Integration Tables

SOME ELEMENTARY FORMS

1. $\int du = u + C$

2. $\int a\,du = au + C$

3. $\int [f(u) + g(u)]\,du = \int f(u)\,du + \int g(u)\,du$

4. $\int u^n\,du = \dfrac{u^{n+1}}{n+1} + C \quad (n \neq -1)$

5. $\int \dfrac{du}{u} = \ln|u| + C$

RATIONAL FORMS CONTAINING $a + bu$

6. $\int \dfrac{u\,du}{a+bu} = \dfrac{1}{b^2}[a + bu - a\ln|a+bu|] + C$

7. $\int \dfrac{u^2\,du}{a+bu} = \dfrac{1}{b^3}\left[\dfrac{1}{2}(a+bu)^2 - 2a(a+bu) + a^2\ln|a+bu|\right] + C$

8. $\int \dfrac{u\,du}{(a+bu)^2} = \dfrac{1}{b^2}\left[\dfrac{a}{a+bu} + \ln|a+bu|\right] + C$

9. $\int \dfrac{u^2\,du}{(a+bu)^2} = \dfrac{1}{b^3}\left[a + bu - \dfrac{a^2}{a+bu} - 2a\ln|a+bu|\right] + C$

10. $\int \dfrac{u\,du}{(a+bu)^3} = \dfrac{1}{b^2}\left[\dfrac{a}{2(a+bu)^2} - \dfrac{1}{a+bu}\right] + C$

11. $\int \dfrac{du}{u(a+bu)} = \dfrac{1}{a}\ln\left|\dfrac{u}{a+bu}\right| + C$

12. $\int \dfrac{du}{u^2(a+bu)} = -\dfrac{1}{au} + \dfrac{b}{a^2}\ln\left|\dfrac{a+bu}{u}\right| + C$

13. $\int \dfrac{du}{u(a+bu)^2} = \dfrac{1}{a(a+bu)} + \dfrac{1}{a^2}\ln\left|\dfrac{u}{a+bu}\right| + C$

FORMS CONTAINING $\sqrt{a + bu}$

14. $\int u\sqrt{a+bu}\,du = \dfrac{2}{15b^2}(3bu - 2a)(a+bu)^{3/2} + C$

15. $\int u^2\sqrt{a+bu}\,du = \dfrac{2}{105b^3}(15b^2u^2 - 12abu + 8a^2)(a+bu)^{3/2} + C$

16. $\int u^n\sqrt{a+bu}\,du = \dfrac{2u(a+bu)^{3/2}}{b(2n+3)} - \dfrac{2an}{b(2n+3)}\int u^{n-1}\sqrt{a+bu}\,du$

17. $\int \dfrac{u\,du}{\sqrt{a+bu}} = \dfrac{2}{3b^2}(bu - 2a)\sqrt{a+bu} + C$

18. $\int \dfrac{u^2\,du}{\sqrt{a+bu}} = \dfrac{2}{15b^3}(3b^2u^2 - 4abu + 8a^2)\sqrt{a+bu} + C$

19. $\int \dfrac{u^n\,du}{\sqrt{a+bu}} = \dfrac{2u^n\sqrt{a+bu}}{b(2n+1)} - \dfrac{2an}{b(2n+1)}\int \dfrac{u^{n-1}\,du}{\sqrt{a+bu}}$

20. $\int \dfrac{du}{u\sqrt{a+bu}} = \begin{cases} \dfrac{1}{\sqrt{a}}\ln\left|\dfrac{\sqrt{a+bu}-\sqrt{a}}{\sqrt{a+bu}+\sqrt{a}}\right| + C & \text{if } a > 0 \\ \dfrac{2}{\sqrt{-a}}\arctan\sqrt{\dfrac{a+bu}{-a}} + C & \text{if } a < 0 \end{cases}$

21. $\int \dfrac{du}{u^n\sqrt{a+bu}} = -\dfrac{\sqrt{a+bu}}{a(n-1)u^{n-1}} - \dfrac{b(2n-3)}{2a(n-1)}\int \dfrac{du}{u^{n-1}\sqrt{a+bu}}$

22. $\int \dfrac{\sqrt{a+bu}\,du}{u} = 2\sqrt{a+bu} + a\int \dfrac{du}{u\sqrt{a+bu}}$

23. $\int \dfrac{\sqrt{a+bu}\,du}{u^n} = -\dfrac{(a+bu)^{3/2}}{a(n-1)u^{n-1}} - \dfrac{b(2n-5)}{2a(n-1)}\int \dfrac{\sqrt{a+bu}\,du}{u^{n-1}}$

FORMS CONTAINING $a^2 \pm u^2$

24. $\int \dfrac{du}{a^2+u^2} = \dfrac{1}{a}\arctan\dfrac{u}{a} + C$

25. $\int \dfrac{du}{a^2-u^2} = \dfrac{1}{2a}\ln\left|\dfrac{u+a}{u-a}\right| + C = \begin{cases} \dfrac{1}{a}\operatorname{arctanh}\dfrac{u}{a} + C & \text{if } |u| < a \\ \dfrac{1}{a}\operatorname{arccoth}\dfrac{u}{a} + C & \text{if } |u| > a \end{cases}$

26. $\int \dfrac{du}{u^2-a^2} = \dfrac{1}{2a}\ln\left|\dfrac{u-a}{u+a}\right| + C = \begin{cases} -\dfrac{1}{a}\operatorname{arctanh}\dfrac{u}{a} + C & \text{if } |u| < a \\ -\dfrac{1}{a}\operatorname{arccoth}\dfrac{u}{a} + C & \text{if } |u| > a \end{cases}$

FORMS CONTAINING $\sqrt{u^2 \pm a^2}$

In formulas 27 through 38, we may replace

$\ln(u + \sqrt{u^2+a^2})$ by $\operatorname{arcsinh}\dfrac{u}{a}$ $\qquad$ $\ln|u + \sqrt{u^2-a^2}|$ by $\operatorname{arccosh}\dfrac{u}{a}$ $\qquad$ $\ln\left|\dfrac{a+\sqrt{u^2+a^2}}{u}\right|$ by $\operatorname{arcsinh}\dfrac{a}{u}$

27. $\int \dfrac{du}{\sqrt{u^2 \pm a^2}} = \ln|u + \sqrt{u^2 \pm a^2}| + C$

28. $\int \sqrt{u^2 \pm a^2}\,du = \dfrac{u}{2}\sqrt{u^2 \pm a^2} \pm \dfrac{a^2}{2}\ln|u + \sqrt{u^2 \pm a^2}| + C$

29. $\int u^2\sqrt{u^2 \pm a^2}\,du = \dfrac{u}{8}(2u^2 \pm a^2)\sqrt{u^2 \pm a^2} - \dfrac{a^4}{8}\ln|u + \sqrt{u^2 \pm a^2}| + C$

(tables continued on back endpapers)

CALCULUS WITH ANALYTIC GEOMETRY

CALCULUS WITH ANALYTIC GEOMETRY

DANIEL J. FLEMING
Department of Mathematics
St. Lawrence University

JAMES J. KAPUT
Department of Mathematics
Southeastern Massachusetts University

HARPER & ROW, PUBLISHERS
New York Hagerstown Philadelphia San Francisco London

Sponsoring Editors: Ann Ludwig, Sharmon Hilfinger
Project Editor: Eva Marie Strock
Production Manager: Laura Argento
Designer: Janet Bollow
Illustrator: John Foster
Cover Designer: Janet Bollow
Cover Illustration: Barbara Hack
Compositor: York Graphic Services
Printer and Binder: Kingsport Press

Library of Congress Cataloging in Publication Data

Fleming, Daniel J 1940–
Calculus with analytic geometry.

Includes index.
1. Calculus. 2. Geometry, analytic. I. Kaput, James J., 1942– joint author. II. Title.
QA303.F5835 515′.15 78-24583
ISBN 0-06-382672-0

CONTENTS

PREFACE

I. THE GENERAL APPROACH

Origins

Five years ago we started developing this text. Since then, preliminary editions have been used at several different colleges and universities in instructional contexts ranging from traditional lecture-discussion classes to independent study situations and most everything in between, including several different versions of PSI courses. We discovered that a text usable in such a wide variety of circumstances required a number of features and a style that would make the book even more effective in traditional lecture courses.

Level and Audience

Our intended audience includes a broad range of students: those in science and engineering, the life and social sciences, and economics and business. We accommodate students with a wide range of backgrounds and abilities by providing coverage of certain precalculus topics (e.g., trigonometry) that is more complete than average and by providing a greater level of detail in our explanations.

Style

Among teachers of mathematics, the word "intuitive" represents almost any attempt to present mathematical ideas in a nonaxiomatic manner, and our text is surely intuitive in this sense. We also move beyond simple informality by being *systematically* intuitive. For example:

1. We control the novelty at each stage, carefully avoiding (whenever possible) introducing new ideas and new notation at the same time.
2. We emphasize how things tie together by extending old ideas into new contexts and by taking old ideas back into familiar contexts.
3. We deliberately examine each major idea from three points of view: *numerical, graphical,* and *algebraic.*

Geometric Growth Theme

Besides applying these three general principles in dealing with such notions as function, limit, derivative, differential, and integral, we also exploit as a basic theme the concrete embodiment of all these ideas in the growth of geometric objects. Each such instance is marked by the symbol ⑤ at the left margin and horizontal rules above and below such discussions. For example, we cover how, as a function of its radius, a circle "grows by circumferences," or a square "grows by the length of the sides in the direction of the growth." Such ideas are discussed first in terms of rate of change, in relation to derivatives (Section 2.3), then in terms of differentials (Section 3.4), and then in terms of the Fundamental Theorem of Calculus, which is seen to generalize all the concrete geometric phenomena discussed to that point (Section 6.3). Similarly, we motivate the proof of the Product Rule geometrically by examining the area of a $u(x)$ by $v(x)$ rectangle (Section 3.1) and develop the method of Lagrange multipliers as a consequence of geometric considerations (Section 19.9).

Rigor

Complete and careful intuitive explanations cannot substitute for precision in formal mathematical statements. We maintain precision in definitions and theorem statements and in formal theorem proofs. In fact, Section 9.3 illustrates, in terms of the functions then available, "Why All the Fuss?"

Applications

In keeping with the expansion of calculus into more and more of the undergraduate curriculum, the current trend in calculus texts is to broaden the areas of applications from science and engineering into other areas. Although we do this, we concentrate our expansion in applications that students can relate to without a great deal of special knowledge and, even more importantly, applications that illustrate how the ideas of calculus relate to everyday life. Such applications are denoted by the symbol ✖ and a vertical rule in the margin and include (among other topics) an examination of

1. Why truckers drive so fast (Section 5.1)
2. Why student apartment rentals can be so costly, even when vacancies exist (Section 5.4)
3. The effects of various sorts of taxes and subsidies on rental costs (Section 5.4)
4. Present value of future income and the effects of inflation on such things as the value of a winning lottery ticket or the real cost of a solar energy installation (Section 8.1)
5. The effects of flashboards on a dam (Section 12.4)
6. The rationale behind government energy taxes (Section 19.9)

Furthermore, we show not merely *that* calculus is used to solve specific problems but *how* calculus is used: *as a conceptual tool* for determining relationships and revealing deeper symmetries. For example, the standard "fence problem" is revisited to expose surprising symmetry (Section 5.1).

Calculators

Calculators are helpful, but not necessary, when using this book. At several important junctures we relate the mathematical idea under consideration to the student's "calculator consciousness." For example, we discuss logical versus numerical accuracy (Section 2.2) and the meaning of infinite limit (Section 4.1) in terms of calculators.

History

We value the history of calculus too much to relegate it to footnotes and portraits. Thus we use the history of calculus at a deeper level, as a guide to the organization of the subject (especially in the choice of geometric motivations) and as the source of an occasional proof (e.g., the proof of L'Hospital's Rule, Section 14.1).

II. ORGANIZATION AND SPECIAL FEATURES

Each chapter begins with an *introduction*, which is an overview of the chapter.

Objectives

Each section begins with a set of *objectives* that tell the student what he or she should be able to do upon completing the section. These objectives are particularly useful when reviewing a section for an exam.

Progress Tests

A very important feature of this text is one to four *progress tests* per section. These tests give the instructor a collection of carefully constructed and straightforward exercises with complete solutions, not just answers, and furthermore, give

1. The student immediate feedback on whether what was just read was understood—*before* the preceding material is combined with other material.
2. The instructor many convenient breakpoints, allowing for classes of different lengths.
3. The instructor flexibility in level of coverage of topics. (Light or quick coverage is achievable by assigning only the Progress Tests, for example.)

PSI Courses

The combination of the objectives and progress tests is especially valuable in any instructional format where students are expected to work through the material without regular lectures. Coupled with the *Instructor's Resource Guide and Test Bank,* the objectives and progress tests eliminate much of the labor in setting up such courses. (See the *Instructor's Resource Guide and Test Bank* for further details.)

Exercises

Each section ends with a collection of *graded exercises.* Each chapter closes with a collection of graded *review exercises,* along with *miscellaneous exercises* and a *self-test.* The progress test problems and the many section exercises provide nearly 7000 exercises to choose from at virtually every level of difficulty. No one is expected to assign or do all the exercises; this huge collection includes enough exercises at any particular level and of any particular type to satisfy the great diversity of needs that arise at different schools and even in a given course. The miscellaneous exercises are more challenging, particularly beyond the early chapters, when more material is available to draw upon. To ensure representative sets of both odds (with answers) and evens (without answers), we frequently pair odd and even exercises of a given type. Students can buy a *Student Supplement* that has complete solutions to odd-numbered section and review exercises and self-tests. A *Solutions Manual* (separate from the *Instructor's Resource Guide and Test Bank*) has answers to evens and solutions to miscellaneous exercises.

III. COVERAGE AND ORDER OF TOPICS

Although any instructor of calculus will recognize from the contents how the chapters interconnect and can be rearranged, the following comments emphasize and point out items not reflected in the contents. Throughout the book we tried to be as flexible as possible in the reordering options because of the special needs of different departments.

1. Student involvement in ϵ–δ work is not requested in the text before Chapter 9, but an instructor wishing to do so can move directly from Section 2.2 to Section 9.1. (Note that Chapter 9 is not a "throwaway theory" chapter, as a glance at the careful development of the idea of limit in Section 9.1 indicates.)

2. Special limits are informally introduced in Section 2.2 only to illustrate ways that limits can fail to exist. Section 4.1 presents more complete treatment in a more natural geometric context as the first step in a systematic approach to describing the behavior of functions.
3. Chapter 6 can be used immediately after Chapter 3 if an early introduction to antiderivatives and integration is desired. The approach to the definite integral is based upon the monotonic case, which is more geometric and less computational than the standard approach. This allows us to establish integrability of all but "infinitely wiggly" functions, an example of which is discussed in Section 9.3.
4. Derivatives of the transcendental functions can be included with the algebraic functions simply by following Chapter 3 with Chapter 7.
5. The exponential and logarithm functions are introduced in Section 7.3 by building on students' earlier experience, and the number e is introduced by examining the slopes of 2^x and 3^x near $x = 0$. We then revisit e as a limit—again as a consequence of geometric considerations. An appendix to Section 7.4 then illustrates how the exponential and logarithm functions can be developed from the integral of $1/t$, thereby illustrating the potency of the Fundamental Theorem of Calculus.
6. Section 11.5 makes a strong case for learning the techniques of integration as a means of making nontrivial use of integration tables.
7. Taylor polynomials are included in their natural context just before power series; power series and Taylor series can be seen as direct generalizations.
8. Computer-generated graphics help illustrate the quadric surfaces in Chapter 17.
9. Green's and Stokes' Theorems are carefully developed in the two-dimensional context to capitalize on students' geometric experience.

Daniel J. Fleming
James J. Kaput

ACKNOWLEDGMENTS

A book in the works as long as this one piles up more than its share of debts. We first wish to thank the many students who helped improve earlier editions and who taught us what is possible. Thanks to Ed Dubinsky, who contributed to an early collection of class notes. We must express appreciation to several generations of typists, including those who married and bore children before we finished the book.

We are also grateful to several generations of reviewers for their many helpful suggestions: W. H. Beyer, The University of Akron; Jerald T. Ball, Chabot Community College, Valley Campus; Robert A. Carman, Santa Barbara City College; Bettye Anne Case, Tallahassee Community College; Arnold Dunn, State University of New York at Potsdam; Garret J. Etgen, University of Houston; William Hemmer, Cuyamaca College; Louis F. Hoelzle, Bucks County Community College; Donald R. Kerr, Jr., Indiana University; Anthony Peressini, University of Illinois at Urbana-Champaign; Evelyn Roane, Northern Virginia Community College; Sister Madeleine Rose, Holy Names College; Dwann E. Veroda, El Camino College; and Bert K. Waits, The Ohio State University. We also thank Arnold Dunn for preparing the *Student Supplement* and the *Solutions Manual.*

In addition, we thank Ted Sundstrom and Doug Kindschi of College IV, Grand Valley State Colleges, for their help and cooperation, and Rick Miller and Gerry Bradshaw of Clarkson College of Technology for their helpful advice. Special acknowledgment to Paul Parente for his patient class testing; he often received portions of text chapter by chapter, sometimes page by page, usually late. Our greatest debt for *direct* help in putting together this book is to our friend, colleague, and most vigorous and perceptive critic, Robert McCabe, of Southeastern Massachusetts University. His constructive influence is absent from few pages indeed.

Finally, we give tremendous thanks to Susan Kaput, our source of greatest indirect help in producing this textbook. She fed, cared for, cleaned up after, and otherwise served three little children (actually, there was only one when we started) and two cantankerous and distracted authors.

TO OUR PARENTS
CHARLES AND LILLIAN FLEMING
AL AND EMILY KAPUT

Chapter 1 ✤

PRELIMINARIES

CONTENTS

This chapter contains a number of ideas and techniques that you need to know to learn calculus. If your math background is strong, you may have seen much of this material earlier. Some items that may be new are: the definition of $|a - b|$ as the distance between a and b, Sec. 1.1; the useful "chart method" for solving certain inequalities, Sec. 1.1; and the discussion of symmetry of parabolas, Sec. 1.3.

We shall use the ideas of this precalculus chapter repeatedly in two ways: (1) as the tools for understanding and applying the calculus, and (2) as a store of examples.

Section 1.1: Numbers, Inequalities, Absolute Value, and Completing the Square

Objectives:

1. Solve and graph inequalities in one variable.

2. Solve and graph equations and inequalities involving absolute value.

3. Complete the square of quadratic polynomials and apply the quadratic formula.

REAL NUMBERS

The set of real numbers, denoted R, contains the integers, the rational numbers, and the irrational numbers. A rational number, a number that can be written as the quotient of two integers, is always expressible by long division as a nonending repeating decimal. On the other hand, an irrational number is a number whose decimal representation is nonending and nonrepeating. For example, $\pi = 3.1415926535\ldots.$

The real numbers do *not* contain the complex numbers. As is common in calculus courses, *we shall confine all our work to the real numbers R*. As a result, expressions such as $\sqrt{x}$ or $\sqrt[6]{x}$ will only be defined for nonnegative x.

IMPOSSIBILITY OF DEFINING DIVISION BY ZERO

The statements "You can't divide by zero" and its equivalent companion "$a/0$ is not defined," which you have heard repeatedly since elementary school, are not made as a matter of taste, simplicity, or convenience, but as a matter of necessity. *We have no choice if we want the other basic rules of arithmetic to hold, because division by zero is inconsistent with these rules.*

To see that this is true, we consider what happens if we attempt to define $a/0$ in the two cases $a \neq 0$ and $a = 0$.

First, suppose $a \neq 0$. If $a/0$ were defined, that is, if $a/0$ were a real number, say b, then $a/0 = b$. But this would imply $a = b \cdot 0 = 0$, contradicting the assumption that $a \neq 0$. Similarly, a contradiction occurs if $a/0$ is defined to be 0. No matter what real number we try to assign the quotient $a/0$, we are led to inevitable contradictions using the basic rules of arithmetic.

ORDER PROPERTIES OF THE REAL NUMBERS

Besides the numerical properties of R associated with addition, subtraction, multiplication, and division, R also has a natural ordering. We use the standard notation for this ordering: $a < b$ means a is less than b, $a \leq b$ means a is less than or equal to b, $a > b$ means a is greater than b, and so on.

1.1.1 **Basic Properties of Order** For any two numbers a and b, either $a < b$, $a = b$, or $a > b$.

1.1.2 $a < b$ implies $(a + d) < (b + d)$ for any number d. That is, adding or subtracting any number to both sides of an inequality leaves the direction of the inequality unchanged.

1.1.3 $a < b$ and $c > 0$ implies $ac < bc$. Multiplying or dividing both sides of an inequality by a *positive* number leaves the direction of the inequality unchanged.

1.1.4 $a < b$ and $c < 0$ implies $ac > bc$. Multiplying or dividing both sides of an inequality by a *negative* number *reverses* the direction of the inequality.

1.1.5 $a < b$ and $ab > 0$ implies $1/a > 1/b$.

1.1.6 $a < b$ and $ab < 0$ implies $1/a < 1/b$.

INTERVALS

We can represent the real numbers by points on a straight line called the *real number line*. We assign the number 0 to one point, which we shall call the *origin*, and the number 1 to some other point to the right of 0, to be regarded as the positive direction. We place an arrow to the right of 1 on this line pointing in the positive direction.

Using the distance from 0 to 1 as our unit of length, we determine the positions of all the integers, which in turn determine the position of all the other real numbers.

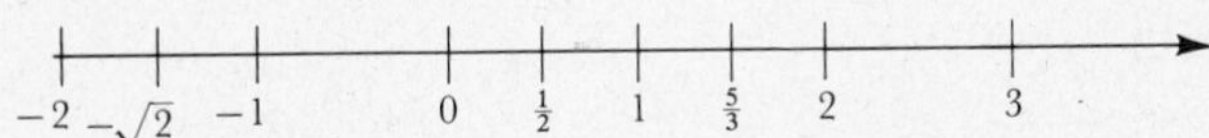

We can now define and graph intervals of real numbers.

1.1.7 **Definition** Suppose $a < b$. Then

(*a*) $[a, b]$ is the set of all real numbers between a and b *including* a and b; that is $[a, b]$ contains all x with $a \leq x \leq b$.
It is called a *closed interval* and its graph is

(*b*) (a, b) is the set of all real numbers between a and b but *not* including a or b. That is, (a, b) contains all x with $a < x < b$. It is called an *open interval* and its graph is

(*c*) Analogously, $[a, b)$ contains its left endpoint but not its right endpoint, and $(a, b]$ contains its right, but not its left, endpoint. These are called *half-open intervals* and their graphs are, respectively,

(*d*) The *infinite interval* $[a, +\infty)$ is the set of all numbers greater than or equal to a. Hence, $[a, +\infty)$ contains all x with $a \leq x$. Its graph is

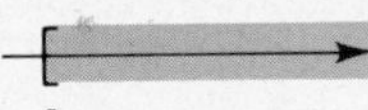

(*e*) $(-\infty, a]$ is the set of all numbers less than or equal to a.
$(a, +\infty)$ and $(-\infty, a)$ are defined similarly.

On occasion we shall refer to R as $(-\infty, +\infty)$.

It is important to remember that the symbols $+\infty$ (plus infinity) and $-\infty$ (minus infinity) do *not* represent real numbers and are used solely as a notational convenience to indicate that the interval is unbounded in either direction.

We use the basic order properties listed earlier to simplify and solve inequalities in much the same way as equations, but with special care when multiplying or dividing by negative numbers. In the following example and henceforth read $\Rightarrow$ as "implies."

Example 1 Solve $-2x - 3 < 5$ and graph the solution.

Solution

$$-2x - 3 < 5$$
$$\Rightarrow -2x < 8 \text{ by (1.1.2)}$$
$$\Rightarrow x > -4 \text{ by (1.1.4)}$$

Thus the solution consists of all x in the infinite interval $(-4, +\infty)$. Its graph is

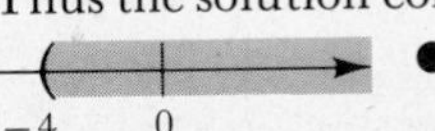

●

PROGRESS TEST 1

1. Test each property (1.1.2) to (1.1.6) using specific numbers.

2. Solve and then graph the solution of:
(*a*) $x + 4 > 3$ (*b*) $-5x \leqslant 10$
(*c*) $3 < 2x - 5$ (*d*) $6 - 5x \leqslant 2x + 9$

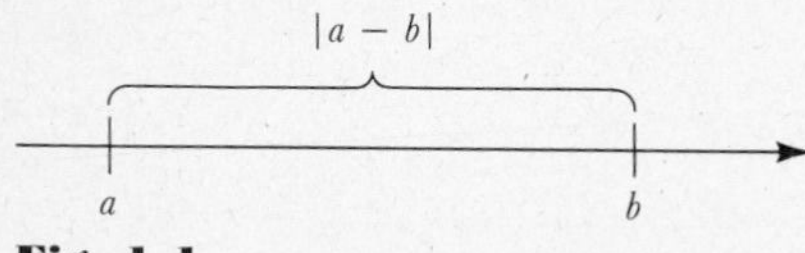

Fig. 1.1

ABSOLUTE VALUE

We base the idea of absolute value on the idea of distance. In particular, we assume that the distance between two points can never be negative.

1.1.8 **Definition** Given two real numbers a and b on the number line, we define $|a - b|$ to be the *distance between a and b*. (See Fig. 1.1.)

1.1.9 **Basic Properties of Absolute Value**

(*a*) $|a + b|$ is the distance between a and $-b$ [because $|a + b| = |a - (-b)|$].

(*b*) $|a| \geq 0$ for all real numbers a, with $|a| = 0$ if and only if $a = 0$. ($|a| = |a - 0|$ is the distance between a and 0, hence is nonnegative; it is zero precisely when $a = 0$, the only number that is zero distance from 0.)

(*c*) $|a| = \begin{cases} a \text{ if } a > 0 \\ 0 \text{ if } a = 0 \\ -a \text{ if } a < 0 \end{cases}$ so, in particular, $|-a| = |a|$

[Note, for example, $|-3| = -(-3) = 3 = |3|$.]

(*d*) $|a \cdot b| = |a|\,|b|$ (This can be proved by examining all combinations of a and b positive, negative, or zero—see Exercise 14, p. 8.)

(*e*) $|a + b| \leq |a| + |b|$ (Often called the *triangle inequality*.)

(*f*) $|a| = \sqrt{a^2}$ [For example, $|-3| = 3 = \sqrt{3^2} = \sqrt{(-3)^2}$.]

(*g*) For $a > 0$, $|x| \leq a$ if and only if $-a \leq x \leq a$, which is true if and only if x is in the interval $[-a, a]$, pictured below.

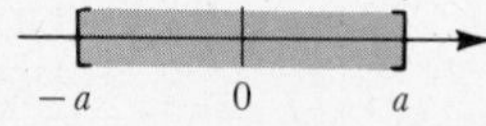

Test the triangle inequality using numbers of your choice.

INEQUALITIES INVOLVING ABSOLUTE VALUE

The key to solving most of the inequalities that we shall consider is to *think in terms of distance on a number line.*

Example 2 Solve $|x - 2| \leq 4$ and graph its solution.

Solution We are looking for the set of numbers x whose distance from 2 is less than or equal to 4. By simply counting off 4 units from either side of 2, we see that x must satisfy $-2 \leq x \leq 6$; that is, the solution consists of all x in the interval $[-2, 6]$. Its graph is

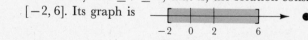

Example 3 Solve $|2x + 3| > 2$ and graph its solution.

Solution We must reduce the given absolute value expression to one of the form $|a - b|$, so that we can interpret it on the number line in terms of distance. By [1.1.9(*d*)]

$$\begin{aligned}|2x+3| &= \left|2\left(x+\frac{3}{2}\right)\right| = |2|\left|x+\frac{3}{2}\right| \\ &= 2\left|x+\frac{3}{2}\right| \qquad \text{by [1.1.9(}c\text{)]} \\ &= 2\left|x-\left(-\frac{3}{2}\right)\right| > 2\end{aligned}$$

Hence, by (1.1.3), $|x-(-3/2)| > 1$, so the solution consists of all numbers x whose distance from $-3/2$ is greater than 1. This includes all x in $(-\infty, -5/2)$ or $(-1/2, +\infty)$. The graph of the solution is

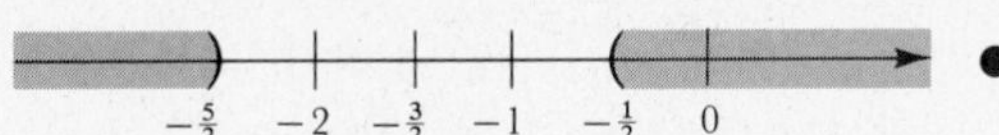

●

PROGRESS TEST 2

Solve the given inequality and then graph its solution.

1. $|x+1| \geq 5$ **2.** $|3x+2| < 6$ **3.** $|2x-6| < 9$ **4.** $|x-1| < 0$

COMPLETING THE SQUARE AND THE QUADRATIC FORMULA

We shall see many occasions where it is useful to complete the square of a quadratic expression. We review the process briefly. Its essence is to take half the coefficient of the x term, square it, and add it to the expression, making compensations to avoid changing the value of the expression.

Example 4 Complete the square of $x^2 + 6x + 11$.

Solution First write $(x^2 + 6x + \quad) + 11$. We then fill in the blank space so that the resulting quadratic trinomial in the parentheses is a perfect square. Take half the coefficient of x, $\frac{6}{2} = 3$, square it, getting 9, and add it in the space provided. Then subtract the 9 from the 11 to avoid changing the value of the total expression:

$$(x^2 + 6x + 9) + 11 - 9 = (x+3)^2 + 2 \quad \bullet$$

If the coefficient of x^2 is not 1, we factor it out before applying the above procedure, as indicated in Example 5.

Example 5 Complete the square of $3x^2 - 6x + 1$.

Solution We first write

$$3x^2 - 6x + 1 = 3(x^2 - 2x + \quad) + 1$$

and then complete the square as before:

$$3(x^2 - 2x + 1) + 1 - 3(1) = 3(x-1)^2 - 2$$

Note that we needed to subtract 3 because we had in fact added $3 \cdot 1$ when we added 1 inside the parentheses. ●

This technique can be applied to the general quadratic expression

$$ax^2 + bx + c$$

to yield the quadratic formula (see Exercise 15).

1.1.10 Quadratic Formula If $ax^2 + bx + c = 0$, with $a \neq 0$, then

$$x = \frac{-b \pm \sqrt{b^2 - 4ac}}{2a} \quad \text{and conversely}$$

We refer to the values of the variable for which an expression is zero as its *roots*. By (1.1.10),

(*a*) If $b^2 - 4ac > 0$, then $ax^2 + bx + c$ has two real roots.
(*b*) If $b^2 - 4ac = 0$, then $ax^2 + bx + c$ has a single repeating root.
(*c*) If $b^2 - 4ac < 0$, then $ax^2 + bx + c$ has no real roots.

Note that our blanket agreement to deal only with *real numbers* leads us to say, in the third case, that "$ax^2 + bx + c$ is never zero," by which we mean it is never zero for real x. The number $b^2 - 4ac$ is usually called the *discriminant* of the quadratic expression $ax^2 + bx + c$.

Shortly, when solving more complicated inequalities, we shall need to determine whether a given quadratic expression can be factored or whether it is always positive or always negative. If we cannot factor the expression by inspection, then we apply the quadratic formula. If $b^2 - 4ac \geq 0$, then we have found the expression's real roots r_1 and r_2 (with $r_1 = r_2$ if $b^2 - 4ac = 0$). In this case we can factor $ax^2 + bx + c = a(x - r_1)(x - r_2)$. This is true because of the fundamental fact from algebra that a polynomial has r as a root if and only if it contains $x - r$ as a factor. On the other hand, if $b^2 - 4ac < 0$, then the expression is never zero, is not factorable, and is always positive or always negative.

Example 6 Factor the following expressions if possible. If the discriminant is negative, determine whether the expression is always positive or always negative.

(*a*) $x^2 - 5x + 1$ (*b*) $2x^2 + x + 2$ (*c*) $3x - x^2 + 4$

Solutions (*a*) We cannot factor $x^2 - 5x + 1$ by inspection, so we apply the Quadratic Formula (1.1.10):

$$x = \frac{+5 \pm \sqrt{25 - 4}}{2} = \begin{cases} \dfrac{5 + \sqrt{21}}{2} \\ \dfrac{5 - \sqrt{21}}{2} \end{cases}$$

Hence

$$x^2 - 5x + 1 = \left[x - \frac{(5 + \sqrt{21})}{2}\right]\left[x - \frac{(5 - \sqrt{21})}{2}\right] \quad \text{(check this by multiplying)}$$

(*b*) Again $2x^2 + x + 2$ is not factorable by inspection, so by (1.1.10):

$$x = \frac{-1 \pm \sqrt{1 - 4 \cdot 2 \cdot 2}}{2 \cdot 2}$$

But $1 - (4 \cdot 2 \cdot 2) = -15 < 0$, so $2x^2 + x + 2$ is not factorable, hence always positive or negative. But substituting $x = 0$ yields a value of 2, so $2x^2 + x + 2$ is always positive.

(*c*) We first put $3x - x^2 + 4$ in "standard form": $-x^2 + 3x + 4$. But this can be factored as $(-x + 4)(x + 1)$. [The same factorization would result if we used (1.1.10); check this.] ●

SOLVING INEQUALITIES—THE CHART METHOD

In applying the calculus it is quite often necessary to determine when an expression is positive or negative. We present a systematic "chart" method by working through a detailed solution of the inequality

$$\frac{x^2 + x}{x - 2} < 0$$

We factor the numerator and denominator of the expression as far as possible:

$$\frac{x^2 + x}{x - 2} = \frac{(x)(x + 1)}{x - 2}$$

and analyze each factor:

1. x is positive when $x > 0$ and negative when $x < 0$.
2. $x + 1$ is positive when $x > -1$ and negative when $x < -1$.
3. $1/(x - 2)$ is positive when $x > 2$ and negative when $x < 2$.

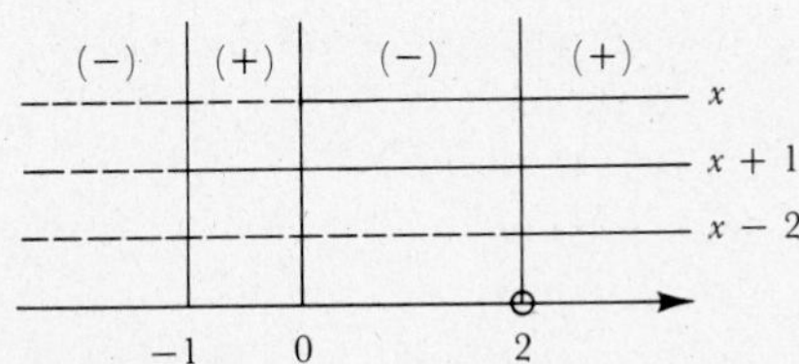

Fig. 1.2

This information is summarized in a chart (Fig. 1.2), which contains a horizontal line for each factor appearing in the numerator or denominator and above the number line marked at each place where a factor changes sign. A vertical line is drawn through each such point. For each factor we draw a horizontal line dashed where the factor is negative, solid where the factor is positive. We also circle points where the denominator is zero since they cannot be part of the solution. [Note we wrote $x - 2$ rather than $1/(x - 2)$.]

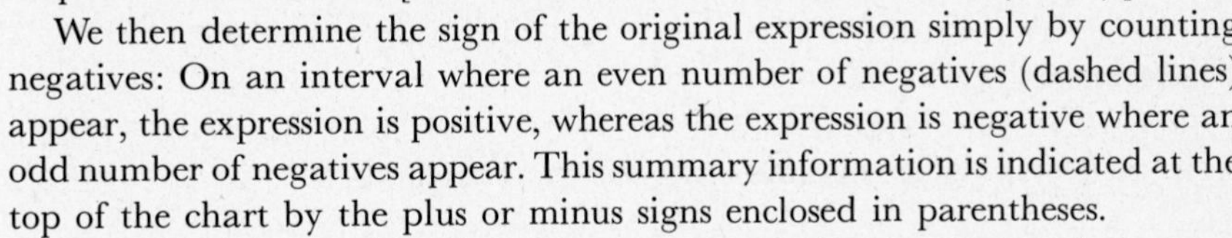

We then determine the sign of the original expression simply by counting negatives: On an interval where an even number of negatives (dashed lines) appear, the expression is positive, whereas the expression is negative where an odd number of negatives appear. This summary information is indicated at the top of the chart by the plus or minus signs enclosed in parentheses.

In the interval $(-\infty, -1)$ we have three negatives, so the product

$$\frac{x(x + 1)}{x - 2} = x(x + 1)\frac{1}{x - 2}$$

is negative. In the interval $(-1, 0)$ we have two negatives, so the product is positive. Similarly the product is negative for x in $(0, 2)$ and positive for x in $(2, +\infty)$. Thus the collection of numbers x for which

$$\frac{x^2 + x}{x - 2} < 0$$

is the set of all x in the intervals $(-\infty, -1)$ and $(0, 2)$.

You should get into the habit of checking your work. In this case a good check is to insert a specific value from each interval and at least one not in these intervals to see whether the result agrees with your answer. For example, in the above problem substitute $x = -2$ and $x = 3$, and observe that you get $-1/2$ and 12, respectively, which is correct.

Example 7 Solve the inequality

$$\frac{6 + x - x^2}{(x^2 + x + 1)(x + 4)(x - 6)^2} \geq 0$$

Solution We factor as far as possible:

$$\frac{(3 - x)(x + 2)}{(x^2 + x + 1)(x + 4)(x - 6)^2}$$

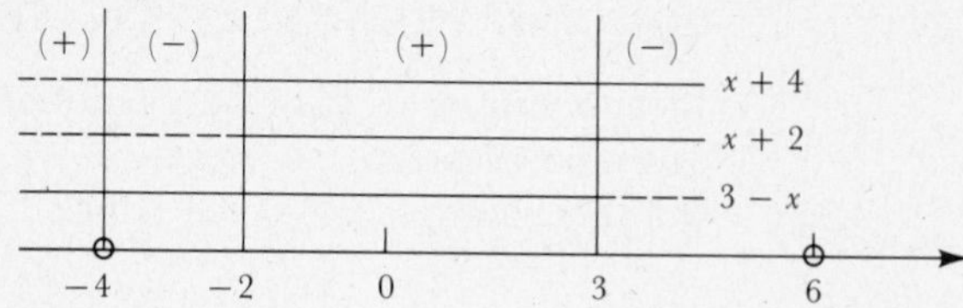

Fig. 1.3

Since the sign of the entire expression is unaffected by any factor that is never negative, a factor such as $(x - 6)^2$ can be omitted from the sign analysis. Note, however, that 6 cannot be part of the solution because the denominator is zero for $x = 6$. Similarly, -4 cannot be part of the solution.

We are unable to factor $x^2 + x + 1$ by inspection; we therefore write its discriminant $1^2 - 4 \cdot 1 \cdot 1 = -3 < 0$. Thus $x^2 + x + 1$ is always positive or always negative. We substitute $x = 0$ and conclude that $x^2 + x + 1$ is always positive. Hence we can ignore $x^2 + x + 1$ in the sign analysis. Thus the original problem reduces to the chart analysis (Fig. 1.3) of

$$\frac{(3 - x)(x + 2)}{x + 4} \qquad \text{with } x \neq 6$$

[Do not be careless. The factor $(3 - x)$ is positive to the *left* of 3 and negative to the right!]

Thus the solution is all x in the intervals $(-\infty, -4)$, $[-2, 3]$. Check this! ●

PROGRESS TEST 3

Solve the following inequalities using the chart method:

1. $x(x - 3)(6 - x) < 0$

2. $\dfrac{(x^2 - 5x + 4)(x + 2)}{(x^2 + 3)(2x + 1)} \geq 0$

SECTION 1.1 EXERCISES

In Exercises 1 to 6, solve the inequality and graph its solution.

1. $|2x + 6| \leq 5$

2. $|x - 8| > \frac{3}{2}$

3. $|5x - 10| \leq 0$

4. $|x - 4| < 4$

5. $|x + 3| \leq 6$

6. $|3x - 1| \geq 0$

7. Solve the equations:
(*a*) $|3x + 4| = 2$ (*b*) $|x - 5| = 0$

In Exercises 8 to 13, solve the given inequality.

8. $\dfrac{(x + 1)(2x - 3)}{x + 5} \geq 0$

9. $\dfrac{(2 - 5x)(x^2 - 3x - 4)}{(x^2 + 2x + 2)(x + 2)} \leq 0$

10. $\dfrac{(x^2 + 2x + 1)(3x + 1)}{(2x^2 - 6x - 20)(1 - x)} \leq 0$

11. $\dfrac{(x^2 + 6x + 9)(x^3 + x^2)}{3x - 5} < 0$

12. $\dfrac{(x^4 + 4)(x^2 - 1)}{(2x + 1)^2(x - 4)} > 0$

13. $\dfrac{(x^2 - 5)^2}{(x^4 - 4)(x + 3)} \geq 0$

14. Prove that $|a \cdot b| = |a|\,|b|$ by showing equality for the four cases involving combinations of positive and negative a and b. (The zero case is trivial.)

15. Derive the Quadratic Formula by completing the square of $ax^2 + bx + c$.

16. As a means of sharpening your algebra skills, prove that (1.1.10) is valid by direct substitution of each of the roots into $ax^2 + bx + c$.

17. Prove that $|a \cdot b| = |a|\,|b|$ using [1.1.9(f)] and the properties of square roots.

18. The double inequality $\frac{1}{2} < |x - 2| < 1$ stands for the two inequalities $\frac{1}{2} < |x - 2|$ and $|x - 2| < 1$. Hence x satisfies the double inequality if it satisfies both inequalities simultaneously.

(*a*) Graph $\frac{1}{2} < |x - 2| < 1$
(*b*) Graph $0 < |x - 2| < 1$
(*c*) Graph $0 < |x - 2| < \frac{1}{4}$

Section 1.2: Review of Coordinate Geometry

Objectives:

1. Use a coordinate system to plot points in a plane and graph linear equations.

2. Determine the distance between a pair of points in a plane.

3. Determine the midpoint of a line segment.

4. Determine the slope of a straight line.

5. Determine an equation for a straight line, given enough information to define the line uniquely.

RECTANGULAR COORDINATES AND THE MEANING OF GRAPHS

In a way similar to how we plot real numbers on a number line, we can plot *ordered pairs* of real numbers in a plane, as indicated in Fig. 1.4. We say *ordered* pair because order *does* make a difference. That is, $(a, b) \neq (b, a)$ unless $a = b$.

In Fig. 1.4 the horizontal line is called the *x axis,* the vertical line is called the *y axis,* and the intersection of these two lines, that is, the point labeled $(0, 0)$, is called the *origin.* The four *quadrants* are labeled I to IV, respectively. We say that x and y are the *coordinates* of the point labeled by (x, y), and the whole scheme is called the *rectangular coordinate system for the plane* or, more simply, the *x-y plane.*

Notice that we use the same notation for the coordinates of a point, say $(-\pi, 2)$, as we do for the open interval $(-\pi, 2)$. The context will inevitably make clear which meaning applies.

Given an equation involving x and y, we can graph the set of points corresponding to the values of x and y that make the equation into a true statement. The graph of this set is called the *graph of the equation,* and such pairs (x, y) are said to *satisfy the equation.*

Example 1 Graph the equation $y = |x|$.

Solution An ordered pair (x, y) is part of the graph if the second coordinate (often called the y coordinate) equals the absolute value of the first coordinate (the x coordinate). Hence the points $(1, 1)$, $(2, 2)$, $(5, 5)$, $(0, 0)$, $(-1, 1)$, $(-2, 2)$, $(-3, 3)$ are all part of the graph, whereas points such as $(1, -1)$, $(0, 1)$, $(-2, -2)$ are not. We plot the points of the graph in Fig. 1.5 and draw an unbroken line through them. ●

DISTANCE AND MIDPOINT

If (x_1, y_1) and (x_2, y_2) are points in the plane as in Fig. 1.6(a), recall that the *distance* between these two points is given by the formula

1.2.1
$$\text{Distance} = \sqrt{(x_1 - x_2)^2 + (y_1 - y_2)^2}$$

This follows from the Pythagorean theorem.

The *midpoint* of the line segment connecting these points [see Fig. 1.6(b)] is the point $(\bar{x}, \bar{y})$, whose coordinates are given by formulas

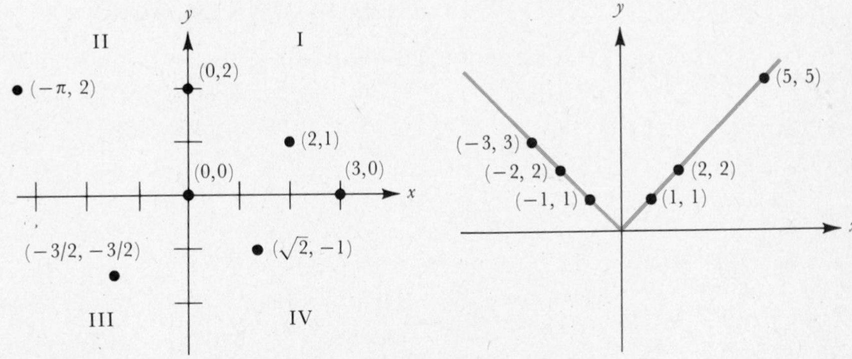

Fig. 1.4 **Fig. 1.5**

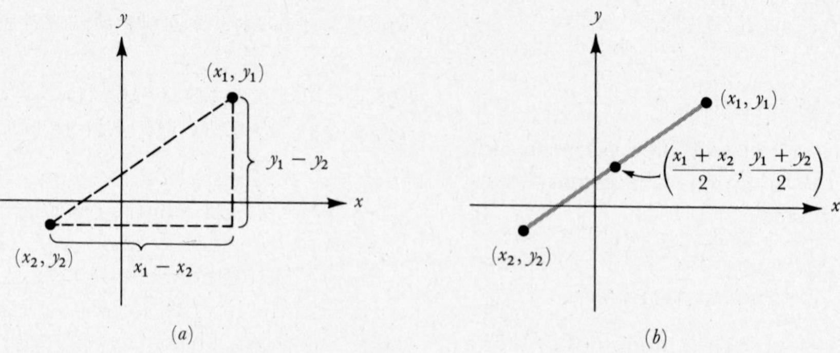

Fig. 1.6

1.2.2

$$\bar{x} = \frac{x_1 + x_2}{2} \qquad \bar{y} = \frac{y_1 + y_2}{2}$$

Example 2 Suppose that $(x_1, 0)$ and $(x_2, 0)$ are two points on the x axis. Show that the distance between these points is the same as the distance used in defining the absolute value $|x_1 - x_2|$.

Solution By (1.2.1) the distance between $(x_1, 0)$ and $(x_2, 0)$ is

$$\sqrt{(x_1 - x_2)^2 + (0 - 0)^2} = \sqrt{(x_1 - x_2)^2}$$
$$= |x_1 - x_2| \qquad \text{by [1.1.9}(f)] \quad \bullet$$

PROGRESS TEST 1

1. Let $A = (1, 3)$ and $C = (3, 5)$. Plot A, C, and the midpoint B of line segment AC. Be sure to give the coordinates of B.
2. Find the distance from A to C.
3. Let $R = (-1, 3)$ and $T = (5, -2)$. Plot R, T, and the midpoint S of line segment RT. Be sure to give the coordinates of S.
4. Find half the distance from R to T.
5. Find the distance from R to S. Does this answer agree with that of Prob. 4?
6. Prove that the distance between the points (x_1, y_1) and (x_2, y_2) is unaffected if we interchange the roles of (x_1, y_1) and (x_2, y_2) in the formula (1.2.1).
7. Prove that, in general, the distance between the midpoint given by (1.2.2) and the point (x_1, y_1) is half the distance between the points (x_1, y_1) and (x_2, y_2).

SLOPES OF STRAIGHT LINES

1.2.3 **Definition** Suppose L is a straight line in the x-y plane that is not parallel to the y axis, and suppose (x_1, y_1) and (x_2, y_2) are two distinct points on L (see Fig. 1.7). Then by the *slope of L* we mean the number m given by

$$m = \frac{y_2 - y_1}{x_2 - x_1}$$

Example 3 Determine the slope m of the straight line passing through the points $(-3, 2)$ and $(6, -1)$.

Solution Letting $(-3, 2) = (x_1, y_1)$ and $(6, -1) = (x_2, y_2)$, as in Fig. 1.8, we find that

$$m = \frac{-1-2}{6-(-3)}$$
$$= \frac{-3}{9} = -\frac{1}{3} \quad \bullet$$

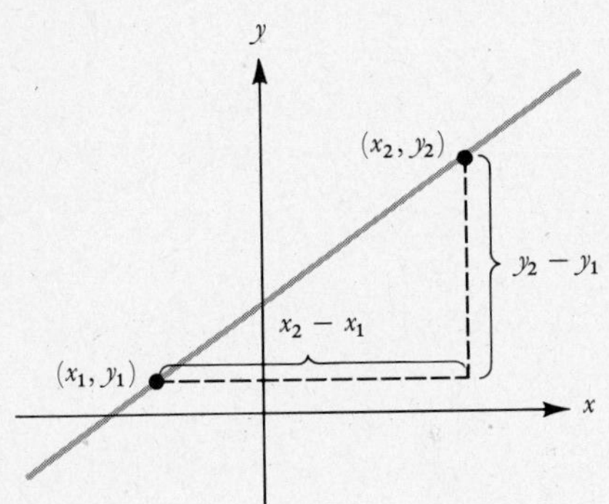

Fig. 1.7

Notice that if we reverse the roles of (x_1,y_1) and (x_2,y_2) in Example 3, letting $(-3, 2) = (x_2,y_2)$ and $(6, -1) = (x_1,y_1)$, then

$$m = \frac{2-(-1)}{-3-6} = \frac{3}{-9} = -\frac{1}{3}$$

as before. It is irrelevant which point is chosen as (x_1,y_1) and which is chosen as (x_2,y_2) *as long as you are consistent in applying (1.2.3)*. This is true because

$$\frac{y_2-y_1}{x_2-x_1} = \frac{-(y_1-y_2)}{-(x_1-x_2)} = \frac{y_1-y_2}{x_1-x_2}$$

The word "straight" here has its intuitive meaning: the slope of a straight line is a fixed number independent of the points used to compute it. (See Exercise 32.) The idea of slope is important in the calculus, so it is important that you get a good feel for what it means geometrically. The line of slope $-1/3$ in Fig. 1.8 slopes downward from left to right; in fact, it decreases $1/3$ unit vertically for each unit increase in x. Put differently, it drops 1 unit whenever x increases by 3 units. In Fig. 1.9 are several graphs, each containing at least two straight lines whose slopes m are

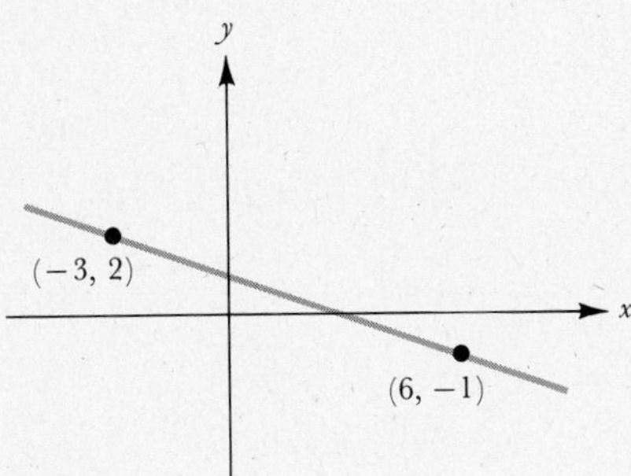

Fig. 1.8

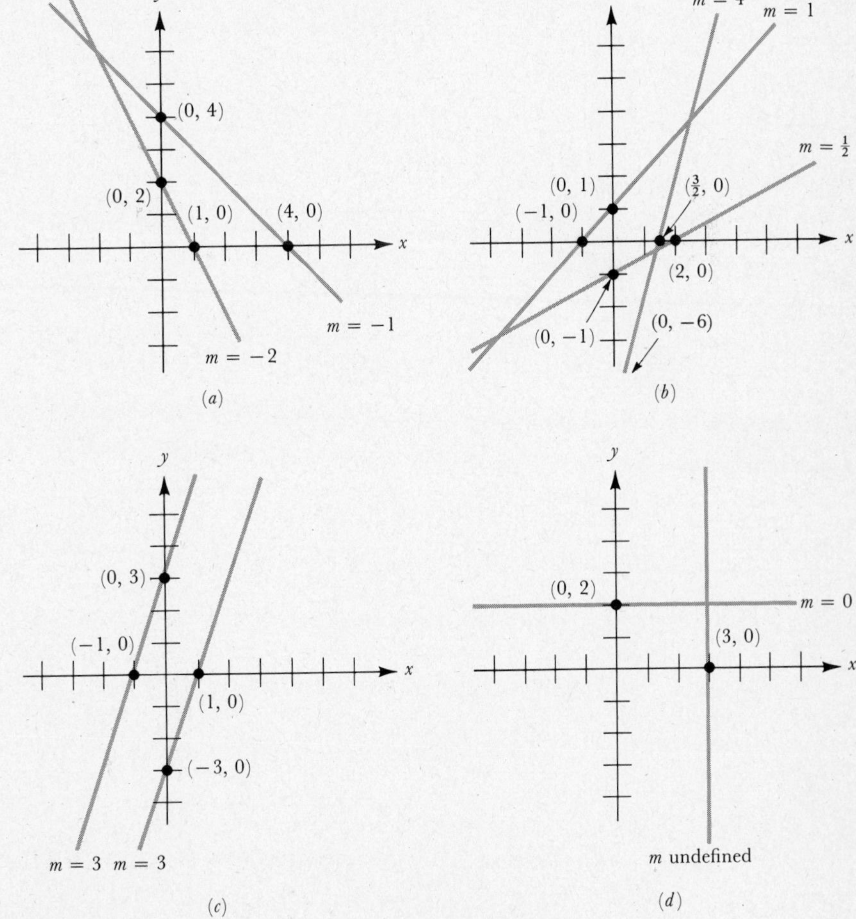

Fig. 1.9

indicated. Also indicated are points where these lines cross the x and y axes. Check the slopes using these points.

1.2.4 Basic Properties of Slope

(*a*) In Fig. 1.9(*a*) we see that *to have a negative slope means that the line slopes down to the right;* the more negative m is, the steeper the downward slope. (Compare with $m = -1/3$ in Fig. 1.8.)

(*b*) In Fig. 1.9(*b*) we see that *to have a positive slope means that the line slopes up to the right;* the larger m is, the steeper the upward slope.

(*c*) In Fig. 1.9(*c*) we see that *two lines with the same slope are parallel.* This is true because for a given change Δx (read "delta x") in x, each line rises (or falls) exactly the same amount Δy, so $\Delta y/\Delta x$ is the same for each line.

(*d*) In Fig. 1.9(*d*) we see that *a horizontal line has zero slope, and conversely.* This is true because, for any two distinct points (x_1, y_1) and (x_2, y_2) on a horizontal line, $y_1 = y_2$. This implies that

$$m = \frac{y_2 - y_1}{x_2 - x_1} = \frac{0}{x_2 - x_1} = 0$$

(*e*) In Fig. 1.9(*d*) we also see that *the slope of a vertical line is undefined.* This is true because for any two distinct points (x_1, y_1) and (x_2, y_2) on a vertical line, $x_1 = x_2$. This implies that the quotient that would define the slope,

$$m = \frac{y_2 - y_1}{x_2 - x_1} = \frac{y_2 - y_1}{0}$$

is not a real number.

(*f*) Suppose that L_1 has slope $m_1 \neq 0$ and L_2 has slope m_2. Then L_2 is *perpendicular to L_1 (written $L_2 \perp L_1$)* if and only if $m_2 = -1/m_1$. (Proof left as Exercise 33.)

PROGRESS TEST 2

Give the slope of each of the straight lines in 1 to 3.

1.

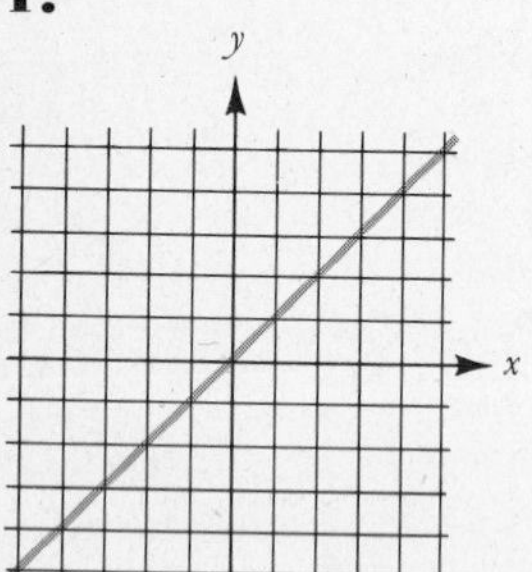

2.

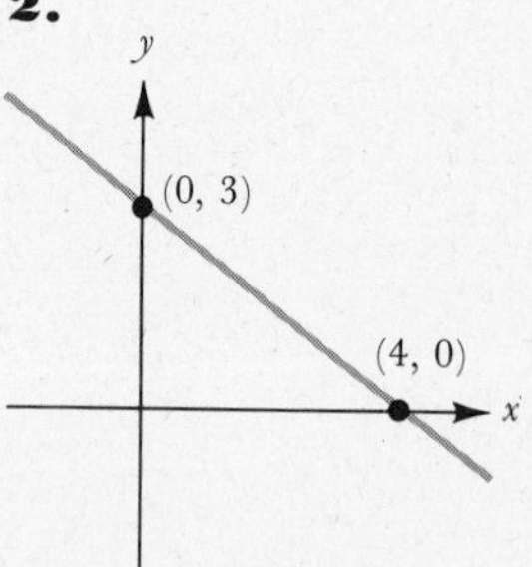

3.

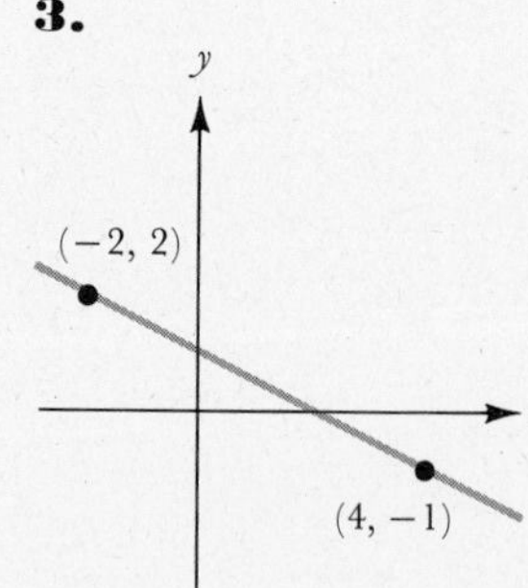

4. Which of the lines in 1 to 3 has the most negative slope?

5. Are any two of the lines in 1 to 3 parallel, and if so, which two?

6. Are any two lines in 1 to 3 perpendicular, and if so, which two?

7. If a line slopes down to the right, with y decreasing at the same rate as x is increasing, what is the slope of this line?

8. Give the slope of a line perpendicular to the line in Prob. 7.

EQUATIONS OF STRAIGHT LINES

We saw that to graph an algebraic equation in x and y is to determine the set of all points in the x-y plane whose coordinates (x, y) satisfy that equation. In plotting these points we obtain a *geometric* picture of the *algebraic* equation. In some cases it is possible to go the other way; that is, given a set of points in the x-y plane, or some geometric information that *determines* a set of points, we may be able to derive an algebraic equation whose graph, in turn, equals the given set of points.

In general, the connection between a graph and its equation is difficult to make, but it is easy for straight lines because of the definition of "straight line": *a line whose slope is always the same no matter which two points are used to compute it.*

As you know from geometry, several kinds of information determine a straight line: for example, a point and a slope, two points, and so forth. We now list the familiar forms of the equation of a straight line L. Each of the forms (a) to (d) corresponds to a different way of uniquely determining a straight line.

1.2.5 Equations of a Straight Line

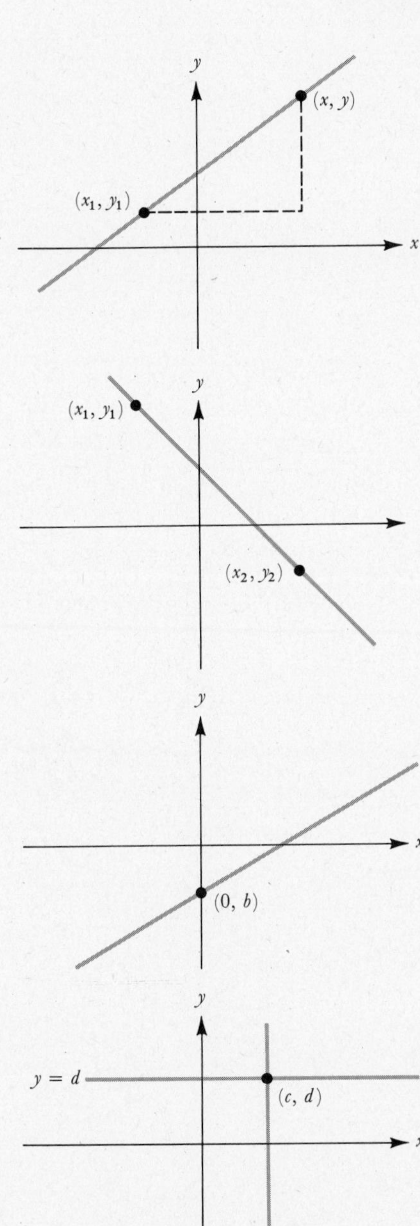

(a) *Point-slope form:* If the line L passes through (x_1, y_1) and has slope m, then L has the equation

$$y - y_1 = m(x - x_1)$$

[If L is to be straight, then the slope determined by (x_1, y_1) and any other point (x, y) must be m; that is $(y - y_1)/(x - x_1) = m$. This equation can be put in the above form.]

(b) *Two-point form:* If the line L passes through (x_1, y_1) and (x_2, y_2), then L has the equation

$$y - y_1 = \frac{y_2 - y_1}{x_2 - x_1}(x - x_1)$$

provided that L is not vertical, that is, provided $x_1 \neq x_2$. [This follows from (a).]

(c) *Slope-intercept form:* If L intercepts the y axis at $(0, b)$ and has slope m, then L has the equation,

$$y = mx + b$$

[This is a special case of (a) with $(x_1, y_1) = (0, b)$.] We shall refer to the point $(0, b)$ as the y intercept. Sometimes we shall simply refer to b as the y intercept.

(d) *Parallel-to-axes (special) forms:* A vertical line through the point (c, d) has the equation

$$x = c$$

[A point (x, y) is on this line if and only if $x = c$, regardless of the values y assumes.] Similarly, a horizontal line through (c, d) has the equation

$$y = d$$

[A point (x, y) is on this line if and only if $y = d$.] Note that the equation of the y axis is $x = 0$, whereas the equation of the x axis is $y = 0$.

(e) *General form:* Every equation that can be put in general form

$$ax + by + c = 0$$

so long as a and b are not both zero, is the equation of a straight line. Conversely, every straight line has such an equation. All the above forms can be put in this general form.

1.2.6 Definition Any equation that can be put in the form $ax + by + c = 0$ is called a *linear equation.*

In Examples 4 to 9 we shall write an equation of a line that is determined by some geometric information, leaving the equation in the general form, clearing of fractions where necessary.

Example 4 Determine an equation of the line with slope 3 passing through $(-2, 7)$.

Solution Using the point-slope form, we obtain $y - 7 = 3[x - (-2)]$, so $3x - y + 13 = 0$. ●

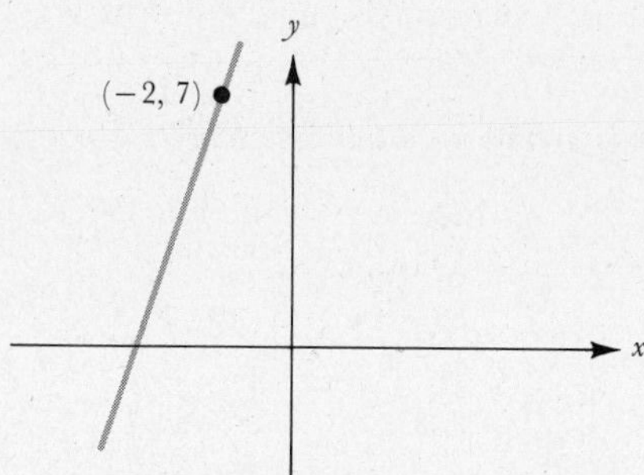

Example 5 Determine an equation of the line passing through $(5, 2)$ and $(-6, 1)$.

Solution Using the two-point form, we obtain

$$y - 2 = \frac{1 - 2}{-6 - 5}(x - 5)$$

so $x - 11y + 17 = 0$. ●

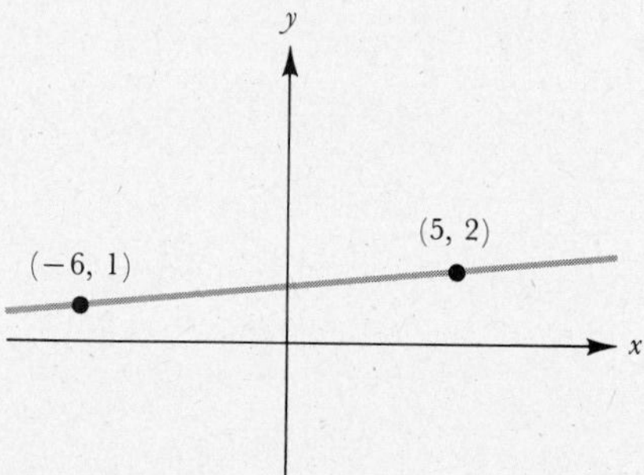

Example 6 Determine an equation of the line through $(6, -1)$ and $(6, 8)$.

Solution Both points have the same x coordinate, so the line is parallel to the y axis, with the equation $x = 6$. (Note that the slope of this line is undefined.) ●

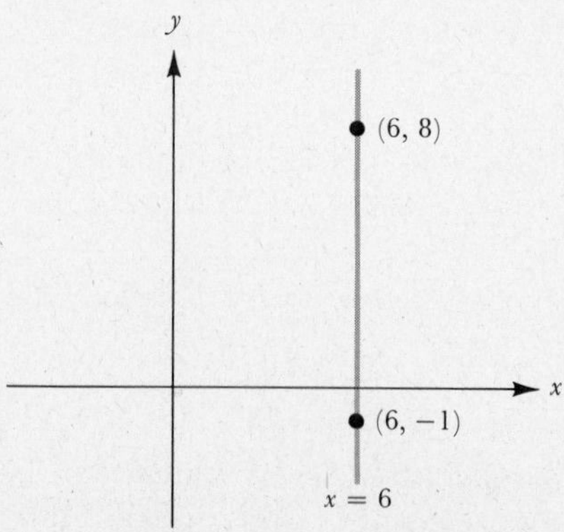

Example 7 Determine an equation of the line through (1, 2) and (7, 2).

Solution Both points have the same y coordinate, so the line is parallel to the x axis, with the equation $y = 2$. (Note that the slope of this line is zero.) ●

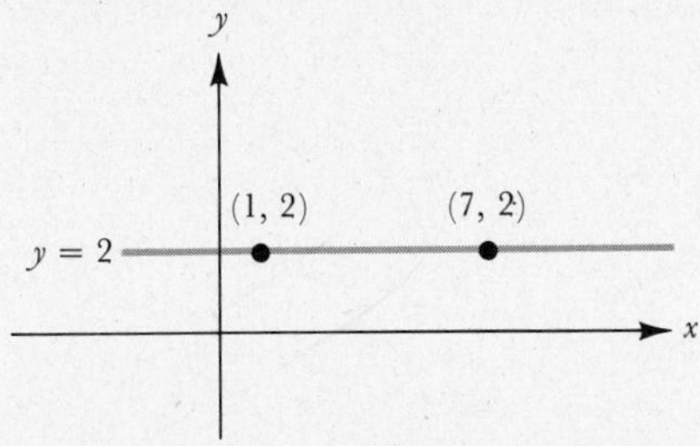

Example 8 Determine an equation of the line through $(1, -4)$ and perpendicular to L with the equation $x + 5y - 3 = 0$.

Solution Solving for y to put L in slope-intercept form

$$y = -\frac{1}{5}x + \frac{3}{5}$$

we see that L has a slope of $-1/5$. Thus the line we are considering has a slope of 5. Using the point-slope formula we obtain $y - (-4) = 5(x - 1)$, or $5x - y - 9 = 0$. ●

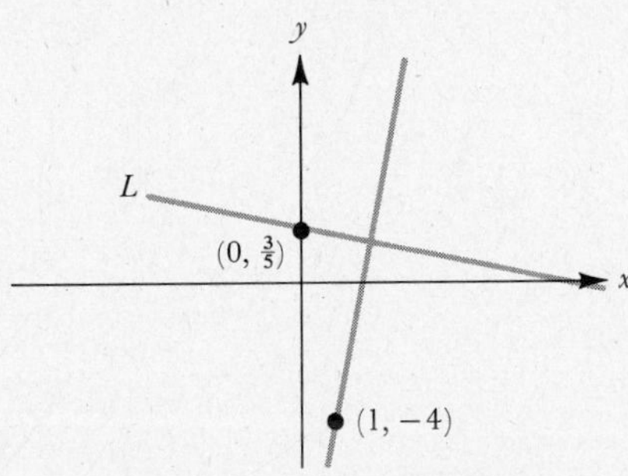

Example 9 Determine an equation of the perpendicular bisector of the line segment joining $(-3, 2)$ and $(5, 6)$.

Solution We begin by sketching graphs of the lines from the given information. Now L_1 has a slope of $(6 - 2)/(5 + 3) = 1/2$, so the perpendicular bisector has a slope of -2. The midpoint of the line segment has coordinates $(\bar{x}, \bar{y})$ where

$$\bar{x} = \frac{5 - 3}{2} = 1 \qquad \text{and} \qquad \bar{y} = \frac{6 + 2}{2} = 4$$

Thus the equation of L_2 is $y - 4 = -2(x - 1)$, or $2x + y - 6 = 0$. ●

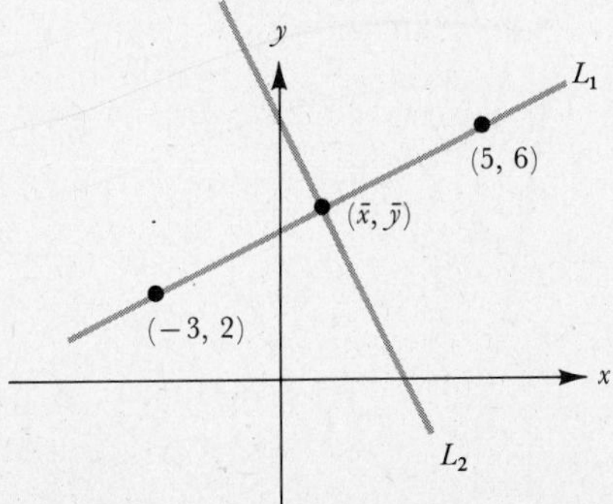

Examples 4 to 9 show that, as long as the given geometric information uniquely determines a line, we can write an equation for that line. Conversely, a given linear equation can be graphed with little difficulty.

Example 10 Graph the equation $5x + 3y = 10$.

Solution We put the equation in slope-intercept form

$$y = -\frac{5}{3}x + \frac{10}{3}$$

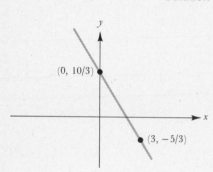

Hence the y intercept is $(0, 10/3)$. Letting $x = 3$, we find the corresponding $y = -5/3 \cdot 3 + 10/3 = -5/3$. We can now draw the graph through the two points $(0, 10/3)$ and $(3, -5/3)$. ●

Sometimes in graphing linear equations, as in Example 10, it is convenient to determine the point where the graph intercepts the x axis by setting $y = 0$ and solving for x. Given $5x + 3y = 10$, we would have $5x = 10$, or $x = 2$, yielding the point $(2, 0)$.

The point where a graph intercepts the x axis is called an x *intercept*.

Another way to graph a linear equation is to put it in slope-intercept form, read off the y intercept, and then use the slope to draw the graph. In the above example, the slope $-5/3$ tells us that whenever x increases by 3, y decreases by 5.

PROGRESS TEST 3

1. Determine equations for the lines with the following properties:
(*a*) Through $(3, 5)$ with slope 4
(*b*) Through the two points $(3, 5)$ and $(4, 9)$
(*c*) With y intercept $(0, 2)$ and slope -2

2. Find the slope and y intercept of the line with equation:
(*a*) $6x - 2y + 5 = 0$
(*b*) $3x = 4y + 8$

3. Graph the following equations:
(*a*) $y = 2x - 3$
(*b*) $2y + 4x + 5 = 0$

4. Find an equation for the line L_1 through $(-5, 3)$ and perpendicular to the line L_2 that passes through $(7, 0)$ and $(-8, 1)$. Leave your answer in point-slope form.

SECTION 1.2 EXERCISES

1. How far apart are:
(*a*) $(\frac{1}{3}, \frac{2}{5})$ and $(-6, -5)$ (*b*) $(0, -2)$ and $(0, 7)$

2. Express your solution to 1(*b*) in terms of absolute value. (Note that the y axis is a number line.)

What is the midpoint of the line segment joining:

3. $(\frac{1}{3}, \frac{2}{5})$ and $(-6, -5)$ **4.** $(5, 0)$ and $(15, 0)$ **5.** $(0, 6)$ and $(6, 0)$ **6.** $(7, 7)$ and the answer to (5)

7. (*a*) What is the slope of the line through the two points in Exercise 5?
(*b*) What is the equation of this line in slope-intercept form?

8. (*a*) What is the slope of the straight-line through $(-1, -1)$ and $(3, 11)$?
(*b*) Does $(-1, -1)$ lie on the straight line through $(-3, -7)$ and $(3, 11)$?
(*c*) What is the slope of the straight line through $(-1, -1)$ and $(3, 15)$?

9. Which is the steepest upward sloping line in Exercise 8?

For Exercises 10 to 14 find the equation of the line as described.

10. With slope 3 through the point $(4, 5)$
11. Through the points $(4, 1)$ and $(5, 5)$
12. With slope -2 and y intercept $(0, 1)$
13. Vertical through $(3, 4)$
14. Horizontal through $(3, 4)$
15. Write your answer to Exercise 11 in:

(*a*) Point-slope form (*b*) Slope-intercept form
(*c*) General form

16. Write the equation of the line with y intercept 3 and parallel to the line with equation $x + 4y - 12 = 0$.

17. Find the equation of the line through (0, 1) with negative slope whose x and y intercepts are the same distance from the origin.

In 18 to 23 graph the equation.

18. In Exercise 10
19. In Exercise 11
20. In Exercise 12
21. In Exercise 17
22. $2x - y = 4$
23. $y - 2x + 1 = 0$

24. Write your answer to Exercise 10 in (*a*) point-slope form, (*b*) slope-intercept form, (*c*) general form.

25. (*a*) Determine an equation of the line through (0, 1) whose y coordinate increases by 1 whenever the x coordinate increases by 1.
(*b*) Are the lines in Exercises 17 and (*a*) perpendicular?
(*c*) Graph each on the same coordinate system.
(*d*) Give the equation in point-slope form of the line that is parallel to that in part (*a*) and passes through the origin.

26. Show that the diagonals of a rectangle bisect each other. (You may assume that one corner is at (0, 0) and two sides are on the axes.)

27. Given the three points $(-3, 4)$, $(3, 2)$, and $(12, -1)$:
(*a*) Show they lie in a straight line using slopes.
(*b*) Show they lie in a straight line using distance.
(*c*) Give an equation for this line.

28. (*a*) The points (2, 8), (10, 2), and (2, 2) form the vertices of a right triangle. Show that the perpendicular bisectors of each side meet in a single point. What is this point?
(*b*) Show that the situation in part (*a*) holds for any right triangle. (You may assume that the two legs lie on the axes.)
(*c*) Does part (*b*) prove that the midpoint of the hypotenuse is equidistant from the vertices?

29. Determine three points on the line segment joining (2, 8) and (10, 2) that subdivide the segment into four pieces of equal length.

30. Prove that the diagonals of a square are perpendicular.

31. The three points $(-1, 7)$, $(2, 8)$ and $(10, 2)$ are three of the vertices of a parallelogram.
(*a*) Name a fourth vertex in the first quadrant.
(*b*) Name another possible fourth vertex in the second quadrant.

32. Prove that the slope of a straight line is independent of the points used to compute it using similar triangles as follows. Let Q_1 and Q_2 and P_1 and P_2 be two pairs of points on a line l.
(*a*) Show that the right triangles whose hypotenuses are the line segments Q_1Q_2 and P_1P_2 are similar.
(*b*) Show that the appropriate ratios (giving the slope) are equal.

33. Prove [1.2.4(f)].

Section 1.3: Circles and Parabolas

Objectives:

1. Graph an equation of a circle, giving its center and radius.

2. Determine an equation for a given circle.

3. Graph an equation of a parabola, giving its axis of symmetry, extreme point, intercepts, and concavity.

CIRCLES

We now know how to write an equation for a straight line and, conversely, which kinds of equations have straight-line graphs. The next simplest kind of geometric figure is the circle. The following example develops an equation for a circle with given center and radius.

Example 1 Determine an equation for the circle with center (4, 7) and radius 3.

Solution As shown in Fig. 1.10, a point (x, y) is on the given circle if and only if its distance from (4, 7) equals 3. That is, by the distance formula (1.2.1),

$$\sqrt{(x-4)^2 + (y-7)^2} = 3$$

Squaring both sides of the equation, we obtain a simplified equation of the circle:

$$(x-4)^2 + (y-7)^2 = 9 \quad \bullet$$

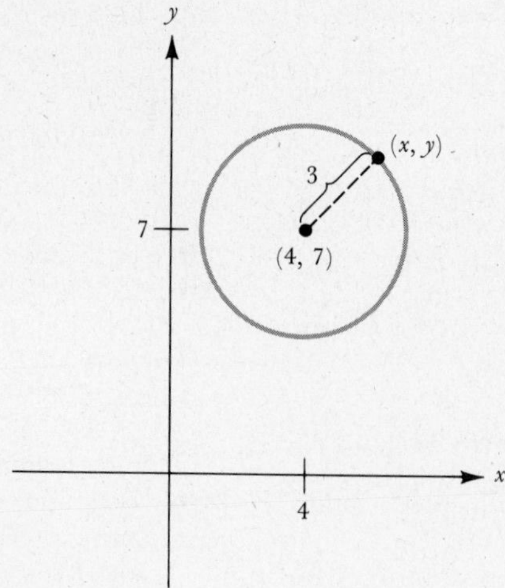

Fig. 1.10

The same method applies in the general case to give the following theorem.

1.3.1 **Theorem** The equation of the circle with center (h, k) and radius $r (r > 0)$ is

$$(x - h)^2 + (y - k)^2 = r^2$$

Note that the equation in Theorem (1.3.1) is *equivalent* to the distance equation

$$\sqrt{(x - h)^2 + (y - k)^2} = r$$

because $r > 0$. [This means that every solution of (1.3.1) is also a solution of the distance equation.]

Given the geometric information describing a circle, we are now able to determine its equation. We can also reverse the process, determining the circle when given the equation.

Example 2 Describe the circle with equation $(y - 1)^2 + (x + 5)^2 = 17$.

Solution Regarding $x + 5$ as $x - (-5)$, we see that this is an equation of the circle with radius $\sqrt{17}$ and center $(-5, 1)$. ●

Often the information determined in Example 2 is less obvious. By multiplying out the factors of the given equation and combining all terms on the left side, we obtain the equivalent equation

$$x^2 + 10x + y^2 - 2y + 9 = 0$$

We *could* have been asked to describe the graph of *this* equation. However, by separately completing the square of the x terms and then the y terms, we get

$$(x^2 + 10x + 25) + (y^2 - 2y + 1) + 9 = 25 + 1$$

(We added 25 and 1 to the right-hand side of the equation to balance what we added when we completed the square on the left.)

Finally, we write the expressions in parentheses as perfect squares and subtract 9 from both sides to recover the original equation

$$(x + 5)^2 + (y - 1)^2 = 17$$

which we know is the equation of the circle with radius $\sqrt{17}$ and center $(-5, 1)$.

This technique is quite limited in scope, giving the equation of a circle *only* if the coefficients of x^2 and y^2 are equal (in which case we can divide through by this coefficient, making the coefficient of the squared terms one) and *only* if the number resulting on the right side of the equation (after completing the squares) is not negative. If the coefficients of x^2 and y^2 were different, then the graph would not be a circle, but an ellipse (if they had the same sign) or a hyperbola (if they had opposite signs). These cases are handled in Chap. 10.

PROGRESS TEST 1

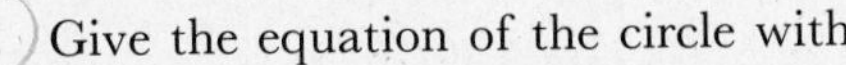

1. Give the equation of the circle with
(*a*) Center $(1, 4)$ and radius 2
(*b*) Center $(-3, 3)$ and radius $\sqrt{5}$
(*c*) Center (m, n) and radius t

2. Describe the graph (if it exists) of the equation
(*a*) $x^2 + 4x + y^2 - 6y + 1 = 0$
(*b*) $y^2 + x^2 + 6x + 12y + 3 = 0$
(*c*) $x^2 + 2x + y^2 - 5y + 40 = 0$

PARABOLAS

Our starting point is the graph of the equation

$$y = x^2$$

In the absence of further information, we form the graph by first filling in a table of data, including sufficient entries to reveal the general shape of the graph. We then plot the data and draw an unbroken curve through the resulting points (Fig. 1.11).

We plotted additional points at $x = \pm 1/2$ near where the graph seemed to be changing direction.

The graph of

$$y = x^2 + c$$

is the graph of $y = x^2$ slid c units vertically (up if $c > 0$, down if $c < 0$). (See Fig. 1.12.) A fact visually obvious in all these graphs is the vertical symmetry about the y axis.

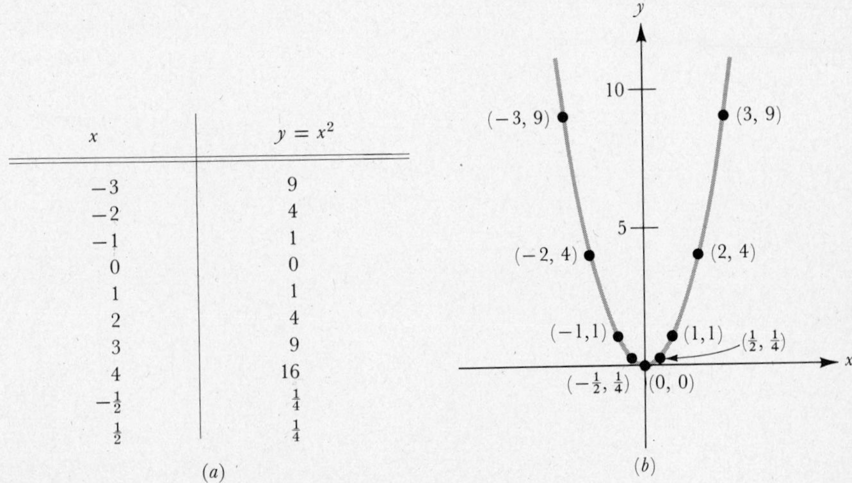

x	$y = x^2$
-3	9
-2	4
-1	1
0	0
1	1
2	4
3	9
4	16
$-\frac{1}{2}$	$\frac{1}{4}$
$\frac{1}{2}$	$\frac{1}{4}$

(*a*)

(*b*)

Fig. 1.11

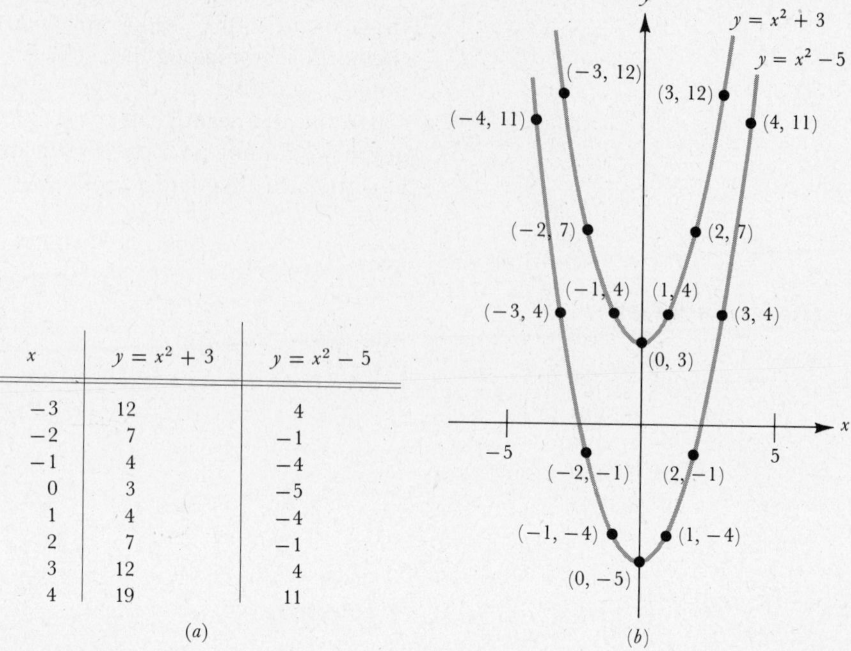

x	$y = x^2 + 3$	$y = x^2 - 5$
-3	12	4
-2	7	-1
-1	4	-4
0	3	-5
1	4	-4
2	7	-1
3	12	4
4	19	11

Fig. 1.12

We now graph $y = 2x^2, y = \frac{1}{10}x^2, y = -x^2$, and $y = -2x^2$, all on the same coordinate axes. (See Fig. 1.13.)

x	$y = 2x^2$	$y = \frac{1}{10}x^2$	$y = -x^2$	$y = -2x^2$
-3	18	$\frac{9}{10}$	-9	-18
-2	8	$\frac{4}{10}$	-4	-8
-1	2	$\frac{1}{10}$	-1	-2
0	0	0	0	0
1	2	$\frac{1}{10}$	-1	-2
2	8	$\frac{4}{10}$	-4	-8
3	18	$\frac{9}{10}$	-9	-18
4	32	$\frac{16}{10}$	-16	-32

(a)

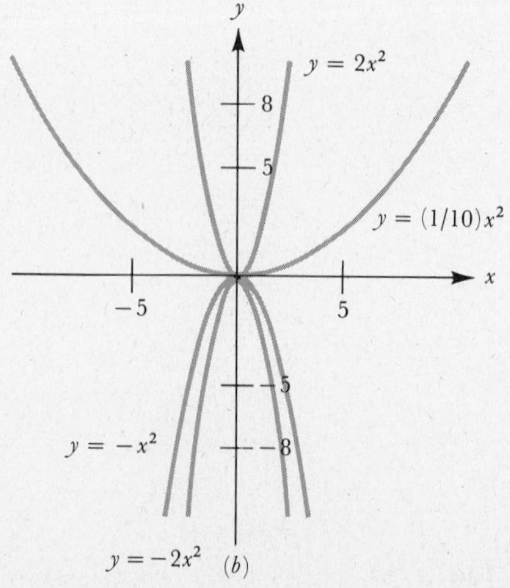

Fig. 1.13

From these graphs, we can see that the graph of $y = ax^2$ opens upward if $a > 0$ and downward if $a < 0$. The graph is flatter when a is small than when a is large. Again, the graph of

$$y = ax^2 + c$$

is merely the graph of $y = ax^2$ slid vertically by c units, and any such graph is symmetric about the y axis (the line $x = 0$).

Algebraically, *symmetry about the y axis means that the same value of y results when x is replaced by $-x$.*

As our last case we consider graphs in the general case

$$y = ax^2 + bx + c$$

Example 3 Sketch the graph of the equation $y = x^2 - 5x + 6$.

Solution The only method of attack at this point is to construct and then plot a table of data. (See Fig. 1.14.)

We plotted the point at $x = 5/2$ because the graph seemed to be turning between $x = 2$ and $x = 3$. ●

x	$y = x^2 - 5x + 6$
-3	30
-2	20
-1	12
0	6
1	2
2	0
3	0
4	2
5	6
6	12
$\frac{5}{2}$	$-\frac{1}{4}$

(*a*)

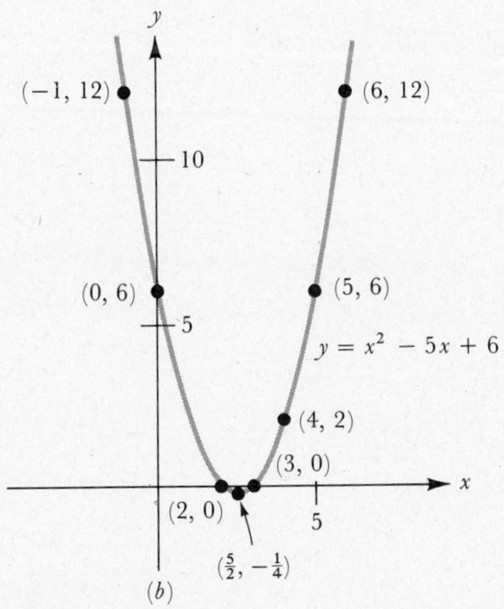

Fig. 1.14

Since the graph of $y = x^2 - 5x + 6$ intercepts the x axis when $y = 0$, we could have determined its x intercepts by solving the quadratic equation $x^2 - 5x + 6 = 0$ [by factoring or applying the quadratic formula (1.1.10)]. Of course we could determine the y intercept by setting $x = 0$. In fact both of these remarks apply to graphs of any equation that can be put in the form

$$y = ax^2 + bx + c$$

However, as yet we have no general technique that would guarantee that the graph of $y = x^2 - 5x + 6$, for example, reaches its minimum at the point $(5/2, -1/4)$. Although the ability to determine and deal with such "critical points" is among the central topics of differential calculus, we can use the idea of completing the square to handle not only this particular case but the general case as follows:

$$\begin{aligned} y = ax^2 + bx + c &= a\left(x^2 + \frac{b}{a}x + \quad\right) + c \\ &= a\left(x^2 + \frac{b}{a}x + \frac{b^2}{4a^2}\right) + \left(c - a \cdot \frac{b^2}{4a^2}\right) \\ &= a\left(x + \frac{b}{2a}\right)^2 + \left(c - \frac{b^2}{4a}\right) \end{aligned}$$

The second term merely slides the curve $y = a[x + b/(2a)]^2$ vertically by $c - b^2/(4a)$ units. The variable term is the product of a with a perfect square $[x + b/(2a)]^2$. For $a > 0$, the nonnegative term $a[x + b/(2a)]^2$ reaches its minimum, namely zero, when $x + b/(2a) = 0$, that is, when $x = -b/(2a)$. For $a < 0$, the nonpositive term $a[x + b/(2a)]^2$ reaches its maximum, namely zero, again when $x = -b/(2a)$. Proved most of 1.3.2:

1.3.2 **Theorem** The graph of the equation $y = ax^2 + bx + c$ reaches its extreme point when $x = -b/(2a)$. This is a minimum if $a > 0$, and a maximum if $a < 0$. In either case, the point (x, y) with $x = -b/(2a)$ is often called the *vertex* of the parabola. Furthermore, the graph is symmetric about the vertical line $x = -b/(2a)$.

Proof: The symmetry statement follows because the graph of $y = a[x + b/(2a)]^2 + [c - b^2/(4a)]$ behaves exactly as does the graph of the equation $Y = aX^2 + [c - b^2/(4a)]$ in the X-Y plane; here we relabel $x + b/(2a)$ as X. ■

Applying (1.3.2) to the equation $y = x^2 - 5x + 6$ of Example 3, we get $-b/(2a) = -(-5)/(2 \cdot 1) = 5/2$, which is the value of x we suspected yielded the minimum point, namely $(5/2, -1/4)$.

1.3.3 **Definition** The graph of any equation that can be put in the form $y = ax^2 + bx + c$ with $a \neq 0$ is called a *parabola*. It is said to be *concave up* if $a > 0$ and *concave down* if $a < 0$.

The line $x = -b/(2a)$ is called the *axis of symmetry* of the parabola.

Notice that the axis of symmetry does not depend on the constant c; the effect of c is only in the vertical height of the graph. Equations of the form $y = ax^2 + c$ have axes of symmetry $x = -0/(2a) = 0$ since $b = 0$.

Example 4 Sketch the graph of $y + x^2 = x + 12$.

Solution First we put the equation in the form $y = -x^2 + x + 12$. The axis of symmetry is $x = -1/[2(-1)] = 1/2$, the y intercept is 12, and the x intercepts (using the Quadratic Formula) are

$$x = \frac{-1 \pm \sqrt{1 + 48}}{2} = \frac{-1 \pm 7}{-2} = -3, 4$$

Finally, the graph (Fig. 1.15) reaches a maximum when $x = 1/2$, so

$$y = -\left(\frac{1}{2}\right)^2 + \frac{1}{2} + 12 = 12\frac{1}{4}$$

We can now sketch the graph. If we want more accuracy, we can plot more points. ●

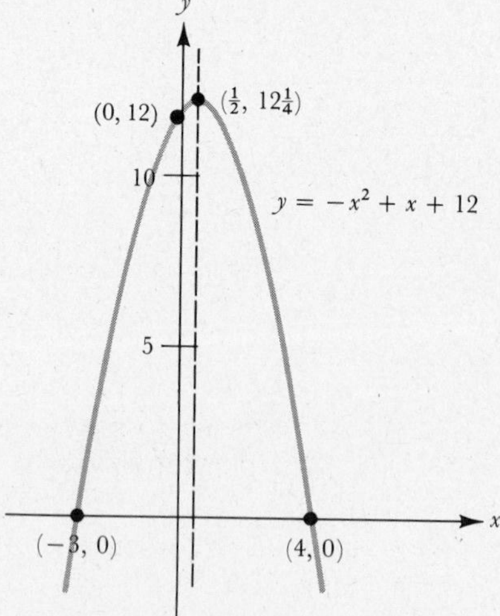

Fig. 1.15

PROGRESS TEST 2

Sketch the graphs of the following equations using the procedure of Example 4, indicating the axis of symmetry with a dashed line and plotting two additional points as a check.

1. $y + 2x^2 = 7x - 3$ **2.** $y = x^2 + 2x - 35$ **3.** $y = 2x^2 + x + 1$

ADDITIONAL REMARKS

Notice in Prob. 3, Progress Test 2, that (1) the graph has no x intercepts, (2) the equation $2x^2 + x + 1 = 0$ has no solutions, (3) the discriminant $1 - 8 = -7$ is negative, and (4) the graph is entirely on one side of the x axis. All four facts about $y = 2x^2 + x + 1$ are equivalent and support one another; they cannot disagree.

The contents of our entire discussion regarding the graphing of parabolas can be "turned on its side" in the sense that the roles of x and y can be interchanged to give an analogous collection of statements.

1.3.4 The graph of $x = ay^2 + by + c$ is a parabola with horizontal axis of symmetry $y = -b/(2a)$, concave right (opening to the right) if $a > 0$ or concave left if $a < 0$, reaching its extreme point when $y = -b/(2a)$, having x intercept c, and y intercepts

$$y = \frac{-b \pm \sqrt{b^2 - 4ac}}{2a}$$

Example 5 Sketch the graph of $x = -y^2 + 3y + 3$, indicating the axis of symmetry with a dashed line.

Solution As shown in Fig. 1.16, the axis of symmetry is $y = -3/2(-1) = 3/2$; the parabola is concave left, since $a = -1 < 0$. The x intercept is 3, and the y intercepts are

$$y = \frac{-3 \pm \sqrt{9 + 12}}{-2} = \frac{3 \mp \sqrt{21}}{2}$$

The extreme point has x coordinate $x = -(3/2)^2 + 3(3/2) + 3 = 21/4$. ●

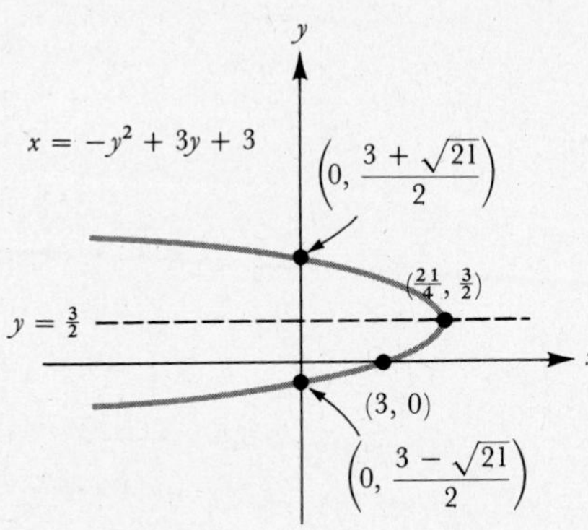

Fig. 1.16

We shall use the phrase "turn . . . on its side" repeatedly in the future in exactly the same way we have just done—to mean that we interchange the roles of x and y.

SECTION 1.3 EXERCISES

Graph Equations 1 to 3 using tables of data.

1. $y = 3x^2 + x - 4$

2. $y = x^2 + 3x + 5$

3. $x = y^2 - 4y$

For Exercises 4 to 13: (a) put the equation into an appropriate form and explain whether its graph is a circle, a parabola, or neither; (b) if it is a circle, give its center and radius, and if it is a parabola, give its axis of symmetry, its concavity (up or down, left or right) and its extreme point (as an ordered pair); (c) if it is a circle or parabola, sketch its graph.

4. $x = y^2 + 2y + 1$

5. $y - x^2 = x - 20$

6. $x^2 + y^2 - 9 = 0$

7. $x^2 - 4x + y^2 + 6y - 12 = 0$

8. $x + y^2 - 4y + 4 = 0$

9. $4x^2 - 12x + 4y - 6 = 0$

10. $y = x^2 + 2x + 2$

11. $x^2 + y^2 + 36 = 0$

12. $y = 4x^2 + 52$

13. $x - y^2 + 2y = 1$

For the equations given in 14 to 21 that do not have graphs, explain why. Otherwise, graph the equation.

14. $y = x^2 - 1$
15. $x = y^2 - 1$
16. $x^2 = 4 - y^2$
17. $(x - 2)^2 + (y + 1)^2 = 49$
18. $x^2 - 4x + y^2 + 2y - 44 = 0$
19. $y - 3x^2 - 11x + 4 = 0$
20. $x^2 + (y - 5)^2 + 20 = 0$
21. $x^2 + y^2 = 0$

CHAPTER EXERCISES

Review Exercises

Solve inequalities 1 to 15 and graph the solution on the number line.

1. $4x - 1 \leq 2$
2. $x + 6 < 7x - 3$
3. $2x + 1 > 2(3x - 10)$
4. $4x - 5 \geq 2(x - 2)$
5. $3x + 7 > -20$
6. $|x - 5| < 5$
7. $|x - 3| > 2$
8. $|2x + 6| \geq 3$
9. $|1 - 2x| \leq 4$
10. $|x + 5| \geq 0$
11. $-|x + 5| \geq 0$
12. $|4x - 7| > 12$
13. $|2 - 5x| \geq 12$
14. $|x - 1| \geq -1$
15. $|x - 1| < 0$

Complete the square of quadratic expressions 16 to 24.

16. $x^2 - 6x + 5$
17. $x^2 + 4x + 8$
18. $x^2 - x - 2$
19. $2x^2 - x - 3$
20. $-x^2 + x + 1$
21. $2x^2 - 2\sqrt{5}x + 6$
22. $8x^2 + 2x - 15$
23. $3x^2 + 2x - 1$
24. $15x^2 - 13x + 2$

Solve inequalities 25 to 30 using the chart method, and graph your solution on a number line.

25. $\dfrac{(x^2 - x - 6)(2x^2 - x)}{(x^2 + 2x + 1)(x + 3)} > 0$
26. $\dfrac{(2x + 1)(x^2 - 1)(x - 4)}{(x - 3)(2x - 1)^2} \geq 0$
27. $\dfrac{(x^3 + 1)(x^2 + 1)(x - 1)}{2x - 3x^2} \leq 0$
28. $\dfrac{(3 - 2x)(-x^2 + 5x - 6)}{x(x^2 - 4)(x + 5)} < 0$
29. $\dfrac{(x^2 + 1)(x^4 + 1)}{(x^2 + 2)(x^6 + 6)} > 0$
30. $\dfrac{x^5(x^2 + 1)(x^2 - 4)^2}{-(x - 1)^2} > 0$

For the pairs of points in 31 to 34, determine (a) the distance between them, (b) the midpoint of the line segment joining them, (c) the slope of the straight line through them, (d) the equation of this straight line in general form, and (e) the equation of the perpendicular bisector of the line segment joining the points.

31. $(3, 6), (5, 1)$
32. $(-4, 5), (5, 7)$
33. $(-4, -2), (-7, -6)$
34. $(-5, -4), (1, 4)$
35. In general form, give the equation of the line
(*a*) With x intercept $(3, 0)$, y intercept $(5, 0)$
(*b*) Through $(3, 0)$ with slope 2
(*c*) Parallel to the line in part (*b*) through $(1, 5)$

In 36 to 39 give the equation of the circle with:

36. Center $(1, 2)$ and radius 4
37. Center $(5, 6)$ and radius $1/3$
38. Center $(-2, -4)$ and through the origin
39. Center $(-2, -4)$ and tangent to the x axis

By completing the squares of the x and y terms in 40 to 44, describe the graphs of the circles with the given equations.

40. $x^2 + y^2 - 8x + 2y - 24 = 0$
41. $x^2 + y^2 + 12x - 16y = 0$
42. $8x^2 + 8y^2 - 72x - 32y + 96 = 0$
43. $2x^2 + 2y^2 - 12x + 16y = 0$
44. $7x = 5y - x^2 - y^2 + 44$

For each equation in 45 to 70, tell whether its graph is a straight line (if so, give its slope and y intercept), a circle (if so, give its center and radius), a parabola (if so, give its axis of symmetry, extreme point, and intercepts), or none of these. If the graph is a straight line, circle or parabola, sketch it.

45. $5x^2 + 2y^2 = 36$
46. $y = -2x + 1$
47. $y = 3x + 2$
48. $y = x(x - 5) + 3$
49. $y^2 + 2y - 6x - 20 = 0$
50. $y^2 - 2y + 8x - 39 = 0$
51. $y - 4 = x + 4$
52. $y = 3x^2 + x - 4$
53. $y = x^2 + 3x + 5$
54. $3 + x^2 = 4 - y^2$
55. $x^2 - 4x - 2y + 10 = 0$
56. $5x^2 + 5y^2 - 7y - 26 = 0$
57. $\dfrac{y - 1}{x - 3} = 2$
58. $y = \frac{1}{2}x - 5$
59. $x = 46.1$
60. $13x - 8y - 1 = 0$
61. $(x - 3)^2 - (y - 2)^2 = 25$
62. $x^2 + y^2 - \frac{13}{3}x + y + 2 = 0$
63. $\dfrac{1}{xy} = y^2$
64. $y = xy + 3$
65. $x^2 = -(y - 2)^2 + 4$

66. $y = 4y^2 - 3x + 1$

67. $3x - y^2 = 6y + 9$

68. $4x - 3 = y$

69. $x^2 + 2y^2 = 49$

70. $4x - y + 2 = 0$

Miscellaneous Exercises

Sketch the graphs of equations 1 to 4:

1. $y = |3x - 2|$

2. $y = 3x - 2$

3. $(x - 3)^2 + (y + 5)^2 - 25 = 0$

4. $y^2 - 4y + x^2 + 12x + 31 = 0$

5. Determine three points on the line segment joining $(1, -1)$ and $(13, 15)$ that divide the segment into four equal parts.

6. What is the smallest number reached by the expression $x^2 - 3x + 5$?

7. Solve: (*a*) $|3x + 8| \geq 9$ (*b*) $|x| < -1$

(*c*) $\dfrac{(x + 2)(x - 1)^2(x^2 + x - 6)^2}{(x - 3)(x^2 + 2x + 2)(x + 9)} \geq 0$

8. Prove that the diagonals of a rhombus are perpendicular bisectors of one another.

9. Assuming that the tangent to a circle at (x_0, y_0) is perpendicular to the radius through (x_0, y_0) (a fact that will be proved later), determine an equation (in point-slope form) for the tangent to the circle $x^2 + y^2 = 25$ at the point $(3, 4)$.

10. Repeat Exercise 9 in more generality. That is, determine an equation for the line tangent to $x^2 + y^2 = r^2$ at (x_0, y_0) (where, of course, $x_0^2 + y_0^2 = r^2$).

11. (*a*) For what value of b does $x + by = 3$ go through the point $(4, 5)$?

(*b*) For what value of c does $x + 2y + c = 0$ go through the point $(4, 5)$?

12. What is the slope of any line parallel to a line perpendicular to the line through $(-3, -7)$ and $(3, 11)$?

13. What is the slope of any line with equation $y = m$ (m a constant)?

14. Where do the lines $y = 2x + 3$ and $2y = 2x + 3$ cross?

15. Do the lines $y = 2x + 3$ and $2y = 4x + 4$ cross? Explain.

SELF-TEST 

1. Solve the following inequalities and graph their solutions:

(*a*) $3(2x - 1) \leq 8x + 5$ (*b*) $|3x - 6| > 10$

(*c*) $\dfrac{x^4(4 - x)(x^2 - x - 6)}{(x + 3)^2(x^2 - 25)} \leq 0$

2. Graph the following equations:

(*a*) $4x - 2 = 7$ (*b*) $y = x^2 + x - 6$

(*c*) $x^2 + 4x + 6y + y^2 - 12 = 0$ (*d*) $4x - 2y + 5 = 0$

3. Let $P = (1, -3)$ and $Q = (3, -1)$.

(*a*) Determine the midpoint of the line segment joining P and Q.

(*b*) Give an equation of the line through P and Q.

(*c*) Give an equation of the line through the origin parallel to that in (*b*).

4. Give the equation of the circle whose center is at $(6, 8)$ and passes through the origin.

Chapter 2 ✗

LIMITS, DERIVATIVES, AND CONTINUITY OF FUNCTIONS

CONTENTS

In this chapter we begin our study of the differential calculus, which is the study of functions and how they change. In Sec. 2.1 we introduce the idea of function. We relate functions to equations, show how they can be described in numerical, geometric, or algebraic terms, and then show various ways of combining functions to get new functions.

Our basic tool in the study of functions is the derivative, which in turn rests on the more fundamental notion of limit. The concept of limit is a subtle and profound idea whose meaning unfolded gradually through the history of mathematics. As you progress through the book, seeing it in its many forms, you too will likely grasp its meaning only gradually. We introduce it in Sec. 2.2 at about the level of precision attained in the eighteenth century, and then go on in Sec. 2.3 to use it to define the derivative as a limit of average rates of change. We interpret the derivative of a function as the instantaneous rate of change of that function (especially as instantaneous velocity), and then in Sec. 2.4 as the slope of the graph of the function. We shall return to a careful exposition of the logically precise definition of limit in Chap. 9, Sec. 9.1. However, this section can be covered any time after Sec. 2.2. The chapter closes with a brief exploration of the idea of continuity and how it relates to the previously introduced ideas.

Section 2.1: Functions: Ways of Describing and Combining Them

Objectives:

1. Describe a function algebraically, geometrically, and numerically.
2. Recognize when a given equation in two variables defines one variable explicitly as a function of the other and when one variable is defined only implicitly as a function of the other.
3. Describe the natural domain and the range of a given function.
4. Form the sum, difference, product, quotient and composition of a pair of functions.

DEFINITION OF FUNCTION

Few mathematicians or users of mathematics would dispute that the most important and useful idea of all mathematics is that of function. To use mathematics is to use functions. A function is used to describe how one quantity depends upon another, ranging from simple verifiable situations (the area A of a square depends upon the length x of its sides according to the formula $A = x^2$), to complex hypothetical situations (the probability P of developing lung cancer depends upon the quantity x of tar taken into the lungs according to some formula). Much of the information (facts, relationships, laws) expressed by any science, be it a physical, biological, or social science, is expressed in the form of functions.

We saw many functions in Chap. 1 given in the form of equations, such as

$$y = x^2 + 2x + 1$$

Here the quantity y depends upon x. However, not all equations describe functions. The following operational definition will clarify the distinction.

2.1.1 **Definition** A *function* is a rule by which each number in a given set of numbers is transformed into another UNIQUELY DETERMINED number. This given set of numbers to which the function applies is called its *domain.* The set of numbers resulting when the function is applied to numbers in the domain is called its *range.*

Consider the following function whose domain is R, the set of all real numbers, and which is defined by the following rule applied to a real number: square it, add twice the original number, and then add 1 to the result.

The number 3 is transformed by this function to the uniquely determined number $(3)^2 + (2) \cdot (3) + 1 = 16$. The number -2 is transformed to the number

$$(-2)^2 + (2) \cdot (-2) + 1 = 1$$

This function transforms any real number x to the unique number

$$x^2 + 2x + 1$$

Obviously, this is the same function mentioned earlier in the form

$$y = x^2 + 2x + 1$$

When a function is given with one variable, say y, equal to an expression involving another variable, say x, we refer to x as the *independent variable* and y as the *dependent variable* (because y *does* depend upon x). Writing a function in this way emphasizes that it *relates* two variables—x and y in this case.

However, as we shall see later, not every equation that gives y in terms of x gives y as a *function* of x. The crucial matter is that y be *uniquely determined* by x.

It is useful to view a function as a *rule for transforming numbers.* In such a case the function is usually given a name, with the most popular single-letter name for functions being simply f.

If f is the name of the function, we can describe f completely by describing what f does to a typical number x in its domain. The usual way of stating that the function f is applied to a number x is "$f(x)$," read "f of x," or "f applied to x." If we call the original function previously introduced f, then the result of applying f to a typical number x in its domain R is given by the equation

$$f(x) = x^2 + 2x + 1$$

The results of applying f to the *specific* numbers we began with are

$$f(3) = 16 \qquad f(-2) = 1$$

We shall often make statements such as "consider the function $f(x) = 4x + 1$ with domain R." Strictly speaking, the name of the function itself is f, and $f(x)$ is the result of applying f to a typical number x in its domain. But since *any function is known by what it does to the numbers in its domain* and since x represents any number in its domain, there should be no confusion in referring to a function by referring to what it does to a typical number x.

It may turn out that functions which look different are really the same. For example,

$$f(x) = x^2 + 2x + 1 \qquad g(x) = (x + 1)^2 \qquad h(w) = (w + 1)^2$$

all with domain R, represent the same rule. Given any real number, the result of applying all three functions is the same. We even could have written

$$f(\ ) = (\ )^2 + 2(\ ) + 1$$

requiring only that the number substituted into the parentheses be a real number. The function is defined by the rule and not by the letters used to represent it.

Example 1 Given that the area of a circle is a function of its radius r with $A(r) = \pi r^2$, show that (*a*) A increases by a factor of four when the radius doubles and (*b*) A increases by a factor of b^2 when the radius increases by a factor of b.

Solutions (*a*) Substitute $2r$ into the area function to get

$$A(2r) = \pi(2r)^2 = \pi 4r^2 = 4A$$

where $A = \pi r^2$.

(*b*) Similarly,

$$\begin{aligned} A(br) &= \pi(br)^2 \\ &= \pi b^2 r^2 \\ &= b^2 A \quad \bullet \end{aligned}$$

Often when a function is given, its domain is not mentioned.

2.1.2 **Definition** The set of all x for which $f(x)$ is a real number is called the *natural domain* of f.

When the domain of f is not mentioned, we shall assume it to be its natural domain. The main issues involved in determining the natural domain of a function are division by zero and even roots of negative numbers.

Example 2 Determine the natural domain of each of the following:

(*a*) $f(x) = \dfrac{1}{x^2 - 4}$ (*b*) $g(x) = \sqrt{9 - x^2}$ (*c*) $h(x) = \dfrac{1}{\sqrt[4]{1 - x^2}}$

Solutions (*a*) The denominator is zero—that is, $x^2 - 4 = 0$—if $x = \pm 2$, so the natural domain of f is the set of all real numbers *except* ± 2.

(*b*) $9 - x^2 \geq 0$ if $9 \geq x^2$ or $-3 \leq x \leq +3$. Hence the natural domain of g is the closed interval $[-3, +3]$.

(*c*) $1 - x^2 > 0$ if $1 > x^2$, so the natural domain of h is the open interval $(-1, 1)$. Note that the numbers ± 1 are not part of the natural domain because $h(\pm 1)$ would lead to division by 0. $\bullet$

When it is necessary to restrict the natural domain of a function artificially, we specify the domain explicitly.

PROGRESS TEST 1

For each of the functions in 1 to 3, compute (a) $f(t)$, *(b)* $f(x+h)$, *(c)* $f(2x)$. *Do not simplify your answers.*

1. $f(x) = 3x^2 - 5x + 2$
2. $f(x) = ax + b$
3. $f(s) = s^3$

4. The circumference C of a circle is given as a function of its radius r by $C = 2\pi r$. What happens to the circumference when the radius doubles?

5. Determine the natural domain of these functions:

(*a*) $f(x) = \sqrt{x^2 - x - 2}$ (*b*) $y = \dfrac{1}{\sqrt[3]{x-8}}$

THE NUMERICAL AND GRAPHICAL DESCRIPTIONS OF A FUNCTION

All the descriptions of functions previously given were algebraic in that in each case the rule has been given explicitly by an algebraic formula. However, since equations can be graphed, and since our functions were given in terms of equations, we can graph functions using what we already know about graphing equations. Recall that an equation in x and y has a graph in the coordinate plane consisting of all pairs (x,y) satisfying the equation. Thus we graph a function $f(x)$ by graphing the points $(x, f(x))$.

2.1.3 **Definition** If f is a function with domain D, then the *graph of* f is the collection of all points $(x, f(x))$ as x varies over D. In particular, to say that the point (x,y) is on the graph of f is equivalent to the statement that $y = f(x)$.

In graphing a function $f(x)$ we have an option in labeling the vertical axis either $f(x)$ or y. It makes little difference, although if we use y, then we really have the graph of the equation $y = f(x)$. We shall refer to the graph of $f(x)$ and the graph of $y = f(x)$ interchangeably.

Often the functions we plot will involve familiar equations, so we shall not need to plot many values of $f(x)$. For example, we know from (1.3.3), p. 22, that the graph of

$$y = f(x) = x^2 + 2x + 1$$

is a parabola, concave up, with vertical axis of symmetry $x = -2/(2 \cdot 1) = -1$, y intercept $(0, 1)$, and so forth.

On the other hand, when confronted with an unfamiliar function, it is most useful to compute several values of the function and compile the results in a table, particularly near points where the behavior of the function may be in doubt. Near such points plot as many values as you need to understand the function.

Example 3 Determine the natural domain and then graph the function

$$f(x) = \frac{x^2 - 9}{x - 3}$$

Solution The function is not defined at $x = 3$ because $(3^2 - 9)/(3 - 3) = 0/0$, which is not a real number. Otherwise, $f(x)$ is defined for all other real numbers. We compute and plot some values near $x = 3$. See Fig. 2.1.

Notice that the points lie in a straight line. (We labeled as many points as is practical.) We indicate that the function is not defined at $x = 3$ by placing an open circle or "hole" in the graph of $f(x)$ at $x = 3$. ●

Obviously, except for the hole, Fig. 2.1(*b*) gives the graph of $g(x) = x + 3$, a fact also apparent from an algebraic manipulation:

x	$f(x) = (x^2 - 9)/(x - 3)$
-1	2
0	3
1	4
2	5
2.5	5.5
2.9	5.9
2.99	5.99
3.01	6.01
3.1	6.1
3.5	6.5
4	7

(*a*)

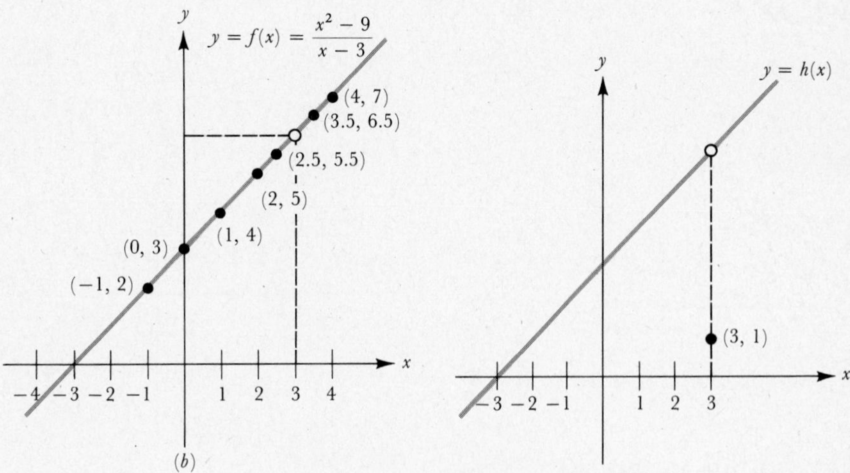

Fig. 2.1

Fig. 2.2

$$f(x) = \frac{x^2 - 9}{x - 3} = \frac{(x - 3)(x + 3)}{x - 3} = 1 \cdot (x + 3) \qquad \text{for } x \neq 3$$

It is important to realize that $(x - 3)/(x - 3) = 1$ only if $x \neq 3$. For $x = 3$, $(x - 3)/(x - 3)$ is not defined. These observations make it clear, however, that no matter how close x is to 3, $f(x)$ *is* defined, and equals $x + 3$. Thus, for example,

$$f(2.99999999) = 5.99999999$$

Nonetheless $f(x) = (x^2 - 9)/(x - 3) \neq g(x) = x + 3$ because of the difference in the domains of $f(x)$ and $g(x)$. In fact, it is worth stating the following general principle:

2.1.4 Two functions f and g are equal if and only if:
(*a*) The domain of f equals the domain of g.
(*b*) $f(x) = g(x)$ for each number x in the common domain of f and g.

We can define yet another function, call it $h(x)$, closely related to $f(x) = (x^2 - 9)/(x - 3)$ and $g(x) = x + 3$ by defining it to be equal to the previous two functions for all $x \neq 3$ and defining $h(3) = 1$, for example. The customary method for defining a function in this way is as follows:

$$h(x) = \begin{cases} x + 3 & \text{if } x \neq 3 \\ 1 & \text{if } x = 3 \end{cases}$$

Thus the domain of $h(x)$ is all of R, and its graph is as shown in Fig. 2.2.

Example 4 Graph the function

$$A(x) = \begin{cases} -1 & \text{if } x < 0 \\ 0 & \text{if } x = 0 \\ 1 & \text{if } x > 0 \end{cases}$$

Solution The solution is given in Fig. 2.3. Notice that the function is always -1 on $(-\infty, 0)$ and is always $+1$ on $(0, +\infty)$. ●

We shall refer to a function that is constant for *all* real numbers as a *constant function.* Thus the function defined by $f(x) = 4$ for all x is a constant function. The graph of $y = f(x)$ is, of course, the horizontal line $y = 4$.

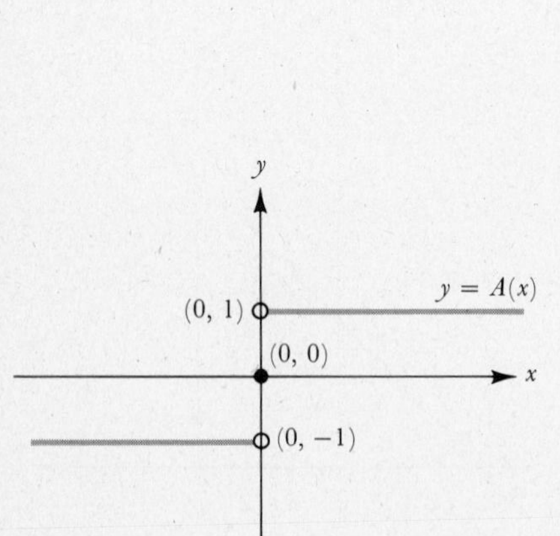

Fig. 2.3

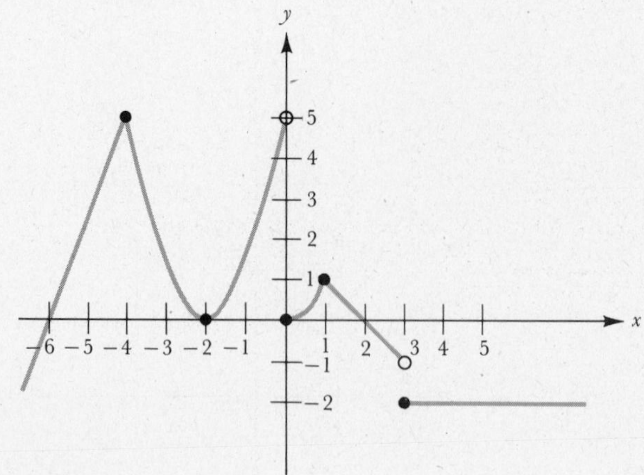

Fig. 2.4

Using the process of patching different algebraic formulas on disjoint intervals, it is possible to construct almost any sort of function.

For example, the function

$$f(x) = \begin{cases} 2x + 12 & \text{for} \quad x \leq -4 \\ x^2 + 4x + 4 & \text{for} \ -4 < x < 0 \\ x^3 & \text{for} \quad 0 \leq x < 1 \\ 2 - x & \text{for} \quad 1 \leq x < 3 \\ -2 & \text{for} \quad 3 \leq x \end{cases}$$

is graphed in Fig. 2.4.

EQUATIONS AND FUNCTIONS

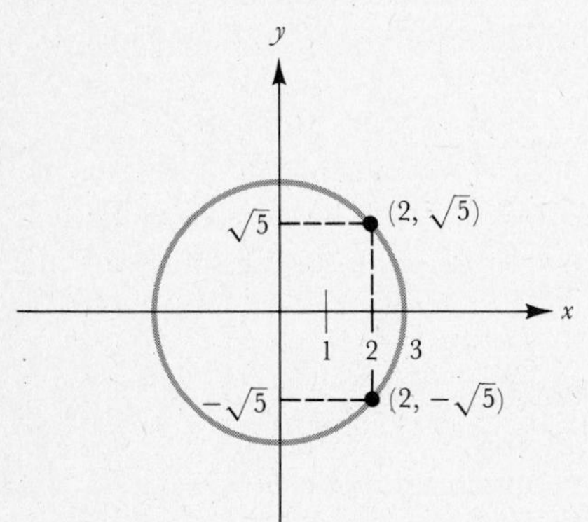

Fig. 2.5

Not all equations in x and y when solved for, say, y automatically yield y as a function of x. By solving

$$x^2 + y^2 = 9$$

for y, we get $y = \pm\sqrt{9 - x^2}$ (see Fig. 2.5). Thus each value of x yields two values of y, so the equation $x^2 + y^2 = 9$ does not define y as a function of x. Letting $x = 2$ yields $y = \sqrt{5}$ and $y = -\sqrt{5}$, or, put geometrically, the vertical line $x = 2$ hits the graph in two places. This suggests the following general rule.

2.1.5 If any vertical line hits the graph of an equation in x and y in more than one point, then the equation does not define y as a function of x (see Fig. 2.6).

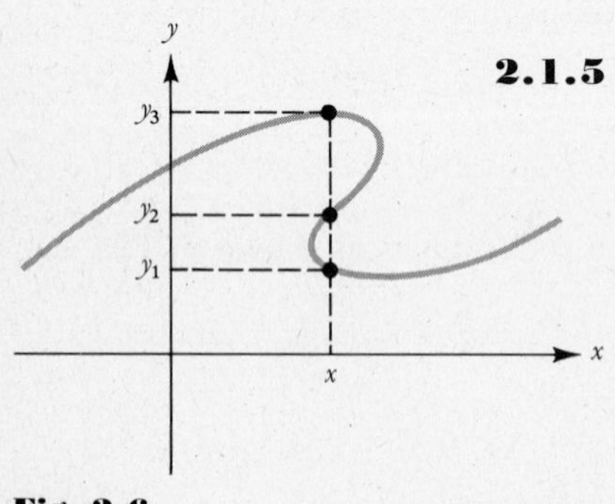

Fig. 2.6

On the other hand, an equation such as $x^2 + y^2 = 9$ (see Fig. 2.5) *does* define more than one function of x *implicitly*. For example, we can define the two functions

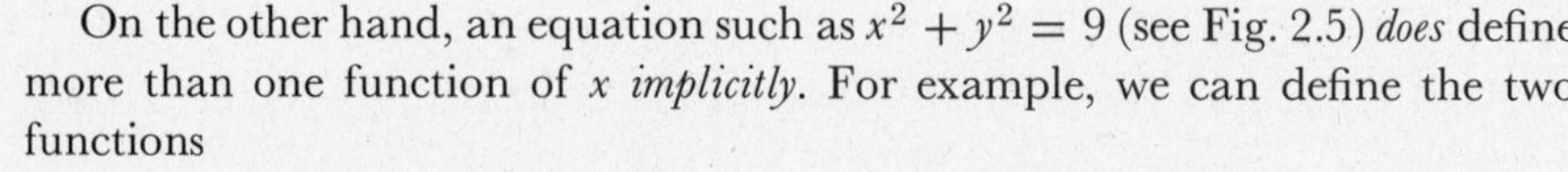

$$U(x) = \sqrt{9 - x^2} \qquad \text{and} \qquad L(x) = -\sqrt{9 - x^2}$$

(see Fig. 2.7). We shall have more to say on implicitly defined functions in Chap. 3.

Occasionally it is necessary or convenient to regard x as a function of y.

Example 5 Use the equation $x + 4 = y^2$ and its graph (a) to define x as a function of y and (b) to define two different functions of x (see Fig. 2.8).

Solution (a) Solving the equation for x in terms of y, we get

$$x = y^2 - 4$$

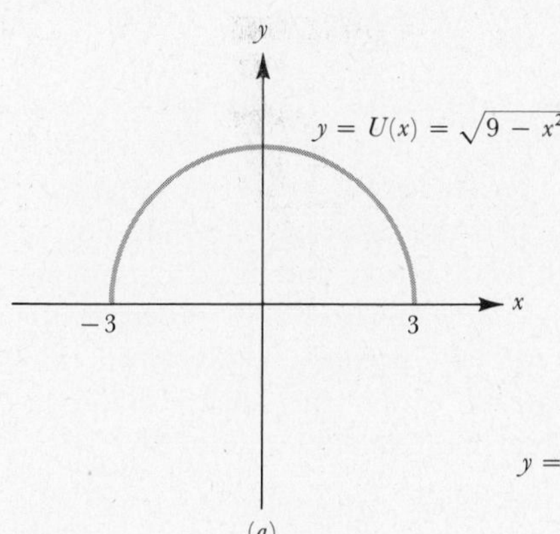

Fig. 2.7

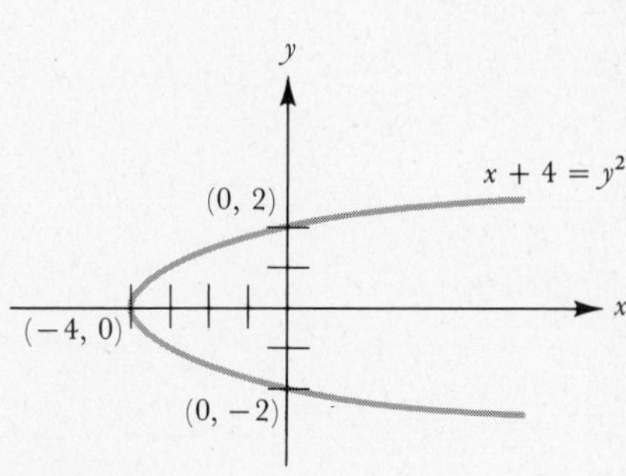

Fig. 2.8

which clearly gives x as a function of y: Each value of y yields a uniquely defined value of x. On the other hand, if we solve for y in terms of x, we do *not* get y as a function of x:

$$y = \pm \sqrt{x + 4}$$

Note also that any vertical line to the right of $x = -4$ hits the graph in two places.
(*b*) For $x \geq -4$, define $s(x) = \sqrt{x + 4}$ and $t(x) = -\sqrt{x + 4}$. ●

PROGRESS TEST 2

1. For $f(x) = (2x - 2)/(x - 1)$ (*a*) determine the natural domain, (*b*) graph the function using a table of data, (*c*) give a simple function that equals the original function everywhere except at one point, and (*d*) define and graph a new function that equals the previously defined functions on their common domain but equals -1 at the point where the original function was undefined.

2. Suppose $2x - 3y + 6 = 0$.
(*a*) Determine whether this equation gives y as a function of x.
(*b*) Determine whether this equation gives x as a function of y.

3. Repeat Prob. 2 for the equation $4y^2 = 8x + 2$.

4. Given the equation $4y^2 = 8x + 2$
(*a*) Determine two distinct functions of x defined implicitly.
(*b*) Give the domain of each.

5. (*a*) Connect the dots in the diagram below with a smooth curve that is to be the graph of a function of x.
(*b*) Repeat (*a*) using a dotted line that will be the graph of a function of y.

6. Sketch the graph of

$$f(x) = \begin{cases} 4 - x^2 & \text{for } x < 1 \\ 2 & \text{for } x = 1 \\ x + 2 & \text{for } x > 1 \end{cases}$$

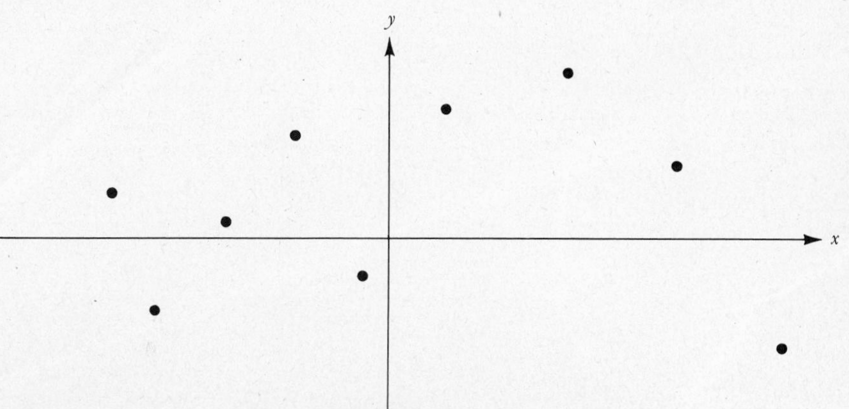

AN ALTERNATIVE DEFINITION OF FUNCTION

We saw that each function f has a graph consisting of all the ordered pairs $(x, f(x))$, where x is in the domain of f. Many authors take this as a starting point, defining a function as a collection of ordered pairs, no two of which have the same first element. Then the set of numbers that appear as first elements of these ordered pairs make up the domain of the function, and those that appear as second elements make up the range. The condition that no two ordered pairs have the same first element corresponds to our uniqueness condition.

The two definitions of function are logically equivalent, but we shall continue to use the rule-based definition except for Exercises 35 and 36.

THE ALGEBRA OF FUNCTIONS

We know how to add, subtract, multiply, and divide algebraic expressions and can use this knowledge to define similar operations on functions.

How should we define the sum of the function $f(x) = x^2$ and the constant function $g(x) = 3$? A natural answer would seem to be the function that transforms any number x into the number $x^2 + 3$. In fact, we shall define the sum of f and g, which is denoted $f + g$, in exactly this way:

$$(f + g)(x) = x^2 + 3$$

(Remember, this is not "$f + g$ *times* x", but "$f + g$ *of* x.") This certainly makes sense in terms of the graphs of f, g, and $f + g$, shown in Fig. 2.9. The effect of adding the constant function g to f is merely to slide the parabola f up by 3 units.

We define $f - g$ similarly:

$$(f - g)(x) = x^2 - 3$$

We can use this example as the basis for a more general definition.

2.1.6 **Definition** The *sum of two functions f and g*, denoted $f + g$, is defined by

$$(f + g)(x) = f(x) + g(x)$$

and the *difference of two functions f and g*, denoted $f - g$, is defined by

$$(f - g)(x) = f(x) - g(x)$$

The domain of $f \pm g$ consists of those numbers common to the domains of f and g.

Graphs of two functions and their sum are shown in Fig. 2.10.

A function such as

$$p(x) = x^4 - 3x^2 + 31$$

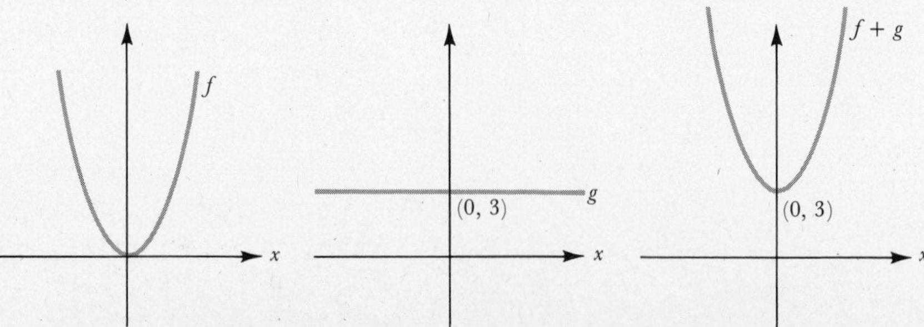

Fig. 2.9

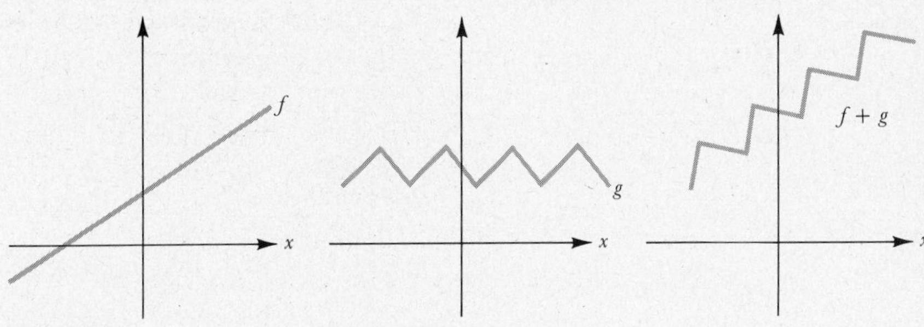

Fig. 2.10

is nothing more than the sum and difference of its separate terms, each of which is a function of x in its own right.

We define the product of two functions f and g similarly.

2.1.7 **Definition** The *product of f and g*, denoted $f \cdot g$, is defined by

$$(f \cdot g)(x) = f(x)g(x)$$

The domain of $f \cdot g$ consists of those numbers common to the domains of f and g.

2.1.8 **Definition** The *quotient of two functions f and g*, denoted f/g, is defined by

$$\left(\frac{f}{g}\right)(x) = \frac{f(x)}{g(x)}$$

The domain of f/g consists of those numbers x that are common to the domains of f and g *and* for which $g(x) \neq 0$.

Unfortunately, unlike the sum and difference, the product and quotient of functions do not lend themselves to easy graphical interpretation.

PROGRESS TEST 3

For each pair of functions f and g in Probs. 1 and 2, find the sum, difference, product, and quotient. Note where any of these is undefined. In Prob. 1 only, sketch the graph of the four resulting functions, compiling a table of data where necessary.

1. $f(x) = 5$, $g(x) = 3x + 1$ **2.** $f(x) = 2x^2$, $g(x) = x$

TYPES OF FUNCTIONS

It is useful to establish a vocabulary for the commonly occurring types of functions that can be built up using the previously discussed operations.

2.1.9 **Definitions**

(*a*) A function of the form $f(x) = mx + b$, where m and b are real numbers, is called a *linear function*.

(*b*) A function of the form $f(x) = ax^2 + bx + c$, where a, b, and c are real numbers ($a \neq 0$), is called a *quadratic function*.

(*c*) A function of the form $f(x) = b_nx^n + b_{n-1}x^{n-1} + \cdots + b_1x + b_0$, where n is a positive integer and $b_0, b_1, \ldots, b_n$ are real numbers, is called a *polynomial function* and is said to be *of degree n* if $b_n \neq 0$.

(*d*) Any quotient of polynomial functions, say $f(x) = p(x)/q(x)$, where $p(x)$ and $q(x)$ are polynomial functions, is called a *rational function.*

(*e*) Any function that can be built up from linear functions by adding, subtracting, multiplying, dividing, or taking *n*th roots is called an *algebraic function.*

By [2.1.9(*e*)],

$$f(x) = \sqrt[3]{\frac{x^4 + \sqrt[5]{x} + 1}{x^2 - \sqrt{x} - 1}}$$

is an algebraic function.

THE COMPOSITION OF FUNCTIONS

We shall now describe a more subtle way of combining functions. Let

$$f(x) = x^3 \qquad \text{and} \qquad g(x) = 2x + 1$$

Earlier we noted that we can substitute various symbols into a function. Thus we can write

$$f(a+b) = (a+b)^3, f(x+h) = (x+h)^3, f(2x) = (2x)^3, \ldots$$

The rule for f requires us to cube any real number put into the parentheses. Why not consider $f(2x + 1)$? According to the rule,

$$f(2x+1) = (2x+1)^3$$

We could have written this as

$$f(g(x)) = (g(x))^3 = (2x+1)^3$$

(Read this as "f of g of x equals g of x cubed," and so on.)

What we have done here is to *apply the function f to the result of applying g.* First g works on x, yielding $2x + 1$, and then f works on $g(x) = 2x + 1$, yielding $(2x + 1)^3$. More generally, we make the following definition:

2.1.10 **Definition** Given two functions f and g, the *composition of f with g*, denoted $f \circ g$, is defined by

$$(f \circ g)(x) = f(g(x))$$

The domain of $f \circ g$ consists of all numbers x in the domain of g for which $g(x)$ is in the domain of f.

Note that the product of f and g is shown by a dot, whereas the composition is indicated by a small circle.

It *does* make a difference whether we say "the composition of f with g" or "the composition of g with f." Given the above two functions

$$f(x) = x^3 \qquad \text{and} \qquad g(x) = 2x + 1$$

the composition of f with g is

$$(f \circ g)(x) = f(g(x)) = f(2x+1) = (2x+1)^3$$

whereas the composition of g with f is

$$(g \circ f)(x) = g(f(x)) = g(x^3) = 2x^3 + 1$$

(The rule given by g is to double the number and add 1. In this latter case the number happens to be the result of applying f, namely x^3.) Obviously $f \circ g$ and $g \circ f$ are not equal.

An algebraic function such as

$$h(x) = \sqrt{4x^2 + 2x + 3}$$

can be regarded as the composition of

$$f(x) = \sqrt{x} \qquad \text{with } g(x) = 4x^2 + 2x + 3$$

Here $h(x) = f(g(x)) = f(4x^2 + 2x + 3) = \sqrt{4x^2 + 2x + 3}$.

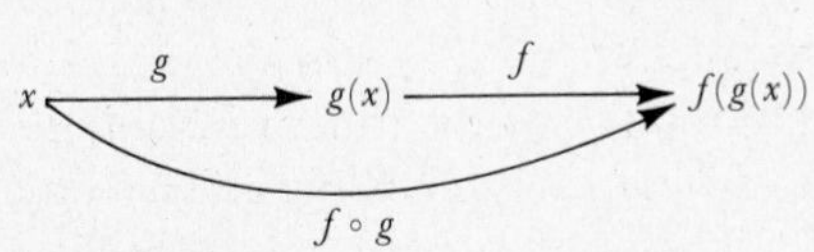

Fig. 2.11

We can think of composition pictorially as in Fig. 2.11. Here x gets transformed by g to $g(x)$, which in turn gets transformed by f to $f(g(x))$. Thus the composition, which we denote by $f \circ g$, transforms x to $f(g(x))$. The composition is really a chain of functions, one following another.

Remember, for the composition of f with g to be *defined* for an x, $g(x)$ must be in the domain of f; that is, $f(g(x))$ must be defined.

Example 6 Let $g(x) = x^2 - 4$, $f(x) = \sqrt{x} + 11$. Determine (*a*) $f \circ g$, (*b*) $g \circ f$, and (*c*) the domain of each composition.

Solutions

(*a*) $(f \circ g)(x) = f(g(x)) = f(x^2 - 4)$
$= \sqrt{x^2 - 4} + 11$

(*b*) $(g \circ f)(x) = g(f(x)) = g(\sqrt{x} + 11)$
$= (\sqrt{x} + 11)^2 - 4$
$= x + 22\sqrt{x} + 117$

(*c*) The domain of $(f \circ g)(x) = \sqrt{x^2 - 4} + 11$ is the set of x such that $x^2 - 4 \geq 0$; that is, $x \leq -2$ or $x \geq 2$. The domain of $(g \circ f)(x) = x + 22\sqrt{x} + 117$ is the set of x such that $x \geq 0$. ●

The domain of composition of $f \circ g$ may sometimes be a proper subset of the natural domain of the *rule* for $f \circ g$. For example, if $f(x) = x^2$ and $g(x) = \sqrt{x}$, then for $x \geqq 0$ $(f \circ g)(x) = x$. This rule has as its natural domain all real numbers, although $f \circ g$ applies only to nonnegative x.

Composition of functions is sometimes usefully expressed in terms of the equation approach to functions. In this case the dependent variable of one equation is substituted into the other equation.

Example 7 Suppose $A = \pi r^2$ and $r = 3t + 1$. Express A as a function of t.

Solution $A = \pi r^2 = \pi(3t + 1)^2$ ●

Example 7 may arise in a physical context. Since $A = \pi r^2$ gives the area of circle as a function of its radius, an interpretation of $r = 3t + 1$ as the radius of a growing circular oil spill as a function of time t yields an expression of the area A as a function of time.

PROGRESS TEST 4

1. For each pair of functions f and g, determine $f \circ g$, $g \circ f$, $f \circ f$, and $g \circ g$.
(*a*) $f(x) = 3x - 2$, $g(x) = 7x$
(*b*) $f(x) = \sqrt[4]{x}$, $g(x) = 2x + 1$
(*c*) $f(x) = \dfrac{1}{x + 1}$, $g(x) = x^2 + 3x + 2$
(*d*) $f(x) = -5$, $g(x) = x^4$

2. Let $f(x) = (x - 2)/3$ and $f^{-1}(x) = 3x + 2$ (read f^{-1} as "f inverse"). Note that f^{-1} does *not* mean $1/f$. Find $f \circ f^{-1}$ and $f^{-1} \circ f$.

3. Given y as a function x: $y = x^2 + 5x - 1$; and x as a function of w: $x = (3w + 2)/w$. Express y as a function of w. Determine y when $w = 1$.

THE THREE POINTS OF VIEW

Besides being a lesson on functions, this section is intended to make another point. A complete understanding of most things must involve more than one point of view. To understand what a human being is, the point of view of chemistry is surely not enough (we have all heard statements such as "the chemical constituents of the human body are worth approximately \$3.50," yet we certainly feel we are worth more than that!) We need to add the points of view of biology, psychology, sociology, anthropology, literature, art, philosophy, and so on to get a more complete idea of what a human being is. The same kind of statement applies to functions.

We have developed three fundamental ways of describing a function:

1. *The symbolic or algebraic description,* usually as a formula
2. *The numerical description,* usually as a table of data
3. *The geometric description,* as given by the graph

In the following work we shall systematically use all three approaches to develop and clarify new ideas as they occur. We hope you too will become actively accustomed to using all three approaches.

SECTION 2.1 EXERCISES

For each function in Exercises 1 to 6, give (a) its natural domain, (b) its range, and (c) its graph, using a table of data where necessary.

1. $f(x) = -10x$ **2.** $g(x) = x^4$

3. $h(x) = \sqrt{16 - x^2}$ **4.** $w(x) = \dfrac{1}{x}$

5. $m(x) = \dfrac{x^2 - 1}{x + 1}$ **6.** $f(x) = \begin{cases} \dfrac{1}{x^2} & \text{if } x \neq 0 \\ 4.5 & \text{if } x = 0 \end{cases}$

In Exercises 7 to 10 give the natural domain of the function.

7. $f(x) = \sqrt{x^3 + 2x^2 - 3x}$ **8.** $g(s) = \sqrt[4]{\dfrac{s^2 - s}{3 - s}}$

9. $h(t) = \sqrt[3]{\dfrac{t + 1}{t - 1}} + \dfrac{1}{\sqrt{t^2 - 5t - 14}}$

10. $w(x) = \sqrt{|x|(x^2 - 1)}$

11. Which of the following descriptions gives a function of x? If a given description fails to give a function, explain why.

(*a*) Double the number x and then add 1 million.

(*b*)

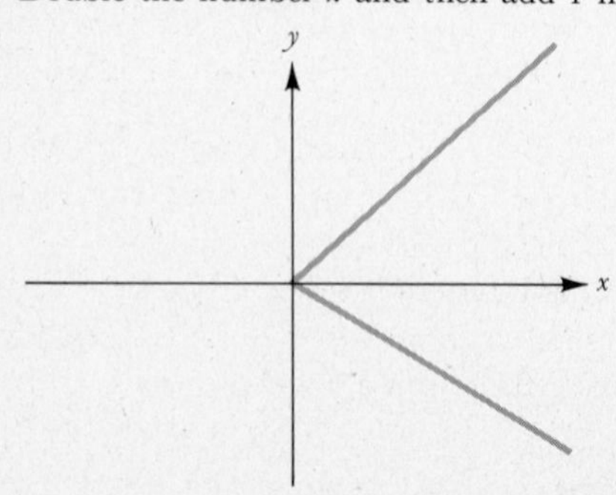

(*c*) $f(x) = 2 - x$ for $x < 0$ and $f(x) = x^2 + 2$ for $x \geq 0$.

(*d*) $g(x) = x^2$ for $x < 2$ and $f(x) = x^3$ for $x > 0$.

(*e*) $f(x) =$ a number that, when raised to the fourth power, yields x.

(*f*)

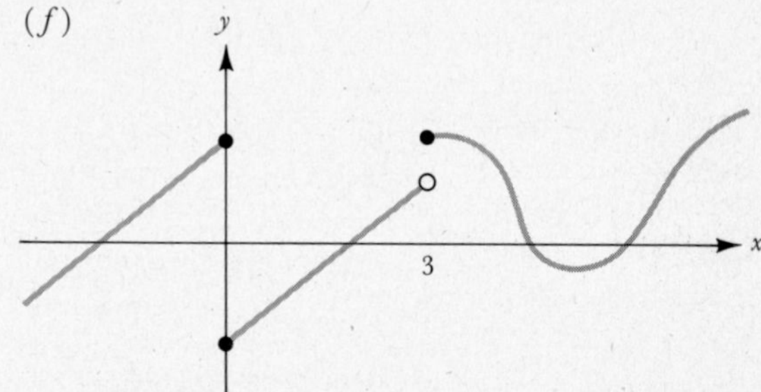

12. Does the equation $x - y^2 - 2y - 4 = 0$

(*a*) Determine y as a function of x?

(*b*) Determine x as a function of y?

13. Determine two distinct functions of x implicitly defined by the equation $16x^2 + y^2 = 25$.

14. Repeat Exercise 13 for the equation $x^3 = y^2$.

15. (*a*) Using a solid curve, connect the dots to give y as a function of x.

(*b*) Repeat (*a*) using a dashed curve to give x as a function of y.

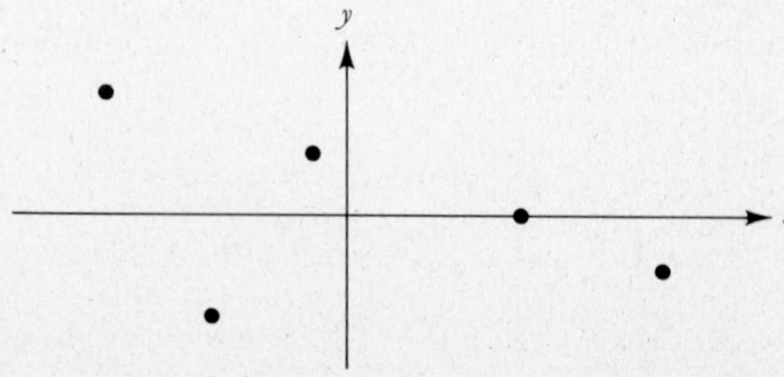

16. Let $f(t) = (t^2 - 2t - 3)/(\sqrt{t} + 1)$. Compute
(a) $f(x)$ (b) $f(2x)$ (c) $f(x + h)$ (d) $f(h)$ (e) $f(3)$

17. Repeat Exercise 16 with $f(r) = (r - 1)/(r + 1) + 2r^2$.

In Exercises 18 to 25 compute, in order,
(a) $f + g$ (b) $f - g$ (c) $f \cdot g$ (d) f/g
(e) $f \circ g$ (f) $g \circ f$ (g) $f \circ f$ (h) $g \circ g$
stating in (a) to (d) which numbers are not *in the natural domains of these functions.*

18. $f(x) = \sqrt{x} + 1$, $g(x) = x^2$

19. $f(x) = x$, $g(x) = 2x + 1$

20. $f(x) = ax + b$, $g(x) = cx + d$

21. $f(x) = 3x - 4$, $g(x) = \dfrac{x + 4}{3}$

22. $f(x) = x$, $g(x) = \dfrac{1}{x}$

23. $f(x) = 2x$, $g(x) = 3x$

24. $f(x) = x^2$, $g(x) = x^3$

25. $f(x) = x^3 + 3x + 1$, $g(x) = 2$

In Exercises 26 to 29 express the function as the composition of two simpler functions. (More than one solution is possible.)

26. $f(x) = (3 - x)^2 - 2(3 - x)$

27. $f(x) = \sqrt{2x + 1}$

28. $f(x) = -(x - 1)^2$

29. $f(x) = 2x + 1$

30. Find $f \circ f$ when $f(x) = 3x + 2$.

31. (a) Graph $f(x) = |x|$. (b) Find $f \circ f$.

32. Graph the function $s(x) =$ largest integer $n \leq x$, sometimes denoted $[x]$. [Note $s(1/2) = s(3/4) = s(.999) = 0$, and $s(1) = 1$.]

33. (a) Graph $f(x) = \begin{cases} x^2 - 5 & \text{for } x < 1 \text{ and } x > 1 \\ 2 & \text{for } x = 1 \end{cases}$
(b) What should $f(1)$ be in part (a) so that the resulting graph can be drawn without lifting your pen from the paper?

34. Graph $f(x) = |x + 1|$ and $g(x) = |x| + 1$ on the same coordinate axes.

35. Give the function $f(x) = x^2$ (domain R) as a set of ordered pairs.

36. Is the set of ordered pairs, the first of which is the absolute value of the second, a function? Explain. Graph this set of ordered pairs.

37. What is the domain of $f \circ g$, where $f(x) = \sqrt{x}$ and $g(x) = -x^2 - 1$?

Section 2.2: Limits: An Informal Introduction

Objectives:

1. Use informal methods, including compiling tables of data and graphs, to determine the behavior of a given function near a given point.
2. Use informal algebraic methods, including careful cancellation and rationalization, to evaluate the limit of a simple function and the limit involving a combination of simple functions.

THE CONCEPT OF LIMIT

Since the most important idea underlying the calculus is that of limit, we shall present it in stages of increasing complexity and formality, beginning with an informal discussion. *All mathematical statements and conclusions in this section will be put on a solid logical foundation in subsequent sections, especially Sec. 9.1, Chap. 9, where we revisit the formal definition of limit in detail.*

Loosely speaking, when dealing with limits, we are concerned with the behavior of a function *near but not at* a particular number. Suppose a function f is defined for all x near a particular number a, but *not necessarily at* $x = a$. *As we choose values of* x *closer and closer to* a, *what happens to the values of* $f(x)$*?*

In some cases your experience and knowledge of certain functions allow you to answer the question immediately.

Example 1 As we choose values of x closer and closer to the number 2, what happens to the corresponding values of $f(x) = x^2$?

Solution Our preliminary approach is to compile a table of data [Fig. 2.12(*a*)] for values of x near 2 and compare these with the graph [Fig. 2.12(*b*)] near $x = 2$.

It is clear from both the table and the graph that, as the values of x approach 2 (from either side), the values of $f(x)$ approach 4. In fact, the values of $f(x)$ can be made as close to 4 as we choose by taking values of x close enough to 2. ●

The standard mathematical shorthand for the conclusion of the previous example is

$$\lim_{x \to 2} x^2 = 4$$

This is read as "the limit as x tends to 2 of x^2 is 4" or "the limit as x approaches 2 of x^2 is 4."

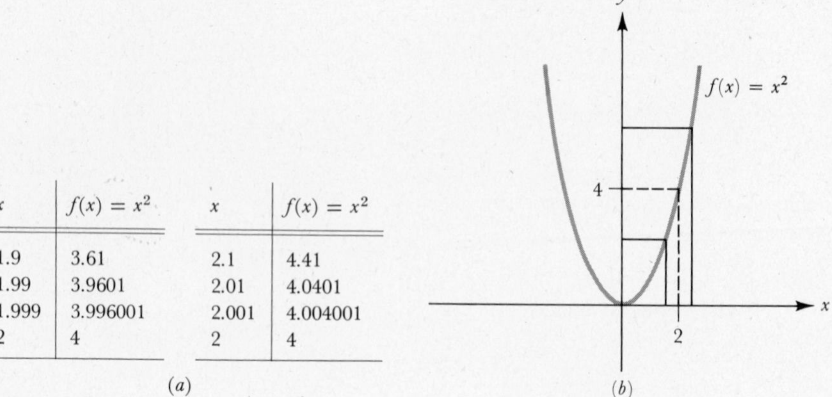

x	$f(x) = x^2$
1.9	3.61
1.99	3.9601
1.999	3.996001
2	4

x	$f(x) = x^2$
2.1	4.41
2.01	4.0401
2.001	4.004001
2	4

Fig. 2.12

Although the value of $f(x) = x^2$ *at* $x = 2$ is 4, we did not use this fact to answer the original question. Often the function is *not* defined at the point in question, as in the following example. (Recall the discussion of holes in the graphs of functions in Sec. 2.1.)

Example 2 As x tends to 2, what happens to the values of $g(x) = (x^3 - 2x^2)/(x - 2)$?

Solution After factoring

$$g(x) = \frac{x^3 - 2x^2}{x - 2} = \frac{x^2(x - 2)}{x - 2}$$

we see that the only difference between $g(x)$ and $f(x) = x^2$ previously examined is that $g(x)$ is not defined at $x = 2$: $g(2) = 0/0$ is not a real number; and the graph of $g(x)$ is a parabola with a hole at $x = 2$ (see Fig. 2.13). Tables of data for g are almost the same as those for f.

Even though $g(2)$ is not defined, we see either from the table of data or the graph that, as x tends to 2, the values of $g(x)$ are tending toward 4. The values of $g(x)$ can be made as close to 4 as we choose by choosing values of x close enough to 2. So again we have

$$\lim_{x \to 2} g(x) = \lim_{x \to 2} \frac{x^3 - 2x^2}{x - 2} = 4$$

even though $g(x)$ is not defined at $x = 2$. ●

x	$g(x) = [x^2(x - 2)]/(x - 2)$
1.9	3.61
1.99	3.9601
1.999	3.996001
2	undefined

x	$g(x) = [x^2(x - 2)]/(x - 2)$
2.1	4.41
2.01	4.0401
2.001	4.004001
2	undefined

(a)

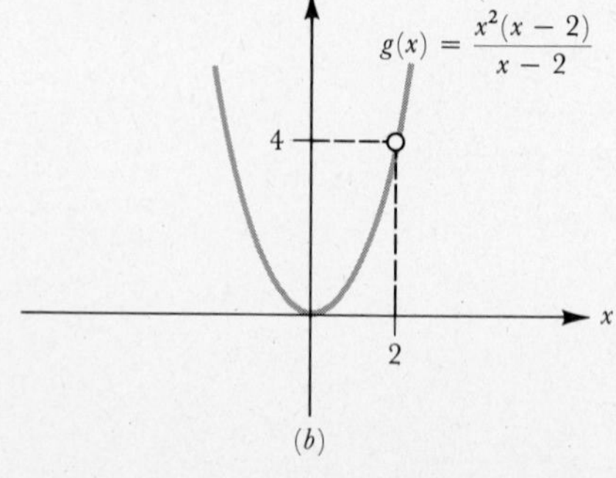

(b)

Fig. 2.13

Writing $g(x)$ as

$$g(x) = x^2 \cdot \frac{x-2}{x-2}$$

we can see why we cannot simply cancel the $x - 2$ factors leaving x^2. For all $x \neq 2$, $g(x)$ is the product of x^2 with $(x-2)/(x-2) = 1$, but at $x = 2$, $(x-2)/(x-2)$ is not defined, hence $g(x)$ is not defined. Thus we *can* cancel provided we are careful in so doing to note that the cancellation does *not* apply when the factors being canceled are zero. We do this by writing

$$g(x) = \frac{x^2(x-2)}{x-2} = x^2 \qquad \text{for } x \neq 2$$

The reason we can then deal with the function that results from canceling is that *in determining the limit we are concerned only with the values of all x* NEAR *the point in question and not* AT *the point* (in this case, the point $x = 2$). The following example illustrates this issue even more dramatically.

Example 3 Determine

$$\lim_{x\to 2} h(x) \text{ for } h(x) = \begin{cases} x^2 & \text{if } x \neq 2 \\ 6 & \text{if } x = 2 \end{cases}$$

Solution *Away* from $x = 2$ this function is the same as f and g of the previous examples. At $x = 2$ it differs from f in that $f(2) = 4$ and $h(2) = 6$. It differs from g because 2 is *not* in the domain of g, whereas it *is* in the domain of h. In particular, then, for x near but not at 2 the table of data and the graph for h (see Fig. 2.14) are the same as those for f and g above. Since the behavior of h for all values of x near but not equal to 2 is the same as that of f and g, the limit as x approaches 2 of $h(x)$ must also be the same. That is, $\lim_{x\to 2} h(x) = 4$ even though $h(2) = 6$. ●

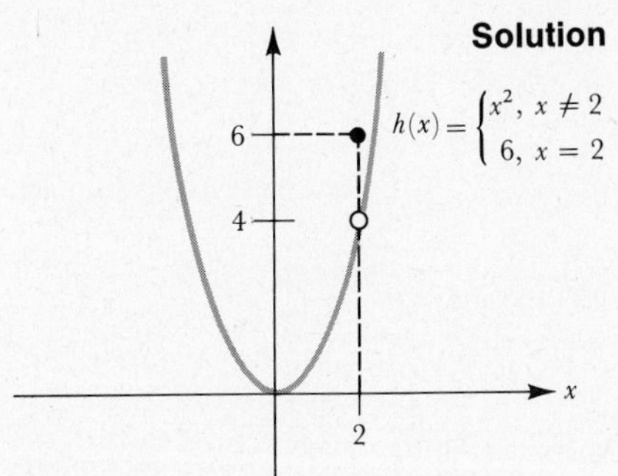

Fig. 2.14

The following example illustrates that the limit concerns the values of x on *both* sides of the point in question.

Example 4 Examine the values of $f(x) = \dfrac{|x|}{x}$ near $x = 0$.

Solution We consider the positive and negative cases separately. For all $x > 0$,

$$f(x) = \frac{|x|}{x} = \frac{x}{x} = 1$$

whereas for all $x < 0$,

$$f(x) = \frac{|x|}{x} = \frac{-x}{x} = -1 \qquad (\text{for } x < 0,\ |x| = -x)$$

On the other hand, $f(x)$ is not defined at $x = 0$. Hence we can sketch the graph, as in Fig. 2.15.

Thus there is no single number which the values of f approach as x approaches 0. In this case we say

$$\lim_{x\to 0} \frac{|x|}{x} \text{ does not exist} \quad ●$$

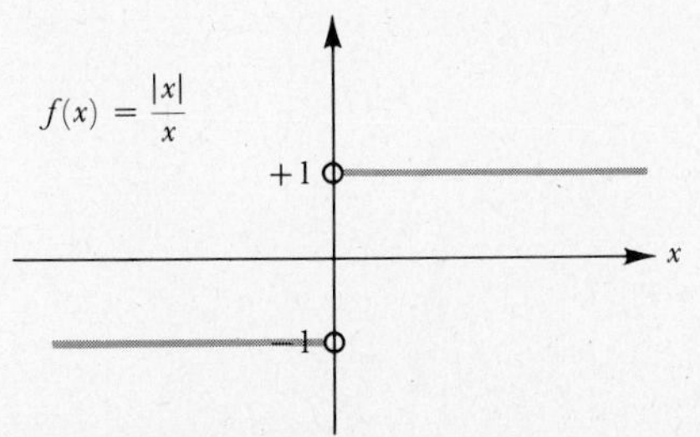

Fig. 2.15

Notice the way this last example differed from the previous three by the way you answer the following question: Given a value of x close to 0, can you tell what $f(x)$ is close to?

Analogous versions of this question for the previous three examples would have "yes" as an answer. (The value of the function would be close to 4.)

ONE-SIDED LIMITS

Up to now we have used the designation

$$\lim_{x\to a} f(x)$$

to denote that we are taking the limit as x tends to a from both sides. Let us then write

$$\lim_{x\to a^+} f(x)$$

to denote that we are taking the limit as x tends to a from the right side. Similarly, we write

$$\lim_{x\to a^-} f(x)$$

to denote that we are taking the limit as x tends to a from the left side. These latter two cases are commonly referred to as *right- and left-hand limits,* respectively.

The following is then certainly reasonable:

2.2.1 Theorem

$$\lim_{x\to a} f(x)$$

exists if and only if $\lim_{x\to a^-} f(x)$ and $\lim_{x\to a^+} f(x)$ both exist and are equal. In this case, of course,

$$\lim_{x\to a^-} f(x) = \lim_{x\to a^+} f(x) = \lim_{x\to a} f(x)$$

We can use the above language to reformulate the conclusions of Example 4 as follows:

$$\lim_{x\to 0^-} \frac{|x|}{x} = -1, \qquad \lim_{x\to 0^+} \frac{|x|}{x} = 1$$

and, since the left- and right-hand limits are unequal,

$$\lim_{x\to 0} \frac{|x|}{x} \text{ does not exist}$$

FUNCTIONS GROWING OR DECREASING WITHOUT BOUND

We now look at another way limits might fail to exist.

Example 5 Determine the values of

$$f(x) = \frac{1}{(x-2)^2}$$

for x near 2.

Solution We construct a table of data and sketch the function (Fig. 2.16). From both the table and the graph we see that as x gets closer to 2, *from both sides,* the values of

x	$1/(x-2)^2$
1.9	100
1.99	10,000
1.999	1,000,000
2	undefined

x	$1/(x-2)^2$
2.1	100
2.01	10,000
2.001	1,000,000
2	undefined

(a)

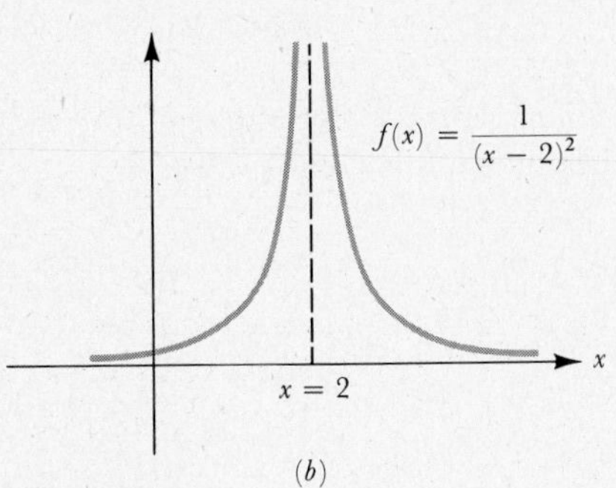

(b)

Fig. 2.16

$f(x)$ are growing without bound and are certainly not approaching any fixed number. Again we say:

$$\lim_{x\to 2} \frac{1}{(x-2)^2} \text{ does not exist} \quad \bullet$$

Once again, in this last example, being told that x is near 2 would not enable you to tell what number $f(x)$ is close to. In both Examples 4 and 5 the limit failed to exist, but for vastly different reasons. In Example 4 $\lim_{x\to 0} |x|/x$ fails to exist because the left- and right-hand limits are unequal. In Example 5 the function fails to have a limit at $x = 2$ because the values of f grow large without bound as x approaches 2. We often express this latter behavior by writing

$$\lim_{x\to 2} \frac{1}{(x-2)^2} = +\infty$$

One must be very careful with this notation. The symbol $+\infty$ does not represent a number, and although we write $= +\infty$, $\lim_{x\to 2} 1/(x-2)^2$ does not in fact equal anything! We are using the shorthand $= +\infty$ here to indicate the *behavior* of $1/(x-2)^2$ for values of x close to 2. And no, we have not found a way to divide by zero. In Example 6 we shall examine the behavior of $f(x) = 1/(x-1)$, which you will see combines the misbehavior of both Examples 4 and 5, decreasing without bound on one side of 1 and increasing without bound on the other. It also indicates that it is not possible to solve the problem of division by zero simply by adopting a fancy new symbol.

We are using the notation $\lim_{x\to 2} f(x) = +\infty$ to mean that, as x tends to a from both sides, $f(x)$ increases without bound. Graphically, this means that the graph of $f(x)$ will cross any horizontal line $y = M$ drawn above a—*no matter how large M is*. Numerically, we can think of the statement to mean that by choosing values of x close enough to a we can get $f(x)$ to exceed the capacity of your calculator—*no matter how large the capacity of your calculator!* (How near to 2 must x be so that $f(x) = 1/(x-2)^2$ will exceed the capacity of *your* calculator?)

We similarly write

$$\lim_{x\to a} f(x) = -\infty$$

if the values of $f(x)$ *decrease without bound* as x tends to a from both sides. Thus, for example,

$$\lim_{x\to 2} \frac{-1}{(x-2)^2} = -\infty$$

Again, we emphasize that this latter notation expresses the behavior of $f(x)$ as x tends to a and does *not* mean that the limit exists in the sense that the values of the function can be made arbitrarily close to some specific real number by choosing x close enough to a.

These ideas extend directly to left- and right-hand limits as illustrated in the following example.

Example 6 Examine the behavior of $f(x) = 1/(x-1)$ near $x = 1$ by determining the behavior as $x \to 1^+$ and as $x \to 1^-$ separately. Sketch the graph of $f(x)$ near $x = 1$.

Solution First we compute some numerical data:

x	$f(x) = 1/(x - 1)$
0	-1
0.5	-2
0.9	-10
0.99	-100
0.999	$-1{,}000$
0.9999	$-10{,}000$
1	undefined

x	$f(x) = 1/(x - 1)$
2	1
1.5	2
1.1	10
1.01	100
1.001	1,000
1.0001	10,000
1	undefined

It is fairly clear that as $x \to 1^+$, the denominator is positive and tends toward zero, so the quotient $1/(x - 1)$ grows without bound. Hence we write

$$\lim_{x\to1^+} \frac{1}{x-1} = +\infty$$

Similarly,

$$\lim_{x\to1^-} \frac{1}{x-1} = -\infty$$

In particular,

$$\lim_{x\to1} \frac{1}{(x-1)}$$

does not exist. Also (see Fig. 2.17),

$$\lim_{x\to1} \frac{1}{x-1} \neq +\infty \qquad \text{and} \qquad \lim_{x\to1} \frac{1}{x-1} \neq -\infty \quad \bullet$$

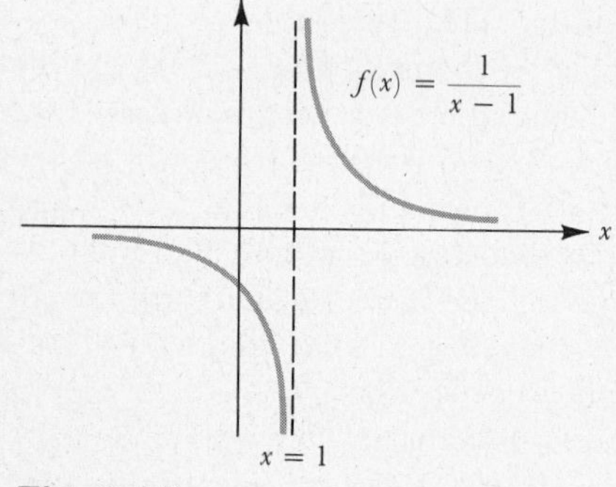

Fig. 2.17

The following example shows that a limit may be difficult to ascertain even *with* a table of values and high accuracy.

Example 7 Determine whether

$$\lim_{x\to2} \frac{(\sqrt{x} - \sqrt{2})}{(x-2)}$$

exists and, if so, determine its value.

Solution First we compute a table of data (using a calculator). From the table below it would be difficult to guess the limit. However, by rationalizing the numerator we can isolate the factor $x - 2$, which can then be carefully canceled in the sense discussed earlier:

$$\frac{\sqrt{x}-\sqrt{2}}{x-2} = \frac{\sqrt{x}-\sqrt{2}}{x-2} \cdot \frac{\sqrt{x}+\sqrt{2}}{\sqrt{x}+\sqrt{2}} = \frac{x-2}{(x-2)(\sqrt{x}+\sqrt{2})}$$

$$= \frac{1}{\sqrt{x}+\sqrt{2}} \qquad \text{if } x \neq 2$$

Now it is surely reasonable to predict that $1/(\sqrt{x} + \sqrt{2})$ will tend to $1/(2\sqrt{2})$ as x tends to 2 from either side. [Note that $1/(2\sqrt{2})$ is approximately 0.3535534, accurate to six decimal places.] $\bullet$

x	$(\sqrt{x} - \sqrt{2})/(x - 2)$
1	0.414214
1.9	0.358087
1.99	0.353996
1.999	0.353598
1.9999	0.353558
2	undefined

x	$(\sqrt{x} - \sqrt{2})/(x - 2)$
3	0.317837
2.1	0.349241
2.01	0.353113
2.001	0.353509
2.0001	0.353549
2	undefined

PROGRESS TEST 1

In Probs. 1 and 2, (*a*) make a table of values of f near $x = a$, (*b*) sketch the graph of f near $x = a$, and (*c*) determine $\lim_{x \to a} f(x)$.

1. $f(x) = 9 - x^2$, $a = 2$

2. $f(x) = \dfrac{x^2 - 9}{x - 3}$, $a = 3$

In Probs. 3 to 6 determine whether $\lim_{x \to a} f(x)$ exists, and if so, determine its value by making a table of values near a (using a calculator or table of square roots) or performing an appropriate algebraic manipulation. When $\lim_{x \to a} f(x)$ does not exist, examine the left- or right-hand limits—if *they* exist.

3. $f(x) = \dfrac{x - 1}{|x - 1|}$, $a = 1$

4. $f(x) = \dfrac{1}{2 - x}$, $a = 2$

5. $f(x) = \dfrac{\sqrt{x} - \sqrt{3}}{x - 3}$, $a = 3$

6. $f(x) = \dfrac{-1}{(x + 3)^4}$, $a = -3$

LIMITS AND CALCULATORS

The previous examples and problems illustrate both the strengths and weaknesses of using a numerical approach to determining limits of functions—in particular, using approximations available through the use of a calculator. By observing the values of $f(x)$ for specific values of x near to the point in question, we can usually get a fairly good idea of what the limit may be; often this amounts to observing the changes in increasingly accurate decimal approximations to some number. However, as Example 7 showed, this is not always sufficient to determine the actual limit without further analysis. But, more important, *no amount of numerical data can provide a logical proof that the limit is a specific number.* A proof requires using the precise definition of limit, found in Sec. 2.5. Thus, no calculator or computer exists—or ever will exist—that can calculate sufficient accuracy to *prove* that $\lim_{x \to a} f(x)$ is some particular number. A proof can be provided only by a human being using precisely defined mathematical concepts and logic. *Numerical* accuracy and *logical* accuracy are entirely different matters. The calculator or computer *can*, however, lead us in the direction of the limit when such exists. We encourage you to use a calculator and perform frequent numerical experiments.

SOME FURTHER EXAMPLES

In all but Example 1 and Prob. 1, Progress Test 1, we determined $\lim_{x \to a} f(x)$ by replacing $f(x)$ by a function $g(x)$ that equaled the original function $f(x)$ everywhere except at $x = a$ itself and hence had the same limit as $x \to a$. We obtained the function $g(x)$ by carefully canceling a factor in $f(x)$ or (in Example 7) by rationalizing and then carefully canceling. We then determined (or "evaluated") $\lim_{x \to a} g(x)$ by direct substitution; that is,

$$\lim_{x \to a} g(x) = g(a)$$

Any such function g, whose limit at a can be determined by direct substitution, is said to be *continuous at* $x = a$. A detailed discussion of continuity appears in Sec. 2.5.

We shall now apply the technique just mentioned to evaluate limits of the type that will appear in the next section, which is on derivatives.

Example 8 Evaluate

$$\lim_{x \to 0} \frac{f(3 + x) - f(3)}{x}$$

where $f(x) = x^2 + 1$.

Solution We shall write out and simplify $[f(3 + x) - f(3)]/x$ before attempting to evaluate the limit. (Recall Sec. 2.1.)

$$\begin{aligned} \frac{f(3 + x) - f(3)}{x} &= \frac{[(3 + x)^2 + 1] - [3^2 + 1]}{x} \\ &= \frac{[9 + 6x + x^2 + 1] - [10]}{x} \\ &= \frac{6x + x^2}{x} \end{aligned}$$

Now, for $x \neq 0$, $(6x + x^2)/x = 6 + x$. In evaluating the limit as x tends to 0, we are interested only in values of x *near* 0, but not *at* 0. For all such x, we know that

$$\frac{f(3 + x) - f(3)}{x} = 6 + x$$

Hence

$$\begin{aligned} \lim_{x \to 0} \frac{f(3 + x) - f(3)}{x} &= \lim_{x \to 0} (6 + x) \\ &= 6 \end{aligned}$$

(Obviously, as x tends to 0, $6 + x$ tends to 6; thus, in effect, we can evaluate this limit by direct substitution.) ●

In the next example the variable in the limit is denoted by h, but our earlier techniques still apply.

Example 9 Evaluate

$$\lim_{h \to 0} \frac{f(1 + h) - f(1)}{h}$$

where $f(x) = 1/x$.

Solution Here $f(1) = 1$ and $f(1 + h) = 1/(1 + h)$. Thus

$$\begin{aligned} \frac{f(1 + h) - f(1)}{h} &= \frac{1/(1 + h) - 1}{h} \\ &= \frac{(1 - 1 - h)/(1 + h)}{h} \\ &= \frac{-h}{(1 + h)h} \qquad \left(\text{Recall } \frac{A/B}{C} = \frac{A}{BC}\right) \end{aligned}$$

Hence, for $h \neq 0$,

$$\frac{f(1 + h) - f(1)}{h} = \frac{-1}{1 + h}$$

Therefore

$$\begin{aligned} \lim_{h \to 0} \frac{f(1 + h) - f(1)}{h} &= \lim_{h \to 0} \frac{-1}{1 + h} \\ &= \frac{-1}{1} = -1 \end{aligned}$$

(Obviously, as h tends to 0, $1 + h$ tends to 1, so the fraction tends to -1.) ●

PROGRESS TEST 2

Evaluate the following limits:

1. $\lim_{h\to 0} \frac{f(-2+h)-f(-2)}{h}$, $f(x) = 3x + 2$

2. $\lim_{h\to 0} \frac{f(1+h)-f(1)}{h}$, $f(x) = \sqrt{x+1}$

Hint: rationalize before taking the limit.

3. $\lim_{t\to 0} \frac{g(t+2)-g(2)}{t}$, $g(t) = t^2 + t$

4. $\lim_{h\to 0} \frac{f(3+h)-f(3)}{h}$, $f(x) = \frac{1}{x^2}$

SECTION 2.2 EXERCISES

Evaluate the following limits in 1 to 12 where possible. In cases where the limit does not seem to exist, compile a table of data and graph the function near the point in question.

1. $\lim_{x\to 0} \frac{5x^3+8x^2}{3x^4-16x^2}$

2. $\lim_{x\to 3} \frac{\sqrt{2x+3}-x}{x-3}$

3. $\lim_{x\to 2} \frac{3x-6}{|x-2|}$

4. $\lim_{x\to 9} \frac{x-9}{\sqrt{x}-3}$

5. $\lim_{x\to 0} \frac{\sqrt{5+x}-\sqrt{5}}{2x}$

6. $\lim_{x\to 2} \frac{x^4-16}{x^2-4}$

7. $\lim_{x\to 1} (2x^3 + x^2 - 9)$

8. $\lim_{t\to 0} \frac{f(2+t)-f(2)}{t}$, where $f(t) = 4t^3$

9. $\lim_{h\to 0} \frac{g(4+h)-g(4)}{h}$, where $g(x) = \frac{1}{x^2+4}$

10. $\lim_{h\to 0} \frac{f(3+h)-f(3)}{h}$, where $f(x) = 16x^2$

11. $\lim_{x\to 3} \frac{f(x)-f(3)}{x-3}$, where $f(x) = 16x^2$

12. $\lim_{x\to 4} \frac{g(x)-g(4)}{x-4}$, where $g(x) = \frac{1}{x^2+4}$

13. Compare the answers to Exercises 10 and 11. Can you explain the correspondence? It may help to examine respective tables of data.

14. Repeat Exercise 13 for the answers to 9 and 12, respectively.

In Exercises 15 to 18 compile a table of data and sketch a graph to determine the behavior of the given function near the given point a.

15. $f(x) = \frac{1}{x+2}$, $a = -2$

16. $g(t) = \frac{1}{(t+1)^3}$, $a = -1$

17. $s(x) = \frac{1}{|2x+1|}$, $a = -\frac{1}{2}$

18. $F(t) = \frac{1}{(2t+1)^2}$, $a = -\frac{1}{2}$

19. For the "largest integer function" $s(x) = [x]$ (see Sec. 2.1, Exercise 32),

(*a*) determine $\lim_{x\to 1^-} s(x)$

(*b*) determine $\lim_{x\to 1^+} s(x)$

(*c*) does $\lim_{x\to 1} s(x)$ exist?

Section 2.3: The Derivative and Rates of Change

Objectives:

1. Compute derivatives of functions using the limit definition.

2. Interpret derivatives of functions in general and in physically related situations as instantaneous rates of change.

AVERAGE AND INSTANTANEOUS RATES OF CHANGE

We remarked in Sec. 2.1 that perhaps the most important and useful mathematical concept is that of function, because it is through functions that we can make precise the quantitative relationships between one thing and another. However, really understanding and using functions usually amount to studying how the quantities related by the functions *change.* This is especially the case if a quantity is given as a function of time. The barometric pressure changes with time, but knowing what the pressure is at any given time is not as valuable as knowing whether and how fast it is rising or falling. The same could be said of the stock or commodity markets. For an outfielder to know the position of a fly ball is not nearly as important as knowing how the position is changing, that is, how the ball is moving. The understanding of physical, economic, social, or biological systems often amounts to the understanding of how they change.

Two basic types of knowledge about rates of change are possible: the *average* rate of change over a given interval and the *instantaneous* rate of change at a given instant. Although it is sometimes sufficient to know the average rate of change over an interval, in other situations the instantaneous rate of change may be more valuable. Being told that the price of wheat would drop from $4.20 per bushel to $3 between July and December (an average loss of 20¢ a month for 6 months) is not particularly valuable, and, in fact, could be downright fraudulent if, during September, the price hit $5. Similarly, knowing that a car hit a tree 1 h after beginning a trip and 20 mi down the road does not tell us how fast the car was going at the instant it hit the tree. All we know is that the car averaged 20 miles per hour (mph) for the previous hour. The driver could have hit the tree at 80 mph while trying to make up time lost while drinking in a pub along the way.

To get a better idea of what is meant by the instantaneous rate of change at a point, we shall examine a more precisely defined situation, one that played an important motivating role in the creation of the calculus in the seventeenth century.

Suppose an object is moving in a straight line and $f(x)$ represents its position at time x. We assume that some positive direction has been assigned, so $f(x)$ can in general be positive, negative, or zero, depending on where the object is located with respect to the origin or reference point.

Now if h represents some change in time, then

$$\frac{f(x+h)-f(x)}{h}$$

gives the *average velocity* over the time interval from x to $x+h$. That is, average velocity is the change in position (displacement) divided by the change in time.

It is well known and can be verified by experiment that, ignoring air resistance, an object falling freely for x seconds will fall approximately

$$f(x) = 16x^2 \text{ ft}$$

(This is a special case of the general gravitation formula $f(x) = \frac{1}{2}gx^2$, where g is the acceleration due to gravity. On earth g is approximately 32 ft/s^2.) In Table 2.1 we show several values of $f(x)$,

x	$f(x) = 16x^2$
0	0
1	16
2	64
3	144
4	256
5	400

Table 2.1

One thing should be apparent: The object falls faster as time passes. It falls 16 ft in the first second and thus averages 16 ft/s for the first second (recall the average rate of velocity is the distance traveled divided by the time elapsed). During the next second (from time $x = 1$ to time $x = 2$) it averages $(64 - 16)/1 = 48$ ft/s. The averages during the third, fourth, and fifth seconds are, respectively, $(144 - 64)/1 = 80$ ft/s, $(256 - 144)/1 = 112$ ft/s, and $(400 - 256)/1 = 144$ ft/s. This certainly agrees with our experience that the longer the object falls, the faster it falls. This is why more people are willing to jump off a chair than off a building.

Computing average velocities is a straightforward application of the basic formula just noted:

$$\text{Average velocity} = \frac{\text{change in position}}{\text{change in time}}$$

x	$f(x) = 16x^2$
4	256
$3\frac{1}{2}$	196
$3\frac{1}{4}$	169
$3\frac{1}{8}$	156.25
3.1	153.76
3.01	144.9616
3.001	144.096016
3.0001	144.00960016
3	144

Table 2.2

But now, can you state how fast the object is falling at the *instant* that 3 s have elapsed, that is, *at* $x = 3$? To answer this question precisely, we need more than arithmetic and algebra; we need to apply the idea of limit. The strategy, however, is simple. We compute the average velocities over smaller and smaller time intervals containing $x = 3$ to see whether these are tending to some fixed number as a limit. If so, we *define* this limit to be the instantaneous velocity at $x = 3$. First, a more detailed table of values for $f(x)$ near $x = 3$ is in order (see Table 2.2).

The average velocity from $x = 3$ to $x = 3\frac{1}{2} = 7/2$ is

$$\frac{f(7/2) - f(3)}{(7/2) - 3} = \frac{196 - 144}{1/2} = 104 \text{ ft/s}$$

Similar computations over smaller and smaller time intervals yield the average velocities in Table 2.3.

We also could have calculated averages for intervals *before* $x = 3$. For example, from $x = 2.9999$ to $x = 3$, the average velocity is

$$\frac{f(2.9999) - f(3)}{2.9999 - 3} = \frac{143.9904002 - 144}{-0.0001} = 95.9984$$

Thus we know that the instantaneous velocity at $x = 3$, say $v(3)$, satisfies

$$95.9984 = v(2.9999) < v(3) < v(3.0001) = 96.0016$$

It certainly appears that as x gets closer and closer to 3 from either side, the average velocities get closer and closer to 96 ft/s. It would surely be reasonable then to state that the *instantaneous velocity* at $x = 3$ is 96 ft/s.

We can use the ideas of the preceding section to clean up things, avoid some of the messy calculations involved here, and make the whole approach more precise. We let h stand for the small positive and negative numbers being added to $x = 3$ and consider the limit as $h \to 0$ of the resulting average.

Thus $3 + h$ can be thought of as a number near $x = 3$. If $h > 0$, say $h = .001$, then $3 + h = 3 + .001 = 3.001$, a number slightly bigger than 3; while if $h < 0$, say $h = -.001$, then $3 + h = 3 - .001 = 2.999$, a number slightly less than 3. We can now calculate the average velocity over a *typical* small time interval between 3 and $3 + h$:

$$\text{Average velocity between 3 and } 3 + h = \frac{f(3 + h) - f(3)}{(3 + h) - 3}$$

$$= \frac{16(3 + h)^2 - 144}{h} = \frac{96h + 16h^2}{h}$$

Time Interval	*Average Velocity (ft/s)*
$x = 3$ to $x = 3\frac{1}{8}$	$\frac{f(3\frac{1}{8}) - f(3)}{3\frac{1}{8} - 3} = \frac{156.25 - 144}{\frac{1}{8}} = 98.00$
$x = 3$ to $x = 3.1$	$\frac{f(3.1) - f(3)}{3.1 - 3} = \frac{153.76 - 144}{.1} = 97.6$
$x = 3$ to $x = 3.01$	$\frac{f(3.01) - f(3)}{3.01 - 3} = \frac{144.9616 - 144}{.01} = 96.16$
$x = 3$ to $x = 3.001$	$\frac{f(3.001) - f(3)}{3.001 - 3} = \frac{144.096016 - 144}{.001} = 96.016$
$x = 3$ to $x = 3.0001$	$\frac{f(3.0001) - f(3)}{3.0001 - 3} = \frac{144.00960016 - 144}{.0001} = 96.0016$

Table 2.3

The idea of *average* velocity means that $h \neq 0$ (average velocities only make sense for nonzero intervals of time!) Hence we know that

$$\frac{96h + 16h^2}{h} = 96 + 16h$$

represents the average velocity during the interval between 3 and $3 + h$.

We define the instantaneous velocity at the instant $x = 3$ to be the limit of such average velocities as $h \to 0$. But, of course,

$$\lim_{h \to 0} (96 + 16h) = 96$$

Example 1 Find the instantaneous velocity under the previous circumstances at the instant $x = 4$.

Solution The average velocity in the interval between time 4 and $4 + h$ equals the distance fallen divided by the time elapsed. The distance fallen is $f(4 + h) - f(4)$, and the time elapsed is $(4 + h) - 4 = h$. Thus the average velocity, with some simplifying, is

$$\frac{f(4 + h) - f(4)}{h} = \frac{128h + 16h^2}{h} = 128 + 16h \qquad (\text{since } h \neq 0)$$

(We are dealing with *average* velocities over a *nonzero* time interval.) Thus the instantaneous velocity at $x = 4$ is

$$\lim_{h \to 0} \frac{f(4 + h) - f(4)}{h} = \lim_{h \to 0} (128 + 16h) = 128 \text{ ft/s} \quad \bullet$$

In determining both of the above limits of average velocities, we knew that $h \neq 0$ and canceled the h factor. Yet we were able to take the limit as $h \to 0$ because, as emphasized in Sec. 2.2, the limit reflects the behavior of the function *near* $h = 0$ and not *at* $h = 0$.

PROGRESS TEST 1

1. With the same "distance fallen" function $f(x) = 16x^2$,
(*a*) Find the average velocity in the interval from $x = 2$ to $x = 2.001$.
(*b*) Find the average velocity in the interval from $x = 1.99$ to $x = 2$.
(*c*) Find the average velocity (in simplified terms) in the interval between $x = 2$ and $x = 2 + h$.

2. Use 1(*c*) to find the instantaneous velocity at $x = 2$.

3. Find the instantaneous velocity at time $x = 1$ using the $\lim_{h \to 0}$ idea.

THE INSTANTANEOUS-VELOCITY FUNCTION

x	*Instantaneous Velocity at Time x*
0	0
1	32
2	64
3	96
4	128
5	160

Table 2.4

Using the techniques just developed we can derive Table 2.4, which gives instantaneous velocities for a freely falling object.

Can you name a function that satisfies the data? When x increases by 1, the instantaneous velocity increases by 32, so the function $g(x) = 32x$ satisfies the Table 2.4. We might be so bold as to conjecture that the function $g(x)$ gives the instantaneous velocity at any instant x. We shall now show that this is the case.

In our previous discussions, to keep things simple and concrete, we calculated some instantaneous velocities at fixed integral values of x. Why not calculate the instantaneous velocity *at any time x*? The same ideas carry through. The average velocity in the time interval between x and $x + h$ is (for $h \neq 0$)

$$\frac{\text{Change in position}}{\text{Change in time}} = \frac{f(x+h) - f(x)}{h} = \frac{16(x+h)^2 - 16x^2}{h}$$
$$= \frac{16(x^2 + 2xh + h^2) - 16x^2}{h} = \frac{32xh + 16h}{h} = 32x + 16h$$

For any fixed x the instantaneous velocity at time x will be the limit as h tends to 0 of the average velocities. But

$$\lim_{h \to 0} (32x + 16h) = 32x$$

just as we expected! We have derived a new function that gives the instantaneous velocity at any time x. It is denoted $f'(x)$ (read "f prime of x") and is called the *derivative of* $f(x) = 16x^2$. We determined (and in effect defined) $f'(x)$ at an unspecified point x by determining

$$f'(x) = \lim_{h \to 0} \frac{f(x+h) - f(x)}{h}$$

Knowing $f'(x) = 32x$ at a typical x, we can now compute the velocity at, for example, $x = 3.76$, by simple substitution:

$$f'(3.76) = 32(3.76) = 120.32 \text{ ft/s}$$

In our previous discussion the motion under consideration was in one direction only (taken to be the positive direction). Consequently, the average velocities computed were actually average speeds, and the instantaneous velocities were instantaneous speeds. For more general straight-line motions *velocity* may be positive or negative. A positive velocity means that the motion is in the positive direction, a negative velocity means that the motion is in the opposite direction, and zero velocity means the object is at rest. In any case *the absolute value of the velocity gives the speed.*

PROGRESS TEST 2

On the moon, because of the moon's smaller size and weaker gravitational pull, the distance an object falls in feet as a function of time x in seconds is approximated by the function $m(x) = 2.6x^2$.

1. Using the technique developed, calculate the derivative function $m'(x)$ that gives the instantaneous velocity at any time x by:
(*a*) Finding the average velocity in the interval from x to $x + h$ and simplifying this expression.
(*b*) Finding the limit of the average velocities as $h \to 0$.

2. How fast is an object falling after 5.9 s have elapsed?

3. Compute $m'(6)$ using the limit definition applied at the particular point $x = 6$.

4. Assuming that the longest safe downward fall on earth involves a "terminal velocity" of 24 ft/s (this is a fall of 3/4 s and a distance of 9 ft), how long is a fall (in seconds) on the moon that gives the same terminal velocity? How many feet is this fall?

DERIVATIVES IN GENERAL

Without regard for any physical interpretation for a function $f(x)$ [where $f(x)$ is not necessarily equal to the function discussed earlier], we can compute the average rate of change of $f(x)$ as x changes from x to $x + h$. This is merely a quotient of the sort already computed several times previously:

$$\frac{f(x+h)-f(x)}{h}$$

We can then go on to form the limit of such quotients as $h \to 0$.

2.3.1 **Definition** The *derivative of f*, denoted f', is a function defined at points in the domain of f by

$$f'(x) = \lim_{h\to 0} \frac{f(x+h)-f(x)}{h}$$

provided that this limit exists. If $f'(x)$ exists at a particular point $x = a$, we refer to $f'(a)$ as the *derivative of $f(x)$ at a* and say *f is differentiable at a*. If f is differentiable for all numbers in some set S, we say f is *differentiable on S*. A function that is differentiable on its entire domain is referred to as a *differentiable function.*

As we saw earlier for the particular functions

$$f(x) = 16x^2 \qquad \text{and} \qquad m(x) = 2.6x^2$$

the derivatives existed for any real number x, and so each is a differentiable function. We also saw that we were able to determine the derivative of f at a particular point in two ways. We found the derivative of f at $x = 3$ by evaluating the limit

$$\lim_{h\to 0} \frac{f(3+h)-f(3)}{h} = 96$$

directly. On the other hand, having found that the derivative of f at an arbitrary point x was $f'(x) = 32x$, later we could compute the derivative at $x = 3$ by simple substitution: $f'(3) = 32 \cdot 3 = 96$. When a function is differentiable at a point $x = a$, its derivative can be determined by either method.

Example 2 For $f(x) = x^2 + 1$, compute (*a*) $f'(5)$ using the definition at the particular point $x = 5$, (*b*) $f'(x)$ for a typical x, (*c*) $f'(5)$ by substitution into $f'(x)$.

Solution

$$\begin{aligned}
(a)\quad f'(5) &= \lim_{h\to 0} \frac{f(5+h)-f(5)}{h} \\
&= \lim_{h\to 0} \frac{[(5+h)^2+1]-[25+1]}{h} \\
&= \lim_{h\to 0} \frac{25+10h+h^2+1-25-1}{h} \\
&= \lim_{h\to 0} \frac{10h+h^2}{h} \\
&= \lim_{h\to 0} (10+h) \qquad \text{since } h \neq 0 \\
&= 10
\end{aligned}$$

$$\begin{aligned}
(b)\quad f'(x) &= \lim_{h\to 0} \frac{f(x+h)-f(x)}{h} \\
&= \lim_{h\to 0} \frac{[(x+h)^2+1]-[x^2+1]}{h} \\
&= \lim_{h\to 0} \frac{x^2+2xh+h^2+1-x^2-1}{h} \\
&= \lim_{h\to 0} \frac{2xh+h^2}{h}
\end{aligned}$$

$$= \lim_{h \to 0} (2x + h) \qquad \text{since } h \neq 0$$

$$= 2x$$

(*c*) Since $f'(x) = 2x$, $f'(5) = 2 \cdot 5 = 10$, which agrees with (*a*) ●

A MATTER OF NOTATION

It is possible to perform the entire previous analysis using other notations. Instead of using h as the notation for the small numbers to be added to x, we can use the symbol Δx (read "delta x"). The symbol Δx is a single indivisible symbol and should be thought of as representing *a small change in x*. Then the quotient

$$\frac{f(x + h) - f(x)}{h}$$

that we have been using takes the form

$$\frac{f(x + \Delta x) - f(x)}{\Delta x}$$

We can take this delta notation a step further by letting Δf denote the change resulting in f when x is changed by the amount Δx. Thus

$$\Delta f = f(x + \Delta x) - f(x)$$

Then our quotient takes the form

$$\frac{f(x + \Delta x) - f(x)}{\Delta x} = \frac{\Delta f}{\Delta x}$$

In addition to the notation $f'(x)$, we denote the derivative by df/dx, so

$$\frac{df}{dx} = \lim_{\Delta x \to 0} \frac{\Delta f}{\Delta x}$$

Sometimes, when a function is given in the form $y = f(x)$ as a functional relationship between two variables, for example

$$y = 16x^2$$

we write Δy to represent the change in y resulting from a change Δx in x. In this case the limit looks like

$$\lim_{\Delta x \to 0} \frac{\Delta y}{\Delta x}$$

and is often denoted as dy/dx or y' instead of df/dx.

In any of these notations, the resulting limit is still called *the derivative with respect to x*. The only difference between the approaches is notation; the idea and the result are exactly the same.

To summarize, if $y = f(x)$, then the derivative with respect to x can be denoted by any of the following:

$$f'(x) \qquad \frac{df}{dx} \qquad \frac{dy}{dx} \qquad y'$$

The whole concept of using different notations to represent the same thing is certainly not foreign to you; for example, 0.5, 1/2, and 2/4 all represent the same number. In our situation each notation has its advantages and disadvantages, and we shall choose among them according to our needs at a particular time, just as we choose the appropriate notation for the number one-half according to the needs of the situation.

Example 3 Compute df/dx for $f(x) = 2x + 3$.

Solution By definition,

$$\begin{aligned}\frac{df}{dx} &= \lim_{\Delta x \to 0} \frac{f(x + \Delta x) - f(x)}{\Delta x} \\ &= \lim_{\Delta x \to 0} \frac{[2(x + \Delta x) + 3] - [2x + 3]}{\Delta x} \\ &= \lim_{\Delta x \to 0} \frac{2x + 2\,\Delta x + 3 - 2x - 3}{\Delta x} \\ &= \lim_{\Delta x \to 0} \frac{2\,\Delta x}{\Delta x} \\ &= \lim_{\Delta x \to 0} (2) \qquad \text{since } \Delta x \neq 0 \\ &= 2 \quad \bullet\end{aligned}$$

Example 4 Given that $y = x^3$, compute dy/dx.

Solution By definition,

$$\begin{aligned}\frac{dy}{dx} = \lim_{\Delta x \to 0} \frac{\Delta y}{\Delta x} &= \lim_{\Delta x \to 0} \frac{[(x + \Delta x)^3] - [x^3]}{\Delta x} \\ &= \lim_{\Delta x \to 0} \frac{x^3 + 3x^2\,\Delta x + 3x(\Delta x)^2 + (\Delta x)^3 - x^3}{\Delta x} \\ &= \lim_{\Delta x \to 0} \frac{3x^2\,\Delta x + 3x(\Delta x)^2 + (\Delta x)^3}{\Delta x} \\ &= \lim_{\Delta x \to 0} [3x^2 + 3x\,\Delta x + (\Delta x)^2] \qquad \text{since } \Delta x \neq 0 \\ &= 3x^2 \quad \bullet\end{aligned}$$

PROGRESS TEST 3

1. Determine df/dx given that $f(x) = 3 - x^2$.
2. Determine $g'(x)$ given that $g(x) = 1/x$. (Combine terms of numerator.)
3. Determine dy/dx given that $y = x^2 + x$.
4. Given $g(x) = 1/x$ (as in Prob. 2), compute
   (*a*) $g'(2)$ applying the limit definition at $x = 2$.
   (*b*) $g'(2)$ by substitution into the result of Prob. 2.

MORE ON INSTANTANEOUS RATES OF CHANGE

When introducing the derivative of a function $f(x)$ in general, we referred to the quotient

$$\frac{f(x + \Delta x) - f(x)}{\Delta x}$$

as the average rate of change of $f(x)$ over the interval between x and $x + \Delta x$. Hence it makes sense to regard $f'(x)$, the limit of such averages as $\Delta x \to 0$, as the instantaneous rate of change of f with respect to x. That this makes sense apart from any particular concrete interpretation for f is the source of its wide usefulness. Whether f describes the distance fallen (so f' gives the instantaneous rate of change of distance with respect to time, that is, the instantaneous velocity) or the profit in a certain economic situation (so f' gives the so-called marginal profit), we rely on the same fundamental idea underlying the interpretation—the instantaneous rate of change of the quantity $f(x)$ with respect to the quantity described by the independent variable x.

Example 5 Given $f(x) = 3$ (constant function), compute and interpret $f'(x)$.

Solution Here, $f(x) = 3 = f(x + \Delta x)$ for any x and any Δx. Therefore the change in f, $\Delta f = f(x + \Delta x) - f(x)$, is equal to zero, so

$$\frac{\Delta f}{\Delta x} = \frac{0}{\Delta x} = 0$$

Hence

$$f'(x) = \lim_{\Delta x \to 0} \frac{\Delta f}{\Delta x} = \lim_{\Delta x \to 0} (0) = 0$$

Naturally enough, $f(x)$, being constant for all x, has zero change Δf, hence zero average change $\Delta f/\Delta x$, and thus zero instantaneous change $f'(x)$. ●

ANOTHER RATE-OF-CHANGE PROBLEM: THE GROWTH OF A CIRCULAR REGION

Material set off by the ⑨ symbol and horizontal rules in the text denotes where the theme of growth of geometric objects is used to build understanding of change and rate of change.

---

⑨ The area A inside a circle is given as a function of its radius by

$$A = \pi r^2$$

Suppose that the length of the radius is increasing. When the circle is small, a given change in radius, Δr, produces a relatively small change in area, ΔA, whereas when the circle is large, the *same* Δr produces a larger increase in area, as can be seen in Fig. 2.18. (Here the delta notation turns out to be convenient.)

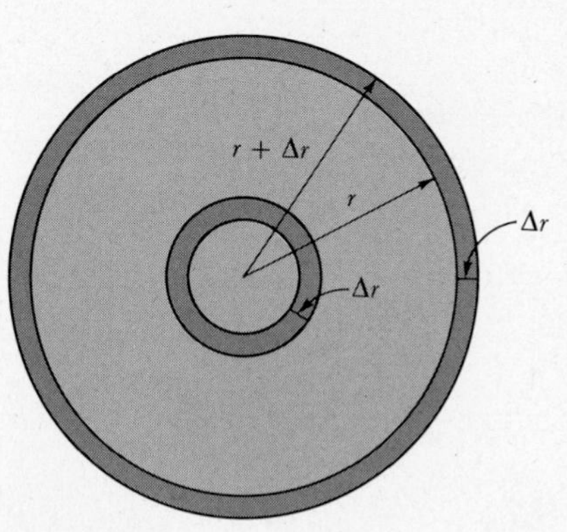

Fig. 2.18

Obviously, the inside strip of width Δr has much smaller area than the outside strip of the *same width*. You have experienced this difference whenever you mowed a lawn in a roughly circular pattern. In that case Δr is the width of the lawn mower and ΔA is the amount you mow in a complete cycle. Certainly it takes much longer to mow the outside strip than the inside one, and the amount you mow is greater. From this we would expect that the rate of change in area, with respect to radius, increases as the radius increases.

To make a more mathematical analysis, we compute

$$\begin{aligned}\Delta A &= A(r + \Delta r) - A(r)\\ &= \pi(r + \Delta r)^2 - \pi r^2\\ &= 2\pi r\,\Delta r + \pi(\Delta r)^2\end{aligned}$$

Notice that ΔA is close to the area of a rectangular strip of width Δr and length $2\pi r$. The additional term $\pi(\Delta r)^2$ represents the amount of area that would theoretically have to be snipped out (in tiny wedges pointing toward the center) to straighten out the circular strip. The *average* rate of change of area with respect to radius is

$$\frac{\Delta A}{\Delta r} = \frac{2\pi r\,\Delta r + \pi(\Delta r)^2}{\Delta r} = 2\pi r + \pi\,\Delta r \qquad (\Delta r \neq 0)$$

Thus the *instantaneous* rate of change of area with respect to radius at radius r is

$$\frac{dA}{dr} = \lim_{\Delta r \to 0} (2\pi r + \pi\,\Delta r) = 2\pi r$$

For $r = 2$, $dA/dr = 2\pi(2) = 4\pi$, and for $r = 30$, $dA/dr = 2\pi(30) = 60\pi$. This mathematically confirms our experience: The larger the radius the larger the rate of change in area. In fact, the rate of change is nothing more than the circumference, which again makes sense: The limit as $\Delta r \to 0$ of the Δr strip *is* a circumference! In effect, the circle "grows by circumferences." In Progress Test 4 and then in the section and chapter exercises we shall see several additional situations analogous to the above. The mathematics together with its physical interpretation provide new insights. Underlying these situations are two fundamental ideas of calculus to be studied later: the idea of differentials and the fundamental theorem of the calculus.

PROGRESS TEST 4

The area A of a square with side x is $A = x^2$.

1. Suppose x increases to $x + \Delta x$. Compute the resulting change in area, ΔA.
2. Sketch the situation, labeling each part of the diagram, noting the area of each section so that each term of ΔA is recognizable in the sketch. (Keep one corner of the square fixed and let all the "growing" take place in the other direction. Shade ΔA.)
3. Compute $\Delta A/\Delta x$ in reduced terms.
4. Compute dA/dx (*a*) in terms of x, (*b*) when $x = 2$, and (*c*) when $x = 200$. (*d*) Discuss the growth of area in terms of dA/dx and the two "edges" in the direction that the growth occurs.

SECTION 2.3 EXERCISES

1. Near the surface of the sun the distance (in feet) fallen by a (rather warm) object in x seconds is approximately $s(x) = 432x^2$.
(*a*) Find the average velocity in the interval between $x = 2$ s and $x = 2\frac{1}{2}$ s.
(*b*) Find the average velocity in the interval between $x = 2$ s and $x = 2\frac{1}{4}$ s.
(*c*) Find the average velocity in the interval between $x = 2$ s and $x = 2.01$ s.
(*d*) Find the average velocity in the interval between $x = 2$ s and $x = (2 + \Delta x)$ s.
(*e*) Find the instantaneous velocity at $x = 2$ s.
(*f*) Find the instantaneous velocity at an arbitrary time x.
(*g*) Find the instantaneous velocity after 5.4 s have elapsed.
(*h*) Given that the largest terminal velocity for a safe jump is 24 ft/s (see Prob. 4, Progress Test 2), what height above the sun gives this terminal velocity and what is the duration of the fall?

In Exercises 2 to 15 use the definition to determine the derivative of the given function at a typical point.

2. $f(x) = 3x - 5$
3. $f(x) = 3 - 2x$
4. $f(x) = 3x^2 - 2$
5. $g(x) = 5x^2 - 3$
6. $g(x) = 6 - 2x^2$
7. $y = 3x^2 - 2x$
8. $y = x$
9. $r(x) = 3x^3$
10. $t(x) = 2x^3 + 5x$
11. $f(w) = w^2 + 3w + 1$
12. $y = 5x^3 - x^2$
13. $y = x^2 - x^3$
14. $f(x) = \dfrac{1}{3x}$
15. $f(x) = \dfrac{1}{3x^2}$
16. Compute and interpret $f'(x)$ for an arbitrary constant function $f(x) = b$.
17. (*a*) Determine $f'(x)$ (using the definition) for $f(x) = \sqrt{x}$ at a typical point x. (You will need to rationalize.)
(*b*) Apply the definition to compute $f'(9)$ and compare with $f'(9)$ obtained by substitution into the result of (*a*).
(*c*) The function $f(x) = \sqrt{x}$ is defined on $[0, +\infty)$. Where is it differentiable?
18. Repeat Exercise 17 for $f(x) = \sqrt{1 - x}$ (defined on $(-\infty, 1]$).

19. ⑨ The volume V of a sphere given as a function of its radius r is $V = \frac{4}{3}\pi r^3$.
(*a*) Compute ΔV (in terms of r and Δr) and interpret geometrically.
(*b*) Determine ΔV for $r = 2$ and $\Delta r = 1$ and compare with ΔV, where $r = 20$ and $\Delta r = 1$. (Thus the amount of blowing required to increase the radius of a *small* balloon by 1 in. is much less than that required to increase the radius of a *large* balloon by 1 in.—ignoring air pressure.)
(*c*) Compute dV/dr.
(*d*) Compare dV/dr and the formula for the surface area of a sphere. Interpret your results.

20. ⑨ Suppose a rectangle has a fixed width of 5 cm and variable length x, so its area A is given by $A = 5x$. Determine and interpret geometrically (drawing a figure): (*a*) ΔA (in terms of x and Δx) and (*b*) dA/dx.

21. The height (in feet) above the ground of a certain object thrown vertically upward from ground level with an initial velocity of 144 ft/s is given as a function of x in seconds by $h(x) = 144x - 16x^2$ (ignoring air resistance).
(*a*) Determine the instantaneous vertical velocity function $h'(x)$.
(*b*) When does the airborne object have zero velocity?
(*c*) At what height is the object at zero velocity?
(*d*) When does the object strike the earth and what is its velocity at that time? (Note that positive height and velocity are in an upward direction.)
(*e*) What is the maximum height reached by the object?
(*f*) Graph the height function.

22. Repeat Exercise 21 for an object thrown vertically with an initial velocity of 96 ft/s, whose distance above ground after x seconds have elapsed is $h(x) = 96x - 16x^2$.

Section 2.4: The Derivative and Slopes of Tangents

Objective:

Relate the derivative of a function to the graph of that function in cases where the derivative is computed using the definition.

DERIVATIVES AND SLOPES

Besides representing the average change of $f(x)$ in the interval between x and $x + \Delta x$, the quotient

$$\frac{f(x + \Delta x) - f(x)}{\Delta x}$$

also represents the slope of the line through the points $P = (x, f(x))$ and $Q = (x + \Delta x, f(x + \Delta x))$; see Fig. 2.19. We simply apply the definition of slope (1.2.3) to the line through P and Q. From the graphical point of view, taking the limit to form the derivative of $f(x)$ amounts to taking the limit of such slopes as $\Delta x \to 0$. Graphically, as Δx tends to 0, the points Q are "moving along" the curve toward P; see Fig. 2.20.

2.4.1 Definition If the derivative of f at the point x exists, that is, if

$$f'(x) = \lim_{\Delta x \to 0} \frac{f(x + \Delta x) - f(x)}{\Delta x}$$

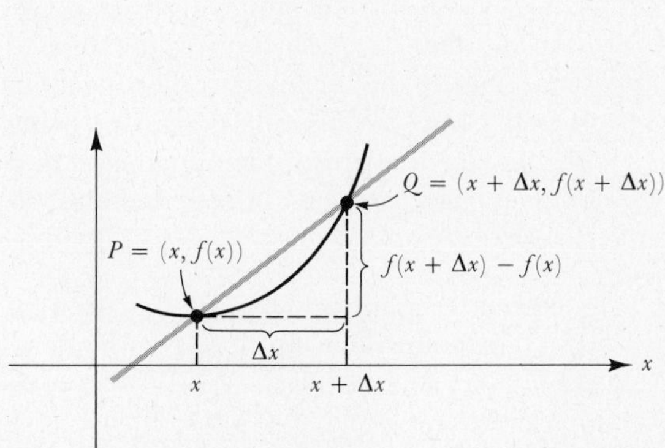

Fig. 2.19

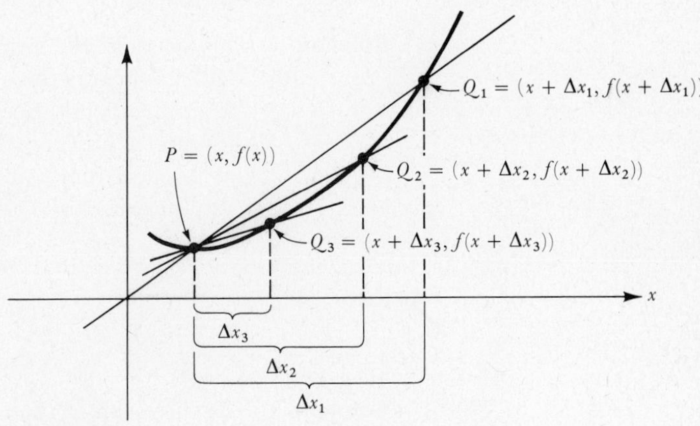

Fig. 2.20

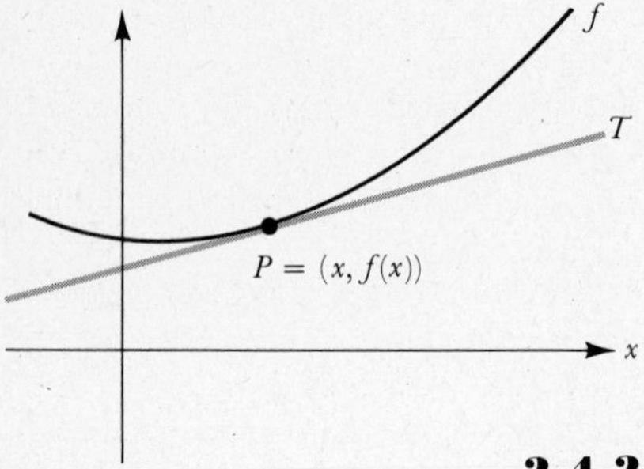

Fig. 2.21

exists, we refer to the unique line through $(x, f(x))$ with slope $f'(x)$ as the *tangent line to the graph of* $f(x)$.

This definition reflects exactly what we see in Figs. 2.19 and 2.20—namely, that as $\Delta x \to 0$, the lines through P and Q are approaching the tangent line to the graph at $(x, f(x))$; see Fig. 2.21. The line T is the tangent line to the graph of $y = f(x)$ at $(x, f(x))$, with a slope of $f'(x)$.

Let us now summarize the result of this discussion.

2.4.2 **Derivatives and Slopes** Wherever $f'(x)$ exists, it gives the slope of the uniquely defined line tangent to the graph of f at $(x, f(x))$. In this case we shall also refer to $f'(x)$ as the *slope of the graph of* $f(x)$, or merely the *slope of* $f(x)$ *at* x.

We shall examine the meaning of this general discussion for the familiar function $f(x) = x^2 + 1$, whose derivative we computed in the preceding section to be

$$f'(x) = 2x$$

In Fig. 2.22 we include a table of data for $f'(x)$ and the graph of $f(x)$ with typical tangents. This new function $f'(x) = 2x$ will give the slope of the tangent to the graph of $y = x^2 + 1$ for any given x.

The table of data fits the picture very well. Note that the tangent line is horizontal, and thus has zero slope, only when $x = 0$.

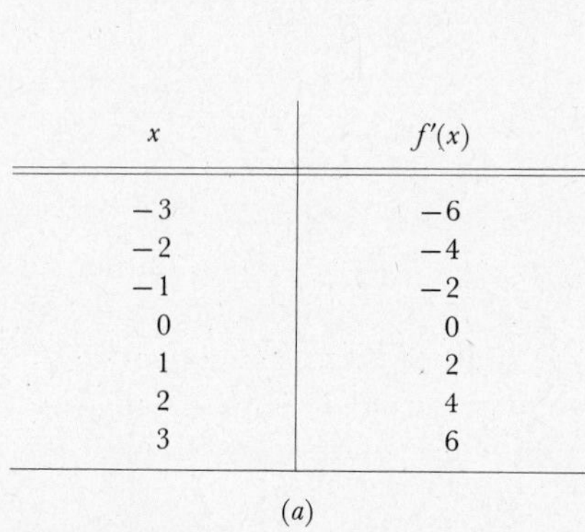

x	$f'(x)$
−3	−6
−2	−4
−1	−2
0	0
1	2
2	4
3	6

(*a*)

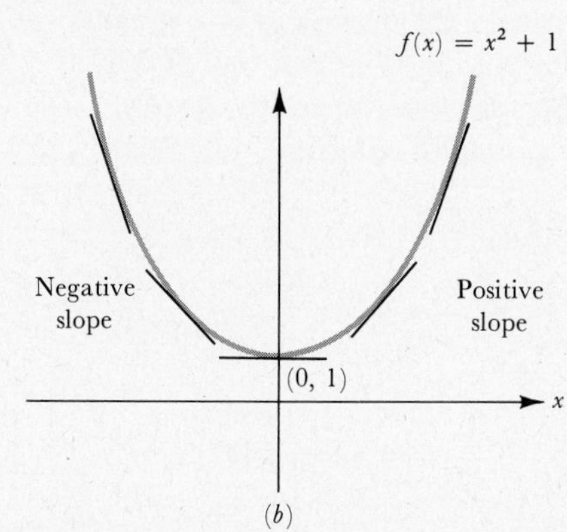

(*b*)

Fig. 2.22

Example 1 Interpret the derivative of $f(x) = 2x + 3$ in terms of its graph. [In Example 3, Sec. 2.3, we showed that $f'(x) = 2$.]

Solution The fact that the graph of $f(x) = 2x + 1$ is a straight line with slope 2 (see Fig. 2.23) is in perfect accord with the fact that $f'(x)$, which gives the slope of the graph at x, is the constant function 2. Notice that in *this* case, the tangent to the graph at any point x is the graph itself. Notice also that the derivative, being the constant function 2, agrees with the fact that this function changes at a constant rate; any change Δx produces a change of $2\,\Delta x$ in the function. ●

Example 2 Interpret the derivative of $y = x^3$ in terms of its graph and give the equation for the line tangent to $y = x^3$ at $x = -1$ in point-slope form. (In Example 4, Sec. 2.3, we found that $dy/dx = 3x^2$.)

Solution The derivative $dy/dx = 3x^2$ tells us that the slope of the graph is positive for all x, except for $x = 0$, where the slope *is* zero. We compute tables of data for y and dy/dx and then sketch the graph. (See Fig. 2.24)

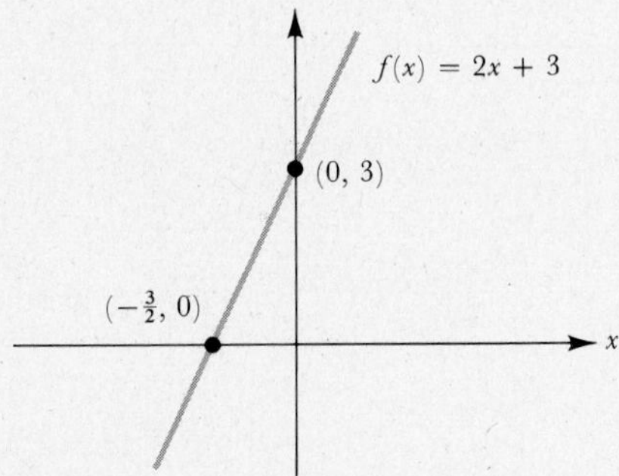

Fig. 2.23

The graph certainly agrees with the derivative: the graph has positive slope, except at $x = 0$, where the derivative $dy/dx = 3x^2$ is zero and the tangent line is horizontal. Furthermore, as $|x|$ increases, the graph gets steeper.

At $x = -1, y = (-1)^3 = -1$ and $dy/dx = 3(-1)^2 = 3$, so the tangent line passes through the point $(-1, -1)$ and has slope 3. Hence its point-slope equation is

$$y - (-1) = 3[x - (-1)]$$

or

$$y + 1 = 3(x + 1) \quad \bullet$$

Notice that the tangent to the graph of $y = x^3$ at $x = 0$ is actually the x axis and in fact crosses the graph at the point of tangency. On the other hand, by putting the equation for the tangent at $(-1, -1)$ into slope-intercept form

$$y = 3x + 2$$

we see that when $x = 2, y = 8$, so this tangent line crosses the curve $y = x^3$ at $(2, 8)$.

It is important to recognize that our definition both of derivative and of tangent is two-sided because, for the defining limit to exist, the left- and right-hand limits

$$\lim_{\Delta x \to 0^-} \frac{f(x + \Delta x) - f(x)}{\Delta x} \quad \text{and} \quad \lim_{\Delta x \to 0^+} \frac{f(x + \Delta x) - f(x)}{\Delta x}$$

must exist and be equal. The following example illustrates that this is not always the case.

Example 3 Show that $g(x) = |x| + 1$ has no derivative, hence no tangent, at $x = 0$, and then interpret this result graphically.

Solution We shall show that

$$\lim_{\Delta x \to 0^-} \frac{g(0 + \Delta x) - g(0)}{\Delta x} \quad \text{and} \quad \lim_{\Delta x \to 0^+} \frac{g(0 + \Delta x) - g(0)}{\Delta x}$$

are unequal. Then we can conclude that the derivative

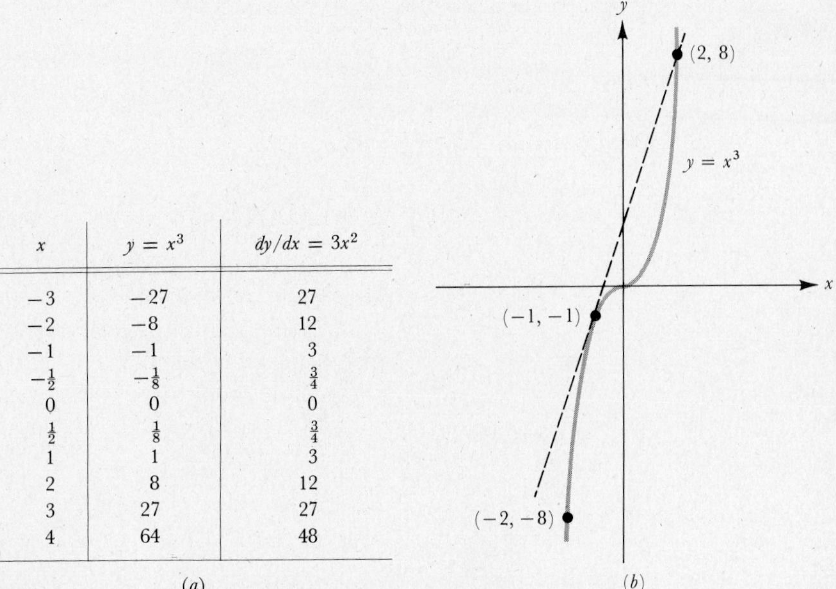

x	$y = x^3$	$dy/dx = 3x^2$
-3	-27	27
-2	-8	12
-1	-1	3
$-\frac{1}{2}$	$-\frac{1}{8}$	$\frac{3}{4}$
0	0	0
$\frac{1}{2}$	$\frac{1}{8}$	$\frac{3}{4}$
1	1	3
2	8	12
3	27	27
4	64	48

(*a*)

(*b*)

Fig. 2.24

$$\lim_{\Delta x \to 0} \frac{g(0 + \Delta x) - g(0)}{\Delta x}$$

does not exist, and hence the tangent at $x = 0$ does not exist.

But

$$\frac{g(0 + \Delta x) - g(0)}{\Delta x} = \frac{[|0 + \Delta x| + 1] - [|0| + 1]}{\Delta x}$$

$$= \frac{|\Delta x| + 1 - 1}{\Delta x}$$

$$= \frac{|\Delta x|}{\Delta x}$$

As we computed in Example 4, Sec. 2.2,

$$\lim_{\Delta x \to 0^-} \frac{|\Delta x|}{\Delta x} = \lim_{\Delta x \to 0^-} \frac{-\Delta x}{\Delta x} \qquad \text{since } |\Delta x| = -\Delta x$$

$$= \lim_{\Delta x \to 0^-} (-1) = -1$$

Also,

$$\lim_{\Delta x \to 0^+} \frac{|\Delta x|}{\Delta x} = \lim_{\Delta x \to 0^+} (1) = 1$$

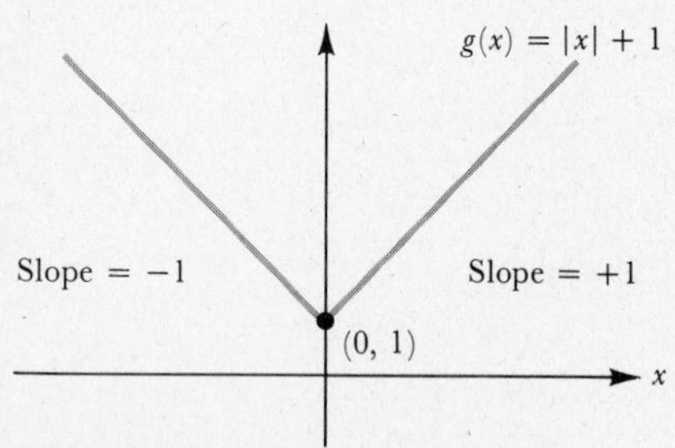

Fig. 2.25

Thus there is no tangent at $x = 0$. Furthermore, for $x > 0$, $g(x) = |x| + 1 = x + 1$ and for $x < 0$, $g(x) = |x| + 1 = -x + 1$, so the graph of g for $x \geq 0$ is the straight line with slope 1 through (0, 1) and for $x \leq 0$, its graph is a straight line with slope -1 through (0, 1). The lack of a well-defined tangent at (0, 1) is reflected in the existence of a sharp corner at (0, 1). (See Fig. 2.25.) ●

In Chap. 4 we shall see that derivatives provide a potent and efficient technique for sketching graphs of functions.

PROGRESS TEST 1

1. For each of the following find $f'(x)$ using the definition and determine an equation for the line tangent to the graph at $(a, f(a))$.

(*a*) $f(x) = 3x^2$ $\quad a = 1$

(*b*) $f(x) = x^2 + x + 1$ $\quad a = 0$

(*c*) $f(x) = 1 - x^3$ $\quad a = 2$

2. Let $f(x) = x^2$ and $g(x) = x^2 + c$, where c is some constant. Show that $f'(x) = g'(x) = 2x$. Interpret this observation graphically.

A REMARK ON OUR APPROACH TO DERIVATIVES

Notice that our introduction to derivatives in Secs. 2.3 and 2.4 has used the three basic ways of viewing functions. We began with the *numerical* point of view, calculating numerical averages based on a physical interpretation. We then switched to the more efficient *algebraic* point of view, computing the limits of algebraic quotients. Finally, we looked at these limits from a *geometric* point of view, recognizing them as slopes of tangent lines.

SECTION 2.4 EXERCISES

In Exercises 1 to 6 compare $f'(x)$ and the graph of $f(x)$, (a) giving a sketch of $f(x)$, (b) describing when the slope is positive, negative, zero, and/or undefined, and (c) giving the equation (in point-slope form) of the line tangent to the graph of f when $x = 1$.

1. $f(x) = 3 - x^2$ **2.** $f(x) = 5 - 2x^2$ **3.** $f(x) = 3x - 5$
4. $f(x) = 2 - 3x$ **5.** $f(x) = 1 - 2x^3$ **6.** $f(x) = x - x^3$

7. (*a*) Show that $g(x) = |x - 2|$ is not differentiable at $x = 2$.
(*b*) Interpret (*a*) in terms of the graph of g.

8. Repeat Exercise 7 for $h(x) = |3x - 9|$ at $x = 3$.

9. (Recall Sec. 2.3, Exercise 17.) For $f(x) = \sqrt{x}$ interpret the behavior of $f'(x)$ near 0 (x not negative) in terms of the graph of $f(x)$ near 0. In particular, examine $\lim_{x\to 0^+} f(x)$ and $\lim_{x\to 0^+} f'(x)$.

10. Repeat Exercise 9 for $f(x) = \sqrt{1 - x}$ near $x = 1$. (Recall Sec. 2.3, Exercise 18.)

11. (*a*) Determine dy/dx, where $y = 1/(x - 1)$.
(*b*) Compare dy/dx near $x = 1$ with the graph of $y = 1/(x - 1)$. (Recall Example 6, Sec. 2.2.)

12. Repeat Exercise 11 for $y = 1/(2 - x)$ near $x = 2$ (Recall Sec. 2.2, Progress Test 1, Prob. 4.)

13. Interpret graphically the fact that $f'(x) = 0$ for any constant function $f(x) = b$.

14. Interpret graphically the fact that $f'(x) = 1$ for $f(x) = x$.

15. Interpret graphically the fact that $f'(x) = 3$ for any function of the form $f(x) = 3x + b$, where b is a constant.

16. Where is the slope of $y = x^2 - 3x + 2$ negative?

17. Where is the slope of $y = x^2 - 6x + 5$ negative?

18. Determine the derivative of the "largest integer function" $s(x) = [x]$ on the open interval $(1/4, 3/4)$. (Recall Sec. 2.1, Exercise 28, and Sec. 2.2, Exercise 18.)

19. *Two types of parallelism:* recall Sec. 2.4, Progress Test 1, Probs. 2 and 3.
(*a*) What is the *vertical* distance between $f(x) = x^2$ and $g(x) = x^2 + 1$ when $x = 0$?
(*b*) Repeat (*a*) for $x = 10$.
(*c*) What is the *horizontal* distance between the graphs of f and g when $y = 1$?
(*d*) Repeat (*c*) for $y = 10$. Compare with (*c*).
(*e*) If the graphs of $f(x)$ and $g(x)$ are defined to be "vertically parallel," suggest an alternative definition of "horizontally parallel."
(*f*) Are the graphs of $f(x) = x^2$ and $h(x) = (x - 1)^2$ horizontally parallel? Are they vertically parallel?

Section 2.5: Limits, Continuity, and Differentiability

Objectives:

1. Apply the basic properties of limits, as expressed in the Limit Theorem, to evaluate limits involving functions that can be built up from quotients of polynomials and *n*th roots.
2. Recognize where functions are continuous.
3. Distinguish between a function that is continuous at a point and a function that has a derivative at that point.

Anyone desiring a complete and detailed discussion of the material up to the following Progress Test 1 may turn directly to Sec. 9.1.

THE LIMIT THEOREM

The following definition states in precise language that

$$\lim_{x\to a} f(x) = L$$

if and only if we can make $f(x)$ stay arbitrarily close to L by choosing x sufficiently close to a.

2.5.1 **Definition** Suppose we are given a function $f(x)$ and numbers a and L, with $f(x)$ defined for all points, except perhaps a itself, of an open interval containing a. We say that

the limit of $f(x)$ as x tends to a is L

and write

$$\lim_{x\to a} f(x) = L$$

if for every positive number ε there is a positive number δ such that

$$0 < |x - a| < \delta \quad \text{implies} \quad |f(x) - L| < \varepsilon$$

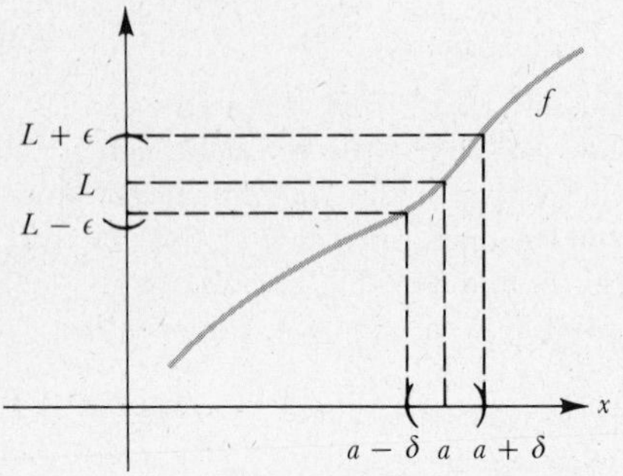

Fig. 2.26

The key element in this definition is that for *every* $\varepsilon > 0$ there must be a $\delta > 0$ such that x being within δ units of a but not equal to a implies that $f(x)$ is within ε units of L. The requirement that $x \neq a$ reflects the remarks made earlier in this chapter that we are not concerned with the behavior of f *at* a, but only *near* a. The limit is unaffected by $f(a)$; in fact, $f(a)$ need not even be defined.

Graphically we are showing that there is an interval about a such that if $x \neq a$ is in this interval, then $f(x)$ is in the (vertical) interval $(L - \varepsilon, L + \varepsilon)$; see Fig. 2.26.

In Chap. 9 we shall prove the following simple but useful fact:

2.5.2 **Theorem** Suppose that $f(x) = mx + b$ is a given function, where m and b are constants and $x = a$ is any real number. Then

$$\lim_{x\to a} f(x) = ma + b$$

That is, $\lim_{x\to a} (mx + b)$ can be evaluated by direct substitution.

As special cases of (2.5.2) we have the following:

2.5.3 **Special Limits**

(*a*) If $f(x) = b$, a constant function ($m = 0$), then $\lim_{x\to a} (b) = b$

(*b*) If $f(x) = x$ ($m = 1, b = 0$), then $\lim_{x\to a} (x) = a$

The next theorem (proved in Chap. 9) justifies the steps required to decompose complicated limits into much simpler limits involving only these special functions, which can then be evaluated by direct substitution.

2.5.4 **The Limit Theorem** Suppose that $\lim_{x\to a} f(x)$ and $\lim_{x\to a} g(x)$ both exist. Then

(*a*) $\lim_{x\to a} [f(x) \pm g(x)] = [\lim_{x\to a} f(x)] \pm [\lim_{x\to a} g(x)]$.

(*b*) $\lim_{x\to a} [f(x) \cdot g(x)] = [\lim_{x\to a} f(x)] \cdot [\lim_{x\to a} g(x)]$. In particular, if b is a constant, then $\lim_{x\to a} [b \cdot g(x)] = b[\lim_{x\to a} g(x)]$ (so constants can be "pulled outside limits").

(*c*) $\lim_{x\to a} [f(x)/g(x)] = [\lim_{x\to a} f(x)]/[\lim_{x\to a} g(x)]$ provided that $\lim_{x\to a} g(x) \neq 0$.

(*d*) $\lim_{x\to a} \sqrt[n]{f(x)} = \sqrt[n]{\lim_{x\to a} f(x)}$, where n is a positive integer. If n is even, then $f(x)$ must be nonnegative.

The best way to appreciate this theorem is to see it at work evaluating limits of much more complicated functions than would be feasible using the definition alone. Although we have stated parts (*a*) and (*b*) for only two functions, they extend directly to sums, differences, and products of any finite number of functions. A similar collection of statements holds for one-sided limits.

Example 1 Determine $\lim_{x\to 4} \sqrt[3]{2x^3 + 3x + 1}$

$$\begin{aligned}
&\lim_{x\to 4} \sqrt[3]{(2x^3 + 3x + 1)} \\
&= \sqrt[3]{\lim_{x\to 4} (2x^3 + 3x + 1)} && \text{Limit Theorem [2.5.4}(d)] \\
&= \sqrt[3]{\lim_{x\to 4} 2x^3 + \lim_{x\to 4} 3x + \lim_{x\to 4} 1} && \text{Limit Theorem [2.5.4}(a)] \\
&= \sqrt[3]{2 \lim_{x\to 4} x^3 + 3 \lim_{x\to 4} x + \lim_{x\to 4} 1} && \text{Limit Theorem [2.5.4}(b)]
\end{aligned}$$

$$= \sqrt[3]{2 \lim_{x\to 4} (x \cdot x \cdot x) + 3 \lim_{x\to 4} x + \lim_{x\to 4} 1}$$

$$= \sqrt[3]{2 \lim_{x\to 4} x \cdot \lim_{x\to 4} x \cdot \lim_{x\to 4} x + 3 \lim_{x\to 4} x + \lim_{x\to 4} 1} \qquad \text{Limit Theorem [2.5.4}(b)]$$

$$= \sqrt[3]{2(4)(4)(4) + 3(4) + 1} \qquad \text{special limits (2.5.3)}$$

$$= \sqrt[3]{141} \quad \bullet$$

The following result is an immediate consequence of the fact that in the definition of $\lim_{x\to a} f(x)$, the behavior of f *at* a is irrelevant. In particular, it records formally our ability to cancel nonzero factors and determine the limit of the "reduced" function. We have applied it dozens of times already.

2.5.5 **Theorem** If $f(x) = g(x)$ for all $x \neq a$ in some open interval containing a, and if $\lim_{x\to a} g(x) = L$, then $\lim_{x\to a} f(x) = L$.

Example 2 Determine

$$\lim_{x\to 2} \frac{x^2 + x - 6}{x^2 - 4}$$

Solution We cannot apply the Limit Theorem [2.5.4(c)] directly because $\lim_{x\to 2} x^2 - 4 = 0$. However, since

$$\frac{x^2 + x - 6}{x^2 - 4} = \frac{(x+3)(x-2)}{(x+2)(x-2)} = \frac{x+3}{x+2} \qquad \text{for } x \neq 2$$

we can first apply Theorem (2.5.5) to conclude that

$$\lim_{x\to 2} \frac{x^2 + x - 6}{x^2 - 4} = \lim_{x\to 2} \frac{x+3}{x+2}$$

Now, since $\lim_{x\to 2} (x + 2) \neq 0$, we have

$$\lim_{x\to 2} \frac{x+3}{x+2} = \frac{\lim_{x\to 2} (x+3)}{\lim_{x\to 2} (x+2)} \qquad \text{by Limit Theorem [2.5.4}(c)]$$

$$= \frac{\lim_{x\to 2} x + \lim_{x\to 2} 3}{\lim_{x\to 2} x + \lim_{x\to 2} 2} \qquad \text{by Limit Theorem [2.5.4}(a)]$$

$$= \frac{2+3}{2+2} \qquad \text{by (2.5.3)}$$

$$= \frac{5}{4} \quad \bullet$$

The Limit Theorem also applies to justify the computations involved in calculating derivatives.

The following theorem will be very useful in some of our later work. The theorem says that if a function f is squeezed between two functions g and h, both of which tend to L as x tends to a, then f itself must tend to L as x tends to a. Its proof is in Chap. 9.

2.5.6 **Squeeze Theorem** If $h(x) \leq f(x) \leq g(x)$ for all $x \neq a$ in some open interval containing a, and if $\lim_{x\to a} h(x) = \lim_{x\to a} g(x) = L$, then $\lim_{x\to a} f(x) = L$.

PROGRESS TEST 1

Determine each of the following three limits, giving reasons for each step.

1. $\lim_{x \to -1} (\sqrt{2}\, x^4 - 2x)$

2. $\lim_{\Delta x \to 0} \dfrac{1/(x + \Delta x) - (1/x)}{\Delta x}$

3. $\lim_{h \to 0} \dfrac{f(x+h) - f(x)}{h}$ where $f(x) = 3x^2 - 2x$ (combine steps—note the limit variable is h, and x plays the role of a constant)

CONTINUITY

We all have a general idea of what a "continuous" process is: It has no sharp breaks or changes. This same general idea applies to functions.

2.5.7 **Definition** We say the function f is *continuous* at a if
(a) $\lim_{x \to a} f(x)$ exists (b) $f(a)$ is defined (c) $\lim_{x \to a} f(x) = f(a)$

Furthermore, we say f *is continuous on a set of numbers* S if f is continuous at each point of S. If S is the set of all real numbers, then we say f is *everywhere continuous* and refer to f as a *continuous function.* If f is not continuous at a, we say f is *discontinuous* at a.

By our definition, a function fails to be continuous at a if any one of the three conditions fails to hold. Since the idea of continuity at a is thus tied to the idea of limit and to whether the function is defined at a, much of our earlier work can be translated directly into the language of continuity.

In particular, a function f may fail to be continuous at a point a for any of the reasons it fails to have a limit at a. For example, if either of the left- or right-hand limits at a fails to exist or equal $f(a)$, then f is discontinuous at a. The following graphically described examples (Figs. 2.27 to 2.31), were discussed on pp. 40–44; $s(x)$, shown in Fig. 2.32, was discussed on p. 60.

The function $g(x)$ fails to satisfy condition (b) [hence (c)] of Definition (2.5.7) at $x = 2$, since $g(2)$ is not defined; see Fig. 2.28.

The function $h(x)$ fails to satisfy condition (c) at $x = 2$, since $\lim_{x \to 2} h(x) = 4$, but $h(2) = 6$; see Fig. 2.29.

The function $D(x)$ fails to satisfy (a) and (b) [hence (c)] at $x = 0$; see Fig. 2.30.

The function $r(x)$ fails to satisfy (a) and (b) [hence (c)] at $x = 2$; see Fig. 2.31.

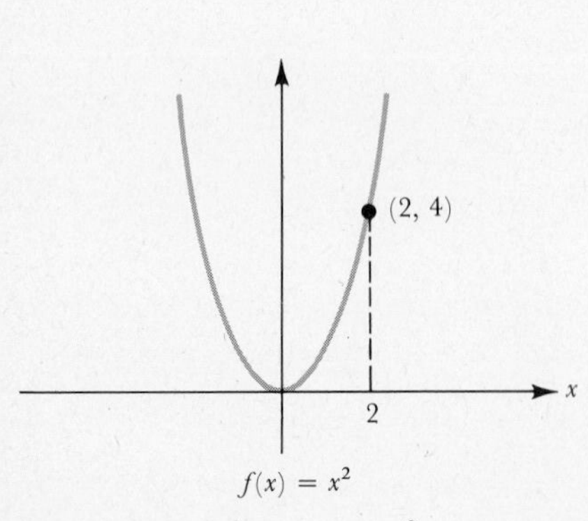

$f(x) = x^2$

continuous at $x = 2$

Fig. 2.27

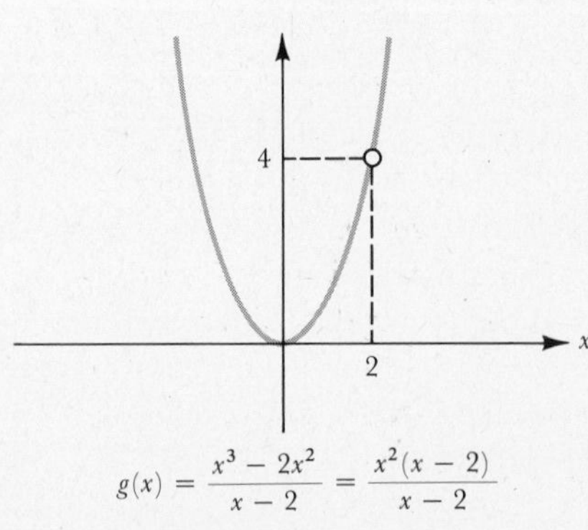

$g(x) = \dfrac{x^3 - 2x^2}{x - 2} = \dfrac{x^2(x-2)}{x-2}$

discontinuous at $x = 2$

Fig. 2.28

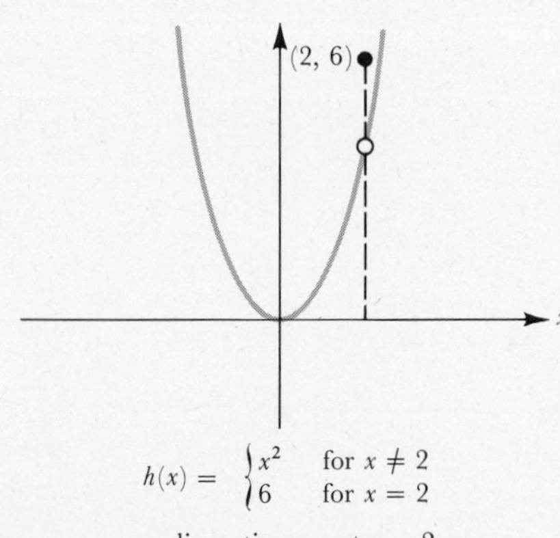

$h(x) = \begin{cases} x^2 & \text{for } x \neq 2 \\ 6 & \text{for } x = 2 \end{cases}$

discontinuous at $x = 2$

Fig. 2.29

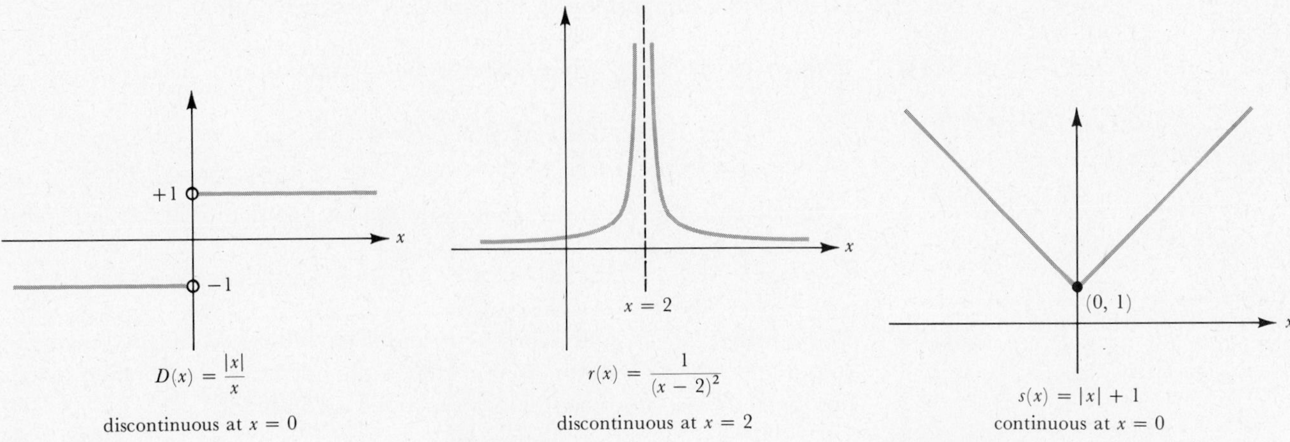

discontinuous at $x = 0$

Fig. 2.30

discontinuous at $x = 2$

Fig. 2.31

continuous at $x = 0$

Fig. 2.32

The function $s(x)$, which has a "corner" at $x = 0$, is nevertheless continuous at $x = 0$; see Fig. 2.32.

Geometrically, as is evident from these illustrations, the idea of continuity at a point has a relatively direct interpretation: The function f is continuous at a if and only if the graph has *no jumps or breaks* at a; that is, it can be traced near a without lifting the pen from the paper.

Algebraically, the idea of continuity at a also has a relatively simple interpretation: the function f is continuous at a if and only if $\lim_{x\to a} f(x)$ *can be evaluated by direct substitution.* We can combine this fact with the Limit Theorem to obtain ways of combining continuous functions to get new continuous functions. For example, if we know that f and g are continuous at a, then $\lim_{x\to a} f(x) = f(a)$ and $\lim_{x\to a} g(x) = g(a)$. By Limit Theorem [2.5.4(*a*)], $\lim_{x\to a} [f(x) \pm g(x)] = [\lim_{x\to a} f(x)] \pm [\lim_{x\to a} g(x)] = f(a) \pm g(a)$, so we can conclude that $f \pm g$ is continuous at a. Similarly, the remaining portions of the following theorem can be deduced.

2.5.8 **Theorem** Suppose f and g are continuous at a. Then $f \pm g$ and $f \cdot g$ are continuous at a, as is f/g if $g(a) \neq 0$. Also, for n a positive integer $\sqrt[n]{f}$ is continuous at a (f must be positive if n is even.)

Now, we know by (2.5.2) that any function of the form $f(x) = mx + b$ is everywhere continuous because any limit of such a function at any point a can be evaluated by direct substitution. Using (2.5.8), we can then apply this fact to show that most familiar algebraic functions are continuous where defined.

The previous discussion overlooked a minor point, which deserves clarification. A function such as $f(x) = \sqrt{x}$ is defined only on $(0, +\infty)$, so, according to our definition of limit, $\lim_{x\to 0} \sqrt{x}$ does not exist, hence $f(x) = \sqrt{x}$ is not continuous at 0. [Recall that (2.5.1) requires $f(x)$ to be defined on an open interval containing 0.] However, $\lim_{x\to 0^+} \sqrt{x}$ *does* exist and equals $f(0) = \sqrt{0} = 0$.

2.5.9 If a function is defined on a closed interval for which a is an endpoint and the appropriate one-sided limit exists and equals the value of the function at a, we shall say that the function is continuous at a.

Hence we say $f(x) = \sqrt{x}$ is continuous at 0, and $g(x) = \sqrt[6]{1-x}$ is continuous at 1. [Here $g(x)$ is defined on $(-\infty, 1]$ and $\lim_{x\to 1^-} \sqrt[6]{1-x} = 0 = g(1)$.]

To handle composition of functions, we first need a limit theorem.

2.5.10 **Theorem** If $\lim_{x \to a} g(x) = b$ and if f is continuous at b, then

$$\lim_{x \to a} f(g(x)) = f(\lim_{x \to a} g(x)) = f(b)$$

In effect, this theorem says that a continuous function preserves limits. An immediate consequence is the following:

2.5.11 **Theorem** If g is continuous at a and f is continuous at $g(a)$, then the composition $f \circ g$ is continuous at a. Hence, provided that the domains match up, the composition of continuous functions is continuous.

Our earlier statement that a continuous function has no jumps or breaks is given formal expression in the following theorem, whose proof rests on the corresponding fact about the real-number line: *it* has no breaks. The proof is beyond the scope of this book.

2.5.12 **The Intermediate-Value Theorem** Suppose f is continuous on the closed interval $[a, b]$ and C is any number between $f(a)$ and $f(b)$. Then there exists at least one number c in $[a, b]$ such that $f(c) = C$.

As is evident from Fig. 2.33(*a*), there may be many c's in $[a, b]$ such that $f(c) = C$. However, if f fails to be continuous at just a single point in $[a, b]$, the conclusion may fail as Example 3 illustrates.

Example 3 Show that the conclusion of the Intermediate-Value Theorem (2.5.12) fails to hold for

$$f(x) = \begin{cases} 2 - x & \text{if } x \le 1 \\ x + 2 & \text{if } x > 1 \end{cases} \qquad \text{on } [0, 3]$$

Solution As shown in Fig. 2.34, $f(0) = 2$ and $f(3) = 5$. If we let $C = 2\frac{1}{2}$, we find that there is no number c between 0 and 3 such that $f(c) = 2\frac{1}{2}$! ●

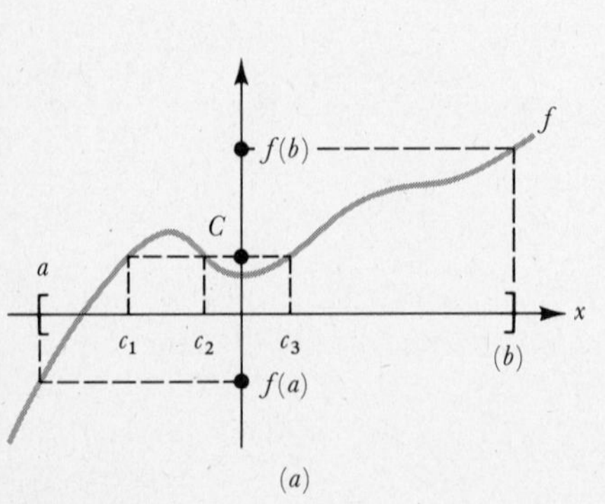

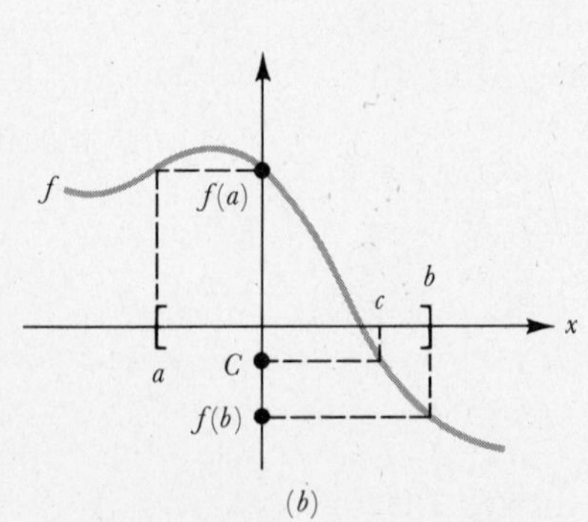

Fig. 2.33

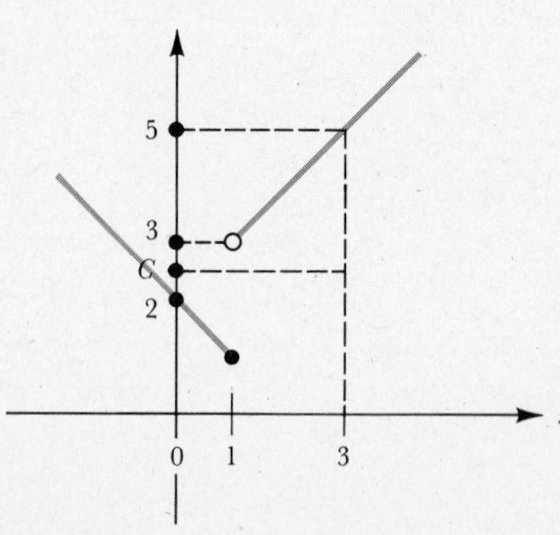

Fig. 2.34

PROGRESS TEST 2

In Probs. 1 to 5 identify the points where the given function is discontinuous and which of the conditions in the definition of continuity fail(s) to hold.

1. $f(x)$ of Prob. 2, Sec. 2.2, Progress Test 1
2. $f(x)$ of Prob. 3, Sec. 2.2, Progress Test 1
3. $f(x)$ of Prob. 4, Sec. 2.2, Progress Test 1
4. $f(x)$ of Prob. 6, Sec. 2.2, Progress Test 1
5. The function in Sec. 2.1, Exercise 11(c)

6. Let $f(x) = \begin{cases} x + 1.95 & \text{for } x < 0 \\ 2 & \text{for } x = 0 \\ x + 2.1 & \text{for } x > 0 \end{cases}$

(a) Show that the conclusion of the Intermediate-Value Theorem fails for this function.
(b) For the same function, name two nonoverlapping closed intervals where the hypothesis (hence conclusion) of the Intermediate-Value Theorem *does* apply.

DIFFERENTIABILITY IMPLIES CONTINUITY

Note the strong contrast between the function $f(x) = x^2 + 1$, whose derivative $f'(x) = 2x$ exists (Sec. 2.3, Example 2) and is zero at $x = 0$, and the function $g(x) = |x| + 1$, whose derivative does not exist at $x = 0$ (Sec. 2.4, Example 3). The graph of $f(x)$ has a horizontal tangent at (0, 1), whereas the graph of $g(x)$ has a sharp corner and *no* uniquely defined tangent at (0, 1). Nonetheless, the graph of neither function has jumps or breaks at (0, 1). Each is continuous at (0, 1), but only $f(x)$ is differentiable at (0, 1). The following discussion establishes the basic relation between differentiability and continuity for a general function f.

The two conditions for a function f on an interval I—(a) f is differentiable on I and (b) f is continuous on I—may be thought of as "good-behavior" conditions on the graph of f. The first says that f has a well-defined tangent at each point in I, which implies that the graph is smooth without sharp corners. The second says that the graph has no breaks or jumps on I. Now, since a graph that cannot even have sharp corners certainly cannot have jumps or breaks, we would expect condition (a) to imply (b) in general. To prove this we shall need some useful equivalent ways of stating continuity at a point. Recall that two statements are equivalent if each implies the other. This means that in practice the two statements can be used interchangeably.

2.5.13 Theorem The following statements are equivalent:

1. $f(x)$ is continuous at $x = a$.
2. $\lim\limits_{x \to a} f(x) = f(a)$.
3. $\lim\limits_{\Delta x \to 0} f(a + \Delta x) = f(a)$.
4. $\lim\limits_{\Delta x \to 0} \Delta f = 0$, where $\Delta f = f(a + \Delta x) - f(a)$.

Proof: The equivalence of statements 1 and 2 is just the definition of continuity at a. We shall show that statements 2 and 3 and statements 3 and 4 are equivalent by systematically rewriting the limit statements. We first show that statements 2 and 3 are equivalent. Suppose $\lim_{x \to a} f(x) = f(a)$ and $x = a + \Delta x$. Then $\Delta x = x - a$, so letting $x \to a$ is equivalent to letting $\Delta x \to 0$. Thus

$$f(a) = \lim_{x \to a} f(x) = \lim_{\Delta x \to 0} f(a + \Delta x)$$

and therefore statements 2 and 3 are equivalent.

To show that 3 and 4 are equivalent, we note that stating

$$f(a + \Delta x) \to f(a) \text{ as } \Delta x \to 0 \qquad \text{(statement 3)}$$

is equivalent to stating

$$\Delta f = [f(a + \Delta x) - f(a)] \to 0 \text{ as } \Delta x \to 0 \qquad \text{(statement 4)} \quad \blacksquare$$

We now can state and prove the promised result.

2.5.14 **Theorem** If f is differentiable at $x = a$, then f is continuous at $x = a$.

Proof: If f is differentiable at $x = a$, then

$$\lim_{\Delta x \to 0} \frac{f(a + \Delta x) - f(a)}{\Delta x}$$

exists. Thus we may write

$$\begin{aligned}
\lim_{\Delta x \to 0} \Delta f &= \lim_{\Delta x \to 0} [f(a + \Delta x) - f(a)] \\
&= \lim_{\Delta x \to 0} \left[\frac{f(a + \Delta x) - f(a)}{\Delta x} \cdot \Delta x\right] \\
&= \left[\lim_{\Delta x \to 0} \frac{f(a + \Delta x) - f(a)}{\Delta x}\right] \cdot [\lim_{\Delta x \to 0} \Delta x] \qquad \text{(Limit Theorem [2.5.4(}b\text{)])} \\
&= f'(a) \cdot 0 = 0
\end{aligned}$$

Thus by the fourth statement of (2.5.13), f is continuous at $x = a$. $\blacksquare$

The results on continuity stated in this section will be used repeatedly throughout the book. In fact, the idea of continuity is basic in calculus. Other important facts relating to continuous functions will be developed as we need them.

SECTION 2.5 EXERCISES

In Exercises 1 to 10 evaluate the given limit, giving reasons for each step.

1. $\lim_{x \to 1} \dfrac{x^2 + x + 3}{3x^2 + 9}$

2. $\lim_{x \to 0} \dfrac{x^2 + x}{2x^2 + 1}$

3. $\lim_{x \to -2} \sqrt{4x^2 + 2}$

4. $\lim_{x \to 2} \dfrac{x^2 - 4}{x + 2}$

5. $\lim_{x \to 1} \dfrac{\sqrt{x} - 1}{x - 1}$

6. $\lim_{x \to 0} \dfrac{(2 + x)^3 - 8}{x}$

7. $\lim_{x \to 3} \dfrac{1/(x + 2) - (1/5)}{x - 3}$

8. $\lim_{h \to 0} \dfrac{\sqrt{4 + h} - 2}{h}$

9. $\lim_{\Delta x \to 0} \dfrac{\sqrt{1 + \Delta x} - 1}{\Delta x}$

10. $\lim_{\Delta x \to 0} \dfrac{(x + \Delta x)^3 - x^3}{\Delta x}$

11. Using $f(x) = |x|/x$ and $g(x) = -|x|/x$, show that $\lim_{x \to a} [f(x) + g(x)]$ may exist when $\lim_{x \to a} f(x)$ and $\lim_{x \to a} g(x)$ do not.

12. Repeat Exercise 11 for $\lim_{x \to a} [f(x) \cdot g(x)]$.

13. Show that $\lim_{x \to a} |f(x)|$ may exist when $\lim_{x \to a} f(x)$ does not. (See Exercise 11.)

14. Assuming that $\lim_{x \to a} f(x) = N$, $\lim_{x \to a} g(x) = M$, $\lim_{x \to a} h(x) = L$, $\lim_{x \to a} i(x) = K$, and that $L + K \neq 0$, give an argument with reasons for each step, proving that

$$\lim_{x \to a} \left[\frac{f(x) + h(x)i(x) - g(x)}{h(x) + i(x)}\right]^{1/5} = \left[\frac{N + LK - M}{L + K}\right]^{1/5}$$

15. Is the combination function in Exercise 14 continuous at $x = a$?

In Exercises 16 to 21 (a) sketch the graph of the functions f; (b) determine those numbers; if any, at which f fails to be continuous; and (c) state the reason(s) why it fails to be continuous.

16. $f(x) = \begin{cases} \dfrac{3}{x - 2} & \text{if } x \neq 2 \\ 7 & \text{if } x = 2 \end{cases}$

17. $f(x) = \begin{cases} \dfrac{x^2 + x - 2}{x - 1} & \text{if } x \neq 1 \\ 5 & \text{if } x = 1 \end{cases}$

18. $f(x) = \begin{cases} \dfrac{x^2 + x - 2}{x - 1} & \text{if } x \neq 1 \\ 3 & \text{if } x = 1 \end{cases}$

19. $f(x) = \dfrac{x^4 - 16}{x^2 - 4}$

20. $f(x) = \dfrac{x - 2}{|x - 2|}$

21. $f(x) = \begin{cases} 2x + 3 & x \leq 1 \\ 8 - 3x & 1 < x < 2 \\ x + 3 & x \geq 2 \end{cases}$

22. Construct a function that is defined everywhere and is discontinuous only at $x = 0$, $x = 2$.

23. Construct a function that is continuous everywhere but at $x = 0$, $x = 2$, where it is undefined.

24. Use the Intermediate-Value Theorem (2.5.12) and Theorem (2.5.9) to prove there is a number a such that $a^3 = 2$.

25. Where is the largest integer function $s(x) = [x]$ discontinuous? (Recall Sec. 2.1, Exercise 32.)

26. Show that the conclusion of the Intermediate-Value Theorem fails for the largest integer function $s(x)$ on $[\frac{1}{2}, \frac{3}{2}]$.

27. Define the functions f and g of Exercise 11 at $x = 0$ in such a way that they can be used to illustrate that the sum of functions discontinuous at a point may be continuous at that point.

28. Repeat Exercise 27 for products.

CHAPTER EXERCISES

Review Exercises

In Exercises 1 to 12 determine the values of the following functions, first at a typical x in the domain, and second at $x = 1$ if 1 is in the domain:

(*a*) $f + g$, (*b*) $f \cdot g$, (*c*) $\dfrac{f}{g}$, (*d*) $f \circ g$, (*e*) $g \circ f$, (*f*) $f \circ f$, (*g*) $g \circ g$.

1. $f(x) = 3x - 1$, $g(x) = x^2$

2. $f(x) = 2x + 3$, $g(x) = 3x^2$

3. $f(x) = \dfrac{1}{x}$, $g(x) = x + 5$

4. $f(x) = \dfrac{1}{x + 2}$, $g(x) = 3x$

5. $f(x) = 3x^4 + 5$, $g(x) = x$

6. $f(x) = x$, $g(x) = x^3 - 2$

7. $f(x) = 4$, $g(x) = x^2 - 3x - 2$

8. $f(x) = x^3 + 3x^2 - 4$, $g(x) = 2$

9. $f(x) = \sqrt{x - 3}$, $g(x) = 2x^2$

10. $f(x) = x^4$, $g(x) = \sqrt[4]{x + 2}$

11. $f(x) = \dfrac{1}{x - 2}$, $g(x) = \dfrac{1}{x + 2}$

12. $f(x) = \frac{1}{3}x - 2$, $g(x) = 3x + 6$

In Exercises 13 to 16 express the given function as a composition of two functions, neither of which is the original function. (*Note that the answer is not unique.*)

13. $f(x) = (x + 1)^3$

14. $g(x) = (x^3)^5 + 3(x^3)^2 + 2(x^3)$

15. $h(x) = 3(x^2)^3 + 4(x^2)^2 + 2(x^2)$

16. $f(x) = \sqrt[3]{3x - 5}$

17. Given that $y = 3x^2 - 2x - 2$, compute y when
(*a*) $x = 2r$ (*b*) $x = 3t + 1$ (*c*) $x = t + \Delta t$
(*d*) $x = w^2$ (*e*) $x = 3u$

18. Repeat Exercise 17 for $y = 5x^2 - 3x + 1$.

Which of the following in 19 to 25 determine(s) y as a function of x? Give a reason for your answer.

19.

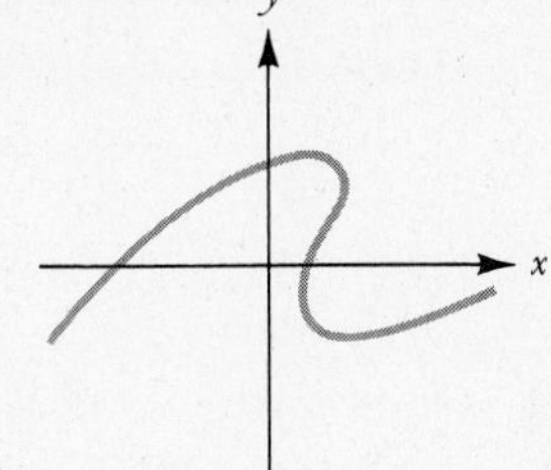

20.

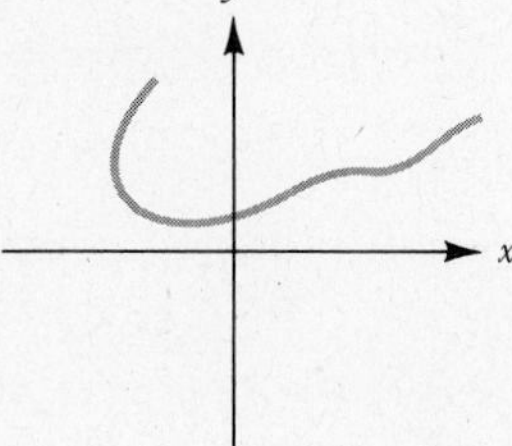

21.

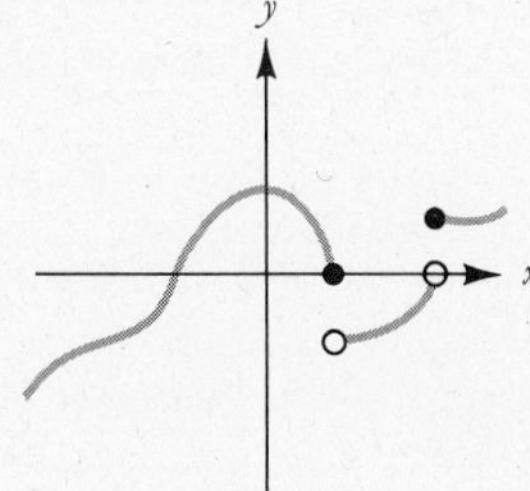

22. $y^2 = x^3$

23. $y^3 = x^2$

24. $\sqrt{y} = x$

25. $1/y = x$

In Exercises 26 to 29, (a) give a linear function that equals h(x) on the natural domain of h(x), and (b) extend the domain of h to include all real numbers in such a way that h(x) equals the linear function for all *x.*

26. $h(x) = \dfrac{x^2 - 2x + 1}{x - 1}$

27. $h(x) = \dfrac{x^2 - 5x}{x}$

28. $h(x) = \dfrac{2x^3 - x^2}{x^2}$

29. $h(x) = \dfrac{x^2 - x - 12}{x - 4}$

In Exercises 30 to 35, (a) determine whether $\lim_{x\to a} f(x)$ exists for the given function at the given point using informal methods (giving reasons for your answer) and (b) if the limit exists, use the Limit Theorem and special limits to justify the steps associated with evaluating the limit.

30. $f(x) = x^2 + 3x - 1,\ a = -3$

31. $g(x) = \dfrac{2x}{\sqrt{x-3}},\ a = 4$

32. $g(x) = \dfrac{2x}{\sqrt{x-3}},\ a = 3$

33. $m(x) = -1/x^2,\ a = 0$

34. $r(x) = 1/x^3,\ a = 0$

35. $f(x) = \begin{cases} 2 - 3x & \text{for } x \neq 1 \\ 2 & \text{for } x = 1 \end{cases},\ a = 1$

36. Is the function in Exercise 35 differentiable at $a = 1$?

37. Which of the functions in Exercises 31, 33, and 35 is continuous at the given point a?

38. Repeat Exercise 37 for the functions in Exercises 30, 32, and 34.

39. Determine $\lim_{x\to a} h(x)$ for $h(x)$ at the point where $h(x)$ is not defined in (*a*) Exercise 27 and (*b*) Exercise 29.

40. Repeat Exercise 39 for functions in (*a*) Exercise 26 and (*b*) Exercise 28.

41. Compute $\lim_{x\to 5^+} \dfrac{2x-10}{|x-5|}$ and $\lim_{x\to 5^-} \dfrac{2x-10}{|x-5|}$.

42. Compute $\lim_{x\to 1^+} \dfrac{1}{2x-2}$ and $\lim_{x\to 1^-} \dfrac{1}{2x-2}$.

43. Give an informal argument to show that $\lim_{x\to 2} (3x - 4) \neq 5$.

In Exercises 44 to 55, (a) using the definition given in (2.3.1), determine the derivative of the given function at a typical number in its domain, and (b) determine the point-slope equation of the line tangent to the graph when the independent variable equals 1.

44. $f(x) = \dfrac{-1}{7x}$

45. $g(x) = x + \dfrac{1}{x}$

46. $h(x) = x - \dfrac{1}{x}$

47. $f(t) = \dfrac{1}{t^2} + 3$

48. $y = 3x^2 + x - 2$

49. $y = \sqrt{x-2}$

50. $f(x) = \dfrac{1}{\sqrt{x}}$

51. $g(x) = \dfrac{1}{x^3}$

52. $m(x) = 5x^2 - 4x + 1$

53. $f(u) = u^3 - 5u^2$

54. $g(w) = \dfrac{1}{w+1}$

55. $h(r) = 6r^2 + 10$

In Exercises 56 to 60 determine the point(s) (if any) where the given function is discontinuous and state why the function is discontinuous there.

56. $f(x) = \begin{cases} -1 & \text{if } x < 2 \\ \dfrac{x}{2} & \text{if } 2 \leq x < 3 \\ \sqrt{x} & \text{if } x > 3 \end{cases}$

57. $f(x) = \dfrac{1}{x^2+1}$

58. $f(x) = \begin{cases} 2x - 1 & \text{if } x < 1 \\ x^2 & \text{if } x \geq 1 \end{cases}$

59. $f(x) = \begin{cases} \dfrac{3}{x-2} & \text{if } x \neq 2 \\ 3 & \text{if } x = 2 \end{cases}$

60. $f(x) = \begin{cases} \dfrac{x^2 - 5x + 6}{x-3} & \text{for } x \neq 3 \\ 1 & \text{for } x = 3 \end{cases}$

In Exercises 61 to 65 sketch the graph of a function that satisfies the given conditions.

61. f is defined for all x in $[0, 2]$, f has a limit at 1, f is not continuous at 1.

62. f is defined for all x in $[0, 2]$, $f(x) < 3$ for all x, $\lim_{x\to 1} f(x) = 3$.

63. f is defined for all x in $[0, 2]$, $\lim_{x\to 1^-} f(x) = -\infty$, $\lim_{x\to 1^+} f(x) = 2$.

64. f is continuous on $[0, 2]$ but is not differentiable at $x = 1$ or $x = 3/2$.

65. f is defined for all x in $[0, 2]$, $\lim_{x\to 1} f(x) = +\infty$.

66. Find $\lim_{x\to 1^-} f(x)$ and $\lim_{x\to 1^+} f(x)$, where

$$f(x) = \begin{cases} x^2 & \text{if } x \geq 1.001 \\ 3x & \text{if } 1 < x < 1.001 \\ 2x + 1 & \text{if } .99 < x < 1 \\ x^2 + 5x & \text{if } x \leq .99 \end{cases}$$

Miscellaneous Exercises

1. As a preparation for Exercise 2, suppose that $l(x) = mx + b$ is a linear function with, say, $m > 0$.

(*a*) Show that $l(x + h)$ is as much above $l(x)$ as $l(x - h)$ is below $l(x)$ for $h > 0$.

(*b*) Interpret part (*a*) graphically.
(*c*) Determine the average of $l(x-h)$ and $l(x+h)$.

2. When examining the "freely falling object" problem numerically, many students recognize that the average velocity between x and $x+h$ for a particular $h>0$ and x is as much above the proposed instantaneous velocity at x as the average velocity between $x-h$ and x is below the proposed instantaneous velocity. See Table 2.3 and subsequent calculations. The students then go on to conjecture that the instantaneous velocity at that point is the "middle" or average of the average velocities on either side of x. (They are often told by the instructor that although this happens to be true here, it is not true in general.)
(*a*) Prove that this conjecture is true for $f(x)=16x^2$ at $x=3$ by comparing

$$\frac{f(3+h)-f(3)}{h} \quad \text{and} \quad \frac{f(3-h)-f(3)}{-h}$$

(*b*) Prove that the conjecture holds for any x with $f(x)=16x^2$.
(*c*) Prove that the conjecture holds for any x with $f(x)=ax^2+bx+c$ any quadratic function (a, b, c are constants).
(*d*) Prove that the conjecture fails for $g(x)=x^3$ at $x=1$.
(*e*) Explain the difference between $g(x)$ and the quadratic function in part (*c*) by regarding the various averages as functions of h and/or noting which are linear and which are not. (See Exercise 1.)
(*f*) Determine those functions for which the conjecture holds and those for which it fails.

---

3. ⑨ The volume V of a right circular cylinder of radius r and height h is given by $V=\pi r^2 h$.
(*a*) Suppose that r is fixed and h is allowed to grow by Δh. Compute ΔV, $\Delta V/\Delta h$, and sketch the situation.
(*b*) Compute dV/dh and relate this result to the area of the "top" of the cylinder.
(*c*) Suppose that h is fixed and r is allowed to grow by Δr. Compute ΔV, $\Delta V/\Delta r$, and sketch the situation.
(*d*) Compute dV/dr and relate this result to the lateral area of a right circular cylinder. (If you do not know the formula, guess it.)

---

4. (An alternative limit notation) If the standard notation for the two-sided limit $\lim_{x\to a} f(x)=L$ is replaced by $\lim_{|x-a|\to 0} f(x)=L$, then give the appropriate analogs for the right- and left-hand limits. (These will not use absolute value notation.)

5. (An alternative definition of derivative) Using arguments similar to those used in proving (2.5.13) and (2.5.14), show that the definition of the derivative of f at $x=a$ can be rewritten as

$$\lim_{x\to a}\frac{f(x)-f(a)}{x-a}$$

6. Using the definition in Exercise 5, determine the derivative of $f(x)=6x^2-3x$ at (*a*) $x=2$ and (*b*) $x=a$.

7. Repeat Exercise 6 for $f(x)=x^2+4$.

8. Determine the point where the line tangent to $y=x^3$ at $x=1$ crosses the curve $y=x^3$.

9. (*a*) Determine where the slope of $f(x)=3x-x^3$ is a maximum, that is, where the graph has steepest positive slope.
(*b*) Determine where the tangent to $f(x)$ is horizontal.
(*c*) Sketch the graph of $f(x)$.

10. Repeat Exercise 9 for $f(x)=x^2-2x^3$.

11. We know that if f is differentiable at a, then f is continuous at a.
(*a*) Is it necessary that f' be differentiable at a when f is?
(*b*) Is it necessary that f' be continuous at a when f' exists at a?

12. (*a*) Where is the "largest integer function" $s(x)=[x]$ differentiable? (Remember that the definition of derivative is two-sided.)
(*b*) What is $s'(x)$ when it exists? Give a geometric argument.

13. Define $s_1(x)=$ smallest integer $n \geqslant x$. [Thus $s_1(1/2)=1=s_1(1)$, with $s_1(1.1)=2$.]
(*a*) Sketch the graph of $s_1(x)$.
(*b*) Where is $s_1(x)$ discontinuous?

14. Refer to Exercise 13.
(*a*) Sketch $s(x)-s_1(x)$.
(*b*) Where is $s(x)-s_1(x)$ continuous?

15. A malfunctioning rocket (weak thrust) is shot vertically upward with an initial velocity of 144 ft/s and is at height $h(x)=144x-3x^2$ after x seconds.
(*a*) Determine the function that gives the instantaneous vertical velocity at time x.
(*b*) When does the airborne rocket have zero velocity?
(*c*) What is its height at this point?
(*d*) When does the rocket crash and what is its downward (negative) velocity at this time?
(*e*) What is the maximum height reached by the rocket?
(*f*) Graph $h(x)$.
(*g*) Compare (*a*) to (*f*) with Sec. 2.3, Exercise 21, to determine the effect of the weak thrust in terms of maximum height and when the rocket strikes the earth.

16. Repeat Exercise 15 for an initial velocity of 96 ft/s and a height function of $h(x)=96x-8x^2$. In part (*g*) compare with Sec. 2.3, Exercise 22.

17. With reference to Exercises 15 and 16, what is the effect of strengthened rocket thrust on the constant a in the function $h(x)=v_0x-ax^2$ (where v_0 is the initial velocity)?

18. (*a*) Does your car have an instantaneous-velocity meter?
(*b*) Which devices on your car would you use to calculate average speeds? (*Hint:* One of these is the clock.)

19. (*a*) Given $f(x)=(1/x)+1$, determine df/dx using the definition.
(*b*) Where does the tangent to the graph of $f(x)$ have a slope of $-1/9$?

20. Given $f(x)=\sqrt{2x+1}$, find $f'(1)$ using the definition.

21. Let $f(x)=mx+b$ be any linear function.
(*a*) What is its tangent at any point?
(*b*) What is the slope of this tangent?

(*c*) Find $f'(x)$ using the definition of derivative.
(*d*) Do your answers to (*a*), (*b*), and (*c*) agree?

22. (*a*) Determine where the slope of $f(x) = x^2 - 2x + 5$ is negative, zero, and positive.
(*b*) Determine the lowest point on the graph of f.

23. (Tangent to a circle)
(*a*) Given $f(x) = \sqrt{1 - x^2}$, whose graph is the upper half of $x^2 + y^2 = 1$, determine $f'(x)$ by rationalizing the numerator of $\Delta f/\Delta x$ before taking the limit as $\Delta x \to 0$.
(*b*) Determine $f'(\sqrt{2}/2)$ and sketch the graph of $f(x)$ and the tangent line.
(*c*) Determine the slope of the radius line through the point of tangency and show that it is perpendicular to the tangent line whose slope is given by $f'(x)$. (This shows that the tangent line determined by the derivative agrees with the classic tangent to a circle.)
(*d*) Show the tangent at $x = 0$ is horizontal.

24. Determine where the function $g(x) = |x^2 - 1|$ is (*a*) continuous and (*b*) differentiable. (Sketch its graph.)

25. By determining left- and right-hand limits separately, determine whether

$$f(x) = \begin{cases} 2x - 1 & \text{for } x < 1 \\ x^2 & \text{for } x \geq 1 \end{cases}$$

is differentiable at $x = 1$.

26. Give reasonable definitions of "left differentiable" and "right differentiable" at $x = a$. (See Exercise 25.)

---

27. ⑨ The volume V of a cube of side x is given by $V = x^3$.
(*a*) Suppose x increases by Δx. Compute ΔV, $\Delta V/\Delta x$, and, if possible, sketch the situation—with one corner fixed and all "growth" in the directions of the other corners of the cube.
(*b*) Compute dV/dx and relate this result to the fact that the growth takes place in the directions of three faces of the cube.

---

28. (Recall Sec. 2.4, Exercise 19.) Let $H(f, g)(x)$ represent the horizontal distance from the point (x, x^2) on the graph of $f(x) = x^2$ to the graph of $g(x) = x^2 + 1$. As x grows without bound, what happens to $H(f, g)(x)$?

29. (See Exercise 28)
(*a*) Given the two functions $f(x) = x^2$ and $g(x) = x^2 + 1$ with $P = (x, x^2)$ $(x \neq 0)$, let Q be the point on the graph of g where the line perpendicular to the tangent to f at P crosses the graph of g. Determine the distance between P and Q.
(*b*) Repeat part (*a*), beginning with the line perpendicular to the tangent at $\bar{Q} = (x, x^2 + 1)$. Let $\bar{P}$ be the point where the perpendicular crosses the graph of f.
(*c*) Compare $H(f, g)(x)$ and the results of (*a*) and (*b*).

SELF-TEST ✣

1. Determine (*a*) $f - g$, (*b*) $f \cdot g$, (*c*) f/g, (*d*) $f \circ g$, (*e*) $g \circ f$, (*f*) $f \circ f$ for $f(x) = 7x - 1$, $g(x) = x^2 + 2x$.

2. Evaluate whichever of the following limits exist:

$$(a)\ \lim_{x \to 1} \frac{x^2 + 2x + 1}{3x + 3} \qquad (b)\ \lim_{x \to 9} \frac{\sqrt{x} - 3}{x - 9}$$

3. Name the point(s) of discontinuity for each of the functions in Prob. 2. Explain your answer.

4. Given that the distance fallen by a certain type of parachute as a function of time is $p(x) = 20x$:
(*a*) Find the average velocity between $x = 3$ and $x = 4$.
(*b*) Find the average velocity between $x = 3$ and $x = 3 + \Delta x$.
(*c*) Find the average velocity between x and $x + \Delta x$.
(*d*) Find the instantaneous velocity at any time x.

5. Does the graph of $f(x) = x^3 + 2x$ have any horizontal tangents? Explain your answer.

6. Determine where on the interval $[a, k]$ the function sketched below is (*a*) continuous, (*b*) differentiable, (*c*) discontinuous [state the reason(s)].

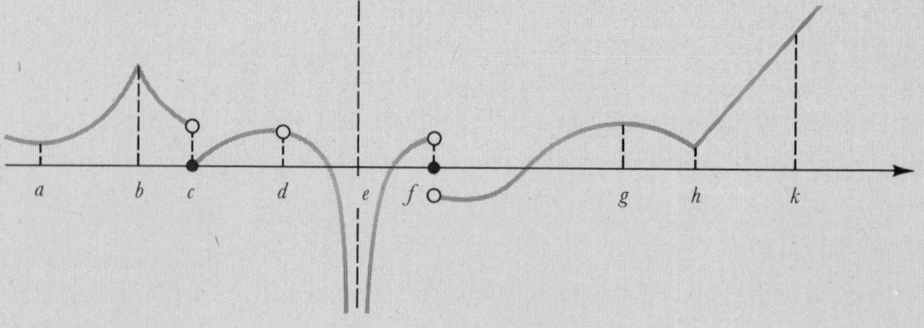

Chapter 3

DIFFERENTIATION

CONTENTS

The purpose of this chapter is to enable you to compute derivatives of any function that can be formed using powers and roots of polynomials. In Secs. 3.1 and 3.2 we develop efficient rules that allow us to calculate derivatives without repeatedly resorting to the basic limit definition. We also extend the application of the derivative to include functions defined implicitly by equations.

In Sec. 3.3 we introduce the differential, which is closely related to the derivative. We interpret differentials algebraically, graphically, and geometrically and show how they can be used in numerical approximations. In Sec. 3.4 we discuss and interpret derivatives of derivatives.

Section 3.1: Basic Rules of Differentiation

Objectives:

Use the rules of differentiation to determine derivatives of sums, products, quotients, and (integral) powers of polynomial functions.

DERIVATIVES OF POLYNOMIAL FUNCTIONS

In Chap. 2 we introduced the derivative of a function $f(x)$ as a means of studying how the values of the function change as x changes. Recall the following.

3.1.1 **Definition** Given that f is defined in some open interval containing x, we define the *derivative of f with respect to x* as

$$\lim_{\Delta x \to 0} \frac{f(x + \Delta x) - f(x)}{\Delta x}$$

and denote it by $f'(x)$, df/dx, or, if $y = f(x)$, y' and dy/dx. The derivative of f with respect to x is also denoted by $D_x f$ or, if $y = f(x)$, $D_x y$. Determining the derivative of a function is commonly referred to as *differentiating* the function.

We have interpreted the derivative as the instantaneous rate of change of f with respect to change in x as well as the slope of the line T tangent to the graph of $f(x)$ at the point $(x, f(x))$. The instantaneous rate of change was defined as the limit of average rates of change over smaller and smaller intervals Δx, whereas the slope of the tangent line was defined as the limit of slopes of "secant lines" through $P = (x, f(x))$ and $Q = (x + \Delta x, f(x) + \Delta x))$, as shown in Fig. 3.1.

Table 3.1 collects for reference some of the results of our examinations of specific functions in Chap. 2. However, direct application of 3.1.1 gives us our first *general* rule (see Progress Test 1, Prob. 9):

3.1.2 **Theorem** The derivative of a constant function is zero.

The pattern evident in functions 3 and 5, Table 3.1, suggests that

$$\text{If } f(x) = x^4, \text{ then } f'(x) = 4x^3$$

We shall prove a general rule that establishes and formalizes this pattern.

3.1.3 **The Power Rule for Positive Integers** If n is a positive integer and $f(x) = x^n$, then $f'(x) = nx^{n-1}$.

Proof: We need to examine $(x + \Delta x)^n$. Although the Binomial Theorem tells us how to expand $(x + \Delta x)^n$ completely, the first three terms tell us all we need to know:

$$(x + \Delta x)^n = x^n + nx^{n-1}(\Delta x) + Ax^{n-2}(\Delta x)^2 + Bx^{n-3}(\Delta x)^3 + \cdots + (\Delta x)^n$$

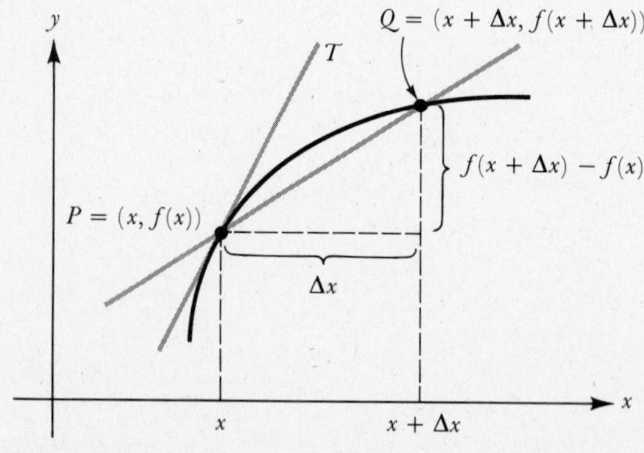

Fig. 3.1

Function	*Derivative*	*Remarks*
1. $f(x) = 3$	$f'(x) = 0$	Zero rate of change, horizontal graph: slope = 0
2. $f(x) = 2x + 3$	$f'(x) = 2$	Constant rate of change, graph is straight line of slope 2
3. $f(x) = x^2$	$f'(x) = 2x$	Parabola concave up, horizontal tangent at $x = 0$
4. $f(x) = x^2 + 1$	$f'(x) = 2x$	Parabola concave up, horizontal tangent at $x = 0$
5. $f(x) = x^3$	$f'(x) = 3x^2$	Slope never negative, with horizontal tangent at $x = 0$
6. $f(x) = 3 - x^2$	$f'(x) = -2x$	Parabola concave down
7. $f(x) = 16x^2$	$f'(x) = 32x$	If f is distance fallen, then f' is instantaneous velocity
8. $f(x) = x^2 + x$	$f'(x) = 2x + 1$	f' is sum of derivatives of x^2 and of x
9. $f(x) = 1/x$	$f'(x) = -1/x^2$	$f(0)$ and $f'(0)$ undefined, but $f'(x) < 0$ otherwise
10. $f(x) = \sqrt{x}$	$f'(x) = 1/(2\sqrt{x})$	$f(x) = x^{1/2}, f'(x) = 1/2x^{-1/2}$

Table 3.1

Here we used the letters A and B, respectively, to denote the coefficients of the third and fourth terms since they are unimportant to the final result. Now, by the definition of the derivative,

$$\begin{aligned} f'(x) &= \lim_{\Delta x \to 0} \frac{f(x + \Delta x) - f(x)}{\Delta x} \\ &= \lim_{\Delta x \to 0} \frac{(x + \Delta x)^n - x^n}{\Delta x} \\ &= \lim_{\Delta x \to 0} \frac{[x^n + nx^{n-1}(\Delta x) + Ax^{n-2}(\Delta x)^2 + \text{terms of higher degree in } \Delta x] - x^n}{\Delta x} \\ &= \lim_{\Delta x \to 0} \frac{nx^{n-1}(\Delta x) + Ax^{n-2}(\Delta x)^2 + \text{terms of higher degree in } \Delta x}{\Delta x} \\ &= \lim_{\Delta x \to 0} (nx^{n-1} + Ax^{n-2}(\Delta x) + \text{terms of higher degree in } \Delta x) \\ &= nx^{n-1} \quad \blacksquare \end{aligned}$$

The last line of the above proof made typical use of the Limit Theorem [2.5.4(*a*)].

The Power Rule agrees with our earlier results; for example, $D_x(x^2) = 2x$. By item 7, Table 3.1, we know

$$D_x(16x^2) = 32x = 16D_x(x^2)$$

The following states more generally that *the derivative of a constant times a function is the constant times the derivative.*

3.1.4 **The Constant-Multiple Rule** If k is a constant and f is differentiable at x, then

$$D_x[kf(x)] = kD_x[f(x)]$$

Proof: By the definition of $D_x[kf(x)]$

$$D_x[kf(x)] = \lim_{\Delta x \to 0} \frac{kf(x + \Delta x) - kf(x)}{\Delta x}$$

$$= \lim_{\Delta x \to 0} \frac{k[f(x + \Delta x) - f(x)]}{\Delta x}$$

$$= k \lim_{\Delta x \to 0} \frac{f(x + \Delta x) - f(x)}{\Delta x} \quad \text{since constants can be pulled out of limits; see [2.5.4(b)]}$$

$$= kD_x[f(x)] \quad \text{by definition of } kD_x[f(x)] \ \blacksquare$$

Combining the Power and Constant-Multiple Rules we have

3.1.5 $$D_x(ax^n) = anx^{n-1} \quad (a \text{ is a constant and } n \text{ a positive integer})$$

All we need now to differentiate a typical polynomial (using rules instead of the definition) is the ability to differentiate term by term. We shall show that *the derivative of a sum is the sum of the derivatives.*

3.1.6 **The Sum Rule** Suppose $f(x) = u(x) + v(x)$, where $u(x)$ and $v(x)$ are differentiable. Then

$$f'(x) = u'(x) + v'(x)$$

Proof: Let us separate the usual limit quotient into a sum of $u(x)$ and $v(x)$ parts and then apply the Limit Theorem [2.5.4(a)].

$$f'(x) = \lim_{\Delta x \to 0} \frac{f(x + \Delta x) - f(x)}{\Delta x}$$

$$= \lim_{\Delta x \to 0} \frac{[u(x + \Delta x) + v(x + \Delta x)] - [u(x) + v(x)]}{\Delta x}$$

$$= \lim_{\Delta x \to 0} \left[\frac{u(x + \Delta x) - u(x)}{\Delta x} + \frac{v(x + \Delta x) - v(x)}{\Delta x}\right]$$

$$= \lim_{\Delta x \to 0} \frac{u(x + \Delta x) - u(x)}{\Delta x} + \lim_{\Delta x \to 0} \frac{v(x + \Delta x) - v(x)}{\Delta x}$$

$$= u'(x) + v'(x) \ \blacksquare$$

Although (3.1.6) is a statement about the sum of *two* differentiable functions, it also applies to the sum of 3, 4, 10, or any number of differentiable functions. Furthermore, it extends directly to differences of functions because

$$\begin{aligned} f(x) &= u(x) - v(x) \\ &= u(x) + (-1)v(x) \end{aligned}$$

Now apply the Sum Rule and then the Constant-Multiple Rule to obtain

$$\begin{aligned} f'(x) &= u'(x) + (-1)v'(x) \\ &= u'(x) - v'(x) \end{aligned}$$

As a result of (3.1.5), (3.1.6), and our discussion, we know that:

3.1.7 **Theorem** Any polynomial function is differentiable.

Example 1 Differentiate $f(x) = 11x^7 + 6x^5 - 3x^2 + 8x - 1$.

Solution We can differentiate term by term and add the results as we go along:

$$\begin{aligned} f'(x) &= 11 \cdot 7x^6 + 6 \cdot 5x^4 - 3 \cdot 2x^1 + 8 \cdot 1 \cdot x^0 - 0 \\ &= 77x^6 + 30x^4 - 6x + 8 \ \bullet \end{aligned}$$

In Exercises 29 and 30 you are asked to interpret the following special cases in terms of graphs and rates of change.

(Linear and Quadratic Functions) Let a, b, c be constants. Then:
(*A*) If $y = ax + b$, then $dy/dx = a$.
(*B*) If $y = ax^2 + bx + c$, then $dy/dx = 2ax + b$.

PROGRESS TEST 1

In Probs. 1 to 10 differentiate the given function using the rules previously developed.

1. $f(x) = x^{41}$
2. $f(x) = x$
3. $f(x) = 4x^3 + 3x^2 + 3x + 2$
4. $g(x) = 51x^{42} + 26x^2 + 432.21$
5. $f(x) = x^8 + 7x^7 + 6x^6 + g(x)$ [where $g(x)$ is differentiable]
6. $s(t) = 4t^3 - 3t/5 + 1$
7. $p(r) = \Pi r^2 + 2\Pi r + \Pi$
8. $f(x) = (x^5 - 3)(x^2 + 4x)$
9. Show that if $f(x) = b$ is a constant function, then $f'(x)$ equals zero.
10. Where does the graph of $f(x) = x^3 - 6x^2$ have a horizontal tangent?

PRODUCTS AND QUOTIENTS

Note that in Prob. 8 of the previous progress test it is necessary to multiply the two factors before differentiating. Attempting to differentiate the product by taking the product of the derivatives leads to an incorrect answer, as can be seen more vividly by treating $f(x) = x^2$ as the product $f(x) = x \cdot x$.

Since the derivative of each factor is 1, the product of the derivatives is 1. But of course $f'(x) = 2x \neq 1$. The derivative of a product is certainly *not* the product of the derivatives!

We shall state the rule for products and then apply it to Prob. 8 before proving it.

3.1.8 **The Product Rule** Suppose $f(x) = u(x)v(x)$, where $u(x)$ and $v(x)$ are differentiable. Then

$$f'(x) = u(x)v'(x) + v(x)u'(x)$$

Thus *the derivative of the product of two functions is the sum of the first function times the derivative of the second function plus the second function times the derivative of the first function.*

Example 2 Use the Product Rule to differentiate $f(x) = (x^5 - 3)(x^2 + 4x)$ and check the result against the previously obtained derivative.

Solution

$$f'(x) = \overbrace{(x^5 - 3)}^{\text{1st}}\overbrace{(2x + 4)}^{\text{Deriv. of 2d}} + \overbrace{(x^2 + 4x)}^{\text{2d}}\overbrace{(5x^4)}^{\text{Deriv. of 1st}}$$
$$= 2x^6 + 4x^5 - 6x - 12 + 5x^6 + 20x^5$$
$$= 7x^6 + 24x^5 - 6x - 12$$

which agrees with the solution obtained to Progress Test 1, Prob. 8. ●

⑤ *Proof of the Product Rule:* In applying the definition of derivatives to the product $f(x) = u(x)v(x)$ it is helpful to regard $u(x)$ and $v(x)$ as the lengths of the sides of a rectangle, so $f(x)$ represents the area of the rectangle. Then, changing x to $x + \Delta x$ changes $f(x)$ to

$$f(x + \Delta x) = u(x + \Delta x)v(x + \Delta x)$$

and $f(x + \Delta x) - f(x)$ may be regarded as the area of the shaded region in Fig. 3.2(a). To simplify notation we label.

$$u(x + \Delta x) - u(x) \text{ as } \Delta u \qquad \text{and} \qquad v(x + \Delta x) - v(x) \text{ as } \Delta v$$

as shown in Fig. 3.2(b). In particular, then, we can regard $f(x + \Delta x) - f(x)$ as the total area of the three regions labeled I, II, and III in Fig. 3.2(b):

$$f(x + \Delta x) - f(x) = \overset{\text{I}}{u(x)\,\Delta v} + \overset{\text{II}}{v(x)\,\Delta u} + \overset{\text{III}}{\Delta u\,\Delta v}.$$

By definition of $f'(x)$ we have

$$f'(x) = \lim_{\Delta x \to 0} \frac{f(x + \Delta x) - f(x)}{\Delta x}$$

(1) $$= \lim_{\Delta x \to 0} \frac{u(x)\,\Delta v + v(x)\,\Delta u + \Delta u\,\Delta v}{\Delta x}$$

$$= \lim_{\Delta x \to 0} \left[u(x)\frac{\Delta v}{\Delta x} + v(x)\frac{\Delta u}{\Delta x} + \frac{\Delta u}{\Delta x}\Delta v \right]$$

$$= \lim_{\Delta x \to 0} \left[u(x)\frac{\Delta v}{\Delta x} \right] + \lim_{\Delta x \to 0} \left[v(x)\frac{\Delta u}{\Delta x} \right] + \lim_{\Delta x \to 0} \left[\frac{\Delta u}{\Delta x}\Delta v \right] \qquad \text{see [2.5.4}(a)]$$

$$= \left[\lim_{\Delta x \to 0} u(x)\right]\left[\lim_{\Delta x \to 0} \frac{\Delta v}{\Delta x}\right] + \left[\lim_{\Delta x \to 0} v(x)\right]\left[\lim_{\Delta x \to 0} \frac{\Delta u}{\Delta x}\right]$$

$$+ \left[\lim_{\Delta x \to 0} \frac{\Delta u}{\Delta x}\right]\left[\lim_{\Delta x \to 0} \Delta v\right] \qquad \text{see [2.5.4}(b)]$$

(2) $$= u(x)v'(x) + v(x)u'(x) + u'(x) \cdot 0$$

$$= u(x)v'(x) + v(x)u'(x)$$

Note: Both $u(x)$ and $v(x)$ are differentiable, so

$$\lim_{\Delta x \to 0} \frac{\Delta v}{\Delta x} = v'(x) \qquad \text{and} \qquad \lim_{\Delta x \to 0} \frac{\Delta u}{\Delta v} = u'(x)$$

Furthermore, $\lim_{\Delta x \to 0} \Delta v = \lim_{\Delta x \to 0} [v(x + \Delta x) - v(x)] = 0$ because $v(x)$, being differentiable, is continuous [by (2.5.14)], and so [2.5.13(d)] applies. ■

Note that although the diagrams of Fig. 3.2 aided in the organization of the above proof, the actual proof is independent of the diagrams. [See Progress Test 2, Prob. 5, for another approach to line (1).]

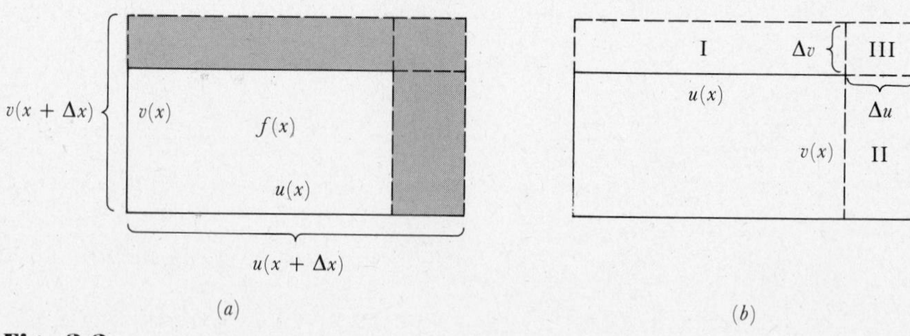

Fig. 3.2

---

⑨ We can apply the Product Rule to $f(x) = x^2$ regarded as the product

$$f(x) = x \cdot x$$

to obtain

$$f'(x) = x \cdot 1 + x \cdot 1 = 2x$$

Here we had $u(x)$ and $v(x) = x$, so in this case the rectangle of Fig. 3.2 reduces to the square associated with Sec. 2.3, Progress Test 4. In fact the idea that a circular disk grows by circumferences or that a square grows by its two sides (the two sides in the direction of the growth) has an interesting extension to the $u(x)v(x)$ rectangle of the Product Rule. The rectangle in Fig. 3.2 grows by the side of length $u(x)$ times the rate of growth in that direction, namely $v'(x)$, plus the side of length $v(x)$ times the rate of growth in the other direction, namely $u'(x)$.

We now turn to quotients of functions, beginning with a simple case.

3.1.9 **Theorem** If $f(x) = 1/v(x)$ for $v(x)$ differentiable and $v(x) \neq 0$, then

$$f'(x) = \frac{-v'(x)}{[v(x)]^2}$$

Proof: As usual, we apply the definitions of $f(x)$ and $f'(x)$:

$$\begin{aligned}
f'(x) &= \lim_{\Delta x \to 0} \frac{f(x + \Delta x) - f(x)}{\Delta x} \\
&= \lim_{\Delta x \to 0} \frac{[1/v(x + \Delta x)] - 1/v(x)}{\Delta x} \\
&= \lim_{\Delta x \to 0} \frac{v(x) - v(x + \Delta x)}{\Delta x(v(x + \Delta x)v(x))} \qquad \left[\frac{(1/A) - (1/B)}{C} = \frac{(B - A)}{CAB}\right] \\
&= \lim_{\Delta x \to 0} -\left[\frac{v(x + \Delta x) - v(x)}{\Delta x}\right]\frac{1}{v(x + \Delta x)v(x)} \\
&= -\lim_{\Delta x \to 0}\left[\frac{v(x + \Delta x) - v(x)}{\Delta x}\right]\lim_{\Delta x \to 0}\frac{1}{v(x + \Delta x)v(x)} \qquad \text{see [2.5.4(}a\text{),(}b\text{)]} \\
(3) \qquad &= -v'(x)\frac{1}{v(x)v(x)} \\
&= \frac{-v'(x)}{[v(x)]^2}
\end{aligned}$$

Note that $v(x)$, being differentiable, is continuous, so [2.5.13(c)] applies to guarantee that

$$\lim_{\Delta x \to 0} v(x + \Delta x) = v(x) \quad \blacksquare$$

Example 3 Use (3.1.9) to differentiate $f(x) = 1/x^2$.

Solution

$$f'(x) = \frac{-2x}{(x^2)^2} = \frac{-2}{x^3} \quad \bullet$$

Note that we can rewrite $1/x$ using a negative exponent as x^{-1} and its derivative (see Table 3.1, function 9) as $-1 \cdot x^{-2}$. Similarly, we can rewrite $1/x^2$ as x^{-2} and its derivative as $-2x^{-3}$.

This suggests that the Power Rule may extend to *all* integers, a fact we shall now formally state and prove.

3.1.10 Power Rule for Any Integer n

If $f(x) = x^n$, then $f'(x) = nx^{n-1}$

Proof: This result already holds for positive integers (3.1.3).

For $n = 0$ and $x \neq 0$, $f(x) = 1$ so $f'(x) = 0$. Hence suppose $f(x) = x^n$, where n is a negative integer. Then $f(x) = 1/x^{-n}$, where $-n$ is a *positive* integer (for example, $x^{-3} = 1/x^{-(-3)} = 1/x^3$). By (3.1.9),

$$\begin{aligned} f'(x) &= \frac{-D_x(x^{-n})}{(x^{-n})^2} \\ &= \frac{-(-n)x^{-n-1}}{x^{-2n}} \\ &= nx^{2n-n-1} \\ &= nx^{n-1} \end{aligned}$$

Note that $D_x(x^{-n}) = -nx^{-n-1}$ by the Power Rule for *positive* integers. ■

We shall now use the Product Rule and the fact that any quotient $u(x)/v(x)$ can be written as a product $u(x)[1/v(x)]$ to prove:

3.1.11 The Quotient Rule Suppose $f(x) = u(x)/v(x)$, where $u(x)$ and $v(x)$ are differentiable and $v(x) \neq 0$. Then

$$f'(x) = \frac{v(x)u'(x) - u(x)v'(x)}{[v(x)]^2}$$

The derivative of a quotient is the denominator times the derivative of the numerator, minus the numerator times the derivative of the denominator, all divided by the denominator squared.

Proof of the Quotient Rule: By the Product Rule applied to $u(x)[1/v(x)]$,

$$\begin{aligned} f'(x) &= u(x)\frac{-v'(x)}{[v(x)]^2} + \frac{1}{v(x)}u'(x) \qquad (3.1.9) \\ &= \frac{-u(x)v'(x) + v(x)u'(x)}{[v(x)]^2} \\ &= \frac{v(x)u'(x) - u(x)v'(x)}{[v(x)]^2} \end{aligned}$$ ■

Example 4 Differentiate $f(x) = \dfrac{3x^4 + x^3 + 2x}{4x^7 + 5x^5}$.

Solution

$$f'(x) = \frac{\overbrace{(4x^7 + 5x^5)}^{\text{den.}}\overbrace{(12x^3 + 3x^2 + 2)}^{\text{deriv. of num.}} - \overbrace{(3x^4 + x^3 + 2x)}^{\text{num.}}\overbrace{(28x^6 + 25x^4)}^{\text{deriv. of den.}}}{\underbrace{[4x^7 + 5x^5]^2}_{\text{den. squared}}}$$ ●

PROGRESS TEST 2

1. Find the derivatives of the following functions. Do not simplify your answers.
   (*a*) $f(x) = (x^3 + 2x)(5x^4 + 20x^3 + 7)$
   (*b*) $f(t) = (t^2 + 9t + 3)(2t + 1)$
   (*c*) $y = \dfrac{x^2 + 1}{x^3 + 1}$
   (*d*) $t(x) = \dfrac{14x^3 + 1}{x^2 - x - 2}$

(e) $f(x) = \dfrac{ax + b}{cx + d}$ all letters in (e) and (f), other than x, denote constants

(f) $u(x) = (ax^2 + bx + c)(mx + d)$

2. Use the Product Rule to differentiate $f(x) = xx^2$.

3. Take the derivative of $f(x) = 7(x^3 + 5x^2 + 2x - 1)$ by means of:

(a) The Constant-Multiple Rule (3.14)

(b) The Product Rule, using 7 as the first factor

4. Keeping in mind that $f(x) = (x^3 + 3x^2 + 4)/x^2 = (x^3 + 3x^2 + 4)(x^{-2})$, differentiate $f(x)$ using (a) the Quotient Rule and (b) the Product Rule. (c) Show that your results are equal.

5. Replacing Δu and Δv by $u(x + \Delta x) - u(x)$ and $v(x + \Delta x) - v(x)$, respectively, in line 1 of the proof of the Product Rule, show that the numerator of the quotient equals $u(x + \Delta x)v(x + \Delta x) - u(x)v(x)$.

MIXING THINGS UP A BIT

Often we need to use the derivative rules in combination. We shall not simplify our following answers unless there is a reason for doing so.

Example 5 Differentiate $y = \dfrac{(3x^{10} + 2x^2 + 1)(4x^3 + x^2)}{2x + 1}$ with respect to x.

Solution We can think of y as a quotient whose numerator is also a product.

$$\frac{dx}{dy} = \frac{\overbrace{(2x + 1)}^{\text{density}}\overbrace{[(3x^{10} + 2x^2 + 1)(12x^2 + 2x) + (4x^3 + x^2)(30x^9 + 4x)]}^{\text{derivative of number}} - \underbrace{\overbrace{[(3x^{10} + 2x^2 + 1)(4x^3 + x^2]}^{\text{number}}}_{\text{density squared}}\overbrace{(2)}^{\text{derivative of density}}}{\underbrace{(2x + 1)^2}_{\text{density squared}}} \quad \bullet$$

Example 6 Determine $D_x f$: $f(x) = (x^3 + 3x^2 + 2)(x^5 + 4x^4 + 5)(3x^{10} + 11x^9 + x)$.

Solution This is a product of three factors, and to take the derivative, we can group the first two together and apply the Product Rule to the resulting two factors:

$$f(x) = \overbrace{[(x^3 + 3x^2 + 2)(x^5 + 4x^4 + 5)]}^{\text{1st factor}}\overbrace{(3x^{10} + 11x^9 + x)}^{\text{2d factor}}$$

$$D_x f = (x^3 + 3x^2 + 2)(x^5 + 4x^4 + 5)(30x^9 + 99x^8 + 1)$$
$$+ (3x^{10} + 11x^9 + x)[(x^3 + 3x^2 + 2)(5x^4 + 16x^3) + (x^5 + 4x^4 + 5)(3x^2 + 6x)]$$

(The derivative of the first aggregate factor is in brackets.) ●

As you may have already realized, most problems of this type are subject to more than one approach. Example 5 could have been done as a product:

$$f(x) = \frac{3x^{10} + 2x^2 + 1}{2x + 1}(4x^3 + x^2)$$

Similarly, we could have grouped the last two factors in Example 6 and proceeded accordingly. Although it usually does not matter which approach you choose in such problems, different approaches may very well lead to answers radically different in appearance.

We present one last example, which illustrates a typical problem.

Example 7 Give an equation of the line tangent to $f(x) = (x + 1)/(2x - 3)$ at $x = 5$.

Solution We shall use the point-slope form of the equation of a line, but first we need the slope at $x = 5$.

$$f'(x) = \frac{(2x-3)(1) - (x+1)(2)}{(2x-3)^2} = \frac{-5}{(2x-3)^2}$$

Thus,

$$f'(5) = \frac{-5}{(2 \cdot 5 - 3)^2} = \frac{-5}{49}$$

is the slope. Now we need the second coordinate of the point on the graph of $f(x)$. Substituting $x = 5$ into the function (into $f(x)$—not $f'(x)$!), we get $f(5) = 6/7$. Thus the point on the curve is $(5, 6/7)$, the slope is $-5/49$, and so the equation of the line is

$$y - \frac{6}{7} = -\frac{5}{49}(x - 5) \quad \bullet$$

SECTION 3.1 EXERCISES

In Exercises 1 to 28 differentiate the given function with respect to the indicated variable using the previous rules.

1. $f(x) = -2x - 3$

2. $f(x) = 5 - 4x^2$

3. $g(x) = x^3 + 5x^2 + 1$

4. $g(t) = t^{10} + t^9 + t^8 + 100$

5. $f(x) = 0.01x - 0.2$

6. $y = \pi r^2$

7. $t(x) = (x^2 + 1)(7x^3 + 3x^2 + x - 3)$

8. $A(t) = (t^5 + 4t^3)(t^2 - 5t - 4)$

9. $f(t) = \dfrac{4t^2 + t}{t^3 + 3t^2} + t$

10. $y(t) = \dfrac{5 - 2t}{1 - t}$

11. $h(s) = \dfrac{1}{s + 1}$

12. $f(r) = \dfrac{r}{r + 1} + \dfrac{r^2}{r^2 + 1}$

13. $y = \dfrac{-7x^2 + 2}{4x^3 + x}$

14. $m(x) = \dfrac{(4x^2 - 2)(3x + 1)}{x - 1}$

15. $f(x) = \dfrac{(x + 1)(x^2 + 2)}{x^2 + 3}$

16. $v = \dfrac{u^2 + 7u^6}{4u^5 - 5} - \dfrac{3}{u^2 + 1}$

17. $g(x) = \dfrac{1}{(x^3 + 3x^2)(7x - 2)}$

18. $f(z) = \dfrac{z}{z + 1} + z^5 - z^3$

19. $m(y) = \dfrac{(y^2 - 5y)(y^4 + 2y - 1)}{6y - 9}$

20. $x(y) = \dfrac{(4y^4 + 3y^3)}{(y - 1)(y^4 + 4y + 2)}$

21. $r(x) = x^{-4} - \dfrac{x^2}{x^3 + 2x + 1}$

22. $x(t) = \dfrac{t}{t^3 + 3} + (t^3 + 2t + 1)(t^2 - 5t - 4)$

23. $u = (x + 1)(x^4 + x^3 + 1)(3x^2 - 7x)$

24. $y = \left[\dfrac{x^2 + 7x}{3x^2 + 1}\right]\left[\dfrac{x^7 + 5x^6 - 3}{x + 2.1}\right]$

25. $y = \dfrac{2x + 4}{x^2 - 3} - \dfrac{2x^3 + 1}{x^2 - x + 1}$

26. $f(w) = (w^5 + 1)(w^2 + 2w + 1) - \dfrac{1}{2w^3}$

27. $f(v) = \dfrac{1}{v^2 - v^3} + \dfrac{6}{v^2}$

28. $f(x) = \dfrac{1}{x^2 + 1} + \dfrac{1}{x^2 + 2} + \dfrac{1}{x^2 + 3}$

29. Interpret remark A before Progress Test 1 (*a*) graphically, (*b*) in terms of rate of change.

30. Repeat Exercise 29 for remark B. Compare with the algebraic approach of (1.3.3). For what x is the tangent line horizontal?

31. (*a*) Find the slope of the line tangent to $f(x) = 3x^2 + 2x^2 + 1$ at $x = 2$.
(*b*) Write the equation of that line in point-slope form.
(*c*) Write the equation of the line perpendicular to the graph of f at $x = 2$.

32. Repeat Exercise 31 for $f(x) = x^4 - 4x^3$ at $x = 1$.

33. Where is the slope of the tangent to the function $f(x) = 1/(x^3 + 1)$ negative?

34. Where is the slope of the tangent to $y = 1/x^2$ negative?

35. An object is thrown upward with initial velocity of 120 ft/s, so its height at time t is given by $h(t) = 120t - 16t^2$. Determine:
(*a*) Its vertical velocity at time t.
(*b*) The maximum height it can attain.
(*c*) The time it hits the ground.

36. Repeat Exercise 35 but with the object thrown from a 96-ft-high platform. Thus its height is given by $H(t) = 120t - 16t^2 + 96$. (Assume that the object does not hit the platform on the way down.)

---

37. ⑨ The surface area of a cube of edge x is composed of six squares of area x^2. Knowing that the rate of change of area of a square with respect to change in its sides is $2x$ tells us that the rate of change of a cube's surface area with respect to x is $12x$ because of at least one of the facts about derivatives. Cite at least one property of derivatives to justify this analysis.

---

38. Let $V = (4/3)\pi r^3$. Find (*a*) dV/dr, (*b*) the derivative of $4\pi r^2$, and (*c*) the derivative of $8\pi r$.

39. Using the Product Rule, assuming that $u(x)$ is differentiable, find a simple formula for the derivative of $f(x) = [u(x)]^2$.

40. Apply the formula developed in Exercise 39 to
(*a*) $f(x) = (x^3 + 3x^2 + 5x + 1)^2$
(*b*) $f(x) = (mx + b)^2$ (m and b constants)

41. Use (3.1.9) and Exercise 39 to find a formula for the derivative of $h(x) = 1/[u(x)]^2$.

42. Apply the formula developed in Exercise 41 to $f(x) = 1/(mx + b)^2$.

43. Apply the formula developed in Exercise 41 to $f(x) = 1/(x^4 + x^3 + x^2 + 3)^2$.

44. Assuming that $f(x) = u(x)/v(x)$ is differentiable, derive the Quotient Rule using the Product Rule applied to $f(x)v(x) = u(x)$.

45. Using the Product Rule, derive a rule for differentiating $f(x) = [g(x)k(x)]m(x)$.

---

46. ⑨ Use a rectangle of height 1 and width $u(x) + v(x)$ to interpret geometrically the numerator of the fraction occurring in the derivation of the Sum Rule. (Let u and v grow in opposite directions.)

47. Use a rectangle of height k and width $f(x)$ to interpret the derivation of (3.1.4).

---

48. Determine df/dx for $f(x) = x^3 + g(x)h(x) - 1/u(x) + 4x$, where g, h, and u are differentiable functions of x.

49. Write the list of differentiation rules in fractional (df/dx) notation.

Section 3.2: Differentiating Composite and Implicitly Defined Functions

Objectives:

Use the Chain Rule to:
1. Differentiate compositions of functions.
2. Differentiate functions raised to rational powers.
3. Differentiate implicitly defined functions.
4. Determine equations of (nonvertical) lines tangent to graphs of equations.

ANOTHER POWER RULE

We can differentiate

$$f(x) = (3x^2 + 5x)^2$$

using the Product Rule:

$$\begin{aligned} f'(x) &= (3x^2 + 5x)(6x + 5) + (3x^2 + 5x)(6x + 5) \\ &= 2(3x^2 + 5x)(6x + 5) \end{aligned}$$

More generally, using the idea of Sec. 3.1, Exercise 39, if $u(x)$ is differentiable and

$$f(x) = [u(x)]^2$$

then

$$\begin{aligned} f'(x) &= u(x)u'(x) + u(x)u'(x) \\ &= 2u(x)u'(x) \end{aligned}$$

If $f(x) = [u(x)]^3$, we may regard $f(x)$ as $[u(x)]^2u(x)$, and apply the Product Rule again using the above result:

$$\begin{aligned} f'(x) &= [u(x)]^2u'(x) + u(x)[2u(x)u'(x)] \\ &= 3[u(x)]^2u'(x) \end{aligned}$$

This is beginning to look like a Power Rule for functions. We now take it one more step to see if the pattern continues. Suppose

$$f(x) = [u(x)]^4$$

Apply the Product Rule to $f(x) = [u(x)]^3u(x)$, using the previous result:

$$\begin{aligned} f'(x) &= [u(x)]^3u'(x) + u(x)(3[u(x)]^2u'(x)) \\ &= 4[u(x)]^3u'(x) \end{aligned}$$

We shall now formalize the pattern that has emerged.

3.2.1 Power Rule for Functions If $u(x)$ is differentiable and

$$f(x) = [u(x)]^n \qquad \text{(where } n \text{ is an integer with } u(x) \neq 0 \text{ for } n \leq 0)$$

then

$$f'(x) = n[u(x)]^{n-1}u'(x)$$

This is a special case of the far more general Chain Rule to be stated and proved shortly, so we shall not prove it here. Of course, the earlier Power Rule for integers is a special case of (3.2.1), with $u(x) = x$, making $u'(x) = 1$: $D_x[(x)^n] = n(x)^{n-1}(1) = nx^{n-1}$.

Example 1 Differentiate $f(x) = (x^4 + 7x^2 + 2x + 3)^5$.

Solution $f'(x) = 5(x^4 + 7x^2 + 2x + 3)^4(4x^3 + 14x + 2)$ ●

Example 2 Differentiate $f(x) = 1/(x^3 + x)^7$, first using the Power Rule for negative exponents and then using the Quotient Rule and (3.2.1) for the positive integer 7.

Solution

$$f(x) = \frac{1}{(x^3 + x)^7} = (x^3 + x)^{-7}$$

Thus $f'(x) = -7(x^3 + x)^{-8}(3x^2 + 1)$.

By the Quotient Rule,

$$\begin{aligned} f'(x) &= \frac{(x^3 + x)^7(0) - (1)7(x^3 + x)^6(3x^2 + 1)}{(x^3 + x)^{14}} \\ &= \frac{(-7)(x^3 + x)^6(3x^2 + 1)}{(x^3 + x)^{14}} = \frac{(-7)(3x^2 + 1)}{(x^3 + x)^8} \end{aligned}$$

which equals the previous result. Using the Power Rule directly yields the quickest answer. ●

Example 3 Differentiate $g(x) = (x^4 + 3x^3)^{11}(7x^2 + 5x + 2)$.

Solution We shall use the Product Rule, using the Power Rule on the first factor.

$$\frac{dg}{dx} = (x^4 + 3x^3)^{11}(14x + 5) + (7x^2 + 5x + 2)11(x^4 + 3x^3)^{10}(4x^3 + 9x^2) \quad ●$$

Example 4 Determine D_th, where

$$h(t) = \left[\frac{4t + 1}{t^2 - 1}\right]^3$$

Solution We use the Power Rule on $[u(t)]^3$ and use the Quotient Rule on $u(t)$:

$$\begin{aligned} D_th &= 3\left[\frac{4t + 1}{t^2 - 1}\right]^2 D_t\left[\frac{4t + 1}{t^2 - 1}\right] \\ &= 3\left[\frac{4t + 1}{t^2 - 1}\right]^2 \frac{(t^2 - 1)(4) - (4t + 1)(2t)}{(t^2 - 1)^2} \quad ● \end{aligned}$$

You are encouraged to compute D_th for $h(t)$ in Example 4 regarded as $h(t) = (4t + 1)^3/(t^2 - 1)^3$.

PROGRESS TEST 1

Differentiate 1 to 6. (You need not simplify your answers.)

1. $f(x) = (x^9 + 2x^4 - x^3 + 41)^7$

2. $g(x) = \dfrac{1}{(x^3 + 7x^2 + 2x + 1)^5}$

3. $v(x) = (x^2 + 2x)(50x^2 + 6x - 3)^4$

4. $f(x) = \dfrac{(x + 4)}{(3x^5 + x^2)^8}$

5. $h(x) = \left[\dfrac{2x + 1}{3x^2 - 4}\right]^6$

6. $g(x) = \dfrac{(x^7 + 6x^5 - x^2)^4}{(5x^2 + 2x)^7}$

THE CHAIN RULE

The function $[u(x)]^n$ is actually the composition of two functions g and u,

$$g(u(x)) \qquad \text{where } g(u) = u^n$$

(You may wish to review the discussion of composition.) The Power Rule for functions says that the derivative of this composition is the derivative of *g with respect to u* [which is $n(u)^{n-1}$], times the derivative of $u(x)$ with respect to x. In other words, the derivative of the composition is

$$g'(u(x))u'(x)$$

which can be written in the easily remembered form

$$\frac{dg}{du}\frac{du}{dx}$$

The following rule gives the derivative of a composition of functions in general, including the above as a special case. Instead of calling it the Composition Rule, we follow tradition and call it the Chain Rule.

3.2.2 **The Chain Rule** Suppose that u is differentiable at x and g is differentiable at $u(x)$. Then the derivative of the composition of g with u, $g(u(x))$, is

$$g'(u(x))u'(x) \qquad \text{or} \qquad \frac{dg}{du}\frac{du}{dx}$$

Before beginning the proof of this most important rule we examine its meaning in a few situations.

Thinking of the derivative in terms of rate of change, the Chain Rule says that *the rate of change of the composition with respect to x is the rate of change of g with respect to u times the rate of change of u with respect to x.*

Consider, for example, the composition of

$$g(u) = 3u + 5 \qquad \text{with } u(x) = 2x + 1$$
$$g(u(x)) = 3[u(x)] + 5 = 3[2x + 1] + 5 = 6x + 8$$

Here u changes twice as fast as x and g changes three times as fast as u, so g changes six times as fast as x. See Exercises 51 and 52 for further numerical illustration of this crucial fact.

⑨ For a more concrete example, recall Sec. 2.1, Example 8. Suppose the radius r of a circular oil spill is increasing at the rate of 3 m/min; that is, the radius as a function of time t in minutes is given by $r(t) = 3t$. The area of the oil spill is πr^2, so as a function of time, the area is

$$A(t) = \pi[r(t)]^2 = \pi[3t]^2$$

The Chain Rule tells us that the rate of increase in area with respect to *time* is the rate of change of area with respect to *radius* times the rate of increase of radius with respect to time; that is,

$$\frac{dA}{dt} = \frac{dA}{dr}\frac{dr}{dt} = 2\pi r(t)r'(t) = 2\pi(3t)(3) = 18\pi t$$

Because the functions are simple, with $A(t) = 9\pi t^2$, the first Power Rule (3.1.2) applies to yield the same result.

To prove the Chain Rule we apply the definition of derivative to the composition $g(u(x))$. A small change Δx in x produces the change Δu in u, which in turn produces the change Δg in g. The rate of change of g with respect to a change in x is the product of the change in g with respect to u, $\Delta g/\Delta u$, and the change in u with respect to x, $\Delta u/\Delta x$. This means that the basic quotient $\Delta g/\Delta x$, whose limit as $\Delta x \to 0$ is dg/dx, may be written as

$$\frac{\Delta g}{\Delta x} = \frac{\Delta g}{\Delta u} \cdot \frac{\Delta u}{\Delta x}$$

This statement makes sense only if $\Delta u \neq 0$ near x, so we divide our proof into two cases.

Case 1: $\Delta u \neq 0$ for small Δx

Case 2: Δu possibly $= 0$

We prove Case 1 here, postponing the more complicated proof of Case 2 to Chap. 9.

By definition,

$$\begin{aligned} \frac{dg}{dx} &= \lim_{\Delta x \to 0} \frac{\Delta g}{\Delta x} \\ &= \lim_{\Delta x \to 0} \frac{\Delta g}{\Delta u} \cdot \frac{\Delta u}{\Delta x} \\ &= \left[\lim_{\Delta x \to 0} \frac{\Delta g}{\Delta u}\right]\left[\lim_{\Delta x \to 0} \frac{\Delta u}{\Delta x}\right] && \text{Limit Theorem [2.5.4}(b)] \\ &= \left[\lim_{\Delta x \to 0} \frac{\Delta g}{\Delta u}\right]\left[\frac{du}{dx}\right] && \text{(because } u \text{ is differentiable with respect to } x) \end{aligned}$$

Looking at $\lim_{\Delta x \to 0} \frac{\Delta g}{\Delta u}$ more closely, we note that u being differentiable means that u is continuous, by (2.5.14). Thus, as $\Delta x \to 0$, we know that $\Delta u \to 0$, by (2.5.13). Therefore,

$$\lim_{\Delta x \to 0} \frac{\Delta g}{\Delta u} = \lim_{\Delta u \to 0} \frac{\Delta g}{\Delta u}$$

But g is differentiable, so by definition

$$\lim_{\Delta u \to 0} \frac{\Delta g}{\Delta u} = \frac{dg}{du}$$

Hence we can conclude that the derivative of the composition of g with u is

$$\begin{aligned} \lim_{\Delta x \to 0} \frac{\Delta g}{\Delta x} &= \lim_{\Delta x \to 0} \frac{\Delta g}{\Delta u}\frac{\Delta u}{\Delta x} \\ &= \frac{dg}{du}\frac{du}{dx} \quad \blacksquare \end{aligned}$$

As we have noted, the Power Rule for Functions is an immediate consequence of the Chain Rule with $g(u) = u^n$. In this case

$$\frac{dg}{dx} = \frac{dg}{du}\frac{du}{dx}$$

$$= n[u(x)]^{n-1}\frac{du}{dx}$$

The general Chain Rule is heavily used in the differentiation of trigonometric, logarithmic, and exponential functions (Chap. 7). Right now we present another application.

IMPLICIT DIFFERENTIATION

Recall from Chap. 2 that an equation in, say, x and y may not be solvable for one of the variables to yield a uniquely defined function of the other. Such equations arise frequently in applications. For this reason it is useful to extend the idea of differentiation to deal with such equations. For example, the equation $x^2 + y^2 = 9$ has a graph, and it is certainly reasonable to seek the slopes of lines tangent to its graph. Yet, as we have seen, solving for $y = \pm\sqrt{9 - x^2}$ does not give y as a function of x. However, the two functions

$$y_1 = \sqrt{9 - x^2} \qquad \text{and} \qquad y_2 = -\sqrt{9 - x^2}$$

are defined *implicitly* by the original equation. (Their graphs are the upper and lower semicircles, respectively.)

For more complicated expressions in x and y it may be very difficult to determine whether there are any functions defined implicitly by the given equation. Determining such functions by actually solving for y in terms of x can be very tedious or even impossible. For example, the equation

$$x^3y^5 + 3y^3 + x^2y - 12 = 0$$

defines many functions $y = f(x)$ implicitly, but to solve for y in terms of x would involve solving a fifth-degree equation in y. The question of whether or when the implicitly defined functions are differentiable is normally addressed in more advanced courses. We shall be content to determine a formula for the derivative of the implicitly defined functions, assuming throughout that the implicitly defined functions are differentiable wherever the formula is defined. We use the Chain Rule to do this. The technique is known as *implicit differentiation.*

Example 5 Assuming that y is implicitly defined as a function of x by the equation $x^2 + y^2 = 9$, determine dy/dx in terms of x and y.

Solution We are assuming that $y = f(x)$ satisfies the given equation, so

$$x^2 + [f(x)]^2 = 9$$

Thus, differentiating both sides with respect to x, we have

$$\frac{d}{dx}[x^2 + [f(x)]^2] = \frac{d}{dx}[9]$$

or

$$\frac{d}{dx}[x^2] + \frac{d}{dx}[f(x)]^2 = 0$$

Now $\frac{d}{dx}[x^2] = 2x$, whereas by the Chain Rule

$$\frac{d}{dx}[f(x)]^2 = 2[f(x)]\frac{df}{dx}$$

Hence we know that

$$2x + 2[f(x)]\frac{df}{dx} = 0$$

so

$$\frac{df}{dx} = \frac{-2x}{2[f(x)]}$$

Since $y = f(x)$, we conclude that

$$\frac{dy}{dx} = \frac{-x}{y} \quad \bullet$$

We replaced y by $f(x)$ in Example 5 to emphasize that y is a function of x and that differentiation is with respect to x. Normally we differentiate each side of the equation in its given form.

Example 6 Determine dy/dx given that y is implicitly defined as a function of x by

$$2x^3 + y^3 + 5y^4x^2 = 19$$

Solution We differentiate each side with respect to x. Using the Sum Rule, we differentiate the left side term by term:

$$\frac{d}{dx}[2x^3] = 6x^2$$

and, by the Chain Rule,

$$\frac{d}{dx}[y^3] = 3y^2\frac{dy}{dx}$$

To differentiate $5y^4x^2$ we must use the Product Rule, regarding $5y^4x^2$ as the product of $5y^4$ and x^2:

$$\frac{d}{dx}[5y^4x^2] = 5y^4(2x) + (x^2)\left(20y^3\frac{dy}{dx}\right)$$

Pulling all these derivatives together, we have

$$6x^2 + 3y^2\frac{dy}{dx} + 10xy^4 + 20x^2y^3\frac{dy}{dx} = 0 = \frac{d(19)}{dx}$$

Hence, solving for dy/dx, we conclude that

$$\frac{dy}{dx} = \frac{-6x^2 - 10xy^4}{3y^2 + 20x^2y^3} \quad \bullet$$

Example 7 Determine the slope of the line tangent to the graph of $2x^3 + 2y^3 - 9xy = 0$ at the point $(1, 2)$. Then give the equation of this tangent line in point-slope form.

Solution First we differentiate implicitly with respect to x,

$$6x^2 + 6y^2y' - 9xy' - 9y = 0$$

and solve for y':

$$y' = \frac{9y - 6x^2}{6y^2 - 9x} = \frac{3y - 2x^2}{2y^2 - 3x}$$

The slope of the tangent line at the point where $x = 1$ and $y = 2$ is found by substituting these values into y':

$$y' = \frac{3(2) - 2(1)^2}{2(2)^2 - 3(1)} = \frac{4}{5}$$

Thus the point-slope form of the equation of the tangent line is

$$y - 2 = \frac{4}{5}(x - 1) \quad \bullet$$

In cases where x is not the independent variable, it is probably best to use the $d\_/d\_$ notation (often called the *Leibniz notation,* after its inventor).

Example 8 Assuming that x is a function of y defined implicitly by the equation

$$x^3y + 3y^2 - 9x^2 = 4$$

find dx/dy in terms of x and y.

Solution The important thing to remember here is that *the differentiation is with respect to* y. Thus $3y^2$ has derivative (with respect to y) $6y$, whereas the derivative of $-9x^2$ is (through the use of the Chain Rule) $-18x(dx/dy)$. The derivative of 4 is 0, and using the Product and Chain Rules, we differentiate x^3y:

$$x^3(1) + y3x^2\frac{dx}{dy} = x^3 + 3x^2y\frac{dx}{dy}$$

Therefore

$$x^3 + 3x^2y\frac{dx}{dy} + 6y - 18x\frac{dx}{dy} = 0$$

So

$$\frac{dx}{dy} = \frac{-x^3 - 6y}{3x^2y - 18x} \quad \bullet$$

Although it seems that we are "differentiating equations," we should note that we are assuming the equations hold for all values of the independent variable in some interval (where, in fact, these equations are "identities"). See Exercise 53.

PROGRESS TEST 2

1. Assuming that y is a function of x defined implicitly by the given equation, determine dy/dx.
(*a*) $xy = 1$ (*b*) $x^2y^2 = 1$
(*c*) $x^2 + y^2 - 2xy + xy^3 = 4$ (*d*) $\dfrac{x+y}{x-y} = 1$

2. For equations (*a*) to (*d*), find dx/dy.

3. The following equation defines x implicitly as a function of t. Find dx/dt.

$$3x^2 + 5t^2 - xt = 3$$

4. Find the point-slope equation of the line tangent to the curve $xy = 1$ at the point $(2, \frac{1}{2})$.

RATIONAL EXPONENTS

We now enlist the help of both the Chain Rule and the principle of implicit differentiation to extend our power rules to rational exponents. Up to now we have dealt only with integer exponents.

Recall that fractional exponents are used merely as another way of writing expressions involving radicals. Thus

$$\sqrt{5} = 5^{1/2}, \qquad \sqrt[3]{5} = 5^{1/3}, \qquad \sqrt[q]{5} = 5^{1/q}, \qquad \sqrt[q]{5^p} = (\sqrt[q]{5})^p = 5^{p/q}$$

Also, $\sqrt[3]{125} = \sqrt[3]{5^3} = (\sqrt[3]{5})^3 = 5$ or, using rational exponents,

$$125^{1/3} = (5^3)^{1/3} = 5^1 = 5$$

Similarly,

$$\frac{1}{\sqrt[3]{(125)^2}} = \frac{1}{(\sqrt[3]{125})^2} = \frac{1}{5} = \frac{1}{5^2}$$

or

$$(125)^{-2/3} = [(125)^{1/3}]^{-2} = 5^{-2} = \frac{1}{25}$$

The same sorts of manipulations apply to algebraic expressions:

$$\sqrt[3]{(1 - x^4)^2} = (1 - x^4)^{2/3}$$
$$\sqrt[5]{(x - 3)^6} = (x - 3)^{6/5} = (x - 3)(x - 3)^{1/5}$$

In general, $u^{p/q}$ is defined to be $\sqrt[q]{u^p}$. When q is even, $u^{p/q}$ is defined only for $u \geq 0$.

By extending the Power Rule to rational exponents we can differentiate expressions involving radicals. We shall state the rule, apply it in a pair of examples where we have obtained the derivative by other means, and then prove it using implicit differentiation.

3.2.3 **The Power Rule for Rational Exponents** Assume that $f(x) = [u(x)]^r$, where $u(x)$ is differentiable and $r = p/q$, with p and q both integers and $q \neq 0$. Then

$$f'(x) = r[u(x)]^{r-1}u'(x)$$

Example 9 Use the Power Rule to differentiate $f(x) = \sqrt{x}$.

Solution Writing $f(x) = x^{1/2}$, we have [with $u(x) = x$]

$$f'(x) = \frac{1}{2}x^{(1/2)-1}(1)$$
$$= \frac{1}{2}x^{-1/2}$$
$$= \frac{1}{2\sqrt{x}} \quad \bullet$$

This result agrees with that obtained using the definition of derivative in Chap. 2 (see Table 3.1). Note that without any overriding reason for a choice, the derivative can be left in either the radical or the rational exponent form.

Example 10 Use the Power Rule to determine dy/dx for $y = \sqrt{9 - x^2}$.

Solution Writing $y = (9 - x^2)^{1/2}$, we have [with $u(x) = 9 - x^2$]

$$\frac{dy}{dx} = \frac{1}{2}(9 - x^2)^{-1/2}(-2x)$$
$$= \frac{-x}{(9 - x^2)^{1/2}}$$
$$= \frac{-x}{\sqrt{9 - x^2}} \quad \bullet$$

Compare with Example 5. There we found that, assuming y as an implicitly defined function of x (one example of which is $y = \sqrt{9 - x^2}$),

$$\frac{dy}{dx} = \frac{-x}{y}$$

This agrees with the result of Example 10, where $y = \sqrt{9 - x^2}$. Notice that dy/dx is undefined at $x = \pm 3$. The tangent lines to the circular graph are vertical at $x = \pm 3$.

Proof of the Power Rule for Rational Exponents: First, note that we need only establish this rule for positive rationals, and then it holds for negative rationals by the usual argument using the Quotient Rule. Furthermore, we need only deal with a rule for $f(x) = x^{p/q}$, because then the result for $[u(x)]^{p/q}$ follows automatically by the Chain Rule.

Assume $y = x^{p/q}$, with $x \neq 0$. Raise both sides of the equation to the qth power to get

$$y^q = (x^{p/q})^q = x^p$$

We now implicitly differentiate with respect to x to obtain

$$qy^{q-1}\frac{dy}{dx} = px^{p-1}$$

We used the Power Rule for *integers,* and on the left side we used the Chain Rule, because y is a function of x.

Solve for dy/dx to get

$$\frac{dy}{dx} = \frac{px^{p-1}}{qy^{q-1}} = \frac{p}{q}\frac{x^{p-1}y}{y^q}$$

But we know that $y = x^{p/q}$ and $y^q = x^p$, so

$$\begin{aligned}\frac{dy}{dx} &= \frac{p}{q}\frac{x^{p-1}(x^{p/q})}{x^p}\\ &= \frac{p}{q}x^{p-1-p}(x^{p/q}) = \frac{p}{q}x^{(p/q)-1}\end{aligned}$$

completing the proof ■

Example 11 Determine $f'(x)$, where $f(x) = \sqrt[3]{x^2}$.

Solution Since $\sqrt[3]{x^2} = x^{2/3}$, we apply (3.2.3):

$$f'(x) = \left(\frac{2}{3}\right)x^{(2/3)-1} = \left(\frac{2}{3}\right)x^{-1/3} = \frac{2}{3x^{1/3}} = \frac{2}{3\sqrt[3]{x}}$$

Notice that $f'(x)$ is undefined if $x = 0$, so $f(x)$ is not differentiable at $x = 0$, even though $f(0)$ is defined ($= 0$). [In fact, $f(x) = x^{2/3}$ is continuous at $x = 0$.] ●

Note that for $x = 0$ and $p/q \geq 1$, the definition of derivative applies to show $dy/dx = 0$. If $x = 0$ and $p/q < 1$, then dy/dx is undefined, as in Example 11.

Example 12 Determine dy/dx if $y = \sqrt{x^5 - 3x^2 + 1}$.

Solution Here $y = (x^5 - 3x^2 + 1)^{1/2}$, so, applying (3.2.3) again,

$$\frac{dy}{dx} = \frac{1}{2}(x^5 - 3x^2 + 1)^{-1/2}(5x^4 - 6x)$$

or

$$\frac{dy}{dx} = \frac{5x^4 - 6x}{2\sqrt{x^5 - 3x^2 + 1}}$$ ●

Example 13 Determine y' when $y = \sqrt[5]{\dfrac{x^2 - 5}{x^3 + 1}}$

Solution We regard $y = \left[\dfrac{x^2 - 5}{x^3 + 1}\right]^{1/5}$. Then

$$y' = \frac{1}{5}\left[\frac{x^2 - 5}{x^3 + 1}\right]^{-4/5} \frac{(x^3 + 1)(2x) - (x^2 - 5)(3x^2)}{(x^3 + 1)^2}$$

or

$$y' = \left[\sqrt[5]{\frac{x^3 + 1}{x^2 - 5}}\right]^4 \left[\frac{-x^4 + 15x^2 + 2x}{5(x^3 + 1)^2}\right] \quad \left[\text{note that } \left(\frac{A}{B}\right)^{-n} = \left(\frac{B}{A}\right)^n\right] \bullet$$

PROGRESS TEST 3

Differentiate the given functions in 1 to 5, leaving your answers in a form using only positive exponents and no radicals.

1. $f(x) = x^{1/2} + x^{1/3}$
2. $f(x) = 3\sqrt{x} + \dfrac{1}{\sqrt[3]{x}}$
3. $g(x) = 4x^{2/3} + (\sqrt[5]{x})^3$
4. $h(x) = \sqrt{x^2 - 3x + 1}$
5. $m(t) = [\sqrt[3]{3t^2 - t}]^{-2}$

SECTION 3.2 EXERCISES

In Exercises 1 to 32, differentiate the given function:

1. $f(x) = 3x^{2/3} + \sqrt{x}$
2. $p(y) = 4/\sqrt{y} - y^{7/5}$
3. $g(t) = \sqrt[4]{2/t}$
4. $q(x) = 2x^{3/7} + 4x\sqrt{x}$
5. $h(z) = 3/z^2 + z^{1/3}/2$
6. $f(t) = (\sqrt[3]{t})^4 + 6/t$
7. $t(w) = 9 - 3w\sqrt[5]{w} + 2/w^2$
8. $g(y) = 2/(\sqrt[3]{y} - 2)$
9. $f(x) = 4\sqrt[3]{(x + 1)^2}$
10. $q(s) = \sqrt{s}/3 + 3/\sqrt{s}$
11. $f(x) = \left[\dfrac{1}{x^2 - 2}\right]^3$
12. $f(x) = (x^2 + 3x + 1)^5$
13. $f(x) = \sqrt{1 - x^2}$
14. $y = \sqrt[3]{1 - x^2}$
15. $g(x) = \sqrt[3]{\dfrac{x^4 + 3x - 2}{5x - 1}}$
16. $f(u) = \sqrt[5]{\dfrac{u^3 - u^2 + 1}{4u - 6}}$
17. $h(x) = \dfrac{1}{\sqrt{x - 1}}$
18. $A = \sqrt{x/(x^2 - 3x + 1)}$
19. $f(z) = (z^4 - 3z - 2)^6$
20. $f(x) = (x^2 - 2x + 5)^2(11x + 4)^{1/2}$
21. $f(y) = (y^3 + 1)^4(5y^2 - y + 2)^7$
22. $f(x) = \sqrt{(5x^6 + x^4 - 2x + 1)^3}$
23. $g(x) = \dfrac{\sqrt[5]{x^2 + 2x}}{\sqrt{x^5 + 5x}}$
24. $y = (x^2 + x + 4)^{-2/3} + \sqrt{x + 2}$
25. $m(t) = \sqrt{t + 2\sqrt[3]{t^2 + t}}$
26. $f(x) = \sqrt{x + \sqrt{5 - x^3}}$
27. $f(h) = \sqrt{\dfrac{h}{h + 1}} - 2\sqrt{\dfrac{h^3 + h}{3h^2 + h}}$
28. $y = \sqrt{(x^2 - 6x + 1)^2 + \sqrt[3]{1 - x}}$
29. $f(t) = (t^4 + 1)^{1/2} - \dfrac{1}{3}(t^3 + t^2 + t + 1)^{-2/3}$
30. $g(x) = \sqrt[3]{1 + \sqrt{x - 1}}$
31. $h(x) = \dfrac{\sqrt{x^5 + 4}}{\sqrt[3]{x^4 + 1}}$
32. $m(r) = \dfrac{\sqrt[4]{r - 1}}{\sqrt[3]{r + 1}}$

In Exercises 33 to 44, using implicit differentiation, (a) find dy/dx and (b) find dx/dy.

33. $x^2 + y^2 = 25$
34. $x^2 + (y - 2)^2 = 9$
35. $3xy + y^3 - x^2 + 2 = 0$
36. $xy = x + y$
37. $x^2y^2 = 4$
38. $\sqrt{x + y} = 1$
39. $\sqrt{xy} + \dfrac{x}{y} = x$
40. $\dfrac{x}{x + y} - \sqrt{x} = \sqrt[3]{y}$
41. $(x + 2)^3 = 3x^2y^2$
42. $(x + y)^4 + \dfrac{x}{y} = 3$
43. $x^2 + 2xy + y^2 = 1$

44. $(x+y)^2 = 1$ (compare with Exercise 43.)

45. Determine dy/dx, given the equation $x = y + 2$, and interpret dy/dx in terms of the graph of $x = y + 2$.

46. (*a*) Determine the equation in point-slope form of the line tangent to the circle of Exercise 33 at the point $(3, 4)$.
(*b*) Is this line perpendicular to the radius through $(3, 4)$?
(*c*) Solve the equation $x^2 + y^2 = 25$ for y and differentiate the positive function with respect to x. Compare with the result of Exercise 33.

47. Repeat Exercise 46(*a*) and (*b*) for the equation $x^2 + y^2 - 6x - 8y = -20$ at $(5, 5)$. Note that you will need to determine the center of the circle by completing the square.

In Exercises 48 and 49 determine dy/dx for the given equation at the given point.

48. $x^2 + xy + y^3 = 11$ at $(1, 2)$

49. $4y^3 - x^2y - 2x + 5y = 0$ at $(0, 0)$

50. Given $3x + 2y - 4 = 0$:
(*a*) Differentiate this equation implicitly to find dy/dx.
(*b*) Solve the original equation for y and find dy/dx explicitly.
(*c*) Interpret dy/dx in terms of the graph of the equation.

51. Suppose $u(x) = 3x + 1$ and $g(u) = u^2 - 5$. Let $x = 1$ and $\Delta x = 1$. Compute Δu and Δg.

52. Repeat Exercise 51 with $u(x) = 2x^2 - 3$, $g(u) = u^2 + 1$, $x = 2$, $\Delta x = 1$.

53. For a a nonzero constant, show by differentiating that $(x+a)^3 = x^3 + 4x^2a + 6xa^2 + a^3$ is not an identity over any interval.

54. For $y = -\sqrt{9 - x^2}$, compute dy/dx using (3.2.3), and compare with the result of Example 5. (Be careful.)

Section 3.3: Differentials

Objectives:

1. Compute and interpret differentials.
2. Use differentials to approximate quantities.

COMPUTING DIFFERENTIALS

Suppose $y = f(x)$ is a differentiable function. We have used the notations dy/dx, df/dx, $D_x y$, $D_x f$, y', and $f'(x)$ interchangeably to denote the derivative of the function with respect to x. Although it is very useful, the fractional Leibniz notation suggests that the derivative is somehow a quotient of two objects, one called dy and the other dx. This interpretation is reinforced by the definition of the derivative as a limit of a genuine quotient and by statements such as the Chain Rule, which written in Leibniz notation is

$$\frac{dy}{dx} = \frac{dy}{du}\frac{du}{dx}$$

It turns out that it *is* possible to define quantities dy and dx in such a way that the quotient of these quantities dy/dx does exist and equals the derivative. These quantities are useful both in calculations of approximations and later in the integral calculus.

3.3.1 **Definition** Let $y = f(x)$ be a differentiable function of x.
(*a*) The *differential of x*, denoted dx, is defined to be any change Δx in x, that is, any real number $\Delta x = dx$.
(*b*) The *differential of y*, denoted dy, is defined by $dy = f'(x)\,dx$. (If y is not mentioned, we use df instead of dy.)

With this definition, if $dx \neq 0$, then the quotient of differentials equals the derivative:

$$\frac{dy}{dx} = \frac{f'(x)\,dx}{dx} = f'(x)$$

We rigged the definition of differentials so that the quotient of differentials, which *looks* just like the Leibniz notation for the derivative, turns out to be *equal* to the derivative when $\Delta x = dx \neq 0$. This duality parallels the situation that occurs for fractional numbers. We can regard $\frac{4}{5}$ as a symbol for a number (dy/dx

as a derivative), or we can regard it as the quotient of two numbers (dy/dx as the quotient of differentials). Our point of view makes no difference, because they are equal! Which point of view we use at any given time depends upon what we happen to be doing at that time.

Calculating differentials amounts to calculating derivatives and multiplying by the appropriate dx. A particular value of dy depends upon two quantities: dx and x.

Example 1 For $y = x^3 + \dfrac{1}{x}$ determine:

(*a*) dy for arbitrary x and dx
(*b*) dy for $x = 2$ and $\Delta x = \frac{1}{2}$
(*c*) dy for $x = 2$ and $\Delta x = 0.01$

Solution (*a*) $dy = \left(3x^2 - \dfrac{1}{x^2}\right) dx$

(*b*) $dy = \left(3 \cdot 2^2 + \dfrac{1}{2^2}\right)\left(\dfrac{1}{2}\right) = 5.875$

(*c*) $dy = \dfrac{47}{4}(0.01) = 0.1175$ ●

Although in Definition (3.3.1) $dx = \Delta x$, dy is not in general equal to Δy, but rather is an approximation to Δy in the following sense. When defining the derivative $f'(x)$, we used Δx to represent a small change in x, which in turn produced a change

$$\Delta y = f(x + \Delta x) - f(x)$$

in $y = f(x)$. We then defined

$$\lim_{\Delta x \to 0} \frac{\Delta y}{\Delta x} = f'(x)$$

But this means that for small Δx,

$$\frac{\Delta y}{\Delta x} \approx f'(x)$$

($\approx$ means "is approximately equal to" or "is close to.") As a result, for small Δx,

$$\Delta y \approx f'(x)\,\Delta x = dy$$

Although the change Δy in the value of a function resulting from a change Δx in its independent variable is often of interest, Δy is usually considerably more difficult to compute than dy. Hence being able to replace Δy by dy for small $dx = \Delta x$ proves to be quite advantageous. Compare the calculation of dy in Example 1(*a*) with just the first two steps necessary to compute Δy for $y = x^3 + 1/x$:

$$\Delta y = \left[(x + \Delta x)^3 + \frac{1}{x + \Delta x}\right] - \left[x^3 + \frac{1}{x}\right]$$

$$= \frac{(x^3 + 3x^2\,\Delta x + 3x(\Delta x)^2 + (\Delta x)^3)(x + \Delta x) + 1}{x + \Delta x} - \frac{x^4 + 1}{x}$$

PROGRESS TEST 1

1. For $y = (x + 1)^2 - (1/x)$ compute:
(*a*) dy at a typical x and dx.
(*b*) dy for $x = 2$ and $dx = 0.1$.
(*c*) dy for $x = 2$ and $dx = 0.01$.
(*d*) Δy for a typical x and Δx. (Do not simplify.)
(*e*) (for calculators) $\Delta y - dy$ for $x = 2$, $dx = \Delta x = 0.01$.

2. For $f(x) = \sqrt{(1 - x^3)/(1 + x^2)}$ compute:
(*a*) df.
(*b*) df for $x = -1$, $dx = 0.01$.

INTERPRETATIONS OF DIFFERENTIALS

We can interpret the approximation of Δy by dy graphically (see Fig. 3.3). T is the line tangent to $y = f(x)$ at the point $P = (x, f(x))$ and $\Delta x = dx$ is a change in x. (We took Δx to be positive in the picture, but could just as well have used $\Delta x < 0$.)

Now (length of $\overline{RS}$)/(length of $\overline{SP}$) is the slope of T, which is $f'(x)$. But the length of $\overline{SP}$ is $\Delta x = dx$, so, substituting, we have

$$\frac{\text{Length of } \overline{RS}}{dx} = f'(x)$$

that is, length of $\overline{RS} = f'(x)\,dx$. But $f'(x)\,dx$ is our definition of dy. Thus we can interpret dy graphically as the change in y *along the tangent line* due to the change $\Delta x = dx$ in x. We already have seen, in interpreting the derivative graphically, that Δy is the change in y *along the curve* due to the same change in x. Thus in Fig. 3.3.

$$\Delta y - dy = \text{length of } \overline{QR}$$

Hence we can readily see that for small Δx, the tangent line is close to the curve, and hence $\Delta y - dy$ is small.

---

⑨ Besides its graphical interpretation, the differential dy has an interpretation in concrete geometric terms for those cases where $y = f(x)$ represents the area or volume of some geometric object. We examined Δy and dy/dx for several such cases in the narrative, progress tests, and exercises of Chap. 2. Perhaps the simplest involves

$$y = x^2$$

regarded as the area of an x-by-x square. (See Chap. 2, Sec. 3, Progress Test 4.) The various constituents of Δy resulting from an increase Δx in x are labeled in the shaded portion of Fig. 3.4(a).

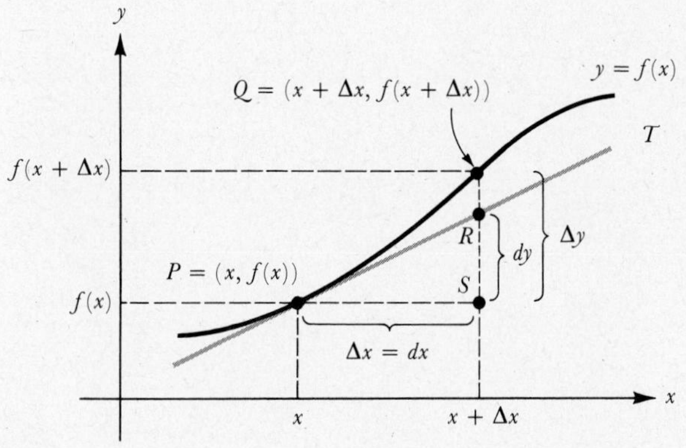

Fig. 3.3

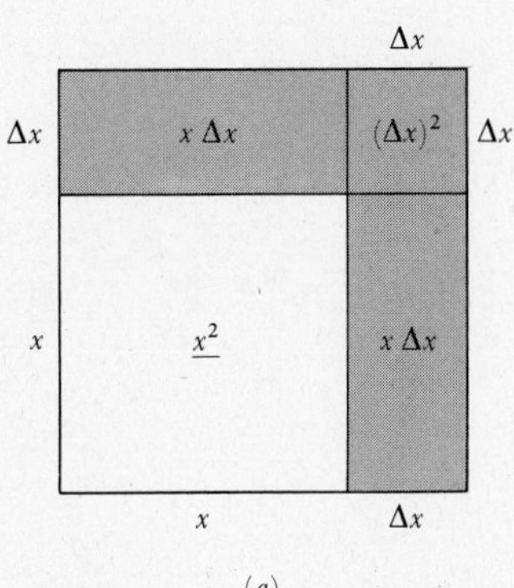

Fig. 3.4

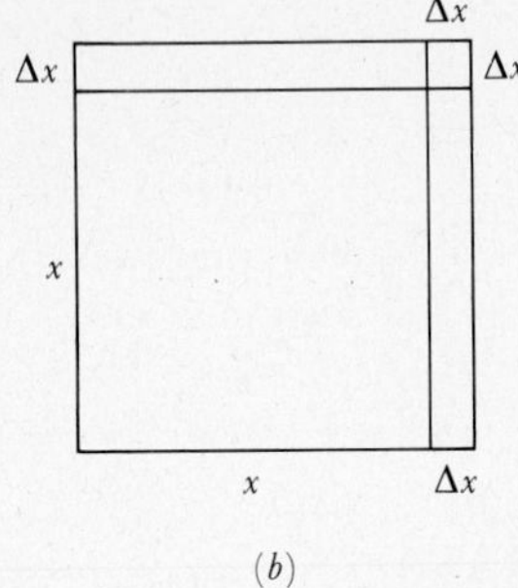

(b)

In particular,

$$\begin{aligned}\Delta y &= (x + \Delta x)^2 - x^2 \\ &= 2x\,\Delta x + (\Delta x)^2\end{aligned}$$

But with $\Delta x = dx$, $2x\,\Delta x = dy$, so

$$\Delta y = dy + (\Delta x)^2$$

As a result,

$$\Delta y - dy = (\Delta x)^2$$

is the area of the upper right corner. This quantity, corresponding to the length of $\overline{QR}$ in Fig. 3.3, is extremely small when Δx is small. For such Δx, the area Δy is made principally of the area of the two long rectangles whose area is $2x\,\Delta x = 2x\,dx = dy$ [Fig. 3.4(*b*)]. That $\Delta y - dy$ approaches 0 much faster than Δx can also be seen numerically in Table 3.2, where x is taken to be 100:

dx	dy	Δy	$\lvert\Delta y - dy\rvert$
1	200	201	1
0.1	20	20.01	0.01
0.01	2	2.0001	0.0001
0.001	0.2	0.200001	0.000001
0.0001	0.02	0.02000001	0.0000001
0.00001	0.002	0.0020000001	0.000000001

Table 3.2

When dx decreases by a factor of $\frac{1}{10}$, $\Delta y - dy$ decreases by a factor of $\frac{1}{100}$.

In this example the "faster than" statement can be written more precisely as

$$\lim_{\Delta x \to 0} \frac{\Delta y - dy}{\Delta x} = 0$$

The statement is true in this case because

$$\frac{\Delta y - dy}{\Delta x} = \frac{(\Delta x)^2}{\Delta x} = \Delta x$$

More important, we can apply this analysis to a differentiable function in general. In so doing, we shall use differentials to clarify a phenomenon that occurred each time we applied the definition to determine the derivative of a specific function $y = f(x)$. In each instance the quotient $\Delta y/\Delta x$ separated into two parts, one of which was the derivative, while the other part $\rightarrow 0$ as $\Delta x \rightarrow 0$. We can write this separation algebraically using differentials:

$$\begin{aligned}\frac{\Delta y}{\Delta x} &= \frac{dy + (\Delta y - dy)}{\Delta x} \\ &= \frac{dy}{\Delta x} + \frac{\Delta y - dy}{\Delta x}\end{aligned}$$

But, by Definition (3.3.1), $dy = f'(x)\,dx$ and $dx = \Delta x$, so $dy/\Delta x = f'(x)$. As a result,

3.3.2
$$\frac{\Delta y}{\Delta x} = f'(x) + \frac{\Delta y - dy}{\Delta x}$$

[Compare (3.3.2) with any derivative computation of Chap. 2!]

We can rewrite (3.3.2) as

$$\frac{\Delta y - dy}{\Delta x} = \frac{\Delta y}{\Delta x} - f'(x)$$

Then

$$\lim_{\Delta x \to 0} \frac{\Delta y - dy}{\Delta x} = \lim_{\Delta x \to 0} \left[\frac{\Delta y}{\Delta x} - f'(x)\right] = 0$$

because $\lim_{\Delta x \to 0} (\Delta y/\Delta x) = f'(x)$ by definition. Thus we have proved in general that $\Delta y - dy \to 0$ "faster than" $\Delta x \to 0$:

3.3.3 Theorem If $y = f(x)$ is differentiable at x, then

$$\lim_{\Delta x \to 0} \frac{\Delta y - dy}{\Delta x} = 0$$

PROGRESS TEST 2

1. Ⓢ The area of a circular disk of radius x is given by $y = f(x) = \pi x^2$. Suppose the radius is increased by Δx. Compute (*a*) Δy, (*b*) dy, and (*c*) $\Delta y - dy$. (*d*) Interpret each of the quantities in (*a*) to (*c*) using a diagram. (*Hint:* The area of a *rectangular* strip of length $2\pi x$ and width Δx is $2\pi x\, \Delta x$. How much does this differ from a strip "bent" to fit around the outside of the circle of radius x? See Chap. 2.)

2. Ⓢ For $x = 10$ in Prob. 1, $dx = 0.01$, calculate (*a*) Δy, (*b*) dy, and (*c*) $\Delta y - dy$, rounding off in the fifth decimal place.

3. Ⓢ The area of an equilateral triangle as a function of the length of its sides x is $A(x) = (\sqrt{3}/4)x^2$, which is the product of half its base, $x/2$, and its height, $(\sqrt{3}/2)x$. Increasing the sides by $dx = \Delta x$ increases the area by ΔA, which is the area of the lower part of the accompanying diagram (partially enclosed by dashed lines). Algebraically, calculate (*a*) ΔA, (*b*) dA, and (*c*) $\Delta A - dA$. Then from the diagram, compute (*d*) the area of rectangle I and (*e*) the sum of the areas of triangles II and III. (*f*) Interpret geometrically the difference between ΔA and dA.

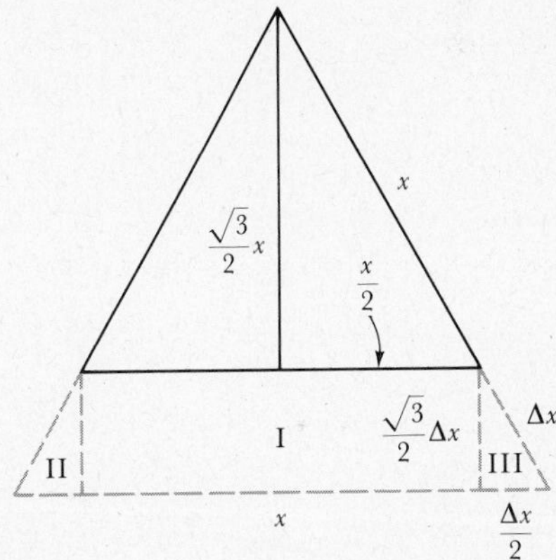

4. In the proof of (3.1.2), that $D_x(x^n) = nx^{n-1}$ for positive integers n, identify the two terms on the right side of (3.3.2).

THE CHAIN RULE FOR DIFFERENTIALS

Each rule for computing derivatives translates directly to a rule for computing differentials because *to compute the differential of a function with respect to some variable, we take the derivative of the function with respect to that variable and then multiply by the differential of that variable.*

The validity of this statement rests on the following:

3.3.4 Chain Rule for Differentials Let $y = g(u)$ and $u(x)$ be differentiable functions of u and x, respectively. Then

$$dy = g'(u(x))u'(x)\,dx \qquad \text{or} \qquad dy = g'(u)\,du$$

The essence of this Chain Rule is that we can regard u both as an *independent* variable (in computing dy) *and* as a *dependent* variable whenever u does depend upon another variable x.

Example 2 If $y = u^5 + 9$, $u = 9x^4 + x^3 - 7$, compute dy.

Solution Now $dy = 5u^4\,du$ and $du = (36x^3 + 3x^2)\,dx$, so

$$dy = 5u^4\,du = 5(9x^4 + x^3 - 7)^4(36x^3 + 3x^2)\,dx \text{ by (3.3.4).} \quad \bullet$$

This Chain Rule is also behind an alternative method for computing implicit derivatives:

Example 3 Use differentials to find dy/dx, given that

$$2x^2y^5 + x^3 - y^4 = 9$$

defines y implicitly as a function of x.

Solution

$$d(2x^2y^5 + x^3 - y^4) = d(9) = 0$$
$$d(2x^2y^5) + d(x^3) - d(y^4) = 0$$
$$2x^2(5y^4)\,dy + y^5(4x)\,dx + 3x^2\,dx - 4y^3\,dy = 0$$

Divide by dx:

$$10x^2y^4\frac{dy}{dx} + 4y^5x + 3x^2 - 4y^3\frac{dy}{dx} = 0$$

Solve for dy/dx:

$$\frac{dy}{dx} = \frac{-4y^5x - 3x^2}{10x^2y^4 - 4y^3} \quad \bullet$$

A common error in using implicit differentiation to find dy/dx is to differentiate a term such as y^4 to get $4y^3$ instead of $4y^3\,dy/dx$, forgetting that y is a function of x. This sort of error can be avoided using differentials because we *always differentiate with respect to whatever variable is present and then multiply by the differential of that variable.*

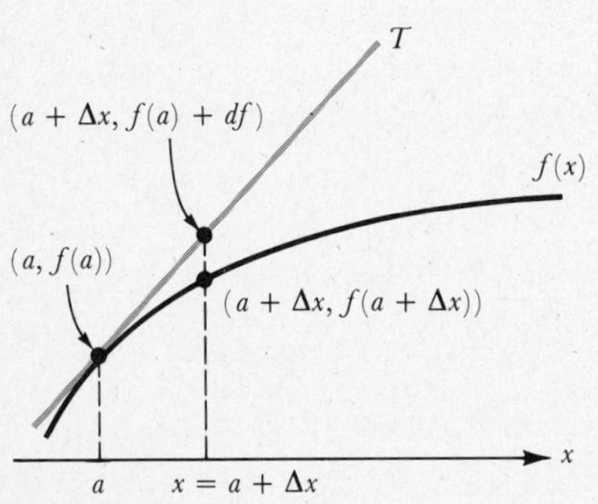

Fig. 3.5

NUMERICAL APPROXIMATIONS

We saw that df provides a "good" approximation to Δf in the sense that not only does $(\Delta f - df) \to 0$ as $\Delta x \to 0$, but $(\Delta f - df)/\Delta x \to 0$ as well. Furthermore, df is easier to compute than Δf. To apply this approximation, we return to the graphical interpretation of df (see Fig. 3.5) for a particular $x = a$ and $\Delta x = x - a$.

The slope of the tangent line T is $f'(a)$, so the point-slope form of the equation of T is

$$y - f(a) = f'(a)(x - a)$$

or

3.3.5 $$y = f(a) + f'(a)\,\Delta x$$

For small Δx, T approximates the graph of f, and since equation (3.3.5) (linear in Δx) can be rewritten as

$$y = f(a) + df$$

the differential df is often said to provide a *linear approximation of f near a.*

In particular, from Fig. 3.5 and (3.3.5) we obtain the useful approximation

3.3.6 $$f(a + \Delta x) \approx f(a) + f'(a)\,\Delta x$$

Example 4 Approximate $\sqrt[5]{33}$ using differentials.

Solution We choose f, a, and $\Delta x = dx$ so that the right-hand side of (3.3.6) is relatively easy to compute and near $\sqrt[5]{33}$, and the left side equals $\sqrt[5]{33}$.

Let $f(x) = x^{1/5}$, $a = 32$, and $\Delta x = dx = 1$. Then $f(a + \Delta x) = f(32 + 1) = f(33) = \sqrt[5]{33}$ is the left-hand side. Now the right-hand side is

$$f(a) + f'(a)\,dx = (32)^{1/5} + \frac{1}{5}(32)^{-4/5}(1)$$

$$= 2 + \frac{1}{5 \cdot 16} = 2 + \frac{1}{80} = 2.0125$$

Hence, by (3.3.6), $\sqrt[5]{33} \approx 2.0125$. ●

The foregoing example embodies the heart of the procedure for approximations using differentials:

To approximate the number b, choose a differentiable function f, a number a, and $\Delta x = dx$ such that:

(*a*) $f(a)$ is relatively easy to compute and near b.

(*b*) $f(a + \Delta x) = b$.

Then $b \approx f(a) + f'(a)\,\Delta x$, by (3.3.6).

Nowdays most calculations such as approximating $\sqrt[5]{33}$ are done by using electronic calculators. (Unless only simple decimals are involved, of course calculators are not really calculators, but "approximators.") Nonetheless, we can use differentials to extend our capacity to approximate *even* in the presence of calculators.

Example 5 Approximate $\sqrt{100.0000001}$.

Solution On most calculators we would get 10 as our "answer," that is, as our approximation. However, by choosing $f(x) = \sqrt{x}$, $a = 100$, and $\Delta x = 0.0000001$

$$\sqrt{100.0000001} \approx \sqrt{100} + \frac{1}{2\sqrt{100}}(0.0000001)$$

$$= 10 + \frac{1}{20}(0.0000001)$$

$$= 10.000000005 \quad ●$$

The important question of the accuracy of differential approximations is addressed first in Chap. 3, Miscellaneous Exercises 1 to 7, and then more completely in Chap. 16. There the differential approximation will be seen as the simplest case of a far more general approximation procedure.

PROGRESS TEST 3

1. Given $y = (4x + 1)^5 + x^3$, find dy.

2. Given $y = 3x^2 - 2x$ and $x = \sqrt[3]{t}$, find dy in terms of t.

3. Given $3w\sqrt{z^2 + 1} + 2z^3 - 9w^4 = 16$, find dw/dz using:

(*a*) Implicit differentiation.

(*b*) Differentials.

4. Determine the differential approximation for

(*a*) $\sqrt{102}$ (*b*) $(\sqrt{0.98})^5$

5. Because of a temperature rise, a $10 \times 10 \times 10$ metal box increases its edge measurement by 0.1 unit. Approximate the resulting increase in volume of the box.

SECTION 3.3 EXERCISES

1. Suppose $f(x) = ax + b$. Let x and $dx = \Delta x$ be given.
(*a*) Find Δf. (*b*) Find df.
(*c*) Show that $\Delta f - df = 0$. (Here the linear approximation is perfect since the function itself is linear.)
(*d*) Show that $dy = dx$ when $y = x$ [so the two parts of Definition (3.3.1) are consistent].

2. For $f(x) = \sqrt{x}$ and a typical $\Delta x = dx$, compute
(*a*) df
(*b*) Δf
and then compute for $x = 1$, $\Delta x = -0.1$:
(*c*) df
(*d*) Δf

3. ⑤ Given a cube whose edge measurement increases from x to $x + \Delta x$, describe, using a diagram, Δy, dy, and $|\Delta y - dy|$, where $y = x^3$ gives the volume of the cube.

In Exercises 4 to 7 determine dy: (a) for a typical x and dx, and (b) for x = 1, dx = 0.1.

4. $y = (3x^2 + 2x - 4)^6 - 2x^3$ **5.** $y = \dfrac{\sqrt{x^3 - x + 2}}{x + 1}$ **6.** $y = 5x$ **7.** $y = \dfrac{1}{x^3}$

8. ⑤ Suppose that the radius x of a circular oil spill is expanding at the rate of 3 m/min.
(*a*) What is the approximate increase in area during the interval from $t = 10$ s to $t = 11$ s?
(*b*) What is the exact increase in area during this time interval?

9. (*a*) Determine df in terms of t and dt, where $f(x) = (\sqrt{x} - 4)/(x^5 - 1)$ and $x(t) = t^2 + 3t - 7$.
(*b*) Compute df when $t = 5$ and $dt = 0.2$.

In Exercises 10 to 17, if the given equations define y implicitly as a function of x, using differentials (a) find dy/dx and (b) find dx/dy.

10. $x^3 - xy + 6y = 0$ **11.** $5x^2 - 2y^2 + 6 = 0$ **12.** $\sqrt{xy} - \sqrt{y} = 0$ **13.** $x^3 + xy^2 - y^3 = 3$

14. $\dfrac{1}{x^2} - \dfrac{1}{y^2} = 1$ **15.** $x^5 - y^5 + x^3y^3 = 34$ **16.** $x^2 + y^2 = 25$ **17.** $xy^2 + \sqrt{xy} + x^3 = 1$

18. Assuming that both x and y in Exercise 16 are functions of t, compute dy/dt in terms of x, y, and dx/dt.

19. Repeat Exercise 18 using the equation in Exercise 17.

Section 3.4: Higher Derivatives

Objectives:

1. Compute higher derivatives.
2. Interpret these higher derivatives graphically and in terms of rates of change.

COMPUTING HIGHER DERIVATIVES

Most differentiable functions can be differentiated again.

If

$$f(x) = 3x^4 + 2x^2 + x - 1$$

then $f'(x) = 12x^3 + 4x + 1$ is itself differentiable. We denote the derivative of $f'(x)$ by $f''(x)$, so

$$f''(x) = 36x^2 + 4$$

Similarly, we denote the derivative of $f''(x)$ by $f'''(x)$

$$f'''(x) = 72x$$

and $f''''(x) = 72$, and $f'''''(x) = 0$. These are called, respectively, the second, third, fourth, and fifth derivatives of f. Beyond the third derivative it is customary to denote the nth derivative by

$$f^{(n)}(x)$$

where the n appears in parentheses to distinguish the nth derivative of f from other operations on f.

Each notation for the derivative extends similarly as indicated in Table 3.3.

Function	First Deriv.	Second Deriv.	Third Deriv.	...	nth Deriv.
$f(x)$	$f'(x)$	$f''(x)$	$f'''(x)$	...	$f^{(n)}(x)$
	$\dfrac{df}{dx}$	$\dfrac{d^2f}{dx^2}$	$\dfrac{d^3f}{dx^3}$	...	$\dfrac{d^nf}{dx^n}$
	$D_xf(x)$	$D_x^2f(x)$	$D_x^3f(x)$	...	$D_x^nf(x)$
$y = f(x)$	$\dfrac{dy}{dx}$	$\dfrac{d^2y}{dx^2}$	$\dfrac{d^3y}{dx^3}$	...	$\dfrac{d^ny}{dx^n}$

Table 3.3

You must be careful to distinguish between

$$\frac{d^ny}{dx^n} \quad \text{and} \quad \left(\frac{dy}{dx}\right)^n$$

The first is the nth derivative with respect to x and the second is the *first* derivative raised to the nth power.

Example 1 Compute

(*a*) $\dfrac{d^4y}{dx^4}$ for $y = x^3$

(*b*) $D_x^2(f)$ for $f(x) = \dfrac{2x-1}{3x+5} + (x^2 - x + 3)^4$

Solution (*a*) $\dfrac{dy}{dx} = 3x^2, \quad \dfrac{d^2y}{dx^2} = 6x, \quad \dfrac{d^3y}{dx^3} = 6, \quad \dfrac{d^4y}{dx^4} = 0$

(*b*) $D_x\left[\dfrac{2x-1}{3x+5} + (x^2 - x + 3)^4\right]$

$$= \frac{(3x+5)(2) - (2x-1)(3)}{(3x+5)^2} + 4(x^2 - x + 3)^3(2x - 1)$$

$$= \frac{13}{(3x+5)^2} + 4(x^2 - x + 3)^3(2x - 1)$$

$$D_x^2(f(x)) = \frac{(3x+5)^2 \cdot 0 - 13(2)(3x+5)(3)}{(3x+5)^4}$$

$$+ 4[(x^2 - x + 3)^3(2) + (2x-1)(3)(x^2 - x + 3)^2(2x - 1)]$$

$$= \frac{-78}{(3x+5)^3} + 8(x^2 - x + 3)^3 + 12(x^2 - x + 3)^2(2x-1)^2 \quad \bullet$$

Example 1 shows that simplification may be helpful between successive computations. Furthermore, (*a*) illustrates that taking the derivative of a polynomial lowers its degree by 1 and taking the nth derivative lowers the degree by n. In particular, if the degree of the polynomial is n, then the nth derivative will be a constant and all subsequent derivatives will be 0.

All the functions we have differentiated in this chapter have nth derivatives for any positive integer n (except at points where denominators are zero). Such functions are said to be *infinitely differentiable.*

We can also find higher derivatives of functions defined implicitly by equations.

Example 2 Given $x^2 - 9y^2 = 9$, find d^2y/dx^2.

Solution Assuming y to be a function of x, we have

$$2x - 18y\frac{dy}{dx} = 0$$

and thus

$$\frac{dy}{dx} = \frac{x}{9y}$$

Then, by the Quotient Rule,

$$\frac{d^2y}{dx^2} = \frac{9y - x(9)\cdot(dy/dx)}{(9y)^2}$$

$$= \frac{9[y - x(x/(9y))]}{(9y)^2} \qquad \left(\text{since } \frac{dy}{dx} = \frac{x}{9y}\right)$$

$$= \frac{9(9y^2 - x^2)}{(9y)^3}$$

Since $x^2 - 9y^2 = 9$, we have

$$\frac{d^2y}{dx^2} = \frac{9(-9)}{(9y)^3} = -\frac{1}{9y^3} \quad \bullet$$

The trick in this sort of implicit differentiation is to simplify the resulting expressions using the original equation. Exercise 30 asks for a computation of d^2y/dx^2 using differentials.

PROGRESS TEST 1

In Probs. 1 to 3 find the first four derivatives of the given function.

1. $f(x) = x^5 + x^4 + x^3 + x^2 + x + 1$
2. $f(x) = x^2 - 5x$
3. $y = 1/x$
4. Find d^2y/dx^2 given that $x^2 + 2xy + 3y^2 = 1$.

INTERPRETATIONS OF HIGHER DERIVATIVES

Since the derivative of a function $f(x)$ with respect to x describes the change in the function with respect to a change in x, the derivative of $f'(x)$ with respect to x must describe the change in $f'(x)$ with respect to x. Similarly, f''' describes the change in f'' with respect to x.

If $s(t)$ is a position function (a function of time t), then $s'(t)$ is the rate of change of position with respect to time, more commonly called "velocity." But then $s''(t)$ is the rate of change of *velocity* with respect to time, more commonly called "acceleration." It is even possible to interpret $s'''(t)$, the rate of change of acceleration with respect to time. This has been called "jerk." (Sitting in a smoothly accelerating automobile we would experience jerk if we were hit from behind by a faster moving vehicle.)

Example 3 Suppose that $s(t) = 50 + 64t - 16t^2$ is a position function. [For example, we can think of $s(t)$ as giving the vertical distance of an object thrown upward at time $t = 0$, where the distance is in feet and time is in seconds. Thus the object is 50 ft above the ground at $t = 0$ (on a platform, perhaps).] Analyze the height, vertical velocity, and acceleration of this object.

Solution Its velocity is given by

$$s'(t) = 64 - 32t$$

Thus, for $t < 2$, $s'(t) > 0$ (the object is moving upward). For $t = 2$, $s'(t) = 0$ (object has stopped—at its highest point). For $t > 2$, $s'(t) < 0$ (object is moving downward; its distance above the ground is decreasing).

Moreover,

$$s''(t) = -32$$

gives the acceleration of the object. This acceleration is, of course, the accelera-tion due to gravity, which when we were introducing the derivative in Chap. 2, was associated with a freely falling object. Notice that $s''(t)$, being negative, means that the acceleration is in a downward direction. ●

Since the first derivative of a function can be interpreted as giving the slopes of the tangent lines to its graph, the second derivative can be interpreted as giving the *change* in the slopes of these lines with respect to x. This tells us something about the "shape" of the graph. We shall examine this idea and its important consequences in the next chapter.

PROGRESS TEST 2

Suppose that the distance between the ground and a particle moving vertically is given by the function $f(t) = t^3 - 6t^2 + 32$.

1. Where is the acceleration positive and where is it negative?

2. What is the velocity when the acceleration is zero?

3. When is the velocity zero?

4. What is the height above the ground when the velocity is zero?

SECTION 3.4 EXERCISES

In Exercises 1 to 12 compute the first three derivatives of the given function.

1. $f(x) = x^3 + x^2$

2. $y = x^8$

3. $g(x) = \dfrac{1}{x^2}$

4. $h(x) = x^{-3}$

5. $f(t) = \sqrt{t}$

6. $f(x) = \pi$

7. $f(x) = 2^3$

8. $g(x) = (3x + 1)^{10}$

9. $h(r) = (r + 4)^5$

10. $y = \dfrac{x^2}{x + 1}$

11. $m(y) = \dfrac{y}{3y + 1}$

12. $A = (x + 11)^{20}$

13. What is the kth derivative of $y = x^n$ for $n > k$? (n a positive integer)

14. What is the kth derivative of $y = x^n$ for $n < k$? (n a positive integer)

15. What is the third derivative of $y = 1/x^n$? (n a positive integer)

16. What is the third derivative of $y = x^{p/q}$? (p and q positive integers and p/q not an integer)

In Exercises 17 to 20 describe the acceleration of a particle moving in a straight line whose directed distance from a given point as a function of time t is given by:

17. $s(t) = 2t^4 - 4t^2 - 5$

18. $g(t) = t^3 - 3t^2 + 1$

19. $f(t) = 4t^3 - 3t^2$

20. $d(t) = 16t$

21. Explain the concavity of the graph of $y = ax^2 + bx + c$ [as defined in (1.3.4)] in terms of the sign of the second derivative of y.

22. What is the millionth derivative of $f(x) = x^{999,999}$?

23. If the velocity of a particle moving in a straight line is given by $v(t) = 36 - 32t$, what is the acceleration?

In Exercises 24 to 29 determine d^2y/dx^2, where y is given implicitly as a function of x by:

24. $x^2 + y - y^3 = 4$

25. $y + x^3y^3 + x^3 = 1$

26. $2y^3 - xy + 5y = 0$

27. $(x + y)^2 = 4x$

28. $y^2 = x$

29. $x^2 - xy + y^2 = 1$

30. Using differentials at each stage, recompute d^2y/dx^2 for x and y in Example 2. Let $dy/dx = y'$ before working on the second derivative. Note that $dy'/dx = d^2y/dx^2$.

31. Repeat Exercise 30 for $x^2 - y^3 = 1$.

CHAPTER EXERCISES

Review Exercises

In Exercises 1 to 35 differentiate the given function with respect to the indicated variable (a, b, c are constants):

1. $f(x) = \sqrt{1 - x^2}$

2. $t(x) = \sqrt[3]{x^5 - 3x^4 - 2x^3 + x^2 - 5x + 7}$

3. $g(t) = (t - 5)^{11}\sqrt[3]{t^7 + 1}$

4. $f(t) = t^{10}(1 - 3t)^5$

5. $f(x) = \dfrac{1}{\sqrt{x + 4}}$

6. $g(x) = \dfrac{x}{x - 1}$

7. $f(x) = \sqrt[5]{\dfrac{x - 1}{x + 1}}$

8. $h(x) = \left(\dfrac{x}{x - 1}\right)^4$

9. $y = (x^2 + 1)^{50}(x^2 - 1)^{10}$

10. $f(x) = \left(\dfrac{x}{x - 1}\right)^4\left(\dfrac{x + 1}{x^2 - 1}\right)^3$

11. $f(x) = (ax^2 + bx + c)^4$

12. $m(x) = \pi^{10} + (4.31)^3$

13. $f(x) = \sqrt{1 + (3x - 5)^6}$

14. $f(x) = (ax + b)^{10}$

15. $h(x) = [(x^3 + 4x - 1)(x^5 - x^2 + x + 2)]^4$

16. $r(s) = \dfrac{s + 1}{s - 1} \cdot \dfrac{s^2 + 1}{s^2 + 2}$

17. $p(s) = (s^4 + s^3 + s^2)^{3/4}$

18. $s(r) = \dfrac{4}{3}\pi r^3$

19. $t(x) = (3.14)^{21}$

20. $f(x) = 3^{-1}$

21. $g(z) = (11z - 4z^2)^{2/3}$

22. $f(x) = (1 + \sqrt{x - 1} + x)^4(x^2 - 7)^5$

23. $t(w) = \dfrac{1 - w}{1 + w^2}$

24. $w(t) = \sqrt{(t^4 + 1)^5 + 6}$

25. $g(y) = \sqrt{y - 1}\sqrt[3]{y + 1}$

26. $f(x) = \sqrt{x + \sqrt{x + 1}}$

27. $g(x) = (x^2 + 1)^5(x^7 + x)^{10}(x^4 + 1)^7$

28. $h(t) = \dfrac{1}{t} + \dfrac{1}{t^2} + \dfrac{1}{t^3}$

29. $t(s) = \sqrt{\dfrac{s + 1}{s - 1}}(s - 4)^9$

30. $f(g) = g^4 + g^3 + g - 6$

31. $f'(x) = 3x^2$

32. $h(x) = x^{.2}$

33. $m(t) = t^{\frac{71}{72}}$

34. $f(x) = \sqrt[3]{1 + \sqrt[3]{1 + x}}$

35. $g(x) = (x + 1)^{10}(x^2 + 1)^5(x^3 + 1)^4$

In Exercises 36 to 45 determine (a) dy/dx and (b) the point-slope equation of the tangent line at the indicated point (where given) using implicit differentiation when y is given implicitly as a function of x by:

36. $xy + x^2y^2 = 1 \quad \left(1, \dfrac{\sqrt{5} - 1}{2}\right)$

37. $3x^2 + 4xy - y^2 = 0 \quad (1, 2(1 + \sqrt{7}))$

38. $\sqrt{x} + \sqrt[3]{y} = 2 \quad (1, 1)$

39. $y^5 + 3x^2y^2 + 7x^2 = 4$ (use differentials), $\left(\sqrt{\dfrac{3}{10}}, 1\right)$

40. $\sqrt{xy} = 1 \quad (1, 1)$

41. $x^2y^2 + x^3y^3 = 1$

42. $x^3 + y^3 = 8 \quad (-1, \sqrt[3]{9})$

43. $\dfrac{x}{y} - 4y = x + 4y \quad \left(4, \dfrac{\sqrt{5} - 1}{2}\right)$

44. $\dfrac{1}{x} + \dfrac{1}{y} = 1 \quad \left(\dfrac{4}{3}, 4\right)$

45. $\dfrac{1}{x^2} + \dfrac{1}{y^2} = 1 \quad \left(\dfrac{2}{\sqrt{3}}, 2\right)$

As Exercises 46 to 55 determine dx/dy using differentials, where x is given implicitly as a function of y by the respective equations in Exercises 36 to 45.

As Exercises 56 to 65 determine d^2y/dx^2 in terms of x and y, where y is given implicitly as a function of x by the respective equations in Exercises 36 to 45.

In Exercises 66 to 75 use differentials to approximate the given quantity.

66. $\sqrt[4]{81.5}$

67. $\sqrt[5]{31}$

68. $(1.01)^{4/5}$

69. $(0.99)^3 + (0.99)^2 + 0.99 - 2$

70. $\sqrt[3]{7.7}$

71. $\sqrt[3]{0.00095}$

72. The empty weight of a spherical water tank 30 ft in diameter, 0.02 ft (approximately 1/4 in.) thick, made of a metal weighing 500 lb/ft^3.

73. The change in $f(x) = x^3 + 5x^2 + x - 4$ when x increases from 2 to 2.01.

74. The change in y where $y = 1/x^2 + \sqrt{x}$ when x decreases from 1 to 0.99.

75. The amount of paint in gallons needed to paint the side of a cylindrical water tank 50 ft tall with a radius of 25 ft. The paint will be about 0.04 in. thick, and 1 gal equals 231 in.^3

76. What is the acceleration of a particle whose directed distance from a given point (as a function of time t) is $s(t) = 3t + 5$?

77. Suppose a particle's directed distance from a fixed point P is given as a function of time t by the function $s(t) = 3t^2 - t^3$.
(*a*) When is its velocity increasing and when is it decreasing?
(*b*) When is the particle traveling away from P and when toward P?
(*c*) What is the maximum positive directed distance from P the particle reaches?

78. Repeat Exercise 77 for $s(t) = 5t - 10t^2$.

79. Determine f' and f'' for $f(x) = \sqrt{\pi^3 + 1}$.

80. Use the definition of derivative to prove the difference rule for derivatives; i.e., if $f(x) = u(x) - v(x)$, then $f'(x) = u'(x) - v'(x)$.

81. Where does the graph of $y = 1/x$ have tangents whose slopes are increasing as we move from left to right, and where are they decreasing?

82. Repeat Exercise 81 for $y = 1/x^2$.

83. An object is hurled upward, and its distance above the ground in feet as a function of time t measured in seconds is given by $h(t) = 144t - 16t^2$.
(*a*) Describe its velocity as a function of time.
(*b*) When does it reach its highest point and what is its maximum height?

84. Suppose that an attorney general announces that the increase in crime is decreasing. What does this mean in terms of the graph of the crime function? Does this mean that crime is decreasing? (The variable here is time.)

85. Given a rocket whose distance above the earth as a function of time t is $s(t) = -(1/3)t^3 + 96t^2 + 100t - 2$, when does the rocket reach maximum *velocity*? (Its acceleration will be zero at that time.)

86. Prove that $D_x[\sqrt{f(x)}] = \dfrac{1}{2\sqrt{f(x)}} D_x[f(x)]$.

87. Prove that $D_x\left[\dfrac{1}{f(x)}\right] = -\dfrac{D_x[f(x)]}{[f(x)]^2}$.

88. Write out all the derivative rules in D_x notation.

Miscellaneous Exercises

(*Differential Approximation Error*) We remarked that for a differentiable function $y = f(x)$, $\Delta f - df$ represents the amount of "bend" in the graph. We now know that the shape of the curve is reflected in the values of $f''(x)$, so the appearance of the second derivative in the following formula, which gives an upper bound on the approximation of Δf by df, should be reasonable. (This formula will be revisited in Chap. 16.) Suppose $f''(x)$ exists and is continuous on $I = [a, a + \Delta x]$ (or $[a + \Delta x, a]$ for $\Delta x < 0$), with $|f''(x)|$ bounded above by M on I. (That is, $|f''(x)| \le M$ for all x in I.) Then

$$|\Delta f - df| \le \frac{M(\Delta x)^2}{2}$$

1. Use this formula to determine an upper bound for the approximation of $\sqrt[5]{33}$ found in Sec. 3.3, Example 4, by:
(*a*) Computing $f''(x)$ for $f(x) = x^{1/5}$.
(*b*) Determining an upper bound M for $f''(x)$ on [32, 33]. (The upper bound will occur at one of the endpoints.)
(*c*) Applying the above formula.

2. Repeat Exercise 1 for the approximation of Review Exercise 66.

3. Repeat Exercise 1 for the approximation of Review Exercise 67.

4. Repeat Exercise 1 for the approximation of Review Exercise 68.

5. Repeat Exercise 1 for the approximation of Review Exercise 69.

6. Repeat Exercise 1 for the approximation of Review Exercise 70.

7. Repeat Exercise 1 for the approximation of Review Exercise 71.

8. Use the Chain rule to show the graph of $f(x) = g(kx)$, for $k > 1$, is k times as steep at each point as is that of $g(x)$.

9. Suppose you attempt to estimate the height of a bridge by measuring the time it takes for a stone to fall to the water below. If you use the distance-fallen formula $s(t) = 16t^2$ (in feet), and the stone takes 5 s to fall, approximate your potential error if your time measurement may be off by as much as 0.2 s.

10. Repeat Exercise 9 with the stone falling only 3 s.

11. Given g a differentiable function of u and u a differentiable function of x with $f(x) = g(u(x))$, determine a formula for d^2f/dx^2. (You will need to use the Product rule.)

12. Given g a differentiable function of u, u a differentiable function of v, and v a differentiable function of x, develop a formula for df/dx where $f(x) = g[u(v(x))]$.

---

13. Ⓢ Given $f(x) = u(x)v(x)$, with u and v differentiable. Write the formula for the differential df of the product and interpret it geometrically. (Recall Fig. 3.2.)

---

14. Ⓢ (Recall Chap. 2, Miscellaneous Exercise 3.) The volume V of a right circular cylinder of radius r and height h is given by $V = \pi r^2 h$.
(*a*) Compute and sketch dV, where r is regarded as a constant and h is the variable.
(*b*) Compute dV, where h is regarded as a constant and r is the variable. What is the difference between dV and ΔV? If ΔV were slit parallel to the axis of the cylinder and flattened out (by distorting its volume), what would be the volume of the nearly rectangular plate resulting?

---

15. (*a*) Assuming that the "curvature" of a given circle should be a constant, show that the second derivative (which measures how the slopes of tangents are chang-

ing) is not a good measure of "curvature" for the upper semicircle $y = \sqrt{r^2 - x^2}$.
(*b*) What geometric figure does have constant "second derivative curvature?"

16. (Recall Chap. 2, Miscellaneous Exercise 23.) Show that the tangent to the upper semicircle $y = \sqrt{r^2 - x^2}$ is perpendicular to the radius for each x in $[-r, r]$. (*Note:* The endpoints need special treatment.)

---

17. (Recall Chap. 2, Miscellaneous Exercise 27.) Interpreting $V = x^3$ as the volume of a cube of edge x, compute dV and interpret $\Delta V - dV$ geometrically. In how many dimensions does $\Delta V - dV$ shrink as $\Delta x \to 0$, and in how many dimensions do the remaining parts of ΔV (these are the constituents of dV) shrink?

---

18. Given that the acceleration of a moving particle is a linear function of t, $a(t) = ct + d$, what sort of function describes the directed distance traveled by this particle?

19. Given that $f(x) = u(x)v(x)$, show that

$$\frac{d^2f}{dx^2} = \frac{d^2u}{dx^2}v(x) + 2\frac{du}{dx}\frac{dv}{dx} + u(x)\frac{d^2v}{dx^2}$$

20. (*a*) Given that $f(x) = u(x)v(x)$, show that

$$\frac{d^3f}{dx^3} = \frac{d^3u}{dx^3}v(x) + 3\frac{d^2u}{dx^2}\frac{dv}{dx} + 3\frac{du}{dx}\frac{d^2v}{dx^2} + u(x)\frac{d^3v}{dx^3}$$

(Use the result of Exercise 19.)
(*b*) Conjecture the nth derivative of $f(x) = u(x)v(x)$.

---

21. (*a*) Generalize Exercise 17 by computing the differential of $f = uvw$, where u, v, and w are all differentiable functions of x. (Introduce Δu, Δv, Δw, as in the proof of the Product Rule.)
(*b*) Interpret $\Delta f - df$ geometrically.

---

22. Compute $D_x[f(x + g(x))]$.

23. Compute $D_x[f((u(x))^2)]$.

24. (Alternative Proof of Power Rule for Rational Exponents) If $f(x) = x^{p/q}$, where p and q are integers, then $f'(x) = (p/q)x^{(p/q)-1}$. We can assume $p = 1$ because the general case follows using the Power rule for integers (since $x^{p/q} = (x^{1/q})^p$).
(*a*) State the derivative quotient, writing Δx as $(x + \Delta x) - x$.
(*b*) Rewrite the identity $(a^q - b^q)/(a - b) = a^{q-1} + a^{q-2}b + a^{q-3}b^2 + \cdots + ab^{q-2} + b^{q-1}$, which holds for any integer q, in order to rewrite the derivative quotient $\Delta f/\Delta x$ in the form $(a - b)/(a^q - b^q)$ for $a = (x + \Delta x)^{1/q}$, $b = x^{1/q}$.
(*c*) Take the limit of the resulting quotient to obtain the derivative.

SELF-TEST ✚

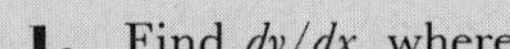

1. Find dy/dx, where

(*a*) $y = \dfrac{\sqrt{x^3 + 4}}{x^7 + x^6}$ (*b*) $y = (x^2 + 1)^{4/5}(x - 2)^8$

(*c*) $x^2y^2 - 3xy + y^2 = 1$ (use differentials)

2. Find d^2y/dx^2 in Prob. 1(*c*).

3. The period P (in seconds) of a pendulum x feet long is given by $P = (1.6)\sqrt{x/2}$. Suppose a large grandfather clock with a pendulum 8 ft long is moved into a cold room, and as a result the pendulum shrinks by 0.06 ft. Using differentials, find the approximate change in the period of the pendulum.

4. Suppose the distance in meters above the ground of an object is given as a function of time x in seconds by $s(x) = 8x - 2x^2$.
(*a*) When is the velocity of this object zero?
(*b*) What is the acceleration of this object?

5. Find the fourth derivative of $g(t) = 1/t^3$.

Chapter 4

APPLICATIONS OF DERIVATIVES TO GRAPHING

CONTENTS

The purpose of this chapter is to enable you to graph functions (and thus describe their behavior) very efficiently. Section 4.1 extends the notion of limit to include cases where the values of a function increase or decrease without bound near a point (vertical asymptotes) and to include descriptions of the behavior of functions when the independent variable increases or decreases without bound. Section 4.2 establishes the connection between the first and second derivatives of a function and its graph. Section 4.3 provides a systematic procedure that pulls together the ideas and techniques of the first two sections and allows you to sketch the graph of a function accurately using relatively few points.

Section 4.4 develops some of the theoretical foundations upon which the practical applications of both this chapter and the next are based. This includes the very important Mean Value Theorem, which will be called upon repeatedly in later chapters to justify basic facts and conclusions.

Section 4.1: Describing the Behavior of Functions Using Special Limits

Objectives:

Determine vertical, horizontal, and oblique asymptotes of the quotient of two polynomials and the quotient of roots of polynomials.

LEFT AND RIGHT LIMITS AND VERTICAL ASYMPTOTES

Recall from Sec. 2.2 that the statement

$$\lim_{x \to a} f(x) = L$$

means that the difference between $f(x)$ and L can be kept as small as we please merely by keeping x close enough to a. Although this statement involves x approaching a from both sides, the analogous *left-hand limit* statement

$$\lim_{x \to a^-} f(x) = L$$

means the same thing *except* that x approaches a only from the left; that is, $x < a$. Similarly, we write the *right-hand* statement

$$\lim_{x \to a^+} f(x) = L$$

if x approaches a only from the right; that is, $x > a$.

Our interest here is in describing the behavior of f near points a where such limits do not seem to exist. In most instances we examine situations where the values of f grow or decrease without bound as x approaches a from one or both sides, as summarized in Fig. 4.1 with graphs accompanied by equivalent algebraic notation. (Recall Chap. 2, pp. 42–44.)

In all the situations described in Fig. 4.1 we can force $f(x)$ to stay larger than any given number M (if $+\infty$ is involved) or less than any given number N (if $-\infty$ is involved) simply by requiring that x be close enough to a. More precisely, we present the following definition.

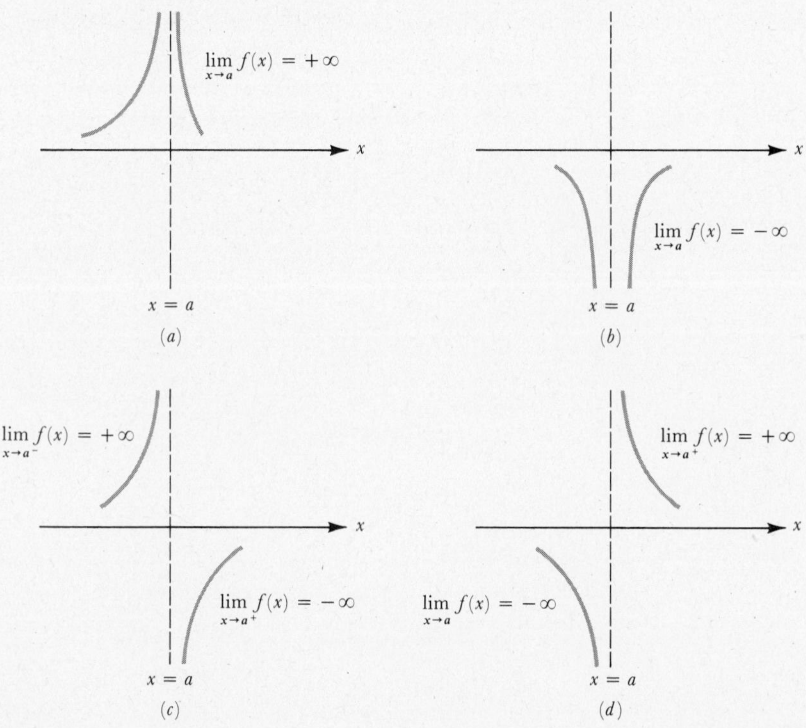

Fig. 4.1

4.1.1 **Definition (Infinite Limits)** Assume that $f(x)$ is defined at all points, except perhaps a, of an open interval containing a. We write

$$\lim_{x \to a} f(x) = +\infty$$

if for every $M > 0$, there is a number $\delta > 0$ such that

$$0 < |x - a| < \delta \text{ implies } f(x) > M$$

Similarly, we write

$$\lim_{x \to a} f(x) = -\infty$$

if for every $N < 0$, there is a number $\delta > 0$ such that

$$0 < |x - a| < \delta \text{ implies } f(x) < N$$

Of course, this definition can be modified to include the one-sided cases illustrated in Fig. 4.1(c) and (d). If any of these situations prevails for a function f at a point a, then we say that f has a *vertical asymptote at* $x = a$.

In practice, as we soon illustrate, showing that f has a vertical asymptote at $x = a$ amounts to writing $f(x)$ as a quotient $u(x)/v(x)$ and then showing that as x tends to a, $u(x)$ tends to some nonzero number k and $v(x)$ tends to zero through either positive or negative values. For $k > 0$, if $v(x)$ tends to zero through positive values, then the quotient is positive and grows without bound. If $v(x)$ tends to zero through negative values, then the quotient is negative and decreases without bound. The reverse conclusions hold, of course, if $k < 0$. Notice that the Limit Theorem for quotients [2.5.4(c)] applies only when $\lim_{x \to a} v(x) \neq 0$, so it does not apply to any of these situations.

Example 1 Show that the following functions have vertical asymptotes at the indicated point a. Construct a table of data for $f(x)$ and sketch the graph of both functions near a.

$$(a)\ f(x) = \frac{2x}{(x-3)^2},\ a = 3 \qquad (b)\ g(x) = \frac{x^2 + x + 1}{2 + x - x^2},\ a = 2$$

Solutions (a) We compute the data for f near $a = 3$; see Table 4.1. Here $\lim_{x \to 3} 2x = 6$, and as $x \to 3^-$, $(x - 3)^2$ approaches 0 through positive values. Hence $\lim_{x \to 3^-} f(x) = +\infty$. Similarly, $\lim_{x \to 3^+} f(x) = +\infty$. Thus $f(x)$ has a vertical asymptote at $a = 3$, with $\lim_{x \to 3} f(x) = +\infty$; see Fig. 4.2($a$).

(b) For $g(x) = \dfrac{x^2 + x + 1}{2 + x - x^2}$, we note that

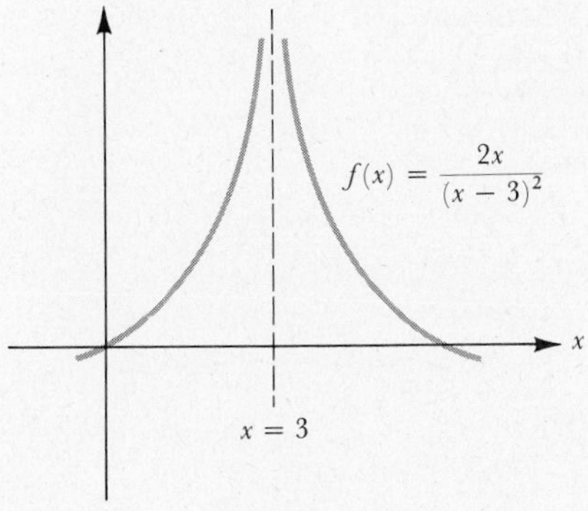

$f(x) = \dfrac{2x}{(x-3)^2}$

$x = 3$

(a)

Fig. 4.2

x	$(x-3)^2$	$f(x) = 2x/(x-3)^2$	x	$(x-3)^2$	$f(x) = 2x/(x-3)^2$
0	9	0	4	1	8
1	4	$\frac{1}{2}$	3.1	0.01	620
2	1	4	3.01	0.0001	60,200
2.5	0.25	20	3.001	0.000001	6,002,000
2.75	0.0625	88			
2.9	0.01	580			
2.99	0.0001	59,800			
2.999	0.000001	5,998,000			

Table 4.1

$$\lim_{x\to 2^-} (x^2 + x + 1) = 7 = \lim_{x\to 2^+} (x^2 + x + 1)$$

while $2 + x - x^2 = (2 - x)(1 + x)$. For x near 2 (on either side), $1 + x$ is near $3 > 0$. However, as x tends to 2^- (through values less than 2), $(2 - x)$ tends toward zero through positive values, hence $(2 - x)(1 + x)$ tends toward zero through positive values, so

$$g(x) = \frac{x^2 + x + 1}{(2 - x)(1 + x)}$$

grows without bound. That is,

$$\lim_{x\to 2^-} g(x) = +\infty$$

On the other hand, as x tends to 2^+ (through values greater than 2), $(2 - x)$ tends to zero through negative values; hence $(2 - x)(1 + x)$ does likewise. As a result,

$$g(x) = \frac{x^2 + x + 1}{(2 - x)(1 + x)}$$

decreases without bound. That is,

$$\lim_{x\to 2^+} g(x) = -\infty$$

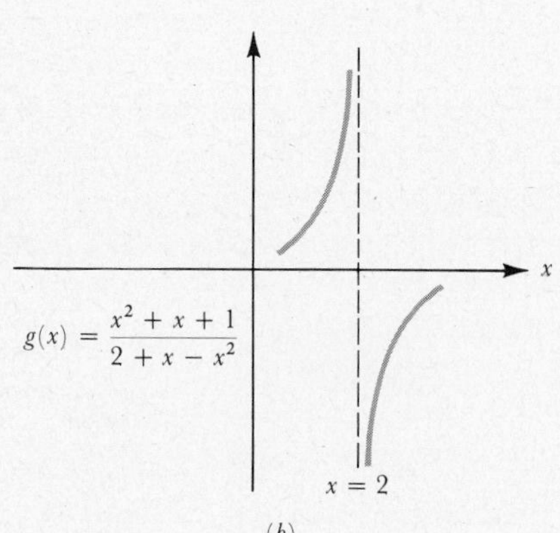

(b)

Hence $g(x)$ has a vertical asymptote at $x = 2$, and we obtain the graph shown in Fig. 4.2(b) to the left. ●

As we know, when examining the behavior of a function f near, but not *at,* a point a, it is possible to perform an algebraic manipulation that replaces f by a function that equals f everywhere except at a. We shall use this idea in the second half of Example 2. Note that the natural place to look for vertical asymptotes is where the expression describing the function has a zero in the denominator.

Example 2 Identify any vertical asymptotes for the functions:

$$(a)\ f(x) = \frac{1}{x} - \frac{1}{x^2} \qquad (b)\ g(x) = \frac{\sqrt{16 - x^2}}{x - 4}$$

Solutions (a) $f(x) = \frac{1}{x} - \frac{1}{x^2} = \frac{x - 1}{x^2}$

The obvious candidate for a vertical asymptote is $x = 0$. Now, as $x \to 0$ through both positive and negative values, $x^2 \to 0$ through positive values. On the other hand,

$$\lim_{x\to 0} (x - 1) = -1$$

Hence $\lim_{x\to 0} f(x) = -\infty$, and $f(x)$ has a vertical asymptote at $x = 0$.

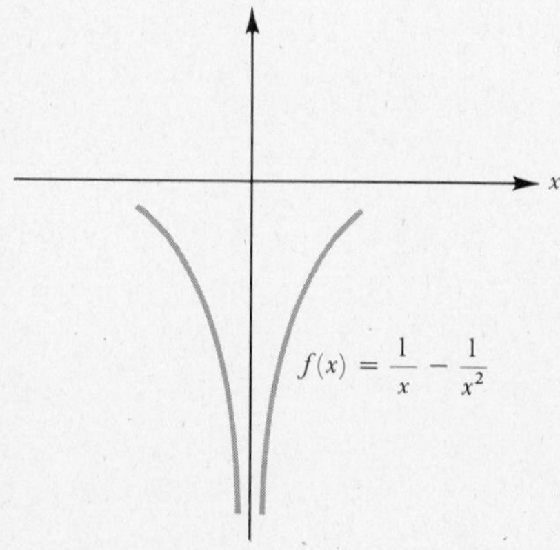

(b) Since $g(x)$ is undefined for $x \geq 4$, we consider what happens to $g(x)$ as x tends to 4 only from the left side. Both the numerator and denominator of

$$g(x) = \frac{\sqrt{16 - x^2}}{x - 4}$$

tend to 0 as $x \to 4^-$, leading to the expression $0/0$, *which has no meaning.* However, rationalizing the numerator and canceling common factors yields

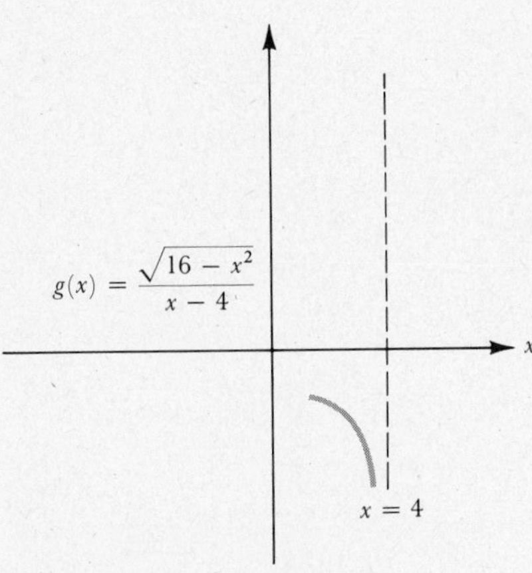

$$\frac{\sqrt{16 - x^2}}{x - 4} \frac{\sqrt{16 - x^2}}{\sqrt{16 - x^2}} = \frac{(4 - x)(4 + x)}{(x - 4)\sqrt{16 - x^2}}$$
$$= \frac{(-1)(4 + x)}{\sqrt{16 - x^2}} \qquad \text{for } x \neq 4$$

Now $\lim_{x \to 4^-}(-1)(4 + x) = -8$. As $x \to 4^-$, $\sqrt{16 - x^2} \to 0$ through positive values, so

$$\lim_{x \to 4^-} g(x) = -\infty \qquad \text{establishing a vertical asymptote at } x = 4$$

Again note that $g(x)$ is not defined for $x \geq 4$. ●

PROGRESS TEST 1

In Probs. 1 to 3 evaluate the given limit.

1. $\lim_{x \to 3} \left[\frac{1}{x - 3} - \frac{5}{(x - 3)^2}\right]$

2. $\lim_{x \to 0^-} \frac{3x - 7}{2x^3}$

3. $\lim_{x \to 1^+} \frac{x^2 - 1}{(x - 1)^2}$

In Probs. 4 and 5, determine any vertical asymptotes for the given function and sketch the function near any vertical asymptotes.

4. $f(x) = \frac{x^6 + x^5 + 2}{x^2 - x - 6}$

5. $g(x) = \frac{2}{x - 1} - \frac{7}{(x - 1)^2}$

6. In Chap. 1 we noted that $\lim_{x \to a} f(x) = +\infty$ if the values of $f(x)$ grow beyond the capacity of your calculator, *regardless of its capacity*. How near must x be to 4 ($x > 4$) for $f(x) = x/(x - 4)$ to exceed the capacity of *your* calculator?

LIMITS AT INFINITY AND HORIZONTAL ASYMPTOTES

Thus far we have examined the behavior of functions $f(x)$ only for relatively small x. We now turn to the question of what happens when x increases or decreases without bound. Consistent with our earlier uses of the symbols $+\infty$ and $-\infty$, we write

$$x \to +\infty \text{ to mean } x \text{ increases without bound}$$

and

$$x \to -\infty \text{ to mean } x \text{ decreases without bound}$$

In these terms we may rephrase our current interest as an investigation of $\lim_{x \to +\infty} f(x)$ and $\lim_{x \to -\infty} f(x)$. For example, given the function

$$f(x) = \frac{1}{x^2} + 5$$

it is fairly clear that as $x \to +\infty$, $1/x^2 \to 0$, so $f(x) \to 5$. That is,

$$\lim_{x \to +\infty} \left(\frac{1}{x^2} + 5\right) = 5$$

We are in effect saying that we can make $1/x^2 + 5$ stay arbitrarily close to 5 by choosing x large enough.

The following general definition expresses this more precisely.

4.1.2 **Definition (Limits at Infinity)**
(*a*) We say that $f(x)$ *tends to L as x increases without bound* and write

$$\lim_{x \to +\infty} f(x) = L$$

if, given any $\varepsilon > 0$, there exists a positive number M such that

$$x > M \text{ implies } |f(x) - L| < \varepsilon$$

(*b*) We say that $f(x)$ *tends to L as x decreases without bound* and write

$$\lim_{x \to +\infty} f(x) = L$$

if, given any $\varepsilon > 0$, there exists a negative number N such that

$$x < N \text{ implies } |f(x) - L| < \varepsilon$$

If either (*a*) or (*b*), or both, applies, we say that $y = L$ is a *horizontal asymptote of the function* $f(x)$.

Part (*a*) says that we can keep the difference $|f(x) - L|$ arbitrarily small (close to zero) by keeping x large enough; (*b*) states that we can keep $|f(x) - L|$ arbitrarily small by keeping $-x$ large enough. Geometrically, (4.1.2) means that the horizontal line $y = L$ is a horizontal asymptote of $f(x)$ if we can keep $f(x)$ close to the line $y = L$ by keeping x far enough to the right or far enough to the left. "Far enough to the left" means that "$-x$ will be large enough" in the sense above. This is the sense in which we shall later use the phrase "large negative x." The limits of (4.1.2) obey a limit theorem exactly analogous to the Basic Limit Theorem (2.5.4), with a replaced throughout by $\pm\infty$. In each following case, determining a horizontal asymptote will eventually reduce to applying the following obvious but useful fact. Its proof is left as Exercise 27.

4.1.3 **Theorem** If p is a positive integer and k is a constant, then

$$\lim_{x \to +\infty} \frac{k}{x^p} = 0 = \lim_{x \to -\infty} \frac{k}{x^p}$$

The key to applying (4.1.3) to the quotient of polynomials $f(x) = p(x)/q(x)$ is the following: *Before taking* $\lim_{x \to \pm\infty}$, *divide the numerator and denominator by* x^n, *where n is the larger of the degrees of* $p(x)$ *or* $q(x)$ [*or equal to the degrees of* $p(x)$ *and* $q(x)$ *if they are equal*].

Since in effect we are replacing f by a function equal to f for all $x \neq 0$, this does not affect the limit, which depends only upon the behavior of the function for x far away from 0.

Example 3 Show that $y = 3$ is a horizontal asymptote of

$$f(x) = \frac{3x^2 + 4}{x^2 + 1}$$

Solution First we divide the numerator and denominator by x^2 since 2 is the degree of both $3x^2 + 4$ and $x^2 + 1$.

Now

$$\lim_{x \to \pm\infty} f(x) = \lim_{x \to \pm\infty} \frac{3x^2/x^2 + 4/x^2}{x^2/x^2 + 1/x^2} \qquad \text{for } x \neq 0$$

$$= \lim_{x \to \pm\infty} \frac{3 + 4/x^2}{1 + 1/x^2} \qquad \text{for } x \neq 0$$

$$= \frac{3 + 0}{1 + 0} = 3 \quad \bullet$$

Notice that the function of this example can be rewritten as

$$f(x) = \frac{1}{x^2 + 1} + 3$$

from which it can also be seen that $\lim_{x\to\pm\infty} f(x) = 3$. Table 4.2 gives some values of $f(x)$ (rounded off in the last place) for large positive and negative x. The graph of $f(x)$ is sketched in Fig. 4.3.

This technique for determining horizontal asymptotes extends to rational functions in the general case, which we now investigate.

4.1.4 Suppose we have

$$f(x) = \frac{a_n x^n + a_{n-1}x^{n-1} + \cdots + a_1 x + a_0}{b_m x^m + b_{m-1}x^{m-1} + \cdots + b_1 x + b_0}$$

where m and n are positive integers and $a_n, b_m \neq 0$. If $n < m$ (so $m - n$ is a positive integer), we divide the numerator and denominator by x^m, to get

$$\lim_{x\to\pm\infty} f(x) = \lim_{x\to\pm\infty} \frac{\dfrac{a_n}{x^{m-n}} + \dfrac{a_{n-1}}{x^{m-n+1}} + \cdots + \dfrac{a_0}{x^m}}{b_m + \dfrac{b_{m-1}}{x} + \cdots + \dfrac{b_0}{x^m}}$$

$$= \frac{0 + 0 + \cdots + 0}{b_m + 0 + \cdots + 0} = 0$$

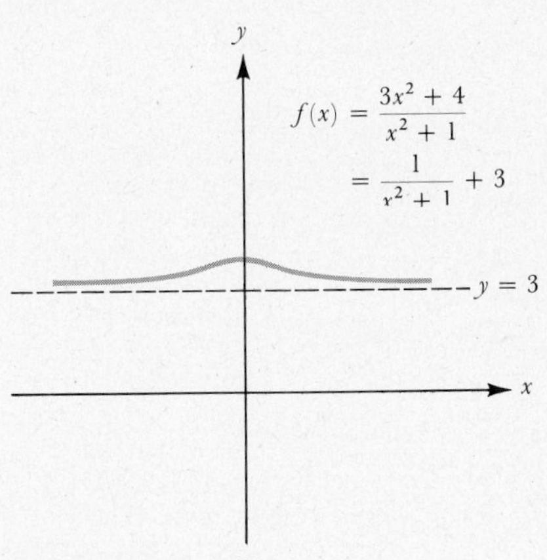

Fig. 4.3

If $n = m$, we then divide numerator and denominator by x^n, as in Example 3, to get

$$\lim_{x\to\pm\infty} f(x) = \frac{a_n}{b_n}$$

Finally, if $n > m$ (so $n - m$ is a positive integer), we divide the numerator and denominator by x^n to get

$$f(x) = \frac{a_n + \dfrac{a_{n-1}}{x} + \cdots + \dfrac{a_0}{x^n}}{\dfrac{b_m}{x^{n-m}} + \dfrac{b_{m-1}}{x^{n-m+1}} + \cdots + \dfrac{b_0}{x^n}}$$

Now, as x grows or decreases without bound, the numerator approaches a_n while the denominator approaches zero. Hence the quotient, $f(x)$, either grows or decreases without bound. This proves the following theorem for $f(x)$, as in (4.1.4):

4.1.5 Theorem

$$\lim_{x\to\pm\infty} f(x) = \begin{cases} 0 \text{ if } n < m & y = 0 \text{ is a horizontal asymptote} \\ a_n/b_n \text{ if } n = m & y = a_n/b_n \text{ is a horizontal asymptote} \\ \pm\infty \text{ if } n > m & \text{no horizontal asymptote} \end{cases}$$

x	$f(x) = 1/(x^2 + 1) + 3$	x	$f(x) = 1/(x^2 + 1) + 3$
10	3.00990099 ···	−10	3.00990099 ···
100	3.0001	−100	3.0001
1000	3.000001	−1000	3.000001
10,000	3.00000001	−10,000	3.00000001

(a)

Table 4.2

You will be asked to apply (4.1.5) in Progress Test 2. The same division-by-powers-of-x technique works for quotients of *roots* of polynomials. Rather than work out another general theorem, however, we shall handle such functions on a case-by-case basis.

An expression consisting of the rth root of a polynomial of degree n will be said to have *relative degree* n/r.

Example 4 Determine any horizontal asymptotes for

$$g(x) = \frac{\sqrt{3x^3 + 2x + 1}}{\sqrt[6]{2x^9 + 1}}$$

Solution The relative degree of the numerator is 3/2, as is the relative degree of the denominator. Hence we divide both the numerator and denominator by $x^{3/2}$. Note that neither $g(x)$ nor $x^{3/2}$ is defined for large negative x; hence we shall examine the behavior of $g(x)$ for large positive x only. Now

$$\begin{aligned}\frac{1}{x^{3/2}}\sqrt{3x^3 + 2x + 1} &= \sqrt{\frac{1}{x^3}}\sqrt{3x^3 + 2x + 1}\\ &= \sqrt{\frac{1}{x^3}(3x^3 + 2x + 1)}\\ &= \sqrt{3 + \frac{2}{x^2} + \frac{1}{x^3}}\end{aligned}$$

and so the limit of the numerator is

$$\lim_{x\to+\infty}\sqrt{3 + \frac{2}{x^2} + \frac{1}{x^3}} = \sqrt{\lim_{x\to+\infty}\left(3 + \frac{2}{x^2} + \frac{1}{x^3}\right)} = \sqrt{3}$$

Also, writing

$$\frac{1}{x^{3/2}} = \frac{1}{x^{9/6}} = \sqrt[6]{\frac{1}{x^9}}$$

we have

$$\begin{aligned}\frac{1}{x^{3/2}}\sqrt[6]{2x^9 + 1} &= \sqrt[6]{\frac{1}{x^9}}\sqrt[6]{2x^9 + 1}\\ &= \sqrt[6]{\frac{2x^9 + 1}{x^9}} = \sqrt[6]{2 + \frac{1}{x^9}}\end{aligned}$$

Thus the limit of the denominator is

$$\lim_{x\to+\infty}\sqrt[6]{2 + \frac{1}{x^9}} = \sqrt[6]{2}$$

and

$$\lim_{x\to+\infty} g(x) = \lim_{x\to+\infty}\frac{\sqrt{3x^3 + 2x + 1}}{\sqrt[6]{2x^9 + 1}} = \frac{\sqrt{3}}{\sqrt[6]{2}}$$

so $g(x)$ has a horizontal asymptote $y = \dfrac{\sqrt{3}}{\sqrt[6]{2}}$. ●

The function $g(x)$ had a horizontal asymptote only in the positive direction and was undefined for large negative x. It is even possible for a function to have two different horizontal asymptotes.

Example 5 Determine the horizontal asymptotes of

$$h(x) = \frac{3x + 2}{\sqrt{4x^2 - 1}}$$

Solution Since $3x + 2$ has degree 1 and $\sqrt{4x^2 - 1}$ has relative degree 1, we shall divide numerator and denominator by x. We shall consider positive x first.

$$\lim_{x\to+\infty} h(x) = \lim_{x\to+\infty} \frac{(1/x)(3x + 2)}{(1/x)\sqrt{4x^2 - 1}}$$

$$= \lim_{x\to+\infty} \frac{3x/x + 2/x}{(1/\sqrt{x^2})\sqrt{4x^2 - 1}}$$

$$= \lim_{x\to+\infty} \frac{3 + 2/x}{\sqrt{4 - 1/x^2}} = \frac{3 + 0}{\sqrt{4 - 0}} = \frac{3}{2}$$

On the other hand, for *negative* x,

$$\frac{1}{x} = -\sqrt{\frac{1}{x^2}} \qquad \left(\text{and } \textit{not } \sqrt{\frac{1}{x^2}}\right)$$

Therefore,

$$\lim_{x\to-\infty} h(x) = \lim_{x\to-\infty} \frac{(1/x)(3x + 2)}{(1/x)\sqrt{4x^2 - 1}}$$

$$= \lim_{x\to-\infty} \frac{3 + 2/x}{-\sqrt{1/x^2}\sqrt{4x^2 - 1}}$$

$$= \lim_{x\to-\infty} \frac{3 + 2/x}{-\sqrt{4 - 1/x^2}} = \frac{3 + 0}{-\sqrt{4 - 0}} = -\frac{3}{2}$$

Thus for positive x, $h(x)$ has the horizontal asymptote $y = 3/2$; for negative x, the horizontal asymptote is $y = -3/2$. ●

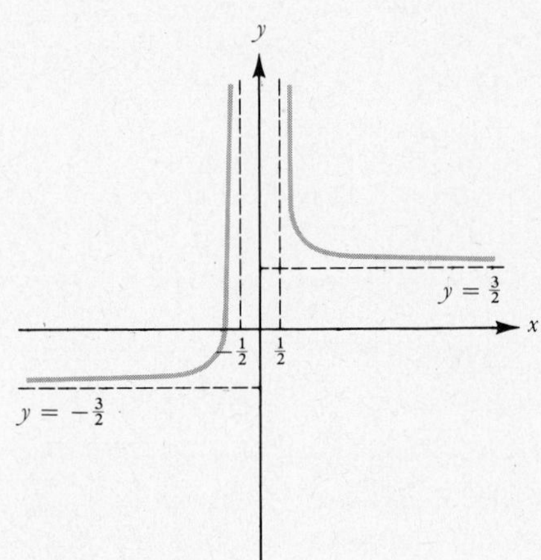

Fig. 4.4

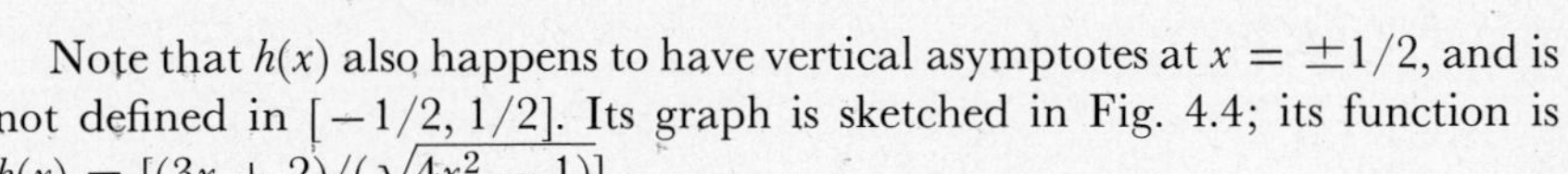

Note that $h(x)$ also happens to have vertical asymptotes at $x = \pm 1/2$, and is not defined in $[-1/2, 1/2]$. Its graph is sketched in Fig. 4.4; its function is $h(x) = [(3x + 2)/(\sqrt{4x^2 - 1})]$.

PROGRESS TEST 2

For the functions in Probs. 1 to 3 use (4.1.5) to determine any horizontal asymptotes.

1. $f(x) = \dfrac{4 - 2x + 7x^2 + x^3}{\sqrt{2x^3 + x - 7}}$

2. $g(x) = \dfrac{4x^3 + x^2 - 9}{8x^2 - x + 1}$

3. $h(x) = \dfrac{4x^5 + x^3 - 9x}{1 + 2x^2 + 3x^7}$

In Probs. 4 to 6 determine any horizontal asymptotes.

4. $f(x) = \dfrac{2x + 1}{\sqrt{3x + 9}}$

5. $f(x) = \dfrac{x^2 + x + 1}{\sqrt[3]{3x^9 + 4x^2 + 1}}$

6. $f(x) = \dfrac{\sqrt{2x^2 + x + 1}}{3x + 2}$

OBLIQUE ASYMPTOTES

If $f(x)$ is a quotient of polynomials

$$f(x) = \frac{p(x)}{q(x)}$$

where the degree of $p(x)$ exceeds that of $q(x)$ by 1, then $f(x)$ has no horizontal asymptotes but *does* approach a line of nonzero slope as $x \to \pm\infty$ in the following sense. By long division,

$$f(x) = cx + d + \frac{r(x)}{q(x)}$$

where $r(x)$, the remainder, has degree less than that of $q(x)$. Now, as $x \to +\infty$, *or* $x \to -\infty$, $r(x)/q(x) \to 0$ by (4.1.5), so the difference between $f(x)$ and the straight line with equation $y = cx + d$

$$f(x) - (cx + d) = \frac{r(x)}{q(x)}$$

approaches zero. If either situation prevails for a function $f(x)$, we say $f(x)$ *has the oblique asymptote* $y = cx + d$. It turns out that c, the slope of the asymptote, is the quotient of the coefficients of the highest degree terms in the numerator and denominator.

Example 6 Determine the oblique asymptote of

$$f(x) = \frac{15x^3 + 5x^2 + 5x + 1}{5x^2 + 1}$$

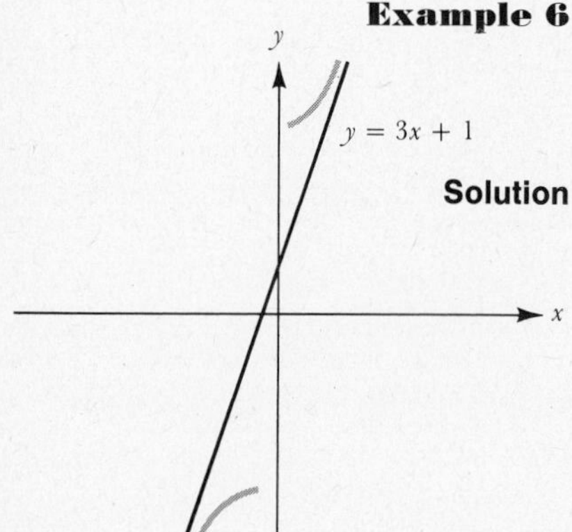

Solution By long division, we have

$$f(x) = 3x + 1 + \frac{2x}{5x^2 + 1}$$

Thus, for large positive and negative x, the graph of $f(x)$ is near the line $y = 3x + 1$. Hence we can describe the graph of $f(x)$ away from the origin as follows. We know that for $x > 0$ the graph of $f(x)$ is above the oblique asymptote [because $2x/(5x^2 + 1) > 0$] and, similarly, for $x < 0$ the graph is below the asymptote. Methods for determining the behavior of $f(x)$ on the remainder of its domain await the systematic use of derivatives to be developed in Sec. 4.2. ●

SECTION 4.1 EXERCISES

For each of the functions in Exercises 1 to 26 determine any vertical, horizontal, or oblique asymptotes.

1. $f(x) = \dfrac{x^2}{x - 2}$

2. $g(x) = \dfrac{x^2}{x^2 - 4}$

3. $y = \dfrac{(x - 1)^2}{2x - 2}$

4. $y = \dfrac{2x^2 - 3x - 1}{x^2 + x - 1}$

5. $f(x) = \dfrac{3x^2 + 2x - 9}{4x^2 + 7}$

6. $u(x) = \dfrac{2x - 1}{\sqrt{2x^2 + 1}}$

7. $v(x) = \dfrac{\sqrt{x^2 - 1}}{3x + 1}$ (Be careful!)

8. $t(x) = \dfrac{(x - 1)^2}{x}$

9. $f(x) = \dfrac{\sqrt{x^2 + 1}}{x + 1}$

10. $g(x) = \dfrac{3x - \sqrt{9x^2 + 2}}{x}$ (treat as the difference of functions)

11. $h(x) = \dfrac{\sqrt{x^4 - 16}}{x - 2}$

12. $f(x) = \dfrac{x^3 - 3x^2}{x^2 + 2x + 1}$

13. $g(x) = \dfrac{x^2 + x + 1}{2x + 9}$

14. $f(x) = \dfrac{3x^2 + 2x + 1}{6x^2 + 3x - 9}$

15. $y = \dfrac{1}{x^2} - \dfrac{1}{x^3}$

16. $f(x) = \dfrac{x^2 + x + 1}{1/x + 1/(2x^2)}$

17. $h(t) = \dfrac{t^2 - 4}{t^4 - t^3 + t^2 - 1}$

18. $m(x) = \dfrac{4x^2}{9 - x^2}$

19. $f(t) = \dfrac{4t^3 - 2t^2 + t}{2t - 1}$

20. $y = \dfrac{11x^2 + 2x - 1}{x - 3}$

21. $f(x) = \dfrac{x^4 - x^3 + 2x - 1}{x^3 + x^2 + x + 1}$

22. $r(x) = \dfrac{x^4 + 3x}{x^2 + 1}$

23. $s(x) = \dfrac{x^6 - x^4 + 5}{(x^2 - 3x - 1)^2}$

24. $f(x) = \dfrac{1/x^2}{1/x^3}$

25. $f(x) = \dfrac{x^2 - 1}{x^2 + 1}$

26. $y = \dfrac{1/x^2}{1/x^4}$

27. Prove (4.1.3)

Section 4.2: Describing the Behavior of Functions Using Derivatives—Part 1

Objectives:

1. Understand the geometric meaning of the following words as they apply to the graphs of functions: increasing, decreasing, relative maximum, relative minimum, concave up, concave down, and point of inflection.

2. Understand the algebraic meaning of the above terms as reflected in the first and second derivatives.

3. Determine relative extremes using both the First- and Second-Derivative Tests.

SOME TERMINOLOGY

To use derivatives efficiently to graph functions, we first establish a working vocabulary. The essence of each definition can be seen in the graphs that accompany it.

4.2.1 Definition We say that the function f is *increasing on the interval I* if

$$x_1 < x_2 \text{ in } I \text{ implies } f(x_1) \leq f(x_2)$$

We say f is *decreasing on I* if

$$x_1 < x_2 \text{ in } I \text{ implies } f(x_1) \geq f(x_2)$$

(See Fig. 4.5.) We use the words *strictly increasing* or *strictly decreasing* when strict inequalities hold in these statements.

These are "left-to-right" definitions in that f is strictly increasing if it goes up as you move from left to right, and strictly decreasing if it goes down as you move from left to right.

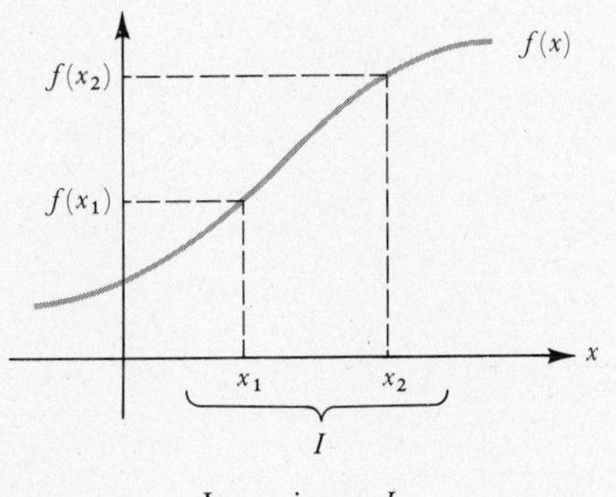

Increasing on I

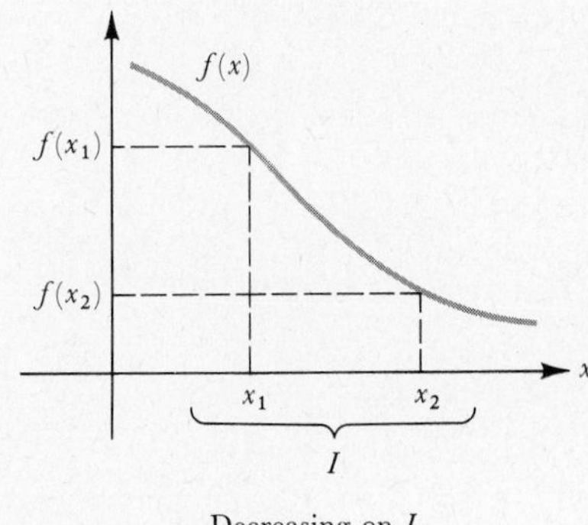

Decreasing on I

Fig. 4.5

4.2.2 Definition We say that the function f *has a relative minimum at* x_0 if $f(x_0)$ is the minimum value of f on some open interval (a, b) containing x_0. That is,

$$f(x_0) \leq f(x) \qquad \text{for all } x \text{ in } (a, b)$$

Similarly, we say that f *has a relative maximum at* x_0 if $f(x_0)$ is the maximum value of f on some open interval (a, b) containing x_0. That is,

$$f(x_0) \geq f(x) \qquad \text{for all } x \text{ in } (a, b)$$

Relative minima and relative maxima are often referred to as *relative extremes* of the function.

In Fig. 4.6 there are *relative minima* at x_0 and x_2, and a *relative maximum* at x_1. Notice that $f(x_0) > f(x_2)$ even though x_0 is a relative minimum of f. This explains the use of the word "relative." There *does* exist an open interval (a, b) containing x_0 (and which does not contain x_2) such that $f(x_0)$ is the smallest value of f *on this* interval. The graph of the function is lower at x_0 than at any other x *near* x_0.

Our definition of relative maximum and relative minimum at x_0 allows the possibility that f is constant on an open interval containing x_0 [so $f(x) = f(x_0)$ there], and our definition of increasing and decreasing functions likewise allows this possibility. However, when graphing functions this is of little importance because, where the function is constant, the special terms are irrelevant: To state the function is constant on some interval and to give the value of that constant completely describes the behavior of the function on that interval.

The next theorem is the first among several to relate extremes of a function to its derivatives.

4.2.3 Theorem If x_0 is a relative extreme of f and $f'(x_0)$ exists, then

$$f'(x_0) = 0$$

That is, the tangent line at x_0 is horizontal.

In Fig. 4.7, $f'(x_0) = 0 = f'(x_1)$ are horizontal tangents, and the relative extremes are at x_0 and x_1.

Caution: The converse of (4.2.3) is not necessarily true! The function $f(x) = x^3 + 1$ shows a case in which $f'(x_0) = 0$ without f having a relative extreme at x_0; see Fig. 4.8.

Proof of (4.2.3): We use the fact that the limit definition of $f'(x_0)$ is two-sided. We consider the case of $f(x_0)$ a relative maximum. (The relative minimum case is similar; see Exercise 17). The key is that for small $|\Delta x|$,

$$f(x_0 + \Delta x) - f(x_0) \leq 0$$

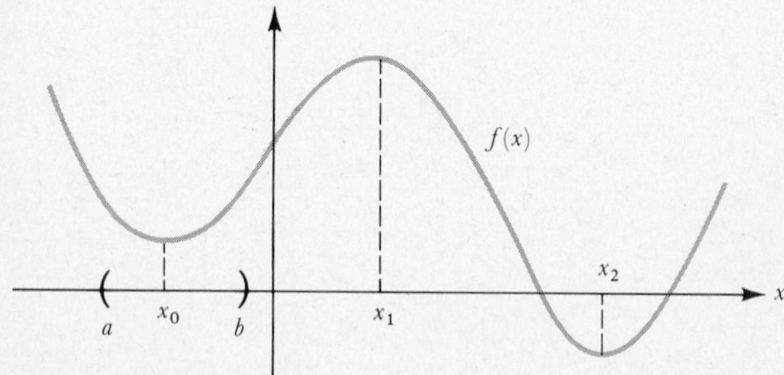

Fig. 4.6

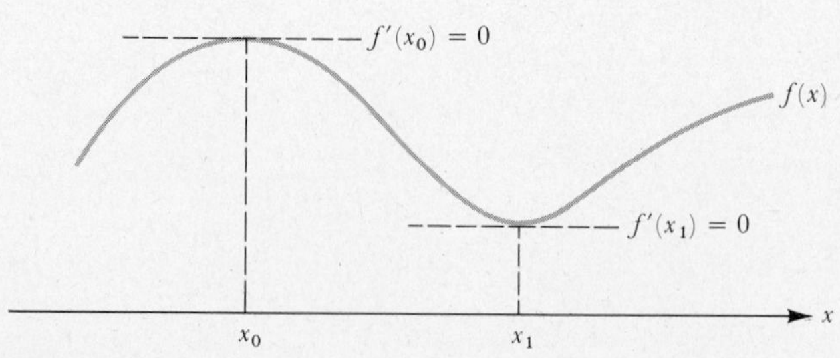

Fig. 4.7

because $f(x_0)$ is a relative maximum. We know that the left and right limits defining $f'(x_0)$ are equal:

4.2.4
$$\lim_{\Delta x \to 0^-} \left[\frac{f(x_0 + \Delta x) - f(x_0)}{\Delta x}\right] = f'(x_0) = \lim_{\Delta x \to 0^+} \left[\frac{f(x_0 + \Delta x) - f(x_0)}{\Delta x}\right]$$

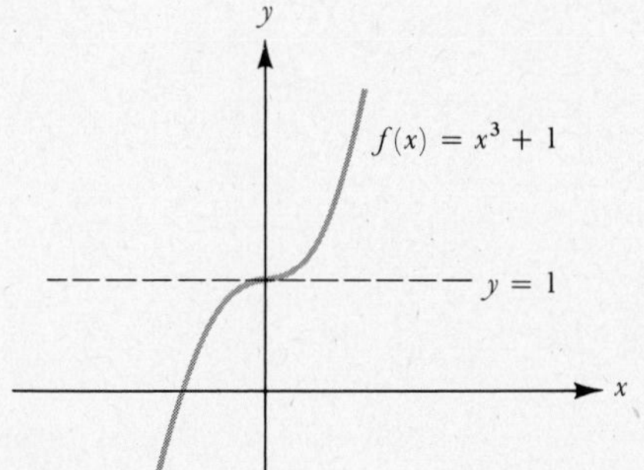

Fig. 4.8

However, since $\Delta x < 0$ on the left side, the bracketed left-hand quotient is positive, and since $\Delta x > 0$ on the right side, the bracketed right-hand quotient is negative. Thus, as $\Delta x \to 0$, the left-hand quotient approaches $f'(x_0)$ through positive values, and the right-hand quotient approaches the *same number* $f'(x_0)$ through negative values. The only way this can happen is if $f'(x_0) = 0$! ■

If $f'(x_0)$ does not exist, f may still have a relative extreme in any of several ways:

(*a*) The left- and right-hand limits appearing in (4.2.4) may exist but fail to be equal, as indicated by the dashed one-sided tangents in Fig. 4.9. Under such circumstances we have a "corner."

(*b*) The left and right limits may grow or decrease without bound, yielding vertical tangents and "cusps" as indicated in Fig. 4.10.

Combinations of the phenomena in (*a*) and (*b*) may also occur.

We shall now examine the *shape* of a graph more closely. Suppose we know that the graph of a function f starts at $(1, 2)$, increases to $(3, 5)$, has a relative maximum at $x = 3$, then decreases to $(4, 0)$. Do we have enough information for a fairly accurate sketch? The continuous functions sketched in Fig. 4.9(*a*) and (*b*) and Fig. 4.10(*a*), all of which satisfy the above, show that the answer to this question is no! (We assume you also thought of the case with f differentiable at $x = 3$.) We need to be more specific about the *way* in which a curve increases or decreases.

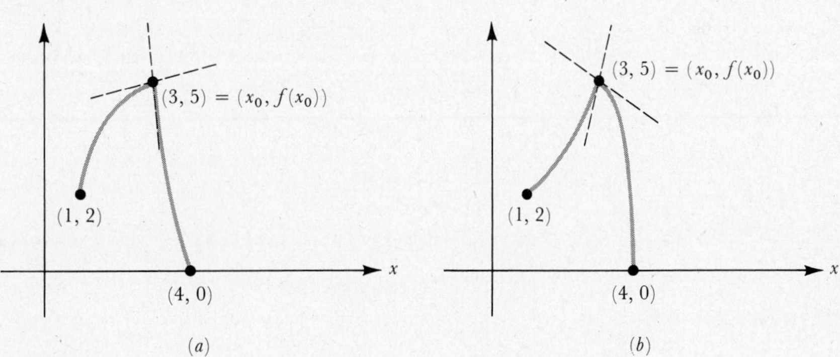

Fig. 4.9

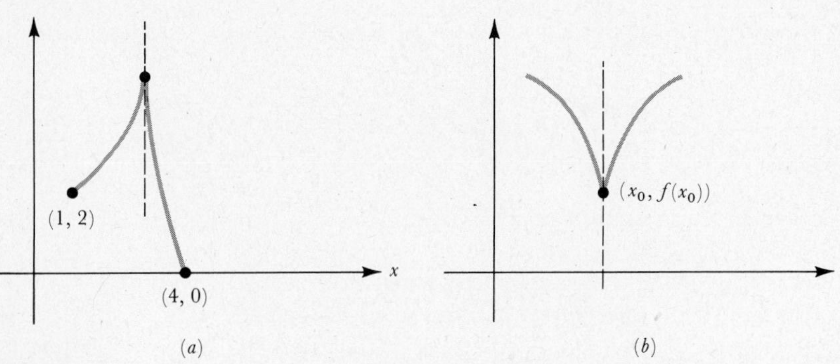

Fig. 4.10

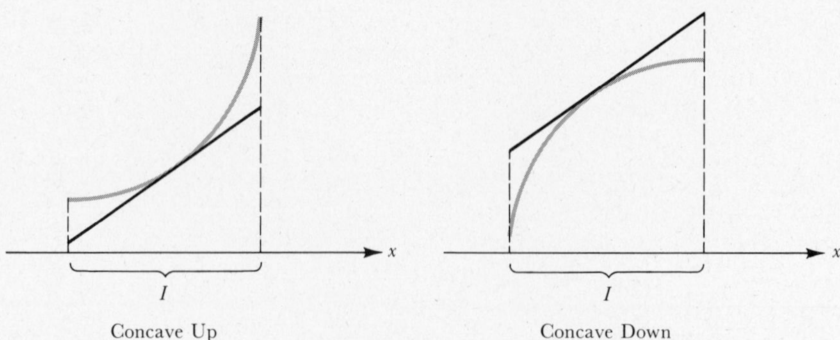

Fig. 4.11

4.2.5 **Definition** We say that the function f is *concave up* on an interval I if the graph of f is above its tangent line at each point of I. Similarly, it is *concave down* on I if its graph lies below its tangent line at each point of I; see Fig. 4.11.

If we return to the previous example and include additional information on the concavity of f, we can in fact get a fairly accurate sketch. For example, suppose f starts at $(1, 2)$, increases in a concave down fashion to $(3, 5)$, and then decreases in a concave up fashion to $(4, 0)$. Then a correct sketch appears in Fig. 4.9(a).

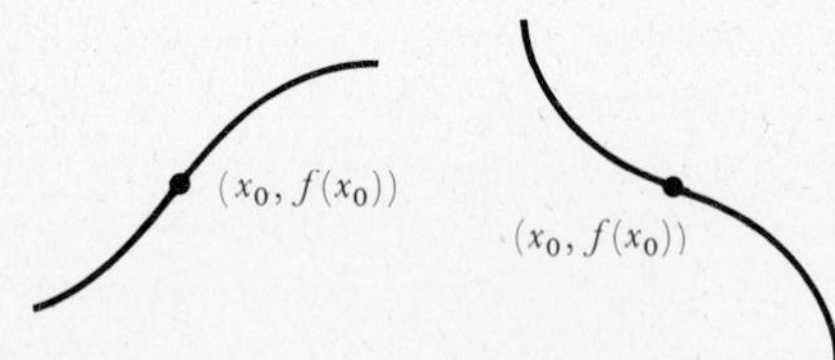

Fig. 4.12

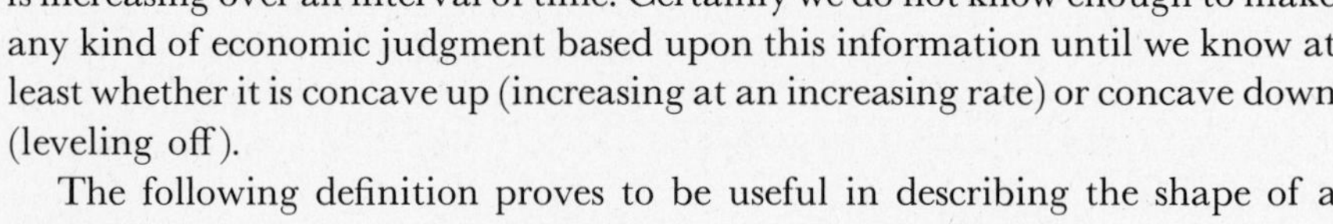

Suppose we are told that a function which describes the unemployment rate is increasing over an interval of time. Certainly we do not know enough to make any kind of economic judgment based upon this information until we know at least whether it is concave up (increasing at an increasing rate) or concave down (leveling off).

The following definition proves to be useful in describing the shape of a graph.

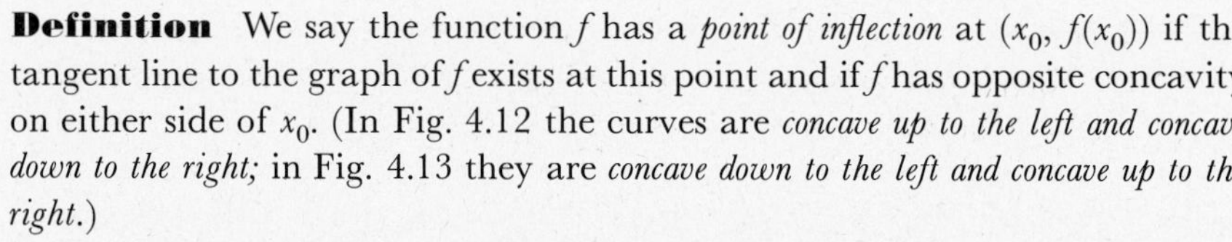

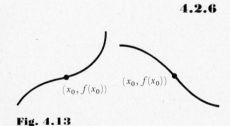

4.2.6 **Definition** We say the function f has a *point of inflection* at $(x_0, f(x_0))$ if the tangent line to the graph of f exists at this point and if f has opposite concavity on either side of x_0. (In Fig. 4.12 the curves are *concave up to the left and concave down to the right;* in Fig. 4.13 they are *concave down to the left and concave up to the right.*)

You should not worry about memorizing these definitions; merely be certain that you can "picture" what they mean.

Note that we sometimes slightly abuse language in cases where confusion does not result. In particular, we often may not distinguish between the function and its graph.

$(x_0, f(x_0))$ $(x_0, f(x_0))$

Fig. 4.13

PROGRESS TEST 1

1. Suppose $f(x)$ has the following graph:

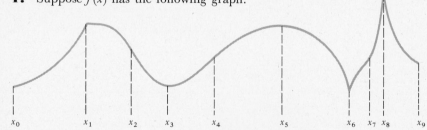

x_0 x_1 x_2 x_3 x_4 x_5 x_6 x_7 x_8 x_9

(a) In which intervals is f decreasing? Increasing?
(b) In which intervals is f concave up? Concave down?
(c) List all relative maxima and minima for f.
(d) At which relative extremes does f have a derivative? What is its value at those points?
(e) List all points of inflection.

2. Sketch the following functions in the interval

$[-2, 5]$, appropriately connecting the given points. (The ordered pairs here refer to points, not intervals.)
(*a*) f increases in a concave down fashion from $(-2, 2)$ to $(2, 4)$, decreases in a concave down fashion to $(3, 0)$, and then decreases in a concave up fashion to $(5, -1)$.
(*b*) f decreases in a concave down fashion from $(-2, 4)$ to $(2, 0)$, increases in a concave down fashion to $(3, 2)$, is constant to $(4, 2)$, and then increases in a concave up fashion to $(5, 4)$.

3. Sketch the graphs of four different functions that satisfy the following conditions: The function decreases from the point $(-2, -1)$ to $(0, -4)$, has a relative minimum at $(0, -4)$, and increases to $(1, -1)$. It is constant on the interval $(1, +\infty)$.

THE MEANING OF THE FIRST DERIVATIVE

In Sec. 4.4, we shall prove the following reasonable and fundamental fact relating a function and its first derivative.

4.2.7 **Increasing-Decreasing Theorem** Suppose f is differentiable on an open interval $I = (a, b)$.
(*a*) If $f'(x) > 0$ for all x in I, then f is strictly increasing on I.
(*b*) If $f'(x) < 0$ for all x in I, then f is strictly decreasing on I.
(*c*) If $f'(x) = 0$ for all x in I, then f is constant on I (that is, its graph is a horizontal straight line).

Since $f'(x)$ is the slope of the tangent to the graph at the point $(x, f(x))$, and since $f'(x) > 0$ means this tangent tilts *up* to the right, the function must be strictly increasing precisely when $f'(x) > 0$. Similarly, it must be strictly decreasing where $f'(x) < 0$ because there the tangent tilts *down* to the right. (See Fig. 4.14.)

The next theorem enables us to use the first derivative to determine relative extremes of a function.

4.2.8 **First-Derivative Test for Relative Extremes** Suppose f is differentiable on an open interval containing the point x_0, but not necessarily *at* x_0. *If f is continuous at x_0 and $f'(x)$ has opposite signs on either side of x_0, then f has a relative extreme at x_0.* In particular,
(*a*) If $f'(x) > 0$ on some interval (a, x_0) and $f'(x) < 0$ on some interval (x_0, b), then f has a relative maximum at x_0.
(*b*) If $f'(x) < 0$ on some interval (a, x_0) and $f'(x) > 0$ on some interval (x_0, b), then f has a relative minimum at x_0.

Figure 4.15 illustrates some of these possibilities. *Note:* We can combine (4.2.3) with (4.2.8) to conclude that if $f'(x_0)$ *does* exist and $f'(x)$ has opposite signs on either side of x_0, then $f'(x_0) = 0$, as in Fig. 4.15(*a*) and (*b*).

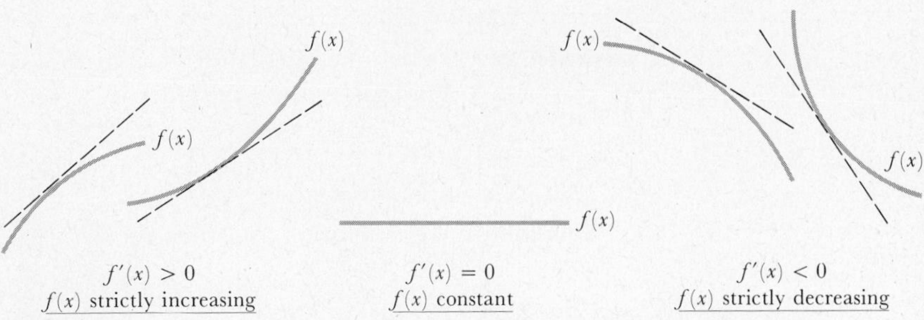

Fig. 4.14

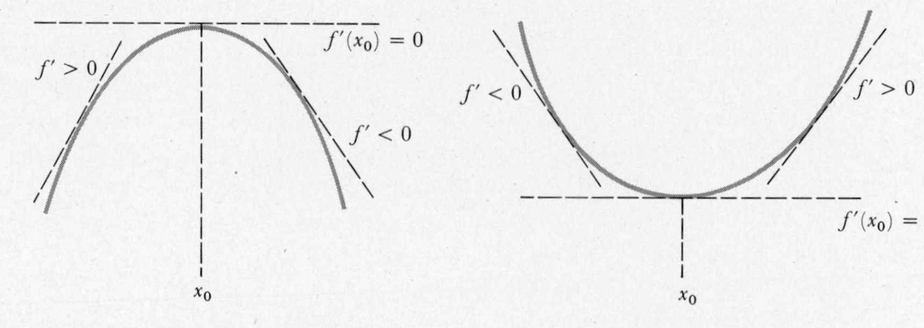

Relative maximum with $f'(x_0) = 0$

(*a*)

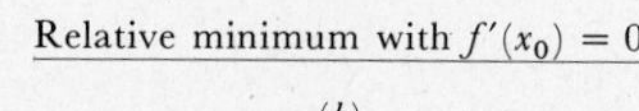

Relative minimum with $f'(x_0) = 0$

(*b*)

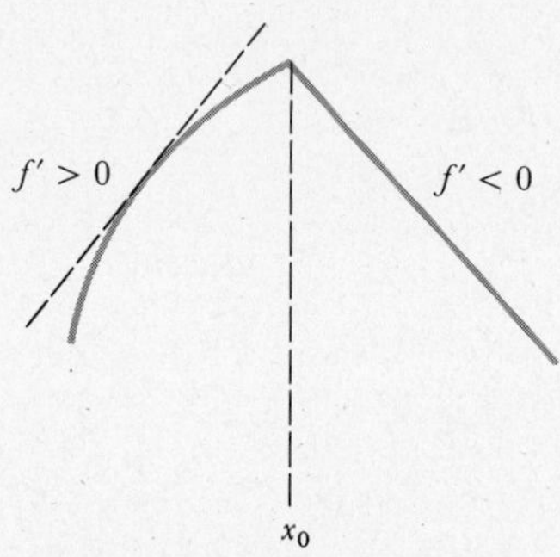

Relative maximum at a corner
$f'(x_0)$ does not exist

(*c*)

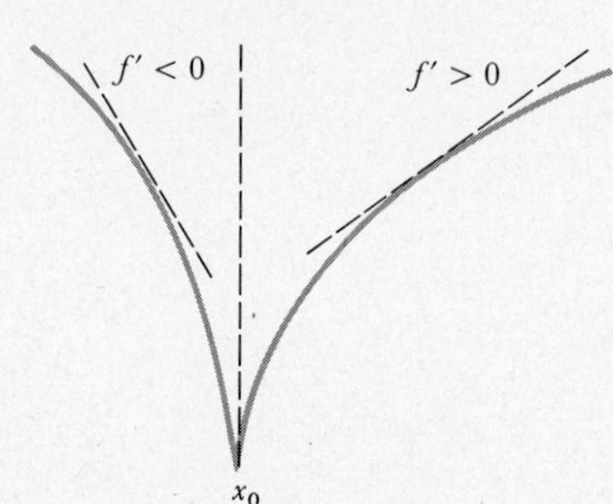

Relative minimum at a cusp
$f'(x_0)$ does not exist

(*d*)

Fig. 4.15

Example 1 Determine any relative extremes of $f(x) = 3x^2 - 6x + 1$.

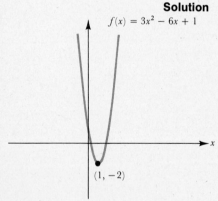

Solution Since $f(x)$ is everywhere differentiable, to apply the First-Derivative Test we determine and then factor the derivative:

$$f'(x) = 6x - 6 = 6(x - 1)$$

Now $f'(x) = 6(x - 1) < 0$ for $x < 1$ and $f'(x) = 6(x - 1) > 0$ for $x > 1$. Hence we have a relative minimum at $x = 1$ with $f(1) = -2$. Of course, $f'(1) = 0$ (horizontal tangent). See Fig. 4.16. ●

$f(x) = 3x^2 - 6x + 1$

$(1, -2)$

Fig. 4.16

For functions with more complicated derivatives it is helpful to determine the sign of the first derivative using the Chart Method developed in Sec. 1.1.

In the following example and henceforth we shall cease using the word "strictly" when dealing with increasing or decreasing functions, treating the constant case explicitly when it arises.

Example 2 Determine where the following function is increasing and where it is decreasing. Find all relative extremes.

$$f(x) = \frac{x^4}{4} - \frac{x^2}{2} + 9$$

Solution We determine and then factor $f'(x)$:

$$\begin{aligned} f'(x) &= x^3 - x \\ &= x(x^2 - 1) \\ &= x(x - 1)(x + 1). \end{aligned}$$

We then use the Chart Method to determine the sign of f'. Notice that to aid in determining extreme points, we marked the axis below the chart using positive slope lines where $f' > 0$ and using negative slope lines where $f' < 0$.

Sign of f':

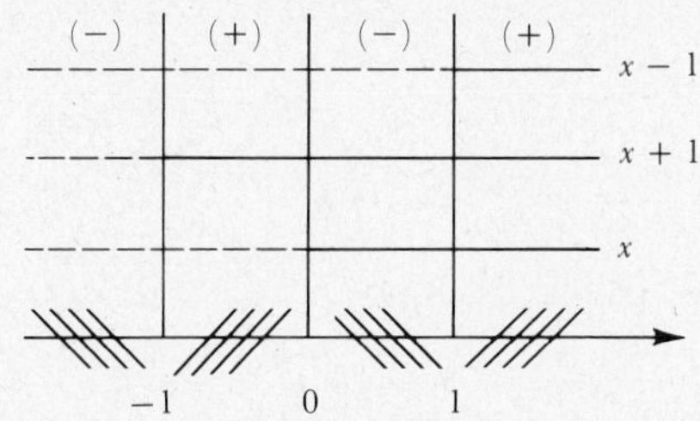

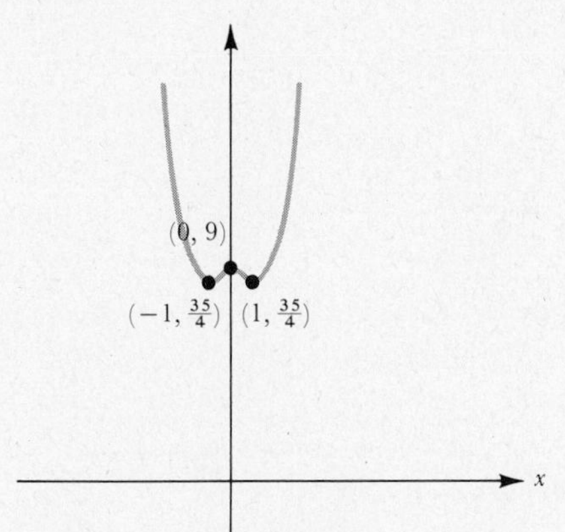

From the Increasing-Decreasing Theorem it follows that f is increasing on $(-1, 0)$ and $(1, +\infty)$ and decreasing on $(-\infty, -1)$ and $(0, 1)$. By the First-Derivative Test, f has relative minima at $x = -1$ and $x = 1$, and has a relative maximum at $x = 0$. We present the accompanying sketch at the left to show the function's behavior near $x = 0$. ●

PROGRESS TEST 2

Find where each of the following functions is increasing and decreasing. Determine all relative extremes. It is not necessary to sketch the graphs.

1. $f(x) = 3x + 1$
2. $f(x) = 3$
3. $f(x) = x^2 - 2x$
4. $f(x) = x^3 + 5x$
5. $f(x) = x^3/3 - x^2 - 3x + 3$ (Use a chart)

THE MEANING OF THE SECOND DERIVATIVE

The derivative of a function tells how that function is changing. Thus, given $f'(x)$, we know the derivative of $f'(x)$, namely $f''(x)$, tells how $f'(x)$ is changing. That is, $f''(x)$ tells how the slopes of lines tangent to $f(x)$ are changing. Contrast the graphs of the two functions in Fig. 4.17, each of which increases from (a, c) to (b, d). Of course, curve A is above the tangent lines and hence concave up, and curve B is concave down. As we move from left to right on A the slopes of the tangent lines are increasing. That is, the first derivative is increasing—which, by (4.2.7), occurs when *its* derivative is positive. Thus *the second derivative of the concave-up function given by curve A is positive.* Similarly, the slopes of tangents of curve B are decreasing, so the first derivative is decreasing. But by (4.2.7), this occurs when *its* derivative is negative. Thus *the second derivative of the concave-down function given by B is negative.* It can be shown that the converses of these observations are true.

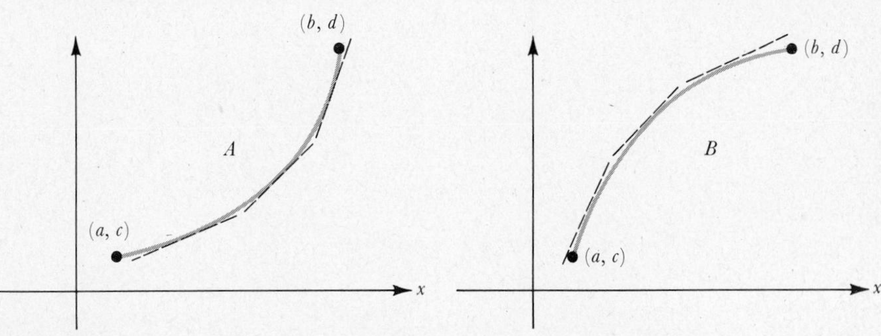

Fig. 4.17

4.2.9 **The Concavity Theorem** Suppose $f''(x)$ exists on an interval I.
(*a*) If $f''(x) > 0$ on I, then f is concave up on I.
(*b*) If $f''(x) < 0$ on I, then f is concave down on I.

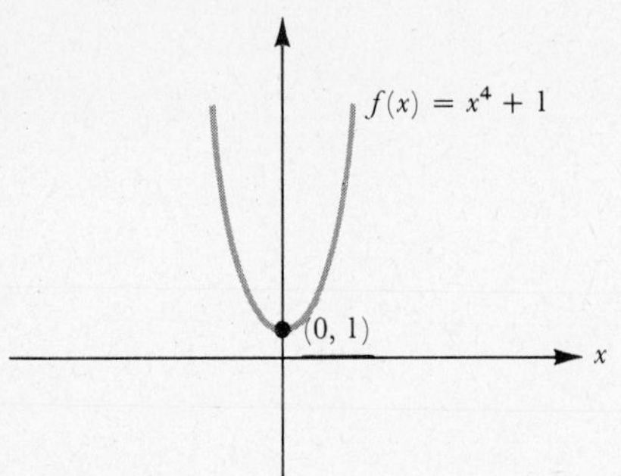

Fig. 4.18

Note: In the case that $f''(x)$ exists and has opposite signs on either side of x_0, then the Concavity Theorem implies that f has a point of inflection at x_0 [see (4.2.6)]. Furthermore, carrying the first-derivative note following (4.2.8) one step further to the second derivative, we can conclude that if $f''(x_0)$ *does* exist at a point of inflection, then $f''(x_0) = 0$. However, just as the first derivative being zero does not imply a relative extreme, the second derivative being zero at x_0 does not imply a point of inflection at x_0. This is illustrated by

$$f(x) = x^4 + 1 \qquad \text{at } x_0 = 0$$

Here $f'(x) = 4x^3$ and $f''(x) = 12x^2$, so $f''(0) = 0$. Yet $f''(x) > 0$ on both sides of 0, and hence there is no change in concavity and thus no inflection point at 0. The graph is concave up for all x. (See Fig. 4.18.)

4.2.10 **Second-Derivative Test for Relative Extremes** Suppose $f'(x_0) = 0$ and $f''(x)$ exists in an open interval containing x_0 (see Fig. 4.19).
(*a*) If $f''(x_0) < 0$, then f has a relative maximum at x_0.
(*b*) If $f''(x_0) > 0$, then f has a relative minimum at x_0.

When the First- and Second-Derivative Tests both apply, we have an excellent chance to check our work for errors.

Example 3 Use the Second-Derivative Test to confirm that

$$f(x) = \frac{x^4}{4} - \frac{x^2}{2} + 9$$

has relative minima at $x = \pm 1$, and show that it does *not* apply to confirm the relative maximum at $x = 0$. (Recall Example 2.)

Solution Since $f'(x) = x^3 - x^2$, we know that

$$\begin{aligned} f''(x) &= 3x^2 - 2x \\ &= x(3x - 2) \end{aligned}$$

Since $f''(0) = 0$, the test does not apply at $x = 0$. On the other hand,

$$f''(1) = (1)(3 - 2) = 1 > 0$$

and

$$f''(-1) = (-1)(3)(-1) - 2 = 5 > 0$$

so [4.2.10(*b*)] *does* apply to confirm the relative minima at $x = \pm 1$. ●

If both the first and second derivatives are zero at x_0, then the Second-Derivative Test does not apply and x_0 may be either a maximum, a minimum, or

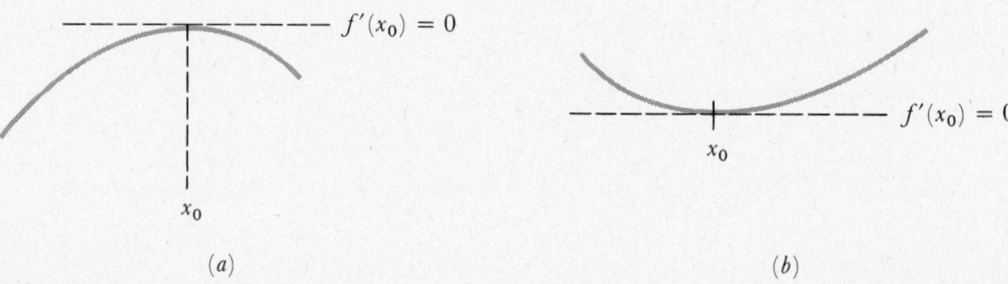

Fig. 4.19

neither. Furthermore, x_0 may or may not be a point of inflection; see Figures 4.8 and 4.18. Note that $f(x) = x^3 + 1$ has a point of inflection at $x = 0$ because $f''(x) = 6x$ changes sign at $x = 0$.

The Chart Method again proves useful in determining concavity and points of inflection for those functions with more complicated second derivatives.

Example 4 Describe the concavity of the following function and determine its points of inflection:

$$f(x) = \frac{x}{x^2 + 1}$$

Solution Since

$$f'(x) = \frac{(x^2 + 1)(1) - x(2x)}{(x^2 + 1)^2} = \frac{1 - x^2}{(x^2 + 1)^2}$$

we know that

$$f''(x) = \frac{(x^2 + 1)^2(-2x) - (1 - x^2)(2)(x^2 + 1)(2x)}{(x^2 + 1)^4}$$

$$= \frac{2x(x^2 + 1)}{(x^2 + 1)^4}[-x^2 - 1 - 2 + 2x^2]$$

$$= \frac{2x}{(x^2 + 1)^3}(x^2 - 3) = \frac{2x(x - \sqrt{3})(x + \sqrt{3})}{(x^2 + 1)^3}$$

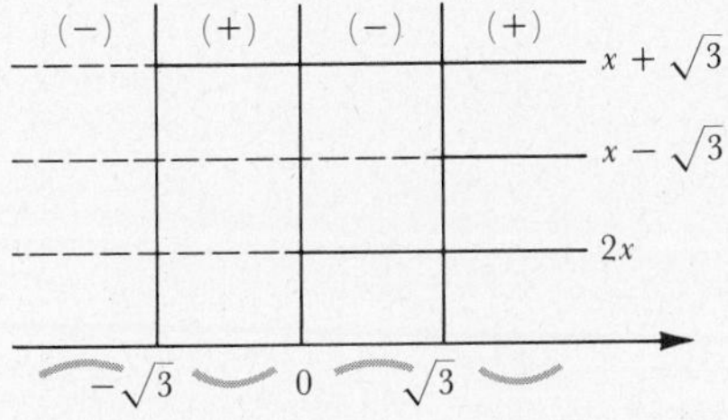

We now construct a chart, noting that $1/(x^2 + 1)^3$ is always positive and thus need not be included in the sign analysis.

From the Concavity Theorem we conclude that f is concave up on $(-\sqrt{3}, 0)$ and $(3, +\infty)$ and is concave down on $(-\infty, -\sqrt{3})$ and $(0, \sqrt{3})$. We note this symbolically below the chart. Furthermore, f has points of inflection at $x = \pm\sqrt{3}$ and $x = 0$. ●

A bit of silliness that may help you remember the significance of the second derivative: "A function is happy to have positive second derivative, so it *smiles* ‿. A function is sad to have negative second derivative, so it *frowns* ⁀."

PROGRESS TEST 3

Determine the concavity of the given functions and find the x coordinates of all points of inflection.

1. $f(x) = \dfrac{x^3}{3} - x^2 + 2$

2. $f(x) = x^4 - 18x^2 + 1$

SECTION 4.2 EXERCISES

1. Sketch the speed as a function of time for a train that moves as follows: At time $t = 0$ it begins moving down the tracks at an increasingly accelerating rate until $t = 3$, when, although it is still increasing speed, the rate of increase slows down, so at $t = 5$ it is traveling at a constant rate. It does so until $t = 6$, when it begins to slow down at an increasing rate, until $t = 7$, when its rate of decrease becomes more gradual until $t = 8$, when it reaches a very slow but constant rate of speed. Identify any points of inflection in the resulting graph.

2. Sketch four different graphs that satisfy the following set of conditions: The function starts at $(2, 0)$ and decreases until it gets to $(4, -2)$. It has a relative minimum at $(4, -2)$ and then increases until it reaches the point $(6, 1)$. It is constant thereafter.

3. Suppose $f(x)$ has the graph shown.

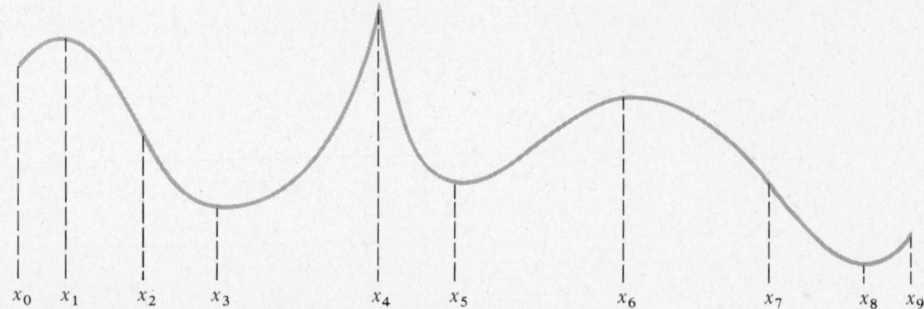

(*a*) On which intervals is f increasing? Decreasing?
(*b*) On which intervals is f concave up? Concave down?
(*c*) List all relative maxima and relative minima.
(*d*) At which relative extremes does f have a derivative?
(*e*) List all but one point of inflection.
(*f*) Where is the unlabeled point of inflection?

4. Explain why the sign chart shown could apply to the derivative of the function in Exercise 3.

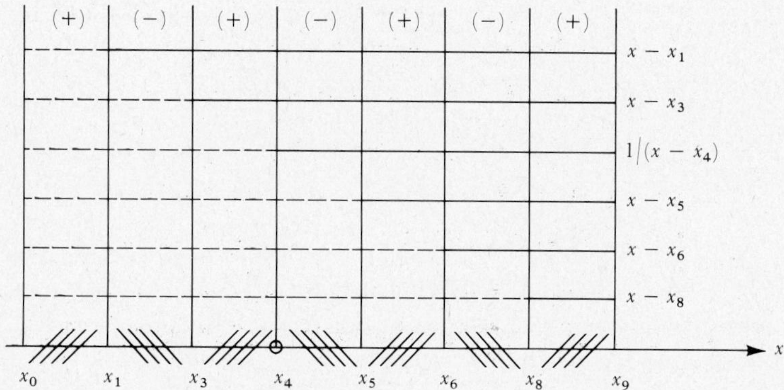

For Exercises 5 to 10, (a) determine where the given function is increasing, (b) determine where it is decreasing, and (c) find all relative extremes and the values of the function at these extremes.

5. $f(x) = 2x^3 + 3x^2 - 12x - 4$

6. $f(x) = 72x^2 - x^4$

7. $f(x) = \dfrac{2x}{\sqrt{4x^2 + 1}}$

8. $f(x) = x(6 - x)^2 + 1$

9. $g(x) = (x^2 - 16)^2$

10. $h(t) = t^3 - 6t^2 + 9t - 5$

For the functions in Exercises 11 to 15 determine the concavity and all points of inflection.

11. $f(x) = 3x^4 - 8x^3 - 12x^2 + 12x + 10$

12. $f(x) = x^4 - 12x^3 + 54x^2 - 108x + 81$

13. $f(x) = 4x + \dfrac{3}{2}(x - 2)^{2/3}$

14. $g(x) = x^{2/3} + 2$

15. $f(x) = \dfrac{1}{3}(x - 5)(x - 2)^2$

16. Where is the *derivative* of $f(x) = (3x^5 - 15x^4 + 20x^3 + 2)/8$ increasing and where is it decreasing?

17. Prove Theorem (4.2.3) for relative minima.

18. Give an example illustrating the necessity for continuity at x_0 in the First Derivative Test.

19. In terms of second derivatives explain why the graph of a linear function has zero concavity (a "straight face").

Section 4.3: Describing the Behavior of Functions Using Derivatives—Part 2

Objectives:

Use the Increasing-Decreasing and Concavity Theorems, First- and Second-Derivative Tests, and asymptotes, to sketch graphs of functions in a systematic way.

A STEP-BY-STEP PROCEDURE FOR GRAPH SKETCHING

We now pull together the information from Secs. 4.1 and 4.2. Of special interest are the points listed in the next definition.

4.3.1 **Definition** A number x_0 in the domain of f is called a *critical number* or *critical point of the function* f if any of the following is true: (*a*) $f'(x_0) = 0$, (*b*) $f'(x_0)$ does not exist, (*c*) $f''(x_0) = 0$, (*d*) $f''(x_0)$ does not exist.

In such circumstances we also refer to $(x_0, f(x_0))$ as a *critical* point.

Critical points are obvious candidates for relative extremes and points of inflection. The following procedure organizes the data we obtain using the machinery now in hand.

PROCEDURE FOR GRAPHING A FUNCTION $f(x)$

1. Note any information from the function itself that may be helpful in sketching the graph (domain, intercepts, symmetry, asymptotes).
2. Determine f' and write it in factored form. Note all critical points determined by f'.
3. Determine f'' and write it in factored form. Note all critical points determined by f''.
4. Using a sign chart for f' if necessary, determine where f is increasing and decreasing and determine relative extremes.
5. Using a sign chart for f'' if necessary, determine concavity and points of inflection, and check your choices of the relative extremes in step 4.
6. Sketch the graph.

We suggest that you use this procedure in your initial attempts at graphing. A few comments on steps 1, 5, and 6 are in order:

Step 1: In most cases the analysis involved in this step is quite simple, but nonetheless can be very helpful in consolidating your information at the end.

For example, suppose that

$$f(x) = \sqrt{x^2 - 1}$$

You would certainly note that f is only defined for values of x satisfying $|x| \geq 1$ and is symmetric with respect to the y axis $[f(x) = f(-x)]$.

Similarly,

$$f(x) = \frac{x-1}{x-2}$$

is defined only for $x \neq 2$ and has a vertical asymptote at $x = 2$. It also has the horizontal asymptote $y = 1$.

Given the function

$$f(x) = x^3 + x^2 - 2x$$

we can factor, getting

$$f(x) = x(x-1)(x+2)$$

and hence $f(0) = 0, f(1) = 0$, and $f(-2) = 0$. Thus the graph meets the x axis at $x = 0$, 1, and -2.

However, for another function, such as

$$f(x) = 3x^5 + x^3 - x + 9,$$

no significant information is provided by step 1.

Step 5: It is important to remember that to use the Second-Derivative Test for relative extremes, the first derivative must be zero at the point. If this *is* the case, then you can check your work in step 4.

Step 6: It is also possible to pick up errors here if your information is not consistent. For example, suppose you have determined that f has a relative maximum at x_0 and that the tangent there is horizontal (that is, $f'(x_0) = 0$). When using the Second-Derivative Test you observe that it is concave down to the left of x_0 and concave up to the right. The concavity results in a graph as in Fig. 4.20(a). On the other hand, $f'(x_0) = 0$ means the graph should be as in Fig. 4.20(b). At this point you would certainly go back and check your work!

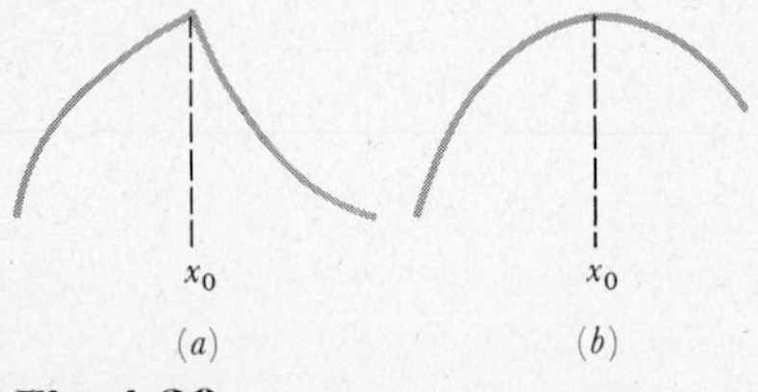

Fig. 4.20

THE GRAPHING PROCEDURE: EXAMPLES

Example 1 Sketch the graph of $f(x) = x^3 - 3x + 5$.

Solution (We shall number our steps in the early examples.)

1. This cubic expression is not easily factorable, so after noting that $f(0) = 5$, we proceed to step 2.
2. $f'(x) = 3x^2 - 3 = 3(x^2 - 1)$
   $= 3(x + 1)(x - 1)$ critical points $x = -1, x = 1$
3. $f''(x) = 6x$ critical point $x = 0$
4. Sign of f':

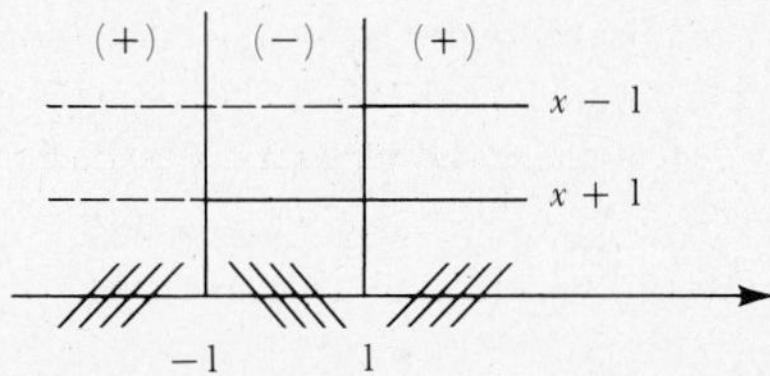
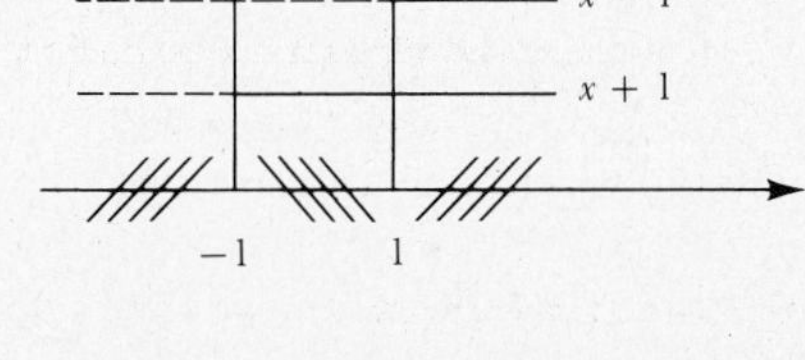

   By the Increasing-Decreasing Theorem, f is increasing on $(-\infty, -1)$ and $(1, +\infty)$, and by the First-Derivative Test, f has a relative maximum at $x = -1$, and relative minimum at $x = 1$. Also, the tangents at these extremes are horizontal since $f'(-1) = f'(1) = 0$ (no cusps or corners).
5. Without using a chart we see that $f''(x) = 6x$ is positive for $x > 0$, so f is concave up in $(0, +\infty)$, and in $(-\infty, 0) f''(x) = 6x$ is negative and hence f is concave down. Furthermore $x = 0$ is a point of inflection. Finally, we check our work in step 4:

$$f''(-1) < 0, \text{ so a relative maximum occurs at } x = -1$$
$$f''(1) > 0, \text{ so a relative minimum occurs at } x = 1$$

   This checks, so we proceed.
6. We first plot our critical points

$$f(-1) = 7, f(0) = 5, f(1) = 3$$

   and then piece together our information in Fig. 4.21. ●

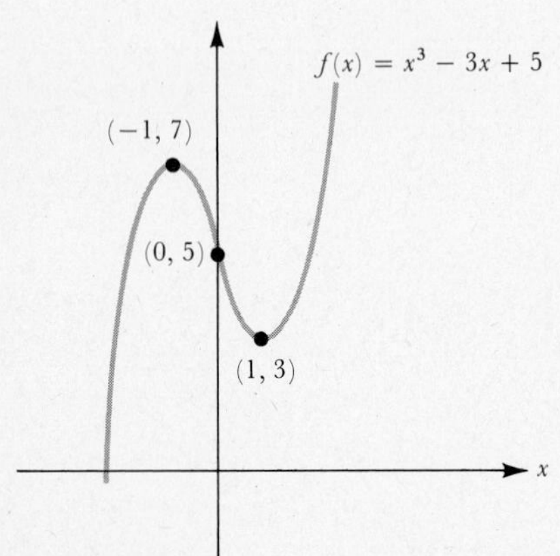

Fig. 4.21

Example 2 Sketch the graph of $f(x) = x^3 + x^2 + 6x - 5$.

Solution

1. Again, little seems to be apparent from the function itself except that $f(0) = -5$.

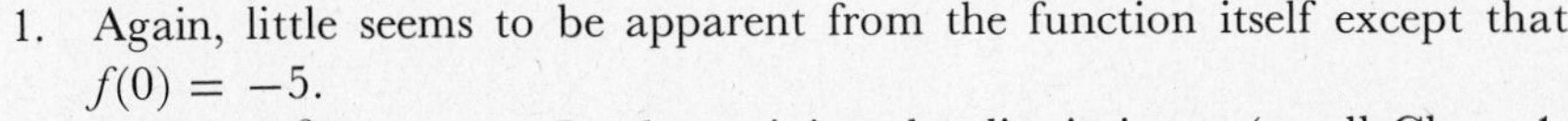

2. $f'(x) = 3x^2 + 2x + 6$. By determining the discriminant (recall Chap. 1, p. 6),
$$b^2 - 4ac = 4 - (4)(3)(6) = -68 < 0$$
we conclude that $f'(x)$ cannot be factored and thus has no real roots. Therefore it is always positive or always negative. By substituting a number into $f'(x)$, for example, $x = 0$, we see that $f'(x)$ is always positive.
3. $f''(x) = 6x + 2 = 6(x + 1/3)$ (critical point $x = -1/3$)
4. Without using a chart, the Increasing-Decreasing Theorem tells us that f is always increasing and has no relative extremes.
5. $f''(x) = 6(x + 1/3) > 0$ for $x > -1/3$
$f''(x) = 6(x + 1/3) < 0$ for $x < -1/3$
so f is concave down to the left of $x = -1/3$, concave up to the right, and has a point of inflection at $x = -1/3$.
6. $f(-1/3) = -1/27 + 1/9 - 6/3 - 5 = -187/27$
$f(0) = -5$ (see Fig. 4.22). ●

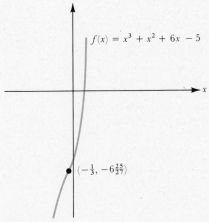

$f(x) = x^3 + x^2 + 6x - 5$

$(-\frac{1}{3}, -6\frac{25}{27})$

Fig. 4.22

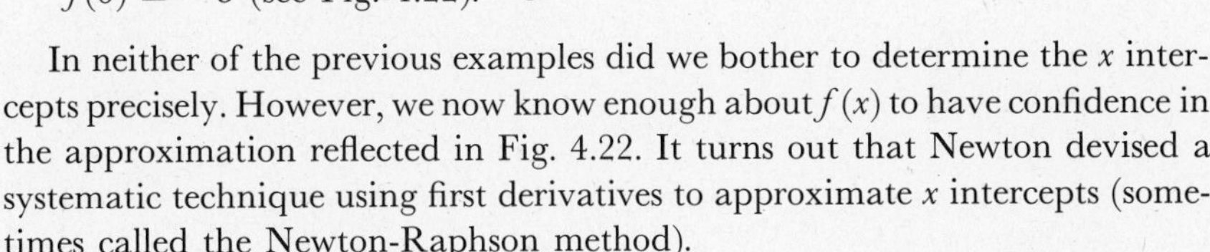

In neither of the previous examples did we bother to determine the x intercepts precisely. However, we now know enough about $f(x)$ to have confidence in the approximation reflected in Fig. 4.22. It turns out that Newton devised a systematic technique using first derivatives to approximate x intercepts (sometimes called the Newton-Raphson method).

Step 1 becomes especially important when asymptotes are involved.

Example 3 Sketch the graph of $f(x) = \dfrac{x^2 - 3x + 6}{x + 2}$.

Solution

1. Here we note a vertical asymptote $x = -2$ with
$$\lim_{x\to-2^-} \frac{x^2 - 3x + 6}{x + 2} = -\infty \quad \text{and} \quad \lim_{x\to-2^+} \frac{x^2 - 3x + 6}{x + 2} = +\infty$$
Also, by long division, $f(x) = \dfrac{x^2 - 3x + 6}{x + 2} = x - 5 + \dfrac{16}{x + 2}$, so
$$\lim_{x\to\pm\infty} [f(x) - (x - 5)] = \lim_{x\to\pm\infty} \frac{16}{x + 2} = 0$$
Hence $y = x - 5$ is an oblique asymptote. Finally, we note that the graph has y intercept $(0, 3)$ and no x intercept. (Discriminant of the quadratic numerator is negative.)
2. $$f'(x) = \frac{(x + 2)(2x - 3) - (x^2 - 3x + 6)(1)}{(x + 2)^2}$$
$$= \frac{x^2 + 4x - 12}{(x + 2)^2} = \frac{(x + 6)(x - 2)}{(x + 2)^2}$$
Critical points are at $x = 2$ and $x = -6$.
3. $$f''(x) = \frac{(x + 2)^2(2x + 4) - (x^2 + 4x - 12)2(x + 2)(1)}{(x + 2)^4} = \frac{32}{(x + 2)^3}$$
There are no new critical points, but f'' changes sign at $x = -2$.

4. Sign of f':

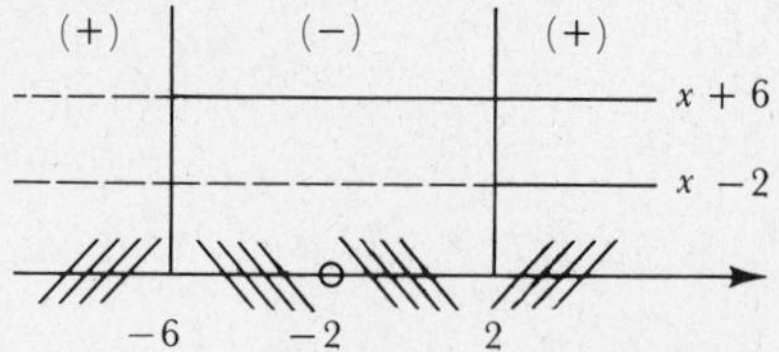

By the First-Derivative Test, f has a relative maximum at $x = -6$ and a relative minimum at $x = 2$. At both these relative extremes f has a horizontal tangent. Note also that f is decreasing on either side of the vertical asymptote $x = -2$.

5. Sign of f'':

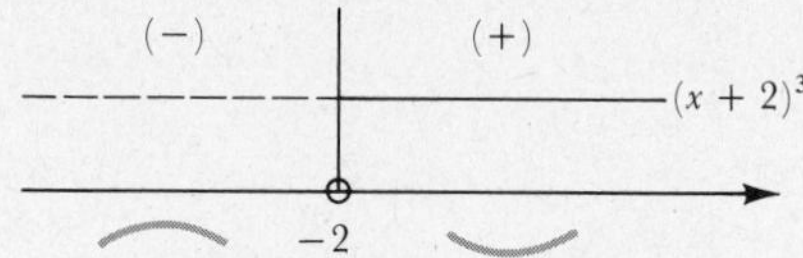

f is concave down to the left of the vertical asymptote $x = -2$ and concave up to the right.

6. $f(2) = 1, f(-6) = -15$; see Fig. 4.23. ●

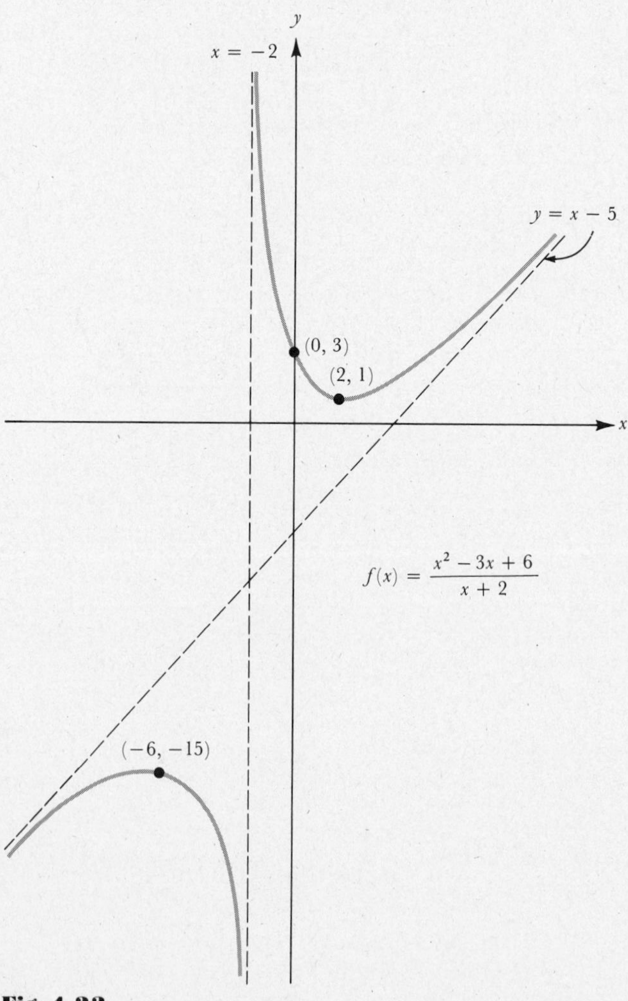

Fig. 4.23

PROGRESS TEST 1

In Probs. 1 and 2 sketch the graph of the given function by following the step-by-step procedure. Indicate clearly all relative extremes and points of inflection.

1. $f(x) = x^4 + \frac{4}{3}x^3 - 4x^2 - \frac{4}{3}$

2. $f(x) = \frac{x}{3x - 1}$

3. Use the first derivative to confirm the sketch of

$$h(x) = \frac{3x + 2}{\sqrt{4x^2 - 1}}$$

in Fig. 4.4.

SOME FURTHER EXAMPLES: FRACTIONAL EXPONENTS

The following pair of examples illustrates two points:

1. Relative extremes can occur where $f'(x)$ does not exist.
2. When dealing with fractional exponents, try to factor, leaving one factor with an integer exponent.

Example 4 Graph $f(x) = 6x^{1/3} + 3x^{4/3}$

Solution

1. $f(x) = 3x^{1/3}(2 + x)$, so $f(0) = f(-2) = 0$
2. $f'(x) = 2x^{-2/3} + 4x^{1/3}$

$$= 2x^{-2/3}(1 + 2x) = \frac{2}{x^{2/3}}(1 + 2x)$$

Critical numbers are $x = -1/2$, $x = 0$ [$f'(0)$ is not defined].

3. $f''(x) = -\frac{4}{3}x^{-5/3} + \frac{4}{3}x^{-2/3}$

$$= \frac{4}{3}x^{-5/3}(-1 + x) = \frac{4}{3x^{5/3}}(-1 + x)$$

Critical numbers are $x = 1$, $x = 0$ [$f''(0)$ is not defined].

4. Sign of f':

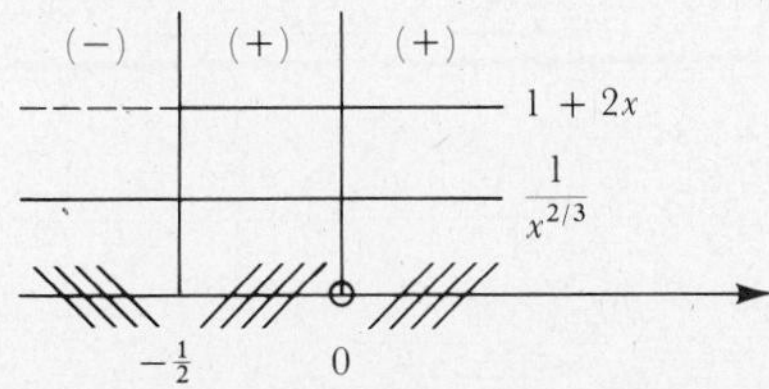

(We circle any points on the number line for which any of the factors is undefined.) Note that the sign of f' is positive on both sides of $x = 0$ since $1/x^{2/3}$ is always positive. Hence f is increasing on both sides of $x = 0$, so the First-Derivative Test does not imply an extreme at $x = 0$ but *does* imply a relative minimum at $x = -1/2$ with a horizontal tangent since $f'(-1/2) = 0$. Because $\lim_{x\to 0} 1/x^{2/3} = +\infty$, we can conclude that the tangent at $x = 0$ is vertical.

5. Sign of f'':

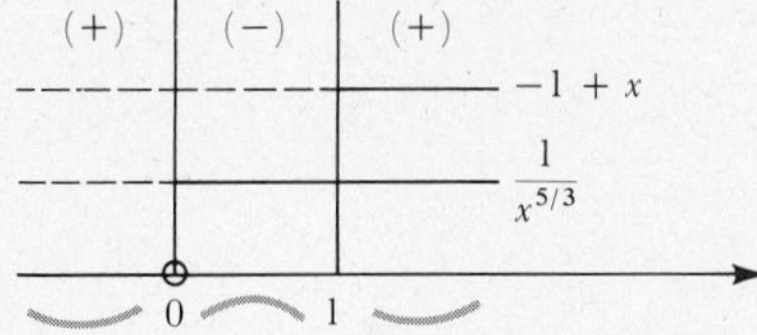

f is concave up for $x < 0$ and $x > 1$ and is concave down in $(0, 1)$. Thus by the sign of f'', we have points of inflection (PI) at $x = 0$ and $x = 1$.

6. See Fig. 4.24. ●

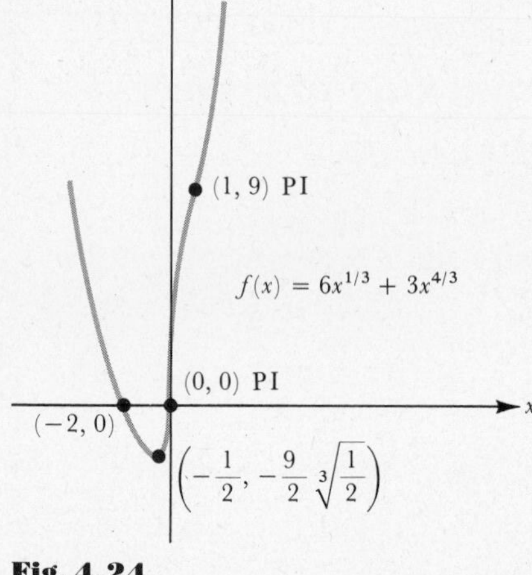

Fig. 4.24

Example 5 Sketch $f(x) = x^{5/3} + 5x^{2/3}$

Solution (We shall not bother to number our steps.)

$$f(x) = x^{5/3} + 5x^{2/3} = x^{2/3}(x + 5) \qquad f(0) = 0, f(-5) = 0$$

$$f'(x) = \frac{5}{3}x^{2/3} + \frac{10}{3}x^{-1/3}$$

$$= \frac{5(x + 2)}{3x^{1/3}} \qquad \text{critical numbers } x = -2,\ x = 0$$

$$f''(x) = \frac{10}{9}x^{-1/3} - \frac{10}{9}x^{-4/3}$$

$$= \frac{10(x - 1)}{9x^{4/3}} \qquad \text{critical numbers } x = 1,\ x = 0$$

Sign of f':

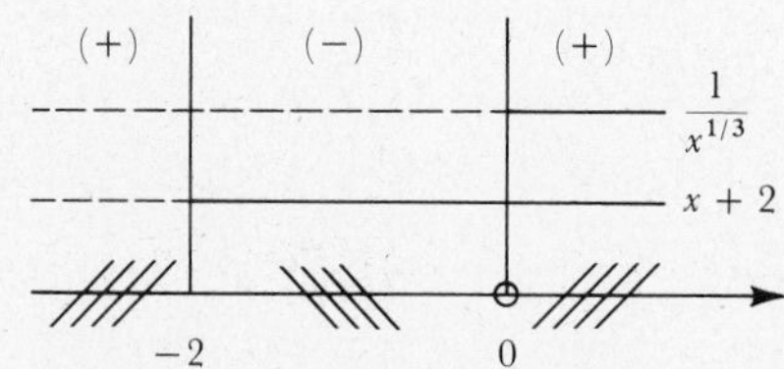

A relative maximum is at $x = -2$ with horizontal tangent since $f'(-2) = 0$. A relative minimum is at $x = 0$ with a vertical tangent (hence cusp) since $\lim_{x\to 0^-} f'(x) = -\infty$ and $\lim_{x\to 0^+} f'(x) = +\infty$.

Sign of f'':

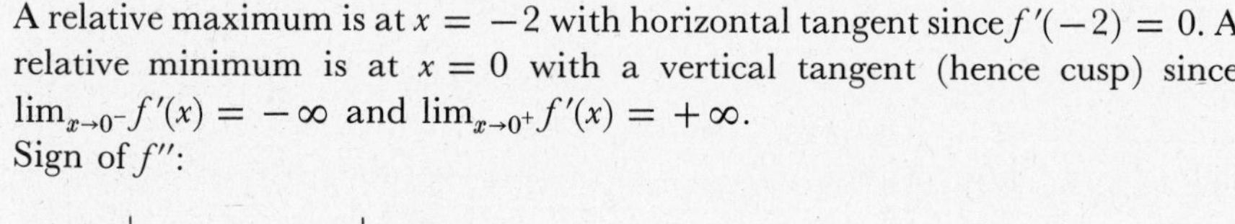

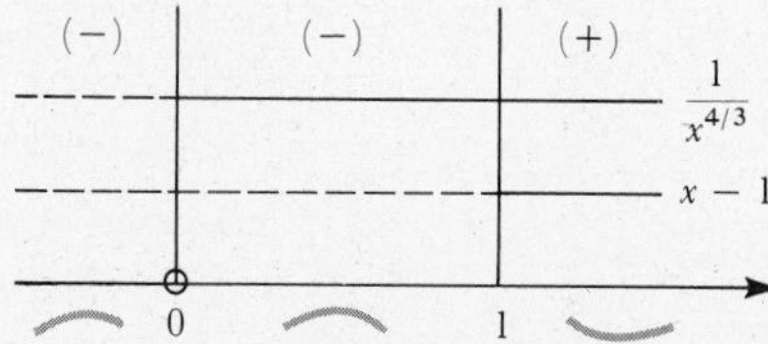

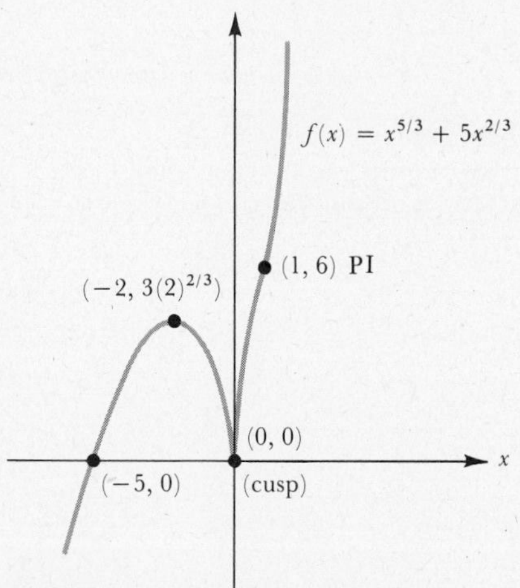

Fig. 4.25

$f''(-2) = (10)(-3)/(9)(\sqrt[3]{-2})^4 < 0$, so there is a relative maximum at $x = -2$ by the Second-Derivative Test. Also, negative concavity on either side of $x = 0$ agrees with the existence of a cusp there. There is a point of inflection at $x = 1$. (See Fig. 4.25.) ●

PROGRESS TEST 2

1. Graph $f(x) = 3x^{5/3} + 15x^{2/3}$, giving the coordinates of all relative extremes and points of inflection.

SECTION 4.3 EXERCISES

Sketch the graphs of the functions in Exercises 1 to 40, giving all pertinent information.

1. $f(x) = x^5 + 1$
2. $f(x) = x^4 - 1$
3. $g(x) = 3 - x - x^2$
4. $h(x) = x^2 + 2x - 1$
5. $f(x) = x^3 + 3x^2 - 9x + 10$
6. $f(x) = x^3 - 6x^2 + 9x - 1$
7. $f(x) = x^3 + 6x^2 + 12x - 5$
8. $f(x) = x^4 - 2x^2 + 1$

9. $f(x) = \dfrac{2x}{x^2 + 1}$ **10.** $f(x) = \dfrac{x + 1}{x^2 + x + 1}$ **11.** $f(x) = \dfrac{1}{\sqrt{x}} + \dfrac{\sqrt{x}}{9}$ **12.** $f(x) = \dfrac{x - 2}{x^2 - 6x + 5}$

13. $f(x) = x^3 + \dfrac{3}{x}$ **14.** $f(x) = \dfrac{2}{\sqrt{x + 1}}$ **15.** $g(x) = \dfrac{2x}{4x - 3}$ **16.** $h(x) = \dfrac{4x}{x^2 + 4}$

17. $y = x^{1/3}$ **18.** $y = x^{2/3}$ **19.** $f(x) = 2x^{6/5} - 12x^{1/5}$ **20.** $g(x) = \sqrt{x^3}$

21. $f(x) = \dfrac{2x^2}{1 - x}$ **22.** $g(x) = \dfrac{x}{\sqrt{2x + 1}}$ **23.** $f(x) = \dfrac{x^3}{x^2 + 1}$ **24.** $f(x) = x\sqrt{4 - x^2}$

25. $h(x) = x^{4/3} + 4x^{1/3}$ **26.** $f(x) = x(x - 1)^5$ **27.** $y = 5x^{2/3} - 3x^{5/3}$ **28.** $f(x) = \dfrac{x}{\sqrt{4 - x^2}}$

29. $f(x) = \sqrt[3]{1 + x^2}$ **30.** $g(x) = \sqrt{1 + x^3}$ **31.** $y = (2 - x)^{1/3}$ **32.** $y = (x + 1)^{1/5}$

33. $f(t) = \sqrt{1 - t^2}$ **34.** $g(t) = t\sqrt{2 - t}$ **35.** $h(t) = 2t^{1/3} + 6t^{-1/3}$ **36.** $f(x) = 12x^{1/3} - 3x^{4/3}$

37. $f(t) = \dfrac{2t^2 - 4}{\sqrt{t^2 - 1}}$ **38.** $f(x) = \dfrac{12 - x^2}{\sqrt{x^2 + 4}}$

39. $g(x) = \dfrac{(x - 1)^2}{x^3}$ **40.** $y = 3(x - 3)^4 - 4(x - 3)^3 - 12(x - 3)^2 + 3$

41. Use first and second derivatives to confirm the sketch of $f(x) = 1/(x^2 + 1) + 3$ in Fig. 4.3.

42. Repeat Exercise 41 for the partial sketch of $f(x) = (x - 1)/x^2$ in Sec. 4.1, Example 2(*a*).

43. Repeat Exercise 41 for the partial sketch of $g(x) = \sqrt{16 - x^2}/(x - 4)$ in Sec. 4.1, Example 2(*b*).

44. Repeat Exercise 41 for the partial sketch of

$$f(x) = \frac{15x^3 + 5x^2 + 5x + 1}{5x^2 + 1} = 3x + 1 + \frac{2x}{5x^2 + 1}$$

in Sec. 4.1, Example 6.

45. Sketch $f(x) = \begin{cases} 4 - x^2 & x \le 2 \\ 3x & x > 2 \end{cases}$

Section 4.4: Extreme Values and the Mean-Value Theorem

Objectives:

1. Determine if and how Rolle's Theorem and the Mean-Value Theorem apply to a given function on a given interval.

2. Begin to appreciate the theoretical depth of the calculus studied to date.

THE EXTREME VALUES OF A FUNCTION ON AN INTERVAL

More often than not, when applying derivatives to solve concrete optimization problems such as those in the next chapter, we are seeking the extreme values of a function on a finite interval. Because the theoretical issues underlying this practical problem have wide-reaching implications throughout the calculus, we take time here to examine them in some detail.

4.4.1 Definition We say that the function f attains a *maximum on a given interval at* c if c is in that interval and $f(c) \ge f(x)$ for all x in that interval, and we say that f attains a *minimum on that interval at* c if c is in that interval and $f(c) \le f(x)$ for all x in that interval. At such points c we say that f attains its *extreme values.*

Thus f attains a maximum on an interval I at c if c is in I and no value of $f(x)$ on I is larger than $f(c)$. A function may attain a maximum or a minimum on I at more than one point of I. Furthermore, such points may or may not yield *relative* extremes and may or may not be endpoints of I. See Fig. 4.26 for some possibilities. In Fig. 4.26(*a*) the function attains maxima at c_1 and c_2 and a minimum at c_3 (all are relative extremes). In Fig. 4.26(*b*) the function attains a maximum at c_1 (a relative maximum) and minima at c_2 (relative) and endpoint b. In Fig. 4.26(*c*) the function attains maxima at infinitely many points comprising the interval $[c_1, c_2]$ and minima at c_3 and endpoint b. In Fig. 4.26(*d*) no extreme values are attained on $[a, b]$.

From Fig. 4.26 (*d*) it is apparent that a function may fail to attain extreme values on an interval, so we need some idea of the conditions guaranteeing that a function *does* attain extreme values on an interval. We take as our starting

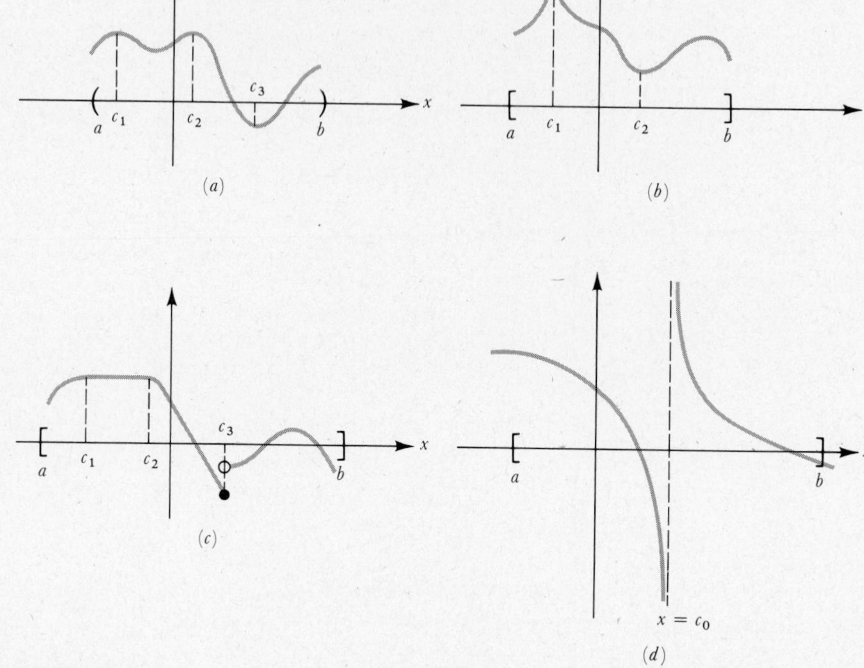

Fig. 4.26

point the Extreme-Value Theorem, which gives us the conditions we need. As with its companion, the Intermediate-Value Theorem for continuous functions, [(2.5.12)], we shall note that although it *can* be proved from more basic principles about the real numbers, the proof is beyond the scope of a first calculus course. (To investigate this further, consult your instructor or any text in advanced calculus.)

4.4.2 **The Extreme Value Theorem** Suppose $[a, b]$ is a closed interval and f is continuous on $[a, b]$. Then f attains both a minimum and a maximum value on $[a, b]$.

Perhaps the most noteworthy thing about this reasonable theorem is that its conclusion may fail if either the interval is not closed or the function is discontinuous on the interval.

For example, if we define

$$f(x) = \begin{cases} \dfrac{1}{(x-1)^2} - \dfrac{1}{2} & \text{for } x \neq 1 \text{ in } [0, 2] \\ 10^{10} & \text{for } x = 1 \end{cases}$$

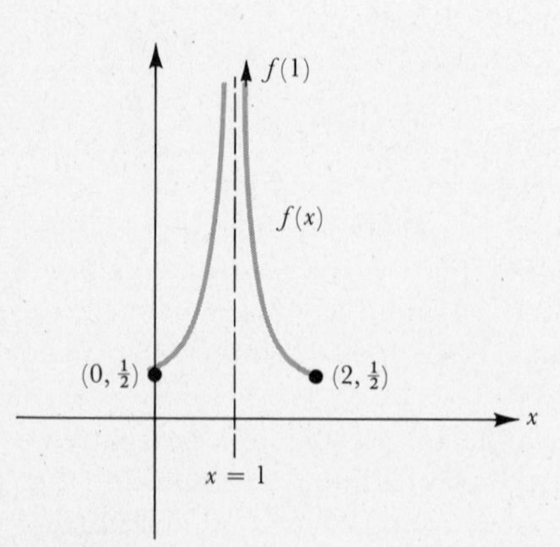

Fig. 4.27

then f is defined on $[0, 2]$ but discontinuous at $x = 1$. (See Fig. 4.27.) Although f attains the minimum value of $1/2$ at each endpoint, $f(x)$ grows without bound as x approaches 1 from either side $[\lim_{x\to 1} f(x) = +\infty]$. No matter how large we attempt to redefine $f(1)$ to be, it is possible to find an x near enough to $x = 1$ so that $f(x) > f(1)$. [For example, $f(1 + 10^{-10}) \approx 10^{20} > 10^{10}$.] In particular, the function f defined above does not attain a maximum on $[0, 2]$. If you still feel that $f(1)$ can be defined so that f attains its maximum on $[0, 2]$ at $x = 1$, name $f(1)$; but remember, $f(1)$ must be a real number!

On the other hand, the function

$$g(x) = 2x + 1$$

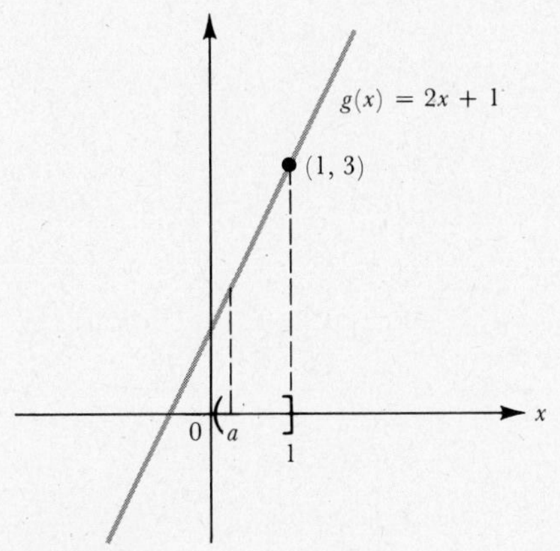

Fig. 4.28

is continuous on the nonclosed interval (0, 1] (see Fig. 4.28) and attains a maximum at $x = 1$; namely, $g(1) = 3$. But $g(x)$ fails to attain a minimum on (0, 1]. Although $g(0) = 1$ and $1 \leq g(x)$ for all x in (0, 1], 0 is not in the interval (0, 1], so it is not true that g attains its minimum on (0, 1] at $x = 0$. Nor could g attain a minimum on (0, 1] at *any* number $a > 0$ in (0, 1] because, for example, $g(a/2) < g(a)$. Thus there is no number in (0, 1] at which g attains a minimum value on (0, 1]. Again, if you think g attains its minimum at some number in (0, 1], name that number!

Of course, it *can* turn out that the conclusion of the theorem may hold even though its hypotheses do not; see Probs. 1(*c*) and (*d*) in Progress Test 1.

PROGRESS TEST 1

1. Sketch examples (an algebraic definition is not necessary) to illustrate each of the following:
(*a*) A continuous function on [1, 4] whose minimum value occurs at two different points.
(*b*) A continuous function on [1, 4] whose maximum and minimum values are distinct and such that each occurs at infinitely many points.
(*c*) A continuous function on (1, 4) that *has* a maximum and minimum value.
(*d*) A discontinuous function on [1, 4] that *has* a maximum and minimum value.

2. Define algebraically a function $f(x)$ that satisfies the given condition.
(*a*) $f(x)$ is continuous on (0, 4) and has a maximum value but no minimum value on (0, 4).
(*b*) $f(x)$ has a minimum value but no maximum value on [0, 4].

THE MEAN-VALUE THEOREM

As we saw in Fig. 4.26(*a*) to (*c*), a function may attain on an interval I a maximum or minimum that is not a relative maximum or relative minimum. If the function is continuous on I (no breaks or jumps) and f attains, say, a maximum value on I at a point c *strictly between* the endpoints of I (whether or not these endpoints are part of I), then there is an open interval containing c and wholly contained in I such that $f(c) \geq f(x)$ for all x in this open subinterval. Thus f has a *relative* maximum at c. Similarly, if f attains a minimum value at d on I where d is *strictly between* the endpoints of I, then f has a *relative* minimum at d. Furthermore, by Theorem (4.2.3) we know that *if f has a derivative at such a relative extreme, then its derivative is zero there.*

Our next step is a natural one: to couple the Extreme-Value Theorem, which guarantees that a function f continuous on $[a, b]$ attains its extreme values there, with Theorem (4.2.3) about zero derivatives at relative extremes to conclude the existence of a point c in the open interval (a, b), where $f'(c) = 0$.

Putting this all together, we have proved Rolle's Theorem, named after the French mathematician Michel Rolle, who published it in 1691.

4.4.3 Rolle's Theorem Suppose that f is continuous on $[a, b]$ and differentiable on (a, b). If $f(a) = f(b)$, then there exists a point c in (a, b) with $f'(c) = 0$.

1. To guarantee that an extreme occurs at c *between* a and b, we assumed that $f(a) = f(b)$.
2. To guarantee that $f'(c)$ exists, we assumed that f is differentiable on the open interval (a, b).

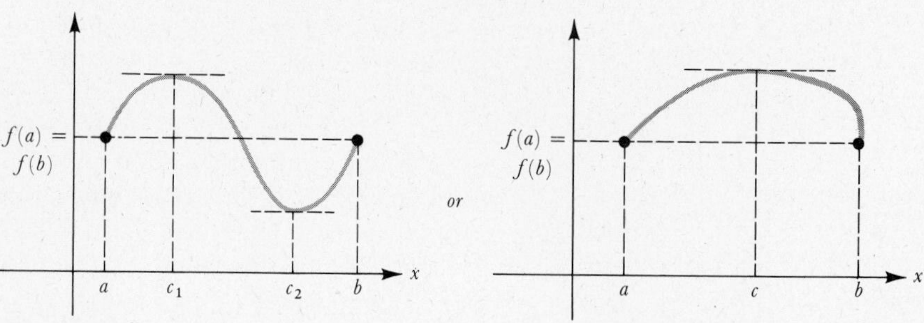

Fig. 4.29

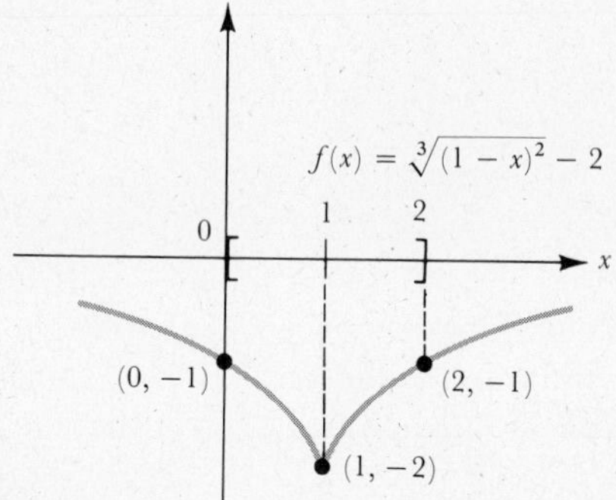

Fig. 4.30

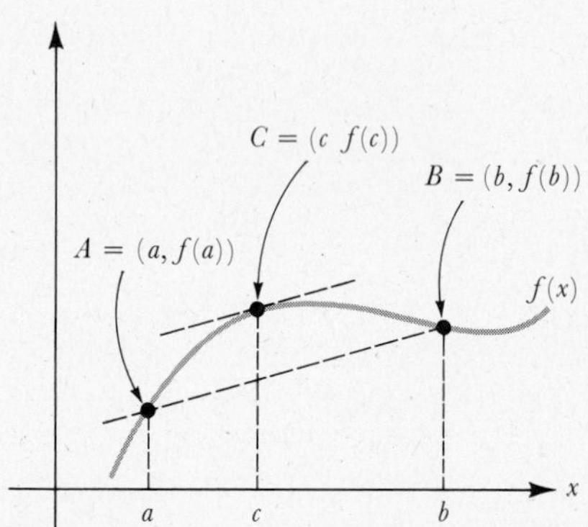

Fig. 4.31

Rephrased geometrically, the conclusion of Rolle's Theorem says there is at least one point on the graph of f between a and b where the tangent line is horizontal, that is, parallel to the line segment joining $(a, f(a))$ and $(b, f(b))$.

If we were to slide the horizontal line $y = f(a) = f(b)$ vertically (either up or down) through the graph of the function, its last contact with the graph would have to be as a tangent line—provided, of course, that the graph is "smooth" enough to *have* a tangent at that point, which *is* the case since f is differentiable on (a, b); see Fig. 4.29. Note that f may have a vertical tangent at an endpoint, as in the second part of Fig. 4.29 at $x = b$.

Rolle's Theorem is also valid in the one case not represented above, where f is constant on $[a, b]$. In that case *every* point c of (a, b) satisfies $f'(c) = 0$.

Figure 4.30 is a sketch of the function

$$f(x) = \sqrt[3]{(1 - x)^2} - 2$$

for which no horizontal tangent line exists on $[0, 2]$ although f reaches a relative minimum of -2 at $x = 1$. Of course, Rolle's Theorem does not apply, because f has a cusp and is not differentiable at $x = 1$. [*Note:* $f'(x) = -2/(3\sqrt[3]{1 - x})$.]

Our main result, the Mean-Value Theorem, is a generalization of Rolle's Theorem to the case where $f(a)$ is not necessarily equal to $f(b)$. Geometrically, its conclusion states that there is at least one point $[c, f(c)]$ on the graph of f between a and b where the tangent line is parallel to the line segment joining $(a, f(a))$ and $(b, f(b))$; see Fig. 4.31.

As with Rolle's Theorem, this conclusion is geometrically clear because if we were to slide the line through A and B vertically (either up or down) without changing its slope $[f(b) - f(a)]/(b - a)$, then its last contact with the graph would have to be as a tangent line parallel to the original.

4.4.4 The Mean-Value Theorem If f is continuous on $[a, b]$ and differentiable on (a, b), then there exists at least one point c in (a, b) such that

$$f'(c) = \frac{f(b) - f(a)}{b - a}$$

Before formally deriving the Mean-Value Theorem (MVT) from Rolle's Theorem, we note that if $f(x)$ (>0) gives distance traveled as a function of time x over the interval from time a to time b, then $b - a$ represents the total time elapsed, and $f(b) - f(a)$ represents the distance traveled during that time interval. Hence

$$\frac{f(b) - f(a)}{b - a}$$

represents the average speed over the time interval. But then the assertion of the MVT is that there is at least one point c in the time interval where the instantaneous speed $f'(c)$ equals the average speed.

As a result, you could be issued a speeding citation for exceeding 55 mph without witnesses and without ever having your speed checked if you were known to have driven more than 55 mi in 1 h. Having had an *average* speed greater than 55 mph, the MVT proves that at least at one point you had an instantaneous speed greater than 55 mph. (Of course you might demand in court that the prosecution prove the MVT from first principles and prove the Extreme-Value Theorem along the way. See Miscellaneous Exercise 3 for an even more subtle case.)

Proof of Mean-Value Theorem: The idea is to subtract an appropriate quantity from the function f so that the resulting function g satisfies the hypotheses of Rolle's Theorem—in particular, $g(a) = g(b)$. We subtract the linear function y whose graph is the line through A and B. The point-slope form of the equation of this line is

$$y - f(a) = \frac{f(b) - f(a)}{b - a}(x - a)$$

so, as a function of x,

$$y = \frac{f(b) - f(a)}{b - a}(x - a) + f(a)$$

Thus the function $g(x) = f(x) - y$ is given by

$$g(x) = f(x) - \frac{f(b) - f(a)}{b - a}(x - a) - f(a)$$

Sure enough,

$$g(a) = f(a) - \frac{f(b) - f(a)}{b - a}(a - a) - f(a) = 0$$

and

$$g(b) = f(b) - \frac{f(b) - f(a)}{b - a}(b - a) - f(a) = 0$$

Furthermore, g is continuous on $[a, b]$ and differentiable on (a, b), with derivative

$$g'(x) = f'(x) - \frac{f(b) - f(a)}{b - a}$$

Hence, by Rolle's Theorem, there exists a c in (a, b) with

$$g'(c) = f'(c) - \frac{f(b) - f(a)}{b - a} = 0$$

But then we have a c in (a, b) such that

$$f'(c) = \frac{f(b) - f(a)}{b - a}$$

which is what we needed to prove. ■

We shall now examine two particular cases to see if and how the MVT applies.

Example 1 Show if and how the MVT applies to $f(x) = x^2 - 4x$ on $[2, 6]$.

Solution Since f is everywhere differentiable, the hypotheses hold. Hence there exists at least one c in $[2, 6]$ with

$$f'(c) = \frac{f(6) - f(2)}{6 - 2} = \frac{16}{4} = 4$$

Now $f'(x) = 2x - 4$, so we know there is a c in $[2, 6]$ such that

$$2c - 4 = 4$$

Of course $c = 4$ works (see Fig. 4.32). ●

Example 2 Show if and how the MVT applies to $f(x) = \sqrt[3]{(1 - x)^2} - 2$ on $[0, 1]$. (See Fig. 4.30 for the graph of f.)

Solution Since $f'(x) = -2/(3\sqrt[3]{1 - x})$, we see that f is differentiable at all points of $[0, 1]$ *except* $x = 1$. However, f is continuous at $x = 1$, so the hypotheses of the MVT are satisfied. (The function need not be differentiable at the endpoints.) Now

$$\frac{f(1) - f(0)}{1 - 0} = \frac{-2 - (-1)}{1} = -1$$

so by the MVT there is a c in $(0, 1)$ such that

$$\frac{-2}{3\sqrt[3]{1 - c}} = -1$$

Solving for c, we find $c = 19/27$. ●

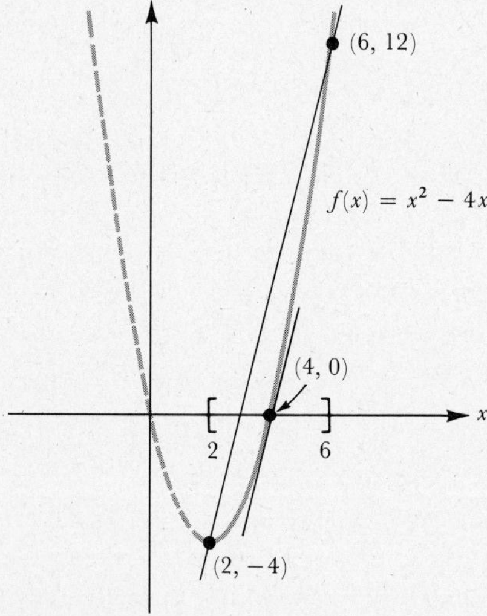

Fig. 4.32

PROGRESS TEST 2

1. Determine whether the Mean Value Theorem applies to the given function on the indicated interval. If so, find all possible values for c. Sketch the situation.

(*a*) $f(x) = \dfrac{x^3}{3} - x - 1$ $\quad [-1, 2]$

(*b*) $f(x) = x^{2/3}$ $\quad [-1, 8]$

(*c*) $f(x) = x^2 + 2x$ $\quad [-2, 1]$

2. Graph the function $f(x) = |2x - 1| - 3$ on $[-1, 2]$. Note that $f(-1) = f(2) = 0$ but there exists no number c where $f'(c) = 0$. Does this contradict Rolle's Theorem?

THEORETICAL APPLICATIONS OF THE MEAN-VALUE THEOREM

Although the MVT has practical applications (for example, in determining bounds on complicated functions over finite intervals), its most important use is as a theoretical tool. The following two theorems are widely used in the remainder of the text.

We already know that the derivative of a constant function is zero. The converse is also true.

4.4.5 Theorem If the derivative of function is zero on an interval (a, b), then the function is constant on (a, b).

4.4.6 Theorem If two functions have the same derivative for all x in an interval (a, b), then the two functions differ by a constant on (a, b).

That is, if $f'(x) = g'(x)$ for all x in (a, b), then there is a constant C such that

$$f(x) = g(x) + C \text{ on } (a, b)$$

Proof of (4.4.5): Suppose that $f'(x) = 0$ for all x in (a, b). It suffices to show that if x_1 and x_2 are any two distinct points in (a, b) with, say, $x_1 < x_2$, then $f(x_1) = f(x_2)$. But f is differentiable on $[x_1, x_2]$, so the MVT applies to guarantee there is a c in (x_1, x_2) such that

$$f'(c) = \frac{f(x_2) - f(x_1)}{x_2 - x_1}$$

But $f'(c) = 0$, so $f(x_2) - f(x_1) = 0$; that is,

$$f(x_2) = f(x_1) \quad \blacksquare$$

Proof of (4.4.6): We apply (4.4.5) to the function

$$h(x) = f(x) - g(x)$$

whose derivative

$$h'(x) = f'(x) - g'(x)$$

is zero on (a, b), to conclude that

$$h(x) = f(x) - g(x) = C \qquad \text{(constant)}$$

on (a, b). Thus

$$f(x) = g(x) + C \text{ on } (a, b) \quad \blacksquare$$

We promised to prove the Increasing-Decreasing Theorem, and now we shall prove part (a). The proof of (b), which is similar, is left as an exercise.

Proof of the Increasing-Decreasing Theorem (4.2.7): We need to show that if $f'(x) > 0$ on (a, b), then f is (strictly) increasing on (a, b). Hence given $x_2 > x_1$ in (a, b), we need to show that

$$f(x_2) > f(x_1)$$

But f is differentiable on $[x_1, x_2]$, so the MVT applies to guarantee c in (x_1, x_2) such that

$$f'(c) = \frac{f(x_2) - f(x_1)}{x_2 - x_1}$$

But $f'(c) > 0$ by hypothesis and $x_2 - x_1 > 0$, so

$$f(x_2) - f(x_1) > 0$$

that is,

$$f(x_2) > f(x_1)$$

as was to be shown. ■

SECTION 4.4 EXERCISES

For each of the functions in Exercises 1 to 5, $f(a) = f(b)$ but there exists no c in (a, b) for which $f'(c) = 0$. Which hypothesis of Rolle's Theorem is violated?

1. $f(x) = \begin{cases} x & 0 \le x \le 1 \\ 2 - x & 1 \le x \le 2 \end{cases} \quad a = 0, b = 2$

2. $f(x) = \begin{cases} 0 & x = 1 \\ x - 2 & 1 < x < 3 \\ 0 & x = 3 \end{cases} \quad a = 1, b = 3$

3. $f(x) = |x| - 2 \quad a = -1, b = 1$

4. $f(x) = |x| \quad a = -1, b = 1$

5. $f(x) = \sqrt{|x|} \quad a = -1, b = 1$

6. (*a*) Which hypothesis of Rolle's Theorem fails for the function $f(x) = |x^2 - 4|$ on $[-3, 3]$?
(*b*) Does the conclusion of Rolle's Theorem hold in this case?
(*c*) Is there a contradiction in this situation?

7. By subtracting the vertical distance between the line with slope $[f(b) - f(a)]/(b - a)$ through the point $(a, f(a))$ (instead of the line through $(b, f(b))$, devise an alternative proof of the MVT.

In Exercises 8 to 15 determine whether the MVT applies for the given function f on the given interval $[a, b]$, and if so, determine each point c such that $f'(c) = [f(b) - f(a)]/(b - a)$.

8. $f(x) = \dfrac{x^3}{3} - x + 1, [-2, 3]$

9. $f(x) = x^3 - 12x + 3, [-3, 3]$

10. $f(x) = x^4, [0, 2]$

11. $f(x) = x^2, [0, 2]$

12. $f(x) = \sqrt{x}, [0, 1]$

13. $f(x) = \sqrt[3]{x}, [-1, 1]$

14. $f(x) = \sqrt[3]{x}, [0, 1]$

15. $f(x) = \sqrt{25 - x^2}, [-5, 5]$

16. Rephrase the proof of the MVT by showing that the function (of x) which gives the vertical distance between the graph of $f(x)$ and the *secant* line through A and B (see Fig. 4.31) has a maximum and satisfies the hypothesis of Rolle's Theorem.

17. Interpret the MVT for linear functions.

CHAPTER EXERCISES

Review Exercises

In Exercises 1 to 9 determine the given limit and describe any asymptotes thus determined.

1. $\lim\limits_{x \to +\infty} \dfrac{\sqrt{x}}{\sqrt{x} + 1}$

2. $\lim\limits_{x \to +\infty} \dfrac{x^2(1 + x)}{1 - x}$

3. $\lim\limits_{x \to 0} \dfrac{x^2 - 4}{2x^2 + x}$

4. $\lim\limits_{x \to +\infty} \dfrac{\sqrt[3]{2x + 4}}{\sqrt{x - 2}}$

5. $\lim\limits_{x \to 2^+} \dfrac{x^3}{x - 2}$

6. $\lim\limits_{x \to 1} \dfrac{x - 2}{(x - 1)^2}$

7. $\lim\limits_{x \to -\infty} \dfrac{3x^3 + 2x}{2x^2 + 2x - 1}$

8. $\lim\limits_{x \to 0} \dfrac{x^2}{x^2 - 1}$

9. $\lim\limits_{x \to 0} \dfrac{x^2(x + 1)}{x - 1}$

In Exercises 10 to 15 determine where the given function is increasing and decreasing and indicate all relative extremes.

10. $f(x) = 12x^3 - 6x^2 + 3x - 1$

11. $g(x) = (1 - x)(1 + x)^3$

12. $f(x) = \left(\dfrac{x - 2}{x + 2}\right)^3$

13. $f(x) = \dfrac{2 - 3x}{2 + x}$

14. $h(x) = \dfrac{x - 3}{x^2 + 4x + 5}$

15. $f(t) = 4t^3 + 27t - 18t^2 - 3$

In Exercises 16 to 19 determine the concavity and points of inflection of the given function.

16. $y = \frac{x^4}{4} - \frac{x^2}{2}$

17. $f(x) = \frac{1 - \sqrt{x}}{1 + \sqrt{x}}$

18. $f(x) = (1 - x)^2(1 + x)^3$

19. $f(x) = 2x^6 - x^4$

In Exercises 20 to 54 graph the given function—include all pertinent information.

20. $f(x) = x^5 - 5x + 2$

21. $y = x^3 - 15x$

22. $f(x) = x^2(1 + x)^3$

23. $f(x) = 3x^4 - 4x^3$

24. $f(x) = 2x^5 - 10x$

25. $y = x^4 - 3x^3 + 3x^2$

26. $g(t) = t^4 + 2t^3 - 3t^2 - 4t$

27. $f(u) = u^4 - 4u^3 - 2$

28. $y = w^4 - 12w^3 + 36w^2 - 10$

29. $f(x) = x^4 - 16x^2$

30. $g(x) = x^3 - 3x^2 - 9x + 10$

31. $f(x) = 2x^6 - 3x^4$

32. $t = 3x^5 - 5x^3$

33. $f(x) = 6x^{1/5} + x^{6/5}$

34. $f(x) = 3x^{5/3} - 15x^{2/3}$

35. $f(x) = \frac{2x}{x^3 + 1}$

36. $y = \frac{3x}{4x - 3}$

37. $y = x^2\sqrt[3]{1 - x}$

38. $f(x) = \frac{x^2}{3x + 1}$

39. $f(x) = \frac{2x^2}{4x - 3}$

40. $g(x) = \frac{2x^2}{1 - x}$

41. $f(x) = x^2(1 + x)^3$

42. $f(x) = x\sqrt[3]{1 - x}$

43. $f(x) = \frac{2x}{(x + 1)^2}$

44. $f(x) = \frac{2x^2}{x^2 + 2}$

45. $f(x) = 3 + \frac{x - 1}{x + 2}$

46. $f(x) = 12x^{1/5} + x^{6/5}$

47. $g(x) = 3x^{4/3} + 6x^{1/3}$

48. $f(x) = 3x^{4/3} - 32x$

49. $f(x) = 2 + (x - 1)^{5/3}$

50. $f(x) = x^2\sqrt[3]{1 - x^2}$

51. $f(x) = 3x^{4/3} + 32x$

52. $f(x) = (x + 2)^{2/3} + 1$

53. $f(x) = \begin{cases} 4 - x & \text{for } x \leq 0 \\ x^2 + 4 & \text{for } x > 0 \end{cases}$

54. $g(x) = \begin{cases} x^2 + 1 & \text{for } x \leq 0 \\ x + 1 & \text{for } x > 0 \end{cases}$

55. $f(t) = 2t^{6/5} + 12t^{1/5}$

56. (*a*) Show that the MVT hypotheses fail for $f(x) = \sqrt[3]{x}$ on $[-1, 8]$.
(*b*) Does its conclusion fail? Explain.

57. Repeat Exercise 56 for $f(x) = \sqrt[3]{1 - x^2}$ on $[-3, 3]$.

58. Do the hypotheses (hence does the conclusion) of the MVT hold for $f(x) = \sqrt{1 - x^2}$ on $[-1, 1]$? If so, what is c?

59. Repeat Exercise 58 for $f(x) = |x^3|$ on $[-1, 1]$. (Be careful at $x = 0$).

Miscellaneous Exercises

1. Use the MVT to determine a linear function that is an upper bound for $f(x) = 2x^3$ on $[0, 1]$.

2. Use the MVT to determine a linear function that is parallel to the upper bound of Exercise 1 but is a *lower* bound for $f(x) = 2x^3$ on $[0, 1]$. (Its graph will be tangent to that of $f(x)$.)

3. Provide the prosecution case that a driver clocked at 45 mph in a 55 mph zone 1 h and 55 mi down the road from a point where he was clocked at 55 mph should be convicted of speeding.

4. Translate the above argument to the case where the initial clocking was at 45 mph and the clocking 1 hr later was at 55 mph.

5. By deriving a contradiction using Rolle's Theorem, show that $p(x) = ax^2 + bx + c$ has at most two real roots. (Assume it has three and work backward.)

6. Show that a cubic polynomial has at most three real roots. (Build on Exercise 5.)

7. Using the MVT and/or its consequences, show that if $f'(x)$ is a polynomial of degree n, then $f(x)$ is a polynomial of degree $n + 1$.

8. Develop some conjectures regarding the degree of a polynomial and the number of relative extremes it can have. (If the graph of a polynomial has as many "peaks and valleys" as your open hand held in front of you, what is the minimum degree of this polynomial?)

9. (Third-Derivative Test for Inflection Points) Suppose $f'''(x)$ exists on an open interval containing $x = a$. Prove that if $f''(a) = 0$ and $f'''(a) \neq 0$, then f has a point of inflection at a. (Apply the Second-Derivative Test to f' and interpret the results.)

10. Show by example that the Third-Derivative Test for Inflection Points may fail if $f'''(a) = 0$.

(The next several exercises extend to higher derivatives the relations among zero derivatives, extremes, and inflection points. This covers cases where the Second- and Third-Derivative Tests fail because the second or third derivative, respectively, is zero.)

11. Suppose that $k \geq 2$ is an integer and $f^{(k)}(x)$ exists in an open interval containing the point $x = a$. Following the given outline, prove that if $f^{(k)}(x)$ has a relative maximum equal to zero at $x = a$ and $f^{(k-1)}(a) = 0$ also, then

$f^{(k-2)}(x)$ also has a relative maximum at $x = a$.
(*a*) Show that $f^{(k)}(x)$ is decreasing in some open interval containing a.
(*b*) Show that $f^{(k-1)}(x)$ changes sign at $x = a$.
(*c*) Show that the conclusion follows.

12. Repeat Exercise 11 for a relative minimum.

13. Suppose the first n derivatives of f exist in an open interval containing the point $x = a$, where $n \geq 2$ is an integer. Prove that, for n an even integer, if $f'(a) = f''(a) = f'''(a) = \cdots = f^{(n-1)}(a) = 0$ and $f^{(n)}(a) \neq 0$, then f has a relative maximum at a if $f^{(n)}(a) < 0$ and a relative minimum at a if $f^{(n)}(a) > 0$. (Apply Exercises 11 and 12 to "back up" to the given conclusion, stepping backward two derivatives at a time.)

14. Under the same conditions as Exercise 13 but with n odd and without the condition that $f'(a) = 0$, prove that f has a point of inflection at $x = a$. (Use same approach as in Exercise 13.)

15. (*a*) Apply Exercise 13 to $f(x) = x^4$.
(*b*) Generalize part (*a*) to any positive even power of x.

16. (*a*) Apply Exercise 14 to $f(x) = x^5$ to prove that f has a point of inflection at $x = 0$.
(*b*) Generalize part (*a*) to any positive odd power of x.

17. Apply Exercises 13 and 14 to determine extreme and inflection points for $f(x) = 5x^6 - 6x^5$. Sketch $f(x)$.

18. Suppose $f(x) = (x - a)^n g(x)$, where $g(a) \neq 0$ and $n > 1$ is an odd integer with $g^{(n)}(x)$ defined in an open interval containing $x = a$. Prove that the graph of $f(x)$ is tangent to the x axis with a point of inflection there.

19. Repeat Exercise 18 but with n even to conclude an extreme point at $x = a$.

20. Apply Exercises 18 and 19 to sketch the graph of $f(x) = (x - 1)^2(x + 2)^3$ near where it touches or crosses the x axis.

21. Repeat Exercise 20 for $f(x) = x^3(x - 1)^2(x + 2)^2$.

22. Prove that $f(x) = 4x^3 + 3x^2 - 6x + 1$ has a root between 0 and 1. (*Hint:* Apply Rolle's Theorem to any function whose derivative is $f(x)$ on $[0, 1]$.)

23. Given that $|f'(x)| \leq 1$ for all numbers x, show that $|f(x) - f(y)| \leq |x - y|$ for all numbers x, y.

24. Let $P(x) = a + a_1x + \cdots + a_nx^n$. Show that between any two consecutive roots of the equation $P'(x) = 0$, there is at most one root of the equation $P(x) = 0$.

SELF-TEST ✣

Sketch the graphs of the functions given in Probs. 1 to 4, indicating all asymptotes, extremes, and points of inflection.

1. $f(x) = x^3 + x^2 + 6x - 5$

2. $f(x) = \dfrac{1}{x^2 - 1}$

3. $f(x) = x - 3 + \dfrac{2}{x + 1}$

4. $f(x) = 3x^{4/3} - 6x^{1/3}$

5. Determine whether the hypotheses of the MVT hold for $f(x) = (x + 2)/(x + 1)$ on $[1, 2]$; if so, determine the numbers c that satisfy the conclusion of the MVT.

Chapter 5

USING DERIVATIVES TO SOLVE PROBLEMS AND UNDERSTAND QUANTITATIVE RELATIONSHIPS

CONTENTS

This chapter illustrates some of the traditional direct applications of derivatives. It is impossible to list all types of applications that are already in use, let alone predict those that might arise in the future. Our goal is that you begin to learn the basic principles involved in using the calculus as a tool to get insights into dynamic relationships and solve specific problems. Generally, the hardest thing to do is to translate a given problem or situation into the appropriate mathematical formulation. This is often referred to as "mathematical modeling" and represents the essence of applied mathematics.

In this chapter modeling a situation means finding appropriate mathematical equations or functions to describe that situation. These situations all involve change, either concrete change, as in the motion of a particle or change in price of an item being manufactured, or abstract change, as in the different possible areas enclosed by a given perimeter. As a result, derivatives are the appropriate tools in attacking such problems. Pay particular attention to the extended examples in Secs. 5.1 and 5.4, where we illustrate the power of the derivative to yield an understanding of relationships that would be difficult to achieve otherwise.

Section 5.1: Optimization Problems

Objectives:

Use derivatives to assist in
1. Determining the extreme values of a function on an interval.
2. Translating certain problems calling for the maximizing or minimizing of a given quantity into appropriate extreme-value problems.

DETERMINING EXTREME VALUES ON AN INTERVAL

We know that a function continuous on a closed interval $[a, b]$ reaches its maximum and minimum values on that interval. If such values occur between a and b, they are also relative extremes, and if the derivative exists at such points, then it is zero there. This suggests the following:

5.1.1 Procedure for Determining Extreme Values of a Function on a Closed Interval $[a, b]$.

(*a*) Determine the values of f at any critical points arising from f' between a and b.
(*b*) Compare these with $f(a)$ and $f(b)$.

If the choice of extreme values in (5.1.1) is not obvious, then apply the graphing machinery of Chap. 4. In the special case that f' exists and does not change sign on $[a, b]$, we know that f is either increasing or decreasing on all of $[a, b]$. Thus:

5.1.2 Theorem If f' exists and does not change sign on $[a, b]$, then f attains its extreme values at the endpoints.

Example 1 Find the extreme values of $f(x) = x/(x + 1)$ on $[-1/2, 1]$.

Solution
$$f'(x) = \frac{(x + 1)(1) - x(1)}{(x + 1)^2} = \frac{1}{(x + 1)^2}$$

Since $f'(x) > 0$ for all x in $[-1/2, 1]$, the function is increasing. It follows that $f(-1/2) = -1$ is the minimum value and $f(1) = 1/2$ is the maximum value. ●

Example 2 Determine any extreme values of $f(x) = x^4 - 2x^2 + 1$ on $[-3, 0]$.

Solution
$$f'(x) = 4x^3 - 4x = 4x(x^2 - 1) = 4x(x - 1)(x + 1)$$

Here it is convenient to use a chart to determine the sign of f'.

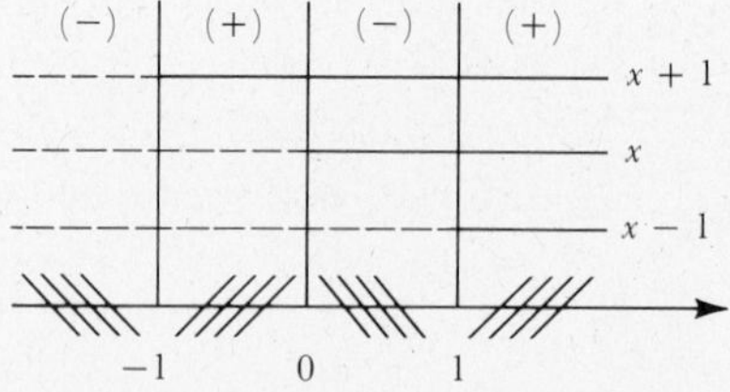

Thus we have relative minima at $x = -1$ and $x = 1$ and a relative maximum at $x = 0$. Since we are only concerned with values of x in $[-3, 0]$ we can disregard the relative minimum at $x = 1$.

Now $f(-1) = 1 - 2 + 1 = 0$, $f(0) = 1$,
and $f(-3) = 81 - 2(9) + 1 = 64$.

The maximum value is 64 and occurs when $x = -3$.

The minimum value is 0 and occurs when $x = -1$. ●

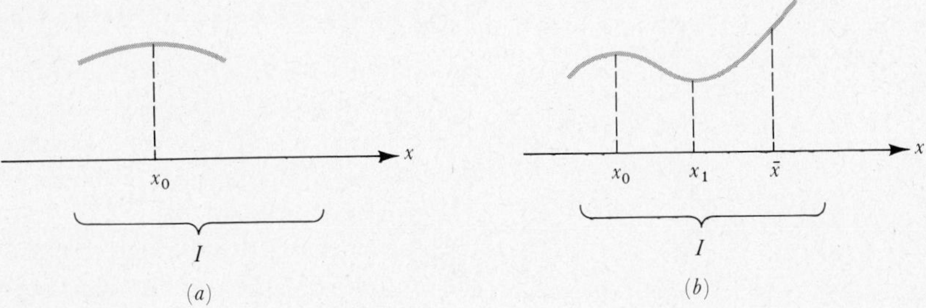

Fig. 5.1

Of course, not all intervals are closed, nor are all functions everywhere continuous. On an arbitrary interval I, one that is not necessarily closed or even bounded, a continuous function need not attain *any* extreme values. However, if f is differentiable on an arbitrary interval I except perhaps at one or more endpoints and has a *single* nonendpoint relative extreme, then this relative extreme is an extreme value on I. Suppose, for example, f has a single relative maximum at x_0 as in Fig. 5.1(*a*). If $f(x_0)$ is not the maximum value of f on I (meaning that $f(\bar{x}) > f(x_0)$ for some other $\bar{x}$ in I) then the graph would have to turn upward elsewhere on I to reach $f(\bar{x})$. This would give rise to *another* relative extreme x_1 on I (a relative minimum, in fact), as in Fig. 5.1(*b*).

More precisely, we have:

5.1.3 **Theorem** Suppose f is differentiable on an interval I, except perhaps at one or more endpoints, and is continuous on all of I. If f has a single relative extreme at x_0 not an endpoint of I, then $f(x_0)$ is an extreme value for f on I.

Example 3 Determine the extreme values if any of the function $f(x) = \pi x^2 + 8/x$ on the interval $(0, +\infty)$.

Solution Since $\lim_{x \to 0^+} f(x) = \lim_{x \to +\infty} f(x) = +\infty$, f has no maximum value on the interval $(0, +\infty)$. Now

$$f'(x) = 2\pi x - \frac{8}{x^2}$$

$$= \frac{2}{x^2}[\pi x^3 - 4]$$

and so the only critical value in the interval is $x = (4/\pi)^{1/3}$.

Since $f''(x) = 2\pi + 16/x^3 > 0$ for all x in $(0, +\infty)$, and in particular $f''((4/\pi)^{1/3}) > 0$, we know by the Second-Derivative Test that f has a relative minimum at $x = (4/\pi)^{1/3}$. It follows from (5.1.3) that f attains its minimum on $(0, +\infty)$ at $(4/\pi)^{1/3}$. (See Fig. 5.2.) ●

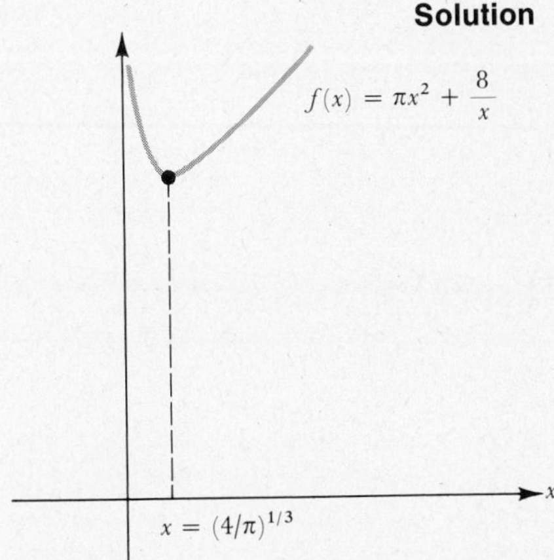

Fig. 5.2

PROGRESS TEST 1

Determine the minimum and maximum values of f on the indicated interval (if they exist).

1. $f(x) = 2x^3 + 3x^2 - 12x + 4$ on $[-3, 3]$

2. $f(x) = x^5 + 2x^3$ on $[0, 3]$

3. $f(x) = \dfrac{2x^2}{1 - x}$ on $(1, +\infty)$

SOLVING SPECIFIC OPTIMIZATION PROBLEMS

Using mathematics to solve concrete problems inevitably involves the practice of "mathematical modeling." Simply stated, mathematical modeling is concerned with translating a real-world problem into a mathematical problem, solving the mathematical problem, and then using the solution in turn to solve the original problem.

The mathematical model for many problems leads to finding the maximum or minimum value of a function on an interval. Such problems often identify themselves by including such superlatives as "least," "greatest," "strongest."

For example, suppose we have a large field adjacent to a river and 100 meters of fence with which we are to enclose part of the field to form a rectangular pasture. Assuming that one side of the pasture will be the unfenced river bank, what are the dimensions of the pasture with largest area that we can enclose? What is this area? This is a maximizing problem; that is, we are maximizing *area* with a fixed amount of fence (perimeter) and a given shape (rectangle). Perhaps the first thing to recognize is that many different pastures will satisfy the given conditions, and it is our job to determine which one will have the largest area. See Fig. 5.3. (The wavy line denotes the river bank.)

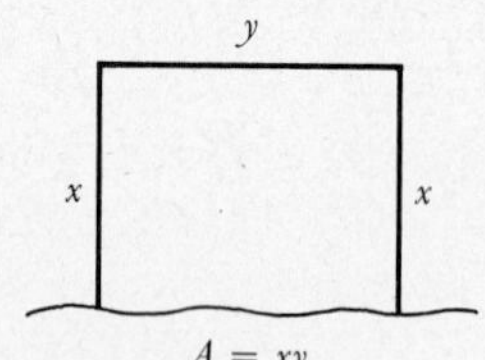

We start by sketching a typical pasture (see figure at left) and labeling the lengths of its sides by x and y. The area A to be maximized is, of course, given by $A = xy$. However, all our differentiation and maximizing techniques apply to functions of one variable. But we can use the given information, that we have only 100 m of fencing, to eliminate one of the variables, say, y. Since we have 100 m of fence to use, we know

$$2x + y = 100$$

Thus $y = 100 - 2x$ and so A may be written as a function of the single variable x:

$$A(x) = x(100 - 2x) = 100x - 2x^2$$

The physical situation gives us natural restrictions on x, namely $0 \leq x \leq 50$. (The two extremes yield the trivial minimum, namely, no area whatsoever!)

Thus our problem is reduced to finding the maximum value of

$$A(x) = 100\,x - 2x^2 \text{ on the interval } [0, 50]$$

But

$$\begin{aligned} A'(x) &= 100 - 4x \\ &= 4(25 - x) \\ &= 0 \text{ when } x = 25 \end{aligned}$$

Thus the only critical point is $x = 25$, and it is a relative maximum. Since the endpoints $x = 0$, $x = 50$ yield $A = 0$, the maximum occurs when $x = 25$. (See

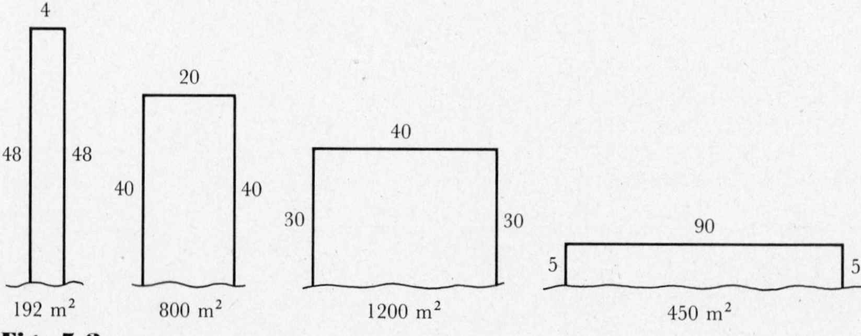

Fig. 5.3

Fig. 5.4 for a graph of $A(x)$ on $[0, 50]$.) Thus the dimensions of the pasture with maximum area are $x = 25$ by $y = 50$, and the area is 1250 m^2.

This example embodies the essence of a general procedure.

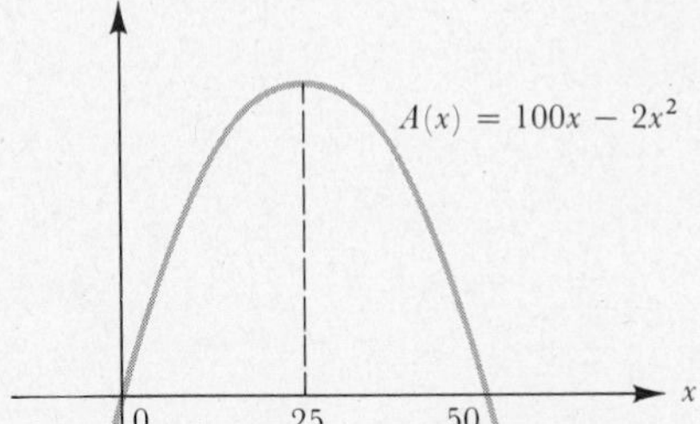

Fig. 5.4

5.1.4 Procedure for Solving Optimization Problems

1. Draw a figure (when appropriate) and assign a letter to each quantity mentioned.
2. Select the quantity to be maximized or minimized and express it in terms of the remaining quantities.
3. Use the given information to eliminate all but one of the quantities in the expression in order to express the quantity to be maximized or minimized as a function of a single variable.
4. Note any restrictions on the domain of this function.
5. Use the methods developed earlier to determine the maximum or minimum.
6. Use the results of step 5 to answer the original question.

Example 4 A cylindrical tin can is to have a volume of 128 cm^3. What are the dimensions that require the least amount of metal?

Solution We assume the can is a right circular cylinder with both a top and bottom.

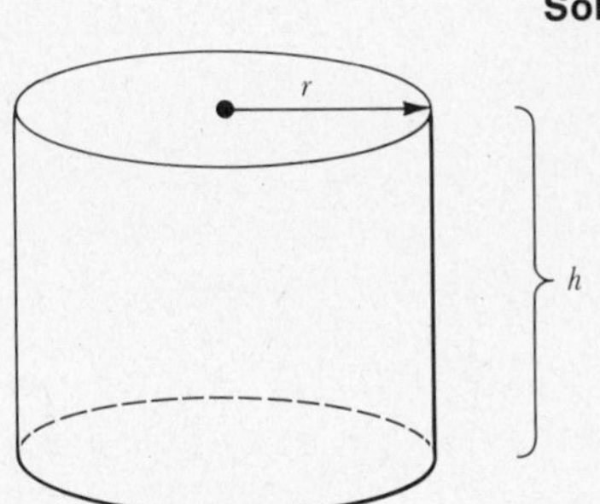

Fig. 5.5

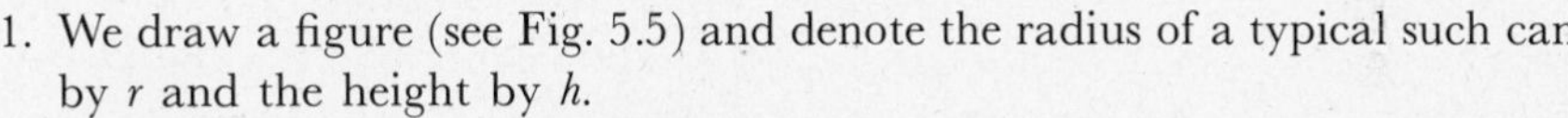

1. We draw a figure (see Fig. 5.5) and denote the radius of a typical such can by r and the height by h.

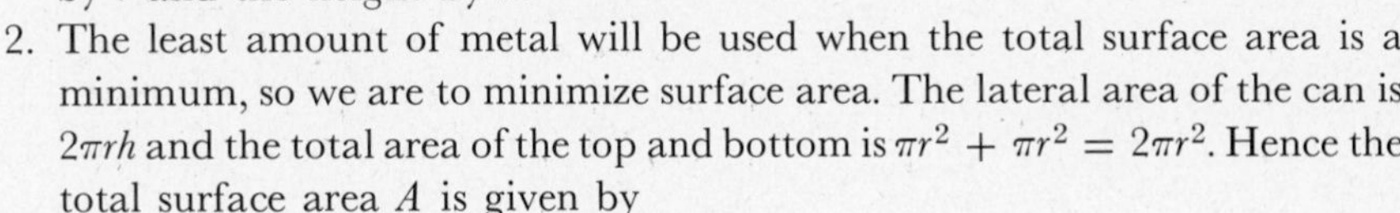

2. The least amount of metal will be used when the total surface area is a minimum, so we are to minimize surface area. The lateral area of the can is $2\pi rh$ and the total area of the top and bottom is $\pi r^2 + \pi r^2 = 2\pi r^2$. Hence the total surface area A is given by

$$A = 2\pi rh + 2\pi r^2$$

3. Now the volume of the can is 128 cm^3, so using the formula for the volume of a right circular cylinder, we have

$$\pi r^2 h = 128$$

Solving for h (which is a bit easier than solving for r), we get

$$h = \frac{128}{\pi r^2}$$

which allows us to express A in terms of the single variable r:

$$A(r) = 2\pi r\left(\frac{128}{\pi r^2}\right) + 2\pi r^2$$

$$= \frac{256}{r} + 2\pi r^2$$

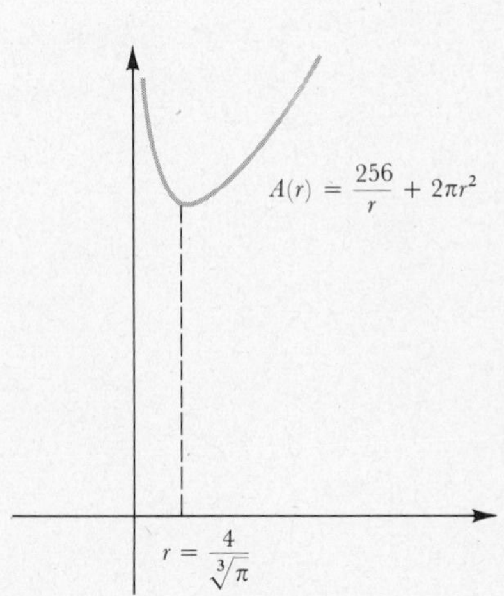

Fig. 5.6

4. The physical restriction on r is simply that $r > 0$. (There is no can with $r = 0$.)
5. Now

$$A'(r) = \frac{-256}{r^2} + 4\pi r$$

$$= \frac{4}{r^2}[\pi r^3 - 64]$$

and so the only critical value on $(0, +\infty)$ is $r = \sqrt[3]{64/\pi} = 4/\sqrt[3]{\pi}$. Now

$$A''(r) = \frac{512}{r^3} + 4\pi > 0$$

for all $r > 0$, so the Second-Derivative Test tells us that $r = 4/\sqrt[3]{\pi}$ yields a relative minimum. Since it is the *only* extreme value of $A(r)$ on $(0, +\infty)$, we can be assured by (5.1.3) that $A(r)$ reaches its minimum value on $(0, +\infty)$ at $r = 4/\sqrt[3]{\pi}$; see Fig. 5.6. Note that as r gets close to zero, $A(r)$ gets large without bound, (an extremely tall can). Also, as r gets large, $A(r)$ gets large without bound (an extremely wide can).

6. Thus the dimensions that use the least amount of metal are

$$r = \frac{4}{\sqrt[3]{\pi}} \approx 2.73 \text{ cm} \qquad \text{and} \qquad h = \frac{128}{\pi r^2} = \frac{128}{\pi(16/\pi^{2/3})} \approx 5.46 \text{ cm} \quad \bullet$$

Example 5 The federal government is to fund both the building of a sewage plant and its connection to towns A and B, which will share the plant. The federal government wishes to locate the plant so as to minimize total connector costs. Town A is 2.5 km downriver from the river port B and 2 km away from the river on the same side as B. (*a*) The larger town A's connecting installation costs are $\$\frac{1}{4}$ million/km and B's costs are $\$\frac{1}{5}$ million/km. Where should the plant be located? (*b*) During the approval procedure a demographic study is made that indicates the population of B will remain stable, but town A will grow, requiring connectors costing $\$\frac{1}{3}$ million/km. Where now should the plant be located?

Solution (*a*) 1. We sketch the situation as in the marginal figure and assign variables to the lengths along the river associated with the plants' location.

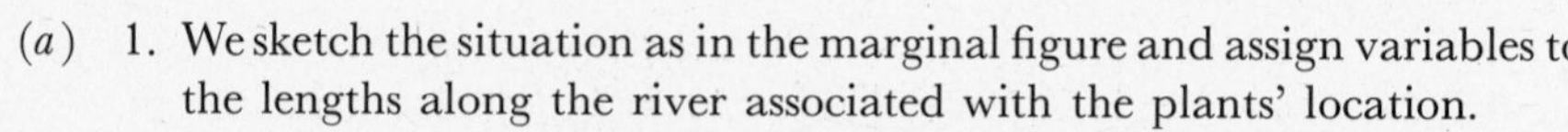

2. We wish to minimize the total cost, which (in millions of dollars) is

$$C(x) = \frac{\sqrt{x^2+4}}{4} + \frac{y}{5}$$

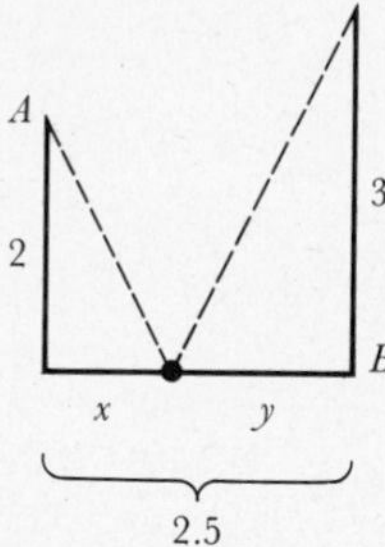

3. Since $x + y = 2.5$, we know that $y = 2.5 - x$ and thus can write C as a function of x:

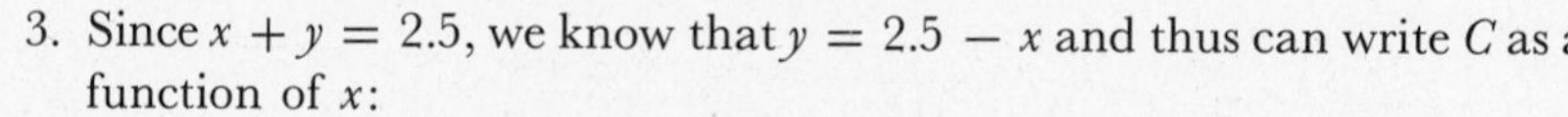

$$C(x) = \frac{\sqrt{x^2+4}}{4} + \frac{2.5-x}{5}$$

4. The situation dictates that $0 \le x \le 2.5$.
5. We compute $C'(x)$ and set it equal to 0:

$$C'(x) = \frac{x}{4\sqrt{x^2+4}} - \frac{1}{5} = 0$$

Solving for x, we obtain $x = \pm(8/3)$. Since neither of these is in the interval $[0, 2.5]$, the minimum must occur at an endpoint. Now $C(0) = \frac{1}{2} + 2.5/5 = 1$ and $C(2.5) = (\sqrt{(2.5)^2+4})/4 \approx 0.8$.

6. Thus the minimum connecting cost occurs when the plant is located at town B.

(*b*) In this case the function to be minimized is

$$C(x) = \frac{\sqrt{x^2+4}}{3} + \frac{2.5-x}{5}, 0 \le x \le 2.5$$

Now

$$C'(x) = \frac{x}{3\sqrt{x^2+4}} - \frac{1}{5} = 0$$

when $x = \pm(3/2)$. We discard $x = -3/2$ since it is not in $[0, 2.5]$. Now

$$C(0) = \frac{2}{3} + \frac{2.5}{5} \approx 1.17$$

$$C\left(\frac{3}{2}\right) = \frac{\sqrt{9/4 + 4}}{3} + \frac{1}{5} \approx 1.03$$

$$C(2.5) = \frac{\sqrt{(2.5)^2 + 4}}{3} \approx 1.07$$

Thus the plant should be located $\frac{3}{2}$ km upriver from A and 1 km downriver from B. ●

As a practical matter, it is wise to convert to decimals while working problems of this type only to assist in comparing magnitudes and perhaps in giving the final answer. Premature punching of the calculator may obscure useful simplifications and clarifying relationships.

PROGRESS TEST 2

We suggest you follow the six-step procedure, especially in Probs. 4 and 5.

1. Given the same field along the river as in our introductory discussion, determine the dimensions and total area of the largest pasture that can be enclosed using 120 m of fence, with the additional condition that the pasture is to be subdivided into two parts (using the same fencing) with a fence perpendicular to the river. Note that the location of the internal fence is irrelevant.

2. Determine the pair of numbers whose product is as large as possible, where the sum of one number and triple the other is fixed at 120. Have you seen this problem before?

3. Repeat Prob. 1 but with the subdividing fence *parallel* to the river.

4. The top and bottom margins of a page in a tiny code book are each 1.5 cm, and the side margins are each 1 cm. If the area of the printed part of the page is fixed at 30 cm^2, what are the dimensions of the page of least area? (Let x and y denote the height and width of the page, respectively.)

5. You are visiting a gold mine where each visitor is given a 12×24-cm piece of cardboard to be made by the visitor into an open-top box by cutting out squares from the corners and folding up the sides. The visitor's box is then filled with gold dust as a free sample to take home. What are the dimensions of the box of largest volume you can construct?

THE DERIVATIVE AS A TOOL FOR DEEPENING INSIGHTS INTO QUANTITATIVE RELATIONSHIPS

You undoubtedly recognized that Probs. 1 and 2, Progress Test 2, were essentially the same; that is, they each translated into the same mathematical formulation: maximize xy, where $3x + y = 120$. Each problem had the *same mathematical model* involving the same mathematical solution based upon the use of first derivatives.

Historically, especially in physics, the derivative model has been used to reveal and express dynamic relationships between variables. Now we shall examine a few situations where money plays an essential role. We use money because it is a quantity which, perhaps not commonly possessed, is commonly understood.

The symbol ✕ *and the vertical rule in the text signal an application designed to illustrate ways that calculus relates to the everyday world.*

✕ **The Fence Problem Revisited** We first consider some variations on the original fence problem (maximize the area of a rectangular pasture one side of which is a river bank—no subdivisions), but instead of being given a fixed 100 m of fence, we are given \$100 to purchase fence. If all the fence costs \$1 per running meter, then this problem has the same mathematical model as the original with the same 25 × 50-m, rectangle solution. Notice that the total length of the fence in one direction is equal to the total length in the other ($2x = 50 = y$).

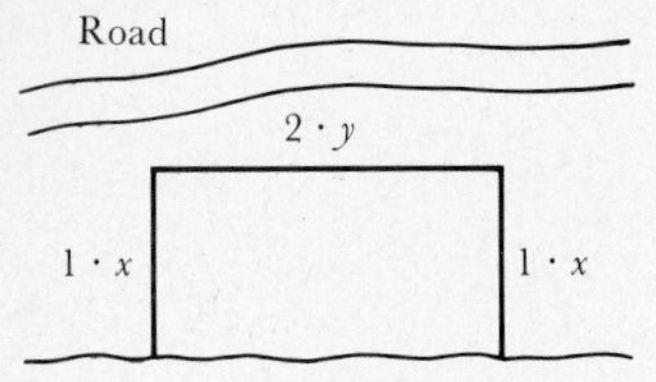

Suppose instead we are given the same \$100 but told that, because there is a road along the side of the fence parallel to the river, a more substantial fence is required on that side, costing \$2/m. What then are the dimensions of the rectangular pasture of largest area that can be enclosed? We proceed as before, drawing a diagram but labeling each side by its total *cost* rather than its length. Of course, the sides perpendicular to the river of length x cost $1 \cdot x = x$ dollars apiece, and the side parallel to the river of length y costs $2y$ dollars. We are still trying to maximize area $A = xy$, but now the given condition (\$100 to spend) translates to

$$2x + 2y = 100$$

We solve for one variable in terms of the other—say, y in terms of x (we *could* do the opposite to get A as a function of y and obtain the same solution)—to obtain

$$y = 50 - x$$

Thus

$$A = A(x) = 50x - x^2 \qquad (0 \leq x \leq 50)$$

Now, if $x = 50$, then all the money is spent on x fence, leaving no money for y fence and zero area. Similarly, $x = 0$ means all the money is spent for y fence, resulting in zero area. Now

$$A'(x) = 50 - 2x = 0 \qquad \text{when } x = 25$$

Since $A''(x) = -2 < 0$ for all x, including $x = 25$, we know that A reaches its maximum at $x = 25$. Then $y = 50 - 25 = 25$, and the largest pasture is a 25 × 25-m square of area 625 m^2.

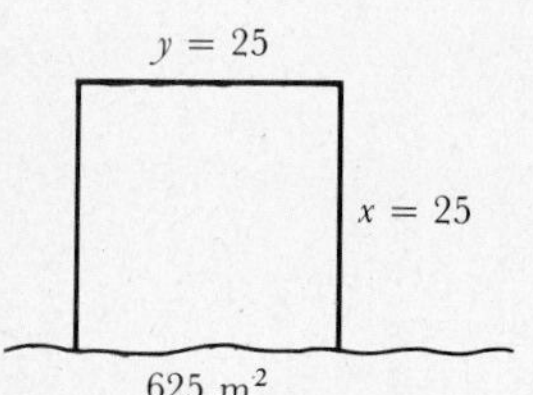

The effect of doubling the cost of the y fence was to cut the area in half and, perhaps even more interesting, to change the *shape* of the largest possible pasture. In effect, the benefit of a "free" fence along the river (which accounted for the lengthening in the direction parallel to the river in the original situation) was neutralized by the doubling of the cost on the opposite side along the road.

Suppose now the cost of the fence perpendicular to the river stays at \$1/m but the fence along the road costs \$5/m and there is still only \$100 for fencing. Then the given condition becomes

$$2x + 5y = 100 \qquad \text{or} \qquad y = 20 - \frac{2}{5}x$$

and

$$A(x) = x\left(20 - \frac{2x}{5}\right) = 20x - \frac{2x^2}{5} \qquad (0 < x < 50)$$

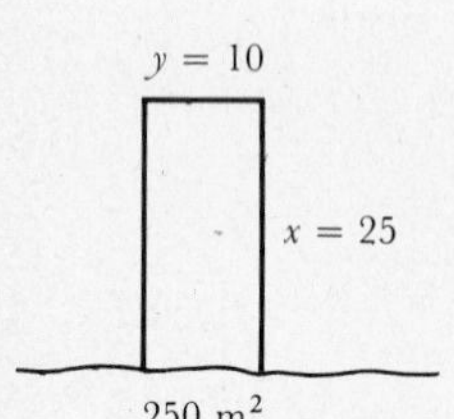

Now

$$A'(x) = 20 - \frac{4}{5}x = 0 \qquad \text{and } A''(x) = \frac{-4}{5} < 0$$

implies a maximum area at $x = 25$ with $y = 20 - (2/5)(25) = 10$ m and $A = (25)(10) = 250$ m^2.

Is it an accident that the length of the x side is 25 m in each of the three cases? One way to check is to let the price of the y fence be k dollars per meter, so the given \$100 constraint is

$$2x + ky = 100$$

so

$$y = \frac{100 - 2x}{k}$$

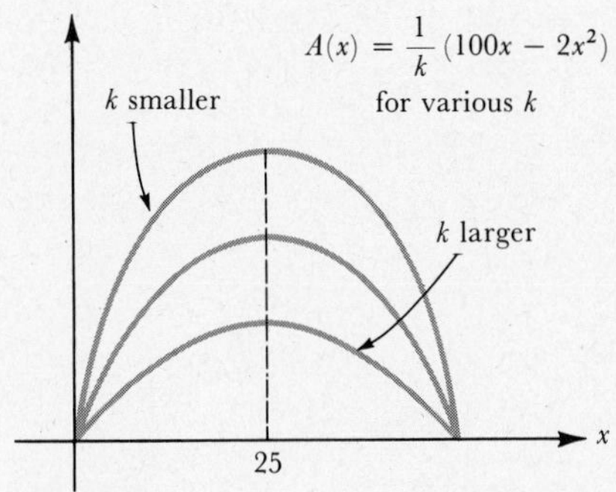

Fig. 5.7

Then

$$A(x) = \frac{1}{k}(100x - 2x^2) \qquad (0 \le x \le 50)$$

and

$$A'(x) = \frac{1}{k}(100 - 4x) = 0 \qquad \text{when } x = 25$$

Sure enough, whatever the cost of the fence along the road, the largest pasture results when each of the other two sides is 25 m. The only effect of k is in the constant $1/k$ in the formula for $A(x)$, which does not affect the location of the relative extreme, as can be seen in Fig. 5.7.

On the other hand, if we also allow the cost of the x fence to vary, letting its cost per meter be j dollars, then the \$100 constraint becomes

$$2jx + ky = 100 \qquad \text{or} \qquad y = \frac{1}{k}(100 - 2jx)$$

Thus

$$A(x) = \frac{1}{k}(100x - 2jx^2) \qquad \left(0 < x \le \frac{50}{j}\right)$$

and

$$A'(x) = \frac{1}{k}(100 - 4jx) = 0 \qquad \text{when } x = \frac{25}{j} \text{ and } y = \frac{50}{k}$$

since $A'' < 0$ and $25/j$ is in $[0, 50/j]$, $x = 25/j$ is a maximum.

Notice that the length of each fence depends only upon *its* cost per meter and not upon the cost of the other! Moreover, the total length of the x fence is $2(25/j) = 50/j$, so the total amount of money spent on the x fence is

$$\frac{50}{j}(j) = 50 \text{ dollars}$$

and the total spent on the y fence is

$$\frac{50}{k}(k) = 50 \text{ dollars}$$

The underlying symmetry of the solution to the original problem (same total length in each direction) prevails here in the general situation: *we spend the same total amount in each direction, namely \$50, no matter what the respective cost per unit length of the fences!* Physically, different relative costs per unit length lead to pastures of different shapes and areas, but our analysis has revealed a deeper underlying symmetry, which, in fact, is based upon the symmetry of the parabolic graphs of the various area functions.

Physical, economic, or even legal restrictions often heighten the importance of endpoints in optimization problems.

× **Minimizing Transportation Costs** A trucking firm estimates that the approximate cost per mile for a given type of truck is $[20 + (1/5)v]$ cents per mile, where v is the velocity in miles per hour and the 20 cents represents "fixed cost" (licenses, insurance, depreciation, and so on). Given that the driver gets \$8/h, what average velocity for a 500-mi trip will minimize the total cost?

The cost per hour of the truck at velocity v is the product of the miles per hour with the cents per mile, or

$$v\left(20 + \frac{v}{5}\right) \text{ cents/h}$$

Adding in the labor cost of 800 cents/h, we conclude that the total hourly cost is

$$20v + \frac{v^2}{5} + 800 \text{ cents/h}$$

Hence the total cost of the trip $C(v)$ is the product of this hourly average with the total hours of the trip

$$\frac{500 \text{ mi}}{v \text{ mph}} = \frac{500}{v} \text{ h}$$

That is,

$$C(v) = \frac{500}{v}\left(20v + \frac{v^2}{5} + 800\right)$$

$$= 500\left(20 + \frac{v}{5} + \frac{800}{v}\right) \text{ cents}$$

Now, we need to minimize $C(v)$ where $v > 0$ and v is less than some reasonable velocity. As usual, we search for relative minima by setting $C'(v) = 0$ and solving for v:

$$C'(v) = 500\left(\frac{1}{5} - \frac{800}{v^2}\right) = 0$$

when

$$v^2 = (5)(800)$$

or

$$v \approx 63.25 \text{ mph}$$

This is a relative minimum by the Second-Derivative Test, so the answer to the original question is: "As close to 63.25 mph as possible."

× Discussion of Truck-Velocity Example—Why Do Truckers Hurry? In terms of United States traffic laws the *legal* interval over which we minimized $C(v)$ is $(0, 55]$, but, as anyone who has driven on major highways knows, this is not the interval in which most truckers drive. The "legal" solution to our problem is 55 mph since $C'(v)$ is positive for $0 < v < 63.25$. Actually, by the Mean-Value Theorem (and common snese), the trucker would need to spend some time at more than 55 mph to average 55 mph. But now we know why truckers, most of whom own their own rigs, prefer higher speeds: to make up for fixed costs. Note that we kept the 500 factored out of our calculations: The length of the trip did not affect the solution. In practical situations, of course, the length of the journey *would* affect the solution because of the regulated length of the driving day for truckers. (The 500-mi trip would take almost 8 h at the optimal speed.)

As we saw, the mathematical analysis of a situation is sometimes less concerned with a particular set of data than with how the elements of the situation relate to one another. Let us determine how changes in fixed costs, especially labor, would affect optimal speed. Suppose we denote the labor costs (in cents per hour) by L. Then we can retrace our earlier solution to see that

$$C(v) = 500\left(20 + \frac{v}{5} + \frac{L}{v}\right)$$

so

$$C'(v) = 500\left(\frac{1}{5} - \frac{L}{v^2}\right)$$

making $C'(v) = 0$ when $v = \sqrt{5L}$—a relative minimum.

Solving this relationship for L, we get $L = v^2/5$. An hourly pay rate corresponding to a 55 mph optimal speed is thus \$6.05. On the other hand, a trucker averaging 65 mph justifies \$8.45/h.

Notice that the 20-cent-per-mile nonlabor fixed costs do *not* enter the optimal velocity issue. However, the $(1/5)v$ term does. It says that the cost per mile increases proportionally with the velocity. If a truck's proportionality constant increased to, say 1/3 (because of increased fuel or tire costs, or a loss of efficiency of the engine), then higher speeds would be more expensive. In fact, at a labor cost of \$8/h, the optimal velocity would then become

$$v = \sqrt{3(800)} \approx 49 \text{ mph}$$

We could even consider a situation where the cost per mile increases with velocity much faster than linearly, as is the case with a tugboat pulling a barge, for example (because of the resistance of the water the tug and barge must plow through). Suppose the cost per mile for our tugboat is

$$20 + \frac{1}{5}v^2$$

Assuming \$8/h labor,

$$C(v) = 500\left(20 + \frac{v^2}{5} + \frac{800}{v}\right)$$

making

$$C'(v) = 500\left(\frac{2}{5}v - \frac{800}{v^2}\right) = 0$$

when

$$v = \sqrt[3]{5/2(800)} \approx 12.6 \text{ mph}$$

Here the increased length of time for the 500-mi trip (almost 40 h) amounts to a decrease in productivity of the driver. This would act to hold wages down—unless, of course, productivity in terms of tonnage-moved increased to compensate, which *is* the case for barges.

PROGRESS TEST 3

1. (*a*) Determine the dimensions and area of the four-sided, rectangular pasture of largest area that can be enclosed by 100 m of fence (same type of fence on all four sides).
(*b*) Compare the symmetry of this example with the original three-sided example discussed on p. 146.
(*c*) How much area can be enclosed if we allow a circular pasture?

2. (*a*) Repeat Prob. 1(*a*), but with the total amount of money available for fence at \$100 and the fence in one direction costing \$2/m and the fence in the perpendicular direction costing \$5/m.
(*b*) How much money is spent on fencing in either direction?

3. Suppose two captives are given 800 m of fence and told to make two separate plots in which they must spend the next 20 years. However, they are not told that they must spend these years in different plots. How should they utilize their fence to achieve maximum area under the given conditions (endpoints are important here)? Explain your result in non-algebraic terms.

SECTION 5.1 EXERCISES

In Exercises 1 to 10 determine the minimum and maximum values of f (if they exist) on the indicated interval.

1. (*a*) $f(x) = \dfrac{x+3}{x-1}$ on $[-3, 0]$

(*b*) Repeat (*a*) for the interval $[-3, 1]$

2. (*a*) $f(x) = x^2 + \dfrac{1}{x}$ on $[\frac{1}{10}, 2]$

(*b*) Repeat (*a*) for $[-1, 2]$

3. $f(x) = \dfrac{x^3}{3} + \dfrac{x^2}{2} - 2x$ on $[-3, 4]$

4. $f(x) = x^4 - 2x^2 + 1$ on $[-3, 0]$

5. $f(x) = 3x + \dfrac{30x}{x-2}$ on $(2, +\infty)$

6. $f(x) = \dfrac{\sqrt{x^2+9}}{2} + \dfrac{7-x}{4}$ on $[0, 4]$

7. $f(x) = x^4 + 5x^3 + 6x^2$ on $[0, 5]$

8. $f(x) = x^{4/3} + 4x^{1/3}$ on $[-2, 5]$

9. $f(x) = x\sqrt{9 - x^2}$ on $[-3, 4]$ (Be careful.)

10. $f(x) = \dfrac{2x}{x^2 + 1}$ on $[-2, 2]$

11. A box with a square base is to have an open top. The area of the material available to make the box is 300 cm^2. What should the dimensions be so that the volume be as large as possible?

12. Repeat Exercise 11 with the conditions that four layers of the material are required for the base of the box.

13. An open-topped, rectangular, square-bottomed aquarium is to hold 1 m^3 of water. Given that the material for the bottom costs twice as much as the sides, what are the dimensions of the least expensive aquarium under these conditions?

14. Repeat Exercise 13, given that material for the bottom costs:
(*a*) four times as much as the sides.
(*b*) k times as much as the sides.

15. Given an aquarium, built under the conditions of Exercise 14(*b*), that is half as tall as it is wide at the base, determine the relative costs of the material for the sides and bottom.

16. A horizontal gutter is to be made from a long piece of sheet iron 80 cm wide by turning up equal widths along the edges into a vertical position. How many centimeters should be turned up at each side to yield a maximum carrying capacity? (The carrying capacity is proportional to cross-sectional area.)

17. At midnight ship B was 90 mi due south of ship A. Ship A sailed east at 15 mph and ship B sailed north at 20 mph. At what time were they closest together? (Draw a diagram giving all appropriate distances t hours after midnight.)

18. (*a*) A cylindrical can that must hold exactly 1000 cm^3 is to be constructed from two materials. The material for the top and bottom costs twice as much as the material to be used for the side. What are the dimensions of the least expensive such can? (*Hint:* Let the cost per square centimeter of the material for the sides be k.)
(*b*) Without working through the details, speculate on the shape of the least expensive can that uses the *same* materials for all parts.

19. The amount of heat received from a heat source is kI/x^2, where I is the intensity, x is the distance from the source, and k is a proportionality constant. (*a*) Where should we put a control panel in the 10-m space between a furnace and a cooling vat (which is not as hot as the furnace) so that the total amount of heat received by the control panel is a minimum, knowing that the heat intensity of the furnace is twice that of the vat? (*b*) Where will the amount of heat received from each source be equal?

20. Repeat Exercise 19 for a furnace:
(*a*) three times as hot as the vat. (*b*) m times as hot as the vat.

21. A piece of wire of length w is cut into two parts, one of which is bent into the shape of a square and the other into the shape of a circle. How should the wire be cut so that the sum of the enclosed area is a minimum?

22. Repeat Exercise 21, but *maximize* area under the conditions that the wire *must* be cut.

23. Find the rectangle of largest area that will fit inside the top semicircle of $x^2 + y^2 = r^2$ with one side on the diameter.

24. Determine an equation of the line tangent to $y = x^3 + 3x^2 + 3$ that has minimum *slope*. (Be careful.)

25. (*a*) A rancher wants to construct a rectangular corral divided into three compartments by two parallel wooden separators that are parallel to one end of the corral. She has \$1200 to spend on this corral. Given that the separators cost \$20/m and the outside fencing costs \$10/m with the requirement that no compartment be less than 8 m long or wide, how should she lay out the corral to achieve maximum total area?
(*b*) Repeat part (*a*) with all conditions the same except with the separators at \$30/m. (Be careful.)
(*c*) Given all conditions as above, at what price of the separators per meter does the 8-m requirement affect the solution?

26. A straight river is 120 m wide. A man standing on one bank wishes to arrive at a house 500 m upriver and on the other bank. If he can run twice as fast as he can swim, how should he proceed to reach the house in the least possible time? (Ignore the current.)

27. The *strength* of a rectangular beam is a constant k (which depends upon the material) times the product of its width and the square of its depth. Determine the dimensions of the strongest such beam that can be cut from a cylindrical log of radius R.

28. The *stiffness* of a rectangular beam is a constant c (which depends upon the material) times the product of its width and the cube of its depth. Determine the dimensions of the stiffest such beam that can be cut from a cylindrical log of radius R.

29. Give the dimensions of the isosceles triangle of largest area that can be inscribed in a circle of radius R.

30. (*a*) Suppose a person travels by boat (bringing along a bicycle) across a 1-mi-wide body of water from A to a landing point L (see diagram), and then bicycles to C at 12 mph. Given that she can row at 3 mph, where should her landing point be to minimize the total traveling time?

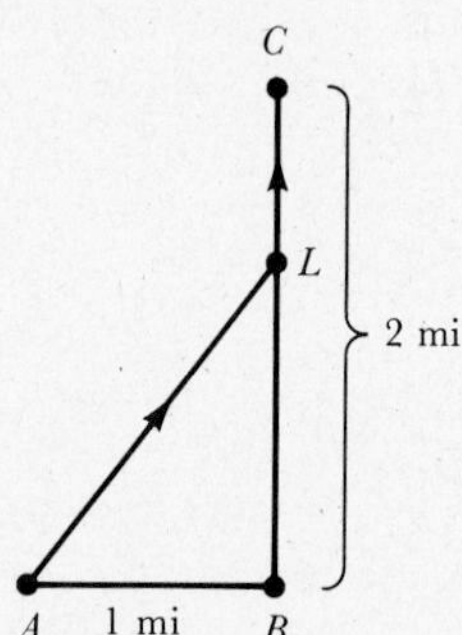

(*b*) Repeat (*a*) for a rowing speed of 5 mph.
(*c*) Repeat (*a*) for a rowing speed of K mph. (Let x be distance from B to L.)

(*d*) Suppose she can get an outboard motor. What speed of the boat is necessary for her to avoid bringing the bicycle?
(*e*) Is there any rowing speed that would justify $L = B$?

31. (*a*) A tank is to have volume 100 m^3 and is to be made in the form of a right circular cylinder with hemispheres attached at each end. If the material for the ends costs three times as much per square meter as that for the sides, find the most economical proportions. (Let c be the cost per square meter for the sides.)
(*b*) Given a tank exactly as in part (*a*) except that the material for the ends costs K times as much as that for the sides, compare the length of the cylindrical part of the tank with its diameter in terms of K.

32. A grain elevator is to be built in the form of a right circular cylinder surmounted by a hemisphere. The material for the floor costs $\$2/m^2$. The material for the sides costs $\$1/m^2$ and the material for the top costs $\$3/m^2$. If the grain elevator is to hold 1000 m^3, what would be the most economical dimensions?

33. Repeat Exercise 32 but with all materials costing the same amount per square meter.

34. Repeat Exercise 32 but with \$1200 to spend on materials and the *volume* to be maximized.

35. Repeat Exercise 32 but with \$1200 to spend on materials, all of which cost the *same* amount per square meter, and with the *volume* to be maximized.

36. What are the relative dimensions of a cylindrical can of fixed surface area S and maximum volume?

37. Find the length of the longest pole that can be carried horizontally around a corner from a corridor 4 m wide. (Note that this length turns out to be the minimum value of certain lengths.)

38. A Norman window is in the shape of a rectangle surmounted by a semicircle. Such a window being built into a church is to be bordered by a narrow strip of gold leaf 12 m in length. Find the dimensions of the Norman window of largest area that can be so bordered.

39. Determine the dimensions of the right circular cone of maximum volume that can be fit inside a sphere of radius 30.

40. If the largest package acceptable to the post office has cross-sectional girth and perimeter totaling at most 72 in., give the volume of the largest cylindrical package that can be mailed.

41. Towns A and B decide to share costs of a water purification plant along a river. Town A is 2.5 mi upriver from B and 2 mi from the river. Town B is 3 mi from the river (on the same side as A). Where along the river should the plant be located to minimize the total pipe involved to connect the towns to the plant?

42. In Exercise 41 it turned out that the plant divided the distance along the river into pieces, the ratio of whose lengths equaled the ratio of the distances of the towns from the river. By letting the distances from A and B to the river be a and b, respectively, determine if this holds in general.

Section 5.2: Related Rates

Objective:

Translate the given information and a given question regarding rates into derivative statements and questions, and then answer the given question by using relations among the given and asked-for rates.

A PROCEDURE FOR SOLVING RELATED RATE PROBLEMS

We now turn from applications involving the extreme values of a function to applications based upon the following fundamental fact: *If x is a quantity that is changing with time, then the instantaneous rate of change of x at time t is given by dx/dt.*

We are concerned with situations in which several quantities, all varying with time, are related by some equation or equations. Given the rate of change of *certain* of these quantities at a particular instant, we shall use the equations relating them to determine the rates of change of *other* quantities at this instant.

In each case the crux of the problem will be to understand that the rates of change, whether given or asked for, are derivatives.

We outline a procedure for handling such "related rate" problems and then illustrate with a few examples.

5.2.1 Procedure for Solving Related Rate Problems

1. Draw a diagram and label any fixed quantities.
2. Express all quantities in terms of time (explicitly or implicitly) and determine any relationships between these quantities.
3. Differentiate with respect to t explicitly or implicitly.
4. Insert the given quantities and solve for the unknown quantity.

Example 1 A balloon is being inflated at the rate of 15 m^3/min. At what rate is the diameter increasing when the diameter has reached 10 m? Assume that the diameter is zero at time zero.

Solution Letting V be the volume at time t and w its diameter at time t, we know $V = (4/3)\pi r^3$, where r is the radius. Since $r = w/2$, we have

$$V = \frac{\pi}{6}w^3$$

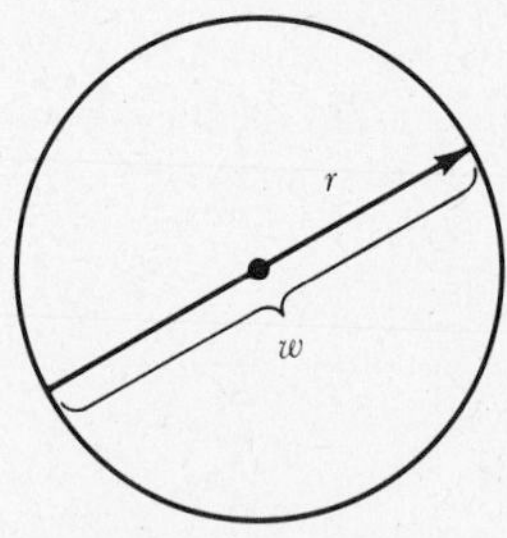

Remembering that V and w are functions of t, we want to interpret information given and asked for in terms of derivatives *with respect to t*. We are given that the rate of change of the volume is 15 m^3/min and we wish to find the rate of change of the diameter when $w = 10$. Thus in terms of derivatives, the problem can be restated: V and w are functions of t related by $V = (\pi/6)w^3$. Given that $dV/dt = 15$, find dw/dt when $w = 10$.

Differentiating with respect to t, we obtain (using the Chain Rule)

$$\frac{dV}{dt} = \frac{\pi}{2}w^2\frac{dw}{dt}$$

Solving for dw/dt, we have

$$\frac{dw}{dt} = \frac{2\,dV/dt}{\pi w^2}$$

Now dw/dt depends upon *both* dV/dt and w. But $dV/dt = 15$ and $w = 10$, so our final answer is

$$\frac{dw}{dt} = \frac{2\,dV/dt}{\pi w^2} = \frac{2(15)}{\pi(10)^2} \approx 0.095 \text{ m/min} \quad \bullet$$

A common error in solving problems of this type is to substitute the given values for t or another quantity into the expressions *before* differentiating. This leads to a dead end because, of course, all the derivatives will be zero. The substitutions should be made *after* differentiating.

Example 2 A conical tank has a depth of 12 ft and a radius at the top of 6 ft. If water is pumped into the tank at the constant rate of 4 ft^3/min, how fast is the radius of the surface of water in the tank changing at the instant that the depth of water in the tank is 6 ft?

Solution Let V be the volume of water in the tank at time t, and let r and h, respectively, denote the radius of the surface and the depth of the water in the tank at time t. Now V, r, h are related by

$$V = \frac{\pi}{3}r^2h$$

We are given that $dV/dt = 4$ and we wish to find dr/dt at the instant that $h = 6$. Since we are neither given nor asked to find the rate at which h is changing, it is convenient to eliminate h from the problem. To do this we examine a cross section of the tank (see the figure) and use similar triangles to conclude that

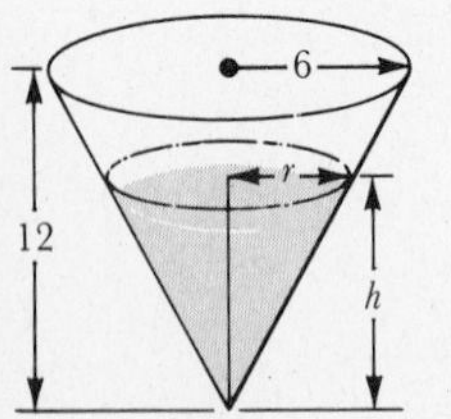

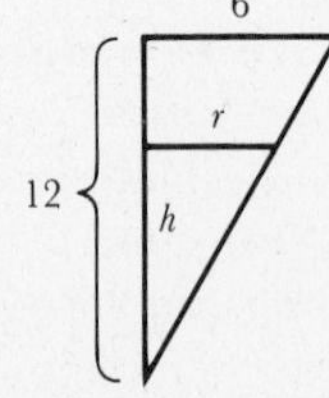

$$\frac{6}{12} = \frac{r}{h} \quad \text{or} \quad h = 2r$$

Thus

$$V = \frac{\pi}{3}r^2(2r) = \frac{2\pi}{3}r^3$$

Since $r = 3$ when $h = 6$, we can restate the problem as: given $V = (2\pi/3)r^3$ and $dV/dt = 4$, find dr/dt when $r = 3$. Now

$$\frac{dV}{dt} = 2\pi r^2 \frac{dr}{dt}$$

so

$$\frac{dr}{dt} = \frac{dV/dt}{2\pi r^2}$$

Substituting the given information, we have

$$\frac{dr}{dt} = \frac{4}{2\pi(3)^2} = \frac{2}{9\pi} \approx 0.071 \text{ ft/min} \quad \bullet$$

Example 3 Two automobiles start from a point A at the same time. One travels west at 80 mph and the other travels north at 45 mph. How fast are the distances between them increasing 3 h later?

Solution Let x be the distance the car traveling west has gone after t hours and y the distance traveled by the car going north (see diagram). Using our right triangle we see that s, x, and y are related by

$$s^2 = x^2 + y^2$$

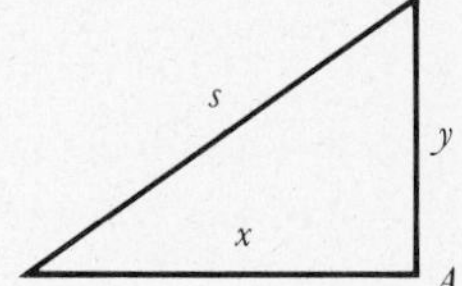

We can differentiate implicitly with respect to t to obtain

$$s\frac{ds}{dt} = x\frac{dx}{dt} + y\frac{dy}{dt}$$

We are given $dx/dt = 80$ and $dy/dt = 45$, and we wish to find ds/dt when $t = 5$. When $t = 5$,

$$x(5) = 80 \cdot 5 = 400$$
$$y(5) = 45 \cdot 5 = 225$$

and

$$s(5) = \sqrt{x(5)^2 + y(5)^2} = \sqrt{210{,}625}$$

Thus

$$\frac{ds}{dt} = \frac{400(80) + 225(45)}{\sqrt{210{,}625}}$$
$$= \frac{42{,}125}{210{,}625} \cdot \sqrt{210{,}625}$$
$$= \frac{(42{,}125)(25)\sqrt{337}}{210{,}625} = 5\sqrt{337} \approx 91.8 \text{ mph} \quad \bullet$$

PROGRESS TEST 1

1. A ladder 15 m tall leans against the side of a house. If the bottom is pulled away at the rate of 4 m/s, determine how fast the top is sliding downward when the bottom is 9 m from the side of the house by following the given outline:

(*a*) Draw a diagram and label the vertical distance y and horizontal distance x.
(*b*) Express all quantities in terms of time and give the relationship between x and y.
(*c*) Differentiate this relationship with respect to time.
(*d*) Insert the given information into the expressions determined in (*b*) and (*c*) to find dy/dt.

2. One airplane flew over an airport at the rate of 5 mi/min. Ten minutes later another flew over at the rate of 4 mi/min. If the first plane was flying west and the second south (at the same altitude), find how fast they were separating 20 min after the second plane flew over.

3. Water is pouring into a conical tank (vertex down) at the rate of 16 m^3/h. The tank is 18 m tall, with a radius on top of 6 m.
How fast is the water level rising when the water is 3 m deep?
(*Hint:* Draw a vertical cross-section diagram and eliminate r from the expression for V using similar triangles.)

WHEN IS A PROBLEM A CALCULUS PROBLEM?

Cover the solution to the following example and attempt its solution yourself.

Example 4 A scientist finds a 200-cm-tall conical stalagmite whose vertical height is 10 times its base diameter. Her evidence indicates that the stalagmite was formed by a constant dripping of a 1 percent limestone solution at the rate of 10,000 cm^3/yr. How old is the stalagmite?

Solution The volume of the stalagmite is

$$\frac{1}{3}\pi\, 10^2(200) \approx 20944 \text{ cm}^3$$

The amount of limestone deposited each year is (0.01)(10,000) = 100 cm^3. Since the rate of increase of volume is constant, the stalagmite is approximately 20944/100 = 209.4 yr old. ●

Despite the appearance of the word "rate," this was not a calculus problem. If all significant rates in the problem (as above) are constant, this is a clue that you can probably solve it without calculus.

SECTION 5.2 EXERCISES

1. Given the conical stalagmite situation of Example 4, find the annual rate of increase in height at the time the measurements were taken. Assume that the shape of the stalagmite remains the same.

2. Repeat Exercise 1 for the same stalagmite when it was only 1 yr old.

3. A tank in the shape of a right circular cone (vertex upward) of radius 7 m and altitude 18 m is being filled with water at the rate of 4 m^3/min.
(*a*) How fast is the water level rising at the instant when the water level is 9 m?
(*b*) How fast is the radius of the surface increasing at the instant when the water level is 15 m?

4. Repeat Exercise 3 with the same data except that the radius of the new tank is only 3 m.

5. At a given instant a small balloon is released from ground level at a point 100 m away from an observer on ground level. If the balloon goes straight up at the rate of 3 m/s, how fast is the balloon receding from the observer after 35 sec?

6. A light is located on the ground 50 m from a building. A man 2 m tall walks from the light toward the building at the rate of 5/3 m/s. How fast is his shadow growing shorter when he is 18 m from the building?

7. (*a*) How fast must a man who is 5 ft 9 in. tall walk out from under a 6-ft-high light bulb so that the leading edge of his shadow (on the ground) moves at 120 mph. (Be careful with units.)
(*b*) Is the shadow's speed constant if his is?

8. At a certain instant the lengths of the legs of a 3-4-5 right triangle are changing. If the longer one is increasing at the rate of 0.4 unit per second, how fast must the shorter one decrease to maintain a constant area? (Of course, it is a 3-4-5 triangle for only an instant.)

9. Repeat Exercise 8, but with the shorter leg increasing at 0.4 unit per second and the longer leg decreasing.

10. When gas is compressed without loss or gain of heat (as in an insulated piston compressor) it obeys the law $pv^{1.4} = k$, where k is a constant, p is the pressure, and v is the volume. This is called "adiabatic compression." Suppose the volume of a certain gas is decreasing at 2 in^3/s while its pressure is 36 lb/in.^{2} and its volume is 54 in.^{3}. At what rate is the pressure changing?

11. If an ideal gas is compressed nonadiabatically, it obeys the law $pv = k$, where k is a constant. Answer the question of Exercise 10 given the same data but obeying the $pv = k$ law.

12. (*a*) The volume of a sphere is increasing at the rate of 1 m^3/min. What rate is its radius increasing when the volume is 27π m^3?
(*b*) At what rate is the radius increasing 9 min later?

13. (*a*) A vertical line slides horizontally across the graph of

$y = x^2$ (for $x \geq 0$) at the rate of 1 unit of distance per unit of time. How fast is the point P of intersection of the line and the graph rising *vertically* when $x = 5$?
(*b*) When $x = 100$?
(*c*) How fast is the point P moving away from the origin at a typical x?

14. (*a*) Given a differentiable function of x and the conditions of Exercise 13(*a*), determine how fast P is rising vertically at a typical x.
(*b*) Determine how fast P is moving away from the origin at a typical x.

15. Suppose the vertical line of Exercise 13(*a*) is moving horizontally at the rate of 3 units per unit of time.
(*a*) How fast is the point P rising vertically at $x = 5$?
(*b*) How fast is the point P rising vertically at a typical x?
(*c*) How fast is P moving away from the origin at a typical x?

16. Suppose the vertical line of Exercise 13(*a*) is moving horizontally at the rate of k units per unit of time. Answer parts (*a*) to (*c*), Exercise 15.

17. Given that the vertical line of Exercise 13(*a*) is moving to the right at $v(t)$ units per unit of time, answer parts (*a*) to (*c*), Exercise 15. (*d*) What rule are you using in your analysis?

18. Suppose there is an opaque vertical band 5 units wide that slides across the coordinate plane horizontally, beginning with its right side at $x = 5$. Let P be the point on the graph of $y = x^2$ that is first to appear on the left side of the band as the band slides along. Answer parts (*a*) and (*b*) of Exercise 13 assuming that x is measured at the right side of the band.
(*c*) How fast is P rising vertically at a typical x?
(*d*) What would happen to the answer of (*c*) if we measured x from the left side of the band?

19. If P (as in Exercise 18) is moving along the curve $y = x^2$, $x \geq 0$ in such a way that its y coordinate is increasing at the rate of 1 unit per unit of time. How fast is its x coordinate increasing at a typical x?

20. Repeat Exercise 19 for the curve $y = x^3$ for $x \geq 0$.

21. (*a*) Repeat Exercise 19 for the curve $y = 5x + 3$ for $x \geq 0$.
(*b*) Suppose $y = mx + b$ is a typical linear function and repeat Exercise 19.

22. Repeat Exercise 19 for the curve $y = x^3 + x$ for $x \geq 0$.

23. At the instant a melting snowball is 16 cm in diameter it is melting at the rate of 2 cm^3/min. How fast is its diameter decreasing at that instant?

24. A 5-m-long horizontal eavestrough has an equilateral triangle cross section 0.1 m across the top.
(*a*) Given that the level of the water is increasing at the rate of 0.01 m/min when the water is at a depth of 0.05 m, at what rate is water flowing into the trough?
(*b*) Later, water is flowing out of the trough at 100 cm^3/min. At what rate is the water level dropping when the level is at 0.08 m?
(*c*) Repeat (*b*) with the water level at 0.02 m.

Section 5.3: Linear Motion

Objectives:

Describe the motion, velocity, and acceleration of a particle moving in a straight line when given a differentiable function describing its position as a function of time.

DESCRIBING THE MOTION OF A PARTICLE

Suppose $s = f(t)$ denotes the position of a particle moving along a straight line as a suitably differentiable function of time t. If $s > 0$, then the particle is s units to the *right* of the origin. If $s < 0$, then the particle is s units to the *left* of the origin; see Fig. 5.8. The point in time chosen as $t = 0$ is merely a reference point. In particular, negative values of t represent time *before* $t = 0$.

For example, suppose the position of the particle at time t is given by

$$s = f(t) = t^2 - 3t - 5$$

When $t = 0, f(0) = -5$, so initially the particle is located 5 units to the left of the origin. When $t = 5$,

$$f(5) = 25 - 15 - 5 = 5$$

and so the particle is 5 units to the right of the origin; see Fig. 5.9.

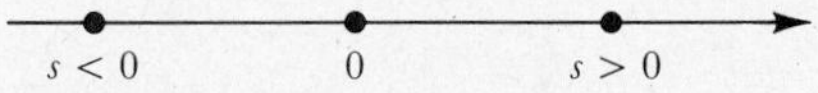

$s < 0$ 0 $s > 0$

Fig. 5.8

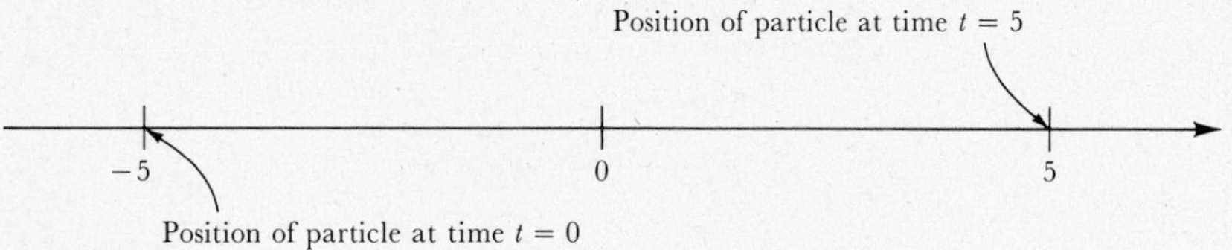

Fig. 5.9

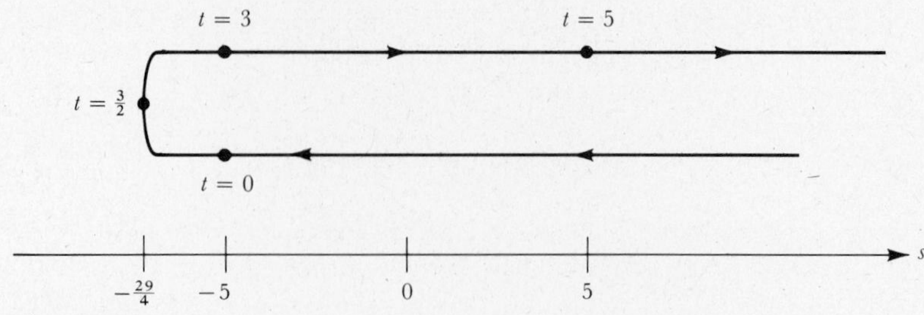

Fig. 5.10

We now investigate what information about the behavior of this particle can be deduced from the first derivative of $s = f(t)$. As we know, $f'(t)$ gives the velocity at time t. This quantity can be either positive or negative. If $f'(t) > 0$, then the particle is moving in the positive direction (to the right), whereas $f'(t) < 0$ means that the particle is moving in the negative direction (to the left). If $f'(t) = 0$, then the particle has stopped. In particular, at the instant at which the particle changes direction its velocity, $f'(t)$ must be zero.

Now consider the previous example. Since $f(t) = t^2 - 3t - 5$, we have

$$f'(t) = 2t - 3 = 2\left(t - \frac{3}{2}\right)$$

Now $2(t - 3/2) < 0$ for $t < 3/2$ and $2(t - 3/2) > 0$ for $t > 3/2$. Thus the particle is moving to the left for time $t < 3/2$, stopped at time $t = 3/2$, and then moving to the right for time $t > 3/2$. Now $f(3/2) = -29/4$, and so the behavior of the particle can be indicated diagramatically on the s axis, as shown in Fig. 5.10.

Acceleration is the rate of change of velocity with respect to time. Thus the derivative of the velocity—that is, the second derivative of $f(t)$—gives the acceleration.

$$s = f(t) \qquad v = f'(t) \qquad a = f''(t)$$

The acceleration tells us whether the particle is speeding up or slowing down. The key is to *compare the sign of the velocity and the acceleration at a given time. If they have the same sign, then the particle is speeding up. If they have opposite signs then the particle is slowing down.* Acceleration in a given direction can be thought of as the effect of a "pull" in that direction (as with gravity, which provides a constant acceleration). Thus when the direction of motion and the direction of acceleration are in agreement, the particle is speeding up in that direction. If these directions are opposite, then the acceleration is in the direction against the motion, so the particle is slowing down. In the above case $f''(t) = 2 > 0$ for all t. Thus $f'(t)$ and $f''(t)$ have opposite signs for $0 < t < 3/2$ and the same sign for $t > 3/2$. Hence the particle slows down from $t = 0$ to $t = 3/2$, and stops at $t = 3/2$. It then accelerates in the positive direction for $t > 3/2$.

The next example illustrates a systematic way to extract and organize information regarding the linear motion of a particle.

Example 1 Describe the motion of a particle whose position s at time t is given by $f(t) = t^3 - 3t^2 - 9t - 2$.

Solution We compute $f'(t)$ and $f''(t)$:

$$f'(t) = 3t^2 - 6t - 9 = 3(t - 3)(t + 1)$$
$$f''(t) = 6t - 6 = 6(t - 1)$$

We use a chart to determine the sign of $f'(t)$ and note the sign of $f''(t)$ directly below the chart and with the same t scale; see Fig. 5.11.

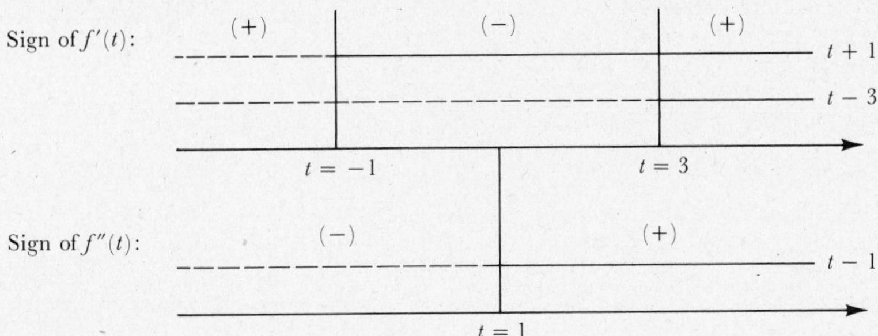

Fig. 5.11

We now evaluate the function at each of the critical values and use $f'(t)$ to diagram the motion parallel to an s axis, as in Fig. 5.12.

$$f(-1) = 3 \qquad f(3) = -29 \qquad f(1) = -13$$

Moving right $t < -1$; stopped at $t = -1$; moving left $-1 < t < 3$; stopped at $t = 3$; and moving right $t > 3$.

When $t < -1, f'(t) > 0$ and $f''(t) < 0$, so the particle is moving to the right and slowing down. It stops at $t = -1$. For $-1 < t < 1$ we have $f'(t) < 0$ and $f''(t) < 0$, so the particle is moving to the left and speeding up. For $1 < t < 3$, $f'(t) < 0$ and $f''(t) > 0$, so the particle continues to the left but is slowing down until it stops when $t = 3$. For $t > 3$ both derivatives are positive, so the particle continues to move right and speed up. ●

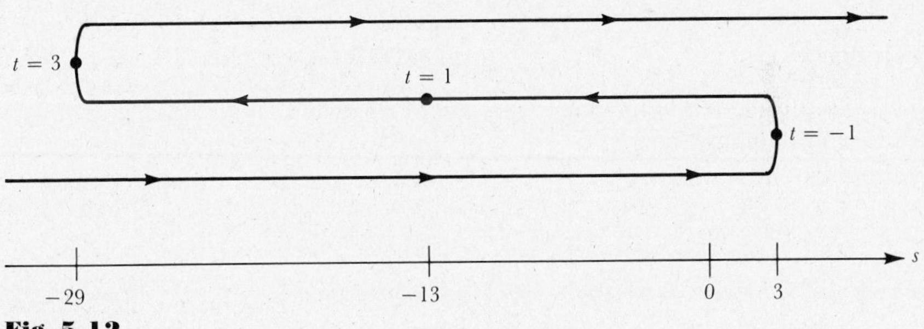

Fig. 5.12

It is important to remember, when using this procedure, that the *time* scale is used in the derivative sign charts and the *distance* scale is used in the motion diagram.

PROGRESS TEST 1

In the following problem $s(t) = t^3 - 9t^2 + 24t$ gives the position of a particle moving along a straight line as a function of t. Describe its motion following the given outline.

1. Find $s'(t)$ and $s''(t)$ and diagram their signs.

2. Evaluate s at each critical point.

3. Diagram the motion of the particle along the s axis.

4. Compare the signs of s' and s'' to determine when it is speeding up and slowing down.

SECTION 5.3 EXERCISES

In Exercises 1 to 12, $s(t)$ gives the position of a particle moving along a straight line. In each case describe in detail the motion of the particle.

1. $s(t) = t^3 - 3t^2 - 24t - 2$

2. $s(t) = 8 - 4t + t^2$

3. $s(t) = \dfrac{4t}{t^2 + 4}$

4. $s(t) = t^4 + 5t^3 + 6t^2$

5. $s(t) = t^4 + 2t^3 - 3t^2 - 4t$

6. $s(t) = 5 - 2t^2 - 3t^3$

7. $s(t) = 7 - 6t^2 + t^4$

8. $s(t) = t + \dfrac{1}{t} \quad (t > 0)$

9. $s(t) = 1 - t - \dfrac{1}{t} \quad (t > 0)$

10. $s(t) = \sqrt{4 + t^2}$

11. $s(t) = \sqrt{9 - t^2}$

12. $s(t) = 3t^5 + 5t^4 + 10t^3 + 30t^2 - 125t$

13. Graph the function of Exercise 1 on the t-s axis, using the sign of s' to determine whether s is increasing or decreasing; use the sign of s'' to determine the graph's concavity.

14. Repeat Exercise 13 for the function of Exercise 4.

15. × Graph your automobile trip home (or to another place along a familiar route) on the t-s axis, measuring t in minutes and s in miles according to the following directions. (Do a careful job on a large sheet of paper. This exercise will be useful as a reference later. You can do this as a "thought experiment" using careful estimates, or with actual measurements, using a clock, speedometer, and odometer.)

(*a*) Label two sets of axes with the appropriate scale for the length of your trip, one above the other with the same t scale. The upper axis will plot velocity against time, and the lower axis, the distance with respect to time. Time t is horizontal.

(*b*) Plot your velocity and distance in either order or simultaneously. (Use pencil because erasing will likely be necessary.) Label each important point in time A, B, C, . . . , (for example, a stoplight), and attach a narrative that is keyed to these labels and explains the entire trip in terms of both derivatives and descriptive language.

(*c*) Check that your three items correspond accurately; this is the key to the exercise.

Section 5.4: Using the Derivative to Help Understand Business and Economics

Objective:

Devise and optimize functions describing business and economics situations.

INTRODUCTION

The differential calculus, which has proved itself to be such a powerful tool for understanding dynamic processes and relationships in the physical sciences and engineering, has for the same reasons become an increasingly important tool in most other sciences as well. We shall illustrate its use by exploring some basic economic concepts, first in general terms, and then more concretely.

In economics and business the word "marginal" means "rate of change." Often the independent variable involves units other than time, such as units of production, sales, price, and so on. We now know that if $P(x) = 400x - x^2 - 5000$ represents profit as a function of x, the number of units produced, then the marginal profit (with respect to units produced) is dP/dx. The derivative arises as naturally here as it does anywhere.

At a given level of production x, we may ask, does it pay to increase production? The answer, of course, depends upon whether $P(x)$ happens to be increasing at that level; that is, whether dP/dx is positive. Here $dP/dx = 2(200 - x)$ is positive for $x < 200$, negative for $x > 200$, and is zero at $x = 200$, which is a critical point. Since we know that $x = 200$ is a relative maximum, the question is answered. (See Fig. 5.13.)

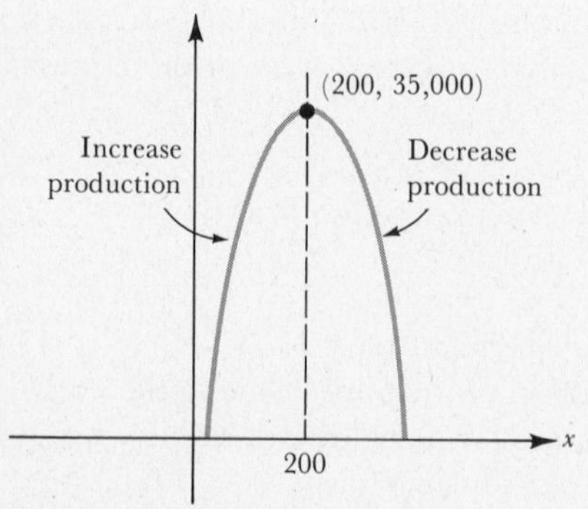

Fig. 5.13

More often than not, the variables in economics and business functions make realistic sense only at integral values (x equals the number of television sets produced, and so on) within some range determined by circumstances. Furthermore, the accounting data and surveys from which the functions are defined represent only samples and approximations. The usual practice is to define a smooth, differentiable function that best fits the data, and round off answers to the appropriate unit. Henceforth we shall assume that functions are defined and differentiable on an appropriate domain and that

marginal (. . .) means the derivative of (. . .)

We now define and examine a few of the more important issues and functions occurring in economics and business. Until further notice, $x \geq 0$ stands for the number of units of whatever is being produced (or sold, manufactured, mined, and so forth). Of course, the following discussion, being elementary, makes several simplifying assumptions to avoid complex issues that would only obscure the role played by the derivative. In particular we avoid complexities involving endpoint extremes for most functions.

MAXIMIZING PROFIT

Denote by $C(x)$ the *cost* of producing x units and by $R(x)$ the *revenue* resulting from the sale (or rental, lease, whatever) of these items. Then, in general,

$$P(x) = R(x) - C(x)$$

is the *profit* function, which normally is to be maximized. As we know, to maximize $P(x)$ we find

$$P'(x) = R'(x) - C'(x)$$

set $P'(x) = 0$, and look for relative maxima. This leads to the equation

$$R'(x) = C'(x)$$

which signifies the classic necessary economic criterion for maximal profit:

marginal revenue equals marginal cost.

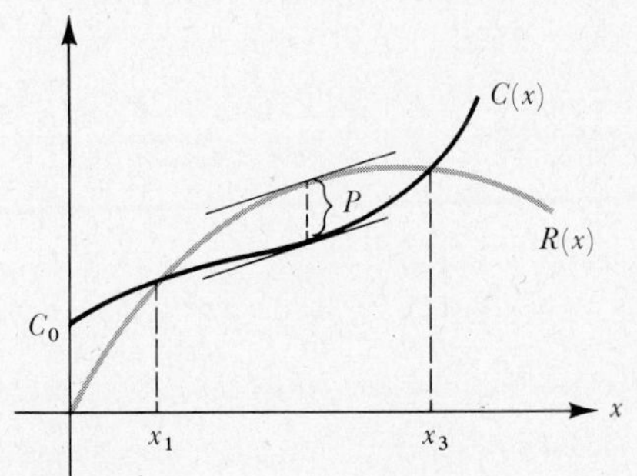

Fig. 5.14

Graphically, a typical situation is given in Fig. 5.14. Profit is the directed vertical distance between the revenue and cost curves, and is a maximum where this distance is greatest.

Now $C(x)$ is normally composed of fixed (nonproduction, overhead) costs C_0 plus variable (production) costs $C_1(x)$. Hence

$$P(x) = R(x) - [C_0 + C_1(x)]$$

so

$$P'(x) = R'(x) - C_1'(x)$$

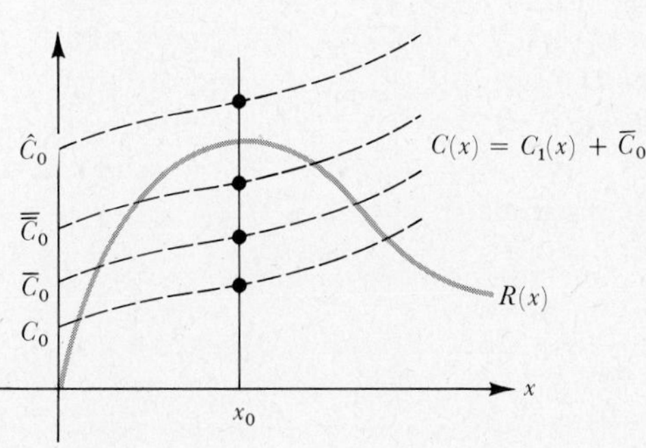

Fig. 5.15

This tells us that fixed costs do not enter into the profit maximizing equation, so actually *marginal revenue will equal marginal production cost.* This means ideally that an increase in overhead costs does not affect the production level decision; it simply means that total profit is decreased by that amount. In Fig. 5.15 we see several different situations, differing only in fixed costs. Optimal production level x_0 remains the same in each case, although with fixed costs $\widehat{C}_0$, the business operates at a loss at all production levels. There x_0 minimizes the loss. Often when making business decisions, it is useful to decompose $C_1(x)$ further when possible into component parts, such as material costs $C_M(x)$ and labor costs $C_L(x)$.

COMPETITION VERSUS MONOPOLY

To examine revenue $R(x)$ more closely, we need to know how $R(x)$ relates to price, which in turn depends upon how much the consumers are willing to pay for the items being produced and how the number of items produced affects the price. As we all know, an increase in price normally leads to a decrease in demand. That is, at a higher price, fewer items can be sold.

If the producer does not sell enough to affect the quantity demanded, as is the case when the producer accounts for only a small part of the market in that product, the demand and price are out of the producer's control; so the price is assumed to be constant, independent of x. Economists normally refer to this as "perfect competition." Letting p be the price, we know that

$$R(x) = px$$

so in a perfect competition situation, the equation $R'(x) = C'(x)$ reduces to

$$C'(x) = p$$

Hence to maximize profit, we find the production level x for which the marginal cost equals the price. This is the production level where the cost of producing one more item is approximately equal to the selling price of that item.

The opposite, monopoly situation, occurs when the producer sets the price $p(x)$. However, even though the price is under control, the demand usually is not, so if the price is too high, the number of units the monopolist will be able to sell will be too low. Under the reasonable assumption that the producer wishes to sell all that is produced, the question of maximizing profit reduces to setting a production level x so that the resulting price $p(x)$, at which all x items are sold, yields a maximum profit

$$P(x) = xp(x) - C(x)$$

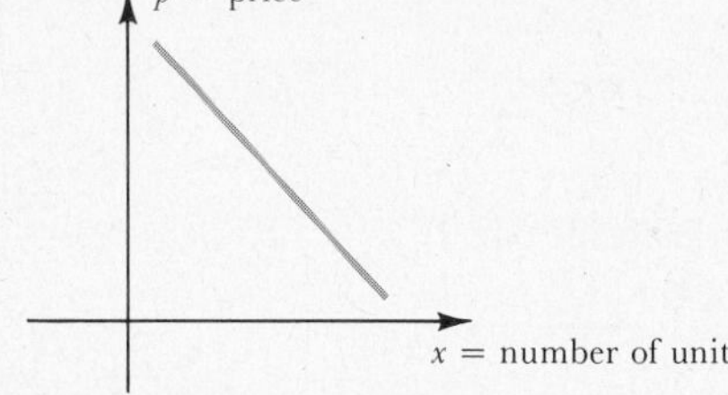

Fig. 5.16

The demand-price function is often assumed to be linear with negative slope within the bounds provided by the situation. Slope is negative because as x increases the price must be lowered to sell the x units produced (see Fig. 5.16).

This general discussion not only makes clear the value of the derivative as a conceptual tool in understanding fundamental economic relationships but also provides a ready means for dealing with specific situations.

Example 1 (Monopoly Situation) Suppose a large coal mining firm produces q carloads of coal per day at a daily cost of $C(q) = q^3 - 150q^2 + 5800q + 500$. If it can sell q carloads at a price of $p(q) = 1000 - 6q$ per carload, how many carloads per day should it produce to (*a*) maximize revenue, (*b*) minimize marginal cost, (*c*) maximize profit? (*d*) What is maximum profit? (The firm is in an area where alternative fuel sources are scarce.)

Solution (*a*) The revenue $R(q) = (1000 - 6q)q$ is maximized when

$$R'(q) = 1000 - 12q = 0 \qquad \text{or when } q = 83\tfrac{1}{3} \text{ carloads}$$

(*b*) The marginal cost is $C'(q) = 3q^2 - 300q + 5800$. $C'(q)$ is minimized when $C''(q) = 6q - 300 = 0$ or when $q = 50$ carloads. We know this is a relative minimum since $C'''(q) = 6 > 0$ for all q. [Note that it does not usually make sense to ask when *costs* are a minimum since in general $C(q)$ will be a strictly increasing function. A quantity more appropriate to minimize is the approximate cost of producing the "next" carload, which, as we remarked before, is approximately equal to $C'(q)$ for q large enough.]

(*c*) The variable price $p(q) = 1000 - 6q$ indicates this is a monopoly situation. We have profit:

$$P(q) = (1000q - 6q^2) - (q^3 - 150q^2 + 5800q + 500)$$

so $$P'(q) = 1000 - 12q - 3q^2 + 300q - 5800 = 0$$

when

$$q^2 - 96q + 1600 = 0$$

or when

$$q \approx 21.5 \text{ and } 74.5$$

Since $P''(q) = -6q + 288$, with $P''(21.5) > 0$ and $P''(74.5) < 0$, we know that maximum profit is achieved when about 74.5 carloads per day are produced. At this production level, the price per carload is \$553.

(*d*) The profit is $P(74.5) = \$41{,}198 - \$13{,}556 = \$27{,}642$. ●

Example 2 (Competitive Situation) Suppose the above coal producer is not a monopolist but faces a purely competitive situation with the going price per carload constant at \$553 and other conditions as in Example 1. What production level yields maximum profit? (Alternative fuel sources are readily available.)

Solution Setting marginal cost equal to \$553, we have

$$3q^2 - 300q + 5800 = 553$$

so

$$q = \frac{100 \pm \sqrt{3004}}{2} \approx 77.4,\ 22.6$$

with $q \approx 77.4$ yielding the maximum profit since $p''\ (77.4) < 0$. ●

Note that it pays to produce a few more carloads per day in the competitive situation, reflecting the ability to sell more coal at \$553 than before.

Progress Test 1 is the base for most of the discussion that follows, so you are urged to work it through in complete detail.

PROGRESS TEST 1 APARTMENT RENTAL AND THE EFFECT OF COLLEGE STUDENTS

× **Part 1 (Lower Demand)** An owner of a 120-unit apartment complex knows from experience that every unit can be rented at \$200/month but that each \$10 increase in rent causes a loss of four tenants.

1. Letting $x =$ the number of units rented, calculate a linear price function $p(x)$ that describes this demand. [*Hint:* Assume $p(120) = 200$, note that $\Delta p = 10$ for $\Delta x = -4$, and use the fact that $p(x)$ is linear.]
2. Calculate a revenue function $R(x)$ for this situation.
3. Determine the
   (*a*) Number of units to be rented to maximize revenue.
   (*b*) Rental price at this point.
   (*c*) Total revenue at this point.
4. Given that the monthly cost of maintaining a rented unit is \$60, the cost of maintaining a vacant unit is almost as high at \$40 (because the landlord must pay utilities on unrented units), and the fixed monthly costs are \$17,000, determine the resulting cost function $C(x)$.

5. (*a*) Determine the profit function $P(x)$ and the number of units to rent to achieve maximum profit.
   (*b*) What *is* the maximum profit?
   (*c*) What is the rental rate?

Part 2 (Higher Demand) Now suppose that a college in the neighborhood expands enrollment and demand for the apartments increases so that all 120 units can be rented at \$240/month and that each \$10 rent increase causes only two vacancies.

1. Calculate the corresponding price function $p_2(x)$.
2. Find the new revenue function $R_2(x)$.
3. Determine the
   (*a*) Number of units to rent to achieve maximum revenue.
   (*b*) Rental price at this point.
   (*c*) Total revenue at this point.
4. Given the same cost function as before, determine the
   (*a*) New profit function $P_2(x)$.
   (*b*) Number of units to rent to maximize profit.
   (*c*) Maximum profit.
   (*d*) New rental rate.

DISCUSSION OF APARTMENT RENTAL

× Although the figures in Progress Test 1 were chosen to exaggerate an effect slightly (especially regarding demand), the point of maximum profit is often less than full occupancy. It can *pay* to have vacant apartments, because the demand situation can change after the buildings are constructed, so that, in effect, the buildings become "too big." In the lower-demand situation the profit at full occupancy is $P(120) = (480)(120) - \frac{5}{2}(120)^2 - 21{,}800 = -200$, a *loss* of \$200. Furthermore, it *can* happen, as in the preceding examples, that a sharp increase in demand (perhaps even leading to a housing shortage) can lead to a shift *downward* in the optimal operating level. This is the sort of thing that can lead to six students sharing a three-room apartment at exorbitant cost.

For the remainder of this discussion we shall assume apartment demand to be at the lower level of part 1. Hence

$$p(x) = 500 - (5/2)x \text{ and } R(x) = 500x - (5/2)x^2$$

We also assume that costs are as before, and thus we obtain the following profit:

$$P(x) = \left(500x - \frac{5}{2}x^2\right) - (20x + 21{,}800)$$
$$= 480x - \frac{5}{2}x^2 - 21{,}800$$

In particular, then, our starting point is an occupancy level of $x = 96$, a rental rate of \$260, and a profit of \$1240. Finally, for simplicity, we shall assume that all quantities are given *per month.*

EFFECTS OF EXCISE VERSUS REAL ESTATE TAXES ON LANDLORD AND TENANTS

× Suppose now the local government decides to levy a 10 percent excise tax on all rentals above \$200. The landlord must now recompute the cost function, adding in the excise tax:

$$C_t(x) = C(x) + \frac{1}{10}\left[\left(500 - \frac{5}{2}x\right) - 200\right]x$$
$$= 50x - \frac{x^2}{4} + 21{,}800$$

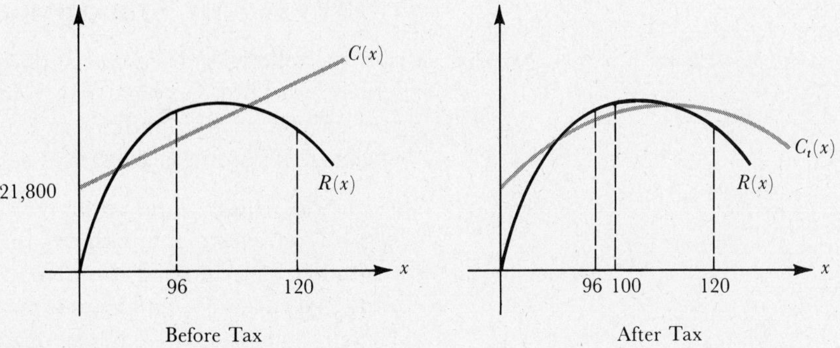

Fig. 5.17

The new profit function becomes

$$P_t(x) = \left(500x - \frac{5}{2}x^2\right) - \left(50x - \frac{x^2}{4} + 21{,}800\right)$$

and, setting $P_t'(x) = 0$, we see that maximum profit of \$700 occurs when $x = 100$ at a rental rate of $p(100) = \$250$; see Fig. 5.17. Here the effect of the excise tax is to change $C(x)$ to $C_t(x)$, which moves the point of maximum profit to the right, from $x = 96$ to $x = 100$. The net return to the government is

$$(100)\left[\frac{1}{10}(250 - 200)\right] = \$500$$

and the loss in profit to the landlord is \$540. From the tenants' point of view, there are four more tenants and the monthly rental rate drops from \$260 to \$250, a net *savings* due to the tax.

From the landlord's point of view, it would have been preferable by \$40 for the government to have raised the \$500 tax revenue through a \$500 increase in real estate tax. This would simply add to fixed costs and decrease profit by that amount, but not change anything else (see Fig. 5.18).

Quite clearly, by adjusting tax type and rate, the government is in a position to influence the housing situation considerably. By raising the excise tax to 20 percent of the rent over \$200, the cost, hence profit, functions change so that the occupancy level that yields maximum profit becomes 105 units and the rental rate \$237.50. The total tax revenue increases to \$787.50, and the landlord's profit drops to \$250.

An interesting (but algebraically complicated) exercise is to calculate the excise tax rate that yields maximum tax revenue.

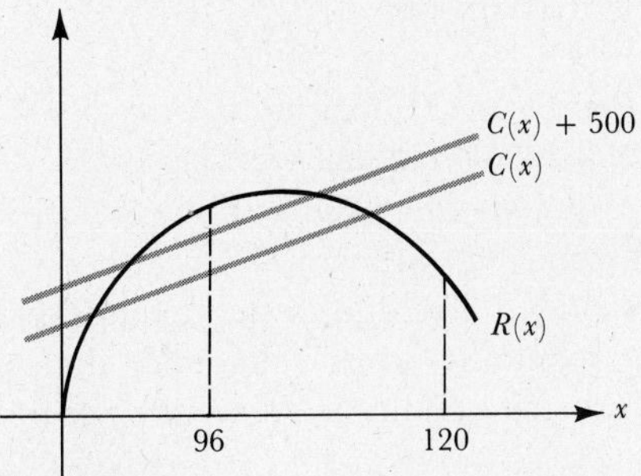

Fig. 5.18

THE EFFECT OF SUBSIDIES ON LANDLORD AND TENANTS

✖ For the following discussion suppose we are at the starting point again with no excise tax, but the local government raises its real estate tax rate, so the increased cost to the landlord is \$1000, bringing fixed costs to \$22,800 and lowering total profit to \$240. We compare two types of subsidies.

Landlord Tax Rebate: The landlord, claiming his \$240 profit to be insufficient return on his investment, petitions the government for a tax rebate based upon the number of vacancies in the apartment complex. In particular, noting his current 24 vacancies as grounds for tax relief, he asks for and receives an agreement guaranteeing a rebate of \$20 for each vacancy. The reasoning in the tax office is that this will lower the \$1000 increase by about (24)(20) = \$480 and increase the landlord's profit by that amount.

However, what the tax office does not take into account is that the demand for apartments is not affected, whereas the landlord's revenue and profit functions are. He will readjust his rental and occupancy rates accordingly so as to maximize profit. The cost function, including the \$1000 real estate tax increase, is $20x + 22{,}800$.

The revenue function increases from

$$500x - \frac{5}{2}x^2$$

to

$$\left(500x - \frac{5}{2}x^2\right) + 20(120 - x)$$

making the new, subsidized, profit function

$$\begin{aligned} P_s(x) &= \left(500x - \frac{5}{2}x^2\right) + 20(120 - x) - (20x + 22{,}800) \\ &= 460x - \frac{5}{2}x^2 - 20{,}400 \end{aligned}$$

Then $P_s'(x) = 460 - 5x = 0$, which implies that only $x = 92$ units should be rented. The rental rate is

$$p(92) = 500 - \frac{5}{2}(92) = \$270$$

and the total rebate from local government is (20)(28) = \$560, so only \$440 of the \$1000 tax increase goes to the city. Furthermore, the new profit is $P_s(92) = \$760$, which is \$520 more than before the rebate.

What is the net result of *this* subsidy? The city loses \$560 in tax revenue, the landlord gains only \$520 in profit, but four fewer units are rented, and the rent for the remaining 92 tenants goes up by \$10, meaning that the tenants end up paying an additional total of \$920. This is obviously an inefficient and unfair sort of subsidy.

Rent Subsidy: Beginning with the same conditions, including the new \$1000 real estate tax increase, suppose a federal rent subsidy program is begun. Although most such apply to lower-income families and are a function of income, we will simplify matters by assuming that all tenants get a \$20 rent subsidy. This affects the demand-price situation as follows.

Assuming that economic, housing, and population conditions are unchanged, the linear demand-price function has the same slope $(-5/2)$ because a \$10 increase in rent still drives out four tenants. With the \$20 subsidy, the landlord can now rent all 120 units at \$220 (rather than \$200 as before). Let p_r denote the new rent linear subsidy price function since $p_r(x)$ has the form

$$p_r(x) = \frac{-5}{2}x + b$$

with $$p_r(120) = 220 = \frac{-5}{2}(120) + b$$

we conclude that $$p_r(x) = 520 - \frac{5}{2}x$$

It is your turn to compute the other effects of the subsidy in Progress Test 2.

PROGRESS TEST 2

1. Determine the new rent subsidy profit function $P_r(x)$.
2. Determine the number of units rented to maximize profit.
3. Determine the new profit.
4. Compute the amount the landlord received for each unit rented.
5. What is the net amount each tenant must pay after the subsidy?
6. Compute the total net subsidy to tenants and compare this with the increase in profit to the landlord.

× RENT SUBSIDY AND RENT CONTROL

As a direct consequence of the rent subsidy, $1000 went to the tenants (a $10 savings per tenant), and the landlord's profit increased by $1960, almost double that amount. Thus to call this a "rent subsidy" is somewhat misleading. (The increase in profit "makes room" for an additional rise in real estate taxes of perhaps $1000. This would leave the landlord with $1200 profit and make the rent subsidy a form of federal revenue sharing.) However, four more units are rented than before. Because the supply-demand situation was altered by the $20 subsidy, the landlord was able to raise rents from $260 to $270, meaning that the $20 subsidy only saved $10 for each tenant.

This naturally raises the issue of whether the supply-demand situation should be controlled in some way. For example, suppose the rent subsidy is coupled with a rent freeze on the landlord, so the $20 goes directly toward the lowering of the rent for the 96 tenants, from $260 to $240. This lower net rent, since demand is still the same, results in attracting more potential tenants. As far as the potential tenants are concerned, the rental price is now $240, so that $p(x) = 240 = 500 - (5/2)x$, which implies $x = 104$. This means that eight additional tenants are willing to move in. However, with the rent freeze, the landlord receives $260 per unit, making the total revenue $(104)(260) =$ $27,040, while costs amount to $(20)(104) + 22{,}800 =$ $24,880. Hence profit is $27{,}040 - 24{,}880 =$ $2160.

Again, the landlord's profit rises almost as much as without the freeze. Furthermore, the total rent subsidy of $(20)(104) =$ $2080 goes toward lowering the rent of the 104 tenants. Hence the landlord and the tenants both have about $2000 more to spend even though the net subsidy was only $2080. What makes the difference between this and the uncontrolled rent subsidy is the increased number of tenants. The net effect of the controls is to make the apartments affordable for eight more tenants, who then contribute, as do all the others, to a share of the profit. This situation contrasts sharply with the tax rebate scheme, which encouraged lower occupancy and higher rents. Note, however, that if the landlord's costs were rising more rapidly than revenue ($260x$) beyond the $x = 96$ point, the additional eight tenants who were able to afford rental at a price of 240 (to them, 260 to the landlord) would not be admitted if the landlord wanted to maximize profit. This same sort of situation can lead to shortages when prices are frozen: Producers strike, refusing to produce beyond the maximal profit point (although the word "strike" is seldom used).

Remarks on Our Simplifications: Besides being rigged to make calculations simple at each turn, the previous discussion assumes a simple linear price-demand function. This makes assumptions about the willingness of tenants to move in response to rent changes. These "elasticity" assumptions are not entirely realistic. Also, we ignored the impact of taxes and subsidies on demand in the larger rental market. Since demand at a particular apartment complex depends upon the larger market, some of our assumptions regarding constancy of demand are also not entirely realistic. Nonetheless, the analysis overall is reasonable enough to reflect the ways that derivatives can be used to analyze the effect of economic policies. Furthermore, a more sophisticated analysis based on a more complicated mathematical model can account for those phenomena ignored in the above discussion. Incidentally, we shall put the apartment complex up for sale in Sec. 8.1.

SECTION 5.4 EXERCISES

1. Suppose a firm can sell any number of its products at \$12 apiece, but it costs $0.001x^3 + 0.3x + 50$ to produce x items.
(*a*) Calculate the revenue function for this situation.
(*b*) Calculate the profit function.
(*c*) Determine which production level yields maximum profit and the actual profit at that level.
(*d*) Determine the level at which *revenue* is maximized (be careful!).

2. Repeat Exercise 1(*a*) to (*c*) for a new selling price of \$15.

3. A company charges \$100 per item for the first 200 items, but in an effort to encourage larger orders and increase revenue it makes a special offer to lower the price by \$0.25 per each item of the entire order for each item ordered beyond 200. What size order will yield maximum revenue for the company? (This is the maximum size order the company would accept under the special offer.)

4. (*a*) Repeat Exercise 3 but with the price of only those items beyond 200 lowered by \$0.25.
(*b*) Repeat part (*a*) with the price lowered by k dollars.
(*c*) When $k = \$0.50$ in part (*b*), the maximum-size order turns out to be 200. Explain. Use a graph.

5. Suppose that a wholesaler has purchased 180 identical television sets for \$150 apiece. She finds she can sell them all at a price of \$160, but for each additional dollar added to the price, one fewer set can be sold.
(*a*) How many sets should she sell to maximize sales revenue?
(*b*) Give the profit function for this situation.
(*c*) How many should she sell to maximize profit?
(*d*) Compare the profit at the point of maximum revenue with the maximum profit.

6. (*a*) Suppose you were hired to sell the television sets of Exercise 5 at a commission of 10 percent. How many would you sell to maximize your commission?
(*b*) What is your maximum commission?

7. Repeat Exercise 6 with a 25 percent commission. Explain your decision.

8. Let Q represent quantity of a product sold. Suppose Q is related to its price p by $Q = k[4 - (1/p)]$, where k is some constant. The government wishes to impose an excise tax on this product. However, if the tax rate t is too high, the quantity sold will fall, and thus tax revenues will diminish.
(*a*) Find the quantity saleable Q_t as a function of tax rate t.
(*b*) Determine the tax revenue function R.
(*c*) At which tax rate t will revenue be maximized? (Note that your answer will involve p and k.)

9. (Average Cost) In certain situations where profit is not important, it is still important to minimize cost per item. (A production run of free samples, a government press run of tax forms, a run of license plates at the prison, and so on.) This average cost is $C(x)/x$.
(*a*) Analyze average cost for $C(x) = 5x + 4$.
(*b*) Analyze average cost for any linear cost function $C(x) = ax + b$.
(*c*) What is the effect of large fixed costs b?
(*d*) Find the point of minimum average cost for $C(x) = 0.00025x^2 + x + 1000$.
(*e*) Note that the average cost is the slope of the line from the origin out to the cost function. Graph the function in part (*c*) accurately and analyze the slope of this line.

10. A monopolistic enterprise has a price (demand) function $p(q) = a - bq$ and a cost function $C(q) = Aq^2 + Bq + C$, where a, b, A, B, C are positive constants with $a > B$. Given that the company will attempt to maximize net profit (after taxes), what rate r should the government charge on each unit q to maximize tax $T = rq$?

11. Given that the company in Exercise 10 will ignore taxes in its cost function and production level decision and merely pass them on to consumers, what should be the government's optimal tax rate? [Note that the tax will change the price-demand function by adding rq to $p(q)$ without changing demand.]

12. Bill Ashley estimates that his August 10 peach crop will be 2000 bushels, with a profit of \$2/bushel. He also estimates that the crop will grow by 80 bushels/day and, because of increasing market glut, his profit per bushel drops by $1\frac{1}{2}$ cents/day. Given a prohibitive frost threat after 30 days, when should he pick the crop to maximize profit?

13. Given the same basic conditions as in Exercise 12, except variable marketing estimates, determine the

(*a*) Daily price drop that would justify picking on August 10.

(*b*) Daily price drop that would justify picking on September 10.

CHAPTER EXERCISES

Review Exercises

In Exercises 1–20 determine (a) the maximum and (b) the minimum values of the given function on the given interval (if these values exist).

1. $f(x) = \frac{x^3}{3} + \frac{x^2}{2} + x + 3$ on $[1, 4]$

2. $f(x) = \frac{(x+3)}{(x+5)}$ on $[-9, -6]$

3. $f(x) = \frac{x^3}{3} - \frac{7x^2}{2} - 18x - 1$ on $[-3, 10]$

4. $f(x) = x(x^2 + 9)$ on $[0, 5]$

5. $g(x) = \frac{x^4}{2} - 3x^3 + 6x^2 - 4x$ on $[-1, 3]$

6. $h(x) = \frac{x^2}{2x+1}$ on $\left[-\frac{1}{4}, 0\right]$

7. $h(x) = \frac{x^2}{2x+1}$ on $[-1, 0]$

8. $y = \frac{3x+2}{\sqrt{4x^2-1}}$ on $\left(\frac{1}{2}, 1\right]$

9. $y = x - \sqrt{x}$ on $[0, +\infty)$

10. $f(x) = x^2 - 2x + 5$ on $[-3, 4]$

11. $f(x) = x^3 - \frac{21}{4x^2} + 9x - 4$ on $[1, 2]$

12. $f(x) = \frac{1}{4}x^4 - \frac{3}{2}x^2$ on $[1, 2]$

13. $f(x) = x^{4/3} - 4x^{1/3}$ on $[-5, 0]$

14. $f(x) = x^3 - x^2 + x - 1$ on $[1, 4]$

15. $f(x) = \frac{4x}{(x^2+4)}$ on $[-3, 4]$

16. $f(x) = x^2 + \frac{1}{x^2}$ on $\left[-2, \frac{-1}{2}\right]$

17. $f(x) = \frac{2x^2}{x^2+1}$ on $[-3, 4]$

18. $f(x) = (x+2)(x-2)^3$ on $[-2, 0]$

19. $y = \frac{x}{(3x+1)^2}$ on $\left[\frac{1}{3}, +\infty\right)$

20. $f(x) = (8 - x)^{2/3}$ on $[0, 16]$

21. The product of two numbers is 16. Determine two such numbers so that the square of one plus the cube of the other is as small as possible.

22. Repeat Exercise 21 with the square of one plus the cube of the other as *large* as possible.

23. A fence 6 m high is 2 m from the side of a house. What is the length of the shortest ladder that will rest on the level ground outside the fence and lean on the side of the house?

24. What are the dimensions of the base of a rectangular box of greatest volume that can be constructed from 100 in.^{2} of cardboard if the base is to be twice as long as it is wide?

25. At noon, ship A, steaming east at 16 mph, is due south of ship B, which is steaming south at 12 mph. They are 100 mi apart at noon. At what time are they closest together?

26. A metal cup in the shape of a right circular cylinder is to hold exactly 100 in.^{3} of liquid. What are the dimensions of such a cup which would be the least expensive to make?

27. A particle is moving along a straight line with the position at time t given by

$$f(t) = \frac{2t-3}{1+t^2}$$

(*a*) Determine in which direction the particle is moving when $t = 0$, $t = 3$.

(*b*) Is the particle slowing down or speeding up when $t = 3$?

28. A particle is moving along a straight line with its position at time t given by

$$f(t) = \frac{t-2}{2t^2+3}$$

(*a*) Determine in which direction the particle is moving when $t = 0$, $t = 2$.

(*b*) Is the particle slowing down or speeding up when $t = 2$?

29. Describe the motion of the particle whose position at time t is given by $f(t) = 2t^3 + 3t^2 - 36t + 1$.

30. Describe the motion of the particle whose position at time t is given by $f(t) = 2t^3 - 12t^2 - 30t + 2$.

31. A conical tank, full of water, is 12 ft high and 20 ft in diameter at the top. If water is let out at the bottom at the rate of 4 ft^3/min, find the rate of change of the depth of the water 10 min after the water starts running out.

32. A man starts walking eastward at 5 ft/s from a point A. Ten seconds later a man starts walking west at 2 ft/s from a point B, 3000 ft north of A. How fast are they separating 2 min after the second man starts?

33. A light is on the ground 40 ft from a building. A man 6 ft tall walks from the light toward the building at 6 ft/s. How rapidly is his shadow on the building growing shorter when he is 20 ft from the building?

34. At the rate of 8 ft^3/min, water flows into an initially empty conical tank that is 10 ft high and 10 ft in diameter at the top. How fast is the water level rising after 4 min?

35. Let the cost of selling x items be $S(x)$, so the average marketing cost per item is $S(x)/x$. Show that the average marketing cost per item is minimized when the marginal sales cost equals the average sales cost.

36. A tank in the shape of a right circular cylinder of radius 6 m is being filled with water at the rate of $1/3\ \text{m}^3/\text{s}$. How fast is the *radius of the surface* increasing at the instant when the depth is 4 m?

37. Repeat Exercise 36, but determine the rate at which the height h is increasing. Is either Exercise 36 or 37 a calculus problem?

38. A man $5\frac{3}{4}$ ft tall walks at the rate of 3 ft/s toward a street light that is 18 ft above the ground. At what rate is his shadow changing when he is 8 ft from the base of the light?

39. Sand is pouring from a spout at the rate of $3\ \text{m}^3/\text{min}$ and forming a conical pile in such a way that the radius at the base is always one-third the altitude. How fast is the altitude of the pile increasing at the instant when the radius of the base is 2 m?

A particle moving along a straight line has its position at time t given by $s(t)$. In Exercises 40 to 44 give a detailed description of the motion of the particle:

40. $s(t) = \dfrac{t^3}{3} - \dfrac{t^2}{2} - 6t$

41. $s(t) = t^4 - 32t + 48$

42. $s(t) = t^3 + 3t^2 + 3t + 2$

43. $s(t) = \dfrac{t}{t^2 + 5}$

44. $s(t) = t^4 - \dfrac{4}{3}t^3$

Miscellaneous Exercises

1. (*a*) We are given a 200-m stone wall and 400 m of fencing. Using the stone wall as one side of the pasture, what are the dimensions and area of the largest rectangular pasture that can be enclosed?
(*b*) Repeat part (*a*), but with the "rectangular" condition weakened into the "trapezoidal" condition indicated in the accompanying diagram.

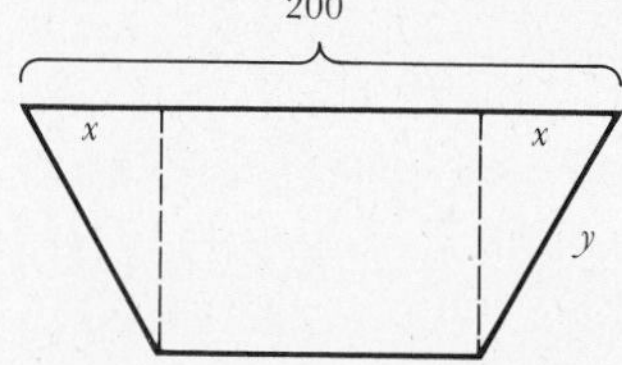

(*c*) What, in terms of area gained, is the value of the weaker "trapezoidal" condition over the "rectangular" condition?

2. A wire of length L is cut into two pieces, one of which is bent to form a circle and the other an equilateral triangle.
(*a*) How should the wire be cut so that the area enclosed by the two pieces will be a maximum?
(*b*) Is there a solution to the corresponding minimum problem?

3. (*a*) Find the point of maximum sales commissions given that x dozen urns can be sold at $50 - x$ dollars apiece, and that the commission rate is 10 percent.
(*b*) Is this point different if the commission rate is 20 percent?

4. Suppose we are producing the urns sold in Exercise 3 and our cost function is $C(x) = x^3 + 2x + 100$ (x is dozens of urns).
(*a*) How many urns would we produce to maximize revenue?
(*b*) How many urns would we produce to maximize profit?
(*c*) How many urns would we produce if we also had to add to our cost the commissions paid in part (*b*) of Exercise 3?
(*d*) Would the production level need to be changed from part (*c*) if the commission rate were lowered from 20 to 10 percent? Explain your answer.

5. A manufacturer will make rectangular open-topped boxes from square sheets of cardboard 1 m on a side by cutting out squares from each corner and folding up the sides.
(*a*) What are the dimensions of such boxes having maximum volume?
(*b*) In an effort to utilize waste, the manufacturer decides to make similar boxes from the squares cut out, in such a way that the total volume of the five boxes constructed from a single sheet is a maximum and the smaller boxes are twice as tall as they are wide. What are the dimensions of each type of box? (Approximate to two decimal places.)
(*c*) Compare the total volume in parts (*a*) and (*b*).

6. Suppose that a seller asks \$100 for an item and a buyer offers \$80. One way for the seller to "split the difference" and settle on a price p is to maximize the product $(100 - p)(p - 80)$.
(*a*) Find the price p that maximizes this product.
(*b*) Discuss the result in (*a*) graphically and determine whether it really "splits the difference."

7. Repeat Exercise 6(*a*) in the general case where the seller's price is p_1 and the buyer's offer is p_2.

8. Suppose the union's last wage request was for \$10/h, management's last offer was \$7/h, and negotiations are stalemated. An arbitrator comes in with a settlement, as she puts it, "to maximize the function $3p^3 - 576p + 10{,}000$."
(*a*) One side claims that this settlement skews the average in favor of its opposite. Which side makes the claim?
(*b*) Suppose you are in the union and the arbitrator offers to let you pick whole numbers m and n to maximize the formula

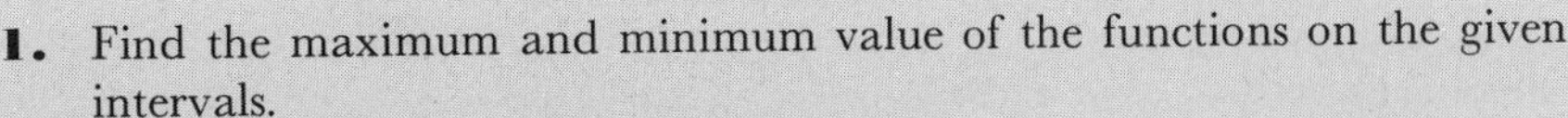

$$(m + n)^2p^3 - (m^2p_1^2 + 2mnp_1p_2 + n^2p_2^2)p + 10{,}000$$

where p_1 is your last offer and p_2 is management's last offer. How would you pick m and n under the restriction that m and n cannot differ by more than 2?

9. What are the dimensions of the rectangular pasture that can be enclosed at a fixed cost C, given that the costs per unit length of the four sides are, respectively, a, b, c, and d? Examine this situation for underlying regularities.

SELF-TEST ✤

1. Find the maximum and minimum value of the functions on the given intervals.
(*a*) $f(x) = x^3 - 3x^2 - 9x + 10$ on $[-2, 4]$
(*b*) $f(x) = \dfrac{x+1}{x+2}$ on $[0, 4]$

2. Describe in detail the motion of the particle whose position at time t is given by $s(t) = 2t^3 - 21t^2 + 36t - 1$.

3. A water trough with vertical cross section in the shape of an equilateral triangle (one vertex down) is being filled at the rate of 4 ft^3/min. If the trough is 12 ft long, how fast is the water level rising when the water reaches a depth of $1\frac{1}{2}$ ft?

4. Your business produces murns, which can be sold at a price of $99 - 2x$ dollars apiece, where x is the thousands of murns sold, and which cost $(1/3)x^3 + 3x + 900$ to produce. (You are a monopolist in the murns market.)
(*a*) How many murns should you produce to maximize your revenue?
(*b*) How many should you produce to maximize profit and what is your selling price?
(*c*) Given your answer to (*b*), how many murns would a purely competitive producer sell at that price level to maximize profit?

5. A rectangular box has a square base and no top. The combined area of the sides and bottom is 48 ft^2. Find the dimensions of the box of maximum volume meeting these specifications.

Chapter 6

INTEGRATION

CONTENTS

Chapters 1 to 5 dealt with the differential calculus and some of its applications. We now begin the study of the integral calculus, whose connection with the differential calculus will be provided by the Fundamental Theorem of Calculus. Just as tangent lines and velocity played an important role in motivating and providing concrete realizations for derivatives, area under a curve will do the same for integrals.

We show in Sec. 6.1 how differentiation can be turned around to yield antidifferentiation, with applications essentially reversed from those for derivatives. (Given the velocity function, we determine the position function, for example.) We then turn to approximating areas in Sec. 6.2, laying the groundwork for the definition and properties of the definite integral in Sec. 6.3. The notation for the definite integral will be very similar to that for the antiderivative, and the actual bridge between the two will again be provided by the Fundamental Theorem.

In Secs. 6.4 to 6.6 we shall turn to some basic applications of integration: averages, areas between curves, and volumes of certain solids, respectively. Although more applications appear in later chapters, these will be prototypes for what follows.

Section 6.1: Antidifferentiation

Objectives:

1. Solve simple differential equations, including those involving acceleration and velocity functions.
2. Use substitution and the Power Rule to determine antiderivatives.

AN INTRODUCTION TO ANTIDIFFERENTIATION

Given a function f, we can in most cases routinely determine its derivative by applying the appropriate formula.

Suppose now we consider the inverse problem: Given a function f, determine a function F whose derivative is f. More precisely, we have:

6.1.1 **Definition** A function F is called *an antiderivative of f* on the interval I, if $F'(x) = f(x)$ for all x in I. Determining such functions F is called *antidifferentiation.*

Consider, for example, the function

$$f(x) = 3x^2$$

defined on $(-\infty, +\infty)$. Since $D_x(x^3) = 3x^2$, we know $F(x) = x^3$ qualifies as an antiderivative of $f(x) = 3x^2$. But so do

$$F_1(x) = x^3 + 5 \qquad F_2(x) = x^3 - \frac{2}{3} \qquad F_3(x) = x^3 + \frac{2\sqrt{5}}{11}$$

It is fairly obvious that any function of the form

$$F(x) = x^3 + C$$

when C is any real number, is an antiderivative of $f(x) = 3x^2$. Using that the derivative of a constant is zero, we can conclude that a similar statement holds for antiderivatives in general.

6.1.2 **Theorem** If F is an antiderivative of a function f on an interval I and if C is any constant, then $F(x) + C$ is also an antiderivative of f on I.

From a geometric point of view this fact is also quite plausible, since the graph of $F(x) + C$ is vertically parallel to that of $F(x)$ in the sense that lines tangent to any two antiderivatives at points with the same x coordinate must be parallel. (See Fig. 6.1.)

Furthermore, as a consequence of the Mean-Value Theorem in Chap. 5, we showed that all antiderivatives of $f(x)$ differ by a constant. This converse to (6.1.2) can be restated as follows:

6.1.3 **Theorem** If $G(x)$ and $F(x)$ are both antiderivatives of $f(x)$ on I, then there is a constant C such that

$$G(x) = F(x) + C \qquad \text{for all } x \text{ in } I$$

Thus, for example, even though $f(x) = 3x^2$ has infinitely many antiderivatives, we can be assured they are *all* of the form $x^3 + C$ for C a real number.

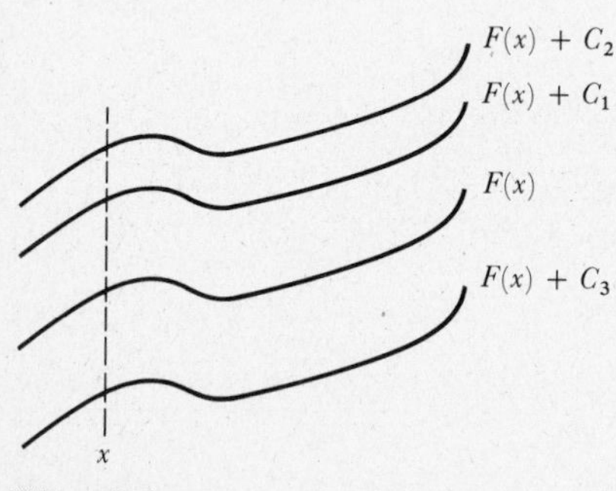

Fig. 6.1

SOME NOTATION AND BASIC ANTIDERIVATIVE RULES

We use the notation

$$\int f(x)\,dx$$

to indicate the process of antidifferentiating the function f. Thus we can write

$$\int 3x^2\,dx = x^3 + C$$

to denote that $x^3 + C$ is a typical antiderivative of $f(x) = 3x^2$.

In effect, $\int(—)\,dx$ indicates the operation of determining an antiderivative in the same way that

$$D_x(—) \qquad \frac{d}{dx}(—) \qquad \text{or} \qquad (—)'$$

all indicated the reverse operation of determining the derivative.

In what follows, when an interval is not mentioned in an antiderivative statement, it is tacitly assumed that the function f and the antiderivative F are both defined on some interval with, of course, $F' = f$ on that interval.

Example 1 Determine a typical antiderivative in each of the following:

(*a*) $\int x^2\,dx$ (*b*) $\int 4x^8\,dx$ (*c*) $\int \sqrt{x}\,dx$

Solutions (*a*) $\displaystyle\int x^2\,dx = \frac{x^3}{3} + C$ because $D_x\left(\frac{x^3}{3} + C\right) = \frac{3x^2}{3} + 0 = x^2$.

(*b*) $\displaystyle\int 4x^8\,dx = \frac{4x^9}{9} + C$ because $D_x\left(\frac{4x^9}{9} + C\right) = 9\frac{4x^8}{9} + 0 = 4x^8$.

(*c*) Here it is helpful to write $\sqrt{x}$ as $x^{1/2}$. Then $\displaystyle\int \sqrt{x}\,dx = \int x^{1/2}\,dx = \frac{x^{3/2}}{3/2} + C$, or $\displaystyle\int \sqrt{x}\,dx = (2/3)\sqrt{x^3} + C$. Here

$$D_x\left(\frac{2}{3}\sqrt{x^3} + C\right) = D_x\left(\frac{2}{3}x^{3/2} + C\right) = \frac{3}{2}\cdot\frac{2}{3}x^{1/2} + 0 = \sqrt{x} \quad \bullet$$

As Example 1 illustrates, antidifferentiation often involves reverse use of derivative formulas, combined with a bit of creative manipulation. To help in this process we shall state some derivative rules in their antiderivative form:

6.1.4 (Basic Antiderivative Rules)

(*a*) $\int kf(x)\,dx = k\int f(x)\,dx$ for k a constant.

(*b*) $\int[f(x) + g(x)]\,dx = \int f(x)\,dx + \int g(x)\,dx$.

(*c*) (*Power Rule*) If $r \neq -1$ is a rational number, then

$$\int x^r\,dx = \frac{x^{r+1}}{r+1} + C$$

Each of these result from applying the corresponding derivative rule. For example, the derivative of the right-hand side of (*b*) is the sum of the derivatives:

(1) $D_x[\int f(x)\,dx + \int g(x)\,dx] = D_x[\int f(x)\,dx] + D_x[\int g(x)\,dx]$

But, by definition of the antiderivative,

$$D_x[\int f(x)\,dx] = f(x) \qquad \text{and} \qquad D_x[\int g(x)\,dx] = g(x)$$

so the right-hand side of (1) is $f(x) + g(x)$. However, again by definition of the antiderivative, this is the derivative of the left side of (*b*), proving (*b*).

Notice that (*a*) means that constants can be pulled in front of an antiderivative sign, as in

$$\int 3x^2\,dx = 3\int x^2\,dx = 3\left[\frac{x^3}{3} + C\right] \qquad \text{(by the Power Rule)}$$

$$= x^3 + 3C$$

Since the constant C added to an antiderivative represents any arbitrary real number, so does $3C$. It is customary to write the arbitrary constant for a typical antiderivative simply as C. In fact, we normally do not write the constant until we reach the last step of the antiderivative process as in the next example.

Example 2 Evaluate: (*a*) $\int (4x^5 + 9x)\,dx$ (*b*) $\int \left(3\sqrt{t} + \frac{2}{t^2}\right) dt$

Solutions Check the result by differentiation.

$$(a)\ \int (4x^5 + 9x)\,dx = \int 4x^5\,dx + \int 9x\,dx \qquad \text{by [6.1.4}(b)\text{]}$$

$$= 4\int x^5\,dx + 9\int x\,dx \qquad \text{by [6.1.4}(a)\text{]}$$

$$= 4\frac{x^6}{6} + 9\frac{x^2}{2} + C \qquad \text{by [6.1.4}(c)\text{]}$$

$$= \frac{2}{3}x^6 + \frac{9}{2}x^2 + C$$

$$\textit{Check: } D_x\left[\frac{2}{3}x^6 + \frac{9}{2}x^2 + C\right] = 4x^5 + 9x$$

$$(b)\ \int \left(3\sqrt{t} + \frac{2}{t^2}\right) dt = \int (3t^{1/2} + 2t^{-2})\,dt$$

$$= 3\int t^{1/2}\,dt + 2\int t^{-2}\,dt$$

$$= 3\frac{t^{3/2}}{3/2} + 2\frac{t^{-1}}{-1} + C$$

$$= 2t^{3/2} - \frac{2}{t} + C$$

$$\textit{Check: } D_t\left[2t^{3/2} - \frac{2}{t} + C\right] = 3t^{1/2} - 2(-1)t^{-2} \qquad \bullet$$

The definition of the antiderivative provides an unusual opportunity for checking potential answers. As usual, the form in which we leave our answer depends upon what, if anything, we plan to do with it. Generally, we shall leave our answers in forms involving positive exponents and simplified fractions.

You may have wondered why the differential appears in antiderivative statements. In fact, the underlined part of

$$\int \underline{f(x)\,dx} = F(x) + C$$

is the differential of $F(x)$ because

$$dF(x) = F'(x)\,dx = f(x)\,dx$$

Hence, taking the antiderivative of both sides, we have

6.1.5
$$\int dF(x) = F(x) + C$$

Although later we shall see theoretical reasons for including differentials in our notation, at this point it is mainly a convenience whose value will become more apparent as we go along.

PROGRESS TEST 1

Evaluate 1 to 4. Check your answer.

1. $\int \frac{3\,dx}{x^3}$

2. $\int (3t^2 + 4\sqrt{t})\,dt$

3. $\int \frac{6z + z^{1/3}}{2z^{3/4}}\,dz$ (break into a sum)

4. $\int (w^2 + 1)^2\,dw$ (expand first)

5. Show that variables *cannot* be pulled in front of antiderivative signs, by showing $\int x^3\,dx \neq x\int x^2\,dx$. (Two antiderivatives are guaranteed to be unequal if their derivatives are unequal.)

APPLICATIONS—DIFFERENTIAL EQUATIONS

A *differential equation* is an equation involving unknown functions and their derivatives. To *solve* a differential equation is to determine a function or class of functions that satisfy the given equation. Usually the solution of a differential equation involves antidifferentiation. For example,

$$\frac{dy}{dx} = 3x^2$$

is a differential equation that we can rewrite as $dy = 3x^2\,dx$. Its solution is determined by antidifferentiation [applying (6.1.5)]:

$$y = \int dy = \int 3x^2\,dx = x^3 + C$$

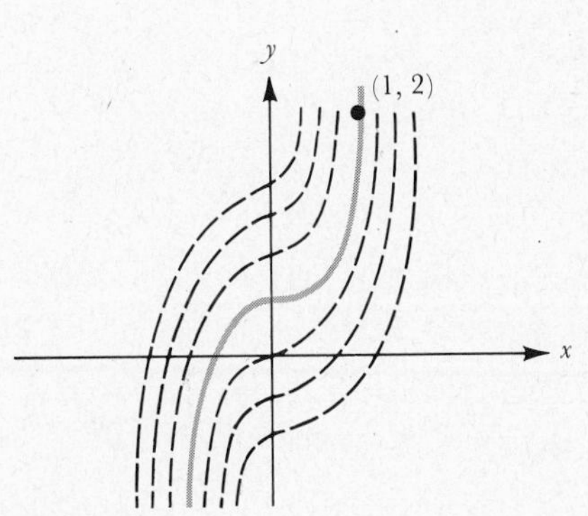

Fig. 6.2

In this case the solution is an infinite class of functions, one for each real number C. If we are given an additional condition, we can usually specify a particular function from among all those satisfying the original equation. If we are told, for example, that $y = 2$ when $x = 1$, then we know that

$$2 = 1^3 + C$$

so $C = 1$. Thus $y = x^3 + 1$ is the *unique* solution of the differential equation $dy/dx = 3x^2$ under the additional condition that $y = 2$ when $x = 1$. Geometrically, functions satisfying $dy/dx = 3x^2$ have graphs such as the family of parallel curves in Fig. 6.2, and the effect of the additional condition is to single out the unique function whose graph (solid curve) passes through (1, 2).

A differential equation arises most frequently as the mathematical expression of some given fact or relationship. We could arrive at the same differential equation just studied by stating in terms of derivatives that "the slope of the graph of the function at the number x is $3x^2$."

Most commonly, differential equations are used in physics and engineering—for example, in describing motion with acceleration due to gravity. Recall that if we measure distance positively in the upward direction, then acceleration and velocity in the downward direction will be negative. In particular, ignoring air resistance, the downward acceleration due to gravity on earth is approximately 32 ft/s^2. Since acceleration is the derivative of velocity v, we have a differential equation

$$\frac{dv}{dt} = -32 \qquad \text{or} \qquad dv = -32\,dt$$

Thus

$$v(t) = \int -32\,dt = -32t + C$$

Hence, knowing the velocity at a particular time t_0, (say, $t_0 = 0$), we have a completely determined function that gives the velocity. Commonly, the velocity at $t = 0$ is called the *initial velocity* and is denoted v_0, so

6.1.6 $$v(t) = v_0 - 32t$$

Example 3 Given that an object is thrown vertically upward at 96 ft/s, determine the velocity at $t = 2$, $t = 3$, and $t = 4$.

Solution By (6.1.6),

$$v(t) = 96 - 32t$$

so

$$v(2) = 32 \text{ ft/s} \qquad v(3) = 0 \text{ ft/s} \qquad v(4) = -32 \text{ ft/s} \quad \bullet$$

Just as we were able to differentiate more than once, we can antidifferentiate more than once. Knowing that (instantaneous) velocity is the derivative ds/dt of the position function, we have the differential equation

$$\frac{ds}{dt} = v_0 - 32t \qquad \text{or} \qquad ds = (v_0 - 32t)\,dt$$

Hence we have the function giving the height above the ground at time t:

$$s(t) = \int (v_0 - 32t)dt$$

$$= v_0 t - 16t^2 + C$$

Again, given the *initial position*, that is $s(0)$, usually denoted s_0, we have the uniquely determined function

6.1.7 $$s(t) = s_0 + v_0 t - 16t^2$$

Example 4 Determine the height above the ground after 2, 3, 4 s, respectively, of an object thrown vertically upward from a platform 10 ft above ground at 96 ft/s.

Solution We are given that $s(0) = s_0 = 10$, so by (6.1.7)

$$s(t) = 10 + 96t - 16t^2$$

Thus

$$s(2) = 138 \text{ ft} \qquad s(3) = 154 \text{ ft} \qquad s(4) = 138 \text{ ft} \quad \bullet$$

We constructed the function in (6.1.7) in two stages, beginning with the constant gravity function. We *could* have presented the original problem as a *second-order* differential equation:

6.1.8 $$\frac{d^2s}{dt^2} = g \qquad (g \text{ is the constant gravitational acceleration})$$

To solve (6.1.8) for a particular function, we need two additional conditions, one on velocity and one on position—usually given as initial conditions.

Example 5 (The acceleration due to gravity in metric units is 9.8 m/s^2.) Suppose an object is thrown upward from 10 m below ground level at an initial velocity of 50 m/s. What is the maximum height in meters reached by the object, and what is its velocity in meters per second when it reaches ground level again?

Solution First we must determine its velocity and height functions.

$$\frac{ds}{dt} = v(t) = \int(-9.8)\,dt = -9.8t + C_0$$

Now $v(0) = (-9.8)(0) + C_0 = 50$, so $v(t) = 50 - 9.8t$. Hence

$$s(t) = \int(50 - 9.8t)\,dt = 50t - 4.9t^2 + C_1$$

and

$$s(0) = C_1 = -10$$

so

$$s(t) = 50t - 4.9t^2 - 10$$

By both common sense and the work of Chap. 5, the object reaches maximum height when $ds/dt = v(t) = 0$.

But $50 - 9.8t = 0$ when $t \approx 5.1$ s, and

$$s(5.1) \approx 117.6 \text{ m}$$

The object reaches ground level when

$$s(t) = 50t - 4.9t^2 - 10 = 0$$

or when

$$t = \frac{50 \pm 48}{9.8} \approx 0.2 \text{ or } 10$$

The object thus reached ground level the second time 10 s after being thrown upward. Its velocity is $v(10) = 50 - (9.8)(10) = -48$ m/s. ●

PROGRESS TEST 2

The downward acceleration on the moon is approximately 5 ft/s.

1. Give a formula for the vertical velocity of an object hurled upward from the surface of the moon with an initial velocity of 20 ft/s.
2. Give the height function for this object.
3. Supposing the greatest initial vertical velocity of a hit baseball is about 150 ft/s, give the minimum roof height of a domed baseball park on the moon (so that pop flies will not hit the roof).
4. (*a*) How long would it take a ball hit vertically with an initial velocity of 150 ft/s to return to the surface of the moon?
   (*b*) What would its velocity be at that time?
5. Solve the second-order differential equation $d^2y/dx^2 = 12x + 2$ given that $y = 3$ when $x = 0$ and $dy/dx = 10$ when $x = 1$.

THE POWER RULE FOR FUNCTIONS; ELEMENTARY SUBSTITUTION

After some reflection, we can compute

$$\int(2x^3 + 5)^{10}6x^2\,dx$$

as follows. We are looking for a function whose derivative is $(2x^3 + 5)^{10}6x^2$, and we know that by the Power Rule for Functions (3.2.1), the derivative of $(2x^3 + 5)^n$ is $n(2x^3 + 5)^{n-1}(6x^2)$, for n an integer. If we take $n = 11$, we shall have solved the problem, provided that we divide through the 11 and add the constant C:

$$\int(2x^3 + 5)^{10}(6x^2)\,dx = \frac{(2x^3 + 5)^{11}}{11} + C$$

In effect we recognized the original antiderivative to be of the form

$$\int u^{10}\,du \qquad \text{with } u = (2x^3 + 5),\ du = 6x^2\,dx$$

From this we concluded that

$$\int u^{10}\,du = \frac{u^{11}}{11} + C = \frac{(2x^3 + 5)^{11}}{11} + C$$

Antiderivative problems are seldom this neat; more typically, we need to solve a problem such as

$$\int x^2(2x^3 + 5)^{10}\,dx$$

However, our procedure remains the same. We must change the form of the function we are dealing with into one that we recognize as the derivative of some other function. Often the best way to do this is to make a substitution and take the differential as follows:

$$\text{Let } u = 2x^3 + 5, \text{ so } du = 6x^2\,dx$$

We then compare du with $x^2\,dx$ by solving $du = 6x^2\,dx$ for $x^2\,dx$ to obtain $x^2\,dx = du/6$. Thus the original antiderivative problem can now be stated as

$$\begin{aligned}\int x^2(2x^3 + 5)^{10}\,dx &= \int u^{10}\frac{du}{6}\\ &= \frac{1}{6}\int u^{10}\,du \qquad \text{by [6.1.4(}a\text{)]}\\ &= \frac{1}{6}\frac{u^{11}}{11} + C \qquad \text{by the Power Rule [6.1.4(}c\text{)]}\\ &= \frac{(2x^3 + 5)^{11}}{66} + C\end{aligned}$$

Check: $D_x\left[\frac{(2x^3 + 5)^{11}}{66} + C\right] = \frac{11(2x^3 + 5)^{10}}{66}(6x^2) + 0 = (2x^3 + 5)^{10}x^2$

In the preceding problem we actually used the Power Rule for Functions (3.2.3), which we now state in its antiderivative form.

6.1.9 **Power Rule for Functions** If u is a differentiable function of x and $r \neq -1$ is a rational number, then

$$\int u^r\,du = \frac{u^{r+1}}{r + 1} + C$$

Most applications of (6.1.9) are of the preceding type: We find a factor of the form

$$(\text{expression in } x)^r$$

inside the antiderivative sign. If such exists, we try the substitution

$$u = \text{expression in } x$$

That is, we let u stand for the expression in x *inside* the parentheses, and compare du with the factors of the integrand other than u^r. *We shall succeed if du turns out to be a constant multiple of these other factors.*

Example 6 Determine $\int \left(\frac{x\,dx}{\sqrt{2x^2 + 9}}\right)$ and check the result.

Solution Rewrite the original problem as $\int(2x^2+9)^{-1/2}x\,dx$. Let $u = 2x^2 + 9$ [*not* $(2x^2+9)^{-1/2}$!!]. Now $du = 4x\,dx$, so $x\,dx = (1/4)\,du$. Thus we can substitute:

$$\begin{aligned}\int (2x^2+9)^{-1/2}x\,dx &= \int u^{-1/2}\frac{1}{4}\,du\\ &= \frac{1}{4}\int u^{-1/2}\,du\\ &= \frac{1}{4}\frac{u^{1/2}}{1/2} + C \qquad \text{by the Power Rule for Functions}\\ &= \frac{1}{2}u^{1/2} + C = \frac{1}{2}(2x^2+9)^{1/2} + C\end{aligned}$$

Check: $$D_x\left[\frac{1}{2}(2x^2+9)^{1/2} + C\right] = \left(\frac{1}{2}\right)\frac{1}{2}(2x^2+9)^{-1/2}(4x) + 0 = \frac{x}{\sqrt{2x^2+9}} \qquad \bullet$$

Note that when using substitution methods you should always give the answer in terms of the original variable. The substitution variable should not appear in your answer.

PROGRESS TEST 3

Using (6.1.9), determine

1. $\int x(1-5x^2)^7\,dx$

2. $\int (x+1)(x^2+2x+7)^{1/3}\,dx$

3. $\displaystyle\int \frac{dv}{\sqrt{(2v+1)^3}}$

4. $\displaystyle\int \left(2+\frac{1}{w}\right)\frac{dw}{w^2}$

MORE ON SUBSTITUTION

Example 7 Determine $\int(x^2+1)^3x^3\,dx$.

Solution Letting $u = x^2 + 1$, we know that $du = 2x\,dx$ so $x^3\,dx = (x^2/2)\,du$. Unfortunately, $x^3\,dx$ and du do *not* differ by a constant factor. However,

$$u = x^2 + 1 \text{ implies } x^2 = u - 1$$

and so we can replace the x^2 factor on the right side of

$$x^3\,dx = \frac{x^2}{2}\,du$$

by $u - 1$, getting $x^3\,dx = (u-1)/2\,du$. Thus

$$\begin{aligned}\int (x^2+1)^3x^3\,dx &= \int u^3\left(\frac{u-1}{2}\right)du\\ &= \frac{1}{2}\int (u^4 - u^3)\,du\\ &= \frac{1}{2}\left(\frac{u^5}{5} - \frac{u^4}{4}\right) + C\\ &= \frac{(x^2+1)^5}{10} - \frac{(x^2+1)^4}{8} + C \qquad \bullet\end{aligned}$$

We hasten to note that manipulations such as that in Example 7 will not always solve an antidifferentiation problem. Consider, for example,

$$\int \sqrt{x^2 - 11}\, dx$$

Here, letting $u = x^2 - 11$, we find $du = 2x\, dx$, so we are missing a factor of x. We *cannot* solve this problem by an elementary substitution. We shall devote Chap. 11 to solving antidifferentiation problems of this and even more complicated types.

SECTION 6.1 EXERCISES

In Exercises 1 to 8 find the indicated general antiderivative (check your answer).

1. $\int (x^2 + x)\, dx$ **2.** $\int (3x - 5)\, dx$ **3.** $\int x^{-2}\, dx$ **4.** $\int \sqrt[3]{x}\, dx$

5. $\int (\sqrt{x} + 5)\, dx$ **6.** $\int \frac{dx}{x^4}$ **7.** $\int \left(\frac{1}{x^3} + x\right) dx$ **8.** $\int \left(1 + \frac{2}{\sqrt{x}} + \sqrt{x}\right) dx$

9. In Exercises 1, 3, 5, and 7 determine the unique antiderivative passing through the points (0, 1) (for Exercise 1), (−1, 3) (for Exercise 3), (1, 20/3) (for Exercise 5), $(2, 5\pi + 3/4)$ (for Exercise 7), respectively.

10. In Exercises 2, 4, 6, and 8 determine the unique antiderivative passing through the points (−1, 8), (1, 3/4), (−2, 1), and (4, 18), respectively.

In Exercises 11 to 14 solve the second-order differential equation using the given information:

11. $\frac{d^2y}{dx^2} = x$; $y = 1$ when $x = 0$, and $\frac{dy}{dx} = 3$ when $x = 2$.

12. $\frac{d^2y}{dx^2} = 1 - 6x$; $y = \frac{5}{2}$ when $x = -1$, and $\frac{dy}{dx} = -5$ when $x = -1$.

13. $\frac{d^2s}{dt^2} - 12t = 0$; $\frac{ds}{dt} = 25$ when $t = 1$, $s(0) = 0$.

14. $\frac{d^2s}{dt^2} + 60t^4 = 12t^2$; $\frac{ds}{dt} = 0$ when $t = 0$, $s(1) = -100$.

15. (*a*) Find the maximum height achieved by an object that has an initial vertical velocity of 100 ft/s at ground level and a vertical acceleration of −10 ft/s² (it is partially self-propelled).
(*b*) Find the length of time needed for the object in (*a*) to reach the ground again.

16. A rock dropped from a bridge hits the water in 4 s. Ignoring air resistance as usual, calculate the height of the bridge above the water.

17. Suppose you are on a large planet where the acceleration due to gravity is −100 ft/s². From the same height as the bridge in Exercise 16, how long would it take the stone to reach the water?

18. Suppose the greatest initial vertical velocity of a hit baseball is about 150 ft/s on the planet with gravitational acceleration of −100 ft/s². (The baseball and bat would be heavier, but the players stronger.)
(*a*) How high should be the roof of a domed stadium?
(*b*) How long would it take the ball to return to the surface after being hit with an initial velocity of 150 ft/s?

In Exercises 19 to 34 determine the indicated antiderivative.

19. $\int (x^5 + 3)^7 x^4\, dx$

20. $\int x^2(6 - x^3)^4\, dx$

21. $\int (x^3 + x^2 + 1)^4(3x^2 + 2x)\, dx$

22. $\int (x^2 - 2x + 1)^{11}(x - 1)\, dx$

23. $\int \sqrt{x^3 - 3x}(x^2 - 1)\, dx$ **24.** $\int \left(\sqrt{x} + \frac{2}{\sqrt{x}}\right) dx$ **25.** $\int \frac{dx}{(2x - 1)^3}$ **26.** $\int x^3\sqrt{9 + x^4}\, dx$

27. $\int \sqrt{x + 1/x}\left(\frac{x^2 - 1}{x^2}\right) dx$ **28.** $\int \frac{9x^2 + 3x}{x^{1/3}}\, dx$ **29.** $\int (x^2 - 1)^2\, dx$ **30.** $\int \sqrt[5]{(2x + 5)^3}\, dx$

31. $\int \frac{(\sqrt{x} + 1)^9}{\sqrt{x}}\, dx$

(*In Exercises 32 to 34 it may be helpful to refer to Example 7.*)

32. $\int (x^3 + 2)^7 x^5\, dx$ **33.** $\int (x^2 - 5)^5 x^3\, dx$ **34.** $\int (x + 1)^3 x\, dx$

35. Determine the function whose graph has slope mx (m a real number) at any x and passes through the point (0, 0).

Section 6.2: Approximating Areas

Objectives:

1. Approximate areas of regions bounded by simple functions using rectangle approximations.

2. Take limits of these approximations to determine exact areas for linear or quadratic functions.

3. Understand the relationships among the various types of approximations and the limits of these approximations.

INTRODUCTION

Suppose R is a bounded region in a plane enclosed by some curve C, as in Fig. 6.3. Although we may not have a ready formula that would give us the area of R, we can be reasonably sure that as long as its boundary is not too wild, it *has* a well-defined area. In fact, we shall *assume* this to be the case and concentrate on *determining* the area. Later, in (6.5.1), we use the theory developed to define the notion of area precisely.

We can determine the area of a straight-sided (polygonal) region relatively easily by decomposing it into disjoint triangles and summing their areas; however, the problem of determining the area of a region with curved sides is much more challenging. The ancient Greeks, especially Archimedes, used what was called the "method of exhaustion." They approximated the curved region using smaller and smaller straight-sided regions—obviously a type of limiting process. Even though the Greeks achieved remarkable accuracy, particularly in special cases such as the circle, more than 2000 years elapsed before the essential idea of *limit* was sufficiently developed to provide the means for a more systematic and complete solution to the area problem.

The solution we shall develop evolved slowly through the seventeenth and eighteenth centuries and was finally put in the form we shall use in the middle of the nineteenth century by the German mathematician Georg Riemann. This approach will use rectangles as the approximating figure.

But why study area? Aside from the fact that the area problem is of considerable intrinsic interest, it provides the concrete realization of mathematical ideas of much wider utility—in much the same way as the notion of instantaneous velocity provided a starting point in our study of instantaneous rates of change and derivatives. In fact, it will turn out that many problems that on the surface have nothing to do with area will have solutions that amount to finding the area of a given region. See p. 189 for further discussion of this issue.

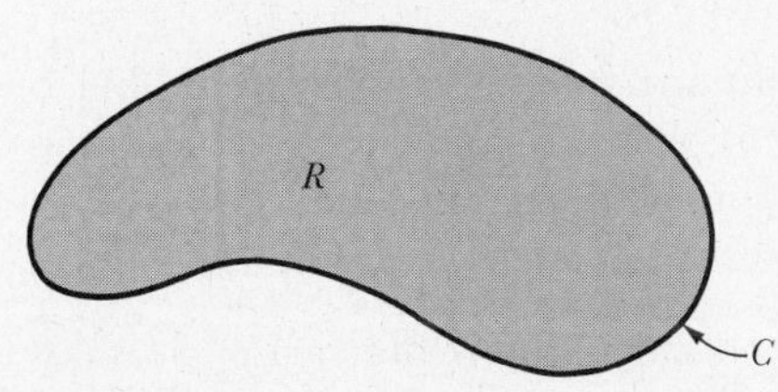

Fig. 6.3

APPROXIMATIONS

To apply our experience with functions and limits, we first assume that the region R, whose area A we seek, is bounded by the graph of a continuous function $f(x)$, the lines $x = a$ and $x = b$ (with $a < b$), and the x axis. (See Fig. 6.4.) Furthermore, to avoid complications with negative quantities, and to ensure that the limit representing the area of R turns out to be nonnegative, we assume that $f(x) \geq 0$ on $[a, b]$. These assumptions are not particularly restrictive since most regions of interest can be broken into subregions of this type.

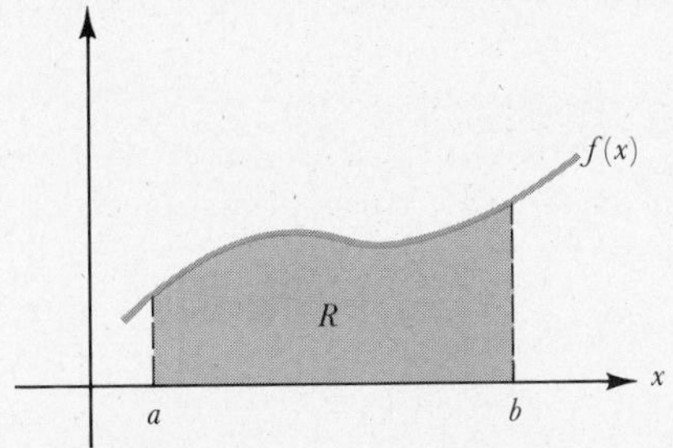

Fig. 6.4

We partition $[a, b]$ into, say, n subintervals using the points

$$x_0 = a, x_1, x_2, \ldots, x_{n-1}, x_n = b$$

and on each subinterval we base a rectangle whose height is determined by some value of $f(x)$ on that subinterval. Since $f(x)$ is continuous on $[a, b]$, it is continuous on each closed subinterval $[x_{i-1}, x_i]$ $(1 \leq i \leq n)$, and so by the Extreme-Value Theorem (4.4.2), $f(x)$ attains a minimum value m_i and a maximum value M_i on $[x_{i-1}, x_i]$. Thus we have two rectangles based upon each such subinterval (see Fig. 6.5), one (shaded) of height m_i and the other of height M_i.

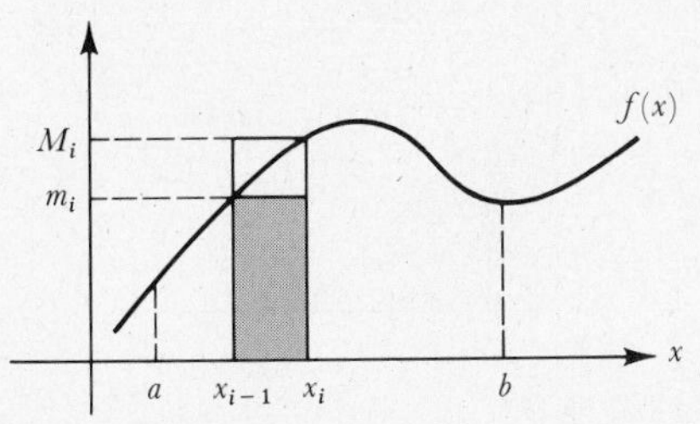

Fig. 6.5

Denoting the width of the rectangle by $\Delta x_i = x_i - x_{i-1}$, the area of the shorter rectangle is $m_i \, \Delta x_i$, and the area of the taller rectangle is $M_i \, \Delta x_i$. If we

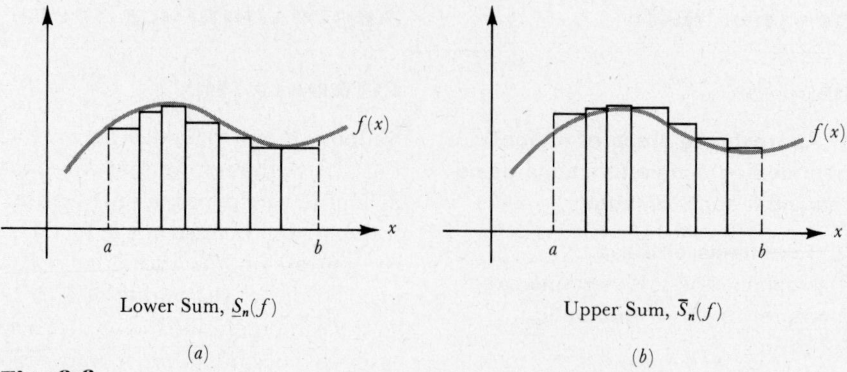

Fig. 6.6

add the areas of all n shorter rectangles, we obtain an approximation $\underline{S}_n(f)$ that *underestimates* A; that is, $\underline{S}_n(f) \leq A$. If we add the areas of all n taller rectangles, we get an approximation $\overline{S}_n(f)$ which *overestimates* A; that is, $A \leq \overline{S}_n(f)$. (See Fig. 6.6.) We refer to $\underline{S}_n(f)$ as a *lower sum* and $\overline{S}_n(f)$ as an *upper sum*. Thus,

$$\underline{S}_n(f) \leq A \leq \overline{S}_n(f)$$

Our next step (following Archimedes) is to improve the approximations by taking narrower and narrower rectangles. In Fig. 6.7, we shade the improvement obtained by replacing one rectangle by two narrower rectangles.

The idea, now, is to *form the limit of such lower and upper sums as the widths of the rectangles approach zero and thus close in on A from both sides:*

6.2.1

$$\underline{S}_n(f) \to A \leftarrow \overline{S}_n(f)$$

Here is where we apply the ideas of limit and function and are thus able to go beyond Archimedes, one of the greatest mathematicians of all time. (Newton once noted, in explaining his own momentous achievements, that he "stood on the shoulders of giants." We are standing on top of a whole pyramid of giants.)

To get a feel for the types of issues and calculations that arise, we shall examine some particular cases. We start by applying our technique to a case where the answer can be confidently obtained by more elementary means. We shall use rectangles to approximate the area of the triangular region R between the x axis and the graph of

$$f(x) = 1 - x \text{ from } x = 0 \text{ to } x = 1 \text{ (see Fig. 6.8)}$$

(R, being a triangle of base 1 and height 1, has area $\frac{1}{2}$.) We shall first partition

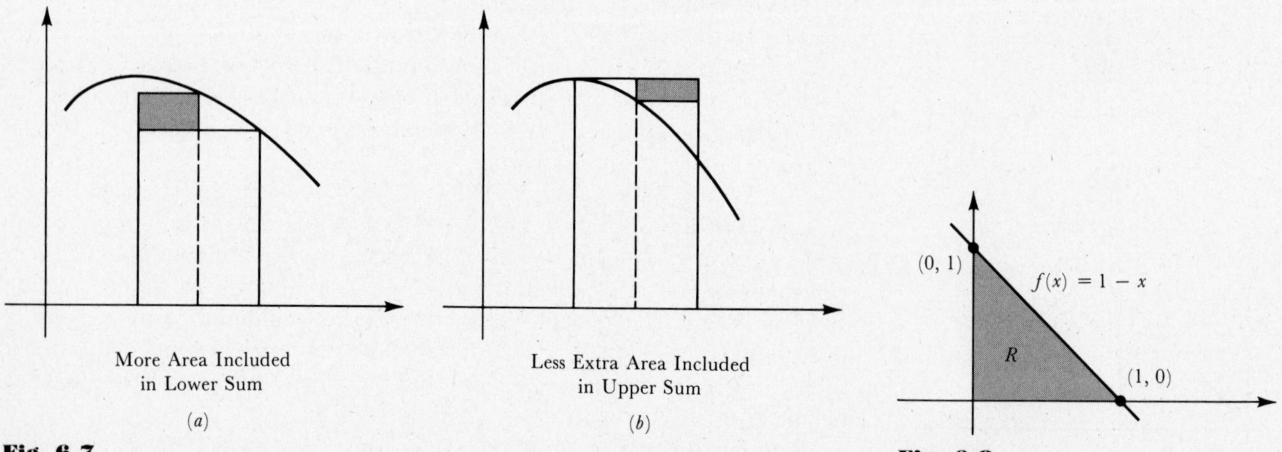

Fig. 6.7

Fig. 6.8

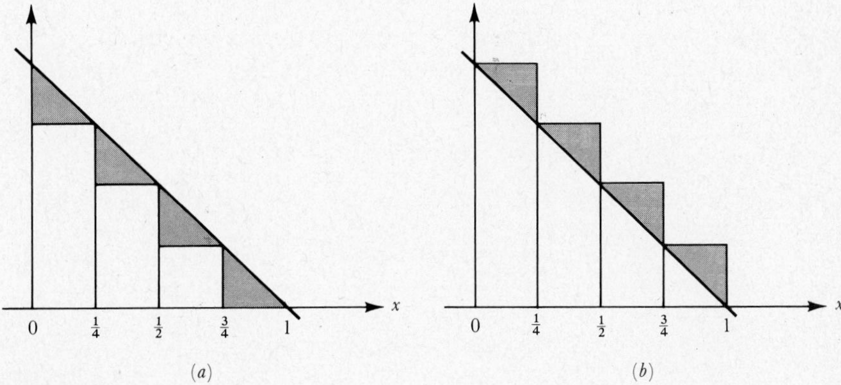

Fig. 6.9

[0, 1] into four equal parts, with $x_0 = 0$, $x_1 = \frac{1}{4}$, $x_2 = \frac{1}{2}$, $x_3 = \frac{3}{3}$, $x_4 = 1$. Since $f(x) = 1 - x$ is decreasing on [0, 1], the maximum value on any subinterval occurs at its left endpoint, and the minimum value occurs at the right endpoint. Hence the lower sum based upon this partition [see Fig. 6.9(*a*)] is

$$\underline{S}_4(f) = f\left(\frac{1}{4}\right)\left(\frac{1}{4}\right) + f\left(\frac{1}{2}\right)\left(\frac{1}{4}\right) + f\left(\frac{3}{4}\right)\left(\frac{1}{4}\right) + f\left(1\right)\left(\frac{1}{4}\right)$$
$$= \frac{3}{4}\cdot\frac{1}{4} + \frac{1}{2}\cdot\frac{1}{4} + \frac{1}{4}\cdot\frac{1}{4} + 0\cdot\frac{1}{4} = \frac{3}{16} + \frac{1}{8} + \frac{1}{16} = \frac{3}{8}$$

The upper sum based on this partition [see Fig. 6.9(*b*)] is

$$\overline{S}_4(f) = f\left(0\right)\left(\frac{1}{4}\right) + f\left(\frac{1}{4}\right)\left(\frac{1}{4}\right) + f\left(\frac{1}{2}\right)\left(\frac{1}{4}\right) + f\left(\frac{3}{4}\right)\left(\frac{1}{4}\right)$$
$$= 1\cdot\frac{1}{4} + \frac{3}{4}\cdot\frac{1}{4} + \frac{1}{2}\cdot\frac{1}{4} + \frac{1}{4}\cdot\frac{1}{4}$$
$$= \frac{1}{4} + \frac{3}{16} + \frac{1}{8} + \frac{1}{16} = \frac{5}{8}$$

The error in each case (the area of the shaded region) is $\frac{1}{8}$.

Taking a finer partition, say 10 subdivisions, with resulting narrower rectangles, should improve on this rough estimate. Let us take the evenly spaced points (for convenience)

$$x_0 = 0 \qquad x_1 = 0.1 \qquad x_2 = 0.2 \qquad x_3 = 0.3 \qquad \ldots \qquad x_9 = 0.9 \qquad x_{10} = 1$$

Again using right endpoints for $\underline{S}_{10}(f)$ and left endpoints for $\overline{S}_{10}(f)$, we have:

$$\underline{S}_{10}(f) = f(.1)\frac{1}{10} + f(.2)\frac{1}{10} + f(.3)\frac{1}{10} + \cdots + f(.9)\frac{1}{10} + f(1)\frac{1}{10}$$
$$= [0.9 + 0.8 + 0.7 + \cdots + 0.1 + 0]\frac{1}{10}$$
$$= [9 + 8 + 7 + \cdots + 1 + 0]\frac{1}{10^2}$$

(We kept the zero term visible to emphasize that we are dealing with a sum of ten terms and put the sum in a form that will be comparable to the general case later.) To total the numbers in the brackets we pair extreme terms together: the first and last, the second and next to last, and so on, working toward the middle. This way, the sum of each pair is the same, and the total is

$$(9 + 0) + (8 + 1) + (7 + 2) + (6 + 3) + (5 + 4) = 5\cdot 9 = 45$$

Hence

$$\underline{S}_{10}(f) = 45 \cdot \frac{1}{10^2} = 0.45$$

Similarly,

$$\begin{aligned}\overline{S}_{10}(f) &= f(0)\frac{1}{10} + f(.1)\frac{1}{10} + f(.2)\frac{1}{10} + \cdots + f(.9)\frac{1}{10} \\ &= [1 + 0.9 + 0.8 + \cdots + 0.1]\frac{1}{10} \\ &= [10 + 9 + 8 + \cdots + 1]\frac{1}{10^2} \\ &= 55 \cdot \frac{1}{10^2} \qquad \text{(by the pairing technique)} \\ &= 0.55\end{aligned}$$

The error of each sum is reduced to 0.05.

To generalize this process and to take a limit, we now partition [0, 1] into n subintervals using points, which for convenience we take to be distance $1/n$ apart along [0, 1]. Hence we have

$$x_0 = 0 \quad x_1 = \frac{1}{n} \quad x_2 = \frac{2}{n} \quad x_3 = \frac{3}{n} \quad \cdots \quad x_{n-1} = \frac{n-1}{n} \quad x_n = \frac{n}{n} = 1$$

Here $\Delta x_i = 1/n$ for each subinterval $[x_{i-1}, x_i]$. We shall parallel the technique used for $n = 10$ exactly. Using right endpoints for the decreasing function $f(x) = 1 - x$, we obtain

$$\begin{aligned}\underline{S}_n(f) &= \left[f\left(\frac{1}{n}\right) + f\left(\frac{2}{n}\right) + f\left(\frac{3}{n}\right) + \cdots + f\left(\frac{n-1}{n}\right) + f\left(\frac{n}{n}\right)\right]\frac{1}{n} \\ &= \left[\left(1 - \frac{1}{n}\right) + \left(1 - \frac{2}{n}\right) + \left(1 - \frac{3}{n}\right) + \cdots + \left(1 - \frac{n-1}{n}\right) + \left(1 - \frac{n}{n}\right)\right]\frac{1}{n} \\ &= \left[\frac{n-1}{n} + \frac{n-2}{n} + \frac{n-3}{n} + \cdots + \frac{1}{n} + 0\right]\frac{1}{n}\end{aligned}$$

6.2.2

$$= [(n-1) + (n-2) + (n-3) + \cdots + 1 + 0]\frac{1}{n^2}$$

Obviously, we need a formula for the sum of n consecutive integers. It is useful to develop a general formula for such sums.

Suppose we have a sum of integers from 1 to m. We add the sum to itself, pairing extreme terms as before but aligning them vertically, as follows:

$$\begin{array}{r} 1 \;+\; 2 \;+\; 3 \;+ \cdots +\; (m-1) \;+\; m \\ + \qquad\qquad\qquad\qquad\qquad\qquad \\ m \;+\; (m-1) + (m-2) + \cdots +\; 2 \;+\; 1 \\ \hline (m+1) + (m+1) + (m+1) + \cdots + (m+1) + (m+1) \end{array}$$

There are m such $(m + 1)$ terms, so the total is $m(m + 1)$. But since this is twice the sum of 1 through m, we can conclude the following:

6.2.3 (Sum of Consecutive Integers)

$1 + 2 + 3 + \cdots + (m - 1) + m = [m(m + 1)]/2$

Returning to $\underline{S}_n(f)$ and (6.2.2) we have the sum (in reverse order) from 1 to $n - 1 = m$, so the total will be

$$\underline{S}_n(f) = \left[\frac{(n-1)(n-1+1)}{2}\right]\frac{1}{n^2}$$
$$= \frac{(n-1)(n)}{2n^2}$$
$$= \frac{1}{2}\left[\frac{n-1}{n}\right] = \frac{1}{2}\left[1 - \frac{1}{n}\right]$$

Thus, $\underline{S}_n(f) < 1/2$ *for all n*. But, more important,

$$\lim_{n\to+\infty} \underline{S}_n(f) = \lim_{n\to+\infty} \frac{1}{2}\left[1 - \frac{1}{n}\right] = \frac{1}{2}$$

Similarly, using left endpoints for the taller rectangles,

$$\overline{S}_n(f) = \left[f(0) + f\left(\frac{1}{n}\right) + f\left(\frac{2}{n}\right) + \cdots + f\left(\frac{n-1}{n}\right)\right]\frac{1}{n}$$
$$= \left[1 + \left(1 - \frac{1}{n}\right) + \left(1 - \frac{2}{n}\right) + \cdots + \left(1 - \frac{n-1}{n}\right)\right]\frac{1}{n}$$
$$= \left[\frac{n}{n} + \frac{n-1}{n} + \frac{n-2}{n} + \cdots + \frac{1}{n}\right]\frac{1}{n}$$
$$= [n + (n-1) + (n-2) + \cdots + 1]\frac{1}{n^2}$$
$$= \left[\frac{n(n+1)}{2}\right]\frac{1}{n^2} \qquad \text{(using the sum formula with } n = m)$$
$$= \frac{1}{2}\left[\frac{n+1}{n}\right] = \frac{1}{2}\left[1 + \frac{1}{n}\right]$$

Thus, $\overline{S}_n(f) > 1/2$ *for all n* and, again,

$$\lim_{n\to+\infty} \overline{S}_n(f) = \lim_{n\to+\infty} \frac{1}{2}\left[1 + \frac{1}{n}\right] = \frac{1}{2}$$

Hence, for this particular example, where we knew $A = 1/2$, we achieved our aim (6.2.1). As the width $1/n$ of the rectangles approaches zero, we were able to conclude that

$$\underline{S}_n(f) = \frac{1}{2}\left[1 - \frac{1}{n}\right] \longrightarrow \frac{1}{2} \longleftarrow \frac{1}{2}\left[1 + \frac{1}{n}\right] = \overline{S}_n(f)$$

BUT WHY AREA?

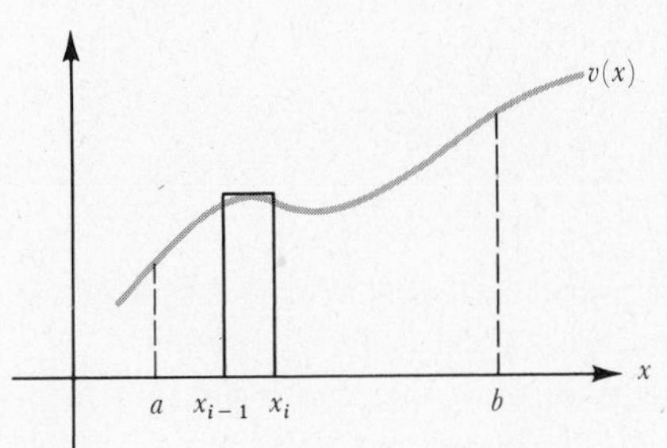

We are now in a position to illustrate how the area of a region might arise in dealing with a problem initially stated without any reference to area. Suppose $v(x) > 0$ gives the instantaneous velocity of some object as a function of time x from $x = a$ to $x = b$, where the graph of $v(x)$ is as shown. A typical rectangle of height $v(x_i)$ and width $x_i - x_{i-1} = \Delta x_i$ has an area $v(x_i)\,\Delta x_i$. But what is the product of a constant velocity $v(x_i)$ and a time interval Δx_i? A distance! (If you travel at 30 mph for $\frac{1}{10}$ h, you cover $(30)(\frac{1}{10}) = 3$ mi.) Hence the area of the rectangle—which, by its flat top, assumes that the velocity is constant—is approximately the distance traveled by the object over the interval from x_{i-1} to x_i. The sum of the areas of such rectangles approximates the total distance traveled from time a to time b. How would we improve this approximation? We would take more and more, but narrower and narrower, rectangles. As we saw when defining derivatives, the smaller the interval, the better the constant velocity approximates the variable velocity. The limit of such approximations is both the area under the graph of $v(x)$ from a to b *and* the total distance traveled by the object from time a to time b.

Thus in setting up the machinery to solve the area problem we also provide the means for solving a much wider class of problems.

PROGRESS TEST 1

1. Suppose that $f(x) = 4 - x^2$ on $[-2, 2]$ is subdivided by the points $x_0 = -2$, $x_1 = 0$, $x_2 = 1$, $x_3 = 3/2$, $x_4 = 2$.
(*a*) Sketch and compute the corresponding lower sum $\underline{S}_4(f)$.
(*b*) Sketch and compute the corresponding upper sum $\overline{S}_4(f)$.

2. Suppose that $f(x) = 2x$ on $[0, 1]$ and $[0, 1]$ is evenly subdivided into 10 subintervals.
(*a*) Compute $\underline{S}_{10}(f)$. (Note that $f(x) = 2x$ is increasing.)
(*b*) Write out $\underline{S}_n(f)$ for $[0, 1]$ partitioned into n subintervals of length $1/n$, and express the sum in brackets as a sum of consecutive integers.

A MORE COMPACT NOTATION

Before we go on to show how this rectangle approximation process applies in much more generality, it is useful to introduce some shorthand notation for sums.

If $a_1, a_2, \ldots, a_n$ is a collection of n numbers, then their sum

$$a_1 + a_2 + \cdots + a_n$$

can be written in the abbreviated form

$$\sum_{i=1}^{n} a_i$$

which is read "the sum from $i = 1$ to $i = n$ of a sub i." We call this *sigma notation* or *summation notation*. Thus, for example, if the typical ith term in the sum is $a_i = 2i^2$, then

$$a_1 + a_2 + a_3 + a_4 + a_5$$

$$= \sum_{i=1}^{5} 2i^2 = 2(1)^2 + 2(2)^2 + 2(3)^2 + 2(4)^2 + 2(5)^2 = 110$$

The earlier sum of consecutive integers from 1 to m has its ith term equal to i; that is, $a_i = i$. Hence we may rewrite (6.2.3) as

$$\sum_{i=1}^{m} i = \frac{m(m+1)}{2}$$

Notice that the role of i in a_i is merely as an index; any other letter would do as well. For example, in the preceding formula,

$$\sum_{i=1}^{m} i = \sum_{j=1}^{m} j = \sum_{k=1}^{m} k = \frac{m(m+1)}{2}$$

and the index letter does not even appear in the final sum. For this reason the index letter is often called a "dummy variable."

We can rewrite the sums appearing earlier in sigma notation. The partition of $[0, 1]$ into n subintervals of equal length may be written with its typical ith point identified as follows:

$$x_0 = 0 \qquad x_1 = \frac{1}{n} \qquad x_2 = \frac{2}{n} \qquad x_3 = \frac{3}{n} \qquad \cdots \qquad x_i = \frac{i}{n}$$

$$\cdots \qquad x_{n-1} = \frac{n-1}{n} \qquad x_n = \frac{n}{n} = 1$$

Then the typical ith subinterval looks like

$$[x_{i-1}, x_i] = \left[\frac{i-1}{n}, \frac{i}{n}\right]$$

Since, as we noted, $f(x) = 1 - x$ is decreasing on $[0, 1]$, it reaches its minima at right endpoints. Thus, on the ith subinterval, the height of the shorter rectangle is

$$f\left(\frac{i}{n}\right) = 1 - \frac{i}{n} = \frac{n-i}{n}$$

Its width Δx_i is $1/n$, so its area is

$$\frac{n-i}{n} \cdot \frac{1}{n} = \frac{n-i}{n^2}$$

Thus, adding these areas from $i = 1$ to $i = n$, we have the lower sum

$$\underline{S}_n(f) = \sum_{i=1}^{n} f(x_i)\,\Delta x_i = \sum_{i=1}^{n} f\left(\frac{i}{n}\right)\frac{1}{n}$$

$$= \sum_{i=1}^{n} \frac{n-i}{n^2} = \sum_{i=1}^{n} (n-i)\frac{1}{n^2}$$

Since the constant factor $1/n^2$ appears in each term and does not involve i, it can be factored out, leaving

$$\left[\sum_{i=1}^{n} (n-i)\right]\frac{1}{n^2}$$

where the sum inside the brackets is merely the sum of consecutive integers from $n - 1$ down to 0:

$$\sum_{i=1}^{n}(n-i) = \underset{i=1}{(n-1)} + \underset{i=2}{(n-2)} + \underset{i=3}{(n-3)} + \underset{\cdots}{\cdots} + \underset{i=n-1}{[n-(n-1)]} + \underset{i=n}{(n-n)}$$

Hence we have arrived at exactly the same sum $\underline{S}_n(f)$ achieved earlier in (6.2.2).

PROGRESS TEST 2

1. Compute $\Sigma_{j=2}^{5}(2j^2 - 1)$ by substituting the four values of j into $2j^2 - 1$.

2. For $f(x) = 2x$ on $[0, 1]$, as in Prob. 2, Progress Test 1:

(*a*) Rewrite $\underline{S}_{10}(f)$ in sigma notation.

(*b*) Determine $\underline{S}_n(f)$ in sigma notation for an even partition of $[0, 1]$ into n subintervals.

MORE ON APPROXIMATING AREAS

We now move to a case where we do not have a ready area formula: Find the area above the x axis and under

$$f(x) = x^2 \qquad \text{from } x = 0 \text{ to } x = 1$$

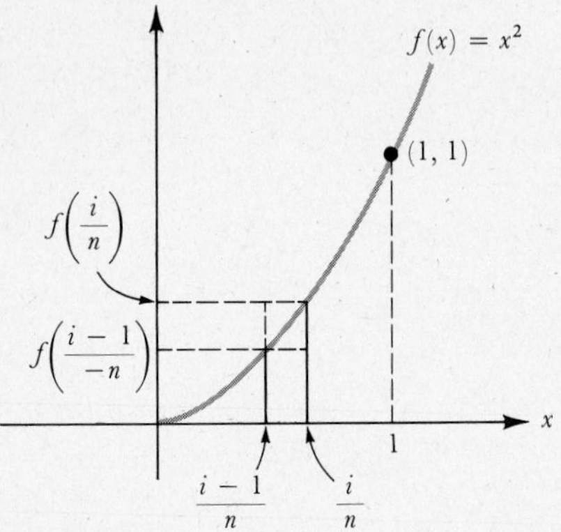

Fig. 6.10

As can be seen in Fig. 6.10, since f is increasing throughout $[0, 1]$, lower sums will require left endpoints and upper sums will require right endpoints.

Using an even partition of $[0, 1]$ into n subintervals, a typical subinterval looks like

$$[x_{i-1}, x_i] = \left[\frac{i-1}{n}, \frac{i}{n}\right]$$

Hence, in sigma notation,

$$\underline{S}_n(f) = \sum_{i=1}^{n} f(x_{i-1})\,\Delta x_i = \sum_{i=1}^{n} f\left(\frac{i-1}{n}\right)\frac{1}{n}$$

$$= \sum_{i=1}^{n} \left(\frac{i-1}{n}\right)^2 \frac{1}{n} = \sum_{i=1}^{n} (i-1)^2 \frac{1}{n^3}$$

$$\overline{S}_n(f) = \sum_{i=1}^{n} f(x_i)\,\Delta x_i = \sum_{i=1}^{n} f\left(\frac{i}{n}\right)\frac{1}{n}$$

$$= \sum_{i=1}^{n} \left(\frac{i}{n}\right)^2 \frac{1}{n} = \sum_{i=1}^{n} i^2 \frac{1}{n^3}$$

It turns out that for $n = 100$ we get

$$\underline{S}_{100}(f) = 0.32835 \qquad \text{and} \qquad \overline{S}_{100}(f) = 0.33835$$

and for $n = 1000$ we get

$$\underline{S}_{1000}(f) = 0.3328335 \qquad \text{and} \qquad \overline{S}_{1000}(f) = 0.3338335.$$

These quantities *seem* to be closing in on the number $\frac{1}{3}$, a hypothesis even more plausible when we note that the average of the upper and lower sums for $n = 1000$ is 0.3333335. To confirm this, we need a formula for the sum of squares of consecutive integers, because all our sums involve such squares. A formula, whose proof is outlined in the exercises, is

6.2.4 Sums of Squares $\displaystyle \sum_{i=1}^{n} i^2 = \frac{n(n+1)(2n+1)}{6}$

Applying this to the upper sum, after factoring out $1/n^3$ (which appears in each term and does not involve i), we have

$$\overline{S}_n(f) = \left[\sum_{i=1}^{n} i^2\right]\frac{1}{n^3}$$

$$= \left[\frac{n(n+1)(2n+1)}{6}\right]\frac{1}{n^3}$$

$$= \frac{2n^3 + 3n^2 + n}{6n^3} = \frac{1}{3} + \frac{1}{2n} + \frac{1}{6n^2}$$

Now, it is clear that, as n grows without bound, the width $\Delta x_i = 1/n$ of the rectangles approaches zero and

$$\overline{S}_n(f) \text{ approaches } \frac{1}{3}$$

The argument that

$$\underline{S}_n(f) = \left[\sum_{i=1}^{n} (i-1)^2\right]\frac{1}{n^3}$$

also approaches $\frac{1}{3}$ involves the same basic idea. However, it is more valuable to examine the *difference between the upper and lower sums.*

We expand the upper and lower sums and align common terms vertically to facilitate the subtraction of $\underline{S}_n(f)$ from $\overline{S}_n(f)$. (Note that each has n terms.)

$$\overline{S}_n(f) = \sum_{i=1}^{n} \frac{i^2}{n^3} = \frac{1^2}{n^3} + \frac{2^2}{n^3} + \frac{3^2}{n^3} + \cdots + \frac{(i-1)^2}{n^3} + \frac{(i)^2}{n^3} + \cdots + \frac{(n-1)^2}{n^3} + \frac{n^2}{n^3}$$

$$\underline{S}_n(f) = \sum_{i=1}^{n} \frac{(i-1)^2}{n^3} = \frac{0^2}{n^3} + \frac{1^2}{n^3} + \frac{2^2}{n^3} + \cdots + \frac{(i-1)^2}{n^3} + \frac{(i)^2}{n^3} + \cdots + \frac{(n-1)^2}{n^3}$$

$$\overline{S}_n(f) - \underline{S}_n(f) = \frac{-0^2}{n^3} + 0 + 0 + \cdots + 0 + 0 + \cdots + 0 + \frac{n^2}{n^3}$$

$$= \frac{n^2}{n^3} = \frac{1}{n}$$

Thus, as $n \to +\infty$,

1. The width of the rectangles approaches 0.
2. The difference between the upper and lower sums likewise approaches 0.
3. One of the sums, namely $\overline{S}_n(f)$, approaches $\frac{1}{3}$.

Hence, combining these facts, we have proved that the area is *exactly* $\frac{1}{3}$. It is of special interest to examine the difference between upper and lower sums, $\overline{S}_n(f) - \underline{S}_n(f)$, in graphical terms. In Fig. 6.11 we shaded each region representing the difference between taller and shorter rectangles. We then slid each shaded region to the left and stacked it along the vertical axis so the total area would be easier to recognize. The regions "stack up" to form a single rectangle of height

$$f(1) - f(0) = 1 - 0 = 1$$

and width $1/n$. This vertical "error rectangle" [whose area is $\overline{S}_n(f) - \underline{S}_n(f)$] thus has area $1/n$. The previously derived algebraic result and its graphical interpretation establish emphatically that

$$\overline{S}_n(f) - \underline{S}_n(f) \to 0$$

as the width of the rectangles $\to$ 0.

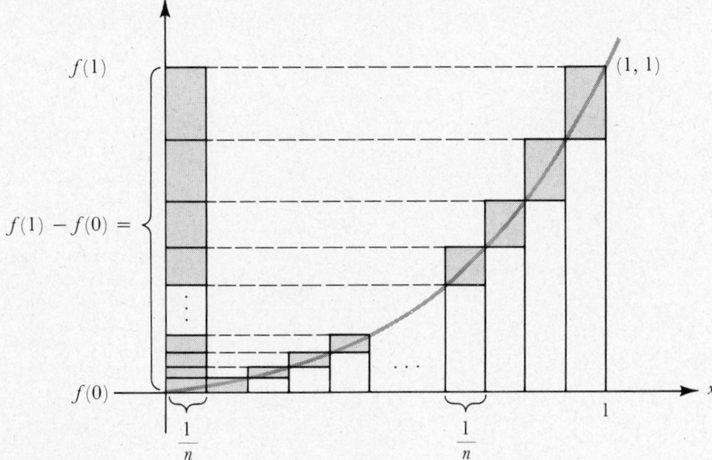

Fig. 6.11

A GENERALIZATION TO MONOTONIC FUNCTIONS

The previous examples point the way to a more general analysis, which will show that the difference between the upper and lower sums, and hence the approximation error, approaches zero as the width of the rectangles approaches zero. In several stages of increasing generality, we shall show that the rectangle-approximation technique applies to regions bounded by a wide class of functions.

If f is either increasing or decreasing on all of $[a, b]$, we shall say that f is *monotonic on* $[a, b]$. Right now, we make the following assumption: *$f(x)$ is a nonnegative function continuous and monotonic on $[a, b]$.*

We shall carry out the examination assuming that $f(x)$ is *increasing* and leave the decreasing case to the next Progress Test. We are approximating the area A of the region R bounded by the graph of f, the x axis, and the lines $x = a$ and $x = b$.

We shall focus our attention on $\overline{S}_n(f) - \underline{S}_n(f)$ because this difference includes both the errors involved in approximating by upper sums as well as errors involved using lower sums. As a result, if we can force $\overline{S}_n(f) - \underline{S}_n(f)$ to zero by taking narrower and narrower rectangles, then we can be guaranteed that the approximation to A, provided by either sum, can be made as accurate as we please. This will establish the validity of the rectangle-approximation technique for regions bounded by the type of function under consideration.

Suppose we have an even partition of $[a, b]$ into n subintervals of length $\Delta x_i = (b - a)/n$ with the points

$$x_0 = a, x_1, x_2, \ldots, x_i, \ldots, x_{n-1}, x_n = b$$

As before, we form the difference between the upper and lower sums, using left endpoints for lower sums and right endpoints for upper sums. We align the sums vertically, and factor out $\Delta x_i = (b - a)/n$ to save space:

$$\overline{S}_n(f) = \sum_{i=1}^{n} f(x_i)\,\Delta x_i = [f(x_1) + f(x_2) + \cdots + f(x_i) + \cdots + f(x_{n-1}) + f(x_n)]\frac{b-a}{n}$$

$$\underline{S}_n(f) = \sum_{i=1}^{\infty} f(x_{i-1})\,\Delta x_i = [f(x_0) + f(x_1) + f(x_2) + \cdots + f(x_{i-1}) + f(x_i) + \cdots + f(x_{n-1})]\frac{b-a}{n}$$

$$\begin{aligned}\overline{S}_n(f) - \underline{S}_n(f) &= [-f(x_0) + 0 + 0 + \cdots + 0 + \cdots + 0 + f(x_n)]\frac{b-a}{n}\\ &= [f(x_n) - f(x_0)]\frac{b-a}{n}\\ &= [f(b) - f(a)]\frac{b-a}{n}\end{aligned}$$

Now, since a and b are fixed, we know that

$$\begin{aligned}\lim_{n\to+\infty} [\overline{S}_n(f) - \underline{S}_n(f)] &= \lim_{n\to+\infty} \frac{[f(b) - f(a)](b-a)}{n}\\ &= 0\end{aligned}$$

Thus, by taking more and more, but narrower and narrower, rectangles we can improve our area estimate to arbitrary accuracy. The graphical interpretation of $\overline{S}_n(f) - \underline{S}_n(f)$ is much the same as before. However, in this case (see Fig. 6.12), the area of the "error rectangle" is

$$[f(b) - f(a)]\left(\frac{b-a}{n}\right)$$

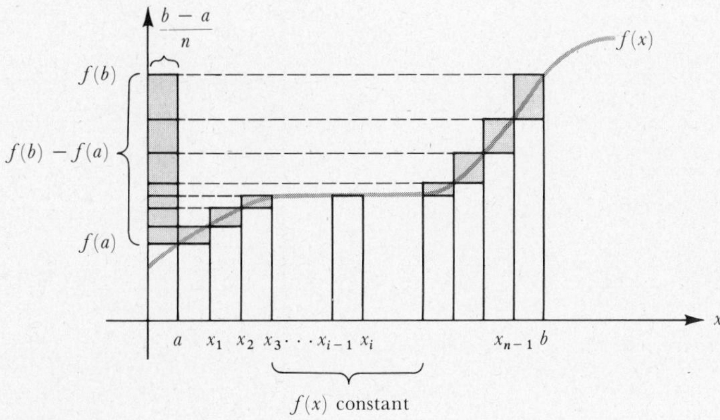

Fig. 6.12

Furthermore, if $f(x)$ happens to be constant on some subinterval of $[a, b]$, the maximum and minimum values of $f(x)$ on that interval are the same, so the "taller" and "shorter" rectangles are equal (see Fig. 6.12). Hence, on such subintervals, zero error is contributed to $\overline{S}_n(f) - \underline{S}_n(f)$, and so our analysis applies to functions that may be constant over some subinterval.

PROGRESS TEST 3

Suppose $f(x) \geq 0$ and is continuous and *decreasing* on $[a, b]$, and $x_0 = a, x_1, x_2, \ldots, x_i, \ldots, x_{n-1}, x_n = b$ is an even partition of $[a, b]$ into subintervals of length $(b - a)/n$.

1. Determine $\overline{S}_n(f) - \underline{S}_n(f)$ algebraically as previously. [Be careful about endpoints, since $f(x)$ is decreasing.]

2. Determine $\lim_{n \to +\infty} [\overline{S}_n(f) - \underline{S}_n(f)]$.

3. Draw a "decreasing" version of Fig. 6.12.

FURTHER GENERALIZATIONS

In all our work we used partitions of $[a, b]$ into equal-length subintervals for convenience. Later, in applications, such even partitions are less convenient.

6.2.5 **Definition** Let P be a partition of $[a, b]$ into n, not necessarily equal-length subintervals, with P consisting of the points

$$a = x_0, x_1, x_2, \ldots, x_i, \ldots, x_{n-1}, x_n = b$$

We call the largest of the numbers $\Delta x_i = x_i - x_{i-1}$ (for $1 \leq i \leq n$) the *norm of* P and denote it by $\|P\|$. (Thus $\|P\|$ is the width of the widest rectangle determined by P.)

We can now make some important observations extending our earlier work. The supporting argument follows.

6.2.6 **Theorem** Assuming that $f(x)$ is nonnegative, continuous, and monotonic on $[a, b]$, then the rectangle-approximation method yields arbitrary accuracy provided that $\|P\|$ is small enough.

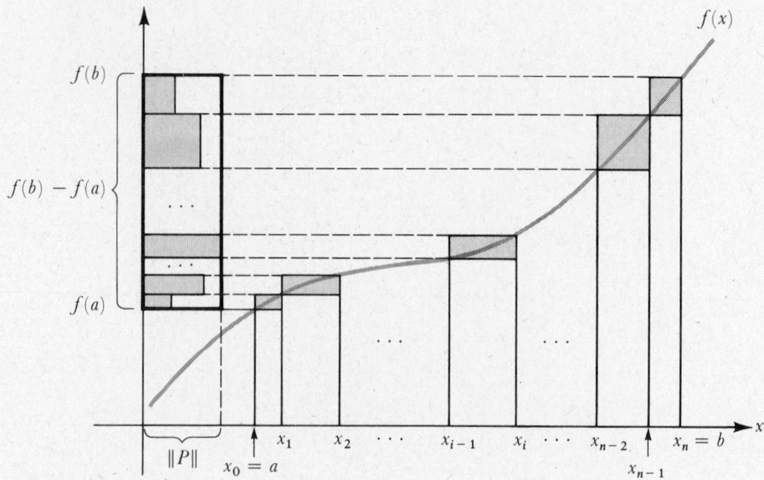

Fig. 6.13

A look back at our earlier calculation of $\overline{S}_n(f) - \underline{S}_n(f)$ shows that, whether or not the partition is even, all terms of the form $f(x_i)\,\Delta x_i$ cancel from $\overline{S}_n(f) - \underline{S}_n(f)$ except for $f(x_0)\,\Delta x_0 = f(a)\,\Delta x_0$ and $f(x_n)\,\Delta x_n = f(b)\,\Delta x_n$. Graphically, we can still stack the differences between upper and lower rectangles along the vertical axis (see Fig. 6.13) except that they do not form an "error-rectangle" as before. Nonetheless, the total error region (shaded) *is* contained within a rectangle (with the heavy black border). We can thus see that by letting $\|P\| \to 0$, we force $\overline{S}_n(f) - \underline{S}_n(f)$ to zero.

Since the upper and lower sums really depend upon the choice of points that make up the partition P (and not on their number), we shall alter our notation slightly and let $\overline{S}_P(f)$ denote the upper sum and $\underline{S}_P(f)$ denote the lower sum. Hence we can restate the above conclusion as

$$\lim_{\|P\|\to 0} [\overline{S}_P(f) - \underline{S}_P(f)] = 0$$

Heretofore we always used endpoints of the subintervals $[x_{i-1}, x_i]$ in our sums. Again, in future applications, the freedom to create approximating sums based upon points *other* than the endpoints will prove to be most valuable. If x_i^* is any point in the subinterval $[x_{i-1}, x_i]$, then $f(x_i^*)$ satisfies

$$m_i \leq f(x_i^*) \leq M_i$$

where m_i and M_i are the minimum and maximum values of $f(x)$ on $[x_{i-1}, x_i]$. As a result, the area of the rectangle of height $f(x_i^*)$ is between that of the shortest and tallest rectangles based upon $[x_{i-1}, x_i]$.

Given such an arbitrary choice of points $x_1^*, x_2^*, \ldots, x_i^*, \ldots, x_n^*$ in the respective subintervals, we shall denote the total area $\Sigma_{i=1}^n f(x_i^*)\,\Delta x_i$ enclosed by rectangles of heights $f(x_i^*)$ by $S_P^*(f)$. Thus we have

$$\underline{S}_P(f) \leq S_P^*(f) \leq \overline{S}_P(f)$$

Consequently, knowing that the error involved in using upper or lower sums to approximate A can be made as small as desired allows us to conclude that the same is true for $S_P^*(f)$. In particular,

$$\lim_{\|P\|\to 0} S_P^*(f) = \lim_{\|P\|\to 0} \sum_{i=1}^{n} f(x_i^*)\,\Delta x_i = A$$

This means that, as long as the width $\|P\|$ of the widest rectangle approaches zero, $S_P^*(f)$ approaches A *independently of the way the points x_i^* are chosen in the subintervals.*

We shall say f is *piecewise monotonic on* $[a, b]$ if there is at most a finite collection of points c in $[a, b]$ at which f goes from strictly increasing to strictly decreasing or vice versa. (See Fig. 6.14 for an example of a piecewise monotonic function.)

With minor extension of our arguments we shall have established the following theorem, which summarizes our achievements on the area problem to date.

6.2.7 **Theorem** Suppose f is a function continuous and nonnegative on $[a, b]$, and A is the area of the region between the graph of f and the x axis from $x = a$ to $x = b$. If P is any partition of $[a, b]$ and x_i^* is any choice of points in the subintervals $[x_{i-1}, x_i]$ $(1 \leq i \leq n)$ determined by P, then

(*a*) $\underline{S}_P(f) \leq A \leq \overline{S}_P(f)$.

(*b*) $\underline{S}_P(f) \leq S_P^*(f) \leq \overline{S}_P(f)$.

If, further, f is piecewise monotonic on $[a, b]$, then

(*c*) $\lim_{\|P\| \to 0} \underline{S}_P(f) = \lim_{\|P\| \to 0} S_P^*(f) = \lim_{\|P\| \to 0} \overline{S}_P(f) = A$.

Proof of (6.2.7): Our earlier arguments, which proved (*a*) and (*b*), involved only continuity and nonnegativity. Conclusion (*c*) has been shown for functions monotonic on $[a, b]$. However, if f is only piecewise monotonic on $[a, b]$, then our earlier argument applies on each of the finite collection of subintervals to show that the difference between the upper and lower sums can be made as small as we please on each subinterval. (There would be four such subintervals for the function in Fig. 6.14.) It follows that the finite sum of these differences—the *total* difference between the upper and lower sums on $[a, b]$—can be made as small as we please by choosing $\|P\|$ sufficiently small. Hence the arguments on the previous several pages apply to show that the indicated limits equal A, and so (*c*) is proved. ∎

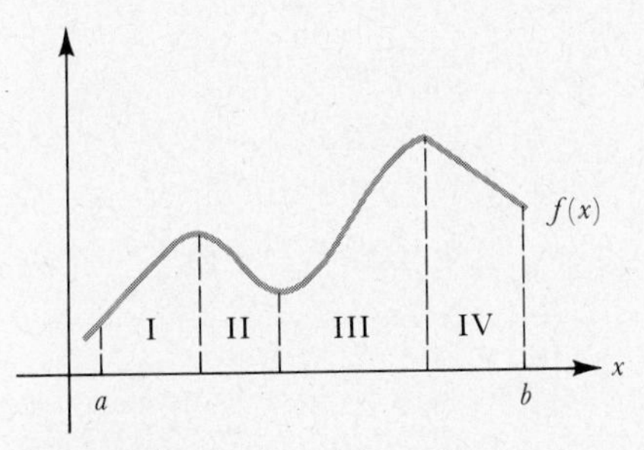

Fig. 6.14

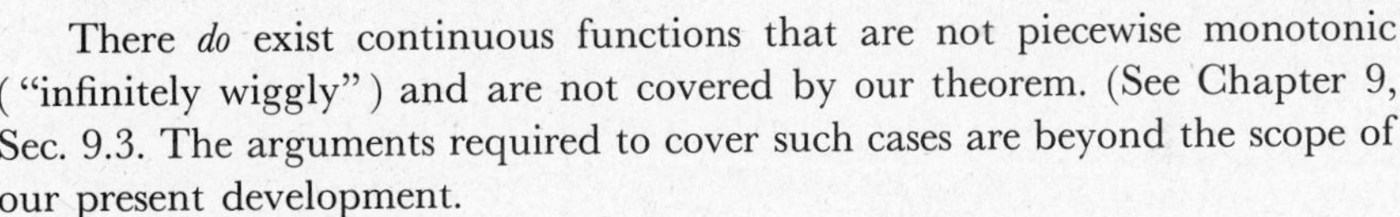

There *do* exist continuous functions that are not piecewise monotonic ("infinitely wiggly") and are not covered by our theorem. (See Chapter 9, Sec. 9.3. The arguments required to cover such cases are beyond the scope of our present development.

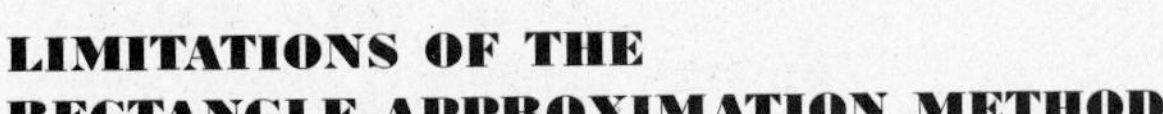

LIMITATIONS OF THE RECTANGLE-APPROXIMATION METHOD

Although we have indeed come a long way in attacking the area problem, we have in Theorem (6.2.7) achieved a classic example of what mathematicians call an "existence theorem." It tells us that all the appropriate limits do exist and do lead to the area of the region in question. *It does* not *give us any ready means for computing the area.*

For functions f satisfying [6.2.7(*c*)] we can exercise the freedoms to choose regular partitions and points $x_i^* = x_i$ to obtain

6.2.8
$$A = \lim_{n \to +\infty} \sum_{i=1}^{n} f\left(a + \frac{i(b-a)}{n}\right) \frac{b-a}{n}$$

However, as is evident in the following exercises, the various techniques needed to determine the sums and evaluate the limits restrict use of (6.2.8) to relatively simple functions.

The *real* solution to the area computation problem is provided by the Fundamental Theorem of Calculus, whose importance is reflected in its title. It will be developed in the next section.

SECTION 6.2 EXERCISES

In Exercises 1 to 10 sketch the function over the indicated interval. Then for the given partition P, (a) compute $\underline{S}_P(f)$ and (b) compute $\overline{S}_P(f)$. For the listed points x_i^* in the subintervals determined by P, (c) compute $S_P^*(f)$ and (d) compute $\|P\|$.

1. $f(x) = x + 1$ on $[0, 2]$; $P = \{0, \frac{1}{2}, 1, \frac{3}{2}, 2\}$; $x_1^* = \frac{1}{4}, x_2^* = \frac{5}{8}, x_3^* = \frac{5}{4}, x_4^* = 1.9$

2. $f(x) = x + 1$ on $[0, 2]$; $P =$ even partition into 10 subintervals. $x_i^* =$ midpoint of subinterval

3. $f(x) = 3 - 2x$ on $[0, 1]$; $P = \{0, \frac{1}{4}, \frac{1}{2}, \frac{3}{4}, 1\}$; $x_1^* = \frac{1}{8}, x_2^* = \frac{3}{8}, x_3^* = \frac{9}{16}, x_4^* = 0.9$

4. $f(x) = 3 - 2x$ on $[0, 1]$; $P =$ even partition into five subintervals; $x_i^* =$ midpoint of ith subinterval

5. $f(x) = 9 - x^2$ on $[-2, 3]$; $P = \{-2, -1, 1, 1\frac{1}{2}, 3\}$; $x_1^* = -\frac{3}{2}, x_2^* = 0, x_3^* = 1, x_4^* = 2$

6. $f(x) = x^2 + 1$ on $[-1, 2]$; $P = \{-1, -\frac{1}{2}, 0, 1, \frac{3}{2}, 2\}$; $x_1^* = -1, x_2^* = -0.1, x_3^* = \frac{1}{2}, x_4^* = \frac{7}{4}$

7. $f(x) = 1 + x^3$ on $[1, 4]$; $P = \{1, 2, 2\frac{1}{2}, 4\}$; $x_1^* = 1, x_2^* = 2\frac{1}{4}, x_3^* = 3$

8. $f(x) = x^2 + x + 1$ on $[-3, 3]$; $P = \{-3, -1, 0, 2, 3\}$; $x_1^* = -2, x_2^* = -1, x_3^* = 1, x_4^* = 3$

9. $f(x) = 1/x$ on $[1, 5]$; $P = \{1, 2, 3, 5\}$; $x_1^* = 1, x_2^* = \frac{5}{2}, x_3^* = 4$

10. $f(x) = \frac{1}{3}x^3 + \frac{1}{2}x^2 - 2x$ on $[-3, 0]$; $P = \{-3, -2, -1, 0\}$; $x_1^* = -3, x_2^* = -1, x_3^* = -\frac{1}{2}$

11. Describe and compare $\underline{S}_P(f)$, $\overline{S}_P(f)$ and $S_P^*(f)$ for a constant function $f(x) = b$ on an interval $[c, d]$ for any partition P of $[c, d]$.

In Exercises 12 to 15 determine the value of the indicated sum.

12. $\sum_{i=1}^{20} i$

13. $\sum_{i=1}^{12} 2i^2$

14. $\sum_{i=1}^{6} (2i^2 - 5)$

15. $\sum_{j=2}^{7} (3j^2 + 1)$

Using methods analogous to those used in this section, determine the areas of the regions R indicated in Exercises 16 and 17. [Take an even partition, determine either $\underline{S}_P(f)$ or $\overline{S}_P(f)$, and take the limit as $\|P\| \to 0$, which is equivalent to the limit as $n \to +\infty$].

16. R bounded by $f(x) = x + 1$, the x axis, $x = 0$ and $x = 2$

17. R bounded by $f(x) = 3 - 2x$, the x axis, $x = 0$, and $x = 1$

In Exercises 18 to 20, use that

$$\sum_{i=1}^{n} c = nc \quad \text{for } c \text{ a constant}, \quad \sum_{i=1}^{n} (a_i \pm b_i) = \left(\sum_{i=1}^{n} a_i\right) \pm \left(\sum_{i=1}^{n} b_i\right), \text{ and } \sum_{i=1}^{n} ka_i = k \sum_{i=1}^{n} a_i$$

to simplify the following sums and to put them in a form to which (6.2.3) and (6.2.4) apply. Then apply these formulas to obtain the sums entirely in terms of n.

18. $\sum_{i=1}^{n} \left(\frac{i-1}{n}\right)^2 \frac{1}{n}$

19. $\sum_{i=1}^{n} \left(\frac{2i+1}{n}\right)^2 \frac{1}{n}$

20. $\sum_{i=1}^{n} \left[1 - \left(\frac{2i}{n}\right)^2\right] \frac{2}{n}$

In Exercises 21 to 31, use (6.2.8) to determine the area between the given function and the given interval [a, b].

21. $f(x) = x^2$, $[0, 2]$

22. $f(x) = x^2 + 1$, $[0, 2]$

23. $f(x) = x^2 + x + 1$, $[0, 2]$

24. $f(x) = 9 - x^2$, $[0, 2]$

25. $f(x) = 9 - x^2$, $[-3, -1]$

26. $f(x) = 9 - x^2$, $[-3, 3]$

27. $f(x) = 2x^2 + x + 3$, $[0, 2]$

28. $f(x) = 2x^2 + 3x$, $[0, 1]$

29. $f(x) = x - x^2$, $[0, 1]$

30. $f(x) = x^2 - 2x + 1$, $[0, 2]$

31. $f(x) = x^2 - 4x + 4$, $[0, 3]$

32. Prove (6.2.4) using the following outline. Let S denote the sum of the squares of the consecutive integers from 1 to n.
(*a*) Prove the identity $n^3 - (n - 1)^3 = 3n^2 - 3n + 1$.
(*b*) Successively substitute the numbers $1, 2, 3, \ldots, n$ into both sides of this identity (but do not evaluate the resulting expressions), aligning the results vertically.
(*c*) Show that all terms but the term n^3 cancel on the left side when the vertical columns are added, whereas on the right side $3S - 3(1 + 2 + 3 + \cdots + n) + n$ results.
(*d*) Solve the resulting equation for S to obtain $S = n^3/3 + n(n + 1)/2 - n/3$ [using (6.2.3) along the way].

33. Prove that $\sum_{i=1}^{n} i^3 = [n(n + 1)/2]^2$ by an argument analogous to that of Exercise 32, but using the identity $n^4 - (n - 1)^4 = 4n^3 - 6n^2 + 4n - 1$.

In Exercises 34 to 37, use the previous formula (together with any others already developed) to determine the area of the region between the given function and the given interval $[a, b]$.

34. $f(x) = x^3$, $[0, 1]$

35. $f(x) = x^3 + 1$, $[0, 2]$

36. $f(x) = -x^3 + 3x^2 - 2x$, $[1, 2]$

37. $f(x) = x^3 + x^2$, $[0, 4]$

38. By examining an n by $n + 1$ rectangle of 1 by 1 squares, devise yet another proof of (6.2.3).

Section 6.3: The Definite Integral and the Fundamental Theorem of Calculus

Objectives:

1. Understand the meaning and properties of the definite integral and its connection with antiderivatives as expressed in the Fundamental Theorem.

2. Use the Fundamental Theorem and the Power Rule to evaluate definite integrals.

DEFINITION AND MEANING OF THE DEFINITE INTEGRAL

In Sec. 6.2 we showed that if $f(x)$ is nonnegative, continuous, and piecewise monotonic on $[a, b]$, then

$$\lim_{\|P\|\to 0} \sum_{i=1}^{n} f(x_i^*)\,\Delta x_i$$

exists and equals the area of the region between $[a, b]$ and the graph of $f(x)$. Furthermore, as long as $\|P\| \to 0$, this limit is independent of the way the partitions P are chosen and of the way the points x_i^* are chosen in the subintervals determined by P.

Although we were ostensibly attacking the "area problem," and so assumed $f(x)$ was nonnegative on $[a, b]$, a careful reexamination of our arguments leading to the limit will reveal that the nonnegativity of $f(x)$ played no essential role. It simply guaranteed that the limit itself would be nonnegative. By relaxing the nonnegativity condition we shall unlock a whole new world of applications to work, energy, and economics and will be able to determine averages and volumes, and solve a whole host of problems currently out of reach. In fact, the applications of this limit are perhaps even more varied than those of the derivative limit.

In the next definition, and often in future work, we shall write

$$\sum_{P} f(x_i^*)\,\Delta x_i \qquad \text{instead of} \qquad \sum_{i=1}^{n} f(x_i^*)\,\Delta x_i$$

to emphasize that the sum depends upon the nature of the partition P and not simply upon the number of subintervals formed by P.

6.3.1 Definition of the Definite Integral Suppose the function f is defined on the closed interval $[a, b]$. If

$$\lim_{\|P\|\to 0} \sum_{P} f(x_i^*)\,\Delta x_i$$

exists and is independent of the choices of x_i^* in the subintervals determined by partitions P of $[a, b]$, we denote it by

$$\int_a^b f(x)\,dx$$

call it the *definite integral* of f from a to b, and say that f is *integrable* on $[a, b]$. We refer to a as the *lower limit of integration*, b as the *upper limit of integration*, and $f(x)$ as the *integrand*. We often express this definition as

$$\int_a^b f(x)\,dx = \lim_{\|P\|\to 0} \sum_{P} f(x_i^*)\,\Delta x_i$$

and refer to the sum appearing on the right as a *Riemann sum* for f.

The precise definition of the limit in (6.3.1) requires that for any $\varepsilon > 0$, there is a $\delta > 0$ such that

$$\|P\| < \delta \qquad \text{implies} \qquad \left| \int_a^b f(x)\,dx - \sum_P f(x_i^*)\,\Delta x_i \right| < \varepsilon$$

for any choices of x_i^* in the subintervals determined by P.

It is apparent from (6.3.1) that a function unbounded on $[a, b]$ cannot be integrable on $[a, b]$, because in such a case it is possible to construct arbitrarily large Riemann sums. In Chap. 14 we shall extend our work to handle the unbounded case.

Despite the apparent complexity of the limit defining $\int_a^b f(x)\,dx$, there is an obvious notational similarity to the *antiderivative* dealt with in Sec. 6.1. Furthermore our remarks in Sec. 6.2, suggesting that by approximating the area under the graph of a velocity function we are in fact approximating the distance traveled over the time interval from a to b, even strengthen the temptation to connect the two ideas. Later the Fundamental Theorem of Calculus will reveal that the connection indeed does exist. However, the two ideas of definite integral $\int_a^b f(x)\,dx$ and antiderivative $\int f(x)\,dx$ are *not* the same. After all, $\int_a^b f(x)\,dx$ is a *number*, whereas $\int f(x)\,dx$ is a *family of functions* (any two of which differ by a constant).

Our earlier work effectively established the integrability of functions that are continuous and piecewise monotonic on $[a, b]$. The arguments required to extend this result to *arbitrary* continuous functions (and thus to cover the "infinitely wiggly" case) are beyond the scope of this text but can be found in any advanced calculus text. Hence we assume the following:

6.3.2 If f is continuous on $[a, b]$, then f is integrable on $[a, b]$. That is, $\int_a^b f(x)\,dx$ exists.

Actually, we can use (6.3.2) to cover cases in which f is defined on $[a, b]$ and has a finite number of jumps or breaks on $[a, b]$ provided there is no point x_0 in $[a, b]$ with $\lim_{x\to x_0} f(x) = \pm\infty$, that is, as long as f is bounded on $[a, b]$. We then know the graph of f will look something like that in Fig. 6.15. Such a function is sometimes said to be *bounded and piecewise continuous* on $[a, b]$.

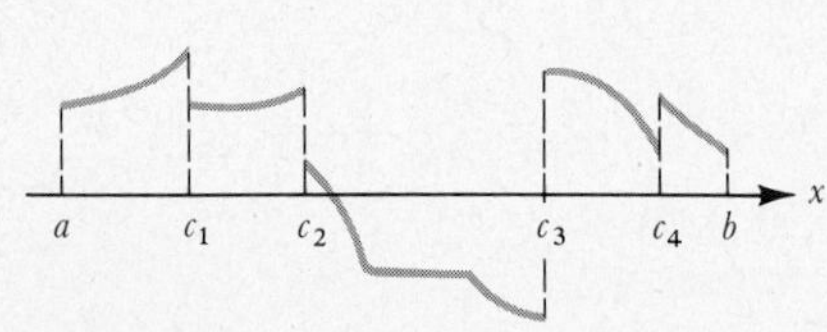

Fig. 6.15

We merely apply (6.3.2) to each subinterval of $[a, b]$ on which f is continuous. In the case of Fig. 6.15 this amounts to applying (6.3.2) separately to the intervals $[a, c_1]$, $[c_1, c_2]$, $[c_2, c_3]$, $[c_3, c_4]$, $[c_4, b]$ and then adding the results.

Our examination of the basic properties of the definite integral will be based upon the interpretation that has carried us this far: area. Thus it is of some importance to interpret $\int_a^b f(x)\,dx$ when $f(x) \le 0$ on $[a, b]$. (See Fig. 6.16.)

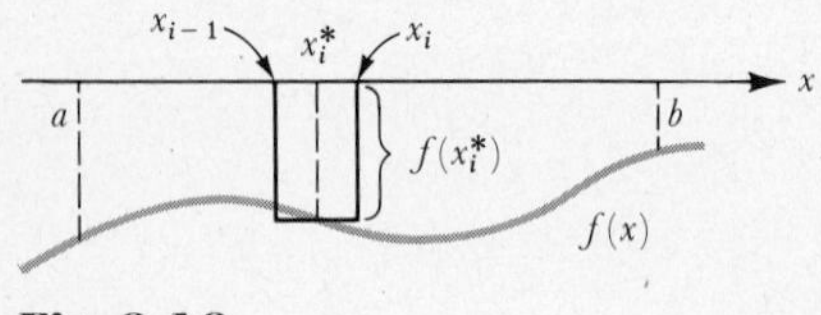

Fig. 6.16

But, of course, for x_i^* any point in a typical subinterval $[x_{i-1}, x_i]$ determined by a partition P of $[a, b]$, $f(x_i^*)\,\Delta x_i$ is merely the negative of the area of the rectangle based on $[x_{i-1}, x_i]$. Passing to the limit of the sum of such negative numbers as $\|P\| \to 0$, we have the following theorem.

6.3.3 **Theorem** If $f(x) \le 0$ and is integrable on $[a, b]$, then $\int_a^b f(x)\,dx = -A(R)$, the negative of the area $A(R)$ of the region R between the graph of f and the interval $[a, b]$.

For example, we can show by techniques used in Sec. 6.2 that

$$\int_1^4 (x-2)\,dx = \frac{3}{2}$$

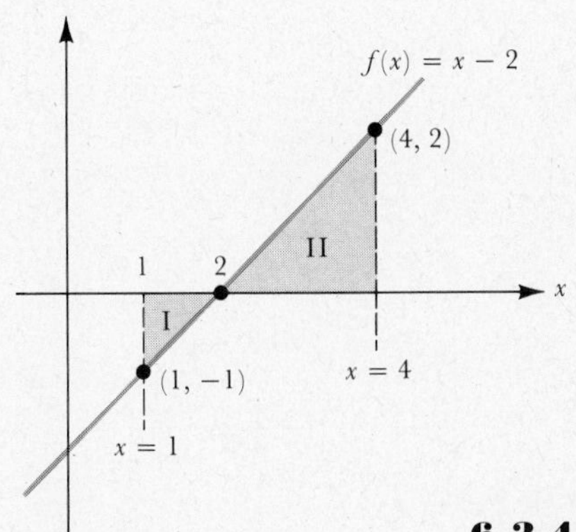

Fig. 6.17

We see in Fig. 6.17 that the area of the triangular region I is $A(\text{I}) = \frac{1}{2}(1)(1) = \frac{1}{2}$ and that the area of triangular region II is $A(\text{II}) = \frac{1}{2}(2)(2) = 2$. Thus

$$\int_1^4 (x - 2)\, dx = A(\text{II}) + (-A(\text{I})) = 2 - \frac{1}{2} = \frac{3}{2}$$

PROPERTIES OF THE DEFINITE INTEGRAL

Our definition of $\int_a^b f(x)\, dx$ presupposes that $a < b$. We shall extend our definitions to cover situations where $a > b$ or $a = b$ as follows. (Unless otherwise noted, *we shall assume that functions are integrable over an appropriate interval I.*)

6.3.4 Definition (*a*) If $b < a$, then we define

$$\int_a^b f(x)\, dx = -\int_b^a f(x)\, dx$$

(*b*) If a is in the domain of f, then we define

$$\int_a^a f(x)\, dx = 0$$

The first definition simply helps things work out in calculations. The second makes sense geometrically because the only "rectangles" that fit over or under the one-point interval $[a, a]$ have zero area.

We summarize important properties of the definite integral in the following.

6.3.5 Properties of the Definite Integral

(*a*) $\displaystyle\int_a^b f(x)\, dx = \int_a^c f(x)\, dx + \int_c^b f(x)\, dx$

(*b*) $\displaystyle\int_a^b kf(x)\, dx = k\int_a^b f(x)\, dx$, where k is any constant

(*c*) $\displaystyle\int_a^b [f(x) \pm g(x)]\, dx = \int_a^b f(x)\, dx \pm \int_a^b g(x)\, dx$

(*d*) If $f(x)$ is bounded above by M on $[a, b]$ and below by m on $[a, b]$, then

$$m(b - a) \le \int_a^b f(x)\, dx \le M(b - a)$$

(*e*) If $f(x) \le g(x)$ on $[a, b]$, $\displaystyle\int_a^b f(x)\, dx \le \int_a^b g(x)\, dx$

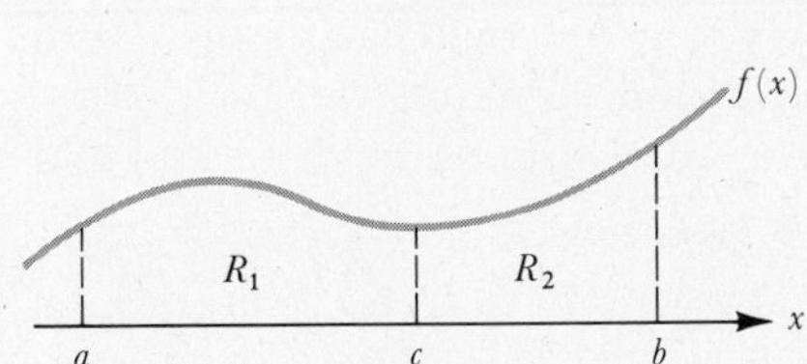

$\int_a^b f(x)\, dx$ = area of R
= area of R_1 + area of R_2
$= \int_a^c f(x)\, dx + \int_c^b f(x)\, dx$

Fig. 6.18

Discussion of (6.3.5) In part (*a*) we make no assumptions about the relative order of a, b, and c. If $a < c < b$ and $f(x) \ge 0$ on $[a, b]$, then it becomes a perfectly reasonable statement about the area of a region R that is broken into two subregions R_1 and R_2 (see Fig. 6.18):

Parts (*b*) and (*c*) follow from the corresponding facts about the sums used in defining the definite integral. Part (*d*) has a straightforward interpretation in terms of area when $f(x) \ge 0$ on $[a, b]$; see Fig. 6.19. The area of the rectangles of width $b - a$ and heights, respectively, m and M clearly satisfy

$$m(b - a) \le \int_a^b f(x)\, dx = A(R) \le M(b - a) \qquad (R \text{ is shaded})$$

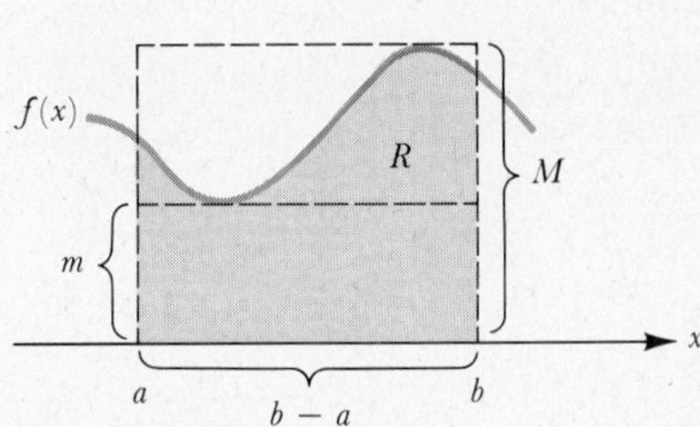

Fig. 6.19

Part (*e*) again has a direct interpretation in terms of area when $0 \le f(x) \le g(x)$ on $[a, b]$. Letting R_f be the region bounded by the graph of $f(x)$, the x axis, and

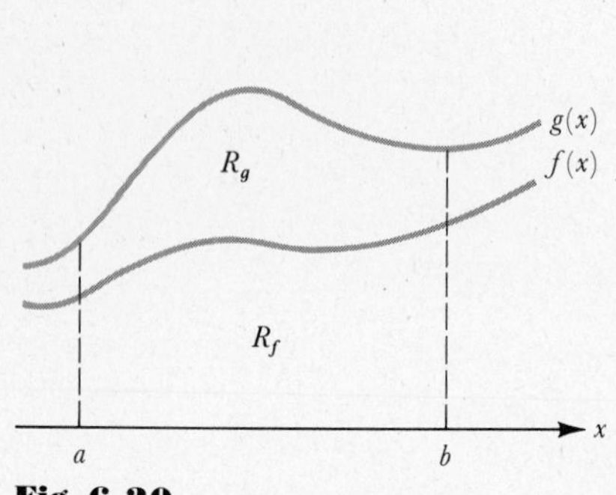

Fig. 6.20

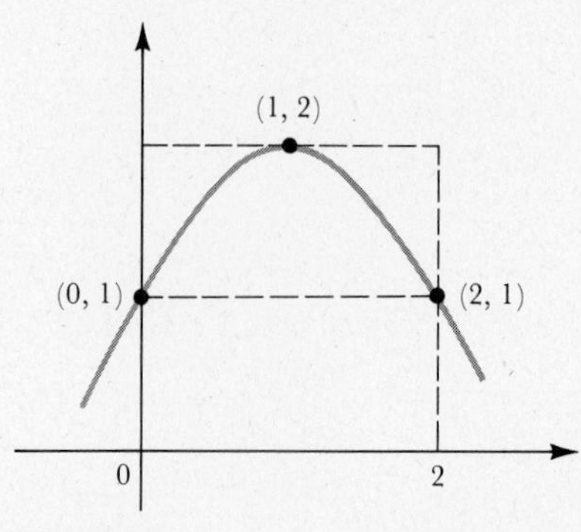

Fig. 6.21

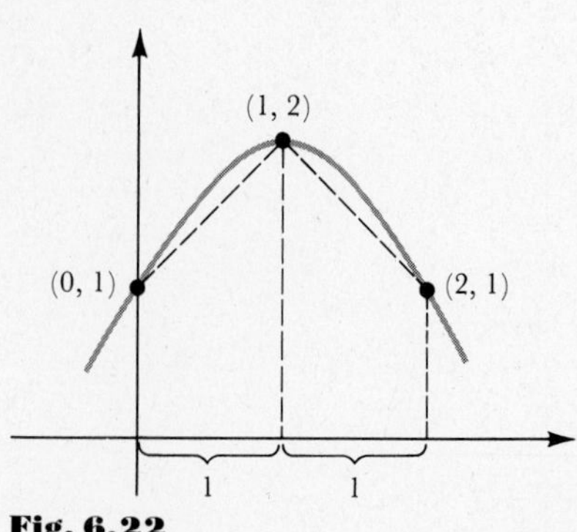

Fig. 6.22

the lines $x = a$ and $x = b$, and letting R_g be the region similarly bounded by the graph of $g(x)$, which contains R_f, it is clear from Fig. 6.20 that

$$\int_a^b f(x)\,dx = A(R_f) \leq A(R_g) = \int_a^b g(x)\,dx$$

We could also interpret part (*e*) if $f(x) \leq g(x) \leq 0$ on $[a, b]$ by using (6.3.3) and negatives of areas.

Example 1 Use [6.3.5(*d*)] to determine upper and lower bounds for

$$\int_0^2 (1 + 2x - x^2)\,dx$$

Solution (See Fig. 6.21.) For $f(x) = 1 + 2x - x^2$ we have the minimum value of f on $[0, 2]$ achieved at both $x = 0$ and $x = 2$ with $f(0) = 1 = f(2)$. The maximum is achieved at $x = 1$ with $f(1) = 2$. Hence, with $b - a = 2 - 0 = 2$, we have

$$2 = (1)(2) \leq \int_0^2 (1 + 2x - x^2)\,dx \leq 2(2) = 4 \quad \bullet$$

We can improve on the lower bound of Example 1 by breaking the region into trapezoidal pieces as in Fig. 6.22 and then determining the areas of the trapezoids to obtain (using the formula for the area of a trapezoid of base b and heights h_1 and h_2, $A = b(h_1 + h_2)/2$. That is,

$$\int_0^2 (1 + 2x - x^2)\,dx \geq 1\left(\frac{1+2}{2}\right) + 1\left(\frac{1+2}{2}\right) = 3$$

In Chap. 11 we shall return to elaborate on this trapezoid-approximation idea.

PROGRESS TEST 1

1. Use [6.3.5(*d*)] to determine upper and lower bounds for $\int_{-3}^{4} (\frac{1}{3}x^3 + \frac{1}{2}x^2 - 2x)\,dx$

2. Determine the following definite integrals, using the interpretation of the definite integral as area where $f(x)$ is nonnegative, using (6.3.3) where $f(x)$ is negative, and using [6.3.5(*a*)] if $f(x)$ changes sign on $[a, b]$. (Draw a sketch when necessary.)

(*a*) $\int_2^7 (-3)\,dx$ (*b*) $\int_0^3 (1 - x)\,dx$

(*c*) $\int_{-2}^1 (x + 1)\,dx$ (*d*) $\int_{-6}^{-1} 4\,dx$

3. Determine the indicated definite integral for the functions sketched in (*a*) and (*b*).

(*a*) $\int_1^8 f(x)\,dx$ (*b*) $\int_0^7 g(x)\,dx$

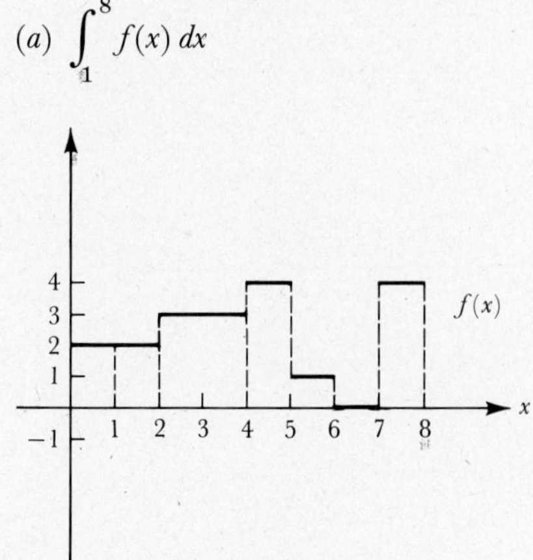

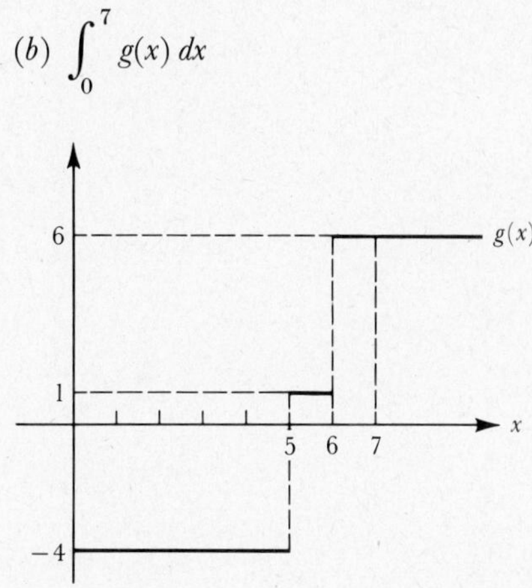

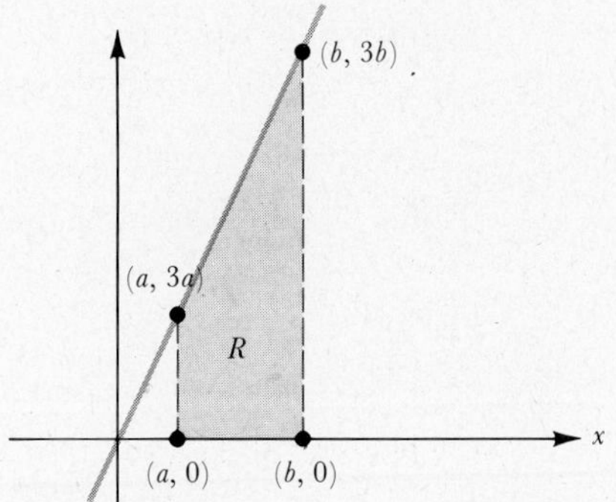

Fig. 6.23

THE FUNDAMENTAL THEOREM: A PRELIMINARY DISCUSSION

So far, we are able to obtain definite integrals without using fancy manipulations of sums and limits only for linear or step functions as in Progress Test 1. But in such cases calculus was not needed to get the answer! This situation resembles that in Chap. 2, when we could compute derivatives only by applying the limit definition. We need simpler rules that will allow us to compute definite integrals such as

$$\int_0^2 \frac{x\,dx}{\sqrt[3]{x^2+4}} \quad \text{or} \quad \int_0^1 (x^2+1)^4 x^3\,dx$$

without resorting to the limit-of-sums definition.

We already remarked on the similarity of the antiderivative notation and the definite integral notation. We shall now pursue this relationship.

Example 2 Determine the area under the graph of $f(x) = 3x$ from $x = a$ to $x = b$ with $0 < a < b$. (See Fig. 6.23.)

Solution Of course the definite integral $\int_a^b (3x)\,dx$ will give the desired area, and this integral could be computed using the definition and the appropriate finite sum formula applied to the resulting Riemann sum. However, the area of this trapezoidal region is just the difference of the areas of two triangles. (See Fig. 6.24.)
Thus

$$\int_a^b (3x)\,dx = A(R) = \underline{\underline{\frac{3b^2}{2} - \frac{3a^2}{2}}} \quad \bullet$$

Now a typical antiderivative of $3x$ is $F(x) = 3x^2/2 + C$ for some constant C.

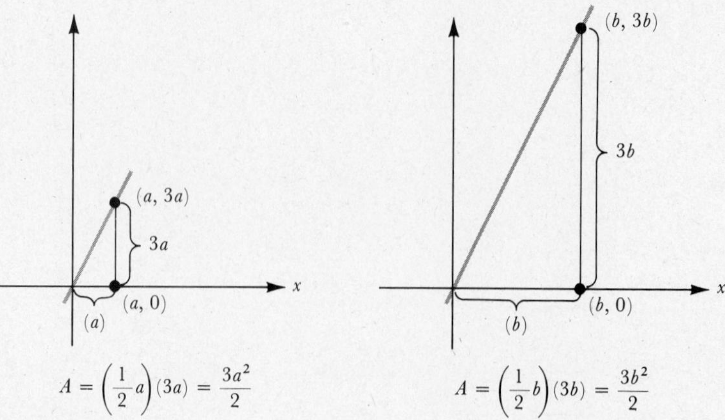

Fig. 6.24

But then

$$F(b) - F(a) = \left(\frac{3b^2}{2} + C\right) - \left(\frac{3a^2}{2} + C\right)$$
$$= \frac{3b^2}{2} - \frac{3a^2}{2}$$

Thus for this example we could have determined the value of $\int_a^b (3x)\,dx$ by *finding any antiderivative* F of $3x$ and *forming the difference* $F(b) - F(a)$*!*

Similarly, we can compute the area under the graph of

$$f(t) = 32t \qquad \text{from } t = 0 \text{ to } t = x$$

using either the definite integral limit-of-sums approach or by using an antiderivative $F(t) = 16t^2$ evaluated at $t = 0$ and $t = x$. Either way we determine the area to be

$$\int_0^x 32t\,dt = 16x^2$$

Now, of course, $f(t) = 32t$ is the familiar function that gives the velocity of a freely falling object after t seconds. We know it is the derivative of the "distance-fallen function" $F(t) = 16t^2$ or, in the terminology of Sec. 6.1, its antiderivative is a distance function. Hence the area $16x^2$ has two interpretations:

1. As the area under the graph of $f(t) = 32t$.
2. As the distance fallen after $t = x$ seconds have elapsed.

Obviously, the relationship between antiderivatives and integrals is intimate, but it needs further investigation.

---

⑨ Why not regard the right-hand side of the region bounded by the graph of $f(t) = 32t$, the t axis, and the line $t = x$, as a *variable,* with the line $t = x$ sweeping out the area under the graph from left to right? Then *the area is a function of x:*

$$\int_0^x 32t\,dt = F(x) = 16x^2$$

Previously, when thinking of $F(x)$ as a distance function, we computed its derivative to obtain the velocity function $F'(x) = 32x$. Let us now compute and interpret the derivative of $F(x)$ where $F(x)$ is regarded as an *area function.*

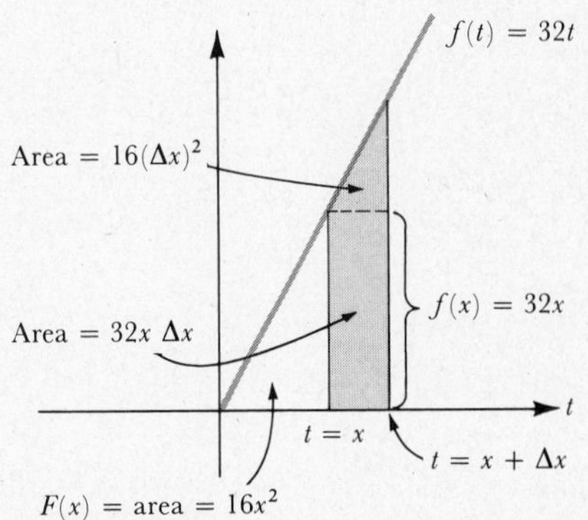

Fig. 6.25

By definition,

$$F(x + \Delta x) - F(x) = 16(x + \Delta x)^2 - 16x^2$$
$$= 32x\,\Delta x + 16(\Delta x)^2$$

the difference between the areas of the larger and smaller triangles, is the area of the shaded vertical strip (see Fig. 6.25). Its two terms correspond to the two parts separated by a dashed line. Now, dividing through by Δx and forming the limit, we have

$$F'(x) = \lim_{\Delta x \to 0} \frac{32x\,\Delta x + 16(\Delta x)^2}{\Delta x}$$
$$= 32x$$

Hence, geometrically, as the width of the strip approaches zero we are left with the length $f(x) = 32x$ of the line sweeping out the region.

Many times before we have observed the connection between the rate of growth of a geometric object and the size of its boundary in the direction of the growth.

The circle, growing by its circumference: $D_x(\pi x^2) = 2\pi x$
The sphere, growing by its surface area: $D_x(\frac{4}{3}\pi x^3) = 4\pi x^2$
The square, growing by the length of the two sides in the direction of the growth: $D_x(x^2) = 2x$
The cube, growing by the area of the three sides in the direction of the growth: $D_x(x^3) = 3x^2$
The cylinder, growing taller by the area of its "roof": $D_h(\pi r^2 h) = \pi r^2$
The cylinder, growing wider by its lateral surface area: $D_r(\pi r^2 h) = 2\pi rh$

In each case *the growth in the size of the geometric object* (volume in three dimensions, area in two dimensions) *is the size of the boundary in the direction of the growth* (an area for a three-dimensional object, a length for a two-dimensional object).

The derivative of $F(x) = 16x^2$ *regarded as an area function is the length of the side in the direction of the growth, namely* $32x$*!* That this result is not an accident peculiar to this case, but rather holds for continuous functions in general, is the extraordinary conclusion of the Fundamental Theorem of Calculus, discussed after Progress Test 2.

PROGRESS TEST 2

1. Evaluate the following definite integrals $\int_a^b f(x)\,dx$ by determining an antiderivative F of f and then computing $F(b) - F(a)$. (Each definite integral appeared earlier, so compare your results with those obtained previously.)

(*a*) $\int_0^1 x^2\,dx$ (*b*) $\int_1^4 (x - 2)\,dx$

(*c*) $\int_0^2 (1 + 2x - x^2)\,dx$

2. What is the area of the region under the graph of the constant function $f(x) = 5$ from $x = a$ to $x = b$ $(a < b)$?

3. ⑨ (*a*) What is the area of the region under the graph of $f(t) = 5$ from $t = 0$ to $t = x$?
(*b*) Denoting your answer to (*a*) by $A(x)$, find $A'(x)$ using the definition of derivative and interpret $A(x + \Delta x)$ and $A(x + \Delta x) - A(x)$ geometrically.

4. Find the area of the trapezoidal region under $f(x) = 3x + 5$ from a to b $(0 < a < b)$ using geometric reasoning and compare your result with $F(b) - F(a)$, where F is any antiderivative of f.

THE FUNDAMENTAL THEOREM OF CALCULUS

We shall now state the theorem for arbitrary continuous functions and then prove it for nonnegative continuous functions to take advantage of area interpretations.

6.3.6 **The Fundamental Theorem of the Calculus, Form I** Suppose f is continuous on $[a, b]$, and for any x in $[a, b]$ suppose F is defined by the following:

$$F(x) = \int_a^x f(t)\,dt$$

Then

$$F'(x) = f(x)$$

In other words, $\int_a^x f(t)\,dt$ is an antiderivative of f on $[a, b]$.

Note that to find the derivative of $\int_a^x f(t)\,dt$ you simply write down the integrand with t replaced by x. Just as we did in the previous discussion, we use one variable in the integrand (t in this case) and another as the upper limit of integration (x in this case). This helps keep separate the two roles played by the function f.

We saw that for $f(t) = 32t$,

$$F(x) = \int_0^x 32t\,dt \qquad \text{implies} \qquad F'(x) = 32x$$

Put more directly, we have

$$D_x\left[\int_0^x 32t\,dt\right] = 32x$$

Similarly,

$$D_x\left[\int_2^x \frac{dt}{\sqrt{1+t^2}}\right] = \frac{1}{\sqrt{1+x^2}}$$

Proof of Fundamental Theorem, Form I, with $f(t) \geq 0$ on $[a, b]$: We know that $F(x)$ represents the area of the region R_x below the graph of $y = f(t)$ and above the interval $[a, x]$; see Fig. 6.26. That is,

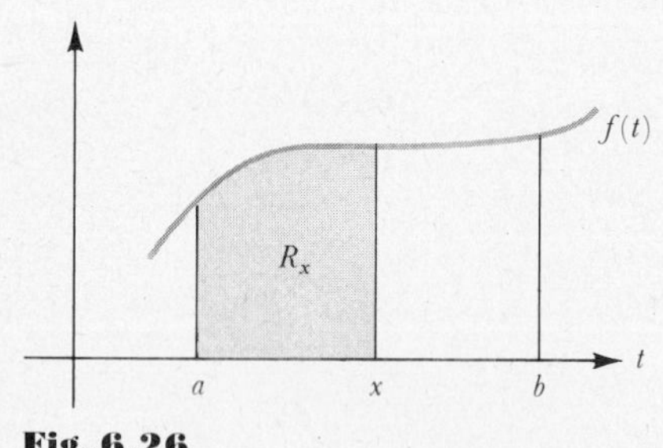

Fig. 6.26

$$F(x) = A(R_x) = \int_a^x f(t)\,dt$$

Notice that $F(a) = \int_a^a f(t)\,dt = 0$ and $F(b) = \int_a^b f(t)\,dt$.

Let us now determine the derivative of this area function,

$$F'(x) = \lim_{\Delta x \to 0} \frac{F(x+\Delta x) - F(x)}{\Delta x}$$

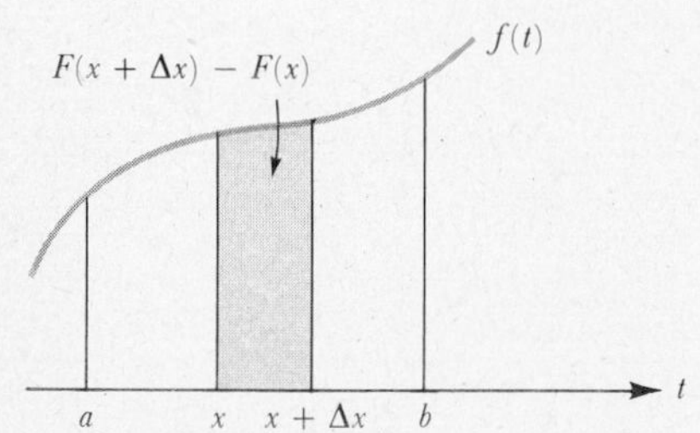

Fig. 6.27

Suppose $\Delta x > 0$. Then, as shown in Fig. 6.27, $F(x + \Delta x)$ is the area under the graph of f out to $x + \Delta x$ and $F(x)$ is the area under the graph of f out to x. Thus $F(x + \Delta x) - F(x)$ is the area of the shaded region.

We now turn our attention to the interval $I = [x, x + \Delta x]$. Now by (4.4.2), the Extreme-Value Theorem, since f is continuous on the closed interval I, f attains a minimum value on I at, say, m_x and a maximum value at, say, M_x. Thus, by [6.3.5(*d*)], with the width of the interval I being Δx, we obtain (see Fig. 6.28)

$$f(m_x)\,\Delta x \leq F(x+\Delta x) - F(x) \leq f(M_x)\,\Delta x$$

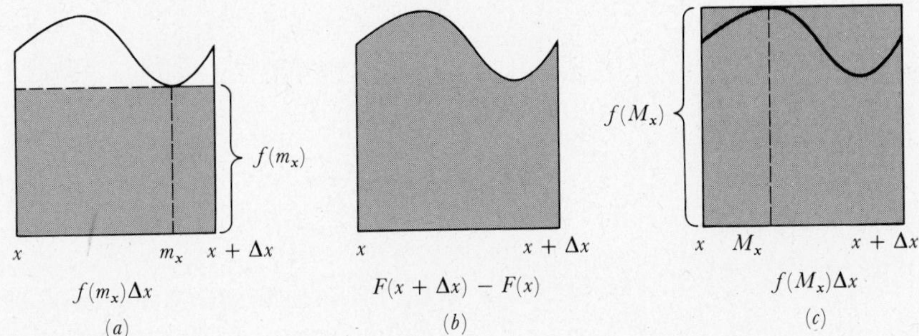

Fig. 6.28

Thus, dividing through by Δx, we have the limit quotient for F trapped:

$$f(m_x) \leq \frac{F(x + \Delta x) - F(x)}{\Delta x} \leq f(M_x)$$

We know that

$$x \leq m_x \leq x + \Delta x \qquad \text{and} \qquad x \leq M_x \leq x + \Delta x$$

so letting $\Delta x \to 0$ forces

$$m_x \to x \qquad \text{and} \qquad M_x \to x$$

But f is continuous, so

$$\lim_{\Delta x \to 0} f(m_x) = f(x) \qquad \text{and} \qquad \lim_{\Delta x \to 0} f(M_x) = f(x)$$

Since in turn, the quotient

$$\frac{F(x + \Delta x) - F(x)}{\Delta x}$$

is always trapped between $f(m_x)$ and $f(M_x)$, we have by the Squeeze Theorem (2.5.6),

$$F'(x) = \lim_{\Delta x \to 0} \frac{F(x + \Delta x) - F(x)}{\Delta x} = f(x)$$

completing the proof for $\Delta x > 0$. A similar argument applies if $\Delta x < 0$. ■

We are now in a position to make the computation of definite integrals much easier and to substantiate the underlying conjecture of Progress Test 2. The Fundamental Theorem, Form I, says that $\int_a^x f(t)\,dt$ is an antiderivative of f on $[a, b]$. Thus any antiderivative F of f on $[a, b]$ is of the form

$$F(x) = \int_a^x f(t)\,dt + C$$

Now

$$F(b) = \int_a^b f(t)\,dt + C$$

and

$$F(a) = \int_a^a f(t)\,dt + C$$
$$= 0 + C = C$$

Thus

$$F(b) - F(a) = \int_a^b f(t)\,dt + C - C$$

$$= \int_a^b f(t)\,dt$$

which gives us:

6.3.7 The Fundamental Theorem of the Calculus, Form II Suppose f is continuous on $[a, b]$ and F is any antiderivative of f on $[a, b]$. Then

$$\int_a^b f(x)\,dx = F(b) - F(a)$$

This remarkable result says that to evaluate $\int_a^b f(x)\,dx$ you simply find an antiderivative of f, substitute the number b and then subtract the result of substituting the number a. The importance of Sec. 6.1 on antidifferentiation should now be clear.

In fact, we shall often refer to an antiderivative as an *indefinite integral,* to $f(x)$ in $\int f(x)\,dx$ as an *integrand* (just as with definite integrals), and to the process of determining either indefinite or definite integrals as *integration.*

We shall use the notation

$$F(x)\Big|_a^b = F(b) - F(a) \qquad \text{or} \qquad \Big[F(x)\Big]_a^b = F(b) - F(a)$$

Thus

$$\int_a^b f(x)\,dx = F(x)\Big|_a^b = F(b) - F(a)$$

where F is any indefinite integral of f.

Example 3 Evaluate $\int_0^2 (x^2 + 4)^3 x\,dx$.

Solution Our first concern is to find an antiderivative

$$\int (x^2 + 4)^3 x\,dx$$

We shall use the substitution technique with $u = x^2 + 4$, so $du = 2x\,dx$, making $x\,dx = (1/2)\,du$. Thus we have

$$\int (x^2 + 4)^3 x\,dx = \int u^3 \frac{1}{2}\,du$$

$$= \frac{1}{2}\frac{u^4}{4} + C \qquad \text{by the Power Rule}$$

$$= \frac{(x^2 + 4)^4}{8} + C$$

Since we are interested in *any* antiderivative, we can take $C = 0$. Hence

$$\int_0^2 (x^2 + 4)^3 x\,dx = \frac{(x^2 + 4)^4}{8}\Big|_0^2$$

$$= \frac{8^4}{8} - \frac{4^4}{8}$$

$$= 512 - 32 = 480 \quad \bullet$$

Two remarks concerning Example 3 have general application:

1. Even if we had not let $C = 0$, the constant C in the antiderivative $(x^2 + 4)^4/8 + C$ would disappear in the evaluation at the limits of integration 0 and 2. Hence we use the simplest antiderivative.

2. It was essential to write the antiderivative in terms of x rather than u. If we had substituted the given limits of integration 0 and 2 into $u^4/8$, we would get an incorrect answer. In the next section we shall devise a Chain Rule for Integration allowing us to switch the given x limits to u limits.

PROGRESS TEST 3

Evaluate the following definite integrals:

1. $\int_{-1}^{1} (4x^3 - 3x^2 - 5)\,dx$ **2.** $\int_{0}^{1} x(x^2 + 1)^2\,dx$ **3.** $\int_{0}^{2} \frac{(x^2 + 1)\,dx}{\sqrt{x^3 + 3x + 1}}$

USES AND LIMITATIONS OF THE FUNDAMENTAL THEOREM

Our main tool in using the Fundamental Theorem to evaluate definite integrals is the Power Rule for antiderivatives. However, we emphasize that our new technique for evaluating definite integrals $\int_a^b f(x)\,dx$ is only as effective as our ability to find an associated antiderivative $\int f(x)\,dx$. For example, since our Power Rule does not hold for $n = -1$, at present we have no way, beyond the definition of the definite integral, to evaluate

$$\int_1^2 \frac{dx}{x}$$

even though we are assured by (6.3.2) that $\int_1^2 dx/x$ exists. In fact, it represents the area of the shaded region in Fig. 6.29. In later chapters we shall resolve the problem of this particular definite integral in several different ways.

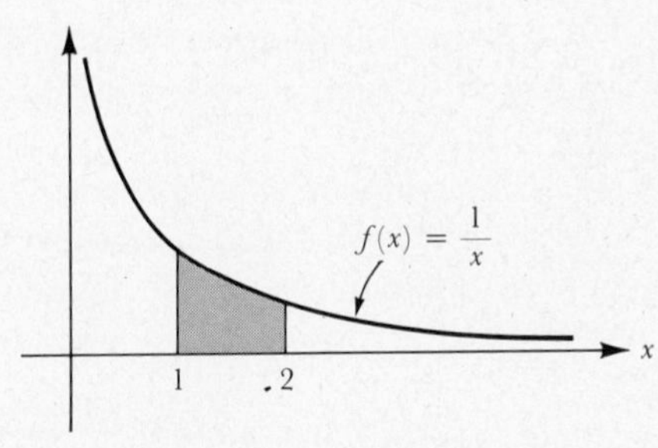

Fig. 6.29

SECTION 6.3 EXERCISES

In Exercises 1 to 12 evaluate the given definite integral using the Fundamental Theorem of Calculus.

1. $\int_0^2 (4 - 2x)\,dx$ **2.** $\int_{-2}^{2} (x - x^3)\,dx$ **3.** $\int_1^{10} (2x^2 + 1)\,dx$ **4.** $\int_{-4}^{-1} x^3\,dx$

5. $\int_0^{1/2} (3x - 9)\,dx$ **6.** $\int_{-3}^{1} 5(x^2 - 1)^4 x\,dx$ **7.** $\int_{1/2}^{1} \sqrt{2x - 1}\,dx$ **8.** $\int_{-2}^{2} \sqrt[3]{4x + 1}\,dx$

9. $\int_0^1 \sqrt{3x^2 + 6x + 1}\,(x + 1)\,dx$ **10.** $\int_{-2}^{2} (x^3 + 6x - 1)^3(x^2 + 2)\,dx$

11. $\int_{-1}^{1} (x^3 + 1)^3 x^5\,dx$ **12.** $\int_{-1}^{2} (x - 1)^3 x\,dx$

In Exercises 13 and 14, use [6.3.5(a)] to evaluate the given definite integral.

13. $\int_0^5 |x^2 - 1|\,dx$ **14.** $\int_{-2}^{2} |x^2 - 4|\,dx$

15. Sketch the region bounded by the curves $f(x) = x^3$, the x axis, $x = -2$, and $x = 2$. Without evaluating, prove that $\int_{-2}^{2} x^3\,dx = 0$. Cite specific theorems.

16. Repeat Exercise 15 for $f(x) = -x$ and $\int_{-2}^{2}(-x)\,dx$.

17. ⑤ Suppose $f(x) \geq 0$ on $[a, b]$ and $c > 0$ is a constant. Show that $\int_a^b f(x)\,dx$ and $\int_a^b (f(x) + c)\,dx$ differ by the area of a certain rectangle.

18. Show that the area under $f(x) = x^n$ for rational $n \geq 0$ from $x = 0$ to $x = b$ is $1/(n + 1) \cdot \text{base} \cdot \text{height}$. Interpret this result for $n = 0$ and $n = 1$.

19. Suppose a ball is thrown upward from ground level at an initial velocity of 96 ft/s. Noting that upward velocity is positive and downward velocity negative, sketch the velocity function on the time axis. Then use a definite integral to determine the total distance traveled by the ball in the air. Interpret your results. (This shows that care must be taken when negative velocities are involved.)

20. Suppose the acceleration $a(t)$ is nonnegative from $t = 0$ to $t = b$. What does the area under the graph of $a(t)$ represent? Explain first in terms of approximating rectangles.

21. Repeat Exercise 20, if $f(x)$ is the marginal profit function and x is the number of items.

22. Differentiate $F(x) = \int_a^x (3t^2 + 2t)\,dt$:
(*a*) Using Form I of the Fundamental Theorem.
(*b*) By first evaluating the definite integral.

23. Repeat Exercise 22 for $G(x) = \int_a^x (4t^3 - 6t + 1)\,dt$.

24. Compute $G'(x)$ for $G(x) = \int_a^b (t^2 + 3t)\,dt$. (Be careful.)

For Exercises 25 and 26, note that properties of the definite integral [see (6.3.5)] were discussed prior to any mention of derivatives and the Fundamental Theorem. Show that the properties of derivatives in combination with application of Form I (to an appropriate variation of $\int_a^x f(t)\,dt$ yield statements consistent with the given property of definite integrals.

25. Property [6.3.5(*b*)]

26. Property [6.3.5(*c*)]

Section 6.4: A Mean-Value Theorem and the Chain Rule for Integrals

Objectives:

1. Determine the average value of a function over a given interval.
2. Evaluate definite integrals using substitution by replacing the original limits with the limits in terms of the substitution variable.
3. Use the Chain Rule for Derivatives in combination with Form I of the Fundamental Theorem to differentiate functions defined by integrals.

AVERAGE (MEAN) VALUE OF A FUNCTION

× Suppose that the function $f(t)$ shown in Fig. 6.30 gives the temperature (in degrees Celsius) on a given afternoon as a function of time t marked off in hours. ($t = 0$ is noon. About 4 P.M. a thunderstorm hit.)

What was the average temperature for the afternoon? It hit a high of 30 and a low of 16, so your first answer might be $(30 + 16)/2 = 23$. But this does not really reflect the behavior of the temperature through the afternoon at all. We might improve on this by taking an hourly average; that is, we average the temperatures at the end of each of the 5 h, to get

6.4.1
$$\begin{aligned}\frac{27 + 27 + 29 + 28 + 16}{5} &= \frac{1}{5}[f(1)(1) + f(2)(1) + f(3)(1) + f(4)(1) + f(5)(1)] \\ &= 25.4\end{aligned}$$

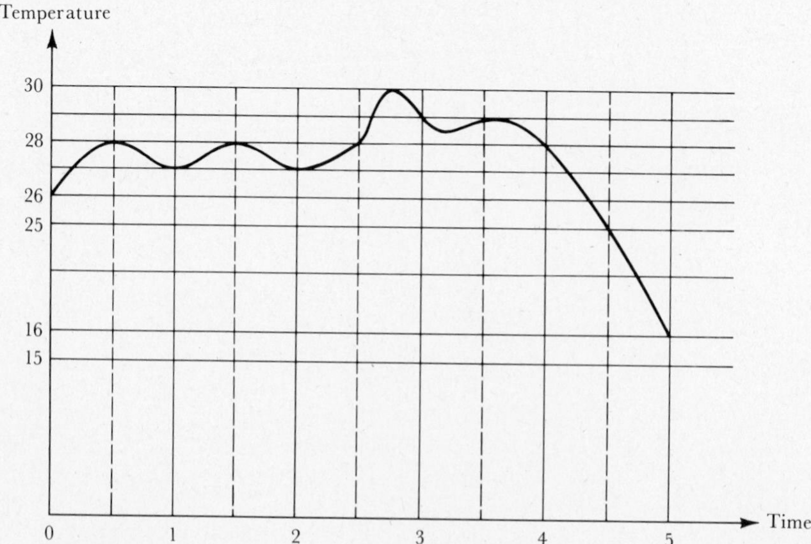

Fig. 6.30

Note that the right-hand side of (6.4.1) is one-fifth of the Riemann sum

$$\sum_{i=1}^{5} f(t_i^*)\,\Delta t_i$$

with $t_i^* = i$, $i = 1, \ldots, 5$ and $\Delta t_i = 1$.

But this average does not quite tell the story either, since the temperature was above 25.4 for almost 90 percent of the afternoon (before the storm). We could improve upon this average by taking a half-hour average (that is, 10 intervals of length 1/2 h):

$$\frac{28 + 27 + 28 + 27 + 28 + 29 + 29 + 28 + 25 + 16}{10} = 26.5$$

We can write the above quotient as

$$\frac{1}{5}\left[28\left(\frac{1}{2}\right) + 27\left(\frac{1}{2}\right) + 28\left(\frac{1}{2}\right) + 27\left(\frac{1}{2}\right) + 28\left(\frac{1}{2}\right) + 29\left(\frac{1}{2}\right) + 29\left(\frac{1}{2}\right) + 28\left(\frac{1}{2}\right) + 25\left(\frac{1}{2}\right) + 16\left(\frac{1}{2}\right)\right]$$

which again exhibits it as one-fifth of a Riemann sum

$$\sum_{i=1}^{10} f(t_i^*)\,\Delta t_i$$

where $t_i^* = i/2$, $i = 1, \ldots, 10$, and $\Delta t_i = 1/2$.

But even this average omits some of the temperature changes during the afternoon; it does not take into account the high temperature of 30, for example. To get the best average, the one that takes into account *all* the fluctuations, we let the length of the intervals approach zero and average more and more temperatures, getting

$$\lim_{\Delta t_i \to 0} \frac{1}{5} \sum_{i=1}^{n} f(t_i^*)\,\Delta t_i = \frac{1}{5} \lim_{\Delta t_i \to 0} \sum_{i=1}^{n} f(t_i^*)\,\Delta t_i = \frac{1}{5} \int_0^5 f(t)\,dt$$

This gives rise to the following definition.

6.4.2 Definition The *average (mean) value of a continuous function f over an interval* $[a, b]$ is defined to be

$$\frac{1}{b - a} \int_a^b f(x)\,dx$$

Example 1 Determine the average value of the function $f(x) = 3x^2 + 1$ on the interval $[0, 4]$.

Solution By (6.4.2) the average is given by

$$\frac{1}{4 - 0} \int_0^4 (3x^2 + 1)\,dx$$

Now

$$\int_0^4 (3x^2 + 1)\,dx = x^3 + x \Big|_0^4 = 68$$

Thus the average value of $f(x) = 3x^2 + 1$ on $[0, 4]$ is $(1/4)(68) = 17$. ●

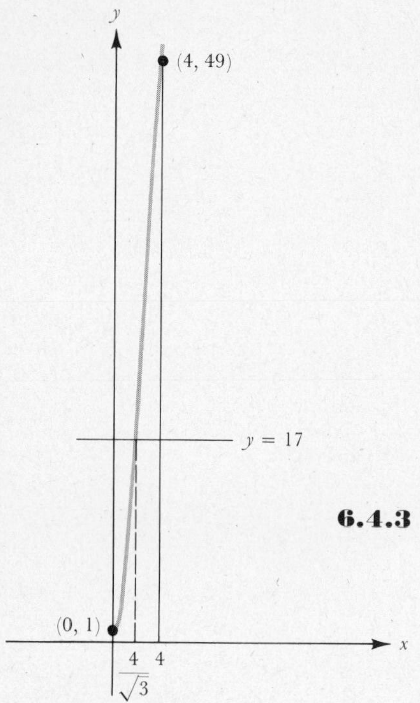

Fig. 6.31

If $f(x) = 3x^2 + 1$ in Example 1 represents the velocity of an accelerating vehicle over the time interval from 0 to 4, then, by definition, the average velocity (in appropriate units) over that interval is 17. Notice that this is *not* the simple arithmetic average of the velocities at $x = 0$ ($f(0) = 1$) and at $x = 4$ ($f(4) = 49$), which is $(49 + 1)/2 = 25$. Nonetheless, there *is* a time c with $0 < c < 4$ such that at that time $f(c) = 17$. This simply says that the graph of $f(x) = 3x^2 + 1$ crosses the horizontal line $y = 17$ between $x = 0$ and $x = 4$ when $3x^2 + 1 = 17$, or when $x = 4/\sqrt{3} \approx 2.3$. (See Fig. 6.31.) Thus the accelerating vehicle attains its average velocity once during the interval.

Similarly, in reference to the temperature in Fig. 6.30, the average temperature is between its low of 16 and its high of 30 and the temperature must have been at that figure at least once during the day. As you may already have guessed, we are working toward a Mean-Value Theorem for Integrals. It is, in fact, the MVT for derivatives stated in integral form.

6.4.3 **The Mean-Value Theorem for Integrals** Suppose f is continuous on $[a, b]$. Then there exists at least one number c between a and b such that

$$f(c) = \frac{1}{b - a}\int_a^b f(x)\,dx$$

Proof: If F is an antiderivative of f on $[a, b]$, then the MVT for derivatives (F is differentiable on $[a, b]$) guarantees at least one number c such that

$$F'(c) = \frac{F(b) - F(a)}{b - a} \tag{1}$$

But F is an antiderivative of f, so $F'(c) = f(c)$, and by the Fundamental Theorem, Form II, we know that

$$\int_a^b f(x)\,dx = F(b) - F(a)$$

Hence we can rewrite (1) as

$$f(c) = \frac{1}{b - a}\int_a^b f(x)\,dx \quad \blacksquare$$

If $f(x) \geq 0$ on $[a, b]$, then we can interpret the MVT for Integrals geometrically. (See Fig. 6.32.) We can rewrite its conclusion in the equivalent form

$$\int_a^b f(x)\,dx = f(c)(b - a)$$

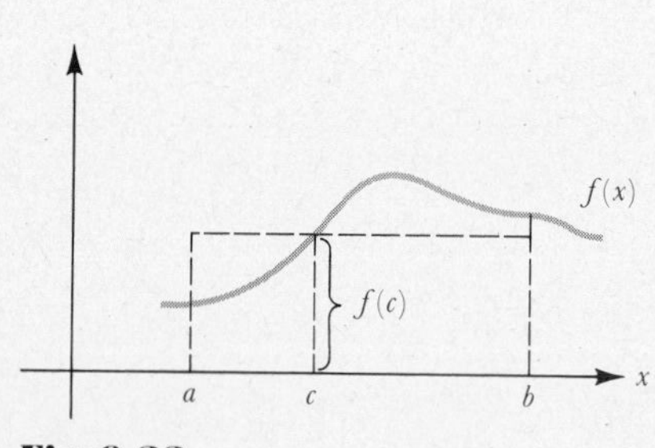

Fig. 6.32

Since $f(c)(b - a)$ is the area of the rectangle based on $[a, b]$ with height $f(c)$, we interpret the conclusion of the MVT for Integrals as stating the existence of a rectangle on $[a, b]$ whose area equals the area between f and $[a, b]$.

PROGRESS TEST 1

1. Find the point on the interval [2, 7] where the function $f(x) = x^2 - 2x + 1$ assumes its average value. (First compute the average value.)

2. The temperature in degrees F on a given afternoon is given by $T = T(x) = -x^2 + 4x + 70$, where x is measured in hours $0 \leq x \leq 6$ (noon to 6 P.M.). What was the average temperature for the afternoon? What was the high and low for the afternoon? (Note that the total day's energy consumption—for air conditioning, heating, and so on—depends upon the true average temperature for that day, and not just the average of the high and low.)

3. (*a*) What is the average velocity, in feet per second, of a stone dropped from a bridge if it takes 5 s for the stone to hit the water below?
(*b*) Compare this with the arithmetic average of the velocities at $t = 0$ and $t = 5$.
(*c*) Interpret the result of (*b*) in terms of the graph of the velocity function.

THE CHAIN RULE FOR INTEGRALS

In Sec. 6.3, Example 3, when computing

$$\int_0^2 (x^2 + 4)^3 x\,dx$$

using the substitution $u = x^2 + 4$, $x\,dx = (1/2)\,du$, we arrived at

$$\frac{1}{2}\int u^3\,du = \frac{u^4}{8}$$

We then remarked that to compute the definite integral using the given limits of integration, we would need to convert the antiderivative to an expression strictly involving x, thereby reducing the problem to

$$\left.\frac{(x^2 + 4)^4}{8}\right|_0^2 = 480$$

Sometimes it will be convenient to switch the given x limits of integration to u limits. In this case, since $u = x^2 + 4$ we know that $u(0) = 4$ and $u(2) = 8$, so we could also evaluate the definite integral without switching back, as follows:

$$\frac{1}{2}\int_4^8 u^3\,du = \left.\frac{u^4}{8}\right|_4^8 = \frac{1}{8}[8^4 - 4^4] = 480$$

Our ability to change limits of integration in this way is based upon the following rule.

6.4.4 **The Chain Rule for Definite Integrals** If $u(x)$ is differentiable on an interval I containing $[a, b]$, then

$$\int_a^b f(u(x))u'(x)\,dx = \int_{u(a)}^{u(b)} f(u)\,du$$

This rule in turn is based upon:

6.4.5 **The Chain Rule for Antiderivatives** If F is an antiderivative of f, then

$$\int f(u(x))u'(x)\,dx = \int f(u)\,du = F(u) + C = F(u(x)) + C$$

In each of the previous statements we are assuming that an antiderivative of f exists on an interval J such that $u(x)$ is in J for x in I, and that f is continuous on J. We now prove (6.4.5), from which (6.4.4) follows.

Proof of (6.4.5): We apply the Chain Rule for derivatives:

$$\begin{aligned} D_x[F(u(x))] &= F'(u(x))u'(x) \\ &= f(u(x))u'(x) \qquad \text{(since } F' = f\text{)} \end{aligned}$$ ■

Example 2 Use the Chain Rule for Definite Integrals to evaluate

$$\int_0^1 (x^2 + 1)^4 x \, dx$$

Solution Let

$$u = x^2 + 1 \qquad \frac{1}{2} du = x \, dx$$

Now

$$x = 0 \quad \text{implies} \quad u = 0^2 + 1 = 1$$

and

$$x = 1 \quad \text{implies} \quad u = 1^2 + 1 = 2$$

Hence

$$\begin{aligned}\int_0^1 (x^2 + 1)^4 x \, dx &= \frac{1}{2} \int_1^2 u^4 \, du \\ &= \frac{1}{2} \left[\frac{u^5}{5} \right]_1^2 \\ &= \frac{1}{2} \left[\frac{32}{5} - \frac{1}{5} \right] = \frac{31}{10}\end{aligned}$$ ●

We can also use the Chain Rule for Derivatives to differentiate functions defined by definite integrals. Form I of the Fundamental Theorem states that

$$D_x \left[\int_a^x f(t) \, dt \right] = f(x)$$

Thus, for example, if

$$F(x) = \int_1^x \frac{dt}{\sqrt{t} + 1}$$

then

$$D_x[F(x)] = D_x \left[\int_1^x \frac{dt}{\sqrt{t} + 1} \right] = \frac{1}{\sqrt{x} + 1}$$

However, given

$$\int_1^{x^2+5} \frac{dt}{\sqrt{t} + 1}$$

we can regard this definite integral as $F(x^2 + 5)$, where

$$F(x) = \int_1^x \frac{dt}{\sqrt{t} + 1}$$

Hence

$$\begin{aligned}D_x \left[\int_1^{x^2+5} \frac{dt}{\sqrt{t} + 1} \right] &= D_x[F(x^2 + 5)] \\ &= F'(x^2 + 5) \cdot D_x(x^2 + 5) \qquad \text{(Chain Rule for } \textit{Derivatives}\text{)} \\ &= \frac{1}{\sqrt{x^2 + 5} + 1} (2x) \\ &= \frac{2x}{\sqrt{x^2 + 5} + 1}\end{aligned}$$

We could even have been asked to differentiate

$$G(x) = \int_{x^2+5}^1 \frac{dt}{\sqrt{t} + 1}$$

The first step would be to apply [6.3.4(*a*)], rewriting $G(x)$ as

$$G(x) = -\int_1^{x^2+5} \frac{dt}{\sqrt{t}+1}$$

We would then proceed as before.

PROGRESS TEST 2

1. Determine the given definite integral using (6.4.4).

$$\int_0^2 \sqrt{x^3+1}\, x^2\, dx$$

In Problems 2 to 4 use Form I of the Fundamental Theorem and any additional necessary properties of the definite integral to find $G'(x)$.

2. $G(x) = \int_1^x \frac{dt}{3t+2}$ **3.** $G(x) = \int_x^1 (t^2+1)\, dt$ **4.** $G(x) = \int_1^{1/x} \frac{dt}{t} \quad x \neq 0$

SECTION 6.4 EXERCISES

In Exercises 1 to 10, (a) determine the average value of the function on the given interval and (b) determine where in the interval f achieves this average value.

1. $f(x) = 2x + 5$, $[0, 10]$
2. $f(x) = x^2 + x + 1$, $[1, 4]$
3. $g(x) = x^3 + x - 2$, $[2, 7]$
4. $h(x) = \sqrt{x} + 1$, $[0, 4]$
5. $y = 1 - x$, $[-2, 3]$
6. $f(x) = 5x^4$, $[0, 2]$
7. $f(x) = 5x^4$, $[-2, 2]$ (compare with Exercise 6)
8. $y = 1 - x^2$, $[-1, 1]$ (sketch)
9. $y = x^2 - 4$, $[-2, 2]$ (sketch)
10. $f(x) = x\sqrt{x^2+6}$, $[0, 3]$

In Exercises 11 to 16 use the Chain Rule for Definite Integrals to evaluate the given definite integral.

11. $\int_0^2 \sqrt{4x+1}\, dx$ **12.** $\int_2^5 \frac{x\, dx}{\sqrt{x^2+5}}$ **13.** $\int_0^1 x(x^2-1)^5\, dx$

14. $\int_1^7 x\sqrt[3]{x^2+15}\, dx$ **15.** $\int_{-1}^1 \sqrt[3]{x-1}\, dx$ **16.** $\int_0^1 (x^3+3x+4)^5(x^2+1)\, dx$

17. Suppose the amount of oil on hand at a storage depot, in thousands of gallons, is given as a function of time t (in weeks) by the graph shown. By determining the number of gallons as an algebraically defined function of t for each convenient interval, determine the average number of gallons on hand during the 10-week interval from $t = 0$ to $t = 10$. *Hint:* $a = 0$, $b = 10$; use [6.3.5(*a*)].

In Exercises 18 to 25 determine $F'(x)$.

18. $F(x) = \int_3^x \sqrt{s^2+s+1}\, ds$

19. $F(x) = \int_1^{3x^2+1} (w^2+2w)\, dw$

20. $F(x) = \int_x^2 \sqrt{t+1}\, dt$

21. $F(x) \int_1^{2x-1} \frac{t+1}{2}\, dt$

22. $F(x) = \int_x^4 \frac{dt}{t^2+1}$

23. $F(x) = \int_{2x}^{x^2} \sqrt{t^2+2}\, dt$ *Hint:* Let c be a number between $2x$ and x^2, and apply [6.3.5(*a*)].

24. $F(x) = \int_{\sqrt{x}}^{5x} (w^2 + 2w)\, dw$ (see Exercise 23)

25. $F(x) = \int_0^5 (t^2 + t)\, dt$

26. Show that the average value of a linear function $f(x) = mx + b$ on an interval $[c, d]$ can be determined arithmetically, without integration.

27. A ball is thrown vertically from ground level at an initial velocity of 96 ft/s. What is its average velocity for the duration of its flight? (Note that positive direction is upward.)

28. Repeat Exercise 27 for the same ball thrown with the same initial velocity but from a 48-ft platform.

29. Show that $\int_2^3 (x^2 - 5x + 10)^3(2x - 5)\, dx = 0$ by using (6.4.4).

Section 6.5: Area Between Graphs of Functions, and Interpretations

Objectives:

1. Use definite integrals to determine the area of regions whose boundaries are the graphs of given equations.

2. Be able to choose the appropriate independent variable for a given problem.

INTRODUCTION

We already established that the area of a region in a plane may in fact represent a wide variety of quantities in problems whose original formulation does not involve area. We shall now concentrate on techniques for computing areas efficiently.

Another payoff from our work with areas will be our ability to move from *approximations* of quantities based upon Riemann sums to computations of *exact* quantities based upon definite integrals. The ideas underlying our work with areas underlie virtually all applications of definite integrals.

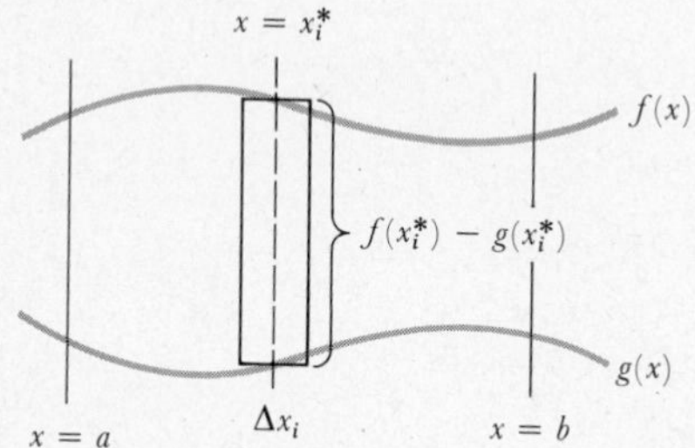

Fig. 6.33

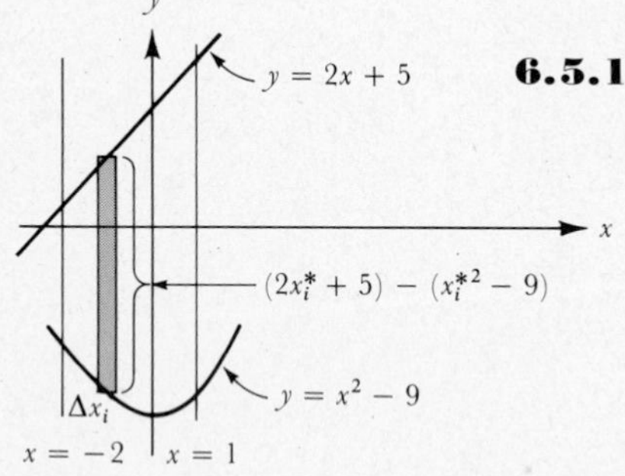

Fig. 6.34

AREA BETWEEN CURVES

Suppose R is the region bounded by $f(x)$ and $g(x)$ (both integrable on $[a, b]$), and the lines $x = a$ and $x = b$ $(a < b)$ with $f(x) \geq g(x)$ on $[a, b]$. To approximate the area of R we subdivide $[a, b]$ with a partition P into n subintervals and sum the areas of the approximating rectangles determined by each subinterval (see Fig. 6.33). We let x_i^* be any point in a typical subinterval $[x_{i-1}, x_i]$ determined by P and sketch in the typical rectangle.

We have

6.5.1
$$A(R) \approx \sum_{i=1}^{n} [f(x_i^*) - g(x_i^*)]\, \Delta x_i$$

a Riemann sum for $f(x) - g(x)$. Letting $\|P\| \to 0$, the approximations approach the *exact* area as a limit. That is,

$$A(R) = \lim_{\|P\| \to 0} \sum_{i=1}^{n} [f(x_i^*) - g(x_i^*)]\, \Delta x_i = \int_a^b (f(x) - g(x))\, dx$$

Of course, with $g(x) = 0$ we have our earlier result for area.

Example 1 Determine the area of the region bounded by the curves $y = x^2 - 9$, $y = 2x + 5$ and the straight lines $x = -2$ and $x = 1$.

Solution First we draw a sketch of the region (see Fig. 6.34). Since $2x + 5 \geq x^2 - 9$ on $[-2, 1]$, we know that $2x + 5$ plays the role of $f(x)$ and $x^2 - 9$ the role of $g(x)$. We then draw in a typical rectangle and form the approximating sum:

$$A(R) \approx \sum_{i=1}^{n} [(2x_i^* + 5) - (x_i^{*2} - 9)]\, \Delta x_i$$

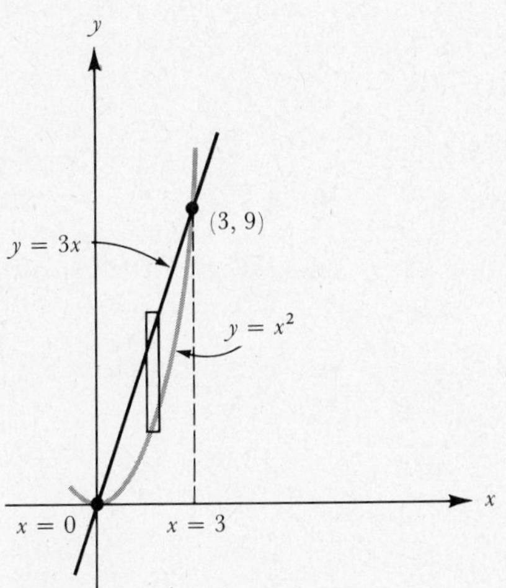

Fig. 6.35

Hence, by (6.5.1),

$$A(R) = \int_{-2}^{1} [(2x + 5) - (x^2 - 9)]\, dx$$
$$= \int_{-2}^{1} (2x - x^2 + 14)\, dx$$
$$= x^2 - \frac{x^3}{3} + 14x \Big|_{-2}^{1}$$
$$= \left(1 - \frac{1}{3} + 14\right) - \left(4 + \frac{8}{3} - 28\right)$$
$$= 36 \quad \bullet$$

In practice we shall apply (6.5.1) in a more streamlined fashion, omitting subscripts and superscripts on x in describing the typical rectangle, and passing directly to the definite integral. The important thing is to identify the region and the typical rectangle clearly. Here and henceforth we shall ignore specific units unless they play a special role in the problem.

Often the straight lines $x = a$ and $x = b$ are not given but are determined by finding the points of intersection of the bounding curves.

Example 2 Find the area of the region enclosed by the curves $y = 3x$ and $y = x^2$.

Solution To find the points of intersection we set $3x = x^2$ and solve, getting $x = 0$ and $x = 3$. Thus the curves intersect at $(0, 0)$ and $(3, 9)$; see Fig. 6.35.

A typical approximating rectangle has a width Δx and height $3x - x^2$, hence an area $(3x - x^2)\,\Delta x$. By (6.5.1), we know that the total *exact* area is given by

$$\int_0^3 (3x - x^2)\, dx = \frac{3x^2}{2} - \frac{x^3}{3}\Big|_0^3$$
$$= \left(\frac{27}{2} - 9\right) - (0) = \frac{9}{2} \quad \bullet$$

PROGRESS TEST 1

Sketch the region enclosed by the indicated curves and find its area.

1. $R: y = x^3, y = 8, x = -1, x = 2$
2. $R: y = \sqrt{x}, y = x^2$

MORE ON AREAS

It is important to realize that to use (6.5.1) we need $f(x) \geq g(x)$ on $[a, b]$. If, for example, the region R were as pictured in Fig. 6.3.6, we would break it up into three subregions and apply (6.5.1) to each, getting

$$A(R) = \int_a^c [f(x) - g(x)]\, dx + \int_c^d [g(x) - f(x)]\, dx + \int_d^b [f(x) - g(x)]\, dx$$

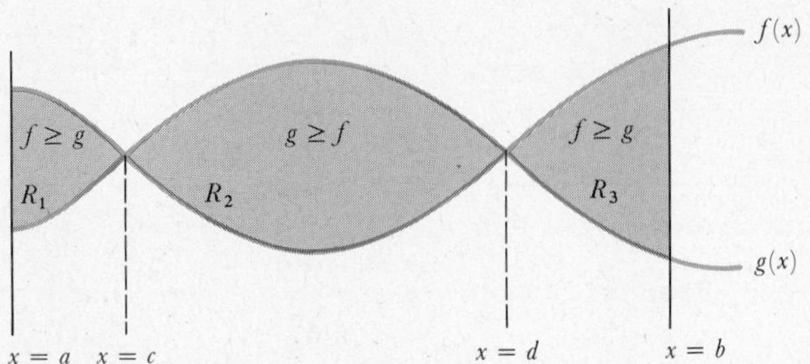

Fig. 6.36 $R = R_1 \cup R_2 \cup R_3$

You can anticipate the possibility of this situation by setting $f(x) = g(x)$ and checking for solutions *between* $x = a$ and $x = b$. Then use a sketch to determine "who's on top."

Example 3 Determine the area of the region bounded by the curves $y = 1 - x^2$ and $y = x^2 - 1$, and the straight lines $x = -2$ and $x = 2$.

Solution Setting $1 - x^2 = x^2 - 1$, we obtain $2x^2 = 2$, so $x = \pm 1$. We sketch the curves (see Fig. 6.37), obtaining three separate regions:

$$\begin{array}{ll} R_1\colon x^2 - 1 \geq 1 - x^2 & \text{on } [-2, -1] \\ R_2\colon 1 - x^2 \geq x^2 - 1 & \text{on } [-1, 1] \\ R_3\colon x^2 - 1 \geq 1 - x^2 & \text{on } [1, 2] \end{array}$$

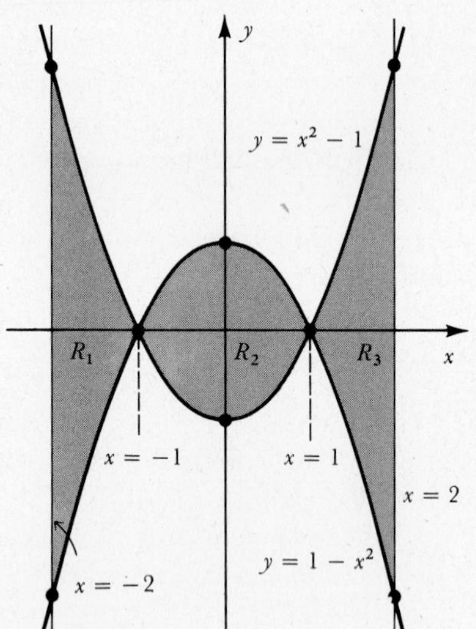

Fig. 6.37

Hence

$$A(R) = \int_{-2}^{-1} [(x^2 - 1) - (1 - x^2)]\,dx + \int_{-1}^{1} [(1 - x^2) - (x^2 - 1)]\,dx$$
$$+ \int_{1}^{2} [(x^2 - 1) - (1 - x^2)]\,dx$$
$$= \int_{-2}^{-1} (2x^2 - 2)\,dx + \int_{-1}^{1} (2 - 2x^2)\,dx + \int_{1}^{2} (2x^2 - 2)\,dx$$
$$= \left[\frac{2}{3}x^3 - 2x\right]_{-2}^{-1} + \left[2x - \frac{2}{3}x^3\right]_{-1}^{1} + \left[\frac{2}{3}x^3 - 2x\right]_{1}^{2}$$
$$= \left[\left(-\frac{2}{3} + 2\right) - \left(\frac{-16}{3} + 4\right)\right] + \left[\left(2 - \frac{2}{3}\right) - \left(-2 + \frac{2}{3}\right)\right]$$
$$+ \left[\left(\frac{16}{3} - 4\right) - \left(\frac{2}{3} - 2\right)\right]$$
$$= -\frac{2}{3} + 2 + \frac{16}{3} - 4 + 2 - \frac{2}{3} + 2 - \frac{2}{3} + \frac{16}{3} - 4 - \frac{2}{3} + 2 = 8 \quad \bullet$$

y = d
Δy_i
$q(y_i^*) - p(y_i^*)$
y = c
$x = p(y)$
$x = q(y)$

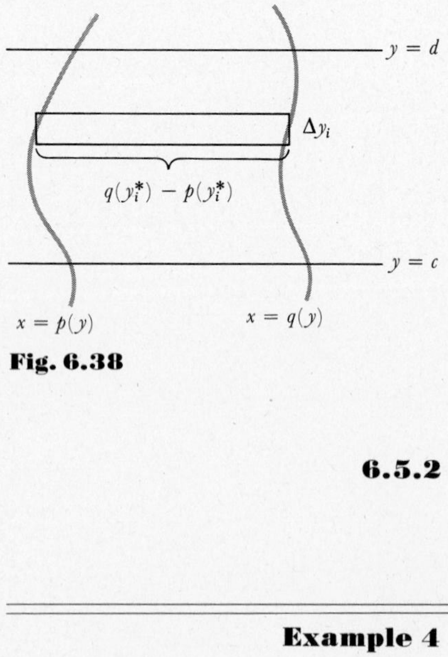

Fig. 6.38

In some situations, when given equations in x and y describing a region, it is convenient to regard x as a function of y and "turn (6.5.1) on its side." (See Fig. 6.38.)

Suppose $q(y) \geq p(y)$ on $[c, d]$. Then the approximating rectangles are horizontal and have area $[q(y_i^*) - p(y_i^*)]\,\Delta y_i$, so we have the Riemann sum

$$\sum_{i=1}^{n} [q(y_i^*) - p(y_i^*)]\,\Delta y_i$$

as an approximation of $A(R)$. Thus we conclude the following:

6.5.2 $$A(R) = \int_{c}^{d} [q(y) - p(y)]\,dy$$

Example 4 Find the area of the region R bounded by the curves $y^2 = x + 1$ and $y = x - 1$.

Solution Considering y as the independent variable, the two functions are

$$x = q(y) = y^2 - 1 \qquad \text{and} \qquad x = p(y) = y + 1$$

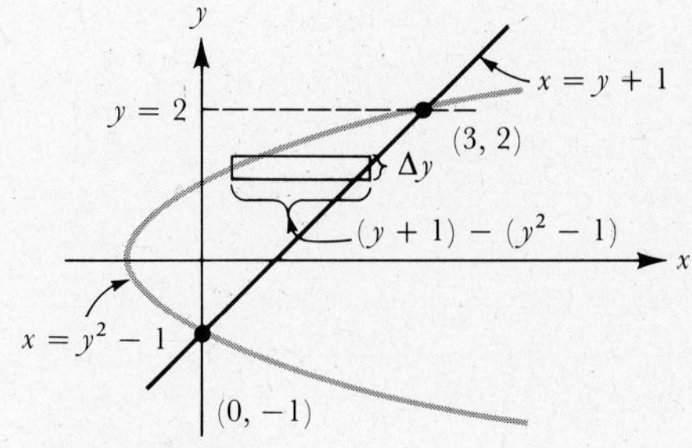

Fig. 6.39

Setting $q(y) = p(y)$, we obtain

$$y^2 - 1 = y + 1 \qquad \text{or} \qquad (y-2)(y+1) = 0$$

Hence the curves intersect when $y = 2$ and $y = -1$. The corresponding values of x are $x = 3$ and $x = 0$, respectively. (See Fig. 6.39.)

Now $y + 1 \geq y^2 - 1$ on $[-1, 2]$, so a typical rectangle has area $[(y + 1) - (y^2 - 1)]\,\Delta y$. By (6.5.2),

$$\begin{aligned} A(R) &= \int_{-1}^{2} [(y+1) - (y^2 - 1)]\,dy \\ &= \int_{-1}^{2} (y - y^2 + 2)\,dy \\ &= \frac{y^2}{2} - \frac{y^3}{3} + 2y \Big|_{-1}^{2} \\ &= \left(2 - \frac{8}{3} + 4\right) - \left(\frac{1}{2} + \frac{1}{3} - 2\right) = \frac{9}{2} \quad \bullet \end{aligned}$$

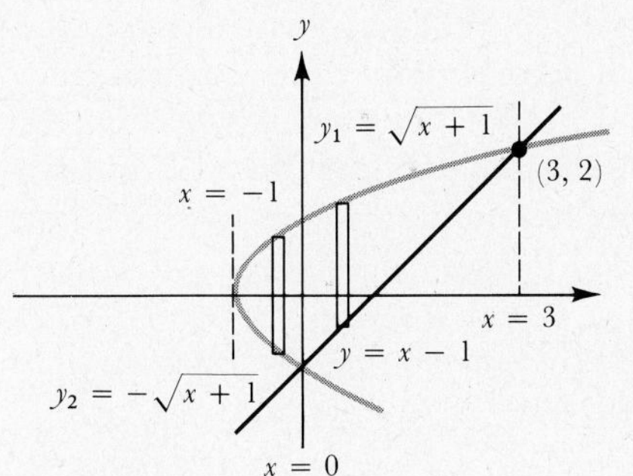

Fig. 6.40

To determine the area of the region in Example 4 (see Fig. 6.39) using x as the independent variable and vertical rectangles, we would need first to express all the boundaries of the region as graphs of functions of x. In practice this usually means solving the given equations for y, being careful to distinguish between the various functions of x defined *implicitly* by a given equation. For example, solving $y^2 = x + 1$ for y yields two functions of x:

$$y_1 = \sqrt{x+1} \qquad \text{above the } x \text{ axis}$$

$$y_2 = -\sqrt{x+1} \qquad \text{below the } x \text{ axis; see Fig. 6.40}$$

It is important to recognize that *there is no single typical vertical rectangle* in this example. Between $x = -1$ and $x = 0$, a typical rectangle has $y_1 = \sqrt{x+1}$ on top and $y_2 = -\sqrt{x+1}$ on the bottom, and hence has area $[\sqrt{x+1} - (-\sqrt{x+1})]\,\Delta x$. Between $x = 0$ and $x = 3$, a typical rectangle has $y_1 = \sqrt{x+1}$ on top and $y = x - 1$ on the bottom, and hence has area $[\sqrt{x+1} - (x-1)]\,\Delta x$. As a result the total area of the region must be given as a sum of two integrals, one for each type of typical rectangle:

$$A(R) = \int_{-1}^{0} [\sqrt{x+1} - (-\sqrt{x+1})]\,dx + \int_{0}^{3} [\sqrt{x+1} - (x-1)]\,dx$$

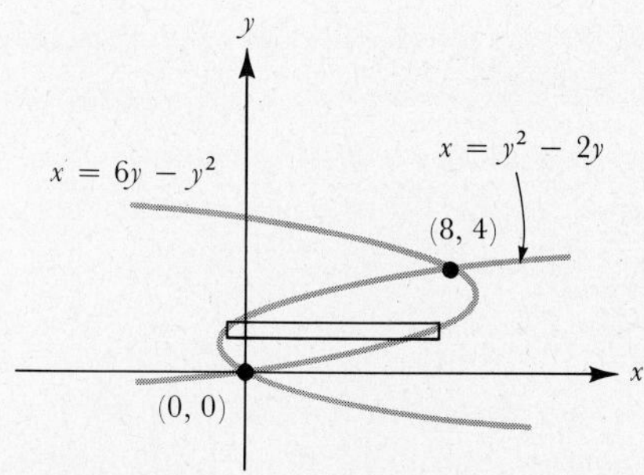

Fig. 6.41

You are invited to show these two integrals yield $A(R) = 9/2$.

To determine the area of the region enclosed by the graphs of

$$x + 2y = y^2 \qquad \text{and} \qquad x + y^2 = 6y$$

we would need to sketch each horizontal-axis parabola, determine where they cross, and then decide whether to use horizontal or vertical rectangles. A glance at Fig. 6.41 tells the outcome. How many integrals would be needed if we used vertical rectangles?

PROGRESS TEST 2

Find the area of the region R bounded by the curves $x = 9 - y^2$ and $x = y + 3$:

1. Treating y as a function of x.

2. Treating x as a function of y.

INTERPRETATIONS OF AREAS IN BUSINESS AND ECONOMICS

The basic idea of a definite integral as a limit of Riemann sums, especially when connected to the area between two curves, applies in business and economics whenever a *total* is sought. Hence the total ______ (cost, revenue, profit, and so forth) associated with producing units from a to b ($a < b$) is the definite integral of the ______ function from a to b (this function being appropriately integrable, of course.)

MAXIMIZING PROFIT OVER TIME

In operations where the revenue, cost, and hence profits all vary considerably but predictably with time, the determination of the maximum profit point requires an analysis of all these factors as functions of *time* rather than as functions of *units produced* (as was the case in Sec. 5.4).

Suppose the coal producer of Sec. 5.4, Example 1, realizes that the local coal reserves are running out, and therefore the amount of coal, hence the rate of revenue flow $R'(t)$ will decrease, the cost rate $C'(t)$ of mining it will increase, and so the profit rate $P'(t) = R'(t) - C'(t)$ will decrease with the passage of time. The major issues for the producer are: (1) How long should the mine be operated to maximize profit? (2) What is the total profit to expect? Under the assumption that production per unit time is constant, the answers to these questions are found by examining the area between the graphs of $R'(t)$ and $C'(t)$; see Fig. 6.42.

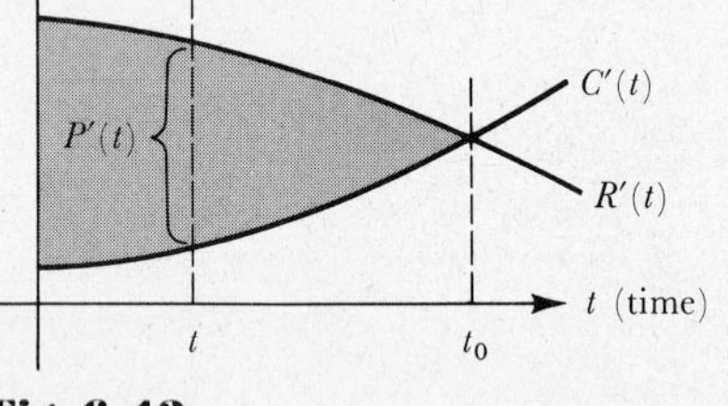

Fig. 6.42

Our experience with area and integration tells us that the total profit is

6.5.3
$$\int_0^{t_0} (R'(t) - C'(t))\, dt$$

where t_0 is the point in time when the marginal revenue equals the marginal cost (marginal in this case with respect to time).

In practice, of course, the rate of change functions are predictions given as average rates per unit time (daily, weekly, monthly, and so on), which we treat as instantaneous rates of change—that is, as derivatives—in order to use calculus techniques.

CONSUMER AND PRODUCER SURPLUS

In Chap. 5 we remarked that to get a consumer to buy a certain quantity x of some commodity, the price must be set accordingly. From this point of view, price is a function of x,

$$p = D(x)$$

This function, inevitably decreasing (the larger the quantity x to be sold, the lower must be the price), is called a *demand function*. Similarly, the price at which a supplier is willing to sell a quantity x is given by the *supply function*. In a competitive situation the market equilibrium occurs when demand equals supply—that is, at (x_0, p_0), where the demand and supply curves intersect (see Fig. 6.43).

However, not all consumers have the same desire or use for a given product, so the market equilibrium price p_0 is a bargain to some consumers who are willing to pay more than p_0 for a given product. The measure of the total bargain to the consumers is called the *consumer surplus* (CS) and is represented by

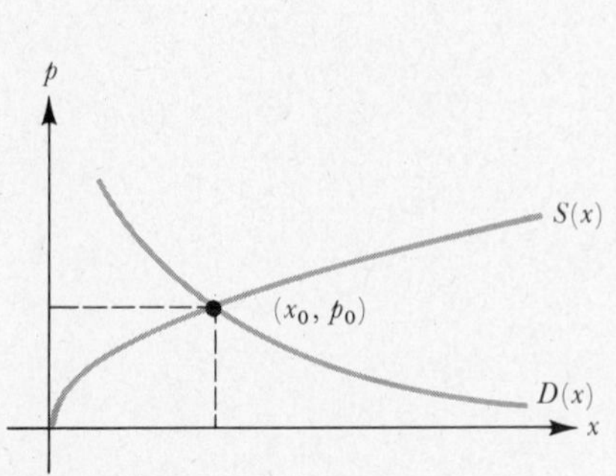

Fig. 6.43

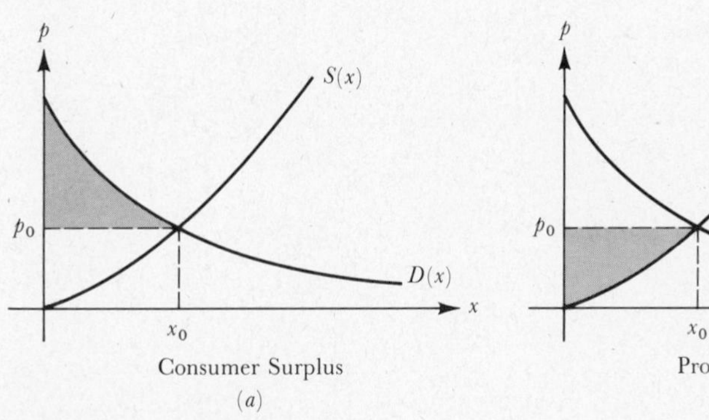

Fig. 6.44

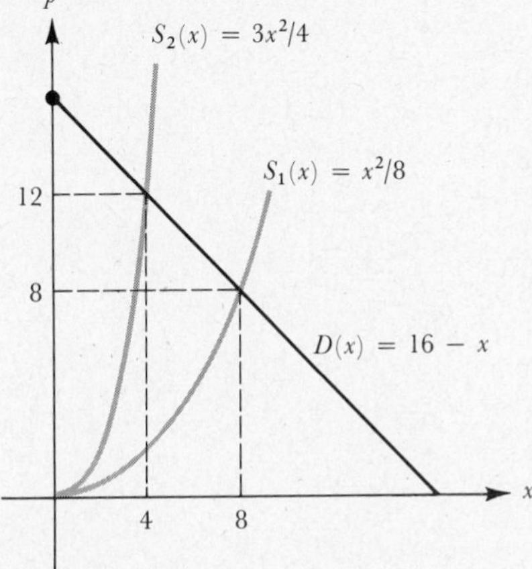

Fig. 6.45

the shaded area in Fig. 6.44(a). (Actually, in practice all the curves involved, hence the regions, are defined only for strictly positive x.)

Similarly, in a typical situation some producers would be willing to produce and sell at a price lower than p_0. The total benefit to producers is called the *producer surplus* (PS) and is represented by the shaded area in Fig. 6.44(b). Hence

6.5.4 $$\text{CS} = \int_0^{x_0} D(x)\,dx - p_0x_0 \qquad \text{PS} = p_0x_0 - \int_0^{x_0} S(x)\,dx$$

From the integrals in (6.5.4) we see that the consumer surplus is the potential total revenue received by producers (if each consumer paid what he/she felt the product was worth) minus the amount that consumers actually spent. Similarly, the producer surplus is the difference between the revenue actually received and the lower revenue that some of the producers would have been willing to accept. CS and PS thus provide a crude measure of the benefits of competition and the lack of a totally auction-based marketplace.

Example 5 Determine CS and PS if the demand function is $D(x) = 16 - x$ and the supply function is: (a) $S_1(x) = x^2/8$ before a merger of two large competitors; and (b) $S_2(x) = 3x^2/4$ after the merger.

Solution We sketch all three functions on the same coordinate axes in Fig. 6.45.

(a) Before the merger, $\text{CS}_1 = \int_0^8(16 - x)\,dx - 8 \cdot 8 = 32$ and $\text{PS}_1 = 64 - \int_0^8 x^2/8\,dx = 42\frac{2}{3}$. ($b$) After the merger, $\text{CS}_2 = \int_0^4(16 - x)^9dx - 4 \cdot 12 = 8$ and $\text{PS}_2 = -\int_0^4 3x^2/4\,dx = 32$. Hence the result of the merger is that the consumer surplus drops by 24, from 32 to 8 (a loss of 75 percent). While the producer surplus drops by $10\frac{2}{3}$, from $42\frac{2}{3}$ to 32 (a loss of 25 percent), the 50 percent increase in price and the lower cost of producing 50 percent fewer units lead to a potentially dramatic increase in profit for the producer. Of course, the consumer pays 50 percent more for each unit. ●

PROGRESS TEST 3

1. Determine the time t_0 to shut down a coal mine given that the revenues are decreasing linearly from a rate (in thousands of dollars) of 50 per month to zero when the coal runs out in 10 years, and given that the monthly cost rate is climbing according to the formula $C'(t) = t^2/24$, where t is measured in months.

2. What is the maximum profit from the mine?

3. Determine the consumer surplus and producer surplus given that the demand function is $D(x) = 10 - x$ and the supply function is $S(x) = x^2/5$.

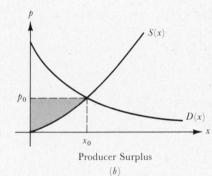

SECTION 6.5 EXERCISES

In Exercises 1 to 20 determine the area of those regions which are bounded by the given graphs.

1. $y = x^2, y = x, x = 1/2, x = 4$

2. $y = x^2 + x + 1, x = 0, x = 3, y = 0$

3. $y = x^3, y = x^2$

4. $x = y^2, x = 1, x = 2$

5. $y = x^2, y = 18 - x^2$

6. $y = \sqrt{x}, y = x^3$

7. $y = x^2 - 2x + 4, y = 0$

8. $x = y^2 - 1, y = x - 1$

9. $y = x^4, y = 2x - x^2$

10. $y^2 = 4x, y = 2x - 4$

11. $y = x^2 + 2x + 1, y + x^2 + 2x = 9, x = 3$

12. $2x = 7 - y, y = x^2 - 4x + 5$

13. $2y - 5 = 4x^2, x = 0, y = 0$

14. $2y = 4x^3 + 5, x = 0, y = 0, x = 1$

15. $2y^2 + y = x + 1, y - x = 1$

16. $y^2 = x - 1, x = 0, y = \pm 1$

17. $y = x^3 - x, y = 0$

18. $2x^2 = 2y + 1, x + 2 = 2y$

19. $y = 1/\sqrt{x - 1}, y = 0, x = 5, x = 2$

20. $xy^2 = 1, x = 1, x = 4$

21. Graph $y^2 = x^2 - x^4$ by following this outline:
(*a*) Note symmetries with respect to each axis.
(*b*) Solve for y in terms of x and note the possible values of x.
(*c*) Determine intercepts and use dy/dx to describe the graph in the first quadrant.
(*d*) Use symmetry to complete the graph.

22. Repeat Exercise 21 for $y^2 = x^2(4 - x^2)$.

23. Determine the total area enclosed by $y^2 = x^2 - x^4$.

24. Determine the total area enclosed by $y^2 = x^2(4 - x^2)$.

In Exercises 25 to 32 determine the area of the region enclosed by the graphs of the given equations.

25. $y^2 = x^3$ and $y = 2x - 3$

26. $y^2 = x^4(1 - x^3)$

27. $y^3 = x, x = y$

28. $x + y^2 = y^3, x = 4y^2$

29. $4y^2 = x^2(4 - x^2)$

30. $3y^2 = x^4(x^3 + 3), y = 0, x = 0, x = 3$

31. $y^2 = (3 - x)^3, x = 0, y = 0, x = 2$

32. $y^4 = x^4(1 + x^3), x = 0, y = 0, x = 1$

In Exercises 33 to 36 assume that vehicle V_1 has velocity v_1 and vehicle V_2 has velocity v_2. In the same way that the area between a nonnegative velocity function and the interval $[a, b]$ represents the distance traveled over the time interval $[a, b]$, the area between two velocity functions $v_1(t)$ and $v_2(t)$ with $v_1(t) \geq v_2(t)$ on the interval $[a, b]$ represents the distance that the vehicle V_1 moves ahead of the vehicle V_2 over the time interval $[a, b]$.

33. Suppose $v_1(t) = 15t$ and $v_2(t) = 10t$ over $[0, 20]$. Compute the distance V_1 pulls ahead of V_2 in the interval $[0, 20]$ using:
(*a*) The idea just described.
(*b*) Elementary (noncalculus) means.

34. Justify the statement made before Exercise 33 using interpretation of the areas of approximating rectangles.

35. Suppose $v_1(t) = 4t^3$ and $v_2(t) = 3t^2$ on $[1, 4]$ with V_1 200 units of distance behind V_2 at $t = 1$. Does V_1 catch up to V_2 over the interval $[1, 4]$?

36. Suppose $v_1(t) = t + 1, v_2(t) = (2/3)\sqrt{t}$ and at $t = 0$, V_1 is 63 units of distance behind V_2.
(*a*) How far apart are they at $t = 4$?
(*b*) How far apart are they at $t = 9$?
(*c*) How long does it take for V_1 to be 145 units ahead of V_2?

37. What is the area of the triangle with vertices $(1, 1)$, $(6, 3)$, $(3, 9)$? Use a definite integral.

38. Repeat Exercise 37 for the triangle with vertices $(-1, 2)$, $(1, 8)$, $(4, -2)$.

39. Suppose the marginal cost for producing the xth item is $2 + 1/(x + 1)^2$ (dollars). What is the total cost of producing the first 100 items?

40. Suppose the marginal revenue and marginal costs are given (x is thousands of items produced) as $R'(x) = 5x + 1 - x^2$ and $C'(x) = 4x - x^2$, determine the total profit for the first 4000 items produced.

41. Given a demand function $D(x) = 2 + \sqrt{9 - x}$ and a supply function

$$S(x) = \begin{cases} x - 1 & \text{for } x \geq 1 \\ 0 & \text{for } 0 \leq x \leq 1 \end{cases}$$

compute the consumer and producer surplus. (Be careful at $x = 1$; separate integrals are required—perhaps.)

42. Suppose that the marginal revenues from an oil field are given by $R'(x) = x^3 - 3x^2 + 2x + 10$ and the marginal costs are given by $C'(x) = -x^3 + 4x^2 - 3x + 10$ (x measured in thousands of barrels per day). Beyond a certain point, when marginal cost exceeds marginal revenue (the shutdown point), the government steps in and pays the difference between marginal cost and marginal revenue until they are equal again.
(*a*) What is the shutdown point and the total profit at that time?
(*b*) What is the total profit during the subsidized period?
(*c*) Sketch $R'(x)$ and $C'(x)$.

Section 6.6: Volumes of Solids of Revolution

Objectives:

Set up and evaluate definite integrals that describe the volumes of solids generated by revolving plane regions about an axis, using either the disk or the shell method, as appropriate.

THE DISK METHOD

We can think of a solid sphere as being generated by revolving the region enclosed by a semicircle about its diameter. Similarly, a solid right circular cone results from revolving the region enclosed by a right triangle about one of its legs (see Fig. 6.46). In this section we shall deal with more general solids generated by revolving a given plane region about a given line. Such a solid is known as a *solid of revolution,* and the line is called the *axis of revolution.* As in our treatment of area problems, we assume the region to be bounded by the graphs of functions.

Let us now determine the volume of the solid generated by revolving about the x axis the region below the graph of $y = f(x)$ from $x = a$ to $x = b$, where $f(x)$ is continuous and nonnegative on $[a, b]$; see Fig. 6.47.

As we know, the area under the graph can be approximated using vertical rectangles of width Δx_i (see Fig. 6.47) based upon a partition P of $[a, b]$. When each rectangle is revolved about the x axis, a *disk* is generated (see Fig. 6.48). The volume of a disk of radius R and thickness h is $\pi R^2 h$, so the volume of a typical disk generated by such a rectangle is

$$\pi [f(x_i)]^2 \, \Delta x_i$$

Since the sum of the rectangular areas approximates the area of the region being revolved, the sum of the volumes of the disks approximates what we shall define as the volume of the solid generated. Thus we say the approximate volume of the solid is

$$\sum_{i=1}^{n} \pi [f(x_i)]^2 \, \Delta x_i$$

But this is a Riemann sum, and if we take finer and finer partitions P of $[a, b]$, we get better and better approximations. Hence it makes sense to define the actual volume V of the solid of revolution to be the limit, as $\|P\| \to 0$, of such Riemann sums. But, by definition, this limit is a definite integral.

6.6.1 Definition $$V = \lim_{\|P\| \to 0} \sum_{i=1}^{n} \pi [f(x_i)]^2 \, \Delta x_i = \int_a^b \pi [f(x)]^2 \, dx$$

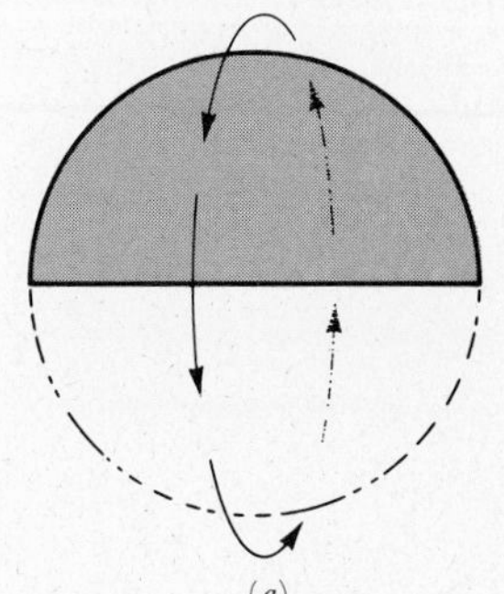

(a)

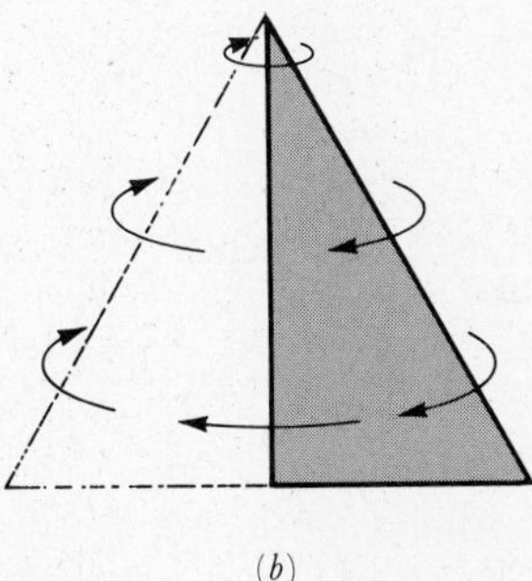

(b)

Fig. 6.46

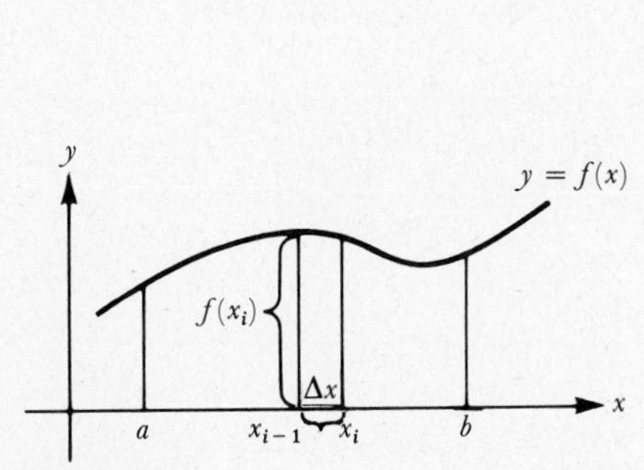

Fig. 6.47

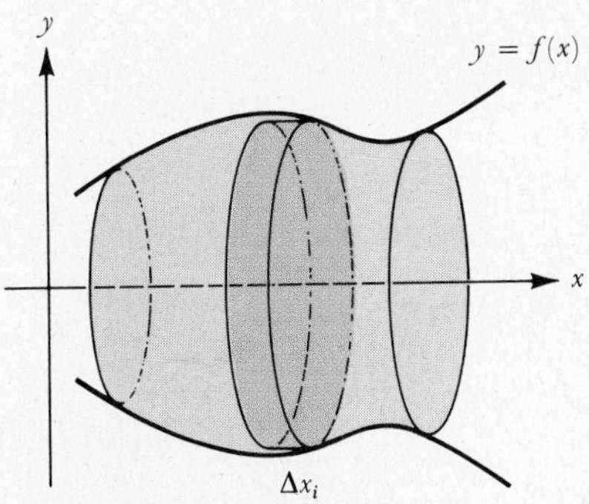

Fig. 6.48

Example 1 Determine the volume of a sphere of radius r generated by rotating the positive semicircle determined by $x^2 + y^2 = r^2$ about the x axis, as shown in the sketch.

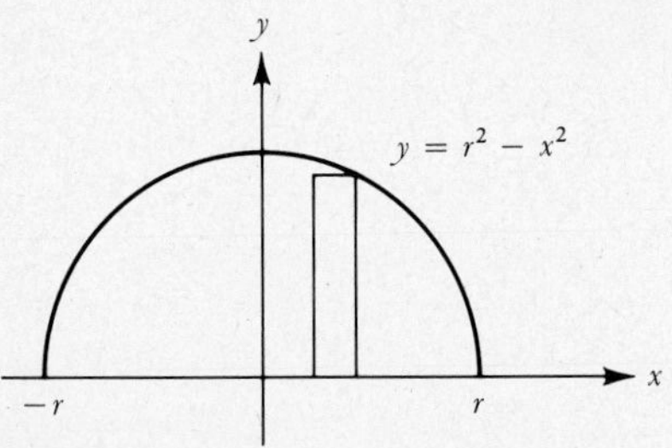

Solution We first solve for (positive) y in terms of x: $y = f(x) = \sqrt{r^2 - x^2}$. By (6.6.1),

$$\begin{aligned} V &= \int_{-r}^{r} \pi[\sqrt{r^2 - x^2}]^2\,dx \\ &= \pi\int_{-r}^{r} (r^2 - x^2)\,dx = \pi\left[r^2x - \frac{x^3}{3}\right]_{-r}^{r} \\ &= \pi\left[\left(r^3 - \frac{r^3}{3}\right) - \left(-r^3 + \frac{r^3}{3}\right)\right] \\ &= \pi\left[\frac{2}{3}r^3 + \frac{2}{3}r^3\right] = \frac{4}{3}\pi r^3 \quad \bullet \end{aligned}$$

The disk approach also works for regions revolved about the vertical axis. In this case the generating rectangles are horizontal, so the given equations will need to be solved for x in terms of y.

Example 2 Determine the volume of the right circular cone generated by revolving the following region R about the y axis. R is enclosed by the x and y axes and the line $x + 2y = 4$.

Solution We first sketch the region and draw in a typical *horizontal* rectangle, as in Fig. 6.49.

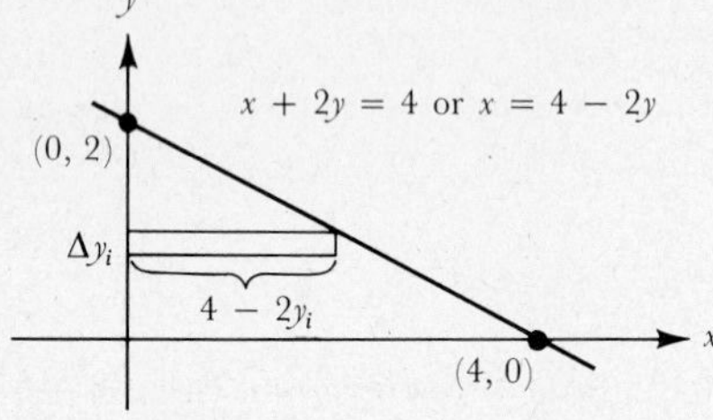

Fig. 6.49

A typical horizontal rectangle has width Δy_i and length $x = 4 - 2y_i$. Revolved about the y axis, it generates a disk of volume $\pi(4 - 2y_i)^2\,\Delta y_i$. The sum of such volumes

$$\sum_{i=1}^{n} \pi(4 - 2y_i)^2\,\Delta y_i$$

is a Riemann sum that leads to a definite integral. Thus, by the horizontal version of (6.6.1), we have

$$\begin{aligned} V &= \int_0^2 \pi(4 - 2y)^2\,dy \\ &= \pi\int_0^2 (16 - 16y + 4y^2)\,dy \\ &= \pi\left[16y - 8y^2 + \frac{4}{3}y^3\right]_0^2 \\ &= \pi\left[32 - 32 + \frac{4}{3}\cdot 8\right] = \frac{32}{3}\pi \quad \bullet \end{aligned}$$

Example 3 shows that the volume of a solid of revolution depends not only on the size of the region revolved but also on the axis of revolution. In part (b) we use disks with holes—that is, washer-shaped solids.

Example 3 Let R be the region bounded by $y = x^2$, $x = 1$, and the x axis. Compare the volumes of the solids generated by revolving R about (a) the x axis, (b) the y axis.

Solution (*a*) Revolving R about the x axis, we can apply (6.6.1) directly (see Fig. 6.50) to get

$$V = \int_0^1 \pi(x^2)^2\,dx = \pi \int_0^1 x^4\,dx = \pi\left[\frac{x^5}{5}\right]_0^1 = \frac{\pi}{5}$$

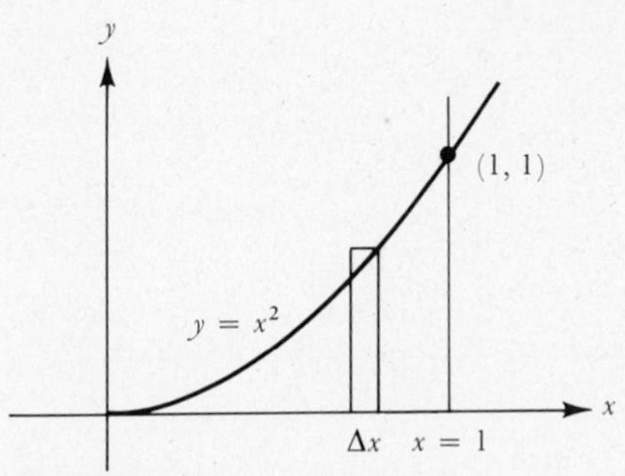

Fig. 6.50

(*b*) When rotated about the y axis, a typical horizontal rectangle has its right extremity on $x = \sqrt{y}$; see Fig. 6.51. It is clear that a typical rectangle generates a washer-shaped solid; see Fig. 6.52. The outside radius is 1 and the radius of the hole is $\sqrt{y_i}$. Now the volume of the washer-shaped disk is the volume of the solid disk minus the volume of the hole:

$$\pi(1)^2\,\Delta y_i - \pi(\sqrt{y_i})^2\,\Delta y_i = \pi(1 - y_i)\,\Delta y_i$$

The sum of such volumes is a Riemann sum:

$$\sum_{i=1}^{n} \pi(1 - y_i)\,\Delta y_i$$

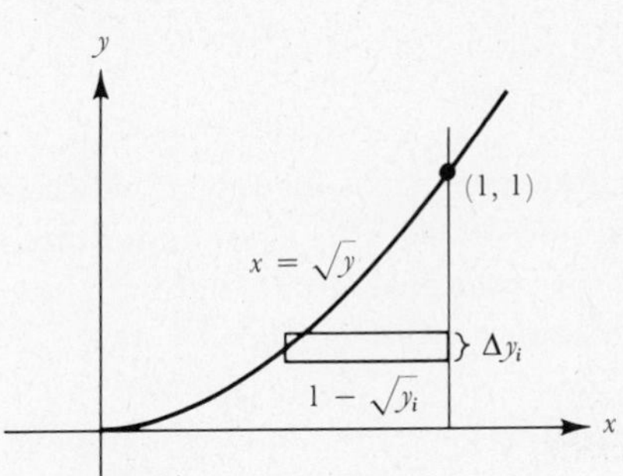

Fig. 6.51

Thus

$$V = \int_0^1 \pi(1 - y)\,dy$$

$$= \pi\left[y - \frac{y^2}{2}\right]_0^1 = \pi\left[1 - \frac{1}{2}\right] = \frac{\pi}{2}$$

The region generates less volume when revolved about the x axis than when revolved about the y axis. The region is—in a sense to be defined precisely using integrals in Chap. 12—"closer" to the x axis than to the y axis. ●

We can also use similar techniques to compute volumes of solids generated by revolving regions about vertical or horizontal lines other than the coordinate axes.

Example 4 Determine the volume of the solid generated by revolving the region of Example 3 about the line $y = -2$.

Solution A typical vertical rectangle generates a washer-shaped solid of outer radius $x_i^2 + 2$ and inner radius 2; see Fig. 6.53. Hence the volume of a typical

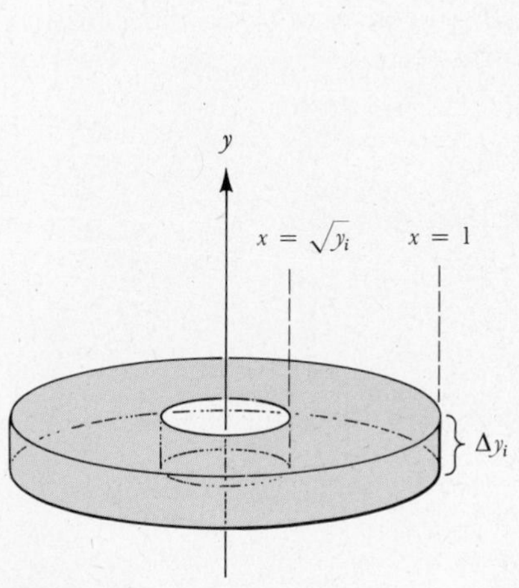

Fig. 6.52

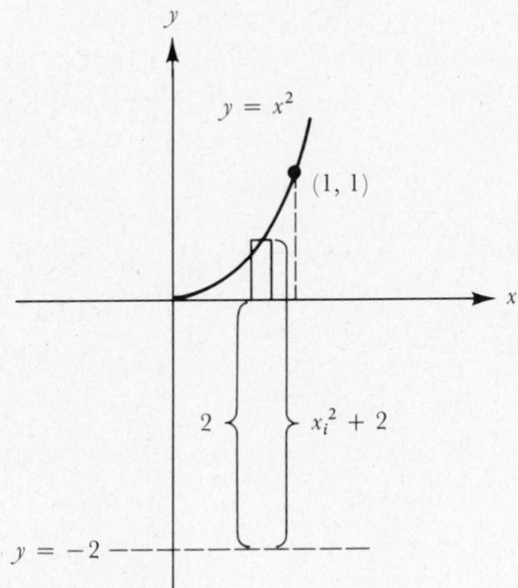

Fig. 6.53

washer is

$$\pi(x_i^2 + 2)^2\,\Delta x_i - \pi(2)^2\,\Delta x_i = \pi[x_i^4 + 4x_i^2]\,\Delta x_i$$

and the total volume of the solid is given by

$$\pi\int_0^1 (x^4 + 4x)\,dx = \pi\left[\frac{x^5}{5} + \frac{4}{3}x^3\right]_0^1 = \frac{23}{15}\pi \quad \bullet$$

PROGRESS TEST 1

1. Let R be the triangular region bounded by the lines $y = (r/h)x$, $x = h$, and the x axis. Determine the volume of the right circular cone of height h and radius r generated by rotating R about the x axis.

2. (*a*) Determine the volume of the solid generated by revolving the following region R about the x axis: R bounded by $2y = 4x^3 + 5$, the x and y axes, and the line $x = 1$.
(*b*) Repeat (*a*) but with R revolved about the line $y = 5$.

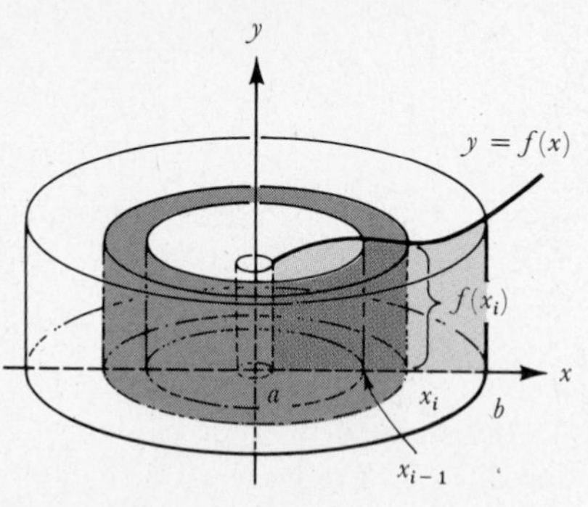

Fig. 6.54

THE SHELL METHOD

Suppose a solid is generated by revolving about the y axis the region between the graph of $y = f(x)$ and the interval $[a, b]$ $(a \geq 0)$, with $f(x) \geq 0$ and continuous on $[a, b]$. Each vertical rectangle in this case generates a right circular cylindrical shell (which is nothing more than a washer-shaped solid); see Fig. 6.54.

But before working the problem in terms of $f(x)$, we note that the volume of a right circular cylindrical shell of height h, inner radius r, and outer radius R is the volume of the outer cylinder minus the volume of the inner cylinder. We write this volume in a way that will eventually lead to a new integral formula for the volume of a solid of revolution:

6.6.2
$$\pi R^2 h - \pi r^2 h = \pi(R + r)(R - r)h = 2\pi\frac{R + r}{2}h(R - r)$$

Now $(R + r)/2$ is the average of the inner and outer radii and $R - r$ is the thickness of the shell. This agrees with the approximate volume we get by cutting the shell parallel to its axis and flattening it out to a rectangular solid whose width is 2π times the average radius of the shell (which is its approximate circumference). The height of the solid is h and its thickness is $R - r$ (see Fig. 6.55).

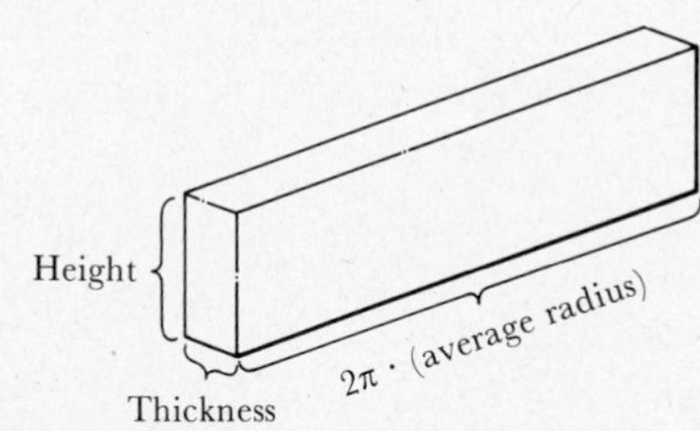

Fig. 6.55

Translating back to our original situation (Fig. 6.54), where as usual we are given a partition P of $[a, b]$, we see that the outer radius is x_i, the inner radius x_{i-1}, the average radius $(x_i + x_{i-1})/2 = x_i^*$, the height $f(x_i^*)$, and the thickness $x_i - x_{i-1} = \Delta x_i$. Hence, by (6.6.2), a typical rectangle generates a shell of volume

6.6.3
$$2\pi x_i^* f(x_i^*)\,\Delta x_i$$

Adding all these volumes, we get the Riemann sum

$$\sum_{i=1}^{n} 2\pi x_i^* f(x_i^*)\,\Delta x_i$$

which approximates the volume of the whole solid. Thus the exact volume is given by

6.6.4
$$V = \lim_{\|P\|\to 0}\sum_{i=1}^{n} 2\pi x_i^* f(x_i^*)\,\Delta x_i = \int_a^b 2\pi x f(x)\,dx$$

We shall now determine the volume of the solid generated by revolving about the y axis the region of Problem 2 in the Progress Test above. (The disk approach would require messy calculations and two separate integrals.)

Example 5 The region R bounded by $2y = 4x^3 + 5$, the x and y axes, and the line $x = 1$ is revolved about the y axis. Using the shell approach, determine the volume of the solid generated. (See the Progress Test solution for a sketch.)

Solution Here $f(x) = y = 2x^3 + 5/2$, $a = 0$, $b = 1$, so

$$V = 2\pi \int_0^1 x\left(2x^3 + \frac{5}{2}\right) dx$$

$$= 2\pi \left[\frac{2x^5}{5} + \frac{5x^2}{4}\right]_0^1$$

$$= \frac{33}{10}\pi \quad \bullet$$

The choice of whether to use the disk or shell approach on a problem will usually be dictated by which is easier—either easier to set up or easier to integrate. In any case, the first step should be to sketch the region to be revolved.

Example 6 Let R be the region enclosed by the curves $y = x^2 - 2x$ and $y + 2x = 9$. Determine the volume of the solid generated by revolving R about the line $x = 4$.

Solution We first sketch R, determining the intersection of the curves by setting $x^2 - 2x = 9 - 2x$. Here we get $x = \pm 3$, so $y = 3$ and 15, respectively. (See Fig. 6.56.)

If we use the disk method, we need two separate integrals because two types of typical rectangles are involved, one above $y = 3$ and one below $y = 3$. Hence we shall use the shell method, which requires a single integral.

Ignoring subscripts and superscripts, a typical rectangle has height $(9 - 2x) - (x^2 - 2x)$ and its center is at a distance $4 - x$ from the axis of revolution. By (6.6.3) it generates a shell of volume

$$2\pi(4 - x)[(9 - 2x) - (x^2 - 2x)]\,\Delta x$$

The sum of volumes of such shells generated by rectangles from $x = -3$ to $x = 3$ approximates the total volume. The exact volume is thus given by

$$\int_{-3}^{3} 2\pi(4 - x)[(9 - 2x) - (x^2 - 2x)]\,dx = 2\pi \int_{-3}^{3} (x^3 - 4x^2 - 9x + 36)\,dx$$

$$= 2\pi\left[\frac{x^4}{4} - \frac{4}{3}x^3 - \frac{9}{2}x^2 + 36x\right]_{-3}^{3}$$

$$= 2\pi\left[\left(\frac{81}{4} - 36 - \frac{81}{2} + 108\right) - \left(\frac{81}{4} + 36 - \frac{81}{2} - 108\right)\right]$$

$$= 2\pi[144] = 288\pi \quad \bullet$$

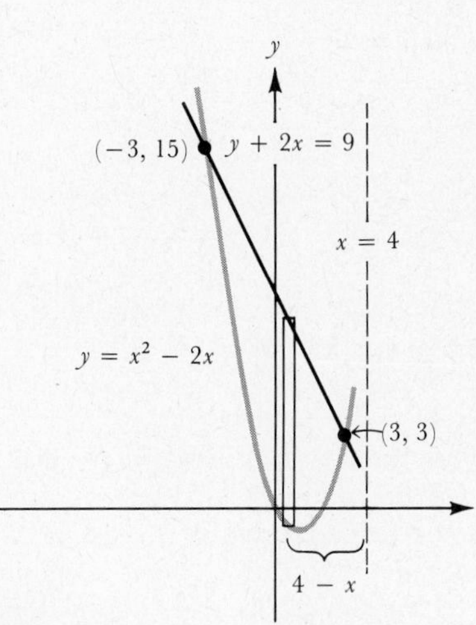

Fig. 6.56

Sometimes the heart of the matter in applications is setting up the region and the axis to generate the appropriate solid. (See Sec. 6.3 Exercises.)

PROGRESS TEST 2

1. Determine the volume of a sphere of radius r using the shell method. Revolve about the x axis.
2. Determine the volume of the solid generated by revolving about the line $x = 3$ the region bounded by $y = x^2 - 2x + 1$ and $y + x^2 + 2x = 9$.

GENERAL REMARKS ON SETTING UP INTEGRALS

Just as with area problems, volume problems offer a choice of approach, which automatically gives us the choice of independent variable. This can sometimes be confusing, but if we take the trouble to use a sketch and include a typical rectangle, the rest will follow naturally. *The width of the typical rectangle determines the independent variable.* For example, if the width happens to be Δy_i, then the dominant variable is y, and the given equations will need to be solved for x in terms of y and the limits of integration determined accordingly.

Another caution: We are using the phrase "typical rectangle" here in a precise way; *all others must be like it in terms of its extreme ends* (upper and lower, or left and right). Of course, some problems may need to be broken into parts with one typical rectangle for each.

SECTION 6.6 EXERCISES

1. Determine the volume of the solid generated by revolving about the x axis the region bounded by the curves $x = y$ and $y = x^2$ using (*a*) the disk method, and (*b*) the shell method.
2. Repeat Exercise 1 revolving about the line $x = -2$.
3. Determine the volume of the solid generated by revolving, about the x axis, the region bounded by $y = x^2 + 7$, $y = 16$, using (*a*) the disk method, and (*b*) the shell method.
4. Repeat Exercise 3 revolving about the line $y = 1$.

In Exercises 5 to 36 determine the volume V of the solid generated by revolving the given region R about the given axis.

5. R bounded by $y = x + 1$ and the x and y axes, about x axis.
6. R as in Exercise 5, about y axis.
7. R bounded by $y = 2x$, $x = 4$, and $y = 0$, about y axis.
8. R as in Exercise 7, about x axis.
9. R as in Exercise 7, about line $x = 5$.
10. R as in Exercise 7, about line $y = 6$.
11. R bounded by $f(x) = x^{-3/2}$, $x = 1$, and $x = 2$, about x axis.
12. R as in Exercise 11, about y axis.
13. R bounded by $y = x$ and $y = 4x - x^2$, about y axis.
14. R as in Exercise 13, about x axis.
15. R as in Exercise 13, about line $x = -5$.
16. R as in Exercise 13, about line $y = -6$.
17. R bounded by $x = y^2$ and $y = x^2$, about y axis.
18. R as in Exercise 17, about x axis.
19. R as in Exercise 17, about line $x = 2$.
20. R as in Exercise 17, about line $x = 200$.
21. R bounded by $y = x^3$, $x = 3$, and $y = 0$, about x axis.
22. R bounded by $y = 4 - x^2$ and x axis, about x axis.
23. R bounded by $y + 4x^2 - 8x = -1$, $x = 0$, $y = 0$, and $x = 1$, about x axis.
24. R as in Exercise 23, about y axis.
25. R bounded by $y = x$ and $2y = x^2$, about x axis.
26. R as in Exercise 25 about y axis.
27. R as in Exercise 25 about line $x = 4$.
28. R bounded by $x^2 = 4y$, $y = 4$, about x axis.
29. R bounded by $y = x^2 - 2x$, y axis, $y = 3$, about y axis.
30. R bounded by $y^2 = x$, $x = y^3$, about x axis.
31. R bounded by $y^2 - 8x = 16$, $y^2 + 16x = 16$, above x axis, about x axis.
32. R bounded by $f(x) = 1/(x - 1)^3$, $x = -1$, coordinate axes, about x axis.
33. R bounded by $y = x^2 + 3x - 6$ and $x + y + 3 = 0$, about x axis.
34. R as in Exercise 33, about line $x = 4$.
35. R bounded by $y = \sqrt{4 + x}$ and x and y axes, about x axis.
36. R bounded by $y = 3x^2 - x^3$ and x axis, about y axis.
37. Determine the volume of the torus generated by revolving the region enclosed by the circle $(x - 5)^2 + y^2 = 9$ about the y axis.
38. Derive a general formula for the volume of a torus of cross-sectional radius r and central radius R. (R is distance from the axis of rotation to the center of generating circle.)
39. Determine the volume remaining of a sphere of radius a with a cylindrical hole of radius r centered on a diameter of the sphere drilled through the sphere ($r < a$).
40. Suppose a sphere of radius r is cut by a plane leaving a "cap-shaped" piece of maximum thickness h. Show that the volume of this cap is $(\pi h^2/3)(3r - h)$. (This formula

is useful in determining the quantity of liquid in a spherical tank.)

41. Determine the volume of the ring generated by revolving the triangular region with vertices (5, 6), (5, 1), and (9, 3) about the y axis.

42. Repeat Exercise 41, revolving about the x axis.

43. Repeat Exercise 41 for the vertices (2, −4), (5, −7), and (10, −4), about the y axis.

44. Repeat Exercise 43, about the x axis.

CHAPTER EXERCISES

Review Exercises

In Exercises 1 to 40 determine the indicated antiderivative.

1. $\int\left(\frac{2}{\sqrt{x}} + x^3 - \sqrt[3]{x}\right) dx$ **2.** $\int\left(2x^2 - \frac{4}{x^5} + 9\right) dx$ **3.** $\int \frac{x^3 + x^2 - \sqrt{x}}{x^5} dx$ **4.** $\int(2 + \sqrt{5})\, ds$

5. $\int(t + 2)(3t - 1)\, dt$ **6.** $\int(2w + 2)^2\, dw$ **7.** $\int \frac{z^3 + 2z - 3}{\sqrt{z}} dz$ **8.** $\int(4x + 1)^{10}\, dx$

9. $\int(x^2 + 2)^{2/3} x\, dx$ **10.** $\int(w + 1)(w^2 + 2w + 6)^5\, dw$

11. $\int \frac{dt}{(3t + 5)^3}$ **12.** $\int(3 - x^2)^4\, 5x\, dx$ **13.** $\int(\pi + 1)^3 t\, dt$ **14.** $\int(x^3 + 1)^{1/3} x^5\, dx$

15. $\int(x^2 - 4x + 4)^{18}\, dx$ **16.** $\int \frac{dx}{\sqrt[3]{(2x + 3)^4}}$ **17.** $\int\left(3 + \frac{6}{w}\right)\frac{dw}{w^2}$ **18.** $\int(\pi x^2 + 3)^3\, 6x\, dx$

19. $\int \frac{(\sqrt{w + 1} + 2)^8\, dw}{\sqrt{w + 1}}$ **20.** $\int 2k^2\, dx$ **21.** $\int(3x - 1)^4\, dx$ **22.** $\int 51\, dx$

23. $\int x^{2/3}(x^3 + \sqrt{x})\, dx$ **24.** $\int \frac{x\, dx}{(x^2 + 1)^4}$ **25.** $\int \frac{dx}{\sqrt[3]{ax + b}}$ **26.** $\int \sqrt{x^4 + x^2}\, dx$

27. $\int t\sqrt[3]{t^2 + 2}\, dt$ **28.** $\int v^2\sqrt{v + 3}\, dv$ **29.** $\int(4x^3 - 5)^5 x^2\, dx$ **30.** $\int \frac{dx}{\sqrt{3x - 7}}$

31. $\int \sqrt[5]{t^3 - t^2 + 1}\left(t^2 - \frac{2}{3}t\right) dt$ **32.** $\int[(w^2 - 2w)^8 w - (w^2 - 2w)^8]\, dw$

33. $\int \sqrt{x + 1/x}\left(1 - \frac{1}{x^2}\right) dx$ **34.** $\int \sqrt{x^4 - 4x^3 + 12x^2 - 1}\,(8x^3 - 24x^2 + 48x)\, dx$

35. $\int \pi^2\, dy$ **36.** $\int\left(\frac{1}{x^2} - \frac{3}{x^3} + x^3\right) dx$ **37.** $\int \frac{(x^3 + 9x^2)\, dx}{\sqrt[3]{x^4 + 12x^3 - 5}}$ **38.** $\int x^3\sqrt[3]{6 + 3x^2}\, dx$

39. $\int \frac{x^5 + x^4 + x^3 - 3}{x^3} dx$ **40.** $\int \frac{x^3\, dx}{\sqrt{3x - 2}}$

41. (*a*) Sketch the graph of $f(x) = x^2 + x + 1$ on [0, 5]. (*b*) Find the Riemann sum associated with the partition $P = \{0, 1, 2, 3, 4, 5\}$ and $(x_i^*) = \{0, 1.5, 3, 3.5, 5\}$. Sketch.
(*c*) Find $\underline{S}_P$ and $\overline{S}_P$ for this partition. Sketch.
(*d*) Use the Fundamental Theorem to determine the area between $f(x)$ and [0, 5].

42. Follow the same format as in Exercise 41 for $f(x) = x^3 - 3x^2 + 2x$ on [0, 4] with $P = \{0, 1, 2, 3, 4\}$ $(x_i^*) = \{\frac{1}{2}, 1, 2, 3.5\}$. What is the area of the region bounded by $y = f(x)$ and $y = 0$?

In Exercises 43 to 46 evaluate the given limits.

43. $\lim_{n\to\infty} \sum_{i=1}^{n} \frac{3i}{n^2}$ **44.** $\lim_{n\to\infty} \sum_{i=1}^{n} \frac{2i^2 + i + 3}{5n^3}$ **45.** $\lim_{n\to\infty} \sum_{i=1}^{n} \frac{4i^3 + i}{2n^4}$ **46.** $\lim_{n\to\infty} \sum_{i=1}^{n} \frac{8i + 6}{n^2}$

In Exercises 47 and 48 use Definition (6.3.1) and the finite-sum formulas to evaluate the given integral.

47. $\int_0^4 (3x^2 + 1)\, dx$ **48.** $\int_{-1}^{1} (1 - 2x^2)\, dx$

In Exercises 49 to 64 use the Fundamental Theorem (and the Chain Rule if necessary) to evaluate the given definite integral.

49. $\int_0^{\pi} 8\,dt$ **50.** $\int_{-1}^{1} x^3\,dx$ **51.** $\int_0^{6} (3x^2 + 2x + 2)\,dx$ **52.** $\int_1^{2} (2z + 1)^2\,dz$

53. $\int_1^{4} \frac{dt}{\sqrt{t}}$ **54.** $\int_0^{5} (x^3 - 3x^2 + 2x)\,dx$ **55.** $\int_0^{4} x\sqrt{x^2 + 7}\,dx$ **56.** $\int_1^{4} \frac{x^2 - x + 2}{x^4}\,dx$

57. $\int_1^{3} \frac{w^2 + 2}{w^2}\,dw$ **58.** $\int_2^{b} (3x + 1)\,dx$ **59.** $\int_0^{5} |x - 2|\,dx$ **60.** $\int_0^{8} x\sqrt{x^2 + 1}\,dx$

61. $\int_{-1}^{1} \frac{x\,dx}{\sqrt{x^2 + 4}}$ **62.** $\int_{-4}^{-1} x\sqrt[3]{x^2 + 3}\,dx$ **63.** $\int_0^{2} x^2\sqrt{x^3 + 1}\,dx$ **64.** $\int_{-4}^{4} x|x - 1|\,dx$

In Exercises 65 to 72 solve the given differential equation with the given conditions on the solution.

65. $\frac{dy}{dx} = x^2 + x + 2 \quad y(0) = 1$

66. $\frac{d^2y}{dx^2} = 3x + 7 \quad y'(0) = 3, y(0) = 4$

67. $\frac{dy}{dx} = \frac{x + 1}{\sqrt{x^2 + 2x + 4}} \quad y(0) = 6$

68. $\frac{d^2y}{dx^2} = 4 \quad y'(1) = 2, y(1) = 3$

69. $\frac{d^2y}{dx^2} = 3x^2 + \sqrt{x} \quad \frac{dy}{dx} = 1$ when $x = 4$, $y = 2$ when $x = 4$

70. $\frac{dy}{dx} = \frac{x}{\sqrt{3x^2 + 1}} \quad y = 2$ when $x = 1$

71. $f'(x) = 3x^2 \quad f'(0) = 1, f(1) = 3/4$

72. $f''(x) = x - 1 \quad f'(0) = 1, f(1) = 3/4$

73. The acceleration of a ball rolling down a certain curved inclined plane is given by $a(t) = 8t - 3t^2$ cm/s^2. Assuming that the distance (in centimeters) is measured positively from the starting point at the top of the plane, how far does the ball roll after (*a*) 2 s, (*b*) 3 s, (*c*) 4 s, and (*d*) 5 s? Verbally describe (*e*) the velocity and (*f*) the distance rolled for this ball.

74. Repeat Exercise 73 for a ball rolling with acceleration $a(t) = 1 + 2t - 6t^2$.

75. Derive a formula (ignoring air resistance) for the height of a bridge as a function of time t in seconds it takes a rock dropped from the bridge to hit the water below.

In Exercises 76 to 79, determine $F'(x)$.

76. $F(x) = \int_{1/2}^{x} \sqrt{t^5 + 1}\,dt$

77. $F(x) = \int_x^{30} (t^3 + 1)^8\,dt$

78. $F(x) = \int_{x^2}^{2x} (t^2 + 1)^3\,dt$

79. $F(x) = \int_0^{7} \frac{dx}{\sqrt{x^4 + x^2 + 9}}$

In Exercises 80 to 85, determine $F'(s)$.

80. $F(s) = \int_0^{s} \frac{dx}{\sqrt{9 - x^2}}$ **81.** $F(s) = \int_{-s}^{2} \sqrt{x^2 + 5}\,dx$ **82.** $F(s) = \int_s^{4} \sqrt[3]{t^2 + 1}\,dt$

83. $F(s) = \int_0^{3} \sqrt{x^4 + x^2}\,dx$ **84.** $F(s) = \int_{s^2+1}^{7} \frac{dx}{\sqrt{x^2 + x + 1}}$ **85.** $\int_{1-s^2}^{s^3} \frac{dt}{\sqrt{t + 1}}$

In Exercises 86 to 89, (a) determine the average value of the given function on the given interval and (b) determine where on that interval the function achieves that average value.

86. $f(x) = x^3 + 1$ on $[-2, 2]$

87. $g(x) = 8 - x^2$ on $[-3, 3]$

88. $h(t) = \sqrt{t}$ on $[1, 9]$

89. $f(t) = 3t^2 - 1$ on $[-1, 3]$

90. (*a*) Sketch the function $f(x) = \sqrt[3]{x^2}$ on $[-1, 1]$.

(*b*) Determine $\int_{-1}^{1} \sqrt[3]{x^2}\,dx$.

(*c*) What is the average value of $f(x)$ on $[-1, 1]$?

(*d*) Where does $f(x)$ achieve this average value?

In Exercises 91 to 98 determine the area of the region bounded by the given curves.

91. $y = x^3, y = 0, x = -1/2, x = 8$

92. $y = |x^3 - 4x|, y = 0$

93. $y = |9 - x^2|, y = x + 3$

94. $y = x^3, y = 3x^2 - 2x$

95. $y = \sqrt{x}, y = x/2$

96. $x = 1 - y^2, x = y^2 - 1$

97. $y^2 = 4 - x, x + 4y = 4$

98. $y = |x|, y = 1 - x^2$

In Exercises 99 to 119 set up and simplify, but do not evaluate, the definite integral giving the requested quantity. The accompanying figure (used in Exercises 99 to 119), involving the lines $x = 0$, $x = 1$, $y = 0$, $y = 1$, $y = x^2$, and $y = x^3$, describes several regions as indicated, including I, II, and III. By the regions I + II and II + III we mean the enclosed regions outside of III and I respectively. The region I + II + III means the whole one-by-one square.

99. Area of: (*a*) I (*b*) III (*c*) II + III (*d*) I + II (*e*) II.

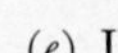

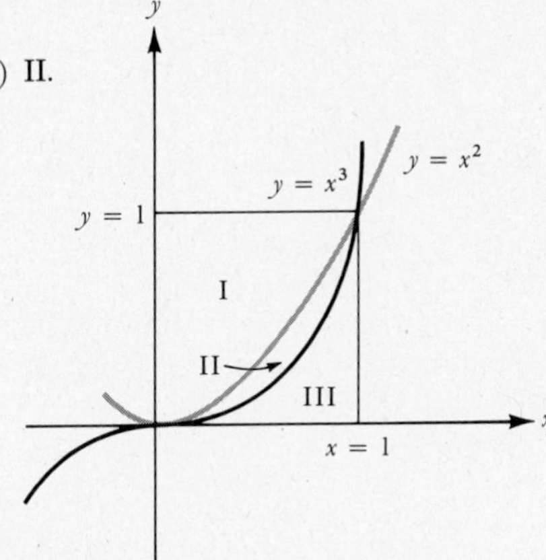

In Exercises 100 to 109, by the disk method, the volume of the solid generated by revolving the given region.

100. I about the x axis

101. II + III about the x axis

102. II about the x axis

103. I + II about the y axis

104. I about the line $y = 1$

105. III about the x axis

106. I + III about the x axis

107. I about the y axis

108. III about the y axis

109. III about the line $y = 1$

In Exercises 110 to 119, by the shell method, the volume of the solid generated by revolving the given region.

110. I about the y axis

111. II + III about the y axis

112. II about the y axis

113. I about $x = 1$

114. I about the x axis (compare with Exercise 100)

115. III about the x axis (compare with Exercise 105)

116. III about the y axis

117. I + II about the y axis

118. III about $x = 1$

119. II about $x = 1$

In Exercises 120 to 125 determine the volume of the solid generated by revolving the given region R about the given axis.

120. R bounded by $4x^2 + 9y^2 = 36$, about the x axis (Be careful not to compute double the volume.)

121. R bounded by $y/2 = \sqrt{x + 2}$, the x and y axes, and $y = 2$, about the y axis.
(*a*) The smaller region R bounded by $x^2 + y^2 = 4$ and $x + y + 2 = 0$ about the y axis.
(*b*) Repeat for the larger region (be careful to avoid duplication).

122. R bounded by $y^2 + 1 = x$, $y + x = 7$, and $x = 10$, about $x = 10$.

123. R bounded by $y^2 = 1/x$, $x = 1$, $x = 2$, and the positive x axis, about $x = 1$.

124. R bounded by $x = y^2 + 1$ and $y + 3 = x$, about the y axis.

125. Same R as Exercise 124, about $y = -1$.

Miscellaneous Exercises

1. If $f(x) = -f(x)$ for all x show that $\int_{-b}^{b} f(x)\,dx = 0$.

2. Given that $f(-x) = f(x)$ for all x and $\int_0^7 f(x)\,dx = 3$, find $\int_{-7}^{7} f(x)\,dx$.

3. Find

$$\lim_{n\to\infty} \sum_{i=1}^{n} \frac{16\sqrt{i}}{(n^{4/3})}$$

(*Hint:* Consider $f(x) = \sqrt[3]{x}$ and express the limit as a definite integral.)

4. A particle is falling with a velocity $v(t) = 1 + 2t$.
(*a*) Find the average velocity with respect to time on the interval $0 \leq t \leq 4$.
(*b*) Assuming that $s(0) = 0$, find the position function $s = s(t)$ of the particle.

5. Suppose the second derivative of a function of x is $6x$ and that the function passes through the point $(1, 4)$ with slope 4. What is the function?

6. Repeat Exercise 5 with second derivative equal to

$-12x + 2$ and the function passing through $(0, 4)$ with slope 1.

7. Given $F(x) = \int_1^{x^2} 5t\,dt$, determine (*a*) $F'(x)$ and (*b*) $F''(x)$ (*c*) Sketch $F(x)$ from $x = 1$ to $x = 4$.

8. Repeat Exercise 7 for $F(x) = \int_1^{\sqrt{x}} (t^4 + t^2)\,dt$.

9. Suppose that a vehicle decelerates at the constant rate of 8 ft/s² from a velocity of 88 ft/s (60 mph).
(*a*) How many seconds are needed to stop?
(*b*) How far does the vehicle travel over this period?

10. Repeat Exercise 9 for a deceleration of 4 ft/s² from 60 ft/s².

11. Repeat Exercise 9 for a nonconstant deceleration of $4t$ ft/s² from 88 ft/s.

12. Suppose vehicle V_1 has acceleration a_1 and vehicle V_2 has acceleration a_2 in appropriate units of distance per second-squared. Given that V_1 is 100 units of distance behind V_2 traveling at the same velocity as V_2 but with acceleration $a_1(t) = 4t$, while $a_2(t) = t$, how long does it take V_1 to overtake V_2?

13. Repeat Exercise 12 but with V_1 traveling at 10 units/s faster than V_2 and $a_1(t) = t = a_2(t)$.

14. Repeat Exercise 12 but with $a_1(t) = t/2$, $a_2(t) = t$. (Be careful.)

15. Interpret the functions and results of Exercises 12 to 14 graphically.

16. Given a uniform plate whose density (mass per square unit) is m and has the same shape and size as the region enclosed by the curves $y = 4 - x^2$ and $y = x^2 - 3x + 2$, what is its total mass?

17. Determine $G''(x)$, where $G(x) = \int_3^{[\int_2^x (t^2+1)\,dt]} \sqrt{1 + s}\,ds$.

18. Determine $F'(x)$, where $F(x) = G(\int_a^{x^3} (2t^2 + 1)\,dt)$.

19. Suppose f is continuous everywhere, with

$$\int_a^b f(x)\,dx = 0$$

for all choices of a and b. Show that $f(x) = 0$ for all x in R.

20. Given

$$g(x) = \int_{x^2}^{1-x} \frac{3 + t}{t^3}\,dx \qquad \text{for } x \geq 1$$

determine (*a*) $g(1)$, (*b*) $g(2)$, (*c*) $g'(x)$, (*d*) $g'(1)$, and (*e*) $g'(2)$.

21. Repeat Exercise 20 for

$$g(x) = \int_{1-x^2}^{4x} (t^2 + 1)\,dt \qquad \text{for all } x$$

22. A tank in the shape of a hemisphere of radius 6 m is being filled with water at the rate of 1/3 m³/s. How fast is the radius of the surface increasing at the instant when the depth is 4 m? (See Sec. 6.6, Exercise 40.)

SELF-TEST ✦

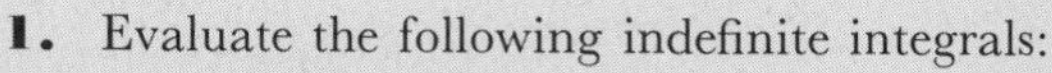

1. Evaluate the following indefinite integrals:

(*a*) $\int \frac{\sqrt{x - 1/x}\,(1 + x^2)}{x^2}\,dx$ (*b*) $\int \frac{(\sqrt{x} + 1)^{3/5}}{\sqrt{x}}\,dx$

2. Use the Fundamental Theorem to evaluate the following definite integrals:

(*a*) $\int_0^4 \frac{3u^2\,du}{\sqrt{2u^3 + 3}}$ (*b*) $\int_{-1}^1 \frac{x\,dx}{\sqrt{x^2 + 1}}$

3. Find the area of the region bounded by the curves $y = x^3 - x^2$ and $y = 2x$.

4. For x in $(-\infty, \infty)$ define $F(x) = \int_0^x \frac{t}{\sqrt{t^4 + 1}}\,dt$.

(*a*) Determine $F'(x)$.
(*b*) Let $g(s) = F(\sqrt{s} + s^2)$. Find dg/ds.

5. An object is thrown upward with an initial velocity of 36 ft/s from a height 80 ft above the ground.
(*a*) How high does the object rise?
(*b*) At what velocity does it strike the ground?

6. (*a*) Determine the average value of the function $f(x) = 3x^2 + 2x + 1$ on the interval $[0, 4]$.
(*b*) Where in $[0, 4]$ does f achieve this average value?

7. Suppose R is the region enclosed by the graphs of $y = 2x^2$, $y - 3 = 5x$, and $x = 0$. Determine the volume of the solid generated by revolving R about (*a*) the x axis, (*b*) the y axis, and (*c*) the line $x = 4$.

Chapter 7

TRIGONOMETRIC, EXPONENTIAL, AND LOGARITHM FUNCTIONS: THEIR DEFINITIONS AND DERIVATIVES

Until now all our work has concerned algebraic functions. We shall now broaden our scope to include certain functions that cannot be expressed in terms of algebraic operations on polynomials. These "transcendental functions" include trigonometric, exponential, and logarithm functions, and have wide practical use in situations where the algebraic functions simply do not apply.

In Sec. 7.1 we briefly review radian measure and the trigonometric functions, and then in Sec. 7.2 we determine the derivatives of these functions. Sections 7.3 and 7.4 repeat the procedure: We first define the exponential and logarithm functions in a way that builds on your earlier experience with them, and then determine their derivatives. To keep our introduction to the exponential functions simple and avoid a number of technical issues, we make some reasonable assumptions regarding their continuity and their fundamental properties, assumptions shown to be valid in the Sec. 7.4 appendix.

Section 7.5 applies the derivatives determined in Secs. 7.2 and 7.4 to problems analogous to those dealt with in Chaps. 4 and 5. This last section can be postponed or omitted in favor of Sec. 8.1, which involves growth and decay. These growth and decay applications involve minimal use of integration, whereas none of Chap. 7 involves integration except a brief appendix to Sec. 7.4.

Section 7.1: Trigonometric Functions—Definitions and Properties

Objectives:

1. Convert degree measure to radian measure and vice versa.

2. Define and graph each of the six trigonometric functions, noting the period and any points where such functions fail to be defined.

MEASURING ANGLES

We take as a starting point the notion of a half-line (we shall call it the *generating line*) fixed at one end and generating an angle by sweeping in a plane from some initial position to some terminal position. For convenience we often assume the initial position is the positive horizontal axis of a given *s*-*t* coordinate system for the plane. We shall call this initial position the *initial side* of the angle and the terminal position the *terminal side* of the angle. Following custom we call a rotation of the generating line in the counterclockwise direction positive and a rotation in the clockwise direction negative. (See Fig. 7.1.)

A *full positive rotation*, one in which the generating line rotates in a counterclockwise direction until it coincides with the initial side, is a unit of angle measure. Another, more common unit of angle measure, the degree, is defined as 1/360 of a full positive rotation. The degree was handed down to us from the Babylonians, who did their calculations in base 60.

Far more important for our and most scientific purposes is *radian* (rad) *measure*. For reasons that will become clear as we go along, radian measure is the most practical measure for the calculus.

To define radian measure, we assume that a unit circle is centered at the origin of the *s*-*t* plane. We shall denote by $P = (s, t)$ the intersection of the generating line and the given unit circle. As the generating line sweeps out an angle, the point P moves along the circle beginning on the initial side at $(1, 0)$.

7.1.1 Definition The directed length of the path traveled by P on the unit circle as the generating line sweeps out the angle is the *radian measure* of that angle. Length along the perimeter of the circle is measured positively from $(0, 1)$ in a counterclockwise direction, and negatively in a clockwise direction.

Since the circumference of the unit circle is 2π, the directed distance traveled by P in a full positive rotation is 2π. Thus a full positive rotation measures $2\pi \approx 6.28$ rad. The directed distance traveled by P in $\frac{1}{8}$ of a full negative rotation is $\frac{1}{8}(-2\pi) = -\pi/4$, so $\frac{1}{8}$ of a full negative rotation measures $-\pi/4 \approx -0.785$ rad. In Fig. 7.2 are some sample angles, with both the radian measure and the degree measure given for each.

If we rotate the generating line $1\frac{1}{4}$ full positive rotations, the directed distance traveled by P along the circle and the radian measure of the angle generated is

$$\frac{5}{4}(2\pi) = \frac{5}{2}\pi \approx 7.85$$

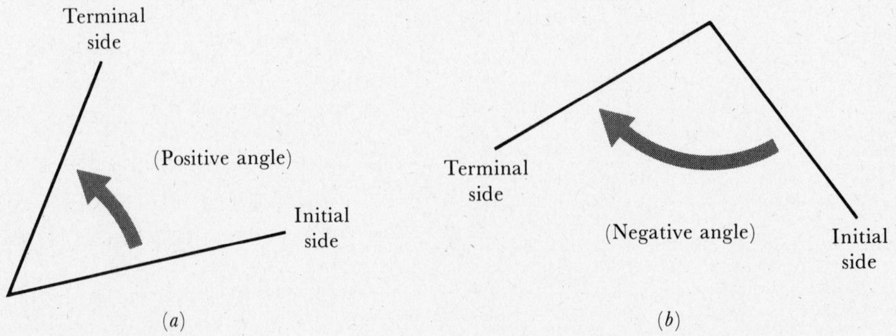

Terminal side

(Positive angle)

Initial side

(*a*)

Terminal side

(Negative angle)

Initial side

(*b*)

Fig. 7.1

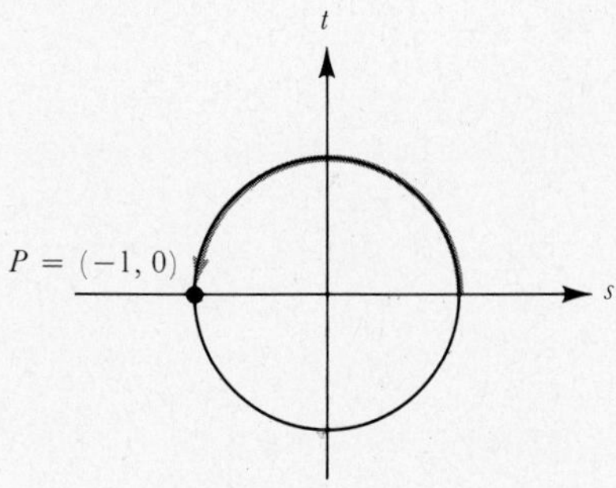

1/2 full positive rotation
radian measure = (1/2)(2π) = π ≈ 3.14
180 degrees

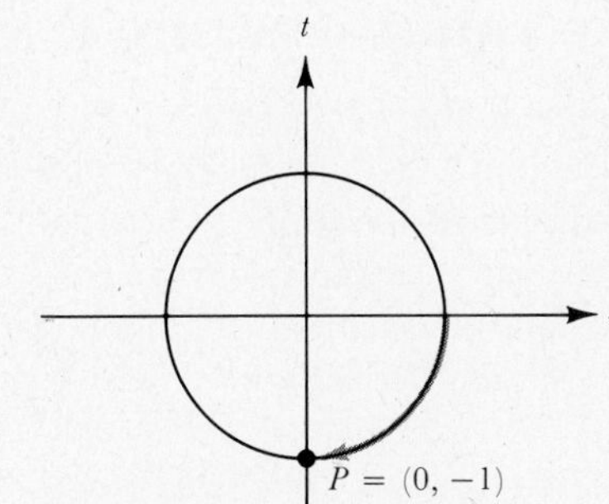

1/4 full negative rotation
radian measure = 1/4 (−2π) = −π/2 ≈ −1.57
−90 degrees

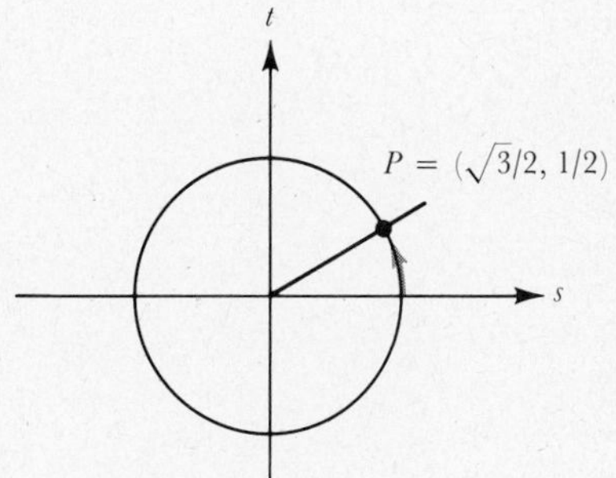

1/12 full positive rotation
radian measure = (1/12)(2π) = π/6 ≈ 0.52
30 degrees

Fig. 7.2

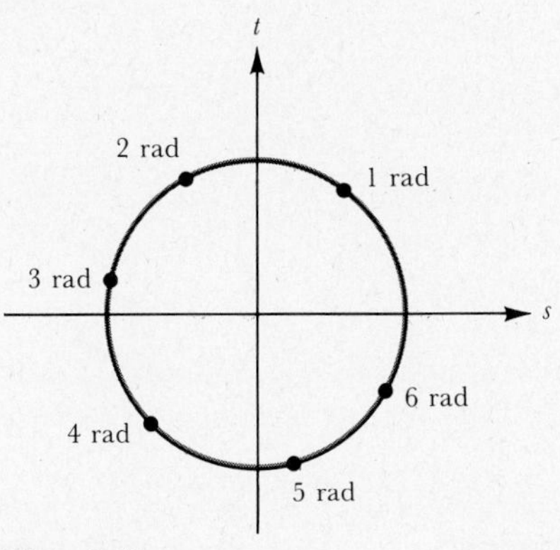

Fig. 7.3

7.1.2

Since there is no restriction on the size and direction of the rotation of the generating line, the directed distance traveled by P and hence the *radian measure of an angle, can be any real number*. In Fig. 7.3 we have marked off angles of measures 1 to 6 rad.

Given any angle, its degree and radian measures must be the same proportion or multiple of the measure of a full positive rotation. Hence, letting D be the degree measure of the angle and R the radian measure, we have

$$\frac{D}{360} = \frac{R}{2\pi}$$

which we can simplify a bit to get

$$\frac{D}{180} = \frac{R}{\pi}$$

This simple relationship lets us convert one measure to the other through simple substitution. For example, given an angle of 65°, $\frac{65}{180} = R/\pi$, so $R = (13/36)\pi$ is its measure in radians. Often the measure of an angle in radians is not reduced to decimal form but is left containing π as a factor. We often identify an angle by its measure and use abbreviated phrases like "consider the angle $\pi/4$" instead of "consider the angle whose radian measure is $\pi/4$."

PROGRESS TEST 1

1. Give the rotation and degree measure of the following angles:

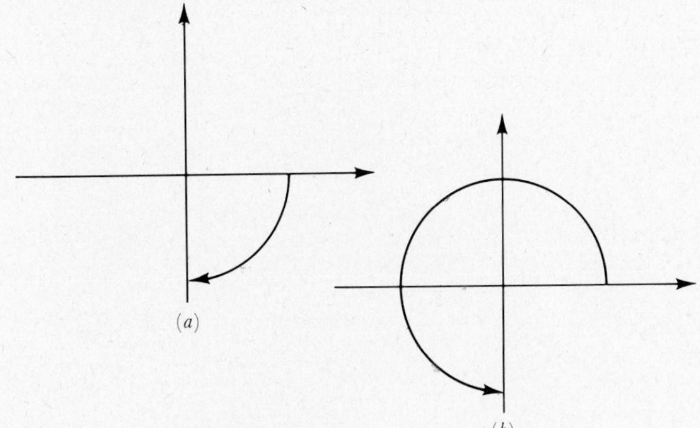

2. Convert your answers to Prob. 1 to radian measure. (Leave π as a factor.)

3. Give the radian and degree measure of the angle consisting of $\frac{5}{2}$ rotations.

4. Convert to radian measure: (*a*) 45°, (*b*) 30°, (*c*) 60°, (*d*) 250°, (*e*) −75°. (Leave π as a factor.)

5. Convert to degree measure: (*a*) $\pi/4$ rad, (*b*) $\pi/12$ rad, (*c*) 0.35 rad, (*d*) −41 rad, (*e*) 62,000 rad. (Put your answer in decimal form.)

THE TRIGONOMETRIC FUNCTIONS

Assume we have an angle of x rad whose terminal side intersects the unit circle at the point (s, t), as in Fig. 7.4, which contains the usual labels for the four quadrants of the plane as well.

We can use the coordinates (s, t) as the basis for the six trigonometric functions of x defined as follows, with their respective standard abbreviations given parenthetically.

7.1.3 Definitions of the Trigonometric Functions

$$\text{sine } (x) = t \quad (\sin x) \qquad \text{cosecant } (x) = \frac{1}{t} \quad (\csc x)$$

$$\text{cosine } (x) = s \quad (\cos x) \qquad \text{secant } (x) = \frac{1}{s} \quad (\sec x)$$

$$\text{tangent } (x) = \frac{t}{s} \quad (\tan x) \qquad \text{cotangent } (x) = \frac{s}{t} \quad (\cot x)$$

We shall now examine and graph the function $\sin x$ for $0 \le x \le 2\pi$. For reference we recall some useful right triangles in Fig. 7.5, each of whose hypotenuse is length 1. (The radian measure of the indicated angle appears inside each triangle.)

In Fig. 7.6 we give the values of $\sin x$ in decimal form, approximating when irrational numbers are involved. Now $\sin \pi/2 = 1$, since P is at $(0, 1)$.

As x continues to increase from $\pi/2$ to π, the point P continues around the circle in the second quadrant, with its t coordinate decreasing from 1 when $x = \pi/2$ and $P = (0, 1)$ to 0 when $x = \pi$ and $P = (-1, 0)$; see Fig. 7.7.

As x continues to grow from π to $\frac{3}{2}\pi$, the t coordinate of P (now in the third quadrant) continues to decrease from 0 at $x = \pi$ to -1 at $x = \frac{3}{2}\pi$; see Fig. 7.8(a). As x continues to grow from $\frac{3}{2}\pi$ to 2π, the t coordinate of P (now in the fourth quadrant) increases from -1 at $x = \frac{3}{2}\pi$ to 0 at $x = 2\pi$; see Fig. 7.8(b).

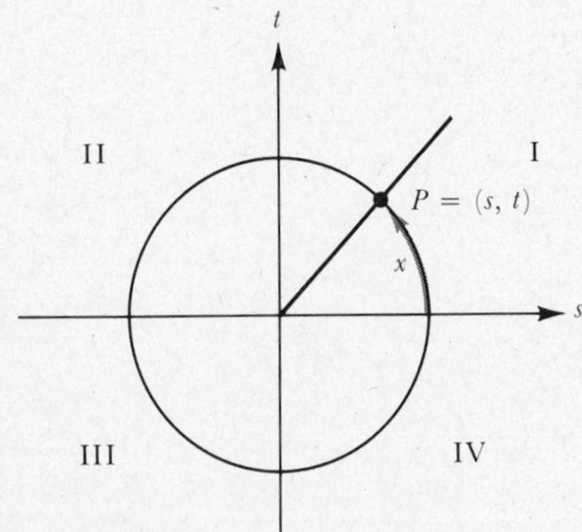

Fig. 7.4

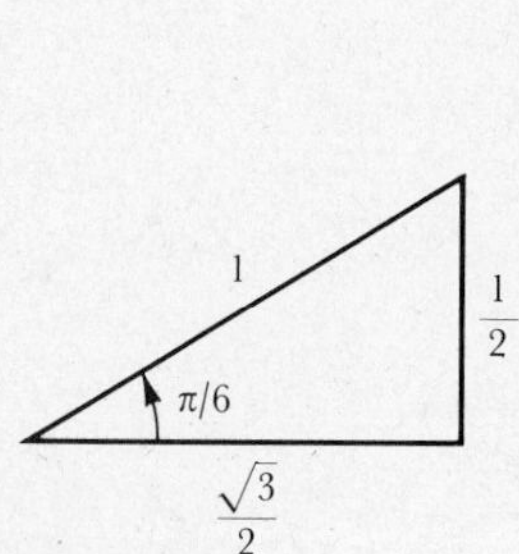

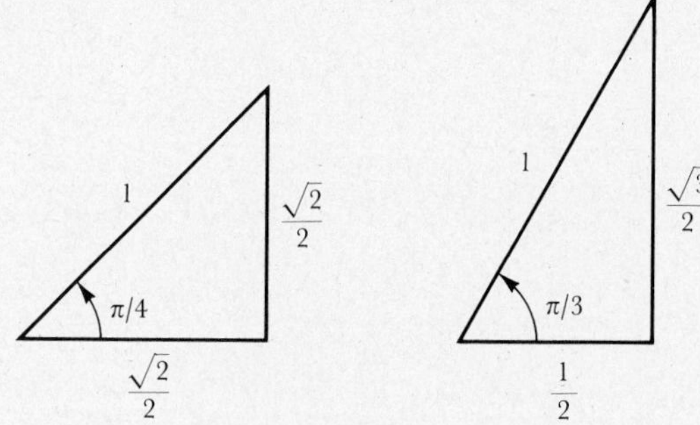

Fig. 7.5

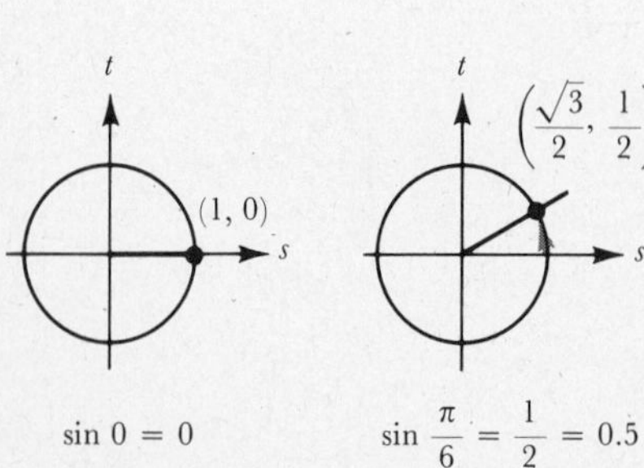

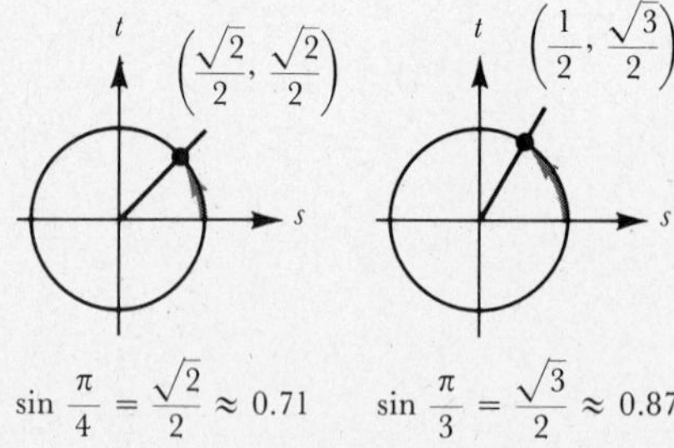

Fig. 7.6

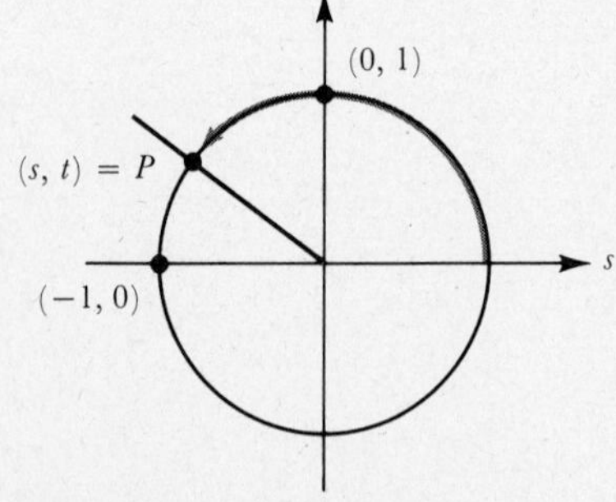

Fig. 7.7

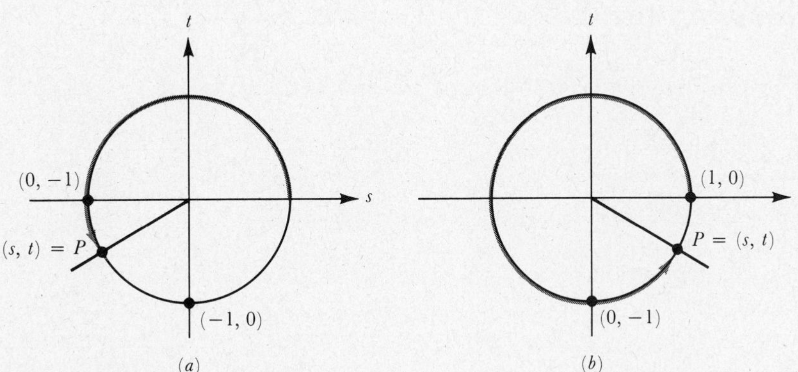

Fig. 7.8

We now plot the function $y = \sin x$ on the x-y coordinate system, labeling certain of these points. (See Fig. 7.9.)

Obviously, as x increases beyond 2π, the sine function begins to repeat itself *because the same* (s, t) *coordinates recur*. The sine function continues the cycle again from $x = 2\pi$ to $x = 4\pi$, and then again from $x = 4\pi$ to $x = 6\pi$, and so on.

Similarly, as x *decreases* from $x = 0$ to $x = -2\pi$, the same (s, t) coordinates occur. For example, $\sin(-\pi/6) = -1/2 = -0.5 = \sin(11\pi/6)$. (See Fig. 7.10.)

Again, $\sin x$ repeats its cycle as x decreases from $x = -2\pi$ to $x = -4\pi$, and so on. Putting all this information together, we get a more complete graph of $\sin x$. (See Fig. 7.11.)

7.1.4 Definition Any function that repeats itself in every interval of fixed length p and in no smaller interval is said to be *periodic of period* p. That is, $f(x + p) = f(x)$ for all x in the domain of f.

As we have seen, the sine function is periodic of period 2π.

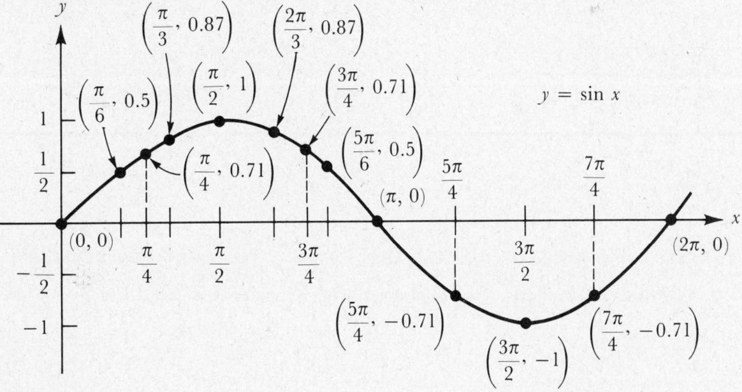

Fig. 7.9

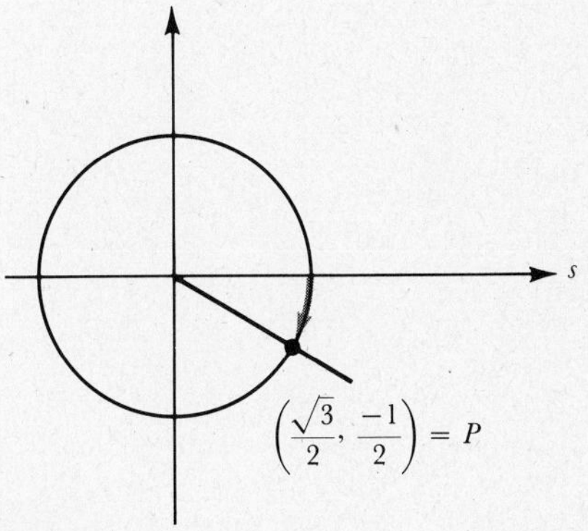

$\left(\frac{\sqrt{3}}{2}, \frac{-1}{2}\right) = P$

Fig. 7.10

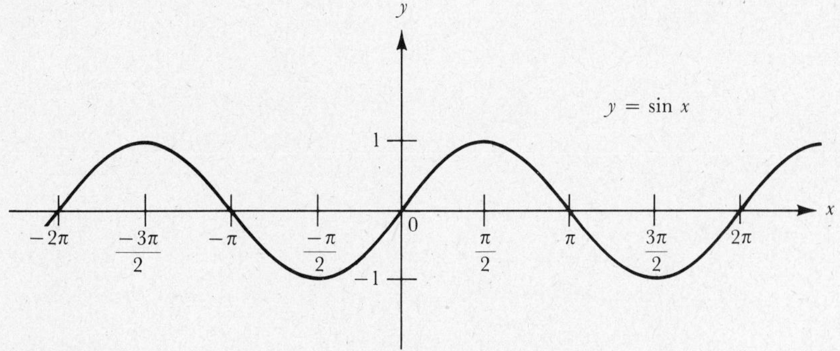

Fig. 7.11

PROGRESS TEST 2

1. Using the familiar right triangles given earlier, fill in the values of $\cos x$ in the table to the right:
2. Graph all the points listed in the table on the x-y plane, draw a smooth curve through them, and extend the graph in either direction beyond the interval $[0, 2\pi]$.
3. What is the period of $\cos x$?

x	$\cos x$	x	$\cos x$	x	$\cos x$
0		$\left(\frac{2}{3}\right)\pi$		$\left(\frac{4}{3}\right)\pi$	
$\frac{\pi}{6}$		$\left(\frac{3}{4}\right)\pi$		$\left(\frac{3}{2}\right)\pi$	
$\frac{\pi}{4}$		$\left(\frac{5}{\pi}\right)\pi$		$\left(\frac{5}{3}\right)\pi$	
$\frac{\pi}{3}$		$\left(\frac{7}{6}\right)\pi$		$\left(\frac{7}{4}\right)\pi$	
$\frac{\pi}{2}$		$\left(\frac{5}{4}\right)\pi$		$\left(\frac{11}{6}\right)\pi$	
				2π	

MORE TRIGONOMETRIC FUNCTIONS

The sine and cosine functions behave very similarly. In fact, one is just the other slid horizontally by $\pi/2$ units; that is, $\cos x = \sin(x + \pi/2)$. (See Fig. 7.12.)

The tangent function, defined by

$$\tan x = \frac{t}{s}$$

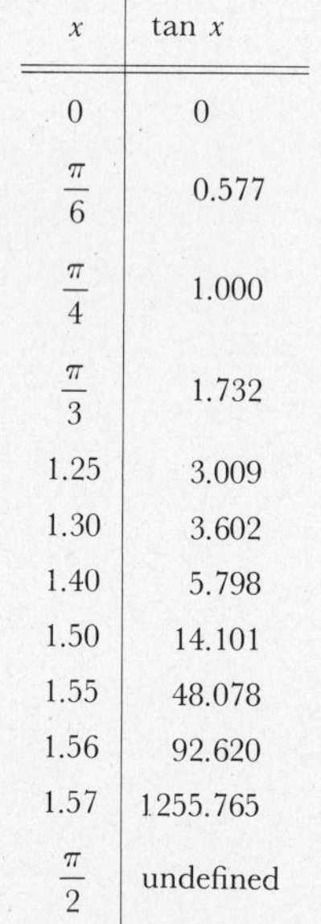

x	$\tan x$
0	0
$\frac{\pi}{6}$	0.577
$\frac{\pi}{4}$	1.000
$\frac{\pi}{3}$	1.732
1.25	3.009
1.30	3.602
1.40	5.798
1.50	14.101
1.55	48.078
1.56	92.620
1.57	1255.765
$\frac{\pi}{2}$	undefined

Table 7.1

behaves somewhat differently, because for certain values of x, namely, whenever the point $P = (s, t)$ is on the t axis, $s = 0$, so $\tan x$ is undefined. In particular this is true for $x = \pm\pi/2$. The tangent function *is* defined for x arbitrarily close to $\pm\pi/2$. We list some values of $\tan x$ for $0 \leq x \leq \pi/2$ in Table 7.1. As $x \to (\pi/2)^-$, $t \to 1$ and $s \to 0^+$. Hence

$$\lim_{x\to(\pi/2)^-} \tan x = +\infty$$

That is, the quotient t/s grows without bound. Similarly, as $x \to (-\pi/2)^+$, $t \to -1$, while $s \to 0^+$. Hence

$$\lim_{x\to-(\pi/2)^+} \tan x = -\infty$$

We also have

$$\lim_{x\to(\pi/2)^+} \tan x = -\infty \qquad \text{and} \qquad \lim_{x\to(3\pi/2)^-} \tan x = +\infty$$

This pattern repeats itself in the intervals $(-\pi/2, \pi/2)$, $(\pi/2, 3\pi/2)$, $(3\pi/2, 5\pi/2)$, and so forth. Clearly $\tan x$ is periodic of period π. (See Fig. 7.13.)

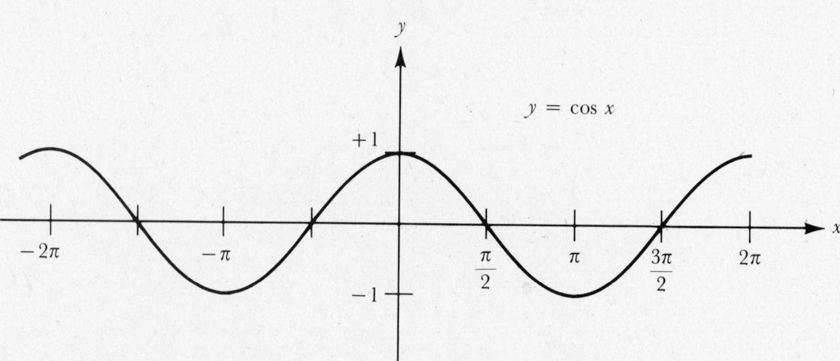

Fig. 7.12

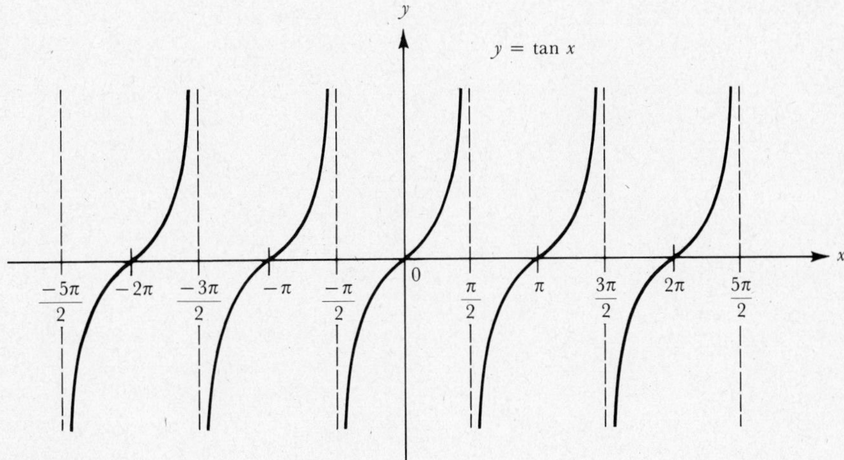

Fig. 7.13

PROGRESS TEST 3

1. Following the procedure above use the definition of the cotangent function, $\cot x = s/t$, to analyze and graph $y = \cot x$ for $0 < x < 2\pi$.
2. What is the period of $\cot x$?
3. Describe all x where $\cot x$ is undefined.
4. In terms of the terminal side of the angle, in which quadrants is $\cot x$ positive and which negative?
5. Describe all x for which $\sec x$ is undefined.
6. Describe all x for which $\csc x$ is undefined.
7. What is the smallest possible value of $|\sec x|$?
8. What is the smallest possible value of $|\csc x|$?

TRIGONOMETRIC IDENTITIES

We shall now recall the basic trigonometric identities but shall not concentrate on proofs.

7.1.5 Trigonometric Identities

$$\tan x = \frac{\sin x}{\cos x} \qquad \cot x = \frac{\cos x}{\sin x} \qquad \sec x = \frac{1}{\cos x} \qquad \csc x = \frac{1}{\sin x}$$

$$\cos^2 x + \sin^2 x = 1 \qquad \tan^2 x + 1 = \sec^2 x \qquad \cot^2 x + 1 = \csc^2 x$$

$$\cos x = \sin\left(\frac{\pi}{2} - x\right) \qquad \sin x = \cos\left(\frac{\pi}{2} - x\right) \qquad \tan x = \cot\left(\frac{\pi}{2} - x\right)$$

$$\sin(a \pm b) = \sin a \cos b \pm \cos a \sin b$$

$$\cos(a \pm b) = \cos a \cos b \mp \sin a \sin b$$

$$\tan(a \pm b) = \frac{\tan a \pm \tan b}{1 \mp \tan a \tan b}$$

$$\sin 2x = 2 \sin x \cos x \qquad \cos 2x = \cos^2 x - \sin^2 x \qquad \tan 2x = \frac{2 \tan x}{1 - \tan^2 x}$$

$$\sin^2 x = \frac{1 - \cos 2x}{2} \qquad \cos^2 x = \frac{1 + \cos 2x}{2}$$

$$\tan x = \frac{\sin 2x}{1 + \cos 2x} = \frac{1 - \cos 2x}{\sin 2x}$$

Some of the above identities, such as

$$\sin^2 x + \cos^2 x = 1$$

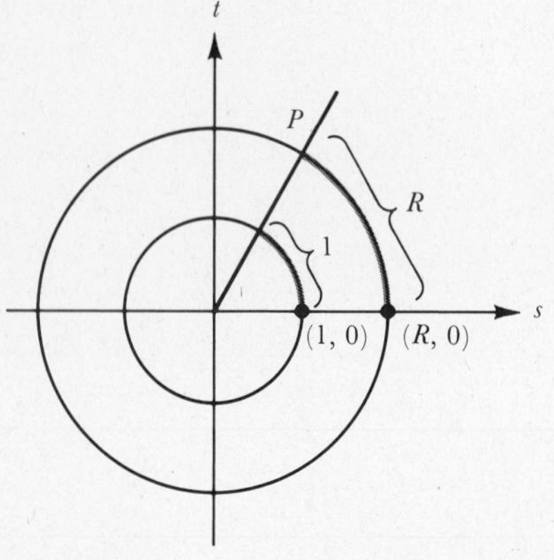

Fig. 7.14

follow from the fact that we used the unit circle to define the trigonometric functions:

$$\sin^2 x + \cos^2 x = t^2 + s^2 = 1$$

Others require much fancier manipulations. [Recall that customarily we write $\sin^2 x$ instead of $(\sin x)^2$, and so on.] Later we shall show how derivatives can be used to relate these identities to one another.

Also, for convenience, we give some values of the sine, cosine, and tangent functions in Table 7.2.

We can use the identities and Table 7.2 to compute additional values of the trigonometric functions.

Example 1 Determine $\sin \pi/12$, accurate through three decimal places.

Solution

$$\sin\frac{\pi}{12} = \sin\left(\frac{\pi}{4} - \frac{\pi}{6}\right)$$

$$= \sin\frac{\pi}{4}\cos\frac{\pi}{6} - \cos\frac{\pi}{4}\sin\frac{\pi}{6}$$

$$= \frac{\sqrt{2}}{2}\frac{\sqrt{3}}{2} - \frac{\sqrt{2}}{2}\frac{1}{2}$$

$$= \frac{\sqrt{2}}{4}(\sqrt{3} - 1) \approx 0.259 \quad \bullet$$

Values of Sin, Cos, Tan at Popular x

x	$\sin x$	$\cos x$	$\tan x$
0	0	1	0
$\frac{\pi}{6}$	$\frac{1}{2}$	$\frac{\sqrt{3}}{2}$	$\frac{1}{\sqrt{3}}$
$\frac{\pi}{4}$	$\frac{\sqrt{2}}{2}$	$\frac{\sqrt{2}}{2}$	1
$\frac{\pi}{3}$	$\frac{\sqrt{3}}{2}$	$\frac{1}{2}$	$\sqrt{3}$
$\frac{\pi}{2}$	1	0	undefined
$\frac{2\pi}{3}$	$\frac{\sqrt{3}}{2}$	$\frac{-1}{2}$	$-\sqrt{3}$
$\frac{3\pi}{4}$	$\frac{\sqrt{2}}{2}$	$\frac{-\sqrt{2}}{2}$	-1
$\frac{5\pi}{6}$	$\frac{1}{2}$	$\frac{-\sqrt{3}}{2}$	$\frac{-1}{\sqrt{3}}$
π	0	-1	0

Table 7.2

ADDITIONAL REMARKS

Although we used the *unit* circle in defining radian measure and the trigonometric functions, we could just as well have used a circle of arbitrary positive radius R. In this case, one radian would be defined as the angle obtained when the point P on the defining circle travels a distance equal to R in the counterclockwise direction. See Fig. 7.14, where we have superimposed a unit circle over the generating line to show that the size of the angle is the same, whether or not we use a unit circle. Analogously, we could define $\sin x = t/R$ instead of $\sin x = t/1$, and so forth. The same functions result.

It is important to remember, however, that the variable x stands for numbers and *not* angles. The domain of each trigonometric function consists of a set of real numbers, which are the radian measures of angles. In fact, we used angles to introduce the trigonometric functions mainly to take advantage of your earlier experience with geometry and trigonometry. In later chapters we shall see how these functions can be defined entirely independently of geometric considerations. Finally, when working with a calculator, *be sure that it is set to the degree mode or radian mode as appropriate.*

SECTION 7.1 EXERCISES

1. Using methods similar to those used to describe $\tan x$, describe and graph $\sec x$ for $-\pi/2 < x < 3\pi/2$.

2. Repeat Exercise 1 for $\csc x$.

3. What is the period of (*a*) $\sec x$ and (*b*) $\csc x$?

In Exercises 4 to 11 convert the given degree measure to radian measure (leaving a factor of π in your answer.)

4. 50 **5.** 15 **6.** 110 **7.** 400

8. -2 **9.** 10^7 **10.** 3.14 **11.** -3.14

In Exercises 12 to 18 convert the given radian measure to degree measure.

12. 1

13. 0.62

14. $\frac{2}{5}\pi$

15. $\frac{\pi}{10}$

16. $\frac{-6}{7}\pi$

17. 200π

18. $\frac{2}{3}$

In Exercises 19 to 29 determine the value of the given function at the indicated number using Table 7.2 and, if necessary, the trigonometric identities.

19. $\sin 200\pi$

20. $\cos\left(-\frac{\pi}{2}\right)$

21. $\cos\left(\frac{3}{4}\right)\pi$

22. $\csc\frac{\pi}{8}$

23. $\sin\left(\frac{-2\pi}{3}\right)$

24. $\cot\left(\frac{-7}{8}\pi\right)$

25. $\sec\frac{-2}{3}\pi$

26. $\tan\left(\frac{5}{6}\pi\right)$

27. $\tan\left(\frac{5}{12}\pi\right)$

28. $\cot\left(\frac{11}{12}\pi\right)$

29. $\cos 50\pi$

30. Determine $\sin(10^{100}\pi) - \sin(\pi/2)$. (Note that your calculator cannot do this one.)

31. Derive the identity $\tan^2 x + 1 = \sec^2 x$ from $\sin^2 x + \cos^2 x = 1$.

32. Repeat Exercise 31 for $\cot^2 x + 1 = \csc^2 x$.

33. Show how the identity $\sin 2x = 2 \sin x \cos x$ follows from the identity for $\sin(a + b)$.

34. Analogously to Exercise 33, derive the identity $\cos 2x = \cos^2 x - \sin^2 x$.

35. Use the identity in Exercise 34 to derive the identity $\sin^2 x = (1 - \cos 2x)/2$.

36. Use the identity in Exercise 34 to derive the identity $\cos^2 x = (1 + \cos 2x)/2$.

Section 7.2: Derivatives of Trigonometric Functions

Objectives:

1. Differentiate each of the six trigonometric functions and combinations of these functions with other differentiable functions.

2. Determine limits involving trigonometric functions.

THE DERIVATIVES OF THE SINE AND COSINE FUNCTIONS

By carefully examining the graphs of the sine and cosine functions in Fig. 7.15 we can learn some important facts about these functions.

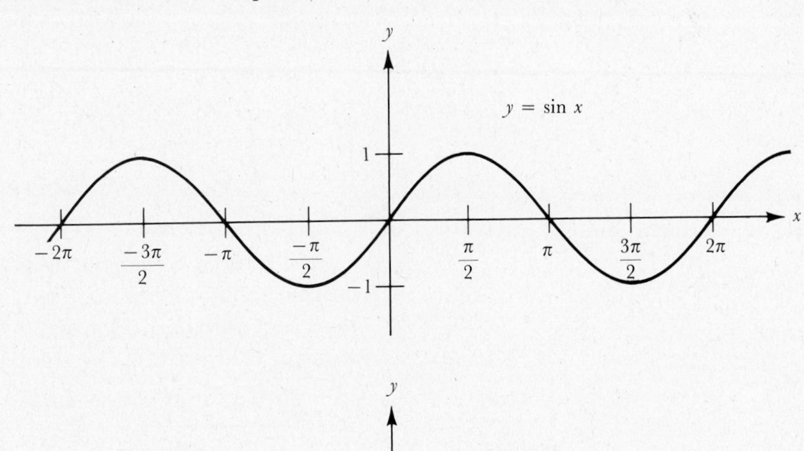

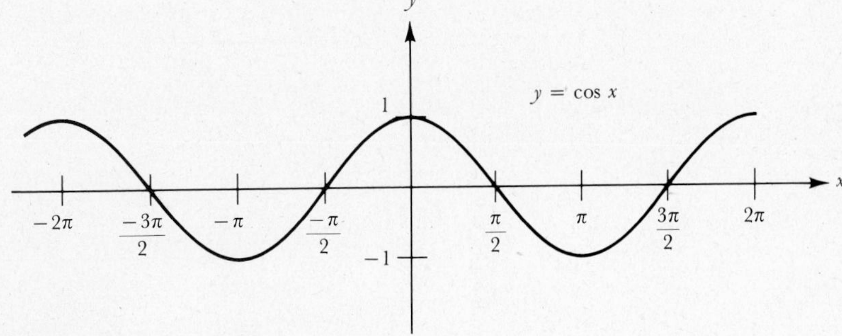

Fig. 7.15

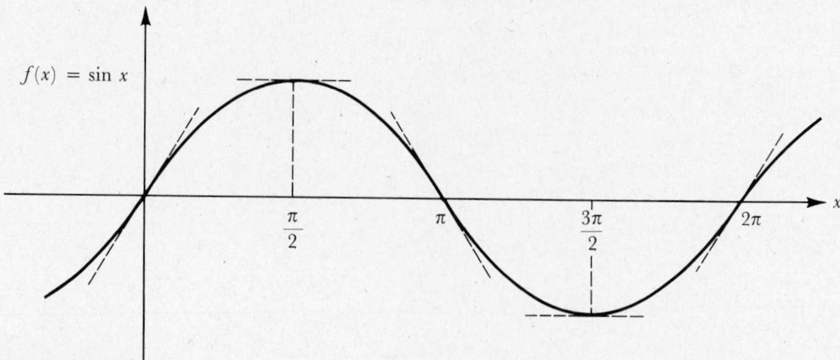

Fig. 7.16

Each function appears to be continuous on $(-\infty, +\infty)$—a fact that can be proved rigorously, but which we shall assume.

The sine function is *odd*, that is, $\sin(-x) = -\sin x$; and the cosine function is *even*, that is, $\cos(-x) = \cos x$.

Each function appears to be differentiable in that the graphs appear to be "smooth"; that is, they have well-defined tangents at each point.

Let us examine the slopes of the tangents to the sine function to see whether we can guess what its derivative might be (see Fig. 7.16). Near $x = 0$ the slope is about $+1$, and it decreases to 0 (horizontal tangent line) at $x = \pi/2$. It then becomes negative and decreases to about -1 at $x = \pi$. It then increases again, reaching 0 (horizontal tangent line) at $x = 3\pi/2$, and continues increasing, reaching about 1 at $x = 2\pi$.

We sketch this *slope function*—that is, the derivative of $\sin x$ between 0 and 2π—in Fig. 7.17. This certainly *looks like* the cosine function between 0 and 2π. Furthermore, this same pattern is repeated with a period of 2π for all x.

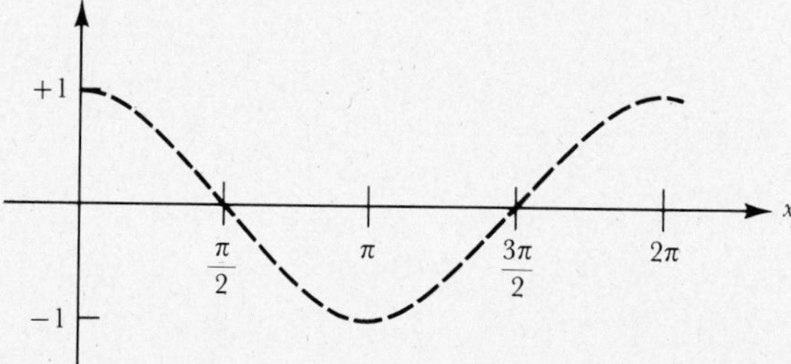

Fig. 7.17

We shall now *prove* that the derivative of $\sin x$ is $\cos x$. We shall handle this derivative carefully because the derivatives of all the other trigonometric functions will be based upon it.

We begin by rewriting the standard derivative quotient:

$$\frac{f(x + \Delta x) - f(x)}{\Delta x} = \frac{\sin(x + \Delta x) - \sin x}{\Delta x}$$

$$= \frac{(\sin x \cos \Delta x + \sin \Delta x \cos x) - \sin x}{\Delta x}$$

Now we break the quotient into two terms and regroup:

$$\frac{\sin \Delta x \cos x}{\Delta x} + \frac{\sin x \cos \Delta x - \sin x}{\Delta x} = \cos x \left[\frac{\sin \Delta x}{\Delta x}\right] + \sin x \left[\frac{\cos \Delta x - 1}{\Delta x}\right]$$

Below, we shall prove the following:

7.2.1 $(a)\ \lim_{\Delta x \to 0} \frac{\sin \Delta x}{\Delta x} = 1$ and $(b)\ \lim_{\Delta x \to 0} \frac{\cos \Delta x - 1}{\Delta x} = 0$

Then using (7.2.1), we can prove:

7.2.2 $D_x(\sin x) = \cos x$

Proof of (7.2.2), using (7.2.1):

$$\begin{aligned} D_x(\sin x) &= \lim_{\Delta x \to 0} \left(\cos x \left[\frac{\sin \Delta x}{\Delta x}\right] + \sin x \left[\frac{\cos \Delta x - 1}{\Delta x}\right]\right) \\ &= \cos x \lim_{\Delta x \to 0} \left[\frac{\sin \Delta x}{\Delta x}\right] + \sin x \lim_{\Delta x \to 0} \left[\frac{\cos \Delta x - 1}{\Delta x}\right] \\ &= (\cos x)(1) + (\sin x)(0) \\ &= \cos x \quad \blacksquare \end{aligned}$$

We shall now derive [7.2.1(b)] from [7.2.1(a)], leaving as our last step the proof of [7.2.1(a)].

Proof that [7.2.1(b)] follows from [7.2.1(a)]:

First,

$$\begin{aligned} \frac{\cos \Delta x - 1}{\Delta x} &= \frac{\cos \Delta x - 1}{\Delta x} \frac{\cos \Delta x + 1}{\cos \Delta x + 1} \\ &= \frac{\cos^2 \Delta x - 1}{\Delta x(\cos \Delta x + 1)} \\ &= \frac{-\sin^2 \Delta x}{\Delta x(\cos \Delta x + 1)} \qquad (\text{since } \sin^2 \Delta x + \cos^2 \Delta x = 1) \\ &= \frac{\sin \Delta x}{\Delta x}\left(\frac{-\sin \Delta x}{\cos \Delta x + 1}\right) \end{aligned}$$

Hence

$$\begin{aligned} \lim_{\Delta x \to 0} \frac{\cos \Delta x - 1}{\Delta x} &= \lim_{\Delta x \to 0} \left[\frac{\sin \Delta x}{\Delta x} \cdot \frac{-\sin \Delta x}{\cos \Delta x + 1}\right] \\ &= \lim_{\Delta x \to 0} \frac{\sin \Delta x}{\Delta x} \cdot \lim_{\Delta x \to 0} \frac{-\sin \Delta x}{\cos \Delta x + 1} \\ &= 1 \cdot \frac{0}{1 + 1} \qquad \text{assuming } [7.2.1(a)] \\ &= 0 \quad \blacksquare \end{aligned}$$

Besides using the Limit Theorem (2.5.4) in these arguments, we also used our assumption that the sine and cosine functions are continuous when we evaluated the limits in the last step by substitution. The same will be true in the proof of [7.2.1(a)].

Thus we have reduced the proof that $D_x(\sin x) = \cos x$ to a proof of [7.2.1(a)]. This argument is of special interest because it will be the first limit argument that does not boil down to the rationalizing and/or canceling routines that sufficed so far. Before moving to the proof, we notice that in Table 7.3 (in which the values were computed using a calculator) and in Fig. 7.18,

$$\sin \Delta x \approx \Delta x$$

for small Δx. (The values in the table are rounded off to the relevant digits.) In Fig. 7.18 we see that *for Δx small, the length of the arc Δx and the length of the straight line segment $\overline{PA}$ are almost equal; hence their ratio $(\sin \Delta x)/\Delta x$ must be very close to 1.* The proof that follows is essentially an elaboration of this observation.

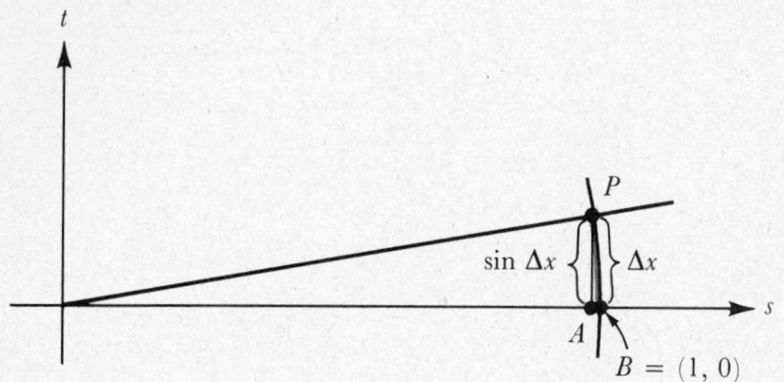

Fig. 7.18

Δx	$\sin \Delta x$	$\dfrac{\sin \Delta x}{\Delta x}$
0.2	0.198	0.993
0.1	0.09983	0.9983
0.01	0.0099998	0.999983
0.001	0.0009999998	0.99999983
0.0001	0.0000999999	0.999999995

Table 7.3

Fig. 7.19

Proof of [7.2.1(a)]: $\lim\limits_{\Delta x \to 0} \dfrac{\sin \Delta x}{\Delta x} = 1$: First, we need only deal with $\Delta x > 0$ because the sine function is odd, making

$$\frac{\sin(-\Delta x)}{-\Delta x} = \frac{-\sin \Delta x}{-\Delta x} = \frac{\sin \Delta x}{\Delta x}$$

Now, suppose we have a circle of radius 1 with an angle of Δx rad, as indicated in Fig. 7.19. Using the geometry of this situation, we trap the quotient in question between 1 and $\cos \Delta x$, and apply the Squeeze Theorem (2.5.6). Now,

$$\text{area of triangle } POB < \text{area of sector } POB < \text{area of triangle } COB$$

Using the facts that the area of a triangle is $\frac{1}{2}$ the base times the height and the area of the Δx sector of the circle is $\Delta x/(2\pi)$ times the area of the whole circle, $\pi \cdot 1^2 = \pi$, the above inequality translates to

$$\frac{1}{2}\overline{OB} \cdot \overline{PA} < \frac{\Delta x}{2\pi} \cdot \pi < \frac{1}{2}\overline{OB} \cdot \overline{CB}$$

Multiplying through by 2 and using that $\overline{OB} = 1$, we get

$$\overline{PA} < \Delta x < \overline{CB}$$

But $\sin \Delta x = \overline{PA}$ by definition, and, by similar triangles,

$$\tan \Delta x = \frac{\overline{PA}}{\overline{OA}} = \frac{\overline{CB}}{\overline{OB}} = \frac{\overline{CB}}{1} = \overline{CB}$$

so we have

$$\sin \Delta x < \Delta x < \tan \Delta x$$

Since $\sin \Delta x > 0$, $(0 < \Delta x < \pi/2)$, we can divide through by $\sin \Delta x$ to get

$$1 < \frac{\Delta x}{\sin \Delta x} < \frac{\tan \Delta x}{\sin \Delta x} = \frac{1}{\cos \Delta x} \qquad \left(\text{using } \tan \Delta x = \frac{\sin \Delta x}{\cos \Delta x}\right)$$

Finally, we can invert each term of the inequality, reversing inequality signs (remember, $2 < 3$ implies $\frac{1}{2} > \frac{1}{3}$) to get the promised "trap" of $(\sin \Delta x)/\Delta x$ between 1 and $\cos \Delta x$:

$$1 > \frac{\sin \Delta x}{\Delta x} > \cos \Delta x$$

Since this holds for *all* Δx with $0 < \Delta x < \pi/2$, we know that

$$1 \geq \lim_{\Delta x \to 0} \frac{\sin \Delta x}{\Delta x} \geq \lim_{\Delta x \to 0} \cos \Delta x = 1$$

Thus, by the Squeeze Theorem, the only possibility is that

$$\lim_{\Delta x \to 0} \frac{\sin \Delta x}{\Delta x} = 1$$

and the proof is completed. ■

Since the derivative of $\sin x$ is $\cos x$, the Chain Rule implies the following:

If $u(x)$ is a differentiable function of x, then

7.2.3 $$D_x[\sin u(x)] = \cos u(x) \cdot u'(x)$$

In stating and applying this rule, we follow custom and use as few parentheses as are necessary to avoid confusion. Hence we have written $\sin u(x)$ instead of $\sin (u(x))$, and $\cos u(x) \cdot u'(x)$ instead of $[\cos (u(x))] \cdot u'(x)$.

Proof of (7.2.3): By the Chain Rule (3.2.2), with g equal to the sine function, we have

$$\begin{aligned} D_x[\sin (u(x))] &= D_x[g(u(x))] \\ &= g'(u(x)) \cdot u'(x) \end{aligned}$$

But g' is the cosine function, so

$$g'(u(x)) \cdot u'(x) = \cos(u(x)) \cdot u'(x) \quad \blacksquare$$

Henceforth, in establishing additional derivative rules, we shall use the phrase "by the Chain Rule" repeatedly as we just did above but not carry out the argument, which is the same in every case.

Example 1 Compute: (*a*) $D_x[\sin(x^2 + 3)]$ and (*b*) $D_x[x^3 \sin(5x + 1)]$.

Solution By (7.2.3):

(*a*) $$\begin{aligned} D_x[\sin(x^2 + 3)] &= (\cos(x^2 + 3))(2x) \\ &= 2x \cos(x^2 + 3) \end{aligned}$$

(Again, it is customary to omit the parentheses and reorder factors to avoid ambiguity.)

(*b*) $$\begin{aligned} D_x[x^3 \sin(5x + 1)] &= x^3(\cos(5x + 1))(5) + (\sin(5x + 1))(3x^2) \\ &= 5x^3 \cos(5x + 1) + 3x^2 \sin(5x + 1) \quad \bullet \end{aligned}$$

We shall also apply (7.2.3) to compute $D_x(\cos x)$. But first we need:

7.2.4 $$(a)\ \cos x = \sin\left(\frac{\pi}{2} - x\right) \qquad (b)\ \sin x = \cos\left(\frac{\pi}{2} - x\right)$$

These statements follow by applying the identities (7.1.5) regarding the sine and cosine of a difference, respectively, with $a = \pi/2$, $b = x$.

Now, by (7.2.3), we have

$$\begin{aligned} D_x[\cos x] &= D_x\left[\sin\left(\frac{\pi}{2} - x\right)\right] && \text{by [7.2.4}(a)\text{]} \\ &= \left(\cos\left(\frac{\pi}{2} - x\right)\right)(-1) && \text{by (7.2.3)} \\ &= (\sin x)(-1) && \text{by [7.2.4}(b)\text{]} \\ &= -\sin x \end{aligned}$$

Hence, by the Chain Rule, for $u(x)$ a differentiable function of x:

7.2.5 $$D_x[\cos u(x)] = -\sin u(x) \cdot u'(x)$$

Example 2 Compute but do not simplify: (*a*) $D_x[\cos(x^3 - 2x + 4)]$ (*b*) $D_x[(\cos x)(1 - \sin 5x)]$ (*c*) $D_x[\sin^3(7x + 1)]$

Solution (*a*) $D_x[\cos(x^3 - 2x + 4)] = -(\sin(x^3 - 2x + 4))(3x^2 - 2)$
(*b*) $D_x[(\cos x)(1 - \sin 5x)] = (\cos x)(-\cos 5x)(5) + (1 - \sin 5x)(-\sin x)$
(*c*) Remember, $\sin^3(7x + 1) = (\sin(7x + 1))^3$.

$$\begin{aligned} D_x[\sin^3(7x + 1)] &= 3\sin^2(7x + 1) \cdot D_x[\sin(7x + 1)] \\ &= 3\sin^2(7x + 1) \cdot (\cos(7x + 1))(7) \end{aligned}$$ ●

In part (*c*) we used both the Power Rule and the Chain Rule.

PROGRESS TEST 1

In Probs. 1 to 8 differentiate the indicated function.

1. $f(x) = \sin(x^4 + 3x^3 + 2x + 1)$
2. $f(x) = \cos(x^5 - 4x^3 - 3x + 1)$
3. $g(x) = \sin^5(3x^2 + x - 1)$
4. $t(x) = \sqrt{\cos(x^2 + x)}$
5. $r(x) = \dfrac{\sin^2(3x + 1)}{\cos^3(5x - 2)}$
6. $s(t) = \cos^4(t^2 + 1)\sin^3(7t) + \sin 4t$
7. $m(p) = 2\sin p \cos p$
8. $f(x) = \sin(\sqrt{4x + 5})$

DERIVATIVES OF THE TANGENT, COTANGENT, SECANT, AND COSECANT FUNCTIONS

We shall now employ the Quotient Rule in combination with (7.2.3) and (7.2.5) and trigonometric identities to differentiate some of the remaining trigonometric functions:

$$\begin{aligned} D_x[\tan x] &= D_x\left[\frac{\sin x}{\cos x}\right] \\ &= \frac{\cos x(\cos x) - (\sin x)(-\sin x)}{\cos^2 x} \\ &= \frac{\cos^2 x + \sin^2 x}{\cos^2 x} = \frac{1}{\cos^2 x} = \sec^2 x \end{aligned}$$

Also,

$$\begin{aligned} D_x[\sec x] &= D_x\left[\frac{1}{\cos x}\right] \\ &= \frac{(-1)(-\sin x)}{\cos^2 x} = \frac{1}{\cos x}\,\frac{\sin x}{\cos x} \\ &= \sec x \tan x \end{aligned}$$

Parallel arguments establish the derivatives of $\cot x$ and $\csc x$. (See Progress Test 2.) Assuming that $u(x)$ is a differentiable function of x, we have by the Chain Rule the following derivative rules:

7.2.6 $$D_x[\tan u(x)] = \sec^2 u(x) \cdot u'(x)$$

7.2.7 $$D_x[\cot u(x)] = -\csc^2 u(x) \cdot u'(x)$$

7.2.8 $$D_x[\sec u(x)] = \sec u(x) \tan u(x) \cdot u'(x)$$

7.2.9 $$D_x[\csc u(x)] = -\csc u(x) \cot u(x) \cdot u'(x)$$

Example 3 Differentiate the following functions:

(*a*) $f(x) = \tan(5x - 3)$
(*b*) $f(x) = \sec(4x^3 - x)$
(*c*) $y = \csc^3(5x)$
(*d*) $g(x) = \sqrt[3]{\cot(x - 1)}$
(*e*) $f(x) = \tan^4(\sin 7x)$

Solution We shall employ a variety of derivative notations.

(*a*) $f'(x) = \sec^2(5x - 3)(5)$

(*b*) $$\frac{df}{dx} = \sec(4x^3 - x)(\tan(4x^3 - x))\, D_x(4x^3 - x)$$
$$= \sec(4x^3 - x)(\tan(4x^3 - x))(12x^2 - 1)$$

(*c*) $$\frac{dy}{dx} = 3\csc^2(5x)\frac{d(\csc(5x))}{dx}$$
$$= 3\csc^2(5x)(-\csc(5x))(\cot(5x))(5)$$
$$= -15\csc^3(5x)\cot(5x)$$

(*d*) $$g'(x) = D_x[(\cot(x - 1))^{1/3}]$$
$$= \frac{1}{3}(\cot(x - 1))^{-2/3}\, D_x[\cot(x - 1)]$$
$$= \frac{1}{3}(\cot(x - 1))^{-2/3}(-\csc^2(x - 1))(1)$$
$$= -\frac{\csc^2(x - 1)}{3\sqrt[3]{\cot^2(x - 1)}}$$

(*e*) $$f'(x) = 4\tan^3(\sin 7x)\, D_x[\tan(\sin 7x)]$$
$$= 4\tan^3(\sin 7x)\sec^2(\sin 7x)\, D_x[\sin 7x]$$
$$= 4\tan^3(\sin 7x)\sec^2(\sin 7x)\cos 7x \cdot 7$$ ●

In part (*e*) we applied first the Power Rule and then the Chain Rule three times. Most errors in differentiating functions of that type (for example, $\sin^3(\cos 5x)$) occur because the Chain Rule is not completely applied.

First and second derivatives of the trigonometric functions are of course useful in graphing. We show here how the graph of the tangent function (Fig. 7.13, p. 239) agrees with the information provided by its derivatives.

Example 4 Show that $f(x) = \tan x$ has the following properties:

(*a*) It is increasing on its natural domain.
(*b*) It is concave up where it is positive and concave down where it is negative.
(*c*) It has a point of inflection at each point where it crosses the x axis.

Solution

(*a*) Now $f'(x) = \sec^2 x$ and $f(x) = \tan x$ have the same natural domain, namely, all x for which $\cos x \neq 0$ [$\sec x = 1/(\cos x)$ and $\tan x = (\sin x)/(\cos x)$]. But since $f'(x) = \sec^2 x$ is positive for all such x, $\tan x$ is increasing on its natural domain.

(*b*) Since $$f''(x) = 2\sec x(\sec x \tan x)$$
$$= 2\sec^2 x \tan x$$
and $2\sec^2 x > 0$, we know that $f(x)$ and $f''(x)$ have the same sign on their (common) natural domain. In particular, assertion (*b*) follows.

(*c*) Where $f(x) = \tan x$ crosses the x axis, it, and hence its second derivative, changes sign, from which assertion (*c*) follows. ●

The successively higher derivatives of certain trigonometric functions tend to repeat or change in a systematic way (see Progress Test 2, Prob. 8 and section exercises 49 to 52).

PROGRESS TEST 2

In Probs. 1 to 5, differentiate the given function. (It is not necessary to simplify your answers.)

1. $y = \csc[(x-4)^5]$

2. $f(x) = \tan^3(3x^7 + 2x^4 + 3x - 1)$

3. $h(x) = \sqrt{\cot(x^3 + 2x)}$

4. $f(x) = \sec(3x+1)\cdot\tan(1-x)$

5. $g(y) = \sec^5[(3y^2 - y)^4]$

6. Prove that $D_x[\cot x] = -\csc^2 x$.

7. Prove that $D_x[\csc x] = -\csc x \cot x$.

8. Describe the 100th derivative of $f(x) = \sin ax$.

MORE LIMITS

Variations of the following limit (established 7.2.1) occur frequently

7.2.10
$$\lim_{t\to 0} \frac{\sin t}{t} = 1$$

Example 5 Evaluate $\lim_{x\to 0} \dfrac{\sin 4x}{x}$.

Solution We wish to use the above limit, so we multiply and divide by 4, getting

$$\lim_{x\to 0} \frac{4\sin 4x}{4x} = 4\lim_{x\to 0}\frac{\sin 4x}{4x}$$

But now letting $t = 4x$, $t \to 0$ as $x \to 0$, so

$$4\lim_{x\to 0}\frac{\sin 4x}{4x} = 4\cdot\lim_{t\to 0}\frac{\sin t}{t}$$
$$= 4\cdot 1 = 4 \quad \bullet$$

This example obviously generalizes to:

7.2.11 For any constant k, $\lim_{x\to 0} \dfrac{\sin kx}{x} = k$.

RADIANS VERSUS DEGREES

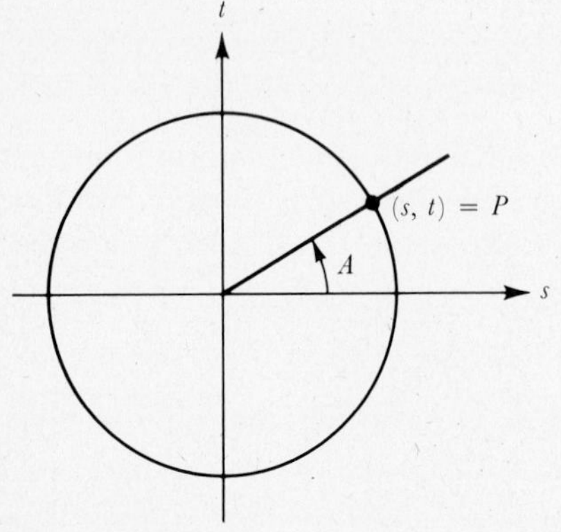

Fig. 7.20

Suppose that at the beginning we had decided to use degree instead of radian measure of angles. An angle A, which yields a point $P = (s, t)$ on the defining circle, has a larger degree measure than radian measure. Nevertheless, we would want the sine of the given angle A to be fixed—namely, as t. For example, suppose A is the angle whose size in radians is $\pi/6$ and whose size in degrees is 30. (See Fig. 7.20.) In terms of radians, we already have $\sin \pi/6 = 0.5$. On the other hand, in terms of degrees we want the sine of 30 to be 0.5 also. Hence we can define a new function $\sin_{\text{deg}}$ with $\sin_{\text{deg}}(30) = 0.5$. This new function $\sin_{\text{deg}}(x)$ has period 360 with, for example,

$$\sin_{\text{deg}}(0) = 0 \qquad \sin_{\text{deg}}(30) = 0.5 \qquad \sin_{\text{deg}}(90) = 1$$

Our new function $\sin_{\text{deg}}$, although defined for numbers (and not angles), accounts for the different way of measuring angles. This is described schematically by the dashed arrow in Fig. 7.21, with the conversion formula connecting the two measures.

The graph of $y = \sin_{\text{deg}}(x)$ in Fig. 7.22, even with the horizontal scale squeezed in drastically, is much "flatter" than our previously defined $\sin x$.

We can translate the degree measure x of an angle into radians, getting $(\pi/180)x$, so that

$$\sin_{\text{deg}}(x) = \sin\left(\frac{\pi}{180}x\right)$$

This can also be seen directly by recognizing $\sin_{\text{deg}}$ to be the composition of sin with $(\pi/180)x$ (see Fig. 7.21). By the Chain Rule,

$$D_x[\sin_{\text{deg}}(x)] = D_x\left[\sin\left(\frac{\pi}{180}x\right)\right] = \cos\left(\frac{\pi}{180}x\right)\left(\frac{\pi}{180}\right)$$
$$= \cos_{\text{deg}}(x)\left(\frac{\pi}{180}\right)$$

In terms of slope this shows again that the graph of $\sin_{\text{deg}}(x)$ should be much flatter than that of the radian-based sine function. Its slope never exceeds $\pi/180 \approx 0.0175$.

These same arguments apply to all the other trigonometric functions and should help explain why we said early in Sec. 7.1 that "radian measure is the most practical measure for the calculus." We do not need to worry about that $\pi/180$ factor every time we take a derivative.

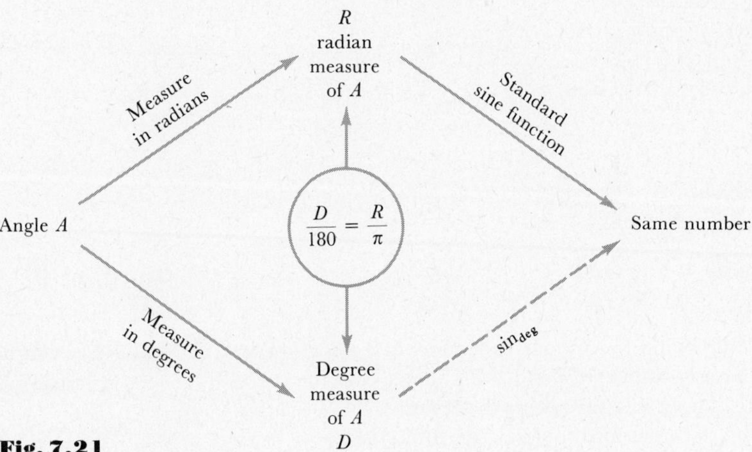

Fig. 7.21

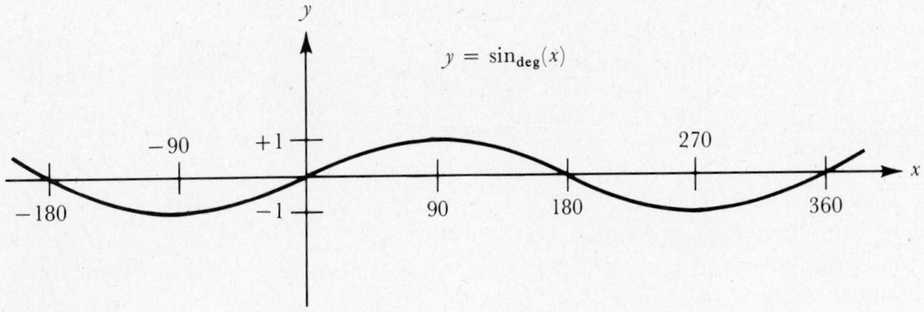

Fig. 7.22

SECTION 7.2 EXERCISES

In Exercises 1 to 30, differentiate the indicated function:

1. $f(x) = \cos(x - 2)$
2. $f(x) = \sin[(x^3 - 3x)^5]$
3. $g(x) = \tan(2x - 1)$
4. $m(x) = \csc(1 - x)$
5. $y = \cot^2(3x - 2)$
6. $y = m \sin^2 x + b$
7. $f(t) = \sin t + \cos t$
8. $h(\theta) = A \sin(a\theta) + B \cos(b\theta)$
9. $f(x) = \sin x \cot^2(3x)$
10. $f(x) = \sin^3(\cos(x))$
11. $f(x) = \sqrt{\dfrac{1 - \cot x}{x}}$
12. $r(t) = \pi(\sin^2 t)(t^3 + 1)$
13. $f(\theta) = a \sin \theta\, b \cos \theta$
14. $g(t) = \sec t \tan t$
15. $f(x) = \sec^2 x$
16. $y = \dfrac{\sqrt{\sec x}}{x}$
17. $u(x) = \tan^2(\sqrt{x+1}) - \sec^2(\sqrt{x+1})$ (simplify)
18. $v(x) = \dfrac{\sin(11x^2 + 3x + 4)}{\cos(11x^2 + 3x + 4)}$ (simplify)
19. $f(x) = \sin(\cos 3x)$
20. $f(x) = \cos(\sin 3x)$
21. $y = \tan^5(x^2 + 1)$
22. $y = \cot^5(x^2 + 1)\tan^5(x^2 + 1)$ (simplify)
23. $g(x) = \sqrt{\sin 4x + 1}\,\sqrt[3]{\csc 3x}$
24. $y = (11x + \csc x)^4$
25. $g(x) = \dfrac{\sin^5(3x + 2)}{\tan(2x - 1)}$
26. $f(x) = (x + \sin^2 3x)(1 - \cos^2 5x)$
27. $l(x) = \dfrac{\sqrt{\tan x}}{\csc^2(4x - 1)}$
28. $f(x) = \left(\dfrac{1}{x}\right) \sin x$
29. $f(x) = x \sin\left(\dfrac{1}{x}\right)$
30. $y = \csc^7(\cot(x^2))$

In Exercises 31 to 34 differentiate the given function with respect to t. Assume that x, y, θ are all functions of t.

31. $f(t) = \cos x + 2t \sin t$
32. $y = \sqrt{x^2 + \theta^2} + \tan x$
33. $f(t) = x \sin \theta$
34. $y = x \cos\theta + y \sin \theta$

In Exercises 35 to 40 determine dy/dx using implicit differentiation.

35. $\sin^2 x + \cos^2 y = 1$
36. $2 \sin x \cos x = y$
37. $\sqrt{\sin 2x + 4} = \tan y$
38. $\sin(x + y) = \cos(x - y)$
39. $\cos(x - y) = x - y$
40. $\tan(x + y) = x + y$

41. By employing the trigonometric identity $\cos(a + b) = \cos a \cos b - \sin a \sin b$ and (7.2.1) prove that the derivative of $\cos x$ is $-\sin x$ using the definition of derivative.

42. Give a formula for the derivative of $f(x) = \tan^n x$ that is entirely in terms of $\tan x$ (n an integer).

43. Give a formula for the derivative of $f(x) = \sec^n x$ (n an integer).

44. What is the equation of the line tangent to $f(x) = \tan(ax)$ at $x = 0$?

45. Given $\cos(x - y) = 1$;
(*a*) Find dy/dx implicitly if it exists (be careful!).
(*b*) Interpret (*a*) and give a sketch of the equation $\cos(x - y) = 1$ anyway. (Note that there is more than one line in the graph.)

46. Prove the standard identity $\sin^2 x + \cos^2 x = 1$ using that a function with zero derivative everywhere must be a constant function.

47. Repeat Exercise 46 for $\tan^2 x + 1 = \sec^2 x$.

48. Repeat Exercise 46 for $\cot^2 x + 1 = \csc^2 x$.

In Exercises 49 to 52 determine the given derivative.

49. $f^{(100)}(x)$, where $f(x) = A \cos x$
50. $f^{(4)}(x)$, where $f(x) = \cos x + \sin x$
51. $f^{(8)}(x)$, where $f(x) = A \sin ax$
52. $d^{15}y/dx^{15}$, where $y = \cos 3x$

53. Use the first and second derivatives to describe the graph of $y = \sec x$, for (*a*) $-\pi/2 < x < \pi/2$ and (*b*) $\pi/2 < x < 3\pi/2$.

54. Repeat Exercise 53 for $y = \csc x$, for (*a*) $0 < x < \pi$ and (*b*) $\pi < x < 2\pi$.

In Exercises 55 to 59, evaluate the given limits:

55. $\lim_{x \to 0} x \sec x$

56. $\lim_{\Delta x \to 0} \frac{\tan \Delta x}{\Delta x}$

57. $\lim_{x \to 0} \frac{\cos 5x}{\cot 3x}$

58. $\lim_{x \to 0} \frac{x^2 + x}{\sin x}$

59. $\lim_{x \to 0} \frac{x \sin x}{1 - \cos x}$

60. Calculate the derivative of $\tan_{\text{deg}}(x)$ using the Chain Rule, expressing the result in terms of "degree functions."

61. Repeat Exercise 60 for $\cot_{\text{deg}}(x)$.

62. By employing an analysis analogous to that involved with Figs. 7.16 and 7.17, show that it is reasonable that $D_x(\cos x) = -\sin x$.

63. (*a*) Using the degree-based approach, show that the crucial limit necessary to compute $D_x[\sin_{\text{deg}}(x)]$ from the definition

$$\lim_{\Delta x \to 0} \frac{[\sin_{\text{deg}}(\Delta x)]}{(\Delta x)} \text{ equals } \lim_{\Delta x \to 0} \frac{\left[\sin\left(\frac{\pi}{180}\Delta x\right)\right]}{(\Delta x)}.$$

(*b*) Evaluate the limit in (*a*) using (7.2.11).

64. Modify the geometric argument used to prove [7.2.1(*a*)] to evaluate the limit in Exercise 63(*a*). (The area of a sector changes.)

Section 7.3: Inverse Functions and Exponential Logarithm Functions

Objectives:

1. Determine inverses of simple functions.
2. Compare graphs of functions and their respective inverses.
3. Graph exponential and logarithm functions.
4. Use the properties of exponential and logarithm functions to manipulate expressions involving these functions.

INVERSES OF FUNCTIONS

The graph of a function that is either strictly increasing or strictly decreasing on I has the property that a horizontal line will hit the graph in at most one point. (See Fig. 7.23.) For such functions, when we are given a value of the function, that is, a number $\bar{x}$ on the vertical axis, we can tell *exactly* which x satisfies $f(x) = \bar{x}$. For the function $f(x) = x^3$, which is increasing on $[-10, 10]$, given a number, say $\bar{x} = 64$, we know exactly which x satisfies

$$f(x) = x^3 = 64$$

namely, $x = 4$. This contrasts sharply with the function $g(x) = x^2$, which changes direction on $[-10, 10]$. Given the number 64, we *cannot* tell which x satisfies

$$g(x) = x^2 = 64$$

because it could be either $x = 8$ or $x = -8$. In the case of $f(x) = x^3$, we can "undo" f by taking the cube root and get a unique answer. In the case of $g(x) = x^2$, there are two possible square roots and no unique answer.

7.3.1 Definition Suppose the function f has a given domain and range. We say f *has an inverse,* denoted f^{-1}, if, for every $\bar{x}$ in the range of f, there is a unique x in the domain of f with $f(x) = \bar{x}$. In this case f^{-1} is a function defined by:

$$f^{-1}(\bar{x}) = x \qquad \text{where } x \text{ is the unique number with } f(x) = \bar{x}$$

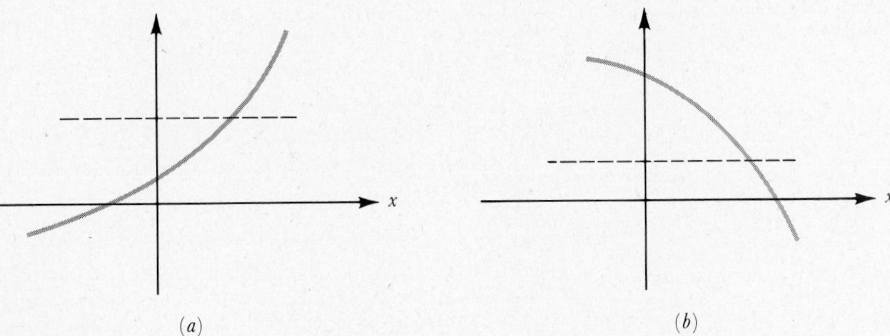

Fig. 7.23

The domain of f^{-1} is the range of f, and the range of f^{-1} is the domain of f. We will consistently use the notation f^{-1} (pronounced "f inverse") for the inverse of f, and so f^{-1} will *not* mean $1/f$.

The following theorem (whose proof is outlined in Exercise 34) gives us conditions on f which guarantee that f^{-1} exists.

7.3.2 **Theorem (Existence of Inverses)** If f is continuous and is either strictly increasing or strictly decreasing on an interval I, then f has an inverse f^{-1} (whose range is I).

Example 1 Given $f(x) = x^3$, determine f^{-1}.

Solution We know f is continuous and increasing on $(-\infty, +\infty)$, so f^{-1} exists. By the definition, $f^{-1}(\bar{x})$ is that unique number for which $f(x) = \bar{x}$; that is, $f^{-1}(\bar{x})$ is that unique number which when cubed is $\bar{x}$. But of course that unique number is $\sqrt[3]{\bar{x}}$, so

$$f^{-1}(\bar{x}) = \sqrt[3]{\bar{x}} \quad \bullet$$

We used $\bar{x}$ in defining f^{-1} both in (7.3.1) and in the example to distinguish it from the variable x that was used to denote numbers in the domain of f. The letter we use as the variable makes no difference since a function is known by what it does. Thus we shall often use an $\bar{x}$ in the construction of f^{-1} from f, but then relabel the resulting function f^{-1} using the more customary x. Hence for $f(x) = x^3$ we can write

$$f^{-1}(x) = \sqrt[3]{x}$$

Example 2 Show that $f(x) = 3x + 2$ has an inverse on R and determine f^{-1}.

Solution $f(x) = 3x + 2$, being a nonconstant linear function, is continuous and strictly increasing on its domain R. Given an $\bar{x}$ in the range, we can algebraically compute the unique x with $f(x) = \bar{x}$ by setting $3x + 2 = \bar{x}$ and solving for x:

$$x = \frac{\bar{x} - 2}{3}$$

Thus $f^{-1}(\bar{x}) = (\bar{x} - 2)/3$, again a linear function, with slope $\frac{1}{3}$ and intercept $(0, -\frac{2}{3})$. Finally, relabeling, we have

$$f^{-1}(x) = \frac{x - 2}{3} \quad \bullet$$

It is useful to think of the inverse function f^{-1} as "undoing" what the original function $f(x)$ "did." The above example, $f(x) = 3x + 2$, triples x and then adds 2, so to undo what $f(x)$ did to x, common sense tells us first to subtract the 2, getting $3x$, and then to divide by 3, getting back to x. (A bit like unwrapping a package—work from the outside in.) But of course, *this is exactly what f^{-1} does:* it subtracts 2 and then divides by 3.

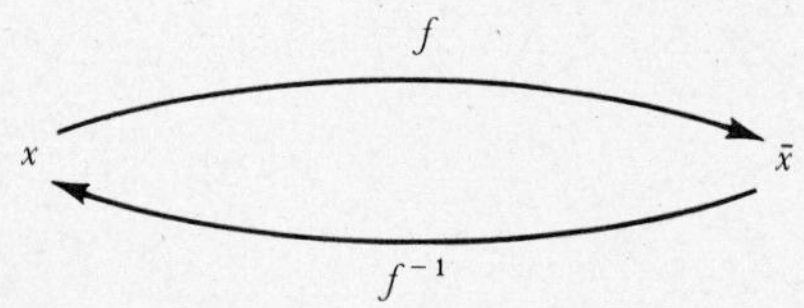

Fig. 7.24

In general, the composition of f^{-1} with f should leave x unchanged. In terms of Fig. 7.24, x gets taken on a round trip back to x.

Algebraically,

$$(f^{-1} \circ f)(x) = f^{-1}(f(x)) = x$$

for each x in the domain of f; and similarly,

$$(f \circ f^{-1})(x) = f(f^{-1}(x)) = x$$

for each x in the domain of f^{-1} (the range of f).

In the special case of $f(x) = 3x + 2$, $f^{-1}(x) = (x - 2)/3$, we get

$$\begin{aligned}(f^{-1} \circ f)(x) &= f^{-1}(f(x)) = f^{-1}(3x + 2)\\ &= \frac{(3x + 2) - 2}{3} = \frac{3x}{3} = x\end{aligned}$$

and

$$\begin{aligned}(f \circ f^{-1})(x) &= f(f^{-1}(x)) = f\left(\frac{x - 2}{3}\right)\\ &= 3\left(\frac{x - 2}{3}\right) + 2 = x - 2 + 2 = x\end{aligned}$$

for all real numbers x.

Example 3 Show that for $a \neq 0$,

$$f(x) = ax + b \qquad (a \neq 0) \qquad \text{implies} \qquad f^{-1}(x) = \frac{x - b}{a}$$

Solution To undo f, we first subtract b, leaving ax, and then divide by a, so $f^{-1}(\bar{x}) = (\bar{x} - b)/a$. We can accomplish this construction algebraically by setting $ax + b = \bar{x}$ and solving for x to get

$$x = \frac{\bar{x} - b}{a}$$

Hence $f^{-1}(\bar{x}) = (\bar{x} - b)/a$, or, using x as the independent variable,

$$f^{-1}(x) = \frac{x - b}{a}$$

Note that this is a linear function of slope $1/a$ through the point $(0, -b/a)$. ●

Here are three simple applications of Example 3 (for all x in R):

(*a*) Given $f(x) = 2x$, $(a = 2, b = 0)$, $f^{-1}(x) = x/2$.

(*b*) Given $f(x) = x + 4$, $(a = 1, b = 4)$, $f^{-1}(x) = (x - 4)/1 = x - 4$.

(*c*) Given $f(x) = -x$, $(a = -1, b = 0)$, $f^{-1}(x) = x/(-1) = -x$.

This last example is its own inverse! But this makes sense: How would you undo a function that reverses the sign of x? You reverse it *again*, getting $-(-x) = x$.

GRAPHS AND DERIVATIVES OF INVERSE FUNCTIONS

The relationship between the graph of f and the graph of f^{-1} both on the same coordinate system is very revealing. In Fig. 7.25 we sketch the examples considered so far, including the diagonal line $y = x$ (dashed). In Fig. 7.25(*a*) we include a pair of tangent lines for future reference.

That all this symmetry about the dashed line $y = x$ is no accident can be seen from the following observation and Fig. 7.26. *The set of ordered pairs determined by* f^{-1} *are precisely those determined by f but with their order reversed.* The diagonal line acts like a mirror with the point (r, s) being "reflected" across the diagonal to the point (s, r). Of course, a point such as $(2, 2)$ *on* the diagonal is its own reflection.

Notice that in Fig. 7.25(e) the graph of the function that was its own inverse, $f(x) = -x = f^{-1}(x)$, is perpendicular to the diagonal; hence it is its own reflection across the diagonal. On the other hand, $f(x) = x + 4$ and $f^{-1}(x) = x - 4$, having the same slope, are parallel, equidistant from the diagonal, as shown in Fig. 2.25(d).

The fact that *the graphs of f and f^{-1} are symmetric across the diagonal* enables us to provide a convincing, but informal, argument for the following important theorem. We leave its formal proof to more advanced courses.

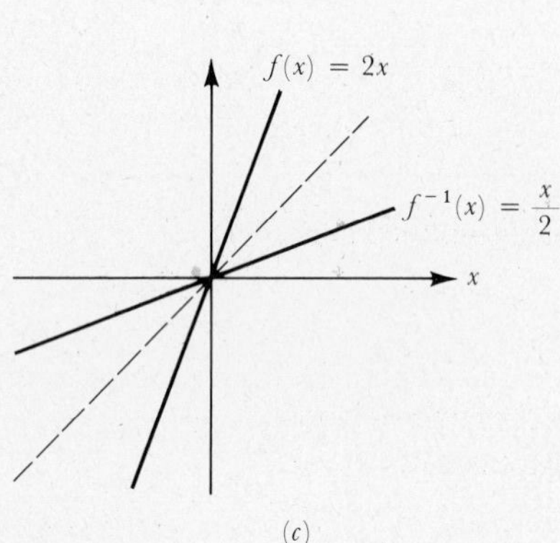

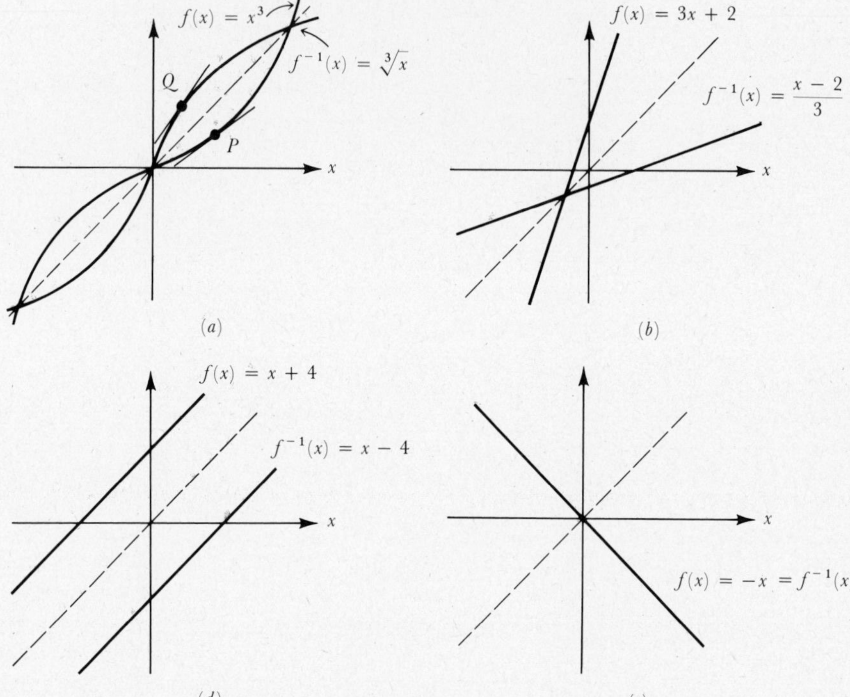

Fig. 7.25

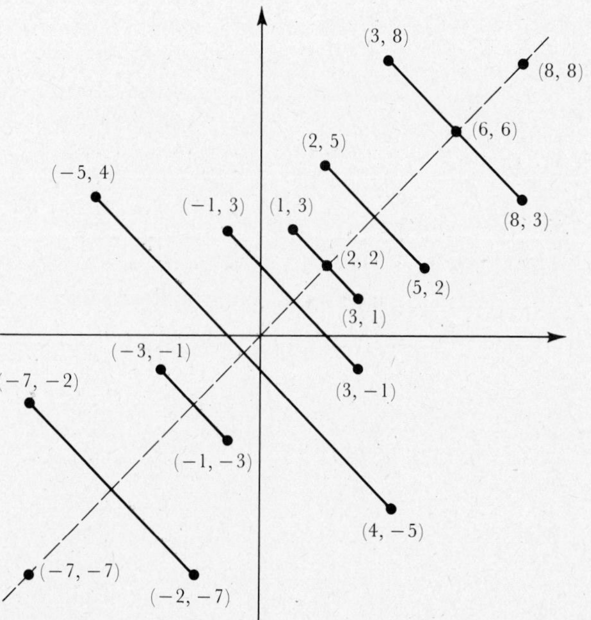

Fig. 7.26

7.3.3 Theorem (Derivatives of Inverse Functions) If f is differentiable at x with $df/dx = a \neq 0$, then df^{-1}/dx exists at $f(x)$ and equals $1/a$.

If the graph of f has a *nonhorizontal* tangent at $P = (x, f(x))$, then the graph of f^{-1} has a *nonvertical* tangent at $Q = (f(x), f^{-1}(f(x))) = (f(x), x)$. [See Fig. 7.25($a$), for example.] Furthermore, if the tangent to f has slope $a \neq 0$, then Example 3 guarantees that the corresponding tangent to f^{-1} has slope $1/a$ since the reflection across the diagonal of the tangent to f at P is the tangent to f^{-1} at Q.

The importance of a *nonhorizontal* tangent to f can be seen graphically in Fig. 7.25(a) by examining the graphs of $f(x) = x^3$ and $f^{-1}(x) = \sqrt[3]{x}$ at their common point (0, 0). The tangent to f is horizontal (slope = 0), and the tangent to f^{-1} is vertical (undefined slope). Algebraically, $D_x[f^{-1}(x)] = D_x(\sqrt[3]{x}) = (1/3)x^{-2/3} = 1/(3\sqrt[3]{x^2})$ obviously is undefined at $x = 0$.

PROGRESS TEST 1

In Probs. 1 and 2 (a) determine f^{-1}, (b) compute $f \circ f^{-1}$ and $f^{-1} \circ f$, and (c) sketch f and f^{-1} on the same coordinate axes, using a dashed line to represent $y = x$.

1. $f(x) = 2x - 1$

2. $f(x) = x^3 - 1$

3. Given $f(x) = x^3 - 1$, what are the coordinates of the point on the graph of $f^{-1}(x)$ where the slope of the tangent line is undefined?

EXPONENTIAL AND LOGARITHM FUNCTIONS

Many phenomena, especially those involving natural growth or decay, are best described by what are called "exponential" functions.

Suppose we monitor the population of an idealized bacteria culture (no deaths, food or oxygen shortages, and so on) in which the bacteria reproduce once every minute. Begin at time $x = 0$ with 1 bacterium. After 1 min it divides in two, yielding 2, each of which divides when $x = 2$ to yield a total of 4, each of which divides when $x = 3$ to yield 8, and so on, yielding successively, 16 at $x = 4$, 32 at $x = 5$, 64 at $x = 6, \ldots$. The number of bacteria present at time x is

$$\underbrace{2 \cdot 2 \cdot \cdots \cdot 2}_{x \text{ repetitions}} = 2^x$$

The function that gives the population as a function of time x is therefore

$$r(x) = 2^x \qquad (x \text{ an integer})$$

Similarly, a quantity that tripled in each unit of time would yield the amounts (beginning with 1 at $x = 0$): 3 at $x = 1$, 9 at $x = 2$, 27 at $x = 3$, 81 at $x = 4, \ldots$. The amount present at time x is

$$g(x) = 3^x$$

On the other hand, if a quantity were cut in half for each unit of time x (with 1 unit of this quantity at the beginning point, $x = 0$), at $x = 1$, $\frac{1}{2}$ unit remains, at $x = 2$ one half of *this* remains, that is, $\frac{1}{4}$ remains; at $x = 3$, $\frac{1}{2} \cdot \frac{1}{4} = \frac{1}{8}$ remains. At time x, the amount remaining is

$$s(x) = \left(\frac{1}{2}\right)^x$$

In general, we shall be dealing with functions of the form

$$f(x) = b^x \qquad \text{where } b > 0$$

Note that for $b < 0$, b^x is defined only for some x and not others, and generally is a colossal mess. In fact, we shall confine our attention mainly to the case $b > 1$, because for $b = 1$, 1^x is a constant function; and for $0 < b < 1$, we know $1 < 1/b$. By (7.3.6), $(1/b)^x = (b^{-1})^x = b^{-x}$, so functions such as $(1/2)^x$ can be handled in terms of 2^{-x}.

7.3.4 **Definition** A function of the form $f(x) = b^x$, where $b > 0$, is called an *exponential function of base b*. More specifically, we make the following definitions:

(*a*) For $x = m$ a positive integer, we define $f(m) = b^m$ to be the product of b with itself m times.

(*b*) For $x = 0$, $f(0) = b^0 = 1$.

(*c*) For $x = m$ a negative integer, $f(m) = b^m = 1/b^{-m}$ (where $-m > 0$, of course).

(*d*) For rational $x = m/n$ (m, n integers, $n \neq 0$), $f(m/n) = b^{m/n} = \sqrt[n]{b^m} = (\sqrt[n]{b})^m$.

The problem of defining b^x for irrational x is somewhat more subtle, involving the use of limits. We shall illustrate the principle involved by examining the case $10^{\sqrt{2}}$ in detail.

The number $\sqrt{2}$, being irrational, has an infinite, nonrepeating decimal representation, which to nine places is

$$\sqrt{2} \approx 1.414213562$$

It is important to recognize that the various decimal approximations of $\sqrt{2}$ are rational numbers: for example, $1.414 = 1 + \frac{414}{1000} = \frac{1414}{1000}$, a quotient of integers. Thus $10^{1.414}$ is already defined. Furthermore, the sequence of decimals

$$(1) \qquad 1.4,\ 1.41,\ 1.414,\ 1.4142,\ 1.41421, \ldots$$

provides an increasingly accurate approximation of $\sqrt{2}$. Now, using the sequence (1), which has $\sqrt{2}$ as its limit, we can obtain the sequence

$$(2) \qquad 10^{1.4},\ 10^{1.41},\ 10^{1.414},\ 10^{1.4142},\ 10^{1.41421}, \ldots$$

each term of which is 10 raised to a rational power and hence is defined in [7.3.4(*d*)]. We *define* $10^{\sqrt{2}}$ to be the number approximated by this sequence. In effect, we know $10^{\sqrt{2}}$ as well as we know $\sqrt{2}$. Furthermore, when $10^{\sqrt{2}}$ is defined in this way, we are guaranteed that as x approaches $\sqrt{2}$, $h(x) = 10^x$ approaches $10^{\sqrt{2}}$, that is,

$$\lim_{x \to \sqrt{2}} h(x) = h(\sqrt{2})$$

Hence $h(x)$ is not only defined at $\sqrt{2}$, it is continuous at $\sqrt{2}$.

The above very informal argument applies to any other irrational x as well as to any other base $b > 0$. It certainly supports the following two basic assumptions:

7.3.5 (*a*) Exponential functions have domain R.

(*b*) Exponential functions are continuous on R.

In addition, we make the following two assumptions familiar from high school algebra:

7.3.6 (*a*) $(b^{x_1})(b^{x_2}) = b^{x_1+x_2}$.

(*b*) $(b^{x_1})^{x_2} = b^{x_1 x_2}$.

In Fig. 7.27 we graphed several exponential functions that will guide us in formulating further properties of the exponential functions.

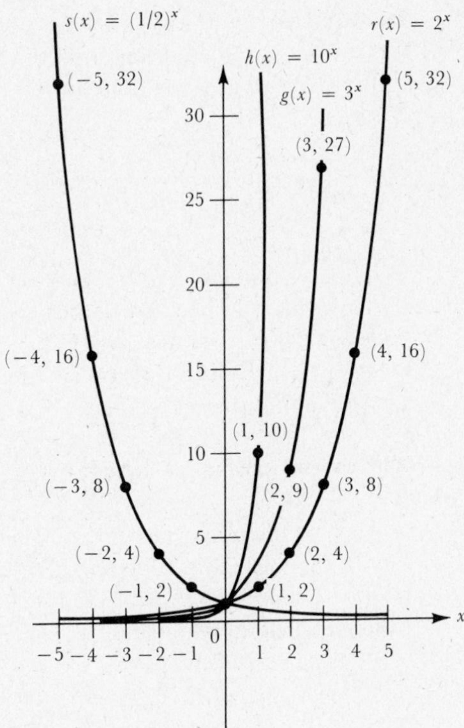

Fig. 7.27

The functions graphed suggest the last two assumptions we shall need for further development (we assume $b \neq 1$):

7.3.7 (*a*) Each exponential function is either strictly increasing ($b > 1$) or strictly decreasing ($b < 1$) with domain R and range $(0, +\infty)$.

(*b*) The larger the base $b > 1$, the steeper the slope of the graph for $x \geq 0$.

As we shall illustrate later in the appendix to Sec. 7.4, these assumptions are used only to provide a convenient starting place that takes advantage of your experience with exponential and logarithm manipulations.

By [7.3.7(*a*)] and (7.3.2) we know the following:

7.3.8 **Theorem** Every exponential function of base $b \neq 1$ has an inverse whose domain is $(0, +\infty)$ and whose range is R.

Consider the function $r(x) = 2^x$, for example. Its inverse is defined as follows:

$$r^{-1}(\overline{x}) \text{ is the unique number } x \text{ with } r(x) = 2^x = \overline{x}$$

Thus

$r^{-1}(8)$ is the unique number x with $2^x = 8$—that is, $r^{-1}(8) = 3$ since $2^3 = 8$

$r^{-1}(4) = 2$, since $2^2 = 4$

$r^{-1}(32) = 5$, since $2^5 = 32$

$r^{-1}\left(\frac{1}{16}\right) = -4$ since $r(-4) = 2^{-4} = \frac{1}{2^4} = \frac{1}{16}$

As you likely know, there is a traditional name for r^{-1}, namely $\log_2$, the "logarithm base 2." More generally, we have:

7.3.9 **Definition** For any base $b > 0$ $(b \neq 1)$ the inverse of the exponential function b^x is denoted $\log_b x$, and is referred to as the *logarithm base b*.

By (7.3.8) $\log_b x$ has domain $(0, +\infty)$ and range R. Also by definition,

$$\log_b(x) \text{ is that unique number } y \text{ with } b^y = x$$

It is convenient to determine $\log_b$ of a number x by answering the question:

"b to what power gives x?"

Thus we find $\log_{10}(100)$ by asking "10 to what power gives 100?" Of course the answer is 2, so $\log_{10}(100) = 2$. Similarly, $\log_{10}(0.0001) = -4$ since $10^{-4} = 0.0001$. We know $\log_b(b^x) = x$ because the answer to "b to what power gives b^x" is obviously x. (Who is buried in Grant's Tomb?) This is just another way of saying that the logarithm and exponential functions are inverses.

7.3.10 **Theorem** For all x, $\log_b(b^x) = x$; and for all $x > 0$, $b^{(\log_b x)} = x$.

Each of the properties of exponential functions implies a companion property of the logarithm functions. For example $\log_b(1) = 0$ because $b^0 = 1$, and $\log_b(b) = 1$ because $b^1 = b$. We list some algebraic properties of the logarithm functions below. We assume x, x_1, x_2 are positive.

7.3.11 (*a*) $\log_b(1) = 0$ and $\log_b(b) = 1$
(*b*) $\log_b(x_1 \cdot x_2) = \log_b(x_1) + \log_b(x_2)$
(*c*) $\log_b(x^p) = p \log_b(x)$
(*d*) $\log_b\left(\dfrac{x_1}{x_2}\right) = \log_b(x_1) - \log_b(x_2)$

These properties are useful in simplifying more complicated expressions.

Example 4 Evaluate $[\log_2(4)]^3 + \log_2(8^3) + \log_2\left(\dfrac{2^{0.1}}{2^5}\right)$.

Solution We examine each term separately:

$$[\log_2(4)]^3 = 2^3 = 8, \ \log_2(8^3) = 3\log_2 8 = 3 \cdot 3 = 9, \text{ and}$$

$$\log_2\left(\frac{2^{0.1}}{2^5}\right) = \log_2(2^{0.1}) - \log_2(2^5) = 0.1 - 5 = -4.9$$

Hence the value of the original expression is

$$8 + 9 - 4.9 = 12.1 \quad \bullet$$

Example 5 Solve for x: $\log_3(x^2) + \log_3(\sqrt{x}) = 6$.

Solution We combine terms on the left using [7.3.11(*b*)] to get

$$\log_3(x^2 \cdot \sqrt{x}) = \log_3(x^{5/2}) = 6$$

Hence $x^{5/2} = 3^6$, by the definition of $\log_3$.
Squaring both sides, $x^5 = 3^{12}$, so

$$x = 3^{12/5} \approx 13.97 \quad \bullet$$

We can use the graphical symmetry of functions and their inverses to help graph the logarithm functions. We shall continue to concentrate on the case

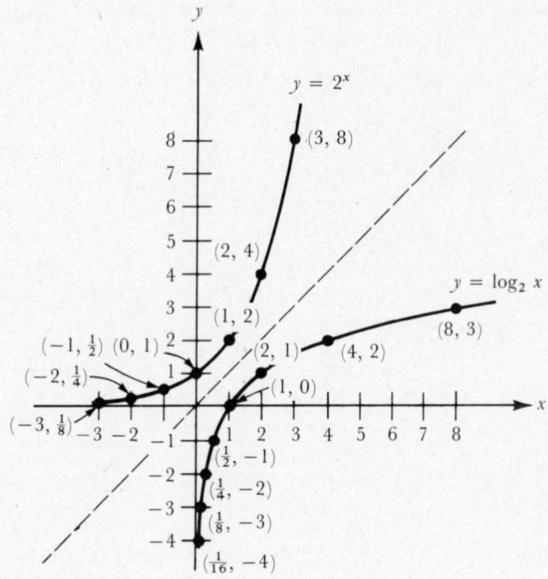

Fig. 7.28

$b > 1$. For base $b > 1$, all the exponential functions have pretty much the same shape, varying only in steepness, so we will graph $y = \log_2 x$ (see Fig. 7.28) knowing that the logarithm functions of other bases $b > 1$ look very similar.

The domain of $\log_2(x)$, being $(0, +\infty)$, means that its graph lies entirely to the right of the vertical axis. Its range is all of R, even though the curve appears to be leveling off as we move to the right. (The domain of 2^x is R.)

In general, $y = \log_b x$ *has no horizontal asymptotes* because, given any y, there exists some $x > 0$ such that $\log_b x = y$. Furthermore, *for* $x \leq 0$, $\log_b x$ *is not defined*. To ask for $\log_b(-3)$ is to ask "b to what power gives -3?" *No* power of $b > 0$ can yield a negative number.

PROGRESS TEST 2

1. Evaluate:
(*a*) $\log_3(9) - \log_3(1/3)^5$ (*b*) $\log_2(1) + \log_3(1)$
(*c*) $\log_2(4^5)$

2. Solve for x:
(*a*) $\log_5(x - 1) + \log_5(x + 1) = 0$
(*b*) $3^{x+1} = 81$
(*c*) $2 \log_{10} x - \log_{10} 1/x = 3$

3. Graph $y = \log_3 x$ and $y = 3^x$ on the same axes.

4. Find $f \circ g$ and simplify to remove fractional exponents wherever possible:

(*a*) $f(x) = \log_b x, g(x) = \dfrac{\sqrt{x^2 + x + 4}(3x + 1)}{(x^2 + 1)^6}$

(*b*) $f(x) = b^x, g(x) = \log_b(4x + 5)$

5. Determine $\log_b x$ in each case:
(*a*) $\log_2(1/8)$ (*b*) $\log_2 64$
(*c*) $\log_2\left(\dfrac{1}{32}\right)$ (*d*) $\log_2 1$
(*e*) $\log_3 27$ (*f*) $\log_5 25$
(*g*) $\log_5(\sqrt{5})$ (*h*) $\log_5(1/5)$
(*i*) $\log_b(1)$ (*j*) $\log_2(8^{10})$
(*k*) $\log_b(b^2)$ (*l*) $\log_3(3^4)$

6. Simplify:
(*a*) $3^{[x \log_3(1+x) - 2x \log_3(1-x)]}$
(*b*) $\log_5(5^x) - \log_5(5^{-x})$
(*c*) $2^{[\log_2(x^2-4) + 3 \log_2(x-1)]}$

SECTION 7.3 EXERCISES

In Exercises 1 to 5 determine f^{-1} and then graph f and f^{-1} on the same coordinate axes. State the natural domain and the range of f and f^{-1}.

1. $f(x) = x - 1$

2. $f(x) = 3x^3$

3. $f(x) = \sqrt[5]{x+1}$

4. $f(x) = \dfrac{1}{x^3}$

5. $f(x) = \dfrac{3x+4}{2}$

In Exercises 6 to 17 evaluate the given numerical expression.

6. $\log_2(4 \cdot 16)$

7. $(\log_3 9)^3$

8. $\log_3(9^3)$

9. $\log_5(5^{-3})$

10. $\log_6 1$

11. $\log_7 49$

12. $\log_7 7$

13. $\log_3 \dfrac{1}{243}$

14. $\log_{10}(10^{10})$

15. $\log_2(256^5)$

16. $\log_2\left(\dfrac{1}{128}\right)$

17. $\log_{16} 2$

In Exercises 18 to 29 solve the given equation for the indicated variable using the definitions and properties of exponential and logarithm functions.

18. $2^{x+1} = 2^{3x-2}$

19. $3^{2x+1} = 3^{4x-5}$

20. $\log_3(8x - 1) = \log_3(3x + 4)$

21. $\log_5(4x + 5) = 2$

22. $4^{x-1} = 2^{3x+1}$

23. $4^{5x-2} = 64$

24. $\log_5(4x + 1) = \log_5(5x - 7)$

25. $\dfrac{1}{2}\log(2^{x-4}) = 0$

26. $\log_b 10 = \dfrac{1}{2}$

27. $\log_b 8 = 1$

28. $\dfrac{1}{2}(\log_b 8 - \log_b 128) = 1$

29. $2^{x^2-5x+6} = 1$

30. Write as a single term: $2 \log_5 4 - \dfrac{1}{2}\log_5 64 - \log_5 2$.

31. Write as a single term: $\log_{10}[\log_3(\log_5 125)]$.

32. Graph $y = \left(\dfrac{1}{2}\right)^x$ and $y = \log_{1/2} x$ in the same coordinate system.

33. Graph $y = \left(\dfrac{1}{3}\right)^x$ and $y = \log_{1/3} x$ in the same coordinate system.

34. Suppose f is continuous and strictly increasing on an interval I. Prove (7.3.2) that f has an inverse f^{-1} whose range is I by:

(*a*) Choosing an arbitrary $\bar{x}$ in the range of f and using the Intermediate-Value Theorem to show $f^{-1}(\bar{x})$ exists in I.

(*b*) Showing there cannot be two different numbers x_1 and x_2 in I with $f^{-1}(\bar{x}) = x_1$ and $f^{-1}(\bar{x}) = x^2$. (Use the strictly increasing hypothesis.)

35. Explain why the graphs of f and f^{-1} can cross *only* on the diagonal $y = x$.

Section 7.4: Derivatives of Exponential and Logarithm Functions

Objectives:

1. Differentiate exponential and logarithm functions of any base $b > 0$.

2. Differentiate combinations of these functions with other differentiable functions.

3. Understand the meaning and importance of the number *e* as a base for the natural exponential and logarithm functions.

DERIVATIVES OF EXPONENTIAL FUNCTIONS

We have all heard statements about the "exponential growth" of energy consumption, population, and what not, with the usual implication that the growth rate is not only large, but increasing. Our best tool for dealing with rates of change is the derivative, so we shall now calculate the derivative of an arbitrary exponential function $f(x) = b^x$ of base $b > 0$. For reasons to become apparent shortly, we first attempt to calculate the derivative at $x = 0$.

$$f'(0) = \lim_{h\to 0} \frac{f(0+h) - f(0)}{h}$$

$$= \lim_{h\to 0} \frac{b^{0+h} - b^0}{h}$$

$$= \lim_{h \to 0} \frac{b^h - 1}{h} \qquad \text{if this limit exists}$$

This, like the $(\sin \Delta x)/\Delta x$ limit, is not obvious. It is reasonable to assume that $f(x) = b^x$ has a tangent at $x = 0$. Thus we now assume that $f'(0)$(the previous limit) exists.

Computing $f'(x)$ at an *arbitrary* x, we have

$$\begin{aligned}
f'(x) &= \lim_{h \to 0} \frac{f(x+h) - f(x)}{h} \\
&= \lim_{h \to 0} \frac{b^{x+h} - b^x}{h} \\
&= \lim_{h \to 0} \frac{b^x b^h - b^x}{h} && \text{by [6.3.6}(a)] \\
&= \lim_{h \to 0} \frac{b^x(b^h - 1)}{h} \\
&= b^x \lim_{h \to 0} \frac{b^h - 1}{h} && \text{since } b^x \text{ does not involve } h \\
&= b^x f'(0) && \text{by our earlier calculation}
\end{aligned}$$

Thus we have:

7.4.1 Theorem If $f(x) = b^x$, then $f'(x) = b^x \cdot f'(0)$.

Geometrically, we have shown that *the derivative of a typical exponential function $f(x) = b^x$ at any x is the product of itself with the slope at $x = 0$.* Since all the exponential functions satisfy $b^0 = 1$ and hence pass through $(0, 1)$, and since the derivative of each depends upon its slope at $(0, 1)$, it is natural now to examine slopes of exponential functions at or near $x = 0$.

We saw that for most scientific purposes, the radian-based trigonometric functions are used because otherwise all derivatives involve a "correction factor" (which is $\pi/180$ if degree-based functions are used). The same is true here: The "correction factor" in this case is $f'(0)$, the slope at $x = 0$. Thus *the simplest derivative will be provided by that base for which $f'(0) = 1$.* Our task then is to determine which base b yields $f'(0) = 1$; that is,

for which base b does $f(x) = b^x$ have slope 1 at $x = 0$?

Recall our earlier statement [7.3.7(b)]:
(1) The larger the base $b > 1$, the steeper the graph of $f(x) = b^x$, $x \geq 0$. Also,
(2) if $f'(0)$ exists and is nonzero, then the graph of $f(x) = b^x$ is concave upward.

The second statement follows from (7.4.1) because $f''(x) = b^x[f'(0)]^2$. [Note $b^x > 0$ for all x so that $f''(x) > 0$.]

As we know, statement (2) means that as x increases, so does the slope. Our search for the "best base" begins with the slopes of

$$r(x) = 2^x \qquad \text{and} \qquad g(x) = 3^x \qquad \text{near } x = 0$$

At this point, we do not know the slopes of these functions at $x = 0$. However, since each is concave upward, the slope of the tangent T_2 to $r(x) = 2^x$ at $x = 0$ $[r'(0)]$ is *less* than the slope of the line L_2 through $(0, 1)$ and $R = (0.1, 2^{0.1})$; see Fig. 7.29(a). Similarly, the slope of the tangent T_3 to $g(x) = 3^x$ at $x = 0$ $[g'(0)]$ is *greater* than that of the line L_3 through $S = (-0.1, 3^{-0.1})$; see Fig. 7.29(b). Using a calculator, we find that the slope of L_2 is:

$$\frac{r(0 + 0.1) - r(0)}{0.1} = \frac{2^{0.1} - 1}{0.1} \approx 0.72$$

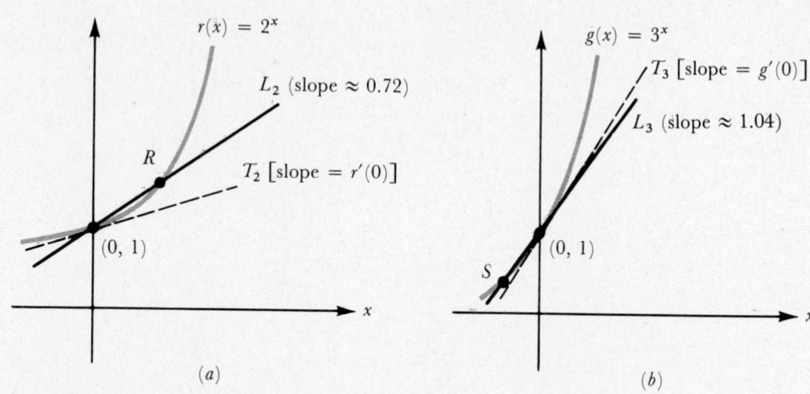

Fig. 7.29

On the other hand, the slope of L_3 is

$$\frac{g[0 + (-0.1)] - g(0)}{-0.1} = \frac{3^{-0.1} - 1}{-0.1} \approx 1.04$$

Thus we know:

$$r'(0) < 0.72 < 1 < 1.04 < g'(0)$$

Hence 3^x is too steep and 2^x is not steep enough at $x = 0$. *Whatever the "best base" is, it must be between* 2 *and* 3.

We now *assume* that such a base between 2 and 3 exists and make the following definition.

7.4.2 **Definition** We define the number e to be that number for which the slope of $f(x) = e^x$ is exactly 1 at $(0, 1)$, that is for which $f'(0) = 1$. The function $f(x) = e^x$ is often called the *natural exponential function* (see Fig. 7.30).

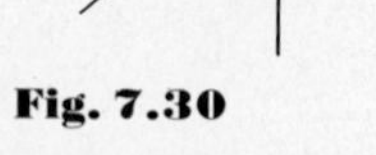

Fig. 7.30

This number is called e after Leonhard Euler, an eighteenth-century Swiss mathematician who first recognized and exploited its wide-ranging significance. It is one of the most important numbers in all mathematics. It is irrational and, to 15 places,

$$e \approx 2.718281828459045$$

We shall see it many times in the coming pages and later show also how it can be defined in terms of a limit (see p. 266).

Now, armed with definition (7.4.2) we recall from (7.4.1) that with $f(x) = e^x$:

$$f'(x) = e^x \lim_{h \to 0} \frac{e^h - e^0}{h} = e^x f'(0) = e^x \cdot 1 = e^x$$

Hence *the derivative of* $f(x) = e^x$ *is itself!* By the Chain Rule, if $u(x)$ is a differentiable function of x, we have

7.4.3 $$D_x[e^{u(x)}] = e^{u(x)} \cdot u'(x)$$

Example 1 Compute $D_x[e^{5x^2+3x}]$.

Solution

$$D_x[e^{5x^2+3x}] = e^{5x^2+3x} \cdot (10x + 3) \quad \bullet$$

Because the natural exponential function has such a simple derivative, we use it to complete the task we began earlier, finding the derivative of an *arbitrary* exponential function. To do this we need:

7.4.4 **Definition** The inverse of the natural exponential function e^x is called the *natural logarithm function* and is denoted $\ln x$ (read "ell n of x").

Hence we write $\ln(x)$ instead of $\log_e x$. This means

$$(3) \qquad \ln(e^x) = x \text{ for all } x \qquad \text{and} \qquad e^{\ln x} = x \text{ for positive } x$$

because e^x and $\ln x$ are inverses. As we know, the graph of the natural logarithm function is the mirror image of the graph of $y = e^x$ across the diagonal $y = x$. From Fig. 7.31 we can see that $\ln x < 0$ for $0 < x < 1$. Moreover,

$$\ln 1 = 0 \qquad \text{and} \qquad \ln e = 1$$

Also, since $y = e^x$ is asymptotic to the x axis, $y = \ln x$ is asymptotic to the y axis ($\lim_{x\to 0^+} \ln x = -\infty$).

Now, suppose we are given an arbitrary exponential function $f(x) = b^x$ for some base $b > 0$. Applying the right side of (3) with $x = b$, we have

$$b = e^{\ln b}$$

Raising each side to the xth power, we have $b^x = (e^{\ln b})^x$, or, by [7.3.6(b)],

7.4.5 $$b^x = e^{(\ln b)x}$$

This means that we can express *any* exponential function in terms of the *natural* exponential function. But then, knowing that b^x is merely e^{kx} for the constant $k = \ln b$, we can apply (7.4.3) to conclude, for $u(x)$ a differentiable function of x, that

$$D_x[b^{u(x)}] = D_x[e^{(\ln b)u(x)}] = e^{(\ln b)u(x)}(\ln b)u'(x) = b^{u(x)}u'(x)\ln b$$

or

7.4.6 $$D_x[b^{u(x)}] = b^{u(x)} \cdot u'(x)\ln b$$

Example 2 Compute $D_x[2^{x^3+5x-7}]$.

Solution $D_x[2^{x^3+5x-7}] = 2^{x^3+5x-7}(3x^2 + 5)\ln 2$ ●

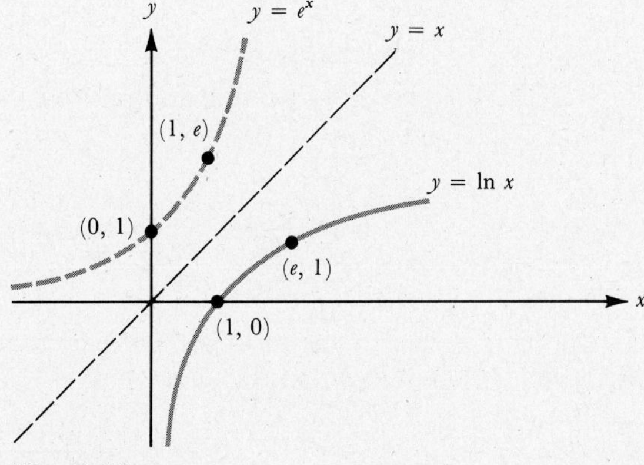

Fig. 7.31

As with the trigonometric functions, successively higher derivatives of exponential functions are often regular. For example, each higher derivative of $y = e^{3x}$ merely introduces a factor of 3:

$$\frac{dy}{dx} = 3e^{3x}, \frac{d^2y}{dx^2} = 3^2e^{3x}, \ldots, \frac{d^ny}{dx^n} = 3^ne^{3x} = 3^ny$$

PROGRESS TEST 1

1. Differentiate the following functions on their natural domains:

(*a*) $f(x) = e^{-x}$ (*b*) $f(x) = e^{2-x^2}$

(*c*) $g(x) = e^{x^2} \cdot (4x^3 + x)$ (*d*) $g(t) = 2^{t^3+2t}$

(*e*) $y = \pi^x$ (*f*) $y = x^\pi$

(*g*) $s(x) = \dfrac{e^x}{2^x}$ (simplify your answer)

(*h*) $t(x) = e^{x-(x \ln 2)}$

2. Compare the answers to Prob. 1(*g*) and (*h*) and explain using (7.4.5).

3. Given that $e^{xy} = x$, determine dy/dx using implicit differentiation.

4. Name the one nonexponential function that is identical to its derivative.

5. Determine the 100th derivative of $g(x) = e^{kx+c}$.

6. (Requires calculator) For $f(x) = b^x$ we know $f'(0) = b^0 \ln b = \ln b$, so we can compare the slopes of various exponential functions at their common point (0, 1). Fill in the following tables to four decimal places. Note that table (*b*) agrees with the value of *e* given earlier.

(*a*)

b	*Slope of* b^x *at* (0,1)
1/2	
1	
2	
3	
5	
10	

(*b*)

b	*Slope of* b^x *at* (0,1)
2.5	
2.6	
2.7	
2.71	
2.72	
e	

DERIVATIVES OF LOGARITHM FUNCTIONS

Since $y = \ln x$ means $y = \log_e x$, we know by definition of logarithm functions (as inverses of exponential functions) that

$$e^y = x$$

Also, since e^x is differentiable with no horizontal tangents, we know by (7.3.3) that $\ln x$ is differentiable. Thus we can differentiate the above equation implicitly to obtain:

$$e^y \frac{dy}{dx} = 1 \quad \text{or} \quad \frac{dy}{dx} = \frac{1}{e^y}$$

But $e^y = x$, so for $y = \ln x$ we have

$$\frac{dy}{dx} = \frac{1}{x}$$

For $u(x)$ differentiable and positive [so $\ln u(x)$ is defined], we apply the Chain Rule to get:

7.4.7
$$D_x[\ln u(x)] = \frac{u'(x)}{u(x)}$$

Example 3 Compute $D_x[\ln(4x^5 + x^2 + 3x + 2)]$.

Solution $D_x[\ln(4x^5 + x^2 + 3x + 2)] = \dfrac{20x^4 + 2x + 3}{4x^5 + x^2 + 3x + 2}$ ●

Example 4 Given $f(x) = \ln(1/x)$, compute $f'(x)$:
(*a*) Directly.
(*b*) Using that $f(x) = \ln(1) - \ln(x)$.

Solution (*a*) $f'(x) = \dfrac{-1/x^2}{1/x} = -\dfrac{1}{x}$.

(*b*) $f(x) = 0 - \ln(x) \qquad [\ln(1) = 0 \text{ because } e^0 = 1]$

Thus $f'(x) = -1/x$. ●

The natural logarithm function is defined only for positive numbers, but $\ln|x|$ is defined for *all* nonzero x. In practice we often need the composition of the natural logarithm function with other, not necessarily positive, functions $u(x)$. So that the composition be defined for those x making $u(x) < 0$, we use $\ln|u(x)|$ instead of $\ln(u(x))$.

7.4.8 **Theorem** If $u(x)$ is differentiable and nonzero, then

$$D_x[\ln|u(x)|] = \frac{u'(x)}{u(x)}$$

Proof: We shall prove the case $u(x) = x$, from which the general case will follow via the Chain Rule.

If $x > 0$, $|x| = x$, and so the derivative has already been computed [in (7.4.7)] to be $1/x$.

If $x < 0$, then $|x| = -x > 0$, so

$$\ln|x| = \ln(-x) = \ln[(-1)x]$$

By (7.4.7),

$$D_x[\ln[(-1)x]] = \frac{(-1)}{-x} = \frac{1}{x}$$

Hence $D_x[\ln|x|] = 1/x$ in both cases. ■

We can repeat for general logarithms the implicit differentiation that gave us the derivative of the natural logarithm:

$$y = \log_b x \text{ implies } b^y = x$$

Hence by (7.4.6), $b^y(dy/dx)\ln b = 1$, so

$$\frac{dy}{dx} = \frac{1}{b^y \ln b} = \frac{1}{x \ln b}$$

Again, by the Chain Rule, if $u(x)$ is a differentiable function of x, then

7.4.9

$$D_x[\log_b|u(x)|] = \frac{u'(x)}{u(x)}\frac{1}{\ln b}$$

Example 5 Given $f(x) = \log_2(x^3 + x)$, find $f'(x)$.

Solution $f'(x) = \dfrac{3x^2 + 1}{x^3 + x}\dfrac{1}{\ln 2}$. ●

PROGRESS TEST 2

In Probs. 1 to 5 differentiate with respect to x.

1. $f(x) = \ln(3x^4 + 2x^2 + x + 1)$

2. $h(x) = \dfrac{\ln x}{x}$

3. $f(x) = \ln(\ln(x))$

4. $r(x) = 2^{(x \ln x)}$

5. $f(x) = \log_3(4x^3 + x^2 - x)$

6. Find y' given $\ln(xy) = 1$.

DERIVATIVES AND THE GRAPHS OF EXPONENTIAL AND LOGARITHM FUNCTIONS

We can reexamine the graphs of exponential and logarithm functions in terms of their derivatives. (Since these graphs were the basis of our assumptions in determining the derivatives, we are merely showing that these results are consistent with the assumptions.)

Example 6 Show that $f(x) = b^x$ is everywhere increasing for $b > 1$, constant if $b = 1$, and decreasing if $0 < b < 1$. Recall Fig. 7.27, p. 257.

Solution Since $f(x) = b^x > 0$ for all x, we know that the sign of $f'(x) = b^x \ln b$ is the same as the sign of $\ln b$. But $\ln b > 0$ for $b > 1$ and $\ln b < 0$ for $0 < b < 1$. For $b = 1$, $\ln b = 0$, so $f'(x) = 0$ and thus $f(x) = 1^x$ is a constant, namely 1. ●

Example 7 For $b > 1$, show $f(x) = \log_b x$ is increasing and concave down on $(0, +\infty)$.

Solution Since both factors of $f'(x) = (1/x)(1/\ln b)$ are positive on $(0, +\infty)$, $f(x)$ is increasing there. Since

$$f''(x) = D_x\left[\frac{1}{(\ln b)\,x}\right]$$

$$= \frac{1}{\ln b} \cdot \frac{-1}{x^2} < 0 \qquad \text{for all } x \text{ in } (0, +\infty)$$

we know that $f(x)$ is concave down there. Recall Fig. 7.28, p. 259. ●

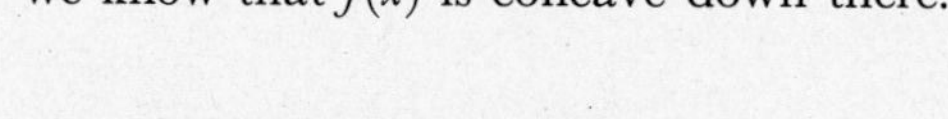

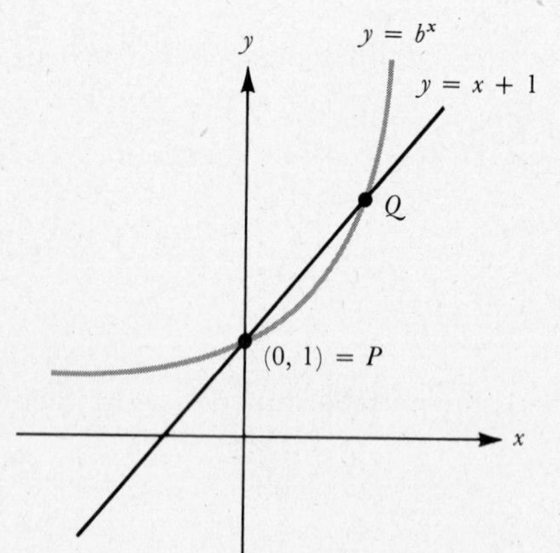

Fig. 7.32

A LIMIT DEFINITION OF e

As we remarked earlier, the number e can be described as a limit. Originally, we searched for a base b such that the line tangent to $y = b^x$ at $x = 0$ would have slope 1. But then the equation of the tangent line at (0, 1) is $y = x + 1$. Suppose for a moment that $b > 1$ is an arbitrary base not equal to e. Then the graphs of

$$y = b^x \qquad \text{and} \qquad y = x + 1$$

intersect when $b^x = x + 1$. We already know they intersect when $x = 0$; but for $b \neq e$, the line $y = x + 1$ is not a tangent line and intersects $y = b^x$ at another point Q. (See Fig. 7.32. For $b < e$, Q is as shown. For $b > e$, Q would be to the left of the y axis.) Raising both sides of $b^x = x + 1$ to the $1/x$ power $(x \neq 0)$ we have

7.4.10 $$b = (x + 1)^{1/x}$$

Figure 7.32 is similar to the figures used in Chap. 2, Sec. 2.4 to interpret the derivative as the slope of a tangent line. There, we took the limit as $x \to 0$ and let the point Q "move down" the *given* curve approaching P as the limit. Here, we want $y = x + 1$ to be the tangent line, so we hold *it* fixed, and consider what happens as b approaches e. By definition of e, as b approaches e, the different curves (for different bases b) intersect the line $y = x + 1$ at points Q that move along $y = x + 1$ closer and closer to (0, 1). This means that the x coordinate of Q approaches 0. But, looking again at (7.4.10), this implies by the definition of e that as the left side approaches e, the right side is the limit of $(1 + x)^{1/x}$ as $x \to 0$. Hence:

7.4.11 $$e = \lim_{x\to 0} (1 + x)^{1/x}$$

This limit can be rewritten using the substitution $u = 1/x$ (so $x = 1/u$). Since the limit in (7.4.11) exists, $\lim_{x\to 0^-}$ and $\lim_{x\to 0^+}$ both equal e. But letting $x \to 0^+$ is equivalent to letting $u \to +\infty$. Hence we can conclude from (7.4.11) that

7.4.12 $$e = \lim_{u\to +\infty} \left(1 + \frac{1}{u}\right)^u$$

We can use this limit representation of e to check the 15-place value of e given earlier and determine e to any specified degree of accuracy. You are invited to fill out and extend Table 7.4.

u	$\left(1 + \frac{1}{u}\right)^u$
1	2
2	2.25
3	2.37037
4	2.44141
5	2.48832
10	2.59374
50	2.69159
100	2.70481
1,000	2.71692
10,000	2.71815
50,000	
100,000	
1,000,000	

Table 7.4

SECTION 7.4 EXERCISES

In Exercises 1 to 36 determine the derivative of the given function.

1. $f(x) = e^{(10x^3+3x-2)}$

2. $f(x) = \ln(4x - 3x^5)$

3. $g(x) = 10^{4x} + 3$

4. $y = \ln(2x - 1) + e^{-x}$

5. $h(x) = e^{x^2/(4x-1)} + 31x$

6. $y = (\ln x)e^x$

7. $y = \ln(e^x)$ (simplify and explain)

8. $y = e^{\ln x}$ (simplify and explain)

9. $y = e^{x^2}$

10. $y = (e^x)^2$ (compare with $D_x[e^{2x}]$)

11. $f(x) = e^{x^2} \cdot \ln(5x + 1)$

12. $f(x) = e^{e^x} + \log_3(x^4 - x)$

13. $g(x) = \dfrac{e^{3x}}{4x^3 + 2x - 3}$

14. $f(t) = e^{t^2} \ln t$

15. $f(x) = \dfrac{\ln(x^3 + x)}{\ln(x - 1)} + 2^{3x}$

16. $f(t) = \sin(e^t)$

17. $f(x) = \ln[\sin^2(5x^2 - 7x)]$

18. $g(x) = \log_{10}(\cot^4 x)$

19. $H(t) = e^{(\sin t + \cos t)}$

20. $f(x) = e^{\pi^2+6\pi}$

21. $m(x) = \ln(\sec x + \tan x)$

22. $y = e^{(e^x)}$

23. $y = (e^e)^x$ (compare with Exercise 22)

24. $f(x) = e^x \cos x$

25. $g(x) = e^{x \cos x}$

26. $m(z) = \ln\left(\dfrac{z + 1}{z - 1}\right)$

27. $n(z) = \ln(z^p)$

28. $N(z) = p \ln z$ (compare with Exercise 27)

29. $f(x) = \sin^2(e^x)$

30. $g(x) = \cos^2(e^x)$

31. $y = \sec(e^{2x})$

32. $y = e^{x^2} \csc x$

33. $f(t) = e^{\cos^2 t}$

34. $g(t) = e^{\sin^2 t}$

35. $y = \dfrac{1}{\ln x}$

36. $y = \ln(\ln x)$

In Exercises 37 to 40 determine the indicated higher-order derivative.

37. $\dfrac{d^ny}{dx^n}$ for $y = 2^x$

38. $\dfrac{d^ny}{dx^n}$ for $y = 3^{2x}$

39. $f^{(100)}(x)$ for $f(x) = \ln x$

40. $g^{(4)}(x)$ for $g(x) = e^{(e^x)}$

In Exercises 41 to 46 determine dy/dx using implicit differentiation.

41. $e^{x+y} = e^x$

42. $x^2e^y = 2$

43. $\ln\left(\frac{x}{y}\right) + xy = 3$

44. $\ln(x^2 + y^2) = 2\ln x + 2\ln y$

45. $(\ln x)^2 + (\ln y)^2 = 1$

46. $e^{\sin y} = x$

47. Where is the line that is tangent to $y = x - \ln x$ horizontal?

48. Where is the curve $y = e^{-x^2}$ concave downward?

49. Graph $y = \ln|x|$.

50. Graph $y = \ln(x^2)$.

SECTION 7.4 APPENDIX AN INTEGRAL APPROACH TO LOGARITHMS

Our informal development of the exponential and logarithm functions began with an informal definition of b^x (especially regarding irrational x), assumed their basic algebraic properties (rules for exponents), and also assumed their continuity and differentiability. We then defined the logarithm functions as inverses of the exponential functions. We shall now outline a more mathematically complete development that is almost the reverse of this and that will illustrate the tremendous power of the Fundamental Theorem of Calculus.

Recall that Form 1 (6.3.6) says: If f is continuous on $[a, b]$ with x in $[a, b]$, then we can define a new function F by $F(x) = \int_a^x f(t)\,dt$. Furthermore, $F'(x) = f(x)$.

7.4.A.1 **Definition** For any $x > 0$, define $\ln x = \displaystyle\int_1^x \frac{dt}{t}$.

Since $1/t$ is continuous on $(0, +\infty)$, we know $\ln x$ has domain $(0, +\infty)$ and is differentiable (hence continuous) with derivative $1/x$. We know $\ln 1 = \int_1^1 (dt)/(t) = 0$ by [6.3.4(*b*)]; it follows that $\ln x$ of (7.4.A.1) is the same as $\ln x$ informally defined in Sec. 7.4 because the two functions have the same domain, the same derivative, and are equal at $x = 1$. All the properties of $\ln x$ (including its graph—using the argument of Example 7) can be established from definition (7.4.A.1). (For $0 < x < 1$, $\ln x$ is in the negative of the area of the shaded region in Fig. 7.33.)

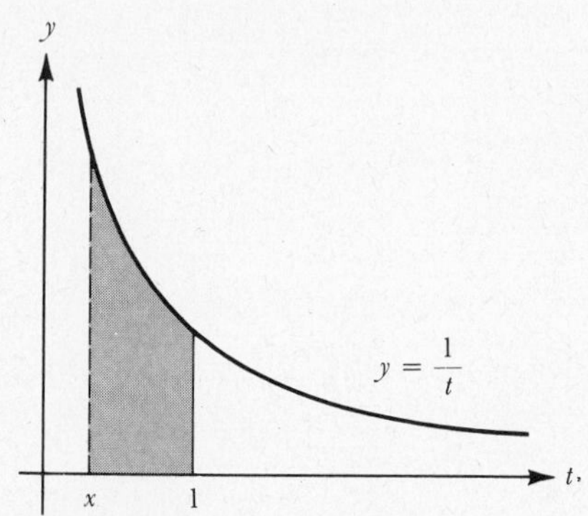

Fig. 7.33

The algebraic properties of $\ln x$ now follow. For example, we shall prove

$$\ln(x_1x_2) = \ln x_1 + \ln x_2 \qquad x_1, x_2 > 0$$

by regarding x_1 as a variable, x_2 as a constant, and differentiating:

$$D_{x_1}[\ln(x_1x_2)] = \frac{x_2}{x_1x_2} = \frac{1}{x_1} = D_{x_1}[\ln x_1]$$

Hence $\ln(x_1x_2) = \ln x_1 + C$ by (4.4.6). But for $x_1 = 1$, $\ln(1 \cdot x_2) = \ln x_2 = \ln 1 + C = 0 + C$, so $C = \ln x_2$, finishing the proof. See the exercises for further development.

Since $\ln x$ is a strictly increasing differentiable function with domain $(0, +\infty)$, range R (see Exercise 1) and no horizontal tangents, we know by (7.3.3) that $\ln x$ has a differentiable inverse. Because $\ln x$ is the same as in Sec. 7.4, we would expect it to have the same inverse, namely, e^x. However, we must still show that this inverse—call it e^x—has the properties attributed to it in Secs. 7.3 and 7.4. But its derivative follows from that of $\ln x$: Set $y = e^x$, so $\ln y = x$. Then $(1/y)(dy/dx) = 1$, which implies $dy/dx = y = e^x$, as expected. The algebraic properties follow from those of $\ln x$ as well. For example, to prove

$$e^{x_1+x_2} = e^{x_1}e^{x_2}$$

we note that $x_1 = \ln y_1$ and $x_2 = \ln y_2$ for some $y_1, y_2 > 0$ (the range of $\ln x$ is

R). Since e^x is the inverse of the logarithm function, $y_1 = e^{x_1}$, $y_2 = e^{x_2}$. Then $e^{x_1+x_2} = e^{(\ln y_1 + \ln y_2)} = e^{\ln(y_1 y_2)} = y_1 y_2 = e^{x_1}e^{x_2}$. Other properties follow similarly.

Finally, we can use the fact that the (natural) exponential function is defined for all real numbers to define b^x to be $e^{x \ln b}$. The properties of b^x now follow from those of e^x.

SECTION 7.4 APPENDIX EXERCISES

1. Prove $\lim_{x\to+\infty} \ln x = +\infty$ as follows (a similar argument shows $\lim_{x\to 0^+} \ln x = -\infty$ from which it follows that the range of ln is R.)
(*a*) Suppose $M > 0$ is given. Show there is a positive integer k such that $k \ln 10 > M$ (note $\ln 10 > 0$).
(*b*) Show that for $x > 10^k$, $\ln x > M$.

2. Using a derivative argument analogous to that in the previous discussion, prove [7.3.11(*c*)] for rational p.

3. Use [7.3.11(*c*)] with $p = -1$ to prove [7.3.11(*d*)].

4. (*a*) Using the Intermediate-Value Theorem (2.5.12) and Exercise 1, prove that there exists a unique number E such that $\ln E = 1$.
(*b*) Use (*a*) and the fact that e^x is the inverse of the strictly increasing natural logarithm function to conclude that $E = e$.

5. (*a*) For $f(x) = \ln x$, compute $f'(1)$ using the definition of derivative $\lim_{h\to 0}[(f(x+h) - f(x))/h]$ at $x = 1$.
(*b*) Use the continuity of $\ln x$, the definition of e^x as the inverse of $\ln x$, and either (7.4.11) or (7.4.12) to give another proof that $E = e$, where e is as defined in Sec. 7.4.

6. (*a*) Using that $x^r = e^{r \ln x}$ by definition for *all* r (including irrational r), prove that $D_x(x^r) = rx^{r-1}$.
(*b*) Use (*a*) and Exercise 2 to prove that [7.3.11(*c*)] holds for *all* p.

7. Use the corresponding property for e^x and the definition of b^x as $e^{x \ln b}$ $(b > 0)$ to prove [7.3.6(*a*)].

8. Repeat Exercise 7 for [7.3.6(*b*)].

9. (*a*) Use (7.4.A.1) and the substitution $u = t/a$ to prove that $\ln b = \int_a^{ab}(dt/t)(a, b > 0)$.
(*b*) Using part (*a*) and $\int_1^{ab}(dt/t) = \int_1^a(dt/t) + \int_a^{ab}(dt/t)$, prove that $\ln(ab) = \ln a + \ln b$.

Section 7.5: Applications of Derivatives of Transcendental Functions

Objectives:

1. **Use logarithmic differentiation to compute derivatives of certain complicated functions.**
2. **Use derivatives of trigonometric, exponential, and logarithm functions to graph transcendental functions, and solve optimization and related rate problems.**

LOGARITHMIC DIFFERENTIATION

Differentiating a function such as

$$f(x) = \frac{(3x^7 - 2x^2 + x)^2\sqrt{x^3+1}}{\sqrt[3]{x^2 - 7x + 1}}$$

involves complex and repeated applications of virtually all the differentiation rules. Just as common logarithms (base 10) are used to simplify complicated numerical computations, natural logarithms can be used to simplify complicated algebraic calculations. We illustrate how to use the properties of natural logarithms to take the derivative of the above function.

First, take the natural logarithm of both sides and break down the right side to a simple sum and difference:

$$\begin{aligned}\ln[f(x)] &= \ln\left[\frac{(3x^7 - 2x^2 + x)^2\sqrt{x^3+1}}{\sqrt[3]{x^2 - 7x + 1}}\right]\\ &= \ln[(3x^7 - 2x^2 + x)^2] + \ln\sqrt{x^3+1} - \ln\sqrt[3]{x^2 - 7x + 1}\\ &= 2\ln(3x^7 - 2x^2 + x) + \frac{1}{2}\ln(x^3+1) - \frac{1}{3}\ln(x^2 - 7x + 1)\end{aligned}$$

Second, differentiate both sides, getting

$$\frac{f'(x)}{f(x)} = 2\cdot\frac{21x^6 - 4x + 1}{3x^7 - 2x^2 + x} + \frac{1}{2}\cdot\frac{3x^2}{x^3+1} - \frac{1}{3}\cdot\frac{2x - 7}{x^2 - 7x + 1}$$

Finally, solve for $f'(x)$ by multiplying by $f(x)$:

$$f'(x) = \left[\frac{42x^6 - 8x + 2}{3x^7 - 2x^2 + x} + \frac{3x^2}{2x^3 + 2} - \frac{2x - 7}{3x^2 - 21x + 3}\right] \cdot \left[\frac{(3x^7 - 2x^2 + x)^2\sqrt{x^3 + 1}}{\sqrt[3]{x^2 - 7x + 1}}\right]$$

This three-step procedure is called *logarithmic differentiation.*

The above function, although messy, could have been differentiated using conventional means, but the following cannot.

Example 1 Differentiate $f(x) = x^x (x > 0)$

Solution

$$\ln f(x) = \ln(x^x) = x \ln x$$

so

$$\frac{f'(x)}{f(x)} = x \cdot \frac{1}{x} + (\ln x) \cdot 1 = 1 + \ln x$$

Hence $f'(x) = x^x(1 + \ln x)$. ●

Example 1 illustrates the following general principle:

To determine the derivative of any function of the form $f(x) = v(x)^{u(x)}$, where $u(x)$ and $v(x)$ are differentiable ($v(x) > 0$) use logarithmic differentiation.

Example 2 Differentiate $f(x) = (3x + 1)^{x^2e^x}$.

Solution $\ln f(x) = (x^2e^x)\ln(3x + 1)$, so, applying the Product Rule twice,

$$\frac{f'(x)}{f(x)} = (x^2e^x) \cdot \frac{3}{3x + 1} + \ln(3x + 1)[x^2e^x + e^x 2x]$$

Thus,

$$f'(x) = \left[\frac{3x^2e^x}{3x + 1} + (2xe^x + x^2e^x)\ln(3x + 1)\right](3x + 1)^{x^2e^x}$$ ●

It turns out that logarithmic differentiation is also handy in proving and extending some of the standard differentiation formulas. We can now confirm that the Power Rule works for *any* real-number exponents, not just rational exponents. The argument is ultimately based upon the assumption that the exponential functions are defined and continuous for all real numbers. This assumption was shown unnecessary in Sec. 7.4, Appendix.

7.5.1 **(Power Rule for any Real Exponent)** Given that $u(x)$ is differentiable and $f(x) = [u(x)]^s$ (s any real number), then

$$f'(x) = s \cdot [u(x)]^{s-1} \cdot u'(x)$$

Proof:

$$\ln f(x) = s \cdot \ln u(x)$$

so

$$\frac{f'(x)}{f(x)} = s \cdot \frac{u'(x)}{u(x)}$$

Hence

$$f'(x) = s \cdot \frac{u'(x)}{u(x)} \cdot [u(x)]^s$$
$$= s[u(x)]^{s-1} \cdot u'(x) \quad ■$$

PROGRESS TEST 1

Find the derivatives of the following three functions:

1. $y = \dfrac{\sqrt{7x - 1}\,\sqrt[3]{x^2 + 5x}}{(x - 11)^7}$

2. $f(x) = \sqrt[3]{\dfrac{(x-5)^5(3x+1)^7}{\sqrt{8x - 1}}}$

3. $f(x) = x^{(x^2)}$

USING DERIVATIVES TO GRAPH TRANSCENDENTAL FUNCTIONS

The derivative can be extremely useful in graphing transcendental functions.

Example 3 Graph $f(x) = \cos(2x) + 2\cos x$ on $[\pi/4, 2\pi]$ and determine its extreme values. Ignore inflection points.

Solution

$$f'(x) = -2\sin 2x - 2\sin x = -2(\sin 2x + \sin x)$$
$$= -2(2\sin x\cos x + \sin x) \quad \text{(by a double-angle formula)}$$
$$= -2\sin x(2\cos x + 1)$$

Hence $f'(x) = 0$ when $\sin x = 0$ and when $\cos x = -1/2$, that is, when $x = 0, \pi, 2\pi, 2\pi/3$, and $4\pi/3$. We ignore $x = 0$ since it is outside $[\pi/4, 2\pi]$. We now chart the sign of f':

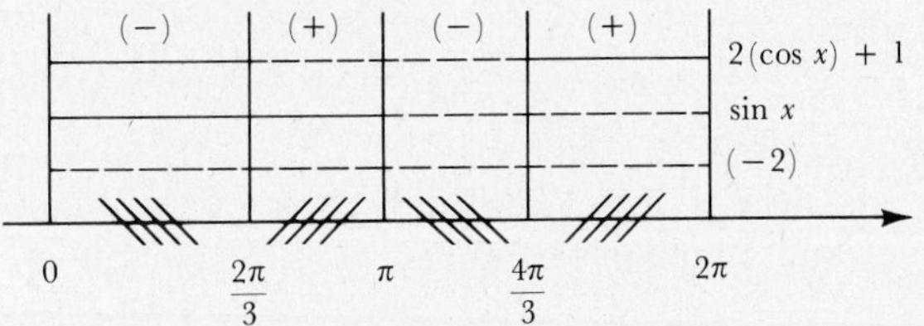

We conclude that there are minima at $x = 2\pi/3, 4\pi/3$, and a maximum at $x = \pi$. $f(2\pi/3) = -3/2, f(\pi) = -1, f(4\pi/3) = -3/2$. Also $f(\pi/4) = \sqrt{2}$ and $f(2\pi) = 3$. Hence the function achieves its maximum at $x = 2\pi$ and its minimum at the two relative minima $x = 2\pi/3, 4\pi/3$; see Fig. 7.34. ●

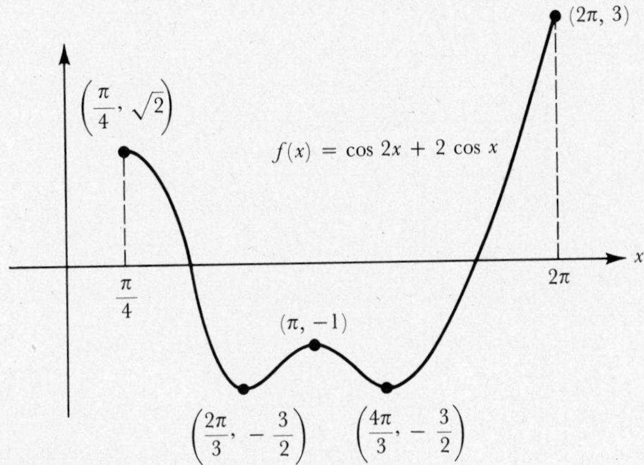

Fig. 7.34

To graph exponential and logarithm functions, it is useful to assume the following pair of limits. These will be proved in more general form in (14.1.5) of Chap. 14.

7.5.2 Theorem (a) $\lim_{x\to+\infty} \frac{\ln x}{x} = 0$ (b) $\lim_{x\to+\infty} \frac{x}{e^x} = 0$

Each limit may be regarded as a statement about the relative growth rates of the functions involved. Notice that, for example, letting $x = 10^{1000}$, we have

$$\frac{\ln x}{x} = \frac{\ln(10^{1000})}{10^{1000}} = \frac{1000 \ln 10}{10^{1000}} = \frac{\ln 10}{10^{997}} = \frac{2.3}{10^{997}} = \underbrace{0.00\ldots0}_{996 \text{ zeros}}23$$

Example 4 Determine the maximum value of $f(x) = (\ln x)/x$.

Solution Note that $f(x)$ is defined only for $x > 0$ and is negative in $(0, 1)$.

$$f'(x) = \frac{x(1/x) - (\ln x)(1)}{x^2} = \frac{1 - \ln x}{x^2}$$

Now, $\ln e = 1$, with

$$1 - \ln x \geq 0 \qquad \text{when } x \leq e$$
$$1 - \ln x \leq 0 \qquad \text{when } x \geq e$$

Hence $f(x)$ is increasing in $(0, e)$ and decreasing in $(e, +\infty)$ with the single relative maximum in $(1, +\infty)$ at $x = e$. By (5.1.3), $f(e) = 1/e$ must be the maximum value of f. ●

PROGRESS TEST 2

1. Graph $f(x) = xe^x$, noting all pertinent data.
2. Determine where $y = x + \sin x$ is increasing and decreasing and locate any and all relative extremes. Then graph $y = x + \sin x$. (We could call this the "step-with-a-thick-rug" function.)

OPTIMIZATION AND RELATED RATES

Transcendental functions arise in optimization problems as naturally as the functions considered in Chap. 5.

Example 5 Suppose a 10-ft-long symmetric V-shaped watering trough is to be made from two boards, each 1 ft wide. What vertex angle between the boards will yield the maximum capacity?

Solution It is most convenient to break the cross section into two right triangles. (See Fig. 7.35.)

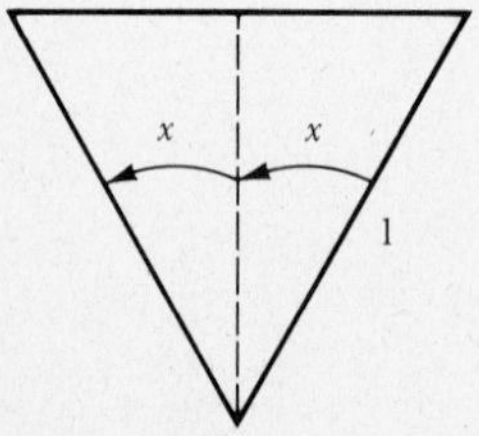

Fig. 7.35

$$\begin{aligned}\text{Volume} &= \text{length times cross-sectional area}\\ &= (10)[(\sin x)(\cos x)] \qquad \left(\sin x = \tfrac{1}{2} \text{ base and } \cos x = \text{height}\right)\\ &= 5 \sin 2x \qquad \text{(double-angle formula)}\end{aligned}$$

Clearly, we are maximizing $V(x) = 5 \sin 2x$, with $0 \leq x \leq \pi/2$. Now $V'(x) = 10 \cos 2x = 0$ when $2x = \pi/2$, that is, when $x = \pi/4$. Since

$V''(x) = -20 \sin 2x$ and $V''(\pi/4) = -20 < 0$, we know that $x = \pi/4$ is a relative maximum.

The two endpoints $x = 0, x = \pi/2$ yield zero volume, so the best vertex angle is $2(\pi/4) = \pi/2$ rad, which yields a volume of 5 ft^3. ●

Trigonometric functions are, of course, useful in situations where related rates involve changing angles, as indicated in Prob. 2, Progress Test 3.

PROGRESS TEST 3

1. A horizontal gutter is to be made from a long piece of sheet metal that is 0.8 m wide, by symmetrically turning up 0.2 m on either side to an angle that maximizes the carrying capacity. If x is the angle the edge makes with the vertical, what x gives the maximum carrying capacity? (Assume that the carrying capacity is directly proportional to the cross-sectional area.)

2. A boat is passing in a straight line 300 ft offshore from an onshore radar installation at the rate of 50 ft/s. At what rate must the radar turn when the boat has traveled 300 ft past its closest point to the radar?

SECTION 7.5 EXERCISES

In Exercises 1 to 13 use logarithmic differentiation to determine the derivative of the given function.

1. $y = \dfrac{\sqrt{(x-5)^2+3}}{(x^2-4)^4}$

2. $y = \dfrac{\sqrt{x^2-1}\,\sqrt[3]{x+1}}{x^4-3x}$

3. $f(x) = (x+1)^x$

4. $f(x) = (2x+1)^x$

5. $k(x) = \sqrt[2]{x^3-x}\,\sqrt{x^5+4x^2} - \sqrt[5]{x^4+4x+4}$

6. $m(x) = \sqrt[5]{\dfrac{(3x-2)^3}{\sqrt{x+1}}}$

7. $t(x) = \dfrac{(x-5)^4(3x^3+9x-2)^7\sqrt{4x+1}}{\sqrt[3]{3x^7-2x+1}}$

8. $g(x) = \dfrac{x^9e^{11x}\sqrt{e^{3x-2x}}}{(3x^7-2x-7)^4}$

9. $f(x) = x^{(x^3)}$

10. $f(x) = (2x)^{(\ln x)}$

11. $g(x) = \dfrac{2x^{(\ln x)}}{e^{5x}}$

12. $h(x) = e^{x^x}$ (use Example 1)

13. $g(x) = (x^x)^2$

14. Determine $D_x[x^{x+1}]$ (*a*) directly and (*b*) using $x^{x+1} = xx^x$ and the Product Rule.

15. Repeat Exercise 14 for $g(x) = (x^x)^2$ using that $(x^x)^2 = x^{2x}$ and the Power Rule.

16. Use logarithmic differentiation to reprove the Product Rule.

17. Use logarithmic differentiation to reprove the Quotient Rule.

18. Determine the concavity of $g(x) = e^{x^2}$.

In Exercises 19 to 30 graph the given function on the indicated interval.

19. $f(x) = x \ln x$ on $(0, +\infty)$

20. $f(x) = e^{-x^2}$ on $(-\infty, +\infty)$

21. $y = (\ln x)^2$ on its natural domain

22. $g(x) = \cos 4x$ on $[0, 2\pi]$

23. $h(x) = 3 \sin(2x - \pi)$ on $[-\pi, 2\pi]$

24. $f(x) = 2x + \sin 2x$ on $[0, 2\pi]$

25. $y = x^x$ on $(0, +\infty)$

26. $y = x^{x^2}$ on $(0, +\infty)$

27. $f(t) = e^{-t} \sin t$ on $(-\infty, +\infty)$

28. $y = \sin \dfrac{2\pi}{t}$ on $(0, 2)$

29. $y = t \cos \dfrac{2\pi}{t}$ on $(0, 4)$

30. $y = e^{\sin t}$ on $[-\pi, 4\pi]$

31. A searchlight sweeps the sky at a rate of 1 rad/min. If the cloud level is 1/4 mi above the ground, how fast (in rad/min) is the searchlight beam sweeping the clouds:
    (*a*) Directly overhead?
    (*b*) 1/4 mi past the point directly overhead?

32. Given a sphere of radius 10 units, find the altitude of the right circular cone of maximum volume that can be inscribed in the sphere. Let $2x$ be the vertex angle of the cone.

33. What is the narrowest alley a 64-ft-long hook-and-ladder truck can turn into from a 27-ft-wide street ignoring the width of the truck?

34. (A rocket tracking problem) A rocket is fired 200 ft exactly downwind from an observer who measures that the angle the rocket makes with the horizontal is increasing at the rate of $\pi/21$ rad/s. The wind is blowing at 20 ft/s. What is the rate of departure of the rocket from its launching point after 7 s have elapsed? (*Note:* To maintain the constant angle increase with the horizontal, the rocket must be accelerating.)

35. Find any extreme values of $f(x) = \dfrac{\sin x}{1 - 2\cos x}$ on $[0, 2\pi]$.

36. Given an 8-ft-high prison wall and a 16-ft ladder, what is the largest horizontal distance an inside ladder can reach outside the wall. (Let θ be the angle between the ladder and the ground.)

37. What is the natural domain of $f(x) = \ln(\sin x)$?

38. If x increases from 100 to 101, what is the approximate change in $f(x) = x^2 \ln x$? (Use differentials.)

39. Find dy in terms of the independent variable:
(*a*) $y = \sec^3(5x + 1)$, where $x = x(\theta) = e^\theta$
(*b*) $y = 4e^{3x^2+2}$, where $x(t) = t^5 + t^4$

CHAPTER EXERCISES

Review Exercises

(Note that Exercises 1 to 21 involve only precalculus ideas.)

In Exercises 1 to 4 determine $f^{-1}(x)$ if it exists. If f does not have an inverse, name two numbers x_1 and x_2 with $f(x_1) = f(x_2)$.

1. $f(x) = x^5 - 7$

2. $f(x) = (x - 5)^3 - 4$

3. $f(x) = 4x^2 + 3x - 1$

4. $f(x) = \sqrt[3]{x - 5}$

5. Which is larger, $(2.5)^3$ or $3^{2.5}$?

6. For which x is $\tan 20x$ undefined?

7. In which quadrant is the terminal side of an angle of -1000 rad?

In Exercises 8 to 13 evaluate the given expression.

8. $\log_3[(\sqrt{x + 1})(x + 1)^5]$ if $x = 2$

9. $5 \log_2\left(\dfrac{\sqrt[3]{4}}{\sqrt[5]{16}}\right)$

10. $\log_2(1/4)^{-3/7}$

11. $\log_3(3\sqrt{3})$

12. $3(2 \log_3(41))$

13. $(\log_2 4)^3 - \log_2(4^3)$

14. If $\log_{10} 51 = a$:
(*a*) What is $\log_{10}(5.1)$?
(*b*) What is $\log(0.51)$?
(*c*) What is $\log_{10}(51{,}000)$?

15. Simplify $\log_b(b(\log_b(x))$.

In Exercises 16 to 21 solve the given equation for x.

16. $\log_3(x + 1) + \log_3(x) = 1$

17. $\log_2(256) = x^3$

18. $\log_4(x - 1) = 2$

19. $\log_2(x + 1) + \log_2(x + 2) = \log_2(12)$

20. $\log_x 9 = 3$

21. $\log_2(\log_2(x)) = 0$

In Exercises 22 to 66 determine the derivative of the function with respect to the indicated independent variable.

22. $g(x) = \sin(x^3 - 7x^2 - 6x + 2) + \cos(x^5 - 4)$

23. $h(x) = e^{x^2+2x+1} - \ln(x^2 - 3x - 5)$

24. $f(z) = (\tan z)\sec(z^2 + 1)$

25. $m(x) = e^{x^3-6x} \cdot \cot\sqrt{x - 1}$

26. $f(x) = \sin^2(2x + 1) \cdot \cos^2(2x - 1)$

27. $s(x) = \dfrac{e^{\tan(4x)}}{\sec^3 x}$

28. $f(t) = [\log_2(t^2 + t)] \cdot 2^{4t}$

29. $f(x) = \dfrac{xe^x}{\cot^3(x^2 + 9x)}$

30. $g(x) = \dfrac{\sqrt{x - 1}}{\sqrt{x + 1}} + \dfrac{\sqrt{x - 1}}{e^{\sqrt{x+1}}}$

31. $h(s) = \sqrt{\csc^3(e^s)}$

32. (a) $f(x) = \sin(\sqrt{3x - 8})$
(b) $y = \sqrt{\sin(3x - 8)}$

33. $y = \tan(x^2 - x) \cdot \cot(x^3 - x^2)$

34. $f(x) = \cot^3(11x^4 + x^3)^5$

35. $g(x) = xe^{x+7} - 3x^4$

36. $f(x) = \dfrac{\sqrt{7x - 4}\sqrt[3]{2x^4 - x^2}(x - 7)^{10}}{\sqrt[5]{2x + 1}\sqrt[4]{\cos x}}$

37. $g(x) = 5^{x^2+x+1} = 2^{x-3}$

38. (a) $f(z) = \sin(\cos^2(z - 1))$
(b) $g(z) = \sin^2(\cos(z - 1))$

39. $f(x) = [\csc^7(x^4 + x^3)] \cdot e^{x^5+1}$

40. $s(x) = (x^2 + 1)^{5x}$

41. $f(x) = \ln(\csc x \cot x)$

42. $f(x) = \sqrt[3]{\dfrac{\sqrt{2x - 1} + 4}{\sqrt{\cot x}}}$

43. $m(x) = \left[\dfrac{\ln(x^3 + 2x - 4)}{x^5 + 4x^4 - 6}\right]^{11/8}$

44. $p(t) = (\sin^3 t)\sqrt{\cos(t^5 - t^3 - 4)}$

45. (a) $r(x) = \sin^5(\cos x)$
(b) $h(x) = \sin(\cos^5 x)$

46. $f(x) = \dfrac{\sqrt[3]{x^3 + x}\sqrt[7]{x - 1}(4x^3 + 7x^2 - 2x - 6)^9(2x + 1)^4}{\sqrt{x + 1}\sqrt[5]{x - 1}}$

47. $g(t) = e^{\log_3(\cos t)}$

48. $m(x) = \tan\sqrt{2x^3 + 1} - \ln(x^2 + 8)$

49. $f(t) = 4e^{\csc^2 t} + \cot^5(t^4 + 1)$

50. $t(v) = \ln(\tan^2(v^4 + 1))$

51. $b(x) = x^{4x-3}$

52. $p(y) = e^{\sin^2(y-4)}$

53. $f(u) = \sqrt{\ln(\sin^4 u) + u^2}$

54. $g(x) = \log_{10}(x^4 - x + 12)$

55. $f(x) = \ln(\csc x + \cot x)$

56. $r(x) = \pi^{e-2^3}$

57. $f(x) = \sqrt[3]{\dfrac{\sin^2(x + 1)}{\log_2(9x - 1)}} \cdot (x^2 - 6)^7$

58. $g(x) = \sin x^2 \cot(x^2 + 1)$

59. $r(x) = (\sin x)^{\cos x}$

60. $f(x) = \dfrac{x + \tan^3 x}{4 - \sin^2 3x}$

61. $y = x^{\cot x}$

62. $f(x) = 7 \sin 2x + 5 \cos 3x$

63. $g(x) = \sqrt{e^x - x^2}$

64. $y(x) = \sqrt[5]{\dfrac{(x^3 - 1)^2(x + 1)^4}{\sqrt{x - 1}\sqrt{x^2 + 1}}}$

65. $f(x) = 5e^{\sin x^2}$

66. $y = \sin^{11}(\sec(5x - 1)^4)$

In Exercises 67 to 78 determine dy/dx, where y is an implicit function of x defined by the given equation.

67. (a) $\sin xy + \sec x = 1$
(b) $e^{\cos y} = x$

68. (a) $e^{x^3y^3} = xy$
(b) $\ln(x^2 + y^2) = e$
(Can you give the graph of this equation?)

69. $e^x + e^y = xe^x$

70. $\sin(x + y) = \cos(x - y)$

71. $x \sin y + y \cos x = 0$

72. $e^{x+y} + x + y = 1$

73. $10^{x+y} = 2^{x-y}$

74. $\tan xy + xy = 0$

75. $\csc(x + y) - \sec(x - y) = x$

76. $\sin(x + y) = y \cos x$

In Exercises 77 to 85 sketch the given function (ignore inflection points).

77. $f(x) = \sin 2x + 2$ on $[0, 2\pi]$

78. $f(x) = x^2e^x$ on $(-\infty, +\infty)$

79. $f(x) = e^{1/x}$ on its natural domain

80. $y = \dfrac{1}{1 - \sin x}$ on $[0, 2\pi]$

81. $y = 4 \cos 8x$ on $[-\pi, \pi]$

82. $y = \ln|\cos x|$ on $[0, 3\pi]$

83. $y = \ln|\sin x|$ on $[0, 2\pi]$

84. $y = e^{\cos x}$ on $[-2\pi, 2\pi]$

85. $y = e^{\sin 2x}$ on $[0, 2\pi]$

86. A conical tent is made from a circular piece of canvas of radius 10 ft by cutting out a sector and sewing up what remains. Under the condition that the sector cut out be at least $\pi/2$ rad (to make a rain flap), what is the largest-volume tent that can be made?

87. Differentiate $g(x) = \ln[\ln(\ln(x))]$.

88. Differentiate $g(x) = e^{(e^x)^2}$.

89. Differentiate $f(x) = \pi e^{-2}$.

90. Find the maximum of $f(x) = e^x \sin x$ on $[0, 6\pi]$.

91. Differentiate $f(x) = (x^9 + 2x)^{1/\pi}$.

92. Determine $g'(x)$ and $g''(x)$, where $g(x) = x^e$.

In Exercises 93 to 95 a particle moving in a straight line has its position at time t given by s(t). Describe the velocity and position of this particle.

93. $s(t) = e^{-t} \sin t$

94. $s(t) = |\sin t|$ (Note especially what happens as multiples of π.)

95. $s(t) = t + \sin t$

96. The force necessary to pull a weight W along a horizontal plane with a cable is given by $F = kW/(\cos\theta + k\sin\theta)$, where W is the weight, k is the "coefficient of friction," and θ is the angle the cable makes with the horizontal.
(*a*) Given that $k = \sqrt{3}$, at what angle is the least force required to pull the weight?
(*b*) For a smaller coefficient of friction, $k = 1/\sqrt{3}$, what is the best angle at which to pull the weight?

97. An airplane is flying at an altitude of 2 mi in a path that will take it directly over an observer. The observer notes that the plane's angle of elevation is $\pi/3$ rad and is increasing at the rate of $(1/120)\pi$ rad/s.
(*a*) How fast is the plane moving at this instant?
(*b*) Suppose instead that the observer finds the same rate of angle change when the plane is directly overhead, what is the plane's speed?

98. Describe the dimensions of the smallest isosceles triangle (in terms of area) that circumscribes a circle of radius 1. (Let 2θ be the angle between the sides of equal length.)

99. (*a*) Give the acceleration function for a particle whose distance from a given point is $s(t) = e^t + e^{-t}$.
(*b*) Show that $s(t)$ satisfies the differential equation $d^{20}s/dt^{20} - s(t) = 0$.

In Exercises 100 to 102, if x increases from 10 to 10.1, determine the approximate change in the given expression.

100. 2^x

101. $\ln x$

102. $\sin(10x^2 + x)$

Miscellaneous Exercises

1. Use logarithmic differentiation to develop a *formula* that gives the derivative of $f(x) = v(x)^{u(x)}$. [Both functions are differentiable and $v(x) \neq 0$.]

2. Compare the derivatives of e^{x+y} and $e^x e^y$ (both with respect to x and under the assumption that y is a function of x).

3. Find the equation of the line tangent to the curve $xy - e^x \sin y = \pi$ at $(1, \pi)$.

4. Repeat Exercise 3 for $f(x) = \ln(\log_2(x))$ at $x = 4$.

5. Given a sphere of radius r, what are the dimensions of the right circular cone of maximum surface area that can be inscribed in this sphere? (Let the cross-sectional vertex angle of the cone be 2θ.)

6. At what rate is the angle between the minute and hour hands of a clock decreasing at 3 o'clock (in radians per hour)?

In Exercises 7 to 12 sketch the given function, ignoring inflection points.

7. $f(x) = \dfrac{x}{1 - \sin x}$ on $[0, 2\pi]$

8. $y = \dfrac{1}{1 - \cos x}$ on $[-\pi, \pi]$

9. $f(x) = \sin(\cos x)$ on $[0, 4\pi]$

10. $f(x) = \cos(\sin x)$ on $[0, 4\pi]$

11. $y = \dfrac{\text{cox } x}{1 - 2\sin x}$

12. $y = \sin(e^x)$ on $(-\infty, 2]$

13. Name at least one function $g(x)$ such that $g'(x) = (1/2)g(x)$.

14. For the measure of an angle in revolutions (so an angle generated by one complete counterclockwise rotation of the generating line has measure 1)
(*a*) Give a formula analogous to (7.1.2) that converts between radian and revolution measure.
(*b*) Define $\sin_{rev}(x)$ analogous to $\sin_{deg}$ and evaluate $\sin_{rev}(x)$ at the measures of the same angles used in illustrating $\sin_{deg} x$.
(*c*) Graph $y = \sin_{rev}(x)$.
(*d*) Determine $D_x[\sin_{rev}(x)]$ using the Chain Rule and the formula in (*a*).
(*e*) Determine $\lim_{x\to 0} \dfrac{\sin_{rev}(x)}{x}$ using (7.2.11) and compare with $D_x[\sin_{rev}(x)]$ at $x = 0$.

15. [Proof that for small $\Delta x > 0$, $\dfrac{\sin(x + \Delta x) - \sin x}{\Delta x} \approx \cos x$.] In the accompanying sketch, suppose that x increases by $\Delta x > 0$, so the intersection of the generating line with the unit circle moves from $B = (\cos x, \sin x)$ to $P = (\cos(x + \Delta x), \sin(x + \Delta x))$. Then $\overline{DP}$ is the growth of the sine function and $\overline{BD}$ the (negative) growth of the cosine function. The previous argument, that $\lim_{\Delta x \to 0} (\sin \Delta x)/\Delta x = 1$ when rotated by x radians, shows in effect that as $\Delta x \to 0$, the difference between the straight-line segment between P and B and the circular segment between P and B approaches zero. For small Δx—that is, $\Delta x \approx 0$—we shall assume that they are approximately equal and thus regard the figure DPB with circular side PB as a triangle and refer to it as "Δ" DPB and refer to the circular side as "PB."

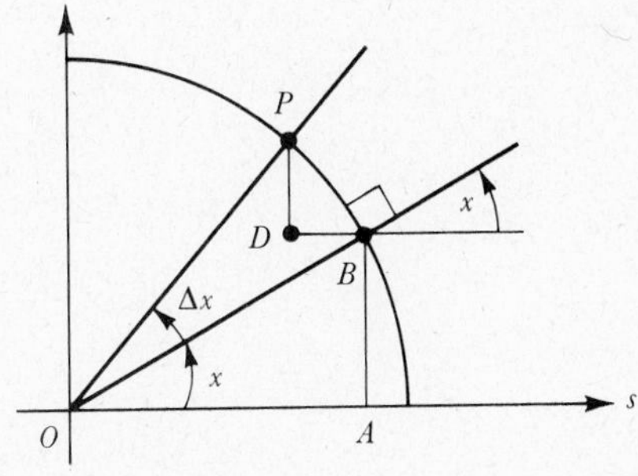

(*a*) Prove "Δ" DPB is similar to ΔABO.

(*b*) Conclude $\dfrac{\sin(x + \Delta x) - \sin x}{\Delta x} \approx \cos x$.

SELF-TEST ✤

1. Differentiate:

(*a*) $f(x) = \dfrac{e^{\tan 4x}}{\sec^3 x}$ (*b*) $g(x) = \ln(\ln(x))$

(*c*) $h(x) = \sin^5(\cos(2x - 1))$ (*d*) $m(x) = \dfrac{\sqrt[3]{x^2 + 1}(x - 4)^{12}(x^3 + 2x)^{9/5}}{\sqrt[4]{x + 1}\sqrt[5]{\sin x}}$

(*e*) $f(x) = \pi e^2$ (*f*) $r(x) = \log_2(x^2 + 1)$

(*g*) $s(x) = \ln(\csc^2 x - 1)$ (*h*) $t(x) = \sec^2 x + c$

2. Sketch the graphs of:

(*a*) $f(x) = e^{x^2}$ on $(-\infty, +\infty)$ (*b*) $g(x) = e^x \cos x$ on $[0, 4\pi]$

3. Find dy/dx given that y is an implicit function of x defined by the equation $\ln(x^3) + e^y = 1$.

4. The range R in feet of a projectile fired with initial velocity v at an angle of x radians with the horizontal is given by $R = (v^2/32) \sin 2x$. Given $v_0 = 64$ ft/s, determine the size of the angle that yields greatest range.

Chapter 8

MORE ON TRANSCENDENTAL FUNCTIONS

CONTENTS

In this chapter we continue our development of transcendental functions. We begin in Sec. 8.1 with the basic application of exponential functions in describing "natural" growth and decay phenomena such as growth of populations, continuously compounded interest, radioactive decay, inflation, return on capital investments (including alternative energy sources), and so on. Sections 8.2 and 8.3 are concerned with integrating exponential, logarithmic, and trigonometric functions and their combinations, including powers and products of trigonometric functions.

In Sec. 8.4 we define inverses for restricted trigonometric functions, and in Sec. 8.5 we show how these inverse trigonometric functions appear in integration problems. In Sec. 8.6 we examine certain frequently occurring combinations of exponential functions called "hyperbolic" functions.

Section 8.1: Growth and Decay of Physical Quantities and Money

Objectives:

1. Define and use exponential functions to describe simple growth and decay phenomena, including continuously compounded interest.

2. Set up and solve differential equations describing growth/decay phenomena in the presence of constraints, including variable constraints.

THE BASIC EQUATIONS

We saw that the derivative of a typical exponential function is the product of itself with a constant, its derivative at zero. If this function is describing a variable quantity A at time t, then $dA/dt = A(t) \cdot A'(0)$. We shall now show conversely that if dA/dt is proportional to $A(t)$, then $A(t)$ is an exponential function. We begin by assuming

the rate of growth (or decay if $dA/dt < 0$) at time t is proportional to the amount on hand at time t

This statement is often expressed as the differential equation

8.1.1
$$\frac{dA}{dt} = kA$$

where k is the proportionality constant. This situation occurs commonly enough to be called *natural growth* (if $k > 0$) or *natural decay* (if $k < 0$). Below, when we use these phrases, we mean that equation (8.1.1) applies. We can solve (8.1.1) by first noting that the derivative dA/dt can be regarded as a quotient of differentials dA and dt. We can thus separate the variables A and t by multiplying both sides by dt and dividing both sides by A to get

8.1.2
$$\frac{dA}{A} = k\,dt$$

The next step is to antidifferentiate, but to antidifferentiate the left side we first rewrite the differentiation formula (7.4.8) in terms of A and t:

$$\frac{d}{dt}[\ln|A|] = \frac{dA/dt}{A}$$

But this tells us that $\ln|A|$ is an antiderivative of

$$\frac{dA/dt}{A}$$

That is,

$$\ln|A| = \int \left(\frac{dA/dt}{A}\right) dt = \int \frac{dA}{A}$$

But $dA/A = k\,dt$ by (8.1.2), so we can conclude that

8.1.3
$$\ln|A| = \int k\,dt = kt + C$$

Then by definition of the natural logarithm,

$$|A| = e^{kt+C} = e^C e^{kt}$$

Hence

$$A = \pm e^C e^{kt}$$

Since $\pm e^C$ is a constant, we can replace $\pm e^C$ by C_0, where $|C_0| = e^C$, to get the natural growth/decay function

8.1.4
$$A = C_0 e^{kt}$$

As with any general solution of a differential equation, this is really an infinite family of solutions, and additional information is required to determine a particular solution. For example, if we know the amount on hand at a particular time, say at $t = 0$, then we can conclude that

$$A(0) = C_0 e^{k \cdot 0} = C_0 \cdot 1 = C_0$$

Hence, in (8.1.4), C_0 is the amount present initially. Of course the amount $A(t)$ present at time t depends not only upon the amount present initially but also upon the rate k. Usually, k must be inferred from other given information.

Example 1 Suppose that the number of bacteria in a culture grows naturally from 10,000 to 800,000 in 20 h. Give the function that describes the number present as a function of t and determine how many bacteria will be present after 30 h.

Solution Knowing that the growth is natural means that (8.1.4) applies, and so the problem reduces to finding C_0 and k. If we take $t = 0$ to be the initial time, at which 10,000 bacteria are present, we know $A(t) = 10{,}000e^{kt}$. Measuring t in hours, we are also given that $A(20) = 800{,}000$, so

$$10{,}000e^{k \cdot 20} = 800{,}000$$

or

$$e^{k \cdot 20} = 80$$

Taking the natural logarithm of both sides, we get $20k = \ln 80$, making

$$k = \frac{\ln 80}{20} \approx \frac{4.382}{20} \approx 0.219$$

Thus we know that

$$A(t) = 10{,}000e^{0.219t}$$

We also know that

$$\begin{aligned} A(30) &= 10{,}000e^{(0.219)(30)} \\ &= 10{,}000e^{6.57} \approx (10{,}000)(713.3698) = 7{,}133{,}698 \end{aligned}$$ ●

Note that Example 1 makes the standard assumption necessary to apply calculus: Although the number of bacteria is an integer, we assume that $A(t)$ is a differentiable function of t. Of course, whenever we make such an assumption, we should be certain that the answer is rounded off to the nearest integer.

If a quantity A is growing naturally, that is, according to the formula $A(t) = C_0 e^{kt}$, it is often of interest to determine *doubling time:* Given

$$A(t_1) = C_0 e^{kt_1} \quad \text{and} \quad A(t_2) = C_0 e^{kt_2} \quad (t_2 > t_1)$$

how large must $t_2 - t_1$ be so that $A(t_2) = 2A(t_1)$? We must solve $C_0 e^{kt_2} = 2C_0 e^{kt_1}$ for $t_2 - t_1$. Divide through by C_0, and take the natural logarithm of both sides:

$$kt_2 = \ln[2e^{kt_1}] = \ln 2 + kt_1$$

Hence

$$kt_2 - kt_1 = \ln 2$$

so

$$t_2 - t_1 = \frac{\ln 2}{k}$$

Because the right side of this equation involves only ln 2 and k, we can conclude that doubling time $t_2 - t_1$ depends only on the *magnitude* of $t_2 - t_1$ and *not* on:

(a) C_0, the amount at $t = 0$

(b) The choice of time t_1 on which the calculation is based (*c*) The quantity $A(t_1)$ at $t = t_1$

Since $\ln 2 \approx 0.69$, we can conclude a useful *rule of thumb for approximating doubling time: Divide 0.69 by k.*

Clearly, tripling time will be given by $t_2 - t_1 = (\ln 3)/k$. This argument suggests the following more general theorem:

8.1.5 **Theorem** For any $b > 0$, if $A = C_0 e^{kt}$, then the amount of time Δt required for A to change by a factor of b is given by

$$\Delta t = \frac{\ln b}{k}$$

What sort of function $A(t) = C_0 e^{kt}$ changes by a factor of b in each unit of time, that is, for $\Delta t = 1$? By (8.1.5) we need $k = \ln b$, from which it follows that

$$A(t) = C_0 e^{(\ln b)t} = C_0 b^t \qquad \text{by (7.4.5)}$$

In other words, *the exponential growth functions we arrived at through the assumption that the growth rate is proportional to the amount on hand, as reflected in the differential equation $dA/A = k$, is exactly the same as that sort of growth we discussed when introducing exponential functions in Sec. 7.3!*

Of course, if the quantity is decaying naturally with $0 < b < 1$, then $\ln b < 0$, so $k < 0$. In such an instance the rate is often expressed in terms of the length of time required for the amount to be cut in half, commonly called its *half-life*.

An important application of this occurs in radioactive decay because the amount of a radioactive element decreases at a rate proportional to the amount remaining.

Example 2 Carbon 14, a radioactive isotope of carbon widely used in dating organic materials, decays naturally with a half-life of 5670 years. Give the decay function for carbon 14.

Solution We know the function has the form $A(t) = C_0 e^{kt}$, where k remains to be determined.

Regardless of the amount of carbon 14 present initially [where $C_0 = A(0)$ and t is measured in years], we know that

$$A(5{,}670) = C_0 e^{k \cdot 5670} = \frac{1}{2} C_0$$

Thus

$$e^{5670k} = \frac{1}{2}$$

making

$$5{,}670k = \ln\left(\frac{1}{2}\right) \approx -0.693$$

so

$$k \approx -0.00012225$$

Hence

$$A(t) = C_0 e^{-.00012225t} \quad \bullet$$

The practice of carbon 14 dating rests on the assumption that the *proportion* of carbon 14 present in any naturally occurring carbon compound is constant (being produced by the impact of cosmic rays on standard carbon). Thus when an organism that has been assimilating carbon with this given proportion of carbon 14 dies and stops assimilating *new* carbon, the amount of carbon 14 in the remains decays to lower and lower fractions of the original proportion.

Example 3 Approximately, how old is a bone that contains only one-tenth the amount of carbon 14 it contained originally?

Solution

$$A(t) = C_0 e^{-0.00012225t}$$

where C_0 is the original carbon 14 (at $t = 0$). We know the amount decreased by a factor of $\frac{1}{10}$, and we know k, so we can use (8.1.5) and solve for the time elapsed directly:

$$\Delta t = \frac{\ln(\frac{1}{10})}{-0.00012225} \approx 18{,}835 \text{ years} \quad \bullet$$

PROGRESS TEST 1

1. Suppose the population of a city is growing exponentially and increasing, from 20,000 twenty years ago to 50,000 today.
(*a*) Give the city's population as a function of time t measured in years with $t = 0$, 20 years ago.
(*b*) When will the population reach 100,000 if it continues to grow according to this formula?

2. Suppose you talk a rich uncle into giving you 1 cent for the first day you do not smoke, 2 cents for the second, 4 cents for the third, 8 for the fourth, and so forth, doubling each additional day you do not smoke.
(*a*) Write the formula $A = 2^{t-1}$ (where $t - 1$ is the number of smokeless days) in the form of (8.1.4).
(*b*) How many smokeless days before you become (i) a millionaire, (ii) a billionaire, (iii) a trillionaire?

3. If $C_0 = 5$, give two descriptions of the natural decay function that corresponds to a half-life of 1 unit of time.

GROWTH AND DECAY OF MONEY

We shall now examine the idea of continuously compounded interest to clarify the meaning of "rate" in reference to rates of natural growth and to arrive at the exponential function in an alternative way, a way compatible with the limit description of e; see (7.4.12), page 267.

Suppose we have a given amount of money (often referred to as "principal") C_0 in a bank on which interest is paid, say at a rate i, where i is expressed as a decimal. Compounded annually, this yields $C_0 + iC_0$ after 1 year. If it is compounded twice a year, then after $\frac{1}{2}$ year has elapsed, the interest earned on C_0 for the first half-year, $(i/2)C_0$, is added to the principal and the total, $C_0 + (i/2)C_0$, earns interest for the second half of the year. The interest for the second half is $(i/2)[C_0 + (i/2)C_0]$, which yields a total principal after one full year of

$$\begin{aligned}\left(C_0 + \frac{i}{2}C_0\right) + \frac{i}{2}\left(C_0 + \frac{i}{2}C_0\right) &= C_0\left[\left(1 + \frac{i}{2}\right) + \frac{i}{2}\left(1 + \frac{i}{2}\right)\right] \\ &= C_0\left(1 + \frac{i}{2}\right)\left(1 + \frac{i}{2}\right) \\ &= C_0\left(1 + \frac{i}{2}\right)^2\end{aligned}$$

A similar analysis applies if interest is compounded n times per year (after each period of length $1/n$), yielding a total principal after one full year of

$$C_0\left(1 + \frac{i}{n}\right)^n$$

If this process is repeated for t years, the resulting total principal is

$$C_0\left(1 + \frac{i}{n}\right)^{tn}$$

(the number of compoundings is merely increased by a factor of t).

If we let the frequency of compoundings grow without bound (monthly, weekly, daily, every hour, every second, and so on), compounding the interest continuously, the total principal at the end of t years is the limit

$$\lim_{n\to+\infty} C_0\left(1 + \frac{i}{n}\right)^{nt} = \lim_{n\to+\infty} C_0\left[\left(1 + \frac{i}{n}\right)^{n}\right]^{t} = C_0\left[\lim_{n\to+\infty}\left(1 + \frac{i}{n}\right)^{n}\right]^{t}$$

The limit inside the brackets is *almost* the same as the limit we arrived at in (7.4.12). In fact, if we let $u = n/i$, we get (omitting the constant C_0)

$$\begin{aligned}\lim_{n\to+\infty}\left(1 + \frac{i}{n}\right)^{n} &= \lim_{n\to+\infty}\left(1 + \frac{1}{n/i}\right)^{(n/i)i}\\ &= \lim_{u\to+\infty}\left(1 + \frac{1}{u}\right)^{ui}\\ &= \left[\lim_{u\to+\infty}\left(1 + \frac{1}{u}\right)^{u}\right]^{i} \qquad (i \text{ does not vary as } u \to +\infty)\\ &= [e]^{i} \qquad \text{by (7.4.12)}\end{aligned}$$

Hence, the resulting total principal becomes, after t years of continuously compounded interest,

$$P(t) = C_0(e^{i})^{t} = C_0e^{it}$$

Of course, this is exactly the same as the formula that results if we assume the money "grows naturally"; in this case the constant of proportionality is i, the interest rate. The rate i is the rate at which *each dollar* is growing at any instant. It is the derivative of $P(t) = C_0e^{it}$ divided by $P(t)$:

$$\frac{C_0e^{it}(i)}{C_0e^{it}} = i$$

However, this rate should be distinguished from the "effective annual rate," which is the amount that each dollar grows over a whole year of compounding. For example, at a rate of $i = 0.06$ compounded continuously, after one year, one dollar grows to

$$P(1) = 1 \cdot e^{(0.06)(1)} \approx \$1.06184$$

The effective annual rate in this case would likely be advertised as 6.18 percent. For optimistic investors, our rule of thumb for doubling time says that money paying at the rate i compounded continuously will double in about $0.69/i$ years. For $i = 0.06$ this is about 11.6 years.

× Inflation can be thought of as a decay of money. Since inflation occurs continuously, it can be thought of mathematically as "natural" decay, acting as a negative interest rate. For example, by (8.1.5) we know that the half-life of money when the inflation rate is 7 percent is:

$$\Delta t = \frac{\ln(1/2)}{-0.07} \approx 9.9 \text{ years}$$

(The general "halving-time" formula is both too obvious and too depressing to state here.)

As expected, the interest rate and inflation rate have opposite effects on net return. If C_0 dollars are invested at an interest rate i during a period when the inflation rate is d, then after t years these dollars are worth (in today's dollars)

$$P(t) = C_0e^{(i-d)t}$$

PROGRESS TEST 2

1. What compound interest rate yields an effective annual return of 100 percent?
2. What compound interest rate will yield an effective annual return of approximately 271.83 percent?
3. What growth rate will cause the population of the United States to double in about 60 years? (This is approximately the current growth rate.)

PRESENT AND FUTURE VALUE OF INVESTMENTS

Using that the future value (neglecting inflation) of C_0 dollars invested at a continuously compounded interest rate i is $P(t) = C_0e^{it}$, we can determine how much money C_0 needs to be so invested now to return P dollars in t years: we merely solve the previous equation for C_0, getting

8.1.6 $$C_0 = Pe^{-it}$$

Economists refer to C_0 as the *present value* of P dollars to be paid in t years *discounted at rate* i.

Example 4 Determine the present value of a paid-up insurance policy guaranteed to pay a lump sum of $50,000 to the beneficiary in 20 years, given that money could be equivalently invested during this period at 7 percent compounded continuously. (Note that the policy's market value or loan collateral value is based upon this present value.)

Solution By (8.1.6)

$$C_0 = (50{,}000)e^{-(0.07)20} = (50{,}000)e^{-1.4} \approx \$12{,}330 \quad \bullet$$

Suppose again that money can be invested at the rate i compounded continuously during a period. However, instead of receiving P dollars in a lump sum b years from now, we are to get the P dollars in n equal installments of P/n at the times $t_0 = 0, t_1, \ldots, t_j, \ldots, t_n = b$, where $t_j - t_{j-1} = \Delta t = b/n$. (Hence $n = b/\Delta t$.) Then the present value of a typical jth payment at time t_j is

$$\frac{P}{n}e^{-it_j} = Pe^{-it_j}\frac{1}{b/\Delta t} = \frac{P}{b}e^{-it_j}\,\Delta t$$

Thus the sum of all such payments has present value

$$\sum_{j=0}^{n} \frac{P}{b}e^{-it_j}\,\Delta t$$

which is clearly a Riemann sum based upon the given partition of $[0, b]$. If we let the frequency of payments increase without bound, with $\Delta t \to 0$, the income approaches a continuous constant stream and so its present value approaches the definite integral [whose value can be determined by (8.2.2), page 289]

$$\int_0^b \frac{P}{b}e^{-it}\,dt$$

In this case the annual rate of payment is a constant P/b dollars per year, but an analogous argument applies when the income stream varies continuously with time. If the annual rate of income at time t is given by the continuous function $f(t)$, then the present value of this income stream over the next b years is

$$\int_0^b f(t)\,e^{-it}\,dt$$

Since integration techniques necessary to handle such integrals will not be available until Chap. 11, we confine ourselves to the constant case.

The market value of income producing property is based upon the present value of the income stream it can be expected to produce.

Example 5 ⨯ **(Recall the Apartment-Rental Discussion of Sec. 5.4.)** The local college, in need of dorm space and tired of hearing complaints of student ripoffs, decides to purchase the apartment building. The battle-weary owner asks \$230,000 for the building. Assuming that the monthly income is \$2000, the expected life of the complex is 20 years, and money could earn 8 percent during this period, is this a fair price?

Solution The annual income rate is (12)(2000) = 24000, so we compute (8.2.2)

$$\int_0^{20} 24000e^{-0.08t}\,dt = \frac{24000}{-0.08}\left[e^{(-0.08)(20)} - e^0\right] \approx \$239{,}431$$

The answer is yes because the value of the income exceeds the asking price. ●

The previous analysis of income-producing investments also applies to other types of investments as well. For example, the "true value" of a solar hot water system is the present value of the "income stream" not spent over the expected life of the system. A similar remark applies to any other money-saving investment; see Exercises 28 and 29.

PROGRESS TEST 3

⨯ Congratulations! You have won a \$1 million lottery, which entitles you to \$50,000 per year for the next 20 years in approximately a continuous stream.

1. Determine the present value of your ticket given that the interest rate over this period will be 8 percent.

2. Determine the actual present value of the after-tax income assuming an income tax of 30 percent.

CONSTRAINED GROWTH—SOME DIFFERENTIAL EQUATIONS

Although the word "natural" is used to describe the growth-decay phenomena studied so far in this section, exponential growth of a population occurs only when growth conditions are perfect. This occurs in nature, if at all, for only a short period of time, after which the growth becomes constrained by such limiting factors as access to nutrients and space, declining fertility, increasing toxicity of the environment due to waste products, etc. We shall now examine two ways such constraints can be incorporated into a more sophisticated and realistic mathematical description of the growth situation.

Suppose that the growth rate $k = k(t)$ is a function of time t (instead of being constant as earlier). The simplest case is $k(t)$ linear, say $k(t) = a + bt$. In this case, separating the variables as before, we get the differential equation

$$\frac{dA}{A} = (a + bt)\,dt$$

that leads to the solution.

8.1.7 $$A = Ce^{at+bt^2/2}$$

In this case, if $a, b > 0$, A is a strictly increasing function whose graph is concave upward, but if a and b are of opposite sign, the graph has a relative extreme at $t = -a/b$ (see Exercise 19).

The second case we examine assumes that the population has an upper bound $\bar{A}$ determined by the limits of the environment to support growth. In this case we assume that *the rate of growth is proportional to the product of the current population A and its difference from the maximum supportable population $\bar{A}$*. That is, we assume that

8.1.8 $$\frac{dA}{dt} = kA(\bar{A} - A)$$

This represents the effects of both forces—the tendency to grow opposed by the environmental limitations. Note that as A gets closer to $\bar{A}$, dA/dt approaches 0. We omit the details proving that the solution of this differential equation is given by:

8.1.9 **The Logistic Function:** $A(t) = C\bar{A}/(C + e^{-k\bar{A}t})$

The behavior of this function and that in (8.1.7) will be explored in the exercises. Notice that by allowing either k or $\bar{A}$, or both, to vary with time, we can develop functions even more sensitive to the subtle influences on population growth.

SECTION 8.1 EXERCISES

1. Suppose an exponentially growing population doubles in 3 years. How long will it take to triple?

2. Suppose an exponentially growing population triples in 3 years. How long did it take to double?

3. (*a*) Suppose a naturally growing population increases from 20,000 to 30,000 in 5 years. How long will it take to increase from 30,000 to 45,000?
(*b*) Compare the function $A(t) = 20{,}000e^{kt}$, where k is as determined in part (*a*), with the function $B(t) = 20{,}000(1.5)^t$.

4. Suppose the population of a city growing exponentially at the rate of 8 percent per year is 100,000. When was the population 75,000?

5. (*a*) If the half-life of radioactive catatonium is 20 years, how long will it take for the substance to decay to one-third of its original amount?
(*b*) What is the "third-life" of catatonium?
(*c*) Does catatonium have a "ninth-life"? And if so, what is it?

6. What is the effective annual rate for money that accumulates interest at a rate of 12 percent per month compounded continuously?

7. Suppose the population of a city grows by 20 percent every 10 years. How long will it take the population to double?

8. A piece of organic material is found to have only 1 percent of its carbon 14 remaining. Approximately how old is it?

9. Repeat Exercise 8 for a piece of material with 2 percent carbon 14.

10. Carbon 14 dating is only accurate up to about 70,000 years. Below approximately what percentage of carbon 14 would the age estimate be undependable?

11. Given \$10,000 to invest at 9 percent compounded continuously for 5 years or 10 percent compounded annually, which situation will yield the larger return?

12. Repeat Exercise 11 for a period of 10 years.

13. (*a*) Suppose that a \$25 U.S. savings bond maturing in 10 years can be purchased for \$18.50. What is the growth rate (interest) of these bonds?
(*b*) What is the growth rate if the price is lowered to \$15?
(*c*) Neglecting patriotism, which is a more profitable 10-year investment, the \$15 savings bond or a bank deposit paying $5\frac{1}{4}$ percent compounded continuously?
(*d*) What would your answer be to part (*c*) if the prevailing bank interest rates look like they are likely to drop soon?

14. Given that the United States population grows exponentially and was 205 million in 1970 and 225 million in 1975:
(*a*) What will the population be in the year 2000?
(*b*) When will the population reach 500 million?

15. Exercise 14 led to a growth rate of about 0.0186. Answer the corresponding questions if the growth rate is reduced to 0.015.

16. (*a*) Given that the half-life of a certain radioactive substance is k years, write a function of the form $A = C_0(1/2)^u$ (where u is a linear function of time t) that describes the amount on hand at time t.

(*b*) Convert your answer in (*a*) to the natural exponential form.

17. Given that the government charges \$650 for a 10-year \$1000 savings bond, what is the implied rate of continuous compounding?

18. (*a*) Which is more profitable, an investment that guarantees to triple your money in 20 years, or a return of $5\frac{1}{2}$ percent continuously compounded?
(*b*) Given a 5 percent inflation rate through the next 20 years, reevaluate your above decision. Explain your conclusion.

19. Graph (8.1.7) for $a = 1$, $b = -4$.

20. (*a*) Where does equation (8.1.9) have a point of inflection?
(*b*) Graph (8.1.9) for $C = 1$, $\bar{A} = 1000$, and $k = 0.05$.

21. Suppose that the air resistance on a falling body is proportional to its downward velocity, so the downward acceleration is 32 ft/s^2 minus a positive constant k times v.
(*a*) Write an appropriate differential equation involving v.
(*b*) Solve this equation.
(*c*) Where does the initial velocity v_0 appear in the solution?
(*d*) What happens to v as t grows large?
(*e*) Describe the effect of a parachute in terms of k.

22. An infectious disease is usually assumed to spread through a susceptible population at a rate proportional to the product of the number infected times the number not yet infected. (The same assumption has been applied to the spread of rumors.)
(*a*) Give a differential equation that models this assumption.
(*b*) Solve this equation.

23. (Newton's Law of Cooling) The temperature T of a body in a medium at temperature T_1 changes at a rate directly proportional to $T - T_1$; that is, $dT/dt = k(T - T_1)$. Show that, as a function of time $T(t) = (T_0 - T_1)e^{kt} + T_1$, where $T_0 = T(0)$, the initial temperature.

24. (See Exercise 23.) Suppose a thermometer reading 80°F is brought into a freezer set at 10°F and its temperature falls to 45°F in 5 min. Determine how long it will take to cool to (*a*) 30°F, (*b*) 15°F. Then (*c*) determine the thermometer's temperature after 1 h in the freezer.

25. (See Exercise 23.) A pot of boiling water (100°C) is removed from the stove in a room at 20°C and cools 5°C in 5 min. With $t = 0$ being the time it was removed from the stove, how long will it take to cool down to 30°C?

26. (See Exercise 23.) (*a*) Repeat Exercise 25 for another pot that cools to 20°C in 20 min.
(*b*) Which pot in Exercises 25 and 26 is better insulated? Or are they equally well insulated?

27. Repeat Example 5 under the assumption that (because of expected inflation) money will return 10 percent during the period.

28. × Assuming that money earns 7 percent over a given period of 10 years (compounded continuously), what is the present value of a solar hot water system that will save \$40/month over these 10 years? What is the actual number of dollars saved by this system in these 10 years?

29. Suppose a door-to-door insulation salesman offers to insulate your house for \$5000. Assuming money earns 8 percent and that an average of \$50/month will be saved over 20 years, is it worth the investment?

30. Give the present value version of the interest-inflation formula immediately preceding Progress 2 [see (8.1.6)]. Then interpret its meaning in the three cases: (*a*) $i > d$, (*b*) $i < d$, (*c*) $i = d$.

Section 8.2: Integrating Transcendental Functions: Basic Formulas

Objectives:

1. Integrate logarithmic, exponential, and trigonometric functions.

2. Integrate combinations of these functions.

LOGARITHMIC AND EXPONENTIAL FUNCTIONS

By the Fundamental Theorem of the Calculus (see Chap. 6) we know that if F is an antiderivative of f (so $F' = f$), then

$$\int_a^b f(x)\,dx = F(b) - F(a)$$

The Power Rule for derivatives led to the Power Rule for antiderivatives:

$$\int u^r\,du = \frac{u^{r+1}}{r+1} + C \qquad \text{for } r \neq -1$$

However, in Chap. 7 we determined derivative formulas for a wide class of functions. Each derivative formula likewise gives rise to an antidifferentiation formula. For example, if $u(x)$ is a differentiable function of x, then

$$D_x[\ln|u(x)|] = \frac{u'(x)}{u(x)}$$

It follows that

$$\int \frac{u'(x)}{u(x)}\, dx = \ln|u(x)| + C$$

Since $1/u(x) = u(x)^{-1}$, we have extended the Power Rule to cover the case $r = -1$:

8.2.1

$$\int u^r\, du = \begin{cases} \dfrac{u^{r+1}}{r+1} + C & \text{for } r \neq -1 \\ \ln|u| + C & \text{for } r = -1 \end{cases}$$

In (8.2.1) and frequently below, to keep the formulas brief we write $u(x)$ simply as u, and (using the definition of differential) we write $u'(x)\, dx$ as du.

Example 1 Evaluate $\int \frac{x^2\, dx}{x^3 + 1}$.

Solution Here we recognize that the numerator is almost the differential of the denominator. That is, letting $u = x^3 + 1$, we know $du = 3x^2\, dx$, so $x^2\, dx = (1/3)\, du$. Hence

$$\int \frac{x^2\, dx}{x^3 + 1} = \int \frac{(1/3)\, du}{u} = \frac{1}{3}\int \frac{du}{u}$$
$$= \frac{1}{3}\ln|u| + C$$
$$= \frac{1}{3}\ln|x^3 + 1| + C$$

We can also write this antiderivative as $\ln|\sqrt[3]{x^3 + 1}| + C$. ●

Henceforth in this section *we assume that $u = u(x)$ is a differentiable function of x.* The derivative formula,

$$D_x[e^{u(x)}] = e^{u(x)}u'(x)$$

implies the antiderivative formula

8.2.2

$$\int e^{u(x)}u'(x)\, dx = e^{u(x)} + C$$

Example 2 Determine $\int_0^1 e^{x^2+2x}(x + 1)\, dx$.

Solution Let $u = x^2 + 2x$, so $du = (2x + 2)\, dx$. Thus $(x + 1)\, dx = \frac{1}{2}\, du$ and

$$\int e^{x^2+2x}(x + 1)\, dx = \int e^u \frac{1}{2}\, du = \frac{1}{2}\int e^u\, du$$
$$= \frac{1}{2}e^u + C = \frac{1}{2}e^{x^2+2x} + C$$

Hence, by the Fundamental Theorem,

$$\int_0^1 e^{x^2+2x}(x+1)\,dx = \left[\frac{1}{2}e^{x^2+2x}\right]_0^1$$
$$= \frac{1}{2}e^3 - \frac{1}{2}e^0 = \frac{1}{2}(e^3 - 1) \quad \bullet$$

In (7.2.3) we concluded that

$$D_x[\sin u(x)] = [\cos u(x)]\, u'(x)$$

from which it follows that $\int \cos u\, du = \sin u + C$. Similarly, each of the other derivative formulas developed in Chap. 7 yields an antiderivative formula. These are now summarized.

8.2.3 (*a*) $\int \cos u\, du = \sin u + C$ (*b*) $\int \sin u\, du = -\cos u + C$

(*c*) $\int \sec^2 u\, du = \tan u + C$ (*d*) $\int \csc^2 u\, du = -\cot u + C$

(*e*) $\int \sec u \tan u\, du = \sec u + C$ (*f*) $\int \csc u \cot u\, du = -\csc u + C$

(*g*) $\int b^u\, du = \dfrac{1}{\ln b} b^u + C$

Example 3 Evaluate $\displaystyle\int \frac{\sec^2 \sqrt{x}}{\sqrt{x}}\,dx$.

Solution Since we have a formula involving $\sec^2 u$ [(8.2.3(*c*)], we try the substitution $u = \sqrt{x} = x^{1/2}$. Then $du = (1/2)x^{-1/2}\,dx = (dx/2)\sqrt{x}$, so $dx/\sqrt{x} = 2\,du$. Hence

$$\int \frac{\sec^2 \sqrt{x}}{\sqrt{x}}\,dx = \int \sec^2 \sqrt{x}\left(\frac{dx}{\sqrt{x}}\right) = \int \sec^2 u(2\,du)$$
$$= 2\int \sec^2 u\,du$$
$$= 2\tan u + C \qquad \text{from [8.2.3(}c\text{)]}$$
$$= 2\tan\sqrt{x} + C \quad \bullet$$

Example 4 Determine $\displaystyle\int \frac{\sin(\ln x)}{x}\,dx$.

Solution Let $u = \ln x$, so $du = dx/x$. Then

$$\int \frac{\sin(\ln x)}{x}\,dx = \int \sin u\,du$$
$$= -\cos u + C = -\cos(\ln x) + C \quad \bullet$$

Example 5 Determine $\displaystyle\int \sqrt{e^x - 1}\, e^x\,dx$.

Solution Note that $\sqrt{e^x - 1} = (e^x - 1)^{1/2}$ and that for $u = e^x - 1$, we have $du = e^x\,dx$. Hence

$$\int \sqrt{e^x - 1}\, e^x\,dx = \int u^{1/2}\,du$$
$$= \frac{u^{3/2}}{3/2} + C = \frac{2}{3}(e^x - 1)^{3/2} + C \quad \bullet$$

PROGRESS TEST 1

In Probs. 1 to 4 evaluate the given integral.

1. $\int \cos 3x\, e^{\sin 3x}\, dx$ **2.** $\int e^{2x} \csc e^{2x} \cot e^{2x}\, dx$ **3.** $\int (1 - \sin^2 x)\cos x\, dx$ **4.** $\int_{-4}^{5} \frac{dx}{x+5}$

5. Determine the area under one arch of the sine curve from 0 to π.

MORE ANTIDERIVATIVE FORMULAS

Among the six basic trigonometric functions, antiderivative formulas for only the sine and cosine appear in (8.2.3). We shall determine the integrals of the four remaining trigonometric functions using (8.2.1).

8.2.4 (*a*) $\int \tan u\, du = \ln|\sec u| + C$

(*b*) $\int \cot u\, du = \ln|\sin u| + C$

(*c*) $\int \sec u\, du = \ln|\sec u + \tan u| + C$

(*d*) $\int \csc u\, du = \ln|\csc u - \cot u| + C$

Proof: We shall derive (*a*) and (*c*). Formulas (*b*) and (*d*) are obtained in a similar way and are left as exercises.

(*a*) $\int \tan x\, dx = \int (\sin x/\cos x)\, dx$. Let $v = \cos x$, then $dv = -\sin x\, dx$, so $-dv = \sin x\, dx$. Hence $\int \tan x\, dx = -\int dv/v = -\ln|v| + C = -\ln|\cos x| + C = \ln|(\cos x)|^{-1} + C = \ln|1/(\cos x)| + C = \ln|\sec x| + C$.
The formula follows from the Chain Rule for Integrals.

(*c*) We shall evaluate $\int \sec x\, dx$ by first rewriting the integrand as a fraction whose numerator is the derivative of the denominator:

$$\sec x = \frac{\sec x(\sec x + \tan x)}{\sec x + \tan x} = \frac{\sec x \tan x + \sec^2 x}{\sec x + \tan x}$$

Now

$$\int \sec x\, dx = \int \frac{(\sec x \tan x + \sec^2 x)\, dx}{\sec x + \tan x} = \int \frac{dv}{v}$$

where $v = \sec x + \tan x$. Thus

$$\int \sec x\, dx = \ln|v| + C = \ln|\sec x + \tan x| + C$$

Again the general formula follows by the Chain Rule. ●

Example 6 Determine $\int \frac{x\, dx}{\cos(x^2)}$.

Solution Since $1/\cos(x^2) = \sec(x^2)$, we know

$$\begin{aligned}\int \frac{x\, dx}{\cos(x^2)} &= \int (\sec x^2)x\, dx \\ &= \int \sec u \frac{1}{2}\, du \qquad (u = x^2,\ du = 2x\, dx) \\ &= \frac{1}{2}\ln|\sec u + \tan u| + C \\ &= \frac{1}{2}\ln|\sec x^2 + \tan x^2| + C = \ln\sqrt{|\sec x^2 + \tan x^2|} + C\end{aligned}$$ ●

PROGRESS TEST 2

Evaluate each integral in Probs. 1 to 3.

1. $\int 2^x \csc 2^x\,dx$

2. $\int \tan^3 x\,dx$ (use $1 + \tan^2 x = \sec^2 x$)

3. $\int e^x \tan e^x\,dx$

4. Prove [8.2.4(*b*)].

SECTION 8.2 EXERCISES

In Exercises 1 to 36 evaluate the given integral.

1. $\int \frac{\sec^2 \sqrt{x+1}}{\sqrt{x+1}}\,dx$

2. $\int \frac{\sqrt{1+\ln x}}{x}\,dx$

3. $\int_0^1 \frac{dx}{2x+1}$

4. $\int_{2/5}^1 \frac{dx}{5x-1}$

5. $\int \frac{x\,dx}{x^2+1}$

6. $\int \frac{x^5\,dx}{x^6-1}$

7. $\int \frac{2^{\sqrt{x}}}{\sqrt{x}}\,dx$

8. $\int \frac{dx}{x \ln x}$

9. $\int \frac{dx}{x\sqrt{\ln x}}$

10. $\int x^2 \cot x^3\,dx$

11. $\int (xe^{x^2+2x} + e^{x^2+2x})\,dx$

12. $\int_0^{\pi/3} e^{\sec x} \sec x \tan x\,dx$

13. $\int x^2 \csc x^3\,dx$

14. $\int \frac{\sin(1/x)}{x^2}\,dx$

15. $\int \frac{x^3 \ln(x^4+5)}{x^4+5}\,dx$

16. $\int e^{4x} \cos e^{4x}\,dx$

17. $\int_0^{\pi/4} \frac{e^{\tan x}\,dx}{\cos^2 x}$

18. $\int \frac{e^{\csc x} \cos x}{\sin^2 x}\,dx$

19. $\int_0^1 \frac{x}{e^{x^2}}\,dx$

20. $\int e^x \tan e^x\,dx$

21. $\int_{-2}^1 e^{x+1}\,dx$

22. $\int \frac{\sec^2 2x}{7 + 3\tan 2x}\,dx$

23. $\int \frac{\sin x\, e^{\sqrt{\cos x}}}{\sqrt{\cos x}}\,dx$

24. $\int \frac{\sin x\, e^{\sec x}}{\cos^2 x}\,dx$

25. $\int_{-1}^1 \frac{e^x}{1+e^x}\,dx$

26. $\int \cos 2x \tan(\sin 2x)\,dx$

27. $\int x\, 3^{x^2+1}\,dx$

28. $\int \sqrt[3]{e^{x-3}+1}\, e^{x-3}\,dx$

29. $\int \frac{\csc(\sqrt{x}+1)\cot(\sqrt{x}+1)\,dx}{\sqrt{x}}$

30. $\int \sin 3x \sqrt[3]{\cos 3x}\,dx$

31. $\int \frac{dx}{e^x}$

32. $\int \frac{dx}{\cos^2 x \sqrt{\tan x}}$

(*In Exercises 33 to 36 divide before integrating.*)

33. $\int \frac{x^3 + 3x^2 + x - 1}{x-5}\,dx$

34. $\int \frac{x^2-1}{2x+1}\,dx$

35. $\int \frac{2x^2+3x+5}{2x+1}\,dx$

36. $\int \frac{4-3x}{8-5x}\,dx$

37. Determine the area under the tangent curve from $x = \pi/6$ to $x = \pi/3$.

38. Determine the area under the curve $y = 1/(x+3)$ from $a = -2$ to $b = 6$.

39. Determine the volume of the solid generated by revolving the region under the curve $y = e^{x^2}$ from $x = 0$ to $x = 1$ about the y axis.

40. Determine the volume of the solid generated by revolving the region bounded by the lines $x = -1$, $x = 1$, $y = e^x$, and the x axis about the x axis.

41. Determine the average value of $f(x) = e^x$ on the interval $[0, 1]$.

42. Determine the average value of $f(x) = (\ln x)/x$ on $[1, e]$.

43. Evaluate $\int \sin x \cos x\,dx$ in two ways: (*a*) by letting $u = \sin x$ and (*b*) by letting $u = \cos x$. (*c*) Explain the difference in the appearance of your answers. (*d*) Evaluate $\int_0^\pi \sin x \cos x\,dx$ by using the two antiderivatives obtained in (*a*) and (*b*) and applying the Chain Rule for Integrals in each case.

44. Evaluate $\int \sec^2 x \tan x\,dx$ in two ways: (*a*) by letting $u = \sec x$ and (*b*) letting $u = \tan x$. (*c*) Explain the difference in the appearance of your answers. (*d*) As in Exercise 43(*d*), evaluate $\int_0^{\pi/4} \sec^2 x \tan x\,dx$ in two ways.

45. Prove [8.2.4(*d*)].

46. Compute $\int e^{\ln x}\,dx$.

47. Compute $\int \ln(e^x)\,dx$.

48. Compute $D_x[\int_1^x e^{\ln t}\,dt]$.

Section 8.3: Integrating Powers and Products of Trigonometric Functions

Objectives:

Evaluate integrals of the following:

1. Certain powers of sines and cosines
2. Products of certain powers of sines and cosines
3. Certain powers of the remaining trigonometric functions
4. Certain products of powers of tangents and secants or cotangents and cosecants

SINES AND COSINES

We shall make frequent use of the following identities.

8.3.1 (*a*) $\sin^2 t + \cos^2 t = 1$ (*b*) $\sin^2 t = \dfrac{1 - \cos 2t}{2}$

(*c*) $\cos^2 t = \dfrac{1 + \cos 2t}{2}$

Our use will generally be one of two types. We shall use either (*a*) in dealing with odd powers of sine and cosine or the double-angle formulas (*b*) and (*c*) in dealing with even powers.

8.3.2 Procedure for Evaluating Integrals of the Form $\int \sin^n u \cos^m u\, du$

Case I: If either m or n is a positive odd integer, use the identity $\sin^2 u + \cos^2 u = 1$ in order to apply the Power Rule to a power of either sines or cosines. (See Example 1.)

Case II: If both m and n are nonnegative even integers, use double-angle formulas, repeatedly if necessary, to lower the powers of sine and cosine.

The simplest situation has either m or n equal to zero.

Example 1 Evaluate $\int \sin^3 x\, dx$.

Solution Here n is odd and $m = 0$, so the Case I procedure applies. Rewrite the integral as indicated:

$$\begin{aligned}
\int \sin^3 x\, dx &= \int \sin^2 x \sin x\, dx \\
&= \int (1 - \cos^2 x) \sin x\, dx \qquad \text{(using [8.3.1}(a)\text{])} \\
&= \int \sin x\, dx - \int \cos^2 x \sin x\, dx \\
&= -\cos x - \int v^2(-dv) \qquad (v = \cos x,\ dv = -\sin x\, dx) \\
&= -\cos x + \frac{v^3}{3} + C = -\cos x + \frac{\cos^3 x}{3} + C \quad \bullet
\end{aligned}$$

Example 2 Evaluate $\int \cos^4 x\, dx$.

Solution Here $n = 0$ and m is even, so the Case II procedure applies. Rewrite the integral as indicated:

$$\begin{aligned}
\int \cos^4 x\, dx &= \int (\cos^2 x)^2\, dx \\
&= \int \left(\frac{1 + \cos 2x}{2}\right)^2 dx \qquad \text{(using [8.3.1}(c)\text{])} \\
&= \frac{1}{4} \int (1 + 2\cos 2x + \cos^2 2x)\, dx \\
&= \frac{1}{4}\left[\int dx + \int (\cos 2x)2\, dx + \int \frac{1 + \cos 4x}{2}\, dx\right] \qquad \text{(using [8.3.1}(c)\text{])} \\
&= \frac{x}{4} + \frac{\sin 2x}{4} + \frac{x}{8} + \frac{1}{8}\int \cos 4x\, dx \\
&= \frac{3x}{8} + \frac{\sin 2x}{4} + \frac{\sin 4x}{32} + C \quad \bullet
\end{aligned}$$

In the last step we did not explicitly write out the substitution $u = 4x$, $du = 4\,dx$. More generally, using the substitution $u = kx$, $du/k = dx$, we see that

$$\int \cos kx\,dx = \int \cos u \frac{du}{k} = \frac{1}{k}\sin u + C = \frac{1}{k}\sin kx + C$$

Similarly,

$$\int \sin kx\,dx = \frac{-1}{k}\cos kx + C$$

In using procedure (8.3.2), where neither n nor m is zero, the computations are very similar to those in Examples 1 and 2.

Example 3 Evaluate $\int \sin^4 x \cos^2 x\,dx$.

Solution Here Case II applies.

$$\begin{aligned}\int \sin^4 x \cos^2 x\,dx &= \int [\sin^2 x]^2 \cdot \cos^2 x\,dx\\ &= \int \left[\frac{1-\cos 2x}{2}\right]^2\left[\frac{1+\cos 2x}{2}\right]dx\\ &= \frac{1}{8}\int (1 - 2\cos 2x + \cos^2 2x)(1+\cos 2x)\,dx\\ &= \frac{1}{8}\int (1 - \cos 2x - \cos^2 2x + \cos^3 2x)\,dx\\ &= \frac{1}{8}\int dx - \frac{1}{8}\int \cos 2x\,dx - \frac{1}{8}\int \cos^2 2x\,dx + \frac{1}{8}\int \cos^3 2x\,dx\\ (1) \qquad &= \frac{1}{8}x - \frac{1}{16}\sin 2x - \frac{1}{8}\int \cos^2 2x\,dx + \frac{1}{8}\int \cos^3 2x\,dx\end{aligned}$$

Now

$$\begin{aligned}\int \cos^2 2x\,dx &= \frac{1}{2}\int [1 + \cos 4x]\,dx\\ &= \frac{1}{2}x + \frac{1}{8}\sin 4x\end{aligned}$$

and

$$\begin{aligned}\int \cos^3 2x\,dx &= \int (1 - \sin^2 2x)\cos 2x\,dx\\ &= \frac{1}{2}\int (1 - u^2)\,du \qquad \left(\text{u} = \sin 2x,\ \frac{du}{2} = \cos 2x\,dx\right)\\ &= \frac{1}{2}\left(u - \frac{u^3}{3}\right) = \frac{1}{2}\sin 2x - \frac{1}{6}\sin^3 2x\end{aligned}$$

Substituting these results into (1) and combining like terms, we obtain

$$\int \sin^4 2x \cos^2 2x\,dx = \frac{1}{16}x - \frac{1}{64}\sin 4x - \frac{1}{48}\sin^3 2x + C \quad \bullet$$

It is important to realize that Case I applies even when one of m or n is not a positive integer, as in Prob. 3, Progress Test 1.

PROGRESS TEST 1

Evaluate each of the following indefinite integrals:

1. $\int \sin^4 x\,dx$ **2.** $\int \cos^5 x\,dx$ **3.** $\int \sqrt{\sin 3x}\cos^3 3x\,dx$ **4.** $\int \sin^2 \frac{x}{2}\cos^2 \frac{x}{2}\,dx$

TANGENTS AND SECANTS

We shall now consider integrals involving the remaining four trigonometric functions. We shall give procedures for tangents and secants, but an analogous analysis applies for cotangents and cosecants. As is explained in the remarks at the end of this section, our procedures are by no means all-inclusive. We shall use the following identities.

8.3.3 (*a*) $1 + \tan^2 t = \sec^2 t$ (*b*) $1 + \cot^2 t = \csc^2 t$

Example 4 Evaluate $\int \tan^2 u\, du$.

Solution Use [8.3.3(*a*)] to rewrite the integral as follows:

$$\int \tan^2 u\, du = \int (\sec^2 u - 1)\, du$$
$$= \tan u - u + C \quad \bullet$$

We can extend the idea of Example 4 to handle integrals of the form

$$\int \tan^m u\, du \qquad \text{where } m > 2 \text{ is positive integer}$$

$$\int \tan^m u\, du = \int \tan^{m-2} u(\sec^2 u - 1)\, du$$
$$= \int \tan^{m-2} u \sec^2 u\, du - \int \tan^{m-2} u\, du$$
$$= \frac{\tan^{m-1} u}{m-1} - \int \tan^{m-2} u\, du$$

We have thus lowered the degree by 2. Repeated applications will reduce the degree to 2 if m is even (so Example 4 applies) or to 1 if m is odd, in which case [8.2.4(*a*)] applies: $\int \tan u\, du = \ln|\sec u| + C$.

We shall now apply the cotangent version of the above procedure.

Example 5 Evaluate $\int \cot^3 u\, du$.

Solution We apply [8.3.3(*b*)] to rewrite the integral as follows:

$$\int \cot^3 u\, du = \int \cot u(\csc^2 u - 1)\, du$$
$$= \int \cot u(\csc^2 u\, du) - \int \cot u\, du$$
$$= \frac{-\cot^2 u}{2} - \ln|\sin u| + C \quad \bullet$$

The first term of the answer is a result of the Power Rule with $v = \cot u$, $-dv = \csc^2 u\, du$; and the second term is a result of applying [8.2.4(*b*)]. Note that $-\ln|\sin u| = \ln|\sin u|^{-1} = \ln|\csc u|$, so the answer could also be presented as

$$\ln|\csc u| - \frac{\cot^2 u}{2} + C$$

We could even rewrite $(\cot^2 u)/2$ as

$$\frac{\csc^2 u - 1}{2} = \frac{\csc^2 u}{2} - \frac{1}{2}$$

and absorb the $-1/2$ into the constant C, leaving our answer in the form

$$\ln|\csc u| - \frac{\csc^2 u}{2} + C$$

Manipulations such as these are common in applications, so you should be on the alert for various forms of an answer.

We now move on to more general situations.

8.3.4 Procedure for Evaluating Integrals of the Form $\int \sec^n u \tan^m u \, du$

Case I: If n is a positive even integer, factor out $\sec^2 u \, du$ and replace the remaining secant factors by tangents using the identity $1 + \tan^2 u = \sec^2 u$.

Case II: If m is a positive odd integer, factor out $\sec u \tan u \, du$ and replace the remaining tangent factors by secants using the identity $\tan^2 u = \sec^2 u - 1$.

To apply the procedure of Case I, it is not necessary that m be an integer.

Example 6 Evaluate $\int \sqrt[3]{\tan x} \sec^4 x \, dx$.

Solution Here $n = 4$, so Case I applies.

$$\begin{aligned}
\int \sqrt[3]{\tan x} \sec^4 x \, dx &= \int \tan^{1/3} x \sec^2 x \sec^2 x \, dx \\
&= \int \tan^{1/3} x \, [1 + \tan^2 x] \sec^2 x \, dx \\
&= \int u^{1/3} [1 + u^2] \, du \qquad (u = \tan x, du = \sec^2 x \, dx) \\
&= \int (u^{1/3} + u^{7/3}) \, du \\
&= \frac{3u^{4/3}}{4} + \frac{3u^{10/3}}{10} + C \\
&= \frac{3 \tan^{4/3} x}{4} + \frac{3 \tan^{10/3} x}{10} + C \quad \bullet
\end{aligned}$$

Again, to apply Case II, n need not be a positive integer.

Example 7 Evaluate $\displaystyle\int \frac{\tan^5 x}{\sqrt{\sec x}} \, dx$.

Solution Here $m = 5$, so Case II applies.

$$\begin{aligned}
\int \frac{\tan^5 x}{\sqrt{\sec x}} &= \int \sec^{-1/2} x \tan^5 x \, dx \\
&= \int \sec^{-3/2} x \tan^4 x \sec x \tan x \, dx \\
&= \int \sec^{-3/2} x [\tan^2 x]^2 \sec x \tan x \, dx \\
&= \int \sec^{-3/2} x [\sec^2 x - 1]^2 \sec x \tan x \, dx \\
&= \int u^{-3/2} [u^2 - 1]^2 \, du \qquad (u = \sec x, du = \sec x \tan x \, dx) \\
&= \int (u^{5/2} - 2u^{1/2} + u^{-3/2}) \, du \\
&= \frac{2u^{7/2}}{7} - \frac{4u^{3/2}}{3} - 2u^{-1/2} + C \\
&= \frac{2 \sec^{7/2} x}{7} - \frac{4 \sec^{3/2} x}{3} - 2 \sec^{-1/2} x + C \quad \bullet
\end{aligned}$$

PROGRESS TEST 2

Evaluate the following integrals:

1. $\int \tan^7 x \, dx$
2. $\int \sec^{2/3} x \tan^3 x \, dx$
3. $\int \csc^4 x \, dx$

A CAUTION

In Secs. 8.2 and 8.3 we cataloged *some* combinations of transcendental functions that can be handled using substitution and trigonometric identities. At this point we cannot handle integrals such as $\int x \sin x\, dx$, $\int xe^x\, dx$, or integrals of odd powers of the secant (or cosecant) function. These all await *integration by parts* in Chap. 11, Sec. 11.1. In Sec. 11.5 we shall show how tables of integrals can also be used to evaluate integrals of trigonometric functions.

Other integrals, *resembling* those dealt with previously—for example, $\int e^{x^2}\, dx$ and $\int \sqrt{\sin x}\cos^2 x\, dx$—can be determined only by using infinite series and numerical techniques (Chap. 16).

SECTION 8.3 EXERCISES

In Exercises 1 to 30 evaluate the given integral.

1. $\int \sin^3 x\, dx$

2. $\int \cos^5 2x\, dx$

3. $\int \sin^2 x \cos^4 x\, dx$

4. $\int (\sec t \tan t)^2\, dt$

5. $\int \cot^7 \theta\, d\theta$

6. $\int \sqrt{\sin(3x + 1)} \cos(3x + 1)\, dx$

7. $\int \tan^2(2x - 5)\, dx$

8. $\int_0^{\pi/3} \sec^4 2x\, dx$

9. $\int_0^{\pi/4} \tan^5 x\, dx$

10. $\int_0^{\pi/6} \cos^4 3x \sin 3x\, dx$

11. $\int (\cos x \sin x)^3\, dx$

12. $\int \sqrt{\cos x} \sin^3 x\, dx$

13. $\int \cot^5 x \csc^2 x\, dx$

14. $\int \sin^2 x \cos^3 x\, dx$

15. $\int \cot^5 x \csc^4 x\, dx$

16. $\int \sin^4 x \cos^4 x\, dx$

17. $\int \frac{\sin^3 u}{\sqrt[3]{\cos^2 u}}\, du$

18. $\int \cot^2 x \csc^4 x\, dx$

19. $\int \tan^3 \theta \sec^3 \theta\, d\theta$

20. $\int \cos^{2/3} t \sin^3 t\, dt$

21. $\int e^x \sin^2 e^x\, dx$

22. $\int \frac{\cot^5(\ln x)}{x}\, dx$

23. $\int \tan^3 x \sqrt[3]{\sec x}\, dx$

24. $\int_{\pi/4}^{\pi/2} \csc^2 x \cot^4 x\, dx$

25. $\int_0^{\pi/8} \sin^6 s\, ds$

26. $\int_0^{\pi/6} \cos^4 x\, dx$

27. $\int_{-\pi/2}^{\pi/2} \sin^3 100x\, dx$

28. $\int_{\pi/6}^{\pi/4} \frac{\sin^3 x}{\cos^2 x}\, dx$

29. $\int_{\pi/4}^{\pi/2} \cot^3 x\, dx$

30. $\int \frac{\sqrt{\sin^5 x}}{\cos x}\, dx$

31. Determine the volume of the solid generated by revolving about the x axis the region bounded by the x axis and one arch of the sine curve.

32. Repeat Exercise 31 for one arch of the cosine curve. Would you expect your answer to be different from that of Exercise 31?

33. Determine the average value of $f(x) = \sin^2 x$ between 0 and $\pi/2$.

34. Determine the average value of $g(x) = \tan x$ on $[-\pi/4, \pi/4]$.

In Exercises 35 to 40 apply the following identities:

$$\sin a \cos b = \frac{1}{2}[\sin(a-b) + \sin(a+b)]$$

$$\sin a \sin b = \frac{1}{2}[\cos(a-b) - \cos(a+b)]$$

$$\cos a \cos b = \frac{1}{2}[\cos(a-b) + \cos(a+b)]$$

35. $\int \sin 5x \cos 2x\, dx$ **36.** $\int \sin 4x \cos 2x\, dx$

37. $\int (\sin 5x + \cos 3x)^2\, dx$ (*Hint:* Expand and then let $a = 5$, $b = 3$ in the first identity.)

38. $\int \sin 3x \sin 10x\, dx$ **39.** $\int \cos 7x \cos 5x\, dx$

40. $\int \cos 3x \cos x \cos 5x\, dx$

41. Evaluate and compare $\int \sin 2x\, dx$ and $2\int \sin x \cos x\, dx$.

42. Evaluate and compare $\int \cos 2x\, dx$ and $\int (\cos^2 x - \sin^2 x)\, dx$.

43. Evaluate and compare $\displaystyle\int \tan 2x\, dx$ and $\displaystyle\int \frac{2\tan x}{1-\tan^2 x}\, dx$.

Section 8.4: Inverse Trigonometric Functions

Objectives:

1. Define and evaluate inverse trigonometric functions.

2. Differentiate inverse trigonometric functions.

3. Use the skills in (1) and (2) in applications and in determining relationships between functions involving inverse trigonometric functions.

INTRODUCTION

Recall that when introducing the idea of inverse function we contrasted the functions $f(x) = x^3$ and $g(x) = x^2$. The first, being strictly increasing on its domain R, had an inverse.

On the other hand, an attempt to define $g^{-1}(x)$ for x in the range of g—the set of nonnegative real numbers—runs into difficulty because there are two possible square roots of x: $+\sqrt{x}$ and $-\sqrt{x}$. That is, a nonzero x in the range of g "came from" two different numbers.

We can avoid this difficulty by trimming our ambitions. Rather than define an inverse for the *whole* function $g(x) = x^2$, we shall define an inverse for *part* of g, that portion of g defined on the nonnegative real numbers. This restricted function, denoted g_p, with domain $[0, +\infty)$ is strictly increasing and thus has an inverse:

$$g_p^{-1}(x) = +\sqrt{x} \qquad \text{for all } x \text{ in range of } g_p$$

which also happens to be $[0, +\infty)$.

Graphically, we formed g_p from g by lopping off the left side of the parabola; see Fig. 8.1(*a*). The graph of g_p^{-1} is the mirror image of the graph of g_p across the (dashed) diagonal; see Fig. 8.1(*b*).

We shall follow this same procedure in determining inverses for the trigonometric functions, none of which is strictly increasing on its entire domain.

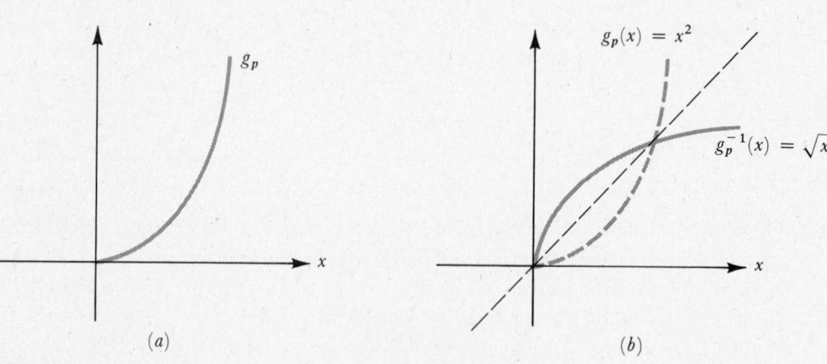

Fig. 8.1

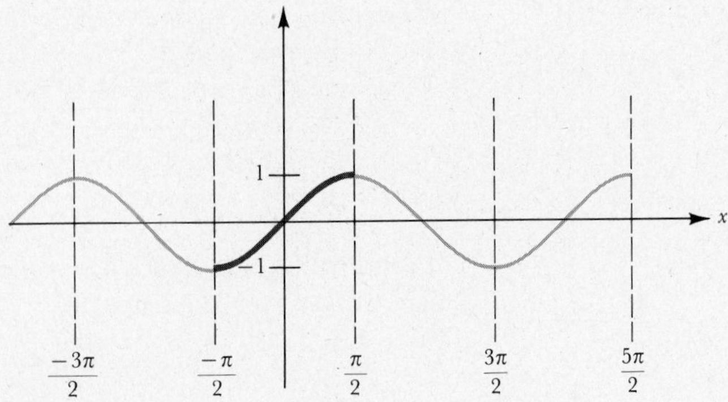

Fig. 8.2

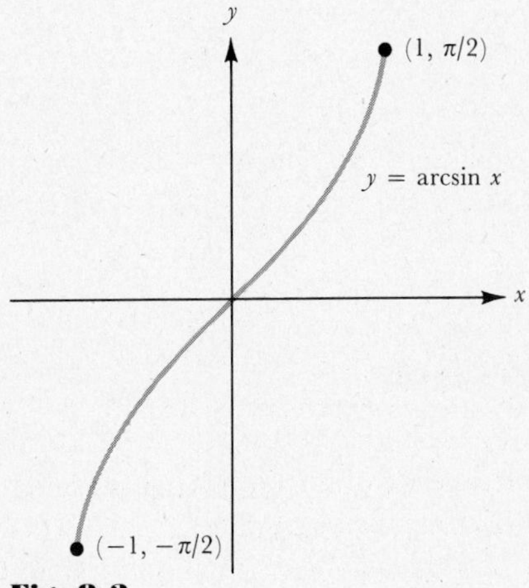

Fig. 8.3

THE INVERSE SINE FUNCTION

As evident from the graph of the sine function (Fig. 8.2), there are many ways we could restrict the domain to get a strictly monotonic piece of the sine function. Following tradition we choose the restriction of $\sin x$ to the interval $[-\pi/2, \pi/2]$. The graph of this restricted sine function is the thickened portion in Fig. 8.2. The function

$$f(x) = \sin x \qquad x \text{ in } [-\pi/2, \pi/2]$$

is strictly increasing on its domain $[-\pi/2, \pi/2]$ and has range $[-1, 1]$. Hence it has an inverse whose domain is $[-1, 1]$ and whose range is $[-\pi/2, \pi/2]$. More precisely, we have:

8.4.1 **Definition** For each x in $[-1, 1]$ we define the *inverse sine of x*, denoted *arcsin x*, to be the unique y in $[-\pi/2, \pi/2]$ such that $\sin y = x$.

Some authors write the inverse sine of x as $\sin^{-1} x$, but we prefer $\arcsin x$. It is easier to say and, more important, it avoids the possible confusion with $1/(\sin x)$. The graph of $y = \arcsin x$, the reflection across the diagonal $y = x$ of our restricted sine function, appears in Fig. 8.3.

Example 1 Compute (*a*) to (*e*):
(*a*) $\arcsin (1/\sqrt{2})$ (*b*) $\arcsin(-1/\sqrt{2})$ (*c*) $\arcsin(\sin \pi/6)$,
(*d*) $\arcsin(\sin 5\pi/6)$ (*e*) $\sin(\arcsin 1/2)$

Solution (*a*) $\arcsin (1/\sqrt{2})$ is that number y with $-\pi/2 \le y \le \pi/2$ such that $\sin y = 1/\sqrt{2}$. Since $\sin \pi/4 = 1/\sqrt{2}$ and $-\pi/2 \le \pi/4 \le \pi/2$, we know that $\arcsin (1/\sqrt{2}) = \pi/4$.

(*b*) $\arcsin(-1/\sqrt{2})$ is that y with $-\pi/2 \le y \le \pi/2$ such that $\sin y = -1/\sqrt{2}$. But $\sin(-\pi/4) = -1/\sqrt{2}$ and $-\pi/2 \le -\pi/4 \le \pi/2$, so $\arcsin(-1/\sqrt{2}) = -\pi/4$.

(*c*) $\arcsin(\sin \pi/6)$ is that y in $[-\pi/2, \pi/2]$ such that $\sin y = \sin \pi/6$. Thus $y = \pi/6$. (Who, we ask again, is buried in Grant's Tomb?)

(*d*) $\arcsin(\sin 5\pi/6)$ is that y in $[-\pi/2, \pi/2]$ such that $\sin y = \sin 5\pi/6$. Now $5\pi/6$ is *not* in $[-\pi/2, \pi/2]$. However, $\sin 5\pi/6 = 1/2 = \sin \pi/6$, and $\pi/6$ *is* in $[-\pi/2, \pi/2]$, so $\arcsin(\sin 5\pi/6) = \arcsin(1/2) = \pi/6$.

(*e*) Now $\arcsin 1/2$ is that y in $[-\pi/2, \pi/2]$ such that $\sin y = 1/2$, so $\arcsin 1/2 = \pi/6$. Thus $\sin(\arcsin 1/2) = \sin(\pi/6) = 1/2$. ●

Recall that, in general, we have the inverse relationships $f^{-1}(f(x)) = x$ for x in the domain of f and $f(f^{-1}(x)) = x$ for x in the domain of f^{-1}. Parts (*c*) and (*e*), Example 1, are special cases, respectively, of:

8.4.2 (*a*) For x in $[-\pi/2, \pi/2]$, $\arcsin(\sin x) = x$.
(*b*) For x in $[-1, 1]$, $\sin(\arcsin x) = x$.

Part (*d*) illustrates that if x is *not* in $[-\pi/2, \pi/2]$, then there is a unique x_1 that *is* in $[-\pi/2, \pi/2]$ such that

$$\arcsin(\sin x) = \arcsin(\sin x_1) = x_1$$

PROGRESS TEST 1

Compute the indicated quantity.

1. $\arcsin 1/2$ **2.** $\arcsin(-1/2)$ **3.** $\arcsin(\sin \pi/8)$ **4.** $\sin(\arcsin 0.012)$
5. $\arcsin(\sin 2\pi/3)$

DEFINITIONS OF THE OTHER INVERSE TRIGONOMETRIC FUNCTIONS

We shall follow the same pattern and define inverses of the remaining trigonometric functions by first appropriately restricting their domains.

We restrict the cosine function to $[0, \pi]$, the heavy part of the graph in Fig. 8.4, where it is strictly decreasing and hence has an inverse.

8.4.3 **Definition** For each x in $[-1, 1]$ we define the *inverse cosine of x*, denoted by *arccos x*, to be the unique y in $[0, \pi]$ such that $\cos y = x$. (See Fig. 8.5.)

To define the arctangent function, we first restrict the tangent function as indicated by the heavy part of the graph in Fig. 8.6. Note that the domain of this strictly increasing restricted tangent function is $(-\pi/2, \pi/2)$, and its range is R.

8.4.4 **Definition** For each real number x we define the *inverse tangent of x*, denoted by *arctan x*, to be the unique y in $(-\pi/2, \pi/2)$ such that $\tan y = x$. (See Fig. 8.7.)

Because its domain is all of R, the arctangent is perhaps the most widely used of the inverse trigonometric functions. As is evident from Fig. 8.7,

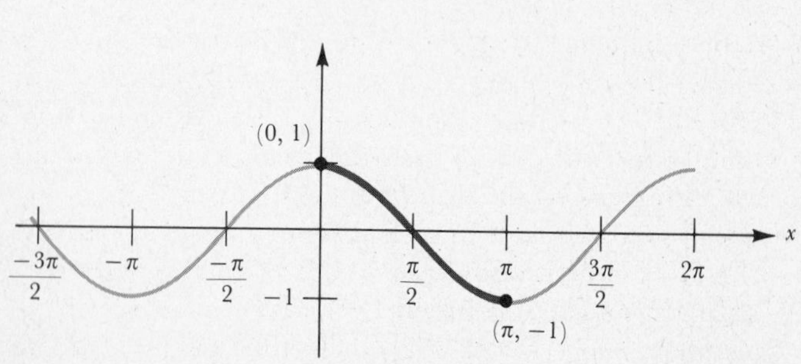

Fig. 8.4

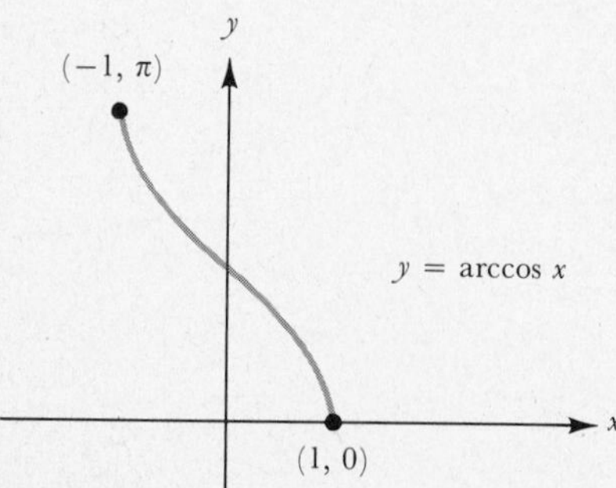

Fig. 8.5

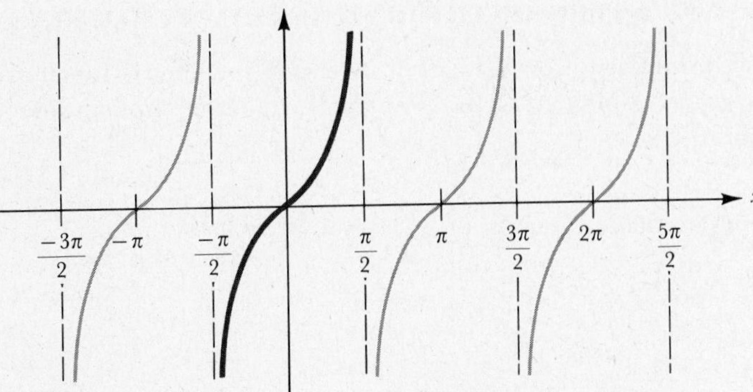

Fig. 8.6

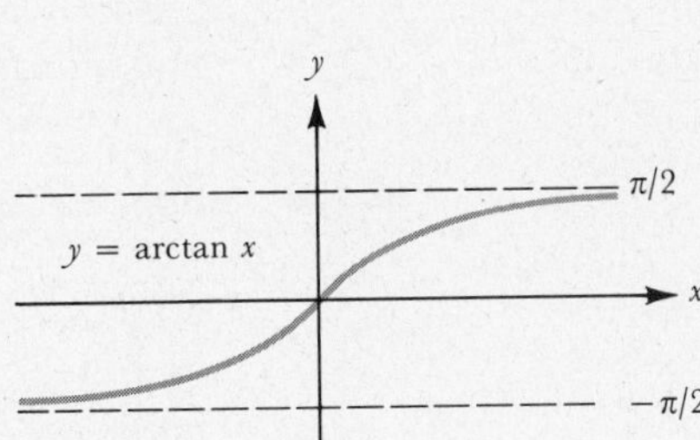

Fig. 8.7

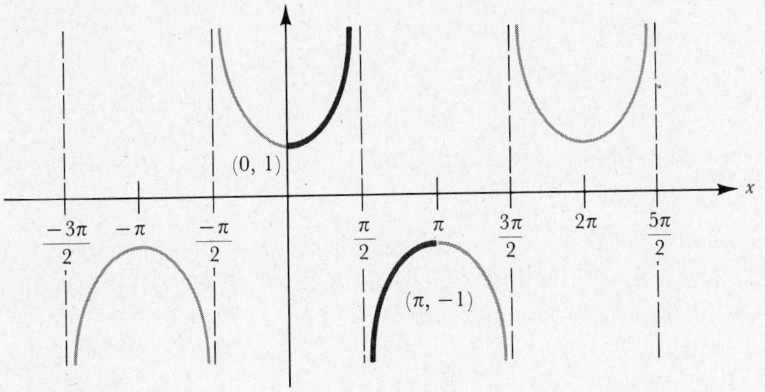

Fig. 8.8

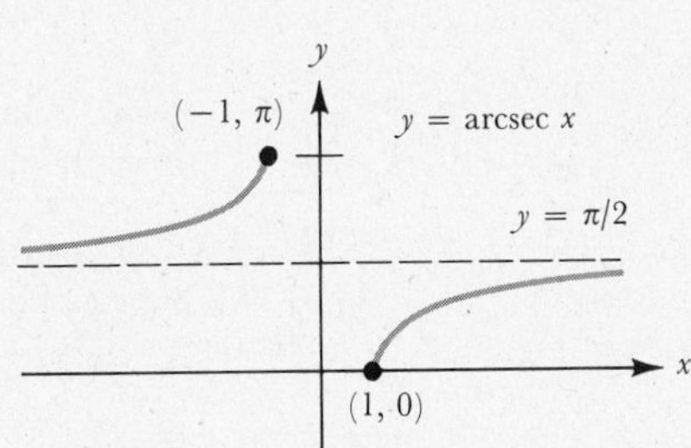

Fig. 8.9

8.4.5 (*a*) $\lim_{x\to+\infty} \arctan x = \pi/2$ (*b*) $\lim_{x\to-\infty} \arctan x = -\pi/2$

The previous definitions of inverse trigonometric functions are based upon rather standard restrictions. No such consensus exists on how to restrict the secant function to obtain a strictly increasing or strictly decreasing function. We choose the restriction to $[0, \pi/2)$ and $(\pi/2, \pi]$, as in Fig. 8.8. The range of this restricted secant function consists of the disjoint intervals $(-\infty, -1]$ and $[1, +\infty)$.

8.4.6 **Definition** For each x in $(-\infty, -1]$ or $[1, +\infty)$, we define the *inverse secant of x*, denoted by *arcsec x,* to be the unique y in $[0, \pi]$ such that $\sec y = x$. (See Fig. 8.9.)

Note that y will never equal $\pi/2$ in (8.4.6) because $\sec \pi/2$ is undefined. From Fig. 8.9 it is apparent that:

8.4.7 (*a*) $\lim_{x\to+\infty} \operatorname{arcsec} x = \pi/2$ (*b*) $\lim_{x\to-\infty} \operatorname{arcsec} x = \pi/2$

The inverses of the cotangent and cosecant functions are seldom used in practice and will be defined later in this section in terms of those inverses already defined.

PROGRESS TEST 2

Compute those of the following that are defined.

1. $\arctan(-\sqrt{3})$ **2.** $\operatorname{arcsec}(-2)$ **3.** $\operatorname{arcsec}[\sec(-\pi/4)]$ **4.** $\arctan[\tan(19\pi/6)]$
5. $\operatorname{arcsec}(1/2)$

DERIVATIVES OF INVERSE TRIGONOMETRIC FUNCTIONS

We shall compute formulas for derivatives of inverse trigonometric functions similarly to the way we computed $D_x[\ln x]$—namely, by using implicit differentiation.

8.4.8 Suppose u is a differentiable function of x. Then

(*a*) $D_x(\arcsin u) = \dfrac{1}{\sqrt{1-u^2}}\dfrac{du}{dx}, \ |u| < 1$

(*b*) $D_x(\arccos u) = -\dfrac{1}{\sqrt{1-u^2}}\dfrac{du}{dx}, \ |u| < 1$

(*c*) $D_x(\arctan u) = \dfrac{1}{1+u^2}\dfrac{du}{dx}$

(*d*) $D_x(\text{arcsec}\, u) = \dfrac{1}{|u|\sqrt{u^2-1}}\dfrac{du}{dx}, \ |u| > 1$

Proof: As usual, we shall prove the formulas for $u(x) = x$, from which the stated formulas follow by the Chain Rule.

(*a*) Let $y = \arcsin x$, $-1 < x < 1$. Then by definition $\sin y = x$. Moreover, by the Theorem on Inverse Functions (7.3.3) dy/dx exists, so we can differentiate implicitly with respect to x, getting

$$\cos y \frac{dy}{dx} = 1 \qquad \text{and} \qquad \frac{dy}{dx} = \frac{1}{\cos y}$$

Since $\sin^2 y + \cos^2 y = 1$, we know that $\cos y = \pm\sqrt{1-\sin^2 y}$. But for y in $[-\pi/2, \pi/2]$, $\cos y \geq 0$, so $\cos y = +\sqrt{1-\sin^2 y}$. Thus

$$\frac{dy}{dx} = \frac{1}{\sqrt{1-\sin^2 y}}$$

$$= \frac{1}{\sqrt{1-x^2}} \qquad \text{since } x = \sin y$$

(*b*) See Exercise 59 below.

(*c*) Let $y = \arctan x$, so $x = \tan y$ and

$$1 = \sec^2 y \frac{dy}{dx} \qquad \text{or} \qquad \frac{dy}{dx} = \frac{1}{\sec^2 y}$$

Since $\sec^2 y = 1 + \tan^2 y = 1 + x^2$, we know that

$$\frac{dy}{dx} = \frac{1}{1+x^2}$$

(*d*) Let $y = \text{arcsec}\, x$, where we know $0 < y < \pi$. Now $\sec y = x$, so

(1) $$\sec y \tan y \frac{dy}{dx} = 1 \qquad \text{or} \qquad \frac{dy}{dx} = \frac{1}{\sec y \tan y}$$

But $\sec y = x$ and $\tan y = \pm\sqrt{\sec^2 y - 1} = \pm\sqrt{x^2-1}$, so

(2) $$\frac{dy}{dx} = \frac{1}{x(\pm\sqrt{x^2-1})}$$

What is the sign of dy/dx? Well, for $0 < y < \pi/2$, $\sec y$ and $\tan y$ are both positive, and for $\pi/2 < y < \pi$, $\sec y$ and $\tan y$ are both negative. Thus the product $\sec y \tan y$ is always positive, and therefore $dy/dx = 1/(\sec y \tan y)$ in (1) must be positive for all values of y. We can guarantee this by writing

$$\frac{dy}{dx} = \frac{1}{|x|\sqrt{x^2 - 1}}$$

[Notice that simply choosing the + sign in (2) would not suffice.] ■

Notice that the derivatives of $y = \arcsin x$, $y = \arctan x$, and $y = \operatorname{arcsec} x$ are all nonnegative on their respective domains, a fact that agrees with their respective graphs: Each function is increasing on its natural domain. On the other hand, $y = \arccos x$, which is decreasing, has a negative derivative.

The derivatives of the remaining inverse trigonometric functions are seldom needed in practice. This will be clearer after we establish a simple relation between $\arcsin x$ and $\arccos x$. Since

$$D_x(\arccos x) = -\frac{1}{\sqrt{1 - x^2}} = -D_x(\arcsin x)$$

we know that

$$D_x(\arccos x + \arcsin x) = D_x(\arccos x) + D_x(\arcsin x) = 0$$

Hence $\arccos x + \arcsin x = C$, a constant [by (4.4.6)]. For $x = 0$, $\arccos 0 + \arcsin 0 = \pi/2 + 0 = C$, so $C = \pi/2$. Thus we can conclude:

8.4.9 Theorem $\arccos x = \pi/2 - \arcsin x$

Following (8.4.9), we define the inverse cotangent and cosecant:

8.4.10 Definition (*a*) $\operatorname{arccot} x = \pi/2 - \arctan x$
(*b*) $\operatorname{arccsc} x = \pi/2 - \operatorname{arcsec} x \qquad (|x| \geq 1)$

Example 2 Differentiate $f(x) = \arcsin(2/x^2)$.

Solution

$$\begin{aligned} f'(x) &= \frac{1}{\sqrt{1 - (2/x^2)^2}} D_x\left(\frac{2}{x^2}\right) \\ &= \frac{1}{\sqrt{1 - (4/x^4)}} \cdot \frac{-4}{x^3} \\ &= \frac{1}{\sqrt{(x^4 - 4)/x^4}} \cdot \frac{-4}{x^3} \\ &= \frac{x^2}{\sqrt{x^4 - 4}} \cdot \frac{-4}{x^3} = \frac{-4}{x\sqrt{x^4 - 4}} \end{aligned}$$ ●

Example 3 Differentiate $f(x) = \operatorname{arcsec}(\ln x)$

Solution

$$\begin{aligned} f'(x) &= \frac{1}{|\ln x|\sqrt{(\ln x)^2 - 1}} D_x(\ln x) \\ &= \frac{1}{x|\ln x|\sqrt{(\ln x)^2 - 1}} \end{aligned}$$ ●

Example 4 A balloon is released (from eye level) 300 m from an observer and rises at the rate of 240 m/min. Find the rate (in radians per minute) at which the angle of elevation of the observer's line of sight is increasing when the balloon is 400 m high.

Solution Let θ be the angle of elevation at time t and x the height of the balloon at time t. Then (see sketch)

$$\theta = \arctan\left(\frac{x}{300}\right)$$

$$\frac{d\theta}{dt} = \frac{1}{1 + (x/300)^2}\frac{d}{dt}\left(\frac{x}{300}\right)$$

$$= \frac{90{,}000}{90{,}000 + x^2}\,\frac{1}{300}\,\frac{dx}{dt}$$

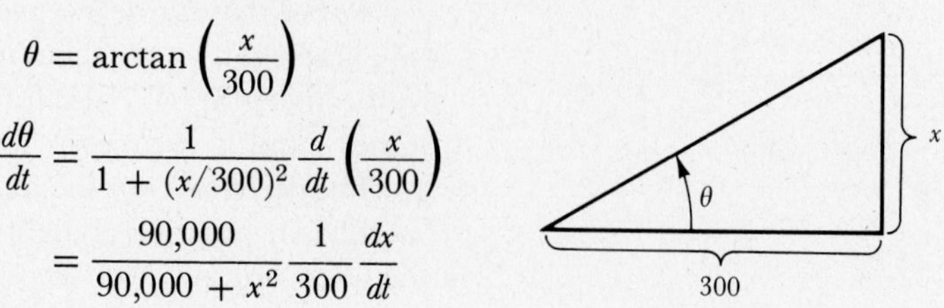

Letting $x = 400$ and $dx/dt = 240$, we have

$$\frac{d\theta}{dt} = \frac{90{,}000}{90{,}000 + 160{,}000}\cdot\frac{1}{300}\cdot 240$$

$$= \frac{72{,}000}{250{,}000} = 0.288 \text{ rad/min} \quad \bullet$$

PROGRESS TEST 3

Find dy/dx:

1. $y = \arcsin(2x^2 - x + 3)$

2. $y = \operatorname{arcsec}(\sqrt{x^2 + 1})$

SECTION 8.4 EXERCISES

In Exercises 1 to 20 evaluate the given expression.

1. $\arctan(-1/\sqrt{3})$

2. $\operatorname{arcsec}(-2/\sqrt{3})$

3. $\arcsin(\sqrt{3}/2)$

4. $\tan[\operatorname{arcsec}(-3)]$

5. $\cot[\arctan(2/3)]$

6. $\arccos 1/2$

7. $\operatorname{arccot} 1$

8. $\operatorname{arccot}(\tan 3\pi/4)$

9. $\operatorname{arccsc}(2)$

10. $\arcsin(\csc \pi/6)$

11. $\csc[\operatorname{arccsc}(-2)]$

12. $\operatorname{arcsec}(-2/\sqrt{3})$

13. $\cot[\arctan(-1)]\cdot$

14. $\sin[\arccos(\sqrt{3}/2)]$

15. $\arctan(\cot \pi/3)$

16. $\arctan[\tan(9\pi/8)]$

17. $\operatorname{arcsec}(\sec 2\pi/3)$

18. $\arcsin[\sin(29\pi/4)]$

19. $\sin[\lim_{x\to+\infty} \operatorname{arcsec}(2x)]$

20. $\csc[\lim_{x\to-\infty} \arctan(x/3)]$

In Exercises 21 to 40 differentiate the given expression.

21. $y = \arcsin\sqrt{x}$

22. $y = \arctan\sqrt{x}$

23. $y = \operatorname{arcsec}\sqrt{x^2 - 1}$

24. $f(x) = x^2 \arcsin(2x)$

25. $f(x) = \arctan(3/x)$

26. $f(x) = x \arctan(x^2)$

27. $f(x) = \operatorname{arcsec}(1/x)$

28. $f(x) = \arcsin(2x/\sqrt{1 + 4x^2})$

29. $g(t) = t \arccos(t^2 + 1)$

30. $g(x) = \sqrt{\arctan\sqrt{x}}$

31. $y = \operatorname{arccsc}(x^2 - 1)$

32. $f(u) = \operatorname{arccot}\sqrt{u^2 - 1}$

33. $h(u) = \arctan(\ln u)$

34. $f(x) = \arcsin(x/\sqrt{x^2 + 1})$

35. $f(x) = e^x \arctan(e^x)$

36. $f(x) = \operatorname{arcsec}[(4x - 5)/7]$

Section 8.6: Hyperbolic Functions and Inverse Hyperbolic Functions

Objectives:

1. Define and differentiate the hyperbolic functions.
2. Extract the hyperbolic functions from combinations of exponential functions.
3. Define inverse hyperbolic functions.
4. Interpret inverse hyperbolic functions as logarithms.
5. Use inverse hyperbolic functions in evaluating integrals.

INTRODUCTION AND DEFINITIONS

Certain combinations of exponential functions occur frequently enough, particularly in applications, to get names of their own. It turns out that the relationships among these functions and among their derivatives parallel relationships among the trigonometric functions. Their names reveal this parallel.

8.6.1 Definition For all x, we define

$$(a)\ \sinh x = \frac{e^x - e^{-x}}{2} \qquad (b)\ \cosh x = \frac{e^x + e^{-x}}{2}$$

These are called the *hyperbolic sine* and *hyperbolic cosine* functions, respectively. (Sinh x is pronounced "cinch x," and cosh x is pronounced the way it looks.)

The graphs of these functions can be obtained by simple addition of $(1/2)e^x$ and $(-1/2)e^{-x}$ in the case of sinh x, and addition of $(1/2)e^x$ and $(1/2)e^{-x}$ in the case of cosh x; see Fig. 8.10.

We define the remaining hyperbolic functions in terms of sinh x and cosh x.

8.6.2

$$(a)\ \tanh x = \frac{\sinh x}{\cosh x} = \frac{e^x - e^{-x}}{e^x + e^{-x}} \qquad (b)\ \coth x = \frac{\cosh x}{\sinh x} = \frac{e^x + e^{-x}}{e^x - e^{-x}}$$

$$(c)\ \operatorname{sech} x = \frac{1}{\cosh x} = \frac{2}{e^x + e^{-x}} \qquad (d)\ \operatorname{csch} x = \frac{1}{\sinh x} = \frac{2}{e^x - e^{-x}}$$

These functions obey identities analogous to the trigonometric identities. We state three and give the computational proof of the first.

8.6.3

$$(a)\ \cosh^2 x - \sinh^2 x = 1 \qquad (b)\ 1 - \tanh^2 x = \operatorname{sech}^2 x$$

$$(c)\ 1 - \coth^2 x = -\operatorname{csch}^2 x$$

Proof of (a):

$$\cosh^2 x - \sinh^2 x = \left(\frac{e^x + e^{-x}}{2}\right)^2 - \left(\frac{e^x - e^{-x}}{2}\right)^2$$

$$= \frac{e^{2x} + 2e^0 + e^{-2x}}{4} - \frac{e^{2x} - 2e^0 + e^{-2x}}{4}$$

$$= \frac{2+2}{4} = 1 \quad ■$$

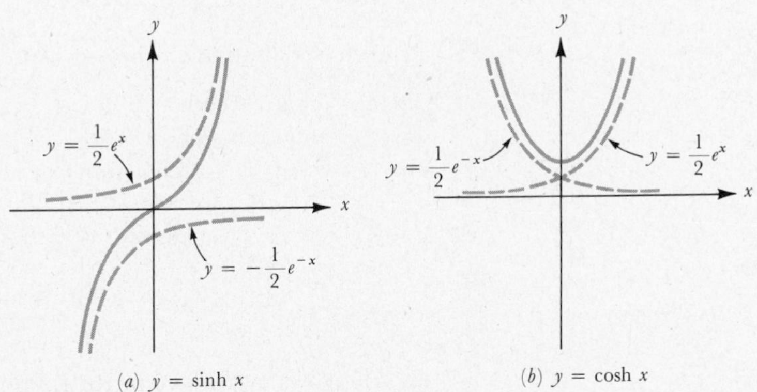

Fig. 8.10

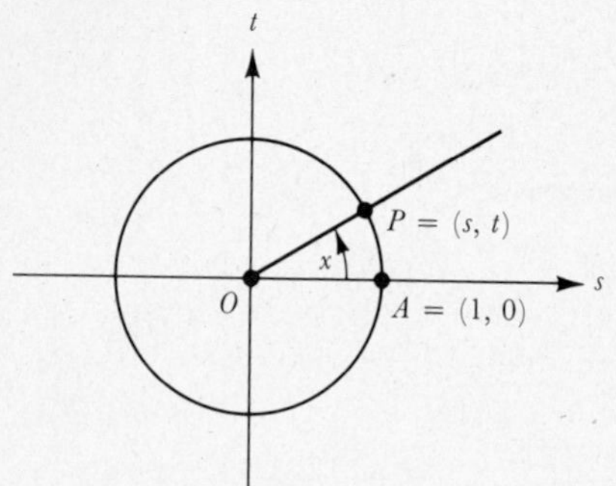

Fig. 8.11

WHY THE WORD "HYPERBOLIC"?

The ordinary trigonometric functions are also often called "circular functions" because of the role of the circle in defining them. Given the unit circle in the s-t plane we defined $\sin x = t$ and $\cos x = s$, where x is the radian measure of the angle illustrated in Fig. 8.11. We then defined the remaining trigonometric functions in terms of this circle. Because $\cos x$ and $\sin x$ are the coordinates of P on the unit circle, we know that $\cos^2 x + \sin^2 x = 1$.

For purposes of continuing and deepening the parallel between trigonometric and hyperbolic functions, we shall reinterpret x in terms of the area of sector OAP. The area of the whole unit circle is $\pi r^2 = \pi$. Since $x/(2\pi)$ is the fraction of the whole 2π radians of a full rotation, the area of sector OAP is the same fraction of the total area π. That is,

$$\text{Area of sector } OAP = \frac{x}{2\pi} \cdot \pi = \frac{x}{2}$$

Thus the variable x for the trigonometric functions is twice the area of sector OAP.

If the generating line rotated in a clockwise direction, we would interpret the area as negative, so the variable x can take on all real values.

But if we begin with the set of points (s, t) satisfying $s^2 - t^2 = 1$, we have the so-called unit hyperbola, pictured in Fig. 8.12. This particular conic section is examined in detail in Sec. 10.2.

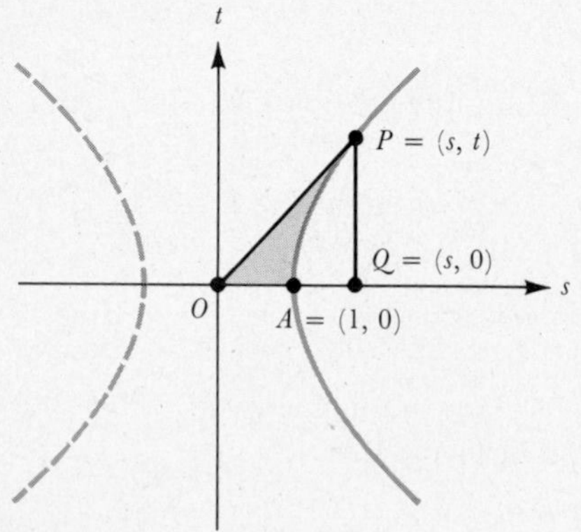

Fig. 8.12

This graph has two branches and we use the right (solid) one. As the point P moves upward and to the right along this branch of the hyperbola, the area of the shaded region OAP increases from 0 without bound, and if P moves downward and to the right below the horizontal axis, we interpret the area as negative. Hence, if we let x be double the area of OAP, x can take on all real values. We could then redefine

$$\sinh x = t \qquad \cosh x = s$$

where $P = (s, t)$, and define the remaining hyperbolic functions in terms of these as before.

In Exercise 64 we shall show that the x serving as the variable is really twice the area of OAP. Notice that since P is on the hyperbola, we know that [8.6.3(a)] holds.

You may wish to examine the behavior of $\cosh x$ and $\sinh x$ in terms of the motion of P. (For example, $\cosh x \geq 1$ since P is on the right-hand branch of the hyperbola.)

DERIVATIVES OF HYPERBOLIC FUNCTIONS

The analogy with trigonometric functions continues among the derivatives of the hyperbolic functions, although we need to be careful with minus signs.

8.6.4 For u a differentiable function of x:

(a) $D_x(\sinh u) = \cosh u \, du/dx$ (b) $D_x(\cosh u) = \sinh u \, du/dx$
(c) $D_x(\tanh u) = \operatorname{sech}^2 u \, du/dx$ (d) $D_x(\coth u) = -\operatorname{csch}^2 u \, du/dx$
(e) $D_x(\operatorname{sech} u) = -\operatorname{sech} u \tanh u \, du/dx$ (f) $D_x(\operatorname{csch} u) = -\operatorname{csch} u \coth u \, du/dx$

Proof of (a):

$$D_x(\sinh x) = D_x\left(\frac{e^x - e^{-x}}{2}\right) = \frac{e^x - e^{-x}(-1)}{2} = \cosh x$$

so (a) follows by the Chain Rule. Proofs of (b) through (f) are left as progress test problems and section exercises. ■

HANGING CABLES

Hyperbolic functions are useful in describing natural phenomena. Whereas the trigonometric functions are periodic and are thus useful for describing cyclic phenomena, the hyperbolic functions are not periodic and thus they describe different sorts of phenomena.

Suppose a homogenous flexible cable is hanging between two points symmetric with respect to the y axis such that its lowest point is $(0, a)$. An analysis of the forces on each segment of cable to determine its shape leads to the differential equation

$$\frac{d^2y}{dx^2} = \frac{1}{a}\sqrt{1 + \left(\frac{dy}{dx}\right)^2}$$

A straightforward calculation shows that $y = a \cosh (x/a)$ satisfies this differential equation, and therefore the shape of this curve (often called a "catenary curve") is the graph of the hyperbolic cosine function. Hence we see graphs of hyperbolic cosines every time we look at telephone or electric cables suspended between poles.

ANTIDIFFERENTIATION OF HYPERBOLIC FUNCTIONS

As we know, every derivative formula gives rise to an antiderivative formula. Here are the six we have so far:

8.6.5
(*a*) $\int \sinh u\, du = \cosh u + C$
(*b*) $\int \cosh u\, du = \sinh u + C$
(*c*) $\int \operatorname{sech}^2 u\, du = \tanh u + C$
(*d*) $\int \operatorname{csch}^2 u\, du = -\coth u + C$
(*e*) $\int \operatorname{sech} u \tanh u\, du = -\operatorname{sech} u + C$
(*f*) $\int \operatorname{csch} u \coth u\, du = -\operatorname{csch} u + C$

The techniques for handling combinations of hyperbolic functions are analogous to those for trigonometric functions, including the use of identities.

Example 1 Evaluate $\int \sinh^3 x\, dx$.

Solution

$$\sinh^3 x = (\sinh x)(\sinh^2 x) = \sinh x(1 + \cosh^2 x)$$

by [8.6.3(*a*)], so

$$\int \sinh^3 x\, dx = \int \sinh x\, dx + \int \cosh^2 x \sinh x\, dx$$
$$= \cosh x + \frac{\cosh^3 x}{3} + C \quad \bullet$$

Identities analogous to double-angle formulas are developed in the section exercises.

PROGRESS TEST 1

In Probs. 1 and 2 prove the given statement.

1. [8.6.3(*b*)]
2. [8.6.4(*b*)]
3. Show that $D_x(\cosh x + \sinh x) = \cosh x + \sinh x$.
4. Given that $\tanh x = 1/2$, determine the values of the remaining hyperbolic functions.
5. Evaluate $\int \operatorname{sech}^4 x\, dx$.

DEFINITIONS OF INVERSE HYPERBOLIC FUNCTIONS

The hyperbolic sine, being strictly increasing on its domain R has an inverse defined on its range R; see Fig. 8.13. On the other hand, we are required to restrict the hyperbolic cosine to its "right half"—the solid part of Fig. 8.14(a).

8.6.6 Definition

(a) For all x, $\operatorname{arcsinh} x$ is that unique y such that $\sinh y = x$.

(b) For all $x \geq 1$, $\operatorname{arccosh} x$ is that unique $y \geq 0$ such that $\cosh y = x$.

Figures 8.13 through 8.18 contain graphs of each hyperbolic function in (a) and the respective inverses in (b). For functions that are not strictly increasing, their restriction to a strictly increasing "piece" is given as the solid part of (a). Asymptotes are dotted. Included is any significant information regarding domains.

Since the hyperbolic functions are defined in terms of exponential functions, it is reasonable to expect that their inverses can be expressed in terms of logarithm functions. For example, we have:

8.6.7 (a) $\operatorname{arcsinh} x = \ln(x + \sqrt{x^2 + 1})$

(b) $\operatorname{arccosh} x = \ln(x + \sqrt{x^2 - 1}) \qquad x \geq 1$

(c) $\operatorname{arctanh} x = \dfrac{1}{2}\ln\left(\dfrac{1 + x}{1 - x}\right) \qquad |x| < 1$

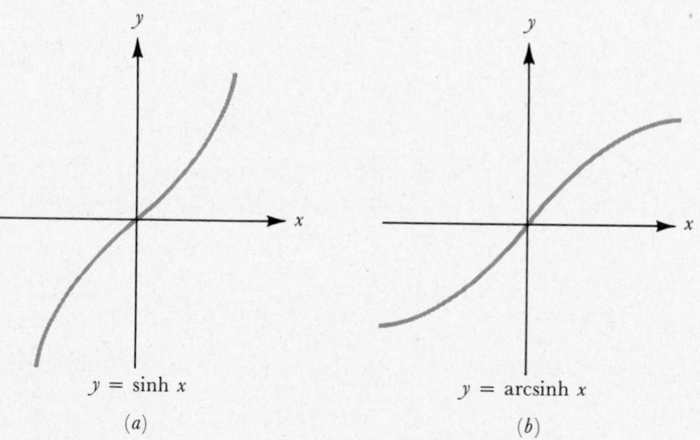

Fig. 8.13

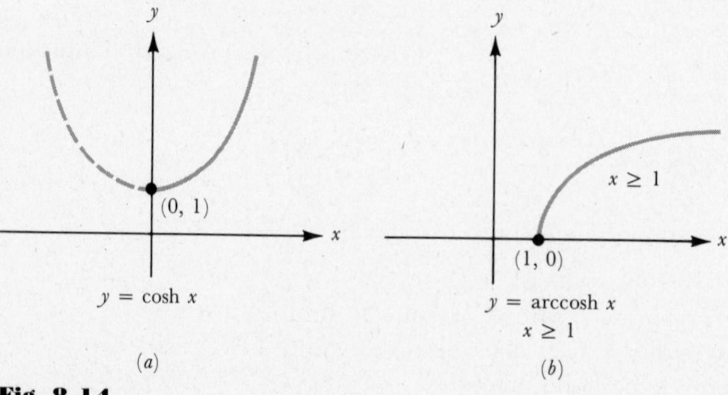

Fig. 8.14

Note that (*a*), for example, can be proved by showing the left and right sides have the same derivative and then noting each side is 0 when $x = 0$. (A standard implicit differentiation argument shows that $D_x(\text{arcsinh}\, x) = 1/\sqrt{x^2 + 1}$.)

Of course, each derivative formula for inverse hyperbolic functions gives rise to an antiderivative formula. Since the antiderivative formulas are more commonly used, we list these only.

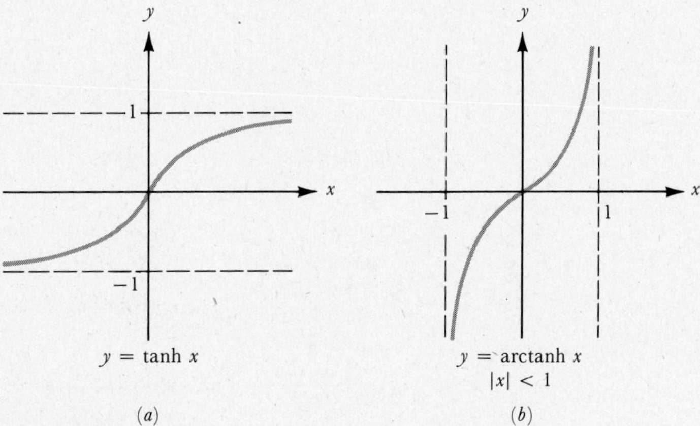

Fig. 8.15

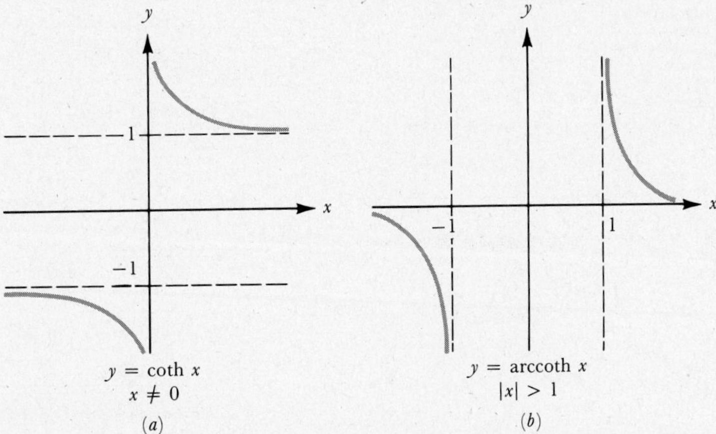

Fig. 8.16

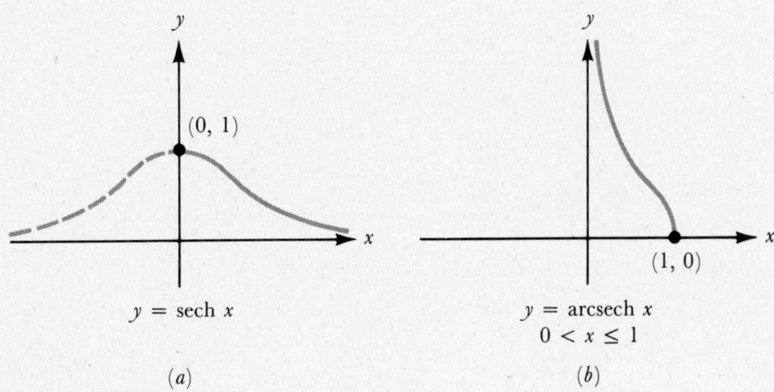

Fig. 8.17

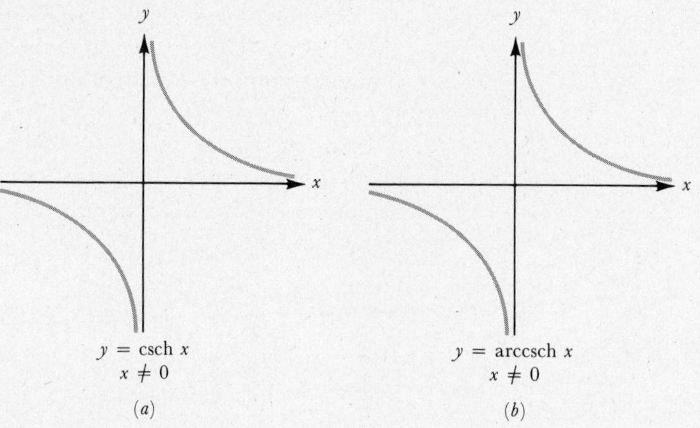

Fig. 8.18

8.6.8 (a) $\displaystyle\int \frac{du}{\sqrt{u^2+a^2}} = \operatorname{arcsinh}\frac{u}{a} + C = \ln(u + \sqrt{u^2+a^2}) + C \qquad a > 0$

(b) $\displaystyle\int \frac{du}{\sqrt{u^2-a^2}} = \operatorname{arccosh}\frac{u}{a} + C = \ln(u + \sqrt{u^2-a^2}) + C \qquad u > a > 0$

(c) $$\int \frac{du}{a^2-u^2} = \begin{cases} \dfrac{1}{a}\operatorname{arctanh}\dfrac{u}{a} + C & \text{if } |u| < a \\ \dfrac{1}{a}\operatorname{arccoth}\dfrac{u}{a} + C & \text{if } |u| > a \end{cases}$$
$$= \frac{1}{2a}\ln\left|\frac{a+u}{a-u}\right| + C \qquad |u| \neq a,\ a \neq 0$$

Example 2 Evaluate $\displaystyle\int \frac{dx}{\sqrt{(x-2)^2+5}}$.

Solution Here we use [8.6.8(a)] with $u = x - 2$, $a = \sqrt{5}$ to get

$$\int \frac{dx}{\sqrt{(x-2)^2+(\sqrt{5})^2}} = \operatorname{arcsinh}\frac{x-2}{\sqrt{5}} + C \quad \bullet$$

The answer to Example 2 could also have been expressed as a natural logarithm.

Virtually all integration problems of this section will yield to other methods developed in Chap. 11, so one should be especially alert to the variety of forms an "answer" can take.

PROGRESS TEST 2

In Probs. 1 to 3 evaluate the given integral. Express your answer to Prob. 1 as a simplified logarithm.

1. $\displaystyle\int \frac{dx}{\sqrt{x^2+6x+3}}$ **2.** $\displaystyle\int \frac{-dx}{9x^2-12x}$ **3.** $\displaystyle\int \frac{\cos x\,dx}{\sqrt{1+\sin^2 x}}$

SECTION 8.6 EXERCISES

1. Given that $\operatorname{csch} x = 2$, determine the values of the remaining hyperbolic functions at $x = 2$.

2. Repeat Exercise 1, given that $\cosh x = 3/2$.

In Exercises 3 to 20 differentiate the given function.

3. $f(x) = \sinh^3(x^4)$

4. $y = \tan^3\sqrt{x+1}$

5. $y = \sqrt{1 + 2\sinh x \cosh x}$

6. $t(x) = [\coth^2(x+1)]\operatorname{csch} x$

7. $u = \tanh(1/x)$

8. $y = \ln(\coth e^x)$

9. $f(x) = \ln(\tanh x - \operatorname{csch} x)$

10. $g(x) = \tanh^2(\operatorname{sech} x)$

11. $y = x^{\sinh x}$

12. $y = x^{\cosh x}$

13. $f(x) = \ln(\sinh e^x)$

14. $y = \operatorname{sech}^3(x-1)$

15. $s(t) = t^2 \cosh 2t$

16. $g(x) = x(\cosh x) - \sqrt{x^2 - 1}$

17. $y = \dfrac{\sinh x}{\arccos x}$

18. $h(x) = \cosh(\csc x)$

19. $y = \arcsin(\sinh x)$

20. $y = \cosh(\cos x)$

In Exercises 21 to 31 prove the given statement.

21. [8.6.3(*c*)]

22. [8.6.4(*d*)]

23. [8.6.4(*e*)]

24. [8.6.4(*f*)]

25. [8.6.7(*b*)]

26. [8.6.7(*c*)]

27. [8.6.7(*a*)]

28. [8.6.4(*c*)]

29. [8.6.7(*f*)]

30. [8.6.8(*a*)], by differentiating.

31. [8.6.8(*b*)], by differentiating.

32. (*a*) Show that $\cosh x + \sinh x = e^x$ and $\cosh x - \sinh x = e^{-x}$.
(*b*) Show that $\sinh(a+b) = (e^a e^b - e^{-a}e^{-b})/2$.
(*c*) Combine (*a*) and (*b*) to prove that $\sinh(a+b) = \sinh a \cosh b + \cosh a \sinh b$.

33. Prove that $\cosh(a+b) = \cosh a \cosh b + \sinh a \sinh b$ (see Exercise 32).

34. Use Exercise 32 to prove that $\sinh 2a = 2\sinh a \cosh a$.

35. Use Exercise 33 to prove that $\cosh 2a = \cosh^2 a + \sinh^2 a$.

36. Prove that $\sinh^2 a = (\cosh 2a - 1)/2$.

37. Prove that $\cosh^2 a = (\cosh 2a + 1)/2$.

38. Determine $\lim_{x\to+\infty} \tanh x$.

39. Determine $\lim_{x\to\pm\infty} \operatorname{sech} x$.

40. Determine $\lim_{x\to\pm\infty} \operatorname{csch} x$.

41. Determine $\lim_{x\to 0} \coth x$.

42. Determine $\lim_{x\to 0} \operatorname{csch} x$.

In Exercises 43 to 63 evaluate the given integral. (Note that some integrals may involve the use of hyperbolic identities, including those in Exercises 33 to 37.)

43. $\displaystyle\int \frac{\operatorname{csch}\sqrt{x}\coth\sqrt{x}}{\sqrt{x}}\,dx$

44. $\displaystyle\int \tanh^2 u \operatorname{sech}^2 u\,du$

45. $\displaystyle\int_{-1}^{1} \cosh^3 x\,dx$

46. $\displaystyle\int_{-1}^{1} \sinh^5 x\,dx$

47. $\displaystyle\int_{0}^{1} \sinh^2 x\,dx$

48. $\displaystyle\int \cosh^2 x\,dx$

49. $\displaystyle\int x\tanh(x^2)\ln(\cosh x^2)\,dx$

50. $\displaystyle\int \operatorname{csch}^4 x\,dx$

51. $\displaystyle\int \frac{dx}{\sqrt{1-3x^2}}$

52. $\displaystyle\int \frac{dx}{\sqrt{2-3x^2}}$

53. $\displaystyle\int \frac{dz}{5-7z^2}$

54. $\displaystyle\int \frac{e^t\,dt}{1-e^{2t}}$

55. $\int \frac{dx}{\sqrt{2 + 3x^2}}$

56. $\int \frac{dx}{\sqrt{9x^2 + 6x + 3}}$

57. $\int \frac{e^x + e^{-x}}{e^x - e^{-x}}\,dx$

58. $\int \frac{dt}{8t + 2 - 4t^2}$

59. $\int_2^3 \frac{dx}{1 - x^2}$

60. $\int \frac{dx}{\sqrt{x^2 - 6x + 3}}$

61. $\int \frac{x\,dx}{x^4 + 16}$

62. $\int \frac{\sin x\,dx}{\sqrt{\cos^2 x + 1}}$

63. $\int_0^1 \frac{dz}{\sqrt{z^2 - 8z + 13}}$

64. Recall Fig. 8.12.
(*a*) What is the area of triangle OQP in terms of sinh and cosh?
(*b*) The area of the shaded region is the answer to (*a*) minus the area of region AQP, this latter area is $\int_0^x \sinh x\, d(\cosh x)$. Evaluate this integral.
(*c*) Show that x is twice the area of OAP.

65. (*a*) A body is falling through air. The air resistance is proportional to the square of the velocity and the proportionality constant is k. Then the distance fallen in feet as a function of time t in seconds is $s = (1/k)\ln(\cosh \sqrt{32kt})$. Find the velocity as a function of time.
(*b*) Show that the velocity tends to a constant, and determine this constant as t becomes large. Interpret your results.
(*c*) Would k be larger for a person with a parachute or without a parachute?
(*d*) Find the acceleration as a function of time.

66. Evaluate the integral $\int \sinh x \cosh x\, dx$ in two ways and show that your results are equal.

67. Prove that $\tanh 2x = \frac{2\tanh x}{1 + \tanh^2 x}$.

68. Reprove [8.6.3(*a*)] using derivatives.

69. Reprove [8.6.3(*b*)] using derivatives.

70. Solve the differential equation $(k^2/32)\, dv/dt = k^2 - v^2$.

71. (*a*) Compute and simplify $D_x[\cosh(\ln x)]$
(*b*) Antidifferentiate the result of (*a*) and then prove that $\cosh(\ln x) = (x + 1/x)/2$.

72. Use first and second derivatives to confirm the graph of $y = \cosh x$, given in Fig. 8.10(*b*).

73. Repeat Exercise 72 for $y = \sinh x$, given in Fig. 8.10(*a*).

74. Repeat Exercise 72 for $y = \tanh x$, given in Fig. 8.15(*a*).

75. Repeat Exercise 72 for $y = \operatorname{csch} x$, given in Fig. 8.18(*a*).

CHAPTER EXERCISES

Review Exercises

1. How long will it take \$100 to grow to \$10,000 if interest is compounded continuously at 10 percent?

2. Repeat Exercise 1, but with the original principal doubled and the interest rate cut in half.

3. (*a*) Suppose the population of a colony of fruit flies doubles every 20 days. Beginning with 10 flies, how many fruit flies will there be in 60 days?
(*b*) How many will there be in 60 days if there were only five flies to begin with?
(*c*) If the population doubled every 10 days and there were 10 flies to begin with, how many fruit flies will be present in 60 days?

In Exercises 4 to 21 compute the given quantity if defined.

4. $\arcsin(\sin 11\pi/4)$

5. $\operatorname{arcsec}(\cos \pi/2)$

6. $\arcsin 1$

7. $\arctan 1$

8. $\operatorname{arcsec} 1$

9. $\arcsin 1/\sqrt{2}$

10. $\arctan 100$

11. $\operatorname{arcsec} 10$

12. $\arcsin(\tan \pi/4)$

13. $\arctan(\sin \pi/2)$

14. $\cos(\arctan 1 + \arctan(1/\sqrt{3}))$

15. $\tan(\arcsin 0)$

16. $\sin(\arctan 0)$

17. $\arcsin(\arctan(-1))$

18. $\operatorname{arcsec}(1/2)$

19. $\operatorname{arcsec} \sqrt{2}$

20. $\operatorname{arcsec}(2/\sqrt{3})$

21. $\arctan(-1)$

In Exercises 22 to 53 differentiate the given function.

22. $f(x) = (\arcsin x)^3$

23. $f(x) = \sqrt{\arctan x}$

24. $g(x) = \arcsin(x^3)$

25. $g(x) = \arctan \sqrt{x}$

26. $h(x) = \ln(\operatorname{arcsec}(x+1))$

27. $r(x) = \cot(\ln(\operatorname{arcsec} x))$

28. $y = e^{x \arctan x}$

29. $m(x) = x^2 \arctan 2x$

30. $y = \dfrac{\arctan x}{\arctan(1/x)}$

31. $p(x) = \arctan(\log_2 x)$

32. $y = \sin(\arcsin x^2)$

33. $g(x) = \arcsin(\cos^3 x)$

34. $f(x) = \arcsin(3x+1)$

35. $f(x) = \arctan(\ln x)$

36. $f(x) = [\arcsin(3x+1)]^4$

37. $f(x) = \dfrac{1}{a^2+x^2} \arctan x$

38. $f(x) = \ln|\arctan x|$

39. $f(x) = \arcsin\left(\dfrac{\cos x + 1}{\sin x}\right)$

40. $f(x) = e^{\operatorname{arcsec} x}$

41. $f(x) = \operatorname{arcsec}(\csc x)$

42. $f(x) = \operatorname{arcsec}(e^x)$

43. $f(x) = \dfrac{1}{\sqrt{a}} \arctan \sqrt{x/a}$

44. $y = x^2 \operatorname{arccosh} x$

45. $g(x) = \dfrac{\cosh x}{1 + \sinh x}$

46. $f(x) = \cosh^2 x - \cosh(x^2)$

47. $y = \cosh(\ln \sec x)$

48. $f(x) = e^{-x} \sinh x$

49. $y = x^{\tanh x}$

50. $h(t) = \operatorname{csch}(\ln x) - \operatorname{sech}(\ln x)$

51. $f(x) = \dfrac{\cosh 2x}{1 - \cosh 2x}$

52. $m(z) = \tanh(z \cot z)$

53. $y = \ln\left[\dfrac{1 + \tanh x}{1 - \tanh x}\right]^{1/2}$

In Exercises 54 to 145 evaluate the given integral.

54. $\int 4 \sin(\pi x)\, dx$

55. $\int x^{-2/3} \sec(1 + \sqrt[3]{x})\, dx$

56. $\displaystyle\int \frac{\sec(e^{-x})\tan(e^{-x})\, dx}{e^x}$

57. $\displaystyle\int \frac{dx}{3x+9}$

58. $\displaystyle\int \frac{\ln(\sin x)}{\sec x}\, dx$

59. $\displaystyle\int \frac{dx}{(3x+9)^2}$

60. $\displaystyle\int \frac{x^4 + 3x^2 - x + 7}{x - 1}\, dx$

61. $\displaystyle\int \frac{3x^2 + 2x + 7}{(3x+9)}\, dx$

62. $\displaystyle\int \frac{dx}{(3x+5)^{1/2}}$

63. $\displaystyle\int_0^{2\pi} |\sin x|\, dx$

64. $\displaystyle\int \frac{dx}{(2x-7)^2}$

65. $\displaystyle\int_1^{e^3} \frac{\sqrt{\ln x}}{x}\, dx$

66. $\displaystyle\int \frac{dx}{4x+3}$

67. $\int \cos^3 3x\, dx$

68. $\displaystyle\int \tan^3\left(\frac{x}{3}\right) dx$

69. $\displaystyle\int \frac{\sec^2 x\, dx}{\sqrt[3]{1 + \tan x}}$

70. $\displaystyle\int \frac{e^{\tan x}}{\cos^2 x}\, dx$

71. $\int x \cdot \pi^{x^2+3}\, dx$

72. $\int \cos^2 x \sin^3 x\, dx$

73. $\displaystyle\int \frac{x\, dx}{\sin(x^2)}$

74. $\displaystyle\int \frac{[\ln(1 + \cos x)]^3}{(\csc x)(1 + \cos x)}\, dx$

75. $\displaystyle\int_{\pi/6}^{2\pi} |\cos x|\, dx$

76. $\int x \csc(3x^2) \cot(3x^2)\, dx$

77. $\int 2x^2 e^{x^3+1}\, dx$

78. $\displaystyle\int \frac{\tan^2(\ln x)\, dx}{x}$

79. $\displaystyle\int \frac{(1+4x)\sqrt{\ln(3x^3 + 9x^4)}}{x + 3x^2}\, dx$

80. $\int e^x \tan(1 + e^x)\, dx$

81. $\displaystyle\int \frac{dx}{x(1 + \ln x)^3}$

82. $\displaystyle\int \frac{\sin^3(3x)\, dx}{\sqrt[3]{\cos(3x)}}$

83. $\displaystyle\int \frac{4^{\sqrt{x+1}}}{\sqrt{x}}\, dx$

84. $\int_0^{\pi} \sin^4 2x \, dx$

85. $\int_0^{\pi} \sin 3x \cos 5x \, dx$

86. $\int (\tan x \sec x)^4 \, dx$

87. $\int x \sin^2(x^2 + 1) \, dx$

88. $\int \cot^5 x \csc^4 x \, dx$

89. $\int \frac{\cos^3 x \, dx}{\sin^2 x}$

90. $\int \cos x \cos 3x \, dx$

91. $\int \sin^3 3x \cos^5 3x \, dx$

92. $\int (\sin^3 x) \sqrt{\cos x} \, dx$

93. $\int \cos^3 x \sqrt{\sin x} \, dx$

94. $\int_{-\pi/4}^{\pi/4} \tan^3 x \cos x \, dx$

95. $\int_{-\pi}^{\pi} \cos^6 \frac{x}{2} \, dx$

96. $\int \cot 2x \csc^4 2x \, dx$

97. $\int (\cot x) \ln(\sin x) \, dx$

98. $\int_{\pi/6}^{\pi/4} \frac{\cos 2x}{\sin^6 2x} \, dx$

99. $\int \cot 3x \csc^4 3x \, dx$

100. $\int \frac{\tan^3 x}{\sec x} \, dx$

101. $\int \tan^{1/2} x \sec^4 x \, dx$

102. $\int \frac{x + 4}{\sqrt{2 + 2x - x^2}} \, dx$

103. $\int \frac{dx}{\sqrt{5 - 3x^2}}$

104. $\int \frac{dx}{\sqrt{e^{-2x} - 1}}$

105. $\int \frac{5 \, dx}{x\sqrt{2x^2 - 1}}$

106. $\int \frac{\ln(\arcsin x)}{(\arcsin x)\sqrt{1 - x^2}} \, dx$

107. $\int_0^1 \frac{dx}{4x^2 + 12x + 14}$

108. $\int_0^2 \frac{x^2 \, dx}{x^2 + 4}$

109. $\int \frac{dx}{(x - 3)\sqrt{x^2 - 6x + 5}}$

110. $\int_{\pi/6}^{\pi/3} \frac{dx}{\sqrt{6x - 1 - x^2}}$

111. $\int \frac{dx}{e^{4x} + e^{-4x}}$

112. $\int \frac{du}{a + u^2}$

113. $\int \frac{dx}{\sqrt{1 - e^{4x}}}$

114. $\int \frac{dx}{\sqrt{1 - 4x^2}}$

115. $\int_1^e \frac{dx}{x\sqrt{3 - (\ln x)^2}}$

116. $\int \frac{3 \, dx}{(3x + 1)\sqrt{9x^2 + 6x}}$

117. $\int \frac{dx}{x^2 \arctan x + \arctan x}$

118. $\int \frac{x \, dx}{2 + 9x^4}$

119. $\int \frac{dx}{x\sqrt{1 - x}}$

120. $\int \frac{dx}{9x^2 + 12x + 8}$

121. $\int \frac{dx}{x^{1/2} + x^{3/2}}$

122. $\int \frac{dx}{\sqrt{1 - x^2} \arcsin x}$

123. $\int_5^7 \frac{\cos x \, dx}{\sqrt{4 - \sin^2 x}}$

124. $\int \frac{(\arctan x)^2}{x^2 + 1} \, dx$

125. $\int \frac{\sin x \, dx}{\sqrt{4 - \cos^2 x}}$

126. $\int \frac{x \, dx}{\sqrt{x^2 + 4}}$

127. $\int \frac{\sec x \tan x \, dx}{1 + \sec^2 x}$

(Most of the integrals in Exercises 128 to 145 involve hyperbolic functions.)

128. $\int \tanh^2 x \, dx$

129. $\int \tanh x \, dx$

130. $\int \frac{\cosh \sqrt{x} \, dx}{\sqrt{x}}$

131. $\int \cosh^3 4x \, dx$

132. $\int \ln \sqrt{x + \sqrt{x^2 + 1}} \, dx$

133. $\int \frac{dt}{(e^t + e^{-t})^2}$

134. $\int \frac{dt}{\sqrt{t^2 - t}}$

135. $\int \frac{dx}{5 - 9x^2 - 12x}$

136. $\int_0^1 \frac{e^t\,dt}{\sqrt{e^{2t}+1}}$

137. $\int \frac{dx}{\sqrt{10x-25x^2}}$

138. $\int \frac{dx}{\sqrt{x^2-2x+26}}$

139. $\int (x^2+1)^{-1/2}\,dx$

140. $\int_1^e \frac{e^{x+1}}{1-e^{2x+2}}\,dx$

141. $\int \frac{dx}{\sqrt{x^2-6x+10}}$

142. $\int x \sinh^3(x^2)\,dx$

143. $\int_{\ln(1/e^2)}^{\ln(1/e)} \cosh 2x\,dx$

144. $\int \coth^2 t\,dt$

145. $\int_0^1 \tanh^3 x\,dx$

146. Prove that $(\cosh x + \sinh x)^3 = \cosh 3x + \sinh 3x$.

147. Solve the differential equation $d^2y/dx^2 = k^2y$.

148. Determine the area of the region bounded by the curve $y = 1/(x^2+1)$, the lines $x = -1$, $x = 1$, and the x axis.

149. Determine the area of the region bounded by $y = 3/(x^2+1)$ and the x axis from $x = 0$ to $x = 1$.

150. Show that $y = e^{ax}\sinh bx$ satisfies the differential equation $y'' - 2ay' + (a^2 - b^2)y = 0$.

Miscellaneous Exercises

1. A painting masterpiece is 4 ft tall and will hang with its bottom edge 8 ft above floor level. How far from the wall on which the painting is hanging should a person stand so that the angle subtended by the top and bottom of the painting is a maximum, assuming that the person's eyes are 5 ft above the floor?

2. Assuming that a cube of ice melts at a rate proportional to its surface area and decreases from 15 cm on an edge to 12 cm in 1 h, how long will it take to melt completely away? (*Note:* The rate of melting is a rate of decrease of volume.)

3. (*a*) The half-life of radium is about 1600 years. How long will it take a chunk of radium to decay to 99 percent of its original mass?
(*b*) Without using a calculator or tables, determine how long it will take the radium to *lose* 75 percent of its mass.

4. Strontium 90, a product of hydrogen bomb explosions, has a half-life of about 27 years. If the amount in a particular place is four times the "safe" level, how long will it be before that place is "safe"?

5. Repeat Exercise 4 if it is discovered that what was originally regarded as "safe" is double the "really safe" level.

6. Graph the equation $y = x - 2\arctan x$.

7. Suppose a telescope is mounted on a 100-m-tall building 100 m away from a 200-m-tall building. This telescope will follow the flight of a freely falling object as it drops to the ground from the top of the 200-m-tall building. At which height of the object is the telescope turning at the fastest rate?

8. Repeat Exercise 7 with the object dropped from the 150-m level instead of 200 m.

9. Prove a "difference formula" for $\cosh(a - b)$.

10. Prove $\arcsin(\tanh x) = \arctan(\sinh x)$.

11. (*a*) Determine the inverse of the function $f(x) = \cosh(3x+1)$.
(*b*) Calculate the derivative of f^{-1} found in part (*a*).

12. Evaluate $\int \cos x \cos 3x \cos 5x\,dx$.

13. Evaluate $\int \sin 2x \sin 4x \sin 6x\,dx$.

14. Evaluate $\int \frac{(\operatorname{sech}^2 \sqrt{x})\sqrt{\tanh \sqrt{x}}\,dx}{\sqrt{x}}$.

15. (*a*) Given a mixture of radioactive substances containing 6 g of A and 3 g of B, where the half-life of A is 10 years and the half-life of B is 20 years, how long will it take for the mixture to contain equal parts of A and B?
(*b*) How long will it take for the mixture to contain twice as much B as A?

16. (*a*) A porous material dries at a rate directly proportional to the amount of moisture remaining (depending upon the material and the relative humidity). If a wet washcloth loses a quarter of its moisture in 30 min, how long will it take to lose 90 percent of its moisture?
(*b*) How long will it take to lose 99 percent of its moisture?

17. (*a*) If the world's population continues to grow exponentially at its current rate of 1.8 percent and was 3.9 billion in 1975, when will it reach 6 billion?
(*b*) When will it reach 10 billion?

18. (*a*) Using the data given in Exercise 17(*a*) and the statement that the current estimate of the upper limit for the human population on this planet is 10 billion, how long will it take to reach 6 billion assuming (8.1.9) holds?
(*b*) When will it reach 8 billion?
(*c*) When will it reach 9 billion?

19. Solve Exercise 18 assuming that the growth rate falls linearly 0.0002 per year beginning in 1975.

20. Solve Exercise 17 assuming that the growth rate falls linearly 0.0002 per year beginning in 1975.

21. Solve Exercise 17 assuming that the growth rate increases linearly 0.0002 per year beginning in 1975.

22. Evaluate $\int \frac{\operatorname{sech}^2 x\,dx}{\sqrt{\tanh^2 x - 4}}$.

23. Prove that $(\cosh x - \sinh x)^n = \cosh nx - \sinh nx$.

24. Evaluate $\int \frac{[\operatorname{arccos}(\coth x)]\operatorname{csch}^2 x}{\sqrt{1 - \coth^2 x}}\, dx$.

25. Prove that $\arctan x = \arcsin\left(\frac{x}{\sqrt{1 + x^2}}\right)$.

26. Prove that $\arcsin x = \arctan\left(\frac{x}{\sqrt{1 + x^2}}\right)$.

27. Determine the volume of the solid generated by revolving about the x axis the region enclosed by $y = \sqrt{\sinh x}$, the x axis, and the line $x = \ln 2$.

In Exercises 28 to 30 graph the given function (be careful with domains).

28. $f(x) = \sin(\arcsin x)$ **29.** $f(x) = \arcsin(\sin x)$

30. $f(x) = \sin(\arccos x)$

31. Sketch the graph of (*a*) $y = \arctan(e^{-x})$ and (*b*) $y = \operatorname{arcsec}(x/2\pi)$ (on its natural domain).

32. Using implicit differentiation, graph the equation $\arcsin(x - y) = 0$. If you are unsuccessful, try using the definition of arcsin and common sense. (Be careful, this problem is trickier than it looks.)

33. (*a*) Set up a differential equation describing the change in temperature T of a hot object brought into a cool room where the temperature is rising at the rate of 1°C every 5 min.
(*b*) Solve this differential equation.

34. When a capacitor discharges, the rate of change of voltage (in volts per second) is proportional to the voltage, with a negative proportionality constant. Give the voltage as a function of time t.

35. Given that $y^2 + x \sinh y + \cosh^2 x = 1$, determine dy/dx.

SELF-TEST ✦

1. (*a*) Determine the initial market value of a \$100 savings bond that matures in 10 years if it is to return 5 percent compounded continuously.
(*b*) What is the above bond worth 5 years from maturity?

2. Set up a differential equation giving the change in temperature T of a hot object brought into a cool room where the temperature R of the room is rising at the rate dR/dt.

In Probs. 3 to 12, evaluate the given integral.

3. $\int (\sin x)\frac{e^{\tan^2 x}}{\cos^3 x}\, dx$

4. $\int \frac{x^2 + x + 1}{x + 1}\, dx$

5. $\int \frac{(\sin x)\sqrt{\ln(\sec x)}}{\cos x}\, dx$

6. $\int \frac{\tan(3 + \sqrt{x})}{\sqrt{x}}\, dx$

7. $\int \cos^4 2x \sin^3 2x\, dx$

8. $\int \tan^3 x \sec^2 x\, dx$

9. $\int \frac{\arctan x\, dx}{x^2 + 1}$

10. $\int \frac{dx}{x\sqrt{x^6 - 5}}$

11. $\int \frac{dx}{(e^x - e^{-x})^2}$

12. $\int \frac{dx}{\sqrt{4x^2 - 4x - 3}}$

In Probs. 13 and 14, differentiate the given function.

13. $f(x) = [\arcsin(1/x)]^2$

14. $g(x) = \frac{\cosh x}{1 - \tanh x}$

Chapter 9

IMPORTANT IDEAS REVISITED

CONTENTS

In this chapter we return to examine some of the ideas treated earlier. Although it is both impossible and unwise to try to prove all the statements we use (*every* theory must start with *some* assumptions), our reexamination will hopefully help illustrate the depth and richness of the ideas underlying our work to date.

We begin in Sec. 9.1 with a careful look at the formal definition of limit, and then in Sec. 9.2 we apply this definition to prove some of the basic theorems on limits that we have used repeatedly in our development. In Sec. 9.3 we examine the behavior of three particular functions to show the need for precision in mathematical statements involving limits, derivatives, and integrals. We show that informally phrased statements provide inadequate descriptions of the behavior of such functions.

Section 9.1: The Formal Definition of Limit

Objectives:

1. **Interpret as "nearness conditions" and graph absolute value inequalities of the form $0 < |x - a| < \delta$ and $|f(x) - L| < \varepsilon$.**

2. **Apply the definition of limit for certain simple, mainly linear, functions.**

THE IDEA OF NEARNESS

In Chap. 2, Sec. 2.2, we repeatedly discussed the idea of limit in terms of "closer and closer," "nearer," and so on. The following function (which first appeared in Sec. 2.5, Progress Test 2) illustrates the inadequacy of these informal statements and the need for the more precise definition of "$\lim_{x\to a} f(x) = L$" given in (2.5.1) and repeated below.

$$f(x) = \begin{cases} x + 1.95 & \text{for } x < 0 \\ 2 & \text{for } x = 0 \\ x + 2.1 & \text{for } x > 0 \end{cases}$$

The graph of $f(x)$ is given in Fig. 9.1. Quite clearly, as x gets nearer and nearer to 0 from either side, $f(x)$ gets nearer and nearer to 2. Just as clearly, however, $\lim_{x\to 0} f(x) \neq 2$.

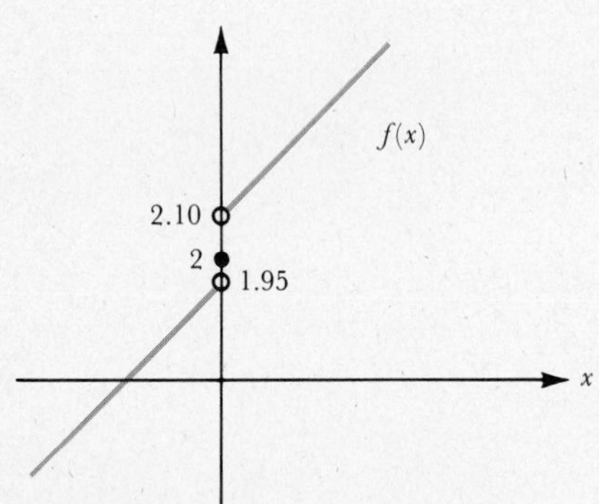

Fig. 9.1

The problem is, that although $f(x)$ *does* get closer and closer to 2, no value of $f(x)$ except $f(0)$ is within 0.05 unit of 2. As a result, no matter how close we take nonzero x to zero, we cannot force the values of $f(x)$ to be within 0.05 unit of 2. We need a more precise definition of limit. But before we give this definition we review the idea of "nearness" as it can be expressed using absolute value.

Recall from Chap. 1, Sec. 1.1, that $|x - a|$ = distance between x and a; that is,

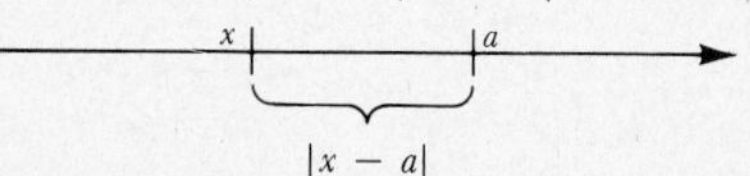

Suppose we wish to insist that x be a number within 0.5 of the number 2. Then we want the distance between x and 2 to be less than 0.5 unit. But the distance between x and 2 is $|x - 2|$, so our condition will be satisfied for any number x for which $|x - 2| < 0.5$. Graphically, x must be in the interval (1.5, 2.5); see Fig. 9.2. Certainly any such x in this interval is less than 0.5 unit from the number 2.

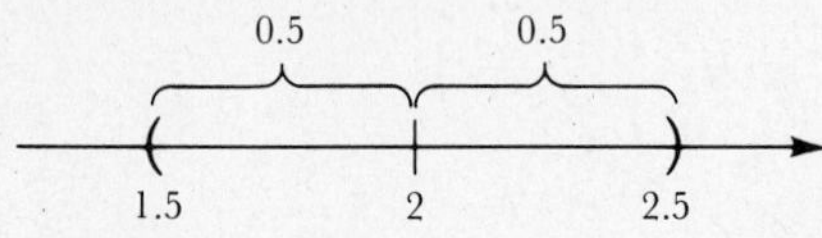

Fig. 9.2

Now suppose δ represents some small positive number (δ is the lowercase Greek letter delta). How do we specify that x be within δ units of the number a?

In light of the above, this is equivalent to specifying that x satisfy the inequality

$$|x - a| < \delta$$

or, again equivalently, that x be in the interval $(a - \delta, a + \delta)$:

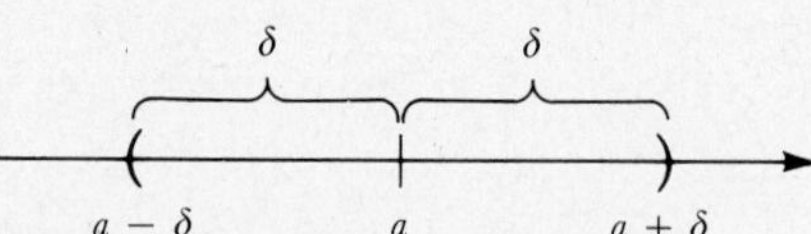

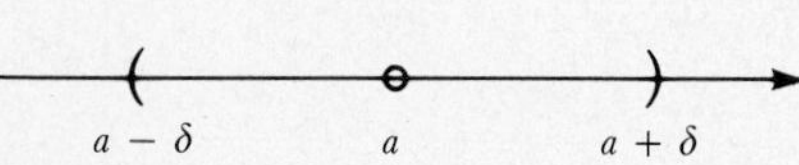

Fig. 9.3

Now, $|x - a| = 0$ only when $x = a$, so the condition that x be within δ units of a, but *not equal* to a, can be given by

$$0 < |x - a| < \delta$$

The left-hand side of the inequality, $0 < |x - a|$, simply guarantees that $x \neq a$, whereas the right-hand side, $|x - a| < \delta$, guarantees that x be within δ units of a. This gives rise to the interval $(a - \delta, a + \delta)$ with the point a removed; see Fig. 9.3. We refer to such as a *deleted interval.*

Similarly, the condition that $f(x)$ be within ε units of the number L ($\varepsilon > 0$) is equivalent to the condition that $f(x)$ satisfy the inequality

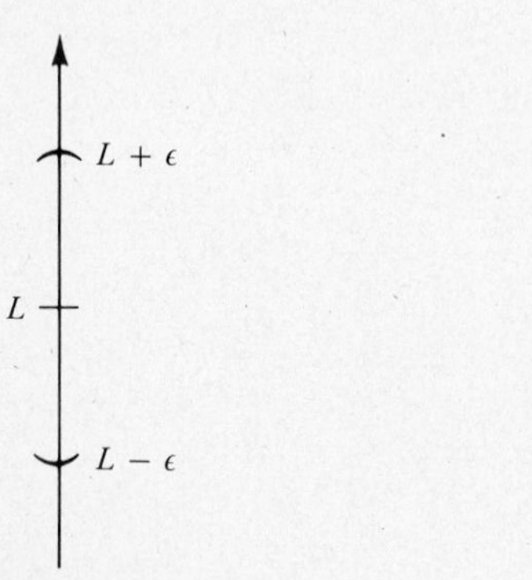

Fig. 9.4

$$|f(x) - L| < \varepsilon$$

or, equivalently, that $f(x)$ be in the interval $(L - \varepsilon, L + \varepsilon)$; see Fig. 9.4. We pictured the interval vertically since values of $f(x)$ are measured on the vertical axis.

Example 1 Show that when x is within 0.01 of 1, it follows that $f(x) = 2x + 3$ is within 0.02 of 5. Sketch the situation.

Solution We wish to show that when $|x - 1| < 0.01$, it follows that $|f(x) - 5| < 0.02$. Evaluating the function and simplifying, we see that

$$\begin{aligned} |f(x) - 5| &= |(2x + 3) - 5| \\ &= |2x - 2| \\ &= 2|x - 1| \\ &< 2(0.01) \qquad \text{(since } |x - 1| < 0.01\text{)} \\ &= 0.02 \end{aligned}$$

Thus we can conclude that when $|x - 1| < 0.01$, $|f(x) - 5| < 0.02$, which is what we wanted to show.

A slightly exaggerated sketch is given in Fig. 9.5. ●

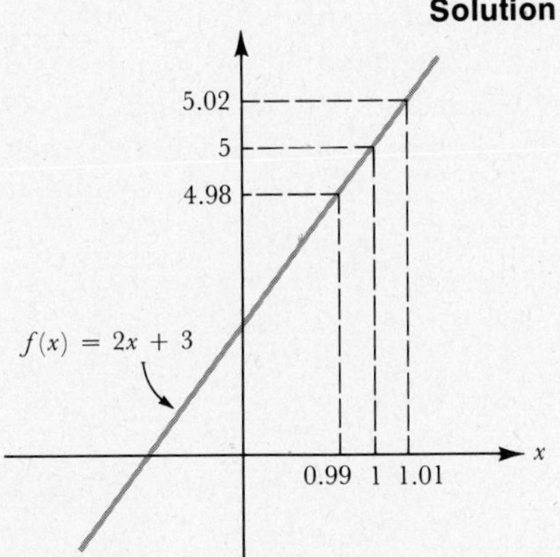

Fig. 9.5

Example 2 shows that the calculations and reasoning of the previous example apply even when the small numbers involved are not specified.

Example 2 Suppose $f(x) = 2x + 3$ and $\varepsilon > 0$ is any small positive number. Show that if x is within $\varepsilon/2$ units of 1, then $f(x)$ is within ε units of 5.

Solution We wish to show that if $|x - 1| < \varepsilon/2$, then $|f(x) - 5| < \varepsilon$. As before,

$$\begin{aligned} |f(x) - 5| &= |(2x + 3) - 5| \\ &= 2|x - 1| \\ &< 2\frac{\varepsilon}{2} \qquad \text{(since } |x - 1| < \varepsilon/2\text{)} \\ &= \varepsilon \end{aligned}$$

so $|f(x) - 5| < \varepsilon$, as we needed to show. ●

Notice that, having done Example 2, we could answer the following question: How near must x be to 1 to guarantee that $f(x)$ be within ε of 5? (*Answer:* x must be within $\delta = \varepsilon/2$ of 1.) Suppose we had not done Example 2 and had been asked: How near must x be to 1 to guarantee that $f(x)$ be within 0.5 of 5? We could answer this question by writing out the stipulated condition and simplifying. We need

$$|f(x) - 5| < 0.5$$

that is,

$$|(2x + 3) - 5| < 0.5$$

or

$$|2x - 2| < 0.5 \qquad \text{or} \qquad 2|x - 1| < 0.5$$

Now the answer can be recognized. Divide both sides of the inequality by 2 to get

$$|x - 1| < \frac{0.5}{2} = 0.25$$

Thus x must be within 0.25 of 1 to guarantee that $f(x)$ be within 0.5 of 5. In exactly the same way we could answer the question: How near must x be to 1 to guarantee that $f(x)$ be within ε of 5? (*Answer:* x must be within $\delta = \varepsilon/2$ of 1; here we replace 0.5 in the above argument by ε.)

PROGRESS TEST 1

1. The inequalities in (*a*) and (*b*) gives rise to an interval. Sketch the interval.
(*a*) $|x-2| < 1/2$ (*b*) $0 < |x-9| < 3$

2. Write down an inequality that has as its solution the given interval.
(*a*) (-1 ... o 2 ...) 5 (deleted interval)
(*b*) $(1.99, 2.01)$
(*c*) (-8 ... o -3 ...) $+2$ (deleted interval)

3. (*a*) Show that if x is within 0.1 unit of 2, then $f(x) = 4x - 1$ is within 0.4 unit of 7.
(*b*) Draw a sketch of the situation in (*a*).

4. For each of the following, determine how near x must be to a to ensure that $f(x)$ will be within ε units of L. Write your answer as an inequality involving absolute value.
(*a*) $f(x) = 3x + 1,\ a = 3,\ \varepsilon = 1,\ L = 10$
(*b*) $f(x) = 2x - 2,\ a = -1,\ \varepsilon = 1/2,\ L = -4$
(*c*) $f(x) = (1/2)x + 1,\ a = 1/4,\ \varepsilon = 1/5,\ L = 9/8$

THE DEFINITION OF LIMIT

We now recall the formal definition of limit.

9.1.1 Definition Suppose we are given a function $f(x)$ and numbers a and L with $f(x)$ defined for all points, except perhaps for a itself, of an open interval containing a. We say

the limit of $f(x)$ as x tends to a is L

and write

$$\lim_{x\to a} f(x) = L$$

if for every positive number ε, there is a positive number δ such that

$$0 < |x - a| < \delta \qquad \text{implies} \qquad |f(x) - L| < \varepsilon$$

This definition requires that for every $\varepsilon > 0$ there must be a $\delta > 0$ such that x being in the *deleted* interval $(a - \delta, a + \delta)$ implies that $f(x)$ is in $(L - \varepsilon, L + \varepsilon)$. The definition formally records our often repeated remarks regarding indifference to what happens *at a* and the concern only with *x near a*. The limit is unaffected by $f(a)$—and, in fact, $f(a)$ need not even be defined.

To prove that $\lim_{x\to a} f(x) = L$ in a particular instance, we assume that $\varepsilon > 0$ is a number out of our control, given in advance. Our task then is to show that there exists a $\delta > 0$ such that $0 < |x - a| < \delta$ implies $|f(x) - L| < \varepsilon$.

The logic of this approach is simple. Having assumed ε to be an *arbitrary* positive number, if we then show that an appropriate $\delta > 0$ exists, we can conclude that such a δ exists for *every* $\varepsilon > 0$, as required by the definition. Graphically we are showing that there is a deleted interval about a such that if x is in this interval, then $f(x)$ is in the (vertical) interval $(L - \varepsilon, L + \varepsilon)$. See Fig. 9.6. Another illustration appears in Fig. 2.26, page 62.

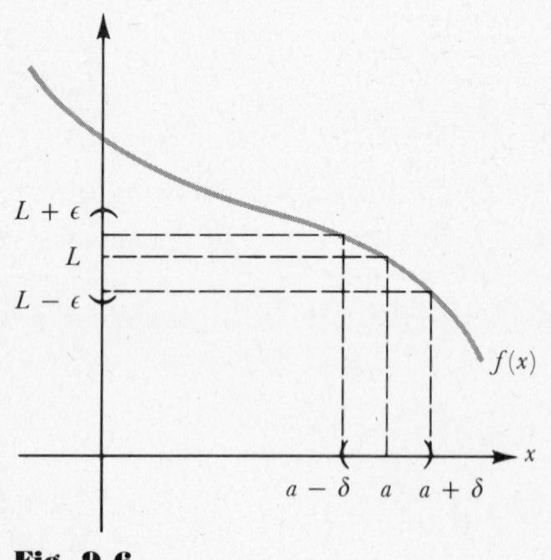

Fig. 9.6

Example 3 Prove, using (9.1.1), that $\lim_{x\to 1} (3x + 1) = 4$.

Solution Here $L = 4$ and $a = 1$. Suppose $\varepsilon > 0$ is given. We must show that there exists a $\delta > 0$ such that

$$0 < |x - 1| < \delta \qquad \text{implies} \qquad |(3x + 1) - 4| < \varepsilon$$

But

$$|(3x + 1) - 4| = |3x - 3|$$
$$= 3|x - 1| < \varepsilon \qquad \text{if } |x - 1| < \varepsilon/3$$

That is, by choosing $\delta = \varepsilon/3$, we know that

$$0 < |x - 1| < \delta \qquad \text{implies} \qquad |(3x + 1) - 4| < \varepsilon$$

Hence, by the definition, $\lim_{x\to 1} (3x + 1) = 4$. ●

In Example 3 the function was defined at a. In Example 4, $f(a)$ is not defined. (The function here is essentially twice the function in Sec. 2.2, Progress Test 1, Prob. 2, page 45.)

Example 4 Prove, using (9.1.1), that $\lim_{x\to 3} \dfrac{2x^2 - 18}{x - 3} = 12$.

Solution Here $L = 12$ and $a = 3$. Suppose $\varepsilon > 0$ is given. It is now our task to show there exists a $\delta > 0$ such that

$$0 < |x - 3| < \delta \qquad \text{implies} \qquad \left|\frac{2x^2 - 18}{x - 3} - 12\right| < \varepsilon$$

However, since $0 < |x - 3|$, we know that $x \neq 3$. For *such* x,

$$\frac{2x^2 - 18}{x - 3} = \frac{2(x - 3)(x + 3)}{(x - 3)} = 2(x + 3)$$

Thus the condition to be satisfied,

$$\left|\frac{2x^2 - 18}{x - 3} - 12\right| < \varepsilon$$

reduces to $|2(x + 3) - 12| = 2|x - 3| < \varepsilon$. But this condition is satisfied when $|x - 3| < \varepsilon/2$. Hence choose $\delta = \varepsilon/2$, so then

$$0 < |x - 3| < \delta = \frac{\varepsilon}{2} \qquad \text{implies} \qquad \left|\frac{2x^2 - 18}{x - 3} - 12\right| < \varepsilon \qquad ●$$

In Example 4 the requirement that x be in the *deleted* interval about 3 enabled us to replace $(2x^2 - 18)/(x - 3)$ by $2(x + 3)$.

We can now apply Definition (9.1.1) to *prove* our contention in the lead-off discussion of this section that the limit as x tends to 0 of the function given there does not equal 2 even though the values of the function get closer and closer to 2 as x gets closer to 0.

Example 5 Suppose $f(x) = \begin{cases} x + 1.95 & \text{for } x < 0 \\ 2 & \text{for } x = 0 \\ x + 2.1 & \text{for } x > 0 \end{cases}$

Prove that

$$\lim_{x\to 0} f(x) \neq 2 \qquad \text{(see Fig. 9.1)}$$

Solution Here $a = 0$ and the alleged $L = 2$. To prove $\lim_{x\to 0} f(x) = 2$, we would need to show that for *every* $\varepsilon > 0$ there exists a $\delta > 0$ such that

$$0 < |x - 0| < \delta \qquad \text{implies} \qquad |f(x) - 2| < \varepsilon$$

Hence to prove $\lim_{x\to 0} f(x) \neq 2$, we need to find an $\varepsilon > 0$ *for which there does not exist such a* δ. A clue to finding such an ε is given by a close-up view of the graph of $f(x)$ near $x = 0$ (see sketch). There are no values of $f(x)$ within 0.05 unit of 2 except $f(0) = 2$. Choose an ε with $0 < \varepsilon < 0.05$, say $\varepsilon = 0.03$. We claim there is no $\delta > 0$ such that

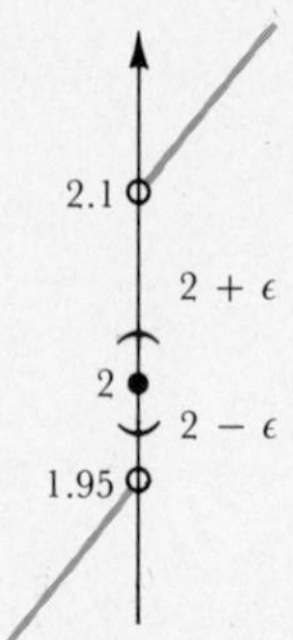

$$0 < |x - 0| = |x| < \delta \qquad \text{implies} \qquad |f(x) - 2| < \varepsilon = 0.03$$

If there were such a δ, we would have $0 < |x| < \delta$, implying that *all* the values of $f(x)$ are within 0.03 unit of 2. But there are *no* values of $f(x)$ within 0.03 unit of 2 (in fact there are none within 0.05 unit of 2) except $f(0) = 2$! But $x \neq 0$ by the condition $0 < |x|$. Thus the statement

$$0 < |x| < \delta \qquad \text{implies} \qquad |f(x) - 2| < 0.03$$

which is supposed to hold *for all such x* [in the deleted interval $(-\delta, +\delta)$], holds *for no such x*.

Hence there is no δ that "works" for $\varepsilon = 0.03$, proving our claim that

$$\lim_{x \to 0} f(x) \neq 2 \quad \bullet$$

ONE-SIDED LIMITS

We frequently needed to use left- and right-hand limits. Our formal definition (9.1.1) is *two*-sided in the sense that the condition

$$0 < |x - a| < \delta$$

determines a *symmetric* deleted interval about a: $(a - \delta, a + \delta)$. We shall break the absolute-value statement into a pair of statements and manipulate these to show how each of the two resulting statements determines "its half" of the deleted interval.

The absolute-value statement $0 < |x - a| < \delta$ is, by (1.1.9), equivalent to the pair of statements

$$0 < (x - a) < \delta \qquad \text{and} \qquad 0 < -(x - a) < \delta$$

But by multiplying the second of this pair through by -1 and appropriately reversing inequality signs, we see that the inequality $0 < |x - a| < \delta$ is equivalent to the pair of inequalities

$$0 < x - a < \delta \qquad \text{and} \qquad -\delta < x - a < 0$$

Finally, we add a to each set of inequalities to get

$$a < x < a + \delta \qquad \text{and} \qquad a - \delta < x < a$$

Hence we can associate the two halves of the absolute-value statement with the two halves of the deleted interval $(a - \delta, a + \delta)$. (See Fig. 9.7.)

We can now make formal definitions of left- and right-hand limits. (We leave the definition of $\lim_{x \to a^+} f(x) = L$ as a Progress Test problem.)

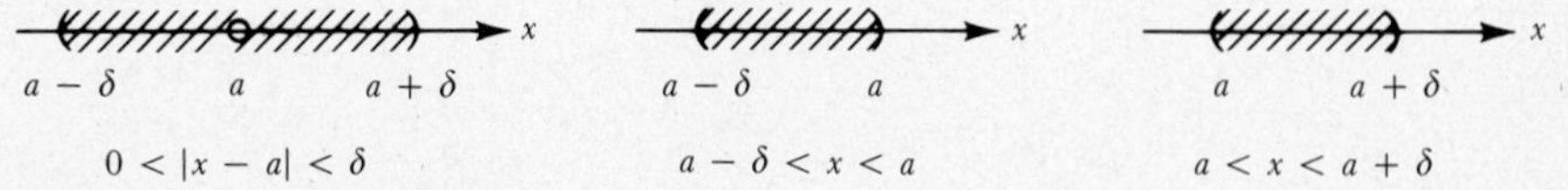

$a-\delta \quad a \quad a+\delta$ $\quad$ $a-\delta \quad a$ $\quad$ $a \quad a+\delta$

$0<|x-a|<\delta$ $\quad$ $a-\delta<x<a$ $\quad$ $a<x<a+\delta$

Fig. 9.7

9.1.2 Definition Suppose $f(x)$ is defined for all x in some open interval (c, a). We say *the limit of $f(x)$ as x tends to a from the left is L*, and write

$$\lim_{x\to a^-} f(x) = L$$

if for every $\varepsilon > 0$, there is a number $\delta > 0$ such that

$$a - \delta < x < a \qquad \text{implies} \qquad |f(x) - L| < \varepsilon$$

Example 6 Show, using (9.1.2), that $\lim_{x\to 0^-} f(x) = 1.95$, where $f(x)$ is as defined in Example 5.

Solution Here $L = 1.95$ and $a = 0$. Suppose $\varepsilon > 0$ is given. It is our task to show that there exists a $\delta > 0$ such that

$$-\delta < x < 0 \qquad \text{implies} \qquad |f(x) - 1.95| < \varepsilon$$

However, for $x < 0$, f is defined by $f(x) = x + 1.95$ (hence is defined in $(c, 0)$ for any negative number c), and thus for such an x

$$|f(x) - 1.95| = |(x + 1.95) - 1.95| = |x|$$

Since any positive number δ for which $-\delta < x < 0$ will satisfy $|x| < \delta$, we take $\delta = \varepsilon$. ●

PROGRESS TEST 2

In Probs. 1 and 2 prove, using (9.1.1), that $\lim_{x\to a} f(x) = L$. Draw a sketch of the function near the point in question.

1. $f(x) = 1 - 2x$, $a = -1$, $L = 3$

2. $f(x) = \dfrac{x^2 - 4}{x - 2}$, $a = 2$, $L = 4$

3. Give a formal definition of right-hand limit analogous to (9.1.2).

4. Use your definition in Prob. 3 to prove that $\lim_{x\to 0^+} f(x) = 2.1$ where $f(x)$ is as defined in Example 5.

THE LIMIT IN A GENERAL CASE

We can now use a generalized version of the above limit arguments to prove the following theorem, which was originally stated as (2.5.2).

9.1.3 Theorem Suppose that $f(x) = mx + b$ is a given function, where m and b are constants and $x = a$ is any real number. Then

$$\lim_{x\to a} f(x) = ma + b$$

Proof: As usual, assume that $\varepsilon > 0$ is given. We must find a $\delta > 0$ such that

$$0 < |x - a| < \delta \qquad \text{implies} \qquad |(mx + b) - (ma + b)| < \varepsilon$$

But

$$\begin{aligned}|(mx + b) - (ma + b)| &= |m(x - a)| \\ &= |m|\,|x - a| < \varepsilon\end{aligned}$$

when

$$|x - a| < \frac{\varepsilon}{|m|} \qquad \text{unless } m = 0$$

If $m = 0$, then any δ works since $f(x)$ is a constant function with $|m|\,|x - a| = 0$. Thus, choosing $\delta = \varepsilon/|m|$, for $m = 0$, or δ arbitrary for $m = 0$, we can conclude that

$$0 < |x - a| < \delta \qquad \text{implies} \qquad |f(x) - (ma + b)| < \varepsilon$$

thereby proving the theorem. ■

All our formal limit arguments to date, including the proof of the last general theorem, have involved linear functions. The ε-δ arguments for nonlinear functions are less routine.

Example 7 Use (9.1.1) to show that $\lim_{x\to 2} 1/x = 1/2$.

Solution We have $L = 1/2$, $a = 2$, with $\varepsilon > 0$ given. We need to show that there exists $\delta > 0$ such that

$$0 < |x - 2| < \delta \qquad \text{implies} \qquad \left|\frac{1}{x} - \frac{1}{2}\right| < \varepsilon \qquad \text{(see Fig. 9.8)}$$

For $\varepsilon < 1/2$, the lines $y = 1/2 + \varepsilon$ and $y = 1/2 - \varepsilon$ intersect the graph of $f(x) = 1/x$ when $1/x_1 = 1/2 + \varepsilon$ and $1/x_2 = 1/2 - \varepsilon$, respectively; that is, when

$$x_1 = \frac{2}{1 + 2\varepsilon} \qquad \text{and} \qquad x_2 = \frac{2}{1 - 2\varepsilon}$$

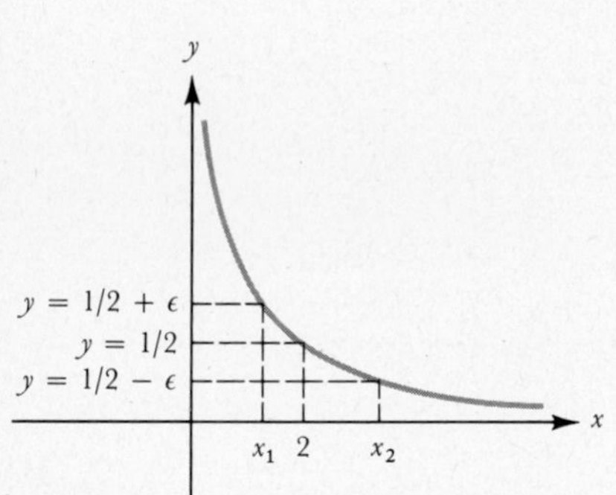

Fig. 9.8

respectively.

We shall select δ to be the smaller of the distances between 2 and x_1 and between 2 and x_2. But

$$|x_1 - 2| = \frac{4\varepsilon}{1 + 2\varepsilon} \qquad \text{and} \qquad |x_2 - 2| = \frac{4\varepsilon}{1 - 2\varepsilon} > |x_1 - 2|$$

so we choose $\delta = 4\varepsilon/(1 + 2\varepsilon)$.

In this argument, if we allow $\varepsilon \geq 1/2$, then $y = 1/2 - \varepsilon$ lies below the graph of $f(x) = 1/x$ for positive x, so $|f(x) - 1/2| < \varepsilon$ for all $x > 2$. Hence the same $\delta = 4\varepsilon/(1 + 2\varepsilon)$ will suffice. ●

After making the effort to show that a particular function has limit L as x tends to a, it is helpful to know that the function cannot tend to any value other than L. That is, limits, if they exist, are unique.

9.1.4 **Theorem (Uniqueness of Limits)** If $\lim_{x\to a} f(x) = L_1$ and $\lim_{x\to a} f(x) = L_2$, then $L_1 = L_2$.

Proof: Assume that $\lim_{x\to a} f(x) = L_1$ and $\lim_{x\to a} f(x) = L_2$, with $L_1 \neq L_2$. We shall derive a contradiction, showing the assumption to be untrue. Let $k = |L_1 - L_2|$, so $k > 0$. Then there exist $\delta_1 > 0$ and $\delta_2 > 0$ such that

$$0 < |x - a| < \delta_1 \qquad \text{implies} \qquad |f(x) - L_1| < \frac{k}{2}$$

and

$$0 < |x - a| < \delta_2 \qquad \text{implies} \qquad |f(x) - L_2| < \frac{k}{2}$$

Let δ be the smaller of the numbers δ_1 and δ_2. Then $0 < |x - a| < \delta$ means that $0 < |x - a| < \delta_1$ *and* $0 < |x - a| < \delta_2$. Thus

$$\begin{aligned} k = |L_1 - L_2| &= |[L_1 - f(x)] + [f(x) - L_2]| \\ &\le |L_1 - f(x)| + |f(x) - L_2| \qquad \text{by [1.1.9}(e)] \\ &< \frac{k}{2} + \frac{k}{2} = k, \text{ a contradiction} \end{aligned}$$

Hence L_1 must equal L_2. ■

SECTION 9.1 EXERCISES

In Exercises 1 to 6 use informal methods to determine L and then use the definition of limit to prove that $\lim_{x\to a} f(x) = L$ *for the given function* $f(x)$ *and point* a.

1. $f(x) = 5x + 2$, $a = 3$

2. $f(x) = 2x + 5$, $a = 1$

3. $f(x) = \dfrac{1}{3}x - 2$, $a = 6$

4. $f(x) = \dfrac{1}{5}x - 1$, $a = 5$

5. $f(x) = \dfrac{x^2 - 9}{x + 3}$, $a = -3$

6. $f(x) = \dfrac{x^2 - x - 6}{x - 3}$, $a = 3$

7. Prove, using the definition of limit, that $\lim_{x\to 3} x^2 \neq 7$.

8. Prove, using the definition of limit, that $\lim_{x\to 2} (|x - 2|)/(x - 2)$ does not exist.

9. Prove, using the definition of limit, that $\lim_{x\to 1} (5x^2 - 5x)/x = 0$.

10. Prove that $\lim_{x\to a} f(x) = L$ if and only if $\lim_{x\to a^+} f(x) = L$ and $\lim_{x\to a^-} f(x) = L$.

11. (*a*) How near to $a = 2$ must x be so that $1/(x + 1)$ be within 0.2 of $L = 1/3$? (Draw a sketch near $a = 2$.)
(*b*) How near to $a = 2$ must x be so that $1/(x + 1)$ be within 0.01 of 1/3?
(*c*) How near to $a = 2$ must x be so that $1/(x + 1)$ be within ε ($\varepsilon > 0$) of 1/3? [Having given your answer, you have proved that $\lim_{x\to 2} 1/(x + 1) = 1/3$.]

12. Repeat Exercise 11 for $f(x) = x^2$ with $a = 2$, $L = 4$.

13. Show that $\lim_{x\to 4} \sqrt{x} = 2$ using the definition of limit.

14. Show that $\lim_{x\to 0} (\sqrt{x} + 1) = 1$ using the definition of limit.

15. Prove, using the definition, that $\lim_{x\to a} f(x) = b$ for any constant function $f(x) = b$.

16. Prove that $\lim_{x\to 0^+} (1/x) \neq L$ for any real number L.

17. Repeat Exercise 16 for $\lim_{x\to 0^-} (1/x)$.

18. Prove that $\lim_{x\to 1^-} s(x) = 0$ for the largest integer function $s(x)$ of Sec. 2.1, Exercise 28.

19. Prove that $\lim_{x\to 1^+} s(x) = 1$. (See Exercise 18.)

In Exercises 20 to 23 show that the indicated limits do not exist.

20. $\displaystyle\lim_{x\to 1} \frac{1}{(x - 1)^2}$

21. $\displaystyle\lim_{x\to 2^+} \frac{1}{x - 2}$

22. $\lim_{x\to 2} f(x)$, where $f(x) = \begin{cases} x^2 & \text{for } x \le 2 \\ x + 1 & \text{for } x > 2 \end{cases}$

23. $\lim_{x\to 1} f(x)$, where $f(x) = \begin{cases} 2x + 1 & \text{for } x < 1 \\ x + 3 & \text{for } x \ge 1 \end{cases}$

24. Show that $\lim_{x\to 3} f(x) = 10$, where

$$f(x) = \begin{cases} 2x + 4 & \text{for } x < 3 \\ x + 7 & \text{for } x > 3 \end{cases} \qquad \text{(sketch)}$$

25. Show $\lim_{x\to 2} f(x) = 1$, where

$$f(x) = \begin{cases} 5 - 2x & \text{for } x < 2 \\ x - 1 & \text{for } x > 2 \end{cases} \qquad \text{(sketch)}$$

26. Using an argument analogous to the proof of (9.1.4), prove using the definition of "limit at infinity" [see (4.1.2)] that $\lim_{x\to +\infty} f(x)$ is unique.

Section 9.2: Limits and Continuity

Objectives:

1. Understand the meaning of abstract limit statements.
2. Use this understanding to prove simple facts about limits.

PROOF OF THE LIMIT THEOREM

Having explored the meaning of $\lim_{x \to a} f(x) = L$, we are now in a position to prove selected portions of the Limit Theorem (2.5.4).

9.2.1 The Limit Theorem Suppose that $\lim_{x \to a} f(x)$ and $\lim_{x \to a} g(x)$ both exist. Then

(*a*) $\lim_{x \to a} [f(x) \pm g(x)] = [\lim_{x \to a} f(x)] \pm [\lim_{x \to a} g(x)]$.

(*b*) $\lim_{x \to a} [f(x) \cdot g(x)] = [\lim_{x \to a} f(x)] \cdot [\lim_{x \to a} g(x)]$

and in particular, if b is a constant, then

$$\lim_{x \to a} [b \cdot g(x)] = b[\lim_{x \to a} g(x)]$$

(*c*) $\lim_{x \to a} \left[\dfrac{f(x)}{g(x)}\right] = \dfrac{\lim_{x \to a} f(x)}{\lim_{x \to a} g(x)}$ provided that $\lim_{x \to a} g(x) \neq 0$

(*d*) $\lim_{x \to a} \sqrt[n]{f(x)} = \sqrt[n]{\lim_{x \to a} f(x)}$

where n is a positive integer. If n is even, then $f(x)$ must be nonnegative.

To facilitate our proofs, we assume that

$$\lim_{x \to a} f(x) = L \qquad \text{and} \qquad \lim_{x \to a} g(x) = M$$

Proof that $\lim_{x \to a} [f(x) + g(x)] = L + M$:

Suppose that $\varepsilon > 0$ is given, so $\varepsilon/2 > 0$ also. Since $\lim_{x \to a} f(x) = L$, there exists a $\delta_1 > 0$ such that

$$0 < |x - a| < \delta_1 \qquad \text{implies} \qquad |f(x) - L| < \frac{\varepsilon}{2}$$

and since $\lim_{x \to a} g(x) = M$, there exists a $\delta_2 > 0$ such that

$$0 < |x - a| < \delta_2 \qquad \text{implies} \qquad |g(x) - M| < \frac{\varepsilon}{2}$$

Letting δ be the smaller of δ_1 and δ_2, we know that $0 < |x - a| < \delta$ means that both $0 < |x - a| < \delta_1$ *and* $0 < |x - a| < \delta_2$. Thus

$$\begin{aligned} &|[f(x) + g(x)] - (L + M)| \\ &= |[f(x) - L] + [g(x) - M]| \\ &\leq |f(x) - L| + |g(x) - M| < \frac{\varepsilon}{2} + \frac{\varepsilon}{2} = \varepsilon \quad \blacksquare \end{aligned}$$

Proof that $\lim_{x \to a} [f(x) \cdot g(x)] = LM$:

We shall first prove that

(i) If $M = 0$, then $\lim_{x \to a} [f(x) \cdot g(x)] = 0$.

Given $\varepsilon > 0$, we need a $\delta > 0$ such that

$$0 < |x - a| < \delta \qquad \text{implies} \qquad |f(x)g(x) - 0| = |f(x)g(x)| < \varepsilon$$

But since $\lim_{x\to a} f(x) = L$, there is a $\delta_1 > 0$ such that

$$0 < |x - a| < \delta_1 \quad \text{implies} \quad |f(x) - L| < 1$$

For $0 < |x - a| < \delta_1$, we thus know

$$|f(x)| - |L| \leq |f(x) - L| < 1$$

so

$$|f(x)| < 1 + |L|$$

Since $\lim_{x\to a} g(x) = 0$, there is a $\delta_2 > 0$ such that

$$0 < |x - a| < \delta_2 \quad \text{implies} \quad |g(x)| < \frac{\varepsilon}{1 + |L|}$$

Letting δ be the smaller of δ_1 and δ_2, we know that $0 < |x - a| < \delta$ means that both $0 < |x - a| < \delta_1$ *and* $0 < |x - a| < \delta_2$. Thus

$$\begin{aligned} |f(x)g(x)| &= |f(x)|\,|g(x)| \\ &< (1 + |L|)\frac{\varepsilon}{1 + |L|} = \varepsilon \end{aligned}$$

which completes the proof of (i).

(ii) We shall now use (i) to prove the general case. (The following argument has a geometric interpretation; see Section Exercise 5.)

Rewrite $f(x)g(x)$ as $(f(x) - L)g(x) + L(g(x) - M) + LM$. Then, by [9.2.1(*a*)],

$$\lim_{x\to a} f(x)g(x) = \lim_{x\to a} [(f(x) - L)g(x)] + \lim_{x\to a} [L(g(x) - M)] + \lim_{x\to a} LM$$

But $\lim_{x\to a} f(x) = L$ and $\lim_{x\to a} g(x) = M$, so $\lim_{x\to a} (f(x) - L) = 0$ and $\lim_{x\to a} (g(x) - M) = 0$. Also, $\lim_{x\to a} (LM) = LM$ by (9.1.3), with $m = 0$. Hence, applying (i),

$$\lim_{x\to a} [f(x)g(x)] = 0 + 0 + LM$$

proving the general case. ■

Of course, if $f(x)$ is the constant function, $f(x) = -1$, we know that $\lim_{x\to a} [-1g(x)] = -1\lim_{x\to a} g(x)$. Using this fact and [9.2.1(*a*)] for sums, we can conclude [9.2.1(*a*)] for differences.

Proof that $\lim \dfrac{f(x)}{g(x)} = \dfrac{L}{M}$, provided that $M \neq 0$:

We shall first prove (assuming $M \neq 0$)

(i) $$\lim_{x\to a} \frac{1}{g(x)} = \frac{1}{M}$$

We know that $|M|/2 > 0$ and $\lim_{x\to a} g(x) = M$, so there is a $\delta_1 > 0$ such that

$$0 < |x - a| < \delta_1 \quad \text{implies} \quad |M - g(x)| < \frac{|M|}{2}$$

But $|M| - |g(x)| \leq |M - g(x)|$, so if $0 < |x - a| < \delta_1$, then

$$|M| - |g(x)| < \frac{|M|}{2}$$

which implies $-|g(x)| < |M|/2 - |M| = -|M|/2$. This in turn implies $|g(x)| > |M|/2$, so $1/|g(x)| < 2/|M|$.

Also, there is a $\delta_2 > 0$ such that

$$0 < |x - a| < \delta_2 \qquad \text{implies} \qquad |M - g(x)| < \frac{\varepsilon M^2}{2}$$

Letting δ be the smaller of δ_1 and δ_2, we know that $0 < |x - a| < \delta$ means that both

$$0 < |x - a| < \delta_1 \qquad \textit{and} \qquad 0 < |x - a| < \delta_2$$

Thus

$$\left|\frac{1}{g(x)} - \frac{1}{M}\right| = \left|\frac{M - g(x)}{M \cdot g(x)}\right| = \frac{1}{|M|} \cdot \frac{1}{|g(x)|} \cdot |M - g(x)|$$
$$< \frac{1}{|M|} \cdot \frac{2}{|M|} \cdot \frac{\varepsilon M^2}{2} = \varepsilon$$

proving (i). Now

$$\lim_{x \to a}\left[\frac{f(x)}{g(x)}\right] = \lim_{x \to a}\left[f(x) \cdot \frac{1}{g(x)}\right]$$
$$= \left[\lim_{x \to a} f(x)\right]\left[\lim_{x \to a} \frac{1}{g(x)}\right] \qquad \text{by (i)}$$
$$= L \cdot \frac{1}{M} = \frac{L}{M} \qquad \blacksquare$$

The general proof that $\lim_{x \to a} \sqrt[n]{f(x)} = \sqrt[n]{L}$ is too difficult to be handled here, so we omit it. In Progress Test 1, however, we outline the proof for $n = 2$ and $f(x) = x$.

PROGRESS TEST 1

1. Using that $||a| - |b|| \leq |a - b|$, prove that $\lim_{x \to a} f(x) = L$ implies that $\lim_{x \to a} |f(x)| = |L|$.
2. Using Prob. 1, prove that if $\lim_{x \to 0} f(x) = 0$, then $\lim_{x \to a} |f(x)| = 0$, and conversely.
3. Prove that $\lim_{x \to a} \sqrt{x} = \sqrt{a}$ $(a \geq 0)$ by following the given steps:
   (*a*) Show that $\sqrt{x} - \sqrt{a} = (x - a)/(\sqrt{x} + \sqrt{a})$.
   (*b*) Given $\varepsilon > 0$, suppose $\delta_1 = a/2$. Show that $0 < |x - a| < \delta_1$ implies that $\sqrt{x} > \sqrt{a/2}$.
   (*c*) Let $k = \sqrt{a/2} + \sqrt{a}$, so that $k < \sqrt{a} + \sqrt{x}$ by (*b*). Use (*a*) to show that $0 < |x - a| < \delta_1$ implies $|\sqrt{x} - \sqrt{a}| < |x - a|/k$.
   (*d*) Let $\delta_2 = \varepsilon k$ and show that $0 < |x - a| < \delta_2$ implies $|\sqrt{x} - \sqrt{a}| < \varepsilon$.
   (*e*) Use the results of (*c*) and (*d*) to conclude that $\lim_{x \to a} \sqrt{x} = \sqrt{a}$.

COMPOSITION AND THE SQUEEZE THEOREM

The following important fact about composition of functions was stated as (2.5.10) but was not proved.

9.2.2 If $\lim_{x \to a} g(x) = b$ and if f is continuous at b, then

$$\lim_{x \to a} f(g(x)) = f\left(\lim_{x \to a} g(x)\right) = f(b)$$

Proof: We need to show that $\lim_{x \to a} f(g(x)) = f(b)$, so let $\varepsilon > 0$ be given. We need to show that there is a $\delta > 0$ such that $0 < |x - a| < \delta$ implies $|f(g(x)) - f(b)| < \varepsilon$. Since f is continuous at b, we know that $\lim_{t \to b} f(t) = f(b)$ by definition (2.5.7). Thus there is a $\delta_1 > 0$ such that

$$|t - b| < \delta_1 \qquad \text{implies} \qquad |f(t) - f(b)| < \varepsilon$$

Letting $t = g(x)$, we conclude that

$$|g(x) - b| < \delta_1 \quad \text{implies} \quad |f(g(x)) - f(b)| < \varepsilon$$

But since $\lim_{x\to a} g(x) = b$, there is a $\delta > 0$ such that $0 < |x - a| < \delta$ implies $|g(x) - b| < \delta_1$, which in turn implies

$$|f(g(x)) - f(b)| < \varepsilon$$

Thus we have shown that the necessary δ exists, completing the proof. ■

The last item on our unfinished-business agenda is the proof of the important Squeeze Theorem, stated as (2.5.6) without proof and then used in the proof of the Fundamental Theorem of the Calculus (6.3.1), and again in showing that $\lim_{\Delta x\to 0} (\sin \Delta x)/\Delta x = 0$.

9.2.3 **Squeeze Theorem** If $h(x) \leq f(x) \leq g(x)$ for all $x \neq a$ in some open interval I containing a, and if $\lim_{x\to a} h(x) = \lim_{x\to a} g(x) = L$, then $\lim_{x\to a} f(x) = L$.

Proof: Suppose $\varepsilon > 0$. We need to show there is a $\delta > 0$ such that $0 < |x - a| < \delta$ implies $|f(x) - L| < \varepsilon$. But since $\lim_{x\to a} h(x) = L$ and $\lim_{x\to a} g(x) = L$, there are $\delta_1 > 0$ and $\delta_2 > 0$, respectively, such that

(i) $\quad 0 < |x - a| < \delta_1 \quad$ implies $\quad |h(x) - L| < \varepsilon$

and

(ii) $\quad 0 < |x - a| < \delta_2 \quad$ implies $\quad |g(x) - L| < \varepsilon$

Now for $x \neq a$ but close enough to a so that x is in I, we know that

(iii) $\quad h(x) \leq f(x) \leq g(x)$

Choosing δ_3 small enough so that $0 < |x - a| < \delta_3$ implies that x is in I, we know that $0 < |x - a| < \delta_3$ implies (iii).

Let δ be the smallest of δ_1, δ_2, δ_3, and assume $0 < |x - a| < \delta$. Then (i) to (iii) all hold. Thus by (ii), $g(x) - L < \varepsilon$, by (iii), $h(x) - L \leq f(x) - L \leq g(x) - L$, and by (i), $-\varepsilon < h(x) - L$. Hence, putting these statements together, we have

$$-\varepsilon < h(x) - L \leq f(x) - L \leq g(x) - L < \varepsilon$$

from which we can conclude that $|f(x) - L| < \varepsilon$, completing the proof. ■

SECTION 9.2 EXERCISES

1. Prove using the definition of limit that if $\lim_{x\to a} f(x) = L$ and $\lim_{x\to a} g(x) = M$, then $\lim_{x\to a} [f(x) - g(x)] = L - M$.

2. Prove using the definition of limit that if $f(x) = b$, a constant function, and $\lim_{x\to a} g(x) = M$, then $\lim_{x\to a} [bg(x)] = bM$.

3. Prove, using the definition of right-hand limit, that if $\lim_{x\to a^+} f(x) = L$ and $\lim_{x\to a^+} g(x) = M$, then $\lim_{x\to a^+} [f(x) + g(x)] = L + M$.

4. Repeat Exercise 3 for $\lim_{x\to a^-}$.

5. Illustrate the proof of the second part of the Limit Theorem, [9.2.1(*b*)], geometrically as follows: (*a*) Draw an $f(x)$ by $g(x)$ rectangle. (*b*) Inside this rectangle and based on one corner, draw an L by M rectangle, where $L < f(x)$ and $M < g(x)$. (*c*) Relate all four rectangular pieces to the proof of [9.2.1(*b*)]. (*d*) Shade all but the L by M part of the rectangle and describe what happens geometrically as $x \to a$.

6. Use a geometric approach analogous to that in Exercise 5 to illustrate the argument requested in Exercise 2.

7. Prove that if $0 < f(x) \leq g(x)$ for all $x \neq a$ in an open interval I containing a, and if $\lim_{x\to a} g(x) = 0$, then $\lim_{x\to a} f(x) = 0$.

8. Prove that if $0 < f(x) \leq g(x)$ for all $x \neq a$ in an open interval I containing a, and if $\lim_{x\to a} f(x) > 0$, then $\lim_{x\to a} g(x) > 0$.

Exercises 9 to 21 involve the application and elaboration of definitions of "infinite limits" (4.1.1) and "limits at infinity" (4.1.2).

9. Prove that $\lim_{x\to 2^+} 1/(x-2) = +\infty$ using Definition (4.1.1) as follows: Let $M > 0$ be given. Show that there is a $\delta > 0$ such that $x - 2 < \delta$ implies $1/(x-2) > M$. (Name such a δ in terms of M.)

10. Prove that $\lim_{x\to 3^+} 2/(x-3) = +\infty$ using (4.1.1) as in Exercise 9.

11. Prove that $\lim_{x\to 3} 5/(x-3)^2 = +\infty$ using (4.1.1).

12. Prove that $\lim_{x\to 2^-} (x+2)/(x^2-4) = -\infty$ using (4.1.1).

13. Prove that $\lim_{x\to +\infty} 1/(x-5) = 0$ using Definition (4.1.2) as follows: Let $\varepsilon > 0$ be given. Show that there is an $M > 0$ such that $x > M$ implies $|1/(x-5)| < \varepsilon$. (M is given in terms of ε.)

14. Prove that $\lim_{x\to +\infty} 2/(x-3) = 0$ using (4.1.2) as in Exercise 13.

15. Prove that $\lim_{x\to +\infty} 4/(x-2)^2 = 0$ using (4.1.2).

16. Prove that $\lim_{x\to -\infty} 2/(x+3)^2 = 0$ using (4.1.2).

17. Prove that if $\lim_{x\to a} |f(x)| = 0$, then $\lim_{x\to a} 1/|f(x)| = +\infty$.

18. Prove that if $\lim_{x\to a} f(x) = 0$ and $f(x) < 0$ for x in $(-\delta_1, +\delta_1)$ for some $\delta_1 > 0$, then $\lim_{x\to a} 1/f(x) = -\infty$.

19. Using an argument analogous to that for [9.2.1(a)], prove that if $\lim_{x\to +\infty} f(x) = A$ and $\lim_{x\to +\infty} g(x) = B$, then $\lim_{x\to +\infty} [f(x) + g(x)] = A + B$.

20. Using an argument analogous to that used in the proof of [9.2.1(b)], prove that if $\lim_{x\to +\infty} g(x) = k \neq 0$, then

$$\lim_{x\to +\infty} \frac{1}{g(x)} = \frac{1}{k}$$

21. Using an argument analogous to the proof of the Squeeze Theorem, prove the following using Definition (4.1.2): If $h(x) \leq f(x) \leq g(x)$ for all $x > \bar{M}$ for some number $\bar{M} > 0$, and $\lim_{x\to +\infty} h(x) = L$ and $\lim_{x\to +\infty} g(x) = L$, then $\lim_{x\to +\infty} f(x) = L$.

Section 9.3: Why All the Fuss?

Objective:

Appreciate the necessity for careful mathematical statements in defining and evaluating limits and in proving theorems.

INTRODUCTION

When confronted with complex-appearing definitions or detailed proofs of theorems that, on the surface, seem obvious, many students have legitimate doubts about their usefulness. This is especially true because often these definitions and theorems are given when the sorts of functions and situations that would render simpler statements inadequate are not yet available for exploration. We shall now briefly examine a few functions whose behavior illustrates the need for care in stating definitions of limit and infinite limit and for care in dealing with derivative quotients and the integrability of functions. Functions such as these literally *forced* mathematicians of the nineteenth century into the definitions we have been using.

For example, let $f(x) = \sin(1/x)$ for $x > 0$, $h(x) = (1/x)\sin(1/x)$ for $x > 0$. (See Fig. 9.9.) We first examine $\lim_{x\to 0^+} f(x)$. As $x \to 0^+$, $1/x \to +\infty$, so $\sin(1/x)$ oscillates between $+1$ and -1 as many times in, say the interval $(0, 1)$, as $\sin x$ oscillates between $+1$ and -1 in the interval $(1, +\infty)$. Notice that *we can make $\sin(1/x)$ as close as we please to 0 by choosing x sufficiently close to 0.* However, we shall show that $\lim_{x\to 0^+} \sin(1/x) \neq 0$: Given any small $\varepsilon > 0$, say $\varepsilon < 1$, and any $\delta > 0$, there are some values of x in $(0, \delta)$ for which $\sin(1/x)$ is within ε of 0 and others for which $\sin(1/x)$ is *not* within ε of 0. In fact, whenever $1/x$ is an odd multiple of $\pi/2$, $\sin(1/x)$ is either $+1$ or -1 and hence outside $(-\varepsilon, \varepsilon)$. No matter how small $\delta > 0$, we can find integers k large enough so that

$$\frac{1}{(2k+1)(\pi/2)} < \delta.$$

(Note that odd integers are those integers of the form $2k + 1$, where k is an integer.) For

$$x = \frac{1}{(2k+1)(\pi/2)}$$

$1/x = (2k+1)(\pi/2)$, so $\sin(1/x) = \pm 1$.

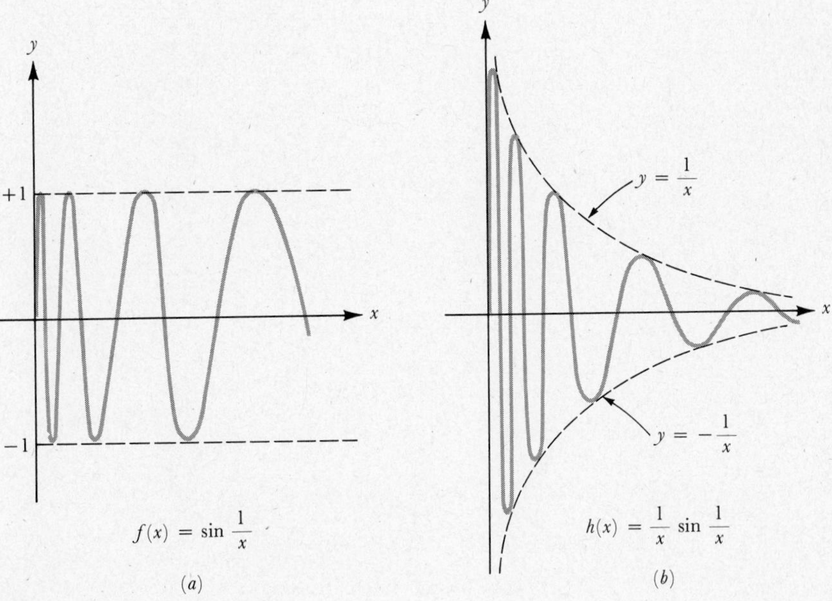

Fig. 9.9

The lesson of this discussion is that the loose paraphrase just emphasized does not suffice as a limit statement.

We can use $h(x) = (1/x)\sin(1/x)$ to make a similar point about infinite limits. *By choosing x sufficiently close to 0, we can make* $h(x) = (1/x)$ $(\sin 1/x)$ *as large as we please.* That is, given any $M > 0$, we can find $\delta > 0$ such that $h(x) > M$ for *some* x in $(0, \delta)$. (See Sec. Exercise 1.) However, no matter how small we choose δ, *there will be other values of x in* $(0, \delta)$ *for which* $h(x) < M$. [See Fig. 9.10(b) and Sec. Exercise 2.] In fact, we can find x in $(0, \delta)$ for which $h(x) = (1/x)\sin(1/x)$ is an arbitrarily large negative number.

Hence $\lim_{x\to 0^+} (1/x)\sin(1/x) \neq +\infty$ and $\lim_{x\to 0^+} (1/x)(\sin 1/x) \neq -\infty$.

THE CHAIN RULE

Recall the formal statement of the Chain Rule:

9.3.1 If u is a differentiable function for all numbers in an open interval containing x, and g is a differentiable function of u in an open interval containing $u(x)$, then the derivative of $g(u(x))$ is $(dg/du)\,(du/dx)$

In Sec. 3.2 we gave a partial proof of (9.3.1) by introducing the expression $(\Delta g/\Delta u)\,(\Delta u/\Delta x)$, where Δx is a small change in x, Δu the corresponding change in u, and Δg the resulting change in g. We assumed, because of the Δu in the denominator above, that $\Delta u \neq 0$. However, for

$$u(x) = \begin{cases} x^2 \sin \dfrac{1}{x} & \text{for } x \neq 0 \\ 0 & \text{for } x = 0 \end{cases}$$

the statement $\Delta u \neq 0$ is not true near $x = 0$. (See Fig. 9.10 and Exercise 3.) Now $u(x)$ is everywhere differentiable (see Exercise 4), but $\Delta u = u(\Delta x) = 0$ whenever $1/\Delta x$ is an integral multiple of π [which is the case for $\Delta x = 1/(k\pi)$, k an integer.] A different proof is required to cover functions such as this particular $u(x)$.

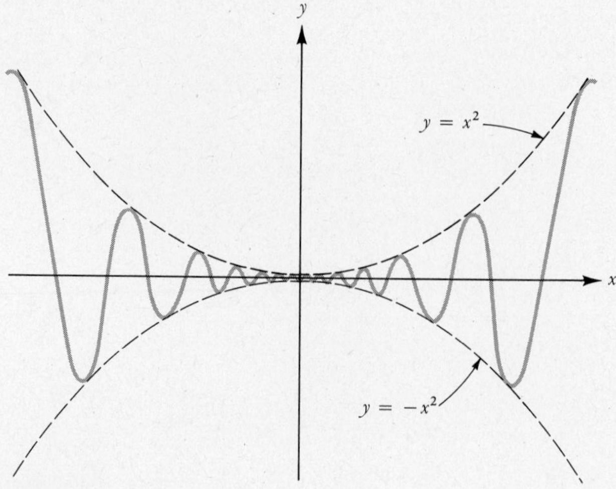

Fig. 9.10

Proof of Chain Rule: [For general g and u, as in (9.3.1.)]

Define $Q = \Delta g/\Delta u - dg/du$ for $\Delta u \neq 0$ and define $Q = 0$ for those Δx making $\Delta u = 0$. Since

$$\lim_{\Delta x \to 0} \Delta u = 0 \qquad (u \text{ is continuous})$$

and since g is differentiable at $u(x)$,

$$\lim_{\Delta u \to 0} Q = \lim_{\Delta x \to 0} Q = 0$$

Rearranging the equation defining Q, we get an equation that is true for *all* Δx, including those for which $\Delta u = 0$:

$$\Delta g = \frac{dg}{du}\,\Delta u + \Delta u\, Q$$

Then upon dividing this equation by Δx and applying the limit theorem, we get

$$\begin{aligned}
\lim_{\Delta x \to 0} \frac{\Delta g}{\Delta x} &= \lim_{\Delta x \to 0} \frac{\dfrac{dg}{du}\,\Delta u + \Delta u\, Q}{\Delta x} \\
&= \lim_{\Delta x \to 0} \frac{dg}{du}\frac{\Delta u}{\Delta x} + \lim_{\Delta x \to 0} \frac{\Delta u}{\Delta x} Q \\
&= \frac{dg}{du}\frac{du}{dx} + \frac{du}{dx} \cdot 0 \qquad (\lim_{\Delta x \to 0} Q = 0) \\
&= \frac{dg}{du} \cdot \frac{du}{dx} \quad \blacksquare
\end{aligned}$$

A DEFINITE INTEGRAL

The function

$$u(x) = \begin{cases} x^2 \sin \dfrac{1}{x} & \text{for } x \neq 0 \\ 0 & \text{for } x = 0 \end{cases} \qquad \text{(see Fig. 9.10)}$$

being differentiable, is also continuous. However, in the terminology of Chap. 6, it is "infinitely wiggly"; that is, there is no closed interval $[a, b]$ containing 0 on which $u(x)$ is monotonic. Recall that in our analysis of area and the definite integral we demonstrated that $\int_a^b f(x)\,dx$ exists for continuous functions that are piecewise monotonic on $[a, b]$, but we had to *assume* that a general (possibly "infinitely wiggly") continuous function is integrable. Hence $u(x)$ is a function to which our earlier analysis did not apply and for which the more general assumption was necessary.

SECTION 9.3 EXERCISES

1. Show that given any $M > 0$, there is a $\delta > 0$ such that $h(x) = (1/x)\sin(1/x) > M$ for some x in $(0, \delta)$.

2. For Exercise 1 show that, for some x in $(0, \delta)$, $h(x) < -M$.

3. Show that for

$$u(x) = \begin{cases} x^2 \sin(1/x) & \text{for } x \neq 0 \\ 0 & \text{for } x = 0 \end{cases}$$

and for any $\delta > 0$, there is a Δx in $(0, \delta)$ with $\Delta u = u(0 + \Delta x) - u(0) = u(\Delta x) = 0$.

4. (*a*) Show that $u(x)$ (as in Exercise 3) is differentiable for $x \neq 0$ by applying the derivative formulas and examining the results.
(*b*) Show that $u(x)$ is differentiable at $x = 0$ by applying the definition of derivative at $x = 0$.
(*c*) Is $u'(x)$ continuous at $x = 0$? Explain.

5. By considering the function $u(x)$ (as in Exercise 3) show that the following does not suffice as a paraphrase of $\lim_{x\to+\infty} u(x) = +\infty$: "We can make $u(x)$ arbitrarily large by choosing x sufficiently large."

Miscellaneous Exercises

1. State and prove a uniqueness theorem for "limits at infinity."

2. Assume that $f(x) \leq 0$ on (a, b) and $a < c < b$. Prove that if $\lim_{x\to c} f(x)$ exists and equals L, then $L \leq 0$.

3. Suppose

$$f(x) = \begin{cases} 0 & \text{if } x \text{ is rational} \\ 1 & \text{if } x \text{ is irrational} \end{cases}$$

Is f continuous at $x = 0$? (*Note:* every open interval contains both rational and irrational numbers.)

4. Suppose

$$g(x) = \begin{cases} 0 & \text{if } x \text{ is rational} \\ x & \text{if } x \text{ is irrational} \end{cases}$$

Is g continuous at $x = 0$?

5. (*a*) Show that

$$f(x) = \begin{cases} \dfrac{-x^2}{2} & \text{if } x < 0 \\ \dfrac{x^2}{2} & \text{if } x \geq 0 \end{cases}$$

has a continuous first derivative at $x = 0$.
(*b*) Show that $f''(0)$ does not exist.

6. Let $[u]$ denote the greatest integer less than u. Sketch $f(x) = [1/x]$ and prove that $\lim_{x\to 0^+} f(x) = +\infty$.

7. Let $g(x) = 1/f(x)$, where $f(x)$ is as in Exercise 6. (*a*) Determine if $\lim_{x\to 0} g(x)$ exists. (*b*) Determine if $g(x)$ is continuous on any interval containing 0 when we define $g(0) = 0$.

SELF-TEST ✚

In Probs. 1 and 2 prove the given statement using the definition of limit.

1. $\lim_{x\to 2} (5 - 3x) = -1$ **2.** $\lim_{x\to 1} \sqrt[3]{x} = 1$

3. Suppose that $\lim_{x\to a} f(x) = L$, $\lim_{x\to a} g(x) = M$, $\lim_{x\to a} h(x) = N$. Prove that if $h(x) < f(x) < g(x)$ for all x in a deleted interval containing a, then $N \leq L \leq M$.

4. Determine $f(x)$, $g(x)$, $h(x)$, and a number a such that the hypotheses of Prob. 3 are satisfied and $N = L = M$.

Chapter 10

A BRIEF REVIEW OF ANALYTIC GEOMETRY

CONTENTS

Curves obtained by intersecting a right circular cone with a plane are called *conic sections* or, more simply, *conics*. The result could be a single point, a straight line, a pair of intersecting straight lines, an ellipse, a parabola, or a hyperbola. We usually refer to the first three as degenerate conics.

We begin by placing the conics in one of the so-called standard positions on the usual rectangular xy-coordinate system. We are then able to translate the geometric definitions into equations in x and y. By making use of a translation and/or rotation of axes we can obtain equations for any conic regardless of its position with respect to the coordinate axes.

We then consider the reverse problem: determining which equations have conic sections as their graphs.

One advantage of this analytic approach to geometry, that is, relating geometric objects to equations in x and y, is that we can eventually bring to bear the machinery of the calculus.

Section 10.1: Translation of Axes and the Parabola

Objectives:

1. Identify and sketch equations whose graphs are parabolas with axes parallel to one of the coordinate axes, using translation of axes if necessary.

2. Determine the focus, vertex, directrix, and axis of a parabola from its equation.

3. Determine the equation of a parabola from geometric information that uniquely determines a parabola.

TRANSLATION OF AXES

We can often simplify the process of graphing an equation in x and y by introducing a new coordinate system. We refer to this process as a *transformation of coordinates.* If the new coordinate system is an XY system, then the effect of transforming the coordinates is to transform the original equation in x and y to a new, in general simpler, equation in X and Y. We do this in such a way that the equation changes but the graph does not. The easiest transformation of coordinates to work with is called a *translation of axes.* In this case the XY-coordinate system is obtained from the xy-coordinate system by a parallel displacement as in Fig. 10.1. In effect, we slide a new coordinate system under the graph.

Let us assume that the origin of the XY system is located at the point with xy coordinates (h, k). Now any point P in the plane has two sets of coordinates (x, y) and (X, Y). From Fig. 10.1 it is not difficult to see that these coordinates are related by the equations:

10.1.1 $$x = X + h \qquad y = Y + k$$

or, equivalently,

10.1.2 $$X = x - h \qquad Y = y - k$$

Given the XY coordinates we can use (10.1.1) to obtain the xy coordinates. To go the other way we use (10.1.2).

Example 1 Identify the graph of the equation

$$16x^2 + 16y^2 - 96x + 128y + 144 = 0$$

and then use a translation of axes to simplify and sketch the equation.

Solution We first regroup the terms,

$$16x^2 - 96x + 16y^2 + 128y = -144$$

divide by 16,

$$(x^2 - 6x + \quad) + (y^2 + 8y + \quad) = -9$$

and complete the square

$$(x^2 - 6x + 9) + (y^2 + 8y + 16) = -9 + 9 + 16$$

This in turn can be written as

$$(x - 3)^2 + (y + 4)^2 = 16$$

Thus the graph is a circle of radius 4 and center $(3, -4)$. Now let $X = x - 3$ and $Y = y + 4$, and the equation becomes

$$X^2 + Y^2 = 16$$

which in the XY system is a circle of radius 4 centered at the *origin.*

By (10.1.2) the substitution $X = x - 3$, $Y = y + 4$ is equivalent to translating the origin of the xy-coordinate system to the point $(3, -4)$. We now sketch the curve $X^2 + Y^2 = 16$ in the XY system and obtain as a consequence the graph of $16x^2 + 16y^2 - 96x + 128y + 144 = 0$ in the xy system. (See Fig. 10.2.) ●

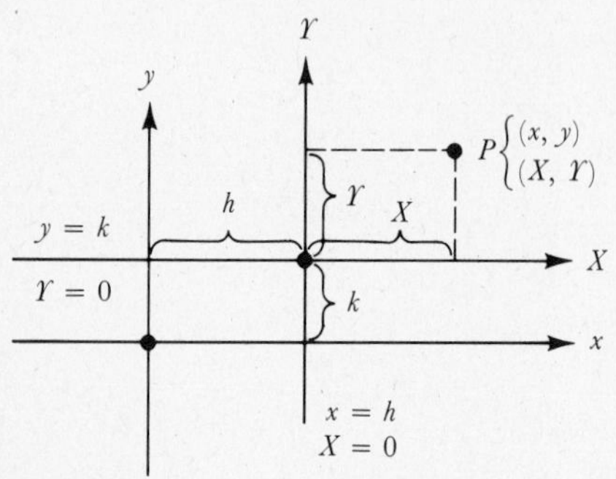

Fig. 10.1

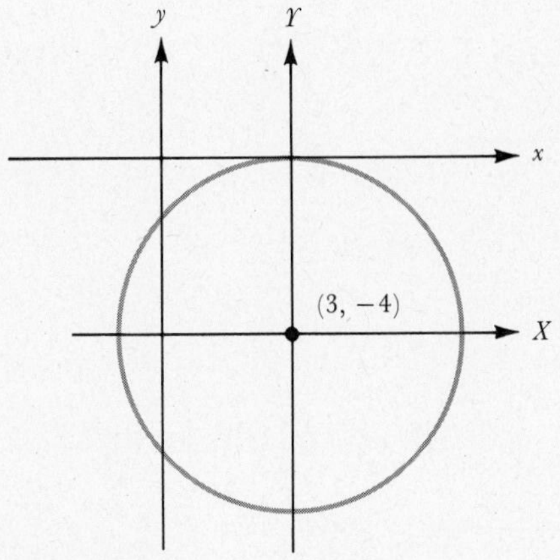

Fig. 10.2

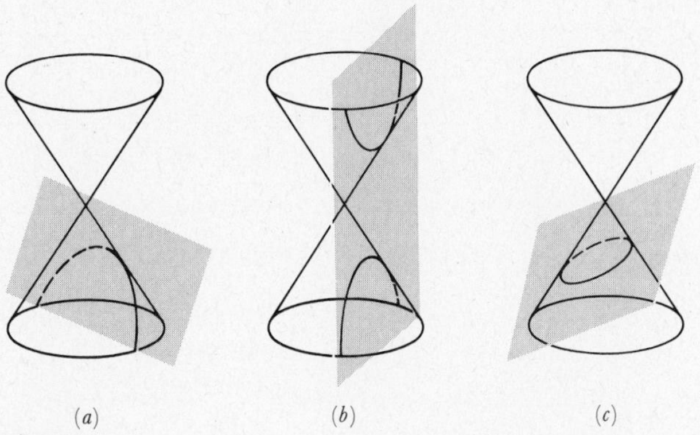

(a) (b) (c)

Fig. 10.3

THE CONIC SECTIONS

The conic sections consist of ellipses, parabolas, and hyperbolas. They are so named because they can be realized as the intersection of a plane with a right circular cone. (See Fig. 10.3.)

In Fig. 10.3(a) a plane parallel to an "element" of the cone (an "element" is a straight line on the cone passing through the vertex) intersects the cone in a parabola. In Fig. 10.3(b) a plane parallel to the axis of the cone intersects the cone in a hyperbola (degenerate if the plane contains the axis). In Fig. 10.3(c) a plane not parallel to either the axis or an element intersects the cone in an ellipse (a circle if the plane is perpendicular to the axis).

THE PARABOLA

10.1.3 **Definition** A *parabola* is the set of all points in a plane that are equidistant from a fixed point F and a fixed line L. The point F is called the *focus* and the line L is called the *directrix*. In addition, the line L' through the focus and perpendicular to the directrix is called the *axis* of the parabola. The point V on the axis midway between the focus and the directrix is called the *vertex*. (See Fig. 10.4.)

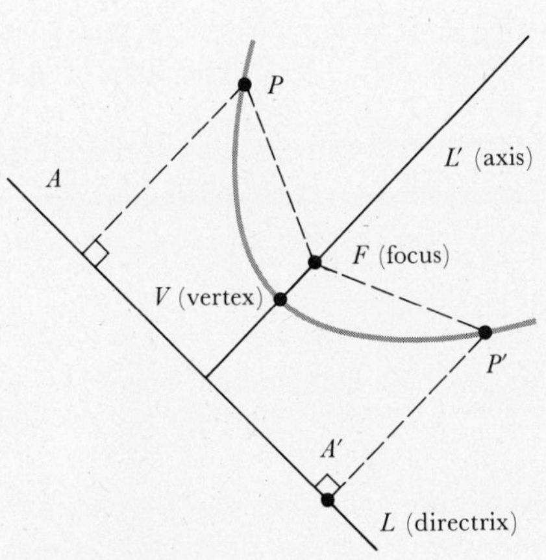

Fig. 10.4

By definition a point P is on the parabola if and only if $|PA| = |PF|$, where A is the base of the perpendicular from P to L. The vertex, being midway between the focus and the directrix, is on the parabola. Furthermore, if P' is the reflection of P through L', then from Fig. 10.4 it is clear that $|P'F| = |P'A'|$, so the parabola is symmetric with respect to its axis.

If a parabola is positioned so that the vertex is at the origin and the axis coincides with one of the coordinate axes, we say the parabola is in a *standard position*. There are four possible standard positions for a given parabola, depending upon whether the focus is located to the right or left of the vertex on the x axis or above or below the vertex on the y axis.

Equations of a Parabola We shall derive an equation of the parabola in one of the standard positions. It will then be an easy task to derive equations for the parabola in any of the other standard positions.

Let $2p$ represent the distance ($p > 0$) from the directrix to the vertex and assume that the axis of the parabola is the x axis with the focus to the right of

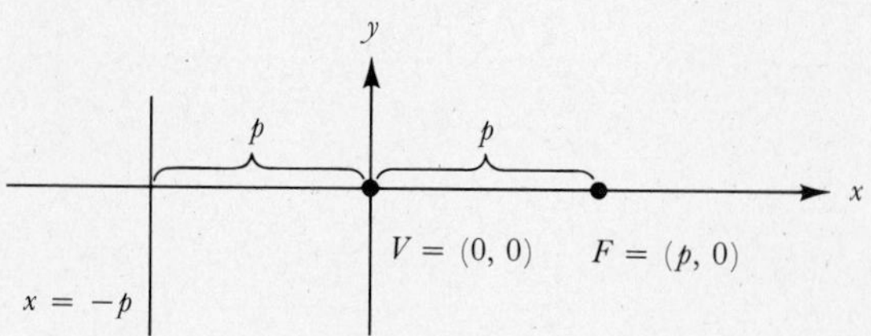

Fig. 10.5

the vertex. (See Fig. 10.5.) It follows that $F = (p, 0)$ and the directrix L has equation $x = -p$.

Let $P = (x, y)$. Then P is on the parabola if and only if $|AP| = |PF|$, where $A = (-p, y)$; see Fig. 10.6. Now

$$|AP| = \sqrt{(x+p)^2 + (y-y)^2} = |x+p|$$

and

$$|PF| = \sqrt{(x-p)^2 + (y-0)^2} = \sqrt{(x-p)^2 + y^2}$$

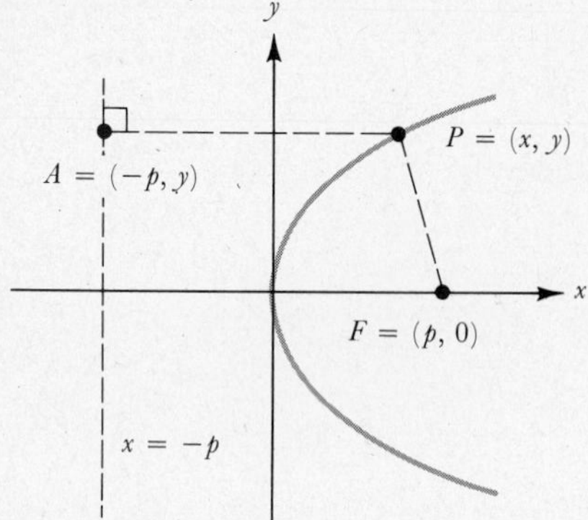

Fig. 10.6

Thus $P = (x, y)$ is on the parabola if and only if

$$\sqrt{(x-p)^2 + y^2} = |x+p|$$

or, equivalently,

$$(x-p)^2 + y^2 = (x+p)^2$$

Squaring the terms $(x-p)^2$ and $(x+p)^2$ and simplifying, we obtain the equivalent equation $y^2 = 4px$.

Thus we have proved the following:

10.1.4 Theorem If $p > 0$, then a necessary and sufficient condition for the point (x, y) to be on the parabola with focus $(p, 0)$ and directrix $x = -p$ is that

$$y^2 = 4px$$

Example 2 Describe the curve $y^2 = 3x$.

Solution Here $3 = 4p$, so $p = 3/4$ and $y^2 = 3x = 4(3/4)x$. Thus $y^2 = 3x$ is a parabola with vertex $(0, 0)$, focus $(3/4, 0)$ and directrix $x = -3/4$. The axis of the parabola is the x axis. (See Fig. 10.7.) ●

If in our development above we had placed the focus at $(-p, 0)$ and taken the line $x = p$ as the directrix, then we would have in effect replaced p by $-p$. This would lead to the equation

$$y^2 = -4px$$

The axis of this parabola is again the x axis, but now the parabola opens to the left.

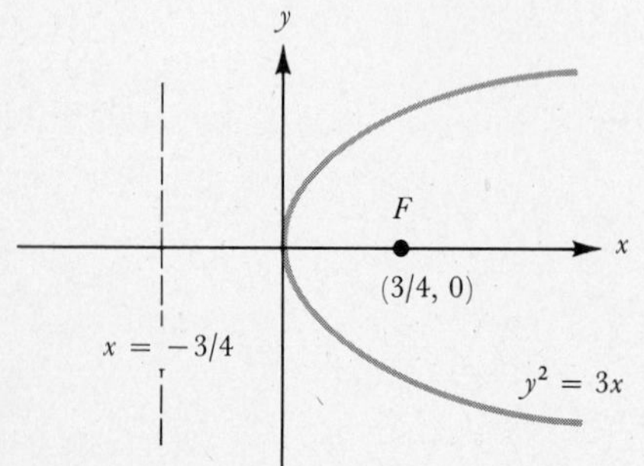

Fig. 10.7

If we interchange the roles of x and y in the previous discussions, we obtain the equation $x^2 = 4py$ for the parabola with focus $(0, p)$ and directrix $y = -p$. Similarly we obtain the equation $x^2 = -4py$ for the parabola with focus $(0, -p)$ and directrix $y = p$. We summarize these results in Table 10.1.

By a translation of axes we can derive the equations for a parabola with a vertical or horizontal axis. For example, suppose the parabola has its vertex at $V = (h, k)$ and has the line $y = k$ as its axis. If the focus is at $(h + p, k)$, then the directrix will be $x = h - p$. Now let $X = x - h$, $Y = y - k$. In the XY-coordinate system $V = (0, 0)$ and the axis is the X axis. Since in this coordinate system

	Equation	Vertex	Focus	Directrix	Axis	Sketch
(a)	$y^2 = 4px$ $p > 0$	(0, 0)	$(p, 0)$	$x = -p$	x axis	$F = (p, 0)$; $x = -p$
(b)	$y^2 = -4px$ $p > 0$	(0, 0)	$(-p, 0)$	$x = p$	x axis	$F = (-p, 0)$; $x = p$
(c)	$x^2 = 4py$ $p > 0$	(0, 0)	$(0, p)$	$y = -p$	y axis	$F = (0, p)$; $y = -p$
(d)	$x^2 = -4py$ $p > 0$	(0, 0)	$(0, -p)$	$y = p$	y axis	$y = p$; $F = (0, -p)$

Table 10.1 Parabolas in Standard Form

the focus is at $(p, 0)$, we can apply Theorem (10.1.4) to conclude that the parabola has the equation

$$X^2 = 4pY$$

Thus in the original xy-coordinate system, the equation becomes

$$(x - h)^2 = 4p(y - k)$$

Example 3 Identify and sketch the graph of the equation $x^2 - 10x - 8y + 53 = 0$.

Solution We begin by putting the x terms on one side and the y term and constant on the other:

$$x^2 - 10x = 8y - 53$$

Complete the square in x:

$$x^2 - 10x + 25 = 8y - 53 + 25 = 8y - 28$$

or

$$(x - 5)^2 = 8\left(y - \frac{28}{8}\right) = 4(2)\left(y - \frac{7}{2}\right)$$

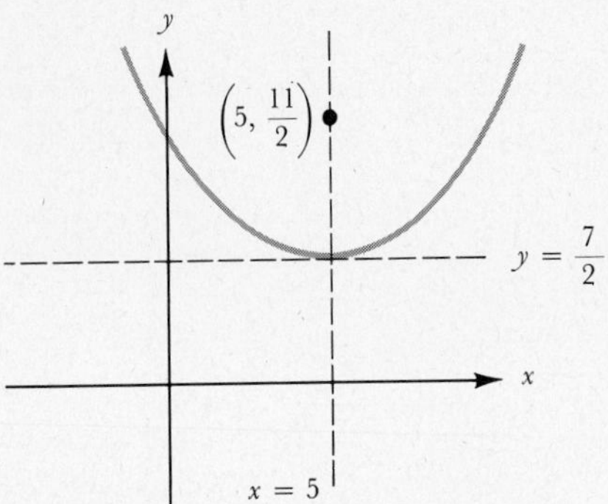

Fig. 10.8

This is of the form $X^2 = 4pY$, which from Table 10.1 has its vertex at the origin of the XY-coordinate system and opens upward. The focus is located p units up on the Y axis, and the directrix is the line $Y = -p$. Since $X = x - 5$, $Y = y - 7/2$, and $p = 2$, in the original xy system we have a parabola with vertex $(5, 7/2)$, focus $(5, 11/2)$, axis $x = 5$, and directrix $y = 3/2$. (See Fig. 10.8.) ●

A reverse version of Example 3 might ask for the equation of the parabola given that its focus was located at $(5, 11/2)$ and that its directrix had equation $y = 3/2$.

PROGRESS TEST 1

In Probs. 1 to 3 determine the vertex, focus, and directrix of the parabolas with the given equations and sketch.

1. $y = 2x^2$ **2.** $4x = y^2$

3. $y^2 - 4y + 8x + 28 = 0$

4. Find an equation for the parabola with vertex at $(4, -3)$ and focus at $(4, 0)$ and sketch.

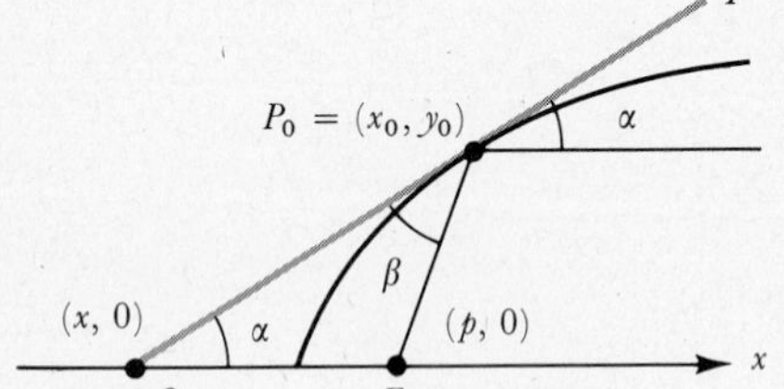

Fig. 10.9

PARABOLIC REFLECTORS

Let $P_0 = (x_0, y_0)$ be a point unequal to the vertex on the parabola $y^2 = 4px$ and let T denote the line tangent to the parabola at P_0. Let α denote the angle between T and the line through P_0 parallel to the x axis. (See Fig. 10.9.) Finally, let β denote the angle P_0F makes with T. We shall show that $\alpha = \beta$, from which we can deduce the reflective properties that make the parabolic shape so useful in applications.

Now α is also the angle between T and the x axis. Since $dy/dx = 2p/y$, T has slope $2p/y_0$ and equation $y - y_0 = (2p/y_0)(x - x_0)$. Letting $y = 0$, we have

$$-y_0 = \frac{2p}{y_0}(x - x_0) \qquad \text{or} \qquad \frac{-y_0^2}{2p} + x_0 = x$$

But (x_0, y_0) is on $y^2 = 4px$, so

$$x = \frac{-4px_0}{2p} + x_0 = -x_0$$

Thus Q has coordinates $(-x_0, 0)$. Now

$$|QF| = p + x_0$$

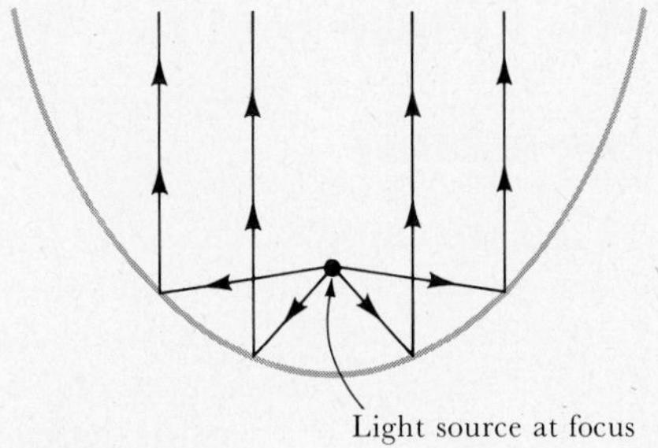

Fig. 10.10

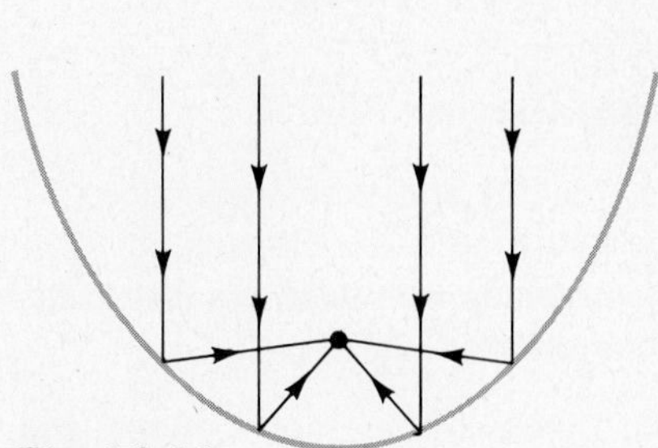
Fig. 10.11

and

$$\begin{aligned} |P_0F| &= \sqrt{(x_0 - p)^2 + y_0^2} \\ &= \sqrt{x_0^2 - 2px_0 + p^2 + y_0^2} \\ &= \sqrt{x_0^2 + 2px_0 + p^2} \qquad \text{since } y_0^2 = 4px_0 \\ &= \sqrt{(x_0 + p)^2} = x_0 + p \end{aligned}$$

Therefore triangle FQP_0 is isosceles, with equal base angles; that is, $\alpha = \beta$.

Now, if a parabola is revolved about its axis, a surface, called a *paraboloid,* is generated whose cross section parallel to the axis is a parabola. When light is reflected from this surface, the angle of incidence equals the angle of reflection, so if a light source is placed at the focus of the parabola generating the paraboloid, then light rays will be reflected outward in rays parallel to the axis, providing a directed beam of light. (See Fig. 10.10.)

The same principle is used in the design of a reflecting telescope, except that now the parabolic surface causes parallel rays coming in to concentrate at the focus. The only difference is the direction of the light rays. (See Fig. 10.11.)

SECTION 1 EXERCISES

In Exercises 1 to 12, (a) determine the vertex, focus, and directrix, and (b) sketch the parabola, identifying these items in the sketch.

1. $y^2 = 20x$
2. $3y^2 - 18x = 0$
3. $y^2 + x = 0$
4. $x^2 - 2y = 0$
5. $5x^2 - 400y = 0$
6. $x^2 - 2x + y = 0$
7. $3y^2 = 4x$
8. $y^2 - 6x + 3x = 0$
9. $x^2 + 10x - 3y = -1$
10. $3x^2 + 12x - 5 + y = 0$
11. $y^2 - 5y - 3 = 10x$
12. $4x^2 - 12x - y = -3$

In Exercises 13 to 22 determine an equation of the parabola in standard position with vertex at (0, 0) and:

13. Focus at $(4, 0)$
14. Focus at $(-6, 0)$
15. Directrix $x = -10$
16. Directrix $y = 5$
17. Focus at $(0, -8)$
18. Focus at $(0, 7)$
19. Passing through $(4, -2)$, opening downward
20. Passing through $(2, 3)$, opening upward
21. Passing through $(3, 5)$; two parabolas are possible—give each
22. Passing through $(-2, -3)$; two parabolas are possible—give each

In Exercises 23 to 30 (a) determine an equation of the indicated parabola, and (b) sketch the parabola.

23. Vertex at $(5, 1)$, focus at $(5, 3)$
24. Vertex at $(2, -3)$, focus at $(2, 1)$
25. Vertex at $(1, 4)$, focus at $(1, 6)$
26. Focus at $(-1, -1)$, directrix $x = -3$
27. Focus at $(4, 6)$, directrix $y = 8$
28. Vertex at $(8, 2)$, directrix $y = 4$
29. Vertex at $(-6, 4)$, directrix $x = 12$
30. Focus at $(5, 8)$, directrix $y = -2$

Section 10.2: The Ellipse and the Hyperbola

Objectives:

1. Identify and sketch equations whose graphs are ellipses or hyperbolas with axes parallel to the coordinate axes.

2. Determine the foci, vertices, and major and minor axes of an ellipse from its equation.

3. Determine the foci, vertices, transverse and conjugate axes, and asymptotes of a hyperbola from its equation.

4. Determine the equation of an ellipse or hyperbola from given information.

THE ELLIPSE

10.2.1 Definition An *ellipse* is the set of all points in a plane the sum of whose distances from two fixed points, F_1 and F_2, is constant. (See Fig. 10.12.) The points F_1 and F_2 are called the *foci* of the ellipse. The straight line through F_1 and F_2 is called the *major axis* of the ellipse, and the perpendicular bisector of the line segment F_1F_2 is called the *minor axis*. The intersection of the major and minor axes is called the *center* of the ellipse.

Let P be an arbitrary point on the ellipse. We denote the sum of the distances $|PF_1|$ and $|PF_2|$ by $2a$. Thus P is on the ellipse if and only if

$$|PF_1| + |PF_2| = 2a$$

If we flip the triangle F_1PF_2 (see Fig. 10.13) about its base F_1F_2, we see that P', the reflection of P through the major axis, is also on the ellipse. By turning the triangle over, we see that P'', the reflection of P through the minor axis, is also on the ellipse. Thus an ellipse is symmetric with respect to both of its axes.

The definition can be used to draw an ellipse. Attach the ends of a string of length $2a$ to the foci. If the tip of a pencil is used to hold the string taut, it will trace out the ellipse as it is moved around since clearly the sum of the distances from the tip to the points F_1 and F_2 will always equal $2a$. (See Fig. 10.13.)

With respect to a given xy-coordinate system, the ellipse is said to be in *standard position* if the center is at the origin and the foci are on either the x or y axis.

To obtain an equation of an ellipse in one of the two standard positions, we let $2c$ denote the distance F_1F_2 and we place the foci at $(-c, 0)$ and $(c, 0)$, so the major axis is the x axis.

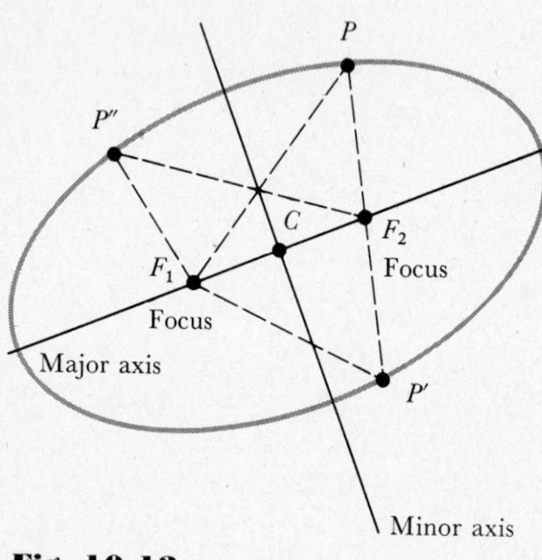

Fig. 10.12

Note from Fig. 10.14 that $2c < 2a$ since $2c$ is the length of the base of triangle F_1PF_2 and $2a = |PF_1| + |PF_2|$ is the sum of the lengths of the other two sides. In particular, $c < a$. Now $P = (x, y)$ is on the ellipse if and only if

$$|PF_1| + |PF_2| = 2a$$

Using the distance formula, we see that this equation is equivalent to

$$\sqrt{(x + c)^2 + y^2} + \sqrt{(x - c)^2 + y^2} = 2a$$

Now if we square both sides, rearrange, and square again, we obtain

$$\left(1 - \frac{c^2}{a^2}\right)x^2 + y^2 = a^2 - c^2$$

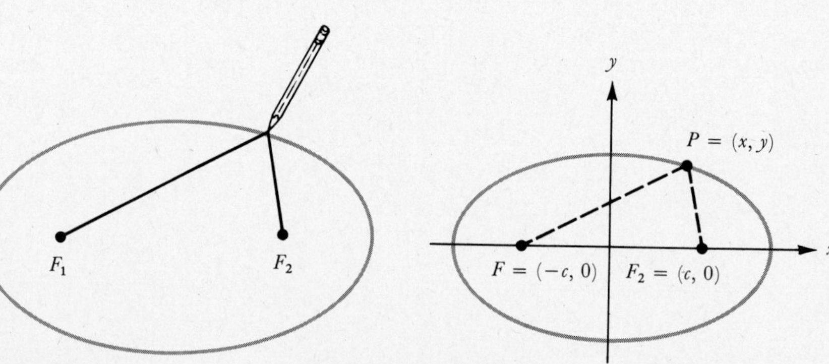

Fig. 10.13 **Fig. 10.14**

or

$$\frac{a^2 - c^2}{a^2}x^2 + y^2 = a^2 - c^2$$

Now $a^2 - c^2 > 0$ since $a > c$. Let $b^2 = a^2 - c^2$ and divide both sides by b^2

$$\frac{x^2}{a^2} + \frac{y^2}{b^2} = 1 \tag{1}$$

This shows that if P is on the ellipse, then the coordinates of P will satisfy (1). Unlike the parabolic case, it does not follow immediately that each previous step is reversible. The content of Exercise 24 is to show that each point (x, y) which satisfies (1) is in turn on the ellipse.

Thus we have proved the following:

10.2.2 **Theorem** Let $\mathcal{E}$ be an ellipse with foci located at $(-c, 0)$ and $(c, 0)$ $(c > 0)$ and let $2a$ $(a > c)$ denote the sum of the distances from any point (x, y) on the ellipse to the foci. Then $\mathcal{E}$ has equation

$$\frac{x^2}{a^2} + \frac{y^2}{b^2} = 1 \qquad \text{where } b^2 = a^2 - c^2$$

From the equation it is algebraically clear that the ellipse is symmetric with respect to the x and y axes and the origin. If $y = 0$, then $x = \pm a$. The points $(-a, 0)$, $(a, 0)$, where the ellipse intersects its major axis, are called the *vertices* of the ellipse. The distance $2a$ between the vertices is called the *length of the major axis*. The ellipse intersects the minor axis at $(0, -b)$ and $(0, b)$. The distance $2b$ between these points is called the *length of the minor axis*.

Since $c \neq 0$ and $b^2 = a^2 - c^2$, a is always larger than b. That is, the ellipse is elongated in the direction of its major axis.

Example 1 Determine an equation for the ellipse with foci at $(-3, 0)$, $(3, 0)$ and $a = 7$.

Solution Since $a = 7$ and $c = 3$, we have $b^2 = a^2 - c^2 = 49 - 9 = 40$, so $b = 2\sqrt{10}$. Thus the ellipse (see Fig. 10.15) has equation

$$\frac{x^2}{49} + \frac{y^2}{40} = 1 \quad \bullet$$

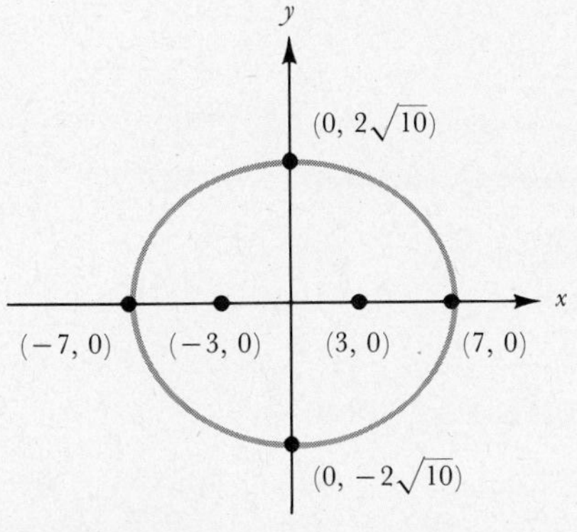

Fig. 10.15

If in our development we had placed the foci at $(0, -c)$ and $(0, c)$, then in effect we would have interchanged the roles of x and y, leading to the equation

$$\frac{x^2}{b^2} + \frac{y^2}{a^2} = 1$$

Again, the major axis (now the y axis) would have length $2a$, and the minor axis (x axis) would have length $2b$. The vertices would be the points $(0, \pm a)$ on the y axis. Since $a > b$, we can recognize from the equations which is the major axis by following these steps:

1. Write the equation in the form

$$\frac{x^2}{(\quad)^2} + \frac{y^2}{[\quad]^2} = 1$$

2. If $(\quad) > [\quad]$, then $(\quad) = a$ and the x axis is the major axis.
3. If $[\quad] > (\quad)$, then $[\quad] = a$ and the y axis is the major axis.

We summarize in Table 10.2.

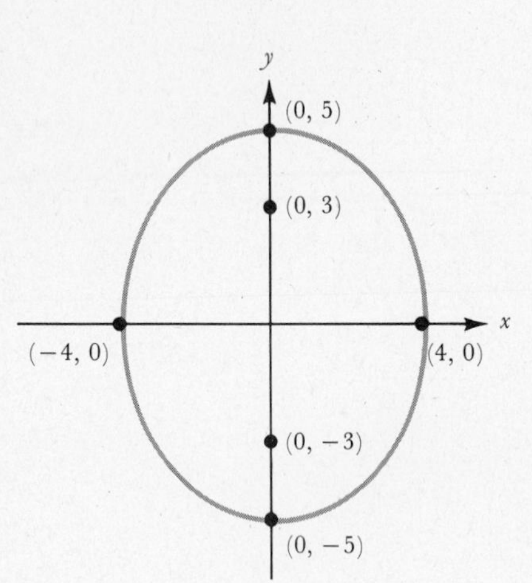

Fig. 10.16

Equation	*Major Axis*	*Minor Axis*	*Foci*	*Vertices*	*Sketch*
$\frac{x^2}{a^2} + \frac{y^2}{b^2} = 1$	x axis	y axis	$(c, 0)$ $(-c, 0)$	$(a, 0)$ $(-a, 0)$	y, $(-a, 0)$, $(a, 0)$, x
$\frac{x^2}{b^2} + \frac{y^2}{a^2} = 1$	y axis	x axis	$(0, c)$ $(0, -c)$	$(0, a)$ $(0, -a)$	y, $(0, a)$, x, $(0, -a)$

Table 10.2 Ellipses in Standard Form ($a > b$, $b^2 = a^2 - c^2$)

Example 2 Sketch the equation $25x^2 + 16y^2 = 400$, labeling foci and vertices.

Solution Divide by 400. Then

$$\frac{25x^2}{400} + \frac{16y^2}{400} = 1 \qquad \text{or} \qquad \frac{x^2}{16} + \frac{y^2}{25} = 1$$

Since $25 > 16$, $a^2 = 25$ and the major axis is the y axis. Now $b^2 = 16$ and so $c^2 = 25 - 16 = 9$. Thus the foci are at $(0, -3)$ and $(0, 3)$; see Fig. 10.16. ●

As with the parabola, it is easy to obtain any translation of an ellipse for which the axes are horizontal and vertical by separately completing the square in x and y.

PROGRESS TEST 1

1. Determine the equation of the ellipse with center at $(0, 0)$, a focus at $(4, 0)$, and major axis of length 8.
2. Repeat Prob. 1, but translated so that the center is at $(8, -6)$.
3. Determine the center, vertices, and foci, and then sketch the ellipse with equation

$$16x^2 - 96x + 25y^2 - 200y = -144.$$

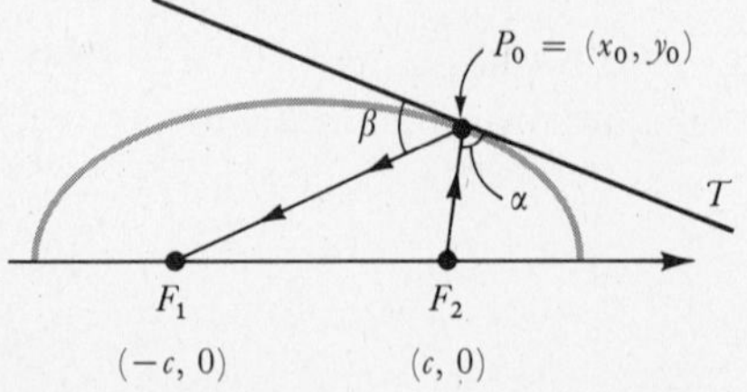

Fig. 10.17

REFLECTIVE PROPERTIES OF THE ELLIPSE

An ellipsoid is obtained by revolving an ellipse about its major axis. Like the paraboloid, the ellipsoid also has an interesting reflective property. In this case a light or sound source emanating from one focus will be reflected to the other focus. (See Fig. 10.17.) We omit its development.

THE HYPERBOLA

10.2.3 **Definition** A *hyperbola* is the set of all points in a plane such that the absolute value of the difference of the distances from these points to two fixed points, F_1 and F_2, is constant. The points F_1 and F_2 are called the *foci* of the hyperbola.

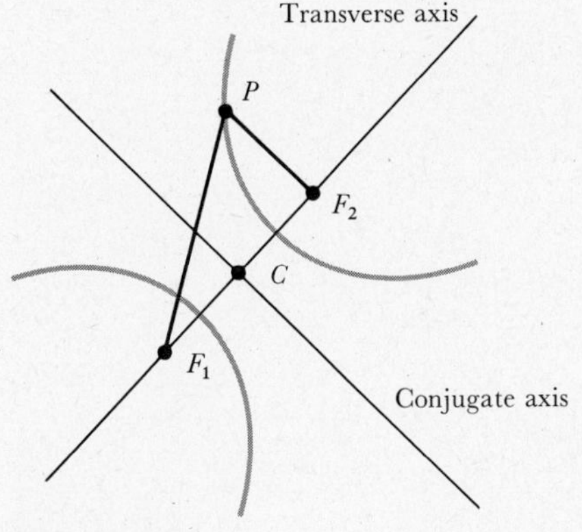

Fig. 10.18

(See Fig. 10.18.) The straight line through F_1 and F_2 is called the *transverse axis* of the hyperbola, and the perpendicular bisector of the segment F_1F_2 is called the *conjugate axis*. The point where the transverse and conjugate axes meet is called the *center* of the hyperbola.

Let P be a point on the hyperbola and let $2a$ denote the absolute value of the difference of the distances $|PF_1|$ and $|PF_2|$. Then P is on the hyperbola if and only if

$$|PF_1| - |PF_2| = \pm 2a$$

We shall derive an equation for the hyperbola in a standard position by placing the center at the origin and the foci on the x axis at $(-c, 0)$ and $(c, 0)$; see Fig. 10.18. Now

$$|PF_2| - |PF_1| = \pm 2a$$

so

$$\sqrt{(x-c)^2 + y^2} - \sqrt{(x+c)^2 + y^2} = \pm 2a$$

Proceeding as with the ellipse, we rearrange, square, simplify, and then square again to get

(1) $$\frac{x^2}{a^2} - \frac{y^2}{c^2 - a^2} = 1$$

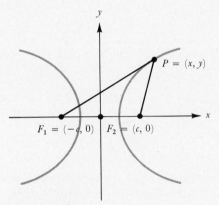

Fig. 10.19

Since the sum of the lengths of two sides of a triangle exceeds the length of the third side, we observe from Fig. 10.19 that

$$2c + |PF_1| > |PF_2| \quad \text{and} \quad 2c + |PF_2| > |PF_1|$$

or

$$2c > |PF_2| - |PF_1| \quad \text{and} \quad 2c > |PF_1| - |PF_2|$$

These latter two inequalities imply

$$2c > \big||PF_1| - |PF_2|\big| = 2a$$

and hence $c > a$. In particular, $c^2 - a^2 > 0$. Thus, letting $b^2 = c^2 - a^2$, we can rewrite (1) as

(2) $$\frac{x^2}{a^2} - \frac{y^2}{b^2} = 1$$

As with the ellipse the steps used in deriving (2) are reversible. Letting $y = 0$ in (2), we see that $x = \pm a$ and thus the hyperbola intersects its transverse axis at $(\pm a, 0)$. These points are called the *vertices* of the hyperbola.

In summary, we have proved the following:

10.2.4 **Theorem** Let $c > a > 0$. The hyperbola with foci $(\pm c, 0)$ and vertices $(\pm a, 0)$ has the equation

$$\frac{x^2}{a^2} - \frac{y^2}{b^2} = 1 \qquad \text{where } b^2 = c^2 - a^2$$

The transverse axis of this hyperbola is the x axis, the conjugate axis is the y axis, and the center is at the origin.

If in the previous development the vertices were on the y axis at $(0, \pm a)$ and the foci at $(0, \pm c)$, then essentially the same computations would yield

$$\frac{y^2}{a^2} - \frac{x^2}{b^2} = 1$$

where again $b^2 = c^2 - a^2$.

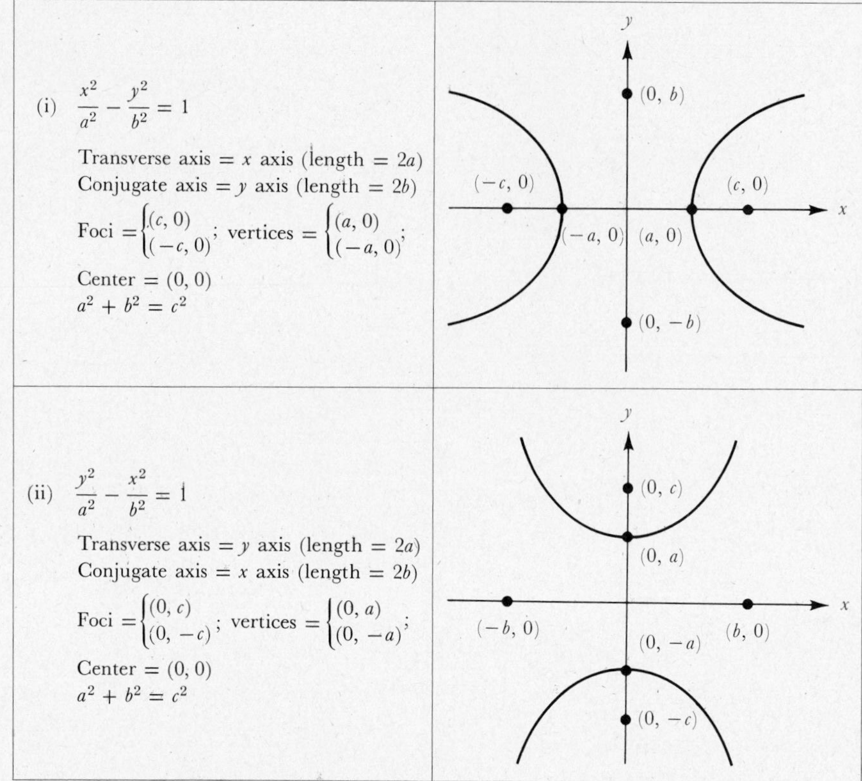

(i) $\dfrac{x^2}{a^2} - \dfrac{y^2}{b^2} = 1$ Transverse axis $= x$ axis (length $= 2a$) Conjugate axis $= y$ axis (length $= 2b$) Foci $= \begin{cases}(c, 0)\\(-c, 0)\end{cases}$; vertices $= \begin{cases}(a, 0)\\(-a, 0)\end{cases}$; Center $= (0, 0)$ $a^2 + b^2 = c^2$	
(ii) $\dfrac{y^2}{a^2} - \dfrac{x^2}{b^2} = 1$ Transverse axis $= y$ axis (length $= 2a$) Conjugate axis $= x$ axis (length $= 2b$) Foci $= \begin{cases}(0, c)\\(0, -c)\end{cases}$; vertices $= \begin{cases}(0, a)\\(0, -a)\end{cases}$; Center $= (0, 0)$ $a^2 + b^2 = c^2$	

Table 10.3 Hyperbolas in Standard Form

Note that, unlike the ellipse, the relative sizes of a and b are not instrumental in determining the axes. However, *the quantity appearing below the variable with the minus sign is always taken to be* b^2.

The distance $2a$ between the vertices is called the *length of the transverse axis.* The number $2b$ is called the *length of the conjugate axis.* Of course it is clear from the equations that a hyperbola does not intersect its conjugate axis. It is equally clear that in either of the standard positions the hyperbola is symmetric with respect to the x and y axes and with respect to the origin.

Table 10.3 gives a summary of the pertinent information and the appropriate sketch.

Example 3 Identify and sketch the graph of the equation $y^2 - 3x^2 = 9$, giving all relevant information.

Solution We divide by 9 to obtain

$$\frac{y^2}{9} - \frac{x^2}{3} = 1$$

Thus $a^2 = 9$, $b^2 = 3$, and $c^2 = a^2 + b^2 = 12$, and so $a = 3$, $b = \sqrt{3}$, $c = 2\sqrt{3}$. Since the equation has the form of (ii) in Table 10.3, the transverse axis is the y axis and the conjugate axis is the x axis. The foci are at $(0, \pm 2\sqrt{3})$ and the vertices are at $(0, \pm 3)$; see Fig. 10.20. ●

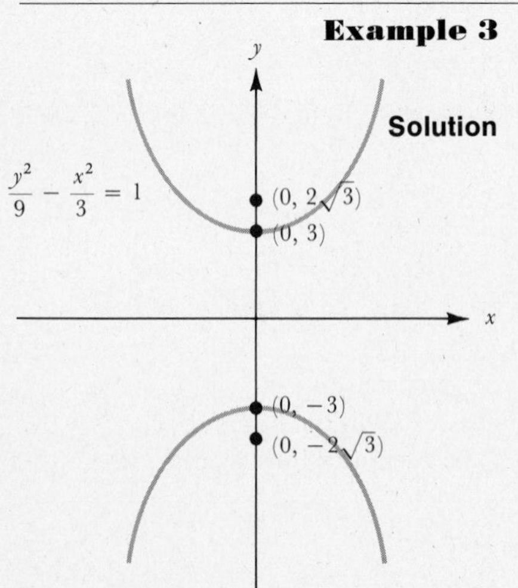

Fig. 10.20

ASYMPTOTES AND THE CENTRAL RECTANGLE

The *central rectangle* of the hyperbola

$$\frac{x^2}{a^2} - \frac{y^2}{b^2} = 1$$

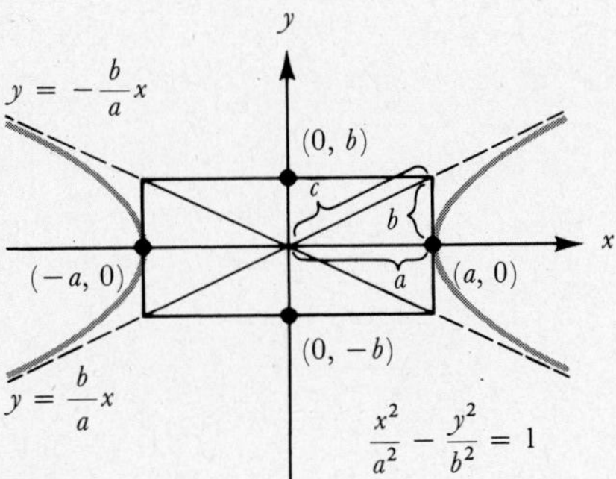

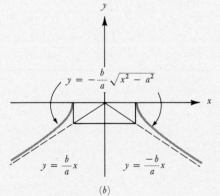

Fig. 10.21

is obtained by drawing parallel lines through the vertices $(\pm a, 0)$ and parallel lines through the points $(0, \pm b)$ on the conjugate axis. (See Fig. 10.21.) The diagonals of the central rectangle lie along the lines $y = \pm(b/a)x$.

Solving for y in terms of x, we have

$$y = \pm\frac{b}{a}\sqrt{x^2 - a^2}$$

Now

$$\lim_{x\to+\infty}\left[\frac{b}{a}\sqrt{x^2 - a^2} - \frac{b}{a}x\right] = \lim_{x\to+\infty}\frac{b}{a}\left[\sqrt{x^2 - a^2} - x\right]$$

$$= \lim_{x\to+\infty}\frac{b}{a}\left[\sqrt{x^2 - a^2} - x\cdot\frac{\sqrt{x^2 - a^2} + x}{\sqrt{x^2 - a^2} + x}\right]$$

$$= \lim_{x\to+\infty}\frac{-ab}{\sqrt{x^2 - a^2} + x} = 0$$

Similarly, one shows

$$\lim_{x\to-\infty}\left[\frac{b}{a}\sqrt{x^2 - a^2} - \left(\frac{-b}{a}x\right)\right] = 0$$

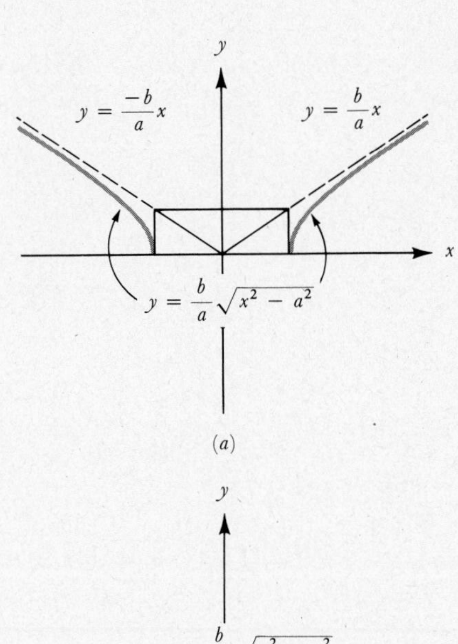

Fig. 10.22

Thus $y = \pm(b/a)x$ are oblique asymptotes to the graph of $y = (b/a)\sqrt{x^2 - a^2}$, the *upper* half of the hyperbola $x^2/a^2 - y^2/b^2 = 1$; see Fig. 10.22(*a*). Analogously, $y = \pm(b/a)x$ are oblique asymptotes of $y = (-b/a)\sqrt{x^2 - a^2}$, the *lower* half of the hyperbola $x^2/a^2 - y^2/b^2 = 1$; see Fig. 10.22(*b*).

For the hyperbola $y^2/a^2 - x^2/b^2 = 1$ a similar argument shows that the lines $y = \pm(a/b)x$ are oblique asymptotes.

As an aid in sketching the graph of a hyperbola it is a good idea first to sketch the central rectangle and then use the extensions of the diagonals as the oblique asymptotes.

Example 4 Sketch the hyperbola

$$16x^2 - 9y^2 = 144$$

Solution Dividing by 144, we have

$$\frac{x^2}{9} - \frac{y^2}{16} = 1$$

Thus $a^2 = 9$, $b^2 = 16$, and $c^2 = 9 + 16 = 25$. The transverse axis is the x axis with the foci at $(\pm 5, 0)$ and the vertices at $(\pm 3, 0)$. The conjugate (y) axis has

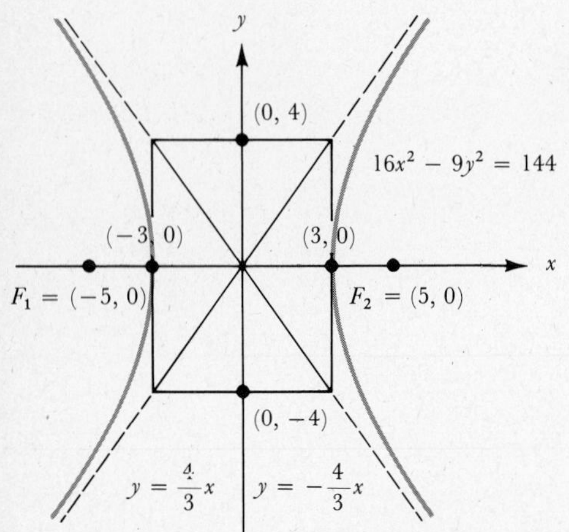

Fig. 10.23

length $2b = 8$. Hence the oblique asymptotes are $y = \pm(4/3)x$, and we can sketch the hyperbola. (See Fig. 10.23.) ●

If a hyperbola in standard position is translated so that its center is at the point (h, k), then, by separately completing the square of the x and y terms, any equation for such a hyperbola can be put in one of the forms

$$\frac{(x-h)^2}{a^2} - \frac{(y-k)^2}{b^2} = 1 \quad \text{or} \quad \frac{(y-k)^2}{a^2} - \frac{(x-h)^2}{b^2} = 1$$

PROGRESS TEST 2

Determine the central rectangle, foci, and vertices and then sketch the hyperbola $9ax^2 - 36x - 4y^2 - 8y = 112$.

SECTION 10.2 EXERCISES

In Exercises 1 to 16 sketch the graph of the given equation, labeling the center, foci, and vertices.

1. $9x^2 + 25y^2 - 18x - 150y + 9 = 0$
2. $16x^2 - 9y^2 - 160x - 72y = -112$
3. $25x^2 + 169y^2 + 100x - 1352y = 69$
4. $x^2 + 4y^2 - 4 = 0$
5. $4x^2 + y^2 - 4 = 0$
6. $x^2 + 4y^2 - 6x + 5 = 0$
7. $9x^2 + 4y^2 + 18x - 8y + 4 = 0$
8. $x^2 + 4y^2 + 2x + 40y = -97$
9. $3x^2 + y^2 + 24x = -2y$
10. $4y^2 - 9x^2 + 16y + 18x = 29$
11. $9x^2 - y^2 - 36x + 8y = 5$
12. $-4x^2 + 9y^2 - 16x - 18y = 43$
13. $9y^2 - x^2 - 36y + 2x = -26$
14. $25x^2 + 4y^2 - 50x - 16y = 59$
15. $x^2 + y^2 - 6y + 10x = -18$
16. $x^2 + 2x - y^2 - 4y + 4 = 0$)

In Exercises 17 to 20 determine the equation of the ellipse satisfying the given conditions.

17. Vertices $(\pm 5, 0)$, center $(0, 0)$, $b = 4$
18. Center $(0, 0)$, one vertex $(10, 0)$, $a = 3$
19. One focus at $(6, 6)$, $b = 2$, center $(4, 6)$
20. Foci $(3, 2)$, $(9, 2)$, $a = 4$

In Exercises 21 to 24 determine the equation of the hyperbola satisfying the given conditions.

21. Vertices $(5, 3)$, $(1, 3)$, $c = 2$, $b = 2$
22. Center $(0, 4)$, foci on y axis $a = 4$, $c = 5$
23. Center $(1, 2)$, foci $(4, 2)$, $(-2, 2)$, $b = 2$
24. Foci $(1, 2)$, $(1, 6)$, $a = 1$.
25. Show that each point (x, y) that satisfies $x^2/a^2 + y^2/b^2 = 1$ is on the ellipse with foci $(\pm c, 0)$.

Section 10.3: Rotation of Axes and the General Quadratic Equation

Objective:

Determine the graph of a general second-degree equation by performing an appropriate rotation, if necessary, to obtain an equivalent equation in standard form.

ROTATION OF AXES

Any equation of a conic section *with axes parallel to the coordinate axes* can now be easily identified. If it is not in standard form, then we appropriately complete the square to reveal its graph as a translation of one of the standard forms. We shall now develop the means for identifying the equation of a conic section whose axes are *not* parallel to the coordinate axes, a conic section that has been rotated through an angle α, as in Fig. 10.24, or rotated and translated, as in

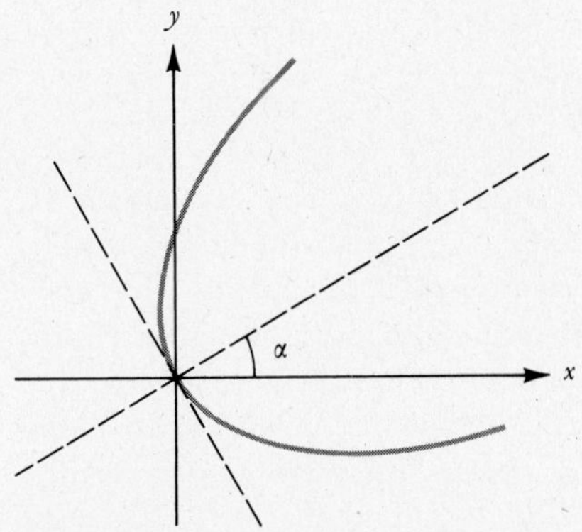

Fig. 10.24

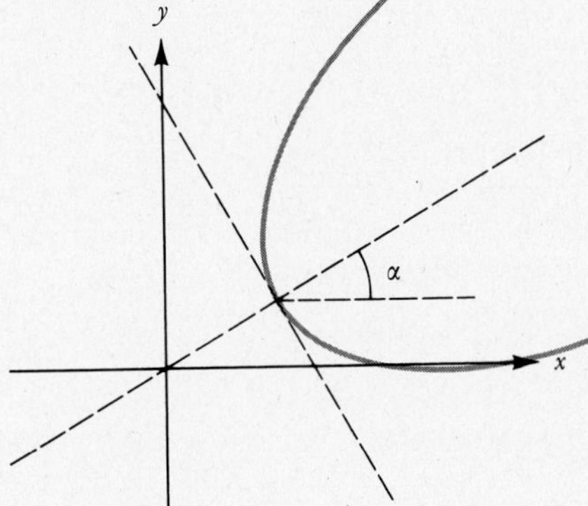

Fig. 10.25

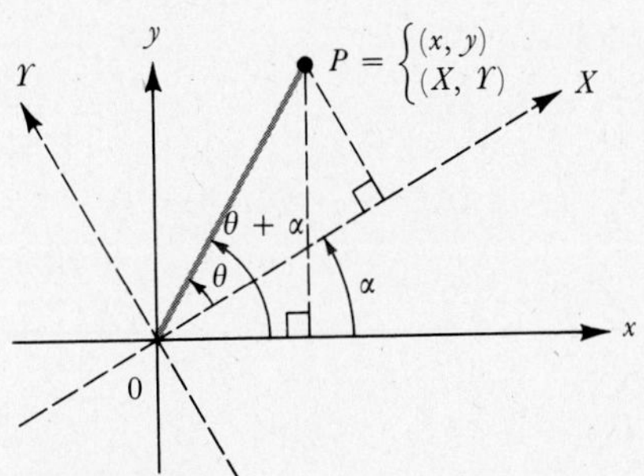

Fig. 10.26

Fig. 10.25. We shall concentrate on a rotation of one of the standard forms, since once this has been obtained, the translation can be handled as before.

Let us assume that the xy-coordinate system has been rotated through an angle α to obtain an XY system. The rotation is counterclockwise if $\alpha > 0$ and clockwise if $\alpha < 0$. If P is any point in the plane, then P has coordinates (x, y) in the xy system and coordinates (X, Y) in the XY system. (See Fig. 10.26.)

Given the coordinates in one system, we would like to derive the coordinates in the other. If θ measures the angle of inclination of the segment OP with the positive X axis, then $\theta + \alpha$ gives the angle of inclination of OP with the x axis. From Fig. 10.26 we know that

(1) $$x = |OP|\cos(\theta + \alpha) \qquad y = |OP|\sin(\theta + \alpha)$$

and

(2) $$X = |OP|\cos\theta \qquad Y = |OP|\sin\theta$$

Using trigonometric identities for $\cos(a + b)$ and $\sin(a + b)$, we have

$$x = |OP|\cos\theta\cos\alpha - |OP|\sin\theta\sin\alpha$$

and

$$y = |OP|\sin\theta\cos\alpha + |OP|\cos\theta\sin\alpha$$

From equations (2) we can thus conclude:

10.3.1
$$x = X\cos\alpha - Y\sin\alpha$$
$$y = X\sin\alpha + Y\cos\alpha$$

Now if a rotation of the xy system through an angle α is used to obtain the XY system, then a rotation of the XY system through an angle $-\alpha$ will yield the xy system. Thus, interchanging x with X, y with Y, and replacing α by $-\alpha$ in (10.3.1), we obtain

$$X = x\cos(-\alpha) - y\sin(-\alpha)$$
$$Y = x\sin(-\alpha) + y\cos(-\alpha)$$

But $\cos(-\alpha) = \cos\alpha$ and $\sin(-\alpha) = -\sin\alpha$, so we have derived the following equations that enable us to obtain the XY coordinates when given the xy coordinates:

10.3.2
$$X = x\cos\alpha + y\sin\alpha$$
$$Y = -x\sin\alpha + y\cos\alpha$$

Example 1 Identify the graph of

$$x^2 + 2\sqrt{3}\,xy - y^2 = 4$$

by deriving its equation in the XY system obtained by rotating the xy system through an angle of $\pi/6$.

Solution Since $\cos\pi/6 = \sqrt{3}/2$ and $\sin\pi/6 = 1/2$ we have from (10.3.1) that

$$x = \frac{\sqrt{3}}{2}X - \frac{1}{2}Y \qquad \text{and} \qquad y = \frac{1}{2}X + \frac{\sqrt{3}}{2}Y$$

Thus in the XY system the equation

$$x^2 + 2\sqrt{3}\,xy - y^2 = 4$$

becomes the equation

$$\left(\frac{\sqrt{3}}{2}X - \frac{1}{2}Y\right)^2 + 2\sqrt{3}\left(\frac{\sqrt{3}}{2}X - \frac{1}{2}Y\right)\left(\frac{1}{2}X + \frac{\sqrt{3}}{2}Y\right) - \left(\frac{1}{2}X + \frac{\sqrt{3}}{2}Y\right)^2 = 4$$

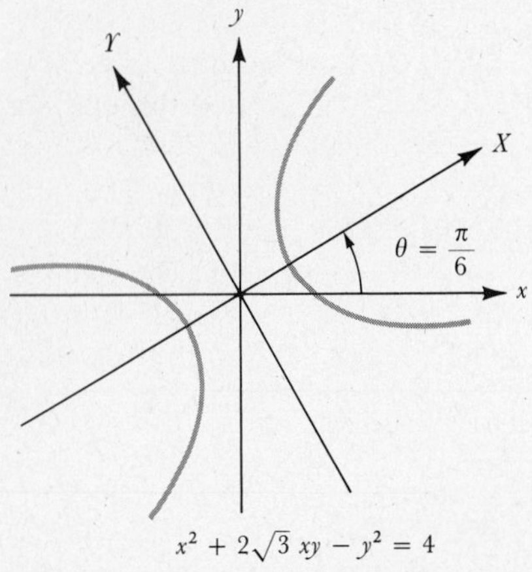

Fig. 10.27

which, if we multiply out and simplify, reduces to

$$\frac{X^2}{2} - \frac{Y^2}{2} = 1$$

This is a hyperbola in the XY-coordinate system with $a^2 = b^2 = 2$ and $c^2 = a^2 + b^2 = 4$. Thus the foci are at $(\pm 2, 0)$ and the vertices are at $(\pm\sqrt{2}, 0)$. The transverse axis is the X axis and the conjugate axis is the Y axis.

Using (10.3.1) with $X = \pm 2$, $Y = 0$, we get that the foci in the xy system are located at $(\sqrt{3}, 1)$ and $(-\sqrt{3}, -1)$. The vertices are located at $(\sqrt{6}/2, \sqrt{2}/2)$ and $(-\sqrt{6}/2, -\sqrt{2}/2)$. Since the transverse axis is the X axis ($Y = 0$), and since $Y = (-1/\sqrt{2})x + (\sqrt{3}/2)y$ in the xy system, the transverse axis is the straight line $x = \sqrt{3}y$. Similarly, the conjugate axis has equation $y = -\sqrt{3}x$. (See Fig. 10.27.) ●

PROGRESS TEST 1

1. Give the new coordinates of the points $(1, 0)$, $(0, 1)$, $(\sqrt{3}, 1)$, $(1, \sqrt{3})$ and $(1, 1)$ if the axes are rotated an angle of $\pi/6$ rad.
2. Give the new equation of the line $(\sqrt{3} - 4)y - 2 = (1 + 4\sqrt{3})x$ after applying the above rotation.
3. Apply the same $\alpha = \pi/6$ rotation to the equation $x^2 - \sqrt{3}xy + 2y = 10$ and simplify the result.

THE GENERAL SECOND-DEGREE EQUATION

The equation

10.3.3 $$Ax^2 + Bxy + Cy^2 + Dx + Ey + F = 0$$

where A and C are not both zero, is called *the general equation of the second degree*. If $B = 0$, then by completing the square (if necessary), we can identify the graph as one of the conic sections (possibly degenerate).

In particular,

$$Ax^2 + Cy^2 + Dx + Ey + F = 0$$

has as its graph:

(*a*) a circle if $A = C$
(*b*) an ellipse if $A, C \neq 0$ and have the same sign
(*c*) a hyperbola if $A, C \neq 0$ and have opposite signs
(*d*) a parabola if A or C is zero

Of course in the degenerate case (*a*) or (*b*) may reduce to a single point or no locus whatsoever, and (*c*) may result in two intersecting straight lines.

We shall now show that if $B \neq 0$, then a rotation of axes will transform (10.3.3) into the equivalent equation

10.3.4 $$A'X^2 + C'Y^2 + D'X + E'Y + F' = 0$$

As before, this can be easily graphed in the XY system and hence in the xy system.

Starting with the original equation

(3) $$Ax^2 + Bxy + Cy^2 + Dx + Ey + F = 0$$

we apply (10.3.1) and let

$$x = X\cos\alpha - Y\sin\alpha \qquad y = X\sin\alpha + Y\cos\alpha$$

where α is to be determined. It is tedious but straightforward to show that the above substitution transforms (3) to

10.3.5
$$A'X^2 + B'XY + C'Y^2 + D'X + E'Y + F' = 0$$

where

$$B' = B(\cos^2\alpha - \sin^2\alpha) + 2(C - A)\sin\alpha\cos\alpha$$

Since

$$\cos 2\alpha = \cos^2\alpha - \sin^2\alpha \qquad \text{and} \qquad \sin 2\alpha = 2\sin\alpha\cos\alpha$$

we can write

$$B' = (\cos 2\alpha)B + (C - A)\sin 2\alpha$$

If $B = 0$, then no rotation is necessary. If $B \neq 0$, then B' will equal zero for any choice of α that makes

$$(\cos 2\alpha)B + (C - A)\sin 2\alpha = 0$$

This is the case if

10.3.6
$$\cot 2\alpha = \frac{A - C}{B}$$

On the interval $(0, \pi)$ the cotangent function has as its range $(-\infty, \infty)$. Thus if $B \neq 0$, there is a unique α with $0 < \alpha < \pi/2$ for which (10.3.6) is satisfied. It is not necessary to remember the formula given in (10.3.5.) because in practice we use (10.3.6) and standard trigonometric identities to determine $\cos\alpha$ and $\sin\alpha$. We then use the transformation given in (10.3.1).

Example 2 Identify and graph the equation

$$8y^2 + 6xy - 9 = 0$$

Solution Now $A = 0$, $B = 6$, and $C = 8$ so, by (10.3.6), $\cot 2\alpha = -8/6 = -4/3$. It follows that $\cos 2\alpha = -4/5$, so

$$\sin\alpha = \sqrt{\frac{1 - \cos 2\alpha}{2}} = \frac{3}{\sqrt{10}} \qquad \text{and} \qquad \cos\alpha = \sqrt{\frac{1 + \cos 2\alpha}{2}} = \frac{1}{\sqrt{10}}$$

Thus $\alpha \approx 71°34'$. With these values of $\sin\alpha$ and $\cos\alpha$ we apply (10.3.1) to obtain $x = (1/\sqrt{10})(X - 3Y)$ and $y = (1/\sqrt{10})(3X + Y)$. Now in the XY system our original equation becomes

$$8\left[\frac{1}{\sqrt{10}}(3X + Y)\right]^2 + 6\left[\frac{1}{\sqrt{10}}(X - 3Y)\right]\left[\frac{1}{\sqrt{10}}(3X + Y)\right] - 9 = 0$$

which simplifies to

$$X^2 - \frac{Y^2}{9} = 1$$

Thus the graph is a hyperbola with $a = 1$, $b = 3$ and $c = \sqrt{10}$. The transverse axis is the X axis, which in the xy-coordinate system has equation $y = 3x$, and the conjugate axis has equation $x = -3y$. The foci are at $(1, 3)$ and $(-1, -3)$, and the vertices are at $(1/\sqrt{10}, 3/\sqrt{10})$ and $(-1/\sqrt{10}, -3/\sqrt{10})$. (See Fig. 10.28.) ●

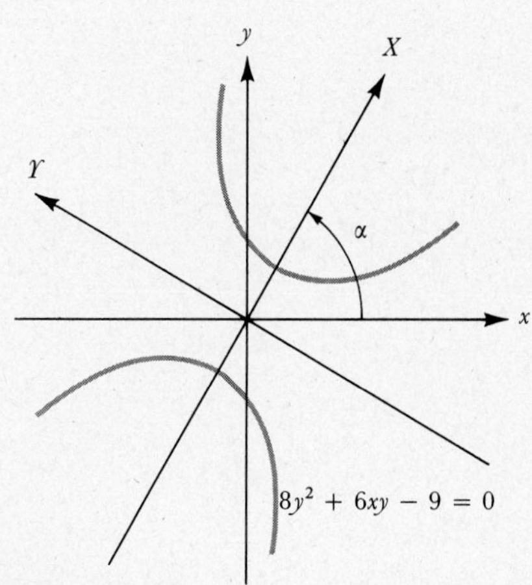

Fig. 10.28

Using the procedure illustrated, one can always obtain a rotation of axes that transforms the general second-degree equation (10.3.3) into one of the form given in (10.3.4). Then, using a translation of axes if necessary, its graph can be identified as a conic.

It is even possible to identify the graph of a second-degree equation directly from the equation itself. We shall give the criteria without proof.

10.3.7 **Theorem** The graph of the equation

$$Ax^2 + Bxy + Cy^2 + Dx + Ey + F = 0$$

is

(*a*) a hyperbola if $B^2 - 4AC > 0$
(*b*) a parabola if $B^2 - 4AC = 0$
(*c*) an ellipse if $B^2 - 4AC < 0$

In this we are using the terms hyperbola, parabola, and ellipse in the loose sense that allows for possible degeneracy.

PROGRESS TEST 2

Reduce $8x^2 + 5y^2 - 4xy - 36 = 0$ to a standard form and sketch.

SECTION 10.3 EXERCISES

Identify the graph of each of the following equations and give its essential features. Check by applying (10.3.7).

1. $x^2 + y^2 - 2xy - 4\sqrt{2}x - 4\sqrt{2}y = 0$
2. $x^2 + xy + y^2 = 6$
3. $x^2 - 2\sqrt{3}xy - y^2 = 1$
4. $x^2 - xy + y^2 - x - y = 3$
5. $3x^2 - 4xy + 8x = 1$
6. $7x^2 + 8xy + y^2 = 3$
7. $4x^2 - 4xy + y^2 = 6$
8. $2x^2 - \sqrt{3}xy + y^2 = 8$
9. $x^2 + xy + 1 = 0$
10. $2x^2 + 4xy - y^2 - 2x + 3y = 12$
11. $6x^2 + 5xy - 6y^2 = 1$
12. $3x^2 + 3y^2 - 2xy - 4\sqrt{6}x - 4\sqrt{6}y = 12$
13. $25x^2 + 14xy + 25y^2 = 288$
14. $9x^2 - 24xy + 16y^2 - 50x - 100y = 175$
15. $5x^2 + 6xy + 5y^2 - 32x - 32y = -32$
16. $3x^2 + 4y^2 + 5 = 0$
17. $x^2 - 9 = 0$
18. $9x^2 - 24xy + 16y^2 - 56x - 92y = -688$
19. $5x^2 + 8xy + 5y^2 = 36$
20. $3x^2 + 10xy + 3y^2 + 8\sqrt{2}x - 8\sqrt{2}y = 0$

CHAPTER EXERCISES

Review Exercises

In Exercises 1 to 30 sketch the graph of the given equation and note all pertinent data (foci, vertices, center, and the like).

1. $9x^2 - y^2 = 9$
2. $y^2 - x^2 = 9$
3. $x^2 + 4y^2 = 2x - 8y - 1$
4. $x^2/2 + 2y^2 = 8$
5. $4x^2 - 9y^2 + 36y + 8x + 4 = 0$
6. $y^2 - 4x^2 = 16$
7. $y^2 - 12 = x$
8. $y^2 = 16 + 4x^2$
9. $y = 40x - 4x^2 - 97$
10. $x^2 = -12y$
11. $x = 8y^2 + 16y + 10$
12. $y^2 = -16x$
13. $9x^2 - 4y^2 - 54x - 16y + 29 = 0$
14. $2x^2 = 4 - y^2$
15. $2x = y^2 + 8y + 22$
16. $(x + 5)^2 + 4(y - 3)^2 = 1$
17. $25x^2 + 4y^2 - 250x = 16y - 541$
18. $x - 2 = \sqrt{(y + 1)^2}$
19. $4x^2 = 2y - y^2$
20. $y + 3x^2 + 6x - 1 = 0$
21. $9x^2 - 8y^2 = -72$
22. $2x + 4y^2 = 24y - 39$

23. $y + 64x = 4x^2 + 254$

24. $3x^2 + 12y^2 = 36$

25. $2x = 4y^2 + 12y + 8$

26. $4(x + 3/2)^2 - 12(2 - y)^2 = 48$

27. $(x - 12)^2 = 4 - (y + 3)^2$

28. $x = -3y^2 - 30y - 75$

29. $2x^2 - 4x = 3y^2 + 6y + 7$

30. $(x/2 + 3)^2 + (3y + 6)^2 = 9$

In Exercises 31 to 40 determine the rotation that eliminates the xy term, rewrite the equation in the XY-coordinate system determined by this rotation, and sketch.

31. $32y^2 - 72xy + 53x^2 - 80 = 0$

32. $x^2 + 3y^2 = 2y - 2\sqrt{3}xy - 2\sqrt{3}x$

33. $5x^2 - 8xy + 5y^2 - 9 = 0$

34. $x^2 + y^2 = 2xy - x - y$

35. $11x^2 + y^2 = 20 + 10\sqrt{3}xy$

36. $x^2 + y^2 = 4 + 3xy$

37. $3x^2 + 4xy = 4$

38. $7x^2 + 7y^2 = 16 + 18xy$

39. $8x^2 + 5y^2 = 36 + 4xy$

40. $16y^2 - 24xy + 9x^2 = 60y + 80x - 100$

In Exercises 41 to 50 use the given information to find an equation for the indicated conic section.

41. Parabola, focus $(0, 4)$, vertex $(0, 0)$

42. Parabola, focus $(2, -1)$, vertex $(2, 3)$

43. Parabola, focus $(1, 1)$, directrix $y = -1$

44. Ellipse, foci $(-2, -1 \pm 3\sqrt{3})$, minor axis has length 6

45. Ellipse, foci $(3 \pm \sqrt{5}, 7)$, major axis has length 6

46. Ellipse, foci $(\pm 1, 0)$, major axis has length 12

47. Ellipse, a focus $(6, -1)$, center $(3, -1)$, major axis has length 10

48. Hyperbola, foci $(\pm 5, 0)$, a vertex $(3, 0)$

49. Hyperbola, a focus $(0, 13)$, a vertex $(0, 12)$, center $(0, 0)$

50. Hyperbola, foci $(15, 3)$, $(-11, 3)$, a vertex $(7, 3)$

Miscellaneous Exercises

1. A line tangent to $y^2 = 4px$ at P meets the directrix at Q. If the focus of the parabola is F, then show $FQ \perp FP$.

2. Given an ellipse $x^2/a^2 + y^2/b^2 = 1$, show that the length of the chord passing through a focus and perpendicular to the major axis has length $2b^2/a$.

3. Is it possible for lines from the foci of an ellipse to a common point P on the ellipse to be perpendicular? Explain.

4. Find an equation for the parabola with directrix $y = (3/4)x$ and focus $(2, 3)$.

5. Show that the parabola $y = ax^2 + bx + c$ has vertex $(-b/2a, (b^2 - 4ac)/4a)$.

6. Show that the tangents to an ellipse and a hyperbola with the same foci are perpendicular at the points of intersection.

7. Let F_1 and F_2 be the foci of a hyperbola. Show that the tangent line to the hyperbola at a point P bisects the angle between F_1P and F_2P.

8. Find the area of the largest rectangle with sides parallel to the coordinate axes that can be inscribed in the ellipse $x^2/a^2 + y^2/b^2 = 1$.

9. Show that any line distinct from but parallel to an asymptote of a hyperbola will intersect the hyperbola in exactly one point.

SELF-TEST ✦

1. (*a*) Identify and sketch the graph of the equation $y^2 + 13 = 4x - 6y$.
(*b*) Find an equation for the parabola with focus $(-2, 4)$ and directrix $x = -6$.

2. (*a*) Identify and sketch the graph of the equation $3x^2 + y^2 = 4y - 6x + 14$.
(*b*) Find an equation for the ellipse with foci at $(-4, 1)$ and $(0, 1)$, given that its major axis has length 6.

3. (*a*) Identify and sketch the graph of the equation $5x^2 - 4y^2 = 10x + 24y + 51$.
(*b*) Find an equation for the hyperbola with foci at $(1, 1)$ and $(-1, -1)$ and with $a = 1$.

4. Identify and sketch the graph of the equation $6x^2 - 12x + 11 = y^2 - 24xy - 26y$.

Chapter 11

TECHNIQUES OF INTEGRATION

CONTENTS

Differentiation usually is the relatively routine procedure of applying a formula. Antidifferentiation seldom involves direct application of a formula and can sometimes involve considerable ingenuity. Our basic use of antiderivatives is as a shortcut in evaluating definite integrals as mandated by the Fundamental Theorem of Calculus, and in this chapter we shall extend considerably our ability to evaluate integrals, both indefinite and definite.

Sections 11.1 through 11.4 provide straightforward illustrations of the most commonly used techniques of integration, and Sec. 11.5 addresses the problem of what to do when confronted with an integral without a context signaling which method to use. We include as part of our general strategy the use of tables of integrals. However, as we point out in Sec. 11.5, to use such a table we must often perform the same sorts of manipulating, substituting, and classifying as is involved in Secs. 11.1 through 11.4 when evaluating integrals *without* tables. Hence the techniques are important whether or not we use integration tables. Lastly, fairly simple continuous functions exist ($\sqrt{1 + x^3}$, for example) whose antiderivatives are not expressible as finite combinations of the functions studied to date and therefore cannot be determined using any of these techniques of integration. Section 11.6 provides two numerical approximation techniques, the Trapezoid Rule and Simpson's Rule, that enable us to approximate definite integrals when we cannot determine the antiderivative and hence cannot use the Fundamental Theorem.

To fit the special needs of this chapter, the end of the chapter is organized somewhat differently than most of the others. We have not distinguished between Review and Miscellaneous exercises. The chapter exercises are especially important because, as a source of context-free integrals, they help you learn the subtle skill of choosing which manipulations and techniques to use on a given integral.

Section 11.1: Integration by Parts

Objective:

Recognize integrands that yield to the method of integration by parts and then apply this method to obtain the indefinite integral.

INTRODUCTION

In Chap. 6 we introduced the substitution technique for evaluating integrals, both definite and indefinite. It is based upon the Chain Rules for Integrals and Antiderivatives, (6.4.4) and (6.4.5), respectively. In Chap. 8 we applied the substitution technique to a wide variety of integrals involving exponential, logarithmic, trigonometric, and inverse trigonometric functions. Nonetheless, a relatively simple integral such as

$$\int x \cos x \, dx$$

fits none of our available formulas no matter what sort of u substitution is tried. (This antiderivative arises, for example, when we use the shell method to determine the volume of the solid formed by revolving about the y axis the region under $y = \cos x$ from $x = 0$ to $x = \pi/2$.) We need a function whose derivative is the product $x \cos x$. A good start toward such an antiderivative is $x \sin x$ because, by applying the Product Rule to $x \sin x$, we get $x \cos x + \sin x$. This is close, but contains the extra term $\sin x$. However, if we add $\cos x$ to the proposed antiderivative, we have $y = x \sin x + \cos x$, and

$$\frac{dy}{dx} = x \cos x + \sin x - \sin x = x \cos x$$

We have found it!

$$\int x \cos x \, dx = x \sin x + \cos x + C$$

A more systematic process for "undoing" the Product Rule is the topic of this section.

THE FORMULA FOR INTEGRATION BY PARTS

Suppose u and v are differentiable functions of x. Then, by the Product Rule,

$$\frac{d}{dx}(uv) = u\frac{dv}{dx} + v\frac{du}{dx}$$

On the other hand, by rearranging terms,

$$u\frac{dv}{dx} = \frac{d}{dx}(uv) - v\frac{du}{dx}$$

and then multiplying through by dx and taking the antiderivative of both sides, we obtain the following:

11.1.1 Formula for Integration by Parts

$$\int u \, dv = uv - \int v \, du$$

We apply this formula to an integral $\int f(x)\,dx$ by thinking of $\int f(x)\,dx$ as $\int u\,dv$ and replacing $\int u\,dv$ by $uv - \int v\,du$ in the hope that the new integral will be easier to evaluate than $\int f(x)\,dx$. Since most integrals $\int f(x)\,dx$ can be written in the form $\int u\,dv$ in several different ways, we have a choice in deciding which part of $f(x)\,dx$ to call u and which part (necessarily containing dx) to call dv. We illustrate the technique by applying it to the integral considered earlier.

Example 1 Use (11.1.1) to determine $\int x \cos x \, dx$.

Solution Here we can let $u = \cos x$ and $dv = x\, dv$, which together imply

$$du = -\sin x \, dx \qquad \text{and} \qquad v = \int x \, dx = \frac{x^2}{2}$$

Then

$$\int x \cos x \, dx = \int u \, dv = uv - \int v \, du$$
$$= (\cos x)\frac{x^2}{2} - \int \frac{x^2}{2}(-\sin x \, dx)$$

But now the integral appearing on the right side, containing the integrand $(x^2/2) \sin x$, is even worse than the original. Let us try instead $u = x$ and $dv = \cos x \, dx$, so

$$du = dx \text{ and } v = \int \cos x \, dx = \sin x$$

Then

$$\int \overset{u}{x} \overbrace{\cos x \, dx}^{dv} = \overset{u}{x} \overbrace{\sin x}^{v} - \int \overbrace{\sin x}^{v} \overset{du}{dx}$$
$$= x \sin x - (-\cos x) + C$$
$$= x \sin x + \cos x + C$$

which is the same result achieved earlier. ●

Two comments on this technique are in order:

1. Although we added an arbitrary constant C at the end of the above computation, we did not do so when integrating to get v from dv. Any constant C_1 added to v will drop out of the final result, as can be seen if we replace v by $v + C_1$ in (11.1.1). (See Exercise 49.)

2. The integral of dv must be computable; otherwise we cannot even write $uv - \int v \, du$.

Example 2 Evaluate $\int \ln x \, dx$.

Solution We have two possible choices of u and dv: (*a*) $u = 1$ and $dv = \ln x \, dx$ or (*b*) $u = \ln x$ and $dv = dx$. However, since we have no formula for $\int \ln x \, dx$, which is the original problem anyway, choice (*a*) is unworkable. For choice (*b*) we have $du = dx/x$ and $v = \int dx = x$; so, by (11.1.1.),

$$\int \ln x \, dx = (\ln x)x - \int x \frac{dx}{x}$$
$$= x \ln x - \int dx$$
$$= x \ln x - x + C \quad ●$$

PROGRESS TEST 1

In Probs. 1 to 3 evaluate the given integral.

1. $\int xe^x \, dx$ **2.** $\int x \sec^2 x \, dx$ **3.** $\int \arcsin x \, dx$

4. Differentiate and simplify to check your result in Prob. 3.

MORE INTEGRATION BY PARTS

Some problems will involve repeated applications of integration by parts.

Example 3 Evaluate $\int (x^2 + 2x)e^x\,dx$

Solution Let $u = x^2 + 2x$, and $dv = e^x\,dx$. Then

$$du = (2x + 2)\,dx \qquad v = e^x$$

so

(1) $$\int (x^2 + 2x)e^x\,dx = (x^2 + 2x)e^x - 2\int (x + 1)e^x\,dx$$

Note that the integral on the right-hand side is similar to the integral we started with, except that the polynomial factor in the integrand has degree one less than the original polynomial factor. Applying the by-parts method to this latter integral with $\bar{u} = (x + 1)$, $d\bar{v} = e^x\,dx$, we obtain

$$\begin{aligned}\int (x + 1)e^x\,dx &= (x + 1)e^x - \int e^x\,dx \\ &= (x + 1)e^x - e^x + C\end{aligned}$$

This in turn can now be substituted into (1) to get

$$\begin{aligned}\int (x^2 + 2x)e^x\,dx &= (x^2 + 2x)e^x - 2[(x + 1)e^x - e^x] + C \\ &= x^2e^x + C \quad \bullet\end{aligned}$$

It should be clear that if $p(x)$ is a polynomial, then any problems of the form

$$\int p(x)e^x\,dx \qquad \text{or} \qquad \int p(x)e^{ax}\,dx$$

can be done in this fashion. The latter case would involve various constant factors involving a. We would, of course, have to repeat the process as many times as the degree of $p(x)$. Also, in each step we would choose u to be the polynomial and dv to be the exponential. Such problems as

$$\int p(x)\sin ax\,dx \qquad \text{and} \qquad \int p(x)\cos ax\,dx$$

can be handled in a similar way.

Sometimes successive integration by parts will result in a constant multiple of the original integral appearing on the right-hand side. In this case we can then combine this multiple with the original integral to solve the problem.

Example 4 Evaluate $\displaystyle\int e^{2x}\sin 3x\,dx$.

Solution Let $u = e^{2x}$ and $dv = \sin 3x\,dx$. Then

$$du = 2e^{2x}\,dx \qquad v = -\frac{1}{3}\cos 3x$$

and

(2) $$\int e^{2x}\sin 3x\,dx = -\frac{1}{3}e^{2x}\cos 3x + \frac{2}{3}\int e^{2x}\cos 3x\,dx$$

Again we apply the by-parts method to the integral on the right, with

$$\bar{u} = e^{2x} \qquad d\bar{v} = \cos 3x\,dx$$

$$d\bar{u} = 2e^{2x}\,dx \qquad \bar{v} = \frac{1}{3}\sin 3x$$

We obtain

$$\int e^{2x}\cos 3x\,dx = \frac{1}{3}e^{2x}\sin 3x - \frac{2}{3}\int e^{2x}\sin 3x\,dx$$

We now substitute this result into (2) to get

$$\int e^{2x}\sin 3x\,dx = -\frac{1}{3}e^{2x}\cos 3x + \frac{2}{9}e^{2x}\sin 3x - \frac{4}{9}\int e^{2x}\sin 3x\,dx$$

This is simply an equation of the form

$$I = A + B - \frac{4}{9}I$$

where I is the original integral. It can be solved for $I = \int e^{2x}\sin 3x\,dx$ by adding $(4/9)\int e^{2x}\sin 3x\,dx$ to both sides:

$$\frac{13}{9}\int e^{2x}\sin 3x\,dx = -\frac{1}{3}e^{2x}\cos 3x + \frac{2}{9}e^{2x}\sin 3x$$

Thus

$$\int e^{2x}\sin 3x\,dx = \frac{-3}{13}e^{2x}\cos 3x + \frac{2}{13}e^{2x}\sin 3x + C \quad \bullet$$

We could have done Example 4 just as easily by choosing u to be the trig factor and dv to be the exponential factor in *each* step. In either case, *once you have decided on your choices for u and dv in the first step, you must then maintain this same format in the second step, or the second step will simply undo the first and no solution will result.* If you do not believe this, try it.

One other piece of practical advice: Write out your solutions completely and systematically. Leave yourself plenty of space. Messy, cramped calculations are not particularly reliable.

PROGRESS TEST 2

Evaluate the following:

1. $\int \sin(\ln x)\,dx$ **2.** $\int (3x^3 + x)\ln x\,dx$ **3.** $\int x^2\sin 2x\,dx$

INTEGRATING ODD POWERS OF SECANT OR COSECANT

In Sec. 8.3 we noted that the technique for evaluating $\int \sec^n u\,du$ or $\int \csc^n u\,du$ for n a positive odd integer required integration by parts.

Example 5 Show that $\displaystyle\int \sec^3 x\,dx = \frac{1}{2}[\sec x\tan x + \ln|\sec x + \tan x|] + C.$

Solution We begin by factoring out a $\sec^2 x\,dx$ to serve as dv.

$$\int \sec^3 x\,dx = \int \sec x\sec^2 x\,dx$$

Now let $u = \sec x$ and $dv = \sec^2 x\,dx$; so

$$du = \sec x\tan x\,dx \qquad \text{and} \qquad v = \tan x$$

Then

$$\begin{aligned}\int \sec^3 x\,dx &= \sec x\tan x - \int \sec x\tan^2 x\,dx\\ &= \sec x\tan x - \int \sec x(\sec^2 x - 1)\,dx\\ &= \sec x\tan x - \int \sec^3 x\,dx + \int \sec x\,dx\end{aligned}$$

Transposing $-\int \sec^3 x\,dx$, we obtain

$$2\int \sec^3 x\,dx = \sec x \tan x + \int \sec x\,dx$$
$$= \sec x \tan x + \ln|\sec x + \tan x|$$

and so

$$\int \sec^3 x\,dx = \frac{1}{2}[\sec x \tan x + \ln|\sec x + \tan x|] + C \quad \bullet$$

SECTION 11.1 EXERCISES

In Exercises 1 to 30 evaluate the given integral.

1. $\int x \sin x\,dx$
2. $\int x \csc^2 x\,dx$
3. $\int_1^2 \arctan x\,dx$
4. $\int_1^e x \ln x\,dx$
5. $\int x2^x\,dx$
6. $\int x^3 e^{x^2}\,dx$
7. $\int \cos(\ln x)\,dx$
8. $\int x^2 \cos x\,dx$
9. $\int x^2 \ln x\,dx$
10. $\int x \operatorname{arcsec} x\,dx$
11. $\int_{\pi/8}^{\pi/6} x \sec 2x \tan 2x\,dx$
12. $\int_0^{\pi/4} x \tan x\,dx$
13. $\int x \arctan x\,dx$
14. $\int x \csc^2 3x\,dx$
15. $\int e^{2x} \cos x\,dx$
16. $\int e^{3x} \sin 2x\,dx$
17. $\int_1^e \frac{\ln x}{x}\,dx$
18. $\int_1^e x^4 \ln x\,dx$
19. $\int e^{4x} \cos \frac{x}{2}\,dx$
20. $\int (x^2 + 1)\sin 3x\,dx$
21. $\int_0^1 (x^2 - 1)e^{2x}\,dx$
22. $\int_1^e (\ln x)^2\,dx$
23. $\int \frac{\ln x}{\sqrt{x}}\,dx$
24. $\int \frac{\arctan \sqrt{x}}{\sqrt{x}}\,dx$
25. $\int 3x \tan^2 2x\,dx$
26. $\int \frac{x^3\,dx}{\sqrt{1 - x^2}}$
27. $\int x \sinh x\,dx$
28. $\int x \cosh x\,dx$
29. $\int \operatorname{arcsinh} x\,dx$
30. $\int \operatorname{arccosh} x\,dx$

In Exercises 31 to 36 determine the trigonometric or hyperbolic integral using integration by parts instead of a trigonometric or hyperbolic identity.

31. $\int \sin^2 x\,dx$
32. $\int \cos^2 x\,dx$
33. $\int \sin 3x \sin x\,dx$
34. $\int \cos 2x \cos x\,dx$
35. $\int \sin 3x \cos 5x\,dx$
36. $\int \cos x \sin 4x\,dx$
37. Evaluate $\int \sec^5 x\,dx$.
38. Evaluate $\int \csc^5 x\,dx$.

39. Given $\int dx/(x \ln x)$, what is wrong with letting $u = 1/(\ln x)$, $dv = dx/x$? Suggest a correct alternative.

40. Use integration by parts to derive the general reduction formula $\int (\ln x)^n\,dx = x(\ln x)^n - n\int(\ln x)^{n-1}\,dx$ for any positive integer n.

41. Use integration by parts to derive a general reduction formula for $\int x^n e^x\,dx$ for n a positive integer.

42. Try two different choices of u and dv for $\int[(\sin x)/x]\,dx$. [Neither will work because $\int[(\sin x)/x]\,dx$ has no elementary antiderivative! See Sec. 11.5.]

43. Determine the volume of the solid generated by revolving about the y axis the region bounded by the x and y axes, $y = e^x$, and $x = 1$.

44. Repeat Exercise 43 for the region bounded by the x axis and $y = \sin x$ from $x = \pi$ to $x = 2\pi$.

45. Determine the area of the region enclosed by $y = \ln x$ from $x = 1$ to $x = e$.

46. Determine the area of the region bounded by $y = x^2 e^x$ from $x = -1$ to $x = 1$.

47. Determine the average value of $f(x) = \ln x$ from $x = 10^{-4}$ to $x = 10^4$.

48. Determine the average value of $f(x) = x^2 \arctan x$ from $x = 0$ to $x = 1$.

49. Show that any constant C_1 added when computing v in integrating by parts drops out of the final integral.

50. × (Recall Sec. 8.1, Example 5.) Suppose the monthly income from an apartment complex grows linearly from \$1500 to \$2500 over the 20-year expected life of the complex during a period when money returns 8 percent. Determine the market value of the complex based upon the present value of its earning power. Compare this value with the \$239,431 based upon a constant \$2000/month income.

51. × (Present Value of Personal Earning Power) In the 30 years after graduation, person A's income grows according to $A(x) = (x^2/30) + 15$, and person B's income grows according to $B(x) = 15 - (x/30)(x - 60)$ (x is the years since graduation, money is measured in thousands of dollars). A and B both start at \$15,000 and finish at \$45,000.

(*a*) Determine the total money earned by each over the 30 years. (Sketch each earning graph.)

(*b*) Assuming that money pays 8 percent compounded continuously over this period, determine the present value of the two income streams.

(*c*) Does the 8 percent present value discounting exaggerate the difference between A and B? Discuss.

Section 11.2: Trigonometric Substitutions

Objective:

Transform integrals involving radicals containing quadratic terms, $\sqrt{u^2 \pm a}$ or $\sqrt{a^2 - u^2}$, into trigonometric integrals by replacing u by either $a \sin \theta$, $a \tan \theta$, or $a \sec \theta$.

INTRODUCTION

Certain frequently occurring integrals involving a radical term of the form

$$\sqrt{a^2 - u^2} \qquad \sqrt{u^2 + a^2} \qquad \text{or} \qquad \sqrt{u^2 - a^2}$$

will not yield to a direct power-rule substitution. Contrast

$$\int x\sqrt{1 - x^2}\, dx \qquad \text{and} \qquad \int \sqrt{1 - x^2}\, dx$$

The first integral, after the insertion of the constant factor 2, is a routine application of the Power Rule:

$$\frac{1}{2}\int \sqrt{1 - x^2}\, 2x\, dx = \frac{1}{2}\int v^{1/2}\, dv = \frac{1}{2}\left(\frac{v^{3/2}}{3/2}\right) + C = \frac{(\sqrt{1 - x^2})^3}{3} + C$$

On the other hand, the integral $\int \sqrt{1 - x^2}\, dx$ is missing an x factor, which cannot be inserted because x is a variable. However, by making the substitution $x = \sin \theta$, so $dx = d(\sin \theta) = \cos \theta\, d\theta$, we arrive at the trigonometric integral

$$\int \sqrt{1 - \sin^2 \theta}\, \cos \theta\, d\theta$$

which can now be handled using the techniques of Chap. 8. In this section we shall systematically examine three substitutions of this type to cover the three types of radicals appearing above.

THE SUBSTITUTION $u = a \sin \theta$

If an integral contains the term

$$\sqrt{a^2 - u^2} \qquad a > 0$$

and does not yield easily to the Power Rule, we make the following substitution and draw the triangle in Fig. 11.1: $u = a \sin \theta$, so $du = a \cos \theta\, d\theta$.

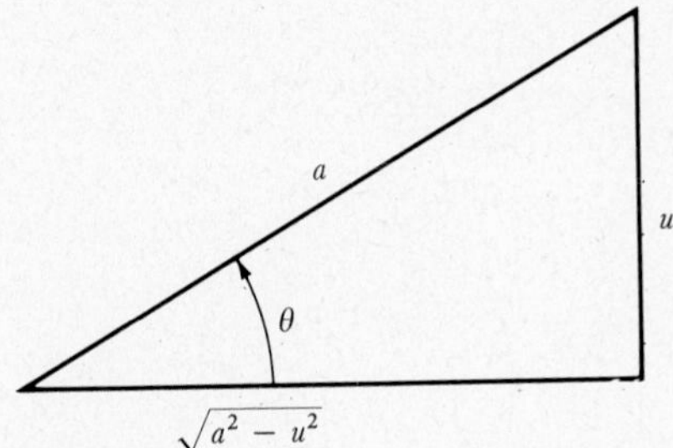

Fig. 11.1

The labeling of the right triangle follows from the Pythagorean Theorem and the fact that $\sin \theta = (u/a)$. In addition, we assume that $-\pi/2 \leq \theta \leq \pi/2$ because, in returning from θ to the original variable after integrating, or in changing limits of integration in definite integrals, we need to regard θ as $\arcsin(u/a)$.

Example 1 Evaluate $\displaystyle\int \frac{x^3\, dx}{\sqrt{9 - x^2}}$.

Solution Clearly the Power Rule does not apply directly, so we let $x = 3 \sin \theta$ and draw the right triangle in Fig. 11.2. Then $dx = 3 \cos \theta\, d\theta$ and

$$\begin{aligned}\sqrt{9 - x^2} = \sqrt{9 - 9\sin^2 \theta} = 3\sqrt{1 - \sin^2 \theta} &= 3\sqrt{\cos^2 \theta} \\ &= 3\,|\cos \theta| \\ &= 3 \cos \theta\end{aligned}$$

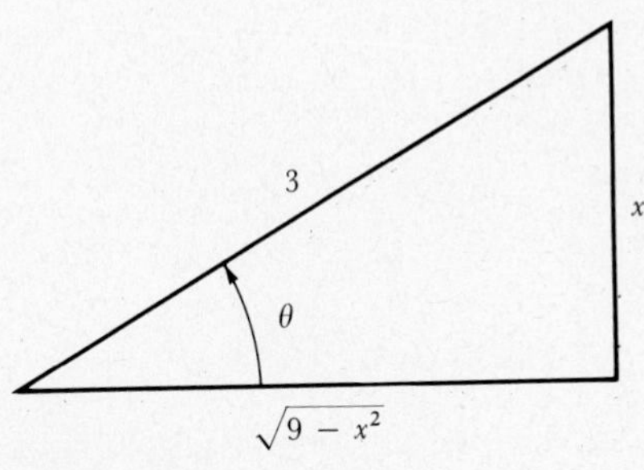

Fig. 11.2

(Note that since $-\pi/2 \leq \theta \leq \pi/2$, $\cos \theta \geq 0$ so $|\cos \theta| = \cos \theta$.) Hence

$$\begin{aligned}\int \frac{x^3\, dx}{\sqrt{9 - x^2}} &= \int \frac{27 \sin^3 \theta (3 \cos \theta\, d\theta)}{3 \cos \theta} \\ &= 27 \int \sin^3 \theta\, d\theta \\ &= 27 \int (1 - \cos^2 \theta) \sin \theta\, d\theta \\ &= 27\left[-\cos \theta + \frac{\cos^3 \theta}{3}\right] + C\end{aligned}$$

To switch back to the original variable x we refer to Fig. 11.2 and notice that $\cos\theta = (\sqrt{9 - x^2})/3$. Thus

$$\int \frac{x^3\,dx}{\sqrt{9-x^2}} = 27\left[\frac{-\sqrt{9-x^2}}{3} + \frac{(\sqrt{9-x^2})^3}{27\cdot 3}\right] + C$$

$$= -9\sqrt{9-x^2} + \frac{(\sqrt{9-x^2})^3}{3} + C \quad \bullet$$

Often, in evaluating the trigonometric integral that results after the substitution, we need to use double-angle formulas. This complicates the switch back to the original variable, as indicated in Example 2.

Example 2 Evaluate $\int \frac{x^2\,dx}{\sqrt{9-x^2}}$.

Solution Except for the degree of x in the numerator, this integral is the same as that in Example 1, so the same substitution and the same triangle (Fig. 11.2) lead to the integral

$$\int \frac{x^2\,dx}{\sqrt{9-x^2}} = \int \frac{9\sin^2\theta(3\cos\theta\,d\theta)}{3\cos\theta}$$

$$= 9\int \sin^2\theta\,d\theta$$

$$= 9\int \frac{1-\cos 2\theta}{2}\,d\theta \quad \text{[double-angle formula (8.3.1), p. 293]}$$

$$= \frac{9}{2}\left[\theta - \frac{1}{2}\sin 2\theta\right] + C$$

Now to express our answer in terms of x using the triangle in Fig. 11.2 we need functions given in terms of θ, *not* 2θ. Thus we use the identity $\sin 2\theta = 2\sin\theta\cos\theta$ to get

$$\frac{9}{2}\left[\theta - \frac{1}{2}\sin 2\theta\right] + C = \frac{9}{2}\left[\theta - \sin\theta\cos\theta\right] + C$$

Since $x = 3\sin\theta$, or $\sin\theta = x/3$, we conclude that $\theta = \arcsin(x/3)$. Also, $\cos\theta = (\sqrt{9-x^2})/3$, so putting all this together, we have

$$\frac{9}{2}[\theta - \sin\theta\cos\theta] + C = \frac{9}{2}\left[\arcsin\frac{x}{3} - \frac{x}{3}\,\frac{\sqrt{9-x^2}}{3}\right] + C$$

$$= \frac{9}{2}\arcsin\frac{x}{3} - \frac{x\sqrt{9-x^2}}{2} + C \quad \bullet$$

Similar care in using identities applies when using definite integrals. We can either switch the limits of integration or express the antiderivative in terms of the original variable and use the original limits of integration.

Example 3 Evaluate $\int_0^{\sqrt{5}/2} \frac{x^4\,dx}{\sqrt{5-x^2}}$

Solution Let $x = \sqrt{5}\sin\theta$, so $dx = \sqrt{5}\cos\theta\,d\theta$ and

$$\sqrt{5-x^2} = \sqrt{5 - 5\sin^2\theta} = \sqrt{5\cos^2\theta} = \sqrt{5}\cos\theta$$

We shall evaluate the integral by expressing the integration limits in terms of θ. Since $\theta = \arcsin(x/\sqrt{5})$, we know that $x = 0$ implies $\theta = \arcsin(0/\sqrt{5}) =$

$\arcsin 0 = 0$. Also, $x = \sqrt{5}/2$ implies $\theta = \arcsin[\sqrt{5}/(2\sqrt{5})] = \arcsin 1/2 = \pi/6$. Returning to the original integral, we have

$$\int_0^{\sqrt{5}/2} \frac{x^4\,dx}{\sqrt{5 - x^2}} = \int_0^{\pi/6} \frac{25 \sin^4\theta \sqrt{5}\cos\theta\,d\theta}{\sqrt{5}\cos\theta}$$

$$= 25\int_0^{\pi/6} \sin^4\theta\,d\theta$$

$$= \frac{25}{4}\int_0^{\pi/6} [1 - \cos 2\theta]^2\,d\theta$$

$$= \frac{25}{4}\int_0^{\pi/6} [1 - 2\cos 2\theta + \cos^2 2\theta]\,d\theta$$

$$= \frac{25}{4}\int_0^{\pi/6} \left[1 - 2\cos 2\theta + \frac{1}{2}(1 + \cos 4\theta)\right] d\theta$$

$$= \frac{25}{4}\int_0^{\pi/6} \left[\frac{3}{2} - 2\cos 2\theta + \frac{1}{2}\cos 4\theta\right] d\theta$$

$$(3) \qquad = \frac{25}{4}\left[\frac{3}{2}\theta - \sin 2\theta + \frac{1}{8}\sin 4\theta\right]_0^{\pi/6}$$

$$= \frac{25}{4}\left[\left(\frac{3\pi}{12} - \sin\frac{\pi}{3} + \frac{1}{8}\sin\frac{2\pi}{3}\right) - \left(\frac{3}{2}\cdot 0 - \sin 0 + \frac{1}{8}\sin 0\right)\right]$$

$$= \frac{25}{4}\left[\frac{\pi}{4} - \frac{7\sqrt{3}}{16}\right] = \frac{25}{64}(4\pi - 7\sqrt{3}) \quad \bullet$$

If we had not changed the limits of integration, we would need first to rewrite the antiderivative appearing in (3) in terms of θ, and then switch to x, a two-step procedure because of the 4θ. The former approach is more efficient.

PROGRESS TEST 1

Evaluate the following:

1. $\displaystyle\int \frac{x^3}{\sqrt{7 - x^2}}\,dx$

2. $\displaystyle\int_0^2 \sqrt{4 - x^2}\,dx$

THE SUBSTITUTIONS $u = a \tan\theta$ AND $u = a \sec\theta$

The next two trigonometric substitutions follow the same pattern as $u = a\sin\theta$, so we present all three in tabular form. In each case it is understood that the Power Rule does not apply directly.

11.2.1 **The Trigonometric Substitutions**

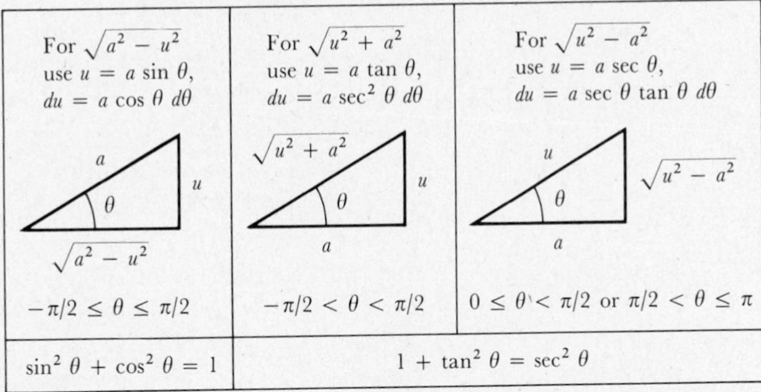

For $\sqrt{a^2 - u^2}$ use $u = a\sin\theta$, $du = a\cos\theta\,d\theta$	For $\sqrt{u^2 + a^2}$ use $u = a\tan\theta$, $du = a\sec^2\theta\,d\theta$	For $\sqrt{u^2 - a^2}$ use $u = a\sec\theta$, $du = a\sec\theta\tan\theta\,d\theta$
hypotenuse a, opposite u, adjacent $\sqrt{a^2 - u^2}$	hypotenuse $\sqrt{u^2 + a^2}$, opposite u, adjacent a	hypotenuse u, opposite $\sqrt{u^2 - a^2}$, adjacent a
$-\pi/2 \le \theta \le \pi/2$	$-\pi/2 < \theta < \pi/2$	$0 \le \theta < \pi/2$ or $\pi/2 < \theta \le \pi$
$\sin^2\theta + \cos^2\theta = 1$	$1 + \tan^2\theta = \sec^2\theta$	

Example 4 Evaluate $\int \frac{x^3\,dx}{\sqrt{x^2+9}}$.

Solution Let $x = 3\tan\theta$. Then $dx = 3\sec^2\theta\,d\theta$ and $\sqrt{x^2+9} = \sqrt{9\tan^2\theta + 9} = \sqrt{9\sec^2\theta} = 3\sec\theta$. Thus

$$\begin{aligned}\int \frac{x^3\,dx}{\sqrt{x^2+9}} &= 27\int \frac{\tan^3\theta\cdot 3\sec^2\theta\,d\theta}{3\sec\theta}\\ &= 27\int \tan^3\theta\sec\theta\,d\theta\\ &= 27\int \tan^2\theta(\tan\theta\sec\theta\,d\theta)\\ &= 27\int(\sec^2\theta - 1)(\tan\theta\sec\theta\,d\theta)\\ &= 9\sec^3\theta - 27\sec\theta + C\\ &= \frac{(x^2+9)^{3/2}}{3} - 9\sqrt{x^2+9} + C \quad \bullet\end{aligned}$$

$\sqrt{x^2+9}$ x θ 3

Example 5 Evaluate $\int \frac{(x^2-4)^{3/2}\,dx}{x}$.

Solution Let $x = 2\sec\theta$, then $dx = 2\sec\theta\tan\theta\,d\theta$ and $\sqrt{x^2-4} = \sqrt{4\sec^2\theta - 4} = \sqrt{4\tan^2\theta} = 2\tan\theta$. Now

$$\begin{aligned}\int \frac{(x^2-4)^{3/2}\,dx}{x} &= \int \frac{(\sqrt{x^2-4})^3\,dx}{x}\\ &= \int \frac{8\tan^3\theta\, 2\sec\theta\tan\theta\,d\theta}{2\sec\theta}\\ &= 8\int \tan^4\theta\,d\theta\\ &= 8\int \tan^2\theta\tan^2\theta\,d\theta\\ &= 8\int(\sec^2\theta - 1)\tan^2\theta\,d\theta\\ &= 8\left[\int \tan^2\theta\sec^2\theta\,d\theta - \int\tan^2\theta\,d\theta\right]\\ &= \frac{8\tan^3\theta}{3} - 8\tan\theta + 8\theta + C\\ &= \frac{(x^2-4)^{3/2}}{3} - 4\sqrt{x^2-4} + 8\,\text{arcsec}(x/2) + C \quad \bullet\end{aligned}$$

x $\sqrt{x^2-4}$ θ 2

As indicated in Progress Test 2, a trigonometric substitution can sometimes be useful even when the integrand does not contain a radical term. Furthermore, when a quadratic expression $ax^2 + bx + c$ appears inside a radical, it is often useful to complete the square and rewrite the expression as a sum or difference of squares so that one of the substitutions in (11.2.1) is appropriate. For example,

$$\int \sqrt{6x - x^2}\,dx = \int \sqrt{9 - (x-3)^2}\,dx = \int \sqrt{9 - u^2}\,du$$

with $u = x - 3$. This latter integral can be evaluated using the trigonometric substitution $u = 3\sin\theta$. Be certain, when using such two-stage substitutions, to return to the *original* variable. (See Prob. 2, Progress Test 2.)

In addition, if we are confronted with a radical of the form $\sqrt{9 - 4x^2}$, for example, we let $u = 2x$, regard $\sqrt{9-4x^2}$ as $\sqrt{9-u^2}$, and make the substitution $2x = 3\sin\theta$. Then $dx = (3/2)\cos\theta\,d\theta$, and so on. Alternatively, we could factor the coefficient of x^2 out of the radical and write $\sqrt{9-4x^2} = 2\sqrt{9/4 - x^2}$. We then let $x = (3/2)\sin\theta$ and proceed as usual. (See Exercises 9 and 10, for example.)

PROGRESS TEST 2

In Probs. 1 and 2 evaluate the given integral.

1. $\int \frac{\sqrt{x^2 - 16}}{x}\,dx$

2. $\int \sqrt{6x - x^2}\,dx$

3. Transform $\int \frac{dx}{(x^2 + a^2)^2}$ into a trigonometric integral.

SECTION 11.2 EXERCISES

In Exercises 1 to 30 evaluate the given integral.

1. $\int \sqrt{25 - x^2}\,dx$

2. $\int \frac{\sqrt{4 - x^2}}{x}\,dx$

3. $\int \frac{dx}{(x^2 + 4)^{3/2}}$

4. $\int \frac{x^2}{\sqrt{x^2 + 4}}\,dx$

5. $\int_{2\sqrt{2}}^{4} \frac{dx}{x\sqrt{x^2 + 16}}$

6. $\int_0^{\sqrt{3}} \frac{x^3\,dx}{(x^2 + 3)^{3/2}}$

7. $\int \frac{x^2\,dx}{\sqrt{x^2 - 4}}$

8. $\int x^3\sqrt{1 - x^2}\,dx$

9. $\int \frac{x^2\,dx}{(4 - 9x^2)^{3/2}}$

10. $\int (4x^2 + 9)^{5/2}\,dx$

11. $\int \sqrt{x^2 - 7}\,dx$

12. $\int \frac{dx}{\sqrt{6 - 4x - 2x^2}}$

13. $\int \frac{\sqrt{5x^2 - 4}}{x}\,dx$

14. $\int_0^1 \frac{x^2\,dx}{\sqrt{4 - x^2}}$

15. $\int_2^4 \frac{x^3\,dx}{\sqrt{4x^2 - 9}}$

16. $\int \sqrt{9 - 8x^2}\,dx$

17. $\int \sqrt{4x^2 - 25}\,dx$

18. $\int \sqrt{x^2 + 5}\,dx$

19. $\int \frac{(x + 5)\,dx}{(x^2 + 2x + 6)^2}$

20. $\int (a^2 - x^2)^{3/2}\,dx$

21. $\int \frac{dx}{(x^2 + 9)^3}$

22. $\int \frac{x^2\,dx}{\sqrt{4x^2 + 9}}$

23. $\int_0^{2\sqrt{3}} \frac{x^3\,dx}{\sqrt{x^2 + 4}}$

24. $\int \frac{(2x - 3)\,dx}{(x^2 + 2x - 3)^{3/2}}$

25. $\int \frac{\sqrt{x^2 - 5}}{x}\,dx$

26. $\int \frac{x\,dx}{\sqrt{3x^2 + 1}}$

27. $\int \frac{2x + 3}{x^2 + x + 1}\,dx$

28. $\int \frac{(x + 3)}{x^2 + 2x + 5}\,dx$

29. $\int \frac{dx}{\sqrt{4x - 4x^2}}$

30. $\int \frac{x^3\,dx}{(6x^2 + 12x - 5)^{3/2}}$

31. Develop a formula (which will be useful in the next section) for

$$\int \frac{(Ax + B)\,dx}{ax^2 + bx + c}$$

(*a*) by completing the square and (*b*) by breaking the integral into parts. (Assume that $b^2 - 4ac < 0$.)

In Exercises 32 to 34 use a trigonometric substitution to re-prove the given formula.

32. $\int \frac{du}{\sqrt{a^2 - u^2}} = \arcsin\frac{u}{a} + C$

33. $\int \frac{du}{a^2 + u^2} = \frac{1}{a}\arctan\frac{u}{a} + C$

34. $\int \frac{du}{|u|\sqrt{u^2 - a^2}} = \frac{1}{a}\operatorname{arcsec}\frac{u}{a} + C$

In Exercises 35 to 37 use the appropriate hyperbolic substitution to prove the given formula.

35. $\int \frac{du}{\sqrt{a^2 + u^2}} = \operatorname{arcsinh}\frac{u}{a} + C$

36. $\int \frac{u^2\,du}{\sqrt{u^2 - a^2}} = \frac{1}{2}u\sqrt{u^2 - a^2} + \frac{a^2}{2}\operatorname{arccosh}\frac{u}{a} + C$

37. $\int \frac{du}{(a^2 - u^2)^{3/2}} = \frac{1}{a^2}\sinh\left(\operatorname{arctanh}\frac{u}{a}\right) + C$

38. Use integration to determine the area inside the upper half of the circle $(x - 2)^2 + (y - 3)^2 = 25$.

39. Determine the area between $y = \sqrt{x^2 + 1}$ and the x axis from $x = 0$ to $x = 2\sqrt{2}$.

40. Determine the area enclosed by the ellipse $9x^2 + 4y^2 = 36$.

41. Repeat Exercise 40 for $x^2 + 9y^2 = 9$.

42. Develop a general formula for the area enclosed by the ellipse $a^2x^2 + b^2y^2 = a^2b^2$.

Section 11.3: Partial Fractions

Objective:

Integrate quotients of polynomials (rational functions) by first breaking them into sums of simple parts—their partial fractions.

INTRODUCTION

Suppose you need to integrate

$$f(x) = \frac{1}{1 - x^2} = \frac{1}{(1 - x)(1 + x)}$$

If you recognize that

$$f(x) = \frac{1/2}{1 - x} + \frac{1/2}{1 + x}$$

then

$$\int f(x)\,dx = \int \frac{1/2}{1 - x}\,dx + \int \frac{1/2}{1 + x}\,dx$$
$$= -\frac{1}{2}\ln|1 - x| + \frac{1}{2}\ln|1 + x| + C$$
$$= \frac{1}{2}\ln\left|\frac{1 + x}{1 - x}\right| + C = \ln\sqrt{\left|\frac{1 + x}{1 - x}\right|} + C$$

If we were confronted with a more difficult rational function to integrate, such as

$$g(x) = \frac{4x^2 + 14x - 2}{x^3 + 3x^2 - x - 3}$$

an analogous decomposition of $g(x)$ into easily integrated "partial fractions" (whose sum is $g(x)$) is harder to come by; but it does exist, as can be seen by combining the following over a common denominator as indicated:

$$\frac{3}{x + 1} + \frac{2}{x - 1} - \frac{1}{x + 3}$$
$$= \frac{3(x - 1)(x + 3) + 2(x + 1)(x + 3) - (x + 1)(x - 1)}{(x + 1)(x - 1)(x + 3)}$$
$$= \frac{4x^2 + 14x - 2}{x^3 + 3x^2 - x - 3} = g(x)$$

As is probably apparent from this calculation, the method for writing the partial fractions of a given quotient of polynomials amounts to undoing what we do when we combine fractions over a common denominator. Since the last step in combining fractions over a common denominator is to multiply the factors of this common denominator, it is reasonable that the first step in going the other way is to factor the denominator. But before we go too far, it is helpful to review some basic algebraic facts and settle on some common terminology.

Recall that $f(x)$ is called a *rational function* if $f(x) = p(x)/q(x)$ for $p(x)$ and $q(x)$ polynomials. Analogous to the distinction between proper and improper fractions for rational *numbers,* we say the *rational function* $f(x) = p(x)/q(x)$ *is proper* if the degree of the numerator is strictly less than that of the denominator. Otherwise $f(x)$ is said to be *improper. All comments in this section regarding procedures for integrating rational functions will apply to proper rational functions only.* Of course by dividing the denominator into the numerator, any improper rational function can be written as a polynomial plus a proper rational function. Since polynomials can be routinely integrated, any method that allows us to integrate proper rational functions automatically extends to improper rational functions by long division. Also, to simplify matters somewhat we shall assume that the denominator $q(x)$ has leading coefficient 1. Nothing is lost by this assumption because if the leading coefficient is $a \neq 1$, we can divide numerator and denominator by a.

Example 1 Express

$$\frac{4x^5 + 3x^4 - 9x^2 + x + 1}{2x^3 + 4x^2 + 3}$$

as a polynomial plus a proper rational function with leading coefficient 1 in the denominator.

Solution By long division we have

$$\frac{4x^5 + 3x^4 - 9x^2 + x + 1}{2x^3 + 4x^2 + 3} = 2x^2 - \frac{5}{2}x + 5 - \frac{(55/2)\,x^2 - x + 14}{2x^3 + 4x^2 + 3}$$

which in turn equals

$$2x^2 - \frac{5}{2}x + 5 - \frac{(55/4)x^2 - (1/2)x + 7}{x^3 + 2x^2 + 3/2} \quad \bullet$$

Now that we have assumed for $f(x) = p(x)/q(x)$ that the degree of $q(x)$ is less than that of $p(x)$ and that the coefficient of the highest power of $q(x)$ is 1, we must factor the polynomial $q(x)$. The Fundamental Theorem of Algebra says that $q(x)$ can be factored into a product of linear factors, such as $(x - a_1)$, $(x - a_2)$, and so on, and irreducible quadratic factors, such as $(x^2 + b_1x + c_1)$, $(x^2 + b_2x + c_2)$, and so forth. By "irreducible" we mean simply that the factor $x^2 + bx + c$ cannot be further factored using real numbers. We saw in Chap. 1 that a general quadratic expression $ax^2 + bx + c$ is irreducible precisely when $b^2 - 4ac < 0$.

If $q(x)$ contains the factor $(x - a)^k$, such as $q(x) = (x - 2)^3(x - 5)$, we say $(x - a)$ is a *k-times-repeated linear factor*. If $k = 1$, as for the factor $(x - 5)$ above, we say $x - a$ is a *nonrepeated linear factor*. Similarly, we refer to the k-times-repeated, or nonrepeated, irreducible *quadratic factor* $x^2 + bx + c$ according to whether $q(x)$ contains $(x^2 + bx + c)^k$ for $k > 1$, or whether $q(x)$ contains $(x^2 + bx + c)^1$, respectively.

One last preliminary fact we need from algebra is the following.

11.3.1 If $p_1(x)$ and $p_2(x)$ are polynomials such that $p_1(x) = p_2(x)$ for all x except possibly $x = a_1, a_2, \ldots, a_n$, then $p_1(x) = p_2(x)$ for *all* x. That is, two polynomials which are known to agree on all but a finite collection of numbers must be identical.

We shall outline our procedures for decomposing $f(x) = p(x)/q(x)$ into a sum of *partial fractions* according to the four types of factorizations of $q(x)$ given in Cases 1–4.

Case 1: $q(x)$ Is a Product of Nonrepeated Linear Factors

We are assuming $q(x)$ can be written in the form

$$q(x) = (x - a_1)(x - a_2) \ldots (x - a_n)$$

where the numbers $a_1, a_2, \ldots, a_n$ are all different.

Then there exist constants $A_1, A_2, \ldots, A_n$ such that

$$\frac{p(x)}{q(x)} = \frac{A_1}{x - a_1} + \frac{A_2}{x - a_2} + \cdots + \frac{A_n}{x - a_n}$$

We are now going to illustrate the method for determining the constants $A_1, \ldots, A_n$ with a specific example before commenting on the general procedure.

Example 2 Determine the partial fraction decomposition for the rational function

$$\frac{x^2 + x + 1}{(x - 1)(x^2 - x - 6)}$$

Solution We first factor $q(x)$:

$$q(x) = (x - 1)(x^2 - x - 6) = (x - 1)(x + 2)(x - 3)$$

We now determine constants A_1, A_2, A_3 such that

$$\frac{x^2 + x + 1}{(x - 1)(x^2 - x - 6)} = \frac{A_1}{x - 1} + \frac{A_2}{x + 2} + \frac{A_3}{x - 3}$$

We now recombine the right-hand side over the common denominator $(x - 1)(x + 2)(x - 3)$ and *leave everything in factored form* to obtain

$$\frac{x^2 + x + 1}{(x - 1)(x^2 - x - 6)} = \frac{A_1(x + 2)(x - 3) + A_2(x - 1)(x - 3) + A_3(x - 1)(x + 2)}{(x - 1)(x + 2)(x - 3)}$$

These two rational functions are defined for all $x \neq 1, -2, 3$. They have the same denominator and hence their polynomial numerators must be equal for all $x \neq 1, -2, 3$. But then, by (11.3.1), they must be identical. Thus

$$x^2 + x + 1 = A_1(x + 2)(x - 3) + A_2(x - 1)(x - 3) + A_3(x - 1)(x + 2)$$

is valid for *all x. Now successively substitute the roots 1, −2, 3 of $q(x)$ into both sides of this equation.* Note that with each substitution all terms except one on the right-hand side reduce to zero. For $x = 1$, $3 = A_1(3)(-2)$; for $x = -2$, $3 = A_2(-3)(-5)$; and for $x = 3$, $13 = A_3(2)(5)$. Thus

$$A_1 = -\frac{1}{2} \qquad A_2 = \frac{1}{5} \qquad A_3 = \frac{13}{10}$$

Hence

$$\frac{x^2 + x + 1}{(x - 1)(x^2 - x - 6)} = -\frac{1}{2} \cdot \frac{1}{x - 1} + \frac{1}{5} \cdot \frac{1}{x + 2} + \frac{13}{10} \cdot \frac{1}{x - 3} \quad \bullet$$

Having achieved the partial-fraction decomposition of

$$\frac{x^2 + x + 1}{(x - 1)(x^2 - x - 6)}$$

in Example 2, we can now integrate routinely as follows:

$$\int \frac{x^2 + x + 1}{(x - 1)(x^2 - x - 6)}\, dx = \int \left[-\frac{1}{2}\frac{1}{x - 1} + \frac{1}{5}\frac{1}{x + 2} + \frac{33}{10}\frac{1}{x - 3} \right] dx$$

$$= -\frac{1}{2} \ln |x - 1| + \frac{1}{5} \ln |x + 2| + \frac{13}{10} \ln |x - 3| + C$$

All the algebraic calculations pay off by changing a rather difficult integral into a routine application of

$$\int \frac{A\, dx}{x - a} = A \ln |x - a| + C$$

The Substitution Procedure used in Example 2 can be outlined as follows:

11.3.2 (*a*) Factor $q(x)$ into distinct linear factors.
(*b*) Write the partial-fraction decomposition for $p(x)/q(x)$ in terms of A_1, A_2, $\ldots$, A_n.

(c) Combine the partial fractions over the common denominator $q(x)$, leaving everything in factored form, and set $p(x)$ equal to this numerator.

(d) Substitute the n distinct roots of $q(x)$ into both sides of the equation determined in step (c) to obtain $A_1, A_2, \ldots, A_n$.

PROGRESS TEST 1

Integrate each of the following by first finding the partial-fraction decomposition.

1. $\displaystyle\int \frac{x^5 - 8x^3 - 8x + 1}{x^3 - 9x}\,dx$ (Be sure to divide!) **2.** $\displaystyle\int \frac{(4x + 2)\,dx}{2x^2 + 4x - 6}$ (Note that the coefficient of x^2 is not 1.)

Case 2: $q(x)$ Is a Product of Linear Factors, Some of Which Are Repeated

In this case each factor $(x - a)^k$ gives rise to k partial fractions of the form

$$\frac{A_1}{x - a}, \frac{A_2}{(x - a)^2}, \ldots, \frac{A_k}{(x - a)^k}$$

Then $p(x)/q(x)$ is equal to the sum of all the partial fractions arising from each of the possibly repeated factors $(x - a_i)^{n_i}$.

Thus, for example, the partial-fraction decomposition of

$$\frac{x^2 + x + 1}{(x - 1)(x + 3)^2(x - 2)^3}$$

is

$$\frac{A_1}{x - 1} + \frac{A_2}{(x + 3)} + \frac{A_3}{(x + 3)^2} + \frac{A_4}{(x - 2)} + \frac{A_5}{(x - 2)^2} + \frac{A_6}{(x - 2)^3}$$

Note that here and always *the number of unknowns to be determined in the partial-fraction decomposition must always equal the degree of* $q(x)$.

Example 3 Evaluate $\displaystyle\int \frac{(x + 1)\,dx}{x(x - 1)^3}$.

Solution First we write

$$\frac{x + 1}{x(x - 1)^3} = \frac{A_1}{x} + \frac{A_2}{x - 1} + \frac{A_3}{(x - 1)^2} + \frac{A_4}{(x - 1)^3}$$

As before, we combine the right-hand side over the common denominator $x(x - 1)^3$ and then equate numerators.

$$x + 1 = A_1(x - 1)^3 + A_2x(x - 1)^2 + A_3x(x - 1) + A_4x \tag{4}$$

Letting $x = 0$ and 1, we obtain $A_1 = -1$ and $A_4 = 2$.

We now have two more unknowns to determine and no more roots to exploit, so we simply substitute *any* two numbers, distinct from the values 0 and 1 already used, into both sides of equation (4). Let $x = -1, x = 2$ (these equal A_1 and A_4 only by coincidence) and we obtain, respectively,

$$0 = -8A_1 - 4A_2 + 2A_3 - A_4$$
$$3 = A_1 + 2A_2 + 2A_3 + 2A_4$$

But $A_1 = -1$ and $A_4 = 2$ so we have the equations

$$-4A_2 + 2A_3 = -6$$
$$A_2 + 2A_3 = 0$$

Solving, we get

$$A_2 = \frac{6}{5} \quad \text{and} \quad A_3 = -\frac{3}{5}.$$

Thus

$$\int \frac{x+1}{x(x-1)^3}\,dx = \int \left[\frac{-1}{x} + \frac{6}{5}\frac{1}{x-1} - \frac{3}{5}\frac{1}{(x-1)^2} + \frac{2}{(x-1)^3}\right] dx$$

$$= -\ln|x| + \frac{6}{5}\ln|x-1| + \frac{3}{5(x-1)} - \frac{1}{(x-1)^2} + C \quad \bullet$$

The only variation in the Substitution Procedure outlined previously is in step (*d*). In this case there will not be n distinct roots to substitute. However, we *can* substitute n distinct numbers, always remembering to include among them *all* roots of $q(x)$. This is what we did in Example 3.

PROGRESS TEST 2

1. Evaluate the following indefinite integral: $\int \frac{x^4+1}{x^3-x^2}\,dx$.

Case 3: $q(x)$ Is a Product of Linear (Perhaps Repeated) and Nonrepeated Irreducible Quadratic Factors

We assume henceforth that the discriminants of the quadratic factors are negative; hence the quadratic factors are irreducible. Suppose $q(x)$ factors as

$$q(x) = (x-a_1)^{n_i} \cdots (x-a_k)^{n_k}(x^2+b_1x+c_1)\cdots(x^2+b_jx+c_j)$$

Each nonrepeated quadratic factor $(x^2+b_ix+c_i)$ of $q(x)$ gives rise to the partial fraction

$$\frac{A_ix+B_i}{x^2+b_ix+c_i}$$

Then $p(x)/q(x)$ is equal to the sum of all the partial fractions arising from each of the factors of $q(x)$.

For example, the partial-fraction decomposition of

$$\frac{x^2+x}{(x-1)^2(x^2+x+1)(x^2-3x+5)}$$

is

$$\frac{A_1}{x-1} + \frac{A_2}{(x-1)^2} + \frac{A_3x+A_4}{x^2+x+1} + \frac{A_5x+A_6}{x^2-3x+5}$$

Again, the number of unknowns in the partial-fraction decomposition equals the degree of $q(x)$. Also, by completing the square in the denominator, any integral of the form

$$\int \frac{Ax+B}{x^2+bx+c}\,dx$$

can be transformed to

$$\int \frac{(A'u+B')\,du}{u^2+d^2}$$

which in turn can be broken into two integrals, the first yielding a logarithm and the second an arctangent.

Example 4 Evaluate $\int \frac{x^2 - 2x - 3}{(x-1)(x^2+2x+2)}\,dx$.

Solution

$$\frac{x^2 - 2x - 3}{(x-1)(x^2+2x+2)} = \frac{A_1}{x-1} + \frac{A_2x + A_3}{x^2+2x+2}$$
$$= \frac{A_1(x^2+2x+2) + (A_2x + A_3)(x-1)}{(x-1)(x^2+2x+2)}$$

Thus

$$x^2 - 2x - 3 = A_1(x^2+2x+2) + (A_2x + A_3)(x-1)$$

For $x = 1$, $-4 = 5A_1$, so $A_1 = -4/5$; for $x = 0$, $-3 = 2A_1 - A_3$, so $A_3 = 7/5$; and for $x = -1$, $0 = A_1 + 2A_2 - 2A_3$, so $A_2 = 9/5$. As a result,

$$\int \frac{(x^2 - 2x - 3)}{(x-1)(x^2+2x+2)}\,dx$$
$$= \int \left[-\frac{4}{5}\frac{1}{x-1} + \frac{1}{5}\frac{9x+7}{x^2+2x+2}\right] dx$$
$$= -\frac{4}{5}\ln|x-1| + \frac{1}{5}\int \frac{9x+7}{(x+1)^2+1}\,dx$$
$$= -\frac{4}{5}\ln|x-1| + \frac{1}{5}\int \frac{(9(u-1)+7)\,du}{u^2+1} \qquad \text{(Let } u = x+1, \text{ so } du = dx \text{ and } x = u-1\text{)}$$
$$= -\frac{4}{5}\ln|x-1| + \frac{9}{5}\int \frac{u\,du}{u^2+1} - \frac{11}{5}\int \frac{du}{u^2+1}$$
$$= -\frac{4}{5}\ln|x-1| + \frac{9}{10}\ln|u^2+1| - \frac{11}{5}\arctan u + C$$
$$= -\frac{4}{5}\ln|x-1| + \frac{9}{10}\ln(x^2+2x+2) - \frac{11}{5}\arctan(x+1) + C \quad \bullet$$

PROGRESS TEST 3

1. Evaluate $\int \frac{(x+1)\,dx}{(x-1)(x^2+x+1)}$.

Case 4: $q(x)$ Is a Product of Linear and Repeated Quadratic Factors

In this case $q(x)$ is of the form

$$q(x) = (x - a_1)^{n_1} \cdots (x - a_k)^{n_k}(x^2 + b_1x + c_1)^{m_1} \cdots (x^2 + b_jx + c_j)^{m_j}$$

Each repeated quadratic factor of the form $(x^2 + bx + c)^k$ gives rise to the partial fractions:

$$\frac{A_1x + A_2}{x^2+bx+c}, \frac{A_3x + A_4}{(x^2+bx+c)^2}, \ldots, \frac{A_{2k-1}x + A_{2k}}{(x^2+bx+c)^k}$$

For example,

$$\frac{x^2+3}{x(x^2+1)^2(x^2+x+1)^3}$$

has the partial-fraction decomposition

$$\frac{A_1}{x} + \frac{A_2x + A_3}{(x^2+1)} + \frac{A_4x + A_5}{(x^2+1)^2} + \frac{A_6x + A_7}{(x^2+x+1)}$$
$$+ \frac{A_8x + A_9}{(x^2+x+1)^2} + \frac{A_{10}x + A_{11}}{(x^2+x+1)^3}$$

Before going on we should consider the problem of evaluating an integral of the form

$$\int \frac{(Ax + B)\,dx}{(x^2 + bx + c)^n}$$

There are essentially three steps:

(*a*) Complete the square of the term $x^2 + bx + c$ and write the integral in the form

$$\int \frac{(Eu + F)\,du}{(u^2 + d^2)^n}$$

(*b*) Break this into two integrals. The first,

$$\int \frac{Eu\,du}{(u^2 + d^2)^n}$$

can be evaluated using the Power Rule.

(*c*) To evaluate

$$\int \frac{F\,du}{(u^2 + d^2)^n}$$

make the trigonometric substitution $u = d\tan\theta$, and the problem will reduce to integrating an integer power of $\cos\theta$.

Example 5 Evaluate $\displaystyle\int \frac{2x^4 + 3x^2 + x + 1}{x(x^2 + 1)^2}\,dx.$

Solution Now

$$\frac{2x^4 + 3x^2 + x + 1}{x(x^2 + 1)^2} = \frac{A_1}{x} + \frac{A_2x + A_3}{x^2 + 1} + \frac{A_4x + A_5}{(x^2 + 1)^2}$$

$$= \frac{A_1(x^2 + 1)^2 + (A_2x + A_3)x(x^2 + 1) + (A_4x + A_5)x}{x(x^2 + 1)^2}$$

Thus

$$2x^4 + 3x^2 + x + 1 = A_1(x^2 + 1)^2 + (A_2x + A_3)x(x^2 + 1) + (A_4x + A_5)x$$

In this example we shall modify the Substitution Procedure for determining the unknowns A_1, A_2, A_3, A_4, A_5. We multiply out the right-hand side and combine like powers of x to get

$$2x^4 + 3x^2 + x + 1$$
$$= (A_1 + A_2)x^4 + A_3x^3 + (2A_1 + A_2 + A_4)x^2 + (A_3 + A_5)x + A_1$$

We now equate coefficients of like powers of x to obtain

$$\begin{aligned} A_1 + A_2 \qquad\qquad &= 2 \\ A_3 \qquad\qquad &= 0 \\ 2A_1 + A_2 \quad + A_4 \qquad &= 3 \\ A_3 \qquad + A_5 &= 1 \\ A_1 \qquad\qquad &= 1 \end{aligned}$$

Thus $A_3 = 0$ and $A_1 = 1$. From the fourth equation, $A_5 = 1 - A_3 = 1 - 0 = 1$. Similarly $A_2 = 2 - A_1 = 2 - 1 = 1$, and $A_4 = 3 - 2A_1 - A_2 = 3 - 2 - 1 = 0$. Therefore

$$\begin{aligned}\int \frac{(2x^4+3x^2+x+1)\,dx}{x(x^2+1)^2} &= \int\left[\frac{1}{x}+\frac{x}{x^2+1}+\frac{1}{(x^2+1)^2}\right]dx\\ &= \ln|x|+\frac{1}{2}\ln(x^2+1)+\int\frac{dx}{(x^2+1)^2} \qquad [x=(1)\tan\theta]\\ &= \ln|x|+\frac{1}{2}\ln(x^2+1)+\int\frac{\sec^2\theta\,d\theta}{\sec^4\theta}\\ &= \ln|x|+\frac{1}{2}\ln(x^2+1)+\int\cos^2\theta\,d\theta \qquad \left(\cos^2\theta=\frac{1+\cos 2\theta}{2}\right)\\ &= \ln|x|+\frac{1}{2}\ln(x^2+1)+\frac{1}{2}(\theta+\sin\theta\cos\theta)+C\\ &= \ln|x|+\frac{1}{2}\ln(x^2+1)+\frac{1}{2}\left(\arctan x+\frac{x}{x^2+1}\right)+C \quad \bullet\end{aligned}$$

In some problems the integrand is already in the form provided by a partial-fraction decomposition. In such cases applying the procedure will lead to constants that merely reproduce the original integral. For example, given

$$\int \frac{3x+1}{(x^2+4x+6)^2}\,dx$$

we would note that x^2+4x+6 is irreducible, and so the integrand is already of the form

$$\frac{Ax+B}{(x^2+bx+c)^n}$$

Thus no partial-fraction decomposition is necessary.

PROGRESS TEST 4

1. Evaluate $\displaystyle\int \frac{x^4+5x^3+10x^2+8x+1}{(x-1)(x^2+2x+2)^2}\,dx$.

SECTION 11.3 EXERCISES

In Exercises 1 to 24 evaluate the given integral.

1. $\displaystyle\int \frac{(2x-1)\,dx}{x^2-x-6}$

2. $\displaystyle\int \frac{(3x+4)\,dx}{x^2-4x-5}$

3. $\displaystyle\int \frac{x\,dx}{x^3+5x^2+6x}$

4. $\displaystyle\int \frac{(x+1)\,dx}{x^3-5x^2+4x}$

5. $\displaystyle\int \frac{(x^2+1)\,dx}{x^3-4x^2+4x}$

6. $\displaystyle\int \frac{(3x^2-2)\,dx}{x^4-6x^3+5x^2}$

7. $\displaystyle\int \frac{x^2\,dx}{(x^2+3x+1)(x-1)}$

8. $\displaystyle\int \frac{(x^3+1)\,dx}{(x^2+1)(x^2+x+1)}$

9. $\displaystyle\int \frac{(x^2+x-4)\,dx}{(x^2+x+1)^2}$

10. $\displaystyle\int \frac{(2x^2-3)\,dx}{(x^2+2)^2(x+1)}$

11. $\displaystyle\int_2^3 \frac{x\,dx}{x^2+x-2}$

12. $\displaystyle\int_2^3 \frac{(x+1)\,dx}{(x+2)(x-1)^2}$

13. $\displaystyle\int_1^2 \frac{dx}{x^3(x+1)}$

14. $\displaystyle\int_0^{\sqrt{5}/5} \frac{dx}{(x^2+5)^2}$

15. $\displaystyle\int \frac{dx}{x^3+x^2+x}$

16. $\displaystyle\int \frac{(3x^4+x^2)\,dx}{x^2+10x+21}$

17. $\displaystyle\int \frac{(x^2+1)\,dx}{x^3+3x^2+2x}$

18. $\displaystyle\int \frac{(2x^2-1)\,dx}{(x+1)^2(x-3)}$

19. $\int \frac{x+1}{(x^2+3x+6)^2}\,dx$

20. $\int \frac{2x+1}{(x^2+x+1)^{50}}\,dx$

21. $\int \frac{(x-1)\,dx}{(x^2+3x+5)(x+1)}$

22. $\int \frac{x+3}{x^4+4x^3+5x^2}\,dx$

23. $\int \frac{(x^4+x)\,dx}{(x-1)(x-2)(x-3)}$

24. $\int \frac{(4x+1)\,dx}{(x^2+1)(x^2-1)^2}$

In Exercises 25 and 26 compute the given integral (a) using partial fractions, and (b) using a trigonometric substitution. Compare your answers.

25. $\int \frac{dx}{a^2-x^2}$

26. $\int \frac{x\,dx}{x^2-a^2}$

27. Name a third (and easiest) method for Exercise 26.

28. Use the partial-fraction method to rederive the formula for $\int \sec x\,dx$ using that

$$\sec x = \frac{1}{\cos x} = \frac{\cos x}{1-\sin^2 x}$$

29. Use a method analogous to that of Exercise 28 to recalculate $\int \sec^3 x\,dx$.

30. Compute $\int_0^{\pi/6} \frac{\cos x\,dx}{\sin^2 x - 5\sin x + 5}$.

31. Compute $\int_{\pi/4}^{\pi/2} \frac{\sin x\,dx}{\cos^2 x - 2\cos x + 1}$.

Section 11.4: Rationalizing Substitutions

Objective:

Transform certain integrands involving radicals and/or transcendental functions into rational functions.

INTEGRALS INVOLVING TERMS $(ax + b)^{p/q}$

The substitution $u = (ax+b)^{1/q}$ can often be used to transform an integral involving $(ax+b)^{p/q}$ (p and q integers, $q > 0$) into a rational function of u. The key is not to compute du directly, but to solve for x, getting

$$x = \frac{u^q - b}{a}$$

from which we can conclude that

$$dx = \frac{qu^{q-1}}{a}\,du$$

Example 1 Evaluate $\int \frac{(2x-1)\,dx}{(x-2)^{2/3}+1}$.

Solution We let $u = (x-2)^{1/3}$; then $u^3 = x - 2$, so $x = u^3 + 2$ and $dx = 3u^2\,du$. Thus

$$\begin{aligned}
\int \frac{(2x-1)\,dx}{(x-2)^{2/3}+1}
&= \int \frac{[2(u^3+2)-1]\,3u^2\,du}{u^2+1} \\
&= \int \frac{(6u^5+9u^2)\,du}{u^2+1} \\
&= \int \left(6u^3 - 6u + 9 + \frac{6u-9}{u^2+1}\right) du \\
&= \frac{3u^4}{2} - 3u^2 + 9u + 3\ln(u^2+1) - 9\arctan u + C \\
&= \frac{3(x-2)^{4/3}}{2} - 3(x-2)^{2/3} + 9(x-2)^{1/3} - 9\arctan[(x-2)^{1/3}] + C
\end{aligned}$$ ●

It may be that the integrand contains two (or more) terms of the form $(ax+b)^{p/q}$, $(ax+b)^{p_1/q_1}$. We can then eliminate both radicals simultaneously by letting $u = (ax+b)^{1/r}$ where r is the least common multiple of q and q_1.

Example 2 Transform $\int \frac{dx}{x^{3/4} - x^{2/3}}$ into an integral of a rational function.

Solution Let $u = x^{1/12}$. Then $u^{12} = x$, $12u^{11}\,du = dx$, $u^3 = x^{1/4}$, and $u^4 = x^{1/3}$, so $u^9 = x^{3/4}$ and $u^8 = x^{2/3}$. As a result,

$$\int \frac{dx}{x^{3/4} - x^{2/3}} = 12 \int \frac{u^{11}\,du}{u^9 - u^8} = 12 \int \frac{u^3\,du}{u - 1} \quad \bullet$$

We shall refer to manipulations such as these that change the integrand into a rational function, as *rationalizations* of the integral. In much of this section we shall concentrate on rationalizing rather than evaluating. Earlier methods apply to the resulting rational integrand.

PROGRESS TEST 1

Rationalize the following integrals:

1. $\int \frac{dx}{x(\sqrt{x+1} - 2)}$

2. $\int \frac{\sqrt{2x+1}}{1 + \sqrt[3]{2x+1}}\,dx$

INTEGRALS INVOLVING A RADICAL TERM $\sqrt{x^2 \pm a^2}$ OR $\sqrt{a^2 - x^2}$ AND x^p, p AN ODD INTEGER

We can rationalize such integrals, where p is positive or negative, by letting u equal the radical term and then taking the differential of u^2. In each case $u\,du$ will be either $x\,dx$ or $-x\,dx$. Thus we shall want to factor an x out of x^p to pair with dx. Since $p - 1$ is even, we can replace x^{p-1} with a polynomial in u, as indicated in Example 3.

Example 3 Rationalize $\int \frac{\sqrt{x^2+9}}{x}\,dx$.

Solution Let $u = \sqrt{x^2 + 9}$. Then $u^2 = x^2 + 9$, so $2u\,du = 2x\,dx$; that is, $u\,du = x\,dx$. As a result,

$$\int \frac{\sqrt{x^2+9}}{x}\,dx = \int \frac{\sqrt{x^2+9}\,(x\,dx)}{x^2} = \int \frac{u(u\,du)}{u^2 - 9}$$

$$= \int \left[1 + \frac{9}{(u-3)(u+3)}\right] du \quad \bullet$$

Notice that evaluating the resulting rationalized integral in Example 3 obviously involves partial fractions. Using the trigonometric substitution $x = 3 \tan \theta$, a rational integrand eventually results, leading again to partial fractions before a final answer is possible.

In examples where you have a choice between a rationalizing substitution and a trigonometric substitution, the particular approach you begin with is not crucial. In practice, if one method seems to lead to a rather complicated integral, you might want to give up on it and try the alternate method. More often than not, the work involved in either approach is the same, and so it becomes a matter of personal preference.

PROGRESS TEST 2

Evaluate the following:

1. $\displaystyle\int \frac{(x^3 - x)\,dx}{\sqrt{2 - x^2}}$

2. $\displaystyle\int x^3\sqrt{x^2 - 5}\,dx$

SOME REMARKS ON SIMILAR SITUATIONS

Other types of integrands not discussed previously also yield rational functions of u if we let u equal the radical term. For example, to evaluate $\int (\sqrt[4]{1 + x^3})/x^4\,dx$ we try $u = \sqrt[4]{1 + x^3}$. Then $u^4 = 1 + x^3$, $4u^3\,du = 3x^2\,dx$, and so

$$\int \frac{\sqrt[4]{1 + x^3}}{x^4}\,dx = \int \frac{\sqrt[4]{1 + x^3}\,(x^2\,dx)}{x^6} = \int \frac{u(4/3)u^3\,du}{(u^4 - 1)^2} = \frac{4}{3}\int \frac{u^4\,du}{(u^4 - 1)^2}$$

This can now be handled by the partial-fraction method.

We may also, on occasion, rationalize an integrand involving an exponential term such as e^x by letting $u = e^x$. Now, rather than computing du directly, we solve for x getting $\ln u = x$, so $du/u = dx$. For example, consider

$$\int \frac{dx}{e^{6x} + 3e^{4x} + e^{2x} + 1}$$

Let $u = e^{2x}$. Then $\ln u = 2x$, so $du/u = 2\,dx$ and

$$\int \frac{dx}{e^{6x} + 3e^{4x} + e^{2x} + 1} = \int \frac{du}{u(u^3 + 3u^2 + u + 1)}$$

which can be handled by earlier methods. (Which?)

INTEGRALS OF RATIONAL FUNCTIONS OF sin *x* AND cos *x*

A somewhat technical but useful substitution

$$u = \tan\frac{x}{2}$$

can be used to transform an integrand which is a rational function of $\sin x$ and $\cos x$ into one which is a rational function of u. Then $x = 2\arctan u$, so $dx = 2\,du/(1 + u^2)$. From Fig. 11.3 we see that $\sin(x/2) = u/\sqrt{u^2 + 1}$ and $\cos(x/2) = 1/\sqrt{u^2 + 1}$. Now, using the identities $\sin 2\theta = 2\sin\theta\cos\theta$ and $\cos 2\theta = \cos^2\theta - \sin^2\theta$ with $\theta = x/2$, we obtain

$$\sin x = 2\sin\frac{x}{2}\cos\frac{x}{2} = 2\frac{u}{\sqrt{u^2 + 1}}\frac{1}{\sqrt{u^2 + 1}} = \frac{2u}{u^2 + 1}$$

and

$$\cos x = \cos^2\left(\frac{x}{2}\right) - \sin^2\left(\frac{x}{2}\right) = \frac{1}{u^2 + 1} - \frac{u^2}{u^2 + 1} = \frac{1 - u^2}{u^2 + 1}$$

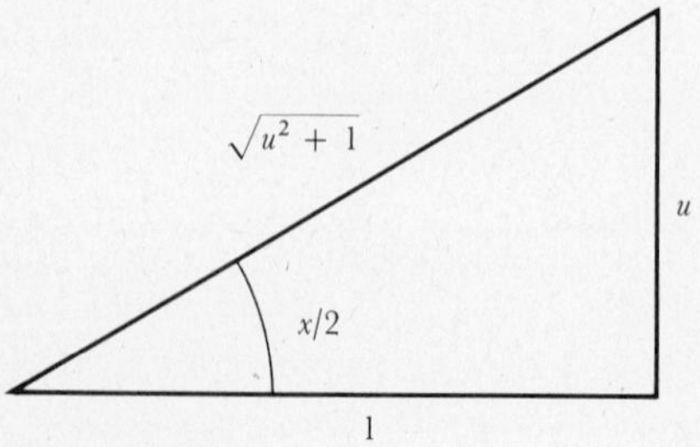

$\sqrt{u^2 + 1}$ u $x/2$ 1

Fig. 11.3

Example 4 Rationalize $\int \frac{\tan x}{1 + \cos x}\,dx$.

Solution Let

$$u = \tan\frac{x}{2},\ dx = \frac{2\,du}{u^2 + 1},\ \sin x = \frac{2u}{u^2 + 1},\ \cos x = \frac{1 - u^2}{u^2 + 1}$$

Then

$$\int \frac{\tan x\,dx}{1 + \cos x} = \int \frac{[\sin x/\cos x]}{1 + \cos x}$$

$$= \int \frac{[(2u/(u^2 + 1))] \cdot [(u^2 + 1)/(1 - u^2)] \cdot [2\,du/(u^2 + 1)]}{1 + (1 - u^2)/(u^2 + 1)} = 2\int \frac{u\,du}{1 - u^2} \quad \bullet$$

PROGRESS TEST 3

Rationalize the following:

1. $\int \frac{dx}{3 - 2\cos x}$
2. $\int \frac{dx}{\sqrt{1 + e^x}}$
3. $\int \frac{\sqrt[3]{3x^5 + 5}}{x}\,dx$

SECTION 11.4 EXERCISES

In Exercises 1 to 30 (a) rationalize the given integral and then (b) evaluate the resulting integral.

1. $\int \frac{x\,dx}{\sqrt[3]{x - 1}}$
2. $\int \frac{\sqrt[3]{1 - x^2}}{x^2}\,dx$
3. $\int \sqrt{3 + \sqrt{x}}\,dx$
4. $\int \frac{\sqrt{x}\,dx}{\sqrt[3]{x} + 1}$
5. $\int \frac{3\sqrt{x + 1} - 2}{\sqrt{x + 1} + 1}\,dx$
6. $\int \frac{2\sqrt[3]{x - 1} + 4}{\sqrt[3]{x - 1} - 1}\,dx$
7. $\frac{dx}{\sqrt[3]{x^2} + \sqrt{x}}$
8. $\int \frac{\sqrt[5]{x} - 1}{\sqrt{x} + 2}\,dx$
9. $\int (x^2 + 3)\sqrt{x^2 - 4}\,dx$
10. $\int \frac{\sqrt[3]{x + 1}}{x^2 - 5}\,dx$
11. $\int \frac{(1 + \sqrt{x})\,dx}{x - x^{3/2}}$
12. $\int \frac{\sqrt{9 - x^2}}{x}\,dx$
13. $\int_0^1 \sqrt[3]{e^{3x} + 8}\,dx$
14. $\int_{\pi/4}^{\pi/3} \frac{dx}{x\sqrt{x^2 + 1}}$
15. $\int \sqrt{1 + \sqrt{1 + \sqrt{x}}}\,dx$
16. $\int \frac{dx}{(e^x + 1)^2}$
17. $\int \frac{dx}{e^x + e^{-x}}$
18. $\int \frac{\sqrt[3]{x + 2}}{x}\,dx$
19. $\int \frac{\sqrt{9 - x^2}}{x^3}\,dx$
20. $\int \frac{1 + x^{1/3}}{1 + x^{2/3}}\,dx$
21. $\int \frac{dx}{1 + \sqrt{2x + 1}}$
22. $\int \frac{dx}{2\sin x - \cos x + 1}$
23. $\int \frac{dx}{1 + \tan x}$
24. $\int_0^1 x^5\sqrt{5 - x^2}\,dx$
25. $\int \frac{e^x\,dx}{e^{3x} + 1}$
26. $\int \frac{dx}{e^x - e^{-x}}$
27. $\int \frac{\sqrt{x} + \sqrt[3]{x}}{\sqrt[5]{x} + \sqrt[6]{x}}\,dx$
28. $\int_0^{\pi/2} \frac{2}{1 + 2\cos x}\,dx$
29. $\int_0^{\pi/2} \frac{dx}{\sin x + \cos x}$
30. $\int \frac{e^x\,dx}{1 - e^{2x}}$
31. Compute $\int \cos x\,dx$ using the substitution $u = \tan(x/2)$.
32. Repeat Exercise 31 for $\int \sin x\,dx$.
33. Repeat Exercise 31 for $\int \cot x\,dx$.
34. Repeat Exercise 31 using the substitution $u = \cot(x/2)$.
35. Repeat Exercise 32 using the substitution $u = \cot(x/2)$.
36. Recompute $\int \sec^3 x\,dx$ using $u = \tan x/2$. (Recall Sec. 11.1, Example 5.)

In Exercises 37 to 40 rationalize the given integral.

37. $\int \frac{(x + \sqrt{x})\,dx}{x^2 + 2x^{1/3}}$
38. $\int \frac{dx}{\sqrt{1 + \sqrt{x}}}$
39. $\int \frac{(\sin^2 x + 2\sin x + 1)\,dx}{3 + \cos x}$
40. $\int \sqrt{\frac{x + 1}{1 - x}}\,dx$

Section 11.5: General Strategies, Including the Use of Integration Tables

Objectives:

1. Establish a general plan when dealing with integrals when the technique to use is not obvious.
2. Transform given integrals into forms appearing in integration tables, and then use the appropriate formulas to evaluate these integrals.

INTRODUCTION

Having studied the basic integration formulas of specific functions in Chap. 8 and the more general techniques of the previous four sections, you have one more integration skill yet to master: what to do when confronted with an integral without a section title indicating which method to use. Generally, the idea is to find some formula that fits the integral. Right now, while your working knowledge of the specific formulas and techniques is fresh, the most efficient approach is to spend some time manipulating the integral into some form recognizable in terms of a basic formula. Later, in months and years to come when time has eroded your memory of the specifics, the table of integrals becomes more important.

A reasonable question, then, is *why learn integration techniques in the first place?* The answer, to be illustrated in a few pages, is that to use integration tables in any but the most routine cases we need to apply virtually every technique of integration studied earlier. Most integrals arising in practice need to be modified before they can be matched to a formula appearing in a table, and it is in this modifying and matching that we need to apply the skills developed previously.

AN OVERVIEW—NONELEMENTARY FUNCTIONS

Before moving on it is useful to put our work with antiderivatives in perspective. Our main interest in searching for antiderivatives is in using these to evaluate the sorts of definite integrals that occur in most applications. Of course our ability to evaluate an integral such as $\int_a^b f(x)\,dx$ by first finding an antiderivative F of f and then computing $F(b) - F(a)$ is based on the Fundamental Theorem of Calculus. Our methods for finding such an antiderivative are based upon relatively simple ideas. On one hand, much of what is involved is merely algebraic manipulation. This includes, for example, completing the square, multiplying by constants, using long division, applying trigonometric identities, and breaking rational functions into their partial fractions. On the other hand, since antidifferentiation is merely differentiation in reverse, the *calculus* part of our techniques can be classified in terms of differentiation ideas. First, each specific integration formula is a derivative formula written in reverse. Moreover, our two basic techniques of integration, elementary substitution and integration by parts, are simply reversals of the Chain Rule and the Product Rule, respectively.

However, even our best attempts to find antiderivatives are doomed to fail in some cases because relatively simple functions exist whose antiderivatives cannot be written as finite combinations of the functions studied to date. Strange, but true! For example, antiderivatives of $\int \sin(x^2)\,dx$ *exist,* but, unfortunately, they cannot be written as a finite algebraic combination or composition of rational, algebraic and transcendental functions considered so far. To write an antiderivative of $\sin(x^2)$ we need an infinite series (see Sec. 16.4).

Finite combinations of the functions we have studied so far are called *elementary functions*. The function whose derivative is $\sin(x^2)$ is a nonelementary function. However, if needed, we can approximate the *definite* integral $\int_0^{\pi/2} \sin(x^2)\,dx$ (the area under $y = \sin(x^2)$ from $x = 0$ to $x = \pi/2$), using its definition as a limit of Riemann sums. Since $\sin(x^2)$ is continuous on $[0, \pi/2]$, we know by (6.3.3) that this limit exists. More efficient techniques for approximating such definite integrals are the subject of Sec. 11.6. Furthermore, no easy criteria exist that separate those functions whose antiderivatives are elementary—and that can thus be determined using techniques of integration or tables of integrals—from those whose antiderivatives are nonelementary.

A GENERAL STRATEGY

When confronted with an integral to evaluate, it is best to proceed deliberately, with a general plan in mind. Try simple approaches first. Does the integral resemble a known formula? If so, how is it different? What kind of manipulation is needed to get the integral into the right form? If a substitution looks promising, try it; *but write it out completely*. For example, $\int (x\,dx)/(1 + x^4)$ involves a quotient of rational functions, but plunging into the messy partial-fraction approach will involve a difficult factorization of $1 + x^4$. Maybe a trig substitution will help. But again, we should look for a simpler approach first. Except for the fourth power of x, the integral resembles the integral leading to the arctangent function

$$\int \frac{dx}{1 + x^2} = \arctan x + C$$

But by letting $u = x^2$, the denominator becomes $1 + u^2$. Now $u = x^2$ implies $du = 2x\,dx$, so $x\,dx = du/2$. Thus we can rewrite $\int (x\,dx)/(1 + x^4)$ as follows:

$$\int \frac{x\,dx}{1 + x^4} = \int \frac{du/2}{1 + u^2} = \frac{1}{2}\int \frac{du}{1 + u^2} = \frac{1}{2}\arctan u + C$$
$$= \frac{1}{2}\arctan x^2 + C$$

By writing out the substitution completely we were able to determine whether the substitution led to an integral that matched a formula *exactly*. *In applying formulas, "close" does not count. There must be an exact fit.*

USING INTEGRATION TABLES

Lists of worked-out integrals exist in various lengths, running from a few dozen formulas as on the front and back endpapers, to hundreds of pages in handbooks. Although virtually every integral appearing in the book to this point could be worked out using such tables, using tables in practice involves much the same sort of classifying, manipulating, and substituting as was involved in the techniques presented earlier.

Example 1 Evaluate $\int \sqrt{3x^2 + 8}\,dx$ using the table of integrals.

Solution We shall apply formula [28], which states that

$$\int \sqrt{u^2 \pm a^2}\,du = \frac{u}{2}\sqrt{u^2 \pm a^2} \pm \frac{a^2}{2}\ln|u + \sqrt{u^2 \pm a^2}|$$

Here we have $u = \sqrt{3}\,x$, $a = \sqrt{8}$, so $du = \sqrt{3}\,dx$ and $dx = du/\sqrt{3}$. Hence

$$\int \sqrt{3x^2 + 8}\,dx = \frac{1}{\sqrt{3}}\int \sqrt{u^2 + (\sqrt{8})^2}\,du$$
$$= \frac{1}{\sqrt{3}}\left[\frac{\sqrt{3}\,x}{2}\sqrt{3x^2 + 8} + \frac{8}{2}\ln|\sqrt{3}\,x + \sqrt{3x^2 + 8}|\right] + C$$
$$= \frac{x}{2}\sqrt{3x^2 + 8} + \frac{4}{\sqrt{3}}\ln|\sqrt{3}\,x + \sqrt{3x^2 + 8}| + C \quad \bullet$$

Many formulas occurring in integral tables take the form of *reduction formulas* that allow us to replace integrals by succeedingly simpler integrals in a systematic way. Most of these reduction formulas are especially useful in cases where repeated integration by parts would be required.

Example 2 Use the integration table to determine $\int x^5 e^{3x}\,dx$.

Solution By formula [98],

$$\int u^n e^{au}\,du = \frac{1}{a}u^n e^{au} - \frac{n}{a}\int u^{n-1}e^{au}\,du$$

Here $n = 5$, $u = 3x$, $dx = du/3$, so we apply the formula repeatedly as follows:

$$\begin{aligned}
\int x^5 e^{3x}\,dx &= \frac{1}{3}x^5 e^{3x} - \frac{5}{3}\int x^4 e^{3x}\,dx = \frac{1}{3}x^5 e^{3x} - \frac{5}{3}\left[\frac{1}{3}x^4 e^{3x} - \frac{4}{3}\int x^3 e^{3x}\,dx\right] \\
&= \frac{1}{3}x^5 e^{3x} - \frac{5}{9}x^4 e^{3x} + \frac{20}{9}\left[\frac{1}{3}x^3 e^{3x} - \frac{3}{3}\int x^2 e^{3x}\,dx\right] \\
&= \frac{1}{3}x^5 e^{3x} - \frac{5}{9}x^4 e^{3x} + \frac{20}{27}x^3 e^{3x} - \frac{20}{9}\left[\frac{1}{3}x^2 e^{3x} - \frac{2}{3}\int x^1 e^{3x}\,dx\right] \\
&= \frac{1}{3}x^5 e^{3x} - \frac{5}{9}x^4 e^{3x} + \frac{20}{27}x^3 e^{3x} - \frac{20}{27}x^2 e^{3x} + \frac{40}{27}\left[\frac{1}{3}x^1 e^{3x} - \frac{1}{3}\int x^0 e^{3x}\,dx\right] \\
&= \frac{1}{3}x^5 e^{3x} - \frac{5}{9}x^4 e^{3x} + \frac{20}{27}x^3 e^{3x} - \frac{20}{27}x^2 e^{3x} + \frac{40}{81}xe^{3x} - \frac{40}{81}\left[\frac{1}{3}e^{3x}\right] + C
\end{aligned}$$ ●

Example 3 Transform $\int \sqrt{\tan x}\,dx$ into a familiar form handled earlier.

Solution A search turns up no formula resembling the given integral, so we try a substitution, $u = \tan x$, $du = \sec^2 x\,dx$. Thus $dx = du/(\sec^2 x) = du/(1 + u^2)$ and $\int \sqrt{\tan x}\,dx = \int(\sqrt{u}/(1 + u^2)\,du$. This last integral does not appear in a table but can be rationalized using $v = \sqrt{u}$, $dv = 1/2\sqrt{u}\,du$, $du = 2v\,dv$ and

$$\int \frac{\sqrt{u}}{1 + u^2}\,du = \int \frac{v}{1 + v^4}(2v\,dv) = 2\int \frac{v^2\,dv}{1 + v^4}$$

We now have a rational integrand, which, again, does not turn up in the table. But by using a partial-fraction decomposition based on the factorization of $1 + v^4 = (v^2 + \sqrt{2}\,v + 1)(v^2 - \sqrt{2}\,v + 1)$ we obtain, finally,

$$2\int \frac{v^2\,dv}{1 + v^4} = -\frac{\sqrt{2}}{2}\int \frac{v\,dv}{v^2 + \sqrt{2}\,v + 1} + \frac{\sqrt{2}}{2}\int \frac{v\,dv}{v^2 - \sqrt{2}\,v + 1}$$

We complete the square and let $w = v \pm \sqrt{2}/2$. Then, for example,

$$\int \frac{v\,dv}{v^2 + \sqrt{2}v + 1} = \int \frac{(w + \sqrt{2}/2)\,dw}{w^2 + (1/\sqrt{2})^2}$$

which breaks down into a natural logarithm and an arctangent (formulas [5] and [24].) ●

Note that Example 3 required completing the square, two substitutions, including a rationalizing substitution, and a partial-fraction decomposition.

Formulas also exist for definite integrals and, in particular, for products of powers of sine and cosine. See Wallis' formulas [125–126].

Example 4 Evaluate $\int_0^{\pi/2} \cos^5 x\, dx$.

Solution By formula [125], with $n = 5$ (odd), we have

$$\int_0^{\pi/2} \cos^5 x\, dx = \frac{(4)(2)}{(5)(3)(1)} = \frac{8}{15} \quad \bullet$$

PROGRESS TEST 1

Using integral tables, evaluate:

1. $\int \frac{dx}{5x\sqrt{3x+2}}$ **2.** $\int \sin^4 x\, dx$ **3.** $\int_0^{\pi/2} \cos^4 x \sin^2 x\, dx$

SECTION 11.5 EXERCISES

(Note: The Chapter Exercises provide exercises relating to the first part of this section.) In Exercises 1 to 30, (a) give the number of an appropriate integration formula in the table of integrals (the answer is not necessarily unique) and (b) apply that formula to evaluate the given integral. (For some exercises a reduction formula may apply and be used more than once, leading to an integral that may be evaluated using another formula.)

1. $\int \frac{dx}{\sqrt{x^2-4}}$ **2.** $\int \sqrt{x^2+4}\, dx$ **3.** $\int \frac{dx}{x^2\sqrt{x^2-9}}$ **4.** $\int \frac{2x\, dx}{1-x^2}$

5. $\int \frac{x\, dx}{\sqrt{5+4x}}$ **6.** $\int \frac{2\, dx}{1-x^2}$ **7.** $\int \sqrt{5-4x^2}\, dx$ **8.** $\int x^{-3}\sqrt{3+x}\, dx$

9. $\int x \cot^2 x^2\, dx$ **10.** $\int x^2 \cos^2 x^3\, dx$ **11.** $\int x^2 e^x\, dx$ **12.** $\int \frac{e^x}{x^3}\, dx$

13. $\int x \arcsin x^2\, dx$ **14.** $\int \frac{\operatorname{arcsec}\sqrt{x-1}}{\sqrt{x-1}}\, dx$ **15.** $\int e^{3x}\sin x\, dx$ **16.** $\int \cos 5x \sin 3x\, dx$

17. $\int x^3 \sin x\, dx$ **18.** $\int x^3 \cos x\, dx$ **19.** $\int \sin^2 x \cos^3 x\, dx$ **20.** $\int \cos^2 x \sin^3 x\, dx$

21. $\int \sin^6 x\, dx$ **22.** $\int \cos^6 x\, dx$ **23.** $\int \frac{dx}{x\sqrt{1-x^2}}$ **24.** $\int (\sqrt{16-x^2})^3\, dx$

25. $\int_0^{\pi/2} \sin^4 x\, dx$ **26.** $\int_0^{\pi/2} \cos^4 x\, dx$ **27.** $\int_0^{\pi/4} \sin^4 2x\, dx$ **28.** $\int_0^{\pi/4} \sin^5 x\, dx$

29. $\int \frac{dx}{\sqrt{5x^2+3x+2}}$ **30.** $\int \frac{dx}{x^2+3x+1}$

In Exercises 31 to 36 some manipulation is required before the given integral can be transformed into a form that fits the table. Evaluate the given integral.

31. $\int \frac{dx}{(x+1)\sqrt{x^2+2x}}$ **32.** $\int \frac{dx}{e^{3x}-e^{-3x}}$

33. $\int \frac{x\, dx}{\sec(x^2)+\tan(x^2)}$ **34.** $\int e^{2x}\sqrt{1+e^{2x}}\, dx$

35. $\int \ln(x^2+1)\, dx$ **36.** $\int \sqrt{\frac{\cos\sqrt{x}}{x}}\, dx$

(Integrate by parts.)

Section 11.6: Numerical Integration

Objectives:

1. Approximate the values of definite integrals using either trapezoidal or parabolic approximations of the integrands.
2. Where appropriate, determine bounds on the errors involved in these approximations.

INTRODUCTION

As noted earlier, if $f(x)$ is continuous on $[a, b]$, then $\int_a^b f(x)\,dx$ exists. However, in some cases the antiderivative $\int f(x)\,dx$ may not be an elementary function and hence not computable using the methods studied in Chaps. 6, 8, or 11. Such integrals occur quite naturally in practice and do not necessarily involve especially complicated functions. For example, $\int_a^b e^{-x^2}\,dx$ arises as the area from $x = a$ to $x = b$ under the widely used probability "normal curve." Although this definite integral exists (it is a real number), it simply cannot be computed using the antiderivative shortcut.

This is a situation analogous to what we confront with an irrational number—say, $\sqrt{2}$. It exists (arising, for example, as the length of the hypotenuse of a 1, 1, $\sqrt{2}$ right triangle), but can only be approximated using decimals. Just as with the *definition* of $\sqrt{2}$ (a number that, when squared, yields 2), the *definition* of definite integral fails to provide a systematic procedure for determining the approximation. The purpose of this section is to introduce two numerical methods for approximating the definite integral $\int_a^b f(x)\,dx$, where we assume that $f(x)$ is continuous on $[a, b]$. We shall describe our methods for *positive* functions, which frames our discussion in the language of area under the graph, but our results hold for any continuous functions. In fact our numerical methods (except for the discussions of approximation error) make sense even when only a finite table of data is given rather than a completely defined function (see the Section Exercises).

Wide availability of computing machines has made numerical integration a common and very useful practice, and many particular techniques are in use. In fact, the study of such numerical techniques is the subject of many books in itself. Here we shall examine two of the basic techniques, the Trapezoid Rule and Simpson's Rule.

Even these rules are available on most programmable calculators, so that your becoming proficient in the actual calculations is not the whole purpose here. Perhaps even more important is your becoming familiar with the sorts of issues that arise in numerical integration: When is it appropriate or necessary? How accurate are the estimates? How does the accuracy relate to the integrand, the limits of integration, or the number of subdivisions? When might one method be preferable to another?

In Sec. 16.4 we briefly examine another method for approximating definite integrals using infinite series.

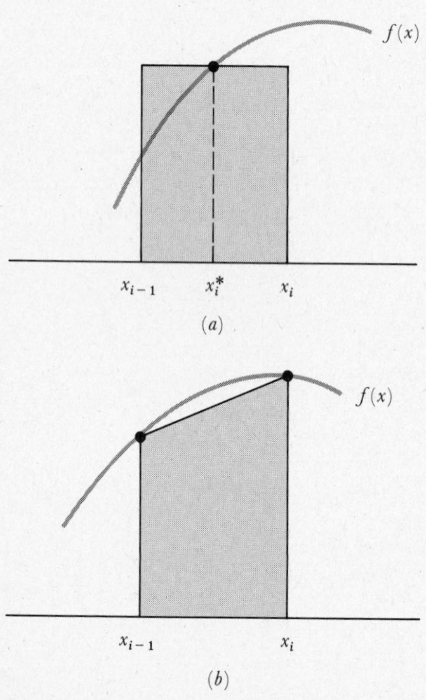

Fig. 11.4

THE TRAPEZOID RULE

Instead of using flat-topped rectangles [Fig. 11.4(a)] as in the definition of definite integral, we use trapezoids [Fig. 11.4(b)].

The area of a trapezoid of base b and heights h_1 and h_2 is $b(h_1 + h_2)/2$. Suppose the interval $[a, b]$ is decomposed into equal subintervals by the points $x_0 = a,\ x_1 = a + \Delta x,\ x_2 = a + 2\,\Delta x,\ \ldots,\ x_i = a + i\,\Delta x,\ \ldots,\ x_n = b$. As shown in Fig. 11.5, the trapezoid associated with a typical ith subinterval has area

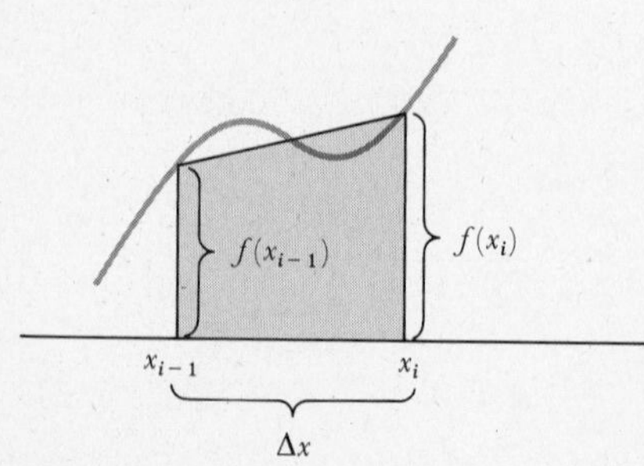

Fig. 11.5

$$\frac{1}{2}[f(x_{i-1}) + f(x_i)]\,\Delta x$$

We simply add these to get an approximation of $\int_a^b f(x)\,dx$; that is,

$$\frac{1}{2}[f(x_0) + f(x_1)]\,\Delta x + \frac{1}{2}[f(x_1) + f(x_2)]\,\Delta x + \cdots$$
$$+ \frac{1}{2}[f(x_{n-2}) + f(x_{n-1})]\,\Delta x + \frac{1}{2}[f(x_{n-1}) + f(x_n)]\,\Delta x$$
$$= \frac{1}{2}[f(x_0) + 2f(x_1) + 2f(x_2) + \cdots + 2f(x_{n-2}) + 2f(x_{n-1}) + f(x_n)]\,\Delta x$$
$$= \frac{b-a}{n}\left[\frac{1}{2}f(x_0) + f(x_1) + f(x_2) + \cdots + f(x_{n-1}) + \frac{1}{2}f(x_n)\right]$$

11.6.1 **The Trapezoid Rule** The nth trapezoidal approximation is given by

$$\int_a^b f(x)\,dx \approx \frac{b-a}{n}\left[\frac{f(a)}{2} + f(x_1) + \cdots + f(x_{n-1}) + \frac{f(b)}{2}\right]$$

where for each integer i between 1 and n, $x_i = a + i(b-a)/n$.

As has become our habit, we shall first apply the rule to a case where we know the answer, $4\int_0^1 dx/(1+x^2)$. (Four times the area under $f(x) = 1/(1+x^2)$ from 0 to 1.) We already know that

$$4\int_0^1 \frac{dx}{1+x^2} = 4 \arctan 1 - \arctan 0 = 4\left(\frac{\pi}{4}\right) = \pi$$

Example 1 Determine the tenth trapezoidal approximation of $\int_0^1 (4\,dx)/(1+x^2)$.

Solution Here $a = 0$, $b = 1$, $n = 10$, and $f(x) = 4/(1+x^2)$ with $\Delta x = 1/10$. In Table 11.1 we shall arrange our calculations in an orderly way, and we suggest that you do likewise. We round off to four decimal places.

$x_0 = 0$	$f(x_0) = \frac{4}{1+0^2}$	$= 4.0000$
$x_1 = \frac{1}{10}$	$f(x_1) = \frac{4}{1+(1/10)^2}$	≈ 3.9604
$x_2 = \frac{2}{10}$	$f(x_2) = \frac{4}{1+(2/10)^2}$	≈ 3.8462
$x_3 = \frac{3}{10}$	$f(x_3) = \frac{4}{1+(3/10)^2}$	≈ 3.6697
$x_4 = \frac{4}{10}$	$f(x_4) = \frac{4}{1+(4/10)^2}$	≈ 3.4483
$x_5 = \frac{5}{10}$	$f(x_5) = \frac{4}{1+(5/10)^2}$	$= 3.2000$
$x_6 = \frac{6}{10}$	$f(x_6) = \frac{4}{1+(6/10)^2}$	≈ 2.9412
$x_7 = \frac{7}{10}$	$f(x_7) = \frac{4}{1+(7/10)^2}$	≈ 2.6846
$x_8 = \frac{8}{10}$	$f(x_8) = \frac{4}{1+(8/10)^2}$	≈ 2.4390
$x_9 = \frac{9}{10}$	$f(x_9) = \frac{4}{1+(9\cdot 10)^2}$	≈ 2.2099
$x_{10} = 1$	$f(x_{10}) = \frac{4}{1+1^2}$	$= 2.0000$

Table 11.1

Thus, from Table 11.1,

$$\int_0^1 \frac{4\,dx}{1+x^2} \approx \frac{1}{10}\left[\frac{4.0000}{2} + f(x_1) + \cdots + f(x_9) + \frac{2.0000}{2}\right]$$
$$\approx 3.1399 \quad \bullet$$

This approximation is in error by

$$\pi - 3.1399 \approx 0.0017 \qquad \text{(again, rounded to four places)}$$

In this example, we knew the antiderivative and hence the value of the definite integral. Consequently we also knew the error. Numerical integration is usually used when this is not the case, and as a result, having some idea of the accuracy is quite important. We now give an upper bound for this error and then discuss it.

11.6.2 **Trapezoid Rule Error** Suppose that f'' exists and is continuous on $[a, b]$ with $|f''(x)| \leq M$ on $[a, b]$. If the nth trapezoidal approximation is denoted T_n, then

$$\left|\int_a^b f(x)\,dx - T_n\right| \leq \frac{(b-a)^3 M}{12n^2}$$

Although we shall not prove this error statement, we note that it makes sense: The error bound increases with the size of the interval $[a, b]$, decreases as the number of subdivisions increases, and depends on the size of the second derivative on $[a, b]$. This last fact occurs because the deviation of $f(x)$ from the slant-line top of the trapezoid depends on the "bend" in the graph, which depends upon how much the slope $f'(x)$ changes, which in turn depends on the size of $|f''(x)|$ on $[a, b]$.

The graph and the error bound both imply that the Trapezoid Rule gives the exact value of the definite integral when $f(x)$ is a linear function. In such a case $f'' \equiv 0$, so the error is 0.

Example 2 Compute an upper bound for the error of the approximation in Example 1 and compare it with the actual error.

Solution Here $b - a = 1$, $n = 10$, so it remains to determine an upper bound for the second derivative of $f(x) = 4/(1 + x^2)$ on $[0, 1]$. Now $f'(x) = -8x/(1 + x^2)^2$ and $f''(x) = 8(3x^2 - 1)/(1 + x^2)^3$. To determine an upper bound for $f''(x)$ on $[0, 1]$ we determine whether it is increasing or decreasing on $[0, 1]$ by checking the sign of *its* derivative. After simplifying, we find

$$f'''(x) = \frac{96x(1-x^2)}{(1+x^2)^4} \geq 0 \qquad \text{on } [0, 1]$$

Thus $f''(x)$ is increasing on $[0, 1]$ and hence reaches its maximum at $x = 1$. By (11.6.2),

$$\left|\int_0^1 \frac{4\,dx}{1+x^2} - 3.1399\right| \leq \frac{(1)^3|f''(1)|}{12(10)^2}$$
$$= \frac{16/8}{12(100)} = \frac{1}{600}$$
$$\approx 0.0017 \qquad \text{(rounded to four places)} \quad \bullet$$

Example 3 Determine the number of subdivisions required in a trapezoidal estimate of $\int_0^1 (4\,dx)/(1 + x^2)$ to guarantee accuracy to within 0.00005 (accuracy through four decimal places).

Solution Using the data accumulated in Example 2, we need to find a positive integer n such that

$$\left| \int_0^1 \frac{4\,dx}{1+x^2} - T_n \right| \leq \frac{(1)^3|f''(1)|}{12(n^2)} = \frac{2}{12n^2} = \frac{1}{6n^2} < 0.00005$$

We have a choice here either to solve the inequality systematically for n, or to use trial and error to determine an n that will suffice. Since the numbers here are easy to deal with, we use the former approach, obtaining

$$\frac{1}{6(0.00005)} < n^2$$

Solved for n, this inequality tells us that $n > 57.7$. Thus any $n \geq 58$ will yield the desired accuracy. ●

PROGRESS TEST 1

1. Using that $\ln x = \int_1^x \frac{dt}{t}$:

(a) Estimate ln 2 using the Trapezoid Rule, with $n = 5$.

(b) Determine an upper bound for the error of this estimate.

(c) Compare the error bound with the actual error through four places.

2. What value of n is necessary to provide accuracy through five decimal places?

3. (a) Estimate $\int_0^2 e^{-x^2}\,dx$ using the Trapezoid Rule, with $n = 10$.

(b) Determine an upper bound for the error of this estimate.

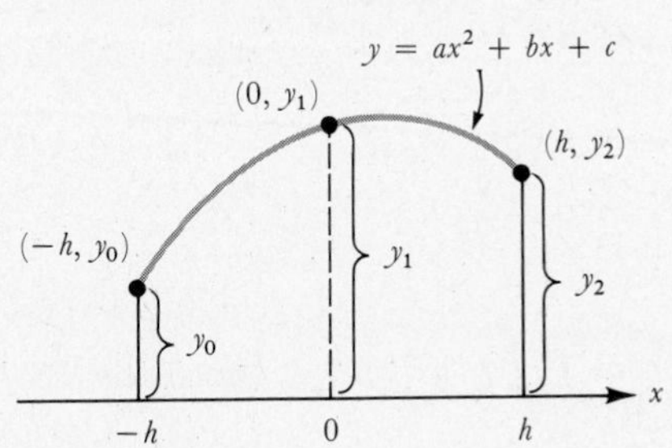

Fig. 11.6

SIMPSON'S RULE

We shall improve on the trapezoidal approximation of a definite integral by replacing the straight-line top of the trapezoid with a parabolic curve that fits the graph of the integrand more accurately. Other types of curves could be—and sometimes are—used, leading to different types of approximations, each of which has its particular advantages and disadvantages. Some of these are explored briefly in the section exercises. Simpson's Rule is based upon the parabolic approximation and has the advantages of being both simple and amenable to machine calculation.

The formula for the area of the region under a typical parabolic curve $y = ax^2 + bx + c$ over the interval $[-h, h]$, labeled as in Fig. 11.6, is

$$A = \frac{2ah^3}{3} + 2ch \qquad \text{(see Exercise 24)}$$

We can rewrite A in terms of the heights y_0, y_1, and y_2 as

11.6.3 $$A = \frac{h}{3}[y_0 + 4y_1 + y_2] \qquad \text{(see Exercise 25)}$$

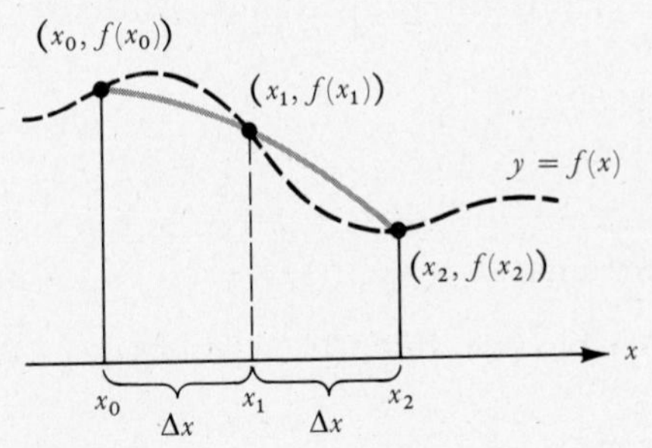

Fig. 11.7

Now, using the fact that three noncollinear points determine a unique parabola, we suppose that $(x_0, f(x_0))$, $(x_1, f(x_1))$, $(x_2, f(x_2))$ are three points on the graph of a continuous function as in Fig. 11.7. By (11.6.3), we know that

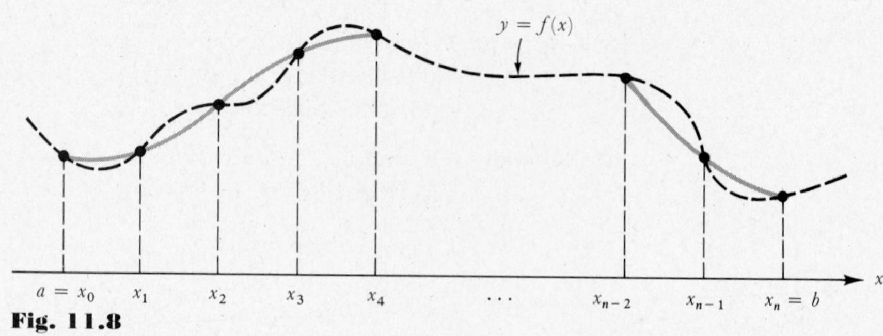

Fig. 11.8

11.6.4
$$A = \frac{\Delta x}{3}[f(x_0) + 4f(x_1) + f(x_2)]$$

We can now use (11.6.4) to approximate $\int_a^b f(x)\,dx$. First we break the interval $[a, b]$ into an even number n of subintervals, each of length $\Delta x = (b - a)/n$. An even number is required because each parabolic arc covers two adjacent intervals. The $n/2$ parabolic arcs cover double intervals, as indicated in Fig. 11.8. We now simply add the areas under the arcs to approximate $\int_a^b f(x)\,dx$:

$$\frac{\Delta x}{3}[f(x_0) + 4f(x_1) + f(x_2)] + \frac{\Delta x}{3}[f(x_2) + 4f(x_3) + f(x_4)] + \cdots + \frac{\Delta x}{3}[f(x_{n-2}) + 4f(x_{n-1}) + f(x_n)]$$

Except for the endpoints, each "even point" appears twice, so, noting that $\Delta x = (b - a)/n$, we recombine the above sum to obtain the following:

11.6.5 **Simpson's Rule** For each even positive integer n,

$$\int_a^b f(x)\,dx \approx \frac{b-a}{3n}[f(a) + 4f(x_1) + 2f(x_2) + \cdots + 2f(x_{n-2}) + 4f(x_{n-1}) + f(b)$$

where, for each $0 \le i \le n$,

$$x_i = a + i\left(\frac{b-a}{n}\right)$$

We call this the nth *parabolic approximation* of $\int_a^b f(x)\,dx$ and denote it by S_n.

A simple way to remember Simpson's Rule is as

$$\frac{b-a}{3n}[(\text{ends}) + 4(\text{odds}) + 2(\text{evens})]$$

We will use Simpson's Rule on the same integral as in Example 1, whose precise value we know to be π.

$x_0 = 0.0$	$f(0) = 4.000000$
$x_1 = 0.1$	$4f(0.1) \approx 15.841584$
$x_2 = 0.2$	$2f(0.2) \approx 7.692308$
$x_3 = 0.3$	$4f(0.3) \approx 14.678899$
$x_4 = 0.4$	$2f(0.4) \approx 6.896552$
$x_5 = 0.5$	$4f(0.5) = 12.800000$
$x_6 = 0.6$	$2f(0.6) \approx 5.882353$
$x_7 = 0.7$	$4f(0.7) \approx 10.738255$
$x_8 = 0.8$	$2f(0.8) \approx 4.878049$
$x_9 = 0.9$	$4f(0.9) \approx 8.839780$
$x_{10} = 1.0$	$f(1) = 2.000000$

Table 11.2

Example 4 Use Simpson's Rule with $n = 10$ to approximate $\int_0^1 (4\,dx)/(1 + x^2)$.

Solution Again, we arrange our calculations in a table (see Table 11.2). Adding the numbers of the right-hand column, we get

$$\int_0^1 \frac{4\,dx}{1 + x^2} \approx \frac{1}{30}(94.247787) = 3.141592667 \quad \bullet$$

Since $|\pi - 3.141592667| \approx 0.000000187$ (rounded off in the last place), this estimate is accurate through six decimal places.

The following formula provides bounds on the error in Simpson's Rule approximations. We omit its proof.

11.6.6 **Simpson's Rule Error** Suppose $f^{(4)}$ exists and is continuous on $[a, b]$ with $|f^{(4)}(x)| \leq M$ on $[a, b]$. Then

$$\left| \int_a^b f(x)\,dx - S_n \right| \leq \frac{(b-a)^5 M}{180n^4}$$

Example 5 Determine an upper bound on the error in Example 4.

Solution We need $f^{(4)}(x)$, where $f(x) = 4/(1 + x^2)$. In computing a bound for the trapezoid-approximation error (Example 2), we found

$$f'''(x) = \frac{96x(1 - x^2)}{(1 + x^2)^4}$$

so

$$f^{(4)}(x) = \frac{96(1 - 10x^2 + 5x^4)}{(1 + x^2)^5}$$

A maximum for $f^{(4)}(x)$ on $[0, 1]$ occurs at $x = 0$ with $f^{(4)}(0) = 96 = M$. (The proof requires using $f^{(5)}(x)$; see Exercise 26.)

Since $b - a = 1$, $n = 10$, we have

$$\left| \int_0^1 \frac{4\,dx}{1 + x^2} - 3.141592667 \right| \leq \frac{(1)^5 \cdot 96}{180 \cdot 10^4} = \frac{8}{15} 10^{-4} \approx 0.0000533 \quad \bullet$$

Notice that (rounded off in the last digit) the possible error for 10 subdivisions with the Trapezoid Rule was 0.0017 compared to a possible error of only 0.0000533 for Simpson's Rule.

Notice that the error bound tells us that Simpson's Rule gives the exact value of the definite integral when $f(x)$ is a polynomial of degree ≤ 3. In such cases $f^{(4)} = 0$, so the error is 0.

PROGRESS TEST 2

(We parallel the problems of Progress Test 1 so that the accuracy of the two approaches can be compared and you can use some calculations done earlier.)

1. Using $\ln x = \int_1^x \frac{dt}{t}$, carrying calculations through five decimals:
(*a*) Estimate ln 2 using Simpson's Rule with $n = 4$.
(*b*) Determine an upper bound for the error of this estimate.
(*c*) Compare the error bound with the actual error.

2. (*a*) What value of n is necessary to provide accuracy through five decimal places?
(*b*) (For calculators) What value of n is necessary to provide accuacy through 15 places (thereby exceeding the accuracy of most hand-held calculators)?

3. (*a*) Estimate $\int_0^2 e^{-x^2}\,dx$ using Simpson's Rule with $n = 10$.
(*b*) Determine an upper bound for the error of this estimate, using that $|f^{(4)}(x)| < 12$ on $[0, 2]$.
(*c*) Compare with the corresponding result for the trapezoid approximation.

SECTION 11.6 EXERCISES

In Exercises 1 to 10, (a) estimate the definite integral using the Trapezoid Rule with n = 4, (b) determine an upper bound for the approximation error, (c) estimate the integral using Simpson's Rule with n = 4, (d) determine an upper bound for the approximation error, and (e) compare the results of (b) and (d) with the actual error.

1. $\int_0^2 3x^2\,dx$ **2.** $\int_0^1 6x^5\,dx$ **3.** $\int_0^{\pi} \sin x\,dx$ **4.** $\int_1^3 \frac{dx}{1+x}$

5. $\int_0^1 e^x\,dx$ **6.** $\int_0^{\pi/2} \cos x\,dx$ **7.** $\int_0^{20} (2x-1)\,dx$ **8.** $\int_0^{20} (4x+1)\,dx$

9. $\int_0^1 (5x^4-4)\,dx$ **10.** $\int_0^1 (x^4-4)\,dx$

For Exercises 11 to 19 follow instructions (a) to (d).

11. $\int_0^1 \sqrt{1+x^4}\,dx$ **12.** $\int_0^{\pi} \sqrt{1+\cos^2 x}\,dx$ **13.** $\int_0^1 \sqrt{1+e^x}\,dx$ **14.** $\int_0^{\pi/2} \sqrt{\sin x}\,dx$

15. $\int_1^2 \frac{\sin x}{x}\,dx$ **16.** $\int_0^1 \sqrt{1+x^3}\,dx$ **17.** $\int_0^1 \frac{2^x}{x+1}\,dx$ **18.** $\int_0^{\pi/2} \sqrt{1-2\sin^2 x}\,dx$

19. $\int_0^2 e^{x^2}\,dx$

20. How many subdivisions are necessary to determine $\int_0^1 \sqrt{\sin x}\,dx$ through four decimal places using (*a*) the Trapezoid Rule and (*b*) Simpson's Rule? (See Exercise 14.)

21. Repeat Exercise 20 for $\int_0^1 \sqrt{1+e^x}\,dx$ (See Exercise 13.)

22. Show that the Trapezoid Rule provides the average of the Riemann sums based, respectively, on left and right endpoints of $[x_{i-1}, x_i]$, associated with the definite integral of a continuous function.

23. How many subdivisions are needed to approximate $\ln(3/2)$ through six decimal places using (*a*) the Trapezoid Rule and (*b*) Simpson's Rule? (*Hint:* $\ln(3/2) = \ln 3 - \ln 2$.)

24. Prove that the area under $y = ax^2 + bx + c$ from $x = -h$ to $x = h$ is $2ah^3/3 + 2ch$. (See Fig. 11.6.)

25. Prove (11.6.3) as follows:

(*a*) Express y_0, y_1, y_2 in terms of h.

(*b*) Use the equations in (*a*) to show that $2ah^2 = y_0 + y_2 - 2y_1$.

(*c*) Use (*b*) to complete the proof.

26. Complete Example 5 by showing that $f^{(4)}(x)$ has a relative minimum at $x = 1/\sqrt{3}$ and then comparing $|f^{(4)}(1/\sqrt{3})|$, $|f^{(4)}(0)|$, and $|f^{(4)}(1)|$.

27. Suppose a motorist noted the following speeds in miles per hour at 5-min intervals of a 1-hr trip: 0, 30, 40, 45, 55, 25, 35, 0, 25, 55, 60, 50, 0. Approximate the total distance traveled using the Trapezoid Rule. (Assume that time is measured in hours, with $x_0 = 0$, $x_1 = 1/12$, $\ldots, x_{12} = 12/12 = 1$ and the total distance is approximately $\int_0^1 f(x)\,dx$ when $f(x)$ is the velocity at time x; here we only know $f(x)$ at the given values.)

28. Using (*a*) the Trapezoid Rule and (*b*) Simpson's Rule, approximate the total area of a cornfield whose dimensions in meters are listed in the accompanying sketch at 10-m intervals.

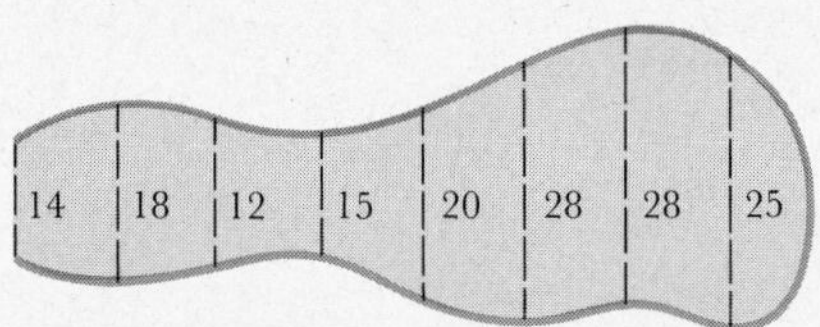

29. Using the Trapezoid Rule, determine the approximate volume of the solid generated by revolving the figure in Exercise 28 about its left end.

30. Suppose a cyclist were clocked at 10-min intervals with the nine speeds resulting corresponding to the numbers in Exercise 28. (*Note:* The ninth is implied to be 0.) Use the Trapezoid Rule to approximate the total distance traveled over that span.

31. Repeat Exercise 30, but instead compute the average velocity over the measured interval.

CHAPTER EXERCISES

In Exercises 1 to 111 evaluate the given integral. Consult with your instructor on whether or not to use tables. These integrals are not graded with respect to difficulty.

1. $\int x\sqrt{1-x^2}\,dx$ **2.** $\int \cot x \csc^3 x\,dx$ **3.** $\int_0^1 x^5 e^{x^2}\,dx$

4. $\int \frac{dx}{\sqrt{x^2-4x+13}}$ **5.** $\int \frac{\sqrt{x^2-5}}{x}\,dx$ **6.** $\int \frac{dx}{(2-3x^2)^{3/2}}$

7. $\int_0^1 x\sqrt{1+x}\,dx$ **8.** $\int \frac{x^3+9}{x^2+x}\,dx$ **9.** $\int \frac{dx}{(x^2+9)^2}$

10. $\int_0^{\pi} \sin 5x \sin x\,dx$ **11.** $\int x \cot x \csc x\,dx$ **12.** $\int e^{-2x} \sin 2x\,dx$

13. $\int \frac{x^3-1}{x^3+x}\,dx$ **14.** $\int \frac{dx}{\sqrt{1-\sin x}}$ **15.** $\int (\arctan x)^2\,dx$

16. $\int \frac{dx}{x^3\sqrt{x^2-9}}$ **17.** $\int \sqrt{\frac{1+x}{1-x}}\,dx$ **18.** $\int \sqrt{2-\sin^2 x}\cos^3 x\,dx$

19. $\int (\tan x + \cot x)\,dx$ **20.** $\int (x \ln x)^2\,dx$ **21.** $\int x\sqrt{3x+1}\,dx$

22. $\int \frac{dx}{2\sin x + 2\cos x + 5}$ **23.** $\int \frac{dx}{1+\cos x}$ **24.** $\int \frac{dx}{\tan x - \sin x}$

25. $\int_0^1 \frac{x+1}{x^2+1}\,dx$ **26.** $\int \sqrt{\tan x}\sec x\,dx$ **27.** $\int x \sin^2 4x\,dx$

28. $\int \frac{x^2\,dx}{\sqrt{2x-x^2}}$ **29.** $\int_1^{\sqrt{2}} (x-\sqrt{x^2-1})\,dx$ **30.** $\int_0^{\pi/8} \frac{\sin^4 x}{\cos^6 x}\,dx$

31. $\int x^2 \sinh x\,dx$ **32.** $\int_1^e (\ln x)^3\,dx$ **33.** $\int \frac{1+\sqrt{x}}{1-\sqrt{x}}\,dx$

34. $\int \frac{\sin x \cos x}{\sin^2 x - \cos^2 x}\,dx$ **35.** $\int \frac{3x^2-2x+1}{x^3-x^2-12x}$ **36.** $\int x \operatorname{arctanh} x\,dx$

37. $\int_0^1 \frac{dx}{(1+x^2)^2}$ **38.** $\int \frac{dx}{x^2+4}$ **39.** $\int (x^2+2)2^x\,dx$

40. $\int_0^1 x^3\sqrt{16-x^2}\,dx$ **41.** $\int \frac{x^5+x^3}{\sqrt{x^2+9}}\,dx$ **42.** $\int_0^{\pi/b} e^{ax}\sin bx\,dx$

43. $\int \frac{(x-2)\,dx}{(x-1)^2(x+1)}$ **44.** $\int \frac{(x^2+x+1)\,dx}{(x^2+x-2)(x^2+3x+2)}$ **45.** $\int (\tan x \sec x)^3\,dx$

46. $\int \frac{dx}{x^2\sqrt{5-x^2}}$ **47.** $\int \sqrt[3]{e^x+1}\,dx$ **48.** $\int \frac{dx}{(x^2+1)(x^2+x+1)}$

49. $\int \frac{\tan x\,dx}{\tan x + \sec x}$ **50.** $\int x \csc x \cot x\,dx$ **51.** $\int \frac{x^3\,dx}{\sqrt{3x^2+7}}$

52. $\int_0^a \sqrt{x^2+a^2}\,dx$ **53.** $\int_0^{\pi/b} e^{ax}\cos bx\,dx$ **54.** $\int_2^9 \frac{(1+x)\,dx}{(x-1)^{2/3}-4}$

55. $\int \frac{(3x^2-1)\,dx}{(x+1)^2(x-2)}$ **56.** $\int \frac{dx}{(x^2+x-2)(x^2+5x-14)}$ **57.** $\int \frac{x\ln x}{\sqrt{x^2-9}}\,dx$

58. $\int_2^4 \frac{x-2}{x^2-x}\,dx$ **59.** $\int \frac{\sqrt[3]{x+1}}{x}\,dx$ **60.** $\int \frac{dx}{x(x^2+3x+4)}$

61. $\int \frac{dx}{2+\sqrt{x+3}}$ **62.** $\int \frac{x^2+1}{(x^2-1)^2}\,dx$ **63.** $\int \frac{dx}{x^2\sqrt{x^2+4}}$

64. $\int \frac{x^2+5}{x^3-x^2}\,dx$ **65.** $\int_{-\pi/4}^{\pi/4} \sec^5 x\,dx$ **66.** $\int_0^a \sqrt{a^2-x^2}\,dx$

67. $\int_a^{2a} \sqrt{x^2-a^2}\,dx$ **68.** $\int x^3\sqrt{2x^2-5}\,dx$ **69.** $\int_1^2 \frac{\sqrt{5+x^2}}{x^3}\,dx$

70. $\int \frac{3x^3+x^2+1}{x^2-5x+4}\,dx$ **71.** $\int \frac{(3x-2)\,dx}{(x-1)(x^2+6x+8)}$ **72.** $\int_0^{\pi/2} x^2\cos x\,dx$

73. $\int_0^{\pi/2} \frac{\sin x\,dx}{(1+\cos x)^2}$ **74.** $\int (\arcsin x)^2\,dx$ **75.** $\int x \tan x \sec x\,dx$

76. $\int \frac{x\,dx}{(x-1)^2(x^2+1)}$

77. $\int \frac{dx}{(x^2+1)^5}$

78. $\int \frac{\sqrt{x^2+9}}{x}\,dx$

79. $\int \frac{dx}{\sqrt{x-1}-(x-1)^{3/2}}$

80. $\int \frac{dx}{(4x^2-24x+27)^{3/2}}$

81. $\int \sin 2x\,\sqrt{\cos^3 x}\,dx$

82. $\int \frac{(2x+3)\,dx}{(x^2+6x+9)^4}$

83. $\int \sqrt{x}\ln x\,dx$

84. $\int \frac{(4-9x^2)^{3/2}}{x^4}\,dx$

85. $\int (4-9x^2)^{3/2}x^4\,dx$

86. $\int \frac{x^3\,dx}{\sqrt{9-x^2}}$

87. $\int \frac{x\,dx}{x^4-1}$

88. $\int x^5\sqrt{x^2-1}\,dx$

89. $\int xe^{x^2}\cos x^2\,dx$

90. $\int \frac{x^2+1}{\sqrt{x^3-1}}\,dx$

91. $\int \frac{dx}{1+\sin x}$

92. $\int \sin^3 x\cos^2 x\,dx$

93. $\int e^{2x}\sin 3x\,dx$

94. $\int \frac{dx}{\sqrt[3]{e^{3x}+1}}$

95. $\int \frac{dx}{\sqrt[3]{1+\sqrt[3]{x}}}$

96. $\int \frac{dx}{1-\cos x+\sin x}$

97. $\int \frac{x}{\sqrt{1+9x}}\,dx$

98. $\int e^{\sqrt[3]{x}}x^{-2/3}\,dx$

99. $\int \frac{dx}{x^2+4x+7}$

100. $\int \cos x\cos 3x\cos 5x\,dx$

101. $\int \frac{\sqrt{4-9x^2}}{x}\,dx$

102. $\int \frac{dx}{\sqrt{(4-x^2)^5}}$

103. $\int x^2 \operatorname{arcsinh} x\,dx$

104. $\int x^2\arctan x^2\,dx$

105. $\int e^{\arcsin x}\,dx$

106. $\int e^{2x+e^x}\,dx$

107. $\int \left(\frac{2}{x}-\frac{1}{x^2}\right)e^{2x}\,dx$

108. $\int \frac{1-x}{1+\sqrt{x}}\,dx$

109. $\int \sinh 3x\cosh 5x\,dx$

110. $\int_0^{\pi/2} e^{-x}\cos x\,dx$

111. $\int \ln(\cos x)\sin x\,dx$

In Exercises 112 to 115, (a) estimate the definite integral using the Trapezoid Rule with $n = 4$, (b) determine an upper bound for the approximation error, (c) estimate the integral using Simpson's Rule with $n = 4$, (d) determine an upper bound for the approximation error, and (e) compare the results of (b) and (d) with the actual error.

112. $\int_0^1 2x\sqrt{x^2+1}\,dx$

113. $\int_0^2 4x^3\,dx$

114. $\int_0^1 e^{-x}\,dx$

115. $\int_0^1 e^{2x}\,dx$

For Exercises 116 to 121 follow instructions (a) to (d) appearing before Exercise 112.

116. $\int_0^2 \frac{1-e^{-x}}{x}\,dx$

117. $\int_1^2 \frac{e^x}{x}\,dx$

118. $\int_0^{\pi} \sqrt{x}\sin x\,dx$

119. $\int_0^2 \frac{dx}{\sqrt{1+x^3}}$

120. $\int_0^{\pi} \sqrt{1+\sin^2 x}\,dx$

121. $\int_1^2 \frac{\cos x}{x}\,dx$

122. (Midpoint Rule) For f continuous on $[a, b]$, let $\overline{x}_i = (x_{i-1} + x_i)/2$ where we have a division of $[a, b]$ into n equal-length subintervals and n is an even number. By the Midpoint Rule, we mean the approximation of $\int_a^b f(x)\,dx$ by

$$\frac{b-a}{n}[f(\overline{x}_1) + f(\overline{x}_2) + \cdots + f(\overline{x}_n)]$$

(*a*) Interpret the Midpoint Rule graphically.
(*b*) Use this rule to approximate ln 2 with $n = 10$ and compare with the earlier results of applying our other two rules.
(*c*) Repeat (*b*) to approximate $\int_0^1 (4\,dx)/(1 + x^2)$ with $n = 10$.

123. (*a*) How can we use the previous calculations of $f(\overline{x}_i)$ to refine a Midpoint Rule approximation when we double the number of subdivisions from n to $2n$?
(*b*) Compare or contrast this situation with that for, first, the Trapezoid Rule and, second, Simpson's Rule.
(*c*) What if we use the average of the highest and lowest values of $f(x)$ on each subinterval $[x_{i-1}, x_i]$?

124. (For calculators) Estimate $\int_1^2 x^x\,dx$ using Simpson's Rule, with $n = 10$.

125. Determine the volume of the solid generated by revolving about the y axis the region bounded by the x and y axes, $x = 1$, and $y = e^{-x}$.

126. Repeat Exercise 125 for the region bounded by the x and y axes, $y = \cos x$, and $x = \pi/2$.

127. Determine the present value of a continuous income stream expected to grow linearly from \$1000 to \$6000/year over the next 10 years, assuming that the prevailing rate of return on money during this period is 8 percent compounded continuously.

128. Repeat Exercise 127 but with the income *dropping* linearly from \$6000 to \$1000 over the next 10 years.

129. Evaluate $\int \ln(x + \sqrt{x^2 - 1})\,dx$ using a hyperbolic substitution.

130. Repeat Exercise 129 for $\int \ln(x + \sqrt{x^2 + 1})\,dx$.

131. Show that area of the sector of a circle of radius r subtended by an angle of θ radians is $(r^2/2)\,\theta$. [*Hint:* Base the sector on the positive x axis, break the region into a (right) triangular piece and a second piece having the circular part of the region as part of its boundary.]

132. Use integration by parts to derive the reduction formula (top, next column)

$$\int \frac{du}{(u^2 + a^2)^n} = \frac{u}{2a^2(n-1)(u^2 + a^2)^{n-1}} + \frac{2n-3}{2a^2(n-1)} \int \frac{du}{(u^2 + a^2)^{n-1}}$$

133. Use Exercise 132 to evaluate $\int \frac{dx}{(x^2 + 2x + 5)^3}$.

134. Derive, for $bc \neq ad$,

$$\int \frac{x\,dx}{(ax + b)(cx + d)} = \frac{1}{bc - ad}\left[\frac{b}{a}\ln(ax + b) - \frac{d}{c}\ln(cx + d)\right].$$

135. Apply the formula in Exercise 134 to evaluate

$$\int \frac{x\,dx}{6x^2 - 7x - 3}.$$

136. Which *one* of the following is elementary?

(*a*) $\int \frac{\ln x\,dx}{x + 1}$ (*b*) $\int \frac{dx}{\ln x}$ (*c*) $\int \frac{dx}{x\sqrt{\ln \sqrt{x}}}$

SELF-TEST ✤

In Probs. 1 to 7 evaluate the given integrals.

1. $\int x \sec^2 x\,dx$

2. $\int x^2 \sqrt{1 - x}\,dx$

3. $\int_0^1 x^3 \sqrt{1 - x^2}\,dx$

4. $\int_0^1 \frac{x^2 - 2}{\sqrt[3]{x^3 - 6x + 1}}\,dx$

5. $\int \frac{x^3 + x^2 + x + 1}{(x^2 - 1)^2(x^2 + 5x + 3)}\,dx$

6. $\int \frac{dx}{x\sqrt{4x^2 + 1}}$

7. $\int e^{x + e^x}\,dx$

8. Use partial fractions to determine $\int \frac{\cos x\,dx}{\sin^2 x + 5\sin x + 4}$.

9. Evaluate $\int \frac{dx}{\sqrt{x} + \sqrt[3]{x}}$

10. Using (*a*) the Trapezoid Rule and then (*b*) Simpson's Rule, approximate $\int_0^2 dx/\sqrt{1 + x^3}$ accurate within 0.005.

11. How many subdivisions are necessary to provide an approximation of $\int_0^1 e^{x^2}\,dx$ accurate through six decimal places using the Trapezoid Rule?

Chapter 12

PARAMETRIC EQUATIONS AND MORE APPLICATIONS OF INTEGRATION

CONTENTS

The same general idea that made the definite integral such a useful means for defining and computing areas, averages, and volumes extends to include other applications as well. These include formulas for computing quantities reflected in the section titles. To make our coverage of lengths of curves and areas of surfaces generated by revolving these curves about axes as inclusive as possible, we introduce a new way of describing curves in Sec. 12.1.

We express the variables x and y both as functions of a third variable called the "parameter," say t. The two equations $x = x(t)$, $y = y(t)$ are called "parametric equations" and will prove to be especially useful in later chapters as an efficient means of describing motion in terms of time.

As with virtually all the applications in this text, the applications are somewhat streamlined, because the basic idea is to learn *how* to use the integral calculus to solve specific problems and gain insight into situations. It is important to learn *when* a definite integral should be used in attacking a problem, and then to construct the appropriate definite integral when it is required.

Section 12.1: Parametric Equations

Objectives:

1. Sketch the graph of a pair of parametric equations either directly by plotting points or by eliminating the parameter.

2. Determine the slope and equation of a line tangent to the graph of a pair of parametric equations.

GRAPHING PARAMETRIC EQUATIONS

Thus far we have considered curves in the plane that arise as graphs of functions of the form $y = f(x)$, $x = g(y)$, or as graphs of equations of the form $F(x,y) = 0$. It is sometimes useful to regard a curve as consisting of points (x,y), where x and y are both functions of a third variable called the parameter. More formally:

12.1.1 Definition If x and y are both functions of the same independent variable t, and if D is a set of real numbers that is common to the domains of both functions, then the equations

$$x = x(t), y = y(t) \qquad t \text{ in } D$$

are called *parametric equations* and t is called the *parameter*. By the *graph* of a pair of parametric equations we mean the collection of all points $(x(t), y(t))$ as t varies over D.

If a pair of parametric equations is given without specifying D, then we shall assume that D is the intersection of the natural domains of $x(t)$ and $y(t)$.

Example 1 Graph the parametric equations $x = \sqrt{t}$, $y = t^{3/2} + 1$.

Solution The intersection of the two domains is all $t \geq 0$. We construct a table showing the points corresponding to a few values of t.

t	0	1/4	1	16/9	2	9/4	4
$x(t)$	0	1/2	1	4/3	$\sqrt{2} \approx 1.41$	3/2	2
$y(t)$	1	9/8	2	91/27	$(\sqrt{2})^3 + 1 \approx 3.83$	35/8	9

Plotting the points $(x(t), y(t))$ and noting that, as $t \to +\infty$, $x(t)$ and $y(t)$ both $\to +\infty$, we sketch the graph in Fig. 12.1. ●

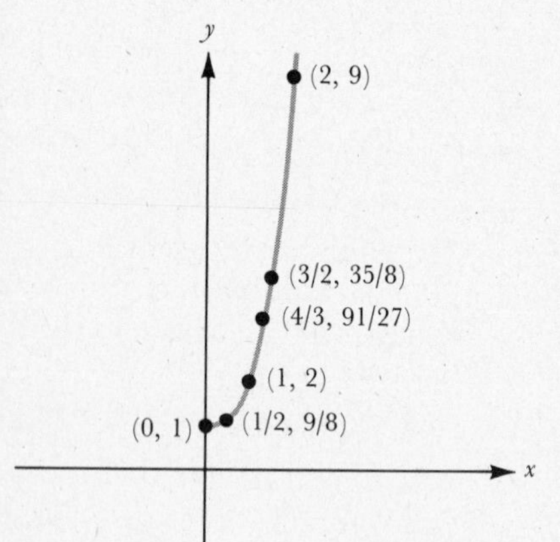

Fig. 12.1

Sometimes, but not always, we can sketch parametric equations by *eliminating the parameter*. In Example 1, if we solve $x = \sqrt{t}$ for t, we get $t = x^2$. We can then substitute $t = x^2$ in the expression for y to get

$$y = t^{3/2} + 1 = (x^2)^{3/2} + 1 = x^3 + 1$$

We must be careful when eliminating the parameter to keep in mind the possible values for x and y as the parameter ranges through its domain. In this particular case, $x = \sqrt{t}$ implies $x \geq 0$, so although we eliminated the parameter to get $y = x^3 + 1$, the graph of the original parametric equations is only that part of the graph of $y = x^3 + 1$ for which $x \geq 0$.

In the special case where $x = x(t) = t$, the parameter amounts merely to another letter used in place of x, so $y = y(t)$ is equivalent to giving y as the same function of x.

Example 2 Compare the graphs of the following three pairs of parametric equations:

(*a*) $x = 2t, y = 3t,$ (*b*) $x = 2t^2, y = 3t^2,$ (*c*) $x = 2 \cos t, y = 3 \cos t.$

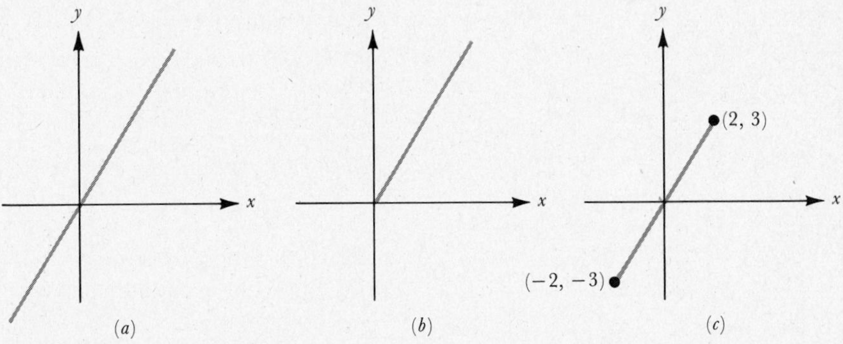

Fig. 12.2

Solution The domain in all three pairs is R. Eliminating the parameter in (a) gives $t = x/2$, so $y = 3(x/2) = (3/2)x$. Since t is any real number, the graph of (a) is the entire straight line $y = (3/2)x$.

In (b), $x \geq 0, y \geq 0$. To eliminate the parameter in (b) we solve for t^2, to get $t^2 = x/2$. Then $y = 3(x/2) = (3/2)x$. Again the graph is a straight line, but this time only $x \geq 0$ is part of the graph.

In (c), we know $-1 \leq \cos t \leq 1$, so $2(-1) \leq x \leq 2(1)$ and $3(-1) \leq y \leq 3(1)$. Here we can eliminate the parameter by solving for $\cos t$; that is, $\cos t = x/2$ and $y = 3(x/2)$ again.

The graphs of (a) to (c) are given in Fig. 12.2. ●

Another way to eliminate the parameter is to exploit some relationship between the functions for x and y.

Example 3 Identify and sketch the graph of $x = 2\cos t, y = 3\sin t, 0 \leq t \leq 2\pi$.

Solution Here it is not particularly convenient to solve one of the equations for t or an expression involving t and then substitute directly into the other. It is easier to use that $\cos^2 t + \sin^2 t = 1$. We solve the two given equations for $\cos t$ and $\sin t$, respectively, getting

$$\cos t = \frac{x}{2} \qquad \text{and} \qquad \sin t = \frac{y}{3}$$

We can then conclude that

$$\left(\frac{x}{2}\right)^2 + \left(\frac{y}{3}\right)^2 = 1 \qquad \text{or} \qquad \frac{x^2}{4} + \frac{y^2}{9} = 1$$

The graph of this latter equation is an ellipse, so it remains to determine which part of this ellipse is included in the graph of the original parametric equations. To do this, we monitor what happens to the points $(2\cos t, 3\sin t)$ as t increases through its domain. As t increases from 0 to $\pi/2$, x steadily decreases from 2 to 0 and y steadily increases from 0 to 3. We tabulate this first-quadrant data as follows:

t	$0 \to \pi/2$
$x(t)$	$2 \to 0$
$y(t)$	$0 \to 3$

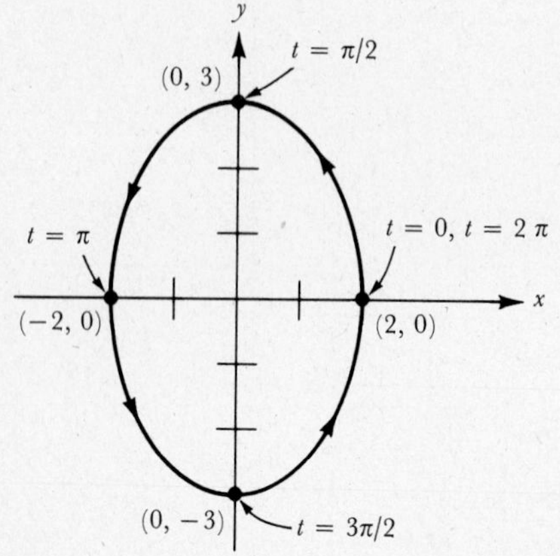

Fig. 12.3

A similar analysis on $[\pi/2, \pi]$, $[\pi, 3\pi/2]$ and $[3\pi/2, 2\pi]$ yields

t	$0 \to \pi/2$	$\pi/2 \to \pi$	$\pi \to 3\pi/2$	$3\pi/2 \to 2\pi$
$x(t)$	$2 \to 0$	$0 \to -2$	$-2 \to 0$	$0 \to 2$
$y(t)$	$0 \to 3$	$3 \to 0$	$0 \to -3$	$-3 \to 0$

Thus, as t increases from 0 to 2π, the points $(2 \cos t, 3 \sin t)$ trace out the entire ellipse in a counterclockwise direction, reaching $(0, 3)$ when $t = \pi/2$, $(-2, 0)$ when $t = \pi$, $(0, -3)$ when $t = 3\pi/2$, and finally returning to $(2, 0)$ when $t = 2\pi$. (See Fig. 12.3.) ●

It should be realized that the foregoing example generalizes.

12.1.2 For $0 \leq t \leq 2\pi$, the graph of $x(t) = a \cos t, y(t) = b \sin t$ is an ellipse centered at the origin. If $a = b$, then the graph is a circle of radius a.

This last example also suggests that we can use parametric equations to describe motion in the plane. This is indeed the case, and we shall return to explore this idea in much more depth when we study vectors in Chaps. 17 and 18.

Lest our eagerness to eliminate parameters in the above graphing examples persuade you that the important thing about parameters is to get rid of them, we now present an application that shows how parameters are useful in describing curves that would be difficult to describe using x and y alone. Furthermore, the parametric equations resulting in this example provide a good illustration of a case in which the elimination of parameters is quite difficult—and not especially valuable. (See Exercise 31.)

Example 4 Determine a pair of parametric equations whose graph is the curve traced out by a fixed point P on the circumference of a circle of radius a as the circle rolls without slipping along the x axis.

Solution We assume that the point P starts at the origin and then moves clockwise as the circle rolls to the right. (See Fig. 12.4.)

Let (x, y) be the coordinates of P and let θ be the angle $\measuredangle ACP$ measured in radians. Now

$$x = \text{length } OB = \text{length } OA - \text{length } PD$$

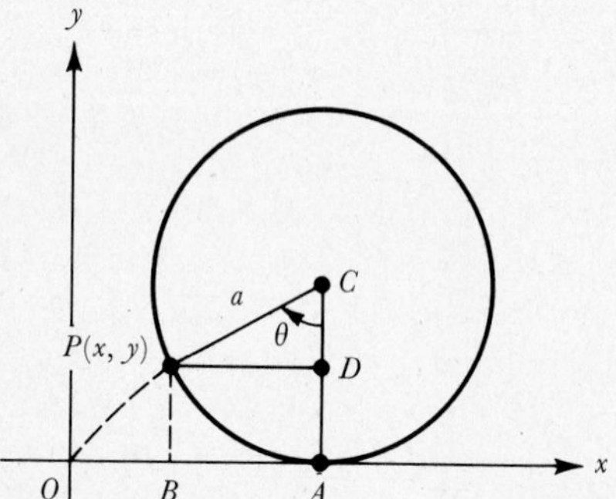

Fig. 12.4

Since there is no slippage, the length OA is the same as the length of the arc $\widehat{PA}$ of the circle. (Think about this!) Now the arc $\widehat{PA}$ has length $a\theta$ and

$$\text{length } PD = (\text{length } PC)\sin\theta = a \sin\theta$$

so

$$x = a\theta - a\sin\theta = a(\theta - \sin\theta)$$

Also

$$\begin{aligned} y &= \text{length } PB \\ &= \text{length } CA - \text{length } CD \\ &= a - a\cos\theta = a(1 - \cos\theta) \end{aligned}$$

Thus the curve has parametric equations

12.1.3 **Cycloid** $x = a(\theta - \sin\theta) \qquad y = a(1 - \cos\theta)$ ●

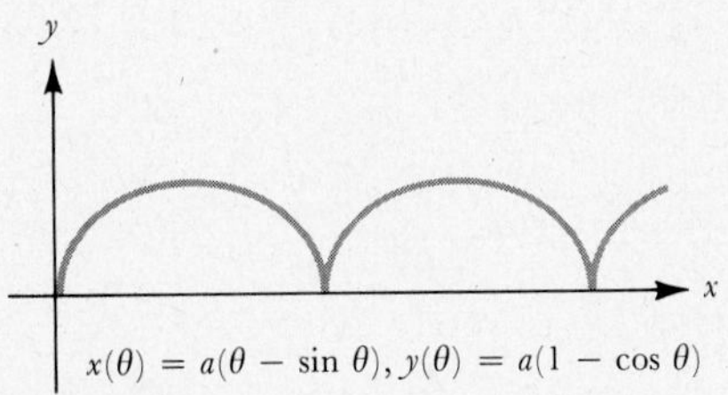

Fig. 12.5

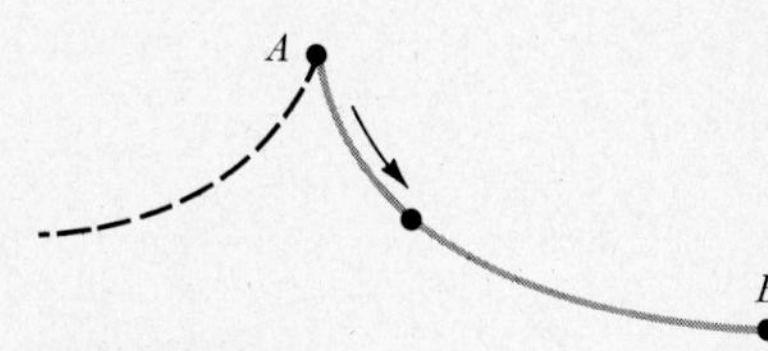

Fig. 12.6

A curve given by (12.1.3), called a *cycloid*, is sketched in Fig. 12.5. Cycloids have fascinating properties. Suppose a frictionless bead is to slide down a wire from A to B, as in Fig. 12.6. What shape wire allows the quickest slide? Not a straight wire, but a wire in the shape of an inverted cycloid with A in the position of the "point" of the cycloid. Furthermore, if B is at the lowest point of the cycloid, then the length of time required for the slide from A to B is the same as the length of time for the slide *from any intermediate point* to B!

Other types of cycloids can be generated by following the path of a point *inside* a rolling wheel or the path of a point outside the rolling circumference of the wheel (as on the flange of a railroad train's wheel).

PROGRESS TEST 1

Graph the given parametric equations.

1. $x = \sqrt{t}, y = t - 4$

2. $x = 2 \sec t, y = 4 \tan^2 t, 0 \leq t \leq \pi$

3. $x = t - 1, y = 2t + 3$

TANGENT LINES AND DERIVATIVES

In Example 2 we saw by eliminating the parameter that the graph of $x = 2t$, $y = 3t$ is the straight line $y = (3/2)x$—a line through the origin with slope 3/2. The slope of this line could be reckoned from the parametric equations themselves. A given change Δt in t induces the change $2(t + \Delta t) - 2t = 2\,\Delta t$ in x and $3(t + \Delta t) - 3t = 3\,\Delta t$ in y. Hence the slope of the line (the y change divided by the x change) is $(3\,\Delta t)/(2\,\Delta t) = 3/2$ *no matter what t is*. Even if we did not know that the graph is a straight line, this last observation that the slope is the same for all t would guarantee it.

Suppose now we have a pair of parametric equations (not necessarily linear)

$$x = x(t), y = y(t) \qquad t \text{ in } I$$

where I is an interval.

We shall now compute the slope m_0 of the line tangent to the graph of these equations at t in I (provided this tangent exists). We proceed as we did in Chap. 2; that is, we examine the slope of the line through $P = (x(t), y(t))$ and $Q = (x(t + \Delta t), y(t + \Delta t))$ as $\Delta t \to 0$. (See Fig. 12.7.)

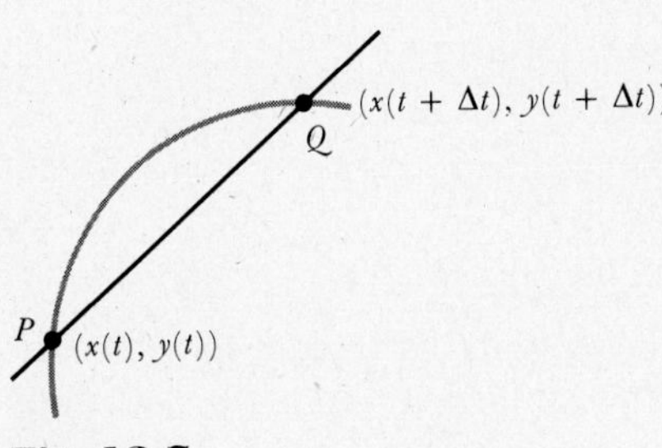

Fig. 12.7

Provided $x(t + \Delta t) - x(t) \neq 0$ for small $|\Delta t|$, the slope of this line is

$$\frac{y(t + \Delta t) - y(t)}{x(t + \Delta t) - x(t)}$$

which we can rewrite as

$$\frac{[y(t + \Delta t) - y(t)]/\Delta t}{[x(t + \Delta t) - x(t)]/\Delta t} = \frac{\Delta y/\Delta t}{\Delta x/\Delta t}$$

Now, if $x(t)$ and $y(t)$ are differentiable at t with $dx/dt \neq 0$, then

12.1.4
$$m_0 = \lim_{\Delta t \to 0} \frac{\Delta y/\Delta t}{\Delta x/\Delta t} = \frac{\lim_{\Delta t \to 0} \Delta y/\Delta t}{\lim_{\Delta t \to 0} \Delta x/\Delta t} = \frac{dy/dt}{dx/dt} = \frac{y'(t)}{x'(t)}$$

Example 5 Determine an equation for the line tangent to the cycloid

$$x = 2(t - \sin t), \qquad y = 2(1 - \cos t)$$

at the point corresponding to $t = \pi/6$.

Solution Now P_0 has coordinates $(\pi/3 - 1, 2 - \sqrt{3})$. Since

$$x'(t) = 2 - 2\cos t \qquad \text{and} \qquad y'(t) = 2 \sin t$$

we know that

$$m_0 = \frac{2\sin(\pi/6)}{2 - 2\cos(\pi/6)} = \frac{1}{2 - \sqrt{3}}$$

Thus, using the point-slope formula, the tangent line T_0 to the curve at P_0 is the graph of the equation

$$y - (2 - \sqrt{3}) = \frac{1}{2 - \sqrt{3}}\left[x - \left(\frac{\pi}{3} - 1\right)\right] \quad \bullet$$

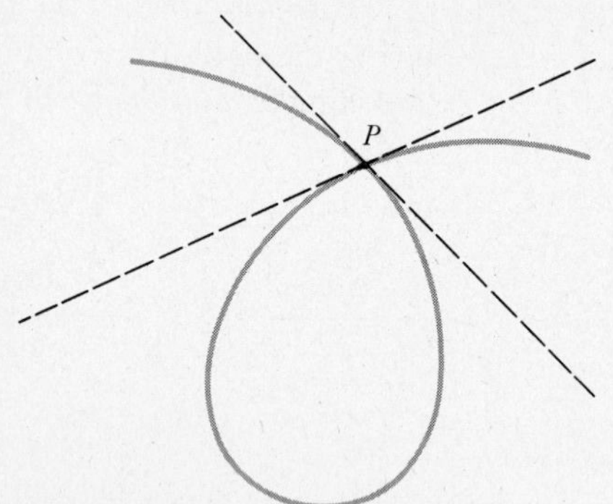

Fig. 12.8

The slope formula in (12.1.4) must be used with care. In places where cusps or corners exist there is no well-defined tangent. On the other hand, if a curve happens to cross itself, with $P = (x(t_0), y(t_0)) = (x(t_1), y(t_1))$ for $t_0 \neq t_1$, then it is possible to have two or more different but well-defined tangents, with different slopes, at P. (See Fig. 12.8.)

We should also like to be able to discuss vertical tangents where such tangents exist, because they occur frequently when parametric equations are graphed. It is not enough simply to require that $x'(t_0)$ be zero because, for example, $y'(t_0)$ may also be zero. To avoid these and other potential problems we shall usually restrict our attention to curves that have at least one well-defined tangent at each point—curves that are "smooth" in the following sense:

12.1.5 **Definition** A curve given parametrically by $x = x(t)$, $y = y(t)$, for t in some interval I, is said to be *smooth on I* if

(*a*) $x'(t)$, and $y'(t)$ exist and are continuous on I, and
(*b*) $[x'(t)]^2 + [y'(t)]^2 \neq 0$ for t in I.

Now from our earlier discussion we know the following:

12.1.6 **Theorem** Suppose C is a smooth curve given parametrically on I by $x = x(t)$, $y = y(t)$.

(*a*) If $x'(t_0) = 0$, then C has a vertical tangent at $(x(t_0), y(t_0))$ with equation $x = x(t_0)$.
(*b*) If $y'(t_0) = 0$, then C has a horizontal tangent at $(x(t_0), y(t_0))$ with equation $y = y(t_0)$.

Example 6 Determine those values of t in $[0, 2\pi]$ and those points in the plane for which the graph of $x = 2\cos t$, $y = 3 \sin t$ has (*a*) vertical tangents and (*b*) horizontal tangents. (See Example 3 and Fig. 12.3.)

Solution $x'(t) = -2\sin t$, $y'(t) = 3\cos t$, so condition [12.1.5(*a*)] holds.

(*a*) $x'(t) = 0$ when $t = 0, \pi, 2\pi$.
For such t, $y'(t) = 3, -3, 3$, respectively, so condition [12.1.5(*b*)] holds. Thus vertical tangents exist at $t = 0, \pi, 2\pi$. Now $x(0) = x(2\pi) = 2$, $x(\pi) = -2$. Also $y(0) = 0 = y(\pi) = y(2\pi)$. Hence there are vertical tangents to the elliptical graph at $(2, 0)$ and $(-2, 0)$.

(*b*) $y'(t) = 0$ when $t = \pi/2, 3\pi/2$.
For such t, $x'(t) = -2$ and $+2$, respectively, so condition [12.1.5(*b*)] holds. Thus horizontal tangents exist at $t = \pi/2, 3\pi/2$. Now $x(\pi/2) = 0 = x(3\pi/2)$ and $y(\pi/2) = 3$, $y(3\pi/2) = -3$. Hence there are horizontal tangents to the graph at $(0, 3)$ and $(0, -3)$. These conclusions that the curve is smooth with the indicated tangents fit Fig. 12.3. ●

PROGRESS TEST 2

1. Determine an equation for the line tangent to the graph of $x = e^t$, $y = e^{-t}$ at $t = \ln 3$ without eliminating the parameter.
2. Determine any values of t for which there are horizontal or vertical lines tangent to $x = 2\cot t$, $y = 2\sin^2 t$ with t in $(0, \pi)$.
3. Show that condition [12.1.5(*a*)] does not hold for the parametric equations in Prob. 2 at $t = 0$.

ADDITIONAL REMARKS

Although the cycloid example given earlier indicates that parametric equations arise naturally in solving specific problems, we have only introduced the topic of parametric equations here. We shall see parameters used in many different contexts in the coming chapters—for example, as angles in polar coordinate equations and as time in motion and projectile problems in Chap. 18.

SECTION 12.1 EXERCISES

In Exercises 1 to 16, (a) eliminate the parameter and (b) sketch the graph.

1. $x = t^2, y = t^2 + t$
2. $x = \dfrac{1}{t}, y = \dfrac{t^2 - 1}{t^2}$
3. $x = t^2 - 1, y = 4t$
4. $x = \dfrac{1}{(t-1)^2}, y = 3t + 1$
5. $x = e^{-t}, y = e^t$
6. $x = \cos 2t, y = \sin 2t, 0 \le t \le \pi$
7. $x = 1 - \dfrac{1}{t}, y = 1 + \dfrac{1}{t}$
8. $x = 3\tan t, y = 4\sec t$
9. $x = \sec t, y = -\tan^2 t, \dfrac{-\pi}{2} < t < \dfrac{\pi}{2}$
10. $x = \sqrt{3}t, y = 2t - 16t^2$
11. $x = 4 - t^4, y = t^2$
12. $x = 2t, y = \sqrt{1 - t^2}, -1 \le t \le 1$
13. $x = 2t, y = \sqrt{1 - t^2}, 0 \le t \le 1$
14. $x = \sqrt{t^2 - 1}, y = \sqrt{t^2 + 1}$
15. $x = \dfrac{2t}{1 + t^2}, y = \dfrac{1 - t^2}{1 + t^2}$
16. $x = 3\cos^2 t, y = 2\sin^2 t$

In exercises 17 to 20, (a) sketch the given equations by plotting a table of data, and (b) check the result in (a) by eliminating the parameter.

17. $x = \cos 2t, y = \cos t, 0 \le t \le 2\pi$
18. $x = \sin t, y = \sin 2t$
19. $x = \sin t, y = 2 + 3\sin t$
20. $x = 1 + \cos t, y = 2\cos t$

In Exercises 21 to 30, without eliminating the parameter (a) determine dy/dx, (b) give an equation of the line tangent to the curve at the given value of t, and (c) determine the values of t where vertical tangents exist and the curve is also smooth.

21. $x = 2t + 1, y = t^2 - 2, t = 2$

22. $x = t^4 + 1, y = t^2 - 1, t = 0$

23. $x = t^3, y = t^9 + 1, t = 1$

24. $x = \tan t, y = \sec t, t = \pi/4$

25. $x = 1 - t, y = e^{-t}, t = 1$

26. $x = e^t, y = e^{2t+1}, t = 1$

27. $x = t^3 - 3t, y = t^2, t = 1$

28. $x = t^3, y = t^2 + 1, t = 2$

29. $x = e^t, y = t^2 + 1, t = \ln 2$

30. $x = \sec 2t, y = \tan 2t, t = \pi/6$

31. Eliminate the parameter from the equations for a cycloid (12.1.3). (Solve the y equation for θ.)

32. Suppose a circular *fixed* spool of radius $a > 0$ is centered at the origin and is wound with thread. Find parametric equations for the curve traced out by the end of the thread as the thread is unwound from the spool. Assume that the end starts at $(a, 0)$ and that the thread is kept taut. (See Fig. *A*.) Use θ as the parameter and note that the segment RP is tangent to the spool at R. This curve is called an *involute* of a circle.

33. Triangle ABP is a rigid equilateral triangle of side a with A sliding on the y axis while B slides on the x axis. (See Fig. *C*.) Determine parametric equations giving the x and y coordinates of P in terms of θ as indicated.

34. Let C be a circle of radius a centered on the y axis at $(0, a)$, and thus tangent to the x axis at the origin. (See Fig. *B*.) Let L_1 be the horizontal line $y = 2a$ and let L be an arbitrary line through the origin. Denote the intersection of L with L_1 by A, and the intersection of L with the circle by B. Let $P = (x, y)$ be the point of intersection of the vertical line through A and the horizontal line through B. Find parametric equations for the curve traced out by such points P using θ, the angle from the x axis to the variable line L, as the parameter. (This curve is called the *witch of Agnesi*.)

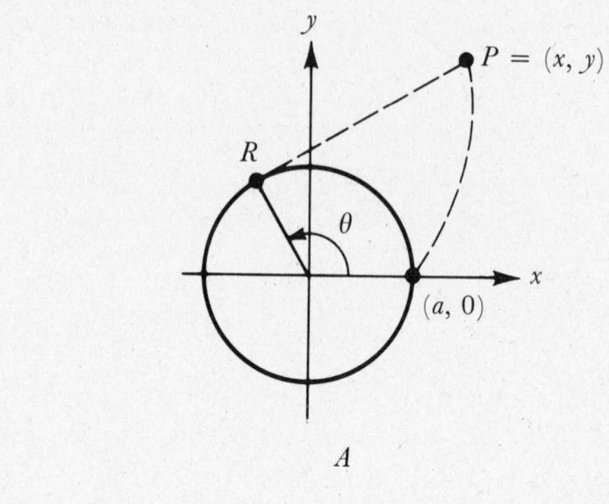

A

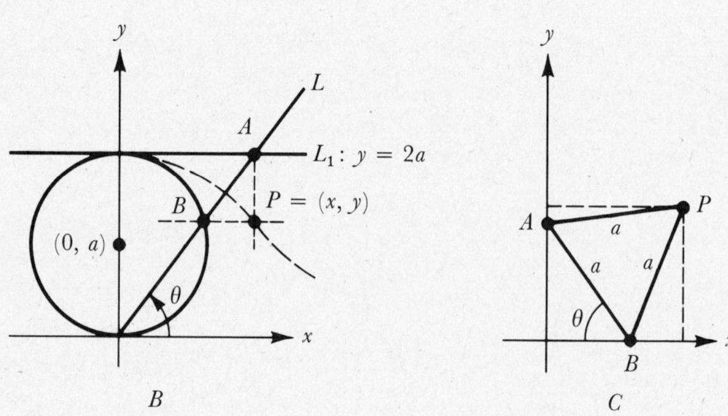

B C

Section 12.2: Arc Length and Area of Surfaces of Revolution

Objectives:

1. Use definite integrals to determine the length of plane curves.

2. Use definite integrals to determine areas of surfaces generated by revolving given plane curves about the coordinate axes.

LENGTH OF CURVES

Suppose we are given a piece of curve—usually called an *arc*—in the *x-y* plane. This arc may be the graph of a function of x, a function of y, or the graph of a pair of parametric equations. Since the parametric case includes $y = f(x)$ and $x = g(y)$ as special cases, we shall first determine the length of an arc that is the graph of

$$x = x(t), y = y(t) \qquad a \le t \le b$$

First, for convenience in our discussion, we assume that this arc does not cross or overlap itself by assuming that

$$(x(t_0), y(t_0)) \neq (x(t_1), y(t_1)) \qquad \text{for } t_0 \neq t_1 \text{ in } [a, b)$$

Such an arc is called *simple*. We *are* allowing the possibility that $(x(a), y(a)) = (x(b), y(b))$. In this case the arc is called *closed*. Our discussion thus applies to such simple closed arcs as circles and ellipses.

One last assumption, stated out of necessity rather than convenience, is that the arc be smooth on $[a, b]$. Our plan for determining the length of this smooth simple arc is:

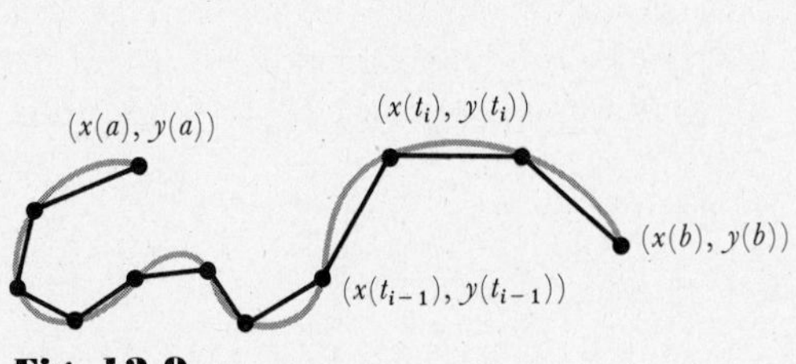

Fig. 12.9

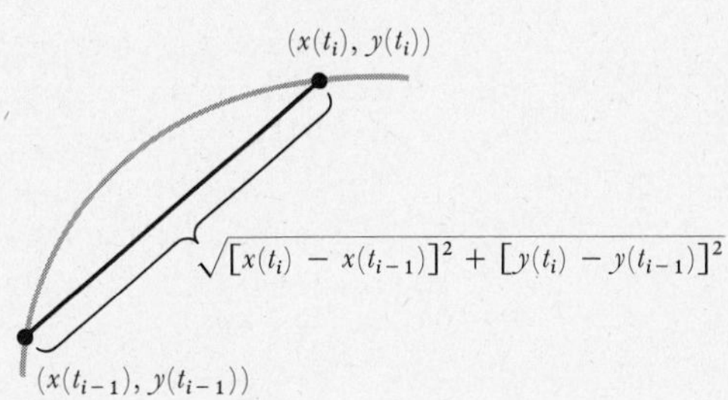

Fig. 12.10

1. Approximate its length using straight-line segments based on a partition of $[a, b]$; see Fig. 12.9.
2. Rewrite the resulting sum to get a Riemann sum.
3. Take the limit as the norm of the partition approaches zero to obtain a definite integral.

Thus we let P be a partition of $[a, b]$ into subintervals

$$[t_0, t_1], \ldots, [t_{i-1}, t_i], \ldots, [t_{n-1}, t_n] \qquad t_0 = a,\ t_n = b$$

Using the distance formula, the length of the curve from $(x(t_{i-1}), y(t_{i-1}))$ to $(x(t_i), y(t_i))$ is approximated (see Fig. 12.10) by

$$\Delta s_i = \sqrt{[x(t_i) - x(t_{i-1})]^2 + [y(t_i) - y(t_{i-1})]^2}$$

We multiply Δs_i by

$$\frac{t_i - t_{i-1}}{\sqrt{(t_i - t_{i-1})^2}} \qquad (= 1)$$

[with $\Delta t = t_i - t_{i-1} > 0$ guaranteeing that $\sqrt{(t_i - t_{i-1})^2} = t_i - t_{i-1}$] to get

$$\begin{aligned}\Delta s_i &= \sqrt{\frac{[x(t_i) - x(t_{i-1})]^2 + [y(t_i) - y(t_{i-1})]^2}{(t_i - t_{i-1})^2}}\,(t_i - t_{i-1}) \\ &= \sqrt{\left[\frac{x(t_i) - x(t_{i-1})}{t_i - t_{i-1}}\right]^2 + \left[\frac{y(t_i) - y(t_{i-1})}{t_i - t_{i-1}}\right]^2}\,\Delta t_i\end{aligned}$$

Each of the functions $x(t)$ and $y(t)$ is differentiable on $[t_{i-1}, t_i]$, so the Mean-Value Theorem (4.4.4) applies to each. Hence there is a $\overline{t}_i$ in (t_{i-1}, t_i) such that

$$\frac{x(t_i) - x(t_{i-1})}{t_i - t_{i-1}} = x'(\overline{t}_i)$$

and a $\widehat{t}_i$ in (t_{i-1}, t_i) such that

$$\frac{y(t_i) - y(t_{i-1})}{t_i - t_{i-1}} = y'(\widehat{t}_i)$$

Thus the length of a typical approximating line segment can be written as

$$\Delta s_i = \sqrt{[x'(\overline{t}_i)]^2 + [y'(\widehat{t}_i)]^2}\,\Delta t_i$$

which then implies that the total length of the arc is approximated by

$$\sum_{i=1}^{n} \Delta s_i = \sum_{i=1}^{n} \sqrt{[x'(\overline{t}_i)]^2 + [y'(\widehat{t}_i)]^2}\,\Delta t_i$$

The next natural step is to take the limit as the norms of the partitions of $[a, b]$ tend to zero and arrive at a definite integral formula for the length of the arc.

Unfortunately, a typical term of the sum may possibly involve distinct points $\bar{t}_i$ and $\hat{t}_i$ from the subinterval $[t_{i-1}, t_i]$. Using arguments beyond the scope of this text (but which can be found in any book on advanced calculus) it can be shown that because $x'(t)$ and $y'(t)$ are continuous, we can let $\bar{t}_i = \hat{t}_i$ and thus arrive at a Riemann sum

$$\sum_{i=1}^{n} \sqrt{[x'(\hat{t}_i)]^2 + [y'(\hat{t}_i)]^2}\, \Delta t_i$$

Furthermore, the continuity of $x'(t)$ and $y'(t)$ guarantees that $\sqrt{[x'(t)]^2 + [y'(t)]^2}$ is continuous on $[a, b]$, so the limit will be the definite integral

$$\int_a^b \sqrt{[x'(t)]^2 + [y'(t)]^2}\, dt$$

Hence we may define the length as follows:

12.2.1 **Definition** The length of a simple smooth arc

$$x = x(t), y = y(t) \qquad a \le t \le b$$

is given by

$$s = \lim_{\|P\| \to 0} \left[\sum_{i=1}^{n} \Delta s_i \right] = \int_a^b \sqrt{[x'(t)]^2 + [y'(t)]^2}\, dt$$

In the case that the curve is the graph of a function $y = f(x)$, $a \le x \le b$, with continuous first derivative on $[a, b]$, we can regard $x = x(t) = t$. Then $x'(t) = 1$ and $y'(t) = f'(t)$, so we have the following:

12.2.2 **Theorem** The length of the graph of a continuously differentiable function $y = f(x)$, $a \le x \le b$, is

$$\int_a^b \sqrt{1 + [f'(x)]^2}\, dx$$

Similarly, we have:

12.2.3 **Theorem** The length of the graph of a continuously differentiable function $x = g(y)$, $a \le y \le b$, is

$$\int_a^b \sqrt{1 + [g'(y)]^2}\, dy$$

Example 1 Use (12.2.1) to determine the circumference of a circle of radius r.

Solution We center the circle at (0, 0) and use (12.1.2) to represent it parametrically by

$$x = r \cos t, y = r \sin t \qquad 0 \le t \le 2\pi$$

Then

$$x'(t) = -r \sin t, y'(t) = r \cos t$$

so

$$\sqrt{[x'(t)]^2 + [y'(t)]^2} = \sqrt{r^2 \sin^2 t + r^2 \cos^2 t} = r\sqrt{1} = r$$

Thus

$$s = \int_0^{2\pi} r\, dt = rt \Big|_0^{2\pi} = 2\pi r \quad \bullet$$

Example 2 Show that using (12.2.2) to find the length of the straight line graph of $f(x) = mx$ from $x = 0$ to $x = b$ gives the same result as the formula for the distance between two points.

Solution $f'(x) = m$, so the length by (12.2.2) is

$$\int_0^b \sqrt{1 + m^2}\, dx = (\sqrt{1 + m^2})x \Big|_0^b = b\sqrt{1 + m^2}$$

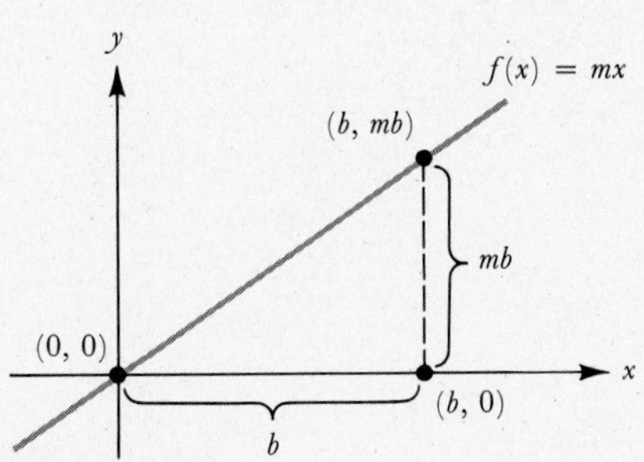

Fig. 12.11

As shown in Fig. 12.11 this equals the result obtainable by the distance formula:

$$\sqrt{b^2 + (mb)^2} = b\sqrt{1 + m^2} \quad \bullet$$

[Notice that if $m = 0$, then our result says that the length of the segment of the x axis from $(0, 0)$ to $(b, 0)$ is $b\sqrt{1 + 0^2} = b$.]

The arc in Fig. 12.12 can be described in many ways, three of which are: (a) $y = x^{2/3}$, $1 \le x \le 8$; (b) $x = y^{3/2}$, $1 \le y \le 4$; and (c) $x = t^3$, $y = t^2$, $1 \le t \le 2$. (Check by eliminating the parameter.)

The length should be the same no matter which arc length formula we use—even though the integrands and the limits of integration will differ in each case. The three integrals are:

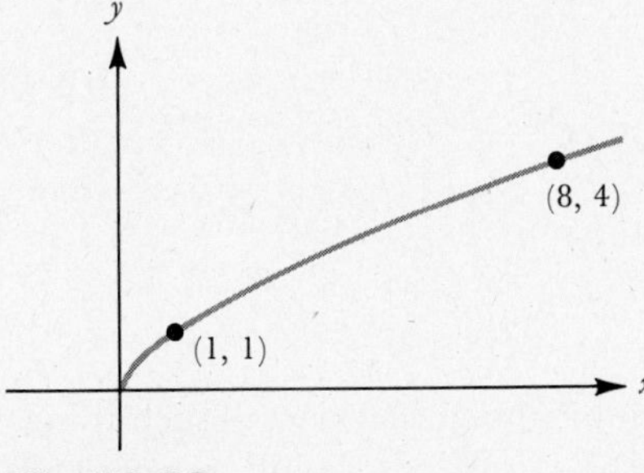

Fig. 12.12

(a) $\displaystyle\int_1^8 \sqrt{1 + \left[\frac{2}{3}x^{-1/3}\right]^2}\, dx$ using 12.2.2

(b) $\displaystyle\int_1^4 \sqrt{1 + \left[\frac{3}{2}y^{1/2}\right]^2}\, dy$ using 12.2.3

(c) $\displaystyle\int_1^2 \sqrt{[3t^2]^2 + [2t]^2}\, dt$ using 12.2.1

We leave the proof that these three integrals are equal (≈ 7.634) as Exercise 44.

PROGRESS TEST 1

Determine the length of the following arcs:

1. $x = e^t \cos t$, $y = e^t \sin t$, $0 \le t \le \pi$

2. $24xy = x^4 + 48$ from $x = 1$ to $x = 3$

3. $x = \dfrac{e^y + e^{-y}}{2}$ from $y = -1$ to $y = 1$ (use a hyperbolic function if you wish)

AREAS OF SURFACES OF REVOLUTION

In Sec. 6.6 we saw that by revolving a *region* about a line a solid of revolution is generated. We shall now examine the outer surface of such a solid.

If a plane curve is revolved about an axis, the curve sweeps out what we shall call a *surface of revolution.* If the curve is a straight-line segment of length L parallel to the axis of revolution at distance r, the surface swept out is the lateral surface of a right circular cylinder of area $2\pi rL$. If the curve is a straight line of length L with one endpoint on the axis of revolution and the other at distance r from the axis of revolution, it sweeps out the lateral surface of a right circular cone of area πrL. (See Fig. 12.13.)

The surface area of both the cylinder and the cone can be made more obvious by slitting either along the generating curve and flattening out the surface. The cylinder yields a rectangle of height $2\pi r$ and width L, whereas the cone yields the sector of a circle of radius L.

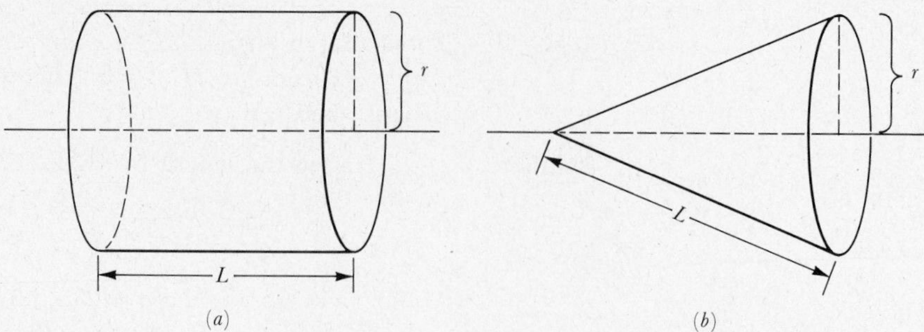

Fig. 12.13

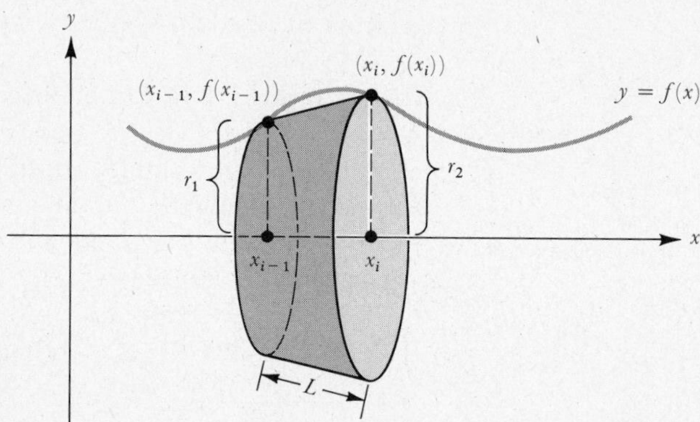

Fig. 12.14

To develop a general technique to determine the surface area of such things as a parabolic reflector, we shall consider the surface of revolution generated when the graph of a function $y = f(x)$ from $x = a$ to $x = b$ is revolved about the x axis. [As we have seen, a given plane curve may also be described as the graph of a pair of parametric equations, and after we have completed the $y = f(x)$ case we shall give analogous formulas for curves given parametrically.] Since our approach to determining the surface area will involve our arc length formulas, we assume that f has a continuous first derivative on $[a, b]$.

Again, we begin by approximating the graph with straight-line segments based on a partition P of $[a, b]$ as in Fig. 12.14.

Each segment of length L, whose endpoints are at distances r_1 and r_2 from the axis, generates the frustum of a cone (a chopped-off cone) with lateral surface area determined by L and the average of the two radii:

12.2.4
$$2\pi\left(\frac{r_1 + r_2}{2}\right)L = \pi(r_1 + r_2)L$$

This formula (a circular version of the area of a trapezoid) can be derived from the earlier formula for the surface area of a complete cone using the cutting and flattening idea (see Exercise 43 below). As with the formula for arc length, a typical segment as pictured in Fig. 12.14 has a length that can be put in the form

$$\sqrt{1 + [f'(x_i^*)]^2}\,\Delta x_i$$

where x_i^* is some number in $[x_{i-1}, x_i]$.

Since the two ends of the segment are at distances $f(x_{i-1})$ and $f(x_i)$ from the x axis (by 12.2.4), this typical segment generates a surface of area

$$\Delta S_i = 2\pi\,\frac{|f(x_{i-1}) + f(x_i)|}{2}\,\sqrt{1 + [f'(x_i^*)]^2}\,\Delta x_i$$

[We use absolute-value signs to allow for the possibility that $f(x) < 0$.] By the Intermediate-Value Theorem (2.5.12) there is an $\overline{x}_i$ in $[x_{i-1}, x_i]$ at which f attains the average of $f(x_{i-1})$ and $f(x_i)$; that is,

$$f(\overline{x}_i) = \frac{f(x_{i-1}) + f(x_i)}{2}$$

Hence

$$\Delta S_i = 2\pi |f(\overline{x}_i)| \sqrt{1 + [f'(x_i^*)]^2}\, \Delta x_i$$

and $\Sigma_{i=1}^n \Delta S_i$ is the total surface area generated by all the line segments based on the partition P of $[a, b]$. As with the arc length development, the detailed arguments leading from $\Sigma_{i=1}^n \Delta S_i$ to a genuine Riemann sum and then to a definite integral are beyond the scope of a first calculus course. Note that a typical term ΔS_i involves possibly distinct numbers $\overline{x}_i$ and x_i^* in each subinterval. Nonetheless, it is geometrically reasonable that such sums provide increasingly accurate approximations to the actual surface area as the norms of the partitions P get smaller. Furthermore, as $\|P\|$ decreases, so does the difference between $\overline{x}_i$ and x_i^* in a typical subinterval, and since both $f(x)$ and $f'(x)$ are continuous, it is reasonable to expect that the sums tend to a definite integral. Hence we make the following:

12.2.5 **Definition** Suppose $f'(x)$ exists and is continuous on $[a, b]$. Then the area of the surface of revolution obtained by revolving the graph of $y = f(x)$, $a \leq x \leq b$, about the x axis is

$$S = \lim_{\|P\| \to 0} \sum_{i=1}^{n} \Delta S_i = 2\pi \int_a^b |f(x)| \sqrt{1 + [f'(x)]^2}\, dx$$

To generalize the formula in (12.2.5) it is helpful to reexamine our arc length formulas in terms of differentials. For a function $y = f(x)$ as in (12.2.5), the arc length formula (12.2.2.) also applies. In particular, if we let u be a dummy variable and x be any number between a and b, we can think of arc length s as a function of x:

$$s = s(x) = \int_a^x \sqrt{1 + [f'(u)]^2}\, du$$

where $s(x)$ is the length of the arc from a to x. But then, by the Fundamental Theorem of Calculus,

$$\frac{ds}{dx} = \sqrt{1 + [f'(x)]^2}$$

so

$$ds = \sqrt{1 + [f'(x)]^2}\, dx$$

It is especially useful to regard ds as the length of a short piece of arc. Let us assume the graph of $y = f(x)$ lies in the upper half plane. We can then reinterpret the surface area formula as

12.2.6(a)
$$S = 2\pi \int_a^b y\, ds$$

which geometrically supports the intuitive idea that the total surface area is the sum of the areas $2\pi y\, ds$ of short "cylinders" of radius y and length ds. (See Fig. 12.15.)

If instead the curve is revolved about the y axis and the graph lies in the right half plane, then a similar argument yields:

12.2.6(b)
$$S = 2\pi \int_a^b x\, ds$$

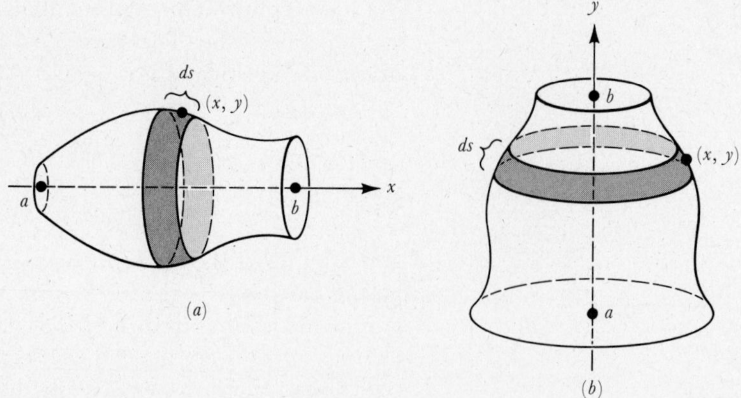

Fig. 12.15

These formulas are equally valid if the curve is given in the form $x = g(y)$, $a \leq y \leq b$. In this case

$$ds = \sqrt{1 + [g'(y)]^2}\, dy$$

and [12.2.6(a)] and [12.2.6(b)] are interpreted to mean, respectively,

$$S = 2\pi \int_a^b y \sqrt{1 + [g'(y)]^2}\, dy$$

and

$$S = 2\pi \int_a^b g(y) \sqrt{1 + [g'(y)]^2}\, dy$$

Finally, if the curve happens to be the graph of the parametric equations $x = x(t), y = y(t)$, where, again, $x'(t)$ and $y'(t)$ exist and are continuous on $[a, b]$, then we can revolve the graph about either the x or the y axis to get two different surfaces.

12.2.7 (*a*) Revolved about the x axis, the graph generates a surface with area

$$2\pi \int_a^b y(t)\, ds \qquad \text{where } ds = \sqrt{[x'(t)]^2 + [y'(t)]^2}\, dt$$

(*b*) Revolved about the y axis, the graph generates a surface with area

$$2\pi \int_a^b x(t)\, ds \qquad \text{where } ds = \sqrt{[x'(t)]^2 + [y'(t)]^2}\, dt$$

In fact, each formula (12.2.5 to 12.2.7) is a special case of the following:

12.2.8 **General Surface Area Formula** The area of the surface generated by revolving the arc C about the axis L is

$$2\pi \int_a^b \mathfrak{D}\, ds$$

where $\mathfrak{D}$ is the distance of the differential of arc length ds from the axis L.

Depending upon the axis L and the form that C is given in, the general formula will be filled out in any of several ways. You are encouraged to start with this formula in setting up surface area integrals. In each case the sum-of-cylinders idea applies.

Example 3 Determine the surface area of a sphere of radius r with its opposite ends sliced off to a depth of ε units, $0 < \varepsilon < r$. (See Fig. 12.16.)

Solution We shall revolve the graph of $f(x) = \sqrt{r^2 - x^2}$ about the x axis from $x = -r + \varepsilon$ to $x = r - \varepsilon$. Now $f(x) > 0$ on $[-r + \varepsilon, r - \varepsilon]$, so $|f(x)| = f(x)$, and $f'(x) = -x/(\sqrt{r^2 - x^2})$, so by (12.2.5) the total surface area is

$$2\pi \int_{-r+\varepsilon}^{r-\varepsilon} \sqrt{r^2 - x^2}\sqrt{1 + \frac{x^2}{r^2 - x^2}}\,dx = 2\pi \int_{-r+\varepsilon}^{r-\varepsilon} \sqrt{r^2 - x^2}\sqrt{\frac{r^2 - x^2 + x^2}{r^2 - x^2}}\,dx$$

$$= 2\pi \int_{-r+\varepsilon}^{r-\varepsilon} r\,dx$$

$$= 2\pi\Big[rx\Big]_{-r+\varepsilon}^{r-\varepsilon} = 4\pi r(r - \varepsilon) \quad \bullet$$

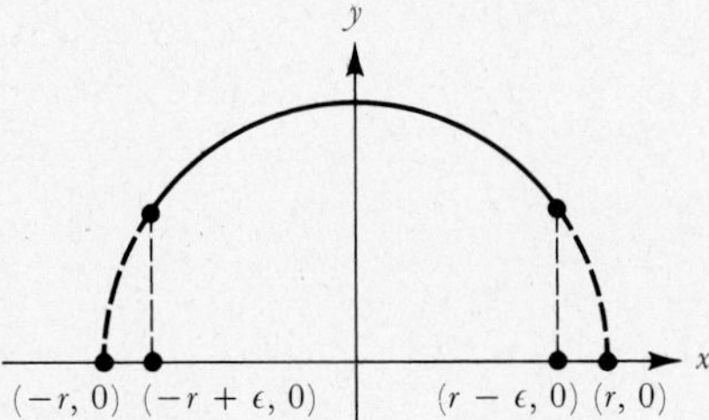

Fig. 12.16

Notice that because $f(x) = \sqrt{r^2 - x^2}$ is not differentiable at $x = \pm r$, we could not use (12.2.5) to determine the surface area of the *entire* sphere generated by revolving the whole semicircle about the x axis.

Example 4 Use (12.2.7) to determine that the surface area of a sphere of radius r is $4\pi r^2$.

Solution We revolve about the x axis the upper semicircle, which is the graph of the parametric equations.

$$x = r\cos t, \; y = r\sin t \qquad \text{with } 0 \le t \le \pi$$

(Note that for t in $[\pi, 2\pi]$, we get the lower semicircle.) Now

$$ds = \sqrt{[r(-\sin t)]^2 + [r\cos t]^2}\,dt = \sqrt{r^2(\sin^2 t + \cos^2 t)}\,dt = r\,dt$$

Thus, by [12.2.7(*a*)] the surface area of the sphere is

$$S = 2\pi \int_0^{\pi} |r\sin t| r\,dt$$

$$= 2\pi r^2 \int_0^{\pi} \sin t\,dt \qquad (\sin t \ge 0 \text{ in } [0, \pi])$$

$$= 2\pi r^2\Big[-\cos t\Big]_0^{\pi} = 2\pi r^2[-(-1) - (-1)] = 4\pi r^2 \quad \bullet$$

Compare the results of Example 4 with that of Example 3 as $\varepsilon \to 0$.

Example 5 Determine the area of the parabolic surface generated by rotating $y = x^2$ $3/4 \le y \le 6$ about the y axis. (See Fig. 12.17.)

Solution First solve for x, using the positive solution: $x = g(y) = \sqrt{y}$. Now $dx/dy = g'(y) = 1/(2\sqrt{y})$ is continuous on $[3/4, 6]$, so by [12.2.6(b)] the area is

$$2\pi \int_{3/4}^{6} \sqrt{y}\sqrt{1 + \left(\frac{1}{2\sqrt{y}}\right)^2}\,dy = 2\pi \int_{3/4}^{6} \sqrt{y}\frac{\sqrt{4y + 1}}{\sqrt{4y}}\,dy$$

$$= \frac{\pi}{4}\int_{3/4}^{6} (4y + 1)^{1/2}\,4\,dy$$

$$= \frac{\pi}{6}\Big[(4y + 1)^{3/2}\Big]_{3/4}^{6}$$

$$= \frac{\pi}{6}[125 - 8] = \frac{117}{6}\pi \quad \bullet$$

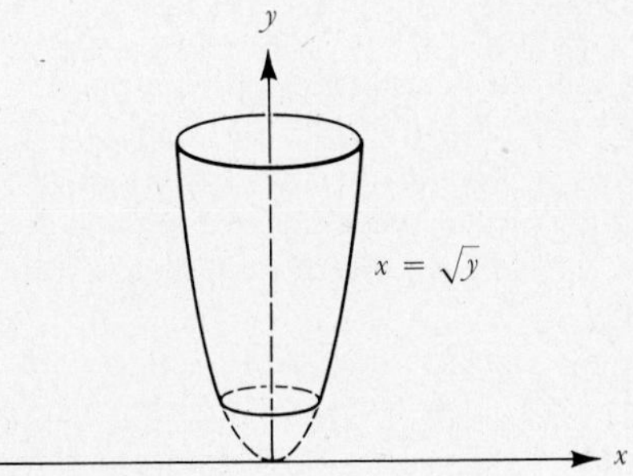

Fig. 12.17

PROGRESS TEST 2

Determine the areas of the surfaces generated by revolving the given curves about the given axes.

1. $y^2 = 4x$, $1 \le x \le 3$, x axis

2. $x = y^3$, $0 \le y \le 2$, y axis

3. $x = 2 + \cos t$, $y = \sin t$, $0 \le t \le \pi$, y axis

SECTION 12.2 EXERCISES

In Exercises 1 to 16, (a) set up a definite integral that equals the length of the given arc and (b) evaluate that integral.

1. $y = (1 - x^{2/3})^{3/2}$, $0 \le x \le 1$

2. $y = \ln x$, $1 \le x \le \sqrt{3}$

3. $x = \ln(\cos y)$, $0 \le y \le 1$

4. $x = \dfrac{y^2}{4} - \dfrac{1}{2}\ln y$, $\dfrac{1}{e} \le y \le 1$

5. $x = t^2$, $y = 3t$, $-2 \le t \le 0$

6. $x = t + \sin t$, $y = 1 - \cos t$, $0 \le t \le \pi$

7. $x = \arcsin \dfrac{t}{4}$, $y = \ln \sqrt{16 - t^2}$, $0 \le t \le 2$

8. $x = 3\sin^3 t$, $y = 3\cos^3 t$, $0 \le t \le \pi$

9. $y = \ln(x + \sqrt{x^2 - 1})$, $3 \le x \le 5$

10. $9y^2 = 4(x - 1)^3$, $1 \le x \le 4$

11. $y = \dfrac{x^4 + 3}{6x}$, $1 \le x \le 2$

12. $y = \ln(\sec x)$, $0 \le x \le \dfrac{\pi}{4}$

13. $y = \dfrac{x^4}{4} + \dfrac{1}{8x^2}$, $1 \le x \le 2$

14. $y = \arcsin(e^{-x})$, $0 \le x \le \sqrt{2}$

15. $x = 3\sin t \cos t$, $y = 3\sin^2 t$, $0 \le t \le \pi$

16. $y = t$, $x = \ln(\sec t)$, $0 \le t \le \pi/4$ (compare with Exercise 12)

In Exercises 17 to 31, (a) set up a definite integral that equals the area of the surface generated by revolving the given curve about the given axis and (b) evaluate that integral.

17. $y = \dfrac{1}{2}(e^x + e^{-x})$ from $x = 0$ to $x = 2$, x axis

18. $y = \sin x$ from $x = \pi/4$ to $x = \pi/2$, x axis (a "ring" shape)

19. $4x = y^2 - 2\ln y$ from $y = 1$ to $y = 2$, y axis

20. $y = 2\sqrt{1 - x}$ from $x = -1$ to $x = 0$, x axis

21. $x = 2\cos^3 t$, $y = 2\sin^3 t$, $0 \le t \le \pi$, x axis

22. $x = 2t$, $y = t^2$, $0 \le t \le 4$, x axis

23. $x = (1/2)t^2 + 1/t$, $y = 4\sqrt{t}$, $1 \le t \le 4$, x axis

24. $x = 3t^2$, $y = 2t^3$, $0 \le t \le 1$, y axis

25. $x^2 - y^2 = 1$ between $(1, 0)$ and $(4, \sqrt{15})$, y axis

26. $18x^2 = y(6 - y)^2$ between $(0, 0)$ to $(0, 6)$, y axis

27. $y = x^{4/3}$, $1 \le x \le 8$, x axis

28. $y = e^{-x}$, $0 \le x \le 1$, x axis

29. $x = (1/6)y^3 + 1/(2y)$, $1 \le y \le 2$, y axis

30. $8y^2 = x^2 - x^4$, $0 \le x \le 1$, x axis

31. $x = t^2/2$, $y = (1/3)(2t + 1)^{3/2}$, $0 \le t \le 4$, y axis

In Exercises 32 to 36 use the general surface area formula (12.2.8) to set up a definite integral giving the area of the surface generated by revolving the given curve about the given axis.

32. $y = x^2/16 - 2\ln x$, $4 \le x \le 8$, y axis

33. $2x = y^2 - 2$, $0 \le y \le 2\sqrt{2}$, x axis

34. $y = x^3/3 + (4x)^{-1}$, $1 \le x \le 2$, the line $y = -1$

35. $9y^2 = x(x - 3)^2$ between $(0, 0)$ and $(3, 0)$, the line $x = 3$

36. $y = x^3/6 + (1/2)x$, $1 \le x \le 2$, y axis

37. Determine the length of one arch of the cycloid $x = t - \sin t$, $y = 1 - \cos t$.

38. Determine the area of the surface generated by revolving one arch of the cycloid in Exercise 37 about the x axis.

39. Repeat Exercise 38, but revolve about the y axis.

40. Use (*a*) the Trapezoid Rule and (*b*) Simpson's Rule with 10 subdivisions to approximate the length of one arch of the sine curve.

41. Repeat Exercise 40 to approximate the length of the arc $y = x^3/6 + x/2$ from $x = 0$ to $x = 1$.

42. Determine the area of the surface generated by revolving $y = \ln x$, $1/e \le x \le e$, about the y axis:
(*a*) First, by using x as the independent variable.
(*b*) Second, by using y as the independent variable. (You will need to invert the equation, determine y limits of integration, and use a trigonometric substitution.)

43. A cone of radius r and slant height l has area πrl, so a frustum of slant height L as pictured has a lateral surface area that is the difference between the area of the whole cone and that of the smaller cone of slant height l and radius r; that is,

$$\pi R(L + l) - \pi rl = \pi(Rl - rl + RL)$$

(*a*) Show that $(L + l)/R = l/r$. (*b*) Use (*a*) to prove (12.2.4). Use (12.2.5) to show that the right circular

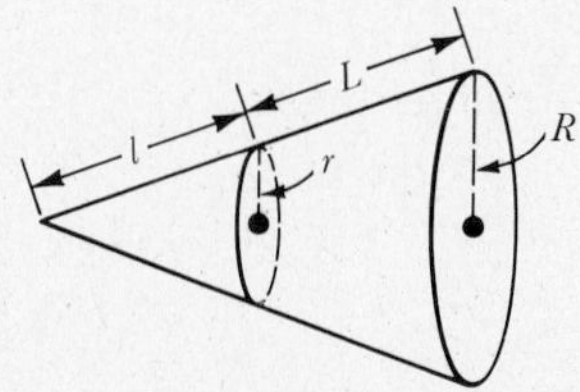

cone of radius r and height h generated by the line $y = (r/h)\,x$ from $x = 0$ to $x = h$ has lateral surface area $\pi r\sqrt{h^2 + r^2}$; here $\sqrt{h^2 + r^2}$ is the slant height. [This shows that (12.2.5) is consistent with the assumption about the area of a cone upon which it was based.]

44. Show that the three definite integrals associated with Fig. 12.12 are equal.

45. (*a*) Given a sphere of radius r generated by revolving $y = +\sqrt{r^2 - x^2}$ about the x axis, a "spherical zone of height h" ($h < 2r$) is generated by revolving the portion of $y = \sqrt{r^2 - x^2}$ that lies above the interval $[-h/2, h/2]$. Determine the surface area of this zone.
(*b*) Show that this surface area is independent of which interval of length h is chosen in $[-r, r]$. (Think of what this means if you pass parallel planes 1 cm apart through the earth and compare the zone near the equator with that near a pole.)

Section 12.3: Centroids of Plane Regions, Curves, and Solids and Surfaces of Revolution

Objectives:

1. Calculate moments and centers of mass for discrete point masses in the plane.

2. Calculate moments and centroids of regions in the plane, plane curves, solids of revolution, and surfaces of revolution.

MOMENTS

Although most of the geometric applications in this section originated in physics and engineering, special knowledge of these areas is unnecessary. Furthermore, we shall deemphasize this aspect of the topic as we progress (often ignoring the units involved, for example). Instead, we shall stress the mathematical essence of these ideas, which will then apply to other areas including population distributions and statistics.

Suppose a uniform bar is oriented along the x axis and balanced on a fulcrum situated at the origin. If we hang a 25-lb mass at $x = -4$, where should we hang a 10-lb mass to balance the bar? The answer, as most any schoolchild would determine (such balance scales are often used in teaching arithmetic) is at $x = 10$. We balanced the tendency of the bar to twist in a counterclockwise direction about its fulcrum.

12.3.1 Definition Given a mass m concentrated at directed distance $\bar{x}$ from a given point P, we define the *moment of that mass with respect to P* as the product $m\bar{x}$. If several masses are placed at different points, then the *total moment* of this system with respect to P is the sum of the moments with respect to P.

In our example we needed to place the mass so that the sum of the moments with respect to the origin was zero. Conversely, we could *begin* with a collection of masses hanging from the bar and ask where the fulcrum should be moved so that the bar balances.

Suppose a 20-lb mass is at $x = -2$ and an 80-lb mass is at $x = 3$ (see Fig. 12.18). Where is the "balance point"? This is the point where all the mass could be concentrated without affecting the total moment of the system with respect to the origin.

The sum of the moments with respect to the origin is

$$(20)(-2) + (80)(3) = 200$$

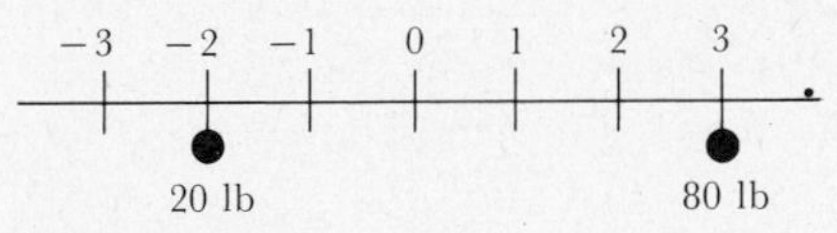

Fig. 12.18

Since the total mass is 100 lb and the moment is the product of mass and directed distance from the origin, we have $100\bar{x} = 200$, so $\bar{x} = 2$, and the system behaves as if all its mass were concentrated at $\bar{x} = 2$. This also means the system could be balanced at $\bar{x} = 2$.

Quite obviously this discussion extends to any finite number of masses placed along the x axis (the role of the bar is superfluous).

12.3.2 Definition Suppose we are given a system of point masses along the x axis. The point $\bar{x}$ at which we could concentrate all the mass of this system without changing the total moment of the system with respect to the origin is called its center of *mass* and is computed by

$$\bar{x} = \frac{\text{total moment}}{\text{total mass}}$$

Example 1 Given the system of masses 5, 25, 50, 10, and 30 at the points -10, -2, 1, 5, and 6, respectively, determine the center of mass of the system.

Solution The total moment is

$$(5)(-10) + (25)(-2) + (50)(1) + (10)(5) + (30)(6) = 180$$

and the total mass is

$$5 + 25 + 50 + 10 + 30 = 120$$

Hence the center of mass is

$$\bar{x} = \frac{180}{120} = \frac{3}{2} \quad \bullet$$

The above discussion regarding masses along a straight *line* generalizes to a system of n masses in the *plane,* whose respective masses and positions are listed:

$$m_1,\ (x_1,y_1);\ m_2,\ (x_2,y_2);\ \ldots\ ;\ m_i,\ (x_i,y_i);\ \ldots\ ;\ m_n,\ (x_n,y_n).$$

The moment of a typical mass m_i [located at (x_i, y_i)] about the y axis is defined to be the product of its mass m_i and its directed distance x_i from the y axis. Similarly, its moment about the x axis is defined to be $m_i y_i$.

12.3.3 Definition Given the system of point masses in the plane as above, we define the total *moment of the system with respect to the y axis* to be

(*a*) $M_y = \sum_{i=1}^{n} m_i x_i$

and the total *moment with respect to the x axis* to be

(*b*) $M_x = \sum_{i=1}^{n} m_i y_i$

Given the total mass $M = \Sigma_{i=1}^{n} m_i$, we define the *center of mass of this system* to be the point $(\bar{x}, \bar{y})$ where

(*c*) $\bar{x} = \dfrac{M_y}{M}$ and $\bar{y} = \dfrac{M_x}{M}$

Just as in the linear case, the center of mass is the balance point—the point at which all the mass could be concentrated without changing the total moment of the system. (A mobile with all its objects in one plane could be hung from its center of mass.)

PROGRESS TEST 1

1. (*a*) Determine the total moment with respect to the origin of the four masses distributed as follows: 4, 6, 7, 9 lb at $x = -6, -2, 0, 8$, respectively.
(*b*) Where is this system's center of mass?
2. Given the same masses in the plane with the same respective x coordinates as in Prob. 1, and with y coordinates -1, 3, 2, and 4, respectively, determine: (*a*) M_y, (*b*) M_x, and (*c*) the center of mass in the plane.
3. Repeat Prob. 1 if a mass of 10 lb is added at $x = 60$.
4. (*a*) Determine the center of mass of a system of four 1-lb masses at the points (1, 4), (2, 5), (4, 7), (5, 16).
(*b*) Repeat (*a*) for a system of four 2-lb masses at the same locations.
(*c*) Repeat (*a*) for a system of four k-lb masses at the same locations.

CENTROID OF A PLANE REGION

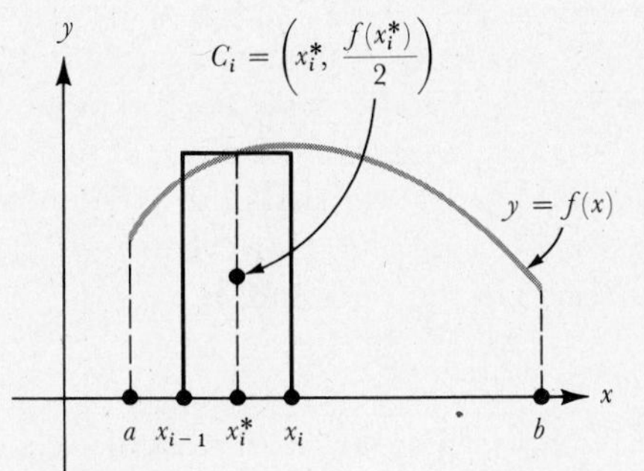

Fig. 12.19

This work with discrete masses extends to thin plates of uniform thickness and density. We assume that a plate is imposed on a region in the x-y plane between the graph of a continuous, nonnegative function $y = f(x)$ and an interval $[a, b]$. We approximate the region with rectangles based on a partition P of $[a, b]$, where in this case a typical rectangle over the subinterval $[x_{i-1}, x_i]$ has height $f(x_i^*)$ with $x_i^* = (x_{i-1} + x_i)/2$, the midpoint of $[x_{i-1}, x_i]$; see Fig. 12.19.

If ρ is the density of the plate in mass per unit area, then the mass of a rectangle is the product of ρ with its area. In this case the area is $f(x_i^*)\,\Delta x_i$, so the mass is $\rho f(x_i^*)\,\Delta x_i$. We assume the mass of a typical rectangle to be concentrated at its center $C_i = (x_i^*, f(x_i^*)/2)$, so the moment of the rectangle with respect to the y axis is $x_i^* \rho f(x_i^*)\,\Delta x_i$. We approximate the total moment of the plate with respect to the y axis by the sum

$$\sum_{i=1}^{n} x_i^* \rho f(x_i^*)\,\Delta x_i = \rho \sum_{i=1}^{n} x_i^* f(x_i^*)\,\Delta x_i$$

As usual, we take the limit as $\|P\| \to 0$ to get a definite integral formula for M_y:

12.3.4 **Definition** $M_y = \rho \int_a^b x f(x)\,dx$

Under the same assumptions, the moment of a typical rectangle with respect to the x axis is the product of the mass $\rho f(x_i^*)\,\Delta x_i$ and the distance $f(x_i^*)/2$ of the center from the x axis, $(f(x_i^*)/2)\,\rho f(x_i^*)\,\Delta x_i$. We approximate the total moment with respect to the x axis with the sum

$$\sum_{i=1}^{n} \frac{f(x_i^*)}{2} \rho f(x_i^*)\,\Delta x_i = \frac{\rho}{2} \sum_{i=1}^{n} [f(x_i^*)]^2\,\Delta x_i$$

and take the limit as $\|P\| \to 0$ to obtain:

12.3.5 **Definition** $M_x = \frac{\rho}{2} \int_a^b [f(x)]^2\,dx$

The total mass M of the plate is the product of the density and area, and the area is $\int_a^b f(x)\,dx$, so

$$M = \rho \int_a^b f(x)\,dx = \rho A$$

Using the above formulas we can now define the center of mass of the plate to be the point $(\overline{x}, \overline{y})$ where:

12.3.6 Definition
$$\overline{x} = \frac{M_y}{\rho A} = \frac{\rho \int_a^b x f(x)\, dx}{\rho \int_a^b f(x)\, dx} = \frac{\int_a^b x f(x)\, dx}{\int_a^b f(x)\, dx}$$

and

$$\overline{y} = \frac{M_x}{\rho A} = \frac{(\rho/2) \int_a^b [f(x)]^2\, dx}{\rho \int_a^b f(x)\, dx} = \frac{(1/2) \int_a^b [f(x)]^2\, dx}{\int_a^b f(x)\, dx}$$

Since the constant density ρ cancels from the above formulas, just as the constant mass did in Problem 4 of Progress Test 1, it is clear that *the point* $(\overline{x}, \overline{y})$ *depends only upon the shape of the plate.* The constant density, hence mass, are unimportant to the discussion. Henceforth in this section, unless otherwise noted, we shall ignore density and mass. (An essentially equivalent assumption is to assume that $\rho = 1$ instead.) Thus the discussion focuses on the plane region R defined by the graph of the function and the role of the physical plate disappears. In these circumstances, and henceforth, we shall use common terminology and speak of $(\overline{x}, \overline{y})$ as the *centroid of the region R*. However, we shall continue to speak of "moments" and use the notation M_y and M_x.

Example 2 Determine the centroid of the region bounded by the positive semicircle $f(x) = \sqrt{r^2 - x^2}$ and the x axis (see sketch).

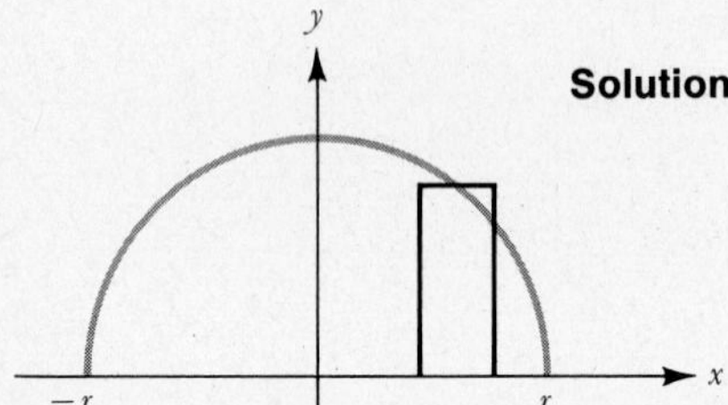

Solution

$$M_y = \int_{-r}^{r} x\sqrt{r^2 - x^2}\, dx$$
$$= \left[\left(\frac{-1}{2}\right)\left(\frac{2}{3}\right)(r^2 - x^2)^{3/2}\right]_{-r}^{r} = 0$$
$$M_x = \int_{-r}^{r} \frac{(r^2 - x^2)}{2}\, dx = \frac{1}{2}\left[r^2 x - \frac{x^3}{3}\right]_{-r}^{r} = \frac{2r^3}{3}$$

The area A of the region is $\pi r^2/2$, so

$$\overline{x} = \frac{0}{A} = 0 \qquad \text{and} \qquad \overline{y} = \frac{2r^3/3}{\pi r^2/2} = \frac{4}{3\pi} r$$

(As we expect from the symmetry of the semicircle, the centroid is on the y axis.) ●

The principles responsible for (12.3.6) can be used to compute centroids of regions for which horizontal rectangles are more convenient.

Example 3 Determine the centroid of the region bounded by an isosceles triangle of base b and height h.

Solution We center the region on the y axis and base it on the x axis as indicated in the accompanying sketch. To avoid breaking each integral into two parts, we use horizontal rectangles with opposite ends on the legs of the triangle. Symmetry with respect to the vertical axis tells us $M_y = \overline{x} = 0$, so it remains to determine M_x.

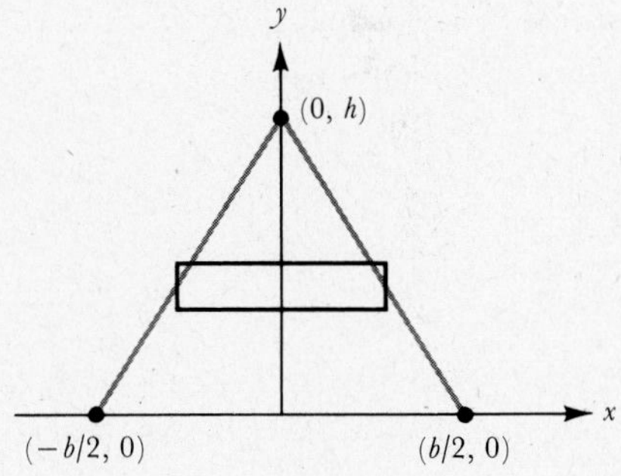

The equation of the right leg is $y_r = (-2h/b)x + h$ which, solved for x, gives $x_r = -by/(2h) + b/2$. The corresponding equation for the left side is $x_l = by/(2h) - b/2$, so the length of a typical rectangle is $x_r - x_l = -by/h + b$. Hence the moment of a typical rectangle with respect to the x axis is $y_i^*(-by_i^*/h + b)\,\Delta y_i$. Thus the total moment is

$$M_x = \int_0^h y\left(-\frac{by}{h} + b\right) dy = b\int_0^h \left(-\frac{y^2}{h} + y\right) dy$$

$$= b\left[-\frac{y^3}{3h} + \frac{y^2}{2}\right]_0^h = b\left[-\frac{h^2}{3} + \frac{h^2}{2}\right] = \frac{bh^2}{6}$$

Also, the area of the triangle is $A = \frac{1}{2}bh$. (This could also be computed using an integral.) Therefore

$$\bar{y} = \frac{M_x}{A} = \frac{bh^2/6}{bh/2} = \frac{bh^2}{6}\cdot\frac{2}{bh} = \frac{h}{3} \quad \bullet$$

You may recall from elementary geometry that the "centroid of a triangle" is the intersection of its medians (a median joins a vertex with the midpoint of its opposite side). You might wish to show that our centroid is in the same place.

PROGRESS TEST 2

1. Determine the centroid of the region R bounded by $y = 4x - x^2$ and the x axis

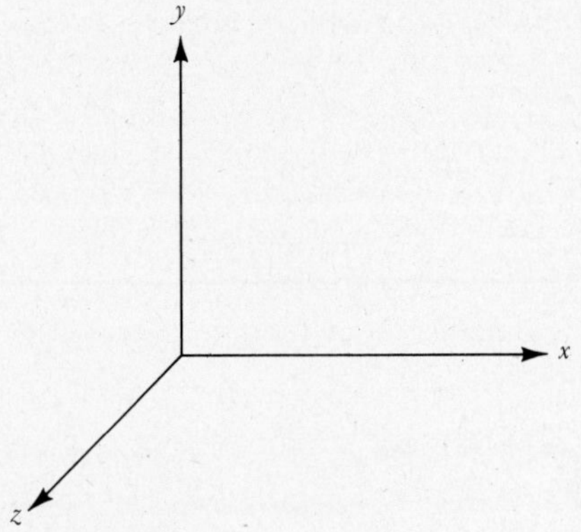

CENTROIDS OF SOLIDS OF REVOLUTION, CURVES, AND SURFACES OF REVOLUTION

Having discussed one-dimensional moments about a point and two-dimensional moments about a line, we use the same ideas to deal with three-dimensional moments about a *plane,* beginning with the simplest type of three-dimensional figures, solids of revolution. We shall define moments of such solids with respect to planes perpendicular to the axis of revolution, which makes it convenient to introduce a third coordinate axis called the z axis. It is perpendicular to the x-y plane, which means it comes out at you from the paper. Since each pair of axes determines a plane, we have the x-y plane, the x-z plane, and the y-z plane. (See Fig. 12.20.)

Fig. 12.20

Now suppose the region R under the graph of $y = f(x)$ from $x = a$ to $x = b$ (where $y = f(x) > 0$ and suitably integrable on $[a, b]$) is revolved about the x axis. The formula for the volume of the solid so generated is developed by first noting that a typical rectangle generates a disk of volume $\pi[f(x_i^*)]^2\,\Delta x_i$. (Here x_i^* is the midpoint of $[x_{i-1}, x_i]$.) Again we assume the solid to be of uniform density, so we can ignore density and mass. We define the moment of this disk with respect to the y-z plane to be

$$x_i^* \cdot \pi[f(x_i^*)]^2\,\Delta x_i$$

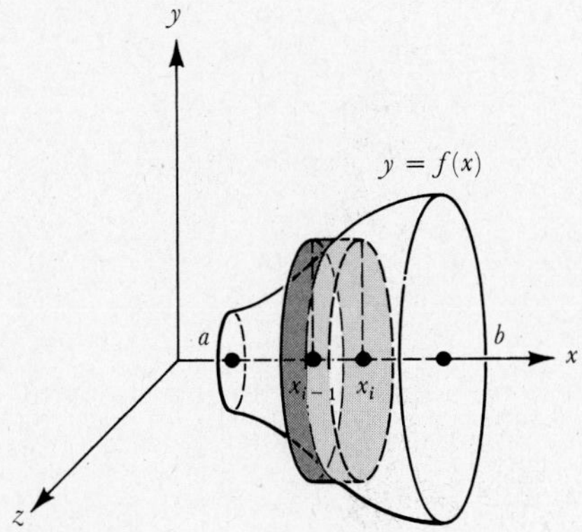

Note that the disk is parallel to the y-z plane while the axis of revolution, the x axis, is perpendicular to the y-z plane. (See Fig. 12.21.)

12.3.7 Definition The total moment of the solid about the y-z plane is

$$M_{yz} = \pi\int_a^b x[f(x)]^2\,dx$$

Fig. 12.21

By symmetry the centroid will lie on the x axis. To determine $\overline{x}$ we need to divide by the volume (instead of area as in two dimensions):

12.3.8 **Definition** $$\overline{x} = \frac{M_{yz}}{V} = \frac{\pi \int_a^b x[f(x)]^2\,dx}{\pi \int_a^b [f(x)]^2\,dx}$$

A similar formula can be developed using the shell approach. (See Exercise 20.)

Example 4 Determine the centroid of the solid hemisphere generated by revolving about the x axis the region under the positive semicircle $f(x) = \sqrt{r^2 - x^2}$ from $x = 0$ to $x = r$.

Solution We can apply (12.3.8) directly:

$$M_{yz} = \pi \int_0^r x(\sqrt{r^2 - x^2})^2\,dx$$

$$= \pi\left[\frac{r^2x^2}{2} - \frac{x^4}{4}\right]_0^r = \frac{\pi}{4}r^4$$

The volume of the hemisphere is $(2/3)\pi r^3$, so

$$\overline{x} = \frac{\pi r^4/4}{2\pi r^3/3} = \frac{3}{8}r \quad \bullet$$

We shall now define the centroid of the plane curve $y = f(x)$ from $x = a$ to $x = b$ where f' exists and is continuous on $[a, b]$. (Think of the curve as a thin wire of uniform thickness and density.) As we saw in Sec. 12.2, a typical approximating chord has length $\sqrt{1 + [f'(x_i^*)]^2}\,\Delta x_i$ where x_i^* is a point in $[x_{i-1}, x_i]$; see Fig. 12.22. We define the moments of this chord about the y and x axes, respectively, to be

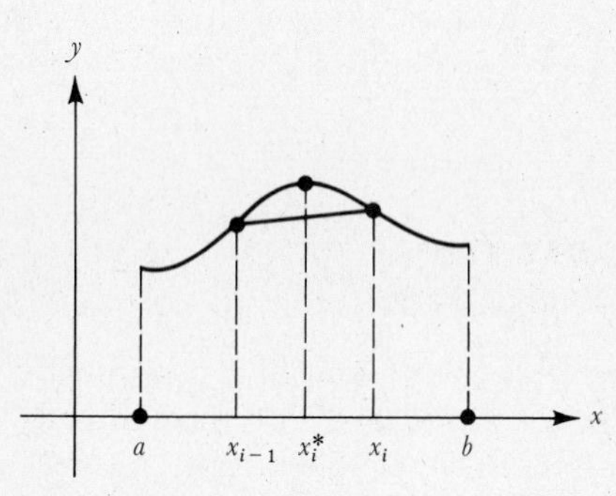

Fig. 12.22

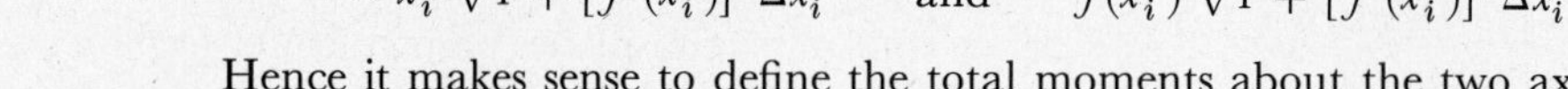

$$x_i^* \sqrt{1 + [f'(x_i^*)]^2}\,\Delta x_i \qquad \text{and} \qquad f(x_i^*)\sqrt{1 + [f'(x_i^*)]^2}\,\Delta x_i$$

Hence it makes sense to define the total moments about the two axes of the curve $y = f(x)$ from $x = a$ to $x = b$ as follows:

12.3.9 **Definition** $$M_y = \int_a^b x\sqrt{1 + [f'(x)]^2}\,dx \qquad M_x = \int_a^b f(x)\sqrt{1 + [f'(x)]^2}\,dx$$

The centroid $(\overline{x}, \overline{y})$ is defined by

$$\overline{x} = \frac{M_y}{s}, \overline{y} = \frac{M_x}{s} \qquad \text{where } s = \int_a^b \sqrt{1 + [f'(x)]^2}\,dx$$

If the same curve is rotated about the x axis, a surface of revolution results, with area

$$A = 2\pi \int_a^b f(x)\sqrt{1 + [f'(x)]^2}\,dx$$

It has its centroid on the axis of revolution so we need only define the moment with respect to the y-z plane—which, as you can surely predict, is

12.3.10 **Definition** $$M_{yz} = 2\pi \int_a^b x \cdot f(x)\sqrt{1 + [f'(x)]^2}\,dx$$

Hence the centroid is at $(\overline{x}, 0)$ where $\overline{x} = M_{yz}/A$.

Example 5 Set up but do not evaluate simplified integrals giving the centroid of the curve $24xy = x^4 + 48$ from $x = 1$ to $x = 3$. (The length of this curve was determined in Sec. 12.2, Progress Test 1, Prob. 2, page 407.)

Solution Examining the solution to the arc length problem, we note that

$$y = \frac{1}{24}x^3 + \frac{2}{x} \qquad \sqrt{1 + \left(\frac{dy}{dx}\right)^2} = \frac{x^2}{8} + \frac{2}{x^2}$$

and thus

$$M_y = \int_1^3 x\left(\frac{x^2}{8} + \frac{2}{x^2}\right) dx$$

and

$$M_x = \int_1^3 \left(\frac{x^3}{24} + \frac{2}{x}\right)\left(\frac{x^2}{8} + \frac{2}{x^2}\right) dx$$

Therefore,

$$\overline{x} = \frac{1}{s}\int_1^3 \left(\frac{x^3}{8} + \frac{2}{x}\right) dx \qquad \text{and} \qquad \overline{y} = \frac{1}{s}\int_1^3 \left(\frac{x^5}{192} + \frac{x}{3} + \frac{4}{x^3}\right) dx$$ ●

Example 6 Set up but do not evaluate simplified integrals giving the centroid of the surface generated by revolving the curve of Example 5 about the x axis.

Solution We can apply (12.3.10) directly, using information noted in Example 5:

$$M_{yz} = 2\pi \int_1^3 x\left(\frac{1}{24}x^3 + \frac{2}{x}\right)\left(\frac{x^2}{8} + \frac{2}{x^2}\right) dx$$

We note from Example 5 that the surface area is $(2\pi)M_x$, so

$$\overline{x} = \frac{1}{M_x}\int_1^3 \left(\frac{x^4}{192} + \frac{x^2}{3} + \frac{4}{x^2}\right) dx$$ ●

PROGRESS TEST 3

In Probs. 1 to 3 we shall use the graph of $hy = rx$ from $x = 0$ to $x = h$, as shown.

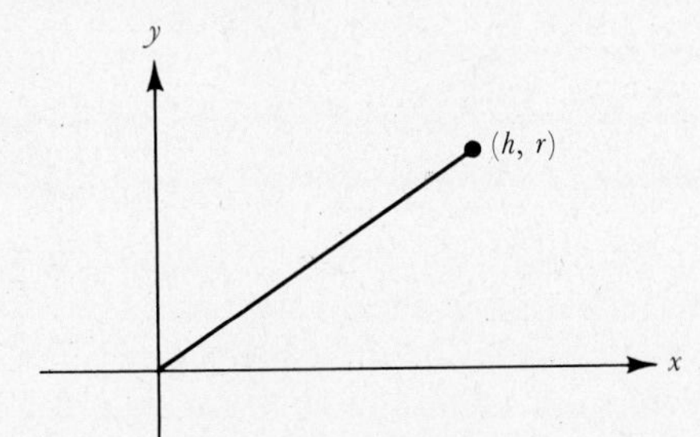

1. Show that the centroid of this line segment is its midpoint (a fact that holds in general).

2. Show that the centroid of the conical solid generated by revolving the region bounded by $hy = rx$ from $x = 0$ to $x = h$ about the x axis is at $\overline{x} = (3/4)h$.

3. Show that the centroid of the conical surface generated by revolving the curve $hy = rx$ from $x = 0$ to $x = h$ about the x axis is $\overline{x} = (2/3)h$.

4. Define the centroid of the surface generated by revolving the graph of a function $x = g(y)$ about the y axis from $y = c$ to $y = d$. (g' is continuous on $[c, d]$.)

PAPPUS' THEOREMS

It was stated earlier that moments and centroids have wide application in science and engineering, in the life and social sciences, and in statistics. Here, however, we shall consider a different sort of application, named after its third-century A.D. Greek discoverer, Pappus of Alexandria.

12.3.11 **Pappus' Theorem for Volumes** Suppose $f(x) \geq 0$ on $[a, b]$ and A is the area of the region R bounded by $f(x)$, the x axis, and the lines $x = a$ and $x = b$.

If $(\bar{x},\bar{y})$ is the centroid of R, then R generates a solid of volume $2\pi\bar{x}A$ when revolved about the y axis, and a solid of volume $2\pi\bar{y}A$ when revolved about the x axis. More generally, if a region R is revolved about some axis, then the volume of the resulting solid is the area of R times the circumference of the circle traveled by the centroid of R.

Proof: The proof of this useful result is surprisingly simple. Consider $2\pi\bar{y}A$. By (12.3.6),

$$\bar{y} = \frac{\frac{1}{2}\int_a^b [f(x)]^2\,dx}{A}$$

so

$$2\pi\bar{y}A = 2\pi\left[\frac{\frac{1}{2}\int_a^b [f(x)]^2\,dx}{A}\right]\cdot A = \pi\int_a^b [f(x)]^2\,dx$$

But this last formula is of course the disk formula (6.6.1) for the volume of the solid generated by revolving R about the x axis. ■

Other versions of Pappus' Theorems are proved similarly. (See the Section and Chapter Exercises.)

Example 7 Determine a formula for the volume of a torus with cross-sectional radius r and central radius R, $r < R$. (See Fig. 12.23.)

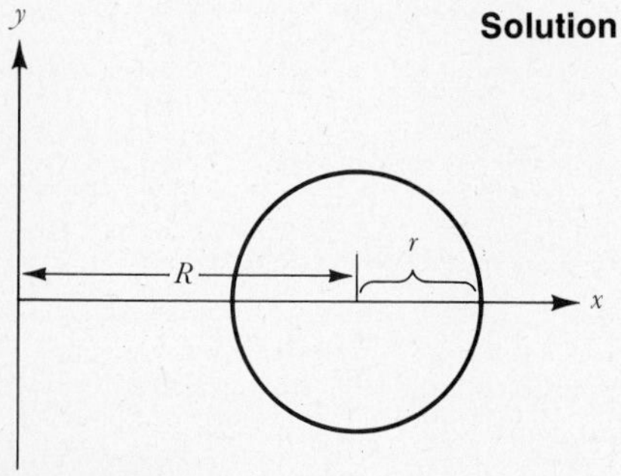

Solution Such a torus can be generated by revolving the circular disk D bounded by $(x - R)^2 + y^2 = r^2$ about the y axis. The centroid of D is at its center $(R, 0)$ at distance R from the y axis. Since the area of D is πr^2 and the centroid travels a circumference of $2\pi R$, the volume of the torus is

$$(2\pi R)(\pi r^2) = 2\pi^2 R r^2$$

(This result agrees with those of Sec. 6.6 Exercises 37 and 38, page 228.) ●

y
R
r
x

Fig. 12.23

Notice that Pappus' Theorem could be used to determine the centroid of a region of known area if we know the volume of the solid it generates when revolved about some axis. (See Exercises 49 to 51.)

MOMENTS OF INERTIA

The product mx of a mass m and a distance x from a given point is sometimes called the "first moment" of that mass about the given point. The product mx^2 is called the *second moment* or *moment of inertia* about the given point. While we are ignoring the units involved, the sum of second moments of a system of masses measures the resistance of that system to rotation about the reference point as well as its resistance to change in velocity when *in* motion. Note that a pair of nonzero masses whose *first* moments add to zero has nonzero *second* moment. (A perfectly balanced barbell resists rotation about the balance point.)

Most of the formulas developed in this section for uniform solids and surfaces have analogs for second moments, with the idea of centroid replaced by *radius of gyration*. However, care must be taken with the density constant, because it does not cancel from all the formulas as it did for centroids. These ideas are explored in Chapter Miscellaneous Exercises 10 to 24. Third and higher moments, used in statistics, are defined analogously. In Chap. 20 we shall also describe situations where the density is a variable.

SECTION 12.3 EXERCISES

1. Given the masses 10, 20, and 30 lb at (2, 3), (5, 9), and (2, 5), respectively, determine (*a*) M_y, (*b*) M_x, and (*c*) the center of mass.

2. Repeat Exercise 1 for the masses 15, 10, 5, and 30 kg at the points $(-3, 2)$, $(4, 12)$, $(1, 1)$, and $(3, -15)$, respectively.

In Exercises 3 to 15 (a) set up integrals and quotients giving the centroid of the region R bounded by the given curves; (b) evaluate these expressions, giving numerical answers.

3. $x + y^2 = 4$ and the y axis

4. $x + y^2 = 4$ and $x = y + 2$

5. $y = x^3 - 3x^2 - x + 3$ and the x axis from $x = -1$ to $x = 1$

6. $y = x^3 - 3x^2 - x + 3$ and the x axis from $x = 1$ to $x = 3$

7. $y = x^3 - 3x^2 - x + 3$ and the x axis

8. $y = x^3$, the x axis, and $x = 1$

9. $y = \ln x$, the x axis, and $x = e$

10. $y = x^2$ and $x = y^2$

11. $y = \sin x$ and the x axis from $x = 0$ to $x = \pi/2$

12. $y^2 = x$ and $2y + 3 = x$

13. $y + 4 = x^2$ and $y + x^2 = 2x$

14. $y = e^x + e^{-x}$ and the x axis from $x = -1$ to $x = 1$.

15. $y = (\ln x)/x$, the x axis, and $x = e$

16. Prove that the centroid of a rectangle is its center. (Let the rectangle be bounded by $x = 0, x = a, y = 0, y = b$.)

17. Prove that the centroid of a circular disk is its center.

18. (*a*) Suppose $f(x) \geq g(x) \geq 0$ on $[a, b]$. Determine a formula for M_x, the moment with respect to the x axis of the region R between the graphs of $f(x)$ and $g(x)$ on $[a, b]$.
(*b*) Determine the moment M_{-4} of the region R of part (*a*) about the line $y = -4$.

19. Determine a formula for the solid generated by revolving the region R of Exercise 18 about the x axis.

20. (Shell Method) Suppose the region bounded by the x axis, $y = f(x)$, $x = a$, and $x = b$ ($f(x) \geq 0$ on $[a, b]$) is revolved about the y axis. Thus a typical vertical rectangle generates a shell whose centroid is at $(0, f(x_i)/2)$ and whose moment with respect to the x-z plane is the product of its volume (ignoring density as usual) and the distance of its centroid from the x-z plane.
(*a*) Determine the total moment of the solid with respect to the x-z plane.
(*b*) Determine the centroid of the solid.

21. Extend the formulas of Exercise 20 to the solid generated by revolving the region R of Exercise 18 about the y axis.

In Exercises 22 to 33 determine the centroid of the solid of revolution generated by revolving the given region about the given axis. You will need to use the results of previous exercises in some cases.

22. R as in Exercise 3 about the y axis

23. R as in Exercise 5 about the x axis

24. R bounded by $y = x^4$ and $y = 2x - x^2$ about the y axis

25. R as in Exercise 24 about the x axis

26. R as in Exercise 5 about the y axis

27. R as in Exercise 6 about the y axis.

28. R bounded by $y = x^2$ and $y = x^3$ about the x axis using the disk method

29. R as in Exercise 28 about the x axis using the shell method (see Exercise 20)

30. R as in Exercise 28 about the y axis using the disk method

31. R as in Exercise 28 about the y axis using the shell method

32. R as in Exercise 4 about the y axis

33. R as in Exercise 5 about the line $y = -3$

34. Determine the centroid of a frustum of a right circular cone with smaller radius r, larger radius R, and height h

In Exercises 35 to 44 determine the centroid of the given curve segment. Exercise numbers refer to Section 12.2 Exercises, page 412.

35. Exercise 1

36. Exercise 2

37. Exercise 3

38. Exercise 4

39. Exercise 11

40. Exercise 12

41. Exercise 13

42. Exercise 14

43. $12xy = 4y^4 + 3$ from $y = 1$ to $y = 3$

44. $x^2 = 4y$ from $(-2, 1)$ to $(2, 1)$

45. Determine a formula for the centroid of the surface generated by revolving the graph of $y = f(x)$ from $x = a$ to $x = b$ about the line $y = k$. Assume that $f(x) \geq k$ and f' is continuous on $[a, b]$.

46. Determine a formula for the centroid of the surface generated by revolving the graph of $y = f(x)$ from $x = a$ ($a > 0$) to $x = b$ about the y axis. Assume f' is continuous and positive on $[a, b]$.

47. Determine the centroid of the surface generated by revolving about the x axis the curve $y = x^4/4 + 1/(8x^2)$ from $x = 1$ to $x = 2$.

48. Repeat Exercise 47 with the curve revolved about the line $y = -2$.

49. Use Pappus' Theorem for Volumes and the fact that the area of a semicircular region of radius r is $(\pi/2)r^2$ to determine its centroid.

50. State and prove a Pappus Theorem for surface areas that will give the surface area in terms of arc lengths and the distance traveled by the centroid of the curve.

51. Use the result of Exercise 50 and the fact that the arc length of a semicircle is πr to determine the centroid of a semicircular curve.

In Exercises 52 to 57 determine the centroid of the surface of revolution whose area is requested in the given exercise from Section 12.2 Exercises.

52. Exercise 18

53. Exercise 19

54. Exercise 20

55. Exercise 27

56. Exercise 28

57. Exercise 29

Section 12.4: WORK AND FLUID PRESSURE

Objectives:

1. Define a definite integral giving the total work done by a continuously variable force acting over an interval.

2. Define a definite integral giving the total force exerted by a fluid on a vertical submerged surface.

3. Use centroids to simplify or avoid the calculations involved in objectives 1 and 2.

WORK

To do work is to use energy, so it is important to have a good quantified definition of "work." A starting point is provided by the following classical definition from physics:

12.4.1 Definition The work W done by a constant force F acting through a distance x along a straight line is defined to be the product Fx.

Generally, the work unit is the product of a distance unit and a force unit. As long as the force is constant, such as in lifting a fixed weight a short distance (compared to the radius of the earth), we do not need calculus to compute the work done; arithmetic suffices. We are interested here in defining and computing work done by a force that *varies* as it acts over some given distance. Hence we assume that the variable force acting from a to b is described by a continuous function $f(x)$, where x varies from a to b. Since the definition (12.4.1) only applies to a *constant* force, we make recourse to the now-standard approach of breaking the interval $[a, b]$ into n subintervals based on a partition P and approximating the force on each subinterval $[x_{i-1}, x_i]$ by a constant $f(x_i)$. Then an approximation to the actual work done on the subinterval is given by

$$W_i = f(x_i)\,\Delta x_i$$

We add these approximations to get an approximation of the total work done over $[a, b]$: $\Sigma_{i=1}^{n} f(x_i)\,\Delta x_i$. The smaller each subinterval, the less the continuous function $f(x)$ can vary on each subinterval, so the better the approximation. Then, as usual, we take the limit of the Riemann sums as $\|P\| \to 0$, resulting in a definite integral.

12.4.2 Definition We define the total work done by the continuously varying force $f(x)$ from $x = a$ to $x = b$ by

$$W = \lim_{\|P\|\to 0} \sum_{i=1}^{n} f(x_i)\,\Delta x_i = \int_a^b f(x)\,dx$$

The simplest and most commonly used illustrative example of a variable force acting over an interval is that required to stretch or compress a simple spring over a fixed distance. Suppose a simple spring is oriented horizontally along the x axis fixed at its left end with its right end at $x = 0$; see Fig. 12.24. The seventeenth century English physicist Robert Hooke discovered that within certain bounds (as long as the spring can return to its natural length), the force necessary to stretch the spring is directly proportional to the distance it is stretched. Thus for some constant k, the *elasticity* of the spring, the force is given by kx.

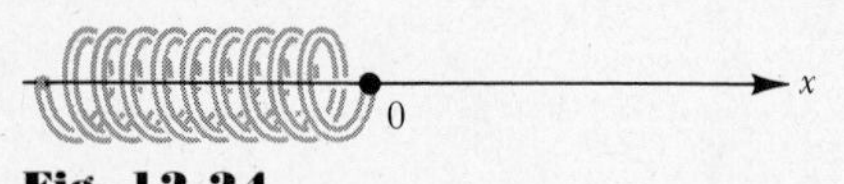

Fig. 12.24

Example 1 Suppose the elasticity of a spring is 4 lb/in. How much work is done in stretching the spring 6 in. beyond its natural length? How much additional work is done in stretching the spring an additional 6 in.?

Solution We orient the spring as in Fig. 12.24. The force necessary to hold the spring stretched at x inches beyond its natural length is $f(x) = 4x$. Hence, by (12.4.2), the work done in stretching it the first 6 in. is

$$\int_0^6 4x\,dx = 2x^2\Big|_0^6 = 72 \text{ in.-lb}$$

while the work done in stretching it the second 6 in. is

$$\int_6^{12} 4x\,dx = 2x^2 \Big|_6^{12} = 216 \text{ in.-lb} \quad \bullet$$

Notice that the elasticity of a spring can be inferred from simple data, such as the fact that it takes a force of 10 lb to hold a spring stretched 2 in. We could conclude that the elasticity is $k = 5$ since we know from Hooke's Law that $10 = k \cdot 2$.

Another sort of work problem involves lifting. The following two examples are intended to illustrate the general approach:

Break the work into simple parts, add these up, and take a limit to define the integral.

The first shows that we can break the work into pieces by breaking *what is being lifted* into pieces, and the second shows that we can break the work into pieces by breaking the *distance* into pieces (as with the spring above).

Example 2 Determine the work done in pumping all the water from a vertical cylindrical tank 6 ft in diameter and 10 ft tall to a height 8 ft above the top of the tank.

Solution The key is to realize that each thin horizontal slab of water must be lifted a different distance. (See Fig. 12.25.)

Measuring x from the top of the tank downward, a typical slab of thickness Δx_i has volume $\pi 3^2\,\Delta x_i$. The density of water is about 62.4 lb/ft^3, so its weight is $62.4(\pi 3^2\,\Delta x_i)$. (In the following Progress Test 1 we shall consider a conical tank, where slabs have varying radius.) This slab needs to be lifted $(x_i + 8)$ ft, so the amount of work needed to do this is about

$$(62.4)(\pi \cdot 3^2\,\Delta x_i)[x_i + 8] = (62.4)9\pi(x_i + 8)\,\Delta x_i$$

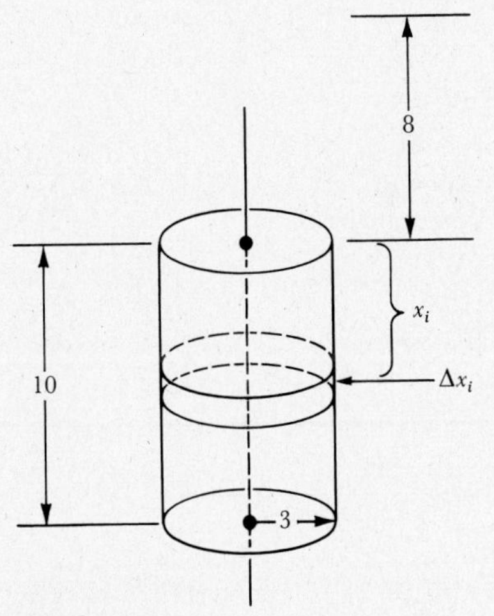

Fig. 12.25

As usual, we add up these "little works" based on the slabs to approximate the total work and then let the thickness of the slabs approach zero to get a definite integral:

$$\int_0^{10} (62.4)9\pi(x + 8)\,dx = (62.4)9\pi\left[\frac{x^2}{2} + 8x\right]_0^{10}$$

$$= (62.4)9\pi(130) \approx 229{,}361.4 \text{ ft-lb}$$

Note that we added slabs from $x = 0$ (the top of the tank) to $x = 10$ (the bottom). Hence the limits of integration 0 and 10. $\bullet$

We should point out that this solution could have been organized a bit differently, but with the same final answer. For example, instead of saying $x = 0$ is at the top of the tank, we could have chosen $x = 0$ to be the destination of the water. Then a typical slab would be lifted x_i feet, but we would add up work based on slabs from $x = 8$ (top of tank) to $x = 18$ (bottom). This yields the integral

$$\int_8^{18} (62.4)9\pi x\,dx = (62.4)9\pi\left[\frac{x^2}{2}\right]_8^{18}$$

$$= (62.4)9\pi[162 - 32]$$

$$= (62.4)9\pi(130)$$

as before.

In general, the way we choose coordinate axes in work problems is unimportant provided we are consistent after making the choice.

Example 3 A 60-ft chain weighing 1.5 lb/ft is hanging freely off a bridge (see sketch). How much work is done in pulling the chain up onto the bridge?

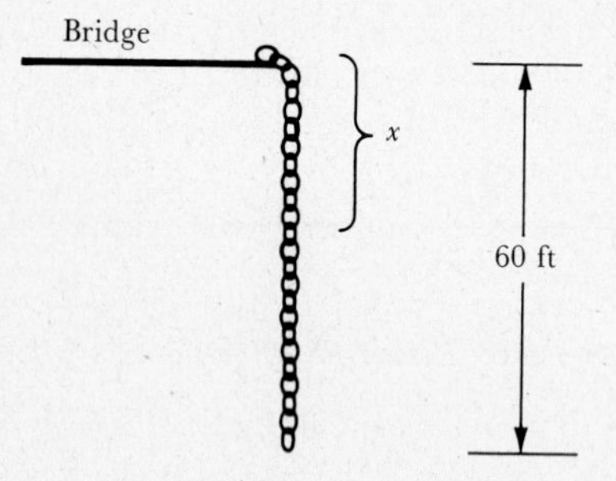

Solution The key is to realize that as the chain is pulled up, the amount of force needed decreases. If x feet are hanging, the force necessary to pull the chain a small distance Δx is about $1.5x$ lb. Hence the work thus accomplished is $1.5x\,\Delta x$. Now $0 \le x \le 60$, so the total work is

$$\int_0^{60} 1.5x\,dx = \frac{3}{4}x^2\Big|_0^{60} = 2700 \text{ ft-lb} \quad \bullet$$

This chain-pulling problem can be analyzed from another point of view to yield the same result. Instead of thinking that we are pulling x feet of chain Δx feet, we can think of the chain as being made of short pieces (which, of course, it is) of length Δx_i. Then such a piece, weighing $(1.5)\,\Delta x_i$ is pulled x_i feet, so the work done in pulling each Δx_i length of chain to the bridge is $1.5x_i\,\Delta x_i$. We then add up such "little works" from $x = 0$ to $x = 60$ to get $\Sigma\, 1.5x_i\,\Delta x_i$, and pass to the definite integral as usual to obtain the same integral as before.

Note that *power* is work per unit time. In United States customary units, for example, 550 ft-lb/s is one unit of power—called the *horsepower* (hp). A unit of power in the metric system is the *watt* (W). One horsepower equals 746 W. See Exercises 8 and 9 for further discussion.

PROGRESS TEST 1

1. Find the work done in stretching a spring from 2 in. to 8 in. if it takes 30 lb to hold it stretched 2 in.

2. By following the indicated steps, determine the work done in pumping the oil out of the top of a conical tank 12 ft in diameter at the top and 18 ft deep. The oil weighs 50 lb/ft^3.
(*a*) Determine the weight of a slab of thickness Δy, which is at distance $18 - y$ from the top of the tank. (See sketch.)
(*b*) Determine the work done in lifting that slab of oil.
(*c*) Determine the total work done in pumping out the tank.

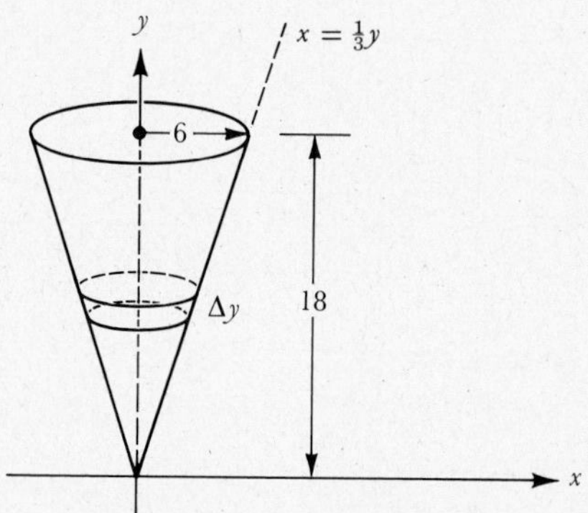

3. Determine the work done in lifting a 400-lb weight 100 ft with a chain that weighs 5 lb/ft.

FLUID PRESSURE

The total force on the horizontal flat bottom of a full 6-ft-deep, 10- by 20-ft rectangular swimming pool is the total weight of the water; that is,

$$(62.4)(10 \cdot 20 \cdot 6) = 74{,}880 \text{ lb}$$

Everything is constant in this problem, so arithmetic, not calculus, suffices to determine the total force. The opposite is the case when we attempt to determine the force on one vertical end of the pool, because the force per square foot varies continuously with the depth. We digress briefly.

Force per unit area is usually called "pressure." It is, in effect, the weight of a column of fluid above a square unit of area. Most liquids, such as water, are incompressible, and thus they have a constant density, say ρ. This means that the pressure at depth d is merely the density ρ times the volume of the column:

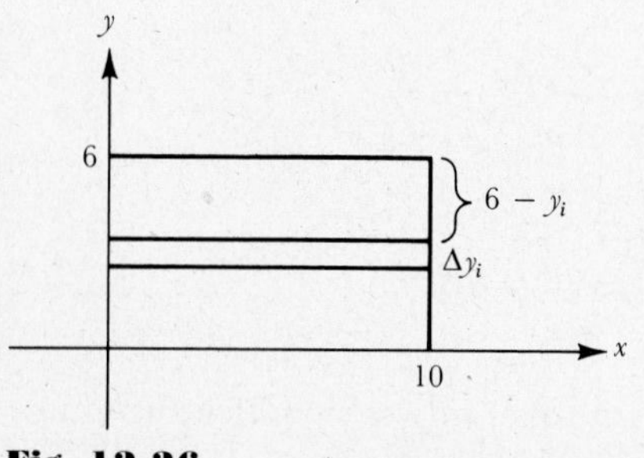

Fig. 12.26

$$\rho(A \cdot d) = \rho(1 \cdot d) = \rho d$$

We shall use one more basic fact about fluid force first made explicit by the seventeenth century French mathematician Blaise Pascal: Within any fluid, the force exerted at each point is equal in all directions.

Returning to the pool, we can now conclude that the pressure at any level on the vertical face of the pool is 62.4 times the depth at that level. As usual we impose a coordinate system on the situation. (See Fig. 12.26.) The pressure at depth $6 - y_i$ is $62.4(6 - y_i)$ and, although the pressure changes with the depth, we can assume that the pressure is constant on a narrow horizontal strip of thickness Δy_i straddling the depth $6 - y_i$. Under this assumption, the total force on the strip is approximately the product of the pressure per unit area and the total area:

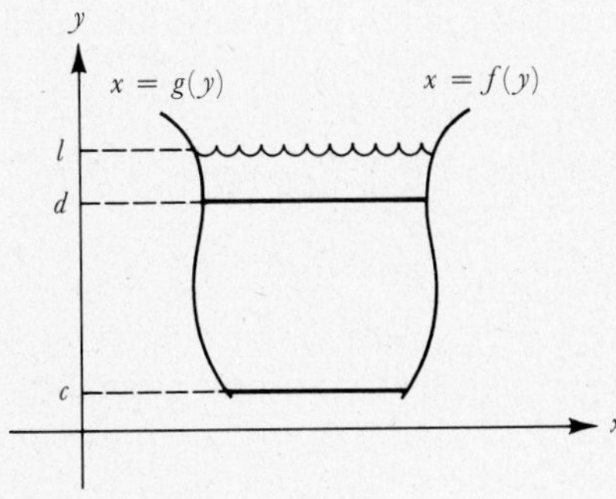

Fig. 12.27

$$[62.4(6 - y_i)][10\, \Delta y_i] = 624(6 - y_i)\, \Delta y_i$$

If we were to add all these approximations based on a partition P of the vertical interval $[0, 6]$ into n such subintervals, we would have an approximation of the total force on the end wall of the pool:

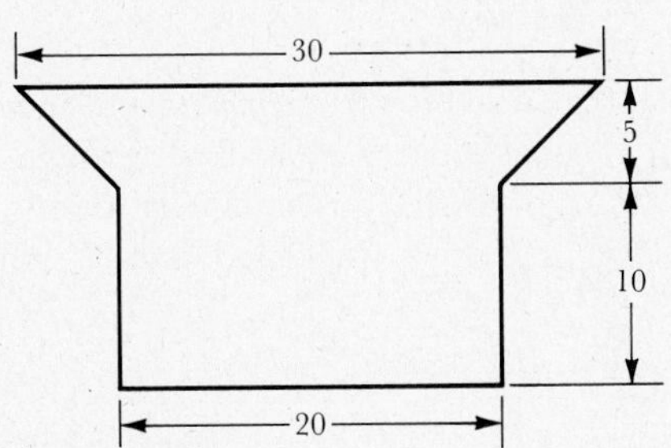

Fig. 12.28

$$\sum_{i=1}^{n} 624(6 - y_i)\, \Delta y_i$$

As usual, we take the limit as $\|P\| \to 0$ to obtain a definite integral that describes the total force exactly:

$$624 \int_0^6 (6 - y)\, dy = 11{,}232 \text{ lb}$$

Notice that this computation involved only the end wall of the pool and the depth of the water, but no other physical features. Hence this could also have been the computation of the force on the vertical face of a dam of the same size at the end of a 100-mi-long canal, or even the computation of the force on a floodgate on top of a much larger dam.

But, dams or pools, we use the same routine: *Take a problem whose crucial element (pressure in this case) is a variable, break it into small enough parts where we can assume the element is constant, add the parts to approximate the whole, and then take the limit as the size of the pieces approaches* 0 *to obtain a precise definite integral description.*

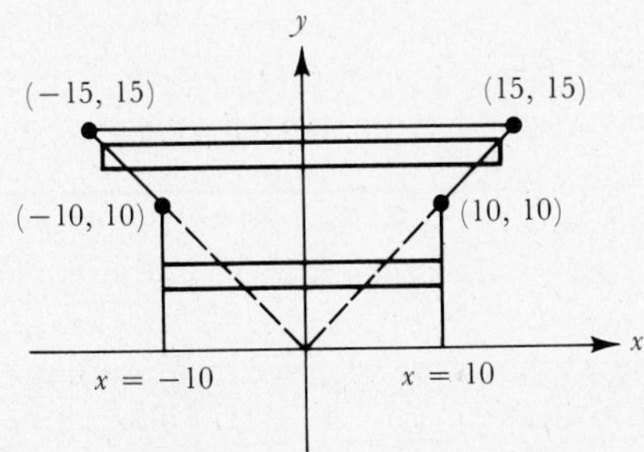

Fig. 12.29

This previous fluid pressure problem generalizes directly to the force exerted by a liquid of constant density ρ on a vertical surface whose shape is defined by functions $x = g(y)$ and $x = f(y)$ from $y = c$ to $y = d$ with $g(y) \leq f(y)$ and continuous on $[c, d]$, as in Fig. 12.27. Assume that the liquid reaches level $l \geq d$ on the superimposed y axis. Then the total force is given by

12.4.3 $$\text{Total force} = \rho \int_c^d [f(y) - g(y)](l - y)\, dy$$

Example 4 Determine the total force on a dam whose inside vertical face is shaped as in Fig. 12.28, with water reaching to its top.

Solution We impose a coordinate system with the bottom of the dam resting on the x axis and with the y axis on the vertical centerline. (See Fig. 12.29.) From $y = 0$ to $y = 10$, the width of a typical horizontal strip is 20. From $y = 10$ to $y = 15$ the boundary at the right has equation $y = x$, so $f(y) = y$, and at the left it has equation $y = -x$, so $g(y) = -y$. Hence a typical strip has area

$$[f(y) - g(y)]\, \Delta y = [y - (-y)]\, \Delta y = 2y\, \Delta y$$

Over both parts of the dam, a typical strip has depth $15 - y$, so the total force is

$$F = 62.4 \int_0^{10} 20(15 - y)\,dy + 62.4 \int_{10}^{15} 2y(15 - y)\,dy$$

$$= 62.4\left[2000 + \frac{875}{3}\right] = 143{,}000 \text{ lb} \quad \bullet$$

In doing problems of this type, we strongly suggest that you not attempt to apply (12.4.3) directly but that you make a fresh approach to each problem, beginning with a sketch and an appropriate choice of coordinates.

PROGRESS TEST 2

1. Determine the total force on the vertical end of a symmetric triangular trough that is 6 ft deep, 4 ft across the top, and filled with a liquid of density 50 lb/ft^3.

2. × Suppose 3-ft flashboards are added to the top of the dam in Example 4, surmounting the dam with a 3- by 30-ft rectangle, and suppose water rises to the top of the flashboards.
(*a*) What is the total force on the flashboards?
(*b*) What is the total force on the whole dam, including the flashboards? [Note that it is *not* the sum of (*a*) and the result of Example 4.]
(*c*) What is the force added to the *original* part of the dam because of the flashboards? What percentage increase is this? (This problem illustrates why lowering the water level behind a weakened dam is so crucial.)

USING CENTROIDS IN WORK AND FLUID PRESSURE PROBLEMS: PAPPUS-TYPE THEOREMS

From our study of centroids—or centers of mass when physical quantities are involved—we know that a thin, homogeneous plate balances on its centroid—or center of mass. A similar statement holds for solids. In essence, the centroid point is a point that can represent the whole plate or solid; it is a kind of "average point" that for some purposes can stand for all the others. Recalling from Sec. 6.4 the relation between integrals and averages, the use of the word and idea of "averages" should not be entirely surprising. It has some useful consequences as well.

Consider the work done in pumping the water from the cylindrical tank in Example 2 to a height 8 ft above the top of the tank. (See Fig. 12.25.) The tank is 10 ft tall, so its centroid is 5 ft above the bottom. The radius of the tank is 3 ft and thus its volume is $\pi 3^2 \cdot 10 = 90\pi$ ft^3, so the total weight of the water is $(62.4)(90\pi)$. If all the water were concentrated at the centroid, then the work done in lifting it the 13 ft to 8 ft above the top of the tank is

$$(62.4)(90\pi)(13) = 229{,}361.4 \text{ ft-lb}$$

which is exactly the result of Example 2.

A similar analysis applies to Example 3, where we determined the work done in pulling a hanging 60-ft chain onto a bridge. We could have regarded all the weight of the chain as being concentrated at its centroid—which happens to be its midpoint in this case: 30 ft below the bridge. The total weight of the chain is $(1.5)(60) = 90$ lb and is thought of as being lifted 30 ft, yielding a total work of $(90)(30) = 2700$ ft-lb, just as before.

We can use similar thinking to determine the total force on one side of a submerged vertical surface. We think of the entire surface as represented by its centroid and compute the pressure at that depth. Then the total force is the product of the total area and the pressure at the centroid.

The force on the 10-ft-wide, 6-ft-deep rectangular end wall of the swimming pool discussed turned out to be 11,232 lb. Using centroids, we note that the area is 60 ft^2, the centroid is at depth 3 ft, and the pressure at 3 ft is (64.2)(3), and so the total force should be, as expected,

$$(62.4)(3)(60) = 11{,}232 \text{ lb}$$

In general, the choice whether or not to use the centroid approach in work or pressure problems depends upon whether the centroid is known, which is the case for most common symmetric shapes.

PROGRESS TEST 3

1. Use the centroid result of Sec. 12.3, Progress Test 3, Prob. 2 to determine the work done in pumping the oil of density 50 lb/ft^3 out the top of an 18-ft-tall conical tank of radius 6 ft (vertex down). Compare with Sec. 12.4, Progress Test 1, Prob. 2.

2. Repeat Prob. 1 with the tank upside down (vertex up).

3. Use the centroid result of Sec. 12.3, Example 3, to determine the total force on the vertical end of a 6-ft-deep symmetric triangular trough 4 ft across the top filled with a liquid of density 50 lb/ft^3. Compare with Sec. 12.4, Progress Test 2, Prob. 1.

SECTION 12.4 EXERCISES

In Exercises 1 to 12 determine the work done in accomplishing the given task.

1. Pumping the contents of a half-full cylindrical water tank to a level 5 ft above the tank. The tank is standing on its end, has a radius of 6 ft and a length of 8 ft.

2. Pumping full of water a conical tank through a hole in the bottom. The tank is 15 ft deep and 10 ft in diameter across the top.

3. Lifting a 200-lb crate a height of 60 ft with a 2-lb/ft chain, 15 ft of which are used in wrapping the crate.

4. Pumping a cylindrical tank full of 60-lb/ft^3 oil through the bottom of the tank. Its radius is 9 ft, its bottom is a hemisphere, and its cylindrical height is 10 ft. (This will require two integrals.)

5. Stretching a spring from 5 in. to 8 in. if it takes 3 lb to hold it stretched 3 in.

6. Filling the tank in Exercise 4 through an opening at the top from a source at the level of the bottom of the tank.

7. Lifting a leaky pail of water 100 ft out of a well with a 2 lb/ft cable. The pail begins with 50 lb of water and leaks at a rate of 1 lb every 4 ft.

8. (*a*) Emptying the water from a full 12-ft-long trough through a pipe 3 ft above the top of the trough. The vertical cross sections are isosceles trapezoids 2 ft wide at the bottom, 6 ft wide at the top, and 2 ft deep.
(*b*) If a 1/10-hp pump does 55 ft-lb of work per second, how long will it take this pump to empty the trough?
(*c*) Assuming a 50 percent efficient electric pump (only half the power fed in is put out by the pump), what power source must be provided (in terms of watts) to empty the trough in 1 min?

9. Repeat Exercise 8 for a hemispherical tank (flat on top) 20 ft in diameter that is to be emptied through a discharge pipe at the top of the tank.

10. Lifting a 500-lb anchor from a depth of 50 ft to a deck 10 ft above the water using a cable which weighs 2 lb/ft out of the water and 1.5 lb/ft in the water. The anchor weighs 400 lb in the water. (Break this into several parts. You can assume that the entire anchor passes through the water surface instantly.)

11. Lifting a 2-lb/ft chain that hangs doubled from a building down to a level of 100 ft from the top by (*a*) pulling on one side of the chain and (*b*) pulling on both sides simultaneously.

12. Stopping a 4000-lb automobile that is traveling at 88 ft/s (60 mi/h). The brakes provide a deceleration of -10 ft/s^2. (*Note:* Force equals the mass of the vehicle times the acceleration.)

In Exercises 13 to 17 determine the total force on the indicated vertical surface.

13. (*a*) A 4-ft-wide, 6-ft-tall rectangular exit gate whose top is 24 ft below the surface of water.
(*b*) Same as (*a*) but with the water lowered by 12 ft.

14. The end of a water-filled trapezoidal-cross-sectional water trough 2 ft wide at the bottom, 4 ft wide at the top, and 3 ft deep.

15. (*a*) The outer surface of a 2-ft-diameter circular submarine porthole whose center is 1000 ft below the surface of the ocean (assume that the density of seawater is 64 lb/ft^3).
(*b*) What is the average pressure per square inch on this surface?
(*c*) What is the total force on the top half of the window?
(*d*) What is the total force on the bottom half of the window? [Use part (*a*).]

16. The end of a water trough, with a triangular (isosceles) cross section, that is 4 ft deep and 4 ft across the top.

17. The end of a symmetric dam as pictured in the sketch.

18. How much must the water behind the dam in Exercise 17 be lowered to cut the force on the dam in half?

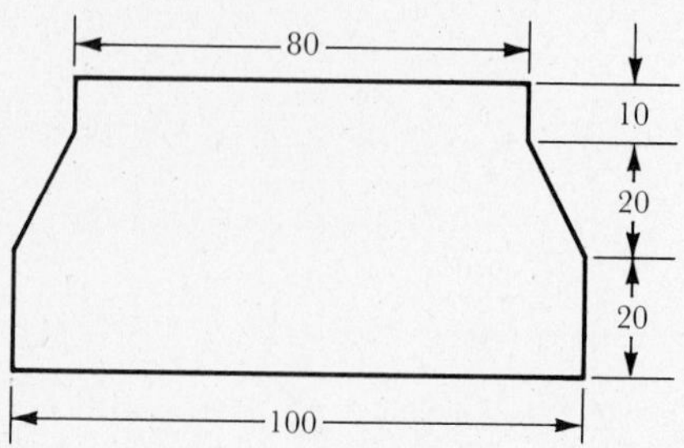

In Exercises 19 to 28, use the centroid approach to solve the corresponding exercise.

19. Exercise 1
20. Exercise 2
21. Exercise 4 (do each part separately and add them)
22. Exercise 6
23. Exercise 13
24. Exercise 14
25. Exercise 15(*a*)
26. Exercise 15(*c*)
27. Exercise 17
28. Exercise 16

29. Ignoring the weight of the tank, determine the work done in standing a cylindrical tank on end if the tank is horizontal, full of 60-lb/ft^3 oil, and 30 ft long, and has a diameter of 10 ft.

30. From Newton's Law of Gravitation it can be shown that the force of gravity exerted on an object which weighs w lb at the earth's surface is $(4000)^2 w/x^2$, where x is the number of miles from the center of the earth. (The radius of the earth is 4000 mi.) Determine the work done (in foot-pounds) in lifting a payload of 10 lb to a height 400 mi above the surface of the earth.

31. Suppose the rocket doing the lifting in Exercise 30 begins with 200 lb of fuel and consumes this linearly (with respect to distance x), exhausting the fuel supply at $x = 4400$. What amount of work was expended in lifting the fuel. (Note $x = 4000$ at liftoff.)

32. Assuming appropriate units, two particles of opposite electrostatic charges e_1 and e_2 attract each other with a force of $e_1 e_2/x^2$, where x is the distance between them. If our particle of 3 units of charge is fixed at the origin, how much work is done in moving a particle of opposite charge of 4 units from $x = 1$ to $x = 10$?

33. What is the total work done by a 200-lb man who climbs a ladder to a 50-ft-high platform carrying the end of an uncoiling rope (of length greater than 50 ft) that weighs 1 lb/ft?

34. Same as Exercise 33, but here the rope is only 30 ft long.

35. Same as Exercise 33, but here the rope is 50 ft long and *dangling from the platform.* He carries the loose end up while the other end is attached to the platform.

36. A vertical a by b rectangular plate has its top edge at the surface of a liquid. Determine how much the plate should be lowered to double the total force on the plate.

CHAPTER EXERCISES

Review Exercises

In Exercises 1 to 17 graph the given pair of parametric equations.

1. $x = 3 + 2 \sin t, y = -2 + 4 \cos t, 0 \le t \le 2\pi$
2. $x = |t|, y = t^2$
3. $x = \sin 2t, y = 2 \sin^2 t, 0 \le t \le 2\pi$
4. $x = 2 + \sec t, y = 3 + 2 \tan t, -\pi/2 < t < \pi/2$
5. $x = 3e^t, y = 1 - e^t$
6. $x = 1 - 3t, y = 2 + 7t$
7. $x = 1 + 1/t, y = t - 1/t$
8. $x = -1 + 3 \sin s, y = 2 + 4 \cos s, 0 \le s \le \pi$
9. $x = 2 + t^2/2, y = -1 + t^3/8$
10. $x = t^2, y = t^4 + 1$
11. $x = t^2 + t + 1, y = t^2/2 + t - 1$
12. $x = t, y = 1/t$
13. $x = 2t + 1, y = \ln(t^2 + 1)$
14. $x = 3 \sec t, y = 2 \tan t, \pi/2 < t < 3\pi/2$
15. $x = t + 1, y = t^3$
16. $x = \ln t, y = e^{4t}$
17. $x = \cos t, y = -3 + \sin t, -\pi/2 \le t \le 3\pi/2$

In Exercises 18 to 21, (a) show that the graph of the given parametric equations is a subset of $y = x^2 - 1$, (b) determine the slope of the line tangent to the graph at $t = 1$, and (c) give the equation of this line in point-slope form.

18. $x = \sec t, y = \tan^2 t$
19. $x = 1/t, y = (1 - t^2)/t^2$
20. $x = e^{-t}, y = e^{-2t} - 1$
21. $x = \sin t, y = -\cos^2 t$

In Exercises 22 to 28 determine an equation for the tangent line to the graph at the indicated point.

22. $x = \sec t, y = -\tan^2 t, t = \pi/4$

23. $x = \cos^3 t, y = \sin^3 t, t = \pi/6$

24. $x = e^t/(1 + e^t), y = 1 + 2e^t, t = 0$

25. $x = 2t^2 + t + 1, y = 1 - 3t, t = 0$

26. $x = 2t/(1 + t^2), y = (1 - t^2)/(1 + t^2), t = 1/2$

27. $x = \cos t, y = 1 + \sin t, t = \pi/4$

28. $x = t^2 + t + 1, y = (t - 1)^3, t = 3$

29. Find all values of t where the curve $x = \cos^4 t, x = \sin^4 t$ has vertical and horizontal tangents, $-2\pi \leq t \leq 2\pi$.

30. Repeat Exercise 29 for the curve $x = \sin^3 t, y = \cos^3 t$, $-2\pi \leq t \leq 2\pi$.

In Exercises 31 to 45 determine the length of the given arc.

31. $x = t, y = 2t^{3/2}, 0 \leq t \leq 4$

32. $x = t^2, y = t - 1, -1 \leq t \leq 1$

33. $x = 2t^3, y = 3t^2, 0 \leq t \leq 1$

34. $x = t, y = \cosh t, -1 \leq t \leq 1$

35. $y = \ln(1 - x^2), 0 \leq x \leq 1/2$

36. $16y = x^2 - 32 \ln x, 4 \leq x \leq 8$

37. $x = e^t \sin t, y = e^t \cos t, 0 \leq t \leq \pi$

38. $x = 2 \sin t, y = 2 \cos t, 0 \leq t \leq \pi$

39. $x = \ln(\sin t), y = t + 1, \pi/4 \leq t \leq \pi/2$

40. $6x = y^3 + 3/y, 1 \leq y \leq 3$

41. $f(y) = \ln(\csc y), \pi/6 \leq y \leq \pi/4$

42. $f(x) = \ln[\coth(x/2)], 1 \leq x \leq 2$

43. $x = t \cos t, y = t \sin t, 0 \leq t \leq \pi$

44. $x = t/2, y = t^4/4, 1 \leq t \leq 2$

45. $x = t^2/4 + (1/2)t, y = 2\sqrt{t}, 1 \leq t \leq 4$

In Exercises 46 to 56 determine the area of the surface generated by revolving the given arc about the given axis.

46. $x = t^2, y = 2t, 0 \leq t \leq \sqrt{3}$, x axis

47. $x = 2 \cos^3 t, y = 2 \sin^3 t, 0 \leq t \leq \pi/2$, x axis

48. The arc in Exercise 36, y axis

49. The arc in Exercise 36, x axis

50. $3y = x^3, 0 \leq x \leq 2$, x axis

51. $3y = x^3, 0 \leq x \leq 2$, y axis

52. $xy = 1$ from $(1, 1)$ to $(2, 1/2)$, x axis

53. $y = \ln x, 1 \leq x \leq 5$, y axis

54. $x = 2 \ln t, y = t^2, 1 \leq t \leq 2$, x axis

55. $x = \cos^2 \theta, y = (1/2) \sin^2 \theta, 0 \leq \theta \leq \pi/4$, x axis

56. $x = 3t^2, y = 2t^3, 0 \leq t \leq 1$, y axis

In Exercises 57 to 87 refer to the diagram shown. Set up, but do not evaluate, integrals giving the requested quantity in Exercises 57 to 81.

57. The length of $y = x^2$ from $x = 0$ to $x = 1$.

58. The length of $y = x^3$ from $x = 0$ to $x = 1$.

The area of the surface generated by revolving [from (0, 0) to (1, 1)]:

59. $y = x^2$ about the x axis

60. $y = x^2$ about the y axis

61. $y = x^3$ about the y axis

62. $y = x^3$ about the x axis

63. $y = x^2$ about the line $y = 1$

64. $y = x^2$ about the line $x = 1$

In Exercises 65 to 81 give each coordinate of the centroid by dividing the integral that gives the appropriate moment by the appropriate symbol A, V, s, *or* S *(area, volume, arc length, or surface area, respectively).*

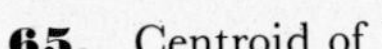

65. Centroid of I

66. Centroid of III

67. Centroid of II + III

68. Centroid of I + II

69. Centroid of II

In Exercises 70 to 75 give the centroid of the solid generated by revolving the given region about the given axis.

70. I about x axis

71. II about x axis

72. III about x axis

73. I about y axis

74. II about y axis

75. III about y axis

In Exercises 76 and 77 give the centroid of the given arc from (0, 0) to (1, 1).

76. $y = x^2$

77. $y = x^3$

In Exercises 78 to 81 give the centroid of the surface of the corresponding exercise.

78. Exercise 60

79. Exercise 61

80. Exercise 62

81. Exercise 59

82. Suppose a tank in the shape of the solid in Exercise 70 is full of a homogeneous liquid of density ρ. Determine the work (in unspecified units) done in pumping out the tank through the top, (*a*) without using centroids and (*b*) using centroids. (*c*) How much work is done by the liquid if it runs out through a hole in the bottom of the tank?

83. Repeat Exercise 82 using the solid in Exercise 71.

84. Repeat Exercise 82 using the solid in Exercise 74.

85. Repeat Exercise 82 using the solid in Exercise 75.

86. Suppose a tank in the shape of the solid in Exercise 73 is divided in half on its vertical cross section by a vertical plate. Determine the total force on one side of the plate, (*a*) without using centroids and (*b*) using centroids.

87. Repeat Exercise 86 using the solid in Exercise 75.

88. If a spring has an unstretched length of 6 in. and requires a force of 4 lb to stretch it to 8 in., determine the work done in stretching the spring to 12 in.

89. Determine the centroid of a typical triangle of base b and height h. (Let its base sit on the x axis and its third vertex on the y axis.)

90. Determine the area inside the ellipse enclosed by the graph of $x = a\cos t$, $y = b\sin t$, $0 \le t \le 2\pi$.

91. Compute the length of the line segment joining (a, b) and (c, d), using (*a*) the distance formula and (*b*) the arc length formula.

92. Determine the length of $9y^2 = (x^2 + 2)^3$ from $x = 0$ to $x = 6$.

Miscellaneous Exercises

1. Determine the work done in turning a conical tank upside down if the tank of height 60 ft and radius 30 ft originally has its point in an upward position and is full of gasoline of density 40 lb/ft^3. Ignore the weight of the tank.

2. Suppose the tank in Exercise 1 is entirely comprised of sheet metal weighing 2 lb/ft^2. What is the work done in turning it on end—ignoring its contents? (*Hint:* Use centroids and treat the flat bottom separately, totaling the two parts of the tank for an answer.)

3. Determine the work done by gravity when one end of a doubled-up 200-ft (in total length) hanging chain is dropped to hang its full length.

4. What is the total force on one side of a vertical divider that divides the tank in Exercise 1 in half: (*a*) before it is turned on end, and (*b*) after it is turned on end?

5. A hemispherical tank (flat side up) is full of a fluid of density ρ. What is the total force on one side of a vertical dividing wall in this tank if the wall is halfway between the center and outside rim of the tank?

6. The average pressure on a vertical surface submerged in a fluid of constant density is the total force divided by its area. Does this idea relate to the Mean-Value Theorem for Integrals?

7. Give a precise statement and a proof analogous to that given for Pappus' Theorem for the fact that we can calculate the total force on a submerged vertical surface by using its centroid, its area, and the pressure at the centroid.

8. Use our formulas to determine the volume of the frustum of a cone whose larger radius is R, whose smaller radius is r, and whose height is h. (This will show that our results are consistent with the hypotheses on which they are based.)

9. An object was dragged 60 ft over uneven ground and the force used was measured at 10-ft intervals, beginning just as the object began to move. The measurements were, respectively 60, 30, 40, 50, 20, 10, and 30 lb.
(*a*) What was the approximate work done if the Trapezoid Rule is used?
(*b*) Repeat (*a*) using Simpson's Rule.

Exercises 10 to 24 concern the development and application of formulas for second and higher moments. Let R *be the region between* $f(x) \ge 0$ *and* $[a, b]$.

10. Suppose a homogeneous plate of constant density ρ has the shape of the region R. Define the total moment of inertia I_y of this plate about the y axis (use a definite integral).

11. Repeat Exercise 10, but give I_x (about the x axis).

12. Define the "radius of gyration" of the plate analogous to the center of mass. (If stuck, see Exercise 20.)

13. Determine the radius of gyration of a homogeneous plate of constant density ρ that has the shape of the region bounded by $y = e^{-x}$ from $x = 0$ to $x = 2$.

14. Define the third moment of a homogeneous plate of constant density ρ with respect to each axis (the plate has shape R as in Exercises 10 to 12.)

15. Determine the third moments of the plate described in Exercise 10.

16. Determine the radius of gyration of a rectangle of width a and height d, oriented with its base on the x axis and lower left corner at $(0, 0)$.

17. Determine the moment of inertia with respect to the yz plane of a solid of density ρ generated by revolving the region R about the x axis (denoted I_{yz}).

18. Suppose f' exists and is continuous on $[a, b]$. Determine the moments of inertia I_y and I_x of a thin wire of uniform density ρ whose graph is in the shape of $y = f(x)$.

19. For $f(x)$ as in Exercise 18, determine the moment of inertia I_{yz} of a uniform surface of constant density ρ generated when $y = f(x)$ is revolved about the x axis.

20. The *radius of gyration* of a region with respect to an axis l is defined to be the number r_l, where $r_l^2 = I_l/A$. Similarly, the radius of gyration of a solid about plane P is a number r_P where $r_P^2 = I_P/V$. Curves and surfaces of revolution are handled similarly. Assume throughout that all densities are constant and equal ρ. Determine the radii of gyration (with respect to the two axes) of a horizontal rectangle whose lower left corner is at $(0, 0)$ and whose upper right corner is at (a, b) $(a, b > 0)$.

21. Determine r_{yz} for the cylinder generated by revolving the above rectangle about the x axis. (See Exercise 20.)

22. Determine the *center of gyration* (using the two radii of gyration) of the region under the first arch of $y = \sin x$ (on $[0, \pi]$).

23. Determine the radius of gyration with respect to a plane containing its base of a right circular cone of radius R and height h.

24. Determine the radius of gyration of a sphere of radius R with respect to a plane a units from its center.

25. Suppose that a vertical plate in the shape of the region R, bounded by the continuous curves $x = g(y)$, $x = f(y)$, $y = c, y = d$, $(g(y) \leq f(y)$ on $[c, d])$, is submerged in a compressible fluid with its top edge l units below the surface. If the density of the fluid increases continuously with the depth according to the function $\rho(y)$, give the total force on the plate (in terms of a definite integral).

26. (*a*) In the sketch shown, $OP = AB$. Give parametric equations for the locus of all points P so determined. (θ is the parameter.)

(*b*) Sketch this curve. It is called the *cissoid of Diocles.*

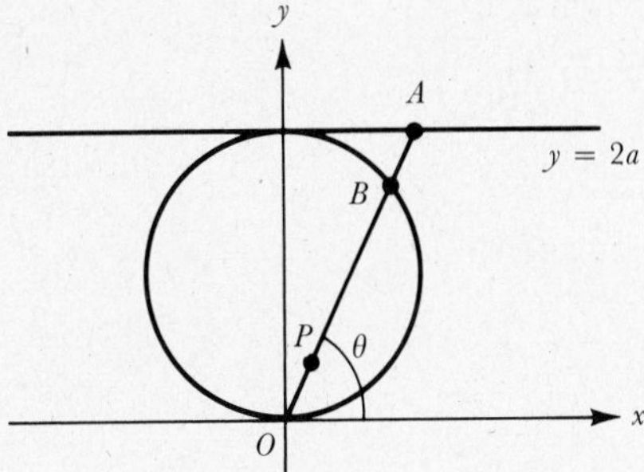

SELF-TEST ✚

1. (*a*) Graph the equations $x = t^{2/3}, y = t + 2$.
(*b*) Determine the point-slope form of the equation for the line tangent to the graph when $t = 1$.

2. Determine the length of the arc $x = t^3, y = (3/2)\,t^2,\ 1 \leq t \leq 3$.

3. Determine the area of the surface generated by revolving the arc in Prob. 2 about the x axis.

In Probs. 4 to 6 set up but do not evaluate definite integrals describing the requested quantities. Let R be the region bounded by the curves $y = 2x^2$, $y = 5x + 3$ and the y axis.

4. The centroid of R

5. The centroid of the solid generated by revolving R about the y axis

6. The centroid of that part of $y = 2x^2$ forming a boundary of R

In Probs. 7 to 9, numerical answers are required.

7. A tank in the shape of a flat-bottomed parabola of revolution generated by revolving $y = 2x^2 - 5$ about the y axis from $y = 0$ to $y = 5$ is full of liquid of density 50 lb/ft^3. Determine the work done in pumping to a height 10 ft above the tank enough contents to lower the level by 3 ft from the full position.

8. Determine the total force on one side of a vertical divider that separates the above tank in two halves: (*a*) before the pumping and (*b*) after the pumping.

9. Determine the volume of a rectangular-cross-section torus whose cross section is 3 ft high and 5 ft wide and whose inner radius is 2 ft.

10. Suppose the line $y = 3x$ from $x = 0$ to $x = 2$ is revolved about the y axis to generate a conical tank and a vertical divider is placed in the center to divide it in half. Suppose now that the tank is filled with a fluid whose density varies continuously with the depth according to the function $x = \rho(y)$—a *nonconstant* density. Determine a definite integral formula that gives the total force on one side of the divider.

Chapter 13

POLAR COORDINATES

CONTENTS

Although it may be possible to drive a nail with a wrench, a hammer is surely the better tool. Similarly, certain curves in the plane are more efficiently dealt with using polar, rather than rectangular, coordinates. Put roughly, polar coordinates for a point P in the plane consist of a pair of numbers, the first of which is a distance from P to the origin and the second of which is the measure of an angle whose initial side is the positive x axis and whose terminal side passes through the origin and P. Relatively simple formulas allow us to transform between polar and rectangular coordinates in either direction. Several of the important ideas considered earlier, such as graphing, derivatives, areas, and arc lengths, for example, can be reinterpreted in this new context.

Besides providing us with a new, intrinsically useful approach to these previously studied ideas, polar coordinates also provide us with a valuable opportunity to deepen our understanding of them by revealing them in a new language. This is especially true for determining the area of a region in the plane bounded by polar curves. Although the formulas are all different, the basic idea is the same: Subdivide into convenient subregions, add their areas to form a Riemann sum approximation, and then pass to a limit given by a definite integral.

Section 13.1: Polar Coordinates

Objectives:

1. Plot points and graphs of equations in polar coordinates.

2. Transform polar coordinates and equations into rectangular coordinates, and vice versa.

INTRODUCTION TO POLAR COORDINATES

We start with a horizontal half-line extending infinitely to the right. This half-line is called the *polar axis,* and its initial point is called the *pole.* (See Fig. 13.1.) We now consider directed angles with initial side the polar axis and vertex at the pole. As usual, positive measures are assigned to angles generated in a counterclockwise direction. The half-line starting at the pole that is the terminal side of the angle θ is called the *ray* θ. Examples appear in Fig. 13.2.

Of course any two angles differing by an even multiple of π determine the same ray (see Fig. 13.3). The ray $(\theta + \pi)$ is referred to as the *opposite* of the ray θ (see Fig. 13.4).

We assume that the polar axis is a copy of the positive real-number line, as is every ray.

13.1.1 **Definition** Suppose P is a point in the plane. We say that *P has polar coordinates* (r, θ) if P is located $|r|$ units out along the ray θ if $r > 0$, and out along the opposite of ray θ if $r < 0$. The pole has polar coordinates $(0, \theta)$ for all real numbers θ.

Pole Polar Axis

Fig. 13.1

Figure 13.5 shows the location of the points with polar coordinates $(2, \pi/4)$, $(3, -\pi/6)$, $(-1, \pi/2)$, and $(-3, -2\pi/3)$. Note that if we add π to θ and replace r by $-r$, we obtain another pair of polar coordinates for the same point. The same is true if we subtract π from θ and replace r by $-r$. Thus the previous four points can also be realized, respectively, as $(-2, 5\pi/4)$, $(-3, 5\pi/6)$, $(1, 3\pi/2)$, and $(3, \pi/3)$; see Fig. 13.6.

With this definition, every pair of real numbers (r, θ) uniquely determines a point in the plane. However, *a given point in the plane does* NOT *have a unique pair of polar coordinates.* To the contrary, there are infinitely many polar coordinates for any point. Since ray θ = ray $(\theta \pm 2n\pi)$ for $n = 0, 1, 2, \ldots$, it follows that the

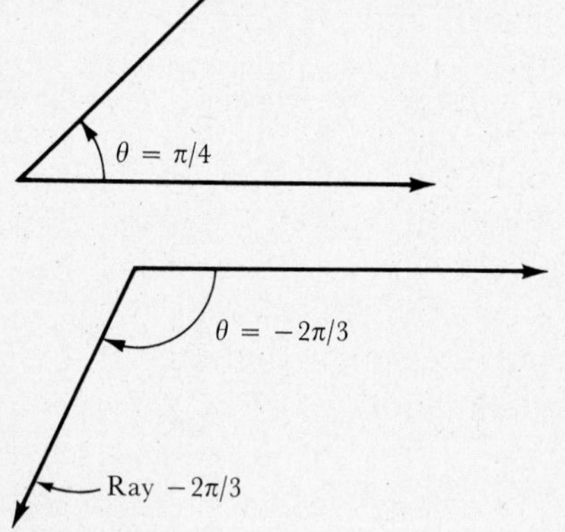

Fig. 13.2

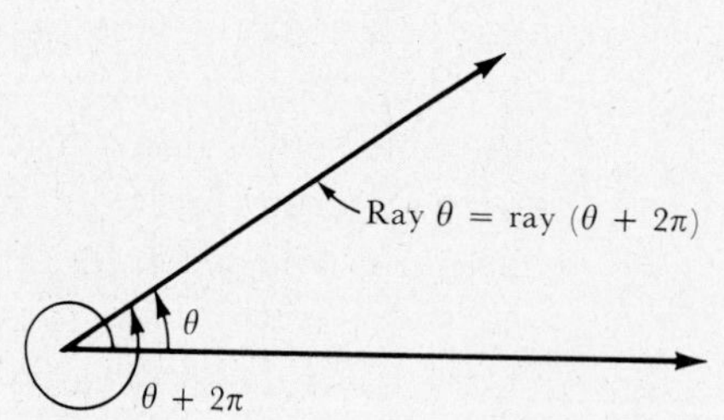

Fig. 13.3

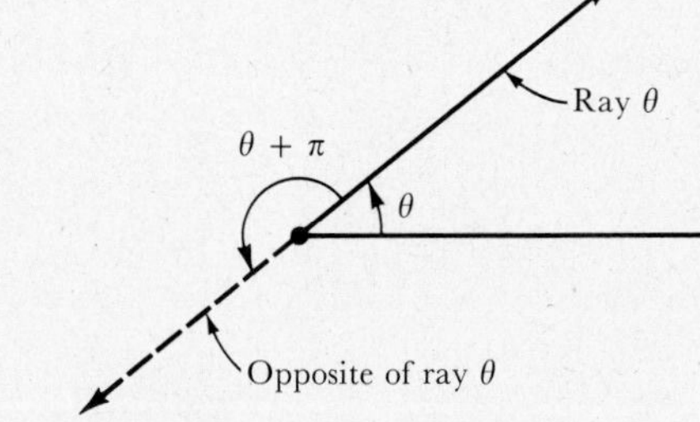

Fig. 13.4

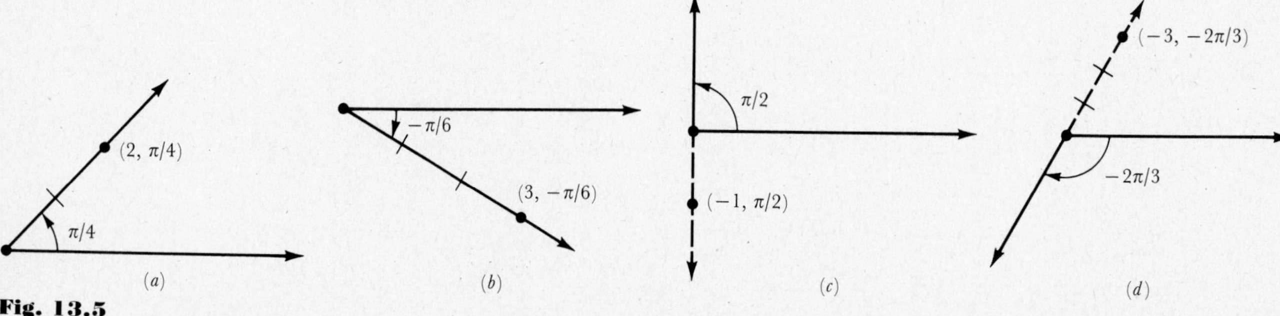

Fig. 13.5

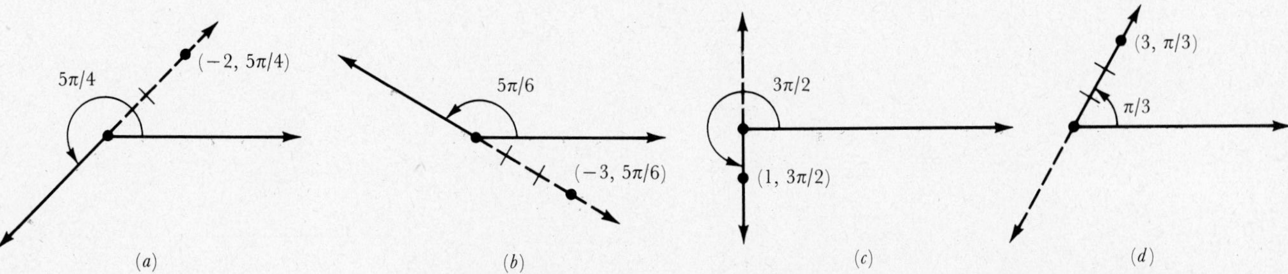

Fig. 13.6

point with polar coordinates (r, θ) will also have polar coordinates $(r, \theta \pm 2n\pi)$. Furthermore, as in the previous examples, if P has polar coordinates (r, θ), then P also has coordinates $(-r, \theta \pm \pi)$. We could also add even multiples of π to $\theta \pm \pi$. Having so many different polar coordinates for the same point may seem confusing, but it turns out to be one of this system's main advantages. It does, however, require that you be careful.

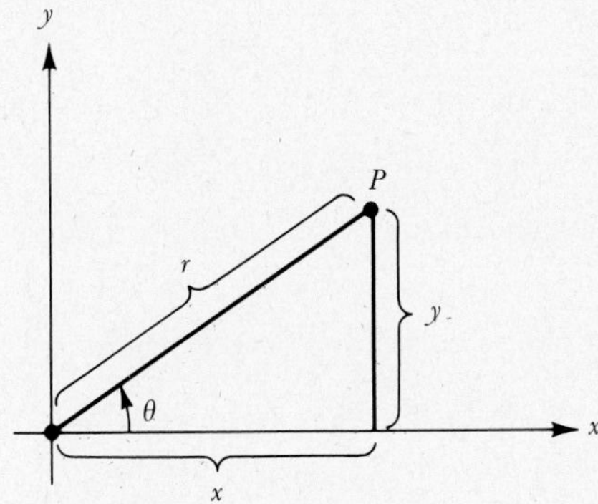

Fig. 13.7

POLAR COORDINATES VERSUS RECTANGULAR COORDINATES

To see the relationship between polar and rectangular coordinates, we begin by letting the pole coincide with the origin and the polar axis with the positive x axis.

Now suppose P is a point in the plane. What is the relationship between the rectangular coordinates (x, y) of P and any pair of polar coordinates (r, θ)?

Let us assume initially that P is in the first quadrant and that (r, θ) is a pair of polar coordinates for P with $r, \theta > 0$. Then from Fig. 13.7, the rectangular coordinates may be given as follows:

13.1.2 Polar to Rectangular

$$x = r\cos\theta \qquad y = r\sin\theta$$

Furthermore, these equations hold for *any other* polar coordinates for P.

Example 1 Determine the rectangular coordinates of the points with polar coordinates
(*a*) $P = (2, \pi/3)$ (*b*) $P = (-5, 11\pi/6)$

Solutions (*a*) Since

$$x = r\cos\theta = 2\cos(\pi/3) = 2(1/2) = 1$$

and

$$y = r\sin\theta = 2\sin(\pi/3) = 2(\sqrt{3}/2) = \sqrt{3}$$

we know that P has rectangular coordinates $(1, \sqrt{3})$; see Fig. 13.8(*a*).

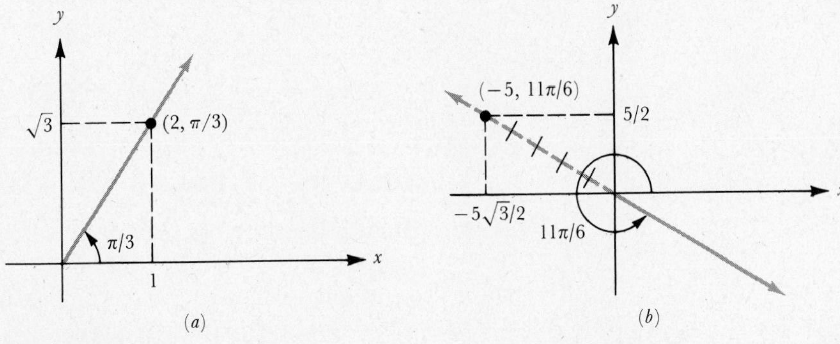

Fig. 13.8

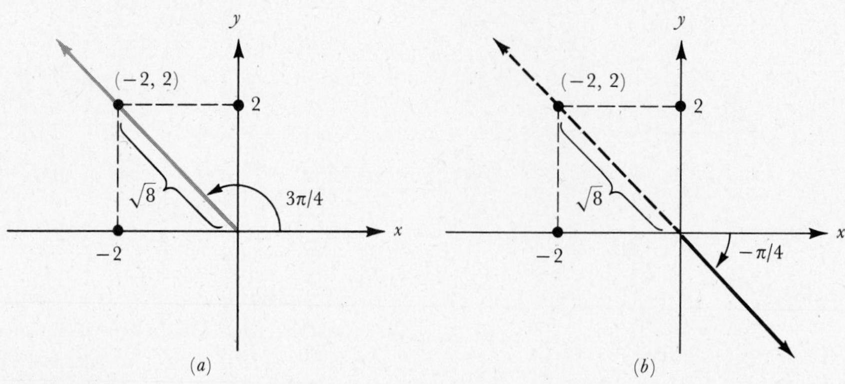

Fig. 13.9

(*b*) Since

$$x = r\cos\theta = -5\cos(11\pi/6) = -5(\sqrt{3}/2) = -5\sqrt{3}/2$$

and

$$y = r\sin\theta = -5\sin(11\pi/6) = -5(-1/2) = 5/2$$

we know that P has rectangular coordinates $(-5\sqrt{3}/2, 5/2)$; see Fig. 13.8(*b*). ●

To transform rectangular coordinates to polar coordinates, we again refer to Fig. 13.7 and observe the following:

13.1.3 Rectangular to Polar

$$x^2 + y^2 = r^2 \qquad \tan\theta = y/x$$

Example 2 Determine a pair of polar coordinates for the point with rectangular coordinates P: $(-2, 2)$

Solution Since $x = -2$ and $y = 2$, we know that $r^2 = x^2 + y^2 = 8$ and $\tan\theta = y/x = -1$. Of course, there are many angles θ with $\tan\theta = -1$. But since $(-2, 2)$ is in the second quadrant, we take $\theta = 3\pi/4$ and hence $r = \sqrt{8}$; see Fig. 13.9(*a*). Therefore one pair of polar coordinates is $(\sqrt{8}, 3\pi/4)$. We could have chosen $\theta = -\pi/4$. In that case we would have to go $\sqrt{8}$ units along the opposite of ray $(-\pi/4)$ to get to P, and so we would choose $r = -\sqrt{8}$, obtaining polar coordinates $(-\sqrt{8}, -\pi/4)$ for P; see Fig. 13.9(*b*). ●

PROGRESS TEST 1

1. Suppose P has polar coordinates $(\sqrt{3}, -\pi/6)$ and Q has polar coordinates $(-2, 7\pi/3)$.
   (*a*) Plot the points P and Q.
   (*b*) Determine four additional polar coordinates for P.
   (*c*) Determine four additional polar coordinates for Q.
   (*d*) Give the rectangular coordinates for P and Q. [Note that the answers to (*b*) and (*c*) are by no means unique.]
2. Determine two pairs of polar coordinates for the point R with rectangular coordinates $(1, \sqrt{3})$.

GRAPHS OF POLAR EQUATIONS

13.1.4 Definition By *the graph of a polar equation in r and θ, F(r, θ) = 0,* we mean the collection of all points that have at least one pair of polar coordinates (r, θ) satisfying the equation. If r can be expressed uniquely as a function of θ, then we write the equation in the form $r = f(\theta)$ and refer to *the graph of the polar function f.*

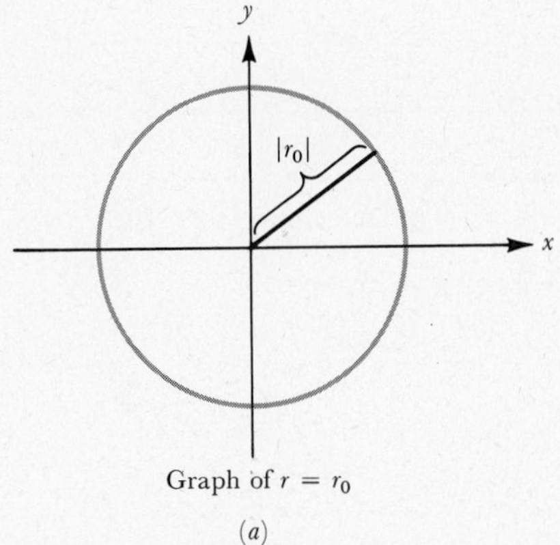

Graph of $r = r_0$

(a)

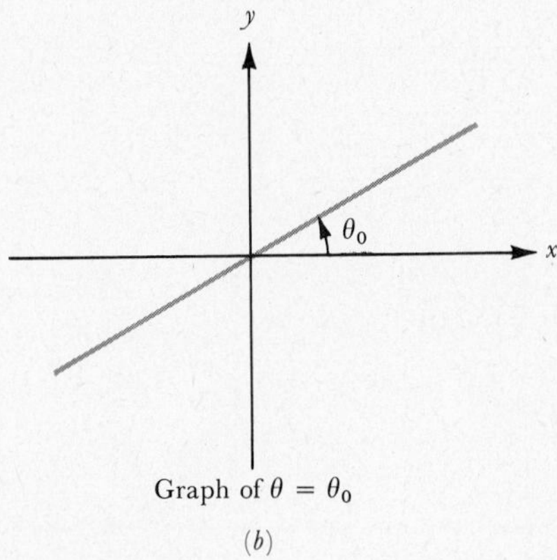

Graph of $\theta = \theta_0$

(b)

Fig. 13.10

The multiplicity of the polar coordinates for a given point may present some problems that were not encountered with rectangular equations. For example, let $F(r, \theta) = r^2 - \cos\theta$. Since $F(1, \pi) = 1^2 - \cos\pi = 1 - (-1) = 2$, the coordinates $(1, \pi)$ do not satisfy the equation $F(r, \theta) = 0$. However, the point with polar coordinates $(1, \pi)$ also has polar coordinates $(-1, 0)$, and $F(-1, 0) = (-1)^2 - \cos 0 = 1 - 1 = 0$. Hence, by definition, the point $(1, \pi)$ *is* on the graph of $F(r, \theta) = 0$ since it has *a* pair of coordinates satisfying the equation.

In rectangular coordinates the basic equations

$$x = x_0 \qquad \text{and} \qquad y = y_0$$

correspond, respectively, to vertical and horizontal straight lines. If r_0 is a constant, then the polar equation

$$r = r_0$$

has as its graph all points of the form (r_0, θ), where θ can be any real number. This is a circle of radius $|r_0|$ centered at the origin; see Fig. 13.10(*a*). Thus we can see how the choice of coordinate systems can dramatically change the form of an equation. Note that if $r_0 = 0$, then the polar coordinates for the pole are of the form $(0, \theta)$ where θ is any real number.

If θ_0 is a constant, then the graph of $\theta = \theta_0$ is all points (r, θ_0) where r can be any real number. This is a straight line through the origin with angle of inclination θ_0; see Fig. 13.10(*b*).

To graph more general polar equations we substitute various values of θ into the equation, determine the corresponding values of r, and plot the resulting points (r, θ). To help simplify matters we use the various properties of the trigonometric functions and note symmetries whenever possible, particularly the following:

13.1.5 **Symmetry** The graph of $F(r, \theta) = 0$ is

(*a*) *symmetric about the origin* if $F(r, \theta) = F(-r, \theta)$
(*b*) *symmetric about the x axis* if $F(r, \theta) = F(r, -\theta)$
(*c*) *symmetric about the y axis* if $F(r, \theta) = F(r, \pi - \theta)$

The validity of (13.1.5) follows from the fact that $(-r, \theta)$, $(r, -\theta)$ and $(r, \pi - \theta)$ are the reflections of (r, θ) through, respectively, the origin, the x axis, and the y axis, as illustrated in Fig. 13.11.

Example 3 Graph the polar equation $r = 1 + \cos\theta$.

Solution First note that the period of $\cos\theta$ is 2π, so we can restrict our attention to $0 \le \theta \le 2\pi$. Since $\cos(-\theta) = \cos(\theta)$, the equation is unchanged if we replace θ by $-\theta$. By [13.1.5(*b*)] the graph is symmetric with respect to the x axis, so we

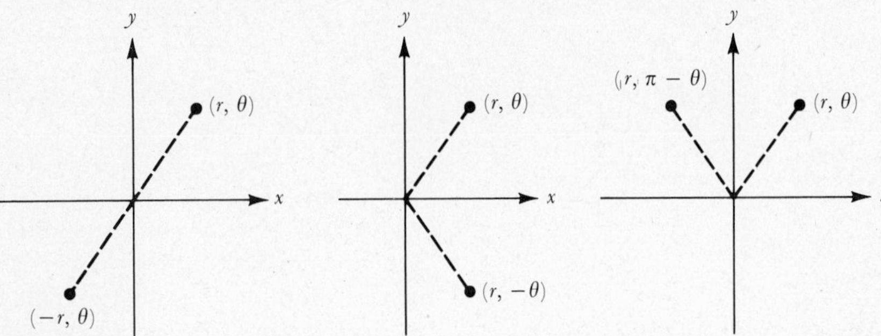

Fig. 13.11

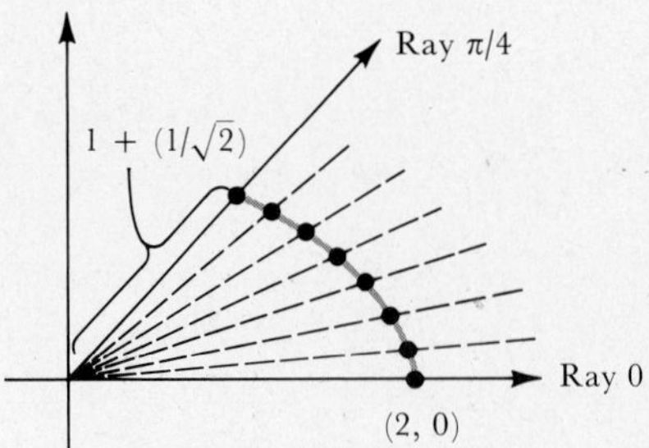

Fig. 13.12

θ:	$0 \to \pi/4$
$\cos\theta$:	$1 \to 1/\sqrt{2}$
$r = 1 + \cos\theta$:	$2 \to 1 + (1/\sqrt{2})$

θ:	$\pi/4 \to \pi/2$	$\pi/2 \to 3\pi/4$	$3\pi/4 \to \pi$
$\cos\theta$:	$1/\sqrt{2} \to 0$	$0 \to -1/\sqrt{2}$	$-1/\sqrt{2} \to -1$
$r = 1 + \cos\theta$:	$1 + 1/\sqrt{2} \to 1$	$1 \to 1 - 1/\sqrt{2}$	$1 - 1/\sqrt{2} \to 0$

Table 13.1

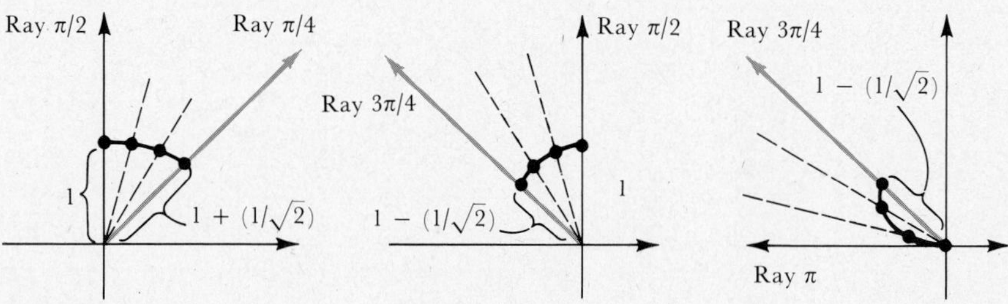

Fig. 13.13

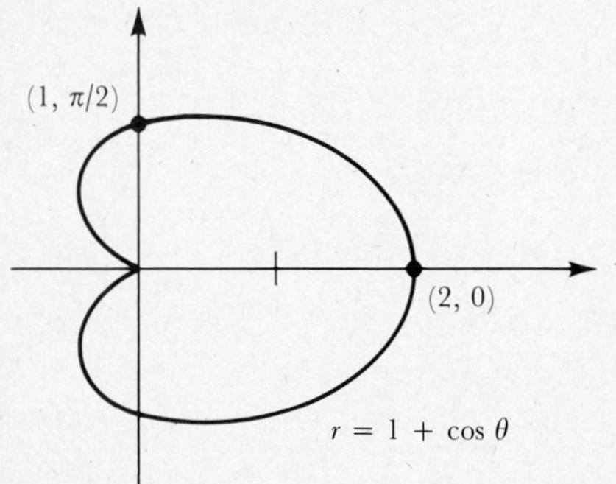

Fig. 13.14

can further restrict our attention to $0 \leq \theta \leq \pi$. Now, as θ increases from 0 to $\pi/4$, $\cos\theta$ steadily decreases from 1 to $1/\sqrt{2}$; thus $r = 1 + \cos\theta$ decreases from 2 to $1 + 1/\sqrt{2}$. On ray 0 we are two units out. As we sweep from ray 0 to ray $\pi/4$, the distances out along the rays decrease from 2 to $1 + 1/\sqrt{2}$. We indicate this piece of the graph in Fig. 13.12, along with the tabulated data.

A similar analysis on the intervals $\pi/4 \leq \theta \leq \pi/2$, $\pi/2 \leq \theta \leq 3\pi/4$, and $3\pi/4 \leq \theta \leq \pi$ yields the data in Table 13.1 and the corresponding graphs (see Fig. 13.13).

Putting the pieces together, we obtain the upper half of the graph in Fig. 13.14, from which the lower half follows by the symmetry established earlier. ●

The graph of $r = 1 + \cos\theta$ is referred to as a *cardioid* because of its heart shape. It is not difficult to see that for a positive constant a, the graphs of

$$r = a(1 \pm \cos\theta) \qquad \text{and} \qquad r = a(1 \pm \sin\theta)$$

are all cardioids (see Fig. 13.15), each of which is a rotation of $r = a(1 + \cos\theta)$. [Note that, for example, $a(1 - \cos\theta) = a(1 - (-\cos(\pi - \theta))) = a(1 + \cos(\pi - \theta))$ and $a(1 + \sin\theta) = a(1 - \cos(\theta - 3\pi/2))$.]

It is important to realize that if any of the three symmetry conditions of (13.1.5) hold, then the corresponding symmetry is guaranteed. However, if one of these conditions fails, the graph may *still* possess that particular symmetry. In the next example we shall graph the polar equation $r = 2 \sin 2\theta$, which has all three symmetries even though none of conditions (a), (b), or (c) in (13.1.5) holds.

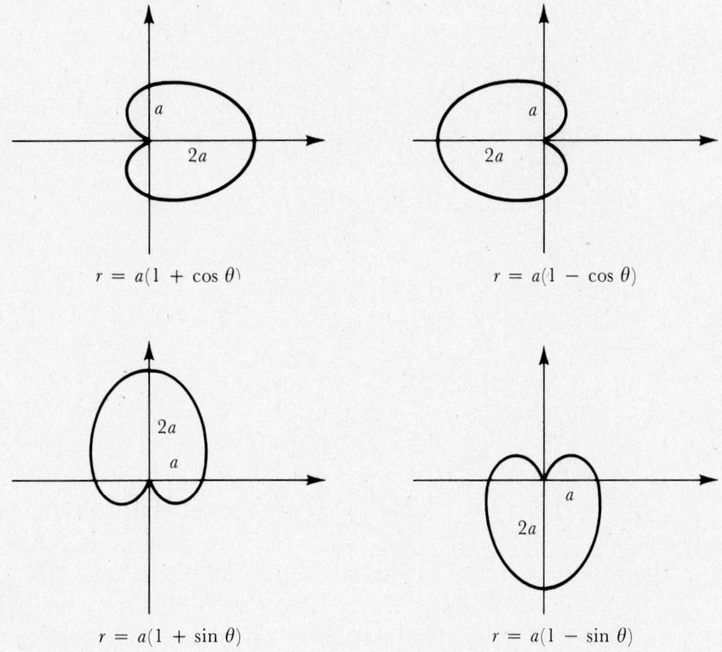

Fig. 13.15

There are, however, other tests for symmetry. For example, the reflection of the point (r, θ) through the x axis can also be realized as $(-r, \pi - \theta)$. Hence, if $F(r, \theta) = F(-r, \pi - \theta)$, the graph of $F(r, \theta) = 0$ will be symmetric about the x axis. Similarly, if $F(r, \theta) = F(-r, -\theta)$, then the graph is symmetric about the y axis; and if $F(r, \theta) = F(r, \theta + \pi)$, the graph will be symmetric about the origin. These latter conditions *are* satisfied by the equation $r = 2 \sin 2\theta$, and hence the graph has all three symmetries.

Example 4 Graph the polar equation $r = 2 \sin 2\theta$.

Solution Since $\sin 2\theta$ has period π, we can restrict our attention to $0 \leq \theta \leq \pi$. As observed, all three symmetries are valid so we will plot the graph for $0 \leq \theta \leq \pi/2$ from the data in Table 13.2 and then use the symmetries to obtain the entire graph. Starting at the origin, we move further out along each ray as we sweep from ray 0 to ray $\pi/8$, to ray $\pi/4$, where we are 1 unit from the origin; see Fig. 13.16(*a*).

As we sweep from ray $\pi/4$ through ray $3\pi/8$ to ray $\pi/2$, we begin to move back along the corresponding rays until we are only out $1/\sqrt{2}$ units at ray $3\pi/8$ and back to the origin at ray $\pi/2$; see Fig. 13.16(*b*). Because of the three symmetries the complete graph has a copy of this loop in each of the four quadrants (see Fig. 13.17). ●

θ:	$0 \to \pi/8$	$\pi/8 \to \pi/4$	$\pi/4 \to 3\pi/8$	$3\pi/8 \to \pi/2$
2θ:	$0 \to \pi/4$	$\pi/4 \to \pi/2$	$\pi/2 \to 3\pi/4$	$3\pi/4 \to \pi$
$\sin 2\theta$:	$0 \to 1/\sqrt{2}$	$1/\sqrt{2} \to 1$	$1 \to 1/\sqrt{2}$	$1/\sqrt{2} \to 0$
$r = 2\sin 2\theta$:	$0 \to 2/\sqrt{2}$	$2/\sqrt{2} \to 2$	$2 \to 2/\sqrt{2}$	$2/\sqrt{2} \to 0$

Table 13.2

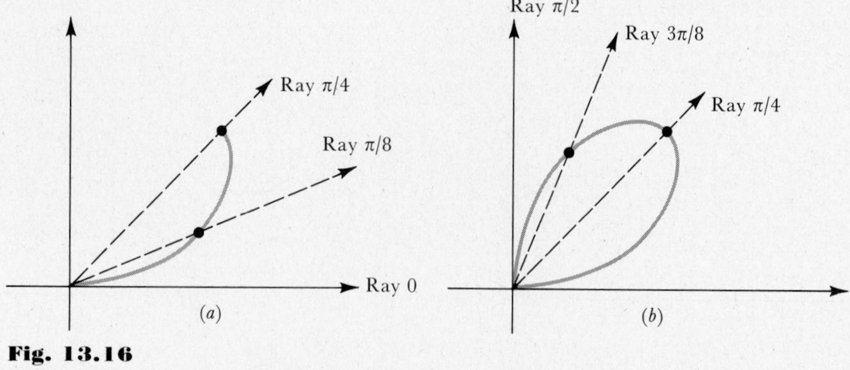

Fig. 13.16

$r = 2 \sin 2\theta$

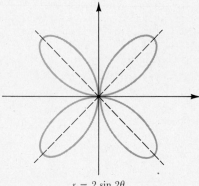

Fig. 13.17

The graph of $r = 2 \sin 2\theta$ is called a *four-leafed rose*. Due to the symmetries, we needed only to plot the graph for values of θ between $\theta = 0$ and $\theta = \pi/2$. However, one should note that the values of θ between $\pi/2$ and π give rise to the leaf in the *fourth* quadrant, whereas the values of θ between $3\pi/2$ and 2π produce the leaf in the second quadrant. This is because $r = 2 \sin 2\theta$ is negative in the intervals $\pi/2 < \theta < \pi$ or $3\pi/2 < \theta < 2\pi$, and so points are plotted along the opposite ray in each case. A similar construction shows that $r = 2 \cos 2\theta$ is also a four-leafed rose, but with the leaves centered on the axes (see Fig. 13.18).

In general one can show that $r = a \sin n\theta$, $r = a \cos n\theta$, $a > 0$, will produce roses with $2n$ leaves if n is an even integer and n leaves if n is an odd integer.

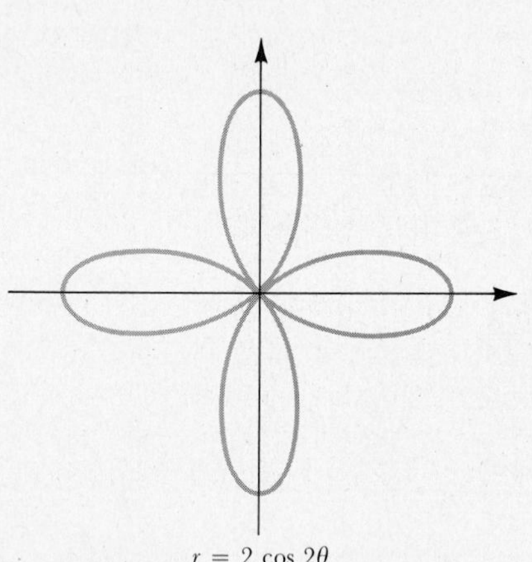

Fig. 13.18

In some cases, transforming from one coordinate system to another can make the graphing easier. For example, to graph $r = 2a \sin \theta$ we can multiply by r to obtain $r^2 = 2ar \sin \theta$.

But $r^2 = x^2 + y^2$ and $y = r \sin \theta$, so the graph of the polar equation $r = 2a \sin \theta$ is the same as the graph of the rectangular equation $x^2 + y^2 = 2ay$. But, completing the square in y as follows,

$$\begin{aligned} x^2 + y^2 = 2ay &\Leftrightarrow x^2 + y^2 - 2ay = 0 \\ &\Leftrightarrow x^2 + y^2 - 2ay + a^2 = a^2 \\ &\Leftrightarrow x^2 + (y - a)^2 = a^2 \end{aligned}$$

we get the rectangular equation of a circle centered on the y axis at $(0, a)$ with radius a; see Fig. 13.19(a). Similarly, $r = 2a \cos \theta$ yields a circle centered on the x axis at $(a, 0)$ with radius a, as in Fig. 13.19(b).

On the other hand, by changing rectangular equations to polar equations, we can sometimes make the graphing easier.

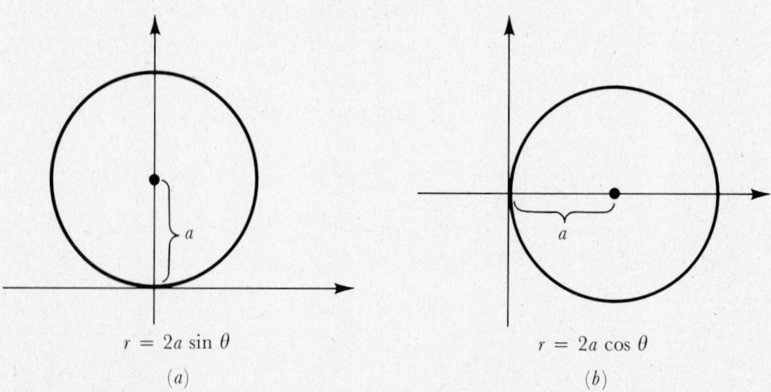

Fig. 13.19

Example 5 Graph the rectangular equation $(x^2 + y^2)^2 = 4(x^2 - y^2)$. (Its graph is sometimes called a *lemniscate of Bernoulli.*)

Solution Since $x^2 + y^2 = r^2$, $x = r\cos\theta$, and $y = r\sin\theta$, we have

$$(r^2)^2 = 4(r^2\cos^2\theta - r^2\sin^2\theta)$$

Dividing by r^2 and using the trigonometric identity $\cos 2\theta = \cos^2\theta - \sin^2\theta$, the equation becomes

$$r^2 = 4\cos 2\theta$$

Since $\cos 2\theta$ has period π, we need only consider $0 \le \theta \le \pi$. Also, we get the same equation if we replace r by $-r$ and θ by $-\theta$, so the graph is symmetric with respect to the origin and the x axis. Hence we shall plot $r = +2\sqrt{\cos 2\theta}$ and then use symmetry. On $[0, \pi/4]$ we have the following data:

θ:	$0 \to \pi/4$
2θ:	$0 \to \pi/2$
$\cos 2\theta$:	$1 \to 0$
$r = 2\sqrt{\cos 2\theta}$:	$2 \to 0$

This leads to the first-quadrant graph in Fig. 13.20, from which the third-quadrant graph (actually the graph of $r = -2\sqrt{\cos 2\theta}$) follows by symmetry with respect to the origin.

By symmetry with respect to the x axis we obtain the dashed portion of the graph in the second and fourth quadrants. Now, for $\pi/4 < \theta < 3\pi/4$, $\cos 2\theta < 0$, and so there are no points that satisfy $r^2 = 4\cos 2\theta$ and we get no additional points. On $[3\pi/4, \pi]$ we have a table similar to the above, with $r = 2\sqrt{\cos 2\theta}$ increasing from 0 to 2 and $r = -2\sqrt{\cos 2\theta}$ decreasing from 0 to -2. This accounts for the second- and fourth-quadrant dashed graph in Fig. 13.20. Thus the polar equation $r^2 = 4\cos 2\theta$ has the graph shown in Fig. 13.21 (a lemniscate), which is also the graph in rectangular coordinates of $(x^2 + y^2)^2 = 4(x^2 - y^2)$. ●

In general, the graph of any equation of the form

$$r = k(a \pm b\cos\theta) \qquad \text{or} \qquad r = k(a \pm b\sin\theta)$$

is called a *limacon*. If $a = b$, the limacon is a cardioid (as in Example 3). If $a > b$, then the basic shape is that of Fig. 13.22(a); if $a < b$, there is an inner loop as in Fig. 13.22(b).

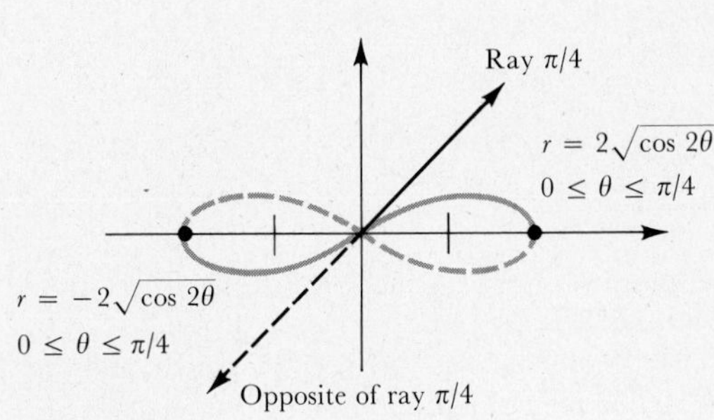

Fig. 13.20

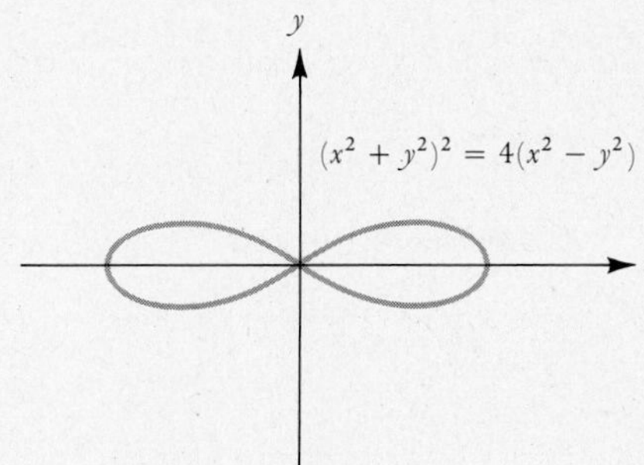

Fig. 13.21

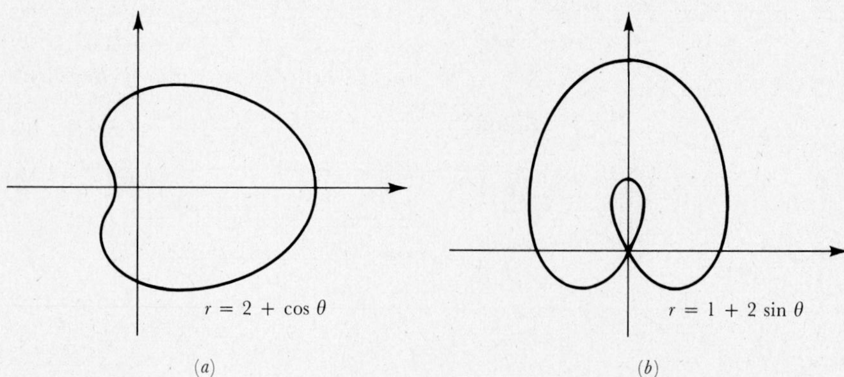

Fig. 13.22

PROGRESS TEST 2

1. Using symmetry and a table, show that the graph of $r = 2 + \cos\theta$ is as in Fig. 13.22(*a*).

2. Determine and sketch a polar equation that has the same graph as the rectangular equation $(x^2 + y^2)^2 = 4xy$.

SECTION 13.1 EXERCISES

In Exercises 1 to 15, (a) plot the point with the given polar coordinates, (b) name a pair of coordinates for the same point that uses the opposite of the original ray, and (c) give a third pair of coordinates for the given point.

1. $(2, \pi/4)$ **2.** $(3, -\pi/4)$ **3.** $(2, 7\pi/6)$
4. $(-2, \pi/6)$ **5.** $(-4, 0)$ **6.** $(1, 400\pi)$
7. $(-2, \pi/3)$ **8.** $(1, 49\pi/6)$ **9.** $(\sqrt{2}, -14\pi/3)$
10. $(2, 0)$ **11.** $(0, \pi/9)$ **12.** $(5, -2\pi/3)$
13. $(1, 2)$ **14.** $(1, -3)$ **15.** $(1, -2)$

In Exercises 16 to 24 determine a pair of polar coordinates for the points with the given rectangular coordinates.

16. $(1, 1)$ **17.** $(-4, 0)$ **18.** $(0, \pi)$
19. $(2\sqrt{3}, 2)$ **20.** $(3, 0)$ **21.** $(3\sqrt{2}, -3\sqrt{2})$
22. $(\sqrt{6}, -\sqrt{2})$ **23.** $(3, -3)$ **24.** $(2, 2)$

In Exercises 25 to 33 find the rectangular coordinates for the points with the given polar coordinates.

25. $(6, 3\pi/4)$ **26.** $(-2, 11\pi/6)$ **27.** $(-3, \pi/3)$
28. $(2, 0)$ **29.** $(2\sqrt{2}, -\pi/4)$ **30.** $(5, 25\pi/4)$
31. $(2, 1)$ **32.** $(1, 2)$ **33.** $(1, 1)$

In Exercises 34 to 42, find an equation in rectangular coordinates for the given polar equation.

34. $r = 4\cos\theta$ **35.** $\theta = \pi/3$ **36.** $r = 2$
37. $r^2 \sin 2\theta = 2$ **38.** $r^3 = 3\cos\theta\sin\theta$ **39.** $r^2 = \cos 2\theta$
40. $r = 1 - 2\sin\theta$ **41.** $r^2 = 1 + \cos\theta$ **42.** $r = \dfrac{2}{2 + \cos\theta}$

43. Show that the graph of $r = 1 + 2\sin\theta$ is as in Fig. 13.22(*b*) by following the given steps:
(*a*) Prove symmetry with respect to the y axis.
(*b*) Give a table for $0 \le \theta \le \pi/2$ and then use (*a*) to complete the graph for $0 \le \theta \le \pi$.
(*c*) Determine for which θ in $[\pi, 3\pi/2]$, $r = 1 + 2\sin\theta = 0$.
(*d*) Compile a table for $\pi \le \theta \le 3\pi/2$, sketch, and then use (*a*) to complete the graph.

In Exercises 44 to 71 sketch the graph of the given polar equations. (You may use earlier remarks and discussion as a guide regarding the general shape and dimensions of the graph.)

44. $\theta = 14\pi/3$
45. $r = -8$
46. $r = 4 \sin 2\theta$
47. $r = \cos 4\theta$
48. $r = 4 \sin 3\theta$
49. $r = 2 \cos 3\theta$
50. $r^2 = 9 \cos 2\theta$
51. $r^2 = 4 \sin 3\theta$
52. $r = 9 \sin \theta$
53. $r = 3 \sin 2\theta$
54. $r = 3 - 2 \sin \theta$
55. $r = 18 \cos \theta$
56. $r = 1 - 2 \sin \theta$
57. $r = 2 + \cos \theta$

(The graphs of the equations in Exercises 58 to 61 are called *conchoids*.)

58. $r = 5 \sec \theta$
59. $r = 2 \csc \theta$
60. $r = 2 \csc \theta - 1$
61. $r = \csc \theta - 2$
62. $r = \theta$ (Archimedian spiral)
63. $r = -3\theta$
64. $r = e^{\theta}$ (logarithmic spiral)
65. $r = 2e^{\theta/(2\pi)}$
66. $r \sin \theta = -1$
67. $r = 5 \cos 3\theta$
68. $r^2 = 9 \cos 2\theta$
69. $r^2 = 9 \sin 2\theta$
70. $r = 2 \sec \theta - 1$
71. $r = \dfrac{2}{2 - \cos \theta}$

In Exercises 72 to 75, transform the given rectangular equation into a polar equation and sketch the graph in polar coordinates superimposed on a rectangular system.

72. $(x^2 + y^2)^2 = 2x^2 + 2y^2 - 3$
73. $(x^2 + y^2)^2 = x^2 - y^2$
74. $x + 2y = 5$
75. $(x^2 + y^2)^3 = 4x^2y^2$
76. (*a*) Show that the graph of the equation $r^2 = 2ar \cos(\theta - \phi) + b$ is a circle with center (a, ϕ), provided $b^2 + c > 0$.
(*b*) Determine the radius of this circle.
77. Transform the linear equation $y = mx$ into polar coordinates.
78. Transform the equation $y = 2$ into polar coordinates.

Exercises 79 to 85 concern the polar equations for conic sections.

79. Show that the graph of $r = 1/[1 - (1/2)\cos \theta]$ is an ellipse.
80. Show that the graph of $r = 1/(1 - \cos \theta)$ is a parabola with a "hole" in the graph. Where is the "hole"?
81. Show that the graph of $r = 1/(1 - 2 \cos \theta)$ is a hyperbola. Describe this hyperbola.
82. Show that the graph of $4r \cos \theta = r - 8$ is a hyperbola. Describe this hyperbola.
83. Describe the conic section that is the graph of $r = 9/(3 + \sin \theta)$.
84. Repeat Exercise 83 for $r = 2/[(1/2) + \sin \theta]$.
85. Determine the polar equation for the hyperbola $x^2 - y^2 = 1$.

Section 13.2: Polar Graphs: Intersections and Tangent Lines

Objectives:

1. Determine intersections between graphs of polar equations.

2. Determine slopes and equations of lines tangent to polar graphs.

INTERSECTIONS OF POLAR GRAPHS

Determining all the points of intersection of two rectangular graphs is the relatively straightforward (although not always easy) task of solving a pair of simultaneous equations. Because a point has many different "names" in polar coordinates, the situation for polar equations is not as straightforward. If we set the polar equations

$$r = \sin \theta \qquad \text{and} \qquad r = \cos \theta$$

equal, we get $\sin \theta = \cos \theta$, or $\tan \theta = 1$. This implies

$$\theta = \pi/4, \pi/4 \pm \pi, \pi/4 \pm 2\pi, \ldots$$

For $0 \leq \theta \leq 2\pi$, we get the two solutions

$$(1/\sqrt{2}, \pi/4) \qquad \text{and} \qquad (-1/\sqrt{2}, 5\pi/4)$$

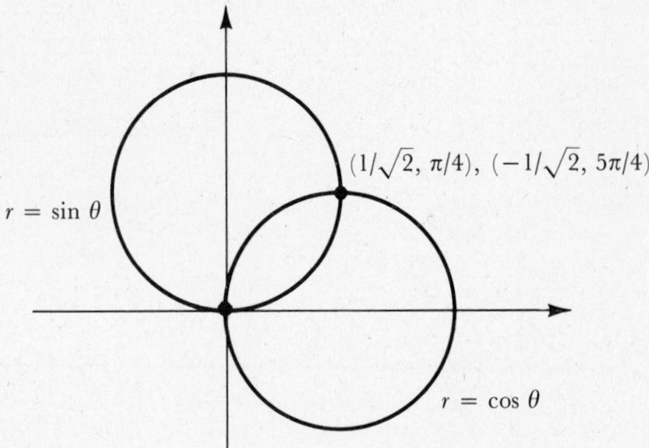

Fig. 13.23

which are polar coordinates for the same point! However, from our work in Section 13.1, we know that the graph of each equation is a circle through the origin, as given in Fig. 13.23.

Thus the simultaneous solution of two polar equations need *not* yield all the points of intersection! The problem is that the polar representations of certain points of intersection may not be the same for both curves. In this case the coordinates of the origin which satisfy $r = \sin\theta$ are of the form $(0, n\pi)$, while for $r = \cos\theta$ they are of the form $(0, \pi/2 + n\pi)$.

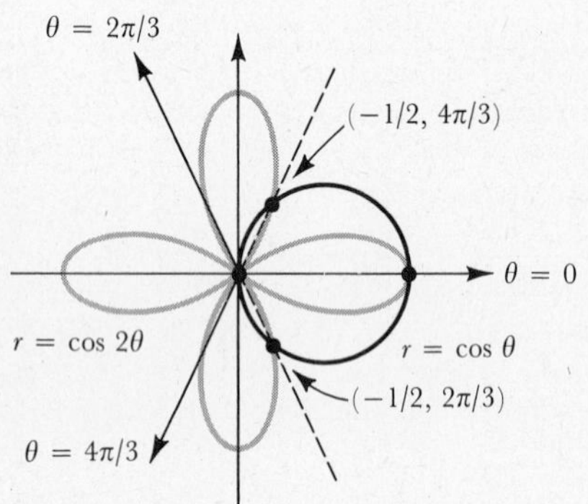

Fig. 13.24

If we think of (r, θ) as the location of a particle moving along the graph at time θ, then simultaneous solutions of $F(r, \theta) = 0$ and $G(r, \theta) = 0$ would arise when the two particles collide—that is, when they are both at the same point at the *same time* θ. However, they may arrive at the same point at different times θ. This would lead to a point of intersection of the two *graphs,* but such a point would not be determined by solving the two equations simultaneously. There are analytic procedures for determining nonsimultaneous solutions, but we shall be content to solve the equations simultaneously and then graph the equations. Any nonsimultaneous solutions can then be readily observed.

Example 1 Determine the points of intersection of the circle $r = \cos\theta$ and the four-leafed rose $r = \cos 2\theta$.

Solution From Fig. 13.24 we can expect four points of intersection. Setting $\cos 2\theta = \cos\theta$ and using the identity $\cos 2\theta = 2\cos^2\theta - 1$, we have

$$2\cos^2\theta - 1 = \cos\theta$$

or

$$2\cos^2\theta - \cos\theta - 1 = 0$$

or

$$(2\cos\theta + 1)(\cos\theta - 1) = 0$$

Hence $\cos\theta = -1/2$ and $\cos\theta = 1$. As a result, $\theta = 0$ gives $(1, 0)$, $\theta = 2\pi/3$ gives $(-1/2, 2\pi/3)$, and $\theta = 4\pi/3$ gives $(-1/2, 4\pi/3)$—all on both curves.

The origin does not arise as a simultaneous solution but is clearly on both graphs. For example, it is on $r = \cos\theta$ as $(0, \pi/2)$, and on $r = \cos 2\theta$ as $(0, \pi/4)$. ●

PROGRESS TEST 1

Determine all points of intersection of the following pairs of polar equations:

1. $r = 1 - \cos\theta$ (cardioid), $r = \cos\theta$.
2. $r = 3/2 - \cos\theta$ (limacon), $\theta = \pi/3$.

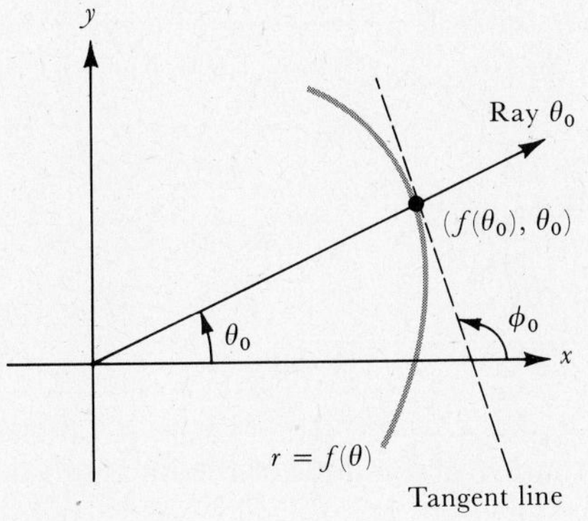

Fig. 13.25

TANGENT LINES TO POLAR GRAPHS

We now turn our attention to equations such as $r = 1 + \cos\theta$, where r is uniquely determined as a *function* of θ. More generally, we suppose that $r = f(\theta)$, with f differentiable with respect to θ, and examine what can be learned about the function and its graph from the derivative. In particular, we ask: *What is the slope of the line tangent to the graph at the point* $(f(\theta_0), \theta_0)$? That is, what is $\tan\phi_0$ (where ϕ_0 is the angle of inclination of the tangent line with the x axis)? (See Fig. 13.25.)

The graph of the constant function $r = f(\theta) = 1$ is the unit circle, and $f'(\theta) = 0$, so the answer to the above question in polar coordinates is certainly *not* provided by $f'(\theta)$ directly. We transform the given function to rectangular coordinates using the equations

$$x = r\cos\theta \qquad \text{and} \qquad y = r\sin\theta$$

This gives the parametric equations

$$x = f(\theta)\cos\theta \qquad \text{and} \qquad y = f(\theta)\sin\theta$$

whose graph is the same as that of $r = f(\theta)$. By (12.1.4), or the Chain Rule, the slope is given by

$$\frac{dy}{dx} = \frac{dy/d\theta}{dx/d\theta} \qquad \text{if } \frac{dx}{d\theta} \neq 0$$

Since

$$\frac{dy}{d\theta} = f(\theta)\cos\theta + f'(\theta)\sin\theta$$

and

$$\frac{dx}{d\theta} = -f(\theta)\sin\theta + f'(\theta)\cos\theta$$

we have proved the following theorem.

13.2.1 **Theorem** If $\left.\dfrac{dx}{d\theta}\right|_{\theta=\theta_0} \neq 0$, then

$$\tan\phi_0 = \frac{f(\theta_0)\cos\theta_0 + f'(\theta_0)\sin\theta_0}{-f(\theta_0)\sin\theta_0 + f'(\theta_0)\cos\theta_0}$$

gives the slope of a polar curve $r = f(\theta)$ at the point $(f(\theta_0), \theta_0)$.

Example 2 Show that the cardioids

$$r = 1 + \sin\theta \qquad \text{and} \qquad r = 1 - \sin\theta$$

are perpendicular at the point $(1, \pi)$; see Fig. 13.15, page 439.

Solution Note that $(1, \pi)$ *is* on both curves. Let $M_0 = \tan\phi_0$ denote the slope of the tangent line to $r = 1 + \sin\theta$ at $(1, \pi)$ and $M_1 = \tan\phi_1$ the slope of the tangent line to $r = 1 - \sin\theta$ at the same point. By (13.2.1),

$$M_0 = \tan\phi_0 = \frac{(1 + \sin\pi)\cos\pi + \cos\pi\sin\pi}{-(1 + \sin\pi)\sin\pi + \cos\pi\cos\pi} = \frac{-1}{1} = -1$$

and

$$M_1 = \tan\phi_1 = \frac{(1 - \sin\pi)\cos\pi + (-\cos\pi)\sin\pi}{-(1 - \sin\pi)\sin\pi + (-\cos\pi)\cos\pi} = \frac{-1}{-1} = 1$$

Since $M_0M_1 = -1$, the tangent lines are perpendicular. ●

Suppose we let ψ_0 (see Fig. 13.26) denote the least positive angle between the ray θ_0 and the tangent line to the graph at $(f(\theta_0), \theta_0)$. Then, from Fig. 13.26,

$$\phi_0 = \theta_0 + \psi_0$$

which, as we shall show, implies the following:

13.2.2
$$\tan \psi_0 = \frac{f(\theta_0)}{f'(\theta_0)}$$

The simplicity of the formula for $\tan \psi_0$ makes it easier to work with the angle between the ray θ_0 and the tangent line than with the slope of the tangent line itself as in rectangular coordinates.

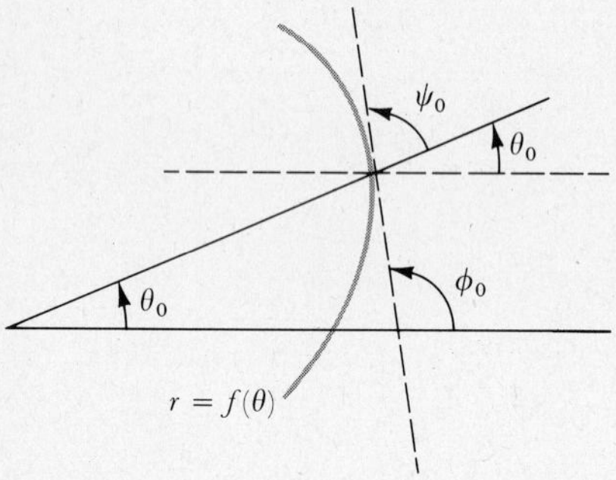

To prove (13.2.2) we divide numerator and denominator of (13.2.1) by $\cos \theta_0$ to get

(1)
$$\tan \phi_0 = \frac{f'(\theta_0)\tan \theta_0 + f(\theta_0)}{f'(\theta_0) - f(\theta_0)\tan \theta_0}$$

Now $\phi_0 = \theta_0 + \psi_0$, so

$$\psi_0 = \phi_0 - \theta_0$$

Taking the tangent of both sides and using the trigonometric identity

$$\tan(a - b) = \frac{\tan a - \tan b}{1 + \tan a \tan b}$$

we obtain

$$\tan \psi_0 = \frac{\tan \phi_0 - \tan \theta_0}{1 + \tan \phi_0 \tan \theta_0}.$$

We then substitute the expression for $\tan \phi_0$ given in (1) and simplify the resulting fraction to get (13.2.2).

ψ_0 θ_0 ϕ_0 θ_0 $r = f(\theta)$

Fig. 13.26

Example 3 Use (13.2.2) and the identity

$$\tan(a + b) = \frac{\tan a + \tan b}{1 - \tan a \tan b}$$

to find the slope of the tangent line to the polar curve $r = \sin^2 3\theta$ at $(1/2, \pi/4)$.

Solution Let ψ_0 be the angle between the tangent line and the ray $\pi/4$. Then $f(\theta) = \sin^2 3\theta$ and $f'(\theta) = 6 \sin 3\theta \cos 3\theta$, so, by (13.2.2),

$$\tan \psi_0 = \frac{\sin^2(3\pi/4)}{6 \sin(3\pi/4)\cos(3\pi/4)} = \frac{-1}{6}$$

Now $\phi_0 = \theta_0 + \psi_0$, where $\theta_0 = \pi/4$ and ϕ_0 is the angle of inclination of the tangent line (see Fig. 13.26). Thus

$$\begin{aligned}\tan \phi_0 &= \tan(\theta_0 + \psi_0)\\ &= \frac{\tan \theta_0 + \tan \psi_0}{1 - \tan \theta_0 \tan \psi_0}\\ &= \frac{\tan(\pi/4) + (-1/6)}{1 - [\tan(\pi/4)](-1/6)} = \frac{5}{7} \quad \bullet\end{aligned}$$

PROGRESS TEST 2

1. Show that the angle ψ between the tangent line to the cardioid $r = 1 - \cos\theta$ and the ray θ is $\theta/2$ for each point (r, θ) on the curve. [Use the identity $\tan(\theta/2) = (1 - \cos\theta)/\sin\theta$.]

2. Determine the slope of the tangent to $r = 1 - \cos\theta$ at $(1, \pi/2)$.

SECTION 13.2 EXERCISES

In Exercises 1 to 20 determine all points of intersection of the graphs of the given equations.

1. $r = 3, \theta = \pi/6$
2. $r = \cos\theta, \theta = \pi/4$
3. $r = \cos\theta, r = 1$
4. $r\cos\theta = 2, r = 4$
5. $r = \cos\theta, r^2 = \cos 2\theta$
6. $r = \cos 2\theta, r = \sin\theta$
7. $r = \cos\theta, r = 1 - \cos\theta$
8. $r = 2, r = 2\cos 2\theta$
9. $r^2 = 2\cos 2\theta, r = 1$
10. $r = \sin 2\theta, r = 1 - \cos 2\theta$
11. $r = 2\sin 2\theta, r = 1$
12. $r = \cos 3\theta, r = \sin 3\theta$
13. $r = \pi/2, r = \theta/2$
14. $r = \pi/2, r = 2\theta$
15. $r = 1 - \sin\theta, r = 1 + \sin\theta$
16. $r = 1 + \cos\theta, r = 1 - \sin\theta$
17. $r = \sin\theta, r = \sec\theta$
18. $r = \cos\theta, r = \csc\theta$
19. $r^2 = \sin 2\theta, r^2 = \cos 2\theta$
20. $r = \tan\theta, r = 2\cos\theta$
21. Determine where the graph of $r = (5/2)\theta$ crosses itself.
22. Determine where the graph of $r = 1 + 4\sin 2\theta$ crosses itself.

In Exercises 23 to 34 determine (a) the slope of the tangent line to the curve at $(f(\theta_0), \theta_0)$ and (b) the tangent of the angle ψ_0 between this tangent line and the ray θ_0.

23. $r = 1 + 2\cos\theta, (2, \pi/3)$
24. $r = 1 + 3\sin\theta, (5/2, \pi/6)$
25. $r = 2 + 2\cos\theta, (2, \pi/2)$
26. $r = 3\cos 2\theta, (3/2, \pi/6)$
27. $r = 2(1 - 2\sin\theta), (2, 0)$
28. $r = 2\cos\theta, (1, \pi/3)$
29. $r = \tan\theta, \theta_0 = \pi/4$
30. $r = \theta, (\pi/3, \pi/3)$
31. $r = \cos 2\theta, \theta_0 = \pi/12$
32. $r = 4\cos 3\theta, \theta_0 = 3\pi/4$
33. $r = 1 - \cos\theta, \theta_0 = \pi/2$
34. $r = 4\sin^2\theta, (1, 5\pi/6)$
35. $r = 1 + 2\sin\theta, (1 - \sqrt{2}, 5\pi/4)$
36. $r = 2 - 4\cos\theta, (4, (2/3)\pi)$
37. $r = 4(1 + \cos\theta), \theta_0 = 5\pi/6$.
38. Determine the lines tangent to the three-leafed rose $r = \cos 3\theta$ at the pole.
39. Repeat Exercise 38 for the four-leafed rose $r = \cos 2\theta$.
40. Show that the angle ψ is a constant for the logarithmic spiral $r = e^{\theta}$.
41. Determine the angle α between the tangents to the graphs of $r = \cos\theta$ and $r = -\sin 2\theta$ at their points of intersection.
42. Repeat Exercise 41 for $r = 1 + \cos\theta$ and $r = \sin\theta$.
43. Use symmetry to determine the points of intersection between $r = 2\cos(\theta/2)$ and $r = 2\sin(\theta/2)$.
44. Prove that the formula for the angle of inclination ϕ of the tangent to the circle $r = 2\sin\theta$ in terms of θ is $\phi = 2\theta$, where $\phi = \theta + \psi$ and where ψ is as in Fig. 13.26.
45. Repeat Exercise 44 to get the formula $\phi = \theta/2$ for $r = 1/(1 - \cos\theta)$.
46. Show that $r = 1 + \cos\theta$ and $r = 1 - \cos\theta$ intersect at right angles.
47. Repeat Exercise 46 for $r = 4\sin\theta$ and $r\sin\theta = 2$.
48. Repeat Exercise 46 for $r = e^{\theta}$ and $r = e^{-\theta}$.
49. (*a*) Give an informal argument that the maximum distance from the pole to a point (r, θ) on a polar graph of $r = f(\theta)$ (where $f'(\theta)$ exists) occurs at a point where the tangent line is perpendicular to the ray θ.
(*b*) Test the statement in (*a*) on the cardioid $r = 2 - 2\cos\theta$.

Section 13.3: Area and Arc Length in Polar Coordinates

Objectives:

1. Determine the area of a region bounded by polar curves.
2. Determine the length of a polar curve.
3. Determine the surface area of the surface obtained by revolving a polar curve about the *x* or *y* axes.

AREA IN POLAR COORDINATES

In Chap. 6 we determined the area of a region bounded by continuous curves first by approximating the region with convenient—namely, rectangular—subregions. Our approach to regions bounded by polar curves is essentially the same except that the "convenient" approximating subregion is a circular sector rather than a rectangle.

A circular sector of a disk of radius r enclosed by an angle of $\Delta\theta$ radians is the fraction $(\Delta\theta)/(2\pi)$ of the entire disk. Hence its area A is the same fraction of the area of the entire disk. Thus $A/\pi r^2 = (\Delta\theta)/(2\pi)$. Solving for A, we get

13.3.1
$$A = \frac{1}{2}r^2\,\Delta\theta$$

Now suppose R is the region bounded by the graph of the polar equation $r = f(\theta)$ and the rays $\theta = \alpha$ and $\theta = \beta$, where $f(\theta)$ is positive and continuous on $[\alpha, \beta]$ and $\beta - \alpha \le 2\pi$. A typical region of this kind is sketched in Fig. 13.27.

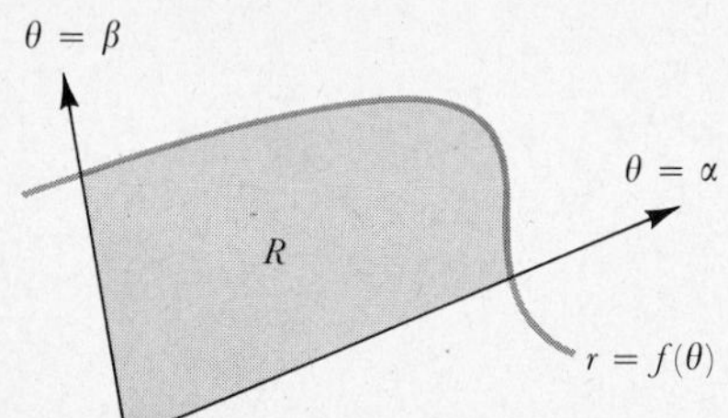

Fig. 13.27

Subdivide $[\alpha, \beta]$ into n distinct subintervals

$$[\theta_0, \theta_1], \ldots, [\theta_{i-1}, \theta_i], \ldots, [\theta_{n-1}, \theta_n]$$

where $\theta_0 = \alpha$, $\theta_n = \beta$, and

$$\Delta\theta_i = \theta_i - \theta_{i-1} \qquad \text{for each } 0 < i \le n$$

Let $\|P\|$ denote the length of the largest subinterval in this partition.

In each subinterval $[\theta_{i-1}, \theta_i]$ choose a point θ_i^*. The circular sector between ray θ_{i-1} and ray θ_i of radius $f(\theta_i^*)$ approximates the region bounded by these rays and the graph of $r = f(\theta)$; see Fig. 13.28. By (13.3.1), the area of this sector is

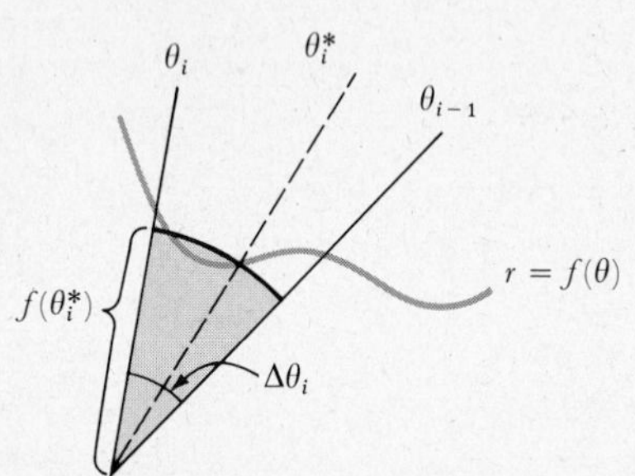

Fig. 13.28

$$\frac{1}{2}[f(\theta_i^*)]^2\,\Delta\theta_i$$

We sum the areas of these n sectors to approximate the total area of R:

$$A(R) \approx \frac{1}{2}\sum_{i=1}^{n}[f(\theta_i^*)]^2\,\Delta\theta_i$$

Figure 13.29 depicts the situation when $n = 4$.

Now as $\|P\| \to 0$, these approximations will tend to $A(R)$. But

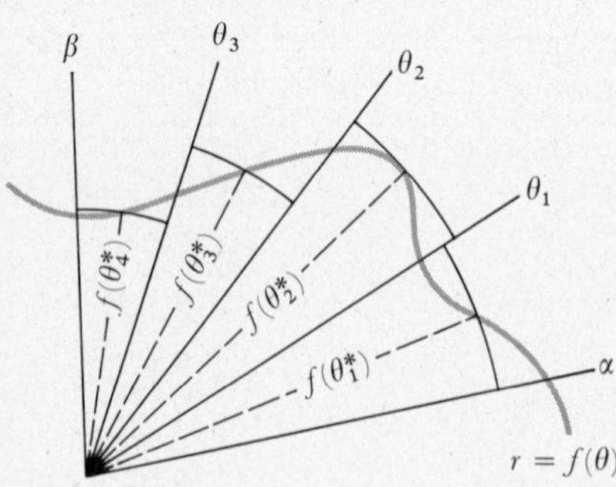

Fig. 13.29

$$\frac{1}{2}\sum_{i=1}^{n}[f(\theta_i^*)]^2\,\Delta\theta_i$$

is a Riemann sum for the continuous function $G(\theta) = (1/2)[f(\theta)]^2$ on $[\alpha, \beta]$, so in the limiting case it will tend to the definite integral

$$\int_\alpha^\beta G(\theta)\,d\theta = \frac{1}{2}\int_\alpha^\beta [f(\theta)]^2\,d\theta$$

To summarize, we have the following:

13.3.2 **Theorem** If $f(\theta)$ is positive and continuous on the interval $[\alpha, \beta]$ of length $\le 2\pi$, then the area of the region bounded by the polar graph of $r = f(\theta)$ and the rays $\theta = \alpha$ and $\theta = \beta$ is given by

$$A(R) = \frac{1}{2}\int_\alpha^\beta [f(\theta)]^2\,d\theta$$

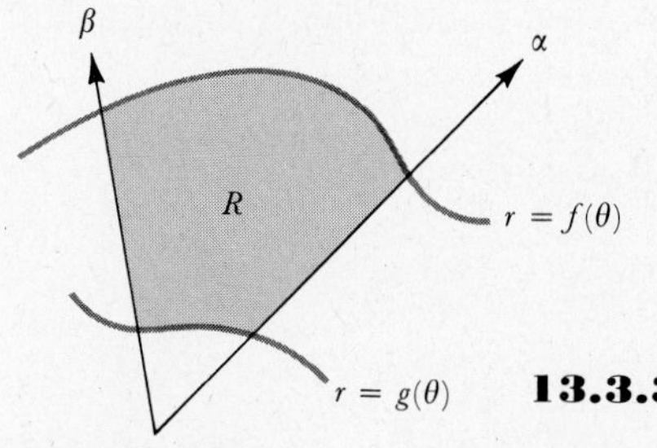

Fig. 13.30

Of course, the area of a more general region such as that in Fig. 13.30 is obtained as a difference of integrals:

$$A(R) = \frac{1}{2}\int_\alpha^\beta [f(\theta)]^2\,d\theta - \frac{1}{2}\int_\alpha^\beta [g(\theta)]^2\,d\theta$$

or

13.3.3

$$A(R) = \frac{1}{2}\int_\alpha^\beta [(f(\theta))^2 - (g(\theta))^2]\,d\theta$$

Example 1 Determine the area of the region enclosed by the four-leafed rose $r = 2\cos 2\theta$ (see Fig. 13.31).

Solution We find the area of the first leaf, which occurs for values of θ in $[-\pi/4, \pi/4]$, and then multiply by 4.

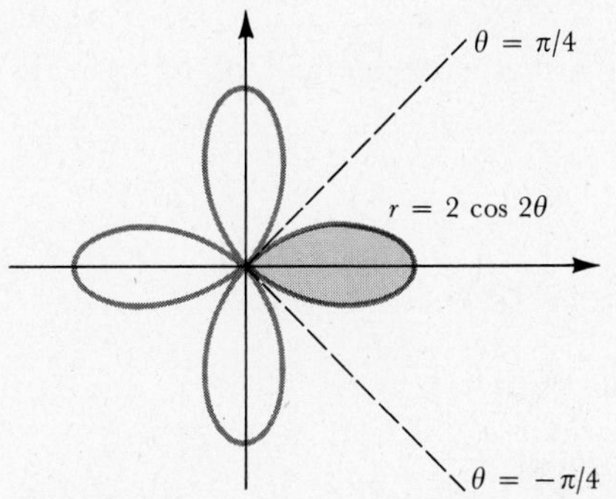

Fig. 13.31

$$\begin{aligned} A(R) &= 4\left[\frac{1}{2}\int_{-\pi/4}^{\pi/4} (2\cos 2\theta)^2\,d\theta\right] \\ &= 2\int_{-\pi/4}^{\pi/4} (4\cos^2 2\theta)\,d\theta \\ &= 8\int_{-\pi/4}^{\pi/4} \left(\frac{1+\cos 4\theta}{2}\right)d\theta \\ &= 4\left(\theta + \frac{\sin 4\theta}{4}\right)\Bigg|_{-\pi/4}^{\pi/4} = 2\pi \quad \bullet \end{aligned}$$

Example 2 Determine the area of each of the following regions:
(*a*) Inside the circle $r = 3\cos\theta$ and outside the cardioid $r = 1 + \cos\theta$.
(*b*) Inside both the circle $r = 3\cos\theta$ and the cardioid $r = 1 + \cos\theta$; see Fig. 13.32(*a*) and (*b*).

Solution Setting $1 + \cos\theta = 3\cos\theta$, we get points of intersection when $\cos\theta = 1/2$, so $\theta = \pm\pi/3$.
(*a*) The desired region lies between the rays $\theta = \pm\pi/3$ and is of the type in Fig. 13.30 with $f(\theta) = 3\cos\theta$ and $g(\theta) = 1 + \cos\theta$. Because of the symmetry with

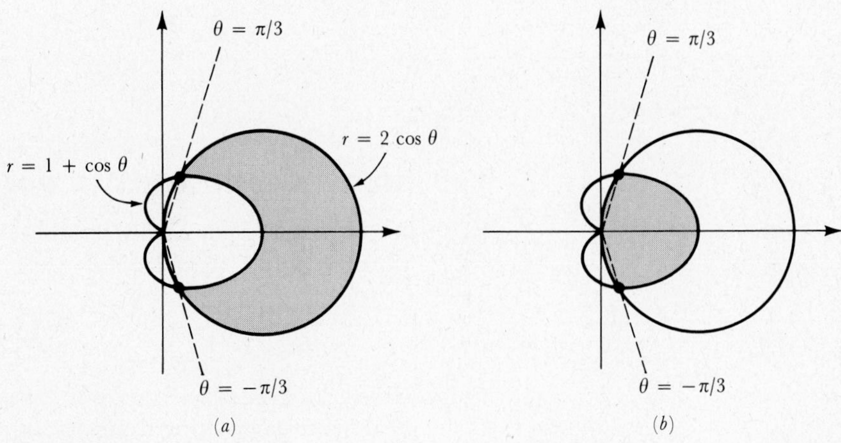

Fig. 13.32

respect to the x axis, we can find the area between the rays $\theta = 0$ and $\theta = \pi/3$ and then multiply by 2. Thus

$$\begin{aligned} A(R) &= 2\left[\frac{1}{2}\int_0^{\pi/3} ([3\cos\theta]^2 - [1+\cos\theta]^2)\right] d\theta \\ &= \int_0^{\pi/3} (9\cos^2\theta - 1 - 2\cos\theta - \cos^2\theta)\, d\theta \\ &= 8\int_0^{\pi/3} \cos^2\theta\, d\theta - 2\int_0^{\pi/3} \cos\theta\, d\theta - \int_0^{\pi/3} d\theta \\ &= 4\int_0^{\pi/3} (1+\cos 2\theta)\, d\theta - 2\int_0^{\pi/3}\cos\theta\, d\theta - \int_0^{\pi/3} d\theta \\ &= 4\left[\theta + \frac{\sin 2\theta}{2}\right]_0^{\pi/3} - \left[2\sin\theta\right]_0^{\pi/3} - \left[\theta\right]_0^{\pi/3} = \pi \end{aligned}$$

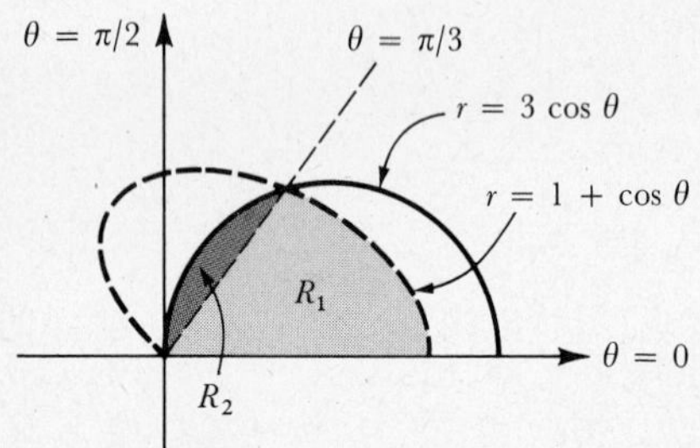

Fig. 13.33

(*b*) Here the desired region is comprised of two regions of the basic type considered in Theorem (13.3.2).

We again use the symmetry and work in the upper half plane (see Fig. 13.33):

$$\begin{aligned} A(R) &= 2[A(R_1) + A(R_2)] \\ &= 2\left[\frac{1}{2}\int_0^{\pi/3} (1+\cos\theta)^2\, d\theta + \frac{1}{2}\int_{\pi/3}^{\pi/2} (3\cos\theta)^2\right] d\theta \\ &= \int_0^{\pi/3} (1 + 2\cos\theta + \cos^2\theta)\, d\theta + 9\int_{\pi/3}^{\pi/2} (\cos^2\theta)\, d\theta \\ &= \theta + 2\sin\theta + \frac{1}{2}\left(\theta + \frac{\sin 2\theta}{2}\right)\Big|_0^{\pi/3} + \frac{9}{2}\left(\theta + \frac{\sin 2\theta}{2}\right)\Big|_{\pi/3}^{\pi/2} \\ &= 5\pi/4 \quad \bullet \end{aligned}$$

PROGRESS TEST 1

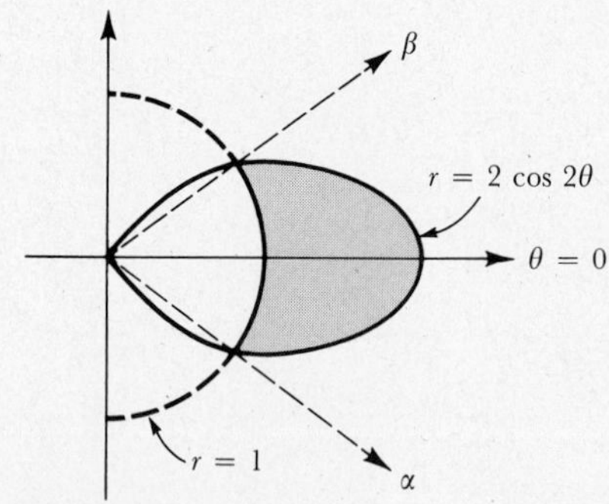

1. Show that formula (13.3.2) gives the expected area of a circle of radius a when the circle is described as $r = f(\theta) = a$, $0 \leq \theta \leq 2\pi$.

2. Determine the area A inside the right-leaf of the four-leafed rose $r = 2\cos 2\theta$ and outside the circle $r = 1$ (see sketch). (First solve for α and β.)

MORE ABOUT AREA IN POLAR COORDINATES

In constructing our earlier formulas, we assumed that $f(\theta) \geq 0$ on $[\alpha, \beta]$. On the other hand, the formula

$$\frac{1}{2}\int_\alpha^\beta [f(\theta)]^2\, d\theta$$

which involves $[f(\theta)]^2$, does not "notice" whether $f(\theta) < 0$. This can sometimes lead to errors resulting from certain areas being counted twice if $f(\theta)$ changes

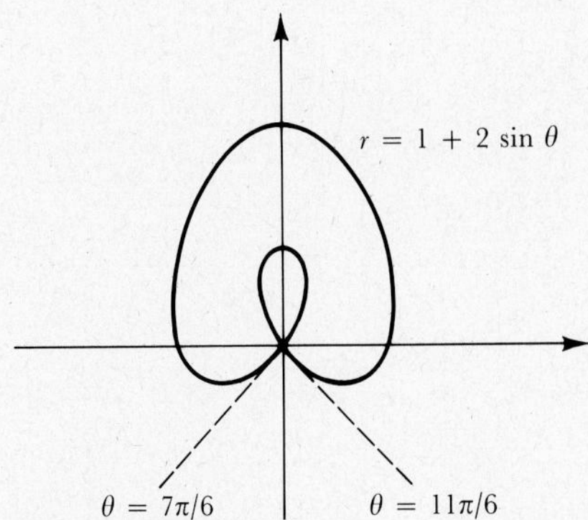

Fig. 13.34

sign on $[\alpha, \beta]$. The inner loop of the limacon $r = 1 + 2\sin\theta$ occurs when $7\pi/6 \leq \theta \leq 11\pi/6$ (see Fig. 13.34 and recall Sec. 13.1, Exercise 43). Hence if we were to compute

$$\frac{1}{2}\int_0^{2\pi}(1+2\sin\theta)^2\,d\theta = \frac{1}{2}\int_0^{7\pi/6}(1+2\sin\theta)^2\,d\theta + \frac{1}{2}\int_{7\pi/6}^{11\pi/6}(1+2\sin\theta)^2\,d\theta + \frac{1}{2}\int_{11\pi/6}^{2\pi}(1+2\sin\theta)^2\,d\theta$$

then the three integrals on the right correspond, respectively, to the areas of the shaded parts of Fig. 13.35(*a*) to (*c*).

Thus the area of the entire limacon is

$$\frac{1}{2}\int_0^{7\pi/6}(1+2\sin\theta)^2\,d\theta + \frac{1}{2}\int_{11\pi/6}^{2\pi}(1+2\sin\theta)^2\,d\theta$$

while $(1/2)\int_0^{2\pi}(1+2\sin\theta)^2\,d\theta$ counts the area of the loop twice.

The moral of this discussion is clear: When computing the area of a region enclosed by the graph of $r = f(\theta)$ on $[\alpha, \beta]$, be certain that $f(\theta) \geq 0$ on $[a, b]$. If $f(\theta) < 0$ for some θ in $[\alpha, \beta]$, break the interval into subintervals where $f(\theta)$ has a constant sign and interpret the associated integrals carefully.

ARC LENGTH AND SURFACE AREA IN POLAR COORDINATES

By (12.2.1), the length of the graph of the parametric equations

$$x = x(t) \qquad \text{and} \qquad y = y(t) \qquad a \leq t \leq b$$

is

$$s = \int_a^b \sqrt{[x'(t)]^2 + [y'(t)]^2}\,dt$$

We can describe any polar curve

$$r = f(\theta) \qquad \alpha \leq \theta \leq \beta$$

parametrically by using $x = r\cos\theta$ and $y = r\sin\theta$. This yields

$$x = f(\theta)\cos\theta \qquad \text{and} \qquad y = f(\theta)\sin\theta \qquad \alpha \leq \theta \leq \beta$$

with the polar angle θ serving as the parameter. Now, assuming f has a continuous derivative on $[\alpha, \beta]$, we have

$$x'(\theta) = -f(\theta)\sin\theta + f'(\theta)\cos\theta$$
$$y'(\theta) = f(\theta)\cos\theta + f'(\theta)\sin\theta$$

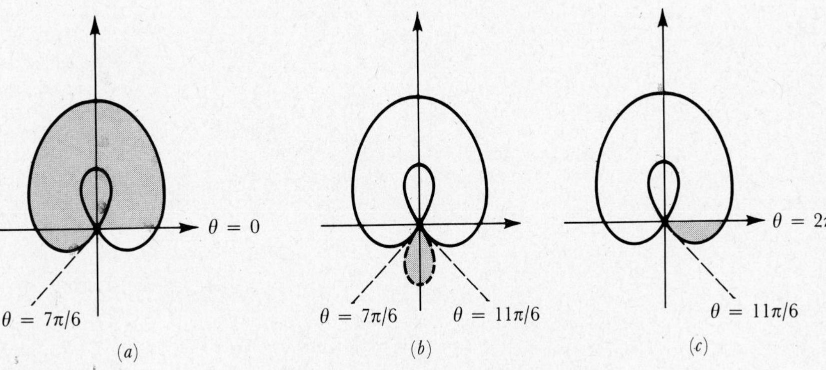

Fig. 13.35

Thus $[x'(\theta)]^2 + [y'(\theta)]^2$

$$\begin{aligned}
&= [f(\theta)]^2\sin^2\theta - 2f(\theta)f'(\theta)\sin\theta\cos\theta + [f'(\theta)]^2\cos^2\theta \\
&\quad + [f(\theta)]^2\cos^2\theta + 2f(\theta)f'(\theta)\sin\theta\cos\theta + [f'(\theta)]^2\sin^2\theta \\
&= [f(\theta)]^2(\cos^2\theta + \sin^2\theta) + [f'(\theta)]^2(\cos^2\theta + \sin^2\theta) \\
&= [f(\theta)]^2 + [f'(\theta)]^2
\end{aligned}$$

This proves the following:

13.3.4 Arc Length $$s = \int_\alpha^\beta \sqrt{[f(\theta)]^2 + [f'(\theta)]^2}\, d\theta$$

That is, $s = \int_\alpha^\beta ds$, where, for polar curves, $ds = \sqrt{[f(\theta)]^2 + [f'(\theta)]^2}\, d\theta$.

Example 3 Find the length of the polar curve $r = \sin^2(\theta/2)$, $0 \le \theta \le \pi$.

Solution Letting $f(\theta) = \sin^2(\theta/2)$, we obtain

$$f'(\theta) = \sin(\theta/2)\cos(\theta/2)$$

Hence

$$\begin{aligned}
ds &= \sqrt{[f(\theta)]^2 + [f'(\theta)]^2}\, d\theta \\
&= \sqrt{\sin^4(\theta/2) + \sin^2(\theta/2)\cos^2(\theta/2)}\, d\theta \\
&= \sqrt{\sin^2(\theta/2)[\sin^2(\theta/2) + \cos^2(\theta/2)]}\, d\theta \\
&= \sqrt{\sin^2(\theta/2)}\, d\theta \\
&= |\sin(\theta/2)|\, d\theta \\
&= \sin(\theta/2)\, d\theta \qquad \text{because } \sin(\theta/2) \ge 0 \text{ on } [0, \pi]
\end{aligned}$$

Therefore

$$s = \int_0^\pi \sin(\theta/2)\, d\theta = -2\cos(\theta/2)\Big|_0^\pi = 2 \quad \bullet$$

In a similar way we can use (12.2.7) to find the area of a surface obtained by revolving the polar curve $r = f(\theta)$, $\alpha \le \theta \le \beta$ about an axis. We replace x by $f(\theta)\cos\theta$, y by $f(\theta)\sin\theta$, and ds by $\sqrt{[f(\theta)]^2 + [f'(\theta)]^2}\, d\theta$.

Example 4 Find the area of the surface obtained by revolving the logarithmic spiral $r = e^\theta$, $0 \le \theta \le \pi/2$, about the x axis.

Solution By (12.2.7), page 410, $S = 2\pi \int_\alpha^\beta y\, ds$, where $y = f(\theta)\sin\theta$, and $ds = \sqrt{(e^\theta)^2 + (e^\theta)^2}\, d\theta = \sqrt{2e^{2\theta}}\, d\theta = \sqrt{2}e^\theta\, d\theta$. Hence

$$\begin{aligned}
S &= 2\pi \int_0^{\pi/2} f(\theta)\sin\theta\, ds \\
&= 2\pi \int_0^{\pi/2} e^\theta(\sin\theta)\sqrt{2}e^\theta\, d\theta \\
&= 2\sqrt{2}\pi \int_0^{\pi/2} e^{2\theta}\sin\theta\, d\theta \\
&= 2\sqrt{2}\pi\left[\frac{1}{5}e^{2\theta}(2\sin\theta - \cos\theta)\right]_0^{\pi/2} \qquad \text{(integration by parts)} \\
&= \frac{2\sqrt{2}\pi}{5}(1 + 2e^\pi) \quad \bullet
\end{aligned}$$

PROGRESS TEST 2

1. Compute the total length of the cardioid $r = 1 + \cos\theta$. (Use x axis symmetry.)

2. Set up, but do not evaluate, a definite integral giving the area of the surface generated by revolving the first quadrant piece of $r = 1 + \cos\theta$ about the y axis. Express the integrand entirely in terms of $\theta/2$.

3. Same as Prob. 2, but about the x axis. Express the integrand in terms of θ.

SECTION 13.3 EXERCISES

In Exercises 1 to 24 determine the area of the region bounded by the graph of the given equations. Use symmetry whenever possible. Figures and exercises given earlier in the chapter may also prove helpful.

1. Inside both loops of $r = 2\sin 2\theta$

2. $r = 2 + \cos\theta$

3. $r = 3 + 2\cos\theta$

4. $r = 5 + \sin\theta$

5. $r = 2 - \sin 2\theta$

6. $r = 2\theta$, with $0 \le \theta \le 1$, $\theta = 0$, $\theta = 1$

7. $r = e^{\theta}$, with $0 \le \theta \le \pi$, $\theta = 0$, $\theta = \pi$

8. $r = 2 + 2\sin 3\theta$, $\theta = 0$, $\theta = \pi/3$

9. Inside the three-leafed rose $r = 2\cos 3\theta$

10. $r = \sqrt{\cos\theta}$, $\theta = 0$, $\theta = \pi/4$

11. $r = 6\sec\theta$, $\theta = 0$, $\theta = \pi/4$

12. $r^2 = 4\sin 2\theta$

13. Inside one loop of $r = a\sin 4\theta$

14. Inside one loop of $r = a\cos 5\theta$

15. Inside $r = 5\cos\theta$ and outside of $r = 2 + \cos\theta$

16. Inside both $r = 5\cos\theta$ and $r = 2 + \cos\theta$

17. Inside both $r = 1 + \cos\theta$ and $r = 1 - \cos\theta$

18. Inside the smaller loop of $r = 1 - 2\cos\theta$

19. Inside the larger loop of $r = 1 - 2\cos\theta$

20. Inside $r = 2(1 + \sin\theta)$ and outside $r = 4\sin\theta$

21. Inside $r^2 = 4\cos 2\theta$ and outside $r = 2\sin\theta$

22. Inside the larger loop of $r = 2 - 4\sin\theta$ and outside the smaller loop

23. Inside the larger loop of $r = a - b\sin\theta$ and outside the smaller loop, where $a > b$ and $\arcsin(a/b) = \theta_0$

24. The area common to the two circles $r = \sin\theta$, $r = \cos\theta$

25. (*a*) Use definite integrals in polar form to compute the area of the circle $r = a\cos\theta$.
(*b*) Compare the answer to (*a*) with $\frac{1}{2}\int_0^{2\pi} a^2\cos^2\theta\, d\theta$ and explain.

In Exercises 26 to 31, (a) set up and (b) evaluate a definite integral giving the length of the indicated arc.

26. $r = 1 + \sin\theta, 0 \le \theta \le 2\pi$

27. $r = \sin^3(\theta/3), 0 \le \theta \le \pi$

28. $r = a\theta, 0 \le \theta \le 2\pi, a > 0$

29. $r = 2(1 - \cos\theta), 0 \le \theta \le 2\pi$

30. $r = 3\sec\theta, 0 \le \theta \le \pi/4$

31. $r = a\sin^4(\theta/4), 0 \le \theta \le \pi/2$

In Exercises 32 to 36, (a) set up and (b) evaluate a definite integral giving the area of the surface generated by revolving the indicated arc about the given axis.

32. $r = \sin 2\theta, 0 \le \theta \le \pi/2$, x axis

33. $r = 2\sin\theta, 0 \le \theta \le \pi/2$, x axis

34. $r^2 = 4\cos 2\theta, 0 \le \theta \le \pi/4$, x axis

35. $r = 1 + \sin\theta, 0 \le \theta \le \pi/2$, y axis

36. $r = e^{\theta}, 0 \le \theta \le \pi/2$, y axis

37. Show that the circumference of a circle of radius $a > 0$ is $2\pi a$ in two ways:
(*a*) By applying (13.3.4) to $r = a$.
(*b*) By applying (13.3.4) to $r = 2a\cos\theta$. (Be careful about counting the same arc twice.)

38. Show that the surface area of a sphere of radius $a > 0$ is $4\pi r^2$ in two ways:
(*a*) Based on $r = a$, $\theta = 0$, $\theta = \pi$.
(*b*) Based on $r = 2a\cos\theta$. (Be careful about counting the same area twice.)

39. (*a*) Show that the graph of $r = a\sin\theta + b\cos\theta$ is a circle and describe that circle.
(*b*) For such a circle prove that ds is proportional to $d\theta$.

40. Show that the area formula (13.3.3) is consistent with the sector formula (13.3.1), on which the former is based, by computing the area of the region enclosed by $r = a$, $r = b$, $\theta = 0$, $\theta = \Delta\theta$, $a < b$, $0 < \Delta\theta < 2\pi$.

CHAPTER EXERCISES

Review Exercises

In Exercises 1 to 16 graph the given polar equation.

1. $\theta = -7\pi/6$

2. $r = -3/2$

3. $r = 2\sec^2(\theta/2)$

4. $r = 2 - \sin\theta$

5. $r \cos \theta = 4$

6. $r = 2 \csc \theta$

7. $r = 2 + \sin 2\theta$

8. $r = \sqrt{1 - \cos \theta}$

9. $r = 1 + \sin^2 \theta$

10. $r^2 = 5 \cos 2\theta$

11. $r^2 = 16 \cos 2\theta$

12. $r = -\theta, \theta > 0$

13. $r = 4/(1 - \cos \theta)$

14. $r = 4 \sin 3\theta$

15. $r(1 + 2 \cos \theta) = -4$

16. $r = 2(2 + \sin \theta)$

17. $r = \cot \theta, -\pi/2 < \theta < \pi/2$

18. $r = 2 \cos(\theta/2)$

In Exercises 19 to 22, (a) transform the given equation into polar form and (b) sketch the graph.

19. $x^2 + y^2 - 4y = 0$

20. $y^2 = 4x + 4$

21. $2xy = k^2$

22. $(x^2 + y^2)^2 = 2a^2xy$

In Exercises 23 to 28 convert the given equation to rectangular coordinates.

23. $r = a \csc \theta$

24. $r = 1/(1 - \sin \theta)$

25. $r = a \sin 2\theta$

26. $r^2 = 2a^2 \cos 2\theta$

27. $r^2 = a^2 \sin^2 \theta$

28. $r = 1/(2 - \cos \theta)$

In Exercises 29 to 36 determine all points of intersection of the graphs of the given equations.

29. $r = 2 \cos \theta, r = \sqrt{3}$

30. $r \cos \theta = 2, r = 4$

31. $r \sin \theta = 2, r = 1$

32. $r = \sin^2 \theta, r = \cos^2 \theta$

33. $r = a \sin \theta, r = a \csc \theta$

34. $r = a \cos \theta, r = a \sec \theta$

35. $r = 3 \sin \theta, r = 12/(3 + \sin \theta)$

36. $r = 1 + \cos \theta, r = \cos(\theta/2)$

In Exercises 37 to 42, (a) determine the slope of the tangent line and (b) determine the tangent of the angle ψ_0 between the line tangent to the curve at $(f(\theta_0), \theta_0)$ and the ray θ_0.

37. $r = 1 - \sin \theta, \theta_0 = \pi/2$

38. $r = 2 - 3 \cos \theta, \theta_0 = \pi/2$

39. $r = a\theta, \theta_0 = \pi/2$

40. $r = a\theta, \theta_0 = \pi$

41. $r = 5 \sin 3\theta, \theta_0 = \pi/4$

42. $r^2 = 8 \cos 2\theta, \theta_0 = \pi/6$

43. Show that $r = \sec^2(\theta/2)$ and $r = 2 \csc^2(\theta/2)$ are perpendicular at their points of intersection.

44. Repeat Exercise 43 for the cardioids $r = 2(1 - \cos \theta)$ and $r = 4(1 + \cos \theta)$.

In Exercises 45 to 64 determine the area of the region bounded by the graphs of the given equations.

45. $r = 2 - \cos \theta, \theta = \pi/4, \theta = \pi$

46. Inside $r = \sqrt{|\cos \theta|}$

47. $r = 3 \cos 2, \theta = 0, \theta = \pi/6$

48. Inside one leaf of $r = \sin n\theta$, n a positive integer

49. Enclosed by $r = 2 \cos 4\theta$

50. Inside $r = 2 \sin \theta$, outside $r = \sin \theta + \cos \theta$

51. Enclosed by $r = \sin 5\theta$

52. Inside $r^2 = 2 \cos 2\theta$, outside $r = 1$

53. Inside $r = 2\sqrt{|\sin 2\theta|}$, outside $r = \sqrt{2}$

54. Inside $r = 2$, outside $r = 2(1 - \cos \theta)$

55. Inside $r = 3$, outside $r = 3(1 - \cos \theta)$

56. Enclosed by $r^2 = 2 \cos 2\theta$

57. Inside $r^2 = 2 \cos 2\theta$, outside $r = 1$

58. Inside $r = 2 \sin \theta$ and $r = 2 \cos \theta$

59. Inside $r = 3 \cos \theta$, outside $r = 1 + \cos \theta$

60. Inside $r = 2 \cos^2(\theta/2)$ and $r = 2 \sin^2(\theta/2)$

61. Inside $r = 2 \cos 3\theta$

62. Inside $r = 2 - \cos \theta$

63. Inside $r = 1 + \sin \theta$ and the lower loop of $r^2 = (1/2)\sin \theta$

64. Inside one leaf of $r = \cos n\theta$, n a positive integer

In Exercises 65 to 72, (a) determine a definite integral that gives the length of the indicated arc and (b) evaluate that integral.

65. $r = e^{-\theta}, 0 \leq \theta \leq 2\pi$

66. $r = 2^{\theta}, 0 \leq \theta \leq \pi$

67. $r = \cos^2(\theta/2), 0 \leq \theta \leq 2\pi$

68. $r = 1 - \sin \theta, 0 \leq \theta \leq 2\pi$

69. $r = 2\theta^2, 0 \leq \theta \leq 2\pi$

70. $r = 1/\theta, \pi/2 \leq \theta \leq \pi$

71. $r = \cos^3(\theta/3), 0 \leq \theta \leq 3\pi/2$

72. $r = 4 - 4 \cos \theta, 0 \leq \theta \leq 2\pi$

In Exercises 73 to 78, (a) determine a definite integral giving the area of the surface generated by revolving the indicated arc about the indicated axis, and (b) evaluate that integral.

73. $r = 6 \sec\theta, 0 \le \theta \le \pi/4, y$ axis

74. $r = 2(1 - \cos\theta), 0 \le \theta \le \pi, x$ axis

75. $r^2 = 4\cos 2\theta, 0 \le \theta \le \pi/6, x$ axis

76. Arc as in Exercise 75 about y axis

77. One loop of $r^2 = 4\cos 2\theta$ about x axis

78. One loop of $r^2 = 4\sin 2\theta$ about y axis

Miscellaneous Exercises

1. Graph $r = -3$

2. Graph the *hyperbolic spiral* $r = 2\pi/\theta$

3. Graph *Freeth's nephroid* $r = 1 + 2\sin(\theta/2)$

4. (*a*) On the same graph, sketch $r = 2 + \cos\theta$, $r = 2 + 2\cos\theta$, $r = 1 + 2\cos\theta$, $r = 1/8 + 2\cos\theta$.
(*b*) Describe the shape of the limacon $r = a + 2\cos\theta$ as $a \to 0$.
(*c*) Use the answer to (*b*) to explain why the circle $r = 2\cos\theta$ is traced out twice as $\theta: 0 \to 2\pi$

5. (*a*) On the same graph, sketch $r = 2 + 4\cos\theta$, $r = 2 + 2\cos\theta$, $r = 2 + \cos\theta$, $r = 2 + (1/8)\cos\theta$.
(*b*) Describe the shape of the limacon $r = 2 + b\cos\theta$ as $b \to 0$.
(*c*) Discuss the answer to (*b*) in terms of the graph of the circle $r = 2$.

6. (Oblique coordinate system) Suppose a rectangular xy coordinate system is given on the plane. We define the "60° oblique uv coordinate system" as follows: Let the u axis coincide with the x axis. The v axis is a line through the origin making a 60° angle with the positive u axis. Then any point P in the plane is on the intersection of a pair of lines $u = u_0$ and $v = v_0$, where $u = u_0$ is a line parallel to the v axis and intersecting the u axis at u_0, and $v = v_0$ is a line parallel to the u axis and intersecting the v axis at v_0. We refer to the pair (u_0, v_0) as the uv coordinates for P (see sketch).

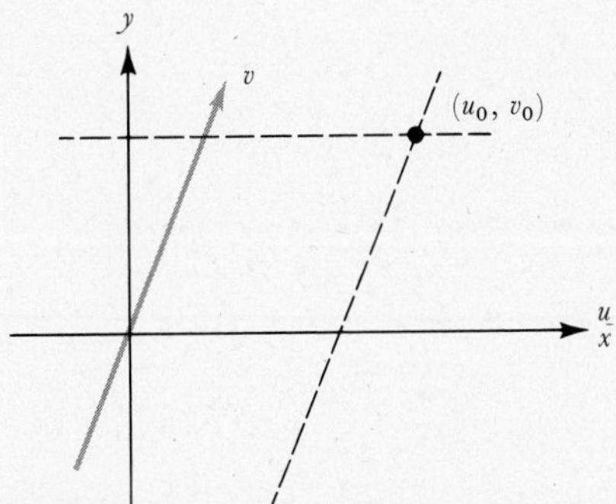

(*a*) Prove that the equations relating xy and uv coordinates are $y = v\sqrt{3}/2$, $x = u + (v/2)$.
(*b*) Determine the xy coordinates of a point with uv coordinates $(1, 1)$.
(*c*) Determine the uv coordinates of a point with xy coordinates $(4, 2)$.
(*d*) Repeat (*c*) for $(4, 0)$.
(*e*) Repeat (*c*) for $(0, \sqrt{3})$
(*f*) Translate the xy circle $x^2 + y^2 = a^2$ into uv coordinates.
(*g*) Translate the xy line $y = mx$ into uv coordinates.
(*h*) Translate the uv line $u = mv$ into xy coordinates.
(*i*) Derive a distance formula for the uv system.
(*j*) Determine the uv distance between the point with xy coordinates $(1, 1)$ and the origin.

7. Describe the graph of $1/r = 3 + 6\sin\theta$.

8. Determine the mean value of $f(\theta) = 1 - \sin\theta$ on $[0, 2\pi]$.

9. Determine the area of the region shaded in the accompanying sketch. (Each circle has radius a.)

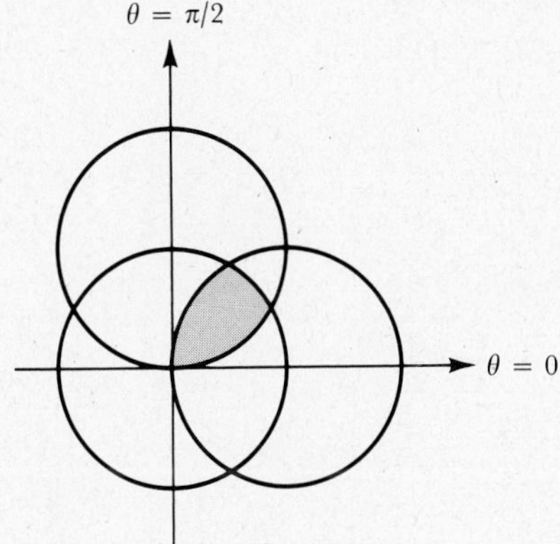

Exercises 10 to 13 are locus problems in polar coordinates.

10. Obtain a polar equation for the cissoid of Chap. 12, Miscellaneous Exercise 36.

11. Suppose F_1 has (polar) coordinates $(a, 0)$, F_2 has (polar) coordinates (a, π). Show that the locus of all points P satisfying $(\overline{PF_1})(\overline{PF_2}) = a^2$ is a lemniscate.

12. (Ovals of Cassini) For F_1, F_2 as in Exercise 11, determine an equation for the locus of all points P such that $(\overline{PF_1})(\overline{PF_2}) = b^2$. Show that for $a = b$ a lemniscate results.

13. Determine the locus of all points P equidistant from the pole and the line $r = \csc\theta$.

14. Determine the angle ψ at an arbitrary point of the *lituus* $r^2\theta = a^2$.

SELF-TEST ✢

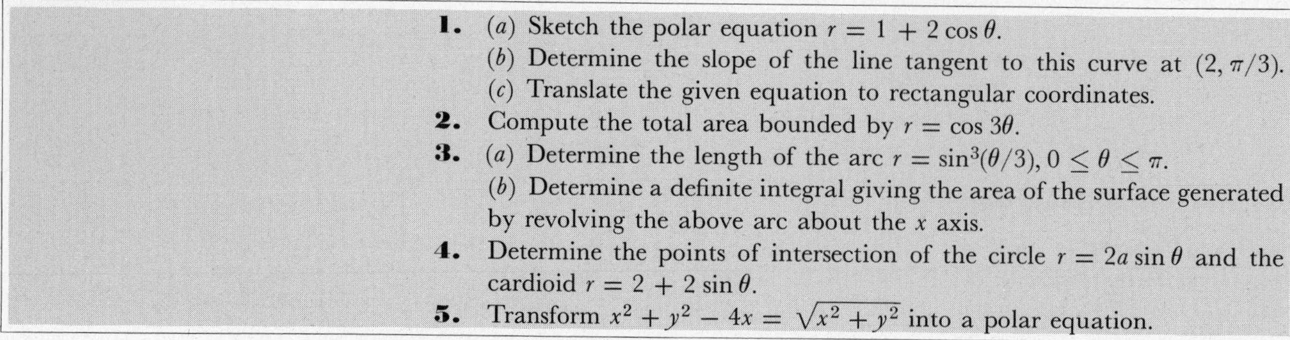

1. (*a*) Sketch the polar equation $r = 1 + 2\cos\theta$.
   (*b*) Determine the slope of the line tangent to this curve at $(2, \pi/3)$.
   (*c*) Translate the given equation to rectangular coordinates.
2. Compute the total area bounded by $r = \cos 3\theta$.
3. (*a*) Determine the length of the arc $r = \sin^3(\theta/3), 0 \leq \theta \leq \pi$.
   (*b*) Determine a definite integral giving the area of the surface generated by revolving the above arc about the x axis.
4. Determine the points of intersection of the circle $r = 2a\sin\theta$ and the cardioid $r = 2 + 2\sin\theta$.
5. Transform $x^2 + y^2 - 4x = \sqrt{x^2 + y^2}$ into a polar equation.

Chapter 14

INDETERMINATE FORMS AND IMPROPER INTEGRALS

CONTENTS

This chapter extends our ability to evaluate limits and integrals. Limits of quotients, where both numerator and denominator either approach zero or grow without bound, will be subject to a systematic and powerful method of evaluation known as L'Hospital's Rule. Previously such limits, as well as others reducible to this type, had to be evaluated on a case-by-case basis.

In Sec. 14.3 we use the idea of limit to define and evaluate integrals whose integrand grows or decreases without bound between the limits of integration, as well as to define and evaluate integrals which have at least one infinite limit of integration.

These ideas will be used frequently in subsequent chapters.

Section 14.1: Indeterminate Forms and L'Hospital's Rules

Objective:

Evaluate limits of quotients when both the numerator and denominator tend to 0 or both tend to $\pm\infty$.

LIMITS OF QUOTIENTS

Earlier in studying limits of quotients

$$\lim_{x\to a} \frac{f(x)}{g(x)}$$

we saw that a variety of things can happen, depending on the behavior of f and g near a. The simplest case occurs when $\lim_{x\to a} f(x) = k$ and $\lim_{x\to a} g(x) = j \neq 0$, because then the Limit Theorem (2.5.4) applies to tell us that

$$\lim_{x\to a} \frac{f(x)}{g(x)} = \frac{k}{j}$$

However, if $g(x)$ approaches 0 through positive values, say, while $\lim_{x\to a} f(x) = k \neq 0$, then $x = a$ is a vertical asymptote, with

$$\lim_{x\to a} \frac{f(x)}{g(x)} = \pm\infty$$

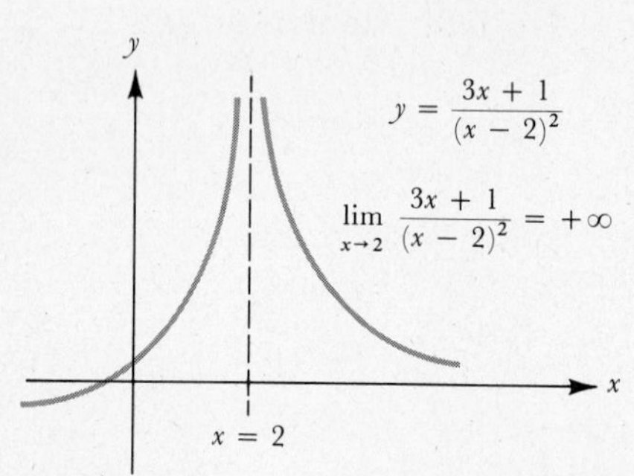

Fig. 14.1

depending on whether $k > 0$ or $k < 0$, respectively. (See, for example, Fig. 14.1.) On the other hand, if $\lim_{x\to a} g(x) = \pm\infty$, and $\lim_{x\to a} f(x) = k$, a real number, then

$$\lim_{x\to a} \frac{f(x)}{g(x)} = 0$$

(You may wish to review Sec. 4.1.)

Of special interest are the "hard" cases where *both* the numerator and denominator tend to 0, or both tend to $\pm\infty$. We have special names for such situations.

14.1.1 **Definition** If $\lim_{x\to a} f(x) = \lim_{x\to a} g(x) = 0$, then the limit of the quotient, $\lim_{x\to a} [f(x)/g(x)]$ is said to have *indeterminate form 0/0*. If $\lim_{x\to a} f(x) = \pm\infty$ and $\lim_{x\to a} g(x) = \pm\infty$, then $\lim_{x\to a} [f(x)/g(x)]$ is said to have *indeterminate form* ∞/∞. These definitions also apply for one-sided limits as well as for $a = \pm\infty$ (limits at infinity).

We shall deal with the 0/0 case first. Actually, this case is not new or rare because, for example, it occurs every time we use the definition to calculate the derivative of a differentiable function:

$$\lim_{\Delta x\to 0} \frac{f(x+\Delta x) - f(x)}{\Delta x}$$

Here $\lim_{\Delta x\to 0} [f(x+\Delta x) - f(x)] = 0$. (Why?)

Through algebraic manipulations we were able to deal with such situations quite handily until we came to the derivative of $\sin x$, when we needed an elaborate geometric argument to show that

$$\lim_{x\to 0} \frac{\sin x}{x} = 1$$

We will now present a collection of techniques which are all versions of what is known as L'Hospital's Rule, named after the seventeenth century French mathematician G. F. de L'Hospital, who included them in his calculus text, the first ever published (1696). Actually, the techniques were invented by his

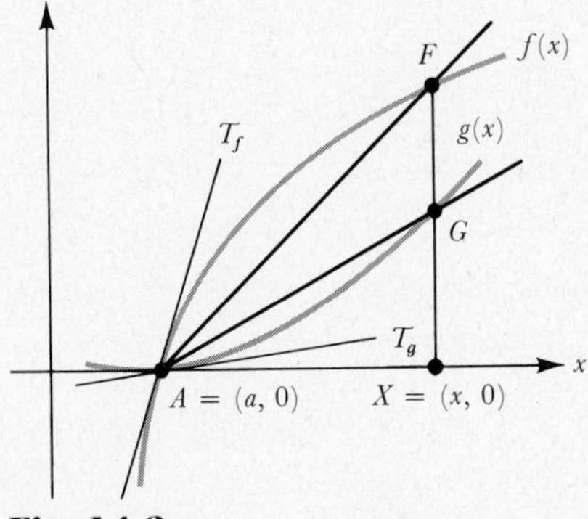

Fig. 14.2

teacher, Johann Bernoulli (the same man who brought us logarithmic differentiation), whose work was the basis for L'Hospital's text.

The basic idea of all the versions of L'Hospital's Rule is to evaluate the limit of the quotient $f(x)/g(x)$ by evaluating the limit of the quotients of the derivatives $f'(x)/g'(x)$. An oversimplified motivation for this approach is that *when an indeterminate form is involved, the limit of the quotient is really determined by the relative rates of change of the numerator and denominator as $x \to a$. How do we compare their rates of change? Using their derivatives, of course!* An informal geometric analysis helps show us what form the rule should take.

Assume that a is a real number with $\lim_{x\to a} f(x) = \lim_{x\to a} g(x) = 0$, so in Fig. 14.2 the graphs of $f(x)$ and $g(x)$ both cross the x axis at $A = (a, 0)$.

By definition of the quotient of two functions, $f(x)/g(x)$ is the quotient of the vertical lengths FX and GX. We know that the respective slopes of the secant lines through A and F and through A and G are FX/XA and GX/XA. We also know that

14.1.2
$$\frac{FX}{GX} = \frac{FX/XA}{GX/XA}$$

Now, as $x \to a$, the point $X \to A$, which means the two secant lines FA and GA approach the respective tangent lines T_f and T_g. Thus the *quotient* of the secant line slopes on the right side of (14.1.2) approaches the *quotient* of the tangent line slopes; that is, they approach the quotient of the derivatives. Hence the quotient on the left side of (14.1.2), which is really $f(x)/g(x)$, must likewise approach the quotient of the derivatives at a. Although this is not the precise conclusion of L'Hospital's Rule, it is the form most often used in practice. We now state the rule in a more precise and general form. A complete proof can be found in any book on advanced calculus.

14.1.3 **L'Hospital's Rule: 0/0** Suppose $f'(x)$ and $g'(x)$ exist near a, and $g'(x) \neq 0$ near a with $\lim_{x\to a} f(x) = \lim_{x\to a} g(x) = 0$. To determine $\lim_{x\to a} [f(x)/g(x)]$ calculate $f'(x)$ and $g'(x)$ separately and evaluate the limit of the quotient $f'(x)/g'(x)$. The conclusion is as follows:

$$\text{If } \lim_{x\to a} \frac{f'(x)}{g'(x)} = L, \text{ then } \lim_{x\to a} \frac{f(x)}{g(x)} = L$$

This rule holds whether a and L are real numbers or $\pm\infty$. It holds for left- and right-hand limits as well.

Note that if a is a real number, the phrase "all x near a" means all x in some open interval containing a but not $x = a$ (often called a "deleted neighborhood of a"). If $a = +\infty$, this phrase means all x larger than some real number N, that is, all x in $(N, +\infty)$. Similarly, for $a = -\infty$, it means all x in $(-\infty, N)$ for some real number N.

We first apply the rule where we are confident of the answer—before moving to cases where the answer is not at all obvious.

Example 1 Compute $\lim_{x\to 0} \dfrac{\sin x}{x}$.

Solution The hypotheses hold since numerator and denominator are everywhere differentiable and $\lim_{x\to 0} \sin x = 0 = \lim_{x\to 0} x$. $D_x(\sin x) = \cos x$ and $D_x(x) = 1$. Now

$$\lim_{x\to 0} \frac{\cos}{1} = 1$$

and therefore, by L'Hospital's Rule,

$$\lim_{x \to 0} \frac{\sin x}{x} = 1$$

as well. ●

Notice that although L'Hospital's Rule gives us an easy means of computing $\lim_{x\to 0} [(\sin x)/x]$, we could not have used it when computing the limit originally, because to use the rule, we needed the derivative of $\sin x$.

It is important to realize that in applying this rule, we differentiate numerator and denominator *separately,* form the quotient of the derivatives, take the limit of this quotient, and *if* this latter limit is L, then the original limit is L. Thus, in phrasing our solutions below, our last step will be to equate L (if it exists) and the original limit in question.

Example 2 Evaluate $\lim_{x \to 1} \dfrac{\ln x}{1 - x}$.

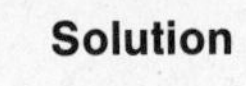

Solution Now $\lim_{x\to 1}(\ln x) = 0 = \lim_{x\to 1}(1 - x)$, and each function is differentiable near $x = 1$, so we use L'Hospital's Rule:

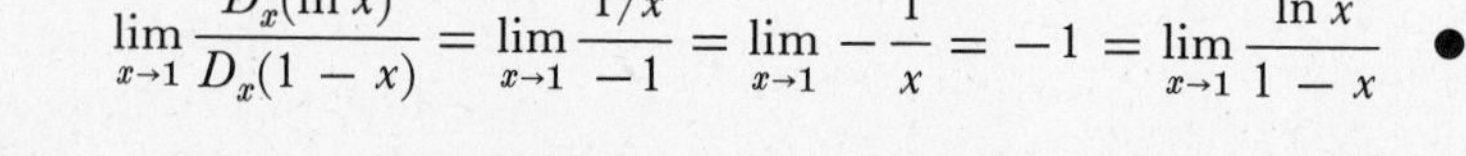

$$\lim_{x \to 1} \frac{D_x(\ln x)}{D_x(1 - x)} = \lim_{x \to 1} \frac{1/x}{-1} = \lim_{x \to 1} -\frac{1}{x} = -1 = \lim_{x \to 1} \frac{\ln x}{1 - x} \quad ●$$

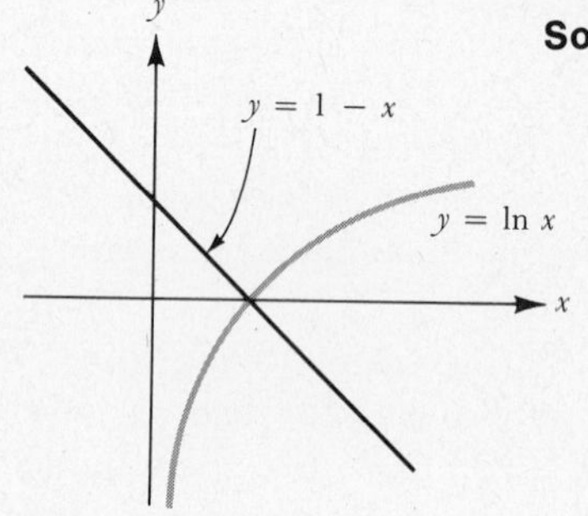

Fig. 14.3

A glance at the graphs of $y = \ln x$ and $y = 1 - x$ confirms the result of Example 2 because the graphs cross at right angles. (See Fig. 14.3.) We would expect the quotient of the derivatives at $x = 1$ to be -1.

Some problems may need repeated use of L'Hospital's Rule before we reach a limit whose value can be recognized.

Example 3 Determine $\lim_{x \to 0} \dfrac{\sin x - x}{x^3}$.

Solution Since $\lim_{x\to 0}(\sin x - x) = \lim_{x\to 0} x^3 = 0$, we take derivatives, the quotient of which is

$$\frac{\cos x - 1}{3x^2}$$

Again, $\lim_{x\to 0}(\cos x - 1) = \lim_{x\to 0} 3x^2 = 0$, so we can reapply the rule. The quotient of the derivatives this time is $(-\sin x)/6x$ and, again, $\lim_{x\to 0} (-\sin x) = \lim_{x\to 0} 6x = 0$, so once again we differentiate numerator and denominator separately and take the limits:

$$\lim_{x \to 0} \frac{-\cos x}{6} = \frac{-1}{6}$$

Retracing the equalities (we shall not bother doing this in the future) we have our answer:

$$-\frac{1}{6} = \lim_{x \to 0} \frac{-\cos x}{6} = \lim_{x \to 0} \frac{-\sin x}{6x} = \lim_{x \to 0} \frac{\cos x - 1}{3x^2} = \lim_{x \to 0} \frac{\sin x - x}{x^3} \quad ●$$

Example 4 Compute $\lim_{x\to+\infty} \dfrac{1-\cos(1/x)}{\arctan(1/x)}$.

Solution As $x \to +\infty$, $1/x \to 0$, $\cos(1/x) \to 1$, so $[1-\cos(1/x)] \to 0$ and $\arctan(1/x) \to \arctan 0 = 0$. We differentiate numerator and denominator, and then take the limit:

$$\lim_{x\to+\infty} \frac{(\sin(1/x))(-1/x^2)}{\dfrac{1}{1+(1/x)^2}(-1/x^2)} = \lim_{x\to+\infty} \frac{\sin(1/x)}{\dfrac{1}{1+(1/x)^2}} = \frac{0}{1}$$

$$= 0 = \lim_{x\to+\infty} \frac{1-\cos(1/x)}{\arctan(1/x)} \quad \bullet$$

PROGRESS TEST 1

In Probs. 1 to 4 evaluate the given limits.

1. $\lim_{x\to 0} \dfrac{2^x - e^x}{x}$

2. $\lim_{x\to 0} \dfrac{1-\cos x}{\sin x}$

3. $\lim_{x\to 0} \dfrac{\sin x - \tan x}{x^2}$

4. $\lim_{x\to+\infty} \dfrac{\ln(1+1/x)}{1/x}$

5. Use L'Hospital's Rule to evaluate the important limit $\lim_{h\to 0}(e^h - 1)/h$ (which gives the slope of $y = e^x$ at $x = 0$).

THE INDETERMINATE FORM ∞/∞

If both $f(x)$ and $g(x)$ increase or decrease without bound as x approaches a, we have the second indeterminate form ∞/∞ defined in (14.1.1). There is another form of L'Hospital's Rule for this case. Its proof can be found in any advanced calculus text.

14.1.4 **L'Hospital's Rule:** ∞/∞ Suppose $f'(x)$ and $g'(x)$ exist near a and $g'(x) \neq 0$ near a with $\lim_{x\to a} f(x) = \pm\infty$ and $\lim_{x\to a} g(x) = \pm\infty$. The conclusion is as follows:

$$\text{If } \lim_{x\to a} \frac{f'(x)}{g'(x)} = L, \text{ then } \lim_{x\to a} \frac{f(x)}{g(x)} = L$$

This rule holds whether a and L are real numbers or $\pm\infty$. It holds for left- and right-hand limits as well.

Example 5 Determine $\lim_{x\to+\infty} \dfrac{x^2}{e^x}$.

Solution The numerator and denominator each grow without bound, so (14.1.4) applies. We differentiate numerator and denominator to get the limit of the quotient $2x/e^x$ as $x \to +\infty$ to which (14.1.4) applies again. We differentiate again, getting

$$\lim_{x\to+\infty} \frac{2}{e^x} = 0$$

It follows that

$$\lim_{x\to+\infty} \frac{x^2}{e^x} = 0 \quad \bullet$$

We can now prove a generalized form of (7.5.2), a very useful pair of limits, the second of which generalizes Example 5.

14.1.5 Theorem

(*a*) If p is any positive number, $\lim_{x\to+\infty} \dfrac{\ln x}{x^p} = 0$.

(*b*) If p is any number, $\lim_{x\to+\infty} \dfrac{x^p}{e^x} = 0$.

Proof:
(*a*) The hypotheses of (14.1.4) hold. (Note that $\ln x$ grows without bound as x does.) Thus

$$\lim_{x\to+\infty} \frac{D_x(\ln x)}{D_x(x^p)} = \lim_{x\to+\infty} \frac{1/x}{px^{p-1}} = \lim_{x\to+\infty} \frac{1}{pxx^{p-1}} = \lim_{x\to+\infty} \frac{1}{px^p}$$

Since $p > 0$, this latter limit is zero; hence the original limit is likewise zero.

(*b*) We consider two cases, (i) $p \leq 0$ and (ii) $p > 0$.

Case (i). If $p \leq 0$, then $-p \geq 0$, so we can conclude that $\lim_{x\to+\infty} 1/(x^{-p}e^x) = 0$ without L'Hospital's Rule.

Case (ii). If $p > 0$, then L'Hospital's Rule applies, so we differentiate numerator and denominator as many times as are necessary to lower the degree of the numerator to zero or below, where case (i) applies. ■

From (14.1,5), we can conclude that $\ln x$ grows *slower* than any positive power of x and e^x grows *faster* than any power of x. (See Exercises 39 and 40.)

Often it is necessary to perform a variety of manipulations of the limit with different indeterminate forms appearing in the same problem.

Example 6 Determine $\lim_{x\to0^+} \dfrac{\ln x}{\cot x}$.

Solution We know that $\lim_{x\to0^+} \ln x = -\infty$ and $\lim_{x\to0^+} \cot x = +\infty$, so (14.1.4) applies. Hence we differentiate and then evaluate

$$\lim_{x\to0^+} \frac{1/x}{-\csc^2 x}$$

But this limit is indeterminate, of the form ∞/∞ again, and the derivatives of the numerator and denominator do not look as if they will help resolve the problem. Hence we write $\csc^2 x = 1/\sin^2 x$ and rearrange;

$$\lim_{x\to0^+} \frac{1/x}{-\csc^2 x} = \lim_{x\to0^+} -\frac{\sin^2 x}{x}$$

Although this latter limit is of indeterminate form $0/0$, we can use (14.1.3) to reduce it to the following:

$$\lim_{x\to0^+} \frac{-2\sin x\cos x}{1} = \frac{0}{1} = 0$$

Hence

$$\lim_{x\to0^+} \frac{\ln x}{\cot x} = 0 \quad \bullet$$

PROGRESS TEST 2

Evaluate the given limits.

1. $\lim_{x\to(\pi/2)^-} \dfrac{1+\sec x}{\tan x}$ **2.** $\lim_{x\to 1^+} \dfrac{\ln(x-1)}{1/(x-1)}$ **3.** $\lim_{x\to+\infty} \dfrac{x+\ln x}{x\ln x}$

A CAUTION

If we use L'Hospital's Rule in an attempt to evaluate

$$\lim_{x\to 0^+} \frac{x}{x^2+\cos x}$$

we obtain

$$\lim_{x\to 0^+} \frac{1}{2x-\sin x} = +\infty$$

However, looking again at the original limit, we find that

$$\lim_{x\to 0^+} \frac{x}{x^2+\cos x} = \frac{0}{0+1} = 0$$

L'Hospital's Rule does not apply, since the original limit is neither of the two basic indeterminate forms! The moral: *Don't use L'Hospital's Rule until you check to see that its hypotheses are satisfied. The original limit must be an indeterminate form.*

SECTION 14.1 EXERCISES

In Exercises 1 to 33 evaluate the given limit.

1. $\lim_{x\to 0} \dfrac{x}{\tan x}$ **2.** $\lim_{x\to 0} \dfrac{e^x-e^{-x}}{\sin x}$ **3.** $\lim_{x\to 0} \dfrac{3x-\tan x}{\sin x-3x}$ **4.** $\lim_{x\to 0} \dfrac{\sin x}{x^3}$

5. $\lim_{x\to 0} \dfrac{\sqrt[3]{x}}{\sin x}$ **6.** $\lim_{x\to 1^-} \dfrac{\sqrt{1-x}}{\ln x}$ **7.** $\lim_{x\to 0} \dfrac{\cos x-1}{x}$ **8.** $\lim_{x\to 0} \dfrac{1-e^x}{x}$

9. $\lim_{x\to 0} \dfrac{3\cos x+x-3}{x^3}$ **10.** $\lim_{x\to+\infty} \dfrac{e^{-x}}{x^2}$ **11.** $\lim_{x\to 0} \dfrac{\ln x}{\csc x}$ **12.** $\lim_{x\to\pi/2} \dfrac{\sec x}{\sec 4x}$

13. $\lim_{x\to 0} \dfrac{xe^x-x}{1-\cos 2x}$ **14.** $\lim_{x\to 0} \dfrac{\sin^2 x}{\sin x^2}$ **15.** $\lim_{x\to 1} \dfrac{\arccos x}{x-1}$ **16.** $\lim_{x\to+\infty} \dfrac{(\ln x)^3}{\sqrt[3]{x}}$

17. $\lim_{x\to 0} \dfrac{e^{2x}-\cos 2x}{2x\sin 2x}$ **18.** $\lim_{x\to a} \dfrac{\ln(x/a)}{x-a}$ **19.** $\lim_{x\to 0} \dfrac{x}{\tan x}$ **20.** $\lim_{x\to 0} \dfrac{\tanh x}{\tanh 3x}$

21. $\lim_{x\to 0} \dfrac{\sinh x}{x}$ **22.** $\lim_{x\to(\pi/2)^+} \dfrac{\ln\tan x}{\ln\sec x}$ **23.** $\lim_{x\to+\infty} \dfrac{\ln(x^3+x)}{\ln(x^3-x)}$ **24.** $\lim_{x\to 5} \dfrac{2^x+3^x}{5x-1}$

25. $\lim_{x\to 0} \dfrac{x-\arctan x}{\tan^3 x}$ **26.** $\lim_{x\to 0} \dfrac{\arcsin x-x}{x^3}$ **27.** $\lim_{x\to 0} \dfrac{\tan x-\sin x}{\tan^3 x}$ **28.** $\lim_{x\to 0} \dfrac{\ln\tan 2x}{\ln\sin 3x}$

29. $\lim_{x\to 0} \dfrac{x-xe^x}{\sin^2 x}$ **30.** $\lim_{x\to+\infty} \dfrac{1/x^2}{\sin^2(1/x)}$ **31.** $\lim_{x\to 0} \dfrac{e^{2x^2}-1}{\sin^2 x}$ **32.** $\lim_{x\to 0} \dfrac{2\cos x-2+x^2}{x^4}$

33. $\lim_{x\to 0} \dfrac{e^x-\sin x-1}{1-\cos 2x}$

34. Determine the limit in Exercise 30 without using L'Hospital's Rule.

35. (*a*) Show that the hypotheses of L'Hospital's Rule (14.1.3) hold for $\lim_{x\to-\infty}(e^x/b^x)$, which is indeterminate of form 0/0 provided $b > 1$.
(*b*) Show that (14.1.3) does not determine this limit.
(*c*) Determine the limit using other means.

36. Draw a sketch analogous to Fig. 14.3 to confirm that $\lim_{x\to 0}(e^x-1)/x = 1$.

37. Use L'Hospital's Rule to determine any horizontal tangents to the cardioid $r = 2(1+\cos\theta)$.

38. Use L'Hospital's Rule to show that the cycloid $x = t-\sin t$, $y = t-\cos t$ has a vertical tangent at $t = 0$. (See Chap. 12, page 401.)

39. Extend [14.1.5(*a*)] to $\lim_{x\to+\infty} \dfrac{(\ln x)^q}{x^p} = 0$ for all positive p and q.

40. Extend [14.1.5(*b*)] to $\lim_{x\to+\infty} \dfrac{x^p}{e^{qx}} = 0$ for any p and any positive q.

Section 14.2: Other Indeterminate Forms

Objectives:

1. Transform indeterminate limits into forms considered in Sec. 14.1.
2. Use the properties of the natural logarithm function to evaluate indeterminate forms involving variable exponents.

THE INDETERMINATE FORMS $0 \cdot \infty$ AND $\infty - \infty$

Consider the following limit of a product:

$$\lim_{x\to 0^+} x\,|\ln x|$$

As $x \to 0^+$, the first factor gets smaller while the second grows without bound. The limit of the product cannot be determined without further analysis.

Similarly, we cannot determine the following without further analysis:

$$\lim_{x\to 0}\left(\frac{1}{x} - \frac{1}{\sin x}\right)$$

14.2.1 **Definition**

(*a*) If $\lim_{x\to a} f(x) = 0$ and $\lim_{x\to a} g(x) = \pm\infty$, then $\lim_{x\to a} [f(x)g(x)]$ is said to be of *indeterminate form* $0 \cdot \infty$.

(*b*) If, as $x \to a$, one of $f(x)$ or $g(x)$ grows without bound while the other decreases without bound, then $\lim_{x\to a} (f(x) + g(x))$ is said to be of *indeterminate form* $\infty - \infty$. (Roughly speaking, $f(x)$ and $g(x)$ pull in opposite directions, and the winner is not obvious.)

Our approach in both of these situations will be to use appropriate manipulations and/or identities to reduce these to indeterminate forms considered in Sec. 14.1. In general, we handle the indeterminate form $0 \cdot \infty$ using

$$f(x) \cdot g(x) = \frac{f(x)}{1/g(x)} \qquad \text{or} \qquad f(x) \cdot g(x) = \frac{g(x)}{1/f(x)}$$

The indeterminate form $\infty - \infty$ may involve a variety of manipulations, ending with a fraction whose limit is indeterminate of form $0/0$ or ∞/∞.

Example 1 Evaluate $\lim_{x\to 0^+} x|\ln x|$.

Solution Consider $\lim_{x\to 0^+} |\ln x|/(1/x)$, which is of form ∞/∞. Hence we apply (14.1.4) to get

$$\lim_{x\to 0^+} \frac{1/x}{-1/x^2} = \lim_{x\to 0^+} (-x) = 0$$

Thus the original limit is 0. (You are invited to try the alternative approach.) ●

Example 2 Evaluate $\lim_{x\to 0}\left[\frac{1}{x} - \frac{1}{xe^{ax}}\right]$ (a is any real number).

Solution A typical technique is to combine terms over a common denominator:

$$\lim_{x\to 0} \frac{e^{ax} - 1}{xe^{ax}}$$

This limit is indeterminate of form $0/0$, so we differentiate. But

$$\lim_{x\to 0} \frac{ae^{ax}}{xae^{ax} + e^{ax}} = \frac{a}{0+1} = a$$

so the original limit is a. ●

It is a mistake to assume that every limit can be handled as routinely as those treated thus far. Sometimes a variety of indeterminate forms may arise in the same problem, requiring a degree of ingenuity. Follow the next example closely.

Example 3 Evaluate $\lim_{x\to 0^+} (\cot x + \ln x)$.

Solution This limit is of indeterminate form $\infty - \infty$, so we rewrite it

$$\text{(1)} \qquad \lim_{x\to 0^+}\left[\frac{\cos x}{\sin x} + \ln x\right] = \lim_{x\to 0^+} \frac{\cos x + \sin x \ln x}{\sin x}$$

We know that the denominator approaches 0 through positive values and $\cos x$ approaches 1. But $\sin x \ln x$ is of indeterminate form $0 \cdot \infty$ and needs special treatment. Now

$$\lim_{x\to 0^+} (\sin x \ln x) = \lim_{x\to 0^+} \frac{\ln x}{1/\sin x}$$

which is of indeterminate form ∞/∞, so we apply (14.1.4), differentiating numerator and denominator to arrive at

$$\lim_{x\to 0^+}\left[\frac{1/x}{(-\cos x/\sin^2 x)}\right] = \lim_{x\to 0^+} \frac{(1/x)\sin^2 x}{-\cos x} = \lim_{x\to 0^+} \frac{[(\sin x)/x]\sin x}{-\cos x}$$

We already know that $\lim_{x\to 0^+}(\sin x)/x = 1$, so

$$\lim_{x\to 0^+} \frac{[(\sin x)/x]\sin x}{-\cos x} = \frac{1 \cdot 0}{-1} = 0.$$

Hence $\lim_{x\to 1^+}(\sin x \ln x) = 0$ also.

Return to the original limit rewritten on the right side of (1). As $x \to 0^+$, the numerator approaches $1 + 0$, while the denominator approaches 0^+, so the fraction grows without bound. Thus

$$\lim_{x\to 0^+}(\cot x + \ln x) = +\infty \quad \bullet$$

PROGRESS TEST 1

Evaluate the given limits.

1. $\lim_{x\to(\pi/2)^-} (\tan x - \sec x)$ (rewrite in terms of sines and cosines)

2. $\lim_{x\to 1^+} \left(\frac{x}{x-1} - \frac{1}{\ln x}\right)$

3. $\lim_{x\to(\pi/2)^-} [\tan x \ln(\sin x)]$ (use $1/\tan x = \cot x$)

THE INDETERMINATE FORMS 0^0, 1^∞, and ∞^0

In some situations a limit of the form

$$\lim_{x\to a} f(x)^{g(x)}$$

may be indeterminate because the base may be approaching 1 while the exponent grows without bound. This was the case in (7.4.11) and (7.4.12), page 267, where we described e as

$$e = \lim_{x\to+\infty}\left(1 + \frac{1}{x}\right)^x \qquad \text{and} \qquad e = \lim_{x\to 0}(1 + x)^{1/x}$$

These are said to be *indeterminate of form 1^∞*. Other similar indeterminate forms are ∞^0 and 0^0.

Our strategy for dealing with such indeterminate forms that arise when evaluating

$$\lim_{x\to a} f(x)^{g(x)}$$

is analogous to our use of logarithmic differentiation in differentiating $f(x)^{g(x)}$: *Take the natural logarithm and then use the properties of the natural logarithm to simplify the situation.* The original indeterminate form will thus be transformed into one that we already know how to handle.

Our first case will be used frequently in later chapters.

14.2.2 Theorem $$\lim_{x\to+\infty}\left(1+\frac{r}{x}\right)^x = e^r$$

Proof: This limit is indeterminate of form 1^∞. Let $y = [1 + (r/x)]^x$ and take natural logarithms. We know that

$$\ln[\lim_{x\to+\infty} y] = \ln\left[\lim_{x\to+\infty}\left(1+\frac{r}{x}\right)^x\right] = \lim_{x\to+\infty}\left[\ln\left(1+\frac{r}{x}\right)^x\right]$$

because the natural logarithm function is continuous. (This enables us to interchange ln and lim as necessary.) Now

$$\lim_{x\to+\infty}\left[\ln\left(1+\frac{r}{x}\right)^x\right] = \lim_{x\to+\infty}\left[x\ln\left(1+\frac{r}{x}\right)\right]$$

and this latter limit is indeterminate of form $\infty \cdot 0$. Rewrite it as

$$\lim_{x\to+\infty}\frac{\ln(1+r/x)}{1/x}$$

which is of form 0/0. Apply L'Hospital's Rule (14.1.3):

$$\lim_{x\to+\infty}\left(\frac{1/(1+r/x)\cdot[-r/x^2]}{-1/x^2}\right) = \lim_{x\to+\infty}\frac{r}{1+r/x} = r$$

Thus

$$\ln[\lim_{x\to+\infty} y] = r$$

Knowing this, we can conclude that

$$\lim_{x\to+\infty} y = e^r \quad \blacksquare$$

The last step of the above proof is important since up to that point we have not found the original limit, but the *logarithm* of that limit. Hence we need apply the exponential function to "undo" the logarithm.

Example 4 Determine $\lim_{x\to0^+} x^{x^2}$.

Solution This limit is of form 0^0. Now

$$\ln[\lim_{x\to0^+} x^{x^2}] = \lim_{x\to0^+}[\ln(x^{x^2})]$$

$$= \lim_{x\to0^+}[x^2\ln x]$$

(which is indeterminate of form $0 \cdot \infty$)

$$= \lim_{x\to0^+}\frac{\ln x}{1/x^2} \qquad \text{(form } \infty/\infty\text{)}$$

Hence we differentiate, getting

$$\lim_{x\to 0^+} \frac{1/x}{-2/x^3} = \lim_{x\to 0^+} \frac{x^2}{-2} = 0$$

Thus $\ln(\lim_{x\to 0^+} x^{x^2}) = 0$, so

$$\lim_{x\to 0^+} (x^{x^2}) = e^0 = 1 \quad \bullet$$

PROGRESS TEST 2

Evaluate the given limits.

1. $\lim_{x\to +\infty} \left(\frac{x+2}{x}\right)^x$ **2.** $\lim_{x\to 0^+} x^{\sin x}$ **3.** $\lim_{x\to 1} x^{1/(1-x^2)}$

AN EXTENSION TO $0^{\pm\infty}$

The logarithm strategy works on limits involving $f(x)^{g(x)}$ even when the limit is not indeterminate, as the proof of the following theorem shows.

14.2.3 **Theorem** Suppose that $\lim_{x\to a} f(x) = 0$ with $f(x) > 0$ for all x near a.

(*a*) If $\lim_{x\to a} g(x) = +\infty$, then $\lim_{x\to a} f(x)^{g(x)} = 0$.

(*b*) If $\lim_{x\to a} g(x) = -\infty$, then $\lim_{x\to a} f(x)^{g(x)} = +\infty$.

Proof: We prove (*a*) and leave (*b*) as Exercise 35.

$$\ln[\lim_{x\to a} f(x)^{g(x)}] = \lim_{x\to a}[\ln(f(x)^{g(x)})] = \lim_{x\to a} [g(x)\ln f(x)]$$

As x approaches a, the first factor $g(x)$ grows without bound while the second factor decreases without bound (since $f(x) \to 0$). Hence

$$\ln[\lim_{x\to a} f(x)^{g(x)}] = -\infty$$

which implies

$$\lim_{x\to a} f(x)^{g(x)} = 0 \qquad \text{(Why?)} \quad \blacksquare$$

SECTION 14.2 EXERCISES

In Exercises 1 to 34 evaluate the given limit.

1. $\lim_{x\to +\infty} x \tan(1/x)$

2. $\lim_{x\to +\infty} x \sin(1/x)$

3. $\lim_{x\to 1} (x-1)\tan(\pi x/2)$

4. $\lim_{x\to 0} \left[\frac{1}{x^2} - \frac{1}{x^2 \sec x}\right]$

5. $\lim_{x\to 0} \left[\frac{1}{e^x - 1} - \frac{1}{\arctan x}\right]$

6. $\lim_{x\to +\infty} e^{-x} \ln x$

7. $\lim_{x\to 0} \left[\frac{1}{\sin x} - \frac{1}{x}\right]$

8. $\lim_{x\to \pi/2} \left[2\sec^2 x - \frac{1}{1-\sin x}\right]$

9. $\lim_{x\to 0} x \csc 2x$

10. $\lim_{x\to 0^+} x^x$

11. $\lim_{x\to 0^+} (\cos 3x)^{1/x}$

12. $\lim_{x\to 0^+} \csc x \arcsin x$

13. $\lim_{x\to 0^-} (e^x + x)^{1/x}$

14. $\lim_{x\to 0^+} (1+x)^{\ln x}$

15. $\lim_{x\to 0^+}(\ln x)^x$

16. $\lim_{x\to 0^+}(1-x)^{\cot x}$

17. $\lim_{x\to 0^+} x^{1/\ln x}$

18. $\lim_{x\to 0^+} x^{(x^x)}$

19. $\lim_{x\to 0^+}(1/x)^{\sin x}$

20. $\lim_{x\to +\infty}(\cos 1/x)^{x^2}$

21. $\lim_{x\to 1^-}(1-x^2)^{1/(1-x)}$

22. $\lim_{x\to 0^+}(\sec x)^{\cot x}$

23. $\lim_{x\to 1}(1+\ln x)^{\cos(1-x)}$

24. $\lim_{x\to 0}(\cos 2x)^{1/x^2}$

25. $\lim_{x\to\pi/2}(\tan^2 x)^{\cos^2 x}$

26. $\lim_{x\to e}(\ln x)^{1/(x-e)}$

27. $\lim_{x\to 0^+}(ax+1)^{\cot x}\quad (a>0)$

28. $\lim_{x\to 0}\left(\dfrac{\cos 2x+\cos 3x}{2}\right)^{1/x^2}$

29. $\lim_{x\to 0^+}[\ln(x+\sqrt{1+x^2})-\ln x]$

30. $\lim_{x\to 0}(\sec x-\cos x)^{\tan 2x}$

31. $\lim_{x\to +\infty}(\cosh x)^{1/x}$

32. $\lim_{x\to b^+}(\log_b x)^{1/(x-b)}\quad (b>1)$

33. $\lim_{x\to 0}[\csc^2 x-1/x^2]$

34. $\lim_{x\to 0^+}(\sec 2x)^{\cot^2 x}$

35. Prove [14.2.3(b)].

36. Compare $\lim_{x\to 0^+}\dfrac{1-2\cot x}{\csc x}$ and $\lim_{x\to 0^-}\dfrac{1-2\cot x}{\csc x}$.

37. Compare $\lim_{x\to 0^+}(e^x+x)^{1/x}$ with the limit in Exercise 13.

38. Prove that $\lim_{x\to 0}[e^{-1/x^2}/x]=0$.

39. Compute $\lim_{x\to 0} 2^{[(1/x)-\csc x]}$

Section 14.3: Improper Integrals

Objectives:

1. Evaluate integrals with infinite limits of integration.
2. Evaluate integrals whose integrands grow or decrease without bound between the limits of integration.

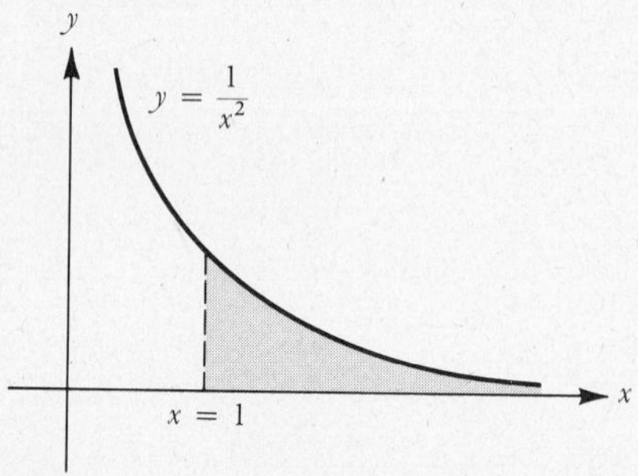

Fig. 14.4

INTEGRALS WITH INFINITE LIMITS OF INTEGRATION

This section deals with two types of integrals commonly called "improper." Our first concern is in evaluating integrals such as

$$\int_1^{+\infty}\frac{dx}{x^2}$$

which may be thought of as representing the area of the region to the right of $x=1$ between the x axis and $y=1/x^2$. (See Fig. 14.4.)

Definite integrals have only been defined over finite intervals, but the basic tool of calculus, namely limit, allows us to extend the definition in a natural way. For this case we have

$$\int_1^{10}\frac{dx}{x^2}=0.9 \qquad \int_1^{100}\frac{dx}{x^2}=0.99 \qquad \int_1^{1000}\frac{dx}{x^2}=0.999 \qquad \int_1^{b}\frac{dx}{x^2}=1-\frac{1}{b}$$

and thus it is natural to define

$$\int_1^{+\infty}\frac{dx}{x^2}=\lim_{b\to+\infty}\int_1^b\frac{dx}{x^2}=\lim_{b\to+\infty}\left(1-\frac{1}{b}\right)=1$$

14.3.1 **Definition**

(*a*) Suppose $f(x)$ is integrable in $[a,b]$ for every b in $[a,+\infty)$. If $\lim_{b\to+\infty}\int_a^b f(x)\,dx$ exists, then we say $\int_a^{+\infty} f(x)$ is *convergent,* and define $\int_a^{+\infty} f(x)\,dx=\lim_{b\to+\infty}\int_a^b f(x)\,dx$.

(*b*) Suppose $f(x)$ is integrable in $[a, b]$ for every a in $(-\infty, b]$. If $\lim_{a\to-\infty} \int_a^b f(x)\,dx$ exists, then we say $\int_{-\infty}^b f(x)\,dx$ is *convergent,* and define $\int_{-\infty}^b f(x)\,dx = \lim_{a\to-\infty} \int_a^b f(x)\,dx$.

(*c*) Suppose $f(x)$ is integrable over all finite intervals $[a, b]$ and c is any real number. If both $\int_{-\infty}^c f(x)\,dx$ and $\int_c^{+\infty} f(x)\,dx$ are convergent, then we define $\int_{-\infty}^{+\infty} f(x)\,dx = \int_{-\infty}^c f(x)\,dx + \int_c^{+\infty} f(x)\,dx$.

If any of the limits in (*a*) to (*c*) fails to exist and be finite, we say that the associated integral is *divergent.*

By [14.3.1(*a*)], we see that $\int_1^{+\infty} dx/x^2$ is convergent and equals 1.

Example 1 Evaluate $\displaystyle\int_4^{+\infty} \frac{dx}{\sqrt{x}}$.

Solution

$$\int_4^{+\infty} \frac{dx}{\sqrt{x}} = \lim_{b\to+\infty} \int_4^b x^{-1/2}\,dx = \lim_{b\to+\infty} \Big[2\sqrt{x}\Big]_4^b$$
$$= \lim_{b\to+\infty} [2\sqrt{b} - 4] = +\infty$$

Hence $\displaystyle\int_4^{+\infty} \frac{dx}{\sqrt{x}}$ is divergent. ●

Example 2 Evaluate $\displaystyle\int_{-\infty}^{+\infty} \frac{dx}{x^2+1}$.

Solution By definition, we must evaluate

$$\int_{-\infty}^{c} \frac{dx}{x^2+1} \quad \text{and} \quad \int_{c}^{+\infty} \frac{dx}{x^2+1}$$

where c is any real number. For convenience we can choose $c = 0$. Now

$$\int_{-\infty}^{0} \frac{dx}{x^2+1} = \lim_{a\to-\infty} \int_a^0 \frac{dx}{x^2+1} = \lim_{a\to-\infty} \Big[\arctan x\Big]_a^0$$
$$= \lim_{a\to-\infty} [\arctan 0 - \arctan a]$$
$$= \lim_{a\to-\infty} (-\arctan a) = -\lim_{a\to-\infty} \arctan a = -\left(-\frac{\pi}{2}\right) = \frac{\pi}{2}$$

(We can expect a nonnegative answer here since the integrand is positive for all x.) On the other hand

$$\int_0^{+\infty} \frac{dx}{x^2+1} = \lim_{b\to+\infty} \int_0^b \frac{dx}{x^2+1}$$
$$= \lim_{b\to+\infty} [\arctan b - \arctan 0]$$
$$= \lim_{b\to+\infty} \arctan b = \frac{\pi}{2}$$

Hence

$$\int_{-\infty}^{+\infty} \frac{dx}{1+x^2} = \frac{\pi}{2} + \frac{\pi}{2} = \pi \quad ●$$

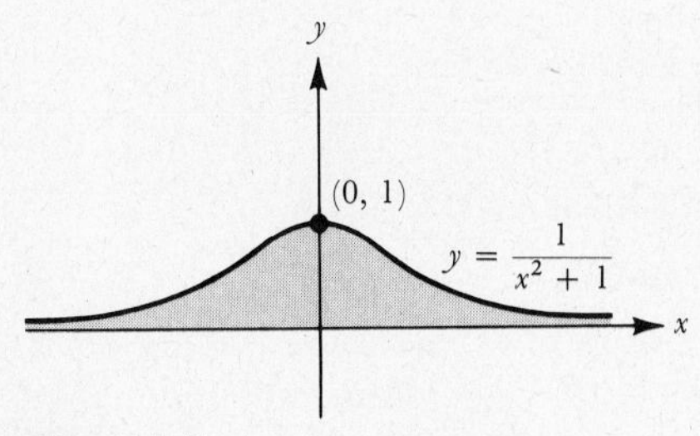

Fig. 14.5

Example (2) gives us the area of the region between the curve $y = 1/(x^2 + 1)$ and the x axis. (See Fig. 14.5.)

Example 3 Evaluate $\int_0^{+\infty} \cos x \, dx$.

Solution

$$\int_0^{+\infty} \cos x \, dx = \lim_{b\to+\infty} \int_0^b \cos x \, dx = \lim_{b\to+\infty} \Big[\sin x\Big]_0^b$$
$$= \lim_{b\to+\infty} [\sin b - \sin 0]$$
$$= \lim_{b\to+\infty} [\sin b]$$

This limit does not exist (oscillates between $+1$ and -1), so $\int_0^{+\infty} \cos x \, dx$ is divergent. ●

Example 4 Compare $\lim_{t\to+\infty} \int_{-t}^{t} 4x^3 \, dx$ and $\int_{-\infty}^{+\infty} 4x^3 \, dx$.

Solution

$$\lim_{t\to+\infty} \int_{-t}^{t} 4x^3 \, dx = \lim_{t\to+\infty} \Big[x^4\Big]_{-t}^{t}$$
$$= \lim_{t\to+\infty} [t^4 - (-t)^4]$$
$$= \lim_{t\to+\infty} 0 = 0$$

On the other hand,

$$\int_{-\infty}^{+\infty} 4x^3 \, dx = \lim_{a\to-\infty} \int_a^0 4x^3 \, dx + \lim_{b\to+\infty} \int_0^b 4x^3 \, dx$$
$$= \lim_{a\to-\infty} \Big[x^4\Big]_a^0 + \lim_{b\to+\infty} \Big[x^4\Big]_0^b$$
$$= \lim_{a\to-\infty} \left(-a^4\right) + \lim_{b\to+\infty} \left(b^4\right)$$

Neither of these limits is a real number, so the integral $\int_{-\infty}^{+\infty} 4x^3 \, dx$ diverges. Since

$$\int_{-\infty}^{+\infty} 4x^3 \, dx \neq \lim_{t\to+\infty} \int_{-t}^{t} 4x^3 \, dx$$

the latter limit *cannot* be used as a shortcut means of evaluating $\int_{-\infty}^{+\infty} 4x^3 \, dx$. ●

PROGRESS TEST 1

Evaluate these improper integrals:

1. $\displaystyle\int_1^{+\infty} \frac{2 \ln x}{x} \, dx$ **2.** $\displaystyle\int_{-\infty}^{0} e^x \, dx$ **3.** $\displaystyle\int_{-\infty}^{+\infty} xe^{-x^2} \, dx$

INTEGRALS WITH INFINITE INTEGRANDS

Can the first quadrant region to the left of $x = 1$ and below $y = 1/x^2$ be regarded as having finite area? (See Fig. 14.4, page 468.) Unfortunately, neither our definition of definite integral based on limits of finite sums nor the Fundamental Theorem of Calculus applies.

The function is not bounded on [0, 1] and is thus not integrable on [0, 1]. However, $y = 1/x^2$ is continuous on $[\varepsilon, 1]$ for any $0 < \varepsilon \leq 1$, so $\int_\varepsilon^1 1/x^2\,dx$ does exist. Again, we can use a limit to answer the above questions:

$$\lim_{\varepsilon\to 0^+}\int_\varepsilon^1 \frac{dx}{x^2} = \lim_{\varepsilon\to 0^+}\left[-\frac{1}{x}\right]_\varepsilon^1 = \lim_{\varepsilon\to 0^+}\left[-1+\frac{1}{\varepsilon}\right] = +\infty$$

Although in this case the region does not have a finite area, the technique used to answer the question makes sense in general.

14.3.2 Definition

(*a*) Suppose $f(x)$ is integrable on $[a+\varepsilon, b]$ for every $a+\varepsilon$ in $(a, b]$ $(\varepsilon > 0)$, and $\lim_{x\to a^+} f(x) = \pm\infty$. If $\lim_{\varepsilon\to 0^+}\int_{a+\varepsilon}^b f(x)\,dx$ exists, then we say $\int_a^b f(x)\,dx$ is *convergent* and define $\int_a^b f(x)\,dx = \lim_{\varepsilon\to 0^+}\int_{a+\varepsilon}^b f(x)\,dx$.

(*b*) Suppose $f(x)$ is integrable on $[a, b-\varepsilon]$ for every $b-\varepsilon$ in $[a, b)$ $(\varepsilon > 0)$, and $\lim_{x\to b^-} f(x) = \pm\infty$. If $\lim_{\varepsilon\to 0^+}\int_a^{b-\varepsilon} f(x)\,dx$ exists, then we say $\int_a^b f(x)\,dx$ is *convergent* and define $\int_a^b f(x)\,dx = \lim_{\varepsilon\to 0^+}\int_a^{b-\varepsilon} f(x)\,dx$.

(*c*) Suppose $\lim_{x\to c^+} f(x) = \pm\infty$ and/or $\lim_{x\to c^-} f(x) = \pm\infty$ for $a < c < b$, and both $\int_a^c f(x)\,dx$ and $\int_c^b f(x)\,dx$ are convergent. Then we say $\int_a^b f(x)\,dx$ is *convergent* and define $\int_a^b f(x)\,dx = \int_a^c f(x)\,dx + \int_c^b f(x)\,dx$.

If any of the limits in (*a*) to (*c*) fails to exist and be finite, then we say the associated integral is *divergent*.

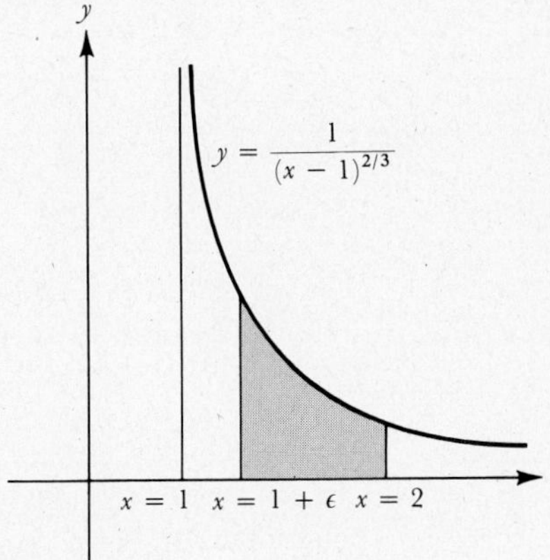

Fig. 14.6

By (14.3.2) and our earlier calculations, $\int_0^1 dx/x^2$ is divergent. A similar calculation shows that $\int_{-1}^0 dx/x^2$ is divergent. Thus, by [14.3.2(*b*)], $\int_{-1}^1 dx/x^2$ is divergent. However, if we had somehow ignored or forgotten the discontinuity at 0 in $[-1, 1]$ and tried to apply the Fundamental Theorem of Calculus, we would get

$$\int_{-1}^1 \frac{dx}{x^2} = -\frac{1}{x}\bigg|_{-1}^1 = -\frac{1}{1} - \left(-\frac{1}{-1}\right) = -1 - 1 = -2$$

which is nonsense because the integral of a positive function could never be negative.

Example 5 Evaluate $\displaystyle\int_1^2 \frac{dx}{(x-1)^{2/3}}$. (See Fig. 14.6.)

Solution Here $\displaystyle\lim_{x\to 1^+}\frac{1}{(x-1)^{2/3}} = +\infty$, so

$$\int_1^2 \frac{dx}{(x-1)^{2/3}} = \lim_{\varepsilon\to 0^+}\int_{1+\varepsilon}^2 \frac{dx}{(x-1)^{2/3}}$$
$$= \lim_{\varepsilon\to 0^+}\left[3(x-1)^{1/3}\right]_{1+\varepsilon}^2$$
$$= \lim_{\varepsilon\to 0^+}\left[3 = 3\varepsilon^{1/3}\right] = 3 \quad \bullet$$

It is best to use two different limit variables in dealing with integrals when the point of discontinuity lies strictly between a and b.

Example 6 Evaluate $\displaystyle\int_0^3 \frac{2x\,dx}{(x^2-1)^{2/3}}$

Solution $\displaystyle\lim_{x\to 1}\frac{2x}{(x^2-1)^{2/3}} = +\infty$, so we must use [14.3.2(*c*)].

$$\lim_{\varepsilon\to 0^+}\int_0^{1-\varepsilon}\frac{2x\,dx}{(x^2-1)^{2/3}}+\lim_{\delta\to 0^+}\int_{1+\delta}^{3}\frac{2x\,dx}{(x^2-1)^{2/3}}$$
$$=\lim_{\varepsilon\to 0^+}\Big[3(x^2-1)^{1/3}\Big]_0^{1-\varepsilon}+\lim_{\delta\to 0^+}\Big[3(x^2-1)^{1/3}\Big]_{1+\delta}^{3}$$
$$=\lim_{\varepsilon\to 0^+}[3(-2\varepsilon+\varepsilon^2)^{1/3}+3]+\lim_{\delta\to 0^+}[6-3(2\delta+\delta^2)^{1/3}]$$
$$=3+6=9 \quad \bullet$$

In some cases it may be necessary to use L'Hospital's Rule in evaluating the limit associated with an improper integral.

Example 7 Evaluate $\int_0^1 \ln x\,dx$.

Solution Since $\lim_{x\to 0^+}(\ln x)=-\infty$,

$$\int_0^1 \ln x\,dx=\lim_{\varepsilon\to 0^+}\int_\varepsilon^1 \ln x\,dx$$
$$=\lim_{\varepsilon\to 0^+}\Big[x\ln x - x\Big]_\varepsilon^1 \qquad \text{(integration by parts)}$$
$$=\lim_{\varepsilon\to 0^+}[-1-\varepsilon\ln\varepsilon+\varepsilon]=-1-\lim_{\varepsilon\to 0^+}(\varepsilon\ln\varepsilon) \qquad (0\cdot\infty)$$

It remains to evaluate the indeterminate form $\lim_{\varepsilon\to 0^+}\frac{\ln\varepsilon}{1/\varepsilon}$. By L'Hospital's Rule,

$$\lim_{\varepsilon\to 0^+}\frac{1/\varepsilon}{-1/\varepsilon^2}=0=\lim_{\varepsilon\to 0^+}\frac{\ln\varepsilon}{1/\varepsilon}$$

Hence

$$\int_0^1 \ln x\,dx=-1 \quad \bullet$$

PROGRESS TEST 2

Evaluate the given improper integrals.

1. $\int_0^2 \frac{dx}{\sqrt{4-x^2}}$ **2.** $\int_0^{33} \frac{dx}{(x-1)^{4/5}}$ **3.** $\int_0^4 \frac{2x\,dx}{x^2-4}$

IMPROPER INTEGRALS INVOLVING POWERS OF X

Improper integrals involving the antiderivative

$$\int\frac{dx}{x^p}$$

occur frequently enough, particularly in the study of infinite series, to warrant systematic treatment. Their straightforward proofs are left as exercises.

14.3.3 Theorem

(*a*) $p\le 1$ implies $\int_1^{+\infty}\frac{dx}{x^p}$ diverges.

(*b*) $p>1$ implies $\int_1^{+\infty}\frac{dx}{x^p}$ converges and equals $\frac{1}{p-1}$.

14.3.4 Theorem

(*a*) $p < 1$ implies $\int_0^1 \frac{dx}{x^p}$ converges and equals $\frac{1}{1-p}$.

(*b*) $p \geq 1$ implies $\int_0^1 \frac{dx}{x^p}$ diverges.

SECTION 14.3 EXERCISES

In Exercises 1 to 40 evaluate the given integral.

1. $\int_0^{+\infty} e^{-x}\,dx$

2. $\int_0^{+\infty} \frac{3x^2}{e^{x^3}}\,dx$

3. $\int_{-\infty}^0 \frac{dx}{e-x}$

4. $\int_{10^{10}}^{+\infty} \frac{dx}{x}$

5. $\int_1^{+\infty} \frac{\ln x}{x}\,dx$

6. $\int_1^{+\infty} \frac{\ln x}{x^2}\,dx$

7. $\int_0^1 \frac{dx}{(1-x)^4}$

8. $\int_0^1 \frac{dx}{\sqrt[3]{x-1}}$

9. $\int_1^2 \frac{dx}{\sqrt{x-1}}$

10. $\int_0^{e^2} \frac{\ln\sqrt{x}}{\sqrt{x}}\,dx$

11. $\int_0^1 x\ln x\,dx$

12. $\int_0^{\pi/2} \sec x\tan x\,dx$

13. $\int_1^{+\infty} \frac{dx}{\sqrt[3]{x^4}}$

14. $\int_{-\infty}^{+\infty} \frac{dx}{\sqrt[3]{x^4}}$

15. $\int_{-\infty}^0 e^{-x}\cos x\,dx$

16. $\int_0^{+\infty} e^{-x}\sin x\,dx$

17. $\int_{-\infty}^{+\infty} \frac{dx}{x^2+4}$

18. $\int_0^2 \frac{dx}{\sqrt{x-1}}$

19. $\int_0^1 \frac{dx}{x^2-1}$

20. $\int_0^{+\infty} \sin x\,dx$

21. $\int_{\pi/3}^{\pi/2} \frac{\sin x}{\sqrt{\cos x}}\,dx$

22. $\int_0^{+\infty} \frac{e^{-x}\,dx}{1+e^{-x}}$

23. $\int_0^1 \frac{(\ln x)^2}{x}\,dx$

24. $\int_{-a}^a \frac{x\,dx}{\sqrt{a^2-x^2}}$

25. $\int_0^{+\infty} xe^{-x}\,dx$

26. $\int_0^{+\infty} x^2e^{-x}\,dx$

27. $\int_{-\infty}^{+\infty} \frac{x\,dx}{x^4+1}$

28. $\int_e^{+\infty} \frac{dx}{x(\ln x)^3}$

29. $\int_1^{+\infty} \frac{dx}{x(\ln x)^3}$

30. $\int_{-\pi/2}^{+\pi/2} \tan x\,dx$

31. $\int_0^{+\infty} \frac{dx}{x^2-2x+10}$

32. $\int_0^{+\infty} \frac{dx}{x^2-6x+11}$

33. $\int_{-\infty}^{+\infty} \frac{dx}{4x^2+12x+13}$

34. $\int_0^{+\infty} e^{-x}\sin 3x\,dx$

35. $\int_0^{\pi/2} \sec x\,dx$

36. $\int_0^{\pi/2} \frac{dx}{1-\sin x}$

37. $\int_1^{+\infty} \frac{dx}{x\sqrt{x^2-1}}$

38. $\int_0^{+\infty} \frac{dx}{x\sqrt{x^2-1}}$

39. $\int_{-1}^{+\infty} \frac{e^{-\sqrt{x}}}{\sqrt{x}}\,dx$ (*Note:* You will need to evaluate three limits.)

40. $\int_{-\infty}^{+\infty} \frac{e^{-\sqrt[3]{x}}}{\sqrt[3]{x^2}}\,dx$ (*Note:* You will need to evaluate four limits.)

41. (*a*) Is $\int_{-1}^1 \frac{\sin\sqrt{x}}{\sqrt{x}}\,dx$ proper or improper?

(*b*) Evaluate the integral in (*a*).

42. Repeat Exercise 41 for $\int_{-\pi/2}^{+\pi/2} \frac{x\,dx}{\csc x^2}$.

43. Prove (14.3.3). **44.** Prove (14.3.4).

45. Generalize (14.3.3) to powers of $x - c$.

46. Generalize (14.3.4) to powers of $x - c$.

47. (*a*) Generalize (14.3.3) by replacing the limit of integration by any $a > 0$.

(*b*) Generalize (14.3.4) by replacing the limit of integration by any $a > 0$.

48. Prove that $\int_0^{+\infty} \frac{dx}{x^p}$ diverges for all real numbers p.

49. Without using parametric equations or polar coordinates, but rather using an improper integral, show that the circumference of a circle of radius r is $2\pi r$.

50. Use the approach of Exercise 49 to show that the surface area of a sphere is $4\pi r^2$.

51. Determine the centroid of the semispherical surface generated by revolving $y = \sqrt{r^2 - x^2}$, $0 \leq x \leq r$, about the y axis.

52. Determine the total first-quadrant area between the x and y axis and $y = e^{-x}$.

53. (A fake paradox) Let R be the region bounded by the line $x = 1$, the positive x axis and the graph of $y = 1/x$, and let S be the solid of revolution generated by rotating R about the x axis.

(*a*) Prove that S has finite volume.

(*b*) Prove that S does not have finite surface area.

(*c*) If S were turned to a vertical position and filled with paint, then we would have a finite amount of paint covering an infinite surface area. Chapter 15 will help clarify this situation. Explain why this is a "fake paradox."

54. Let R be the region bounded by $y = x^{-2/3}$, $x = 1$, and $y = 0$.

(*a*) Show that R has infinite area.

(*b*) Show that the volume of the solid generated by revolving R about the x axis is finite.

55. Determine the length of the spiral $r = e^{-\theta}$ for $\theta \geq 0$.

56. Compute the area between the curves $y = \pm 1/\sqrt{x}$ to the right of $x = 4$.

57. (*a*) Prove that $\int_0^{+\infty} x^k e^{-x}\,dx = k\int_0^{+\infty} x^{k-1}e^{-x}\,dx$.

(*b*) Use (*a*) to show that for n a positive integer, $\int_0^{+\infty} x^n e^{-x}\,dx = n!$

58. × Suppose the apartment complex of Example 5, Sec. 8.1, is assumed to have infinite life during which it continues to return \$2000/month income with money returning 8 percent. What is its market value (the present value of its unending income stream)? (Note, of course, that the total income is infinite.)

59. × (See Exercise 58 and Sec. 11.1, Exercise 50.) Suppose the same apartment complex of Exercise 58 returns an unending monthly income stream that grows linearly from \$0/month at a rate of \$1 per year. What now is its market value?

CHAPTER EXERCISES

Review Exercises

In Exercises 1 to 36 evaluate the given limit.

1. $\lim_{x\to -1} \dfrac{\sin \pi x}{1+x}$

2. $\lim_{x\to 0} \dfrac{x\cos x - \sin x}{x}$

3. $\lim_{x\to 0} \dfrac{\sec x - 1}{x^2}$

4. $\lim_{x\to 0} \dfrac{\sec x \tan x}{2x}$

5. $\lim_{x\to 0^+} \dfrac{\sin x}{\sqrt{x}}$

6. $\lim_{x\to 0} \dfrac{\cot 2x}{\cot 3x}$

7. $\lim_{x\to 0} \dfrac{\ln \sin x}{\ln \tan x}$

8. $\lim_{x\to +\infty} x \tan \dfrac{2}{x}$

9. $\lim_{x\to 0} (e^x + 3x)^{1/x}$

10. $\lim_{x\to 0^+} \sin x \ln x$

11. $\lim_{x\to \pi} \dfrac{\cos mx}{\cos nx}$ (m,n odd integers)

12. $\lim_{x\to 0} \dfrac{\ln x^2}{\cot x^2}$

13. $\lim_{x\to 0} \dfrac{x}{\cot x}$

14. $\lim_{x\to 0} \dfrac{e^x + 2e^{-x} - 3}{1 - \cos 3x}$

15. $\lim_{x\to 0} \dfrac{\sin x + 3x}{2x}$

16. $\lim_{x\to 1} \cot x \ln \cos x$

17. $\lim_{x\to 0} (1 + 2x)^{1/(5x)}$

18. $\lim_{x\to 0} \dfrac{x\ln(1+x)}{1 - \cos x}$

19. $\lim_{x\to 0} (\cos x)^{\cot^2 x}$

20. $\lim_{x\to +\infty} (1 - x^2)e^{-x^2}$

21. $\lim_{x\to 0} \left(\dfrac{1}{x} - \dfrac{1}{e^x - 1}\right)$

22. $\lim_{x\to 1^-} \left(\dfrac{1}{x^2 - 1}\right)^{x-1}$

23. $\lim_{x\to \pi/2} \dfrac{\sin x + \cos x - 1}{1 + \cos x - \sin x}$

24. $\lim_{x\to 0} \dfrac{1 - \cos x - (x^3/6)}{x^4}$

25. $\lim_{x\to 0} \dfrac{\sin x - \sinh x}{x^2}$

26. $\lim_{x\to (\pi/2)^-} (\sin x)^{\sec x}$

27. $\lim_{x\to 0^+} (\sin x \ln x)$

28. $\lim_{x\to 0} \dfrac{\sin x^2}{x \sin x}$

29. $\lim_{x\to +\infty} \dfrac{3^x - 2^x}{x}$

30. $\lim_{x\to 0} \left[\dfrac{\ln(1+x)}{x^2} - \dfrac{1}{x(1+x)}\right]$

31. $\lim_{x\to 0} \dfrac{3^x - 2^x}{x}$

32. $\lim_{x\to 0^+} \dfrac{\tan x}{x}$

33. $\lim_{x\to \pi/2} \dfrac{\sin mx}{\sin nx}$ [m,n even integers]

34. $\lim_{x\to \pi/2} (\sin x)^{\tan^2 x}$

35. $\lim_{x\to 0^+} \dfrac{\ln \tan x}{\ln \tan 3x}$

36. $\lim_{x\to 0} \dfrac{e^x - 2\sin x - e^{-x}}{\sin x - x}$

In Exercises 37 to 56 evaluate the given integral.

37. $\int_0^{\pi/2} \tan^2 x\,dx$

38. $\int_{-1}^{2} \dfrac{dx}{\sqrt{2 + x - x^2}}$

39. $\int_0^{\pi/2} \dfrac{1 - \cos x}{\sin^2 x}\,dx$

40. $\int_{-\infty}^{+\infty} \dfrac{dx}{e^x + e^{-x}}$

41. $\int_1^{+\infty} \dfrac{\ln x}{x^2}\,dx$

42. $\int_4^{5} \dfrac{dx}{\sqrt{x - 4}}$

43. $\int_1^{e} \dfrac{dx}{x \ln x}$

44. $\int_0^{\pi/2} (\sec x \tan x - \sec^2 x)\,dx$

45. $\int_0^{+\infty} x^2 e^{-x}\,dx$

46. $\int_0^{+\infty} \dfrac{dx}{x + 1}$

47. $\int_0^{1} (\ln x)^2\,dx$

48. $\int_{-1}^{1} \sqrt{\dfrac{1+x}{1-x}}\,dx$

49. $\int_{1/2}^{1} \frac{dx}{x^4\sqrt{1-x^2}}$

50. $\int_0^{\pi/2} \sec x\, dx$

51. $\int_{-2}^{+\infty} \frac{dx}{(x+1)^4}$

52. $\int_1^{+\infty} \frac{dx}{x\sqrt{1-x^2}}$

53. $\int_0^{+\infty} e^{-x} \sin x\, dx$

54. $\int_0^{+\infty} e^{-2x} \cosh x\, dx$

55. $\int_0^1 \frac{\ln x}{\sqrt{x}}\, dx$

56. $\int_{-a}^{a} \frac{x\, dx}{\sqrt{a^2 - x^2}}$

57. Compare $\lim_{x\to 1^-} (x-1)^{1/(x^2-1)}$ with the limit in Exercise 22 above.

58. Compare $\lim_{x\to 0^+} x^{1/x}$ and $\lim_{x\to +\infty} x^{1/x}$.

59. Evaluate the following limit

$$\lim_{x\to\sqrt{2}} \frac{2x^4 - 10x^2 + 12}{x^4 + x^2 - 6}$$

(*a*) Using L'Hospital's Rule and (*b*) by some other means.

60. Find the area between the graph of $y^2 = x^2/(4 - x^2)$ and the lines $x = \pm 2$. (Sketch the graph.)

61. (*a*) Find the area of the region bounded by the x axis, the negative y axis, and the graph of $y = \ln x$.
(*b*) Find the volume of the solid of revolution generated by revolving this region about the y axis.
(*c*) Find the surface area of this solid.

62. (*a*) Find the area between the graph of $y = x/(x^2+1)^2$ and the positive x axis.
(*b*) What is the volume of the solid of revolution generated by rotating this region about the x axis?
(*c*) What is the surface area of this solid?

63. (*a*) Is $\int_{-1}^{1} \frac{\sin x}{x}\, dx$ improper?
(*b*) Use the Trapezoid Rule with eight subdivisions to get an estimate for the integral in (*a*).

64. (*a*) Evaluate $\int_0^1 \frac{dx}{e^x - e^{-x}}$. (*b*) Evaluate $\int_1^{+\infty} \frac{dx}{e^x - e^{-x}}$.

65. Compare (*a*) $\lim_{x\to 0^+} (\cos x)^{\csc x}$ and (*b*) $\lim_{x\to 0^-} (\cos x)^{\csc x}$.

66. Compute $\lim_{x\to 0} \frac{\sin ax}{\sin bx}$.

67. Compute $\lim_{x\to 0} \frac{\tan ax}{\tan bx}$.

68. Suppose R is the region bounded by $y = 1/\sqrt{x}$, the positive y axis, and the line $x = 1$. Determine the area of R thinking in terms of (*a*) vertical rectangles and (*b*) horizontal rectangles. (The two answers should agree, but two different types of improper integrals occur.)

Miscellaneous Exercises

1. Prove that the integral $\int_e^{+\infty} dx/(x(\ln x)^p)$ converges if $p > 1$ and diverges if $p \le 1$.

2. What value of x makes the vertical line through it divide the following region into pieces of equal area? It is bounded by the curve $y = 1/(1+x^2)$ and the positive x axis.

3. The "gamma function" is defined by $\Gamma(x) = \int_0^{+\infty} t^{x-1}e^{-t}\, dt$. (See Sec. 14.3, Exercise 57.)
(*a*) Show that $\Gamma(x)$ is defined (integral converges) for $x > 0$.
(*b*) Show that $\Gamma(x)$ is undefined (integral diverges) for $x \le 0$.
(*c*) Using integration by parts, demonstrate that $\Gamma(x+1) = x\Gamma(x)$.
(*d*) Using part (*c*), show that $\Gamma(n) = (n-1)!$ for n a positive integer. (Hence the gamma function generalizes the "factorial" operation to any positive number.)

4. ($\int_a^x t^u\, dt$ near $u = -1$) Let $f(u) = \int_a^x t^u\, dt$, where $0 < a < x$.
(*a*) Evaluate the integral and compute $\lim_{u\to -1} f(u)$, using L'Hospital's Rule.
(*b*) Compare your result with $\int_a^x t^{-1}\, dt$.

5. (*a*) Using the opposite of the "take logarithms" strategy, evaluate

$$\lim_{x\to 0} \left[\ln \frac{\sin x}{x}\right]$$

(*b*) What hypotheses did you use in your argument?

6. Repeat Exercise 5 for $\lim_{x\to 0} \left[\ln \frac{1 - e^{-x}}{x}\right]$.

7. Prove that $\lim_{x\to 0} \frac{e^{-1/x^2}}{x} = 0$.

8. (*a*) Prove that if k is a positive integer, then $\lim_{x\to 0} (e^{-1/x^2}/x^k) = 0$.
(*b*) Use part (*a*) to show that

$$f(x) = \begin{cases} e^{-1/x^2} & \text{for } x \ne 0 \\ 0 & \text{for } x = 0 \end{cases}$$

has continuous higher derivatives of all orders at $x = 0$. (This is clearly true for $x \ne 0$.)

9. Compute (*a*) $\lim_{x\to 0^+} \frac{\int_0^x e^{t^2}\, dt}{\int_0^x e^t\, dt}$ and (*b*) $\lim_{x\to +\infty} \frac{\int_0^x e^t\, dt}{\int_0^x e^{t^2}\, dt}$.

Exercises 10 to 18 concern the use of Comparison Theorems in showing convergence or divergence of $\int_a^{+\infty} f(x)\, dx$ and in establishing bounds on approximation error when the antiderivative is nonelementary.

10. Suppose $f(x)$ and $g(x)$ are continuous with $0 \le f(x) \le g(x)$ on $[a, +\infty)$.
(*a*) Prove that if $\int_a^{+\infty} g(x)\, dx$ converges, then $\int_a^{+\infty} f(x)\, dx$ converges. [Use a Squeeze Theorem on $\lim_{b\to+\infty} \int_a^b f(x)\, dx$.]
(*b*) Prove that if $\int_a^{+\infty} f(x)\, dx$ diverges, then $\int_a^{+\infty} g(x)\, dx$ diverges.

11. Use Exercise 10 to show that $\int_1^{+\infty} e^{-x^2}\, dx$ converges. (Note that for $x \ge 1$, $1/e^{x^2} \le 1/e^x$.)

12. Use Exercise 10 to show that $\int_1^{+\infty} \frac{dx}{\ln(1+x)}$ diverges.

13. Using the Trapezoid Rule with $n = 10$ in Sec. 11.6, Progress Test 1, we estimated $\int_0^2 e^{-x^2}\,dx \approx 0.88186$ with an error less than 0.000066.
(*a*) Prove that $\int_2^{+\infty} e^{-x^2}\,dx \le \int_2^{+\infty} e^{-2x}\,dx$.
(*b*) Using (*a*) to show that $\int_0^{+\infty} e^{-x^2}\,dx \le \int_0^2 e^{-x^2}\,dx + \int_2^{+\infty} e^{-2x}\,dx$ and, computing $\int_2^{+\infty} e^{-2x}\,dx$, conclude that $\int_0^{+\infty} e^{-x^2}\,dx$ is within $(-1/2)(-e^{-4}) \approx 0.0091578$ of $\int_0^2 e^{-x^2}\,dx$.
(*c*) Use (*a*) and (*b*) to give and justify an estimate of $\int_0^{+\infty} e^{-x^2}\,dx$ accurate to within 0.01.

14. Use the idea of Exercise 13, but using an appropriate c in $\int_0^c e^{-x^2}\,dx$ and $\int_c^{+\infty} e^{-cx}\,dx$, to approximate $\int_0^{+\infty} e^{-x^2}\,dx$ accurate through four decimal places.

15. $\int_0^{+\infty} dx/\sqrt{1+x^4}$ can be broken into a sum $\int_0^c dx/\sqrt{1+x^4} + \int_c^{+\infty} dx/\sqrt{1+x^4}$. The second interval is bounded above by $\int_c^{+\infty} dx/x^2$. What value of c is required to ensure an approximation of $\int_0^{+\infty} dx/\sqrt{1+x^4}$ accurate within 0.005? (*Hint:* Use trial and error, beginning with an analysis of $\int_c^{+\infty} dx/x^2$.)

16. How large must b be in order that we can replace $\int_b^{+\infty} dx/\sqrt{1+x^3}$ by $\int_b^{+\infty} dx/x^{3/2}$ as a means of approximating $\int_0^{+\infty} dx/\sqrt{1+x^3}$ within 0.01?

17. A probability density function $f(x)$ must be everywhere defined and nonnegative with $\int_{-\infty}^{+\infty} f(x)\,dx = 1$. Show that

$$f(x) = \begin{cases} (1/2)e^{-x/2} & \text{for } x \ge 0 \\ 0 & \text{for } x < 0 \end{cases}$$

is a probability density function.

18. (See Exercise 17.) Use the idea of Exercise 13 to estimate $\int_{-\infty}^{+\infty} (1/\sqrt{\pi})e^{-x^2/2}\,dx$ within 0.01. (This limit is actually 1, and $(1/\sqrt{\pi})e^{-x^2/2}$ is called the "unit normal density function." The closely related function $\operatorname{erf}(t) = (2/\sqrt{\pi})\int_0^t e^{-x^2/2}\,dx$ is called the "error function." Both are widely used in probability and statistics.)

19. M_y, the first moment of the region under a probability density function, is also called the "expected value" for that function and is denoted μ. Compute the expected value for the unit normal density function.

20. Compute the expected value (see Exercise 19) for the function in Exercise 17.

21. The standard deviation associated with a probability density function is defined to be the square root of $\int_{-\infty}^{+\infty} (x-\mu)^2 f(x)\,dx$. (It is actually a second moment.) Compute the standard deviation for the unit normal function in Exercises 18 and 19.

22. Compute the standard deviation for the density function in Exercises 17 and 20.

SELF-TEST ✣

1. Evaluate the given limit:

(*a*) $\lim_{x\to 0} \frac{xe^{\cos x}}{1 + \sin x - \cos x}$ (*b*) $\lim_{x\to+\infty} \frac{\ln(\ln(\ln x))}{\ln(\ln x)}$

(*c*) $\lim_{x\to 1^-} x^{1/(1-x)}$ (*d*) $\lim_{x\to 0^+} (\sin x)^{\sin x}$

2. Evaluate $\int_{-\infty}^{+\infty} e^{(x-e^x)}\,dx$. (Factor the integrand.)

3. Which of (*a*) $\int_{-\infty}^{+\infty} \frac{dx}{x^2+1}$ and (*b*) $\int_{-\infty}^{+\infty} \frac{dx}{x^2+2x+1}$ is "more improper"?

4. Evaluate both integrals in Prob. 3.

5. Is $\int_{-1}^{1} \frac{1-\cos x}{x^2}\,dx$ improper?

6. Is $\int_0^1 x \ln x\,dx$ improper?

Chapter 15

CONVERGENCE OF SEQUENCES AND SERIES

CONTENTS

We shall define the convergence of an infinite sequence of numbers $a_1, a_2, a_3, \ldots, a_n, \ldots$ analogously to the convergence of $\lim_{x\to+\infty} f(x)$. For example, the sequence 1, 1/4, 1/9, . . . , $1/n^2$, . . . converges to 0 (as $n \to +\infty$). We then use this limit idea to extend the notion of addition (which applies only to a finite collection of numbers) to an infinite series $a_1 + a_2 + a_3 + \cdots + a_n + \cdots$. The sum will be defined to be the limit (as $n \to +\infty$) of the finite sums $a_1 + a_2 + a_3 + \cdots + a_n$, when this limit exists.

The bulk of this chapter is devoted to methods for determining when the limits of infinite sequences and infinite series exist. These methods will have their main application in the next chapter, where we use infinite series to approximate functions. As a result our attention here is on the techniques themselves and not on their use in solving specific problems. In this way the style of this chapter is similar to Chap. 11, Techniques of Integration. As was the case then, learning *how* to use a particular technique is probably not as difficult or as important as learning *when* to use it. A collection of miscellaneous exercises provides practice in choosing the technique in the absence of a context that would suggest which technique to use.

Section 15.1: Sequences

Objectives:

1. Translate between the general term of a sequence and its elements.

2. Determine whether a given sequence is increasing or decreasing, bounded or unbounded.

3. Determine, in simple cases, whether a given sequence is convergent or divergent and, if convergent, determine its limit.

INTRODUCTION

Unending or "infinite" sequences of numbers such as the following occur throughout mathematics:

(i) $-1, +1, -1, +1, \ldots, (-1)^n, \ldots$

(ii) $-\frac{1}{2}, \frac{1}{4}, -\frac{1}{8}, \frac{1}{16}, \ldots, \left(-\frac{1}{2}\right)^n, \ldots$

(iii) $\frac{3}{10}, \frac{3}{10}+\frac{3}{100}, \frac{3}{10}+\frac{3}{100}+\frac{3}{1000}, \ldots, \frac{3}{10}+\frac{3}{100}+\cdots+\frac{3}{10^n}, \ldots$

Since all our work in this and the next chapter involves sequences, we give a formal definition.

15.1.1 Definition A *sequence* is a function whose domain is the set of positive integers. Instead of writing the value of the function at a typical integer n as $f(n)$, we write it as a_n. The numbers $a_1, a_2, \ldots, a_n, \ldots$ are called the *elements* of the sequence, and the number a_n is called a *general element* of the sequence.

A sequence can be described by giving (1) its first several elements, with the understanding that the implied pattern is to continue, or (2) a general element (usually as an algebraic formula).

Example 1 Determine the first four elements of each of the following sequences.

(*a*) $a_n = \dfrac{(-1)^{n-1}}{3^n}$ (*b*) $b_n = 1 + \dfrac{1}{2} + \cdots + \dfrac{1}{n}$

Solutions

(*a*) $a_1 = \dfrac{(-1)^{1-1}}{3^1} = \dfrac{(-1)^0}{3} = \dfrac{1}{3}$

$a_2 = \dfrac{(-1)^{2-1}}{3^2} = \dfrac{(-1)^1}{3^2} = \dfrac{-1}{9}$

$a_3 = \dfrac{(-1)^{3-1}}{3^3} = \dfrac{(-1)^2}{3^3} = \dfrac{1}{27}$

$a_4 = \dfrac{(-1)^{4-1}}{3^4} = \dfrac{(-1)^3}{3^4} = \dfrac{-1}{81}$

(*b*) $b_1 = 1$

$b_2 = 1 + \dfrac{1}{2} = \dfrac{3}{2}$

$b_3 = 1 + \dfrac{1}{2} + \dfrac{1}{3} = \dfrac{11}{6}$

$b_4 = 1 + \dfrac{1}{2} + \dfrac{1}{3} + \dfrac{1}{4} = \dfrac{50}{24}$ ●

The problem of finding the general element when given the first few elements can be considerably more complex. For example, suppose you were presented with the sequence in Example 1(*a*) displayed as

$$1/3, -1/9, 1/27, -1/81, \ldots$$

The denominators are increasing powers of three and thus are given by $3^n, n = 1, 2, \ldots$, whereas the numerators are alternating plus and minus ones and thus can be given by $(-1)^{n-1}, n = 1, 2, \ldots$. Hence $a_n = (-1)^{n-1}/3^n$.

Usually the form of a general element is not unique. For example, we could also have given the numerators above as $(-1)^{n+1}$. Other helpful facts in dealing with sequences are the following (test these on $n = 1, 2, 3$):

15.1.2 (*a*) The positive even integers may be written as $2n$, $n = 1, 2, \ldots$.
(*b*) The positive odd integers may be written as $2n - 1$, $n = 1, 2, \ldots$.
(*c*) The positive integers beginning with $k + 1$ (k a nonnegative integer) may be written as $n + k$, $n = 1, 2, \ldots$.

Example 2 Determine a general element for the sequence $1/2, -3/4, 5/6, -7/8, \ldots$.

Solution Temporarily ignoring the signs, we note that the numerators consist of successive odd integers $2n - 1$, and the denominators consist of successive even integers $2n$. The signs alternate $+1, -1, +1, -1$, so

$$a_n = \frac{(-1)^{n+1}(2n-1)}{2n} \qquad n = 1, 2, \ldots \quad \bullet$$

PROGRESS TEST 1

1. Write out the first four elements of each of the following sequences:

(*a*) $a_n = (1 - 1/n)^n$ (*b*) $a_n = 1 + (-1)^n$

(*c*) $a_n = \begin{cases} 0 & n \text{ even} \\ \dfrac{n+1}{2} & n \text{ odd} \end{cases}$ (*d*) $a_n = \dfrac{2^n - 1}{2^n}$

2. Find the general element of each of the following sequences:
(*a*) $4/3, 5/4, 6/5, \ldots$
(*b*) $1, 1/2, 1, 1/4, 1, 1/6, \ldots$
(*c*) $1/4, 3/16, 5/36, 7/64, \ldots$

LIMITS OF SEQUENCES

Sequence (iii), given on page 478, when rewritten as

$$0.3, 0.33, 0.333, 0.3333, \ldots$$

clearly represents a sequence of decimals that provides an increasingly accurate approximation of the number $1/3$.

By going far enough out in the sequence we can guarantee that a_n is within any specified distance of $1/3$—that is, $\lim_{n\to+\infty} a_n = 1/3$.

15.1.3 **Definition** We say that the sequence a_n tends to L as n grows without bound and write

$$\lim_{n\to+\infty} a_n = L$$

if given any $\varepsilon > 0$, there exists a positive integer K such that

$$n > K \text{ implies } |a_n - L| < \varepsilon$$

We shall also write this as "$a_n \to L$ as $n \to +\infty$" or, even more simply, as "$a_n \to L$," and say the sequence is *convergent*. If $\lim_{n\to+\infty} a_n$ fails to exist, then we say a_n is *divergent*.

This definition closely resembles that of $\lim_{x\to+\infty} f(x) = L$ in (4.1.2) because by definition a sequence is itself a function defined on the subset of $[1, +\infty)$ consisting of the positive integers. Limits of sequences obey most of the laws with which we are now quite familiar. For example, the limit of a sequence, when it exists, is unique (see Sec. 15.1, Exercise 81). For the record we state:

15.1.4 Limit Theorem for Sequences Suppose $\lim_{n\to+\infty} a_n = A$, $\lim_{n\to+\infty} b_n = B$.

(*a*) $\lim_{n\to+\infty} (a_n \pm b_n) = A \pm B$

(*b*) $\lim_{n\to+\infty} (a_n b_n) = AB$ and $\lim_{n\to+\infty} (ca_n) = cA$ (c a constant)

(*c*) $\lim_{n\to+\infty} \left(\dfrac{a_n}{b_n}\right) = \dfrac{A}{B}$ ($b_n \neq 0$ and $B \neq 0$)

(*d*) $\lim_{n\to+\infty} \sqrt[k]{a_n} = \sqrt[k]{A}$, k a positive integer. If k is even, then $a_n \geq 0$.

(*e*) $\lim_{n\to+\infty} |a_n| = |A|$.

(*f*) If f is continuous at A, then $\lim_{n\to+\infty} f(a_n) = f(A)$. Continuous functions preserve limits of sequences.

The proofs are left as Sec. 15.1, Exercises 82–86.

To take advantage of our earlier work with limits at infinity, we often regard a sequence as the restriction of a function defined on the entire interval $[1, +\infty)$, even assuming, if necessary, that this function is differentiable. In particular, if $\lim_{x\to+\infty} f(x) = L$ and $f(n) = a_n$, then $\lim_{n\to+\infty} a_n = L$.

Example 3 Evaluate (*a*) $\lim_{n\to+\infty} \dfrac{1 + \sqrt{n}}{1 - \sqrt{n}}$ and (*b*) $\lim_{n\to+\infty} \dfrac{n^2 + n + 1}{e^n}$.

Solutions (*a*) As in Sec. 4.1, we multiply numerator and denominator of a typical element by $1/\sqrt{n}$ and then evaluate the limit of the resulting expression:

$$\lim_{n\to+\infty} \frac{(1/\sqrt{n}) + 1}{(1/\sqrt{n}) - 1} = \frac{0 + 1}{0 - 1} = -1$$

(*b*) We shall apply L'Hospital's Rule to a differentiable function of which the given sequence is a restriction, $f(x) = (x^2 + x + 1)/e^x$ (∞/∞).

$$\lim_{x\to+\infty} \frac{x^2 + x + 1}{e^x} = \lim_{x\to+\infty} \frac{2x + 1}{e^x}$$

We apply L'Hospital's Rule again to get

$$\lim_{x\to+\infty} \frac{2}{e^x} = 0$$

But then

$$\lim_{x\to+\infty} \frac{x^2 + x + 1}{e^x} = 0$$

from which we can conclude that

$$\lim_{n\to+\infty} \frac{n^2 + n + 1}{e^n} = 0. \quad \bullet$$

In part (*a*) of Example 3 we used, without mention, the Limit Theorem for Sequences as well as the fact that $\lim_{n\to+\infty} 1/\sqrt{n} = 0$ (which follows from the fact that $\lim_{x\to+\infty} 1/\sqrt{x} = 0$.)

The following Squeeze Theorem, whose proof is left as an exercise, will be valuable in using sequences with known limits to compute limits of new sequences.

15.1.5 **Squeeze Theorem** If $b_n \leq a_n \leq c_n$ for each n, and if $\lim_{n\to+\infty} b_n = \lim_{n\to+\infty} c_n = L$, then

$$\lim_{n\to+\infty} a_n = L$$

Theorem (15.1.5) is most often applied when b_n is the constant sequence, all of whose elements are zero. In this case $b_n \to 0$ trivially, so showing that $c_n \to 0$ allows us to conclude that $a_n \to 0$.

Example 4 Determine $\lim_{n\to+\infty} a_n$, where

$$a_1 = \frac{1}{4},\ a_2 = \frac{1 \cdot 3}{4 \cdot 8},\ a_3 = \frac{1 \cdot 3 \cdot 5}{4 \cdot 8 \cdot 12}, \ldots$$

Solution We first note that a_n is a quotient whose numerator is the product of the odd integers while the denominator consists of a product of multiples of 4. Hence

$$a_n = \frac{1 \cdot 3 \cdot 5 \cdot \cdots \cdot (2n-1)}{4 \cdot 8 \cdot 12 \cdot \cdots \cdot 4n} = \frac{1}{4} \cdot \frac{3}{8} \cdot \frac{5}{12} \cdot \cdots \cdot \frac{(2n-1)}{4n}$$

In this latter form we see that a_n is a product of fractions, each of which is less than 1, so the product should be less than 1. In fact each element is less than its predecessor. We rewrite the first few elements as follows, hoping to apply the Squeeze Theorem:

$$a_1 = \frac{1}{4} = 1 \cdot \frac{1}{4},\ a_2 = \frac{1}{4} \cdot \frac{3}{8} = 1 \cdot \frac{3}{4} \cdot \frac{1}{8} < \frac{1}{8}$$

$$a_3 = \frac{1}{4} \cdot \frac{3}{8} \cdot \frac{5}{12} = 1 \cdot \frac{3}{4} \cdot \frac{5}{8} \cdot \frac{1}{12} < \frac{1}{12}$$

In general,

$$a_n = \frac{1}{4} \cdot \frac{3}{8} \cdot \frac{5}{12} \cdot \cdots \cdot \frac{2n-1}{4n} = 1 \cdot \frac{3}{4} \cdot \frac{5}{8} \cdot \cdots \cdot \frac{2n-1}{4(n-1)} \cdot \frac{1}{4n} < \frac{1}{4n}$$

because the initial factors indicated are each less than one, so their product is as well. Consequently, for each n,

$$0 \leq a_n \leq \frac{1}{4n}$$

Since $\lim\limits_{n\to+\infty} 0 = \lim\limits_{n\to+\infty} \frac{1}{4n} = 0$, we know that

$$\lim_{n\to+\infty} a_n = 0$$

by the Squeeze Theorem. ●

It is worthwhile to collect some of the more common limits that we shall encounter in the pages to come. The first six follow from limits determined in Chaps. 4 and 14.

Recall that $n!$ (read "n factorial") is shorthand for

$$n \cdot (n-1)(n-2) \cdot \cdots \cdot 3 \cdot 2 \cdot 1$$

the product of all positive integers less than or equal to n.

15.1.6 Theorem Special Limits

(*a*) If $p(n)$ and $q(n)$ are polynomials whose highest degree terms are $p_k n^k$ and $q_j n^j$, respectively, then

$$\lim_{n\to+\infty} \frac{p(n)}{q(n)} = \begin{cases} 0 & \text{if } k < j \\ p_k/q_j & \text{if } k = j \\ \pm\infty & \text{if } k > j \end{cases}$$

(*b*) If $p > 0$, $\lim_{n\to+\infty} \frac{1}{n^p} = 0$.

(*c*) For all p, $\lim_{n\to+\infty} \frac{n^p}{e^n} = 0$.

(*d*) If $p > 0$, $\lim_{n\to+\infty} \frac{\ln n}{n^p} = 0$.

(*e*) If r is any real number,

$$\lim_{n\to+\infty} \left(1 + \frac{r}{n}\right)^n = e^r \qquad \text{and} \qquad \lim_{n\to+\infty} \left(\frac{n}{n+r}\right)^n = e^{-r}$$

(*f*) If k is any real number, $\lim_{n\to+\infty} n^{k/n} = 1$.

(*g*) If $|r| < 1$, then $\lim_{n\to+\infty} r^n = 0$.

(*h*) If r is any real number, $\lim_{n\to+\infty} \frac{r^n}{n!} = 0$.

Proof:

(*a*) The statement follows from (4.1.5), p. 113

(*b*) The statement follows from (4.1.4), p. 113

(*c*) The statement follows from |14.1.5(*b*)|, p. 462

(*d*) The statement follows from |14.1.5(*a*)|, p. 462

(*e*) The first limit follows from (14.2.2) p. 466 and the second is its reciprocal.

(*f*) (*f*) follows from (*d*) since $\ln(n^{k/n}) = k(\ln n)/n \to 0$. Hence $n^{k/n} \to e^0 = 1$.

(*g*) We shall show that $\lim_{n\to+\infty} |r|^n = 0$, from which it follows that $\lim_{n\to+\infty} r^n = 0$ by [15.1.4(*e*)]. Suppose some small $\varepsilon > 0$ is given, say $\varepsilon < 1$. By definition (15.1.3), we must show there is an integer K such that $n \geq K$ implies $|r|^n < \varepsilon$. But, taking the natural logarithm of both sides of this last inequality, we have

$$\ln|r|^n < \ln \varepsilon \qquad \text{or} \qquad n \ln|r| < \ln \varepsilon$$

(Note that the natural logarithm, being an increasing function, preserves the direction of inequalities. Convince yourself of this!) Dividing through by the negative quantity $\ln|r|$ (recall that $|r| < 1$ so $\ln|r| < 0$), we get

$$n > \frac{\ln \varepsilon}{\ln|r|}$$

Thus, by choosing $n \geq K$, where K is an integer greater than $(\ln \varepsilon)/(\ln |r|)$, we know that the original inequality $|r|^n < \varepsilon$ holds, as required.

(*h*) We omit the Squeeze Theorem-based proof. ■

Applications of these limits appear in Progress Test 2. Although we have concentrated on those limits that *do* exist, we should note that limits of sequences can fail to exist in essentially the same ways that limits at infinity can fail to exist.

Example 5 Show that the following sequences diverge:

(*a*) $a_n = (-1)^n$ (*b*) $a_n = \begin{cases} 1 & n \text{ odd} \\ 1/n & n \text{ even} \end{cases}$ (*c*) $a_n = \dfrac{n}{\ln(n+1)}$

Solutions (*a*) This sequence alternates between -1 (for n odd) and $+1$ (for n even) and hence approaches no single number as $n \to +\infty$.

(*b*) The even elements approach 0 whereas the odd elements are all 1. Choosing $\varepsilon < 1/2$, we see that neither 0 nor 1 can be the limit of this sequence because no matter how large the integer K, there will be even integers $n > K$ such that both $|a_n - 0| < \varepsilon$ and $|a_n - 1| > \varepsilon$.

(*c*) We know (using L'Hospital's Rule, for example) that

$$\frac{\ln(x+1)}{x} \to 0^+ \text{ as } x \to +\infty$$

Hence $\dfrac{\ln(n+1)}{n} \to 0^+$ as $n \to +\infty$. But then

$$\frac{1}{[\ln(n+1)]/n} = \frac{n}{\ln(n+1)} \to +\infty \text{ as } n \to +\infty \quad \bullet$$

PROGRESS TEST 2

Evaluate the following limits:

1. $\lim_{n\to+\infty} \left(\dfrac{n}{n-\pi}\right)^n$
2. $\lim_{n\to+\infty} \dfrac{(-1)^n 2^n}{n!}$
3. $\lim_{n\to+\infty} \dfrac{(-1)^n 3^{n+2}}{4^n}$
4. $\lim_{n\to+\infty} \dfrac{\ln n}{n^{.01}}$
5. $\lim_{n\to+\infty} \dfrac{2n^2 - 6n + 1}{3n^3 + n + 2}$
6. $\lim_{n\to+\infty} \sin n\pi$

BOUNDED AND MONOTONIC SEQUENCES

15.1.7 Definition

(*a*) A sequence a_n is said to be *increasing* if $a_{n+1} \geq a_n$ for all n.

(*b*) A sequence a_n is said to be *decreasing* if $a_{n+1} \leq a_n$ for all n.

(*c*) A sequence that is either increasing or decreasing is called *monotonic*.

We use the modifier "strictly" when strict inequalities hold in the previous three statements.

Example 6 illustrates two standard techniques for determining whether a sequence is monotonic.

Example 6 Are the following sequences monotonic?

(*a*) $a_n = \dfrac{1 \cdot 3 \cdot 5 \cdot \ldots \cdot (2n-1)}{n!}$ (*b*) $a_n = \dfrac{\ln(n+3)}{n+3}$

Solutions (*a*) Note that for positive term sequences the conditions $a_{n+1} \geq a_n$ and $a_{n+1} \leq a_n$ are equivalent, respectively, to $a_{n+1}/a_n \geq 1$ and $a_{n+1}/a_n \leq 1$.

Now

$$a_{n+1} = \frac{1 \cdot 3 \cdot 5 \cdot \cdots \cdot (2n-1)[2(n+1)-1]}{(n+1)!}$$
$$= \frac{1 \cdot 3 \cdot 5 \cdot \cdots \cdot (2n-1)(2n+1)}{n!(n+1)}$$

Thus

$$\frac{a_{n+1}}{a_n} = \frac{1 \cdot 3 \cdot 5 \cdot \cdots \cdot (2n-1)(2n+1)}{n!(n+1)} \cdot \frac{n!}{1 \cdot 3 \cdot 5 \cdot \cdots \cdot (2n-1)}$$
$$= \frac{2n+1}{n+1} > 1 \qquad \text{for all } n$$

Hence a_n is a strictly increasing sequence.

(*b*) An approach similar to (*a*) leads to

$$\frac{a_{n+1}}{a_n} = \frac{\ln(n+4)}{n+4} \cdot \frac{n+3}{\ln(n+3)}$$

but it is not clear how this quotient compares to 1, so we examine the original sequence $a_n = [\ln(n+3)]/(n+3)$. We shall know whether a_n is increasing or decreasing if we can determine the sign of the derivative of $f(x) = [\ln(x+3)]/(x+3)$. Now

$$f'(x) = \frac{(x+3)\dfrac{1}{x+3} - \ln(x+3)(1)}{(x+3)^2} = \frac{1 - \ln(x+3)}{(x+3)^2}$$

Thus $f'(x) < 0$ if and only if $1 - \ln(x+3) < 0$ or $\ln(x+3) > 1 = \ln e$. But since the natural logarithm function is strictly increasing, $\ln(x+3) > \ln e$ is equivalent to $x + 3 > e$, which is certainly true for all $x \geq 1$. Hence $f(x) = [\ln(x+3)]/(x+3)$ is decreasing on $[1, +\infty)$, so the sequence $a_n = [\ln(n+3)]/(n+3)$ is likewise decreasing. ●

15.1.8 Definition

(*a*) A sequence a_n is said to be *bounded above* if there exists a number U such that $a_n \leq U$ for all n.

(*b*) A sequence a_n is said to be *bounded below* if there exists a number L such that $L \leq a_n$ for all n.

(*c*) A sequence that is both bounded above and below is called *bounded*.

Recall, for example, the increasing sequence of rational numbers

$$1, 1.4, 1.41, 1.414, 1.4142, 1.41421, 1.414213, \ldots$$

which provide, at each stage, an increasingly accurate approximation of the irrational number $\sqrt{2}$. This sequence is bounded above by 1.5, for example.

We need to assume the next fundamental theorem to guarantee, in effect, that any sequence of real numbers that *should* converge actually *does* converge (to a real number).

15.1.9 Theorem If a_n is increasing and bounded above, or decreasing and bounded below, then a_n converges.

The proof of (15.1.9) rests on the basic "completeness" of the real numbers; the real-number line has no "holes." In this respect the set of real numbers is essentially different from its subset of rational numbers: the above sequence of rational numbers, although bounded above, does not converge to a rational number because $\sqrt{2}$ is irrational.

Although (15.1.9) tells us that certain limits exist, it does not provide a means for identifying the actual limit.

Example 7 Show that $a_n = \dfrac{n!}{n^n}$ converges.

Solution

$$\frac{a_{n+1}}{a_n} = \frac{(n+1)!}{(n+1)^{n+1}} \cdot \frac{n^n}{n!} = \frac{(n+1)n!}{(n+1)(n+1)^n} \cdot \frac{n^n}{n!}$$

$$= \frac{n^n}{(n+1)^n} = \left(\frac{n}{n+1}\right)^n < 1 \qquad \text{for all } n$$

Thus a_n is decreasing. Since all the elements of a_n are positive, a_n is bounded below (by 0 for example). Thus a_n is convergent, by (15.1.9). ●

We shall show in Sec. 15.3 that $a_n = n!/n^n$ converges to 0.

When discussing convergence of sequences or using theorems regarding convergence, it is usually possible to disregard erratic behavior of early elements in the sequence, because convergence depends only on the later elements—the a_n, for $n \geq j$, where j is a fixed integer. Thus, for example, $a_n = 10^n/2^{n^2}$ is not decreasing, because $a_1 < a_2$. However, for $n \geq 2$, $a_n > a_{n+1}$. Since each element is positive, (15.1.9) tells us that a_n converges. (See Exercise 62 next page.)

SECTION 15.1 EXERCISES

In Exercises 1 to 14 write the first four elements of the given sequence.

1. $a_n = \dfrac{2n+1}{n+3}$

2. $a_n = \dfrac{(-1)^{n+1}}{n+1}$

3. $a_n = ne^{-n}$

4. $a_n = \dfrac{n+3}{(n-1)!}$

5. $a_n = 1 \cdot 3 \cdot 5 \cdot \cdots \cdot (2n-1)$

6. $a_n = \dfrac{n}{\ln(n+1)}$

7. $a_n = \sin(3\pi/n)$

8. $a_n = \dfrac{n!}{1 \cdot 3 \cdot 5 \cdot \cdots \cdot (2n-1)}$

9. $a_n = \dfrac{2 \cdot 4 \cdot \cdots \cdot (2n)}{2^n n!}$

10. $a_n = (-1)^{n+1} - 1$

11. $a_n = n^2 + \dfrac{(-1)^n}{n+1}$

12. $a_n = \begin{cases} n^2 + n & n \text{ even} \\ 4n - 1 & n \text{ odd} \end{cases}$

13. $a_n = \dfrac{1-(-1)^n}{2}$

14. $a_n = (-1)^n\left(1 - \dfrac{(-1)^n}{2^n}\right)$

In Exercises 15 to 26 give a general element for each of the following sequences. (The form of an answer is not unique.)

15. 2/5, −4/25, 8/125, −16/625, ...

16. 1/2, −3/4, 5/8, −7/16, ...

17. 3/5, 9/7, 27/9, 81/11, ...

18. 5/7, −7/9, 9/11, −11/13, ...

19. 6, 5, 6, 5, ...

20. 3, 1, 3, 1, ...

21. 1, 1/2, 1/6, 1/24, 1/120, ...

22. 1/2, 5/4, 9/8, 13/16, ...

23. 2/22, −3/32, 4/42, −5/52, ...

24. $2 \cdot 4, 4 \cdot 6, 6 \cdot 8, 8 \cdot 10, \ldots$

25. $-1 \cdot 3, 3 \cdot 5, -5 \cdot 7, 7 \cdot 9, \ldots$

26. 1, 1/4, 1/27, 1/64, 1/3125, ...

In Exercises 27 to 40 determine whether the given sequence is increasing, decreasing, or not monotonic.

27. $a_n = \dfrac{3^n}{4^n + 1}$

28. $a_n = \dfrac{2^n + 1}{3^n}$

29. $a_n = \dfrac{(n+1)^2}{e^n}$

30. $a_n = \dfrac{\ln(n+1)}{\sqrt{n}}$

31. $a_n = \dfrac{2^n(1 \cdot 3 \cdot 5 \cdot \cdots \cdot (2n-1))}{2 \cdot 4 \cdot 6 \cdot \cdots \cdot (2n)}$

32. $a_n = \dfrac{n^2}{n!}$

33. $a_n = \ln\left(\dfrac{n+1}{2n}\right)$

34. $a_n = \dfrac{n+1}{n-1}$

35. $a_n = \dfrac{n}{n+1}$

36. $a_n = \dfrac{3n}{4n+1}$

37. $a_n = \dfrac{\sqrt{n}+1}{n}$

38. $a_n = n^{1/n}$

39. $a_n = \ln\left(\dfrac{n+2}{n+1}\right)$

40. $a_n = \cos n\pi$

In Exercises 41 to 50 determine whether the given sequence is bounded above or bounded below.

41. $a_n = \dfrac{2^n}{2 \cdot 4 \cdot 6 \cdot \cdots \cdot (2n)}$

42. $a_n = \dfrac{3^{n+1}}{3 \cdot 6 \cdot 9 \cdot \cdots \cdot (3n)}$

43. $a_n = \dfrac{2n^2+n}{3n+5}$

44. $a_n = \dfrac{3^n}{1+3^n}$

45. $a_n = ne^{-n}$

46. $a_n = \ln\left(\dfrac{3n}{n+1}\right)$

47. $a_n = \dfrac{(2n)!}{n^n}$

48. $a_n = \dfrac{(2n)!}{3 \cdot 5 \cdot 7 \cdot \cdots \cdot (2n+1)}$

49. $a_n = (2^n + 3^n)^{1/n}$

50. $a_n = (1 + 2/n)^{2/n}$

In Exercises 51 to 78 determine the limit, if it exists, of the given sequence. Give reasons for your conclusions.

51. $a_n = \dfrac{3n^3+2n^2-1}{1+2n-n^3}$

52. $a_n = \dfrac{2n^3+6n+9}{3n^2+5}$

53. $a_n = \left[\dfrac{n}{n-\sqrt{2}}\right]^n$

54. $a_n = \dfrac{(-1)^n 5^n}{(n-1)!}$

55. $a_n = \dfrac{2^n+5}{3^n}$

56. $a_n = \dfrac{\sqrt{2n+5}}{n+6}$

57. $a_n = \dfrac{\sin n}{n}$

58. $a_n = \dfrac{e^n}{e^n-1}$

59. $a_n = \left(1+\dfrac{3}{n}\right)^2$

60. $a_n = \dfrac{\sqrt{n}}{2+3\sqrt{n}}$

61. $a_n = \tan\left(\dfrac{\sqrt{\pi^2 n^2 - 8}}{4n}\right)$

62. $a_n = \dfrac{10^n}{2^{n^2}}$

63. $a_n = \cos(n + \pi/2)$

64. $a_n = \sin(n\pi/2)$

65. $a_n = \left(\dfrac{n+e}{n}\right)^n$

66. $a_n = \dfrac{\sqrt[3]{3n^6+5}}{2n^2+9}$

67. $a_n = \dfrac{2n^2+n-1}{4n^4+7}$

68. $a_n = \left(\dfrac{n+1}{n+2}\right)^{n+1}$ (*Hint:* let $m = n + 1$)

69. $a_n = (-1)^n \dfrac{2^{n+3}}{3^{n-1}}$

70. $a_n = n \sin\left(\dfrac{1}{n}\right)$

71. $a_n = n^{7/n}$

72. $a_n = \ln\left(\dfrac{n}{2n+1}\right)$

73. $a_n = \left(1-\dfrac{7}{n}\right)^n$

74. $a_n = \dfrac{\log_2 n}{n^{.0001}}$

75. $a_n = \dfrac{\sqrt{n}}{e^n}$

76. $a_n = \dfrac{1 \cdot 3 \cdot 5 \cdot \cdots \cdot (2n-1)}{n!}$

77. $a_n = \dfrac{1}{n}\displaystyle\int_0^n e^{t^2}\,dt$

78. $a_n = \dfrac{n \arctan n}{2n^2+5}$

79. Show that $a_n = (1 + 2^n)^{1/n}$ is convergent.

80. Show that $a_n = \dfrac{(2n)!}{2^{2n}(n!)^2}$ is convergent.

81. Prove that if $\lim_{n\to+\infty} a_n = L$ and $\lim_{n\to+\infty} a_n = L'$, then $L = L'$.

82. Prove [15.1.4(*a*)] for sums.

83. Prove [15.1.4(*b*)]. (Translate the idea of the corresponding proof for standard limits of functions [9.2.1(*b*)].

84. Prove [15.1.4(*c*)].

85. Prove [15.1.4(*e*)].

86. Prove [15.1.4(*f*)].

87. Prove the Squeeze Theorem.

Section 15.2: Infinite Series—An Introduction

Objectives:

Use the definition of convergence and divergence and basic facts about infinite series, including geometric series, to determine convergence or divergence of certain series and to determine the sum of convergent series.

INTRODUCTION AND DEFINITION OF INFINITE SERIES

Although we know how to add any *finite* collection of numbers, we have no definition for the "infinite sum"

$$1 + \frac{1}{2} + \left(\frac{1}{2}\right)^2 + \left(\frac{1}{2}\right)^3 + \cdots$$

However, we can certainly form the "partial sums" as follows:

$$s_1 = 1 + \frac{1}{2},\ s_2 = 1 + \frac{1}{2} + \left(\frac{1}{2}\right)^2 = 1 + \frac{3}{4},$$

$$s_3 = 1 + \frac{1}{2} + \left(\frac{1}{2}\right)^2 + \left(\frac{1}{2}\right)^3 = 1 + \frac{7}{8},$$

$$s_4 = 1 + \frac{1}{2} + \left(\frac{1}{2}\right)^2 + \left(\frac{1}{2}\right)^3 + \left(\frac{1}{2}\right)^4 = 1 + \frac{15}{16},$$

$$s_5 = 1 + \frac{15}{16} + \left(\frac{1}{2}\right)^5 = 1 + \frac{31}{32}$$

Each new term seems to cut the distance to 2 in half. In fact it appears that for each positive integer n,

15.2.1 $$s_n = 1 + \frac{1}{2} + \left(\frac{1}{2}\right)^2 + \left(\frac{1}{2}\right)^3 + \cdots + \left(\frac{1}{2}\right)^n = 1 + \frac{2^n - 1}{2^n} = 2 - \frac{1}{2^n}$$

This can be verified as follows:

$$\begin{aligned}\frac{1}{2}s_n &= s_n - \frac{1}{2}s_n\\ &= \left[1 + \left(\frac{1}{2}\right) + \left(\frac{1}{2}\right)^2 + \cdots + \left(\frac{1}{2}\right)^n\right]\\ &\quad -\frac{1}{2}\left[1 + \left(\frac{1}{2}\right) + \left(\frac{1}{2}\right)^2 + \cdots + \left(\frac{1}{2}\right)^n\right]\\ &= 1 + \left(\frac{1}{2}\right) + \left(\frac{1}{2}\right)^2 + \cdots + \left(\frac{1}{2}\right)^n - \left(\frac{1}{2}\right) - \left(\frac{1}{2}\right)^2 - \left(\frac{1}{2}\right)^3 - \cdots - \left(\frac{1}{2}\right)^{n+1}\\ &= 1 - \left(\frac{1}{2}\right)^{n+1}\end{aligned}$$

Hence

$$s_n = 2\left[1 - \left(\frac{1}{2}\right)^{n+1}\right] = 2 - \left(\frac{1}{2}\right)^n$$

Arithmetic and algebra brought us this far, but, as was the case with instantaneous rates of change in Chap. 2, to complete the analysis we need the notion of limit. We now define

$$1 + \frac{1}{2} + \left(\frac{1}{2}\right)^2 + \left(\frac{1}{2}\right)^3 + \cdots = \lim_{n\to+\infty} s_n = \lim_{n\to+\infty}\left[2 - \left(\frac{1}{2}\right)^n\right]$$

$$= 2 \qquad \text{by } [15.1.6(g)]$$

This gives a meaning for the "infinite sum" based on the notion of limit of a sequence. It is consistent with our earlier observation that the partial sums are tending toward 2.

We shall use this same approach to define "infinite sums" in general.

It will be convenient to use the shorter sigma notation for sums (introduced to write Riemann sums in Sec. 6.1). We denote the sum

$$a_1 + a_2 + \cdots + a_n$$

by $\Sigma_{k=1}^{n} a_k$. The index letter k serves as a variable, usually referred to as a "dummy variable" because it need not appear when the sum is written out. For example,

$$\sum_{k=1}^{4} (3k^2 - 1) = (3 \cdot 1^2 - 1) + (3 \cdot 2^2 - 1) + (3 \cdot 3^2 - 1) + (3 \cdot 4^2 - 1) = 86$$

Here $a_k = (3k^2 - 1)$, but exactly the same sum could also be written as $\Sigma_{i=1}^{4} (3i^2 - 1)$.

For j an integer, the sum

$$a_j + a_{j+1} + a_{j+2} + \cdots + a_n$$

is written as $\Sigma_{k=j}^{n} a_k$. Thus we can write the sum s_4 appearing earlier as

$$s_4 = \sum_{k=0}^{4}\left(\frac{1}{2}\right)^k = \left(\frac{1}{2}\right)^0 + \left(\frac{1}{2}\right)^1 + \left(\frac{1}{2}\right)^2 + \left(\frac{1}{2}\right)^3 + \left(\frac{1}{2}\right)^4$$

15.2.2 Definition We refer to

$$\sum_{k=1}^{\infty} a_k = a_1 + a_2 + a_3 + \cdots$$

as an *infinite series,* or more simply as a *series,* and refer to the numbers a_k as its *terms.* Associated with each infinite series is its sequence of *partial sums*

$$s_1 = a_1,\ s_2 = a_1 + a_2, \ldots,\ s_n = a_1 + a_2 + \cdots + a_n, \ \cdots$$

We say that the series *converges* to S, and write

$$\sum_{k=1}^{\infty} a_k = S \qquad \text{or} \qquad a_1 + a_2 + a_3 + \cdots = S$$

if $\lim_{n\to+\infty}(s_n = S)$. In this case we often refer to $\Sigma_{k=1}^{\infty} a_k$ as a *convergent series* and say that S is its *sum.*

If $\lim_{n\to+\infty} s_n$ does not exist, then we say that $\Sigma_{k=1}^{\infty} a_k$ is a *divergent series.* We also refer to

$$\sum_{k=j}^{\infty} a_k = a_j + a_{j+1} + a_{j+2} + \cdots \qquad (j \text{ an integer})$$

as an infinite series, with partial sums, convergence, and divergence defined analogously.

Just as in the introductory example, $\Sigma_{k=0}^{\infty} (1/2)^k$, we have defined the numerical value of a general infinite series, if it exists, to be the limit of its sequence of partial sums. Thus, by our earlier discussion $\Sigma_{k=0}^{\infty} (1/2)^k$, is convergent with sum equal to 2.

On the other hand,

$$(1) \qquad \sum_{k=1}^{\infty} k = 1 + 2 + 3 + \cdots$$

is divergent, because, by (6.2.3), page 188,

$$\lim_{n\to+\infty} s_n = \lim_{n\to+\infty} \sum_{k=1}^{n} k = \lim_{n\to+\infty} \frac{n(n+1)}{2} = +\infty$$

As is evident in the definition, an infinite series can be written either in sigma form or in expanded form. Depending upon circumstances, we may or may not choose to include a general term in the expanded form. Thus we could also have written the series in (1) as

$$\sum_{k=1}^{\infty} k = 1 + 2 + 3 + \cdots + k + \cdots$$

The last three dots (ellipsis) indicate that the terms of the series continue without end.

Note that we use separate indices, one for a typical kth term of the series and another for the nth partial sum—the sum as k runs from 1 to n. As usual, it is not especially important which letters are used.

Example 1 Determine whether the series

$$\sum_{k=1}^{\infty} \frac{1}{k^2 + k}$$

converges or diverges. If it converges, find its sum.

Solution Note that

$$\sum_{k=1}^{\infty} \frac{1}{k^2 + k} = \sum_{k=1}^{\infty} \frac{1}{k(k+1)} = \sum_{k=1}^{\infty} \left[\frac{1}{k} - \frac{1}{k+1}\right]$$

Thus

$$\begin{aligned} s_n &= \frac{1}{1\cdot 2} + \frac{1}{2\cdot 3} + \frac{1}{3\cdot 4} + \cdots + \frac{1}{(n-1)n} + \frac{1}{n(n+1)} \\ &= \left(1 - \frac{1}{2}\right) + \left(\frac{1}{2} - \frac{1}{3}\right) + \left(\frac{1}{3} - \frac{1}{4}\right) \\ &\qquad + \cdots + \left(\frac{1}{n-1} - \frac{1}{n}\right) + \left(\frac{1}{n} - \frac{1}{n+1}\right) \end{aligned}$$

The interior terms cancel one another in pairs as indicated, and we are left with

$$s_n = 1 - \frac{1}{n+1}$$

Since

$$\lim_{n\to+\infty} \left(1 - \frac{1}{n+1}\right) = 1,$$

we conclude that

$$\sum_{k=1}^{\infty} \frac{1}{k^2 + k} = 1$$

That is, the series converges to 1. ●

A series such as that in Example 1 is often referred to as a *telescoping series* because of the way its interior terms collapse. (You may wish to compute the first several partial sums of Example 1 numerically—its convergence to 1 becomes quite convincing.)

Example 2 Investigate convergence or divergence of the series $\sum_{i=1}^{\infty} (-1)^{i+1}$.

Solution

$s_1 = (-1)^{1+1} = (-1)^2 = 1$

$s_2 = (-1)^{1+1} + (-1)^{2+1} = (-1)^2 + (-1)^3 = 0$

$s_3 = (-1)^{1+1} + (-1)^{2+1} + (-1)^{3+1} = (-1)^2 + (-1)^3 + (-1)^4 = 1$

It is not difficult to see that in general

$$s_n = \begin{cases} 1 & \text{if } n \text{ is odd} \\ 0 & \text{if } n \text{ is even} \end{cases}$$

As a result $\lim_{n\to+\infty} s_n$ does not exist, so $\Sigma_{i=1}^{\infty} (-1)^{i+1}$ diverges. ●

Examples 1 and 2 show that determining convergence or divergence of a series often amounts to manipulating its partial sums s_n into a form that allows us to recognize their limit as $n \to +\infty$.

PROGRESS TEST 1

Determine the convergence or divergence of the given series. For convergent series, find the sum.

1. $\sum_{k=1}^{\infty} \frac{1}{(k+4)(k+5)}$ (Use partial fractions.)
2. $\sum_{k=1}^{\infty} (1 + (-1)^k)$ (Write out the first six or seven partial sums. It is not necessary to give the nth partial sum in general.)
3. $\sum_{j=1}^{\infty} \ln\left(\frac{j}{j+1}\right)$

BASIC FACTS ABOUT SERIES

Theorem 15.2.3 is obtained by combining the Limit Theorem for Sequences with basic facts about ordinary addition.

15.2.3 Theorem Suppose $\sum_{k=1}^{\infty} a_k$ and $\sum_{k=1}^{\infty} b_k$ converge and c is any number. Then

(*a*) $\sum_{k=1}^{\infty} (a_k \pm b_k)$ converges with $\sum_{k=1}^{\infty} (a_k \pm b_k) = \sum_{k=1}^{\infty} a_k \pm \sum_{k=1}^{\infty} b_k$

(*b*) $\sum_{k=1}^{\infty} ca_k$ converges with $\sum_{k=1}^{\infty} ca_k = c\sum_{k=1}^{\infty} a_k$

Note that the above arithmetic is valid only for *convergent* series. However, we do have:

15.2.4 Theorem

(*a*) If $c \neq 0$ and $\sum_{k=1}^{\infty} a_k$ diverges, then $\sum_{k=1}^{\infty} ca_k$ diverges.

(*b*) If $\sum_{k=1}^{\infty} a_k$ converges and $\sum_{k=1}^{\infty} b_k$ diverges, then $\sum_{k=1}^{\infty} (a_k + b_k)$ diverges.

If we add or subtract a finite number of terms from an infinite series, we will not affect the convergence or divergence of the series, because the convergence or divergence of its sequence of partial sums is unaffected by the "early" elements. If the series converges, this will, of course, change the sum by the sum of the affected terms. (See Sec. 15.2, Exercise 52.)

15.2.5 **Theorem** Let j be a positive integer. Then the series $\Sigma_{k=1}^{\infty} a_k$ and $\Sigma_{k=j+1}^{\infty} a_k$ either both converge or both diverge. Furthermore, if $\Sigma_{k=1}^{\infty} a_k = L$, then

$$\sum_{k=j+1}^{\infty} a_k = L - (a_1 + a_2 + \cdots + a_j)$$

Example 3 Find the sum of the series $\sum_{i=3}^{\infty} 1/2^i$.

Solution We know that $\Sigma_{i=0}^{\infty} 1/2^i = 2$; thus $1 + 1/2 + 1/4 + \Sigma_{i=3}^{\infty} 1/2^i = 2$, and so $\Sigma_{i=3}^{\infty} 1/2^i = 2 - (1 + 1/2 + 1/4) = 2 - 7/4 = 1/4$. ●

Converge or diverge? That is the question to be asked of any series. Our first armament in attacking it is simple enough.

15.2.6 ***k*th-Term Test for Divergence** If $\lim_{k\to+\infty} a_k \neq 0$, then $\sum_{k=1}^{\infty} a_k$ diverges.

Proof: We shall show the equivalent statement that if $\Sigma_{k=1}^{\infty} a_k$ is to converge, then $\lim_{k\to+\infty} a_k = 0$.

Suppose, then, that $\sum_{k=1}^{\infty} a_k = L$ and $k > 1$. Thus

$$\begin{aligned} a_k &= (a_1 + a_2 + \cdots + a_k) - (a_1 + a_2 + \cdots + a_{k-1}) \\ &= s_k - s_{k-1} \end{aligned}$$

But

$$\begin{aligned} \lim_{k\to+\infty} a_k &= \lim_{k\to+\infty} (s_k - s_{k-1}) \\ &= \lim_{k\to+\infty} s_k - \lim_{k\to+\infty} s_{k-1} \qquad [15.1.4(a)] \\ &= L - L = 0 \quad ■ \end{aligned}$$

This theorem is sometimes more briefly called "the *k*th-Term Test." It also applies when $\lim_{k\to+\infty} a_k$ does not exist. Thus, for example, we know that $\Sigma_{k=1}^{\infty} (-1)^k$ diverges.

Example 4 Determine convergence or divergence of

(*a*) $\displaystyle\sum_{m=1}^{\infty} \cos\left(\frac{m}{m^2+1}\right)$ (*b*) $\displaystyle\sum_{n=1}^{\infty} \left(1 + \frac{1}{n}\right)^n$

Solution (*a*) $\displaystyle\lim_{m\to+\infty} \cos\left(\frac{m}{m^2+1}\right) = \cos 0 = 1$, so

$$\sum_{m=1}^{\infty} \cos\left(\frac{m}{m^2+1}\right) \qquad \text{diverges}$$

(*b*) $\lim_{n\to+\infty}\left(1+\frac{1}{n}\right)^n = e$, by [15.1.6(*e*)], so

$$\sum_{n=1}^{\infty}\left(1+\frac{1}{n}\right)^n \quad \text{diverges} \quad \bullet$$

Beware: The above divergence test causes trouble in two ways:

1. It encourages a confusion between sequences and series. But remember, even when we know that $\Sigma_{k=1}^{\infty} a_k$ is convergent, its value is determined not by a limit of a_k's but as a limit of the sequence of partial sums s_n. Thus part (*b*), Example 4 tells us not that the series sums to e but that the series is divergent.
2. It tempts using the converse—which is not true, as Example 5 will illustrate. If $\lim_{k\to+\infty} a_k = 0$, then nothing can be concluded regarding the convergence or divergence of $\Sigma_{k=1}^{\infty} a_k$.

Example 5 (Harmonic Series) Show that $\sum_{k=1}^{\infty}\frac{1}{k}$ diverges.

Solution We shall show that even though $\lim_{k\to+\infty} a_k = \lim_{k\to+\infty} 1/k = 0$, the sequence of partial sums grows without bound, so $\Sigma_{k=1}^{\infty} 1/k$ is divergent. To simplify the arithmetic we shall confine our attention to the partial sums $s_1, s_2, s_4, s_8, \ldots$. Since the entire sequence of partial sums is increasing, showing *these* sums grow without bound implies that the entire sequence of partial sums does likewise.

But consider the following sequence of inequalities. (We have underlined those fractions that replace a larger fraction and thus yield the inequality.)

$$s_1 = 1 > 1\cdot\frac{1}{2}$$

$$s_2 = 1+\frac{1}{2} > \underline{\frac{1}{2}}+\frac{1}{2} = 2\cdot\frac{1}{2}$$

$$s_4 = 1+\frac{1}{2}+\left(\frac{1}{3}+\frac{1}{4}\right) > \underline{\frac{1}{2}}+\frac{1}{2}+\left(\underline{\frac{1}{4}}+\frac{1}{4}\right) = 3\cdot\frac{1}{2}$$

$$s_8 = 1+\frac{1}{2}+\left(\frac{1}{3}+\frac{1}{4}\right)+\left(\frac{1}{5}+\frac{1}{6}+\frac{1}{7}+\frac{1}{8}\right)$$

$$> \frac{1}{2}+\frac{1}{2}+\left(\underline{\frac{1}{4}}+\frac{1}{4}\right)+\left(\underline{\frac{1}{8}}+\underline{\frac{1}{8}}+\underline{\frac{1}{8}}+\frac{1}{8}\right) = 4\cdot\frac{1}{2}$$

$$s_{16} > 5\cdot\frac{1}{2} \qquad \text{(Check this!)}$$

$$s_{2^{n-1}} > n\cdot\frac{1}{2}$$

Since the partial sums can be made larger than any given number, $\Sigma_{k=1}^{\infty} 1/k$ diverges. ●

The above series, $\Sigma_{k=1}^{\infty} 1/k$, is called the *harmonic series*. It illustrates how common sense may be misleading when dealing with series. Although successive partial sums are formed by adding ever smaller additional terms, the sum nevertheless grows without bound. (2^{15} terms are needed before the sum exceeds $10 = 20/2$. How many terms are needed to exceed 100?)

PROGRESS TEST 2

Determine the convergence or divergence of the following series. If convergent, find the sum.

1. $\sum_{k=3}^{\infty} \frac{1}{k^2+k}$ (see Example 1)

2. $\sum_{n=2}^{\infty} \left(\frac{\sqrt{n}}{\ln n}\right)^2$

3. $\sum_{n=1}^{\infty} \frac{n-1}{2n+5}$

4. $\sum_{n=2}^{\infty} \frac{\pi}{2^n}$

GEOMETRIC SERIES

15.2.7 **Definition** For any numbers a and r, a series of the form

$$\sum_{k=0}^{\infty} ar^k = a + ar + ar^2 + \cdots + ar^k + \cdots$$

is called a *geometric series.*

Thus geometric series are those series whose ratio of each term with its predecessor is a constant. Our introductory series $\Sigma_{k=0}^{\infty}(1/2)^k$ was obviously a geometric series with $a = 1$, $r = 1/2$. The proof of the following theorem is a direct generalization of the argument that showed $\Sigma_{k=0}^{\infty}(1/2)^k = 2$.

15.2.8 **Theorem** Suppose $\sum_{k=0}^{\infty} ar^k$ is a geometric series.

(*a*) If $|r| < 1$, then $\sum_{k=0}^{\infty} ar^k = \frac{a}{1-r}$.

(*b*) If $|r| \geq 1$, then $\sum_{k=0}^{\infty} ar^k$ diverges.

Since the behavior of geometric series is so completely determined by the above theorem, a standard strategy for analyzing a series involving powers of constants is to manipulate it into a form that either reveals the given series to be geometric or allows direct comparison with a geometric series.

Example 6 Determine the convergence or divergence of the following series. If convergent find the sum.

(*a*) $\sum_{n=2}^{\infty} \frac{2^{n-1}+4^{n-1}}{3^{n-1}}$ (*b*) $\sum_{n=3}^{\infty} \frac{(-1)^n 3^{n+1}}{4^{n-2}}$

Solution (*a*) $\sum_{n=2}^{\infty} \frac{2^{n-1}+4^{n-1}}{3^{n-1}} = \sum_{n=2}^{\infty}\left[\left(\frac{2}{3}\right)^{n-1} + \left(\frac{4}{3}\right)^{n-1}\right]$

By (15.2.5), *where* we start the summation index cannot affect the convergence or divergence of the series. Since $\Sigma_{n=2}^{\infty}(2/3)^{n-1}$ converges ($|r| = 2/3 < 1$), and $\Sigma_{n=2}^{\infty}(4/3)^{n-1}$ diverges ($|r| = 4/3 > 1$), the given series diverges, by [15.2.4(*b*)]. Alternatively, we could have observed that since $(2^{n-1} + 4^{n-1})/3^{n-1} > 1$, the general term does not tend to zero, and thus the series diverges, by (15.2.6).

(*b*) We first rewrite the general term in order to reveal that the given series is geometric:

$$\frac{(-1)^n 3^{n+1}}{4^{n-2}} = \frac{(-1)^n 3^n \cdot 3}{4^{n-2}}$$
$$= \frac{4^2 \cdot 3 \cdot (-1)^n 3^n}{4^2 \cdot 4^{n-2}}$$
$$= \frac{48 \cdot (-3)^n}{4^n} = 48\left(\frac{-3}{4}\right)^n$$

Thus

$$\sum_{n=3}^{\infty} \frac{(-1)^n 3^{n+1}}{4^{n-2}} = \sum_{n=3}^{\infty} 48\left(\frac{-3}{4}\right)^n$$

Now

$$\sum_{n=0}^{\infty} 48\left(\frac{-3}{4}\right)^n = \frac{48}{1-(-3/4)} \qquad \text{by [15.2.8}(a)\text{], with } r = -3/4,\ a = 48$$
$$= \frac{192}{7}$$

so

$$\sum_{n=3}^{\infty} 48\left(\frac{-3}{4}\right)^n = \frac{192}{7} - 48\left[\left(\frac{-3}{4}\right)^0 + \left(\frac{-3}{4}\right)^1 + \left(\frac{-3}{4}\right)^2\right] \qquad \text{by (15.2.5)}$$
$$= \frac{-81}{7} \quad \bullet$$

We use an infinite series every time we use a decimal because a decimal *is* an infinite series:

$$.a_1a_2a_3 \cdots = a_1 \cdot 10^{-1} + a_2 \cdot 10^{-2} + a_3 \cdot 10^{-3} + \cdots$$

where a_k is an integer between 0 and 9, inclusive. Furthermore, an infinite *repeating* decimal (which thus represents a rational number) is a geometric series. This fact, coupled with [15.2.8(*a*)], can be used to express a repeated decimal as a quotient of integers in reduced terms.

Example 7 Find the rational fraction in lowest terms for the repeating decimal 0.6121212. . . .

Solution 0.6121212 . . .

$$= \frac{6}{10} + \frac{12}{1{,}000} + \frac{12}{100{,}000} + \frac{12}{10{,}000{,}000} + \cdots$$
$$= \frac{6}{10} + \frac{12}{1{,}000}\left[1 + \frac{1}{100} + \left(\frac{1}{100}\right)^2 + \cdots\right]$$
$$= \frac{6}{10} + \frac{12}{1{,}000}\left[\sum_{n=0}^{\infty}\left(\frac{1}{100}\right)^n\right]$$
$$= \frac{6}{10} + \frac{12}{1{,}000} \cdot \frac{1}{1 - 1/100} \qquad \text{by [15.2.8}(a)\text{], with } r = \frac{1}{100}$$
$$= \frac{6}{10} + \frac{12}{1{,}000} \cdot \frac{100}{99}$$
$$= \frac{6}{10} + \frac{12}{10 \cdot 99} = \frac{6 \cdot 99 + 12}{10 \cdot 99}$$
$$= \frac{606}{10 \cdot 99} = \frac{101}{5 \cdot 33} = \frac{101}{165} \quad \bullet$$

PROGRESS TEST 3

In Probs. 1 to 3 determine the convergence or divergence of the given series. Find the sum of the convergent series.

1. $\sum_{n=1}^{\infty} \frac{3}{(3-\pi)^{n-1}}$ **2.** $\sum_{n=4}^{\infty} \frac{5^{n+2}}{3^n}$ **3.** $\sum_{n=1}^{\infty} \frac{2^n}{3^{n+1}}$

4. Find the rational fraction in lowest terms for the repeating decimal 0.32424 . . . .

SECTION 15.2 EXERCISES

In Exercises 1 to 31 determine the convergence or divergence of the given series. If convergent, find the sum.

1. $\sum_{k=1}^{\infty} \frac{1}{k^2-2k}$ **2.** $\sum_{k=1}^{\infty} \frac{1}{2k^2+k}$ **3.** $\sum_{k=2}^{\infty} \frac{3}{k^2-1}$ **4.** $\sum_{n=4}^{\infty} \frac{6}{(n+2)(n+1)}$

5. $\sum_{k=1}^{\infty} \sin\left(\frac{\pi}{2}-\frac{1}{k^2}\right)$ **6.** $\sum_{n=1}^{\infty} (-1)^{n^2}$ **7.** $\sum_{j=1}^{\infty} (-1)^{2j+1}$ **8.** $\sum_{i=1}^{\infty} \frac{1}{1+\sqrt{i}}$

9. $\sum_{m=1}^{\infty} \frac{3m+1}{5m-1}$ **10.** $\sum_{k=1}^{\infty} \frac{(-1)^k e^k}{k^4}$ **11.** $\sum_{i=1}^{\infty} \frac{i^2+i+2}{\ln(i+1)}$ **12.** $\sum_{j=2}^{\infty} \frac{2j-1}{j^2(j-1)^2}$

13. $\sum_{n=1}^{\infty} \ln\left(\frac{2n+1}{n+2}\right)$ **14.** $\sum_{n=1}^{\infty} \frac{12}{5^{n-1}}$ **15.** $\sum_{n=2}^{\infty} \frac{\sqrt{n}}{\ln n}$ **16.** $\sum_{n=1}^{\infty} \sin\left(\frac{2n+7}{n+1}\right)$

17. $\sum_{k=1}^{\infty} \frac{3^{k-1}+4^k}{5^{k+2}}$ **18.** $\sum_{k=3}^{\infty} \frac{(-1)^k 3^k}{4^k}$ **19.** $\sum_{k=1}^{\infty} \left(2^k+\frac{1}{2^k}\right)$ **20.** $\sum_{k=0}^{\infty} \frac{1}{2^{k+2}}$

21. $\sum_{n=1}^{\infty} \frac{1}{(n+2)(n+3)}$ **22.** $\sum_{n=1}^{\infty} \frac{\pi^n}{(3.1)^{n+1}}$ **23.** $\sum_{n=1}^{\infty} (\sqrt{n+1}-\sqrt{n})$ **24.** $\sum_{n=1}^{\infty} \left(\frac{n+2}{n}\right)^n$

25. $\sum_{n=1}^{\infty} \frac{1}{(2n-1)(2n+1)}$ **26.** $\sum_{n=1}^{\infty} \left(\frac{n}{n-3}\right)^n$ **27.** $\sum_{n=1}^{\infty} n^{2/n}$ **28.** $\sum_{k=1}^{\infty} 3\left[\frac{1}{k(k+2)}+\left(\frac{2}{3}\right)^k\right]$

29. $\sum_{k=1}^{\infty} 2\left[\frac{1}{2^k}+(-1)^k\right]$ **30.** $\sum_{k=1}^{\infty} \left[\frac{3}{k^2-5k}-\frac{4^{k-1}}{7^{k+1}}\right]$ **31.** $\sum_{k=1}^{\infty} \left(\frac{2^{k-2}}{3^{k+1}}-\frac{2}{k^2-4}\right)$

32. For any series $\Sigma_{k=1}^{\infty} a_k$ and $k > 1$, $s_k = s_{k-1} + a_k$, with $s_1 = a_1$. Hence a given sequence S_n can be regarded as the sequence of partial sums of a series. Compute the series $\Sigma_{k=1}^{\infty} a_k$ that has $S_n = 1 - 1/4^n$ as its sequence of partial sums.

33. Repeat Exercise 32 for $S_n = \frac{n+1}{n+2}$.

34. Repeat Exercise 32 for $S_n = \ln(n+2)$. (Treat $k = 1$ separately.)

35. Repeat Exercise 32 for $S_n = \ln[2n/(n+1)]$.

36. Repeat Exercise 32 for $S_n = n/(3n+1)$.

37. Repeat Exercise 32 for $S_n = n^2$.

38. (*a*) Show that $\Sigma_{n=1}^{\infty} 1/(2^n + n)$ converges by showing that its partial sums are bounded above.
(*b*) Determine the sum of $\Sigma_{n=1}^{\infty} 1/(2^n + n)$.

In Exercises 39 to 41 express the given repeating decimal in the form p/q.

39. 0.2171717 . . . **40.** 0.123123 . . .

41. 0.120120 . . .

42. Is the decimal representation of π a geometric series?

43. A ball is dropped from a height of 2 m, and each time it strikes the floor it bounces to two-thirds its previous height. What is the total distance traveled by the ball before it finally stops bouncing? Or does it *ever* stop bouncing? Is there a *last* bounce? Discuss this with someone.

44. Repeat Exercise 43 for a ball dropped from a height of 2 m and which bounces only to one-third its previous height.

45. (*a*) A (real) honey wagon is being drawn by a horse traveling at 5 mi/h back to the aviary from a market 10 mi away. A busy bee, determined to get its honey back, travels back and forth between the wagon (which has some drippings on it) and its hive at 30 mi/h. Assuming that the rest time at either end of the bee's trip is negligible, what is the total distance traveled by the bee as the wagon returns home?
(*b*) Redo the problem assuming that you had never heard of infinite series. (*Hint:* The bee flies for as long as the horse walks. Don't be upset; John von Neumann, one of the cleverest mathematicians of the twentieth century, is said to have solved the problem using series.)

46. Banks create capital by loaning out a proportion of their deposits, a portion of which are then deposited in banks again. Then a proportion is loaned out and a portion redeposited, and so on. The federal government requires banks to hold a certain proportion of their deposits. What is the total amount of capital created if (*a*) \$1000 is deposited in a bank, (*b*) the banks are required to hold 50 percent of deposits, and (*c*) all the money loaned at each stage finds its way back into bank deposits?

47. Repeat Exercise 46 if holding requirements are lowered to 40 percent. How much more capital per initial \$1000 is created by the government's action?

48. Repeat Exercise 46 but with hypothesis (*c*) changed to reflect the fact that only 75 percent of the money loaned at each stage returns to banks as deposits.

49. At what time between 1 and 2 o'clock is the minute hand of a clock exactly on top of the hour hand? [*Hint:* The hour hand moves 1/12 as fast as the minute hand, so the hour hand reaches $1 + 1/12$ when the minute hand travels from straight up to 1. When the minute hand reaches $1 + 1/12$, the hour hand goes to $(1 + 1/12) + (1/12 \cdot 1/12)$.]

50. (See Exercise 49) When is the *next* time the minute hand catches up with the hour hand?

51. (See Exercises 49 and 50)
(*a*) How many times strictly between noon and midnight does the minute hand coincide with the hour hand?
(*b*) If you have not solved (*a*), pretend you are 10 years old and try again.

52. Prove [15.2.3(*b*)].

53. Prove (15.2.5).

Section 15.3: Positive-Term Series

Objectives:

1. **Apply the Comparison and Limit Comparison Tests, and the Integral, Ratio, and Root Tests to determine the convergence or divergence of series of positive terms.**

2. **Choose the appropriate test(s) to determine convergence or divergence of a given series.**

INTRODUCTION

Although we were able to put the partial sums of telescoping and geometric series into a form whose limit was then straightforward to compute, in general this is not so easy to accomplish. We shall now develop various tests designed to tell us whether or not a given series converges. Determining the actual sums of series will be among the topics of Chap. 16.

Unless otherwise noted, we shall assume that *the terms of all series* $\Sigma_{k=1}^{\infty} a_k$ *dealt with in this section are nonnegative.* For simplicity we shall refer to such as "positive-term series." Of course, any theory and technique devoted to convergence will also extend to series that may have a finite number of negative terms, but whose terms a_k are nonnegative for all $k \geq j$ for some positive integer j.

For positive-term series,

$$s_{n+1} = s_n + a_{n+1} \geq s_n \qquad \text{for all } n$$

which implies that the sequence of partial sums is increasing. Then by (15.1.9) we have the following:

15.3.1 Theorem If the sequence of partial sums of the positive term series $\Sigma_{k=1}^{\infty} a_k$ is bounded above, then $\Sigma_{k=1}^{\infty} a_k$ converges. Otherwise $\Sigma_{k=1}^{\infty} a_k$ diverges.

THE INTEGRAL TEST

The next test links our work with convergence of improper integrals to convergence of series.

15.3.2 The Integral Test Suppose $f(x)$ is nonnegative, continuous, and decreasing on $[1, +\infty)$. Then the series $\Sigma_{k=1}^{\infty} f(k)$ converges or diverges according as the improper integral $\int_1^{+\infty} f(x)\,dx$ converges or diverges.

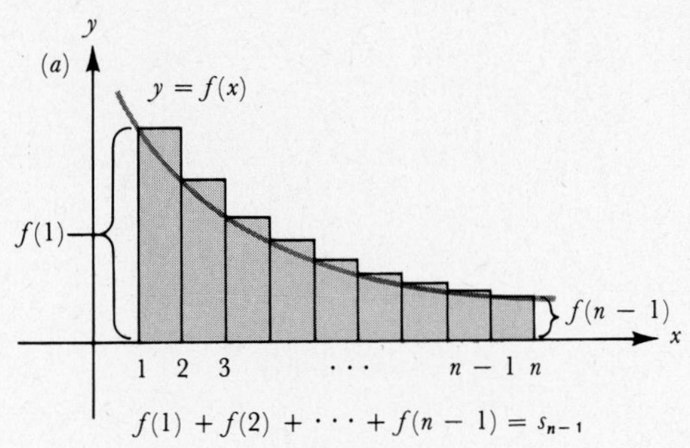

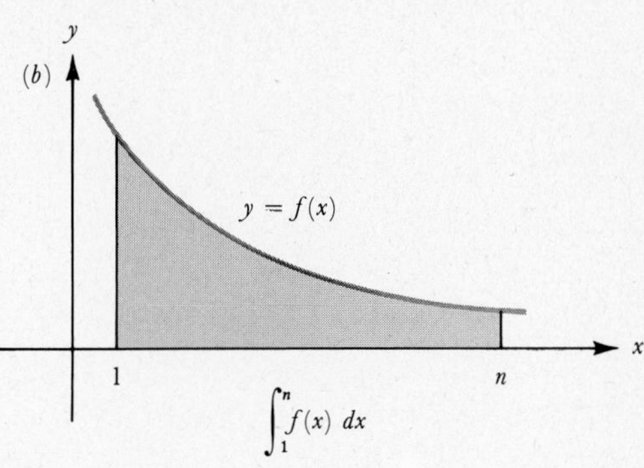

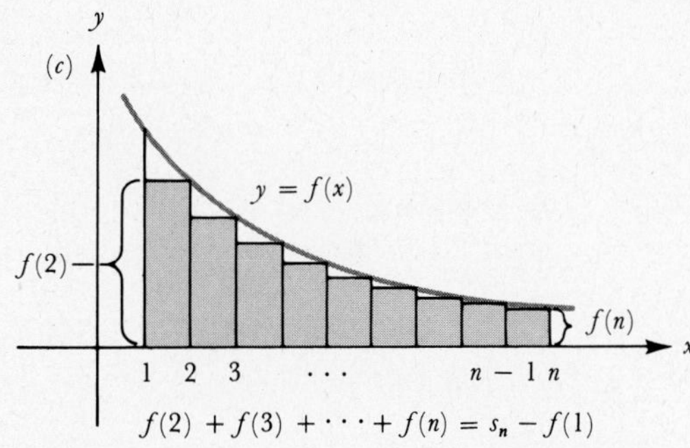

Fig. 15.1

Proof: Since the function involved is nonnegative, we can phrase our argument in terms of the areas in Fig. 15.1, from which, for each $n > 1$,

$$s_n - f(1) \le \int_1^n f(x)\,dx \le s_{n-1} \le s_n \tag{1}$$

(Note that $s_{n-1} \le s_n$ because $f(n) \ge 0$ for all n. Also, each rectangle has width 1.)

Suppose first that $\int_1^{+\infty} f(x)\,dx$ converges with, say,

$$\lim_{n\to+\infty} \int_1^n f(x)\,dx = L$$

Then the sequence of partial sums s_n is bounded above by $L + f(1)$, and hence $\sum_{k=1}^{\infty} f(k)$ converges, by (15.3.1); see Fig. 15.1(*c*).

On the other hand, if $\int_1^{+\infty} f(x)\,dx$ diverges, then the partial sums s_n are unbounded, by the last inequality in (1), so $\sum_{k=1}^{\infty} f(k)$ diverges, by (15.3.1). ■

We apply the Integral Test to a positive term series $\sum_{k=j}^{\infty} a_k$ (j a fixed positive integer) by associating with this given series a function $f(x)$ with $f(k) = a_k$ for all $k \ge j$.

Example 1 Determine the convergence or divergence of the series $\displaystyle\sum_{k=2}^{\infty} \frac{1}{k \ln k}$.

Solution Let $f(x) = 1/(x \ln x)$; then $f(k) = 1/(k \ln k)$ and f is positive, continuous, and decreasing on $[2, +\infty)$.

$$\int_2^{+\infty} \frac{dx}{x \ln x} = \lim_{n\to+\infty} \int_2^n \frac{dx}{x \ln x}$$
$$= \lim_{n\to+\infty} \int_{x=2}^{x=n} \frac{du}{u} \qquad \left(u = \ln x,\, du = \frac{dx}{x}\right)$$
$$= \lim_{n\to+\infty} \Big[\ln u\Big]_{x=2}^{x=n}$$
$$= \lim_{n\to+\infty} \Big[\ln(\ln x)\Big]_2^n$$
$$= \lim_{n\to+\infty} [\ln(\ln n) - \ln(\ln 2)] = +\infty$$

Thus $\sum_{k=2}^{\infty} \frac{1}{k \ln k}$ diverges by the Integral Test. ●

Notice that $\Sigma_{k=2}^{\infty} 1/(k \ln k)$ diverges even though the kth term tends to zero as $k \to +\infty$.

15.3.3 **Definition** A series of the form $\Sigma_{k=1}^{\infty} 1/k^p$, p any real number, is called a p *series*. The p series for $p = 1$ is the *harmonic series*.

By Example 5 in Sec. 15.2, we know that the harmonic series diverges. The Integral Test will be used to prove the following more general fact.

15.3.4 **Theorem** The p series $\Sigma_{k=1}^{\infty} 1/k^p$ converges if $p > 1$ and diverges if $p \le 1$.

Proof: Assume $p < 0$. Then $1/k^p = k^{-p} = k^q (q > 0)$, and so $\lim_{k\to+\infty} 1/k^p = \lim_{k\to+\infty} k^q = +\infty$. If $p = 0$, then $1/k^p = 1/k^0 = 1$. Thus for $p \le 0$ the p series diverges by the kth-Term Test.

Now assume $p > 0$. Then $f(x) = 1/x^p$ is continuous, positive, and decreasing on $[1, +\infty)$, and $\Sigma_{k=1}^{\infty} f(k) = \Sigma_{k=1}^{\infty} 1/k^p$. By the Integral Test, the convergence or divergence of $\Sigma_{k=1}^{\infty} 1/k^p$ depends on $\int_1^{+\infty} dx/x^p$. But, by (14.3.3), page 472, the improper integral, hence the series, diverges for $p \le 1$ and converges for $p > 1$. ■

Thus, for example, by (15.3.4),

$$\sum_{k=1}^{\infty} \frac{1}{\sqrt{k}} \qquad \sum_{n=10^{10}}^{\infty} \frac{1}{n} \qquad \sum_{j=1}^{\infty} \frac{1}{j^{.01}}$$

are divergent p series, and

$$\sum_{n=1}^{\infty} \frac{1}{n^2} \qquad \sum_{i=1}^{\infty} \frac{1}{i^{3/2}} \qquad \sum_{m=1}^{\infty} \frac{1}{m^{1.001}}$$

are convergent p series.

Naturally, the Integral Test will be most useful on series $\Sigma_{k=1}^{\infty} a_k$ where the function $f(x)$ such that $f(k) = a_k$ is relatively easy to antidifferentiate.

PROGRESS TEST 1

Determine the convergence or divergence of each of the following series:

1. $\sum_{k=2}^{\infty} \frac{1}{k(\ln k)^2}$ **2.** $\sum_{k=1}^{\infty} \frac{k}{e^{k^2}}$ **3.** $\sum_{m=1}^{\infty} \frac{\ln m}{m}$ **4.** $\sum_{k=1}^{\infty} \cos(k^2 + 1)$

COMPARISON TESTS

Now that we have a store of series whose convergence or divergence is known, we shall develop tests that allow us to determine the behavior of new series by comparing them with the known series.

15.3.5 **The Comparison Test** Suppose $\Sigma_{k=1}^{\infty} a_k$ and $\Sigma_{k=1}^{\infty} b_k$ are positive-term series.
(*a*) If $a_k \leq b_k$ for all k and if $\Sigma_{k=1}^{\infty} b_k$ converges, then $\Sigma_{k=1}^{\infty} a_k$ converges.
(*b*) If $a_k \geq b_k$ for all k and if $\Sigma_{k=1}^{\infty} b_k$ diverges, then $\Sigma_{k=1}^{\infty} a_k$ diverges.

Proof: Let s_n and t_n denote the partial sums of $\Sigma_{k=1}^{\infty} a_k$ and $\Sigma_{k=1}^{\infty} b_k$, respectively. Since we are dealing with positive-term series, s_n and t_n are increasing sequences.
(*a*) If $a_k \leq b_k$ for all k, then

$$s_n = a_1 + a_2 + \cdots + a_n \leq b_1 + b_2 + \cdots + b_n = t_n$$

for all n. If $\Sigma_{k=1}^{\infty} b_k$ converges, then $\lim_{n\to+\infty} t_n = L$, so the increasing sequence of partial sums s_n for $\Sigma_{k=1}^{\infty} a_k$ is bounded above by L. Hence $\Sigma_{k=1}^{\infty} a_k$ converges, by (15.3.1).
(*b*) If $a_k \geq b_k$ for all k, then $s_n \geq t_n$ for all n. If $\Sigma_{k=1}^{\infty} b_k$ diverges, then $\lim_{n\to+\infty} t_n = +\infty$, so the sequence of partial sums s_n for $\Sigma_{k=1}^{\infty} a_k$ cannot be bounded above. Hence $\Sigma_{k=1}^{\infty} a_k$ diverges by (15.3.1). ■

To apply the Comparison Test to a series $\Sigma_{k=1}^{\infty} a_k$, you must compare it with a series $\Sigma_{k=1}^{\infty} b_k$, whose convergence or divergence you already know. Such a series is often referred to as a "test series," with the most frequently used test series being geometric series and p series.

Since we know by (15.2.5) that for a fixed integer j, $\Sigma_{k=j}^{\infty} a_k$ converges if and only if $\Sigma_{k=1}^{\infty} a_k$ converges, it suffices to prove the appropriate inequality $a_k \leq b_k$ or $a_k \geq b_k$ only for all $k \geq j$ for a fixed integer j.

Example 2 Test the following series for convergence or divergence:

(*a*) $\displaystyle\sum_{k=1}^{\infty} \frac{1}{3^k + \sqrt{k}}$ (*b*) $\displaystyle\sum_{k=1}^{\infty} \frac{\ln k}{\sqrt{k}}$

Solutions Note that the general terms of both series tend to zero, so the kth-Term Test does not apply.
(*a*) As our test series we take the convergent geometric series $\Sigma_{k=1}^{\infty} 1/3^k$. Now $3^k + \sqrt{k} > 3^k$. Thus $1/(3^k + \sqrt{k}) < 1/3^k$, so the series $\Sigma_{k=1}^{\infty} 1/(3^k + \sqrt{k})$ is, term by term, less than the known convergent series $\Sigma_{k=1}^{\infty} 1/3^k$ and hence converges by the Comparison Test.
(*b*) If $k \geq 3$, then $\ln k \geq \ln 3 > 1$ and hence $(\ln k)/\sqrt{k} > 1/\sqrt{k}$. But $\Sigma_{k=1}^{\infty} 1/\sqrt{k}$ is a divergent p-series, and hence $\Sigma_{k=1}^{\infty} (\ln k)/\sqrt{k}$ diverges by the Comparison Test. ●

We should point out that the Integral Test would be difficult to apply in the above example, especially part (*a*).

The following elaboration of the Comparison Test is useful because it replaces the chore of establishing an inequality with the often simpler task of evaluating a limit. The comparison idea is still the same except that we focus attention on the behavior of *quotients* of respective terms.

15.3.6 **The Limit Comparison Test** Suppose $\Sigma_{k=1}^{\infty} a_k$ and $\Sigma_{k=1}^{\infty} b_k$ are (strictly) positive-term series.

(*a*) If $\lim_{k\to+\infty} a_k/b_k = c > 0$, then either both series converge or both diverge.
(*b*) If $\lim_{k\to+\infty} a_k/b_k = 0$ and $\Sigma_{k=1}^{\infty} b_k$ converges, then $\Sigma_{k=1}^{\infty} a_k$ converges.
(*c*) If $\lim_{k\to+\infty} a_k/b_k = +\infty$ and $\Sigma_{k=1}^{\infty} b_k$ diverges, then $\Sigma_{k=1}^{\infty} a_k$ diverges.

Proof: [We prove (*a*). Parts (*b*) and (*c*) are similar and are left as exercises.] If the sequence a_k/b_k converges to $c > 0$, then, by definition (15.1.3), for any $\varepsilon > 0$, there is a positive integer K such that $k > K$ implies

$$\left|\frac{a_k}{b_k} - c\right| < \varepsilon$$

This absolute-value inequality is equivalent to the double inequality

$$c - \varepsilon < \frac{a_k}{b_k} < c + \varepsilon$$

or

$$(c - \varepsilon)b_k < a_k < (c + \varepsilon)b_k \qquad (2)$$

Now if $\Sigma_{k=1}^{\infty} b_k$ converges, then so does $\Sigma_{k=1}^{\infty} (c + \varepsilon)b_k$, by [15.2.3(*b*)]. Hence $\Sigma_{k=1}^{\infty} a_k$ converges, by the Comparison Test applied to the right side of (2). On the other hand, if $\Sigma_{k=1}^{\infty} b_k$ diverges, then so does $\Sigma_{k=1}^{\infty} (c - \varepsilon)b_k$, and thus $\Sigma_{k=1}^{\infty} a_k$ diverges as well. ■

The key to using any comparison test is to recognize an appropriate test series $\Sigma_{k=1}^{\infty} b_k$. In the case of quotients of polynomials, a good test series is $\Sigma_{k=1}^{\infty} 1/k^p$, where p is the degree of the denominator minus the degree of the numerator.

Example 3 Test $\displaystyle\sum_{k=1}^{\infty} \frac{2k^2 + k - 1}{3k^5 + k^4 - k^2 + 5}$ for convergence or divergence.

Solution We use the convergent p-series $\Sigma_{k=1}^{\infty} 1/k^3$ as the $\Sigma_{k=1}^{\infty} b_k$ test series. Then

$$\frac{2k^2 + k - 1}{3k^5 + k^4 - k^2 + 5} \div \frac{1}{k^3} = \frac{2k^5 + k^4 - k^3}{3k^5 + k^4 - k^2 + 5}$$

But the limit of this quotient, by [15.1.6(*a*)], is $2/3 > 0$, so the given series converges by the Limit Comparison Test. ●

A similar approach works when roots of polynomials are involved, except that we use *relative* degrees. For example, to test the series $\Sigma_{k=1}^{\infty} (\sqrt[3]{k} + 1/k + 2)$ we would use as our test series $\Sigma_{k=1}^{\infty} 1/k^{2/3}$.

Arguments parallel to those used in the previous example yield the following theorem:

15.3.7 **Theorem** If, for each k, a_k is either a quotient of polynomials or roots of polynomials, then $\Sigma_{k=1}^{\infty} a_k$ converges if the degree or relative degree of the denominator exceeds that of the numerator by more than one. Otherwise the series diverges.

PROGRESS TEST 2

Test each of the following for convergence or divergence.

1. $\sum_{k=1}^{\infty} \frac{\ln k}{k^{3/2}}$ (Use $\sum_{k=1}^{\infty} 1/k^{5/4}$ as a limit-test series.)

2. $\sum_{k=1}^{\infty} \frac{\arctan k}{\sqrt{k}+1}$ (Use $\sum_{k=1}^{\infty} 1/\sqrt{k}$ as a limit-test series.)

3. $\sum_{k=1}^{\infty} \frac{k+2}{\sqrt{3k+5}}$

THE RATIO AND ROOT TESTS

Our last two tests for positive-term series are based on comparing a series with a geometric series.

15.3.8 **The Ratio Test** Suppose $\sum_{k=1}^{\infty} a_k$ is a positive term series and $\lim_{k\to+\infty} \frac{a_{k+1}}{a_k} = L$ or $\lim_{k\to+\infty} \frac{a_{k+1}}{a_k} = +\infty = L$

(*a*) If $L < 1$, then $\sum_{k=1}^{\infty} a_k$ converges.

(*b*) If $L > 1$ or $L = +\infty$, then $\sum_{k=1}^{\infty} a_k$ diverges.

(*c*) If $L = 1$, then the test is inconclusive and $\sum_{k=1}^{\infty} a_k$ may either converge or diverge.

Proof:

(*a*) Suppose $L < 1$. Choose r with $L < r < 1$. Since $\lim_{k\to+\infty} a_{k+1}/a_k = L$, we can find K such that, for $k \geq K$,

$$\frac{a_{k+1}}{a_k} < r$$

That is,

$$a_{k+1} < r \cdot a_k$$

In particular,

$$a_{K+1} < ra_K$$
$$a_{K+2} < ra_{K+1} < r(ra_K) = r^2 a_K$$
$$a_{K+3} < ra_{K+2} < r(r^2 a_K) = r^3 a_K$$

and, in general,

$$a_{K+j} < r^j a_K \qquad \text{for each } j = 1, 2, \ldots$$

But $\Sigma_{j=1}^{\infty} r^j a_K = a_K \Sigma_{j=1}^{\infty} r^j$ is a convergent geometric series since $|r| = r < 1$. Thus by the Comparison Test,

$$\sum_{j=1}^{\infty} a_{K+j} = a_{K+1} + a_{K+2} + \cdots$$

converges and hence $\Sigma_{k=1}^{\infty} a_k$ converges.

(*b*) If $L > 1$ or $L = +\infty$, then in either case we can find K such that $k \geq K$ implies

$$\frac{a_{k+1}}{a_k} > 1$$

that is, $a_{k+1} > a_k$. And since the a_k's are all positive, the sequence of a_k's cannot tend to zero. Thus $\Sigma_{k=1}^{\infty} a_k$ diverges by the kth-Term Test.

(*c*) Consider $\Sigma_{k=1}^{\infty} 1/k$ and $\Sigma_{k=1}^{\infty} 1/k^2$. For each series $a_{k+1}/a_k \to 1$ as $k \to +\infty$, but the first diverges and the second converges. Hence if $L = 1$, anything can happen. ■

Warning: Be sure you form the ratio in the correct order. Replace k everywhere in a_k by $k + 1$ and then divide by a_k.

Example 4 Test the series $\displaystyle\sum_{k=1}^{\infty} \frac{2^k(k+1)}{k!}$.

Solution

$$\frac{a_{k+1}}{a_k} = \frac{2^{k+1}[(k+1)+1]}{(k+1)!} \cdot \frac{k!}{2^k(k+1)}$$

$$= \frac{2^{k+1}(k+2)k!}{2^k(k+1)k!(k+1)} = \frac{2(k+2)}{(k+1)^2}$$

Thus

$$\lim_{k\to+\infty} \frac{a_{k+1}}{a_k} = \lim_{k\to+\infty} \frac{2(k+2)}{(k+1)^2} = 0 < 1$$

and so the series converges. ●

Example 5 Test the series $\displaystyle\sum_{k=1}^{\infty} \frac{k^k}{k!}$.

Solution

$$\frac{a_{k+1}}{a_k} = \frac{(k+1)^{k+1}}{(k+1)!} \cdot \frac{k!}{k^k}$$

$$= \frac{(k+1)(k+1)^k}{k!(k+1)} \cdot \frac{k!}{k^k}$$

$$= \frac{(k+1)^k}{k^k} = \left(\frac{k+1}{k}\right)^k = \left(1 + \frac{1}{k}\right)^k$$

Then $\lim_{k\to+\infty} a_{k+1}/a_k = e$ by [15.1.6(*e*)], so the series diverges. ●

Knowing *how* to use a particular test is not as easy as knowing *when* to use it. As with techniques of integration, for some examples several different tests can be applied successfully. However, it can be shown that if a_k involves only quotients of polynomials, roots of polynomials, or logarithms of polynomials, then the Ratio Test will yield 1 and hence give no information. Thus, for example, the Ratio Test fails when applied to any of the following:

$$\sum_{k=1}^{\infty} \frac{k^2+k+1}{4k^5+k^2} \qquad \sum_{k=1}^{\infty} \frac{\sqrt{k+1}}{3k+9} \qquad \sum_{n=1}^{\infty} \frac{\ln(n+1)}{\sqrt{n^2+4n}}$$

As a rule of thumb (as always there are exceptions), the Ratio Test can best be applied to series involving factorial or factorial-like terms and/or terms of the form a^k.

15.3.9 **The Root Test** Suppose $\Sigma_{k=1}^{\infty} a_k$ is a positive-term series and $\lim_{k\to+\infty} (a_k)^{1/k} = L$ or $\lim_{k\to\infty} (a_k)^{1/k} = +\infty = L$.

(*a*) If $L < 1$, then $\Sigma_{k=1}^{\infty} a_k$ converges.
(*b*) If $L > 1$, or $L = +\infty$, then $\Sigma_{k=1}^{\infty} a_k$ diverges.
(*c*) If $L = 1$, then the test is inconclusive and $\Sigma_{k=1}^{\infty} a_k$ may either converge or diverge.

The proof, involving a comparison with geometric series, resembles that of the Ratio Test and we leave it as Exercises 57 to 59. The Root Test is particularly useful if the kth term of the series $\Sigma_{k=1}^{\infty} a_k$ is itself a kth power.

Example 6 Test the series $\displaystyle\sum_{k=1}^{\infty} \left(\frac{k}{3k+2}\right)^k$.

Solution

$$a_k^{1/k} = \frac{k}{3k+2} \to \frac{1}{3} < 1 \qquad \text{as } k \to +\infty,$$

so the series $\displaystyle\sum_{k=1}^{\infty} \left(\frac{k}{3k+2}\right)^k$ converges. ●

It should be clear by now that your ability to use successfully the various series tests is directly related to your ability to compute limits of sequences. When working the exercises below, refer whenever necessary to the special limits developed in Sec. 15.1, especially those in (15.1.6), page 482.

PROGRESS TEST 3

Test each of the following series for convergence or divergence:

1. $\displaystyle\sum_{k=1}^{\infty} \frac{2^k}{k^k}$ **2.** $\displaystyle\sum_{k=1}^{\infty} \frac{k!}{3 \cdot 6 \cdot 9 \cdot \cdots \cdot 3k}$ **3.** $\displaystyle\sum_{n=1}^{\infty} \frac{2^n(n+1)}{(n-1)!\,3^n}$ **4.** $\displaystyle\sum_{n=1}^{\infty} \left(\frac{n}{n+2}\right)^n$

SECTION 15.3 EXERCISES

In Exercises 1 to 54 test the given series for convergence or divergence.

1. $\displaystyle\sum_{k=1}^{\infty} \frac{k+1}{k3^k}$ **2.** $\displaystyle\sum_{k=1}^{\infty} \frac{k2^{-k}}{k^2+k+1}$ **3.** $\displaystyle\sum_{k=1}^{\infty} \frac{5}{k^2+9}$

4. $\displaystyle\sum_{k=3}^{\infty} \frac{(\ln k)^{-1/3}}{k}$ **5.** $\displaystyle\sum_{k=2}^{\infty} \frac{1}{k(\ln k)^2}$ **6.** $\displaystyle\sum_{k=1}^{\infty} \frac{\ln k}{k}$

7. $\displaystyle\sum_{k=1}^{\infty} \frac{\sin(\pi/2k)}{k^2}$ **8.** $\displaystyle\sum_{k=1}^{\infty} \tan\left(\frac{k\pi}{2k+1}\right)$ **9.** $\displaystyle\sum_{k=1}^{\infty} k^4 e^{-k^5}$

10. $\displaystyle\sum_{k=2}^{\infty} k^{-2}(\ln k)^{-1/2}$ **11.** $\displaystyle\sum_{k=1}^{\infty} \frac{(5000)^k}{k!}$ **12.** $\displaystyle\sum_{k=1}^{\infty} \frac{2^k}{2!}$

13. $\displaystyle\sum_{k=2}^{\infty} \frac{1}{\sqrt{k^2-1}}$ **14.** $\displaystyle\sum_{k=1}^{\infty} \frac{(k!)^2}{(k^2)!}$ **15.** $\displaystyle\sum_{k=1}^{\infty} \frac{1}{\sqrt{k!}}$

16. $\displaystyle\sum_{k=1}^{\infty} \frac{k!}{3 \cdot 5 \cdot 7 \cdot \cdots \cdot (2k+1)}$ **17.** $\displaystyle\sum_{k=1}^{\infty} \frac{3k+1}{2^k}$ **18.** $\displaystyle\sum_{k=1}^{\infty} \frac{k}{(2k)!}$

19. $\sum_{k=1}^{\infty} \frac{1}{k^{(k+1)/k}}$

20. $\sum_{k=1}^{\infty} \frac{1 \cdot 3 \cdot 5 \cdot \dots \cdot (2k-1)}{2 \cdot 4 \cdot 6 \cdot \dots \cdot (2k)}$

21. $\sum_{k=1}^{\infty} \frac{\sqrt{k+1}}{k^{5/2}+k^2}$

22. $\sum_{k=1}^{\infty} \frac{k^2}{3k^3+4}$

23. $\sum_{k=1}^{\infty} \frac{1 \cdot 4 \cdot 7 \cdot \dots \cdot (3k-2)}{1 \cdot 5 \cdot 9 \cdot \dots \cdot (4k-3)}$

24. $\sum_{k=1}^{\infty} \frac{\ln k}{k^{2/3}}$

25. $\sum_{k=1}^{\infty} k^2 \sin(1/k)$

26. $\sum_{k=1}^{\infty} \frac{k^2+k+1}{\sqrt{2k+5}}$

27. $\sum_{k=1}^{\infty} k/2^{k^2}$

28. $\sum_{k=1}^{\infty} k/\pi^k$

29. $\sum_{k=1}^{\infty} \frac{\sin(1/k)}{k}$

30. $\sum_{k=2}^{\infty} \frac{1}{k(\ln k)^{4/3}}$

31. $\sum_{k=1}^{\infty} \frac{1 \cdot 3 \cdot 5 \cdot \dots \cdot (2k-1)}{k!}$

32. $\sum_{k=1}^{\infty} \frac{k}{4k^3+1}$

33. $\sum_{n=1}^{\infty} \frac{n^2+n+1}{4n^5+n^3+3}$

34. $\sum_{n=1}^{\infty} \frac{1}{n^n}$

35. $\sum_{j=2}^{\infty} \frac{1}{\ln j \ln(j+1)}$

36. $\sum_{i=1}^{\infty} i \sin(1/i)$

37. $\sum_{i=1}^{\infty} \frac{e^{1/i}}{i^2}$

38. $\sum_{j=1}^{\infty} \frac{2^j(j^2+1)}{(j-1)!}$

39. $\sum_{m=1}^{\infty} \frac{3^m a^m}{m!}$ (a a constant)

40. $\sum_{n=1}^{\infty} \frac{\ln n}{n^{1.01}}$

41. $\sum_{n=1}^{\infty} \frac{n \arctan n}{n+1}$

42. $\sum_{n=1}^{\infty} \frac{\arcsin(1/n)}{\sqrt{n^2+1}}$

43. $\sum_{n=1}^{\infty} \frac{\sqrt{n^3+1}}{\sqrt[3]{4n^8+n^2}}$

44. $\sum_{n=1}^{\infty} \frac{n+1}{3n+2}$

45. $\sum_{n=1}^{\infty} \left(\frac{n}{2n^2+1}\right)^n$

46. $\sum_{n=1}^{\infty} \frac{2^n(n!)^2}{(3n)!}$

47. $\sum_{n=1}^{\infty} \frac{3}{2+5\ln n}$

48. $\sum_{k=1}^{\infty} \left(1+\frac{4}{k}\right)^k$

49. $\sum_{m=1}^{\infty} \left|\sin\left(\frac{m+1}{m^2+4}\right)\right|^m$

50. $\sum_{n=1}^{\infty} \frac{(\ln n)2^n}{5^n}$

51. $\sum_{k=1}^{\infty} \frac{\ln k}{k^2}$

52. $\sum_{k=1}^{\infty} \frac{1+\sqrt{k+1}}{2^k}$

53. $\sum_{k=2}^{\infty} \frac{1 \cdot 3 \cdot 5 \cdot \dots \cdot (2k-3)}{2 \cdot 4 \cdot 6 \cdot \dots \cdot (2k-2)} \left(\frac{2}{3}\right)^k$

54. $\sum_{k=1}^{\infty} \frac{1}{3^k - \cos k}$

55. Use the Integral Test on $\Sigma_{k=1}^{\infty} 1/k^4$ to determine an upper bound on the error involved in using the first nine terms to approximate the sum. (*Hint:* Compare

$$\sum_{k=1}^{\infty} \frac{1}{k^4} - \sum_{k=1}^{10} \frac{1}{k^4} = \sum_{k=11}^{\infty} \frac{1}{k^4}$$

and the appropriate integral.)

56. Repeat Exercise 55 for the first 10 terms of $\Sigma_{k=2}^{\infty} 1/k^3$.

57. Prove [15.3.9(*a*)].

58. Prove [15.3.9(*b*)].

59. Prove [15.3.9(*c*)].

Section 15.4: Series with Alternating Positive and Negative Terms

Objectives:

1. Determine convergence of series of alternating positive and negative terms.

2. Determine convergence or divergence of $\Sigma_{k=1}^{\infty} |a_k|$.

3. Approximate to any given accuracy the sum of a convergent series of alternating positive and negative terms.

CONVERGENCE OF SERIES WITH NEGATIVE TERMS

Except for the kth-Term Divergence Test, we have yet to develop tests for convergence or divergence of series with negative terms. However, given any series $\Sigma_{k=1}^{\infty} a_k$, we can take the absolute value of its terms to get the positive-term series $\Sigma_{k=1}^{\infty} |a_k|$, to which the tests of the last section apply.

15.4.1 Definition If $\Sigma_{k=1}^{\infty} |a_k|$ converges, we say that the series $\Sigma_{k=1}^{\infty} a_k$ is *absolutely convergent.*

Example 1 Show that the following series are absolutely convergent:

(*a*) $\displaystyle\sum_{k=1}^{\infty} \frac{\sin k}{k^2}$ (*b*) $\displaystyle\sum_{k=1}^{\infty} \frac{(-1)^k(k+1)}{2^k}$

Solutions (*a*) Since $|\sin k| \leq 1$ for all k, we have $|(\sin k)/k^2| \leq 1/k^2$. But $\Sigma_{k=1}^{\infty} 1/k^2$ is a convergent p-series. Thus the positive-term series $\Sigma_{k=1}^{\infty} |(\sin k)/k^2|$ converges by the Comparison Test, and so the original series $\Sigma_{k=1}^{\infty} (\sin k)/k^2$ is absolutely convergent.

(*b*) $|a_k| = \left|\dfrac{(-1)^k(k+1)}{2^k}\right| = \dfrac{k+1}{2^k}$. Thus

$$\left|\frac{a_{k+1}}{a_k}\right| = \frac{k+2}{2^{k+1}} \cdot \frac{2^k}{k} = \frac{k+2}{2k} \to \frac{1}{2} < 1$$

as $k \to +\infty$. Hence

$$\sum_{k=1}^{\infty} \left|\frac{(-1)^k(k+1)}{2^k}\right| = \sum_{k=1}^{\infty} \frac{k+1}{2^k}$$

converges by the Ratio Test, so the original series is absolutely convergent. ●

15.4.2 Theorem If $\Sigma_{k=1}^{\infty} |a_k|$ converges, then $\Sigma_{k=1}^{\infty} a_k$ converges. That is, *an absolutely convergent series is convergent.*

Proof: Assume that $\Sigma_{k=1}^{\infty} |a_k|$ converges. Now

$$-|a_k| \leq a_k \text{ implies } 0 \leq |a_k| + a_k$$

But $a_k \leq |a_k|$ in turn implies

(1) $$|a_k| + a_k \leq |a_k| + |a_k| = 2|a_k|$$

Since $\Sigma_{k=1}^{\infty} |a_k|$ converges, we know that $\Sigma_{k=1}^{\infty} 2|a_k|$ converges. Hence, by (1) and the Comparison Test, the series of nonnegative terms $\Sigma_{k=1}^{\infty} (|a_k| + a_k)$ converges. Finally, since $a_k = (a_k + |a_k|) - |a_k|$, and since $\Sigma_{k=1}^{\infty} (a_k + |a_k|) - |a_k|$ is the difference of convergent series, and hence converges by [15.2.3(*a*)], we can conclude that $\Sigma_{k=1}^{\infty} a_k$ converges. ■

By (15.4.2), our work in Example 1 showed that the two given series are convergent.

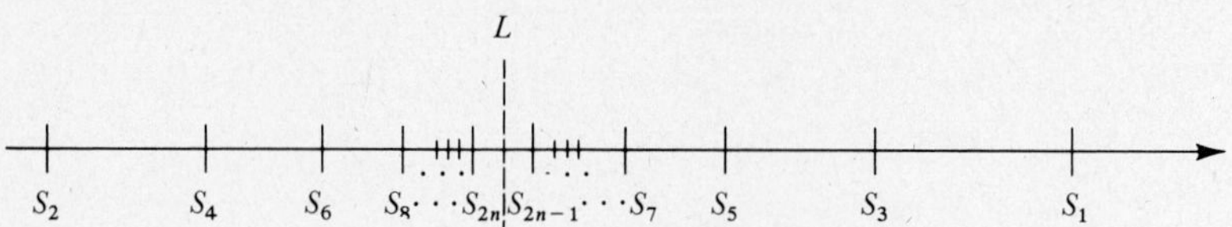

Fig. 15.2

15.4.3 **Definition** A series whose successive terms alternate in sign is called an *alternating series.* Such a series is of the form $\Sigma_{k=1}^{\infty} (-1)^k a_k$ or $\Sigma_{k=1}^{\infty} (-1)^{k+1} a_k$ where $a_k > 0$ for all k. (In the first instance the initial term is negative and in the second instance, positive.)

The following test is our main tool in dealing with convergence of alternating series. It expresses the notion that a certain type of alternating series may converge because its positive and negative terms partly cancel one another enough so that the sequence of partial sums converges.

15.4.4 **Alternating-Series Test** Suppose that for an alternating series $\Sigma_{k=1}^{\infty} (-1)^{k+1} a_k$ both

$$\lim_{k\to+\infty} a_k = 0 \qquad \text{and } a_k > a_{k+1} \qquad \text{for all } k$$

Then $\Sigma_{k=1}^{\infty} (-1)^{k+1} a_k$ converges.

The essence of the justification is to be found in Fig. 15.2, where some partial sums are plotted. At each stage as n increases, a term that is smaller in absolute value than the previous term is added or subtracted from the previous partial sum, so the partial sums jump back and forth, taking shorter jumps each time. The length of the jumps approaches 0 as $n \to +\infty$. As a result, all the odd partial sums are on one side of L, converging to L, and the even sums are on the opposite side, likewise converging to L. We leave the algebraic details to more advanced courses.

The Alternating Series Test likewise applies to alternating series whose first term is negative as well as to series for which its hypotheses hold for all $k \geq j$ for a fixed integer j.

Example 2 Show that $\displaystyle\sum_{k=1}^{\infty} \frac{(-1)^{k+1}}{k}$ converges.

Solution The hypotheses of the Alternating-Series Test obviously hold with $a_k = 1/k$, so $\Sigma_{k=1}^{\infty} (-1)^{k+1}/k$ converges. ●

Notice that

$$\sum_{k=1}^{\infty} \left| \frac{(-1)^{k+1}}{k} \right| = \sum_{k=1}^{\infty} \frac{1}{k}$$

which is the harmonic series and is divergent. Thus Example 2 provides an example of a series that converges but is not absolutely convergent, showing that *the converse of (15.4.2) is false.* The situation in Example 2 occurs often enough to merit a definition.

15.4.5 Definition If $\Sigma_{k=1}^{\infty} a_k$ converges but $\Sigma_{k=1}^{\infty} |a_k|$ diverges, we say that $\Sigma_{k=1}^{\infty} a_k$ is *conditionally convergent*.

Thus

$$\sum_{k=1}^{\infty} \frac{(-1)^{k+1}}{k}$$

is conditionally convergent.

PROGRESS TEST 1

For each of the following, determine whether the given series is absolutely convergent, conditionally convergent, or divergent.

1. $\displaystyle\sum_{k=1}^{\infty} \frac{(-1)^n(k+2)}{3k+5}$

2. $\displaystyle\sum_{k=1}^{\infty} \frac{(-1)^{k+1}\ln k}{k^{10/9}}$ (*Hint:* Use Limit Comparison Test for absolute convergence, with $b_k = 1/k^{19/18}$.)

3. $\displaystyle\sum_{m=1}^{\infty} \frac{(-1)^m \ln m}{m}$ (See Sec. 15.3, Progress Test 1, Prob. 3, page 498.)

APPROXIMATING ALTERNATING SERIES WITH PARTIAL SUMS

If $\Sigma_{k=1}^{\infty} (-1)^{k+1} a_k$ satisfies the hypotheses of (15.4.4)—and thus converges to, say, L—then, as noted earlier, the odd partial sums decrease toward L and the even partial sums increase toward L. Thus for n even, $s_n < L < s_{n+1}$, and for n odd, $s_{n+1} < L < s_n$. In either case,

$$|L - s_n| < |s_{n+1} - s_n| = |(-1)^{n+2}a_{n+1}| = a_{n+1}$$

This provides us with the following useful error estimate.

15.4.6 First-Term-Neglected Estimate for Alternating Series Suppose that $\Sigma_{k=1}^{\infty} (-1)^{k+1}a_k = L$ with $\lim_{k\to+\infty} a_k = 0$ and $a_k > a_{k+1}$ for all k. Then

$$|L - \sum_{k=1}^{n} (-1)^{k+1}a_k| \le a_{n+1}$$

That is, the sum of the first n terms is within a_{n+1} (the first term neglected) of the actual sum of the series.

Example 3 How many terms of $\Sigma_{k=1}^{\infty} (-1)^{k+1}/k!$ should be added to obtain an approximation of the sum accurate through three decimal places? (The error must be less than 0.0005.)

Solution Since

$$\frac{a_{k+1}}{a_k} = \frac{k!}{(k+1)!} = \frac{1}{k+1} \to 0 < 1$$

as $k \to +\infty$, we know by the Ratio Test that the given series is absolutely convergent.

By (15.4.6), the absolute value of the error in using the first n terms is bounded by $a_{n+1} = 1/(n+1)!$. Hence we need n large enough so that

$$\frac{1}{(n+1)!} < 0.0005$$

For $n = 5$,

$$\frac{1}{(n+1)!} = \frac{1}{720} \approx 0.001389 > 0.0005$$

For $n = 6$,

$$\frac{1}{(n+1)!} = \frac{1}{5040} \approx 0.000198 < 0.0005$$

Hence the first six terms of the series will yield sufficient accuracy. ●

PROGRESS TEST 2

1. How many terms of the series $\Sigma_{k=1}^{\infty} (-1)^{k+1}/4^{2k+1}$ are required to ensure an approximation of the sum of the series within 10^{-4}?

A GENERAL PROCEDURE FOR CONVERGENCE TESTING

When confronted with a series $\Sigma_{k=1}^{\infty} a_k$ whose convergence or divergence is not obvious by inspection, the following three-step procedure may prove useful in organizing your approach to the problem.

1. Try the kth-Term Divergence Test. If $\lim_{k\to+\infty} a_k = 0$, then:
2. Apply an appropriate positive term series test to $\Sigma_{k=1}^{\infty} |a_k|$. If $\Sigma_{k=1}^{\infty} |a_k|$ converges, you have finished since $\Sigma_{k=1}^{\infty} a_k$ converges absolutely. If $\Sigma_{k=1}^{\infty} |a_k|$ diverges and if $\Sigma_{k=1}^{\infty} a_k$ is an alternating series, then:
3. Apply the Alternating-Series Test to determine whether $\Sigma_{k=1}^{\infty} a_k$ is conditionally convergent.

Series exist (but not in this text) that do not yield to any of our tests. For example, if $\Sigma_{k=1}^{\infty} |a_k|$ diverges but $\Sigma_{k=1}^{\infty} a_k$ is not alternating (but contains negative terms a_k for $k > j$ for any fixed j), then step 3 does not apply. Such cases will yield to approaches developed in more advanced courses.

A collection of positive-term series to test for convergence or divergence in the absence of context is provided in Miscellaneous Exercises 1 to 34.

SECTION 15.4 EXERCISES

In Exercises 1 to 11, (a) show that the given alternating series is convergent, and (b) determine the error involved in using the first five terms as an approximation to the series.

1. $\displaystyle\sum_{k=1}^{\infty} \frac{(-1)^{k+1}k}{k^2+1}$

2. $\displaystyle\sum_{k=1}^{\infty} \frac{(-1)^{k+1}\log_{10} k}{k}$

3. $\displaystyle\sum_{k=1}^{\infty} \frac{(-1)^{k+1}}{\ln(k+1)}$

4. $\displaystyle\sum_{k=1}^{\infty} \frac{(-1)^{k+1}}{\sqrt{k+1}}$

5. $\displaystyle\sum_{k=1}^{\infty} \frac{(-1)^{k+1}}{k!2^k}$

6. $\displaystyle\sum_{k=1}^{\infty} \frac{(-1)^{k+1}k}{3^k}$

7. $\displaystyle\sum_{k=1}^{\infty} \frac{(-1)^{k+1}}{k^2}$

8. $\displaystyle\sum_{k=1}^{\infty} \frac{(-1)^{k}2^{2k-2}}{2^2 \cdot 4^2 \cdot 6^2 \cdot \cdots \cdot (2^k-2)^2}$

9. $\displaystyle\sum_{k=1}^{\infty} \frac{k+1}{k}\left(\frac{-1}{2}\right)^k$

10. $\displaystyle\sum_{k=3}^{\infty} (-1)^{k+1}\frac{3^k}{k!}$

11. $\displaystyle\sum_{k=3}^{\infty} \frac{(-1)^{k+1}}{\ln(2k-3)}$

In Exercises 12 to 39 determine whether the given series converges absolutely, converges conditionally, or diverges.

12. $\sum_{k=1}^{\infty} \frac{(-1)^{k+1}}{\sqrt[3]{k}}$

13. $\sum_{k=1}^{\infty} (-1/n)^3$ (Be careful!)

14. $\sum_{k=1}^{\infty} \frac{(-1)^k}{\sqrt{k^3+1}}$

15. $\sum_{k=2}^{\infty} (-1)^k \frac{2k+1}{k-1}$

16. $\sum_{k=2}^{\infty} \frac{(-1)^{k+1}}{k \ln k}$

17. $\sum_{k=1}^{\infty} \frac{(-1)^k k}{\ln(k+1)}$

18. $\sum_{n=1}^{\infty} \frac{(-1)^n n}{2n+1}$

19. $\sum_{n=1}^{\infty} \frac{(-1)^n n}{e^n}$

20. $\sum_{k=2}^{\infty} \frac{(-1)^k}{k(\ln k)^2}$

21. $\sum_{m=1}^{\infty} \frac{(-1)^{m+1}(3m^2+1)}{m^3+1}$

22. $\sum_{j=1}^{\infty} \frac{(-1)^j(j+1)}{\ln(2^j)}$

23. $\sum_{n=1}^{\infty} \frac{(-1)^n \sqrt{n}}{n^3+9n}$

24. $\sum_{n=1}^{\infty} \frac{(-1)^{n+1}(n-2)}{\ln(n+1)}$

25. $\sum_{n=1}^{\infty} \frac{(-1)^n n!}{n^n}$

26. $\sum_{n=1}^{\infty} \frac{(-1)^{n+1} n^2}{2^n}$

27. $\sum_{n=1}^{\infty} \frac{(-1)^n (3n+1)}{\sqrt{n^2+5}}$

28. $\sum_{n=1}^{\infty} \frac{\sin n}{n^2}$

29. $\sum_{n=1}^{\infty} \frac{(-1)^{n+1} \cdot 2 \cdot 4 \cdot \cdots \cdot (2n)}{3 \cdot 5 \cdot \cdots \cdot (2n+1)}$

30. $\sum_{n=1}^{\infty} \frac{(-1)^n 4^n}{3^{n+2}}$

31. $\sum_{n=1}^{\infty} \frac{(-1)^{n+1} \arctan n}{n}$

32. $\sum_{k=1}^{\infty} \sin\left(\frac{k\pi}{4}\right)$

33. $\sum_{n=1}^{\infty} \frac{(-1)^{n+1} 2^n}{(n^2+1)(n+1)!}$

34. $\sum_{k=1}^{\infty} \sin\left(\frac{\pi}{4k^2}\right)$

35. $\sum_{n=1}^{\infty} \frac{(-1)^{n+1} n!}{3 \cdot 5 \cdot \cdots \cdot (2n+1)}$

36. $\sum_{k=2}^{\infty} \frac{(-1)^k}{\ln(\ln k)}$

37. $\sum_{k=1}^{\infty} (-1)^k \left(\frac{\pi}{2} - \arctan k\right)$

38. $\sum_{k=1}^{\infty} (-1)^k \sin\left(\frac{1}{k}\right)$

39. $\sum_{k=1}^{\infty} (-1)^k \left[1 - \cos\left(\frac{1}{k}\right)\right]$

40. By rearranging terms we can make the sum of a conditionally convergent series any number we wish. By following the given outline, show that $\Sigma_{k=1}^{\infty} (-1)^{k+1}/k$ can be rearranged to yield 2 as its sum.
(*a*) Separate the given series into two (divergent) series $\mathcal{O}$ and $\mathcal{E}$, one consisting of the odd terms and the other of the even terms.
(*b*) Choose just enough terms from $\mathcal{O}$ so that the sum exceeds 2.
(*c*) Add just enough terms from $\mathcal{E}$ to bring the sum under 2.
(*d*) Add just enough terms from $\mathcal{O}$ to exceed 2 again.
(*e*) Show that a series constructed in this way converges to 2.

41. Show by the procedure given in Exercise 40 that $\Sigma_{k=1}^{\infty} (-1)^{k+1}/3k$ can be rearranged to yield a sum of -1.8.

42. If $\Sigma_{k=1}^{\infty} a_k$ is absolutely convergent and we form another series, each of whose partial sums $\overline{S}_n$ is a rearrangement of the original partial sum S_n, does $\lim_{n\to+\infty} \overline{S}_n = \Sigma_{k=1}^{\infty} a_k$? State your result as a theorem.

43. Give an example of a convergent series $\Sigma_{k=1}^{\infty} a_k$ such that $\Sigma_{k=1}^{\infty} \sqrt{a_k}$ diverges ($a_k > 0$ for all k).

44. Give an example of a divergent series $\Sigma_{k=1}^{\infty} a_k$ such that $\Sigma_{k=1}^{\infty} a_k^{\,4}$ converges.

45. (Note Exercises 43 and 44.) Prove that if $\Sigma_{k=1}^{\infty} a_k$ converges absolutely, then for any $p \geq 1$, $\Sigma_{k=1}^{\infty} a_k^{\,p}$ converges.

46. Suppose that $S = 1 - 1/2 + 1/3 - 1/4 + 1/5 - 1/6 + 1/7 - \cdots$ and therefore $(1/2)S = 1/2 - 1/4 + 1/6 - 1/8 + \cdots$. Then $S + (1/2)S = 1 + 1/3 - 1/2 + 1/5 + 1/7 - 1/4 + \cdots = S$ (with its terms rearranged). But $S + (1/2)S = S$ implies $S = 0$, whereas $S = (1 - 1/2) + (1/3 - 1/4) + (1/5 - 1/6) + \cdots > 0$. Where is the flaw that leads to the contradiction?

CHAPTER EXERCISES

Review Exercises

In Exercises 1 to 8 determine a_1, a_2, a_3, a_4, for each of the given sequences.

1. $a_n = n \sin(\pi/n)$

2. $a_n = (-1)^{n+1}(n+1)^{1/n}$

3. $a_n = \dfrac{2n+1}{4n-3}$

4. $a_n = \dfrac{2 \cdot 4 \cdot \cdots \cdot (2n)}{3n+5}$

5. $a_n = \dfrac{(-1)^n n!}{2^{n+1}}$

6. $a_n = \sqrt{\dfrac{n+1}{2n-1}}$

7. $a_n = \begin{cases} 1/2^n & n \text{ even} \\ (n+1)/(n+3) & n \text{ odd} \end{cases}$

8. $a_n = 3 + (-1)^n 3$

In Exercises 9 to 18 determine a general element a_n for the given sequence.

9. $1, 1/2, 1, 1/4, 1, 1/6, \ldots$

10. $2/5, 5/2, 4/7, 7/4, \ldots$

11. $1, -1/4, 1/16, -1/64, \ldots$

12. $1, 1 - 1/2, 1 - 1/2 + 1/4, 1 - 1/2 + 1/4 - 1/8, \ldots$

13. $3 \cdot 7, -5 \cdot 9, 7 \cdot 11, -9 \cdot 13, \ldots$

14. $\dfrac{1}{2}, -\dfrac{1 \cdot 3}{2 \cdot 4}, \dfrac{1 \cdot 3 \cdot 5}{2 \cdot 4 \cdot 6}, -\dfrac{1 \cdot 3 \cdot 5 \cdot 7}{2 \cdot 4 \cdot 6 \cdot 8}, \ldots$

15. $-1/3, 1/27, -1/243, 1/2187, \ldots$

16. $1/2, 1/(2\sqrt{5}), 1/(3\sqrt{6}), 1/(4\sqrt{7}), \ldots$

17. $1/3, 1/3, 3/11, 1/5, 5/37, 3/35, \ldots$ (replace canceled factors)

18. $1, 1/4, 1/27, 1/256, \ldots$

In Exercises 19 to 24 determine whether the given sequence is monotonic or bounded in any way.

19. $a_n = n^2 e^{-n}, n = 2, 3, \ldots$

20. $a_n = (-1)^n \sqrt{n+1}$

21. $a_n = \dfrac{4n}{n^2+1}$

22. $a_n = \dfrac{\ln n}{n^2} n = 3, 4, 5, \ldots$

23. $a_n = |1 - n^2|$

24. $a_n = (2^n + 3^n)^{1/n}$

In Exercises 25 to 62 determine the limit, if it exists, of the given sequence.

25. $a_n = \dfrac{2n+1}{4n-2}$

26. $a_n = \dfrac{\ln n}{\sqrt[5]{n+5}}$

27. $a_n = \dfrac{3n^2+n+1}{n+2}$

28. $a_n = \dfrac{n^2+2n+1}{e^n}$

29. $a_n = \dfrac{2n^2+2n+4}{3n^3+n^2+1}$

30. $a_n = (-1)^n \dfrac{7^n}{10^{n+2}}$

31. $a_n = \dfrac{\sqrt{2n+1}}{\sqrt[3]{n^2+n}}$

32. $a_n = (n+3)^{1/n}$

33. $a_n = \dfrac{\sqrt[3]{4n^3+5}}{3n+1}$

34. $a_n = (-1)^n(\pi/4)^{n+1}$

35. $a_n = \left(\dfrac{3+n}{n}\right)^n$

36. $a_n = \sqrt{\dfrac{3n+5}{2+12n}}$

37. $a_n = (1 - 5/n)^n$

38. $a_n = \begin{cases} (-1)^n/2n & n \text{ even} \\ (\ln n)/n & n \text{ odd} \end{cases}$

39. $a_n = \left(\dfrac{n}{n-2}\right)^n$

40. $a_n = \begin{cases} 1/2 & n \text{ even} \\ n/\sqrt{4n^2+1} & n \text{ odd} \end{cases}$

41. $a_n = \left(\dfrac{n+1}{n+3}\right)^{n+1}$

42. $a_n = \dfrac{\ln n}{e^n}$

43. $a_n = \dfrac{(\ln n)^3}{n}$

44. $a_n = \left(\dfrac{n+1}{n-1}\right)^n$

45. $a_n = n^2 - \sqrt{n^4 - n^2 + 2}$

46. $a_n = n \sin(1/n)$

47. $a_n = \dfrac{(-1)^n 3^n}{n!}$

48. $a_n = \ln\left(\dfrac{n^2+n-1}{n+3}\right)$

49. $a_n = (n+1)^{1/n}$

50. $a_n = \dfrac{\sin n}{n}$

51. $a_n = (1/n)^{n^2}$

52. $a_n = \sin(n)$

53. $a_n = \dfrac{(-1)^n 3^{n+4}}{5^{n-2}}$

54. $a_n = (e^n + n)^{2/n}$

55. $a_n = \left(1 + \dfrac{1}{2n}\right)^{n^2}$

56. $a_n = \left(\dfrac{n+1}{n+3}\right)^{n+1}$

57. $a_n = \sqrt{n} - \sqrt{n+2}$

58. $a_n = \dfrac{n^2 + n + 1}{e^n}$

59. $a_n = \dfrac{(-1)^n 4^n}{(1.1)^n n!}$

60. $a_n = (-1)^n \sqrt{n}$

61. $a_n = \dfrac{3n^2 + n - 6}{\sqrt{4n + 1}}$

62. $a_n = \dfrac{\sqrt{n^2 + 2n + 1}}{\sqrt[3]{8n^3 + n^2 + 1}}$

For those of the following series in Exercises 63 to 77 that converge, determine the sum.

63. $\sum_{k=1}^{\infty} \dfrac{5}{3^{k-1}}$

64. $\sum_{k=0}^{\infty} \dfrac{4}{3^k}$

65. $\sum_{k=1}^{\infty} \dfrac{5}{3^{k+2}}$

66. $\sum_{k=1}^{\infty} \dfrac{2^k + 5^{k-1}}{3^{k-1}}$

67. $\sum_{k=1}^{\infty} \dfrac{(-1)^k 3^{k-1} + 4^{k-1}}{5^k}$

68. $\sum_{k=1}^{\infty} \dfrac{e^k}{\pi^{k+1}}$

69. $\sum_{k=1}^{\infty} (-1)^k \dfrac{4^{k+1}}{3^k}$

70. $\sum_{k=3}^{\infty} \dfrac{1}{k^2 - k}$

71. $\sum_{k=2}^{\infty} \dfrac{3}{k^2 - 1}$

72. $\sum_{k=0}^{\infty} \dfrac{(-1)^k 4^k + 2^k}{5^k}$

73. $\sum_{k=1}^{\infty} \dfrac{1}{(k + 7)(k + 8)}$

74. $\sum_{k=2}^{\infty} \dfrac{3}{k^2 + 5k + 6}$

75. $\sum_{n=1}^{\infty} \left[\sin\left(\dfrac{1}{n}\right) - \sin\left(\dfrac{1}{n + 1}\right)\right]$

76. $\sum_{n=1}^{\infty} [\arctan n - \arctan (n + 1)]$

77. $\sum_{k=1}^{\infty} \ln\left(\dfrac{k + 1}{k + 2}\right)$

(See Miscellaneous Exercises 1 to 34 for practice in determining convergence or divergence of positive-term series.)

In Exercises 78 to 93 determine whether the given series is absolutely convergent, conditionally convergent, or divergent.

78. $\sum_{k=1}^{\infty} \dfrac{(-1)^{k+1} 2^k}{(k + 1)!}$

79. $\sum_{k=1}^{\infty} \dfrac{(-1)^k \cdot 3 \cdot 6 \cdot 9 \cdot \cdots \cdot (3k)}{4 \cdot 8 \cdot 12 \cdot \cdots \cdot (4k)}$

80. $\sum_{k=1}^{\infty} \dfrac{(-1)^{k+1}(k + 1)}{3k^2 + 5}$

81. $\sum_{k=1}^{\infty} \sin\left(\dfrac{\pi k + 3}{4 + 2k}\right)$

82. $\sum_{n=1}^{\infty} \dfrac{(-1)^n (n + 1)}{2^n \ln(n + 1)}$

83. $\sum_{n=1}^{\infty} (-1)^{n+1} \ln\left(\dfrac{2n + 5}{4n + 6}\right)$

84. $\sum_{k=1}^{\infty} \dfrac{(-1)^n \sin k}{k^2}$

85. $\sum_{k=1}^{\infty} \dfrac{(-1)^{k+1} 2^k}{(k^2 + 1)k!}$

86. $\sum_{k=1}^{\infty} \dfrac{(-1)^k \sqrt{k}}{2k^2 + 9}$

87. $\sum_{k=1}^{\infty} \dfrac{(-1)^k (3k - 2)}{(1 + 2k)}$

88. $\sum_{k=2}^{\infty} \dfrac{(-1)^k \log_2 k}{\sqrt{k + 1}}$

89. $\sum_{k=1}^{\infty} \dfrac{(-1)^k k^k}{(2k + 1)^k}$

90. $\sum_{k=1}^{\infty} \dfrac{(-1)^{k+1} 2^k}{3^k k}$

91. $\sum_{k=1}^{\infty} \dfrac{(-1)^{k+1}}{(k + 1)!}$

92. $\sum_{k=1}^{\infty} \dfrac{(-1)^{k+1} k}{k^2 + 1}$

93. $\sum_{k=1}^{\infty} \dfrac{(-1)^{k+1} \ln k}{k^2 + 1}$

In Exercises 94 to 97 find the series $\Sigma_{k=1}^{\infty} a_k$ that has as its sequence of partial sums s_n the given sequence.

94. $s_n = \dfrac{n + 2}{n + 3}$

95. $s_n = \dfrac{2n + 1}{3n + 2}$

96. $s_n = \ln(n + 1)$

97. $s_n = \dfrac{n}{6n + 36}$

Miscellaneous Exercises

In Exercises 1 to 34 determine whether the given series converges or diverges, citing the appropriate test(s) and comparison test series where appropriate.

1. $\sum_{k=1}^{\infty} \dfrac{1 + 2k}{4k + 2}$

2. $\sum_{k=1}^{\infty} \dfrac{\sqrt{2k + 5}}{k^2 + k + 1}$

3. $\sum_{k=1}^{\infty} \dfrac{\arctan k}{k^2 + 1}$

4. $\sum_{k=1}^{\infty} \dfrac{\sqrt[5]{k^4 + 1}}{k^3 + k^2 + 1}$

5. $\sum_{k=1}^{\infty} \dfrac{2^{k+1}(k + 1)!}{(k + 1)^{k+1}}$

6. $\sum_{k=1}^{\infty} \dfrac{k \arctan k}{k + 1}$

7. $\sum_{k=1}^{\infty} \left(\dfrac{k}{2k^2 + 1}\right)^k$

8. $\sum_{k=1}^{\infty} \dfrac{1}{2^k \ln (k + 1)}$

9. $\sum_{k=1}^{\infty} \cos(1/k)$

10. $\sum_{k=1}^{\infty} \frac{(k+5)!}{2^k(k+2)!}$

11. $\sum_{k=1}^{\infty} \frac{1}{\ln(k+1)\ln(k+2)}$

12. $\sum_{k=1}^{\infty} \frac{\log_3 k}{k^{14/13}}$

13. $\sum_{k=1}^{\infty} \frac{(k!)^2}{2^k k!}$

14. $\sum_{k=1}^{\infty} \ln\left(\frac{k+2}{k+3}\right)$

15. $\sum_{k=1}^{\infty} \left(\frac{k+1}{k+2}\right)^k$

16. $\sum_{k=1}^{\infty} \frac{4^k(1 \cdot 3 \cdot 5 \cdot \cdots \cdot (2k-1))}{(k+1)!}$

17. $\sum_{k=2}^{\infty} \frac{1}{e^k \ln k}$

18. $\sum_{k=1}^{\infty} \left(\frac{3k+2}{\pi k+4}\right)^k$

19. $\sum_{k=1}^{\infty} \frac{\ln k}{k!}$

20. $\sum_{k=1}^{\infty} \frac{2^k \ln(k+1)}{k \cdot 3^k}$

21. $\sum_{j=1}^{\infty} \frac{3^j(j+2)}{1 \cdot 3 \cdot 5 \cdot \cdots \cdot (2j-1)}$

22. $\sum_{j=1}^{\infty} \frac{2^j j!}{j^j}$

23. $\sum_{n=7}^{\infty} \frac{\ln n}{\sqrt{n+1}}$

24. $\sum_{n=1}^{\infty} \frac{\ln n}{n^{9/8}+1}$

25. $\sum_{k=1}^{\infty} \left(\frac{k+1}{1+3k^2}\right)^k$

26. $\sum_{k=1}^{\infty} k^{-1/2} \arctan k$

27. $\sum_{i=3}^{\infty} \frac{\sqrt{2i^2-1}}{1+i^{5/2}}$

28. $\sum_{i=2}^{\infty} \left(1-\frac{3}{i}\right)^i$

29. $\sum_{n=1}^{\infty} \frac{3^n(2n^2+1)}{(n+1)!}$

30. $\sum_{n=1}^{\infty} \frac{1}{\ln(n+1)\ln(n+2)}$

31. $\sum_{k=1}^{\infty} \frac{|\sin k|}{3k+5}$

32. $\sum_{k=1}^{\infty} \frac{(3k+5)^k}{(7k+6)^k}$

33. $\sum_{k=1}^{\infty} \frac{\sqrt{k^2+2k+1}}{\sqrt[3]{k^6+k^3+1}}$

34. $\sum_{k=1}^{\infty} \frac{2^k(2k+5)}{5^k(k+1)}$

35. Show that $\sum_{k=2}^{\infty} \frac{1}{k(\ln k)^p}$ converges if $p \geq 1$ and diverges if $p < 1$.

36. Show that $\sum_{k=1}^{\infty} \frac{\ln k}{k^p}$ converges if $p > 1$ and diverges if $p \leq 1$.

37. Show that if $\sum_{k=1}^{\infty} a_k$ converges, then $\sum_{k=1}^{\infty} 1/a_k$ diverges.

38. Show that if $\lim_{k\to+\infty} a_n$ exists, then $\sum_{k=2}^{\infty} (a_k - a_{k-1})$ converges.

39. The sequence a_n is said to be *Cauchy* if for every $\varepsilon > 0$, there is a positive integer K such that $n,m \geq K$ implies $|a_n - a_m| < \varepsilon$. Prove that $a_n = 1/n$ is Cauchy. (Choose K so that $n,m \geq K$ implies $1/n < \varepsilon/2$ and $1/m < \varepsilon/2$.)

40. Prove that if a_n converges, then a_n is Cauchy. (See Exercise 39.) (The converse, which is true, is considerably harder to prove, and reflects the fact mentioned earlier in this chapter that every sequence that *should* converge does converge; the real-number line is "complete" and has no holes.)

41. An *infinite product* is an expression $\Pi_{k=1}^{\infty}(1+a_k) = (1+a_1)(1+a_2)(1+a_3)\cdots$ whose convergence is defined as the convergence of its sequence of partial products $p_n = (1+a_1)(1+a_2)(1+a_3)\cdots(1+a_n)$. If $a_k \geq 0$ for all $k \geq 1$, prove that $p_n \geq 1 + a_1 + a_2 + \cdots + a_n$.

42. (See Exercise 41.) For $a_k \geq 0$, prove that the convergence of $\Pi_{k=1}^{\infty}(1+a_k)$ implies the convergence of $\Sigma_{k=1}^{\infty} a_k$.

43. (See Exercise 41.) Prove the converse of Exercise 42. (Consider $\ln p_n$.)

44. (Generalized Ratio Test) Suppose $a_k \geq 0$ for all $k \geq 1$ and that there is a number $R < 1$ and an integer K such that $k \geq K$ implies $a_{k+1}/a_k \leq R$. (Note that $\lim_{k\to+\infty} a_{k+1}/a_k$ need not exist.) Show that $\Sigma_{k=1}^{\infty} a_k$ converges.

45. Apply the result of Exercise 44 to prove that

$$1 + 1/2 + 1/3 + 1/6 + 1/9 + 1/18 + 1/27 + \cdots$$

converges. (Note that the Ratio Test fails.)

46. Let $S_n = \Sigma_{k=1}^{n} 1/k$. Using an upper Riemann sum for $f(x) = 1/x$ based on the integer partition of $[1, +\infty)$ into unit width rectangles, show that $S_n > \ln(n+1)$. (Draw a diagram.)

47. (See Exercise 46.) By "stacking" the rounded-base triangular errors along the vertical axis inside a rectangle of area 1 (as we did in Chapter 6), show that the increasing sequence $a_n = S_n - \ln(n+1)$ is bounded above and hence converges. (It is a transcendental number, approximately equal to 0.577, and is denoted by γ. It is called the *Euler constant.*)

48. Prove that if $\lim_{n\to+\infty} a_n = A$ and $\lim_{n\to+\infty} b_n = B$, then $\lim_{n\to+\infty}(a_n b_n) = AB$. (*Hint:* See the earlier arguments for standard limits.)

SELF-TEST ✚

In Probs. 1 to 5 determine the limit of the given sequence.

1. $a_n = \left(\dfrac{n}{n-3}\right)^n$

2. $a_n = \dfrac{3^n}{2^n\, n!}$

3. $a_n = n^{-b/n}$

4. $a_n = \dfrac{\ln n}{\sqrt[5]{n}}$

5. $a_n = \dfrac{(-1)^n 5^n}{(2\pi)^n}$

In Probs. 6 and 7 determine the convergence or divergence of the given series. If convergent, determine the sum.

6. $\displaystyle\sum_{k=2}^{\infty} \frac{(-1)^k 2^k + 3^{k+1}}{4^{k-1}}$

7. $\displaystyle\sum_{k=1}^{\infty} \frac{1}{k^2 + 7k + 12}$

In Probs. 8 to 11 test the given series for convergence or divergence. State which test(s) you are using.

8. $\displaystyle\sum_{n=1}^{\infty} \frac{\sqrt{2n+1}}{n^2 + 3n + 1}$

9. $\displaystyle\sum_{n=1}^{\infty} \left(\frac{n+3}{n^2 + n + 1}\right)^n$

10. $\displaystyle\sum_{n=1}^{\infty} \frac{2^n n!}{n^n}$

11. $\displaystyle\sum_{n=1}^{\infty} \frac{1}{n(\ln n + 1)^3}$

12. Is the following series absolutely convergent, conditionally convergent, or divergent? Explain.

$$\sum_{k=1}^{\infty} \frac{(-1)^{k+1} k^2}{k^3 + 2k}$$

13. How many terms of the series

$$\sum_{n=1}^{\infty} \frac{(-1)^{n+1}}{5^n n!}$$

should you sum in order to approximate the series to within 10^{-5}?

Chapter 16

POWER SERIES AND TAYLOR SERIES

CONTENTS

The ideas of this chapter are at the heart of calculus. Although they appear rather late in this and most calculus texts, they were developed historically as part of, or even in advance of, many of the fundamental notions appearing in earlier chapters. Power series provide a new way of writing virtually all the functions we have dealt with to date as well as a way of writing new types of functions, "nonelementary functions," which we first saw in Sec. 11.5 as antiderivatives of familiar functions.

Basically, a power series can be thought of as an "infinite-degree polynomial." In Sec. 16.1 we set the stage for power series *representations* of functions by examining genuine polynomial *approximations* of functions. Section 16.2 defines the notion of power series and relates power series to the infinite series of Chap. 15. In Sec. 16.3 we develop conditions under which we can form limits of the polynomial approximations of functions in Sec. 16.1 to achieve *exact* power series representations of those functions. We also study techniques for differentiating, integrating, and otherwise manipulating power series. Section 16.4 illustrates some applications of power series, showing how they yield powerful new approximation and integration techniques.

The facts developed in this chapter are applied almost everywhere calculus is applied, including both concrete applications and applications in further mathematics, especially differential equations.

Section 16.1: Taylor Polynomials

Objectives:

1. Construct Taylor Polynomial approximations for a given function.
2. Determine upper bounds on the error of such approximations.

TAYLOR POLYNOMIALS

Because calculations involving polynomial functions are relatively easy, polynomials are often used to approximate other functions. To evaluate a polynomial we need only add and multiply numbers. On the other hand, how would we evaluate e^x at $x = 0.1$? Use a calculator? A table? The values available from such sources are only approximations; more important, such approximations are usually achieved by using polynomials to approximate the original functions.

In (3.3.5) we used the differential $df = f'(x)\,dx$ to approximate the function $f(x)$. There we approximated $f(x)$ near $(a, f(a))$ by the linear function

$$f(a) + f'(a)(x - a) \approx f(x) \qquad \text{(Recall Fig. 3.5, page 98)}$$

But $f(a) + f'(a)(x - a)$ is merely a first-degree polynomial in x whose graph is the straight line tangent to the graph of $f(x)$ at $(a, f(a))$. For x near a, $f(x)$ is near the tangent line.

We can improve on this approximation by using a *second*-degree polynomial whose graph is not only tangent to that of f at $(a, f(a))$ but also has the same concavity at that point; that is, it "bends" the same way that f does at $x = a$. One such polynomial is

$$P_2(x) = f(a) + f'(a)(x - a) + \frac{f''(a)}{2}(x - a)^2$$

It passes through the point $(a, f(a))$ because

$$P_2(a) = f(a) + 0 + 0 = f(a)$$

Its first derivative

$$P_2'(x) = f'(a) + \frac{f''(a)}{2} \cdot 2 \cdot (x - a)$$

evaluated at $x = a$ is

$$P_2'(a) = f'(a) + f''(a)(a - a) = f'(a)$$

which is the same as that of f. Finally, its concavity, given by

$$P_2''(x) = f''(a)$$

agrees with that of f at $x = a$.

A third-degree polynomial would probably be an even better approximation if *its* derivatives equaled those of f at $x = a$. Then the rate of change of *its* second derivative would be the same as that of f. One such polynomial is

$$P_3(x) = f(a) + f'(a)(x - a) + \frac{f''(a)}{2}(x - a)^2 + \frac{f'''(a)}{3 \cdot 2}(x - a)^3$$

Each of the following derivatives evaluated at $x = a$ equals the respective derivative of f at $x = a$ since all but the appropriate term are zero when $x = a$:

$$P_3'(x) = 0 + f'(a) + \frac{f''(a)}{2} \cdot 2 \cdot (x - a) + \frac{f'''(a)}{3 \cdot 2} \cdot 3(x - a)^2$$
$$P_3''(x) = 0 + 0 + f''(a) + f'''(a)(x - a)$$
$$P_3'''(x) = 0 + 0 + 0 + f'''(a)$$

The same reasoning carried n steps would give us the polynomial

$$P_n(x) = f(a) + \frac{f'(a)}{1!}(x - a) + \frac{f''(a)}{2!}(x - a)^2 + \frac{f'''(a)}{3!}(x - a)^3 + \cdots + \frac{f^{(n)}(a)}{n!}(x - a)^n$$

Note that the nth derivative of $P_n(x)$ is given by

$$P_n^{(n)}(x) = \frac{f^{(n)}(a)}{n!} \cdot n! = f^{(n)}(a) \qquad \text{(Check this!)}$$

16.1.1 **Definition** We say that

$$P_n(x) = f(a) + \frac{f'(a)}{1!}(x - a) + \frac{f''(a)}{2!}(x - a)^2 + \cdots + \frac{f^{(n)}(a)}{n!}(x - a)^n$$

is the *nth Taylor polynomial for f at a.*

(The English mathematician Brook Taylor worked with many of the ideas in this chapter during the early 1700s in the generation after Newton.)

Example 1 Compute $P_4(x)$ at $a = 1$ for $f(x) = \ln x$.

Solution We must first compute the first four derivatives and evaluate each at $a = 1$.

$$f'(x) = \frac{1}{x} \qquad f'(1) = 1$$

$$f''(x) = \frac{-1}{x^2} \qquad f''(1) = -1$$

$$f'''(x) = \frac{2}{x^3} \qquad f'''(1) = 2$$

$$f^{(4)}(x) = \frac{-3 \cdot 2}{x^4} \qquad f^{(4)}(1) = -6$$

Hence

$$\begin{aligned} P_4(x) &= \ln 1 + \frac{1}{1!}(x - 1) + \frac{(-1)}{2!}(x - 1)^2 + \frac{2}{3!}(x - 1)^3 + \frac{(-6)}{4!}(x - 1)^4 \\ &= (x - 1) - \frac{1}{2}(x - 1)^2 + \frac{1}{3}(x - 1)^3 - \frac{1}{4}(x - 1)^4 \quad \bullet \end{aligned}$$

APPROXIMATIONS OF SIN X USING TAYLOR POLYNOMIALS

We shall examine Taylor polynomials for $\sin x$ at $a = 0$ algebraically, graphically, and numerically. In so doing, we shall see how a typical trigonometric table can be generated using Taylor polynomials. But first we need derivatives at 0:

$$\begin{aligned} f(0) &= \sin 0 = 0 \\ f'(0) &= \cos 0 = 1 \\ f''(0) &= -\sin 0 = 0 \\ f'''(0) &= -\cos 0 = -1 \\ f^{(4)} &= \sin 0 = 0 \end{aligned}$$

Obviously, this cycle of four derivatives repeats indefinitely, so in particular, every even derivative of $f(x)$ is 0. Thus, for example:

$$P_5(x) = f(0) + f'(0)(x-0) + \frac{f''(0)}{2!}(x-0)^2 + \frac{f'''(0)}{3!}(x-0)^3 + \frac{f^{(4)}(0)}{4!}(x-0)^4 + \frac{f^{(5)}(0)}{5!}(x-0)^5$$

$$= x - \frac{x^3}{3!} + \frac{x^5}{5!}$$

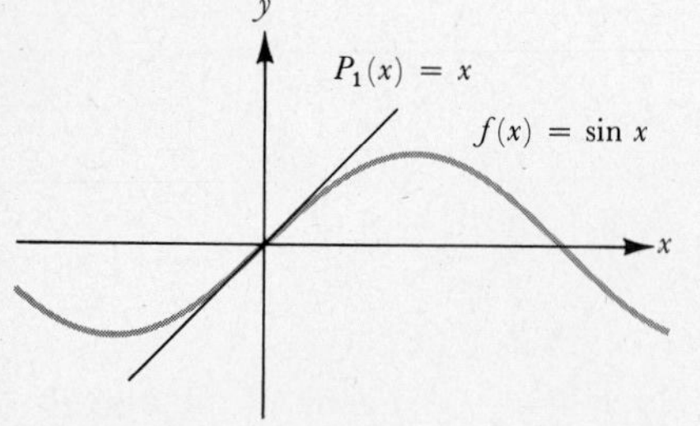

Fig. 16.1

and

$$P_9(x) = x - \frac{x^3}{3!} + \frac{x^5}{5!} - \frac{x^7}{7!} + \frac{x^9}{9!}$$

We dealt with $P_1(x) = x$ before, when examining $\lim_{x\to 0}(\sin x)/x$. We noted then that for small x, $\sin x \approx x$, making $(\sin x)/x \approx 1$. Geometrically this means that the graph of the first Taylor polynomial is close to that of $f(x) = \sin x$ for x near 0; it is the tangent line at $(0, 0)$; see Fig. 16.1.

In Figs. 16.2 and 16.3 we give sketches of

$$P_3(x) = x - \frac{x^3}{6} \qquad \text{and} \qquad P_5(x) = x - \frac{x^3}{6} + \frac{x^5}{120}$$

respectively. Between $-\pi/2$ and $+\pi/2$ we exaggerated the difference between the approximation and $f(x) = \sin x$ so that the graphs could be visibly distinguished. Even so, the increasing accuracy of $P_n(x)$ as n increases is quite evident. Further information regarding accuracy is provided by a close inspection of Table 16.1, where all values are rounded off in the sixth decimal place. The R_5 column on the right gives $P_5(x) - \sin x$, that is, the error involved in using $P_5(x)$ as an approximation of $\sin x$.

For x close to 0—say, x in $[0, 0.1]$—all the approximations are quite good. However, as reflected in the graphs, for larger x, higher degree polynomials are

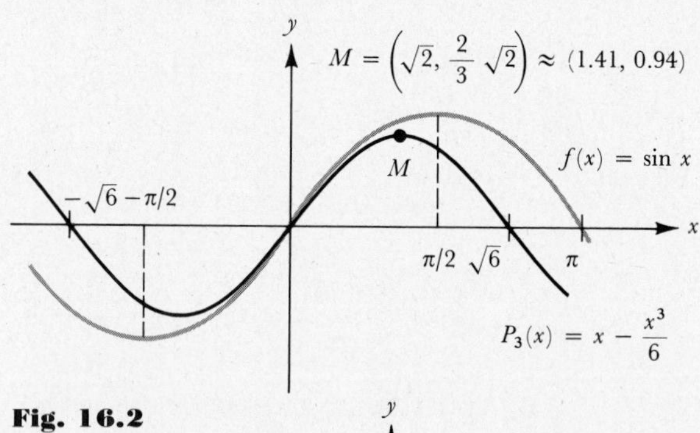

Fig. 16.2

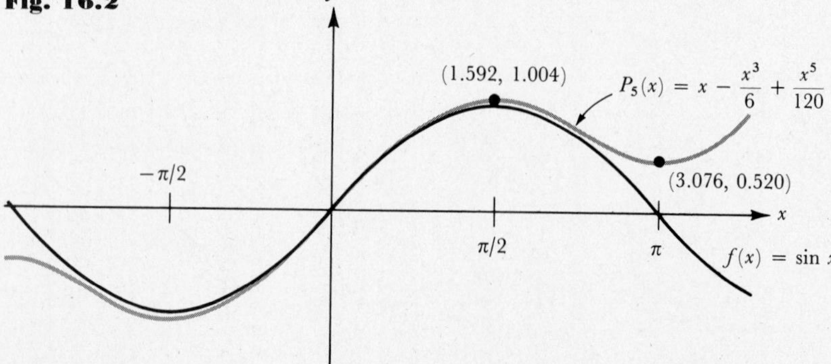

Fig. 16.3

$P_1 = x$	P_3	P_5	P_7	P_9	$\sin x$	R_5
0.000000	0.000000	0.000000	0.000000	0.000000	0.000000	0.000000
0.100000	0.099833	0.099833	0.099833	0.099833	0.099833	0.000000
0.200000	0.198667	0.198670	0.198670	0.198670	0.198670	0.000000
0.400000	0.389333	0.389418	0.389418	0.389418	0.389418	0.000000
0.500000	0.479167	0.479427	0.479426	0.479426	0.479426	−0.000001
$0.785398 \approx \frac{\pi}{4}$	0.704652	0.707143	0.707107	0.707107	0.707107	−0.000036
0.900000	0.778500	0.783421	0.783326	0.783326	0.783327	−0.000094
1.000000	0.833333	0.841667	0.841468	0.841471	0.841471	−0.000196
1.300000	0.933833	0.964774	0.963529	0.963559	0.963558	−0.001216
1.400000	0.942667	0.987485	0.985393	0.985450	0.985450	−0.002035
1.500000	0.937500	1.000781	0.997391	0.997497	0.997495	−0.003286
$1.570796 \approx \frac{\pi}{2}$	0.924832	1.004525	0.999843	1.000004	1.000000	−0.004525
2.000000	0.666667	0.933333	−0.907937	0.909347	0.909297	−0.024036
3.000000	−1.500000	0.525000	0.091071	0.145313	0.141120	−0.383880
4.000000	−6.666667	1.866667	−1.388413	−0.661728	−0.756802	−2.623469
6.000000	−30.000000	34.800000	−20.742857	7.028571	−0.279415	−35.079416

Table 16.1

required in order to give accurate approximations. For x as large as 0.4 (about 23 degrees) each polynomial of degree 5 or more is accurate through six decimal places, while $P_7(x)$ yields this accuracy for x as large as $\pi/4$. The best approximation shown is that of $P_9(x)$, which gives five-place accuracy for x as large as $\pi/2$ and is within 0.1 even at $x = 4$. To get this close at $x = 6$ would require $P_{15}(x)$.

In a standard trigonometric table, values of sine (as well as the other functions) are only needed for x in $[0, \pi/4]$, so $P_5(x)$ can be used to give a four-place table for the sine function.

PROGRESS TEST 1

1. Determine the eighth Taylor polynomial at $a = 0$ for $f(x) = \cos x$.
2. (*a*) Graph $P_1(x)$ (a constant function) and $\cos x$ on the same coordinate axes.
   (*b*) Repeat for $P_2(x)$.
   (*c*) Repeat for $P_4(x)$.
3. Compute $P_5(x)$ at $a = 0$ for $f(x) = e^x$.

TAYLOR'S REMAINDER FORMULA

How accurate are the approximations provided by Taylor polynomials? Before we answer this question we establish some convenient notation.

16.1.2 **Definition:** If $P_n(x)$ is the nth Taylor polynomial approximation of $f(x)$ at a, we denote by $R_n(x)$ the difference between $f(x)$ and $P_n(x)$ and call it the *nth remainder term.* Thus $f(x) = P_n(x) + R_n(x)$.

In Table 16.1 the right-hand column consists of the fifth remainder term for the sine function's Taylor polynomials at $a = 0$. The following important theorem describes $R_n(x)$ in terms of the $(n + 1)$st derivative of f and will thus enable us to give an upper bound for the error associated with a Taylor polynomial approximation.

16.1.3 **Taylor's Formula:** Suppose f is a function whose $(n + 1)$st derivative exists on an open interval I containing a. Then for any number $x \neq a$ in I, there exists a number c strictly between a and x such that

$$R_n(x) = \frac{f^{(n+1)}(c)}{(n+1)!}(x-a)^{n+1}$$

or equivalently

$$f(x) = f(a) + \frac{f'(a)}{1!}(x-a) + \frac{f''(a)}{2!}(x-a)^2 + \cdots + \frac{f^{(n)}(a)}{n!}(x-a)^n + \frac{f^{(n+1)}(c)}{(n+1)!}(x-a)^{n+1}$$

We adopt the common convention that $f^{(0)}(a) = f(a)$. This will simplify matters when summation notation is used.

Notice that if we take $x = b$, with $a < b$, and $n = 0$ in the above theorem, then its conclusion states that there is a c strictly between a and b such that

$$f(b) = f(a) + \frac{f'(c)}{1!}(b-a)$$

This equation can be rewritten as

$$f'(c) = \frac{f(b) - f(a)}{b-a}$$

Hence in this particular case the conclusion of (16.1.3) is the conclusion of the Mean-Value Theorem (4.4.4). For this reason (16.1.3) is also referred to as the **Generalized Mean-Value Theorem.** As with the original Mean-Value Theorem, its proof is an application of Rolle's Theorem to a specially designed function.

Proof of (16.1.3): We assume that x is a fixed number with $a < x$; arguments for $a > x$ are similar. For each t in $[a, x]$, define a function of t as follows:

$$F(t) = f(x) - f(t) - f'(t)(x-t) - \frac{f''(t)}{2!}(x-t)^2 - \cdots - \frac{f^{(n)}(t)}{n!}(x-t)^n - \frac{R_n(x)}{(x-a)^{n+1}}(x-t)^{n+1}$$

Now $F(a) = f(x) - P_n(x) - R_n(x) = 0$ and $F(x) = 0$. Furthermore, $F(t)$, being the product and difference of functions differentiable on (a, x), is itself a differentiable function of t on (a, x). Hence, Rolle's Theorem applies to guarantee that there is a c in (a, x) such that $F'(c) = 0$. But thus we need to compute $F'(t)$. We shall use the Product Rule on all but the first two terms (remember, F is a function of t and x is fixed):

$$\begin{aligned} F'(t) = 0 - f'(t) &- [f'(t)(-1) + (x-t)f''(t)] \\ &- \left[\frac{f''(t)}{2!}2(x-t)(-1) + (x-t)^2\frac{f'''(t)}{2!}\right] \\ &- \left[\frac{f'''(t)}{3!}3(x-t)^2(-1) + (x-t)^3\frac{f^{(4)}(t)}{3!}\right] \\ &- \cdots - \left[\frac{f^{(n)}(t)}{n!}n(x-t)^{n-1}(-1) + (x-t)^n\frac{f^{(n+1)}(t)}{n!}\right] \\ &- \frac{R_n(x)(n+1)}{(x-a)^{n+1}}(x-t)^n(-1) \\ &= \frac{-f^{(n+1)}(t)}{n!}(x-t)^n + \frac{R_n(x)(n+1)}{(x-a)^{n+1}}(x-t)^n \end{aligned}$$

Notice that the $f'(t)$ terms canceled and that the second term in each bracketed pair canceled out the first term in the next pair.

By Rolle's Theorem then, there is a c in (a, x) such that

$$F'(c) = \frac{-f^{(n+1)}(c)}{n!}(x - c)^n + \frac{R_n(x)}{(x - a)^{n+1}}(n + 1)(x - c)^n = 0$$

Dividing through by $(x - c)^n$ (since $x \neq c$), we conclude that

$$R_n(x) = \frac{f^{(n+1)}(c)(x - a)^{n+1}}{(n + 1)n!} = \frac{f^{(n+1)}(c)}{(n + 1)!}(x - a)^{n+1} \quad \blacksquare$$

An immediate consequence of Taylor's Formula is based on the simple fact that if the approximation error equals $[f^{(n+1)}(c)/(n + 1)!](x - a)^{n+1}$ for some c strictly between a and x, then its absolute value can be no larger than $[M/(n + 1)!]\,|x - a|^{n+1}$, where M is an upper bound on the possible numbers $|f^{(n+1)}(c)|$.

16.1.4 **Error Bound for Taylor Approximations:** If the hypotheses of (16.1.3) hold and if $|f^{(n+1)}(t)| \leq M$ on $[a, x]$ (or $[x, a]$ if $x < a$), then

$$|R_n(x)| < \frac{M}{(n + 1)!}|x - a|^{n+1}$$

For $n = 1$, (16.1.4) reduces to an upper bound for the differential approximation of Chap. 3. (See Chap. 3 Miscellaneous Exercises 1 to 7.) For $n = 1$, the upper bound for $|R_1(x)|$ involves the magnitude of the second derivative of f near a, which should make sense by the following argument. $P_1(x)$ is the tangent line approximation to the graph at $(a, f(a))$. The error of this approximation depends on the amount of "bend" in the graph near $(a, f(a))$, which in turn depends upon the rate at which the slopes of the tangent lines change near $(a, f(a))$. But these slopes are given by $f'(x)$ near $(a, f(a))$ and, of course, the rate at which $f'(x)$ changes is given by $f''(x)$ near $(a, f(a))$.

We now revisit the approximation of $\sqrt[5]{33}$ originally given in Sec. 3.3, Example 4 (accurate to three decimal places). Note that "n-place accuracy" means an error less than $5 \times 10^{-(n+1)}$.

Example 2 Estimate $\sqrt[5]{33}$ through six decimal places.

Solution Let $f(x) = x^{1/5}$ and compute Taylor polynomials for $f(x)$ at $a = 32$. We shall attempt a third Taylor approximation. Now

$$f'(x) = \frac{1}{5}x^{-4/5} \qquad f''(x) = \frac{1}{5} \cdot \frac{-4}{5}x^{-9/5} \qquad f'''(x) = \frac{1}{5} \cdot \frac{-4}{5} \cdot \frac{-9}{5}x^{-14/5}$$

$$f^{(4)}(x) = \frac{1}{5} \cdot \frac{-4}{5} \cdot \frac{-9}{5} \cdot \frac{-14}{5}x^{-19/5}$$

We know that

$$|R_4(33)| \leq \frac{M(1)^4}{4!} = \frac{M}{24}$$

where M is an upper bound for

$$|f^{(4)}(t)| = \left|\frac{-504}{(625)t^{19/5}}\right|$$

on [32, 33]. But $|f^{(4)}(t)|$ reaches its maximum on [32, 33] at 32 because $504/(625t^{19/5})$ is decreasing on [32, 33]. Hence

$$|f^{(4)}(t)| \leq \frac{504}{(625)(32)^{19/5}} = \frac{504}{625} \cdot \frac{1}{2^{19}} \approx 0.00000154$$

Choosing $M = 0.00000154$, we know that

$$|R_4(33)| \leq \frac{M}{24} \approx 0.0000000641$$

Hence the third Taylor polynomial yields the desired accuracy. Now

$$f(32) = 2 \qquad f'(32) = \frac{1}{80} \qquad f''(32) = -\frac{1}{3200} \qquad f'''(32) = \frac{9}{512{,}000}$$

and therefore

$$f(33) = 2 + \frac{1}{80}(33 - 32) - \frac{1}{3200} \cdot \frac{1}{2!}(33 - 32)$$
$$+ \frac{9}{512{,}000} \cdot \frac{1}{3!} \cdot (33 - 32) \approx 2.012347 \quad \bullet$$

Notice that in finding M in Example 1 we found an upper bound for $|f^{(4)}(t)|$, not $f^{(4)}(t)$. If we had not reached the desired accuracy with the third Taylor polynomial, we would merely compute additional terms as necessary.

Before continuing, we gather together three of the Taylor approximations computed at $a = 0$ so far (see Progress Test 1).

16.1.5 (*a*) $\sin x \approx x - \frac{x^3}{3!} + \frac{x^5}{5!} - \frac{x^7}{7!} + \cdots + \frac{(-1)^n x^{2n+1}}{(2n+1)!}$

(*b*) $\cos x \approx 1 - \frac{x^2}{2!} + \frac{x^4}{4!} - \frac{x^6}{6!} + \cdots + \frac{(-1)^n x^{2n}}{(2n)!}$

(*c*) $e^x \approx 1 + x + \frac{x^2}{2!} + \frac{x^3}{3!} + \cdots + \frac{x^n}{n!}$

It is important to realize that this polynomial approximation for $\sin x$ is not $P_n(x)$ but $P_{2n+1}(x)$. [$P_{2n+1}(x)$ happens to equal $P_{2n+2}(x)$—show this!] Similarly, the previous polynomial approximation for $\cos x$ is $P_{2n}(x)$ [which equals $P_{2n+1}(x)$]. The given polynomial for e^x *is* $P_n(x)$.

APPROXIMATING *e*

Using the earlier limit representation of e given in (7.4.12), page 267

$$e = \lim_{u \to +\infty} \left(1 + \frac{1}{u}\right)^u$$

$u = 10{,}000$ is required to reach an estimate of e accurate through three decimal places. As is often the case, Taylor polynomials provide a much more efficient approximation method. By taking $x = 1$ in [16.1.5(*c*)] we know that

$$e = e^1 = 1 + 1 + \frac{1^2}{2!} + \frac{1^3}{3!} + \cdots + \frac{1^n}{n!} + R_n(1)$$
$$= 1 + 1 + \frac{1}{2!} + \frac{1}{3!} + \cdots + \frac{1}{n!} + R_n(1)$$

where $R_n(1)$ is the error involved in using the sum of the previous terms as an approximation of e. By (16.1.4) we know that

$$|R_n(x)| \leq \frac{M|x^{n+1}|}{(n+1)!}$$

where M is an upper bound for $|f^{(n+1)}(t)| = |e^t| = e^t$ on $[0, x]$. Since e^t [the $(n+1)$st derivative of e^t] is increasing on $[0, x]$, it reaches a maximum *at* x. In particular then for $x = 1$, we know that the $(n+1)$st derivative reaches a maximum of $e^1 = e$ on $[0, 1]$. *But we are trying to approximate e*, so we cannot use $M = e$ as an upper bound on the error of the approximation! However, we *do* know that $e < 3$ (recall the original definition of e), so we can use $M = 3$ as our upper bound. Hence

$$|R_n(1)| \leq \frac{3|1^{n+1}|}{(n+1)!} = \frac{3}{(n+1)!}$$

For example, with $n = 6$ we get

$$e \approx P_6(1) \approx 1 + 1 + \frac{1}{2} + \frac{1}{6} + \frac{1}{24} + \frac{1}{120} + \frac{1}{720}$$

$$\approx 2.718056 \qquad \text{and} \qquad |R_6(1)| \approx 0.000595$$

and with $n = 10$ we get

$$e \approx P_{10}(1) \approx 2.7182818 \qquad \text{and} \qquad |R_{10}(1)| \approx 0.0000000751$$

This means that $P_{10}(1)$ provides six-place accuracy. In fact $n = 17$ yields the 15-place accuracy of the estimate given earlier and $n = 39$ yields 45-place accuracy.

BOUNDS ON THE APPROXIMATION ERROR FOR SIN X

Determining an upper bound for $|f^{(n+1)}(t)|$ on an interval, which can then be used to yield an upper bound for $|R_n(x)|$, is sometimes a creative art. In Example 2 we used the value of $|f^{(n+1)}(t)|$ at an endpoint, whereas in approximating e the endpoint value turned out to be the number e we were approximating.

In approximating $f(x) = \sin x$ using $P_5(x) = x - (x^3/3!) + (x^5/5!)$ we used prior knowledge of $\sin x$ to compute $R_5(x)$, from which we concluded that $P_5(x)$ gives four-place accuracy on $[0, \pi/4]$ (see Table 16.1). We can now use (16.1.4) to verify this fact. However, instead of using an upper bound for $|R_5(x)|$ we use that

$$\begin{aligned} P_6(x) &= f(0) + f'(0)x + \frac{f''(0)}{2!}x^2 + \frac{f'''(0)}{3!}x^3 + \frac{f^{(4)}(0)}{4!}x^4 \\ &\qquad + \frac{f^{(5)}(0)}{5!}x^5 + \frac{f^{(6)}(0)}{6!}x^6 \\ &= P_5(x) \end{aligned}$$

because $f^{(6)}(0) = -\sin 0 = 0$. Hence we are entitled to use an upper bound for $|R_6(x)|$ and so we use $n = 6$ in (16.1.4). Now $|f^{(6+1)}(t)| = |-\cos t| \leq 1$ on $[0, \pi/4]$. Thus

$$|R_6(x)| \leq \frac{1 \cdot |\pi/4|^7}{7!} \approx 0.0000366$$

which says that $P_5(x)$ gives the four-place accuracy claimed. Note that $P_7(x)$ gives six-place accuracy on $[0, \pi/4]$.

PROGRESS TEST 2

1. Determine $\sqrt[3]{65}$ to within seven decimal places (using $f(x) = x^{1/3}$).

2. (*a*) Determine $e^{0.1}$ to within six decimal places.
(*b*) Determine $\sqrt[3]{e}$ to within four decimal places.
(*c*) What degree Taylor polynomial for e^x at $a = 0$ is necessary to approximate e^5 to within four decimal places?

3. Calculate $P_{n+1}(x)$ for $f(x) = \ln(1 + x)$ at $a = 0$.

SECTION 16.1 EXERCISES

In Exercises 1 to 11, part (a), compute $P_n(x)$ for the given $f(x)$, a, and n.

1. (*a*) $f(x) = e^{-x}$, $a = 0$, $n = 4$
(*b*) Give $P_n(x)$ for arbitrary n.

2. (*a*) $f(x) = \ln(1 - x)$, $a = 0$, $n = 4$
(*b*) Give $P_n(x)$ for arbitrary n.

3. (*a*) $f(x) = \ln(\cos x)$, $a = 0$, $n = 3$
(*b*) Use (*a*) to approximate $\ln(\cos 0.1)$.
(*c*) Determine an upper bound on the error of the approximation in (*b*).

4. (*a*) $f(x) = \ln(\sin x)$, $a = \pi/2$, $n = 3$
(*b*) Use (*a*) to approximate $\ln(\sin \pi/3)$.
(*c*) Determine an upper bound on the error of the approximation in (*b*).

5. (*a*) $f(x) = \sqrt{x + 1}$, $a = 0$, $n = 3$
(*b*) Use (*a*) to approximate $\sqrt{0.9}$.
(*c*) Use (*a*) to approximate $\sqrt{2}$.

6. (*a*) $f(x) = 2^x$, $a = 0$, $n = 3$
(*b*) Use (*a*) to approximate $\sqrt{2}$.
(*c*) Compare part (*b*) with Exercise 5(*c*).
(*d*) Determine an upper bound on the error 6(*b*).

7. (*a*) $f(x) = \ln\left(\dfrac{1 + x}{1 - x}\right)$, $a = 0$, $n = 4$
(Use properties of ln before differentiating.)
(*b*) Compare (*a*) with Exercise 2(*a*).

8. (*a*) $f(x) = \arcsin x$, $a = 0$, $n = 2$
(*b*) Use (*a*) to approximate $\pi/6$.

9. (*a*) $f(x) = \arctan x$, $a = 0$, $n = 2$
(*b*) Use (*a*) to approximate $\pi/4$.

10. (*a*) $f(x) = \cosh x$, $a = 0$, $n = 4$
(*b*) Give $P_n(x)$ for arbitrary n.

11. (*a*) $f(x) = \sinh x$, $a = 0$, $n = 4$
(*b*) Give $P_n(x)$ for arbitrary n.

12. (*a*) Compute $P_3(x)$ and $P_4(x)$ for $f(x) = x^3 - x^2 - 9x + 9$ at $a = 0$.
(*b*) Repeat (*a*) at $a = 2$, denoting the results by $\bar{P}_3(x)$ and $\bar{P}_4(x)$, respectively.
(*c*) Compare $P_3(x)$ and $\bar{P}_3(x)$ by expanding and simplifying $\bar{P}_3(x)$.
(*d*) Use (*b*) to compute $f(2.01)$.

13. (*a*) Compute $P_4(x)$ and $P_5(x)$ for $f(x) = (x^4/4) - (x^3/3) - x^2 + 1$ at $a = 0$.
(*b*) Repeat (*a*) at $a = 1$, denoting the results by $\bar{P}_4(x)$ and $\bar{P}_5(x)$
(*c*) Compare $P_4(x)$ and $\bar{P}_4(x)$ by expanding and simplifying $\bar{P}_4(x)$.
(*d*) Use *b* to compute $f(1.01)$.

14. For $f(x) = e^x$ and $a = 0$, graph (*a*) $P_1(x)$, (*b*) $P_2(x)$, (*c*) $P_3(x)$, (*d*) $P_4(x)$.

In Exercises 15 to 20 compute the given quantity to the given accuracy using Taylor polynomials and (16.1.4) to guarantee the accuracy. (You may use results of previous exercises and leave your answer as a finite numerical sum.)

15. ln 2 accurate within 10^{-4}

16. $\ln(3/2)$ accurate within 10^{-3}

17. $1/e$ accurate within 10^{-4}

18. e^2 accurate with 10^{-4}

19. $\sqrt{15}$ accurate through four decimal places

20. $\sqrt{80}$ accurate through four decimal places

21. Using $a = \pi/8$ instead of $a = 0$, which Taylor polynomials yield six-place accuracy for $\sin x$ on $[0, \pi/4]$?

22. Repeat Exercise 21 for $f(x) = \cos x$.

23. Compute $P_6(x)$ for $f(x) = \sin^2 x$ by first using a double-angle formula.

24. (*a*) If $f(x) = \dfrac{1}{1 - x}$, show that at $a = 0$, $P_n(x) = 1 + x + x^2 + \cdots + x^n$.
(*b*) Using (*a*) show that $R_n(x) = \dfrac{x^{n+1}}{1 - x}$.

25. Use Exercise 24 to show that

$$\frac{1}{1 + x^2} = 1 - x^2 + x^4 - x^6 + \cdots + (-1)^n x^{2n} + \frac{(-1)^{n+1} x^{2n+2}}{1 + x^2}$$

Section 16.2: Power Series

Objectives:

Determine the radius of convergence and interval of convergence of a given power series.

INTRODUCTION

In the previous section we saw how a (finite-degree) polynomial $P_n(x)$ can be used to approximate a given function. What happens if we attempt to achieve perfect accuracy in the approximation by letting n grow without bound? The answer to this question will couple the ideas developed in Chap. 15 regarding infinite series with the notion of "infinite-degree polynomial," whose formal definition and official name are given in the following:

16.2.1 **Definition** An expression of the form

$$b_0 + b_1(x - a) + b_2(x - a)^2 + \cdots + b_n(x - a)^n + \cdots$$

where $b_0, b_1, \ldots, b_n, \ldots$ and a are constants, is called a *power series in* $(x - a)$. If $a = 0$, then the series

$$b_0 + b_1x + b_2x^2 + \cdots + b_nx^n + \cdots$$

is called a *power series in* x.

We shall often denote the above series, respectively, using summation notation:

$$\sum_{n=0}^{\infty} b_n(x - a)^n \qquad \text{and} \qquad \sum_{n=0}^{\infty} b_nx^n$$

with the convention that $(x - a)^0 = 1$ and $x^0 = 1$—even when $x = a$ and $x = 0$. (This convention enables us to avoid the cumbersome notation necessary to separate out the $n = 0$ term, which technically would be required to avoid 0^0, an indeterminate form.)

That a power series is different from an infinite series can be seen by asking the question of either power series in (16.2.1): Does this power series converge?

Since convergence has only been defined for series of numbers, the question makes no sense. However, for each number substituted for x in a power series we *do* get an infinite series. For example, in (16.2.1) if $b_n = 1$ for all n, then we get $\Sigma_{n=0}^{\infty} x^n$, which is a geometric series for each x and converges, to $1/(1 - x)$, for all $|x| < 1$, by (15.2.8). The appropriate question to ask of a power series is: *For which values of x does it converge?* Notice that $\Sigma_{n=0}^{\infty} b_n(x - a)^n$ always converges at $x = a$. It will turn out that if there are other points at which the power series converges, then the set of these points will always comprise an interval.

THE RADIUS OF CONVERGENCE AND INTERVAL OF CONVERGENCE OF A POWER SERIES

As Example 1 shows, the basic tool in determining the numbers x for which a given power series converges is the Ratio Test (15.3.8), page 501.

Example 1 Determine those x for which $\displaystyle\sum_{n=0}^{\infty} \frac{nx^n}{2^n}$ converges.

Solution We regard $nx^n/2^n$ as u_n and examine the series $\Sigma_{n=0}^{\infty} |u_n|$. We form the ratio of successive terms

$$\left|\frac{u_{n+1}}{u_n}\right| = \left|\frac{(n + 1)x^{n+1}}{2^{n+1}} \cdot \frac{2^n}{nx^n}\right| = \frac{1}{2}\left(\frac{n + 1}{n}\right)|x|$$

Since

$$\lim_{n\to+\infty}\left[\frac{1}{2}\left(\frac{n+1}{n}\right)|x|\right]=\frac{1}{2}|x|$$

we know by the Ratio Test that the series converges whenever $(1/2)|x|<1$, that is, when $|x|<2$. It diverges when $(1/2)|x|>1$, that is, when $|x|>2$. At $x=\pm 2$ the Ratio Test fails, and so we examine each case separately. If $x=2$, the power series becomes

$$\sum_{n=0}^{\infty}\frac{n}{2^n}\cdot 2^n=\sum_{n=0}^{\infty} n$$

which diverges [nth-Term Test, (15.2.6)]. If $x=-2$, the power series becomes

$$\sum_{n=0}^{\infty}\frac{n}{2^n}(-2)^n=\sum_{n=0}^{\infty}(-1)^n n$$

which again diverges. Hence $\Sigma_{n=0}^{\infty} nx^n/2^n$ converges for all x in $(-2, 2)$. ●

The approach of the above example suggests a proof of the following theorem:

16.2.2 **Theorem** Suppose $\sum_{n=0}^{\infty} b_n(x-a)^n$ is a power series, with

$$\lim_{n\to+\infty}\left|\frac{b_{n+1}}{b_n}\right|=L$$

(*a*) If $L=0$, then $\Sigma_{n=0}^{\infty} b_n(x-a)^n$ converges absolutely for all x.

(*b*) If $0<L<+\infty$, then $\Sigma_{n=0}^{\infty} b_n(x-a)^n$ converges absolutely for all x such that $|x-a|<1/L$, diverges for all x such that $|x-a|>1/L$, and may either converge or diverge when $|x-a|=1/L$.

(*c*) If $L=+\infty$, then $\Sigma_{n=0}^{\infty} b_n(x-a)^n$ converges only for $x=a$.

Proof: We prove (*b*), leaving (*a*) and (*c*) as exercises. From the limit of the ratio of successive terms as in Example 1:

$$\lim_{n\to+\infty}\left|\frac{b_{n+1}(x-a)^{n+1}}{b_n(x-a)^n}\right|=\lim_{n\to+\infty}\left|\frac{b_{n+1}}{b_n}\right||x-a|=L|x-a|$$

By the Ratio Test, the series converges absolutely when

$$L|x-a|<1 \qquad \text{or} \qquad |x-a|<\frac{1}{L}$$

It diverges when $L|x-a|>1$, or $|x-a|>1/L$. The Ratio Test fails when $L|x-a|=1$, and the examples will illustrate that the series can either converge or diverge when $|x-a|=1/L$. ■

16.2.3 **Definition** Suppose $\Sigma_{n=0}^{\infty} b_n(x-a)^n$ is as in (16.2.2). If condition (*a*) holds, we say that $\Sigma_{n=0}^{\infty} b_n(x-a)^n$ has an *infinite radius of convergence* and *interval of convergence* $(-\infty, +\infty)$. If condition (*b*) holds, we say that $\Sigma_{n=0}^{\infty} b_n(x-a)^n$ has *radius of convergence* $r=1/L$. In this case we say that the *interval of convergence* is the set of all x for which $\Sigma_{n=0}^{\infty} b_n(x-a)^n$ converges. (This interval may or may not include endpoints.) If condition (*c*) holds, we say that $\Sigma_{n=0}^{\infty} b_n(x-a)^n$ has *zero radius of convergence* and an *interval of convergence* consisting of the single point a.

Note that the radius of convergence of the power series in Example 1 is 2, and the interval of convergence is $(-2, 2)$.

We should point out that although the hypothesis in the previous theorem and definition holds for all power series in this text, it does not hold for all power series. Nonetheless, it can be shown that the set of x for which *any* power series converges is an interval of one of the above types. Notice also that, in effect, the above theorem says that convergence at the endpoints of the interval of convergence must be handled on a case-by-case basis. (See, however, Sec. 16.2, Exercises 41 and 42).

Example 2 Find the radius of convergence r and interval of convergence I for each of the given power series:

(a) $\sum_{n=0}^{\infty} n!x^n$ (b) $\sum_{n=1}^{\infty} \frac{x^n}{n^n}$

(c) $\sum_{n=1}^{\infty} \frac{n!x^n}{1 \cdot 3 \cdot 5 \cdot \cdots \cdot (2n-1)}$ (d) $\sum_{n=1}^{\infty} \frac{x^n}{\ln(n+1)}$

Solutions We apply the Ratio Test to the series $\Sigma_{n=0}^{\infty} |u_n|$ where $u_n = b_n x^n$.

(a) $$\left|\frac{u_{n+1}}{u_n}\right| = \left|\frac{(n+1)!x^{n+1}}{n!x^n}\right| = (n+1)|x|$$

If $x \neq 0$, then $\lim_{n\to+\infty}(n+1)|x| = +\infty$, and thus the series diverges by the Ratio Test for all $x \neq 0$. Hence $r = 0$ and $I = \{0\}$.

(b) $$\left|\frac{u_{n+1}}{u_n}\right| = \left|\frac{x^{n+1}}{(n+1)^{n+1}} \cdot \frac{n^n}{x^n}\right|$$
$$= \frac{n^n}{(n+1)^{n+1}}|x| = \frac{n_n}{(n+1)^n} \cdot \frac{1}{(n+1)}|x|$$
$$= \left(\frac{n}{n+1}\right)^n \cdot \frac{1}{(n+1)}|x|$$

Now

$$\lim_{n\to+\infty}\left(\frac{n}{n+1}\right)^n = e^{-1} \quad \text{and} \quad \lim_{n\to+\infty}\frac{1}{n+1} = 0$$

Thus, for all x,

$$\lim_{n\to+\infty}\left|\frac{u_{n+1}}{u_n}\right| = e^{-1} \cdot 0 \cdot |x| = 0$$

Since

$$\lim_{n\to+\infty}\left|\frac{u_{n+1}}{u_n}\right| < 1$$

for all x, the series converges absolutely for all x. Hence $r = +\infty$ and $I = (-\infty, +\infty)$.

(c) $$\left|\frac{u_{n+1}}{u_n}\right| = \left|\frac{(n+1)!x^{n+1}}{1 \cdot 3 \cdot 5 \cdots (2n-1)(2n+1)} \cdot \frac{1 \cdot 3 \cdot 5 \cdot \cdots \cdot (2n-1)}{n!x^n}\right|$$
$$= \frac{(n+1)|x|}{2n+1}$$

Now

$$\lim_{n\to+\infty}\frac{(n+1)|x|}{2n+1} = \frac{|x|}{2} < 1$$

for $|x| < 2$, so $r = 2$. Letting $x = \pm 2$ we obtain the series

$$\sum_{n=1}^{\infty} \frac{2^n n!}{1 \cdot 3 \cdot 5 \cdot \cdots \cdot (2n-1)} \quad \text{and} \quad \sum_{n=1}^{\infty} \frac{(-1)^n 2^n n!}{1 \cdot 3 \cdot 5 \cdot \cdots \cdot (2n-1)}$$

But

$$\frac{2^n n!}{1 \cdot 3 \cdot 5 \cdot \cdots \cdot (2n-1)} = \frac{2 \cdot 4 \cdot 6 \cdot \cdots \cdot (2n)}{1 \cdot 3 \cdot 5 \cdot \cdots \cdot (2n-1)} > 1$$

since each of the factors

$$\frac{2}{1}, \frac{4}{3}, \frac{6}{5}, \ldots, \frac{2n}{2n-1}$$

is bigger than 1. Thus both of the endpoint series diverge by the nth-Term Test, making $I = (-2, 2)$.

(*d*) $$\left|\frac{u_{n+1}}{u_n}\right| = \frac{\ln(n+1)}{\ln(n+2)}|x| \to |x|$$

as $n \to +\infty$, making $r = 1$. When $x = 1$, the power series becomes

$$\sum_{n=1}^{\infty} \frac{1}{\ln(n+1)}$$

Now $\ln(n+1) < n + 1$. Thus

$$\frac{1}{\ln(n+1)} > \frac{1}{n+1}$$

so the series diverges by comparison with the divergent series $\Sigma_{n=1}^{\infty} 1/(n+1)$. When $x = -1$, we obtain

$$\sum_{n=1}^{\infty} \frac{(-1)^n}{\ln(n+1)}$$

which converges by the Alternating-Series Test. Hence $I = [-1, 1)$. ●

Example 3 Find the interval I and radius of convergence of

$$\sum_{n=0}^{\infty} \frac{n(x+2)^n}{3^n(n+1)}$$

Solution Now

$$\left|\frac{u_{n+1}}{u_n}\right| = \left|\frac{(n+1)(x+2)^{n+1}}{3^{n+1}(n+2)} \cdot \frac{3^n(n+1)}{n(x+2)^n}\right| = \frac{1}{3}\left[\frac{n^2+2n+1}{n^2+2n}\right]|x+2|$$

Therefore

$$\lim_{n \to +\infty} \left|\frac{u_{n+1}}{u_n}\right| = |x+2| \lim_{n \to +\infty} \frac{1}{3}\left(\frac{n^2+2n+1}{n^2+2n}\right) = \frac{|x+2|}{3}$$

Thus by the Ratio Test the series converges absolutely when $|x+2|/3 < 1$, that is, when $|x+2| < 3$. It diverges when $|x+2| > 3$, so $r = 3$. Since $a = -2$ and $r = 3$, we must examine the endpoints of the interval $(-2-3, -2+3) = (-5, 1)$. Substituting $x = -5$ and $x = 1$ we obtain, successively,

$$\sum_{n=0}^{\infty} \frac{(-1)^n n}{n+1} \quad \text{and} \quad \sum_{n=0}^{\infty} \frac{n}{n+1}$$

both of which diverge, by the nth-Term Test. Thus $I = (-5, 1)$. ●

PROGRESS TEST 1

Find the radius r and interval I of convergence for each of the following power series.

1. $\sum_{n=0}^{\infty} \frac{(-1)^n n x^n}{3^n}$ **2.** $\sum_{n=1}^{\infty} \frac{x^n}{n(n+1)}$ **3.** $\sum_{n=1}^{\infty} \frac{(x-4)^n}{n}$

SECTION 16.2 EXERCISES

In Exercises 1 to 18 determine the radius of convergence r and the interval of convergence I for the given power series.

1. $\sum_{n=0}^{\infty} \frac{(-1)^n 3n x^n}{2^n}$ **2.** $\sum_{n=0}^{\infty} \frac{(-1)^n 2n x^n}{3^{n-1}}$ **3.** $\sum_{n=1}^{\infty} n^n x^n$ **4.** $\sum_{n=1}^{\infty} \frac{\ln n}{n} x^n$

5. $\sum_{n=0}^{\infty} \frac{n^n}{n+1} x^n$ **6.** $\sum_{n=1}^{\infty} \frac{3^n x^{2n}}{n^2}$ **7.** $\sum_{n=1}^{\infty} \frac{(x-5)^n}{3n}$ **8.** $\sum_{n=0}^{\infty} \frac{(n+1)(x+1)^n}{n+2}$

9. $\sum_{n=0}^{\infty} n!(x+3)^n$ **10.** $\sum_{n=0}^{\infty} \frac{3^n x^n}{\ln(n+2)}$ **11.** $\sum_{n=0}^{\infty} \frac{(2x)^n}{n^2+1}$ **12.** $\sum_{n=1}^{\infty} \frac{(x+2)^{2n+1}}{n2^n}$

13. $\sum_{n=0}^{\infty} \frac{(3x)^n}{n!}$ **14.** $\sum_{n=0}^{\infty} \frac{(2n)! x^n}{2^{2n}(n!)^2}$

15. $\sum_{n=1}^{\infty} \frac{n!(x-2)^n}{3^n \cdot 1 \cdot 3 \cdot 5 \cdot \cdots \cdot (2n-1)}$ **16.** $\sum_{n=1}^{\infty} \frac{2^n(x-3)^n}{n^2(n+1)}$

17. $\sum_{n=0}^{\infty} \frac{(-1)^n(x+7)^{2n}}{(2n)!}$ **18.** $\sum_{n=0}^{\infty} \frac{(-1)^n 4^{n+1}(x-3)^n}{3^{2n}}$

19. For the power series $\sum_{n=1}^{\infty} \frac{n! x^n}{n^n}$

(*a*) Show that $r = e$.
(*b*) Show that the series diverges at $x = e$ by showing that the ratio a_{n+1}/a_n always exceeds 1, where $a_n = n!e^n/n^n$. [*Hint:* Show that

$$\left(\frac{n+1}{n}\right)^n \leq e$$

by taking the natural logarithm and using the fact that $\ln(1 + 1/n) \leq 1/n$.]
(*c*) Show how the argument in (*b*) applies when $x = -e$.

20. For $\sum_{n=1}^{\infty} \frac{1 \cdot 3 \cdot 5 \cdot \cdots \cdot (2n-1)x^n}{2^n n!}$

(*a*) Show that $r = 1$.
(*b*) Show divergence at $x = 1$ by comparison with the divergent series $\sum_{n=1}^{\infty} 1/(2n)$.
(*c*) Apply the Alternating-Series Test for the series at $x = -1$ by following the given outline:
(i) Show that $a_n > a_{n+1}$ for all n, where $a_n = 1 \cdot 3 \cdot 5 \cdot \cdots \cdot (2n-1)/(2^n n!)$
(ii) Show that $a_n^2 \to 0$ as $n \to +\infty$ (from which it follows that $a_n \to 0$.) *Hint:* Show that $a_n^2 < (2n-1)/(2n)^2$ and apply the Squeeze Theorem.

In Exercises 21 to 36, (a) determine r and (b) determine I for the given power series.

21. $\sum_{n=1}^{\infty} \frac{(2x+3)^n}{2^n(2n+1)}$ **22.** $\sum_{n=2}^{\infty} \frac{(-1)^n x^n}{n \ln n}$

23. $\sum_{n=1}^{\infty} \frac{(-1)^n \pi^n (x-1)^n}{n^2+n}$ **24.** $\sum_{n=0}^{\infty} \frac{(-1)^n(x+2)^n}{4^{2n}}$

25. $\sum_{n=0}^{\infty} \frac{(-1)^n}{(n!)^2}\left(\frac{x}{2}\right)^n$ **26.** $\sum_{n=0}^{\infty} \frac{(-1)^n}{n!(n+1)!}\left(\frac{x}{2}\right)^n$

27. $\sum_{n=1}^{\infty} \frac{(n^2+n)(x+2)^n}{\sqrt{n}\, 2^n}$ **28.** $\sum_{n=0}^{\infty} \frac{(2-x)^{n+1}}{\sqrt{n}}$

29. $\sum_{n=1}^{\infty} \frac{x^{3n}}{\sqrt{3n}}$

30. $\sum_{n=1}^{\infty} \frac{(-1)^n x^{5n}}{\sqrt{5n}}$

31. $\sum_{n=2}^{\infty} \frac{x^n}{(\ln n)^n}$ [Use the Root Test (15.3.9).]

32. $\sum_{n=1}^{\infty} \frac{2^n x^n}{n^4}$

33. $\sum_{n=1}^{\infty} \frac{n^n x^n}{n!}$ (See Exercise 19.)

34. $\sum_{n=1}^{\infty} n^n (x-1)^n$

35. $\sum_{n=0}^{\infty} [1 + (-1)^n] x^n$

36. $\sum_{n=1}^{\infty} \frac{[(-1)^n + 1] x^n}{n}$

In Exercises 37 to 40 determine the set of x for which the given series converges. (Note these are not power series in x, but nonetheless the question makes sense and can be answered. Note also that the set of such x need not be an interval.)

37. $\sum_{n=1}^{\infty} \frac{(\sin x)^n}{n^2}$

38. $\sum_{n=1}^{\infty} \frac{(\cos x)^n}{n}$

39. $\sum_{n=1}^{\infty} (\ln \sqrt[2n]{x})^n$

40. $\sum_{n=1}^{\infty} (x^2 = 2x + 1/4)^n$ (Consider first $\sum_{n=1}^{\infty} u^n$, where $u = x^2 - 2x + 1/4$ and then use a sketch of $x^2 - 2x + 1/4$.)

41. Show that if $\Sigma_{n=0}^{\infty} b_n x^n$ converges absolutely at one endpoint of its finite interval of convergence, then it does likewise at the other.

42. Show that if a power series $\Sigma_{n=0}^{\infty} b_n x^n$ has a half-open interval of convergence, then convergence is conditional at that endpoint.

43. (Root Test) Prove that if $\lim_{n\to+\infty} |b_n|^{1/n} = L$, then the radius of convergence of $\Sigma_{n=0}^{\infty} b_n x^n$ is $1/L$.

44. Show that the Root Test approach to the radius of convergence (see Exercise 43) applies to $\Sigma_{n=1}^{\infty} x^n/10^{n-(-1)^n}$, whereas the Ratio Test fails.

Section 16.3: Differentiation and Integration of Power Series—Taylor Series

Objectives:

Compute a power series $\Sigma_{n=0}^{\infty} b_n(x-a)^n$ that equals a given infinitely differentiable function on an interval $(a-r, a+r)$, $r > 0$, by using differentiation, integration, or other algebraic manipulation of a given power series, or by using Taylor polynomial coefficients.

DIFFERENTIATION OF POWER SERIES

If $\Sigma_{n=0}^{\infty} b_n x^n$ has a nonzero radius of convergence r, then for each x in $(-r, r)$, $\Sigma_{n=0}^{\infty} b_n x^n$ converges to a number. If we call this number $f(x)$, then $\Sigma_{n=0}^{\infty} b_n x^n$ defines a function on $(-r, r)$ given by

$$f(x) = \sum_{n=0}^{\infty} b_n x^n \qquad \text{for } x \text{ in } (-r, r)$$

In such a circumstance we shall say $\Sigma_{n=0}^{\infty} b_n x^n$ **represents** $f(x)$ on $(-r, r)$. When introducing power series we noted that for each x in $(-1, 1)$, $\Sigma_{n=0}^{\infty} x^n$ is a geometric series converging to $1/(1-x)$. Thus $\Sigma_{n=0}^{\infty} x^n$ represents $f(x) = 1/(1-x)$ on $(-1, 1)$ because

$$f(x) = \frac{1}{1-x} = \sum_{n=0}^{\infty} x^n \qquad \text{for } x \text{ in } (-1, 1)$$

We can ask, in general, what are the properties of functions that can be represented by power series? As remarked earlier, a power series can be thought of as a generalized ("infinite-degree") polynomial, and polynomials have very convenient mathematical properties. They can be differentiated and integrated term-by-term as many times as we wish. It turns out that similar properties hold for power series, provided we stay in $(-r, r)$. Furthermore, as the following theorem asserts, the derivative of the power series that represents $f(x)$ will represent $f'(x)$ on $(-r, r)$. We leave its rather involved proof to more advanced courses.

16.3.1 **Differentiation of Power Series** Suppose $\Sigma_{n=0}^{\infty} b_n x^n$ has a radius of convergence $r > 0$ and $f(x) = \Sigma_{n=0}^{\infty} b_n x^n$ on $(-r, r)$. Then f is differentiable on $(-r, r)$ and

$$f'(x) = \sum_{n=1}^{\infty} n b_n x^{n-1} \qquad \text{for each } x \text{ in } (-r, r)$$

This theorem makes assertions about the *radius* of convergence and not the *interval* of convergence, which may or may not contain any of its endpoints. For the most part we shall confine our attention to the open interval and ignore the case-by-case analysis of endpoints. A similar theorem holds for powers of $(x - a)$, $a \neq 0$.

As illustrated in the next example, the best way to differentiate a power series is to write out its first several terms before differentiating. It is not advisable to work only with the general term because confusions regarding the indices are quite likely.

Example 1 Prove that $\displaystyle\sum_{n=0}^{\infty} \frac{x^n}{n!} = e^x$ for all real numbers x.

Solution By the Ratio Test the series $\Sigma_{n=0}^{\infty} \dfrac{x^n}{n!}$ converges for all x. (Check this!) If

$$(1) \qquad f(x) = \sum_{n=0}^{\infty} \frac{x^n}{n!} = 1 + x + \frac{x^2}{2!} + \frac{x^3}{3!} + \cdots + \frac{x^n}{n!} + \cdots$$

for all x, then by (16.3.1),

$$\begin{aligned} f'(x) &= 1 + \frac{2x}{2!} + \frac{3x^2}{3!} + \cdots + \frac{nx^{n-1}}{n!} + \cdots \\ &= 1 + x + \frac{x^2}{2!} + \cdots + \frac{x^{n-1}}{(n-1)!} + \cdots \\ &= 1 + x + \frac{x^2}{2!} + \cdots + \frac{x^{n-1}}{(n-1)!} + \frac{x^n}{n!} + \cdots \\ &= \sum_{n=0}^{\infty} \frac{x^n}{n!} = f(x) \end{aligned}$$

Which function equals its derivative? We showed in Section 8.1 that if $dy/dx = y$, then $y = ce^x$. Hence $f(x) = ce^x$ for some constant c. But since $f(0) = 1$ in (1), we can conclude that $c = 1$ and $f(x) = e^x$. ●

In Sec. 16.1 we constructed $P_n(x)$, the nth Taylor polynomial for $f(x) = e^x$ at $a = 0$, using the first n derivatives of e^x, getting

$$\begin{aligned} e^x \approx P_n(x) &= f(0) + \frac{f'(0)}{2!} + \frac{f''(0)}{2!} x^2 + \cdots + \frac{f^{(n)}(0)}{n!} x^n \\ &= 1 + \frac{x}{1!} + \frac{x^2}{2!} + \cdots + \frac{x^n}{n!} \end{aligned}$$

Since e^x has derivatives of all orders, the same procedure that gives the Taylor polynomials can be used to generate the power series in Example 1.

16.3.2 **Definition** Let f be a function which has derivatives of all orders at $x = a$. The series

$$\sum_{n=0}^{\infty} \frac{f^{(n)}(a)}{n!}(x-a)^n$$

is called the *Taylor series for f at a* or the *Taylor series expansion of f about a*. If $a = 0$, then the Taylor series

$$\sum_{n=0}^{\infty} \frac{f^{(n)}(0)}{n!}x_n$$

is often referred to as the *Maclaurin series*. (The British mathematician Colin Maclaurin was a younger contemporary of Taylor in the 1700s.)

In effect the limit, as $n \to +\infty$, of the Taylor *polynomial approximations* of e^x is the Taylor *series* for e^x. Furthermore, and most important, the power series represents e^x *exactly* for each x. Recall that (16.1.3) states that

$$f(x) = P_n(x) + R_n(x)$$

where $R_n(x)$, the nth Taylor remainder, is the error involved in using $P_n(x)$ to approximate $f(x)$. If a function f has derivatives of all orders in an open interval I containing a, then it is always possible to form the Taylor series for f at a. However, Example 8 will show that *the Taylor series for f need not represent f on I*, even when the power series converges for all x. The key is whether or not $R_n(x) \to 0$ as $n \to +\infty$.

16.3.3 **Taylor's Theorem** Suppose f has derivatives of all orders on $(a - r, a + r)$ for $r > 0$. Then

$$f(x) = \sum_{n=0}^{\infty} \frac{f^{(n)}(a)}{n!}(x-a)^n$$

if and only if $\lim_{k \to +\infty} R_k(x) = 0$.

Proof: For each positive integer n we know that $f(x) - P_k(x) = R_k(x)$ on $(a - r, a + r)$.

$$f(x) = \sum_{n=0}^{\infty} \frac{f^{(n)}(a)}{n!}(x-a)^n \quad \text{if and only if}$$

$$f(x) = \lim_{k \to +\infty} P_k(x) \quad \text{if and only if}$$

$$\lim_{k \to +\infty} [f(x) - P_k(x)] = 0 \quad \text{if and only if}$$

$$\lim_{k \to +\infty} R_k(x) = 0 \quad \blacksquare$$

Example 2 Show that the Maclaurin series for $\sin x$ is given by

$$\sin x = x - \frac{x^3}{3!} + \frac{x^5}{5!} - \cdots + (-1)^n \frac{x^{2n+1}}{(2n+1)!} + \cdots$$

$$= \sum_{n=0}^{\infty} \frac{(-1)^n x^{2n+1}}{(2n+1)!} \quad \text{for all } x$$

Solution We have already shown in [16.1.5(*a*)] that the Taylor polynomials for $\sin x$ are the partial sums for the series indicated above for all x. Hence it remains to show that as $n \to +\infty$, the remainder tends to 0. But, by (16.1.3),

$$R_n(x) = \frac{f^{(n+1)}(c_n)}{(n+1)!}x^{n+1}$$

for some c_n with $|c_n| < |x|$. But the only possibilities for $f^{(n+1)}(c_n)$ are $\pm \sin c_n$ or $\pm \cos c_n$. In any of these cases $|f^{(n+1)}(c_n)| \le 1$, so

$$0 \le |R_n(x)| = \frac{|f^{(n+1)}(c_n)|\,|x|^{n+1}}{(n+1)!} \le \frac{|x|^{n+1}}{(n+1)!}$$

However, by the Special Limit Theorem [15.1.6(*h*)], for any real number r, $\lim_{n\to\infty} r^n/n! = 0$. Hence

$$0 \le \lim_{n\to+\infty} |R_n(x)| \le \lim_{n\to+\infty} \frac{|x|^{n+1}}{(n+1)!} = 0 \qquad \text{for all } x$$

It follows by the Squeeze Theorem (15.1.5) that $\lim_{n\to+\infty} |R_n(x)| = 0$, so $\lim_{n\to+\infty} R_n(x) = 0$. The conclusion follows by Taylor's Theorem. ●

Although we could develop the Maclaurin series for $\cos x$ similarly, we choose instead to differentiate the series for $\sin x$ term by term. By the Differentiation Theorem for Power Series (16.3.1), the derivative of the sine series will give the cosine series:

$$D_x[\sin x] = D_x\left[x - \frac{x^3}{3!} + \frac{x^5}{5!} - \frac{x^7}{7!} + \cdots + \frac{(-1)^n x^{2n+1}}{(2n+1)!} + \cdots\right]$$

$$(n = 0, 1, 2, \ldots)$$

so

$$\begin{aligned}\cos x &= 1 - \frac{3x^2}{3!} + \frac{5x^4}{5!} - \frac{7x^6}{7!} + \cdots + \frac{(-1)^n(2n+1)x^{2n}}{(2n+1)!} + \cdots \\ &= 1 - \frac{x^2}{2!} + \frac{x^4}{4!} - \frac{x^6}{6!} + \cdots + \frac{(-1)^n x^{2n}}{(2n)!} + \cdots \\ &= \sum_{n=0}^{\infty} \frac{(-1)^n x^{2n}}{(2n)!} \qquad \text{for all } x\end{aligned}$$

Of course, the partial sums of this series are precisely the Taylor polynomials for $\cos x$ recorded in [16.1.5(*b*)], page 522.

PROGRESS TEST 1

1. Show that the Maclaurin series for $f(x) = 1/(1-x)$ is $\Sigma_{n=0}^{\infty} x^n$ for x in $(-1, 1)$. [Note that earlier statements confirm that this power series represents $f(x) = 1/(1-x)$ on $(-1, 1)$.]
2. Use Prob. 1 to give a power series representation for $g(x) = 1/(1-x)^2$ on $(-1, 1)$.

COMPUTATIONS OF POWER SERIES REPRESENTATIONS

Problems 1 and 2 of Progress Test 1 suggests that power series may arise or be computed in more than one way. The following theorem will be very useful in dealing with power series representations for functions because it guarantees, in effect, that *any power series representation of a function on an open interval is its Taylor series*—no matter how that series is computed.

16.3.4 Uniqueness of Power Series Representations If $f(x) = \sum_{n=0}^{\infty} b_n(x-a)^n$ for $|x-a| < r$, then $b_n = f^n(a)/n!$ for each $n = 0, 1, 2, \ldots$. Hence if

$$\sum_{n=0}^{\infty} b_n(x-a)^n = \sum_{n=0}^{\infty} c_n(x-a)^n \qquad \text{on } (a-r, a+r)$$

then

$$b_n = c_n = \frac{f^{(n)}(a)}{n!} \qquad \text{for each } n = 0, 1, 2, \ldots$$

Any two power series representations of a function f on an interval (a − r, a + r) must be the same.

Proof: We apply the Differentiation Theorem (16.3.1) repeatedly:

$$\begin{aligned}
f(x) &= b_0 + b_1(x-a) + b_2(x-a)^2 + b_3(x-a)^3 + b_4(x-a)^4 + \cdots \\
f'(x) &= b_1 + 2b_2(x-a) + 3b_3(x-a)^2 + 4b_4(x-a)^3 + \cdots \\
f''(x) &= 2b_2 + 3 \cdot 2b_3(x-a) + 4 \cdot 3b_4(x-a)^2 + \cdots \\
f'''(x) &= 3 \cdot 2b_3 + 4 \cdot 3 \cdot 2b_4(x-a) + 5 \cdot 4 \cdot 3 \cdot 2b_5(x-a)^2 + \cdots \\
&\vdots \\
f^{(n)}(x) &= n!b_n + (n+1)!b_{n+1}(x-a) + (n+2)!b_{n+2}(x-a)^2 + \cdots
\end{aligned}$$

Substituting $x = a$, we obtain

$$f^{(n)}(a) = n!b_n$$

so

$$b_n = \frac{f^{(n)}(a)}{n!} \quad \blacksquare$$

Notice that in the above proof we used the fact, implied by (16.3.1), that if f has a power series representation on $(a-r, a+r)$, then f has derivatives of all orders on $(a-r, a+r)$. Some authors refer to a function with such a power series representation as *analytic on* $(a-r, a+r)$.

We are now free to compute power series representations for functions without resorting to the usually difficult procedure of computing the Taylor series directly via nth derivatives and the often more difficult chore of showing that $R_n(x) \to 0$. Notice, for example, how difficult it would be to compute higher derivatives in the following example.

Example 3 Compute the Maclaurin series for $f(x) = xe^x$.

Solution By Example 1, we know that $e^x = \sum_{n=0}^{\infty} x^n/n!$ for all x. Hence

$$xe^x = x\left[\sum_{n=0}^{\infty} \frac{x^n}{n!}\right] = \sum_{n=0}^{\infty} \frac{x^{n+1}}{n!} \qquad \text{for all } x$$

(Note that x acts as a constant with respect to the series $\sum_{n=0}^{\infty} x^n/n!$, so [15.2.3(*b*)] page 490, applies.) ●

Another way of constructing power series representations amounts to a "composition" with a known series. In such cases it is important to note how the radius of convergence is affected. In general, if the original representation is valid for $|x-a| < r$ and we replace x by $u(x)$ in this representation, then the new representation is valid for all x such that $|u(x) - a| < r$. Usually we solve this last inequality for $|x-a|$.

Example 4 Construct power series representations for

(*a*) $\dfrac{1}{1+x}$ (*b*) $\dfrac{1}{1+4x^2}$ (*c*) $\dfrac{1}{(1-x^2)^2}$

Solution (*a*) We regard $1/(1+x)$ as $1/(1-(-x))$ and replace x by $-x$ in the power series for $1/(1-x)$ at $a=0$:

$$\frac{1}{1+x} = \sum_{n=0}^{\infty}(-x)^n = \sum_{n=0}^{\infty}(-1)^n x^n = 1 - x + x^2 - x^3 + \cdots$$

The original radius of convergence is 1. But $|-x| < 1$ if and only if $|x| < 1$, so the radius of convergence of $\Sigma_{n=0}^{\infty}(-1)^n x^n$ is also 1.

(*b*) We use the series constructed in (*a*) as the starting point and replace x by $4x^2$, getting

$$\begin{aligned}\frac{1}{1+4x^2}\\ &= \sum_{n=0}^{\infty}(-1)^n(4x^2)^n = 1 - 4x^2 + 4^2(x^2)^2 - 4^3(x^2)^3 + 4^4(x^2)^4 - \cdots\\ &= \sum_{n=0}^{\infty}(-1)^n 4^n x^{2n}\end{aligned}$$

The original series converges for $|x| < 1$, so the new series converges when $|4x^2| < 1$, that is, when $|x| < 1/2$.

(*c*) In Progress Test 1, we differentiated the series for $1/(1-x)$ to get the series for $1/(1-x)^2$:

$$\frac{1}{(1-x)^2} = \sum_{n=1}^{\infty} nx^{n-1} = 1 + 2x + 3x^2 + \cdots \qquad \text{for } x \text{ in } (-1,1)$$

We now replace x by x^2 to get the desired power series:

$$\begin{aligned}\frac{1}{(1-x^2)^2} &= \sum_{n=1}^{\infty} n(x^2)^{n-1} = 1 + 2(x^2) + 3(x^2)^2 + 4(x^2)^3 + \cdots\\ &= \sum_{n=1}^{\infty} n(x^{2n-2})\end{aligned}$$

The original series converges for $|x| < 1$, and since $|x^2| < 1$ if and only if $|x| < 1$, we know that the latter series has a radius of convergence equal to 1. ●

INTEGRATION OF POWER SERIES

The Differentiation Theorem for power series (16.3.1) has an equally useful analog for integration. [We omit its proof, which is an application of (16.3.1) and the Fundamental Theorem of Calculus.]

16.3.5 **Integration of Power Series** Suppose $\Sigma_{n=0}^{\infty} b_n x^n$ has a radius of convergence $r > 0$ and $f(x) = \Sigma_{n=0}^{\infty} b_n x^n$. Then f is integrable on $(-r, r)$ and

$$\int_0^x f(t)\,dt = \sum_{n=0}^{\infty} \frac{b_n x^{n+1}}{n+1}$$

This theorem allows us to integrate power series term by term on the interior of the interval convergence and to construct Taylor or Maclaurin series that could be difficult to construct using derivatives. A similar theorem holds for power series in $(x - a)$ with 0 replaced by a in the lower limit of integration.

Example 5 Determine a power series representation for

$$f(x) = \arctan x \qquad \text{for } |x| < 1$$

Solution Using Example 4(*a*), we replace x by x^2 in

$$\frac{1}{1+x} = \sum_{n=0}^{\infty} (-1)^n x^n$$

to obtain

$$\frac{1}{1+x^2} = \sum_{n=0}^{\infty} (-1)^n (x^2)^n = \sum_{n=0}^{\infty} (-1)^n x^{2n} \qquad \text{for } |x| < 1$$

But now, applying (16.3.5) to this latter series, we have

$$\arctan x = \int_0^x \frac{dt}{1+t^2} = \sum_{n=0}^{\infty} \frac{(-1)^n x^{2n+1}}{2n+1} \qquad \text{for } |x| < 1 \quad \bullet$$

Notice that while arctan x is defined for all real numbers, the power series representation is not. It can be shown using Taylor's Remainder Theorem that the representation is valid on $[-1, 1]$. However, for $|x| > 1$, the power series diverges.

Our techniques also extend to power series at nonzero a.

Example 6 Determine the Taylor series for $\ln x$ at $a = 1$.

Solution Writing

$$\frac{1}{x} = \frac{1}{1 + (x-1)}$$

we then replace x by $x - 1$ in the power series representation of $1/(1 + x)$; that is, $\Sigma_{n=0}^{\infty} (-1)^n x^n$ [see Example 4(*a*)]. We obtain

$$\frac{1}{x} = \frac{1}{1+(x-1)} = \sum_{n=0}^{\infty} (-1)^n (x-1)^n \qquad |x - 1| < 1$$

We now apply the Integration Theorem to get

$$\ln x = \int_1^x \frac{dt}{t} = \sum_{n=0}^{\infty} \frac{(-1)^n (x-1)^{n+1}}{n+1} \qquad |x - 1| < 1$$

(Compare with the Taylor polynomials for $\ln x$ in Sec. 16.1, page 517.) $\bullet$

PROGRESS TEST 2

Determine power series for each function given at $a = 0$. Note the radius of convergence for this power series.

1. xe^{-x}

2. $\sin x^2$

3. $\ln\left(\frac{1+x}{1-x}\right)$, $(|x| < 1)$

(In Prob. 3, integrate the series for $1/(1 + x)$ and $1/(1 - x)$, respectively, and then subtract, using the fact that the difference of convergent series is convergent.)

THE BINOMIAL SERIES

Most of our earlier manipulations with series have used the geometric series as a starting point. We now introduce another "starting-point" series, which is a generalization of the Binomial Theorem.

16.3.6 **Binomial Theorem** If m is a positive integer, and a and b are real numbers, then

$$(a+b)^m = a^m + \binom{m}{1}a^{m-1}b + \binom{m}{2}a^{m-2}b^2 + \cdots + \binom{m}{m-1}ab^{m-1} + b^m$$

where

$$\binom{m}{n} = \frac{m!}{n!(m-n)!} = \frac{m(m-1)(m-2)\cdots(m-n+1)}{n!}$$

Our particular interest is in expanding $(1+x)^m$, where m is not necessarily a positive integer. We shall define an appropriate series, show that its radius of convergence is 1, and assert that it *does* represent the function $(1+x)^m$.

16.3.7 **Definition** The series

$$1 + mx + \frac{m(m-1)}{2!}x^2 + \cdots + \frac{m(m-1)\cdots(m-n+1)}{n!}x^n + \cdots$$

$$= 1 + \sum_{n=1}^{\infty}\binom{m}{n}x^n$$

for m any real number, is called the *binomial series*. The numbers

$$\binom{m}{n} = \frac{m(m-1)\cdots(m-n+1)}{n!}$$

are called the *binomial coefficients*.

We can use the Ratio Test to determine the radius of convergence of the binomial series:

$$\left|\frac{\binom{m}{n+1}x^{n+1}}{\binom{m}{n}x^n}\right|$$

$$= |x|\left|\frac{m(m-1)\cdots(m-n+1)(m-n)}{(n+1)!}\cdot\frac{n!}{m(m-1)\cdots(m-n+1)}\right|$$

$$= |x|\frac{|m-n|}{(n+1)}$$

Since $\lim_{n\to+\infty}|x|\,|m-n|/(n+1) = |x|$, we know that the radius of convergence is 1.

For m a positive integer, the binomial series reduces to the polynomial expansion of $(1+x)^m$ given by the Binomial Theorem. (Check this.) In general we have the following:

16.3.8 **Theorem** For m any real number and any x in $(-1, 1)$,

$$(1+x)^m = 1 + \sum_{n=1}^{\infty}\binom{m}{n}x^n$$

This theorem can be proven either using Taylor series or manipulations of the derivatives of $(1 + x)^m$ and the binomial series. An outline of the proof is given in Exercise 45. In any case, the Uniqueness Theorem for power series (16.3.4) tells us that the traditional binomial coefficients are also the Taylor coefficients $f^{(n)}(0)/n!$.

Example 7 Determine the Maclaurin series representation for $\sqrt{1+x}$.

Solution Let $m = 1/2$. Then, by (16.3.8),

$$(1 + x)^{1/2} = 1 + \sum_{n=1}^{\infty} \binom{1/2}{n} x^n \qquad \text{for } |x| < 1$$

Now

$$\binom{1/2}{1} = \frac{1}{2}$$

$$\binom{1/2}{2} = \frac{1/2(1/2 - 1)}{2!} = \frac{(1/2)(-1/2)}{2!} = \frac{-1}{2^2 2!}$$

$$\binom{1/2}{3} = \frac{1/2(1/2 - 1)(1/2 - 2)}{3!} = \frac{(1/2)(-1/2)(-3/2)}{3!} = \frac{1 \cdot 3}{2^3 \cdot 3!}$$

and, in general

$$\begin{aligned}\binom{1/2}{n} &= \frac{1/2(1/2 - 1)(1/2 - 2) \cdots (1/2 - n + 1)}{n!} \\ &= \frac{1/2(-1/2)(-3/2) \cdots ((3 - 2n)/2)}{n!} \\ &= \frac{(-1)^{n-1}(1 \cdot 3 \cdot 5 \cdot \cdots \cdot (2n - 3))}{2^n n!}\end{aligned}$$

Hence

$$\sqrt{1+x} = 1 + \sum_{n=1}^{\infty} \frac{(-1)^{n-1}(1 \cdot 3 \cdot 5 \cdot \cdots \cdot (2n - 3))}{2^n \cdot n!} x^n \quad \bullet$$

PROGRESS TEST 3

1. Determine a power series representation of $f(x) = 1/\sqrt{1-x}$ for x in $(-1, 1)$.

MACLAURIN SERIES FOR e^{-1/x^2}

The following example promised earlier shows that a Taylor series developed from a given infinitely differentiable function can converge for all x yet fail to represent that function.

Example 8 Show that the Maclaurin series for

$$f(x) = \begin{cases} e^{-1/x^2} & \text{if } x \neq 0 \\ 0 & \text{if } x = 0 \end{cases}$$

converges everywhere but represents f only when $x = 0$.

Solution First, $f(0) = 0$. To determine $f'(0)$ and subsequent derivatives we shall use the definition of derivative:

$$f'(0) = \lim_{x\to 0} \frac{f(x) - f(0)}{x - 0} = \lim_{x\to 0} \frac{f(x)}{x}$$
$$= \lim_{x\to 0} \frac{e^{-1/x^2}}{x} = \lim_{x\to 0} \frac{1/x}{e^{1/x^2}}$$

Since $\lim_{x\to 0} |1/x| = \lim_{x\to 0} |e^{1/x^2}| = +\infty$, we can apply l'Hospital's Rule (∞/∞) to obtain

$$f'(0) = \lim_{x\to 0} \frac{1/x}{e^{1/x^2}} = \lim_{x\to 0} \frac{-1/x^2}{e^{1/x^2}(-2/x^3)}$$
$$= \lim_{x\to 0} \frac{x}{2e^{1/x^2}} = 0$$

$$\text{For } x \neq 0, D_x(e^{-1/x^2}) = \frac{2}{(x^3)}(e^{-1/x^2})$$

and therefore

$$f'(x) = \begin{cases} \dfrac{2e^{-1/x^2}}{x^3} & \text{if } x \neq 0 \\ 0 & \text{if } x = 0 \end{cases}$$

Now

$$f''(0) = \lim_{x\to 0} \frac{f'(x) - f'(0)}{x - 0} = \lim_{x\to 0} \frac{f'(x)}{x}$$
$$= \lim_{x\to 0} \frac{2e^{-1/x^2}}{x^4} = \lim_{x\to 0} \frac{2/x^4}{e^{1/x^2}} \qquad (\infty/\infty)$$
$$= \lim_{x\to 0} \frac{-8/x^5}{e^{1/x^2}(-2/x^3)} = \lim_{x\to 0} \frac{4}{x^2 e^{1/x^2}} = 0$$

Similarly, $f^{(n)}(0) = 0$ for all n. Thus the Maclaurin series for f is

$$0 + 0x + \frac{0x^2}{2!} + \cdots + \frac{0x^n}{n!} + \cdots = 0$$

But $f(x) \neq 0$ if $x \neq 0$ and thus, although the Maclaurin series converges for all values of x, it only represents f at $x = 0$. By Taylor's Theorem, $R_n(x) \not\to 0$ except for $x = 0$. ●

SOME STANDARD MACLAURIN SERIES

We list for reference the Maclaurin series that have been developed to date or can be determined in a similar fashion.

16.3.9 (*a*) $\sin x = \sum_{n=0}^{\infty} \frac{(-1)^n x^{2n+1}}{(2n+1)!}$

(*b*) $\cos x = \sum_{n=0}^{\infty} \frac{(-1)^n x^{2n}}{(2n)!}$

(*c*) $e^x = \sum_{n=0}^{\infty} \frac{x^n}{n!}$

(*d*) $\ln(1+x) = \sum_{n=0}^{\infty} \frac{(-1)^n x^{n+1}}{n+1} \qquad |x| < 1$

$$(e)\quad \arctan x = \sum_{n=0}^{\infty} \frac{(-1)^n x^{2n+1}}{2n+1} \qquad |x| < 1$$

$$(f)\ (1+x)^m = 1 + \sum_{n=1}^{\infty} \frac{m(m-1)\cdots(m-n+1)}{n!} x^n \qquad |x| < 1$$

$$(g)\quad \sinh x = \sum_{n=0}^{\infty} \frac{x^{2n+1}}{(2n+1)!}$$

$$(h)\quad \cosh x = \sum_{n=0}^{\infty} \frac{x^{2n}}{(2n)!}$$

SECTION 16.3 EXERCISES

In Exercises 1 to 14 give a power series representation (at $a = 0$ unless otherwise indicated) for the given function on the indicated interval. If the interval is not given, identify the radius of convergence. You may use representations developed earlier unless specific directions call for another approach.

1. $f(x) = \dfrac{1}{(1-x)^3}$, $|x| < 1$

2. $f(x) = e^{3x}$, all x

3. $f(x) = x^2 e^{-x^2}$, all x

4. $f(x) = (\sin x)/x$, all $x \neq 0$

5. $f(x) = (x-1)\ln x$, at $a = 1$

6. $f(x) = \sin x$, at $a = \pi/4$

7. $f(x) = (1 - \cos x)/x$, $x \neq 0$

8. $f(x) = \ln(1 + x^2)$

9. $f(x) = x \ln(1 + x^4)$, $|x| < 1$

10. $f(x) = \dfrac{\arctan(x^3)}{x}$

11. $f(x) = \displaystyle\int_0^x \frac{dt}{1-4t^2}$, $|x| < 1/2$

12. $f(x) = \displaystyle\int_0^x \ln\left(\frac{1+t}{1-t}\right) dt$, $|x| < 1$

13. $f(x) = \displaystyle\int_0^x \frac{t\,dt}{4+t^4}$, $|x| < \sqrt{2}$

14. $f(x) = \displaystyle\int_0^x e^{-t^2}\,dt$

15. (*a*) Use the definition of Maclaurin series to prove [16.3.9(*g*)].
(*b*) Derive [16.3.9(*g*)] from [16.3.9(*h*)].

16. Derive [16.3.9(*g*)] using the definition of sinh x.

17. Derive [16.3.9(*h*)] using the definition of cosh x.

18. Use integration and Example 7 to give a power series for $\sqrt{(1+x)^3}$.

19. Determine a power series representation for $f(x) = (1 - 2x^2)/(1 + 2x^2)^2$ by using the following outline:
(*a*) Replace x in the geometric series [which represents $1/(1-x)$] by $-2x^2$ and note the radius of convergence of the resulting series.
(*b*) Multiply this series by x and differentiate both the function represented and the series to obtain a series representation for $f(x)$.

20. Determine a power series expansion for $\int_0^x \arctan t^2\,dt$, $|x| < 1$, by the following given outline:
(*a*) Make an appropriate substitution into [16.3.9(*e*)].
(*b*) Integrate.

21. Determine a power series expansion for $f(x) = x/(x^2 + 4x + 6)$ in powers of $x + 2$ by the following given outline:
(*a*) Complete the square and break the fraction into two terms by writing its numerator as $(x + 2) - 2$.
(*b*) Substitute $(-x^2/2)$ in the geometric series and determine the radius of convergence.
(*c*) Replace x in the resulting representation by $x + 2$ and multiply both sides by $(x + 2)/2$ to obtain a representation of the first fraction term of (*a*).
(*d*) Subtract the representation for $2/[(x^2 + 2) + 2]$ that occurred in (*c*) and combine the resulting series by writing out the first several terms.

22. Use the idea of Exercise 21 to determine a power series representation of $f(x) = \ln(x^2 + 4x + 13)$ in powers of $x + 2$.

23. (*a*) Determine a power series for $f(x) = \arcsin x$ by making an appropriate substitution into the series in Progress Test 3, Prob. 1, and integrating.
(*b*) Write out and simplify the first four terms of the series in (*a*).

24. Determine a power series representation of $f(x) = 1/(x^2 - 4x + 5)$ in terms of $x - a$ for an appropriate a. State the radius of convergence.

25. Repeat Exercise 24 for $f(x) = \arctan\left(\dfrac{x-2}{3}\right)$.

26. Repeat Exercise 24 for $f(x) = \ln(2 - x - x^2)$. (*Hint:* Factor $2 - x - x^2$.)

In Exercises 27 to 38 compute the Maclaurin series (except Exercise 28) for the given function using earlier results where appropriate. Give the radius of convergence. (Note that in some exercises trigonometric identities may be useful.)

27. $\sin x \cos x$

28. $(x - 1)e^{-x^2+2x-1}$, at $a = 1$

29. $\sin^2(2x)$

30. $\int_0^x \ln(1-t)\,dt$

31. $\dfrac{x}{(1-x)^2}$

32. $\dfrac{\sin(x^2)}{x}$

33. $\int_0^x u^3\cos(\sqrt{u})\,du$

34. $\cos^2(x/3)$

35. $\sqrt{6+3x}$

36. $\dfrac{x}{\sqrt{1-x^2}}$

37. $\sqrt[3]{1+x}$

38. $\dfrac{x}{\sqrt[3]{1-x^2}}$

39. Suppose that

$$f(x) = \sum_{n=0}^{\infty} \frac{(-1)^n x^{2n+1}}{(2n+1)!}$$

for all x. Show that $y = f(x)$ satisfies the differential equation $y'' + y = 0$.

40. Use the result of Exercise 4 to give another argument that $\lim_{x\to 0}[(\sin x)/x] = 1$.

41. Prove that $f(x) = \ln(1+x)$ is represented by its Maclaurin series on $(0, 1]$ by following the given outline. (This fact has, in effect, been proved earlier, and you may use some of these earlier calculations as a check on your work.)

(*a*) Determine $f^{(n)}(x)$ and then $f^{(n+1)}(x)/(n+1)!$.
(*b*) Use (*a*) to determine $R_n(x)$ for x in $(0, 1]$.
(*c*) Show that $x/(1+c_n) < 1$.
(*d*) Use (*c*) to prove that $R_n(x) \to 0$ as $n \to +\infty$. [$R_n(x)$ also tends to 0 on $(-1, 0)$, but this is more difficult to show.]

42. Using the Taylor series for $\ln x$ at $a = 1$ developed in Example 6, show directly that $\lim_{n\to+\infty} R_n(x) = 0$ for x in $(1, 2]$. [$R_n(x)$ also tends to 0 for x in $(0, 1]$, but this is more difficult to show.]

43. Find the Taylor series for $1/\sqrt{x}$ at $a = 1$ and show directly that $\lim_{n\to+\infty} R_n(x) = 0$ for x in $[1, 2)$.

44. The differential equation

$$x\frac{d^2y}{dx^2} + \frac{dy}{dx} + xy = 0$$

is called **Bessel's equation of order zero.** For $x > 0$ define

$$J_0(x) = \sum_{n=0}^{\infty} \frac{(-1)^n x^{2n}}{(n!)^2 2^{2n}}$$

Show that $x(J_0(x))'' + (J_0(x))' + x(J_0(x)) = 0$.

45. Let $f(x) = 1 + \Sigma\binom{m}{n}x^n$, $-1 < x < 1$.
(*a*) Show that $(1+x)f'(x) = mf(x)$.
(*b*) Show that $f(x) = (1+x)^m$ by showing that $\ln(f(x)) = \ln[(1+x)^m]$.

Section 16.4: Miscellaneous Applications of Power Series

Objectives:

1. Use power series to approximate the values of the functions they represent.
2. Use power series to represent nonelementary antiderivatives.
3. Use power series to approximate definite integrals.

NUMERICAL APPROXIMATIONS

Our work with Taylor polynomials in Sec. 16.1 gave us methods to approximate the value of a function, with bounds on $|R_n(x)|$ providing accuracy estimates. In some cases the calculations required in using $R_n(x)$ are quite complicated, but can be circumvented using (15.4.6), the First-Term-Neglected Estimate for an alternating series $\sum_{n=0}^{\infty}(-1)^n a_n$ whose terms are decreasing in absolute value (recall page 507):

> the sum of the terms $a_0 - a_1 + a_2 - \cdots + a_k$ is within $|a_{k+1}|$ of the sum of the entire series

Example 1 Approximate $\ln(3/2)$ through three decimal places using the power series representation of $\ln(1+x)$, $|x| < 1$.

Solution Using [16.3.9(*d*)], for $|x| < 1$, we have

$$\ln(1+x) = \sum_{n=0}^{\infty} \frac{(-1)^n x^{n+1}}{n+1}$$

Thus, with $3/2 = 1 + 1/2$,

$$\ln(3/2) = \sum_{n=0}^{\infty} \frac{(-1)^n(1/2)^{n+1}}{n+1} = \sum_{n=0}^{\infty} \frac{(-1)^n}{2^{n+1}(n+1)}$$

This is an alternating series with $1/[2^{n+1}(n+1)]$ decreasing, so the sum

$$\sum_{n=0}^{k} \frac{(-1)^n}{(2^{n+1})(n+1)}$$

is within $1/[2^{k+2}(k+2)]$ of the sum of the entire series, namely, $\ln(3/2)$.

If $k = 5$, the error will be bounded by

$$\frac{1}{2^7 \cdot 7} = \frac{1}{896} \approx 0.001$$

If $k = 6$, the error will be bounded by

$$\frac{1}{2^8 \cdot 8} = \frac{1}{2048} \approx 0.000488$$

yielding the desired accuracy (error < 0.0005). Hence

$$\ln(3/2) \approx \frac{1}{2} - \frac{1}{4 \cdot 2} + \frac{1}{8 \cdot 3} - \frac{1}{16 \cdot 4} + \frac{1}{32 \cdot 5} - \frac{1}{64 \cdot 6} + \frac{1}{128 \cdot 7} \approx 0.406 \quad \bullet$$

In the previous example we used "error" when, to be precise, we meant the *absolute value* of the error. We shall use "error" in this way throughout this section.

Notice that the above technique is preferable to a Taylor polynomial approach *when* it applies, because we merely compute additional terms of the series to improve our accuracy and examine the size of the first term not added. This is in contrast to the Taylor polynomial approach, where computation of bounds on the $(k + 1)$st derivative can be cumbersome and need be repeated for each improvement in accuracy.

If we can manipulate a known series to fit a given situation, we can then avoid calculating derivatives and compute numerical values directly. In Sec. 16.1, we revisited the approximation of $\sqrt[5]{33}$ using Taylor polynomials for $f(x) = x^{1/5}$ at $a = 32$, computing an approximation accurate through six decimal places. Using the binomial series, only five terms yield an approximation accurate through eight places. (See Sec. 16.4, Exercise 8.)

APPLICATIONS OF POWER SERIES TO INTEGRATION

In Sec. 11.6 we developed two methods, the Trapezoid Rule and Simpson's Rule, for approximating a definite integral for which the antiderivative was not an elementary function, a function expressible as a *finite* combination of functions studied up to then (powers, roots, quotients and compositions of polynomials, trig functions, or log and exponential functions). We now have the means to express such nonelementary functions more explicitly. By manipulating such power series representations, we have yet another way to approximate specific definite integrals. However, the following example accomplishes more than we were able to do in Section 11.6, namely, represent an actual nonelementary *function.*

Example 2 Compute an antiderivative of e^{-x^2}.

Solution (It was noted in Sec. 11.5 that such an antiderivative is not an elementary function.) Substituting $-x^2$ in [16.3.9(*c*)], we have

$$e^{-x^2} = \sum_{n=0}^{\infty} \frac{(-x^2)^n}{n!} = \sum_{n=0}^{\infty} \frac{(-1)^n x^{2n}}{n!} \qquad \text{for all } x$$

Thus, applying the Integration Theorem for power series (16.3.5), we obtain

$$\begin{aligned}\int_0^x e^{-t^2}\,dt &= \int_0^x \left[\sum_{n=0}^{\infty} \frac{(-1)^n t^{2n}}{n!}\right] dt \\ &= \sum_{n=0}^{\infty} \left[\frac{(-1)^n}{n!} \int_0^x t^{2n}\,dt\right] \\ &= \sum_{n=0}^{\infty} \frac{(-1)^n x^{2n+1}}{n!(2n+1)}\end{aligned}$$

This gives us a power series representation of $\int_0^x e^{-t^2}\,dt$, valid for all x. ●

As Prob. 3 of both progress tests in Sec. 11.6 we estimated $\int_0^2 e^{-x^2}\,dx$ using 10 subdivisions and used derivatives to determine bounds on the errors involved; actually, we used an informal bound on the fourth derivative for Simpson's Rule because the fifth derivative was too complicated to analyze directly. Three-place accuracy resulted using either rule.

Example 3 (*a*) Estimate $\int_0^2 e^{-x^2}\,dx$ using

$$\sum_{n=0}^{k} \frac{(-1)^n x^{2n+1}}{n!(2n+1)} \qquad \text{for } k = 10$$

(*b*) Give a bound for the error in (*a*).
(*c*) Give a bound for the error when $k = 20$.

Solution (*a*) We merely substitute into the power series:

$$\begin{aligned}\int_0^2 e^{-x^2}\,dx &\approx 2 + \frac{(-1)2^3}{1!\cdot 3} + \frac{(-1)^2 2^5}{2!\cdot 5} + \frac{(-1)^3 2^7}{3!\cdot 7} + \cdots + \frac{(-1)^{10}2^{21}}{10!\cdot 21} \\ &= 2 - \frac{2^3}{1!\cdot 3} + \frac{2^5}{2!\cdot 5} - \frac{2^7}{3!\cdot 7} + \cdots + \frac{2^{21}}{10!\cdot 21} \approx 0.889045\end{aligned}$$

(*b*) We have an alternating series, so it remains to show that its terms are decreasing in absolute value. This can be done by showing that

$$\frac{|a_n|}{|a_{n+1}|} > 1 \qquad \text{for } |a_n| = \left|\frac{(-1)^n\, 2^{2n+1}}{n!(2n+1)}\right|$$

But

$$\begin{aligned}\frac{|a_n|}{|a_{n+1}|} &= \frac{2^{2n+1}}{(n!)(2n+1)} \cdot \frac{(n+1)!(2n+3)}{2^{2n+3}} \\ &= \frac{2n+3}{2n+1} \cdot \frac{(n+1)}{2^2} > \frac{n+1}{4} > 1\end{aligned}$$

for $n > 3$. Thus, the terms beyond a_3 are decreasing in value, so the error is bounded by

$$|a_{11}| = \frac{|(-1)^{11}2^{23}|}{11!\cdot 23} \approx 0.00914$$

(*c*) For $k = 20$, the error is bounded by

$$|a_{21}| = \frac{2^{43}}{21! \cdot 43} = 0.000000004 \qquad \text{(eight-place accuracy)} \quad \bullet$$

Note that although we have leaned heavily on the alternating-series error bound, constructions comparable to the above series approximations apply for other types of series. However, determining bounds on the error involved may be more difficult, perhaps requiring use of Taylor's Remainder.

PROGRESS TEST 1

1. Here we improve on the Trapezoid Rule and Simpson's Rule approximations of ln 2 that occurred as Prob. 1 of both progress tests in Sec. 11.6. There, the Trapezoid Rule with five subdivisions gave ln 2 accurate through two places, Simpson's Rule with four subdivisions gave $\ln 2 \approx 0.693$ accurate in the third decimal place. To achieve five-place accuracy the Trapezoid Rule required 183 subdivisions and Simpson's Rule required 14 subdivisions.
(*a*) Determine a number x between 0 and 1 such that $(1 + x)/(1 - x) = 2$.
(*b*) Using the series for $\ln[(1 + x)/(1 - x)]$ developed in Progress Test 2, Sec. 16.3, determine how many terms are required to reach an approximation accurate in the third decimal place, given the previous information.
(*c*) If $\ln 2 \approx 0.693147$ is accurate in the sixth decimal place, how many terms are required to exceed the accuracy of the Trapezoid Rule with 183 subdivisions and Simpson's Rule with 14 subdivisions?

2. (*a*) Compute the (nonelementary) antiderivative $\int_0^x \sin(t^2)\,dt$.
(*b*) Use (*a*) to determine $\int_0^1 \sin(t^2)\,dt$ to within 10^{-6}.

SECTION 16.4 EXERCISES

In the problems below you may leave your approximation as a finite numerical sum.

1. (*a*) Use the Maclaurin series for arctan x with $n = 3$ to approximate π.
(*b*) Repeat (*a*) with $n = 5$.
(*c*) What value of n is required for 10-place accuracy?
(*d*) Compare the above results with the approximations of π given in Sec. 11.6 using the Trapezoid Rule and Simpson's Rule.

2. Compute $\displaystyle\int_0^1 e^{-x^3}\,dx$ within 10^{-6}.

3. Compute $\displaystyle\int_0^{1/2} \frac{dx}{1 + x^4}$ within 10^{-6}.

4. Using Sec. 16.3, Exercise 4, determine $\int_0^{1/2} (\sin x)/x\,dx$ within 10^{-6}.

5. Use the binomial series to approximate $\sqrt[3]{1.1}$ to five-place accuracy.

6. Use five terms of the appropriate binomial series to approximate $\sqrt[3]{28}$.

7. Approximate $\int_0^1 \sqrt{4 - x^2}\,dx$ through four decimal places.

8. By following the given outline, use the first five terms of the binomial series for $[1 + (x/32)]^{1/5}$ to approximate $\sqrt[5]{33}$ and give a bound for the error involved.
(*a*) Change the form of the given expression to $(1/2)(32 + x)^{1/5}$ and show that the radius of convergence of its binomial series is 32.
(*b*) Write out and add the first five terms of this binomial series for $x = 1$.
(*c*) Show that the nth term a_n of the binomial series is of the form $a_n = u/(n!5^n2^{5n-1})$, where u is the product of the numbers $5(k - 1) - 1$ for $k = 2, 3, \ldots, n$.
(*d*) It can be shown that the $|a_n|$ are decreasing. Use this fact to give a bound on the error involved in using the first five terms of the series as an approximation.
(*e*) Why does the series for $(1 + x)^{1/5}$ fail to yield an estimate for $\sqrt[5]{32}$?

9. (Another approximation of ln 2) (*a*) Show that the Maclaurin series for $\ln(1 - x)$ on $(-1, 1)$ converges at its left-hand endpoint and represents ln 2 there.
(*b*) Use (*a*) to give another series for ln 2.
(*c*) Use (*b*) to approximate ln 2 with $n = 2$.
(*d*) Repeat (*c*) with $n = 5$.

10. (*a*) Use the power series for $-\ln(1 - x)$ with $x = 1/2$ to give another series for ln 2. [*Note:* $-\ln(1/2) = \ln 2$.]
(*b*) Use (*a*) to approximate ln 2 with $n = 2$.
(*c*) Repeat (*b*) with $n = 5$.

11. Compare the three series approximations for ln 2 from Progress Test 1, Prob. 1 and Exercises 9 and 10, and the two integration approximations from Sec. 11.6, and rank them in order of convergence efficiency.

12. (*a*) Compute $\displaystyle\int_0^x \frac{dt}{1 + t^3}$.
(*b*) Estimate $\displaystyle\int_0^1 \frac{dx}{1 + x^3}$ within 10^{-6}.

In Exercises 13 to 17 evaluate the given integral to within 10^{-4}.

13. $\int_0^1 \sqrt{1+x^4}\,dx$ **14.** $\int_0^1 e^{-x^3}\,dx$ **15.** $\int_0^{1/2} \frac{dx}{1+x^4}$ **16.** $\int_0^{1/3} \frac{dx}{1+x^3}$

17. $\int_0^{1/2} \arctan(x^2)\,dx$

18. (*a*) Determine a formal alternating series for $\int_0^x (1-e^{-t})/t\,dt$ (the "exponential integral") without regard to convergence.
(*b*) Show that

$$\lim_{\varepsilon \to 0^+} \int_\varepsilon^x \frac{1-e^{-t}}{t}\,dt = \int_0^x \frac{1-e^{-t}}{t}\,dt$$

as defined in (*a*).
(*c*) Compute $\int_0^1 \frac{1-e^{-t}}{t}\,dt$ to within 10^{-6}.

CHAPTER EXERCISES

Review Exercises

In Exercises 1 to 9, (a) determine the fourth Taylor polynomial for the given function at the given point, (b) determine an upper bound on the error involved in using this polynomial to approximate the quantity given in brackets.

1. e^x, $a = 1$; $[\sqrt{e}]$
2. $\sin x$, $a = \pi/6$; $[\sin(\pi/4)]$
3. $\cos x$, $a = \pi/4$; $[\cos(\pi/3)]$
4. $\tan x$, $a = \pi/4$; $[\tan(\pi/6)]$
5. $\csc x$, $a = \pi/4$; $[\csc(\pi/6)]$
6. 2^x, $a = 0$; $[\sqrt[3]{2}]$
7. $x \sin x$, $a = \pi/2$; $[(\sin 0.1)/10]$
8. $x^2 \ln x$, $a = 1$; $[e^2]$
9. $e^x + e^{2x}$, $a = 0$; $[e + e^2]$

10. Use Taylor polynomials to approximate $\sqrt{15}$ accurate through four decimal places.

11. Expand the polynomial $f(x) = x^4 + 2x^3 + 2x^2 + 4x + 1$ in powers of $x - 2$.
(*b*) Use part (*a*) to compute $f(2.1)$.

12. (*a*) Determine the nth Taylor polynomial for $f(x) = 1/\sqrt{1-x}$ at $a = 0$.
(*b*) Use the result of part (*a*) to estimate $(0.9)^{-1/2}$ through six decimal places.

In Exercises 13 to 34, (a) determine the radius of convergence and (b) determine the interval of convergence for the given power series.

13. $\sum_{n=0}^{\infty} (x-4)^n$ **14.** $\sum_{n=0}^{\infty} \frac{(-1)^n n x^n}{2^n(n^2+1)}$ **15.** $\sum_{n=0}^{\infty} (-1)^n x^{3n}$ **16.** $\sum_{n=1}^{\infty} \frac{(-1)^n n! x^n}{n^4 3^n}$

17. $\sum_{n=1}^{\infty} \frac{x^{n+2}}{3n}$ **18.** $\sum_{n=0}^{\infty} \frac{nx^n}{(n+1)^2}$ **19.** $\sum_{n=2}^{\infty} \frac{(-1)^n x^n}{n(\ln n)^2}$ **20.** $\sum_{n=0}^{\infty} \frac{n(3x-4)^n}{(n+1)^2}$

21. $\sum_{n=0}^{\infty} \frac{(-1)^n (x-3)^n}{2^{3n+1}(n^2+1)}$ **22.** $\sum_{n=0}^{\infty} \frac{(-1)^n \sqrt{n}(x-1)^n}{e^n(n+1)}$ **23.** $\sum_{n=0}^{\infty} \frac{\arctan n}{n+1}(x-2)^n$ **24.** $\sum_{n=0}^{\infty} \frac{3^n n(x+2)^n}{2n+5}$

25. $\sum_{n=1}^{\infty} \frac{n! x^n}{3^{n^2+n}}$ **26.** $\sum_{n=1}^{\infty} \frac{(-1)^n 2^n (x+4)^n}{4^n n^2}$ **27.** $\sum_{n=1}^{\infty} \frac{(x+6)^n}{n^n}$ **28.** $\sum_{n=0}^{\infty} \ln(n^2+1) x^n$

29. $\sum_{n=0}^{\infty} \frac{(-1)^n x^n}{3n+1}$ **30.** $\sum_{n=1}^{\infty} \frac{(-1)^n 3^n (x-1)^{2n}}{n^2}$ **31.** $\sum_{n=0}^{\infty} \frac{(-1)^n x^n}{3^n+1}$ **32.** $\sum_{n=0}^{\infty} \frac{(-1)^n 2^n x^n}{n!}$

33. $\sum_{n=0}^{\infty} \frac{(-1)^n n! 2^n x^{n+1}}{4^{n+1}(n+2)!}$ **34.** $\sum_{n=1}^{\infty} \frac{(x-1)^n}{2 \cdot 5 \cdot 8 \cdot \cdots \cdot (3n-1)}$

In Exercises 35 to 50 using any appropriate technique, determine a Maclaurin series representation for the given function. Indicate the open interval on which this representation is valid.

35. $x \arctan(2x)$ **36.** $x^2 e^{-x/3}$ **37.** $x \sin 4x^2$ **38.** $\frac{\cos(\pi x)}{x}$

39. $\sqrt{5+x}$ **40.** $\sqrt{2-x^3}$ **41.** $x^2 \cos \sqrt{x}$ **42.** $\frac{x}{(1+x^2)^2}$

43. $\dfrac{x^2+1}{x-1}$

44. $x^2\sin(3x^2)$

45. $\sqrt[3]{8+x}$

46. $\ln(2+x^2)$

47. $F(x)=\int_0^x \dfrac{t\,dt}{1-t^4}$

48. $F(x)=\int_0^x \dfrac{t\,dt}{3+t^3}$

49. $F(x)=\int_0^x t^2\arctan(4t^2)\,dt$

50. $\sinh(x^2)$

In Exercises 51 to 54 approximate (as a finite numerical sum) the given quantity to within the indicated degree of accuracy.

51. $\int_0^1 \cos\sqrt{x}\,dx \quad (10^{-3})$

52. $\int_0^{1/2} x\sqrt{1-x^2}\,dx \quad (10^{-4})$

53. $\int_0^{1/3} (1-x^2)^{-3/2}\,dx \quad (10^{-4})$

54. $\int_0^{1/2} x^2\sinh x\,dx \quad (10^{-4})$

Miscellaneous Exercises

1. Show that if $f(x)$ has a continuous $(n+1)$st derivative on an open interval containing a, then

$$\lim_{x\to a}\frac{f(x)-P_n(x)}{(x-a)^n}=0$$

2. Suppose $P_n(x)$ is the nth Taylor polynomial for f at a where f has a continuous nth derivative on an open interval containing a and x. If $F(x)=\int_a^x f(t)\,dt$, then show that the nth Taylor polynomial for $F(x)$ is given by $\bar{P}_n(x)=\int_a^x P_n(t)\,dt$.

3. Compute $P_n(x)$ for $f(x)=(b+x)^m$ where m is an integer and b a constant, and compare the coefficients with the usual binomial coefficients.

In Exercises 4 to 14, (a) determine the radius of convergence, and (b) the interval of convergence for the given power series.

4. $\sum_{n=1}^{\infty}(4x)^{2n}$

5. $\sum_{n=0}^{\infty}\dfrac{(-1)^n x^n}{\sqrt{3n+1}}$

6. $\sum_{n=0}^{\infty}\dfrac{(-1)^n}{n!(n+p)!}\left(\dfrac{x}{2}\right)^{2n+p}$ (p a positive integer)

7. $\sum_{n=0}^{\infty}\dfrac{(-1)^n x^n}{2^{(n^2)}}$

8. $\sum_{n=0}^{\infty}\dfrac{(2n^2+n)x^n}{3^n\sqrt{n+1}}$

9. $\sum_{n=1}^{\infty}\dfrac{(-1)^n(x-3)^{2n}}{2^{2n+2}}$

10. $\sum_{n=0}^{\infty} b_n x^n$ where $b_n=\sum_{k=0}^{n}\dfrac{1}{2^k}$

11. $\sum_{n=1}^{\infty}\dfrac{(-1)^n\,3\cdot5\cdot7\cdot\dots\cdot(2n+1)}{2\cdot4\cdot6\cdot\dots\cdot(2n)}x^n$

12. $\sum_{n=1}^{\infty}\left(\dfrac{n+1}{n}\right)^n x^n$

13. $\sum_{n=1}^{\infty}\dfrac{n!}{3\cdot6\cdot9\cdot\dots\cdot(3n)}x^n$

14. $\sum_{n=2}^{\infty}\dfrac{1\cdot3\cdot5\cdot\dots\cdot(2n-3)}{4\cdot6\cdot\dots\cdot(2n)}x^n$

In Exercises 15 to 19 determine a power series representation for each of the given functions and give the open interval on which this representation is valid.

15. $\dfrac{\cos^2 x}{x}$

16. $F(x)=\int_0^x \ln\left(\dfrac{2+t}{2-t}\right)dt$

17. $F(x)=\int_0^x \sqrt{1-t^3}\,dt$

18. $\operatorname{arcsinh} x$

19. $F(x)=\int_0^x \ln(3+2t)\,dt$

In Exercises 20 to 23 approximate the following to within the indicated degree of accuracy.

20. $\int_0^{1/2}\sqrt{1-x^3}\,dx \quad (10^{-4})$

21. $\int_0^{1/2} x\arctan(x^2)\,dx \quad (10^{-4})$

22. $\int_0^1 x^2e^{-x^2}\,dx \quad (10^{-5})$

23. $\int_0^{1/3}\dfrac{dx}{1+x^4} \quad (10^{-4})$

24. Let i be such that $i^2=-1$ and use the Maclaurin series for e^x to show Euler's formula $e^{i\theta}=\cos\theta+i\sin\theta$.

SELF-TEST ✚

1. Find the interval of convergence of each of the following series.

(*a*) $\displaystyle\sum_{n=1}^{\infty} \frac{(-1)^n n(x+1)^{2n}}{2^n(n+1)}$ (*b*) $\displaystyle\sum_{n=1}^{\infty} \frac{(-1)^n 3^n x^n}{n^2+2n}$

2. Find a Maclaurin series for $f(x) = x/(5+x)^2$ valid for $|x| < 5$.

3. Use the first four terms of an appropriate power series to approximate $\displaystyle\int_0^{1/2} \sqrt{1+x^4}\,dx$. How accurate is this approximation?

Chapter 17

VECTORS AND SOLID GEOMETRY

CONTENTS

Many physical quantities, such as force, velocity, torque, and momentum, have associated with them both a magnitude and a direction. In dealing with these quantities mathematically it is useful to introduce the notion of a vector. The general definition of a vector does not involve any geometric considerations. However, the types of vectors used in the calculus can be presented either geometrically or analytically. We shall use both approaches, choosing the interpretation as the situation warrants.

We begin this chapter with a discussion of coordinates in three-space and a study of some simple surfaces. This solid geometry will play much the same role in our later study of functions of two variables that plane geometry played in our study of functions of one variable. In fact, this chapter is, in many respects, the "Chapter 1" for calculus of more than one variable.

Section 17.1: Coordinates in Three-Space

Objectives:

1. Plot points in Cartesian coordinates.

2. Recognize the equation of a sphere.

3. Determine the distance between two points in space.

4. Determine the midpoint of a line segment joining two points in space.

5. Recognize the equation of a vertical or horizontal plane.

CARTESIAN COORDINATES

In Chap. 1 we introduced rectangular coordinates and noted the one-to-one correspondence between points in the plane and ordered pairs of real numbers. This coordinate system has been an indispensable tool in our study of the calculus to date. To further this study we shall introduce a three-dimensional coordinate system that will allow us to identify points in space with ordered triples of real numbers in a similar way.

We start with a horizontal plane which has the usual x-y rectangular coordinate system positioned so that the positive y axis extends to the right and the positive x axis extends toward us; see Fig. 17.1(a). A third real number line, called the z axis, is placed perpendicularly through the origin of the x-y plane with its positive direction upward and its zero point at the origin of the x-y plane. We refer to these intersecting number lines arranged in this fashion as the *Cartesian coordinate system.* If you were to grasp the z axis with your right hand and curl your fingers from the positive x axis to the positive y axis, then your thumb, if extended, would point in the direction of the positive z axis, as in Fig. 17.1(b). For this reason our system is often referred to as a *right-handed system.*

Each pair of axes determines a *coordinate plane.* These three mutually perpendicular planes are called the x-y plane, the x-z plane, and the y-z plane.

If P is a point in space, then the unique plane through P parallel to the y-z plane intersects the x axis at some point with one-dimensional coordinate x_0. We refer to this number as the x *coordinate* of the point P. In a similar way one obtains the y *coordinate* y_0 and the z *coordinate* z_0 by considering the intersection with the y and z axes of the planes through P parallel, respectively, to the x-z plane and the x-y plane. We thus label the point P as (x_0, y_0, z_0) and refer to this triple of real numbers as the *Cartesian coordinates of P* (see Fig. 17.2). With this convention, points on the x axis are of the form $(x, 0, 0)$; those on the y and z axes have the form $(0, y, 0)$ and $(0, 0, z)$, respectively.

Conversely, any ordered triple (x_0, y_0, z_0) of real numbers uniquely determines a point P as the intersection of three planes: the plane through $(x_0, 0, 0)$ parallel to the y-z plane, the plane through $(0, y_0, 0)$ parallel to the x-z plane, and the plane through $(0, 0, z_0)$ parallel to the x-y plane. Figure 17.3 shows the location of some typical points in space.

In light of this identification we refer to the set of all ordered triples of real numbers as *Cartesian three-space.* If the point P has Cartesian coordinates (x, y, z), then we will often write $P = (x, y, z)$.

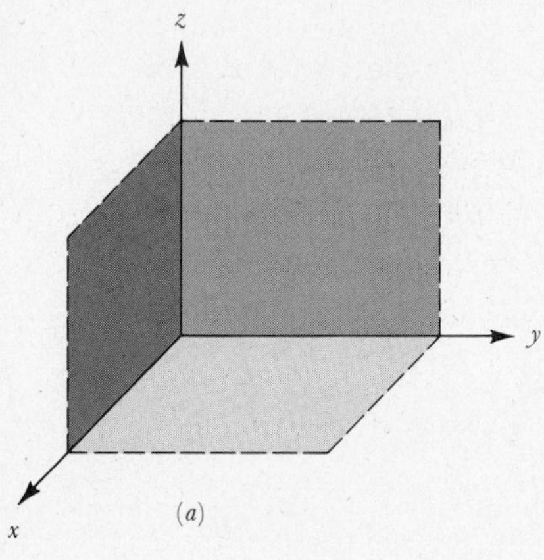

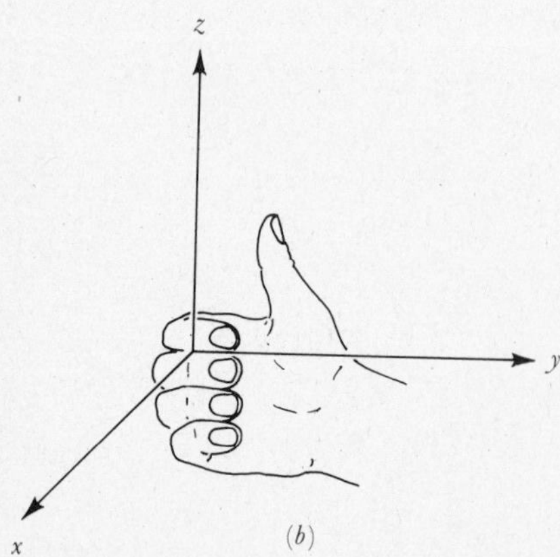

Fig. 17.1

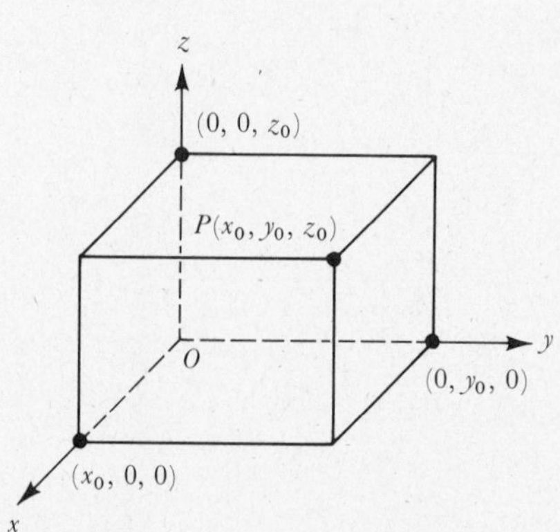

Fig. 17.2

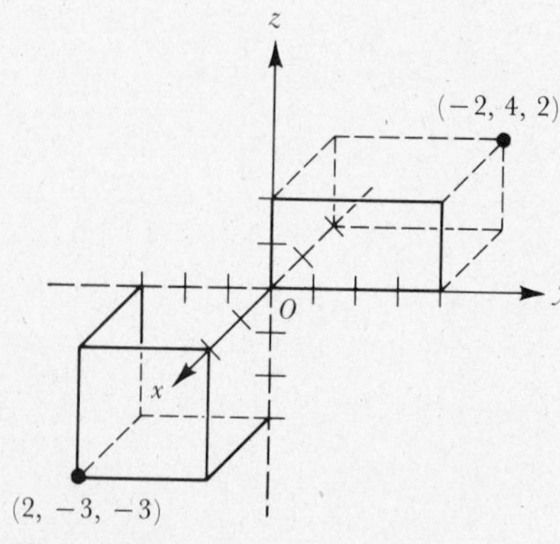

Fig. 17.3

The three coordinate planes divide Cartesian three-space into eight subregions called *octants*. The *first octant* consists of those points with nonnegative x, y, and z coordinates. There is no standard numbering for the remaining octants.

THE GRAPH OF AN EQUATION

17.1.1 **Definition** The *graph of an equation* in x, y, and z consists of all those points whose coordinates satisfy the given equation.

It is not difficult to determine whether a given point is on the graph of a particular equation since one need only check that the coordinates satisfy the equation. For example, the point $(2, 3, 8)$ is on the graph of $x^2 + y^2 - z = 5$ since $(2)^2 + (3)^2 - (8) = 4 + 9 - 8 = 5$; but the point $(1, 2, -1)$ is not on the graph since $(1)^2 + (2)^2 - (-1) = 6 \neq 5$. The more general problem of sketching the graph of an equation $F(x, y, z) = 0$ will be dealt with in more detail in the next section and in subsequent chapters.

The context plays an important role in graphing equations. For example, in rectangular coordinates, the equation $x = y$ has as its graph a straight line through the origin; see Fig. 17.4(a). In Cartesian coordinates we interpret the equation $x = y$ as an equation in x, y, and z where the z variable does not appear. A point with coordinates (x, y, z) will be on the graph if and only if $x = y$, and hence z can be any real number. The graph is a *plane* perpendicular to the x-y plane passing through the *line* in the x-y plane with rectangular equation $x = y$; see Fig. 17.4(b).

An even simpler class of equations is of the form $x = k$, $y = k$, or $z = k$, whose graphs are planes parallel to the coordinate planes (see Fig. 17.5). As we

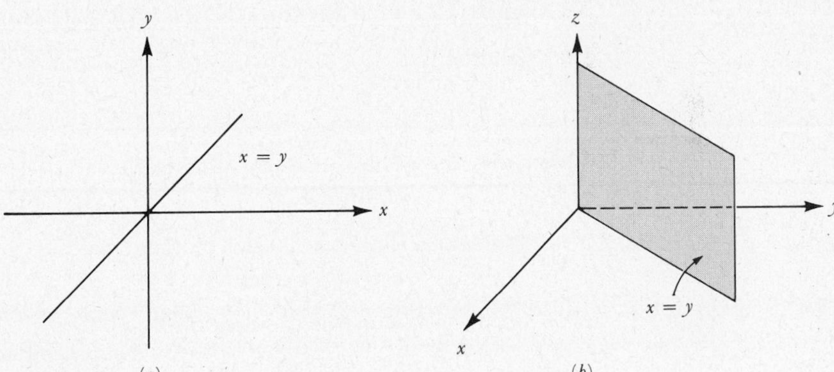

Fig. 17.4

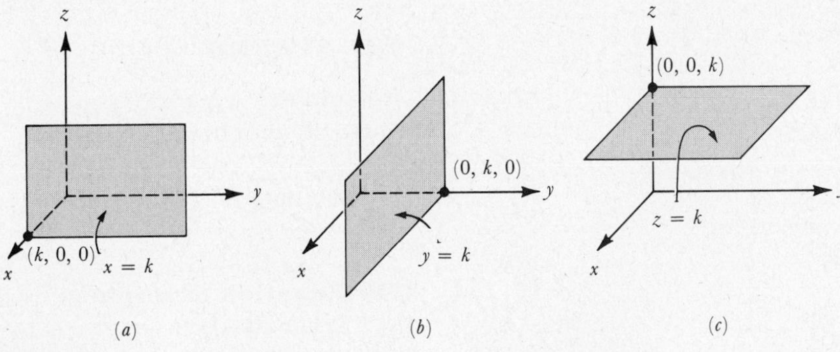

Fig. 17.5

shall see in the next section the chore of graphing a more complex equation can often be simplified by examining its intersections with planes of this type.

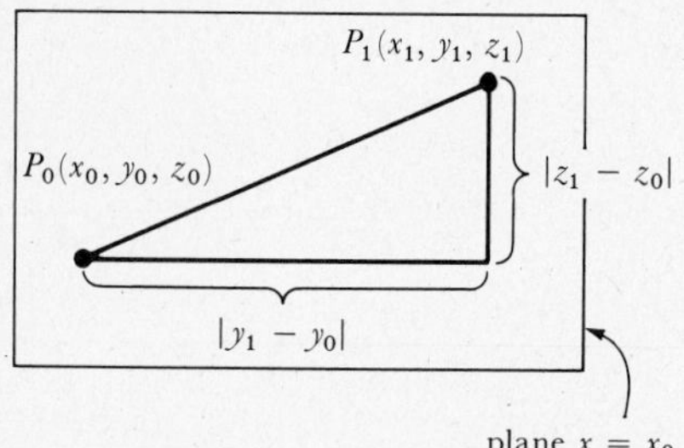

Fig. 17.6

THE DISTANCE FORMULA

17.1.2 The Distance Formula If $P_0 = (x_0, y_0, z_0)$ and $P_1 = (x_1, y_1, z_1)$ are points in space, then the distance between them is given by

$$d(P_0, P_1) = \sqrt{(x_1 - x_0)^2 + (y_1 - y_0)^2 + (z_1 - z_0)^2}$$

Proof: If both points lie in a plane parallel to one of the coordinate planes, then, as was the case for the distance formula in the plane, the Pythagorean theorem applies directly. Figure 17.6 depicts the situation when both points lie in the plane $x = x_0$. Now

$$d(P_0, P_1)^2 = |y_1 - y_0|^2 + |z_1 - z_0|^2$$

so

$$d(P_0, P_1) = \sqrt{(y_1 - y_0)^2 + (z_1 - z_0)^2}$$

which gives the result since $x_1 - x_0 = 0$ in this case.

If P_0 and P_1 do not lie in a plane parallel to one of the coordinate planes, then we consider the rectangular parallelepiped (box) with opposite vertices P_0 and P_1, as in Fig. 17.7. Applying the Pythagorean theorem to the triangle P_0SQ, we see that P_0Q has length $\sqrt{(x_1 - x_0)^2 + (y_1 - y_0)^2}$. Now apply the theorem again to the triangle P_0QP_1 to obtain

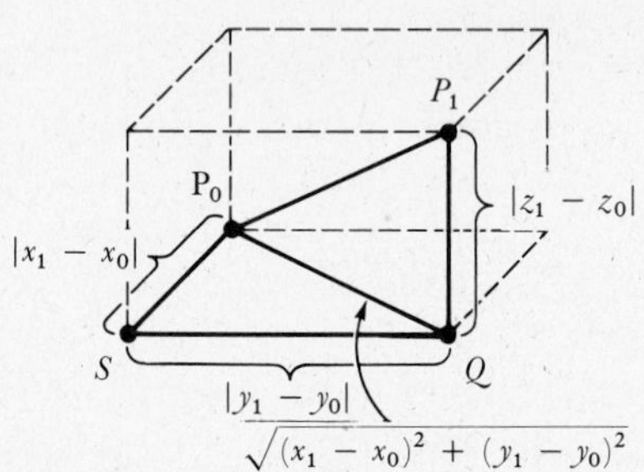

Fig. 17.7

$$\begin{aligned} d(P_0, P_1) &= \sqrt{[\sqrt{(x_1 - x_0)^2 + (y_1 - y_0)^2}]^2 + (|z_1 - z_0|)^2} \\ &= \sqrt{(x_1 - x_0)^2 + (y_1 - y_0)^2 + (z_1 - z_0)^2} \quad \blacksquare \end{aligned}$$

Example 1 Find the distance between the points $P_0 = (-1, 2, 3)$ and $P_1 = (3, -2, 4)$.

Solution

$$d(P_0, P_1) = \sqrt{(3 - (-1))^2 + (-2 - 2)^2 + (4 - 3)^2} = \sqrt{33} \quad \bullet$$

Example 2 Show that the points $P = (3, -1, 2)$, $Q = (4, 1, 6)$, and $R = (5, 3, 10)$ are collinear.

Solution The points are pictured in Fig. 17.8. It suffices to show that $d(P, Q) + d(Q, R) = d(P, R)$

which is the case since

$$\begin{aligned} d(P, Q) &= \sqrt{(4 - 3)^2 + (1 - (-1))^2 + (6 - 2)^2} = \sqrt{21} \\ d(Q, R) &= \sqrt{(5 - 4)^2 + (3 - 1)^2 + (10 - 6)^2} = \sqrt{21} \\ P(P, R) &= \sqrt{(5 - 3)^2 + (3 - (-1))^2 + (10 - 2)^2} = 2\sqrt{21} \quad \bullet \end{aligned}$$

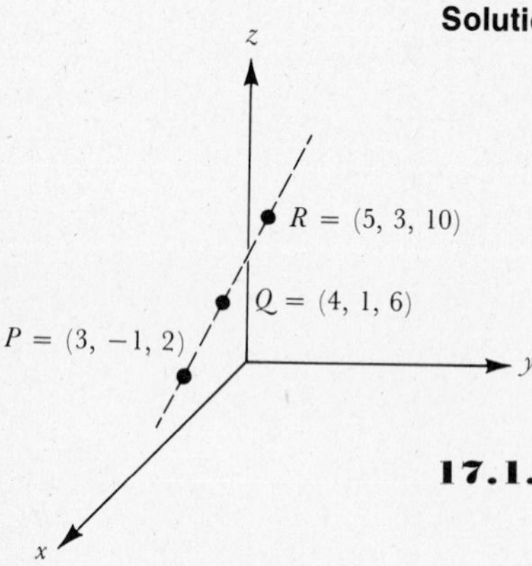

Fig. 17.8

THE SPHERE

17.1.3 Definition A *sphere of radius* r ($r > 0$) and center (a, b, c) is the set of all points whose distance from (a, b, c) is r.

As an immediate consequence of the distance formula (see Exercise 21) we have:

17.1.4 The Equation of a Sphere Let $r > 0$. The sphere of radius r and center (a, b, c) has equation $(x - a)^2 + (y - b)^2 + (z - c)^2 = r^2$. Conversely, every such equation has as its graph a sphere.

Example 3 Show that $x^2 + y^2 + z^2 + 4x - 2y + 6z + 5 = 0$ is the equation of a sphere. Determine its center and radius.

Solution Our objective is to show that this equation can be written in the form (17.1.4). To do this we first regroup the terms $(x^2 + 4x \quad) + (y^2 - 2y \quad) + (z^2 + 6z \quad) = -5$; complete the square in x, y, and z; $(x^2 + 4x + 4) + (y^2 - 2y + 1) + (z^2 + 6z + 9) = -5 + 4 + 1 + 9$; and rewrite the term as $(x + 2)^2 + (y - 1)^2 + (z + 3)^2 = 9$.
It follows that the equation has as its graph a sphere of radius 3 centered at $(-2, 1, -3)$. ●

As was the case for circles in the plane, it is not difficult to see that any second-degree equation of the form

$$Ax^2 + By^2 + Cz^2 + Dx + Ey + Fz + G = 0 \text{ with } A = B = C\ (\neq 0)$$

can be written in the equivalent form $(x - a)^2 + (y - b)^2 + (z - c)^2 = k$ and thus will either have no graph $(k < 0)$ or will have as its graph a sphere $(k > 0)$ or a single point $(k = 0)$.

THE MIDPOINT OF A LINE SEGMENT

Using an argument similar to the two-dimensional case it is routine to derive:

17.1.5 **The Midpoint Formula** The midpoint of the line segment joining $P_0 = (x_0, y_0, z_0)$ and $P_1 = (x_1, y_1, z_1)$ has coordinates

$$\left(\frac{x_0 + x_1}{2}, \frac{y_0 + y_1}{2}, \frac{z_0 + z_1}{2}\right)$$

Example 4 Give an equation for the sphere with diameter PQ, where $P = (-1, 2, 4)$ and $Q = (3, 1, 6)$.

Solutions The radius of the sphere is

$$r = (1/2)\,d(P, Q) = (1/2)\sqrt{(3 - (-1))^2 + (1 - 2)^2 + (6 - 4)^2} = \sqrt{21}/2$$

The center C of the sphere is located at the midpoint of the diameter

$$C = \left(\frac{-1 + 3}{2}, \frac{2 + 1}{2}, \frac{4 + 6}{2}\right) = (1, 3/2, 5)$$

From (17.1.4) an equation for this sphere is

$$(x - 1)^2 + \left(y - \frac{3}{2}\right)^2 + (z - 5)^2 = \frac{21}{4} \quad ●$$

PROGRESS TEST 1

1. Draw a Cartesian coordinate system and plot the points $P = (-2, 1, 3)$, $Q = (1, -3, -5)$ and $R = (3, 2, 1)$.
2. Show that the points $P = (-2, 3, 4)$, $Q = (1, -1, 5)$, and $R = (-4, 2, 6)$ are the vertices of a right triangle and find its area.
3. Identify the graphs of the following equations:
   (a) $x^2 + y^2 + z^2 - 4x + 6y - 2z + 14 = 0$.
   (b) $4x^2 + 4y^2 + 4z^2 - 4x - 24y + 8z + 5 = 0$.

SECTION 17.1 EXERCISES

1. Plot each of the following points:
(*a*) (1, 3, 6) (*b*) (1, −5, 2)
(*c*) (−2, −1, 3) (*d*) (−2, 1, 5)
(*e*) (2, 3, −4) (*f*) (3, −2, −1)
(*g*) (−4, −1, −5) (*h*) (2, −1, −3)

In Exercises 2 to 7, P and Q are given to be the opposite vertices of a rectangular parallelepiped with faces parallel to the coordinate planes. Give the remaining vertices.

2. $P = (2, 1, 6)$, $Q = (3, 2, 4)$
3. $P = (-3, 2, 1)$, $Q = (4, -6, 2)$
4. $P = (4, -1, 2)$, $Q = (3, 0, 9)$
5. $P = (-2, -1, -2)$, $Q = (0, 4, 1)$
6. $P = (3, -1, -2)$, $Q = (-1, -2, 3)$
7. $P = (-4, -1, 1)$, $Q = (4, 6, -8)$
8. Determine the length and midpoint of each of the diagonals PQ in Exercises 2 to 7.

In Exercises 9 to 12 determine whether the three points are collinear.

9. (2, 3, −2), (6, 2, −3/2), (10, 1, −1)
10. (−3, 4, 5), (−2, 2, 8), (−5, 8, −1)
11. (1/2, −1, 0), (−1, 1, 3), (−4, 1, 6)
12. (1, 3, 2), (2, 5, 1), (0, 1, 3)

In Exercises 13 to 16 determine which of the triangles having the given vertices P, Q, R are right triangles.

13. $P = (1, 2, 8)$, $Q = (2, 1, 3)$, $R = (1, 4, 5)$
14. $P = (1, -1, 2)$, $Q = (3, -1, 6)$, $R = (-2, 1, 3)$
15. $P = (5, 0, -3)$, $Q = (2, -1, 4)$, $R = (6, 4, -2)$
16. $P = (0, 1, 2)$, $Q = (2, 1, 3)$, $R = (1, 3, 0)$

In Exercises 17 to 20 identify the graphs (if they exist) of the given equations.

17. $x^2 - 2x + y^2 + 4y + z^2 - 8z = 123$
18. $x^2 + y^2 + z^2 - x + 2y + 6z = 23/4$
19. $3x^2 + 3y^2 + 3z^3 + 6y - 12z = -39$
20. $x^2 + y^2 + z^2 - 12z = -38$
21. Prove Theorem (17.1.4).
22. Prove Theorem (17.1.5).

Sketch the graph in Cartesian three-space of each of the following equations.

23. $x = -2$
24. $y = 3$
25. $z = 3/2$
26. $3x + y = 1$
27. $x - 2y = 4$
28. $2y + 3z = 5$

Section 17.2: Surfaces

Objectives:

Identify and sketch the graphs of equations of (1) planes, (2) cylinders, (3) surfaces of revolution, and (4) quadric surfaces.

INTRODUCTION

By a *surface* we shall mean the graph of any equation in three variables. We shall write such an equation symbolically as $F(x, y, z) = 0$. The purpose of this section is to establish certain basic surfaces and their equations, namely, planes, cylinders, surfaces of revolution, and the quadric surfaces. Later we shall frequently use these surfaces as a source of examples, so it is crucial that you be able to recognize and sketch them. We shall assume that you are familiar with the equations in two variables for the conic sections (see Chap. 10).

SYMMETRIES

Symmetries are even more useful in graphing surfaces than in graphing curves. In addition to symmetries with respect to the origin and the axes, we are also concerned with symmetries with respect to the coordinate planes. Table 17.1 summarizes the various types of symmetries possible for two points P and Q:

I	II
(a) origin	$Q = (-a, -b, -c)$
(b) x axis	$Q = (a, -b, -c)$
(c) y axis	$Q = (-a, b, -c)$
(d) z axis	$Q = (-a, -b, c)$
(e) x-y plane	$Q = (a, b, -c)$
(f) x-z plane	$Q = (a, -b, c)$
(g) y-z plane	$Q = (-a, b, c)$

Table 17.1 Symmetries Let $P = (a, b, c)$. We say P and Q satisfy the symmetry indicated in column I if and only if Q has the coordinates indicated in column II.

symmetry with respect to the origin, one of the coordinate axes, or one of the coordinate planes. Some illustrations appear in Fig. 17.9.

A surface S is said to be *symmetric with respect to a line* L if and only if for every point P on S, the point Q that is symmetric to P with respect to L is also on S. A similar statement holds for the symmetry of a surface with respect to a point or a plane. Using Table 17.1 it is easy to state necessary and sufficient conditions for the graph of an equation $F(x, y, z) = 0$ to obey a particular symmetry. For example, the graph of $F(x, y, z) = 0$ will be symmetric to the x-y plane if and only if $F(x, y, z) = F(x, y, -z)$. That is, the equation remains unchanged when z is replaced by $-z$.

Example 1 Identify the symmetries of the graphs of the given equation.
(a) $x^2 - 2y + y^3 - 6 = 0$
(b) $z = \sqrt{x^2 + y^2}$

Solutions (a) If we replace x by $-x$ in equation (a), the equation is unchanged. From Table 17.1(g) it follows that the graph of this equation is symmetric with respect to the y-z plane. None of the other symmetries holds.

(b) Replacing either x by $-x$, or y by $-y$, or both, will not change equation (b), and hence the graph is symmetric with respect to the y-z plane, the x-z plane, and the z axis. None of the other symmetries holds. ●

SKETCHING SURFACES

In general the intersection of two surfaces will result in a curve. Of particular help in sketching the graph of an equation are its curves of intersection with

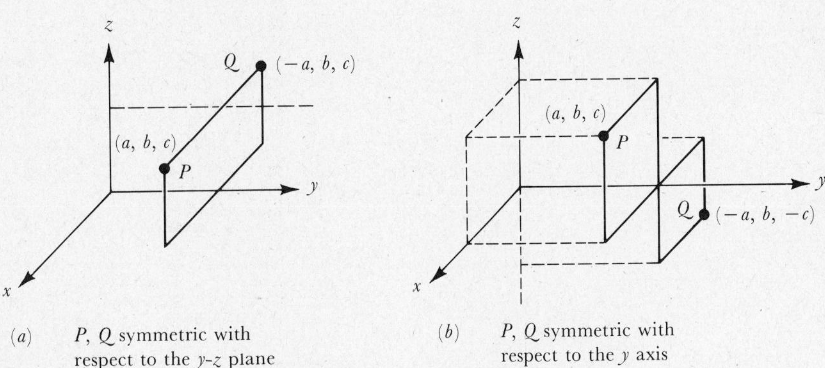

(a) P, Q symmetric with respect to the y-z plane

(b) P, Q symmetric with respect to the y axis

Fig. 17.9

either the coordinate planes or planes parallel to the coordinate planes. We refer to these curves as *traces* of the surface. We begin with the simplest type of surface, a plane.

17.2.1 **Definition** An equation of the form $Ax + By + Cz + D = 0$, where at least one of the coefficients A, B, C is not zero, is called a *linear equation in the variables x, y, and z.*

In Sec. 17.5 we shall show that the graph of a linear equation is a plane, and conversely, every plane has an equation of this form. In this section let us assume this and concentrate on graphing such equations.

Example 2 Sketch the graph of the plane $3x + 2y + 6z - 18 = 0$.

Solution Letting $y = z = 0$, we see that the plane intersects the x axis at the point (6, 0, 0). Similarly, the y and z intercepts are (0, 9, 0) and (0, 0, 3). Joining these points with straight-line segments, we obtain a sketch of the portion of the plane in the first octant (see Fig. 17.10). ●

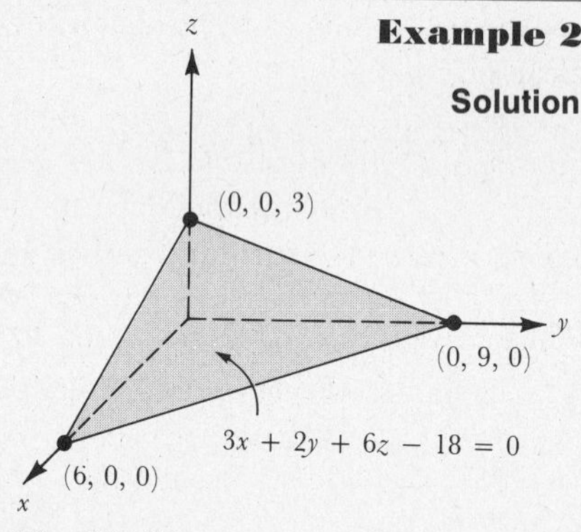

Fig. 17.10

If $D \neq 0$ in (17.2.1), then we can find the x, y, and z intercepts as in Example 2. Of course, the relevant portion of the plane that we sketch need not be in the first octant [see Example 3(*a*)]. If $D = 0$, then the plane will pass through the origin, and in this case we use the traces in the coordinate planes to obtain the sketch [see Example 3(*b*)].

Example 3 Sketch the graphs of the following planes:
(*a*) $-x + 2y + z = 6$
(*b*) $3x + 2y - z = 0$

Solutions (*a*) The intercepts are (−6, 0, 0), (0, 3, 0) and (0, 0, 6). Joining these points with straight line segments, we sketch the portion of the plane in the octant directly behind the first octant; see Fig. 17.11(*a*).

(*b*) The plane goes through the origin. The trace in the x-z plane is the straight line $3x - z = 0$, and the trace in the y-z plane is the line $2y - z = 0$. Using the origin and points P, Q on these traces, we sketch a portion of the plane in Fig. 17.11(*b*). ●

We now examine a slightly more complex example.

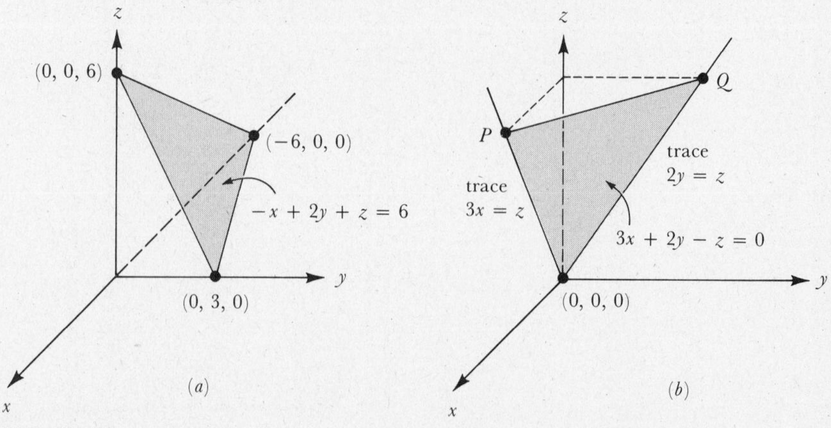

Fig. 17.11

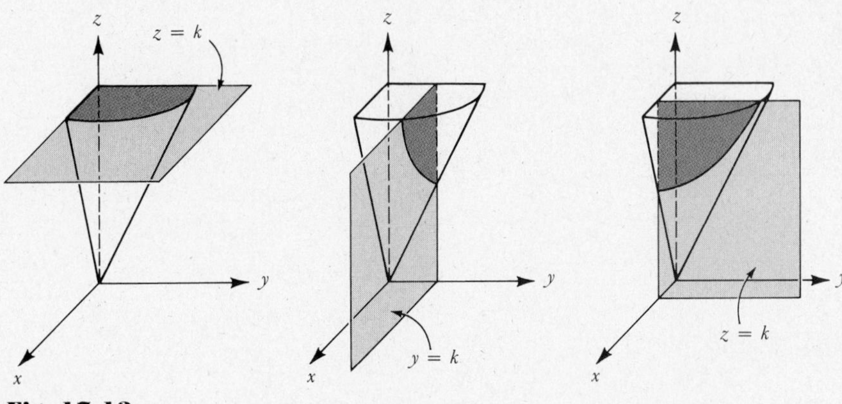

Fig. 17.12

Example 4 Sketch the surface $z = \sqrt{x^2 + y^2}$ by first examining its traces.

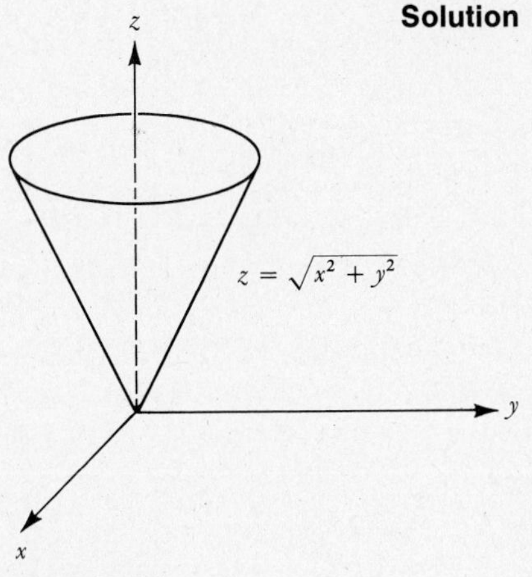

Fig. 17.13

Solution We noted in Example 1(*b*) that the graph will be symmetric with respect to the x-z and y-z planes. It is clear that $z \geq 0$, and hence it suffices to sketch the graph in the first octant and then use the symmetric properties to complete the sketch. If $z = 0$, then we have $\sqrt{x^2 + y^2} = 0$, which implies $x = y = 0$. Thus the trace in the x-y plane is the single point $(0, 0, 0)$. If $z = k > 0$, then $\sqrt{x^2 + y^2} = k$ is equivalent to $x^2 + y^2 = k^2$. This is the equation of a circle of radius k centered at the "origin." Since we are in the plane $z = k$, the "origin" in this case means the point $(0, 0, k)$. (See the first part of Fig. 17.12 above.)

The trace in the x-z plane ($y = 0$) is $z = |x|$. In the planes $y = k$ parallel to the x-z plane, we have $z = \sqrt{x^2 + k^2}$, which is the upper branch of the hyperbola $z^2 - x^2 = k^2$. Similarly, in the y-z plane the trace is $z = |y|$, and in the planes $x = k$ we obtain the upper branch of the hyperbola $z^2 - y^2 = k^2$. The resulting surface is the upper nappe of a right circular cone. See Fig. 17.12 for the remaining first octant traces. The graph is sketched in Fig. 17.13. ●

PROGRESS TEST 1

1. Sketch the graph of the equation $x^2 + z^2 = 4y$. Note relevant symmetries and traces.

2. Sketch the graphs of the planes with equations
(*a*) $3x - 2y + 2z = 6$
(*b*) $-2x + 3y + z = 0$

CYLINDERS

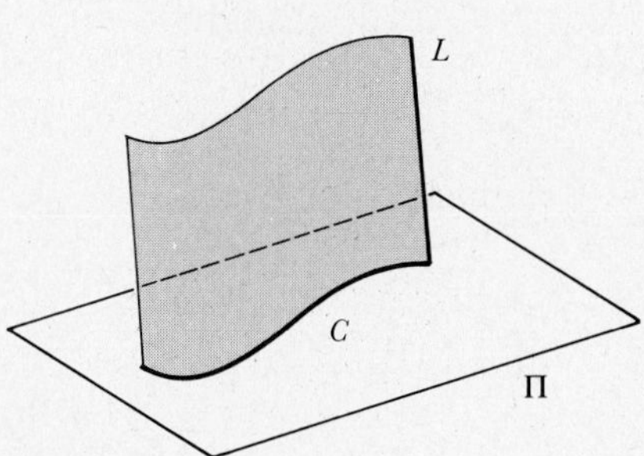

Fig. 17.14

17.2.2 Definition Suppose C is a curve that lies in a plane Π, with L a line not parallel to Π that passes through some point on C. (See Fig. 17.14.) The collection of all points on all lines through C and parallel to L is called a *cylinder*. The curve C is called the *directrix* of the cylinder. The line L and all lines through C parallel to L are called *rulings* or *generators* of the cylinder.

Cylinders with the directrix in one of the coordinate planes and rulings perpendicular to the plane of the directrix are particularly easy to recognize.

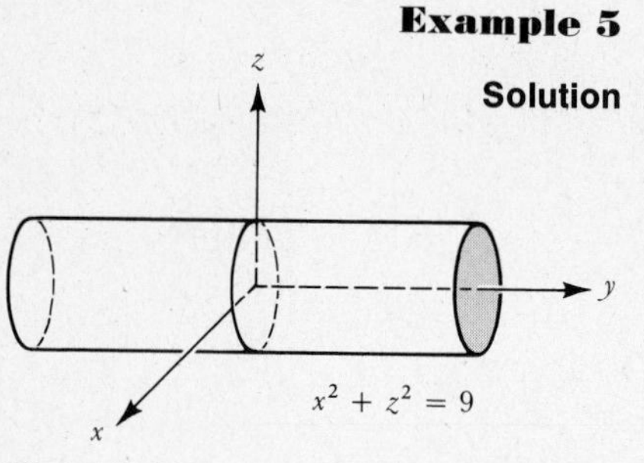

Fig. 17.15

Example 5 Sketch the surface S with equation $x^2 + z^2 = 9$.

Solution A point $P = (x, y, z)$ is on S if and only if $x^2 + z^2 = 9$. In particular, if C is the circle in the x-z plane of radius 3 centered at the origin, then we know $P = (x, y, z)$ is on S if and only if $(x, 0, z)$ is on the circle C. Thus S is a cylinder with rulings parallel to the y axis, whose directrix is the circle C in the x-z plane. Of course S is an example of the familiar right circular cylinder. (See Fig. 17.15.) ●

The proof of the following resembles the solution of Example 5 and so is omitted.

17.2.3 **Theorem** The graph of an equation $F(x, y, z) = 0$ where one of the variables does not appear is a cylinder. The rulings of the cylinder are parallel to the axis of the absent variable. The directrix lies in the plane of the variables that *do* appear in the equation and is the graph of $F(x, y, z) = 0$ considered as an equation in *two* variables.

Example 6 Graph the surfaces with the following equations.
(*a*) $y = 2x^2$ (*b*) $x^2 + 2x + 4z^2 = 15$

Solutions (*a*) The directrix is the parabola in the x-y plane with equation $y = 2x^2$. The surface is a parabolic cylinder with rulings parallel to the z axis; [see Fig. 17.16(*a*)].
(*b*) We complete the square and rewrite equation (*b*) as $(x + 1)^2/16 + z^2/4 = 1$. In the x-z plane this is the equation of an ellipse. The surface is an elliptic cylinder with rulings parallel to the y axis; see Fig. 17.16(*b*). ●

SURFACES OF REVOLUTION

A surface obtained by revolving a plane curve C about a line L in the plane of the curve is called a **surface of revolution.** The line L is called the **axis** of the surface of revolution generated by the curve C (Fig. 17.17). The intersection of a surface of revolution with a plane perpendicular to its axis is *always* a circle.

Suppose the plane curve $z = f(y)$ in the y-z plane is revolved about the y axis to generate a surface of revolution S; see Fig. 17.18. Let $P = (x, y, z)$ be an arbitrary point and consider the plane through P perpendicular to the y axis.

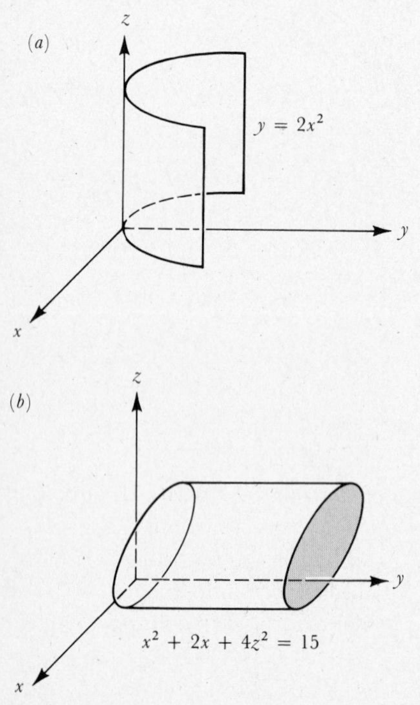

Fig. 17.16

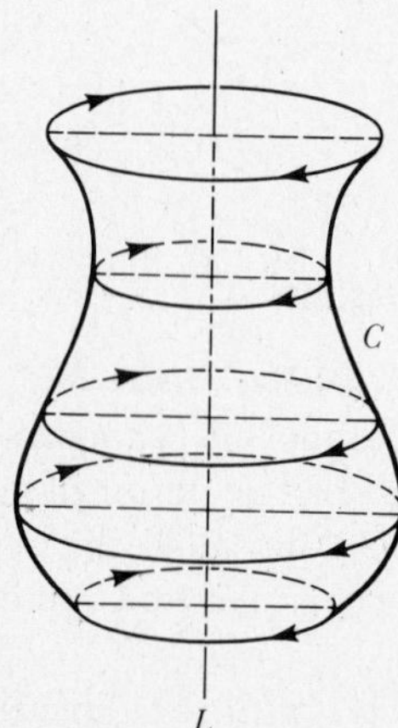

Fig. 17.17

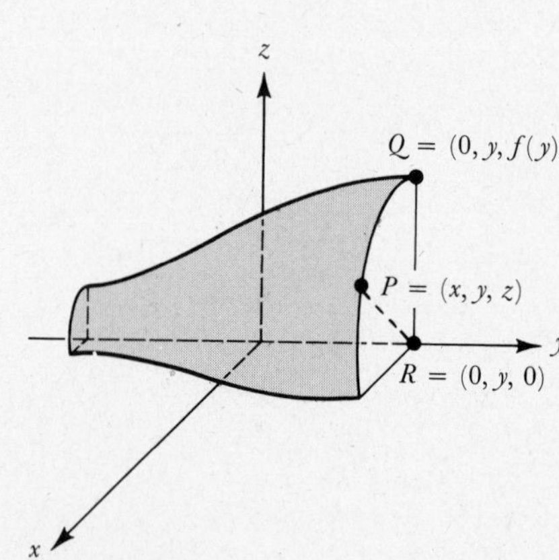

Fig. 17.18

Plane of Generating Curve	Equation of Generating Curve	Axis of Revolution	Equation of Surface of Revolution
(*a*) xy (*b*) xz	$y = f(x)$ $z = f(x)$	x axis	$y^2 + z^2 = [f(x)]^2$
(*c*) xy (*d*) yz	$x = g(y)$ $z = g(y)$	y axis	$x^2 + z^2 = [g(y)]^2$
(*e*) xz (*f*) yz	$x = h(z)$ $y = h(z)$	z axis	$x^2 + y^2 = [h(z)]^2$

Table 17.2 Surfaces of Revolution

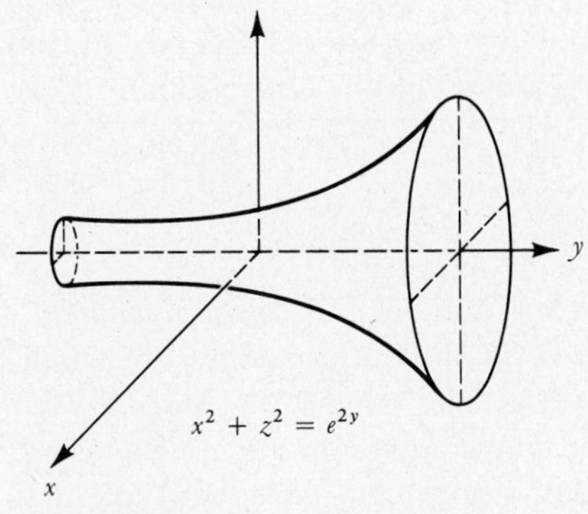

Fig. 17.19

The circular cross section of S with this plane has its center on the y axis at $R = (0, y, 0)$. The point $Q = (0, y, f(y))$ in the y-z plane is on this cross section. Now P will be on the surface if and only if it is on this circular cross section, that is, if and only if $|PR|^2 = |QR|^2$. But $|PR|^2 = x^2 + z^2$ and $|QR|^2 = [f(y)]^2$. Thus P is on S if and only if $x^2 + y^2 = [f(y)]^2$.

A similar equation can be obtained for any other surface of revolution whose axis of revolution is one of the coordinate axes. These results are summarized in Table 17.2. Note that in each case *the equation of the generating curve is expressed as a function of the variable associated with the axis of revolution.*

Example 7 Determine an equation for the surface obtained by revolving the curve $y = \ln x$ in the x-y plane about the y axis; see Fig. 17.19.

Solution Since the axis of revolution is the y axis, we solve $y = \ln x$ for x in terms of y, getting $x = e^y$. Now from (*c*) in Table 17.2 we obtain $x^2 + z^2 = e^{2y}$ as the desired equation. ●

Example 8 Identify and sketch the surface with equation $9x^2 + y^2 + z^2 = 36$.

Solution We write the equation in the form $y^2 + z^2 = 36 - 9x^2 = [f(x)]^2$, where $y = f(x) = 3\sqrt{4 - x^2}$. The graph of $y = 3\sqrt{4 - x^2}$ in the x-y plane is the upper half of the ellipse $x^2/4 + y^2/36 = 1$; see Fig. 17.20(*a*). Thus the surface is the surface of revolution obtained by revolving this semiellipse $y = 3\sqrt{4 - x^2}$ about the x axis; see Fig. 17.20(*b*). ●

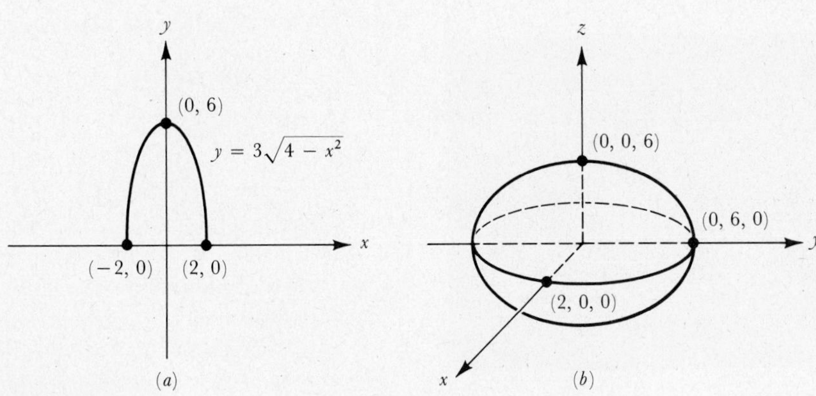

Fig. 17.20

We refer to the graph of the equation in Example 8 as an *ellipsoid of revolution*. It could also have been interpreted as the surface obtained by revolving the semiellipse $z = 3\sqrt{4 - x^2}$ in the x-z plane about the x axis. Many surfaces of revolution can be generated in more than one way.

PROGRESS TEST 2

Identify and sketch each of the following surfaces:

1. $z = \ln x$

2. $x^2 - 4y^2 = 16$

3. $3x^2 + 3y^2 = 12z$

4. Find an equation for the surface obtained by revolving the curve $z^2 = x^3$ about the z axis.

QUADRIC SURFACES

The graph of a second-degree equation in x, y, and z is called a *quadric surface*. We shall not attempt a complete discussion of quadric surfaces, but rather shall consider those quadric surfaces that are positioned symmetrically with respect to the coordinate axes and have relatively simple equations. Of course, using the material already developed in this section one can recognize the graph of a second-degree equation that is a cylinder, surface of revolution, or a sphere. We assume in the following discussions that a, b, and c are *positive constants*.

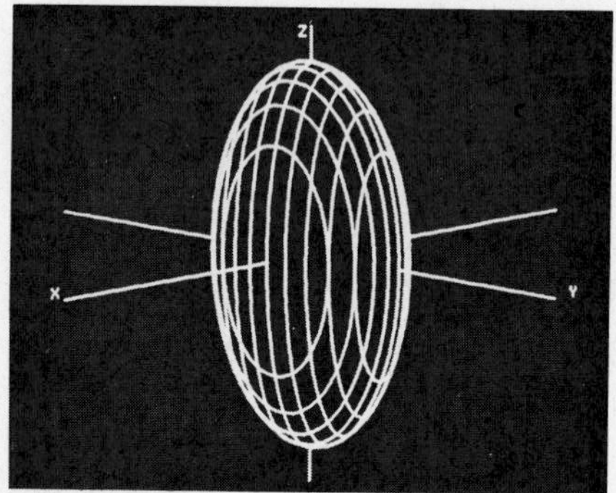

Fig. 17.21 (computer graphics for Figs. 17.21 to 17.28 by Dr. Lee Estes, Southeastern Massachusetts University).

THE ELLIPSOID

The graph of the equation

17.2.4
$$\frac{x^2}{a^2} + \frac{y^2}{b^2} + \frac{z^2}{c^2} = 1$$

is called an *ellipsoid* (Fig. 17.21). The ellipsoid has all the symmetries noted in Table 17.1. Each trace in the respective coordinate planes is an ellipse, as is each trace in the planes $x = k$, $(|k| < a)$, $y = k$ $(|k| < b)$ and $z = k$ $(|k| < c)$.

If any two of the constants are equal, then the ellipsoid is an *ellipsoid of revolution*. For example, $x^2/a^2 + y^2/a^2 + z^2/c^2 = 1$ can be written as $x^2 + y^2 = [(a/c)\sqrt{c^2 - z^2}]^2$, which is the surface obtained by revolving the semiellipse $y = (a/c)\sqrt{c^2 - z^2}$ in the y-z plane about the z axis. If $a = b = c$, then the ellipsoid is a sphere.

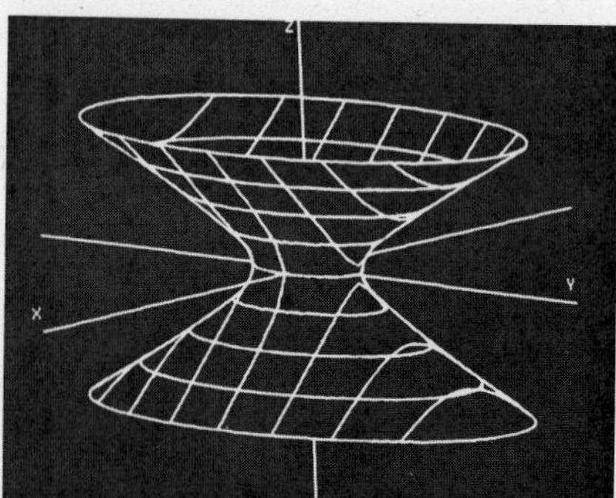

Fig. 17.22

THE HYPERBOLOID OF ONE SHEET

The graph of the equation

17.2.5
$$\frac{x^2}{a^2} + \frac{y^2}{b^2} - \frac{z^2}{c^2} = 1$$

is called a *hyperboloid of one sheet;* see Fig. 17.22. It has all the symmetries of Table 17.1. The trace in the x-y plane is the ellipse $x^2/a^2 + y^2/b^2 = 1$ and the traces in the x-z and y-z planes are the hyperbolas $x^2/a^2 - z^2/c^2 = 1$ and $y^2/b^2 - z^2/c^2 = 1$, respectively. The traces in the planes $z = k$ yield ellipses that increase in size as $|k|$ increases. The traces in the planes $x = k$ and $y = k$ are hyperbolas.

If $a = b$, then the surface is a *hyperboloid of revolution,* obtained by revolving the hyperbola $x^2/a^2 - z^2/c^2 = 1$ about the z axis (the conjugate axis).

Equations of the form $x^2/a^2 - y^2/b^2 + z^2/c^2 = 1$ and $-x^2/a^2 + y^2/b^2 + z^2/c^2 = 1$ also represent hyperboloids of one sheet, which envelop the y and x axes, respectively, rather than the z axis as does (17.2.5).

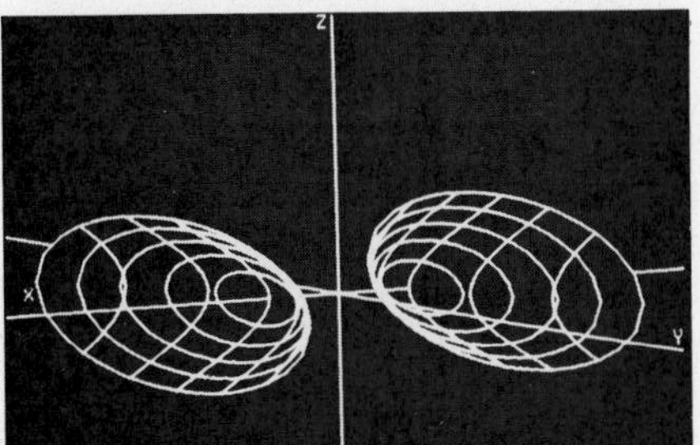

Fig. 17.23

THE HYPERBOLOID OF TWO SHEETS

The graph of the equation

17.2.6
$$\frac{x^2}{a^2} - \frac{y^2}{b^2} - \frac{z^2}{c^2} = 1$$

is called a *hyperboloid of two sheets;* see Fig. 17.23. The traces in the planes $y = k$ and $z = k$ are hyperbolas. There are no traces in the planes $x = k$ if $|k| < a$, and the traces in the planes $x = k$ when $|k| > a$ are ellipses that increase in size as $|k|$ increases.

If $b = c$, then the equation is that of a *hyperboloid of revolution,* obtained by revolving the hyperbola $x^2/a^2 - z^2/c^2 = 1$ about the x axis (the transverse axis).

The equations $-x^2/a^2 - y^2/b^2 + z^2/c^2 = 1$ and $-x^2/a + y^2/b^2 - z^2/c^2 = 1$ are also hyperboloids of two sheets. The two separated pieces are situated on the z axis for the former and on the y axis for the latter.

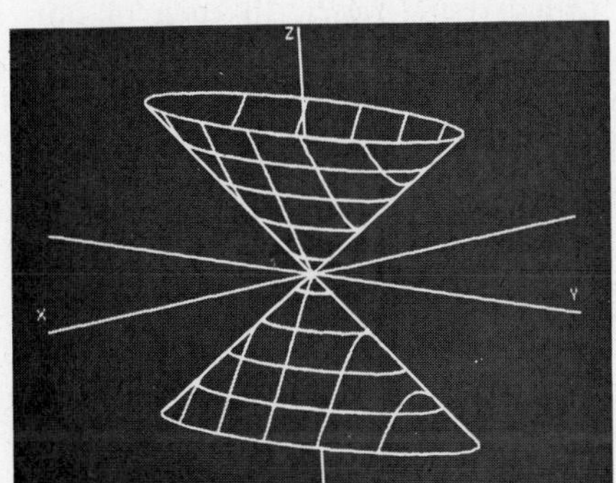

Fig. 17.24

THE ELLIPTIC CONE

The graph of the equation

17.2.7
$$\frac{x^2}{a^2} + \frac{y^2}{b^2} - \frac{z^2}{c^2} = 0$$

is called an *elliptic cone;* see Fig. 17.24. The trace in the *x-y* plane is the origin, whereas the traces in the planes $z = k$ are ellipses that increase in size as $|k|$ does. The traces in the planes $x = k$ and $y = k$ are hyperbolas that reduce to intersecting straight lines in the coordinate planes $x = 0$ and $y = 0$.

If $a = b$, then the cone is called a *right circular cone,* and it can be obtained by revolving the line $x = (a/c)z$ in the *x-z* plane about the z axis. The equations $x^2/a^2 - y^2/b^2 + z^2/c^2 = 0$ and $-x^2/a^2 + y^2/b^2 + z^2/c^2 = 0$ are also elliptic cones, which envelop the y and x axes, respectively.

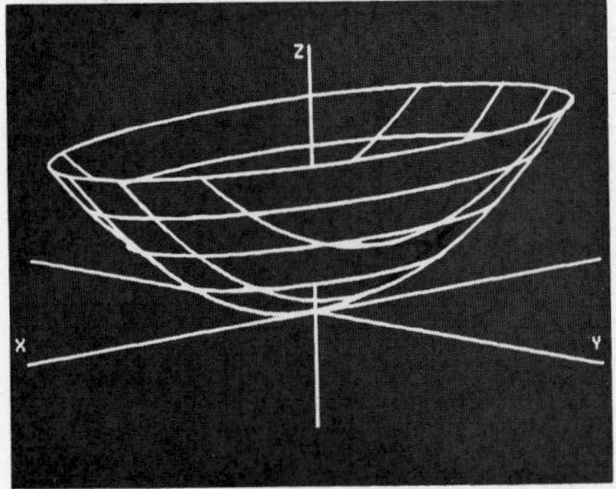

Fig. 17.25

THE ELLIPTIC PARABOLOID

The graph of the equation

17.2.8
$$\frac{x^2}{a^2} + \frac{y^2}{b^2} = \frac{z}{c}$$

is called an *elliptic paraboloid;* see Fig. 17.25. The trace in the planes $z = k$ are ellipses if $k > 0$. No trace exists if $k < 0$. The traces in the planes $x = k$ and $y = k$ are parabolas. The graph is symmetric with respect to the *x-z* and *y-z* planes.

If $a = b$, the equation is that of a *paraboloid of revolution* obtained by revolving the parabola $y^2 = (a^2/c)z$ about the z axis. The equations $x^2/a^2 + z^2/b^2 = y/c$ and $y^2/a^2 + z^2/b^2 = x/c$ are also elliptic paraboloids, which open about the positive y and x axes, respectively. If in the above discussion we allow c to be negative, then the graphs will open about the appropriate negative coordinate axis.

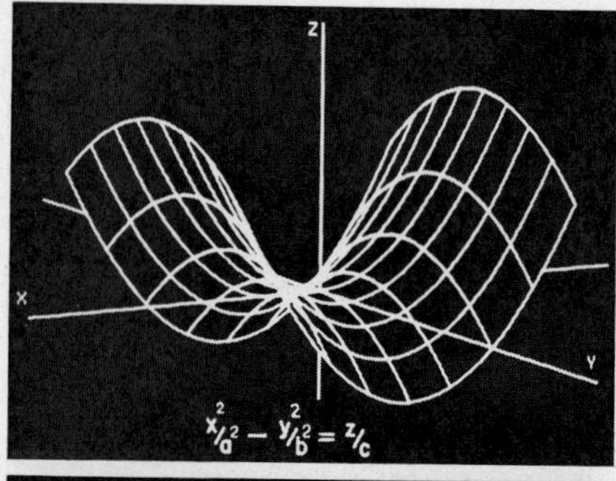

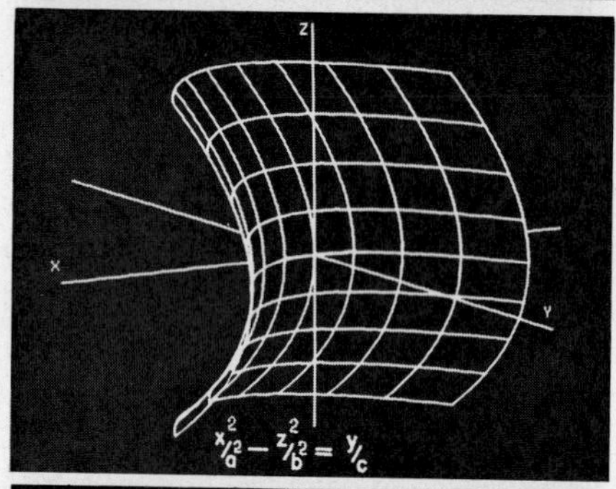

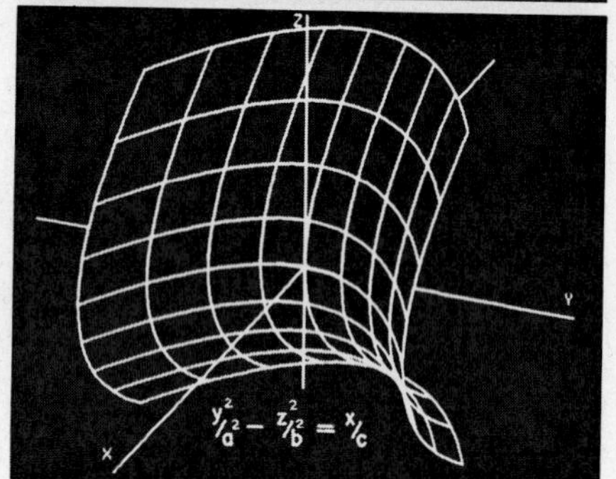

Fig. 17.26

THE HYPERBOLIC PARABOLOID

The graph of the equation

17.2.9 $$\frac{x^2}{a^2} - \frac{y^2}{b^2} = \frac{z}{c}$$

is called a *hyperbolic paraboloid;* see Fig. 17.26. The graph is symmetric with respect to the x-z and y-z planes. The trace in the x-z plane is a parabola opening upward, whereas the trace in the y-z plane is a parabola opening downward. The trace in the x-y plane is a pair of intersecting straight lines. The traces in the planes $z = k$ are hyperbolas with transverse axes parallel to x axis if $k > 0$ and parallel to the y axis if $k < 0$. The traces in the planes $x = k$ and $y = k$ are parabolas opening downward in the first case and upward in the second.

Equations of the form $x^2/a^2 - z^2/b^2 = y/c$ and $y^2/a^2 - z^2/b^2 = x/c$ also represent hyperbolic paraboloids, as do any of the above with $c < 0$. Because of their saddle shape, they are the most difficult of the quadric surfaces to sketch. The three situations, with $c > 0$, are sketched in the three parts of Fig. 17.26.

SUMMARY AND EXAMPLES

Ellipsoids, hyperboloids, and cones are called *central conics.* In each of our preceding discussions the *center* was the origin. Elliptic and hyperbolic paraboloids are examples of *noncentral conics.* In our previous development the origin is called the vertex of the elliptic paraboloid; for the hyperbolic paraboloid it is called the *saddle point.* If in any of the equations (17.2.4) to (17.2.9) or their variations the variables x, y, and z are replaced, respectively, by $x - h, y - k$, and $z - l$, then the graph will be the same surface except translated so that its center (vertex or saddle point) is located at the point (h, k, l).

Example 9 Identify and sketch the graph of:
(*a*) $9x^2 - 16y^2 + z^2 - 144 = 0$
(*b*) $x^2 - 2x + 4y^2 - 16y + 1 = 16z$

Solutions
(*a*) Equation (*a*) can be rewritten as $x^2/16 - y^2/9 + z^2/144 = 1$, which has the form of (16.2.5) with z and y interchanged. Thus the graph is a hyperboloid of one sheet; see Fig. 17.27 (left).
(*b*) We first complete the square in x and y, obtaining $(x^2 - 2x + 1) + 4(y^2 - 4y + 4) + 1 = 16z + 1 + 16$ or, equivalently, $(x - 1)^2/16 + (y - 2)^2/4 = z + 1$. This has the form of (17.2.8) with x, y, z replaced by $x - 1, y - 2$ and $z + 1$, respectively. The graph is an elliptic paraboloid opening upward with vertex at $(1, 2, -1)$; see Fig. 17.27 (right). ●

Being able to sketch the various quadric surfaces in space will be a great aid to your learning and using the calculus of several variables which follows. Of course, the only way to learn is to practice. However, more general surfaces can be extremely difficult to sketch. An example of such is

$$z = \frac{|\sin(\pi/8)(x - 16)|}{|x - 16|} + \frac{|\sin(\pi/8)(y - 16)|}{|y - 16|}$$

which is sketched in Fig. 17.28.

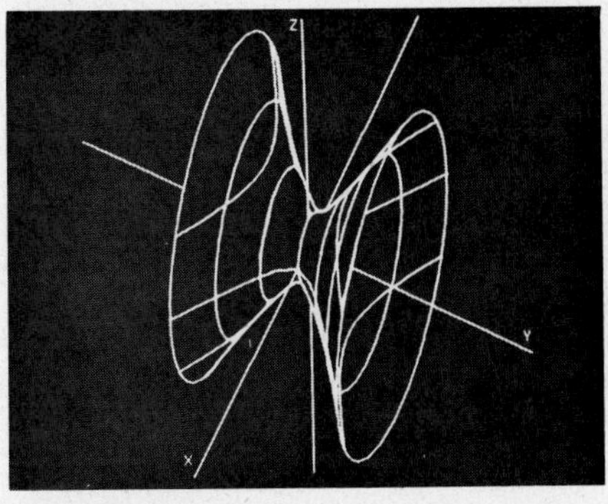

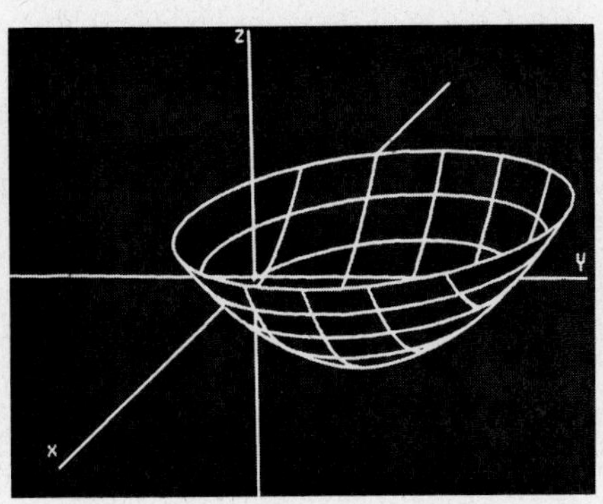

Fig. 17.27

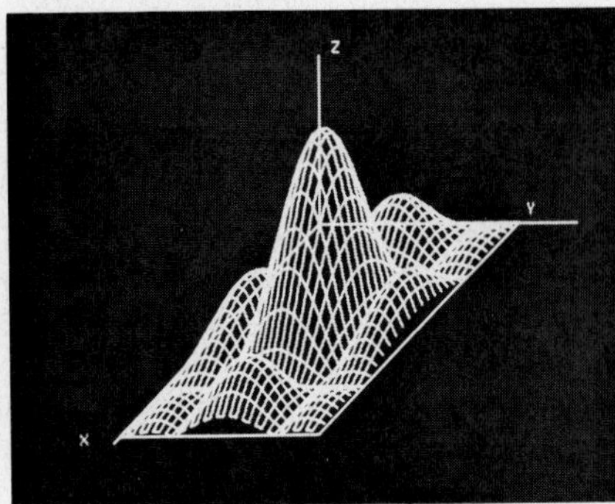

Fig. 17.28

PROGRESS TEST 3

Identify and sketch the graph of:

1. $x^2 - 4y^2 + z^2 = 16$ **2.** $y^2 - x^2 = 2z$ **3.** $x^2 + z^2 - 2x - 2z - 36y + 38 = 0$

SECTION 17.2 EXERCISES

In Exercises 1 to 30 identify and sketch the graph of the given equation.

1. $2x - 4y = 6$
2. $x - 2y + 3z = 4$
3. $x = z$
4. $3y^2 + 3z^2 - [\ln(x - 1)]^2 = 0$
5. $x^2 + z^2 = y^2$
6. $\frac{(y-1)^2}{3} - \frac{(x+1)^2}{5} = z + 3$
7. $x^2 - 8z = 0$
8. $5x^2 + 25y^2 + 10z^2 = 50$
9. $2y^2 - 4x^2 - 12z^2 = 12$
10. $x^2/4 + 2z^2 = 8y$
11. $4(x - 1)^2 + y^2 + 2y + z^2 = 4$
12. $x^2 - 4z^2 = 4y^2$
13. $x^2 + y^2 - (z - 1)^4 = 0$
14. $|x| + |z| = 1$
15. $x^2 - 5x + 2y^2 + 4y = 2$
16. $z - 4x = 1$
17. $2x + 4y - 8z = 0$
18. $x^2 + z^2 = |y|$
19. $y = \cos z$
20. $4x^2 - 2z^2 + 4y^2 = 0$
21. $5x^2 - y^2 - 5z^2 = 10$
22. $x^2 - y^2/3 + z^2 = 0$
23. $3x^2 + y^2 - 6z^2 = 36$
24. $x^2 + 3y^2 - 9z^2 = 0$
25. $2x^2 - 4y^2 = 8z$
26. $x^2 + y^2 + 5y = 2z$
27. $x^2 - y^2 - z^2 + 4x - 6y = 9$
28. $x^2 + y^2 - 2y + z^2 - 2z = 0$
29. $2z^2 + 2x^2 - [\arctan y]^2 = 0$
30. $z^2 = x^2 + 4y^2 - 2x + 8y + 4z$

In Exercises 31 to 35 determine an equation for the surface obtained by revolving the given plane curve about the indicated axis.

31. $y = 3x$, y axis **32.** $y = \sqrt[3]{z + 4}$, y axis **33.** $x^2 - z^2 = 1$, z axis **34.** $y^2 = z^3$, z-axis
35. $y = \ln(z - 1)$, y axis **36.** Find an equation for the torus obtained by revolving circle $(x - 2)^2 + z^2 = 1$ about the z axis.

Section 17.3: Vectors in the Plane and in Space

Objectives:

1. Add, subtract and form scalar multiples of vectors.

2. Interpret the operations of addition, subtraction, and scalar multiplication of vectors geometrically.

INTRODUCTION

Many physical quantities, such as velocity, acceleration, force, or momentum, possess both a magnitude *and* a direction. Such quantities are called *vector quantities* or *vectors* and are quite different from such quantities as mass or temperature, which can be completely described by means of a real number. These latter quantities are usually referred to as *scalar* quantities.

Vectors of the type mentioned above can be represented either geometrically or analytically. The geometric approach has the advantage of visual clarity and is very intuitive. However, actual computations with vectors are more readily done using an analytic representation. Consequently, we shall discuss both approaches. Our emphasis will be on the analytic approach, but we shall use the geometric interpretation for illustrative purposes and in our applications.

PLANE VECTORS

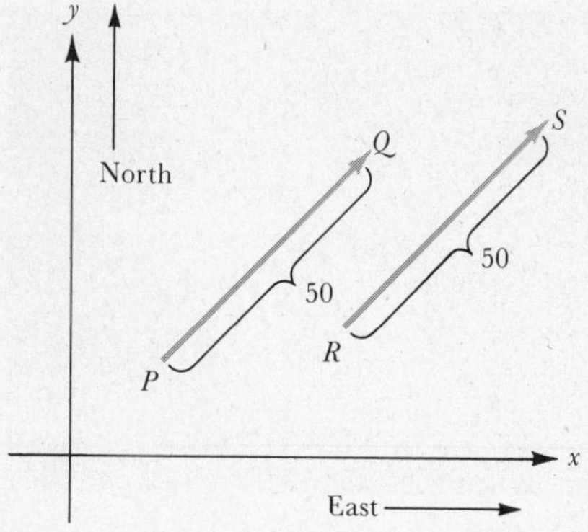

Fig. 17.29

If P and Q are distinct points in the plane, then the directed line segment from P to Q is called a *geometric vector*. We denote it by $\overrightarrow{PQ}$ and refer to P as the *initial point* (also *base* or *tail*) of the geometric vector and Q as the *terminal point* (or *head*). Vectors can be *represented* by means of geometric vectors. For example, a force of 50 lb in the northeasterly direction applied to an object located at P can be represented as in Fig. 17.29, where $\overrightarrow{PQ}$ has length 50 and points toward the northeast. The *same* force applied to a point R would be represented by a geometric vector $\overrightarrow{RS}$, which also has length 50 and points to the northeast. With this in mind we make the convention of identifying geometric vectors $\overrightarrow{PQ}$ and $\overrightarrow{RS}$ that have the same length and the same direction, and we write $\overrightarrow{PQ} = \overrightarrow{RS}$.

In print we usually denote a vector by a single boldface letter such as **A** (when writing by hand we instead write $\vec{A}$). If $\overrightarrow{PQ}$ is a geometric vector which represents **A**, then we refer to $\overrightarrow{PQ}$ as a **representative** of **A** and write $\mathbf{A} = \overrightarrow{PQ}$. Figure 17.30 shows several representatives of a vector **A**. The particular representative of **A** that has its base at the origin is called the **standard representative of A.**

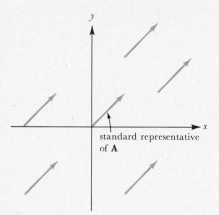

Fig. 17.30

Using standard representatives, we can define a one-to-one correspondence between vectors and points in the plane by letting **A** correspond to the point R, which is the head of its standard representative. In particular, if **A** has standard representative $\overrightarrow{OR}$, where $R = (a_1, a_2)$, then the representative of **A** with base $P = (x, y)$ will be $\overrightarrow{PQ}$, where $Q = (x + a_1, y + a_2)$; see Fig. 17.31. Note that the length of this vector is $\sqrt{a_1{}^2 + a_2{}^2}$. If α is the angle of inclination with the positive x axis $(0 \leq \alpha < 2\pi)$, then

$$\cos\alpha = \frac{a_1}{\sqrt{a_1{}^2 + a_2{}^2}} \qquad \text{and} \qquad \sin\alpha = \frac{a_2}{\sqrt{a_1{}^2 + a_2{}^2}}$$

That is, *the length and direction of* **A** *are completely determined by the numbers* a_1 *and* a_2, the coordinates of the head of its standard representative. This observation leads to our analytic definition of a vector.

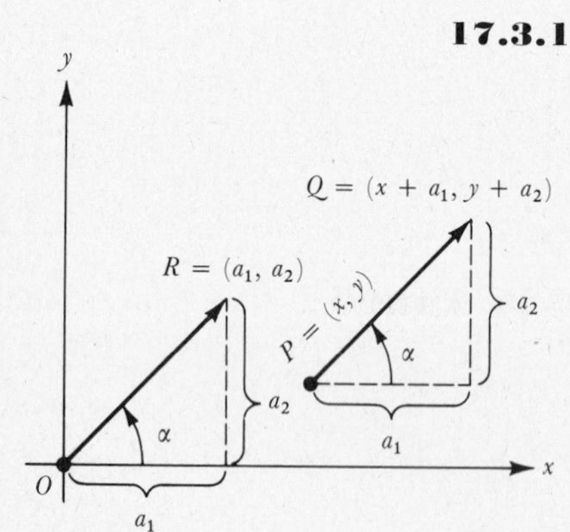

Fig. 17.31

17.3.1 **Definition** A *two-dimensional* (plane) *vector* is an ordered pair of real numbers $\langle a_1, a_2 \rangle$. The numbers a_1 and a_2 are called respectively, the *x* and *y components* or *coordinates* of the vector. Two vectors are equal if and only if they have the same x component and the same y component. That is, $\langle a_1, a_2 \rangle = \langle b_1, b_2 \rangle$ if and only if $a_1 = b_1$ *and* $a_2 = b_2$. The collection of all plane vectors is denoted by R_2.

As noted, we shall use boldface letters to denote vectors. We usually use the same letters (though always lowercase) with the subscripts 1 and 2 to denote the respective components. Thus we write $\mathbf{A} = \langle a_1, a_2 \rangle$, $\mathbf{B} = \langle b_1, b_2 \rangle$, and so on. The use of pointed parentheses to denote vectors allows us to maintain a useful distinction between points in the plane and vectors.

17.3.2 **Definition** The vector $\langle 0, 0 \rangle$ is called the *zero vector* and is denoted by **0**.

17.3.3 **Definition** The *length* of a vector $\mathbf{A} = \langle a_1, a_2 \rangle$, denoted $\|\mathbf{A}\|$, is given by

$$\|\mathbf{A}\| = \sqrt{a_1{}^2 + a_2{}^2}$$

$\|\mathbf{A}\|$ is also referred to as the *magnitude* or *norm* of **A**. If $\mathbf{A} \neq \mathbf{0}$, then the *direction* of **A** is the unique angle α, $0 \leq \alpha < 2\pi$, such that

$$\cos\alpha = \frac{a_1}{\|\mathbf{A}\|} \quad \text{and} \quad \sin\alpha = \frac{a_2}{\|\mathbf{A}\|}$$

We assign no direction to the zero vector.

In the future we shall simply refer to vectors using whichever interpretation is most convenient. As we go along, new definitions concerning operations on vectors will be given in terms of the components (that is, analytically), but we shall follow with the appropriate geometric interpretation. If **A** is given in the form $\overrightarrow{PQ}$, then you will need to determine the components of **A** before applying any of the definitions. To help in this we have:

17.3.4 **Theorem** If $\mathbf{A} = \overrightarrow{PQ}$, where $P = (x_1, y_1)$ and $Q = (x_2, y_2)$, then

$$\mathbf{A} = \langle x_2 - x_1, y_2 - y_1 \rangle$$

Proof: The idea is, simply, to check that $\overrightarrow{OR}$, the standard representative of **A** given in Fig. 17.32, has the same length and direction as $\overrightarrow{PQ}$. But $\|\overrightarrow{OR}\| = \sqrt{(x_2 - x_1)^2 + (y_2 - y_1)^2} = \|\overrightarrow{PQ}\|$ and

$$\cos\alpha_1 = \cos\alpha_2 = \frac{x_2 - x_1}{\sqrt{(x_2 - x_1)^2 + (y_2 - y_1)^2}}$$

$$\sin\alpha_1 = \sin\alpha_2 = \frac{y_2 - y_1}{\sqrt{(x_2 - x_1)^2 + (y_2 - y_1)^2}}$$

Thus $\overrightarrow{OR}$ and $\overrightarrow{PQ}$ have the same length and direction, and hence they are equal. ■

Example 1 Let $\mathbf{A} = \overrightarrow{PQ}$ where $P = (1, 2)$ and $Q = (-3, -2)$.
(*a*) Determine the components of **A**.
(*b*) Determine the length and direction of **A**.
(*c*) Sketch $\overrightarrow{PQ}$ and the standard representative of **A**.

Solutions (*a*) By (17.3.4), $a_1 = -3 - (1) = -4$, $a_2 = -2 - (2) = -4$. Thus $\mathbf{A} = \langle -4, -4 \rangle$.

(*b*) By (17.3.3), $\|\mathbf{A}\| = \sqrt{(-4)^2 + (-4)^2} = \sqrt{32} = 4\sqrt{2}$ and

$$\cos\alpha = \frac{a_1}{\|\mathbf{A}\|} = \frac{-4}{4\sqrt{2}} = \frac{-1}{\sqrt{2}} = \sin\alpha$$

Since $0 \le \alpha \le 2\pi$, we must have $\alpha = 5\pi/4$.

(*c*) See Fig. 17.33. ●

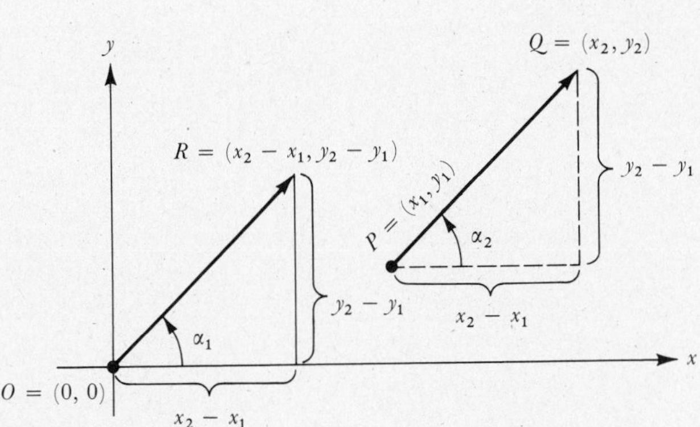

Fig. 17.32

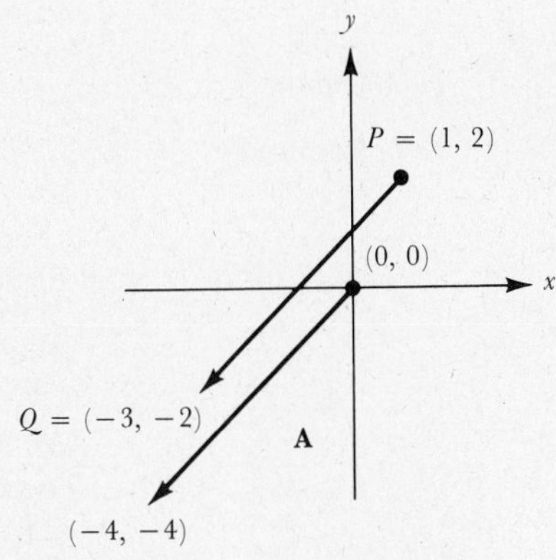

Fig. 17.33

PROGRESS TEST 1

1. Let $\mathbf{A} = \langle -2, 3\rangle$. If $P = (1, -4)$, find Q such that $\mathbf{A} = \overrightarrow{PQ}$. Sketch.
2. Let $P = (1, -3), Q = (-6, 7), R = (-4, -3)$. Find S such that $\overrightarrow{PQ} = \overrightarrow{RS}$.
3. Find the length and direction of $\mathbf{A} = \langle 1, \sqrt{3}\rangle$.

OPERATIONS ON VECTORS

17.3.5 **Definition** Let $\mathbf{A} = \langle a_1, a_2\rangle$ and $\mathbf{B} = \langle b_1, b_2\rangle$ be two vectors. Then their *sum* $\mathbf{A} + \mathbf{B}$ is defined by

$$\mathbf{A} + \mathbf{B} = \langle a_1 + b_1, a_2 + b_2\rangle$$

That is, to add two vectors, simply add componentwise.

Example 2 Compute $\mathbf{A} + \mathbf{B}$ when
(*a*) $\mathbf{A} = \langle -1, 0\rangle$, $\mathbf{B} = \langle 3, 7\rangle$.
(*b*) $\mathbf{A} = \overrightarrow{PQ}$, $\mathbf{B} = \overrightarrow{RS}$, where $P = (1, -1)$, $Q = (3, 7)$, $R = (-3, 4)$, and $S = (8, 1)$.

Solutions (*a*) $\mathbf{A} + \mathbf{B} = \langle -1, 0\rangle + \langle 3, 7\rangle = \langle -1 + 3, 0 + 7\rangle = \langle 2, 7\rangle$

(*b*) Before applying (17.3.5) we must first use (17.3.4) to determine the components of $\mathbf{A}$ and $\mathbf{B}$.

$$\mathbf{A} = \overrightarrow{PQ} = \langle 3 - 1, 7 - (-1)\rangle = \langle 2, 8\rangle$$
$$\mathbf{B} = \overrightarrow{RS} = \langle 8 - (-3), 1 - 4\rangle = \langle 11, -3\rangle$$

and thus

$$\mathbf{A} + \mathbf{B} = \langle 2, 8\rangle + \langle 11, -3\rangle = \langle 13, 5\rangle \quad \bullet$$

17.3.6 **Definition** If $\mathbf{A} = \langle a_1, a_2\rangle$, then the *negative* of $\mathbf{A}$, denoted $-\mathbf{A}$, is defined by $-\mathbf{A} = \langle -a_1, -a_2\rangle$. The *difference* $\mathbf{A} - \mathbf{B}$ is defined by $\mathbf{A} - \mathbf{B} = \mathbf{A} + (-\mathbf{B})$.

Since

$$\mathbf{A} + (-\mathbf{B}) = \langle a_1, a_2\rangle + \langle -b_1, -b_2\rangle = \langle a_1 - b_1, a_2 - b_2\rangle$$

we see that subtraction of vectors is also componentwise.

Example 3 Find $\mathbf{A} - \mathbf{B}$ if $\mathbf{A} = \langle 2, -7\rangle$ and $\mathbf{B} = \langle -3, 4\rangle$.

Solution

$$\mathbf{A} - \mathbf{B} = \langle 2, -7\rangle - \langle -3, 4\rangle = \langle 2, -7\rangle + \langle 3, -4\rangle = \langle 5, -11\rangle \quad \bullet$$

The following properties of vector addition and subtraction are an immediate consequence of the corresponding properties of real numbers and the componentwise nature of the definitions.

17.3.7 **Theorem** Let $\mathbf{A}$, $\mathbf{B}$, and $\mathbf{C}$ be vectors. Then
(*a*) $\mathbf{A} + \mathbf{B} = \mathbf{B} + \mathbf{A}$
(*b*) $\mathbf{A} + (\mathbf{B} + \mathbf{C}) = (\mathbf{A} + \mathbf{B}) + \mathbf{C}$

(*c*) $\mathbf{A} + \mathbf{0} = \mathbf{A}$
(*d*) $\mathbf{A} - \mathbf{A} = \mathbf{0}$

Proof: We illustrate by proving (*a*). The proofs of (*b*), (*c*), and (*d*) are similar.

$$\begin{aligned}\mathbf{A} + \mathbf{B} &= \langle a_1, a_2\rangle + \langle b_1, b_2\rangle \\ &= \langle a_1 + b_1, a_2 + b_2\rangle \\ &= \langle b_1 + a_1, b_2 + a_2\rangle \\ &= \langle b_1, b_2\rangle + \langle a_1, a_2\rangle = \mathbf{B} + \mathbf{A} \quad \blacksquare\end{aligned}$$

When dealing with vectors it is traditional to refer to a real number as a *scalar.*

17.3.8 **Definition** If $\mathbf{A} = \langle a_1, a_2\rangle$ is any vector and k is any scalar, then $k\mathbf{A} = \langle ka_1, ka_2\rangle$. We refer to $k\mathbf{A}$ *as a scalar multiple* of $\mathbf{A}$.

17.3.9 **Theorem** Let $\mathbf{A}$ and $\mathbf{B}$ be vectors and k_1 and k_2 be scalars. Then
(*a*) $k_1\mathbf{0} = \mathbf{0}$
(*b*) $0\mathbf{A} = \mathbf{0}$
(*c*) $(k_1k_2)\mathbf{A} = k_1(k_2\mathbf{A})$
(*d*) $k_1(\mathbf{A} + \mathbf{B}) = k_1\mathbf{A} + k_1\mathbf{B}$
(*e*) $(k_1 + k_2)\mathbf{A} = k_1\mathbf{A} + k_2\mathbf{A}$
(*f*) $1\mathbf{A} = \mathbf{A}$
(*g*) $(-1)\mathbf{A} = -\mathbf{A}$

Proof: We shall prove (*d*):

$$\begin{aligned}k_1(\mathbf{A} + \mathbf{B}) &= k_1(\langle a_1, a_2\rangle + \langle b_1, b_2\rangle) \\ &= k_1\langle a_1 + b_1, a_2 + b_2\rangle \\ &= \langle k_1(a_1 + b_1), k_1(a_2 + b_2)\rangle \\ &= \langle k_1a_1 + k_1b_1, k_1a_2 + k_1b_2\rangle \\ &= \langle k_1a_1, k_1a_2\rangle + \langle k_1b_1, k_1b_2\rangle \\ &= k_1\langle a_1, a_2\rangle + k_1\langle b_1, b_2\rangle = k_1\mathbf{A} + k_1\mathbf{B} \quad \blacksquare\end{aligned}$$

An important relationship between the length of $k\mathbf{A}$ and the length of $\mathbf{A}$ is given by the following:

17.3.10 **Theorem** Let $\mathbf{A} = \langle a_1, a_2\rangle$ and let k be any scalar. Then

$$\|k\mathbf{A}\| = |k|\,\|\mathbf{A}\|$$

Proof:

$$\begin{aligned}\|k\mathbf{A}\| &= \|\langle ka_1, ka_2\rangle\| \\ &= \sqrt{(ka_1)^2 + (ka_2)^2} \\ &= \sqrt{k^2({a_1}^2 + {a_2}^2)} \\ &= \sqrt{k^2}\sqrt{{a_1}^2 + {a_2}^2} = |k|\,\|\mathbf{A}\| \quad \blacksquare\end{aligned}$$

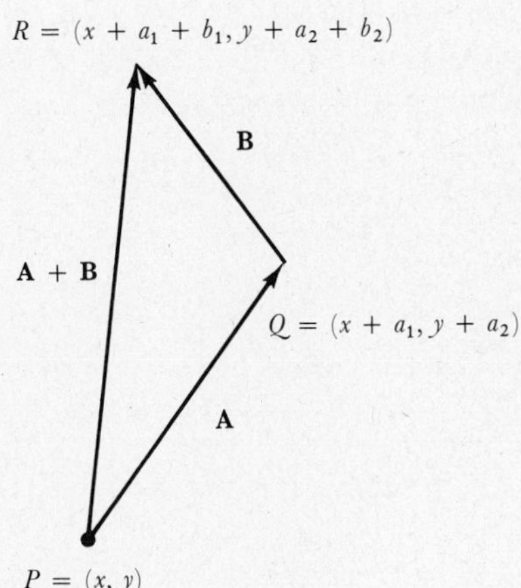

Fig. 17.34

THE GEOMETRY OF VECTOR OPERATIONS

Vector Addition: Let $\mathbf{A} = \langle a_1, a_2\rangle$, $\mathbf{B} = \langle b_1, b_2\rangle$, and let $P = (x, y)$ be any point in the plane. The representative of $\mathbf{A}$ with base P is $\overrightarrow{PQ}$, where $Q = (x + a_1, y + a_2)$; and the representative of $\mathbf{B}$ with base Q is $\overrightarrow{QR}$, where $R = ((x + a_1) + b_1, (y + a_2) + b_2)$; see Fig. 17.34. Now $\mathbf{A} + \mathbf{B} = \langle a_1 + b_1, a_2 + b_2\rangle$, and so the representative of $\mathbf{A} + \mathbf{B}$ with base P is $\overrightarrow{PS}$, where $S = (x + (a_1 + b_1), y + (a_2 + b_2))$. But $x + (a_1 + b_1) = (x + a_1) + b_1$ and $y + (a_2 + b_2) = (y + a_2) + b_2$, so $R = S$. Thus we have:

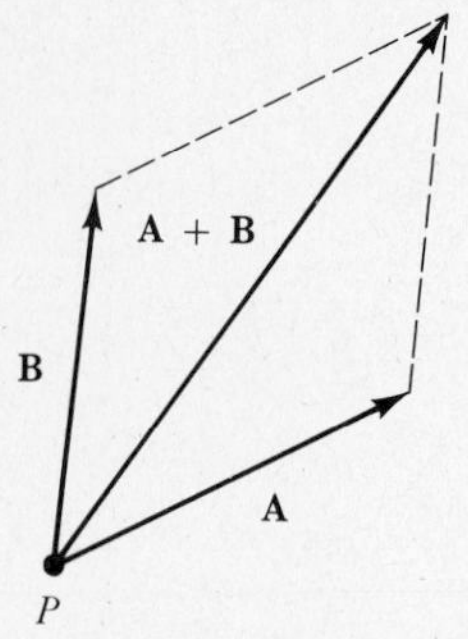

Fig. 17.35

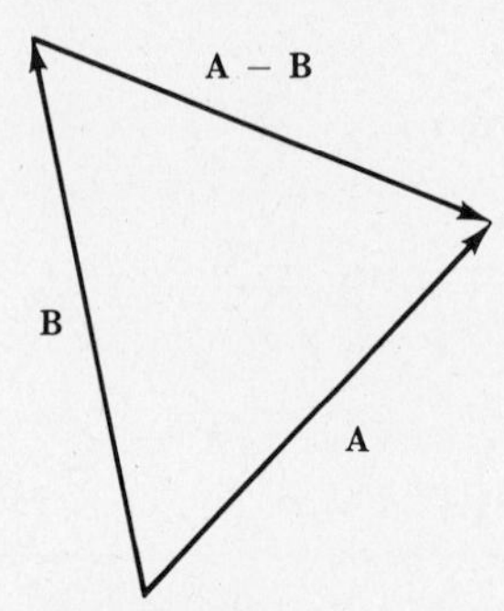

Fig. 17.36

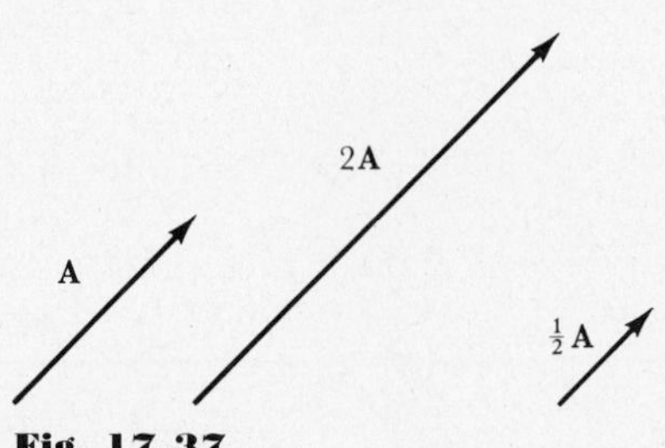

Fig. 17.37

17.3.11 **Tail-to-Head Addition** A representative for A + B can be obtained by placing a representative of **B** with its base at the head of some representative of A and taking as the representative of A + B the vector from the base of A to the head of **B**.

Since representatives of **A**, **B**, and **A** + **B** comprise the sides of a triangle, we can use the fact that the length of one side of a triangle is less than or equal to the sum of the lengths of the remaining sides to obtain the following:

17.3.12 **Triangle Inequality** $\|\mathbf{A} + \mathbf{B}\| \leq \|\mathbf{A}\| + \|\mathbf{B}\|$

An equivalent realization of the vector sum can be obtained by taking representatives of A and **B** with the *same* initial point P and then forming the parallelogram with these representatives as adjacent sides (see Fig. 17.35). The diagonal with base P is then a representative of **A** + **B**. This is referred to as the "parallelogram law of addition."

If, in Fig. 17.35, **A** and **B** represent forces applied at P, then **A** + **B** is the *resultant force*. That is, the single force **A** + **B** applied at P will have the same effect as the two combined forces **A** and **B**.

Vector Subtraction: Since **A** − **B** is the vector that when added to **B** yields A, we can use the tail-to-head method to obtain a geometric realization.

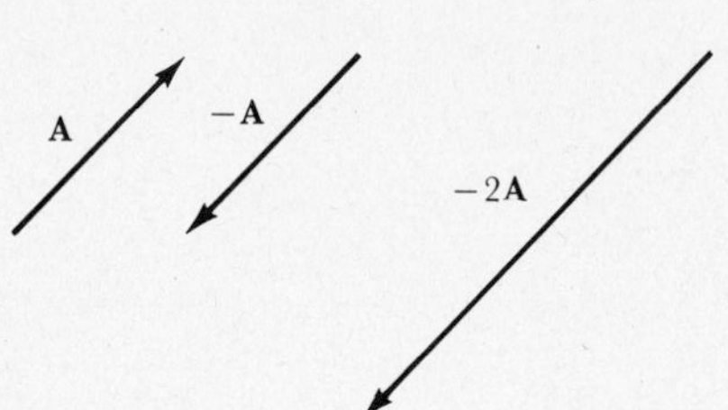

Fig. 17.38

17.3.13 **Vector Subtraction** If A and **B** have the same initial point, then the vector from the head of **B** to the head of **A** is a representative of **A** − **B**. Thus **A** − **B** "points" from **B** to A; see Fig. 17.36.

Scalar Multiplication: Since $\|k\mathbf{A}\| = |k|\,\|\mathbf{A}\|$, we know that if $k > 0$, then k**A** will have k times the length of **A** and the same direction; see Fig. 17.37. If $k < 0$, then k**A** will have $|k|$ times the length of **A** but the *opposite direction;* see Figs. 17.37 and 17.38 and the discussion on page 569.

PROGRESS TEST 2

1. Let $\mathbf{A} = \langle 2, -4 \rangle$, $\mathbf{B} = \langle -3, 1 \rangle$, $\mathbf{C} = \langle -2, 3 \rangle$. Determine $(\mathbf{A} + \mathbf{B}) + \mathbf{C}$, $\mathbf{A} - \mathbf{B}$, and $\mathbf{C} - \mathbf{A}$, and illustrate each geometrically.
2. Let $\mathbf{A} = \langle -3, 1 \rangle$, $\mathbf{B} = \langle 4, -2 \rangle$, $\mathbf{C} = \langle -1, -1 \rangle$. Determine
(*a*) $3\mathbf{A} + 2\mathbf{B} - \mathbf{C}$ (*b*) $\|\mathbf{C}\|(\mathbf{A} + \mathbf{B})$
(*c*) $\|3\mathbf{A} - 2\mathbf{B}\|$

UNIT VECTORS

17.3.14 **Definition** A vector of length 1 is called a *unit vector.*

17.3.15 **Theorem** Let $\mathbf{A} \neq \mathbf{0}$. Then

(*a*) $\dfrac{\mathbf{A}}{\|\mathbf{A}\|}$ is a unit vector having the same direction as $\mathbf{A}$.

(*b*) $\mathbf{A} = \|\mathbf{A}\|\langle\cos\alpha, \sin\alpha\rangle$, where α is the direction of $\mathbf{A}$.

Proof:
(*a*) By (17.3.10) and (17.3.8),

$$\left\|\frac{\mathbf{A}}{\|\mathbf{A}\|}\right\| = \left\|\left(\frac{1}{\|\mathbf{A}\|}\right)\mathbf{A}\right\| = \left|\frac{1}{\|\mathbf{A}\|}\right| \|\mathbf{A}\| = \frac{1}{\|\mathbf{A}\|}\|\mathbf{A}\| = 1$$

(*b*) This is an immediate consequence of (17.3.3). ■

From the above it follows that if $\|\mathbf{A}\| = 1$, then the x component of $\mathbf{A}$ is $\cos\alpha$ and the y component of $\mathbf{A}$ is $\sin\alpha$, where α is the direction of $\mathbf{A}$. For this reason a unit vector is often referred to as a **direction vector**. In particular, we have:

17.3.16 **Theorem:** If $\mathbf{A}$ is a direction vector, then $k\mathbf{A}$ has length $|k|$. The direction of $k\mathbf{A}$ will be the *same* as $\mathbf{A}$ if $k > 0$, and *opposite* to that of $\mathbf{A}$ if $k < 0$.

Example 4 Determine a vector $\mathbf{B}$ of length 3/2 that has the same direction as $\mathbf{A} = \overrightarrow{PQ}$, where $P = (-1, 2)$, $Q = (3, 4)$.

Solution $\mathbf{A} = \langle 3 - (-1), 4 - 2\rangle = \langle 4, 2\rangle$ and $\|\mathbf{A}\| = \sqrt{16 + 4} = 2\sqrt{5}$. Thus

$$\frac{\mathbf{A}}{\|\mathbf{A}\|} = \left\langle \frac{2}{\sqrt{5}}, \frac{1}{\sqrt{5}} \right\rangle$$

is a unit vector in the direction of $\mathbf{A}$. Therefore

$$\mathbf{B} = \frac{3}{2}\frac{\mathbf{A}}{\|\mathbf{A}\|} = \left\langle \frac{3}{\sqrt{5}}, \frac{3}{2\sqrt{5}} \right\rangle$$

is a vector of length 3/2 that has the same direction as $\mathbf{A}$. ●

17.3.17 **Definition** The vectors $\mathbf{i} = \langle 1, 0\rangle$ and $\mathbf{j} = \langle 0, 1\rangle$ are called the *unit coordinate vectors.*

Note that $\mathbf{i}$ and $\mathbf{j}$ are unit vectors in the directions of the positive x and y axes, respectively. By our rules for addition and scalar multiplication, for an arbitrary vector $\mathbf{A}$ we have

$$\begin{aligned} \mathbf{A} &= \langle a_1, a_2\rangle \\ &= \langle a_1, 0\rangle + \langle 0, a_2\rangle \\ &= a_1\langle 1, 0\rangle + a_2\langle 0, 1\rangle = a_1\mathbf{i} + a_2\mathbf{j} \end{aligned}$$

If $\mathbf{A}$ and $\mathbf{B}$ are vectors and α and β are scalars, then the vector $\alpha\mathbf{A} + \beta\mathbf{B}$ is called a *linear combination* of $\mathbf{A}$ and $\mathbf{B}$. Thus *every vector can be written* (*uniquely*) *as a linear combination of the unit coordinate vectors* $\mathbf{i}$ *and* $\mathbf{j}$. This is illustrated in Fig. 17.39.

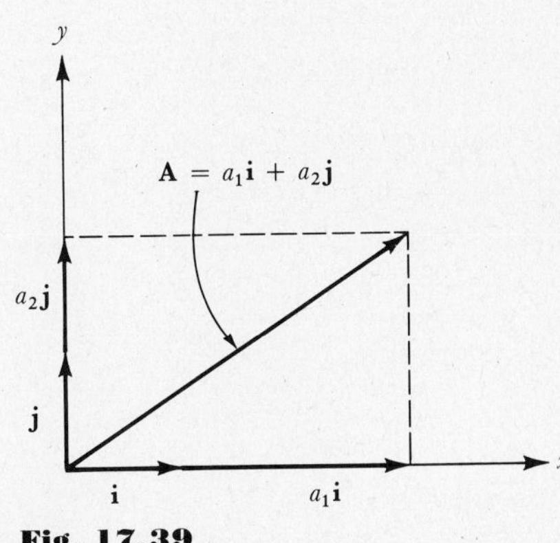

Fig. 17.39

VECTORS IN SPACE

The concept of a three-dimensional vector (3-vector) is a direct extension of the notion of a plane vector. That is, a 3-*vector* is an ordered *triple* of real numbers. If

$\mathbf{A} = \langle a_1, a_2, a_3 \rangle$ is such a vector, then the numbers a_1, a_2, and a_3 are called, respectively, the x, y, and z *components* or *coordinates* of $\mathbf{A}$. As with plane vectors (2-vectors), the operations are defined componentwise. Specifically, if $\mathbf{A} = \langle a_1, a_2, a_3 \rangle$ and $\mathbf{B} = \langle b_1, b_2, b_3 \rangle$ are vectors, and if k is a scalar, then

$$\mathbf{A} + \mathbf{B} = \langle a_1 + b_1, a_2 + b_2, a_3 + b_3 \rangle$$
$$\mathbf{A} - \mathbf{B} = \langle a_1 - b_1, a_2 - b_2, a_3 - b_3 \rangle$$
$$k\mathbf{A} = \langle ka_1, ka_2, ka_3 \rangle$$

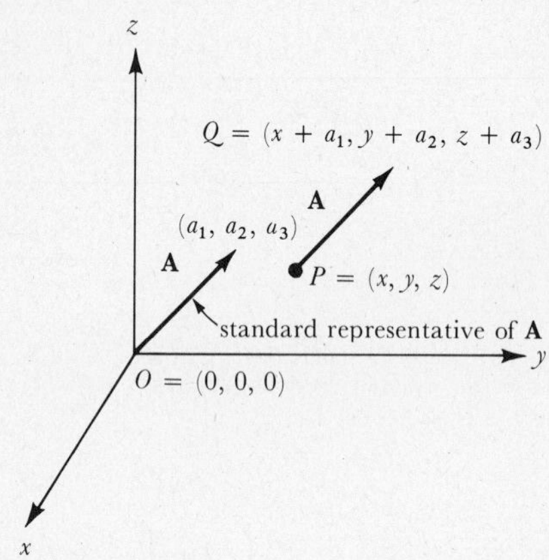

Fig. 17.40

The properties of vector addition, subtraction, and scalar multiplication given in Theorems (17.3.7) and (17.3.9) extend directly to 3-vectors by taking into account the additional component. The collection of all 3-vectors is denoted by R_3.

If $\mathbf{A} = \langle a_1, a_2, a_3 \rangle$ and if $P = (x, y, z)$ is any point, then the directed line segment from P to $Q = (x + a_1, y + a_2, z + a_3)$, denoted $\overrightarrow{PQ}$, is called a *representative* of $\mathbf{A}$. As with plane vectors, this gives us a geometric realization of a vector in R_3. Again, we do not distinguish between representatives of the same vector. Thus if $\overrightarrow{PQ}$ is a representative of $\mathbf{A}$, we write $\mathbf{A} = \overrightarrow{PQ}$. The notions of **base, head,** and **standard representative** of $\mathbf{A}$ are as for 2-vectors; see Fig. 17.40.

As with vectors in R_2, we can obtain the components of a vector given in the form $\overrightarrow{PQ}$ by subtracting the coordinates of P from those of Q.

17.3.18 Theorem If $\mathbf{A} = \overrightarrow{PQ}$, where $P = (x_1, y_1, z_1)$ and $Q = (x_2, y_2, z_2)$, then

$$\mathbf{A} = \langle x_2 - x_1, y_2 - y_1, z_2 - z_1 \rangle$$

The *norm* or *length* of $\mathbf{A} = \langle a_1, a_2, a_3 \rangle$ is again the length of any representative of $\mathbf{A}$. Thus, from the distance formula (17.1.2), we have

$$\|\mathbf{A}\| = \sqrt{a_1^2 + a_2^2 + a_3^2}$$

As before, $\|k\mathbf{A}\| = |k|\,\|\mathbf{A}\|$ and $\mathbf{A}/\|\mathbf{A}\|$ is a unit vector.

For 3-vectors the *unit coordinate vectors* are given by

$$\mathbf{i} = \langle 1, 0, 0 \rangle \qquad \mathbf{j} = \langle 0, 1, 0 \rangle \qquad \mathbf{k} = \langle 0, 0, 1 \rangle$$

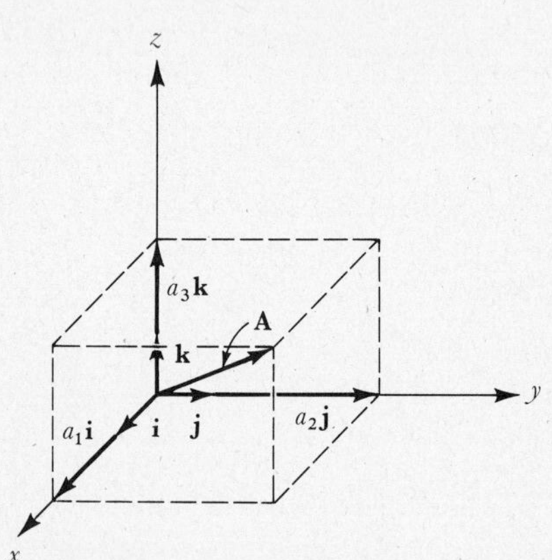

Fig. 17.41

and every 3-vector can be uniquely expressed as a *linear combination* of $\mathbf{i}$, $\mathbf{j}$, and $\mathbf{k}$ by means of

$$\langle a_1, a_2, a_3 \rangle = a_1\mathbf{i} + a_2\mathbf{j} + a_3\mathbf{k}$$

It is often convenient to think of R_2 as being a subset of R_3 by identifying a typical vector $\langle a_1, a_2 \rangle$ in R_2 with the vector $\langle a_1, a_2, 0 \rangle$ in R_3. Thus, in effect, when dealing with linear combinations of the form $a_1\mathbf{i} + a_2\mathbf{j}$, we can always treat $\mathbf{i}$ and $\mathbf{j}$ as being the unit coordinate vectors in R_2.

SECTION 17.3 EXERCISES

In Exercises 1 to 10, (a) determine Q such that $\overrightarrow{PQ} = \mathbf{A}$, (b) determine $\|\mathbf{A}\|$, and (c) sketch the standard representative of $\mathbf{A}$.

1. $\mathbf{A} = \langle -2, 3 \rangle$, $P = (4, -1)$
2. $\mathbf{A} = \langle -3/2, 1/3 \rangle$, $P = (3, -2)$
3. $\mathbf{A} = 4\mathbf{i} + 2\mathbf{j}$, $P = (2, 1/2)$
4. $\mathbf{A} = \langle -3, -8 \rangle$, $P = (-2, 9)$
5. $\mathbf{A} = -2\mathbf{i} + 6\mathbf{j}$, $P = (-11, -6)$
6. $\mathbf{A} = \langle -2, -1, -\sqrt{2} \rangle$, $P = (1, 1, 1)$
7. $\mathbf{A} = \langle 5, -2, 1 \rangle$, $P = (-2, 1, 2)$
8. $\mathbf{A} = \langle -1, -2, 3 \rangle$, $P = (3, 1, 4)$
9. $\mathbf{A} = \langle 3, -1, 0 \rangle$, $P = (2, -1, 1)$
10. $\mathbf{A} = \langle -2, 1, 1 \rangle$, $P = (2, -1, 2)$

In Exercises 11 to 20 write $\mathbf{A} = \overrightarrow{PQ}$ in component form.

11. $P = (-3, 5)$, $Q = (9, -3)$
12. $P = (4/3, 1/2)$, $Q = (0, -2)$
13. $P = (\sqrt{2}, 1 + \sqrt{3})$, $Q = (-1, 1)$
14. $P = (-1/2, 7/9)$, $Q = (7, -2/9)$

15. $P = (3, 7)$, $Q = (-6, 0)$
16. $P = (2, 1, -1)$, $Q = (4, -1, 7)$
17. $P = (3, \sqrt{2}, -1)$, $Q = (-1, 2, 4)$
18. $P = (-2, 4, 13)$, $Q = (7, -1, 12)$
19. $P = (\pi, 1, e)$, $Q = (-2, \sqrt{2}, 1)$
20. $P = (7, -1, \sqrt{3})$, $Q = (3, 2, -1)$

In Exercises 21 to 30 compute $\mathbf{A} + \mathbf{B}$ *and* $\mathbf{A} - \mathbf{B}$, *and illustrate geometrically.*

21. $\mathbf{A} = \langle -2, 1\rangle$, $\mathbf{B} = \langle 3, 5\rangle$
22. $\mathbf{A} = 3\mathbf{i} - 2\mathbf{j}$, $\mathbf{B} = \langle -1, 4\rangle$
23. $\mathbf{A} = \langle -7, -3\rangle$, $\mathbf{B} = \langle -2, -1\rangle$
24. $\mathbf{A} = \langle 11, 3\rangle$, $\mathbf{B} = -8\mathbf{i} + 2\mathbf{j}$
25. $\mathbf{A} = \langle 6, -2, -1\rangle$, $\mathbf{B} = \langle 2, 1, 1\rangle$
26. $\mathbf{A} = \langle 5, -1, 2\rangle$, $\mathbf{B} = \langle 3, 2, -1\rangle$
27. $\mathbf{A} = \langle 8, -4, 2\rangle$, $\mathbf{B} = \langle 4, -4, 0\rangle$
28. $\mathbf{A} = \mathbf{i} + 2\mathbf{j} - \mathbf{k}$, $\mathbf{B} = 3\mathbf{i} + 2\mathbf{j} + 2\mathbf{k}$
29. $\mathbf{A} = \overrightarrow{PQ}$, $P = (1, 6)$, $Q = (-3, 2)$; $\mathbf{B} = 7\mathbf{i} + 2\mathbf{j}$
30. $\mathbf{A} = \overrightarrow{PQ}$, $P = (-1, 4, 1)$, $Q = (-3, -5, 2)$; $\mathbf{B} = 4\mathbf{i} - 2\mathbf{j} + \mathbf{k}$

In Exercises 31 to 35 determine the indicated vector or scalar quantity where $\mathbf{A} = \langle 1, 0, -2\rangle$, $\mathbf{B} = \langle -3, 1, 17\rangle$, $\mathbf{C} = \langle 6, 1, -5\rangle$.

31. $3\mathbf{A} + \mathbf{B} - \mathbf{C}$
32. $\|\mathbf{A}\|\,\mathbf{B} + \|\mathbf{B}\|\,\mathbf{A}$
33. $\|\mathbf{A} - \mathbf{C}\|\,(\mathbf{A} + 2\mathbf{B})$
34. $\|3\mathbf{A} - 2\mathbf{B}\|$
35. $2\mathbf{A}/\|\mathbf{A}\| + 4\mathbf{C}/\|\mathbf{B}\|$

In Exercises 36 to 41 determine a plane vector $\mathbf{B}$ *with the same direction as* $\mathbf{A}$ *and with the indicated length.*

36. $\mathbf{A} = \langle -3, 4\rangle$, $\|\mathbf{B}\| = 6$
37. $\mathbf{A} = \langle 1, \sqrt{8}\rangle$, $\|\mathbf{B}\| = 2$
38. $\mathbf{A} = \langle 2, -3\rangle$, $\|\mathbf{B}\| = 1$
39. $\mathbf{A} = 4\mathbf{i} - 4\mathbf{j}$, $\|\mathbf{B}\| = 3$
40. $\mathbf{A} = \overrightarrow{PQ}$; $P = (1, -2)$, $Q = (3, 6)$; $\|\mathbf{B}\| = 1/2$
41. $\mathbf{A} = \overrightarrow{PQ}$; $P = (-1, -3)$, $Q = (-1, 6)$; $\|\mathbf{B}\| = 9$
42. Prove Theorem (17.3.9), part (*e*).
43. Prove Theorem (17.3.9), part (*g*).

Section 17.4: The Dot and Cross Products

Objectives:

1. Evaluate the dot product of two vectors.
2. Use the dot product to determine (*a*) the angle between two vectors, (*b*) scalar and vector projections, and (*c*) the work done by a force in moving an object along a straight line.
3. Evaluate cross products and apply the various geometric properties to compute areas and volumes.

INTRODUCTION

In this section we discuss two different types of vector multiplication. The first, called the *dot product,* is defined for vectors in R_2 or R_3. The second, called the *cross product,* is only defined for vectors in R_3. Both products are very useful in many geometric and physical applications.

THE DOT PRODUCT

17.4.1 Definition If $\mathbf{A} = \langle a_1, a_2, a_3\rangle$ and $\mathbf{B} = \langle b_1, b_2, b_3\rangle$, then the *dot product* of $\mathbf{A}$ and $\mathbf{B}$, denoted $\mathbf{A}\cdot\mathbf{B}$, is defined by

$$\mathbf{A}\cdot\mathbf{B} = a_1b_1 + a_2b_2 + a_3b_3$$

The dot product is also referred to as the *scalar product* or the *inner product.*

It is important to note that *the result of taking the dot product of two vectors is a scalar.* Of course, if $\mathbf{A}$, $\mathbf{B}$ are in R_2, then the dot product is given by

$$\begin{aligned}\mathbf{A}\cdot\mathbf{B} &= \langle a_1, a_2\rangle\cdot\langle b_1, b_2\rangle\\ &= a_1b_1 + a_2b_2\end{aligned}$$

Example 1 Determine $\mathbf{A}\cdot\mathbf{B}$ if $\mathbf{A} = \langle 2, 1, -3\rangle$, $\mathbf{B} = \langle -4, 5, -2\rangle$.

Solution

$$\begin{aligned}\mathbf{A}\cdot\mathbf{B} &= \langle 2, 1, -3\rangle\cdot\langle -4, 5, -2\rangle\\ &= 2(-4) + 5(1) + (-3)(-2) = -8 + 5 + 6 = 3.\end{aligned}$$ ●

We summarize some properties of the dot product that follow directly from the definition.

17.4.2 Theorem Suppose **A**, **B**, **C** are vectors in R_3 (or R_2) and k is any scalar, then
(*a*) $\mathbf{A}\cdot\mathbf{B} = \mathbf{B}\cdot\mathbf{A}$
(*b*) $\mathbf{A}\cdot(\mathbf{B}+\mathbf{C}) = \mathbf{A}\cdot\mathbf{B} + \mathbf{A}\cdot\mathbf{C}$
(*c*) $k(\mathbf{A}\cdot\mathbf{B}) = (k\mathbf{A})\cdot\mathbf{B} = \mathbf{A}\cdot(k\mathbf{B})$
(*d*) $\mathbf{0}\cdot\mathbf{A} = 0$
(*e*) $\mathbf{A}\cdot\mathbf{A} = \|\mathbf{A}\|^2$

To explore a useful geometric interpretation of the dot product, we first define the angle between two vectors. Notice that even when the vectors are in R_3, two vectors with a common base point lie in a plane.

17.4.3 Definition Suppose **A** and **B** are nonzero vectors. If **A** is not a scalar multiple of **B**, then by the *angle* θ *between* **A** *and* **B** we mean the least positive angle between representatives with the same initial point. If $\mathbf{A} = k\mathbf{B}$, then we define the angle between **A** and **B** to be 0 if $k > 0$ and π if $k < 0$ (see Fig. 17.42).

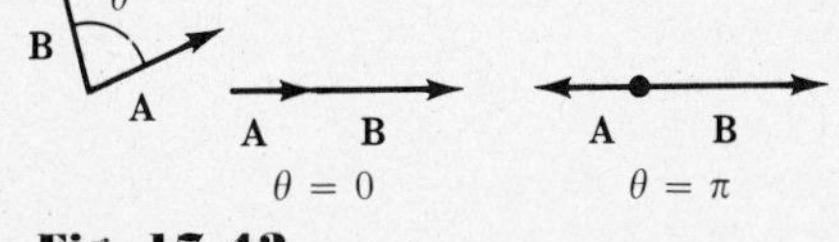

Fig. 17.42

Note that the angle between two vectors is by definition between 0 and π. Further, this angle is directionless in the sense that the angle between **A** and **B** is the same as the angle between **B** and **A**.

We say **A** and **B** are *parallel* and write $\mathbf{A}\|\mathbf{B}$ if $\theta = 0$ or $\theta = \pi$. If $\theta = \pi/2$, then we say that **A** and **B** are *perpendicular* or *orthogonal* and write $\mathbf{A} \perp \mathbf{B}$. It is convenient to define the zero vector to be both parallel and orthogonal to every vector. With this definition $\mathbf{A}\|\mathbf{B}$ if and only if $\mathbf{A} = k\mathbf{B}$ for some scalar k.

We now give the geometric significance of the dot product, which is a consequence of the Law of Cosines.

17.4.4 Theorem Let θ be the angle between the nonzero vectors **A** and **B**. Then

$$\cos\theta = \frac{\mathbf{A}\cdot\mathbf{B}}{\|\mathbf{A}\|\,\|\mathbf{B}\|}$$

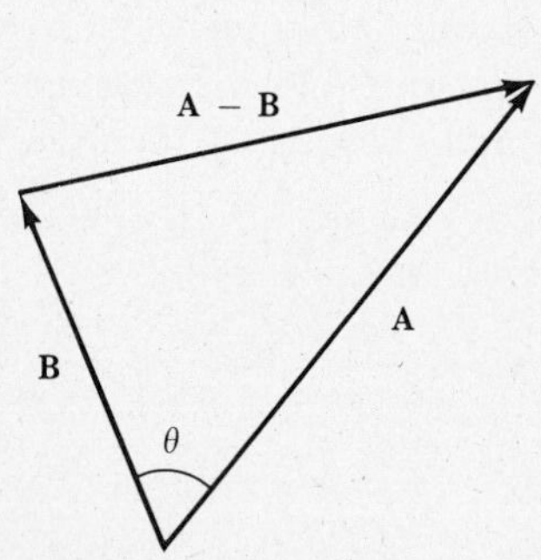

Fig. 17.43

Proof: Choose representatives of **A** and **B** with the same initial point and consider the triangle that has these representatives and a representative of $\mathbf{A} - \mathbf{B}$ as its three sides (see Fig. 17.43). By the law of cosines

$$\|\mathbf{A}-\mathbf{B}\|^2 = \|\mathbf{A}\|^2 + \|\mathbf{B}\|^2 - 2\|\mathbf{A}\|\,\|\mathbf{B}\|\cos\theta$$

Now, by [17.4.2(*e*)],

$$\begin{aligned}\|\mathbf{A}-\mathbf{B}\|^2 &= (\mathbf{A}-\mathbf{B})\cdot(\mathbf{A}-\mathbf{B})\\ &= (\mathbf{A}-\mathbf{B})\cdot\mathbf{A} + (\mathbf{A}-\mathbf{B})\cdot(-\mathbf{B}) \qquad \text{by [17.4.2}(b)]\\ &= \mathbf{A}\cdot\mathbf{A} - \mathbf{B}\cdot\mathbf{A} - \mathbf{A}\cdot\mathbf{B} + \mathbf{B}\cdot\mathbf{B} = \|\mathbf{A}\|^2 + \|\mathbf{B}\|^2 - 2\mathbf{A}\cdot\mathbf{B}\end{aligned}$$

Comparing this with the above, we conclude that

$$-2\mathbf{A}\cdot\mathbf{B} = -2\,\|\mathbf{A}\|\,\|\mathbf{B}\|\cos\theta$$

and hence

$$\cos\theta = \frac{\mathbf{A}\cdot\mathbf{B}}{\|\mathbf{A}\|\,\|\mathbf{B}\|} \quad \blacksquare$$

As an immediate consequence of (17.4.4) we have:

17.4.5 Theorem $\mathbf{A} \perp \mathbf{B}$ if and only if $\mathbf{A}\cdot\mathbf{B} = 0$.

Example 2 Let $\mathbf{A} = \langle 2, -1 \rangle$ and $\mathbf{B} = \langle k, 3 \rangle$. Determine k such that $\mathbf{A} \perp \mathbf{B}$.

Solution

$$\mathbf{A} \cdot \mathbf{B} = \langle 2, -1 \rangle \cdot \langle k, 3 \rangle = 2k - 3$$

Thus $\mathbf{A} \cdot \mathbf{B} = 0$ if and only if $k = 3/2$, so $\mathbf{B} = \langle 3/2, 3 \rangle$. ●

Example 3 Determine the angle between the vectors $\mathbf{A} = \langle -1, 3, 4 \rangle$ and $\mathbf{B} = \langle 1, 2, -2 \rangle$.

Solution

$$\cos\theta = \frac{\langle -1, 3,4 \rangle \cdot \langle 1, 2, -2 \rangle}{\|\langle -1, 3, 4 \rangle\| \, \|\langle 1, 2, -2 \rangle\|} = -1/\sqrt{26} \approx -0.1961$$

Since $0 \le \theta \le \pi$, we have $\theta \approx 1.768$, or $\theta \approx 101.3°$; see Fig. 17.44. ●

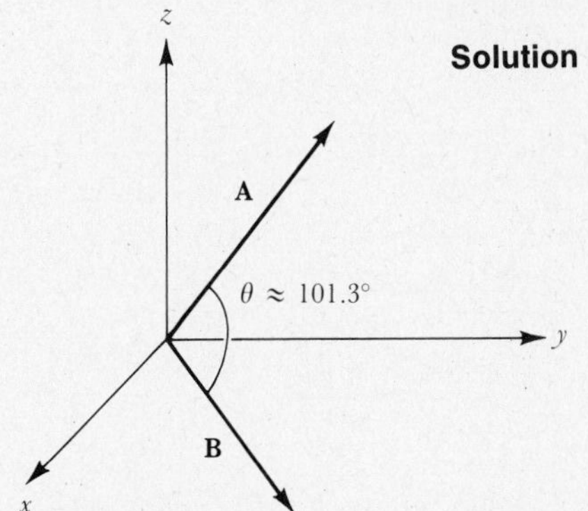

Fig. 17.44

If neither $\mathbf{A}$ nor $\mathbf{B}$ is zero, then from (17.4.4) we have

$$\mathbf{A} \cdot \mathbf{B} = \|\mathbf{A}\| \, \|\mathbf{B}\| \cos\theta$$

Since $|\cos\theta| \le 1$, we conclude:

17.4.6 Cauchy-Schwarz Inequality $\quad |\mathbf{A} \cdot \mathbf{B}| \le \|\mathbf{A}\| \, \|\mathbf{B}\|$

PROGRESS TEST 1

1. Compute the cosine of the angle between the vectors $\mathbf{A}$ and $\mathbf{B}$.
(*a*) $\mathbf{A} = \langle 2, -1 \rangle$, $\mathbf{B} = \langle 5, 10 \rangle$
(*b*) $\mathbf{A} = \langle 2, 1, -6 \rangle$, $\mathbf{B} = \langle 3, -1, 1 \rangle$

2. For which choices of k will $\mathbf{A} = \langle k, 3, -4 \rangle$ be perpendicular to $\mathbf{B} = \langle 4k, -1, k \rangle$?

3. Show that $\|\mathbf{A} + \mathbf{B}\|^2 = \|\mathbf{A}\|^2 + 2\mathbf{A} \cdot \mathbf{B} + \|\mathbf{B}\|^2$

SCALAR AND VECTOR PROJECTIONS

Suppose $\mathbf{A}$ and $\mathbf{B}$ are nonzero vectors with α the angle between them. Consider representatives of $\mathbf{A}$ and $\mathbf{B}$ with the same initial point P and let L be the straight line through this representative of $\mathbf{B}$. Drop a perpendicular from the head of $\mathbf{A}$ to the line L and let Q denote the point where this perpendicular meets L (see Fig. 17.45). The vector with representative $\overrightarrow{PQ}$ is called the *vector projection* of $\mathbf{A}$ on $\mathbf{B}$ or the *vector component of* $\mathbf{A}$ *in the direction* of $\mathbf{B}$ and is denoted by

$$\mathbf{Proj_B\, A}$$

Note that $\mathbf{Proj_B\, A}$ has the direction of $\mathbf{B}$ if $0 \le \alpha < \pi/2$, whereas it has the opposite direction of $\mathbf{B}$ if $\pi/2 < \alpha \le \pi$. If $\alpha = \pi/2$ then $\mathbf{Proj_B\, A} = \mathbf{0}$.

The *scalar projection* of $\mathbf{A}$ on $\mathbf{B}$ or the *component* of $\mathbf{A}$ in the *direction* of $\mathbf{B}$ is defined by

$$\mathrm{Proj}_{\mathbf{B}}\, \mathbf{A} = \|\mathbf{A}\| \cos\alpha$$

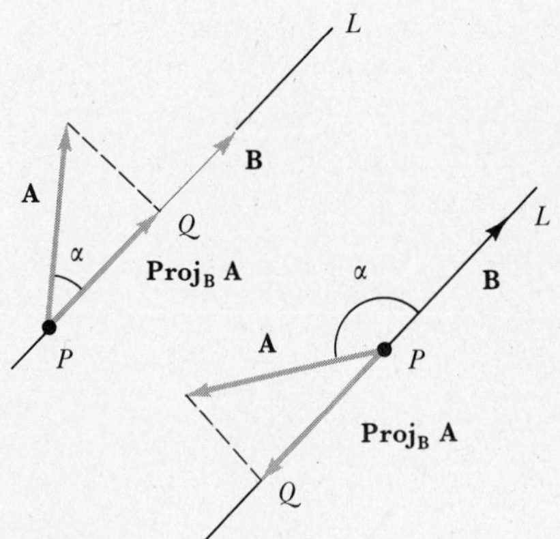

Fig. 17.45

From Fig. 17.45 we note that if $0 \le \alpha < \pi/2$, then $\|\mathbf{Proj_B\, A}\| = \|\mathbf{A}\| \cos\alpha = \mathrm{Proj}_{\mathbf{B}}\, \mathbf{A}$. On the other hand, if $\pi/2 < \alpha \le \pi$, then $\|\mathbf{Proj_B\, A}\| = \|\mathbf{A}\| \cos(\pi - \alpha) = -\|\mathbf{A}\| \cos\alpha = -\mathrm{Proj}_{\mathbf{B}}\, \mathbf{A}$. Thus the absolute value of the *scalar* projection is equal to the length of the *vector* projection. Since the sign of the scalar projection reflects whether $\mathbf{Proj_B\, A}$ has the direction of $\mathbf{B}$ or $-\mathbf{B}$, and since $\mathbf{B}/\|\mathbf{B}\|$ is a unit vector in the direction of $\mathbf{B}$, by (17.3.15) we know that

17.4.7
$$\mathbf{Proj_B\ A} = (\mathrm{Proj_B\ A})\frac{\mathbf{B}}{\|\mathbf{B}\|}$$

The following gives an easy method for determining vector and scalar projections using dot products.

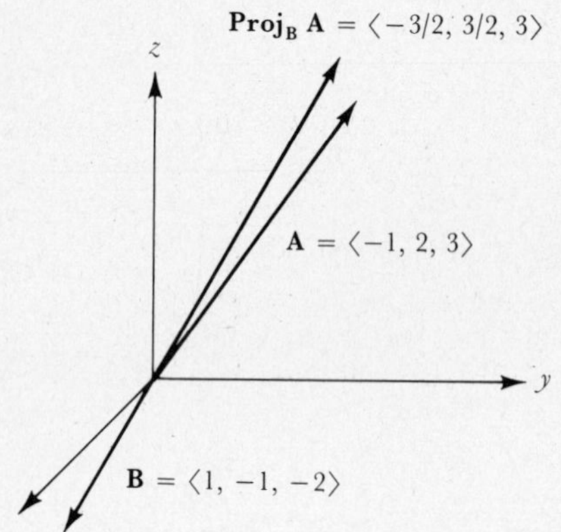

Fig. 17.46

17.4.8 **Projections and Dot Products** Suppose **A** and **B** are nonzero vectors. Then

(*a*) $\mathrm{Proj_B\ A} = \dfrac{\mathbf{A}\cdot\mathbf{B}}{\|\mathbf{B}\|}$.

(*b*) $\mathbf{Proj_B\ A} = \left(\dfrac{\mathbf{A}\cdot\mathbf{B}}{\mathbf{B}\cdot\mathbf{B}}\right)\mathbf{B}$.

Proof: By (17.4.4), $\cos\alpha = \dfrac{\mathbf{A}\cdot\mathbf{B}}{\|\mathbf{A}\|\,\|\mathbf{B}\|}$, and so

$$\mathrm{Proj_B\ A} = \|\mathbf{A}\|\cos\alpha = \frac{\|\mathbf{A}\|\mathbf{A}\cdot\mathbf{B}}{\|\mathbf{A}\|\,\|\mathbf{B}\|} = \frac{\mathbf{A}\cdot\mathbf{B}}{\|\mathbf{B}\|}$$

Also,

$$\mathbf{Proj_B\ A} = (\mathrm{Proj_B\ A})\frac{\mathbf{B}}{\|\mathbf{B}\|} = \frac{\mathbf{A}\cdot\mathbf{B}}{\|\mathbf{B}\|}\frac{\mathbf{B}}{\|\mathbf{B}\|} = \left(\frac{\mathbf{A}\cdot\mathbf{B}}{\|\mathbf{B}\|^2}\right)\mathbf{B} = \left(\frac{\mathbf{A}\cdot\mathbf{B}}{\mathbf{B}\cdot\mathbf{B}}\right)\mathbf{B} \quad \blacksquare$$

Example 4 Determine $\mathbf{Proj_B\ A}$ and illustrate geometrically, where $\mathbf{A} = \langle -1, 2, 3\rangle$ and $\mathbf{B} = \langle 1, -1, -2\rangle$.

Solution Referring to Fig. 17.46 above,

$$\mathbf{Proj_B\ A} = \left(\frac{\langle -1, 2, 3\rangle\cdot\langle 1, -1, -2\rangle}{\langle 1, -1, -2\rangle\cdot\langle 1, -1, -2\rangle}\right)\langle 1, -1, -2\rangle$$
$$= \frac{-9}{6}\langle 1, -1, -2\rangle = \langle -3/2, 3/2, 3\rangle \quad \bullet$$

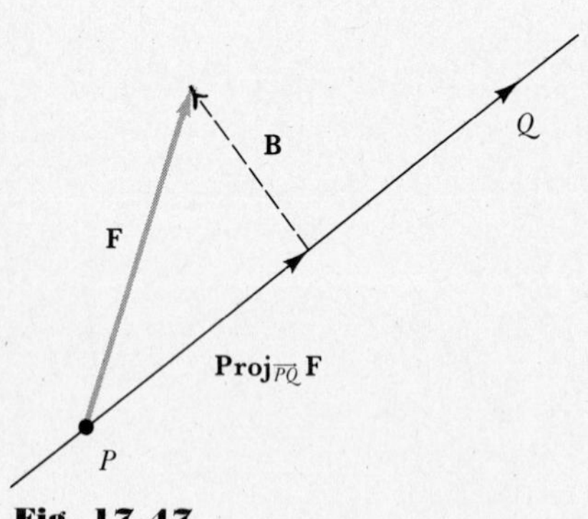

Fig. 17.47

The notion of a scalar projection allows an interesting physical interpretation of the dot product. In our prior discussion of work (Sec. 12.4) we always assumed that the constant force was applied along the line of the motion. Because of this, the direction was irrelevant to the calculations and so we were able to represent the force as a scalar. In this case, work was defined as force times displacement. More generally, a constant force is a vector, and the direction in which the force is applied need not be in the direction of the motion. Specifically, suppose a constant force is applied to move an object from a point P to a point Q along a straight line. Note in Fig. 17.47 that $\mathbf{F} = \mathbf{Proj}_{\overrightarrow{PQ}}\mathbf{F} + \mathbf{B}$. However, since **B** is perpendicular to the line of motion, it will make no contribution to the work done. Thus, we define the work done by **F** in moving the object from P to Q to be the component of **F** in the direction of the motion times the displacement. That is,

17.4.9 **Definition-Work**
$$W = (\mathrm{Proj}_{\overrightarrow{PQ}}\mathbf{F})\|\overrightarrow{PQ}\|$$

By (17.4.8), $W = (\mathrm{Proj}_{\overrightarrow{PQ}}\mathbf{F})\|\overrightarrow{PQ}\| = \dfrac{\mathbf{F}\cdot\overrightarrow{PQ}}{\|\overrightarrow{PQ}\|}\|\overrightarrow{PQ}\| = \mathbf{F}\cdot\overrightarrow{PQ}$. Thus we have another interpretation of the dot product of two vectors: It is the work done by the first vector (considered as a force) as the point of application moves along the length of the second.

PROGRESS TEST 2

1. Let $A = \langle a_1, a_2, a_3 \rangle$. Show that the components of A in the direction of the positive x, y, and z axes are, respectively, the x, y, and z components of A.
2. Determine the scalar and vector projections of $A = \langle 2, 1, -6 \rangle$ into $\mathbf{B} = \langle 3, -1, 1 \rangle$.
3. A constant force of magnitude 10 lb and in the direction of the vector $\langle 1, 1 \rangle$ is applied to move an object 30 ft along a straight line. Find the work done.

SOME GEOMETRY

We have not yet specifically defined what we mean by the "direction" of a vector in R_3. To do this we introduce the notion of *direction angles*. These will be useful in our later study of lines in space.

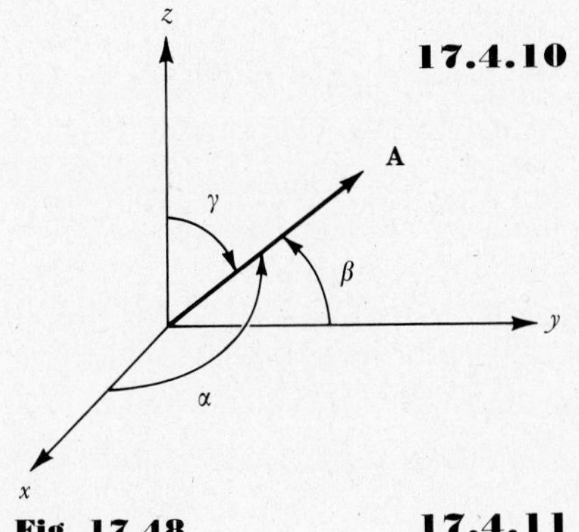

Fig. 17.48

17.4.10 **Definition** Suppose A is a nonzero vector in R_3. The *direction angles* of A, denoted α, β, and γ, are, respectively, the angles between **A** and the coordinate vectors **i**, **j**, and **k**. The numbers $\cos\alpha$, $\cos\beta$, and $\cos\gamma$ are called the *direction cosines* of **A** (see Fig. 17.48).

If A has direction angles α, β, and γ, then $-\mathbf{A}$ will have direction angles $\pi - \alpha$, $\pi - \beta$, and $\pi - \gamma$ (check this). Since $\cos(\pi - \theta) = -\cos\theta$, it follows that the direction cosines of $-\mathbf{A}$ are precisely the negatives of the direction cosines of A. From (17.4.4) we conclude the following:

17.4.11

$$\cos\alpha = \frac{\mathbf{A}\cdot\mathbf{i}}{\|\mathbf{A}\|\,\|\mathbf{i}\|} = \frac{\langle a_1, a_2, a_3\rangle \cdot \langle 1, 0, 0\rangle}{\|\mathbf{A}\|} = \frac{a_1}{\|\mathbf{A}\|}$$

$$\cos\beta = \frac{\mathbf{A}\cdot\mathbf{j}}{\|\mathbf{A}\|\,\|\mathbf{j}\|} = \frac{\langle a_1, a_2, a_3\rangle \cdot \langle 0, 1, 0\rangle}{\|\mathbf{A}\|} = \frac{a_2}{\|\mathbf{A}\|}$$

$$\cos\gamma = \frac{\mathbf{A}\cdot\mathbf{k}}{\|\mathbf{A}\|\,\|\mathbf{k}\|} = \frac{\langle a_1, a_2, a_3\rangle \cdot \langle 0, 0, 1\rangle}{\|\mathbf{A}\|} = \frac{a_3}{\|\mathbf{A}\|}$$

As a consequence of these observations, we have:

17.4.12 **Theorem** If A is a nonzero vector with direction cosines $\cos\alpha$, $\cos\beta$, and $\cos\gamma$, then

(*a*) $\mathbf{A} = \|\mathbf{A}\|\langle\cos\alpha, \cos\beta, \cos\gamma\rangle$ (*b*) $\cos^2\alpha + \cos^2\beta + \cos^2\gamma = 1$

Proof:

(*a*) $\|\mathbf{A}\|\langle\cos\alpha, \cos\beta, \cos\gamma\rangle$

$= \langle \|\mathbf{A}\|\cos\alpha, \|\mathbf{A}\|\cos\beta, \|\mathbf{A}\|\cos\gamma\rangle$

$= \langle \|\mathbf{A}\|\,a_1/(\|\mathbf{A}\|), \|\mathbf{A}\|\,a_2/(\|\mathbf{A}\|), \|\mathbf{A}\|\,a_3/(\|\mathbf{A}\|)\rangle = \langle a_1, a_2, a_3\rangle = \mathbf{A}$

(*b*) $\cos^2\alpha + \cos^2\beta + \cos^2\gamma$

$$= \frac{a_1^2}{\|\mathbf{A}\|^2} + \frac{a_2^2}{\|\mathbf{A}\|^2} + \frac{a_3^2}{\|\mathbf{A}\|^2} = \frac{a_1^2 + a_2^2 + a_3^2}{\|\mathbf{A}\|^2} = 1$$

since $\|\mathbf{A}\|^2 = \mathbf{A}\cdot\mathbf{A} = a_1^2 + a_2^2 + a_3^2$. ■

Note that [17.4.12(*a*)] emphasizes again the fact that *a vector is completely determined by its length* ($\|\mathbf{A}\|$) *and its direction* (*the direction cosines*). It follows from (17.4.11) that if **A** is a unit vector, then the components of **A** are the direction cosines of **A**.

In particular, we have:

17.4.13 **Theorem** If $\mathbf{A} \neq 0$, then $\mathbf{A}/\|\mathbf{A}\|$ is a unit vector with the same direction as **A**.

Example 5 Find the direction cosines for the vector $\mathbf{A} = \overrightarrow{PQ}$, where $P = (1, 2, -1)$ and $Q = (-1, 4, 6)$.

Solution We first use (17.3.18) to determine the components of $\mathbf{A}$:

$$\mathbf{A} = \langle -1 - 1, 4 - 2, 6 - (-1) \rangle = \langle -2, 2, 7 \rangle$$

Now $\|\mathbf{A}\| = \sqrt{4 + 4 + 49} = \sqrt{57}$, so $\mathbf{A}/\|\mathbf{A}\| = \langle -2/\sqrt{57}, 2/\sqrt{57}, 7/\sqrt{57} \rangle$ is a unit vector having the same direction as $\mathbf{A}$. Thus $\cos\alpha = -2/\sqrt{57}$, $\cos\beta = 2/\sqrt{57}$, and $\cos\gamma = 7/\sqrt{57}$. ●

Example 6 Prove that a parallelogram is a rectangle if and only if its diagonals have the same length.

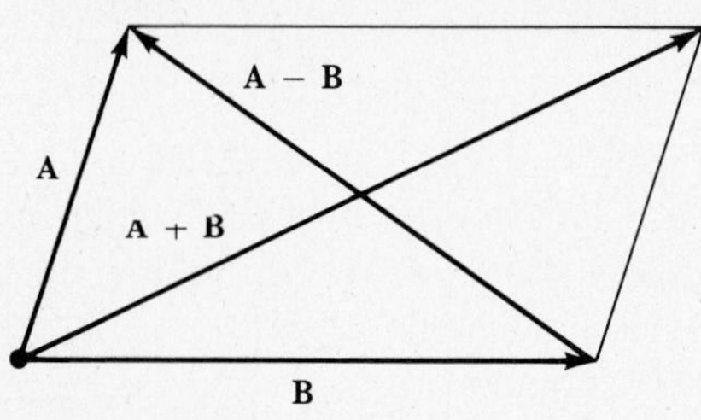

Fig. 17.49

Solution Let P, Q, R, and S denote the vertices of the parallelogram in Fig. 17.49. Let $\mathbf{A} = \overrightarrow{PQ}$ and $\mathbf{B} = \overrightarrow{PS}$. Now by our geometric realization of vector addition and subtraction, the diagonal $\overrightarrow{PR}$ is a representative of $\mathbf{A} + \mathbf{B}$ and the diagonal $\overrightarrow{SQ}$ is a representative of $\mathbf{A} - \mathbf{B}$. Thus the parallelogram has diagonals of equal length if and only if $\|\mathbf{A} + \mathbf{B}\| = \|\mathbf{A} - \mathbf{B}\|$ or, equivalently, $\|\mathbf{A} + \mathbf{B}\|^2 = \|\mathbf{A} - \mathbf{B}\|^2$. Now $\|\mathbf{A} + \mathbf{B}\|^2 = (\mathbf{A} + \mathbf{B}) \cdot (\mathbf{A} + \mathbf{B}) = \mathbf{A} \cdot \mathbf{A} + 2\mathbf{A} \cdot \mathbf{B} + \mathbf{B} \cdot \mathbf{B}$ and $\|\mathbf{A} - \mathbf{B}\|^2 = (\mathbf{A} - \mathbf{B}) \cdot (\mathbf{A} - \mathbf{B}) = \mathbf{A} \cdot \mathbf{A} - 2\mathbf{A} \cdot \mathbf{B} + \mathbf{B} \cdot \mathbf{B}$. Equating these, we obtain $\mathbf{A} \cdot \mathbf{B} = 0$; that is, $\mathbf{A} \perp \mathbf{B}$. Thus the diagonals have equal length if and only if $\overrightarrow{PQ} \perp \overrightarrow{PS}$, which is in turn true if and only if the parallelogram is a rectangle. ●

The problem of finding the distance from a point to a line can be realized as a minimization problem and solved using the methods of Chap. 5. A much simpler solution can be obtained using vector methods.

Example 7 Use vector methods to show that the distance from a point (x_0, y_0) to a line L represented by $ax + by = c$ is given by

$$d = \frac{|ax_0 + by_0 - c|}{\sqrt{a^2 + b^2}}$$

Solution Let $\mathbf{N} = \langle a, b \rangle$. Then $\mathbf{N}$ is perpendicular to the line L (see Progress Test 3, Prob. 3). Suppose $P_1 = (x_1, y_1)$ is any point on L and place a representative of $\mathbf{N}$ with its base at P. Then the desired distance is $\|\mathbf{Proj_N}\, \overrightarrow{P_1Q}\|$ (see Fig. 17.50). Note that this is the case regardless of which side of L the representative of $\mathbf{N}$ happens to lie on. Now

$$\begin{aligned}
\|\mathbf{Proj_N}\, \overrightarrow{P_1Q}\| &= |\operatorname{Proj_N} \overrightarrow{P_1Q}| \\
&= \frac{|\mathbf{N} \cdot \overrightarrow{P_1Q}|}{\|\mathbf{N}\|} = \frac{|\langle a, b \rangle \cdot \langle x_0 - x_1, y_0 - y_1 \rangle|}{\sqrt{a^2 + b^2}} \\
&= \frac{|a(x_0 - x_1) + b(y_0 - y_1)|}{\sqrt{a^2 + b^2}} \\
&= \frac{|ax_0 + by_0 - (ax_1 + by_1)|}{\sqrt{a^2 + b^2}} = \frac{|ax_0 + by_0 - c|}{\sqrt{a^2 + b^2}}
\end{aligned}$$

In this last step we were able to use the fact that $ax_1 + by_1 = c$ since (x_1, y_1) was assumed to be on L. ●

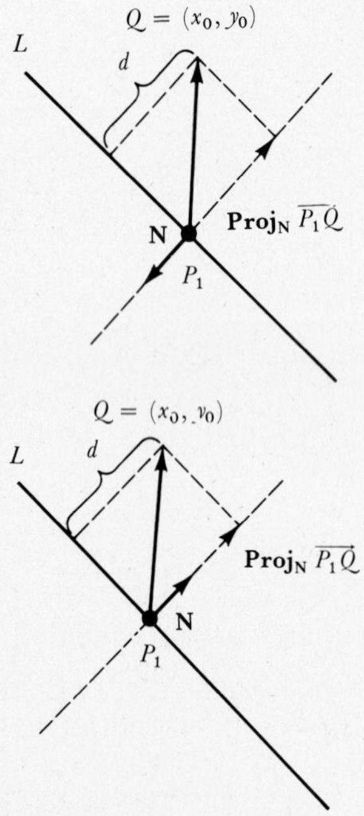

Fig. 17.50

PROGRESS TEST 3

1. Determine the vector **B** of length 2 with the same direction as $\mathbf{A} = \langle -2, 1, 3\rangle$.
2. Compute the direction angles of the vector $\mathbf{A} = \langle 1, -1, \sqrt{2}\rangle$.
3. Show that $\mathbf{N} = \langle a, b\rangle$ is perpendicular to the line $ax + by = c$. (*Hint:* Choose $P_1 = (x_1, y_1)$, $P_2 = (x_2, y_2)$ on L. Then $\overrightarrow{P_1P_2} \| L$, so it suffices to show that $\overrightarrow{P_1P_2} \perp \mathbf{N}$.)

THE CROSS PRODUCT

The cross product of two vectors differs markedly from the dot product in that (1) the cross product of two vectors is itself a vector, and (2) some of the usual rules of algebra are not valid for cross products.

17.4.14 **Definition** Let $\mathbf{A} = \langle a_1, a_2, a_3\rangle$ and $\mathbf{B} = \langle b_1, b_2, b_3\rangle$ be two vectors. The *cross product* of **A** and **B**, denoted $\mathbf{A} \times \mathbf{B}$, is defined by

$$\mathbf{A} \times \mathbf{B} = \langle a_2b_3 - a_3b_2, a_3b_1 - a_1b_3, a_1b_2 - a_2b_1\rangle.$$

A simple way to remember (17.4.14) is to express it in terms of the first row expansion of a determinant.

17.4.15
$$\begin{aligned}\mathbf{A} \times \mathbf{B} &= (a_2b_3 - a_3b_2)\mathbf{i} + (a_3b_1 - a_1b_3)\mathbf{j} + (a_1b_2 - a_2b_1)\mathbf{k} \\ &= \begin{vmatrix} a_2 & a_3 \\ b_2 & b_3 \end{vmatrix}\mathbf{i} - \begin{vmatrix} a_1 & a_3 \\ b_1 & b_3 \end{vmatrix}\mathbf{j} + \begin{vmatrix} a_1 & a_2 \\ b_1 & b_2 \end{vmatrix}\mathbf{k} \\ &= \begin{vmatrix} \mathbf{i} & \mathbf{j} & \mathbf{k} \\ a_1 & a_2 & a_3 \\ b_1 & b_2 & b_3 \end{vmatrix}\end{aligned}$$

Example 8 Determine $\mathbf{A} \times \mathbf{B}$ for $\mathbf{A} = \langle 2, 1, -3\rangle$, $\mathbf{B} = \langle -4, 2, 1\rangle$.

Solution
$$\begin{aligned}\mathbf{A} \times \mathbf{B} = \begin{vmatrix} \mathbf{i} & \mathbf{j} & \mathbf{k} \\ 2 & 1 & -3 \\ -4 & 2 & 1 \end{vmatrix} &= \begin{vmatrix} 1 & -3 \\ 2 & 1 \end{vmatrix}\mathbf{i} - \begin{vmatrix} 2 & -3 \\ -4 & 1 \end{vmatrix}\mathbf{j} + \begin{vmatrix} 2 & 1 \\ -4 & 2 \end{vmatrix}\mathbf{k} \\ &= (1 + 6)\mathbf{i} - (2 - 12)\mathbf{j} + (4 + 4)\mathbf{k} \\ &= \langle 7, 10, 8\rangle\end{aligned}$$ ●

Although the following properties of the cross product are direct consequences of the definition [see the proof of part (*d*), for example], a real understanding of cross products will develop from the geometric interpretation.

17.4.16 **Theorem** Let **A**, **B**, and **C** be vectors in R_3 and let k be a scalar. Then
(*a*) $\mathbf{A} \times \mathbf{B} = -(\mathbf{B} \times \mathbf{A})$
(*b*) $k(\mathbf{A} \times \mathbf{B}) = (k\mathbf{A}) \times \mathbf{B} = \mathbf{A} \times (k\mathbf{B})$
(*c*) $\mathbf{A} \times (\mathbf{B} + \mathbf{C}) = (\mathbf{A} \times \mathbf{B}) + (\mathbf{A} \times \mathbf{C})$
(*d*) $\mathbf{A} \times \mathbf{A} = \mathbf{0}$
(*e*) $\mathbf{A} \times \mathbf{0} = \mathbf{0} \times \mathbf{A} = \mathbf{0}$
(*f*) $\mathbf{A} \cdot (\mathbf{B} \times \mathbf{C}) = (\mathbf{A} \times \mathbf{B}) \cdot \mathbf{C}$
(*g*) $\|\mathbf{A} \times \mathbf{B}\|^2 = \|\mathbf{A}\|^2 \|\mathbf{B}\|^2 - (\mathbf{A} \cdot \mathbf{B})^2$

Proof of (d): By the definition,
$$\mathbf{A} \times \mathbf{A} = (a_2a_3 - a_3a_2)\mathbf{i} + (a_3a_1 - a_1a_3)\mathbf{j} + (a_1a_2 - a_2a_1)\mathbf{k} = \mathbf{0}$$ ■

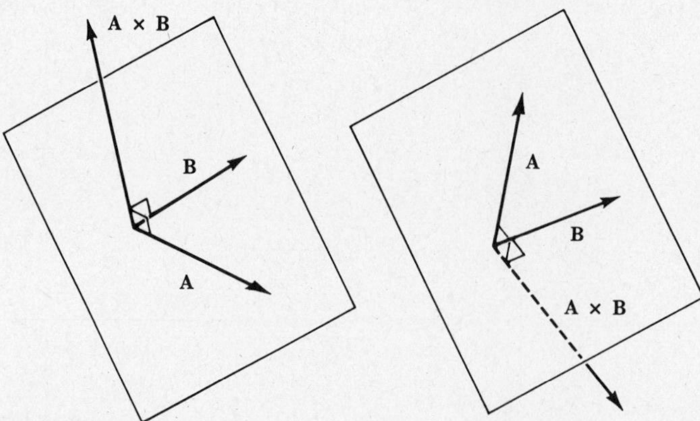

Fig. 17.51

We now turn to the geometric significance of the cross product.

17.4.17 **Theorem** The vector $\mathbf{A} \times \mathbf{B}$ is perpendicular to both $\mathbf{A}$ and $\mathbf{B}$.

Proof: Now, using the properties (d) and (f) from (17.4.16), we have $\mathbf{A}\cdot(\mathbf{A} \times \mathbf{B}) = (\mathbf{A} \times \mathbf{A})\cdot\mathbf{B} = \mathbf{0}\cdot\mathbf{B} = 0$ and hence $\mathbf{A} \perp \mathbf{A} \times \mathbf{B}$. A similar argument shows $\mathbf{B} \perp \mathbf{A} \times \mathbf{B}$. ■

If $\mathbf{A}$ and $\mathbf{B}$ are not parallel, then representatives of $\mathbf{A}$ and $\mathbf{B}$ with the same initial point determine a plane. From (17.4.17), $\mathbf{A} \times \mathbf{B}$ is perpendicular to this plane. It can be shown that the correct direction for $\mathbf{A} \times \mathbf{B}$ is the one that makes A, B, $\mathbf{A} \times \mathbf{B}$ a right-hand triple, as depicted in Fig. 17.51. One easily checks that this is the case with $\mathbf{i}$, $\mathbf{j}$, $\mathbf{k}$ since

$$\mathbf{i} \times \mathbf{j} = \langle 1, 0, 0\rangle \times \langle 0, 1, 0\rangle = \begin{vmatrix} \mathbf{i} & \mathbf{j} & \mathbf{k} \\ 1 & 0 & 0 \\ 0 & 1 & 0 \end{vmatrix}$$

$$= \begin{vmatrix} 0 & 0 \\ 1 & 0 \end{vmatrix}\mathbf{i} - \begin{vmatrix} 1 & 0 \\ 0 & 0 \end{vmatrix}\mathbf{j} + \begin{vmatrix} 1 & 0 \\ 0 & 1 \end{vmatrix}\mathbf{k} = \mathbf{k}$$

17.4.18 **Theorem** Suppose $\mathbf{A}$ and $\mathbf{B}$ are nonzero vectors in R_3 and θ is the angle between them. Then $\|\mathbf{A} \times \mathbf{B}\| = \|\mathbf{A}\|\,\|\mathbf{B}\| \sin\theta$.

Proof: Now $\mathbf{A}\cdot\mathbf{B} = \|\mathbf{A}\|\,\|\mathbf{B}\| \cos\theta$. Thus, from [17.4.16($g$)],

$$\begin{aligned} \|\mathbf{A} \times \mathbf{B}\|^2 &= \|\mathbf{A}\|^2\,\|\mathbf{B}\|^2 - (\mathbf{A}\cdot\mathbf{B})^2 \\ &= \|\mathbf{A}\|^2\,\|\mathbf{B}\|^2 - \|\mathbf{A}\|^2\,\|\mathbf{B}\|^2 \cos^2\theta \\ &= \|\mathbf{A}\|^2\,\|\mathbf{B}\|^2 (1 - \cos^2\theta) \\ &= \|\mathbf{A}\|^2\,\|\mathbf{B}\|^2 \sin^2\theta \end{aligned}$$

Since $\sin\theta \geq 0$, the theorem follows by taking square roots. ■

As an immediate consequence we have the following generalization of [17.4.16(d)]:

17.4.19

$$\mathbf{A} \parallel \mathbf{B} \text{ if and only if } \mathbf{A} \times \mathbf{B} = \mathbf{0}$$

If $\mathbf{A}$ and $\mathbf{B}$ are not parallel, then $\|\mathbf{A} \times \mathbf{B}\|$ has an interesting geometrical realization. Suppose $\mathbf{A}$ and $\mathbf{B}$ have the same initial point and suppose $\mathcal{P}$ is the parallelogram with coterminal sides **A** and **B** (see Fig. 17.52). Then the height of $\mathcal{P}$ is $\|\mathbf{B}\| \sin \theta$, and the base has length **A**. Hence the area of $\mathcal{P}$ is given by:

17.4.20 Area of a Parallelogram

$$A(\mathcal{P}) = \|\mathbf{A}\|\,\|\mathbf{B}\| \sin \theta = \|\mathbf{A} \times \mathbf{B}\|$$

Example 9 Compute the area of the parallelogram having the standard representatives of $\mathbf{A} = \langle 3, 2, 1\rangle$ and $\mathbf{B} = \langle -1, 4, 3\rangle$ as coterminal sides; see Fig. 17.53.

Solution

$$\mathbf{A} \times \mathbf{B} = \begin{vmatrix} \mathbf{i} & \mathbf{j} & \mathbf{k} \\ 3 & 2 & 1 \\ -1 & 4 & 3 \end{vmatrix} = \langle 2, -10, 14\rangle$$

Thus $A(\mathcal{P}) = \|\mathbf{A} \times \mathbf{B}\| = \sqrt{4 + 100 + 196} = 10\sqrt{3}$. ●

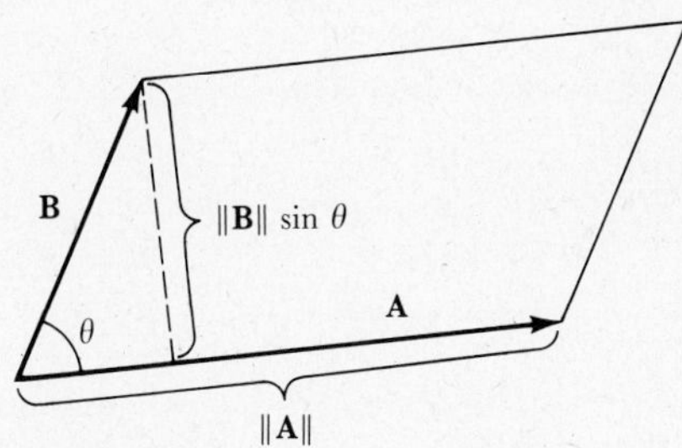

Fig. 17.52

There is also an interesting geometric interpretation of the *triple scalar product* $(\mathbf{A} \times \mathbf{B}) \cdot \mathbf{C}$. Suppose **A**, **B**, and **C** form a right-hand triple and consider representatives with the same initial point. Let $\mathcal{P}$ be the parallelepiped with these representatives as coterminal edges (see Fig. 17.54). The face containing **A** and **B** is a parallelogram with area $\|\mathbf{A} \times \mathbf{B}\|$. Since the volume of $\mathcal{P}$ is given by the product of the area of its base and its height h, we have

$$V(\mathcal{P}) = \|\mathbf{A} \times \mathbf{B}\| h$$

Now the height h is the scalar projection of **C** onto $\mathbf{A} \times \mathbf{B}$; see Fig. 17.54(*b*). Thus

$$\begin{aligned} V(\mathcal{P}) &= \|\mathbf{A} \times \mathbf{B}\| \operatorname{Proj}_{\mathbf{A} \times \mathbf{B}} \mathbf{C} \\ &= \|\mathbf{A} \times \mathbf{B}\| \left(\frac{(\mathbf{A} \times \mathbf{B}) \cdot \mathbf{C}}{\|\mathbf{A} \times \mathbf{B}\|} \right) = (\mathbf{A} \times \mathbf{B}) \cdot \mathbf{C} \end{aligned}$$

By the same argument, but using different faces to represent the base, we obtain:

17.4.21 Volume of a Parallelepiped

$$V(\mathcal{P}) = (\mathbf{A} \times \mathbf{B}) \cdot \mathbf{C} = (\mathbf{B} \times \mathbf{C}) \cdot \mathbf{A} = (\mathbf{C} \times \mathbf{A}) \cdot \mathbf{B}$$

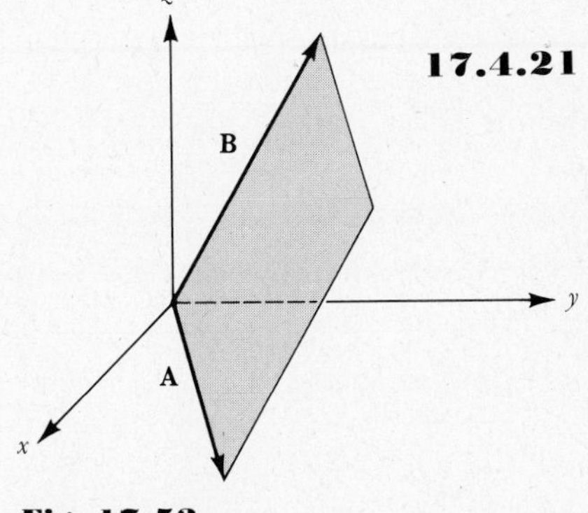

Fig. 17.53

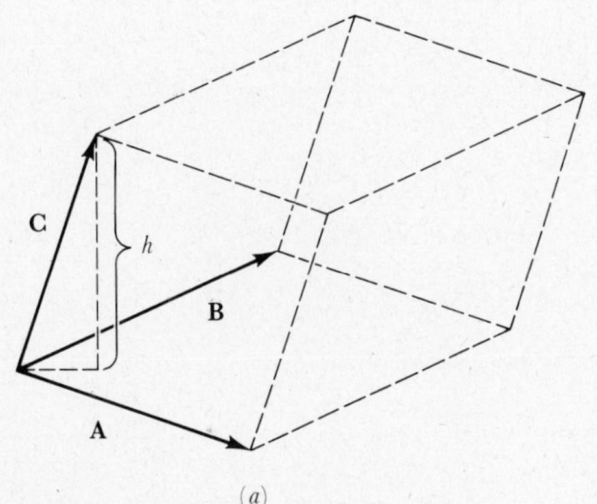

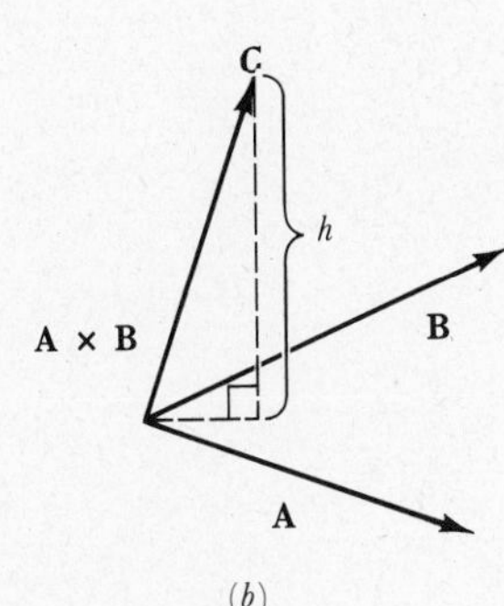

Fig. 17.54

17.4.22 Theorem The vectors **A**, **B**, **C** form a right-hand triple if and only if

$$(\mathbf{A} \times \mathbf{B}) \cdot \mathbf{C} > 0$$

Note that **C** lies in the plane of **A** and **B** if and only if $\mathbf{A} \times \mathbf{B}$ is perpendicular to **C**—that is, if and only if $(\mathbf{A} \times \mathbf{B}) \cdot \mathbf{C} = 0$. Thus we have

17.4.23 Theorem If **A**, **B**, and **C** are vectors with the same initial point, then they are all in the same plane ("coplanar") if and only if

$$(\mathbf{A} \times \mathbf{B}) \cdot \mathbf{C} = 0$$

Example 10 The standard representatives of the vectors $\mathbf{A} = \langle 1, 2, -1 \rangle$, $\mathbf{B} = \langle 3, -1, 4 \rangle$ and $\mathbf{C} = \langle 7, 5, -3 \rangle$ form the coterminal edges of a parallelepiped $\mathcal{P}$. Determine the volume of $\mathcal{P}$.

Solution

$$V(\mathcal{P}) = |(\mathbf{A} \times \mathbf{B}) \cdot \mathbf{C}|$$

$$= \left| \left\langle \begin{vmatrix} 2 & -1 \\ -1 & 4 \end{vmatrix}, -\begin{vmatrix} 1 & -1 \\ 3 & 4 \end{vmatrix}, \begin{vmatrix} 1 & 2 \\ 3 & -1 \end{vmatrix} \right\rangle \cdot \langle 7, 5, -3 \rangle \right|$$

$$= |\langle 7, -7, -7 \rangle \cdot \langle 7, 5, -3 \rangle|$$

$$= |49 - 35 + 21| = 35 \quad \bullet$$

In the previous argument we wrote $h = \text{Proj}_{\mathbf{A} \times \mathbf{B}}\, \mathbf{C}$, which uses the fact that **C** and $\mathbf{A} \times \mathbf{B}$ are on the same side of the plane determined by **A** and **B**. In other words, **A**, **B**, and **C** form a right-hand triple. In general, if **A** and **B** are not parallel and if **C** does not lie in the plane of **A** and **B**, then $V(\mathcal{P}) = |(\mathbf{A} \times \mathbf{B}) \cdot \mathbf{C}|$. In fact, the sign of the triple scalar product tells us whether **A**, **B**, and **C** form a right-hand triple.

PROGRESS TEST 4

Let $\mathbf{A} = \langle 1, 2, -3 \rangle$, $\mathbf{B} = \langle 2, -4, 5 \rangle$, $\mathbf{C} = \langle 7, 2, -4 \rangle$.

1. Find the area of the parallelogram with coterminal sides **A** and **B**.

2. Find the volume of the parallelepiped with coterminal edges **A**, **B**, and **C**.

3. Is **A**, **B**, **C** a right-hand triple?

4. Determine a vector **D** of length 2 which is perpendicular to the plane of **A** and **C** and such that **A**, **C**, **D** forms a right-hand triple.

SECTION 17.4 EXERCISES

In Exercises 1 to 10, (a) determine the dot product of the given vectors and (b) find the cosine of the angle between them.

1. $\langle 3, -17 \rangle, \langle 4, 2 \rangle$

2. $\langle 0, 6 \rangle, \langle -9, 3 \rangle$

3. $\langle 7, \pi \rangle, \langle 2\pi, -5 \rangle$

4. $\mathbf{i} - 6\mathbf{j}, 3\mathbf{i} + \mathbf{j}$

5. $2\mathbf{i} + 5\mathbf{j}, 6\mathbf{i} - 1/2\mathbf{j}$

6. $\langle 1, 1, -1 \rangle, \langle 2, 0, 4 \rangle$

7. $\langle 2, 4, 4 \rangle, \langle -4, 2, -4 \rangle$

8. $\langle 1, 2, -1 \rangle, \langle 2, 1, 0 \rangle$

9. $\langle 3, -1, 1 \rangle, \langle -2, 1, -4 \rangle$

10. $\mathbf{i} + 2\mathbf{j} + 3\mathbf{k}, -\mathbf{i} + 6\mathbf{j} - 2\mathbf{k}$

*In Exercises 11 to 20 find (a) $Proj_{\mathbf{B}}$ **A** and (b) $\mathbf{Proj_B}$ **A**.*

11. $\mathbf{A} = \langle 2, 3 \rangle, \mathbf{B} = \langle -4, -1 \rangle$

12. $\mathbf{A} = \langle 4, -3 \rangle, \mathbf{B} = 2\mathbf{i} - \mathbf{j}$

13. $\mathbf{A} = \langle 8, 1 \rangle, \mathbf{B} = \langle -1, 1 \rangle$

14. $\mathbf{A} = \langle 3, 2 \rangle, \mathbf{B} = \langle -7, 1 \rangle$

15. $\mathbf{A} = \mathbf{i} - \mathbf{j}, \mathbf{B} = 5\mathbf{j}$

16. $\mathbf{A} = \langle 1, -1, 2 \rangle, \mathbf{B} = \langle 2, 1, 5 \rangle$

17. $\mathbf{A} = \langle 3, 6, -2 \rangle, \mathbf{B} = \langle 0, -2, 3 \rangle$

18. $\mathbf{A} = \langle 4, -7, 1 \rangle, \mathbf{B} = \langle 6, 1, 5 \rangle$

19. $\mathbf{A} = 3\mathbf{i} + 9\mathbf{j} - \mathbf{k}, \mathbf{B} = -2\mathbf{i} + 9\mathbf{j} + \mathbf{k}$

20. $\mathbf{A} = \mathbf{j} + 2\mathbf{k}, \mathbf{B} = 4\mathbf{i} + 2\mathbf{j}$

In Exercises 21 to 24 find k so that the angle between the given vectors is $\pi/4$.

21. $\mathbf{A} = \langle 3, 2\rangle, \mathbf{B} = \langle 2, k\rangle$

22. $\mathbf{A} = 3\mathbf{i} + 2\mathbf{j}, \mathbf{B} = k\mathbf{i} - \mathbf{j}$

23. $\mathbf{A} = \langle 2, 1\rangle, \mathbf{B} = \langle k, -3\rangle$

24. $\mathbf{A} = k\mathbf{i} + \mathbf{j}, \mathbf{B} = 2\mathbf{i} - \mathbf{j}$

25. If **A** and **B** are perpendicular and of unit length, and if $\mathbf{C} = \alpha\mathbf{A} + \beta\mathbf{B}$, show that $\alpha = \mathbf{C}\cdot\mathbf{A}$, $\beta = \mathbf{C}\cdot\mathbf{B}$.

26. Show that $\mathbf{A} \perp \mathbf{B}$ if $\|\mathbf{A} + \mathbf{B}\| = \|\mathbf{A} - \mathbf{B}\|$.

27. Let $\mathbf{A} \neq 0$ and suppose $\mathbf{A}\cdot\mathbf{B} = \mathbf{A}\cdot\mathbf{C}$. Does it follow that $\mathbf{B} = \mathbf{C}$?

In Exercises 28 to 31 use vector methods.

28. Show that the diagonals of a rectangle are equal.

29. Show that the medians of a triangle meet in a point.

30. Show that the sum of the squares of two sides of a triangle is equal to the square of the third side if and only if the triangle is a right triangle.

31. Find the angle between the diagonal of a cube and one of its edges.

32. Show that if **A**, **B**, **C** are nonzero and mutually perpendicular in R_3 and if α, β, γ are scalars such that $\alpha\mathbf{A} + \beta\mathbf{B} + \gamma\mathbf{C} = 0$, then necessarily $\alpha = \beta = \gamma = 0$.

In Exercises 33 and 34 find the work done by the illustrated constant force of magnitude 30 lb as the point of application moves from P to Q.

33.

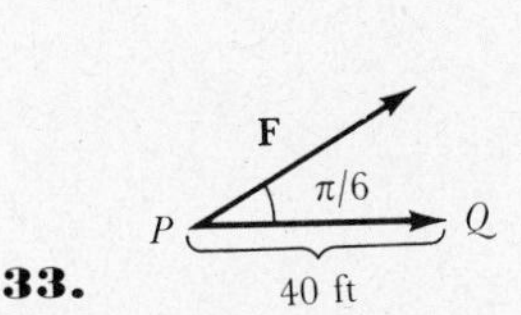

34.

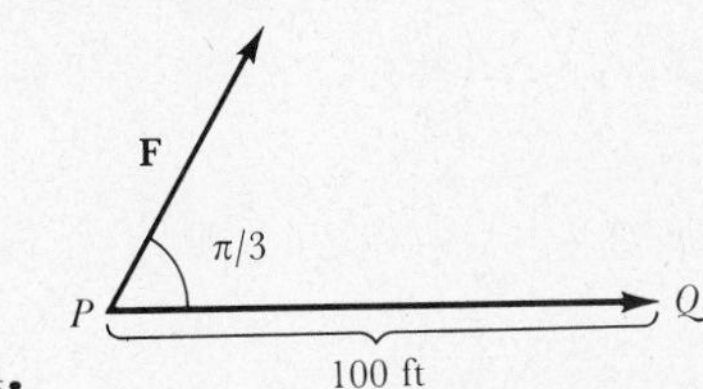

35. Prove Theorem 17.4.1.

In Exercises 36 to 49, suppose $\mathbf{A} = \langle 1, 0, 2\rangle$, $\mathbf{B} = \langle 2, 1, -3\rangle$, $\mathbf{C} = \langle -1, 4, 6\rangle$ *and* $\mathbf{D} = \langle 1, 3, 0\rangle$ *and determine*

36. $\mathbf{A} \times \mathbf{B}$

37. $\mathbf{D} \times \mathbf{C}$

38. $(\mathbf{A} - \mathbf{B}) \times (\mathbf{A} + \mathbf{B})$

39. $\mathbf{A} \times (\mathbf{B} \times \mathbf{C})$

40. $\mathbf{A} \times \mathbf{D}$

41. $(\mathbf{A} \times \mathbf{B})\cdot(\mathbf{C} \times \mathbf{D})$

42. $(\mathbf{A} \times \mathbf{B}) \times \mathbf{C}$

43. $\mathbf{A}\cdot(\mathbf{B} \times \mathbf{C})$

44. The area of the parallelogram with coterminal sides **A** and **B**.

45. The volume of the parallelepiped with coterminal edges **A**, **B**, and **C**.

46. A unit vector perpendicular to the plane of **B** and **D**.

47. A vector of length 3/2 that is perpendicular to both **B** and **C**.

48. The area of the triangle with sides **B**, **C**, and **B** − **C**.

49. Two distinct unit vectors perpendicular to the plane through the three points (1, 0, 1), (2, 2, −4), and (−1, 3, 2).

50. Show that, for any 3-vectors **A** and **B**, $(\mathbf{A} - \mathbf{B}) \times (\mathbf{A} + \mathbf{B}) = 2(\mathbf{A} \times \mathbf{B})$.

51. Suppose that representatives of the 3-vectors **A**, **B**, **C**, **D** all lie in the same plane. What can you say about $(\mathbf{A} \times \mathbf{B}) \times (\mathbf{C} \times \mathbf{D})$?

52. Show that the line segment joining the midpoints of two sides of a triangle is parallel to, and one-half the length of, the third side.

Section 17.5: Lines and Planes

Objectives:

1. Determine parametric equations of a uniquely determined line.
2. Give the equation of a uniquely determined plane.
3. Determine whether lines intersect, are parallel, or are skew.
4. Determine whether planes are parallel and, if not, determine the equation of their intersection.
5. Use the skills 1 to 4 in combination.

LINES

In this section, if $P = (x, y, z)$ denotes a *point* in three-space, then $\mathbf{P} = \langle x, y, z\rangle$ will always represent the *vector* whose components are the same as the coordinates of P. Thus, geometrically, the standard representative of the *vector* **P** has its head at the *point* P (see Fig. 17.55).

Note that if $P = (x, y, z)$ and $Q = (x_1, y_1, z_1)$ are arbitrary points in space, then (see Fig. 17.56)

$$\begin{aligned}\overrightarrow{PQ} &= \langle x_1 - x, y_1 - y, z_1 - z\rangle \\ &= \langle x_1, y_1, z_1\rangle - \langle x, y, z\rangle = \mathbf{Q} - \mathbf{P}\end{aligned}$$

Let L be a line in space, $P = (x_1, y_1, z_1)$ a point on L, and $\mathbf{M} = \langle a, b, c\rangle$ a vector parallel to L. If $R = (x, y, z)$, then a necessary and sufficient condition for the point R to be on L is that the vector $\overrightarrow{PR}$ be parallel to **M** (see Fig. 17.57). But

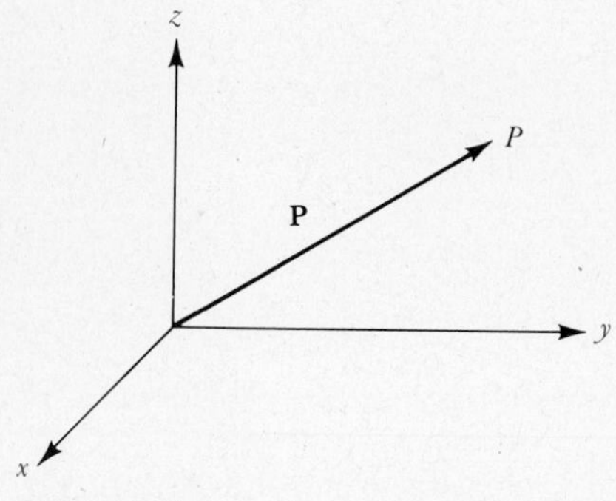

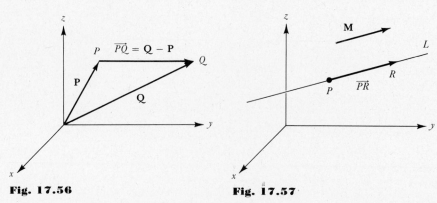

Fig. 17.55

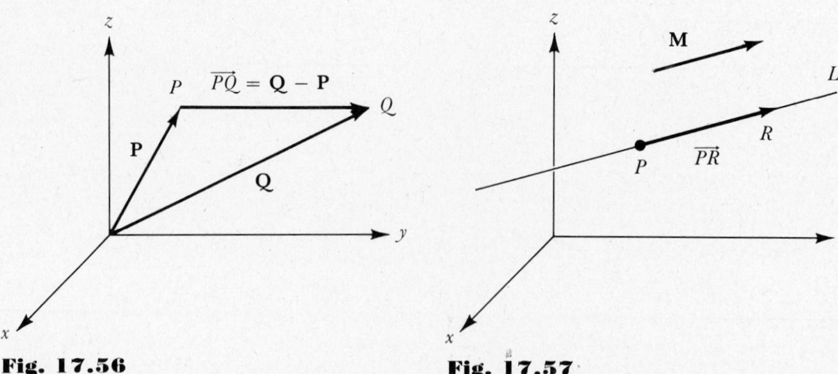

Fig. 17.56

Fig. 17.57

two vectors are parallel if and only if one is a scalar multiple of the other. Thus R is on L if and only if $\overrightarrow{PR} = t\mathbf{M}$, for t some scalar, or $\mathbf{R} - \mathbf{P} = t\mathbf{M}$; that is,

$$\mathbf{R} = \mathbf{P} + t\mathbf{M}$$

Thus we have a vector equation for the straight line through the point P and parallel to the vector $\mathbf{M}$. It can be thought of as a three-dimensional analog of the $y = mx + b$ equation of a straight line in the plane; the vector $\mathbf{M}$ is the three-dimensional "slope."

Since $\mathbf{R} = \langle x, y, z\rangle$ and $\mathbf{P} + t\mathbf{M} = \langle x_1 + at, y_1 + bt, z_1 + ct\rangle$, by equating components we see that the vector equation $\mathbf{R} = \mathbf{P} + t\mathbf{M}$ is equivalent to the parametric equations

$$x = x_1 + at \qquad y = y_1 + bt \qquad z = z_1 + ct$$

Recall from Sec. 12.1 (you may wish to review Sec. 12.1) that the graph of a pair of parametric equations $x = x(t), y = y(t)$ is a curve that consists of all points in the plane of the form $(x(t), y(t))$. Similarly, a set of three parametric equations $x = x(t), y = y(t), z = z(t)$ will have as its graph a *space curve* consisting of all points of the form $(x(t), y(t), z(t))$. In this particular instance, by our construction, the graph will be the straight line through P and parallel to L. In summary, we have:

17.5.1 Theorem The straight line through $P = (x_1, y_1, z_1)$ and parallel to $\mathbf{M} = \langle a, b, c\rangle$ has vector equation

$$(a)\ \mathbf{R}(t) = \mathbf{P} + t\mathbf{M}$$

and parametric equations

$$(b)\ x = x_1 + at, \qquad y = y_1 + bt, \qquad z = z_1 + ct$$

Fig. 17.58

The components of the vector $\mathbf{M}$ are called *direction numbers* for L. If we choose any other vector parallel to L, we obtain a different, but equivalent, set of parametric equations for L. In particular, any set of numbers proportional to the direction numbers for a line are themselves direction numbers, so direction numbers are not unique.

Example 1 Determine parametric equations for the line L that passes through the points $P_1 = (1, 2, -1)$ and $P_2 = (3, 4, 2)$.

Solution Since L goes through the points P_1 and P_2, it is parallel to $\overrightarrow{P_1P_2} = \langle 2, 2, 3\rangle$; see Fig. 17.58. Taking this vector as $\mathbf{M}$ and using P_1 as the point on L, we apply [17.5.1(b)] to obtain $x = 1 + 2t, y = 2 + 2t, z = -1 + 3t$. ●

If the direction numbers of L—say, a, b, and c—are all nonzero, then we can eliminate t in [17.5.1(b)] to obtain *symmetric equations* for L.

17.5.2

$$\frac{x - x_1}{a} = \frac{y - y_1}{b} = \frac{z - z_1}{c}$$

In the plane, a pair of straight lines will either be parallel or intersect. In space, nonparallel lines need not intersect. In this case we say the lines are *skew*. We can use the above algebraic descriptions of lines to determine whether a given pair of lines intersect or are skew.

Example 2 Show that the following lines L_1 and L_2 are skew.

$$L_1: \quad \frac{x-1}{2} = \frac{y-2}{1} = \frac{z-3}{-4}$$

$$L_2: \quad x = 2 + 4t, y = -1, z = 3 + 3t$$

Solution L_1 is parallel to $\mathbf{M}_1 = \langle 2, 1, -4 \rangle$, and L_2 is parallel to $\mathbf{M}_2 = \langle 4, 0, 3 \rangle$. Since $\mathbf{M}_1$ is not a scalar multiple of $\mathbf{M}_2$, the lines L_1 and L_2 are not parallel. L_1 has parametric equations $x = 1 + 2s, y = 2 + s, z = 3 - 4s$. They will intersect if and only if there is a value for s and a value for t yielding the same point—that is, s and t such that $1 + 2s = 2 + 4t, 2 + s = -1, 3 - 4s = 3 + 3t$. From the second equation, $s = -3$. Substitution of $s = -3$ into the first equation yields $t = -7/4$, and substitution of $s = -3$ into the third equation yields $t = 4$. Thus there are no values of s and t that yield the same point. (The equations are inconsistent.) Hence L_1 and L_2 do not intersect. ●

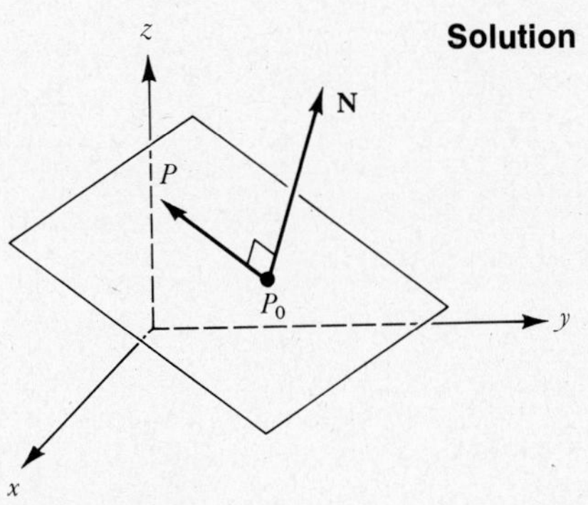

Fig. 17.59

EQUATIONS OF PLANES

17.5.3 **Definition** Suppose $\mathbf{N} = \langle a, b, c \rangle$ is a nonzero vector and $P_0 = (x_0, y_0, z_0)$ is a point. The collection of all points $P = (x, y, z)$ such that $\overrightarrow{P_0P}$ is perpendicular to $\mathbf{N}$ is called the *plane* through P_0 with *normal vector* $\mathbf{N}$.

If $\overrightarrow{P_0P}$ is perpendicular to $\mathbf{N}$, then it is perpendicular to any nonzero scalar multiple of $\mathbf{N}$, and conversely. Hence any nonzero scalar multiple of a normal vector to a plane is *itself* a normal vector to that plane (see Fig. 17.59).

17.5.4 **Theorem** The plane Π through $P_0 = (x_0, y_0, z_0)$ with normal $\mathbf{N} = \langle a, b, c \rangle$ has the equation $a(x - x_0) + b(y - y_0) + c(z - z_0) = 0$.

Proof: By definition $P = (x, y, z)$ is on Π if and only if $\mathbf{N} \cdot \overrightarrow{P_0P} = 0$. The equation follows since

$$\begin{aligned}\mathbf{N} \cdot \overrightarrow{P_0P} &= \langle a, b, c \rangle \cdot \langle x - x_0, y - y_0, z - z_0 \rangle \\ &= a(x - x_0) + b(y - y_0) + c(z - z_0) \quad \blacksquare\end{aligned}$$

Some authors call the components of any vector normal to a plane its *attitude numbers*.

Example 3 Determine an equation for the plane through $(2, -1, 2)$ with normal $\langle 3, -2, 4 \rangle$.

Solution By (17.5.4), the plane has equation $3(x - 2) - 2(y + 1) + 4(z - 2) = 0$, or, more simply, $3x - 2y + 4z = 16$. The x, y, and z intercepts are, respectively,

(16/3, 0, 0), (0, −8, 0), and (0, 0, 4). A portion of the plane is sketched in Fig. 17.60. ●

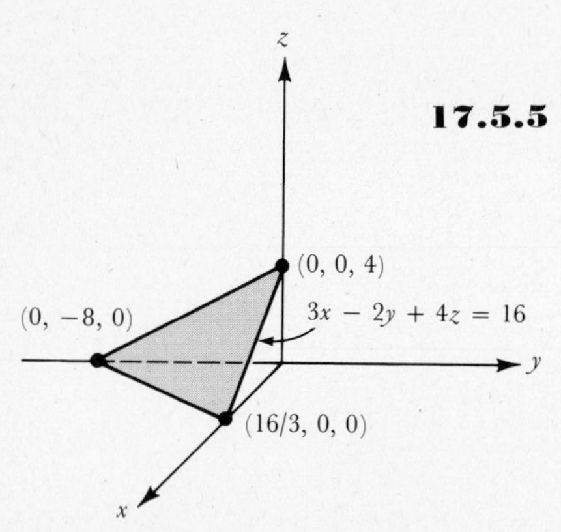

Fig. 17.60

The following connection between planes and their equations was promised in Sec. 17.2, page 556.

17.5.5 **Theorem** Every plane has an equation of the form $ax + by + cz = d$ where a, b, and c are not all zero. Conversely, the graph of an equation of this form is a plane with normal $\mathbf{N} = \langle a, b, c\rangle$.

Proof: From (17.5.4), every plane has an equation $a(x - x_0) + b(y - y_0) + c(z - z_0) = 0$, or alternatively, $ax + by + cz = d$, where $d = ax_0 + by_0 + cz_0$.

Conversely, let $P_0 = (x_0, y_0, z_0)$ be a fixed point on the graph of $ax + by + cz = d$ and let $P = (x, y, z)$ be any point on the graph. Then $ax_0 + by_0 + cz_0 = d$ and $ax + by + cz = d$.

Subtracting these equations we have $a(x - x_0) + b(y - y_0) + c(z - z_0) = d - d = 0$, which, by (17.5.4), is the equation of a plane through P_0 with normal $\mathbf{N} = \langle a, b, c\rangle$. ■

Example 4 A straight line L passes through the point (3, 1, 2) and is perpendicular to the plane Π: $x + 2y + 2z = 6$. Find the point where this line intersects the plane.

Solution By (17.5.5), $\mathbf{N} = \langle 1, 2, 2\rangle$ is normal to Π. Since L is also perpendicular to Π, it follows that $\mathbf{N}$ is parallel to L. Thus L goes through (3, 1, 2) and has direction numbers 1, 2, 2. From (17.5.1), L has parametric equations $x = 3 + t$, $y = 1 + 2t$, $z = 2 + 2t$. If the point $(3 + t, 1 + 2t, 2 + 2t)$ on L is also to be on Π, then we must have $3 + t + 2(1 + 2t) + 2(2 + 2t) = 6$. Solving for t, we have $t = -1/3$, which, when substituted into the equations for L, gives (8/3, 1/3, 4/3) as the desired point of intersection. ●

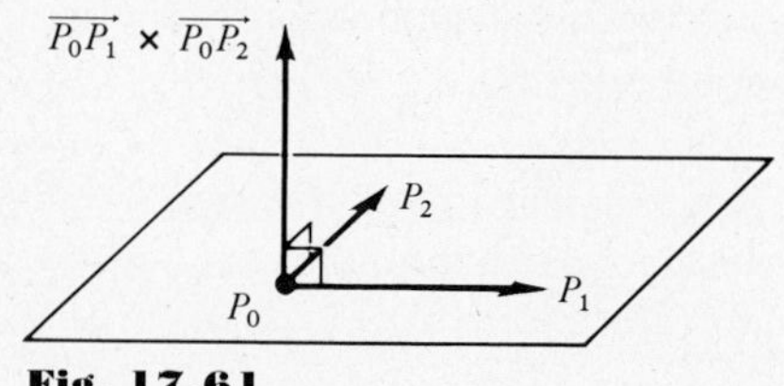

Fig. 17.61

PLANES AND CROSS PRODUCTS

Suppose the points P_0, P_1, P_2 are not collinear. Then $\overrightarrow{P_0P_1}$ and $\overrightarrow{P_0P_2}$ will not be parallel, and hence the cross product $\mathbf{N} = \overrightarrow{P_0P_1} \times \overrightarrow{P_0P_2}$ will be a vector perpendicular to both $\overrightarrow{P_0P_1}$ and $\overrightarrow{P_0P_2}$. By definition (17.5.3), the points P_1 and P_2 are on the plane determined by P_0 and $\mathbf{N}$. As we would expect, three noncollinear points determine a plane (see Fig. 17.61).

Example 5 Find an equation for the plane through the points $P_0 = (1, 2, -1)$, $P_1 = (3, 4, 2)$, and $P_2 = (-1, 0, 4)$.

Solution By the discussion above, $\mathbf{N} = \overrightarrow{P_0P_1} \times \overrightarrow{P_0P_2}$ is a normal to the plane. But

$$\mathbf{N} = \langle 2, 2, 3\rangle \times \langle -2, -2, 5\rangle$$
$$= \left\langle \begin{vmatrix} 2 & 3 \\ -2 & 5 \end{vmatrix}, -\begin{vmatrix} 2 & 3 \\ -2 & 5 \end{vmatrix}, \begin{vmatrix} 2 & 2 \\ -2 & -2 \end{vmatrix} \right\rangle = \langle 16, -16, 0\rangle$$

By (17.5.4), the plane has equation $16(x - 1) - 16(y - 2) + 0(z + 1) = 0$, or $y - x = 1$. [In applying (17.5.4) we used the point P_0. We could just as well have used P_1 or P_2.] ●

We define *planes* to be *parallel* or *perpendicular* according to whether their *normals* are parallel or perpendicular, respectively.

Example 6 Determine an equation for the plane Π through the point $(2, 1, -6)$ and perpendicular to both of the planes Π_1: $x + 2y + z = 1$ and Π_2: $2x - 3y + z = 4$.

Solution The normals for Π_1 and Π_2 are, respectively, $\mathbf{N}_1 = \langle 1, 2, 1\rangle$ and $\mathbf{N}_2 = \langle 2, -3, 1\rangle$. Any vector **N** that is perpendicular to both $\mathbf{N}_1$ and $\mathbf{N}_2$ will be a suitable normal for Π. Thus we take

$$\mathbf{N} = \mathbf{N}_1 \times \mathbf{N}_2 = \left\langle \begin{vmatrix} 2 & 1 \\ -3 & 1 \end{vmatrix}, -\begin{vmatrix} 1 & 1 \\ 2 & 1 \end{vmatrix}, \begin{vmatrix} 1 & 2 \\ 2 & -3 \end{vmatrix} \right\rangle = \langle 5, 1, -7\rangle$$

The equation for Π is $5(x - 2) + (y - 1) - 7(z + 6) = 0$, or $5x + y - 7z = 53$. ●

PROGRESS TEST 1

1. Find parametric equations for the following lines:
   (*a*) Through $(7, -2, 5)$ and parallel to $\langle 3, 1/2, -1\rangle$.
   (*b*) Through $(2, 1, 6)$ and $(-5, 3, 7)$.
2. Do the following lines intersect? If not, determine if they are parallel or skew. $x = 1 - 3t$, $y = 7 + t$, $z = 11 + 4t$; $(x - 3)/1 = (y - 9)/(-1) = (z - 4)/3$.
3. Find an equation for the following planes Π:
   (*a*) Through $(2, 6, -4)$ with normal $\langle 4, -1, 2\rangle$.
   (*b*) Through $(1, 2, -3)$, $(2, 4, 6)$, $(-2, -1, 1)$.

SOME ADDITIONAL GEOMETRY

Using vector methods it is relatively simple to determine the (perpendicular) distance from a point $P_0 = (x_0, y_0, z_0)$ to a plane Π: $ax + by + cz = d$. Let $P_1 = (x_1, y_1, z_1)$ be a point on Π and place a representative of $\mathbf{N} = \langle a, b, c\rangle$ with its base at P_1. From Fig. 17.62 it is apparent that the desired distance D is the absolute value of the scalar projection of $\overrightarrow{P_1P_0}$ onto **N**. Note that the absolute value is necessary since the representative of **N** need not be on the same side of Π as the point P_0.

Example 7 Show that the distance from the point $P_0 = (x_0, y_0, z_0)$ to the plane Π: $ax + by + cz = d$ is given by

$$D = \frac{|ax_0 + by_0 + cz_0 - d|}{\sqrt{a^2 + b^2 + c^2}}$$

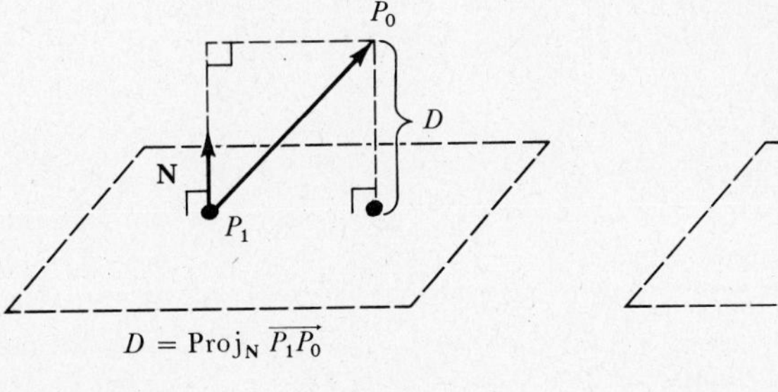

Fig. 17.62

Solution From our previous discussion,

$$D = |\text{Proj}_{\mathbf{N}}\, \overrightarrow{P_1P_0}|$$
$$= \frac{|\langle x_0 - x_1, y_0 - y_1, z_0 - z_1\rangle \cdot \langle a, b, c\rangle|}{\|\langle a, b, c\rangle\|}$$
$$= \frac{|ax_0 + by_0 + cz_0 - (ax_1 + by_1 + cz_1)|}{\sqrt{a^2 + b^2 + c^2}} = \frac{|ax_0 + by_0 + cz_0 - d|}{\sqrt{a^2 + b^2 + c^2}}$$

The last step follows because $P_1 = (x_1, y_1, z_1)$ is on Π and hence satisfies $ax_1 + by_1 + cz_1 = d$. ●

Example 8 Suppose the straight line L has the vector equation $\mathbf{R}(t) = \mathbf{P} + t\mathbf{U}$, where $\mathbf{U}$ is a unit vector. Let Q be any point. Determine the point S on L that is closest to Q.

Solution The situation is diagramed in Fig. 17.63. Since S is on L, we know

(1) $\mathbf{S} = \mathbf{P} + t\mathbf{U}$

for some t. Now $(\mathbf{S} - \mathbf{Q}) \perp \mathbf{U}$, so

$$(\mathbf{S} - \mathbf{Q}) \cdot \mathbf{U} = 0$$

so

$$[(\mathbf{P} + t\mathbf{U}) - \mathbf{Q}] \cdot \mathbf{U} = 0$$

so

$$(\mathbf{P} - \mathbf{Q} \cdot \mathbf{U} + t\mathbf{U} \cdot \mathbf{U} = 0$$

But $\mathbf{U} \cdot \mathbf{U} = \|\mathbf{U}\|^2 = 1$, so we can solve this latter equation for t, getting

$$t = (\mathbf{Q} - \mathbf{P}) \cdot \mathbf{U}$$

Substituting in (1), we have $\mathbf{S} = \mathbf{P} + [(\mathbf{Q} - \mathbf{P}) \cdot \mathbf{U}]\mathbf{U}$. ●

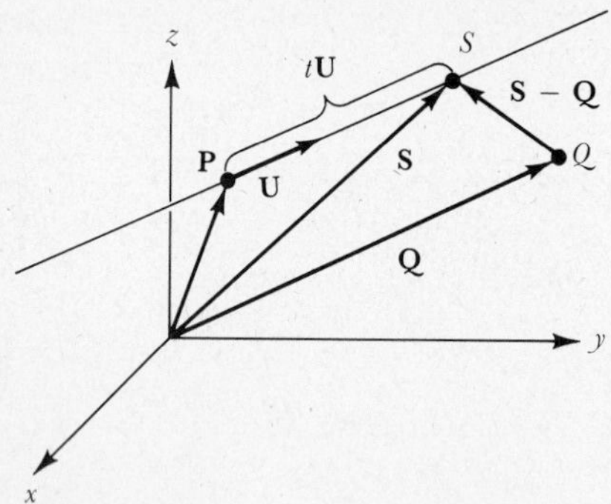

Fig. 17.63

SECTION 17.5 EXERCISES

In Exercises 1 to 8 compute parametric equations for the line satisfying the given conditions.

1. Through $(1, 2, -6)$ and $(-3, 4, 2)$.

2. Through $(8, -1, 4)$ and $(-2, 0, 7)$.

3. Through $(-2, -6, 3)$ and parallel to $\langle 6, 1, 2\rangle$.

4. Through $(1/2, -8, e)$ and parallel to $\langle 0, 6, 9\rangle$.

5. Through $(0, -2, 6)$ and parallel to both the planes $2x - y + 7z = 8$ and $3x - 2y - z = 3$.

6. The intersection of the planes $x + 2y - 2z = 5$ and $5x - 2y - z = 2$.

7. The intersection of the planes $x + y - 2z = 3$ and $4x - y + z = 6$.

8. Through the point $(2, 1, -1)$ and perpendicular to the line $x = 1 + 3t, y = 2 - t, z = 3 + 2t$.

In Exercises 9 to 16 find an equation for the plane satisfying the given conditions.

9. Through $(-6, 3, 1/2)$ with normal $\langle 2, 1, -4\rangle$.

10. Through $(4, -6, -1)$ with normal $\langle 3, -2, -6\rangle$.

11. Through $(1, 2, -6)$ $(3, 7, 9)$ and $(4, -1, 6)$.

12. Through $(-3, -4, 6)$ $(2, -8, 1)$ and $(3, 7, -4)$.

13. Through $(6, 1, -7)$ and containing the line $x = 2 - t$, $y = 4 + 3t$, $z = 8 + 7t$.

14. Containing the intersecting lines $x = 1 + 2t, y = 4 - t$, $z = 3 + 4t$ and $(x - 3)/(-6) = (y - 7)/(-1) = (z - 8)/(-13)$.

15. Containing the parallel lines $x = 2 + t$, $y = 3 - 6t$, $z = -1 + t$ and $x = -4 + 2t$, $y = 8 - 12t$, $z = 3 + 2t$.

16. Through $(2, -1, 7)$ and parallel to the plane $4x - 8y + 6z = 27$.

17. Find the distance from the point $(2, 1, -6)$ to the plane $2x + 3y - z = 4$.

18. Let $\mathbf{N}$ be a unit normal for the plane Π, which passes through the point $P_0 = (x_0, y_0, z_0)$. Show that the point S on Π closest to the point Q satisfies $\mathbf{S} = \mathbf{P}_0 \cdot \mathbf{N} - \mathbf{Q} \cdot \mathbf{N})\mathbf{N}$.

19. Find the point on the plane $x + y - z = 3$ closest to the point $(2, 1, -3)$.

20. The line L is obtained by "projecting" the line L_1: $x = 1 + 2t$, $y = 3 - t$, $z = 2 + t$ onto the plane

$2x + y - z = 7$. That is, through each point of L one drops a perpendicular to the plane $2x + y - z = 7$. The line L is made up of the points where these perpendiculars meet the plane. Find equations for L.

21. Find equations for the line L of intersection of the planes $x + y + 2z = 6$ and $2x - y - z = 4$.

22. Find the distance from the plane $3x + y - z = 2$ to the point $(2, -1, 5)$.

23. Find the distance from the point $Q = (1, 2, -1)$ to the line L: $x = 3 + t, y = -2 + 2t, z = 1 + 2t$.

CHAPTER EXERCISES

Review Exercises

In Exercises 1 to 35 identify (plane, cylinder, surface of revolution, particular quadric surface) and sketch the surface with the given Cartesian equation.

1. $x^2 + y^2 + z^2 - 2x + 6y - 4z = 11$
2. $9x^2 + y^2 - 18x - 4y + 4 = 0$
3. $9x^2 + 9y^2 - 4z^2 - 18x + 18y - 16z + 2 = 0$
4. $9z^2 + 18y^2 - 4x^2 = 0$
5. $y^2 + z^2 = e^x$
6. $x^2 + y^2 + z^2 + 4x + 8y + 11 = 0$
7. $y^2 - 2y = z^2 - 6z + 8$
8. $4y^2 - 8y - 9x^2 - 36x - 32 = 36z$
9. $4x^2 - 9z^2 = 36y$
10. $3x^2 + 3z^2 = y^{2/3}$
11. $4x^2 + 4y^2 + 4z^2 + 4x + 12y - 4z = 5$
12. $y = 4z^2 - 24z + 36$
13. $9x^2 + y^2 + 72x - 4y + 148 = 9z$
14. $y^2 + 4z^2 = 4x$
15. $x^2 + y^2 = (\arcsin z)^2$
16. $x^2 + y^2 + 18y + z^2 = 0$
17. $x^2 + 4x + y^2 - 10y + z^2 - 16x + 92 = 0$
18. $9z^2 + 54z + 4y^2 - 72y = -369$
19. $4x^2 - y^2 + 2y - 2z^2 - 8z = 14$
20. $27x^2 + 12y^2 - 9z^2 + 216x - 48y - 18z + 363 = 0$
21. $12x^2 + 4y^2 + 5z^2 - 24x + 56y - 40z + 228 = 0$
22. $9y^2 - 4x^2 - 36z^2 = 36$
23. $3x - 2y + 7z = 0$
24. $9z^2 + 36x^2 - 4y^2 = 36$
25. $x + 2y + 3z = 18$
26. $27x^2 + 12y^2 + 9z^2 = 108$
27. $x = y + 3z$
28. $x^2 + 3y^2 = 9$
29. $x^2 = z^{1/4} - y^2$
30. $z = \sqrt{x}$
31. $y + 4z = 12$
32. $|x| + |y| = 4$
33. $z = \ln y$
34. $x^2 + y^2 = (\ln z)^2$
35. $x = z^3$

In Exercises 36 to 45 find an equation for the surface of revolution obtained by revolving the given plane curve about the indicated axis.

36. $x = 4z^2$, x axis
37. $x = 4z^2$, z axis
38. $x^2/9 + y^2/16 = 1$, x axis
39. $x^2/9 + y^2/16 = 1$, y axis
40. $z^2 = y^3$, z axis
41. $z^2 = y^3$, y axis
42. $z = e^y$, z axis
43. $z = e^y$, y axis
44. $yz = 1$, y axis
45. $yz = 1$, z axis

In Exercises 46 to 55 find (a) the length of **A**, *(b) a unit vector in the direction of* **A** *and (c) the cosine of the direction of* **A**.

46. $\mathbf{A} = \langle 4, -3\rangle$
47. $\mathbf{A} = \langle \sqrt{10}, \sqrt{15}\rangle$
48. $\mathbf{A} = \langle 3, 1/2\rangle$
49. $\mathbf{A} = \langle -2, -6\rangle$
50. $\mathbf{A} = \overrightarrow{PQ}, P = (1, -1), Q = (2, 4)$
51. $\mathbf{A} = \langle -1, 0\rangle$
52. $\mathbf{A} = \langle -1, -3\rangle$
53. $\mathbf{A} = \langle -2/3, 1/4\rangle$
54. $\mathbf{A} = \langle 0, -2\rangle$
55. $\mathbf{A} = \overrightarrow{PQ}, P = (-2, -4), Q = (1, 5)$

In Exercises 56 to 60 determine (a) $\mathbf{A} \times \mathbf{B}$ and *(b) the area of the parallelogram with coterminal sides* **A** *and* **B**.

56. $\mathbf{A} = \langle 2, 1, -1\rangle, \mathbf{B} = \langle 3, 0, 4\rangle$
57. $\mathbf{A} = \langle 0, -2, 3\rangle, \mathbf{B} = \langle 2, 3, 0\rangle$
58. $\mathbf{A} = \langle 1, 2, -6\rangle, \mathbf{B} = \langle -2, 1, 3\rangle$
59. $\mathbf{A} = \langle 3, 0, 1\rangle, \mathbf{B} = \langle 0, 2, -4\rangle$
60. $\mathbf{A} = \langle 0, 2, 0\rangle, \mathbf{B} = \langle 5, 0, 0\rangle$

In Exercises 61 to 63 find the volume of the parallelepiped with coterminal edges **A**, **B**, *and* **C**.

61. $\mathbf{A} = \langle 1, 1, 0\rangle$, $\mathbf{B} = \langle 0, 2, 3\rangle$, $\mathbf{C} = \langle 4, 0, 1\rangle$
62. $\mathbf{A} = \langle 1, 2, -1\rangle$, $\mathbf{B} = \langle 0, 2, 0\rangle$, $\mathbf{C} = \langle 3, 4, 0\rangle$
63. $\mathbf{A} = \langle 3, -2, -1\rangle$, $\mathbf{B} = \langle 1, 3, -2\rangle$, $\mathbf{C} = \langle 1, 1, 1\rangle$

In Exercises 64 to 70 find an equation for the plane satisfying the given conditions:

64. Through $(1, 2, -1)$, $(1, 3, 2)$, $(2, -1, 5)$.

65. Through $(2, -3, 0)$, $(0, 5, 7)$, $(6, 0, -4)$.

66. Through $(-2, -1, -4)$ and parallel to the plane $3x + 4y - 2z = 4$.

67. Through $(6, 2, -4)$ and perpendicular to the line $x = 2 - t, y = 3 + 4t, z = 8t$.

68. Through $(2, 1, -2)$ and containing the line $x = 4 + t$, $y = 2 - 3t$, $z = 4 + 2t$.

69. Containing the intersecting lines $x = 3 + t, y = 2 - 6t$, $z = 4 + 2t$, and $x = 7 - 4t, y = 7 - 5t, z = -4 + 8t$.

70. Containing the parallel lines $x = 2 - t$, $y = 4 + 3t$, $z = 7 - 8t$, and $x = 4 - 2t, y = 3 + 6t, z = -1 - 16t$.

In Exercises 71 to 74 find equations for the straight line satisfying the given conditions:

71. Through $(2, 1, -8)$ and $(3, 2, 5)$.

72. Through $(-5, 1, 4)$ and $(0, 8, -9)$.

73. Through $(5, 6, 0)$ and parallel to $\langle -2, 1, -8\rangle$.

74. Through $(2, 1, 1)$ and perpendicular to the line $x = 4 - t, y = 5 + 7t, z = -1 + 3t$.

75. Find the distance from the point $(-1, 2, 4)$ to the plane $x + 3y - z = 5$.

76. Find the distance from the point $(0, 2, -8)$ to the plane $3x - y + z = 0$.

77. Use vector methods to show that the altitudes of a triangle meet in a point.

78. Use vector methods to show that the midpoints of the sides of a quadrilateral form the vertices of a parallelogram.

79. Use vector methods to show that the diagonals of a parallelogram bisect each other.

In Exercises 80 to 83 find the work done as the constant force with the given magnitude and direction is applied to move an object in the plane along a straight line from P to Q.

80. $\|\mathbf{F}\| = 30$ lb, $\alpha = \pi/6$, $P = (1, -1)$, $Q = (3, 5)$

81. $\|\mathbf{F}\| = 21$ lb, $\alpha = \pi/4$, $P = (2, 1)$, $Q = (4, 7)$

82. $\|\mathbf{F}\| = 2$ lb, $\alpha = \pi/3$, $P = (3, 5)$, $Q = (5, 10)$

83. $\|\mathbf{F}\| = 40$ lb, $\alpha = \pi/6$, $P = (4, -2)$, $Q = (8, -6)$

84. Force $\mathbf{F}_1$ has magnitude $\|F_1\| = 30$ lb and direction $\alpha = \pi/3$. Force $\mathbf{F}_2$ has magnitude $\|F_2\| = 20$ lb and direction $\alpha = \pi/6$. Find the work done by the resultant force $\mathbf{F}_1 + \mathbf{F}_2$ in moving an object in the plane along the x axis from $(0, 0)$ to $(40, 0)$.

Miscellaneous Exercises

1. Find an equation for the sphere that is tangent to the plane $2x + 3y + 2z = 12$ and has its center at $(3, 2, 1)$.

2. Find an equation for the sphere that has its center on the y axis and passes through the points $(0, 0, 5)$ and $(1, -2, 0)$.

3. An elliptical-cross-section torus is a surface of revolution obtained by revolving an ellipse about a line that does not intersect the ellipse. Find an equation for the elliptical torus generated by revolving the ellipse $(x - c)^2/a^2 + y^2/b^2 = 1$ in the x-y plane about the y axis ($c > a$).

4. A surface is said to be *ruled* if through every point on the surface there passes a straight line all of whose points are on the surface. Complete the argument sketched below to show that the hyperboloid of one sheet, $x^2/a^2 + y^2/b^2 - z^2/c^2 = 1$, is a ruled surface.

(*a*) Show that for every choice of θ the straight line $x = a\cos\theta + (a\sin\theta)t$, $\quad y = b\sin\theta - (b\cos\theta)t$, $z = t$ lies on the hyperboloid.

(*b*) Let $A = (x_1, y_1, z_1)$ be an arbitrary point on the hyperboloid. It suffices to show that there exists a θ such that the straight line in part (*a*) goes through A. Since t must equal z, one can do this by showing that $x_1 = a\cos\theta + (a\sin\theta)z_1, y_1 = b\sin\theta - (b\cos\theta)z_1$ can be solved for θ. (*Hint:* Solve these equations for $\cos\theta$ and $\sin\theta$ and show that this suffices to determine θ.)

5. Suppose the xyz coordinate system is rotated through an angle θ about the z axis to obtain an XYZ system:

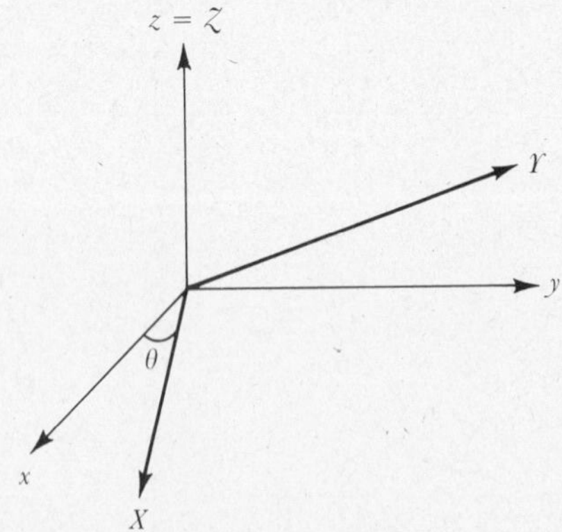

(*a*) Show that if a point P has coordinates (X, Y, Z) in this latter system, then it has coordinates (x, y, z) in the former, where $x = X\cos\theta - Y\sin\theta$, $y = X\sin\theta + Y\cos\theta$, and $z = Z$.

(*b*) Identify the quadric surface $z = xy$ by first rotating through an angle of $\pi/4$ radians.

6. Find the angle between the diagonal of a cube and the diagonal of one of its faces.

7. Let L_1 and L_2 be lines parallel to the vectors $\mathbf{v}_1$ and $\mathbf{v}_2$, respectively, and let α be the angle between $\mathbf{v}_1$ and $\mathbf{v}_2$. The angle θ between L_1 and L_2 is defined by $\theta = \alpha$ if $0 \leq \alpha \leq \pi/2$ and by $\theta = \pi - \alpha$ if $\pi/2 < \alpha \leq \pi$. Determine the angle between each of the following pairs of lines.

(*a*) $x = 2 - t, y = 3 + 2t, z = 4t$;
$x = 3 + 6t, y = -2 - 4t, z = 1 + 3t$.

(b) $x = -2t, y = 4 + t, z = -1 - 5t$;
$(x - 6)/1 = (y - 5)/3 = (z + 8)/(-7)$.

By the angle between two planes we mean the angle between lines normal to these planes. In Exercises 8 and 9 determine the angle between each of the following pairs of planes.

8. $3x + y - z = 4$, $2x + 4y + 3z = 12$

9. $x - y - 5z = 1$, $2x + y - z = 3$

12. Show that three points P, Q, R in space are collinear if and only if $\overrightarrow{PQ} \times \overrightarrow{PR} = \mathbf{0}$.

14. Let Π be a plane with normal $\mathbf{N} = \langle a, b, c\rangle$, and let $X = (x, y, z)$ be an arbitrary point on Π. Show that there exists a real number d such that $\mathbf{X} \cdot \mathbf{N} = d$ for all X.

16. Let L be the straight line with equation $\mathbf{X} = \mathbf{P}_0 + t\mathbf{M}$ and let Π be the plane with equation $\mathbf{X} \cdot \mathbf{N} = d$, where $\|\mathbf{N}\| = 1$. If $\mathbf{N} \cdot \mathbf{M} \neq 0$, show that L intersects Π at the head of the standard representative for

$$\mathbf{P}_0 + \frac{[d - (\mathbf{P}_0 \cdot \mathbf{N})]}{\mathbf{N} \cdot \mathbf{M}} \mathbf{M}$$

18. Show that the distance from the point A to the line through the points B and C is given by

$$d = \frac{\|\overrightarrow{BA} \times \overrightarrow{BC}\|}{\|\overrightarrow{BC}\|}$$

20. Show that if $\mathbf{A}$ is perpendicular to $\mathbf{B}$ and $\mathbf{C} = \alpha\mathbf{A} + \beta\mathbf{B}$, then $\alpha = \mathbf{C} \cdot \mathbf{A}$ and $\beta = \mathbf{C} \cdot \mathbf{B}$.

22. Show that $\|\mathbf{A} + \mathbf{B}\|^2 - \|\mathbf{A} - \mathbf{B}\|^2 = 4\mathbf{A} \cdot \mathbf{B}$.

24. Show that $\|\mathbf{A}\|\mathbf{B} + \|\mathbf{B}\|\mathbf{A}$ is perpendicular to $\|\mathbf{A}\|\mathbf{B} - \|\mathbf{B}\|\mathbf{A}$.

26. When does $|\mathbf{A} \cdot \mathbf{B}| = \|\mathbf{A}\|\|\mathbf{B}\|$? Justify.

28. Show that $(\mathbf{A} + \mathbf{B}) \cdot (\mathbf{A} - \mathbf{B}) = \|\mathbf{A}\|^2 - \|\mathbf{B}\|^2$.

10. Determine an equation for the plane through the points $(1, 2, 5)$ and $(3, 6, -2)$ and parallel to the line of intersection of the planes $x + y - 3z = 2$ and $2x - y + 2z = 4$.

11. Determine an equation for the plane through the point $(3, -1, 4)$ and perpendicular to the line of intersection of the planes $x - y + 5z = 8$ and $2x - 3y + z = 4$.

13. Show that four points P, Q, R, S in space are coplanar if and only if $\overrightarrow{PS} \cdot (\overrightarrow{PQ} \times \overrightarrow{QR}) = 0$.

15. Prove the converse of Exercise 14. That is, if $\mathbf{N} = \langle a, b, c\rangle$ and if d is any real number, then the collection of all $X = (x, y, z)$ such that $\mathbf{X} \cdot \mathbf{N} = d$ represents a plane.

17. Let L_1, L_2 be skew straight lines through the points P, Q and R, S respectively. Show that the distance from L_1 to L_2 is given by

$$d = \frac{|\overrightarrow{PS} \cdot (\overrightarrow{PQ} \times \overrightarrow{RS})|}{\|\overrightarrow{PQ} \times \overrightarrow{RS}\|}$$

19. Show that the standard representative of

$$\frac{\|\mathbf{B}\|\mathbf{A} + \|\mathbf{A}\|\mathbf{B}}{\|\mathbf{A}\|\|\mathbf{B}\|}$$

bisects the angle between the standard representatives of $\mathbf{A}$ and $\mathbf{B}$.

21. Give examples with $\mathbf{A} \neq \mathbf{0}$ to show that $\mathbf{A} \cdot \mathbf{C} = \mathbf{A} \cdot \mathbf{B}$ need not imply $\mathbf{C} = \mathbf{B}$.

23. Let $\mathbf{A}$ and $\mathbf{B}$ be nonzero vectors. Determine the scalar k such that $\mathbf{A} - k\mathbf{B}$ is perpendicular to $\mathbf{B}$.

25. When does $\|\mathbf{A} + \mathbf{B}\| = \|\mathbf{A}\| + \|\mathbf{B}\|$? Justify.

27. Show that $\|\mathbf{A} - \mathbf{B}\| \geq \|\mathbf{A}\| - \|\mathbf{B}\|$. [*Hint:* Consider the equation $\mathbf{A} = \mathbf{B} + (\mathbf{A} - \mathbf{B})$.]

SELF-TEST ✣

1. Identify and sketch the surface whose equation in Cartesian coordinates is given by $x^2 + 4y^2 - 2x - 8y - 4z + 13 = 0$.

2. Find an equation for the surface obtained by revolving the graph of the plane curve $y = x^3 + 9$ about the y axis.

3. Find an equation for the plane through the points $(1, 2, 5)$, $(3, 3, 6)$, and $(2, 4, 8)$. What is the distance from this plane to the origin?

4. Find parametric equations for the line through $(2, 2, 6)$ and parallel to $\langle 1, -1, 2\rangle$. Does this line intersect the line with equations $x = 5 - 2t$, $y = 1 + t$, $z = 6 - t$?

5. Let $\mathbf{A} = \langle 2, 1\rangle$, $\mathbf{B} = -3\mathbf{i} + 4\mathbf{j}$, $\mathbf{C} = \overrightarrow{PQ}$, where $P = (1, 2)$ and $Q = (-1, 6)$.

(a) Find the vector projection of $\mathbf{C} + 4\mathbf{A}$ onto $\mathbf{B}$.

(b) Find a vector in the direction of $\mathbf{A} + \mathbf{B}$ that has the same length as $2\mathbf{B} - 3\mathbf{C}$.

6. Let $\mathbf{A} = \langle 1, 2, -1\rangle$, $\mathbf{B} = \langle 3, -4, 5\rangle$, $\mathbf{C} = \langle 1, 7, -2\rangle$.

(a) Find the area of the parallelogram with coterminal sides $\mathbf{A}$ and $\mathbf{B}$.

(b) Find the volume of the parallelepiped with coterminal edges $\mathbf{A}$, $\mathbf{B}$, and $\mathbf{C}$.

CHAPTER 18 ✖

VECTOR VALUED FUNCTIONS

CONTENTS

In this chapter we introduce and study the idea of a vector valued function. Unlike "scalar" functions, which transform real numbers into real numbers, vector functions transform real numbers into vectors. Since parametric equations relate real numbers to points in the plane or space, vector functions are closely related to parametric equations.

In Sec. 18.1 the concepts of limit, continuity, differentiability, and integrability are extended—essentially coordinatewise—to vector functions. In Sec. 18.2 we study the geometry of their graphs, paying particular attention to the important concept of curvature, the measure of the amount of "bend" in a curve. Section 18.3 is concerned with using vector functions to describe motion in the plane and in space. We generalize earlier work with straight-line motion (describable using scalar functions) to much more general motion in the plane or space.

Section 18.1: Vector Functions

Objectives:

1. Sketch the graphs of vector functions.

2. Differentiate and integrate vector functions.

INTRODUCTION

Up to this point we have studied the calculus of real functions of a real variable, that is, functions that transform real numbers into real numbers. Henceforth we shall refer to such functions as *scalar functions*. Since many physical quantities of interest are in fact vector quantities, it is important to consider vector valued functions of a real variable or, more briefly, *vector functions*.

Vector functions have the form

18.1.1
$$\mathbf{F}(t) = \langle x(t), y(t), z(t)\rangle$$
$$= x(t)\mathbf{i} + y(t)\mathbf{j} + z(t)\mathbf{k}$$

where the *component functions* $x(t)$, $y(t)$ and $z(t)$ are scalar functions defined on some common domain D. (**i**, **j**, and **k** are the unit coordinate vectors.) If the domain of a vector function is not specified, then we assume it is the intersection of the natural domains of the component functions.

The graph of (18.1.1), called a *space curve*, is by definition the graph of the parametric equations

$$x = x(t),\ y = y(t),\ z = z(t) \qquad t \in D$$

These are precisely the points traced out by the head of the standard representatives of $\mathbf{F}(t)$. When graphing a space curve, we usually place arrowheads on the graph to indicate the direction that the points are traced out as t increases (see Fig. 18.1).

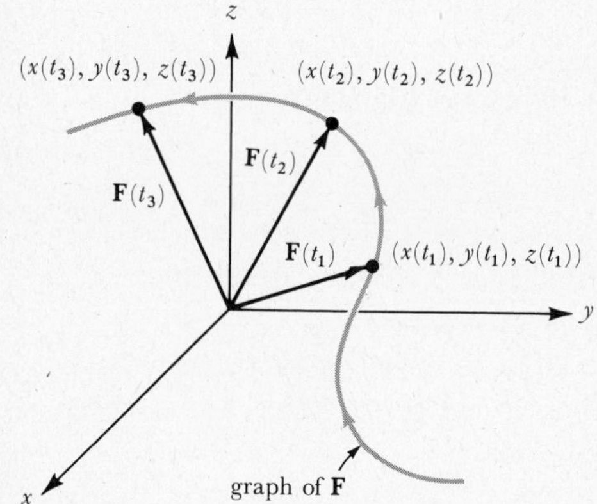

Fig. 18.1

If $z(t) = 0$ for all t in D, then the graph lies in the x-y plane. In this case we usually write

$$\mathbf{F}(t) = \langle x(t), y(t)\rangle = x(t)\mathbf{i} + y(t)\mathbf{j}$$

where now $\mathbf{i} = \langle 1, 0\rangle$ and $\mathbf{j} = \langle 0, 1\rangle$. We refer to the graph of **F** in this latter case as a *plane curve*. In either situation we often refer to "the curve **F**," when in effect we mean the curve that is the graph of the vector function **F**.

Vector functions are particularly useful in describing the motion of an object in space (or in the plane). To accomplish this we assume that the mass of the object is located at its center of mass and refer to it as a *particle*. If t denotes time, then the head of the standard representative of $\mathbf{F}(t)$ gives the position of the particle at time t. In this setting we often refer to **F** as the *position function* of the motion and $\mathbf{F}(t)$ as a *position vector* (see Fig. 18.2). On occasion we shall make a distinction between the position *vector* $\mathbf{F}(t)$ and the *point* $F(t)$ located at the head of the standard representative of $\mathbf{F}(t)$.

z
location of the particle at time t
path of the particle
position vector $\mathbf{F}(t)$
y
x

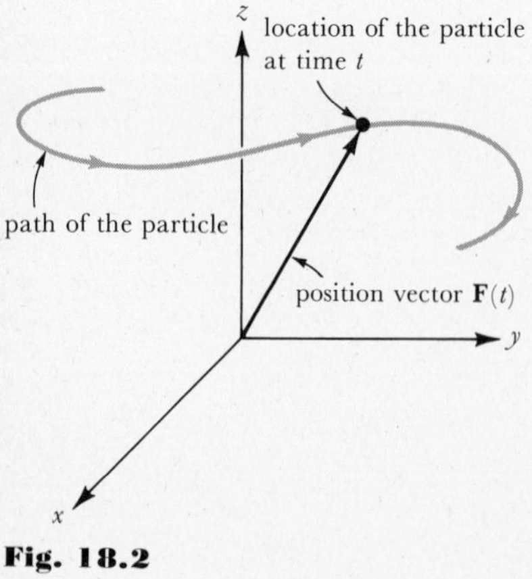

Fig. 18.2

SOME EXAMPLES

Several plane curves were discussed in Sec. 12.1 in the context of parametric equations. In particular, $\mathbf{F}(t) = \langle 2\cos t, 3\sin t\rangle$ has as its graph an ellipse (Sec. 12.1, Example 3, page 399). The graph of $\mathbf{F}(t) = \langle a(t - \sin t), a(1 - \cos t)\rangle$ is a cycloid (Sec. 12.1, Example 4, page 400). Similarly, we know from work in Sec. 17.5, pages 581 to 583, that the graph of any vector function of the form

$$\mathbf{F}(t) = \langle at + c_1, bt + c_2, ct + c_3\rangle$$

is a straight line in space.

Example 1 Sketch the graph C of the vector function

$$\mathbf{F}(t) = \langle a \cos t,\ b \sin t,\ t \rangle \qquad a,\ b > 0$$

Solution The parametric equations are

$$x = a \cos t,\ y = b \sin t,\ z = t$$

We first eliminate the parameter from the first two equations. Since $x/a = \cos t$ and $y/b = \sin t$, we have

$$\frac{x^2}{a^2} + \frac{y^2}{b^2} = \cos^2 t + \sin^2 t = 1$$

It follows that C lies on the elliptical cylinder (parallel to the z axis) $x^2/a^2 + y^2/b^2 = 1$. As t increases, the points move steadily up and around the cylinder making a complete revolution around the cylinder on each interval of length 2π. We refer to such a curve as an *elliptical helix* or, if $a = b$, a *circular helix*. A table of values on $[0, 2\pi]$ and a sketch is given in Fig. 18.3. ●

Example 2 Sketch the portion of the space curve $\mathbf{F}(t) = \langle \cos t, \sin t, 2 \sin t \rangle$ that lies in the first octant.

Solution First we eliminate the parameter from x and y and from x and z: $x = \cos t$, $y = \sin t$, so

$$x^2 + y^2 = 1 \tag{1}$$

Also, $z = 2 \sin t$, so $\sin t = z/2$ and hence

$$x^2 + \left(\frac{z}{2}\right)^2 = 1 \tag{2}$$

Equation (1) is a circular cylinder parallel to the z axis and equation (2) is an elliptical cylinder parallel to the y axis. The graph of $\mathbf{F}$ must lie on both, so it must be the intersection of these cylinders. The portion in the first octant sketched in Fig. 18.4 corresponds to the values $0 \le t \le \pi/2$. ●

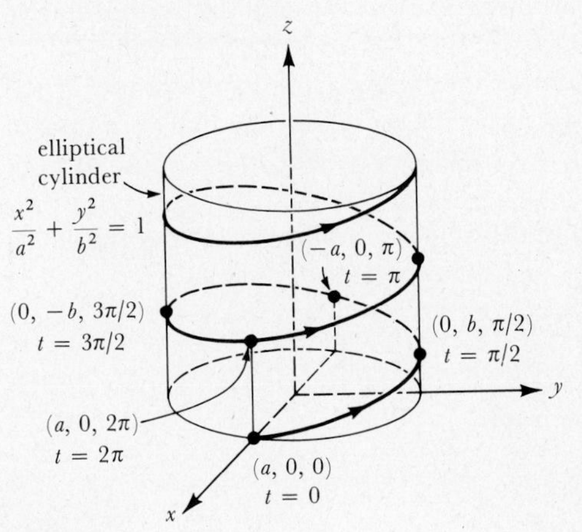

Fig. 18.3

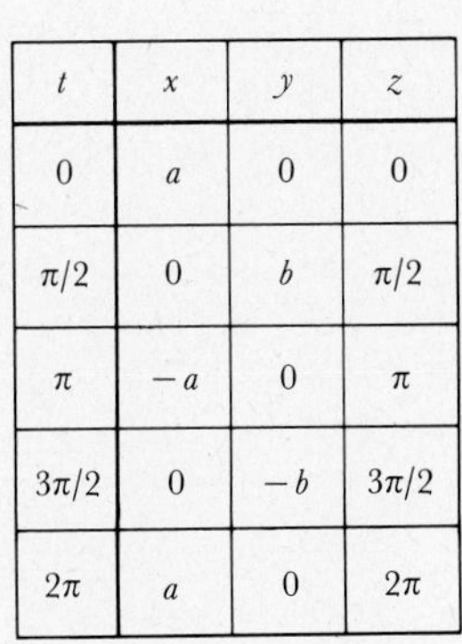

t	x	y	z
0	a	0	0
$\pi/2$	0	b	$\pi/2$
π	$-a$	0	π
$3\pi/2$	0	$-b$	$3\pi/2$
2π	a	0	2π

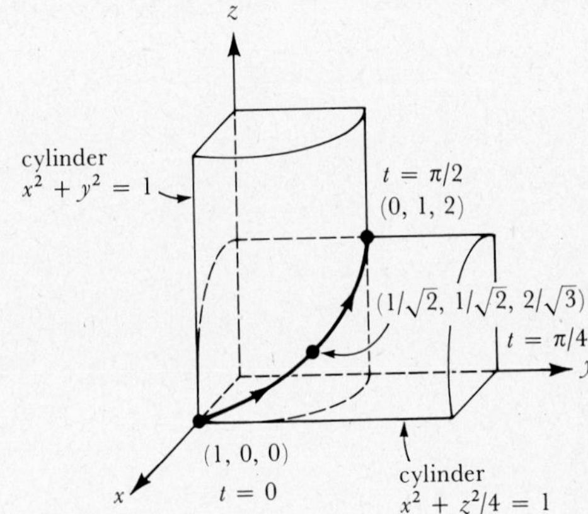

Fig. 18.4

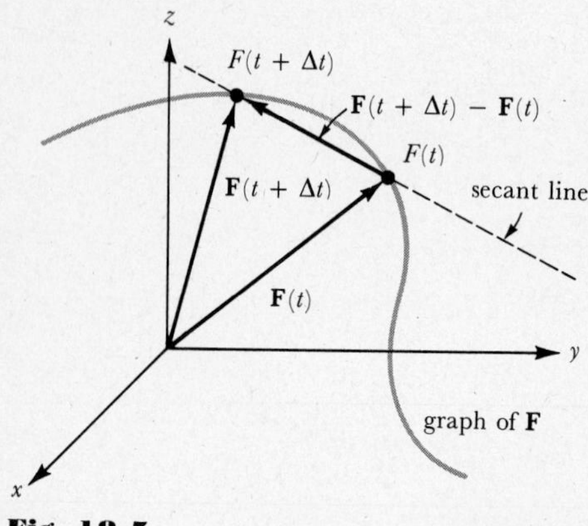

Fig. 18.5

THE CALCULUS OF VECTOR FUNCTIONS

We define limits of vector functions componentwise. That is, we define

$$\lim_{t \to t_0} \langle x(t), y(t), z(t) \rangle = \langle a, b, c \rangle$$

if and only if

$$\lim_{t \to t_0} x(t) = a, \ \lim_{t \to t_0} y(t) = b, \text{ and } \lim_{t \to t_0} z(t) = c$$

Our definition of continuity of a vector function is a direct analog of continuity for scalar functions.

18.1.2 **Definition** The vector function **F** is *continuous at* t_0 if $\mathbf{F}(t_0)$ is defined and $\lim_{t \to t_0} \mathbf{F}(t) = \mathbf{F}(t_0)$. If **F** is continuous at each point in a set of numbers I, then we say **F** is *continuous on* I. A function **F** continuous on its domain will simply be referred to as *continuous*.

With this definition, **F** is continuous at t_0 if and only if each of the component functions is continuous at t_0.

We define the derivative of a vector function using the familiar limits.

18.1.3 **Definition** The derivative of a vector function **F**, denoted by any of $\mathbf{F}'$, $D_t\mathbf{F}$, or $d\mathbf{F}/dt$, is the vector function defined by

$$\mathbf{F}'(t) = \lim_{\Delta t \to 0} \frac{\mathbf{F}(t + \Delta t) - \mathbf{F}(t)}{\Delta t}$$

at all t for which this limit exists.

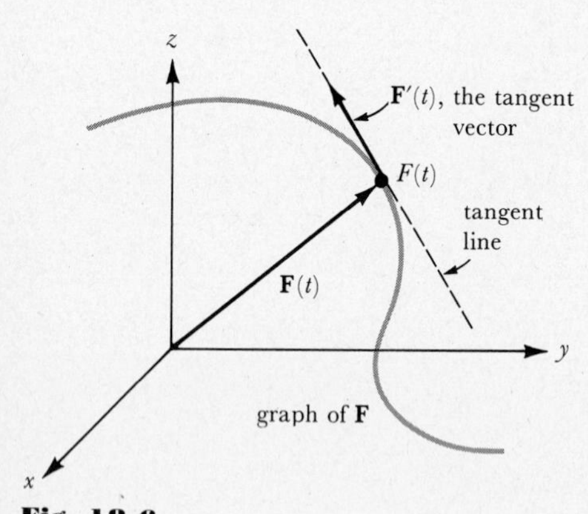

Fig. 18.6

Let us analyze (18.1.3) geometrically. For small Δt, $\mathbf{F}(t + \Delta t) - \mathbf{F}(t)$ has as a representative the short vector from the head of the standard representative for $\mathbf{F}(t)$ to the head of the standard representative for $\mathbf{F}(t + \Delta t)$; see Fig. 18.5. For Δt small (say, positive)

$$\frac{\mathbf{F}(t + \Delta t) - \mathbf{F}(t)}{\Delta t} = \frac{1}{\Delta t}[\mathbf{F}(t + \Delta t) - \mathbf{F}(t)]$$

is an elongation of this vector. In particular, the line through this representative is a secant line.

As $\Delta t \to 0$ these secant lines tend to the tangent line to the graph at the point $F(t)$. Thus $\mathbf{F}'(t)$ is a tangent vector in the sense that the representative of $\mathbf{F}'(t)$ whose base is at the point $F(t)$, lies in the tangent line to the graph at $F(t)$; see Fig. 18.6.

It should not be surprising that the differentiability of **F** is completely determined by the differentiability of its components.

18.1.4 **Theorem** Suppose $\mathbf{F}(t) = \langle x(t), y(t), z(t) \rangle$. Then $\mathbf{F}'(t)$ exists if and only if each of $x'(t)$, $y'(t)$, and $z'(t)$ exists. Furthermore,

$$\mathbf{F}'(t) = \langle x'(t), y'(t), z'(t) \rangle$$

Proof:

$$\mathbf{F}(t + \Delta t) - \mathbf{F}(t) = \langle x(t + \Delta t) - x(t), y(t + \Delta t) - y(t), z(t + \Delta t) - z(t) \rangle$$

Thus

$$\lim_{\Delta t \to 0} \frac{\mathbf{F}(t + \Delta t) - \mathbf{F}(t)}{\Delta t}$$
$$= \lim_{\Delta t \to 0} \left\langle \frac{x(t + \Delta t) - x(t)}{\Delta t}, \frac{y(t + \Delta t) - y(t)}{\Delta t}, \frac{z(t + \Delta t) - z(t)}{\Delta t} \right\rangle$$
$$= \left\langle \lim_{\Delta t \to 0} \frac{x(t + \Delta t) - x(t)}{\Delta t}, \lim_{\Delta t \to 0} \frac{y(t + \Delta t) - y(t)}{\Delta t}, \lim_{\Delta t \to 0} \frac{z(t + \Delta t) - z(t)}{\Delta t} \right\rangle$$

These latter limits exist and equal the derivatives of the component functions if and only if the component functions are differentiable. ■

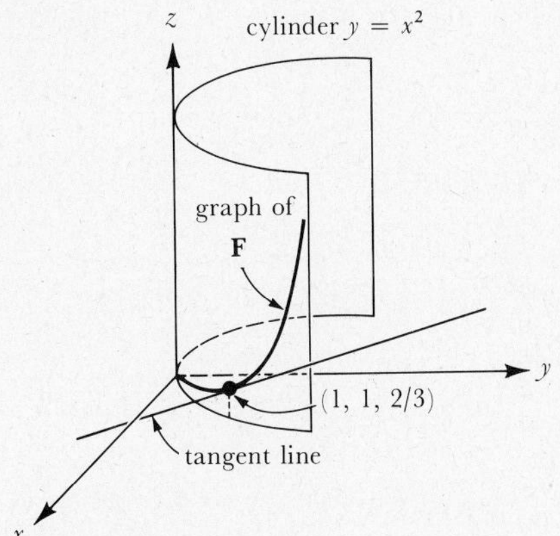

Fig. 18.7

Repeated applications of (18.1.4) yield higher order derivatives of **F**. Thus, for example,

$$\mathbf{F}''(t) = \langle x''(t), y''(t), z''(t) \rangle$$

provided, of course, that the second derivative of each component function exists.

Example 3 Determine parametric equations for the line tangent to the graph of the *twisted cubic* $\mathbf{F}(t) = \langle t, t^2, (2/3)t^3 \rangle$ at the point $P = (1, 1, 2/3)$.

Solution The point $P = (1, 1, 2/3)$ corresponds to the value $t = 1$. Since $\mathbf{F}'(t) = \langle 1, 2t, 2t^2 \rangle$, $\mathbf{F}'(1) = \langle 1, 2, 2 \rangle$ is a tangent vector to the curve at P. By (17.5.1), the tangent line, sketched in Fig. 18.7, has equations

$$x = 1 + t, y = 1 + 2t, z = 2/3 + 2t \quad \bullet$$

PROGRESS TEST 1

Let $\mathbf{F}(t) = \langle e^t \cos t, e^t \sin t, e^t \rangle$

1. Sketch the graph of **F**. **2.** Compute $\mathbf{F}'(t)$.

3. Give equations for the line tangent to the graph of **F** at the point $(1, 0, 1)$.

RULES FOR DIFFERENTIATING VECTOR FUNCTIONS

18.1.5 **Theorem** If **F** and **G** are differentiable vector functions and h is a differentiable scalar function, then:

(*a*) $D_t [\mathbf{F}(t) + \mathbf{G}(t)] = \mathbf{F}'(t) + \mathbf{G}'(t)$

(*b*) $D_t [h(t)\mathbf{F}(t)] = h(t)\mathbf{F}'(t) + h'(t)\mathbf{F}(t)$

(*c*) $D_t [\mathbf{F}(h(t))] = \mathbf{F}'(h(t))h'(t)$

(*d*) $D_t [\mathbf{F}(t) \cdot \mathbf{G}(t)] = \mathbf{F}(t) \cdot \mathbf{G}'(t) + \mathbf{F}'(t) \cdot \mathbf{G}(t)$

(*e*) $D_t [\mathbf{F}(t) \times \mathbf{G}(t)] = \mathbf{F}(t) \times \mathbf{G}'(t) + \mathbf{F}'(t) \times \mathbf{G}(t)$

Proof: All these rules follow in a straightforward fashion. We shall illustrate by proving (*c*). Since

$$\mathbf{F}(h(t)) = \langle x(h(t)), y(h(t)), z(h(t)) \rangle$$

we have, by (18.1.4),

$$\begin{aligned} D_t\mathbf{F}(h(t)) &= \langle D_t[x(h(t))], D_t [y(h(t))], D_t [z(h(t))] \rangle \\ &= \langle x'(h(t))h'(t), y'(h(t))h'(t), z'(h(t))h'(t) \rangle \\ &= \langle x'(h(t)), y'(h(t)), z'(h(t)) \rangle h'(t) \\ &= \mathbf{F}'(h(t))h'(t) \quad ■ \end{aligned}$$

One must be cautious when working with cross products since the order is important. Keep this in mind when using [18.1.5(e)].

Example 4 Let $\mathbf{F}(t)$ be a differentiable vector function on the interval I and suppose $\|\mathbf{F}(t)\| = k$ for all t in I. Show that $\mathbf{F}(t) \cdot \mathbf{F}'(t) = 0$ for each t in I.

Solution By (17.4.2), $\|\mathbf{F}(t)\|^2 = \mathbf{F}(t) \cdot \mathbf{F}(t) = k^2$, so

$$D_t[\mathbf{F}(t) \cdot \mathbf{F}(t)] = D_t(k^2) = 0$$

By [18.1.5(d)],

$$\begin{aligned} D_t[\mathbf{F}(t) \cdot \mathbf{F}(t)] &= \mathbf{F}(t) \cdot \mathbf{F}'(t) + \mathbf{F}'(t) \cdot \mathbf{F}(t) \\ &= 2\mathbf{F}(t) \cdot \mathbf{F}'(t) \qquad \text{by (17.4.2)} \end{aligned}$$

Therefore $2\mathbf{F}(t) \cdot \mathbf{F}'(t) = 0$, which implies $\mathbf{F}(t) \cdot \mathbf{F}'(t) = 0$. ●

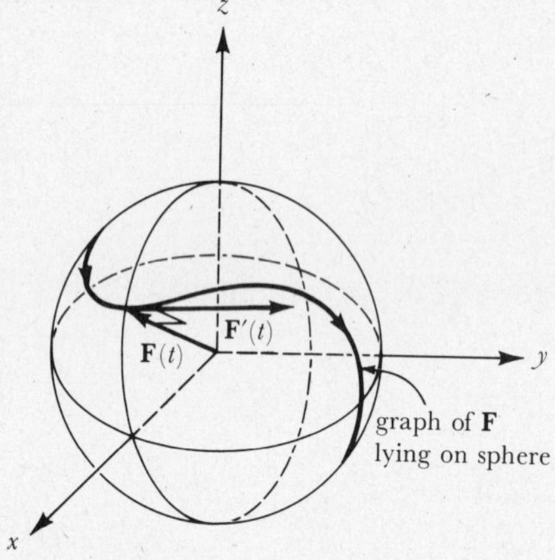

Fig. 18.8

Example 4 has a nice geometric interpretation. If $\mathbf{F}(t) = \langle x(t), y(t), z(t)\rangle$ is a curve such that $\|\mathbf{F}(t)\| = k$, then $[x(t)]^2 + [y(t)]^2 + [z(t)]^2 = k^2$ for each t in I. That is, the graph of $\mathbf{F}$ lies on the sphere of radius k centered at the origin. The reasonable conclusion of Example 4 is that the tangent vector $\mathbf{F}'(t)$ is perpendicular to the position vector $\mathbf{F}(t)$ for each t (see Fig. 18.8).

Part (c) of Theorem (18.1.5) is called the *Chain Rule* for vector functions and will be used extensively in the remainder of this chapter. For emphasis we restate it in Leibnitz notation.

18.1.6 **Chain Rule** If $\mathbf{F} = \mathbf{F}(s)$ and $s = h(t)$, then

$$\frac{d\mathbf{F}}{dt} = \frac{d\mathbf{F}}{ds}\frac{ds}{dt}$$

INTEGRALS OF VECTOR FUNCTIONS

18.1.7 **Definition** If $\mathbf{F}(t) = \langle x(t), y(t), z(t)\rangle$ has integrable components on $[a, b]$, then we define

(a) $$\int_a^b \mathbf{F}(t)\,dt = \left\langle \int_a^b x(t)\,dt, \int_a^b y(t)\,dt, \int_a^b z(t)\,dt \right\rangle$$

Similarly, if each component has an antiderivative we define

(b) $$\int \mathbf{F}(t)\,dt = \left\langle \int x(t)\,dt, \int y(t)\,dt, \int z(t)\,dt \right\rangle$$

Example 5 Let $\mathbf{F}(t) = \langle \sin t, \cos t, 2t\rangle$. Determine

(a) $\int \mathbf{F}(t)\,dt$ (b) $\int_0^{\pi/4} \mathbf{F}(t)\,dt$

Solutions

(a) $$\begin{aligned} \int \mathbf{F}(t)\,dt &= \langle \textstyle\int \sin t\,dt, \int \cos t\,dt, \int 2t\,dt\rangle \\ &= \langle -\cos t + c_1, \sin t + c_2, t^2 + c_3\rangle \\ &= \langle -\cos t, \sin t, t^2\rangle + \langle c_1, c_2, c_3\rangle \\ &= \langle -\cos t, \sin t, t^2\rangle + \mathbf{C} \qquad \text{where } \mathbf{C} = \langle c_1, c_2, c_3\rangle \end{aligned}$$

(b) $$\begin{aligned} \int_0^{\pi/4} \mathbf{F}(t)\,dt &= \left\langle \int_0^{\pi/4} \sin t\,dt, \int_0^{\pi/4} \cos t\,dt, \int_0^{\pi/4} 2t\,dt \right\rangle \\ &= \left\langle -\cos t\,\Big|_0^{\pi/4}, \sin t\,\Big|_0^{\pi/4}, t^2\,\Big|_0^{\pi/4} \right\rangle \\ &= \left\langle 1 - \frac{1}{\sqrt{2}}, \frac{1}{\sqrt{2}}, \frac{\pi^2}{16} \right\rangle \end{aligned}$$ ●

Note that whereas the indefinite integral of a scalar function involves an arbitrary scalar constant, the indefinite integral of a vector function involves an arbitrary *vector* constant. The following result is an immediate consequence of (18.1.7) and the corresponding result for scalar functions.

18.1.8 **Theorem** If $D_t[\mathbf{F}(t)] = \mathbf{G}(t)$, then $\int \mathbf{G}(t)\,dt = \mathbf{F}(t) + \mathbf{C}$.

Example 6 Given that $\mathbf{F}'(t) = \langle \sin t, e^t, 2t \rangle$ and that $\mathbf{F}(0) = \langle 2, 1, 2 \rangle$, find $\mathbf{F}(t)$.

Solution Since $\mathbf{F}'(t) = \langle \sin t, e^t, 2t \rangle$, by (18.1.8),

$$\begin{aligned}\mathbf{F}(t) &= \int \langle \sin t, e^t, 2t \rangle\,dt + \mathbf{C}\\ &= \left\langle \int \sin t\,dt, \int e^t\,dt, \int 2t\,dt \right\rangle + \langle c_1, c_2, c_3 \rangle\\ &= \langle -\cos t, e^t, t^2 \rangle + \langle c_1, c_2, c_3 \rangle\end{aligned}$$

Let $t = 0$, so

$$\begin{aligned}\langle 2, 1, 2 \rangle = \mathbf{F}(0) &= \langle -\cos 0, e^0, (0)^2 \rangle + \langle c_1, c_2, c_3 \rangle\\ &= \langle -1, 1, 0 \rangle + \langle c_1, c_2, c_3 \rangle\end{aligned}$$

Thus

$$\langle c_1, c_2, c_3 \rangle = \langle 2, 1, 2 \rangle - \langle -1, 1, 0 \rangle = \langle 3, 0, 2 \rangle$$

and

$$\mathbf{F}(t) = \langle -\cos t, e^t, t^2 \rangle + \langle 3, 0, 2 \rangle = \langle 3 - \cos t, e^t, t^2 + 2 \rangle \quad \bullet$$

PROGRESS TEST 2

1. Determine $\mathbf{F}(t)$ if $\mathbf{F}''(t) = \langle 3t^2, 8t^3, 2 \rangle$, $\mathbf{F}(0) = \langle 0, 0, 0 \rangle$, and $\mathbf{F}'(0) = \langle 4, 6, 8 \rangle$.
2. Suppose $\mathbf{F}$ is differentiable and $\mathbf{F}(t) \neq 0$ for all t. Show that

$$D_t(\|\mathbf{F}(t)\|) = \frac{\mathbf{F}(t) \cdot \mathbf{F}'(t)}{\|\mathbf{F}(t)\|}$$

ARCS AND ARC LENGTH

A vector function $\mathbf{F}$ whose domain consists of a closed interval $I = [a, b]$ is called an *arc*. Following the language of Chap. 12, an arc $\mathbf{F}$ is said to be *smooth* if the components of $\mathbf{F}$ have continuous derivatives on I and $\|\mathbf{F}'(t)\| \neq 0$ for all t in (a, b). If $\mathbf{F}$ is smooth, then the graph of $\mathbf{F}$ will have no breaks, corners, or cusps. If $\mathbf{F}(t_1) \neq \mathbf{F}(t_2)$ whenever t_1, t_2 are in $[a, b)$ with $t_1 \neq t_2$, then the arc is said to be *simple*. Geometrically, the graph of a simple arc does not cross itself. A simple arc for which $\mathbf{F}(a) = \mathbf{F}(b)$ is said to be *closed*. These properties of an arc are illustrated in Fig. 18.9.

Definition (12.2.1) on the lengths of plane arcs can be routinely extended to the three-dimensional setting.

18.1.9 **Definition** The *length* s of a *smooth arc* $\mathbf{F}(t) = \langle x(t), y(t), z(t) \rangle$, $a \leq t \leq b$, is given by

$$s = \int_a^b \|\mathbf{F}'(t)\|\,dt = \int_a^b \sqrt{[x'(t)]^2 + [y'(t)]^2 + [z'(t)]^2}\,dt$$

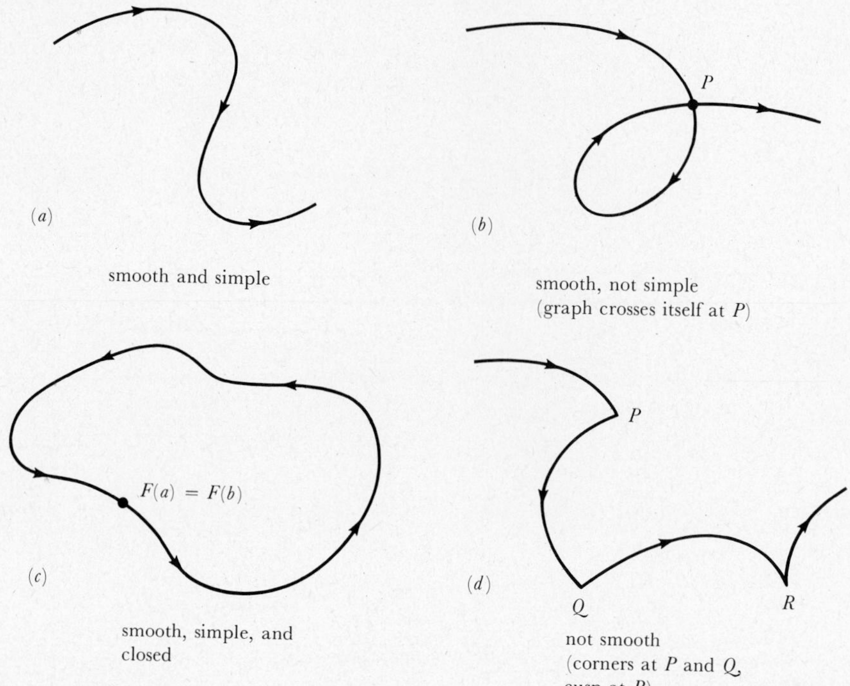

Fig. 18.9

F(a)
F(b)
positive direction
along **F**

F(a)
F(b)
negative direction
along **F**

Fig. 18.10

If **F** is a smooth arc, then by the *positive direction* along the graph of **F** we mean the direction that the head of the standard representative of $\mathbf{F}(t)$ moves along the graph as t increases. The opposite direction is referred to as the *negative direction.* If **F** is simple, then the positive direction can be unambiguously indicated by placing arrow heads along the curve; see Fig. 18.10.

Suppose $\mathbf{F}(t)$, $a \le t \le b$, is a smooth simple arc and t_0 is a fixed number in $[a, b]$. We define the *arc length function* on $[a, b]$ by

18.1.10
$$s(t) = \int_{t_0}^{t} \|\mathbf{F}'(w)\| \, dw$$

If $t > t_0$, then $s(t)$ represents the length of the curve between the points $F(t_0)$ and $F(t)$; see Fig. 18.11. If we think of **F** as describing the motion of a particle, then $s(t)$ represents the distance traveled (along the graph of **F**) by the particle from time t_0 to time t. From (18.1.10) and the Fundamental Theorem of the Calculus, we have that

18.1.11
$$\frac{ds}{dt} = \|\mathbf{F}'(t)\|$$

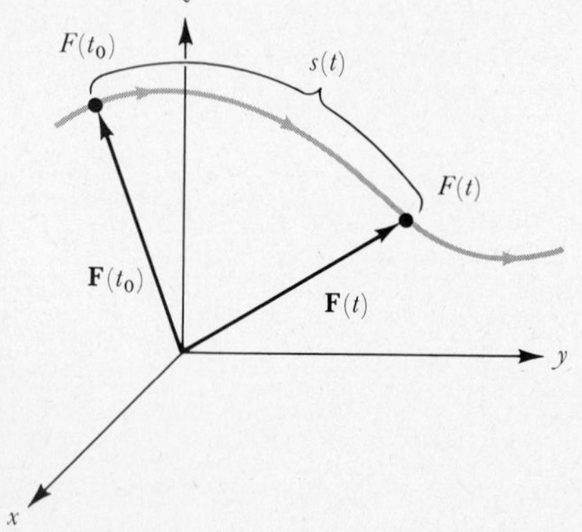

Fig. 18.11

In particular, since **F** is smooth, $ds/dt > 0$, so s is a strictly increasing function of t. Consequently the function s is invertible. Hence, if $t(s)$ denotes the inverse, we can express the given function **F** in terms of arc length as follows:

$$\mathbf{G}(s) = \mathbf{F}(t(s))$$

The calculations involved in expressing a vector function in terms of arc length are, in general, quite difficult. Nonetheless, as we shall see, it is of theoretical value that it can be done. In particular, an interesting consequence of using arc length as a parameter is that the derivative with respect to s is always a *unit* tangent vector.

18.1.12 **Theorem** Let $\mathbf{F}(t)$ be a smooth vector function and let s represent arc length along the graph of $\mathbf{F}$. Then $d\mathbf{F}/ds$ is a unit tangent vector.

Proof: By the Chain Rule,

$$\frac{d\mathbf{F}}{dt} = \frac{d\mathbf{F}}{ds}\frac{ds}{dt}$$

Now

$$\frac{ds}{dt} = \|\mathbf{F}'(t)\| = \left\|\frac{d\mathbf{F}}{dt}\right\|$$

and, since $\mathbf{F}$ is smooth,

$$\left\|\frac{d\mathbf{F}}{dt}\right\| \neq 0$$

Solving the above equation for $d\mathbf{F}/ds$, we have

$$\frac{d\mathbf{F}}{ds} = \frac{\dfrac{d\mathbf{F}}{dt}}{\left\|\dfrac{d\mathbf{F}}{dt}\right\|}$$

Clearly $d\mathbf{F}/ds$ has unit length, and since it is a positive scalar multiple of the tangent vector, it is itself a tangent vector. ■

VELOCITY AND ACCELERATION

As noted earlier, one of the most useful applications of vector functions is in the study of motion. If $\mathbf{G}(t) = \langle x(t), y(t), z(t)\rangle$ is the position function of a particle moving in space, then $\mathbf{G}(0)$ is called the *initial position*. Many motions of interest take place in a plane—including, for example, the motion of a planet about the sun. In cases such as these our position function has the form $\mathbf{G}(t) = \langle x(t), y(t)\rangle$.

When using arc length as a parameter, we usually take $s(0) = 0$. Thus $s(t)$ measures the distance traveled *along the path* of the motion from the point $(x(0), y(0), z(0))$. Of course, we measure time from some convenient starting point, which may allow for negative t and hence negative s. For example, in describing the plane path of a boat being tracked by radar, we may take $t = 0$ seconds to represent the instant it is directly in front of the radar installation. Then $G(-2)$, the head of the position vector $\mathbf{G}(-2)$, represents its position 2 s prior to its arrival in front of the installation; see Fig. 18.12.

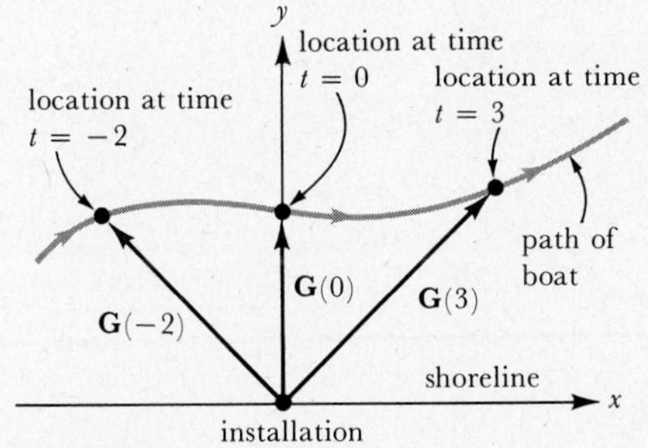

Fig. 18.12

We shall assume that the position functions describing motions are twice differentiable.

18.1.13 **Definition** Suppose $\mathbf{G}(t)$ is the position vector of a particle moving in space. (*a*) The vector $\mathbf{V}(t) = \mathbf{G}'(t)$ is called the *velocity vector* of the particle at time t. (*b*) The vector $\mathbf{A}(t) = \mathbf{V}'(t) = \mathbf{G}''(t)$ is called the *acceleration vector*.

By (18.1.11), $\|\mathbf{V}(t)\| = ds/dt$ is the *speed* of the particle. Thus the velocity vector points in the direction of the motion and has magnitude equal to the speed of the particle. This is completely analogous to the situation in Sec. 5.3 when we studied straight-line motion. In that setting the absolute value of the scalar velocity gives the speed, and the sign of the velocity determines the direction. The analogy between the acceleration vector $\mathbf{A}(t)$ for a motion in the plane or in space and the scalar acceleration $a(t)$ for a straight-line motion is not as direct. Both, of course, are obtained by differentiating the velocity. But, in general, $\|\mathbf{A}(t)\|$ will *not* give d^2s/dt^2, the scalar acceleration in the path. A more detailed description of motion and vector acceleration will be given in Sec. 18.3.

SECTION 18.1 EXERCISES

In Exercises 1 to 10 sketch the graph of the given vector function.

1. $F(t) = \langle 2 \sec t, 4 \tan^2 t\rangle$

2. $F(t) = \langle t^2, t^2 + t\rangle$

3. $F(t) = \langle -2 + \cos t, 3 + \sin t\rangle$

4. $F(t) = (2t + 1)\mathbf{i} + (3t - 6)\mathbf{j}$

5. $F(t) = (4 - 4t^4)\mathbf{i} + t^2\mathbf{j}$

6. $F(t) = \langle 2 \cos t, 2 \sin t, t\rangle$

7. $F(t) = \langle t \cos t, t \sin t, t\rangle$

8. $F(t) = \langle 2t^2, 4t, t^3\rangle$

9. $F(t) = (t + 2)\mathbf{i} + t\mathbf{j} + t^2\mathbf{k}$

10. $F(t) = e^t\mathbf{i} + e^{-t}\mathbf{j} + \sqrt{2}t\,\mathbf{k}$

In Exercises 11 to 13 determine whether $\mathbf{F}$ *is continuous at* $t = t_0$.

11. $F(t) = \begin{cases} \langle t^2 - 1, 2t + 7\rangle & 0 \le t \le 1 \\ \langle \pi/2 - \arcsin t, 9t\rangle & t > 1 \end{cases};\ t_0 = 1$

13. $F(t) = \begin{cases} \langle 4 \arctan t, 2t^2 - t + 1, 2\rangle & t \le 1 \\ \langle \pi/2 + \arcsin t, 2\sqrt{t}, 2t\rangle & t > 1 \end{cases};\ t_0 = 1$

12. $F(t) = \begin{cases} \langle (\sin t)/t, t \ln t\rangle & t > 0 \\ \langle 1, 0\rangle & t = 0 \end{cases};\ t_0 = 0$

In Exercises 14 to 20 determine $\mathbf{F}'(t)$ *and* $\mathbf{F}''(t)$.

14. $F(t) = \langle 2\sqrt{t}, t^3 + 6t\rangle$

15. $F(t) = \langle \sin \sqrt{t}, e^{t^2}\rangle$

16. $F(t) = \langle t \ln t, e^{3t}, t^2\rangle$

17. $F(t) = \langle \cos^2 t, 2 \tan t, t/(t + 1)\rangle$

18. $F(t) = \langle \csc^3 t, t^2 \cos t\rangle$

19. $F(t) = \ln(\sec t)\mathbf{i} + 5\mathbf{j}$

20. $F(t) = (t - \cos t)\mathbf{i} + t\mathbf{j} - (1/\sqrt{t})\mathbf{k}$

The line through $F(t_0)$ *and parallel to* $\mathbf{F}'(t_0)$ *is called the* tangent line *to* $\mathbf{F}$ *at* $F(t_0)$. *The plane through* $F(t_0)$ *with normal* $\mathbf{F}'(t_0)$ *is called the* normal plane *to* $\mathbf{F}$ *at* $F(t_0)$. *In Exercises 21 to 24 find equations for the tangent line and normal plane at the indicated point.*

21. $F(t) = \langle e^t, e^{-t}, \sqrt{2} + t\rangle$, $(1, 1, \sqrt{2})$

22. $F(t) = \langle t^2 + 1, 2 - t, t^2 - 2t\rangle$, $(2, 1, -1)$

23. $F(t) = \langle t \sin t, t \cos t, \sqrt{3}\, t\rangle$, $(\pi/2, 0, \sqrt{3}\pi/2)$

24. $F(t) = \langle (2/3)t^{3/2}, t, t\rangle$, $(18, 9, 9)$

In Exercises 25 to 30 determine $\mathbf{F}(t)$ *from the given information.*

25. $F'(t) = \langle 3t + 1, t^2 + 2\rangle$, $F(1) = \langle 3, -1\rangle$

26. $F'(t) = \langle e^t \sin t, e^t \cos t\rangle$, $F(0) = \langle 1, -1\rangle$

27. $F'(t) = \langle \cos t, \sin t, 4\pi t\rangle$, $F(\pi/2) = \langle 2, 0, \pi^3/3\rangle$

28. $F''(t) = \langle 6t, 12t^2 + 1\rangle$, $F(0) = \langle -2, 1\rangle$, $F'(0) = \langle 3, 2\rangle$

29. $F''(t) = \langle 0, -32\rangle$, $F(0) = \langle 0, 0\rangle$, $F'(0) = \langle 150{,}150\sqrt{3}\rangle$

30. $F''(t) = \langle 12t^2 + 8, 20t^3 + 2, 12\rangle$, $F(0) = \langle 0, 1, 0\rangle$, $F'(0) = \langle 16, 0, 4\rangle$

31. Suppose that $\mathbf{F}$ is three times differentiable. Show that

$$D_t[\mathbf{F} \cdot (\mathbf{F}' \times \mathbf{F}'')] = \mathbf{F} \cdot (\mathbf{F}' \times \mathbf{F}''')$$

32. Show that if a particle moves in space with constant speed, then the velocity and acceleration vectors must be perpendicular.

33. A particle moves in space in such a way that the position vector is always perpendicular to the velocity vector. Show that the particle moves along a sphere centered at the origin.

34. Find a vector function whose graph is the intersection of the surfaces $z = \sqrt{x^2 + y^2}$ and $y + z = 2$.

35. Suppose that the graphs of the two smooth curves $\mathbf{F}$ and $\mathbf{G}$ do not intersect. If $\|\mathbf{F} - \mathbf{G}\|$ has a minimum at $t = t_0$, show that $\mathbf{F}(t_0) - \mathbf{G}(t_0)$ is perpendicular to both $\mathbf{F}'(t_0)$ and $\mathbf{G}'(t_0)$.

Section 18.2: The Unit Tangent and Normal Vectors; Curvature

Objectives:

1. Determine the unit tangent and unit normal vectors for a vector function.

2. Determine the curvature for a plane or space curve.

INTRODUCTION

We saw in Chap. 4 the tremendous usefulness of first and second derivatives in graphing scalar functions. In Sec. 5.3 we used these derivatives to describe the motion of a particle traveling in a straight line. We now wish to extend each of these endeavors (*a*) to describe the geometrical characteristics of plane and space curves and (*b*) to describe *nonlinear* motion of particles in a plane or in space. In this section we concentrate on (a) and in the next section we concentrate on (b). We begin with the simplest case, describing curves in the plane.

PLANE CURVES: UNIT TANGENT AND NORMAL VECTORS

Given a plane curve, our aim is to construct a pair of perpendicular unit vectors *based on the curve*—in effect, to act as coordinate vectors. These will better enable us to describe the curve near that point. *We assume throughout this discussion that the vector function* $\mathbf{F}$ *is smooth with smooth derivatives.*

18.2.1 **Definition** If $\mathbf{F}$ is a vector function, then for each t the *unit tangent vector* at $\mathbf{F}(t)$, denoted $\mathbf{T}(t)$, is defined by

$$\mathbf{T}(t) = \frac{\mathbf{F}'(t)}{\|\mathbf{F}'(t)\|}$$

Note that $\mathbf{T}$ actually *is* a unit vector, that is, $\|\mathbf{T}\| = 1$. Also, since $\mathbf{T}$ is a positive scalar multiple of the tangent vector $\mathbf{F}'$, it is itself a tangent vector. As noted in (18.1.12), if s represents arc length along the curve, then

$$\frac{d\mathbf{F}(t)}{ds} = \mathbf{T}(t)$$

18.2.2 **Definition** For each t the *unit normal vector* at $\mathbf{F}(t)$, denoted $\mathbf{N}(t)$, is defined by

$$\mathbf{N}(t) = \frac{\mathbf{T}'(t)}{\|\mathbf{T}'(t)\|}$$

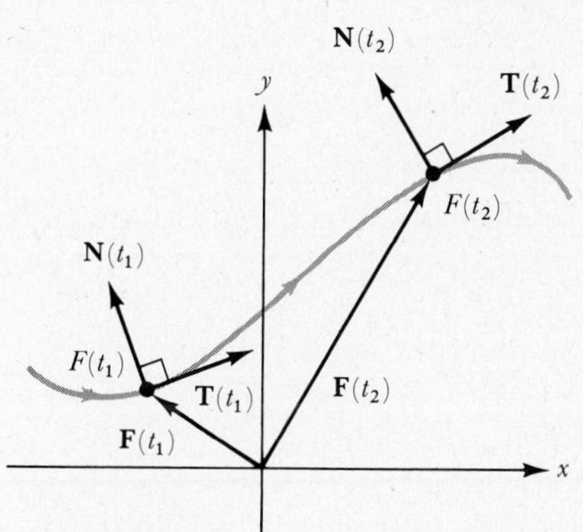

Fig. 18.13

Since we are assuming that $\mathbf{F}'$ has a smooth derivative, it follows that $\mathbf{T}$ is smooth. In particular, $\mathbf{T}'(t)$ exists and is $\neq \mathbf{0}$ for all t, and hence $\mathbf{N}(t)$ is defined for all t. Also, since $\|\mathbf{T}\| = 1$, we know from Example 4 of Sec. 18.1 that $\mathbf{T} \perp \mathbf{T}'$. But $\mathbf{N}$ has the same direction as $\mathbf{T}'$, and so we conclude:

18.2.3 $\qquad \mathbf{T}(t) \perp \mathbf{N}(t) \qquad$ for all t

When sketching $\mathbf{T}(t)$ and $\mathbf{N}(t)$ we use representatives whose base is on the curve at the point $F(t)$. Thus $\mathbf{T}$ and $\mathbf{N}$ are orthogonal unit vectors that move along the curve as t moves through the domain of $\mathbf{F}$ (see Fig. 18.13).

Example 1 Compute $\mathbf{T}(t)$ and $\mathbf{N}(t)$ for the vector function $\mathbf{F}(t) = \langle t^2, 2t\rangle$ and sketch their representatives at the points corresponding to $t = 0$ and $t = 1$.

Solution We have $\mathbf{F}'(t) = \langle 2t, 2\rangle$, so $\|\mathbf{F}'(t)\| = 2\sqrt{t^2+1}$ and hence

$$\mathbf{T}(t) = \left\langle \frac{t}{\sqrt{t^2+1}}, \frac{1}{\sqrt{t^2+1}} \right\rangle$$

Similarly,

$$\mathbf{T}'(t) = \left\langle \frac{1}{(t^2+1)^{3/2}}, \frac{-t}{(t^2+1)^{3/2}} \right\rangle$$

and

$$\|\mathbf{T}'(t)\| = \frac{1}{t^2+1}$$

so

$$\mathbf{N}(t) = \left\langle \frac{1}{\sqrt{t^2+1}}, \frac{-t}{\sqrt{t^2+1}} \right\rangle$$

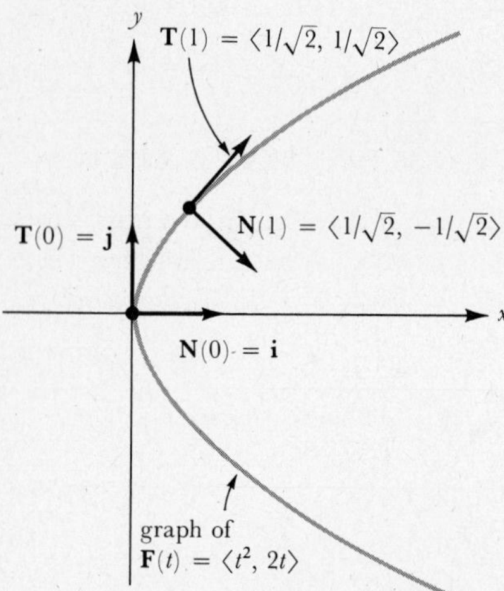

Fig. 18.14

Thus $\mathbf{F}(0) = \langle 0, 0\rangle$, $\mathbf{T}(0) = \langle 0, 1\rangle = \mathbf{j}$, $\mathbf{N}(0) = \langle 1, 0\rangle = \mathbf{i}$, and $\mathbf{F}(1) = \langle 1, 2\rangle$, $\mathbf{T}(1) = \langle 1/\sqrt{2}, 1/\sqrt{2}\rangle$, $\mathbf{N}(1) = \langle 1/\sqrt{2}, -1/\sqrt{2}\rangle$. These representatives are sketched in Fig. 18.14. ●

THE DIRECTION OF THE UNIT NORMAL

Since $\mathbf{T}(t)$ is a unit vector, we know from (17.3.15) that

$$(1) \qquad \mathbf{T}(t) = \langle \cos\phi(t), \sin\phi(t)\rangle$$

where $\phi(t)$ is the direction of $\mathbf{T}(t)$. Figure 18.15(a) shows a typical situation where $\phi(t)$ is increasing and Fig. 18.15(b) shows a situation where $\phi(t)$ is decreasing.

Differentiating (1), we have

$$\begin{aligned}\mathbf{T}'(t) &= \langle(-\sin\phi(t))\phi'(t), (\cos\phi(t))\phi'(t)\rangle \\ &= \phi'(t)\langle -\sin\phi(t), \cos\phi(t)\rangle\end{aligned}$$

Now since $\|\langle -\sin\phi(t), \cos\phi(t)\rangle\| = 1$, we know that $\|\mathbf{T}'(t)\| = |\phi'(t)|$. Let

$$\mathbf{u}(t) = \langle -\sin\phi(t), \cos\phi(t)\rangle$$

Then

$$(2) \qquad \mathbf{N}(t) = \frac{\mathbf{T}'(t)}{\|\mathbf{T}'(t)\|} = \frac{\phi'(t)\mathbf{u}(t)}{|\phi'(t)|} = \begin{cases}\mathbf{u}(t) & \text{if } \phi'(t) > 0 \\ -\mathbf{u}(t) & \text{if } \phi'(t) < 0\end{cases}$$

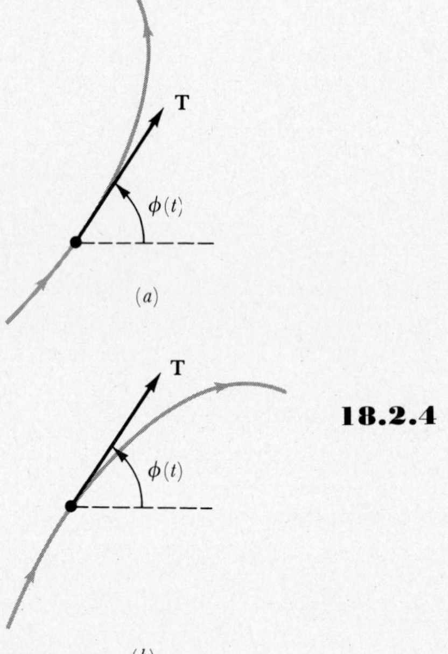

Fig. 18.15

Since

$$-\sin\phi(t) = \cos(\phi(t) + \pi/2) \qquad \text{and} \qquad \cos\phi(t) = \sin(\phi(t) + \pi/2)$$

we can conclude that $\mathbf{u}$ is always a unit vector 90° in advance of $\mathbf{T}$. That is, a counterclockwise rotation of $\mathbf{T}$ through 90° will coincide with $\mathbf{u}$. It follows that $-\mathbf{u}$ is 90° in retreat of $\mathbf{T}$. Thus from (2) we can conclude:

18.2.4 **Theorem** Suppose $\mathbf{F}$ is a vector function and $\phi(t)$ denotes the direction of $\mathbf{T}(t)$. Then
(a) $\phi'(t) > 0$ implies $\mathbf{N}(t)$ has direction $\phi(t) + \pi/2$.
(b) $\phi'(t) < 0$ implies $\mathbf{N}(t)$ has direction $\phi(t) - \pi/2$.

Theorem (18.2.4) is illustrated in Fig. 18.16. By applying (18.2.4) in a variety of situations, it can be shown that the representative of $\mathbf{N}$ always lies on the concave side (inside) of the curve. And the representatives of $\mathbf{N}$ and $\mathbf{T}$ are always on opposite sides of the curve. The situation at $F(t_2)$ in Fig. 18.13 thus could not occur.

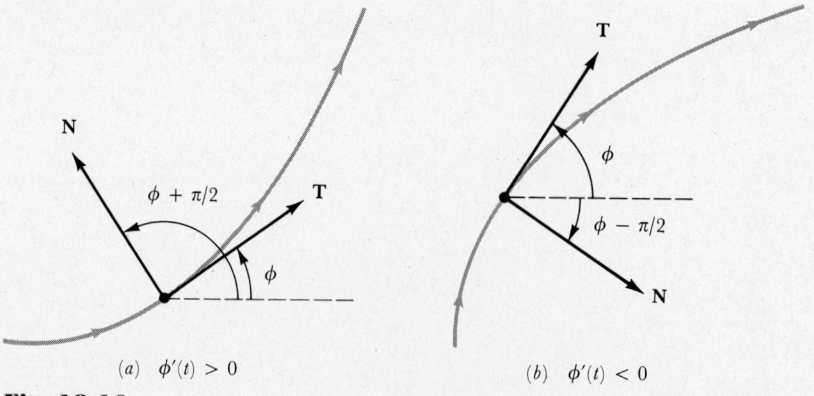

Fig. 18.16

PROGRESS TEST 1

Determine $\mathbf{T}(t)$ and $\mathbf{N}(t)$ for the given vector functions and sketch their representatives at the point corresponding to the indicated value of t.

1. $\mathbf{F}(t) = \langle \cos 2t, \sin 2t \rangle,\ t = \pi/8$

2. $\mathbf{F}(t) = \langle t - \sin t, 1 - \cos t \rangle,\ t = \pi$

CURVATURE

If an object is in motion and if no force acts on this object, then according to Newton's first law of motion its velocity is a constant; that is, it will move at a constant speed in a straight line. Therefore, for a particle to have nonconstant velocity, that is, to have either nonconstant speed or to move in other than a straight line, it must be subject to a force. We are especially interested in relating the *shape* of the path traveled by a particle and the forces necessary to keep the particle on that path. We shall give a quantitative definition of what we intuitively regard as the "sharpness" of a curve: the rate at which its direction changes.

18.2.5 **Definition** If $\mathbf{F}$ is a plane vector function, then the *curvature* of $\mathbf{F}$, $k(t)$, is defined by

$$k(t) = \left| \frac{d\phi}{ds} \right|$$

where $\phi(t)$ is the direction of the tangent vector $\mathbf{T}(t)$ and s represents arc length along the graph of $\mathbf{F}$.

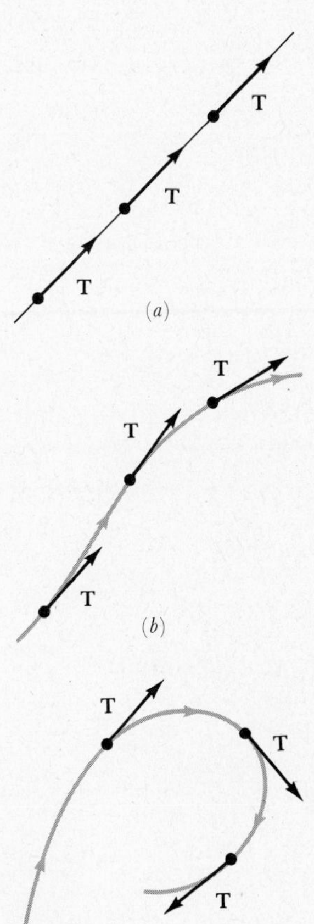

Fig. 18.17

This definition expresses exactly what we would wish: It measures the change in direction of the curve with respect to *distance along the curve* rather than with respect to t. This ensures that we are measuring a property of the curve itself rather than properties having to do with the way we move along the curve.

Now $\mathbf{T} = \langle \cos \phi, \sin \phi \rangle$, so by the Chain Rule

$$\frac{d\mathbf{T}}{ds} = \langle -\sin \phi, \cos \phi \rangle \frac{d\phi}{ds}$$

and hence

$$\left\| \frac{d\mathbf{T}}{ds} \right\| = \|\langle -\sin \phi, \cos \phi \rangle\| \left| \frac{d\phi}{ds} \right| = \left| \frac{d\phi}{ds} \right|$$

Thus an alternate formulation of curvature is given by

18.2.6
$$k(t) = \left\| \frac{d\mathbf{T}(t)}{ds} \right\|$$

If $\mathbf{F}$ is a straight line, then the curvature is zero and the unit tangent does not change direction as you move along the curve; see Fig. 18.17(a). If a curve is turning gradually, then $\mathbf{T}$ changes direction gradually, so k will be small; see Fig. 18.17(b). However, if the curve is turning sharply, then a small change in arc length produces a large change in the direction of $\mathbf{T}$ and thus k will be large; see Fig. 18.17(c).

The next formulas for $k(t)$ are in general easier to compute than (18.2.5) or (18.2.6) because they use the derivatives of $\mathbf{F}$ with respect to t.

18.2.7 **Theorem** Suppose C is the graph of $\mathbf{F}(t) = \langle x(t), y(t) \rangle$. Then

(a) $k(t) = \dfrac{\|\mathbf{T}'(t)\|}{\|\mathbf{F}'(t)\|}$

(b) $k(t) = \dfrac{|x'(t)y''(t) - y'(t)x''(t)|}{[(x'(t))^2 + (y'(t))^2]^{3/2}}$

Proof: (a) By the Chain Rule

$$\frac{d\mathbf{T}(t)}{ds} = \frac{d\mathbf{T}(t)}{dt}\frac{dt}{ds} = \frac{\mathbf{T}'(t)}{ds/dt} = \frac{\mathbf{T}'(t)}{\mathbf{F}'(t)}$$

and hence

$$k(t) = \left\|\frac{d\mathbf{T}(t)}{ds}\right\| = \frac{\|\mathbf{T}'(t)\|}{\|\mathbf{F}'(t)\|}$$

(b) Since ϕ gives the direction of the tangent vector, $\tan\phi$ gives the slope of the curve. Thus

$$\tan\phi = \frac{dy}{dx} = \frac{dy/dt}{dx/dt} = \frac{y'(t)}{x'(t)}$$

Differentiating with respect to t, we have

(3) $$(\sec^2\phi)\frac{d\phi}{dt} = \frac{x'(t)y''(t) - y'(t)x''(t)}{(x'(t))^2}$$

But $\sec^2\phi = 1 + \tan^2\phi = 1 + \left[\dfrac{y'(t)}{x'(t)}\right]^2 = \dfrac{[x'(t)]^2 + [y'(t)]^2}{[x'(t)]^2}$.

Substituting for $\sec^2\phi$ in (3) and solving for $d\phi/dt$, we get

(4) $$\frac{d\phi}{dt} = \frac{x'(t)y''(t) - y'(t)x''(t)}{[x'(t)]^2 + [y'(t)]^2}$$

Now

$$\frac{d\phi}{ds} = \frac{d\phi/dt}{ds/dt} = \frac{d\phi/dt}{\sqrt{[x'(t)]^2 + [y'(t)]^2}}$$

so we substitute $d\phi/dt$ from (4) into this latter expression. After simplifying and taking absolute values, we have

$$k(t) = \left|\frac{d\phi}{ds}\right| = \frac{|x'(t)y''(t) - y'(t)x''(t)|}{[(x'(t))^2 + (y'(t))^2]^{3/2}} \quad \blacksquare$$

By [18.2.7(*a*)] and definition (18.2.2),

$$k\mathbf{N} = \frac{\|\mathbf{T}'\|}{\|\mathbf{F}'\|}\frac{\mathbf{T}'}{\|\mathbf{T}'\|} = \frac{\mathbf{T}'}{\|\mathbf{F}'\|} = \frac{d\mathbf{T}/dt}{ds/dt} = \frac{d\mathbf{T}}{dt}\frac{dt}{ds} = \frac{d\mathbf{T}}{ds}$$

and thus we have:

18.2.8 $$\frac{d\mathbf{T}(t)}{ds} = k(t)\mathbf{N}(t)$$

Although [18.2.7(*b*)] has the most complicated appearance of the various formulas for curvature, it is usually the simplest to work with.

Example 2 Show that the curvature of a circle of radius r is constant and equals $1/r$.

Solution Since the position of the circle is irrelevant, we take as our vector function

$$\mathbf{F}(t) = \langle r\cos t, r\sin t\rangle$$

which has as its graph the circle of radius r centered at the origin. Now

$$x'(t) = -r\sin t \qquad y'(t) = r\cos t$$
$$x''(t) = -r\cos t \qquad y''(t) = -r\sin t$$

so, by [18.2.7(*b*)],

$$k(t) = \frac{|(-r\sin t)(-r\sin t) - (r\cos t)(-r\cos t)|}{[(-r\sin t)^2 + (r\cos t)^2]^{3/2}}$$
$$= \frac{|r^2(\sin^2 t + \cos^2 t)|}{[r^2(\sin^2 t + \cos^2 t)]^{3/2}} = \frac{r^2}{(r^2)^{3/2}} = \frac{1}{r} \quad \bullet$$

Example 2 suggests the following:

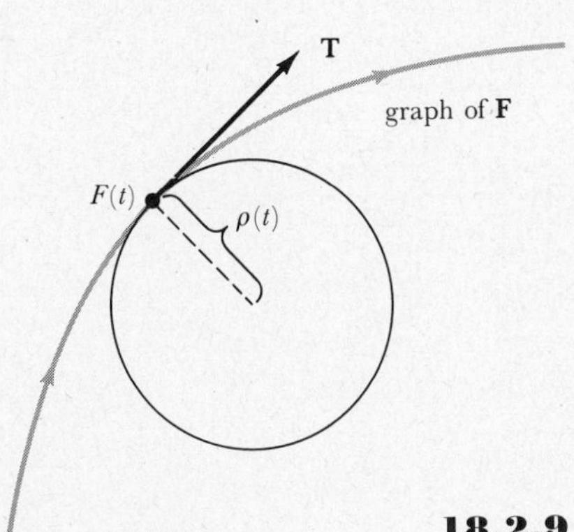

Fig. 18.18

18.2.9 **Definition** For $k(t) \neq 0$, we define the *radius of curvature* $\rho(t)$ by $\rho(t) = 1/k(t)$. The circle of radius $\rho(t)$ on the concave side of the graph of **F** and tangent to the curve at $F(t)$ is called the *circle of curvature*.

The circle of curvature may be thought of as the circle that "best fits" the curve at the given point (see Fig. 18.18).

Example 3 Find $k(t)$ for the vector function $\mathbf{F}(t) = \langle 2t, t^2\rangle$ and sketch the circle of curvature at the point $(0, 0)$.

Solution By [18.2.7(*b*)],

$$k(t) = \frac{|2(2) - 2t(0)|}{[(2)^2 + (2t)^2]^{3/2}} = \frac{1}{2\sqrt{1 + t^2}}$$

Therefore $k(0) = 1/2$ so $\rho(0) = 2$. The center of curvature is thus 2 units in the direction determined by $\mathbf{N}(0) = \mathbf{j}$, that is, at the point $(0, 2)$ on the y axis (see Fig. 18.19). $\bullet$

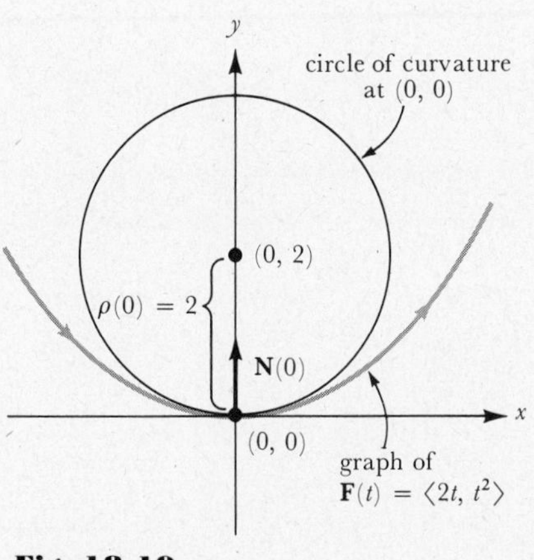

Fig. 18.19

If a curve is given in scalar functional form $y = f(x)$ [or $x = g(y)$], then we can apply [18.2.7(*b*)] to obtain the curvature as a function of the independent variable. (See Exercise 29.)

18.2.10 **Theorem** If f'' exists, then the curvature of the graph of f at the point $(x, f(x))$ is given by

$$k(x) = \frac{|f''(x)|}{[1 + (f'(x))^2]^{3/2}}$$

In Exercise 30 it is noted that the second derivative itself is an inappropriate measure of curvature of the graph of f. In Exercise 31 it is illustrated that the rate of change of the angle that the tangent line makes with the x axis is likewise an inadequate curvature measure.

For engineering purposes it is often of interest to know where a given curve has its sharpest "bend," that is, its maximum curvature.

Example 4 Determine the point on the graph of $y = \ln x$ where the curvature is a maximum.

Solution Let $y = \ln x$. Then $y' = 1/x$ and $y'' = -1/x^2$, so

$$k(x) = \frac{|-1/x^2|}{[1 + (1/x)^2]^{3/2}} = \left(\frac{x^2}{1 + x^2}\right)^{3/2} \frac{1}{x^2} = \frac{x}{(1 + x^2)^{3/2}}$$

and

$$k'(x) = \frac{(1 + x^2)^{3/2} - x(3/2)(1 + x^2)^{1/2} 2x}{(1 + x^2)^3}$$

$$= \frac{(1 + x^2)^{1/2}[1 + x^2 - 3x^2]}{(1 + x^2)^3} = \frac{2[(1/\sqrt{2}) - x][(1/\sqrt{2}) + x]}{(1 + x^2)^{5/2}}$$

An analysis of the sign of $k'(x)$ shows that a relative maximum occurs at $x = 1/\sqrt{2}$. Since this is the only extreme in $(0, +\infty)$, then $(1/\sqrt{2}, \ln(1/\sqrt{2}))$ must be the point where the maximum value of $k(x)$ occurs. ●

PROGRESS TEST 2

1. Find the curvature for the function $\mathbf{F}(t) = \langle t^3, t^2 \rangle$.
2. Determine the point on the curve $y = e^x$ where the curvature is a maximum.

SPACE CURVES AND THE MOVING TRIHEDRAL

As in two dimensions, the unit tangent and unit normal vectors for space curves are given, respectively, by

$$\mathbf{T}(t) = \frac{\mathbf{F}'(t)}{\|\mathbf{F}'(t)\|} \quad \text{and} \quad \mathbf{N}(t) = \frac{\mathbf{T}'(t)}{\|\mathbf{T}'(t)\|}$$

Since $\mathbf{T}$ and $\mathbf{N}$ are perpendicular, the angle between them is $\pi/2$. Thus $\mathbf{B} = \mathbf{T} \times \mathbf{N}$ is also a unit vector since, by (17.4.18),

$$\|\mathbf{T} \times \mathbf{N}\| = \|\mathbf{T}\| \, \|\mathbf{N}\| \sin\frac{\pi}{2} = 1$$

The vector $\mathbf{B}$ is called the *unit binormal*. The vectors $\mathbf{T}$, $\mathbf{N}$, and $\mathbf{B}$ are mutually orthogonal of unit length and form a right-hand triple. We refer to them as the *moving trihedral*. In effect they yield a coordinate system that is similar to that of the $\mathbf{i}, \mathbf{j}, \mathbf{k}$ vectors but "moves" along the curve.

Example 5 Determine $\mathbf{T}$, $\mathbf{N}$, and $\mathbf{B}$ for the circular helix $\mathbf{F}(t) = \langle a \cos t, a \sin t, bt \rangle$, and sketch their representatives at the points corresponding to $t = 0$ and $t = \pi/2$.

Solution Now $\mathbf{F}'(t) = \langle -a \sin t, a \cos t, b \rangle$, so

$$\|\mathbf{F}'(t)\| = \sqrt{a^2 \sin^2 t + a^2 \cos^2 t + b^2} = \sqrt{a^2 + b^2}$$

Thus

$$\mathbf{T}(t) = \left\langle \frac{-a \sin t}{\sqrt{a^2 + b^2}}, \frac{a \cos t}{\sqrt{a^2 + b^2}}, \frac{b}{\sqrt{a^2 + b^2}} \right\rangle$$

Similarly,

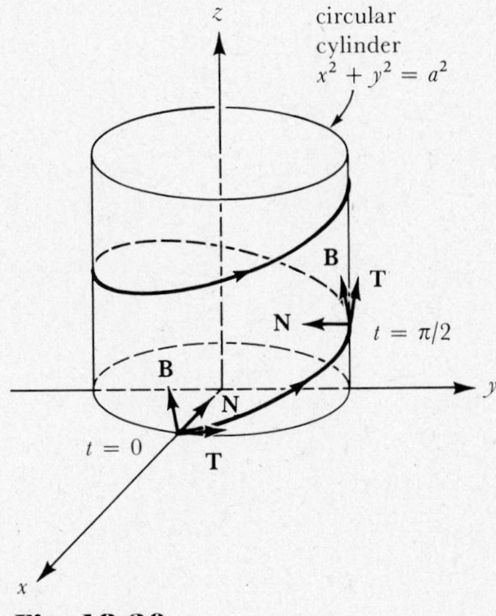

Fig. 18.20

$$\mathbf{T}'(t) = \left\langle \frac{-a\cos t}{\sqrt{a^2+b^2}}, \frac{-a\sin t}{\sqrt{a^2+b^2}}, 0 \right\rangle$$
$$= \frac{-a}{\sqrt{a^2+b^2}}\langle \cos t, \sin t, 0\rangle$$

so

$$\|\mathbf{T}'(t)\| = \frac{a}{\sqrt{a^2+b^2}} \qquad \text{and} \qquad \mathbf{N}(t) = \langle -\cos t, -\sin t, 0\rangle$$

Finally,

$$\mathbf{B}(t) = \mathbf{T}(t) \times \mathbf{N}(t) = \begin{vmatrix} \mathbf{i} & \mathbf{j} & \mathbf{k} \\ \dfrac{-a\sin t}{\sqrt{a^2+b^2}} & \dfrac{a\cos t}{\sqrt{a^2+b^2}} & \dfrac{b}{\sqrt{a^2+b^2}} \\ -\cos t & -\sin t & 0 \end{vmatrix}$$
$$= \left\langle \frac{b\sin t}{\sqrt{a^2+b^2}}, \frac{-b\cos t}{\sqrt{a^2+b^2}}, \frac{a}{\sqrt{a^2+b^2}} \right\rangle$$

The representatives at $t = 0$ and $t = \pi/2$ are sketched in Fig. 18.20. ●

We defined the curvature for *plane* curves as the absolute value of the rate of change of the direction of the tangent vector with respect to arc length. Since the direction of a vector in three dimensions cannot be expressed in terms of a single angle, this definition is no longer convenient. However, in (18.2.6) we showed that this definition is equivalent to $\|d\mathbf{T}/ds\|$. We use this formulation to define curvature in R_3.

18.2.11 **Definition** The *curvature* k of a space curve $\mathbf{F}$ is defined by

$$k(t) = \left\| \frac{d\mathbf{T}(t)}{ds} \right\|$$

As before, we have

18.2.12
$$\frac{d\mathbf{F}(t)}{ds} = \mathbf{T}(t) \qquad \text{and} \qquad \frac{d\mathbf{T}(t)}{ds} = k(t)\mathbf{N}(t)$$

The following gives us a computational version of curvature for a space curve. We omit its proof.

18.2.13 **Theorem** Suppose $\mathbf{F}(t)$ is a space curve. Then

$$k(t) = \frac{\|\mathbf{T}'(t)\|}{\|\mathbf{F}'(t)\|} = \frac{\|\mathbf{F}'(t) \times \mathbf{F}''(t)\|}{\|\mathbf{F}'(t)\|^3}$$

Example 6 Compute the curvature for the circular helix $\mathbf{F}(t) = \langle a\cos t, a\sin t, bt\rangle$.

Solution

$$\mathbf{F}'(t) \times \mathbf{F}''(t) = \langle -a\sin t, a\cos t, b\rangle \times \langle -a\cos t, -a\sin t, 0\rangle$$
$$= \langle ab\sin t, -ab\cos t, a^2\rangle$$

Thus

$$k(t) = \frac{\|\langle ab\sin t, -ab\cos t, a^2\rangle\|}{\|\langle -a\sin t, a\cos t, b\rangle\|^3}$$
$$= \frac{\sqrt{a^2b^2+a^4}}{(\sqrt{a^2+b^2})^3} = \frac{|a|\sqrt{a^2+b^2}}{(\sqrt{a^2+b^2})^3} = \frac{|a|}{a^2+b^2} \quad ●$$

Example 7 Find $k(t)$ for the vector function $\mathbf{F}(t) = \langle e^t \cos t, e^t \sin t, e^t \rangle$.

Solution

$$\mathbf{F}'(t) = \langle e^t(\cos t - \sin t), e^t(\sin t + \cos t), e^t \rangle$$
$$\mathbf{F}''(t) = \langle -2e^t \sin t, 2e^t \cos t, e^t \rangle$$

and

$$\mathbf{F}' \times \mathbf{F}'' = \begin{vmatrix} \mathbf{i} & \mathbf{j} & \mathbf{k} \\ e^t(\cos t - \sin t) & e^t(\sin t + \cos t) & e^t \\ -2e^t \sin t & 2e^t \cos t & e^t \end{vmatrix}$$
$$= e^{2t}\langle \sin t - \cos t, -\cos t - \sin t, 2 \rangle \quad \text{(Check!)}$$

Thus

$$\frac{\|\mathbf{F}' \times \mathbf{F}''\|}{\|\mathbf{F}'\|^3} = \frac{\|e^{2t}\langle \sin t - \cos t, -\cos t - \sin t, 2\rangle\|}{\|e^t\langle \cos t - \sin t, \sin t + \cos t, 1\rangle\|^3} = \left(\frac{\sqrt{2}}{3}\right)e^{-t} \quad \bullet$$

SECTION 18.2 EXERCISES

In Exercises 1 to 10:

(*a*) Find $\mathbf{T}(t)$ and $\mathbf{N}(t)$.
(*b*) Sketch a portion of the curve near $t = t_0$.
(*c*) Sketch representatives of $\mathbf{T}(t_0)$ and $\mathbf{N}(t_0)$ at $F(t_0)$.

1. $\mathbf{F}(t) = (t \cos t)\mathbf{i} + (t \sin t)\mathbf{j}$, $t_0 = 0$
2. $\mathbf{F}(t) = (3 - 2t^2)\mathbf{i} + (4 + 5t)\mathbf{j}$, $t_0 = 2$
3. $\mathbf{F}(t) = (2 \sin 3t)\mathbf{i} + (4 + 5t)\mathbf{j}$, $t_0 = \pi$
4. $\mathbf{F}(t) = \langle t - \sin t, 1 - \cos t\rangle$, $t_0 = \pi/4$
5. $\mathbf{F}(t) = \langle t - \sin t, 1 - \cos t\rangle$, $t_0 = \pi/6$
6. $\mathbf{F}(t) = \langle t^3, 3t\rangle$, $t_0 = 2$
7. $\mathbf{F}(t) = \langle 2t, e^t + e^{-t}\rangle$, $t_0 = 0$
8. $\mathbf{F}(t) = \langle \ln(\sin t), \ln(\cos t)\rangle$, $t_0 = \pi/4$
9. $\mathbf{F}(t) = \langle t^3/3 - t, t^2\rangle$, $t_0 = 1$
10. $\mathbf{F}(t) = \langle 2 - \cos t, 1 - \sin t\rangle$, $t_0 = \pi$

In Exercises 11 to 15:

(*a*) Find $\mathbf{T}(t)$, $\mathbf{N}(t)$ and $\mathbf{B}(t)$.
(*b*) Sketch a portion of the curve near $t = t_0$.
(*c*) Sketch representatives of $\mathbf{T}(t_0)$, $\mathbf{N}(t_0)$, and $\mathbf{B}(t_0)$ at $F(t_0)$.

11. $\mathbf{F}(t) = \langle e^t \cos t, e^t \sin t, e^t\rangle$, $t_0 = \pi/2$
12. $\mathbf{F}(t) = \langle 3t \cos t, 3t \sin t, 4t\rangle$, $t_0 = \pi/2$
13. $\mathbf{F}(t) = \langle t^2/2, t^3/3, t\rangle$, $t_0 = 1$
14. $\mathbf{F}(t) = (\cos t)\mathbf{i} + e^t\mathbf{j} + (t^2 + 2)\mathbf{k}$, $t_0 = 0$
15. $\mathbf{F}(t) = 4(\sin)2t\mathbf{i} + (2t - \sin 2t)\mathbf{j} + \cos(2t)\mathbf{k}$, $t_0 = \pi/2$

In Exercises 16 to 25 find the curvature. The results of earlier exercises may be useful.

16. $\mathbf{F}(t) = \langle \cos 2t, \sin 2t\rangle$
17. $\mathbf{F}(t) = \langle e^t \sin t, e^t \cos t\rangle$
18. $\mathbf{F}(t) = \langle t^3, 3t\rangle$
19. $\mathbf{F}(t) = \langle t^3, t^2\rangle$
20. $\mathbf{F}(t) = \langle t^3/3 - t, t^2\rangle$
21. $\mathbf{F}(t) = \langle a \cos b, b \sin t\rangle$
22. $\mathbf{F}(t) = \langle e^t, 2t, e^{-t}\rangle$
23. $\mathbf{F}(t) = \langle 1/t, t^3/3, t\rangle$
24. $f(x) = \sin x$
25. $g(y) = y^2/3$

26. Show that the curvature for a polar curve $r = f(\theta)$ is given by

$$k(\theta) = \frac{|2[f'(\theta)]^2 - f(\theta)f''(\theta) + [f(\theta)]^2|}{([f(\theta)]^2 + [f'(\theta)]^2)^{3/2}}$$

[*Hint:* Express $r = f(\theta)$ parametrically with parameter θ.]

27. Use Exercise 26 to find the curvature of the polar curves.
(*a*) $r = 2 + \sin\theta$ (*b*) $r = 4 \cos 2\theta$

28. For each of the following determine the point where the maximum curvature occurs.
(*a*) $y = x^2$ (*b*) $y = \sin x$, $-\pi \le x \le \pi$
(*c*) $\mathbf{F}(t) = \langle a \sin t, b \cos t\rangle$, $0 \le t \le 2\pi$

29. Prove (18.2.10). [*Hint:* Express $y = f(x)$ parametrically with $x = t$.]

30. Show that $|f''(x)|$ does not provide an adequate measure of curvature. [Consider $y = f(x) = ax^2 + bx + c$.]

31. (*a*) Let $\theta = \theta(x)$ be the angle of inclination of the tangent line to $y = f(x)$. Show

$$\left|\frac{d\theta}{dx}\right| = \frac{|f''(x)|}{\sqrt{1 + [f'(x)]^2}}$$

(*b*) Show that $d\theta/dx$ is not an adequate measure of curvature. [Consider $y = x^2$ at $(1, 1)$ and $y = -\sqrt{x}$ at $(1, -1)$; the curvature should be the same at these points.]

In Exercises 32 to 36 sketch the circle of curvature at the indicated point.

32. $F(t) = \langle 8t, 2 - 2t^2 \rangle, t = 1$

33. $F(t) = \langle t^2/2, t^3/3 \rangle, t = 1$

34. $F(t) = \langle e^t \cos t, e^t \sin t \rangle, t = \pi/2$

35. $y = \sin x, x = \pi/4$

36. $y = \ln x, x = 1$

As the circle of curvature moves along the graph of the vector function **F** its center traces out a curve called the *evolute* of **F**. By (18.2.9), the evolute of **F** is the graph of the vector function **C**, where

$$C(t) = F(t) + \rho(t)N(t)$$

37. Determine and sketch the evolute of the parabola $F(t) = \langle t, t^2 \rangle$.

38. Using the computations from Progress Test 1, show that the evolute of the cycloid $F(t) = \langle t - \sin t, 1 - \cos t \rangle$ is the cycloid $C(t) = \langle t - \sin t, \cos t - 1 \rangle$.

Section 18.3: Motion

Objectives:

1. Describe the motion of a particle when given the position function.
2. Determine the position function of a projectile from information about its velocity and acceleration.
3. Determine the tangential and normal components of acceleration.

INTRODUCTION

As mentioned when introducing vector functions, such functions provide an excellent means for describing the motion of a particle in the plane or in space. The position function of such a particle will be denoted by **G** and, as usual, the velocity and acceleration are denoted by **V** and **A** respectively. The graph of **G** is usually referred to as the *path* or *trajectory* of the particle. From Sec. 18.1 we recall

18.3.1 $$V(t) = G'(t)$$

18.3.2 $$A(t) = G''(t) = V'(t)$$

When using arc length, we shall assume it is measured from the initial position $G(0)$. That is,

18.3.3 $$s(t) = \int_0^t \|V(w)\|\, dw$$

Thus for $t \geq 0$, $s(t)$ acts as an odometer, giving the distance that the particle has traveled in its path at time t. Furthermore,

18.3.4 $$\frac{ds}{dt} = \|V(t)\|$$

gives the instantaneous speed of the particle. Now $T = V/\|V\|$ is a unit vector that points in the direction of the motion at any instant. By writing the velocity vector in the form

18.3.5 $$V = \left(\frac{ds}{dt}\right)T$$

we emphasize that the velocity is completely determined by the speed and direction of the particle.

Example 1 The position function of a particle moving in the plane is given by $G(t) = \langle 2t^2, t^4/4 - 1 \rangle$. Sketch the path and sketch the representatives of $V(1)$ and $A(1)$.

Solution Eliminating the parameter, we have $y = x^2/16 - 1$ and so the path is along this parabola, starting at $(0, -1)$ and moving up the right branch as t increases. The velocity and acceleration are given, respectively, by

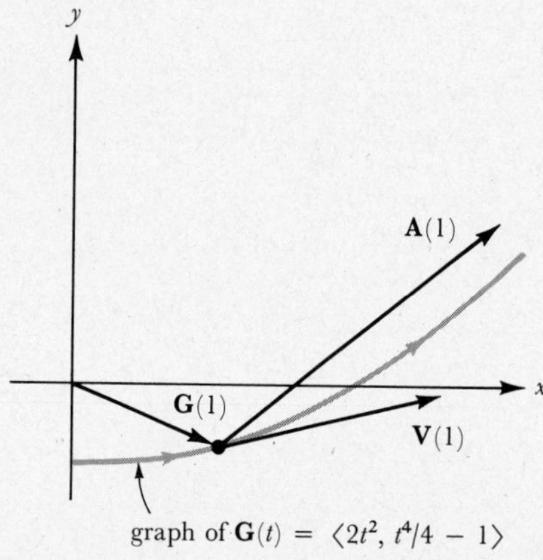

Fig. 18.21

$$\mathbf{V}(t) = \langle 4t, t^3 \rangle \qquad \text{and} \qquad \mathbf{A}(t) = \langle 4, 3t^2 \rangle$$

Thus $\mathbf{V}(1) = \langle 4, 1 \rangle$ and $\mathbf{A}(1) = \langle 4, 3 \rangle$.

The path and representatives of $\mathbf{V}(1)$ and $\mathbf{A}(1)$ are sketched in Fig. 18.21. ●

According to Newton's *second law of motion* the resultant (sum) of all the forces acting on a particle of mass m that is moving along some path in space is related to the acceleration of the particle by the vector equation

18.3.6 $$\mathbf{F} = m\mathbf{A}$$

Example 2 A particle of mass m has position function

$$\mathbf{G}(t) = \langle a \cos \omega t, b \sin \omega t \rangle$$

where a, b, and ω are positive constants.

(*a*) Describe the trajectory.

(*b*) Determine the velocity, speed, and acceleration.

(*c*) Determine the force acting on the particle.

Solutions (*a*) The path is the ellipse $x^2/a^2 + y^2/b^2 = 1$. The particle starts at $(a, 0)$ and travels in the counterclockwise direction.

(*b*) Now $\mathbf{V}(t) = \langle -a\omega \sin \omega t, b\omega \cos \omega t \rangle$ and $\mathbf{A}(t) = \langle -a\omega^2 \cos \omega t, -b\omega^2 \sin \omega t \rangle = -\omega^2 \mathbf{G}(t)$. The speed is given by

$$\|\mathbf{V}(t)\| = \omega \sqrt{a^2 \sin^2 \omega t + b^2 \cos^2 \omega t}$$

(*c*) By (18.3.6), $\mathbf{F} = m\mathbf{A} = -m\omega^2 \mathbf{G}$ ●

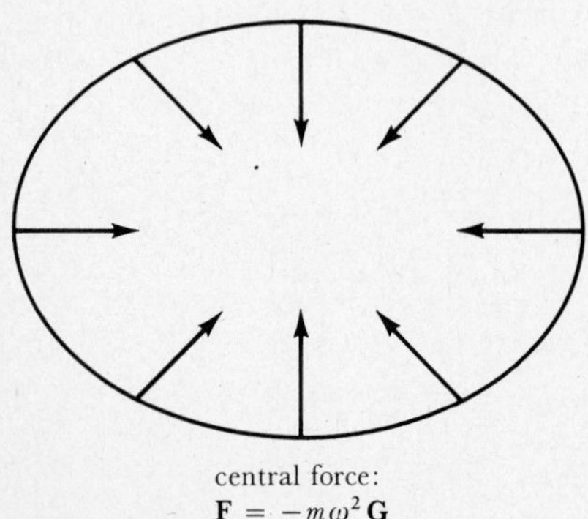

central force: $\mathbf{F} = -m\omega^2 \mathbf{G}$

Fig. 18.22

In the above example the force is proportional to the position vector. Any force of this type is called a *central force*. If, as in Example 2, the constant of proportionality is negative, then the force is called a *centripetal* or *attractive* force. The centripetal force of Example 2 is illustrated in Fig. 18.22. If $a = b$ (a circular trajectory), then this is the sort of force exerted on a ball being whipped in a circular path at the end of a piece of string held fixed at one end.

PROGRESS TEST 1

1. Deduce from (18.3.6) Newton's first law of motion: A particle subject to no forces will remain at rest or will travel in a straight line with constant velocity.

2. Show that the force acting on the particle with position function $\mathbf{G}(t) = \langle a \cosh \omega t, a \sinh \omega t \rangle$, for a, $\omega > 0$, is a central force.

THE PROJECTILE PROBLEM

Suppose a projectile of mass m is fired with an angle of elevation of α radians and with an initial speed of v_0 feet per second. We wish to determine the path of the projectile under the assumptions that the path lies in a plane and that the only force acting on the projectile is the force due to gravity. (We are neglecting forces due to wind and air resistance.) This situation generalizes the vertical velocity explored in Sec. 6.1.

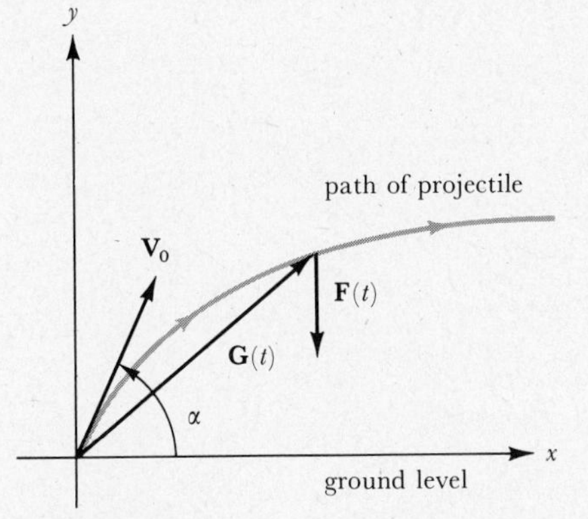

Fig. 18.23

We choose coordinates so that the x axis coincides with ground level. Suppose that $\mathbf{G}(t) = \langle x(t), y(t) \rangle$ denotes the position vector of the projectile with $\mathbf{G}(0) = \mathbf{0}$. That is, the projectile is fired at ground level from the origin. Our object is to determine $\mathbf{G}(t)$. If $\mathbf{V}(t)$ denotes the velocity of the projectile, then $\mathbf{V}_0 = \mathbf{V}(0)$ is the initial velocity. Since the initial speed is v_0 ft/s and the initial direction is α, by (17.3.15), page 569, we know that

$$\mathbf{V}(0) = \langle v_0 \cos \alpha, v_0 \sin \alpha \rangle$$

The force of gravity $\mathbf{F}(t)$ acting on the projectile is directed vertically downward, in the direction of $-\mathbf{j}$. It has magnitude mg, where g is the gravitational constant. Thus $\mathbf{F}(t) = -mg\mathbf{j} = \langle 0, -mg \rangle$; see Fig. 18.23.

By (18.2.6),

$$-mg\mathbf{j} = m\mathbf{A}$$

so

(1) $$\mathbf{A} = -g\mathbf{j}$$

But $\mathbf{A} = \mathbf{G}''$, so this is a vector differential equation that can be solved for $\mathbf{G}$. Integrating (1) we have

$$\mathbf{V}(t) = -gt\mathbf{j} + \mathbf{C}_1$$

Letting $t = 0$, we conclude that $\mathbf{C}_1 = \mathbf{V}(0) = \mathbf{V}_0$, so

(2) $$\mathbf{V}(t) = -gt\mathbf{j} + \mathbf{V}_0$$

Integrating again, we obtain

$$\mathbf{G}(t) = \frac{-gt^2}{2}\mathbf{j} + \mathbf{V}_0 t + \mathbf{C}_2$$

But $\mathbf{G}(0) = \mathbf{0} = \mathbf{C}_2$, so

(3) $$\mathbf{G}(t) = \frac{-gt^2}{2}\mathbf{j} + \mathbf{V}_0 t$$

We know that $\mathbf{V}_0 = \langle v_0 \cos \alpha, v_0 \sin \alpha \rangle$, so we can write (2) and (3) in component form to obtain:

18.3.7 $$\mathbf{G}(t) = \langle (v_0 \cos \alpha)t, (v_0 \sin \alpha)t - (g/2)t^2 \rangle$$

and

18.3.8 $$\mathbf{V}(t) = \langle v_0 \cos \alpha, v_0 \sin \alpha - gt \rangle$$

Suppose that $\alpha = \pi/2$. Then physically we are describing a vertically projected object. In such a case

$$\begin{aligned} \mathbf{G}(t) &= \langle (v_0)(0)t, (v_0)(1)t - (g/2)t^2 \rangle \\ &= \langle 0, v_0 t - (g/2)t^2 \rangle \end{aligned}$$

Hence $\mathbf{G}(t)$ is equivalent to a scalar function. In fact it is equivalent both in form and interpretation to the scalar *position-above-the-ground* function in (6.1.7). Similarly, the velocity function reduces to

$$\mathbf{V}(t) = \langle 0, v_0 - gt \rangle$$

which again agrees completely with the vertical velocity function (6.1.6). Henceforth, unless otherwise noted, we shall assume $0 < \alpha < \pi/2$.

Example 3 Show that the path of the projectile given by (18.3.7) is parabolic.

Solution By (18.3.7), $x(t) = (v_0 \cos \alpha)t$ and $y(t) = (v_0 \sin \alpha)t - (g/2)t^2$. Since $\cos \alpha \neq 0$, we can solve for t in terms of x,

$$t = \frac{x}{v_0 \cos \alpha}$$

and conclude that

$$\begin{aligned} y &= v_0 \sin \alpha \left(\frac{x}{v_0 \cos \alpha} \right) - \frac{g}{2} \left(\frac{x}{v_0 \cos \alpha} \right)^2 \\ &= (\tan \alpha)x - \left(\frac{g}{2v_0^2 \cos^2 \alpha} \right) x^2 \end{aligned}$$

Since α, v_0 and g are constants, this function has the form $y = Ax - Bx^2$, which is the equation of a concave-down parabola. ●

Example 4 A projectile is fired from ground level with an initial speed of 1000 ft/s and with an angle of elevation of $\pi/6$ rad. Determine: (*a*) the time when the projectile strikes the ground, (*b*) the range of the projectile, and (*c*) the speed of the projectile at the instant of impact.

Solution Since $v_0 = 1000$ and $\alpha = \pi/6$, by (18.3.7),

$$\mathbf{G}(t) = \langle 500\sqrt{3}t, 500t - (g/2)t^2 \rangle$$

(*a*) The time of impact will occur when

$$y(t) = t[500 - (g/2)t] = 0$$

But $t = 0$ corresponds to the initial position, so the projectile strikes the ground after $t = 1000/g = 31.25$ seconds; see Fig. 18.24.

(*b*) The range of the projectile is the horizontal distance traveled during the flight. Since the projectile will not strike the ground until $t = 31.25$, the range is

$$x(31.25) = 500\sqrt{3}(31.25) = 15{,}625\sqrt{3} \approx 27{,}063.3 \text{ ft}$$

(*c*) $\mathbf{V}(t) = \langle 500\sqrt{3}, 500 - gt \rangle$, so the velocity at impact is

$$\begin{aligned} \mathbf{V}(31.25) &= \langle 500\sqrt{3}, 500 - 32(31.25) \rangle \\ &= \langle 500\sqrt{3}, -500 \rangle \end{aligned}$$

The speed at impact is

$$\|\mathbf{V}(31.25)\| = \sqrt{(500\sqrt{3})^2 + (-500)^2} = 1000 \text{ ft/s}$$ ●

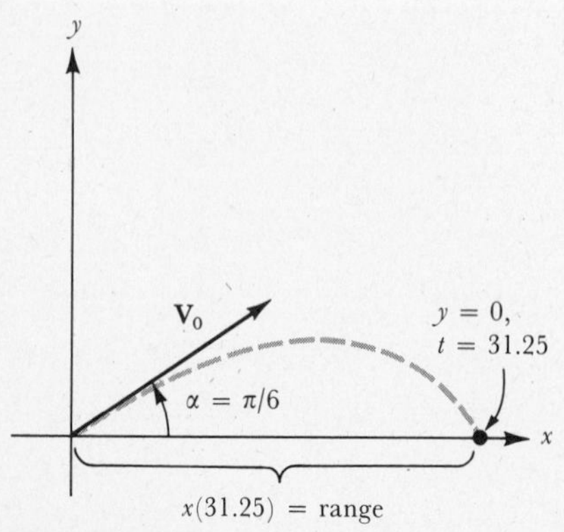

Fig. 18.24

PROGRESS TEST 2

1. A projectile is fired from ground level with initial speed v_0 feet per second at an angle of elevation of α radians.
   (*a*) Determine the time of impact.
   (*b*) Show that the terminal speed equals the initial speed.
   (*c*) Determine the maximum height the projectile attains.
   (*d*) If v_0 is fixed and α is allowed to vary, then determine the angle of elevation that will yield the maximum range.

TANGENTIAL AND NORMAL COMPONENTS OF ACCELERATION

Since acceleration is the derivative of velocity, it must reflect any changes in either the direction or the speed of the object. The next result shows very vividly how the acceleration is affected by such changes. Since $\mathbf{F} = m\mathbf{A}$, these changes must also be reflected in the force.

18.3.9 **Theorem** If $\mathbf{G}(t)$ denotes the position function of a particle moving in space, then

$$\mathbf{A}(t) = \left(\frac{d^2s}{dt^2}\right)\mathbf{T}(t) + k(t)\left(\frac{ds}{dt}\right)^2\mathbf{N}(t)$$

Proof: From (18.3.5), $\mathbf{V}(t) = (ds/dt)\mathbf{T}(t)$. Thus, differentiating with respect to t, we have

$$\mathbf{A}(t) = \left(\frac{d^2s}{dt^2}\right)\mathbf{T} + \frac{ds}{dt}\mathbf{T}'(t) \tag{4}$$

But

$$\mathbf{T}'(t) = \frac{d\mathbf{T}}{dt} = \frac{d\mathbf{T}}{ds}\frac{ds}{dt}$$
$$= k(t)\mathbf{N}(t)\frac{ds}{dt} \qquad \text{(by 18.2.12)}$$

Substitution of this latter expression into (4) gives the desired result. ■

The scalar functions

18.3.10
$$a_{\mathrm{T}} = \frac{d^2s}{dt^2} \qquad \text{and} \qquad a_{\mathrm{N}} = k\left(\frac{ds}{dt}\right)^2$$

are referred to as the *tangential* and *normal* components of acceleration, respectively.

Similarly, the vector functions

18.3.11
$$a_{\mathrm{T}}\mathbf{T} \qquad \text{and} \qquad a_{\mathrm{N}}\mathbf{N}$$

are called, respectively, the *tangential* and *normal acceleration.* The *tangential and normal forces* are obtained by multiplying each function in (18.3.11) by m.

Since $\mathbf{A} = a_{\mathrm{T}}\mathbf{T} + a_{\mathrm{N}}\mathbf{N}$ we have

$$\|\mathbf{A}\|^2 = (a_{\mathrm{T}}\mathbf{T} + a_{\mathrm{N}}\mathbf{N}) \cdot (a_{\mathrm{T}}\mathbf{T} + a_{\mathrm{N}}\mathbf{N})$$
$$= a_{\mathrm{T}}^2\mathbf{T}\cdot\mathbf{T} + 2a_{\mathrm{T}}a_{\mathrm{N}}\mathbf{T}\cdot\mathbf{N} + a_{\mathrm{N}}^2\mathbf{N}\cdot\mathbf{N}$$

But $\mathbf{T} \cdot \mathbf{T} = \mathbf{N} \cdot \mathbf{N} = 1$ and $\mathbf{T} \cdot \mathbf{N} = 0$, so $\|\mathbf{A}\|^2 = a_{\mathrm{T}}^2 + a_{\mathrm{N}}^2$, and hence

18.3.12

$$a_{\mathrm{N}} = \sqrt{\|\mathbf{A}\|^2 - a_{\mathrm{T}}^2}$$

Equation (18.3.12) is an excellent aid in computing a_{T} and a_{N}.

Example 5 Determine $a_{\mathrm{T}}(t)$ and $a_{\mathrm{N}}(t)$ for the motion with position function $\mathbf{G}(t) = \langle e^t \sin t, e^t \cos t\rangle$.

Solution Since $\mathbf{G}(t) = e^t\langle \sin t, \cos t\rangle$, we know that

$$\begin{aligned}\mathbf{V}(t) = \mathbf{G}'(t) &= e^t\langle \cos t, -\sin t\rangle + e^t\langle \sin t, \cos t\rangle \\ &= e^t\langle \cos t + \sin t, \cos t - \sin t\rangle\end{aligned}$$

and

$$\frac{ds}{dt} = \|\mathbf{V}(t)\| = |e^t|\sqrt{(\cos t + \sin t)^2 + (\cos t - \sin t)^2} = \sqrt{2}e^t$$

Thus

$$a_{\mathrm{T}}(t) = \frac{d^2s}{dt^2} = \sqrt{2}e^t$$

Now

$$\begin{aligned}\mathbf{A}(t) &= \mathbf{V}'(t) \\ &= e^t\langle -\sin t + \cos t, -\sin t - \cos t\rangle \\ &\quad + e^t\langle \cos t + \sin t, \cos t - \sin t\rangle = e^t\langle 2\cos t, -2\sin t\rangle\end{aligned}$$

Hence

$$\|\mathbf{A}(t)\| = |e^t|\sqrt{4\cos^2 t + 4\sin^2 t} = 2e^t$$

By (18.3.12),

$$a_{\mathrm{N}}(t) = \sqrt{(2e^t)^2 - (\sqrt{2}e^t)^2} = \sqrt{4e^{2t} - 2e^{2t}} = \sqrt{2}e^t \quad \bullet$$

× **Good Driving: Understanding the Laws of Motion** Our bodies experience the above tangential and normal force vectors riding in an automobile whenever we change our speed or direction. As the automobile speeds up, the back of our seat pushes us forward with a tangential force $ma_{\mathrm{T}}\mathbf{T}$, where m is our mass. Similarly, as we turn, the interior of the car forces us in the direction of the turn with a normal force $ma_{\mathrm{N}}\mathbf{N}$.

Let us now turn our attention to the normal force

$$\mathbf{F} = Ma_{\mathrm{N}}\mathbf{N} = Mk\left(\frac{ds}{dt}\right)^2\mathbf{N} \qquad (M = \text{mass of car})$$

acting on the car as we travel around a curve at constant speed. Assuming the car does not skid, this force acts on the car via friction where the tires meet the road. If the curve is sharp, then k and hence $\mathbf{F}$ must be large. If the speed ds/dt is large, then again $\mathbf{F}$ must be large. If both the curvature and the speed are large, then $\mathbf{F}$ will be very large. However, if the normal force necessary to turn the car exceeds the force available through friction, then the car skids. In particular, if the available force decreases because of rain, snow, ice, or bald tires, then, for a given curvature k, ds/dt must be kept small to prevent skidding.

Whether or not we choose to obey the statutory driving laws, we have no choice but to obey the laws of motion.

POLAR COMPONENTS OF ACCELERATION

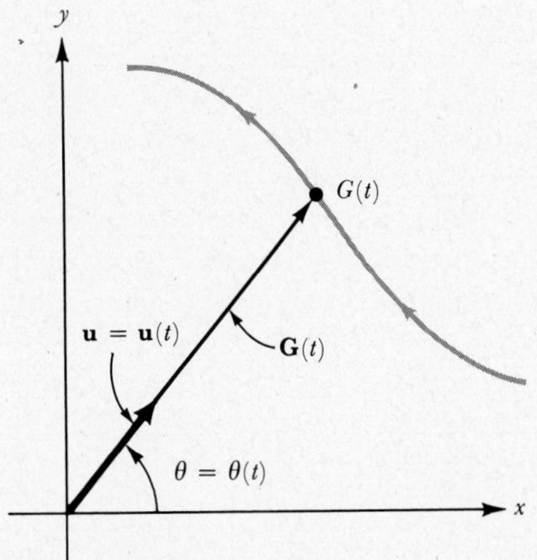

Fig. 18.25

Theorem (18.3.9) gives a decomposition of the acceleration into components that are tangent and normal to the path, that is, components that are parallel and perpendicular to the velocity vector. In the case of a plane motion, by working in polar coordinates we can obtain an equally useful decomposition of **A** into a pair of components, respectively parallel and perpendicular to the *position vector.*

Suppose $\mathbf{G}(t) = \langle x(t), y(t)\rangle$, and let $(r(t), \theta(t))$ denote polar coordinates for $(x(t), y(t))$. Then $\mathbf{u}(t) = \langle \cos\theta(t), \sin\theta(t)\rangle$ is a unit vector parallel to $\mathbf{G}(t)$. Since $\|\mathbf{G}(t)\| = |r(t)|$, we can write **G** in polar form as $\mathbf{G}(t) = r(t)\mathbf{u}(t)$; see Fig. 18.25.

If $\mathbf{v}(t) = \langle -\sin\theta(t), \cos\theta(t)\rangle$, then $\mathbf{v}(t)$ is a unit vector perpendicular to $\mathbf{u}(t)$ that points in the direction of advancing θ. Figure 18.26 shows the situation with representatives of **u, v** on the path. Now

$$\frac{d\mathbf{u}}{d\theta} = \langle -\sin\theta, \cos\theta\rangle = \mathbf{v} \quad \text{and} \quad \frac{d\mathbf{v}}{d\theta} = \langle -\cos\theta, -\sin\theta\rangle = -\mathbf{u}$$

so

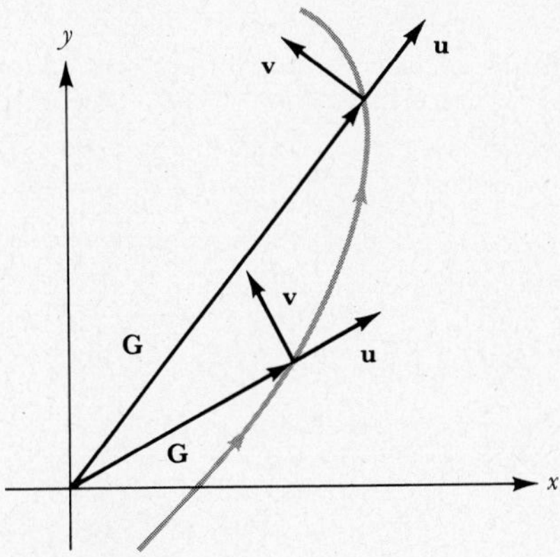

Fig. 18.26

$$\text{(5)} \qquad \frac{d\mathbf{u}}{dt} = \frac{d\mathbf{u}}{d\theta}\frac{d\theta}{dt} = \mathbf{v}\frac{d\theta}{dt} \quad \text{and} \quad \frac{d\mathbf{v}}{dt} = \frac{d\mathbf{v}}{d\theta}\frac{d\theta}{dt} = -\mathbf{u}\frac{d\theta}{dt}$$

Since $\mathbf{G} = r\mathbf{u}$,

$$\begin{aligned}\mathbf{V} = \frac{d\mathbf{G}}{dt} &= r\frac{d\mathbf{u}}{dt} + \frac{dr}{dt}\mathbf{u} \\ &= r\frac{d\theta}{dt}\mathbf{v} + \frac{dr}{dt}\mathbf{u} \qquad \text{[by (5)]}\end{aligned}$$

Differentiating and again, using (5), we have

$$\begin{aligned}\mathbf{A} &= r\frac{d^2\theta}{dt^2}\mathbf{v} + \frac{dr}{dt}\frac{d\theta}{dt}\mathbf{v} + r\frac{d\theta}{dt}\frac{d\mathbf{v}}{dt} + \frac{d^2r}{dt^2}\mathbf{u} + \frac{dr}{dt}\frac{d\mathbf{u}}{dt} \\ &= r\frac{d^2\theta}{dt^2}\mathbf{v} + \frac{dr}{dt}\frac{d\theta}{dt}\mathbf{v} - r\left(\frac{d\theta}{dt}\right)^2\mathbf{u} + \frac{d^2r}{dt^2}\mathbf{u} + \frac{dr}{dt}\frac{d\theta}{dt}\mathbf{v}\end{aligned}$$

Grouping terms, we obtain

18.3.13
$$\mathbf{A} = \left[r\frac{d^2\theta}{dt^2} + 2\frac{dr}{dt}\frac{d\theta}{dt}\right]\mathbf{v} + \left[\frac{d^2r}{dt^2} - r\left(\frac{d\theta}{dt}\right)^2\right]\mathbf{u}$$

KEPLER'S LAWS OF PLANETARY MOTION

Early in the seventeenth century the German astronomer Johann Kepler used a detailed study of data collected by the Danish astronomer Tycho Brahe to conclude that:

I. The planets move in planar elliptical orbits about the sun, with the sun located at a focus of the ellipse.

II. A planet's position vector based at the sun sweeps out area at a constant rate. (See Fig. 18.27, where the areas of the shaded regions are equal. This implies that the planet takes the same time in going from P to Q as from R to S.)

III. The ratio $r_a{}^3/T^2$ is the same for all planets, where r_a is the average of a planet's shortest and longest distances from the sun and T is the time necessary to complete an orbit.

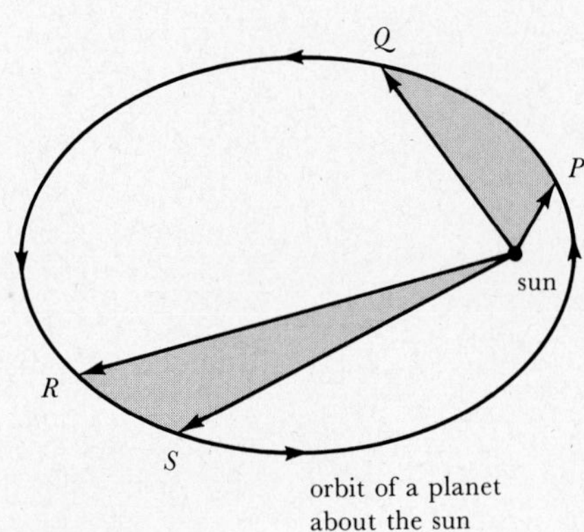

Fig. 18.27

In the previous century the Polish astronomer Copernicus had switched the roles of the earth and the sun as the center of the universe, thereby overthrowing the 1500-year-old Ptolemaic view that the sun revolved about the earth. However, it was Kepler who saw clearly (but without a telescope) the elliptical motion of the planets. Then, approximately 60 years later, Newton followed Kepler's giant step by perhaps an even greater step when he showed how these laws follow from a universal inverse square law of gravitational attraction. We shall show how (18.3.13) and the inverse square law imply conclusion II above.

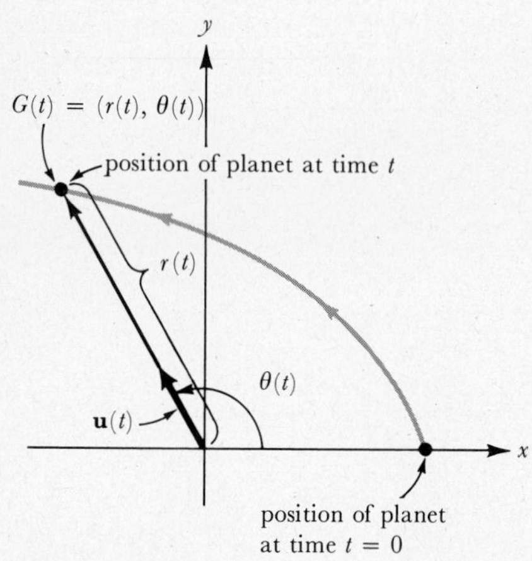

Fig. 18.28

We set up our coordinate system so that the sun is at the origin. Let $\mathbf{G}(t)$ denote the position of the planet and assume $\theta(0) = 0$. Then, as above, we have $\mathbf{G} = r\mathbf{u}$ (see Fig. 18.28). Now Newton's law of gravitational attraction asserts that the force acting on the planet is a central force. Thus $\mathbf{F} = c\mathbf{G} = cr\mathbf{u}$. But $\mathbf{F} = m\mathbf{A}$, so $\mathbf{A}$ is a scalar times $\mathbf{u}$. By (18.3.13), the only way this can happen is if the coefficient of $\mathbf{v}$ is zero. That is,

(6) $$r^2\frac{d^2\theta}{dt^2} + 2\frac{dr}{dt}\frac{d\theta}{dt} = 0$$

Now let $w = d\theta/dt$. Then (6) becomes

$$r\frac{dw}{dt} + 2w\frac{dr}{dt} = 0$$

so

$$r\,dw + 2w\,dr = 0$$

and

$$\frac{dw}{w} = \frac{-2dr}{r}$$

Integrating, we have

$$\ln|w| = -2\ln|r| + \ln|k|$$

so

$$\ln|r^2w| = \ln|k|$$

It follows that

(7) $$r^2w = r^2\frac{d\theta}{dt} \qquad \text{is constant}$$

Now if $A = A(t)$ denotes the area (see Fig. 18.29) swept out by the position vector from time $t = t_0$ to time t and $\theta_0 = \theta(t_0)$, $\theta = \theta(t)$, then, by (13.3.2), our formula for area in polar coordinates, page 448, we have

$$A = \frac{1}{2}\int_{\theta_0}^{\theta} r^2\,d\theta$$

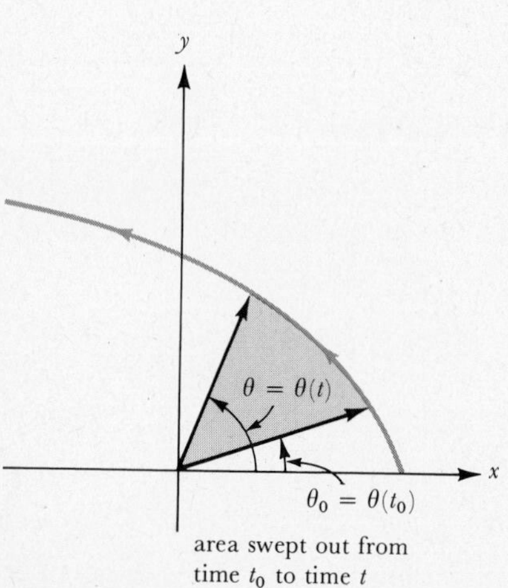

Fig. 18.29

But $dA/d\theta = \frac{1}{2}r^2$ by the Fundamental Theorem of Calculus, so by the Chain Rule

$$\frac{dA}{dt} = \frac{dA}{d\theta}\frac{d\theta}{dt} = \frac{1}{2}r^2\frac{d\theta}{dt}$$

which, by (7), is constant. Since dA/dt is the derivative with respect to time of the area swept out, Kepler's second law is established.

SECTION 18.3 EXERCISES

In Exercises 1 to 10 find the velocity, acceleration, and speed of the particle whose position function is given by $\mathbf{G}(t)$.

1. $\mathbf{G}(t) = \langle t^2 + 1, 2t - 3, t \rangle$

2. $\mathbf{G}(t) = \langle e^{-t}, e^{2t} \rangle$

3. $\mathbf{G}(t) = \langle \ln \sin t, \ln \cos t \rangle$

4. $\mathbf{G}(t) = \langle 2 \sin 2t, 2 \cos 2t, t \rangle$

5. $\mathbf{G}(t) = \langle 3 \sin t, 4 \cos t, t \rangle$

6. $\mathbf{G}(t) = \langle t \cos t, t \sin t \rangle$

7. $\mathbf{G}(t) = \langle e^{-t} \sin t, e^{-t} \cos t \rangle$

8. $\mathbf{G}(t) = \langle t \cos t + \sin t, \sin t - t \cos t \rangle$

9. $\mathbf{G}(t) = \langle e^t + e^{-t}, e^t - e^{-t} \rangle$

10. $\mathbf{G}(t) = \langle 3t, 7t - t^2, 2t \rangle$

In Exercises 11 to 15 determine the position function $\mathbf{G}(t)$ *from the given information.*

11. $\mathbf{V}(t) = \langle 3 - 2t, 4 + 3t \rangle$, $\mathbf{G}(0) = \langle 1, -1 \rangle$

12. $\mathbf{V}(t) = \langle t^2 + 1, 3t^3 \rangle$, $\mathbf{G}(0) = \langle -4, 1 \rangle$

13. $\mathbf{A}(t) = \langle 0, -8 \rangle$, $\mathbf{V}(0) = \langle 3, 1 \rangle$, $\mathbf{G}(0) = \langle 0, 0 \rangle$

14. $\mathbf{A}(t) = \langle 3 \sin t, 3 \cos t \rangle$, $\mathbf{V}(0) = \langle 1, -1 \rangle$, $\mathbf{G}(0) = \langle 3, 2 \rangle$

15. $\mathbf{A}(t) = \left\langle \dfrac{1}{(t + e)^2}, 12t^2 \right\rangle$, $\mathbf{V}(0) = \langle 2, 2 \rangle$, $\mathbf{G}(0) = \langle 1, 3 \rangle$

In Exercises 16 to 20 find the normal and tangential components of acceleration for the given motion.

16. $\mathbf{G}(t) = \langle 3 \sin t, 3 \cos t \rangle$

17. $\mathbf{G}(t) = \langle \sin(t^2), \cos(t^2) \rangle$

18. $\mathbf{G}(t) = \langle 4t, t^2, 2t^2 \rangle$

19. $\mathbf{G}(t) = \langle e^{-t}, e^t \rangle$

20. $\mathbf{G}(t) = \langle e^t \cos t, e^t \sin t, t \rangle$

21. A projectile is fired from a point k feet above the ground with an initial speed of v_0 ft/s and at an angle of elevation α with the horizontal. Determine the position function for the projectile.

22. J. N. is an excellent golfer. When he swings full with any iron he imparts an initial speed of approximately 176 ft/s on the golf ball. When he uses a one-iron the angle of elevation at which the ball leaves the ground is approximately 30°. How many yards does J. N. hit a one-iron?

23. Higher number irons are known to impart a higher angle of elevation. J. N. (see Exercise 22) can hit a wedge (10-iron) 120 yd. What angle of elevation does his wedge impart the golf ball?

24. Dumbo, the human cannonball, is fired from his cannon with an initial speed of $80(\sqrt{10/3})$ ft/s. He has high hopes that he will land on a special cushion located 200 ft down range. What angles of elevation are possible to ensure Dumbo's safe return?

25. A building used to house Dumbo's circus (see Exercise 24) has a flat ceiling 75 ft high. Dumbo still wishes to land on his cushion 200 ft away, but he has no desire to go through the ceiling in the process. What angle of elevation should he instruct his helpers to use?

26. On the 4th of July a special firecracker with an adjustable time fuse is fired from the top of a building 100 ft high with an initial velocity of 480 ft/s and at an angle of elevation of $\pi/6$ rad with the horizontal. The fuse is to be set so that the firecracker will explode at the instant it reaches its highest point. For how many seconds should the fuse be set? If the fuse malfunctions and the firecracker does not explode, how many feet from the base of the building will it land?

In Exercises 27 to 30 find the polar components of acceleration for the motion of the particle with position function $\mathbf{G}(t) = \langle r(t), \theta(t) \rangle$.

27. $r(t) = 2 \cos 4t, \theta(t) = 2t$

28. $r(t) = 1 - \cos 3t, \theta(t) = 3t$

29. $r(t) = 2 \sin 2t^2, \theta(t) = t^2$

30. $r(t) = e^{2t}, \theta(t) = 2t$

CHAPTER EXERCISES

Review Exercises

In Exercises 1 to 10 determine $\mathbf{F}'(t)$ *and* $\mathbf{F}''(t)$.

1. $\mathbf{F}(t) = \langle t^2 - \ln t, t - 3t^3 \rangle$

2. $\mathbf{F}(t) = \langle te^{-t}, \cos(t^2) \rangle$

3. $\mathbf{F}(t) = \langle 4t - \sqrt{t}, (\ln t)^2 \rangle$

4. $\mathbf{F}(t) = [t/(t^2 + 1)]\mathbf{i} + (\arctan t)\mathbf{j}$

5. $\mathbf{F}(t) = (\arcsin t)\mathbf{i} - \csc(t^2)\mathbf{j}$

6. $\mathbf{F}(t) = \langle \sin 2t, \cos \sqrt{t}, t^2 \rangle$

7. $\mathbf{F}(t) = \langle \sqrt{t^2 + 1}, 1/t, e^{t^2} \rangle$

8. $\mathbf{F}(t) = \langle t + 1/\sqrt{t}, 4 - 3/t, 4t^3 \rangle$

9. $\mathbf{F}(t) = \langle \sqrt{\sin t}, \ln(t^2 + 2), e^{\sqrt{t}} \rangle$

10. $\mathbf{F}(t) = (t^2 - 4/t)\mathbf{i} + (1/t^2 - \sqrt{t})\mathbf{k}$

In Exercises 11 to 20 determine $\mathbf{F}(t)$ *from the given information.*

11. $\mathbf{F}'(t) = \langle 12t^2 - 4t, -25t^4\rangle$, $\mathbf{F}(0) = \langle -3, 5\rangle$

12. $\mathbf{F}'(t) = \langle 1/\sqrt{1-t^2}, t\sqrt{1-t^2}\rangle$, $\mathbf{F}(0) = \langle \pi/4, 2/3\rangle$

13. $\mathbf{F}'(t) = \langle 1/(t+1), e^t\rangle$, $\mathbf{F}(0) = \langle 4, -3\rangle$

14. $\mathbf{F}''(t) = \langle 12t, 24(1-t^2)\rangle$, $\mathbf{F}(0) = \langle 3, 1\rangle$, $\mathbf{F}'(0) = \langle 2, 0\rangle$

15. $\mathbf{F}''(t) = \langle \sin t, 18t^2\rangle$, $\mathbf{F}(0) = \langle 2, -1\rangle$, $\mathbf{F}'(0) = \langle 4, 3\rangle$

16. $\mathbf{F}'(t) = \langle 3t^2, \sin t, \sec t\rangle$, $\mathbf{F}(\pi/4) = \langle 0, 1, 2\rangle$

17. $\mathbf{F}'(t) = \langle 4t+2, t^2-3t, 4\rangle$, $\mathbf{F}(0) = \langle 2, 1, 4\rangle$

18. $\mathbf{F}'(t) = \langle 1/(t^2+1), 1/\sqrt{1-t^2}, \sqrt{t^2+1}\rangle$, $\mathbf{F}(0) = \langle 3, 4, 7\rangle$

19. $\mathbf{F}''(t) = \langle 18t, \sin t, 12t^2\rangle$, $\mathbf{F}(0) = \langle 0, 0, 0\rangle$, $\mathbf{F}'(0) = \langle 2, 3, 1\rangle$

20. $\mathbf{F}''(t) = \langle 3t^2 - 2t, 1/(t+1)^2, 4t\rangle$, $\mathbf{F}(0) = \langle 0, 0, 0\rangle$, $\mathbf{F}'(0) = \langle 2, 1, -1\rangle$

In Exercises 21 to 30 sketch the graph of $\mathbf{F}$.

21. $\mathbf{F}(t) = \langle t^2, 3t-6\rangle$

22. $\mathbf{F}(t) = \langle t^2, t^6\rangle$

23. $\mathbf{F}(t) = \langle e^t \sin t, e^t \cos t\rangle$, $0 \le t \le 2\pi$

24. $\mathbf{F}(t) = \langle \cos^3 t, \sin^3 t\rangle$, $0 \le t \le 2\pi$

25. $\mathbf{F}(t) = \langle \sec t, \tan^2 t\rangle$, $-\pi/2 < t < \pi/2$

26. $\mathbf{F}(t) = \langle 4\sin 2t, 3\cos 2t, t\rangle$, $0 \le t \le 2\pi$

27. $\mathbf{F}(t) = \langle 2\cos t, 4\sin t, 2t\rangle$, $0 \le t \le 2\pi$

28. $\mathbf{F}(t) = \langle t\cos t, t, t\sin t\rangle$, $0 \le t \le 2\pi$

29. $\mathbf{F}(t) = 2t\mathbf{i} + t^3\mathbf{j} + 3t^2\mathbf{k}$, $-1 \le t \le 1$

30. $\mathbf{F}(t) = 2(\sin t)\mathbf{i} + e(\cos t)\mathbf{j} + 4(\sin t)\mathbf{k}$, $0 \le t \le \pi/2$

In Exercises 31 to 35 determine $\mathbf{T}(t)$ *and* $\mathbf{N}(t)$ *and sketch their representatives at* $t = t_0$.

31. $\mathbf{F}(t) = \langle 3\cos 2t, 3\sin 2t\rangle$, $t_0 = \pi/4$

32. $\mathbf{F}(t) = \langle 4-3t, t^2\rangle$, $t_0 = 1$

33. $\mathbf{F}(t) = \langle \sqrt{t}, t^2\rangle$, $t_0 = 1$

34. $\mathbf{F}(t) = \langle t^2/2, t^3/3\rangle$, $t_0 = 0$

35. $\mathbf{F}(t) = \langle e^{-t}, e^t\rangle$, $t_0 = 0$

In Exercises 36 to 40 determine $\mathbf{T}(t)$, $\mathbf{N}(t)$, *and* $\mathbf{B}(t)$ *and sketch their representatives at* $t = t_0$. *Determine* $k(t_0)$.

36. $\mathbf{F}(t) = 6(\sin 2t)\mathbf{i} + 6(\cos 2t)\mathbf{j} + 5t\mathbf{k}$, $t_0 = \pi/2$

37. $\mathbf{F}(t) = (2-3t)\mathbf{i} + 4t\mathbf{j} + 2t^{3/2}\mathbf{k}$, $t_0 = 1$

38. $\mathbf{F}(t) = \langle e^t, e^t\cos t, e^t \sin t\rangle$, $t_0 = \pi$

39. $\mathbf{F}(t) = \langle 4\sin t, 2t - \sin 2t, \cos 2t\rangle$, $t_0 = \pi/2$

40. $\mathbf{F}(t) = \langle t^2, t^3, t\rangle$, $t_0 = 0$

In Exercises 41 to 50 determine $k(t)$.

41. $\mathbf{F}(t) = \langle \cos 3t, \sin 3t\rangle$

42. $\mathbf{F}(t) = \langle 3\cos t, 5\sin t\rangle$

43. $\mathbf{F}(t) = \langle 2t+4, t^2-2\rangle$

44. $\mathbf{F}(t) = \langle 2t\sin t, 2t\cos t\rangle$

45. $\mathbf{F}(t) = \langle \cos t + t\sin t, \sin t - t\cos t\rangle$

46. $y = x^4/4 + 1/(8x^2)$

47. $y = \sqrt{x}$

48. $y = e^{-x}$

49. $y = 1/x$

50. $y = 2x^2 + x - 1$

In Exercises 51 to 57 determine the tangential and normal components of acceleration.

51. $\mathbf{G}(t) = \langle 3\cos t, 4\sin t\rangle$

52. $\mathbf{G}(t) = \langle 2t+4, t^2-2\rangle$

53. $\mathbf{G}(t) = \langle t^2, \sqrt{t}\rangle$

54. $\mathbf{G}(t) = \langle t\sin t, t\cos t\rangle$

55. $\mathbf{G}(t) = \langle \cos t + t\sin t, \sin t - t\cos t\rangle$

56. $\mathbf{G}(t) = \langle t, t^2, t^3\rangle$

57. $\mathbf{G}(t) = \langle 4\cos t, 9\sin t, t\rangle$

In Exercises 58 to 62 sketch the circle of curvature at the indicated point.

58. $y = x^4/4$, $x = 1$

59. $y = x^3/3$, $x = 1$

60. $\mathbf{F}(t) = \langle 2t^2, 2t\rangle$, $t = 0$

61. $\mathbf{F}(t) = \langle 1/t, t\rangle$, $t = 1$

62. $\mathbf{F}(t) = \langle 2\cos t, 3\sin t\rangle$, $t = \pi/2$

In Exercises 63 to 65 determine the evolute of the indicated vector function. (For a definition of the evolute of a function see the discussion following Exercise 36 in Sec. 18.2).

63. $\mathbf{F}(t) = \langle 2t^2, 2t\rangle$

64. $\mathbf{F}(t) = \langle 1/t, t\rangle$

65. $\mathbf{F}(t) = \langle 2\cos t, 3\sin t\rangle$

Miscellaneous Exercises

1. Show that if $\mathbf{F}$ is twice differentiable, then $[\mathbf{F}\cdot\mathbf{F}']' = \|\mathbf{F}\|^2 + \mathbf{F}\cdot\mathbf{F}''$.

2. Suppose that $\mathbf{F}$, $\mathbf{G}$, $\mathbf{H}$ are differentiable. Show that

$$[\mathbf{F}\cdot(\mathbf{G}\times\mathbf{H})]' = \mathbf{F}'\cdot(\mathbf{G}\times\mathbf{H}) + \mathbf{F}\cdot(\mathbf{G}'\times\mathbf{H}) + \mathbf{F}\cdot(\mathbf{G}\times\mathbf{H}')$$

3. Suppose that the graph of a smooth curve does not go through the origin. Show that $\mathbf{F}(t_0) \perp \mathbf{F}'(t_0)$, where $F(t_0)$ is the point on the graph closest to the origin.

4. Express the curve

$$F(t) = \langle 2(\cos t + t \sin t), 2(\sin t - t \cos t)\rangle$$

using arc length as a parameter measured from the point corresponding to $t = 0$.

5. The motion of a particle in the plane is given by $F(t) = \langle \cos^3 t, \sin^3 t\rangle$, $0 \leq t \leq \pi/2$. After how many seconds will the particle have traveled 3/8 of a unit along the curve?

6. At what point on the curve $y = 6x - x^2$ is the curvature a minimum?

7. A hill has angle of elevation $\pi/6$ with the ground. A projectile is fired with initial speed v_0 feet per second from the base of the hill (toward the hill) at an angle of elevation $\alpha(> \pi/6)$. How far down range will the projectile have traveled when it strikes the hill?

8. At what angle of elevation should the projectile in Exercise 7 be fired in order to maximize the range?

9. A particle moves along a curve in such a way that the tangent vector at each point on the curve is always perpendicular to the standard representative to that point. Describe the curve.

10. The vector function $\mathbf{F}(t) = \langle t \cos t, t \sin t, 9 - t^2\rangle$ lies on a particular quadric surface. Use this observation to sketch the graph of **F**.

11. The space curve $\mathbf{F}(t) = \langle 3 \cos t, 3 \cos 2t, 3 \sin t\rangle$ can be realized as the intersection of two quadric surfaces. Use this observation to sketch the portion of the graph of **F** in the first octant.

12. Find parametric equations for and sketch the space curve joining the points (0, 0, 9) and (1, 3, −1) obtained by intersecting the surfaces with equations $x^2 + y^2 + z = 9$ and $y = 3x$.

13. Find **T**, **N**, and **B** at any point on the space curve $F(t) = \langle 2 \sin t, 2 \cos t, -2 \ln(\cos t)\rangle$, $|t| < \pi/2$.

14. Show that the acceleration vector of a particle moving in space has length

$$\left[\left(\frac{d^2s}{dt^2}\right)^2 + (k(t))^2\left(\frac{ds}{dt}\right)^4\right]^{1/2}$$

15. A projectile is designed to curve leftward with an acceleration of L feet per second squared. Use a three-dimensional analysis to describe the motion of such a projectile.

16. Let **F** be a vector function. Show that $\lim_{t\to t_0} \mathbf{F}(t) = \mathbf{L}$ if and only if for every $\varepsilon > 0$ there exists a $\delta > 0$ such that $\|\mathbf{F}(t) - \mathbf{L}\| < \varepsilon$ whenever $0 < |t - t_0| < \delta$.

The representatives of $\mathbf{T}(t)$ *and* $\mathbf{N}(t)$ *with their bases at* $F(t)$ *on the graph of a space curve* **F** *determine a plane called the osculating plane at* $\mathbf{F}(t)$. *In Exercises 17 to 19 determine the osculating plane at the indicated point.*

17. $F(t) = \langle e^t, e^t \cos t, e^t \sin t\rangle$, $t = \pi/2$

18. $F(t) = \langle \cos t, -\sin t, t\rangle$, $t = \pi/2$

19. $F(t) = \langle 2 \cos t, 2 \sin t, t\rangle$, $t = \pi/6$

20. Discuss the behavior of the circle of curvature near a point of inflection on a plane curve.

SELF-TEST ✢

1. Let $\mathbf{F}(t) = \langle t, t^2 - 2t\rangle$.
(*a*) Sketch the graph of **F**.
(*b*) Determine $\mathbf{N}(t)$ and $\mathbf{T}(t)$ and sketch their representatives at the point corresponding to $t = 1$.
(*c*) Compute the curvature of **F**.

2. The position function of a particle moving in the plane is given by $\mathbf{G}(t) = \langle e^{-t} \cos t, e^{-t} \sin t\rangle$.
(*a*) Find $\mathbf{V}(t)$ and $\mathbf{A}(t)$.
(*b*) Find the tangential and normal components of acceleration.

3. Determine the curvature for the space curve $\mathbf{F}(t) = \langle t^2, t + 1, 2t^2\rangle$.

4. A projectile has a range of 64 miles. What is its initial speed?

5. The position function of a particle is given by

$$F(t) = \langle 3t^2, 2t^3\rangle \qquad t \geq 0$$

How long does it take the particle to move $20\sqrt{10} - 2$ units along the curve from the origin?

Chapter 19 ✤

FUNCTIONS OF SEVERAL VARIABLES

CONTENTS

The next two chapters are devoted to a systematic study of the calculus of functions of several variables. As with functions of one variable this study is broken into two parts: the differential calculus and the integral calculus. This chapter is concerned with the former, and Chap. 20 is devoted to the latter. The first few sections of this chapter lay the groundwork for such a study by extending the fundamental notions of limit, continuity, and differentiability to this more general setting. The later sections build on this foundation with an emphasis on how the differential calculus of functions of several variables can be applied to solve specific problems.

Being able to deal with phenomena involving two, three, or more variables greatly increases the variety of applications that are possible. On the other hand, our increase in versatility is accompanied by an increase in complexity. Although the ideas of limit, continuity, and differentiability are direct extensions of the respective single-variable ideas, the presence of more than one variable leads to complications both theoretically and computationally. We shall concentrate on the basics, illustrating where and how some of the more important complications occur, but leaving the more subtle aspects of the theory to more advanced courses.

Section 19.1: Functions of Several Variables

Objectives:

1. Determine the domain of functions of two or three variables.

2. Sketch the graphs of functions of two variables.

3. Sketch the level curves of a function of two variables and the level surfaces of a function of three variables.

FUNCTIONS OF TWO VARIABLES

Often in geometry and in the physical, biological, and social sciences, because of the inherent interconnectedness of phenomena, one quantity depends upon more than one other quantity. We shall extend our definition of function to cover such situations.

As before, R denotes the real numbers, R_2 denotes the set of ordered pairs (x, y) of real numbers, and R_3 denotes the set of all ordered triples (x, y, z) of real numbers. We assume the usual identification between R_3 and points in space. Note that in this chapter we are interested in ordered pairs and ordered triples considered as points rather than as vectors. We begin with functions of two variables.

19.1.1 **Definition** A *function f of two variables* is a rule that assigns to each point (x, y) in some subset D of R_2 a *unique* real number denoted by $f(x, y)$. The set D of points to which f applies is called the *domain of f*, and the set of all numbers $f(x, y)$ resulting when f is applied to these points [as (x, y) varies over D] is called the *range of f*. We shall sometimes denote the range and domain of f by $R(f)$ and $D(f)$, respectively.

In practice we often do not specify $D(f)$, but rather assume that f has its natural domain. That is, $D(f)$ is the set of all points (x, y) that yield a real number when substituted into the rule for f.

Example 1 Determine the domain of the function

$$f(x, y) = \sqrt{144 - 16x^2 - 9y^2}$$

Solution A point (x, y) is in $D(f)$ if and only if $16x^2 + 9y^2 \leq 144$, or equivalently,

$$\frac{x^2}{9} + \frac{y^2}{16} \leq 1$$

Thus $D(f)$ consists of all points inside or on the ellipse with equation

$$\frac{x^2}{9} + \frac{y^2}{16} = 1 \qquad \text{(see Fig. 19.1)} \quad \bullet$$

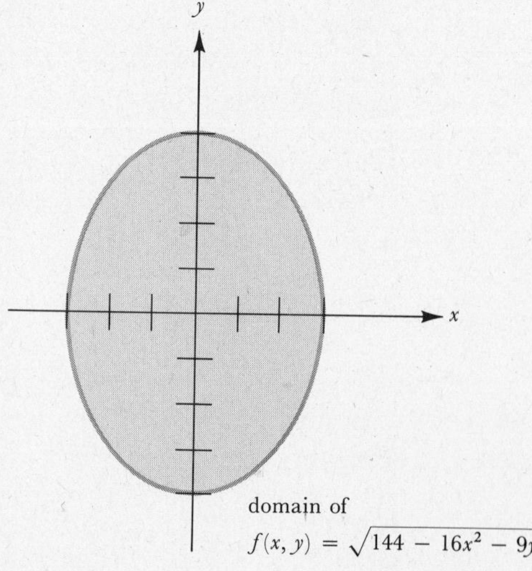

Fig. 19.1

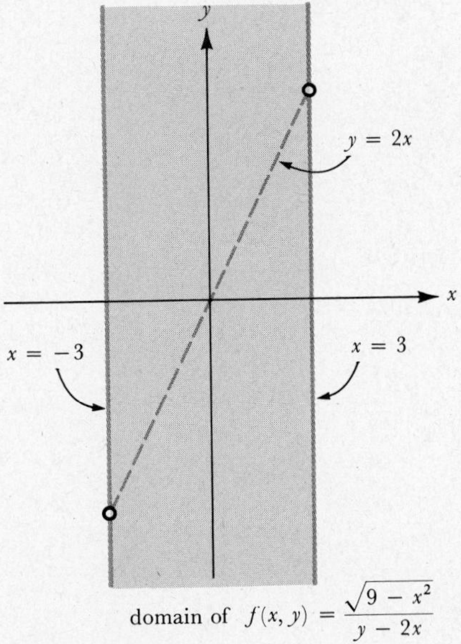

Fig. 19.2

Example 2 Determine the domain of the function

$$f(x,y) = \frac{\sqrt{9 - x^2}}{y - 2x}$$

Solution For $f(x,y)$ to be defined we must have $9 - x^2 \geq 0$ and $y - 2x \neq 0$. Now $9 - x^2 \geq 0$ if and only if $|x| \leq 3$. Thus the points in $D(f)$ must lie in the vertical strip extending between and including the points on the lines $x = -3$ and $x = 3$. In addition, since $y - 2x \neq 0$, we must discard fom this set all points on the line $y = 2x$. A sketch of the domain is given in Fig. 19.2. We have used dashes to indicate that the points along $y = 2x$ are not to be included in $D(f)$. ●

GRAPHS

19.1.2 **Definition** The *graph* of a function f of two variables consists of all the points $(x,y,f(x,y))$ in R_3 as (x,y) varies over $D(f)$ in R_2.

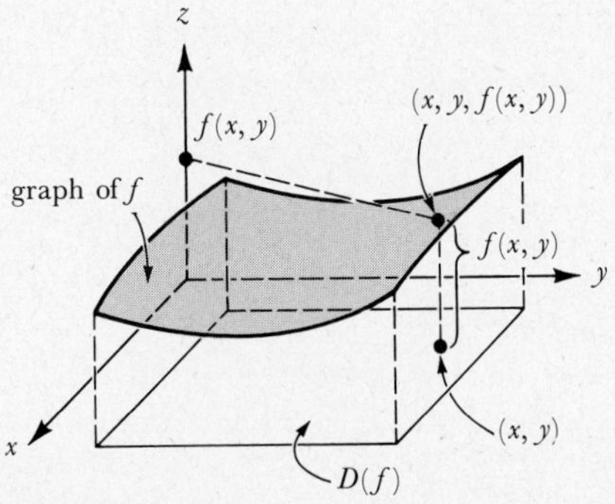

Note that the graph of the function f is the same as the graph of the equation $z = f(x,y)$. In this setting the domain $D(f)$ will be some subset of the x-y plane and the graph of f will, in general, be a surface. If (x,y) is in $D(f)$, then $f(x,y)$ gives the directed distance from the x-y plane to the point $(x,y,f(x,y))$ on the graph of f. Figure 19.3 shows a typical situation where $f(x,y) \geq 0$ for (x,y) in $D(f)$.

When we write a function in the form $z = f(x,y)$, we refer to x and y as *independent* variables and z as the *dependent* variable. In many instances the problem of graphing a function of two variables is simplified by using our knowledge of equations in R_3 developed in Chap. 17, especially Sec. 17.2.

Fig. 19.3

Example 3 Sketch the graph of the function $f(x,y) = \sqrt{x^2 + y^2}$.

Solution Letting $z = \sqrt{x^2 + y^2}$ and squaring both sides, we have

$$z^2 = x^2 + y^2$$

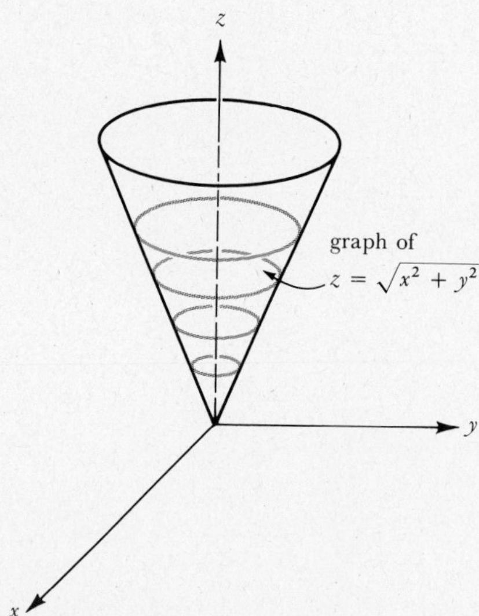

Fig. 19.4

which from (17.2.7) or Sec. 17.2, Example 4, we know is the equation of a circular cone whose axis is the z axis. The graph of $z = \sqrt{x^2 + y^2}$ is the upper nappe of this cone (see Fig. 19.4). ●

The following example does not lead to one of the familiar quadric surfaces, but the techniques for obtaining the graph are much the same.

Example 4 Sketch the portion of the graph of $f(x,y) = e^x + y$ that lies in the first octant.

Solution Restricting our attention to the first octant is equivalent to restricting the domain of f to the first quadrant in the x-y plane. Hence the graph of f is the graph of the equation $z = e^x + y$, $x, y \geq 0$. The traces in the x-z and y-z planes ($y = 0$ and $x = 0$, respectively) are the curves $z = e^x$ and $z = y + 1$, respectively. Since $x \geq 0$, we know $e^x \geq 1$, and hence there is no trace in the planes $z = k$ unless $k \geq 1$. Some typical traces are given in Fig. 19.5. Using these we obtain a sketch of the graph of f in Fig. 19.6. ●

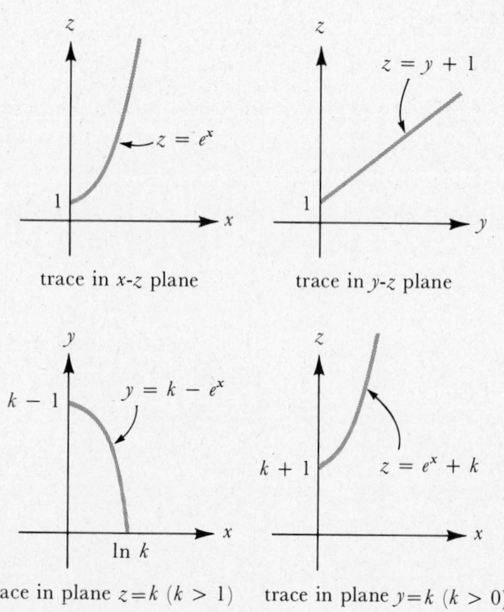

Fig. 19.5

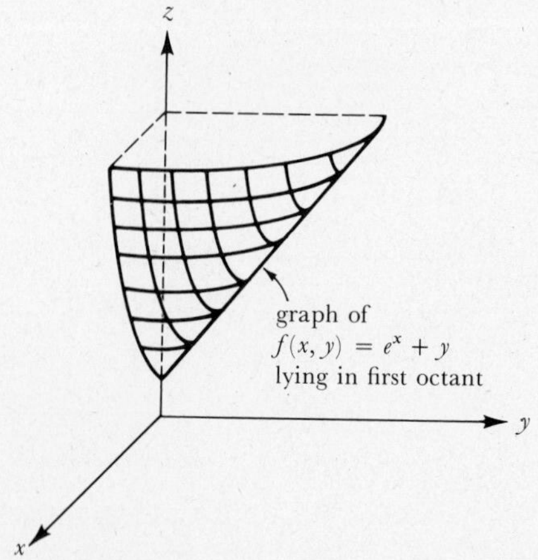

Fig. 19.6

PROGRESS TEST 1

1. Determine and sketch the domain of
$f(x,y) = \sqrt{xy - 1}$

2. Sketch the graphs of the following functions.
(*a*) $f(x,y) = 4 - 2x - y$
(*b*) $f(x,y) = \sqrt{3x^2 + 4y^2 - 12}$

FUNCTIONS OF SEVERAL VARIABLES

Definition (19.1.1) easily extends to functions of three, four, or, for that matter, any finite number of variables. For the most part we shall be concerned with functions of two or three variables. A function of three variables will be denoted by

$$w = f(x, y, z)$$

where the domain is now a subset of R_3. In this setting, w is the dependent variable and x, y, and z are the independent variables. For example, the function

$$w = \frac{xyz}{\sqrt{4 - x^2 - y^2 - z^2}}$$

is well defined if and only if

$$x^2 + y^2 + z^2 < 4$$

Hence the domain of f is the set of all points inside the sphere of radius 2 centered at the origin. Note that none of the points on the sphere itself is in $D(f)$ because any such points will result in a zero denominator.

Whenever we have occasion to refer to functions of four or more variables, we denote the independent variables by x_i and write

$$w = f(x_1, x_2, \ldots, x_n)$$

The domain in this case will consist of some collection of ordered n-tuples of real numbers. That is, $D(f)$ is a subset of R_n, where R_n denotes the collection of *all* ordered n-tuples of real numbers $(x_1, x_2, \ldots, x_n)$. Of course, for $n \geq 4$, $D(f)$ has no geometric realization.

You have at your command rather elaborate machinery for sketching graphs of functions of *one* variable. However, sketching the graph of functions of *two* variables is considerably more difficult. For functions of three or more variables it cannot be done since the graph will be a subset of R_n, where $n \geq 4$.

The fact that we cannot "picture" a function of three variables often leaves students rather ill at ease when dealing with them. Nonetheless, they are a traditional part of the elementary calculus and arise very naturally in applications. Notice, for example, that the density of a nonhomogeneus substance varies depending on location, and hence is a function of the three variables necessary to determine the location.

LEVEL CURVES AND LEVEL SURFACES

As noted, the graph of a function of two variables is in general a surface in three-space. We can obtain information about this surface (and hence about the function) by examining the curves that result when we slice the surface with planes parallel to the *x-y* plane. Such curves, which are sketched by projecting them into the *x-y* plane, are called *level curves* of the function. More precisely we have:

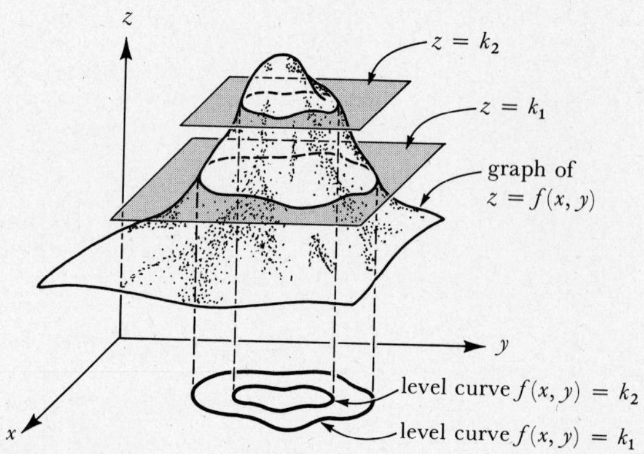

Fig. 19.7

19.1.3 Definition Let $f(x, y)$ be a function of two variables and let k be a constant. The graph in the x-y plane of the equation $f(x, y) = k$ is called a *level curve of f* (see Fig. 19.7).

Level curves are what cartographers use in drawing contour maps of a region. Think of the terrain that makes up the region as a surface. The level curves (*contour curves*) that appear on the map are marked to indicate the altitude. The curves marked 3000 represent the curve of intersection of the terrain (considered as a surface) with a horizontal plane of altitude 3000 ft. Level curves appearing very close together on the map indicate steep terrain, and level curves that are far apart indicate flatter terrain. A typical situation is pictured in Fig. 19.8(a) and (b). In both cases the difference in altitude is 500 ft. The curves in (a) reflect the steepness of the mountain while those in (b) reflect a more gradually sloping hill.

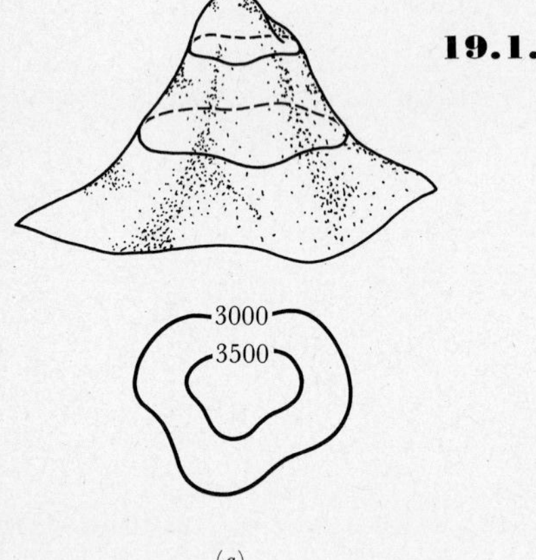

Fig. 19.8

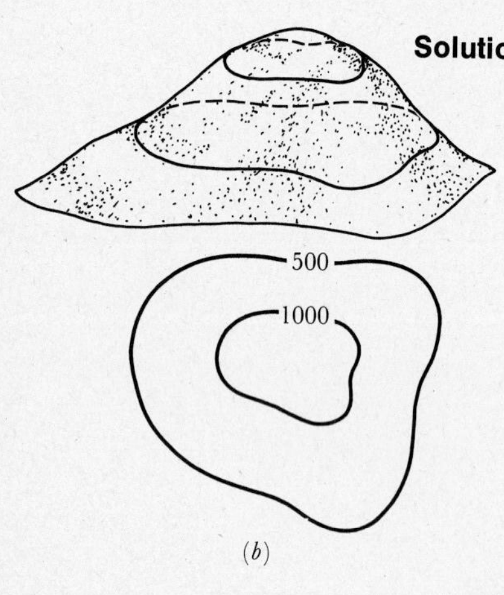

Example 5 Sketch some level curves of the surface

$$f(x, y) = y^2 - 2x$$

Solution The curves $y^2 - 2x = c$ are parabolas opening to the right, and since $y^2 = 2(x + \frac{1}{2}c)$, the vertices are on the x axis at $(-\frac{1}{2}c, 0)$. Some of the level curves corresponding to various values of c are shown in Fig. 19.9. ●

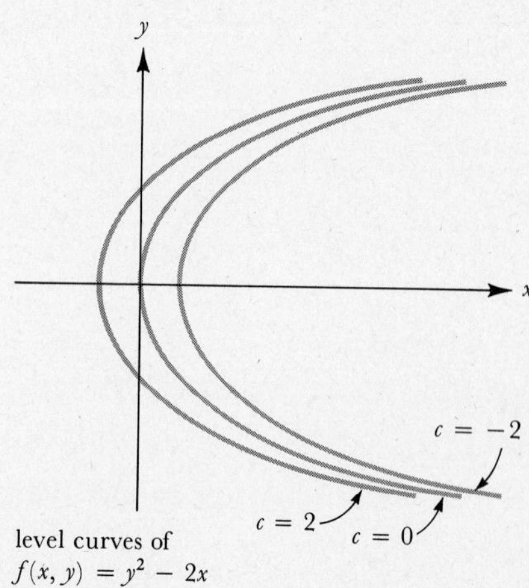

level curves of $f(x, y) = y^2 - 2x$
Fig. 19.9

Suppose a metal plate occupies a region in the x-y plane, and suppose the function $T(x, y)$ gives the temperature distribution of the plate at a given instant. That is, at this particular instant the temperature at the point on the plate with coordinates (x, y) is $T(x, y)$. For a given c, the level curve $T(x, y) = c$ identifies those points on the plate having the same temperature c; these curves are called *isotherms,* or *isothermal curves.* Similarly, the surface pressure (corrected for different altitudes) over the United States varies as a function of location, yielding a function of two variables whose level curves, commonly called *isobars,* are the curves one normally sees on daily weather maps. These curves consist of points having equal barometric pressure. The closer together the isobars, the greater the differences in pressure, and, since air moves in such a way as to equalize pressure, the greater the wind.

If a *solid* metal object is heated, then the temperature distribution is a function of *three* variables, and we speak of *isothermal surfaces* $T(x, y, z) = k$. In general, for functions of three variables we have:

19.1.4 **Definition** If $f(x, y, z)$ is a function of three variables and if k is a constant, then the surface with equation

$$f(x, y, z) = k$$

is called a *level surface* of f.

Example 6 Examine the level surfaces of the function

$$f(x, y, z) = \frac{z}{x^2 + y^2}$$

Solution If $k = 0$, then the level surface is $z = 0$, the x-y plane. For $k \neq 0, f(x, y, z) = k$ becomes

$$z = k(x^2 + y^2)$$

which has as its graph a paraboloid with vertex at the origin. The paraboloid opens about the positive z axis if $k > 0$ and the negative z axis if $k < 0$. The larger $|k|$, the steeper the paraboloid. Some typical level surfaces are sketched in Fig. 19.10. ●

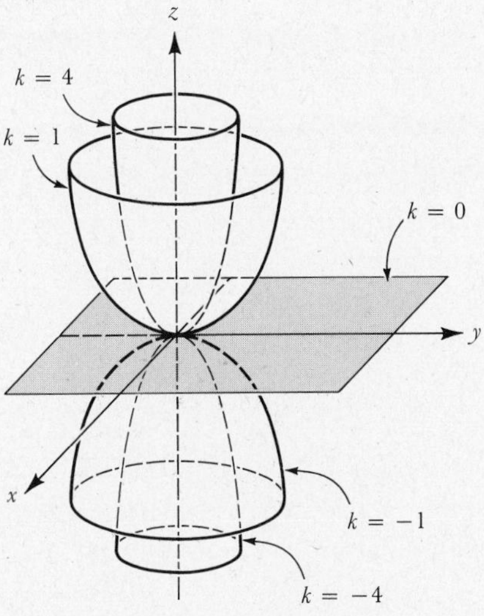

level surfaces of $f(x, y, z) = \frac{z}{x^2 + y^2}$

Fig. 19.10

PROGRESS TEST 2

1. Determine the domain of the function $w = \cos x + \sin y + \arcsin\sqrt{z^2 - 1}$. Describe the domain geometrically.

2. Sketch the level curves for the function $z = x/y^2$ corresponding to $c = 0, 1, -1$.

3. Sketch the level surfaces for the function $w = z^2 - x^2 - y^2$ corresponding to $c = 0, 1, -1$.

SECTION 19.1 EXERCISES

In Exercises 1 to 10 find the domain of the given functions.

1. $z = \sqrt{3x^2 + 5y^2 - 15}$
2. $z = \dfrac{y^2 - x^2}{y - x}$
3. $z = \dfrac{x}{y} - 1$
4. $z = \sqrt{xy}$
5. $z = \ln(x + y)$
6. $z = \arcsin\left(\sqrt{\dfrac{y - x}{y + x}}\right)$
7. $z = \ln(4x - y)$
8. $z = \sqrt{12 - x^2 - y}$
9. $w = \dfrac{x}{y - z}$
10. $w = \dfrac{x^2}{y^2 - z^2}$

In Exercises 11 to 20 sketch the graph of the given function.

11. $z = x^2 + 2y^2$
12. $z = 9 - x^2$
13. $z = x^3 + y$
14. $z = xy$
15. $z = \sqrt{x^2 + 9y^2}$
16. $z = -\sqrt{9 - x^2 - y^2}$
17. $z = y^2 - x^2$
18. $z = \sqrt{x^2 + 2y^2 - 1}$
19. $z = \sqrt{x^2 + y^2 + 1}$
20. $z = 2x + e^y$

In Exercises 21 to 25 describe the level curves.

21. $z = x^2 + 9y^2$
22. $z = \sqrt{9 - 3x^2 - 2y^2}$
23. $z = xy$
24. $z = x^2 + y$
25. $z = \dfrac{x - 2}{y + 1}$

In Exercises 26 to 30 sketch the level surfaces corresponding to the indicated values of c.

26. $w = x^2 + y^2 + z^2$; $c = 0, 1, 4$
27. $w = x + y + 2z$; $c = \pm 1, \pm 4$
28. $w = x^2 + y^2 - 2z$; $c = 0, 1, -1$
29. $w = x^2 + 2y^2 + 3z^2$; $c = 0, 1, -1$
30. $w = xyz$; $c = 1$

Section 19.2: Limits and Continuity

Objectives:

1. Determine limits of functions of two variables.

2. Determine points where a function of two variables is continuous

LIMITS

Before introducing the concept of a limit for functions of several variables, it is instructive to revisit the single-variable setting. Suppose the domain of f contains a deleted open interval about the point a. We recall that

$$\lim_{x \to a} f(x) = L$$

if for every $\varepsilon > 0$, there exists a $\delta > 0$ such that

$$(1) \qquad 0 < |x - a| < \delta \text{ implies } |f(x) - L| < \varepsilon$$

Now in R, $|x - y|$ is the distance between x and y. If we denote this distance by $d(x, y)$, then we can rephrase (1) as

$$(2) \qquad 0 < d(x, a) < \delta \text{ implies } d(f(x), L) < \varepsilon$$

Since we also have a notion of distance between points in R_2 given by

$$d[(x_1, y_1), (x_2, y_2)] = \sqrt{(x_2 - x_1)^2 + (y_2 - y_1)^2}$$

it is fairly simple to formulate (2) for functions of two variables. The only problem that remains is to specify the two-dimensional analog of the statement that "f contains a deleted open interval about the point a."

19.2.1 Definition Suppose $P = (a, b)$ is a point in R_2 and r is a positive number. The *open disk* $S_r(P)$ is the collection of all points (x, y) such that $d[(x, y), (a, b)] < r$. A set of the form $S_r(P)$ with the point P removed is called a *deleted open disk* about (a, b) (see Fig. 19.11).

19.2.2 Definition Suppose f is a function of two variables whose domain contains some deleted open disk about (a, b). Let L be some real number. We say

$$\lim_{(x,y)\to(a,b)} f(x, y) = L$$

if, given any $\varepsilon > 0$, there exists a $\delta > 0$ such that

$$0 < d[(x, y), (a, b)] < \delta \text{ implies } d[f(x, y), L] < \varepsilon$$

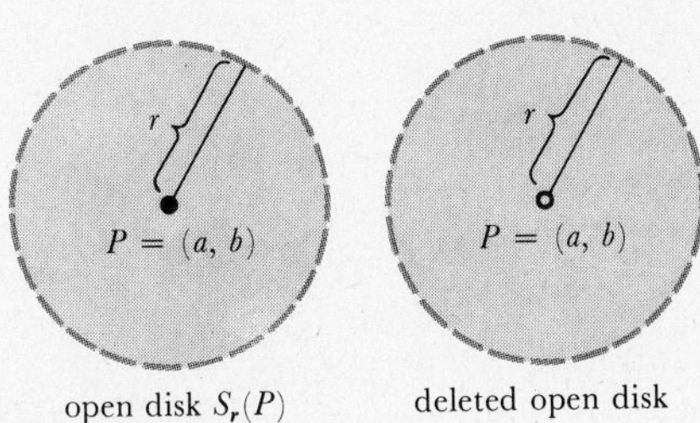

Fig. 19.11

As before, if

$$\lim_{(x,y)\to(a,b)} f(x, y) = L$$

then we say $f(x, y)$ *has limit* L *or tends to* L *as* (x, y) *tends to* (a, b). We sometimes denote this by $f(x, y) \to L$ as $(x, y) \to (a, b)$.

Graphically, the function f has limit L as (x, y) tends to (a, b) if, given any open interval $(L - \varepsilon, L + \varepsilon)$ on the z axis, one can exhibit a deleted open disk $S_\delta(a, b)$ about (a, b) in the x-y plane with the property that every point in this disk gets sent by f into the interval $(L - \varepsilon, L + \varepsilon)$; see Fig. 19.12.

Example 1 Show that

$$\lim_{(x,y)\to(0,0)} \frac{x^2 y}{x^2 + y^2} = 0$$

Solution Since we are taking the limit at $(0, 0)$ we can assume that $(x, y) \neq (0, 0)$. For all such points, $x^2/(x^2 + y^2) \leq 1$, and hence

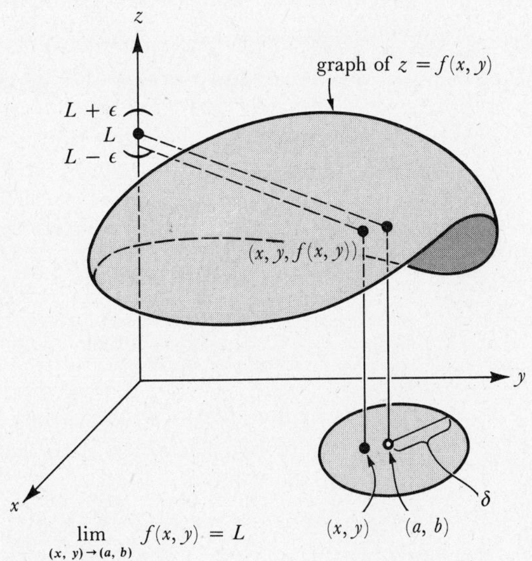

Fig. 19.12

$$d[(f(x,y), 0] = \left|\frac{x^2y}{x^2+y^2} - 0\right| = \frac{x^2}{x^2+y^2}|y| \leq |y|$$

Now $|y| = \sqrt{y^2} \leq \sqrt{x^2+y^2} = d[(x,y),(0,0)]$, and thus if we take $\delta = \varepsilon$, then

$$0 < d[(x,y),(0,0)] < \delta = \varepsilon$$

implies

$$d[(f(x,y), 0] \leq |y| \leq d[(x,y),(0,0)] < \varepsilon$$

Hence, by Definition (19.2.2),

$$\lim_{(x,y)\to(0,0)} \frac{x^2y}{x^2+y^2} = 0 \quad \bullet$$

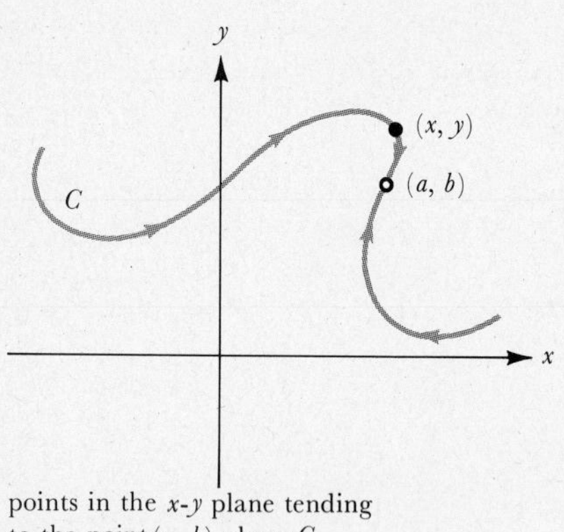

points in the x-y plane tending to the point (a, b) along C

Fig. 19.13

If $\lim_{(x,y)\to(0,0)} f(x,y) = L$ and if C is *any* smooth curve through (a, b), then it can be shown that $f(x,y)$ will tend to L as (x,y) tends to (a, b) *along* C (see Fig. 19.13). In particular, if letting (x,y) tend to (a, b) along *two distinct curves* through (a, b) leads to *different values* for L, then $\lim_{(x,y)\to(a,b)} f(x,y)$ *does not exist.* (This corresponds to the one-dimensional criterion that if $\lim_{x\to a^+} f(x) \neq \lim_{x\to a^-} f(x)$, then $\lim_{x\to a} f(x)$ does not exist.)

Example 2 Show that

$$\lim_{(x,y)\to(0,0)} \frac{x^2-y^2}{x^2+y^2}$$

does not exist.

Solution Let L_m be the straight line through the origin with slope m. Then (x,y) is on L_m if and only if $y = mx$. Also, $(x, mx) \to (0,0)$ along L_m as $x \to 0$ (see Fig. 19.14). Since we are investigating the limit at $(0,0)$, we can assume $x \neq 0$. Now

$$f(m, mx) = \frac{x^2 - m^2x^2}{x^2+m^2x^2} = \frac{1-m^2}{1+m^2}$$

and so clearly $f(x, mx) \to (1-m^2)/(1+m^2)$ as $x \to 0$. Since this leads to different results for different choices of m, f does not have a limit at $(0,0)$. $\bullet$

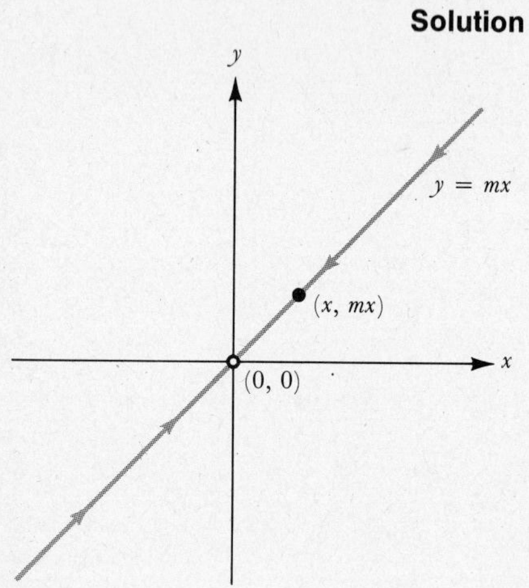

Fig. 19.14

The usual limit theorems remain valid for functions of two variables. That is, if

$$\lim_{(x,y)\to(a,b)} f(x,y) = L \qquad \text{and} \qquad \lim_{(x,y)\to(a,b)} g(x,y) = M$$

then

$$\lim_{(x,y)\to(a,b)} [f(x,y) + g(x,y)] = L + M$$

$$\lim_{(x,y)\to(a,b)} kf(x,y) = kL \qquad (k \text{ a constant})$$

$$\lim_{(x,y)\to(a,b)} f(x,y)g(x,y) = LM$$

$$\lim_{(x,y)\to(a,b)} f(x,y)/g(x,y) = L/M \qquad (\text{provided } M \neq 0)$$

If f is a function of a single variable whose domain contains the range of the function g of two variables, then we can form the *composition function* $f \circ g$ by means of the rule

$$(f \circ g)(x,y) = f(g(x,y))$$

As in (2.5.10), we have

19.2.3 Theorem If f is continuous at $x = C$ and if $\lim_{(x,y)\to(a,b)} g(x,y) = C$, then

$$\lim_{(x,y)\to(a,b)} (f \circ g)(x,y) = f(C)$$

PROGRESS TEST 1

1. Let $f(x,y) = \dfrac{x^2y}{x^4 + y^2}$.
   (a) Determine the limit of $f(x,y)$ as (x,y) tends to $(0,0)$ along any straight line through the origin.
   (b) Determine the limit of $f(x,y)$ as (x,y) tends to $(0,0)$ along the parabola $y = x^2$.
   (c) Does $\lim_{(x,y)\to(0,0)} f(x,y)$ exist?

2. Use definition (19.2.2) to prove that

$$\lim_{(x,y)\to(a,b)} (x) = a$$

CONTINUITY

A point P is said to be an *interior point* of a set D if D contains some open disk about P. If all the points of a set are interior points, then the set is called *open.* A point P with the property that *every* open disk about P contains at least one point in D and one point not in D is called a *boundary point* of D, and the collection of all such points is called the *boundary* of D. Boundary points of a set need not be elements of the set. If a set *does* contain all of its boundary points, then it is said to be *closed.* We illustrate several of the possible situations in Fig. 19.15, where we used dashes for those portions of the boundary that are not in the set.

19.2.4 Definition Suppose f is a function of two variables and (a, b) is an interior point of $D(f)$, the domain of f. Then f is said to be *continuous at* (a, b) if

$$\lim_{(x,y)\to(a,b)} f(x,y) = f(a,b)$$

A function f is said to be *continuous on an open set* D if it is continuous at each point of D.

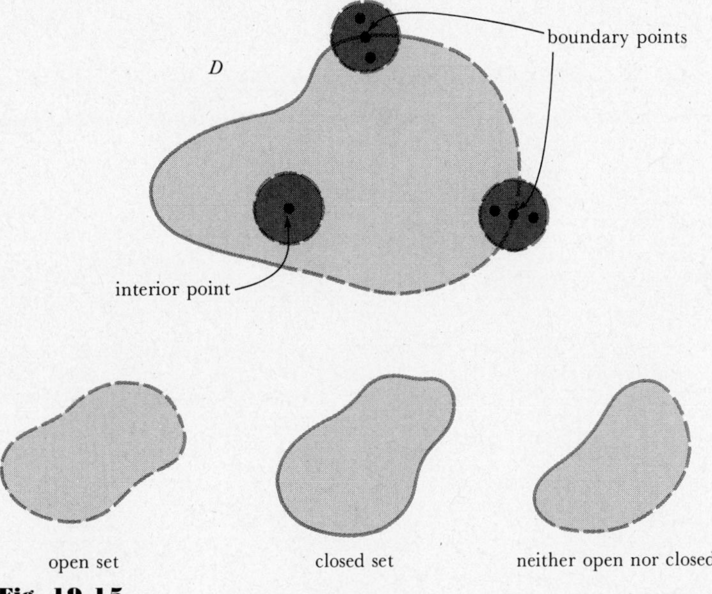

Fig. 19.15

Note that this definition is essentially the same as that for the one-variable case (2.5.7). That is, the function is continuous at a point if it has a limit at this point that equals the value of the function at the point. The graph of a continuous function of two variables is a surface that has no holes or breaks.

From the definition and the limit properties it can be shown that sums, differences, scalar multiples, products, quotients (where the denominator is not zero), and compositions of continuous functions will themselves be continuous. In particular, *polynomial functions* of two variables, such as $f(x,y) = 4x^4y^2 + 2xy^3 + 6y^2 + 9$, are everywhere continuous. *A rational function* of two variables (that is, a quotient of polynomial functions) is continuous at all points where the denominator is not zero. As a result, a function such as

$$f(x,y) = \cos\left(\frac{x^2 + 2xy - y^2}{x^2 + y^2 + 1}\right)$$

is continuous everywhere since it is the composition of the continuous cosine function with the continuous rational function $(x^2 + 2xy - y^2)/(x^2 + y^2 + 1)$. In particular, limits of such functions can be obtained by direct evaluation. Thus

$$\lim_{(x,y)\to(0,0)} \cos\left(\frac{x^2 + 2xy - y^2}{x^2 + y^2 + 1}\right) = \cos(0) = 1$$

FUNCTIONS OF THREE VARIABLES

All statements of this section can be routinely extended to functions of three variables. In this setting, distance between points is given by

$$d[(x_1, y_1, z_1), (x_2, y_2, z_2)] = \sqrt{(x_2 - x_1)^2 + (y_2 - y_1)^2 + (z_2 - z_1)^2}$$

and if $P = (a, b, c)$, then $S_\varepsilon(P)$, called the *open ball about P,* consists of all points that are less than ε units from P.

SECTION 19.2 EXERCISES

In Exercises 1 and 2 use the definition to establish the given limit.

1. $\displaystyle\lim_{(x,y)\to(1,-2)} (3x - 4y + 1) = 12$

2. $\displaystyle\lim_{(x,y)\to(0,0)} \frac{x^2y^2}{x^2 + y^2} = 0$

In Exercises 3 to 6 show that $\lim_{(x,y)\to(a,b)} f(x,y)$ *does not exist by exhibiting distinct curves through* (a, b) *with the property that* $f(x,y)$ *tends to distinct limits as* $(x,y) \to (a,b)$ *along these curves.*

3. $f(x,y) = \dfrac{y}{x + y}$, $(a, b) = (0, 0)$

4. $f(x,y) = \dfrac{x^5 + y}{x^4 + y^2}$, $(a, b) = (0, 0)$

5. $f(x,y) = \dfrac{x^2 + y}{x^2 + y^2}$, $(a, b) = (0, 0)$

6. $f(x,y) = y^2 \ln(x^2 + y^2)$, $(a, b) = (0, 0)$

In Exercises 7 to 12 use the limit theorems to evaluate the given limits.

7. $\displaystyle\lim_{(x,y)\to(1,1)} 4xy^2 + 2x - 9$

8. $\displaystyle\lim_{(x,y)\to(0,0)} \frac{x^2y + 9x + 6}{3 + x^2 + y^4}$

9. $\displaystyle\lim_{(x,y)\to(-1,0)} \frac{x^2 - y^2}{x^2 + y^2}$

10. $\displaystyle\lim_{(x,y)\to(3,1)} 3x^2 + x/y$

11. $\displaystyle\lim_{(x,y)\to(2,-1)} \frac{8xy^2 + 2y}{x + 3y}$

12. $\displaystyle\lim_{(x,y)\to(0,0)} \frac{4x^2 + 5}{xy + 9}$

13. Show how (19.2.3) can be used to establish the continuity of $f(x,y) = \sin(x + y)$ at $(0, 0)$.

In Exercises 14 and 15 determine whether $f(x,y)$ is continuous at (a, b).

14. $f(x,y) = \begin{cases} \dfrac{\sin(x+y)}{x+y} & (x,y) \neq (0,0) \\ 0 & (x,y) = (0,0) \end{cases}$ $\quad (a,b) = (0,0)$

15. $f(x,y) = \begin{cases} \dfrac{x+y}{|x|+|y|} & (x,y) \neq (0,0) \\ 0 & (x,y) = (0,0) \end{cases}$ $\quad (a,b) = (0,0)$

16. Show that $\lim_{(x,y)\to(2,5)} (3x + 4y) = 26$.

Hint:
(i) Use the triangle inequality for the absolute value to show that

$$d[(3x+4y), 26] \leq 3|x-2| + 4|y-5|$$

(ii) Note that

$$|x-2| = \sqrt{(x-2)^2} \leq \sqrt{(x-2)^2 + (y-5)^2} = d[(x,y), (2,5)]$$

Obtain a similar inequality involving $|y - 5|$ and combine with (i) to show that

$$d[(3x+4y), 26] \leq 7d[(x,y), (2,5)]$$

17. Show that open disks are open sets.

18. Give a definition of continuity at the point (a, b, c) for the three-variable function $f(x,y,z)$. [Parallel (19.2.4).]

19. Give a definition of limit for functions of three variables paralleling (19.2.2).

20. Use Exercise 19 to prove that $\lim_{(x,y,z)\to(a,b,c)} (y) = b$.

Section 19.3: Partial Derivatives

Objectives:

1. **Determine the partial derivatives of a function of two or three variables.**

2. **Determine equations for the tangent plane and normal line to the graph of a function of two variables at a point.**

INTRODUCTION

For a function of a single variable the derivative measures the rate of change of the function with respect to its independent variable.

For a function of two variables it would thus seem reasonable to consider *two* derivatives to measure the rate of change of the function with respect to each of its independent variables. Such derivatives, called *partial derivatives,* are defined as follows:

19.3.1 **Definition** Suppose (a, b) is a point in the domain of the function f. *The partial derivative of f with respect to x at (a, b),* denoted $f_x(a, b)$, is defined by

(*a*) $$f_x(a,b) = \lim_{\Delta x \to 0} \frac{f(a+\Delta x, b) - f(a,b)}{\Delta x}$$

provided that this limit exists. Similarly, *the partial derivative of f with respect to y at (a, b),* denoted $f_y(a, b)$, is defined by

(*b*) $$f_y(a,b) = \lim_{\Delta y \to 0} \frac{f(a, b+\Delta y) - f(a,b)}{\Delta y}$$

Note that these definitions give rise to two new functions f_x and f_y, called the *partial derivatives of f.* These are given by

19.3.2 (*a*) $$f_x(x,y) = \lim_{\Delta x \to 0} \frac{f(x+\Delta x, y) - f(x,y)}{\Delta x}$$

(*b*) $$f_y(x,y) = \lim_{\Delta y \to 0} \frac{f(x, y+\Delta y) - f(x,y)}{\Delta y}$$

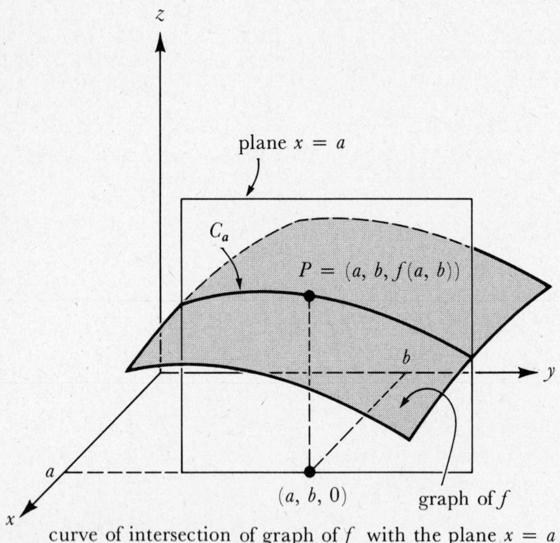

Fig. 19.16

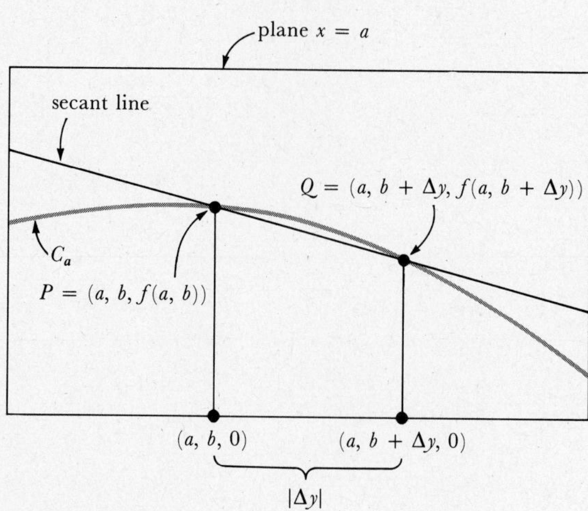

Fig. 19.17

The domains of these functions consist of all those points (x, y) in the domain of f for which their respective defining limits exist.

THE GEOMETRIC INTERPRETATION OF PARTIAL DERIVATIVES

Suppose that $f_y(a, b)$ exists and let C_a denote the curve obtained by intersecting the vertical plane $x = a$ with the graph of f (see Fig. 19.16).

We can compute the slope of the tangent line to C_a at the point $P = (a, b, f(a, b))$ by working in the plane $x = a$ and proceeding as in Sec. 2.4. If Δy denotes a change in y, then the secant line through P and $Q = (a, b + \Delta y, f(a, b + \Delta y))$ has slope

$$\frac{f(a, b + \Delta y) - f(a, b)}{\Delta y} \qquad \text{(see Fig. 19.17)}$$

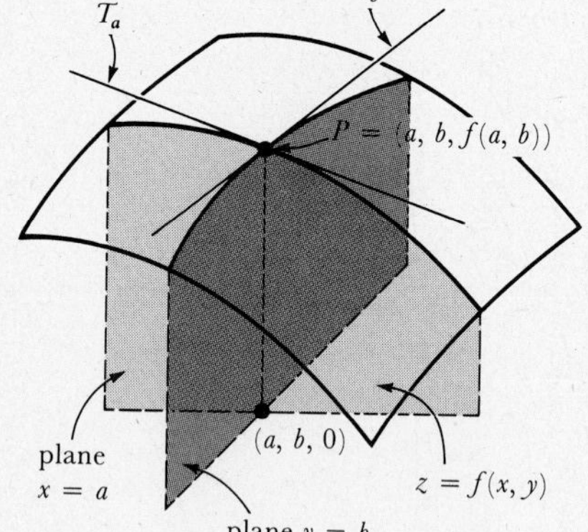

Fig. 19.18

As $\Delta y \to 0$, the points Q slide along the curve C_a toward P. In turn, the corresponding secant lines tend to the line T_a tangent to C_a at P. The slope of T_a is given by

$$m_a = \lim_{\Delta y \to 0} \frac{f(a, b + \Delta y) - f(a, b)}{\Delta y}$$

which is, by definition, equal to $f_y(a, b)$. Similarly, if C_b denotes the curve obtained by intersecting the graph of f with the plane $y = b$, then a completely analogous argument shows that $f_x(a, b)$ represents the slope of the line T_b tangent to C_b at P (see Fig. 19.18).

COMPUTING PARTIAL DERIVATIVES

If y is held fixed, then $f(x,y)$ can be considered as a function of the single variable x, and [19.3.2(a)] amounts to the derivative of this single-variable function with respect to x in the sense of (2.3.1). Thus in most instances we can use our knowledge of ordinary derivatives from Chaps. 2 and 3 to find f_x. *We simply treat y as a constant and differentiate with respect to x.* Similarly, *to compute f_y we differentiate with respect to y, holding x fixed.*

Example 1 Determine f_x and f_y for the function $f(x,y) = 2x^3y^2 - 3x^4 + 9y^2 + 6$ and find their values at the point $(1, 2)$.

Solution To find f_x we treat y as if it were a constant. Thus $f_x(x,y) = 6x^2y^2 - 12x^3$. Similarly, holding x constant and differentiating with respect to y, we have

$$f_y(x,y) = 4x^3y + 18y$$

Hence

$$f_x(1,2) = 12, \qquad f_y = (1,2) = 44. \quad \bullet$$

Alternative derivative notations Alternative notations for f_x and f_y are

$$f_x(x,y) = f_1(x,y) = \frac{\partial f}{\partial x} \qquad \text{and} \qquad f_y(x,y) = f_2(x,y) = \frac{\partial f}{\partial y}$$

The latter notation is reminiscent of the Leibniz notation for functions of a single variable, and we shall understand $\frac{\partial}{\partial x}$ to mean "take the partial derivative with respect to x" in much the same way that $\frac{d}{dx}$ was understood to mean "take the derivative with respect to x." Thus, for example,

$$\frac{\partial}{\partial x}(3x^2y^4 + 9x) = 6xy^4 + 9 \qquad \text{and} \qquad \frac{\partial}{\partial y}(3x^2y^4 + 9x) = 12x^2y^3$$

If the function is given in the form $z = f(x,y)$, then

$$\frac{\partial z}{\partial x} = z_x \qquad \text{and} \qquad \frac{\partial z}{\partial y} = z_y$$

are also used to denote f_x and f_y, respectively.

Example 2 Determine $\frac{\partial f}{\partial x}$ and $\frac{\partial f}{\partial y}$ for the function $f(x,y) = y^2 \sin(x^3 + xy)$.

Solution One must be careful in this example to use the appropriate derivative rules.

$$\begin{aligned}
\frac{\partial f}{\partial x} &= \frac{\partial}{\partial x}[y^2 \sin(x^3 + xy)] \\
&= y^2 \frac{\partial}{\partial x}[\sin(x^3 + xy)] \\
&= y^2[\cos(x^3 + xy)]\frac{\partial}{\partial x}(x^3 + xy) \\
&= y^2[\cos(x^3 + xy)](3x^2 + y) = (3x^2y^2 + y^3)\cos(x^3 + xy)
\end{aligned}$$

Note that in the second step we needed to use the Chain Rule.

$$\begin{aligned}
\frac{\partial f}{\partial y} &= \frac{\partial}{\partial y}[y^2 \sin(x^3 + xy)] \\
&= y^2 \frac{\partial}{\partial y}[\sin(x^3 + xy)] + [\sin(x^3 + xy)]\frac{\partial}{\partial y}(y^2) \\
&= y^2[\cos(x^3 + xy)]\frac{\partial}{\partial y}(x^3 + xy) + [\sin(x^3 + xy)]2y \\
&= y^2[\cos(x^3 + xy)](x) + 2y \sin(x^3 + xy) \\
&= y[xy \cos(x^3 + xy) + 2 \sin(x^3 + xy)]
\end{aligned}$$

In this latter case the differentiation was with respect to y and hence we were required to treat f as the product of two functions. ●

In some cases it may be necessary to resort to the limit definition to determine the existence or nonexistence of a partial derivative at a particular point.

Example 3 Let $f(x,y) = \begin{cases} \dfrac{x^2y}{x^2+y^2} & \text{if } (x,y) \neq (0,0) \\ 0 & \text{if } (x,y) = (0,0) \end{cases}$

Determine $f_x(x,y)$.

Solution If $(x,y) \neq (0,0)$, then

$$f(x,y) = \frac{x^2y}{x^2+y^2}$$

so

$$f_x(x,y) = \frac{(x^2+y^2)(2xy) - x^2y(2x)}{(x^2+y^2)^2} = \frac{2xy^3}{(x^2+y^2)^2}$$

If $(x,y) = (0,0)$, then

$$f_x(0,0) = \lim_{\Delta x \to 0} \frac{f(0+\Delta x, 0) - f(0,0)}{\Delta x}$$

$$= \lim_{\Delta x \to 0} \frac{f(\Delta x, 0) - 0}{\Delta x} = \lim_{\Delta x \to 0} \frac{0-0}{\Delta x} = 0$$

[Since we were taking the limit as $\Delta x \to 0$, we assumed $\Delta x \neq 0$. Thus $(\Delta x, 0) \neq (0,0)$ and so, by the definition of $f(x,y)$,

$$f(\Delta x, 0) = \frac{(\Delta x)^2 \cdot 0}{(\Delta x)^2 + 0^2} = 0]$$ ●

PROGRESS TEST 1

In Probs. 1 and 2 determine f_x and f_y.

1. $f(x,y) = 3x^2y^4 + x^3y^2$

2. $f(x,y) = xy \ln(x^2+y^2)$

3. Determine $f_x(0,0)$ for

$$f(x,y) = \begin{cases} \dfrac{3x^3+5y^3}{x^3+y^2} & \text{if } (x,y) \neq (0,0) \\ 0 & \text{if } (x,y) = (0,0) \end{cases}$$

PARTIAL DERIVATIVES AS RATES OF CHANGE

Suppose at a particular instant the temperature of a heated metal plate located in the plane is given by $T(x,y) = 9 - x^2 + y^4$, where (x,y) represents a typical point on the plate. The partials

$$\frac{\partial T}{\partial x} = -2x \qquad \text{and} \qquad \frac{\partial T}{\partial y} = 4y^3$$

measure the rate of change of the temperature with respect to a horizontal and vertical change in direction, respectively. In particular, if a particle located at

$(-1, 2)$ (where $\partial T/\partial x = 2$, $\partial T/\partial y = 32$) were to be moved slightly to the right ($\Delta x > 0$), the temperature would increase by 2° per unit of displacement, and if it were to move slightly downward ($\Delta y < 0$), the temperature would decrease by 32° per unit of displacement.

---

⑨ For another example, recall that the volume of a right circular cylinder is given as a function of two variables by $V = \pi r^2 h$, where r is the radius and h is the height. Thus

$$\frac{\partial V}{\partial r} = 2\pi rh \qquad \frac{\partial V}{\partial h} = \pi r^2$$

Note that $\partial V/\partial r$ is the formula for the lateral surface area of a cylinder of radius r and height h, and $\partial V/\partial h$ is the formula for the area of a circle of radius r. This is, of course, what we saw in Chaps. 2 and 3 when we treated V as functions of the height and radius separately. We review the situation in our new terminology. Now $\partial V/\partial r$ measures the rate of change of the volume with respect to the radius. Suppose you leave the height of the cylinder fixed and make a small change Δr in the radius. The thin shells pictured in Fig. 19.19 represent the change in the volume ΔV induced by this change in the radius. For Δr small, $\Delta V/\Delta r$ is approximately equal to the lateral surface area of the cylinder ($\Delta V/\Delta r = 2\pi rh + \pi h\,\Delta r$).

Similarly, if r is fixed and h is changed by Δh units, then the resulting change in the volume ΔV is represented by a thin disk of radius r (see Fig. 19.20). In fact, $\Delta V/\Delta h$ equals the area of a circle of radius r; that is, $\Delta V/\Delta h = (\pi r^2\,\Delta h)/\Delta h$.

Thus we see that in the limiting cases ($\Delta r \to 0$ or $\Delta h \to 0$), if h is fixed, then the rate of growth of the volume with respect to r is "by lateral surface areas." If r is fixed, the rate of growth of the volume is "by areas of circular disks."

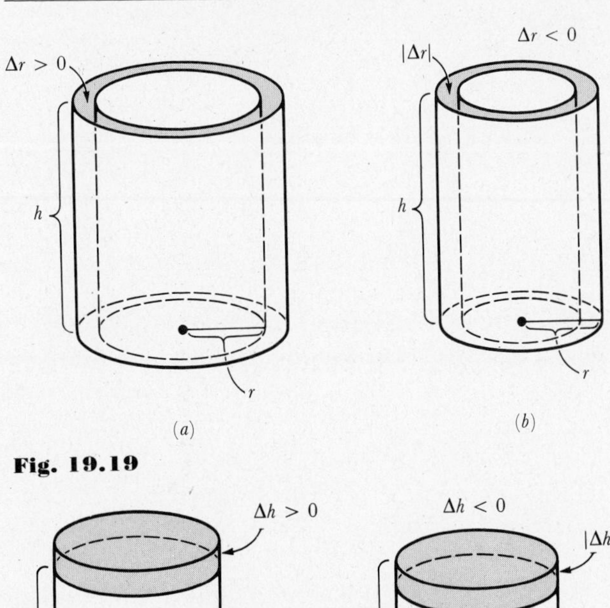

Fig. 19.19

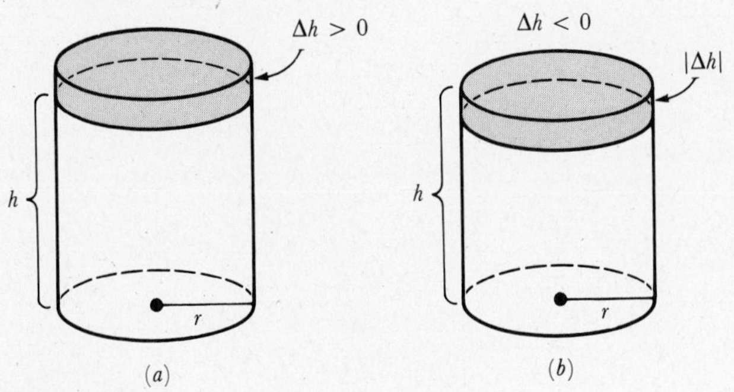

Δh > 0

Δh < 0

|Δh|

h

h

r

r

(a)

(b)

Fig. 19.20

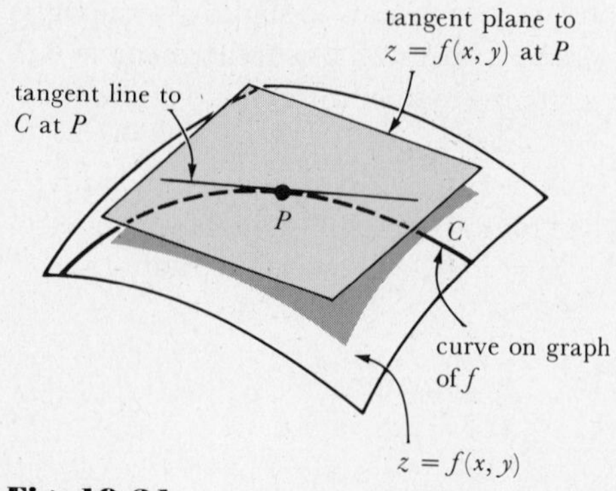

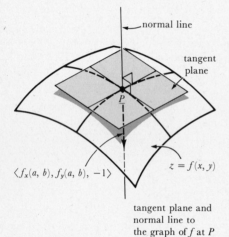

Fig. 19.21

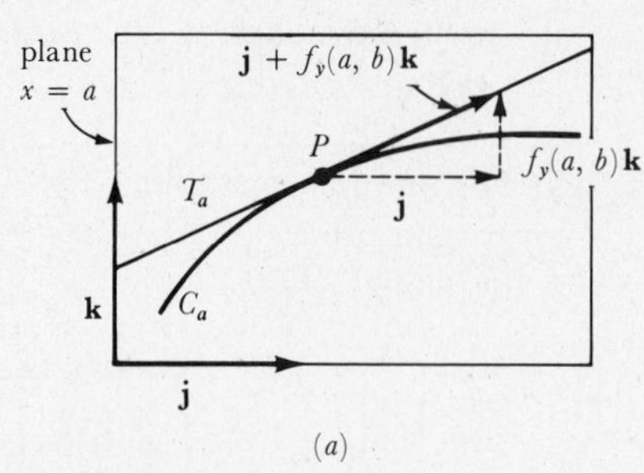

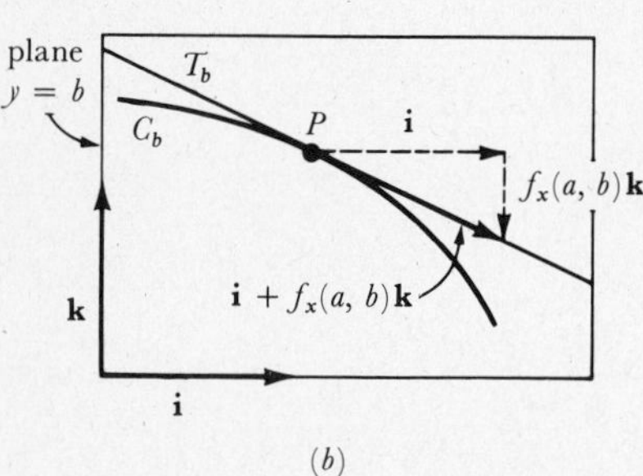

Fig. 19.22

normal line
tangent plane
P
$\langle f_x(a, b), f_y(a, b), -1\rangle$
$z = f(x, y)$

tangent plane and normal line to the graph of f at P

Fig. 19.23

TANGENT PLANES

The analog of the tangent line to the graph of a function of one variable is the notion of a *tangent plane* to the graph of a function of two variables. Let S denote the graph of $z = f(x, y)$ and let C be a curve lying on S and passing through the point $P = (a, b, f(a, b))$. The *tangent plane to S at P*, denoted Π_T, should contain the tangent line to C at P for any such curve C (see Fig. 19.21). In particular, if the tangent plane exists it will contain the lines T_a and T_b. Now T_a lies in the plane $x = a$ and in this plane has slope $f_y(a, b)$. Hence $\mathbf{j} + f_y(a, b)\mathbf{k}$ is a vector parallel to T_a; see Fig. 19.22(*a*). Similarly, $\mathbf{i} + f_x(a, b)\mathbf{k}$ is parallel to T_b; see Fig. 19.22(*b*).

A normal to Π_T can be obtained by crossing these vectors

$$\begin{aligned}\mathbf{N}_T &= (\mathbf{j} + f_y(a, b)\mathbf{k}) \times (\mathbf{i} + f_x(a, b)\mathbf{k}) \\ &= \langle 0, 1, f_y(a, b)\rangle \times \langle 1, 0, f_x(a, b)\rangle \\ &= \begin{vmatrix} \mathbf{i} & \mathbf{j} & \mathbf{k} \\ 0 & 1 & f_y(a, b) \\ 1 & 0 & f_x(a, b) \end{vmatrix} = \langle f_x(a, b), f_y(a, b), -1\rangle\end{aligned}$$

Since Π_T goes through $(a, b, f(a, b))$, we can use (17.5.4), page 583, to give its equation.

19.3.3 **Theorem** Suppose f is a function of two variables whose tangent plane Π_T exists at $(a, b, f(a, b))$. Then the vector

$$\langle f_x(a, b), f_y(a, b), -1\rangle$$

is a normal to Π_T, and Π_T has the equation

$$f_x(a, b)(x - a) + f_y(a, b)(y - b) = z - f(a, b)$$

If $z = f(x, y)$ has a tangent plane at $P = (a, b, f(a, b))$, then the line L_N through P and perpendicular to the tangent plane is called the *normal line to f at P* (see Fig. 19.23).

Since $\langle f_x(a, b), f_y(a, b), -1\rangle$ is normal to Π_T, we conclude the following:

19.3.4 **Theorem** The normal line to $z = f(x, y)$ at $(a, b, f(a, b))$ has parametric equations

$$x = a + f_x(a, b)t \qquad y = b + f_y(a, b)t \qquad z = f(a, b) - t$$

Example 4 Find equations for the tangent plane and normal line to the surface

$$z = 3x^4y^2 - 2xy^3 \text{ at the point } (-1, 3, 81)$$

Solution Now $f_x(x,y) = 12x^3y^2 - 2y^3$ and $f_y(x,y) = 6x^4y - 6xy^2$, so $f_x(-1,3) = -162$ and $f_y(-1,3) = 72$. Thus

$$\Pi_T: \quad -162(x+1) + 72(y-3) = z - 81$$

and

$$L_N: \quad x = -1 - 162t, y = 3 + 72t, z = 81 - t \quad \bullet$$

It is possible for $f_x(a,b)$ and $f_y(a,b)$ to exist and for the tangent plane at $P = (a, b, f(a,b))$ to *fail* to exist. Such a situation would occur, for example, if you intersect the graph with a plane (distinct from $x = a$ or $y = b$) perpendicular to the x-y plane and get a resulting curve of intersection that either has no tangent line at P or has a tangent line that does not lie in the plane determined by T_a and T_b. It can be shown that if the partial derivatives are continuous, then this will not happen.

PROGRESS TEST 2

1. Determine equations for the tangent plane and normal line to the graph of the function $f(x,y) = \sqrt{x^2+y^2}$ at the point $(3, 4, 5)$.
2. The temperature distribution of a heated plate in a plane is given by $T(x,y) = 12 - 3x^2 + 6y^5$. Determine the rate of change of the temperature in the direction of the positive x axis at $(3, -1)$.

FUNCTIONS OF THREE VARIABLES

Partial derivatives of functions of three variables are defined in an analogous way. In this setting, two variables are held fixed and the function is differentiated with respect to the remaining variable.

19.3.5 **Definition** If $w = f(x,y,z)$, then the *partial derivatives* f_x, f_y, and f_z are defined by:

(a) $f_x(x,y,z) = \lim_{\Delta x\to 0} \dfrac{f(x+\Delta x, y, z) - f(x,y,z)}{\Delta x}$

(b) $f_y(x,y,z) = \lim_{\Delta y\to 0} \dfrac{f(x, y+\Delta y, z) - f(x,y,z)}{\Delta y}$

(c) $f_z(x,y,z) = \lim_{\Delta z\to 0} \dfrac{f(x, y, z+\Delta z) - f(x,y,z)}{\Delta z}$

whenever the appropriate limits exist.

As before, we shall on occasion use the alternate notation

$$f_x = \frac{\partial f}{\partial x} \qquad f_y = \frac{\partial f}{\partial y} \qquad f_z = \frac{\partial f}{\partial z}$$

Example 5 Determine f_x, f_y, and f_z for the function $f(x,y,z) = y^2\sin(xyz) + ze^{xy}$.

Solution Holding y and z constant and differentiating with respect to x, we have

$$f_x(x,y,z) = y^2[\cos(xyz)]\,yz + ze^{xy}\cdot y = y^3z\cos(xyz) + yze^{xy}$$

Similarly, using the Product and Chain Rules on the first term,

$$f_y(x,y,z) = xyz^2\cos(xyz) + 2y\sin(xyz) + xze^{xy}$$

and

$$f_z(x,y,z) = xy^3\cos(xyz) + e^{xy} \quad \bullet$$

HIGHER-ORDER PARTIALS

If f is a function of the two variables x and y, then the partial derivatives f_x and f_y are themselves functions of two variables. The partial derivatives of these functions (if they exist) are called *second partials* of f. In our notation they are

$$(f_x)_x,\ (f_x)_y,\ (f_y)_x,\ (f_y)_y$$

where, for example, $(f_x)_y$ means the partial derivative of the function f_x with respect to y. It is customary to drop the parentheses and write them as

$$f_{xx}, f_{xy}, f_{yx}, f_{yy}$$

where the order of differentiation is read from left to right. In the "∂" notation the four second partials are

$$\frac{\partial}{\partial x}\left(\frac{\partial f}{\partial x}\right), \frac{\partial}{\partial y}\left(\frac{\partial f}{\partial x}\right), \frac{\partial}{\partial x}\left(\frac{\partial f}{\partial y}\right), \frac{\partial}{\partial y}\left(\frac{\partial f}{\partial y}\right)$$

which we collapse to

$$\frac{\partial^2 f}{\partial x^2}, \frac{\partial^2 f}{\partial y\,\partial x}, \frac{\partial^2 f}{\partial x\,\partial y}, \frac{\partial^2 f}{\partial y^2}$$

respectively. In this latter notation we must be careful to read the order of differentiation *from right to left*.

Example 6 Determine the four second partials for the function $f(x,y) = x^2y^3 - 3xy^2$.

Solution Since $f_x(x,y) = 2xy^3 - 3y^2$ and $f_y(x,y) = 3x^2y^2 - 6xy$, we have

$$f_{xx}(x,y) = 2y^3 \qquad f_{xy}(x,y) = 6xy^2 - 6y$$
$$f_{yx}(x,y) = 6xy^2 - 6y \qquad f_{yy}(x,y) = 6x^2y - 6x \quad \bullet$$

The partial derivatives f_{xy} and f_{yx} are called *mixed partials* of f. In the above example the mixed partials are equal. This is in fact the case for most of the functions that we will encounter. The theorem telling us the conditions under which mixed partials are equal follows. Its proof is usually given in more advanced courses.

19.3.6 **Theorem** If f_{xy} and f_{yx} are continuous at (a, b), then $f_{xy}(a,b) = f_{yx}(a,b)$.

Examples of functions whose mixed partials are *not* equal are given in Exercises 51 and 52.

Third-, fourth-, and higher-order partials are defined in a completely analogous way. For example,

$$f_{yxyy} = \frac{\partial^4 f}{\partial y^2\, \partial x\, \partial y}$$

indicates the fourth-order partial derivative obtained by differentiating first with respect to y, then x, and then twice more with respect to y.

Example 7 Determine $\dfrac{\partial^3 f}{\partial x\, \partial y\, \partial x}$ for $f(x, y) = x^3y + xe^y$.

Solution

$$\frac{\partial f}{\partial x} = 3x^2y + e^y$$

$$\frac{\partial^2 f}{\partial y\, \partial x} = \frac{\partial}{\partial y}(3x^2y + e^y) = 3x^2 + e^y$$

$$\frac{\partial^3 f}{\partial x\, \partial y\, \partial x} = \frac{\partial}{\partial x}(3x^2 + e^y) = 6x \quad \bullet$$

If all partial derivatives of a given order are continuous, then repeated application of (19.3.6) shows that the order of differentiation is irrelevant. Thus, for example,

$$f_{xyx} = f_{xxy} = f_{yxx}$$

In particular, each of these derivatives for the function in Example 7 would equal $6x$ (check this!). In many instances the independence of the order of differentiation can simplify our computations.

Example 8 Let $f(x, y) = xy^2 \sec^2(y^3)$. Determine $\dfrac{\partial^5 f}{\partial x^2\, \partial y^3}$.

Solution If we differentiate using the indicated order, then we must begin by differentiating three times with respect to y, which, in view of the products involved, is rather tedious. However, since derivatives of f of any order will result in continuous functions, we have

$$\frac{\partial^5 f}{\partial x^2 \partial y^3} = \frac{\partial^5 f}{\partial y^3 \partial x^2}$$

Now $\partial f/\partial x = y^2 \sec^2(y^3)$, so $\partial^2 f/\partial x^2 = 0$, and hence

$$\frac{\partial^5 f}{\partial y^3\, \partial x^2} = 0 \quad \bullet$$

Higher-order partials for functions of three variables are denoted in the same way, and the corresponding theorem on mixed partials carries over. In fact, the definitions and theorems hold for functions of more than three variables as well.

Example 9 Compute the mixed second partials for the function

$$f(x, y, z) = xe^y + ye^z + ze^x$$

Solution $f_x = e^y + ze^x, f_y = xe^y + e^z, f_z = ye^z + e^x$. Thus

$$f_{xy} = e^y = f_{yx} \qquad f_{xz} = e^x = f_{zx} \qquad f_{yz} = e^z = f_{zy}$$

(We used the continuity of the second partials to cut the number of computations in half.) $\bullet$

PROGRESS TEST 3

1. Determine f_x, f_y, and f_z for the function $f(x, y, z) = xz^2 \ln(x^2 + y^2 + z^2)$.

2. Determine the indicated partial derivative:
(*a*) $f(x, y) = y^3 \sin(xy)$, f_{xyx}
(*b*) $f(x, y) = x \tan(y^3)$, f_{yyxx}
(*c*) $f(x, y, z) = xy^2z^3 - 2/(xy)$, f_{xyz}

SECTION 19.3 EXERCISES

In Exercises 1 to 20 determine the indicated partial derivatives.

1. $f(x,y) = x^2\sqrt{x + y^3} - xy^2; f_x, f_y$

2. $f(x,y) = \arcsin\left(\dfrac{x}{\sqrt{x^2 + y^2}}\right); f_x, f_y$

3. $f(x,y) = e^{x/y^2} \ln(y/x^2); f_x, f_y$

4. $f(x,y) = x^2 \tan(\pi y^3); f_x, f_y$

5. $f(x,y) = (x^2 - y^2)/\sqrt{x^3 + xy^2}; f_x, f_y$

6. $f(x,y) = 3\pi x^2 - \cos(x^2/y^3); f_x, f_y$

7. $f(x,y) = \sqrt[3]{x^2y + 1} - 1/\sqrt[3]{y^2x + 1}; f_x, f_y$

8. $f(x,y) = x^2\cot(x,y) - y^2 \csc(x^2\sqrt{y}); f_x, f_y$

9. $f(x,y,z) = \ln(\sqrt{x^2 + y^2 + z^2}); f_x, f_y$

10. $f(x,y,z) = xyz^2 + x^2y^4/(3x^2 + yz); f_x, f_y, f_z$

11. $f(x,y,z) = \cos^3(\sqrt{xz^4} + yx^3); f_x, f_y, f_z$

12. $f(r,\theta) = r^2 \cos(2r - \theta) + 3\theta^2r^3; f_r, f_\theta$

13. $f(s,t) = 4s^2\sqrt{t^2s + 9} + 1/t^2 + s; f_s, f_t$

14. $f(u,v,w,z) = 3u^4wz^2 - 9w^3v^2 + 8uvw; f_u, f_v, f_w, f_z$

15. $f(x,y,z,t) = tz^3\sqrt{xy^2} + xzt; f_x, f_y, f_z, f_t$

16. $f(r,s,t,u) = e^{(s+t)/u} + t^2r^3 \ln(u/s); f_r, f_s, f_t, f_u$

17. $f(x,y,z,w) = w \sin(x^2y^3z); f_x, f_y, f_z, f_w$

18. $f(w,x) = e^{\sqrt{w^2 + x}} - \ln(x/w^2); f_w, f_x$

19. $f(r,h) = \sec^2(r^2 + 2rh); f_r, f_h$

20. $f(V,I,Z) = V(V^2 + I^3Z) - \sqrt{V^2 + I^2 + Z^2}; f_V, f_I, f_Z$

In Exercises 21 to 25 determine equations for the tangent plane and normal line to the surface at the given point.

21. $z = 4 - x^2 - y^2; P = (2, 1, -1)$

22. $z = x^2 + y + 3\ln(xy); P = (1, 1, 2)$

23. $z = e^y \cos \pi x; P = (1, 0, -1)$

24. $z = \sqrt{x^2 + 2y^2}; P = (3, 4, \sqrt{41})$

25. $z = 3x^2y^3 - 2x^2; P = (-1, 2, 22)$

In Exercises 26 to 29 determine the value of the indicated partial derivative at the given point.

26. $f(x,y) = \begin{cases} \dfrac{3xy^2}{x^2 + y^2} & (x,y) \neq (0,0) \\ 0 & (x,y) = (0,0); \end{cases} \quad f_x(0,0)$

27. $f(x,y) = \begin{cases} \dfrac{4x^4}{x^3 + y^2} & (x,y) \neq (0,0) \\ 0 & (x,y) = (0,0); \end{cases} \quad f_x(0,0)$

28. $f(x,y) = \begin{cases} \dfrac{x^2 + y^2 - 2y}{x + y - 1} & (x,y) \neq (0,1) \\ 0 & (x,y) = (0,1); \end{cases} \quad f_y(0,1)$

29. $f(x,y) = \begin{cases} \dfrac{\sin(x + y^2)}{y^2 + 2y} & (x,y) \neq (0,0) \\ 0 & (x,y) = (0,0); \end{cases} \quad f_x(0,0)$

30. Let $f(x,y) = \begin{cases} \dfrac{xy}{x^2 + y^2} & \text{if } (x,y) \neq (0,0) \\ 0 & \text{if } (x,y) = (0,0) \end{cases}$

(*a*) Show that $f(x,y)$ is not continuous at $(0,0)$.
(*b*) Show that $f_x(0,0)$ and $f_y(0,0)$ exist.
(Thus the existence of partial derivatives at a point need not imply the continuity of the function at that point!)

31. Show that $f(x,y) = 4x^4 - 2x^2y^2 + 3y^4$ satisfies the partial differential equation

$$x\frac{\partial f}{\partial x} + y\frac{\partial f}{\partial y} = 4f(x,y)$$

32. Show that $w = x^2y + zy^2 + xz^2$ satisfies the partial differential equation $w_x + w_y + w_z = (x + y + z)^2$.

33. The temperature distribution of a heated plate in the plane is given by $T(x,y) = (12 - 3x^2 + 6y^5)^2$. What change in temperature will result from a small upward movement from the point $(3, -1)$?

34. Determine f_x, f_y, and f_z if

$$f(x,y,z) = \int_{xz^4}^{x^2y} \sin(t^2)\, dt$$

In Exercises 35 to 44 determine all distinct second partials.

35. $f(x,y) = x^2\sqrt{y} + 3x/y$

36. $f(x,y) = (x-y)^2/(y^2-x)$

37. $f(x,y) = x^2 \arctan(xy)$

38. $f(x,y) = x \ln(x/y)$

39. $f(x,y) = \sin(xe^y)$

40. $f(x,y) = \arcsin(xy)$

41. $f(x,y) = x/\sqrt{x^2+y^2}$

42. $f(x,y) = \sqrt{x^3+y^2}$

43. $f(x,y,z) = \sqrt{x^2+y^2+z^2}$

44. $f(x,y,z) = x^2 \ln(xy/z)$

In Exercises 45 to 50 determine the indicated partial of f.

45. $f(x,y) = x\tan(y^3)$, f_{yyxx}

46. $f(x,y) = x^2y^2 - x^4y + 3$, $\partial^3 f/\partial x \partial y \partial z$

47. $f(u,v,w) = \ln(u^2 + 3v - w^3)$, $\partial^3 f/\partial u^2\, \partial v$

48. $f(P,I) = \ln[\sin(P-I)]$, f_{PPI}

49. $f(r,\theta) = r^2e^{r\theta}$, $\partial^4 f/\partial\theta^4$

50. $f(x,y,z) = xe^y + ye^z + ze^x$, $\partial^4 f/\partial x\, \partial y^2\, \partial z$

51. Let f be defined by

$$f(x,y) = \begin{cases} (xy)\left(\dfrac{x^2-y^2}{x^2+y^2}\right) & \text{if } (x,y) \neq (0,0) \\ 0 & \text{if } (x,y) = (0,0) \end{cases}$$

(*a*) Use definition [19.3.2(*a*)] to show that $f_x(0,y) = -y$ for all y.

(*b*) Use definition [(19.3.2(*b*)] to show that $f_y(x,0) = x$ for all x.

(*c*) Using (*a*) and the fact that

$$f_{xy}(0,0) = \lim_{\Delta y \to 0} \frac{f_x(0,\Delta y) - f_x(0,0)}{\Delta y}$$

show that $f_{xy}(0,0) = -1$.

(*d*) Using (*b*) and the fact that

$$f_{yx}(0,0) = \lim_{\Delta x \to 0} \frac{f_y(\Delta x,0) - f_y(0,0)}{\Delta x}$$

show that $f_{yx}(0,0) = 1$.

52. Using an argument similar to that in Exercise 51 show that $f_{xy}(0,0) \neq f_{yx}(0,0)$ for the function $f(x,y)$

$$= \begin{cases} x^2\arctan(y/x) - y^2\arctan(x/y) & \text{if } xy \neq 0 \\ 0 & \text{if } xy = 0 \end{cases}$$

53. Verify that the function $f(x,y) = \arctan(x/y)$ satisfies $\partial^2 f/\partial x^2 + \partial^2 f/\partial y^2 = 0$.

54. Determine $\partial^2 F/\partial x\, \partial y$ if

$$F(x,y) = \int_0^{xy^2} \arctan(t)\, dt$$

55. Show that $f(x,y,z) = (x^2+y^2+z^2)^{-1/2}$, $(x^2+y^2+z^2 \neq 0)$, satisfies $f_{xx} + f_{yy} + f_{zz} = 0$. This latter equation is often called *Laplace's equation in three dimensions*.

Section 19.4: Differentiability and the Total Differential

Objectives:

1. Understand geometrically and algebraically the notion of differentiability of functions of two variables in relation to the corresponding notion for functions of one variable.

2. Use differentials to approximate the value of a two- or three-variable function near a given point.

INTRODUCTION

How should we define differentiability for a function of two or three variables? One might be tempted to define it in terms of the existence of the partial derivatives. However, in Exercise 30, Sec. 19.3, the function

$$f(x,y) = \begin{cases} \dfrac{xy}{x^2+y^2} & (x,y) \neq (0,0) \\ 0 & (x,y) = (0,0) \end{cases}$$

was shown to possess both first partials at $(0,0)$ and yet be discontinuous at that point! Certainly, differentiability should imply continuity. Figure 19.24 illustrates another weakness of the existence of the first partials as a condition for differentiability. The curves of intersection with the planes $x = 0$ and $y = 0$ have horizontal tangent lines at $(0,0)$. That is, $f_x(0,0) = f_y(0,0) = 0$. However, the curve of intersection with the plane $y = x$ has a corner at $(0,0)$ and hence no tangent line there. Thus this function has no tangent plane at $(0,0)$. Any definition of differentiability for a function of two variables at a point should also ensure the existence of a tangent plane at that point.

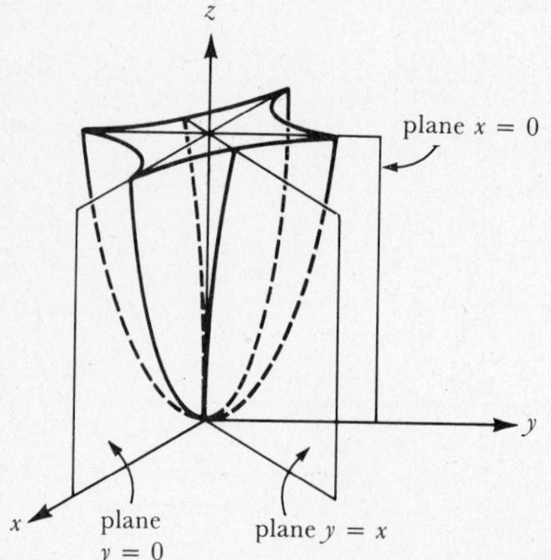

Fig. 19.24

In Sec. 3.3 it was shown that if a function of a single variable is differentiable at x, then

$$\lim_{\Delta x \to 0} \frac{\Delta f - df}{\Delta x} = 0 \tag{1}$$

where $\Delta f = f(x + \Delta x) - f(x)$ is the change in f resulting from the change Δx in the independent variable. It is this intimate relationship between the increment in f and the differential of f that we wish to preserve. We begin by extending these notions to functions of several variables.

THE TOTAL DIFFERENTIAL

19.4.1 Definition If f is a function of two variables, then the *total differential of f,* denoted df, or $df(x,y)$ is defined by

$$df = f_x(x,y)\,\Delta x + f_y(x,y)\,\Delta y$$

where Δx and Δy are arbitrary real numbers and (x,y) is a point at which f possesses first partials. If f is given in the form $z = f(x,y)$, then we also denote the total differential by dz.

Note that df is a function of the four variables $x, y, \Delta x$, and Δy. As in the single-variable case, we define the differentials of the independent variables by

19.4.2 $$dx = \Delta x \qquad \text{and} \qquad dy = \Delta y$$

If Δx and Δy are considered as increments in x and y, then the change from (x,y) to $(x + \Delta x, y + \Delta y)$ produces a change Δf (or Δz) in the function f, given by

19.4.3 $$\Delta f = f(x + \Delta x, y + \Delta y) - f(x,y) \qquad \text{(see Fig. 19.25)}$$

Example 1 Compare the values of Δf and df for the function $f(x,y) = x^2 + xy^2$ if $x = 1$, $y = 3$, $\Delta x = 0.01$, and $\Delta y = -0.02$.

Solution By (19.4.1),

$$\begin{aligned} df &= f_x(x,y)\,\Delta x + f_y(x,y)\,\Delta y \\ &= (2x + y^2)\,\Delta x + (2xy)\,\Delta y \end{aligned}$$

Letting $x = 1, y = 3, \Delta x = 0.01, \Delta y = -0.02$, we have

$$df = 11(0.01) + 6(-0.02) = -0.01$$

From (19.4.3),

$$\begin{aligned} \Delta f &= f(1 + 0.01, 3 - 0.02) - f(1, 3) \\ &= f(1.01, 2.98) - f(1, 3) \\ &= (1.01)^2 + (1.01)(2.98)^2 - 10 = -0.010696 \quad \bullet \end{aligned}$$

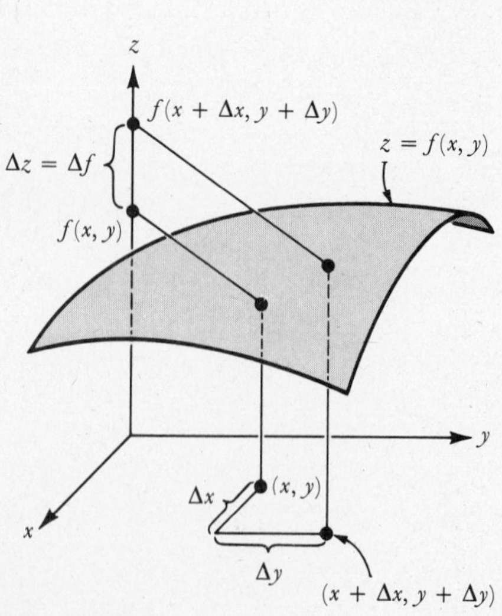

Fig. 19.25

Example 1 illustrates that the differential df of a function of two variables is a useful approximation of Δf in the same way that it was a useful approximation for functions of one variable. In particular: (*a*) The computations involved in computing df are considerably less tedious than those in computing Δf. (*b*) For small Δx and Δy, df is close to Δf. In this case df is within 0.000696 of Δf.

The following theorem gives a condition on the partial derivatives of f that ensures this "good approximation" property. It will also guide us in formulating a definition of differentiability for a function of two variables.

19.4.4 **Theorem** Suppose the function $f(x, y)$ has continuous first partials in some open disk S about (a, b). Then

$$\Delta f(a, b) = df(a, b) + G_1\,\Delta x + G_2\,\Delta y$$

where G_1 and G_2 are both functions of Δx and Δy, satisfying

$$\lim_{(\Delta x, \Delta y)\to(0,0)} G_1(\Delta x, \Delta y) = \lim_{(\Delta x, \Delta y)\to(0,0)} G_2(\Delta x, \Delta y) = 0$$

Proof: We choose Δx and Δy small enough that the rectangle centered at (a, b) of height $2|\Delta y|$ and width $2|\Delta x|$ lies inside A. Now

$$\Delta f = f(a + \Delta x, b + \Delta y) - f(a, b)$$

or

(2) $\Delta f = [f(a + \Delta x, b + \Delta y) - f(a, b + \Delta y)] + [f(a, b + \Delta y) - f(a, b)]$

Now fix Δx and Δy (both > 0) and let $p(x) = f(x, b + \Delta y)$, $q(y) = f(a, y)$. Then p and q are differentiable on the intervals $[a, a + \Delta x]$ and $[b, b + \Delta y]$, respectively. Applying the Mean-Value Theorem (4.4.4) to each of p and q, we have (see Fig. 19.26)

(3) $$p(a + \Delta x) - p(a) = p'(d)\,\Delta x \qquad a < d < a + \Delta x$$

and

(4) $$q(b + \Delta y) - q(b) = q'(c)\,\Delta y \qquad b < c < b + \Delta y$$

Now $p'(x) = f_x(x, b + \Delta y)$ and $q'(y) = f_y(a, y)$, so (3) and (4) are equivalent to

(5) $$f(a + \Delta x, b + \Delta y) - f(a, b + \Delta y) = f_x(d, b + \Delta y)\,\Delta x$$

and

(6) $$f(a, b + \Delta y) - f(a, b) = f_y(a, c)\,\Delta y$$

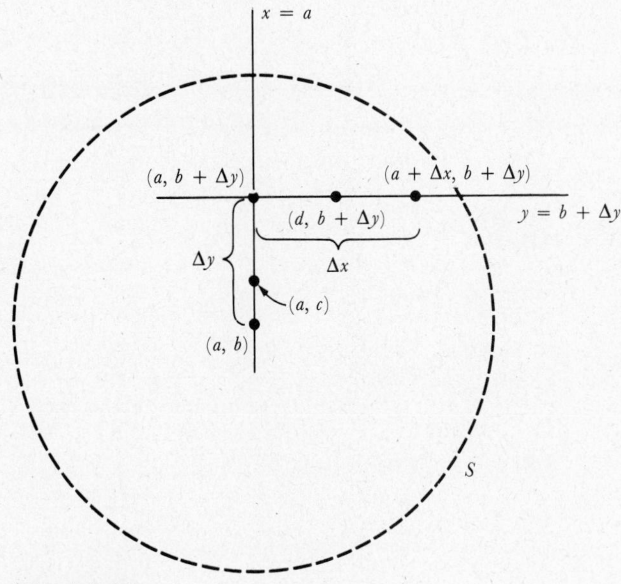

Fig. 19.26

respectively. Substituting (5) and (6) into (2), we have

$$\Delta f = f_x(d, b + \Delta y)\,\Delta x + f_y(a, c)\,\Delta y$$

Let

$$G_1 = f_x(d, b + \Delta y) - f_x(a, b) \qquad \text{and} \qquad G_2 = f_y(a, c) - f_y(a, b)$$

Then

$$\begin{aligned}\Delta f &= [f_x(a, b) + G_1]\,\Delta x + [f_y(a, b) + G_2]\,\Delta y \\ &= f_x(a, b)\,\Delta x + f_y(a, b)\,\Delta y + G_1\,\Delta x + G_2\,\Delta y \\ &= df + G_1\,\Delta x + G_2\,\Delta y\end{aligned}$$

Now as Δx and Δy tend to zero, we conclude that $d \to a$, $c \to b$, and $b + \Delta y \to b$. Since f_x and f_y are continuous, it follows that

$$G_1 = [f_x(d, b + \Delta y) - f_x(a, b)] \to [f_x(a, b) - f_x(a, b)] = 0$$

and

$$G_2 = [f_y(a, c) - f_y(a, b)] \to [f_y(a, b) - f_y(a, b)] = 0$$

The argument is essentially the same if we remove the restriction Δx, $\Delta y > 0$. ■

DIFFERENTIABILITY

Note that the conclusion of (19.4.4) ensures that $df \approx \Delta f$ for Δy and Δx small. With this in mind we provide the following definition.

19.4.5 **Definition of Differentiability** Suppose that f is a function of two variables whose first partials exists in some open disk about (a, b). We say *f is differentiable at* (a, b) if there exist functions G_1 and G_2 such that

(*a*)
$$\Delta f(a, b) = df(a, b) + G_1\Delta x + G_2\Delta y$$

and

(*b*)
$$\lim_{(\Delta x,\Delta y)\to(0,0)} G_1(\Delta x, \Delta y) = \lim_{(\Delta x,\Delta y)\to(0,0)} G_2(\Delta x, \Delta y) = 0$$

If f is differentiable at (a, b), then $\Delta f - df = G_1\,\Delta x + G_2\,\Delta y$. Thus

$$\begin{aligned}0 &\le \left|\frac{\Delta f - df}{|\Delta x| + |\Delta y|}\right| \\ &\le |G_1|\frac{|\Delta x|}{|\Delta x| + |\Delta y|} + |G_2|\frac{|\Delta y|}{|\Delta x| + |\Delta y|} \le |G_1| + |G_2|\end{aligned}$$

Since this latter sum tends to zero as $(\Delta x, \Delta y) \to (0, 0)$, we have:

19.4.6
$$\lim_{(\Delta x,\Delta y)\to(0,0)} \frac{\Delta f - df}{|\Delta x| + |\Delta y|} = 0$$

This is the two-variable analog of Theorem (3.3.3) mentioned in (1) above. In fact, condition (19.4.6) is *equivalent* to f being differentiable at (a, b); see Exercise 22.

Finally, to complete our analogy to the one-variable case, we have:

19.4.7 **Theorem** If a function f is differentiable at (a, b), then it is continuous at (a, b).

Proof: From (19.4.5), $\Delta f = df + G_1\,\Delta x + G_2\,\Delta y$. That is,

$$f(a + \Delta x, b + \Delta y) - f(a, b) = f_x(a, b)\,\Delta x + f_y(a, b)\,\Delta y + G_1\,\Delta x + G_2\,\Delta y$$

Thus

$$\lim_{(\Delta x,\Delta y)\to(0,0)} [f(a+\Delta x, b+\Delta y) - f(a,b)] = 0$$

or, equivalently,

$$\lim_{(\Delta x,\Delta y)\to(0,0)} f(a+\Delta x, b+\Delta y) = f(a,b)$$

Let $x = a + \Delta x$, $y = b + \Delta y$. Then $(x,y) \to (a,b)$ if and only if $(\Delta x, \Delta y) \to (0,0)$, and hence

$$\lim_{(x,y)\to(a,b)} f(x,y) = f(a,b) \quad \blacksquare$$

As might be expected, it is in general rather difficult to establish differentiability from either the definition or the equivalent condition (19.4.6). However, (19.4.4) implies the next theorem, which provides a condition guaranteeing differentiability and is much easier to check.

19.4.8 **Sufficient Condition for Differentiability** If $f_x(x,y)$ and $f_y(x,y)$ exist and are continuous in some open disk about (a,b), then f is differentiable at (a,b).

THE TOTAL DIFFERENTIAL AND THE TANGENT PLANE

The geometric interpretation of the total differential of a function of two variables again directly parallels the one-variable case. There, the fact that $df \approx \Delta f$ for Δx small was reflected geometrically by the fact that the tangent line is near the curve—provided you are near the point of tangency. Furthermore, as indicated in Fig. 19.27, df is the change in the height of the tangent line as we move from a to $a + \Delta x$, whereas Δf is the corresponding change in the height of the graph *of the function itself.*

Suppose now that f is a function of two variables. It can be shown that the differentiability of f at (a,b) is equivalent to the existence of a tangent plane to the graph of $z = f(x,y)$ at $P = (a, b, f(a,b))$. We shall assume this result. By (19.3.3), the tangent plane has the equation

(7) $$z - z_0 = f_x(a,b)(x-a) + f_y(a,b)(y-b)$$

where $z_0 = f(a,b)$.

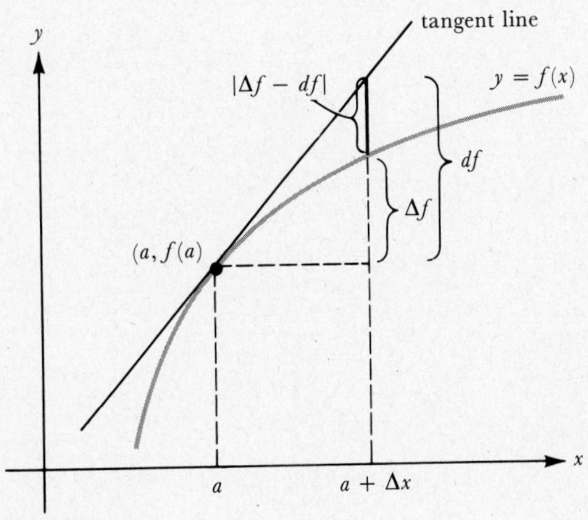

Fig. 19.27

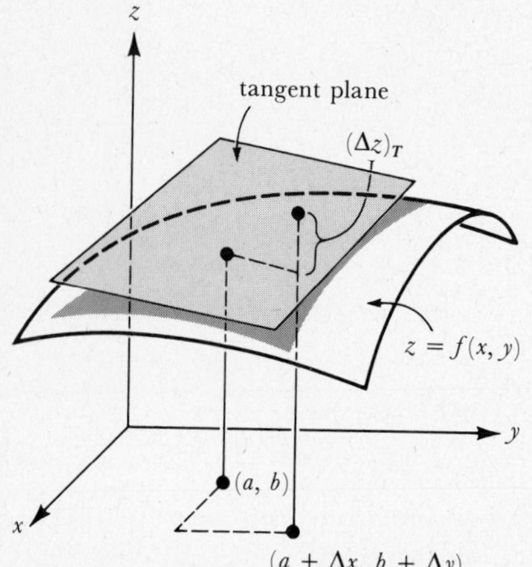

Fig. 19.28

Let $x = a + \Delta x$, $y = b + \Delta y$. Then (7) becomes

$$z - z_0 = f_x(a, b)\,\Delta x + f_y(a, b)\,\Delta y \tag{8}$$

Now the left side of equation (8) represents the change $(\Delta z)_T$ in the z coordinate *of the tangent plane* as you move from the point (a, b) to $(a + \Delta x, b + \Delta y)$; see Fig. 19.28. On the other hand the right side of equation (8) is by definition dz, so we have

$$(\Delta z)_T = dz$$

Now since $z = f(x, y)$, we know that

$$\Delta z = f(a + \Delta x, b + \Delta y) - f(a, b)$$

represents the change in the z coordinate *of the function itself* as we make the same move from (a, b) to $(a + \Delta x, b + \Delta y)$ in the x-y plane (Fig. 19.25). And since f is differentiable at (a, b), $\Delta z \approx dz$ for Δx and Δy small. But $dz = (\Delta z)_T$, so $\Delta z \approx (\Delta z)_T$. In Fig. 19.29, $|\Delta z - (\Delta z)_T|$ measures the vertical distance from the point on the surface above $(a + \Delta x, b + \Delta y)$ to the point on the tangent plane above $(a + \Delta x, b + \Delta y)$. Thus the fact that $\Delta z \approx dz$ is realized geometrically by the fact that the tangent plane is near the surface—provided you are near the point of tangency.

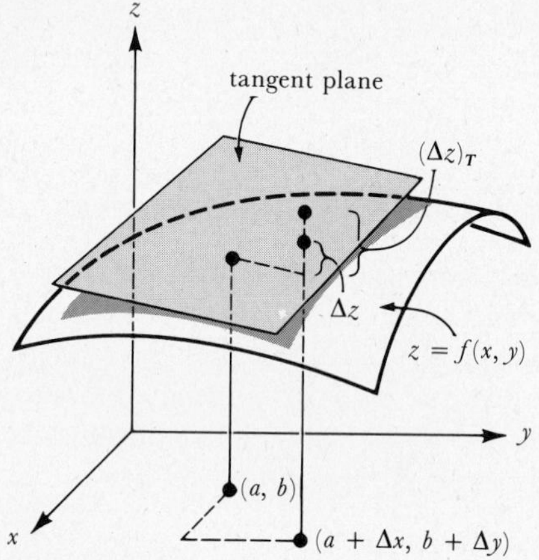

Fig. 19.29

If $w = f(x, y, z)$ is a function of three variables, then we extend (19.4.1) in a natural way and define the *total differential* df by

19.4.9 $$df = f_x(x, y, z)\,\Delta x + f_y(x, y, z)\,\Delta y + f_z(x, y, z)\,\Delta z$$

In turn the increment of f is given by

$$\Delta f = f(a + \Delta x, b + \Delta y, c + \Delta z) - f(a, b, c)$$

and *f is differentiable at (a, b, c)* if the first partials exist in some open ball about (a, b, c) and

$$\Delta f = df + G_1\,\Delta x + G_2\,\Delta y + G_3\,\Delta z$$

for functions G_1, G_2, and G_3 that tend to zero as $(\Delta x, \Delta y, \Delta z) \to (0, 0, 0)$. In this case there is no simple geometric realization for the differential, but we still have $df \approx \Delta f$ for Δx, Δy, and Δz small. As with functions of two variables, the continuity of the first partials ensures differentiability, and differentiability at a point ensures continuity at that point.

In many computations we use the approximation $\Delta f \approx df$ in the form

$$f(a + \Delta x, b + \Delta y) \approx f(a, b) + df$$

or

$$f(a + \Delta x, b + \Delta y, c + \Delta z) \approx f(a, b, c) + df$$

Example 2 Use the total differential to approximate $\sqrt{(3.02)^2 + (3.96)^2}$.

Solution Let $f(x, y) = \sqrt{x^2 + y^2}$, with $a = 3$, $b = 4$, $\Delta x = 0.02$, and $\Delta y = -0.04$. Then

$$f(a + \Delta x, b + \Delta y) = \sqrt{(3.02)^2 + (3.96)^2}$$

Now

$$f_x(x, y) = \frac{x}{\sqrt{x^2 + y^2}} \quad \text{and} \quad f_y(x, y) = \frac{y}{\sqrt{x^2 + y^2}},$$

so $f_x(3, 4) = 3/5$ and $f_y(3, 4) = 4/5$. By (19.4.6),

$$\sqrt{(3.02)^2 + (3.96)^2} \approx 5 + \frac{3}{5(0.02)} - \frac{4}{5(0.04)} = 4.98 \quad \bullet$$

PROGRESS TEST 1

1. Let $z = 2x^2 - 3xy + y^2$. Determine $|\Delta z - dz|$ if $x = -1, y = 3, \Delta x = -0.01$, and $\Delta y = 0.02$.
2. Approximate $\sqrt{(3.01)^2 + (2.97)^3}$.

SECTION 19.4 EXERCISES

In Exercises 1 to 10 determine $|\Delta f - df|$ for the indicated values of x, y, Δx, and Δy.

1. $f(x,y) = 2xy^2$; $x = -2, y = 3, \Delta x = 0.01, \Delta y = 0.04$
2. $f(x,y) = xy + x^3$; $x = 1, y = 4, \Delta x = -0.03, \Delta y = 0.3$
3. $f(x,y) = xy^2 - 2x^2$; $x = 4, y = 5, \Delta x = 0.001, \Delta y = 0.04$
4. $f(x,y) = 3x^2 + xy - 4y^2$; $x = 1, y = -3, \Delta x = 0.06, \Delta y = 0.08$
5. $f(x,y) = e^{-xy}$; $x = 2, y = 1.5, \Delta x = 0.03, \Delta y = -0.02$
6. $f(x,y) = \ln(y/x)$; $x = 4, y = 3, \Delta x = 0.07, \Delta y = -0.04$
7. $f(x,y) = xe^{xy}$; $x = 2, y = -2, \Delta x = 0.2, \Delta y = -0.1$
8. $f(x,y) = 4x^2 - 2xy + y^2$; $x = 1, y = -2, \Delta x = -0.02, \Delta y = 0.03$
9. $f(x,y) = xy + \ln(xy)$; $x = 1, y = 5, \Delta x = -0.02, \Delta y = 0.01$
10. $f(x,y) = x - y/x + y$; $x = 3, y = 5, \Delta x = 0.01, \Delta y = 0.02$

In Exercises 11 to 15 use (19.4.6) to show that the functions are differentiable for all (x, y).

11. $f(x,y) = x^2 + 4xy$
12. $f(x,y) = 3x^2 + 5y^2$
13. $f(x,y) = y^2x - 2xy$
14. $f(x,y) = x + y/x \quad (x \neq 0)$
15. $f(x,y) = y^2/x \quad (x \neq 0)$

16. Let $f(x,y) = \begin{cases} \dfrac{x^2y^2}{x^4 + y^4} & \text{if } (x,y) \neq (0,0) \\ 0 & \text{if } (x,y) = (0,0) \end{cases}$

(*a*) Determine f_x and f_y for $(x,y) \neq (0,0)$
(*b*) Use the definitions of f_x and f_y to find $f_x(0,0)$ and $f_y(0,0)$.
(*c*) Use Theorem (19.4.8) to show that f is differentiable at $(0,0)$.

17. When measuring the dimensions of a right circular cylinder, maximum errors of 1 percent and 0.5 percent are possible in measuring the radius and height, respectively. Approximate the maximum possible error in calculating the volume.

18. Approximate the length of the third side of a triangle with adjacent sides of lengths 15.8 cm and 30.2 cm if the angle between these sides is 58°.
19. Approximate $\sqrt{(0.98)^2 + (1.97)^2 + (2.02)^2}$.
20. Approximate $\sqrt{(297)^2 + (402)^2}$
21. Approximate $\sin 46° \cos 29°$.
22. Let f be a function of two variables such that

$$\lim_{(\Delta x, \Delta y) \to (0,0)} \frac{\Delta f(a,b) - df(a,b)}{|\Delta x| + |\Delta y|} = 0$$

Show that f is differentiable at (a, b).

23. A rectangular box is measured and found to have base 3 cm by 7 cm and height 18 cm. Approximate the maximum error in the volume measurement if the maximum error in measuring the dimensions was 0.02 cm.

Section 19.5: The Chain Rule

Objectives:

Use the appropriate version of a Chain Rule to compute partial (or ordinary) derivatives of the composition of two functions.

INTRODUCTION

The familiar Chain Rule for functions of a single variable enables us to differentiate the composition of two functions using the derivatives of the functions being composed. The situation for functions of two or more variables is more complex since several different types of compositions can be formed.

For example, if $z = f(x,y)$ and if in turn $x = h(r,s)$ and $y = g(r,s)$, then we can compose f with the functions h and g to yield z as a function of r and s:

$$z = f(h(r,s),\ g(r,s))$$

[Of course, for such a composition to make sense each (r,s) must be in the domains of both h and g and the resulting pairs $(h(r,s), g(r,s))$ must be in the domain of f. We shall assume throughout this section that all functions considered have domains so that the compositions being discussed are well defined.]

On the other hand, if we compose the functions $x = p(t)$ and $y = q(t)$ with f, then we have z as a function of the single variable t:

$$z = f(p(t), q(t))$$

In the former case a Chain Rule is in order to express the partial derivatives of z with respect to r and s in terms of the partials of f, h, and g. In the latter case we would obtain the (ordinary) derivative of z with respect to t in terms of *partial* derivatives of f and (ordinary) derivatives of p and q.

Even if we limit our attention to functions of one, two, or three variables, there are several different ways that compositions can be formed, and so several versions of a Chain Rule are needed. We shall begin by giving the statement and proof of one particular version of the Chain Rule. The general Chain Rule that encompasses all possibilities follows.

THE CHAIN RULE—A SPECIAL CASE

19.5.1 **Theorem** Let $z = f(x, y)$ be differentiable and suppose that the functions $x = x(r, s)$ and $y = y(r, s)$ are continuous and possess partial derivatives with respect to r and s. Then z considered as a function of r and s has partial derivatives with respect to these variables given by

$$(a) \qquad \frac{\partial z}{\partial r} = \frac{\partial z}{\partial x}\frac{\partial x}{\partial r} + \frac{\partial z}{\partial y}\frac{\partial y}{\partial r}$$

and

$$(b) \qquad \frac{\partial z}{\partial s} = \frac{\partial z}{\partial x}\frac{\partial x}{\partial s} + \frac{\partial z}{\partial y}\frac{\partial y}{\partial s}$$

Proof: Since the proofs of (a) and (b) are similar, we shall prove (a). Now z, as a function of r and s, is given by

$$z = f(x(r, s), y(r, s))$$

By definition,

$$(1) \qquad \frac{\partial z}{\partial r} = \lim_{\Delta r \to 0} \frac{f(x(r + \Delta r, s), y(r + \Delta r, s)) - f(x(r, s), y(r, s))}{\Delta r}$$

provided that this limit exists.

Now since s is fixed, we take $\Delta s = 0$, and so

$$(2) \qquad \Delta x = x(r + \Delta r, s) - x(r, s) \qquad \Delta y = y(r + \Delta r, s) - y(r, s)$$

Thus

$$x(r + \Delta r, s) = x(r, s) + \Delta x = x + \Delta x$$

and

$$y(r + \Delta r, s) = y(r, s) + \Delta y = y + \Delta y$$

Substituting in (1), we have

$$\frac{\partial z}{\partial r} = \lim_{\Delta r \to 0} \frac{f(x + \Delta x, y + \Delta y) - f(x, y)}{\Delta r} = \lim_{\Delta r \to 0} \frac{\Delta z}{\Delta r}$$

Since f is differentiable, by definition we know that

$$(3) \qquad \Delta z = \frac{\partial z}{\partial x}\Delta x + \frac{\partial z}{\partial y}\Delta y + G_1 \Delta x + G_2 \Delta y$$

where $G_1, G_2 \to 0$ as $(\Delta x, \Delta y) \to (0, 0)$.

We can assume G_1, G_2 are zero if $\Delta x = \Delta y = 0$ because, if not, we can simply redefine them to be zero at $(0, 0)$ as done in the proof of the single-variable Chain Rule (9.3.1).

Now dividing (3) by Δr, we have

$$\frac{\Delta z}{\Delta r} = \frac{\partial z}{\partial x}\frac{\Delta x}{\Delta r} + \frac{\partial z}{\partial y}\frac{\Delta y}{\Delta r} + G_1\frac{\Delta x}{\Delta r} + G_2\frac{\Delta y}{\Delta r} \tag{4}$$

Let $\Delta r \to 0$. From (2) and the continuity of $x(r, s)$ and $y(r, s)$, it follows that Δx and $\Delta y \to 0$, and hence G_1, $G_2 \to 0$. Thus, from (4),

$$\begin{aligned}\frac{\partial z}{\partial r} &= \lim_{\Delta r \to 0} \frac{\Delta z}{\Delta r}\\ &= \frac{\partial z}{\partial x}\lim_{\Delta r \to 0}\frac{\Delta x}{\Delta r} + \frac{\partial z}{\partial y}\lim_{\Delta r \to 0}\frac{\Delta y}{\Delta r} + \lim_{\Delta r \to 0}\left[G_1\frac{\Delta x}{\Delta r} + G_2\frac{\Delta y}{\Delta r}\right]\\ &= \frac{\partial z}{\partial x}\frac{\partial x}{\partial r} + \frac{\partial z}{\partial y}\frac{\partial y}{\partial r} \quad \blacksquare\end{aligned}$$

Example 1 Let $z = x^2y + xy$, $x = re^s$, $y = re^{-s}$. Determine $\partial z/\partial r$ and $\partial z/\partial s$.

Solution By (19.5.1.a),

$$\begin{aligned}\frac{\partial z}{\partial r} &= \frac{\partial z}{\partial x}\frac{\partial x}{\partial r} + \frac{\partial z}{\partial y}\frac{\partial y}{\partial r}\\ &= (2xy + y)e^s + (x^2 + x)e^{-s}\\ &= (2r^2 + re^{-s})e^s + (r^2e^{2s} + re^s)e^{-s} = 3r^2e^s + 2r\end{aligned}$$

Similarly, by (19.5.1.b),

$$\begin{aligned}\frac{\partial z}{\partial s} &= \frac{\partial z}{\partial x}\frac{\partial x}{\partial s} + \frac{\partial z}{\partial y}\frac{\partial y}{\partial s}\\ &= (2xy + y)(re^s) + (x^2 + x)(-re^{-s})\\ &= (2r^2 + re^{-s})(re^s) + (r^2e^{2s} + re^s)(-re^{-s}) = r^3e^s \quad \bullet\end{aligned}$$

In the previous example if we express z as a function of r and s directly, we have $z = r^3e^s + r^2$. Thus

$$\frac{\partial z}{\partial r} = 3r^2e^s + 2r \qquad \text{and} \qquad \frac{\partial z}{\partial s} = r^3e^s$$

which agrees, of course, with the results of Example 1.

THE GENERAL CHAIN RULE

We can generalize (19.5.1) to deal with the situation where

$$z = f(x_1, x_2, \ldots, x_n)$$

is a function of the n variables $x_1, x_2, \ldots, x_n$ and where for each i, $1 \le i \le n$,

$$x_i = x_i(u_1, u_2, \ldots, u_m)$$

is a function of the m variables $u_1, u_2, \ldots, u_m$.

In this setting the x_i's are called *intermediate* variables and the u_j's are called *independent* variables. The following general version of the Chain Rule, which has essentially the same proof as (19.5.1), tells us how to find the partials with respect to the independent variables in terms of the partials with respect to the intermediate variables.

19.5.2 **The General Chain Rule** Let $z = f(x_1, x_2, \ldots, x_n)$ be differentiable and suppose that for each $i = 1, 2, \ldots, n$, $x_i = x_i(u_1, u_2, \ldots, u_m)$ is continuous and possesses partial derivatives with respect to each of the variables $u_1, u_2, \ldots, u_m$. Then, for each $j = 1, 2, \ldots, m$,

$$\frac{\partial z}{\partial u_j} = \frac{\partial z}{\partial x_1}\frac{\partial x_1}{\partial u_j} + \frac{\partial z}{\partial x_2}\frac{\partial x_2}{\partial u_j} + \cdots + \frac{\partial z}{\partial x_n}\frac{\partial x_n}{\partial u_j}$$

Hence to find the partial derivative of z with respect to one of the independent variables, say u_j, we must:

(i) Compute the partial derivatives of z with respect to *all* of the *intermediate* variables.
(ii) Compute the partial derivative of all of the intermediate variables with respect to the *given* u_j.
(iii) Multiply the results from (i) and (ii) term by term and add.

Notice that there will always be one term in the sum for each intermediate variable. It is understood in (19.5.2) that any partial derivatives become ordinary derivatives whenever the function being differentiated at any stage is a function of a single variable.

For emphasis we state the conclusion of (19.5.2) in some of the commonly occurring situations.

19.5.3 $$z = f(x, y),\ x = x(t),\ y = y(t)$$

$$\frac{dz}{dt} = \frac{\partial z}{\partial x}\frac{dx}{dt} + \frac{\partial z}{\partial y}\frac{dy}{dt}$$

19.5.4 $$w = f(x, y, z),\ x = x(t), y = y(t), z = z(t)$$

$$\frac{dw}{dt} = \frac{\partial w}{\partial x}\frac{dx}{dt} + \frac{\partial w}{\partial y}\frac{dy}{dt} + \frac{\partial w}{\partial z}\frac{dz}{dt}$$

19.5.5 $$w = f(x, y, z),\ x = x(r, s),\ y = y(r, s),\ z = z(r, s)$$

(*a*) $$\frac{\partial w}{\partial r} = \frac{\partial w}{\partial x}\frac{\partial x}{\partial r} + \frac{\partial w}{\partial y}\frac{\partial y}{\partial r} + \frac{\partial w}{\partial z}\frac{\partial z}{\partial r}$$

(*b*) $$\frac{\partial w}{\partial s} = \frac{\partial w}{\partial x}\frac{\partial x}{\partial s} + \frac{\partial w}{\partial y}\frac{\partial y}{\partial s} + \frac{\partial w}{\partial z}\frac{\partial z}{\partial s}$$

19.5.6 $$z = f(u),\ u = u(x, y)$$

(*a*) $$\frac{\partial z}{\partial x} = \frac{dz}{du}\frac{\partial u}{\partial x}$$

(*b*) $$\frac{\partial z}{\partial y} = \frac{dz}{du}\frac{\partial u}{\partial y}$$

It is probably worthwhile for you to formulate the remaining five versions that one can encounter when dealing with functions of one, two, or three variables.

Example 2 Show that if f is differentiable, then $z = f(x/y)$ satisfies the partial differential equation

$$x\frac{\partial z}{\partial x} + y\frac{\partial z}{\partial y} = 0$$

Solution Let $z = f(u)$ and $u = u(x,y) = x/y$. Then, by (19.5.6),

$$\frac{\partial z}{\partial x} = \frac{dz}{du}\frac{\partial u}{\partial x} = \frac{dz}{du}\left(\frac{1}{y}\right)$$

and

$$\frac{\partial z}{\partial y} = \frac{dz}{du}\frac{\partial u}{\partial y} = \frac{dz}{du}\left(\frac{-x}{y^2}\right)$$

Thus

$$x\frac{\partial z}{\partial x} + y\frac{\partial z}{\partial y} = \frac{x}{y}\frac{dz}{du} - \frac{x}{y}\frac{dz}{du} = 0 \quad \bullet$$

Example 3 A plate exposed to a heat source maintains the temperature distribution $T(x,y) = 9 - 3x^2 - 5y^2 - xy$ over the time interval $[0, a]$. A particle moving along a certain path on the surface of the plate has velocity $\langle -4, 3\rangle$ at the instant $t = t_0$ $(0 < t_0 < a)$ that it passes through the point $(-1, 1)$. Assuming that the particle has the temperature of its location, how fast is the temperature of the particle changing at the instant the particle is at $(-1, 1)$?

Solution Let $\mathbf{F}(t) = \langle x(t), y(t)\rangle$ be the vector function that describes the motion of the particle. Now $\mathbf{V}(t) = \langle dx/dt, dy/dt\rangle$, and since $\mathbf{V}(t_0) = \langle -4, 3\rangle$ we have

$$\left.\frac{dx}{dt}\right|_{t=t_0} = -4 \quad \text{and} \quad \left.\frac{dy}{dt}\right|_{t=t_0} = 3$$

By the Chain Rule in (19.5.3),

$$\begin{aligned}\frac{dT}{dt} &= \frac{\partial T}{\partial x}\frac{dx}{dt} + \frac{\partial T}{\partial y}\frac{dy}{dt}\\ &= (-6x - y)\frac{dx}{dt} + (-10y - x)\frac{dy}{dt}\end{aligned}$$

Since the particle is located at the point $(-1, 1)$ at time t_0, we have $x = -1, y = 1$ at time t_0. Thus

$$\left.\frac{dT}{dt}\right|_{t=t_0} = (6 - 1)(-4) + (-10 + 1)(3) = -47$$

Hence if T is measured in degrees Celsius and t in seconds, then the temperature is decreasing at the rate of 47°C per second at the instant the particle passes through the point $(-1, 1)$. $\bullet$

PROGRESS TEST 1

1. Use the Chain Rule to find the indicated partial derivative.

(a) $\dfrac{\partial z}{\partial r}$, where $z = \dfrac{xy}{\sqrt{x^2 + y^2}}$, $x = r\cos\theta$, $y = r\sin\theta$

(b) $\dfrac{dz}{dt}$, where $z = xy^2 \ln\left(\dfrac{w}{x}\right)$, $x = \dfrac{1}{t}$, $y = t$, $w = t^2 + t$

2. The height of a right circular cone is decreasing at the rate of 4 cm/s and the radius is increasing at the rate of 2 cm/s. How fast is the volume changing at the instant that the radius is 12 cm and the height is 14 cm?

IMPLICIT DIFFERENTIATION

In Sec. 3.2 we developed a method for finding dy/dx if y is defined implicitly as a differentiable function of x by the equation $F(x, y) = k$. Using the Chain Rule we can arrive at a more direct means for determining dy/dx. Let $z = F(t, y)$, where $t = x$ and $y = y(x)$. Then z is a function of x alone, so if F is differentiable, then

$$\begin{aligned}\frac{dz}{dx} &= \frac{\partial F}{\partial t}\frac{dt}{dx} + \frac{\partial F}{\partial y}\frac{dy}{dx}\\ &= \frac{\partial F}{\partial t} + \frac{\partial F}{\partial y}\frac{dy}{dx}\\ &= \frac{\partial F}{\partial x} + \frac{\partial F}{\partial y}\frac{dy}{dx}\end{aligned}$$

Thus if $\partial F/\partial y \neq 0$, we have

19.5.7
$$\frac{dy}{dx} = \frac{-\partial F/\partial x}{\partial F/\partial y} = \frac{-F_x}{F_y}$$

In a completely analogous way, if the equation $F(x, y, z)$ defines z implicitly as a differentiable function of x and y, then

19.5.8
$$\frac{\partial z}{\partial x} = \frac{-F_x}{F_z}, \qquad \frac{\partial z}{\partial y} = \frac{-F_y}{F_z}$$

provided that $F_z \neq 0$.

Example 4 Assuming that the equation $x^3y^2 - 2xy^4 = x^2 + 1$ defines y as a differentiable function of x, find dy/dx.

Solution Let $F(x, y) = x^3y^2 - 2xy^4 - x^2 - 1$. Then $F_x = 3x^2y^2 - 2y^4 - 2x$ and $F_y = 2x^3y - 8xy^3$. Thus, by (19.5.7),

$$\frac{dy}{dx} = -\frac{3x^2y^2 - 2y^4 - 2x}{2x^3y - 8xy^3} = \frac{-3x^2y^2 + 2y^4 + 2x}{2x^3y - 8xy^3} \quad \bullet$$

As indicated in the above example, the idea is to bring all terms on one side of the equation, regard that side of the equation as giving a function of whatever variables appear, and then take partials in order to apply (19.5.7) or (19.5.8).

THE TOTAL DIFFERENTIAL AND THE CHAIN RULE

If $z = f(x, y)$, then the total differential is given by

$$dz = \frac{\partial z}{\partial x}dx + \frac{\partial z}{\partial y}dy \tag{5}$$

In this case $dx = \Delta x$ and $dy = \Delta y$ are arbitrary real numbers. Suppose now that $x = x(u, v)$, $y = y(u, v)$. Then z is a function of u and v, so the total differential is

$$dz = \frac{\partial z}{\partial u}du + \frac{\partial z}{\partial v}dv \tag{6}$$

If, in addition we agree to interpret dx and dy in (5) as total differentials when x and y are intermediate variables, then

$$dx = \frac{\partial x}{\partial u}\,du + \frac{\partial x}{\partial v}\,dv$$

and

$$dy = \frac{\partial y}{\partial u}\,du + \frac{\partial y}{\partial v}\,dv$$

Substituting into (5), we get

$$\begin{aligned} dz &= \frac{\partial z}{\partial x}\left[\frac{\partial x}{\partial u}\,du + \frac{\partial x}{\partial v}\,dv\right] + \frac{\partial z}{\partial y}\left[\frac{\partial y}{\partial u}\,du + \frac{\partial y}{\partial v}\,dv\right] \\ &= \left[\frac{\partial z}{\partial x}\frac{\partial x}{\partial u} + \frac{\partial z}{\partial y}\frac{\partial y}{\partial u}\right]du + \left[\frac{\partial z}{\partial x}\frac{\partial x}{\partial v} + \frac{\partial z}{\partial y}\frac{\partial y}{\partial v}\right]dv \\ &= \frac{\partial z}{\partial u}\,du + \frac{\partial z}{\partial v}\,dv \qquad \text{by the Chain Rule} \end{aligned}$$

Since we can also derive equation (5) from equation (6) by simply reversing the previous steps, we conclude that (5) and (6) are equivalent. In light of this we shall follow the convention of treating dx and dy as total differentials in (5) whenever x and y are intermediate variables. A similar convention applies when $w = f(x, y, z)$.

SECTION 19.5 EXERCISES

In Exercises 1 to 10 find the indicated partial (or ordinary) derivative (a) by using the Chain Rule and (b) by expressing w as a function of the indicated independent variables before differentiating.

1. $w = \dfrac{x}{x^2 + y^2}$, $x = r\cos\theta$, $y = r\sin\theta$; $\partial w/\partial\theta$

2. $w = 3x^2y - y/x$, $x = t^2$, $y = 3t + 4$; dw/dt

3. $w = x^2 + y^2 - 3z$; $x = r\cos\theta$, $y = r\sin\theta$, $z = r^3$; $\partial w/\partial r$

4. $w = \arcsin x$, $x = \dfrac{rst}{\sqrt{r^2 + s^2 + t^2}}$; $\partial w/\partial t$

5. $w = \dfrac{xy}{x^2 + y^2}$, $x = \sin t$, $y = \cos t$; dw/dt

6. $w = (x^2 + y^2 + z^2)^{3/2}$, $x = e^r\cos\theta$, $y = e^r\sin\theta$, $z = e^r$; $\partial w/\partial r$

7. $w = e^{xy}\sin(x + 2y)$, $x = r + 3s - t$, $y = 2r + s + t$; $\partial w/\partial r$

8. $w = \arctan\left(\dfrac{uv}{x}\right)$, $u = y + 2z$, $v = y^2$, $x = z^2y$; $\partial w/\partial z$

9. $w = \dfrac{x^2yz^3}{u} + \dfrac{2u^2y}{z}$, $x = t$, $y = \dfrac{1}{t}$, $z = t^2$, $u = 2t$; dw/dt

10. $w = \ln(x^2 + y^2)$, $x = e^r\cos\theta$, $y = e^r\sin\theta$; $\partial w/\partial r$, $\partial w/\partial\theta$

In Exercises 11 to 20 compute the indicated partial (or ordinary) derivatives.

11. $z = e^{x-y}/e^{x+y}$, $x = \ln(s/t)$, $y = \ln st$; $\partial z/\partial s$, $\partial z/\partial t$

12. $z = x^2 + y^2$, $x = \sin(s - t)$, $y = \cos(s + t)$; $\partial z/\partial s$, $\partial z/\partial t$

13. $z = y^3/x^2$, $x = r^2 - s$, $y = 2rs$; $\partial z/\partial r$, $\partial z/\partial s$

14. $z = e^{xy^2}$, $x = rs$, $y = r/\sqrt{1 + s^2}$; $\partial z/\partial r$, $\partial z/\partial s$

15. $w = e^{-x}y^2\cos z$, $x = t$, $y = 2t$, $z = 4t$; dw/dt

16. $z = \arcsin xy$, $x = s + t$, $y = s - t$; $\partial z/\partial s$, $\partial z/\partial t$

17. $z = e^x + y$, $x = \arctan(u/v)$, $y = \ln(u + v)$; $\partial z/\partial u$, $\partial z/\partial v$

18. $w = x^2 + y^2 + z^2 + v^2$, $x = r\cos t$, $y = r\sin t$, $z = rt$, $v = r$; $\partial w/\partial t$, $\partial w/\partial r$

19. $z = x^y$, $x = 1 + s^2$, $y = e^t$; $\partial z/\partial s$, $\partial z/\partial t$

20. $w = x^{yz}$, $x = r$, $y = r + s$, $z = \cos(rs)$; $\partial w/\partial r$, $\partial w/\partial s$

In Exercises 21 to 25 assume that the given equation defines y implicitly as a differentiable function of x and determine dy/dx.

21. $x^2y^3 - 3xy^4 + x^2 = 9$

22. $x^4 + 5x^2y^3 + y^4 = x$

23. $2x^{1/3} + 4y^{2/3} = 2$

24. $x\cos(xy^2) + y^3 = 3.$

25. $x^2 + \sqrt{xy} = y^2 + 1$

In Exercises 26 to 30 assume that the given equation defines z implicitly as a differentiable function of x and y and determine $\partial z/\partial x$, $\partial z/\partial y$.

26. $xz^4 - y^2xy = z^3 + 4$

27. $\sin(xyz^2) = xz + y$

28. $xe^{yz} - ze^{xy} = 5$

29. $x\sqrt{x^2 + y^2 + z^2} = z + 2$

30. $x^2yz^5 - xy^2z^3 = x^2y + 9$

31. Let $w = f(x^2 - y^2, y^2 - x^2)$, where $f(u, v)$ is differentiable. Show that

$$y\frac{\partial w}{\partial x} + x\frac{\partial w}{\partial y} = 0$$

32. Let $w = f(x,y)$ be differentiable and let $x = r\cos\theta$, $y = r\sin\theta$. Show that

$$\left(\frac{\partial w}{\partial r}\right)^2 + \frac{1}{r^2}\left(\frac{\partial w}{\partial \theta}\right)^2 = \left(\frac{\partial w}{\partial x}\right)^2 + \left(\frac{\partial w}{\partial y}\right)^2$$

33. Let $z = f(x^2 - y)$, where f is differentiable. Show that

$$y\frac{\partial z}{\partial x} + 2xy\frac{\partial z}{\partial y} = \frac{yz}{x}$$

34. If f is differentiable and $w = z^3 f(xy, xz)$, show that

$$x\frac{\partial w}{\partial x} - \frac{\partial w}{\partial y} - z\frac{\partial w}{\partial z} = 3w$$

35. Let F be differentiable and let $w = F(x_1 - x_2, x_2 - x_3, x_3 - x_4, x_4 - x_5)$. Show that

$$\sum_{i=1}^{5}\frac{\partial w}{\partial x_i} = 0$$

36. Let $w = f(\sqrt{x^2 + y^2 + z^2})$, where $w = f(t)$ is differentiable. Show that

$$\left(\frac{\partial w}{\partial x}\right)^2 + \left(\frac{\partial w}{\partial y}\right)^2 + \left(\frac{\partial w}{\partial z}\right)^2 = \left(\frac{df}{dt}\right)^2$$

37. Suppose that the temperature at the point (x, y, z) in space is given by $T(x, y, z) = x^2 + y^2 + z^2 - xyz$. A particle moving along the space curve $F(t) = \langle x(t), y(t), z(t)\rangle$ passes through the point $(1, -1, 2)$ at time $t = t_0$ with velocity $\langle 3, -2, 4\rangle$. How fast is the temperature changing at that instant?

38. The radius of a right circular cylinder is increasing at the rate of 4 cm/s, and the height is decreasing at the rate of 2 cm/s. How fast is the volume changing at the instant when the radius is 20 cm and the height is 40 cm?

39. A particle in the x-y plane passes through the point (3, 4) with velocity $\langle -2, 5\rangle$. Another particle travels along the surface $z = x^2 + y^2$ in such a way that it is always directly above the particle in the x-y plane. How fast is the height above the x-y plane of this latter particle changing at the instant that it passes through the point (3, 4, 5)?

40. Let $f(x, y) = \int_0^{(x^2+y^2)^{1/2}} \frac{dt}{\sqrt{t^2+1}}$. Find $\frac{\partial f}{\partial x}$ and $\frac{\partial f}{\partial y}$.

41. Let $f(x, y) = \int_{h(x)}^{g(y)} k(t)\,dt$. Find $\frac{\partial f}{\partial x}$ and $\frac{\partial f}{\partial y}$.

Section 19.6: The Gradient and the Directional Derivative

Objectives:

1. Determine the gradient of a function of several variables.

2. Determine and interpret the directional derivative of a function of several variables.

INTRODUCTION

If f is a function of two variables, then f_x measures the rate of change of f with respect to a change in distance in the x direction, that is, in the direction of the unit vector **i**. Similarly, f_y measures the rate of change of f in the y direction, the direction of the unit vector **j**. In this section we shall define a "directional derivative," $D_{\mathbf{u}}f$, which measures the rate of change of f with respect to a change in distance in the **u** direction, where **u** is *any* unit vector.

If we place a representative of **u** with its base at (a, b), then the plane through **u** and perpendicular to the x-y plane intersects the surface in a curve that goes through the point $P = (a, b, f(a, b))$. Geometrically, $D_{\mathbf{u}}f$, when evaluated at (a, b), should yield the slope of the tangent line to this curve at P. In particular, the special cases where $\mathbf{u} = \mathbf{i}$ or $\mathbf{u} = \mathbf{j}$ will then yield the familiar geometric interpretations of $f_x(a, b)$ and $f_y(a, b)$. These situations are pictured in Fig. 19.30.

THE DIRECTIONAL DERIVATIVE

Suppose **u** is a unit vector in the x-y plane. Then $\mathbf{u} = \langle\cos\theta, \sin\theta\rangle$, where θ is the direction of **u**. If $L_{\mathbf{u}}$ denotes the line in the x-y plane through (a, b) parallel to **u**, then $L_{\mathbf{u}}$ has parametric equations.

$$x = a + (\cos\theta)t \qquad y = b + (\sin\theta)t$$

We can now regard $L_{\mathbf{u}}$ as a t axis, where $t = 0$ corresponds to the point (a, b) and the positive direction along $L_{\mathbf{u}}$ is in the direction of **u** (see Fig. 19.31).

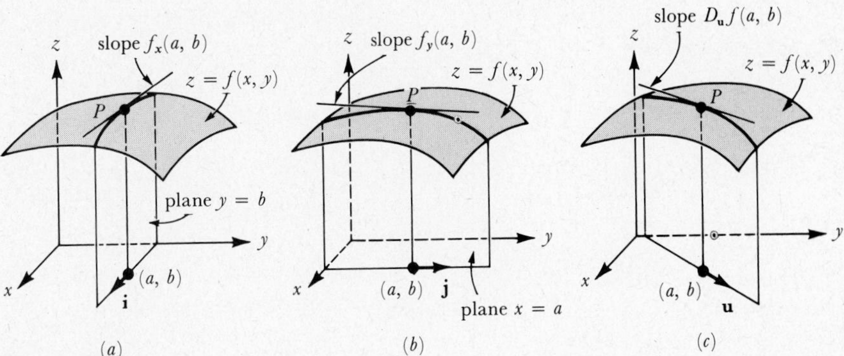

Fig. 19.30

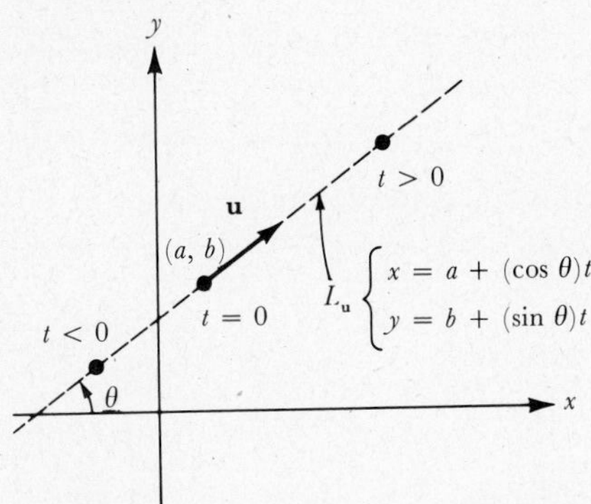

Fig. 19.31

$Q =$
$(a + t\cos\theta, b + t\sin\theta, f(a + t\cos\theta, b + t\sin\theta))$

C

$P = (a, b, f(a, b))$

secant line

line $L_{\mathbf{u}}$

t

(a, b)

plane through L perpendicular to x-y plane

$(a + t\cos\theta, b + t\sin\theta)$

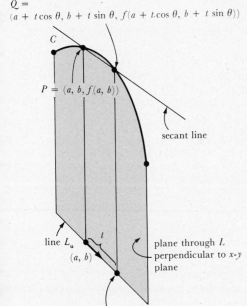

Fig. 19.32

Let C denote the curve of intersection of the surface $z = f(x, y)$ with the plane through $L_{\mathbf{u}}$ and perpendicular to the x-y plane. A point (x, y, z) is on C if and only if

$$x = a + (\cos\theta)t \qquad y = b + (\sin\theta)t \qquad z = f(a + (\cos\theta)t, b + (\sin\theta)t)$$

To determine the slope of the tangent line to C at P we consider first a secant line joining the points $P = (a, b, f(a, b))$ and $Q = (a + (\cos\theta)t, b + (\sin\theta)t, f(a + (\cos\theta)t, b + (\sin\theta)t))$, on C (see Fig. 19.32). This line has slope in the plane given by

$$(1) \qquad \frac{f(a + (\cos\theta)t, b + (\sin\theta)t) - f(a, b)}{t}$$

Furthermore, as $t \to 0$, the points Q tend to P along C. Thus the quotients in (1) will tend to the slope of the tangent line at P if it exists. Motivated by this discussion we make the following definition.

19.6.1 **Definition of Directional Derivative** Let f be a function of two variables. The *directional derivative* of f at (a, b) *in the direction of the unit vector* $\mathbf{u} = \langle\cos\theta, \sin\theta\rangle$ is defined by

$$D_{\mathbf{u}}f(a, b) = \lim_{t \to 0} \frac{f(a + (\cos\theta)t, b + (\sin\theta)t) - f(a, b)}{t}$$

provided this limit exists.

Notice that since t represents a change in distance in the $\mathbf{u}$ direction, $D_{\mathbf{u}}f$ can be interpreted as the rate of change of f in the $\mathbf{u}$ direction.

The following theorem gives us an easy means for computing directional derivatives.

19.6.2 **Theorem** Suppose $\mathbf{u} = \langle u_1, u_2 \rangle$ is a unit vector and $f(x, y)$ is differentiable at (a, b). Then

$$D_{\mathbf{u}}f(a, b) = f_x(a, b)u_1 + f_y(a, b)u_2$$

Proof: Since $\mathbf{u}$ is a unit vector, $u_1 = \cos\theta$ and $u_2 = \sin\theta$, where θ is the direction of $\mathbf{u}$. Let $h(t) = f(a + u_1t, b + u_2t)$. Then, by the definition of the derivative for a function of one variable,

$$\begin{aligned} h'(0) &= \lim_{t\to 0} \frac{h(t) - h(0)}{t} \\ &= \lim_{t\to 0} \frac{f(a + u_1t, b + u_2t) - f(a, b)}{t} \\ &= D_{\mathbf{u}}f(a, b) \qquad \text{by (19.6.1)} \end{aligned}$$

But $h(t)$ is the composition of $z = f(x, y)$ with the functions $x = x(t) = a + u_1t$ and $y = y(t) = b + u_2t$, so by the Chain Rule (19.5.3)

$$\begin{aligned} h'(t) &= \frac{dz}{dt} = \frac{\partial f}{\partial x}\frac{dx}{dt} + \frac{\partial f}{\partial y}\frac{dy}{dt} \\ &= f_x(a + u_1t, b + u_2t)u_1 + f_y(a + u_1t, b + u_2t)u_2 \end{aligned}$$

Letting $t = 0$, we obtain the desired result:

$$D_{\mathbf{u}}f(a, b) = h'(0) = f_x(a, b)u_1 + f_y(a, b)u_2 \quad \blacksquare$$

Note that, as expected, if $\mathbf{u} = \mathbf{i} = \langle 1, 0 \rangle$, then

$$D_{\mathbf{i}}f(a, b) = f_x(a, b)$$

and if $\mathbf{u} = \mathbf{j} = \langle 0, 1 \rangle$, then

$$D_{\mathbf{j}}f(a, b) = f_y(a, b)$$

If $\mathbf{v}$ is any nonzero vector, then we define $D_{\mathbf{v}}f(a, b)$, *the directional derivative* of f in *the direction* of $\mathbf{v}$, by

19.6.3 $$D_{\mathbf{v}}f(a, b) = D_{\mathbf{u}}f(a, b) \qquad \text{where } \mathbf{u} = \frac{\mathbf{v}}{\|\mathbf{v}\|}$$

Example 1 Determine the directional derivative at $(-1, 2)$ in the direction of $\mathbf{v} = \langle 3, 4 \rangle$ for the function $f(x, y) = x^2 + 2xy - y^2$.

Solution Let $\mathbf{u} = \mathbf{v}/\|\mathbf{v}\| = \langle 3/5, 4/5 \rangle$. Then $\mathbf{u}$ is a unit vector in the direction of $\mathbf{v}$. Now $f_x(x, y) = 2x + 2y$ and $f_y(x, y) = 2x - 2y$. Thus

$$\begin{aligned} D_{\mathbf{v}}f(-1, 2) &= D_{\mathbf{u}}f(-1, 2) \\ &= f_x(-1, 2)(3/5) + f_y(-1, 2)(4/5) \\ &= 2(3/5) + (-6)(4/5) = -18/5 \quad \bullet \end{aligned}$$

Suppose a thin plate occupies a region in the x-y plane and suppose $T(x, y)$ is the temperature distribution for the plate, where T is measured in degrees Celsius, and x and y are measured in centimeters. Then $D_{\mathbf{u}}T(a, b)$ gives the rate, in deg C/cm, at which the temperature changes at (a, b) in the direction of $\mathbf{u}$. Suppose a heat-seeking particle is located at position (a, b) on the plate. In what direction will the particle travel? Certainly it would want to head in the direction of maximal increase in temperature. Thus, among all possible directions $\mathbf{u}$, the answer is provided by the direction that yields the largest value for $D_{\mathbf{u}}T(a, b)$. Similarly, a heat-fleeing particle would travel in the direction that yields the least value for $D_{\mathbf{u}}T(a, b)$. As an aid in answering questions of this type it is useful to interpret the conclusion of Theorem (19.6.2) as a dot product.

THE GRADIENT

19.6.4 **Definition** If $z = f(x, y)$, the *gradient of f at (x, y)*, denoted ∇f or $\nabla f(x, y)$, is given by

$$\nabla f = \langle f_x(x, y), f_y(x, y) \rangle$$

The symbol ∇f, using an upside-down delta or "nabla," is read as "del *f*."

Note that if $\mathbf{u} = \langle u_1, u_2 \rangle$ is a unit vector, then

$$\begin{aligned} \nabla f(a, b) \cdot \mathbf{u} &= \langle f_x(a, b), f_y(a, b) \rangle \cdot \langle u_1, u_2 \rangle \\ &= f_x(a, b)u_1 + f_y(a, b)u_2 \end{aligned}$$

But this latter expression is, by (19.6.2), just $D_{\mathbf{u}}f(a, b)$. Thus we can restate (19.6.2) as

19.6.5 **Theorem** Suppose f is differentiable at (a, b) with $\mathbf{u}$ a unit vector. Then

$$D_{\mathbf{u}}f(a, b) = \nabla f(a, b) \cdot \mathbf{u}$$

If α is the angle between $\nabla f(a, b)$ and $\mathbf{u}$, then from one of the basic properties of the dot product (17.4.4), we know that

$$\begin{aligned} D_{\mathbf{u}}f(a, b) &= \nabla f(a, b) \cdot \mathbf{u} \\ &= \|\nabla f(a, b)\| \, \|\mathbf{u}\| \cos \alpha \end{aligned}$$

But $\mathbf{u}$ is a unit vector, so $\|\mathbf{u}\| = 1$, and we conclude that

$$D_{\mathbf{u}}f(a, b) = \|\nabla f(a, b)\| \cos \alpha$$

If (a, b) is fixed and $\mathbf{u}$ (hence α) is allowed to vary, then this expression will be maximized when $\cos \alpha = 1$. Since by definition $0 \le \alpha \le \pi$, the maximum occurs when $\alpha = 0$, that is, when $\mathbf{u}$ has the same direction as $\nabla f(a, b)$. The *minimum* value for $D_{\mathbf{u}}f(a, b)$ will occur when $\cos \alpha = -1$ or $\alpha = \pi$. In this case $\mathbf{u}$ has the opposite direction of $\nabla f(a, b)$, that is, the direction of $-\nabla f(a, b)$. Note that the actual maximum and minimum values for $D_{\mathbf{u}}f(a, b)$ are $\|\nabla f(a, b)\|$ and $-\|\nabla f(a, b)\|$, respectively. We summarize these observations in the following:

19.6.6 **Theorem** Suppose f is differentiable at (a, b). Then

(*a*) The maximum value of a directional derivative at (a, b) is $\|\nabla f(a, b)\|$, and it occurs in the direction of $\nabla f(a, b)$.

(*b*) The minimum value of a directional derivative at (a, b) is $-\|\nabla f(a, b)\|$, and it occurs in the direction of $-\nabla f(a, b)$.

Example 2 Determine the direction that yields the fastest rate of increase of f at $(1, 1)$ where $f(x, y) = x^3y - xy^2$.

Solution Since $f_x = 3x^2y - y^2$ and $f_y = x^3 - 2xy$, $\nabla f(x, y) = \langle 3x^2y - y^2, x^3 - 2xy \rangle$. Also, since $D_{\mathbf{u}}f(1, 1)$ measures the rate of change of f at $(1, 1)$ in the direction of $\mathbf{u}$, the fastest rate of increase will occur in the direction that maximizes $D_{\mathbf{u}}f(1, 1)$. By (19.6.6), this is the direction of $\nabla f(1, 1) = \langle 2, -1 \rangle$. ●

Example 3 The temperature distribution of a thin plate is given by $T(x,y) = 10 - x^2 - 2y^2$. A heat-seeking particle is located at the point $(-1, 3)$ on the plate. Determine the path C along which the particle will travel as it moves towards the origin, the hottest point.

Solution We solve the problem by finding a parameterization for C and then eliminating the parameter to obtain an equation for C. If $x = x(t), y = y(t)$ are parametric equations for C, then the vector $\langle x'(t), y'(t)\rangle$ points in the direction of the motion of the particle. By (19.6.6) this vector should always have the direction of ∇T. This will be the case if

$$\langle x'(t), y'(t)\rangle = \nabla T = \langle -2x, -4y\rangle$$

Setting components equal, we have the two differential equations

$$\frac{dx}{dt} = -2x \qquad \text{and} \qquad \frac{dy}{dt} = -4y$$

which are equivalent to

$$\frac{dx}{x} = -2dt \qquad \text{and} \qquad \frac{dy}{y} = -4dt$$

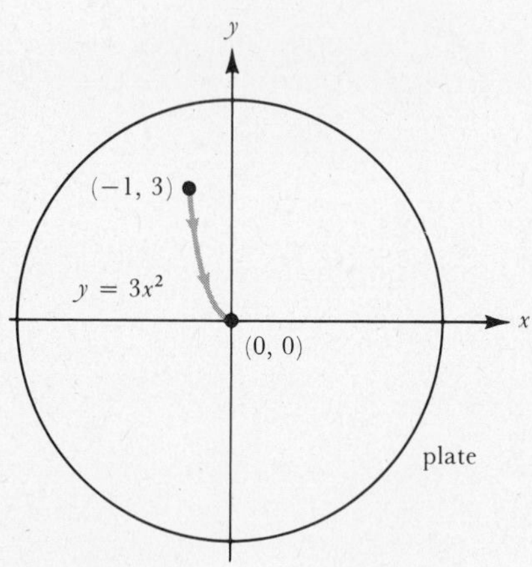

Fig. 19.33

respectively. From our work in Sec. 8.1 we know these have solutions

$$x = c_1 e^{-2t} \qquad \text{and} \qquad y = c_2 e^{-4t}$$

respectively. Since the particle starts at $(-1, 3)$, we know that

$$-1 = x(0) = c_1 \qquad \text{and} \qquad 3 = y(0) = c_2$$

Hence $x = -e^{-2t}$ and $y = 3e^{-4t}$ provide a pair of parametric equations for C. Eliminating the parameter, we observe that the particle will travel along the parabola $y = 3x^2$ from $(-1, 3)$ to $(0, 0)$; see Fig. 19.33. ●

PROGRESS TEST 1

1. Determine $D_{\mathbf{v}}f(a, b)$ for each of the following:

(a) $f(x,y) = 3x^2 + 4y^2$, $\mathbf{v} = \langle 1, 2\rangle$, $(a, b) = (3, -4)$

(b) $f(x,y) = \dfrac{x}{\sqrt{x^2 + y^2}}$, $\mathbf{v} = \langle \sqrt{3}, -1\rangle$, $(a, b) = (3, 4)$

2. The directional derivative of the function $f(x,y)$ at the point $(3, -2)$ in the direction of the vector $\langle 1, 3\rangle$ has value 2 while the directional derivative at this same point in the direction of $\langle 1, 0\rangle$ has value 4. Determine the value of the directional derivative of f at $(3, -2)$ in the direction toward the origin.

3. A particle is located on the surface $z = 3x^2y^2 + 10$ above the point $(2, 3)$ in the x-y plane. In what horizontal direction should the particle initially head in order to traverse the steepest path?

AN EXTENSION TO FUNCTIONS OF THREE VARIABLES

Let $w = f(x, y, z)$ and let $\mathbf{u} = \langle u_1, u_2, u_3\rangle$ be a unit vector.

19.6.7 **Definition** The *directional derivative of* f at (a, b, c) *in the direction of* $\mathbf{u}$ is defined by

$$D_{\mathbf{u}}f(a, b, c) = \lim_{t\to 0} \frac{f(a + u_1 t, b + u_2 t, c + u_3 t) - f(a, b, c)}{t}$$

provided that this limit exists.

As before, $D_{\mathbf{u}}f(a, b, c)$ gives the rate of change of f at (a, b, c) with respect to a change in distance in the **u** direction.

For functions of three variables, we define the gradient by

19.6.8
$$\nabla f(x, y, z) = \langle f_x(x, y, z), f_y(x, y, z), f_z(x, y, z)\rangle$$

Theorems (19.6.5) and (19.6.6) can be routinely extended. Thus if f is differentiable at (a, b, c) and if **u** is a unit vector, then

19.6.9
$$D_{\mathbf{u}}f(a, b, c) = \nabla f(a, b, c) \cdot \mathbf{u}$$

Furthermore, the maximal rate of increase of f is $\|\nabla f(a, b, c)\|$ and occurs when **u** has the direction of $\nabla f(a, b, c)$. Similarly, the minimum rate of $-\|\nabla f(a, b, c)\|$ occurs when **u** has the direction of $-\nabla f(a, b, c)$.

Example 4 Determine the directional derivative of $f(x, y, z) = x^2y - xz + yz^2$ at $(-1, 1, 2)$ in the direction of the vector $\mathbf{v} = \langle 2, 1, 2\rangle$.

Solution We first find a unit vector in the direction of **v**:

$$\mathbf{u} = \frac{\mathbf{v}}{\|\mathbf{v}\|} = \langle 2/3, 1/3, 2/3\rangle$$

Now

$$\nabla f(x, y, z) = \langle 2xy - z, x^2 + z^2, -x + 2yz\rangle$$

so

$$\nabla f(-1, 1, 2) = \langle -4, 5, 5\rangle$$

Thus

$$\begin{aligned} D_{\mathbf{v}}f(-1, 1, 2) &= D_{\mathbf{u}}f(-1, 1, 2) \\ &= \nabla f(-1, 1, 2) \cdot \mathbf{u} \\ &= \langle -4, 5, 5\rangle \cdot \langle 2/3, 1/3, 2/3\rangle \\ &= -8/3 + 5/3 + 10/3 = 7/3 \end{aligned}$$ ●

Example 5 In what direction will the function $f(x, y, z) = x^2 + 2yz - z^2$ experience the greatest rate of increase at the point $(2, -1, 3)$?

Solution Now $\nabla f(x, y, z) = \langle 2x, 2z, 2y - 2z\rangle$, so $\nabla f(2, -1, 3) = \langle 4, 6, -8\rangle$. Thus the maximal rate of increase in f will occur in the direction of the vector $\langle 4, 6, -8\rangle$. ●

SECTION 19.6 EXERCISES

In Exercises 1 to 10 determine the directional derivative of f at the given point in the direction of **v**.

1. $f(x, y) = x\sqrt{x^2 + y^2}$; $(-3, 4)$; $\mathbf{v} = \langle -4, 8\rangle$

2. $f(x, y) = e^x \cos y$; $(0, 0)$; $\mathbf{v} = \langle -\sqrt{3}, 1\rangle$

3. $f(x, y) = \dfrac{x}{\sqrt{x^3 + y^3}}$; $(2, 8)$; $\mathbf{v} = \langle -2, 3\rangle$

4. $f(x, y) = \ln(\sqrt{x^2 + y^2})$; $(1, \sqrt{3})$; $\mathbf{v} = \langle 4, -3\rangle$

5. $f(x, y) = 4x^2e^{2y} - xy^2$; $(3, 0)$; $\mathbf{v} = \langle 1, 1\rangle$

6. $f(x, y, z) = x^2 - 2xy + y^3$; $(1, 0, -3)$ $\mathbf{v} = \langle 2, 1, -6\rangle$

7. $f(x, y, z) = \ln(\sqrt{x^2 + y^2 + z^2})$; $(1, 1, \sqrt{2})$; $\mathbf{v} = \langle 2, -1, 2\rangle$

8. $f(x, y, z) = e^{xy}\cos(yz)$; $(2, 0, 0)$; $\mathbf{v} = \langle 4, 1, 7\rangle$

9. $f(x, y, z) = \dfrac{x}{\sqrt{x^2 + y^2 + z^2}}$; $(2, -2, 1)$; $\mathbf{v} = \langle 3, 2, -4\rangle$

10. $f(x, y, z) = x \arctan\dfrac{y}{z}$; $(3, 0, -2)$; $\mathbf{v} = \langle 4, -1, 2\rangle$

In Exercises 11 to 15 determine the direction that will yield (a) the maximum and (b) the minimum rate of increase in f at the given point.

11. $f(x,y) = x^3y - 3y^2$; $(3, -2)$

12. $f(x,y) = e^x \arctan y$; $(0, 1)$

13. $f(x,y) = \sqrt{x^2 + y^2}$; $(-8, 2)$

14. $f(x,y,z) = x^2 + 3xy - z^2$; $(1, -6, 3)$

15. $f(x,y,z) = x^2y + xy^2z^3$; $(2, 1, -1)$

16. Let C be a smooth curve given by $\mathbf{F}(s) = \langle x(s), y(s), z(s) \rangle$, where s represents arclength and let $F(s_0) = (a, b, c)$. Show that if $f(x,y,z)$ is differentiable at (a, b, c), then

$$\left.\frac{df}{ds}\right|_{s=s_0} = D_{\mathbf{T}} f(a, b, c)$$

where $\mathbf{T}$ denotes the unit tangent to C at $s = s_0$.

In Exercises 17 and 18 use the result of Exercise 16 to determine the solution.

17. Find the rate of change with respect to distance along C of the function $f(x,y) = x^2 + 3xy^2$ at the point $(2, 2)$, where C is the curve with equation $y^3 = 2x^2$.

18. Let C be the space curve given by $F(t) = \langle e^t, e^{-t}, \sqrt{2}t \rangle$. Find the rate of change with respect to distance along C for the function $f(x,y,z) = 3x^2 - 2y^2 + z^2$ at the point $(1, 1, 0)$.

19. The function $z = f(x,y)$ has, at the point $(1, 2)$, directional derivatives that are equal to 2 in the direction toward $(2, 2)$ and -2 in the direction toward $(1, 1)$. What is the directional derivative at $(1, 2)$ in the direction toward $(4, 6)$?

20. The temperature distribution of a metal plate located in the x-y plane is given by $T(x,y) = 8 - 3x^2 - 4y^2$. Find an equation for the curve that a heat-seeking particle starting at $(3, -1)$ will travel along toward the origin.

21. A metal plate in the shape of a circular disk of radius 5 in. is located in the x-y plane with its center at the origin. The temperature distribution for the plate is given by $T(x,y) = 5 + 2x + 4y - x^2 - y^2$. A heat-fleeing particle starts at the origin. At what point on the boundary of the disk will the particle exit the disk? (Use the fact that the differential equation $du/dt + k_1u = k_2$ has solution $u = k_2/k_1 + Ce^{-k_1t}$).

Section 19.7: Tangent Planes and Normal Lines

Objectives:

1. **Determine the tangent plane and normal line to a surface $F(x, y, z) = k$ at a point P.**
2. **Interpret the gradient as a normal to the level curves or level surfaces of a function of two or three variables.**

MORE ABOUT GRADIENTS

In Sec. 19.6 we showed that the gradient of a function f of two or three variables always points in the direction of maximal increase in f. We shall now show that the gradient is always perpendicular to the level curves of a function of two variables, or perpendicular to the level *surfaces* of a function of three variables. We shall assume throughout that the functions under consideration have continuous first partials that are not all zero at the point in question.

Functions of two variables Suppose $P = (a, b)$ is a point in the domain of $f(x,y)$ and C is a smooth level curve of f through (a, b). Then C has an equation of the form

$$f(x,y) = k$$

for some constant k. Let

$$x = x(t) \qquad \text{and} \qquad y = y(t)$$

be a parameterization for C, with the point P corresponding to $t = t_0$. It follows that

$$f(x(t), y(t)) = k$$

and so

$$D_t[f(x(t), y(t))] = 0 \tag{1}$$

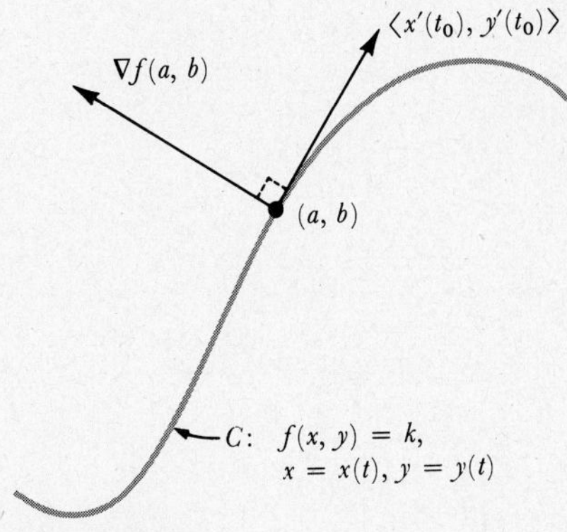

Fig. 19.34

By applying the Chain Rule (19.5.3) to the left side of (1), we have

$$f_x(x,y)x'(t) + f_y(x,y)y'(t) = 0$$

or, equivalently,

$$\nabla f(x,y) \cdot \langle x'(t), y'(t)\rangle = 0$$

When $t = t_0$, $\nabla f(x(t_0),y(t_0)) = \nabla f(a, b)$, and thus we have

(2) $$\nabla f(a, b) \cdot \langle x'(t_0), y'(t_0)\rangle = 0$$

Now, by definition, a vector is perpendicular to a curve at a point if it is perpendicular to any tangent vector to the curve at that point. Since $\langle x'(t_0), y'(t_0)\rangle$ is a tangent vector to C at (a, b), equation (2) implies the following:

19.7.1 **Theorem** If $f(x,y)$ has continuous first partials at (a, b) and if $\nabla f(a, b) \neq \mathbf{0}$, then $\nabla f(a, b)$ is perpendicular to the level curve of f passing through (a, b) (see Fig. 19.34).

Example 1 Sketch the level curve of the function $f(x,y) = 4x^2 + 9y^2$ passing through the point $P = (1, 4\sqrt{2}/3)$ and sketch the representative of $\nabla f(1, 4\sqrt{2}/3)$ at P.

Solution The desired level curve C has the form

$$4x^2 + 9y^2 = k$$

Since the point P is on C, it follows that $k = 4(1)^2 + 9(4\sqrt{2}/3)^2 = 36$. Thus C has equation

$$4x^2 + 9y^2 = 36$$

or, equivalently,

$$\frac{x^2}{9} + \frac{y^2}{4} = 1$$

Now $\nabla f = \langle 8x, 18y\rangle$, so $\nabla f(1, 4\sqrt{2}/3) = \langle 8, 24\sqrt{2}\rangle$. A sketch is given in Fig. 19.35. ●

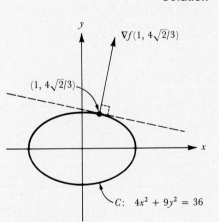

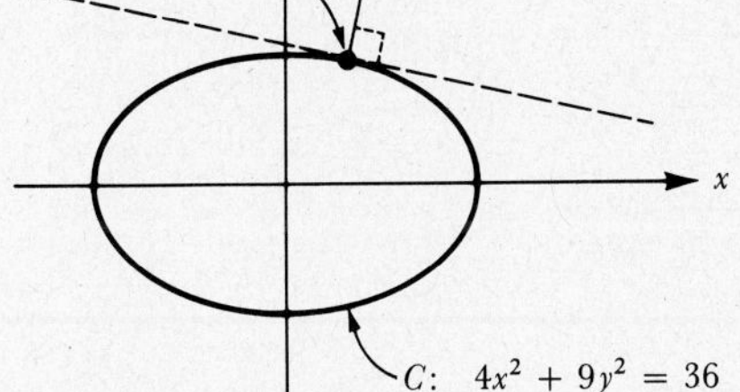

Fig. 19.35

Functions of three variables Now let $w = f(x, y, z)$ and let $P = (a, b, c)$ be a point on the level surface

$$S: \; f(x, y, z) = k$$

Suppose C is any smooth space curve lying on the surface S and passing through P, with

$$x = x(t), \qquad y = y(t), \qquad z = z(t)$$

a parameterization of C. Since C lies on S, we know that

$$f(x(t), y(t), z(t)) = k$$

so

$$D_t[f(x(t), y(t), z(t))] = 0$$

As above, by the Chain Rule (19.5.3), this is equivalent to

$$\nabla f(x(t), y(t), z(t)) \cdot \langle x'(t), y'(t), z'(t)\rangle = 0$$

In particular,

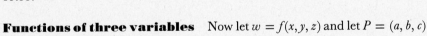

$$\nabla f(a, b, c) \cdot \langle x'(t_0), y'(t_0), z'(t_0)\rangle = 0$$

where t_0 is the value of the parameter corresponding to the point P. Thus $\nabla f(a, b, c)$ is perpendicular to the surface S at (a, b, c) because it is perpendicular to every curve lying on S and passing through (a, b, c); see Fig. 19.36. This establishes the following:

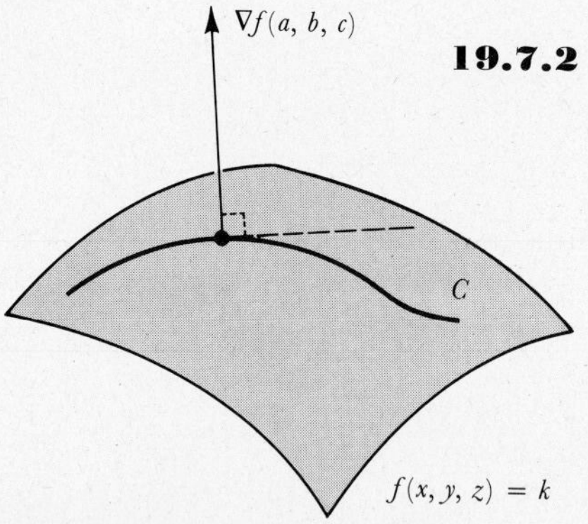

19.7.2 **Theorem** If $f(x, y, z)$ has continuous first partials at (a, b, c) and if $\nabla f(a, b, c) \neq \mathbf{0}$, then $\nabla f(a, b, c)$ is perpendicular to the level surface of f passing through (a, b, c).

Fig. 19.36

In Example 3, Sec. 19.6, we determined the path followed by a heat-seeking particle on a plate whose temperature at a point (x, y) is given by $T(x, y) = 10 - x^2 - 2y^2$. We found that if the particle starts at $(-1, 3)$, it follows the (dark) path $y = 3x^2$ to the origin, the hottest point (see Fig. 19.37). In terms of level curves, as mandated by (19.7.1), *the particle travels in such a way as to cross any level curves at right angles.* At each instant, beginning on the level curve $T(x, y) = -9$, it travels in the direction of the gradient. Also illustrated in Fig. 19.37 are paths followed by heat-seeking particles originating at other points. (In Exercise 31 you are asked to prove that all such paths not originating on the x axis are of the form $y = ax^2$.) There are many other situations occurring in nature that serve to illustrate theorems (19.7.1) and (19.7.2). For example, a charged particle moving in an electromagnetic field will cross the equipotential surfaces at right angles; water flowing freely downhill will seek a path that runs perpendicular to the contour curves of the terrain; and most insects moving toward the source of a sex-attractive pheromone will travel in the direction of the pheromone concentration gradient.

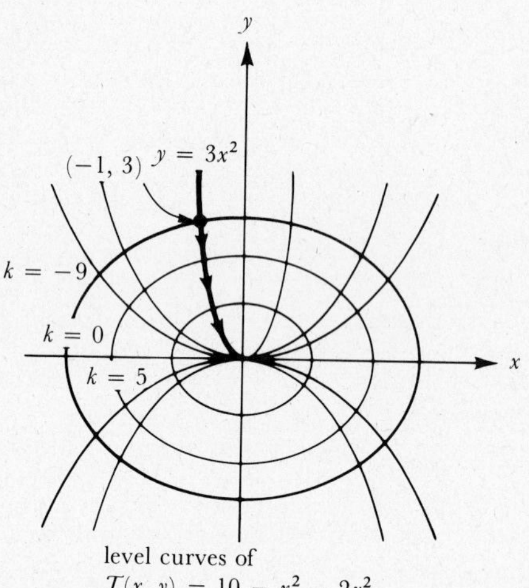

Fig. 19.37

TANGENT PLANES AND NORMAL LINES

In Sec. 19.3 we developed a means for finding the tangent plane to the graph of $z = f(x, y)$ at a point $(a, b, f(a, b))$. Suppose now we are given a surface S with equation

$$F(x, y, z) = k$$

This can be considered as a level surface of the function $w = F(x, y, z)$. If $P = (a, b, c)$ is a point on S, and if F has continuous first partials, then $\nabla F(a, b, c)$ is normal to the surface at P, that is, normal to the tangent plane to S at P.

19.7.3 **Theorem** If $F(x, y, z)$ has continuous first partials and if (a, b, c) is a point on the surface $F(x, y, z) = k$ with $\nabla F(a, b, c) \neq \mathbf{0}$, then the tangent plane to the surface at this point has the equation

$$F_x(a, b, c)(x - a) + F_y(a, b, c)(y - b) + F_z(a, b, c)(z - c) = 0$$

and the normal line has equations

$$x = a + F_x(a, b, c)t, \quad y = b + F_y(a, b, c)t, \quad z = c + F_z(a, b, c)t$$

Example 2 Find the equations for the tangent plane and normal line to the surface

$$\frac{z^2 \sin xy}{\pi^2} + 3y \cos z + 2 = 0$$

at the point $(\pi/2, 1, \pi)$.

Solution Let

$$F(x, y, z) = \frac{z^2 \sin xy}{\pi^2} + 3y \cos z$$

Then the desired surface is the level surface $F(x,y,z) = -2$. Now

$$F_x = \frac{z^2 y \cos xy}{\pi^2}, F_y = \frac{z^2 x \cos xy}{\pi^2} + 3 \cos z,$$

$$F_z = \frac{2z \sin xy}{\pi^2} - 3y \sin z$$

and therefore $F_x(\pi/2, 1, \pi) = 0$, $F_y(\pi/2, 1, \pi) = -3$, and $F_z(\pi/2, 1, \pi) = 2/\pi$.

By (19.7.3), the tangent plane has equation $-3(y-1) + (2/\pi)(z-\pi) = 0$ and the normal line has equations $x = \pi/2, y = 1 - 3t$, and $z = \pi + (2/\pi)t$. ●

Example 3 Show that the formulas given in Theorems (19.3.3) and (19.3.4) for the tangent plane and normal line to the graph of a function $z = f(x,y)$ at a point $(a, b, f(a, b))$ are a special case of Theorem (19.7.3).

Solution If we let $F(x,y,z) = f(x,y) - z$, then the graph of the level surface $F(x,y,z) = 0$ of F is the same as the graph of f. Now

$$F_x = \frac{\partial}{\partial x}[f(x,y) - z] = f_x(x,y)$$

$$F_y = \frac{\partial}{\partial y}[f(x,y) - z] = f_y(x,y)$$

$$F_z = \frac{\partial}{\partial z}[f(x,y) - z] = -1$$

Letting $c = f(a, b)$ and applying (19.7.3), we find that the tangent plane has equation

$$f_x(a,b)(x-a) + f_y(a,b)(y-b) - (z - f(a,b)) = 0$$

which agrees with (19.3.3), page 638. Similarly, the normal line has equations $x = a + f_x(a,b)t$, $y = b + f_y(a,b)t$, $z = f(a,b) - t$, which agrees with (19.3.4), page 638. ●

PROGRESS TEST 1

1. Suppose $f(x,y) = x^2 - 4y^2$. Sketch the level curves of f that pass through the points $(3, \sqrt{5}/2)$ and $(-2\sqrt{10}, 1)$, respectively. Sketch ∇f at each of these points.
2. Find equations for the tangent plane and normal line to the surface $x^3 + 4y^3 + z^3 = 5xyz$ at the point $(2, 1, 2)$.
3. Given the surfaces $y - x^2 + z^2 + 2 = 0$ and $x^2 + y^2 - z = 8$, which intersect in a curve C, find equations for the tangent line to C at the point $(-2, 2, 0)$.

SECTION 19.7 EXERCISES

In Exercises 1 to 6 sketch the level curve of f that passes through the point $P = (a, b)$ and sketch the representative of $\nabla f(a, b)$ at P.

1. $f(x,y) = xy$, $P = (1, 1)$
2. $f(x,y) = y^2 - x^2$, $P = (2, 1)$
3. $f(x,y) = x^2 + 4y^2$, $P = (2, 1)$
4. $f(x,y) = y^2 - x^3$, $P = (1, 1)$
5. $f(x,y) = x^2y$, $P = (3, 1/9)$
6. $f(x,y) = 2x^2 + y^2 + 1$, $P = (1, 1)$
7. Show that the tangent line at (a, b) to the level curve of $f(x,y)$ passing through (a, b) has the equation

$$f_x(a,b)(x-a) + f_y(a,b)(y-b) = 0$$

(Assume that f has continuous first partials and $\nabla f(a, b) \neq \mathbf{0}$.)

In Exercises 8 to 16 find an equation for the tangent line at P to the level curve of f that passes through P. (Use the result of Exercise 7.)

8. $f(x,y) = xy$, $(P = (1, 2)$

9. $f(x,y) = y^2 - 2xy + x^2 - 2x$, $P = (2, 4)$

10. $f(x,y) = 3x^2 - 2xy + y^2$, $P = (1, 1)$

11. $f(x,y) = x^3 + y^3 - 2xy$, $P = (1, 1)$

12. $f(x,y) = x^2 + 4y^2$, $P = (2, 1)$

13. $f(x,y) = xe^{xy}$, $P = (1, 0)$

14. $f(x,y) = x^2y$, $P = (3, 1/9)$

15. $f(x,y) = x^5 - 2y^3 + y^5$, $P = (1, 1)$

16. $f(x,y) = x^3 + 2xy + y^3$, $P = (1, -1)$

17. Show that the level curves of $f(x,y) = x^3 + y^3$ are perpendicular to the level curves of $g(x,y) = xy$ for all $(a, b) \neq (0, 0)$.

18. Let C be a level curve of f that does not go through the origin and let (a, b) be the point on C closest to the origin. Show that the vector $\langle a, b \rangle$ is perpendicular to the tangent line to C at (a, b).

In Exercises 19 to 25 find equations for (a) the tangent plane and (b) normal line to the given surface at the indicated point.

19. $3x^2 - 2y^2 + z^2 = 14$, $P = (2, 1, 2)$

20. $x = \ln(3y/4z)$, $P = (0, 4, 3)$

21. $3x^2y - xz^2 = 14$, $P = (-2, 1, 1)$

22. $\ln(x^2 + y^2 - z^2) = 0$, $P = (1, -1, 1)$

23. $xy = yz - xz^2 = 1$, $P = (1, 1, 1)$

24. $\sin(xz^2) = zy$, $P = (\pi/6, 1/2, 1)$

25. $ye^{xy} = ze^{xz}$, $P = (2, -1, -1)$

26. Let $P = (a, b, c)$ be a point on the sphere $x^2 + y^2 + z^2 = r^2$. Show that the normal line to the sphere at P passes through the point $(-a, -b, -c)$.

27. Find equations for the tangent line to the curve of intersection of the surfaces $x \sin yz = 1$, $ze^{y^2 - x^2} = \pi/2$ at the point $(1, 1, \pi/2)$.

28. Find the points on the surface

$$\frac{x^2}{9} + \frac{y^2}{4} + \frac{z^2}{36} = 1$$

where the tangent plane is parallel to the plane $2x - 3y + z = 4$.

29. Prove that if $f(x,y)$ and $g(x,y)$ are differentiable functions of two variables, then $\nabla(f + g) = \nabla f + \nabla g$.

30. Suppose $\nabla f(a, b) = \nabla g(a, b) \neq \mathbf{0}$ and $\nabla f(a, b) \cdot \nabla g(a, b) = 0$. What can we conclude about the level curves at (a, b)?

31. Show that any path traveled by a heat-seeking particle on a plate with temperature distribution $T(x,y) = 10 - x^2 - 2y^2$ is a parabola of the form $y = ax^2$ provided that the path does not originate on the x axis (see Fig. 19.37, page 664).

Section 19.8: Extremes of Functions of Two Variables

Objectives:

1. Determine relative extremes of a function of two variables.

2. Determine absolute extremes of a continuous function of two variables on a closed and bounded set.

3. Use objectives 1 and 2 to solve applied maximum-minimum problems.

INTRODUCTION

In Chap. 5 we used derivatives to develop powerful and very useful techniques for determining relative and absolute extremes of single-variable functions. Although, as we might expect, the situation is somewhat more complex for functions of two variables, many of the underlying ideas and examples extend our earlier one-variable work.

19.8.1 Definition We say that $f(x,y)$ has a *relative maximum at* (a, b) if the domain of f contains an open disk D about (a, b) such that $f(a, b) \geq f(x,y)$ for all (x,y) in D.

19.8.2 Definition If (a, b) is in a subset S of the domain of f, then we say that f has an *absolute maximum on S at* (a, b) if $f(a, b) \geq f(x,y)$ for all (x,y) in S.

Relative and *absolute minimum* are defined similarly.

DETERMINING RELATIVE EXTREMES

In (4.2.3) we showed that if a function of one variable has a derivative at a relative extreme, then this derivative is zero there. That is, the tangent line at a relative extreme must be horizontal. This fact implies a similar theorem for functions of two variables.

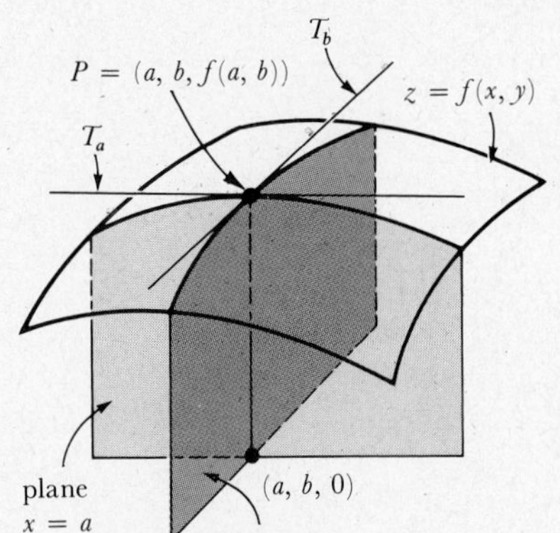

Fig. 19.38

19.8.3 Theorem If f has a relative extreme at (a, b) and if the first partials of f exist at (a, b), then $f_x(a, b) = f_y(a, b) = 0$ or, equivalently, $\nabla f(a, b) = \mathbf{0}$.

Proof: We show $f_x(a, b) = 0$ for a relative maximum. The other cases are similar. Hold $y = b$ constant and differentiate the one variable function $f(x, b)$:

$$D_x[f(x, b)] = f_x(x, b)$$

But f has a relative maximum at (a, b), so $f(a, b) \geq f(x, b)$ for all (x, b) near (a, b). Thus $g(x) = f(x, b)$ has a relative maximum at $x = a$ and so, by (4.2.3), $g'(a) = f_x(a, b) = 0$. ■

Geometrically, the conclusion of (19.8.3) implies that the planes $x = a$ and $x = b$ intersect the graph of f in curves that have horizontal tangent lines at $(a, b, f(a, b))$ (see Fig. 19.38). In particular, if the tangent plane exists at such a point it will be horizontal.

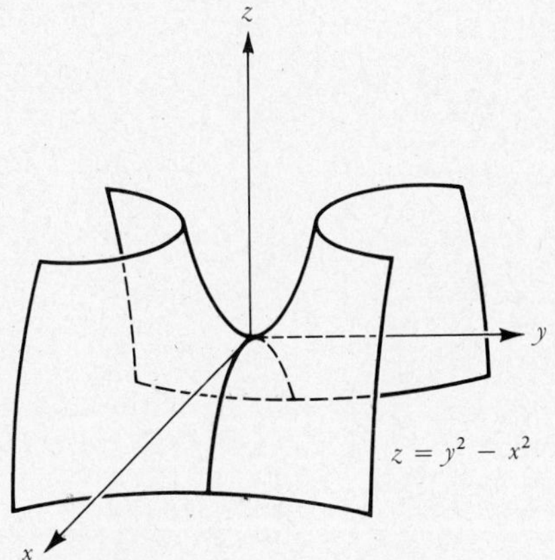

Fig. 19.39

The function $f(x, y) = y^2 - x^2$ shows that the converse of (19.8.3) is false since $\nabla f = \langle -2x, 2y \rangle = \mathbf{0}$ when $x = y = 0$, but f does not have an extreme at $(0, 0)$. To verify this we note that $f(x, 0) = -x^2 < 0$ for all $x \neq 0$. Since $f(0, 0) = 0$, f cannot have a relative minimum at $(0, 0)$. Similarly, $f(0, y) = y^2 > 0$ for all $y \neq 0$, so f cannot have a relative maximum at $(0, 0)$. The sketch in Fig. 19.39 shows that the graph of f has a saddle shape at the origin.

From our work in Chap. 4 we know that a function of one variable can have a relative extreme at a point where the derivative fails to exist. Similarly, $f(x, y) = \sqrt{x^2 + y^2}$ has a relative minimum at $(0, 0)$, but neither

$$f_x(x, y) = \frac{x}{\sqrt{x^2 + y^2}} \qquad \text{nor} \qquad f_y(x, y) = \frac{y}{\sqrt{x^2 + y^2}}$$

exists at $(0, 0)$. The conical graph (see Fig. 19.40) has a sharp point at $(0, 0)$.

Despite its limitations, (19.8.3) is quite helpful in identifying *candidates* for relative extremes.

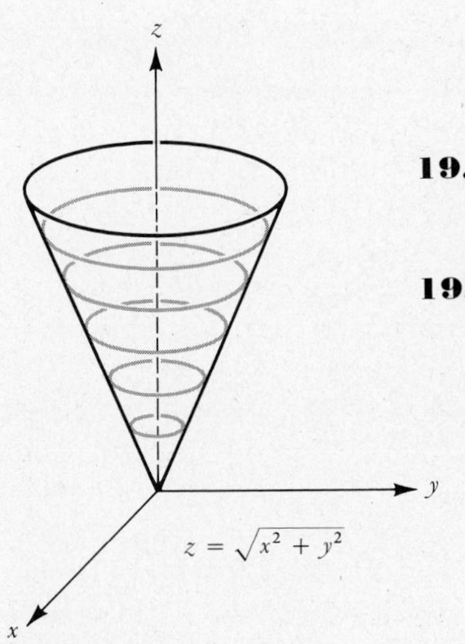

Fig. 19.40

19.8.4 Definition A point (a, b) in the domain of $f(x, y)$ is called a *critical point of* f if either $\nabla f(a, b) = \mathbf{0}$ or $\nabla f(a, b)$ fails to exist.

19.8.5 Definition A critical point of f which is not a relative extreme of f is called a *saddle point of* f.

For the most part we shall deal with differentiable functions, so critical points can be identified by simultaneous solution of the equations

$$f_x(x, y) = 0 \qquad \text{and} \qquad f_y(x, y) = 0$$

The problem, then, is to distinguish saddle points from relative extremes. Our starting point is analogous to the Second-Derivative Test for functions of one variable, except that for functions of two variables the underlying issue of concavity is necessarily more complex.

19.8.6 **Second-Derivative Test for Relative Extremes** Suppose f has continuous second partials in an open disk centered at (a, b) and $\nabla f(a, b) = \mathbf{0}$. At (a, b):

(*a*) If $f_{xx}f_{yy} - f_{xy}^2 > 0$, then f has a relative extreme, which is
  (i) a relative maximum if $f_{xx} < 0$
  (ii) a relative minimum if $f_{xx} > 0$.

(*b*) If $f_{xx}f_{yy} - f_{xy}^2 < 0$, then f has a saddle point.

(*c*) If $f_{xx}f_{yy} - f_{xy}^2 = 0$, then the test is inconclusive.

A proof of (19.8.6) is outlined in Miscellaneous Exercise 17. The idea of the proof is based on examining concavity of curves formed by intersecting vertical planes with the graph of f at (a, b).

To apply (19.8.6) we must first determine the critical points by simultaneously solving the equations $f_x = 0$ and $f_y = 0$.

Example 1 Determine the relative extremes, if any, for the function $f(x, y) = x^3 + y^3 + 3xy$.

Solution We first simultaneously solve

$$f_x = 3x^2 + 3y = 0 \qquad \text{and} \qquad f_y = 3y^2 + 3x = 0$$

From the first we get $y = -x^2$, which, when substituted in the second, yields $3x^4 + 3x = 3x(x^3 + 1) = 0$. Hence $x = 0$ or $x = -1$. The critical points are thus $(0, 0)$ and $(-1, -1)$. Now $f_{xx} = 6x$, $f_{xy} = 3$, and $f_{yy} = 6y$, so at $(0, 0)$

$$f_{xx}f_{yy} - f_{xy}^2 = -(3)^2 < 0$$

Hence by [19.8.6(*b*)], f has a saddle point at $(0, 0)$.

At $(-1, -1)$, $f_{xx}f_{yy} - f_{xy}^2 = (-6)(-6) - (3)^2 = 27 > 0$. By [19.8.6(*a*)], f has a relative maximum at $(-1, -1)$ since $f_{xx}(-1) = -6 < 0$. ●

When the Second-Derivative Test does not apply at a critical point (a, b), we must resort to other methods, one of which is simply to examine the values of f near (a, b). The technique is to suppose that h and k are small numbers and then examine $f(a + h, b + k) - f(a, b)$.

Example 2 Determine the relative extremes, if any, for the function

$$f(x, y) = y^4 - x^2$$

Solution Solving $f_x = -2x = 0$ and $f_y = 4y^3 = 0$, we obtain the single critical point $(0, 0)$. Now

$$f_{xx}f_{yy} - f_{xy}^2 = (-2)(12y^2) - (0)^2 = -24y^2$$

which is 0 at $(0, 0)$. Hence the Test is inconclusive. We examine

$$f(0 + h, 0 + k) - f(0, 0) = f(h, k) = k^4 - h^2$$

Now, on the x axis we have $f(h, 0) = -h^2 \leq 0$, while on the y axis we have $f(0, k) = k^4 \geq 0$. Thus f has opposite sign for various (h, k) near, but not at, $(0, 0)$. Since $f(0, 0) = 0$, we conclude that f attains neither a relative maximum nor a relative minimum at $(0, 0)$, and hence $(0, 0)$ must be a saddle point. ●

PROGRESS TEST 1

Determine and classify all critical points for the given functions.

1. $f(x,y) = x^2 + y^2 + x^2y + 4$

2. $f(x,y) = x^4 + y^4$

ABSOLUTE EXTREMES

The Extreme-Value Theorem for functions of a single variable (4.4.2) guarantees an absolute maximum and minimum for a continuous function on a closed interval $[a, b]$. Determining such extremes amounts to comparing any relative extremes in the interior of the interval with the values of the function at the endpoints—the "boundary" of the interval. The situation for functions of two variables is similar, but a bit more subtle. The definition of continuity on an open set in the plane, (19.2.4), can be extended to the boundary of such a set, yielding a definition of continuity on a *closed* set. A set in the plane is said to be *bounded* if it can be enclosed by a circle of large enough radius. Now, given a function f that is continuous on a closed and bounded set in the plane, an Extreme-Value Theorem for functions of two variables guarantees that f attains an absolute maximum and minimum on D. Such a theorem—and in fact a corresponding theorem for functions of n variables—is normally developed in more advanced courses. We shall assume that the extreme values we need actually exist and concentrate on a procedure for finding them.

19.8.7 Procedure for Determining Absolute Extreme Values of a Continuous Function f on a Closed and Bounded Set D

1. Sketch D and determine any critical points on the interior of D.
2. Describe the boundaries algebraically and describe the restrictions of the function f on the separate pieces of the boundary.
3. Determine any extreme values of f on the respective boundary pieces.
4. Compare the results of 1 and 3 to determine the extreme values of f on D.

As illustrated in the next example, step 3 often consists of applying optimization techniques from Sec. 5.1.

Example 3 Determine the maximum and minimum values of

$$f(x,y) = x^2 + 4xy + y^2 + 6x + 1$$

on the closed triangular region

$$D = \{(x,y)/0 \le x \le 4, x - 4 \le y \le 0\} \qquad \text{(see Fig. 19.41)}$$

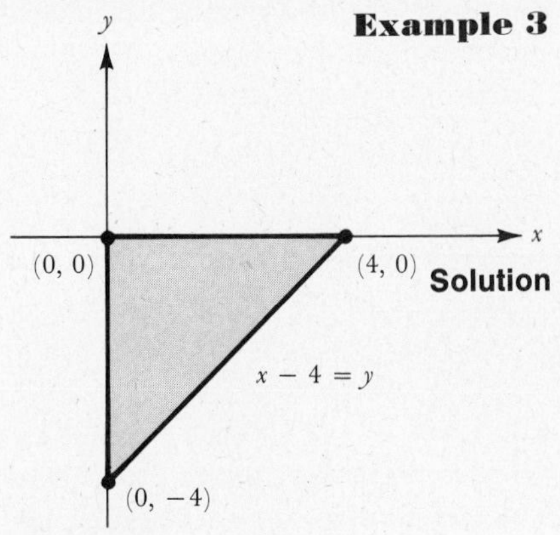

Fig. 19.41

Solution 1. We solve the equations

$$f_x = 2x + 4y + 6 = 0 \qquad \text{and} \qquad f_y = 4x + 2y = 0$$

to obtain the single critical point $(1, -2)$.

2. The boundary of D consists of the three line segments

$$C_1\colon\ x = 0,\ -4 \le y \le 0$$
$$C_2\colon\ y = 0,\ 0 \le x \le 4$$
$$C_3\colon\ y = x - 4,\quad 0 \le x \le 4$$

On C_1: $f(0,y) = y^2 + 1 = g(y)$, $-4 \leq y \leq 0$.
On C_2: $f(x,0) = x^2 + 6x + 1 = h(x)$, $0 \leq x \leq 4$.
On C_3: $f(x, x-4) = x^2 + 4x(x-4) + (x-4)^2 + 6x + 1$
$= 6x^2 - 18x + 17 = k(x)$, $0 \leq x \leq 4$.

3. Now $g'(y) = 2y \leq 0$ for $-4 \leq y \leq 0$, so the extremes of g occur at the endpoints $y = -4$ and $y = 0$. The corresponding points on C_1 are $(0, -4)$ and $(0, 0)$.

Similarly, $h'(x) = 2x + 6 = 0$ when $x = -3$. But $0 \leq x \leq 4$, so $h(x)$ reaches its extremes at $x = 0$ and $x = 4$. The corresponding points on C_2 are $(0, 0)$ and $(4, 0)$. Finally, $k'(x) = 12x - 18 = 0$ when $x = 3/2$, so the extremes of $k(x)$ can occur when $x = 0$, $x = 3/2$, or $x = 4$. Since $y = x - 4$, the corresponding points on C_3 are, respectively, $(0, -4)$, $(3/2, -5/2)$, and $(4, 0)$.

4. From steps 1 and 3 we have the points $(1, -2)$, $(0, 0)$, $(0, -4)$, $(4, 0)$ and $(3/2, -5/2)$. We now compare:

$$f(1,-2) = 1 - 8 + 4 + 6 + 1 = 4$$
$$f(0,0) = 1$$
$$f(4,0) = 16 + 24 + 1 = 41$$
$$f(0,-4) = 16 + 1 = 17$$
$$f(3/2,-5/2) = 9/4 - 60/4 + 25/4 + 18/2 + 1 = 7/2$$

The maximum value is 41 and occurs at $(4, 0)$. The minimum value of 1 occurs at $(0, 0)$. ●

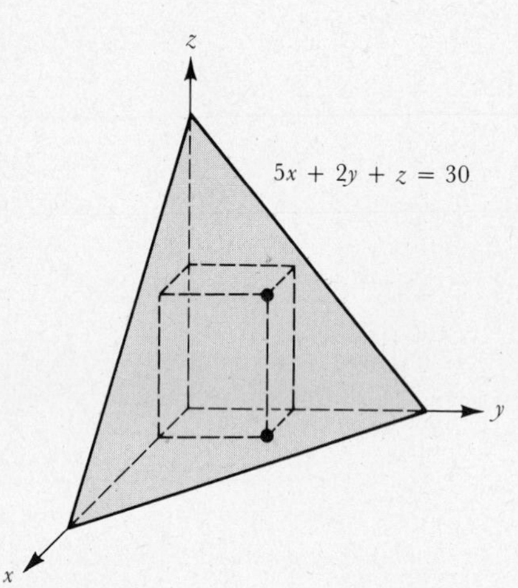

Fig. 19.42

Example 4 An automotive engineer is attempting to fit a rectangular map compartment of maximal volume into a chopped-off corner under the dashboard. She measures and then models the available space as in Fig. 19.42, where the constricting plane has equation $5x + 2y + z = 30$. What are the dimensions and volume of the rectangular map compartment of largest volume that can be built?

Solution We let the dimensions of the compartment be x, y, and z, so its volume is given by $V = xyz$. Since $z = 30 - 5x - 2y$, we can write V as a function of x and y:

$$V(x,y) = 30xy - 5x^2y - 2xy^2$$

We are attempting to maximize V on the triangular region in the x-y plane bounded by the axes and the intersection of the plane $5x + 2y + z = 30$ with the x-y plane (see Fig. 19.43). However, both physical and algebraic considerations imply that $V(x,y) = 0$ for (x,y) on the boundary of this region. Hence it remains to determine a positive relative maximum on the interior.

$$V_x = 30y - 10xy - 2y^2 = 0 \quad \text{and} \quad V_y = 30x - 5x^2 - 4xy = 0$$

Since we are assuming (x,y) is not on the boundary, we know $x \neq 0, y \neq 0$, so we can divide the equations through by y and x, respectively, to get

$$30 - 10x - 2y = 0 \quad \text{and} \quad 30 - 5x - 4y = 0$$

These equations yield the critical point $(2, 5)$. We use the Second-Derivative Test: $V_{xx}V_{yy} - V^2_{xy} = (-10y)(-4x) - (30 - 10x - 4y)^2 > 0$ at $(2, 5)$, with $V_{xx}(2, 5) = -50 < 0$. Hence $(2, 5)$ is a relative maximum for V. Hence the dimensions of the compartment are 2 by 5 by 10 and its volume is 100 in.^{3}. ●

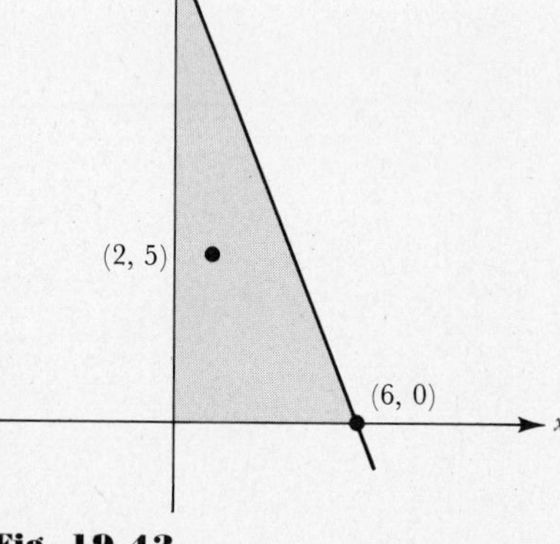

Fig. 19.43

More applications will appear in the next section.

PROGRESS TEST 2

1. The temperature distribution on a metal disk of radius 1 centered at the origin is given by

$$T(x,y) = 2y^2 + 4x^2 + 2y + 1$$

Determine the hottest and coldest points on this disk. (Use y as the independent variable for T on the boundary, describing the right and left halves separately.)

SECTION 19.8 EXERCISES

In Exercises 1 to 20 identify all critical points of the given function and state whether each is a relative maximum, relative minimum, or saddle point.

1. $f(x,y) = 2y^2 - xy + x^2$

2. $f(x,y) = x^2 + xy + y^2 + 3x - 9y$

3. $f(x,y) = 21x^2 - 4x^3 + 2y^2 - 4y^3 + 18xy$

4. $f(x,y) = x^2 + y^2 - 4xy$

5. $f(x,y) = 8x^2 + 4y^2 + 4y + 3$

6. $f(x,y) = 6x^2 + 6y^2 + y^3$

7. $f(x,y) = x^2 - 24xy + 8y^2$

8. $f(x,y) = x^2 + y^3 - 4xy - 8x + 13y + 1$

9. $f(x,y) = x^3 + y^2 - 6xy + 6x + 3y + 2$

10. $f(x,y) = x^4y^2 + x^3y^3 - 12x^3y^2$

11. $f(x,y) = 2x^3 - 3y^4 - 6xy^2$

12. $f(x,y) = x^3y^2 - x^4y^2 - x^3y^3$

13. $f(x,y) = x^4 - 4xy + y^4$

14. $f(x,y) = x^4 - x^2y^2 + y^4$

15. $f(x,y) = e^x \cos y$

16. $f(x,y) = e^{-x} \sin y$

17. $f(x,y) = \dfrac{x}{y} - \dfrac{y}{x}$

18. $f(x,y) = 2x^2 + 2xy + 5y^2 + 4x$

19. $f(x,y) = (y + 2x)e^{-(x^2+y^2)}$

20. $f(x,y) = e^{2y}(x^2 + x + y)$

In Exercises 21 to 31 determine the absolute maximum and minimum values obtained by the function f on the closed domain R bounded by the given curves.

21. $f(x,y) = 2x^2 + y^2 - y$; $y = 1 - x^2$, x axis.

22. $f(x,y) = x^2 - xy + 2y^2 + 7y$; $x = \pm 3$, $y = \pm 3$.

23. $f(x,y) = x^2 + 3y$; x axis, y axis, $x + y = 2$

24. $f(x,y) = x^2 - xy + y^2 - 3x - 4$; x axis, y axis, $x + y = 4$

25. $f(x,y) = x^2 + 4x - 3xy + y^2 - 2y$; x axis, y axis, $x = 2$, $y = 2$

26. $f(x,y) = x^2 + y^2 - xy - x$; $x = \pm 1, y = \pm 1$

27. $f(x,y) = 8x + 2xy - x^2 - 3y^2 - 2$; x axis, y axis, $x = 10$, $y = 3$

28. $f(x,y) = e^{x^2+y^2+x}$; $x = \pm 1, y = \pm 1$

29. $f(x,y) = y^2 - x^2 + 4x^2y^2$; $x^2 + y^2 = 1$

30. $f(x,y) = e^{y-3x}$; $x^2 + y^2 = 1$

31. $f(x,y) = 2x^2 + 4y^2 + 2x + 1$; x axis, $y = 9 - x^2$

32. A rectangular, open-topped box is to hold 256 m^3. Determine the dimensions of such a box that can be constructed using a minimum amount of material.

33. A rectangular box is to hold 36 ft^3. What dimensions will yield the least expensive box if the material used to construct the top and bottom cost three times as much as that used to construct the sides?

34. Show that among all rectangular boxes having a fixed surface area, the cube will have the maximum volume.

35. Find the dimensions of the rectangular box of maximum volume if the sum of the lengths of the edges is 24.

36. What is the volume of the largest rectangular box that can be inscribed in the ellipsoid $x^2/a^2 + y^3/b^2 + z^2/c^2 = 1$? (*Hint:* Maximize the square of the volume.)

37. Use the methods of this section to find the distance from the origin to the plane $ax + by + cz = d$. (*Hint:* Minimize the square of the distance.)

In Exercises 38 and 39 determine the dimensions of the rectangular box of largest volume situated with its bottom on the x-y plane, one corner at the origin, sides parallel to the axes, and its opposite corner on the given plane.

38. $6x + 4y + 3z = 36$

39. $9x + 9y + 3z = 18$

In Exercises 40 and 41 determine the points on the given surface where the distance from the origin to this surface is a minimum.

40. $z^2 - xy = 8$

41. $x^2 - yz = 8$

42. Give another version of [19.8.6(*a*)] with the sign of $f_{xx}(a,b)$ replaced by that of $f_{yy}(a,b)$ and f_{xy} replaced by f_{yx}. Why does this version follow from the original?

43. (*a*) Show that (19.8.6) does not apply to $f(x,y) = x^4 - 3x^2y + 2y^2$ at the critical point $(0,0)$.
(*b*) Show that $(0,0)$ is a saddle point.

44. Repeat Exercise 43 for $f(x,y) = 4x^4 - 5x^2y + y^2$.

Section 19.9: Optimization Under Constraints and Lagrange Multipliers, with Applications in Economics

Objectives:

1. Determine extreme values of functions of more than one variable under constraints using level curves and using Lagrange multipliers.

2. Apply these techniques to problems in economics and elsewhere.

INTRODUCTION

Most applied extreme-value problems involve the determining of extreme values of a multivariable function subject to some constraint. For functions of two variables $f(x,y)$, the constraint often takes the form of an equation in two variables $g(x,y) = C$. Solving a maximizing problem, then, amounts to finding the maximum value of $f(x,y)$ where the points (x,y) *are on the curve* $g(x,y) = C$. Since each value k of f is associated with a unique level curve $f(x,y) = k$ in the plane, the idea is to search among the level curves of f for a level curve providing a maximum value for f on $g(x,y) = C$. As we shall see, this level curve will be tangent to the curve $g(x,y) = C$ in at least one point.

We can interpret the classic "fence problem" in these terms: What are the dimensions of the rectangular pasture of largest area that can be enclosed by a fixed perimeter C? Letting x and y be the dimensions of the pasture, the function $A(x,y) = xy$ is being maximized under the constraint $2x + 2y = C$. The graph of the constraint curve is a straight line of slope $= -1$, while for each possible area k, there is a unique hyperbolic level curve $A(x,y) = xy = k$. The level curve that meets the constraint curve $2x + 2y = C$ and is associated with largest area k is clearly that which is *tangent* to the constraint curve at (x_0, y_0); see Fig. 19.44. The fact that $y_0 = x_0$ (yielding a square x_0 by y_0 pasture) agrees with our earlier work in Sec. 5.1.

On the other hand, if a fence is not needed on one side, say a "y side," then the constraint is given by $y + 2x = C$, so the constraint curve $y = C - 2x$ is tangent as indicated in Fig. 19.45 at a point whose y coordinate is double its x coordinate. (Coincidentally, the rectangle with opposite corners $(0,0)$ and (x_0, y_0) in each case serves as a sketch of the optimal pasture.)

More generally, for a function $f(x,y)$ and constraint $g(x,y) = C$, the maximum occurs where the appropriate level curve of f is tangent to the graph of

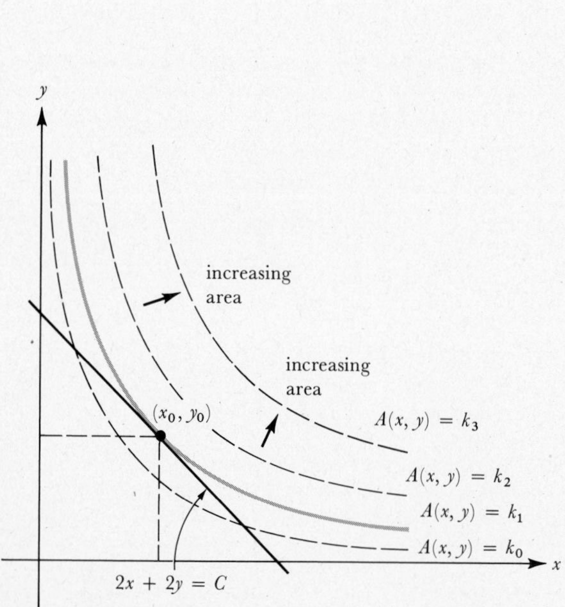

Fig. 19.44

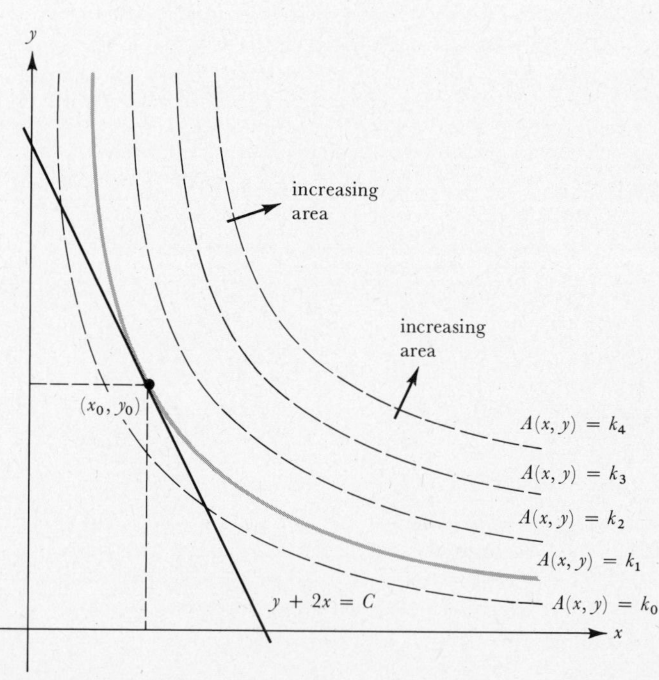

Fig. 19.45

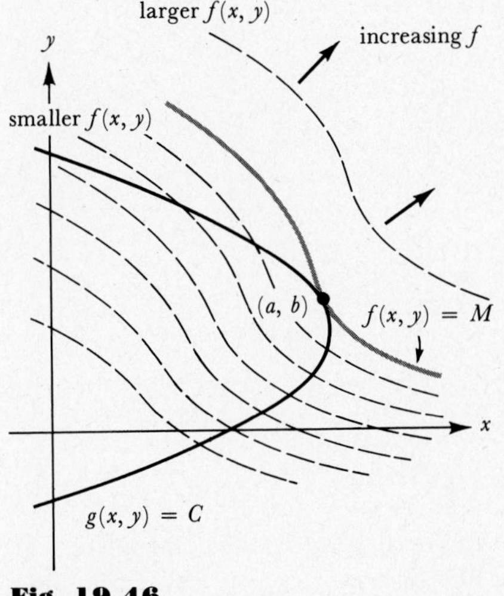

Fig. 19.46

$g(x,y) = C$—provided that the curves in question *have* tangents and we are not on the boundary of $D(f)$. (See Fig. 19.46.)

PREFERENCE, UTILITY, AND INDIFFERENCE

The previous ideas are widely used in economics, where, as anyone who has drawn up a budget can attest, optimization problems inevitably involve constraint conditions. Economists assume that we spend our money in ways that maximize our satisfaction—or *utility,* as they call it. They further assume that our preferences, at least our relative preferences, are roughly quantifiable so that a measure of total utility U is possible.

✖ Suppose on a given Saturday night that a student will divide available funds between x hamburgers and y beers. Depending on personal preferences and relative prices, this student will choose to buy a certain number x of hamburgers and a certain number y of beers in order to maximize utility $U(x,y)$. Given fixed prices, say \$1 for a hamburger and 50¢ for a beer, different choices could quite likely lead to equal utility. For example, it is possible that different combinations of hamburgers and beer yield equal utility. Each person's utility function has level curves, called *indifference curves* by economists. In Fig. 19.47 the student achieves the same utility U_1 from 1 hamburger and 5 beers as with 4 hamburgers and 1 beer. The cost of the former combination is $(1) + (\frac{1}{2})(5) = 3(\frac{1}{2})$ dollars and the cost of the latter is $(1)(4) + (\frac{1}{2})1 = 4\frac{1}{2}$ dollars. However, given a fixed amount to spend, say 3 dollars, the student will attempt to maximize $U(x,y)$ according to the constraint $(1)x + (\frac{1}{2})y = 3$. The optimal solution, as illustrated in Fig. 19.47, is 2 hamburgers and 2 beers. However, given \$5 to spend, the constraint curve moves out so that the optimal point of 2 hamburgers and $5\frac{1}{2}$ beers is reached on the indifference curve U_3 (see Fig. 17.47). This person obviously likes beer. However, if the price of beer goes up to 75¢ then, given the same \$5, this person will choose to have 3 hamburgers and 3 beers on the U_2 indifference curve because the *slope* of the constraint curve changes from -2 to $-4/3$ (see Fig. 19.47). Furthermore, a different person will have different indifference curves. Does the person, one of whose indifference curves appears in Fig. 19.48, have a drinking problem? Note that, given the original hamburger-beer prices, this person will spend \$5 by buying 1 hamburger and 8 beers. It takes 10 hamburgers to make this person as satisfied with one beer as with 8 beers and 1 hamburger.

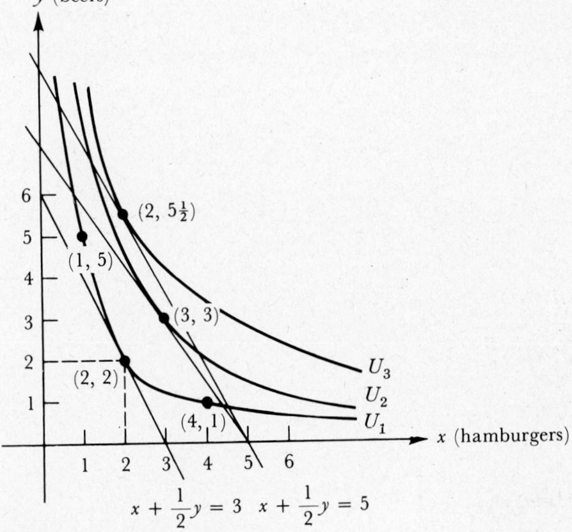

Fig. 19.47

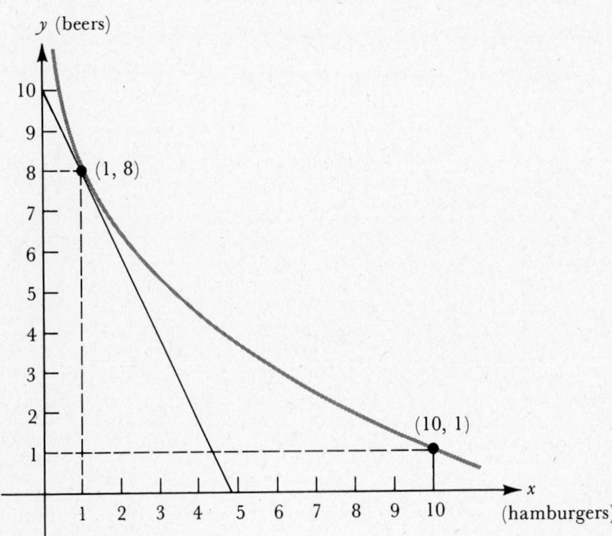

Fig. 19.48

PROGRESS TEST 1

1. Draw a plausible pair of spinach-hamburger indifference curves for a 250-lb 12-year-old. (Plot hamburgers horizontally.)
2. Using the indifference curves in Prob. 1 and assuming that a plate of spinach and a hamburger each cost \$1, illustrate graphically how this 12-year-old might divide \$6 between spinach and hamburgers.

TAXES ON ENERGY CONSUMPTION

× In recent years the federal government has attempted to use a combination of energy taxes and rebates in order to lower energy consumption. The basic idea is to discourage consumption by raising the visible prices via specific taxes. But to avoid overburdening consumers financially and to maintain comparable utility, the government rebates much of the tax for people of average income. To analyze this policy for a typical consumer, we let x be the annual consumption of units of energy, say gallons of gasoline to be specific, and let y be the quantity of all other goods purchased annually.

Before the tax a typical person is operating at the point of maximum utility (x_0, y_0) where that person's budget constraint curve B is tangent to his/her indifference curve of highest utility U_1 (see Fig. 19.49). The imposition of the tax raises the price of gasoline relative to all other goods y, making gasoline less affordable and making other goods appear preferable—as reflected in the more negatively sloped constraint curve B'. Now, to return this typical consumer (who is also a voter) to the original level of utility (satisfaction), the government rebates the tax at income tax time. This rebate has the effect of sliding the budget constraint curve out to B'', allowing a tangent at (x_2, y_2) on the original indifference curve U_1. The net effect of the tax is to lower gasoline consumption from x_0 to x_2 gallons while maintaining equivalent utility; the consumer buys more of y (everything else) to compensate.

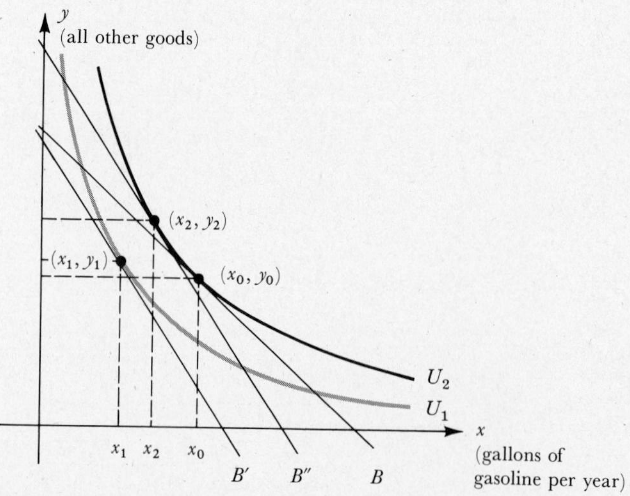

Fig. 19.49

THE LAGRANGE MULTIPLIER TECHNIQUE

Up to now our discussions have focused on the geometric aspects of optimization of $f(x,y)$ under the constraint $g(x,y) = C$. We can move to a powerful algebraic technique by using the fact that if $g(x,y) = C$ is tangent to a level curve of f (and all appropriate partials exist), then the normals to the respective curves must be parallel at the point (a, b) of tangency (recall Fig. 19.46). Since, by (19.7.1), the normals have the directions $\nabla f(a, b)$ and $\nabla g(a, b)$, we can conclude that one gradient is a multiple of the other. That is, $\nabla f(a, b) = \lambda \nabla g(a, b)$. Assuming throughout that neither gradient is **0**, this condition leads to the three equations

19.9.1 $$f_x = \lambda g_x \qquad f_y = \lambda g_y \qquad g(x,y) = C$$

which can be solved simultaneously for x, y, and λ. This will yield *candidates* for extremes of $f(x,y)$, as Fig. 19.50 indicates, from which we then choose the appropriate point(s).

The variable λ in (19.9.1) is called a *Lagrange multiplier* (after the French mathematician Joseph Lagrange, who first developed this technique in the late eighteenth century).

Example 1 Determine the minimum, if it exists, of the function

$$f(x,y) = 2x^2 + 3x + 2xy + 2y + y^2$$

subject to the condition $y^2 - x = 1$.

Solution Let $g(x,y) = y^2 - x$. Then

(1) $$f_x = 4x + 3 + 2y = \lambda g_x = \lambda(-1)$$
(2) $$f_y = 2x + 2 + 2y = \lambda g_y = \lambda(2y)$$

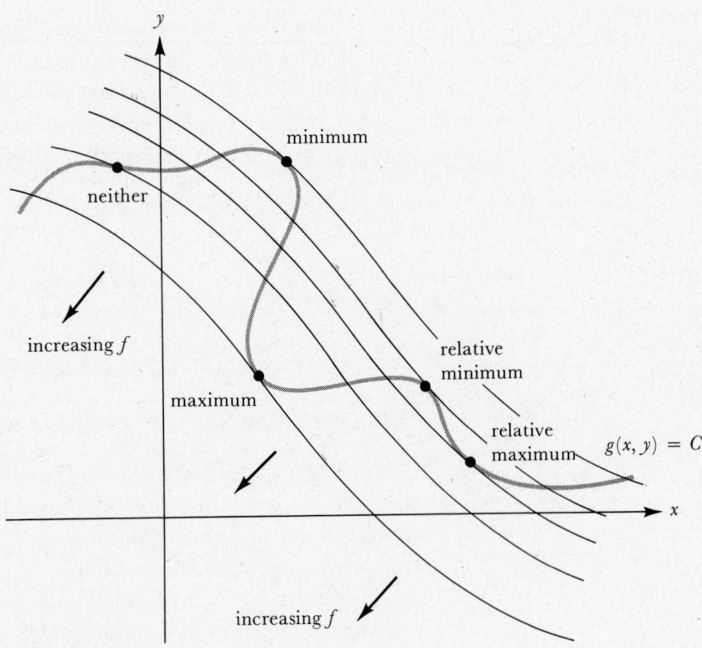

Fig. 19.50

Set $x = y^2 - 1$ in (1) and (2) to obtain the simultaneous equations

$$4y^2 - 4 + 3 + 2y + \lambda = 0 \quad \text{and} \quad 2y^2 - 2 + 2 + 2y - 2y\lambda = 0$$

or

$$4y^2 + 2y - 1 + \lambda = 0 \quad \text{and} \quad y(y + 1 - \lambda) = 0$$

The equation on the right tells us that either $y = 0$ or $y = \lambda - 1$. We use $y = \lambda - 1$ in the equation on the left to obtain a quadratic in λ, which solves to yield $\lambda = 1/4, 1$. Now $\lambda = 1/4$ implies $y = -3/4$, which yields $x = -7/16$ and the point $(-7/16, -3/4)$. Also, $\lambda = 1$ implies $y = 0$ and $x = -1$, which yields the point $(-1, 0)$.

Since $f(-1, 0) = -1$ and $f(-7/16, -3/4) = -155/128$, we conclude that the minimum value of f is $-155/128$ and occurs at $(-7/16, -3/4)$. ●

To determine extreme values of a function on a *region* bounded by a constraint curve $g(x, y) = C$, we can use Lagrange multipliers on the boundary in combination with the techniques of Sec. 19.8 on the interior.

Example 2 Determine the minimum value of $f(x, y) = 2x^2 + 3x + 2xy + 2y + y^2$ on the set of all (x, y) such that $y^2 - x \leq 1$.

Solution Example 1 has given us a minimum value at $(-7/16, -3/4)$ *on* the curve $y^2 - x = 1$, so we now identify the region consisting of points (x, y) satisfying $y^2 - x < 1$. We solve this inequality for x to obtain $x > y^2 - 1$, whose graph consists of all points "inside" the parabola $x = y^2 - 1$ as shaded in Fig. 19.51. Looking for critical points, we set

$$f_x = 4x + 3 + 2y = 0 \qquad f_y = 2x + 2 + 2y = 0$$

which solve to yield the critical point $(-1/2, -1/2)$, which *is* inside the region. By the Second-Derivative Test, $(-1/2, -1/2)$ is a relative minimum. Since $f(-1/2, -1/2) = -5/4$ $(= -160/128)$ and $f(-7/16, -3/4) = -155/128$, we conclude that f reaches its minimum of $-5/4$ at $(-1/2, -1/2)$. ●

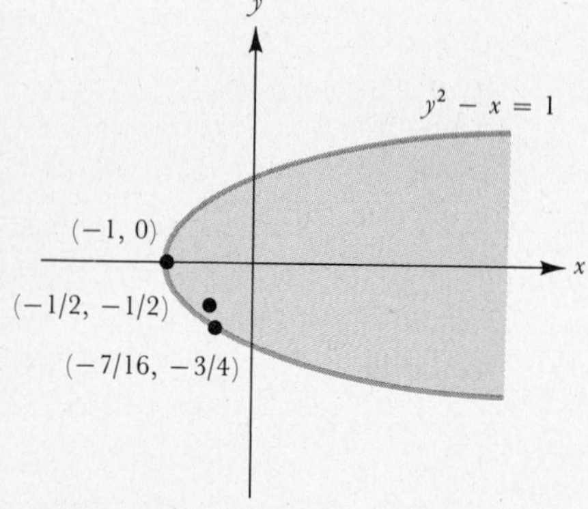

Fig. 19.51

Although we have concentrated on functions of two variables, the Lagrange technique extends directly, via a higher dimensional interpretation of the equation $\nabla f = \lambda \nabla g$, to functions of three or more variables. The key, of course, is that there be enough equations to determine solutions. These extensions are treated in the section and chapter exercises.

PROGRESS TEST 2

1. Determine the extreme values of $f(x, y) = x^2 y$ on the curve $2x^2 + y^2 = 12$.

THE MEANING OF THE LAGRANGE MULTIPLIER

Suppose you are a division vice president asking the company president for an increase in your division's operating budget to be spent on labor costs x and material costs y in order to increase productivity $f(x, y)$. You must prove that a modest budget increase will spur a large productivity increase.

Under your current budget constraint $g(x, y) = C$, you determined the point (a, b) of maximum productivity M [so $f(a, b) = M$] by finding λ so that the following equations all hold:

$$f_x(a, b) = \lambda g_x(a, b) \tag{1}$$
$$f_y(a, b) = \lambda g_y(a, b) \tag{2}$$
$$g(a, b) - C = 0 \tag{3}$$

We shall now show that λ is more than a technical help in computations and can provide the basis for your argument for a larger budget. For *you*, C, (a, b), λ, and M all have been determined—in effect, by the president. But for the *president*, the budget C and hence (a, b), λ, and M are all *variables*.

By (3) we can write $\lambda(g(a, b) - C) = 0$, so

$$M = f(a, b) - \lambda(g(a, b) - C) \tag{4}$$

We shall compute dM/dC, the measure of the rate of change of maximal productivity with respect to budget. Since a, b, and λ all depend on C, we will need the Chain Rule as well as the Product Rule in order to differentiate the right side of (4). We differentiate the two terms on the right of (4) separately:

$$\frac{df(a, b)}{dC} = f_x(a, b)\frac{da}{dC} + f_y(a, b)\frac{db}{dC}$$

and

$$\frac{d}{dC}[\lambda g(a, b) - \lambda C] = \lambda\left(g_x(a, b)\frac{da}{dC} + g_y(a, b)\frac{db}{dC}\right) + g(a, b)\frac{d\lambda}{dC} - \left(\lambda \cdot 1 + C\frac{d\lambda}{dC}\right)$$
$$= \lambda g_x(a, b)\frac{da}{dC} + \lambda g_y(a, b)\frac{db}{dC} + C\frac{d\lambda}{dC} - \lambda - C\frac{d\lambda}{dC}$$

[since $g(a, b) = C$]

$$= \lambda g_x(a, b)\frac{da}{dC} + \lambda g_y(a, b)\frac{db}{dC} - \lambda$$

Therefore, combining these derivatives we have

$$\frac{dM}{dC} = f_x(a, b)\frac{da}{dC} + f_y(a, b)\frac{db}{dC} - \lambda g_x(a, b)\frac{da}{dC} - \lambda g_y(a, b)\frac{db}{dC} + \lambda$$
$$= (f_x(a, b) - \lambda g_x(a, b))\frac{da}{dC} + (f_y(a, b) - \lambda g_y(a, b))\frac{db}{dC} + \lambda$$
$$= \lambda \qquad \text{by equations (1) and (2)}$$

Hence *the Lagrange multiplier is the rate of change of maximal output with respect to change in budget*. If, at a particular optimal operating point (a, b), λ is large and positive, then it makes sense for the president to increase your budget, whereas, if λ is small, then the budget probably should be held constant.

Since the mathematics of the above argument does not depend on the particular interpretation given, we have proved the following theorem.

19.9.2 Theorem Suppose that $f(x,y)$ yields a maximum M at (a, b) under the constraint $g(x,y) = C$ when $\nabla f(a, b) = \lambda \nabla g(a, b)$. Then $\lambda = dM/dC$.

SECTION 19.9 EXERCISES

1. (Reversed Fence Problem) Draw several level curves and the constraint curve that illustrates the solution of the following: Give the dimensions of the three-sided rectangular pasture with an area of 10,000 m^2 requiring the least amount of fence.

2. Repeat Exercise 1 for (*a*) a four-sided rectangular pasture and (*b*) a circular pasture.

3. (*a*) Sketch a pair of plausible milkshake-hamburger indifference curves for the 250-lb 12-year-old of Progress Test 1. (Plot hamburgers horizontally.)

(*b*) Given that hamburgers cost \$1 and milkshakes cost 50¢, use your solutions to (*a*) to estimate how this 12-year-old might spend \$5 on hamburgers and milkshakes.

4. Sketch a hamburger-salad indifference curve for the above-mentioned 250-lb 12-year-old who has decided to lose weight. (Plot hamburgers horizontally.)

In Exercises 5 to 12 use Lagrange multipliers to determine the indicated extreme value for the given functions and constraints.

5. $f(x,y) = xy$, $x^2 + y^2 = 1$, maximum

6. $f(x,y) = xy^2$, $x^2 + y^2 = 12$, all extreme values

7. $f(x,y) = 2x^2 + y^2 - x - y + 1$, $x^2 + y^2 = 1$, maximum

8. $f(x,y) = x^2 + 3y^2$, $4x + y = 6$, minimum

9. $f(x,y) = x^2 + 2y^2 + 10x - 15y$, $2x + 2y + 150 = 0$, minimum

10. $f(x,y) = 4x^2 - 4xy + y^2$, $x^2 + y^2 = 1$, maximum

11. $f(x,y) = x^2 + y^2$, $x^4 + y^4 = 1$, all extreme values

12. $f(x,y) = 5x^2 - 2xy + 3y^2$, $2x^2 + y^2 = 12$, all extreme values

In Exercises 13 to 20 use Lagrange multipliers [solve the four equations resulting from $\nabla f(x,y,z) = \lambda \nabla g(x,y,z)$ *and* $g(x,y,z) = C$*] to determine the indicated extreme value for the given functions and constraints.*

13. $f(x,y,z) = 4xyz$, $x + 2y + 3z = 54$, maximum

14. $f(x,y,z) = x + y + 2z$, $x^2 + y^2 + z^2 = 54$, minimum

15. $f(x,y,z) = x^2 + y^2 + z^2$, $x + 3y - 2z = 12$, minimum

16. $f(x,y,z) = xy^2z^3$, $x + y + z = 12$, maximum

17. $f(x,y,z) = x^3 + y^3 + z^3$, $x + y + z = 1$, maximum

18. $f(x,y,z) = xyz$, $4x^2 + 9y^2 + 16z^2 = 432$, maximum

19. $f(x,y,z) = \sqrt{x^2 + y^2 + z^2}$, $x - 4y - z - 100 = 0$, minimum

20. $f(x,y,z) = x^4 + y^4 + z^4$, $x^2 + y^2 + z^2 = 1$, all extreme values

In Exercises 21 to 25 use Lagrange multipliers to determine the minimum distance between the given surface and the indicated entity. It suffices to minimize the square of the distance.

21. The surface $x^2 - z^2 = 9$ and the origin.

22. The plane $2x - 2y + 4z = 5$ and the origin.

23. The plane $x + 2y + 3z = 73$ and the point $(1, 2, 4)$.

24. The surface $x^2 - y^2 - 2z = -2$ and the origin.

25. The intersection of the planes $x + y + z = 1$, $3x + 2y + z = 1$, and the origin.

26. Use Lagrange multipliers to show that the rectangle of maximum area and fixed perimeter C must be a square.

In Exercises 27 to 32 use the methods of this section to solve the given extreme-value problem.

27. Determine the extreme values of $f(x,y) = (3x^2 + 2y^2)/e^{x^2+y^2}$ on the disk $x^2 + y^2 \leq 1$.

28. Determine the dimensions of a right circular can of volume 16 in.^{3} of minimum surface area.

29. A rectangular aquarium is to have an open top. Given that the material for the bottom costs 2¢/in.^{2} and the material for the sides costs 1¢/in.^{2}, what are the dimensions of the least expensive aquarium that holds 1728 in.^{3}?

30. Determine the rectangle of maximum area and sides parallel to the axes inscribed in the ellipse $4x^2 + y^2 = 1$.

31. Repeat Exercise 30, but maximize the perimeter.

32. Using that the volume of the ellipsoid $x^2/a^2 + y^2/b^2 + z^2/c^2 = 1$ is $(4/3)\pi abc$, determine the ellipsoid $x^2/a^2 + y^2/b^2 + z^2/c^2 = 1$ with minimum volume that passes through $(3, 1, 2)$.

33. A manufacturer produces $f(x,y)$ units of a given product where x and y represent, respectively, the labor and capital costs of the product. If the labor and capital per unit are A and B dollars, respectively, and the total budget is C dollars, prove that at the level (a, b) of maximum production, $f_x(a,b)/A = f_y(a,b)/B = \lambda$, where λ is the Lagrange multiplier. This equation is called the *law of equimarginal productivity*.

34. Suppose that a company has a labor cost x, capital y, production function (in thousands) $f(x,y) = 12x^{3/2}y$ where the labor cost per unit is \$2 and the capital cost per unit is \$3. Suppose also that the total budget is 120 (in thousands of dollars).

(*a*) Determine how to allocate the budget to maximize the production level.

(*b*) What *is* the maximal production level?

(*c*) If the budget is increased by 1 (thousand dollars), what is the effect on the production level? Should the budget be increased? [See (19.9.2).]

35. In Examples 1 and 2 we determined the minimum value of $f(x,y) = 2x^2 + 3x + 2xy + 2y + y^2$ for $y^2 - x \leq 1$.

(*a*) Use the result of Example 2 to adjust the constraint to $y^2 - x \leq a$ for an appropriate a so that the absolute minimum occurs on the boundary.

(*b*) Compare the necessary adjustment with the λ used to compute the minimum on the original boundary.

(*c*) Use Lagrange multipliers to recompute the minimum value of f on the boundary and compare the λ resulting in this solution with that suggested by the remarks preceding (19.9.2).

CHAPTER EXERCISES

Review Exercises

In Exercises 1 to 10 determine and sketch the region in R_2 or R_3 that is the domain of the given function.

1. $f(x,y) = \sqrt{4 - x^2 - y^2}$

2. $f(x,y) = \sqrt{y^2 - x}/xy$

3. $f(x,y) = \left[\dfrac{x + 2 - y}{y - x^2}\right]^{1/4}$

4. $f(x,y) = \dfrac{x^2 + y^2}{xy - x^2}$

5. $f(x,y) = \ln(3 - xy)$

6. $f(x,y) = \arcsin(x + y)$

7. $f(x,y) = \sqrt[4]{x + y} + \sqrt{y - 2x^2}$

8. $f(x,y,z) = \ln(z - x^2 - y^2)$

9. $f(x,y,z) = \sqrt{x^2 + 2x + y^2 + z^2 - 15}$

10. $f(x,y,z) = \arcsin x + \arctan y + \ln z$

In Exercises 11 to 15 show that the indicated limits do not exist.

11. $\displaystyle\lim_{(x,y)\to(0,0)} \frac{2y^2}{x^2 + y^2}$

12. $\displaystyle\lim_{(x,y)\to(0,0)} \frac{4y^2 - x^2}{x^2 + 4y^2}$

13. $\displaystyle\lim_{(x,y)\to(0,0)} \frac{x^4y^4}{x^6 + y^6}$

14. $\displaystyle\lim_{(x,y)\to(0,0)} \frac{x^3 + y^3}{x + y^2}$

15. $\displaystyle\lim_{(x,y)\to(1,2)} \frac{(x - 1)^2}{(x - 1)^2 + (y - 2)^2}$

16. Let $f(x,y) = \begin{cases} \dfrac{x^2y^2}{x^3 + y^3} & \text{if } (x,y) \neq (0,0) \\ 0 & \text{if } (x,y) = (0,0) \end{cases}$

(*a*) Is f continuous at $(0,0)$?
(*b*) Determine if f_x or f_y exist at $(0,0)$.

In Exercises 17 to 21, (a) sketch the graph of f, (b) sketch several level curves of f, (c) sketch the curves of intersection with the planes $x = a$ and $x = b$, and (d) find equations for the tangent lines to the curves in (c) at the point $(a, b, f(a, b))$.

17. $f(x,y) = x^2 + y$; $a = 2, b = 1$

18. $f(x,y) = 4 - x^2$; $a = 1, b = 2$

19. $f(x,y) = 9 - x^2 - y^2$; $a = 1, b = 2$

20. $f(x,y) = x^2 + y^2$; $a = -1, b = 2$

21. $f(x,y) = ye^x$; $a = 1, b = 1$

In Exercises 22 to 28 determine f_x and f_y.

22. $f(x,y) = e^{x^2}\cos(xy)$

23. $f(x,y) = y^2 \arcsin(y/x)$

24. $f(x,y) = x^2 \ln(x^2 + y^2)$

25. $f(x,y) = \cos^2(x + \ln(xy))$

26. $f(x,y) = \arctan(x/y^2)$

27. $f(x,y) = x^2ye^{y/x^2}$

28. $f(x,y) = \displaystyle\int_x^{xy} e^{t^2}\,dt$

In Exercises 29 to 37 determine the first partials with respect to each of the independent variables.

29. $f(x,y,z) = xye^{x^2+y^2+z^2}$

30. $f(u,v,w) = w^2 \arctan(u/v)$

31. $f(r,\theta,z) = \left(\dfrac{r - \theta^2}{z - r^2}\right)^2$

32. $f(r,\theta,t) = \theta^2 \ln(r/\theta) + t/r^3$

33. $f(w,z,p) = 4p^2z + p/w - w/z$

34. $f(r,s,t) = \dfrac{rt - st}{r^2 + t^2}$

35. $f(p,q) = q \arcsin(qp^2)$

36. $f(A,C,D) = D^3\sqrt{AC + C^3}$

37. $f(G,I,P) = P^2 \displaystyle\int_I^{G^3} \frac{dt}{\sqrt{t^3 + 1}}$

In Exercises 38 through 43 determine the indicated partial derivative.

38. $f(x,y) = \arcsin(x^2 + y^2)$, f_{xy}

39. $f(x,y) = x^2ye^{x/y}$, f_{xx}

40. $f(x,y) = y \arcsin(x^2)$, f_{xxyy}

41. $f(x,y,z) = x^2e^y + z^2e^x + e^z$, f_{xyz}

42. $f(x,y,z) = \arctan(xyz)$, f_{xx}

43. $f(x,y) = x\,\text{arcsec}(\sqrt{x})$, f_{xyy}

In Exercises 44 to 46 determine equations for the tangent plane and normal line to the graph of f at the given point.

44. $f(x,y) = x \arcsin(x/(2y))$, $P = (1, 1, \pi/6)$

45. $f(x,y) = y \ln(x^2 + y^2) + e^x$, $P = (0, 1, 1)$

46. $f(x,y) = \dfrac{5x}{\sqrt{x^2 + y^2}}$, $P = (3, 4, 3)$

In Exercises 47 to 49 determine equations for the tangent plane and normal line to the given surface at the indicated point.

47. $x^{1/2} + y^{1/2} + z^{1/2} = 5$, $P = (1, 1, 9)$

48. $xy \cos(y^2z) = \sqrt{\pi}/2$, $P = (1, \sqrt{\pi}, 1/3)$

49. $x \arctan(y/z) - y \arcsin(x/z^2) = \pi/12$, $P = (1, 1, \sqrt{2})$

In Exercises 50 to 54 use the total differential to approximate the given quantity.

50. $f(-3.04, 3.95)$, where $f(x,y) = x/\sqrt{x^2 + y^2}$

51. $\sqrt[3]{99.01 + 25.16}$

52. $\sin 44° \cos 61°$

53. The force of attraction between two particles of masses m and n located r units apart is given by $F = gmn/r^2$. If there is a maximum possible error of 2 percent in the measurements of the masses and 3 percent in the measurement of r, then what is the approximate maximum percentage error in the measurement of F?

54. A can is measured and is found to have a height of 12 in. and a radius of 4 in. If there was a possible maximum error of 0.02 in. in each of these measurements, then what is the approximate error if the volume is computed using these measurements?

55. Let $z = f(x/y, y/z)$, where f is differentiable. Show that

$$\frac{x}{y}\frac{\partial z}{\partial x} + \frac{\partial z}{\partial y} = 0$$

56. Let $z = f(\sqrt{x^2 + y^2})$, where f is differentiable. Show that

$$y\frac{\partial z}{\partial x} - x\frac{\partial z}{\partial y} = 0$$

57. The temperature in a particular fluid at the point (x, y, z) is given by $T(x,y,z) = x^2y - yz + y^2x$. The velocity at time $t = t_0$ of a particle moving along a path through the fluid is given by $\mathbf{V}(t_0) = \langle -3, 4, 7\rangle$. How fast is the temperature of the particle changing at that instant? (The particle assumes the temperature of the fluid at each point.)

58. Let $z = \sqrt{x^2 + y^2}$. Show that $z_{xx}z_{yy} = (z_{xy})^2$.

In Exercises 59 to 63 determine the directional derivative of f at the given point in the direction of the vector **v**.

59. $f(x,y) = 3x^2y - xy^3$; $P = (1, -2)$, $\mathbf{v} = \langle 3, 4\rangle$

60. $f(x,y) = xy/\sqrt{x^2 + y^2}$; $P = (-3, 4)$, $\mathbf{v} = \langle -2, 5\rangle$

61. $f(x,y) = e^x \sin y$; $P = (0, \pi/6)$, $\mathbf{v} = \langle 4, -3\rangle$

62. $f(x,y,z) = x^2 \arcsin(z/y)$; $P = (-1, 2, 0)$, $\mathbf{v} = \langle 1, 1, 1\rangle$

63. $f(x,y,z) = xyz/\sqrt{x^2 + y^2 + z^2}$; $P = (-1, 2, -2)$, $\mathbf{v} = \langle 1, 1, \sqrt{2}\rangle$

In Exercises 64 to 68 determine the maximum and minimum value of the directional derivative of f at the given point.

64. $f(x,y) = x^2\sqrt{x^2 + y^2}$, $P = (3, -4)$

65. $f(x,y) = x^2y^3 - 3x/y$, $P = (1, 1)$

66. $f(x,y) = e^{\sqrt{x^2+y^2}}$, $P = (\sqrt{2}, -\sqrt{2})$

67. $f(x,y,z) = x^2yz^3 - 3zx^2$, $P = (1, -1, 2)$

68. $f(x,y,z) = \ln(\sqrt{x^2 + y^2 + z^2})$, $P = (-1, 1, \sqrt{2})$

In Exercises 69 to 73 identify the critical points of f and state whether each is a relative maximum, relative minimum, or saddle point.

69. $f(x,y) = x^2 + y^2 + 3y + xy + 1$

70. $f(x,y) = y^3 - x^3 + 3xy$

71. $f(x,y) = xy^2 - y - x^2y + x$

72. $f(x,y) = x^4 - 4xy + y^4$

73. $f(x,y) = x \sin y$

74. Find the absolute maximum and minimum values of the function $f(x,y) = x^2 - xy + y^2 - 3x - 4$ on the closed triangular region with vertices (0, 0), (4, 0), and (0, 4).

75. Find the absolute maximum and minimum values of the function $f(x,y) = 2x^2 + y^2 + x^2y + 3x$ on the closed region bounded above by the parabola $y = 1 - x^2$ and below by the line $y = -3$. (*Note:* This problem will involve solving a cubic that clearly has as a solution $x = -1$.)

76. Use the method of Lagrange multipliers to find the extreme values of the function $f(x,y) = 2x^2 + 6y^3$ on the ellipse $x^2 + 3y^2 = 1$.

77. Use the method of Lagrange multipliers to find the extreme values of the function $f(x,y) = y^3 + 2x^3$ on the circle $x^2 + y^2 = 1$.

78. Use the method of Lagrange multipliers to find the maximum value of the function $f(x,y,z) = (xyz)^2$ on the sphere $x^2 + y^2 + z^2 = 1$.

79. Determine the points nearest to and farthest from the origin and that are on the ellipsoid $36x^2 + 9y^2 + 4x^2 = 36$.

80. Suppose a manufacturer has the production function $f(x,y) = 60xy$, where x is labor cost and y is material cost. Suppose that the unit cost of labor is 2 and the unit cost of material is 1 (each in thousands). What will be the increase in production due to increasing the budget from \$600,000 to \$601,000? [Use (19.9.2), noting that the budget constraint is in thousands of dollars.]

81. (*a*) Use Lagrange multipliers to obtain the maximum value of $f(x,y) = x - 2y$ on $y = \cos x$, $-\pi/2 \leq x \leq \pi/2$.
(*b*) Sketch the appropriate level curve tangent to $y = \cos x$.
(*c*) Check the solution to (*a*) using the single-variable maximizing technique of Sec. 5.1.

82. Determine the maximum value of $f(x,y) = 9xy - x^3 - y^3$ for $x + y \leq 6$ in the first quadrant.

83. Find three positive numbers that add up to 12 and such that the sum of their squares is a minimum.

84. Find three positive numbers that add up to 12 and such that their product is a maximum.

85. Find the points on the surface $x^2 - yz = 8$ that are closest to the origin.

Miscellaneous Exercises

1. Let $f(x,y)$ be differentiable at (a,b), and let $\mathbf{F}(t) = \langle x(t), y(t), z(t) \rangle$ be a space curve whose graph lies on the surface $z = f(x,y)$ and goes through the point $P = (a, b, f(a,b))$. Show that the tangent line to $\mathbf{F}$ at P lies in the tangent plane to the surface $z = f(x,y)$ at P.

2. A function $f(x,y)$ has as its trace in the x-z plane the curve $z = x^2 + 2$, and its trace in the plane $x = 1$ is $z = 3 + y$. Assuming that f is differentiable at $(1,0)$, find an equation for the tangent plane at the point on the graph of f above the point $(1,0)$ in the x-y plane.

3. The tangent plane to the graph of a function $f(x,y)$ at the point $(1,2,-1)$ has equation $2x + 3y + 4z = 4$. Find $f_x(1,2)$ and $f_y(1,2)$.

4. Let $f(x,y) = \begin{cases} \dfrac{\sin(x^2+y^2)}{x^2+y^2} & \text{if } (x,y) \neq (0,0) \\ 0 & \text{if } (x,y) = (0,0) \end{cases}$

Determine $f_x(0,0)$ and $f_y(0,0)$.

5. Let $f(x,y) = \begin{cases} \dfrac{2xy}{x^2+y^2} & \text{if } (x,y) \neq (0,0) \\ 0 & \text{if } (x,y) = (0,0) \end{cases}$

(*a*) Show that f is not continuous at $(0,0)$.
(*b*) Show that $f_x(0,0)$ and $f_y(0,0)$ exist.
(*c*) Are f_x and f_y continuous at $(0,0)$?

6. Use the definition of the limit to show that
$$\lim_{(x,y)\to(1,2)} (x^2 + y^2) = 5$$
(*Hint:* $x^2 + y^2 - 5 = (x-1)^2 + (y-2)^2 + 2(x-1) + 4(y-2)$.)

7. Use the definition of the limit to show that
$$\lim_{(x,y)\to(-2,1)} (3x + 5y) = -1$$

8. Is the function $f(x,y) = \begin{cases} \dfrac{\sin(xy)}{xy} & (x,y) \neq (0,0) \\ 1 & (x,y) = (0,0) \end{cases}$

continuous at $(x,y) = (0,0)$?

9. Use the definition to find $f_x(2,3)$ and $f_y(2,3)$, where $f(x,y) = \ln(2x + 3y)$.

10. A function $f(x,y)$ is said to be *positively homogeneous of degree n* if $f(tx, ty) = t^n f(x,y)$ for all $t \geq 0$. Show that if f is differentiable and positively homogeneous of degree n, then $xf_x(x,y) + yf_y(x,y) = nf(x,y)$. (*Hint:* First determine $\partial/\partial t[f(tx,ty)]$ and $\partial/\partial t[t^n f(x,y)]$; then let $t = 1$.)

11. Show that if $f(x,y)$ is differentiable and positively homogeneous of degree n, then $f_x(x,y)$ and $f_y(x,y)$ are positively homogeneous of degree $n-1$ (see Exercise 10).

12. Let
$$f(x,y) = \int_0^{x^2y^2} g(z)\, dz$$
and let $x = t^2 + 2t + 2$ and $y = t^3 - 2t + 2$. Find
$$\left.\frac{df}{dt}\right|_{t=0} \quad \text{given that } g(4) = 2.$$

13. Let $f(t) = \int_2^{t^3} h(w)\, dw$ and let $t = \sqrt{x^2 + y^2 + z^2}$. Find $\partial f/\partial x$ at the point $(1, 1, \sqrt{2})$ given that $h(8) = -4$.

14. The height and length of a rectangular box with square base are measured and these measurements are used to compute the volume. If there was a maximum possible error of 1 percent in measuring the length and 2 percent in measuring the height, then what is the approximate maximum percentage error induced by using these measurements to compute the volume?

15. Formulate a Chain Rule appropriate to compute dz/dt given that $z = f(x,y)$, $x = x(u,v)$, $y = y(u,v)$, $u = u(t)$, and $v = v(t)$.

16. Formulate a Chain Rule that gives $\partial z/\partial r$ given that $z = f(x,y)$, $x = x(u,v)$, $y = y(u,v)$, $u = u(r,s)$, and $v = v(r,s)$.

17. Prove part (i) of [19.8.6(*a*)] by following the given outline: Let $\mathbf{u} = \langle \cos\alpha, \sin\alpha \rangle$ be a unit vector of arbitrary direction, $\Pi_{\mathbf{u}}$ a plane through (a,b) parallel to $\mathbf{u}$ and perpendicular to the x-y plane, and $C_{\mathbf{u}}$ the intersection of $\Pi_{\mathbf{u}}$ with the graph of $z = f(x,y)$.
(*a*) Show that $D_{\mathbf{u}}^2 f(x,y) = [D_{\mathbf{u}} f(x,y)]_x \cos\alpha + [D_{\mathbf{u}} f(x,y)]_y \sin\alpha$.
(*b*) Use that $f_{xy} = f_{yx}$ (why is this true?) to show that $D_{\mathbf{u}}^2 f(x,y) = f_{xx}\cos^2\alpha + 2f_{xy}\sin\alpha\cos\alpha + f_{yy}\sin^2\alpha$.
(*c*) For $\alpha \neq 0, \pi$, show that $(1/\sin^2\alpha) D_{\mathbf{u}}^2 f(a,b)$ is a quadratic polynomial in $\cot\alpha$ with no roots if and only if $f_{xx}(a,b) f_{yy}(a,b) - f_{xy}^2(a,b) > 0$. Show that $D_{\mathbf{u}}^2 f(a,b)$ is likewise positive for $\alpha = 0$ and $\alpha = \pi$.
(*d*) Interpret the sign of $D_{\mathbf{u}}^2 f(a,b)$ and $f_{xx}(a,b)$ in terms of concavity to conclude part (i) of [19.8.6(*a*)].

18. At a particular instant the radius of a right circular cylinder is 12 cm and is increasing at the rate of 5 cm/s and the height is 64 cm and is decreasing at the rate of 3 cm/s. How fast is the surface area changing at this instant? Is it increasing or decreasing?

19. A particle is moving along the curve of intersection of the sphere $x^2 + y^2 + z^2 = 17$ with the plane $2x + y + 3z = 13$. Find the speed at which the particle is traveling at the instant it is at the point $(2,3,2)$ if $dx/dt = 4$ at that instant.

20. Suppose that $z = f(x,y)$, $x = e^r \cos\theta$, $y = e^r \sin\theta$. Show that

$$\left(\frac{\partial z}{\partial r}\right)^2 + \left(\frac{\partial z}{\partial \theta}\right)^2 = e^{2r}\left[\left(\frac{\partial z}{\partial x}\right)^2 + \left(\frac{\partial z}{\partial y}\right)^2\right]$$

21. Show that the normal line at any point on the cone $x^2 + y^2 = z^2$ intersects the z-axis.

22. Use the method of Lagrange multipliers to find the point on the plane $x + 3y - 2z = 4$ which is closest to the origin.

23. Use the method of Lagrange multipliers to determine the points on the plane $x + y + z = 4$ with the property that the sum of the cubes of their coordinates is a maximum.

24. A parallelepiped is to be inscribed in the ellipsoid $x^2/a^2 + y^2/b^2 + z^2/c^2 = 1$ so that the edges are parallel to the coordinate axes. Determine the dimensions if the parallelepiped is to have maximum volume.

25. Determine three positive numbers with the property that their sum is a minimum and the sum of their reciprocals is equal to 1.

26. (*The Method of Least Squares*) Let (x_1,y_1), (x_2,y_2), $\ldots$, (x_n,y_n) be n points in the plane. The line $y = mx + b$ with the property that the sum of the squares of the vertical distances from the points to the line is called the *least-squares line* for the points. Since the vertical distance from (x_i,y_i) to the line $y = mx + b$ is $|y_i - (mx_i + b)|$, the problem is to find the value of m and b that minimizes the function

$$D(m,b) = \sum_{i=1}^{n} (y_i - mx_i - b)^2$$

Show that these values are given by

$$m = \frac{\begin{vmatrix} \sum_{i=1}^{n} y_i & n \\ \sum_{i=1}^{n} x_i y_i & \sum_{i=1}^{n} x_i \end{vmatrix}}{\begin{vmatrix} \sum_{i=1}^{n} x_i & n \\ \sum_{i=1}^{n} x_i^2 & \sum_{i=1}^{n} x_i \end{vmatrix}} \qquad b = \frac{\begin{vmatrix} \sum_{i=1}^{n} x_i & \sum_{i=1}^{n} y_i \\ \sum_{i=1}^{n} x_i^2 & \sum_{i=1}^{n} x_i y_i \end{vmatrix}}{\begin{vmatrix} \sum_{i=1}^{n} x_i & n \\ \sum_{i=1}^{n} x_i^2 & \sum_{i=1}^{n} x_i \end{vmatrix}}$$

SELF-TEST ✚

1. Let $f(x,y) = x^2y^3 - x^2 - y^2 + \ln(xy)$

(*a*) Determine f_x and f_y.

(*b*) Show by direct computation that $f_{xy} = f_{yx}$.

(*c*) Determine equations for the tangent plane and normal line to f at $(1, 1, -1)$.

2. Let $z = f(x/(x + y))$, where $f(u)$ is a differentiable function of one variable. Show that

$$x\frac{\partial z}{\partial x} - y\frac{\partial z}{\partial y} = 0$$

3. Let $f(x,y) = xy\sqrt{x^2 + y^2 + z^2}$.

(*a*) Find the directional derivative of f at $(-\sqrt{2}, 1, 1)$ in the direction of $\langle 2, -1, 2\rangle$.

(*b*) Find the maximum and minimum value of the directional derivative of f at $(-\sqrt{2}, 1, 1)$.

4. Determine the absolute maximum and minimum value of the function $f(x,y) = x^2 + 2y^2 + x + 1$, on the closed disk $x^2 + y^2 \leq 1$.

5. Let

$$f(x,y) = \begin{cases} \dfrac{3x^2 - y^2}{x^2 + y^2} & \text{if } (x,y) \neq (0,0) \\ 0 & \text{if } (x,y) = (0,0) \end{cases}$$

Is f continuous at $(0, 0)$?

Chapter 20 ✖

MULTIPLE INTEGRATION

CONTENTS

This chapter introduces the integral calculus for functions of more than one variable. We begin with the double integral of a function of two variables over a plane region. It is the two-dimensional analog of the definite integral of a (single variable) function over an interval. Just as we used the notion of the area of the region between the graph of a positive function and an interval to introduce the definite integral, we use the volume of the solid between the graph of a positive function of two variables (a surface) and a plane region to introduce the double integral. Once defined, however, we *compute* the double integral using a pair of "iterated integrals." Computing each of the iterated integrals is similar to computing partial derivatives in reverse—we hold one variable fixed and integrate with respect to the other.

In Sec. 20.3 we move up yet one more dimension to define the triple integral. Again, the triple integral once defined is evaluated using iterated integrals. In Sec. 20.4 we revisit these double and triple integrals—in the context of polar coordinates in two dimensions and in the context of cylindrical and spherical coordinates for three dimensions.

Given the new flexibility inherent in using functions of more than one variable, we are able to compute volumes, masses, moments, and other geometric and physical quantities with far fewer restrictions than was the case in Chaps. 6 and 12.

Section 20.1: Double and Iterated Integrals

Objectives:

1. **Understand the definition of the double integral as a limit of Riemann sums.**
2. **Express double integrals in terms of iterated integrals.**
3. **Evaluate iterated integrals.**

INTRODUCTION

Our introduction of the definite integral in Chap. 6 was based on determining the area under the graph of a positive function and above some interval on the x axis. We now consider the analog of this problem for functions of two variables. Specifically, let R be a region in the x-y plane and let $f(x, y)$ be a positive function defined on R. Our object is to determine the volume (if it exists) of the solid region S below the graph of f and above the region R in the x-y plane (see Fig. 20.1). *We assume throughout* two reasonable restrictions on the region R:

(1) R is bounded.

(2) The boundary of R is a closed arc that is composed of a finite collection of smooth arcs.

Some typical such regions R are sketched in Fig. 20.2. In our following discussion we further assume that the volume of S exists and concentrate on how to determine its value. Later we shall discard this assumption and use our development to *define* the volume.

VOLUMES

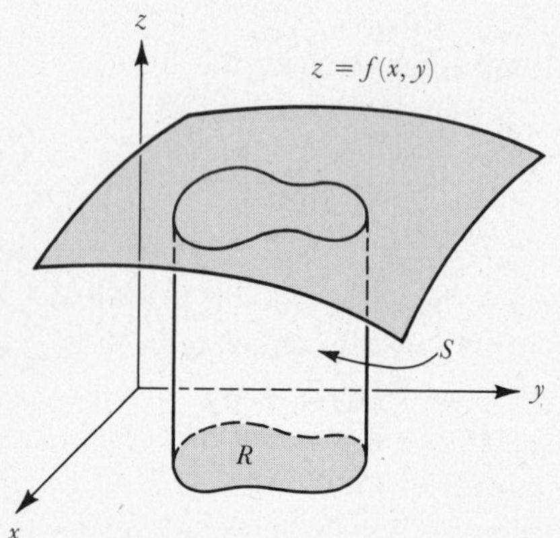

Fig. 20.1

Since the region R is bounded, we can contain it in a rectangle D whose sides are parallel to the coordinate axes. We begin by superimposing a *grid* over D consisting of a finite number of lines parallel to the x and y axes. Consider all the rectangles determined by this grid that lie *wholly within* the region R. Since there is a finite number of such rectangles (say n), we can order them, getting $R_1, R_2, \ldots, R_n$. We refer to this collection as an *inner partition* of R and denote it by P (see Fig. 20.3). By the norm of P, denoted $\|P\|$, we shall mean the length of the largest diagonal of the R_i's.

Now suppose (x_i, y_i) is an arbitrary point in R_i and consider the parallelepiped (rectangular box) whose base is the rectangle R_i and whose height is $f(x_i, y_i)$. This parallelepiped has volume V_i given by

$$V_i = f(x_i, y_i)\,\Delta A_i$$

where ΔA_i denotes the area of R_i. Furthermore, this volume approximates the volume of the solid region above R_i and below the graph of f (see Fig. 20.4). If

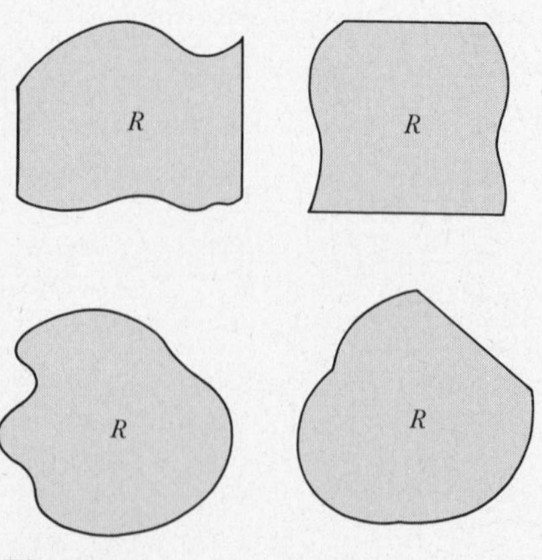

Fig. 20.2

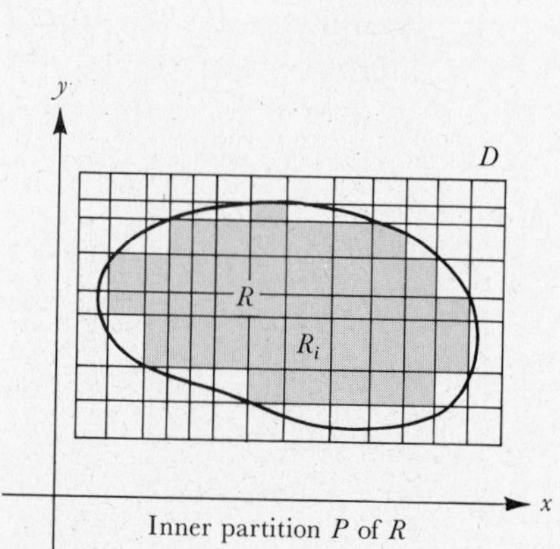

Fig. 20.3

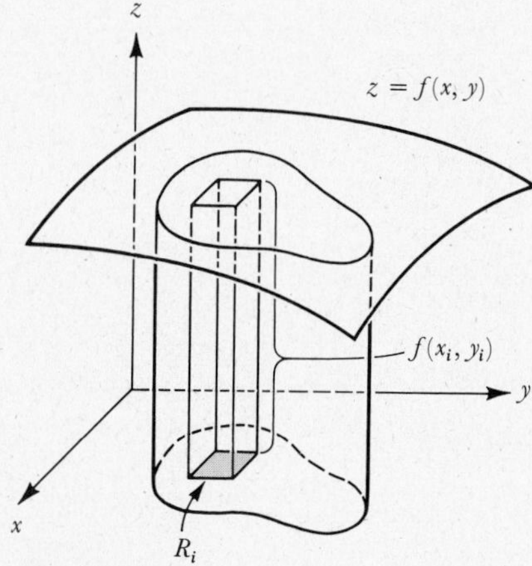

Fig. 20.4

we now sum the volumes of each of the n parallelepipeds obtained in this fashion, we shall have an approximation to $V(S)$, the volume of S. That is,

$$V(S) \approx \sum_{i=1}^{n} f(x_i, y_i)\, \Delta A_i$$

To improve this approximation we can refine the grid by inserting more lines parallel to the x and y axes. This has the effect of increasing n, the number of rectangles in the inner partition P. We continue in this fashion, letting n grow without bound. If we do this in such a way that $\|P\| \to 0$, then the corresponding sums will tend to $V(S)$ independently of how the (x_i, y_i) are chosen in each R_i. We denote this by writing

$$V(S) = \lim_{\|P\| \to 0} \sum_{i=1}^{n} f(x_i, y_i)\, \Delta A_i$$

Just as we dropped the condition that f be positive in Chap. 6 to define the definite integral of a function of a single variable independently of area, we now define the double integral independently of volume.

THE DOUBLE INTEGRAL

Suppose $f(x, y)$ is defined on the region R and suppose P is an inner partition of R consisting of the rectangles $R_1, R_2, \ldots, R_n$.

20.1.1 **Definition** Any sum of the form $\sum_{i=1}^{n} f(x_i, y_i)\, \Delta A_i$, where (x_i, y_i) is in R_i and ΔA_i is the area of R_i, is called a *Riemann sum* of f over P.

Now, as in the single-variable case, f is said to be *integrable* (over R) if the Riemann sums tend to some fixed number. More precisely, we have:

20.1.2 **Definition** We say f is *integrable* over R if there exists a real number I with the following property: Given any $\varepsilon > 0$, there exists a $\delta > 0$ such that whenever P is an inner partition of R with $\|P\| < \delta$, then

$$\left| \sum_{i=1}^{n} f(x_i, y_i)\, \Delta A_i - I \right| < \varepsilon$$

for all choices of (x_i, y_i) in R_i. If f is integrable over R, then the number I (which is unique) is called the *double integral* of f over R and is denoted by

$$\iint_R f(x, y)\, dA$$

We express (20.1.2) symbolically by writing

20.1.3
$$\lim_{\|P\| \to 0} \sum_{i=1}^{n} f(x_i, y_i)\, \Delta A_i = \iint_R f(x, y)\, dA$$

We shall assume the following result, which is usually proved in advanced calculus.

20.1.4 **Theorem:** If f is continuous on R, then f is integrable over R.

In light of our earlier discussion, if f is positive and continuous on R, then we define the *volume of the solid region S lying above R and below the graph of f* by

20.1.5
$$V(S) = \iint_R f(x,y)\,dA$$

PROPERTIES OF THE DOUBLE INTEGRAL

The double integral satisfies many properties similar to those of the definite integral. We shall state these without proof, assuming throughout that f is integrable over the given regions.

20.1.6 **Theorem:** Suppose R_1 and R_2 are two regions in the x-y plane having no points in common except possibly boundary points. Then

$$\iint_{R_1 \cup R_2} f(x,y)\,dA = \iint_{R_1} f(x,y)\,dA + \iint_{R_2} f(x,y)\,dA$$

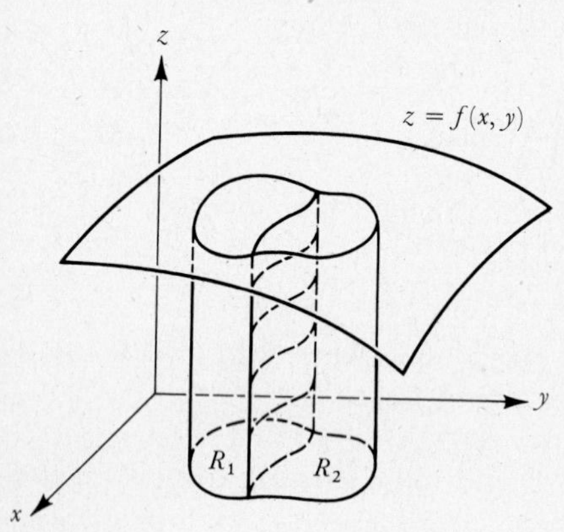

Fig. 20.5

If f is positive on R_1 and R_2, then (20.1.6) is reasonable, since the volume of the region S over $R_1 \cup R_2$ should certainly be the sum of the volumes of the pieces of S that lie above R_1 and R_2, respectively (see Fig. 20.5). Of course (20.1.5) can be extended to any finite collection $R_1, R_2, \ldots, R_n$ of disjoint (except for boundary points) regions in the x-y plane.

The remaining basic properties of the double integral are given by:

20.1.7
$$\iint_R kf(x,y)\,dA = k\iint_R f(x,y)\,dA \qquad k \text{ a constant}$$

20.1.8
$$\iint_R [f(x,y) \pm g(x,y)]\,dA = \iint_R f(x,y)\,dA \pm \iint_R g(x,y)\,dA$$

If $f(x,y) \geq g(x,y)$ on R, then

20.1.9
$$\iint_R f(x,y)\,dA \geq \iint_R g(x,y)\,dA$$

Our restrictions on the types of regions R in the x-y plane that we are considering guarantees that each has a well-defined area. This area could be determined using the methods from Chap. 6. However, we note that if $f(x,y) = 1$ for all (x,y) in R, then

$$\sum_{i=1}^{n} f(x_i, y_i)\,\Delta A_i = \sum_{i=1}^{n} \Delta A_i$$

Since this latter sum is just the sum of the areas of all the rectangles in a typical inner partition P (see Fig. 20.3), it is reasonable that in the limit we shall obtain the area of R,

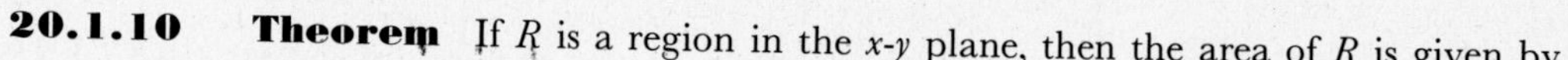

20.1.10 **Theorem** If R is a region in the x-y plane, then the area of R is given by

$$A(R) = \iint_R dA$$

The evaluation of a double integral *from the definition* is in general a difficult chore. Fortunately, we can draw upon our knowledge gained in Chaps. 6 and 12 to aid in this task. We begin by restricting our attention to special types of regions in the x-y plane.

TYPE I AND TYPE II REGIONS

20.1.11 Definition If h and g have continuous derivatives on $[a, b]$, then a region of the form

$$R = \{(x, y) \mid a \leq x \leq b, h(x) \leq y \leq g(x)\}$$

is called a *type I* region. Similarly, if p and q have continuous derivatives on $[c, d]$, then

$$R = \{(x, y) \mid c \leq y \leq d, p(y) \leq x \leq q(y)\}$$

is said to be *type II* region.

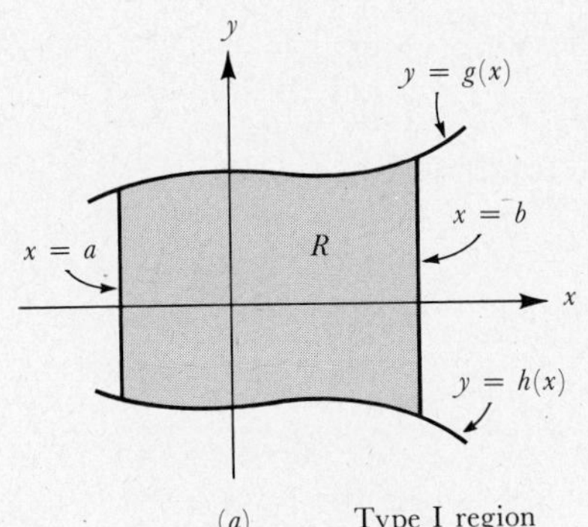

(a) Type I region

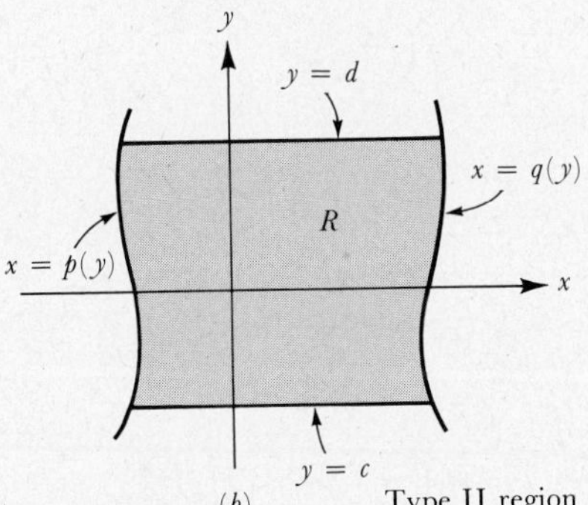

(b) Type II region

Fig. 20.6

Typical type I and II regions are sketched in Fig. 20.6. By (12.2.2), page 406, regions of either type will have boundaries of finite length. Any region R satisfying our original restrictions (1) and (2) can always be decomposed into a finite collection of type I and/or type II regions. Thus if we can develop a means for evaluating a double integral over type I and II regions, we can employ (20.1.6) to handle the general case.

EVALUATION OF DOUBLE INTEGRALS

Suppose S denotes the solid region below the graph of the positive continuous function $f(x, y)$ and above the type I region R (Fig. 20.7). *Fix* x between a and b and let Π_x denote the plane through $(x, 0, 0)$ and parallel to the y-z plane. Then Π_x slices a cross section from S in the form of a plane region below the trace of $z = f(x, y)$ in Π_x and above the interval $[h(x), g(x)]$ (see Fig. 20.8). Since x is fixed, $z = f(x, y)$ is a positive continuous function of the *single* variable y. We know from Chap. 6 that the area of this cross-sectional slice is given by

(3) $$A(x) = \int_{h(x)}^{g(x)} f(x, y)\, dy$$

If we now let x vary over $[a, b]$, then (3) yields a function that gives the areas of the cross sections of S. It can be shown that A is a continuous function of x on $[a, b]$. If we subdivide $[a, b]$ in the usual way, then

$$A(x_i)\, \Delta x_i$$

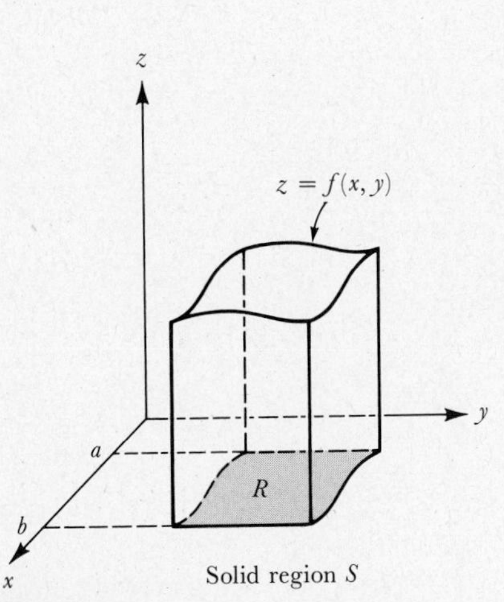

Solid region S

Fig. 20.7

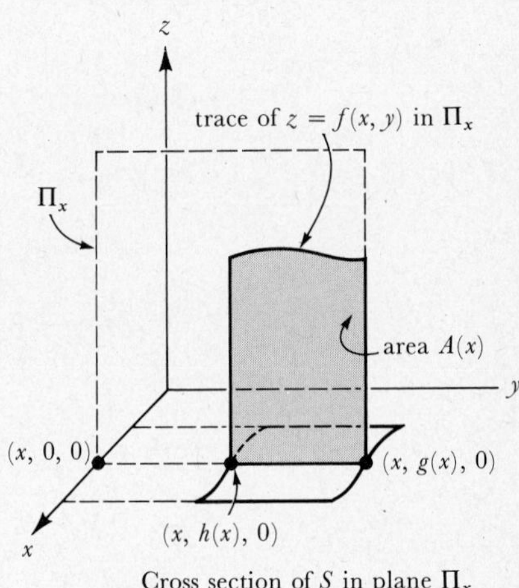

Cross section of S in plane Π_x

Fig. 20.8

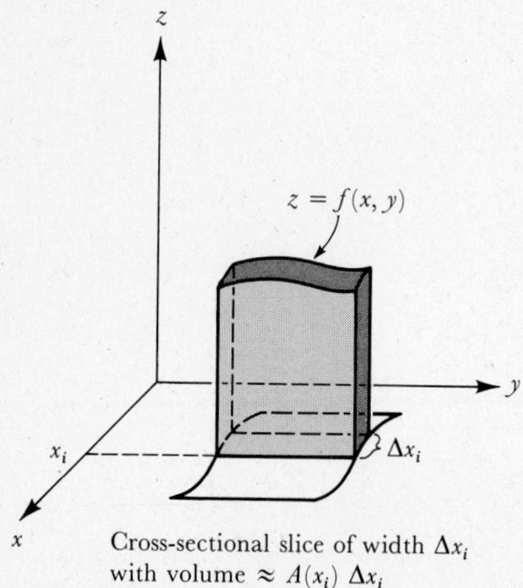

Cross-sectional slice of width Δx_i with volume $\approx A(x_i)\,\Delta x_i$

Fig. 20.9

can be interpreted as an approximation to the volume of the cross-sectional slice of thickness Δx_i (at x_i) as depicted in Fig. 20.9. Adding these volumes we have

$$V(S) \approx \sum_{i=1}^{n} A(x_i)\,\Delta x_i$$

which is an ordinary Riemann sum of the type considered in Chapter 6. Since A is continuous on $[a, b]$, we conclude that

$$V(S) = \int_a^b A(x)\,dx = \int_a^b \left[\int_{h(x)}^{g(x)} f(x,y)\,dy\right] dx$$

In light of (20.1.5), we have

$$\text{(4)} \qquad \iint_R f(x,y)\,dA = \int_a^b \left[\int_{h(x)}^{g(x)} f(x,y)\,dy\right] dx$$

The integral that appears on the right side of (4) is called an *iterated integral.* There are two types of iterated integrals, and they can be defined independently of any volume consideration.

20.1.12 **Definition** If R is a region of type I and f is continuous on R, then *the iterated integral of f over R,* denoted

$$\int_a^b \int_{h(x)}^{g(x)} f(x,y)\,dy\,dx$$

is defined by

$$\int_a^b \int_{h(x)}^{g(x)} f(x,y)\,dy\,dx = \int_a^b \left[\int_{h(x)}^{g(x)} f(x,y)\,dy\right] dx$$

In (20.1.12) it is understood that one first evaluates the inside "partial integral"

$$\int_{h(x)}^{g(x)} f(x,y)\,dy$$

by holding x constant. The result will be a continuous function of x, which is then integrated over $[a, b]$.

In an analogous way we define the iterated integral over a type II region by

20.1.13 **Definition** $\displaystyle\int_c^d \int_{p(y)}^{q(y)} f(x,y)\,dx\,dy = \int_c^d \left[\int_{p(y)}^{q(y)} f(x,y)\,dx\right] dy$

Example 1 Evaluate the iterated integral

$$I = \int_0^2 \int_{-1}^1 (2x + 2y)\,dy\,dx$$

Solution By (20.1.12),

$$\begin{aligned} I &= \int_0^2 \left[\int_{-1}^1 (2x + 2y)\,dy\right] dx \\ &= \int_0^2 \Big[2xy + y^2\Big]_{-1}^{1} dx \\ &= \int_0^2 \Big[(2x(1) + (1)^2) - (2x(-1) + (-1)^2)\Big]\,dx \\ &= \int_0^2 4x\,dx = 2x^2\Big|_0^2 = 8 \quad \bullet \end{aligned}$$

Example 2 Evaluate the iterated integral

$$I = \int_{-1}^{1}\int_{0}^{y+2} (2x + 2y)\,dx\,dy$$

Solution By (20.1.13),

$$\begin{aligned} I &= \int_{-1}^{1}\left[\int_{0}^{y+2} (2x + 2y)\,dx\right] dy \\ &= \int_{-1}^{1}\left[x^2 + 2xy\right]_{0}^{y+2} dy \\ &= \int_{-1}^{1}\left[((y + 2)^2 + 2(y + 2)y) - 0\right] dy \\ &= \int_{-1}^{1} (3y^2 + 8y + 4)\,dy = y^3 + 4y^2 + 4y\Big|_{-1}^{1} = 10 \end{aligned}$$ ●

PROGRESS TEST 1

Evaluate the given iterated integrals:

1. $I = \int_{0}^{1}\int_{0}^{y} (12x^2 + 4y)\,dx\,dy$
2. $I = \int_{1/2}^{1}\int_{0}^{3x} \sin(\pi x^2)\,dy\,dx$

FURTHER EXAMPLES

As suggested in our earlier discussion, we can compute double integrals using iterated integrals:

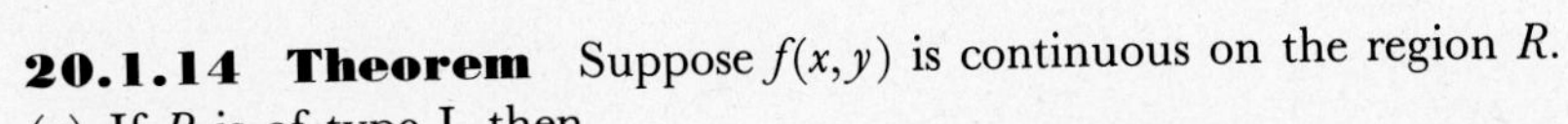

20.1.14 Theorem Suppose $f(x,y)$ is continuous on the region R.
(*a*) If R is of type I, then

$$\iint_R f(x,y)\,dA = \int_a^b\int_{h(x)}^{g(x)} f(x,y)\,dy\,dx$$

(*b*) If R is of type II, then

$$\iint_R f(x,y)\,dA = \int_c^d\int_{p(y)}^{q(y)} f(x,y)\,dx\,dy$$

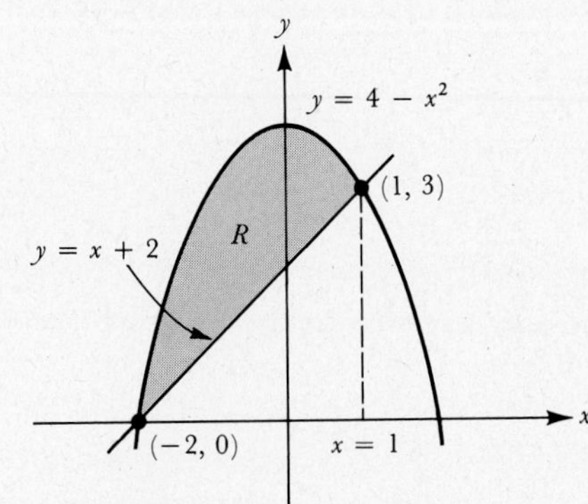

Fig. 20.10

Example 3 Evaluate $\iint_R(2x^2y + 3x)\,dA$, where R is the region bounded by $y = 4 - x^2$ and $y = x + 2$.

Solution Setting $4 - x^2 = x + 2$ and solving, we obtain $x = -2$, $x = 1$. Thus the graphs intersect in the points $(-2, 0)$ and $(1, 3)$; see Fig. 20.10. Now R is a type I region with description

$$R = \{(x,y)\,|\,-2 \le x \le 1, x + 2 \le y \le 4 - x^2\}$$

so, by [20.1.14(*a*)],

$$\int_{-2}^{1}\int_{x+2}^{4-x^2} (2x^2y + 3x)\,dy\,dx$$
$$= \int_{-2}^{1}\left[\int_{x+2}^{4-x^2} (2x^2y + 3x)\,dy\right] dx$$
$$= \int_{-2}^{1}\left[(x^2y^2 + 3xy)\Big|_{y=x+2}^{y=4-x^2}\right] dx$$
$$= \int_{-2}^{1} [x^2(4 - x^2)^2 + 3x(4 - x^2) - x^2(x + 2)^2 - 3x(x + 2)]\,dx$$
$$= \int_{-2}^{1} (x^6 - 9x^4 - 7x^3 + 9x^2 + 6x)\,dx$$
$$= \frac{x^7}{7} - \frac{9x^5}{5} - \frac{7x^4}{4} + 3x^3 + 3x^2 \Big|_{-2}^{1} = \frac{-2481}{140} \quad \bullet$$

It is worth noting that the region R in Example 3 is equal to $R_1 \cup R_2$, where [see Fig. 20.11(a)]

$$R_1 = \{(x,y)\,|\,0 \le y \le 3, -\sqrt{4-y} \le x \le y - 2\}$$

and

$$R_2 = \{(x,y)\,|\,3 \le y \le 4, -\sqrt{4-y} \le x \le \sqrt{4-y}\}$$

Thus

$$\iint_R (2x^2y + 3x)\,dA$$
$$= \iint_{R_1 \cup R_2} (2x^2y + 3x)\,dA$$
$$= \iint_{R_1} (2x^2y + 3x)\,dA + \iint_{R_2} (2x^2y + 3x)\,dA \quad \text{[by (20.1.6)]}$$

(5) $$= \int_0^3\int_{-\sqrt{4-y}}^{y-2} (2x^2y + 3x)\,dx\,dy + \int_3^4\int_{-\sqrt{4-y}}^{\sqrt{4-y}} (2x^2y + 3x)\,dx\,dy$$

These integrals can in turn be evaluated using (20.1.13) and the results added to obtain the value of the double integral. Of course this latter approach is less efficient than the former.

A HELPFUL MNEMONIC DEVICE

By thinking of an iterated integral as a limit of a double summing process, many students are better able to keep track of the variables and limits of integration when performing iterated integration. Although it is possible to develop iterated integrals in such a way as to make this notion precise, we shall merely steal its style and use it as a mnemonic device. In reference to Example 3 and the type I description of R, we can think of

$$\int_{-2}^{1}\int_{x+2}^{4-x^2} (2x^2y + 3x)\,dy\,dx$$

as the limit of a double summing process that first "adds" the values of $2x^2y + 3x$ along elements of the region in the y direction, "adding values along a vertical strip of dy's" between the lower and upper boundary of R, as in Fig. 20.11(b). Since the next variable of integration is x, it is to be expected that the lower and upper limits on the dy's should be expressed in terms of x. The second iteration can then be thought of as "summing the values on the resulting strips

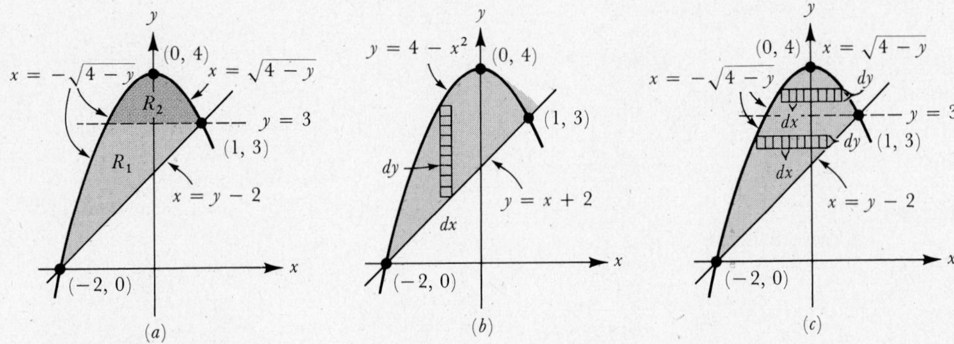

Fig. 20.11

of thickness dx" in the x direction from left to right—from the lower limit $x = -2$ to the upper limit $x = 1$.

In a similar way, we can view the second (type II) approach to the same double integral as "summing the values of $2x^2y + 3x$ along a horizontal strip of dx's" in the x direction from left to right, and then "summing the values on the dy strips" in the y direction from bottom to top. This point of view makes obvious the need in this case to use two separate pairs of iterated integrals when "adding values on dx's first": We get two *different* kinds of horizontal strips depending on whether we are below or above the line $y = 3$; see Fig. 20.11(c). Below $y = 3$, we add dx's along horizontal strips bounded on the left by $x = -\sqrt{4-y}$ and on the right by $x = y - 2$. We then add over such strips (of "thickness dy") in the y direction from $y = 0$ to $y = 3$, getting the first iterated integral in (5). Above $y = 3$, we "add values on dx's along horizontal strips" bounded on the left by $x = -\sqrt{4-y}$ and on the right by $x = \sqrt{4-y}$; see Fig. 20.11(c). We then add over such strips in the y direction from $y = 3$ to $y = 4$, getting the second iterated integral in (5).

Thus in general we can think of the type I approach as generating and then summing over vertical strips and the type II approach as generating and then summing over horizontal strips.

REVERSING THE ORDER OF ITERATION

By switching the order of integration, we can sometimes simplify a problem greatly.

Example 4 Evaluate the iterated integral

$$I = \int_0^1 \int_y^1 e^{x^2}\, dx\, dy$$

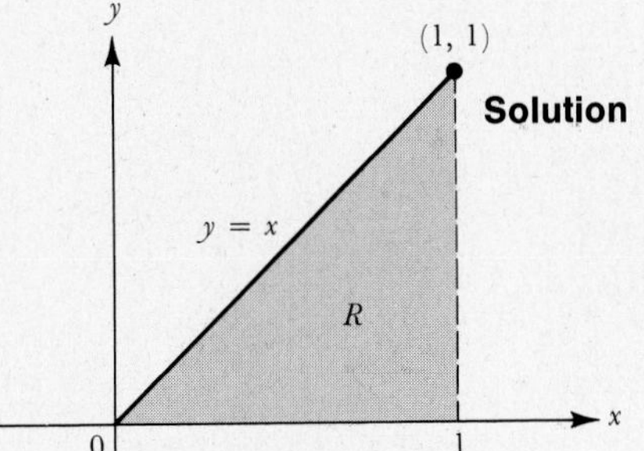

Fig. 20.12

Solution To apply (20.1.13) directly (horizontal strips) we must first determine the indefinite integral $\int e^{x^2}\, dx$. However, this is a nonelementary integral! Suppose instead we use (20.1.14). By [20.1.14(b)] we can regard the given iterated integral as a double integral:

$$\int_0^1 \int_y^1 e^{x^2}\, dx\, dy = \iint_R e^{x^2}\, dA \tag{6}$$

where R is the type II region given by

$$R = \{(x, y) \mid 0 \le y \le 1, y \le x \le 1\} \qquad \text{(see Fig. 20.12)}$$

Now R can also be described as a type I region:

$$R = \{(x,y) \mid 0 \le x \le 1, 0 \le y \le x\}$$

Thus, by [20.1.14(*a*)],

$$(7) \qquad \iint_R e^{x^2}\,dA = \int_0^1 \int_0^x e^{x^2}\,dy\,dx \qquad \text{(vertical strips)}$$

Combining (6) and (7), we obtain

$$\int_0^1 \int_y^1 e^{x^2}\,dx\,dy = \int_0^1 \int_0^x e^{x^2}\,dy\,dx$$
$$= \int_0^1 \left[e^{x^2} y\right]_0^x dx$$
$$= \int_0^1 x e^{x^2}\,dx = \frac{1}{2} e^{x^2} \Big|_0^1 = \frac{e-1}{2} \quad \bullet$$

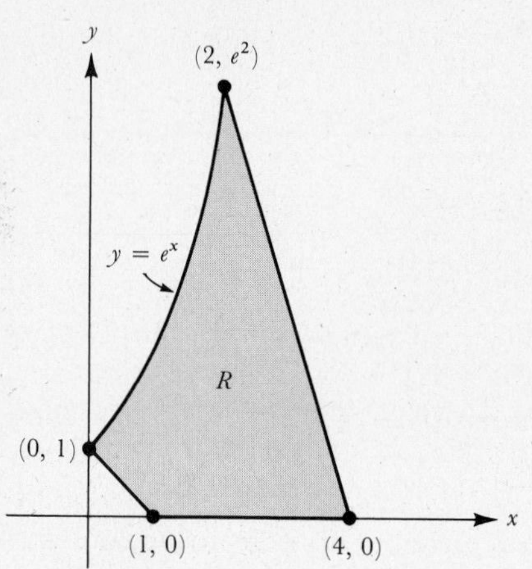

Fig. 20.13

Whether or not we use the mnemonic device involving strips, it should be clear that the process of reversing the order of integration is not simply a matter of interchanging differentials and renaming variables! The form of the new iterated integral depends on the new description of the region involved. Always sketch the region.

PROGRESS TEST 2

1. Suppose R is the region bounded by the graphs of $y = x^2$ and $y = x$.
(*a*) Sketch and describe R as a region of type I.
(*b*) Find the associated iterated integral I of $f(x,y) = 4x + 2y$ over R (vertical strips).

2. Sketch the region R over which the iterated integral is evaluated and give an equivalent iterated integral (or integrals) in which the order of integration is reversed.

(*a*) $I = \int_0^{\pi/2} \int_0^{\sin x} f(x,y)\,dy\,dx$

(*b*) $I = \int_{-3}^{4} \int_{y^2-9}^{y+3} f(x,y)\,dx\,dy$

3. Set up an iterated integral (or integrals) equal to $\iint_R f(x,y)\,dA$, where R is the region depicted in Fig. 20.13.

SECTION 20.1 EXERCISES

In Exercises 1 to 14 evaluate the given iterated integral.

1. $\int_0^2 \int_0^{x^2+1} (60x - 40y)\,dy\,dx$

2. $\int_{-1}^3 \int_{x^2+1}^{2x+4} (2x^2y - 1)\,dy\,dx$

3. $\int_{-1}^2 \int_3^4 \frac{3y}{2}\sqrt{x^2+y^2}\,dy\,dx$

4. $\int_0^{\pi/2} \int_0^y y \sin x\,dx\,dy$

5. $\int_0^1 \int_x^0 \frac{dy\,dx}{4+x^2}$

6. $\int_0^{\pi/4} \int_0^{\sin x} e^y \cos x\,dy\,dx$

7. $\int_0^{\pi/2} \int_0^{\cos\theta} r^3 \cos\theta\,dr\,d\theta$

8. $\int_0^1 \int_0^t \frac{s\,ds\,dt}{(4-s^2)^{3/2}}$

9. $\int_1^2 \int_0^y \frac{dx\,dy}{x^2+y^2}$

10. $\int_1^4 \int_x^{x^{2/3}} 3y^2\sqrt{x}\,dy\,dx$

11. $\int_0^2 \int_0^{\sqrt{y}} 3x^3\,dx\,dy$

12. $\int_0^{\pi/3} \int_0^{\sec x \tan x} y^3 \cos^4 x\,dy\,dx$

13. $\int_{\pi/2}^{3\pi/2} \int_{2y}^0 y \sin(x-y)\,dx\,dy$

14. $\int_0^1 \int_0^x (\sin x + \sin y)\,dy\,dx$

In Exercises 15 to 26, (a) express $\iint_R f(x,y)\,dA$ as an appropriate iterated integral (or integrals) and (b) evaluate the resulting iterated integral.

15. $f(x,y) = 2x + 3y$; R is the rectangle bounded by $x = -1$, $x = 5$, $y = -3$, and $y = 2$.

16. $f(x,y) = x^2y$; R is the region bounded by $y = x^2$ and $y = 4x$.

17. $f(x,y) = xy/(x^2 + y^2)$; R is bounded by $x = 0$, $y = 1$, $y = 2$, and $x = y$.

18. $f(x,y) = 2$; R is the disk centered at the origin of radius 9.

19. $f(x,y) = \cos(\pi x^2)$; R is the region bounded by $x = 1/2$, $x = 1$, $y = 0$, and $y = 2x$.

20. $f(x,y) = x\sin(y/2)$; R is the region bounded by $y = 0$, $y = \pi/2$, $x = 0$, and $x = 2y$.

21. $f(x,y) = \sin^3 x \sin y$; R is bounded by $y = 0$, $y = \pi$, $x = \arctan(1/y)$, $x = \pi/2$.

22. $f(x,y) = x^3y - 2y^3$; R is the region above $y = x^2 + 4$ and below $y = 12$.

23. $f(x,y) = \sqrt{x/y}$; R is the region below $y = 2x$, above $y = x$, and between $x = 1$ and $x = 4$.

24. $f(x,y) = 1/(x+y)^{3/2}$; R is bounded by $y = 1$, $y = 4$, $x = 0$, and $x = 4y$.

25. $f(x,y) = \sin x + \cos y$; R is bounded by $y = 0$, $y = \pi/2$, $x = 0$, $x = y$.

26. $f(x,y) = 10x + 10y$; R is bounded by $y = x^2$ and $y = \sqrt{x}$.

In Exercises 27 to 32 evaluate the given integral by first reversing the order of integration.

27. $\int_0^1\int_x^1 \sin y^2\,dy\,dx$

28. $\int_0^2\int_{2x}^4 e^{y^2}\,dy\,dx$

29. $\int_0^2\int_{x/2}^{x} 4e^{-y^2}\,dy\,dx$

30. $\int_0^3\int_{x/3}^9 \sqrt{4-x^2}\,dx\,dy$

31. $\int_0^3\int_{-x-2}^{4-x^2} (3x+2y)\,dy\,dx$

32. $\int_1^e\int_1^x y^2\ln(x^4)\,dx\,dy$

33. Compute $\iint_R (x - 2y)\,dA$, where R is as in the given figure.

34. Compute $\iint_R (12xy)\,dA$, where R is as in the given figure.

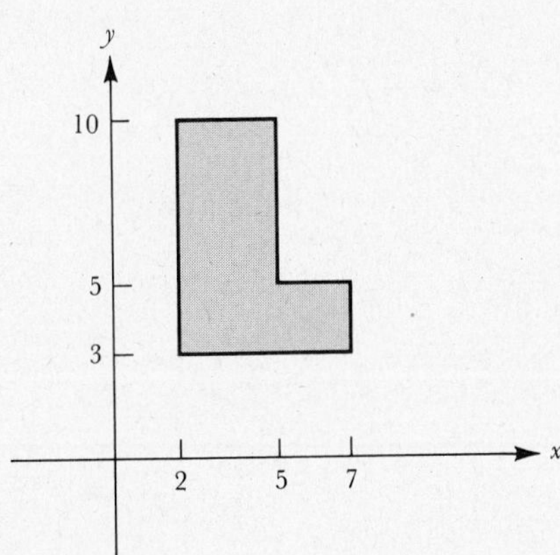

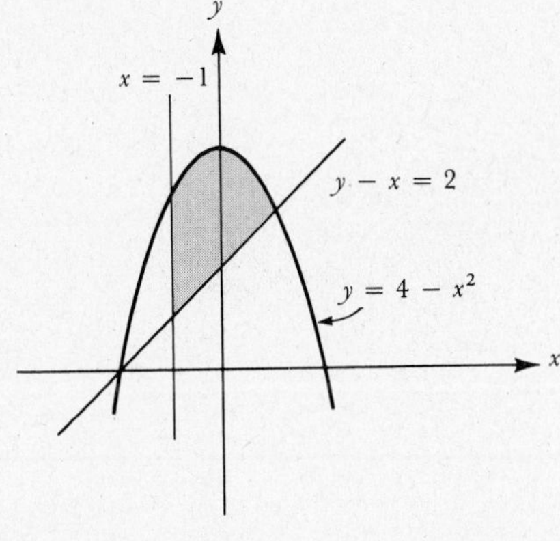

Section 20.2: Applications of Double Integrals

Objectives:

Use double integrals to determine such quantities as area, volume, first and second moments, centers of mass, and forces of attraction.

AREA AND VOLUME

As in Sec. 20.1 we shall restrict our attention to regions R in the x-y plane that can be decomposed into a finite number of subregions of type I or type II. From (20.1.10) the area of such a region is given by

20.2.1
$$A(R) = \iint_R dA$$

We shall in this and subsequent sections make frequent use of the integration tables.

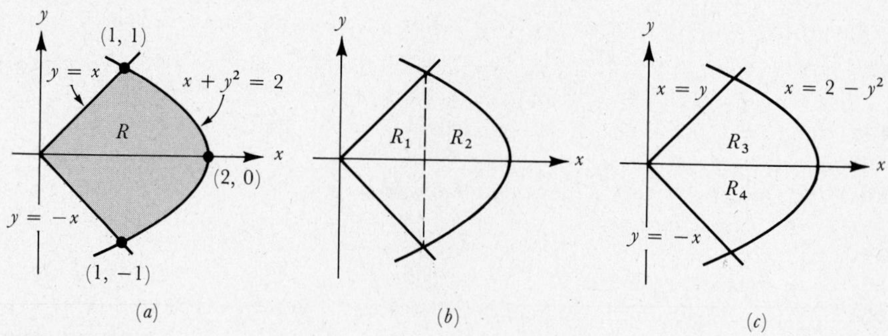

Fig. 20.14

Example 1 Determine the area of the region R bounded by the curves $x = |y|$ and $x + y^2 = 2$.

Solution The region R is sketched in Fig. 20.14. We can decompose R into the two type I regions R_1, R_2 or the two type II regions R_3, R_4. By symmetry $A(R_3) = A(R_4)$ and thus we elect to work with this latter decomposition. Now, by inspecting Fig. 20.14(c), we conclude that

$$R_3 = \{(x,y) \,|\, 0 \le y \le 1, y \le x \le 2 - y^2\}$$

so

$$\begin{aligned} A(R) = 2A(R_3) &= 2\iint_{R_3} dA \\ &= 2\int_0^1 \int_y^{2-y^2} dx\, dy \\ &= 2\int_0^1 x \Big|_y^{2-y^2} dy \\ &= 2\int_0^1 (2 - y^2 - y)\, dy \\ &= 2\left[2y - \frac{y^3}{3} - \frac{y^2}{2}\right]_0^1 = \frac{7}{3} \end{aligned}$$ ●

Example 2 Determine the area of the region R bounded by the graphs of

$$h(x) = \ln x \qquad \text{and} \qquad g(x) = \frac{1}{e-1}(x-1).$$

Solution Since $h(1) = g(1) = 0$ and $h(e) = g(e) = 1$, the curves intersect at the points $(1, 0)$ and $(e, 1)$; see Fig. 20.15. As a region of type I, R has the description

$$R = \left\{(x,y) \,|\, 1 \le x \le e, \frac{1}{e-1}(x-1) \le y \le \ln x\right\}$$

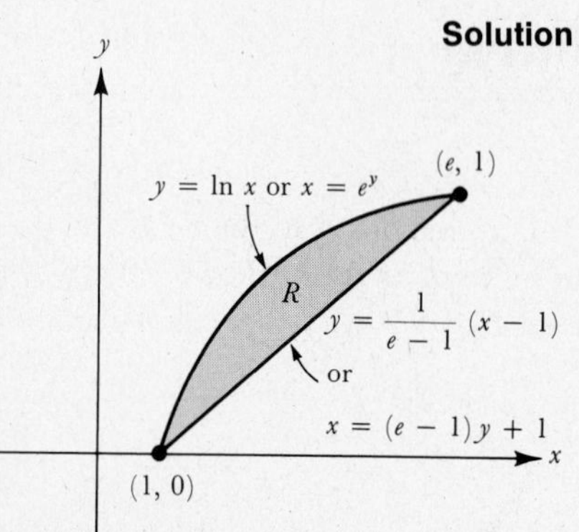

Fig. 20.15

Now R is also a region of type II. Solving

$$y = \ln x \qquad \text{and} \qquad y = \frac{1}{e-1}(x-1)$$

for x in terms of y we have $x = e^y$, $x = (e-1)y + 1$. Thus

$$R = \{(x,y) \,|\, 0 \le y \le 1, e^y \le x \le (e-1)y + 1\}$$

Working with this latter description of R we obtain

$$
\begin{aligned}
A(R) = \iint_R dA &= \int_0^1 \int_{e^y}^{(e-1)y+1} dx\, dy \\
&= \int_0^1 x \Big|_{e^y}^{(e-1)y+1} dy \\
&= \int_0^1 [(e-1)y + 1 - e^y]\, dy \\
&= \frac{(e-1)}{2} y^2 + y - e^y \Big|_0^1 = \frac{3-e}{2} \quad \bullet
\end{aligned}
$$

Also, from (20.1.5), if $f(x,y)$ is continuous and positive on R, then the volume of the solid region S lying above R and below the graph of f has volume given by:

20.2.2

$$V(S) = \iint_R f(x,y)\, dA$$

Example 3 Find the volume of the solid region S in the first octant bounded on the sides by the cylinder $x^2 + y^2 = 4$ and on the top by the plane $3z = 12 - 2x - 2y$.

Solution The region S is sketched in Fig. 20.16. Note that it sits above the quarter-circle R in the x-y plane and below the plane $z = (12 - 2x - 2y)/3$. Now

$$R = \{(x,y) \mid 0 \le x \le 2, 0 \le y \le \sqrt{4 - x^2}\}$$

so

$$
\begin{aligned}
V(S) &= \iint_R \left[\frac{12 - 2x - 2y}{3}\right] dA \\
&= \frac{1}{3} \int_0^2 \int_0^{\sqrt{4-x^2}} (12 - 2x - 2y)\, dy\, dx \\
&= \frac{1}{3} \int_0^2 \left[12y - 2xy - y^2\right]_0^{\sqrt{4-x^2}} dx \\
&= \frac{1}{3} \int_0^2 \left[12\sqrt{4 - x^2} - 2x\sqrt{4 - x^2} - (4 - x^2)\right] dx \\
&= \frac{1}{3} \left[(6x\sqrt{4 - x^2} + 24 \arcsin(x/2)) + \frac{2(4 - x^2)^{3/2}}{3} - 4x + \frac{x^3}{3}\right]_0^2 \\
&= 4\pi - 32/9 \qquad \text{(We used integral formula 40.)} \quad \bullet
\end{aligned}
$$

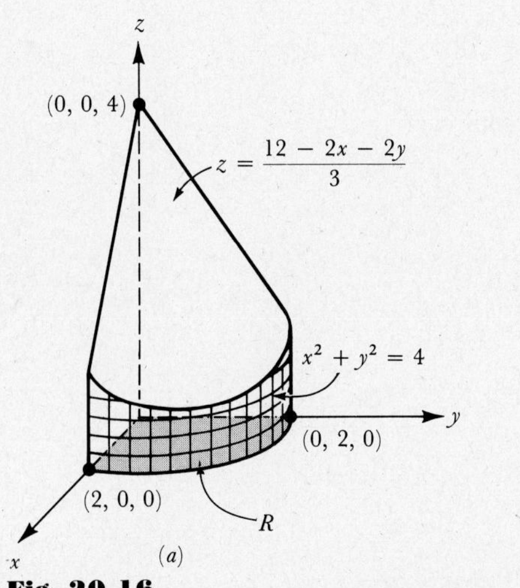

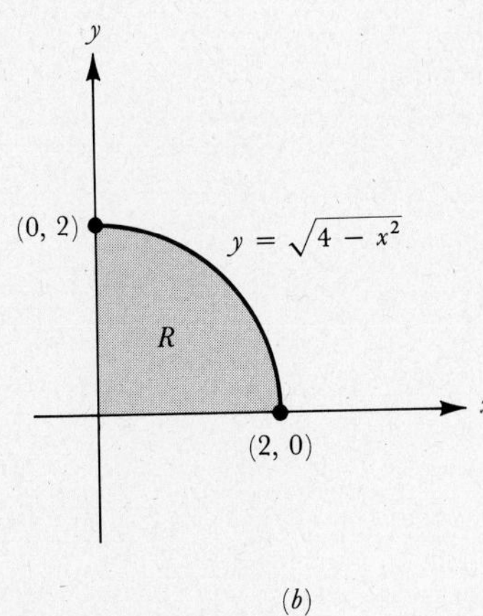

Fig. 20.16

Example 4 Find the volume of the solid region S inside the paraboloid $z = 4x^2 + 9y^2$ and below the plane $z = 36$.

Solution A sketch of the solid S is given in Fig. 20.17. We first determine the region R in the x-y plane above which this solid lies. Setting $4x^2 + 9y^2 = 36$, we see that R is an ellipse centered at the origin. By the symmetry of the problem we can find the volume in the first octant and multiply by 4. Let R_1 be the quarter of the ellipse in the first quadrant. Then

$$R_1 = \left\{(x, y) \,\middle|\, 0 \le x \le 3, 0 \le y \le \frac{\sqrt{36 - 4x^2}}{3}\right\}$$

Now $V(S)$ is the difference of the volumes V_1 and V_2 where V_1 is the volume of the solid region above R_1 and below the plane $z = 36$, and V_2 is the volume of the solid above R_1 and below the paraboloid $z = 4x^2 + 9y^2$. Thus

$$\begin{aligned}
V(S) &= 4\iint_{R_1} 36\, dA - 4\iint_{R_1} (4x^2 + 9y^2)\, dA \\
&= 4\iint_{R_1} [36 - 4x^2 - 9y^2]\, dA \\
&= 4\int_0^3 \int_0^{\sqrt{36-4x^2}/3} [36 - 4x^2 - 9y^2]\, dy\, dx \\
&= 4\int_0^3 \left[36y - 4x^2y - 3y^3 \Big|_0^{\sqrt{36-4y^2}/3}\right] dx \\
&= 4\int_0^3 \left[12\sqrt{36 - 4x^2} - (4/3)x^2\sqrt{36 - 4x^2} - (1/9)(36 - 4x^2)^{3/2}\right] dx \\
&= 4\int_0^3 (1/9)\sqrt{36 - 4x^2}\,(108 - 12x^2 - 36 + 4x^2)\, dx \\
&= \frac{4}{9}\int_0^3 (72 - 8x^2)\sqrt{36 - 4x^2}\, dx \\
&= \frac{32}{9}\int_0^3 (9 - x^2)\sqrt{36 - 4x^2}\, dx \\
&= \frac{64}{9}\int_0^3 (9 - x^2)^{3/2}\, dx \\
&= \frac{64}{9}\left[\frac{x(9 - x^2)^{2/3}}{4} + \frac{27x\sqrt{9 - x^2}}{8} + \frac{243}{8}\arcsin\left(\frac{x}{3}\right)\right]_0^3 = 108\pi
\end{aligned}$$

(We used integral formula 47.) ●

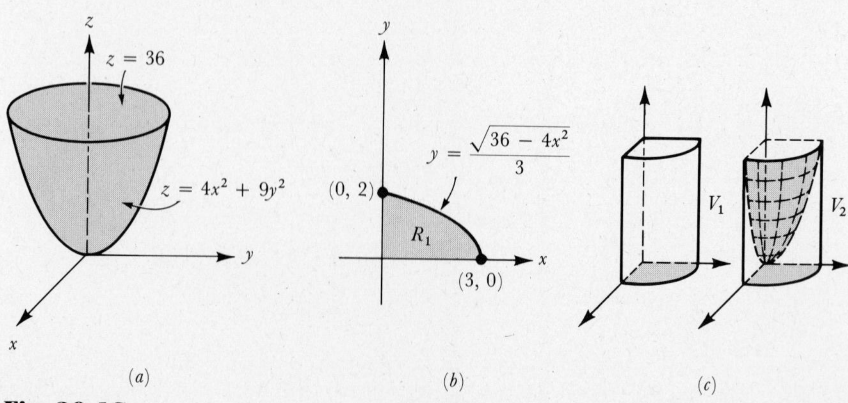

Fig. 20.17

PROGRESS TEST 1

1. Determine the area of the region bounded by the curves $y = x^2$ and $y = x$.
2. Determine the volume of the solid region S in the first octant below the graph of the cylinder $z^2 + y^2 = 1$ and above the triangular region R in the x-y plane bounded by $y = x$, $y = 1$, and $x = 0$.

MOMENTS AND CENTER OF MASS

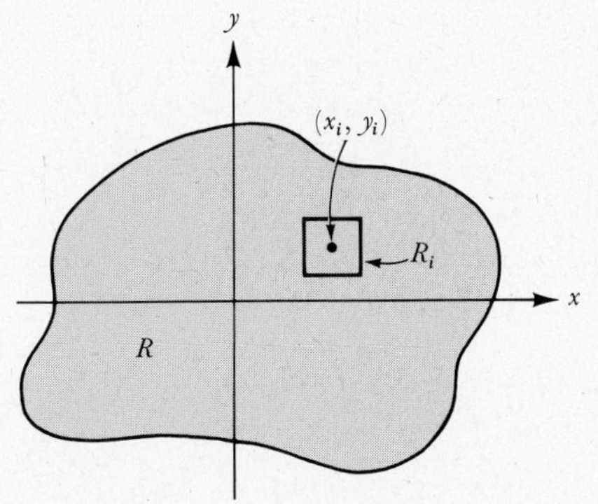

Lamina L occupying the region R

Fig. 20.18

In Chap. 12 we used the definite integral as an aid in giving mathematical definitions for such physical quantities as the center of mass of a lamina of constant density. By a "lamina" we mean a very thin flat sheet, and by "density" we mean area density, that is, mass per unit area. We consider now a lamina L with *nonconstant* density. If R is the region that L occupies in the x-y plane, then we denote the density at (x, y) by $\rho(x, y)$(see Fig. 20.18). We shall assume throughout that ρ is continuous on R.

We first examine the problem of determining M, the mass of L. We begin by considering an inner partition of R into n rectangular subregions $R_1, R_2, \ldots, R_n$. Now the mass of that part of L occupying R_i is approximated by

(1) $$M_i \approx \rho(x_i, y_i)\,\Delta A_i$$

where (x_i, y_i) is some point in R_i. Note that (1) is only an approximation since $\rho(x, y)$ need not be constantly equal to $\rho(x_i, y_i)$ on the rectangle R_i. However, if $\|P\|$ is quite small, then the continuity of ρ ensures that $\rho(x, y)$ will not vary much over R_i. Thus the total mass M is approximated by:

$$M \approx \sum_{i=1}^{n} \rho(x_i, y_i)\,\Delta A_i$$

Since this latter sum is a Riemann sum for the continuous function ρ over R, we *define* the mass of L by

20.2.3 $$M = \iint_R \rho(x, y)\,dA$$

Suppose we wish in turn to determine the center of mass $(\overline{x}, \overline{y})$ for L. Our first chore would be to determine the first moments of L about the axes. To do this we think of each portion of L occupying the rectangle R_i in the partition P as having its mass concentrated at (x_i, y_i). Then (see Fig. 20.19)

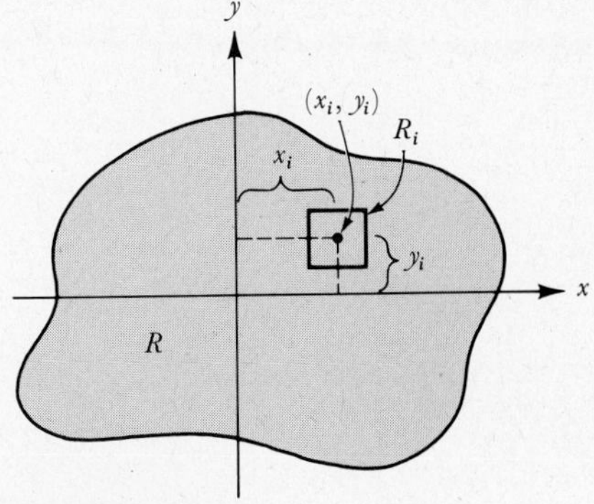

Fig. 20.19

$$(M_x)_i \approx y_i \rho(x_i, y_i)\,\Delta A_i \qquad (M_y)_i \approx x_i \rho(x_i, y_i)\,\Delta A_i$$

where $(M_x)_i$ and $(M_y)_i$ represent the first moments of R_i about the x and y axes, respectively.

Since the moment of the sum is equal to the sum of the moments, we can approximate the first moments of L by

$$M_x \approx \sum_{i=1}^{n} y_i \rho(x_i, y_i)\,\Delta A_i \qquad M_y \approx \sum_{i=1}^{n} x_i \rho(x_i, y_i)\,\Delta A_i$$

Taking limits, we define M_x and M_y by:

20.2.4 $$M_x = \iint_R y\rho(x, y)\,dA$$

20.2.5
$$M_y = \iint_R x\rho(x,y)\,dA$$

It follows that the center of mass is given by $(\bar{x},\bar{y})$, where

20.2.6
$$\bar{x} = \frac{M_y}{M} \qquad \bar{y} = \frac{M_x}{M}$$

Although the above work led to double integrals, we of course use iterated integrals to evaluate the double integrals.

Example 5 A lamina L occupies the region in the first quadrant bounded by $y = \sqrt{x}$ and $x = 4$. Given that the density at each point is proportional to the square of the distance from the origin, find the center of mass of L.

Solution The lamina is sketched in Fig. 20.20. Now $\rho(x,y) = k(x^2 + y^2)$, so

$$M = \iint_R k(x^2 + y^2)\,dA$$

where R is the region given by

$$R = \{(x,y)\,|\,0 \le x \le 4, 0 \le y \le \sqrt{x}\}$$

Fig. 20.20

Thus

$$\begin{aligned} M &= \int_0^4 \int_0^{\sqrt{x}} k(x^2 + y^2)\,dy\,dx \\ &= k\int_0^4 \left[x^2y + \frac{y^3}{3}\right]_0^{\sqrt{x}} dx \\ &= k\int_0^4 \left(x^{5/2} + \frac{x^{3/2}}{3}\right) dx \\ &= k\left[\frac{2x^{7/2}}{7} + \frac{2x^{5/2}}{15}\right]_0^4 = \frac{4288}{105}k \end{aligned}$$

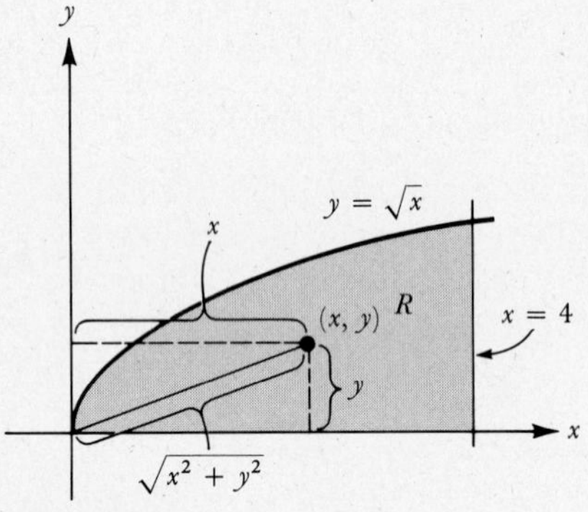

Now

$$\begin{aligned} M_x &= k\int_0^4 \int_0^{\sqrt{x}} y(x^2 + y^2)\,dy\,dx \\ &= k\int_0^4 \left[\frac{x^2y^2}{2} + \frac{y^4}{4}\right]_0^{\sqrt{x}} dx \\ &= k\int_0^4 \left(\frac{x^3}{2} + \frac{x^2}{4}\right) dx \\ &= k\left[\frac{x^4}{8} + \frac{x^3}{12}\right]_0^4 = \frac{112}{3}k \end{aligned}$$

and

$$\begin{aligned} M_y &= k\int_0^4 \int_0^{\sqrt{x}} x(x^2 + y^2)\,dy\,dx \\ &= k\int_0^4 \left[x^3y + \frac{xy^3}{3}\right]_0^{\sqrt{x}} dx \\ &= k\int_0^4 \left(x^{7/2} + \frac{x^{5/2}}{3}\right) dx \\ &= k\left[\frac{2x^{9/2}}{9} + \frac{2x^{7/2}}{21}\right]_0^4 = \frac{7936}{63}k \end{aligned}$$

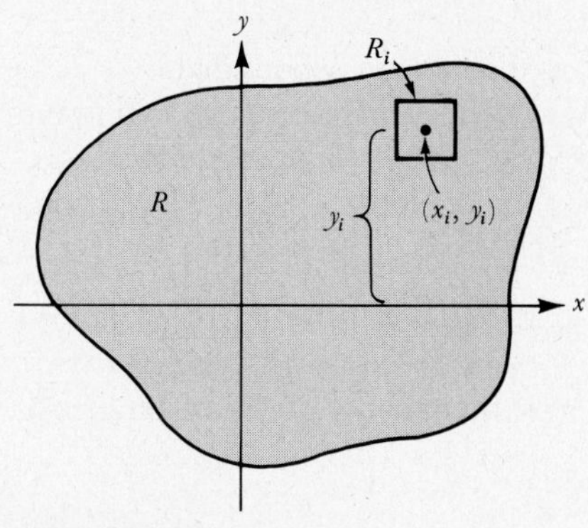

Fig. 20.21

Thus

$$\bar{x} = \frac{7936k}{63} \times \frac{105}{4288k} \approx 3.085 \qquad \text{and} \qquad \bar{y} = \frac{112k}{3} \times \frac{105}{4288k} \approx 0.914 \quad \bullet$$

In a similar way we can obtain defining integrals for moments of inertia. For example, consider the portion of L occupying the rectangle R_i, whose mass we consider as being concentrated at (x_i, y_i). Then $(I_x)_i$, the moment of inertia of this portion of L about the x axis (see Fig. 20.21), is approximated by

$$\begin{aligned}(I_x)_i &= (\text{distance from } x \text{ axis})^2 \times \text{mass} \\ &\approx {y_i}^2 \rho(x_i, y_i)\, \Delta A_i\end{aligned}$$

Hence I_x, the moment of inertia of L about the x axis, is approximated by

$$I_x \approx \sum_{i=1}^{n} {y_i}^2 \rho(x_i, y_i)\, \Delta A_i$$

Thus the defining integral for I_x is given by

20.2.7 $$I_x = \iint_R y^2 \rho(x, y)\, dA$$

In an analogous way the moments of inertia about the y axis and the origin (the *polar moment*) are defined by, respectively,

20.2.8 $$I_y = \iint_R x^2 \rho(x, y)\, dA$$

and

20.2.9 $$I_0 = \iint_R (x^2 + y^2) \rho(x, y)\, dA$$

Example 6 Find the polar moment of the lamina occupying the region R bounded by $y = x^2, y = 4$, and $x = 0$ if the density is inversely proportional to the square of the distance from the origin.

Solution The region is sketched in Fig. 20.22. Now

$$\rho(x, y) = \frac{k}{x^2 + y^2} \qquad \text{and} \qquad R = \{(x, y) \mid 0 \le x \le 2, x^2 \le y \le 4\}$$

Thus

$$\begin{aligned}I_0 &= \iint_R (x^2 + y^2) \frac{k}{(x^2 + y^2)}\, dA \\ &= k \int_0^2 \int_{x^2}^4 dy\, dx = k \int_0^2 (4 - x^2)\, dx = \frac{16k}{3} \quad \bullet\end{aligned}$$

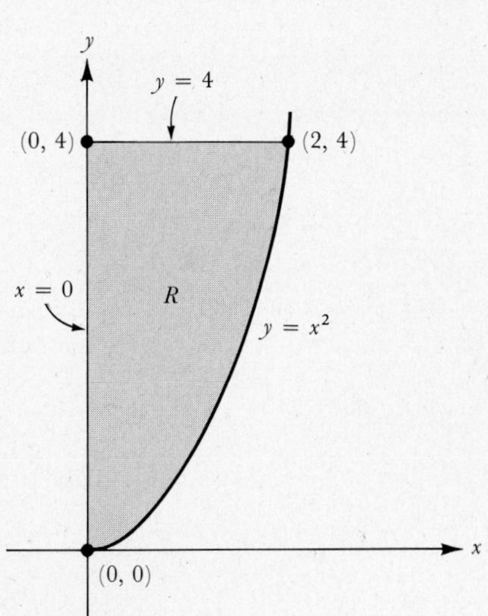

Fig. 20.22

GRAVITATIONAL ATTRACTION

A particle of mass m_1 located at the point Q exerts a force of gravitational attraction $\mathbf{F}$ on a particle of mass m_2 located at P given by

$$\mathbf{F} = \frac{g m_1 m_2 \overrightarrow{PQ}}{\|\overrightarrow{PQ}\|^3}$$

where g is the gravitational constant. (This is a version of Newton's "inverse-square law" of gravitational attraction.)

Example 7 Assume L is the lamina occupying the triangular region R bounded by $x = 1$, $x = 2, y = 0$, and $y = x - 1$, and assume that the density is proportional to the distance from the y axis. Set up iterated integrals to give the components of the force of gravitational attraction **F** exerted by the lamina L on a particle of mass μ located at the origin.

Solution A sketch of the lamina is given in Fig. 20.23. As usual, we partition R into rectangular subregions. We approximate the mass of each of these and treat each as if its mass were concentrated at a single point (x_i, y_i) in the rectangle R_i. The mass of this ith piece is given by

$$m_i \approx kx_i \, \Delta A_i$$

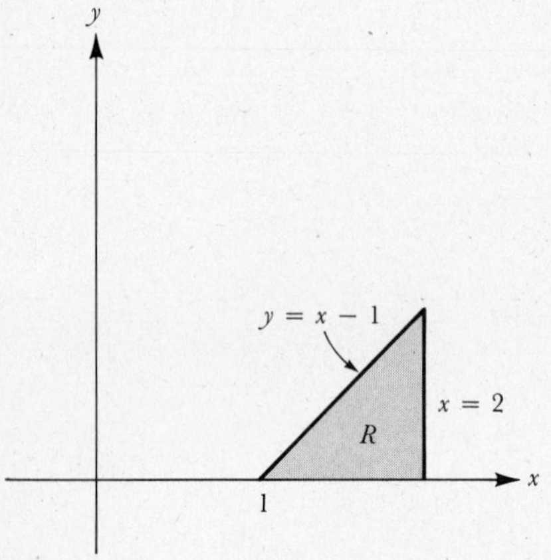

Thus

$$\mathbf{F}_i \approx \frac{g\mu kx_i \, \Delta A_i \langle x_i - 0, y_i - 0 \rangle}{\|\langle x_i - 0, y_i - 0 \rangle\|^3}$$

gives the force of attraction that this portion of the lamina exerts on the particle located at the origin. The components of $\mathbf{F}_i$ are, respectively,

$$\frac{g\mu kx_i^2 \, \Delta A_i}{\sqrt{x_i^2 + y_i^2}} \quad \text{and} \quad \frac{g\mu kx_i y_i \, \Delta A_i}{\sqrt{x_i^2 + y_i^2}}$$

If we add these and pass to the limit, we get the components of **F**, the force of attraction of L on the particle located at the origin. These are, respectively,

$$g\mu k \iint_R \frac{x^2}{\sqrt{x^2 + y^2}} \, dA \quad \text{and} \quad g\mu k \iint_R \frac{xy \, dA}{\sqrt{x^2 + y^2}}$$

Hence

$$\mathbf{F} = \left\langle g\mu k \int_1^2 \int_0^{x-1} \frac{x^2}{\sqrt{x^2 + y^2}} \, dA, \, g\mu k \int_1^2 \int_0^{x-1} \frac{xy \, dA}{\sqrt{x^2 + y^2}} \right\rangle \bullet$$

Fig. 20.23

PROGRESS TEST 2

1. Determine the first moment about the y axis of the lamina occupying the region bounded by $x = 0$, $y = 0$, and $y = 1 - x$ if the density is proportional to the distance from the x axis.

2. Determine the moment of inertia about the x axis of the lamina occupying the region bounded by $y = x^2$, $y = 0$, and $x = 2$ if the density is given by $\rho(x, y) = x + 2$.

SECTION 20.2 EXERCISES

In Exercises 1 to 6 determine the area of the region R bounded by the given curves.

1. $y = x^3, y = x^2$

2. $y = x^2, y = 18 - x^2$

3. $y^2 = 4x, y = 2x - 4$

4. $2y^2 + y = x + 1, y - x = 1$

5. $2x^2 = 2y + 1, x + 2 = 2y$

6. $y = x^3 - x, y = 0$

In Exercises 7 to 12 determine the volume of the indicated solid region S.

7. The solid in the first octant below the graph of $2x + y + 3z = 6$.

8. The solid inside the sphere $x^2 + y^2 + z^2 = 4$ and above $z = 1$.

9. The solid in the first octant bounded by the cylinders $x^2 + y^2 = 9$ and $x^2 + z^2 = 9$.

10. The solid bounded by the planes $x + y + z = 2$, $z = 2$, and the cylinders $x = y^2$ and $y = x^2$.

11. The solid in the first octant bounded by $x^2 + y^2 = 4$, $z = y$, $z = 0$, and $x = 0$.

12. The solid region bounded above by $x^2 + y^2 + z^2 = 1$ and below by $z = 0$.

In Exercises 13 to 18 determine the mass and the center of mass of the given lamina.

13. L is the triangular lamina in the first quadrant bounded by the line $x + y = 2$, given that the density function is $\rho(x,y) = x^2 - 2y$.

14. L occupies the region bounded by $y = 4 - x^2$ and $y + x + 2 = 0$; the density is proportional to the distance from the line $y = 5$.

15. L occupies the region bounded by $x + y^2 = 1$ and $y = x + 1$; the density is proportional to the distance from $x = 7$.

16. L occupies the region bounded by $x = 4 - 3y^2$ and $x = y^2$; the density is constant.

17. L has the shape of an isosceles right triangle and the density is proportional to the distance from the vertex opposite the hypotenuse.

18. L occupies the region bounded by $y = x^2$ and $y = x$; the density is proportional to the distance from the x axis.

In Exercises 19 to 23 determine the moment of inertia of the given lamina about the indicated line.

19. L occupies the region bounded by $y = x^2$, $y = x$, $\rho(x,y) = kx$; about the x axis.

20. L occupies the region bounded by $x = y^2 + 1$ and $y - x + 3 = 0$, and $\rho(x,y) = ky$; about the line $x = -2$.

21. L occupies the quarter-disk in the first quadrant bounded by $x^2 + y^2 = 16$, and $\rho(x,y) = 2x + 2y$; about the line $y = 6$.

22. L occupies the region bounded by $x = y^2$ and $x = y$, and the density is proportional to the sum of the distances from the x and y axes; about the line $y = 4$.

23. L occupies the region bounded by $y = x$, $y = 1$, $y = 4$, and $x = 0$, and the density is proportional to the distance from the x axis; about the y axis.

24. Set up (but do not evaluate) an iterated integral that gives the first moment of the lamina occupying the region bounded by $y^2 = x - 1$ and $y = x - 3$ about the line $y = x + 3$ if the density is given by $\rho(x,y) = xy + x$. (*Hint:* Use the formula for the distance from a point to a line developed in Chap. 17, page 585.)

Section 20.3: Triple Integrals, with Applications

Objectives:

1. Evaluate triple iterated integrals.

2. Apply the theory of the triple integral to solve physical and geometric problems.

INTRODUCTION

If $w = f(x,y,z)$ is defined on some bounded solid region S, then we can define the *triple integral* of f over S, denoted

$$\iiint_S f(x,y,z)\,dV$$

in much the same way as we defined the double integral. Since S is bounded, we can contain it in a parallelepiped D whose faces are parallel to the coordinate planes. We begin by superimposing a *grid* over D consisting of a finite number of planes parallel to the faces of D. This has the effect of subdividing D into a collection of smaller parallelepipeds (boxes), a finite number of which lie wholly within S (see Fig. 20.24). We denote these by $S_1, S_2, \ldots, S_n$ and refer to them as an *inner partition* P of S. A few typical elements from an inner partition are sketched in Fig. 20.25. By $\|P\|$, the *norm* of P, we shall mean the length of the largest diagonal of the S_i's.

Now for each i let (x_i, y_i, z_i) be an arbitrary point in S_i and let ΔV_i denote the volume of S_i. A sum of the form

$$\sum_{i=1}^{n} f(x_i, y_i, z_i)\,\Delta V_i$$

is called a *Riemann sum* of f over P. If we now let $n \to +\infty$ in such a way that $\|P\| \to 0$, and if these Riemann sums tend to some fixed number independently of how the (x_i, y_i, z_i) are chosen in S_i, then we say f is *integrable* over S and we write

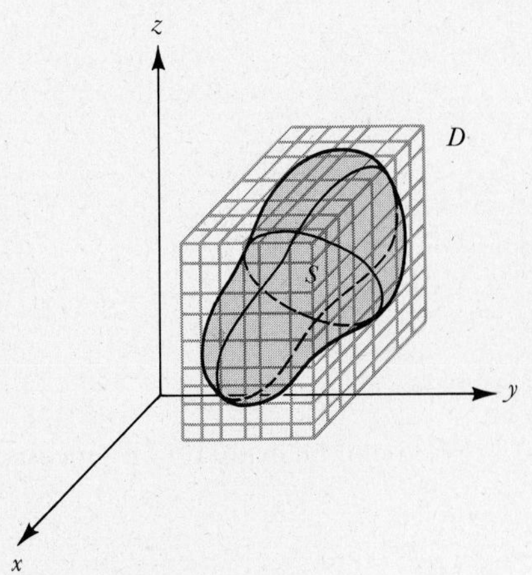

Fig. 20.24

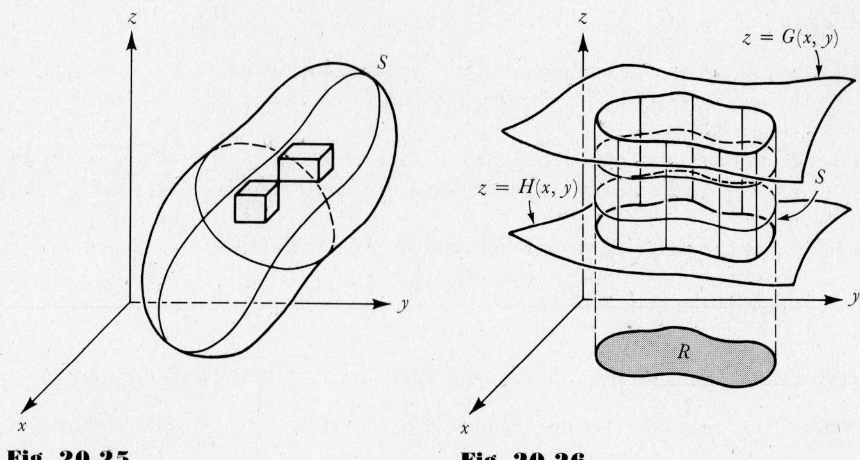

Fig. 20.25

Fig. 20.26

20.3.1 Definition

$$\iiint_S f(x,y,z)\,dV = \lim_{\|P\|\to 0} \sum_{i=1}^{n} f(x_i, y_i, z_i)\,\Delta V_i$$

By a *regular region* in three-space we mean a region of the form

(1) $$S = \{(x,y,z) \mid (x,y) \in R,\ H(x,y) \leq z \leq G(x,y)\}$$

where R is a plane region of the type considered in Section 20.1 and where H and G have continuous partials on R (see Fig. 20.26).

Similarly, regular regions can be of the form

(2) $$S = \{(x,y,z) \mid (x,z) \in R,\ P(x,z) \leq y \leq Q(x,z)\}$$

or

(3) $$S = \{(x,y,z) \mid (y,z) \in R,\ U(y,z) \leq x \leq V(y,z)\}$$

where R is a region in the x-z or y-z plane and the functions P, Q, U, V have continuous first partials. (See Fig. 20.27 for a sketch of the former.)

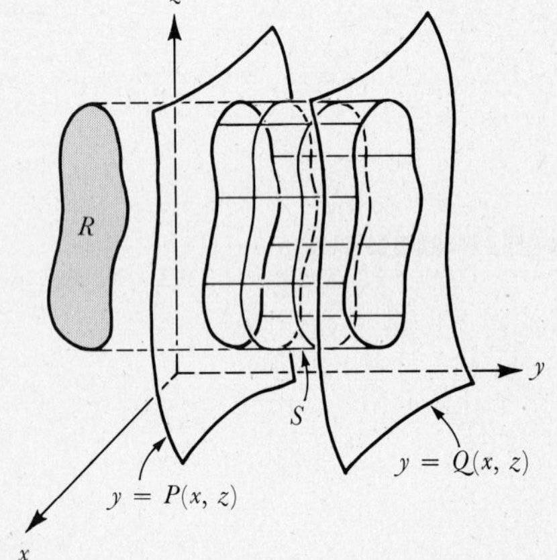

Fig. 20.27

We shall assume the following:

20.3.2 Theorem If $f(x,y,z)$ is continuous in the regular region S, then $\iiint_S f(x,y,z)\,dV$ exists.

If $f(x,y,z) = 1$, then the Riemann sums reduce to $\Sigma_{i=1}^{n} \Delta V_i$, which approximates the volume of S; and thus we have another way to realize the volume of a solid:

20.3.3 $$V(S) = \iiint_S dV$$

EVALUATION OF TRIPLE INTEGRALS

It can be shown that a triple integral over a region can be evaluated by means of iterated integrals. In particular, if

$$S = \{(x,y,z) \mid a \leq x \leq b,\ h(x) \leq y \leq g(x),\ H(x,y) \leq z \leq G(x,y)\}$$

then

20.3.4
$$\iiint_S f(x, y, z)\, dV = \int_a^b \int_{h(x)}^{g(x)} \int_{H(x,y)}^{G(x,y)} f(x, y, z)\, dz\, dy\, dx$$

The iterated integral is obtained by first holding x and y constant and evaluating

$$\int_{H(x,y)}^{G(x,y)} f(x, y, z)\, dz$$

where $H(x,y)$ and $G(x,y)$ are z limits. This reduces the triple iterated integral to a double iterated integral of the form

$$\int_a^b \int_{h(x)}^{g(x)} A(x, y)\, dy\, dx$$

which is evaluated as in Sec. 20.1.

Example 1 Evaluate

$$\int_1^2 \int_0^x \int_1^{xy} 8xy\, dz\, dy\, dx$$

Solution

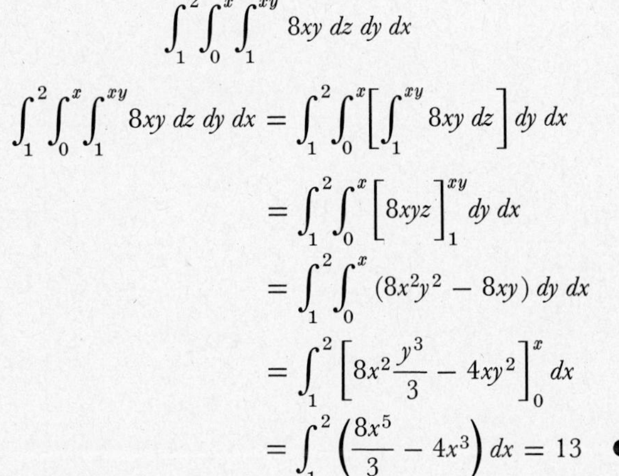

$$\begin{aligned}\int_1^2 \int_0^x \int_1^{xy} 8xy\, dz\, dy\, dx &= \int_1^2 \int_0^x \left[\int_1^{xy} 8xy\, dz\right] dy\, dx \\ &= \int_1^2 \int_0^x \Big[8xyz\Big]_1^{xy} dy\, dx \\ &= \int_1^2 \int_0^x (8x^2y^2 - 8xy)\, dy\, dx \\ &= \int_1^2 \left[8x^2\frac{y^3}{3} - 4xy^2\right]_0^x dx \\ &= \int_1^2 \left(\frac{8x^5}{3} - 4x^3\right) dx = 13\end{aligned}$$ ●

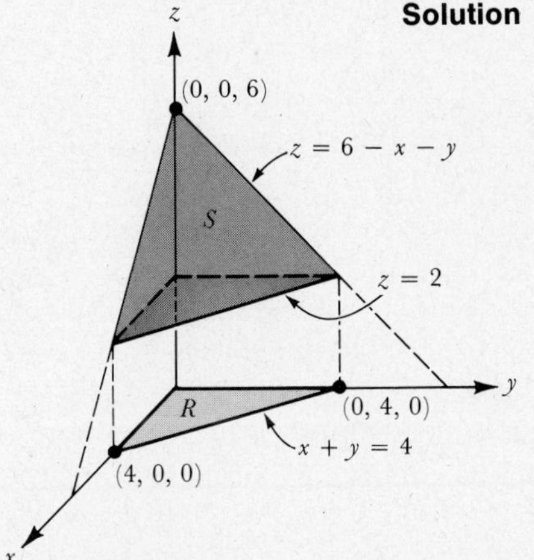

Fig. 20.28

Most triple iterated integrals arise directly from geometric or physical applications, the simplest of which is the determination of a solid's volume.

Example 2 Determine the volume of the solid region S in the first octant bounded below by $z = 2$ and above by the plane $x + y + z = 6$.

Solution By (20.3.3), $V(S) = \iiint_S dV$. To evaluate this integral we shall first express S in the form (1) and then use (20.3.4) to express the triple integral as an iterated integral. A sketch of S is given in Fig. 20.28. Eliminating z from the two equations, we see that S sits over the region R in the first quadrant of the x-y plane bounded by the x and y axes and the straight line $x + y = 4$. Thus

$$R = \{(x, y) \mid 0 \le x \le 4,\ 0 \le y \le 4 - x\}$$

Now (x, y, z) is in S if and only if (x, y) is in R and

$$2 \le z \le 6 - x - y$$

so

$$S = \{(x, y, z) \mid 0 \le x \le 4,\ 0 \le y \le 4 - x,\ 2 \le z \le 6 - x - y\}$$

From this description of S and the order of the differentials, $dV = dz\,dy\,dx$, we can now apply (20.3.4) to evaluate $\iiint_S dV$:

$$\begin{aligned} V(S) &= \int_0^4 \int_0^{4-x} \int_2^{6-x-y} dz\,dy\,dx \\ &= \int_0^4 \int_0^{4-x} \Big[z \Big]_{z=2}^{z=6-x-y} dy\,dx \\ &= \int_0^4 \left[4y - xy - \frac{y^2}{2} \right]_0^{4-x} dx \\ &= \int_0^4 \left[4y - xy - \frac{y^2}{2} \right]_0^{4-x} dx \\ &= \int_0^4 \left(8 - 4x + \frac{x^2}{2} \right) dx = \frac{32}{3} \end{aligned}$$ ●

We evaluated the triple integral in Example 2 using an iterated integral whose order of iteration and associated limits are dictated by (20.3.4). However, this order of integration is by no means unique. Since there are six ways to order the three variables x, y, and z, there are, in general, six possible iterations; each is associated with a particular description of S. For example, if a solid region S were described as

$$S = \{(x, y, z)\,|\, a \leq x \leq b,\ h(x) \leq z \leq g(x),\ H(x, z) \leq y \leq G(x, z)\}$$

then instead of (20.3.4), we would write

$$\iiint_S f(x, y, z)\,dV = \int_a^b \int_{h(x)}^{g(x)} \int_{H(x,z)}^{G(x,z)} f(x, y, z)\,dy\,dz\,dx$$

The inside integral would be obtained by holding x and z fixed and integrating with respect to y.

On the other hand, given a particular triple iterated integral, we can, as with the double iterated integral, read off from the limits of integration the boundaries of the associated solid, paying particular attention to the order of the iteration. The outside differential tells you which variable is being limited by the constant limits on the outside integral, the middle differential indicates the variable being limited by the curves on the middle integral, and the inside differential indicates the variable being limited by the surfaces on the inside integral. For example, the solid S associated with the following iterated integral

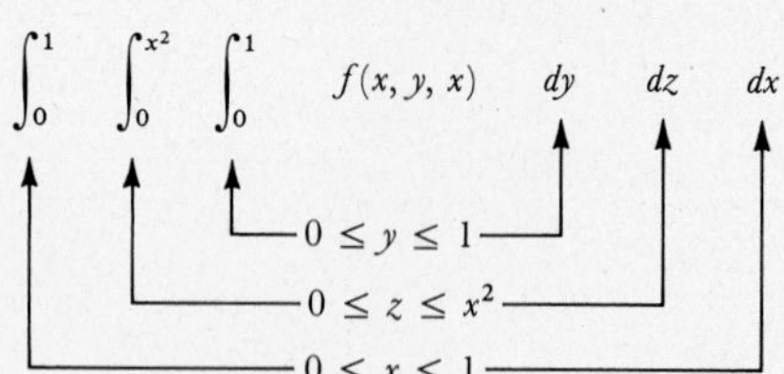

may be described as (see Fig. 20.29)

$$S = \{(x, y, z)\,|\,0 \leq x \leq 1,\ 0 \leq z \leq x^2,\ 0 \leq y \leq 1\}$$

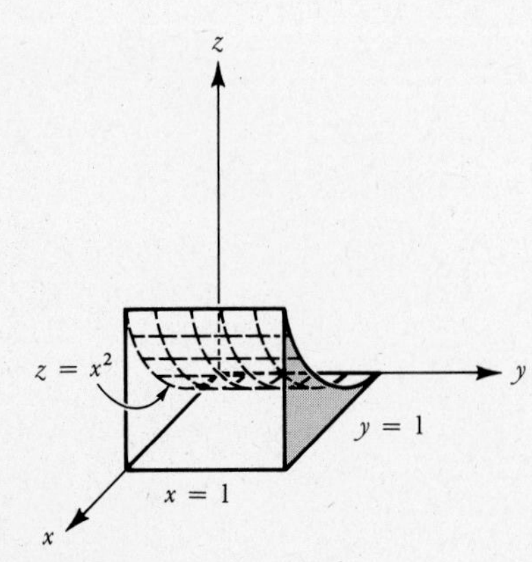

Fig. 20.29

The different iterated integrals that can be used to compute a given triple integral $\iiint_S f(x, y, z)\,dV$ require different descriptions of S. The next example illustrates how to form these descriptions using the projections of S into the coordinate planes.

Example 3 Give the six iterated integrals for the triple integral of Example 2.

Solution Our approach is to sketch the projections of the solid S, as in Fig. 20.28, in each of the coordinate planes; see Fig. 20.30(a). Each such projection has two descriptions—in effect, a type I and a type II description. The projection R_1 in the x-y plane is obtained by eliminating z from the equations of the top and bottom of S, $z = 6 - x - y$ and $z = 2$. This yields one of the boundaries, namely, $y = 4 - x$. The others are $x = 0, y = 0$; thus [see Fig. 20.30(b)]

$$R_1 = \{(x, y) | 0 \le x \le 4,\ 0 \le y \le 4 - x\}$$
$$= \{(x, y) | 0 \le y \le 4, 0 \le x \le 4 - y\}$$

Similarly, by letting $y = 0$, we have the descriptions for R_2:

$$R_2 = \{(x, z) | 0 \le x \le 4,\ 2 \le z \le 6 - x\}$$
$$= \{(x, z) | 2 \le z \le 6,\ 0 \le x \le 6 - z)\} \qquad \text{[see Fig. 20.30(}c\text{)]}$$

Finally, we let $x = 0$ to get R_3:

$$R_3 = \{(y, z) | 0 \le y \le 4,\ 2 \le z \le 6 - y\}$$
$$= \{(y, z) | 2 \le z \le 6,\ 0 \le y \le 6 - z\} \qquad \text{[see Fig. 20.30(}d\text{)]}$$

Now, in turn, (x, y, z) is in S if and only if:

(x, y) is in R_1 and $2 \le z \le 6 - x - y$

or

(x, z) is in R_2 and $0 \le y \le 6 - x - z$

or

(y, z) is in R_3 and $0 \le x \le 6 - y - z$

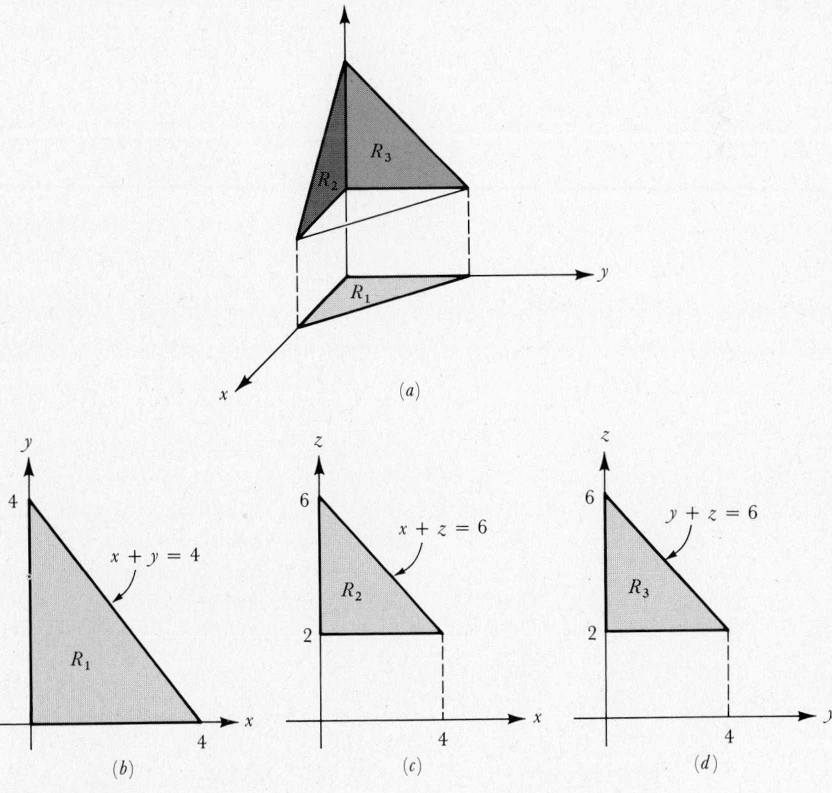

Fig. 20.30

Thus, putting all this information together, we have the following six iterated integrals, given in the order corresponding to the previous descriptions:

$$\iiint_S dV = \int_0^4 \int_0^{4-x} \int_2^{6-x-y} dz\, dy\, dx$$

$$= \int_0^4 \int_0^{4-y} \int_2^{6-x-y} dz\, dx\, dy$$

$$= \int_0^4 \int_2^{6-x} \int_0^{6-x-z} dy\, dz\, dx$$

$$= \int_2^6 \int_0^{6-z} \int_0^{6-x-z} dy\, dx\, dz$$

$$= \int_0^4 \int_2^{6-y} \int_0^{6-y-z} dx\, dz\, dy$$

$$= \int_2^6 \int_2^{6-z} \int_0^{6-y-z} dx\, dy\, dz \quad \bullet$$

We express the pattern developed in Example 3 to assist in more complicated situations.

20.3.5 Guidelines for Setting up Triple Iterated Integrals Over a Solid S

(*a*) The outer two differential variables determine the coordinate plane in which we project the solid S (forming a plane region R).

(*b*) The limits of the outer two integrals are then determined by the boundaries of R, and depend on whether R is given a type I or type II description.

(*c*) The upper and lower limits on the inner integral are the boundary surfaces of S in the direction of the inner differential variable expressed in terms of the other two.

A MNEMONIC DEVICE FOR TRIPLE ITERATED INTEGRALS

As with double iterated integrals, some students find it helpful to view the triple iterated integral as the limit of a triple summing process over the solid S where the values of the function $f(x, y, z)$ being integrated are added in three stages, with the order depending on how the solid region is described. We imagine the values of f being "added" over the solid S in a three-stage geometric process. We begin by "adding" over boxes stacked in the direction of the inside differential variable. This yields a long prism-shaped box between the boundaries of S in the direction of that inside differential variable. It is necessarily perpendicular to the plane of the outer two differential variables in which S is projected. We then "add" over such prisms in the direction of the intermediate differential variable to form a slab that is perpendicular to the direction of the outer differential variable. Finally we "add" over such slabs between the limits of the outer integral.

Figure 20. 31 illustrates the process for

$$\int_a^b \int_{h(x)}^{g(x)} \int_0^{G(x,y)} f(x, y, z)\, dz\, dy\, dx$$

when S is a solid regular region based on the x-y plane. The values of f are added first in the boxes in the z direction from the bottom surface ($z = H(x, y) = 0$ in

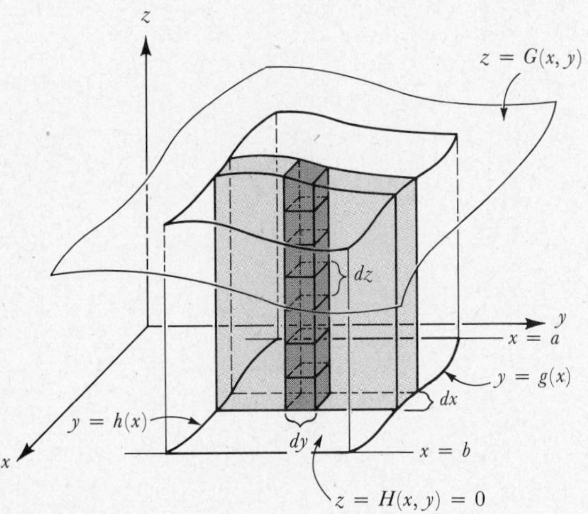

Fig. 20.31

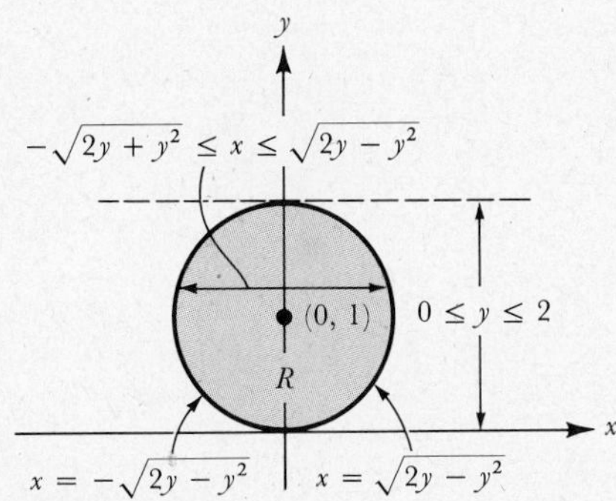

Fig. 20.32

this case) to the top, $G(x, y)$. The values of f are then added on the resulting prisms across a slab in the y direction from $y = h(x)$ to $y = g(x)$. Finally, the values of f are added on slabs ranging in the x direction from $x = a$ to $x = b$.

CHANGING THE ORDER OF ITERATION

The next example involves switching the order of iteration, which requires that the solid S first be recovered from the original iterated integral.

Example 4 Given that

$$I = \iiint_S f(x, y, z)\, dV = \int_0^2 \int_{-\sqrt{2y-y^2}}^{\sqrt{2y-y^2}} \int_{2y}^{x^2+y^2} f(x, y, z)\, dz\, dx\, dy$$

(*a*) sketch the region S, and (*b*) express I as an iterated integral(s) with $dV = dy\, dx\, dz$.

Solution (*a*) From the iterated integral, we have

$$S = \{(x, y, z) \,|\, 0 \leq y \leq 2,\ -\sqrt{2y - y^2} \leq x \leq \sqrt{2y - y^2},\ 2y \leq z \leq x^2 + y^2\}$$

To assist in sketching S, which we need do before applying [20.3.5(*a*)], we project S into the x-y plane ($z = 0$) getting

$$R = \{(x, y) \,|\, 0 \leq y \leq 2,\ -\sqrt{2y - y^2} \leq x \leq \sqrt{2y - y^2}\}$$

If $x = \pm\sqrt{2y - y^2}$, then $x^2 = 2y - y^2$, or $x^2 + y^2 - 2y = 0$. Completing the square in y, we have $x^2 + (y - 1)^2 = 1$, which is a circle of radius 1 centered on the y axis at (0, 1); see Fig. 20.32. Now (x, y, z) is in S if and only if (x, y) is in R and $2y \leq z \leq x^2 + y^2$. Thus S is the solid region bounded below by the paraboloid $z = x^2 + y^2$ and above by the plane $z = 2y$ (see Fig. 20.33).

(*b*) Since the order of integration is $dy\, dx\, dz$, by [20.3.5(*a*)] we need the projection R_1 of the region S in the x-z plane. To do this we eliminate y from the equations

$$(4) \qquad z = 2y \quad \text{and} \quad z = x^2 + y^2$$

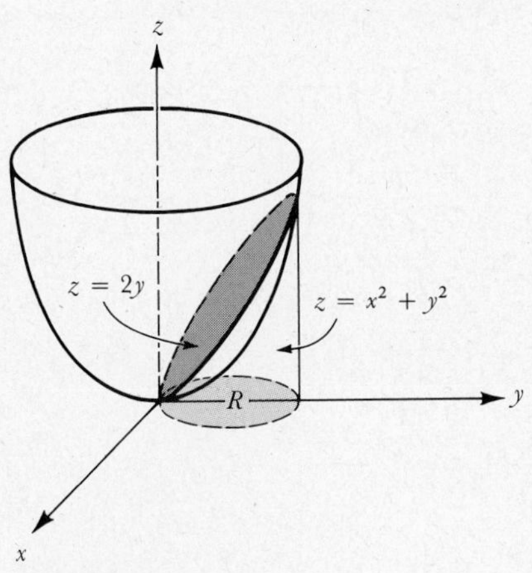

Fig. 20.33

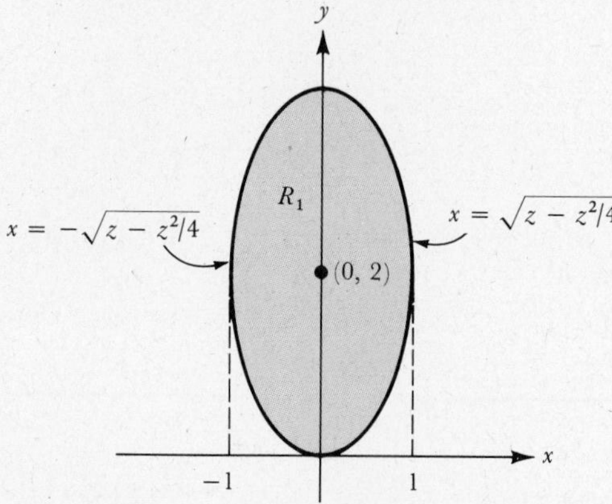

Fig. 20.34

obtaining

$$z = x^2 + \frac{z^2}{4}$$

or, equivalently, by completing the square in z and rearranging, we have

$$\frac{(z-2)^2}{4} + x^2 = 1$$

Thus R_1 is an ellipse in the x-z plane centered on the z axis (see Fig. 20.34). We express x in terms of z to get

$$x = \pm\sqrt{z - z^2/4}$$

and so

$$R_1 = \{(x, z)|0 \leq z \leq 2, -\sqrt{z - z^2/4} \leq x \leq \sqrt{z - z^2/4}\}$$

Hence we know that

$$I = \int_0^2 \int_{-\sqrt{z-z^2/4}}^{+\sqrt{z-z^2/4}} \int_{-}^{-} f(x, y, z)\, dy\, dx\, dz$$

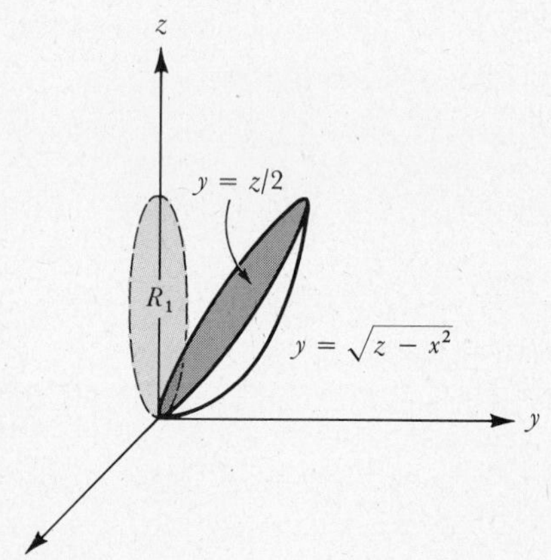

Fig. 20.35

where the inner limits remain to be determined. By [20.3.5(c)] it remains to solve the equations in (4) for y in terms of x and z. We get $y = z/2$ and $y = \pm\sqrt{z - x^2}$. Since S lies to the right of the x-z plane, we choose the plus sign in the latter equation, and conclude that $z/2 \leq y \leq \sqrt{z - x^2}$; see Fig. 20.35. Hence

$$I = \int_0^2 \int_{-\sqrt{z-z^2/4}}^{\sqrt{z-z^2/4}} \int_{z/2}^{\sqrt{z-x^2}} f(x, y, z)\, dy\, dx\, dz \quad \bullet$$

PROGRESS TEST 1

Evaluate the following integrals:

1. $\int_{-1}^{1} \int_0^x \int_0^{xz} xz\, dy\, dz\, dx$

2. $\iiint_S z\, dV$, where S is the solid region in the first octant inside the cylinders $x^2 + y^2 = 1$ and $x^2 + z^2 = 1$. (Sketch S carefully.)

APPLICATIONS OF THE TRIPLE INTEGRAL

The principle involved in applications of the triple integral is the same as for the double integral: Obtain an approximation of the desired quantity as a Riemann sum and then pass to a limit that is a triple integral and describes that quantity exactly. Throughout we shall have occasion to use the fact that the mass of a substance of density k and volume V is kV. Because of the triple integral, our ability to determine mass and moments of solids will extend greatly our capability developed in Sec. 12.4, where we were limited to shapes and densities that could be described using only a single variable.

Example 5 Find the mass M of a cylindrical solid S of radius a and height h if the density at any point is proportional to the distance from the base of the solid.

Solution A right circular cylinder of radius a, whose axis is the z axis, has equation $x^2 + y^2 = a^2$. Thus

$$S = \{(x, y, z) \mid -a \leq x \leq a, -\sqrt{a^2 - x^2} \leq y \leq \sqrt{a^2 - x^2}, 0 \leq z \leq h\}$$

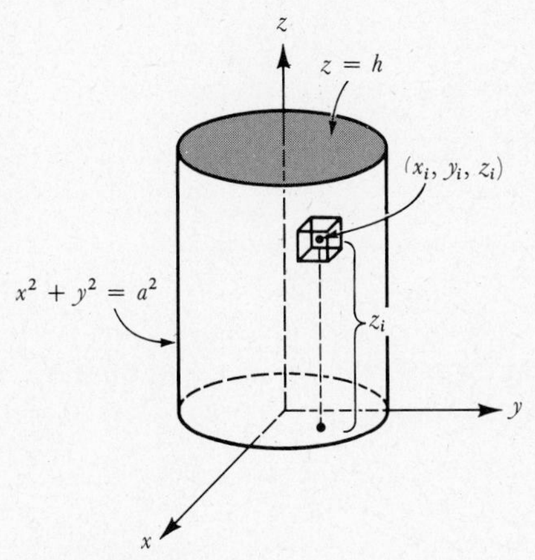

Fig. 20.36

If we impose a grid on S, then a typical subregion S_i has volume ΔV_i. If (x_i, y_i, z_i) is a point in S_i and k is the constant of proportionality, then the density at this point is kz_i and so the mass of S_i is approximated by $M_i \approx kz_i\, \Delta V_i$; see Fig. 20.36. Since

$$M \approx \sum_{i=1}^{n} M_i \approx \sum_{i=1}^{n} kz_i\, \Delta V_i$$

we can pass to the limit to obtain

$$\begin{aligned} M &= \iiint_S kz\, dV \\ &= \int_{-a}^{a} \int_{-\sqrt{a^2-x^2}}^{\sqrt{a^2-x^2}} \int_0^h kz\, dz\, dy\, dx \\ &= \frac{kh^2}{2} \int_{-a}^{a} \int_{-\sqrt{a^2-x^2}}^{\sqrt{a^2-x^2}} dy\, dx \\ &= kh^2 \int_{-a}^{a} \sqrt{a^2 - x^2}\, dx \\ &= kh^2 \left[\frac{a^2}{2} \arcsin\left(\frac{x}{a}\right) + \frac{x\sqrt{a^2 - x^2}}{2}\right]_{-a}^{a} = \frac{kh^2 a^2 \pi}{2} \end{aligned}$$ ●

Example 6 Set up an iterated integral that gives the moment of inertia I_z about the z axis of the homogeneous solid S bounded below by the upper nappe of the cone $3z^2 = x^2 + y^2$ and above by the plane $z = 1$.

Solution A sketch of the solid S with a typical subregion S_i is given in Fig. 20.37. The distance from the point (x_i, y_i, z_i) in S_i to the z axis is $\sqrt{x_i^2 + y_i^2}$ and the mass of S_i is approximated by $k\, \Delta V_i$. Thus the moment of inertia about the z axis of this subregion is approximated by

$$(I_z)_i \approx (\sqrt{x_i^2 + y_i^2})^2 (k\, \Delta V_i)$$

and hence

$$I_z \approx \sum_{i=1}^{n} k(x_i^2 + y_i^2)\, \Delta V_i$$

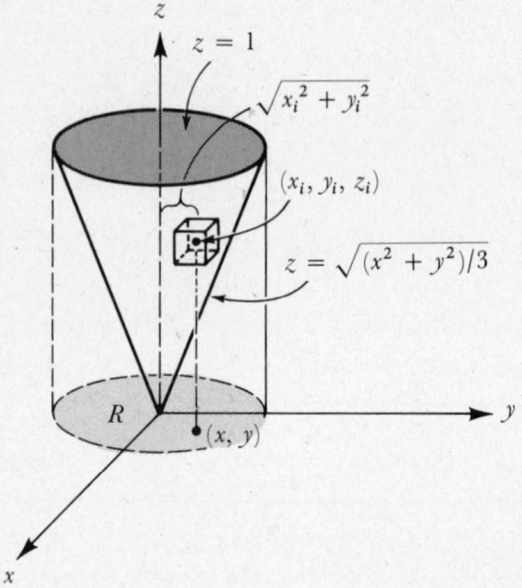

Fig. 20.37

Passing to the limit, we have

$$I_z = \iiint_S k(x^2 + y^2)\, dV$$

To obtain the projection of S in the x-y plane (and thus use dz as the inner differential), we eliminate z from

$$z = 1 \qquad \text{and} \qquad 3z^2 = x^2 + y^2$$

getting

$$x^2 + y^2 = 3$$

a circle centered at the origin of the x-y plane of radius $\sqrt{3}$. This has the description

$$R = \{(x,y) \mid -\sqrt{3} \le x \le \sqrt{3},\ -\sqrt{3 - x^2} \le y \le \sqrt{3 - x^2}\}$$

Now, as indicated in Fig. 20.37, (x, y, z) is in the cone S if and only if (x, y) is in R and

$$\sqrt{(x^2 + y^2)/3} \le z \le 1$$

Thus

$$S = \{(x, y\ z) \mid -\sqrt{3} \le x \le \sqrt{3},\ -\sqrt{3 - x^2} \le y \le \sqrt{3 - x^2},\ \sqrt{(x^2 + y^2)/3} \le z \le 1\}$$

so

$$I_z = \int_{-\sqrt{3}}^{\sqrt{3}} \int_{-\sqrt{3-x^2}}^{\sqrt{3-x^2}} \int_{\sqrt{(x^2+y^2)/3}}^{1} k(x^2 + y^2)\, dz\, dy\, dx \quad \bullet$$

Although the integral in Example 6 can be evaluated, it is not particularly easy to do so. We shall give a technique in the next section that simplifies this chore considerably by switching coordinate systems.

PROGRESS TEST 2

1. Find the first moment about the y-z plane of the homogeneous solid in the first octant bounded by the cylinder $y = x^2$ and the planes $x = 1$ and $z = y$.

2. Set up but do not evaluate a triple iterated integral for the second moment about the z axis of the solid bounded above by $z = 12 - x^2 - y^2$ and below by $z = 3$, where the density at any point is inversely proportional to the distance from the origin.

SECTION 20.3 EXERCISES

In Exercises 1 to 12 evaluate the indicated iterated integral.

1. $\int_0^6 \int_{-3}^2 \int_{-1}^1 36y\, dy\, dz\, dx$

2. $\int_0^5 \int_{-1}^1 \int_0^2 20x\, dx\, dy\, dz$

3. $\int_0^{12} \int_0^{\sqrt{144-x^2}} \int_0^{20-x-y} dz\, dy\, dx$

4. $\int_0^1 \int_0^{\sqrt{y}} \int_0^{\sqrt{x}} x^2 z\, dz\, dx\, dy$

5. $\int_0^1 \int_0^y \int_0^z e^x \, dx \, dz \, dy$

6. $\int_0^1 \int_0^{y^2} \int_0^y 12x(x^2 + y^2) \, dz \, dx \, dy$

7. $\int_0^{\pi/2} \int_0^z \int_x^{z+x} \cos(y + z) \, dy \, dx \, dz$

8. $\int_0^1 \int_0^{2-z} \int_0^{2-y-z} 24z \, dx \, dy \, dz$

9. $\int_0^{\pi/2} \int_0^1 \int_0^{y^2} 12y \cos x \, dz \, dy \, dx$

10. $\int_1^2 \int_{y/2}^{y/\sqrt{2}} \int_0^{\ln(xy)} \frac{e^z \, dz \, dx \, dy}{\sqrt{y^2 - x^2}}$

11. $\int_{\pi/3}^{\pi/2} \int_0^{\sin x} \int_0^{y \sin x} y \cos x \, dz \, dy \, dx$

12. $\int_0^{\pi/3} \int_z^{\pi/3} \int_0^{yz} \sin\left(\frac{x}{z}\right) dx \, dy \, dz$

In Exercises 13 to 17 use triple integration to find the volume of the given solid.

13. The solid region in the first octant bounded above by the plane $x + y + z = 12$ and on the sides by the cylinder $x^2 + y^2 = 9$.

14. The solid region above the plane $z = 2$ and below the hemisphere $z = \sqrt{18 - x^2 - y^2}$.

15. The solid region in the first octant above the plane $z = 1$ and below the plane $2x + 3y + 4z = 12$.

16. The solid region below the paraboloid $z = 9 - (x^2 + y^2)$ and above the x-y plane.

17. The solid region bounded on the sides by the cylinders $y^2 = x$ and $x^2 = y$, below by the plane $x + y + z = 2$, and above by the plane $z = 2$.

In Exercises 18 to 21 the iterated integral represents the volume of a region S. Sketch S and give the limits of integration on the indicated integrals.

18. $V(S) = \int_0^1 \int_0^{3-3x} \int_2^{(24-12x-4y)/6} dz \, dy \, dx$

(*a*) $\iiint_S dV$, where $dV = dy \, dz \, dx$

(*b*) $\iiint_S dV$, where $dV = dx \, dy \, dz$

19. $V(S) = \int_0^3 \int_0^{\sqrt{9-x^2}} \int_0^{2y} dz \, dy \, dx$

(*a*) $\iiint_S dV$, where $dV = dx \, dz \, dy$

(*b*) $\iiint_S dV$, where $dV = dy \, dz \, dx$

20. $V(S) = \int_0^2 \int_0^{2-x} \int_0^{2-x-y} dz \, dy \, dx$

(*a*) $\iiint_S dV$, where $dV = dy \, dz \, dx$

(*b*) $\iiint_S dV$, where $dV = dx \, dy \, dz$

21. $V(S) = \int_1^2 \int_0^{1/x} \int_0^{x^2} dy \, dx \, dz$

(*a*) $\iiint_S dV$, where $dV = dx \, dy \, dz$

(*b*) Explain why $\iiint_S dV$ cannot be expressed as a single iterated integral if $dV = dz \, dy \, dx$.

22. Find the moment of inertia about the z axis of the homogeneous solid inside the cylinder $x^2 + y^2 = 9$, above the plane $z = 0$, and below the plane $z = 3y$.

23. Find the moment of inertia about the y axis of the solid in the first octant inside the cylinder $x^2 + y^2 = 1$ and below the paraboloid $z = x^2 + y^2$, where the density is proportional to the distance from the z axis.

24. Determine the mass of the solid region above the x-y plane and below the paraboloid $z = 8 - x^2 - y^2$ if the density at any point is proportional to the distance from the base.

25. Determine the mass of the homogeneous solid bounded above by $z = 2$, below by $z = 1$, and on the sides by the paraboloid $z = x^2 + y^2$.

26. Determine the mass of the solid region bounded on the sides by the cylinder $y = x^2$ and the plane $y = 2x$, on the bottom by the x-y plane, and on the top by the plane $z = x + y$. The density is proportional to the distance from the x-z plane.

27. Determine (*a*) the mass and (*b*) the center of mass $(\bar{x}, \bar{y}, \bar{z})$ of the solid S in the first octant bounded by the plane $x = 2$ and the cylinder $z^2 + y^2 = 9$, where the density at a point is proportional to its distance from the x axis. (You will need to find three moments.)

Section 20.4: MULTIPLE INTEGRALS IN OTHER COORDINATE SYSTEMS

Objectives:

1. Evaluate double integrals in polar coordinates.

2. Evaluate triple integrals in cylindrical and spherical coordinates.

3. Transform a multiple integral in one coordinate system to an equivalent integral in another coordinate system.

THE DOUBLE INTEGRAL IN POLAR COORDINATES

(You may find it helpful to review polar coordinates at this point.) Our objective is to define the *polar double integral*

(1) $$\iint_R f(r, \theta)\, dA$$

where R is a polar region of type I or type II as depicted in Fig. 20.38. As in the rectangular case, we shall assume that the bounding curves have continuous derivatives. We begin by imposing a *polar grid* over R consisting of concentric circles centered at the pole and half-lines emanating from the pole (see Fig. 20.39). This in turn determines an inner partition P of R consisting of a finite number (say n) of "polar rectangles" $R_1, R_2, \ldots, R_n$ that lie wholly within R. Again we let $\|P\|$ denote the length of the longest diagonal of the R_i's. A *Riemann sum* of f over P is any sum of the form

(2) $$\sum_{i=1}^{n} f(r_i, \theta_i)\, \Delta A_i$$

where (r_i, θ_i) is an arbitrary point in R_i and ΔA_i denotes the area of R_i. Now if

(3) $$\lim_{\|P\| \to 0} \sum_{i=1}^{n} f(r_i, \theta_i)\, \Delta A_i$$

exists independently of how the (r_i, θ_i) are chosen in R_i, then we say f is *integrable* over R and we denote the value of the limit by $\iint_R f(r, \theta)\, dA$. As before, we assume the following:

20.4.1 Theorem If $f(r, \theta)$ is continuous on the polar region R, then f is integrable over R.

If $f(r, \theta) = 1$ on R, then the sum (2) becomes $\Sigma_{i=1}^{n} \Delta A_i$, the sum of the areas of the polar rectangles in the inner partition P of R. Thus, in the limit, we obtain the area of R.

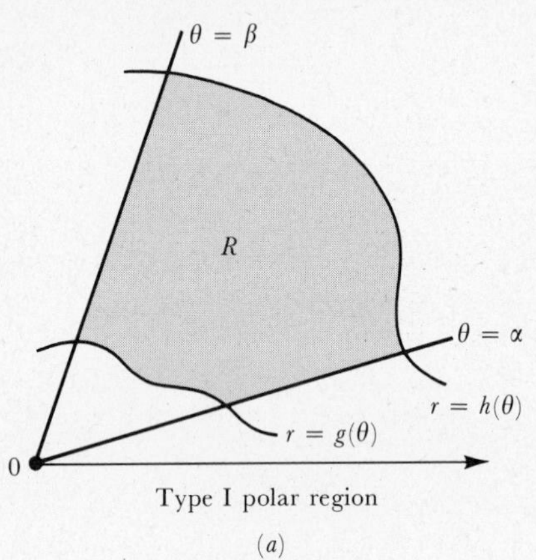

Type I polar region

(a)

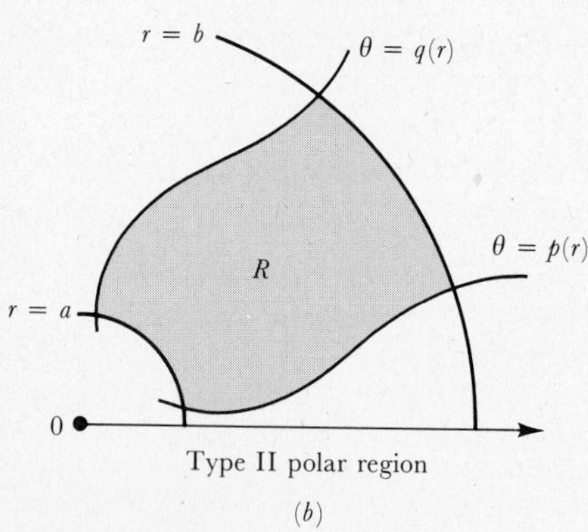

Type II polar region

(b)

Fig. 20.38

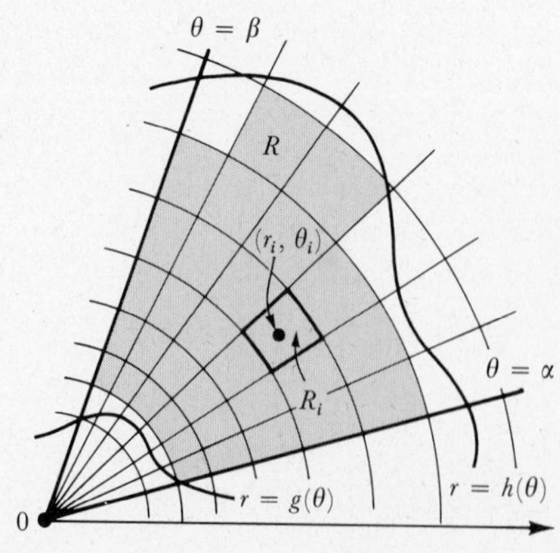

Inner partition P of R

Fig. 20.39

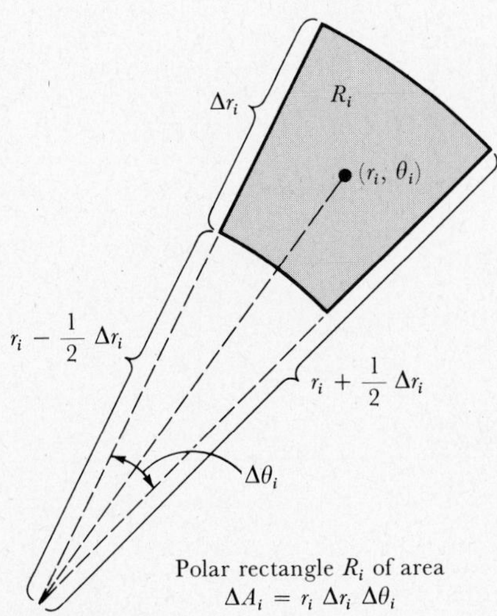

Polar rectangle R_i of area $\Delta A_i = r_i\, \Delta r_i\, \Delta\theta_i$

Fig. 20.40

20.4.2
$$A(R) = \iint_R dA$$

To gain some insight into how we can express the polar integral in (1) as an *iterated* integral, we examine ΔA_i. Suppose (r_i, θ_i) is chosen as the center of R_i (see Fig. 20.40). Now ΔA_i is the difference of the areas of two circular sectors of "angle width" $\Delta\theta_i$. The larger sector has radius $r_i + \Delta r_i/2$, and the smaller has radius $r_i - \Delta r_i/2$. Thus, applying the formula for the area of a circular sector,

$$\begin{aligned}\Delta A_i &= \frac{1}{2}\left(r_i + \frac{1}{2}\Delta r_i\right)^2 \Delta\theta_i - \frac{1}{2}\left(r_i - \frac{1}{2}\Delta r_i\right)^2 \Delta\theta_i \\ &= \frac{1}{2}\left[r_i^2 + r_i\,\Delta r_i + \frac{1}{4}(\Delta r_i)^2\right]\Delta\theta_i - \frac{1}{2}\left[r_i^2 - r_i\,\Delta r_i + \frac{1}{4}(\Delta r_i)^2\right]\Delta\theta_i \\ &= r_i\,\Delta r_i\,\Delta\theta_i\end{aligned}$$

The factor r_i in the expression for ΔA_i leads to a factor of r in the associated iterated integrals—one iteration order for each of the two ways of describing R.

20.4.3 **Theorem** Suppose $f(r, \theta)$ is continuous on the region R.
(*a*) If $R = \{(r, \theta) \mid \alpha \le \theta \le \beta, g(\theta) \le r \le h(\theta)\}$, then

$$\iint_R f(r, \theta)\, dA = \int_\alpha^\beta \int_{g(\theta)}^{h(\theta)} f(r, \theta)\, r\, dr\, d\theta \qquad \text{[type I]}$$

(*b*) If $R = \{(r, \theta) \mid a \le r \le b, p(r) \le \theta \le q(r)\}$, then

$$\iint_R f(r, \theta)\, dA = \int_a^b \int_{p(r)}^{q(r)} f(r, \theta)\, r\, d\theta\, dr \qquad \text{[type II]}$$

Example 1 Determine the area of the region R in the first quadrant that is inside the cardioid $r = 2(1 + \cos\theta)$ and outside the circle $r = 2$.

Solution R is sketched in Fig. 20.41. It can be described as a type I region:

$$R = \{(r, \theta) \mid 0 \le \theta \le \pi/2, 2 \le r \le 2 + 2\cos\theta\}$$

Hence, by (20.4.2) and [20.4.3(*a*)],

$$\begin{aligned}A(R) &= \iint_R dA = \int_0^{\pi/2} \int_2^{2+2\cos\theta} r\, dr\, d\theta \\ &= \frac{1}{2}\int_0^{\pi/2} r^2 \Big|_{r=2}^{r=2+2\cos\theta} d\theta \\ &= \frac{1}{2}\int_0^{\pi/2} (8\cos\theta + 4\cos^2\theta)\, d\theta = 4 + \pi/2 \quad \bullet\end{aligned}$$

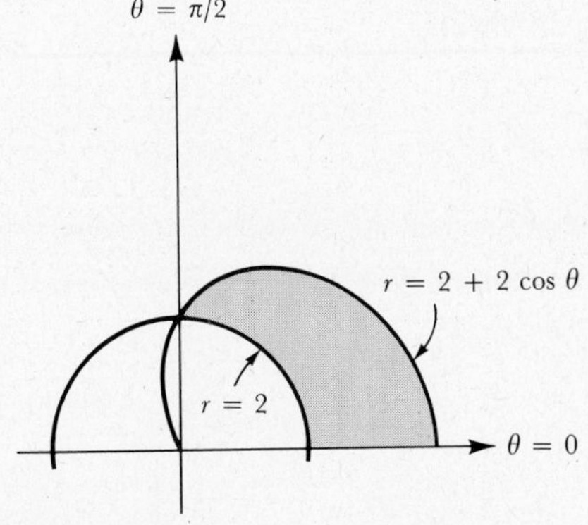

Fig. 20.41

The next example illustrates that certain multiple integral problems posed in terms of rectangular coordinates are simpler to solve using polar coordinates.

Example 2 A lamina L occupies the annular region R in the first quadrant between the semicircles $y = \sqrt{1 - x^2}$ and $y = \sqrt{16 - x^2}$. Find the moment of inertia of L about the x axis if the density at any point of L is inversely proportional to its distance from the origin; see Fig. 20.42(*a*).

Solution In polar coordinates the two semicircles have equations $r = 1$ and $r = 4$, respectively. Thus a description for R in polar coordinates is provided by

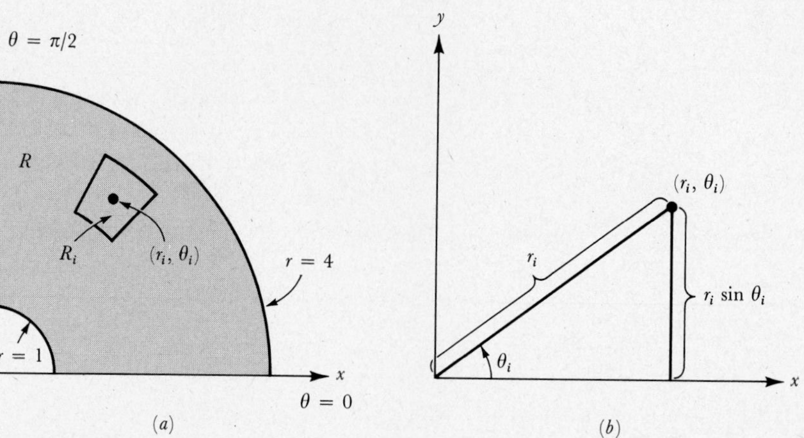

Fig. 20.42

$$R = \{(r, \theta) \,|\, 0 \leq \theta \leq \pi/2, 1 \leq r \leq 4\}$$

(which can be regarded as either type I or type II.)

If R_i is a typical polar rectangle in an inner partition of R, then the distance from (r_i, θ_i) in R to the x axis is $r_i \sin \theta_i$; see Fig. 20.42(*b*). Since r_i measures the distance from the origin, the mass of the piece of L occupying R_i is approximated by

$$\frac{k}{r_i} \Delta A_i$$

Hence the moment of inertia of this piece about the x axis is approximated by

$$\begin{aligned}(I_x)_i &\approx (r_i \sin \theta_i)^2 \left(\frac{k}{r_i} \Delta A_i\right) \\ &= k r_i \sin^2 \theta_i \, \Delta A_i\end{aligned}$$

Thus

$$I_x \approx \sum_{i=1}^{n} k r_i \sin^2 \theta_i \, \Delta A_i$$

Taking the limit as $\|P\| \to 0$, we have

$$\begin{aligned}I_x &= k \iint_R r^2 \sin^2 \theta \, dA \\ &= k \int_0^{\pi/2} \int_1^4 r^2 \sin^2 \theta \, r \, dr \, d\theta \\ &= k \int_0^{\pi/2} \int_1^4 r^3 \sin^2 \theta \, dr \, d\theta \\ &= \frac{k}{4} \int_0^{\pi/2} r^4 \sin^2 \theta \, \Big|_{r=1}^{r=4} d\theta \\ &= \frac{255k}{4} \int_0^{\pi/2} \sin^2 \theta \, d\theta = \frac{255k\pi}{16} \quad \bullet\end{aligned}$$

If we had tried to solve Example 2 in rectangular coordinates, then I_x would have the form

$$I_x = \iint_R \frac{ky^2}{\sqrt{x^2 + y^2}} \, dA$$

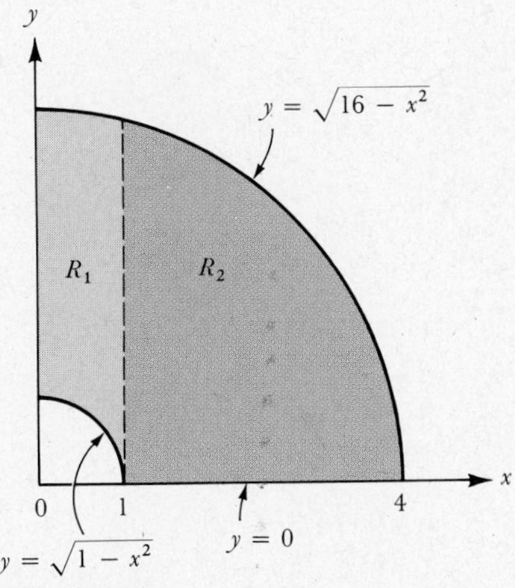

Fig. 20.43

To express this as an iterated integral, we must first subdivide R into the regions R_1 and R_2 as in Fig. 20.43. Thus

$$I_x = \int_0^1 \int_{\sqrt{1-x^2}}^{\sqrt{16-x^2}} \frac{ky^2}{\sqrt{x^2+y^2}}\, dy\, dx + \int_1^4 \int_0^{\sqrt{16-x^2}} \frac{ky^2}{\sqrt{x^2+y^2}}\, dy\, dx$$

Clearly, in this case the polar coordinate approach is to be preferred!

More generally, it is often possible to transform an iterated integral in rectangular coordinates to an equivalent iterated integral in polar coordinates. For example, if the region R has the description

$$R = \{(x, y) \mid a \le x \le b, h(x) \le y \le g(x)\}$$

in rectangular coordinates, and if the *same* region has description

$$R = \{(r, \theta) \mid \alpha \le \theta \le \beta, p(\theta) \le r \le q(\theta)\}$$

in polar coordinates, then

$$\int_a^b \int_{h(x)}^{g(x)} f(x, y)\, dy\, dx = \int_\alpha^\beta \int_{p(\theta)}^{q(\theta)} f(r\cos\theta, r\sin\theta) r\, dr\, d\theta$$

There are several other possibilities, but the procedure in each case is to describe the region R in polar coordinates, replace $f(x, y)$ by $f(r\cos\theta, r\sin\theta)$, replace $dy\, dx$ (or $dx\, dy$) by $r\, dr\, d\theta$ (or $r\, d\theta\, dr$) and form the new iterated integral(s) using the polar description of R. Of course, the order in which the differentials appear will depend on the description of R.

Example 3 Evaluate the iterated integral

$$I = \int_0^4 \int_0^{\sqrt{16-y^2}} (x^2 + y^2)\, dx\, dy$$

by first transforming to an equivalent integral in polar coordinates.

Solution From the iterated integral we read off the region over which the integral is evaluated:

$$R = \{(x, y) \mid 0 \le y \le 4, 0 \le x \le \sqrt{16 - y^2}\}$$

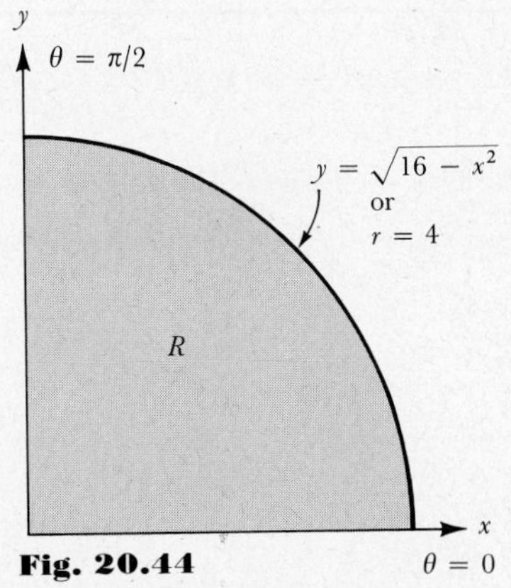

Since R is just the portion of the disk $x^2 + y^2 \le 16$ lying in the first quadrant (see Fig. 20.44), it has polar description

$$R = \{(r, \theta) \mid 0 \le \theta \le \pi/2, 0 \le r \le 4\}$$

Now

$$\begin{aligned} f(r\cos\theta, r\sin\theta) &= (r\cos\theta)^2 + (r\sin\theta)^2 \\ &= r^2(\cos^2\theta + \sin^2\theta) = r^2 \end{aligned}$$

and thus

$$\begin{aligned} I &= \int_0^{\pi/2} \int_0^4 r^2\, r\, dr\, d\theta \\ &= \frac{1}{4} \int_0^{\pi/2} r^4 \Big|_0^4\, d\theta = 64 \int_0^{\pi/2} d\theta = 32\pi \end{aligned}$$ ●

Fig. 20.44

Example 4 Determine the volume of the solid S bounded below by the x-y plane, on the sides by the cylinder $x^2 + y^2 = 1$, and on top by the sphere $x^2 + y^2 + z^2 = 9$ (see Fig. 20.45).

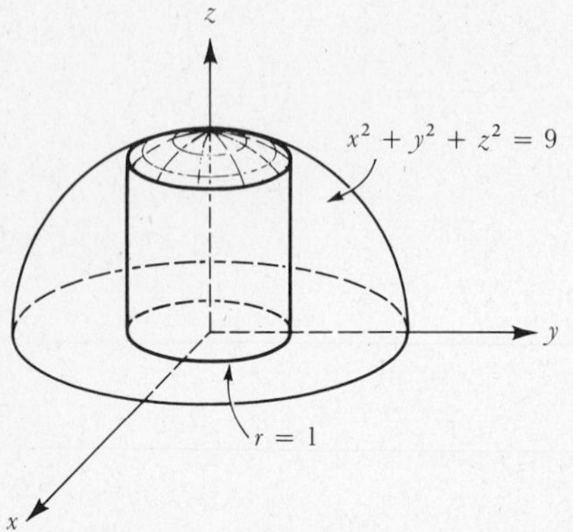

Fig. 20.45

Solution By (20.1.5),

$$V(S) = \iint_R \sqrt{9 - x^2 - y^2}\, dA$$

Now in polar coordinates, R is just the unit disk

$$R = \{(r, \theta) | 0 \le r \le 1, 0 \le \theta \le 2\pi\}$$

so

$$V(S) = \iint_R \sqrt{9 - (x^2 + y^2)}\, dA$$

$$= \int_0^1 \int_0^{2\pi} \sqrt{9 - r^2}\, r\, d\theta\, dr$$

$$= \int_0^1 2\pi \sqrt{9 - r^2}\, r\, dr = 2\pi\left(9 - \frac{16\sqrt{2}}{3}\right) \quad \bullet$$

PROGRESS TEST 1

1. Determine the area of the region R enclosed by one leaf of the four-leafed rose $r = \cos 2\theta$.
2. Determine the volume of the solid region bounded below by the upper nappe of the cone $z = \sqrt{x^2 + y^2}$ and above by the plane $z = 1$.
3. Evaluate $\iint_R e^{x^2+y^2}\, dA$, where
$$R = \{(x, y) | 0 \le x \le 1,\ 0 \le y \le \sqrt{1 - x^2}\}$$

CYLINDRICAL COORDINATES

We have seen that certain plane curves are more naturally expressed in terms of polar coordinates rather than rectangular coordinates. In three dimensions there are two commonly used generalizations of polar coordinates. The first of these, called *cylindrical coordinates,* are particularly useful in physical or geometric situations involving symmetry about the z axis.

20.4.4 **Definition** A point P in space is said to have *cylindrical coordinates* (r, θ, z) if (r, θ) are polar coordinates for Q, the projection of P in the x-y plane, and z is the directed distance from the x-y plane to P. (See Fig 20.46.)

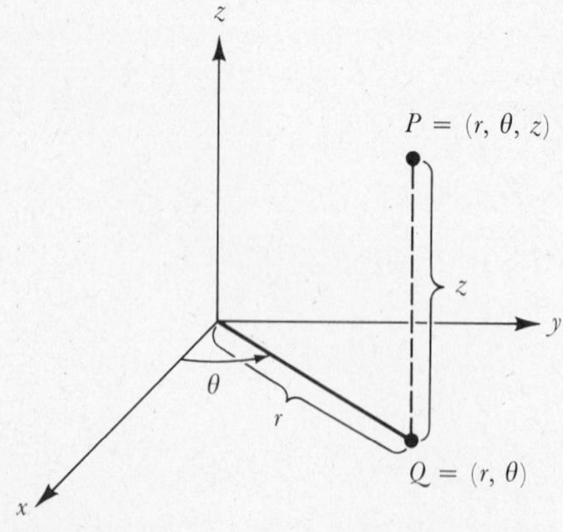

Fig. 20.46

In effect the cylindrical coordinate system is obtained by placing the z axis perpendicularly through the polar plane at the pole. The Cartesian and cylindrical coordinates are related by the equations

20.4.5 $$x = r\cos\theta, \quad y = r\sin\theta, \quad z = z$$

and

20.4.6 $$x^2 + y^2 = r^2, \quad \tan\theta = y/x, \quad z = z$$

The graphs of the basic equations $r = k$, $\theta = k$, and $z = k$ are shown in Fig. 20.47. Using (20.4.5) and (20.4.6) we can transform the equations for more general surfaces from one coordinate system to another system.

Example 5 (*a*) Determine an equation in cylindrical coordinates for the surface with Cartesian equation $x^2 = y^2 + 2z^2$.
(*b*) Determine an equation in Cartesian coordinates for the surface with cylindrical equation $r = 4\cos\theta$.

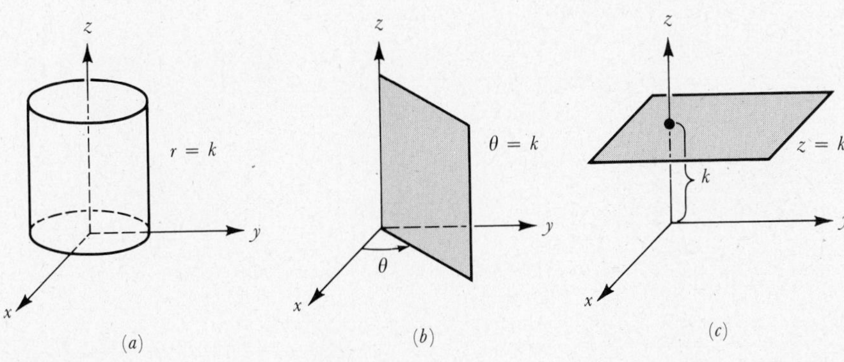

Fig. 20.47

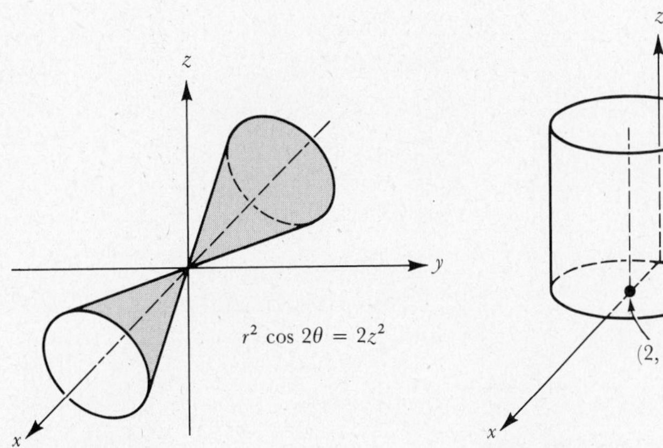

Fig. 20.48

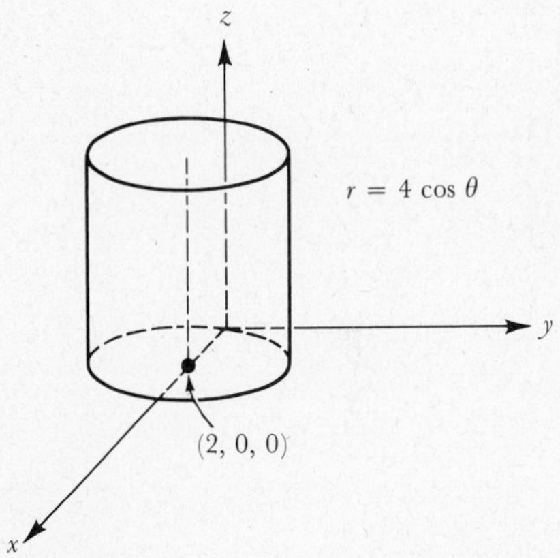

Fig. 20.49

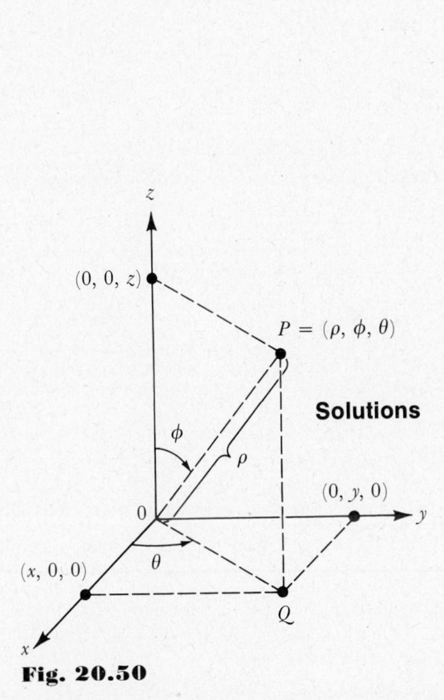

Fig. 20.50

Solutions

(*a*) From Chap. 17 we know that $x^2 = y^2 + 2z^2$ has as its graph an elliptic cone with its axis the x axis. By (20.4.5), $x^2 - y^2 = r^2\cos^2\theta - r^2\sin^2\theta = r^2(\cos^2\theta - \sin^2\theta) = r^2\cos 2\theta$. Thus the cylindrical equation for this cone is $r^2\cos 2\theta = 2z^2$. (See Fig. 20.48.)

(*b*) Multiplying $r = 4\cos\theta$ by r, we have $r^2 = 4r\cos\theta$. But $r^2 = x^2 + y^2$ and $r\cos\theta = x$, so the Cartesian equation is $x^2 + y^2 = 4x$. Upon completing the square in x we get $(x-2)^2 + y^2 = 4$. Thus the graph is a right circular cylinder of radius 2 whose axis is the vertical line through (2, 0, 0). (See Fig. 20.49.) ●

PROGRESS TEST 2

1. Find cylindrical coordinates for the points with Cartesian coordinates
(*a*) $(-1, \sqrt{3}, 4)$ (*b*) $(1, 1, \sqrt{2})$

2. Find cylindrical equations for the surfaces with Cartesian equations
(*a*) $x^2 - y^2 + z^2 = 1$ (*b*) $x^2 + y^2 = 9$

SPHERICAL COORDINATES

The second generalization of polar coordinates to three dimensions is particularly useful when working with problems involving a center of symmetry.

20.4.7 **Definition** A point P in space is said to have *spherical coordinates* (ρ, ϕ, θ) (see Fig. 20.50) if (*a*) $\rho = \|\overrightarrow{OP}\|$, (*b*) ϕ is the angle between the vertical coordinate vector $\mathbf{k}$ and $\overrightarrow{OP}$, and (*c*) θ is any polar angle for the projection of P in the x-y plane.

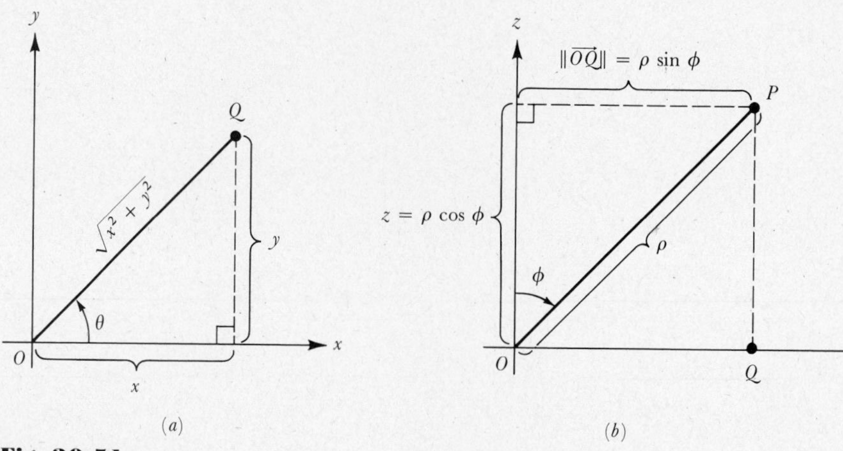

Fig. 20.51

Note that, by definition, ρ is always nonnegative and ϕ is always between 0 and π.

If Q is the projection of P into the x-y plane, then, referring to Fig. 20.51(a), we have

(4) $$x = \|\overrightarrow{OQ}\| \cos \theta, \quad y = \|\overrightarrow{OQ}\| \sin \theta$$

Figure 20.51(b) depicts the plane containing the z-axis and $\overrightarrow{OP}$, from which we observe that

(5) $$\|\overrightarrow{OQ}\| = \rho \sin \phi, \quad z = \rho \cos \phi$$

Combining (4) and (5) we obtain equations for transforming points from spherical to Cartesian coordinates.

20.4.8 $$x = \rho \sin \phi \cos \theta, y = \rho \sin \phi \sin \theta, z = \rho \cos \phi$$

The equations in the other direction are given by

20.4.9 $$\rho = \sqrt{x^2 + y^2 + z^2}, \quad \tan \theta = y/x, \quad \phi = \arccos\left(\frac{z}{\sqrt{x^2 + y^2 + z^2}}\right)$$

As with cylindrical coordinates we can gain some insight into this new system by examining the graphs of the basic equations $\rho = k$, $\phi = k$, and $\theta = k$.

The graph of $\rho = k > 0$ is a sphere of radius k centered at the origin, while $\rho = 0$ has as its graph the origin itself. The graph of $\theta = k$ is the same as in cylindrical coordinates (a vertical plane). The graphs of $\phi = 0$ and $\phi = \pi$ are the positive and negative z axes, respectively, and the graph of $\phi = \pi/2$ is the x-y plane. Otherwise, $\phi = k$ has as its graph the upper or lower nappe of a cone depending on whether $0 < k < \pi/2$ or $\pi/2 < k < \pi$, respectively. These various graphs are shown in Fig. 20.52.

Example 6 Find the Cartesian coordinates for the point with spherical coordinates $(6, \pi/6, 3\pi/4)$.

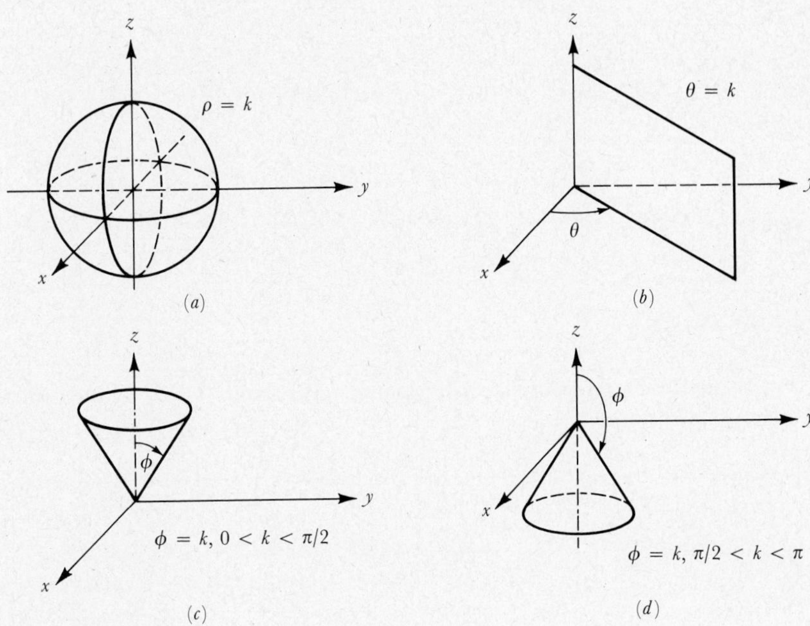

Fig. 20.52

Solution From (20.4.8),

$$x = \rho \sin \phi \cos \theta = 6 \sin(\pi/6)\cos(3\pi/4) = -3/\sqrt{2}$$
$$y = \rho \sin \phi \sin \theta = 6 \sin(\pi/6)\sin(3\pi/4) = 3/\sqrt{2}$$
$$z = \rho \cos \phi = 6 \cos(\pi/6) = 3\sqrt{3}$$

so the point has Cartesian coordinates $(-3/\sqrt{2}, 3/\sqrt{2}, 3\sqrt{3})$. ●

Example 7 Find spherical coordinates for the point P with Cartesian coordinates $(1, -1, \sqrt{2})$.

Solution By (20.4.9),

$$\rho = \sqrt{x^2 + y^2 + z^2} = \sqrt{1 + 1 + 2} = 2$$
$$\tan \theta = y/x = -1$$
$$\phi = \arccos\left(\frac{z}{\sqrt{x^2 + y^2 + z^2}}\right) = \arccos(\sqrt{2}/2) = \pi/4$$

In the x-y plane the projection of P, $(1, -1)$, is in the fourth quadrant, so we can choose $\theta = 5\pi/4$. Thus a set of spherical coordinates is $(2, \pi/4, 5\pi/4)$. ●

Although all texts define spherical coordinates in the same way, some books list the coordinates in the order (ρ, θ, ϕ), rather than (ρ, ϕ, θ) as we have done. This will not affect the graph of a spherical equation, but it *is* important to know which order is being used when you attempt to plot a point with a given set of coordinates.

As before we can use (20.4.8) and (20.4.9) to transform equations for a surface. For example, a sphere of radius 4 centered at $(0, 0, 5)$ has the Cartesian equation $x^2 + y^2 + (z - 5)^2 = 16$ or, equivalently, $x^2 + y^2 + z^2 - 10z = -9$. But $x^2 + y^2 + z^2 = \rho^2$ and $z = \rho \cos \phi$, so the corresponding equation in spherical coordinates is $\rho^2 = 10\rho \cos \phi - 9$.

Conversely, given an equation in spherical coordinates, we can obtain the corresponding Cartesian equation. For example, the equation $\rho \sin \phi \tan \phi \sin 2\theta = 1$ can be written as

$$(\rho \sin \phi \sin \theta)(\rho \sin \phi \cos \theta)2 = \rho \cos \phi$$

by multiplying both sides by $\rho \cos \phi$ and using the identities $\tan \phi = (\sin \phi)/(\cos \phi)$ and $\sin 2\theta = 2 \sin \theta \cos \theta$. Now, by (20.4.9), the corresponding Cartesian equation is $2xy = z$.

PROGRESS TEST 3

1. Find spherical coordinates for the points with Cartesian coordinates
(a) $(-1, \sqrt{3}, 4)$ *(b)* $(1, 1, \sqrt{2})$

2. Find spherical equations for the surfaces with Cartesian equations
(a) $x^2 - y^2 + z^2 = 1$ *(b)* $x^2 + y^2 = 9$

THE TRIPLE INTEGRAL IN CYLINDRICAL AND SPHERICAL COORDINATES

In Cartesian coordinates it is natural to use inner partitions made up of parallelepipeds of the form

$$S_i = \{(x, y, z) \mid x_{i-1} \leq x \leq x_i, y_{i-1} \leq y \leq y_i, z_{i-1} \leq z \leq z_i\}$$

with volume $\Delta V_i = \Delta x_i \, \Delta y_i \, \Delta z_i$. The corresponding integral $\iiint_S f(x, y, z) \, dV$, when expressed as an iterated integral, leads to the differential of volume

$$dV = dx \, dy \, dz$$

When working in cylindrical or spherical coordinates, the natural inner partitions consist, respectively, of "cylindrical wedges"

$$S_i = \{(r, \theta, z) \mid r_{i-1} \leq r \leq r_i, \theta_{i-1} \leq \theta \leq \theta_i, z_{i-1} \leq z \leq z_i\}$$

and "spherical wedges"

$$S_i = \{(\rho, \phi, \theta) \mid \rho_{i-1} \leq \rho \leq \rho_i, \phi_{i-1} \leq \phi \leq \phi_i, \theta_{i-1} \leq \theta \leq \theta_i\}$$

The basic Cartesian, cylindrical, and spherical subregions are sketched in Fig. 20.53.

The corresponding Riemann sums in these latter coordinate systems have the forms

$$\sum_{i=1}^{n} h(\bar{r}_i, \bar{\theta}_i, \bar{z}_i) \, \Delta V_i \qquad \text{or} \qquad \sum_{i=1}^{n} g(\bar{\rho}_i, \bar{\phi}_i, \bar{\theta}_i) \, \Delta V_i$$

Δz_i

Δx_i

Δy_i

Parallelepiped

(a)

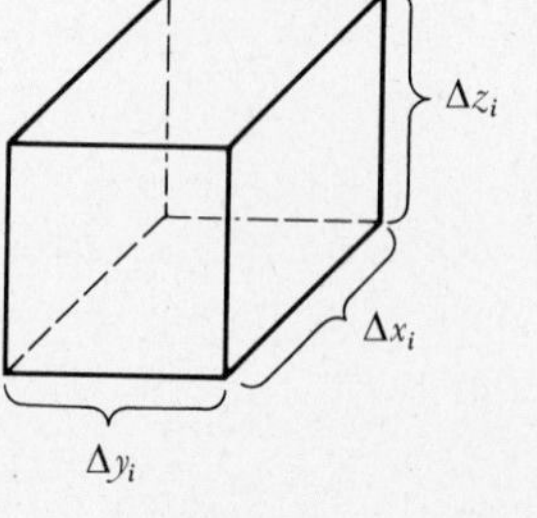

Cylindrical wedge

(b)

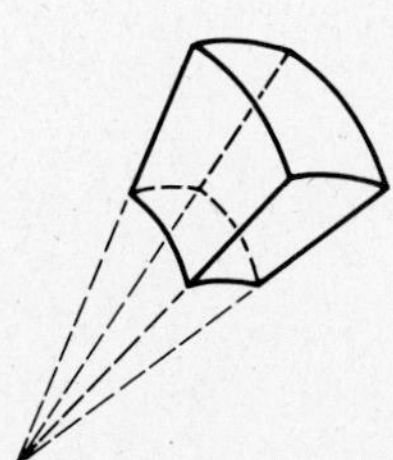

Spherical wedge

(c)

Fig. 20.53

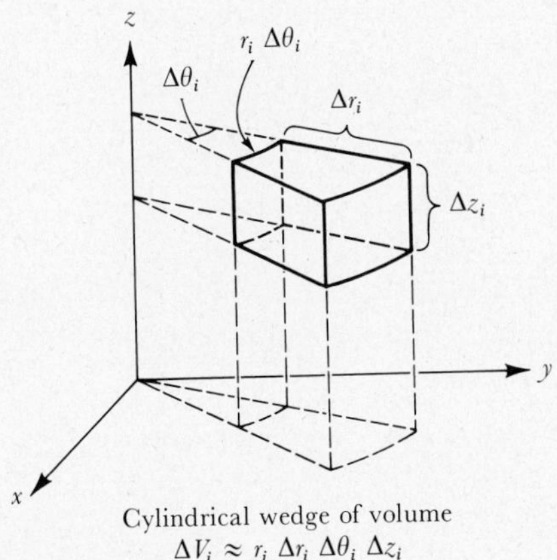

Cylindrical wedge of volume
$\Delta V_i \approx r_i\,\Delta r_i\,\Delta\theta_i\,\Delta z_i$

Fig. 20.54

where $(\bar{r}_i, \bar{\theta}_i, \bar{z}_i)$ and $(\bar{\rho}_i, \bar{\phi}_i, \bar{\theta}_i)$ are arbitrary points in the respective wedges S_i, and ΔV_i represents their respective volumes. If we now refine the partitions in such a way that $\|P\| \to 0$ ($\|P\|$ = length of largest of the diagonals of the S_i's), and if these sums tend to some number independently of how the points are chosen in each S_i, then they give rise to triple integrals in the respective coordinate systems. That is, we define:

20.4.10 $$\iiint_S h(r, \theta, z)\,dV = \lim_{\|P\| \to 0} \sum_{i=1}^{n} h(\bar{r}_i, \bar{\theta}_i, \bar{z}_i)\,\Delta V_i$$

and

20.4.11 $$\iiint_S g(\rho, \phi, \theta)\,dV = \lim_{\|P\| \to 0} \sum_{i=1}^{n} g(\bar{\rho}_i, \bar{\phi}_i, \bar{\theta}_i)\,\Delta V_i$$

It can be shown that if h and g are continuous, and if the functions determining the bounded regions S are continuously differentiable, then these limits will exist. That is, continuous functions are integrable over S.

To determine how to express these integrals as iterated integrals, we must first determine ΔV_i in each case. Now from Fig. 20.54 we see that a typical cylindrical wedge has volume approximately equal to the volume of the parallelepiped with edges of lengths $r_i\,\Delta\theta_i$, Δr_i, and Δz_i. Thus in cylindrical coordinates

$$\Delta V_i \approx r_i\,\Delta r_i\,\Delta\theta_i\,\Delta z_i$$

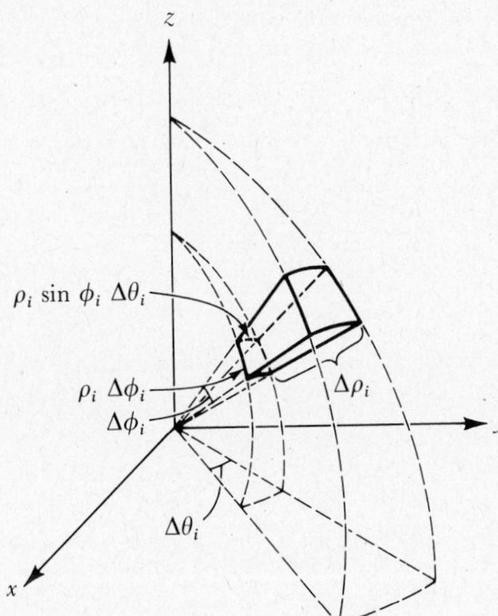

Fig. 20.55

Similarly, the volume of a spherical wedge is approximated by the volume of a parallelepiped with dimensions $\rho_i \sin\theta_i\,\Delta\theta_i$, $\rho_i\,\Delta\phi_i$, and $\Delta\rho_i$, so in spherical coordinates (see Fig. 20.55)

$$\Delta V_i \approx \rho_i^2 \sin\phi_i\,\Delta\rho_i\,\Delta\phi_i\,\Delta\theta_i$$

It can be shown that the corresponding iterated integrals will involve differentials of volume of the form

20.4.12 $$dV = r\,dr\,d\theta\,dz \qquad \text{(cylindrical)}$$

and

20.4.13 $$dV = \rho^2 \sin\phi\,d\rho\,d\phi\,d\theta \qquad \text{(spherical)}$$

Of course, the order in which the differentials of the basic variables appear will depend on how the integral is expressed as an iterated integral and how the solid is described. For example, if S is described in cylindrical coordinates by

$$S = \{(r, \theta, z) \mid \alpha \le \theta \le \beta, f(\theta) \le r \le g(\theta), F(r, \theta) \le z \le G(r, \theta)\}$$

as in Fig. 20.56, then

$$\iiint_S h(r, \theta, z)\,dV = \int_{\alpha}^{\beta} \int_{f(\theta)}^{g(\theta)} \int_{F(r,\theta)}^{G(r,\theta)} k(r, \theta, z)\,r\,dz\,dr\,d\theta$$

$F(r, \theta) \le z \le G(r, \theta)$
$f(\theta) \le r \le g(\theta)$
$\alpha \le \theta \le \beta$

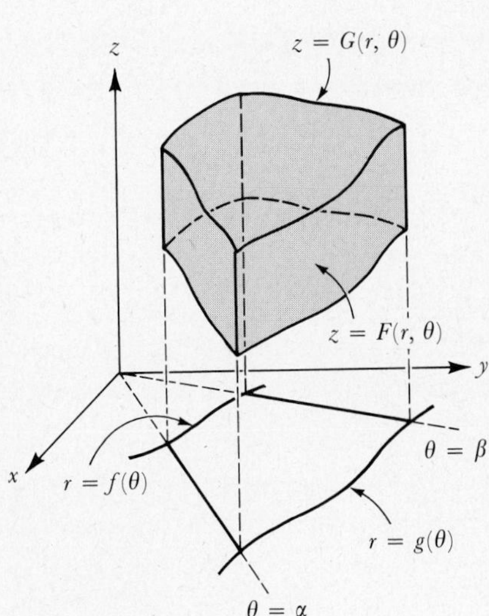

Fig. 20.56

As with Cartesian coordinates, the order in which the differentials appear allows you to read off which variable is being limited by the respective limits of integration.

Example 8 Given that

$$\iiint_S f(r, \theta, z)\, dV = \int_0^1 \int_0^{\pi/2} \int_0^{\sqrt{4-r^2}} f(r, \theta, z) r\, dz\, d\theta\, dr$$

describe and sketch S.

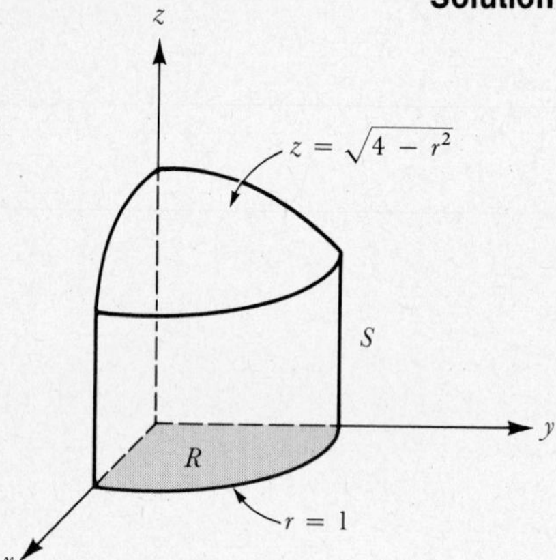

Fig. 20.57

Solution From the order in which the differentials appear, we know that the outside limits of integration are limits on r, the middle limits are on θ, and the inside limits are on z. Thus

$$S = \{(r, \theta, z) | 0 \le r \le 1, 0 \le \theta \le \pi/2, 0 \le z \le \sqrt{4 - r^2}\}$$

Now

$$R = \{(r, \theta) | 0 \le r \le 1, 0 \le \theta \le \pi/2\}$$

describes a quarter circle of radius 1 centered at the origin. Thus

$$\{(r, \theta, z) | (r, \theta) \text{ in } R\}$$

is one-quarter of a vertical right circular cylinder of infinite extent. Since $z \ge 0$, S is bounded below by the x-y plane. Now $z = \sqrt{4 - r^2}$ implies $r^2 + z^2 = 4$. But since $x^2 + y^2 = r^2$, the graph of $r^2 + z^2 = 4$ is a sphere of radius 2 centered at the origin. Since $z \le \sqrt{4 - r^2}$, S is bounded above by the upper half of this sphere. A sketch of S is given in Fig. 20.57. ●

Example 9 Find the volume of the region S bounded below by the cone $\phi = \pi/4$ and bounded above by the sphere $\rho = 2$.

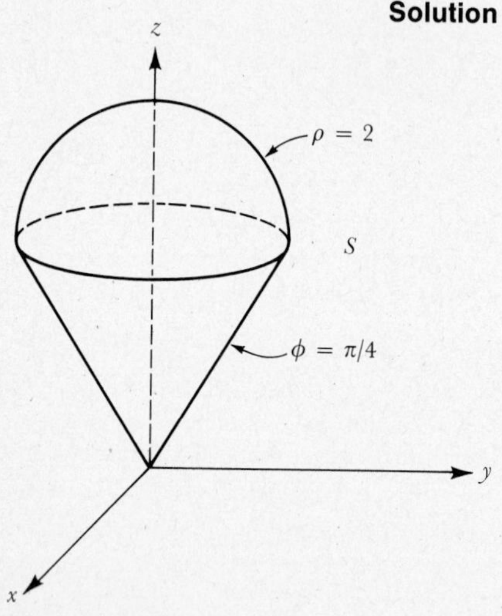

Fig. 20.58

Solution As before, $V(S) = \iiint_S dV$. The region S is sketched in Fig. 20.58. We shall use the symmetry of its ice cream cone shape and multiply by 4 the volume in the first octant. In this octant $0 \le \theta \le \pi/2$. Denote this portion of the solid by $S_{1/4}$. Since

$$S_{1/4} = \{(\rho, \phi, \theta) | 0 \le \theta \le \pi/2,\ 0 \le \phi \le \pi/4,\ 0 \le \rho \le 2\}$$

we have

$$\begin{aligned} V(S) = 4V(S_{1/4}) &= 4 \int_0^{\pi/2} \int_0^{\pi/4} \int_0^{2} \rho^2 \sin \phi\, d\rho\, d\phi\, d\theta \\ &= \frac{32}{3} \int_0^{\pi/2} \int_0^{\pi/4} \sin \phi\, d\phi\, d\theta \\ &= \frac{32}{3} \int_0^{\pi/2} \Big[-\cos \phi \Big]_0^{\pi/4} d\theta \\ &= \frac{32}{3} \left(1 - \frac{1}{\sqrt{2}}\right) \int_0^{\pi/2} d\theta = \frac{16\pi}{3} \left(1 - \frac{1}{\sqrt{2}}\right) \end{aligned}$$ ●

Example 10 Set up, but do not evaluate, an iterated integral giving the moment of inertia about the z axis of the homogeneous region bounded below by the upper nappe of the cone $3z^2 = x^2 + y^2$ and above by the plane $z = 1$.

Solution Letting $x = 0$, we see that the cone meets the y-z plane in the intersecting straight lines $z = \pm(1/\sqrt{3})y$. It follows [see Fig. 20.59(a)] that the cone has equation

$$\phi = \arctan\left(\frac{\sqrt{3}}{1}\right) = \frac{\pi}{3}$$

in spherical coordinates.

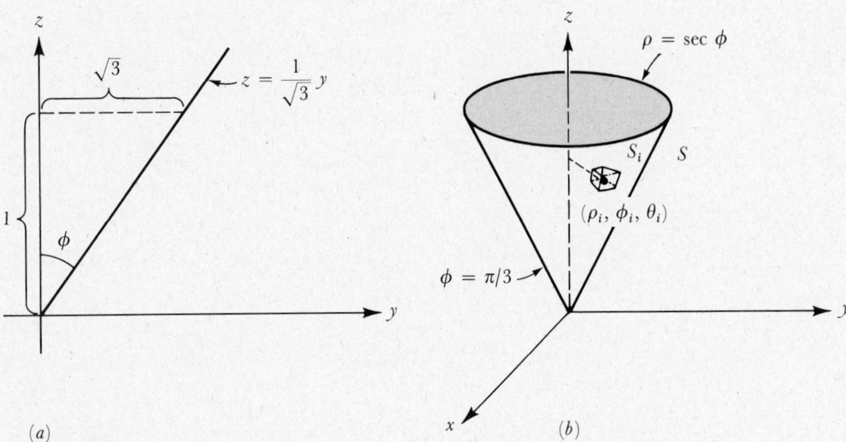

Fig. 20.59

In the equations relating Cartesian and spherical coordinates, $z = \rho \cos \phi$, so the plane $z = 1$ has equation $\rho \cos \phi = 1$ or, equivalently, $\rho = \sec \phi$. Thus a description of S in spherical coordinates is given by

$$S = \{(\rho, \phi, \theta) \mid 0 \leq \theta \leq 2\pi,\ 0 \leq \phi \leq \pi/3, 0 \leq \rho \leq \sec \phi\}$$

Suppose now that S has been broken up into small subregions consisting of spherical wedges, and let $(\rho_i, \phi_i, \theta_i)$ denote a point in a typical subregion S_i. The mass of S_i is given by

20.4.14 $$M_i = k\,\Delta V_i \qquad (k \text{ is the density})$$

Since the distance from $(\rho_i, \phi_i, \theta_i)$ to the z axis is $\rho_i \sin \phi_i$ [see Fig. 20.51(b)], the second moment of this piece about the z axis is

$$(I_z)_i = (\rho_i \sin \phi_i)^2 k\,\Delta V_i$$

Thus

$$I_z \approx \sum_{i=1}^{n} k\rho_i{}^2 \sin^2 \phi_i\,\Delta V_i$$

and hence

$$\begin{aligned} I_z &= \iiint_S k\rho^2 \sin^2 \phi\, dV \\ &= \int_0^{2\pi} \int_0^{\pi/3} \int_0^{\sec \phi} k\rho^2 \sin^2 \phi\, \rho^2 \sin \phi\, d\rho\, d\phi\, d\theta \\ &= k \int_0^{2\pi} \int_0^{\pi/3} \int_0^{\sec \phi} \rho^4 \sin^3 \phi\, d\rho\, d\phi\, d\theta \quad \bullet \end{aligned}$$

TRANSFORMING INTEGRALS FROM CARTESIAN TO CYLINDRICAL OR SPHERICAL COORDINATES

In many instances we can simplify the evaluation of an integral

$$\iiint_S f(x, y, z)\, dV$$

in Cartesian coordinates by transforming it to an equivalent integral in another coordinate system.

From our equations relating the coordinate systems, we have

20.4.15 $$\iiint_S f(x, y, z)\, dV = \iiint_{\hat{S}} f(r \cos \theta, r \sin \theta, z)\, dV$$

where $\widehat{S}$ represents the *same* region S described in *cylindrical coordinates.* In any iteration of this latter integral, dV will be of the form $dV = r\,dr\,d\theta\,dz$ with the order of the differentials depending on the description of $\widehat{S}$. Further, $\widehat{S}$ should be described in such a way that $r \geq 0$ and θ varies over an interval of length $\leq 2\pi$.

The corresponding transformation to spherical coordinates is given by

20.4.16 $$\iiint_S f(x, y, z)\, dV = \iiint_{\bar{S}} f(\rho \sin\phi \cos\theta, \rho \sin\phi \sin\theta, \rho \cos\phi)\, dV$$

where in this case $\bar{S}$ represents the region S described in *spherical coordinates* and dV is some permutation of $\rho^2 \sin\phi\, d\rho\, d\phi\, d\theta$.

Example 11 Express the integral

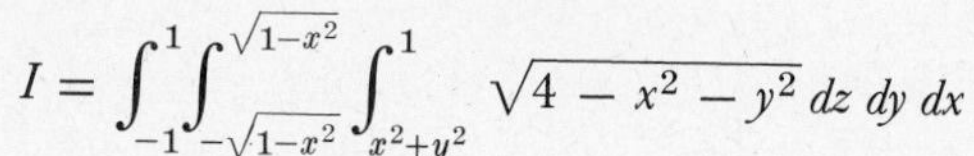

$$I = \int_{-1}^{1}\int_{-\sqrt{1-x^2}}^{\sqrt{1-x^2}}\int_{x^2+y^2}^{1} \sqrt{4 - x^2 - y^2}\, dz\, dy\, dx$$

as an equivalent iterated integral in cylindrical coordinates.

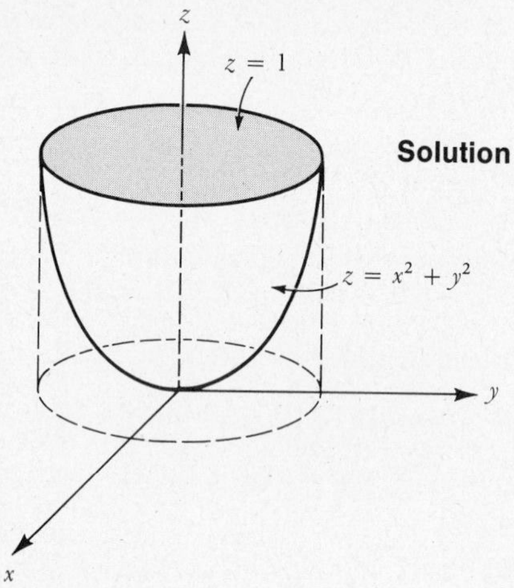

Solution

$$I = \iiint_S \sqrt{4 - x^2 - y^2}\, dV$$

where

$$S = \{(x, y, z) \mid -1 \leq x \leq 1, -\sqrt{1 - x^2} \leq y \leq \sqrt{1 - x^2}, x^2 + y^2 \leq z \leq 1\}$$

Thus S is the solid region above the paraboloid $z = x^2 + y^2$ and below the plane $z = 1$ (see Fig. 20.60). In cylindrical coordinates, S can be described as

$$S = \{(r, \theta, z) \mid 0 \leq \theta \leq 2\pi, 0 \leq r \leq 1, r^2 \leq z \leq 1\}$$

and since $\sqrt{4 - x^2 - y^2} = \sqrt{4 - r^2}$, we have

$$I = \int_0^{2\pi}\int_0^1\int_{r^2}^1 \sqrt{4 - r^2}\, r\, dz\, dr\, d\theta \quad \bullet$$

z = 1
z = x² + y²

Fig. 20.60

Example 12 Express the integral

$$I = \int_0^1\int_0^{\sqrt{1-x^2}}\int_0^{\sqrt{1-x^2-y^2}} (x^2 + y^2 + z^2)^{3/2}\, dz\, dy\, dx$$

as an equivalent iterated integral in spherical coordinates.

Solution

$$I = \iiint_S (x^2 + y^2 + z^2)^{3/2}\, dV$$

where S is the quarter-hemisphere (see Fig. 20.61)

$$S = \{(x, y, z) \mid 0 \leq x \leq 1, 0 \leq y \leq \sqrt{1 - x^2}, 0 \leq z \leq \sqrt{1 - x^2 - y^2}\}$$

In spherical coordinates, S is given by

$$S = \{(\rho, \phi, \theta) \mid 0 \leq \theta \leq \pi/2, 0 \leq \phi \leq \pi/2, 0 \leq \rho \leq 1\}$$

and since

$$(x^2 + y^2 + z^2)^{3/2} = (\rho^2)^{3/2} = \rho^3$$

we have

$$I = \int_0^{\pi/2}\int_0^{\pi/2}\int_0^1 \rho^3 \cdot \rho^2 \sin\phi\, d\rho\, d\phi\, d\theta$$

$$= \int_0^{\pi/2}\int_0^{\pi/2}\int_0^1 \rho^5 \sin\phi\, d\rho\, d\phi\, d\theta \quad \bullet$$

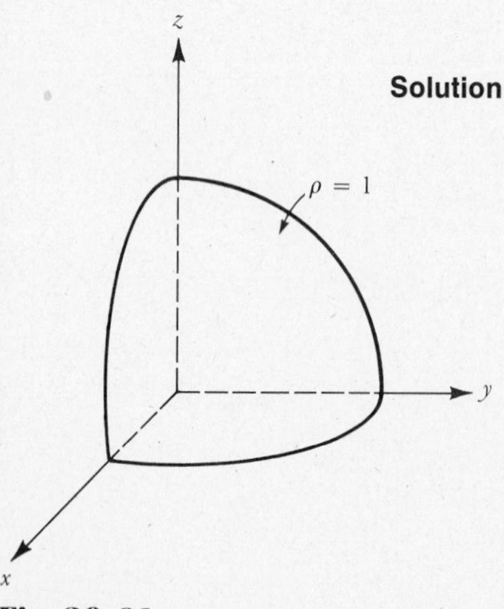

Fig. 20.61

PROGRESS TEST 4

1. Use cylindrical coordinates to find the mass of the homogeneous solid bounded on the sides by $x^2 + y^2 = 1$, on the bottom by the *x*-*y* plane, and on the top by $x^2 + y^2 + z^2 = 2$.

2. Use spherical coordinates to find the mass of the solid in the first octant between the spheres $x^2 + y^2 + z^2 = a^2$ and $x^2 + y^2 + z^2 = b^2$ $(a > b)$ if the density at any point is inversely proportional to its distance from the origin.

SECTION 20.4 EXERCISES

In Exercises 1 to 5, (a) transform the given iterated integral into an equivalent iterated integral in polar coordinates, and (b) evaluate the resulting polar integral.

1. $\int_0^2 \int_{y/\sqrt{3}}^{\sqrt{4-y^2}} \sqrt{x^2 + y^2}\, dx\, dy$

2. $\int_0^{1/\sqrt{2}} \int_y^{\sqrt{1-y^2}} x\, dx\, dy$

3. $\int_0^4 \int_0^{\sqrt{4-x^2}} y^2\, dy\, dx$

4. $\int_0^3 \int_{-\sqrt{9-x^2}}^{\sqrt{9-x^2}} (x^2 + y^2)^{3/2}\, dy\, dx$

5. $\int_1^4 \int_0^x \frac{dy\, dx}{(x^2 + y^2)^{1/2}}$

In Exercise 6 to 10 use a double integral (or integrals) in polar coordinates to determine the indicated area.

6. The area of the region between the concentric circles $x^2 + y^2 = 9$ and $x^2 + y^2 = 1$.

7. The area of the region inside the circle $r = 3 \cos \theta$ and outside the cardioid $r = 1 + \cos \theta$.

8. The total area enclosed by the lemniscate $r^2 = 2 \cos 2\theta$.

9. The area of one leaf of the three-leafed rose $r = 4 \sin 3\theta$.

10. The area enclosed by the smaller loop of the limacon $r = 1 - 2 \sin \theta$.

In Exercises 11 to 15 determine the volume of the indicated solid region.

11. The solid inside the sphere $x^2 + y^2 + z^2 = 4$ that is also inside the cylinder $x^2 + y^2 = 1$.

12. The solid bounded above by the cone $z = r$, below by the *x*-*y* plane, and on the sides by the cylinder $r = 4 \cos \theta$.

13. The solid bounded by the paraboloid $z = 4 - x^2 - y^2$, the *x*-*y* plane, and the cylinder $x^2 + y^2 = 1$.

14. The solid bounded above by the cone $z = \sqrt{x^2 + y^2}$, on the sides by the cylinder $x^2 + y^2 - 3y = 0$, and below by the *x*-*y* plane.

15. The solid inside the ellipsoid $4x^2 + 4y^2 + z^2 = 4$.

In Exercises 16 to 20 find the indicated moment for the lamina occupying the region R and having the given density.

16. R is the region inside the cardioid $r = 1 + \cos \theta$; the density is proportional to the distance from the pole; find M_x.

17. R is the region inside the circle $x^2 + y^2 = 4$ and outside the circle $x^2 + y^2 = 1$; the density is proportional to the distance from the x axis; find M_y.

18. R is the region inside the circle $x^2 - 2x + y^2 = 0$; the lamina is homogeneous; find I_y.

19. R is the region inside the circle $x^2 + y^2 = 9$; the lamina is homogeneous; find I_L, where L is the line $y = 3$.

20. R is the region bounded by the graph of $r = \sin \theta$; the density is proportional to the distance from the y axis; find M_x.

21. A lamina that occupies the region inside the cardioid $r = 2 + 2 \cos \theta$ and outside the circle $r = 2$ has density inversely proportional to the square of the distance from the pole. Find the force of attraction exerted by this lamina on a particle of mass μ located at the origin.

In Exercises 22 to 26 evaluate the given integral by first transforming into an equivalent integral in cylindrical or spherical coordinates (whichever is more convenient).

22. $\int_{-2}^2 \int_{-\sqrt{4-y^2}}^{\sqrt{4-y^2}} \int_0^{\sqrt{16-x^2-y^2}} dz\, dx\, dy$

23. $\int_0^1 \int_{-\sqrt{1-x^2}}^{\sqrt{1-x^2}} \int_{\sqrt{x^2+y^2}}^{\sqrt{2-x^2-y^2}} dz\, dy\, dx$

24. $\int_{-\sqrt{2}}^{\sqrt{2}} \int_{-\sqrt{2-x^2}}^{\sqrt{2-x^2}} \int_0^4 (x^2 + y^2) z\, dz\, dy\, dx$

25. $\int_0^1 \int_0^1 \int_0^{\sqrt{1-x^2}} dy\, dx\, dz$

26. $\int_0^a \int_0^{\sqrt{a^2-x^2}} \int_{h\sqrt{x^2+y^2}/a}^{h} dz\, dy\, dx$

27. Find the first moment about the x-y plane of the solid bounded above by the plane $z = 4$ and below by the paraboloid $z = x^2 + y^2$ where the density at any point is proportional to the distance from the x-y plane.

28. Find the moment of inertia about the z axis of the solid in Exercise 27.

29. Find the moment of inertia about the z axis of the region between the concentric spheres $x^2 + y^2 + z^2 = a^2$ and $x^2 + y^2 + z^2 = b^2$ $(a > b)$ if the density at a point is proportional to its distance from the origin.

30. Find the mass of the homogeneous solid bounded below by the cone $\phi = \pi/6$ and above by the sphere $\rho = 4$.

31. Find the mass of a spherical solid of radius a if the density is proportional to the distance from the center.

32. Find the mass of a right circular cone of height h and radius at the top a, if the density is proportional to the distance from the axis of the cone.

33. Find the force of attraction of the region in the first octant between the concentric spheres $x^2 + y^2 + z^2 = a^2$ and $x^2 + y^2 = b^2$ $(a > b)$ on a particle of mass m located at the origin if the density of this region is inversely proportional to the distance from the origin.

34. Find the moment of inertia about the x axis of the region inside the sphere $x^2 + y^2 + z^2 = 18$ and the cylinder $x^2 + y^2 = 9$ if the density is proportional to the distance from the z axis.

35. Find the mass of the homogeneous solid region bounded by the cylinder $x^2 + y^2 = ax$ and the cone $z^2 = x^2 + y^2$.

CHAPTER EXERCISES

Review Exercises

In Exercises 1 to 10 each of the iterated integrals represents an area or a volume of some region. (a) Sketch the region (in the appropriate coordinate system). (b) Compute the iterated integral.

1. $\int_0^{\pi/4}\int_{\sin x}^{\cos x} dy\,dx$

2. $\int_{-1}^{2}\int_{y^2/2}^{(y+2)/2} dx\,dy$

3. $\int_0^{3/\sqrt{2}}\int_{12-x^2}^{x^2+3} dy\,dx$

4. $\int_0^{\pi}\int_0^{1+\cos\theta} r\,dr\,d\theta$

5. $\int_0^{\pi}\int_{\theta}^{(\theta+\pi)/2} r\,dr\,d\theta$

6. $\int_0^{\pi/3}\int_1^{2\cos\theta} r\,dr\,d\theta$

7. $\int_0^1\int_0^{\sqrt{1-y^2}}\int_0^1 dz\,dx\,dy$

8. $\int_{-2}^{2}\int_{-\sqrt{4-x^2}}^{\sqrt{4-x^2}}(4 - x^2 - y^2)\,dy\,dx$

9. $\int_0^{2\pi}\int_0^4\int_0^{4-r\sin\theta} r\,dz\,dr\,d\theta$

10. $\int_0^{\pi/4}\int_0^{2\pi}\int_0^{-4\cos\phi} \rho^2\sin\phi\,d\rho\,d\phi\,d\theta$

In Exercises 11 to 15 evaluate $\iint_R f(x, y)\,dA$, where

11. $f(x,y) = xy$, $R = \{(x,y)|0 \le x \le 1, x^2 \le y \le \sqrt{x}\}$

12. $f(x,y) = x^2 + 2y$, $R = \{(x,y)|0 \le x \le 1, x^3 \le y \le x^2\}$

13. $f(x,y) = 2x - y$, $R = \{(x,y)|2 \le y \le 3, 1 + y \le x \le \sqrt{y}\}$

14. $f(x,y) = x^2y$, $R = \{(x,y)|1 \le y \le 2, 0 \le x \le y\}$

15. $f(x,y) = \dfrac{x}{\sqrt{1 + x^2 + y^2}}$, $R = \{(x,y)|0 \le x \le 2, 0 \le y \le x^2/2\}$

In Exercises 16 to 20, evaluate the indicated iterated integral.

16. $\int_0^{\sqrt{\pi/2}}\int_0^{x^2} x\sin y\,dy\,dx$

17. $\int_0^1\int_{y^{3/2}}^{\sqrt{y}} \dfrac{x\,dx\,dt}{1 + y^4}$

18. $\int_0^{\pi/4}\int_0^{\tan x} 3y^2\sec x\,dy\,dx$

19. $\int_0^1\int_0^{\cos x} e^{\sin x}\,dy\,dx$

20. $\int_1^{\sqrt{3}}\int_0^{x} \dfrac{x\,dy\,dx}{x^2 + y^2}$

In Exercises 21 to 25, (a) sketch the region over which the integral is being evaluated, (b) write an equivalent integral with the order of integration reversed, and (c) evaluate this latter integral.

21. $\int_0^1\int_y^1 \sqrt{2 + x^2}\,dx\,dy$

22. $\int_0^1\int_{\sqrt[3]{x}}^1 \sqrt{1 + y^4}\,dy\,dx$

23. $\int_0^1\int_{1-\sqrt{1-y}}^{1+\sqrt{1-y}} y\,dx\,dy$

24. $\int_0^1\int_{\sqrt{y}}^1 e^{y/x}\,dx\,dy$

25. $\int_0^1\int_{\arcsin y}^{\pi/2} e^{\cos x}\,dx\,dy$

In Exercises 26 to 30 determine the area of the given region R.

26. R is bounded by $y = 2x^2 - 2x - 1$, $y = x^2 - 2x + 3$.

27. R is bounded by $y = \ln x$, $y = e^x$, $y = 1$, $y = 2$.

28. R is bounded by $y^2 = 4x$, $2x^2 = y$.

29. R is bounded by $y = x$, $y = 3x$, and $x = 2$.

30. R is in the first quadrant and bounded by $y = 0$, $y = x$, and $x^2 + y^2 = 4$.

In Exercises 31 to 35 use the double integral to determine the volume of the given region S.

31. S is bounded by the coordinate planes and $x + y + z = 1$.

32. S is bounded below by the x-y plane and above by the paraboloid $z = 4 - x^2 - y^2$.

33. S is bounded below by the x-y plane, on the sides by the cylinder $x^2 + y^2 = 1$, and above by the plane $z = 3 - x$.

34. S is bounded by the cylinder $x^2 + z^2 = 4$ and by the planes $y = 0$ and $x + y + z = 3$.

35. S is bounded above by $z = 18 - x^2 - y^2$ and below by $z = x^2 + y^2$.

In Exercises 36 to 40 determine the area of the given polar region R.

36. R is enclosed by $r = 4 - \cos\theta$.

37. R is enclosed by $r = 2\sin 3\theta$.

38. R is the region inside both the curves $r = 2\cos\theta$ and $r = 1$.

39. R is the region inside $r = 2\cos\theta$ and outside $r = 1$.

40. R is enclosed by $r^2 = \sin 2\theta$.

41. Find the center of mass of the lamina bounded by $y = x^2 + 2x$ and $y = x + 2$ if the density is proportional to the distance from the x axis.

42. Find the moment of inertia about the y axis of the homogeneous lamina bounded by $r = 1 + \cos\theta$.

43. Find the moment of inertia about the origin of the lamina bounded by $y^2 = x$ and $y = x$ if the density is given by $\rho(x,y) = xy$.

44. Find the moment of inertia about the origin of the homogeneous lamina bounded by one loop of $r^2 = \cos 2\theta$.

45. A lamina occupies the annular region in the first quadrant between $r = 1$ and $r = 4$. Given that the density is inversely proportional to the distance from the origin, determine the force of attraction exerted by this lamina on a particle of mass μ located at the origin.

In Exercises 46 to 49 determine and evaluate an equivalent iterated integral in polar coordinates.

46. $\displaystyle\int_0^4\int_0^{\sqrt{16-x^2}} dy\,dx$

47. $\displaystyle\int_0^2\int_{-\sqrt{4-y^2}}^{\sqrt{4-y^2}} \frac{dx\,dy}{\sqrt{16 - x^2 - y^2}}$

48. $\displaystyle\int_0^4\int_0^{\sqrt{4y-y^2}} (x^2 + y^2)\,dx\,dy$

49. $\displaystyle\int_0^{1/\sqrt{2}}\int_0^{\sqrt{1-x^2}} \frac{dy\,dx}{e^{x^2+y^2}}$

In Exercises 50 to 53 find the volume of the indicated region S by using an appropriate double integral in polar coordinates.

50. S is the region inside both of the surfaces $x^2 + y^2 + z^2 = 16$ and $x^2 + y^2 = 4$.

51. S is bounded below by the upper nappe of the cone $b^2(x^2 + y^2) = a^2z^2$ and above by the plane $z = b$ $(a, b > 0)$.

52. S is bounded above by $z = r$ and below by $z = r^2/2$.

53. S is inside the cylinder $x^2 + y^2 - 4y = 0$ and bounded above and below by the cone $z^2 = x^2 + y^2$.

In Exercises 54 to 63 evaluate the given integral.

54. $\displaystyle\int_0^{\pi/2}\int_0^{\cos\theta}\int_{r^2}^{r\cos\theta} r\,dz\,dr\,d\theta$

55. $\displaystyle\int_0^1\int_0^{2-2y}\int_y^{2-x-y} (y + z)\,dz\,dx\,dy$

56. $\displaystyle\int_0^1\int_0^{1-x^2}\int_0^y 3x\,dz\,dy\,dx$

57. $\displaystyle\int_0^1\int_x^{\sqrt{4-x^2}}\int_1^{e^z} \frac{x}{y}\,dy\,dz\,dx$

58. $\displaystyle\int_0^{\pi}\int_0^2\int_{r^2}^4 r^2\,dz\,dr\,d\theta$

59. $\displaystyle\int_0^{2\pi}\int_0^{\pi/4}\int_0^{\sec\phi} \rho^4\sin^2\phi\,d\rho\,d\phi\,d\theta$

60. $\displaystyle\int_{-1}^2\int_{y^2}^{y+2}\int_{(x^2+y^2)/4}^{y+3} y\,dz\,dx\,dy$

61. $\displaystyle\int_0^1\int_0^x\int_0^{x-y} x\,dz\,dy\,dx$

62. $\displaystyle\int_0^{\pi/2}\int_0^{\cos\theta}\int_0^r r^2\,dz\,dr\,d\theta$

63. $\displaystyle\int_0^{\pi/4}\int_0^{\cos\theta}\int_0^{2\pi} \rho^2\sin\phi\,d\rho\,d\phi\,d\theta$

In Exercises 64 to 68 determine the volume of the solid S.

64. S is bounded by $z = x^2$, $x^2 + y^2 = 4$, and $z = 0$.

65. S is bounded by $x^2 + y^2 = 4z$ and $z = y$.

66. S is bounded above by the paraboloid $z = r^2$, on the bottom by the x-y plane, and on the sides by the cylinder $r = \cos\theta$.

67. S is bounded by the cylinder $x^2 + y^2 = 2x$ and the sphere $x^2 + y^2 + z^2 = 4$.

68. S is bounded below by $\phi = \pi/4$ and above by $\rho = 2\cos\phi$.

69. Give all six iterations for the integral $\iiint_S dV$, where S is the region in the first octant bounded below by $z = 4$ and above by $2x + 3y + 4z = 24$.

70. Give the remaining 5 iterations for the integral

$$\int_0^4 \int_0^z \int_0^x dy\, dx\, dz$$

71. Find the moment of inertia about the z axis for the solid region between the spheres $x^2 + y^2 + z^2 = 1$ and $x^2 + y^2 + z^2 = 4$ if the density is inversely proportional to the distance from the origin.

72. Find the center of mass of the homogeneous solid bounded below by the x-y plane and above by $z = 1 - r^2$.

73. Find the moment of inertia about the x axis of the solid region bounded below by $z = 2x$, above by $z = 4x$, and on the sides by $y = 0$ and $y = 1$ if the density is given by $\rho(x, y) = xy$.

Miscellaneous Exercises

1. Express

$$\int_0^1 \int_1^{\pi/2} dx\, dy + \int_1^{\pi^2/4} \int_{\sqrt{y}}^{\pi/4} dx\, dy$$

as a single iterated integral.

2. Express

$$\int_2^4 \int_2^y dx\, dy + \int_4^6 \int_2^4 dx\, dy + \int_6^{12} \int_{y/3}^4 dx\, dy$$

as a single iterated integral.

3. Express the integral

$$\int_0^2 \int_0^{\sqrt{4-x^2}} \int_0^{(12-2x-2y)/3} dz\, dy\, dx$$

as a sum of iterated integrals with $dV = dx\, dz\, dy$.

4. Express the integral

$$\int_{\pi/6}^{\pi/3} \int_0^1 r\, dr\, d\theta$$

as an equivalent iterated integral (or integrals) in rectangular coordinates.

5. Evaluate

$$\int_0^{\sqrt{3}/2} \int_{1/\sqrt{3}}^{\sqrt{1-y^2}} \ln(x^2 + y^2)\, dx\, dy$$

6. The solid region in the first octant between the spheres $x^2 + y^2 + z^2 = 1$ and $x^2 + y^2 + z^2 = 4$ has a density that is inversely proportional to the distance from the origin. Set up an iterated integral that gives the first moment of the region about the plane $x + y + z = 16$.

7. The density of a bounded solid that does not contain the origin is given by

$$f(x, y, z) = (x^2 + y^2 + z^2)^{3/2}$$

Show that the force of attraction exerted by this solid on a particle of unit mass located at the origin is directed toward the centroid of the region occupied by the solid.

8. Prove the *parallel-axis theorem:* The moment of inertia of a lamina about any axis L in its plane equals $I_p + Mr^2$, where M is the mass of the lamina, r is the distance from L to the center of mass C of the plate, and I_p is the moment of inertia of the lamina about an axis parallel to L through C.

9. Evaluate $\iint_R dA/\sqrt{x - y}$, where R is the triangle bounded by $y = 0$, $y = x$, and $x = 1$. (Be careful.)

10. Show that the volume of the elliptic cylinder $x^2/a^2 + y^2/b^2 = 1$ $(a, b > 0)$ between the x-y plane and the plane $z = cx + dy + h$ does not depend on the constants c and d. (*Hint:* Use a polarlike substitution for x and y.)

11. Use cylindrical coordinates and the fact that $V(S) = \iiint_S dV$ to develop the ("shell method") formula $V = \int_a^b 2\pi x f(x)\, dx$ for the volume of the solid of revolution generated by revolving about the y axis the plane region between the graph of $y = f(x)$ and the interval $[a, b]$, for $f(x) \geq 0$ on $[a, b]$. (You may wish to switch axes.)

12. Find the area outside the parabola $r(1 + \cos\theta) = 2a$ and inside the cardioid $r = 2a(1 + \cos\theta)$.

13. Determine $\iint_R (\sqrt{x} + \sqrt{y})\, dx\, dy$, where R is bounded by $x = 0$, $y = 0$, and $\sqrt{x} + \sqrt{y} = 1$. (*Hint:* Use a parametric representation of $\sqrt{x} + \sqrt{y} = 1$ to obtain a polarlike substitution for x and y.)

14. Suppose C is a solid right circular cylinder of radius a, length h, and uniform density ρ. Determine the gravitational attraction on a particle of unit mass on the axis of C at a distance b from one end of C. (*Hint:* First approximate the attraction of a cross-sectional slice of thickness Δx_i. To do this, use a polar representation of an element of the disk.)

15. Use the idea of Exercise 14 to show that the attraction of a uniform-density solid sphere on a particle of unit mass at distance b from its center ($b >$ radius) is the same as if the total mass were concentrated at the center.

SELF-TEST ✤

1. Evaluate the following:

   (a) $\int_0^1 \int_0^{\sqrt{x}} xy\, dy\, dx$

   (b) $\int_0^{\pi/2} \int_0^{2 \sin \theta} r\, dr\, d\theta$

   (c) $\int_0^1 \int_0^{x^2} \int_0^1 (xy + z)\, dy\, dz\, dx$

   (d) $\int_0^{\pi/2} \int_0^{\pi/4} \int_0^{\sec \phi} \rho^2 \sin \phi\, d\rho\, d\phi\, d\theta$

   (e) $\int_0^{2\pi} \int_0^2 \int_0^{4-r^2} r\, dz\, dr\, d\theta$

2. Sketch the region over which each of the integrals in Prob. 1 is evaluated.
3. Evaluate the following by either transforming to a more convenient coordinate system or using a different order of integration.

   (a) $\int_0^1 \int_0^{\sqrt{1-x^2}} (x^2 + y^2)^{3/2}\, dy\, dx$

   (b) $\int_{-1}^1 \int_{\sqrt{1-y^2}}^{\sqrt{1-y^2}} \int_{\sqrt{x^2+y^2}}^{\sqrt{2-x^2-y^2}} dz\, dx\, dy$

   (c) $\int_0^1 \int_y^1 (x^2 + 1)^{2/3}\, dx\, dy$

4. Find the first moment about the x axis of the lamina occupying the region bounded by $x = y^2 - 1$ and $y + 1 = x$ if the density at any point is proportional to the distance from the line $y = 2$.

Chapter 21 

THE LINE INTEGRAL

CONTENTS

So far our study of the calculus has dealt with several different types of functions. In Chaps. 1 to 16 we were concerned only with functions that transform real numbers into real numbers. In Chap.18 we introduced functions that transform numbers into vectors, and in Chap. 19 we studied functions that send points in R_n into numbers. Thus we have considered functions from R to R, R to R_n, and R_n to R. In this chapter we complete the picture by introducing functions that send vectors into vectors, that is, functions from R_m into R_n. Such vector functions are usually referred to as *vector fields*, and they arise quite naturally in many applied areas such as physics, mechanics, electrostatics, and fluid dynamics. Our approach for the most part is limited to vector fields from R_2 into R_2 and only scratches the surface of a very useful and exciting area of mathematics.

A more advanced course in complex analysis presents two-dimensional vector fields in their natural setting. More general vector fields are usually studied in advanced calculus or mathematical physics. The key concept in this chapter is that of the integral of a vector field over a curve, traditionally called a *line integral*. The main result is Green's Theorem, which relates the line integral of a vector field over a curve to a certain double integral of the type studied in Chap. 20. We study the physical consequences of this result in the context of problems involving the velocity field of a fluid flow. The mathematics developed applies as well to vector fields associated with electromagnetic and gravitational fields, among others.

Section 21.1: Vector Fields

Objectives:

1. Determine when a vector field is conservative (the gradient of some function).

2. Find the potential functions for a conservative vector field.

INTRODUCTION

Suppose a fluid is flowing in a stream whose surface occupies the region D in R_2. If the velocity of a particle in the stream depends only on its location and *not* on time, then the flow is called *steady state*. If, in addition, the flow is horizontal and independent of the depth, then we can regard it as a *two-dimensional fluid flow*. We can now define a function, called a *velocity field*, that assigns to each point (x, y) in D the vector that gives the velocity $\mathbf{v}(x, y)$ of the fluid at that point. The formal "graph" of such a function would be a subset of R_4 and hence could not be sketched. However, we can choose selected points (x, y) in D and sketch the representative of $\mathbf{v}(x, y)$ whose base is at (x, y). A typical such velocity field is sketched in Fig. 21.1.

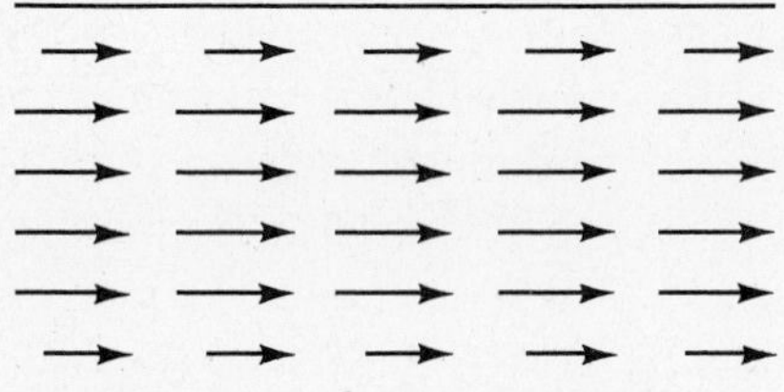

Velocity field of a stream

Fig. 21.1

VECTOR FIELDS

Functions of the above type arise quite frequently in applications ranging from fluid flow to gravitational or electromagnetic fields, whenever it is necessary to associate a vector quantity to each point in a given region. Such functions are referred to as *vector fields*. If the vectors under consideration represent velocities (or forces), then the vector field is called a velocity field (or force field). Specifically, we have:

21.1.1 **Definition** Suppose D is a subset of R_n. A function $\mathbf{F}$ that assigns to each point $(x_1, x_2, \ldots, x_n)$ in D a unique vector $\mathbf{F}(x_1, x_2, \ldots, x_n)$ in R_m is called a *vector field*. Such a vector field is often denoted by writing $\mathbf{F}: D \to R_m$.

The above steady-state fluid flow gave rise to a vector field from a subset of R_2 into R_2. If the flow were not steady state, then the velocity of a particle in the surface of the stream would depend on both the location *and* time. If we let

$$S = \{(x, y, t) \mid (x, y) \text{ in } D,\ a \leq t \leq b\}$$

then $\mathbf{v}(x, y, t)$ would denote the velocity of the particle located at (x, y) at time t. Such a *time-dependent velocity field* is an example of a vector field from a subset of R_3 into R_2.

In this text we shall deal, for the most part, with two-dimensional vector fields, that is, vector fields from a subset of R_2 into R_2. Many of our conclusions, with appropriate modifications, can be extended to three-dimensional vector fields. Several of these extensions will be noted in the exercises.

If $\mathbf{F}: D \to R_2$ is a two-dimensional vector field, then $\mathbf{F}$ can be expressed in *component form* as

$$\mathbf{F}(x, y) = \langle M(x, y), N(x, y) \rangle$$

or, equivalently,

$$\mathbf{F}(x, y) = M(x, y)\mathbf{i} + N(x, y)\mathbf{j}$$

where M and N are scalar functions of two variables defined on D.

Since the points (x, y) in D can be identified with the position vectors $\mathbf{X} = \langle x, y \rangle$, we can also consider $\mathbf{F}$ as a vector function of a vector variable. In this case we write $\mathbf{F}$ in the form

$$\mathbf{F}(\mathbf{X}) = \langle M(x, y), N(x, y) \rangle$$

As usual, we shall use whichever notation best fits the particular situation.

Example 1 Suppose a particle of unit mass is located at the origin. Determine and sketch the force field that gives at each point $X = (x, y)$ the force of attraction exerted by this particle on a typical particle of unit mass located at (x, y).

Solution In Chap. 20 we noted that Newton's inverse-square law states that

$$\mathbf{F} = \frac{gm_1m_2}{\|\overrightarrow{PQ}\|^3}\overrightarrow{PQ}$$

where $\mathbf{F}$ is the force of attraction that a particle of mass m_1 located at Q will exert on a particle of mass m_2 located at P. In our setting, $m_1 = m_2 = 1$, $Q = (0, 0)$, and $P = X = (x, y)$. Thus $\overrightarrow{PQ} = \langle 0 - x, 0 - y\rangle = -\mathbf{X}$ and

$$\|\overrightarrow{PQ}\|^3 = \|-\mathbf{X}\|^3 = \|\mathbf{X}\|^3 = (x^2 + y^2)^{3/2}$$

Thus the force field is given by

$$\mathbf{F}(x, y) = \frac{-g}{\|\mathbf{X}\|^3}\mathbf{X}$$

or, in component form,

$$\mathbf{F}(x, y) = \left\langle \frac{-gx}{(x^2 + y^2)^{3/2}}, \frac{-gy}{(x^2 + y^2)^{3/2}} \right\rangle$$

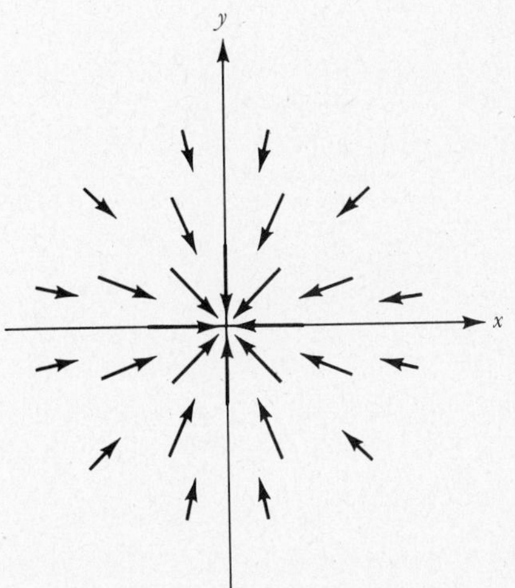

Gravitational force field

Fig. 21.2

The negative sign means that the forces are directed towards the origin. Also, since

(1) $$\|\mathbf{F}(x, y)\| = \frac{g\|\mathbf{X}\|}{\|\mathbf{X}\|^3} = \frac{g}{\|\mathbf{X}\|^2}$$

the magnitude of the force is small for points far from the origin ($\|\mathbf{X}\|$ large) and is large for points near the origin ($\|\mathbf{X}\|$ small). Equation (1) explains why we use the terminology "inverse-square law." A sketch of this gravitational force field is given in Fig. 21.2. ●

GRADIENT FIELDS—CONSERVATION OF ENERGY

An important class of vector fields arises from considering the gradient of a scalar function of several variables. Specifically, if f is a function of two variables and if D denotes the set of points in R_2 at which the first partials of f exist, then the *gradient field*, grad f, denoted by $\nabla f\colon D \to R_2$, is given in component form by

$$\nabla f(x, y) = \langle f_x(x, y), f_y(x, y)\rangle$$

21.1.2

Definition Suppose $\mathbf{F}\colon D \to R_2$ is a two-dimensional vector field. If there exists a scalar function f, differentiable on D, such that

$$\nabla f(x, y) = \mathbf{F}(x, y)$$

for each (x, y) in D, then $\mathbf{F}$ is called a *conservative vector field* and f is called a *potential function* for $\mathbf{F}$ on D.

The genesis of the terms "conservative" and "potential" in (21.1.2) is in physics—in the law of conservation of energy. Specifically, suppose a particle of mass m is moving along the plane curve $\mathbf{X}(t)$ subject to the conservative force $\mathbf{F} = \nabla f$. As usual, $\mathbf{V}(t) = \mathbf{X}'(t)$ and $\mathbf{A}(t) = \mathbf{X}''(t)$ denote the velocity and acceleration of the particle, respectively. By Newton's second law of motion,

$$\mathbf{F}(\mathbf{X}(t)) = m\mathbf{A}(t)$$

But $\mathbf{F} = \nabla f$, so we have

$$m\mathbf{A}(t) - \nabla f(\mathbf{X}(t)) = 0 \tag{2}$$

If we dot both sides of equation (2) with **V**, we obtain

$$m\mathbf{A}(t) \cdot \mathbf{V}(t) - \nabla f(\mathbf{X}(t)) \cdot \mathbf{V}(t) = 0 \tag{3}$$

Now

$$D_t[\|\mathbf{V}(t)\|^2] = D_t[\mathbf{V}(t) \cdot \mathbf{V}(t)] = 2\mathbf{V}'(t) \cdot \mathbf{V}(t) = 2\mathbf{A}(t) \cdot \mathbf{V}(t)$$

so

$$\mathbf{A}(t) \cdot \mathbf{V}(t) = D_t[\tfrac{1}{2}\|\mathbf{V}(t)\|^2]$$

Also, by the Chain Rule,

$$\begin{aligned} D_t[f(\mathbf{X}(t))] &= D_t[f(x(t), y(t))] \\ &= f_x(x(t), y(t))x'(t) + f_y(x(t), y(t))y'(t) \\ &= \nabla f(\mathbf{X}(t)) \cdot \mathbf{V}(t) \end{aligned}$$

Thus we can rewrite equation (3) as

$$D_t[\tfrac{1}{2}m\|\mathbf{V}(t)\|^2 - f(\mathbf{X}(t))] = 0$$

which implies

$$\tfrac{1}{2}m\|\mathbf{V}(t)\|^2 - f(\mathbf{X}(t)) = C \tag{4}$$

for some constant C.

The quantity $(1/2)m\|\mathbf{V}(t))\|^2$ is called the *kinetic energy* of the particle and $-f(\mathbf{X}(t))$ is called its *potential energy*. The sum of its kinetic and potential energies is called its *total energy*. Thus equation (4) is a version of the law of conservation of energy, which states that if the force field is conservative, then the total energy must be constant.

Because of the above, if f is a potential function for the conservative force field **F**, then $-f$ is often referred to as a *potential-energy function* for **F**.

The problem of determining necessary and sufficient conditions for a vector field to be conservative is a significant one that will require some preliminary groundwork. However, in most specific instances we simply *assume* **F** is conservative and then systematically proceed as below either to determine the potential functions or to derive a contradiction.

Example 2 Show that the vector field

$$\mathbf{F}(x, y) = \langle 2xy^3 - 4x, 3x^2y^2 + 2y\rangle$$

is conservative [for *all* (x, y)].

Solution To show that **F** is conservative we must exhibit a scalar function $f(x, y)$ whose first partials exist for all (x, y) such that

$$f_x(x, y) = 2xy^3 - 4x \quad \text{and} \quad f_y(x, y) = 3x^2y^2 + 2y \tag{5}$$

Since f_x is obtained from f by holding y constant and differentiating with respect to x, we can attempt to recapture f by integrating f_x with respect to x—again holding y constant. In this case the "constant of integration" will be a function of y. (Think about this!) Thus

$$\begin{aligned} f(x, y) &= \int f_x(x, y)\, dx + C(y) \\ &= \int (2xy^3 - 4x)\, dx + C(y) = x^2y^3 - 2x^2 + C(y) \end{aligned} \tag{6}$$

All that remains is to determine, if we can, $C(y)$. Now by differentiating the above with respect to y, we know that

$$f_y(x,y) = \frac{\partial}{\partial y}[x^2y^3 - 2x^2 + C(y)] = 3x^2y^2 + C'(y)$$

Our original assumption (5) was that $f_y(x,y) = 3x^2y^2 + 2y$, so, setting these two expressions for f_y equal, we obtain

$$3x^2y^2 + C'(y) = 3x^2y^2 + 2y$$

Hence $C'(y) = 2y$, and so $C(y) = y^2 + k$. Substituting this latter expression in (6), we conclude that any function of the form

$$f(x,y) = x^2y^3 - 2x^2 + y^2 + k \qquad (k \text{ a constant})$$

is a potential function for **F**, and so **F** is conservative. Of course, we can easily check our computations by taking the gradient of f to see if it equals the original vector field **F**. (Check it.) ●

Thus, in general, to determine the potential functions of a conservative vector field

$$\mathbf{F}(x,y) = \langle M(x,y), N(x,y) \rangle$$

we proceed as follows:
(*a*) Set $f_x(x,y) = M(x,y)$.
(*b*) Integrate with respect to x, getting

$$f(x,y) = \int M(x,y)\,dx + C(y)$$

(*c*) Differentiate with respect to y, getting

$$f_y(x,y) = \frac{\partial}{\partial y}\left[\int M(x,y)\,dx\right] + C'(y)$$

(*d*) Set this latter expression equal to $N(x,y)$ and solve for $C'(y)$.
(*e*) Integrate to determine $C(y)$ and hence obtain, from (*a*), the potential functions f.

Dually, we could just as well start by setting $f_y(x,y) = N(x,y)$, then integrate with respect to y to obtain $f(x,y) = \int N(x,y)\,dy + K(x)$, and so on. If in step (*d*) we do not obtain $C'(y)$ as a function of y alone [or $K'(x)$ as a function of x alone with the dual procedure], then we can conclude that the vector field is not conservative.

Example 3 Show that the vector field $\mathbf{F}(x,y) = \langle y, -x \rangle$ is not conservative.

Solution Assume **F** is conservative and let f be any potential function. Then $f_x(x,y) = y$ and $f_y(x,y) = -x$. Integrating the first of these equations with respect to x and the latter with respect to y, we have

$$f(x,y) = xy + C(y) = -xy + K(x)$$

Thus

$$K(x) = 2xy + C(y)$$

Now

$$K'(x) = \frac{\partial}{\partial x}[2xy + C(y)] = 2y + 0 = 2y$$

and so

$$0 = \frac{\partial}{\partial y}[K'(x)] = \frac{\partial}{\partial y}[2y] = 2$$

Thus the assumption that **F** is conservative leads to the contradiction $0 = 2$. ●

The following criterion is useful in determining when a vector field is *not* conservative. It is an immediate consequence of Theorem (19.3.6) on the equality of mixed partials.

21.1.3 **Theorem** Suppose D is an open subset of R_2 and $\mathbf{F}: D \to R_2$ is the vector field given by

$$\mathbf{F}(x,y) = \langle M(x,y), N(x,y) \rangle$$

where M and N have continuous first partials on D. If $\mathbf{F}$ is conservative, then

$$\frac{\partial M}{\partial y}(x,y) = \frac{\partial N}{\partial x}(x,y)$$

for each (x,y) in D.

Proof: Since $\mathbf{F}$ is conservative, there exists a scalar function f such that $\mathbf{F} = \nabla f$ on D. But $\mathbf{F} = \langle M, N \rangle$ and $\nabla f = \langle f_x, f_y \rangle$, so $M = f_x$ and $N = f_y$. Now M and N are continuous with continuous first partials, and hence f has continuous first and second partials. By Theorem (19.3.6), page 640, $f_{xy} = f_{yx}$ on D. The conclusion follows since $M_y = (f_x)_y = f_{xy}$ and $N_x = (f_y)_x = f_{yx}$. ■

Since under the above hypotheses, conservative implies that $M_y = N_x$, it is logically equivalent that $M_y \neq N_x$ implies that $\mathbf{F}$ is *not* conservative.

It turns out that *with a suitable adjustment on the hypotheses,* the condition $M_y = N_x$ is both a necessary *and* sufficient condition for $\mathbf{F} = \langle M, N \rangle$ to be conservative. We will discuss this further following Green's Theorem in Section 21.4. Nonetheless, one should certainly check to see whether $M_y = N_x$ when determining if $\mathbf{F}$ is conservative. If not, there is no need to proceed any further; the field is not conservative. If $M_y = N_x$, then even though there is nothing to guarantee that $\mathbf{F}$ is conservative, you can utilize the procedure outlined earlier to attempt to find a potential function.

PROGRESS TEST 1

Identify those vector fields that are conservative and determine all potential functions.

1. $\mathbf{F}(x,y) = \langle x^2y - 2x, 4xy^2 + y \rangle$, $D = R_2$.
2. The inverse-square field of Example 1. [D consists of all $(x,y) \neq (0,0)$.]
3. $\mathbf{F}(x,y) = \langle 2x \ln(2 + y^2) - 6x^2, 2x^2y/(2 + y^2) + 2y \rangle$, $D = R_2$.

SECTION 21.1 EXERCISES

In Exercises 1 to 6 sketch the given vector field.

1. $\mathbf{F}(x,y) = \langle 0, 1 \rangle$
2. $\mathbf{F}(x,y) = \langle 1, 0 \rangle$
3. $\mathbf{F}(x,y) = \langle y/3, -x/3 \rangle$
4. $\mathbf{F}(x,y) = \langle 1/\sqrt{x^2 + y^2}, 1/\sqrt{x^2 + y^2} \rangle$
5. $\mathbf{F}(x,y) = \langle x, x \rangle$
6. $\mathbf{F}(x,y) = \langle x^2, -y \rangle$

In Exercises 7 to 20 identify those vector fields that are conservative and determine the potential functions. Unless otherwise stated, assume $D = R_2$.

7. $\mathbf{F}(x,y) = \langle 4x - 2y, 5y - 2x \rangle$
8. $\mathbf{F}(x,y) = \langle 2xy + 3, x^2 + 8y \rangle$
9. $\mathbf{F}(x,y) = \langle x^2y - 3x, 4xy^2 + 9y \rangle$
10. $\mathbf{F}(x,y) = \langle (y^2 + 1) \cos x, 2y \sin x \rangle$
11. $\mathbf{F}(x,y) = \langle y^2e^x - y^2, 2ye^x - 2xy + 3y^2 \rangle$
12. $\mathbf{F}(x,y) = \left\langle \dfrac{1 - 2x}{y}, \dfrac{x^2 - x}{y^2} \right\rangle$, $D = \{(x,y) \mid y \neq 0\}$
13. $\mathbf{F}(x,y) = \langle xe^{x^2}y, e^{x^2}y^2 + x \rangle$
14. $\mathbf{F}(r,s) = \langle 2s \sin r \cos r + s^2 \sin r, \sin^2 r - 2s \cos r \rangle$

15. $F(u, v) = \langle 2u \sin v + e^u \cos v, u^2 \cos v - e^u \sin v\rangle$

16. $F(r, \theta) = \langle \theta - 3r^2, r - \theta^2 + 2\theta\rangle$

17. $F(s, t) = \langle e^s(s^2 + t^2 + 2s), 2te^s\rangle$

18. $F(x, y) = \left\langle \dfrac{6xy^2}{3 + x^2} - y^3 + 2x, 6y \ln(3 + x^2) - 3xy^2 \right\rangle$

19. $F(x, y) = \langle 2xe^{x^2} \cos y + 2x, -e^{x^2} \sin y\rangle$

20. $F(x, y) = \langle x^2y \sinh(xy) + 2x \cosh(xy) - 4x^3, x^3 \sinh(xy) + 2y\rangle$

21. A *three-dimensional vector field* F: $D \to R_3$ is called *conservative* if $F = \nabla f$ on D for some differentiable scalar function $f(x, y, z)$. Suppose D is an open subset of R_3 and $F(x, y, z) = \langle M(x, y, z), N(x, y, z), P(x, y, z)\rangle$ is conservative. Show that $M_y = N_x$, $N_z = P_y$, and $P_x = M_z$.

In Exercises 22 to 25 use the following procedure to determine the potential functions for the given three-dimensional conservative vector fields.

(*a*) Set $f_x = M$.
(*b*) Integrate with respect to x, getting $f(x, y, z) = \int M\,dx + C(y, z)$.
(*c*) Differentiate the result in (*b*) with respect to y and set equal to N, obtaining an expression for $C_y(y, z)$.
(*d*) Integrate C_y with respect to y to obtain $C = \int C_y(y, z)\,dy + K(z)$.
(*e*) Substitute (*d*) into (*b*), differentiate with respect to z, and equate with P.

22. $F(x, y, z) = \langle 6x^2y^2 - 3yz^2, 4x^3y - 3xz^2, -6xyz\rangle$

23. $F(x, y, z) = \langle y \sin z, x \sin z, xy \cos z\rangle$

24. $F(x, y, z) = \langle y^2e^z + 3z^2, 2xye^z + 2y, xy^2e^z + 6xy\rangle$

25. $F(x, y, z) = \langle \sin y^2 + 2xz \cos x^2, 2xy \cos y^2, \sin x^2\rangle$

Section 21.2: The Line Integral

Objectives:

1. **Evaluate the line integral of a vector field over a curve.**
2. **Determine the work done by a force field in moving an object along a curve.**

CURVES

If $\mathbf{X}(t)$, $a \le t \le b$, is a simple arc, then we can order the points on the graph by insisting that $X(t_1)$ precede $X(t_2)$ if and only if $t_1 < t_2$, where t_1 and t_2 are allowed to vary over $[a, b)$ if the arc is closed, and over $[a, b]$ otherwise. When the graph of a simple arc is given this ordering, we say it has *the orientation of increasing parameter*. If $\mathbf{X}(t)$ is *continuous*, then the graph together with this ordering is called a *curve* and $\mathbf{X}(t)$ is called a *parameterization of the curve*. We usually denote a curve by C. If we wish to emphasize the particular parameterization we write:

$$C\colon\ \mathbf{X}(t),\ a \le t \le b$$

It is important to note that two quite distinct vector functions can determine the *same* curve C. For example, the simple continuous arcs

$$\mathbf{X}(t) = \langle t, t^2\rangle,\ 0 \le t \le 1$$

and

$$\mathbf{Y}(t) = \langle \sin(t - \pi/2), \sin^2(t - \pi/2)\rangle,\ \pi/2 \le t \le \pi$$

both have as their graphs the piece of the parabola $y = x^2$, $0 \le x \le 1$. Furthermore, in both cases the orientation of increasing parameter provides the same ordering to this graph. That is, the same points precede one another in each ordering. The curve having these parameterizations is sketched in Fig. 21.3, where as in Chap. 18 we have indicated the ordering by placing arrows along the graph.

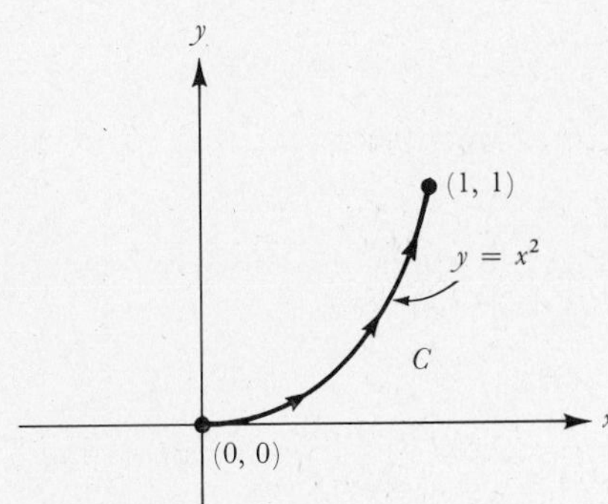

Fig. 21.3

It can be shown that if a curve C has a smooth parameterization, then *any* parameterization of C will be smooth. Hence we shall refer to a curve with a smooth parameterization as a *smooth curve*. A curve C is said to be *piecewise smooth* if it can be formed by piecing together, end to end, a finite collection of smooth curves $C_1, C_2, \ldots, C_n$. In this case we write the following equation:

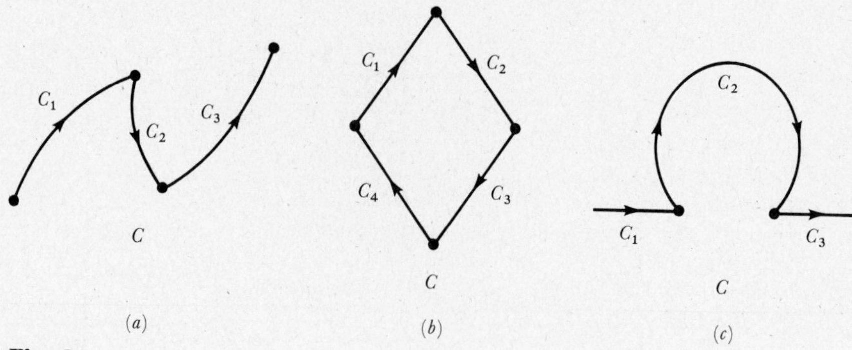

Fig. 21.4

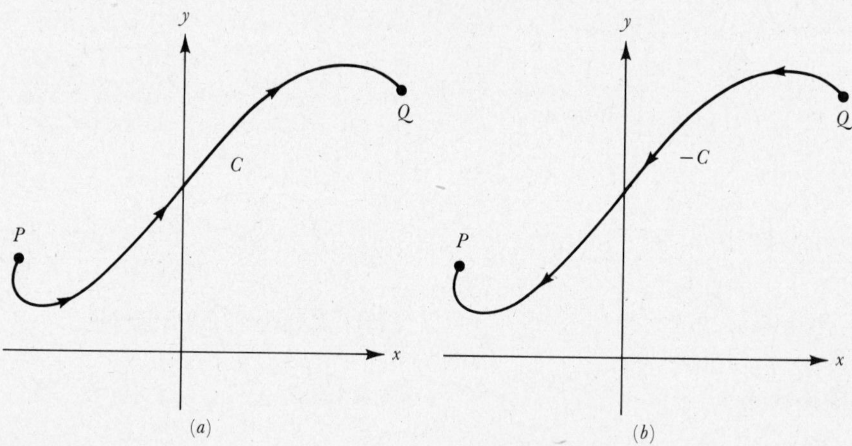

Fig. 21.5

$$C = \bigcup_{i=1}^{n} C_i$$

Some typical piecewise smooth curves are sketched in Fig. 21.4.

Finally, if C is a curve, then by $-C$ we shall mean the curve that has the same graph as C but the *reverse orientation* (see Fig. 21.5). It is worth noting that if C has parameterization

$$C\colon \quad \mathbf{X}(t),\ a \leq t \leq b$$

then a particular parameterization for $-C$ is given by

21.2.1 $$-C\colon \quad \mathbf{X}(a+b-s),\ a \leq s \leq b$$

(Note that as s varies from a to b, $a + b - s$ varies from b to a.)

WORK

In Chap. 17 we noted that if a constant force $\mathbf{F}$ is applied to move an object along a straight line from P to Q, then the work done by this force is given by

(1) $$W = (\text{Proj}_{\overrightarrow{PQ}}\,\mathbf{F})\|\overrightarrow{PQ}\|$$

We now wish to consider a much more general problem. Suppose that $\mathbf{F}\colon D \to R_2$ is a force field defined on some open subset D of R_2, and C is a

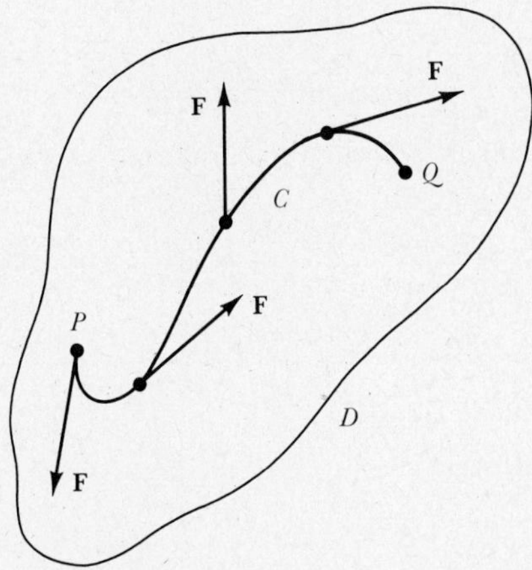

Fig. 21.6

smooth curve contained in D joining the point P to the point Q (see Fig. 21.6). *How much work is done by this variable force in moving an object from P to Q along C?*

To answer this question we shall apply what by now should be a familiar process: We subdivide the curve C into small pieces, approximate the work done on each of these segments of C, add the result, and then take the appropriate limit to obtain a definite integral.

Now suppose C is given parametrically by the simple smooth arc

$$\mathbf{X}(t) = \langle x(t), y(t)\rangle \qquad a \le t \le b$$

Since C is smooth, $\mathbf{V}(t) = \langle x'(t), y'(t)\rangle$ is continuous on $[a, b]$. Let P be a partition of $[a, b]$ into n subintervals

$$[t_0, t_1], \ldots, [t_{i-1}, t_i], \ldots, [t_{n-1}, t_n]$$

where $t_0 = a$, $t_n = b$, $\Delta t_i = t_i - t_{i-1}$, and $\|P\| = \max \Delta t_i$. Let $t_i^* \epsilon [t_{i-1}, t_i]$ for each i, and assume that $\mathbf{F}$ takes on the constant value $\mathbf{F}_i = \mathbf{F}(\mathbf{X}(t_i^*))$ on the segment of C joining $X(t_{i-1})$ to $X(t_i)$. An approximation to the work done by $\mathbf{F}$ along this segment is obtained by taking the product of the scalar projection of $\mathbf{F}_i$ in the direction of the motion with the displacement. That is,

$$W_i \approx (\text{Proj}_{\mathbf{T}_i} \mathbf{F}_i)\, \Delta s_i$$

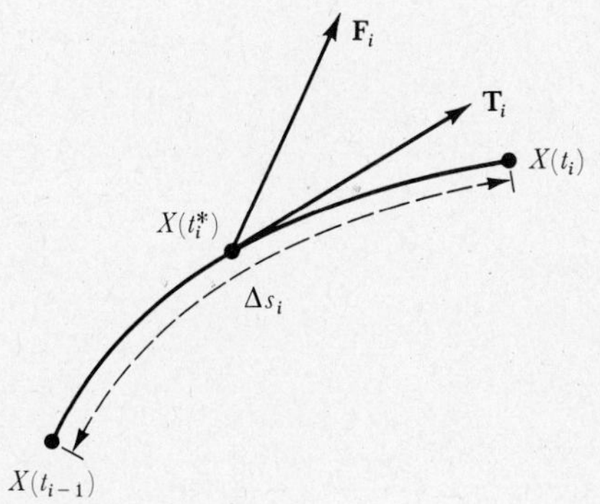

Fig. 21.7

where $\mathbf{T}_i = \mathbf{T}(t_i^*)$ is a unit vector in the direction of the motion and Δs_i is the length of the segment (see Fig. 21.7). Now recall from (18.3.4) that if s represents arc length along C, then $ds = \|\mathbf{V}(t)\|\, dt$, so $\Delta s_i \approx \|\mathbf{V}_i\|\, \Delta t_i$, where $\mathbf{V}_i = \mathbf{V}(t_i^*)$. Thus

$$\begin{aligned} W_i &\approx (\text{Proj}_{\mathbf{T}_i} \mathbf{F}_i)\, \Delta s_i \\ &\approx \left(\frac{\mathbf{F}_i \cdot \mathbf{T}_i}{\|\mathbf{T}_i\|}\right) \|\mathbf{V}_i\|\, \Delta t_i \\ &= (\mathbf{F}_i \cdot \mathbf{T}_i) \|\mathbf{V}_i\|\, \Delta t_i \qquad (\text{since } \|\mathbf{T}_i\| = 1) \\ &= \left(\mathbf{F}_i \cdot \frac{\mathbf{V}_i}{\|\mathbf{V}_i\|}\right) \|\mathbf{V}_i\|\, \Delta t_i = (\mathbf{F}_i \cdot \mathbf{V}_i)\, \Delta t_i \end{aligned}$$

Hence

$$W \approx \sum_{i=1}^{n} (\mathbf{F}_i \cdot \mathbf{V}_i)\, \Delta t_i = \sum_{i=1}^{n} \mathbf{F}(\mathbf{X}(t_i^*)) \cdot \mathbf{V}(t_i^*)\, \Delta t_i$$

If $\mathbf{F}$ has continuous components, then this latter sum is a Riemann sum on $[a, b]$ for the continuous scalar function $\mathbf{F}(\mathbf{X}(t)) \cdot \mathbf{V}(t)$. Thus as $n \to +\infty$ and $\|P\| \to 0$, by (6.3.2),

$$\sum_{i=1}^{n} \mathbf{F}(\mathbf{X}(t_i^*)) \cdot \mathbf{V}(t_i^*)\, \Delta t_i \to \int_a^b \mathbf{F}(\mathbf{X}(t)) \cdot \mathbf{V}(t)\, dt$$

21.2.2 **Definition** Suppose $\mathbf{F}$ is a force field with continuous components on the open set D and suppose C is a smooth curve in D given parametrically by the simple smooth arc $\mathbf{X}(t)$, $a \le t \le b$. The work done by $\mathbf{F}$ in moving an object from $X(a)$ to $X(b)$ along C is defined by

$$W = \int_a^b \mathbf{F}(\mathbf{X}(t)) \cdot \mathbf{V}(t)\, dt$$

It can be shown (see Exercise 16) that the value of the integral in (21.2.2) does not depend on the particular parameterization of the curve C.

Example 1 Find the work done by the force $\mathbf{F}(x, y) = \langle x^2 + 2y, x + y^2 \rangle$ as the point of application moves from the point $(0, 0)$ to $(1, 1)$ along the parabola $y = x^2$.

Solution Let C denote the directed curve from $(0, 0)$ to $(1, 1)$ along the parabola $y = x^2$. A parametric representation of C is given by $\mathbf{X}(t) = \langle x(t), y(t) \rangle = \langle t, t^2 \rangle$, $0 \le t \le 1$. Now $\mathbf{F}(\mathbf{X}(t)) = \mathbf{F}(x(t), y(t)) = \langle t^2 + 2t^2, t + t^4 \rangle = \langle 3t^2, t^4 + t \rangle$ and $\mathbf{V}(t) = \langle x'(t), y'(t) \rangle = \langle 1, 2t \rangle$. Thus

$$\begin{aligned} W &= \int_0^1 \mathbf{F}(\mathbf{X}(t)) \cdot \mathbf{V}(t)\, dt \\ &= \int_0^1 \langle 3t^2, t^4 + t \rangle \cdot \langle 1, 2t \rangle\, dt \\ &= \int_0^1 [3t^2(1) + (t^4 + t)(2t)]\, dt \\ &= \int_0^1 (5t^2 + 2t^5)\, dt = \frac{5t^3}{3} + \frac{t^6}{3} \Big|_0^1 = 2 \end{aligned}$$

(If the magnitude of $\mathbf{F}$ is measured in newtons and distances are measured in meters, then W is measured in joules.) ●

Example 2 Show that the work done by a variable force as the point of application moves from P to Q along the smooth curve C is equal to the change in the kinetic energy of the particle at the end points of C.

Solution Let $\mathbf{X}(t)$, $a \le t \le b$, be a parameterization of C and let m be the mass of the particle on which the force $\mathbf{F}$ is being exerted. By Newton's second law of motion, $\mathbf{F} = m\mathbf{A}$. As noted in Sec. 21.1,

$$D_t\left[\frac{1}{2}\|\mathbf{V}(t)\|^2\right] = \mathbf{A}(t) \cdot \mathbf{V}(t)$$

and hence

$$\begin{aligned} W &= \int_a^b \mathbf{F}(\mathbf{X}(t)) \cdot \mathbf{V}(t)\, dt \\ &= \int_a^b m\mathbf{A}(t) \cdot \mathbf{V}(t)\, dt \\ &= m\int_a^b D_t\left[\frac{1}{2}\|\mathbf{V}(t)\|^2\right] dt \\ &= \frac{1}{2}m\|\mathbf{V}(t)\|^2 \Big|_a^b \\ &= \frac{1}{2}m\|\mathbf{V}(b)\|^2 - \frac{1}{2}m\|\mathbf{V}(a)\|^2 = K_b - K_a \end{aligned}$$

where K_b is the kinetic energy of the particle at Q and K_a is the kinetic energy at P. ●

PROGRESS TEST 1

1. Suppose the force $\mathbf{F}(x, y) = \langle 3x^2 + 10y, 3y^2 - 5x^2 \rangle$ moves a particle from $(-1, -2)$ to $(1, 2)$ along the cubic curve C given parametrically by $\mathbf{X}(t) = \langle t, 2t^3 \rangle$, $-1 \le t \le 1$. Determine the work done.

THE LINE INTEGRAL

In Chap. 6 our attempts at finding the area of a plane region led us to the definition of the definite integral, which in turn proved to be a very fruitful concept in many contexts quite distinct from the original area problem. In much the same way, our above development of an integral definition for work leads us to the fundamental and fruitful notion of a line integral. We merely replace the force field by a general vector field.

Suppose $\mathbf{F}: D \to R_2$ is the vector field

$$\mathbf{F}(x,y) = \langle M(x,y), N(x,y)\rangle$$

where D is an open subset of R_2. Let C be a smooth curve in D with the parameterization C: $\mathbf{X}(t)$, $a \le t \le b$, and let P, $\|P\|$ and t_i^* be as in the development of work on page 739.

21.2.3 **Definition** The *line integral* of $\mathbf{F}$ over C, denoted by

$$\int_C \mathbf{F}(\mathbf{X})\cdot d\mathbf{X} \qquad \text{or} \qquad \int_C M\,dx + N\,dy$$

is defined by

$$\int_C \mathbf{F}(\mathbf{X})\cdot d\mathbf{X} = \lim_{\substack{n\to+\infty \\ \|P\|\to 0}} \sum_{i=1}^{n} \mathbf{F}(\mathbf{X}(t_i^*))\cdot\mathbf{X}'(t_i^*)\,\Delta t_i$$

provided this limit exists independently of how the t_i^*'s are chosen in $[t_{i-1}, t_i]$.

As noted in our discussion of work, if $\mathbf{F}$ has continuous components on D, then the sums in (21.2.3) are Riemann sums for continuous scalar functions on $[a, b]$. This observation, together with (6.3.2), enables us to evaluate $\int_C \mathbf{F}(\mathbf{X})\cdot d\mathbf{X}$.

21.2.4 **Theorem** Suppose $\mathbf{F}(x,y) = \langle M(x,y), N(x,y)\rangle$, where M and N are continuous on the open subset D of R_2. Let C be a smooth curve in D with the parameterization $\mathbf{X}(t)$, $a \le t \le b$. Then $\int_C \mathbf{F}(\mathbf{X})\cdot d\mathbf{X}$ exists and

$$\int_C \mathbf{F}(\mathbf{X})\cdot d\mathbf{X} = \int_a^b \mathbf{F}(\mathbf{X}(t))\cdot\mathbf{X}'(t)\,dt$$

Furthermore, the value of this integral is independent of the particular parameterization of C.

Our notations become clearer when we write the integrands in component form:

$$\begin{aligned}\mathbf{F}(\mathbf{X}(t))\cdot\mathbf{X}'(t)\,dt = \langle M, N\rangle\cdot\langle x'(t), y'(t)\rangle\,dt &= Mx'(t)\,dt + Ny'(t)\,dt \\ &= M\,dx + N\,dy\end{aligned}$$

This yields the computationally useful formula:

21.2.5
$$\int_C M\,dx + N\,dy = \int_a^b M(x(t),y(t))x'(t)\,dt + N(x(t),y(t))y'(t)\,dt$$

If a smooth curve C is decomposed into pieces $C_1, \ldots, C_n$, as in Fig. 21.8(a), then the additivity property of definite integrals implies that

21.2.6
$$\int_C M\,dx + N\,dy = \sum_{i=1}^{n} \int_{C_i} M\,dx + N\,dy$$

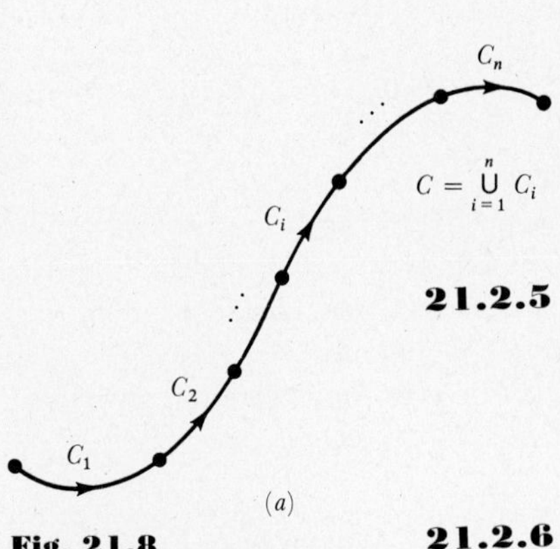

Fig. 21.8

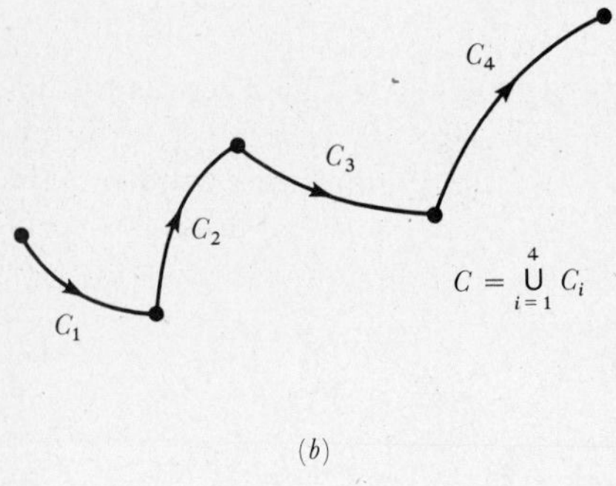

(b)

Hence we extend the notion of a line integral to piecewise smooth curves $C = \cup_{i=1}^{n} C_i$ by insisting the line integral over C be defined as the sum of the integrals over the smooth curves C_i; see Fig. 21.8(b) to the left.

If C has the parameterization $\mathbf{X}(t)$, $a \le t \le b$, then, by (21.2.1), $-C$ has the parameterization $\mathbf{Y}(s)$, $a \le s \le b$, where $\mathbf{Y}(s) = \mathbf{X}(a + b - s)$. By the Chain Rule, $\mathbf{Y}'(s) = \mathbf{X}'(a + b - s)(-1)$. Thus

$$\int_{-C} M\,dx + N\,dy = \int_a^b \mathbf{F}(\mathbf{Y}(s)) \cdot \mathbf{Y}'(s)\,ds$$

$$= \int_a^b \mathbf{F}(\mathbf{X}(a + b - s)) \cdot \mathbf{X}'(a + b - s)(-1)\,ds$$

Now if $t = a + b - s$, then $dt = -ds$. Furthermore, $s = a$ implies $t = b$, and $s = b$ implies $t = a$. With this substitution we have

$$\int_{-C} M\,dx + N\,dy = \int_b^a \mathbf{F}(\mathbf{X}(t)) \cdot \mathbf{X}'(t)\,dt$$

$$= -\int_a^b \mathbf{F}(\mathbf{X}(t)) \cdot \mathbf{X}'(t)\,dt = -\int_C M\,dx + N\,dy$$

Thus we conclude

21.2.7
$$\int_{-C} M\,dx + N\,dy = -\int_C M\,dx + N\,dy$$

Example 3 Evaluate $\int_C xy\,dx + y^2\,dy$, where

(a) C: $\mathbf{X}(t) = \langle t, t^2 \rangle$, $0 \le t \le 1$.
(b) C: $\mathbf{X}(t) = \langle t, t \rangle$, $0 \le t \le 1$.

Solutions (a) Since $x(t) = t, y(t) = t^2$, we have $dx = x'(t)\,dt = dt$ and $dy = y'(t)\,dt = 2t\,dt$. Thus

$$\int_C xy\,dx + y^2\,dy = \int_0^1 [t(t^2)\,dt + (t^2)^2 2t\,dt]$$

$$= \int_0^1 (t^3 + 2t^5)\,dt = \frac{t^4}{4} + \frac{t^6}{3}\bigg|_0^1 = \frac{7}{12}$$

(b) Now $x(t) = y(t) = t$, so $dx = dy = dt$, and

$$\int_C xy\,dx + y^2\,dy = \int_0^1 (t^2\,dt + t^2\,dt)$$

$$= \int_0^1 2t^2\,dt = \frac{2t^3}{3}\bigg|_0^1 = \frac{2}{3} \quad \bullet$$

Note that in Example 3 the curves are, respectively, the piece of the parabola $y = x^2$ joining $(0, 0)$ to $(1, 1)$ and the straight-line segment joining the *same* two points along $y = x$; see Fig. 21.9(a) and (b), respectively, top of page 743.

In situations of this type, where the curve C is given by $y = f(x)$, $a \le x \le b$ [or $x = g(y)$, $c \le y \le d$], one can work directly with the functional equation. For example, in part (a) of Example 3,

$$C\colon\quad y = f(x) = x^2 \qquad 0 \le x \le 1$$

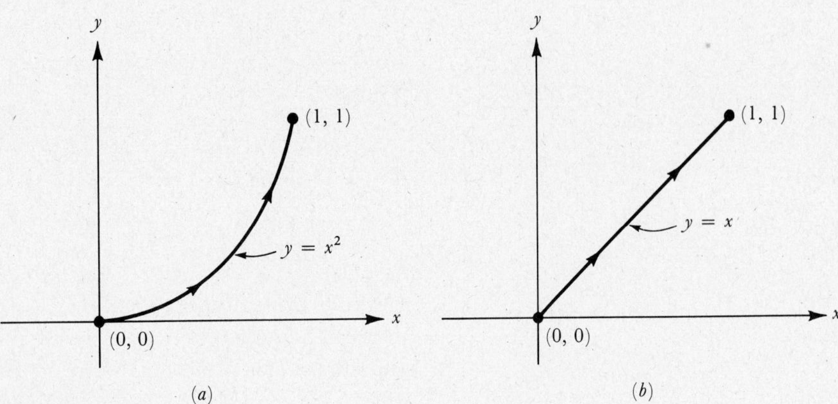

Fig. 21.9

thus $dy = f'(x)\,dx = 2x\,dx$, and so

$$\int_C xy\,dx + y^2\,dy = \int_0^1 [x \cdot x^2\,dx + (x^2)^2\,2x\,dx] = \int_0^1 (x^3 + 2x^5)\,dx = \frac{7}{12}$$

Example 4 Evaluate $\int_C 2xy\,dx + x^2\,dy$, where

(*a*) C is the curve obtained by traversing the x axis from $(-1, 0)$ to $(1, 0)$ and then counterclockwise along the parabola $y = 1 - x^2$ to $(0, 1)$.

(*b*) C is the directed straight-line segment joining $(-1, 0)$ to $(0, 1)$.

Solutions (*a*) C is a piecewise smooth curve consisting of a directed line segment C_1 and a parabolic arc C_2 (see Fig. 21.10). Now C_1 has the parameterization

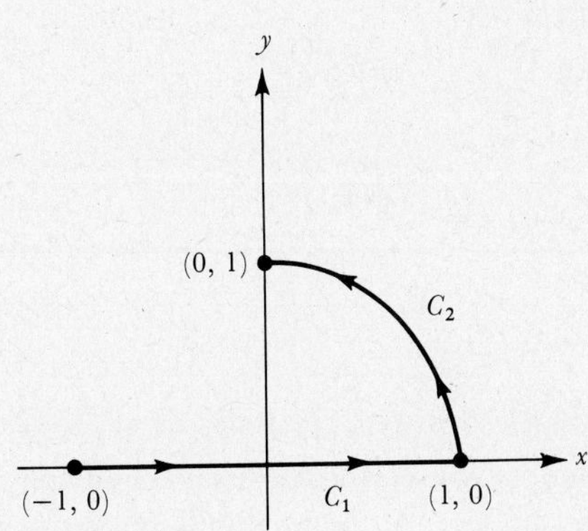

Fig. 21.10

$$C_1\colon\ \langle t, 0\rangle,\ -1 \le t \le 1$$

If we try to parameterize C_2 by letting $x = t, y = 1 - t^2, 0 \le t \le 1$, we obtain the wrong orientation. We can use (21.2.1) to obtain

$$C_2\colon\ \langle 1 - t, 1 - (1 - t)^2\rangle = \langle 1 - t, 2t - t^2\rangle, 0 \le t \le 1$$

Alternatively, we can work with

$$-C_2\colon\ \langle t, 1 - t^2\rangle,\ 0 \le t \le 1$$

and employ (21.2.7). We shall elect the latter course. Thus

$$\begin{aligned}\int_C 2xy\,dx + x^2\,dy &= \int_{C_1} 2xy\,dx + x^2\,dy + \int_{C_2} 2xy\,dx + x^2\,dy\\ &= \int_{C_1} 2xy\,dx + x^2\,dy - \int_{-C_2} 2xy\,dx + x^2\,dy\\ &= \int_{-1}^1 (2t \cdot 0\,dt + t^2 \cdot 0) - \int_0^1 [2t(1 - t^2)\,dt + t^2(-2t\,dt)]\\ &= -\int_0^1 (2t - 4t^3)\,dt = t^4 - t^2\Big|_0^1 = 0\end{aligned}$$

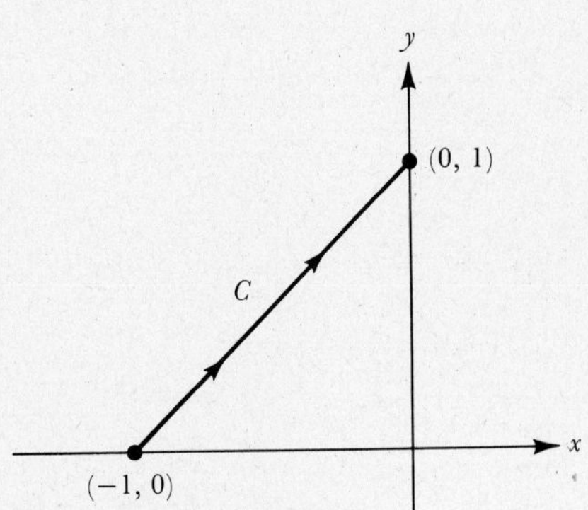

Fig. 21.11

(*b*) By (17.5.1), a directed line segment joining P to Q (see Fig. 21.11) can always be parameterized by the vector equation

$$\mathbf{X}(t) = \mathbf{P} + t\,\overrightarrow{PQ}, 0 \le t \le 1$$

In this case $P = (-1, 0)$ and $Q = (0, 1)$, so $\overrightarrow{PQ} = \langle 1, 1\rangle$, and hence

$$C: \quad \mathbf{X}(t) = \langle -1, 0\rangle + t\langle 1, 1\rangle = \langle -1 + t, t\rangle,\ 0 \le t \le 1$$

Thus

$$\int_C 2xy\,dx + x^2\,dy = \int_0^1 [2(-1+t)t\,dt + (-1+t)^2\,dt]$$
$$= \int_0^1 (3t^2 - 4t + 1)\,dt = t^3 - 2t^2 + t\Big|_0^1 = 0 \quad \bullet$$

Since both of the curves in Example 3 join (0, 0) to (1, 1), we see that the value of a line integral over a curve joining two points will in general depend on the curve. On the other hand, in Example 4 the integrals over two distinct curves joining (−1, 0) to (0, 1) have the same value. It appears that integrals of *certain* vector fields might be independent of the path that one chooses to join two points. We shall investigate this question thoroughly later.

We close this section by giving an example of a line integral over a closed curve.

Example 5 Evaluate

$$\int_C \frac{-y}{x^2+y^2}\,dx + \frac{x}{x^2+y^2}\,dy$$

where C is the unit circle oriented in the counterclockwise direction.

Solution Now C: $\langle \cos t, \sin t\rangle$, $0 \le t \le 2\pi$, so $dx = -\sin t\,dt$ and $dy = \cos t\,dt$. Thus

$$\int_C \frac{-y}{x^2+y^2}\,dx + \frac{x}{x^2+y^2}\,dy$$
$$= \int_0^{2\pi} \left[\frac{-\sin t(-\sin t\,dt)}{\sin^2 t + \cos^2 t} + \frac{\cos t(\cos t\,dt)}{\sin^2 t + \cos^2 t}\right]$$
$$= \int_0^{2\pi} \left[\frac{\sin^2 t\,dt}{1} + \frac{\cos^2 t\,dt}{1}\right] = \int_0^{2\pi} 1 \cdot dt = 2\pi \quad \bullet$$

PROGRESS TEST 2

1. Evaluate $\int_C x^2y\,dx + 2x\,dy$, where C is the square with vertices (0, 0), (0, 2), (2, 2), (2, 0) oriented in the clockwise direction.

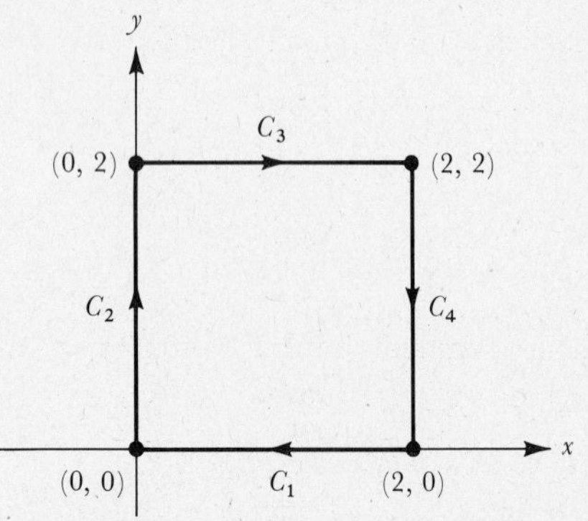

2. Evaluate $\int_C (x - y^2)\,dx + (x^2 + y)\,dy$, where C is the piecewise smooth curve obtained by traversing the line $y = x$ from (0, 0) to (1, 1), and then clockwise along $y = 2 - x^2$ to $(\sqrt{2}, 0)$.

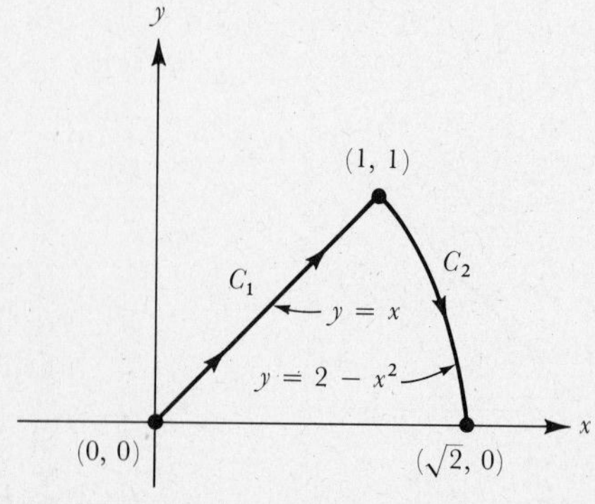

SECTION 21.2 EXERCISES

In Exercises 1 to 10 evaluate the line integral over the given curve.

1. $\int_C (3x^2y - 2x)\,dx + (x^3 + 3y^2)\,dy$; C: $\mathbf{X}(t) = \langle 2t+1, t^2\rangle$, $0 \le t \le 1$

2. $\int_C (3x^2y - 2x)\,dx + (x^3 + 3y^2)\,dy$; C is the directed line segment joining $(1, 0)$ to $(3, 1)$.

3. $\int_C (3x^2y - 2x)\,dx + (x^3 + 3y^2)\,dy$; C is the triangle with vertices $(-2, 0)$, $(0, 1)$, and $(3, 0)$ oriented in the counterclockwise direction.

4. $\int_C \sin y\,dx + x\cos y\,dy$; C is the directed line segment joining $(1, \pi/3)$ to $(2, \pi/2)$.

5. $\int_C (x + y)\,dx + (y + x^2)\,dy$; C is the curve joining $(-3, -3)$ to $(0, 0)$ along the line $y = x$ and then to $(1, 1)$ along $y = x^3$.

6. $\int_C (2xy^2 + 1)\,dx + (2x^2y - 1)\,dy$; C is the rectangle with vertices $(-2, 0)$, $(1, 0)$, $(1, 1)$, and $(-2, 1)$ oriented in the counterclockwise direction.

7. $\int_C (xe^y)\,dx + (x^2e^y)\,dy$; C: $\mathbf{X}(t) = \langle t, t\rangle$, $0 \le t \le 1$

8. $\int_C x^2\cos y\,dx - y^2\sin x\,dy$; C: $\mathbf{X}(t) = \langle t, t\rangle$, $0 \le t \le 1$

9. $\displaystyle\int_C \frac{-y\,dx}{x^2 + y^2} + \frac{x\,dy}{x^2 + y^2}$; C is the ellipse $\dfrac{x^2}{4} + \dfrac{y^2}{9} = 1$ oriented in the counterclockwise direction.

10. $\displaystyle\int_C \frac{2\,dx}{y} + \frac{x\,dy}{y^2}$; C is the straight line segment joining the point $(0, -3)$ to the point $(1, -1)$.

11. Let $\mathbf{F}(x, y, z) = \langle M(x, y, z), N(x, y, z), P(x, y, z)\rangle$ be a three-dimensional vector field defined on some open subset D of R_3 and let C be a smooth curve in D with the parameterization $\mathbf{X}(t) = \langle x(t), y(t), z(t)\rangle$, $a \le t \le b$. Using (21.2.3) as a model, define $\int_C M\,dx + N\,dy + P\,dz$. State the analog of (21.2.5).

In Exercises 12 to 15 evaluate the line integral over the given curve (see Exercise 11).

12. $\int_C xy\,dx + yz\,dy + xz\,dz$; C: $\mathbf{X}(t) = \langle t, t^2, t^3\rangle$, $0 \le t \le 1$

13. $\int_C x^2yz\,dx - xy^2\,dy + z^3\,dz$; C: $\mathbf{X}(t) = \langle t, 1 - t, t^2\rangle$, $0 \le t \le 1$

14. $\int_C x^2\,dx + y^2\,dy + z^2\,dz$; C: $\mathbf{X}(t) = \langle \cos t, \sin t, t\rangle$, $0 \le t \le \pi/2$

15. $\int_C e^x\,dx + e^y\,dy + e^z\,dz$; C: $\mathbf{X}(t) = \langle t^2, t, t^2\rangle$, $0 \le t \le 1$

16. If $\mathbf{X}(t)$, $a \le t \le b$, and $\mathbf{Y}(s)$ $c \le s \le d$, are both parameterizations for the same smooth curve C, then it can be shown that there exists a scalar function $t = h(s)$ defined on $[a, b]$ with range $[c, d]$ such that $h(c) = a$ and $h(d) = b$, $h'(s) > 0$ on $[c, d]$, and $\mathbf{Y}(s) = \mathbf{X}(h(s))$, $c \le s \le d$. Using this information, show that $\int_C \mathbf{F}(\mathbf{X})\cdot d\mathbf{X}$ is independent of the parameterization of C.

In Exercises 17 to 20 use this definition to evaluate the given integral. Let $f(x, y)$ be continuous on D and let C: $\langle x(t), y(t)\rangle$, $a \le t \le b$, be a smooth curve contained in D. The line integral of the scalar function f over C with respect to arc length is defined by $\int_C f(x, y)\,ds = \int_a^b f(x(t), y(t))\sqrt{[x'(t)]^2 + [y'(t)]^2}\,dt$

17. $\int_C x^2y^3\,ds$; C: $\langle \sin t, \cos t\rangle$, $0 \le t \le \pi/2$

18. $\int_C x^2y^2\,ds$; C: $\langle \cos t, \sin t\rangle$, $0 \le t \le \pi/4$

19. $\int_C (y^3 + x)\,ds$; C: $\langle t^3, t\rangle$, $0 \le t \le 1$

20. $\int_C xy^{2/3}\,ds$; C: $\langle t, t^{3/2}\rangle$, $0 \le t \le 1$

A thin wire is assumed to occupy the curve C. Let $f(x, y)$ denote the density (mass per unit length) of the wire at the point (x, y) on C. In Exercises 21 to 25 derive an integral definition for the given quantity. (See the discussion preceding Exercises 17 to 20.)

21. M, the mass of the wire.

22. M_x, the first moment of the wire about the x axis.

23. M_y, the first moment of the wire about the y axis.

24. I_x, the second moment of the wire about the x axis.

25. I_y, the second moment of the wire about the y axis.

Section 21.3: Independence of the Path

Objectives:

1. Determine when a line integral is independent of the path.

2. Evaluate a path-independent line integral using the Fundamental Theorem for Line Integrals.

INTRODUCTION: CONNECTED REGIONS

In this section we shall show that the line integral of a *conservative* vector field over a curve C depends *only* on the endpoints of C and not on the curve itself. This implies the reasonable conclusion that the work done by a conservative force field in moving an object from a point A to a point B depends only on the points and not on the path the object travels in going from A to B. To state these results precisely we shall need:

21.3.1 **Definition** An open subset D of R_2 is *connected* if any two points in D can be joined by a piecewise smooth curve that lies wholly within D.

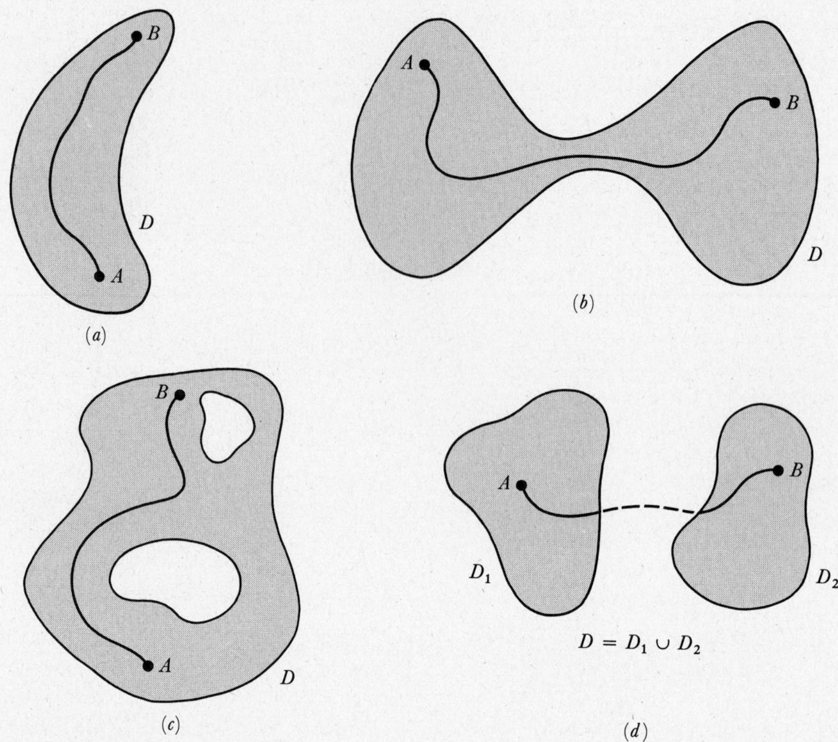

Fig. 21.12

This states mathematically what we *normally* mean by a connected set—one which does not consist of disjoint pieces. Each of the sets in Fig. 21.12(*a*), (*b*), and (*c*) is connected, whereas the region in Fig. 21.12(*d*) is not.

In dealing with connected regions it is often necessary to distinguish between those in (*a*) and (*b*) in Fig. 21.12, which do *not* have "holes," and (*c*), which does.

21.3.2 **Definition** An open connected subset D of R_2 is *simply connected* if whenever C is a simple *closed* piecewise smooth curve lying in D, then all points enclosed by C also lie in D.

The region sketched in Fig. 21.12(*c*) is connected but not simply connected.

21.3.3 **Definition** Suppose D is an open connected subset of R_2 and $\mathbf{F} = \langle M, N \rangle$ is a vector field defined on D. We say that the line integral of $\mathbf{F}$ is *independent of the path* in D if, given any P and Q in D, then $\int_C M\,dx + N\,dy$ is the same for all piecewise smooth curves C joining P to Q and lying in D.

If $\int_C M\,dx + N\,dy$ is independent of the path in D and if P and Q are in D, then we shall often write

$$\int_P^Q M\,dx + N\,dy$$

to denote the common value of the line integral over any curve joining P to Q.

INDEPENDENCE OF THE PATH

By the second form of the Fundamental Theorem of the Calculus (6.3.7),

$$\int_a^b f'(t)\,dt = f(b) - f(a)$$

This means that if the integrand is a derivative, then the value of the definite integral *depends only on the endpoints of the interval*. The following is an analogous version for line integrals.

21.3.4 **The Fundamental Theorem of Calculus for Line Integrals** Suppose $\mathbf{F} = \nabla f$, where f has continuous first partials on the open set D. If P and Q are any two points in D and if C is any piecewise smooth curve in D joining P to Q, then

$$\int_C \nabla f \cdot d\mathbf{X} = f(Q) - f(P)$$

Proof: We shall assume that C is smooth. The extension to piecewise smooth curves is routine (see Exercise 17). Suppose C has the parameterization $\mathbf{X}(t)$, $a \le t \le b$. Then

$$\begin{aligned}
\int_C \nabla f \cdot d\mathbf{X} &= \int_a^b \nabla f(\mathbf{X}(t)) \cdot \mathbf{X}'(t)\,dt \\
&= \int_a^b [f_x(\mathbf{X}(t))x'(t) + f_y(\mathbf{X}(t))y'(t)]\,dt \\
&= \int_a^b D_t[f(\mathbf{X}(t))]\,dt \qquad \text{by the Chain Rule (19.5.3)} \\
&= f(\mathbf{X}(b)) - f(\mathbf{X}(a)) \qquad \text{by (6.3.7)} \\
&= f(\mathbf{Q}) - f(\mathbf{P}) \quad \blacksquare
\end{aligned}$$

We are now in a position to state several equivalent conditions to the path independence of a line integral. First we assume:
1. D is an open connected subset of R.
2. $\mathbf{F} = \langle M, N \rangle$, where M, N have continuous first partials in D.

21.3.5 **Theorem** The line integral $\int_C M\,dx + N\,dy$ is independent of the path in D if and only if $\mathbf{F}$ is conservative.

Proof: Assume $\mathbf{F}$ is conservative with, say, $\mathbf{F} = \nabla f$. Then by (21.3.4), $\int_C M\,dx + N\,dy$ depends only on the endpoints of C and not C itself. Thus $\int_C M\,dx + N\,dy$ is independent of the path in D.

Conversely, assume that the integral of $\mathbf{F}$ is independent of the path in D. Let (a, b) be some fixed point in D, and for each (x, y) in D define

$$f(x, y) = \int_C M\,dx + N\,dy$$

where C is any piecewise smooth curve joining (a, b) to (x, y). Since D is open, there exists some open disk about (x, y) lying entirely within D. It follows that we can choose curves C and $\hat{C}$ in D, as in Fig. 21.13. Now

$$\begin{aligned}
f(x, y) &= \int_C M\,dx + N\,dy \\
&= \int_{C_1} M\,dx + N\,dy + \int_{C_2} M\,dx + N\,dy \\
&= \int_{(a,b)}^{(x_0,y)} M\,dx + N\,dy + \int_{(x_0,y)}^{(x,y)} M\,dx + N\,dy
\end{aligned}$$

Now the first integral depends only on y, so

$$\frac{\partial}{\partial x}\left[\int_{(a,b)}^{(x_0,y)} M\,dx + N\,dy\right] = 0$$

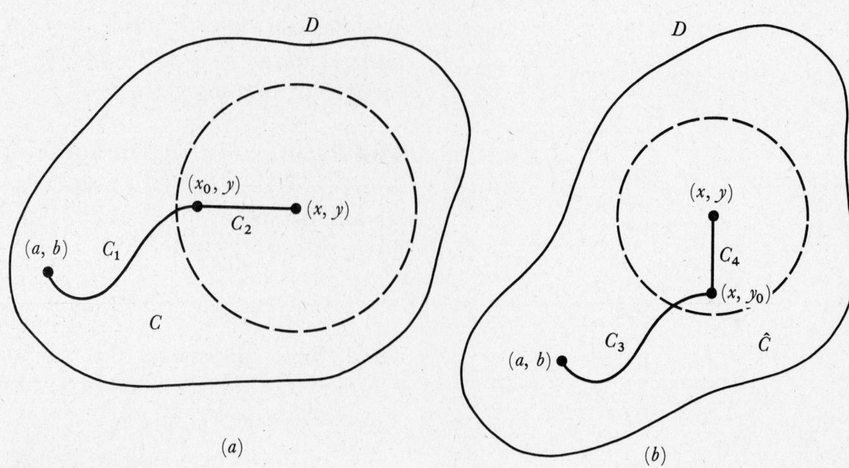

Fig. 21.13

Also, on C_2, $dy = 0$, so

$$\frac{\partial}{\partial x}\left[\int_{(x_0,y)}^{(x,y)} M\,dx + N\,dy\right] = \frac{\partial}{\partial x}\left[\int_{(x_0,y)}^{(x,y)} M\,dx\right]$$
$$= \frac{\partial}{\partial x}\left[\int_{x_0}^{x} M(t,y)\,dx\right] = M(x,y)$$

Thus $f_x = M$. By a similar argument we can work with $\widehat{C}$ to conclude that $f_y = N$. It follows that $\mathbf{F} = \nabla f$, so $\mathbf{F}$ is conservative. ■

21.3.6 **Theorem** The line integral $\int_C M\,dx + N\,dy$ is independent of the path in D if and only if $\int_K M\,dx + N\,dy = 0$ for every piecewise smooth closed curve K contained in D.

Proof: Let P, Q be any two points in D and let C_1, C_2 be any smooth curves joining P to Q; see Fig. 21.14(a). Then $K = C_1 \cup -C_2$ is a piecewise smooth and closed curve [see Fig. 21.14(b)] and so

$$0 = \int_K M\,dx + N\,dy$$
$$= \int_{C_1 \cup -C_2} M\,dx + N\,dy$$
$$= \int_{C_1} M\,dx + N\,dy + \int_{-C_2} M\,dx + N\,dy = \int_{C_1} M\,dx + N\,dy - \int_{C_2} M\,dx + N\,dy$$

Thus

$$\int_{C_1} M\,dx + N\,dy = \int_{C_2} M\,dx + N\,dy$$

and so the integral is independent of the path. The proof of the converse is the content of Exercise 18. ■

In Sec. 21.1 we noted that if $\mathbf{F} = \langle M, N\rangle$ is conservative, then $M_y = N_x$. Now if $\mathbf{F}$ happens to be the vector field

$$\mathbf{F}(x,y) = \left\langle \frac{-y}{x^2+y^2}, \frac{x}{x^2+y^2}\right\rangle$$

and if $D = \{(x,y)\,|\,(x,y) \neq (0,0)\}$, then M_y and N_x are continuous in D and $M_y = N_x = (y^2 - x^2)/(x^2+y^2)^2$. If C is the closed unit circle, then by Example

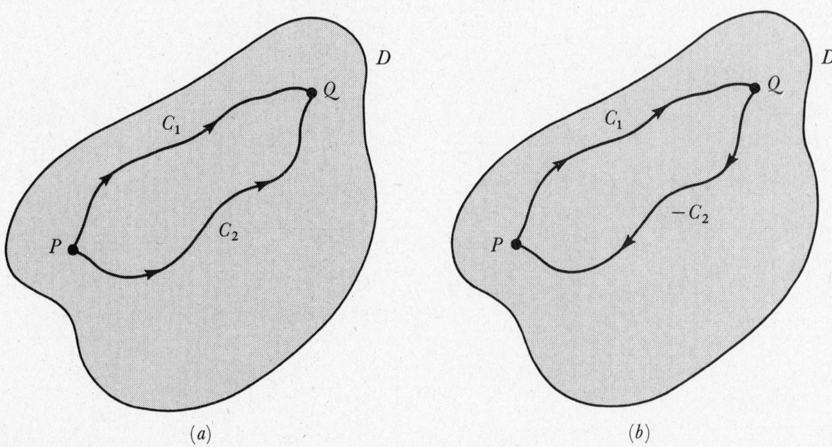

Fig. 21.14

5, Sec. 21.2, page 744, $\int_C M\,dx + N\,dy = 2\pi \neq 0$. Hence, by (21.3.6) and (21.3.5), **F** is *not* conservative in D.

It can be shown that if we strengthen our original condition (1) and insist that D be *simply connected,* then the condition $M_y = N_x$ *is* sufficient to ensure independence of the path.

21.3.7 If D is simply connected, then $\int_C M\,dx + N\,dy$ is independent of the path in D if and only if $M_y = N_x$.

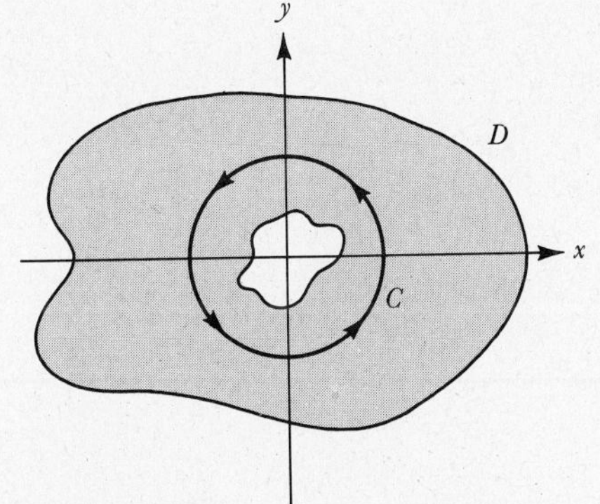

Fig. 21.15

The problem with the above integral of **F** around the unit circle C is that to ensure the continuity of the component functions (and their partials), we must exclude $(0, 0)$ from any connected region containing C. Since C encircles the origin, any such region will have a hole and hence will *not* be simply connected (see Fig. 21.15).

Suppose we let $\widehat{D}$ be the simply connected open subset of R_2 consisting of the x-y plane with the negative x axis and the origin removed. Now, we noted above, if $M = -y/(x^2 + y^2)$, $N = x/(x^2 + y^2)$, then $M_y = N_x$. Furthermore, M and N are continuous in $\widehat{D}$, so by (21.3.7) and (21.3.5), $\mathbf{F} = \langle M, N \rangle$ is a gradient. If we proceed formally as in Sec. 21.1 to find a function f such that $\mathbf{F} = \nabla f$, we obtain $f(x,y) = \arctan(y/x)$. However, f is not continuous on $\widehat{D}$! To see this, take any point on the positive y axis, say $(0, 1)$, and let $P = (x,y)$ tend to $(0, 1)$ along the curves indicated in Fig. 21.16(a). Along one curve $y/x \to +\infty$, so $f(x,y) \to \pi/2$, while along the other $y/x \to -\infty$, so $f(x,y) \to -\pi/2$. In Mis-

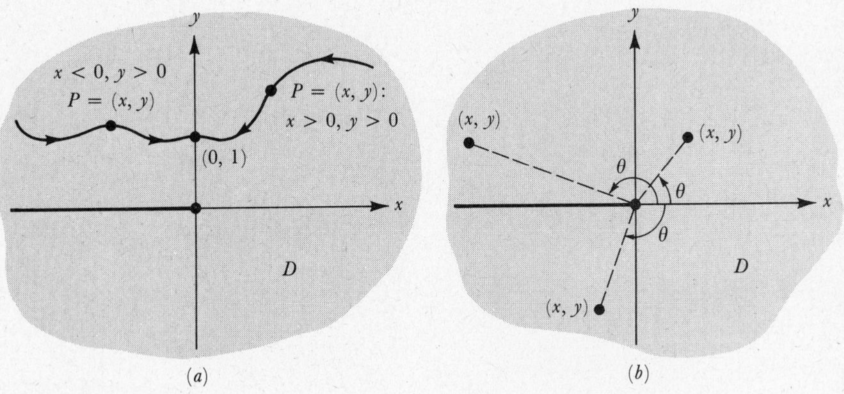

Fig. 21.16

cellaneous Exercise 2 you are asked to show that the correct choice of f is given by

$$f(x,y) = \begin{cases} \arctan(y/x) & x > 0 \\ \pi/2 & x = 0, y > 0 \\ \pi + \arctan(y/x) & x < 0, y > 0 \\ -\pi/2 & x = 0, y < 0 \\ -\pi + \arctan(y/x) & x < 0, y < 0 \end{cases}$$

Although this function may look gruesome, it is merely the function that assigns to each (x, y) its unique polar angle θ between $-\pi/2$ and $\pi/2$; see Fig. 21.16(b).

PROGRESS TEST 1

Evaluate each of the following:

1. $I = \int_{(0,0)}^{(1,1)} [(3x^2y + y^2 - 2x)\,dx + (x^3 + 2xy)\,dy]$

2. $I = \int_{(0,0)}^{(0,\pi/2)} [(e^x \cos y)\,dx - (e^x \sin y)\,dy]$

3. $I = \int_{(e^2,0)}^{(0,1)} \left[\frac{x}{x^2 + y^2}\,dx + \frac{y}{x^2 + y^2}\,dy\right]$

SECTION 21.3 EXERCISES

Evaluate each of the integrals in Exercises 1 to 10.

1. $\int_{(1,-1)}^{(0,2)} (4x^3y^2 + 3x^2)\,dx + (2x^4y - 3y^2)\,dy$

2. $\int_{(0,1)}^{(1,0)} (2xe^{2x}(x + 1) - \cos y)\,dx + (x \sin y + 2y)\,dy$

3. $\int_{(0,0)}^{(\pi,\pi/2)} (\sin y - y \cos x)\,dx + (x \cos y - \sin x)\,dy$

4. $\int_{(0,1)}^{(1,2)} \frac{x}{\sqrt{x^2 + y^2}}\,dx + \left[\frac{y}{\sqrt{x^2 + y^2}} - \frac{1}{y^2}\right] dy$

5. $\int_{(1,4)}^{(2,7)} \frac{y}{(x + y)^2}\,dx - \frac{x}{(x + y)^2}\,dy$

6. $\int_{(1,1)}^{(2,4)} \left(\frac{1}{y} + 2x\right) dx - \frac{x}{y^2}\,dy$

7. $\int_{(0,0)}^{(\pi/6,\pi/3)} (2 \sec x \tan x \tan y + 2x)\,dx + (\sec^2 x \sec^2 y)\,dy$

8. $\int_{(0,0)}^{(1,1)} \frac{(x - 1)\,dx}{(x - 1)^2 + y^2} + \frac{y\,dy}{(x - 1)^2 + y^2}$

9. $\int_{(1,1)}^{(2,\ln 2)} (-y^2e^{-xy})\,dx + (1 - xy)e^{-xy}\,dy$

10. $\int_{(0,0)}^{(1,0)} (y^2 \cosh x - 2x \cosh y)\,dx + (2y \sinh x - x^2 \sinh y)\,dy$

11. Show that

$$\int_C \left[\frac{x}{(x^2 + y^2)^{3/2}}\,dx + \frac{y}{(x^2 + y^2)^{3/2}}\,dy\right]$$

is independent of the path in $D = \{(x,y) \mid (x,y) \neq (0,0)\}$.

In Exercises 12 to 16 show that the integrals are not independent of the path in D using either (21.3.7) or (21.3.6).

12. $\int_C (3x^2 - 2y)\,dx + (2y + 2x)\,dy$; $D = R_2$

13. $\int_C (y^2 + y)\,dx + (2xy - e^y)\,dy$; $D = R_2$

14. $\int_C \frac{-1}{x^2y}\,dx + \frac{x + 1}{xy^2}\,dx$; $D = \{(x,y) \mid (x,y) \neq (0,0)\}$

15. $\int_C (y - y^2)\,dx + (2xy - x)\,dy$; $D = R_2$

16. $\int_C \frac{-y}{(x - 1)^2 + y^2}\,dx + \frac{x - 1}{(x - 1)^2 + y^2}\,dy$; $D = \{(x,y) \mid (x,y) \neq (1,0)\}$

17. Prove Theorem (21.3.4) for piecewise smooth curves.

18. Show that if $\int_C M\,dx + N\,dy$ is independent of the path in D, then $\int_K M\,dx + N\,dy = 0$ for every piecewise smooth closed curve K contained in D.

Section 21.4: Green's Theorem

Objectives:

Understand and use Green's Theorem to aid in evaluating line integrals or double integrals.

INTRODUCTION

Green's Theorem relates the value of a line integral over the boundary of a region in the plane to a certain double integral over that region. It has several interesting physical interpretations, some of which will be discussed in Sec. 21.5. As we shall see shortly, it is also a multipurpose tool for evaluating integrals. We begin our development by discussing the types of regions for which Green's Theorem applies.

REGULAR REGIONS

We shall assume throughout this section that *all curves are piecewise smooth and simple.*

21.4.1 **Definition** A plane region D is called *regular* if it is closed and bounded, and if the boundary of D consists of a finite number of nonintersecting closed curves.

If D is a regular region, then we denote the boundary of D by ∂D. The boundary is said to have the *standard orientation* if each of the curves C_i making up the boundary is oriented so that the region is always on the left as one moves along C_i in the direction of increasing parameter. Some examples of regular regions with their standard oriented boundaries are shown in Fig. 21.17.

Note that the standard orientation condition forces the outer boundary curve to have the counterclockwise orientation, whereas all inner boundary curves are oriented in the clockwise direction.

If D is a regular region with the *standard orientation,* then we shall denote the sum of the integrals of each of the closed curves that make up ∂D by

$$\oint_{\partial D} M\,dx + N\,dy$$

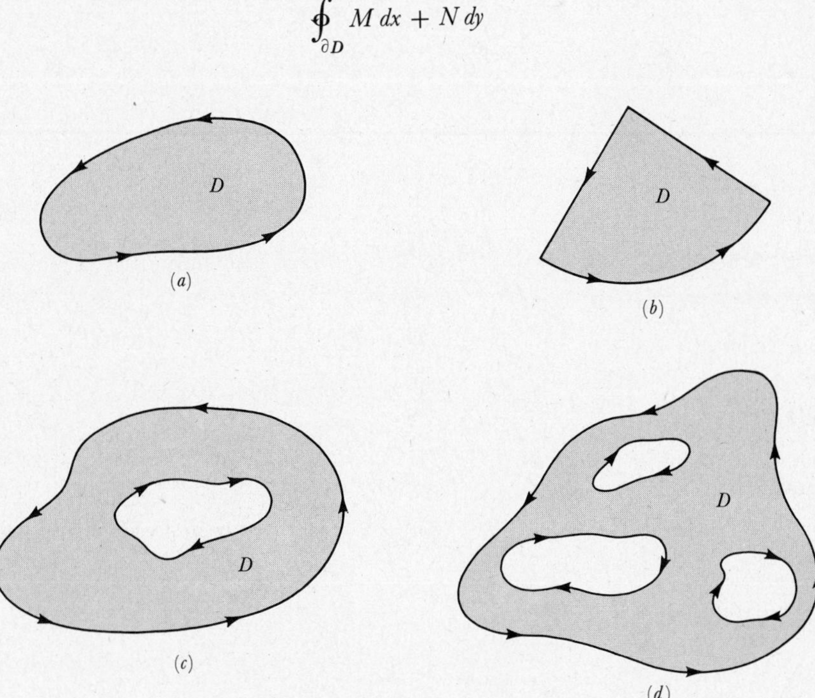

Fig. 21.17

GREEN'S THEOREM AND SOME EXAMPLES

21.4.2 **Green's Theorem** Suppose that M and N have continuous first partials in the regular region D. Then

$$\oint_{\partial D} M\,dx + M\,dy = \iint_D [N_x - M_y]\,dA$$

We shall defer a discussion of the proof of Green's Theorem until after we have had an opportunity to see how it is used in evaluating integrals. In particular, the following examples illustrate situations where Green's Theorem is used to:

1. Transform the problem of evaluating a line integral to that of evaluating a simpler double integral;
2. Transform the problem of evaluating a double integral to that of evaluating a simpler line integral;
3. Transform the problem of evaluating a line integral over a closed curve C to that of evaluating a line integral over a simpler closed curve.

Example 1 Evaluate

$$\oint_{\partial D} (x^2 - y^3)\,dx + (y^2 + x^3)\,dy$$

where D is the annular region of Fig. 21.18 with inner radius r_1 and outer radius r_2.

Solution Since D is a regular region, we apply Green's Theorem with

$$M(x,y) = x^2 - y^3 \qquad N(x,y) = y^2 + x^3$$
$$M_y(x,y) = -3y^2 \qquad N_x(x,y) = 3x^2$$

Thus

$$\oint_{\partial D} (x^2 - y^3)\,dx + (y^2 + x^3)\,dy = \iint_D 3(x^2 + y^2)\,dA$$

Evaluating this latter integral using polar coordinates, we have

$$\iint_D 3(x^2 + y^2)\,dA = 3\int_0^{2\pi}\int_{r_1}^{r_2} r^2 r\,dr\,d\theta = \frac{3\pi}{2}(r_2{}^4 - r_1{}^4) \quad \bullet$$

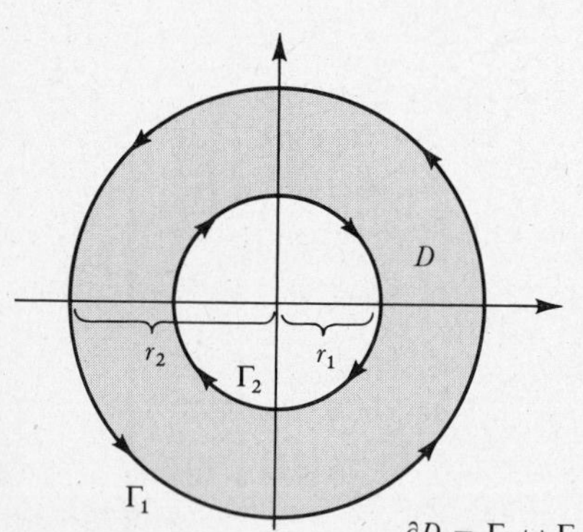

Fig. 21.18

If D is a region in the x-y plane, then we know that the area of D is given by

$$A(D) = \iint_D dA$$

In many instances, when D is regular, determining $A(D)$ can be simplified by Green's Theorem.

Example 2 Show that the area of a regular region D is given by

$$A(D) = \frac{1}{2}\oint_{\partial D} (-y\,dx + x\,dy)$$

Solution Let $M(x,y) = -y$ and $N(x,y) = x$. Then $N_x - M_y = 2$ and so, by Green's Theorem,

$$\frac{1}{2}\oint_{\partial D} (-y\,dx + x\,dy) = \frac{1}{2}\iint_D 2\,dA = \iint_D dA = A(D) \quad \bullet$$

Example 3 Use the result of Example 2 to find the area of the region enclosed by the ellipse $x^2/a^2 + y^2/b^2 = 1$.

Solution Let D be the region enclosed by the ellipse. Then D is a regular region and a counterclockwise parameterization for ∂D is given by

$$\langle a \cos t, b \sin t \rangle, \; 0 \le t \le 2\pi$$

Thus $dx = -a \sin t\, dt$, $dy = b \cos t\, dt$, and so

$$\begin{aligned} A(D) &= \frac{1}{2} \oint_{\partial D} [-y\, dx + x\, dy] \\ &= \frac{1}{2} \int_0^{2\pi} [(-b \sin t)(-a \sin t\, dt) + (a \cos t)(b \cos t\, dt)] \\ &= \frac{ab}{2} \int_0^{2\pi} (\sin^2 t + \cos^2 t)\, dt = \frac{ab}{2} \int_0^{2\pi} dt = \pi ab \quad \bullet \end{aligned}$$

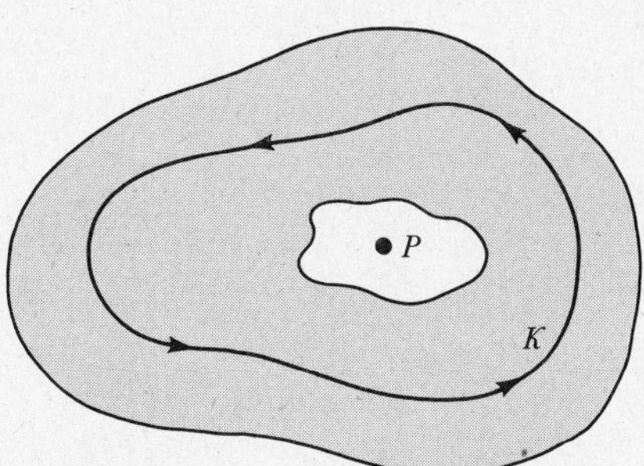

Region containing K but excluding P

Fig. 21.19

If K is a piecewise smooth simple closed curve, then the region inside K is simply connected. Hence if $M_y = N_x$ in this region, then by Theorems (21.3.7) and (21.3.6), we know that $\int_K M\, dx + N\, dy = 0$. However, if K encircles a point at which the partial derivatives are not continuous, then (as noted in Section 21.3) even though $M_y = N_x$ for all *other* points, we can no longer conclude that the line integral over K is zero. The problem, of course, is that any region containing K and excluding the bad point P will necessarily *not* be simply connected, and so we cannot apply Theorem 21.3.7 (see Fig. 21.19). Such a region will, however, be regular and we can use Green's Theorem to transform the problem of integrating over the curve K to the often simpler problem of integrating over a circle.

Specifically, let P be a point encircled by K at which the partials of M or N are discontinuous. Center a circle Γ about P of small enough radius to avoid intersecting K. Then D, the region inside K and outside Γ, is regular, so by Green's Theorem

$$\iint_D [N_x - M_y]\, dA = \oint_{\partial D} M\, dx + N\, dy = \int_K M\, dx + N\, dy + \int_\Gamma M\, dx + N\, dy$$

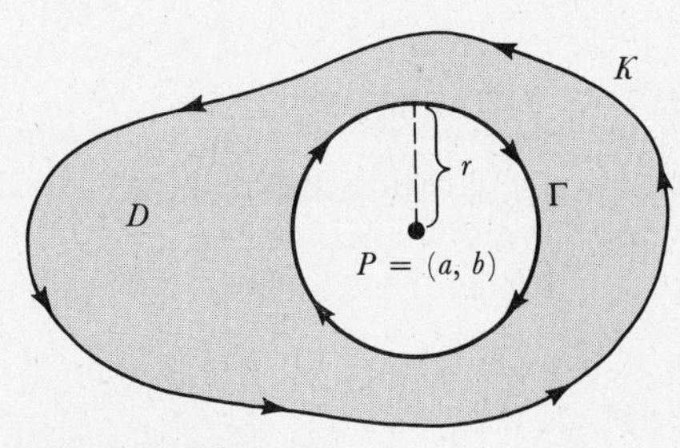

Fig. 21.20

where K has the counterclockwise orientation and Γ has the clockwise orientation. If $M_y = N_x$ in D, then $\iint_D [N_x - M_y]\, dA = 0$, so

$$\int_K M\, dx + N\, dy = -\int_\Gamma M\, dx + N\, dy = \int_{-\Gamma} M\, dx + N\, dy$$

where $-\Gamma$ now has the counterclockwise orientation. In particular, if $P = (a, b)$ and Γ has radius r, then

$$-\Gamma: \; \langle a + r \cos t, b + r \sin t \rangle \qquad 0 \le t \le 2\pi$$

Example 4 Evaluate

$$\int_K \frac{-y}{x^2 + y^2}\, dx + \frac{x}{x^2 + y^2}\, dy$$

where K is a closed curve oriented in the counterclockwise direction consisting of the piece of the ellipse $x^2/9 + y^2/16 = 1$ in the third quadrant joining $(-3, 0)$ to $(0, -4)$, the straight-line segment from $(0, -4)$ to $(3, 0)$, and the semicircle $y = \sqrt{9 - x^2}$ from $(3, 0)$ to $(-3, 0)$ as in Fig. 21.21.

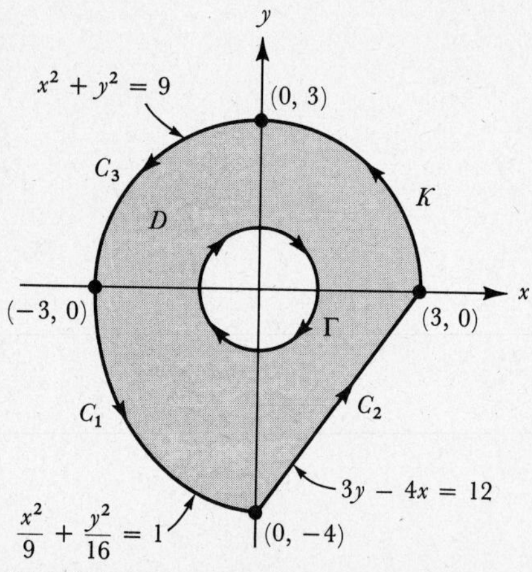

Fig. 21.21

Solution We note that $M_y = N_x = (y^2 - x^2)/(x^2 + y^2)^2$ for all $(x,y) \neq (0,0)$. Let Γ be the unit circle centered at the origin with the clockwise orientation. Then

$$-\Gamma: \quad \langle \cos t, \sin t \rangle \qquad 0 \leq t \leq 2\pi$$

and so, arguing as above and referring to Fig. 21.21, we have

$$\begin{aligned}\int_K \frac{-y}{x^2+y^2}\,dx + \frac{x}{x^2+y^2}\,dy &= \int_{-\Gamma} \frac{-y}{x^2+y^2}\,dx + \frac{x}{x^2+y^2}\,dy \\ &= \int_0^{2\pi} \left[\frac{-\sin t(-\sin t\,dt)}{\cos^2 t + \sin^2 t} + \frac{(\cos t)\cos t\,dt}{\cos^2 t + \sin^2 t}\right] \\ &= \int_0^{2\pi} dt = 2\pi \quad \bullet\end{aligned}$$

One can appreciate the advantage of the above approach by attempting a direct calculation of the line integral in Example 4 using the definition.

PROGRESS TEST 1

1. Let $A(R)$ denote the area of the regular region R. Show that

$$A(R) = \oint_{\partial R} x\,dy = -\oint_{\partial R} y\,dx$$

2. Use Green's Theorem to evaluate

$$\int_K \left(\frac{xy^3}{3} + 4y - e^{2x}\right) dx + (xy^2 + 2\sin y)\,dy$$

where K is the rectangle with vertices $(1,1)$, $(2,1)$, $(1,4)$, $(2,4)$ oriented in the counterclockwise direction.

3. Evaluate

$$\int_K \frac{-y}{(x+1)^2+y^2}\,dx + \frac{(x+1)}{(x+1)^2+y^2}\,dy$$

where K is the counterclockwise curve from $(0,3)$ to $(0,-3)$ along $x = y^2 - 9$ and then to $(0,3)$ along the y axis.

GREEN'S THEOREM: A DISCUSSION OF THE PROOF

We shall prove Green's Theorem for a special type of region and then indicate how the theorem can be extended to more general regular regions. Specifically, we assume that D is simultaneously of Type I and Type II, as discussed in Chap. 20. Thus D can be expressed in the form

$$D = \{(x,y) \mid a \leq x \leq b, f(x) \leq y \leq h(x)\} \tag{1}$$

and in the form

$$D = \{(x,y) \mid c \leq y \leq d, p(y) \leq x \leq q(y)\} \tag{2}$$

Such a region is sketched in Fig. 21.22(a) and (b).

Now assume M and N have continuous partials in D. It suffices to show that

$$\oint_{\partial D} M\,dx = -\iint_D M_y\,dA \tag{3}$$

and

$$\oint_{\partial D} N\,dy = \iint_D N_x\,dA \tag{4}$$

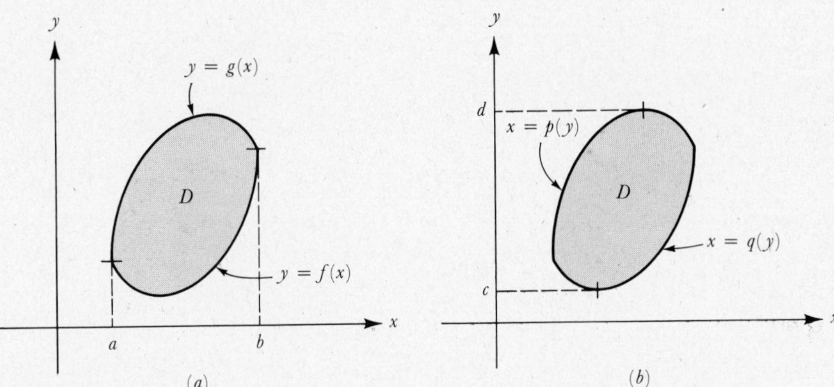

Fig. 21.22

for then we can add these integrals to obtain the conclusion of Green's Theorem. To establish (3) we work with D in the form (1), where C_1 and C_2 are as depicted in Fig. 21.23. Note that C_1 has the parameterization $\langle x, f(x)\rangle$, $a \leq x \leq b$, and $-C_2$ has the parameterization $\langle x, h(x)\rangle$, $a \leq x \leq b$. Now

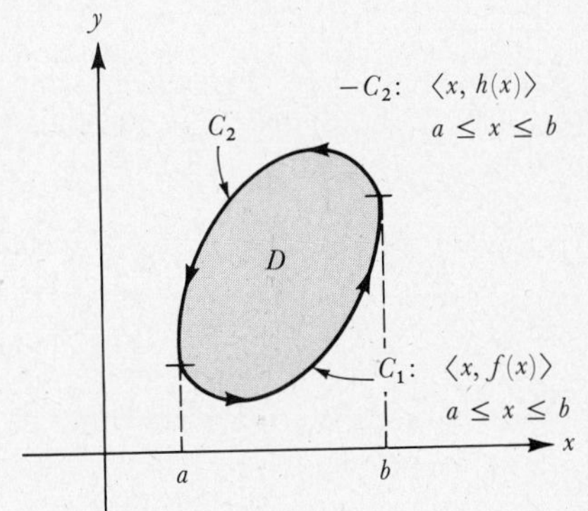

Fig. 21.23

$$-\iint_D M_y\,dA = -\int_a^b \int_{f(x)}^{h(x)} M_y(x,y)\,dy\,dx$$

$$= -\int_a^b \Big[M(x,y)\Big]_{f(x)}^{h(x)} dx \qquad \text{by (6.3.7)}$$

$$= -\int_a^b [M(x, h(x)) - M(x, f(x))]\,dx$$

$$= \int_a^b [M(x, f(x)) - M(x, h(x))]\,dx$$

Also,

$$\oint_{\partial D} M\,dx = \int_{C_1} M(x,y)\,dx + \int_{C_2} M(x,y)dx$$

$$= \int_{C_1} M(x,y)\,dx - \int_{-C_2} M(x,y)\,dx$$

$$= \int_a^b M(x, f(x))\,dx - \int_a^b M(x, h(x))\,dx$$

$$= \int_a^b [M(x, f(x)) - M(x, h(x))]\,dx$$

Thus (3) is established. The proof of (4) is essentially the same if one works with the characterization (2) for D (see Exercise 17).

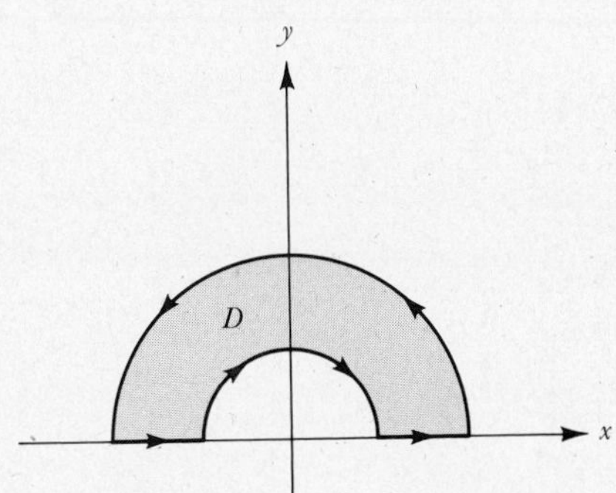

Fig. 21.24

We can gain some insight into how the proof of Green's Theorem could be extended to more general regions by considering the semiannular region D in Fig. 21.24. Now D is not a region of the type considered in our proof, but D can be broken into the two regions D_1 and D_2, which *are* of this type (see Fig. 21.25).

Applying Green's Theorem to D_1 and D_2, we have

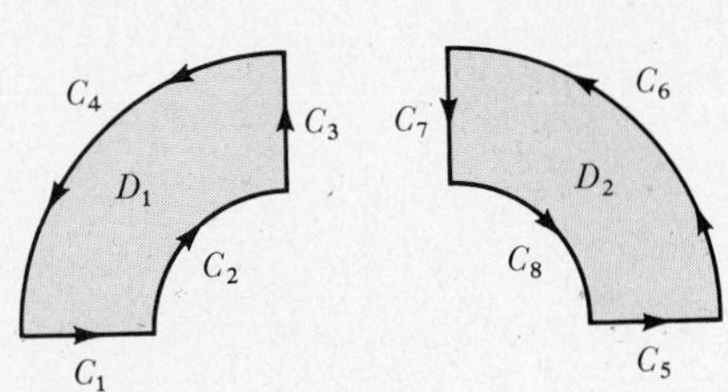

Fig. 21.25

$$\iint_{D_1} [N_x - M_y]\,dA = \sum_{i=1}^{4} \int_{C_i} M\,dx + N\,dy$$

and

$$\iint_{D_2} [N_x - M_y]\,dA = \sum_{i=5}^{8} \int_{C_i} M\,dx + N\,dy$$

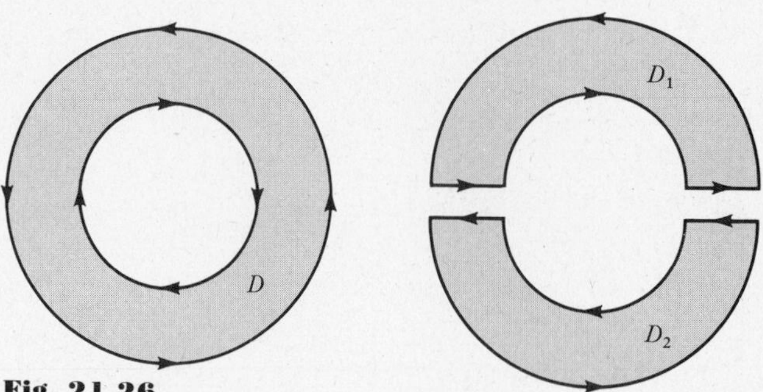

Fig. 21.26

Now

$$\iint_D [N_x - M_y]\,dA = \iint_{D_1} [N_x - M_y]\,dA + \iint_{D_2} [N_x - M_y]\,dA$$
$$= \sum_{i=1}^{8} \int_{C_i} M\,dx + N\,dy$$

However, $-C_3 = C_7$, and so

$$\int_{C_3} M\,dx + N\,dy + \int_{C_7} M\,dx + N\,dy = \int_{C_3} M\,dx + N\,dy - \int_{C_3} M\,dx + N\,dy = 0$$

Thus

$$\iint_D [N_x - M_y]\,dA = \sum_{\substack{i=1,\\ i\neq 3, i\neq 7}}^{8} \int_{C_i} M\,dx + N\,dy = \oint_{\partial D} M\,dx + N\,dy$$

We can in turn now obtain a version of Green's Theorem for an annular region (Fig. 21.26) by breaking it into two semiannular regions and proceeding much as above (see Exercise 18). In fact, a general proof of Green's Theorem can be constructed by breaking up a regular region into carefully chosen subregions and proceeding in a like fashion.

SECTION 21.4 EXERCISES

In Exercises 1 to 5 verify Green's Theorem by directly evaluating both $\oint_D M\,dx + N\,dy$ and $\iint_D [N_x - M_y]\,dA$.

1. $M(x,y) = xy$, $N(x,y) = -x$; $D = \{(x,y) \mid 0 \leq x \leq 2, 0 \leq y \leq 5\}$

2. $M(x,y) = y$, $N(x,y) = 0$; D is the annular region inside the circle $x^2 + y^2 = 9$ and outside the circle $x^2 + y^2 = 1$.

3. $M(x,y) = e^x \sin y$, $N(x,y) = e^x \cos y$; $D = \{(x,y) \mid 0 \leq x \leq \pi/4, 0 \leq y \leq \pi/4\}$

4. $M(x,y) = 3x + 2y$, $N(x,y) = 2x + 4y$; D is the region enclosed by the ellipse $x^2 + (y^2/4) = 1$.

5. $N(x,y) = 3x^2y^4/4 - x^2y$, $N(x,y) = x^3y^3$; $D = \{(x,y) \mid 0 \leq x \leq 2, 0 \leq y \leq 2x\}$

In Exercises 6 to 10 use Green's Theorem to evaluate the given line integral.

6. $\oint_{\partial K} x^2y\,dx + 3x^2y^2\,dy$, where K is the rectangle with vertices $(0,0)$, $(0,3)$, $(2,3)$, $(2,0)$.

7. $\oint_{\partial K} (x^2 + y^2)\,dx + (y^2 - x^2)\,dx$, where K is the circle $x^2 + y^2 = 4$.

8. $\oint_{\partial K} e^x \sin y\,dx + e^x \cos y\,dy$, where K is the circle $(x-1)^2 + (y+5)^2 = 9$.

9. $\oint_{\partial K} (x^2 + y^2)\,dx - x^2\,dy$, where K is the circle $x^2 + y^2 = 9$.

10. $\oint_{\partial K} (x^2y^4 + 3y - \sin x)\,dx + (xy^3 - 2e^y)\,dy$; K is the rectangle with vertices $(0,0)$, $(3,0)$, $(3,4)$, $(0,4)$.

In Exercises 11 to 14, use Green's Theorem to compute the area of the given region (see Example 2 and Prob. 1, Progress Test 1).

11. The triangle with vertices (0, 0), (4, 0), and (4, 3).

12. The region bounded by the curves $y = x^2$ and $y = x + 2$.

13. The region inside the curve $x^{2/3} + y^{2/3} = 4$.

14. The region bounded by the curves $y = x^2 - 1$ and $y = 0$.

15. Let C be a piecewise smooth simple closed curve that does not pass through the point $(2, -1)$. Show

$$\oint_C \frac{(y+1)\,dx}{(x-2)^2+(y+1)^2} + \frac{(2-x)\,dx}{(x-2)^2+(y+1)^2} = \begin{cases} 2\pi & \text{if } (2,-1) \text{ is inside } C \\ 0 & \text{if } (2,-1) \text{ is outside } C \end{cases}$$

16. Show that

$$\oint_K \frac{x\,dx}{x^2+y^2} - \frac{y\,dy}{x^2+y^2} = 0$$

for every piecewise smooth closed curve K that does not pass through the origin.

17. Let D be a region of the form $D = \{(x,y) \mid c \le y \le d,\ p(y) \le x \le q(y)\}$. Show that if M and N have continuous partials in D, then

$$\oint_{\partial D} N\,dy = \iint_D N_x\,dA$$

18. Show how the argument that established Green's Theorem for a semiannular region can be extended to yield the conclusion of Green's Theorem on an annular region. (See Fig. 21.26.)

Section 21.5: Applications of Green's Theorem: Divergence and Curl

Objectives:

1. **Determine the flux of a vector field over a closed curve.**
2. **Determine the circulation of a vector field over a closed curve.**

INTRODUCTION

Green's Theorem has some interesting physical consequences depending on the types of vector fields under consideration. Among the most common are inverse-square fields and velocity fields. The former arise in problems from electrostatics; the latter arise naturally in the study of fluid dynamics. We shall discuss a two-dimensional steady-state fluid flow. This is a classical problem whose importance has been revitalized in recent years as the study of fluids, especially blood, has come under serious mathematical study. Our work is only a first step because of the simplifying assumptions made. Subsequent steps take the study into three dimensions and take into account such additional variables as friction, viscosity, and time-dependent flow (as with a pulse).

Specifically, suppose a fluid of constant density δ is flowing in a stream whose surface occupies the region D in the plane. As in Sec. 21.1 we assume the flow is horizontal and independent of the depth. In effect, this allows us to treat the problem as two dimensional. We further assume the flow is steady state. In this setting the velocity field of the flow has the form

$$\mathbf{v}(x,y) = \langle M(x,y), N(x,y)\rangle$$

A sketch of a typical such steady-state velocity field is given in Fig. 21.27.

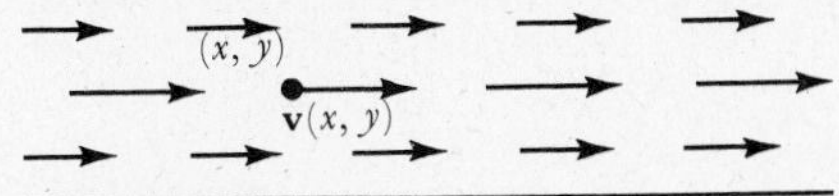

Fig. 21.27

Consider a region R in D whose boundary consists of a piecewise smooth simple closed curve K. We wish to discuss two measures of the flow. One is called the *flux* of $\delta\mathbf{v}$ over K and measures the rate of flow of the mass of the fluid *out* of the region R. The other is called the *circulation* of $\mathbf{v}$ around K and, in a sense to be made more precise, it measures the net tendency of the fluid to flow along K. We assume throughout that K has the counterclockwise orientation and has parameterization

$$K:\ \mathbf{X}(t) = \langle x(t), y(t)\rangle,\ a \le t \le b$$

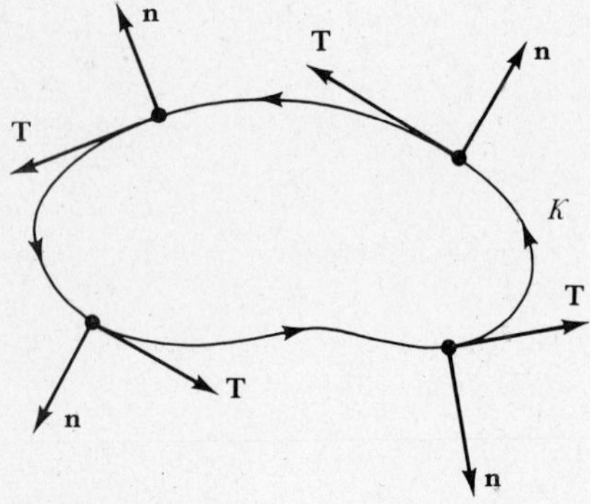

Fig. 21.28

As an aid in determining the flux and circulation we shall use the unit tangent $\mathbf{T}$ to K and $\mathbf{n}$ the *outer* unit normal to K. From Chap. 18 we know that

(1) $$\mathbf{T}(t) = \frac{\mathbf{X}'(t)}{\|\mathbf{X}'(t)\|} = \left\langle \frac{x'(t)}{\|\mathbf{X}'(t)\|}, \frac{y'(t)}{\|\mathbf{X}'(t)\|} \right\rangle$$

or equivalently [since $d\mathbf{X} = \mathbf{X}'(t)\,dt$ and $ds/dt = \|\mathbf{X}'(t)\|$],

(2) $$\mathbf{T} = \frac{d\mathbf{X}}{ds} = \left\langle \frac{dx}{ds}, \frac{dy}{ds} \right\rangle$$

where s represents arc length along K. Since K has the counterclockwise orientation, $\mathbf{n}$ is obtained by rotating $\mathbf{T}$ 90° in the clockwise direction (see Fig. 21.28). Thus $\mathbf{n}$ is given by

(3) $$\mathbf{n} = \left\langle \frac{y'(t)}{\|\mathbf{X}(t)\|}, \frac{-x'(t)}{\|\mathbf{X}'(t)\|} \right\rangle$$

or

(4) $$\mathbf{n} = \left\langle \frac{dy}{ds}, \frac{-dx}{ds} \right\rangle$$

We shall develop an integral giving the flux by summing flow in the direction of $\mathbf{n}$ and then develop an integral giving the circulation by summing flow in the direction of $\mathbf{T}$.

THE FLUX INTEGRAL

Since the dot product of two vectors has the same sign as the cosine of the angle between them (17.4.4), we know that $\mathbf{v} \cdot \mathbf{n} > 0$ at points on K where the fluid is leaving R. Similarly, $\mathbf{v} \cdot \mathbf{n} < 0$ at points where the fluid is entering R. If $\mathbf{v} \cdot \mathbf{n} = 0$, then the flow is parallel to the tangent and so the fluid is neither entering nor leaving R at that point (see Fig. 21.29).

To determine the flux we shall subdivide K into n pieces $C_1, \ldots, C_n$ having arc lengths $\Delta s_1, \ldots, \Delta s_n$, respectively. We let $\mathbf{v}_i$ and $\mathbf{n}_i$ denote the values of $\mathbf{v}$ and $\mathbf{n}$ at some point on C_i and treat them as being constant throughout C_i. Since $\mathbf{n}_i$ is fixed, we are in effect treating C_i as if it were a straight line segment. In Fig. 21.30(a) we see the actual situation, whereas Fig. 21.30(b) shows the idealized approximation. We further assume, for now, that the flow is out of R over C_i as in Fig. 21.30.

If $\mathbf{n}_i$ is parallel to $\mathbf{v}_i$, then the area A_i of the fluid that passes through C_i in time Δt_i is equal to the area of the rectangle with sides of lengths Δs_i and

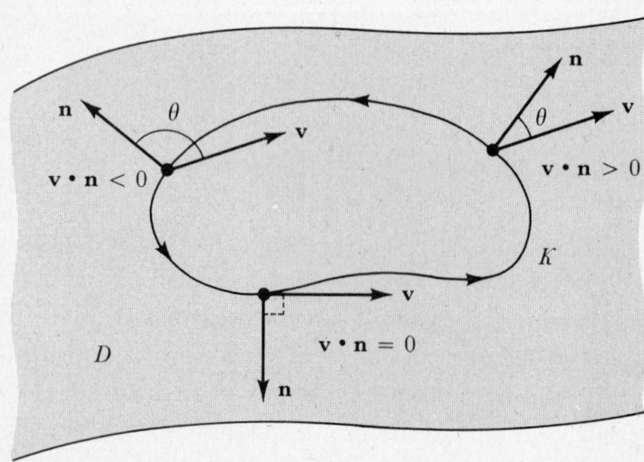

Fig. 21.29

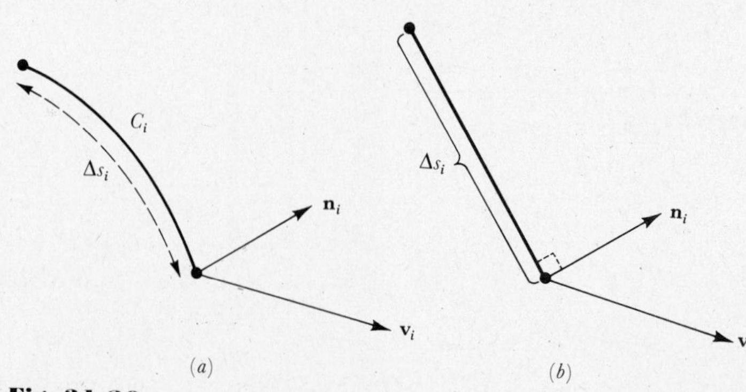

Fig. 21.30

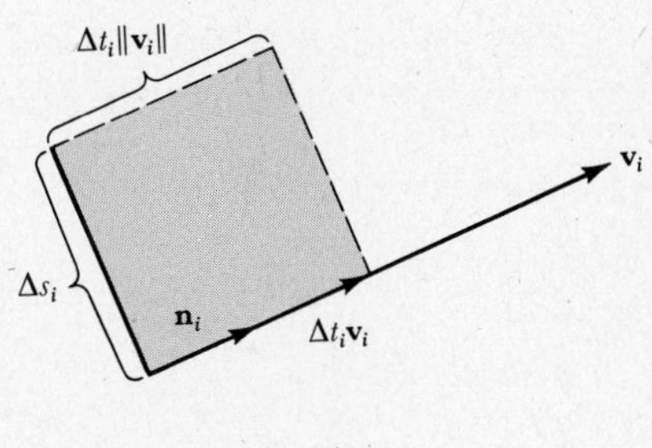

(a)

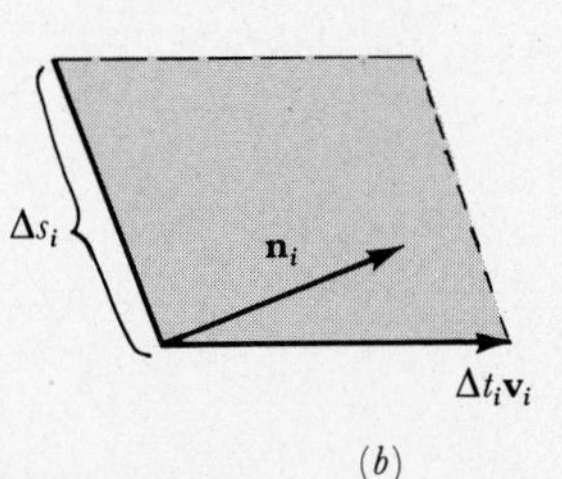

(b)

Fig. 21.31

$\Delta t_i \|\mathbf{v}_i\|$; see Fig. 21.31(*a*). If $\mathbf{n}_i$ is not parallel to $\mathbf{v}_i$, then the area of the amount of fluid that passes through C_i is equal to the area of the parallelogram whose base has length Δs_i and whose height is the scalar projection of $\Delta t_i \mathbf{v}_i$ onto $\mathbf{n}_i$ (see Fig. 21.31). In either case,

$$A_i = (\text{Proj}_{\mathbf{n}_i} \Delta t_i \mathbf{v}_i)\Delta s_i = \left[\frac{\Delta t_i \mathbf{v}_i \bullet \mathbf{n}_i}{\|\mathbf{n}_i\|}\right]\Delta s_i = \Delta t_i \mathbf{v}_i \bullet \mathbf{n}_i \, \Delta s_i$$

The amount of mass flowing through C_i in time Δt_i is approximated by $M_i = \delta A_i$, so the rate of flow is approximately

$$\frac{M_i}{\Delta t_i} = \delta \mathbf{v}_i \bullet \mathbf{n}_i \, \Delta s_i \tag{5}$$

Since we are trying to measure the net *outward* rate of flow, we regard $-A_i$ as the flow *into* R over C_i. However, as noted earlier, in this case $\mathbf{v}_i \bullet \mathbf{n}_i$ would be negative and so our approximation in (5) would still be valid. We now sum up the contributions over all such segments to obtain our approximation to the flux:

$$\sum_{i=1}^{n} \delta \mathbf{v}_i \bullet \mathbf{n}_i \, \Delta s_i \tag{6}$$

This suggests the following definition of the flux of $\delta \mathbf{v}$ over K.

21.5.1 Definition $\quad \text{Flux} = \int_K \delta \mathbf{v} \bullet \mathbf{n} \, ds$

By (4),

$$\mathbf{v} \bullet \mathbf{n} \, ds = \langle M, N \rangle \bullet \left\langle \frac{dy}{ds}, \frac{-dx}{ds} \right\rangle ds$$
$$= \langle M, N \rangle \bullet \langle dy, -dx \rangle = -N\,dx + M\,dy$$

and thus we have

21.5.2 Theorem $\quad \text{Flux} = \int_K \delta \mathbf{v} \bullet \mathbf{n} \, ds = \delta \int_K -N\,dx + M\,dy$

Example 1 Determine the flux of $\mathbf{v}$ over the unit circle $x^2 + y^2 = 1$ where $\mathbf{v} = \langle y^2, x^2 \rangle$.

Solution A counterclockwise parameterization for K, the unit circle, is given by

$$K: \ \langle \cos t, \sin t \rangle \qquad 0 \le t \le 2\pi$$

Thus, by (21.5.2),

$$\text{Flux} = \int_K \delta \mathbf{v} \bullet \mathbf{n} \, ds = \delta \int_K -N\,dx + M\,dy$$
$$= \delta \int_K -x^2\,dx + y^2\,dy$$
$$= \delta \int_0^{2\pi} -\cos^2 t(-\sin t\,dt) + \sin^2 t(\cos t\,dt)$$
$$= \delta \left[\frac{\cos^3 t}{3} + \frac{\sin^3 t}{3}\right]_0^{2\pi} = 0 \quad \bullet$$

It is customary to define the notion of flux for *any* vector field (not merely one associated with a fluid flow).

21.5.3 Definition Suppose $\mathbf{F}(x,y) = \langle M(x,y), N(x,y)\rangle$ is a vector field and K is a piecewise smooth simple closed curve with counterclockwise orientation. The *flux* of $\mathbf{F}$ over K is defined by

$$\text{Flux} = \int_K \mathbf{F}\cdot\mathbf{n}\,ds = \int_K -N\,dx + M\,dy$$

where $\mathbf{n}$ represents the outer unit normal to K.

THE DIVERGENCE THEOREM

There are two types of derivatives associated with a vector field, the "divergence" and the "curl." The first of these is related to the flux integral via Green's Theorem.

21.5.4 Definition The *divergence* of the vector field $\mathbf{F} = \langle M, N\rangle$, denoted div $\mathbf{F}$, is defined by div $\mathbf{F} = M_x + N_y$.

21.5.5 Divergence Theorem Suppose K is a piecewise smooth simple closed curve and that $\mathbf{F} = \langle M, N\rangle$ has continuous first partials throughout the region R enclosed by K. Then

$$\int_K \mathbf{F}\cdot\mathbf{n}\,ds = \iint_R \operatorname{div}\mathbf{F}\,dA$$

where $\mathbf{n}$ represents the outer unit normal to K, and K has the counterclockwise orientation.

Proof: The region R is certainly a regular region, and since K has the counterclockwise orientation, it yields the standard orientation for ∂R. Thus

$$\begin{aligned}\int_K \mathbf{F}\cdot\mathbf{n}\,ds &= \int_K -N\,dx + M\,dy\\ &= \oint_{\partial R} -N\,dx + M\,dy\\ &= \iint_R (M_x - (-N_y))\,dA \qquad \text{(by Green's Theorem)}\\ &= \iint_R (M_x + N_y)\,dA = \iint_R \operatorname{div}\mathbf{F}\,dA \quad \blacksquare\end{aligned}$$

Example 2 Determine the flux of the vector field $\mathbf{F}(x,y) = \langle x^3, y^3\rangle$ over the unit circle K.

Solution By the Divergence Theorem

$$\begin{aligned}\text{Flux} &= \int_K \mathbf{F}\cdot\mathbf{n}\,ds = \iint_D \operatorname{div}\mathbf{F}\,dA \qquad (D \text{ is the unit disk})\\ &= \iint_D (M_x + N_y)\,dA\\ &= 3\iint_D (x^2 + y^2)\,dA\\ &= 3\int_0^{2\pi}\int_0^1 r^2(r\,dr\,d\theta)\\ &= 3\int_0^{2\pi}\int_0^1 r^3\,dr\,d\theta = \frac{3\pi}{2} \quad \bullet\end{aligned}$$

In the case where **F** is a velocity field, div $\mathbf{F}(x,y)$ measures the rate of mass flow per unit area from a point (x,y). A point where div $\mathbf{F} > 0$ is said to be a *source* and a point where div $\mathbf{F} < 0$ is called a *sink*. If div $\mathbf{F} = 0$ throughout the stream, then, reasonably enough, the fluid is said to be *incompressible*. A general vector field for which div $\mathbf{F} = 0$ is said to be *solenoidal*. Note that by the Divergence Theorem, if the flux across K is zero, then either div $\mathbf{F} = 0$ or else the sources and sinks in R must cancel one another.

PROGRESS TEST 1

1. Let $\mathbf{F}(x,y) = \langle x^3 + y^3, y^3 - x^3 \rangle$ and let K be the curve $x^2 + y^2 = 4$ oriented in the counterclockwise direction. Use the Divergence Theorem to determine the flux of **F** over K.

THE CIRCULATION INTEGRAL AND STOKES' THEOREM

Suppose we again consider the two-dimensional flow problem, only now we compare the velocity of the fluid along K with **T**, the direction of K. Now $\mathbf{v} \cdot \mathbf{T}$ measures how close the directions of **v** and **T** match one another; see Fig. 21.32. If we envision the points on K as small beads strung along K, then along a small piece of curve of length Δs the flow of the fluid will tend to move the beads in the direction of K when $\mathbf{v} \cdot \mathbf{T} > 0$ and in the opposite direction when $\mathbf{v} \cdot \mathbf{T} < 0$. The beads will remain stationary at any points where $\mathbf{v} \cdot \mathbf{T} = 0$. If we sum terms of the form $\mathbf{v} \cdot \mathbf{T}\,\Delta s$ and pass to a limit, we will obtain an integral of the form

$$I = \int_K \mathbf{v} \cdot \mathbf{T}\, ds$$

called the *circulation* of **v** around K.

Suppose now we return to our bead analogy and think of the beads as being fixed to the curve. If $I > 0$, then the flow will tend to rotate the curve in the counterclockwise direction. If $I < 0$, the fluid will tend to rotate K clockwise. If $I = 0$, then K will not rotate. It is in this sense that we say the circulation measures the "tendency of the fluid to flow in the direction of K."

As with flux, we define circulation for an arbitrary vector field. Note from (2), page 758, that $\mathbf{T}\, ds = \langle dx, dy \rangle$.

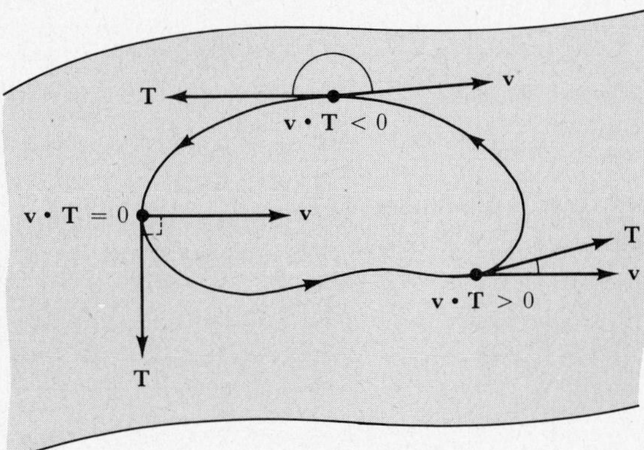

Fig. 21.32

21.5.6 Definition Suppose $\mathbf{F} = \langle M, N \rangle$ is a vector field and K is a piecewise smooth simple closed curve with counterclockwise orientation. The *circulation* of $\mathbf{F}$ around K is defined by

$$\text{Circulation} = \int_K \mathbf{F} \cdot \mathbf{T}\, ds = \int_K M\, dx + N\, dy.$$

STOKES' THEOREM

21.5.7 Definition The *curl* of the vector field $\mathbf{F} = \langle M, N \rangle$, denoted curl $\mathbf{F}$, is defined by

$$\text{curl } \mathbf{F} = (N_x - M_y)\mathbf{k}$$

When working with two-dimensional vector fields, one can just as well define the curl of $\mathbf{F}$ to be the scalar quantity $N_x - M_y$. In higher dimensions it is essential that curl $\mathbf{F}$ be defined as a vector quantity and so we elect this convention (see Exercise 16).

Again, Green's Theorem provides a connection between the curl of $\mathbf{F}$ and the circulation of $\mathbf{F}$ around K. This connection is usually referred to as *Stokes' Theorem.* As before, K has the counterclockwise orientation and R denotes the region enclosed by K.

21.5.8 Stokes' Theorem Suppose that K is a piecewise smooth simple closed curve and $\mathbf{F} = \langle M, N \rangle$ has continuous partials throughout the region R. Then

$$\int_K \mathbf{F} \cdot \mathbf{T}\, ds = \iint_R (\text{curl } \mathbf{F}) \cdot \mathbf{k}\, dA$$

Proof:

$$\begin{aligned}\int_K \mathbf{F} \cdot \mathbf{T}\, ds &= \int_K M\, dx + N\, dy \\ &= \iint_R [N_x - M_y]\, dA \qquad \text{(by Green's Theorem)} \\ &= \iint_R \langle 0, 0, N_x - M_y \rangle \cdot \langle 0, 0, 1 \rangle\, dA = \iint_R (\text{curl } \mathbf{F}) \cdot \mathbf{k}\, dA \quad \blacksquare\end{aligned}$$

If $\mathbf{F}$ is a velocity field, then curl $\mathbf{F}$ measures the tendency of the flow to rotate about $\mathbf{k}$. Stokes' Theorem provides a relationship between the circulation of $\mathbf{F}$ around the *boundary* of a region to the curl of $\mathbf{F}$ on the *interior* of the region. If curl $\mathbf{F} = 0$ throughout, then $\mathbf{F}$ is said to be *irrotational.* Note that if $\mathbf{F}$ is a force field, then the circulation integral $\int_K \mathbf{F} \cdot \mathbf{T}\, ds = \int_K M\, dx + N\, dy$ is just the work done by $\mathbf{F}$ in moving an object along the closed curve K.

PROGRESS TEST 2

1. For $\mathbf{F}$ and K as in Progress Test 1, use Stokes' Theorem to determine the circulation of $\mathbf{F}$ around K.

SECTION 21.5 EXERCISES

In Exercises 1 to 10 determine the flux of $\mathbf{F}$ *over* K *and the circulation of* $\mathbf{F}$ *around* K. *Assume that* K *has the counterclockwise orientation.*

1. $\mathbf{F}(x,y) = \langle x^2 - 2y, y^2 + 4x \rangle$; K is the rectangle with vertices $(0, 0)$, $(3, 0)$, $(3, 4)$, $(0, 4)$.

2. $\mathbf{F}(x,y) = \langle x^2 + \sin y, y^2 - \cos x \rangle$; K is the rectangle with vertices $(0, 0)$, $(\pi/2, 0)$, $(\pi/2, \pi/4)$, $(0, \pi/4)$.

3. $F(x,y) = \langle y^3 + 2y, 3y^2x \rangle$; K is the circle $x^2 + y^2 = 9$.

4. $F(x,y) = \langle \ln(x^2 + y^2), \ln(x^2 + y^2) \rangle$; K is the boundary of the semiannular region between $y = \sqrt{4 - x^2}$ and $y = \sqrt{1 - x^2}$.

5. $F(x,y) = \langle xy, y^2 - x^2 \rangle$; K is the triangle with vertices $(0,0)$, $(3,0)$, $(2,2)$.

6. $F(x,y) = \langle xy^2, x^2y \rangle$; K is the circle $x^2 + y^2 = 25$.

7. $F(x,y) = \langle 4x + 6y, 2x - 9y \rangle$; K is a rectangle with area 42 square units.

8. $F(x,y) = \langle -y/(x^2 + y^2), x/(x^2 + y^2) \rangle$; K is any closed curve that does not encircle or contain $(0,0)$.

9. $F(x,y) = \langle e^x \sin y, e^x \cos y \rangle$; K is any circle centered at $(0,0)$.

10. $F(x,y) = \langle x^2y + y^3, -xy^2 - y^3 \rangle$; K is the circle $x^2 + y^2 = 1$.

11. Let $\mathbf{F}_1 = \langle M_1, N_1 \rangle$, $\mathbf{F}_2 = \langle M_2, N_2 \rangle$ be vector fields and let f be a scalar function of two variables and k any real number. Show that

(*a*) $\operatorname{div}(\mathbf{F}_1 + \mathbf{F}_2) = \operatorname{div} \mathbf{F}_1 + \operatorname{div} \mathbf{F}_2$

(*b*) $\operatorname{div}(k\mathbf{F}_1) = k \operatorname{div} \mathbf{F}_1$

(*c*) $\operatorname{div}(f\mathbf{F}_1) = f \operatorname{div} \mathbf{F}_1 + \nabla f \cdot \mathbf{F}_1$

12. Let $\mathbf{F}$ be an inverse-square field of the form $\mathbf{F}(\mathbf{X}) = -\mathbf{X}/\|\mathbf{X}\|^3$. Show that $\mathbf{F}$ is solenoidal in the region $\|\mathbf{X}\| > r$, where $r > 0$.

13. Let $\mathbf{F} = \langle M, N \rangle$ be a vector field and f a scalar function of two variables. Show that

$$[\operatorname{curl}(f\mathbf{F})] \cdot \mathbf{k} = f(\operatorname{curl} \mathbf{F}) \cdot \mathbf{k} + [\operatorname{curl}\langle f, f \rangle] \cdot \mathbf{F}$$

14. If f has continuous first and second partials, show that $\operatorname{curl} \nabla f = \mathbf{0}$.

15. If $\mathbf{F} = \langle M, N \rangle$, where M and N have continuous first and second partials, show $\operatorname{div}(\operatorname{curl} \mathbf{F}) = 0$.

16. If $\mathbf{F} = \langle M, N, P \rangle$, curl $\mathbf{F}$ is defined by curl $\mathbf{F} = \langle P_y - N_z, M_z - P_x, N_x - M_y \rangle$. This formula is more easily recalled by means of the mnemonic device

$$\operatorname{Curl} \mathbf{F} = \begin{vmatrix} \mathbf{i} & \mathbf{j} & \mathbf{k} \\ \dfrac{\partial}{\partial x} & \dfrac{\partial}{\partial y} & \dfrac{\partial}{\partial z} \\ M & N & P \end{vmatrix}$$

where the expansion is by elements of the first row and $\partial/\partial x$ "times f" is understood to be f_x. Assuming that all partial derivatives that occur exist and are continuous, show that

(*a*) $\operatorname{curl}(\nabla f) = \mathbf{0}$

(*b*) $\operatorname{curl}(\mathbf{F} \times \mathbf{G}) = [(\operatorname{curl} \mathbf{F}) \cdot \mathbf{G}] - [\mathbf{F} \cdot (\operatorname{curl} \mathbf{G})]$

(*c*) $\operatorname{curl}(f\mathbf{F}) = f(\operatorname{curl} \mathbf{F}) + \nabla f \times \mathbf{F}$

CHAPTER EXERCISES

Review Exercises

In Exercises 1 to 10 evaluate the given line integral over the indicated curve using (21.2.5).

1. $\int_C y\,dx - x\,dy$; C is the portion of the parabola $y = 1 - x^2$ from $(0,1)$ to $(1,0)$.

2. $\int_C e^x y\,dx + x^2\,dy$; C is the curve consisting of the line segment from $(0,0)$ to $(1,1)$ followed by the horizontal line segment to $(2,1)$.

3. $\int_C x \sin y\,dx + xy\,dy$; C is the portion of the parabola $y = x^2$ from $(0,0)$ to $(1,1)$ followed by the horizontal line segment to $(-2,1)$.

4. $\int_C x\,dx - y\,dy$; C is the portion of the graph of $y = \sin x$ joining $(0,0)$ to $(\pi/2, 1)$.

5. $\int_C 2xy\,dx + (x^2 + 2y)\,dy$; C is the line segment joining $(1,1)$ to $(2,-1)$.

6. $\int_C xy^2\,dx - x\,dy$; C is the triangle with vertices $(0,-1)$, $(2,0)$, $(0,3)$ oriented counterclockwise.

7. $\int_C (x - y)\,dx + xy\,dy$; C is the closed curve obtained by traversing $y = x^3$ from $(0,0)$ to $(1,1)$ and then in a straight line back to the origin.

8. $\int_C x\,dx - y\,dy$; C is the unit circle centered at the origin oriented counterclockwise.

9. $\int_C y\,e^{xy}\,dx + x\,e^{xy}\,dy$; C is the straight line segment from $(0,0)$ to $(1,1)$.

10. $\int_C (x - y^2)\,dx + x\,dy$; C is the rectangle with vertices $(0,0)$, $(3,0)$, $(3,2)$, $(0,2)$ oriented clockwise.

In Exercises 11 to 20 identify those vector fields that are conservative (on R_2) and determine the potential functions.

11. $F(x,y) = \langle e^{xy}(xy + 1), x^2e^{xy} \rangle$

12. $F(x,y) = \langle 6xy, 3x^2 + 2y \rangle$

13. $F(x,y) = \langle e^x \cos y^2 + 1, 1 - 2y\,e^x \sin y^2 \rangle$

14. $F(x,y) = \langle 2xy/(1 + y^4), \arctan(y^2) \rangle$

15. $F(x,y) = \langle ye^x \cos(e^x), -\sin(e^x) \rangle$

16. $F(x,y) = \langle 2xe^y + y^2e^x, x^2e^y + y^2e^x \rangle$

17. $F(x,y) = \langle \sec x \tan x \tan y - \cos y, \sec x \sec^2 y + \sin y \rangle$

18. $F(x,y) = \langle 1 - 2xy, x^2 - 1 \rangle$

19. $F(x,y) = \langle y^2e^x/(1 + e^x) - 2x, 2y \ln(1 + e^x) \rangle$

20. $F(x,y) = \langle \sin xy + xy \cos xy - 2xy, x^2(\cos xy - 1) \rangle$

In Exercises 21 to 25 show that the given integral is independent of the path and find its value.

21. $\displaystyle\int_{(1,2)}^{(3,-1)} (y^2 - 2x)\,dx + (2xy + 2y)\,dy$

22. $\displaystyle\int_{(0,1)}^{(2,-1)} (2xe^{x^2}y)\,dx + e^{x^2}\,dy$

23. $\displaystyle\int_{(0,1)}^{(1,0)} \frac{ye^x}{(1 + e^x)^2}\,dx - \frac{1}{1 + e^x}\,dy$

24. $\displaystyle\int_{(0,0)}^{(1,1)} (3x^2y - 2xy^3)\,dx + (x^3 - 3x^2y^2 + 2y)\,dy$

25. $\displaystyle\int_{(0,0)}^{(\pi/3,\pi/6)} (\sin y + y^2 \cos x)\,dx + (x \cos y + 2y \sin x)\,dx$

In Exercises 26 to 30 show that the given integral is independent of the path in an appropriate region containing C and evaluate the integral over C.

26. $\int_C \left(2x - \frac{y}{x^2}\right) dx + \frac{dy}{x}$; C: $\mathbf{X}(t) = \langle t, t^2 \rangle, 1 \le t \le 4$

27. $\int_C \frac{x\,dx}{\sqrt{x^2+y^2}} + \frac{y\,dy}{\sqrt{x^2+y^2}}$; C: $\mathbf{X}(t) = \langle \cos t, \sin t \rangle, 0 \le t \le 3\pi/2$

28. $\int_C \left(\frac{1}{y} - \frac{1}{x^2 y}\right) dx - \left(\frac{x^2+1}{xy^2}\right) dy$; C: $\mathbf{X}(t) = \langle t, t^2+1 \rangle, 1 \le t \le 2$

29. $\int_C \left(\frac{1}{y}\cos\frac{x}{y} + 2x\right) dx - \left(\frac{1}{y^2}\cos\frac{x}{y}\right) dy$; C: $\mathbf{X}(t) = \langle \pi t/2, 1+t \rangle, 0 \le t \le 1$

30. $\int_C \frac{-y}{(x+3)^2+y^2}\,dx + \frac{x+3}{(x+3)^2+y^2}\,dy$; C: $\mathbf{X}(t) = \langle \sin t, \cos t \rangle, 0 \le t \le 2\pi$

In Exercises 31 to 35 use one of the area formulas developed in Sec. 21.4, Example 3, or Progress Test 1 to determine the area of the given region.

31. The region enclosed by the curve C: $\langle \cos^3 t, \sin^3 t \rangle$, $0 \le t \le 2\pi$.

32. The region bounded above by $y = 4 - x^2$ and below by $y + x + 2 = 0$.

33. The region inside the ellipse $\frac{x^2}{9} + \frac{y^2}{12} = 1$.

34. The region bounded by the curve $y = x^2$, the x axis, $x = 1$, and $x = 4$.

35. The region above the x axis and below one arch of the cycloid $x = t - \sin t$, $y = 1 - \cos t$.

In Exercises 36 to 40 use Green's Theorem to evaluate the given line integral.

36. $\int_C 3y\,dx + 7x\,dy$; C: $\langle \cos t, \sin t \rangle$, $0 \le t \le 2\pi$

37. $\oint_D (2x^3 - y^3)\,dx + (x^3 + y^3)\,dy$; D is the annular region between $x^2 + y^2 = 1$ and $x^2 + y^2 = 4$.

38. $\int_C (xy + y^2)\,dx + (2xy + 3x^2)\,dx$; C: $\langle \cos t, \sin t \rangle$, $0 \le t \le 2\pi$

39. $\int_C e^x \sin y\,dx + e^x \cos y\,dy$; C: $\langle \cos^3 t, \sin^3 t \rangle$, $0 \le t \le 2\pi$

40. $\int_C x^2 y\,dy + 3xy\,dy$; C is the rectangle with vertices $(0, 0)$, $(2, 0)$, $(2, 3)$, $(0, 3)$ oriented counterclockwise.

In Exercises 41 to 45 determine div **F** *and curl* **F** *for the given vector fields.*

41. $\mathbf{F}(x,y) = \langle 2x, 3y \rangle$

42. $\mathbf{F}(x,y) = \langle x^2 y, -yx^2 \rangle$

43. $\mathbf{F}(x,y) = \left\langle \frac{x}{\sqrt{x^2+y^2}}, \frac{y}{\sqrt{x^2+y^2}} \right\rangle$

44. $\mathbf{F}(x,y) = \left\langle \frac{x}{x^2+y^2}, \frac{y}{x^2+y^2} \right\rangle$

45. $\mathbf{F}(x,y) = \langle x \sin(x/y), y^2 \cos(xy) \rangle$

In Exercises 46 to 55 determine (a) the circulation of **F** *around the curve K and (b) the flux of* **F** *over K where K has the counterclockwise orientation.*

46. $\mathbf{F}(x,y) = \langle 3x^2 + y, x + 5y^2 \rangle$; K is the rectangle with vertices $(0, 0)$, $(2, 0)$, $(2, 3)$, $(0, 3)$.

47. $\mathbf{F}(x,y) = \langle 2x - y, x + 2y \rangle$; K is the rectangle with vertices $(-1, 0)$, $(1, 0)$, $(1, 2)$, $(-1, 2)$.

48. $\mathbf{F}(x,y) = \langle y^2 - x^2, xy \rangle$; K is the rectangle with vertices $(-1, 1)$, $(2, 1)$, $(2, 3)$, $(-1, 3)$.

49. $\mathbf{F}(x,y) = \langle xy^2 + x^2, y^3 - x \rangle$; K is the rectangle with vertices $(2, 5)$, $(5, 5)$, $(5, 7)$, $(2, 7)$.

50. $\mathbf{F}(x,y) = \langle x^2 y^2, 1 - 2y^2 \rangle$; K is the rectangle with vertices $(-3, 1)$, $(2, 1)$, $(2, 3)$, $(-3, 3)$.

51. $\mathbf{F}(x,y) = \langle -y/(x^2+y^2), x/(x^2+y^2) \rangle$; K is the circle $(x - 16)^2 + (y - 9)^2 = 1$.

52. $\mathbf{F}(x,y) = \langle y^3, x^3 \rangle$; K is the unit circle centered at the origin.

53. $\mathbf{F}(x,y) = \langle e^x \sin y, e^x \cos y \rangle$; K is the unit circle centered at the origin.

54. $\mathbf{F}(x,y) = \langle x, y \rangle$; K is the unit circle centered at the origin.

55. $\mathbf{F}(x,y) = \langle e^x \sin y, e^x \cos y \rangle$; K is the triangle with vertices $(0, 0)$, $(1, 1)$, $(0, 1)$.

In Exercises 56 to 60 use Green's Theorem to evaluate the given line integral.

56. $\int_C \frac{x\,dx}{x^2+y^2} + \frac{y\,dy}{x^2+y^2}$; C is the closed curve obtained by traversing $y^2 = x + 16$ counterclockwise from $(0, 4)$ to $(0, -4)$ and then counterclockwise along the ellipse $x^2/9 + y^2/16 = 1$.

57. $\int_C \frac{(x-1)\,dx}{x^2 - 2x + 1 + y^2} + \frac{y\,dy}{x^2 - 2x + 1 + y^2}$; C is the ellipse $(x - 1)^2/16 + y^2/9 = 1$ with the counterclockwise orientation.

58. $\int_C \frac{(x-a)\,dx}{(x-a)^2+(y-b)^2} + \frac{(y-b)\,dy}{(x-a)^2+(y-b)^2}$; C is any simple smooth curve that encloses the point (a, b).

59. $\int_C \frac{-y\,dx}{x^2-4x+4+y^2} + \frac{(x-2)\,dy}{x-4+4+y^2}$; C is the ellipse $(x-2)^2/42 + y^2/18 = 1$ with the counterclockwise orientation.

60. $\int_C \frac{-(y-b)\,dx}{(x-a)^2+(y-b)^2} + \frac{(x-a)\,dy}{(x-a)^2+(y-b)^2}$; C is any simple smooth curve that encloses the point (a, b).

Miscellaneous Exercises

1. Let D be the region inside the circle C_1: $x^2+y^2=100$ and outside *both* circles C_2: $(x-2)^2+y^2=1$ and C_3: $(x+2)^2+y^2=1$. Suppose M_y and N_x are continuous and equal in D. If

$$\oint_{C_3} M\,dx + N\,dy = 4 \qquad \text{and} \qquad \oint_{C_4} M\,dx + N\,dy = 7$$

where C_4: $(x-2)^2+y^2=2$, find

$$\oint_{C_1} M\,dx + N\,dy$$

2. Let $\widehat{D}$ be the simply connected region consisting of the x-y plane with the negative x axis and the origin removed, and let $\mathbf{F}(x,y) = \langle -y/(x^2+y^2),\ x/(x^2+y^2)\rangle$. Show that the function in the discussion following (21.3.7), page 750, satisfies $\nabla f(x,y) = \mathbf{F}(x,y)$ on all of $\widehat{D}$.

3. Show that $\int_C x^2\,dx + y\,x^4\,dy = 0$, where C is *any* ellipse centered at the origin. Is this integral zero over *every* piecewise smooth closed curve that encircles the origin?

4. Let C be a smooth closed curve and let D be the region enclosed by C. If f has continuous second partials, show that

$$\iint_D \left[\frac{\partial^2 f}{\partial x^2} + \frac{\partial^2 f}{\partial y^2}\right] dA = \oint_C \nabla f \cdot \mathbf{n}\,ds$$

where $\mathbf{n}$ is the unit outer normal to C.

5. Let f and g have continuous first partials in R_2. Show that

$$\int_C [f\nabla g + g\nabla f] \cdot d\mathbf{X} = 0$$

for every smooth closed curve in R_2.

6. Let f and g have continuous partials in R_2 and let C be a smooth closed curve in R_2. Show that

$$\iint_D (\nabla f \times \nabla g) \cdot \mathbf{k}\,dA = \oint_C f\nabla g \cdot d\mathbf{X}$$

where D is the region enclosed by C.

In Probs. 7 to 9 let C be a piecewise smooth curve (not necessarily simple). The winding number of C about P = (a, b) is defined by

$$w = \frac{1}{2\pi}\int_C \frac{-(y-b)\,dx}{(x-a)^2+(y-b)^2} + \frac{(x-a)\,dy}{(x-a)^2+(y-b)^2}$$

7. If C is a simple curve, show that the winding number of C about P is either 0, 1, or -1.

8. Use Exercise 7 to give a precise definition of what it means for a simple smooth curve to have a counterclockwise or clockwise orientation.

9. What is the winding number of the curve C: $\langle \cos t, \sin t\rangle$, $0 \le t \le 2n\pi$?

10. A spring has the shape of the circular helix $\langle \cos t, \sin t, t\rangle$, $0 \le t \le 2\pi$. Find the center of mass of this spring if the density at any point is proportional to the square of the distance from the origin. (*Hint:* First define what $\int_C f(x,y,z)\,ds$ means, where f is a scalar function and s is the arc length along C.)

SELF-TEST ✚

1. Evaluate $\int_C (x^2-y^2)\,dx + x\,dy$, where C is the counterclockwise oriented curve obtained by going from $(-4, 0)$ to $(0, -16)$ along $y = x^2 - 16$ and then along a straight line to $(16, 0)$.

2. Use Green's Theorem to evaluate $\int_K (4xy - x^3)\,dx + (2x + y^2)\,dy$, where K is the counterclockwise boundary of the region bounded by the curves $y = x^2$ and $y = x$.

3. Use the Fundamental Theorem for Line Integrals to evaluate $\int_C (y - x^2)\,dx + (x + y^2)\,dy$ where C: $\langle 2t+1, t^2-6t\rangle$, $0 \le t \le 1$.

4. Determine the flux of the vector field $\mathbf{F}(x,y) = \langle x^2 - y^3, 2y^2 - x^4\rangle$ over K, where K is the counterclockwise boundary of the region bounded by the curves $y = 9 - x^2$ and $y = 0$.

5. Determine the circulation of the vector field $\mathbf{F}(x,y) = \langle y - x^4, 2x + y^3\rangle$ around the counterclockwise oriented cardioid $r = 1 + \cos\theta$.

Appendix A

AN INTRODUCTION TO DIFFERENTIAL EQUATIONS

CONTENTS

Virtually every discipline that uses mathematical models expresses important quantitative relationships in equations involving functions and derivatives, that is, differential equations. We saw examples of simple differential equations in Chaps. 6 (objects thrown vertically and being acted upon by the force of gravity), 8 (natural growth and decay of populations, radioactive elements, and money) and 18 (projectile problems). This Appendix provides a brief glimpse of the methods used to "solve" differential equations—to determine functions that satisfy the differential equations. Extensions of some of these methods appear in the exercises.

Section A.1: First Order Differential Equations

Objectives:
Recognize and solve a first order differential equation that is (*a*) separable, (*b*) homogeneous, (*c*) exact, or (*d*) linear.

INTRODUCTION

Any equation involving functions and their derivatives is called a *differential equation.* In particular, if F is a function of $n + 2$ variables, then

$$(1) \quad F(x,y,y',y'',\ldots,y^{(n)}) = 0$$

is called an *ordinary differential equation of order n.* A function f is said to be a *solution* of (1) if

$$F(x,f(x),f'(x),f''(x),\ldots,f^{(n)}(x)) = 0$$

for all x in some interval I. For example, the equation

$$(2) \quad y'' + y = 0$$

is a second order ordinary differential equation. One can easily check that $f(x) = \sin x$ is a solution.

If $g(x,y) = 0$ defines implicitly some function that solves a differential equation, then we refer to $g(x,y) = 0$ as an *implicit solution* of the equation. For example, $x^2 + y^2 = 9$ is an implicit solution of the differential equation

$$y^2y' + xy = 0$$

since upon differentiating $x^2 + y^2 = 9$ implicitly, we have

$$y' = -\frac{x}{y}$$

and so

$$y^2y' + xy = y^2\left(-\frac{x}{y}\right) + xy = -xy + xy = 0$$

It can be shown that any solution of Eq. (2) is of the form

$$(3) \quad y = c_1 \sin x + c_2 \cos x$$

for suitably chosen constants c_1 and c_2. Conversely, every y of the form of (3) is a solution of Eq. (2). For this reason (3) is called the *general solution* of the differential equation. In most instances the general solution of an nth order differential equation will involve n arbitrary constants $c_1, c_2, \ldots, c_n$. If we assign n conditions to y and its derivatives, then we can select from all solutions the *particular solution* that satisfies those conditions. Specifically, if $A_0, A_1, \ldots, A_{n-1}$ are fixed constants, then the problem of finding the particular solution of

$$F(x,y,y',\ldots,y^{(n)}) = 0$$

that satisfies the *initial conditions*

$$y(x_0) = A_0, y'(x_0) = A_1, \ldots, y^{(n-1)}(x_0) = A_{n-1}$$

is called an *initial value problem.*

FIRST ORDER DIFFERENTIAL EQUATIONS

We shall limit our attention to first order equations, which can be written in *derivative form* as

$$\frac{dy}{dx} = f(x,y)$$

or in *differential form* as

$$M(x,y)\,dx + N(x,y)\,dy = 0$$

SEPARABLE DIFFERENTIAL EQUATIONS

A.1.1 **Definition** A first order differential equation is said to be *separable* if it can be written in the form

$$(4) \quad M(x)\,dx + N(y)\,dy = 0$$

where M and N are continuous.

To solve (4) we first write it in the equivalent form

$$M(x) + N(y)y' = 0$$

Now if $y = f(x)$ is a solution, then

$$M(x) + N(f(x))f'(x) = 0$$

which, upon integrating, yields

$$\int M(x)\,dx + \int N(f(x))f'(x)\,dx = C$$

or equivalently,

A.1.2 $$\int M(x)\,dx + \int N(y)\,dy = C$$

Example 1 Solve the differential equation

$$(y^2 + y)\,dx + x\,dy = 0$$

Solution Dividing through by $x(y^2 + y)$, we can write the equation in the equivalent form

$$\frac{dx}{x} + \frac{dy}{y(y+1)} = 0$$

Thus

$$\int \frac{dx}{x} + \int \frac{dy}{y(y+1)} = C$$

Now

$$\int \frac{dx}{x} = \ln|x|$$

and

$$\int \frac{dy}{y(y+1)} = \int \left[\frac{1}{y} - \frac{1}{y+1}\right] dy = \ln\left|\frac{y}{y+1}\right|$$

If we denote C by $\ln|k|$, then our solution can be written in the form

$$\ln|x| + \ln\left|\frac{y}{y+1}\right| = \ln|k|$$

so

$$\ln\left|\frac{xy}{y+1}\right| = \ln|k|$$

and

$$\left|\frac{xy}{y+1}\right| = |k|$$

or equivalently,

$$xy = k(y+1) \quad \bullet$$

Example 2 A tank contains 30 gal of brine in which 5 lb of salt have been dissolved. Pure water flows in the tank at the rate of 2 gal/min. The mixture, which is kept uniform by stirring, runs out of the tank at the same rate as the pure water flows in. How much salt remains in the tank after 10 min?

Solution Let x denote the amount of salt in the tank at time t. Since the rate of inflow is the same as the rate of outflow, the tank contains 30 gal of mixture at any time t. The concentration of salt at time t is $\frac{x}{30}$ lb/gal, and since this mix is flowing out at the rate of 2 gal/min, we have

$$(5)\quad \frac{dx}{dt} = -\frac{x}{50}(2) = -\frac{x}{25}\text{ lb/min}$$

Since the tank contained 5 lb of salt initially, we have the initial condition $x(0) = 5$. Separating the variables from (5), we have the initial value problem

$$\frac{dx}{x} + \frac{dt}{25} = 0 \qquad x(0) = 5$$

Integrating and simplifying, we have

$$x = 5e^{-t/25}$$

Thus after 10 min the tank contains $5e^{-2/5} \approx 3.35$ lb of salt. ●

HOMOGENEOUS EQUATIONS

A.1.3 **Definition** A first order differential equation is said to be *homogeneous* if it can be written in the form

$$(6)\quad \frac{dy}{dx} = f\left(\frac{y}{x}\right)$$

To solve (6), we make the substitution $y = vx$. Then

$$\frac{dy}{dx} = v + x\frac{dv}{dx} \quad\text{and}\quad \frac{y}{x} = v$$

so (6) becomes

$$v + x\frac{dv}{dx} = f(v),\text{ or, equivalently, } \frac{dv}{f(v) - v} = \frac{dx}{x}$$

Since this latter equation is separable in the variables x and v, we can solve it by integration and then replace v by y/x.

Example 3 Solve the differential equation

$$\frac{dy}{dx} = \frac{y^2 + 2xy}{x^2}$$

Solution We note that

$$\frac{dy}{dx} = \left(\frac{y}{x}\right)^2 + 2\left(\frac{y}{x}\right)$$

Thus, letting $y = vx$, we have

$$v + x\frac{dv}{dx} = v^2 + 2v$$

or

$$\frac{dv}{v^2 + v} = \frac{dx}{x}.$$

We integrate (using partial fractions on the left side) to obtain

$$\ln\left|\frac{v}{v+1}\right| = \ln|x| + \ln|c|$$

or equivalently,

$$\frac{v}{v+1} = cx$$

Substituting $v = y/x$ and simplifying, we have

$$y = \frac{cx^2}{1-cx} \quad \bullet$$

A function $F(x,y)$ of two variables is said to be *homogeneous of degree n* if

$$F(tx, ty) = t^n F(x,y)$$

The following theorem is a simple way of identifying a homogeneous differential equation given in differential form. An outline of the proof is in Exercise 22.

A.1.4 **Theorem** The first order differential equation

$$M(x,y)\,dx + N(x,y)\,dy = 0$$

is homogeneous if and only if the functions M and N are both homogeneous of the same degree.

PROGRESS TEST 1

Classify and solve the following differential equations:

1. $3xy\,dx + (x^2+4)\,dy = 0$

2. $(y + \sqrt{x^2+y^2})\,dx - x\,dy = 0$

EXACT DIFFERENTIAL EQUATIONS

A.1.5 **Definition** The differential equation

$$(7) \quad M(x,y)\,dx + N(x,y)\,dy = 0$$

is said to be *exact* in the region R if there exists a function $f(x,y)$ such that $f_x(x,y) = M(x,y)$ and $f_y(x,y) = N(x,y)$ for all (x,y) in R.

Note that the exactness of (7) is equivalent to the vector field $\langle M, N\rangle$, being conservative, and so by (21.3.5) and (21.3.7) we have:

A.1.6 **Theorem** Let R be a simply connected region in the plane. The differential equation

$$M(x,y)\,dx + N(x,y)\,dy = 0$$

is exact in R if and only if $M_y = N_x$.

If $M\,dx + N\,dy = 0$ is exact, then it can be written in the equivalent form $f_x\,dx + f_y\,dy = 0$, where f is such that $f_x = M$ and $f_y = N$. But $f_x\,dx + f_y\,dy = df$, the total differential of f, and so this equation is equivalent to the equation $df = 0$, whose solution is $f(x,y) = C$. Thus solving an exact differential equation is exactly the same as finding a potential function for the conservative vector field $\langle M, N\rangle$.

Example 4 Solve the differential equation

$$(3x^2\ln(3+y^2) - y^4)\,dx + \left(\frac{2x^3y}{3+y^2} - 4xy^3 + 2y\right)dy = 0$$

Solution Since $M_y = \dfrac{6x^2y}{3+y^2} - 4y^3 = N_x$, the equation is exact. Set

$$f_x = 3x^2\ln(3+y^2) - y^4$$

Then, integrating with respect to x, we have

$$f(x,y) = x^3 \ln(3 + y^2) - y^4 x + C(y)$$

Thus

$$f_y(x,y) = \frac{2x^3 y}{3 + y^2} - 4y^3 x + C'(y)$$

Setting $f_y = N_x$, we conclude $C'(y) = 2y$, so $C(y) = y^2$. Thus $f(x,y) = x^3 \ln(3 + y^2) - y^4 x + y^2$, and the general solution of the equation is

$$x^3 \ln(3 + y^2) - y^4 x + y^2 = C \quad \bullet$$

FIRST ORDER LINEAR DIFFERENTIAL EQUATIONS

A.1.7 **Definition** A first order differential equation is said to be *linear* if it can be written in the form

$$(8) \quad \frac{dy}{dx} + P(x)y = Q(x)$$

where P and Q are continuous.

If in Eq. (8) $Q = 0$, then we can write the equation in the separable form

$$\frac{dy}{y} + P(x)\,dx = 0$$

which can be solved to give

$$\ln y + \int P(x)\,dx = C$$

or, equivalently,

$$y\,e^{\int P(x)\,dx} = C$$

Note that, by the Product and Chain Rules,

$$(9) \quad D_x(y\,e^{\int P(x)\,dx}) = \frac{dy}{dx}\,e^{\int P(x)\,dx} + P(x)y\,e^{\int P(x)\,dx}$$

Thus if $Q(x)$ is not the zero function, we can multiply (8) through by $e^{\int P(x)\,dx}$ to obtain

$$\frac{dy}{dx}\,e^{\int P(x)\,dx} + P(x)y\,e^{\int P(x)\,dx} = Q(x)\,e^{\int P(x)\,dx}$$

which from (9) is equivalent to

$$D_x(y\,e^{\int P(x)\,dx}) = Q(x)\,e^{\int P(x)\,dx}$$

Integrating, we obtain,

$$y\,e^{\int P(x)\,dx} = \int Q(x)\,e^{\int P(x)\,dx} + C$$

or, equivalently,

A.1.8 $$y = e^{-\int P(x)\,dx}\left[\int Q(x)\,e^{\int P(x)\,dx} + C\right]$$

The next example, although similar to Example 2, results in a linear rather than a separable differential equation.

Example 5 A tank contains 60 gal of brine in which 15 lb of salt have been dissolved. Brine containing 2 lb of salt per gallon flows into the tank at the rate of 4 gal/min. The mixture, which is kept uniform by stirring, simultaneously flows out of the tank at the rate of 2 gal/min. How much salt is in the tank after 30 min?

Solution As in Example 2, we let $x(t)$ denote the amount of salt in the tank at time t. The rate at which salt is flowing into the tank is

$$\left(\frac{2 \text{ lb}}{\text{gal}}\right)\left(\frac{4 \text{ gal}}{\text{min}}\right) = 8 \text{ lb/min}$$

The concentration of salt in the tank at time t is $x(t)$ divided by $60 + (4 - 2)t$, and so the rate at which salt is flowing out of the tank is

$$\left(\frac{x(t)}{60 + 2t}\ \frac{\text{lb}}{\text{gal}}\right)\left(\frac{2\ \text{gal}}{\text{min}}\right) = \frac{x(t)}{30 + t}\ \text{lb/min}$$

Now $\frac{dx}{dt}$, the rate of change of the amount of salt in the tank at time t, is "rate in" minus "rate out," so

$$\frac{dx}{dt} = 8 - \frac{x}{30 + t}$$

or, equivalently,

$$\frac{dx}{dt} + \left(\frac{1}{30 + t}\right)x = 8$$

This equation is linear, with $P(t) = \frac{1}{30 + t}$ and $Q(t) = 8$. Now $\int P(t)\,dt = \ln(30 + t)$, so $e^{\int P(t)\,dt} = e^{\ln(30+t)} = 30 + t$. Thus the general solution is

$$x(t)(30 + t) = \int 8(30 + t)\,dt + C = \frac{8(30 + t)^2}{2} + C$$

or

$$x(t) = 4(30 + t) + \frac{c}{30 + t} \quad \bullet$$

Since the initial amount of salt in the tank was 15 lb, we have

$$15 = x(0) = 120 + \frac{C}{30}$$

so $C = -3150$ and the amount of salt in the tank at time t is

$$x(t) = 120 + 4t - \frac{3150}{30 + t}$$

Hence $x(30) = 240 - \frac{3150}{60} = 187.5$ lb. $\bullet$

PROGRESS TEST 2

Identify and solve the following differential equations.

1. $dy/dx + (\tan x)y = \cos x$

2. $(2x \tan y - y \sec x \tan x + 2x)\,dx + (x^2 \sec^2 y - \sec x)\,dy = 0$

SECTION A.1 EXERCISES

In Exercises 1 to 10 find the general solution of the given differential equation.

1. $x \csc y\,dx + (x^2 - x - 2)\,dy = 0$

2. $(y^2 + 9)\,dx + (x^2 + 5)\,dy = 0$

3. $(xy + y + x + 1)\,dx + y\,dy = 0$

4. $(x + y)\,dx - x\,dy = 0$

5. $(x\,e^{y/x} - y)\,dx + x\,dy = 0$

6. $(3x^2y^4 - 6x - y)\,dx + (4x^3y^3 + 18y - x)\,dy = 0$

7. $\left(\frac{2xy}{1 + x^2} - y^3 \cos x\right)dx + (\ln(1 + x^2) - 3y^2 \sin x + 1)\,dy = 0$

8. $y\,e^{xy}\,dx + (x\,e^{xy} + 1)\,dy = 0$

9. $dy/dx + 2xy/(x^2 + 1) = 4x$

10. $dy/dx + 2y = x^2 e^{-2x}$

In Exercises 11 to 18 solve the given initial value problem.

11. $(2x \cos y - y^2 \cos x)dx + (1 - 2y \sin x - x^2 \sin y)dy = 0$ $[y(0) = 2]$

12. $(3x^2y^2 - 3x^2)\,dx + (2yx^3 + 4y^3)\,dy = 0$ $[y(0) = 3]$

13. $(y e^x + 2e^x + y^2)\,dx + (e^x + 2xy)\,dy = 0\ [y(0) = 1]$

14. $(y + 3)\,dx + y(x + 2)\,dy = 0\ [y(0) = 2]$

15. $2xy\,dx + (x^2 + 9)\,dy = 0\ [y(0) = 4]$

16. $dy/dx = \dfrac{2xy}{x^2 - y^2}\ [y(0) = 1]$

17. $(x \tan(y/x) + y)\,dx - x\,dy = 0\ [y(2/\pi) = 1]$

18. $\dfrac{dy}{dx} + \dfrac{y}{x^2} = \dfrac{1}{x^2}\quad [x(1) = 2e]$

19. If $\phi(x,y)M(x,y)\,dx + \phi(x,y)N(x,y)\,dy = 0$ is exact, then $\phi(x,y)$ is called an *integrating factor* for the differential equation $M(x,y)\,dx + N(x,y)\,dy = 0$. Solve each of the following by first showing ϕ is an integrating factor and then solving the equivalent equation $\phi M\,dx + \phi N\,dy = 0$.
(*a*) $(2xy^4 - y^3)\,dx + (3x^2y^3 - 2xy^2 + 4y^4)\,dy = 0$ $[\phi(x,y) = y^{-1}]$
(*b*) $(x + y)\,dx + dy = 0\ [\phi(x,y) = e^x]$
(*c*) $(2xy^2 + y)\,dx + (2y^3 - x)\,dy = 0\ [\phi(x,y) = y^{-2}]$

20. A differential equation of the form $\dfrac{dy}{dx} + P(x)y = Q(x)y^n$ is called a *Bernoulli equation.*
(*a*) Show that if $n = 0$ or $n = 1$, then the Bernoulli equation is linear.
(*b*) Show that the substitution $v = y^{1-n}$ reduces the Bernoulli equation to the linear equation

$$\frac{dv}{dx} + [(1 - n)P(x)]v = [(1 - n)Q(x)]$$

21. Use the result of Exercise 20(*b*) to solve
(*a*) $\dfrac{dy}{dx} + y = xy^4$ (*b*) $\dfrac{dy}{dx} + \dfrac{y}{2x} = \dfrac{x}{y^3}$

22. Prove Theorem (A.1.4). (*Hint:* If M and N are both homogeneous of degree n, then let $t = 1/x$).

23. A large tank contains 50 gal of brine in which 8 lb of salt have been dissolved. Brine containing 3 lb of dissolved salt per gallon is flowing into the tank at the rate of 5 gal/min. The mixture, which is kept uniform by stirring, runs out of the tank at the rate of 3 gal/min. How much salt is in the tank after 17 min?

24. By a chemical reaction, chemical A is converted into chemical B. The rate at which A converts into B is proportional to the amount of A present at any time. After 10 min, 60 g of A remains; at the end of 40 min, 15 g of A remains. How many grams of A were present initially? How much of A will remain after 1 h?

Section A.2: Second Order Linear Differential Equations

Objectives:
1. Solve a second order linear homogeneous differential equation with constant coefficients.
2. Use the method of undetermined coefficients or the method of variation of parameter to solve a second order linear nonhomogeneous differential equation with constant coefficients.

INTRODUCTION

A second order differential equation is said to be *linear with constant coefficients* if it can be written in the form

$$(1)\quad \frac{d^2y}{dx^2} + a_1\frac{dy}{dx} + a_2 y = f(x)$$

where $f(x)$ is continuous on some interval I and a_1, a_2 are constants. If $f(x) = 0$ on I, then the equation is called *homogeneous,* and we refer to

$$(2)\quad \frac{d^2y}{dx^2} + a_1\frac{dy}{dx} + a_2 y = 0$$

as the *associated homogeneous equation* of (1). (This is a use of the term "homogeneous," different from that used for first order equations in Sec. A.1.)

We shall concentrate first on solving the homogeneous equation. The following theorem, whose proof can be found in any differential equation text, reduces the problem to that of finding just two solutions.

A.2.1 **Theorem** If $y_1(x)$ and $y_2(x)$ are solutions of (2), and if y_1 is not a scalar multiple of y_2, then the general solution of (2) is given by

$$y = c_1y_1 + c_2y_2$$

where c_1 and c_2 are arbitrary constants.

For example, one easily checks that $y_1(x) = \sin x$ and $y_2(x) = \cos x$ are solutions of $y'' + y = 0$. Since $\sin x$ and $\cos x$ are clearly not scalar multiples of one another, the general solution is given by

$$y = c_1 \sin x + c_2 \cos x$$

THE CHARACTERISTIC EQUATION

A.2.2 **Definition** The polynomial

$$(3) \quad c(m) = m^2 + a_1 m + a_2$$

is called the *characteristic polynomial* of the homogeneous differential Eq. (2). The equation $c(m) = 0$ is called the *characteristic equation.*

If $c(m)$ is an irreducible quadratic, then $c(m) = 0$ has no real roots. Otherwise $c(m) = 0$ has either a twice-repeated real root m_1 or it has two distinct real roots, m_1 and m_2. By applying the quadratic formula we can identify the three possibilities as follows:

(4) If $a_1^2 - 4a_2 > 0$, then $c(m) = 0$ has the distinct real roots.

$$m_1 = \frac{-a_1 + \sqrt{a_1^2 - 4a_2}}{2} \qquad m_2 = \frac{-a_1 - \sqrt{a_1^2 - 4a_2}}{2}$$

(5) If $a_1^2 - 4a_2 = 0$, then $c(m) = 0$ has the repeated real root $m_1 = -a_1/2$

(6) If $a_1^2 - 4a_2 < 0$, then $c(m)$ is irreducible.

We shall show that the solution of (2) will depend on which case occurs. As a first step in that direction we have:

A.2.3 **Theorem** If m_1 is a real root of the characteristic equation of (2), then $y_1(x) = e^{m_1 x}$ is the solution of (2).

Proof: Let $y_1(x) = e^{m_1 x}$. Then $y_1'(x) = m_1 e^{m_1 x}$ and $y_1''(x) = m_1^2 e^{m_1 x}$. Thus $y_1'' + a_1 y_1' + a_2 y_1 = m_1^2 e^{m_1 x} + a_1 m_1 e^{m_1 x} + a_2 e^{m_1 x} = e^{m_1 x}(m_1^2 + a_1 m_1 + a_2) = e^{m_1 x} c(m_1) = 0.$ ■

If $m_1 \neq m_2$, then certainly $e^{m_1 x}$ is not a scalar multiple of $e^{m_2 x}$, and so by (A.2.1) and (A.2.2), we have:

A.2.4 **Theorem** If $c(m) = 0$ has the distinct real roots m_1, m_2, then (2) has general solution

$$y = c_1 e^{m_1 x} + c_2 e^{m_2 x}$$

Example 1 Solve the differential equation $y'' + 2y' - 4y = 0$.

Solution The characteristic equation is $c(m) = m^2 + 2m - 4$. Now $c(m)$ has the distinct real roots $m_1 = -1 + \sqrt{5}$ and $m_2 = -1 - \sqrt{5}$. Thus the general solution is

$$\begin{aligned} y &= c_1 \exp(-1 + \sqrt{5})x + c_2 \exp(-1 - \sqrt{5})x \\ &= e^{-x}[c_1 \exp \sqrt{5}x + c_2 \exp - \sqrt{5}x] \end{aligned}$$ ●

A.2.5 **Theorem** If $c(m) = 0$ has the repeated real root m_1, then (2) has general solution

$$y = c_1 e^{m_1 x} + c_2 x e^{m_1 x}$$

Proof: By (A.2.3) $e^{m_1 x}$ is a solution of (2), and certainly $e^{m_1 x}$ and $x\,e^{m_1 x}$ are not scalar multiples of one another. Thus the result will follow from (A.2.1) if we can show $x\,e^{m_1 x}$ is a solution of (2). Let $y = x\,e^{m_1 x}$. Then

$$\begin{aligned} y'' + a_1 y' + a_2 y &= 2m_1 e^{m_1 x} + m_1^2 x\,e^{m_1 x} + a_1(e^{m_1 x} + m_1 x\,e^{m_1 x}) + a_2 x\,e^{m_1 x} \\ &= (m_1^2 + a_1 m_1 + a_2)x\,e^{m_1 x} + (2m_1 + a_1)\,e^{m_1 x} \\ &= c(m_1)x\,e^{m_1 x} + (2m_1 + a_1)\,e^{m_1 x} \end{aligned}$$

But $c(m_1) = 0$, and since by (5), $m_1 = -a_1/2$, we know $2m_1 + a_1 = 0$. Hence

$$y'' + a_1 y' + a_2 = 0$$ ■

Example 2 Solve the differential equation

$$\frac{d^2y}{dx^2} - 6\frac{dy}{dx} + 9y = 0$$

Solution Since $c(m) = m^2 - 6m + 9 = (m - 3)^2$, the general solution is given by

$$y = c_1 e^{3x} + c_2 x\, e^{3x} \quad \bullet$$

A.2.6 **Theorem** If $c(m)$ is irreducible, then (2) has general solution

$$y = e^{ax}[c_1 \sin bx + c_2 \cos bx]$$

where $a = -a_1/2$ and $b = \sqrt{4a_2 - a_1^2}/2$.

Proof: A direct calculation shows that if $y = e^{ax} \sin bx$, then

$$y'' + a_1 y' + a_2 = e^{ax}[(a^2 - b^2 + a_1 a + a_2) \sin bx + (2ab + a_1 b) \cos bx]$$

Now $a = -a_1/2$, so $a_1 = -2a$, and hence $2ab + a_1 b = 2ab - 2ab = 0$. Similarly,

$$a^2 - b^2 + a_1 a + a_2 = 0$$

Thus $e^{ax} \sin bx$ is a solution of (2). A similar argument shows that $e^{ax} \cos bx$ is a solution, and so by (A.2.1) the general solution is as stated. ■

Example 3 Solve the differential equation $y'' + y' + y = 0$.

Solution Now $c(m) = m^2 + m + 1$ is irreducible. Since $a_1 = a_2 = 1$, we have $a = -1/2$ and $b = \sqrt{3}/2$, and so the general solution is

$$y = e^{-x/2}\left[c_1 \sin \frac{\sqrt{3}x}{2} + c_2 \cos \frac{\sqrt{3}x}{2}\right] \quad \bullet$$

PROGRESS TEST 1

Solve the following differential equations:

1. $y'' + 4y' - 12y = 0$ **2.** $y'' - 10y' + 25y = 0$ **3.** $y'' + y' + 3y = 0$

FINDING PARTICULAR SOLUTIONS FOR A NONHOMOGENEOUS DIFFERENTIAL EQUATION

Since our efforts to date allow us to determine the solution of the associated homogeneous equation (2) more or less by inspection, the following reduces the problem of solving (1) to that of finding any *particular* solution.

A.2.7 **Theorem** Let y_p be any particular solution of (1) and let $y_c = c_1 y_1 + c_2 y_2$ be the general solution of the associated homogeneous equation. Then (1) has general solution

$$y = y_p + y_c$$

Proof: Let y be any solution of (1) and let $z = y - y_p$. Then $y = y_p + z$ and $z'' + a_1 z' + a_2 z = (y - y_p)'' + a_1(y - y_p)' + a_2(y - y_p) = (y'' + a_1 y' + a_2 y) - (y_p'' + a_1 y_p' + a_2 y_p) = f(x) - f(x) = 0$. Thus z is a solution of the associated homogeneous equation, and so $z = y_c$ for some appropriate choice of c_1 and c_2. Thus $y = y_p + y_c$. Conversely, let $y = y_p + z$, where z is any solution of the associated homogeneous equation. Then

$$\begin{aligned} y'' + a_1 y' + a_2 &= (y_p + z)'' + a_1(y_p + z)' + a_2(y_p + z) \\ &= (y_p'' + a_1 y_p' + a_2 y_p) + (z'' + a_1 z' + a_2 z) \\ &= f(x) + 0 = f(x) \end{aligned}$$

and so any function of the form $y_p + y_c$ is a solution of (1). ■

Example 4 Show that $y_p(x) = 2x^2 + 6x + 7$ is a solution of the differential equation

$$(7)\quad y'' - 3y' + 2y = 4x^2$$

and find the general solution.

Solution Since

$$\begin{aligned} y_p'' - 3y_p' + 2y_p &= 4 - 3(4x + 6) + 2(2x^2 + 6x + 7) \\ &= 4 - 12x - 18 + 4x^2 + 12x + 14 = 4x^2 \end{aligned}$$

y_p is a solution of (7). Now the associated homogeneous equation has characteristic polynomial $c(m) = m^2 - 3m + 2 = (m - 1)(m - 2)$. Thus $y_c = c_1e^x + c_2e^{2x}$, and so by (A.2.7) the general solution of (7) is

$$y = c_1e^x + c_2e^{2x} + 4x^2 \quad \bullet$$

VARIATION OF PARAMETERS

Let $y_c = c_1y_1(x) + c_2y_2(x)$ be the general solution of (2). In the method of *variation of parameters* we seek a particular solution to (1) having the form

$$(8)\quad y_p = v_1y_1 + v_2y_2$$

where v_1 and v_2 are functions to be determined. Since there are two functions to be determined, we should expect to use two conditions. The first, of course, is that y_p be a solution of (1). The second, which helps to minimize the computations, is that

$$(9)\quad v_1'y_1 + v_2'y_2 = 0$$

Now

$$y_p' = v_1'y_1 + v_2'y_2 + v_1y_1' + v_2y_2' = v_1y_1' + v_2y_2'$$

and

$$y_p'' = v_1'y_1' + v_2'y_2' + v_1y_1'' + v_2y_2''$$

It is routine to show that

$$y_p'' + a_1y_p' + a_2y_1' = v_1[y_1'' + a_1y_1' + a_2y_1] + v_2[y_2'' + a_1y_2' + a_2y_2] + v_1'y_1' + v_2'y_2'$$

But y_1 and y_2 are solutions of (2), and so $y_1'' + a_1y_1' + a_2y_1 = 0$ and $y_2'' + a_1y_2' + a_2y_2 = 0$. Hence y_p is a solution of (1) if and only if

$$(10)\quad v_1'y_1' + v_2'y_2' = f(x)$$

Solving (9) and (10) simultaneously for v_1' and v_2', we obtain

A.2.8

$$v_1' = \frac{-y_2f}{y_1y_2' - y_2y_1'} \qquad v_2' = \frac{y_1f}{y_1y_2' - y_2y_1'}$$

We can now integrate to determine v_1 and v_2 and then use (8) to form y_p.

Example 5 Solve the differential equation

$$y'' + y = \csc x$$

Solution Since $c(m) = m^2 + 1$, we have $a_1 = 0$, $a_2 = 1$, and from (A.2.6) $a = 0$, $b = 1$. Thus the associated homogeneous equation has general solution

$$y_c = c_1 \sin x + c_2 \cos x$$

Let $y_1 = \sin x$, $y_2 = \cos x$. Then $y_1y_2' - y_2y_1' = (\sin x)(-\sin x) - (\cos x)(\cos x) = -(\sin^2 x + \cos^2 x) = -1$. Thus

$$v_1' = \frac{-\cos x \csc x}{-1} = \cot x \qquad v_2' = \frac{\sin x \csc x}{-1} = -1$$

Thus $v_1 = \int \cot x\, dx = \ln|\sin x|$ and $v_2 = \int -dx = -x$ (we take the constants of integration to be zero), so

$$y_p = (\ln|\sin x|)\sin x - x\cos x$$

and hence the general solution of the equation is

$$\begin{aligned} y &= c_1 \sin x + c_2 \cos x + (\ln|\sin x|)\sin x - x\cos x \\ &= (c_1 + \ln|\sin x|)\sin x + (c_2 - x)\cos x \quad \bullet \end{aligned}$$

UNDETERMINED COEFFICIENTS

The method of *undetermined coefficients* can be applied to find a particular solution of (1) when $f(x)$ is either

(11) e^{kx}, x^n, $\sin cx$, $\cos cx$

(12) a finite product of functions from (11)

(13) a sum of constant multiples of functions from (11) or (12)

Thus a typical such function f would be

$$(14) \quad f(x) = 3x^2e^{2x} + x\sin 4x + 3x^2 - x + 5$$

The method consists of looking for a solution to (1) that has the form

$$y_p = \sum_{i=1}^{n} A_i v_i$$

where the A_i's are constant coefficients to be determined and the v_i's are all distinct functions of the form (11) or (12) that appear either in f or in any derivative of f. For example, in (14) the v_i's consist of x^2e^{2x}, xe^{2x}, e^{2x}, $x\sin 4x$, $x\cos 4x$, $\sin 4x$, $\cos 4x$, x^2, x, 1. The A_i's are then determined by substituting y_p into the differential equation and equating coefficients. To be assured of solving for the A_i's any v_i that is a solution of the associated homogeneous equation must be replaced by x^jv_i, where j is the least positive integer such that x^jv_i is *not* a solution of the homogeneous equation.

Example 6 Solve the differential equation

$$y'' - 4y' + 3y = 2e^{3x} + x^2 - 5$$

Solution The associated homogeneous equation has the characteristic polynomial $c(m) = m^2 - 4m + 3 = (m-1)(m-3)$, so $y_c = c_1e^x + c_2e^{3x}$. Since $f(x) = 2e^{3x} + x^2 - 5$, the v_i's are e^{3x}, x^2, x, 1. However, e^{3x} solves the associated homogeneous equation (take $c_1 = 0$, $c_2 = 1$), so we replace it by xe^{3x}, giving us $y_p = Axe^{3x} + Bx^2 + Cx + D$. Thus $y_p' = 3Axe^{3x} + Ae^{3x} + 2Bx + C$ and $y_p'' = 6Ae^{3x} + 9Axe^{3x} + 2B$. Therefore $y_p'' - 4y_p' + 3y_p = (9A - 12A + 3A)xe^{3x} + (6A - 4A)e^{3x} + (3B)x^2 + (3C - 8B)x + (2B - 4C + 3D)$. Since $9A - 12A + 3A = 0$, the first term drops out. Equating coefficients, we have $2A = 2$, $3B = 1$, $3C - 8B = 0$, $2B - 4C + 3D = -5$, and hence $A = 1$, $B = 1/3$, $C = 8/9$, $D = -19/9$.

Thus

$$y_p = xe^{3x} + \frac{1}{3}x^2 + \frac{8}{9}x - \frac{19}{9}$$

and the general solution is given by

$$y = c_1e^x + c_2e^{3x} + xe^{3x} + \frac{1}{3}x^2 + \frac{8}{9}x - \frac{19}{9} \quad \bullet$$

PROGRESS TEST 2

1. Use the method of variation of parameters to solve

$$y'' - 4y' + 4y = x^2e^{2x}$$

2. Use the method of undetermined coefficients to solve

$$y'' - 4y' + 4y = e^{2x} + 1$$

SECTION A.2 EXERCISES

In Exercises 1 to 12 find the general solution of the given homogeneous differential equation.

1. $y'' - 4y = 0$
2. $y'' + y' - 3y = 0$
3. $y'' - y' - 30y = 0$
4. $y'' + 8y = 0$
5. $y'' + 4y' + 3y = 0$
6. $y'' - 2\sqrt{2}y' + 2y = 0$
7. $y'' - 8y' + 16y = 0$
8. $y'' - 2y' - 6y = 0$
9. $y'' - 2y' + 4y = 0$
10. $y'' - 12y' + 36y = 0$
11. $y'' + 2y' + 8y = 0$
12. $y'' + 4y' + 4y = 0$

In Exercises 13 to 17 use the method of variation of parameters to solve the given equation.

13. $y'' + y = \tan x$
14. $y'' + y = \sec^3 x$
15. $y'' - y = e^x \sin x$
16. $y'' - 2y' + y = e^x/x, x > 0$
17. $y'' - y = \cos x$

In Exercises 18 to 22 use the method of undetermined coefficients to solve the given equation.

18. $y'' + y = x + 3e^{-x}$
19. $y'' - y = e^x \sin x$
20. $y'' - y' - 2y = x^2 + xe^{-x}$
21. $y'' - 2y' + y = xe^x$
22. $y'' - y = x \sin x$

Review Exercises

In Exercises 1 to 10 (a) identify all categories among S separable, H homogeneous, E exact, L linear into which each of the differential equations falls and (b) use any method to solve the differential equation.

1. $x(y^2 + y)\,dx + y(x^2 - 1)\,dy = 0$
2. $(y + e^x)\,dx + (x + e^y)\,dy = 0$
3. $x\dfrac{dy}{dx} + y = x^5$
4. $\dfrac{dy}{dx} - 3y = e^{2x}$
5. $(2x \tan y + y^2 \cos x + 2x)\,dx + (x^2 \sec^2 y + 2y \sin x)\,dy = 0$
6. $\dfrac{dy}{dx} + xy + y = x + 1$
7. $x \cos y\,dy - \sin y\,dx = 0$
8. $(\sec^2 y + 2n \ln y)\,dx + (2x \sec^2 y \tan y + x^2/y)\,dy = 0$
9. $xy' - y = x^3\sqrt{1 - x^2}$
10. $(x - y \tan(y/x))\,dx + (x \tan(y/x))\,dy = 0$

In Exercises 11 to 20 solve the given initial value problem.

11. $xy' + 2y = 6, y(1) = 7$
12. $y^2\,dx - (xy + x^2)\,dy = 0, y(1) = 3$
13. $y' + (1/x)y = 1/e^{x^2}, y(1) = 2$
14. $2xy\,dx = (x^2 + y^2)\,dy, y(2) = 4$
15. $(y + xy)\,dx + x\,dy = 0, y(1) = 3$
16. $(1 - \cos x)y' + y \sin x = 0, y(\pi/4) = 1$
17. $xy' - y = 2x^2, y(2) = 3$
18. $\left(\dfrac{2x - 1}{y}\right)dx + x\left(\dfrac{1 - x}{y^2}\right)dy = 0, y(-1) = 3$
19. $(2xy - 3)\,dx + (x^2 + 4y)\,dy = 0, y(1) = 2$
20. $(x^2 + y^2)\,dx - xy\,dy = 0, y(1) = 2$

In Exercises 21 to 30 determine the general solution of the given second order differential equation.

21. $y'' + 8y = 0$
22. $y'' + 3y' + y = 0$
23. $y'' + 3y' - y = 0$
24. $y'' + 2y' - 48y = 0$
25. $y'' - 2\sqrt{5}y' + 5y = 0$
26. $2y'' + 4y' + 8y = 0$
27. $y'' - 16y' + 64y = 0$
28. $y'' + 3y = 0$
29. $4y'' + 7y = 0$
30. $2y'' + 3y' + 8y = 0$

In Exercises 31 to 35 solve the given second order initial value problem.

31. $y'' - 4y' + 4y = 0; y(2) = 3, y'(2) = 0$
32. $y'' + 4y' + 4y = 0; y(0) = 3, y'(0) = 6$
33. $y'' - 5y' + 6y = 0; y(1) = y'(1) = 1$
34. $y'' - 6y' + 25y = 0; y(0) = y'(0) = -3$
35. $y'' + y = 0; y(\pi/2) = y'(\pi/2) = 1$

In Exercises 36 to 40 use the method of undetermined coefficients to solve the given differential equation.

36. $y'' - y' - 6y = e^x \sin x$
37. $y'' + 2y' + 4y = \cos 4x$
38. $y'' + y = x + e^x$
39. $y'' - 4y' + 4y = xe^{2x}$
40. $y'' + y = 3 - 4 \cos x$

In Exercises 41 to 45 use the method of variation of parameter to solve the given differential equation.

41. $y'' - 3y' + 2y = e^x/(e^x + 1)$

42. $y'' + y = \tan^2 x$

43. $y'' + y = \sec^3 x$

44. $y'' - 4y' + 4y = x^2e^{2x}$

45. $y'' + y = \sec x \tan x$

46. A large tank contains 50 gal of brine in which 25 lb of salt have been dissolved. Pure water flows into the tank at the rate of 1.5 gal/min. The mixture, which is kept uniform by stirring, runs out of the tank at the rate of 1 gal/min. After how much time will the amount of salt in the tank be reduced to 20 lb?

47. A large tank contains 200 gal of brine into which 30 lb of salt have been dissolved. Brine containing 3 lb of salt per gallon flows into the tank at the rate of 4 gal/min. The mixture, which is kept uniform by stirring, flows out of the tank at the rate of 3 gal/min. How much salt is in the tank after 20 min?

48. An object that weighs 15 lb falls from rest in a straight line toward the earth. The force due to air resistance is $4\,v$ pounds, where v is the speed in feet per second. Find the speed and distance fallen by the object after 7 s.

49. By a chemical reaction, chemical A is converted into chemical B. The rate at which the conversion takes place is proportional to the amount of A present at any time. After 2 h, 25 g of chemical A remains; after 3 h only 10 g is left. How much of chemical A was present initially? How much of chemical A will be left after 5 h?

Miscellaneous Exercises

1. Show that if $1/N[M_y - N_x]$ is a function of x alone [say $= p(x)$], then

$$\mu(x)\,M(x,y)\,dx + \mu(x)\,N(x,y)\,dy = 0$$

2. Show that if $1/M[N_x - M_y]$ is a function of y alone [say $= q(y)$], then

$$\nu(y)\,M(x,y)\,dx + \nu(y)\,N(x,y)\,dy = 0$$

is exact where $\nu(y) = e^{\int q(y)\,dy}$

In Exercises 3 to 6 use the results of Exercises 1 and 2 to solve the given differential equation.

3. $(2x^2 + y)\,dx + (x^2y - x)\,dy = 0$

4. $(x^2 + 2y^2)\,dx + xy\,dy = 0$

5. $(x^4 + y^4)\,dx - xy^3\,dy = 0$

6. $(y^2x + y^2 + y)\,dx + (2xy + 1)\,dy = 0$

7. Solve the differential equation $x\dfrac{d^2y}{dx^2} + \dfrac{dy}{dx} = 0$ by first making the substitutions $p = \dfrac{dy}{dx}, \dfrac{db}{dx} = \dfrac{d^2y}{dx^2}$ and solving the resulting first order equation.

8. Solve the differential equation $\dfrac{d^2y}{dx^2} + y\dfrac{dy}{dx} = 0$ by first making the substitutions $p = dy/dx$, $\dfrac{d^2y}{dx^2} = p\dfrac{dP}{dy}$, and solving the resulting first order equation.

9. (*Power Series Solution*) Solve the differential equation (i) $y'' + xy' + y = 0$ using the following procedure:

(a) Assume the series has a power series solution of the form $y(x) = \Sigma_{n=0}^{\infty} C_nx^n$. Substitute this series into (i) and adjust indices to conclude

$$\sum_{n=0}^{\infty} [(n+2)(n+1)C_{n+2} + (n+1)C_n]x^n = 0$$

(b) Using the fact that $\Sigma_{n=0}^{\infty} a_nX^n = 0$ only if $a_n = 0$ for all n, arrive at the "recurrence relation"

$$\text{(ii)}\quad C_{n+2} = \frac{-C_n}{n+2},\ n = 0, 1, 2, \ldots$$

(c) Letting C_0 and C_1 be arbitrary, use (ii) to show that

$$C_{2n} = (-1)^n\frac{C_0}{2^nn!} \qquad C_{2n+1} = (-1)^n\frac{2^nn!C_1}{(2n+1)!}$$

(d) Conclude (i) has solution

$$y(x) = C_0\left[\sum_{n=0}^{\infty}\frac{(-1)^nx^{2n}}{2^nn!}\right] + C_1\left[\sum_{n=0}^{\infty}\frac{(-1)^n2^nn!x^{2n+1}}{(2n+1)!}\right]$$

In Exercises 10 to 15 use a procedure similar to that in Exercise 9 to solve the following differential equations.

10. $y'' - xy = 0$ (Airy's equation)

11. $y'' + xy' + 2y = 0$

12. $y'' - xy' + y = 0$

13. $y'' - xy' - 2y = 0$

14. $y'' + y = 0$

15. $y'' - 2xy = 0$

16. Show that the solution of Exercise 14 obtained using the power series method agrees with the solution obtained using (A.2.6).

17. (*Orthogonal Trajectories*) An *orthogonal trajectory* to a family of plane curves $F(x, y, c) = 0$ is any curve that intersects every curve in the family at right angles. To determine the orthogonal trajectories, proceed as follows:

(*a*) Differentiate the equation $F(x, y, c) = 0$ implicitly with respect to x.

(*b*) Eliminate the parameter c from the two equations in (*a*) to arrive at a differential equation $dy/dx = f(x, y)$.

(*c*) Since two curves are perpendicular if and only if their slopes are negative reciprocals, the general solution of

$$\frac{dy}{dx} = -1/f(x, y)$$

will yield the orthogonal trajectories.

Use this procedure to determine the orthogonal trajectories of the following families:

17-1: $y = cx^2$
17-2: $x^2 + y^2 = c^2$
17-3: $y = cx^3$

18. A heated metal plate has temperature distribution $T(x, y) = 10 - x^2 - 2y^2$. By (19.6.6) and (19.7.1) a heat-seeking particle will travel along a trajectory orthogonal to the isotherms $T(x, y) = C$. Determine the trajectory a heat-seeking particle starting at $(-1, 3)$ will travel along to get to the origin (compare with Sec. 19.6, Example 3).

APPENDIX A SELF-TEST

1. Solve the following first order differential equations:

(*a*) $(xy + 2x)\,dx + 3(x^2 - 2)\,dy = 0$

(*b*) $(\sin^2 y + 2x\cos y)\,dx + (x \sin 2y - x^2 \sin y)\,dy = 0$

(*c*) $\dfrac{dy}{dx} + (\tan x)y = \cos^3 x$

2. Solve the initial value problem

$$\frac{dy}{dx} = \frac{xy - y^2}{x^2}, \; y(1) = 1$$

3. Solve the differential equation $y'' - 4y = x^2 + e^x$

4. Solve the initial value problem $y'' + y = 2\csc x$; $y(\pi/2) = 0$, $y'(\pi/2) = 2$

Appendix B

TABLES

Table 1 Trigonometric Functions

Degrees	*Radians*	*Sin*	*Cos*	*Tan*	*Cot*		
0	0.0000	0.0000	1.0000	0.0000		1.5708	90
1	0.0175	0.0175	0.9998	0.0175	57.290	1.5533	89
2	0.0349	0.0349	0.9994	0.0349	28.636	1.5359	88
3	0.0524	0.0523	0.9986	0.0524	19.081	1.5184	87
4	0.0698	0.0698	0.9976	0.0699	14.301	1.5010	86
5	0.0873	0.0872	0.9962	0.0875	11.430	1.4835	85
6	0.1047	0.1045	0.9945	0.1051	9.5144	1.4661	84
7	0.1222	0.1219	0.9925	0.1228	8.1443	1.4486	83
8	0.1396	0.1392	0.9903	0.1405	7.1154	1.4312	82
9	0.1571	0.1564	0.9877	0.1584	6.3138	1.4137	81
10	0.1745	0.1736	0.9848	0.1763	5.6713	1.3963	80
11	0.1920	0.1908	0.9816	0.1944	5.1446	1.3788	79
12	0.2094	0.2079	0.9781	0.2126	4.7046	1.3614	78
13	0.2269	0.2250	0.9744	0.2309	4.3315	1.3439	77
14	0.2443	0.2419	0.9703	0.2493	4.0108	1.3265	76
15	0.2618	0.2588	0.9659	0.2679	3.7321	1.3090	75
16	0.2793	0.2756	0.9613	0.2867	3.4874	1.2915	74
17	0.2967	0.2924	0.9563	0.3057	3.2709	1.2741	73
18	0.3142	0.3090	0.9511	0.3249	3.0777	1.2566	72
19	0.3316	0.3256	0.9455	0.3443	2.9042	1.2392	71
20	0.3491	0.3420	0.9397	0.3640	2.7475	1.2217	70
21	0.3665	0.3584	0.9336	0.3839	2.6051	1.2043	69
22	0.3840	0.3746	0.9272	0.4040	2.4751	1.1868	68
23	0.4014	0.3907	0.9205	0.4245	2.3559	1.1694	67
24	0.4189	0.4067	0.9135	0.4452	2.2460	1.1519	66
25	0.4363	0.4226	0.9063	0.4663	2.1445	1.1345	65
26	0.4538	0.4384	0.8988	0.4877	2.0503	1.1170	64
27	0.4712	0.4540	0.8910	0.5095	1.9626	1.0996	63
28	0.4887	0.4695	0.8829	0.5317	1.8807	1.0821	62
29	0.5061	0.4848	0.8746	0.5543	1.8040	1.0647	61
30	0.5236	0.5000	0.8660	0.5774	1.7321	1.0472	60
31	0.5411	0.5150	0.8572	0.6009	1.6643	1.0297	59
32	0.5585	0.5299	0.8480	0.6249	1.6003	1.0123	58
33	0.5760	0.5446	0.8387	0.6494	1.5399	0.9948	57
34	0.5934	0.5592	0.8290	0.6745	1.4826	0.9774	56
35	0.6109	0.5736	0.8192	0.7002	1.4281	0.9599	55
36	0.6283	0.5878	0.8090	0.7265	1.3764	0.9425	54
37	0.6458	0.6018	0.7986	0.7536	1.3270	0.9250	53
38	0.6632	0.6157	0.7880	0.7813	1.2799	0.9076	52
39	0.6807	0.6293	0.7771	0.8098	1.2349	0.8901	51
40	0.6981	0.6428	0.7660	0.8391	1.1918	0.8727	50
41	0.7156	0.6561	0.7547	0.8693	1.1504	0.8552	49
42	0.7330	0.6691	0.7431	0.9004	1.1106	0.8378	48
43	0.7505	0.6820	0.7314	0.9325	1.0724	0.8203	47
44	0.7679	0.6947	0.7193	0.9657	1.0355	0.8029	46
45	0.7854	0.7071	0.7170	1.0000	1.0000	0.7854	45
		Cos	*Sin*	*Cot*	*Tan*	*Radians*	*Degrees*

Table 2 Exponential Functions

x	e^x	e^{-x}	x	e^x	e^{-x}
0.00	1.0000	1.000000	**0.50**	1.6487	0.606531
0.01	1.0101	0.990050	0.51	1.6653	.600496
0.02	1.0202	.980199	0.52	1.6820	.594521
0.03	1.0305	.970446	0.53	1.6989	.588605
0.04	1.0408	.960789	0.54	1.7160	.582748
0.05	1.0513	0.951229	**0.55**	1.7333	0.576950
0.06	1.0618	.941765	0.56	1.7507	.571209
0.07	1.0725	.932394	0.57	1.7683	.565525
0.08	1.0833	.923116	0.58	1.7860	.559898
0.09	1.0942	.913931	0.59	1.8040	.554327
0.10	1.1052	0.904837	**0.60**	1.8221	0.548812
0.11	1.1163	.895834	0.61	1.8404	.543351
0.12	1.1275	.886920	0.62	1.8589	.537944
0.13	1.1388	.878095	0.63	1.8776	.532592
0.14	1.1503	.869358	0.64	1.8965	.527292
0.15	1.1618	0.860708	**0.65**	1.9155	0.522046
0.16	1.1735	.852144	0.66	1.9348	.516851
0.17	1.1853	.843665	0.67	1.9542	.511709
0.18	1.1972	.835270	0.68	1.9739	.506617
0.19	1.2092	.826959	0.69	1.9937	.501576
0.20	1.2214	0.818731	**0.70**	2.0138	0.496585
0.21	1.2337	.810584	0.71	2.0340	.491644
0.22	1.2461	.802519	0.72	2.0544	.486752
0.23	1.2586	.794534	0.73	2.0751	.481909
0.24	1.2712	.786628	0.74	2.0959	.477114
0.25	1.2840	0.778801	**0.75**	2.1170	0.472367
0.26	1.2969	.771052	0.76	2.1383	.467666
0.27	1.3100	.763379	0.77	2.1598	.463013
0.28	1.3231	.755784	0.78	2.1815	.458406
0.29	1.3364	.748264	0.79	2.2034	.453845
0.30	1.3499	0.740818	**0.80**	2.2255	0.449329
0.31	1.3634	.733447	0.81	2.2479	.444858
0.32	1.3771	.726149	0.82	2.2705	.440432
0.33	1.3910	.718924	0.83	2.2933	.436049
0.34	1.4049	.711770	0.84	2.3164	.431711
0.35	1.4191	0.704688	**0.85**	2.3396	0.427415
0.36	1.4333	.697676	0.86	2.3632	.423162
0.37	1.4477	.690734	0.87	2.3869	.418952
0.38	1.4623	.683861	0.88	2.4109	.414783
0.39	1.4770	.677057	0.89	2.4351	.410656
0.40	1.4918	0.670320	**0.90**	2.4596	0.406570
0.41	1.5068	.663650	0.91	2.4843	.402524
0.42	1.5220	.657047	0.92	2.5093	.398519
0.43	1.5373	.650509	0.93	2.5345	.394554
0.44	1.5527	.644036	0.94	2.5600	.390628
0.45	1.5683	0.637628	**0.95**	2.5857	0.386741
0.46	1.5841	.631284	0.96	2.6117	.382893
0.47	1.6000	.625002	0.97	2.6379	.379083
0.48	1.6161	.618783	0.98	2.6645	.375311
0.49	1.6323	.612626	0.99	2.6912	.371577
0.50	1.6487	0.606531	**1.00**	2.7183	0.367879

Table 2 Exponential Functions (Continued)

x	e^x	e^{-x}	x	e^x	e^{-x}
1.00	2.7183	0.367879	**1.50**	4.4817	0.223130
1.01	2.7456	.364219	1.51	4.5267	.220910
1.02	2.7732	.360595	1.52	4.5722	.218712
1.03	2.8011	.357007	1.53	4.6182	.216536
1.04	2.8292	.353455	1.54	4.6646	.214381
1.05	2.8577	0.349938	**1.55**	4.7115	0.212248
1.06	2.8864	.346456	1.56	4.7588	.210136
1.07	2.9154	.343009	1.57	4.8066	.208045
1.08	2.9447	.339596	1.58	4.8550	.205975
1.09	2.9743	.336216	1.59	4.9037	.203926
1.10	3.0042	0.332871	**1.60**	4.9530	0.201897
1.11	3.0344	.329559	1.61	5.0028	.199888
1.12	3.0649	.326280	1.62	5.0531	.197899
1.13	3.0957	.323033	1.63	5.1039	.195930
1.14	3.1268	.319819	1.64	5.1552	.193980
1.15	3.1582	0.316637	**1.65**	5.2070	0.192050
1.16	3.1899	.313486	1.66	5.2593	.190139
1.17	3.2220	.310367	1.67	5.3122	.188247
1.18	3.2544	.307279	1.68	5.3656	.186374
1.19	3.2871	.304221	1.69	5.4195	.184520
1.20	3.3201	0.301194	**1.70**	5.4739	0.182684
1.21	3.3535	.298197	1.71	5.5290	.180866
1.22	3.3872	.295230	1.72	5.5845	.179066
1.23	3.4212	.292293	1.73	5.6407	.177284
1.24	3.4556	.289384	1.74	5.6973	.175520
1.25	3.4903	0.286505	**1.75**	5.7546	0.173774
1.26	3.5254	.283654	1.76	5.8124	.172045
1.27	3.5609	.280832	1.77	5.8709	.170333
1.28	3.5966	.278037	1.78	5.9299	.168638
1.29	3.6328	.275271	1.79	5.9895	.166960
1.30	3.6693	0.272532	**1.80**	6.0496	0.165299
1.31	3.7062	.269820	1.81	6.1104	.163654
1.32	3.7434	.267135	1.82	6.1719	.162026
1.33	3.7810	.264477	1.83	6.2339	.160414
1.34	3.8190	.261846	1.84	6.2965	.158817
1.35	3.8574	0.259240	**1.85**	6.3598	0.157237
1.36	3.8962	.256661	1.86	6.4237	.155673
1.37	3.9354	.254107	1.87	6.4483	.154124
1.38	3.9749	.251579	1.88	6.5535	.152590
1.39	4.0149	.249075	1.89	6.6194	.151072
1.40	4.0552	0.246597	**1.90**	6.6859	0.149569
1.41	4.0960	.244143	1.91	6.7531	.148080
1.42	4.1371	.241714	1.92	6.8210	.146607
1.43	4.1787	.239309	1.93	6.8895	.145148
1.44	4.2207	.236928	1.94	6.9588	.143704
1.45	4.2631	0.234570	**1.95**	7.0287	0.142274
1.46	4.3060	.232236	1.96	7.0993	.140858
1.47	4.3492	.229925	1.97	7.1707	.139457
1.48	4.3929	.227638	1.98	7.2427	.138069
1.49	4.4371	.225373	1.99	7.3155	.136695
1.50	4.4817	0.223130	**2.00**	7.3891	0.135335

Table 2 Exponential Functions (Continued)

x	e^x	e^{-x}	x	e^x	e^{-x}
2.00	7.3891	0.135335	**2.50**	12.182	0.082085
2.01	7.4633	.133989	2.51	12.305	.081268
2.02	7.5383	.132655	2.52	12.429	.080460
2.03	7.6141	.131336	2.53	12.554	.079659
2.04	7.6906	.130029	2.54	12.680	.078866
2.05	7.7679	0.128735	**2.55**	12.807	0.078082
2.06	7.8460	.127454	2.56	12.936	.077305
2.07	7.9248	.126186	2.57	13.066	.076536
2.08	8.0045	.124930	2.58	13.197	.075774
2.09	8.0849	.123687	2.59	13.330	.075020
2.10	8.1662	0.122456	**2.60**	13.464	0.074274
2.11	8.2482	.121238	2.61	13.599	.073535
2.12	8.3311	.120032	2.62	13.736	.072803
2.13	8.4149	.118837	2.63	13.874	.072078
2.14	8.4994	.117655	2.64	14.013	.071361
2.15	8.5849	0.116484	**2.65**	14.154	0.070651
2.16	8.6711	.115325	2.66	14.296	.069948
2.17	8.7583	.114178	2.67	14.440	.069252
2.18	8.8463	.113042	2.68	14.585	.068563
2.19	8.9352	.111917	2.69	14.732	.067881
2.20	9.0250	0.110803	**2.70**	14.880	0.067206
2.21	9.1157	.109701	2.71	15.029	.066537
2.22	9.2073	.108609	2.72	15.180	.065875
2.23	9.2999	.107528	2.73	15.333	.065219
2.24	9.3933	.106459	2.74	15.487	.064570
2.25	9.4877	0.105399	**2.75**	15.643	0.063928
2.26	9.5831	.104350	2.76	15.800	.063292
2.27	9.6794	.103312	2.77	15.959	.062662
2.28	9.7767	.102284	2.78	16.119	.062039
2.29	9.8749	.101266	2.79	16.281	.061421
2.30	9.9742	0.100259	**2.80**	16.445	0.060810
2.31	10.074	.099261	2.81	16.610	.060205
2.32	10.176	.098274	2.82	16.777	.059606
2.33	10.278	.097296	2.83	16.945	.059013
2.34	10.381	.096328	2.84	17.116	.058426
2.35	10.486	0.095369	**2.85**	17.288	0.057844
2.36	10.591	.094420	2.86	17.462	.057269
2.37	10.697	.093481	2.87	17.637	.056699
2.38	10.805	.092551	2.88	17.814	.056135
2.39	10.913	.091630	2.89	17.993	.055576
2.40	11.023	0.090718	**2.90**	18.174	0.055023
2.41	11.134	.089815	2.91	18.357	.054476
2.42	11.246	.088922	2.92	18.541	.053934
2.43	11.359	.088037	2.93	18.728	.053397
2.44	11.473	.087161	2.94	18.916	.052866
2.45	11.588	0.086294	**2.95**	19.106	0.052340
2.46	11.705	.085435	2.96	19.298	.051819
2.47	11.822	.084585	2.97	19.492	.051303
2.48	11.941	.083743	2.98	19.688	.050793
2.49	12.061	.082910	2.99	19.886	.050287
2.50	12.182	0.082085	**3.00**	20.086	0.049787

Table 2 Exponential Functions (Continued)

x	e^x	e^{-x}	x	e^x	e^{-x}
3.00	20.086	0.049787	**3.50**	33.115	0.030197
3.01	20.287	.049292	3.51	33.448	.029897
3.02	20.491	.048801	3.52	33.784	.029599
3.03	20.697	.048316	3.53	34.124	.029305
3.04	20.905	.047835	3.54	34.467	.029013
3.05	21.115	0.047359	**3.55**	34.813	0.028725
3.06	21.328	.046888	3.56	35.163	.028439
3.07	21.542	.046421	3.57	35.517	.028156
3.08	21.758	.045959	3.58	35.874	.027876
3.09	21.977	.045502	3.59	36.234	.027598
3.10	22.198	0.045049	**3.60**	36.598	0.027324
3.11	22.421	.044601	3.61	36.966	.027052
3.12	22.646	.044157	3.62	37.338	.026783
3.13	22.874	.043718	3.63	37.713	.026516
3.14	23.104	.043283	3.64	38.092	.026252
3.15	23.336	0.042852	**3.65**	38.475	0.025991
3.16	23.571	.042426	3.66	38.861	.025733
3.17	23.807	.042004	3.67	39.252	.025476
3.18	24.047	.041586	3.68	39.646	.025223
3.19	24.288	.041172	3.69	40.045	.024972
3.20	24.533	0.040764	**3.70**	40.447	0.024724
3.21	24.779	.040357	3.71	40.854	.024478
3.22	25.028	.039955	3.72	41.264	.024234
3.23	25.280	.039557	3.73	41.679	.023993
3.24	25.534	.039164	3.74	42.098	.023754
3.25	25.790	0.038774	**3.75**	42.521	0.023518
3.26	26.050	.038388	3.76	42.948	.023284
3.27	26.311	.038006	3.77	43.380	.023052
3.28	26.576	.037628	3.78	43.816	.022823
3.29	26.843	.037254	3.79	44.256	.022596
3.30	27.113	0.036883	**3.80**	44.701	0.022371
3.31	27.385	.036516	3.81	45.150	.022148
3.32	27.660	.036153	3.82	45.604	.021928
3.33	27.938	.035793	3.83	46.063	.021710
3.34	28.219	.035437	3.84	46.525	.021494
3.35	28.503	0.035084	**3.85**	46.993	0.021280
3.36	28.789	.034735	3.86	47.465	.021068
3.37	29.097	.034390	3.87	47.942	.020858
3.38	29.371	.034047	3.88	48.424	.020651
3.39	29.666	.033709	3.89	48.911	.020445
3.40	29.964	0.033373	**3.90**	49.402	0.020242
3.41	30.265	.033041	3.91	49.899	.020041
3.42	30.569	.032712	3.92	50.400	.019840
3.43	30.877	.032387	3.93	50.907	.019644
3.44	31.187	.032065	3.94	51.419	.019448
3.45	31.500	0.031746	**3.95**	51.935	0.019255
3.46	31.817	.031430	3.96	52.457	.019063
3.47	32.137	.031117	3.97	52.985	.018873
3.48	32.460	.030807	3.98	53.517	.018686
3.49	32.786	.030501	3.99	54.055	.018500
3.50	33.115	0.030197	**4.00**	54.598	0.018316

Table 2 Exponential Functions (Continued)

x	e^x	e^{-x}	x	e^x	e^{-x}
4.00	54.598	0.018316	**4.50**	90.017	0.011109
4.01	55.147	.018133	4.51	90.922	.010998
4.02	55.701	.017953	4.52	91.836	.010889
4.03	56.261	.017774	4.53	92.759	.010781
4.04	56.826	.017597	4.54	93.691	.010673
4.05	57.397	0.017422	**4.55**	94.632	0.010567
4.06	57.974	.017249	4.56	95.583	.010462
4.07	58.577	.017077	4.57	96.544	.010358
4.08	59.145	.016907	4.58	97.514	.010255
4.09	59.740	.016739	4.59	98.494	.010153
4.10	60.340	0.016573	**4.60**	99.484	0.010052
4.11	60.947	.016408	4.61	100.48	.009952
4.12	61.559	.016245	4.62	101.49	.009853
4.13	62.178	.016083	4.63	102.51	.009755
4.14	62.803	.015923	4.64	103.54	.009658
4.15	63.434	0.015764	**4.65**	104.58	0.009562
4.16	64.072	.015608	4.66	105.64	.009466
4.17	64.715	.015452	4.67	106.70	.009372
4.18	65.366	.015299	4.68	107.77	.009279
4.19	66.023	.015146	4.69	108.85	.009187
4.20	66.686	0.014996	**4.70**	109.95	0.009095
4.21	67.357	.014846	4.71	111.05	.009005
4.22	68.033	.014699	4.72	112.17	.008915
4.23	68.717	.014552	4.73	113.30	.008826
4.24	69.408	.014408	4.74	114.43	.008739
4.25	70.105	0.014264	**4.75**	115.58	0.008652
4.26	70.810	.014122	4.76	116.75	.008566
4.27	71.522	.013982	4.77	117.92	.008480
4.28	72.240	.013843	4.78	119.10	.008396
4.29	72.966	.013705	4.79	120.30	.008312
4.30	73.700	0.013569	**4.80**	121.51	0.008230
4.31	74.440	.013434	4.81	122.73	.008148
4.32	75.189	.013300	4.82	123.97	.008067
4.33	75.944	.013168	4.83	125.21	.007987
4.34	76.708	.013037	4.84	126.47	.007907
4.35	77.478	0.012907	**4.85**	127.74	0.007828
4.36	78.257	.012778	4.86	129.02	.007750
4.37	79.044	.012651	4.87	130.32	.007673
4.38	79.838	.012525	4.88	131.63	.007597
4.39	80.640	.012401	4.89	132.95	.007521
4.40	81.451	0.012277	**4.90**	134.29	0.007477
4.41	82.269	.012155	4.91	135.64	.007372
4.42	83.096	.012034	4.92	137.00	.007299
4.43	83.931	.011914	4.93	138.38	.007227
4.44	84.775	.011796	4.94	139.77	.007155
4.45	85.627	0.011679	**4.95**	141.17	0.007083
4.46	86.488	.011562	4.96	142.59	.007013
4.47	87.357	.011447	4.97	144.03	.006943
4.48	88.235	.011333	4.98	145.47	.006874
4.49	89.121	.011221	4.99	146.94	.006806
4.50	90.017	0.011109	**5.00**	148.41	0.006738

Table 2 Exponential Functions (Continued)

x	e^x	e^{-x}	x	e^x	e^{-x}
5.00	148.41	0.006738	**5.50**	244.69	0.0040868
5.01	149.90	.006671	5.55	257.24	.0038875
5.02	151.41	.006605	5.60	270.43	.0036979
5.03	152.93	.006539	5.65	284.29	.0035175
5.04	154.47	.006474	5.70	298.87	.0033460
5.05	156.02	0.006409	**5.75**	314.19	0.0031828
5.06	157.59	.006346	5.80	330.30	.0030276
5.07	159.17	.006282	5.85	347.23	.0028799
5.08	160.77	.006220	5.90	365.04	.0027394
5.09	162.39	.006158	5.95	383.75	.0026058
5.10	164.02	0.006097	**6.00**	403.43	0.0024788
5.11	165.67	.006036	6.05	424.11	.0023579
5.12	167.34	.005976	6.10	445.86	.0022429
5.13	169.02	.005917	6.15	468.72	.0021335
5.14	170.72	.005858	6.20	492.75	.0020294
5.15	172.43	0.005799	**6.25**	518.01	0.0019305
5.16	174.16	.005742	6.30	544.57	.0018363
5.17	175.91	.005685	6.35	572.49	.0017467
5.18	177.68	.005628	6.40	601.85	.0016616
5.19	179.47	.005572	6.45	632.70	.0015805
5.20	181.27	0.005517	**6.50**	665.14	0.0015034
5.21	183.09	.005462	6.55	699.24	.0014301
5.22	184.93	.005407	6.60	735.10	.0013604
5.23	186.79	.005354	6.65	772.78	.0012940
5.24	188.67	.005300	6.70	812.41	.0012309
5.25	190.57	0.005248	**6.75**	854.06	0.0011709
5.26	192.48	.005195	6.80	897.85	.0011138
5.27	194.42	.005144	6.85	943.88	.0010595
5.28	196.37	.005092	6.90	992.27	.0010078
5.29	198.34	.005042	6.95	1043.1	.0009586
5.30	200.34	0.004992	**7.00**	1096.6	0.0009119
5.31	202.35	.004942	7.05	1152.9	.0008674
5.32	204.38	.004893	7.10	1212.0	.0008251
5.33	206.44	.004844	7.15	1274.1	.0007849
5.34	208.51	.004796	7.20	1339.4	.0007466
5.35	210.61	0.004748	**7.25**	1408.1	0.0007102
5.36	212.72	.004701	7.30	1480.3	.0006755
5.37	214.86	.004654	7.35	1556.2	.0006426
5.38	217.02	.004608	7.40	1636.0	.0006113
5.39	219.20	.004562	7.45	1719.9	.0005814
5.40	221.41	0.004517	**7.50**	1808.0	0.0005531
5.41	223.63	.004472	7.55	1900.7	.0005261
5.42	225.88	.004427	7.60	1998.2	.0005005
5.43	228.15	.004383	7.65	2100.6	.0004760
5.44	230.44	.004339	7.70	2208.3	.0004528
5.45	232.76	0.004296	**7.75**	2321.6	0.0004307
5.46	235.10	.004254	7.80	2440.6	.0004097
5.47	237.46	.004211	7.85	2565.7	.0003898
5.48	239.85	.004169	7.90	2697.3	.0003707
5.49	242.26	.004128	7.95	2835.6	.0003527
5.50	244.69	0.004087	**8.00**	2981.0	0.0003355

Table 2 Exponential Functions (Continued)

x	e^x	e^{-x}	x	e^x	e^{-x}
8.00	2981.0	0.0003355	**9.00**	8103.1	0.0001234
8.05	3133.8	.0003191	9.05	8518.5	.0001174
8.10	3294.5	.0003035	9.10	8955.3	.0001117
8.15	3463.4	.0002887	9.15	9414.4	.0001062
8.20	3641.0	.0002747	9.20	9897.1	.0001010
8.25	3827.6	0.0002613	**9.25**	10405	0.0000961
8.30	4023.9	.0002485	9.30	10938	.0000914
8.35	4230.2	.0002364	9.35	11499	.0000870
8.40	4447.1	.0002249	9.40	12088	.0000827
8.45	4675.1	.0002139	9.45	12708	.0000787
8.50	4914.8	0.0002036	**9.50**	13360	0.0000749
8.55	5166.8	.0001935	9.55	14045	.0000712
8.60	5431.7	.0001841	9.60	14765	.0000677
8.65	5710.0	.0001751	9.65	15522	.0000644
8.70	6002.9	.0001666	9.70	16318	.0000613
8.75	6310.7	0.0001585	**9.75**	17154	0.0000583
8.80	6634.2	.0001507	9.80	18034	.0000555
8.85	6974.4	.0001434	9.85	18958	.0000527
8.90	7332.0	.0001364	9.90	19930	.0000502
8.95	7707.9	.0001297	9.95	20952	.0000477
9.00	8103.1	0.0001234	**10.00**	22026	0.0000454

Table 3 Hyperbolic Functions

x	$\sinh x$	$\cosh x$	$\tanh x$
0	0.00000	1.0000	0.00000
0.1	.10017	1.0050	.09967
0.2	.20134	1.0201	.19738
0.3	.30452	1.0453	.29131
0.4	.41075	1.0811	.37995
0.5	.52110	1.1276	.46212
0.6	.63665	1.1855	.53705
0.7	.75858	1.2552	.60437
0.8	.88811	1.3374	.66404
0.9	1.0265	1.4331	.71630
1.0	1.1752	1.5431	.76159
1.1	1.3356	1.6685	.80050
1.2	1.5095	1.8107	.83365
1.3	1.6984	1.9709	.86172
1.4	1.9043	2.1509	.88535
1.5	2.1293	2.3524	.90515
1.6	2.3756	2.5775	.92167
1.7	2.6456	2.8283	.93541
1.8	2.9422	3.1075	.94681
1.9	3.2682	3.4177	.95624
2.0	3.6269	3.7622	.96403
2.1	4.0219	4.1443	.97045
2.2	4.4571	4.5679	.97574
2.3	4.9370	5.0372	.98010
2.4	5.4662	5.5569	.98367
2.5	6.0502	6.1323	.98661
2.6	6.6947	6.7690	.98903
2.7	7.4063	7.4735	.99101
2.8	8.1919	8.2527	.99263
2.9	9.0596	9.1146	.99396
3.0	10.018	10.068	.99505
3.1	11.076	11.122	.99595
3.2	12.246	12.287	.99668
3.3	13.538	13.575	.99728
3.4	14.965	14.999	.99777
3.5	16.543	16.573	.99818
3.6	18.285	18.313	.99851
3.7	20.211	20.236	.99878
3.8	22.339	22.362	.99900
3.9	24.691	24.711	.99918
4.0	27.290	27.308	.99933
4.1	30.162	30.178	.99945
4.2	33.336	33.351	.99955
4.3	36.843	36.857	.99963
4.4	40.719	40.732	.99970
4.5	45.003	45.014	.99975
4.6	49.737	49.747	.99980
4.7	54.969	54.978	.99983
4.8	60.751	60.759	.99986
4.9	67.141	67.149	.99989
5.0	74.203	74.210	.99991

Table 4 Natural Logarithms

N	0	1	2	3	4	5	6	7	8	9
1.0	0000	0100	0198	0296	0392	0488	0583	0677	0770	0862
1.1	0953	1044	1133	1222	1310	1398	1484	1570	1655	1740
1.2	1823	1906	1989	2070	2151	2231	2311	2390	2469	2546
1.3	2624	2700	2776	2852	2927	3001	3075	3148	3221	3293
1.4	3365	3436	3507	3577	3646	3716	3784	3853	3920	3988
1.5	4055	4121	4187	4253	4318	4383	4447	4511	4574	4637
1.6	4700	4762	4824	4886	4947	5008	5068	5128	5188	5247
1.7	5306	5365	5423	5481	5539	5596	5653	5710	5766	5822
1.8	5878	5933	5988	6043	6098	6152	6206	6259	6313	6366
1.9	6419	6471	6523	6575	6627	6678	6729	6780	6831	6881
2.0	6931	6981	7031	7080	7129	7178	7227	7275	7324	7372
2.1	7419	7467	7514	7561	7608	7655	7701	7747	7793	7839
2.2	7885	7930	7975	8020	8065	8109	8154	8198	8242	8286
2.3	8329	8372	8416	8459	8502	8544	8587	8629	8671	8713
2.4	8755	8796	8838	8879	8920	8961	9002	9042	9083	9123
2.5	9163	9203	9243	9282	9322	9361	9400	9439	9478	9517
2.6	9555	9594	9632	9670	9708	9746	9783	9821	9858	9895
2.7	9933	9969	1.0006	0043	0080	0116	0152	0188	0225	0260
2.8	1.0296	0332	0367	0403	0438	0473	0508	0543	0578	0613
2.9	0647	0682	0716	0750	0784	0818	0852	0886	0919	0953
3.0	1.0986	1019	1053	1086	1119	1151	1184	1217	1249	1282
3.1	1314	1346	1378	1410	1442	1474	1506	1537	1569	1600
3.2	1632	1663	1694	1725	1756	1787	1817	1848	1878	1909
3.3	1939	1969	2000	2030	2060	2090	2119	2149	2179	2208
3.4	2238	2267	2296	2326	2355	2384	2413	2442	2470	2499
3.5	1.2528	2556	2585	2613	2641	2669	2698	2726	2754	2782
3.6	2809	2837	2865	2892	2920	2947	2975	3002	3029	3056
3.7	3083	3110	3137	3164	3191	3218	3244	3271	3297	3324
3.8	3350	3376	3403	3429	3455	3481	3507	3533	3558	3584
3.9	3610	3635	3661	3686	3712	3737	3762	3788	3813	3838
4.0	1.3863	3888	3913	3938	3962	3987	4012	4036	4061	4085
4.1	4110	4134	4159	4183	4207	4231	4255	4279	4303	4327
4.2	4351	4375	4398	4422	4446	4469	4493	4516	4540	4563
4.3	4586	4609	4633	4656	4679	4702	4725	4748	4770	4793
4.4	4816	4839	4861	4884	4907	4929	4951	4974	4996	5019
4.5	1.5041	5063	5085	5107	5129	5151	5173	5195	5217	5239
4.6	5261	5282	5304	5326	5347	5369	5390	5412	5433	5454
4.7	5476	5497	5518	5539	5560	5581	5602	5623	5644	5665
4.8	5686	5707	5728	5748	5769	5790	5810	5831	5851	5872
4.9	5892	5913	5933	5953	5974	5994	6014	6034	6054	6074
5.0	1.6094	6114	6134	6154	6174	6194	6214	6233	6253	6273
5.1	6292	6312	6332	6351	6371	6390	6409	6429	6448	6467
5.2	6487	6506	6525	6544	6563	6582	6601	6620	6639	6658
5.3	6677	6696	6715	6734	6752	6771	6790	6808	6827	6845
5.4	6864	6882	6901	6919	6938	6956	6974	6993	7011	7029

Table 4 Natural Logarithms (Continued)

N	0	1	2	3	4	5	6	7	8	9
5.5	1.7047	7066	7084	7102	7120	7138	7156	7174	7192	7210
5.6	7228	7246	7263	7281	7299	7317	7334	7352	7370	7387
5.7	7405	7422	7440	7457	7475	7492	7509	7527	7544	7561
5.8	7579	7596	7613	7630	7647	7664	7681	7699	7716	7733
5.9	7750	7766	7783	7800	7817	7834	7851	7867	7884	7901
6.0	1.7918	7934	7951	7967	7984	8001	8017	8034	8050	8066
6.1	8083	8099	8116	8132	8148	8165	8181	8197	8213	8229
6.2	8245	8262	8278	8294	8310	8326	8342	8358	8374	8390
6.3	8405	8421	8437	8453	8469	8485	8500	8516	8532	8547
6.4	8563	8579	8594	8610	8625	8641	8656	8672	8687	8703
6.5	1.8718	8733	8749	8764	8779	8795	8810	8825	8840	8856
6.6	8871	8886	8901	8916	8931	8946	8961	8976	8991	9006
6.7	9021	9036	9051	9066	9081	9095	9110	9125	9140	9155
6.8	9169	9184	9199	9213	9228	9242	9257	9272	9286	9301
6.9	9315	9330	9344	9359	9373	9387	9402	9416	9430	9445
7.0	1.9459	9473	9488	9502	9516	9530	9544	9559	9573	9587
7.1	9601	9615	9629	9643	9657	9671	9685	9699	9713	9727
7.2	9741	9755	9769	9782	9796	9810	9824	9838	9851	9865
7.3	9879	9892	9906	9920	9933	9947	9961	9974	9988	*0001
7.4	2.0015	0028	0042	0055	0069	0082	0096	0109	0122	0136
7.5	2.0149	0162	0176	0189	0202	0215	0229	0242	0255	0268
7.6	0281	0295	0308	0321	0334	0347	0360	0373	0386	0399
7.7	0412	0425	0438	0451	0464	0477	0490	0503	0516	0528
7.8	0541	0554	0567	0580	0592	0605	0618	0630	0643	0656
7.9	0669	0681	0694	0707	0719	0732	0744	0757	0769	0782
8.0	2.0794	0807	0819	0832	0844	0857	0869	0882	0894	0906
8.1	0919	0931	0943	0956	0968	0980	0992	1005	1017	1029
8.2	1041	1054	1066	1078	1090	1102	1114	1126	1138	1150
8.3	1163	1175	1187	1199	1211	1223	1235	1247	1258	1270
8.4	1282	1294	1306	1318	1330	1342	1353	1365	1377	1389
8.5	2.1401	1412	1424	1436	1448	1459	1471	1483	1494	1506
8.6	1518	1529	1541	1552	1564	1576	1587	1599	1610	1622
8.7	1633	1645	1656	1668	1679	1691	1702	1713	1725	1736
8.8	1748	1759	1770	1782	1793	1804	1815	1827	1838	1849
8.9	1861	1872	1883	1894	1905	1917	1928	1939	1950	1961
9.0	2.1972	1983	1994	2006	2017	2028	2039	2050	2061	2072
9.1	2083	2094	2105	2116	2127	2138	2148	2159	2170	2181
9.2	2192	2203	2214	2225	2235	2246	2257	2268	2279	2289
9.3	2300	2311	2322	2332	2343	2354	2364	2375	2386	2396
9.4	2407	2418	2428	2439	2450	2460	2471	2481	2492	2502
9.5	2.2513	2523	2534	2544	2555	2565	2576	2586	2597	2607
9.6	2618	2628	2638	2649	2659	2670	2680	2690	2701	2711
9.7	2721	2732	2742	2752	2762	2773	2783	2793	2803	2814
9.8	2824	2834	2844	2854	2865	2875	2885	2895	2905	2915
9.9	2925	2935	2946	2956	2966	2976	2986	2996	3006	3016

Use In 10 = 2.30259 and the properties of $\ln x$ to find logarithms of numbers greater than 10 or less than 1.

Table 5 Common Logarithms

N	0	1	2	3	4	5	6	7	8	9
10	0000	0043	0086	0128	0170	0212	0253	0294	0334	0374
11	0414	0453	0492	0531	0569	0607	0645	0682	0719	0755
12	0792	0828	0864	0899	0934	0969	1004	1038	1072	1106
13	1139	1173	1206	1239	1271	1303	1335	1367	1399	1430
14	1461	1492	1523	1553	1584	1614	1644	1673	1703	1732
15	1761	1790	1818	1847	1875	1903	1931	1959	1987	2014
16	2041	2068	2095	2122	2148	2175	2201	2227	2253	2279
17	2304	2330	2355	2380	2405	2430	2455	2480	2504	2529
18	2553	2577	2601	2625	2648	2672	2695	2718	2742	2765
19	2788	2810	2833	2856	2878	2900	2923	2945	2967	2989
20	3010	3032	3054	3075	3096	3118	3139	3160	3181	3201
21	3222	3243	3263	3284	3304	3324	3345	3365	3385	3404
22	3424	3444	3464	3483	3502	3522	3541	3560	3579	3598
23	3617	3636	3655	3674	3692	3711	3729	3747	3766	3784
24	3802	3820	3838	3856	3874	3892	3909	3927	3945	3962
25	3979	3997	4014	4031	4048	4065	4082	4099	4116	4133
26	4150	4166	4183	4200	4216	4232	4249	4265	4281	4298
27	4314	4330	4346	4362	4378	4393	4409	4425	4440	4456
28	4472	4487	4502	4518	4533	4548	4564	4579	4594	4609
29	4624	4639	4654	4669	4683	4698	4713	4728	4742	4757
30	4771	4786	4800	4814	4829	4843	4857	4871	4886	4900
31	4914	4928	4942	4955	4969	4983	4997	5011	5024	5038
32	5051	5065	5079	5092	5105	5119	5132	5145	5159	5172
33	5185	5198	5211	5224	5237	5250	5263	5276	5289	5302
34	5315	5328	5340	5353	5366	5378	5391	5403	5416	5428
35	5441	5453	5465	5478	5490	5502	5514	5527	5539	5551
36	5563	5575	5587	5599	5611	5623	5635	5647	5658	5670
37	5682	5694	5705	5717	5729	5740	5752	5763	5775	5786
38	5798	5809	5821	5832	5843	5855	5866	5877	5888	5899
39	5911	5922	5933	5944	5955	5966	5977	5988	5999	6010
40	6021	6031	6042	6053	6064	6075	6085	6096	6107	6117
41	6128	6138	6149	6160	6170	6180	6191	6201	6212	6222
42	6232	6243	6253	6263	6274	6284	6294	6304	6314	6325
43	6335	6345	6355	6365	6375	6385	6395	6405	6415	6425
44	6435	6444	6454	6464	6474	6484	6493	6503	6513	6522
45	6532	6542	6551	6561	6571	6580	6590	6599	6609	6618
46	6628	6637	6646	6656	6665	6675	6684	6693	6702	6712
47	6721	6730	6739	6749	6758	6767	6776	6785	6794	6803
48	6812	6821	6830	6839	6848	6857	6866	6875	6884	6893
49	6902	6911	6920	6928	6937	6946	6955	6964	6972	6981
50	6990	6998	7007	7016	7024	7033	7042	7050	7059	7067
51	7076	7084	7093	7101	7110	7118	7126	7135	7143	7152
52	7160	7168	7177	7185	7193	7202	7210	7218	7226	7235
53	7243	7251	7259	7267	7275	7284	7292	7300	7308	7316
54	7324	7332	7340	7348	7356	7364	7372	7380	7388	7396

Table 5 Common Logarithms (Continued)

N	0	1	2	3	4	5	6	7	8	9
55	7404	7412	7419	7427	7435	7433	7451	7459	7466	7474
56	7482	7490	7497	7505	7513	7520	7528	7536	7543	7551
57	7559	7566	7574	7582	7589	7597	7604	7612	7619	7627
58	7634	7642	7649	7657	7664	7672	7679	7686	7694	7701
59	7709	7716	7723	7731	7738	7745	7752	7760	7767	7774
60	7782	7789	7796	7803	7810	7818	7825	7832	7839	7846
61	7853	7860	7868	7875	7882	7889	7896	7903	7910	7917
62	7924	7931	7938	7945	7952	7959	7966	7973	7980	7987
63	7993	8000	8007	8014	8021	8028	8035	8041	8048	8055
64	8062	8069	8075	8082	8089	8096	8102	8109	8116	8122
65	8129	8136	8142	8149	8156	8162	8169	8176	8182	8189
66	8195	8202	8209	8215	8222	8228	8235	8241	8248	8254
67	8261	8267	8274	8280	8287	8293	8299	8306	8312	8319
68	8325	8331	8338	8344	8351	8357	8363	8370	8376	8382
69	8388	8395	8401	8407	8414	8420	8426	8432	8439	8445
70	8451	8457	8463	8470	8476	8482	8488	8494	8500	8506
71	8513	8519	8525	8531	8537	8543	8549	8555	8561	8567
72	8573	8579	8585	8591	8597	8603	8609	8615	8621	8627
73	8633	8639	8645	8651	8657	8663	8669	8675	8681	8686
74	8692	8698	8704	8710	8716	8722	8727	8733	8739	8745
75	8751	8756	8762	8768	8774	8779	8785	8791	8797	8802
76	8808	8814	8820	8825	8831	8837	8842	8848	8854	8859
77	8865	8871	8876	8882	8887	8893	8899	8904	8910	8915
78	8921	8927	8932	8938	8943	8949	8954	8960	8965	8971
79	8976	8982	8987	8993	8998	9004	9009	9015	9020	9025
80	9031	9036	9042	9047	9053	9058	9063	9069	9074	9079
81	9085	9090	9096	9101	9106	9112	9117	9122	9128	9133
82	9138	9143	9149	9154	9159	9165	9170	9175	9180	9186
83	9191	9196	9201	9206	9212	9217	9222	9227	9232	9238
84	9243	9248	9253	9258	9263	9269	9274	9279	9284	9289
85	9294	9299	9304	9309	9315	9320	9325	9330	9335	9340
86	9345	9350	9355	9360	9365	9370	9375	9380	9385	9390
87	9395	9400	9405	9410	9415	9420	9425	9430	9435	9440
88	9445	9450	9455	9460	9465	9469	9474	9479	9484	9489
89	9494	9499	9504	9509	9513	9518	9523	9528	9533	9538
90	9542	9547	9552	9557	9562	9566	9571	9576	9581	9586
91	9590	9595	9600	9605	9609	9614	9619	9624	9628	9633
92	9638	9643	9647	9652	9657	9661	9666	9671	9675	9680
93	9685	9689	9694	9699	9703	9708	9713	9717	9722	9727
94	9731	9736	9741	9745	9750	9754	9759	9763	9768	9773
95	9777	9782	9786	9791	9795	9800	9805	9809	9814	9818
96	9823	9827	9832	9836	9841	9845	9850	9854	9859	9863
97	9868	9872	9877	9881	9886	9890	9894	9899	9903	9908
98	9912	9917	9921	9926	9930	9934	9939	9943	9948	9952
99	9956	9961	9965	9969	9974	9978	9983	9987	9991	9996

Table 6 Powers and Roots

n	n^2	$\sqrt{n}$	n^3	$\sqrt[3]{n}$	n	n^2	$\sqrt{n}$	n^3	$\sqrt[3]{n}$
1	1	1.000	1	1.000	51	2,601	7.141	132,651	3.708
2	4	1.414	8	1.260	52	2,704	7.211	140,608	3.732
3	9	1.732	27	1.442	53	2,809	7.280	148,877	3.756
4	16	2.000	64	1.587	54	2,916	7.348	157,464	3.780
5	25	2.236	125	1.710	55	3,025	7.416	166,375	3.803
6	36	2.449	216	1.817	56	3,136	7.483	175,616	3.826
7	49	2.646	343	1.913	57	3,249	7.550	185,193	3.848
8	64	2.828	512	2.000	58	3,364	7.616	195,112	3.871
9	81	3.000	729	2.080	59	3,481	7.681	205,379	3.893
10	100	3.162	1,000	2.154	60	3,600	7.746	216,000	3.915
11	121	3.317	1,331	2.224	61	3,721	7.810	226,981	3.936
12	144	3.464	1,728	2.289	62	3,844	7.874	238,328	3.958
13	169	3.606	2,197	2.351	63	3,969	7.937	250,047	3.979
14	196	3.742	2,744	2.410	64	4,096	8.000	262,144	4.000
15	225	3.873	3,375	2.466	65	4,225	8.062	274,625	4.021
16	256	4.000	4,096	2.520	66	4,356	8.124	287,496	4.041
17	289	4.123	4,913	2.571	67	4,489	8.185	300,763	4.062
18	324	4.243	5,832	2.621	68	4,624	8.246	314,432	4.082
19	361	4.359	6,859	2.668	69	4,761	8.307	328,509	4.102
20	400	4.472	8,000	2.714	70	4,900	8.367	343,000	4.121
21	441	4.583	9,261	2.759	71	5,041	8.426	357,911	4.141
22	484	4.690	10,648	2.802	72	5,184	8.485	373,248	4.160
23	529	4.796	12,167	2.844	73	5,329	8.544	389,017	4.179
24	576	4.899	13,824	2.884	74	5,476	8.602	405,224	4.198
25	625	5.000	15,625	2.924	75	5,625	8.660	421,875	4.217
26	676	5.099	17,576	2.962	76	5,776	8.718	438,976	4.236
27	729	5.196	19,683	3.000	77	5,929	8.775	456,533	4.254
28	784	5.291	21,952	3.037	78	6,084	8.832	474,552	4.273
29	841	5.385	24,389	3.072	79	6,241	8.888	493,039	4.291
30	900	5.477	27,000	3.107	80	6,400	8.944	512,000	4.309
31	961	5.568	29,791	3.141	81	6,561	9.000	531,441	4.327
32	1,024	5.657	32,768	3.175	82	6,724	9.055	551,368	4.344
33	1,089	5.745	35,937	3.208	83	6,889	9.110	571,787	4.362
34	1,156	5.831	39,304	3.240	84	7,056	9.165	592,704	4.380
35	1,225	5.916	42,875	3.271	85	7,225	9.220	614,125	4.397
36	1,296	6.000	46,656	3.302	86	7,396	9.274	636,056	4.414
37	1,369	6.083	50,653	3.332	87	7,569	9.327	658,503	4.431
38	1,444	6.164	54,872	3.362	88	7,744	9.381	681,472	4.448
39	1,521	6.245	59,319	3.391	89	7,921	9.434	704,969	4.465
40	1,600	6.325	64,000	3.420	90	8,100	9.487	729,000	4.481
41	1,681	6.403	68,921	3.448	91	8,281	9.539	753,571	4.498
42	1,764	6.481	74,088	3.476	92	8,464	9.592	778,688	4.514
43	1,849	6.557	79,507	3.503	93	8,649	9.643	804,357	4.531
44	1,936	6.633	85,184	3.530	94	8,836	9.695	830,584	4.547
45	2,025	6.708	91,125	3.557	95	9,025	9.747	857,375	4.563
46	2,116	6.782	97,336	3.583	96	9,216	9.798	884,736	4.579
47	2,209	6.856	103,823	3.609	97	9,409	9.849	912,673	4.595
48	2,304	6.928	110,592	3.634	98	9,604	9.899	941,192	4.610
49	2,401	7.000	117,649	3.659	99	9,801	9.950	970,299	4.626
50	2,500	7.071	125,000	3.684	100	10,000	10.000	1,000,000	4.642

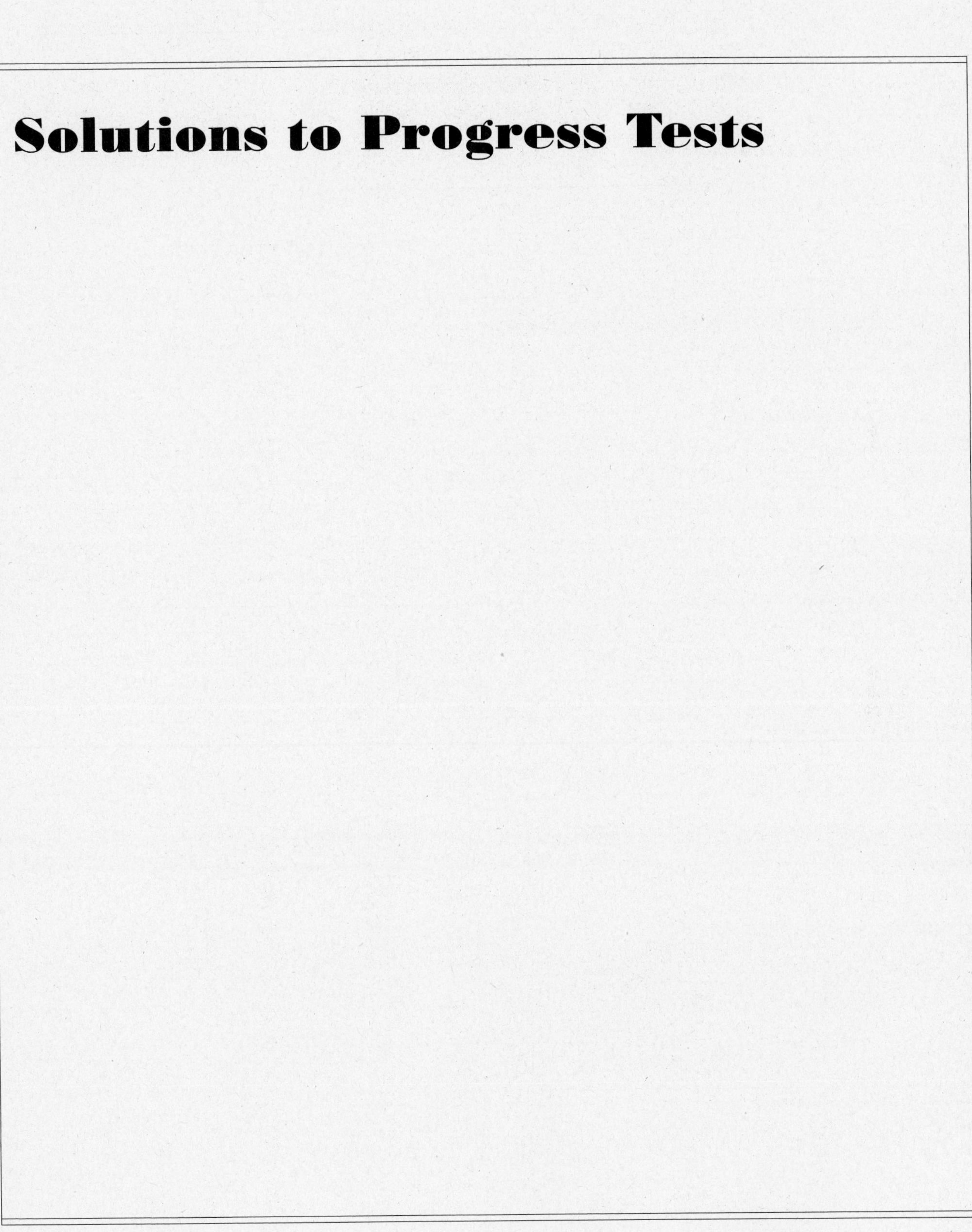

Solutions to Progress Tests

Chapter 1: Solutions to Progress Tests

Sec. 1.1: Progress Test 1

2. (*a*) $x > -1$ (*b*) $x \geq -2$ (*c*) $4 < x$ (*d*) $x \geq -\frac{3}{7}$

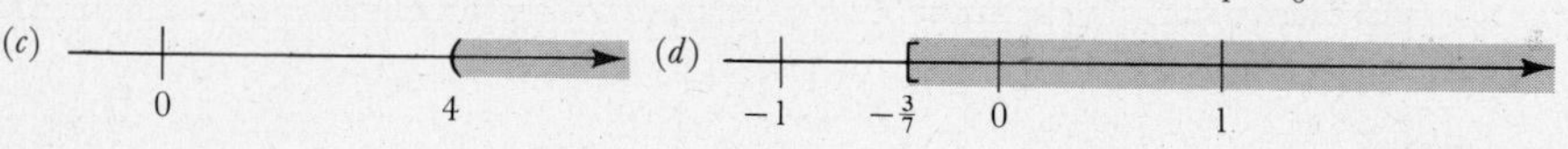

Sec. 1.1: Progress Test 2

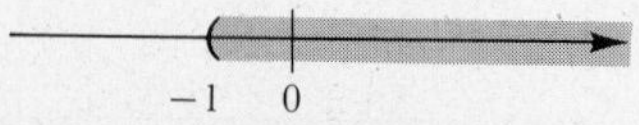

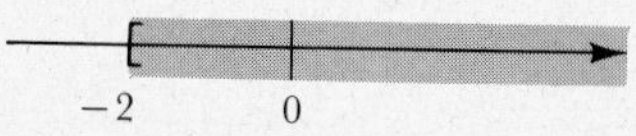

1. All x whose dist. from -1 is ≥ 5: $(-\infty, -6]$, $[4, +\infty)$

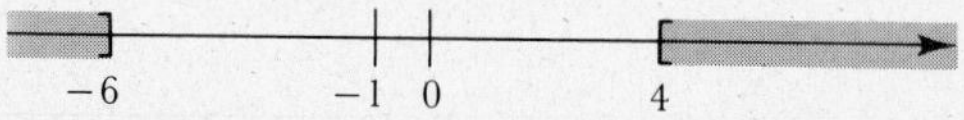

2. $3|x + \frac{2}{3}| < 6 \Rightarrow |x + \frac{2}{3}| < 2$. All x whose dist. from $-\frac{2}{3}$ is < 2: $(-\frac{8}{3}, \frac{4}{3})$.

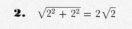

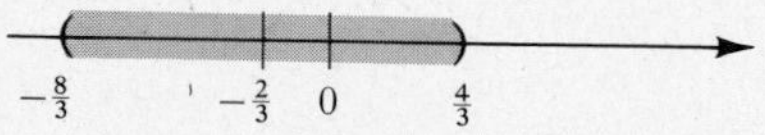

3. $2|x - 3| < 9 \Rightarrow |x - 3| < \frac{9}{2}$. All x whose dist. from 3 is $< \frac{9}{2}$. $(-\frac{3}{2}, \frac{15}{2})$

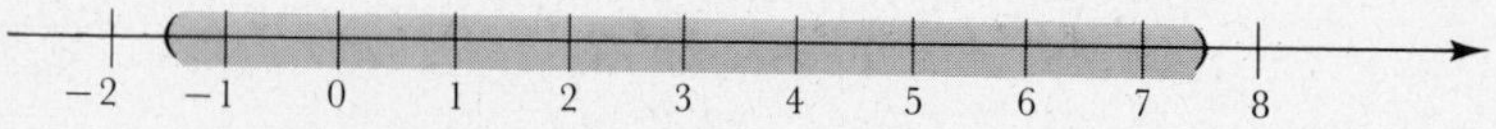

4. No x can have negative dist. from 1, so no x satisfy $|x - 1| < 0$.

Sec. 1.1: Progress Test 3

1. $(0, 3)$, $(6, +\infty)$

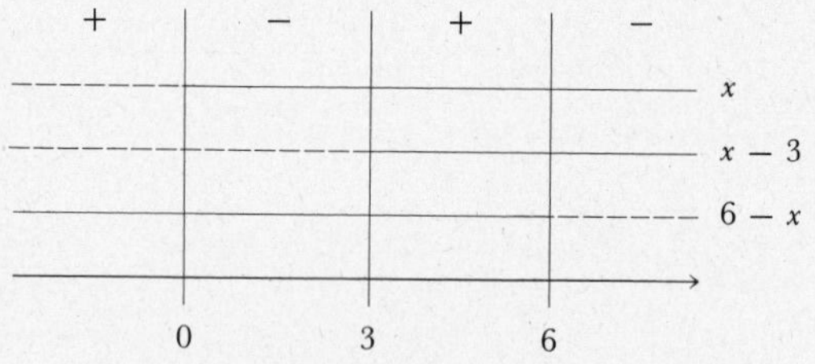

2. $(-\infty, -2]$, $(-\frac{1}{2}, 1]$, $[4, +\infty)$ (*Note:* $x^2 + 3$ does not affect solution and $-\frac{1}{2}$ makes den. $= 0$.)

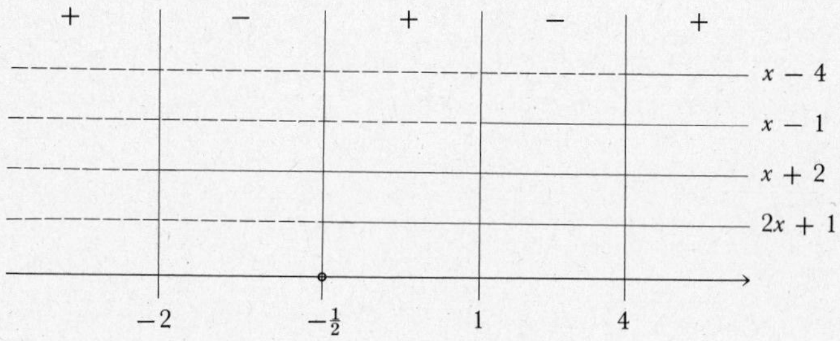

Sec. 1.2: Progress Test 1

1. $C = (3, 5)$, $B = (2, 4)$, $A = (1, 3)$

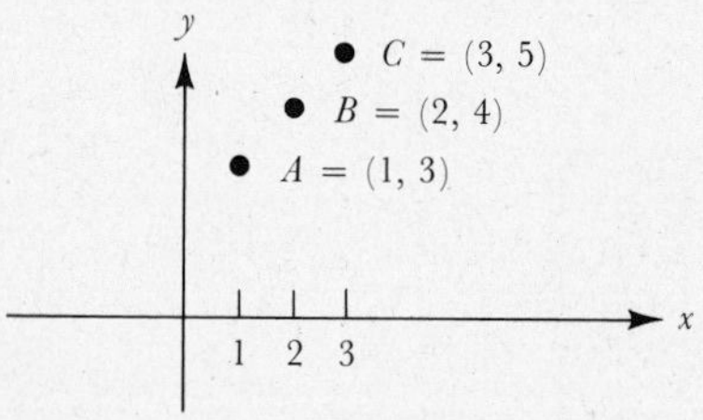

2. $\sqrt{2^2 + 2^2} = 2\sqrt{2}$

3.

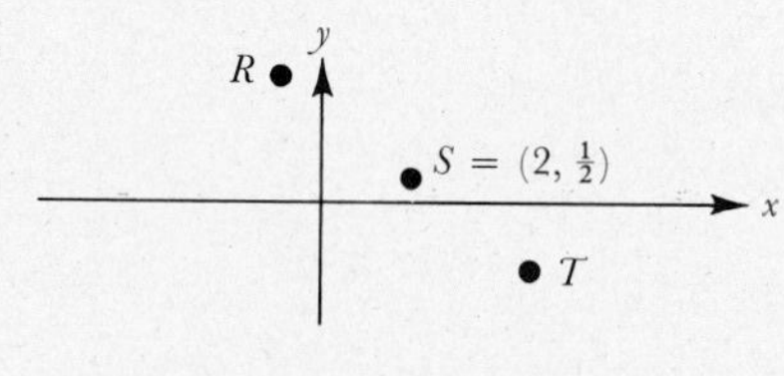

4. $\sqrt{61}/2$. **5.** $\sqrt{9 + 25/4} = \sqrt{61}/2$.

6. $\sqrt{(x_1 - x_2)^2 + (y_1 - y_2)^2} = \sqrt{[(-1)(x_2 - x_1)]^2 + [(-1)(y_2 - y_1)]^2} = \sqrt{(-1)^2(x_2 - x_1)^2 + (-1)^2(y_2 - y_1)^2} = \sqrt{(x_2 - x_1)^2 + (y_2 - y_1)^2}$.

7.
$$\sqrt{\left(x_1 - \frac{x_1 + x_2}{2}\right)^2 + \left(y_1 - \frac{y_1 + y_2}{2}\right)^2} = \sqrt{\left(\frac{2x_1 - x_1 - x_2}{2}\right)^2 + \left(\frac{2y_1 - y_1 - y_2}{2}\right)^2}$$
$$= \sqrt{\frac{(x_1 - x_2)^2}{4} + \frac{(y_1 - y_2)^2}{4}} = \frac{1}{2}\sqrt{(x_1 - x_2)^2 + (y_1 - y_2)^2}.$$

Sec. 1.2: Progress Test 2

1. $+1$. **2.** $\dfrac{3 - 0}{0 - 4} = \dfrac{-3}{4}$. **3.** $\dfrac{2 - (-1)}{-2 - 4} = \dfrac{3}{-6} = -\dfrac{1}{2}$. **4.** 2. **5.** No. **6.** No. **7.** -1 (Draw a picture.)

8. $\dfrac{-1}{(-1)} = 1$.

Sec. 1.2: Progress Test 3

1. (a) $y - 5 = 4(x - 3)$; (b) $y - 5 = \dfrac{9-5}{4-1}(x-3)$ or $y - 9 = 4(x - 4)$; (c) $y = -2x + 2$. **2.** (a) $y = 3x + \frac{5}{2} \Rightarrow m = 3,\ b = \frac{5}{2}$; (b) $y = \frac{3}{4}x - 2 \Rightarrow m = \frac{3}{4},\ b = -2$.

3. (a)

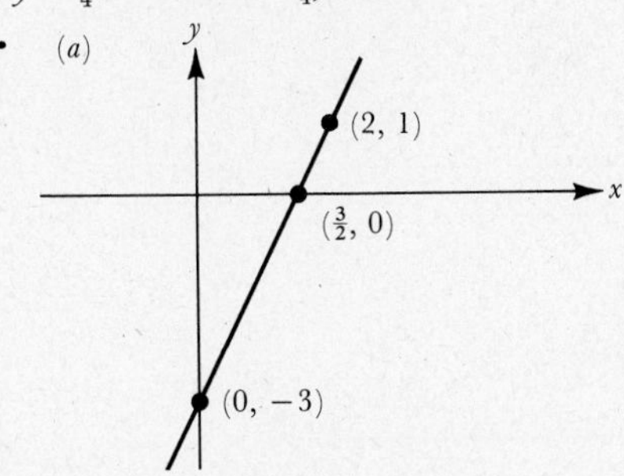

4. Slope of L_2: $\dfrac{1}{-8-7} = -\dfrac{1}{15}$; L_1: $y - 3 = 15(x + 5)$.

Sec. 1.3: Progress Test 1

1. (a) $(x-1)^2 + (y-4)^2 = 4$; (b) $(x+3)^2 + (y-3)^2 = 5$; (c) $(x-m)^2 + (y-n)^2 = t^2$.
2. (a) $(x+2)^2 + (y-3)^2 = -1 + 13$; center $(-2, 3)$, $r = \sqrt{12} = 2\sqrt{3}$.
(b) $(x+3)^2 + (y+6)^2 = -3 + 45$; center $(-3, -6)$, $r = \sqrt{42}$. (c) $(x+1)^2 + (y-\frac{5}{2})^2 = -40 + 1 - \frac{25}{4} = -\frac{131}{4} < 0$, so no graph.

Sec. 1.3: Progress Test 2

1.

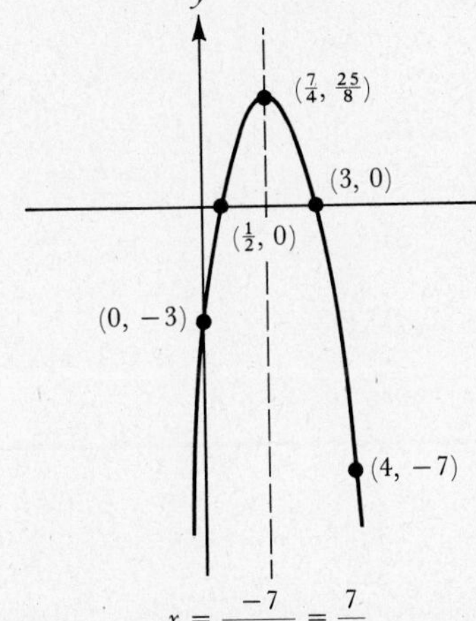

$x = \dfrac{-7}{2(-2)} = \dfrac{7}{4}$

$y = (-2x + 1)(x - 3)$

2.

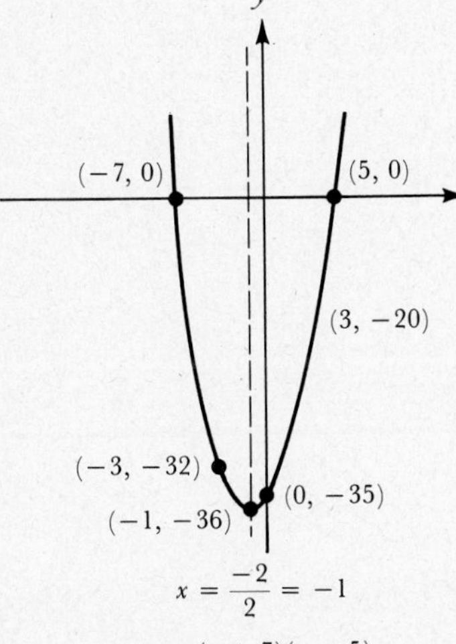

$x = \dfrac{-2}{2} = -1$

$y = (x + 7)(x - 5)$

3.

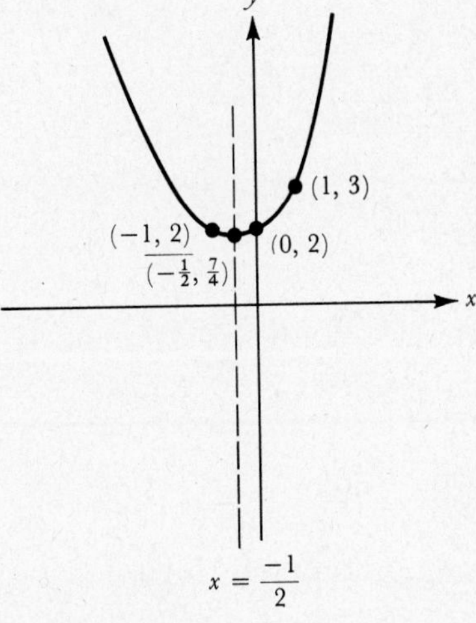

$x = \dfrac{-1}{2}$

discriminant $= 1 - 8 = -7 < 0$

Chapter 2: Solutions to Progress Tests

Sec. 2.1: Progress Test 1

1. (a) $f(t) = 3t^2 - 5t + 2$; (b) $f(x+h) = 3(x+h)^2 - 5(x+h) + 2$; (c) $f(2x) = 3(2x)^2 - 5(2x) + 2$. **2.** (a) $f(t) = at + b$; (b) $f(x+h) = a(x+h) + b$; (c) $f(2x) = a(2x) + b$. **3.** (a) $f(t) = t^3$; (b) $f(x+h) = (x+h)^3$; (c) $f(2x) = (2x)^3$.
4. $C(2r) = 2\pi(2r) = 4\pi r = 2C$; it doubles also.
5. (a) $x^2 - x - 2 = (x+1)(x-2) \geq 0$ Domain contains all x in $(-\infty, -1]$, $[2, +\infty)$ (b) all $x \neq 8$.

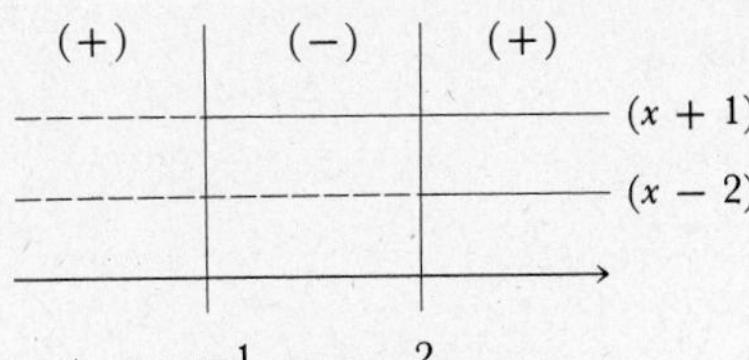

Sec. 2.1: Progress Test 2

1. (*a*) All $x \neq 1$; (*b*)

x	$f(x)$
-1	2
0	2
$\frac{1}{2}$	2
.9	2
.99	2
1	undefined
1.01	2
1.1	2
2	2

(*c*) The function constantly 2 equals $f(x)$ except at $x = 1$. (*d*) $\bar{f}(x) = \begin{cases} 2 \text{ for } x \neq 1 \\ -1 \text{ for } x = 1 \end{cases}$

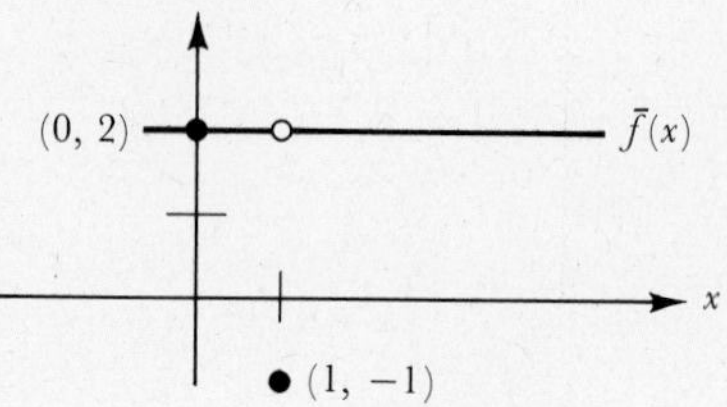

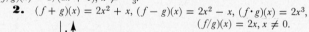

2. (*a*) $y = \frac{2}{3}x + 2$, yes; (*b*) $x = \frac{3}{2}y - 3$, yes [graph is straight line, slope $\frac{2}{3}$, y int. (0, 2)].

3. (*a*) $y = \pm\sqrt{2x + \frac{1}{2}}$, no (*b*) $x = \dfrac{2y^2 - 1}{4}$, yes. **4.** (*a*) $y_1 = \sqrt{2x + \frac{1}{2}}$, $y_2 = -\sqrt{2x + \frac{1}{2}}$; (*b*) all $x \geq -\frac{1}{4}$ (for both).

5.

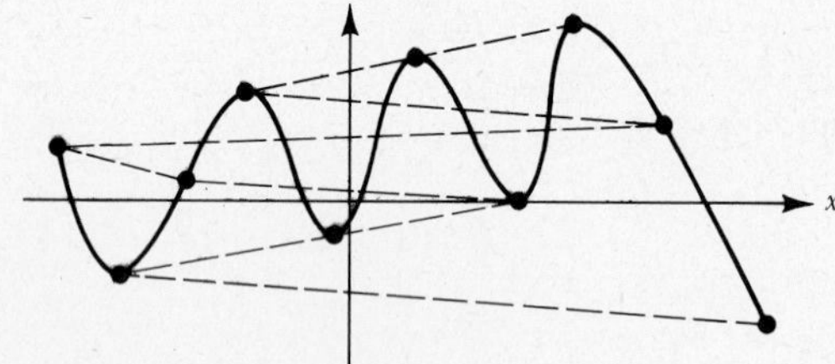

6.

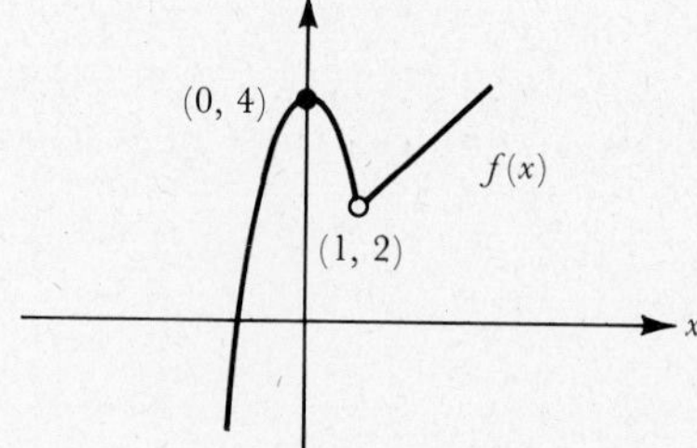

Sec. 2.1: Progress Test 3

1. $(f + g)(x) = 3x + 6$, $(f - g)(x) = 4 - 3x$, $(f \cdot g)(x) = 15x + 5$, $(f/g)(x) = 5/(3x + 1)$, $x \neq -\frac{1}{3}$.

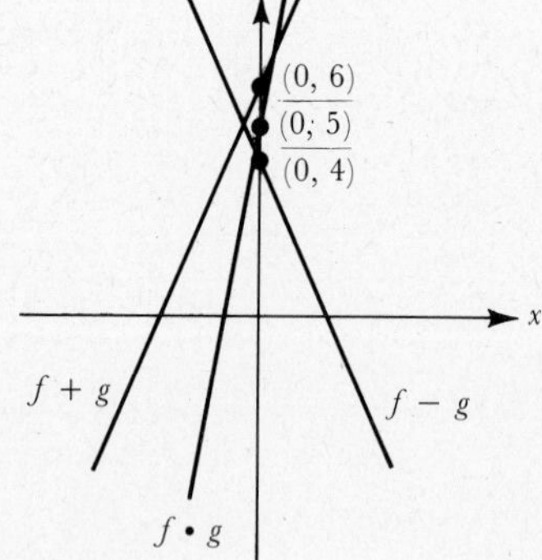

2. $(f + g)(x) = 2x^2 + x$, $(f - g)(x) = 2x^2 - x$, $(f \cdot g)(x) = 2x^3$, $(f/g)(x) = 2x$, $x \neq 0$.

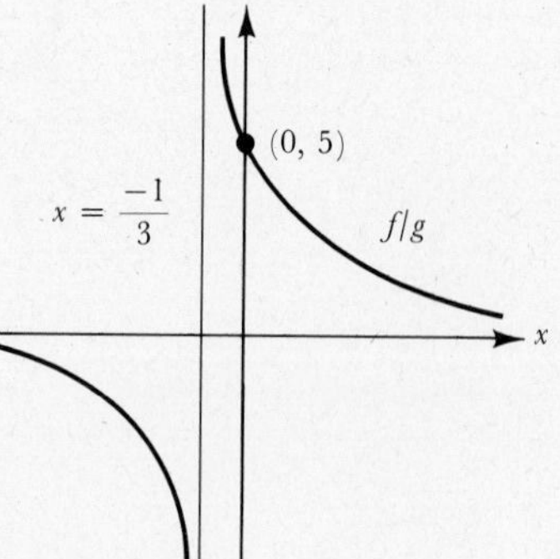

Sec. 2.1: Progress Test 4

1. (*a*) $(f \circ g)(x) = 21x - 2$, $(g \circ f)(x) = 21x - 14$, $(f \circ f)(x) = 9x - 8$, $(g \circ g)(x) = 49x$; (*b*) $(f \circ g)(x) = \sqrt[4]{2x + 1}$, $(g \circ f)(x) = 2\sqrt[4]{x} + 1$, $(f \circ f)(x) = \sqrt[16]{x}$, $(g \circ g)(x) = 4x + 3$; (*c*) $(f \circ g)(x) = \dfrac{1}{x^2 + 3x + 3}$, $(g \circ f)(x) = \dfrac{1}{(x + 1)^2} + \dfrac{3}{x + 1} + 2$, $(f \circ f)(x) = \dfrac{x + 1}{x + 2}$, $(g \circ g)(x) = x^4 + 6x^3 + 16x^2 + 21x + 12$; (*d*) $(f \circ g)(x) = -5$, $(g \circ f)(x) = 625$, $(f \circ f)(x) = -5$, $(g \circ g)(x) = x^{16}$.

2. $(f \circ f^{-1})(x) = x$ and $(f^{-1} \circ f)(x) = x$. **3.** $y = \left(\dfrac{3w + 2}{w}\right)^2 + 5\left(\dfrac{3w + 2}{w}\right) - 1$. When $w = 1$, $y = 49$.

Sec. 2.2: Progress Test 1

1. (*a*)

x	$9 - x^2$
1	8
1.59	6.75
1.9	5.39
1.99	5.0399
1.999	5.003999

x	$9 - x^2$
2.1	4.59
2.01	4.9599
2.001	4.995999

(*b*)

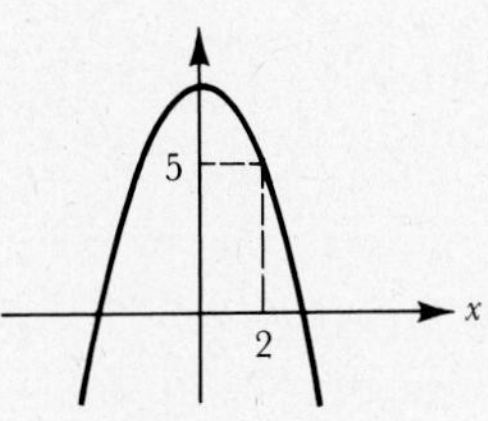

(*c*) $\lim_{x\to 2}(9 - x^2) = 5$.

2. (*a*)

x	$\dfrac{x^2-9}{x-3}$
2.5	5.5
2.9	5.9
2.99	5.99
2.999	5.999
3	undefined

x	$\dfrac{x^2-9}{x-3}$
3.1	6.1
3.01	6.01
3.001	6.001
3	undefined

(*b*)

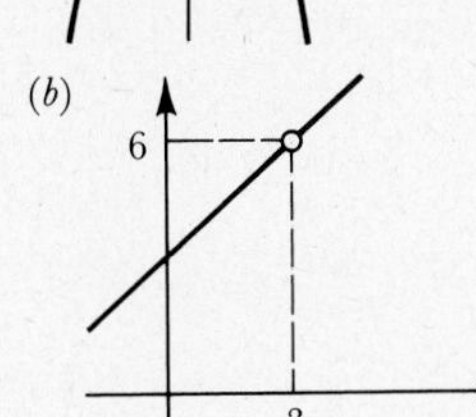

(*c*) $\lim_{x\to 3}\dfrac{x^2-9}{x-3} = 6$

3.

x	$\dfrac{x-1}{\lvert x-1\rvert}$
0.9	-1
0.99	-1
0.999	-1
1	undefined

x	$\dfrac{x-1}{\lvert x-1\rvert}$
1.1	1
1.01	1
1.001	1
1	undefined

$\lim_{x\to 1}\dfrac{x-1}{|x-1|}$ does not exist; $\lim_{x\to 1^-}\dfrac{x-1}{|x-1|} = \lim_{x\to 1^-}\dfrac{x-1}{-(x-1)} = -1$; $\lim_{x\to 1^+}\dfrac{x-1}{|x-1|} = \lim_{x\to 1^+}\dfrac{x-1}{(x-1)} = 1$.

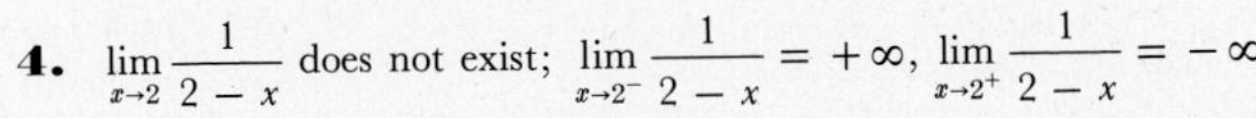

4. $\lim_{x\to 2}\dfrac{1}{2-x}$ does not exist; $\lim_{x\to 2^-}\dfrac{1}{2-x} = +\infty$, $\lim_{x\to 2^+}\dfrac{1}{2-x} = -\infty$.

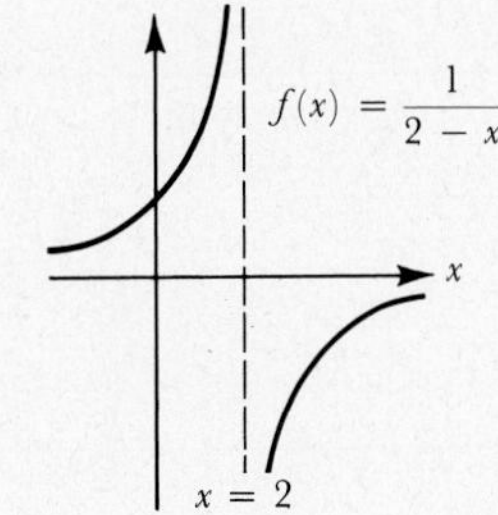

5. $\lim_{x\to 3}\dfrac{\sqrt{x}-\sqrt{3}}{x-3} = \lim_{x\to 3}\dfrac{\sqrt{x}-\sqrt{3}}{(\sqrt{x}+\sqrt{3})(\sqrt{x}-\sqrt{3})} = \lim_{x\to 3}\dfrac{1}{\sqrt{x}+\sqrt{3}}$

[(since $x \neq 3$)] $= \dfrac{1}{2\sqrt{3}} \approx 0.288675$ (this could also have been obtained by rationalizing the numerator).

6. Since $(x+3)^4 > 0$ for all $x \neq -3$, $\dfrac{-1}{(x+3)^4} < 0$ for all $x \neq -3$ and $\lim_{x\to -3}\dfrac{-1}{(x+3)^4} = -\infty$.

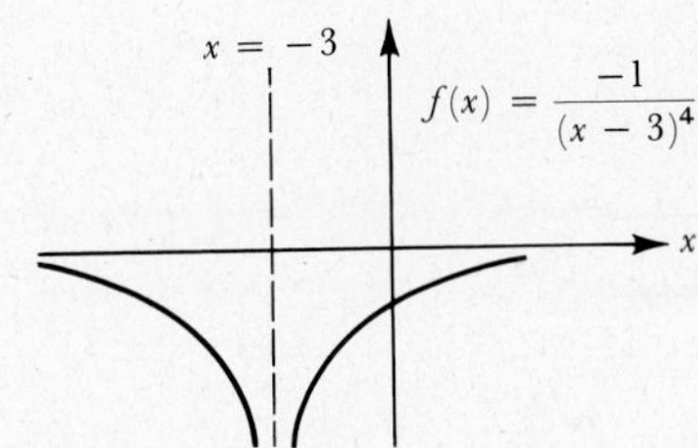

Sec. 2.2: Progress Test 2

1. $\lim_{h\to 0}\dfrac{[3(-2+h)+2]-[3(-2)+2]}{h} = \lim_{h\to 0}\dfrac{3h}{h}$. But $h \neq 0$, so $\dfrac{3h}{h} = 3$, and $\lim_{h\to 0} 3 = 3$ (3 is unaffected by h).

2. $\dfrac{\sqrt{(1+h)+1}-\sqrt{1+1}}{h} = \dfrac{\sqrt{2+h}-\sqrt{2}}{h}\cdot\dfrac{\sqrt{2+h}+\sqrt{2}}{\sqrt{2+h}+\sqrt{2}} = \dfrac{2+h-2}{h(\sqrt{2+h}+\sqrt{2})} = \dfrac{h}{h(\sqrt{2+h}+\sqrt{2})} = \dfrac{1}{\sqrt{2+h}+\sqrt{2}}$ since $h \neq 0$.

Thus $\lim_{h\to 0}\dfrac{f(1+h)-f(1)}{h} = \lim_{h\to 0}\dfrac{1}{\sqrt{2+h}+\sqrt{2}} = \dfrac{1}{2\sqrt{2}}$.

3. $\lim_{t\to 0}\dfrac{[(t+2)^2+(t+2)]-[2^2+2]}{t} = \lim_{t\to 0}\dfrac{t^2+5t}{t} = \lim_{t\to 0}(t+5) = 5$.

4. $\lim_{h\to 0}\dfrac{\dfrac{1}{(3+h)^2}-\dfrac{1}{3^2}}{h} = \lim_{h\to 0}\dfrac{9-(3+h)^2}{h\cdot 9\cdot(3+h)^2} = \lim_{h\to 0}\dfrac{-6h-h^2}{9h(3+h)^2} = \lim_{h\to 0}\dfrac{-6-h}{9(3+h)^2} = \dfrac{-6}{81} = \dfrac{-2}{27}$ $\left(\text{recall } \dfrac{a/b}{h} = \dfrac{a}{hb}\right)$.

Sec. 2.3: Progress Test 1

1. (*a*) $\dfrac{64.064016 - 64}{.001} = 64.016$; (*b*) $\dfrac{63.3616 - 64}{-.001} = 63.84$; (*c*) $\dfrac{16(2+h)^2 - 16(2)^2}{h} = 64 + 16h$. **2.** $64 \text{ ft/s} = \lim_{h\to 0}(64 + 16h)$.

3. $32 \text{ ft/s} = \lim\limits_{h\to 0}(32 + 16h) = \lim\limits_{h\to 0}\dfrac{16(1+h)^2 - 16(1)^2}{h}$.

Sec. 2.3: Progress Test 2

1. (*a*) $\dfrac{m(x+h) - m(x)}{h} = \dfrac{2.6(x+h)^2 - 2.6x^2}{h} = 5.2x + 2.6h$; (*b*) $m'(x) = \lim_{h\to 0}(5.2x + 2.6h) = 5.2x$.

2. $m'(5.9) = (5.2)(5.9) = 30.68$ ft/s. **3.** $m'(6) = \lim\limits_{h\to 0}\dfrac{m(6+h) - m(6)}{h} = \lim\limits_{h\to 0}\dfrac{2.6(36 + 12h + h^2) - (2.6)(36)}{h} = \lim\limits_{h\to 0}\dfrac{31.2h + 2.6h^2}{h} =$ $\lim_{h\to 0}(31.2 + 2.6h) = 31.2$. **4.** $m'(x) = 5.2x = 24$, so $x = 4.6$ s (approx.), $m(4.6) = 2.6(4.6)^2 = 55.0$ ft (approx.).

Sec. 2.3: Progress Test 3

1. $\dfrac{df}{dx} = \lim\limits_{\Delta x\to 0}\dfrac{f(x+\Delta x) - f(x)}{\Delta x} = \lim\limits_{\Delta x\to 0}\dfrac{[3 - (x+\Delta x)^2] - [3 - x^2]}{\Delta x} = \lim\limits_{\Delta x\to 0}\dfrac{-2x\,\Delta x - (\Delta x)^2}{\Delta x} = \lim\limits_{\Delta x\to 0}(-2x - \Delta x)$ [(since $\Delta x \neq 0$)] $= -2x$.

2. $g'(x) = \lim\limits_{\Delta x\to 0}\dfrac{g(x+\Delta x) - g(x)}{\Delta x} = \lim\limits_{\Delta x\to 0}\dfrac{1/(x+\Delta x) - 1/x}{\Delta x} = \lim\limits_{\Delta x\to 0}\dfrac{\dfrac{x - (x+\Delta x)}{x(x+\Delta x)}}{\Delta x} = \lim\limits_{\Delta x\to 0}\dfrac{-\Delta x}{x(x+\Delta x)\,\Delta x} = \lim\limits_{\Delta x\to 0}\dfrac{-1}{x(x+\Delta x)}$ [(since $\Delta x \neq 0$)] $= \dfrac{-1}{x\cdot x} = \dfrac{-1}{x^2}$. **3.** $\dfrac{dy}{dx} = \lim\limits_{\Delta x\to 0}\dfrac{[(x+\Delta x)^2 + (x+\Delta x)] - [x^2 + x]}{\Delta x} = \lim\limits_{\Delta x\to 0}\dfrac{2x\,\Delta x + (\Delta x)^2 + \Delta x}{\Delta x} = \lim\limits_{\Delta x\to 0}(2x + \Delta x + 1)$ [(since $\Delta x \neq 0$)] $= 2x + 1$. **4.** (*a*) $g'(2) = \lim\limits_{\Delta x\to 0}\dfrac{g(2+\Delta x) - g(2)}{\Delta x} = \lim\limits_{\Delta x\to 0}\dfrac{1/(2+\Delta x) - 1/2}{\Delta x} = \lim\limits_{\Delta x\to 0}\dfrac{-\Delta x}{2(2+\Delta x)(\Delta x)} = \lim\limits_{\Delta x\to 0}\dfrac{-1}{2(2+\Delta x)} = \dfrac{-1}{4}$;

(*b*) $g'(2) = \dfrac{-1}{(2)^2} = -\dfrac{1}{4}$.

Sec. 2.3: Progress Test 4

1. $\Delta A = (x+\Delta x)^2 - (x)^2 = 2x\,\Delta x + (\Delta x)^2$. **2.**

	x	Δx
Δx	$x\ \Delta x$	$(\Delta x)^2$
x	x^2	$x\ \Delta x$

3. $\dfrac{\Delta A}{\Delta x} = \dfrac{2x\,\Delta x + (\Delta x)^2}{\Delta x} = 2x + \Delta x$.

4. (*a*) $\dfrac{dA}{dx} = \lim\limits_{\Delta x\to 0}(2x + \Delta x) = 2x$; (*b*) When $x = 2$, $\dfrac{dA}{dx} = 2\cdot 2 = 4$; (*c*) When $x = 200$, $\dfrac{dA}{dx} = 2\cdot 200 = 400$; (*d*) As $\Delta x \to 0$, the growth in A, ΔA, approaches growth in the two edges of length x. The same Δx for x small produces a smaller ΔA than when x is large.

Sec. 2.4: Progress Test 1

1. (*a*) $f'(x) = \lim\limits_{\Delta x\to 0}\dfrac{3(x+\Delta x)^2 - 3x^2}{\Delta x} = \lim\limits_{\Delta x\to 0}(6x + 3\,\Delta x) = 6x$, $y - 3 = 6(x - 1)$;

(*b*) $f'(x) = \lim\limits_{\Delta x\to 0}\dfrac{[2(x+\Delta x)^2 + (x+\Delta x) + 1] - [2x^2 + x + 1]}{\Delta x} = \lim\limits_{\Delta x\to 0}(4x + 2\,\Delta x + 1) = 4x$, $y - 1 = x$;

(*c*) $f'(x) = \lim\limits_{\Delta x\to 0}\dfrac{[1 - (x+\Delta x)^3] - [1 - x^3]}{\Delta x} = \lim\limits_{\Delta x\to 0}(-3x^2 - 3x\,\Delta x - \Delta x^2) = -3x^2$, $y + 7 = -12(x - 2)$.

2. $g'(x) = \lim\limits_{\Delta x\to 0}\dfrac{g(x+\Delta x) - g(x)}{\Delta x} = \lim\limits_{\Delta x\to 0}\dfrac{[(x+\Delta x)^2 + c] - [x^2 + c]}{\Delta x} = \lim\limits_{\Delta x\to 0}(2x + \Delta x) = 2x$, which equals the result obtained in Sec. 2.3,

Progress Test 4, for $f(x) = x^2$. Thus the graph of $g(x)$ is merely the graph of $f(x) = x^2$ slid up (if $c > 0$) or down (if $c < 0$) by c units. They are parallel in the sense that for each x, their respective tangents are parallel. For example, if $c = 3$, we have the following:

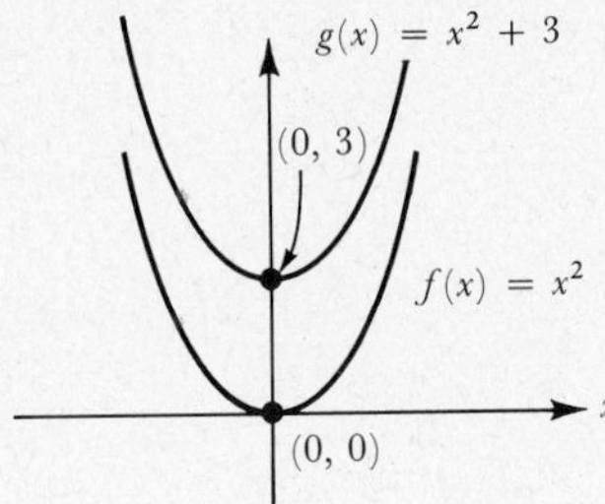

Sec. 2.5: Progress Test 1

[Here LT(a) means Limit Theorem part (a), and SL means special limit.)] **1.** $\lim_{x\to-1}(\sqrt{2}x^4 - 2x) = \lim_{x\to-1}(\sqrt{2}x^4) - \lim_{x\to-1}(2x)$ [LT(a)] $= \sqrt{2}(\lim_{x\to-1} x)(\lim_{x\to-1} x)(\lim_{x\to-1} x)(\lim_{x\to-1} x) - 2(\lim_{x\to-1} x)$ [LT(b)] $= (\sqrt{2})(-1)(-1)(-1)(-1) - (2)(-1)$ [(SL)] $= \sqrt{2} + 2$.
2. $\lim_{\Delta x\to 0} \frac{1/(x+\Delta x) - 1/x}{\Delta x} = \lim_{\Delta x\to 0} \frac{-\Delta x}{x(x+\Delta x)\,\Delta x}$ [(algebra)] $= \lim_{\Delta x\to 0} \frac{-1}{x(x+\Delta x)}$ [$(\Delta x \neq 0)$] $= \lim_{\Delta x\to 0} (-1)/[\lim_{\Delta x\to 0} x(x+\Delta x)]$ [LT(c)] $=$ $(-1)/[x \lim_{\Delta x\to 0}(x+\Delta x)]$ [SL(a), LT(b)] $= (-1)/[x(\lim_{\Delta x\to 0} x + \lim_{\Delta x\to 0} \Delta x)]$ [LT(a)] $= (-1)/(x\cdot x)$ [SL(b)—note x is const., the variable is Δx].
3. $\lim_{h\to 0} \frac{[3(x+h)^2 - 2(x+h)] - [3x^2 - 2x]}{h} = \lim_{h\to 0} \frac{(6xh + 3h^2 - 2h)}{h}$ [(algebra)] $= \lim_{h\to 0}(6x + 3h - 2)$[$(h \neq 0)$] $=$ $\lim_{h\to 0} 6x + \lim_{h\to 0} 3h - \lim_{h\to 0} 2$ [LT(a)] $= (6x - 2) + 3\lim_{h\to 0} h$ [SL(a), LT(b)] $= 6x - 2$ [SL(b)].

Sec. 2.5: Progress Test 2

1. $x = 3$ [(b), (c)]. **2.** $x = 1$ [(a) to (c)]. **3.** $x = 2$ [(a) to (c)]. **4.** $x = -3$ [(a) to (c)]. **5.** cts. everywhere—including $x = 0$. **6.** (a) For $C = 2.02$, there is no c in $[-1, 1]$ with $f(c) = 2.02$, yet $f(-1) = 0.95$, $f(1) = 3.1$ with $f(-1) < C < f(1)$. Note $f(0) = 2$, and for x in $[-1, 0)$, $f(x) < 1.95$, and for x in $(0, 1]$, $f(x) > 2.1$. (b) $[-3, -1]$, $[2, 5]$. (Any closed intervals not containing $x = 0$ suffice.)

Chapter 3: Solutions to Progress Tests

Sec. 3.1: Progress Test 1

1. $f'(x) = 41x^{40}$. **2.** $f'(x) = 1$. **3.** $f'(x) = 12x^2 + 6x + 3$. **4.** $g'(x) = 2142x^{41} + 52x$. **5.** $f'(x) = 8x^7 + 49x^6 + 36x^5 + g'(x)$. **6.** $s'(x) = 12t^2 - \frac{3}{5}$. **7.** $p'(r) = 2\pi r + 2\pi$. **8.** (Need to multiply out before differentiating) $f(x) = x^7 + 4x^6 - 3x^2 - 12x$. Thus $f'(x) = 7x^6 + 24x^5 - 6x - 12$.
9. $\lim_{\Delta x\to 0} \frac{f(x+\Delta x) - f(x)}{\Delta x} = \lim_{\Delta x\to 0} \frac{b-b}{\Delta x} = \lim_{\Delta x\to 0} \frac{0}{\Delta x} = \lim_{\Delta x\to 0} (0) = 0$ by (2.5.3). **10.** $f'(x) = 3x^2 - 12x = 0$ (zero slope) when $x = 0$, $x = 4$.

Sec. 3.1: Progress Test 2

1. (a) $f'(x) = (x^3 + 2x)(20x^3 + 60x^2) + (5x^4 + 20x^3 + 7)(3x^2 + 2)$; ($b$) $f'(t) = (t^2 + 9t + 3)(2) + (2t + 1)(2t + 9)$; ($c$) $y' = \frac{(x^3+1)(2x) - (x^2+1)(3x^2)}{(x^3+1)^2}$; ($d$) $t'(x) = \frac{(x^2 - x - 2)(42x^2) - (14x^3 + 1)(2x - 1)}{(x^2 - x - 2)^2}$; ($e$) $f'(x) = \frac{(cx+d)a - (ax+b)c}{(cx+d)^2}$; ($f$) $u'(x) = (ax^2 + bx + c)(m) + (mx + d)(2ax + b)$. **2.** $df/dx = x(2x) + (x^2)(1) = 3x^2$. **3.** ($a$) $f'(x) = 7(3x^2 + 10x + 2)$; ($b$) $f'(x) = 7(3x^2 + 10x + 2) + (x^3 + 5x^2 + 2x - 1)(0)$. **4.** ($a$) $f'(x) = \frac{x^2(3x^2 + 6x) - (x^3 + 3x^2 + 4)2x}{[x^2]^2}$; ($b$) $f'(x) = (x^3 + 3x^2 + 4)(-2)x^{-3} + (x^{-2})(3x^2 + 6x)$; ($c$) the answer to ($a$) simplifies as follows: $\frac{x(3x^2 + 6x) - (x^3 + 3x^2 + 4)2}{x^3} = x(3x^2 + 6x)x^{-3} + (-2)(x^3 + 3x^2 + 4)x^{-3} = (3x^2 + 6x)x^{-2} + (-2)(x^3 + 3x^2 + 4)x^{-3}$, which equals the answer to (b). **5.** Let $u(x + \Delta x) = \bar{u}$, $v(x + \Delta x) = \bar{v}$, $u(x) = u$, $v(x) = v$. Then line (1) is $u(\bar{v} - v) + v(\bar{u} - u) + (\bar{u} - u)(\bar{v} - v) = u\bar{v} - uv + v\bar{u} - vu + \bar{u}\bar{v} - \bar{u}v - u\bar{v} + uv = \bar{u}\bar{v} - uv$, as to be shown.

Sec. 3.2: Progress Test 1

1. $f'(x) = 7(x^9 + 2x^4 - x^3 + 41)^6(9x^8 + 8x^3 - 3x^2)$. **2.** $g'(x) = -5(x^3 + 7x^2 + 2x + 1)^{-6}(3x^2 + 14x + 2)$.
3. $v'(x) = (x^2 + 2x)(4)(50x^2 + 6x - 3)^3(100x + 6) + (50x^2 + 6x - 3)^4(2x + 2)$.

4. $f'(x) = \dfrac{(3x^5 + x^2)^8(1) - (x + 4)(8)(3x^5 + x^2)^7(15x^4 + 2x)}{(3x^5 + x^2)^{16}}$. **5.** $h'(x) = 6\left[\dfrac{2x + 1}{3x^2 - 4}\right]^5 \dfrac{(3x^2 - 4)(2) - (2x + 1)(6x)}{(3x^2 - 4)^2}$.

6. $g'(x) = \dfrac{(5x^2 + 2x)^7[4(x^7 + 6x^5 - x^2)^3(7x^6 + 30x^4 - 2x)] - (x^7 + 6x^5 - x^2)^4[7(5x^2 + 2x)^6(10x + 2)]}{(5x^2 + 2x)^{14}}$.

Sec. 3.2: Progress Test 2

1. (*a*) $x\dfrac{dy}{dx} + y(1) = 0$, so $\dfrac{dy}{dx} = -\dfrac{y}{x}$; (*b*) $x^2 2y\dfrac{dy}{dx} + y^2 2x = 0$, so $\dfrac{dy}{dx} = -\dfrac{y}{x}$; (*c*) $2x + 2y\dfrac{dy}{dx} - 2x\dfrac{dy}{dx} - 2y + x3y^2\dfrac{dy}{dx} = 0$, so $\dfrac{dy}{dx} = \dfrac{-2x + 2y - y^3}{2y - 2x + 3xy^2}$; (*d*) $\dfrac{(x - y)(1 + dy/dx) - (x + y)(1 - dy/dx)}{(x - y)^2} = 0$, so $\dfrac{dy}{dx} = \dfrac{y}{x}$. **2.** (*a*) $x \cdot 1 + y\dfrac{dx}{dy} = 0$, so $\dfrac{dx}{dy} = -\dfrac{x}{y}$; (*b*) $x^2 2y + y^2 2x\dfrac{dx}{dy} = 0$, so $\dfrac{dx}{dy} = -\dfrac{x}{y}$; (*c*) $2x\dfrac{dx}{dy} + 2y - 2x - 2y\dfrac{dx}{dy} + x3y^2 + y^3\dfrac{dx}{dy} = 0$, so $\dfrac{dx}{dy} = \dfrac{2y - 2x + 3xy^2}{-2x + 2y - y^3}$; (*d*) $\dfrac{(x - y)(dx/dy + 1) - (x + y)(dx/dy - 1)}{(x - y)^2} = 0$, so $\dfrac{dx}{dy} = \dfrac{x}{y}$. **3.** $6x\dfrac{dx}{dt} + 10t - x - t\dfrac{dx}{dt} = 0$, so $\dfrac{dx}{dt} = \dfrac{x - 10t}{6x - t}$. **4.** $x\dfrac{dy}{dx} + y = 0$, so $\dfrac{dy}{dx} = \dfrac{-y}{x}$ and at $x = 2$, $\dfrac{dy}{dx} = \dfrac{-1/2}{2} = -\dfrac{1}{4}$. Thus $y - \dfrac{1}{2} = -\dfrac{1}{4}(x - 2)$.

Sec. 3.2: Progress Test 3

1. $f'(x) = \dfrac{1}{2x^{1/2}} + \dfrac{1}{3x^{2/3}}$. **2.** $f'(x) = \dfrac{3}{2x^{1/2}} - \dfrac{1}{3x^{4/3}}$. **3.** $g'(x) = \dfrac{8}{3x^{1/3}} + \dfrac{3}{5x^{2/5}}$. **4.** $h'(x) = \dfrac{2x - 3}{2(x^2 - 3x + 1)^{1/2}}$.

5. $m'(t) = \dfrac{2 - 12t}{3(3t^2 - t)^{5/3}}$.

Sec. 3.3: Progress Test 1

1. (*a*) $dy = \left[2(x + 1) + \dfrac{1}{x^2}\right]dx$; (*b*) $dy = \left(6 + \dfrac{1}{4}\right)(0.1) = 0.625$; (*c*) $dy = 0.0625$; (*d*) $\Delta y = \left[(x + \Delta x + 1)^2 - \dfrac{1}{x + \Delta x}\right] - \left[(x + 1)^2 - \dfrac{1}{x}\right]$; (*e*) $\Delta y - dy = \left[\left((3.01)^2 - \dfrac{1}{2.01}\right) - \left(3^2 - \dfrac{1}{2}\right)\right] - [0.0625] \approx 0.0000876$ (rounded off in last place).

2. (*a*) $df = \dfrac{1}{2}\sqrt{\dfrac{1 + x^2}{1 - x^3}}\left[\dfrac{(1 + x^2)(-3x^2) - (1 - x^3)(2x)}{(1 + x^2)^2}\right]dx$; (*b*) $df = \dfrac{1}{2}\sqrt{\dfrac{2}{2}}\left[\dfrac{(2)(-3) - (2)(-2)}{2^2}\right](0.01) = -0.0025$.

Sec. 3.3: Progress Test 2

1. (*a*) $\Delta y = 2\pi x\,\Delta x + \pi(\Delta x)^2$; (*b*) $dy = 2\pi x\,\Delta x$;

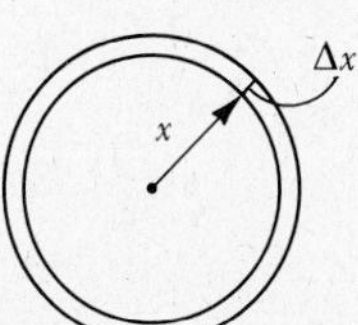

(*c*) $\Delta y - dy = \pi(\Delta x)^2$; Δx 2πx

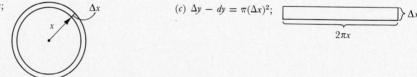

(*d*) area of outside strip is Δy. To bend rectangular strip of width Δx and length $2\pi x$ (circumference of inside circle of radius x), little "wedges" would have to be cut in the strip (try it!) $\Delta y - dy$ is area of all the "wedges" that would need to be cut in rectangular strip in order to bend it around circle (impossible in practice). **2.** (*a*) $0.2\pi + 0.0001\pi \approx 0.6286$; (*b*) $0.2\pi \approx 0.62832$; (*c*) $0.0001\pi \approx 0.00031$.

3. (*a*) $\Delta A = \dfrac{\sqrt{3}}{4}(x + \Delta x)^2 - \dfrac{\sqrt{3}}{4}x^2 = \dfrac{\sqrt{3}}{2}x(\Delta x) + \dfrac{\sqrt{3}}{4}(\Delta x)^2$; (*b*) $dA = \dfrac{\sqrt{3}}{2}x(\Delta x)$; (*c*) $\dfrac{\sqrt{3}}{4}(\Delta x)^2$; (*d*) $\dfrac{\sqrt{3}}{2}x(\Delta x)$; (*e*) $\dfrac{\Delta x}{4}\dfrac{\sqrt{3}}{2}\Delta x + \dfrac{\Delta x}{4}\dfrac{\sqrt{3}}{2}\Delta x = \dfrac{\sqrt{3}}{4}(\Delta x)^2$; (*f*) difference between dA and ΔA is area of triangles II and III. **4.** $f'(x) = nx^{n-1}$ and $\dfrac{\Delta y - dy}{\Delta x}$ includes all other terms—including $\dfrac{x^n}{\Delta x}$.

Sec. 3.3: Progress Test 3

1. $dy = (20(4x + 1)^4 + 3x^2)\,dx$. **2.** $dy = \left(6\sqrt[3]{t}\dfrac{1}{3}\dfrac{1}{\sqrt[3]{t^2}} - (2)\dfrac{1}{3}\dfrac{1}{\sqrt[3]{t^2}}\right)dt$. **3.** (*a*) $3\dfrac{dw}{dz}\sqrt{z^2 + 1} + 3w\dfrac{2z}{2\sqrt{z^2 + 1}} + 6z^2 - 36w^3\dfrac{dw}{dz} = 0$, so solving for $\dfrac{dw}{dz}$, $\dfrac{dw}{dz} = \dfrac{wz/\sqrt{z^2 + 1} + 2z^2}{12w^3 - \sqrt{z^2 + 1}}$; (*b*) $3w\dfrac{(1)\,2z\,dz}{(2)\,\sqrt{z^2 + 1}} + 3\sqrt{z^2 + 1}\,dw + 6z^2\,dz - 36w^3\,dw = 0$, so dividing through by dz, we get: $\dfrac{3wz}{\sqrt{z^2 + 1}} + 3\sqrt{z^2 + 1}\dfrac{dw}{dz} + 6z^2 - 36w^3\dfrac{dw}{dz} = 0$. Then solve for $\dfrac{dw}{dz}$ to get same result as (*a*). **4.** (*a*) Choose $f(x) = \sqrt{x}$, $a = 100$,

$\Delta x = dx = 2$. Then $f'(x) = \dfrac{1}{2\sqrt{x}}$, so $\sqrt{102} = f(100 + 2) \approx f(100) + f'(100)(2) = 10 + \dfrac{1}{10} = 10.1$; (*b*) choose $f(x) = x^{5/2}$, $a = 1$, $\Delta x = dx = -0.02$. Then $f'(x) = (5/2)\,x^{3/2}$, so $(0.98)^{5/2} = f(1 + (-0.02)) \approx f(1) + (5/2)(1)^{3/2}(-0.02) = 1 - (0.05) = 0.95$.
5. $f(x) = x^3$, $a = 10$, $\Delta x = dx = 0.1$, $f'(x) = 3x^2$, so change in volume, $f(10 + 0.1) - f(10) \approx f'(10)(0.1) = (300)(0.1) = 30$ cubic units.

Sec. 3.4: Progress Test 1

1. $f'(x) = 5x^4 + 4x^3 + 3x^2 + 1$, $f''(x) = 20x^3 + 12x^2 + 6x + 2$, $f'''(x) = 60x^2 + 24x + 6$, $f^{(4)}(x) = 120x + 24$.
2. $f'(x) = 2x - 5$, $f''(x) = 2$, $f'''(x) = 0 = f^{(4)}(x)$. **3.** $f'(x) = \dfrac{-1}{x^2}$, $f''(x) = \dfrac{2}{x^3}$, $f'''(x) = \dfrac{-6}{x^4}$, $f^{(4)}(x) = \dfrac{24}{x^5}$.
4. $\dfrac{dy}{dx} = -\dfrac{x + y}{x + 3y}$, $\dfrac{d^2y}{dx^2} = -\left[\dfrac{2y - 2x\,dy/dx}{(x + 3y)^2}\right] = (-2)\dfrac{x^2 + 2xy + 3y^2}{(x + 3y)^3} = \dfrac{-2}{(x + 3y)^3}$.

Sec. 3.4: Progress Test 2

1. $f''(t) = 6(t - 2) < 0$ for $t < 2$, $f''(t) > 0$ for $t > 2$. **2.** Acceleration is zero when $t = 2$, $f'(x) = -12$ is velocity.
3. $f'(t) = 3t(t - 4) = 0$ when $t = 0, 4$. **4.** $f(0) = 32$, $f(4) = 0$.

Chapter 4: Solutions to Progress Tests

Sec. 4.1: Progress Test 1

1. $\lim\limits_{x\to 3} \dfrac{(x - 3) - 5}{(x - 3)^2} = -\infty$. **2.** $+\infty$. **3.** $\lim\limits_{x\to 1^+} \dfrac{x + 1}{x - 1} = +\infty$. **4.** Let $Q = \dfrac{x^6 + x^5 + 2}{(x - 3)(x + 2)}$, so $\lim\limits_{x\to 3^-} Q = \lim\limits_{x\to -2^-} Q = -\infty$ and $\lim\limits_{x\to 3^+} Q = \lim\limits_{x\to -2^+} Q = +\infty$.

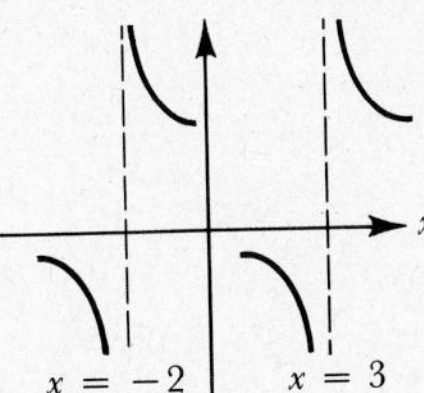

5. Let $\dfrac{2(x - 2) - 7}{(x - 2)^2} = \dfrac{2x - 11}{(x - 2)^2} = P$, so $\lim\limits_{x\to 2} P = -\infty$.

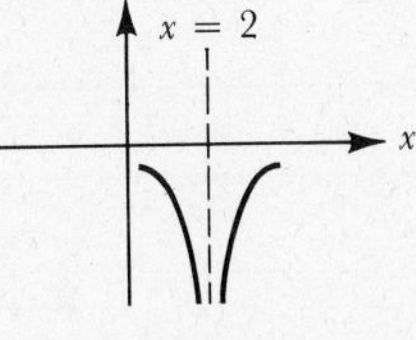

Sec. 4.1: Progress Test 2

1. $y = 1/\sqrt{2}$. **2.** No horiz. asym. **3.** $y = 0$. **4.** No horiz. asym. Divide num. and den. by $\sqrt{x^2}$.
5. $y = 0$. Divide num. and den. by $x^3 = \sqrt[3]{x^9}$. **6.** $y = \sqrt{2}/3$. Divide num. and den. by $\sqrt{x^2}$.

Sec. 4.2: Progress Test 1

1. (*a*) Decr. (x_1, x_3), (x_5, x_6), (x_8, x_9); incr. (x_0, x_1), (x_3, x_5), (x_6, x_8); (*b*) C.U. (x_0, x_1), (x_2, x_4), (x_7, x_8), (x_8, x_9); C.D. (x_1, x_2), (x_4, x_6), (x_6, x_7); (*c*) rel. max. x_1, x_5, x_8; rel. min. x_3, x_6; (*d*) At x_3 and x_5, $f' = 0$; (*e*) x_2, x_4, x_7 (no tan. at x_1).
2. (These are sample solutions)

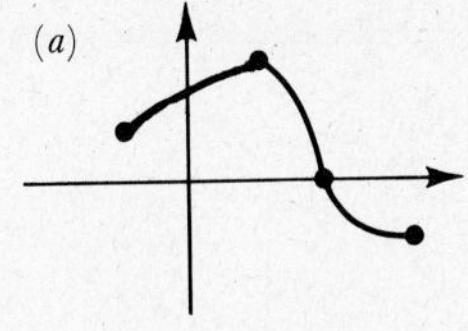

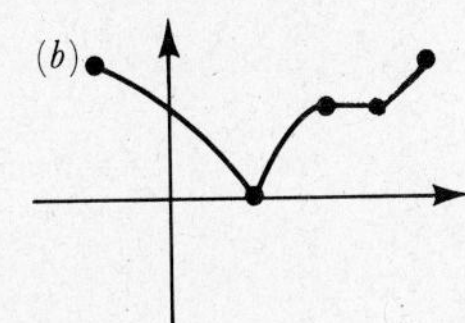

(Other graphs are possible.)

3. Again, these are samples.

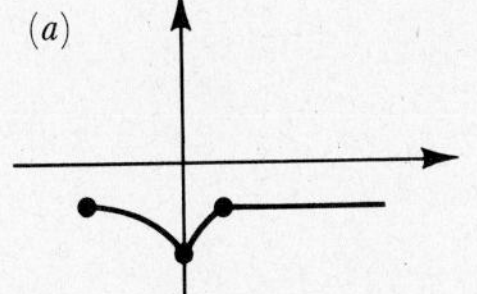

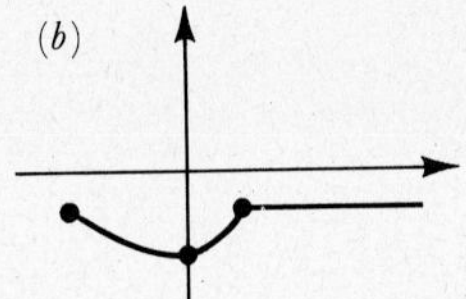

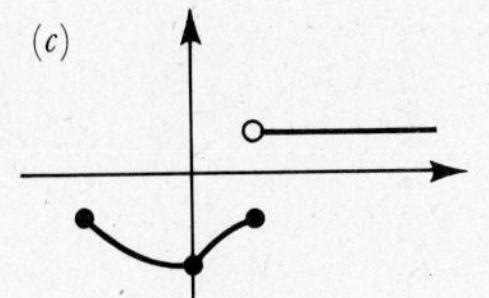

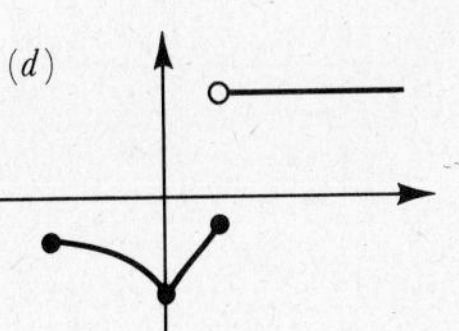

Sec. 4.2: Progress Test 2

1. Linear of slope 3, everywhere increasing. **2.** Const. fn., zero slope everywhere. **3.** $f'(x) = 2(x - 1) < 0$ on $(-\infty, 1)$ and > 0 on $(1, +\infty)$. Rel. min. at $x = 1$. **4.** Increasing everywhere since $f'(x) = 3x^2 + 5 > 0$ for all x.

5. $f'(x) = (x - 3)(x + 1)$ incr. on $(-\infty, 1)$ and $(3, +\infty)$, decr. on $(-1, 3)$. Rel. max. at $x = -1$, rel. min at $x = 3$.

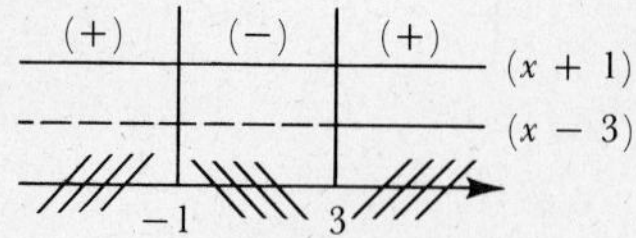

Sec. 4.2: Progress Test 3

1. $f'(x) = x^2 - 2x$, $f''(x) = 2(x - 1)$ changes sign at $x = 1$, a pt. of infl. C.D. on $(-\infty, 1)$, C.U. on $(1, +\infty)$.

2. $f'(x) = 4x^3 - 36x$, $f''(x) = 12(x + \sqrt{3})(x - \sqrt{3})$ See chart for f'': P.I. at $x = \pm\sqrt{3}$

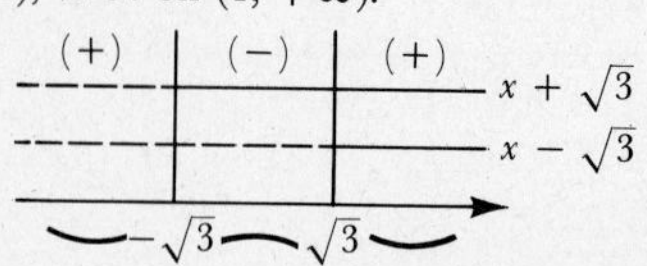

Sec. 4.3: Progress Test 1

1. (1) $f(0) = \frac{4}{3}$; (2) $f'(x) = 4(x)(x + 2)(x - 1) = 0$ when $x = 0, -2, 1$ (crit. pts.); (3) $f''(x) = 4(3x^2 + 2x - 2) = 0$ when $x = -1/3 \pm \sqrt{7}/3$ (crit. pts.); (4)

(−) (+) (−) (+) $(x + 2)$, $(x - 1)$, (x); -2, 0, 1

Rel. max. $x = 0$. Rel. min. $x = -2, 1$. (See incr.-decr. on f' chart.)

(5) (+) (−) (+) $\left(x + \frac{1}{3} - \frac{\sqrt{7}}{3}\right)$, $\left(x + \frac{1}{3} + \frac{\sqrt{7}}{3}\right)$; $-\frac{1}{3} - \frac{\sqrt{7}}{3}$, 0, $-\frac{1}{3} + \frac{\sqrt{7}}{3}$

See concavity on f'' chart. P.I. at $x = -\frac{1}{3} \pm \frac{\sqrt{7}}{3}$.

(6)

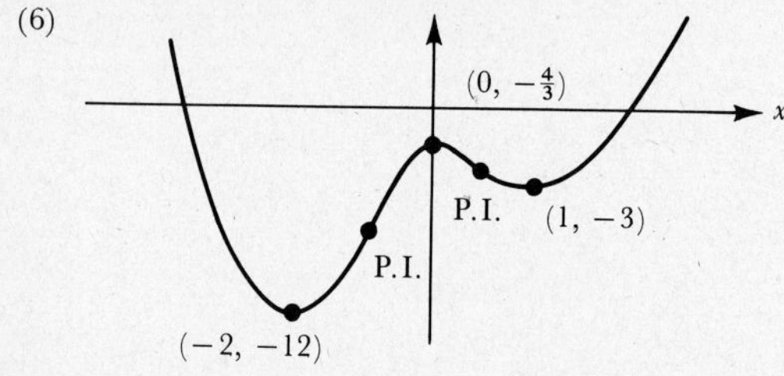

2. (1) $\lim_{x\to\pm\infty} \frac{x}{3x - 1} = \frac{1}{3}$, so $y = \frac{1}{3}$ is horiz. asym. $\lim_{x\to 1/3^-} \frac{x}{3x - 1} = -\infty$ and $\lim_{x\to 1/3^+} \frac{x}{3x - 1} = +\infty$, so vert. asym. at $x = \frac{1}{3}$; $f(0) = 0$; (2) $f'(x) = \frac{-1}{(3x - 1)^2} < 0$ for all $x \neq 1/3$; (3) $f''(x) = \frac{+6}{(3x - 1)^3} > 0$ for $x \neq 1/3$, and $f''(x)$ never 0; (4) $f'(x) < 0$ for all x where defined, so f is decr. for all $x \neq 1/3$; (5) $f''(x) < 0$, hence f C.D. for x in $(-\infty, -1/3)$; $f''(x) > 0$, hence f C.U. for x in $(1/3, +\infty)$;

(6)

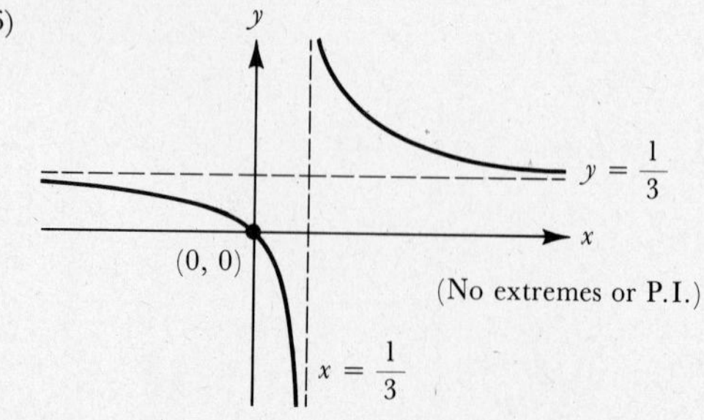

3. $f'(x) = \left((3)\sqrt{4x^2 - 1} - (3x + 2)\frac{8x}{2\sqrt{4x^2 - 1}}\right)/(4x^2 - 1) = \frac{3(4x^2 - 1) - (12x^2 + 8x)}{(4x^2 - 1)\sqrt{4x^2 - 1}} = -\frac{8x + 3}{(4x^2 - 1)^{3/2}}$ defined only in $(-\infty, -1/2)$, $(1/2, +\infty)$. $f'(x) > 0$ hence f incr. for $x < -1/2$, and $f'(x) < 0$, hence f decr. for $x > 1/2$. Finally, $\lim_{x\to -1/2^-} f'(x) = +\infty$, $\lim_{x\to 1/2^+} f'(x) = -\infty$ (vertical asymptotes). No extremes.

Sec. 4.3: Progress Test 2

1. $f(x) = 3x^{2/3}(x + 5), f'(x) = 5x^{-1/3}(x + 2), f''(x) = (10/3)x^{-4/3}(x - 1)$.

Note $f'(0)$ and $f''(0)$ are undef. (cusp.) with $f(0) = 0$. Also $f(-5) = 0$.

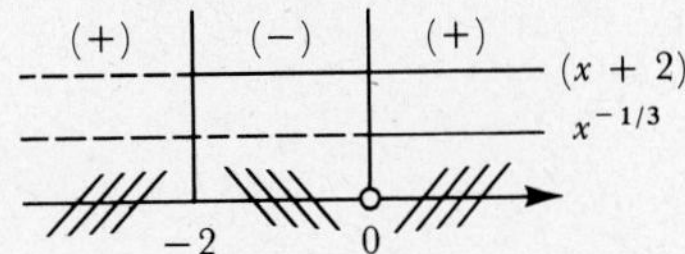

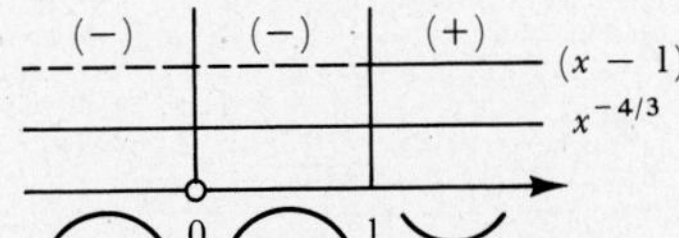

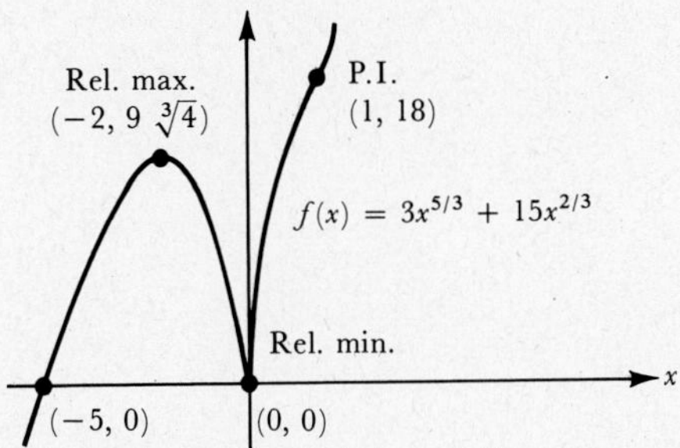

Sec. 4.4: Progress Test 1

(These are sample solutions—of course there are many others.)

1.

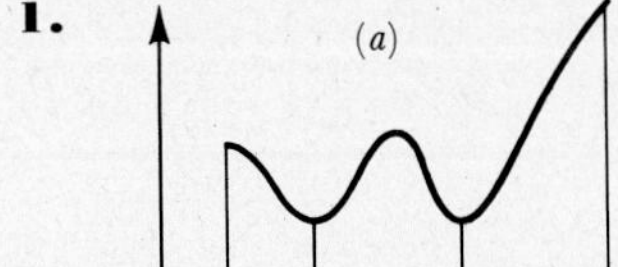

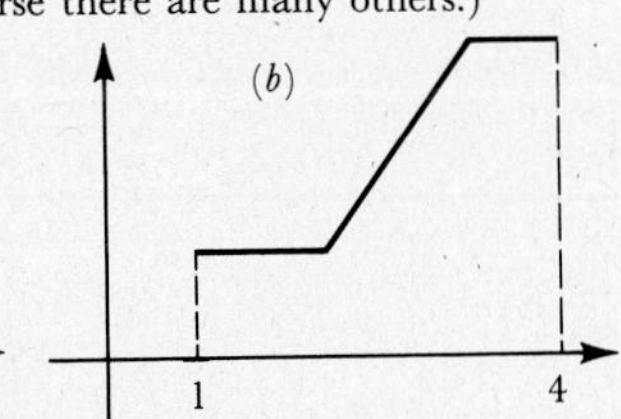

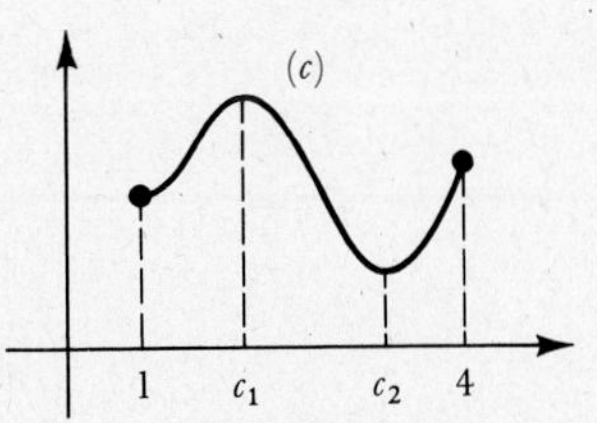

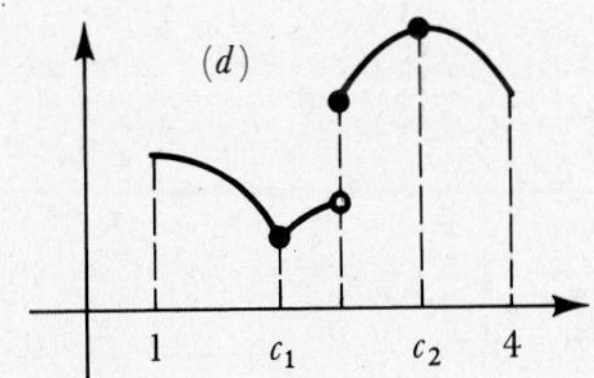

2. (*a*) $f(x) = 4x - x^2$. Max. value 4 occurs at $x = 2$

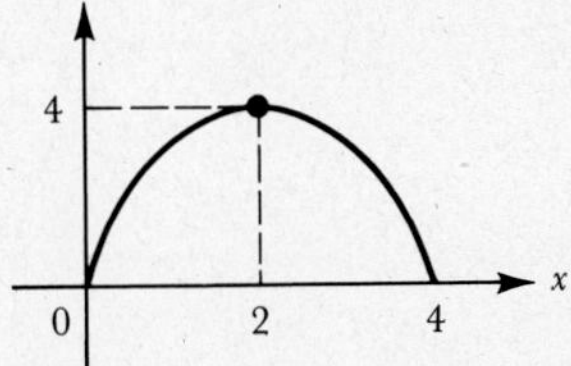

(*b*) $f(x) = \begin{cases} \dfrac{1}{4 - x} & \text{for } x < 4 \\ 1000 & \text{for } x = 4 \end{cases}$

Min. value 1/4 occurs at $x = 0$.

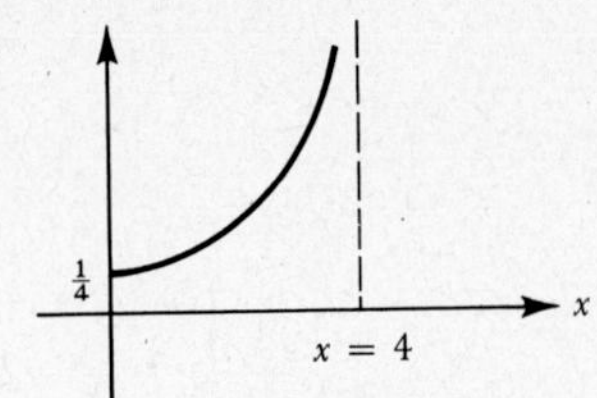

Sec. 4.4: Progress Test 2

1. (*a*) $f(-1) = f(2) = -1/3$ so, actually, Rolle's Theorem applies since f is everywhere diff. $f'(x) = x^2 - 1 = 0$ when $x = \pm 1$. (*b*) Hyp. fail since $f'(0)$ is undef.

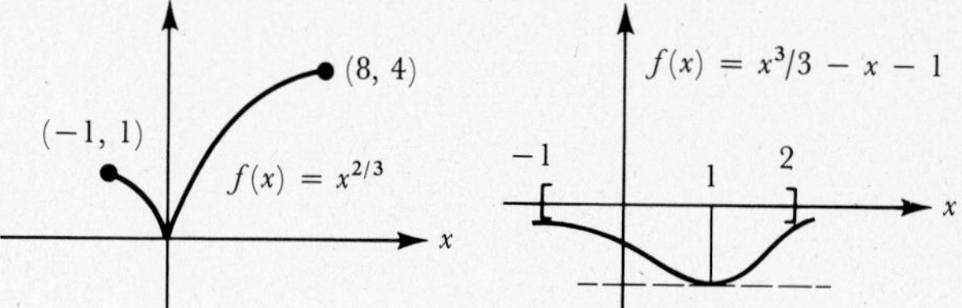

(*c*) f is everywhere diff. $f(1) = 3$, $f(-2) = 0$, $f'(x) = 2x + 2 = \dfrac{3 - 0}{1 - (-2)} = 1$ when $x = -1/2$.

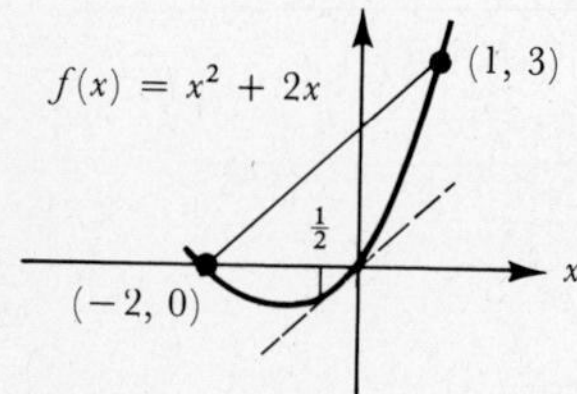

2.

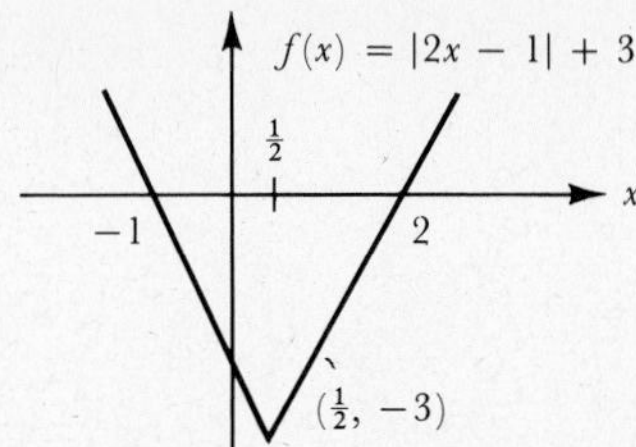

Note $f(x) = \begin{cases} 2x - 4 & \text{for } x > 1/2 \\ -(2x - 1) - 3 = -2x - 2 & \text{for } x < 1/2 \end{cases}$

$f'(x)$ not def. at $x = 1/2$. No.

Chapter 5: Solutions to Progress Tests

Sec. 5.1: Progress Test 1

(Extreme values are underlined.)

1. $f'(x) = 6(x + 2)(x - 1) = 0$ when $x = -2, 1$. $f''(x) = 12x + 6$, so $f''(-2) < 0$, $f''(1) > 0$, making $x = -2$ a rel. max. and $x = 1$ a rel. min. $\underline{f(1) = -3}$, $f(-2) = 24$, $f(-3) = 13$, $\underline{f(3) = 49}$. **2.** $f'(x) = 5x^4 + 6x^2 \geq 0$ on $[0, 3]$, so $f(x)$ is increasing. Hence min. is $\underline{f(0) = 0}$ and max. is $\underline{f(3) = 297}$. **3.** $f'(x) = \dfrac{4(x - x^2) + 2x^2}{(1 - x)^2} = \dfrac{2x(2 - x)}{(1 - x)^2} = 0$ when $x = 0, 2$, so a single rel. extreme on $(1, +\infty)$ at $x = 2$. Note $f'(x) > 0$ for $x < 2$, $f'(x) < 0$ for $x > 2$, so a rel. max. Hence $\underline{f(2) = -8}$ is max. for f on $(1, +\infty)$. Since $\lim_{x \to 1^+} f(x) = -\infty$, no min. value.

Sec. 5.1: Progress Test 2

1. $3x + y = 120$ so $y = 120 - 3x$. $A = xy = x(120 - 3x)$. $dA/dx = 120 - 6x = 0$ when $x = 20$, a rel. max. $A(x)$ def. on $[0, 40]$, so $x = 20$, $y = 60$, $A = 1200$.

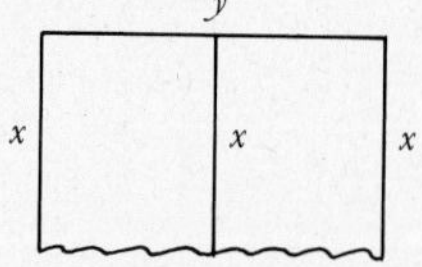

2. Let one no. be x and the other y. Maximize product $x \cdot y = P$. Now $3x + y = 120$, so this is same problem as 1. Hence $x = 20$, $y = 60$, $P = 1200$. (Note that $P > 0$ implies the condition $0 \leq x \leq 40$.) If roles of x and y are reversed (with $3y + x = 120$), same solution results. **3.** $2x + 2y = 120$, so $y = 60 - x$ and $A = xy = x(60 - x)$, defined on $[0, 60]$. $dA/dx = 60 - 2x = 0$ when $x = 30$, a rel. max. $y = 30$ and $A = 900$ (square pasture 30×30).

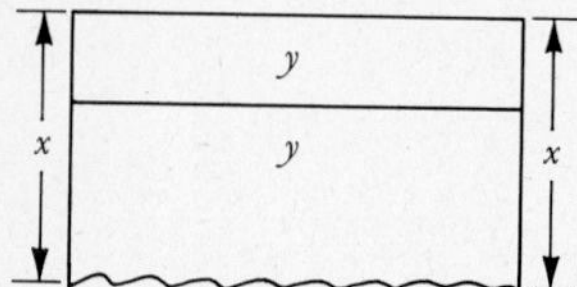

4. (2) Total area $= A = xy$. (3) Area of printed p. is $(x - 3)(y - 2) = 30$, so $y = \dfrac{30}{x - 3} + 2$ and $A = A(x) = x\left(\dfrac{30}{x - 3} + 2\right)$.

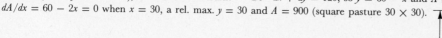

(4) Only restrict. is $x > 3$, so interval is $(3, +\infty)$.

(5) $A'(x) = \dfrac{(x-3)(30) - 30x(1)}{(x-3)^2} + 2 = \dfrac{2(x-3)^2 - 90}{(x-3)^2} = \dfrac{2x^2 - 12x - 72}{(x-3)^2}$

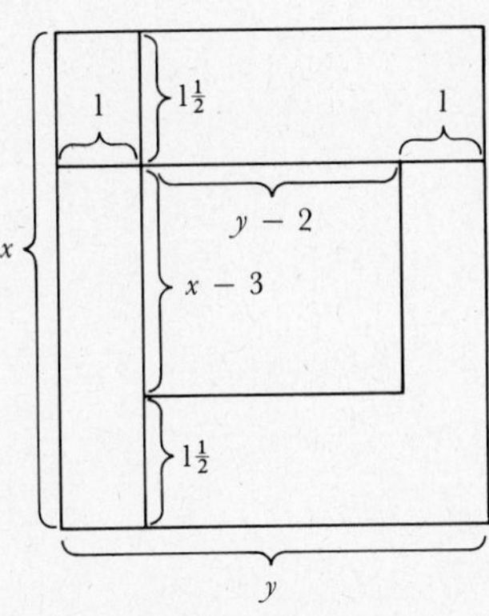

Hence, there is a rel. extreme at $x = 3 + 3\sqrt{5}$ (we disregard $x = 3 - 3\sqrt{5} < 0$). Since $\lim_{x\to+\infty}[A(x)] = +\infty$, we know this is a rel. min. on $(3, +\infty)$. (6) Thus $y = \dfrac{30}{3\sqrt{5}} + 2 = 2 + 2\sqrt{5}$, making dimensions of entire page $(3 + 3\sqrt{5}) \approx 9.71$ cm by $(2 + 2\sqrt{5}) \approx 6.47$ cm.

5. (2) Maximize vol. V of box, which is $V(x) = x(12 - 2x)(24 - 2x) = 288x - 72x^2 + 4x^3$.

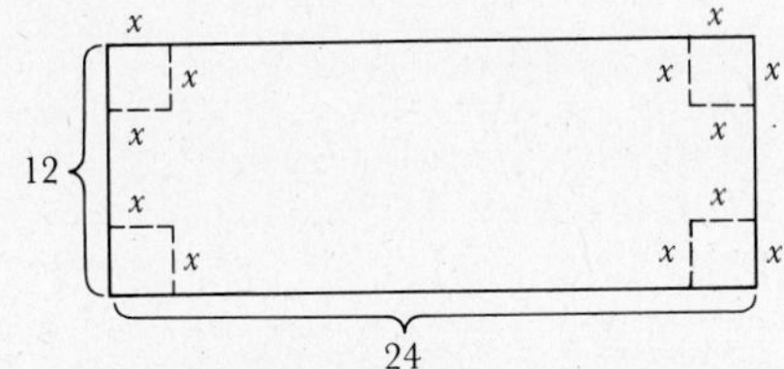

(3) (We *could* have given dimensions using more than 1 variable and then used given horiz. and vert. dimensions to eliminate all but 1.) (4) $0 \leq 2x \leq 12$, so the interval is $[0, 6]$. (5) $V'(x) = 12(24 - 12x + x^2) = 0$ in $[0, 6]$ when $x = 6 - 2\sqrt{3}$. (6) Box will have dimensions $6 - 2\sqrt{3} \approx 2.54$ cm by $4\sqrt{3} \approx 6.93$ cm by 18.93 cm.

Sec. 5.1: Progress Test 3

1. (*a*) Max. $A = xy$ with $2x + 2y = 100$, so $y = 50 - x$ and $A(x) = x(50 - x)$. $dA/dx = 50 - 2x = 0$ when $x = 25$, $0 \leq x \leq 50$. $y = 25$ and $A = 625$ m^2. (*b*) A square pasture, perfectly symmetric, with all earlier symmetries trivially satisfied. (*c*) $100 = 2\pi r$ implies $r = 100/2\pi$ and $A = \pi r^2 = (100/2\pi)^2 = 5000/\pi \approx 1591.5$ m^2. (This is largest possible pasture of any shape.) **2.** (*a*) Max. $A = xy$ with $2(5x) + 2(2y) = 100$, so $y = (50 - 5x)/2$. $A = x(50 - 5x)/2$ and $dA/dx = 25 - 5x = 0$ implies $x = 5$, $y = 25/2$. (*b*) \$50. in each dir. **3.** Perimeter $P = 4x + 4y = 800$. Area $A = x^2 + y^2 = x^2 + (200 - x)^2$.

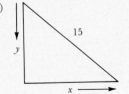

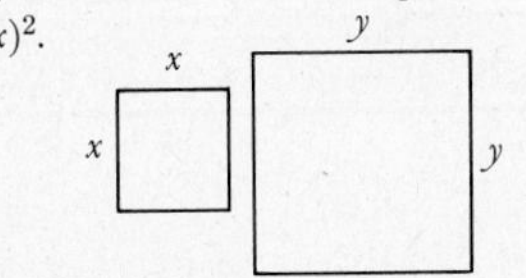

$A'(x) = 4(x - 100) = 0$ at $x = 100$. But $A''(x) = 4 > 0$ means $x = 100$ is a rel. min. [$A(100) = 20{,}000$ m^2 with 2 equal plots]. But $A(x)$ is defined on $(0, 200)$—endpts. are excluded by given cond. But $A(0) = A(200) = 40{,}000$, so max. is achieved at endpts. Hence, to comply with instructions, captives should make 1 square plot as small as possible and use remaining fence to obtain an area as close as possible to 40,000 m^2. One plot is better than two: Put the two plots together along a common side and note the 2 "wasted" sides are eliminated when 1 plot is used.

Sec. 5.2: Progress Test 1

1. (*a*) (*b*) Given $dx/dt = 4$, need find dy/dt when $x = 9$, given that $x^2 + y^2 = 225$.

(*c*) $x\,dx/dt + y\,dy/dt = 0$. (*d*) When $x = 9$, $y = \sqrt{225 - 81} = 12$, so $dy/dt = (-x/y)(dx/dt) = (-9/12)(4) = -3$ m/s. **2.** Let $t = 0$ be time when plane flies over.

5(t + 10) 4t s

$s = \sqrt{16t^2 + (5t + 50)^2} = \sqrt{41t^2 + 500t + 2500}$. $\dfrac{ds}{dt} = \dfrac{41t + 250}{\sqrt{41t^2 + 500t + 2500}} \approx 6.29$ m/min when $t = 20$.

3. $\dfrac{r}{h} = \dfrac{6}{18}$ implies $r = \dfrac{h}{3}$. $V = \dfrac{\pi}{3}r^2h = \dfrac{\pi}{3}\left(\dfrac{h}{3}\right)^2 h = \dfrac{\pi}{27}h^3$. $\dfrac{dV}{dt} = \dfrac{\pi}{9}h^2\dfrac{dh}{dt}$. Given $\dfrac{dV}{dt} = 16$, need $\dfrac{dh}{dt}$. But $\dfrac{dh}{dt} = \dfrac{dV/dt}{\pi/9\ h^2} = \dfrac{9 \cdot 16}{\pi h^2} = \dfrac{9 \cdot 16}{\pi \cdot 3^2} \approx 5.1$ m/s when $h = 3$.

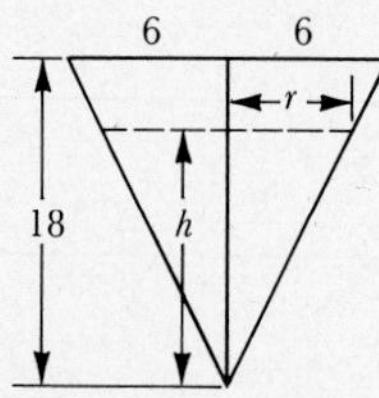

Sec. 5.3: Progress Test 1

1. $s'(t) = 3(t - 2)(t - 4)$, $s''(t) = 6(t - 3)$.

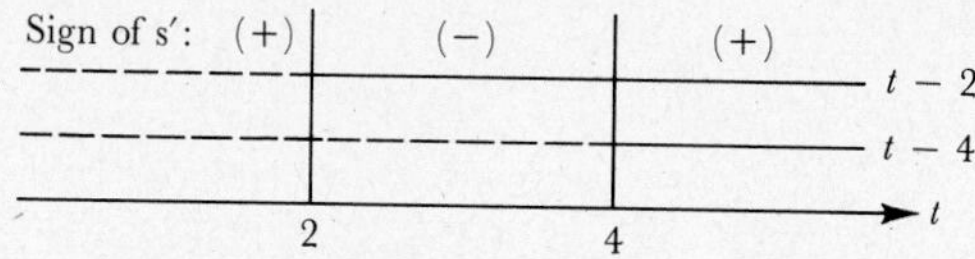

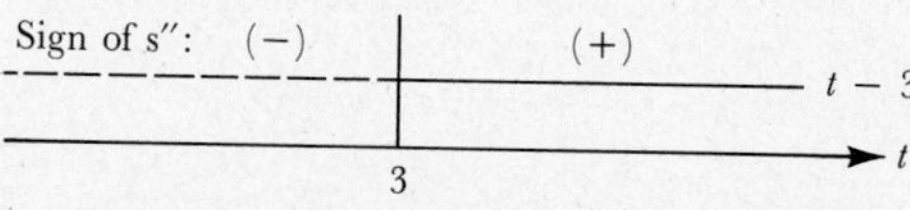

2. $s(2) = 20$, $s(3) = 18$, $s(4) = 16$. **3.**

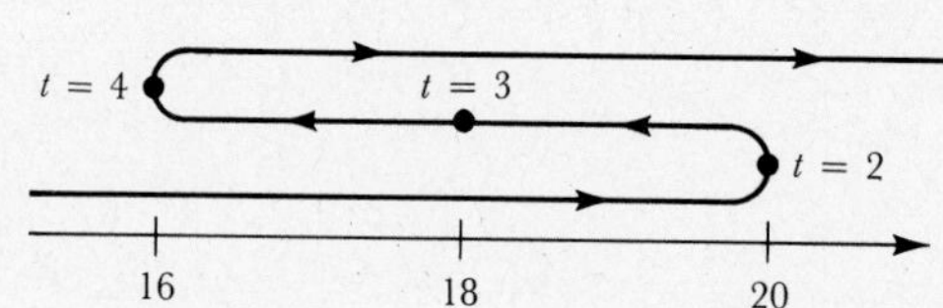

4. $t < 2$, moving right and slowing down; $t = 2$, stopped; $2 < t < 3$, moving left and speeding up; $3 < t < 4$, moving left and slowing down; $t = 4$, stopped; $t > 4$, moving right and speeding up.

Sec. 5.4: Progress Test 1—Part 1 (lower demand)

1. $p(x) = mx + b$, with $\Delta p/\Delta x = m$ (since p is linear). We have $\Delta p/\Delta x = 10/-4 = -2/5 = m$. Since $p(120) = 200$, $b = 500$, so $p(x) = 500 - (5/2)x$. **2.** $R(x) = x(500 - (5/2)x)$. **3.** (*a*) $R'(x) = 500 - 5x = 0$ implies $x = 100$, a rel. max. since $R''(x) = -5 < 0$; (*b*) $p(100) = 500 - (5/2)(100) = \$250$; (*c*) $R(100) = \$25{,}000$. **4.** $C(x) = 60x + (120 - x)(40) + 17{,}000 = 20x + 17{,}000$. **5.** (*a*) $P(x) = (500x - (5/2)x^2) - (20x + 17{,}000) = 480x - (5/2)x^2 - 21{,}800$. $P'(x) = 480 - 5x = 0$ implies $x = 96$, a rel. max. since $P''(x) = -5 < 0$; (*b*) $P(96) = \$1240$; (*c*) $p(96) = \$260$.

Part 2 (higher demand)

1. $\Delta p_2/\Delta x = 10/-2 = -5$ and $p_2(120) = 240 = (-5)(120) + b$ implies $p_2(x) = 840 - 5x$. **2.** $R_2(x) = x(840 - 5x)$. **3.** (*a*) $R_2'(x) = 840 - 10x = 0$ implies $x = 84$; (*b*) $p_2(84) = \$420$; (*c*) $R_2(84) = \$35{,}280$. **4.** (*a*) $P_2(x) = (840x - 5x^2) - (20x + 21{,}800) = 820x - 5x^2 - 21{,}800$; (*b*) $P_2'(x) = 820 - 10x = 0$ implies $x = 82$; (*c*) $P_2(82) = \$11{,}820$; (*d*) $p_2(82) = \$430$.

Sec. 5.4: Progress Test 2

1. $P_r(x) = 500x - (5/2)x^2 - 22{,}800$. **2.** $P_r'(x) = 500 - 5x = 0$ when $x = 100$, a max. value. **3.** $P_r(100) = \$2200$. **4.** $p_r(100) = 520 - (5/2)100 = \$270$. **5.** $270 - 20 = \$250$. **6.** 100 tenants get $(100)(10) = 1000$ and profit increase is $2200 - 240 = \$1960$.

Chapter 6: Solutions to Progress Tests

Sec. 6.1: Progress Test 1

1. $\int \frac{3\,dx}{x^3} = 3\int x^{-3}\,dx = \frac{3x^{-2}}{-2} + C = \frac{-3}{2}\frac{1}{x^2} + C.$ **2.** $\int(3t^2 + 4\sqrt{t})\,dt = 3\int t^2\,dt + 4\int t^{1/2}\,dt = \frac{3t^3}{3} + \frac{4t^{3/2}}{3/2} + C = t^3 + \frac{8}{3}t^{3/2} + C.$

3. $\int \frac{6z + z^{1/3}}{2z^{3/4}}\,dz = \int \frac{6z}{2z^{3/4}}\,dz + \int \frac{z^{1/3}}{2z^{3/4}}\,dz = 3\int z^{1/4}\,dz + (1/2)\int z^{-5/12}\,dz = \frac{3z^{5/4}}{5/4} + (1/2)\frac{z^{7/12}}{7/12} + C = \frac{12}{5}z^{5/4} + \frac{6}{7}z^{7/12} + C.$

4. $\int(w^2 + 1)^2\,dw = \int(w^4 + 2w^2 + 1)\,dw = \int w^4\,dw + 2\int w^2\,dw + \int dw = \frac{w^5}{5} + \frac{2w^3}{3} + w + C.$ **5.** $\int x^3\,dx = \frac{x^4}{4} + C,$

$x\int x^2\,dx = x\left[\frac{x^3}{3} + C\right] = \frac{x^4}{3} + Cx.$ But $D_x\left[\frac{x^4}{3} + Cx\right] \neq x^3.$

Sec. 6.1: Progress Test 2

1. $a(t) = -5$, $v(t) = \int(-5)\,dt = -5t + 20.$ **2.** $h(t) = \int(-5t + 20)\,dt = -\frac{5}{2}t^2 + 20t + C_1$, but $C_1 = 0$ since $h(0) = 0$ (object is hurled from the *surface* of the moon). **3.** $v(t) = -5t + 150 = 0$ when $t = 30$, and $h(30) = -\frac{5}{2}(30)^2 + 150(30) = 2250$ ft (more than $\frac{1}{2}$ mi). **4.** (*a*) $h(t) = -\frac{5}{2}t^2 + 150t = 0$ implies $t = 0$ and $t = 60$. Ball takes 1 full min to return to moon's surface. (*b*) $v(60) = 150 - 5(60) = -150$ ft/s (downward). **5.** $dy/dx = 6x^2 + 2x + C_1 = 10$ when $x = 1$, so $C_1 = 2$. Also, $y = 2x^3 + x^2 + 2x + C_2 = 3$ when $x = 0$, so $C_2 = 3$.

Sec. 6.1: Progress Test 3

1. Let $u = 1 - 5x^2$, so $du = -10x\,dx$ and $x\,dx = -du/10$. $\int x(1 - 5x^2)^7\,dx = \int u^7\left(\frac{-du}{10}\right) = -\frac{1}{10}\int u^7\,du = -\frac{u^8}{80} + C = -\frac{(1 - 5x^2)^8}{80} + C.$

2. Let $u = x^2 + 2x + 7$, so $du = (2x + 2)\,dx = 2(x + 1)\,dx$. Thus $\frac{1}{2}du = (x + 1)\,dx$ and $\int(x^2 + 2x + 7)^{1/3}(x + 1)\,dx = \int u^{1/3}\frac{1}{2}du =$ $\frac{1}{2}\int u^{1/3}\,du = \frac{3}{8}u^{4/3} + C = \frac{3}{8}(x^2 + 2x + 7)^{4/3} + C.$ **3.** $\int \frac{dv}{\sqrt{(2v + 1)^3}} = \int(2v + 1)^{-3/2}\,dv$. Let $u = 2v + 1$, so $du = 2\,dv$. Hence $\frac{1}{2}du = dv$. Thus we have $\int u^{-3/2}\frac{1}{2}du = \frac{1}{2}\int u^{-3/2}\,du = -u^{-1/2} + C = \frac{-1}{\sqrt{2v + 1}} + C.$ **4.** Let $u = 2 + \frac{1}{w}$, so $du = \frac{-1}{w^2}\,dw$. Hence $\frac{dw}{w^2} = -du.$

Thus $\int\left(2 + \frac{1}{w}\right)\frac{dw}{w^2} = \int u(-du) = -\int u\,du = \frac{-u^2}{2} + C = \frac{-1}{2}\left(2 + \frac{1}{w}\right)^2 + C.$

Sec. 6.2: Progress Test 1

1. (*a*) $\underline{S}_4(f) = f(-2)(0 - (-2)) + f(1)(1 - 0) + f(3/2)(3/2 - 1) + f(2)(2 - 3/2) = (0)(2) + (3)(1) + (7/4)(1/2) + (0)(1/2) = 31/8;$

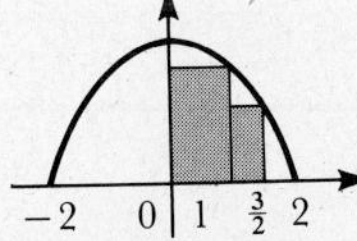

(*b*) $\overline{S}_4(f) = f(0)(0 - (-2)) + f(0)(1 - 0) + f(1)(3/2 - 1) + f(3/2)(2 - 3/2) = (4)(2) + (4)(1) + (3)(1/2) + (7/4)(1/2) = 115/8.$

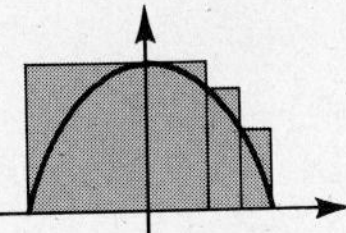

2. (*a*) $f(x) = 2x$ is increasing on $[0, 1]$, so lower sums based on left side of intervals. $\underline{S}_{10}(f) =$ $f(0)\,(1/10) + f(0.1)\,(1/10) + \cdots + f(0.9)(1/10) = (1/10)\,[0 + 0.2 + 0.4 + 0.6 + 0.8 + 1 + 1.2 + 1.4 + 1.6 + 1.8] = 5/10 = 1/2.$

(*b*) $\underline{S}_n(f) = \frac{1}{n}\left[f(0) + f\left(\frac{1}{n}\right) + f\left(\frac{2}{n}\right) + f\left(\frac{3}{n}\right) + \cdots + f\left(\frac{n-1}{n}\right)\right] = \frac{1}{n}\left[\frac{2\cdot 0}{n} + \frac{2\cdot 1}{n} + \frac{2\cdot 2}{n} + \frac{2\cdot 3}{n} + \cdots + \frac{2(n-1)}{n}\right] =$ $\frac{2}{n^2}[0 + 1 + 2 + 3 + \cdots + (n - 1)].$

Sec. 6.2: Progress Test 2

1. We successively substitute 2, 3, 4, 5 into $2j^2 - 1$ and add:

$[2(2)^2 - 1] + [2(3)^2 - 1] + [2(4)^2 - 1] + [2(5)^2 - 1] = 7 + 17 + 31 + 49 = 104.$ **2.** (*a*) $\underline{S}_{10}(f) = \sum_{i=1}^{10} f\left(\frac{i-1}{10}\right)\left(\frac{1}{10}\right) = \sum_{i=1}^{10}\frac{2(i-1)}{10}\frac{1}{10};$

(*b*) $\underline{S}_n(f) = \sum_{i=1}^{n} f\left(\frac{i-1}{n}\right)\left(\frac{1}{n}\right) = \sum_{i=1}^{n} 2\left(\frac{i-1}{n}\right)\left(\frac{1}{n}\right) = \frac{2}{n^2}\sum_{i=1}^{n}(i - 1).$

Sec. 6.2: Progress Test 3

1. Maxima are achieved at left endpoints, minima at right endpoints. $\overline{S}_n(f) - \underline{S}_n(f) = \sum_{i=1}^{n} f(x_{i-1})\left(\frac{b-a}{n}\right) - \sum_{i=1}^{n} f(x_i)\left(\frac{b-a}{n}\right) = [f(x_0) - f(x_n)]\frac{b-a}{n} = \frac{[f(a) - f(b)](b-a)}{n}$. **2.** $\lim_{n\to+\infty} \frac{[f(a) - f(b)](b-a)}{n} = 0$ since numerator does not depend on n.
3. Essentially Fig. 6.12 inverted.

Sec. 6.3: Progress Test 1

1. $M = f(4) = 64/3$, $m = f(1) = -7/6$, $b - a = 7$: $(-7/6)(7) \le \int_{-3}^{4}((1/3)x^3 + (1/2)x^2 - 2x)\,dx \le (64/3)(7)$. **2.** (*a*) $(-3)(5) = -15$; (*b*) $\int_0^3(1 - x)\,dx = \int_0^1(1 - x)\,dx + \int_1^3(1 - x)\,dx = \frac{1}{2} + (-\frac{1}{2})(2)(2) = \frac{-3}{2}$,
(*c*) $\int_{-2}^{1}(x + 1)\,dx = \int_{-2}^{-1}(x + 1)\,dx + \int_{-1}^{1}(x + 1)\,dx = -\frac{1}{2} + \frac{1}{2}(2)(2) = \frac{3}{2}$; (*d*) $\int_{-6}^{-1} 4\,dx = (4)(5) = 20$ (a rectangle of width 5, height 4).
3. (*a*) $(2)(2) + (2)(3) + (1)(4) + (1)(1) + (1)(0) + (1)(4) = 19$; (*b*) $(-4)(5) + (1)(1) + (1)(6) = -13$.

Sec. 6.3: Progress Test 2

1. (*a*) $\int x^2\,dx = x^3/3 + C = F(x)$: $F(1) - F(0) = (1/3 + C) - (0/3 + C) = 1/3$; (*b*) $\int(x - 2)\,dx = x^2/2 - 2x + C = F(x)$: $F(4) - F(1) = (8 - 8 + C) - (1/2 - 2 + C) = 3/2$; (*c*) $\int(1 + 2x - x^2)\,dx = x + x^2 - x^3/3 + C = F(x)$: $F(2) - F(0) = (2 + 4 - 8/3 + C) - (0 + 0 - 0 + C) = 10/3$. **2.** $5b - 5a$. **3.** (*a*) $5x - 5(0) = 5x$; (*b*) $A(x) = 5x$. $A'(x) = \lim_{\Delta x\to 0} \frac{5x + 5\,\Delta x - 5x}{\Delta x} = \lim_{\Delta x\to 0} \frac{5\,\Delta x}{\Delta x} = 5$. Here $A(x + \Delta x) - A(x)$ is the area of the smaller rectangle shown.

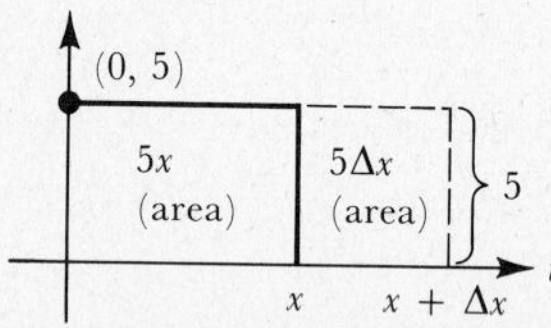

4. The area of the larger trapezoid determined by $f(x)$, $x = 0$ and $x = b$ is the area of its triangular part (above the dashed line) $(\frac{1}{2}b)(3b + 5 - 5) = \frac{3}{2}b^2$ plus the area of its rectangular part (below the dashed line), $5b$. Similarly, the area of the smaller trapezoid determined by $x = 0$ and $x = a$ is $\frac{3}{2}a^2 + 5a$. Hence the area we are looking for is $(\frac{3}{2}b^2 + 5b) - (\frac{3}{2}a^2 + 5a)$, which turns out to be $F(b) - F(a)$, where F is any antiderivative of $f(x) = 3x + 5$ since $\int(3x + 5)\,dx = (3/2)x^2 + 5x + C$.

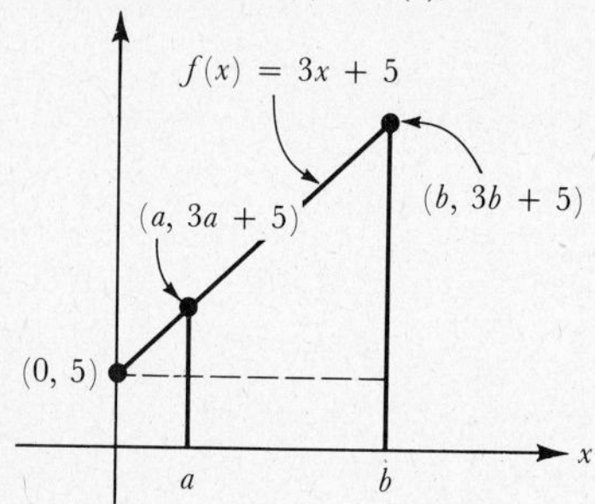

Sec. 6.3: Progress Test 3

1. $\int_{-1}^{1}(4x^3 - 3x^2 - 5)\,dx = [x^4 - x^3 - 5x]_{-1}^{1} = (1 - 1 - 5) - (1 + 1 + 5) = -12$.
2. $\frac{1}{2}\int_0^1 (x^2 + 1)^2(2x\,dx) = \frac{1}{2}\,\frac{(x^2 + 1)^3}{3}\Big|_0^1 = \frac{8}{6} - \frac{1}{6} = \frac{7}{6}$. $\left(\text{Let } u = x^2 + 1, \text{ so } du = 2x\,dx \text{ and } x\,dx = \frac{du}{2}.\right)$
3. $\int_0^2(x^3 + 3x + 1)^{-1/2}(x^2 + 1)\,dx = \frac{1}{3}\int_0^2 (x^3 + 3x + 1)^{-1/2}(3x^2 + 3)\,dx = \frac{1}{3}\,\frac{(x^3 + 3x + 1)^{1/2}}{1/2}\Big|_0^2 = \frac{2}{3}[(8 + 6 + 1)^{1/2} - (0 + 0 + 1)^{1/2}] = \frac{2}{3}$.
$\left[\text{Let } u = x^3 + 3x + 1,\ du = (3x^2 + 3)\,dx, \text{ so } (x^2 + 1)\,dx = \frac{du}{3}.\right]$

Sec. 6.4: Progress Test 1

1. By the MVT for Integrals there is a c between 2 and 7 such that $f(c) = \frac{1}{b - a}\int_a^b f(x)\,dx =$ average value. Now $\int_2^7 (x^2 - 2x + 1)\,dx = \frac{x^3}{3} - x^2 + x\Big|_2^7 = \frac{215}{3}$ and $f(c) = c^2 - 2c + 1$, so $c^2 - 2c + 1 = \frac{1}{5}\left(\frac{215}{3}\right) = \frac{43}{3}$. Thus $3c^2 - 6c - 40 = 0$, which implies $c = \frac{3 \pm \sqrt{129}}{3}$. Of these numbers $\frac{3 + \sqrt{129}}{3}$ is in [2, 7], so f assumes its mean value at $x = \frac{3 + \sqrt{129}}{3} \approx 4.79$. **2.** The average temperature is $\frac{1}{6-0}\int_0^6 T(x)\,dx = \frac{1}{6}\int_0^6 (-x^2 + 4x + 70)\,dx = \frac{1}{6}(420) = 70°$. Now $T'(x) = -2x + 4$, so only critical point on [0, 6] occurs at $x = 2$. Since $T(0) = 70$, $T(2) = 74$, $T(6) = 58$. The high was 74°F and the low was 58°F. **3.** (*a*) $v(t) = 32t$: $\frac{1}{5}\int_0^5 32t\,dt = \frac{1}{5}[16t^2]_0^5 = 80$; (*b*) $\frac{v(0) + v(5)}{2} = \frac{0 + 160}{2} = 80$; (*c*) The ave. vel. is one-fifth area of triangle, and the arithmetic average in (*b*) is half the height of the triangle. The velocity grows linearly so reaches its average halfway through the interval ($32t = 80$ when $t = 5/2$.)

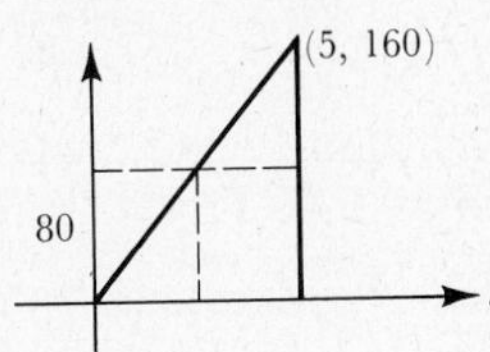

Sec. 6.4: Progress Test 2

1. $u = x^3 + 1$, $du = 3x^2\,dx$, $x^2\,dx = du/3$. When $x = 2$, $u = 9$, and when $x = 0$, $u = 1$.
$\int_0^2 \sqrt{x^3 + 1}\,x^2\,dx = \int_1^9 \sqrt{u}\,\frac{du}{3} = \frac{1}{3}\int_1^9 u^{1/2}\,du = \frac{1}{3}\frac{u^{3/2}}{3/2}\Big|_1^9 = \frac{2}{9}[27 - 1] = \frac{52}{9}$. **2.** $G'(x) = \frac{1}{3x + 2}$ by F.T. **3.** $G(x) = -\int_1^x (t^2 + 1)\,dt$ by (6.3.4). So $G'(x) = -(x^2 + 1)$ by F.T. **4.** $G(x) = F(1/x)$ where $F(x) = \int_1^x dt/t$. So by the Chain Rule
$G'(x) = F'(1/x)\frac{d}{dx}(1/x) = F'(1/x)(-1/x^2)$. But $F'(x) = 1/x$ by the F.T. So $G'(x) = \frac{1}{1/x}(-1/x^2) = -1/x$.

Sec. 6.5: Progress Test 1

1.

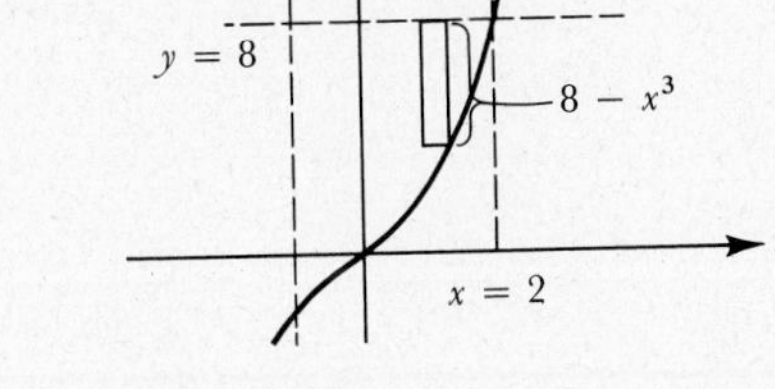

$8 \geq x^3$ on $[-1, 2]$ $A(R) = \int_{-1}^2 (8 - x^3)\,dx = 8x - \frac{x^4}{4}\Big|_{-1}^2 = 20\ 1/4$ **2.** $\sqrt{x} = x^2$ when $x = 0$

2. $\sqrt{x} = x^2$ when $x = 0$ and $x = 1$. $\sqrt{x} \geq x^2$ on [0, 1] $A(R) = \int_0^1 (\sqrt{x} - x^2)\,dx = \frac{2x^{3/2}}{3} - \frac{x^3}{3}\Big|_0^1 = \frac{1}{3}$

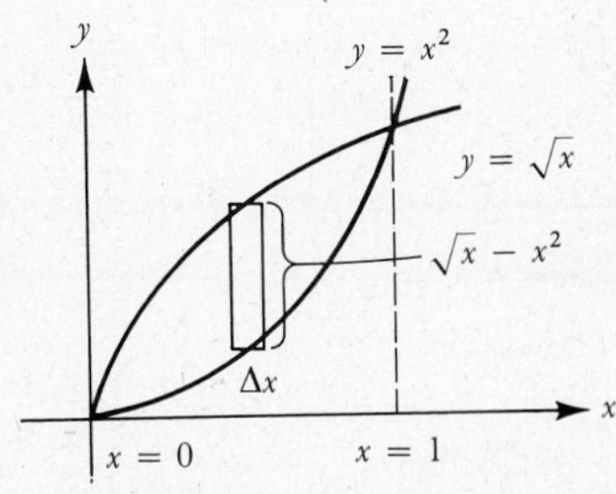

Sec. 6.5: Progress Test 2

1. $9 - y^2 = y + 3$ implies $(y + 3)(y - 2) = 0$, so curves intersect when $y = -3, 2$.

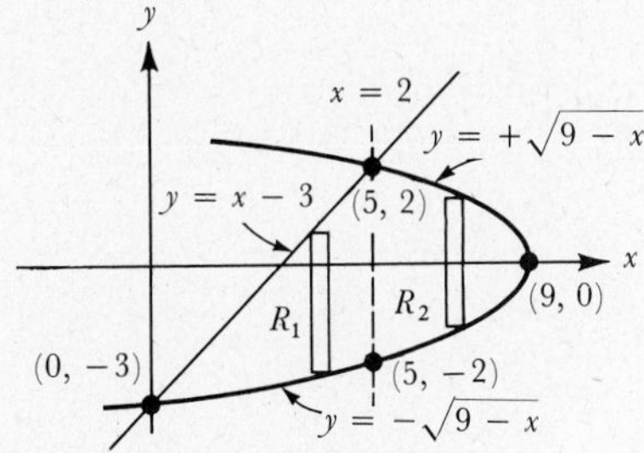

Now $x - 3 \geq -\sqrt{9-x}$ on $[0, 5]$, so $A(R_1) = \int_0^5[(x-3) - (-\sqrt{9-x})]\,dx = \frac{x^2}{2} - 3x - \frac{2}{3}(9-x)|_0^5 = \frac{61}{6}$ and $\sqrt{9-x} \geq -\sqrt{9-x}$ on $[5, 9]$, so $A(R_2) = \int_5^9[\sqrt{9-x} - (-\sqrt{9-x})]\,dx = 2\int_5^9 \sqrt{9-x}\,dx = 2(-\frac{2}{3})(9-x)^{3/2}|_5^9 = \frac{64}{6}$ Thus $A(R) = A(R_1) + A(R_2) = 20\ 5/6$.

2. $A(R) = \int_{-3}^{2}[(9-y^2) - (y+3)]\,dy = \int_{-3}^{2}(6 - y - y^2)\,dy = 20\ 5/6$

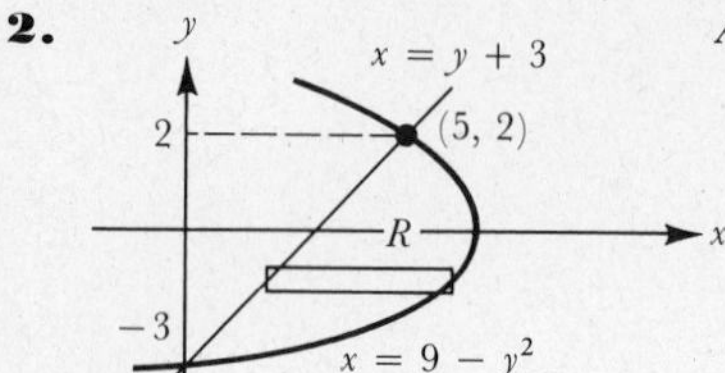

Sec. 6.5: Progress Test 3

1. $F'(t) = 50 - \frac{5}{12}t = \frac{t^2}{24}$, so $t^2 + 10t - 1200 = 0$ and $t = 30$ months or $2\frac{1}{2}$ years. **2.** $\int_0^{30}\left[\left(50 - \frac{5}{12}t\right) - \left(\frac{t^2}{24}\right)\right]dt = 937.5$: profit is $937,500. **3.** $S(x) = D(x)$ when $x = 5$ with $D(5) = 5 = S(5)$. C.S. $= \int_0^5(10 - x - 5)\,dx = 25/2$. P.S. $= \int_0^5\left(5 - \frac{x^2}{5}\right)dx = 50/3$.

Sec. 6.6: Progress Test 1

1. $\pi\int_0^h\left(\frac{r}{h}x\right)^2 dx = \pi\left[\frac{r^2x^3}{3h^2}\right]_0^h = \frac{\pi}{3}r^2h.$

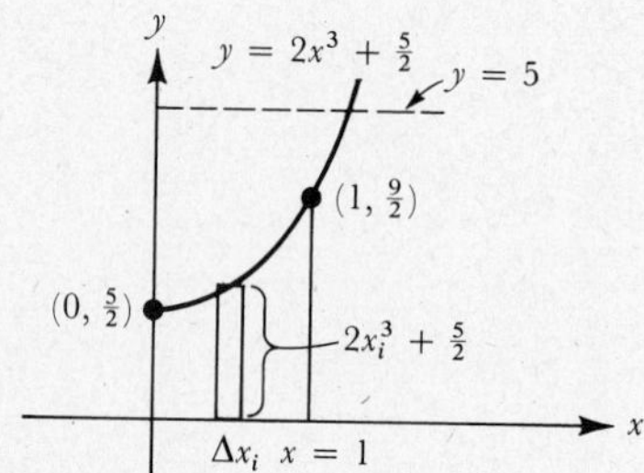

2. (*a*) $\pi\int_0^1\left(2x^3 + \frac{5}{2}\right)^2 dx = \pi\left[\frac{4}{7}x^7 + \frac{5}{2}x^4 + \frac{25}{4}x\right]_0^1 = \frac{261}{28}\pi$;

(*b*) outer radius is 5, inner radius is $5 - \left(2x_i^3 + \frac{5}{2}\right) = \frac{5}{2} - 2x_i^3$.

$$V = \pi\int_0^1\left[(5)^2 - \left(\frac{5}{2} - 2x^3\right)^2\right]dx = \pi\int_0^1\left[\frac{75}{4} + 10x^3 - 4x^6\right]dx = \frac{439}{28}\pi.$$

Sec. 6.6: Progress Test 2

1. Assume the sphere is generated by revolving the top hemisphere of $x^2 + y^2 = r^2$ about the x axis. We shall double the volume generated by the first quadrant portion of $x = \sqrt{r^2 - y^2}$. Hence the volume is $4\pi\int_0^r y\sqrt{r^2 - y^2}\,dy = 4\pi\left[\left(-\frac{1}{2}\right)\left(\frac{2}{3}\right)(r^2 - y^2)^{3/2}\right]_0^r = \frac{4}{3}\pi r^3$.

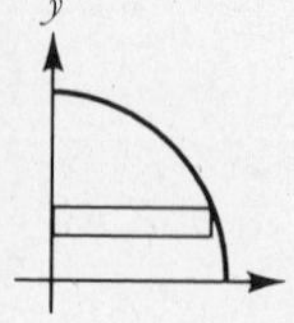

2. First sketch the region and determine its boundaries. The disk method would require three separate integrals, so we use shells. $2\pi\int_{-2}^{2}(3 - x)[(-x^2 - 2x + 9) - (x^2 - 2x + 1)]\,dx = 2\pi\int_{-2}^{2}(2x^3 - 6x^2 - 8x + 24)\,dx = 2\pi\left[\frac{x^4}{2} - 2x^3 - 4x^2 + 24x\right]_{-2}^{2} = 128\pi$.

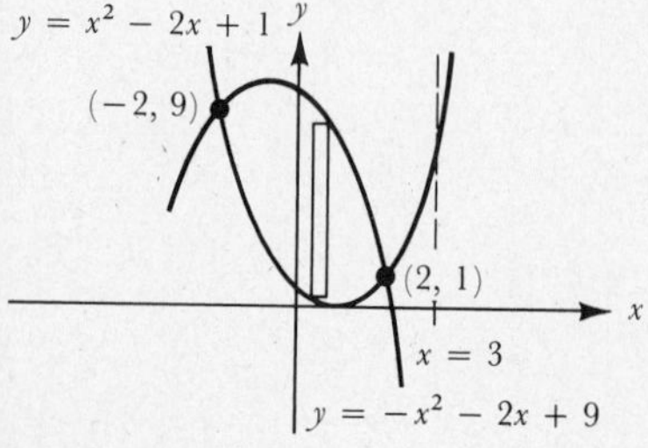

Chapter 7: Solutions to Progress Tests

Sec. 7.1: Progress Test 1

1. (*a*) $-1/4$ rotation, $-90°$; (*b*) $3/4$ rotation, $270°$. **2.** (*a*) $-\pi/2$ rad; (*b*) $(3/2)\pi$ rad. **3.** 5π rad, $900°$. **4.** (*a*) $\pi/4$ rad; (*b*) $\pi/6$ rad; (*c*) $\pi/3$ rad; (*d*) $(25/18)\pi$ rad; (*e*) $(-5/12)\pi$ rad. **5.** (*a*) $45°$; (*b*) $15°$; (*c*) $20.05°$; (*d*) $-2349.13°$; (*e*) $3{,}552{,}338.33°$.

Sec. 7.1: Progress Test 2

1.

x	$\cos x$	x	$\cos x$	x	$\cos x$
0	1	$\frac{2}{3}\pi$	-0.5	$\frac{4}{3}\pi$	-0.5
$\pi/6$	0.87	$\frac{3}{4}\pi$	-0.71	$\frac{3}{2}\pi$	0
$\pi/4$	0.71	$\frac{5}{6}\pi$	-0.87	$\frac{5}{3}\pi$	0.5
$\pi/3$	0.5	π	-1	$\frac{7}{4}\pi$	0.71
$\pi/2$	0	$\frac{7}{6}\pi$	-0.87	$\frac{11}{6}\pi$	0.87
		$\frac{5}{4}\pi$	-0.71	2π	1

2. Fig. 7.12, p. 238. **3.** 2π

Sec. 7.1: Progress Test 3

1.

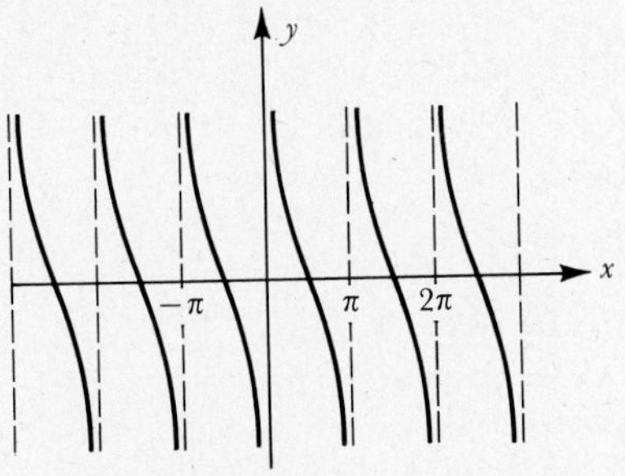

y

x

−π π 2π

2. π **3.** All integral multiples of π. **4.** 2d, 4th

5. All odd multiples of $\pi/2$. **6.** All multiples of π. **7.** 1. **8.** 1.

Sec. 7.2: Progress Test 1

1. $f'(x) = [\cos(x^4 + 3x^3 + 2x + 1)](4x^3 + 9x^2 + 2)$. **2.** $df/dx = [-\sin(x^5 - 4x^3 - 3x + 1)](5x^4 - 12x^2 - 3)$.
3. $g'(x) = 5\,[\sin^4(3x^2 + x - 1)](\cos(3x^2 + x - 1))(6x + 1)$. **4.** $t'(x) = (1/2)(\cos(x^2 + x))^{-1/2}(-\sin(x^2 + x))(2x + 1)$.

5. $$dr/dx = \frac{(\cos^3(5x - 2))\cdot 2\sin(3x + 1)(\cos(3x + 1))\cdot 3 - (\sin^2(3x + 1))\cdot 3(\cos^2(5x - 2))(-\sin(5x - 2))\cdot 5}{\cos^6(5x - 2)}$$

$$= \frac{6\cos(5x - 2)\sin(3x + 1)\cos(3x + 1) + 15\sin^2(3x + 1)\sin(5x - 2)}{\cos^4(5x - 2)}.$$

6. $ds/dt = \cos^4(t^2 + 1)3\sin^2(7t)(\cos 7t)(7) + \sin^3(7t)4\cos^3(t^2 + 1)(-\sin(t^2 + 1))(2t) + (\cos 4t)(4)$. **7.** $m'(p) = 2[(\sin p)(-\sin p) + \cos p \sin p]$. **8.** $\dfrac{df}{dx} = \left[\cos(\sqrt{4x + 5})\right]\dfrac{1}{2\sqrt{4x + 5}}(4)$.

Sec. 7.2: Progress Test 2

1. $y' = -\csc[(x - 4)^5]\cdot\cot[(x - 4)^5]\cdot 5(x - 4)^4$.

2. $f'(x) = 3\tan^2(3x^7 + 2x^4 + 3x - 1)\cdot\sec^2(3x^7 + 2x^4 + 3x - 1)\cdot(21x^6 + 8x^3 + 3)$. **3.** $\dfrac{dh}{dx} = \dfrac{[-\csc^2(x^3 + 2x)](3x^2 + 2)}{2\sqrt{\cot(x^3 + 2x)}}$.

4. $f'(x) = (\sec(3x + 1))(\sec^2(1 - x))(-1) + (\tan(1 - x))(\sec(3x + 1))(\tan(3x + 1))(3)$.

5. $g'(y) = 5\sec^4(3y^2 - y)^4\sec[(3y^2 - y)^4]\tan[(3y^2 - y)^4]\cdot 4(3y^2 - y)^3\cdot(6y - 1)$.

6. $D_x\left[\dfrac{\cos x}{\sin x}\right] = \dfrac{\sin x(-\sin x) - \cos x\cos x}{\sin^2 x} = \dfrac{-(\sin^2 x + \cos^2 x)}{\sin^2 x} = -\csc^2 x$.

7. $D_x\left[\dfrac{1}{\sin x}\right] = \dfrac{(-1)\cos x}{\sin^2 x} = \dfrac{-1}{\sin x}\,\dfrac{\cos x}{\sin x} = -\csc x\cot x$. **8.** Each deriv. of $\sin ax$ introduces an a factor and each deriv. of order a multiple of 4 is $a^n \sin ax$. Since 100 is a multiple of 4, $D_x^{(100)}[\sin ax] = a^{100}\sin ax$.

Sec. 7.3: Progress Test 1

1. (*a*) $f^{-1}(x) = \dfrac{x+1}{2}$; (*b*) $f(f^{-1}(x)) = f\left(\dfrac{x+1}{2}\right) = 2\left(\dfrac{x+1}{2}\right) - 1 = x$, $f^{-1}(f(x)) = f^{-1}(2x-1) = \dfrac{(2x-1)+1}{2} = x$.

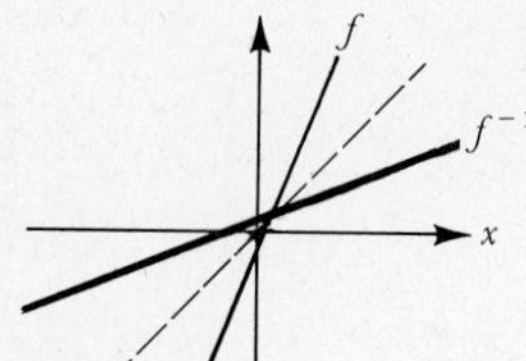

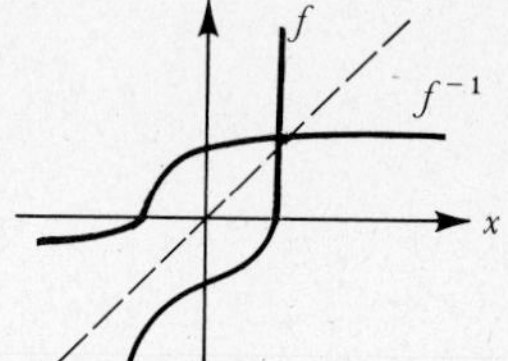

2. (*a*) $f^{-1}(x) = \sqrt[3]{x+1}$; (*b*) $f(f^{-1}(x)) = f(\sqrt[3]{x+1}) = (\sqrt[3]{x+1})^3 - 1 = x + 1 - 1 = x$, $f^{-1}(f(x)) = f^{-1}(x^3 - 1) = \sqrt[3]{(x^3-1)+1} = \sqrt[3]{x^3} = x$. **3.** $(-1, 0)$.

Sec. 7.3: Progress Test 2

1. (*a*) 7; (*b*) 0; (*c*) 10. **2.** (*a*) $\log_5(x^2 - 1) = 0$, so $x^2 - 1 = 1$, $x = \pm\sqrt{2}$; (*b*) $x + 1 = \log_3(81) = 4$, so $x = 3$; (*c*) $\log_{10} x^2 + \log_{10} x = 3$, so $\log_{10} x^3 = 3$. Thus $x^3 = 10^3$, and so $x = 10$. **3.**

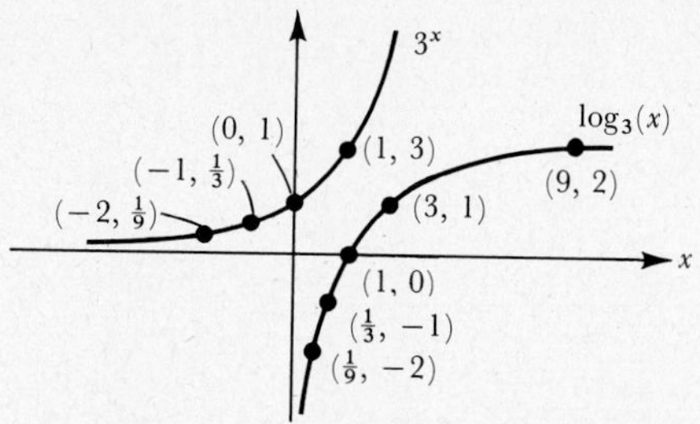

4. (*a*) $(f \circ g)(x) = \log_b\left[\dfrac{(x^2+x+4)^{1/2}(3x+1)}{(x^2+1)^6}\right] = \dfrac{1}{2}\log_b(x^2+x+4) + \log_b(3x+1) - 6\log_b(x^2+1)$;
(*b*) $(f \circ g)(x) = b^{(\log_b(4x+5))} = 4x + 5$. **5.** (*a*) -3; (*b*) 6; (*c*) -5; (*d*) 0; (*e*) 3; (*f*) 2; (*g*) 1/2; (*h*) -1; (*i*) 0; (*j*) 30; (*k*) 2; (*l*) 4. **6.** (*a*) $3\left(\log_3 \frac{(1+x)^x}{(1-x)^{2x}}\right) = \left[\dfrac{1+x}{(1-x)^2}\right]^x$; (*b*) $\log_5(5^{2x}) = 2x$; (*c*) $2^{\log_2[(x^2-4)(x-1)^3]} = (x^2-4)(x-1)^3$.

Sec. 7.4: Progress Test 1

1. (*a*) $f'(x) = -e^{-x}$; (*b*) $f'(x) = e^{2-x^2}(-2x)$; (*c*) $g'(x) = e^{x^2}(12x^2+1) + (4x^3+x)e^{x^2}(2x)$; (*d*) $g'(t) = 2^{t^3+2t} \cdot (3t^2+2) \cdot \ln 2$; (*e*) $\dfrac{dy}{dx} = \pi^x \ln \pi$; (*f*) $\dfrac{dy}{dx} = \pi x^{\pi-1}$ (the Power Rule!); (*g*) $s'(x) = \dfrac{2^x e^x - e^x 2^x \ln 2}{(2^x)^2} = \dfrac{e^x 2^x(1 - \ln 2)}{(2^x)^2} = \dfrac{e^x}{2^x}(1 - \ln 2)$; (*h*) $t'(x) = e^{x - x\ln 2}(1 - \ln 2)$.
2. $s'(x) = t'(x)$ since $s(x) = t(x)$. Note $e^x/2^x = e^x/e^{x\ln 2} = e^{x - x\ln 2}$. **3.** $e^{xy}(x\, dy/dx + y) = 1$, so $dy/dx = (e^{-xy} - y)/x$ $(x \neq 0)$.
4. $f(x) = 0$ for all x. (Special case of ke^x with $k = 0$). **5.** $g^{(100)}(x) = k^{100}\, e^{kx+c}$ **6.**

b	Slope of b^x at (0, 1)	b	Slope of b^x at (0, 1)
$\frac{1}{2}$	−0.6931	2.5	0.9163
1	0	2.6	0.9555
2	0.6931	2.7	0.9933
3	1.0986	2.71	0.9969
5	1.6094	2.72	1.0006
10	2.3026	e	1

Sec. 7.4: Progress Test 2

1. $f'(x) = \dfrac{12x^3 + 4x + 1}{3x^4 + 2x^2 + x + 1}$. **2.** $h'(x) = \dfrac{x(1/x) - \ln x}{x^2} = \dfrac{1 - \ln x}{x^2}$. **3.** $f'(x) = \dfrac{1/x}{\ln x} = \dfrac{1}{x \cdot \ln x}$.

4. $r'(x) = (\ln 2)2^{x\ln x} \cdot (1 + \ln x)$. **5.** $f'(x) = \dfrac{12x^2 + 2x - 1}{4x^3 + x^2 - x}\dfrac{1}{\ln 3}$ **6.** $\dfrac{xy' + y}{xy} = 0$, so $y' = -\dfrac{y}{x}$ $(x \neq 0)$.

Sec. 7.5: Progress Test 1

1. $\ln y = \dfrac{1}{2}\ln(7x - 1) + \dfrac{1}{3}\ln(x^2 + 5x) - 7\ln(x - 11)$. Thus $y' = y\left[\dfrac{7}{2(7x-1)} + \dfrac{2x+5}{3(x^2+5x)} - \dfrac{7}{x - 11}\right]$.

2. $\ln f(x) = \frac{1}{3}\left[5\ln(x-5) + 7\cdot\ln(3x+1) - \frac{1}{2}\ln(8x-1)\right]$. Thus $f'(x) = \frac{f(x)}{3}\left[5\cdot\frac{1}{x-5} + 7\cdot\frac{3}{3x+1} - \frac{1}{2}\cdot\frac{8}{8x-1}\right]$.

3. $\ln f(x) = x^2 \ln x$, so $f'(x)/f(x) = x^2(1/x) + (\ln x)2x = x + x(2\ln x) = x(1+\ln x^2)$. Hence $f'(x) = x^{(x^2)}(x)(1+\ln x^2) = x^{(x^2+1)}(1+\ln x^2)$.

Sec. 7.5: Progress Test 2

1. $f'(x) = e^x(x+1) = 0$ when $x = -1$ and $f''(x) = e^x(x+2) = 0$ when $x = -2$. $f''(-1) = e > 0$, so we have a relative minimum at $x = -1$. The sign of $f''(x)$ changes from minus on the left of $x = -2$ to plus on the right. Hence $f(x)$ is concave down to the left and concave up to the right of $x = -2$, making $x = -2$ a point of inflection. Now $f(-1) = -1/e$ and $f(-2) = -2/e^2$. For $x < -1$, $f(x)$ is decreasing; that is, as x *decreases,* $f(x)$ increases. $f(x)$, being the product of a negative and a positive number, is bounded above by 0 in $(-\infty, 0)$ and is asymptotic to the negative x axis.

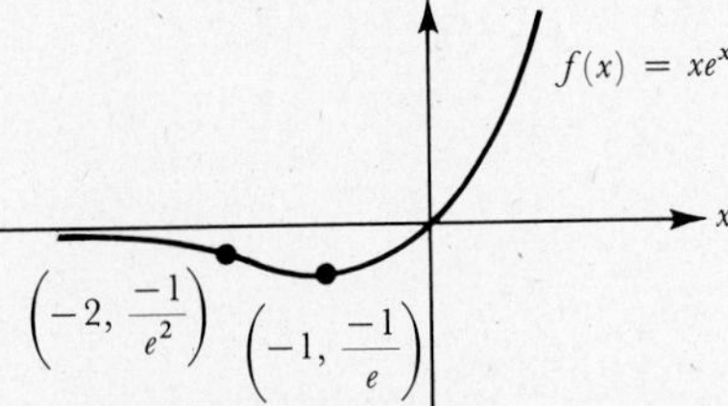

2. $f'(x) = 1 + \cos x \geq 0$ for all x with $f'(x) = 0$ when $\cos x = -1$, that is, when $x = \pi, 3\pi, -\pi$, in fact all odd multiples of π. These are "steps," but not relative extremes since f is never decreasing. For such values of x, $x = (2k+1)\pi$, $f(x) = (2k+1)\pi$ for any integer k.

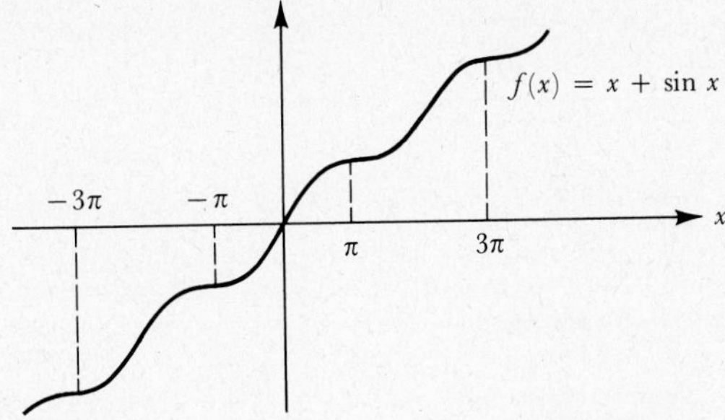

Sec. 7.5: Progress Test 3

1. Cross-sectional area equals $c(x) = (0.4)(0.2\cos x) + 2[\frac{1}{2}(0.2\sin x)(0.2\cos x)] = 0.04(2\cos x + \sin x\cos x)$
$c'(x) = 0.04(-2\sin x - \sin^2 x + \cos^2 x) = 0.04(-2\sin^2 x - 2\sin x + 1) = 0$ when

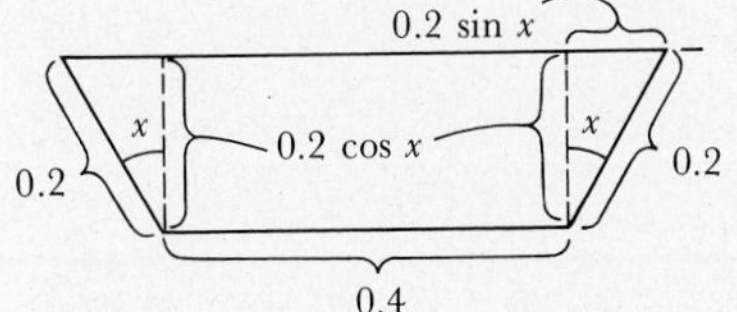

$\sin x = (\sqrt{3}-1)/2 \approx 0.366$ (ignore negative solution since $0 \leq x \leq \pi/2$). Thus $x \approx 0.3747$ rad $[c''(0.3747) < 0]$. **2.** Given $dx/dt = 50$ when $x = 300$. Need $d\theta/dt$. We know $x/300 = \tan\theta$. Diff. implicitly with respect to t: $(1/300)\,dx/dt = \sec^2\theta\,(d\theta/dt)$. When $x = 300$, $\theta = \pi/4$, and $\sec\theta = 2/\sqrt{2}$. Hence $d\theta/dt = (1/300)(50)(1/2) = 1/12 \approx 0.0833$ rad/s.

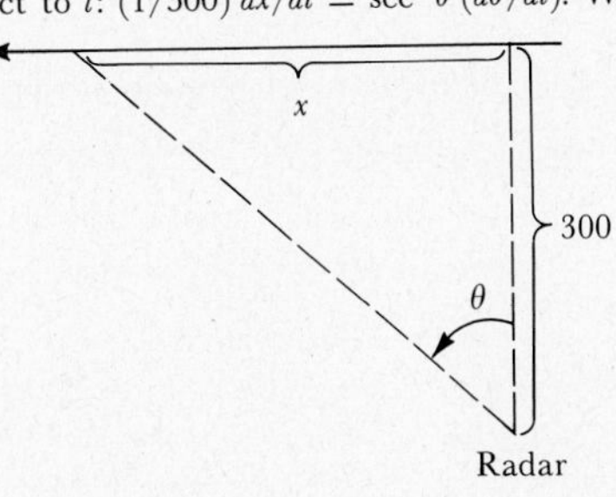

Chapter 8: Solutions to Progress Tests

Sec. 8.1: Progress Test 1

1. (*a*) $20{,}000\,e^{20k} = 50{,}000$ implies $k = (\ln 2.5)/20 \approx 0.0458$, so $A(t) = 20{,}000\,e^{0.0458t}$; (*b*) $20{,}000\,e^{0.0458t} = 100{,}000$ implies $t \approx 35.14$ years.

2. (*a*) $A = 2^{t-1} = e^{(\ln 2)(t-1)} \approx e^{0.693(t-1)}$; (*b*) (*i*) converting dollars to cents, solve $100{,}000{,}000 = e^{0.693(t-1)}$ for t, getting $t = [\ln(10^8)]/0.693 + 1 = (8\ln 10)/0.693 + 1 \approx 26.58 + 1$ day—you become a millionaire after the 27th day; (*ii*) solve $10^{11} = e^{0.693t-1}$ for t, getting $t \approx 37.55$ days; (*iii*) $t \approx 47.5$ days. **3.** $A(t) = 5e^{(\ln 1/2)t} = 5(\frac{1}{2})^t$.

Sec. 8.1: Progress Test 2

1. $P(1) = 2C_0 = C_0 e^{i-1}$ implies $i = \ln 2 \approx 0.69$ (69 percent). **2.** $i = 1$ yields $P(t) = C_0 e^t \approx C_0(2.7183)^t$, with $P(1)$ as needed. **3.** $60 = (\ln 2)/k$ implies $k \approx 1.155$ percent.

Sec. 8.1: Progress Test 3

1. $\displaystyle\int_0^{20} 50000e^{(-0.08)t}\,dt = \frac{50000}{-0.08}[e^{-1.6} - 1] \approx \$498{,}815.$ **2.** $\displaystyle\int_0^{20} 35000e^{(-0.08)t}\,dt \approx \$209{,}833.$

Sec. 8.2: Progress Test 1

1. Let $u = \sin 3x$, $du = 3\cos 3x\,dx$; then $\int \cos 3x\, e^{\sin 3x}\,dx = \frac{1}{3}\int e^u\,du = \frac{1}{3}e^u + C = \frac{1}{3}e^{\sin 3x} + C.$ **2.** Let $u = e^{2x}$, $du = 2e^{2x}\,dx$. Then integral becomes $\int \csc u \cot u\,du/2 = (-\csc u)/2 + C = (-\csc e^{2x})/2 + C.$ **3.** Let $u = \sin x$, $du = \cos x$. $\int(1 - \sin^2 x)\cos x\,dx = \int(1 - u^2)\,du = u - u^3/3 + C = \sin x - (\sin^3 x)/3 + C.$ **4.** Let $u = x + 5$, $du = dx$. Then $\int dx/(x+5) = \int du/u = \ln|u| + C = \ln|x + 5|$. Thus $\int_{-4}^{5} dx/(x+5) = \ln 10 - \ln 1 = \ln 10.$
5. $A = \int_0^{\pi} \sin x\,dx = -\cos x|_0^{\pi} = -\cos \pi - (-\cos 0) = 2.$

Sec. 8.2: Progress Test 2

1. Let $u = 2^x$, $du = (\ln 2)2^x\,dx$, $2^x\,dx = \dfrac{du}{\ln 2}$; $\displaystyle\int(\csc 2^x)\,dx = \int \csc u \frac{du}{\ln 2} = \frac{\ln|\csc 2^x - \cot 2^x|}{\ln 2} + C.$

2. $\displaystyle\int \tan x(\sec^2 x - 1)\,dx = \frac{\tan^2 x}{2} - \ln|\sec x| + C.$ **3.** Let $u = e^x$, $du = e^x\,dx$: $\int \tan u\,du = \ln|\sec u| + C = \ln|\sec e^x| + C.$

4. $\displaystyle\int \cot x\,dx = \int \frac{\cos x}{\sin x}\,dx = \int \frac{dv}{v}$ (with $v = \sin x$) $= \ln|v| + C = \ln|\sin x| + C$. General result follows by Chain Rule.

Sec. 8.3: Progress Test 1

1. Case II. $\displaystyle\int\left(\frac{1 - \cos 2x}{2}\right)^2 dx = \frac{1}{4}\int(1 - 2\cos 2x + \cos^2 2x)\,dx = \frac{1}{4}\int\left[1 - 2\cos 2x + \frac{1}{2}(1 + \cos 4x)\right]dx = \frac{3}{8}x - \frac{1}{4}\sin 2x + \frac{1}{32}\sin 4x + C.$

2. Case I. $\displaystyle\int(1 - \sin^2 x)^2 \cos x\,dx = \sin x - \frac{2\sin^3 x}{3} + \frac{\sin^5 x}{5} + C.$

3. Case I. $\displaystyle\int(\sin 3x)^{1/2}(1 - \sin^2 3x)(\cos 3x\,dx)\left(v = \sin 3x, \frac{dv}{3} = \cos 3x\,dx\right) =$

$\displaystyle\int v^{1/2}(1 - v^2)\frac{dv}{3} = \frac{2v^{3/2}}{3\cdot 3} - \frac{2v^{7/2}}{7\cdot 3} + C = \frac{2}{9}\sin^{3/2} 3x - \frac{2}{21}\sin^{7/2} 3x + C.$

4. $\displaystyle\int \frac{1 - \cos x}{2}\cdot\frac{1 + \cos x}{2}\,dx = \frac{1}{4}\int(1 - \cos^2 x)\,dx = \frac{1}{4}\int\left(1 - \frac{1}{2}[1 + \cos 2x]\right)dx = \frac{1}{8}x - \frac{1}{4}\sin 2x + C.$

Sec. 8.3: Progress Test 2

1. $\displaystyle\int \tan^5 x(\sec^2 x - 1)\,dx = \int \tan^5 x \sec^2 x - \int \tan^5 x\,dx = \frac{\tan^6 x}{6} - \int \tan^3 x(\sec^2 x - 1)\,dx =$

$\displaystyle\frac{\tan^6 x}{6} - \int \tan^3 x \sec^2 x + \int \tan^3 x\,dx = \frac{\tan^6 x}{6} - \frac{\tan^4 x}{4} + \int \tan x \sec^2(x - 1)\,dx = \frac{\tan^6 x}{6} - \frac{\tan^4 x}{4} + \frac{\tan^2 x}{2} + \ln|\sec x| + C.$

2. $\displaystyle\int \sec^{-1/3} x \tan^2 x(\sec x \tan x\,dx) = \int \sec^{-1/3} x(\sec^2 x - 1)\sec x \tan x\,dx = \frac{3\sec^{8/3} x}{8} - \frac{3\sec^{2/3} x}{2} + C.$

3. $\displaystyle\int(1 + \cot^2 x)\csc^2 x\,dx = -\cot x - \frac{\cot^3 x}{3} + C.$

Sec. 8.4: Progress Test 1

1. $\pi/6$ is in $[-\pi/2, \pi/2]$ and $\sin \pi/6 = 1/2$. Thus $\arcsin 1/2 = \pi/6$. **2.** $-\pi/6$ is in $[-\pi/2, \pi/2]$ and $\sin(-\pi/6) = -1/2$, thus $\arcsin(-1/2) = -\pi/6$. **3.** Since $\pi/8$ is in $[-\pi/2, \pi/2]$, $\arcsin(\sin \pi/8) = \pi/8$. **4.** $\sin(\arcsin 0.012) = 0.012$. **5.** Note that $2\pi/3$ is not in $[-\pi/2, \pi/2]$, but $\pi/3$ is, and $\sin(2\pi/3) = \sin \pi/3$; thus $\arcsin(\sin 2\pi/3) = \arcsin(\sin \pi/3) = \pi/3$.

Sec. 8.4: Progress Test 2

1. $-\pi/3$ is in $(-\pi/2, \pi/2)$ and $\tan(-\pi/3) = -\sqrt{3}$, so $\arctan(-\sqrt{3}) = -\pi/3$. **2.** Need y such that $\sec y = (-2)$ and y is in $[0, \pi]$. But $\sec 2\pi/3 = -2$ and y is in $[0, \pi]$, so $\text{arcsec}(-2) = 2\pi/3$. **3.** Now $\sec(-\pi/4) = \sec \pi/4 = \sqrt{2}$, and so $\text{arcsec}(\sec(-\pi/4)) = \text{arcsec}(\sec \pi/4) = \pi/4$. **4.** Now $\tan(19\pi/6) = \tan \pi/6$, so $\arctan(\tan 19\pi/6) = \arctan(\tan \pi/6) = \pi/6$.
5. $\text{arcsec}\, x$ is def. only for $|x| \geq 1$, so $\text{arcsec}(\frac{1}{2})$ is undef. (There is no y such that $\sec y = 1/2$ since $|\sec x| \geq 1$ for all x.)

Sec. 8.4: Progress Test 3

1. $\dfrac{dy}{dx} = \dfrac{D_x(2x^2 - x + 3)}{\sqrt{1 - (2x^2 - x + 3)^2}} = \dfrac{4x - 1}{\sqrt{-4x^4 + 4x^3 - 13x^2 + 6x - 8}}.$

2. $\dfrac{dy}{dx} = \dfrac{1}{|\sqrt{x^2+1}|\sqrt{(\sqrt{x^2+1})^2 - 1}} D_x(\sqrt{x^2+1}) = \dfrac{1}{\sqrt{x^2+1}\sqrt{x^2}} \cdot \dfrac{x}{\sqrt{x^2+1}} = \dfrac{x}{|x|(x^2+1)}.$

Sec. 8.5: Progress Test 1

1. $\displaystyle\int \frac{dx}{|x|\sqrt{2x^2 - 3}} = \frac{1}{\sqrt{2}}\int \frac{dx}{|x|\sqrt{x^2 - \frac{3}{2}}} = \frac{1}{\sqrt{2}} \cdot \frac{1}{\sqrt{\frac{3}{2}}} \operatorname{arcsec}\left(\frac{x}{\sqrt{\frac{3}{2}}}\right) + C = \frac{1}{\sqrt{3}} \operatorname{arcsec}\left(\frac{\sqrt{2}\,x}{\sqrt{3}}\right) + C.$ **2.** $1 - 2x - x^2 = 2 - (x+1)^2$, so $\displaystyle\int \frac{dx}{\sqrt{1 - 2x - x^2}} = \int \frac{dx}{\sqrt{2 - (x+1)^2}}$. Let $u = x + 1$, $du = dx$; then

$\displaystyle\int \frac{dx}{\sqrt{1 - 2x - x^2}} = \int \frac{du}{\sqrt{2 - u^2}} = \arcsin\left(\frac{u}{\sqrt{2}}\right) + C = \arcsin\left(\frac{x+1}{\sqrt{2}}\right) + C.$

3. $\displaystyle\int_{-3}^{3/2} \frac{dx}{x^2 + 9} = \frac{1}{3}\arctan\left(\frac{x}{3}\right)\Big|_{-3}^{3/2} = \frac{1}{3}\arctan\left(\frac{1}{2}\right) - \frac{1}{3}\arctan(-1) = \frac{1}{3}\arctan\left(\frac{1}{2}\right) - \frac{1}{3}\left(\frac{-\pi}{4}\right) = \frac{1}{3}\arctan\left(\frac{1}{2}\right) + \frac{\pi}{12}.$

Sec. 8.5: Progress Test 2

1. Let $u = e^{2x}$, $du = 2e^{2x}\,dx$. Then $\displaystyle\frac{1}{2}\int \frac{du}{\sqrt{5 - u^2}} = \frac{1}{2}\arcsin\frac{u}{\sqrt{5}} + C = \frac{1}{2}\arcsin\frac{e^{2x}}{\sqrt{5}} + C.$

2. $\displaystyle\int(x^3 + 3x^2 - 5x - 17)\,dx + \int \frac{37x + 120}{x^2 + 7}\,dx = x^4/4 + x^3 - \frac{5x^2}{2} - 17x + (37/2)\ln(x^2 + 7) + (120/\sqrt{7})\arctan(x/\sqrt{7}) + C.$

Sec. 8.6: Progress Test 1

1. $1 - \tanh^2 x = 1 - \left(\dfrac{\sinh x}{\cosh x}\right)^2 = \dfrac{\cosh^2 x - \sinh^2 x}{\cosh^2 x} = \left(\dfrac{1}{\cosh x}\right)^2 = \operatorname{sech}^2 x.$ **2.** $D_x(\cosh x) = D_x\left(\dfrac{e^x + e^{-x}}{2}\right) = \dfrac{e^x - e^{-x}}{2} = \sinh x$, so [8.6.4(*b*)] follows by Chain Rule. **3.** Result follows since $\cosh x + \sinh x = e^x$. **4.** $\coth x = \dfrac{1}{1/2} = 2$, $\operatorname{sech}^2 x = 1 - (1/2)^2 = 3/4$, so $\operatorname{sech} x = \sqrt{3}/2$. (Note $\operatorname{sech} x > 0$ for all x.) $\cosh x = \dfrac{1}{\sqrt{3}/2} = \dfrac{2}{\sqrt{3}}$, $\sinh x = \sqrt{\cosh^2 x - 1} = \sqrt{1/3}$. (Note that $\tanh x = 1/2$ implies $x > 0$, so $\sinh x > 0$.) **5.** $\displaystyle\int(1 - \tanh^2 x)\operatorname{sech}^2 x\,dx = \tanh x - \frac{\tanh^3 x}{3} + C.$

Sec. 8.6: Progress Test 2

1. $\displaystyle\int \frac{dx}{\sqrt{x^2 + 6x + 3}} = \int \frac{dx}{\sqrt{(x+3)^2 - (\sqrt{6})^2}} = \operatorname{arccosh}\frac{x+3}{\sqrt{6}} + C = \ln(x + 3 + \sqrt{x^2 + 6x + 3}) + C'.$

2. $$\int \frac{-dx}{9x^2 - 12x} = \int \frac{dx}{4 - (3x-2)^2} = \begin{cases} \dfrac{1}{2}\operatorname{arctanh}\dfrac{3x-2}{2} + C \text{ if } 3x - 2 < 2 \\ \dfrac{1}{2}\operatorname{arccoth}\dfrac{3x-2}{2} + C \text{ if } 3x - 2 > 2 \end{cases}.$$

3. Here $u = \sin x$, $du = \cos x\,dx$, $a = 1$. $\displaystyle\int \frac{du}{\sqrt{1 + u^2}} = \operatorname{arcsinh}(\sin x) + C.$

Chapter 9: Solutions to Progress Tests

Sec. 9.1: Progress Test 1

1. (*a*) (*b*)

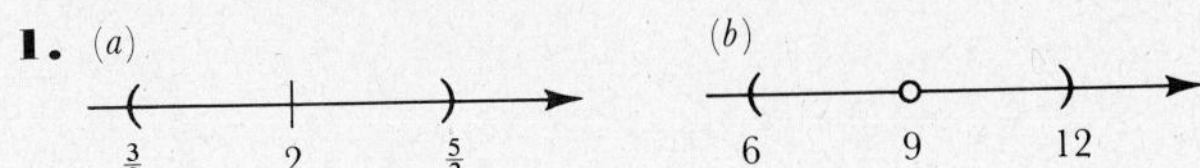

2. (*a*) $0 < |x - 2| < 3$; (*b*) $|x - 2| < 0.01$; (*c*) $0 < |x + 3| < 5$.
3. (*a*) If $|x - 2| < 0.1$, then $|(4x - 1) - 7| = |4x - 8| = 4|x - 2| < 4(0.1) = 0.4$; (*b*)

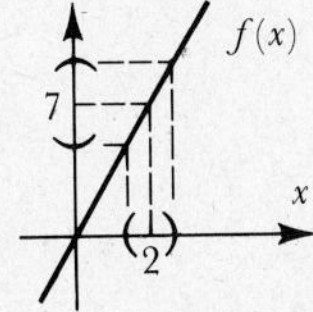

4. (a) $|f(x) - 10| = |3x + 1 - 10| = 3|x - 3| < 1$ if $|x - 3| < \frac{1}{3}$;
(b) $|f(x) - (-4)| = |2x - 2 + 4| = 2|x + 1| < \frac{1}{2}$ if $|x + 1| < \frac{1}{4}$; (c) $|f(x) - \frac{9}{8}| = |\frac{1}{2}x + 1 - \frac{9}{8}| = \frac{1}{2}|x - \frac{1}{4}| < \frac{1}{5}$ if $|x - \frac{1}{4}| < \frac{2}{5}$.

Sec. 9.1: Progress Test 2

1. Given $\varepsilon > 0$, we need $\delta > 0$ such that $0 < |x - (-1)| < \delta$ implies $|(1 - 2x) - 3| < \varepsilon$. But $|(1 - 2x) - 3| = 2|x + 1| = 2|x - (-1)| < \varepsilon$ if $|x - (-1)| < \varepsilon/2$. Choose $\delta = \varepsilon/2$. (See sketch 1.)

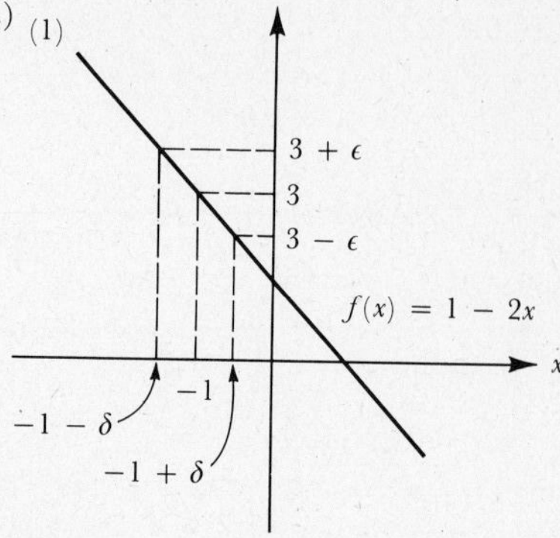

2. Suppose $\varepsilon > 0$ is given. Then we need $\delta > 0$ such that $0 < |x - 2| < \delta$ implies $|(x^2 - 4)/(x - 2) - 4| < \varepsilon$. But $0 < |x - 2|$ implies $x \neq 2$; for such x, $(x^2 - 4)/(x - 2) = x + 2$, so we need $\delta > 0$ such that $|(x + 2) - 4| < \varepsilon$. But $|(x + 2 - 4)| = |x - 2|$, so $\delta = \varepsilon$ works. (See sketch 2.)

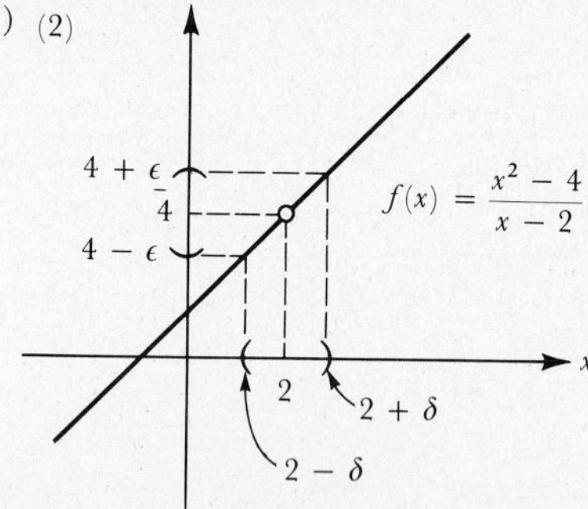

3. Suppose $f(x)$ is defined for all x in some open interval (a, c). We say the limit of $f(x)$ as x tends to a from the right is L, and we write $\lim_{x\to a^+} f(x) = L$, if for every $\varepsilon > 0$, there is a number $\delta > 0$ such that $a < x < a + \delta$ implies $|f(x) - L| < \varepsilon$. **4.** For $x > 0$, $f(x) = x + 2.1$. If $\varepsilon > 0$, then we need $\delta > 0$ such that $0 < x < \delta$ implies $|(x + 2.1) - 2.1| < \varepsilon$. But $|(x + 2.1) - 2.1| = |x|$, so $\delta = \varepsilon$ works.

Sec. 9.2: Progress Test 1

1. Let $\varepsilon > 0$ be given. Since $\lim_{x\to a} f(x) = L$, there is $\delta > 0$ such that $0 < |x - a| < \delta$ implies $|f(x) - L| < \varepsilon$. But $||f(x)| - |L|| \leq |f(x) - L|$, so $0 < |x - a| < \delta$ implies $||f(x)| - |L|| < \varepsilon$ also. **2.** By (1) $\lim_{x\to a} f(x) = 0$ implies $\lim_{x\to a} |f(x)| = 0$. Conversely, assume $\lim_{x\to a} |f(x)| = 0$ and let $\varepsilon > 0$ be given. Then there is $\delta > 0$ such that $0 < |x - a| < \delta$ implies $||f(x)| - 0| < \varepsilon$. But then $||f(x)| - 0| = ||f(x)|| = |f(x)| < \varepsilon$ also, so $\lim_{x\to a} f(x) = 0$. **3.** (a) $(\sqrt{x} - \sqrt{a})\left(\dfrac{\sqrt{x} + \sqrt{a}}{\sqrt{x} + \sqrt{a}}\right) = \dfrac{x - a}{\sqrt{x} + \sqrt{a}}$; (b) distance between x and a less than $\delta_1 = a/2$ implies (i) $x > a/2$, which implies $\sqrt{x} > \sqrt{a/2}$; (c) $k < \sqrt{a} + \sqrt{x}$ implies $\dfrac{1}{k} > \dfrac{1}{\sqrt{a} + \sqrt{x}}$, so $\dfrac{|x - a|}{k} > \dfrac{|x - a|}{\sqrt{a} + \sqrt{x}} = |\sqrt{x} - \sqrt{a}|$; (d) $0 < |x - a| < \delta_2 = \varepsilon k$ implies: (ii) $|\sqrt{x} - \sqrt{a}| < \dfrac{\varepsilon k}{k} = \varepsilon$; (e) Choose δ = smaller of δ_1 and δ_2 so both (i) and (ii) hold, proving $\lim_{x\to a} \sqrt{x} = \sqrt{a}$.

Chapter 10: Solutions to Progress Tests

Sec. 10.1: Progress Test 1

1. $y = 2x^2 \Leftrightarrow x^2 = \frac{1}{2}y \Leftrightarrow x^2 = 4(\frac{1}{8})y$. From Table 1(c) $p = \frac{1}{8}$, vertex = (0, 0), focus = $(0, \frac{1}{8})$, directrix: $y = -\frac{1}{8}$.

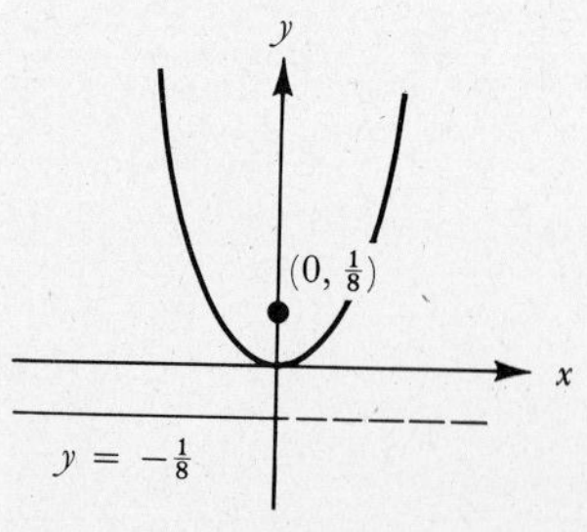

2. $y^2 = 4x = 4(1)x$. From Table 1(*a*) $p = 1$, vertex $= (0, 0)$, focus $= (1, 0)$, directrix: $x = -1$.

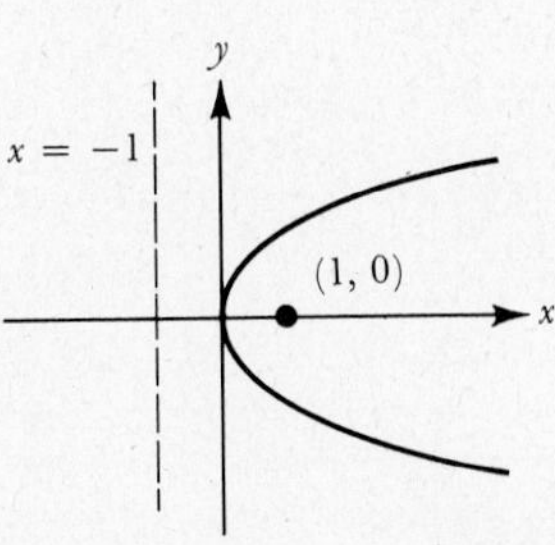

3. $y^2 - 4y = -8x - 28 \Leftrightarrow y^2 - 4y + 4 = -8x - 24 \Leftrightarrow (y - 2)^2 = -8(x + 3) \Leftrightarrow (y - 2)^2 = -4(2)(x + 3) \Leftrightarrow Y^2 = -4(2)X$, where $Y = y - 2$, $X = x + 3$. From Table 1(*b*) $p = 2$ and in *X-Y* system, vertex $= (0, 0)$, focus $= (-2, 0)$, directrix: $X = 2$. Thus in *X-Y* system, vertex $= (-3, 2)$, focus $= (-5, 2)$, directrix: $x = 0$.

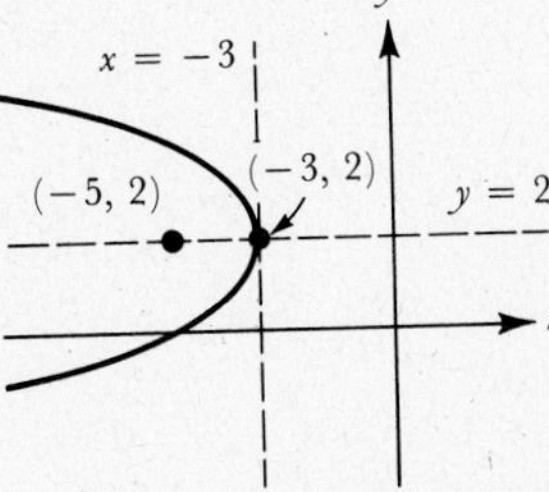

4. Since the focus is 3 units above the vertex on the line $x = 4$, the directrix is 3 units below the vertex and so is the line $y = -6$. Since $p = 3$ and the parabola opens upward, its equation is $X^2 = 4(3)Y$, where $X = x - 4$ and $Y = y + 3$. That is, $(x - 4)^2 = 12(y + 3)$.

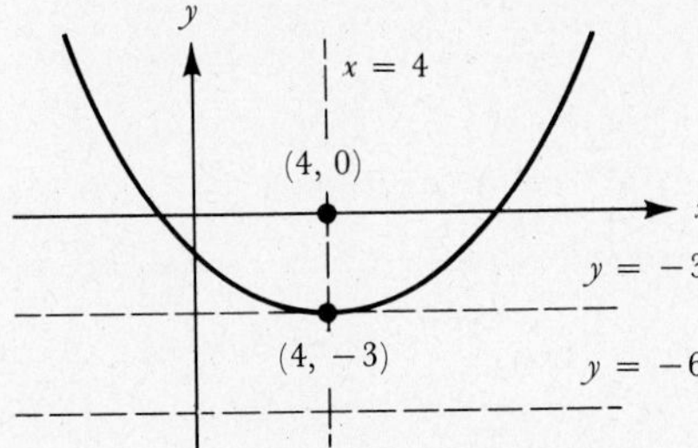

Sec. 10.2: Progress Test 1

1. Since the length of major axis is 16, $2a = 16 \Rightarrow a = 8$ and since $c = 4$, $b^2 = 64 - 16 = 48$. Hence equation is $x^2/64 + y^2/48 = 1$.
2. $(x - 8)^2/64 + (y + 6)^2/48 = 1$. **3.** Factor out 16 and 25 and then complete square as indicated:
$16(x^2 - 6x + 9) + 25(y^2 - 8y + 16) = 106 + 16 \cdot 9 + 25 \cdot 16 = 400$, or $(x = 3)^2/5^2 + (y - 4)^2/4^2 = 1$. Hence center $(3, 4)$, foci at $(6, 4)$, $(0, 4)$. (See sketch.)

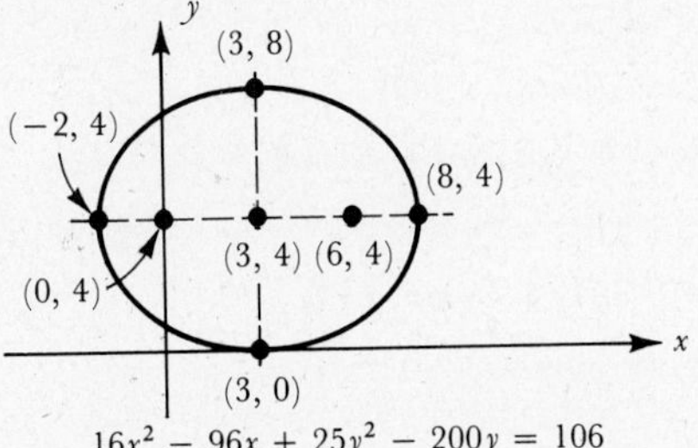

Sec. 10.2: Progress Test 2

$9x^2 - 36x - 4y^2 - 8y = 112 \Leftrightarrow 9(x^2 - 4x \quad\quad) - 4(y^2 - 2y \quad\quad) = 112 \Leftrightarrow 9(x^2 - 4x + 4) - 4(y^2 - 2y + 1) = 112 + 36 - 4 = 144 \Leftrightarrow (x - 2)^2/16 - (y - 1)^2/36 = 1$. Thus $a = 4$, $b = 6$, and $c = \sqrt{a^2 + b^2} = \sqrt{52} = 2\sqrt{13}$. Center $(2, 1)$, foci $(2 + 2\sqrt{13}, 1)$, $(2 - 2\sqrt{13}, 1)$, vertices $(6, 1)$, $(-2, 1)$. See sketch next page.

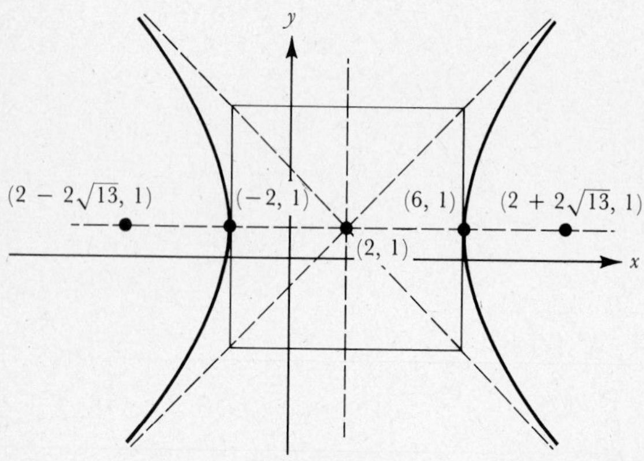

Sec. 10.3: Progress Test 1

1. $(\sqrt{3}/2), (-1/2), (\frac{1}{2}, \sqrt{3}/2), (2, 0), (\sqrt{3}, 1), ((\sqrt{3}+1)/2, (\sqrt{3}-1)/2)$. **2.** $Y - 2 = 4X$. **3.** $Y^2 - \sqrt{3}XY + X + \sqrt{3}Y = 10$.

Sec. 10.3: Progress Test 2

Since $A = 8$, $B = -4$, and $C = 5$, we apply (10.3.6.) to obtain $\cot 2\alpha = -3/4$, $\cos 2\alpha = -3/5$, so $\sin \alpha = \sqrt{\dfrac{1 + 3/5}{2}} = \dfrac{2}{\sqrt{5}}$ and $\cos \alpha = \sqrt{\dfrac{1 - 3/5}{2}} = \dfrac{1}{\sqrt{5}}$. Let $x = (1/\sqrt{5})\,(X - 2Y)$, $y = (1/\sqrt{5})\,(2X + Y)$, substitute and simplify to get $4X^2 + 9Y^2 = 36$, or $X^2/9 + Y^2/4 = 1$. The graph is an ellipse with major axis on the X axis.

Chapter 11: Solutions to Progress Tests

Sec. 11.1: Progress Test 1

1. $u = x,\ dv = e^x\,dx;\ du = dx,\ v = e^x$: $\int xe^x\,dx = xe^x - \int e^x\,dx = xe^x - e^x + C$.

2. $u = x,\ dv = \sec^2 x\,dx;\ du = dx,\ v = \tan x$: $\int x \sec^2 x\,dx = x\tan x - \int \tan x\,dx = x \tan x - \ln|\sec x| + C$.

3. $u = \arcsin x,\ dv = dx;\ du = \dfrac{dx}{\sqrt{1-x^2}},\ v = x$: $\displaystyle\int \arcsin x\,dx = x \arcsin x - \int x\frac{dx}{\sqrt{1-x^2}} = x \arcsin x + \sqrt{1-x^2} + C$.

Sec. 11.1: Progress Test 2

1. $u = \sin(\ln x),\ dv = dx;\ du = \dfrac{\cos(\ln x)}{x}\,dx,\ v = x$: $\int \sin(\ln x)\,dx = x\sin(\ln x) - \int \cos(\ln x)\,dx$; $\bar{u} = \cos(\ln x),\ d\bar{v} = dx;\ d\bar{u} = -\dfrac{\sin(\ln x)\,dx}{x}$; $\bar{v} = x$: $\int \sin(\ln x)\,dx = x\sin(\ln x) - [x\cos(\ln x) + \int \sin(\ln x)\,dx] = x\sin(\ln x) - x\cos(\ln x) - \int \sin(\ln x)\,dx$. Thus $\int \sin(\ln x)\,dx = \dfrac{x}{2}[\sin(\ln x) - \cos(\ln x)] + C$. **2.** $u = \ln x,\ dv = 3x^3 + x;\ du = \dfrac{dx}{x},\ v = \dfrac{3x^4}{4} + \dfrac{x^2}{2}$. $\displaystyle\int(3x^3 + x)\ln x\,dx = \left[\frac{3x^4}{4} + \frac{x^2}{2}\right]\ln x - \int\left(\frac{3x^3}{4} + \frac{x}{2}\right)dx = \left[\frac{3x^4}{4} + \frac{x^2}{2}\right]\ln x - \frac{3x^4}{16} - \frac{x^2}{4} + C$. **3.** $u = x^2,\ dv = \sin 2x\,dx;\ du = 2x\,dx$, $v = -\dfrac{1}{2}\cos 2x$. $\int x^2 \sin 2x\,dx = \dfrac{-x^2}{2}\cos 2x + \dfrac{2}{2}\int x\cos 2x\,dx$. $\bar{u} = x,\ d\bar{v} = \cos 2x\,dx;\ d\bar{u} = dx,\ \bar{v} = \dfrac{1}{2}\sin 2x$. $\int x^2\sin 2x\,dx = -\dfrac{x^2}{2}\cos 2x + \dfrac{x}{2}\sin 2x - \dfrac{1}{2}\int \sin 2x\,dx = -\dfrac{x^2}{2}\cos 2x + \dfrac{x}{2}\sin 2x + \dfrac{1}{4}\cos 2x + C$.

Sec. 11.2: Progress Test 1

1. Let $x = \sqrt{7}\sin\theta$, $dx = \sqrt{7}\cos\theta\,d\theta$, and $\sqrt{7 - x^2} = \sqrt{7 - 7\sin^2\theta} = \sqrt{7\cos^2\theta} = \sqrt{7}\cos\theta$. $\displaystyle\int\frac{x^3\,dx}{\sqrt{7-x^2}} = \int\frac{7\sqrt{7}\sin^3\theta\,\sqrt{7}\cos\theta\,d\theta}{\sqrt{7}\cos\theta} = 7\sqrt{7}\int\sin^3\theta\,d\theta = 7\sqrt{7}\int(1-\cos^2\theta)\sin\theta\,d\theta = 7\sqrt{7}\left(\frac{\cos^3\theta}{3} - \cos\theta\right) + C = 7\sqrt{7}\left(\frac{(7-x^2)^{3/2}}{21\sqrt{7}} - \frac{\sqrt{7-x^2}}{\sqrt{7}}\right) + C = \frac{(7-x^2)^{3/2}}{3} - 7\sqrt{7-x^2} + C$.

$\sqrt{7}$ $\quad x \quad \theta \quad \sqrt{7 - x^2}$

2. Let $x = 2\sin\theta$, $dx = 2\cos\theta\,d\theta$, and $\sqrt{4 - x^2} = \sqrt{4 - 4\sin^2\theta} = \sqrt{4\cos^2\theta} = 2\cos\theta$. $x = 0$ implies

$\theta = \arcsin(0/2) = 0$ and $x = 2$ implies $\theta = \arcsin(2/2) = \pi/2$. Thus $\int_0^2 \sqrt{4 - x^2}\, dx = \int_0^{\pi/2} 2\cos\theta \cdot 2\cos\theta\, d\theta = 4\int_0^{\pi/2} \cos^2\theta\, d\theta =$ $4/2 \int_0^{\pi/2} (1 + \cos 2\theta)\, d\theta = 2[\theta + (1/2)\sin 2\theta]_0^{\pi/2} = 2(\pi/2 + (1/2)\sin\pi) - (0 + (1/2)\sin 0) = \pi$.

2 x θ $\sqrt{4 - x^2}$

Sec. 11.2: Progress Test 2

1. Let $x = 4\sec\theta$, $dx = 4\sec\theta\tan\theta\, d\theta$, $\sqrt{x^2 - 16} = 4\tan\theta$. $\int \frac{\sqrt{x^2 - 16}}{x}\, dx = \int \frac{4\tan\theta\ 4\sec\theta\tan\theta\, d\theta}{4\sec\theta} = 4\int \tan^2\theta\, d\theta = 4(\tan\theta - \theta) +$ $C = 4\left[\frac{\sqrt{x^2 - 16}}{4}\right] - \operatorname{arcsec}\left(\frac{x}{4}\right) + C.$

x $\sqrt{x^2 - 16}$ θ 4

2. $\int \sqrt{6x - x^2}\, dx = \int \sqrt{9 - (x - 3)^2}\, dx[(u = x - 3)] = \int \sqrt{9 - u^2}\, du\ [(u = 3\sin\theta,\ du = 3\cos\theta\, d\theta,$ $\sqrt{9 - u^2} = 3\cos\theta)] = 9\int \cos^2\theta\, d\theta = \frac{9}{2}\int [1 + \cos 2\theta]\, d\theta = \frac{9}{2}[\theta + \frac{1}{2}\sin 2\theta] + C = \frac{9}{2}[\theta + \sin\theta\cos\theta] + C =$ $\frac{9}{2}\arcsin\left(\frac{u}{3}\right) + \frac{u\sqrt{9 - u^2}}{9} + C = \frac{9}{2}\left[\arcsin\left(\frac{x - 3}{3}\right) + \frac{(x - 3)\sqrt{6x - x^2}}{9}\right] + C.$

3 u θ $\sqrt{9 - u^2}$

3. Let $x = a\tan\theta$, $dx = a\sec^2\theta\, d\theta$, then $x^2 + a^2 = a^2\tan^2\theta + a^2 = a^2(\tan^2\theta + 1) = a^2\sec^2\theta$.
$\int \frac{dx}{(x^2 + a^2)^2} = \int \frac{a\sec^2\theta\, d\theta}{(a^2\sec^2\theta)^2} = \frac{1}{a^3}\int \frac{d\theta}{\sec^2\theta}\cos^2\theta\, d\theta.$

Sec. 11.3: Progress Test 1

1. $\frac{x^5 - 8x^3 - 8x + 1}{x^3 - 9x} = x^2 + 1 + \frac{x + 1}{x(x + 3)(x - 3)}$. Now $\frac{x + 1}{x(x + 3)(x - 3)} = \frac{A_1}{x} + \frac{A_2}{x + 3} + \frac{A_3}{x - 3} =$ $\frac{A_1(x + 3)(x - 3) + A_2x(x - 3) + A_3x(x + 3)}{x(x + 3)(x - 3)}$, so $x + 1 = A_1(x + 3)(x - 3) + A_2x(x - 3) + A_3x(x + 3)$. Now $x = 0 \Rightarrow 1 = -9A_1 \Rightarrow A_1 = -1/9$, $x = 3 \Rightarrow 4 = 18A_3 \Rightarrow A_3 = 2/9$, $x = -3 \Rightarrow -2 = 18A_2 \Rightarrow A_2 = -1/9$. Thus $\int (x^2 + 1)\, dx + \int \left[-\frac{1}{9}\cdot\frac{1}{x} - \frac{1}{9}\cdot\frac{1}{x + 3} + \frac{2}{9}\cdot\frac{1}{x - 3}\right] dx = \frac{x^3}{3} + x - \frac{1}{9}\ln|x| - \frac{1}{9}\ln|x + 3| + \frac{2}{9}\ln|x - 3| + C.$

2. $\frac{4x + 2}{2x^2 + 4x - 6} = \frac{2x + 1}{x^2 + 2x - 3} = \frac{2x + 1}{(x + 3)(x - 1)} = \frac{A_1}{x + 3} + \frac{A_2}{x - 1} = \frac{A_1(x - 1) + A_2(x + 3)}{(x + 3)(x - 1)}$, so $2x + 1 = A_1(x - 1) + A_2(x + 3)$ and, letting $x = -3, +1$, we get $-5 = 3A_1$, $-1 = 4A_2$, so $A_1 = 5/4$, $A_2 = 3/4$, with $\int \left[\frac{5}{4}\frac{1}{x + 3} + \frac{3}{4}\cdot\frac{1}{x - 1}\right] dx = \frac{5}{4}\ln|x + 3| + \frac{3}{4}\ln|x - 1| + C.$

Sec. 11.3: Progress Test 2

1. $\frac{x^4 + x}{x^3 - x^2} = x + \frac{x^2 + 1}{x^2(x - 1)}$. Now $\frac{x^2 + 1}{x^2(x - 1)} = \frac{A_1}{x} + \frac{A_2}{x^2} + \frac{A_3}{x - 1} = \frac{A_1x(x - 1) + A_2(x - 1) + A_3x^2}{x^2(x - 1)}$, so $x^2 + 1 = A_1x(x - 1) + A_2(x - 1) + A_3x^2$. Then $x = 0 \Rightarrow A_2 = -1$, and $x = 1 \Rightarrow A_3 = 2$. Then for $x = -1$ (a number $\neq 0, \neq 1$), we get $2 = 2A_1 - 2A_2 + A_3 = 2A_1 + 4$, so $A_1 = -1$, $\int \frac{x^2 + 1}{x^2(x - 1)}\, dx = \int x\, dx + \int \left[\frac{-1}{x} - \frac{1}{x^2} + \frac{2}{x - 1}\right] dx = \frac{x^2}{2} - \ln|x| + \frac{1}{x} + 2\ln|x - 1| + C.$

Sec. 11.3: Progress Test 3

1. $\frac{x + 1}{(x - 1)(x^2 + x + 1)} = \frac{A_1}{x - 1} + \frac{A_2x + A_3}{x^2 + x + 1} = \frac{A_1(x^2 + x + 1) + (A_2 x + A_3)(x - 1)}{(x - 1)(x^2 + x + 1)}$, so $x + 1 = A_1(x^2 + x + 1) + (A_2x + A_3)(x - 1)$. $x = 1$ implies $2 = 3A_1$, so $A_1 = 2/3$, with two unknowns yet to determine. Substitute any two numbers $\neq 1$, say $x = 0$, $x = -1$, to obtain $1 = A_1 - A_3$, $0 = A_1 + 2A_2 - 2A_3$. Now $A_1 = 2/3$ implies $A_3 = A_1 - 1 = 1/3$, and $A_2 = (2A_3 - A_1)/2 = -2/3$. Thus
$\int \frac{(x + 1)\, dx}{(x - 1)(x^2 + x + 1)} = \int \left[\frac{2}{3}\cdot\frac{1}{x - 1} - \frac{1}{3}\cdot\frac{2x + 1}{x^2 + x + 1}\right] dx = \frac{2}{3}\ln|x - 1| - \frac{1}{3}\ln(x^2 + x + 1) + C.$

Sec. 11.3: Progress Test 4

1. $\dfrac{x^4 + 5x^3 + 10x^2 + 8x + 1}{(x-1)(x^2+2x+2)^2} = \dfrac{A_1}{x-1} + \dfrac{A_2x + A_3}{x^2+2x+2} + \dfrac{A_4x + A_5}{(x^2+2x+2)^2} =$

$\dfrac{A_1(x^2+2x+2)^2 + (A_2x + A_3)(x-1)(x^2+2x+2) + (A_4x + A_5)(x-1)}{(x-1)(x^2+2x+2)^2}$. Multiplying out and equating numerators: $x^4 + 5x^2 + 10x^2 + 8x + 1 = (A_1 + A_2)x^4 + (4A_1 + A_2 + A_3)x^3 + (8A_1 + A_3 + A_4)x^2 + (8A_1 - 2A_2 - A_4 + A_5)x + (4A_1 - 2A_3 - A_5)$. Hence

$$\begin{aligned}
A_1 + A_2 \qquad\qquad\qquad &= 1\\
4A_1 + A_2 + A_3 \qquad\qquad &= 5\\
8A_1 \qquad + A_3 + A_4 \qquad &= 10\\
8A_1 - 2A_2 \qquad - A_4 + A_5 &= 8\\
4A_1 \qquad - 2A_3 \qquad - A_5 &= 1.
\end{aligned}$$

This system solves to yield $A_1 = 1, A_2 = 0, A_3 = 1, A_4 = 1, A_5 = 1$, so $\displaystyle\int \frac{(x^4 + 5x^3 + 10x^2 + 8x + 1)\,dx}{(x-1)(x^2+2x+2)^2} =$

$\displaystyle\int\left[\frac{1}{x-1} + \frac{1}{x^2+2x+2} + \frac{x+1}{(x^2+2x+2)^2}\right]dx = \ln|x-1| + \int\frac{dx}{(x+1)^2+1} + \int\frac{x+1}{(x^2+2x+2)^2}\,dx = \ln|x-1| +$ $\arctan(x+1) - \dfrac{1}{2}\dfrac{1}{x^2+2x+2} + C.$

Sec. 11.4: Progress Test 1

1. Let $u = \sqrt{x+1}$; then $u^2 = x + 1$, so $2u\,du = dx$ and $x = u^2 - 1$. $\displaystyle\int\frac{dx}{x(\sqrt{x+1} - 2)} = \int\frac{2u\,du}{(u^2-1)(u-2)} = \int\frac{2u\,du}{(u-1)(u+1)(u-2)}.$

2. Let $u = (2x+1)^{1/6}$; then $u^6 = 2x + 1$ and $6u^5\,du = 2\,dx$. Also, $u^3 = (2x+1)^{3/6} = \sqrt{2x+1}$ and $u^2 = (2x+1)^{2/6} = \sqrt[3]{2x+1}$. Thus $\displaystyle\int\frac{\sqrt{2x+1}\,dx}{1 + \sqrt[3]{2x+1}} = \int\frac{u^3(3u^5)\,du}{1+u^2} = 3\int\frac{u^8\,du}{u^2+1}.$

Sec. 11.4: Progress Test 2

1. Let $u = \sqrt{2 - x^2}$. Then $u^2 = 2 - x^2$ and $-u\,du = x\,dx$; $\displaystyle\int\frac{(x^3 - x)\,dx}{\sqrt{2-x^2}} = \int\frac{(x^2-1)\,x\,dx}{\sqrt{2-x^2}} = \int\frac{[(2-u^2)-1](-u\,du)}{u} =$ $\displaystyle\int(u^2 - 1)\,du = \frac{u^3}{3} - u + C = \frac{(2-x^2)^{3/2}}{3} - \sqrt{2-x^2} + C.$ **2.** Let $u = \sqrt{x^2 - 5}$. Then $u^2 = x^2 - 5$ and $u\,du = x\,dx$

$\displaystyle\int x^3\sqrt{x^2-5}\,dx = \int x^2\sqrt{x^2-5}\,x\,dx = \int(u^2+5)(u)(u\,du) = \int(u^4 + 5u^2)\,du = \frac{u^5}{5} + \frac{5u^3}{3} + C = \frac{(x^2-5)^{5/2}}{5} + \frac{5}{3}(x^2-5)^{2/3} + C.$

Sec. 11.4: Progress Test 3

1. Let $u = \tan(x/2)$. Then $dx = \dfrac{2\,du}{u^2+1}$ and $\cos x = \dfrac{1-u^2}{1+u^2}$, so $\displaystyle\int\frac{dx}{3 - 2\cos x} = \int\frac{\frac{2\,du}{u^2+1}}{3 - 2\left(\frac{1-u^2}{1+u^2}\right)} = \int\frac{2\,du}{5u^2+1} = \frac{2}{5}\int\frac{du}{u^2 + 1/5}.$

2. Let $u = \sqrt{1 + e^x}$. Then $u^2 = 1 + e^x$, $u^2 - 1 = e^x$, so $\ln(u^2 - 1) = x$ and $dx = \dfrac{2u\,du}{u^2-1}$. Thus $\displaystyle\int\frac{dx}{\sqrt{1+e^x}} = \int\frac{1}{u}\cdot\frac{2u\,du}{u^2-1} = \int\frac{du}{u^2-1}.$

3. Let $u = \sqrt{3x^5 + 5}$. Then $u^3 = 3x^5 + 5$, so $3u^2\,du = 15x^4\,dx$ and $\displaystyle\int\frac{\sqrt[3]{3x^5+5}}{x}\,dx = \int\frac{\sqrt[3]{3x^5+5}\,x^4\,dx}{x^5} = \int\frac{u\cdot 1/5\,u^2\,du}{(u^3-5)/3} = \frac{3}{5}\int\frac{u^3\,du}{u^3-5}.$

Sec. 11.5: Progress Test 1

Numbers in brackets refers to the appropriate formula. **1.** [20] with $a = 2$, $b = 3$: $\dfrac{1}{5}\cdot\dfrac{1}{\sqrt{2}}\ln\left|\dfrac{\sqrt{3x+2} - \sqrt{2}}{\sqrt{3x+2} + \sqrt{2}}\right| + C.$ **2.** [73] with $n = 4$, $u = x$: $\displaystyle\frac{-\sin^3 x\cos x}{4} + \frac{3}{4}\int\sin^2 x\,dx = \frac{-\sin^3 x\cos x}{4} + \frac{3}{4}\left(\frac{-\sin x\cos x}{2} + \frac{1}{2}\int dx\right) = \frac{-\sin^3 x\cos x}{4} - \frac{3\sin x\cos x}{8} + \frac{3}{8}x + C.$

3. [126] with $m = 2$, $n = 4$: $\dfrac{(1)(3)(1)}{(6)(4)(2)}\dfrac{\pi}{2} = \dfrac{\pi}{32}.$

Sec. 11.6: Progress Test 1

1. (*a*)

t_i	$f(t_i)$
1.0	1.0000
1.2	0.8333
1.4	0.7143
1.6	0.6250
1.8	0.5556
2.0	0.5000

$\ln 2 = \int_1^2 \frac{dt}{t} \approx \frac{1}{5}\left[\frac{1.0000}{2} + 0.8333 + 0.7143 + 0.6250 + 0.5556 + \frac{0.50000}{2}\right] = \frac{1}{5}[3.4782] = 0.6956$; (*b*) $f''(t) = \frac{2}{t^3}$ reaches its max. on [1, 2] at $t = 1$, with $f''(1) = 2$, so $|\ln 2 - T_5| \le \frac{(1)^3 \cdot 2}{12 \cdot 25} \approx 0.0067$; (*c*) to 4 places, $\ln 2 = 0.6931$, so $|\ln 2 - T_5| = 0.0025$ $\left[\text{Note } f(t) = \frac{1}{t} \text{ is concave up, so } T_5 \text{ is an } \textit{over}\text{estimate of } \ln 2.\right]$ **2.** $M = 2$, $b - a = 1$, so we need a positive integer n such that $\frac{1 \cdot 2}{12n^2} \le 0.000005$, or $\frac{1}{0.00003} \le n^2$, which implies $33333.3 \le n^2$, or $182.6 \le n$. Hence we need $n \ge 183$.

3. (*a*)

x_i	0.0	0.2	0.4	0.6	0.8	1.0	1.2	1.4	1.6	1.8	2.0
$f(x_i)$	1.0000	0.9608	0.8521	0.6977	0.5273	0.3679	0.2369	0.1409	0.0773	0.0392	0.0183

$\int_0^2 e^{-x^2}\,dx \approx \frac{2}{10}[4.4093] = 0.88186$; (*b*) $f''(x) = -2e^{-x^2}(1 - 2x^2)$ does not have an obvious max. on [0, 2], so we must use *its* derivative: $f'''(x) = 4xe^{-x^2}(3 - x^2)$. Since $f'''(x) > 0$ on $(0, \sqrt{3})$ and $f'''(x) > 0$ on $(\sqrt{3}, 2)$, f'' reaches a rel. max. at $x = \sqrt{3}$, with $|f''(\sqrt{3})| = |10e^{-3}| \approx 0.49787$. Now $b - a = 2$, $n = 10$, so $\left|\int_0^2 e^{-x^2}\,dx - T_{10}\right| \le \frac{2^3 \cdot (0.09957)}{12 \cdot 10^2} \approx 0.00332$.

Sec. 11.6: Progress Test 2

1. (*a*) Here $f(t) = 1/t$, $x_0 = 1$, $x_1 = 1.25$, $x_2 = 1.5$, $x_3 = 1.75$, $x_4 = 2$, so $\ln 2 \approx 1/(3 \cdot 4)\left[1 + 4 \cdot \frac{4}{5} + 2 \cdot \frac{2}{3} + 4 \cdot \frac{4}{7} + \frac{1}{2}\right] \approx \frac{1}{12}[8.31905] = 0.69325$; (*b*) $f^{(4)} = \frac{24}{t^5}$ is decreasing on [1, 2], so $M = f^{(4)}(1) = 24$. By (11.6.6), $|\ln 2 - 0.69325| \le \frac{1^5 \cdot 24}{180 \cdot 4^4} = \frac{1}{15 \cdot 2 \cdot 4^3} \approx 0.00053$; (*c*) (through 5 places) $|\ln 2 - 0.69325| = |-0.00010| < 0.00053$.

2. (*a*) $b - a = 1$, $M = 24$, so we need $\frac{24}{180n^4} < 0.000005$, or $\frac{2}{0.000075} < n^4$, or $26666.7 < n^4$. This implies $n > 12.8$. Since we need an even integer, choose $n = 14$ (considerably less than the $n = 183$ for the TR). (*b*) We need $\frac{24}{180n^4} < 5 \cdot 10^{-16}$, or $\frac{2 \cdot 10^{16}}{15 \cdot 5} < n^4$, or $(2.66)10^{14} < n^4$ or $\sqrt[4]{\frac{2}{75}} \cdot 10^4 < n$, which makes $n > (0.4041)10^4$, or $n > 4041$. Choose $n = 4042$. **3.** (*a*) We use the $f(x_i)$ from Sec. 11.6, Progress Test 1, to obtain $\int_0^2 e^{-x^2}\,dx \approx \frac{2}{30}[13.2387] \approx 0.88258$; (*b*) error $\le \frac{12 \cdot 2^5}{180 \cdot 10^4} \approx 0.000213$, which implies 3-place accuracy; (*c*) The TR guarantees 2-place accuracy. Although $f(x) = e^{-x^2}$ has a pt. of infl. in [0, 2], so much of the TR error cancels, SR still yields greater accuracy.

Chapter 12: Solutions to Progress Tests

Sec. 12.1: Progress Test 1

1. $t = x^2$, so $y = x^2 - 4$ (a parabola). Since $t \ge 0$, we know $x \ge 0$, so graph is rt. branch of parabola. **2.** $y = 4\tan^2 t = 4(\sec^2 t - 1)$ and $\sec t = x/2$, so $y = 4[(x/2)^2 - 1] = x^2 - 4$. Now $y = 4\tan^2 t \ge 0$ for all t. As t increases from $t = 0$ toward $\pi/2$, x and y both increase without bound from (2, 0). At $t = \pi/2$, x and y are not defined, but as t increases beyond $\pi/2$ to $t = \pi$, x and y decrease, with (x, y) moving down the left side of the parabola to $(-2, 0)$. **3.** $t = x + 1$, so $y = 2(x + 1) + 3 = 2x + 5$. No restrictions on t, so graph is entire straight line.

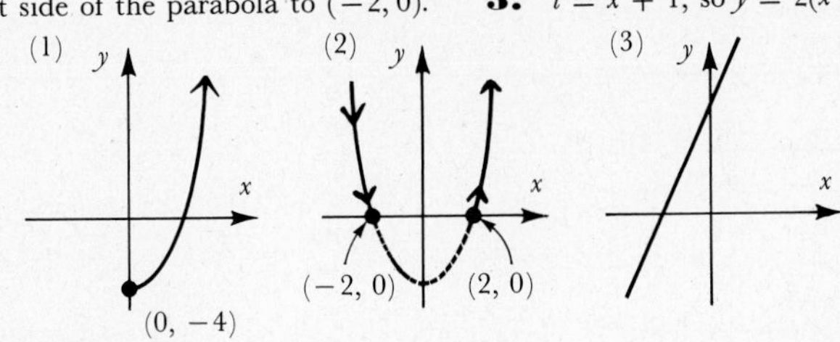

Sec. 12.1: Progress Test 2

1. $x'(t) = e^t, y'(t) = -e^{-t}, (x(\ln 3), y(\ln 3)) = (e^{\ln 3}, e^{-\ln 3}) = (3, 1/3)$. $m_0 = \dfrac{y'(\ln 3)}{x'(\ln 3)} = \dfrac{-e^{-\ln 3}}{e^{\ln 3}} = \dfrac{-1/3}{3} = \left(-\dfrac{1}{9}\right)$. Equation of tan. line is $y - 1/3 = -(1/9)(x - 3)$. **2.** $dx/dt = -2\csc^2 t$, $dy/dt = 4\sin t\cos t = 2\sin 2t$. Now dx/dt is never zero, so no vert. tangents, but $dy/dt = 0$ when $2t$ is a multiple of π. Since $0 < t < \pi$, $t = \pi/2$. Now $x'(\pi/2) = -2$, so a horiz. tangent at $t = \pi/2$, at the pt. $(0, 2)$. **3.** At $t = 0$ $x(t)$, hence $x'(t)$, does not exist.

Sec. 12.2: Progress Test 1

1. $x'(t) = -e^t\sin t + e^t\cos t, y'(t) = e^t\cos t + e^t\sin t$; $[x'(t)]^2 + [y'(t)]^2 = e^{2t}\sin^2 t - 2e^t\sin t\cos t + e^{2t}\cos^2 t + e^{2t}\cos^2 t + 2e^t\sin t\cos t + e^{2t}\sin^2 t = 2e^{2t}(\sin^2 t + \cos^2 t) = 2e^{2t}$. Therefore $s = \sqrt{2}\int_0^\pi e^t\,dt = \sqrt{2}\,e^t|_0^\pi = \sqrt{2}(e^\pi - 1)$.

2. $\displaystyle\int_1^3 \sqrt{1 + \left(\frac{x^2 - 16}{8x^2}\right)^2}\,dx = \int_1^3 \sqrt{\left(\frac{x^4 + 16}{8x^2}\right)^2}\,dx = \int_1^3 \left(\frac{x^2}{8} + \frac{2}{x^2}\right)dx = \frac{29}{12}$. **3.** $\displaystyle\int_{-1}^1 \sqrt{1 + \left[\frac{1}{2}(e^y + e^{-y})\right]^2}\,dy = \frac{1}{2}\int_{-1}^1 \sqrt{4 + (e^{2y} - 2 + e^{-2y})}\,dy = \frac{1}{2}\int_{-1}^1 \sqrt{(e^y + e^{-y})^2}\,dy = \frac{1}{2}[e^y - e^{-y}]_{-1}^1 = e - \frac{1}{e}$ (or $\int_{-1}^1 \sqrt{1 + \sinh^2 y}\,dy = \int_{-1}^1 \cosh y\,dy = [\sinh y]_{-1}^1$).

Sec. 12.2: Progress Test 2

1. $\displaystyle 2\pi\int_1^3 2\sqrt{x}\sqrt{1 + \left(\frac{1}{\sqrt{x}}\right)^2}\,dx = \frac{2}{3}4\pi[(x + 1)^{3/2}]_1^3 = \frac{8\pi}{3}(8 - \sqrt{8})$.

2. $\displaystyle 2\pi\int_0^2 y^3\sqrt{1 + (3y^2)^2}\,dy = 2\pi\cdot\frac{1}{36}\cdot\frac{2}{3}[(1 + 9y^4)^{3/2}]_0^2 = \frac{\pi}{27}[(\sqrt{145})^3 - 1]$.

3. $2\pi\int_0^\pi (2 + \cos t)\sqrt{\sin^2 t + \cos^2 t}\,dt = 2\pi[2t + \sin t]_0^\pi = 4\pi^2$.

Sec. 12.3: Progress Test 1

1. (*a*) Total moment $= 36$; (*b*) $\bar{x} = 36/26 = 18/13$. **2.** (*a*) $M_y = 36$ [from 1.(a)]; (*b*) $M_x = 64$; (*c*) $\bar{x} = 18/13$ [from 1.(b)], $\bar{y} = 64/26 = 32/13$. **3.** (*a*) $36 + 600 = 636$; (*b*) $636/26 = 318/13$. **4.** (*a*) $M_y = (1)(1) + (1)(2) + (1)(4) + (1)(5) = 12$, $M_x = (1)(4) + (1)(5) + (1)(7) + (1)(16) = 32$, $M = 4$, $\bar{x} = 12/4 = 3$, $\bar{y} = 32/4 = 8$; (*b*) $M_y = (2)(1) + (2)(2) + (2)(4) + (2)(5) = 2\cdot 12 = 24$, $M_x = 2\cdot 32 = 64$, $M = 8$, $\bar{x} = 24/8 = 3$, $\bar{y} = 64/8 = 8$; (*c*) $M_y = k\cdot 12$, $M_x = k\cdot 32$, $M = 4k$, $\bar{x} = \dfrac{k\cdot 12}{4k} = 3$, $\bar{y} = \dfrac{k\cdot 32}{4k} = 8$.

Sec. 12.3: Progress Test 2

1. $\displaystyle M_y = \int_0^4 x(4x - x^2)\,dx = 64/3$; $\displaystyle M_x = \frac{1}{2}\int_0^4 (4x - x^2)^2 = \frac{512}{30}$; $\displaystyle A = \int_0^4 (4x - x^2)\,dx = \frac{32}{3}$. Hence $\bar{x} = \dfrac{64/3}{32/3} = 2$, $\bar{y} = \dfrac{512/30}{32/3} = \dfrac{8}{5}$.

Sec. 12.3: Progress Test 3

1. $\displaystyle\frac{dy}{dx} = \frac{r}{h}$, $\displaystyle M_y = \int_0^h \frac{x\sqrt{h^2 + r^2}}{h}\,dx = \frac{\sqrt{h^2 + r^2}}{h}\frac{x^2}{2}\Big|_0^h = \frac{h}{2}\sqrt{h^2 + r^2}$. $\displaystyle M_x = \int_0^h \frac{r}{h}\frac{x\sqrt{h^2 + r^2}}{h}\,dx = \frac{r}{2}\sqrt{h^2 + r^2}$, $\displaystyle s = \int_0^h \frac{\sqrt{h^2 + r^2}}{h}\,dx = \sqrt{h^2 + r^2}$. Hence $\bar{x} = \dfrac{M_y}{s} = \dfrac{h}{2}$, $\bar{y} = \dfrac{M_x}{s} = \dfrac{r}{2}$. **2.** $\displaystyle M_{y-z} = \pi\int_0^h x\left(\frac{r}{h}x\right)^2 dx = \frac{\pi r^2}{h^2}\left[\frac{x^4}{4}\right]_0^h = \frac{\pi}{4}r^2h^2\cdot V = \frac{\pi}{3}r^2h$. Thus $\bar{x} = \dfrac{(\pi/4)\,r^2h^2}{(\pi/3)\,r^2h} = \dfrac{3}{4}h$. **3.** $\displaystyle M_{y-z} = 2\pi\int_0^h x\frac{r}{h}x\frac{\sqrt{h^2 + r^2}}{h}\,dx = \frac{2\pi r\sqrt{h^2 + r^2}}{h^2}\int_0^h x^2\,dx = \frac{2}{3}\pi rh\sqrt{h^2 + r^2}$.

$\displaystyle A = 2\pi\int_0^h \frac{r}{h}x\frac{\sqrt{h^2 + r^2}}{h}\,dx = 2\pi\frac{r}{2}\sqrt{h^2 + r^2}$. Thus $\bar{x} = \dfrac{2\pi\frac{r}{3}h\sqrt{h^2 + r^2}}{2\pi\frac{r}{2}\sqrt{h^2 + r^2}} = \dfrac{2}{3}h$. (Note $\bar{x}$ is different for the conical surface and solid.)

4. $\displaystyle\bar{y} = \frac{M_{x-z}}{s} = \frac{1}{s}[2\pi\int_c^d g(y)\sqrt{1 + [g'(y)]^2}\,dy]$. $\bar{x} = 0$ by symmetry.

Sec. 12.4: Progress Test 1

1. $k = 30/2 = 15$. $W = \int_2^8 15x\,dx = 450$ in.-lb. **2.**

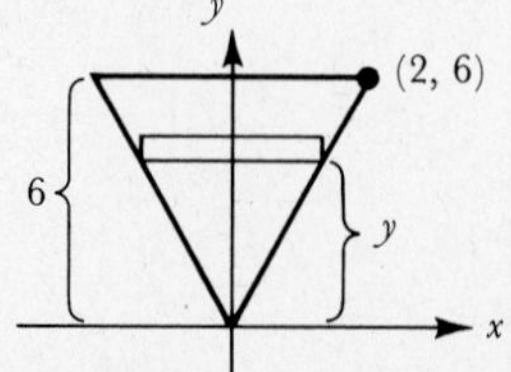

(*a*) The right side boundary of the cross section has equation $y = 3x$, so $x = y/3$ is the radius of a typical disk, which thus weighs $50 \cdot \frac{y^2}{9}\,\Delta y$; (*b*) $\left(50\pi\frac{y^2}{9}\,\Delta y\right)(18 - y)$; (*c*) $W = \frac{50}{9}\pi\int_0^{18} y^2(18 - y)\,dy = 50\pi(972) = 152{,}681.4$ ft-lb.
3. $\int_0^{100}(5x + 400)\,dx = 6500$ ft-lb. [With x feet remaining to lift, $(5x + 400)\,\Delta x$ is approximately work necessary to lift chain and weight Δx feet.]

Sec. 12.4: Progress Test 2

1. Equation of rt. side is $y = 3x$, so $x = f(y) = y/3$, and left side has equation $y = -3x$, so $x = g(y) = -y/3$. Hence total force is $50\int_0^6 (y/3 - (-y/3))(6 - y)\,dy = 50(24) = 1200$ lb. **2.** (*a*) $62.4\int_{15}^{18} 30(18 - y)\,dy = 62.4(135) = 8424$ lb; (*b*) key is new depth 18 instead of 15 ft. $F = 62.4[\int_0^{10} 20(18 - y)\,dy + \int_{10}^{15} 2y(18 - y)\,dy + \int_{15}^{18} 30(18 - y)\,dy] = 62.4[2600 + \frac{2000}{3} + 135] = 212{,}264$; (*c*) $212{,}264 - (143{,}000 + 8424) = 60{,}840$ lb, almost 43% increase.

Sec. 12.4: Progress Test 3

1. The total wt. is 50 times volume $= 50\left[\frac{\pi}{3}6^2 \cdot 18\right] = 50(216\pi)$. The centroid is $(3/4)h = 3/4 \cdot 18 = 27/2$, so oil is lifted 9/2 ft, making total work $= (9/2)(50)(216\pi) \approx 152{,}681.4$ ft-lb.
2. Centroid is at $(1/4)18 = 9/2$, so oil must be lifted $18 - 9/2 = 27/2$ ft. Total work $= (27/2)(50)(216\pi) \approx 458{,}044.2$ ft-lb.
3. Area $= 12$ ft^2, centroid is at depth $(1/3)(6) = 2$ ft. Pressure at 2 ft is $(50)(2) = 100$, so total force is 1200 lb.

Chapter 13: Solutions to Progress Tests

Sec. 13.1: Progress Test 1

1. (*a*)

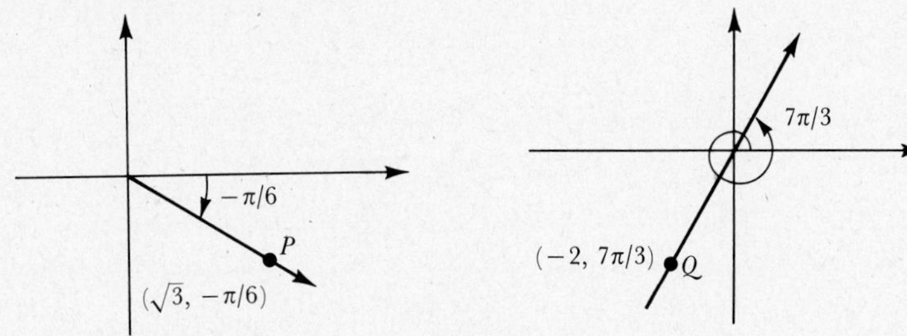

(*b*) $(\sqrt{3}, -\pi/6) = (\sqrt{3}, -\pi/6 \pm 2n\pi) = (-\sqrt{3}, 5\pi/6 \pm 2n\pi)$;
(*c*) $(-2, 7\pi/3) = (-2, \pi/3 \pm 2n\pi) = (2, 4\pi/3 \pm 2n\pi)$; (*d*) $x = \sqrt{3}\cos(-\pi/6) = 3/2$, $y = \sqrt{3}\sin(-\pi/6) = -\sqrt{3}/2$ and $x = -2\cos(7\pi/3) = -1$, $y = -2\sin(7\pi/3) = -\sqrt{3}$. So P has $(3/2, -\sqrt{3}/2)$ and Q has $(-1, -\sqrt{3})$ as r.c. **2.** Since $x = 1$, $y = \sqrt{3}$, we have $r^2 = x^2 + y^2 = 4$ and $\tan\theta = y/x = \sqrt{3}$. Since $(1, \sqrt{3})$ is in 1st quad., take $\theta = \pi/3$, and so $r = 2$. Thus R has p.c. $(2, \pi/3)$, or $(-2, 4\pi/3)$.

Sec. 13.1: Progress Test 2

1. Since $\cos(-\theta) = \cos\theta$, graph is symmetric about x axis and period of $\cos\theta$ is 2π, so only need consider $0 \le \theta \le \pi$.

θ:	$0 \to \pi/4$	$\pi/4 \to \pi/2$	$\pi/2 \to 3\pi/2$	$3\pi/2 \to \pi$
$\cos\theta$:	$1 \to 1/\sqrt{2}$	$1/\sqrt{2} \to 0$	$0 \to -1/\sqrt{2}$	$-1/\sqrt{2} \to -1$
$r = 2 + \cos\theta$:	$3 \to 2 + 1/\sqrt{2}$	$2 + 1/\sqrt{2} \to 2$	$2 \to 2 - 1/\sqrt{2}$	$2 - 1/\sqrt{2} \to 1$

2. Let $x^2 + y^2 = r^2$, $x = r\cos\theta$, $y = r\sin\theta$ and $(x^2 + y^2)^2 = 4xy$ becomes $r^4 = 4r^2\cos\theta\sin\theta$. Divide by r^2 and use identity $\sin 2\theta = 2\sin\theta\cos\theta$ to obtain $r^2 = 2\sin 2\theta$. This is graphed in a manner similar to Example 5.

(Symmetric about origin, no graph for $\pi/2 < \theta < \pi$.)

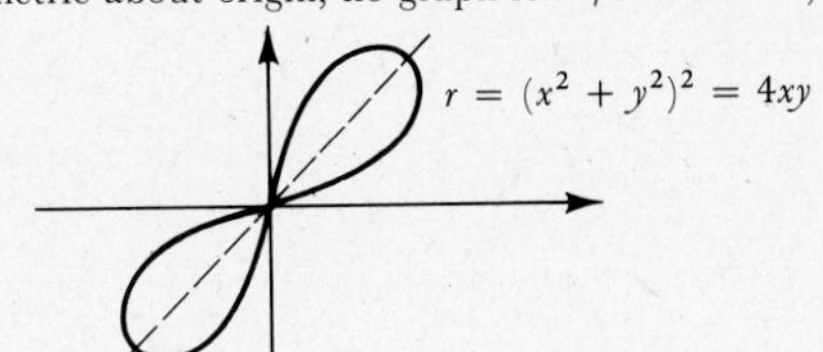

Sec. 13.2: Progress Test 1

1.

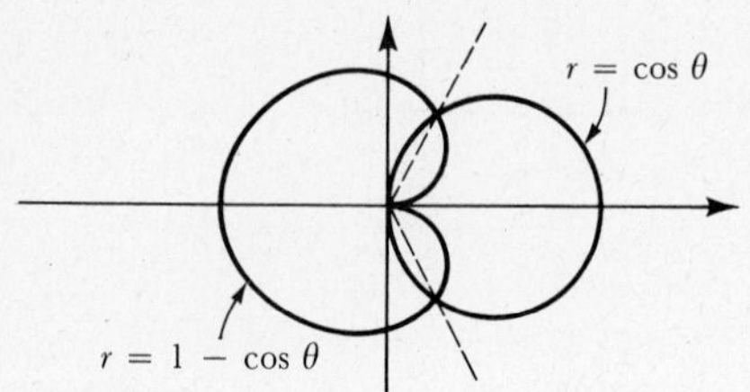

Setting $1 - \cos\theta = \cos\theta \Rightarrow 2\cos\theta = 1 \Rightarrow \cos\theta = 1/2 \Rightarrow \theta = \pm\pi/3$. Thus $(1/2, \pi/3)$, $(1/2, -\pi/3)$ are pts. of intersection. Also from graphs, curves go through origin.

2.

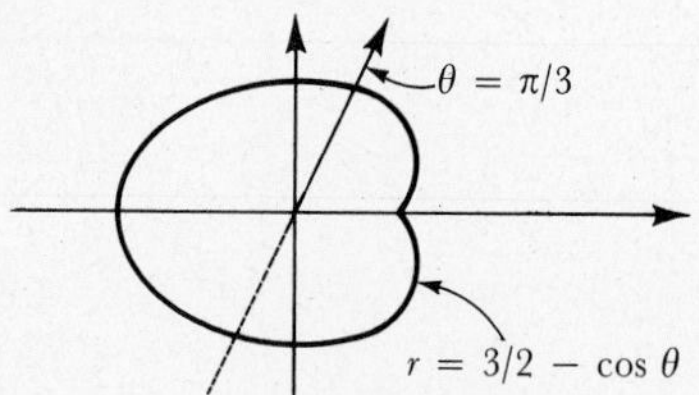

If $\theta = \pi/3$, then $r = 3/2 - \cos\pi/3 = 3/2 - 1/2 = 1$. Thus $(1, \pi/3)$ is on intersection. The pt. $(-2, \pi/3)$ is on $\theta = \pi/3$, but these coords. do not satisfy $r = 3/2 - \cos\theta$. However, $(-2, \pi/3)$ also has coordinates $(2, 4\pi/3)$ and $2 = 3/2 - \cos 4\pi/3$, so this pt. is on both graphs.

Sec. 13.2: Progress Test 2

1. $\tan\psi = (1 - \cos\theta)/(\sin\theta) = \tan\theta/2$. **2.** If ψ_0 is angle between tangent line at $(1, \pi/2)$ and ray $\pi/2$. By 1, $\psi_0 = \theta_0/2 = \pi/4$, so $\phi_0 = \theta_0 + \psi_0 = \pi/2 + \pi/4 = 3\pi/4$ and $\tan\psi_0 = \tan 3\pi/4 = -1$.

Sec. 13.3: Progress Test 1

1. $\frac{1}{2}\int_0^{2\pi} a^2\,d\theta = \frac{a^2}{2}[\theta]_0^{2\pi} = \frac{a^2}{2}(2\pi) = \pi a^2$. **2.** Set $2\cos 2\theta = 1$ so $\cos 2\theta = 1/2$. In 1st and 4th quadrants we have $2\theta = \pm\pi/3$, so $\alpha = -\pi/6$ and $\beta = +\pi/6$. Using x axis symmetry, we double area between 0 and $\pi/6$. By (13.3.3.),

$$A = 2\cdot\frac{1}{2}\int_0^{\pi/6}[(2\cos 2\theta)^2 - (1)^2]\,d\theta = \int_0^{\pi/6}(4\cos^2 2\theta - 1)\,d\theta = \int_0^{\pi/6}\left[4\left(\frac{1+\cos 4\theta}{2}\right) - 1\right]d\theta = \int_0^{\pi/6}(2\cos 4\theta - 1)\,d\theta = \sqrt{3}/2 - \pi/6.$$

Sec. 13.3: Progress Test 2

1. (See Fig. 13.6 for cardioid.) $f(\theta) = 1 + \cos\theta, f'(\theta) = -\sin\theta$, so $ds = \sqrt{(1+\cos\theta)^2 + (-\sin\theta)^2}\,d\theta = \sqrt{2 + 2\cos\theta}\,d\theta$. Now $(1 + \cos\theta)/2 = \cos^2(\theta/2)$, so $2 + 2\cos\theta = 4\cos^2(\theta/2)$. Thus $s = 2\int_0^\pi ds = 2\int_0^\pi\sqrt{4\cos^2(\theta/2)}\,d\theta = 4\int_0^\pi\cos(\theta/2)\,d\theta$ [since $\cos(\theta/2) \geq 0$ in $[0, \pi]$—check!] $= 8[\sin(\theta/2)]_0^\pi = 8$. **2.** $x = f(\theta)\cos\theta = (1 + \cos\theta)\cos\theta$. Using 1, $S = 2\pi\int_0^{\pi/2}x\,ds = 2\pi\int_0^{\pi/2}(1 + \cos\theta)\cos\theta\,2\cos(\theta/2)\,d\theta = 2\pi\int_0^{\pi/2}[2\cos^2(\theta/2)][2\cos^2(\theta/2) - 1][2\cos(\theta/2)]\,d\theta$ (applied $\cos^2(\theta/2) = (1 + \cos\theta)/2$).
3. $y = f(\theta)\sin\theta = (1 + \cos\theta)\sin\theta$, $ds = \sqrt{2 + 2\cos\theta}\,d\theta$, so $S = 2\pi\int_0^{\pi/2}y\,ds = 2\pi\int_0^{\pi/2}(1 + \cos\theta)\sin\theta\sqrt{2 + 2\cos\theta}\,d\theta$.

Chapter 14: Solutions to Progress Tests

Sec. 14.1: Progress Test 1

[The hypotheses of (14.1.2.) hold in each case.] **1.** $\lim_{x\to 0}\frac{2^x\ln 2 - e^x}{1} = \frac{1\ln 2 - 1}{1} = \ln 2 - \ln e = \ln(2/e)$. **2.** $\lim_{x\to 0}\frac{\sin x}{\cos x} = \frac{0}{1} = 0$.

3. Since $\lim_{x\to 0}(\cos x - \sec^2 x) = \lim_{x\to 0} 2x = 0$, use (14.1.2.) again: $\lim_{x\to 0}\frac{-\sin x - 2\sec x\sec x\tan x}{2} = \frac{0}{2} = 0$.

4. $\lim_{x\to+\infty}\frac{(-1/x^2)/(1 + 1/x)}{-1/x^2} = \lim_{x\to+\infty}\frac{1}{1 + 1/x} = 1$. **5.** $\lim_{h\to 0}\frac{e^h}{1} = 1$.

Sec. 14.1: Progress Test 2

[(14.1.4.) applies initially to each problem.] **1.** $\lim_{x\to(\pi/2)^-}\frac{\sec x\tan x}{\sec^2 x} = \lim_{x\to(\pi/2)^-}\cos x\frac{\sin x}{\cos x} = 1$. **2.** [Note $\ln(x - 1) \to -\infty$ as $x \to 1^+$.]

Differentiate, getting $\lim_{x\to 1^+}\frac{1/(x - 1)}{-1/(x - 1)^2} = \lim_{x\to 1^+}[-(x - 1)] = 0$. **3.** $\lim_{x\to+\infty}\frac{1 + 1/x}{1 + \ln x} = 0$ since numerator $\to 1$ and denominator $\to +\infty$.

Sec. 14.2: Progress Test 1

1. $(\infty - \infty)$: Consider $\lim_{x\to(\pi/2)^-}\left(\frac{\sin x}{\cos x} - \frac{1}{\cos x}\right) = \lim_{x\to(\pi/2)^-}\frac{\sin x - 1}{\cos x}$; $(0/0)$: $\lim_{x\to(\pi/2)^-}\frac{\cos x}{(-\sin x)} = \frac{0}{1} = 0$. **2.** $(\infty - \infty)$: Consider $\lim_{x\to1^+}\frac{x\ln x - x + 1}{(x-1)\ln x}$, $(0/0)$. Differentiate numerator and denominator to get $\lim_{x\to1^+}\frac{\ln x}{(x - 1 + x\ln x)/x} = \lim_{x\to1^+}\frac{x\ln x}{x - 1 + x\ln x}$, again $(0/0)$. Differentiate numerator and denominator to get $\lim_{x\to1^+}\frac{1+\ln x}{2+\ln x} = \frac{1}{2} = \lim_{x\to1^+}\left(\frac{x}{x-1} - \frac{1}{\ln x}\right)$. **3.** $(\infty\cdot 0)$. Write as $\lim_{x\to(\pi/2)^-}\frac{\ln(\sin x)}{\cot x}$, $(0/0)$. Differentiate to get $\lim_{x\to(\pi/2)^-}\frac{(\cos x)/(\sin x)}{-\csc^2 x} = \lim_{x\to(\pi/2)^-}(-\cos x\sin x) = 0\cdot 1 = 0 = \lim_{x\to(\pi/2)^-}\tan x\ln(\sin x)$.

Sec. 14.2: Progress Test 2

The limits are indeterminate of forms 1^∞, 0^0, 1^∞ respectively.

1. $\lim_{x\to+\infty}\left[x\ln\left(\frac{x+2}{x}\right)\right] = \lim_{x\to+\infty}\frac{\ln(1+2/x)}{1/x}\ [(0/0)] = \lim_{x\to+\infty}\frac{1/(1+2/x)(-2/x^2)}{-1/x^2} = \lim_{x\to+\infty}\frac{2}{1+2/x} = 2$, so $\lim_{x\to+\infty}\left(\frac{x+2}{x}\right)^x = e^2$.

2. $\lim_{x\to0^+}(\sin x\ln x) = \lim_{x\to0^+}\frac{\ln x}{1/\sin x}\ [(\infty/\infty)]\ \lim_{x\to0^+}\frac{1/x}{\cos x/\sin^2 x} = \lim_{x\to0^+}\frac{(\sin^2 x)/x}{\cos x} = 0$, so $\lim_{x\to0^+}x^{\sin x} = e^0 = 1$.

3. $\lim_{x\to1}\left[\frac{1}{1-x^2}\ln x\right][(0/0)]\ \lim_{x\to1}\frac{1/x}{-2x} = -\frac{1}{2}$, so $\lim_{x\to1}x^{1/(1-x^2)} = e^{-1/2} = 1/\sqrt{e}$.

Sec. 14.3: Progress Test 1

1. $\lim_{b\to+\infty}\int_1^b\frac{2\ln x}{x} = \lim_{b\to+\infty}[(\ln x)^2]_1^b = \lim_{b\to+\infty}(\ln b)^2 = +\infty$. **2.** $\lim_{a\to-\infty}\int_a^0 e^x\,dx = \lim_{a\to-\infty}[e^0 - e^a] = 1 - \lim_{a\to-\infty}e^a = 1 - 0 = 1$.

3. $\lim_{a\to-\infty}\left[-\frac{e^{-x^2}}{2}\right]_a^0 + \lim_{b\to+\infty}\left[-\frac{e^{-x^2}}{2}\right]_0^b = \lim_{a\to-\infty}\left[-\frac{1}{2} + \frac{e^{-a^2}}{2}\right] + \lim_{b\to+\infty}\left[-\frac{e^{-b^2}}{2} + \frac{1}{2}\right] = \frac{1}{2}\lim_{a\to-\infty}(e^{-a^2}) - \frac{1}{2}\lim_{b\to+\infty}(e^{-b^2}) =$ $\frac{1}{2}\lim_{a\to-\infty}\frac{1}{e^{a^2}} - \frac{1}{2}\lim_{b\to+\infty}\frac{1}{e^{b^2}} = 0 - 0 = 0$.

Sec. 14.3: Progress Test 2

1. $\lim_{\varepsilon\to0^+}\left[\arcsin\frac{x}{2}\right]_0^{2-\varepsilon} = \lim_{\varepsilon\to0^+}\left[\arcsin\frac{2-\varepsilon}{2}\right] = \frac{\pi}{2}$. **2.** $\lim_{\varepsilon\to0^+}\int_0^{1-\varepsilon}\frac{dx}{(x-1)^{4/5}} + \lim_{\delta\to0^+}\int_{1+\delta}^{33}\frac{dx}{(x-1)^{4/5}} = \lim_{\varepsilon\to0^+}5[(x-1)^{1/5}]_0^{1-\varepsilon} +$ $\lim_{\delta\to0^+}5[(x-1)^{1/5}]_{1+\delta}^{33} = 5\lim_{\varepsilon\to0^+}[-\varepsilon^{1/5} - (-1)^{1/5}] + 5\lim_{\delta\to0^+}[32^{1/5} - \delta^{1/5}] = 5(1) + 5(2) = 15$.

3. $\lim_{\varepsilon\to0^+}[\ln(4\varepsilon - \varepsilon^2) - \ln 4] + \lim_{\delta\to0^+}[\ln 12 - \ln(4\delta + \delta^2)]$. Neither limit is finite, so integral is divergent.

Chapter 15: Solutions to Progress Tests

Sec. 15.1: Progress Test 1

1. (*a*) 0, 1/4, 8/27, 81/256; (*b*) 0, 2, 0, 2; (*c*) 1, 0, 2, 0; (*d*) 1/2, 3/4, 7/8, 15/16. **2.** (*a*) Apply [15.1.2(*c*)] with $k = 3$ in numerator and $k = 2$ in denominator to get $a_n = \frac{n+3}{n+2}$; (*b*) each odd-numbered elt. is 1 and even numbered elt. is the reciprocal of the number designating the elt., so $a_n = \begin{cases}1 \text{ for } n \text{ odd}\\ 1/n \text{ for } n \text{ even}\end{cases}$; (*c*) the numerators consist of the odd integers $2n - 1$, and denominators consist of squares of the even integers $(2n)^2$, so $a_n = \frac{2n-1}{(2n)^2}$.

Sec. 15.1: Progress Test 2

1. By [15.1.6(*e*)] with $r = -\pi$ we have $\left(\frac{n}{n-\pi}\right)^n \to e^{-(-\pi)} = e^{\pi}$. **2.** By [15.1.6(*h*)] with $r = -2$ we have $\frac{(-1)^n 2^n}{n!} \to 0$.

3. $\frac{(-1)^n\,3^{n+2}}{4^n} = \frac{(-1)^n\,3^n\cdot 3^2}{4^n} = 9\left(\frac{-3}{4}\right)^n$. By [15.1.6(*g*)] with $r = -3/4$, $\left(\frac{-3}{4}\right)^n \to 0$. Hence $\frac{(-1)^n\,3^{n+2}}{4^n} = 9\left(\frac{-3}{4}\right)^n \to 9\cdot 0 = 0$.

4. By [15.1.6(*d*)] with $p = 0.01 > 0$, we have $\frac{\ln n}{n^{.01}} \to 0$. **5.** [15.1.6(*a*)] applies with $k = 2 < 3 = j$, so limit is 0.

6. $\sin n\pi = 0$ for all positive integers n. Thus the seq. is constantly 0 and has limit 0.

Sec. 15.2: Progress Test 1

1. $\frac{1}{(n+4)(n+5)} = \frac{1}{n+4} - \frac{1}{n+5}$; $s_1 = \left(\frac{1}{5} - \frac{1}{6}\right)$; $s_2 = \left(\frac{1}{5} - \frac{1}{6}\right) + \left(\frac{1}{6} - \frac{1}{7}\right) = \frac{1}{5} - \frac{1}{7}$;
$s_3 = \left(\frac{1}{5} - \frac{1}{6}\right) + \left(\frac{1}{6} - \frac{1}{7}\right) + \left(\frac{1}{7} - \frac{1}{8}\right) = \frac{1}{5} - \frac{1}{8}$; and $s_n = \left(\frac{1}{5} - \frac{1}{6}\right) + \left(\frac{1}{6} - \frac{1}{7}\right) + \left(\frac{1}{7} - \frac{1}{8}\right) + \cdots + \left(\frac{1}{n+4} - \frac{1}{n+5}\right) = \frac{1}{5} - \frac{1}{n+5}$.
Thus $\sum_{k=1}^{\infty} \frac{1}{(k+4)(k+5)} = \lim_{n\to+\infty} s_n = \lim_{n\to+\infty}\left(\frac{1}{5} - \frac{1}{n+5}\right) = \frac{1}{5}$. **2.** $s_1 = 1 + (-1) = 0$;
$s_2 = (1 + (-1)) + (1 + 1) = 2$; $s_3 = (1 + (-1)) + (1 + 1) + (1 + (-1)) = 2$; $s_4 = (1 + (-1)) + (1 + 1) + (1 + (-1)) + (1 + 1) = 4$.
0, 2, 2, 4, 4, 6, 6, ... Since $\lim_{n\to+\infty} s_n = +\infty$, the series $\sum_{k=1}^{\infty} (1 + (-1)^k)$ diverges. **3.** $\ln\left(\frac{j}{j+1}\right) = \ln j - \ln(j+1)$;
$s_1 = \ln 1 - \ln 2 = -\ln 2$; $s_2 = (\ln 1 - \ln 2) + (\ln 2 - \ln 3) = \ln 1 - \ln 3 = -\ln 3$; and in general
$s_n = (\ln 1 - \ln 2) + (\ln 2 - \ln 3) + \cdots + (\ln n - \ln(n+1)) = \ln 1 - \ln(n+1) = -\ln(n+1)$. Since $\lim_{n\to+\infty} s_n = \lim_{n\to+\infty} (-\ln(n+1) = -\infty$,
$\sum_{n=1}^{\infty} \ln\left(\frac{n}{n+1}\right)$ diverges.

Sec. 15.2: Progress Test 2

1. We know $\sum_{k=1}^{\infty} \frac{1}{k^2+k} = 1$ by Example 1, so $1 = \sum_{k=1}^{\infty} \frac{1}{k^2+k} = \frac{1}{2} + \frac{1}{6} + \sum_{k=3}^{\infty} \frac{1}{k^2+k}$. Thus $\sum_{k=3}^{\infty} \frac{1}{k^2+k} = 1 - \frac{1}{2} - \frac{1}{6} = \frac{1}{3}$. **2.** Since
$\lim_{n\to+\infty} \frac{\ln n}{n^p} = 0$ for $p > 0$, we have $\lim_{n\to+\infty} \frac{n^p}{\ln n} = +\infty$. Hence $\lim_{n\to+\infty} \left(\frac{\sqrt{n}}{\ln n}\right)^2 = +\infty$, so $\sum_{n=2}^{\infty} \left(\frac{\sqrt{n}}{\ln n}\right)^2$ diverges by (15.2.6.). **3.** $\lim_{n\to+\infty} \frac{n-1}{2n+5} = \frac{1}{2} \neq 0$,
so $\sum_{n=1}^{\infty} \frac{n-1}{2n+5}$ diverges by (15.2.6.). **4.** Since $\sum_{n=0}^{\infty} \frac{1}{2^n} = 2$, $\sum_{n=2}^{\infty} \frac{1}{2^n} = 2 - 1 - \frac{1}{2} = \frac{1}{2}$. Thus $\sum_{n=2}^{\infty} \frac{\pi}{2^n} = \pi \sum_{n=2}^{\infty} \frac{1}{2^n} = \frac{\pi}{2}$.

Sec. 15.2: Progress Test 3

1. $\sum_{n=1}^{\infty} \frac{3}{(3-\pi)^{n-1}} = \sum_{n=1}^{\infty} 3\left(\frac{1}{3-\pi}\right)^{n-1}$. Series diverges. ($|r| = 1/(\pi - 3) > 1$). **2.** $\sum_{n=4}^{\infty} \frac{5^{n+2}}{3^n} = \sum_{n=4}^{\infty} 25\left(\frac{5}{3}\right)^n$. Series diverges. ($|r| = 5/3 > 1$).

3. $\sum_{n=1}^{\infty} \frac{2^n}{3^{n+1}} = \frac{1}{3}\sum_{n=1}^{\infty} \left(\frac{2}{3}\right)^n = \frac{1}{3}\left[\left(\frac{2}{3}\right) + \left(\frac{2}{3}\right)^2 + \cdots\right] = \frac{1}{3}\left[\frac{1}{1-2/3} - 1\right] = \frac{2}{3}$. ($|r| = 2/3 < 1$).

4. $0.3242424\ldots = \frac{3}{10} + \frac{24}{1000} + \frac{24}{100{,}000} + \frac{24}{10{,}000{,}000} + \cdots = \frac{3}{10} + \frac{24}{1000}\left[\sum_{n=0}^{\infty} \left(\frac{1}{100}\right)^n\right] = \frac{3}{10} + \frac{24}{1000} \cdot \frac{100}{99} = \frac{107}{330}$.

Sec. 15.3: Progress Test 1

1. Let $f(x) = \frac{1}{x(\ln x)^2}$. Then f is pos., cont., and decr. on $[2, +\infty)$ and $\int_2^{+\infty} \frac{dx}{x(\ln x)^2} = \lim_{n\to+\infty} \int_2^n \frac{dx}{x(\ln x)^2} = \lim_{n\to+\infty} \int_{x=2}^{x=n} \frac{du}{u^2} =$
$\lim_{n\to+\infty} \frac{-1}{\ln x}\Big|_2^n \left(\begin{matrix} u = \ln x \\ du = dx/x \end{matrix}\right) = \lim_{n\to+\infty} \left[\frac{-1}{\ln n} + \frac{1}{\ln 2}\right] = \frac{1}{\ln 2}$. Therefore $\sum_{k=2}^{\infty} \frac{1}{k(\ln k)^2}$ conv. by Int. Test.

2. Let $f(x) = x/e^{x^2} = xe^{-x^2}$. Clearly f is pos. and cont. on $[1, +\infty)$. Since $f'(x) = e^{-x^2}(1 - x^2) \leq 0$ for $x \geq 1$, f is decr. on $[1, +\infty)$.
$\int_1^{+\infty} xe^{-x^2}\,dx = \lim_{n\to+\infty} \int_1^n xe^{-x^2}\,dx = \lim_{n\to+\infty} \left[\frac{-1}{2e^{x^2}}\right]_1^n = \lim_{n\to+\infty} \left[\frac{-1}{2e^{n^2}} + \frac{1}{2e}\right] = \frac{1}{2e}$, and so $\sum_{k=1}^{\infty} \frac{k}{e^{k^2}}$ conv. by Int. Test.

3. Let $f(x) = \frac{\ln x}{x}$. Then f is pos. and cont. on $(1, +\infty)$. Since $f'(x) = \frac{1 - \ln x}{x^2} < 0$ for $x > e$, f is decr. on $[e, +\infty)$ and hence decr. on
$[3, +\infty)$. $\int_3^{\infty} \frac{\ln x}{x}\,dx = \lim_{n\to+\infty} \int_3^n \frac{\ln x}{x}\,dx = \lim_{n\to+\infty} \left[\frac{(\ln x)^2}{2}\right]_3^n = \lim_{n\to+\infty} \left[\frac{(\ln n)^2}{2} - \frac{(\ln 3)^2}{2}\right] = +\infty$. Thus $\sum_{m=3}^{\infty} \frac{\ln m}{m}$ div. by Int. Test.

4. The Int. Test does not apply. But since $\lim_{k\to+\infty} \cos(k^2 + 1)$ does not exist, $\sum_{k=1}^{\infty} \cos(k^2 + 1)$ div. by kth Term Test.

Sec. 15.3: Progress Test 2

1. $\dfrac{\ln k}{k^{3/2}} \Big/ \left(\dfrac{1}{k^{5/4}}\right) = \dfrac{\ln k}{k^{1/4}} \to 0$ as $k \to +\infty$ by [15.1.6(*d*)]. Hence $\sum_{k=1}^{\infty} \dfrac{\ln k}{k^{3/2}}$ conv. by [15.3.6(*b*)]. **2.** $\dfrac{\arctan k}{\sqrt{k}+1} \Big/ (1/\sqrt{k}) =$ $\arctan k\left(\dfrac{\sqrt{k}}{\sqrt{k}+1}\right) \to \dfrac{\pi}{2}\cdot 1$ as $k \to +\infty$. Hence $\sum_{k=1}^{\infty} \dfrac{\arctan k}{\sqrt{k}+1}$ div. by [15.3.6(*a*)]. **3.** Div. by (15.3.7.) or by kth Term Test.

Sec. 15.3: Progress Test 3

1. $a_k^{1/k} = 2/k \to 0 < 1$. The series conv. **2.** $\dfrac{a_{k+1}}{a_k} = \dfrac{(k+1)!}{3\cdot 6\cdot 9 \ldots 3k(3k+3)} \cdot \dfrac{3\cdot 6\cdot 9 \ldots 3k}{k!} = \dfrac{k+1}{3k+3} \to \dfrac{1}{3} < 1$. The series conv.

3. $\dfrac{a_{n+1}}{a_n} = \dfrac{2^{n+1}(n+2)}{n!\ 3^{n+1}} \cdot \dfrac{(n-1)!\ 3^n}{2^n(n+1)} = \dfrac{2}{3}\cdot\dfrac{n+2}{n(n+1)} \to 0 < 1$. The series conv. **4.** $a_n^{1/n} = \dfrac{n}{n+2} \to 1$. Root Test inconclusive. However, $\lim_{n\to+\infty} \left(\dfrac{n}{n+2}\right)^n = e^{-2} \neq 0$ by [15.1.6(*e*)], so the series div. by nth Term Test.

Sec. 15.4: Progress Test 1

1. $|a_k| = \dfrac{k+2}{3k+5} \to \dfrac{1}{3} \neq 0$, hence $\lim_{k\to+\infty} \dfrac{(-1)^k(k+2)}{3k+5} \neq 0$, and so series div. by kth Term Test.

2. Consider $|a_k|/|b_k| = \dfrac{\ln k}{k^{20/18}} \Big/ \dfrac{1}{k^{19/18}} = \dfrac{\ln k}{k^{1/18}} \to 0$ as $k \to +\infty$ {by [15.1.6(*d*)]}. Hence orig. series conv. absolutely by Limit Comp. T.

3. We know $(\ln m)/m$ is decreasing for $m \geq 3$. Hence the Alternating Series Test applies to show $\sum_{m=1}^{\infty} \dfrac{(-1)^m \ln m}{m}$ conv. However, comparison to the (divergent) harmonic series shows $\sum_{m=1}^{\infty} \dfrac{\ln m}{m}$ div. Hence $\sum_{m=1}^{\infty} \dfrac{(-1)\ln m}{m}$ is conditionally conv.

Sec. 15.4: Progress Test 2

1. $|a_{k+1}| = \dfrac{1}{4^{2(k+1)+1}} = \dfrac{1}{4^{2k+3}}$. We want $\dfrac{1}{4^{2k+3}} < 10^{-4}$, or equivalently, $4^{2k+3} > 10^4 = 10{,}000$. Now $k = 1$ yields $4^5 = 1024 < 10^4$, but $k = 2$ yields $4^7 = 16{,}384 > 10^4$. Thus the sum of the first two terms of this series is within 10^{-4} of the actual sum of the series.

Chapter 16: Solutions to Progress Tests

Sec. 16.1: Progress Test 1

1. $P_8(x) = 1 - \dfrac{x^2}{2!} + \dfrac{x^4}{4!} - \dfrac{x^6}{6!} + \dfrac{x^8}{8!}$.

2. *Note:* $P_4(x) = 1 - \dfrac{x^2}{2} + \dfrac{x^4}{24} = 0$ for $x \approx \pm 1.59$ and $x \approx \pm 3.08$.

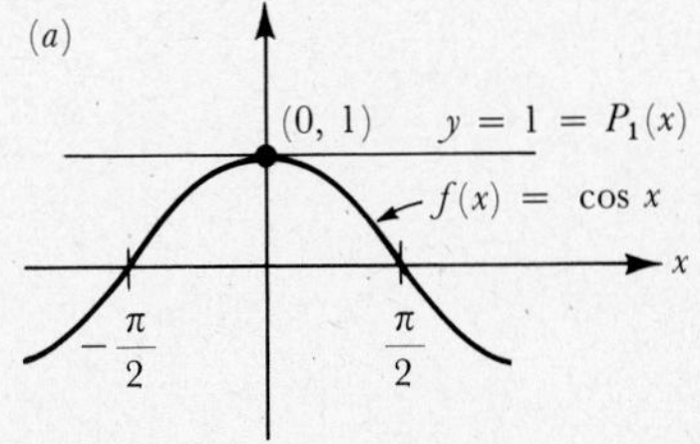

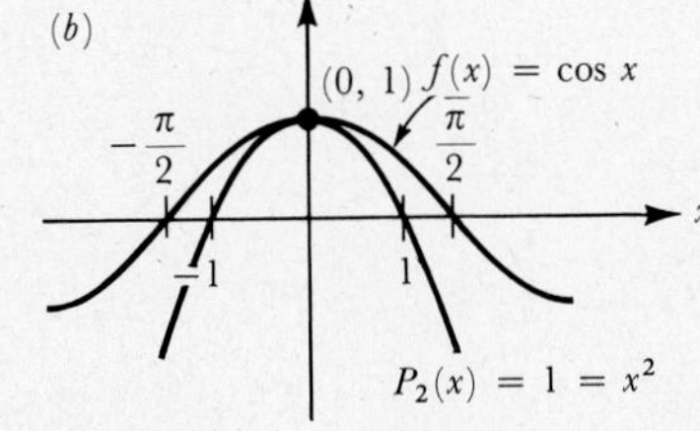

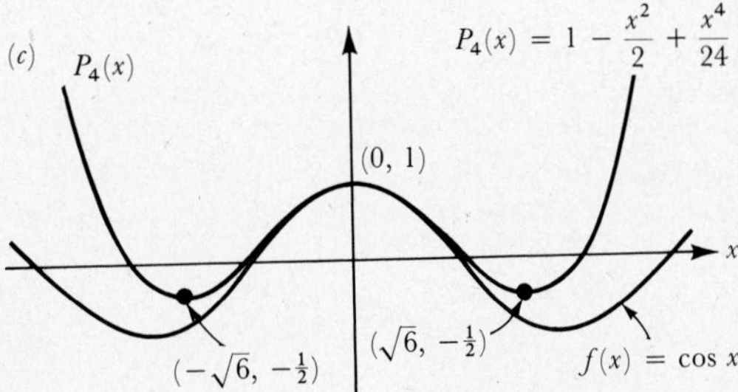

3. $P_5(x) = 1 + x + \dfrac{x^2}{2!} + \dfrac{x^3}{3!} + \dfrac{x^4}{4!} + \dfrac{x^5}{5!}$.

Sec. 16.1: Progress Test 2

1. Choose $n = 3$. $|f^{(4)}(x)| = \left|\left(\frac{1}{3}\right)^4(-2)(-5)(-8)x^{-11/3}\right| \approx 0.0000002354$ at $a = 64$. $|f^{(4)}(x)| \leq |f^{(4)}(64)| = M$ on $[64, 65]$. Then $|R_3(65)| \leq \frac{M(1)^4}{4!} \approx 9.8 \cdot 10^{-9}$ and $\sqrt[3]{65} \approx P_3(65) = 4 + \frac{1}{3}\cdot\frac{1}{16} + \frac{-2}{9}\cdot\frac{1}{1024} + \frac{10}{27}\cdot\frac{1}{65536} \approx 4.0207258$. **2.** (*a*) Let $n = 4$, $M = 3$; then $e^{0.1} = 1 + 0.1 + \frac{(0.1)^2}{2} + \frac{(0.1)^3}{6} + \frac{(0.1)^4}{24} + R_4(0.1) \approx 1.105171 + R_4(0.1)$, where $|R_4(0.1)| \leq \frac{3(0.1)^5}{120} \approx 0.00000025$. (*b*) $|R_4(0.3333)| \leq \frac{3(0.3333)^5}{120} \approx 0.000103$. Will try $n = 5$. $|R_5(0.3333)| \leq \frac{3(0.3333)^6}{720} \approx 0.0000057$. Hence use $P_5\left(\frac{1}{3}\right) = 1 + \frac{1}{3} + \left(\frac{1}{2}\right)\left(\frac{1}{3}\right)^2 + \frac{1}{6}\left(\frac{1}{3}\right)^3 + \frac{1}{24}\left(\frac{1}{3}\right)^4 + \frac{1}{120}\left(\frac{1}{3}\right)^5 \approx 1.3956$. (*c*) $n = 20$ yields $|R_{20}(5)| < 0.000028 < 0.00005$.
3. $P_{n+1}(x) = x - \frac{x^2}{2} + \frac{x^3}{3} - \frac{x^4}{4} + \cdots + \frac{(-1)^{n-1}x^n}{n} + \frac{(-1)^n x^{n+1}}{n+1}$.

Sec. 16.2: Progress Test 1

1. $\lim_{n\to+\infty}\left|\frac{(-1)^{n+1}(n+1)x^{n+1}}{3^{n+1}}\cdot\frac{3^n}{(-1)^n n x^n}\right| = \lim_{n\to+\infty}\left[\frac{n+1}{3n}|x|\right] = \frac{|x|}{3}$, so $r = 3$. At $x = -3$: $\sum_{n=0}^{\infty}\frac{(-1)^n n(-3)^n}{3^n} = \sum_{n=0}^{\infty} n$ diverges. At $x = 3$, $\sum_{n=0}^{\infty}\frac{(-1)^n n\, 3^n}{3^n} = \sum_{n=0}^{\infty}(-1)^n n$ diverges (nth term test). $I = (-3, 3)$. **2.** $\lim_{n\to+\infty}\left|\frac{x^{n+1}}{(n+1)(n+2)}\cdot\frac{n(n+1)}{x^n}\right| = \lim_{n\to+\infty}\frac{n}{n+2}|x| = |x|$, so $r = 1$. At $x = -1$: $\sum_{n=1}^{\infty}\frac{(-1)^n}{n(n+1)}$ conv. absolutely by comparison to p series for $p = 2$. Similarly, conv. absolutely at $x = 1$, so $I = [-1, 1]$.

3. $\lim_{n\to+\infty}\left|\frac{(x-4)^{n+1}}{n+1}\cdot\frac{n}{(x-4)^n}\right| = \lim_{n\to+\infty}\frac{n}{n+1}|x-4| = |x-4|$, so $r = 1$. Examine endpts. of $(4-1, 4+1) = (3, 5)$. At $x = 3$, $\sum_{n=1}^{\infty}\frac{(-1)^n}{n}$ conv. conditionally. At $x = 5$: $\sum_{n=1}^{\infty}\frac{1}{n}$ div., so $I = [3, 5)$.

Sec. 16.3: Progress Test 1

1. $f^{(n)}(x) = \frac{n!}{(1-x)^{n+1}}$, so $f^{(n)}(0) = n!$ **2.** $g(x) = D_x\left[\frac{1}{1-x}\right]$, so (16.3.1) implies $g(x) = \sum_{n=1}^{\infty} nx^{n-1}$.

Sec. 16.3: Progress Test 2

1. Replace x by $-x$ in p.s. for e^x and then multiply through by x: $xe^{-x} = x\left[\sum_{n=0}^{\infty}\frac{(-x)^n}{n!}\right] = x\left[\sum_{n=0}^{\infty}\frac{(-1)^n x^n}{n!}\right] = \sum_{n=0}^{\infty}\frac{(-1)^n x^{n+1}}{n!}$. Rad. of conv. is infinite. **2.** Replace x by x^2 in p.s. for sin x: $\sin(x^2) = \sum_{n=0}^{\infty}\frac{(-1)^n(x^2)^{2n+1}}{(2n+1)!} = \sum_{n=0}^{\infty}\frac{(-1)^n x^{4n+2}}{(2n+1)!}$. Rad. of conv. is infinite.
3. Note $\ln(1-x) = \int_0^x \frac{-dt}{1-t}$, so $\ln\left(\frac{1+x}{1-x}\right) = \ln(1+x) - \ln(1-x) = \int_0^x \frac{dt}{1+t} + \int_0^x \frac{dt}{1-t} = \sum_{n=0}^{\infty}\frac{(-1)^n x^{n+1}}{n+1} + \sum_{n=0}^{\infty}\frac{x^{n+1}}{n+1} = \left[x - \frac{x^2}{2} + \frac{x^3}{3} - \cdots + \frac{(-1)^n x^{n+1}}{n+1} + \cdots\right] + \left[x + \frac{x^2}{2} + \frac{x^3}{3} + \cdots + \frac{x^{n+1}}{n+1} + \cdots\right] = 2x + \frac{2x^3}{3} + \frac{2x^5}{5} + \cdots + \frac{2x^{2n+1}}{2n+1} + \cdots = 2\sum_{n=0}^{\infty}\frac{x^{2n+1}}{2n+1}$, $|x| < 1$.

Sec. 16.3: Progress Test 3

1. $f(x) = (1-x)^{-1/2} = 1 + \sum_{n=1}^{\infty}\binom{-1/2}{n}(-1)^n x^n$ [(for $|x| < 1$)] $= 1 + (-1/2)(-1)x + \frac{(-1/2)(-7/2)}{2!}x^2 + \frac{(-1/2)(-3/2)(-5/2)(-1)}{3!}x^3 +$

$\cdots + \dfrac{(-1/2)(-3/2)\ldots(1/2-n)(-1)^n}{n!}x^n + \cdots = 1 + \dfrac{x}{2} + \dfrac{1\cdot 3}{(2^2)(2!)}x^2 + \dfrac{1\cdot 3\cdot 5}{(2^3)3!}x^3 + \cdots + \dfrac{1\cdot 3\ldots(2n-1)}{(2^n)(n!)}x^n + \cdots$ [(*note:* number of negative factors is always even in each term)] $= 1 + \sum_{n=1}^{\infty}\dfrac{1\cdot 3\ldots(2n-1)}{(2^n)(n!)}x^n$.

Sec. 16.4: Progress Test 1

1. (*a*) $x = \dfrac{1}{3}$ (*b*) $\ln 2 = 2\cdot\sum_{n=0}^{\infty}\dfrac{(1/3)^{2n+1}}{2n+1} = 2\left\{\dfrac{1}{3} + \dfrac{1}{3}\left(\dfrac{1}{3}\right)^3 + \dfrac{1}{5}\left(\dfrac{1}{3}\right)^5 + \dfrac{1}{7}\left(\dfrac{1}{3}\right)^7 + \cdots\right\}$. Now $2\cdot\sum_{n=0}^{\infty}\dfrac{1}{(2n+1)3^{2n+1}} = 2\left\{\dfrac{1}{3} + \dfrac{1}{3\cdot 3^3} + \dfrac{1}{5\cdot 3^5}\right\} \approx 2\left\{0.34650\right\} \approx 0.69300$. Hence three terms are required. (*c*) $n = 5$ gives 0.693147, accurate through six places.

2. (*a*) $\sin(x^2) = \sum_{n=0}^{\infty}\dfrac{(-1)^n x^{4n+2}}{(2n+1)!}$, so $\int_0^x \sin(t^2)\,dt = \sum_{n=0}^{\infty}\dfrac{(-1)^n x^{4n+3}}{(4n+3)(2n+1)!}$. (*b*) Hence we have an alternating series whose terms decrease in absolute value. For $n+1=4$, $\dfrac{1}{19\cdot 9!} = \dfrac{1}{6{,}894{,}720} \approx 0.00000015 < 10^{-6}$. Hence $\int_0^1 \sin(t^2)\,dt \approx \dfrac{1}{3} - \dfrac{1}{7\cdot 3!} + \dfrac{1}{11\cdot 5!} - \dfrac{1}{15\cdot 7!} \approx 0.310268$.

Chapter 17: Solutions to Progress Tests

Sec. 17.1: Progress Test 1

1.

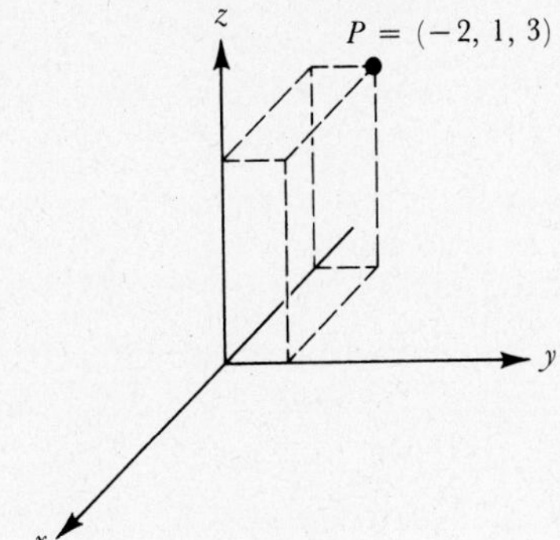

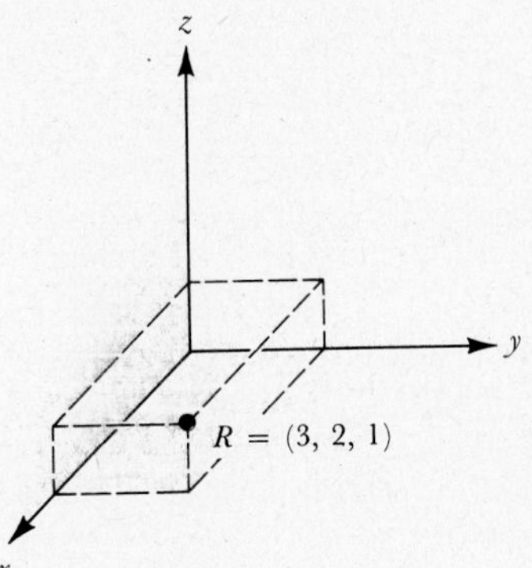

z
y
x
Q = (1, −3, −5)

2.

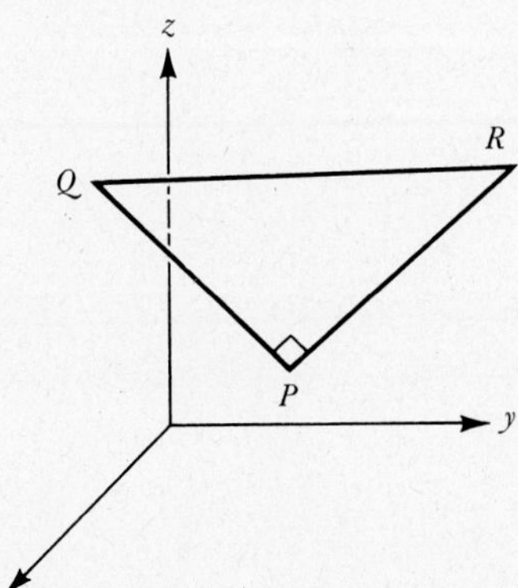

$d(P,Q) = \sqrt{26}$, $d(P,R) = \sqrt{9} = 3$, $d(Q,R) = \sqrt{35}$. Hence $d(P,Q)^2 + d(P,R)^2 = d(Q,R)^2$, so PQR is rt. triangle. $A = (1/2)\,bh = (1/2)\,d(P,Q)\,d(P,R) = 3\sqrt{26}/2$.

3. (*a*) $(x-2)^2 + (y+3)^2 + (z-1)^2 = 0$. Graph is single pt. $(2, -3, 1)$.
(*b*) Divide by 4 and complete sq.: $(x-1/2)^2 + (y-3)^2 + (z+1)^2 = 9$. Sphere centered at $(1/2, 3, -1)$, radius 3.

Sec. 17.2: Progress Test 1

1. Sym. with respect to y-z and x-y planes. Since $y \geq 0$, traces in planes $y = k > 0$ are circles centered on y axis of radii $2\sqrt{k}$. Traces in planes $x = k$ and $z = k$ are parabolas $z^2 = 4y - k^2$ and $x^2 = 4y - k^2$, respectively.

2.

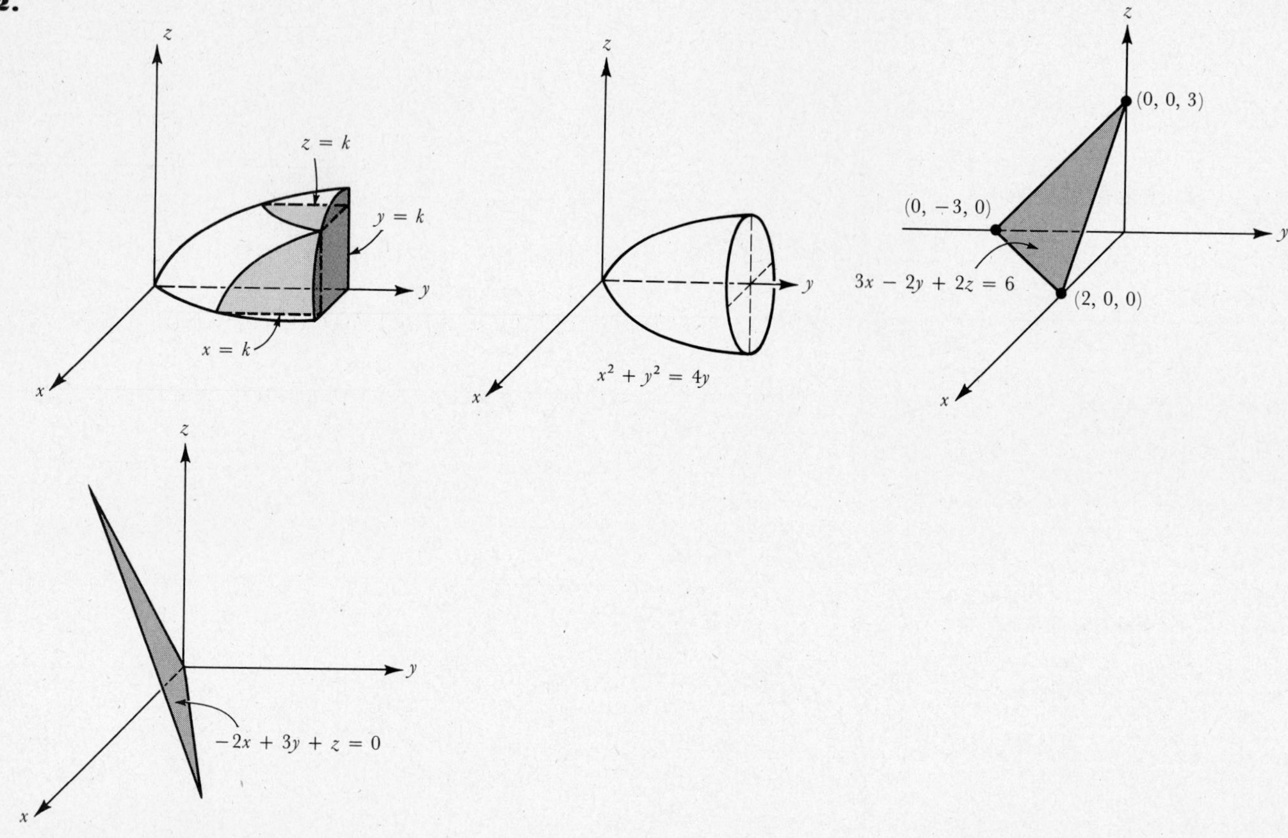

Sec. 17.2: Progress Test 2

1. Cyl. with rulings ∥ to y axis and directrix the plane curve in x-z plane with equation $z = \ln x$.
2. Cyl. with rulings ∥ to z axis and directrix the hyperbola in x-y plane with equation $x^2/16 - y^2/4 = 1$.
3. Surf. of rev.: revolve plane curve $x = 2\sqrt{z}$ in x-z plane about z axis.

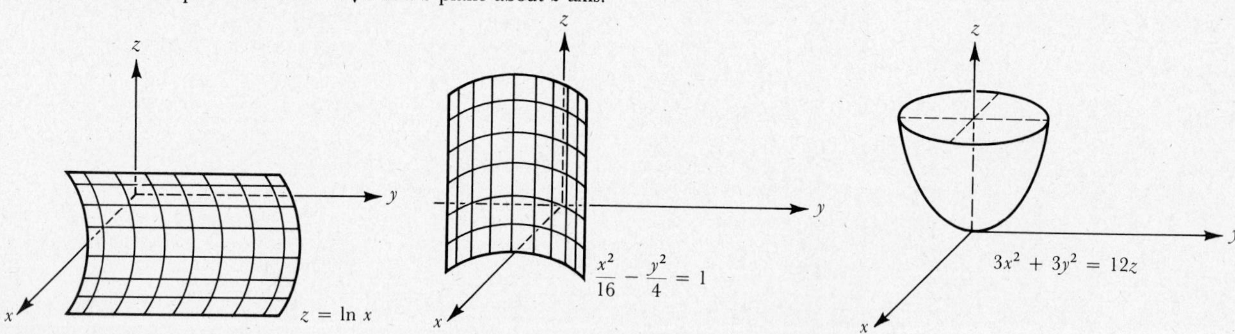

4. Solve for x in terms of z: $x = h(z) = z^{2/3}$. By Table 17.2*e*, equation of surf. is $x^2 + y^2 = [z^{2/3}]^2 = z^{4/3}$.

Sec. 17.2: Progress Test 3

1. $x^2/16 - y^2/4 + z^2/16 = 1$ is of form (17.2.5) with y and z interchanged, thus a hyperboloid of one sheet enveloping y axis, in fact, hyperboloid of rev.

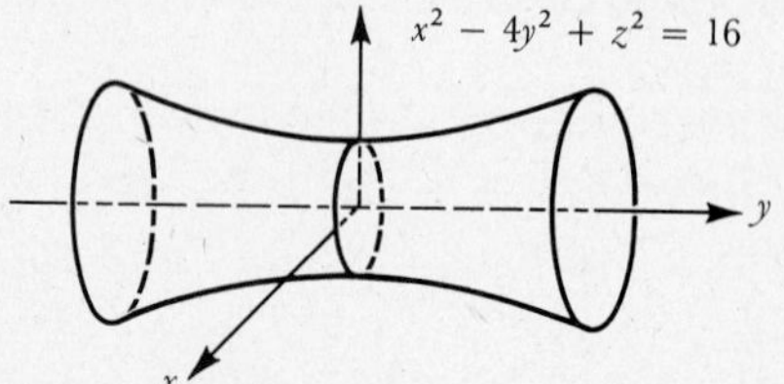

2. Has form of (17.2.9), with $c < 0$, a hyperbolic paraboloid.

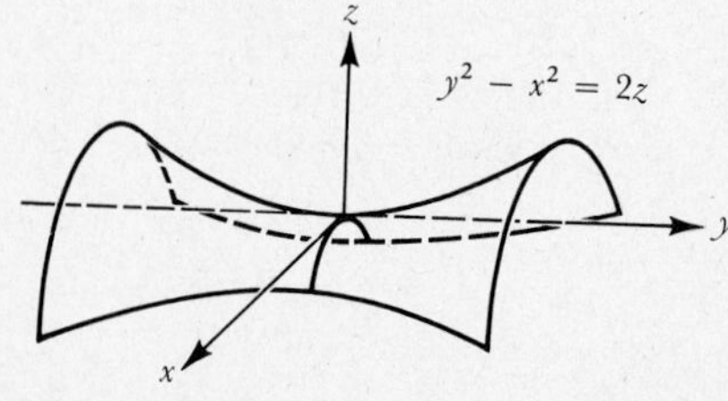

3. Comp. sq. to get $(x-1)^2+(z-1)^2=36(y-1)$, a translated variation of (17.2.8), an elliptic paraboloid with vertex (1, 1, 1), opening upward.

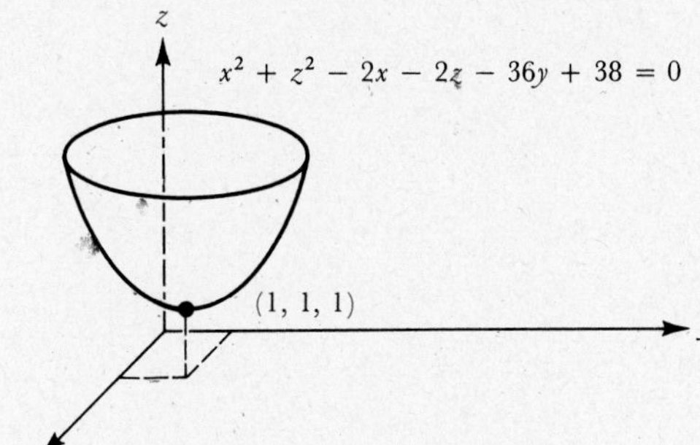

Sec. 17.3: Progress Test 1

1. Let $Q=(q_1,q_2)$. Then $\mathbf{A}=\overrightarrow{PQ}$ if and only if $\langle -2,3\rangle=\langle q_1-1,q_2-(-4)\rangle$. Thus $q_1-1=-2$, $q_2+4=3$; hence $q_1=-1$, $q_2=-1$, so $Q=(-1,-1)$.

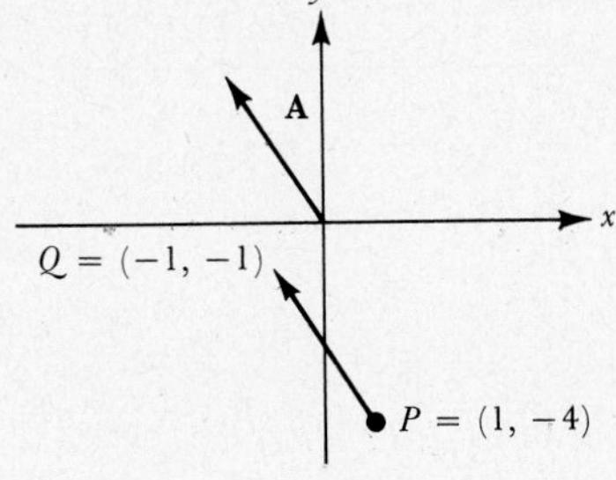

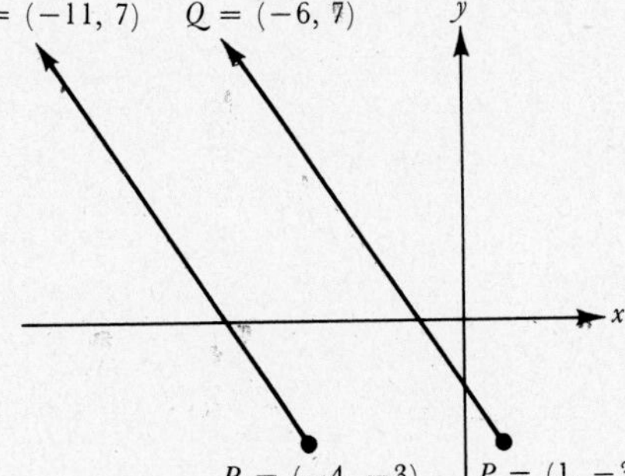

2. Let $S=(s_1,s_2)$. Then $\overrightarrow{PQ}=\langle -6-1,7-(-3)\rangle=\langle -7,10\rangle$ and $\overrightarrow{RS}=\langle s_1+4,s_2+3\rangle$. So $s_1+4=-7$ and $s_2+3=10$; thus $S=(-11,7)$. **3.** $\|\mathbf{A}\|=\sqrt{(1)^2+(\sqrt{3})^2}=\sqrt{4}=2$ and $\cos\alpha=1/2$. Since $0\le\alpha\le\pi/2$, $\alpha=\pi/3$.

Sec. 17.3: Progress Test 2

1. $(\mathbf{A}+\mathbf{B})+\mathbf{C}=(\langle 2,-4\rangle+\langle -3,1\rangle)+\langle -2,3\rangle=\langle -3,0\rangle$; $\mathbf{A}-\mathbf{B}=\langle 2,-4\rangle-\langle -3,1\rangle=\langle 5,-5\rangle$; $\mathbf{C}-\mathbf{A}=\langle -2,3\rangle-\langle 2,-4\rangle=\langle -4,7\rangle$.

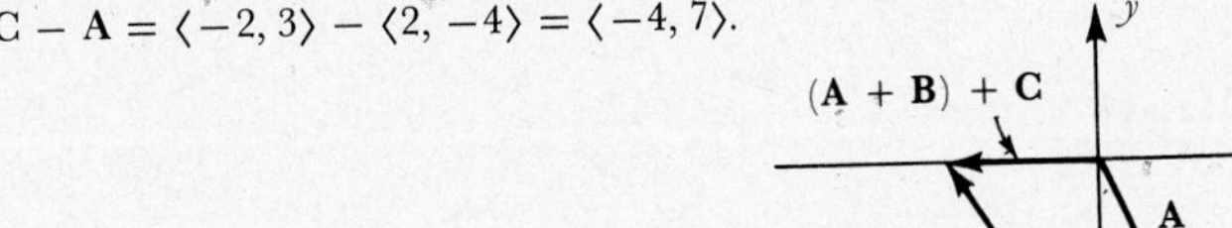

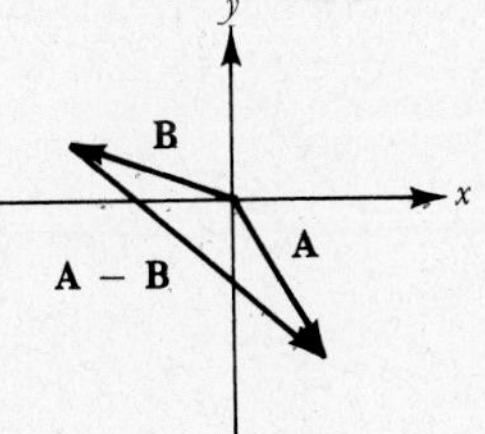

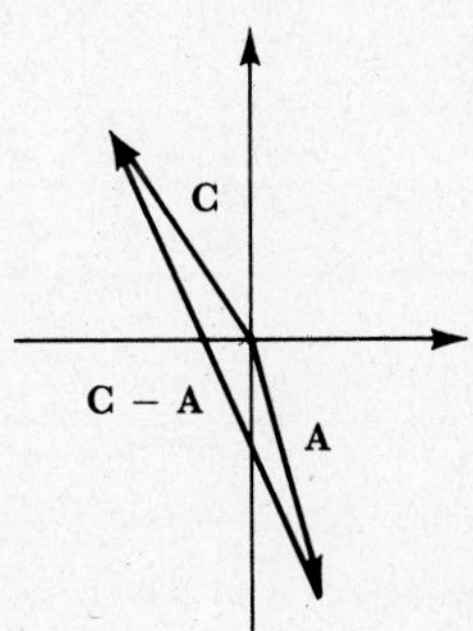

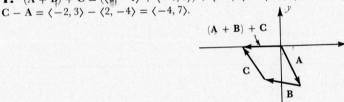

2. (*a*) $3\langle -3,1\rangle+2\langle 4,-2\rangle-\langle -1,-1\rangle=\langle 0,0\rangle$; (*b*) $\|\mathbf{C}\|(\langle -3,1\rangle+\langle 4,-2\rangle)=\langle\sqrt{2},-\sqrt{2}\rangle$; (*c*) $\|3\langle -3,1\rangle-2\langle 4,-2\rangle\|=\|\langle -17,7\rangle\|=\sqrt{289+49}=\sqrt{338}$.

Sec. 17.4: Progress Test 1

1. (*a*) $\cos\theta=0$; (*b*) $\cos\theta=-1/(\sqrt{41}\sqrt{11})\approx-0.0471$. **2.** $\mathbf{A}\cdot\mathbf{B}=4k^2-4k-3=0\Leftrightarrow k=-1/2$ or $k=3/2$.
3. $\|\mathbf{A}+\mathbf{B}\|^2=(\mathbf{A}+\mathbf{B})\cdot(\mathbf{A}+\mathbf{B})=\mathbf{A}\cdot(\mathbf{A}+\mathbf{B})+\mathbf{B}\cdot(\mathbf{A}+\mathbf{B})=\mathbf{A}\cdot\mathbf{A}+\mathbf{A}\cdot\mathbf{B}+\mathbf{B}\cdot\mathbf{A}+\mathbf{B}\cdot\mathbf{B}=\|\mathbf{A}\|^2+2\mathbf{A}\cdot\mathbf{B}+\|\mathbf{B}\|^2$.

Sec. 17.4: Progress Test 2

1. Since **i**, **j**, and **k** are unit vectors in dir. of pos. x, y, and z axes, we apply [17.4.8(*a*)]; $\text{Proj}_{\mathbf{i}}\,\mathbf{A}=(\mathbf{A}\cdot\mathbf{i})/\|\mathbf{i}\|=a_1$, $\text{Proj}_{\mathbf{j}}\,\mathbf{A}=(A\cdot\mathbf{j})/\|\mathbf{j}\|=a_2$, $\text{Proj}_{\mathbf{k}}\,\mathbf{A}=(\mathbf{A}\cdot\mathbf{k})/\|\mathbf{k}\|=a_3$. **2.** $\text{Proj}_{\mathbf{B}}\,\mathbf{A}=\dfrac{\langle 2,1,-6\rangle\cdot\langle 3,-1,1\rangle}{\|\langle 3,-1,1\rangle\|}=\dfrac{-1}{\sqrt{11}}$ and

$\mathbf{Proj_B\,A}=\left(\dfrac{\langle 2,1,-6\rangle\cdot\langle 3,-1,1\rangle}{\|\langle 3,-1,1\rangle\|^2}\right)\langle 3,-1,1\rangle=\langle -3/11,1/11,-1/11\rangle$. **3.** $\langle 1/\sqrt{2},1/\sqrt{2}\rangle$ is a unit vector in dir. of $\langle 1,1\rangle$, and so $10\langle 1/\sqrt{2},1/\sqrt{2}\rangle=\langle 10/\sqrt{2},10/\sqrt{2}\rangle=\mathbf{F}$. Thus $W=\mathbf{F}\cdot\langle 30,0\rangle=300/\sqrt{2}$ ft-lb.

Sec. 17.4: Progress Test 3

1. The vector $\mathbf{A}/\|\mathbf{A}\| = \langle -2/\sqrt{14}, 1/\sqrt{14}, 3/\sqrt{14} \rangle$ has same dir. as **A** but length 1. Thus desired vector **B** is $2\mathbf{A}/\|\mathbf{A}\| = \langle -4/\sqrt{14}, 2/\sqrt{14}, 6/\sqrt{14} \rangle$. **2.** $\|\mathbf{A}\| = \sqrt{1+1+2} = 2$, so $\cos\alpha = 1/2$, $\cos\beta = -1/2$, and $\cos\gamma = \sqrt{2}/2$. Thus $\alpha = \pi/3$, $\beta = 2\pi/3$, $\gamma = \pi/4$. **3.** Suffices to show $\mathbf{N} \cdot \overrightarrow{P_1P_2} = 0$. But $\mathbf{N} \cdot \overrightarrow{P_1P_2} = \langle a, b \rangle \cdot \langle x_2 - x_1, y_2 - y_1 \rangle = a(x_2 - x_1) + b(y_2 - y_1) = ax_2 + by_2 - (ax_1 + by_1)$. Since P_1, P_2 are on L, their coords. satisfy $ax_1 + by_1 = c$, $ax_2 + by_2 = c$, and so $\mathbf{N} \cdot \overrightarrow{P_1P_2} = ax_2 + by_2 - (ax_1 + by_1) = c - c = 0$.

Sec. 17.4: Progress Test 4

1. Area $= \|\mathbf{A} \times \mathbf{B}\| = \left\| \left\langle \begin{vmatrix} 2 & -3 \\ -4 & 5 \end{vmatrix}, -\begin{vmatrix} 1 & -3 \\ 2 & 5 \end{vmatrix}, \begin{vmatrix} 1 & 2 \\ 2 & -4 \end{vmatrix} \right\rangle \right\| = \|\langle -2, -11, -8 \rangle\| = 3\sqrt{21}$.

2. Volume $= |(\mathbf{A} \times \mathbf{B}) \cdot \mathbf{C}| = |\langle -2, -11, -8 \rangle \cdot \langle 7, 2, -4 \rangle| = |-4| = 4$. **3.** $(\mathbf{A} \times \mathbf{B}) \cdot \mathbf{C} = -4 < 0$, so **A**, **B**, **C** is not a rt.-hand triple. **4.** Let $\mathbf{D}' = \mathbf{A} \times \mathbf{C} = \left\langle \begin{vmatrix} 2 & -3 \\ 2 & -4 \end{vmatrix}, -\begin{vmatrix} 1 & -3 \\ 7 & -4 \end{vmatrix}, \begin{vmatrix} 1 & 2 \\ 7 & 2 \end{vmatrix} \right\rangle = \langle -2, -17, -12 \rangle$. Then **A**, **C**, **D**′ is a rt.-hand triple and **D**′ is ⊥ to plane of **A** and **C**. Let $\mathbf{D} = (2\mathbf{D}')/\|\mathbf{D}'\| = \langle -4/\sqrt{437}, -34/\sqrt{437}, -24/\sqrt{437} \rangle$. Then **D** has length 2, and since it is a pos. mult. of **D**′, it is ⊥ to plane of **A**, **C**; and so **A**, **C**, **D** form a rt.-hand triple.

Sec. 17.5: Progress Test 1

1. (*a*) $x = 7 + 3t$, $y = -2 + (1/2)t$, $z = 5 - t$; (*b*) $x = 2 - 7t$, $y = 1 + 2t$, $z = 6 + t$. **2.** We have $x = 1 - 3t$, $y = 7 + t$, $z = 11 + 4t$; $x = 3 + s$, $y = 9 - s$, $z = 4 + 3s$. Equating $1 - 3t = 3 + s$ and $7 + t = 9 - s$, we solve to obtain $t = -2$, $s = 4$. Substituting into respective z coordinates, we obtain $z = 3$ and $z = 16$, respectively. Thus lines do not intersect. Since $\langle -3, 1, 4 \rangle$, $\langle 1, -1, 3 \rangle$ are not scalar multiples of one another, lines are not parallel and hence are skew. **3.** (*a*) $4(x - 2) - (y - 6) + 2(z + 4) = 0$ or $4x - y + 2z = -6$; (*b*) let $P_0 = (1, 2, -3)$, $P_1 = (2, 4, 6)$, $P_2 = (-2, -1, 1)$. Then $\overrightarrow{P_0P_1} = \langle 1, 2, 9 \rangle$, $\overrightarrow{P_0P_2} = \langle -3, -3, 4 \rangle$. Then $\mathbf{N} = \overrightarrow{P_0P_1} \times \overrightarrow{P_0P_2} = \langle 35, -31, 3 \rangle$ is normal to Π. Since plane contains $(1, 2, -3)$, equation is $35(x - 1) - 31(y - 2) + 3(z + 3) = 0$ or $35x - 31y + 3z = -36$.

Chapter 18: Solutions to Progress Tests

Sec. 18.1: Progress Test 1

1. $x^2 + y^2 = e^{2t}(\cos^2 t + \sin^2 t) = e^{2t} = z^2$. Hence graph of **F** lies on cone $x^2 + y^2 = z^2$. Since $z = e^t > 0$, it lies on upper nappe.

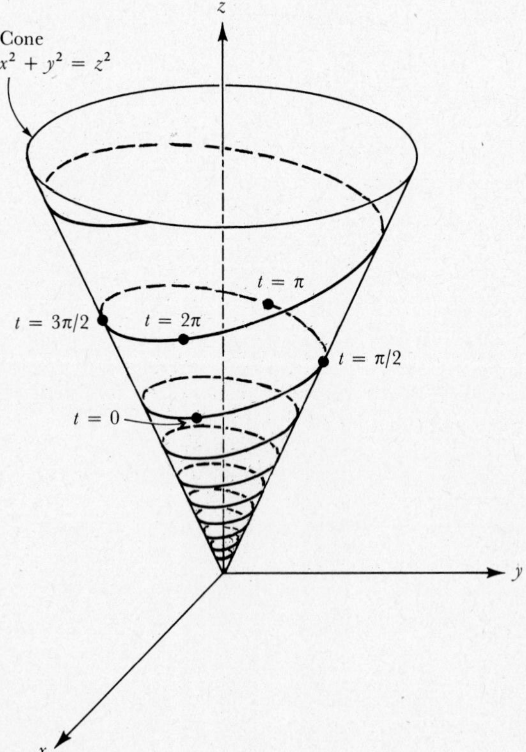

2. $F'(t) = \langle e^t(\cos t - \sin t), e^t(\cos t + \sin t), e^t\rangle$. **3.** Tan. line is parallel to $\mathbf{F}'(0) = \langle 1, 1, 1\rangle$ and goes through $\mathbf{F}(0) = (1, 0, 1)$. Thus parametric equations are $x = 1 + t, y = t, z = 1 + t$.

Sec. 18.1: Progress Test 2

1. $\mathbf{F}'(t) = \langle t^3, 2t^4, 2t\rangle + \langle c_1, c_2, c_3\rangle$. Since $\mathbf{F}(0) = \langle 4, 6, 8\rangle$, $c_1 = 4$, $c_2 = 6$, $c_3 = 8$. Now $\mathbf{F}(t) = \langle t^4/4 + 4t, 2t^5/5 + 6t, t^2 + 8t\rangle + \langle d_1, d_2, d_3\rangle$, and since $\mathbf{F}(0) = \mathbf{0}$, $d_1 = d_2 = d_3 = 0$. Thus $\mathbf{F}(t) = \langle t^4/4 + 4t, 2t^5/5 + 6t, t^2 + 8t\rangle$. **2.** $\|\mathbf{F}(t)\| = \sqrt{\mathbf{F}(t) \cdot \mathbf{F}(t)}$, so
$D_t(\|\mathbf{F}(t)\|) = \frac{1}{2}(\mathbf{F}(t) \cdot \mathbf{F}(t))^{-1/2}[\mathbf{F}(t) \cdot \mathbf{F}'(t) + \mathbf{F}'(t) \cdot \mathbf{F}(t)] = \frac{1}{2}(\mathbf{F}(t) \cdot \mathbf{F}(t))^{-1/2}[2\,\mathbf{F}(t) \cdot \mathbf{F}'(t)] = \dfrac{\mathbf{F}(t) \cdot \mathbf{F}'(t)}{\|\mathbf{F}(t)\|}$.

Sec. 18.2: Progress Test 1

1. $\mathbf{F}'(t) = \langle -2\sin 2t, 2\cos 2t\rangle$, $\|\mathbf{F}'(t)\| = 2$, so $\mathbf{T}(t) = \langle -\sin 2t, \cos 2t\rangle$. Now $\mathbf{T}'(t) = \langle -2\cos 2t, -2\sin 2t\rangle$ and $\|\mathbf{T}'(t)\| = 2$, so $\mathbf{N}(t) = \langle -\cos 2t, -\sin 2t\rangle$. Thus $\mathbf{T}(\pi/8) = \langle -1/\sqrt{2}, 1/\sqrt{2}\rangle$, $\mathbf{N}(\pi/8) = \langle -1/\sqrt{2}, -1/\sqrt{2}\rangle$.

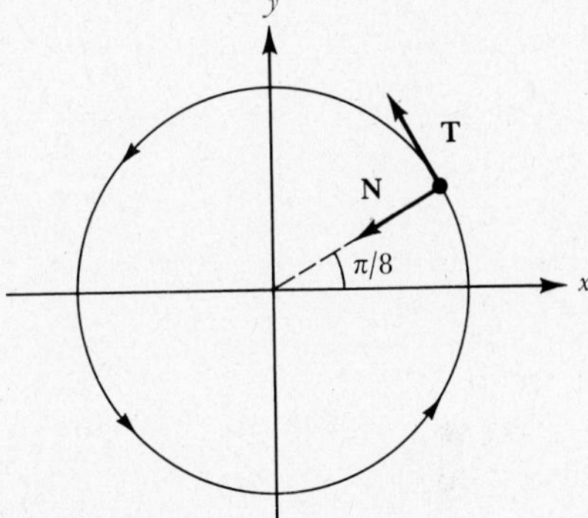

2. $\mathbf{F}'(t) = \langle 1 - \cos t, \sin t\rangle$, $\|\mathbf{F}'(t)\| = \sqrt{2 - 2\cos t}$, so $\mathbf{T}(t) = \left\langle \dfrac{1 - \cos t}{\sqrt{2 - 2\cos t}}, \dfrac{\sin t}{\sqrt{2 - 2\cos t}} \right\rangle = \left\langle \sqrt{\dfrac{1 - \cos t}{2}}, \dfrac{\sin t}{\sqrt{2 - 2\cos t}} \right\rangle$.
$\mathbf{T}'(t) = \left\langle \dfrac{\sin t}{2\sqrt{2 - 2\cos t}}, \dfrac{-\cos^2 t + 2\cos t + 1}{(2 - 2\cos t)^{3/2}} \right\rangle = \left\langle \dfrac{\sin t}{2\sqrt{2 - 2\cos t}}, -\dfrac{1}{2}\sqrt{\dfrac{1 - \cos t}{2}} \right\rangle$. Thus $\|\mathbf{T}'(t)\| = 1/2$ and
$\mathbf{N}(t) = \left\langle \dfrac{\sin t}{\sqrt{2 - 2\cos t}}, -\sqrt{\dfrac{1 - \cos t}{2}} \right\rangle$, and so $\mathbf{T}(\pi) = \langle 1, 0\rangle$, $\mathbf{N}(\pi) = \langle 0, -1\rangle$.

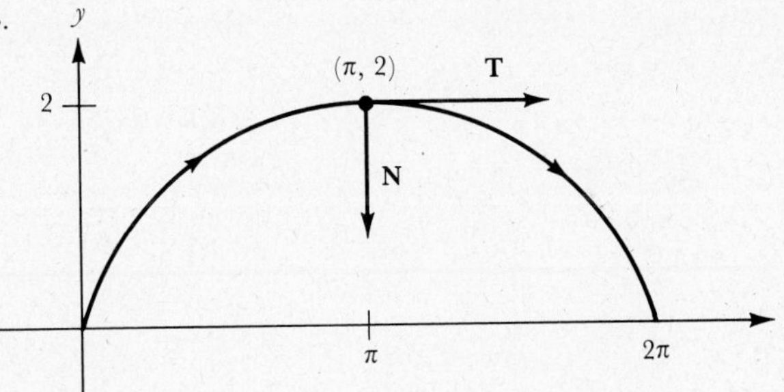

Sec. 18.2: Progress Test 2

1. $x' = 3t^2$, $x'' = 6t$; $y' = 2t$, $y'' = 2$. Thus $k(t) = \dfrac{|3t^2(2) - 2t(6t)|}{[(3t^2)^2 + (2t)^2]^{3/2}} = \dfrac{6t^2}{[9t^4 + 4t^2]^{3/2}}$. **2.** $k(x) = \dfrac{|e^x|}{[1 + (e^x)^2]^{3/2}} = \dfrac{e^x}{[1 + e^{2x}]^{3/2}}$.
$k'(x) = \dfrac{e^x[1 - 2e^{2x}]}{[1 + e^{2x}]^{5/2}}$; max. value occurs at $x = \ln(1/\sqrt{2})$.

Sec. 18.3: Progress Test 1

1. Since $\mathbf{F} = m\mathbf{A}$, if $\mathbf{F} = \mathbf{0}$, then $\mathbf{A} = \mathbf{0}$. Integrating twice, we have $\mathbf{V}(t) = \mathbf{C}_1$ and $\mathbf{G}(t) = \mathbf{C}_1 t + \mathbf{C}_2$, which has as its graph a single pt. if $\mathbf{C}_1 = \mathbf{0}$ and a straight line if $\mathbf{C}_1 \neq 0$. **2.** $\mathbf{V}(t) = \omega\langle a\sinh \omega t, a\cosh \omega t\rangle$ and $\mathbf{A}(t) = \omega^2\langle a\cosh \omega t, a\sinh \omega t\rangle = \omega^2\mathbf{G}(t)$. Thus $\mathbf{F}(t) = m\omega^2\mathbf{G}(t)$, and so is a central (repulsive) force.

Sec. 18.3: Progress Test 2

1. (*a*) $y(t) = (v_0 \sin\alpha)t - (g/2)t^2 = t[v_0\sin\alpha - (g/2)t]$. Thus $y(t) = 0$ when $t = 0$ and when $t = (2v_0\sin\alpha)/g$. Since $t = 0$ corresponds to initial position, impact occurs after $[(2v_0\sin\alpha)/g]$ s. (*b*) By (*a*), flight terminates after $t = [(2v_0\sin\alpha)/g]$ s. Thus terminal speed is $\|\mathbf{V}((2v_0\sin\alpha)/g)\| = \|\langle v_0\cos\alpha, v_0\sin\alpha - g((2v_0\sin\alpha)/g)\rangle\| = \|\langle v_0\cos\alpha, -v_0\sin\alpha\rangle\| = \sqrt{v_0^2(\cos^2\alpha + \sin^2\alpha)} = v_0$.

(*c*) $y(t) = (v_0 \sin \alpha)t - (g/2)t^2$ gives height at time t. Max. ht. occurs when $y'(t) = v_0 \sin \alpha - gt = 0$ (that is, when vert. component of **V** is 0), or $t = (v_0 \sin \alpha)/g$. Thus max. ht. is $y((v_0 \sin \alpha)/g) = (v_0^2 \sin^2 \alpha)/(2g)$. (*d*) By (*a*), time of impact is $(2v_0 \sin \alpha)/g$, and so range is $x(2v_0 \sin \alpha)/g) = v_0 \cos \alpha(2v_0 \sin \alpha)/g = v_0^2(2 \sin \alpha \cos \alpha)/g = (v_0^2 \sin 2\alpha)/g$. Thus max. range occurs when $\sin 2\alpha = 1$; that is, $\alpha = \pi/4$.

Chapter 19: Solutions to Progress Tests

Sec. 19.1: Progress Test 1

1. The pt. (x, y) is in $D(f) \Leftrightarrow xy \geq 1$. Thus for $x > 0, y \geq 1/x$, and for $x < 0, y \leq 1/x$.

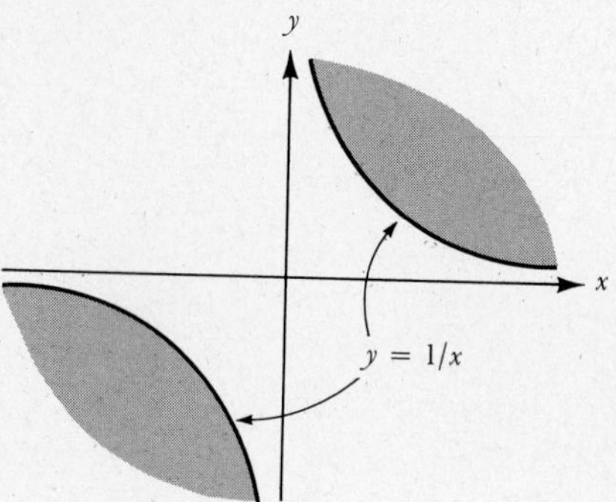

2. (*a*) Graph is a plane with x, y, and z intercepts $(2, 0, 0)$, $(0, 4, 0)$ and $(0, 0, 4)$, respectively;

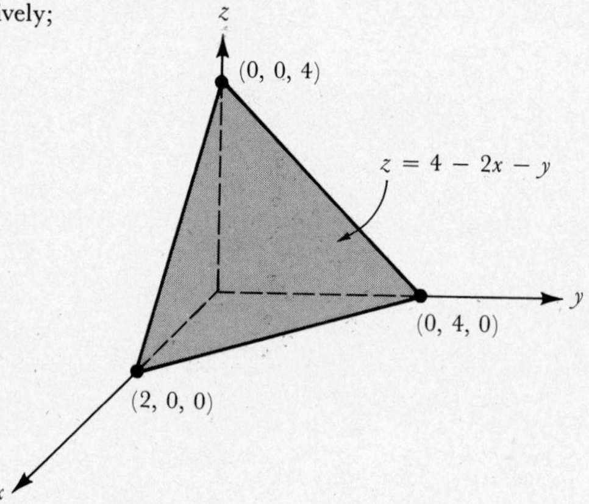

(*b*) $z = \sqrt{3x^2 + 4y^2 - 12} \Rightarrow x^2/4 + y^2/3 - z^2/12 = 1$, whose graph is a hyperboloid of one sheet. Graph of f is the part lying above x-y plane.

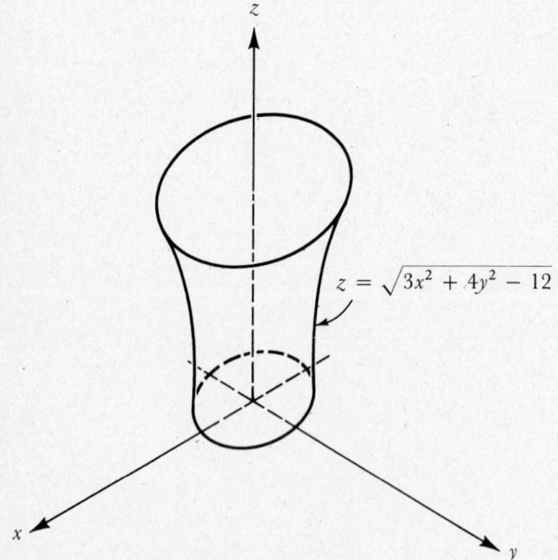

Sec. 19.1: Progress Test 2

1. Pt. (x, y, z) is in $D(f) \Leftrightarrow \sqrt{z^2 - 1} \leq 1$ and $z^2 - 1 \geq 0$. This is true $\Leftrightarrow z$ is in $[-\sqrt{2}, -1]$ or $[1, \sqrt{2}]$. Thus domain consists of two horizontal plates of thickness $\sqrt{2} - 1$ and of infinite extent, one above x-y plane between the planes $z = 1$ and $z = \sqrt{2}$ and one below x-y plane between the planes $z = -1$ and $z = -\sqrt{2}$.

2.

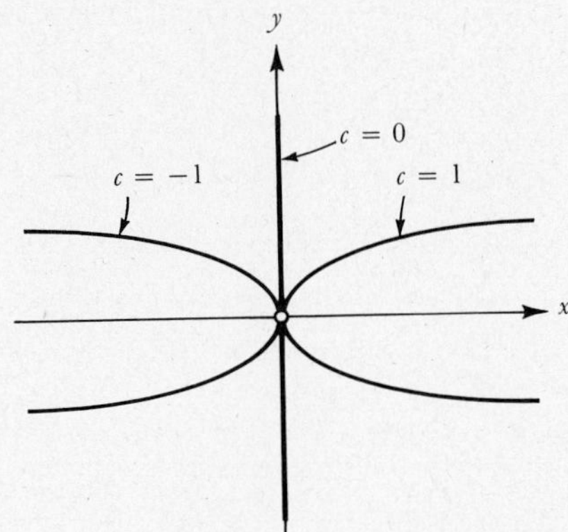

3.

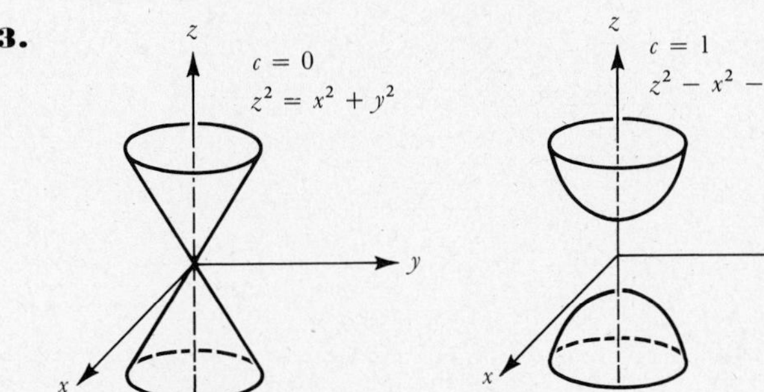

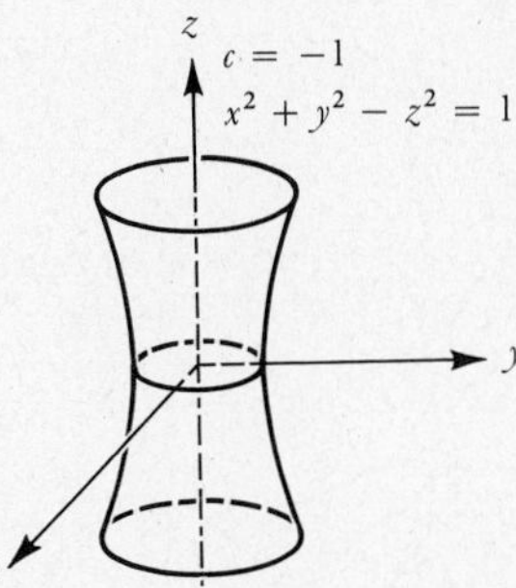

Sec. 19.2: Progress Test 1

1. (*a*) Let L_m be the straight line with equation $y = mx$. Then $(x, mx) \to (0, 0)$ along L_m as $x \to 0$. Now $f(x, mx) = mx/(x^2 + m^2)$, which tends to 0 as $x \to 0$. (*b*) Pts. $(x, x^2) \to (0, 0)$ along $y = x^2$ as $x \to 0$. Since $f(x, x^2) = 1/2$ for $x \neq 0$, $f(x, y) \to 1/2$ as $(x, y) \to (0, 0)$ along $y = x^2$. (*c*) No! For lim to exist, lim would have to be the same along any smooth curves through (0, 0). **2.** Let $\varepsilon > 0$. Since $\sqrt{(x-a)^2} \leq \sqrt{(x-a)^2 + (y-b)^2}$, we choose $\delta = \varepsilon$. Then $0 < d[(x, y), (a, b)] = \sqrt{(x-a)^2 + (y-b)^2} < \delta$ implies $d[f(x, y), f(a, b)] = d[x, a] = \sqrt{(x-a)^2} \leq \sqrt{(x-a)^2 + (y-b)^2} < \delta = \varepsilon$.

Sec. 19.3: Progress Test 1

1. $f_x = 6xy^4 + 3x^2y^2, f_y = 12x^2y^3 + 2x^3y$. **2.** $f_x = 2x^2y/(x^2 + y^2) + y\ln(x^2 + y^2), f_y = 2xy^2/(x^2 + y^2) + x\ln(x^2 + y^2)$.

3. $f_x(0, 0) = \lim_{\Delta x \to 0} \dfrac{f(\Delta x, 0) - f(0, 0)}{\Delta x} = \lim_{\Delta x \to 0} \dfrac{3(\Delta x)^3}{(\Delta x)^3} = 3$.

Sec. 19.3: Progress Test 2

1. $f_x = x/\sqrt{x^2 + y^2}, f_y = y/\sqrt{x^2 + y^2}$, so $f_x(3, 4) = 3/5$ and $f_y(3, 4) = 4/5$. Thus Π_T: $(3/5)(x - 3) + (4/5)(y - 4) = z - 5$ and N: $x = 3 + (3/5)t, y = 4 + (4/5)t, z = 5 - t$. **2.** $T_x(x, y) = -6x$, and so $T_x(3, -1) = -18$. Temp. is changing at the rate of $-18°$ per unit of length.

Sec. 19.3: Progress Test 3

1. $f_x = z^2\ln(x^2 + y^2 + z^2) + 2x^2z^2/(x^2 + y^2 + z^2), f_y = 2xyz^2/(x^2 + y^2 + z^2), f_z = 2xz\ln(x^2 + y^2 + z^2) + 2xz^3/(x^2 + y^2 + z^2)$.
2. (*a*) $f_x = y^4\cos(xy), f_{xx} = -y^5\sin(xy), f_{xxy} = -y^5x\cos(xy) - 5y^4\sin(xy)$. By equality of mixed partials, this also equals f_{xyx}.
(*b*) $f_{yyxx} = f_{xxyy}, f_x = \tan(y^3), f_{xx} = 0$, so $f_{xxyy} = 0$. (*c*) $f_x = y^2z^3 + 2/(x^2y), f_{xy} = 2yz^3 - 2/(x^2y^2)$, so $f_{xyz} = 6yz^2$.

Sec. 19.4: Progress Test 1

1. $\Delta z = f(-1.01, 3.02) - f(-1, 3) = 2(-1.01)^2 - 3(-1.01)(3.02) + (3.02)^2 - 20 = 0.3112$.
$dz = f_x(-1, 3)(-0.01) + f_y(-1, 3)(0.02) = (-13)(-0.01) + 9(0.02) = 0.31$. Thus $|\Delta z - dz| = 0.0012$. **2.** Let $f(x, y) = \sqrt{x^2 + y^3}$, $x = 3$, $y = 3$, $\Delta x = 0.01$, and $\Delta y = -0.03$. Then $f(x + \Delta x, y + \Delta y) = \sqrt{(3.01)^2 + (2.97)^3}$. Now $f_x = x/\sqrt{x^2 + y^3}$ and $f_y = 3y^2/(2\sqrt{x^2 + y^3})$, so $\sqrt{(3.01)^2 + (2.97)^3} \approx 6 + (1/2)(.01) + (9/4)(-.03) = 5.9375$.

Sec. 19.5: Progress Test 1

1. (*a*) $z_r = z_x x_r + z_y y_r = [y^3/(x^2+y^2)^{3/2}]\cos\theta + [x^3/(x^2+y^2)^{3/2}]\sin\theta = \dfrac{r^3\sin^3\theta\cos\theta}{r^3} + \dfrac{r^3\cos^3\theta\sin\theta}{r^3} = (\cos^2\theta + \sin^2\theta)\cos\theta\sin\theta = \cos\theta\sin\theta$; (*b*) $dz/dt = z_x\,dx/dt + z_y\,dy/dt + z_w\,dw/dt = (y^2\ln(w/x) - y^2)(-1/t^2) + (2xy\ln(w/x))(1) + (xy^2/w)(2t+1) = \ln(t^3+t^2) + (3t+2)/(t+1)$. **2.** $V = (\pi/3)r^2h$, so $dv/dt = V_r(dr/dt) + V_h(dh/dt) = (2\pi rh/3)(dr/dt) + (\pi r^2/3)(dh/dt)$. Thus at given instant $dV/dt = (2\pi\cdot 12\cdot 14/3)(2) + (\pi(12)^2/3)(-4) = 32\pi$ cm/s.

Sec. 19.6: Progress Test 1

1. (*a*) $\nabla f(x,y) = \langle 6x, 8y\rangle$, $\mathbf{u} = \mathbf{v}/\|\mathbf{v}\| = \langle 1/\sqrt{5}, 2/\sqrt{5}\rangle$, so $D_\mathbf{v}f(3,-4) = D_\mathbf{u}f(3,-4) = \nabla f(3,-4)\cdot\mathbf{u} = -46/\sqrt{5}$; (*b*) $\nabla f(x,y) = \langle y^2/(x^2+y^2)^{3/2}, -xy/(x^2+y^2)^{3/2}\rangle$, $\mathbf{u} = \mathbf{v}/\|\mathbf{v}\| = \langle\sqrt{3}/2, -1/2\rangle$, so $D_\mathbf{v}f(3,4) = D_\mathbf{u}f(3,4) = \nabla f(3,4)\cdot\mathbf{u} = (16\sqrt{3}+12)/250$. **2.** $D_{\langle 1,3\rangle}f(3,-2) = \nabla f(3,-2)\cdot\langle 1/\sqrt{10}, 3/\sqrt{10}\rangle = 2$. $D_{\langle 1,0\rangle}f(3,-2) = \nabla f(3,-2)\cdot\langle 1,0\rangle = 4$. Thus $f_x(3,-2)(1/\sqrt{10}) + f_y(3,-2)(3/\sqrt{10}) = 2$ and $f_x(3,-2)(1) + f_y(3,-2)(0) = 4$; solving these, we have $f_x(3,-2) = 4$ and $f_y(3,-2) = (2\sqrt{10}-4)/3$. Now $\mathbf{u} = \langle -3/\sqrt{13}, 2/\sqrt{13}\rangle$ is a unit vector that points from $(3,-2)$ toward $(0,0)$. Thus $D_\mathbf{u}f(3,-2) = \nabla f(3,-2)\cdot\mathbf{u} = (4\sqrt{10}-44)/(3\sqrt{13})$. **3.** It should proceed in the direction of $\nabla z(2,3) = \langle 108, 72\rangle$.

Sec. 19.7: Progress Test 1

1. Since $f(3,\sqrt{5}/2) = 4$ and $f(-2\sqrt{10},1) = 36$, level curves have equations $x^2 - 4y^2 = 4$ and $x^2 - 4y^2 = 36$. $\nabla f(3,\sqrt{5}/2) = \langle 6, -4\sqrt{5}\rangle$, $\nabla f(-2\sqrt{10},1) = \langle -4\sqrt{10}, -8\rangle$.

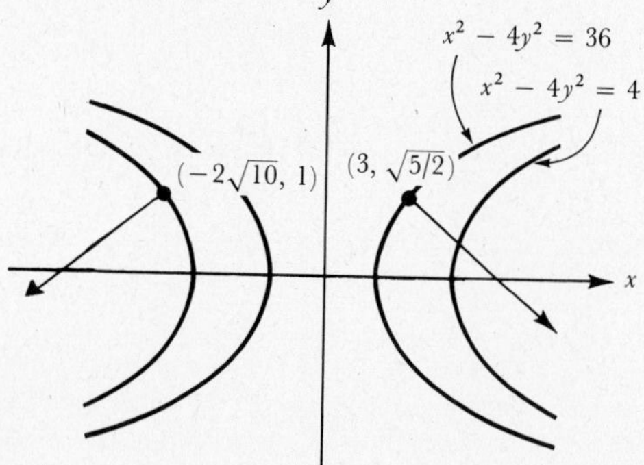

2. Let $F(x,y,z) = x^3 + 4y^3 + z^3 - 5xyz$. Then $F_x = 3x^2 - 5yz$, $F_y = 12y^2 - 5xy$, $F_z = 3z^2 - 5xy$, so $F_x(2,1,2) = 2$, $F_y(2,1,2) = -8$, $F_z(2,1,2) = 2$. Tangent plane: $2(x-2) - 8(y-1) + 2(z-2) = 0$; normal line: $x = 2+2t$, $y = 1-8t$, $z = 2+2t$. **3.** Let $F(x,y,z) = y - x^2 + z^2 + 2$, $G(x,y,z) = x^2 + y^2 - z - 8$. Then $\nabla F(-2,2,0) = \langle 4,1,0\rangle$, $\nabla G(-2,2,0) = \langle -4,4,-1\rangle$. Tan. line to C at $(-2,2,0) = P$ lies in tan. plane to both surfaces at P. But ∇F and ∇G are normals to these planes, so tan. line is parallel to $\nabla F\times\nabla G = \langle -1,4,20\rangle$ and hence has equations $x = -2-t$, $y = 2+4t$, $z = 20t$.

Sec. 19.8: Progress Test 1

1. $f_x = 2x + 2xy = 0$, $f_y = 2y + x^2 = 0$ have simultaneous solutions $(0,0)$, $(\sqrt{2},-1)$, $(-\sqrt{2},-1)$. Since $f_{xx} = 2+2y$, $f_{xy} = 2x$, $f_{yy} = 2$, we have $f_{xx}f_{yy} - f_{xy}{}^2 = 4 + 4y - 4x^2$, which is neg. $(= -8)$ at $(\pm\sqrt{2},-1)$ (saddle pts.) and is pos. $(=4)$ at $(0,0)$. Since $f_{xx}(0,0) = 2 > 0$, $(0,0)$ is rel. min. **2.** $f_x = 4x^3 = 0$, $f_y = 4y^3 = 0$ yield crit. pt. at $(0,0)$ with $f(0,0) = 0$. Test fails, but $f(0+h, 0+k) = h^4 + k^4 > 0$ for all h, $k \neq 0$. Hence rel. min. at $(0,0)$.

Sec. 19.8: Progress Test 2

1. $T_x = 8x = 0$, $T_y = 4y + 2 = 0$ have solution $(0,-1/2)$. Let C_1: $x = \sqrt{1-y^2}$, $-1 \le y \le 1$;

y
C_1: $x = \sqrt{1-y^2}$
max
max
x
C_2: $x = -\sqrt{1-y^2}$
$(0,-1)$ − min

C_2: $x = -\sqrt{1-y^2}$, $-1 \le y \le 1$. Then $g(y) = T(\pm\sqrt{1-y^2}, y) = -2y^2 + 2y + 5$ and $g'(y) = -4y + 2 = 0$ when $y = 1/2$, a rel. max. for g. Pts. on boundary corresponding to $y = 1/2$ and $y = \pm 1$ (endpts.) are, respectively, $(\pm\sqrt{3}/2, 1/2)$ and $(0, \pm 1)$. Now $T(0, 1) = 5$, $T(0, -1) = 1$, $T(0, 1/2) = 5/2$, $T(\pm\sqrt{3}/2, 1/2) = 11/2$. Hence min. temp. $(= 1)$ occurs at $(0, -1)$ and max. $[=(11/2)]$ at $(\pm\sqrt{3}/2, 1/2)$.

Sec. 19.9: Progress Test 1

1. Solution curves should have this general shape and location.

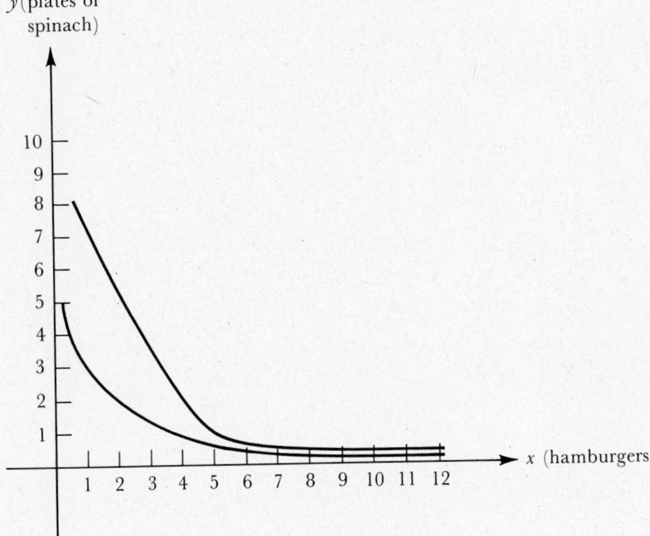

2.

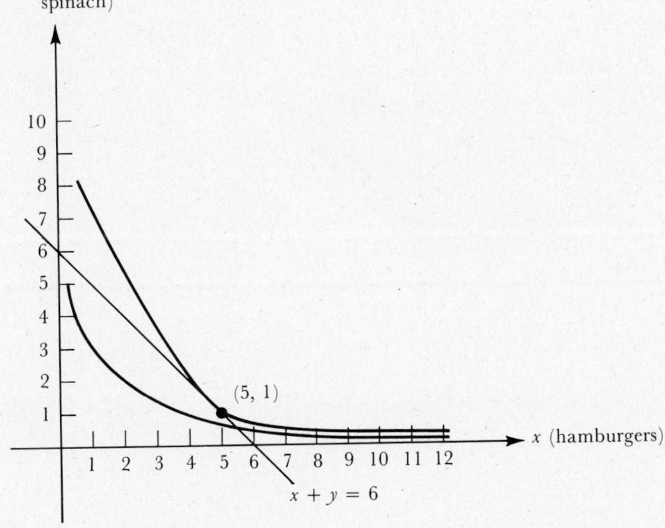

Sec. 19.9: Progress Test 2

1. (i) $f_x = 2xy = \lambda 4x$, (ii) $f_y = x^2 = \lambda(2y)$, (iii) $2x^2 + y^2 = 12$. From (i) $y = 2\lambda$ [$x \neq 0$ by (iii)], so $x^2 = y^2$ by (ii). Hence $x = \pm 2$ by (iii). Hence extremes at $(2, 2)$, $(-2, 2)$, $(-2, -2)$, $(2, -2)$. Computing f at these pts. yields maximum of 8 at $(2, 2)$, $(-2, 2)$ and minimum of -8 at $(-2, -2)$, $(2, -2)$.

Chapter 20: Solutions to Progress Tests

Sec. 20.1: Progress Test 1

1. $I = \int_0^1 \left[\int_0^y (12x^2 + 4y)\,dx\right] dy = \int_0^1 [4x^3 + 4xy]_0^y\,dy = \int_0^1 (4y^3 + 4y^2)\,dy = \dfrac{7}{3}$.

2. $I = \int_{1/2}^1 \left[\int_0^{3x} \sin(\pi x^2)\,dy\right] dx = \int_{1/2}^1 [y\sin(\pi x^2)]_0^{3x}\,dx = \int_{1/2}^1 3x\sin(\pi x^2)\,dx = \dfrac{3}{2\pi}\left[1 + \dfrac{1}{\sqrt{2}}\right]$.

Sec. 20.1: Progress Test 2

1. (a) $R = \{(x, y)/0 \le x \le 1, x^2 \le y \le x\}$;

(b) $I = \int_0^1 \left[\int_{x^2}^{x} (4x + 2y)\, dy\right] dx = \int_0^1 [4xy + y^2]_{x^2}^{x}\, dx = \int_0^1 (5x^2 - 4x^3 - x^4)\, dx = 7/15.$

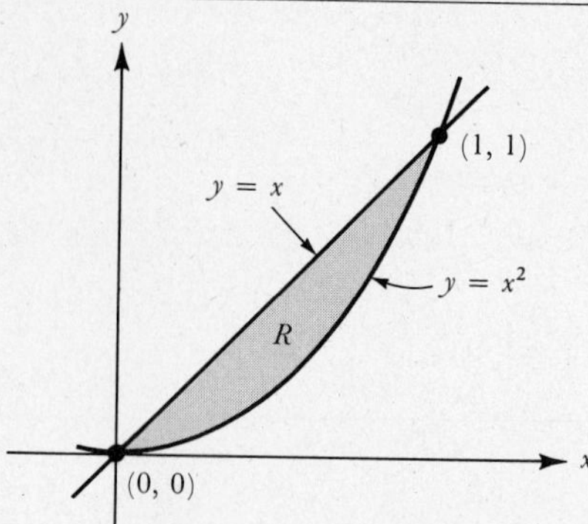

2. (a)

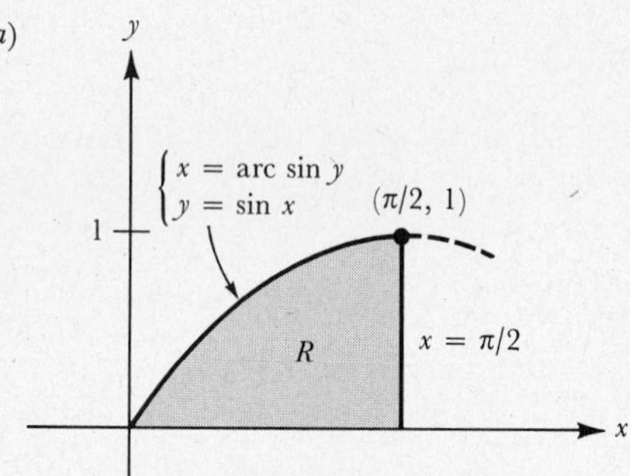

$$I = \int_0^1 \int_{\arcsin y}^{\pi/2} f(x, y)\, dx\, dy$$

(b)

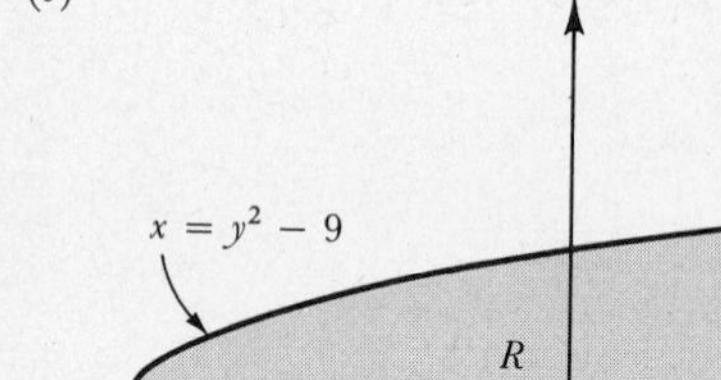
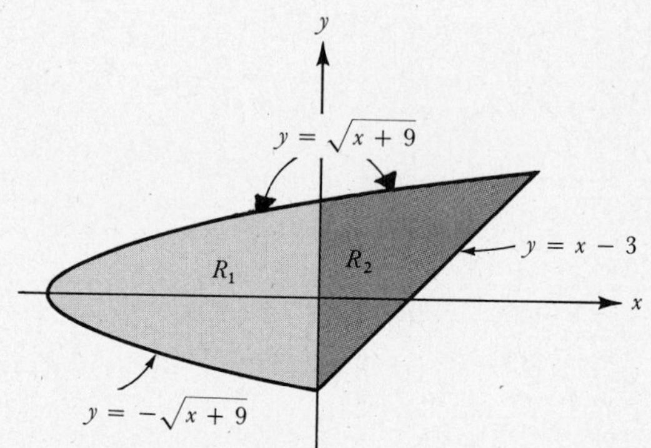

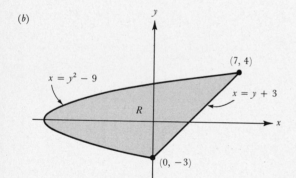

$$I = \int_{-9}^{0} \int_{-\sqrt{x+9}}^{\sqrt{x+9}} f(x, y)\, dy\, dx + \int_0^7 \int_{x-3}^{\sqrt{x+9}} f(x, y)\, dy\, dx.$$

3.

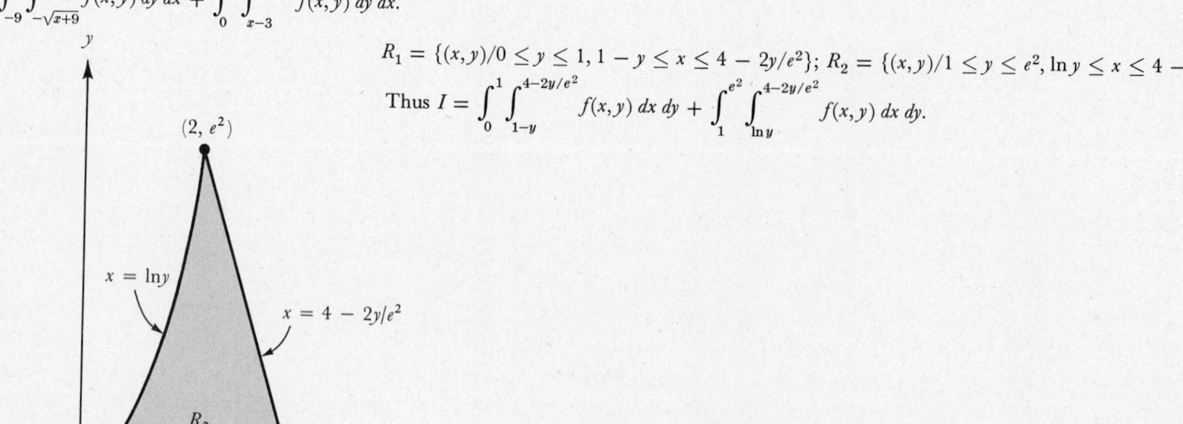

$R_1 = \{(x, y)/0 \le y \le 1, 1 - y \le x \le 4 - 2y/e^2\}$; $R_2 = \{(x, y)/1 \le y \le e^2, \ln y \le x \le 4 - 2y/e^2\}$.

Thus $I = \int_0^1 \int_{1-y}^{4-2y/e^2} f(x, y)\, dx\, dy + \int_1^{e^2} \int_{\ln y}^{4-2y/e^2} f(x, y)\, dx\, dy.$

Sec. 20.2: Progress Test 1

1.

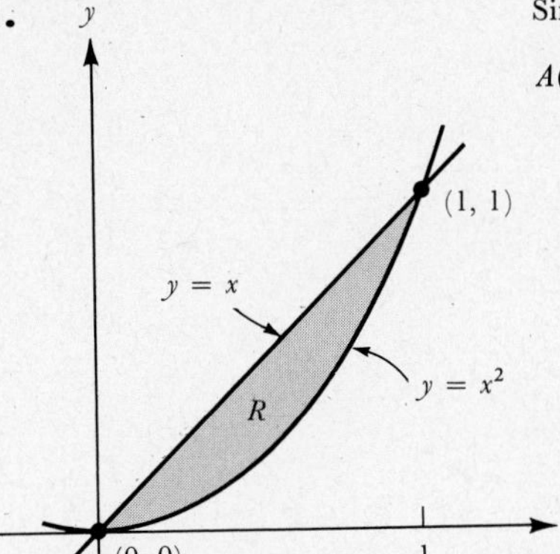

Since $R = \{(x,y)/0 \le x \le 1,\ x^2 \le y \le x\}$,

$$A(R) = \iint_R dA = \int_0^1 \int_{x^2}^{x} dy\,dx = \int_0^1 [x - x^2]\,dx = x^2/2 - x^3/3|_0^1 = 1/6.$$

2.

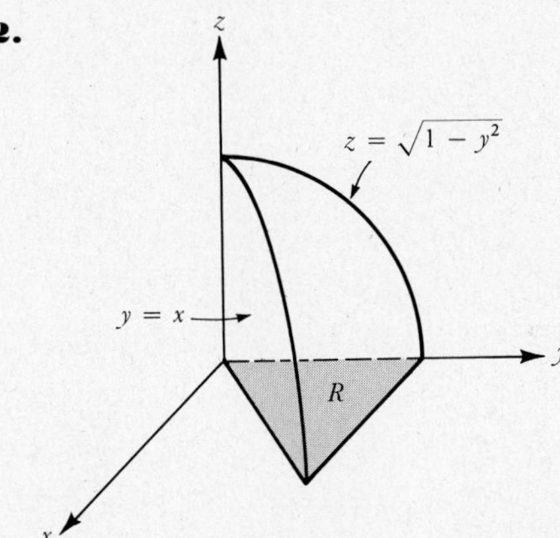

Since $R = \{(x,y)/0 \le y \le 1, 0 \le x \le y\}$, $V(S) = \iint_R \sqrt{1-y^2}\,dA = \int_0^1 \int_0^y \sqrt{1-y^2}\,dx\,dy =$

$\int_0^1 x\sqrt{1-y^2}\,\Big|_0^y dy = \int_0^1 y\sqrt{1-y^2}\,dy = \dfrac{-(1-y^2)^{3/2}}{3}\,\Big|_0^1 = 1/3$. (*Note:* working with the type I description of R leads to a more difficult integration problem.)

Sec. 20.2: Progress Test 2

1.

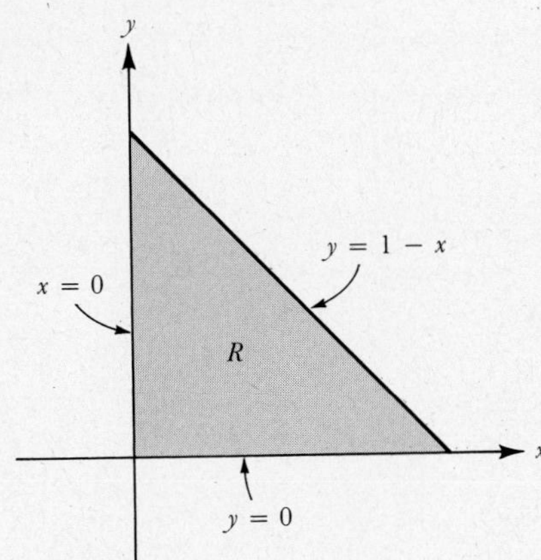

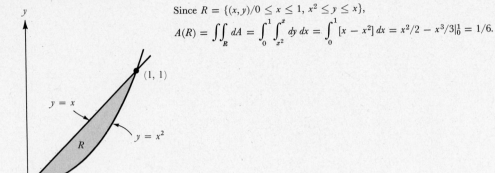

Now $\rho(x,y) = ky$, and since $R = \{(x,y)/0 \le x \le 1, 0 \le y \le 1 - x\}$,
$M_y = \iint_R x\,\rho(x,y)\,dA = \int_0^1\int_0^{1-x} x\,ky\,dy\,dx = k\int_0^1 xy^2/2|_0^{1-x}\,dx$
$= (k/2)\int_0^1(x^3 - 2x^2 + x)\,dx = k/24$.

2.

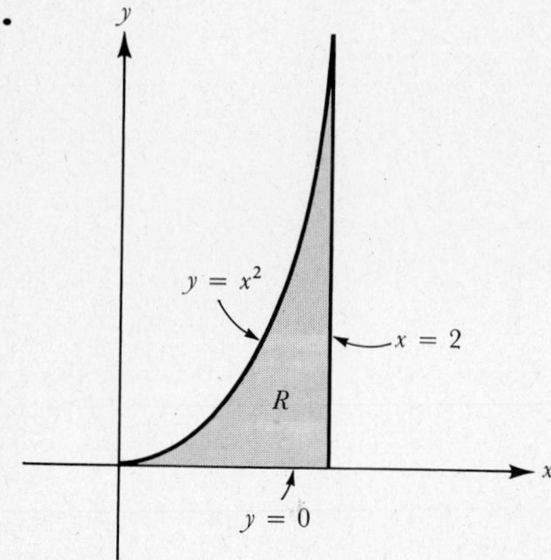

Since $R = \{(x,y)/0 \le x \le 2, 0 \le y \le x^2\}$, $I_x = \iint_R y^2\rho(x,y)\,dA = \int_0^2\int_0^{x^2} y^2(x+2)\,dy\,dx = \int_0^2[y^3(x+2)/3]_0^{x^2}\,dx = (1/3)\int_0^2(x^7 + 2x^6)\,dx = \frac{3840}{168}$.

Sec. 20.3: Progress Test 1

1. $\displaystyle\int_{-1}^{1}\int_0^x\int_0^{xz} xz\,dy\,dz\,dx = \int_{-1}^{1}\int_0^x xyz\Big|_0^{xz} dz\,dx = \int_{-1}^{1}\int_0^x x^2z^2\,dz\,dx = \frac{1}{3}\int_{-1}^{1} x^2z^3\Big|_0^x dx = \frac{1}{3}\int_{-1}^{1} x^5\,dx = 0.$

2.

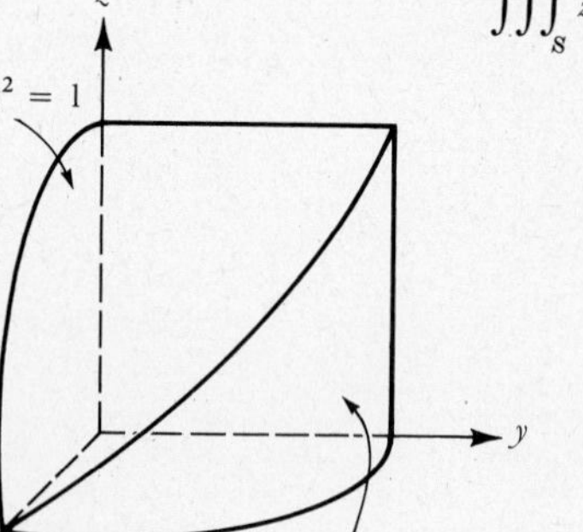

$\displaystyle\iiint_S z\,dV = \int_0^1\int_0^{\sqrt{1-x^2}}\int_0^{\sqrt{1-x^2}} z\,dz\,dy\,dx = \frac{1}{2}\int_0^1\int_0^{\sqrt{1-x^2}} (1-x^2)\,dy\,dx = \frac{1}{2}\int_0^1 (1-x^2)^{3/2}\,dx = 3\pi/32.$

Sec. 20.3: Progress Test 2

1.

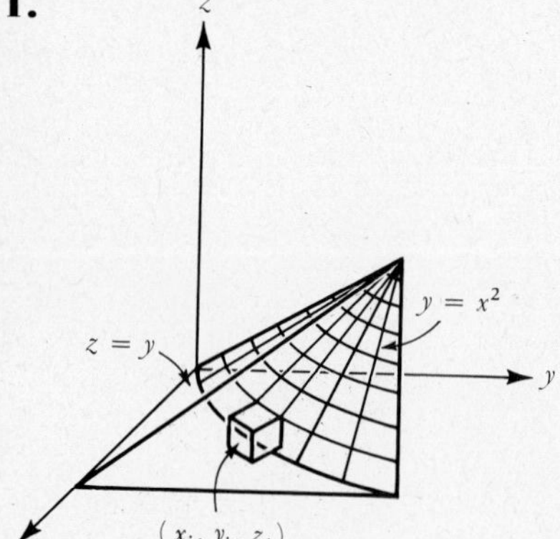

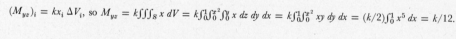

$(M_{yz})_i = kx_i\,\Delta V_i$, so $M_{yz} = k\iiint_S x\,dV = k\int_0^1\int_0^{x^2}\int_0^y x\,dz\,dy\,dx = k\int_0^1\int_0^{x^2} xy\,dy\,dx = (k/2)\int_0^1 x^5\,dx = k/12.$

2.

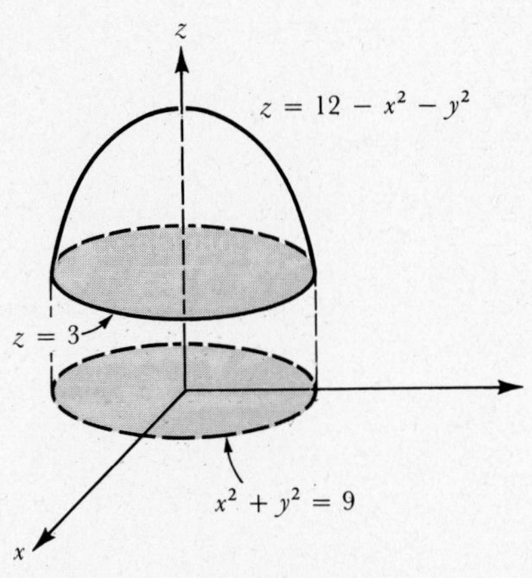

$$(I_z)_i \equiv (\sqrt{x_i^2 + y_i^2})^2\left[\frac{k}{\sqrt{x_i^2 + y_i^2 + z_i^2}}\right]\Delta V_i, \text{ so } I_z = \iiint_S \frac{k(x^2 + y^2)\,dV}{\sqrt{x^2 + y^2 + z^2}} = \int_{-3}^{3}\int_{-\sqrt{9-x^2}}^{\sqrt{9-x^2}}\int_{3}^{12-x^2-y^2} \frac{k(x^2 + y^2)\,dz\,dy\,dx}{\sqrt{x^2 + y^2 + z^2}}.$$

Sec. 20.4: Progress Test 1

1.

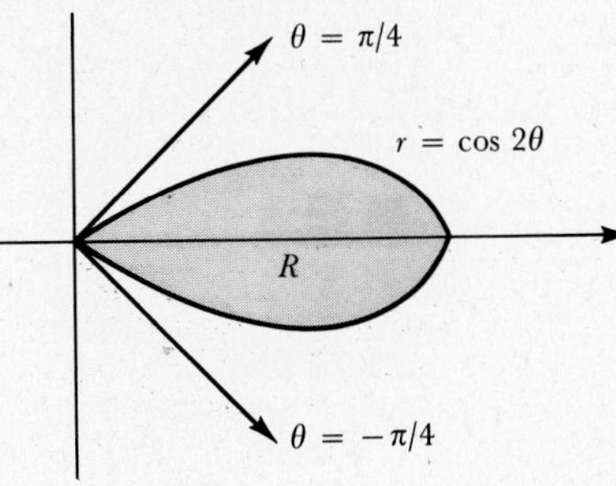

$$A(R) = \int_{-\pi/4}^{\pi/4}\int_0^{\cos 2\theta} r\,dr\,d\theta = \int_{-\pi/4}^{\pi/4}\left[\frac{r^2}{2}\right]_0^{\cos 2\theta} d\theta = \frac{1}{2}\int_{-\pi/4}^{\pi/4} \cos^2 2\theta\,d\theta = \pi/8.$$

2.

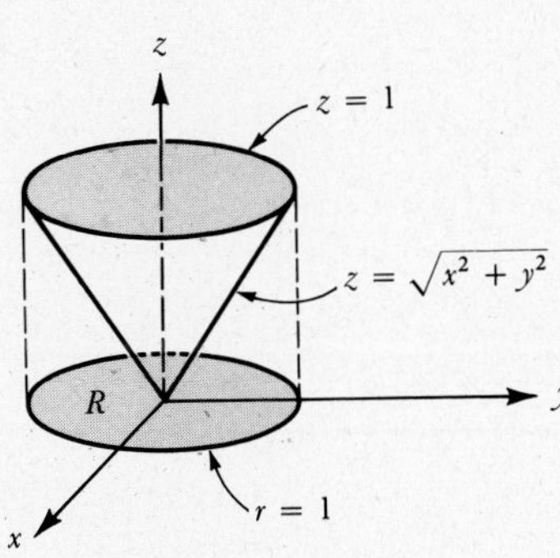

$V(S) = \int_{-1}^{1}\int_{-\sqrt{1-x^2}}^{\sqrt{1-x^2}}[1 - \sqrt{x^2 + y^2}]\,dy\,dx = \iint_R [1 - \sqrt{x^2 + y^2}]\,dA$. In polar coords. $R = \{(r, \theta)/0 \le \theta \le 2\pi,\ 0 \le r \le 1\}$ and $\sqrt{x^2 + y^2} = r$, so $V(S) = \int_0^{2\pi}\int_0^1 (1 - r)r\,dr\,d\theta = \frac{1}{6}\int_0^{2\pi} d\theta = \pi/3$.

3.

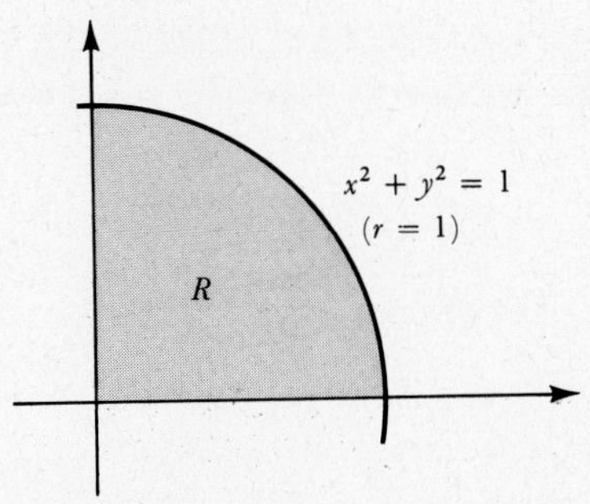

Since $x^2 + y^2 = r^2$ and $R = \{(r, \theta)/0 \le \theta \le \pi/2,$ $0 \le r \le 1\}$ $\iint_R e^{x^2+y^2}\,dA = \int_0^{\pi/2}\int_0^1 e^{r^2}\,r\,dr\,d\theta = (\pi/4)(e - 1)$.

Sec. 20.4: Progress Test 2

1. (*a*) $r = \sqrt{1 + 3} = 2$, $\tan\theta = -\sqrt{3}$, and since $(-1, \sqrt{3})$ is a pt. in the 2d quad. of the *x-y* plane, we choose $\theta = 2\pi/3$. Thus cyl. coords. are $(2, 2\pi/3, 4)$. (*b*) $r = \sqrt{1 + 1} = \sqrt{2}$, $\tan\theta = 1$; thus, since $(1, 1)$ is in 1st quad., $\theta = \pi/4$; so cyl. coords. are $(\sqrt{2}, \pi/4, \sqrt{2})$.
2. (*a*) $x^2 - y^2 = r^2(\cos^2\theta - \sin^2\theta) = r^2\cos 2\theta$. Thus cyl. equation is $r^2\cos 2\theta + z^2 = 1$. (*b*) $x^2 + y^2 = r^2$, so equation $x^2 + y^2 = 9$ becomes $r^2 = 9$ in cyl. coord., which is equivalent to $r = 3$.

Sec. 20.4: Progress Test 3

1. (*a*) $\rho = (1 + 3 + 16)^{1/2} = 2\sqrt{5}$, $\tan\theta = -\sqrt{3}$, so $\theta = 2\pi/3$, $\phi = \arccos(2/\sqrt{5})$. Thus sph. coord. are $(2\sqrt{5}, \arccos(2/\sqrt{5}), 2\pi/3)$. (*b*) $\rho = (1 + 1 + 2)^{1/2} = 2$, $\tan\theta = 1$, so $\theta = \pi/4$ and $\phi = \arccos(\sqrt{2}/2) = \pi/4$. Thus sph. coord. are $(2, \pi/4, \pi/4)$.
2. (*a*) $x^2 - y^2 + z^2 = \rho^2[\sin^2\phi(\cos^2\theta - \sin^2\theta) + \cos^2\phi] = \rho^2[\sin^2\phi\cos 2\theta + \cos^2\phi]$. So sph. equation is $\rho^2[\sin^2\phi\cos 2\theta + \cos^2\phi] = 1$. (*b*) $x^2 + y^2 = \rho^2\sin^2\phi[\cos^2\theta + \sin^2\theta] = \rho^2\sin^2\phi$. Thus sph. equation is $\rho^2\sin^2\phi = 9$. But $\rho \geq 0$ and $\sin\phi \geq 0$, so this is equivalent to $\rho\sin\phi = 3$.

Sec. 20.4: Progress Test 4

1.

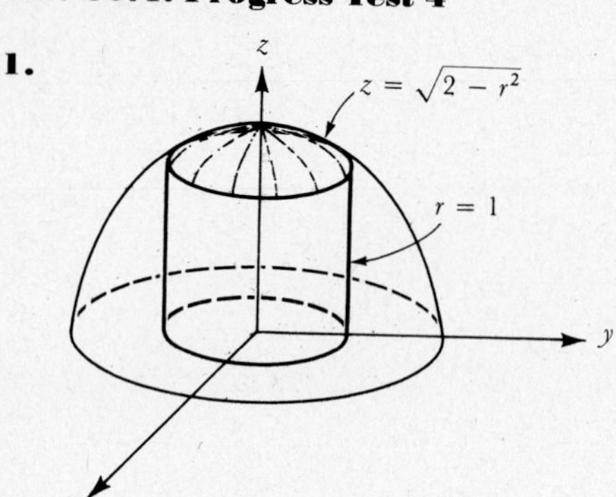

$$M = \int_0^{2\pi}\int_0^1\int_0^{\sqrt{2-r^2}} r\,dz\,dr\,d\theta = \int_0^{2\pi}\int_0^1 r\sqrt{2-r^2}\,dr\,d\theta = \int_0^{2\pi}\frac{2\sqrt{2}-1}{3}\,dr = \frac{2\pi}{3}(2\sqrt{2}-1).$$

2.

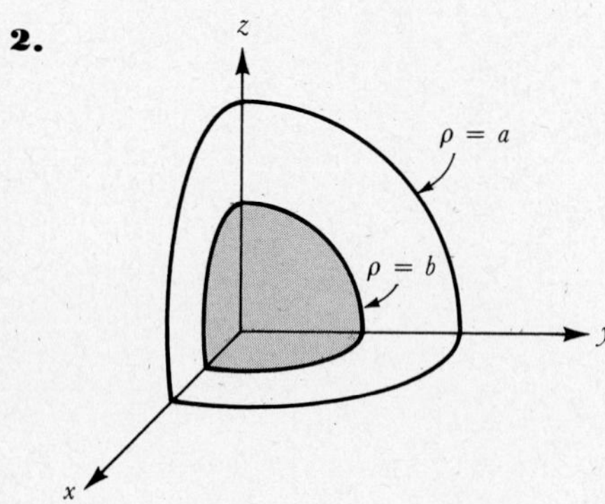

$$\begin{aligned} M &= \int_0^{\pi/2}\int_0^{\pi/2}\int_b^a \frac{k}{\rho}\rho^2\sin\phi\,d\rho\,d\phi\,d\theta \\ &= \frac{k(a^2-b^2)}{2}\int_0^{\pi/2}\int_0^{\pi/2}\sin\phi\,d\phi\,d\theta \\ &= \frac{k(a^2-b^2)}{2}\int_0^{\pi/2} d\theta = \frac{k\pi(a^2-b^2)}{4}. \end{aligned}$$

Chapter 21: Solutions to Progress Tests

Sec. 21.1: Progress Test 1

1. $M_y = x^2 \neq 4y^2 = N_x$. By (21.1.3), **F** is not conservative. **2.** $M_y = N_x = 3gxy(x^2 + y^2)^{-5/2}$. Let $f_x = -gx(x^2 + y^2)^{-3/2}$. Then $f = g(x^2 + y^2)^{-1/2} + C(y)$, so $f_y = -gy(x^2 + y^2)^{-3/2} + C'(y)$. Setting $f_y = N \Rightarrow C'(y) = 0$, so $C(y) = k$, a constant. Thus $\mathbf{F} = \nabla f$, where $f(x, y) = g(x^2 + y^2)^{-1/2} + k$. **3.** $M_y = N_x = 4xy/(2 + y^2)$. Let $f_x = 2x\ln(2 + y^2) - 6x^2$. Then $f = x^2\ln(2 + y^2) - 2x^3 + C(y)$, so $f_y = 2x^2y/(2 + y^2) + C'(y)$. Set $f_y = N$ to conclude $C'(y) = 2y$, so $C(y) = y^2 + k$. Thus $\mathbf{F} = \nabla f$, where $f(x, y) = x^2\ln(2 + y^2) - 2x^3 + y^2 + k$.

Sec. 21.2: Progress Test 1

1. By (21.2.2) $W = \int_{-1}^{1}\mathbf{F}(t, 2t^3)\cdot\langle 1, 6t^2\rangle\,dt = \int_{-1}^{1}\langle 3t^2 + 20t^3, 12t^6 - 5t^2\rangle\cdot\langle 1, 6t^2\rangle\,dt, = \int_{-1}^{1}(3t^2 + 20t^3 + 72t^8 - 30t^4)\,dt = t^3 + 5t^4 + 8t^9 - 6t^5|_{-1}^{1} = 6.$

Sec. 21.2: Progress Test 2

1. C_1: $\langle 2-2t, 0\rangle$, $0 \le t \le 1$, C_2: $\langle 0, 2t\rangle$, $0 \le t \le 1$, C_3: $\langle 2t, 2\rangle$, $0 \le t \le 1$, C_4: $\langle 2, 2-2t\rangle$, $0 \le t \le 1$.

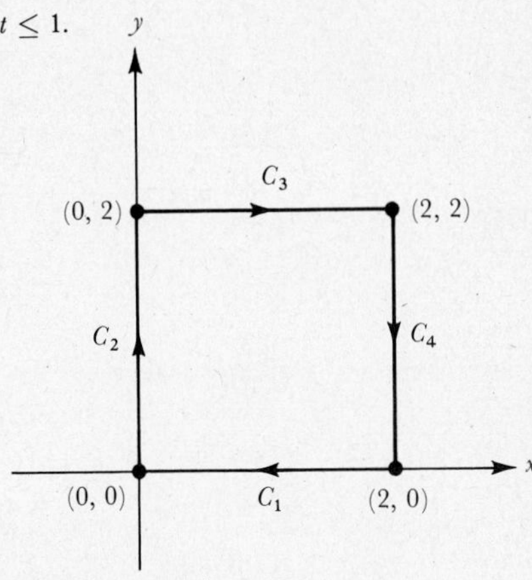

$\int_{C_1} x^2y\,dx + 2x\,dy = 0$ since $y = 0$ and $dy = 0$, $\int_{C_2} x^2y\,dx + 2x\,dy = 0$ since $x = 0$, $\int_{C_3} x^2y\,dx + 2x\,dy = \int_0^1 16t^2\,dt = 16/3$, $\int_{C_4} x^2y\,dx + 2x\,dy = \int_0^1(-8)\,dt = -8$; thus $\int_C x^2y\,dx + 2x\,dy = 16/3 - 8 = -8/3$. **2.** C_1: $\langle t, t\rangle$, $0 \le t \le 1$, C_2: $\langle t, 2-t^2\rangle$, $1 \le t \le \sqrt{2}$. $\int_{C_1}(x-y^2)\,dx + (x^2+y)\,dy = \int_0^1 2t\,dt = 1$, $\int_{C_2}(x-y^2)\,dx + (x^2+y)\,dy = \int_1^{\sqrt{2}}(-t^4 + 4t^2 - 3t - 4)\,dt =$

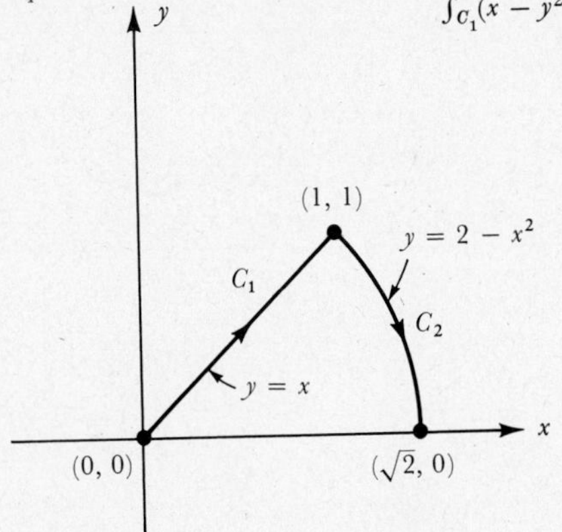

$(41 - 64\sqrt{2})/30$. Thus $\int_C(x-y^2)\,dx + (x^2+y)\,dy = (71 - 64\sqrt{2})/30$.

Sec. 21.3: Progress Test 1

1. $M_y = N_x = 3x^2 + 2y$ (cont. on all of R_2). A potential function is $f(x,y) = x^3y + xy^2 - x^2$. Thus $I = f(1,1) - f(0,0) = 1$. **2.** $M_y = N_x = -e^x \sin y$ (cont. on all of R_2). A potential function is $f(x,y) = e^x \cos y$. Thus $I = f(0, \pi/2) - f(0,0) = -1$. **3.** Let D be any simply connected open set containing $(e^2, 0)$ and $(0, 1)$ but *not* $(0, 0)$. Then $M_y = N_x = -2xy/(x^2+y^2)$; M_y, N_x cont. on D. A potential function is $f(x,y) = (1/2)\ln(x^2+y^2)$. Thus $I = f(0,1) - f(e^2, 0) = (-1/2)\ln(e^4) = -2$.

Sec. 21.4: Progress Test 1

1. Let $M = 0$, $N = x$. Then by Green's Theorem $\oint_{\partial R} x\,dy = \iint_R[N_x - M_y]\,dA = \iint_R dA = A(R)$. Similarly, if $M = -y$ and $N = 0$, then $\oint_{\partial R} y\,dx = \iint_R[N_x - M_y]\,dA = \iint_R dA = A(R)$. **2.** Let $M = xy^3/3 + 4y - e^{2x}$, $N = xy^2 + 2y$, and let D be the region inside the rectangle. Then $\int_K M\,dx + N\,dy = \iint_D[N_x - M_y]\,dA = \iint_D(y^2 - xy^2 - 4)\,dA = \int_1^4\int_{-1}^2(y^2 - xy^2 - 4)\,dx\,dy = -45/2$. **3.** Let D be the regular region outside the circle Γ of radius 1/2 centered at $(-1, 0)$ (clockwise) and inside K. Since $M_y = N_x$,

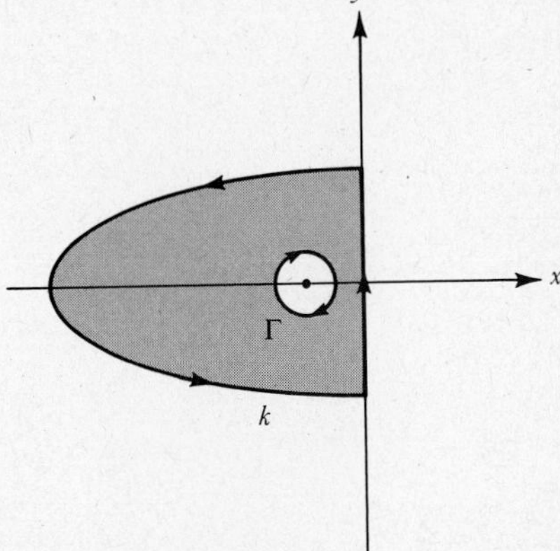

$\oint_{\partial D}(-y/((x+1)^2+y^2)\,dx + (x+1)/((x+1)^2+y^2)\,dy = \iint_D 0\,dA = 0$. Now $\oint_{\partial D} = \int_K + \int_\Gamma = 0 \Rightarrow \int_K = -\int_\Gamma = \int_{-\Gamma}$, where $-\Gamma$ has parameterization $\langle -1+(1/2)\cos t, (1/2)\sin t\rangle$, $0 \le t \le 2\pi$. Thus $\int_K = \int_{-\Gamma} = \int_0^{2\pi}(\sin^2 t + \cos^2 t)\,dt = 2\pi$.

Sec. 21.5: Progress Test 1

1. $\int_K \mathbf{F}\cdot\mathbf{n}\,ds = \iint_R \operatorname{div}\mathbf{F}\,dA = \iint_R 3(x^2+y^2)\,dA = 3\int_0^{2\pi}\int_0^2 r^3\,dr\,d\theta = 24\pi$.

Sec. 21.5: Progress Test 2

1. $\int_K \mathbf{F}\cdot\mathbf{T}\,ds = \iint_R (\operatorname{curl}\mathbf{F}\cdot\mathbf{k}\,dA = -\iint_R[-3(x^2+y^2)]\,dA = -3\int_0^{2\pi}\int_0^2 r^3\,dr\,d\theta = -24\pi$.

Appendix A: Solutions to Progress Tests

Sec. A.1: Progress Test 1

1. (Separable) $\dfrac{3x}{x^2+4}\,dx + \dfrac{dy}{y} = 0 \Leftrightarrow (3/2)\ln(x^2+4) + \ln|y| = \ln|c| \Leftrightarrow 3y(x^2+4) = 2c$. **2.** (Homogeneous) $\dfrac{dy}{dx} = \dfrac{y}{x} + \sqrt{1+\left(\dfrac{y}{x}\right)^2}$. Let $y = vx$ and equation becomes (after simplification) $\dfrac{dv}{\sqrt{v^2+1}} = \dfrac{dx}{x} \Leftrightarrow \ln|v+\sqrt{v^2+1}| = \ln|x| + \ln|c| \Leftrightarrow v + \sqrt{v^2+1} = cx \Leftrightarrow y + \sqrt{x^2+y^2} = cx^2$.

Sec. A.1: Progress Test 2

1. (Linear) $P(x) = \tan x$, so $e^{\int P(x)dx} = \sec x$ and soln. is $y\sec x = \int \cos x \sec x\,dx + c = \int dx + c = x + c$.
2. (Exact) $M_y = N_x = 2x\sec^2 y - \sec x\tan x$. Set $f_x = 2x\tan y - y\sec x\tan x + 2x \Rightarrow f = x^2\tan y - y\sec x + x^2 + \phi(y) \Rightarrow f_y = x^2\sec^2 y - \sec x + \phi'(y)$. Thus $\phi'(y) = 0$, so $\phi(y) = C_1$. The soln. is $x^2\tan y - y\sec x + x^2 + C_1 = C$.

Sec. A.2: Progress Test 1

1. $c(m) = m^2 + 4m - 12 = (m-2)(m+6)$, so $y_c = c_1e^{2x} + c_2e^{-6x}$. **2.** $c(m) = m^2 - 10m + 25 = (m-5)^2$, so $y_c = c_1e^{5x} + c_2xe^{5x}$.
3. $c(m) = m^2 + m + 3$ is irreducible. Now $a_1 = 1$, $a_2 = 3$, so $a = -a_1/2 = -1/2$ and $b = \sqrt{4a_2 - a_1{}^2}/2 = \sqrt{12-1}/2 = \sqrt{11}/2$. Thus $y_c = e^{-x/2}\left[c_1\sin\dfrac{\sqrt{11}\,x}{2} + c_2\cos\dfrac{\sqrt{11}\,x}{2}\right]$.

Sec. A.2: Progress Test 2

1. $c(m) = m^2 - 4m + 4 = (m-2)^2$, so $y_c = c_1e^{2x} + c_2xe^{2x}$. Let $y_1 = e^{2x}$ and $y_2 = xe^{2x}$. Then $y_1y_2' - y_2y_1' = e^{2x}(e^{2x} + 2xe^{2x}) - xe^{2x}(2e^{2x}) = e^{4x}$. Thus $v_1' = \dfrac{-xe^{2x}(x^2e^{2x})}{e^{4x}} = -x^3$ and $v_2' = \dfrac{e^{2x}(x^2e^{2x})}{e^{4x}} = x^2$. Hence $v_1 = \dfrac{-x^4}{4}$ and $v_2 = x^3/3$. Consequently, $y_p = \dfrac{-x^4}{4}(e^{2x}) + \dfrac{x^3}{3}(xe^{2x}) = \dfrac{x^4e^{2x}}{12}$. The general solution of the equation is $y = c_1e^{2x} + c_2xe^{2x} + \dfrac{x^4e^{2x}}{12}$. **2.** As in Prob. 1, $y_c = c_1e^{2x} + c_2xe^{2x}$. The v_i's are e^{2x}, 1, and since e^{2x} and xe^{2x} satisfy the associated equation, we replace e^{2x} by x^2e^{2x}. Thus $y_p = Ax^2e^{2x} + B$; $y_p' = 2Axe^{2x} + 2Ax^2e^{2x}$; $y_p'' = 4Ax^2e^{2x} + 8Axe^{2x} + 2Ae^{2x}$ and $y_p'' - 4y_p' + 4y_p = (4A - 8A + 4A)x^2e^{2x} + (8A - 8A)xe^{2x} + 2Ae^{2x} + 4B = 2Ae^{2x} + 4B$. Thus $2A = 1$, $4B = 1$, so $A = 1/2$, $B = 1/4$. The general solution is $y = c_1e^{2x} + c_2xe^{2x} + (1/2)x^2e^{2x} + 1/4$.

INDEX

Page numbers in parentheses and *italic* refer to the exercise(s) on that page.

30. $\int \frac{\sqrt{u^2+a^2}\,du}{u} = \sqrt{u^2+a^2} - a \ln\left|\frac{a+\sqrt{u^2+a^2}}{u}\right| + C$

31. $\int \frac{\sqrt{u^2-a^2}\,du}{u} = \sqrt{u^2-a^2} - a\,\text{arcsec}\left|\frac{u}{a}\right| + C$

32. $\int \frac{\sqrt{u^2\pm a^2}\,du}{u^2} = -\frac{\sqrt{u^2\pm a^2}}{u} + \ln|u+\sqrt{u^2\pm a^2}| + C$

33. $\int \frac{u^2\,du}{\sqrt{u^2\pm a^2}} = \frac{u}{2}\sqrt{u^2\pm a^2} - \frac{\pm a^2}{2}\ln|u+\sqrt{u^2\pm a^2}| + C$

34. $\int \frac{du}{u\sqrt{u^2+a^2}} = -\frac{1}{a}\ln\left|\frac{a+\sqrt{u^2+a^2}}{u}\right| + C$

35. $\int \frac{du}{u\sqrt{u^2-a^2}} = \frac{1}{a}\,\text{arcsec}\left|\frac{u}{a}\right| + C$

36. $\int \frac{du}{u^2\sqrt{u^2\pm a^2}} = -\frac{\sqrt{u^2\pm a^2}}{\pm a^2 u} + C$

37. $\int (u^2\pm a^2)^{3/2}\,du = \frac{u}{8}(2u^2\pm 5a^2)\sqrt{u^2\pm a^2}$

$$+\frac{3a^4}{8}\ln|u+\sqrt{u^2\pm a^2}| + C$$

38. $\int \frac{du}{(u^2\pm a^2)^{3/2}} = \frac{u}{\pm a^2\sqrt{u^2\pm a^2}} + C$

FORMS CONTAINING $\sqrt{a^2-u^2}$

39. $\int \frac{du}{\sqrt{a^2-u^2}} = \arcsin\frac{u}{a} + C$

40. $\int \sqrt{a^2-u^2}\,du = \frac{u}{2}\sqrt{a^2-u^2} + \frac{a^2}{2}\arcsin\frac{u}{a} + C$

41. $\int u^2\sqrt{a^2-u^2}\,du = \frac{u}{8}(2u^2-a^2)\sqrt{a^2-u^2} + \frac{a^4}{8}\arcsin\frac{u}{a} + C$

42. $\int \frac{\sqrt{a^2-u^2}\,du}{u} = \sqrt{a^2-u^2} - a\ln\left|\frac{a+\sqrt{a^2-u^2}}{u}\right| + C$

$$= \sqrt{a^2-u^2} - a\,\text{arccosh}\,\frac{a}{u} + C$$

43. $\int \frac{\sqrt{a^2-u^2}\,du}{u^2} = -\frac{\sqrt{a^2-u^2}}{u} - \arcsin\frac{u}{a} + C$

44. $\int \frac{u^2\,du}{\sqrt{a^2-u^2}} = -\frac{u}{2}\sqrt{a^2-u^2} + \frac{a^2}{2}\arcsin\frac{u}{a} + C$

45. $\int \frac{du}{u\sqrt{a^2-u^2}} = -\frac{1}{a}\ln\left|\frac{a+\sqrt{a^2-u^2}}{u}\right| + C = -\frac{1}{a}\,\text{arccosh}\,\frac{a}{u} + C$

46. $\int \frac{du}{u^2\sqrt{a^2-u^2}} = -\frac{\sqrt{a^2-u^2}}{a^2u} + C$

47. $\int (a^2-u^2)^{3/2}\,du = -\frac{u}{8}(2u^2-5a^2)\sqrt{a^2-u^2} + \frac{3a^4}{8}\arcsin\frac{u}{a} + C$

48. $\int \frac{du}{(a^2-u^2)^{3/2}} = \frac{u}{a^2\sqrt{a^2-u^2}} + C$

FORMS CONTAINING $2au-u^2$

49. $\int \sqrt{2au-u^2}\,du = \frac{u-a}{2}\sqrt{2au-u^2} + \frac{a^2}{2}\arccos\left(1-\frac{u}{a}\right) + C$

50. $\int u\sqrt{2au-u^2}\,du = \frac{2u^2-au-3a^2}{6}\sqrt{2au-u^2}$

$$+\frac{a^3}{2}\arccos\left(1-\frac{u}{a}\right) + C$$

51. $\int \frac{\sqrt{2au-u^2}\,du}{u} = \sqrt{2au-u^2} + a\arccos\left(1-\frac{u}{a}\right) + C$

52. $\int \frac{\sqrt{2au-u^2}\,du}{u^2} = -\frac{2\sqrt{2au-u^2}}{u} - \arccos\left(1-\frac{u}{a}\right) + C$

53. $\int \frac{du}{\sqrt{2au-u^2}} = \arccos\left(1-\frac{u}{a}\right) + C$

54. $\int \frac{u\,du}{\sqrt{2au-u^2}} = -\sqrt{2au-u^2} + a\arccos\left(1-\frac{u}{a}\right) + C$

55. $\int \frac{u^2\,du}{\sqrt{2au-u^2}} = -\frac{(u+3a)}{2}\sqrt{2au-u^2} + \frac{3a^2}{2}\arccos\left(1-\frac{u}{a}\right) + C$

56. $\int \frac{du}{u\sqrt{2au-u^2}} = -\frac{\sqrt{2au-u^2}}{au} + C$

57. $\int \frac{du}{(2au-u^2)^{3/2}} = \frac{u-a}{a^2\sqrt{2au-u^2}} + C$

58. $\int \frac{u\,du}{(2au-u^2)^{3/2}} = \frac{u}{2\sqrt{2au-u^2}} + C$

FORMS CONTAINING TRIGONOMETRIC FUNCTIONS

59. $\int \sin u\,du = -\cos u + C$

60. $\int \cos u\,du = \sin u + C$

61. $\int \tan u\,du = \ln|\sec u| + C$

62. $\int \cot u\,du = \ln|\sin u| + C$

63. $\int \sec u\,du = \ln|\sec u + \tan u| + C = \ln|\tan(\frac{1}{4}\pi + \frac{1}{2}u)| + C$

64. $\int \csc u\,du = \ln|\csc u - \cot u| + C = \ln|\tan\frac{1}{2}u| + C$

65. $\int \sec^2 u\,du = \tan u + C$

66. $\int \csc^2 u\,du = -\cot u + C$

67. $\int \sec u\tan u\,du = \sec u + C$

68. $\int \csc u\cot u\,du = -\csc u + C$

69. $\int \sin^2 u\,du = \frac{1}{2}u - \frac{1}{4}\sin 2u + C$

70. $\int \cos^2 u\,du = \frac{1}{2}u + \frac{1}{4}\sin 2u + C$

71. $\int \tan^2 u\,du = \tan u - u + C$

72. $\int \cot^2 u\,du = -\cot u - u + C$

73. $\int \sin^n u\,du = -\frac{1}{n}\sin^{n-1}u\cos u + \frac{n-1}{n}\int \sin^{n-2}u\,du$

74. $\int \cos^n u\,du = \frac{1}{n}\cos^{n-1}u\sin u + \frac{n-1}{n}\int \cos^{n-2}u\,du$